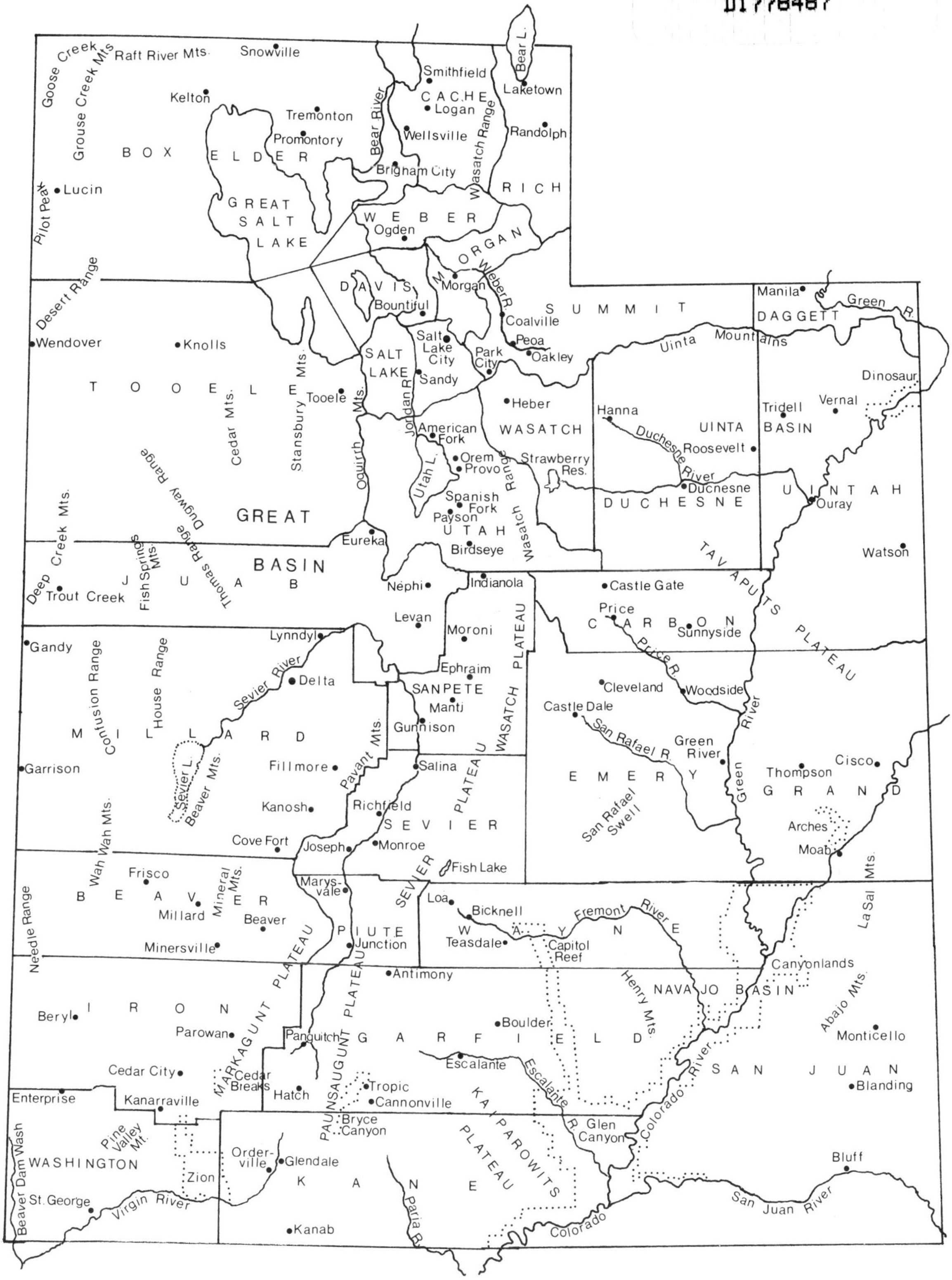
Goose Creek
Grouse Creek Mts.
Raft River Mts.
Snowville
Kelton
Tremonton
Promontory
Smithfield
CACHE
Logan
Wellsville
Bear River
Bear L.
Laketown
Randolph
Wasatch Range
Brigham City
RICH
BOX ELDER
Pilot Peak
Lucin
GREAT SALT LAKE
WEBER
Ogden
MORGAN
Morgan
Weber R.
DAVIS
Bountiful
Desert Range
Wendover
Knolls
SUMMIT
Coalville
Peoa
Oakley
Manila
Green R.
DAGGETT
Uinta Mountains
Salt Lake City
SALT LAKE
Park City
Sandy
Jordan R.
TOOELE
Tooele
Stansbury Mts.
Oquirrh Mts.
Cedar Mts.
Heber
WASATCH
Hanna
Duchesne River
UINTA BASIN
Roosevelt
Tridell
Vernal
Dinosaur
American Fork
Orem
Provo
Utah L.
Strawberry Res.
Spanish Fork
Payson
UTAH
Birdseye
Duchesne
DUCHESNE
UINTAH
Ouray
Watson
Dugway Range
Deep Creek Mts.
GREAT BASIN
Eureka
Fish Springs Mts.
Thomas Range
Trout Creek
JUAB
Nephi
Indianola
Castle Gate
TAVAPUTS PLATEAU
Price
CARBON
Sunnyside
Price R.
Levan
Moroni
Lynndyl
Gandy
Confusion Range
House Range
Sevier River
Delta
Ephraim
SANPETE
Manti
Gunnison
WASATCH PLATEAU
Cleveland
Woodside
Castle Dale
San Rafael R.
Green River
MILLARD
Sevier L.
Beaver Mts.
Garrison
Fillmore
Pavant Mts.
Salina
Thompson
Cisco
GRAND
EMERY
San Rafael Swell
Kanosh
Richfield
SEVIER
Arches
Moab
Wah Wah Mts.
Cove Fort
Joseph
Monroe
Fish Lake
Frisco
Mineral Mts.
BEAVER
Millard
Beaver
Marysvale
Loa
Bicknell
Fremont River
WAYNE
Teasdale
Capitol Reef
La Sal Mts.
Needle Range
Minersville
PIUTE
Junction
Antimony
Canyonlands
Henry Mts.
NAVAJO BASIN
Abajo Mts.
Monticello
Beryl
IRON
Parowan
MARKAGUNT PLATEAU
Panguitch
PAUNSAUGUNT PLATEAU
Boulder
GARFIELD
Escalante
Escalante R.
SAN JUAN
Blanding
Colorado River
Cedar City
Cedar Breaks
Hatch
Tropic
Cannonville
KAIPAROWITS PLATEAU
Enterprise
Kanarraville
Bryce Canyon
Glen Canyon
Pine Valley Mt.
Beaver Dam Wash
WASHINGTON
Orderville
Glendale
Zion
KANE
Bluff
St. George
Virgin River
Paria R.
San Juan River
Kanab
Colorado

GREAT BASIN NATURALIST MEMOIRS

Number 9   Brigham Young University   1987

A Utah Flora

©1987, Brigham Young University
ISBN 0-8425-2260-3

Methodology

Each taxonomic treatment was written in its entirety by its author or authors, beginning with a checklist of taxa known or thought to occur within the state. Keys were constructed based on specimens taken from the herbarium and from concepts in literature. Many have been improved by comparison with living plants in the field. However, most were written in their entirety from herbarium specimens, and their similarities to published treatments are coincidental. Where keys were derived in part from publications, the source is included in literature citations, typically listed following the generic description.

Descriptions are likewise based on specimens, except where those were inadequate for the purpose (some taxa are known from few and sometimes imperfect specimens), and the circumscriptions then came, at least partially, from literature, especially from original descriptions. Format for the descriptions is essentially that standard for much of taxonomic literature. Duration, root, stem (including habit and aspect), leaves, inflorescence, flowers, and fruits are described in essentially that order. Chromosome number is included at the end of descriptions, where that information has been available in literature. We have not attempted to do routine counting of chromosomes as a part of the basic research.

Type specimens of most taxa described from Utah have been examined by one or more of the authors. Decisions on critical groups have been based on the type specimens.

Authors of taxa are cited routinely for each taxon of the rank of family or below. The author names are abbreviated following the list by Chatterley, Welsh, and Welsh (1982) and supplemented by the work of Halliday, Meikle, Story, and Wilkinson (1980). We have chosen not to use all the abbreviations in the latter work because of brevity; e.g., M. E. Jones is cited merely as Jones in the Utah flora and A. Gray is Gray. A complete list of author abbreviations and their given names and surnames, along with year of birth, year of death, or year of publication, is appended to this work.

Discussions following the description provide clarification in many instances, but they routinely include plant community data (typically from low to high elevations), elevational range (in meters), counties from which the taxon is known in Utah (arranged in alphabetical order), distribution beyond Utah (or whether they are endemic to the state), the number of specimens seen from Utah during the study (Arabic numerals) and the number collected in Utah by the author of the particular group (Roman numerals). The number examined is an index to both the familiarity of the author with the taxon and its abundance; the number collected by the author provides an index to that worker's understanding of the taxon in the field.

Taxa known to be adventive or introduced purposefully are so noted in the discussion, and the region from which the plants originated is indicated.

Flowering and fruiting dates have been routinely omitted, because practically everything in Utah flowers in either spring (extending to early summer) or (late summer to) autumn; and many plants that flower in springtime again flower in autumn. Only when the situation is unusual has the flowering date been included.

Midsummer is a poor time to collect plants in Utah, except in high mountains, where it is still springtime. Spring begins in Utah at the lowest and southernmost elevations in Washington, Kane, and San Juan counties as early as January in some years, but more commonly in March and April. As the relationship of the earth to the sun moves from the vernal toward the autumnal equinox, flowering and fruiting of plants move both northward and upward in elevation. Spring arrives at some of the high mountain summits only to meet autumn moving down. Summer at high elevations is a fleeting condition that hardly exists in some years. The autumn-flowering plants reverse the trend of flowering time, finally closing out the season in the lowermost elevations of the southernmost portions of the state typically in October and November (December). There are plants that begin to flower in southern Utah in November and December and continue flowering to spring of the following year (e.g., *Thamnosma* and *Garrya*). For most of Utah spring arrives in April and May, but there are areas where frost can (and often does) occur at all months of the year.

Such ecological data as slopes and valleys (or rocky or dry) have been omitted because practically all plants grow in such sites. The term *saline* has been used in most cases in place of the familiar *alkaline*, since one can observe salinity in the field but must test for alkalinity.

Measurements

Plant parts are measured in metric units. Parts measuring less than 1 cm long are always cited in millimeters (mm) and tenths of millimeters. Millimeter accuracy is routine for parts up to about 25 mm in length or breadth, and then the unit of measurement is changed to centimeters (cm). Decimeters (dm) are used for plants or plant parts that generally exceed 50 centimeters. Meters (m) are used for plants that exceed about 1.5 meters and for all elevational measurements. Kilometers are used for long lineal measurements (except those cited in type locality information where miles are used).

Plant height is taken as the length of the plant above ground level and includes the inflorescence and flowers. Leaf measurements are from the petiole base to the blade apex, unless otherwise indicated. Leaf breadth is taken at the widest position on the blade. Petioles are measured from point of insertion on the stem to the blade base. Stipules are measured from the point of insertion on the stem or petiole base to the apex. Peduncles are measured from the uppermost bract to the first flower or branch of a compound inflorescence. Total flower length is taken from the point of insertion of the pedicel (peduncle) to the apex of the longest petal. Calyx measurements are taken from the point of insertion of the pedicel to the tip of the longest calyx teeth, except in legumes, and then the measurement is to the length of the longest lateral tooth. Calyx teeth and corolla lobe measurements are from the base of the sinus to an adjacent lobe or tooth apex. Petal length is determined from the point of insertion of the petal to its apex. Fruit length measurements include from the base of the ovary (point of insertion on the receptacle) to the tip, excluding style and stigma measurements, unless otherwise noted. Gynophores are excluded from ovary measurements but stipes are not.

Nomenclature

Names given for taxa in this work are consistent with stipulations of the International Code of Botanical Nomenclature (Voss 1983). The principle of priority is the overriding consideration, except where expressly limited by mandate of the Code. Species and generic concepts are mainly conservatively interpreted, with the definition of species considered multifaceted because of differences in breeding mechanisms and point in evolutionary time. The concept of subspecies is herein restricted to those species with usually large geographical and morphological representation. They typically consist of two or more varieties. Varietal designation is more often used alone as the principal category below the species level, especially in less widely ranging and less complex species wherein designation of subspecies is unwarranted. However, no effort has been made to routinely propose new nomenclatural transfers of all subspecies to varietal level. Such a course is not indicated in this provincial treatment that cannot encompass the variation in all taxa beyond the boundaries of Utah.

The names utilized for genera and species are based on the best information available to us and on our understanding of the groups as they occur throughout their natural and adoptive ranges. Major checklists, which attempt to categorize and standardize generic and specific names, are followed only in part (USDA Soil Conservation Service 1982). The choice of generic and specific concepts are best left to competent taxonomists, not legislated by adoption of standardized lists. Workers on specific areas of a country as large and diverse as the United States are certain to know more about the plants of their area than are workers who labor more casually on a broader problem. However, that tendency of more intimate knowledge does not relieve the worker in a limited area from reviewing all the information available from whatever source. We hope we have reviewed the information from all sources carefully, and our taxonomic and nomenclatural decisions result from that review. The final word is not to be found, however, in the following pages.

As additional collections are made, and as summary monographs and revisions become available, the views presented here will be subject to change. The flora of the region and its nomenclature still require much understanding and great labor.

Common Names

Common names have been taken, with modification, from the USDA Forest Service General Technical Report INT-38 and from general nonpublished sources. No attempt has been made to provide common names throughout, especially when they are not known or where they are not in general usage.

Classification

This book is dedicated to the many students of botany who will use it as a reference. Much of its use will be to find names of genera and species of examples in hand. Because of that use and the need for ease and economy of time, a phylogenetic arrangement is carried only to the class level. The book begins with a key to the divisions, and is followed by the fern allies (Lycopodiophyta and Equisetophyta), ferns (Polypodiophyta), conifers (Pinophyta or gymnosperms), and finally the flowering plants (Magnoliophyta), with Magnoliopsids (dicots) first, followed by the Liliopsids (monocots). Within each major group the families are arranged in alphabetical order, as are the genera and species within each family. A list of the taxonomic categories arranged in a phylogenetic sequence follows (the arrangement of the flowering plant families is modified from Cronquist 1981):

GROUP TRACHEOPHYTA

Approximation of a phylogenetic arrangement of families.

LYCOPODIOPHYTA

Family name	Author	Genera	Species	Infra
1. Isoetaceae	slw	2	4	
2. Selaginellaceae	hig	1	5	

EQUISETOPHYTA

Family name	Author	Genera	Species	Infra
1. Equisetaceae	hig	1	4	

POLYPODIOPHYTA

Family name	Author	Genera	Species	Infra
1. Ophioglossaceae	hig	1	4	
2. Polypodiaceae	hig	15	35	
3. Marsileaceae	hig	1	2	
4. Salviniaceae	hig	1	1	

PINOPHYTA

Family name	Author	Genera	Species	Infra
1. Ginkgoaceae	slw	1	1 (intro 1)	
2. Taxaceae	slw	1	2 (intro 2)	
3. Pinaceae	ndaslw	6	23 (intro 12)	
4. Cupressaceae	slwnda	5	14 (intro 10)	
5. Taxodiaceae	slw	3	3 (intro 3)	
6. Ephedraceae	hig	1	4	

MAGNOLIOPHYTA

Magnoliopsida

Family name	Author	Genera	Species	Infra
1. Magnoliaceae	slw	2	4 (intro 4)	
2. Saururaceae	hig	1	1	
3. Nymphaeaceae	slw	2	2 (intro 1)	
4. Ceratophyllaceae	slw	1	1	
5. Ranunculaceae	slw	13	61 (intro 7)	
6. Berberidaceae	slw	2	9 (intro 6)	
7. Papaveraceae	slw	7	15 (intro 6)	
8. Fumariaceae	slw	3	4 (intro 1)	
9. Platanaceae	slw	1	2 (intro 2)	
10. Hamamelidaceae	slw	1	1 (intro 1)	
11. Ulmaceae	slw	3	7 (intro 6)	
12. Cannabaceae	slw	2	2 (intro 1)	
13. Moraceae	slw	3	4 (intro 4)	
14. Urticaceae	slw	2	2	
15. Juglandaceae	slw	2	5 (intro 5)	
16. Fagaceae	slw	2	12 (intro 7)	
17. Betulaceae	slw	5	8 (intro 1)	
18. Nyctaginaceae	slw	6	17 (intro 1)	
19. Aizoaceae	hig	3	3 (intro 1)	
20. Cactaceae	slw	9	28	14
21. Chenopodiaceae	slw	16	58 (intro 13)	13
22. Amaranthaceae	slw	4	11 (intro 6)	
23. Portulacaceae	higslw	6	18	
24. Caryophyllaceae	slw	16	56 (intro 14)	5
25. Polygonaceae	slw	8	89 (intro 3)	25
26. Paeoniaceae	slw	1	1	
27. Elatinaceae	slw	2	3	

A flora of the Uinta Basin has been prepared by Sherel Goodrich and Elizabeth Neese, and its publication is anticipated in 1986. An atlas of distibution maps of Utah plant species has been prepared by Beverly Albee and Leila Shultz, and it is expected to be published shortly.

Philosophical Basis and Species Concept

The underlying principle in this work is reality and its representation as nearly as possible. Reality is interpreted differently by each worker, of course, and one's concept of reality changes as new information is forthcoming. Taxonomic concepts in this work are derived from the study of the plants themselves, both in the herbarium and in the field. All contributors to the taxonomic text are familiar with the plants as they occur in nature, and that familiarity is transferred to the plants as they exist on herbarium sheets. In the natural context plant taxa occur in correlation with geography, geology, and climatic or positional factors. The correlations are important in determination of the nature of the taxon, and it is to these correlations that we have looked for understanding. The juxtaposition of taxa in the field (or the lack thereof) forms another parameter that must be determined, because many of the taxa form morphological intermediates when in contact. Hybridization and its potential must be regarded in taxonomic decisions. Presence of cleistogamy and apomixis within a group or complex must be considered. Other representations of reality, the floras, revisions, and monographs must be studied to supplement the information gained from field and herbarium studies. Thus, taxonomic concepts in this work are derived both from the study of the plants themselves and from the works of other students of the flora.

All indigenous plant species known to occur in Utah are included in the flora. Introduced plants are also covered, but not so intensively as are the native ones. Plant taxonomists have been conditioned to work with the indigenous plants alone, often to the exclusion of introduced ones. The introduced taxa are, however, a portion of the flora of the state and cannot be ignored. Many are established within native or human-induced habitats. Others are merely waifs that grow occasionally in the state. Cultivated plants offer still another group that has been largely neglected by botanists.

This flora attempts to present coverage of all established species, all common adventive taxa, and many of the commonly grown cultivated species. No attempt has been made to provide coverage for all the cultivated plants. There are a great many of them, and they are typically poorly represented in herbaria. Future workers should focus more attention on the cultivated flora. A more reasonable taxonomy of those plants will then be possible.

A taxon is based on the totality of its characters, but the character does not make the taxon. When workers have become intrigued with the value of a character they have been led to absurdity. Within this century fruit differences were used by one worker to separate the species of *Astragalus* into numerous genera, often with closely related and otherwise practically identical species placed in separate genera (e. g., *A. beckwithii* in *Phacomene* and *A. oophorus* in *Phaca*). The character must be considered in context with other features of the plant, with its geography, geological substrates, elevational range, features of slope, sympatry or allopatry, chromosome condition, phytochemistry, breeding mechanisms, and other items important to its interpretation.

Simple recombinant types should not be taken for taxa necessarily. Such unrelated examples might well appear again and again within a group, and keys written to such plants will provide means of identification of these recombinant types, but do they represent taxa? Therein lies perhaps the greatest dilemma of plant classification, i.e., the taxonomic character should reflect relationship, not merely a haphazard recombination of trivial features, but the pursuit of the truly reflective characteristics is often merely an ideal.

Ideally, taxonomic treatments represent a translation of the reality of nature into graphic representation. Difficulty arises, and taxonomists are held in disrepute, when they try to reverse the situation by imposing a system of their own creation onto the plants as they are in the real world.

And, the flora is only imperfectly understood. Many of the plant taxa are known from a few collections only, and it is impossible to represent them beyond the limits of the information available. Many taxa have been discovered recently and we know nothing of their chromosome compliments, breeding mechanisms, and other features. They occur in the treatments presented below as best as possible, but modification and reinterpretation will follow the accumulation of other data. Other workers will evaluate the same data differently, resulting in the presentation of restructured taxonomic opinions. Some will wish to dissect the genera into ever smaller units, based on various tangible and intangible features, as if such restructuring was of tremendous importance. All presentations will be based in fact, as the worker will interpret that fact, but the weighting of the data and the biases of the worker will determine, in some part, the presentation followed.

Prior to the present manuscript, which is acknowledged beforehand as imperfect and stained with the biases of its authors, routine identification of a collection of plants from anywhere in Utah has involved the use of as many as a dozen or more floras and monographs. The summary works of Tidestrom (1925) and Welsh and Moore (1974) functioned merely to tantalize the serious student of Utah plants. The local floras by Garrett (1909 and subsequent), Holmgren (1948, 1957) and Arnow et al. (1977, 1980), served the local areas more or less adequately, but they left students of surrounding areas grasping for tools for identification.

Literature used for determination of plant taxa depended on the locality where the plant was collected. Those taken from southern Utah were identified using the *Arizona Flora* (Kearney and Peebles 1951), those from eastern Utah were determined by use of the *Manual of the Plants of Colorado* (Harrington 1951) and others (Munz 1969; Correll & Johnston 1970; Cronquist et al. 1972, 1977, 1982; Dorn 1977; Hitchcock et al. 1955–1969). Specific groups were checked against the most modern revisionary works and monographs. Serious workers soon acquired a library of books and separates if their determinations were to be at least partially adequate. Now, for the first time, there will be one work that can be used for all but an unknown number of plants taken from within the state of Utah. A general bibliography of works used in identification of Utah plants previously and on which concepts of this flora are based in part is presented below. Specific citations are included with the taxonomic treatments.

GREAT BASIN NATURALIST MEMOIRS

A Utah Flora

Stanley L. Welsh[1], N. Duane Atwood[2], Sherel Goodrich[3], and Larry C. Higgins[4], editors

No. 9   Brigham Young University, Provo, Utah   1987

ABSTRACT.— A comprehensive treatment of the vascular flora of Utah is presented. Keys are provided to families, genera, species, and infraspecific taxa (when present). Taxa are described, ecological data is given, and geographical information is provided. County distribution in Utah is given for each species and infraspecific taxon. General geographical information is given for taxa that extend beyond the boundaries of Utah. Chromosome numbers are provided for each taxon, where that information was available in literature. Indigenous taxa include some 2,572 species and 355 infraspecific taxa, or a total of 2,927. Introduced species number some 580, and the total taxa treated in the flora is 3,507. New nomenclatural proposals include *Mertensia lanceolata* var. *nivalis* (Wats.) Higgins stat. nov.; *Chrysothamnus nauseosus* (Pallas) H. & C. var. *uintahensis* (L. C. Anderson) Welsh stat. nov.; *Machaeranthera canescens* (Pursh) Gray var. *latifolia* (A. Nels.) Welsh stat. nov., and var. *aristatus* (Eastw.) Turner comb. nov.; *Viguiera longifolia* (Robins. & Greenm.) Blake var. *annua* (Jones) Welsh comb. nov.; *Dudleya pulverulenta* (Nutt.) Britt. & Rose var. *arizonica* (Rose) Welsh stat. nov.; *Cordylanthus kingii* Wats. var. *densiflorus* (Chuang & Heckard) Atwood stat. nov.; *Cymopterus acaulis* (Pursh) Raf. var. *fendleri* (Gray) Goodrich comb. nov.

INTRODUCTION

Stanley L. Welsh

Historical Basis

Study of Utah flora began in September 1843, when John Charles Fremont took the first plant collections from the region later included within the state (Fremont 1845). In the late 1860s Sereno Watson collected in Utah and Nevada and authored the most important contribution to the understanding of Utah flora in the 19th century (Watson 1871). His work was used by such important Utah botanists as Marcus E. Jones, who published an important summary treatment of his own 1894 fieldwork in 1895 (Welsh 1982a). A. O. Garrett (1909 and subsequent) published the *Spring Flora of the Wasatch Region* early in the 20th century, which was used by generations of students to as late as midcentury. His work stimulated the subsequent publication of other local floras, including the *Handbook of the Vascular Plants of the Northern Wasatch* by Holmgren (1948, 1957), and the *Flora of the Central Wasatch Front, Utah* by Arnow, Albee, and Wycoff (1977, 1980).

Tidestrom (1925) wrote a summary treatment of the flora of Utah and Nevada, which has long been out of print and badly dated by changes in nomenclature and new information. Welsh and Moore (1973) attempted a more comprehensive treatment of the Utah flora. Their work involved some 2,500 taxa and included all of Utah.

Thus, the present work is not only a fruition of the years of labor by its editors and contributors, but it represents the culmination of more than a century and four decades of exploration and collection of the plants of Utah. Beginning with Fremont, the exploration has gone forward to the present, not in a steady rate but at a varied one as the personnel involved have waxed and waned. Hundreds of collectors have contributed specimens that form the basis of this flora. The few specimens taken by Fremont represent only a very small but significant part of the tremendous number of collections made in the intervening years (Welsh 1986).

Perhaps 400,000 specimens from Utah reside within the herbaria of the state, and possibly almost that many more examples from Utah are present in herbaria elsewhere. More than a thousand type specimens have been taken from within Utah (Welsh 1982a). There has been a renaissance within classical plant taxonomy in Utah during the past three decades. That rebirth of taxonomic emphasis was involved with the improvement in transportation and accessibility of Utah, in the increased number of personnel, in the devotion of those persons, in the environmental movement aided by federal laws, and in an increased emphasis for production of state and regional floras.

---

[1]Life Science Museum and Department of Botany and Range Science, Brigham Young University, Provo, Utah 84602.
[2]USDA Forest Service, Intermountain Region, Ogden, Utah 84401.
[3]Intermountain Forest and Range Experiment Station, Forest Service, U.S. Department of Agriculture, Ogden, Utah 84401. Present address: Vernal Ranger District, Ashley National Forest, Vernal, Utah 84078.
[4]Herbarium, Department of Biology, West Texas State University, Canyon, Texas. 79016.

CONTENTS

INTRODUCTION 1
Historical basis 1
Philosophical basis and species
concept 2
Methodology 3
Measurements 3
Nomenclature 3
Common names 4
Classification 4
Abbreviations 5
Terminology 5

DESCRIPTION OF THE ENVIRONMENT 6
Plant communities 7
Phytogeography 7
Modern setting and impacts 9

ACKNOWLEDGMENTS 10

GENERAL REFERENCES 10

GROUP TRACHEOPHYTA 12
Division **Lycopodiophyta** 13
Division **Equisetophyta** 14
Division **Polypodiophyta** 15
Division **Pinophyta** 24
Division **Magnoliophyta** 35

Class Magnoliopsida 35
Class Liliopsida 647

AUTHOR ABBREVIATIONS 824

GLOSSARY 834

INDEX 846

#	Family name	Author	Genera	Species	Intro	Infra
28.	Guttiferae	hig	1	2		
29.	Tiliaceae	slw	1	7	(intro 7)	
30.	Malvaceae	slw	10	25	(intro 7)	1
31.	Violaceae	slw	1	11	(intro 4)	
32.	Tamaricaceae	slw	1	3	(intro 3)	
33.	Frankeniaceae	slw	1	1	(intro 1)	
34.	Passifloraceae	slw	1	1	(intro 1)	
35.	Cucurbitaceae	hig	7	12	(intro 10)	
36.	Loasaceae	kht	3	20		
37.	Salicaceae	sg	2	35	(intro 10)	
38.	Capparaceae	slw	3	5		
39.	Cruciferae	slw	46	165	(intro 42)	39
40.	Resedaceae	slw	1	1	(intro 1)	
41.	Ericaceae	slw	5	12		
42.	Pyrolaceae	slw	4	9		
43.	Ebenaceae	slw	1	1		
44.	Primulaceae	slw	6	17	(intro 1)	
45.	Saxifragaceae	sg	13	41	(intro 4)	
46.	Crassulaceae	slw	3	6		
47.	Rosaceae	slw	33	111	(intro 45)	9
48.	Leguminosae	slw	46	244	(intro 51)	65
49.	Elaeagnaceae	slw	2	5	(intro 1)	
50.	Haloragaceae	slw	1	2		
51.	Lythraceae	higslw	3	4	(intro 1)	
52.	Punicaceae	hig	1	1	(intro 1)	
53.	Onagraceae	slw	10	59		9
54.	Cornaceae	slw	1	1		
55.	Garryaceae	hig	1	1		
56.	Santalaceae	slw	1	1		
57.	Viscaceae	nda	2	8		
58.	Celastraceae	slw	4	9	(intro 6)	
59.	Aquifoliaceae	slw	1	2	(intro 2)	
60.	Euphorbiaceae	higslw	6	28	(intro 5)	
61.	Rhamnaceae	slw	3	8	(intro 3)	
62.	Vitaceae	slw	2	5	(intro 3)	
63.	Linaceae	slw	1	6		
64.	Polygalaceae	slw	1	3		
65.	Krameriaceae	slw	1	2		
66.	Sapindaceae	slw	1	1		
67.	Hippocastanaceae	kht	1	4	(intro 4)	
68.	Aceraceae	slw	1	12	(intro 9)	
69.	Anacardiaceae	slw	5	8	(intro 5)	
70.	Simaroubaceae	hig	1	1		
71.	Meliaceae	hig	1	1	(intro 1)	
72.	Rutaceae	higslw	2	2		1
73.	Zygophyllaceae	slw	4	4	(intro 1)	
74.	Oxalidaceae	slw	1	2	(intro 2)	
75.	Geraniaceae	slw	3	9	(intro 1)	
76.	Limnanthaceae	slw	1	1		
77.	Tropaeolaceae	slw	1	1	(intro 1)	
78.	Araliaceae	slw	3	4	(intro 3)	
79.	Umbelliferae	sg	25	70	(intro 5)	10
80.	Gentianaceae	hig	5	18		
81.	Apocynaceae	hig	5	10	(intro 3)	1
82.	Asclepiadaceae	hig	3	20	(intro 1)	
83.	Solanaceae	hig	11	33	(intro 14)	
84.	Convolvulaceae	hig	5	8	(intro 4)	
85.	Cuscutaceae	gib	1	12	(intro 1)	
86.	Menyanthaceae	slw	1	1		
87.	Polemoniaceae	slw	13	67	(intro 3)	10
88.	Hydrophyllaceae	nda	9	49		6
89.	Boraginaceae	hig	17	97		
90.	Verbenaceae	hig	3	8		
91.	Labiatae	hig	24	43	(intro 14)	
92.	Hippuridaceae	slw	1	1		
93.	Callitrichaceae	slw	1	3		
94.	Plantaginaceae	hig	1	7	(intro 2)	
95.	Buddlejaceae	hig	1	2	(intro 1)	
96.	Oleaceae	slw	6	19	(intro 13)	27
97.	Scrophulariaceae	nda/ecn	18	137	(intro 12)	
98.	Orobanchaceae	hig	1	5		
99.	Pedaliaceae	hig	1	1		
100.	Bignoniaceae	hig	3	5	(intro 4)	
101.	Lentibulariaceae	slw	1	2		
102.	Campanulaceae	slw	6	10	(intro 3)	
103.	Rubiaceae	higslw	4	12	(intro 1)	1
104.	Caprifoliaceae	slw	5	14	(intro 5)	
105.	Adoxaceae	slw	1	1		
106.	Valerianaceae	hig	4	7	(intro 2)	
107.	Dipsacaceae	hig	1	1	(intro 1)	
108.	Compositae	slw	113	481	(intro 45)	96

Liliopsida

#	Family name	Author	Genera	Species		Infra
1.	Alismaceae	nda	3	4		
2.	Hydrocharitaceae	nda	1	2		
3.	Juncaginaceae	slw	1	3		
4.	Potamogetonaceae	hig	1	19		
5.	Rupiaceae	slw	1	1		
6.	Najadaceae	nda	1	3		
7.	Zannichelliaceae	slw	1	1		
8.	Palmae	slw	2	2		
9.	Lemnaceae	nda	3	8		
10.	Commelinaceae	nda	1	1		
11.	Juncaceae	sg	2	28		2
12.	Cyperaceae	sg	8	120		2
13.	Gramineae	laa	83	254	(intro 67)	8
14.	Sparganiaceae	nda	1	4		
15.	Typhaceae	nda	1	3		
16.	Liliaceae	slw	27	54	(intro 16)	
17.	Amaryllidaceae	slw	3	6	(intro 6)	
18.	Iridaceae	slw	4	11	(intro 6)	
19.	Agavaceae	hig	3	11		2
20.	Orchidaceae	hig	9	23		3
	Totals		765	3152	(intro 580)	355

In the present work there are 2,572 species and 355 infraspecific taxa, or a total of some 2,927 indigenous taxa. Introduced species treated in the work include some 580 species. The total for the flora is 3,507 taxa.

Abbreviations

Abbreviations have been used sparingly, the exceptions mainly being the names of authors of the various families or genera in the text. Those author abbreviations in the list above are ecn (Elizabeth Chase Neese), gib (Garry Innes Baird), hig (Larry Charles Higgins), kht (Kaye Hugie Thorne), laa (Lois A. Arnow), nda (Nephi Duane Atwood), sg (Sherel Goodrich), and slw (Stanley Larson Welsh).

Abbreviations include those for metric units of measure cited above and the following: mi (mile), ca (circa, about, almost), U. S. (United States), U. S. S. R. (Soviet Union), nom. illeg. (illegitimate name), sens. lat. (sensu lato, in a broad sense), sens. latissimo (in the broadest sense), sens. str. (sensu stricto, in a strict sense), ssp. (subspecies), var. (variety), and f. (forma, except following the name of an author of a taxon, in which case it stands for filial or junior). Synonyms are shortened by abbreviation of generic names, except where such abbreviation would be misleading. In some instances the specific and infraspecific names have been abbreviated, especially where they merely repeat the name in current use.

Terminology

Descriptive terminology is mainly standard for the groups involved. Terms are defined in an appended glossary.

DESCRIPTION OF THE ENVIRONMENT

Utah is situated largely between 37 and 41 degrees north latitude and 108 to 114 degrees west longitude. It occupies some 219,990 square km. Elevation ranges between about 665 and 4,126 m, with the lowest elevation being in Beaver Dam Wash in Washington County and the highest at the summit of Kings Peak in the Uinta Mountains (astride the Duchesne-Summit County line).

The state is divided roughly into two halves by the mountain-plateau axis formed by the mountains of the Wasatch Range and the plateaus of central and southern Utah. West of the mountain and plateau system lies the Great Basin, a portion of the Basin and Range Province, whose drainages have no outlet to the sea. East of the dividing line, water falls ultimately to the Colorado River through a large number of tributaries. The Weber, Bear, and Sevier rivers arise to the east of the northeast-southwest trending axis but fall ultimately to the west through canyons that cut the major ranges. The two portions of the state are also roughly divided by the margin of the overthrust belt that extends into Utah from Idaho and exits the state into Nevada along a line that marks essentially the southern edge of the Great Basin. The geology is fundamentally different between the two halves.

The Basin and Range Province is marked with numerous, north-south trending, essentially parallel ranges. The ranges are the result of block faulting and expose strata of sandstone or quartzite, limestone, and dolomite that vary from Precambrian to Permian age (Hintze 1972). The mountain ranges stand with their bases buried in their own detritus. Valleys are filled with the products of erosion. The valley bottoms are frequently occupied by saline pans, salt flats, or fresh to saline lakes or ponds. Slopes above the pans to the mountains proper consist of complex assemblages of alluvial fans coalesced into gently to moderately sloping bajadas. Alluvium is graded in size down slope, with the coarser materials near the montane sources and finer accumulating in the bottoms. Salt is likewise carried to the valley bottoms, where it is abandoned by the moisture that served to transport it there.

Sand is sorted by the prevailing westerly winds and tends to stack along the east sides of valleys. Dunes are present especially in the Sevier Desert (the Lynndyl Dunes are the largest), but they occur also in small portions of many of the western valleys.

Valleys that drain from the Great Basin ranges tend to be short on the west side and drop steeply to the valley floor. Those on the east sides tend to fall less steeply and to be longer. Perennial streams are few, and those frequently disappear into the valley bottoms. Springs are relatively numerous but small. Exceptions are found especially in Snake Valley and the Sevier Desert. Large lakes tend to be saline (Great Salt Lake) or fresh (Utah Lake). Sevier Dry Lake contains water only during exceptionally wet weather cycles.

Because of the closed hydrological system in the Great Basin, at least as far as surface flow is concerned, the geological and topographic influence on vegetative cover is significantly different from that of the Colorado Drainage Basin. The Colorado drainage is an open system, where water flows from the basin to the sea, carrying with it alluvium that would have otherwise persisted as valley fill, alluvial fans, and bajadas. Old pedimental surfaces, consisting of rounded cobble and gravel, persist for a time, perched on ancient erosional surfaces now abandoned by contemporary drainages cut below them. An exception to the low amount of valley fill in the Colorado drainage system is found where that drainage impinges upon the Basin and Range Province, as it does in the southwestern portion of Utah in the Beaver Dam Wash vicinity. There great accumulations of fanglomerate are now exposed along the entrenched drainage of Beaver Dam Creek.

Geological strata are exposed over vast areas of the Colorado Plateau, either with the raw stratum or weathered colluvium as the surficial material. Large alluvial deposits are not typical of the landscape. Each particle removed by erosion is being moved down river to the sea.

Exposed geological strata in the Colorado drainage system consist of shale, mudstone, siltstone, sandstone, and conglomerate sequences. They vary in age from Precambrian to Tertiary, with late Tertiary, Pleistocene, and Recent (Hintze 1972) alluvial deposits being either perched or along drainage bottoms in valleys. Sand dunes are present, especially on the foot of the eastern slope of the San Rafael Swell, near Kanab, and west of Hurricane.

Plant species distributions tend to be correlated with geology, especially where the geological strata are exposed essentially unmodified at the surface. The distributions are likewise secondarily influenced by the presence of alluvium and soils, their texture, salinity, and derivation. Hence, it is not possible to understand plant taxonomy without consideration of the substrate occupied by the particular plant being investigated. The raw and colluvial surfaces available for plants in the Colorado drainage system present an entirely different array of possibilities than in the Great Basin. The basic stratigraphies of the two areas are different—the Great Basin gray and mainly calciferous, the Colorado Basin varicolored and of greater diversity of substrate. Much of the vegetation of the Great Basin is on alluvium, whereas most of that in the Colorado system is directly on the geological strata. Mountains proper in the Great Basin offer exposures of strata, but often the valleys do not.

Large areas of the Great Basin and mountains and high plateaus of Utah bear surficial deposits of igneous material, either intrusive or extrusive in origin. These igneous outcrops offer a not so subtle control of the vegetative cover. Especially is this control demonstrated on the complex igneous outcrops of the Marysvale volcanic centrum, which displays a great thickness and diversity of substrates from low to high elevations. Further, portions of the igneous rocks have been again modified by thermal activities, and this thermal modification adds to the number of habitat types.

Survivors on the raw or colluvial surfaces in the Colorado system tend to be specialists, those plant taxa that can survive on the harsh substrates without serious competition. Plants of a similar nature in the Great Basin tend to occupy the mountain ranges proper, with exceptions mainly occurring on low-elevation exposures. Alluvium tends to be occupied by generalists, those plants that can tolerate a wider degree of habitat variation and more persistent competition by other generalists. Even a thin cover of detritus or alluvium over geological strata in the Colorado Basin, especially, is sufficient to allow for vegetative composition to change abruptly. Where alluvial fans thin toward their edges, they are as effective in insulating the underlying geological strata from plants as where thicker toward the mouth of a draw or canyon.

Insulation of plants from parental materials of geological strata can result from soil development or from burying with alluvium. Soil development is a function, at least in part, of the amount of water available. In Utah, alluvium is produced at all elevations, but water from precipitation increases with elevation. And, since temperature tends to decrease at a uniform rate (approximately 1.8 degrees C per 305 m), evaporation is lower at the higher elevations and more moisture is available for both plant growth and soil development. Soils, like alluvium, tend to be occupied by generalists. However, despite the occurrence of high precipitation at higher elevations, there are places where the larger amounts of water are ineffective. Rock outcrops, cliffs, margins of plateaus, and ridge tops, because of lack of penetration, tend not to be well watered in spite of the large amount. Endemism in Utah is correlated with the availability of raw or colluvial substrates (Welsh 1978a), and these are present in broad, low- to high-elevation expanses in the Colorado Basin and plateaus—but mainly in the Great Basin they are montane.

Plant Communities

Plant community descriptions, for the purposes of this work, have been simplified. Communities have been determined subjectively by the plant collector, who recorded that information on specimen labels, and by the individual author of a specific group. Generally the plant communities are characterized by one or more obvious, typically woody species, growing in any given area. Grasslands, those areas dominated mainly by grass species, occur widely in the state, but the total area occupied by them, when compared to that dominated by woody and herbaceous species together, is not large. In some instances it has been useful to give a general designation to groups of communities within a given elevational and broad vegetational type. In general, the following groupings of plants and some of their most obvious habitat species are treated, starting at low elevation and moving upward.

WARM DESERT SHRUB.—Creosote bush, Joshua tree, blackbrush, cholla, desert willow, burro-bush, sand (old man) sagebrush, and *Hilaria rigida*. Hot, dry slopes, rock outcrops, and drainages of Washington, Kane, and San Juan counties support most of the warm desert shrub vegetative type in Utah.

CHAPARRAL.—Turbinella live-oak, *Ceanothus greggii*, *Arctostaphylos pungens*, and *Garrya flavescens*. Chaparral is mainly developed in the Pine Valley and Bull Valley mountains of Washington County.

SALT DESERT SHRUB.—Shadscale, mat saltbush, Castle Valley saltbush, greasewood, seepweed, ephedra, *Eriogonum corymbosum*, enceliopsis, zuckia, salt grass, and arctic rush. Saline valleys and geological strata, mainly at lower elevations are occupied by this community type.

RIPARIAN COMMUNITIES.—Cottonwood, birch, alder, box elder, red-osier dogwood, horsetail, willow, tamarix, *Forestieria*, rabbitbrush, greasewood, common reed, saltgrass, Nebraska sedge, and arctic rush. Riparian communities occur at practically all elevations in Utah.

HANGING GARDENS.—Maidenhair ferns, Eastwood monkeyflower, cardinal monkeyflower, cave primrose, helleborine orchid, golden sedge, and many others. Hanging gardens occur along exposed water sources, mainly in sandstone along canyon walls in the southern portion of the state.

SAND DESERT SHRUB.—Sand (old man) sagebrush, *Vanclevea*, *Eriogonum leptocladon*, *Sporobolus* species, *Yucca* species, Indian ricegrass, *Psoralidium junceum*, *P. lanceolatum*, and *Amsonia tomentosa*. Sandy areas are mainly at lower elevations in the valleys of western and southeastern Utah.

COOL DESERT SHRUB.—Big sagebrush, black sagebrush, winterfat, rabbitbrush, blue grama, galleta, Indian ricegrass, and dropseed species.

PINYON-JUNIPER OR JUNIPER-PINYON.—Pinyon pine, Utah juniper, big sagebrush, black sagebrush, mutton grass, and needle and thread. This vegetative type occurs over vast areas of the state.

MOUNTAIN BRUSH.—Gambel oak, bigtooth maple, service berry, mountain mahogany, and big sagebrush. Large areas are occupied by mountain brush, mainly on foothills, but also in canyons at lower elevations.

PONDEROSA PINE.—Ponderosa pine, manzanita, and aspen. This community type is present on the drier sites in lower aspen communities and more moist portions of the mountain brush vegetative type. Often it occurs on acidic substrates.

FIR.—White fir, Douglas fir, mixed mountain brush, and aspen. The fir community is mainly restricted to north-facing slopes at moderate elevations in mountains.

ASPEN.—Aspen, mixed mountain brush, big sagebrush, silver sagebrush, and transitional to the next. The aspen vegetative type is very important in the mountains of the state, often forming openings or parklands dominated by grass, grass-forb, and tall forb communities.

SPRUCE-FIR.—Engelmann spruce, alpine fir, lodgepole pine, grass-forb, and sedge-forb. Large areas of spruce-fir forest are present in the mountains of the state, often forming patches around openings dominated by grass-forb, sedge-forb, or tall forb communities. Timberline or tree line occurs at the upper margin of the spruce-fir type.

ALPINE TUNDRA.—Various grass (*Festuca*, *Poa*), forb, (*Geum rossii*, *Arenaria*, *Stellaria*, *Silene*, *Dryas*), sedge (*Carex*, *Kobresia*), and shrub (*Salix*, *Vaccinium*) species at or above timberline. Alpine tundra consists of low-growing plants present in most high-elevation sites in Utah, mainly above 3,355 m elevation. The sites are cold, dry, and exposed to long light periods. They are exposed to proportionally greater amounts of ultraviolet radiation than are the other communities, which are buried beneath progressively more atmosphere downward in elevation.

Phytogeography

Phytogeographic considerations in Utah involve the concepts of floras previous to the recent past, migrational pathways, and development of species in place (the neoendemics). Evolution of the floras of the Intermountain Region have been reviewed by Tidwell, Rushforth, and Simper (1972) and will not be considered further here.

Migration is a principal means of enrichment or depauperation of a flora. Discussion of depauperation through outward migration is essentially moot, unless there is some evidence of a species having existed within an area in the first place. Enrichment is more easily demonstrated because the plants now exist within the area under review. Taxa tend not to move as floras, but to move as individuals. Each individual (if not the taxon) evidently has a set of limits that controls where it can grow and a usually finite means of dispersal. Hence, plants tend to move within the habitat available to them along the lines of least resistance. Where such lines of least resistance are available to several or many taxa, as determined by coincidence of distribution patterns, migration routes are judged to exist.

In Utah there are several main patterns of distribution, each correlated in some degree with one or more migration pathways. Hot dry canyon slopes along the major river systems serve as routes for movement of propagules, especially upward along the lower portion of the canyons from the dry desert country to the south. The redbud (*Cercis occidentalis*) is such a plant, and the pattern can be designated as the redbud pathway. Others with the redbud pattern include numerous Mohavean plants. Two main prongs of this migration pathway occur in Utah, one along the Virgin River and another along Glen Canyon. The arms of this pathway are separated by the highland between the Hurricane fault and the East Kaibab monocline at the Cockscomb (the so-called Dixie corridor). The Utah agave (*Agave utahensis*) and *Thamnosma montana* represent slightly shorter extensions of the redbud type, both occurring on the canyons in both Washington County and in or near Kane County along the canyons of the Colorado. Other Mohavean representation is that characterized by Joshua tree, creosote bush, stinking gourd, and their associates of lower elevations in Washington County. Above them elevationally is developed a chaparral type, dominated by turbinella live oak (*Quercus turbinella*), *Arctostaphylos pungens*, and *Ceanothus greggii* that is unmatched elsewhere in Utah. Both the oak and the *Ceanothus* (and the stinking gourd) are also sparingly represented along canyons of the Colorado in San Juan County. More broadly distributed plants of Mohavean representation include the cholla cacti as a group, and *Opuntia whipplei* more especially. This plant occurs sparingly along the drainages of the San Juan and Colorado rivers as far north as Grand County, and to southern Millard County in the Great Basin.

This latter extension is matched, at least in part, by the extension of *Prunus fasciculata* and extended beyond by *Lycium andersonii*, both principally Mohavean in distribution. Other plants with this extension into the southern Great Basin from the Colorado River Drainage, but not necessarily Mohavean in general distribution, include Bigelow sagebrush (*Artemisia bigelovii*), *Astragalus mollissimus* var. *thompsonae*, and *Astragalus lonchocarpus*. *Nitrophila occidentalis* and *Solidago spectabilis* are other extensions of the Mohavean type that occur in the southern Great Basin portion of Utah. The migration pathway for these plants, designated as the *whipplei* extension, seems not to have been directly across the intervening land between the hot deserts of Washington and Clark counties but up the drainages leading northward from the Virgin, especially up Meadow Valley Wash in Nevada and Beaver Dam Wash of Arizona, Utah, and Nevada.

Evidence of movement of plants from higher elevations to lower elevations along the canyon systems is supported by the presence of plants typical of high elevations in the canyon systems far below their usual region of growth. Aspen, Douglas fir, Rocky Mountain juniper, Utah juniper, and many other species of mountains occur along the canyons. Some of these instances are supported by lone waifs that are away from the main body of the species. However, the canyons tend to act as inverted mountains, being shaded for a much longer portion of the day and having, as a result of the cooling shade, more available water than the surrounding arid lands; the habitats simulate those of the mountains.

Habitat for hydrophytes along the waterways of Utah demonstrate the ease with which propagules move from place to place along them. Birds and other animals, including man, have moved along the waterways because of the ease the low-elevation corridors provide.

The Great Basin, despite its north-south trending ranges that make movement for people difficult in an east-west direction, is essentially open to movement of propagules at low elevations. *Astragalus callithrix*, *A. uncialis*, and *Swertia gypsicola* are examples of low-elevation Great Basin endemics with representation in both Utah and Nevada. Movement at higher elevations is fraught with difficulties involving the transfer of propagules across broad, low-elevation, xeric valleys, and most of the mountain ranges support one or more narrow endemics.

Influencing the northern portion of the Great Basin is a route for movement of plants from the Snake River region of the Pacific Northwest. *Populus trichocarpa* is an example of the Snake River Plains extension into Utah. Plants of the Raft River Mountains tend to share affinities with the Pacific Northwest and belong to the *trichocarpa* example. An expansion of the *trichocarpa* type of distribution is found in *Astragalus purshii*, which occurs from the west and northwest into Box Elder, Rich, and Uintah counties. *Astragalus lentiginous* var. *platyphyllidius* has about that same distribution pattern but is missing from the Uinta Basin proper (the only place in Utah where *A. lentiginosus* sens. lat. is not known to occur).

The mountain and plateau axis formed by the Wasatch Range and its mountain chains, the Wasatch Plateau, and the Sevier-Paunsaugunt Plateau form an elevated route along which boreal plants have migrated. The boreal forest trees are examples of plants with access along the mountain ranges. Other examples include *Chamaerhodos erecta*, which reaches its southern limits in the Tushar and Thousand Lake mountains vicinity. *Astragalus australis* and *Oxytropis viscida* show a similar pattern, but with more disjunct colonies northward along the mountain-plateau axis almost to the Uinta Mountains. *Dryas octopetala* is disjunct in the Uinta Mountains from its usually more northern distribution. The forest trees, especially, have evidently been able to move across the distances between elevated ranges of the Great Basin and the isolated mountains of the Navajo Basin in the Colorado Plateau. Evidence of movement at high elevation from the main ranges of the Rocky Mountains is indicated by ranges of plants in the La Sal and less commonly in the Abajo mountains (e. g., *Draba spectabilis*). *Erigeron vagus* shares a distribution that includes not only the La Sal Mountains but portions of the Sevier-Paunsaugunt Plateau.

Several steppe and prairie elements show migrational pathways from the high cold steppes of Wyoming and from New Mexico and Arizona. Both *Astragalus adsurgens* and *A. spatulatus* show a pattern from Daggett County to the high plateaus of the Sanpete-Emery divide.

Westward, the steppe flora (in depauperate form) sweeps to the crest of the Bear River Range. The diminution of species is marked, even in the Utah portion of the steppe flora. *Astragalus gilviflorus* gives way at a much lower elevation than does *A. spatulatus*, east of the Bear River divide in Rich County. And, possibly *Bouteloua gracilis* in the Uinta Basin shares the *spatulatus-adsurgens* route from the cold steppes.

Elsewhere in the state, blue grama grass appears to have southern affinities, as do other prairie grasses (e.g., *Sorghastrum nutans*, *Panicum virgatum*, *Andropogon gerardii*, and *Schizachyrium scoparium*). For the most part, the tall prairie grasses tend to occur (probably relictually) in mesic to hydric canyon bottoms, or they cluster around seeps and springs (often in or near hanging gardens).

Hanging garden plants show diverse phytogeographic relationships. The cave primrose apparently has affinities with boreal species, far to the north of the distribution of this uniquely adapted species. The main area of distribution of saw grass (*Cladium californicum*) is in warm temperate and subtropical portions of southern North America and central America (Welsh & Toft 1981).

Thus, the flora of Utah has alliances with many portions of the world. Its understanding must be based on the relationship with those other floras.

Modern Setting and Impacts

The occupation of Utah by mankind has not been without cost to the plant life present in the region. The inroads of modern civilization, mainly in the past 140 years, have been great indeed. Indigenous plants have been displaced by agriculture, buildings, roads, cities and towns, and other appurtenances of this civilization. Valley bottoms and other lands near to water were occupied at once following initial ingress into this area. Very little is known, as a consequence, of the original vegetational composition, although the species present are generally understood through examination of protected relicts.

Meadowlands were similarly occupied and many of them have been modified through agronomic and pastoral processes. Much of the meadowland, that still existing, has been subjected to intensive grazing for more than a century. Also, bottom and meadowland early became private property. The combination of private ownership and heavy utilization for grazing and other land use did much to discourage study and collection of plant species. These lands are still only imperfectly known from a botanical standpoint.

Lands considered as suitable for grazing included most of Utah, and much of the state was finally subjected to heavy grazing. Despite the utilization by grazing animals, with their significant impact on spring, seep, and meadowlands, the most important changes in the flora of Utah due to such use appear to belong to two categories—the displacement of species (with attendant reduction in area) and the increase of less palatable plants.

Crop and row agriculture provided another impact on the indigenous vegetation. The native species were eradicated from large areas (about 4.5%) of Utah and replaced by cultivated plants and numerous weeds. Introductions such as cheatgrass and Russian thistle are familiar components of our flora, occupying huge areas in the valleys and foothills over much of Utah. Bindweed and creeping or Canada thistle are pestiferous weeds of consequence that were introduced from the Old World. The poisonous plant halogeton was first discovered in Utah in the mid-1930s, and it now covers much of the sheep winter range in the state.

Like the crop and row agriculture, the physical appurtenances of civilization—the cities, towns, highways, railways, and industrial structures—tended to be placed in valley bottoms, on alluvium or valley fill, thereby reducing the total area occupied by indigenous plant communities. Although little information of a descriptive nature exists in the historical records, it can be inferred that the plant species displaced in these regions were generalists, plant species that still exist in large areas of Utah.

Certain kinds of industrial development, especially ferrous and nonferrous metal mines, tend to be placed in regions where plant specialists, the endemics, might be expected. Except for the open pit operations, those begun prior to any attempt at searching for rare plants, the surface operations for underground mines occupy only small areas, and their total impacts are relatively small. This is true, fortunately, for coal mines generally in Utah. Only small areas have been subjected to open pit or strip mining techniques. Strip mining should be considered as a last-resort operation, since the strippable land is mainly in arid, low-elevation sites that potentially support endemic plant species.

Recent urban sprawl, especially of cities in the Colorado drainage system, has been forced on to colluvia or raw geological substrates where plant specialists occur. An example is to be found in Washington County where urban sprawl and its attendant human activities impinge directly upon the very limited habitat of the bearclaw poppy.

Other modern industrial expansion that might harmfully influence the narrowly restricted specialists includes electrical transmission lines and structures, substations, highways, pipelines, well sites, and military facilities. These features tend to cross geological outcrops indiscriminately, and their impingement on the nonrandomly placed rare plants is thereby possible.

Federal law, especially the Endangered Species Act of 1973, as amended, provides stipulations for protection of restricted plant specialists. Enforcement of those stipulations should guarantee minimal impact and continuity of the numerous restricted taxa in Utah.

Despite all the impacts of civilization, there is no substantiated account of the extinction of any plant species in Utah during its first 140 years. It is hoped that the same will be said following the next century.

Nevertheless, the net results are cumulative, and the flora of Utah has given way to the impingements of humanity. No part of the state is unchanged, and the potential for increased reduction of native flora is great. And, waiting in the Old World are dozens, if not hundreds, of aggressive, weedy species that could be introduced to Utah inadvertently. During each of the past several decades new and costly importations have reached Utah. Musk or nodding thistle, winged or Scotch thistle, and star-thistle are examples of plants that have spread over Utah in the period from 1950 to the present. More will assuredly follow.

ACKNOWLEDGMENTS

This flora project, begun at Brigham Young University in 1960, has a larger counterpart in the Intermountain Flora Project of the New York Botanical Garden. Rather than being competitive, although exclusive in personnel and organization, the projects have been mutually collaborative and beneficial. Drs. Arthur Cronquist, Noel Holmgren, Patricia Holmgren, James Reveal, and Rupert C. Barneby have aided in identification and verification of critical material, have provided information on nomenclature, have given locality data and information on new state records, allowed prior publication of nomenclatural combinations and new taxa, and have given encouragement and help at all times. Exchange specimens, both those sent and those received, have been useful to both projects. We thank them for their help.

Specimens in critical groups have been routinely sent to specialists for verification and/or identification. Groups and specialists are as follows:

Compositae	Arthur Cronquist
Cirsium	Gerald B. Ownbey
Boraginaceae	Larry C. Higgins
Leguminosae (general)	Duane Isely
Astragalus (and other critical legumes)	Rupert C. Barneby
Lupinus	David B. Dunn
Allium	Dale B. McNeal
Chrysothamnus	Loran Anderson
Penstemon	Noel Holmgren
Umbelliferae	Lincoln Constance
Various groups	John Thomas Howell
Eriogonum	James L. Reveal
Loasaceae	H. J. Thompson
Gramineae	Lois A. Arnow
Cruciferae	Reed C. Rollins
Hydrophyllaceae	N. Duane Atwood
Cyperaceae	Sherel Goodrich

Special thanks are hereby tendered to Drs. Dorald L. Allred and Stephen L. Wood of the Life Science Museum at Brigham Young University for their help and encouragement. Drs. Bruce Smith and Jerran Flinders gave financial assistance and allowed time for curation and collection of materials. Many people have assisted with collection and identification of plant materials. Among them are Mont Lewis, Kaye Thorne, Sherel Goodrich, Drs. Glen Moore, Joseph R. (Dick) Murdock, Bertrand F. Harrison, Walter P. Cottam, Arthur Holmgren, Bassett Maguire, Elizabeth Neese, N. Duane Atwood, L. C. Higgins, Steven L. Clark, and Richard Shaw.

Kaye Thorne is here singled out for special thanks. Her duties at the herbarium of Brigham Young University, huge by any standard, were not an obstacle to helping with the completion of the flora. She wrote two family treatments, arranged for loans, provided illustrations of new taxa discovered, and did myriads of other tasks.

Many of the early collectors and avid botanists have passed from the scene. Included are Seville Flowers, Albert Osbun Garrett, Per Axel Rydberg, Marcus Eugene Jones, Sereno Watson, and John Charles Fremont. Their contributions are here gratefully aknowledged.

The writers are especially grateful to the new generation of collectors, including Gary Baird, Merton A. (Ben) Franklin, Joel Tuhy, Betsy Neeley, and Ronald Kass. Beverly Albee is here given special thanks for the loan of distribution maps showing specimens from the herbaria of the University of Utah and Utah State University. The curators of the various herbaria are acknowledged for their help and suggestions.

Collaboration of officials of the various national parks and monuments, who so kindly granted collecting permits, is also acknowledged. The assistance of Larry Hays at Zion National Park, Jim Holland, Vic Vierra, and Ron Sutton at Glen Canyon National Recreation Area, and numerous people, including Superintendent Pete Perry of Canyonlands National Park was particularly helpful. The former superintendent of Arches, Natural Bridges, and Canyonlands, the late Bates Wilson, gave help and encouragement during the early years of fieldwork leading to this flora. Eleanor Inskip of the Canyonlands Natural History Association arranged for float trips through Cataract Canyon in collaboration with the National Park Service. Numerous federal and state agencies have collaborated in allowing and encouraging the collection of critical material. Donna House of the Navajo Heritage program aided in exploration of Navajo lands in Utah.

Thanks are given to Jody Chandler, subject matter librarian at the M. L. Bean Life Science Museum, branch library, for her search of literature for chromosome numbers and for the citation of type locality information. Kathryn Taylor Mastin read the manuscript for typographical errors and uniformity, especially of author abbreviations, and she assisted with preparation of the author list. The glossary was written by N. D. Atwood and supplemented by all other contributors. The index of scientific names was formulated initially by N. D. Atwood and S. Goodrich. Common names were extracted from the text by Wendy Jefferies.

GENERAL REFERENCES

References of a specific nature accompany many of the families and genera in the taxonomic treatment. Many other references have been used in a general way, and their use has contributed to the value of the flora. These general references, many not cited in the text directly, are presented here.

ARNOW, L., B. ALBEE, AND A. WYCKOFF. 1977, 1980. Flora of the central Wasatch Front, Utah. Utah Mus. Nat. Hist., Salt Lake City. 662 pp.

ATWOOD, N. D., S. GOODRICH, AND S. L. WELSH. 1984. New *Astragalus* (Leguminosae) from the Goose Creek drainage, Utah-Nevada. Great Basin Nat. 44: 263–264.

ATWOOD, N. D. AND S. L. WELSH. 1985. New species of *Talinum* (Portulacaceae) from Utah. Great Basin Nat. 45: 485–487.

BAILEY, L. H. 1951. Manual of cultivated plants. The MacMillan Company, New York. 1116 pp.

BAILEY, L. H. AND E. Z. BAILEY. 1976. Hortus third. (Revised and expanded by the staff of the Liberty Hyde Bailey Hortorium). MacMillan Publishing Co., Inc., New York. 1290 pp.

BARNEBY, R. C. AND S. L. WELSH. 1985. New species of *Astragalus* (Leguminosae) from southeastern Utah. Great Basin Nat. 45: 551–552.

BENSON, L. 1979. Plant classification. 2nd ed. D. C. Heath and Co., Lexington, Massachusetts. 901 pp.

BOR, N. L. 1960. The grasses of Burma, Ceylon, India, and Pakistan. Pergamon Press, London. 767 pp.

BROTHERSON, J. D., L. A. SZYSKA, AND W. E. EVENSON. 1980. Poisonous plants of Utah. Great Basin Nat. 40: 229–253.

CHATTERLEY, M., B. T. WELSH, AND S. L. WELSH. 1982. Preliminary index of authors of Utah plant names. Great Basin Nat. 42: 385–394.

CHRISTENSEN, E. M. 1967. Bibliography of Utah botany and wildland conservation. Brigham Young Univ. Sci. Bull. Biol. Ser. 9(1): 1–136.

CHRISTENSEN, E. M. AND S. L. WELSH. 1963. Presettlement vegetation of the valleys of western Summit and Wasatch counties, Utah. Proc. Utah Acad. 40: 200–201.

CORRELL, D. S. AND M. C. JOHNSTON. 1970. Manual of the vascular plants of Texas. Texas Research Foundation, Renner, Texas. 1881 pp.

CRONQUIST, A. 1981. An integrated system of classification of flowering plants. Columbia University Press, New York. 1262 pp.

CRONQUIST, A., A. H. HOLMGREN, N. H. HOLMGREN, AND J. L. REVEAL. 1972. Intermountain flora. Vol. 1. Hafner Publishing Company, Inc., New York. 270 pp.

CRONQUIST, A., A. H. HOLMGREN, N. H. HOLMGREN, J. L. REVEAL, AND P. K. HOLMGREN. 1977. Intermountain flora. Vol. 6. Columbia Univ. Press, New York. 584 pp.

DORN, R. D. 1977. Manual of the vascular plants of Wyoming. Garland Publishing, Inc., New York. Vol. 1 and 2. 1498 pp.

FREMONT, J. C. 1845. Report of exploring expedition to the Rocky Mountains in the year 1843, and to Oregon and North California in the years 1843–44. Washington D.C.

GARRETT, A. O. 1936. Spring flora of the Wasatch region. Ed. 5. Stevens & Wallis, Inc., Salt Lake City. 240 pp.

GLEASON, H. A. AND A. CRONQUIST. 1963. Manual of the vascular plants of the northeastern United States and adjacent Canada. Van Nostrand Reinhold Co., New York. 810 pp.

GOODRICH, S. AND S. L. WELSH. 1983. New variety of *Stephanomeria tenuifolia* (Compositae) from Utah. Great Basin Nat. 43: 373–374.

GOULD, F. W. 1951. Grasses of the southwestern United States. Univ. Ariz.. Biol. Sci. Bull. 7: 1–352.

_____. 1975. The grasses of Texas. Texas A & M Press. College Station, Texas. 653 pp.

GOULD, F. W., AND R. MORAN. 1981. The grasses of Baja California, Mexico. San Diego Society of Natural History, San Diego. 140 pp.

GOULD, F. W. AND R. B. SHAW. 1983. Grass systematics. 2d ed. Texas A & M Press, College Station. Texas. 397 pp.

GRAHAM, E. H. 1937. Botanical studies in the Uinta Basin of Utah and Colorado. Ann. Carnegie Mus. 26: 1–432.

HALLIDAY, P., R. D. MEIKLE, J. STORY, AND H. WILKINSON. 1980. Draft index of author abbreviations compiled at the herbarium, Royal Botanic Gardens, Kew. Her Majesty's Stationer Office. 249 pp.

HANSON, A. A. 1972. Grass varieties in the United States. U.S. Dept. Agric. Handbook No. 170: 1–124.

HARRINGTON, H. D. 1954. Manual of the plants of Colorado. Sage Books, Denver. 666 pp.

HARRISON, B. F., S. L. WELSH, AND G. MOORE. 1964. Plants of Arches National Monument. Brigham Young Univ. Sci. Bull. Biol. Ser. 5 (1): 1–23.

HINTZE, L. F. 1972. Geologic history of Utah. Brigham Young University, Dept. of Geology. Provo. 181 pp.

HITCHCOCK, A. S. 1935. Manual of the grasses of the United States. U.S. Dept. Agric. Misc. Publ. 200: 1–1040.

_____. 1951. Manual of the grasses of the United States. 2nd ed., revised by Agnes Chase. U.S. Dept Agric. Misc. Publ. 200: 1–1051.

HITCHCOCK, C. L., A. CRONQUIST, M. OWNBEY, AND J. W. THOMPSON. 1955. Vascular plants of the Pacific Northwest. Part 5: Compositae. Univ. Washington Publ. Bot. 17(5): 1–343.

_____. 1959. Vascular plants of the Pacific Northwest. Part 4: Ericaceae through Campanulaceae. Univ. Washington Publ. Bot. 17(4): 1–510.

_____. 1961. Vascular plants of the Pacific Northwest. Part 3: Saxifragaceae to Ericaceae. Univ. Washington Publ. Bot. 17(3): 1–614.

_____. 1964. Vascular plants of the Pacific Northwest. Part 2: Salicaceae to Saxifragaceae. Univ. Washington Publ. Bot. 17(2): 1–597.

_____. 1969. Vascular plants of the Pacific Northwest. Univ. Washington Publ. Bot. 17(1): 1–914.

HOLMGREN, A. H. 1948, 1957. Handbook of the vascular plants of the northern Wasatch. Lithotype Process Company, San Francisco. 202 pp.

HOWELL, J. T., E. MCCLINTOCK, AND COLLABORATORS. 1960. Supplement. pp. 1035–1076. In: Arizona flora. University of California Press, Berkeley and Los Angeles.

HUBBARD, C. E. 1968. Grasses: A guide to their structure, identification, uses, and distribution in the British Isles. Revised ed. Penguin Books). Harmondsworth. 463 pp.

HULTEN, E. 1964. The circumboreal plants. I. Almquist and Wiksell. Stockholm. 275 pp.

KASS, R. J. AND S. L. WELSH. 1985. New species of *Primula* (Primulaceae) from Utah. Great Basin Nat. 45: 548–550.

KEARNEY, T. H., R. H. PEEBLES, AND COLLABORATORS. 1969. Arizona flora. University of California Press, Berkeley and Los Angeles. 1032 pp.

KINGSBURY, J. M. 1964. Poisonous plants of the United States and Canada. Prentice-Hall, Inc., Englewood Cliffs, New Jersey. 626 pp.

KNOBLOCH, I. W. 1963. Hybridization in the Gramineae. Darwiniana 12: 624–628.

MCVAUGH, R. 1983. Flora Novo-Galiciana. Vol. 14. Gramineae. Ed. W. R. Anderson. Univ. Mich. Press. Ann Arbor. 436 pp.

MUNZ, P. A. 1968. Supplement to a California flora. University of California Press, Berkeley. 224 pp.

_____. 1970. A California flora. University of California Press, Berkeley. 1681 pp.

MURDOCK, J. R. AND S. L. WELSH. 1971. Land use in Wah Wah and Pine valleys, western Utah. Brigham Young Univ. Sci. Bull. Biol. Ser. 12(4): 1–25.

NEESE, E. AND S. L. WELSH. 1983. A new species of *Penstemon* (Scrophulariaceae) from the Uinta Basin, Utah. Great Basin Nat. 43: 373–374.

NICORA, E. G. 1978. Flora Patagonica, Part 3. Gramineae. Coleccion Cientifica del I. N. T. A., Buenos Aires, Argentina.

NORTHSTROM, T. E. AND S. L. WELSH. 1970. Revision of the *Hedysarum boreale* complex. Great Basin Naturalist 30: 109–130.

REHDER, A. 1949. Bibliography of cultivated trees and shrubs. Arnold Arboretum of Harvard University, Jamaica Plain, Massachusetts. 825 pp.

SCOGGAN, H. J. 1978. The flora of Canada. Part 2. Natl. Museum Canada, Ottawa. 545 pp.

THORNBURG, A. A. 1982. Plant materials for use on surface-mined lands in arid and semiarid regions. U.S. Dept. Agric., SCS-TP-157. 88 pp.

TIDESTROM, I. 1925. Flora of Utah and Nevada. Contr. U.S. Natl. Herb. 25: 1–665.

TIDWELL, W. D., S. R. RUSHFORTH, AND D. SIMPER. 1972. Evolution of floras in the Intermountain Region. pp. 19–39. In: Intermountain Flora. Vol. 1. Hafner Publishing Company, New York.

TRESHOW, M., S. L. WELSH, AND G. MOORE. 1964. Guide to the woody plants of Utah. Ed. 2. Purett Press, Boulder, Colorado. 160 pp.

TSVELEV, N. N. 1984. Grasses of the Soviet Union. Parts 1 and 2. Russian Translation Series 8. A. A. Balkema, Rotterdam. 1196 pp.

TUTIN, T. G., V. H. HEYWOOD, N. A. BURGES, D. M. MOORE, D. H. VALENTINE, S. M. WALTERS, AND D. A. WEBB. 1964–1980. Flora Europaea. Vol. 1–5. Cambridge University Press, Cambridge.

USDA Soil Conservation Service. 1982. National list of scientific plant names. SCS-TP-159. Vol. 1. U.S. Government Printing Office. 416 pp.
Vallentine, J. F. 1961. Important Utah range grasses. Utah State Univ. Ext. Circ. 281: 1–48.
Voss, E. G. 1972. Michigan flora. Part 1. Cranbrook Institute of Science and University of Michigan Herbarium, Bloomfield Hills. 488 pp.
_____. ed. 1983. International code of botanical nomenclature. Adopted by the 13th International Botanical Congress, Sydney, August 1981. Bohn, Scheltema & Holkema, Utrecht. 472 pp.
Watson, S. 1871. Botany. In C. King, Report of the geological exploration of the fortieth parallel, Vol 5. Washington: Government Printing Office. 525 pp.
Weber, W. A. 1976. Rocky Mountain Flora. Colorado Assoc. Univ. Press, Boulder. 479 pp.
Welsh, S. L. 1957. An ecological survey of the vegetation of the Dinosaur National Monument, Utah. Unpublished Thesis, Brigham Young University. 86 pp.
_____. 1970a. New and unusual plants from Utah. Great Basin Nat. 30: 16–22.
_____. 1970b. An undescribed species of *Astragalus* (Leguminosae) from Utah. Rhodora 72: 189–193.
_____. 1971. Description of a new species of *Dalea* (Leguminosae) from Utah. Great Basin Nat. 31: 90–92.
_____. 1974a. Anderson's flora of Alaska. Brigham Young Univ. Press, Provo, Utah. 724 pp.
_____. 1974b. Utah plant novelties in *Astragalus* and *Yucca*. Great Basin Nat. 34: 305–310.
_____. 1975. Utah plant novelties in *Cymopterus* and *Penstemon*. Great Basin Nat. 35: 377–378.
_____. 1978a. Problems of plant endemism on the Colorado Plateau. Memoirs Great Basin Nat. 2: 191–195.
_____. 1978b. Endangered and threatened plants of Utah: a reevaluation. Great Basin Nat. 38: 1–18.
_____. 1981. New taxa of western plants—in tribute. Brittonia 33: 294–303.
_____. 1982a. Utah plant types—Historical perspective 1840 to 1981—Annotated list and bibliography. Great Basin Nat. 42: 1–44.
_____. 1982b. New taxa of thistles (*Cirsium*, Asteraceae) in Utah. Great Basin Nat. 42: 199–202.
_____. 1982c. A new species of *Cryptantha* (Boraginaceae) dedicated to the memory of F. Creutzfeldt. Great Basin Nat. 42: 203–204.
_____. 1983. A bouquet of daisies (*Erigeron*, Compositae). Great Basin Nat. 43: 365–368.
_____. 1983b. New taxa in *Thelesperma* and *Townsendia* (Compositae) from Utah. Great Basin Nat. 43: 369–370.
_____. 1986. History of botanical exploration in Utah: The Fremont era. Utah Acad. Sci. (in press).
Welsh, S. L. and N. D. Atwood. 1977. An undescribed species of *Thelypodiopsis* (Brassicaceae) from the Uinta Basin, Utah. Great Basin Nat. 37: 95–96.
Welsh, S. L., N. D. Atwood, S. Goodrich, E. Neese, K. H. Thorne, and B. Albee. 1981. Preliminary index of Utah vascular plant names. Great Basin Nat. 41: 1–108.
Welsh, S. L., N. D. Atwood, and J. R. Murdock. 1978. Kaiparowits flora. Great Basin Nat. 38: 125–179.
Welsh, S. L., N. D. Atwood, and J. L. Reveal. 1976. Endangered threatened, extinct, endemic, and rare or restricted Utah plants. Great Basin Nat. 35: 327–376.
Welsh, S. L., and R. C. Barneby. 1981. *Astragalus lentiginosus* (Fabaceae) revisited — a unique new variety. Isleya 2(1): 1–2.
Welsh, S. L., and S. Goodrich. 1980. Miscellaneous plant novelties from Alaska, Nevada, and Utah. Great Basin Nat. 40: 78–88.
Welsh, S. L., and G. Moore. 1968. Plants of Natural Bridges National Monument. Proc. Utah Acad. 45: 220–248.
_____. 1973. Utah plants: Tracheophyta. Brigham Young University Press, Provo, Utah. 473 pp.
Welsh, S. L., and E. Neese. 1983a. New species in *Hymenoxys* and *Perityle* (Compositae). Great Basin Nat. 43: 369–370.
_____. 1983b. New variety of *Opuntia basilaris* (Cactaceae) from Utah. Great Basin Nat. 43: 700.
Welsh, S. L., and B. W. Olsen. 1969. A living, prehistoric lumber tree. Proc. Utah Acad. 46: 149–159.
Welsh, S. L., and J. L. Reveal. 1968. A new species of *Townsendia* from Utah. Brittonia 20: 375–377.
Welsh, S. L., and F. G. Smith. 1983. New *Haplopappus* variety in Utah (Compositae). Great Basin Nat. 43: 371–372.
Welsh, S. L., and K. H. Thorne. 1978. Illustrated manual of proposed endangered and threatened plants of Utah. U. S. Fish and Wildlife Publication, Denver, Colordo. 318 pp.
Welsh, S. L., and C. A. Toft. 1981. Biotic communities of hanging gardens in southeastern Utah. National Geographic Society Research Reports 13: 663–682.

GROUP TRACHEOPHYTA - VASCULAR PLANTS

Plants with a well developed vascular system (xylem and phloem) in the sporophyte generation, leaves (either macrophylls or microphylls), and roots; reproduction by spores, or by microspores and megaspores, and by seeds (in most), the latter borne in cones or flowers.

Key to the Divisions and Classes.

1. Plants with small scalelike leaves, usually with a single vein (microphylls); reproduction by means of spores; flowers or woody cones lacking 2
— Plants with large leaves, usually with more than a single vein (macrophylls), if scalelike, as occasionally, otherwise different from above; reproduction by spores or seeds, the latter borne in flowers or cones 3
2(1). Stems jointed, fluted, and hollow in the internodes; leaves not green, reduced to a whorl of connate scales at the nodes; plants neither grass- nor mosslike ....
.... Equisetophyta, p. 9
— Stems not jointed; leaves green and imbricated, not whorled or forming a sheath at the nodes; plants either aquatic and grasslike or terrestrial and mosslike .....
.... Lycopodiophyta, p. 13
3(1). Plants fernlike and with broad leaves or free-floating aquatics with small overlapping leaves; reproduction by spores; flowers and woody cones lacking ....
.... Polypodiophyta, p. 15
— Plants neither fernlike nor free-floating aquatics (except in Lemnaceae); reproduction by spores and seeds, these borne in flowers or cones 4
4(3). Seeds not borne enclosed by ripening carpels, but naked and situated on the surface of a scale, these borne crowded together on an axis and forming a cone; flowers not developed; leaves typically needle- or scalelike ..
.... Pinophyta, p. 24
— Seeds borne in ripening carpels; plants with flowers; leaves mainly not needle- or scalelike (Magnoliophyta) .. 5
5(4). Flower parts mainly 4- or 5-merous; leaves typically net veined; stems increasing in diameter by means of a cambium between the xylem and phloem; cotyledons typically 2 .... Magnoliopsida, p. 35
— Flower parts typically 3-merous; leaves typically parallel veined; stems usually lacking a cambium or, if present, producing entire vascular bundles; cotyledon 1
.... Liliopsida, p. 647

DIVISION LYCOPODIOPHYTA

Clubmosses

Perennial herbs with alternation of generations, the generations differentiated and ultimately independent; sporophyte well developed, with roots, stems, and microphylls; vascular system protostelic, without leaf gaps; leaves typically alternate and often spirally arranged, either scale- or grasslike, with a single, unbranched vascular bundle; sporangia solitary, subtended by a sporophyll, the sporophylls aggregated into a definite or indefinite strobilus; spores dimorphic (megaspores and microspores) in Utah materials.

Key to the Families

1. Plants aquatic, submerged in ponds or lakes, or occasionally growing on exposed mud, grasslike; leaves long and slender, from a broadly clasping base; sporangia at base of leaves .... **Isoetaceae**, p. 13
— Plants terrestrial, growing in dry, rocky situations; leaves small and scalelike; sporangia in terminal strobili .... **Selaginellaceae**, p. 13

ISOETACEAE Reichenb.

Quillwort Family

Plants perennial, aquatic, amphibious, or sometimes terrestrial herbs; stems cormlike, with leaves (microphylls) clustered in a close spiral at the summit of the stem; leaves simple, elongate, dilated basally, the blade hollow and transversely septate, the outermost sterile, the next innermost bearing megasporangia, and the next innermost bearing microsporangia; sporangia solitary, enclosed in a cavity on the ventral side of the leaf base; ligule (a small flap of tissue) borne above the sporangial cavity; spores dimorphic, of microspores and megaspores; $x = 21$.

Isoetes L.

Stems very short; leaves more or less cylindrical, elongate, the peripheral tissues often containing longitudinal strands of sclerenchyma; sporangia borne at the base of the leaves, usually covered by a velum or thin flap of tissue.

Pfeiffer, N. E. 1922. Monograph of Isoetaceae. Ann. Missouri Bot. Garden 9: 79–232.

1. Megaspores obscurely tuberculate or ridged (as viewed at high magnification) 2
— Megaspores spinose, crested, or ridged (as viewed at high magnification) 3
2(1). Leaves mostly less than 15 cm long; hyaline sporophyll margins less than 1 cm long above the sporangium; sclerenchymatous strands essentially lacking . *I. bolanderi*
— Leaves mostly more than 15 cm long; hyaline sporophyll margins mainly 1–5 cm long above the sporangium; sclerenchymatous strands obvious .... *I. howellii*
3(1). Megaspores mostly less than 0.5 mm wide, spinose; leaves subulate .... *I. echinospora*
— Megaspores mostly more than 0.5 mm wide, crested or ridged; leaves linear in the lower portion at least .... *I. lacustris*

Isoetes bolanderi Engelm. Bolander Quillwort. Leaves mostly 2–12 cm long, gradually tapering from the broad base, slender, soft; hyaline margins extending to ca 1 cm above the sporangium; velum covering to ca 1/2 of the sporangium; megaspores 0.3–0.5 mm wide, obscurely tuberculate, ridged, or wrinkled. Ponds, lake margins, and sometimes on mud, at 1310 to 3205 m in Duchesne, Garfield, Salt Lake, Sevier, Summit, and Utah counties; British Columbia to Montana, south to California, Arizona, and Colorado; 14 (i).

Isoetes echinospora Durieu Spiny Quillwort. Leaves mostly 2.5–10 cm long, gradually tapering from the broad base, slender, soft, straight or curved; hyaline margins extending mostly 1–4 (5) cm above the sporangium; velum covering less than 1/2 of the sporangium; megaspores 0.3–0.6 mm wide, more or less spiny with blunt, truncate, or bifid spines; $n = 21$. Ponds, lake margins, and in mud at 2300 to 3085 m in Duchesne and Summit counties; widely distributed in North America; circumboreal; 3 (i).

Isoetes howellii Engelm. Howell Quillwort. Leaves mostly 15–25 (30) cm long, linear, very slender, firm; hyaline margins extending 1–5 cm above the sporangium; velum covering less than 1/2 of the sporangium; megaspores 0.3–0.5 mm wide, with low tubercles, ridges, or wrinkles. Pond margin (Dry Lake) at ca 1735 m in Cache County; Washington to Montana, south to California; 3 (0).

Isoetes lacustris L. Lake Quillwort. Leaves linear, mostly 3–12 cm long, slender, firm; velum covering less than 1/2 of the sporangium; megaspores 0.5–0.8 mm wide, with crests or ridges; $2n = 110$. Reported for the Uinta Mts. (Intermountain Flora 1: 184. 1972); widely disjunct in North America; circumboreal; 0 (0).

SELAGINELLACEAE Reichenb.

Spikemoss Family

Plants low, creeping, forming loose mats or dense tufts among rocks, mosslike in habit and appearance; stems branched, slender, erect or prostrate; leaves (microphylls) numerous, small, oblong to lanceolate, to 3 mm long, sessile and imbricate, all alike and spirally arranged; heterosporous, the sporophylls green, ovate-triangular, slightly larger than the vegetative leaves, arranged in 4 ranks, sharply keeled and forming a 4-angled terminal strobilus often not much different than the vegetative stem; microsporangia and megasporangia axillary and randomly disposed in the strobilus, orange or yellowish; microspores numerous; megaspores 3–4 in each megasporangium, orange; prothalia minute, retained in the spore wall; $x = 7, 8, 9$.

Selaginella Beauv.

Evergreen herbs with dichotomously to monopodially branched stems; leaves numerous, imbricate, small; strobili bisexual, the lower sporophylls usually producing megasporangia and the upper ones microsporangia.

Flowers, S. 1944. Ferns of Utah. Bull. Univ. Utah 35(7): 1–87.

1. Plants loosely matted, the branches distant, long and spreading 2
— Plants densely tufted or matted, the branches short 3

2(1). Stems 2–3 mm thick; leaves gradually tapered to the apex, loosely imbricate, 2.5–3 mm long, the bristle tip 0.3–0.9 mm long.................... *S. underwoodii*

— Stems ca 1 mm thick; leaves abruptly contracted at the apex, appressed and closely imbricate, 1 mm long, a bristle tip lacking or minute .... *S. mutica*

3(1). Leaves tapering to the apex, the bristle tip long and slender, white, 1–2 mm long; plants of the La Sal and Uinta mts................................ *S. densa*

— Leaves abruptly acute, the bristle tip shorter, yellowish green, 0.1–0.5 mm long or obsolete 4

4(3). Terminal setae of leaves evident, mostly 0.2–0.5 mm long; our most common and most widespread species .... *S. watsonii*

— Terminal setae of leaves very short (rarely to 0.2 mm long) or obsolete; plants rare in Kane and Washington counties .... *S. utahensis*

Selaginella densa Rydb. Rydberg Spikemoss. [*S. rupestris* var. *densa* (Rydb.) Clute; *S. scopulorum* Maxon]. Plants caespitose, densely tufted, the stems becoming 10–12 cm long, creeping, with numerous short compact and ascending branches; leaves densely imbricate, 2–3 mm long, 0.2–0.4 mm wide, pale green, brownish below, lanceolate to linear-oblong, tapering toward the apex, rounded and boat-shaped at back, narrowly grooved dorsally, short-ciliate marginally (often sparingly so), erect, 1–2.5 cm long or longer, sharply 4-angled; sporophylls triangular-ovate, 1.5–2 mm long, the bristle tip ca 1 mm long; megaspores ca 0.4 mm thick, more or less distinctly roughened; $2n = 18$. Rocky ledges and talus slopes in pinyon-juniper, sagebrush, spruce-fir, krummholz, and alpine tundra communities at 2700 to 4300 m in Daggett, Duchesne, Grand, San Juan, and Summit counties; Alaska to California, east to Manitoba, the Dakotas, New Mexico, and Arizona; 16 (0).

Selaginella mutica D. C. Eaton Awnless Spikemoss. Plants very slender in widely spreading mats; stems 10–40 cm long, distantly and somewhat pinnately branched, 1 mm thick; leaves closely imbricate in 6 distinct rows, 1 mm long, 0.2–0.3 mm wide, oblong to oblong-ovate, obtuse, the upper ones with very short hyaline points, the margins with spreading cilia; strobili slightly broader than the vegetative branches, sharply 4-angled, long and slender, 1–3 cm long, often curved; sporophylls ovate-triangular, 1.5–1.8 mm long, concave and keeled, the margin ciliate, the apex shortly bristle-tipped; megaspores ca 0.3 mm thick, undulate to nearly smooth or somewhat roughened. Rocky crevices, often in sandstone and shale, in sagebrush, pinyon-juniper, mixed desert shrub, ponderosa pine, and Douglas fir communities at 1730 to 2330 m in Emery, Garfield, San Juan, Uintah, and Wayne counties; Colorado to Arizona, New Mexico, and Texas; 13 (0).

Selaginella underwoodii Hieron. Underwood Spikemoss. [*S. fendleri* (Underw.) Hieron.]. Plants in spreading tufts or mats; stems slender, becoming 20–30 cm long, creeping, the branches long and distant, spreading, to 8 cm long, prostrate or ascending; leaves rather loosely imbricate, dark green, 2–3 mm long, lanceolate, oblong-lanceolate, to triangular-lanceolate, tapering at the apex, tipped with a pale yellowish bristle 0.4–0.9 mm long, the margins shortly and distantly ciliate; strobili ascending or erect, to 3 cm long; sporophylls triangular-ovate to -lanceolate, 1.5–2.5 mm long, the apex shortly mucronate, the margins sparingly ciliate; megaspores ca 0.3 mm thick, somewhat roughened; $n = 7, 14, 18, 21$. Rocky ledges and crevices in sagebrush, mountain brush, and ponderosa pine communities, mainly on Navajo Sandstone, at 1650 to 2500 m in Kane and Washington counties; Arizona, Colorado, Wyoming, New Mexico, and Texas; 4 (0).

Selaginella utahensis Flowers Utah Spikemoss. Plants very similar to *S. watsonii*, but differing in the leaves which typically have a very short white point to 0.1 mm long or less, or occasionally with a setum, but this seldom over 0.2 mm long, or the point obsolete and the leaves wholly muticous. Ledges and crevices in Navajo Sandstone in sagebrush, oakbrush, pinyon-juniper, and ponderosa pine communities at 1060 to 2350 m in Kane and Washington (type from Zion Canyon) counties; Nevada; 4 (i).

Selaginella watsonii Underw. Watson Spikemoss. Plants in dense tufts or somewhat matted; stems 5–15 cm long, creeping; branches erect or ascending, to 4 cm long and 2 mm thick; leaves crowded, imbricate, dark green, brownish below, oblong-lanceolate, 2–3 mm long, 0.5–0.7 mm wide, concave, boat-shaped at back and with a narrow groove dorsally, the margins sparingly ciliate, the apex with a yellowish green bristle 0.2–0.4 mm long; strobili erect or diverging from the stem tips, sharply 4-angled, to 2.5 cm long (often much shorter); sporophylls triangular-lanceolate to ovate-lanceolate, sharply keeled, 2 mm long, 1 mm wide at the base, the margins smooth or finely ciliate; megaspores ca 0.4 mm thick, somewhat roughened. Ledges or talus slopes in mountain brush, ponderosa pine, aspen, spruce-fir, lodgpole pine, krummholz, and alpine tundra communities at 1290 to 4250 m in Beaver, Box Elder, Duchesne, Garfield, Juab, Millard, Piute, Salt Lake (type from Cottonwood Canyon), Sevier, Summit, Tooele, Uintah, Utah, Washington, and Wayne counties; California to Nevada, Oregon and Montana; 70 (iii).

DIVISION EQUISETOPHYTA
Horsetails

Perennial herbs with alternation of generations, both ultimately independent; sporophyte with roots, stems, and whorled scalelike microphylls; stems photosynthetic (or sometimes dimorphic and the fertile ones lacking chlorophyll), longitudinally ribbed and grooved, jointed, and usually hollow in the internodes, simple or with whorled branches through the sheathing leaf bases; sporangia borne beneath stalked peltate scales (sporangiophores) closely grouped in whorls, forming a terminal strobilus; spores alike (homosporous), with the exine forming hygroscopic elaters; $x = 108$.

EQUISETACEAE Michx.
Horsetail Family

Perennial, the stems annual or perennial, typically hollow, jointed, longitudinally ribbed; leaves microphyllous, whorled, small, and scalelike; strobili spikelike, bearing numerous stalked, peltate scales with sporangia on the lower surface; spores numerous, spherical, with a thick perispore consisting of 4 spirally wound bands (elaters), these hygroscopic.

***Equisetum* L.**

Plants rhizomatous perennials; stems annual or perennial and evergreen, with silicified cell walls; strobili borne on photosynthetic stems or on specialized non-photosynthetic stems.

1. Stems annual, typically dimorphic, the fertile ones usually without chlorophyll, the sterile commonly with regular whorls of branches; cones with at least some peduncles much surpassing the subtending sheath, rounded apically .... *E. arvense*
— Stems perennial or annual, all alike, typically unbranched or, if so, lacking regular whorls of branches; cones with peduncles seldom exceeding the subtending sheath, apiculate 2

2(1). Stems slender, 1.5–4 mm thick, 1–3 dm tall, 5- to 12–ridged; central cavity less than half the diameter of the stem; leaves and teeth not sharply differentiated, the teeth persistent .... *E. variegatum*
— Stems more robust, mostly 5–10 mm thick, 3–15 dm tall, 14- to 40–ridged; central cavity more than half the diameter of the stem; leaves and teeth sharply differentiated, the teeth usually deciduous 3

3(2). Stems overwintering; sheaths about as broad as long, finally ash colored and with 2 dark bands; cones evidently apiculate .... *E. hyemale*
— Stems not overwintering; sheaths longer than broad, typically green and with 1 dark band only; cones inconspicuously apiculate .... *E. laevigatum*

***Equisetum arvense* L.** Meadow Horsetail. Stems annual, of 2 types, the sterile ones (5) 10–50 (60) cm tall, 1–5 mm thick, 10- to 12–ridged, the ridges with minute bumps and cross-ridges, the central cavity ca 1/4 the stem diameter, the stomates in 2 broad bands, not sunken, the sheaths 5–10 mm long, greenish, with teeth 1–3 mm long, persistent, separate or some united, brown or blackish, the margins sometimes pale and hyaline; branches in regular whorls, 3- to 4–ridged, solid, usually not branched again; fertile stems whitish, pinkish, brownish, or yellowish, borne in springtime, soon withered, 0.6–3 dm tall, 3–8 mm thick, the sheaths 10–20 mm long, with teeth 5–9 (11) mm long, some connate; strobili 5–35 mm long or more, with peduncles much longer than the subtending sheath, blunt apically; $n = 108$. Moist to somewhat dry places in sagebrush, mountain brush, pinyon-juniper, aspen, and fir communities at 1300 to 3200 m in Beaver, Daggett, Davis, Duchesne, Emery, Garfield, Grand, Iron, Juab, Kane, Morgan, Salt Lake, San Juan, Sanpete, Sevier, Summit, Tooele, Uintah, Utah, Wasatch, Washington, Wayne, and Weber counties; widely distributed in North America; circumboreal; 61 (v).

***Equisetum hyemale* L.** Common Scouringrush. Stems perennial, evergreen, all alike, commonly 2–10 dm tall or more, 4–10 mm thick or more, with (14) 16–20 ridges or more, the ridges with 2 rows of tubercles or 1 row of transverse ridges, the central cavity ca 3/4 the stem diameter, the stomates in 2 rows in each groove, not sunken; sheaths 3–10 (15) mm long, usually with 2 black bands separated by a grayish band at maturity, the teeth 2–4 mm long, deciduous, black, hyaline-margined, jointed to the sheath; strobilus 10–25 (30) mm long, subsessile or with peduncles subequal to the subtending sheath, stoutly apiculate. Streambanks, seeps, and marshes in sagebrush, riparian, mountain brush, ponderosa pine, and aspen-fir communities at 1230 to 2850 m in Beaver, Box Elder, Cache, Daggett, Duchesne, Emery, Garfield, Grand, Juab, Kane, Millard, Piute, Salt Lake, San Juan, Tooele, Utah, Wasatch, Washington, and Weber counties; widespread in North America; Eurasia; 35 (ii).

***Equisetum laevigatum* A. Br.** Smooth Scouringrush. [*E. kansanum* Schaffner; *E. funstonii* A. A. Eaton]. Stems annual, all alike, commonly 2–10 dm tall, 2–8 mm thick, with (14) 16–30 ridges, the ridges smooth or commonly with low transverse wrinkles; central cavity about 2/3–3/4 the diameter of the stem; sheaths widened upward, the upper ones green with an apical dark band; leaves keeled below, the teeth usually scarious-margined, 1–2 mm long, articulated, and soon deciduous; cones short-pedunculate or nearly sessile, 10–25 mm long, rather blunt or inconspicuously apiculate. Riparian and other moist habitats in blackbrush, sagebrush, greasewood, pinyon-juniper, mountain brush, aspen, spruce-fir, and lodgepole pine communities at 1320 to 3350 m in all except Iron, Piute, and Sevier counties; British Columbia to Baja California, east to Ontario and Texas; 90 (iv).

***Equisetum variegatum* Schleicher** Variegated Scouringrush. Stems perennial, evergreen, all alike, commonly (0.5) 1–4 dm tall, 1–2 (4) mm thick, with 5–12 ridges, each ridge with 2 rows of tubercles, the central cavity 1/4–1/3 the diameter of the stem, the stomates in 2 rows in each groove, sunken below the epidermis; sheaths (1) 2–4 mm long, the base not easily distinguished, flared, black or blackish apically, the teeth 1–2 (3) mm long, with conspicuous white-hyaline margins; strobilus (3) 7–10 mm long, subsessile or shortly pedunculate, prominently apiculate. Wet meadows and along streams in aspen, spruce-fir, and alpine tundra communities at 2850 to 3700 m in Beaver, Garfield, Iron, Kane, and Salt Lake counties; Alaska to the Atlantic and south to Washington, Illinois, and Pennsylvania; circumboreal; 4 (0).

DIVISION POLYPODIOPHYTA

Ferns

Perennial herbs with alternation of generations, these ultimately independent; sporophyte with roots, stems, and macrophylls (typically with more than one vein or with branched veins); stele with leaf gaps; stems mostly rhizomatous; leaves typically alternate and large, sometimes reduced; sporangia borne on foliage or modified leaves, typically in sori, or in some borne in specialized sporocarps representing modified leaf segments; sporophylls not aggregated into a strobilus; spores alike (homosporous) or dissimilar (heterosporous).

Key to the families

1. Spores borne in sporangia on green, aerial leaves; plants terrestrial 2
— Spores borne in sporocarps (these usually below ground or water level); plants aquatic or amphibious, often free floating 3

2(1). Spore-bearing leaves strikingly different from the vegetative leaves; sporangium without an annulus, opening by a transverse, gaping slit .... **Ophioglossaceae**, p. 16

— Spore-bearing leaves much like the vegetative ones, at most having narrower segments; sporangium with an annulus .... **Polypodiaceae**, p. 17

3(2). Leaves palmately divided into 4 leaflets, cloverlike with long petioles; plants rooting in mud, often in shallow ponds or lakes, the leaves floating .... **Marsiliaceae**, p. 16

— Leaves entire or 2-lobed, sessile; plants small, branched, 0.5–2 cm long, floating on water ....
.... **Salviniaceae**, p. 24

MARSILEACEAE R. Br.

Pepperwort Family

Plants herbaceous, creeping, rooting in mud, with slender branching rhizomes; leaves erect, filiform or with long 2- to 4-foliolate blades; leaflets cuneate-obovate, the veins dichotomous; sporocarps hard and bony, globose to ellipsoid, pilose to glabrous, pedunculate, borne on the rhizome near the petiole base or on the petiole; sori solitary within the compartments, each producing archegonia.

Marsilea L.

Plants small, cloverlike; leaves long-petiolate, the blades 4-foliolate, cruciform; sporocarps subglobose to ellipsoid, mostly with 2 teeth near the base, splitting into 2 valves at maturity and producing numerous sori on a gelatinous receptacle; sori including both megasporangia and microsporangia.

Johnson, D. M. 1986. Systematics of the New World species of *Marsilea* (Marsileaceae) Syst. Bot. Monogr. 11: 1–87.

1. Superior tooth of sporocarps typically more than 0.4 mm long, acute .... *M. vestita*
— Superior tooth of sporocarp typically (when present) less than 0.4 mm long, blunt. .... *M. oligospora*

Marsilea oligospora Goodding. Rhizomes long-creeping, hairy at the apex; petioles 3–15 cm long; leaflets cuneate, 6–15 mm long, entire, hairy; peduncles distinct from petioles, short; sporocarp solitary, 5–6 mm long, early densely hairy, finally glabrate. Muddy shores of ponds and lakes in Cache (Dry Lake) County; Washington to Montana, south to California and Wyoming; 3 (0).

Marsilea vestita Hook. & Grev. Pepperwort; Waterclover. [*M. mucronata* A. Br.]. Rhizomes long-creeping, densely hairy at the nodes; petioles 2–18 cm long, with broadly cuneate leaflets 5–15 mm long, entire, hairy; peduncles distinct from the petiole, short; sporocarps solitary, 4–8 mm long, early densely hairy, later more sparsely so. Sinks, mudflats, pools, and other moist places in willow, tamarix, cottonwood, and mixed forb communities at 1470 to 2000 m in Emery, Millard, Rich, Salt Lake, Sevier, Uintah, Washington, and Weber counties; British Columbia to Minnesota, south to Arkansas, Texas, Arizona, and California; 14 (0).

OPHIOGLOSSACEAE R. Br.

Adderstongue Family

Plants more or less succulent, with short tuberous erect rhizomes; leaves 1 per stem, green, simple or compound, nodding in bud but not circinate, the fertile portion of the frond distinct, borne erect, arising from the stipe; sporangia borne naked on the fertile spikelike or branched segment; indusia lacking; sporangia without an annulus, opening by a slit; spores tetrahedral, numerous; gametophyte subterranean, threadlike; x = 45, 60.

Botrychium Swartz

Plants from clustered fleshy roots; stem simple, erect, fleshy, surrounded by a sheath of brown scaly leaf bases; leaf solitary with a common stalk bearing one sterile blade and a fertile one, the latter bearing a spicate or branched cluster of globose sporangia in 2 rows; leaves entire to pinnately or palmately lobed, the veins not netted; fertile segment pinnate or bipinnate; x = 45.

Flowers, S. 1944. Ferns of Utah. Bull. Univ. Utah. 35(7): 1–87.

1. Leaves with distinct petioles, tending to be ternate-pinnate, the basal pair of pinnae enlarged, attached near ground level............................ *B. simplex*
— Leaves sessile or nearly so, attached near the middle of the stem 2

2(1). Sterile blade once-pinnatifid, the pinnae fan-shaped and mostly overlapping....................... *B. lunaria*
— Sterile blade mostly bipinnate, the pinnae ovate to oblong, not or only slightly overlapping 3

3(2). Blades deltoid or deltoid-ovate in outline; lower pinnae distant, conspicuously larger and longer than the upper ones .... *B. lanceolatum*
— Blades ovate to oblong in outline; lower pinnae approximate, not conspicuously larger or longer than the upper ones, broadly ovate to obovate-oblong, obtuse ..*B. boreale*

Botrichium boreale Milde Northern Grapefern. [*B. crassinervium* var. *obtusilobum* Rupr.]. Plants stout and fleshy, 8–13 (25) cm tall; common stalk 1/3–1/2 above ground, 3–7 cm long, the sterile leaf blade nearly sessile, yellowish green, to 3 cm long, ovate to oblong in outline, once-pinnate to bipinnate; pinnae approximate or slightly overlapping, 6–10, sessile or decurrent, rhombic, ovate, or elliptical, 4–12 mm long, lobed or divided, the tips obtuse, rounded; fertile segment longer than the sterile, racemose to paniculate, to 3 cm long; 2n = 180. Open moist or wet grassy meadows at ca 3330 m in Summit County; Alaska to Washington, Montana, and Oregon; Siberia; 1 (0).

Botrichium lanceolatum (S. G. Gmel.) Angstrom Lanceleaf Grapefern. [*Osmunda lanceolata* S. G. Gmel.]. Plants stout and fleshy, 6–40 cm tall, the common stalk 4–15 cm long; sterile leaf blade sessile or nearly so, inserted near the top of the plant, broadly deltoid in outline, 1–6 cm long, 1–8 cm wide at the base, the apex acute, pinnately divided, the lower pinnae or segments distant, lanceolate, conspicuously longer and larger than the upper ones that are oblong-lanceolate to ovate, variously lobed, cleft, or divided, the lobes bluntly acute; fertile segment with a short stalk ca 1 cm long or shorter, paniculate, 1–5 cm long. Meadows, open woods, and slopes, at ca 3330 m in Juab County; Labrador to Maine, west to Alaska and south to Washington and Colorado; Greenland, Iceland, and Eurasia; 1 (0).

Botrichium lunaria (L.) Swartz Moonwort. [*Osmunda lunaria* L.; *B. onondagense* Underw.; *B. lunaria* var. *onondagense* (Underw.) House; *B. lunaria* f. *onondagense* (Underw.) Butters & Abbe; *B. minganense* Victorin; *B. lunaria* var. *minganense* (Vict.) Dole; *B. lunaria* ssp. *minganense* (Vict.) Calder & Taylor]. Plants 3–28 cm tall, the common stalk short, sheathed with scaly leaf bases, the sterile blades sessile, from a long sheathing base, diverging below the middle of the plant, the blade 2–10 cm long, ovate-oblong to oblong in outline, once-pinnately divided, the segments 5–15, lunately reniform or fan-shaped, mostly overlapping, entire or the upper ones crenate to incised; fertile segment exceeding the sterile leaf, long-stalked, the fruiting portion simple, racemose, or paniculate; n = 45. Grassy meadows or edges of woods at 2350 to 3500 m in Beaver, Cache, Daggett, Duchesne, Iron (?), Juab, Salt Lake, and Summit counties; Alaska to Labrador, south to California, Arizona, Colorado, and Maine; Greenland, Eurasia, South America, New Zealand, and Australia; 8 (0).

Botrichium simplex E. Hitchc. Little Grapefern. [*B. virginianum* var. *simplex* (E. Hitchc.) Gray]. Plants 3–10 (16) cm tall, the common stalk 0.5–4 cm long, the sterile leaf diverging from at or near the ground level, the petiole to 4 cm long, half as long as the blade or more, the blade 0.3–4 cm long, ovate to ovate-oblong, simple to pinnately divided, the segments fan-shaped, oblong, or rhomboidal, broadly inserted and decurrent on the indeterminate rachis, entire, crenate, or divided above, the fertile segment on a long stalk usually exceeding the sterile leaf, simple or 1– or 2–pinnately divided. Moist to somewhat dry woods and open slopes at 2300 to 3500 m in Kane and Salt Lake counties; British Columbia to Newfoundland and south to California, New Mexico, Indiana, and New Jersey; circumboreal; 2 (0).

POLYPODIACEAE R. Br.
Common Fern Family

Plants with scaly or hairy, creeping rhizomes; leaves coiled in the bud (circinate), forming fiddleheads, petiolate, simple or more commonly compound or decompound (pinnate or ternate-pinnate), often hairy or scaly; fertile and sterile leaves alike or dissimilar; sporangia grouped into sori on the lower leaf surface, these naked or covered by an indusium or by the recurved leaf margin; sporangia with an annulus; spores all alike; prothalia flat, green, aerial; x = 29, 30, 34, 36, 40, 41, 42.

Arnow, L., B. Albee, and A. Wyckoff. 1980. Ferns. pp 485–493. In: Flora of the central Wasatch Front, Utah. University of Utah Printing Services. Salt Lake City.
Cronquist, A. et al. 1972. Ferns. pp 192–220. In: Intermountain Flora. Vol. 1. Hafner Publishing, New York.
Flowers, S. 1944. Ferns of Utah. Bull. Univ. Utah 35: 21–67.

1.	Rhizome and leaves hairy, but not scaly; sori linear, marginal, confluent; stipe bundles several; blades large, coarse, subternately compound; plants of broad distribution	*Pteridium*
—	Rhizome and often the leaves scaly and also hairy; plants of various distribution	2
2(1).	Sori marginal or nearly so, or borne in lines along the veins and then lacking an indusium; stipe bundle solitary; spores tetrahedral	3
—	Sori dorsal on the veins, with or without an indusium, the indusium, when present, not formed by the leaf margin; stipe bundles 2 or more at base; spores bilateral	9
3(2).	Sporangia borne on veins on the underside of a reflexed marginal lobe; plants of the Wasatch Mts. and the southern half of Utah	*Adiantum*
—	Sporangia not borne on the under side of a marginal lobe, but on the leaf surface beneath a marginal lobe; plants of various distribution	4
4(3).	Sporangia borne in lines on the veins; blades pentagonal in outline; plants known from Washington County	*Pityrogramma*
—	Sporangia submarginal; blades various; plants of various distribution	5
5(4).	Leaves dimorphic or not, but the fertile ones always with glabrous, narrow, and elongate ultimate segments 1–3 (5) cm wide	6
—	Leaves usually not dimorphic, the fertile with ultimate segments usually much broader	7
6(5).	Leaves strongly dimorphic, the sterile ones well developed, but obviously shorter than the fertile; stipes greenish or greenish stramineous, at least distally; plants rather broadly distributed	*Cryptogramma*
—	Leaves weakly, if at all, dimorphic, either all alike and fertile, or with a few sterile ones and those not much different than the fertile ones; stipes dark brown; plants of montane sites in Salt Lake and Utah counties	*Aspidotus*
7(5).	Margins of fertile pinnae or pinnules not inrolled, or only slightly and irregularly so; plants mostly of Washington County	*Notholaena*
—	Margins of the fertile pinnae or pinnules conspicuously inrolled, forming a continuous indusial flap; plants rather broadly distributed	8
8(7).	Leaf blades glabrous or nearly so, not conspicuously woolly or scaly, the ultimate segments usually well over 5 mm in length	*Pellaea*
—	Leaf blades conspicuously woolly or scaly or both, at least beneath, the ultimate segments all shorter than 5 mm in length	*Cheilanthes*
9(2).	Indusium none; leaves simple, deeply pinnatifid, the lobes mostly oblong and finely serrate to entire; plants of Washington County	*Polypodium*
—	Indusium present, but often inferior and concealed or soon withering; plants rather broadly distributed	10
10(9).	Indusium peltate, attached by a central stalk, spreading over the sorus; leaves evergreen, often spiny-toothed or sharply serrate	*Polystichum*
—	Indusium not peltate; leaves not spiny-toothed	11
11(10).	Sori oblong; indusia elongate, straight or curved	12
—	Sori round or nearly so; indusia somewhat horseshoe-shaped or circular (in *Woodsia* splitting from the top into hairlike segments)	13
12(11).	Leaves usually less than 3 dm long, including the petiole; blades once-pinnate; pinnae less than twice as long as wide, toothed to subentire; sori straight, attached to the outer side of the vein; plants known from the Uinta Mts. and Washington County	*Asplenium*
—	Leaves 3–10 dm long; blades 2–3 times pinnate; pinnae more than twice as long as wide and pinnate; sori straight or more often curved across the veins; plants of the Uintah and Wasatch mts. and from Washington County	*Athyrium*
13(11).	Indusium horseshoe- or kidney-shaped, attached along the inner, notched margin; leaf blades 3–10 dm long; plants known from Garfield, Kane, and Washington counties	*Dryopteris*
—	Indusium not as above; leaf blades to 3, rarely 4, dm long; plants broadly distributed	14

14(13). Indusium inferior, attached under the sori, splitting at the top into slender or scalelike, somewhat beaded segments; leaves moderately thick, the veins obscure, the margins often irregularly recurving; rhizomes bearing marcescent petiole bases ... *Woodsia*

— Indusium attached on one side only, hoodlike, commonly reflexed at maturity; leaves thin, the veins distinct, the margins usually flat; rhizomes without marcescent leaf bases .... *Cystopteris*

Adiantum L.

Delicate, small ferns with slender, scaly, creeping rhizomes, the scales concolorous, long-attenuate, and rusty; fronds widely spreading, compound; stipe and rachis black or red black, smooth and shiny; blades membranaceous, compound, glabrous, scaleless; pinnules leafletlike, with main veins along one margin, the veinlets dichotomous; sori marginal at the tips of lobes; indusium false, formed by the strongly reflexed tips of the leaflet lobes.

1. Blades longer than broad; stipe not dichotomously branched, continous with the flexuous rachis; leaflets flabellate or rhomboid; rhizome scales minute, less than 0.5 mm wide; plants widespread ...... *A. capillus-veneris*

— Blades broader than long; stipes dichotomously branched apically; leaflets subrectangular; rhizome scales 1–2 mm wide; plants of restricted distribution .... *A. pedatum*

Adiantum capillus-veneris L. Maidenhair Fern. [*A. modestum* Underw.; *A. capillus-veneris* var. *modestum* (Underw.) Fern.; *A. rimicola* Slosson, type from San Juan County; *A. capillus-veneris* var. *modestum* f. *rimicola* Fern.]. Rhizomes creeping, brown scaly; leaves 3–40 cm long, including stipes, pinnately compound, the rachis flexuous; stipes scaly at the base, black below, brown above, shiny and smooth; pinnules broadly obovate-cuneate to rhomboidal, often oblique at the base, the main vein marginal, the margins lobed or cleft, the lobes in turn serrate-dentate; indusium often transversely elongated or curved, greenish, becoming brown in age, the margin thin and hyaline, entire or undulate-erose; n = 30, 60. Seeps and hanging gardens, commonly in sandstone or limestone, at 830 to 1850 m in Emery, Grand, Garfield, Kane, Millard, San Juan and Washington counties; British Columbia to South Dakota, Missouri, Virginia, Florida, and south to Texas, Arizona, California; subtropics of both hemispheres; 69 (xii).

Adiantum pedatum L. Northern Maidenhair Fern. [*A. pedatum* var. *aleuticum* Rupr.; *A. pedatum* ssp. *aleuticum* (Rupr.) Calder & Taylor]. Rhizome thick, beset with brown scales; leaves solitary or few, mostly 20–70 cm long, the petioles purplish black, 10–40 cm long, the blade glabrous, with main pinnae dichotomously branched, shorter than the petioles and more or less parallel with the ground, 3–60 cm long and about as wide; pinnae commonly 8 or more, with the central ones the longest; pinnules broader than long, with a straight lower margin and curved, lobed, upper margin bearing the sori; 2n = 58, 60. Shaded wet cliffs, crevices, and streamsides at 2400 to 3860 m in Garfield, Salt Lake, and Washington counties; Alaska to the Atlantic, south to California, Oklahoma, and Georgia; Asia; 31 (0). Our material is assignable to **var. *aleuticum* Rupr.**

Aspidotus Copel.

Small, mesophytic rockferns with short scaly rhizomes; scales of rhizome narrow, attenuate; leaves glabrous, evergreen, only slightly dimorphic, if at all, 2–4 times pinnate, with free veins; pinnules often confluent; stipes slender and wiry, brown; margins of fertile pinnules reflexed and abruptly scarious-margined, the margin forming an indusium for the submarginal sori.

Aspidotus densa (Brack.) Lellinger [*Cheilanthes siliquosa* Maxon; *Cryptogramma densa* (Brack.) Diels in Engler & Prantl]. Leaves evergreen, glabrous, 6–25 cm long, including the stipe, densely tufted; stipes dark reddish brown, shiny, 4–18 cm long, usually longer than the blades, scaly at the base and on the rhizome, with minute, firm, narrow, acute and almost black scales; blades mostly monomorphic and fertile or a few of them sterile, ovatetriangular to oblong in outline, 2.5–6 cm long, tripinnately compound; pinnae few, 4–8 in offset pairs, the basal ones broadly triangular and longer than the upper ones, the ultimate segments numerous, confluent, linear to linear-elliptic, 3–12 mm long, abruptly tapering to a firm, mucronate tip, the margins strongly revolute, the recurved portion abruptly white-membranous; sori borne on the white indusial leaf margin at its junction with the green portion; n = 30. Rock crevices and moist slopes at 2830 to 3400 m in Salt Lake and Utah counties; British Columbia to Gaspe, south to Oregon, Idaho, and Montana; 2 (0).

Asplenium L.

Evergreen, small to medium ferns; rhizome short, with numerous roots; scales of rhizome dark, long, and slender; leaves 1- to 3-pinnate, or irregularly divided; veins free, simple or forked and not extending to the margin; petiole slender, wiry, green to brown or black; sori laminar, elongate, each borne on a veinlet; indusium hyaline, flaplike, attached along the vein that bears the sorus and opening along the side toward the midline of the segment; spores bilateral. Note: This is an obscure genus in Utah, consisting of small, delicate, easily overlooked ferns. Their rarity in the state together with their inconspicuous nature accounts for the paucity of collections.

1. Leaf blades irregularly forked, with a few narrow segments, each segment 1–2 cm long, entire or with a few teeth; sori very elongate; plants known from the Uinta and La Sal mts. .... *A. septentrionale*

— Leaf blades with many pinnae, not narrowly linear; sori not very elongate 2

2(1). Fronds bipinnate to tripinnate, ovate-deltoid in outline; plants known from Zion National Park .... *A. adiantum-nigrum*

— Fronds once pinnate; plants of various distribution 3

3(2). Stipes brown below, the rachis green or greenish; leaves soft, not evergreen; plants of the Uinta and Wasatch mts. .... *A. viride*

— Stipes and rachis purple brown to nearly black; leaves firm, evergreeen 4

4(3). Stipes and rachis purple brown; leaf divisions without a small, earlike lobe at base, broadest above the base; plants of the Wasatch Mts. .... *A. trichomanes*

— Stipe and rachis blackish; leaf divisions with a small, earlike lobe at base, broadest above the base; plants of Washington County .... *A. resiliens*

Asplenium adiantum-nigrum L. Black Spleenwort. [*A. andrewsii* A. Nels.]. Fronds tufted or a few together, 1–3 dm long; stipes chestnut brown to blackish below and greenish above; blades ovate-deltoid to elongate deltoid, 3–15 cm long, 2.5–7.6 cm wide, bipinnate or ternate, pinnae deltoid below to lanceolate above, the segments ovate-lanceolate, incised-serrate; sori short but almost connected in a continous chain on the pinnae; indusia straight, entire or nearly so; n = 72. Shaded, mesic cliffs of Navajo Sandstone, in the mountain brush and ponderosa pine community at 1750 m in Washington County; Arizona and Colorado; Europe, Asia, and Africa; 4 (i). Our material does not appear to differ in any major respects from the Old World specimens.

Asplenium resiliens **Kunze** Ebony Spleenwort. Fronds clustered, mostly 5–25 cm long; blades subcoriaceous, linear-oblong to narrowly oblanceolate, pinnate; stipe short, slender, shining, black or purplish black, rounded beneath, flattened and with 2 narrow wing-angles above; rachis colored like the stipe; pinnae 15–35, opposite or slightly offset pairs, irregularly crenate-serrulate, asymmetrically attached at the base; sori usually several on each of the fertile pinnae, 1–2 mm long. Shady crevices of sandstone in warm desert shrub to pinyon-juniper and mountain brush communities, usually in shaded canyons, at 1300 to 1800 m in San Juan and Washington counties; Colorado to Pennsylvania, south to Mexico; South America; 5 (i).

Asplenium septentrionale **(L.) Hoffm.** [*Acrostichum septentrionale* L.]. Fronds densely tufted, 5–20 cm long; stipes slender, brown purple at the base, naked for some distance; blades irregularly forking, with 2–5 narrowly linear, grasslike, rather rigid segments 1–2 cm long, these tapering from the middle, entire or with a few long narrow teeth near the apex; sori elongate, 2–3 per segment, usually in pairs near each margin; indusium continuous just within the margin on each side, entire or sparingly short ciliate; n = 36, 72. Rock crevices in sagebrush, pinyon-juniper, mountain brush, and spruce-fir communities at 1600 to 2715 m in Daggett, Grand, San Juan, and Uintah counties; Oregon to South Dakota, south to Baja California, New Mexico, and Oklahoma; 4 (0).

Asplenium trichomanes **L.** Maidenhair Spleenwort. Rhizomes very short; leaves clustered, 5–20 (25) cm long, glabrous, evergreen, associated with persistent leaf bases of previous seasons; petioles slender, curved, shining, dark reddish brown; blade oblong to linear in outline, the rachis colored like the stipe; pinnae mostly 12–35 opposite or offset pairs, 2–9 mm long and 1–7 mm broad, toothed; sori elongate, with conspicuous indusium; n = 36, 72, 81. On rock and crevices of cliffs or talus slopes, in mountain brush, aspen, and spruce-fir communities, at 1750 to 3000 m in Salt Lake and Utah counties; Alaska to the Atlantic, south to Oregon, Arizona, Texas, Alabama, and Georgia; Eurasia; 3 (0).

Asplenium viride **L.** Green Spleenwort. Rhizomes short; leaves clustered, 2–15 cm long, glabrous or sparsely glandular-hairy, usually not evergreen, commonly with persistent leaf bases of previous seasons; petioles slender, brown to purplish brown at base, becoming greenish upwards; blade oblong in outline, the rachis green or greenish; pinnae mostly 3–20 opposite or subopposite pairs, 2–9 mm long and 1–6 mm wide, crenate; sori elongate, with conspicuous indusia. Rock crevices, mostly on limestone, in spruce-fir and alpine tundra communities at 3550 to 3980 m in Cache, Duchesne, Salt Lake, Uintah, and Utah counties; Alaska to the Atlantic, south to Washington, Nevada, Colorado, Wisconsin, and New York; 14 (0).

Athyrium Roth

Medium to large sized, mesophytic, deciduous ferns; rhizomes short, scaly, ascending, clothed with persistent petiole bases of preceding seasons; leaves 2–4 times pinnately compound, the pinnae once to thrice compound or pinnatifid, the veins reaching the margin; petiole coarse, flattened and black basally becoming herbaceous upwards, scaly, with 2 vascular bundles at the base, these anastomosing upwards; sori round to elongate; indusium hyaline, flaplike, attached along the vein on the side of the sorus toward the margin of the segment, or lacking.

1. Leaf blades finely dissected, the pinnules narrowly lanceolate in outline, acute apically; indusium lacking; plants uncommon .... *A. distentifolium*
— Leaf blades rather coarsely dissected, the pinnules lance-oblong to lanceolate in outline, more or less rounded apically to acute; indusium present; plants locally common .... *A. filix-femina*

Athyrium distentifolium **Tausch ex Opiz** Alpine Lady-fern. Rhizomes short, with leaves arranged in a vaselike tuft; leaves mostly 20–80 cm long, glabrous, deciduous, usually borne with persistent leaf bases of previous seasons; petioles coarse, blackish and scaly basally, becoming greenish or straw colored and sparsely scaly upwards; blade 15–60 cm long, lance-elliptic in outline, 2–4 times pinnately compound or pinnatifid; pinnae mostly 15–25 pairs; pinnules numerous, lanceolate in outline, acute; sori round, less than 1 mm wide; indusium lacking. Stream margins and other moist sites at 2700 to 3300 m in Cache and Weber counties; Alaska to California, Nevada, and Colorado, and disjunctly in northeastern North America; circumboreal; 4 (0). Our material belongs to var. *americanum* (**Butters**]) **Cronq.** [*A. alpestre* var. *americanum* Butters; *A. distentifolium* ssp. *americanum* (Butters) Hulten].

Athyrium filix-femina **(L.) Roth** Lady-fern. [*Polypodium filix-femina* L.]. Rhizomes short, with leaves arranged in a vaselike tuft; leaves mostly (20) 30–130 cm long, glabrous, deciduous, usually with persistent leaf bases of the previous seasons; petioles coarse, blackish and scaly basally, becoming greenish or straw colored and sparsely scaly upward; blade (15) 25–100 cm long, lanceolate to lance-elliptic in outline, 2–3 times pinnately compound or pinnatifid; pinnae mostly 20–35 pairs; pinnules numerous, oblong to lanceolate in outline, rounded to acute; sori oblong to horseshoe-shaped; indusium straight or curved, often toothed; 2n = 80. Moist sites in mountain brush, ponderosa pine, aspen, and spruce-fir communities at 2250 to 3200 m in Daggett, Juab, Salt Lake, Sevier, Summit, Uintah, Utah, Washington, and Weber counties; Alaska to the Atlantic, south to California, Texas, and Florida; circumboreal; 20 (v). Our plants have been assigned to var. *cyclosorum* (**Rupr.**) **Ledeb.** [*A. cyclosorum* Rupr.], which is separable only with difficulty from the typical European materials.

Cheilanthes Swartz

Small, evergreen ferns with scaly short to widely creeping, much branched rhizomes; fronds not dimorphic, rigidly erect-spreading, mostly tomentose to glandular or paleaceous; stipes wiry, blackish to reddish brown or stramineous; blades pinnate to bipinnate-pinnatifid or further decompound, with the veins free and thickened at the tips; sori borne at the thickened vein tips, marginal, roundish, and distinct or narrowly confluent; indusia formed by the thin reflexed margin of the ultimate segments.

1. Leaf blades with dense, white or brown scales only; rhizomes elongate; plants of Kane and Washington counties .... *C. covillei*
— Leaf blades with distinct hairs, with scales often present also; rhizomes short and much branched 2
2(1). Leaves densely tomentose, lacking scales, ovate to oblong-lanceolate, ultimate segments oblong-oval to obovate; plants widely distributed .... *C. feei*
— Leaves with both hairs and scales, the blades lanceolate to oblong-lanceolate; plants of restricted distribution ... 3
3(2). Ultimate segments oblong, distant; plants known from Cache County .... *C. gracillima*
— Ultimate segments rounded or obovate, approximate; plants of San Juan (?) County .... *C. eatonii*

Cheilanthes covillei Maxon Coville Lipfern. Loosely or densely tufted ferns 10–30 cm tall; rhizomes short, white-scaly when young, the scales becoming brown in age, each scale with a dark hard stripe; petioles 3–12 cm long, dark brown, scaly; blades 3–4 times pinnate, ovate-lanceolate to lanceolate in outline, 3–12 cm long, 2–6 cm wide, green and glabrous above, densely scaly beneath with whitish to brownish, attenuate, appressed scales; pinnae of 6–15 pairs; ultimate segments very small, rounded to obovate, beadlike, green above, densely scaly beneath. Dry rock crevices in warm desert shrub and pinyon-juniper communities at 850 to 1450 m in Kane and Washington counties; Arizona, Nevada, and California; 8 (i).

Cheilanthes eatonii Baker Eaton Lipfern. Loosely to densely tufted ferns 8–36 cm tall; rhizomes short, much branched, covered with pale brown scales, these with dark hardened centers; petioles 7–18 cm long, tan to dark purplish brown, villous and scaly with tawny scales; blade 3–4 times pinnate, lanceolate to oblong-lanceolate in outline, 5–20 cm long, 2–5 cm wide, tomentose and rusty-scaly, the scales mostly beneath on the rachis and costa; pinnae 8–20, the lowest ones remote; ultimate segments small, roundish or obovate-spatulate, the margins strongly inrolled and with a pale, scarious edge. Dry rock crevices or on talus in mixed desert shrub communities at 1400 to 1800 m in San Juan (?) County; Texas to Oklanoma, Colorado, New Mexico, and Arizona; 0 (0). The species is included here on the basis of a specimen cited as having been taken in Utah (Intermountain Flora 1: 204. 1972).

Cheilanthes feei Moore Fee Lipfern. Tufted ferns 6–15 (30) cm tall; rhizomes short, covered with brown scales, these with dark hard centers; petioles 3–8 cm long, brown to purplish-brown, sparsely hairy, with tawny, spreading pubescence; blade 3–4 times pinnate, ovate-lanceolate to triangular in outline, 3–13 cm long, 1.5–4 cm wide, densely white or brownish tomentose beneath, green or only sparsely long-hairy above; pinnae 6–12 pairs; ultimate segments small, 1–1.5 mm long, rounded, the margins loosely inrolled, the sporangia spread over the entire surface of the segment; $2n = 87$. Crevices and talus slopes in sagebrush, pinyon-juniper, mountain brush, ponderosa pine, and Douglas fir communities at 1200 to 3000 m in Beaver, Cache, Carbon, Daggett, Emery, Garfield, Grand, Iron, Kane, Millard, Salt Lake, San Juan, Sevier, Uintah, Utah, Washington, and Wayne counties; British Columbia to Illinois, south to California, Texas, and Mexico; 100 (ix).

Cheilanthes gracillima D. C. Eaton Lacefern. Loosely to densely tufted ferns 10–20 (30) cm tall; rhizomes short, covered with brown scales, these with hard centers; petioles 3–12 cm long, dark brown, glabrous to sparsely clothed with scattered long hairs or scales or both; blade mostly 2–3 times pinnate, linear-lanceolate to lanceolate in outline, 4–12 cm long, 1–2 cm wide, green and glabrous above; scaly beneath with long tawny or brownish scales, also villous-puberulent as well; pinnae 9–20 on each side, crowded; ultimate segments small, mostly 1–3 mm long, 1–1.5 mm wide, oblong to obovate, green above, rusty woolly beneath, also sometimes with scattered narrow scales, the margins widely recurved. Crevices in mountain brush, aspen, spruce-fir, and alpine tundra communities at 2060 to 3700 m in Cache County; British Columbia to Montana, south to California and Nevada; 2 (0).

Cryptogramma R. Br.

Small, mesophytic, deciduous or evergreen ferns; rhizomes short or somewhat elongate, scaly; leaves dimorphic, fertile and vegetative, mostly (1) 2 or 3 times pinnate or pinnatifid, the veins reaching the margin, at least on vegetative leaves; petiole slender, green, straw colored or purplish black, scaly, with a single vascular bundle; fertile leaves longer than the vegetative ones, with relatively long, narrow pinnules; sori more or less continuous along the margin of pinnules, the margin revolute, forming a false indusium.

1. Rhizomes short, densely leafy and often clothed with old leaf bases; leaves tufted, evergreen; plants of the Uinta and Wasatch mts. and elsewhere in the high plateaus and mountains of southern Utah .... *C. crispa*
— Rhizomes elongate, sparsely leafy, lacking persistent leaf bases; leaves deciduous; stipes dark brown; plants of the Wasatch Mts. .... *C. stelleri*

Cryptogramma crispa (L.) R. Br. Parsley-fern; Rockbrake. [*Osmunda crispa* L.]. Rhizomes short, compactly branched, clothed with scales, persistent leaf bases, and tufted leaves; vegetative leaves (3) 7–25 cm long, with scaly, straw colored to greenish petioles 1.5–17 cm long and ovate to ovate-lanceolate blades 2–3 times pinnately compound, the pinnae commonly 5–11, twice pinnate, the ultimate segments toothed; fertile leaves longer than the vegetative ones, the fertile pinnae about as many as the vegetative, the ultimate segments linear to narrowly oblong; sori more or less continuous, covered by the revolute margin of the ultimate segments; $2n = 60$. Crevices and talus in ponderosa pine, lodgepole pine, spruce-fir, and alpine tundra at 3200 to 3800 m in Cache, Beaver, Duchesne, Grand, Iron, Piute, Salt Lake, San Juan, Sevier, Summit, Utah, Wasatch, and Wayne counties; Alaska east to the Great Lakes, south to California, Mexico, and Nebraska; 28 (ix). Our material belongs to var. **acrostichoides** (R. Br.) C. B. Clarke [*C. acrostichoides* R. Br.].

Cryptogramma stelleri (S. G. Gmel.) Prantl Slender Cliff-brake. [*Pteris stelleri* S. G. Gmel.]. Rhizomes slender, creeping, scaly, with leaves scattered along the length of the rhizome; vegetative leaves 5–15 cm long, with scaly, purplish brown petioles 1–9 cm long and ovate to oblong blades 1 or 2 times pinnate, the pinnae commonly 3–5, these usually merely pinnatifid, the ultimate segments crenately toothed; fertile leaves commonly surpassing the vegetative ones, mostly 6–19 cm long, with petioles 3–9 cm long, these colored much like the vegetative petioles, the ultimate segments lanceolate to lance-linear; sori continuous, the revolute margin membranous, translucent. Moist shaded outcrops at 2500 to 3500 m in Utah County; Alaska to the Atlantic, south to Washington, Nevada, Colorado, Iowa, and West Virginia; Asia; 8 (0).

Cystopteris Bernh.

Small to medium, mesophytic, delicate, deciduous ferns; rhizomes short or elongate, scaly; leaves clustered or scattered, all about alike, mostly 2–4 times pinnate or ternate-pinnate or pinnatifid, the veins reaching the margin; petioles slender, scaly, brown to green or straw colored, not jointed, with 2 vascular bundles; sori round, borne along veins on the lower side of the blade; indusium attached under the sorus, the free tip hoodlike, arching over the sorus, often pushed back as the sorus enlarges, soon withering.

Blasdell, R. F. 1963. A monographic study of the fern genus Cystopteris. Mem. Torrey Bot. Club 21(4): 1–102.

1. Leaves 3–40 cm long, ovate-lanceolate to oblong-lanceolate, the lowest pair of pinnae the shortest; plants common and widespread .... C. fragilis
— Leaves 30–80 cm long, lanceolate to triangular-lanceolate, the lowest pinnae pair the longest; plants restricted and uncommon .... C. bulbifera

Cystopteris bulbifera (L.) Bernh. Bulblet Bladderfern. [*Polypodium bulbiferum* L.]. Leaves tufted, 30–80 cm long, the blades lanceolate to elongate triangular-lanceolate in outline, tripinnate; pinnae deltoid-lanceolate to oblong-lanceolate; pinnules oblong, broadly decurrent, pinnatifid or deeply incised, the segments more or less obtuse and sparingly dentate, minutely glandular on the lower surface; stipe flat in front below, becoming grooved above and with traces of lateral grooves, the rachis and midribs often bearing bulblets on the undersides; $2n = 42, 84$. Crevices and talus slopes up to 3000 m in Salt Lake, San Juan, and Washington counties; Manitoba to Newfoundland, south to Arizona, New Mexico, Arkansas, and Georgia; 3 (0).

Cystopteris fragilis (L.) Bernh. Brittle-fern. [*Polypodium fragile* L.]. Plants loosely tufted, from short, creeping rhizomes; leaves 3–40 cm long (including stipe), oblong-lanceolate to ovate-lanceolate in outline, tripinnate, thin; stipes brown below, yellowish above, smooth, with a groove in front, except at the base, and with 2 lateral grooves; pinnae ovate to oblong-lanceolate; pinnules or segments glabrous, oblong, broadly decurrent, confluent above, dentate to incised, the lower ones often pinnatifid, the tips obtuse or a few acute; indusia small, attached to one side and arching backward at maturity like a hood, free margin rounded or elongate, erose, soon withering; $2n = 84, 126, 168, 252$. Crevices, talus, and in other damp or shady places at 1600 to 4000 m in all Utah counties; Alaska to Newfoundland, south to California, Arizona, New Mexico, and Texas; 284 (xxi). This is the most widespread and most common fern species in Utah.

Dryopteris Adanson

Deciduous ferns from thick rhizomes; leaves 2–3 times pinnate; stipes continuous with the rhizome, sheathed basally with large, fimbriate scales, the upper part often scaly and glandular; pinnae mostly lanceolate; pinnules numerous, the veins pinnate and forked; sori round in outline, borne on the veins, in 2 rows per pinnule; indusium attached at one side as indicated by a deep notch, lunate or horseshoe-shaped, large and conspicuous.

Dryopteris filix-mas (L.) Schott Male Fern. [*Polypodium filix-mas* L.]. Leaves 2–10 dm long, tufted, on short, stout, densely scaly rhizomes; stipes coarse, pale or brownish at base, densely scaly with long, thin, brown scales; blades oblong-lanceolate in outline, bipinnate; pinnae numerous, narrowly lanceolate, to 15 cm long and 4 cm wide, the largest near the middle, the ultimate segments oblong, rounded, finely serrate, the rachis and often the lower surface of segments with elongate, hairlike scales; sori round; indusium conspicuous; $2n = 82, 164$. Shaded moist sites in mountain brush and ponderosa pine communities at 1350 to 1850 m in Garfield, Salt Lake, Sanpete, and Washington counties; British Columbia to Newfoundland, south to California, Texas, Oklahoma, and Vermont; 29 (i).

Notholaena R. Br.

Small xeric ferns; rhizomes short, branched, with long, slender, brownish scales; leaves evergreen, firm, 2–5 times pinnate, the ultimate segments quite small, often waxy or woolly, margins slightly recurved or nearly flat, not forming a false indusium over the sporangium; sori submarginal at the ends of veins.

Tyron, R. 1956. A revision of the American species of Notholaena. Contr. Gray Herb. 179: 1–106.

1. Leaves densely white or brownish tomentose on both surfaces .... N. parryi
— Leaves glabrous or mealy with a waxy powder, not woolly 2
2(1). Leaves glabrous, green, not conspicuously waxy; segments few and large .... N. jonesii
— Leaves abundantly mealy beneath, the segments small and numerous 3
3(2). Rachis of leaf straight or nearly so .... N. limitanea
— Rachis of leaf sharply flexuous; plants not definitely known from Utah, but to be expected in San Juan County .... N. fendleri Kuntze

Notholaena jonesii Maxon Jones Cloak-fern. [*Pellaea jonesii* (Maxon) Morton]. Leaves tufted from a short, conspicuously scaly rhizome, the scales thin, brown, and long-attenuate; petiole and rachis reddish brown, glabrous; blades oblong-ovate to narrowly triangular, bipinnate or partly tripinnate, 3–10 cm long; pinnae few, 4–10 pairs; ultimate segments small, 1.5–4.5 mm long, broadest near the subtruncate to shallowly cordate base, glabrous; margins often slightly inrolled but not covering the sporangia. Crevices in cresote bush, blackbrush, and other warm desert shrub communities, and in sagebrush and pinyon-juniper communities at 830 to 1200 m in Washington County; Arizona, Nevada, and California; 2 (i).

Notholaena limitanea Maxon Border Cloak-fern. [*Pellaea limitanea* (Maxon) Morton]. Leaves tufted, from a short, decumbent woody rhizome 1–4 cm long, conspicuously scaly, the scales wholly scarious, pale castaneous, linear-attenuate; petioles purplish black, 4–14 cm long, glabrous; blades 4–5 times pinnate, deltoid-ovate, 5–15 cm long, 4–11 cm wide, the rachis delicate, with the smallest ones almost capillary, purplish black; pinnae 4–13 pairs, ascending to spreading; ultimate segments sessile or nearly so, 2–3 mm long, linear-oblong to ovate-oblong, the lower surface conspicuously white-mealy, the upper glabrous; margins slightly inrolled but not totally covering the sporangia. Crevices at 1200 to 1450 m in Grand and San Juan Counties; Arizona, New Mexico, Texas, and Mexico; 9 (0).

Notholaena parryi **D. C. Eaton** Parry Cloak-fern. [*Cheilanthes parryi* (D. C. Eaton) Domin]. Densely tufted, woolly ferns, 8–23 cm tall; rhizomes short, covered with scarious, reddish brown scales with dark hard midstripes; petioles blackish purple, 3–10 cm long, loosely viscid-villous; blades ovate-oblong to oblong-lanceolate in outline, bipinnate to pinnatifid, the pinnae in 5–12 pairs; ultimate segments rounded, often obscured by the dense, long, woolly hairs on both sides, but especially so beneath, grayish above, rusty beneath; margins crenate and widely recurved, becoming flattened with age; 2n = 60. Crevices in warm desert shrub at 850 to 1350 m in Washington (type from near St. George) County; Arizona, Nevada, and California; 29 (v).

Pellaea Link

Small xeric ferns; rhizomes short, branched, with dense, long, narrow, brown scales giving a woolly appearance; leaves evergreen, pinnate, firm, the veins free, the surface glabrous or sparsely hairy; petioles green to reddish brown or blackish purple, slender, wiry, breaking off above ground level, leaving a persistent base; sori marginal, confluent, covered by the reflexed leaf margin, this forming a continuous false indusium.

Tryon, A. F. 1957. A revision of the fern genus *Pellaea*, section *Pellaea*. Ann. Missouri bot. Gard. 44: 125–193.

1. Scales of rhizome uniformly colored; petiole and rachis terete or elliptical 2
— Scales of rhizome bicolored, with a hard central stripe; petiole and rachis sulcate, convex, or plane on the upper surface 3
2(1). Rhizomes thick, with numerous, compressed, short, articulated petiole bases; middle and lower pinnae asymmetrical, mitten-shaped .... *P. breweri*
— Rhizomes rather stout, with few petioles persisting, these usually not articulated; pinnae seldom mitten-shaped, the lower ones often with 1 or 2 pairs of pinnules .... *P. glabella*
3(1). Basal pinnae usually less than twice as long as broad, entire or divided into 3–11 segments; rachis of pinnae to 2 cm long .... *P. ternifolia*
— Basal pinnae usually more than twice as long as broad, divided into 9–21 segments; rachis of pinnae to 7 cm long .... *P. truncata*

Pellaea breweri **D. C. Eaton** Brewer Cliff-brake. Rhizome compact, large, with many compressed bases of articulated petioles; scales of rhizome entangled and matted, uniformly rusty brown, lustrous, acicular, the margins sinuate, the apex attenuate; blades 2.5–21 cm long, curved; petiole and rachis terete, brownish, shining, with prominent lines of articulation; blade 1.5–16 cm long, 0.5–3.5 cm wide, linear-oblong, once pinnate or pinnate-pinnatifid, the upper pinnae entire, sessile, the lower asymetrical, deeply 2–lobed, unilateral or mitten-shaped, 0.5–2.5 cm long, 0.3–1 cm wide, the margin narrow, whitish, crenulate. Rocky hillsides, outcrops, and talus in sagebrush, mountain brush, aspen, pine, and spruce-fir communities at 2500 to 3400 m in Beaver, Box Elder, Cache, Grand, Juab, Millard, Salt Lake, Sevier, Summit, Tooele, Uintah, Utah, Wasatch, and Weber counties; Washington to Montana, south to California and Wyoming; 48 (ii).

Pellaea glabella **Mett. ex Kuhn** Suksdorf Cliff-brake. [*P. suksdorfiana* Butters]. Rhizome thickish, compact; scales of rhizome usually not matted, uniformly rusty brown, shining, linear, flexuous, attenuate; leaves 1–36 cm long, usually lax; petiole and rachis terete, glabrous or nearly so, brownish black, shining, the lines of articulation lacking or nearly so; blades 0.7–21 cm long, 0.5–8 cm wide, linear or ovate-lanceolate, 1–2 pinnate, usually bluish green, somewhat glaucous, the upper pinnae entire or auriculate, sessile or subsessile, the lower pinnae entire, 2– to 5–lobed or pinnate with 3–7 segments, the segments 0.5–3.5 cm long, 0.3–1 cm broad, oblong to linear-ovate. Crevices and hanging gardens at 1200 to 2800 m in Emery, Grand, Kane, Sevier, Uintah, and Washington counties; British Columbia to Colorado, Arizona, and New Mexico; 11 (iv).

Pellaea ternifolia **(Cav.) Link** Ternate Cliff-brake. [*Pteris ternifolia* Cav.]. Rhizome thickish, elongate, decumbent; scales of rhizome free, brownish or with tan tips, bicolorous with a narrow sclerotic stripe, subulate, the margins irregularly dentate or erose, the apex attenuate, filiform; leaves 4–50 cm long, erect, straight, stiff; petiole and rachis convex or plane to sulcate, rarely terete, glabrous or rarely pubescent, brownish black, becoming black in age, breaking irregularly, without articulation lines; blade 3–32 cm long, 0.5–6 cm wide, linear, lanceolate, or elongate triangular, 1–2 pinnate, ternate, or with 3–11 segments, the segments 0.5–4 cm long, 0.5–1 cm wide, lanceolate to narrowly oblong, entire or ternately divided; n = 58. Crevices in sandstone at ca 1500 m in Washington County (Zion National Park); Arizona to Texas, Mexico, South America, and Hawaii; 4 (i). Specimens examined fall within the limits of this species as outlined in Tryon's (1957) revision. The species is closely allied to *P. truncata*, but the ternate pinnae are distinctive.

Pellaea truncata **Gooding** Spiny Cliff-brake. [*P. longimucronata* Hook., nom. illeg.]. Rhizome thickish, elongate, decumbent; scales of rhizome appressed, brown, bicolorous, with a broad, hard stripe usually broader than the border, subulate, the margin and apex erose-dentate; leaves 12–38 cm long, erect, straight, stiff, alike, the lower pinnae often sterile, the upper ones fertile; petiole and rachis sulcate, glabrous or nearly so, usually glaucous, castaneous, becoming darker with age, the stipe breaking irregularly, without articulation lines; blade 8–22 cm long, 4–12 cm broad, triangular, bipinnate, rarely tripinnate, grayish green, the segments 0.3–1.5 cm long, 0.1–1 cm wide, narrowly oblong to oval, entire, mucronate apically. Crevices in warm desert shrub, oak, pinyon-juniper, and ponderosa pine communities at 980 to 2000 m in Iron, Kane, San Juan, and Washington counties; Arizona, Colorado, New Mexico, Nevada, and Mexico; 37 (iv).

Pityrogramma Link.

Mesic to xeric evergreen (or deciduous) ferns; rhizome short, clothed with slender scales; leaves tufted; petiole elongate, slender, dark, shining, the base scaly, glabrous above, the blades 3- to 5-angled in outline, rather broad, pinnate to ternately pinnately dissected, white or yellow mealy beneath, otherwise glabrous; sori continuous on the veins, from the midvein to the margins; indusium lacking.

Pityrogramma triangularis (Kaulf.) Maxon Goldback Fern. [*Gymnogramma triangulare* Kaulf.]. Rhizome short, covered with thick scales with hard midstripe; leaves tufted; petioles brown to purplish brown, 6–23 cm long, considerably longer than the blades, these glabrous above, densely white or yellow mealy beneath, 4–10 cm long and about as wide, ternate-pinnately compound; pinnae few, the lowest pair the largest; leaf margins narrowly revolute; n = 30, 60. Crevices and in soil on shaded rocky slopes in warm desert shrub and ponderosa pine communities at 1175 to 1800 m in Washington County; British Columbia south to California and Nevada; 9 (i).

Polypodium L.

Small to medium ferns from creeping, nodulose, branched, scaly rhizomes; leaves scattered, evergreen, firm, pinnatifid, with toothed or entire segments; veins free; sori round in outline, borne on the veins; indusium wanting.

Polypodium hesperium Maxon Western Polypody. Rhizome densely brown scaly; leaves oblong-ovate to oblong-lanceolate in outline, 8–25 cm long including petioles, these 1–12 cm long, yellowish or green, the base brown; blades 4–18 cm long, 2–5 cm wide, deeply parted or pinnatifid; segments oblong to linear-oblong, the margins serrate to nearly entire; sori midway between the margins of the segments and the midveins; indusium lacking; n = 74. Crevices and other mesic sites in mountain brush, ponderosa pine, aspen, and spruce-fir communities at 1500 to 3000 m in Salt Lake, Utah, and Washington counties; British Columbia to Montana, south to California, Arizona, New Mexico, and South Dakota; 28 (i).

Polystichum Roth

Small to medium, mesophytic, evergreen ferns; rhizomes short, stout, scaly; petiole shorter than the blade, conspicuously scaly toward the base with brownish, dimorphic scales, some broad and toothed, others narrow and hairlike; blade coarse, pinnate, oblong to lanceolate in outline, scaly on the rachis and costae, the margin often with pungent teeth; sori laminar, round, prominent, borne on the veins in one or more distinct rows on each side of the midvein; indusium attached at the center on a projection and spreading peltately over the sorus, the margin fringed.

1. Pinnae undivided, the base asymmetrical, auricled on the upper side, the margin long-spinulose dentate or serrate .... *P. lonchitis*
— Pinnae or some of them deeply cleft toward the base ... 2

2(1). Principal pinnae mostly less than 1.5 (1.8) cm long, mostly 1–2 times as long as wide, the spinulose points of the teeth prominent and tending to spread or incurve; plants of Box Elder County .... *P. krukebergii*
— Principal pinnae mostly more than 1.5 cm long, usually 2–3 times as long as wide, the lobes and teeth mostly ascending or incurved, with tip merely callous-mucronate; plants of other distribution ...... *P. scopulinum*

Polystichum krukebergii W. L. Wagner Krukeberg Holly-fern. Leaves 1–4 dm long; petiole scaly at the base, less so above; blade linear-lanceolate or oblong-lanceolate in outline; pinnae 20–40 on each side of rachis, often closely crowded, ovate in outline, 0.8–1.5 (1.8) cm long, 0.5–1 cm wide, 1–2 times as long as wide, commonly with 1 or rarely 2 prominent lobes near the base, the margin toothed or shallowly lobed, with spinulose, spreading or widely incurved teeth; sori borne on the middle and upper pinnae; indusium erose-dentate; 2n = 82. Crevices and other mesic sites at ca 2900 m in Box Elder County; British Columbia to Idaho, south to California; 3 (0).

Polystichum lonchitis (L.) Roth Holly-fern. [*Polypodium lonchitis* L.]. Leaves 2–5 dm long; petiole usually rather short or almost none, yellowish green, scaly; blade elliptic-lanceolate to oblong-lanceolate in outline; pinnae 25–50 per side of rachis, triangular to oblong-lanceolate, falcate, the larger pinnae usually near or above the middle of the leaf, 1–4.5 cm long, 3–13 mm wide, the margins spinulose serrate, the base auricled on the upper side; lower pinnae gradually reduced in size and more distant, nearly triangular, up to 1.5 cm long and nearly as wide; sori borne on the middle and upper pinnae; indusium erose-dentate; n = 41, 82. Talus and crevices in mountain brush, aspen, and spruce fir communities at 2000 to 3350 m in Cache, Duchesne, Iron, Salt Lake, Sanpete, Tooele, and Utah counties; Alaska to Newfoundland, south to California and New Mexico; 53 (ii).

Polystichum scopulinum (D. C. Eaton) Maxon Rock Holly-fern. [*Aspidium aculeatum* var. *scopulinum* D. C. Eaton]. Leaves 1–4 dm long; petiole scaly at the base, less so above, or nearly glabrous; blade ovate-lanceolate to linear-lanceolate in outline; pinnae 20–40 per side of rachis, closely crowded, ovate to lanceolate, (0.8) 1.5–3.2 cm long, 0.5–1 cm wide, 2–3 times as long as wide, usually with 2 or rarely 3 prominent basal lobes, the margin toothed or shallowly lobed, the lobes or teeth sharply crenate-serrate with ascending or incurved callous-mucronate tips; sori borne mostly on the middle and upper pinnae; indusium erose-dentate. Crevices and rocky hillsides at 1950 to 3500 m in Millard, Salt Lake, and Washington counties; Washington to Quebec, south to California and Arizona; 11 (0).

Pteridium Scop.

Medium-sized to rather large mesophytic ferns; rhizomes hairy, deep-seated, elongate, branching, often forming large colonies; leaves deciduous, scattered; petiole coarse, erect and stemlike; blade firm, spreading, three times pinnately or ternately-pinnately compound, the ultimate segments numerous, crowded, sessile and often confluent; sori confluent, protected by the narrowly inrolled indusial leaf margin, and a delicate inner hyaline indusium.

Pteridium aquilinum (L.) Kuhn Bracken Fern. [*Pteris aquilina* L.]. Clone-forming ferns from hairy rhizomes lacking scales; leaves scattered, 6–8 (20) dm long including petiole; petiole shorter than the blades, coarse, green or yellowish; blades firm, broadly triangular in outline, 2–12 dm long, glabrous to short-hairy above, variously long hairy beneath, 2–3 times pinnate, the ultimate segments oblong and entire to toothed or lobed, widely spreading; sori developed along the leaf margins, confluent, protected by the narrowly recurved leaf margin and by a delicate, concealed, inner indusium; n = 52. In moist or dry wooded areas or clearings or on open slopes in oak-mountain brush, sagebrush, pine, aspen and spruce-fir communities at 1700 to 2850 m in Cache, Grand, Juab, Kane, Rich, San Juan, Sanpete, Uintah, Utah, Wasatch, Washington and Weber counties; cosmopolitan; 64 (vii). Our material belongs to the western **var. pubescens** Underw.

Woodsia R. Br.

Small to medium-sized ferns; rhizomes short and thick, covered with brown scales; leaves 2–3 times pinnate, oblong-lanceolate to lanceolate in outline; pinnae pinnately divided, triangular to oblong; pinnules crenate or dentate, the margins flat or recurved; sori round, borne on the veins of the pinnules; indusium arising from below the sorus and splitting at maturity into slender segments which become inconspicuous in age.

1. Leaves glabrous, or merely glandular .... *W. oregana*
— Leaves glandular and with glandless septate hairs, at least on the lower surface .... *W. scopulina*

Woodsia oregana **D. C. Eaton** Oregon Woodsia. Leaves tufted, several to numerous, glabrous to glandular but without non-glandular hairs, 7–25 cm long including the petioles; petioles 2–10 cm long, dark reddish brown near the base, lighter above, the blade 4–15 cm long, 1–4.5 cm wide; pinnae ovate-oblong or triangular, pinnatifid, obtuse apically; pinnules crenate, often lobed or cleft near the base, the margins slightly recurved; indusium platelike, with slender hairlike segments that appear beaded, mostly hidden below the sorus. In moist cliffs or dry shaded places or talus slopes with sagebrush, pinyon-juniper, mountain brush, ponderosa pine and white fir communities at 1850 to 3200 m in Beaver, Box Elder, Cache, Daggett, Emery, Grand, Juab, Kane, Millard, Salt Lake, San Juan, Tooele, and Washington counties; Quebec to British Columbia, south to California, New Mexico, Oklahoma, Wisconsin, and Vermont; 56 (vii).

Woodsia scopulina **D. C. Eaton.** Rocky Mountain Woodsia. Tufted, rather similar in appearance to *W. oregana*; leaves several to numerous, glandular and with longer septate non-glandular hairs, 8–35 cm long including the petiole; petiole dark brown at the base, lighter above and hairy; the blade 5–22 cm long, 1–7 cm wide; pinnae narrow, oblong-lanceolate, pinnatifid, acute, glandular hairy on the lower surface; pinnules oblong, crenate-serrate to lobed, obtuse, the base broadly decurrent, not at all or only slightly contracted; indusia splitting into narrow, beaded hairlike segments at maturity. Rocky ledges, crevices, and talus slopes with mountain brush, aspen, spruce-fir, and alpine tundra communities at 2700 to 3800 m in Beaver, Cache, Duchesne, Garfield, Salt Lake, San Juan, Sevier, Summit, Uintah, Utah, and Wayne counties; Alaska to Quebec, south to California, New Mexico, Oklahoma, Tennessee, and North Carolina; 34 (i).

SALVINIACEAE Reichb.

Waterfern Family

Plants small aquatic free-floating or growing on mud; rhizome branched, with simple roots; leaves 2–ranked or in whorls, opposite or alternate, simple or lobed; sporocarps soft, thin-walled, borne singly or 2 or more on a common stalk at the base of the leaves, 1–loculed, each containing a central often branched receptacle bearing microsprangia or megasporangia.

Azolla Lam.

Small plants with pinnately branched reddish or green free-floating stems covered with minute imbricate 2–lobed leaves and producing rootlets beneath; sporocarps borne in pairs beneath the stem, dimorphic, the small ones ovoid and bearing 1 megaspore, the larger globose and containing numerous microspores.

***Azolla mexicana* Presl** Waterfern; Mosquitofern. Plants forming floating mats to 2 cm across; leaves crowded, the upper lobe commonly less than 1 mm long, tinged with purple, papillose, with narrow hyaline cellular-papillose margins; microsporocarps slightly more than 1 mm thick; glochidia with scattered septae; megaspores 0.25–0.3 mm long, the rounded part minutely pitted. Floating on the surface of lakes and ponds or sloughs and ditches at 1480 to 2050 m in Cache, Salt Lake, and Utah counties; British Columbia to Wisconsin, south to northern South America; 6 (0).

DIVISION PINOPHYTA

Gymnosperms

Shrubs or trees with alternation of generations, the gametophyte generation greatly reduced (micro- and megagametophytes, equivalent to pollen and embryo sac respectively); sporophytes with roots, stems, and leaves; leaves needle- or scalelike or broad, spirally arranged, whorled, or opposite; microsporophylls aggregated on an axis, forming a soft strobilus (the male cone); megasporophylls solitary, paired, or aggregated into compound woody (or fleshy) strobili (female or ovulate cones); ovules typically paired (1 to many) on the surface of a scale, the micropyle exposed during pollination; seeds typically large, and often winged.

Key to the families.

1. Stems jointed; leaves scalelike, typically brownish or blackish, opposite or in whorls of 3; branches green and photosynthetic .... **Ephedraceae**, p. 28
— Stems not jointed; leaves various, needle or scalelike, or broad but, if scalelike, closely overlapping, typically green; branches not green 2
2(1). Leaves broad, fan-shaped, dichotomously veined, deciduous; cones drupelike, covered with an aril .... **Ginkgoaceae**, p. 29
— Leaves needlelike, linear, or scalelike, solitary or in fascicles, persistent or deciduous; cones usually not provided with an aril 3
3(2). Leaves alternate or subopposite, arranged in 2 ranks; plants shrubs or small trees, evergreen; cones with a brightly colored aril .... **Taxaceae**, p. 34

— Leaves opposite, whorled, or spirally arranged (if alternate, as in *Taxodium*, not evergreen and cones not arilate), sometimes 2-ranked; plants shrubs or trees, evergreen or deciduous; cones not with a brightly colored aril 4

4(3). Leaves scalelike or awl shaped; female cones mainly 5-25 mm long, globose or oblong, the scales few and decussate (fleshy in *Juniperus*) ..... **Cupressaceae**, p. 25

— Leaves needlelike, linear, oblong, or less commonly scalelike or awl shaped; female cones typically more than 25 mm long or the scales peltate 5

5(4). Leaves apparently 2-ranked (actually spirally arranged) and flattened, deciduous with the branchlets, or scale- or awllike and persistent .... **Taxodiaceae**, p. 34

— Leaves spirally arranged or fascicled and evidently opposite or whorled, persistent or, if scalelike, papery .. .. **Pinaceae**, p. 29

CUPRESSACEAE Bartling

Cypress Family

Monoecious trees or shrubs; leaves evergreen, opposite or whorled, scalelike or awl-shaped and needlelike; staminate cones small, terminal or axillary, the microsporophylls decussate; ovulate cones terminal and commonly with 2-12 opposite or whorled scales, each bearing 1 to several ovules, dry or fleshy at maturity; x = 11, 12.

1. Branchlets not arranged in flat sprays, the twigs extending more or less in all directions; plants indigenous and cultivated 2

— Branchlets arranged in flat sprays, the twigs more or less in a single plane; plants all cultivated 3

2(1). Cones mostly 15 mm thick or more, dry at maturity, the scales woody and finally separating; seeds numerous under each scale; plants cultivated and long-persisting .... *Cupressus*

— Cones berrylike, the scales fleshy at maturity (drying later), not opening; seed 1 or few; plants indigenous or cultivated .... *Juniperus*

3(1). Cones subglobose, the scales shield-shaped; seeds few under each scale .... *Chamaecyparis*

— Cones oblong, the scales imbricated or valvate; seeds 2 per scale 4

4(3). Bark of trunk exfoliating in plates; cone scales 4 or 6; leaves appearing in whorls of 4 .... *Calocedrus*

— Bark of trunk shredded, not exfoliating in plates; cone scales 8 or more (rarely 6); leaves obviously paired . *Thuja*

Calocedrus Kurz

Monoecious aromatic trees; bark scaly, exfoliating in plates; branchlets compressed, forming flat sprays; leaves scalelike, imbricate, 4-ranked, decurrent at the base, dimorphic, the upper and lower ones flattened or rounded, the lateral ones folded; staminate cones cylindroid; ovulate cones cylindroid, with usually 3 pairs of scales, the middle ones fertile and each with 2 ovules, maturing in one season; seeds winged.

Florin, R. 1956. Nomenclatural notes on genera of living gymnosperms. Taxon 5: 191-192.
Li, Hui-Lin. 1953. A reclassification of *Libocedrus* and Cupressaceae. J. Arnold Arbor. 34: 17-36.

Calocedrus decurrens (**Torr.**) **Florin** Incense Cedar. [*Libocedrus decurrens* Torr.; *Heyderia decurrens* (Torr.) K. Koch; *Thuja decurrens* (Torr.) Voss]. Trees, mainly 5-15 m tall, the trunk to 5 dm thick or more; bark scaly, the exfoliating surface reddish to brownish, finally furrowed; leaves appearing to be in whorls of 4, the tips incurved, 4-6 mm long, greenish to yellowish; staminate cones 4-7 mm long, yellow; ovulate cones 15-25 mm long, cylindroid, reddish brown, pendulous, the scales mucronate; seeds typically 4, 8-10 mm long, the longest wing subequal to the scale. Uncommonly grown ornamental and specimen tree in lower elevation portions of Utah; Oregon, California, and Nevada; 7 (0).

Chamaecyparis Spach

Monoecious fragrant trees with opposite decussate scalelike leaves, the lateral ones folded, the upper and lower ones flattened or rounded; branchlets arranged in more or less flattened sprays; staminate cones ellipsoid to ovoid; ovulate cones subglobose, with 3-6 pairs of peltate scales, each bearing 2-5 ovules; cones dehiscent, woody to leathery, dark to gray brown, glaucous.

Chamaecyparis lawsoniana (**Murray**) **Parl.** Port Orford Cedar. [*Cupressus lawsoniana* Murray]. Trees, mainly 5-15 m tall, with base enlarged and the crown spirelike; branchlets horizontal or pendulous, bearing slender flattened branchlets; bark smooth on young trees, finally thick and divided into broad rounded ridges; leaves green above, glaucous below, mainly 1-1.5 mm long (to 6 mm long on vigorous shoots); male cones oblong, reddish; female cones globose, ca 8 mm thick, reddish brown, more or less glaucous, the scales each with an apical conical projection; seeds ovoid, slightly flattened, 3-4 mm long, narrowly wing-margined; n = 11. Uncommonly grown ornamental and specimen tree in lower elevation portions of Utah; Oregon and California; 7 (0). The closely related **C. nootkatensis** (**D. Don**) **Spach** is reported to be grown in Utah also. It differs in having branchlets scarcely flattened and leaves not or obscurely glandular dorsally.

Cupressus L.

Monoecious fragrant resinous evergreen trees; buds naked; branchlets slender, 4-angled; bark finally scaly; leaves opposite, small, trimorphic—linear and prickly, scalelike, or elongate (on juvenile shoots); male cones 4-sided to subcylindrical, with 5-10 pairs of opposite scales; ovulate cones subglobose, maturing in 2 years, with thickened peltate scales; seeds many on the fertile scales, narrowly winged.

Cupressus arizonica **Greene** Arizona Cypress. Spreading to columnar trees, mainly 5-15 m tall; trunk mainly 1-5 dm thick; twigs stout, 4-angled, 1.5-2 mm thick, branching at nearly right angles; bark finally scaly or even furrowed, usually grayish; leaves scalelike, triangular-ovate, sharply pointed, blue green, commonly glaucous, ca 2 mm long; cones globose, 15-25 mm thick, short-stalked, hard and woody, gray to purplish, often glaucous, with 6-8 flattened scales bearing a hard point in the center, remaining attached for several years, finally opening; seeds ca 2 mm long, purplish brown. Rather commonly grown ornamental in Utah (Sanpete, Utah, and Washington counties); Texas to Arizona and Mexico; 11 (ii). The Arizona cypress persists following introduction, often in apparently inhospitable habitats such as at Castle Cliffs in Washington County. The trees grow rapidly from seed, reaching fence-post size in only a few years.

Juniperus L.

Shrubs or trees with opposite or whorled scalelike or awl-shaped leaves; branchlets subterete or angular, not in flattened sprays; staminate cones subcylindrical to subglobose; ovulate cones subglobose, with mostly 3–8 opposite or whorled scales, becoming fleshy and berrylike at maturity.

1.	Plants spreading to prostrate shrubs	2
—	Plants erect or ascending trees or shrubs	5
2(1).	Leaves needlelike or awl-shaped, jointed to the twig, in whorls of 3; plants indigenous or cultivated ...... *J. communis*	
—	Leaves scalelike or, if needlelike, decurrent on the twig, opposite or in whorls of 3; plants cultivated	3
3(2).	Shrubs spreading, open, to 2 m high or more; main branches ascending, the secondary ones spreading to ascending .... *J. chinensis*	
—	Shrubs low, compact, often less than 6 dm high; main branches prostrate to ascending, the secondary branchlets often strongly ascending	4
4(3).	Leaves dark green, obtuse or acutish, with a strong disagreeable odor when bruised .... *J. sabina*	
—	Leaves bluish green to steel blue (often pinkish in winter), acute or cuspidate, the odor not especially disagreeable when bruised .... *J. horizontalis*	
5(1).	Leaves needlelike, jointed at the base, the upper surface with a white band	6
—	Leaves mostly scalelike, decurrent at the base, not jointed, the upper surface usually lacking a white band	7
6(5).	Leaves concave above, rounded or slightly keeled below; plants trees or spreading shrubs, usually without pendulous branches .... *J. communis*	
—	Leaves narrowly grooved above, conspicuously keeled below; trees with pendulous branches .... *J. rigida*	
7(5).	Branchlets usually coarse, mostly 1–2 mm thick; leaf margins typically denticulate	8
—	Branchlets slender, mainly less than 1 mm thick; leaf margins entire	9
8(7).	Scalelike leaves with a conspicuous resin gland on the back; ovulate cones juicy, dark blue to blackish at maturity; plants rare in Utah .... *J. monosperma*	
—	Scalelike leaves not with a conspicuous resin gland on the back; ovulate cones not juicy, becoming fibrous, and usually brownish to purplish when mature; plants common and widespread .... *J. osteosperma*	
9(7).	Scalelike leaves obtuse; needlelike leaves usually in whorls of 3; plants cultivated .... *J. chinensis*	
—	Scalelike leaves acute; needlelike leaves usually opposite	10
10(9).	Mature scale leaves overlapping those directly above; cones ripening the first season, sweet and juicy; plants introduced .... *J. virginiana*	
—	Mature scale leaves rarely overlapping those directly above, or only slightly so; cones ripening the second season, fibrous and of poor flavor; plants indigenous or cultivated .... *J. scopulorum*	

Juniperus chinensis L. Chinese Juniper; Pfitzer Juniper. Spreading shrubs to columnar trees, mainly 1–10 m tall, with trunks several or solitary; needlelike leaves usually in 3's, the scalelike ones opposite, decurrent on the twig; ovulate cones mostly 5–7 mm in diameter, finally brown to brownish purple, commonly glaucous, maturing in 2 years, usually 2- or 3-seeded; $2n = 22, 44$. Commonly grown ornamentals in much of Utah; introduced from eastern Asia; 19 (0). More than 50 cultivars are known for this species, and many of them are grown in Utah. The maze of horticultural forms is difficult to interpret. The commonly grown Pfitzer juniper becomes massive, and is ill-suited for most plantings around homes. It is less commonly grown than formerly.

Juniperus communis L. Common Juniper. [*J. sibirica* Burgsd.; *J. communis* var. *sibirica* (Burgsd.) Rydb.; *J. communis* var. *montana* Ait., sensu authors]. Spreading shrubs to columnar trees, mainly 0.3–6 m tall, with trunks several or solitary; leaves jointed to the stem, mostly in whorls of 3, awl-shaped, 3–10 (15) mm long, spinulose-tipped, usually marked with a white band on the upper surface, dark green on the lower surface; cones maturing the second season, green, ripening bluish black, 5–10 mm thick, usually 1-seeded; seeds 4–5 mm long; $2n = 22, 44$. Aspen, spruce-fir, and less commonly other plant communities, at 1615 to 3375 m in all Utah counties, except Davis and Rich; Alaska and Yukon, east to the Atlantic, south to California, New Mexico, and Georgia; circumboreal; 78 (xv). The species is represented by indigenous spreading shrubs and by erect or spreading cultivated, mostly Old World, selections. Indigenous plants belong to **var. *depressa* Pursh.** The status of the cultivated material has not been determined.

Juniperus horizontalis Moench Creeping Juniper. Stems commonly decumbent to procumbent or prostrate, mostly 0.5–1.5 m long; leaves decurrent, opposite, scalelike or awl-shaped, 1–4 (6) mm long, acute to spinulose-tipped, lacking a white band on the upper surface; cones maturing the first season, green, ripening blue purple or blue black, glaucous, 5–10 mm thick, mostly 3- to 5-seeded. Cultivated ground cover and ornamental, generally grown in planting strips along curbs and sidewalks, in much of Utah; Alaska and Yukon, east to the Atlantic, south to British Columbia, Colorado, Nebraska, Illinois, and New York; 8 (0).

Juniperus monosperma (Engelm.) Sarg. One-seed Juniper. [*J. occidentalis* var. *monosperma* Engelm.; *J. mexicana* var. *monosperma* (Engelm.) Cory]. Shrubs or small trees, typically with several branches from the ground, mainly 2–4 m tall; bark thin, fibrous and ultimately shredded; branchlets stout, ca 2 mm thick; leaves typically opposite, sometimes in 3's, the scalelike ones 1–3 mm long, yellowish green, the tips often spreading, denticulate marginally, typically with a resin gland on the dorsal side; juvenile leaves awl-shaped, sharp, to 5 mm long, decurrent; staminate cones yellowish brown, 3–4 mm long; ovulate cones subglobose, 4–7 mm thick, dark blue to blue purple at maturity, glaucous, succulent but of bad flavor when fresh; seeds 1 (or 2) per cone. Pinyon-juniper and mixed grass-shrub communities at ca 1750 m in Kane (?), San Juan (?), and Washington (?) counties; Arizona to Oklahoma and Texas, south to Mexico; 1 (0). All identifications of this species for Utah are tentative, and probably they are based on equivocal specimens of *J. osteosperma*.

Juniperus osteosperma (Torr.) Little Utah Juniper; Utah Cedar. [*J. tetragona* var. *osteosperma* Torr.; *J. californica* var. *utahensis* Vasey, type from the Little

Wasatch Mts.; *J. californica* var. *utahensis* Engelm., type from near St. George; *J. occidentalis* var. *utahensis* (Engelm.) Veitch; *J. utahensis* (Englem.) Lemmon]. Shrubs or small trees, with single stem or several branches from the ground, mainly 2–4 m tall; bark thin, fibrous and ultimately shredded; branchlets stout, ca 2 mm thick; leaves typically opposite, sometimes in 3's, the scalelike ones (0.5) 1–3 mm long, yellowish green, denticulate on the margins, the dorsal resin gland not usually apparent; juvenile leaves awl-shaped, sharp, 2–8 mm long, decurrent; staminate cones yellowish brown, 3–4 mm long; ovulate cones subglobose, 6–12 mm thick or more, brownish or blue to blue purple at maturity, glaucous, succulent but of bad flavor when fresh; seeds 1 (or 2) per cone. Riparian, mixed warm and cool desert shrub, sagebrush, mountain brush, juniper, pinyon-juniper, ponderosa pine, and aspen communities at 850 to 2440 m in all Utah counties; Montana and Wyoming south to California, Arizona, and New Mexico; 134 (xxiii). This is the other of the pair of pinyon-juniper or juniper-pinyon components of a vast woodland typically placed between the lower elevation, more xeric, cool desert shrub, dominated by sagebrush, and the higher elevation, more mesic, mountain brush or ponderosa pine communities. Juniper is the more xeric of the duo, often serving as nurse trees for pinyon in well developed forests. Pinyon tends to increase proportionally to juniper at higher elevations, and finally is the main component. Trees of both species occur often along drainages and on dune sands below their usual elevational ranges. Utah juniper increases under grazing, and has spread from the thin substrates of ridges and mountain slopes into deeper valley soils. Reclamation attempts to restore the balance has been successful only in part, with reestablishment of juniper taking only a few years in many instances. Devastation of juniper and pinyon on thin substrates, where it is not seral, seems not to have been a reasonable reclamation option. Utah juniper has been exploited for "cedar" posts and firewood.

Juniperus rigida Sieb. & Zucc. Needle Juniper. Trees, mainly 3–8 m tall, the crown pyramidal, with drooping branchlets; leaves all needlelike, linear, 8–24 mm long, sharply pointed, green, with a broad white band above; staminate cones yellowish, 2–4 mm long; ovulate cones brown to black, glaucous, 6–8 mm thick, 1-seeded (?); 2n = 22. Sparingly grown ornamental of much beauty, planted in lower elevation portions of the state; introduced from eastern Asia; 2 (0). This species should be more widely grown.

Juniperus sabina L. Savin. Shrubs, typically with widely spreading decumbent to procumbent or ascending branches, and ascending branchlets; branchlets ca 1 mm thick; leaflets needlelike and spreading or scalelike and overlapping; staminate cones yellowish, 2–4 mm long; ovulate cones brownish blue, glaucous, mainly 6–8 mm thick, usually 2-seeded. Widely and commonly grown ornamental and ground cover, especially in planting strips, over much of lower elevation portions of Utah; introduced from Eurasia; 4 (0). The cultivar "tamaricifolia," which has leaves usually all needlelike, is the commonly grown phase in Utah.

Juniperus scopulorum Sarg. Rocky Mountain Juniper. [*J. virginiana* var. *scopulorum* (Sarg.) Lemmon; *J. virginiana* var. *montana* Vasey, type from Wasatch Mts.]. Trees, typically 3–6 m tall, with a conical to pyramidal or less commonly rounded crown; bark thin, ultimately fibrous and shredded, or less commonly fissured and breaking into platy scales; branchlets ca 1 mm thick or less; leaves typically opposite, sometimes in 3's, the scalelike leaves 0.5–3 (4) mm long, green or blue green, sometimes with an apparent dorsal resin gland; juvenile leaves needlelike, 3–8 mm long; staminate cones 2–3 mm long, brownish; ovulate cones subglobose, mainly 4–6 mm thick, becoming bluish to purplish at maturity, glaucous, maturing in 2 years, succulent but of bad flavor; seeds typically 2; 2n = 22. Pinyon-juniper, mountain brush, ponderosa pine, Douglas fir, white fir, and aspen communities at 1525 to 2830 m in all Utah counties; British Columbia to the Dakotas, south to Nevada, Arizona, New Mexico, and Texas; 116 (xii). Plants of this species are typically montane, but often occur below their usual elevational range in moist valley bottoms where cold air drainage simulates environmental conditions of the mountains. The wood is of excellent quality and has been used for "cedar" chests and fence posts. The heartwood is red and highly aromatic. The famous Jardine juniper in Logan Canyon, aged at 3600 years, belongs to this species. Horticultural selections are available commercially, and the species is commonly grown in Utah.

Juniperus virginiana L. Virginia Juniper; Red Cedar. Trees, typically 3–6 m tall, with a conical to pyramidal or rounded crown; bark thin, ultimately fibrous and shredded; branchlets ca 1 mm thick or less; leaves typically opposite, sometimes in 3's, the scalelike leaves 0.5–2 (3) mm long, green or blue green; juvenile leaves needlelike, 2–5 mm long; staminate cones mainly 2–3 mm long, yellowish or brownish; ovulate cones subglobose, 5–6 mm thick, becoming blue black at maturity, maturing in one season, succulent and sweet; seeds 1 or 2; 2n = 22. Commonly grown ornamental in much of Utah; northeastern U. S., south to Florida and west to Missouri and Texas; 8 (0). This species is difficult to distinguish from the indigenous, and less commonly cultivated, *J. scopulorum*. It differs in minor, but apparently significant ways, as outlined in the key. Cedar chests and other insect resistant drawers and closets are typically lined with the red heartwood of this species, and the ovulate cones (berries) are used to flavor certain alcoholic beverages.

Thuja L.

Trees or shrubs with opposite decussate scalelike leaves, the lateral leaves folded, the upper and lower rounded or flattened; branchlets arranged in distinctive flattened sprays; staminate cones subglobose; ovulate cones erect or reflexed, oblong-ellipsoid, with 4–6 pairs of laterally attached scales, the middle scales each bearing 2 or 3 ovules; cones dehiscent, woody or fleshy, green or turning tan or brownish at maturity.

1. Branchlets typically in vertical sprays; cone scales fleshy, strongly curved, greenish or yellowish at maturity .... *T. orientalis*
— Branchlets typically in horizontal or at least not vertical sprays; cone scales not fleshy, brown at maturity .... *T. occidentalis*

Thuja occidentalis L. American Arborvitae; White Cedar. Trees (sometimes shrubby) mainly 2–6 m tall; branchlets arranged into flattened sprays, these variously disposed; leaves dark green above, yellowish green beneath, glandular; ovulate cones 10–12 mm long, with 4 or 5 pairs of scales, narrowly ovoid-ellipsoid, finally cylindroid and brown at maturity. Commonly grown ornamental trees or shrublike trees in much of lower elevation Utah (Cache, Uintah, Utah, and Washington counties); introduced from the eastern U. S. and Canada; 20 (0). The western red cedar, *T. plicata* Donn ex G. Don, might also occur in cultivation in Utah. It differs from *T. occidentalis* in the more regularly arranged sprays, glossy green above and with definite whitish marks beneath.

Thuja orientalis L. Oriental Arborvitae. [*Platycladus orientalis* (L.) Franco]. Mainly shrubby small trees, 1–3 m tall; branchlets arranged into flattened sprays, these typically vertically disposed; leaves bright green to yellowish green, glandular; ovulate cones mainly 15–20 mm long, fleshy, green, and greenish to yellowish at maturity, ovoid. Commonly cultivated ornamentals in much of lower elevation regions of Utah (Cache, Davis, Sanpete, and Utah counties); introduced from China and Korea; 8 (0). Several horticultural phases are planted.

EPHEDRACEAE Dumort.

Ephedra Family

Dioecous shrubs; branches green to olive green, opposite or whorled, striate; leaves scalelike, opposite or whorled, more or less connate; male cones compound, borne at the nodes or terminal, with 2–8 microsporophylls, these free or with stalks united, with a calyxlike involucre surrounding the stalks; female cones solitary or whorled, sessile or peduncled, subtended by firm or scarious bracts; seeds 1–3, hard, somewhat angled to almost terete.

Ephedra L.

The stems simulate those of an *Equisetum*, especially in being green and striate; the differences are obvious, however.

Cutler, H. C. 1939. Monograph of the North American species of the genus *Ephedra*. Ann. Missouri Bot. Gard. 26: 373–428.
Benson, L. 1943. Revision of the status of southwestern trees and shrubs. I. *Ephedra*. Amer. J. Bot. 30: 230–233.

1. Leaves and bracts 3 per node; branches whorled; bracts of female cone scales clawed, 6–10 mm wide, scarious .... *E. torreyana*
— Leaves and bracts 2 per node; branches initially opposite; bracts of female cone scales not clawed, 3–5 mm wide, only the margins scarious 2

2(1). Seeds usually solitary, grayish to pale brown, vertically wrinkled; plants of Washington County ... *E. fasciculata*
— Seeds usually 2, dark brown to almost black, smooth ... 3

3(2). Leaf bases gray, deciduous; branchlets gray green, glaucous, divergent .... *E. nevadensis*
— Leaf bases brown, persistent; branches green or yellowish green, erect, broomlike .... *E. viridis*

Ephedra fasciculata A. Nels. Mohave Ephedra. Low often prostrate shrubs 3–10 dm tall; branches flexible to somewhat rigid, terete, pale green, smooth to somewhat roughened, becoming yellowish in age; leaves opposite, 1–3 mm long, with a hyaline somewhat persistent white sheath, the remainder deciduous; male cones 2 to several, obovoid, 4–8 mm long, sessile, with 4–8 pairs of obovate bracts 2–3 mm long, membranous, pale yellow; female cones sessile, ellipsoidal, 6–13 mm long, with 4–7 whorls of elliptic bracts 3–7 mm long, the margins hyaline, the remainder pale brown to green; seeds usually solitary, longitudinally furrowed, pale brown, 5–12 mm long. Dry wash bottoms and rocky slopes in creosote bush, blackbrush, and desert almond communities at 800 to 1100 m in Washington County; Arizona, Nevada, and California; 9 (ii).

Ephedra nevadensis Wats. Nevada Ephedra. Erect shrubs, mainly 3–15 dm tall (or more); branches pale green, glaucous, almost smooth, the older ones grayish, widely divergent; leaves paired, 2–5 mm long, thickened medially on the dorsal side, soon deciduous, leaving gray bases; male cones 1 to several, ellipsoid, 4–8 mm long, sessile or short-pedunculate, with 5–9 pairs of membranous obovate bracts 3–4 mm long; female cones pedunculate, roundish, 5–11 mm long, with 3–5 pairs of ovate bracts 4–8 mm long, pale brown to yellowish green; seeds paired, smooth, brown, 4–9 mm long. Creosote bush, blackbrush, mixed desert shrub, sagebrush, pinyonjuniper, and rabbitbrush communities at 850 to 2150 m in Beaver, Box Elder, Carbon, Daggett, Duchesne, Emery, Juab, Millard, Piute, Sanpete, Sevier, Uintah, and Washington counties; Arizona to Oregon and California; 51 (ii). This plant is consistently hedged back by sheep, to whom it is a valuable sourse of browse.

Ephedra torreyana Wats. Torrey Ephedra. Erect shrubs, 2–10 dm tall (rarely more); branches blue green to olive green, sometimes glaucous, appearing smooth but with many small longitudinal furrows, rigid, terete, to 3.5 mm thick, solitary or whorled at the nodes; leaves ternate or whorled, 2–5 mm long, dorsimedially thickened, connate for nearly 2/3 their length, at maturity the lobes spreading or recurved, somewhat persistent; male cones solitary to several in a whorl, ovate, sessile, 6–8 mm long; bracts ternate in 5–6 whorls, obovate, clawed, scarious except in the center and at the base; female cones solitary to several at the nodes, ovoid, 9–13 mm long, sessile, the bracts in 3's in whorls of 5 or 6, obovate, clawed, 6–9 mm long, scarious, the margins minutely toothed and undulate; seeds solitary or 2, pale brown to yellow green, scabrous, 7–10 mm long. Dry sandy or rocky hillsides in creosote bush, blackbrush, salt desert shrub, mountain brush, and pinyon-juniper communities at 850 to 2330 m in Duchesne, Emery, Garfield, Grand, Kane, San Juan, Uintah, Washington, and Wayne counties; Texas to Arizona, Nevada, and Colorado; 119 (vii). This plant is eaten by livestock, especially by sheep.

Ephedra viridis Cov. Green Ephedra; Mormon Tea; Brigham Tea. Shrubs 1–15 dm tall, spreading to erect; branches rigid to flexible, bright green to yellow green or less commonly olive to gray green, initially opposite, in some finally falsely whorled, typically fastigiate and broomlike; leaves opposite, 1.5–4 mm long, thickened dorsimedially, deciduous and leaving a thickened persistent brown base; male cones 2 or more, obovoid, sessile, 5–7 mm long, the bracts opposite, 2–4 mm long, membranous, pale yellow, ovate; female cones obovoid, 6–10 mm long, sessile or pedunculate, with 4–8 pairs of ovate bracts 4–7 mm long; seeds paired, brown, trigonal, smooth, 5–8 mm long.

1. Female cones sessile or nearly so; stems not viscid .....
   .... *E. viridis* var. *viridis*
— Female cones pedunculate; stems often viscid ....
   .... *E. viridis* var. *viscida*

Var. *viridis* Blackbrush, salt desert shrub, sagebrush, mountain brush, pinyon-juniper, rabbitbrush, and mountain brush communities at 900 to 2950 m in Beaver, Carbon, Daggett, Duchesne, Emery, Garfield, Grand, Juab, Kane, Millard, San Juan, Sanpete, Sevier, Tooele, Uintah, Utah, Washington, and Wayne counties; Wyoming to Colorado, Arizona, Nevada, Oregon, and California; 150 (iv). This is the source of Mormon tea (also known by myriad other names), a yellowish drink made by steeping the branchlets in hot water. The plant is not so severely hedged by browsing animals as some of the other species, but is still of considerable importance.

Var. *viscida* (Cutler) L. Benson [*E. coryi* var. *viscida* Cutler; *E. cutleri* Peebles]. Mostly in sandy areas with blackbrush, mixed desert shrub, mixed grass, rabbitbrush, and pinyon-juniper communities at 900 to 1950 m in Emery, Garfield, Grand, Kane, San Juan, Uintah, Washington, and Wayne counties; Colorado, New Mexico, and Arizona; 40 (iii). In contemporary floras this plant has been regarded at specific rank as *E. cutleri*. However, the length of the stalks of the ovulate cones and the viscid condition of the stems forms a continuum with *E. viridis* in a strict sense, especially where the two grow together. This is the phase of the species that forms stands in extensive sandy grasslands in southeastern Utah, often with only the tips of the stems protruding, almost grasslike, from the sand. Much of the plant is buried within the sandy substrate, which it helps to stabilize.

GINKGOACEAE Engler in Engler & Prantl

Ginkgo Family

Dioecious, deciduous, resinous trees; leaves alternate or clustered on spur shoots, fan-shaped, bilobed, dichotomously veined, petiolate; cones on spur shoots in leaf or bract axils; male cones catkinlike; female cones typically of 2 ovules, on a long peduncle, with usually only one maturing; seeds with a fleshy aril, plumlike; x = 12.

Ginkgo biloba L. Maidenhair Tree; Ginkgo. Trees to 20 m tall or more, the trunk to 1 m thick; bark deeply furrowed and gray in age, pale when young and smoothish; leaf blades mainly 2–6 cm long and 2–8 cm wide or more; cones 2–3 cm thick, 1.5–2 cm thick, green and glaucous, long-pecunculed, ripening and finally fetid; 2n = 24. Commonly grown specimen tree, especially on institutional grounds, in much of Utah; introduced from China; 6 (0). The plant has been cultivated for millenia in China, and is now widely grown in temperate portions of the world.

PINACEAE Lindl.

Pine Family

Plants monoecious (rarely dioecious) trees or shrubs; leaves evergreen or deciduous, needlelike, linear to oblong, borne singly or in clusters of 2–5 or densely aggregated on short lateral shoots; cones solitary, axillary or terminal; male cones small, soft, with spirally arranged microsporophylls; ovulate cones small to large, the several to many scales spirally arranged, each subtended by a bract and bearing 2 ovules; cones woody to leathery or papery; x = 12, 13.

1. Leaves borne in dense clusters on short spur branches; cones variously persistent or deciduous 2
— Leaves borne singly or in clusters of 2–5, persistent; cones typically persistent (except in *Abies*) 3
2(1). Leaves persistent; cones falling apart at maturity . *Cedrus*
— Leaves deciduous; cones persistent .... *Larix*
3(2). Leaves 2–5 per cluster and fitting together to form a cylinder or, if solitary, terete .... *Pinus*
— Leaves borne singly, either angled or flattened but not terete 4
4(3). Branchlets rough where needles have fallen, the leaves deciduous above the persistent base, typically 4-angled and sharply pointed .... *Picea*
— Branches smooth where needles have fallen, the leaves deciduous to the base, typically flattened and bluntly pointed 5
5(4). Cones erect, not persistent, the scales falling from the central axis, with subtending bracts not apparent on the cone surface; terminal buds resinous .... *Abies*
— Cones pendulous, persistent, the scales not deciduous from the central axis, with subtending 3-toothed bracts apparent on the cone surface; terminal buds not resinous .... *Pseudotsuga*

Abies Miller

Evergreen spirelike or conical trees with thin, grayish bark often bulged by resin vesicles when young, dark gray and thick in age; leaves borne singly, spirally arranged, flat, blunt, narrowed to a short stout petiole, wholly deciduous, the leaf scar nearly circular; winter buds blunt, resin covered; male cones catkinlike, cylindrical; ovulate cones stiffly erect, cylindrical, maturing in one season, the scales shed singly at maturity, the slender axis persistent on the branches for several years.

1. Branchlets pubescent; male cones bluish; mature leaves mostly 1.5–1.8 mm wide and 2–3 cm long; trees spirelike or dwarf and shrublike at timberline .... *A. lasiocarpa*
— Branchlets glabrous or nearly so; male cones rose to dark red; mature leaves mostly 1.8–2.5 mm wide and 2.5–5.5 cm long; trees with a conic to rounded crown .. *A. concolor*

***Abies concolor* (Gord. & Glend.) Lindl.** White fir. [*Picea concolor* Gord. & Glend.; *A. grandis* var. *concolor* (Gord. & Glend.) Murray]. Trees to 80 m tall, with the crown conic to rounded; bark thick, smooth when young except for resin blisters, strongly furrowed and dark gray in age; leaves solitary, flat in cross-section, bluish to yellow green, resin ducts lateral and just beneath the epidermis, the stomata in vertical rows on both surfaces, blunt to rounded apically, the leaf scar circular-depressed; male cones rose to dark red (rarely yellowish), less than 15 mm long; ovulate cones yellowish green to greenish purple, erect, oblong-cylindric, 7–12 cm long, the scales fan-shaped, broader than long; seeds broadly winged, 7–12 mm long. Mountain brush, aspen, Douglas fir, and fir communities at 1525 to 3050 m in all Utah counties except Daggett and Rich; Oregon to Wyoming, south to California, Arizona, New Mexico, and Mexico; 78 (xv).

***Abies lasiocarpa* (Hook.) Nutt.** Subalpine Fir. [*Pinus* (*Abies*) *lasiocarpa* Hook.; *A. balsamea* ssp. *lasiocarpa* (Hook.) J. Boivin; *A. subalpina* Engelm. in Ward, type from Wasatch Plateau]. Trees to 40 m tall, with a spirelike crown or dwarfed and shrubby at timberline; bark smooth except for resin blisters when young, becoming furrowed and dark gray in age; branchlets pubescent; leaves 1.2–3 (3.2) cm long, 1–2 mm wide, the upper surface nearly flat, the lower with a prominent ridge, blunt to acute and rigid, solitary, dark blue-green, the stomata on both surfaces, the upper in one more or less continuous band, the lower in 2 bands separated by the ridge, the leaf scar circular; staminate cones bluish, ca 9 mm long; ovulate cones mostly deep purple (rarely brownish green), 6–10 cm long; seeds 6–7 mm long, winged, the wings much larger than the seed, over 1 cm long. Aspen, spruce, spruce-fir, or fir communities, often growing with Engelmann spruce, at (2000) 2470 to 3355 m (often forming krummholz at timberline), in all Utah counties, except Davis and Millard; Alaska, Yukon, and British Columbia, south to Oregon, Arizona, and New Mexico; 68 (v).

Cedrus Trew

Monoecious or dioecious evergreen conifers; bark smooth and gray at first, but furrowed and dark gray in age; leaves alternate and single on young shoots, fascicled and numerous on spurs on older stems, quadrangular, stiff; male cones erect, cylindric; ovulate cones erect, ovoid to cylindroid, with closely imbricated scales that fall apart at maturity; seeds with membranous wings.

Cedrus atlantica (**Endl.**) **Manetti ex Carr.** Atlas Cedar. [*Pinus atlantica* Endl.; *C. libani* ssp. *atlantica* (Endl.) Franco]. Trees to 30 m tall or more, and with trunks to 1 m thick or more; branchlets densely pubescent; leaves 6–25 mm long, bluish green, glaucous, quadrangular, sharply pointed, borne in fascicles of 20–50 on spur branches, singly on shoots of the season; male cones 2.3–5 cm long, the scales tan to brown, tardily deciduous; ovulate cones 6.5–10 cm long, 4.5–7 cm thick, ovoid to obovoid, strongly resinous, the scales deciduous from the persistent rachis, broadly fan-shaped, the seeds each with a fan-shaped yellowish brown wing. Uncommonly, but widely, grown specimen tree in the lower valleys of Utah; introduced from Algeria; 4 (0). The closely related species, *C. lebani* **A. Rich.** or cedar of Lebanon, with which the Atlas cedar has been combined, is sometimes grown here also, but is apparently more frost sensitive. It is distinguished by having longer leaves and larger cones.

Larix Miller

Deciduous trees with scaly bark; leaves dimorphic, thin and scalelike hair-tipped bracteate ones at the tips of short branches, and green needles, these latter borne in clusters on short spur shoots or spirally arranged on long shoots, all deciduous; winter buds blunt; male cones cylindrical and catkinlike; ovulate cones curved upward, ovoid, maturing in one season, the entire cone more or less persistent, not falling apart at maturity.

1. Leaves, at least the longest, (25) 35–50 mm long; male cones yellow; ovulate cone scales much shorter than the long-attenuate, subtending bracts .... *L. occidentalis*
— Leaves mainly less than 25 mm long; male cones brown; ovulate cone scales much longer than the slender subtending bracts .... *L. decidua*

Larix decidua **Miller** European Larch. [*L. europaea* DC.]. Trees to 25 m tall or more, the trunk to 1 m thick or more; young twigs glabrous; bark ultimately thick and furrowed; needles typically 15–35 per spur, pale green, stiffish, mostly 8–25 mm long, flattened, the upper surface almost flat, the lower surface ridged in the middle; male cones brown, 1–2 cm long; ovulate cones mostly 2.5–3.5 cm long, the scales brown, somewhat hairy on the outer surface, truncate to broadly emarginate apically, longer than the attenuate subtending bracts; seeds ca 3 mm long, the wing oblique, ca 7 mm long, brownish; 2n = 24. Rather commonly grown specimen tree in much of lower elevation portions of Utah; introduced from Europe; 8 (i).

Larix occidentalis **Nutt.** Western, Montana, or Mountain Larch; Tamarack, Hakmatack; Western Tamarack. Trees to 50 m tall or more, the trunk to 1 m thick or more; young twigs glabrous or more or less pubescent; bark ultimately thick and furrowed; needles typically 15–30 per spur, pale green, stiffish, mostly (25) 35–50 mm long, rather broadly triangular in section, the upper surface flat or nearly so, the lower one ridged in the middle; male cones yellow, ca 1 cm long; ovulate cones mostly 2.5–3.5 cm long, the scales reddish brown to brownish, usually hairy on the outer surface, truncate to broadly emarginate apically, much surpassed by the long-attenuate subtending bracts; seeds ca 3 mm long, the wings ca 6 mm long, brownish to reddish. Established in a planting in Big Cottonwood Canyon, Salt Lake County; British Columbia to Oregon, east Montana and Idaho; 1 (0).

Picea A. Dietr.

Evergreen trees with scaly bark; needles borne singly, spirally arranged, quadrangular and square in cross section or more or less flattened, often sharply acute, deciduous above the base (quickly deciduous from cut branches), the twigs roughened with persistent peglike leaf bases; winter buds blunt; staminate cones cylindrical, catkinlike; ovulate cones curved downward, ovoid to cylindrical, maturing in one season, the cones more or less persistent, not falling apart at maturity.

1. Branchlets ordinarily pendulous along main branches; cones 10–18 cm long; trees cultivated .... *P. abies*
— Branchlets ordinarily spreading along main branches; cones less than 10 cm long; trees indigenous or cultivated 2
2(1). Twigs (or leaf bases) pubescent; leaves flexible, not sharply pointed, the apex blunt to acute; ovulate cones 3–6 cm long, deciduous following seed maturity .... *P. engelmannii*
— Twigs (and leaf bases) glabrous; leaves rigid, sharply pointed; ovulate cones 6–10 cm long, persistent following seed maturity .... *P. pungens*

Picea abies (**L.**) **Karsten** Norway Spruce. [*Pinus abies* L.]. Trees to 25 m tall or more, the trunk to 1 m thick or more; bark scaly, grayish; branches spreading to curved-ascending, with pendulous branchlets on mature branches; branchlets pubescent to subglabrous or glabrous; needles mainly 10–20 mm long, quadrangular; ovulate cones mainly 10–18 cm long, pendulous, persisting on the tree for one or more years; 2n = 24. Commonly grown ornamental tree in much of Utah; introduced from Europe; 1 (0). The white spruce, *P. glauca* (**Moench**) **Voss**, with cones 2.5–6.5 cm long is uncommonly grown in Utah. The branchlets are glabrous and the needles are 6–15 mm long.

Picea engelmannii Parry ex Engelm. Engelmann Spruce. [*Abies engelmannii* Parry; *P. glauca* ssp. *engelmannii* (Parry) Taylor]. Trees to 4 m tall, or shrubby and low at timberline; bark thin, scaly, reddish brown or cinnamon; twigs (or leaf bases) pubescent or rarely glabrous; leaves glaucous to deep bluish green, 13–25 (30) mm long, blunt to acute apically but not sharply pointed and pungent; staminate cones yellow brown, 7–15 mm long; ovulate cones purplish brown to brown, oblong-cylindric, 3–5.5 cm long, persistent for ca 1 year; scales entire to incised apically, the bracts shorter than the scales; seeds winged, the wing long, pale brown. Spruce, spruce-fir, and lodgepole pine communities at (2285) 2440 to 3420 m in probably all Utah counties; British Columbia to Alberta, south to California, Arizona, and New Mexico; 43 (viii). This tree is often a codominant with subalpine fir, and less commonly with lodgepole pine. The trees are harvested for lumber.

Picea pungens Engelm. Blue Spruce. Trees to 3 m tall, often with a dense conical crown, or with an open and pyramidal crown in age; bark gray to brownish, thickish, scaly, sometimes furrowed; twigs (and leaf bases) glabrous, shiny; leaves bluish green, rigid, 12–30 mm long, tapered to a pungent tip; male cones yellow, 10–15 mm long; ovulate cones purplish brown when young, light brown or stramineus when mature, 6–12 cm long, persistent for at least 2 years; scales thin, the apex erose and undulate, surpassing the bracts; seeds 2–3 mm long, winged, the wing longer than the seed. Riparian and other moist habitats with willow, cottonwood, and other mesophytes at 1830 to 2870 m in Beaver, Box Elder, Carbon, Duchesne, Emery, Garfield, Grand, Kane, Iron, Piute, Salt Lake, San Juan, Sanpete, Sevier, Summit, Uintah, Utah, Wasatch, and Washington counties; Idaho and Wyoming to Arizona and New Mexico; 34 (iii). The blue spruce is the state tree for both Utah and Colorado. The trees are used for lumber.

Pinus L.

Evergreen trees with scaly bark; leaves dimorphic, of thin scalelike ones subtending the base of short spur branches and of green needles borne singly (and terete) or in clusters of 2–5 on spur branches, these spirally arranged on the twigs; winter buds acutish, resinous; male cones ovoid to cylindrical; ovulate cones variously arranged but not erect in age, ovoid to lance-ovoid, maturing in 1 or 2 seasons, more or less persistent, woody, not falling apart at maturity.

1.	Needles borne in bundles of 5	2
—	Needles solitary (and terete) or in clusters of 2 or 3	6
2(1).	Needles entire along the margins; trees indigenous	3
—	Needles minutely serrulate marginally; trees introduced	4
3(2).	Needles 3–7 cm long, the sheaths early deciduous; cones 7–20 cm long, unarmed; plants widely distributed	*P. flexilis*
—	Needles 2–4 cm long, the sheaths persisting for 2–3 years; cones 5–9 cm long, armed with spines; plants local, mainly in southern Utah	*P. longaeva*
4(2).	Ovulate cones typically 15–25 cm long; needles mainly 5–10 cm long; plants grown in Big Cottonwood Canyon, Salt Lake County	*P. monticola*
—	Ovulate cones of various length but, if 15–25 cm long, the needles more than 10 cm long	5
5(4).	Needles less than 12 cm long, slender but not drooping	*P. strobus*
—	Needles 12–20 cm long, slender, drooping	*P. griffithii*
6(1).	Needles solitary (rarely 2 per bundle), cylindrical, terete; trees of the Great and Virgin basins	*P. monophylla*
—	Needles in bundles of 2 or 3	7
7(6).	Needles typically borne in bundles of 3 (at least some)	8
—	Needles typically in bundles of 2	9
8(7).	Cones 15–30 cm long; bark with odor of vanilla; buds not covered with resin droplets; plants introduced, cultivated	*P. jeffreyi*
—	Cones 7–15 cm long; bark with odor of turpentine; buds often with resin droplets; plants indigenous, sometimes cultivated	*P. ponderosa*
9(7).	Trees small to moderately sized; sheath at base of needles deciduous; seeds large, edible; plants mainly of the Colorado Plateau	*P. edulis*
—	Trees small to large; sheath at base of needles persistent; seeds small to moderate, not typically eaten; plants introduced or, if indigenous, the distribution various or otherwise	10
10(9).	Needles 8 cm long or more	11
—	Needles 8 cm long or less	12
11(10).	Trees typically less than 5 m tall; bark scaly, cinnamon; cones 3–5 cm long	*P. densiflora*
—	Trees typically much more than 5 m tall; bark finally furrowed and gray, not cinnamon; cones 8–10 cm long	*P. nigra*
12(10).	Trees with scaly cinnamon colored bark (at least above); cones short stalked; plants introduced	*P. sylvestris*
—	Trees with bark various, but not cinnamon colored; cones sessile or subsessile	13
13(12).	Plants typically shrubby and less than 2.5 m tall, introduced	*P. mugo*
—	Plants typically much over 2.5 m tall; indigenous in northern Utah and uncommonly cultivated	*P. contorta*

Pinus contorta Dougl. ex Loudon Lodgepole Pine; Black Pine. [*P. murrayana* Balf. in Murray; *P. contorta* ssp. *murrayana* (Balf.) Critchf.; *P. contorta* var. *murrayana* (Balf.) Engelm.]. Trees mostly 10–35 m tall, usually in dense stands with small bare trunks and in more open stands with branch covered bases; bark thin, orange brown to gray, scaly, not ridged or furrowed except in large trees in open stands; leaves 2 per cluster, (2) 3–9 cm long, stout, often twisted; staminate cones orange red, 8–10 mm long; ovulate cones borne on the upper branches, subsessile, serotinous, opening and shedding seeds in 2 or more years, 3–6 cm long, the scales narrow, thickened at the end, the umbo dorsal and with a short sharp subpersistent prickle; seeds 3–4 mm long, reddish brown, with a prominent wing 10–12 mm long; $2n = 24$. Aspen, lodgepole pine, and spruce-fir or spruce communities at 2135 to 3355 m in Cache, Daggett, Duchesne, Summit, Uintah, and Wasatch counties; Alaska to Saskatchewan, south to California and Colorado; 24 (i). A specimen taken in the Stansbury Mts., Tooele County, appears to be from a cultivated tree. This is the most common conifer in the Uinta Mts. Our plants belong to var. *latifolia* Engelm. ex Wats. (type from the Uinta Mts.) [*P. contorta* ssp. *latifolia* (Engelm.) Critchf.; *P. divari-*

cata var. *latifolia* (Engelm.) J. Boivin]. Insect infestations of lodgepole pine in recent years have resulted in wholesale death of post- to log-sized trees. Thousands of acres of trees have been killed.

Pinus densiflora Sieb. & Zucc. Japanese Red Pine. Trees small, mainly 4–5 m tall; bark thin, scaly, cinnamon colored (at least above); needles in bundles of 2, mainly 6–11 cm long, bluish green; ovulate cones 2.5–4 cm long, ovoid to cylindroid-ellipsoid, tardily opening, the seed 3–4 mm long, with an oblique wing 7–9 mm long, this stramineous to brownish. Sparingly grown ornamental pine of charm and beauty in at least Utah County; introduced from Japan; 5 (0). Our material apparently belongs to the dwarf horticultural phase passing as **var. umbraculifera** Tanyosho.

Pinus edulis Engelm. Pinyon; Two-needle Pinyon. [*P. monophylla* var. *edulis* (Engelm.) Jones; *P. cembroides* var. *edulis* (Engelm.) Voss]. Small to moderate sized trees with pyramidal to rounded crown, mainly 5–15 (20) m tall, the trunk short; bark thin and scaly, yellowish brown to reddish brown, finally furrowed and grayish; needles 2 (1–3) per cluster, 1.5–5 cm long, rigid, sharply pointed, the sheath deciduous; staminate cones 3–6 mm long; ovulate cones ovoid, (2) 3–5 cm long, short-stalked, yellowish brown to brown, resinous, the scales thickened apically but without a prickle, the umbo dorsal, inconspicuous; seeds brown, mainly 8–16 mm long, ovoid-ellipsoid to ellipsoid, thick-shelled, wingless, averaging smaller than in *P. monophylla* (q.v.); n = 12. Pinyon-juniper and pinyon or sagebrush and lower aspen communities at 1220 to 2745 m in Beaver, Cache, Daggett, Duchesne, Emery, Garfield, Grand, Kane, Iron, Juab, Millard, Piute, Rich, San Juan, Sanpete, Sevier, Uintah, Utah, and Washington counties; Wyoming to Arizona, New Mexico, Oklahoma, and Mexico; 92 (xiv). The two-needle pinyon, usually in association with Utah juniper (which often serves as a nurse plant), forms extensive woodlands in many areas of the state, except in western Utah where it is replaced by *P. monophylla*. The latter plant has a more open crown and more bluish green aspect, in addition to the key characters. Besides being valuable for the edible seeds, the wood is used as a premier firewood. Huge areas of pinyon-juniper woodland have been devastated in attempts (not always wise) at improvement of range forage conditions, especially for grazing by domestic livestock and wildlife.

Pinus flexilis James Limber Pine. [*Apinus flexilis* (James) Rydb.]. Trees, mainly 8–20 m tall, with pyramidal or more commonly rounded crown, sometimes shrubby; mature bark dark brown to blackish, furrowed, and with rectangular scaly plates; young branchlets puberulent; needles 5 per cluster, (2.2) 3.5–7 cm long, rigid, dark green, not serrulate; staminate cones reddish, 7–10 mm long; ovulate cones ovoid to subcylindric-ellipsoid, 6–14 cm long, short-stalked, the scales thickened at the tip, the umbo terminal, unarmed; seeds 7–12 mm long, dark brown, the wing vestigial. Typically on ridge crests in Douglas fir, ponderosa pine, aspen, and spruce-fir or high elevation grass-sagebrush communities at 1830 to 3450 m in all Utah counties, except Uintah; British Columbia and Alberta, south to California, Arizona, and Texas; 80 (xi). Trees of this species are seldom common, but stand flaglike along windswept ridgecrests in many plant communities. The seeds are sufficiently large to be used as human food, but are seldom harvested.

Pinus jeffreyi Balf. in Murray Jeffrey Pine. Trees, mainly 10–40 m tall, the crown finally rounded; mature bark furrowed, with large irregular plates, finally purplish to reddish brown, the furrows exposing red fresh bark, with the odor of vanilla; buds without resin droplets; needles (2) 3 per bundle, bluish green, mainly 12–25 cm long; sheaths of needles persistent; staminate cones 2–3.5 cm long, purplish; ovulate cones 15–25 cm long, ovoid to ovoid-ellipsoid, short-stalked, the scales thickened apically, the dorsal umbo armed with a prickle; seeds 10–12 mm long, the wing prominent, mainly 2–3 cm long. Cultivated ornamental and specimen tree in Utah County (at least); Oregon, California, and Nevada; 2 (0).

Pinus longaeva D. K. Bailey Western Bristlecone. Bushy trees to 20 m tall, or gnarled and shrublike; bark grayish to reddish brown; branches pendulous, twisted and very elongated; leaves 5 per cluster, persistent, 2–4 cm long, rigid, dark green, not serrulate; staminate cones 10–12 mm long; ovulate cones 5.5–8.5 cm long, broadest at the base, reddish brown, the scales thickish, ridged, the umbo dorsal and armed with a slender incurved prickle 4–6 mm long; seeds pale brown, 6–8 mm long, the wings slightly longer. Ridges and open slopes, mainly on limestone or dolomitic substrates, in ponderosa pine, spruce, and spruce-fir communities at 2195 to 3265 m in Beaver, Carbon, Duchesne, Emery, Garfield, Iron, Juab, Kane, Millard, Sanpete, Tooele, Utah, Wayne, and Washington counties; Nevada and California; 45 (ix). This species is distinguished from *P. aristata* Engelm., the bristlecone pine, in having 2 resin ducts per leaf, these not or rarely producing a resin dot at the leaf apex. The western bristlecone is among the oldest living trees in the world, with some known to have lived for 5000 years.

Pinus monophylla Torr. & Frem. Singleleaf Pinyon. [*P. edulis* var. *monophylla* (Torr. & Frem.) Torr. in Ives; *P. cembroides* var. *monophylla* (Torr. & Frem.) Voss]. Trees to 15 (20) m tall, with rounded to flat-topped crown in age; mature bark redddish brown, with narrow flat ridges; needles mostly solitary, rigid, incurved, pale green, sharply pointed, 2.5–3.5 cm long, the sheaths deciduous; staminate cones yellowish, 5–6 mm long; ovulate cones 3.5–5.5 cm long, broadly ovoid, brown, the scales thick, especially at the tip, 4–seeded; seeds mostly 10–17 mm long, wingless, brown, moderately thin-shelled, edible. Pinyon-juniper, pinyon, sagebrush, and lower aspen communities at 820 to 2535 (2960) m in Beaver, Box Elder, Cache, Iron, Juab, Kane, Millard, Tooele, Utah, Wasatch, and Washington counties; Arizona, Nevada, California, and Baja; 37 (iv). The presence within the Great Basin of trees with 2 or even 3 needles instead of the usually rounded, solitary needle has led some botanists to view the unusual plants as hybrids between this and the two-needle pinyon of the Colorado drainage. Trees having more than the usual number of needles are almost exclusively on mesic sites and the condition seems to be ecologically induced instead of indicating genetic interaction. This is the principal nut-pine of commerce, with great quantities of the seeds being harvested in some years for sale as "pine nuts." Stands of this species and Utah juniper in the Great Basin have been devastated similarly to those of the two-needle pinyon to improve rangelands for grazing purposes.

Pinus monticola Dougl. ex D. Don Western White Pine. [*P. strobus monticola* (Dougl.) Nutt.]. Trees to 25 m tall or more; trunks to 1 m in diameter or more; bark thin, gray, smooth at first, finally furrowed into scaly plates; branchlets puberulent; buds ovoid, acute; leaves in bundles of 5, persisting for 2–4 years, glaucous, (4) 5–10 cm long, more or less serrulate along the margins; sheaths deciduous; male cones 7–10 mm long, yellow; ovulate cones typically 15–25 cm long, narrowly oblong, pale brown, pendulous, on stalks 1–2.5 cm long, the umbo terminal, inconspicuous; seeds 6–10 mm long, the wings 20–25 mm long. Known from a planting in Big Cottonwood Canyon (Spruces), Salt Lake County; British Columbia and Alberta to California, Idaho, and Montana; 1 (0).

Pinus mugo Turra Mountain Pine; Mugo Pine. [*P. montana* Miller]. Shrubby low trees mainly less than 6 m tall; trunk seldom readily apparent; bark scaly, gray, not becoming cinnamon above; leaves 2 per bundle, mainly 4.5–7 cm long, green to blue green, persisting for some years, the sheaths more or less persistent; staminate cones yellowish, 8–10 mm long; ovulate cones 4–6 (7) cm long, the umbo dorsal, unarmed; seeds 3–5 mm long, the wing about twice as long as the seed, scarious, shiny; $2n = 24$. Rather commonly grown ornamental in much of lower elevation portions of the State (Cache, Davis, and Utah counties); introduced from Europe; 9 (0).

Pinus nigra Arnold Austrian Pine. Trees, mainly 20–30 m tall, with a rounded crown; bark ultimately thick, gray, and furrowed, not becoming cinnamon upward; needles in bundles of 2, mainly 12–16 cm long, blue green, the sheaths persistent; staminate cones yellowish, 25–35 mm long; ovulate cones 5–8 cm long, tan to brownish, the scale tips brownish, the umbo dorsal, unarmed; seeds 5–7 mm long, the wings 2–3 times as long as the seeds, brownish, oblique; $2n = 24$. Commonly grown ornamental and shade tree of lower elevation portions of Utah; introduced from Eurasia; 6 (0). Austrian pine is frequently mistaken in cultivation for ponderosa pine, which usually has 3 needles per cluster and larger armed cones.

Pinus ponderosa Lawson Ponderosa Pine; Western Yellow Pine; Yellow Pine. Trees, mainly 10–30 m tall; bark ultimately thick, deeply furrowed, and reddish brown, forming polygonal plates, with cinnamon colored furrows; leaves mainly in clusters of 3 (less commonly in 2's), 8–10 cm long, yellow green, persisting at branch ends, the sheaths persistent; male cones yellow to purple, 2–3 cm long; ovulate cones 7–15 cm long, reddish brown, the scale tips yellowish brown and with a stout prickle on the umbo; seeds 6–7 mm long, the wing brownish purple, 2–4 times as long as the seed; $2n = 24$. Mountain brush, ponderosa pine, and aspen communities (less commonly with spruce-fir and lodgepole pine communities) at 1585 to 2685 m in all Utah counties except Box Elder, Cache, Davis, Morgan, Rich, Salt Lake, and Wasatch (?); British Columbia to the Dakotas and Nebraska, south to California and New Mexico; 69 (vi). This is one of the premier lumber trees of Utah, producing cylindrical large boles, often without major branches in the lower 5–10 meters. The wood is used both in rough construction and in cabinet work.

Pinus strobus L. White Pine; Eastern White Pine. Slender trees, mainly 10–25 m tall; trunk to 1 m thick or more; bark scaly, not deeply furrowed; needles in clusters of 5, very slender, serrulate, yellow green to blue green, the sheaths soon deciduous; staminate cones yellowish, 8–10 mm long; ovulate cones 6–25 cm long, on stalks to 2 cm long or more, broadest near the middle, the scales thickish, brown to reddish brown, the umbo terminal, unarmed; seeds 4–6 mm long, the wings 2–4 times as long as the seeds; $2n = 24$. Cultivated ornamental and specimen trees in lower elevation portions of the state; Manitoba to Newfoundland, south to Iowa, Illinois, Tennessee, and Georgia; 4 (0). This pine is only sparingly grown in Utah, but was probably the most important soft wood used in colonial America.

Pinus sylvestris L. Scots Pine. Trees mainly 15–25 m tall, finally with an open spreading crown; bark thin, ultimately furrowed and dark gray below, becoming scaly and cinnamon upward; needles 2 per cluster, 4.5–9 cm long, yellow green, the sheaths persisting; male cones 8–12 mm long, yellowish; ovulate cones 4–7 cm long, conic to ovoid, the umbo dorsal, unarmed; seeds 3–4 mm long, the wing much longer than the seeds; $2n = 24$. Commonly grown ornamental, windbreak, and specimen tree in much of lower elevation Utah (Cache, Utah, and Weber counties); introduced from Europe; 11 (0).

Pinus wallichiana A. B. Jackson Himalayan White Pine. [*P. griffithii* authors, not Parl.]. Slender trees, mainly 10–25 m tall; bark smooth, at least when young; branchlets glabrous, glaucous; leaves in bundles of 5, mainly 18–22 cm long, very slender, serrulate, yellowish to bluish green and glaucous; staminate cones yellowish; ovulate cones 12–25 cm long, yellowish brown, the scales yellowish apically, the umbo subterminal, mucronate, but not sharply awned; seeds 5–6 mm long, the wing several times longer than the seed, brown. Uncommonly grown specimen tree in lower elevation portions of Utah; introduced from the Himalayas; 3 (0).

Pseudotsuga Carr.

Evergreen coniferous trees; bark ultimately blackish to dark gray and deeply furrowed; branchlets with slightly raised oval leaf scars; winter buds reddish brown, non-resinous, pointed; leaves solitary, spirally arranged, flat, more or less petiolate; staminate cones solitary in leaf axils, the microsporphylls expanded apically; ovulate cones terminal or apparently so, pendent, maturing in one season, not falling apart at maturity, the scales rounded, surpassed by subtending, conspicuously 3–lobed bracts; seeds oblong, shorter than the membranous wings.

Pseudotsuga menziesii (Mirbel) Franco Douglas Fir; Red Pine. [*Abies menziesii* Mirb.; *A. taxifolia* Poir. in Lam., not Du Tour 1803 or Desf. 1804; *P. taxifolia* (Poir.) Britt.]. Trees mainly 20–30 m tall; bark smooth when young, finally deeply furrowed and blackish to dark gray in age; branches mainly alternate, usually pubescent for some years; leaves mainly 15–35 mm long, flat, obtuse to acutish, bluish green; staminate cones 5–8 mm long, orange red; ovulate cones 4–6 cm long (exclusive of bracts), ovoid-cylindroid, pendulous, brown to reddish brown, soon deciduous, the scales rounded, the subtending 3–lobed bracts prominently exserted; seeds 5–6 mm long, the wing ca twice as long as the seed; $2n = 26, 27$. White fir, mountain brush, aspen, and spruce-fir communities at 1525 to 3050 m in all Utah counties; British Columbia and Alberta, south to California, New Mexico, Texas, and Mexico; 127 (xxvi). Our material belongs to var. **glauca** (Beissner) Mayr [*Tsuga douglasii* var. *glauca* Beissner. in Jager & Beissner; *P. douglasii* var. *glauca* (Beissner) Mayr; *P. globosa* Flous, type from Big Cottonwood Creek, Salt Lake County]. This species was exploited for lumber, as red pine, by pioneer sawyers. The wood is hard, close-grained, and very durable. Sites logged of this species a century ago still have not grown back trees of usable size.

TAXACEAE S. F. Gray
Yew Family

Evergreen dioecious shrubs or trees; leaves needlelike, flattened, alternate or rarely opposite, often 2-ranked; plants dioecious; staminate cones small, globular, axillary; ovulate cone much reduced, the ovules solitary, each surrounded by a fleshy disc (aril), the aril ripening and brightly colored at maturity; x = 11.

Taxus L.

Treated is a single genus with features of the family.

1. Leaves gradually acuminate; winter bud scales not keeled .... *T. baccata*
— Leaves abruptly acute; winter bud scales keeled .... *T. cuspidata*

Taxus baccata L. English Yew. Trees, mainly 3–7 m tall; winter buds obtuse, the scales persistent at base of branches; leaves mainly 2–3 cm long, gradually acuminate, glossy and dark green above, pale beneath; aril nearly globose, to ca 12 mm thick. Sparingly cultivated ornamental in Utah; introduced from Eurasia; 8 (0). This and other species of yew are poisonous to livestock and people. The poisonous principal is reputedly alkaloidal.

Taxus cuspidata L. Japanese Yew. Shrubs or trees, mainly 0.5–2 m tall; winter buds ovoid-oblong, acute, the scales keeled; leaves mainly 2–3 cm long, abruptly acute, dull green above, with 2 yellowish bands beneath; aril ellipsoid, to ca 12 mm thick. Rather commonly grown ornamental in Utah; introduced from eastern Asia; 6 (0).

TAXODIACEAE Warm.
Taxodium Family

Evergreen or deciduous trees; leaves spirally arranged, though often 2-ranked; male cones terminal or axillary, with spirally arranged microsporophylls, each with several microsporangia; ovulate cones woody, with thickened widely spreading scales, each bearing 2–9 ovules; x = 11.

1. Leaves opposite or apparently so; branchlets deciduous with the leaves in autumn .... *Metasequoia*
— Leaves alternate or apparently so, mostly spirally arranged; branchlets and leaves deciduous or evergreen .. 2
2(1). Leaves scalelike, subulate and rigid, evergreen .... *Sequoiadendron*
— Leaves needlelike, flat, not subulate and rigid, deciduous .... *Taxodium*

Metasequoia Miki

Trees with deciduous branchlets; leaves decussate or apparently 2-ranked; male cones sessile, in long drooping clusters; ovulate cones subglobose, pendulous on naked peduncles; scales 10–15 pairs; seeds 5–8 per scale, 2-winged.

Metasequoia glyptostroboides Hu & Cheng Dawn Redwood. Dioecious (or monoecious) trees to 15 m tall or more; trunks buttressed; bark thin, shredded, or ultimately thickened and furrowed; leaves mainly 5–27 mm long, flattened, 1–1.6 mm wide, deciduous with the branchlet in autumn; ovulate cones 2–2.5 cm long, subglobose, becoming quadrangular upon opening, borne on stalks 1.5–6 cm long, the scales more or less peltate, green, finally becoming brown to tan; 2n = 22. Uncommonly grown specimen and shade tree in Utah; introduced from China; 5 (ii). The trees grow rapidly, with one example reaching a height of more than 10 meters and a basal diameter of 5 dm in 25 years. The wood is light and brittle when dry, but flexible and strong when growing. The species should be more widely planted.

Sequoiadendron Buchholz

Evergreen trees; leaves scalelike; male cones sessile; ovulate cones oblong-ovoid, green, persistent on the tree, woody, with numerous wedge-shaped scales, terminating in a long terete spine, each with 3–12 or more ovules; seeds maturing in 2 years, with 2 thin lateral wings.

Sequoiadendron giganteum (Lindl.) Buchholz Big Tree; Giant Sequoia. [*Wellingtonia gigantea* Lindl.]. Trees to 40 m tall or more; trunks buttressed, to 1 m thick or more, often without branches for some distance above the base in age; crown spirelike to rounded and open in age; bark finally thick; leaves decurrent, thickly set, appressed or with the tips spreading, 3–6 mm long, or 3–12 mm long on leading shoots, blue green; staminate cones 5–6 mm long; ovulate cones oblong-ovoid, 5–8 cm long, greenish, becoming red brown, the scales abruptly dilated into grooved disks; seeds ca 6 mm long, the wings broader than the body; 2n = 22. Sparingly to commonly grown specimen tree and botanical curiosity from Utah County south to Washington County; California; 9 (iii). Trees of this species planted at Brigham Young University by B. F. Harrison in the mid 1940's are now more than 15 m tall, some with basal diameter more than 8 dm. A specimen from the east side of the Pine Valley Mountains, Washington County, planted sometime following the turn of the 20th century, now overtops the ponderosa pine forest in which it grows and has a basal diameter greater than one meter. Specimens of this species in California are among the largest and oldest trees on earth.

Taxodium Rich.

Trees with deciduous branchlets; leaves appearing 2-ranked, alternate, linear; male cones many, small, in catkinlike clusters at branch ends; ovulate cones globose, the scales many, thick, shield-shaped; seeds 2 per scale.

Taxodium distichum (L.) Rich. Bald Cypress; Swamp Cypress. [*Cupressus distica* L.]. Trees to 15 m or more; bark thin, gray brown, rough; leaves mostly 10–16 mm long, distichous, apparently alternate, on slender deciduous twigs 5–12 cm long; male cones 2 mm thick, borne in slender, drooping panicles; ovulate cones subglobose, 2–2.5 cm thick, the scales shield-shaped, tan to brownish. Uncommonly grown shade and specimen tree in Utah; introduced from the southeastern U. S.; 4 (0). Trees planted at the margin of the Botany Pond, Brigham Young University, in the mid 1940's by B. F. Harrison are now more than 12 m tall and have huge buttressed trunks. The trees should be more widely grown in the state.

DIVISION MAGNOLIOPHYTA
The Flowering Plants

Herbs, subshrubs, shrubs, and trees with an alternation of generations; gametophyte generation greatly reduced, represented by microgametophytes (ultimately the pollen grains) and megagametophytes (the embryo sacs); fertilization double; sporophytes with roots, stems, leaves and flowers; xylem typically with vessels; flowers of spirally arranged parts, or of 1–4 whorls of parts cyclically arranged, usually with sepals, petals, stamens, and a pistil (or these variously lacking or modified); sepals typcially enclosing the other parts in bud; petals typically as many as the sepals and borne alternate with them; stamens (microsporophylls) typically as many or twice as many and alternate with the petals (opposite the sepals); pistil variously simple (of one carpel or megasporophyll) and solitary or few to many, or consisting of connate carpels (hence compound); ovule solitary or few to many, borne on placentae by a funiculus; ovary ripening to form a fruit.

CLASS MAGNOLIOPSIDA
The Dicots

Annual, biennial, or perennial herbs, or shrubs or trees; leaves alternate, opposite, or whorled, simple or variously compound, typically net-veined; flower parts typically 4– or 5–merous; embryo typically with 2 cotyledons.

Key to the families.

1.	Perianth consisting of a single whorl, arbitrarily called sepals, or none	Key 1, p. 35
—	Perianth consisting of 2 whorls (sepals and petals)	2
2(1).	Corolla of separate petals	3
—	Corolla of united (connate) petals, at least near the base	Key 4, p. 39
3(2).	Stamens numerous, more than twice as many as the petals	Key 2, p. 37
—	Stamens few, not more than twice as many as the petals	Key 3, p. 37

Key 1.
Perianth consisting of a single whorl.

1).	Plants parasitic on the branches of trees or shrubs, rooting in the host, usually yellow green .... **Viscaceae**, p. 643	
—	Plants not parasitic on branches of trees, rooting in soil	2
2(1).	Plants trees, shrubs, or vines	3
—	Plants herbaceous	29
3(2).	Leaves opposite	4
—	Leaves typically alternate, at least the uppermost ...	11
4(3).	Plants trailing vines	5
—	Plants trees or shrubs	6
5(4).	Leaves compound; stamens and pistils many; flowers showy .... **Ranunculaceae** (*Clematis*), p. 503	
—	Leaves simple, deeply palmately lobed; stamens and pistils 5 or fewer; flowers not showy .... **Cannabaceae** (*Humulus*), p. 96	
6(4).	Ovary superior; fruit a samara, capsule, achene, or drupe	7
—	Ovary inferior (apparently so in Elaeagnaceae); fruit a drupe or a berry	9
7(6).	Flowers perigynous; plants shrubs, mainly less than 1 m tall .... **Rosaceae** (*Coleogyne*), p. 519	
—	Flowers hypogynous; plants trees or shrubs, mainly over 1.5 m tall	8
8(7).	Fruit a double samara; leaves palmately lobed or 3– to 5–foliolate .... **Aceraceae**, p. 41	
—	Fruit a simple samara (1–winged), capsule, or drupe; leaves pinnately compound or simple and typically pinnately veined .... **Oleaceae**, p. 431	
9(6).	Flowers in pendulous catkins; leaves leathery, evergreen; plants of Washington County .... **Garryaceae**, p. 307	
—	Flowers variously arranged but not in catkins; leaves not both leathery and evergreen	10
10(9).	Flowers in corymbose cymes, perfect; ovary usually 2–loculed; stamens typically 5 ...... **Cornaceae**, p. 241	
—	Flowers solitary or in axillary clusters, often imperfect; ovary 1–loculed; stamens 4–8 .... **Elaeagnaceae**, p. 294	
11(3).	Leaves compound	12
—	Leaves simple	15
12(11).	Leaflets armed with spinulose teeth; plants evergreen .... **Berberidaceae**, p. 55	
—	Leaflets entire or serrate, but not spinulose, deciduous	13
13(12).	Leaf rachis very narrowly winged; fruit a thin-fleshed drupe; trees cultivated and established in Washington County .... **Anacardiaceae**, p. 46	
—	Leaf rachis not winged; fruit not a drupe; trees or shrubs of various distribution and persistence	14
14(13).	Leaflets entire; fruit a legume; flowers conspicuous, brightly colored or white .... **Leguminosae**, p. 336	
—	Leaflets toothed or entire; fruit drupaceous; flowers inconspicuous, greenish .... **Juglandaceae**, p. 326	
15(11).	Plants trailing vines .... **Polygonaceae**, p. 470	
—	Plants trees or shrubs	16
16(15).	Flowers of one or both sexes in catkins (aments); plants monoecious or dioecious	17
—	Flowers not in catkins, perfect or imperfect	22
17(16).	Perianth lacking	18
—	Perianth present	20
18(17).	Plants desert shrubs of saline soils; leaves entire, terete or nearly so; fruit a utricle .... **Chenopodiaceae** (*Sarcobatus*), p. 116	
—	Plants of moist situations in a variety of soils; leaves mostly not entire, the blades flat; fruit a nutlet, nut, or capsule	19
19(18).	Plants monoecious; staminate flowers attached to bract of catkin .... **Betulaceae**, p. 57	
—	Plants dioecious; staminate flowers attached to axis of catkin .... **Salicaceae**, p. 546	
20(17).	Fruit a multiple or a syconium, fleshy and elongated, obovoid, or globose and woody ..... **Moraceae**, p. 425	
—	Fruit an acorn or a nut	21
21(20).	Fruit a nut, enclosed in a leafy involucre or in cone-like catkins; leaves serrate or doubly so; plants usually of moist situations, if native .... **Betulaceae**, p. 57	
—	Fruit an acorn, chestnut, or beechnut, with a basal cup, spiny bur, or subtended by bracts; leaves typically lobed; plants seldom of moist situations, unless cultivated .... **Fagaceae**, p. 304	

22(16).	Ovary inferior or apparently so .. **Elaeagnaceae**, p. 294	39(38).	Plants aquatic, typically more or less submerged ... 40
—	Ovary superior 23	—	Plants terrestrial, sometimes growing in moist or wet soil 41
23(22).	Ovary 1-carpellate; fruit an achene or berry 24	40(39).	Leaves entire, opposite, often crowded into terminal rosettes .... **Callitrichaceae**, p. 84
—	Ovary 2- or more-carpellate; fruit an achene or otherwise, but not a berry 25	—	Leaves dichotomously dissected, whorled .... **Ceratophyllaceae**, p. 115
24(23).	Plants armed with thorns (modified leaves), introduced or indigenous shrubs with small but showy perianth; fruit a berry .... **Berberidaceae**, p. 55	41(39).	Perianth lacking entirely (in some Euphorbiaceae, the involucre has petaloid bracts) 42
—	Plants unarmed native shrubs, the perianth showy or not; fruit a plumose achene .... **Rosaceae**, p. 519	—	Perianth present 43
25(23).	Plants typically trees; fruit a samara 26	42(41).	Inflorescence spicate, subtended by a conspicuous petaloid involucre, the whole resembling a single flower .... **Saururaceae**, p. 557
—	Plants normally shrubs; fruit an achene or utricle ... 27		
26(25).	Leaves palmately veined and lobed; fruit of woody capsules borne in heads ..... **Hamamelidaceae**, p. 314	—	Inflorescence a cyathium (a cupshaped involucre enclosing a cluster of staminate flowers each consisting of single pedicellate stamens, and pistillate flowers, consisting of solitary, 3-lobed pistil), which often bear petaloid appendages and glands .... **Euphorbiaceae**, p. 299
—	Leaves pinnately veined, serrate; fruit a samara ... . **Ulmaceae**, p. 611		
27(25).	Plants with milky juice; inflorescence subtended by colored foliose bracts .... **Euphorbiaceae**, p. 299		
—	Plants with watery juice; inflorescence not as above . 28	43(41).	Flowers epigynous .... **Araliaceae**, p. 50
28(27).	Flowers perfect, subtended by a cuplike involucre; stamens 6-9; fruit an achene .... **Polygonaceae**, p. 470	—	Flowers perigynous or hypogynous 44
		44(43).	Flowers perigynous, the ovary enclosed in or seated in a hypanthium 45
—	Flowers usually imperfect, not subtended by a cuplike involucre; stamens 1-5; fruit a utricle .... **Chenopodiaceae**, p. 116	—	Flowers hypogynous, the ovary not enclosed in a hypanthium 48
29(2).	Ovary inferior 30	45(44).	Hypanthium campanulate, the 4 sepals apparent as triangular lobes; plants often aquatic or growing in mud, seldom terrestrial .... **Lythraceae**, p. 417
—	Ovary superior 38		
30(29).	Plants aquatic; leaves entire and in whorls or alternate and dissected 31	—	Hypanthium various, the sepals 4 or 5; plants not aquatic or growing in mud, terrestrial 46
—	Plants terrestrial; leaves various but not as above ... 32	46(45).	Stipules present; leaves alternate .... **Rosaceae**, p. 519
31(30).	Leaves alternate and finely dissected .... **Haloragaceae**, p. 314	—	Stipules none; leaves opposite 47
—	Leaves whorled, entire .... **Hippuridaceae**, p. 315	47(46).	Stamens 3-5; fruit indehiscent . **Nyctaginaceae**, p. 426
32(30).	Ovary 2-loculed, with 1 ovule per locule; fruit 2-seeded 33	—	Stamens many; fruit dehiscent by a circumscissile lid .... **Aizoaceae**, p. 43
—	Ovary 1-loculed, 1- or 2-ovuled, sometimes with several carpels, but with a single ovule developing (1- to 5-loculed and many-seeded in some Aizoaceae); fruit 1- or 2-seeded 34	48(44).	Perianth showy; sepals and petals both present in bud, but the sepals caducous and represented only by scars during anthesis; plants with milky juice .... **Papaveraceae**, p. 450
33(32).	Perianth of connate segments; leaves opposite or whorled; flowers usually in cymes ... **Rubiaceae**, p. 543	—	Perianth showy or not; plants various but not as above 49
—	Perianth segments distinct; leaves alternate or basal; flowers in umbels .... **Umbelliferae**, p. 613	49(48).	Styles and stigmas single 50
		—	Styles and stigmas typically more than 1 51
34(32).	Plants typically fleshy and prostrate; fruit a typically circumscissile capsule .... **Aizoaceae**, p. 43	50(49).	Perianth neither tubular nor corolla-like .... **Urticaceae**, p. 637
—	Plants usually not fleshy, of various habit but sometimes prostrate; fruit not a circumscissile capsule ... 35	—	Perianth tubular, corolla-like ... **Nyctaginaceae**, p. 426
35(34).	Flowers sessile, borne in involucrate heads; anthers united into a tube around the style . **Compositae**, p. 131	51(49).	Leaves mostly deeply palmately 5- to 7-lobed or with 5-7 leaflets; flowers imperfect .... **Cannabaceae**, p. 96
—	Flowers not sessile or in involucrate heads; anthers distinct 36	—	Leaves mostly entire or shallowly lobed; flowers perfect or imperfect 52
36(35).	Leaves alternate .... **Santalaceae**, p. 556	52(51).	Ovary with more than 1 locule 53
—	Leaves opposite 37	—	Ovary with a single locule 54
37(36).	Leaves pinnately parted or compound, at least the cauline ones sheathing at the base; plants dioecious .... **Valerianaceae**, p. 638	53(52).	Leaves opposite .... **Aizoaceae**, p. 43
		—	Leaves alternate, at least above . **Euphorbiaceae**, p. 299
—	Leaves simple, not sheathing at the base; plants with perfect flowers .... **Nyctaginaceae**, p. 426	54(52).	Leaves opposite; ovules and seeds more than 1; fruit a capsule 55
38(29).	Pistils several to many per flower; stamens usually 10 or more .... **Ranunculaceae**, p. 503	—	Leaves alternate or opposite; ovules and seeds solitary; fruit an achene or utricle 56
—	Pistils 1 (3 or 4 in Saururaceae) per flower; stamens 1 to many, but usually 10 or less 39	55(54).	Capsule opening by means of a circumscissile lid .. .. **Aizoaceae**, p. 43
		—	Capsule opening by valves ... **Caryophyllaceae**, p. 101

56(54).	Perianth typically of 6 segments; stamens 3, 6, or 9; stipular sheaths often present (lacking in several genera) .... **Polygonaceae**, p. 470		—	Sepals caducous; plants not succulent .... **Papaveraceae**, p. 450
—	Perianth with 1, 4, or 5 segments; stamens 1, 4, or 5; stipular sheaths not present 57		16(14).	Filaments united into a tube around the pistil .... **Malvaceae**, p. 419
57(56).	Bracts subtending flowers scarious, typically awn-tipped; plants not scurfy .... **Amaranthaceae**, p. 44		—	Filaments not united into a tube, distinct or in clusters and united at the base 17
—	Bracts subtending flowers not scarious, not awned; plants often scurfy .... **Chenopodiaceae**, p. 116		17(16).	Maturing ovary open apically, the seeds exposed; flowers irregular .... **Resedaceae**, p. 518
			—	Maturing ovary closed apically, the seeds not exposed; flowers usually regular 18

Key 2.
Corolla of separate petals; stamens more than twice as many as the petals.

1.	Ovary inferior or partly so 2		18(17).	Stamens attached to the margin of a hypanthium ... 19
—	Ovary superior 7		—	Stamens attached at base of ovary 20
2(1).	Petals numerous; stems thick, succulent, armed with spines or glochids; leaves lacking or caducous .... **Cactaceae**, p. 85		19(18).	Hypanthium 8– to 12–ribbed; stems usually 4–angled .... **Lythraceae**, p. 417
—	Petals few; stems not thick and succulent, though occasionally spiny; leaves present and conspicuous during growing season 3		—	Hypanthium not ribbed; stems not as above .... **Rosaceae**, p. 519
3(2).	Ovary only partly inferior 4		20(18).	Leaves opposite .... **Guttiferae**, p. 314
—	Ovary wholly inferior 5		—	Leaves alternate or basal, but still alternate 21
4(3).	Leaves opposite; fruit a capsule .. **Saxifragaceae**, p. 557		21(20).	Plants robust, glabrous, glaucous perennials; sepals leathery; flowers often 2–4 cm wide or more .... **Paeoniaceae**, p. 450
—	Leaves alternate; fruit a pome .... **Rosaceae**, p. 519			
5(3).	Flowers bright orange-crimson; fruit a leathery berry with fleshy seeds .... **Punicaceae**, p. 501		—	Plants various, but not as above in all respects 22
—	Flowers variously colored; fruit not as above 6		22(21).	Leaves compound, with 3–5 leaflets; ovary short-stipitate .... **Capparaceae** (*Polanisia*), p. 97
6(5).	Plants woody; fruit fleshy .... **Rosaceae**, p. 519		—	Leaves various, but not as above; ovary sessile .... **Ranunculaceae**, p. 503
—	Plants herbaceous; fruit a capsule ... **Loasaceae**, p. 413			

Key 3.
Corolla of separate petals; stamens few, not more than twice as many as the petals.

7(1).	Plants aquatic, with floating leaves .... **Nymphaeaceae**, p. 430		1.	Flowers with more than 1 pistil or the pistil so deeply lobed as to appear so 2
—	Plants terrestrial, sometimes growing in moist or wet soils, but leaf blades not floating 8		—	Flowers with a single pistil 7
8(7).	Plants trees or shrubs 9		2(1).	Plants succulent, fleshy .... **Crassulaceae**, p. 242
—	Plants herbaceous, or woody at base only 14		—	Plants not succulent 3
9(8).	Flowers in cymes, these borne on a wing-like bract; stamens borne in fascicles .... **Tiliaceae**, p. 610		3(2).	Plants monoecious trees, the flowers borne in globose, unisexual heads; leaves palmately lobed .... **Platanaceae**, p. 355
—	Flowers variously arranged, but not on a winglike bract; stamens not in fascicles 10		—	Plants herbs, shrubs, or trees, typically with perfect flowers, these not borne in globose heads; leaves various 4
10(9).	Shrubs or small trees with recurved thorns and pinnate leaves; fruit a legume .... **Leguminosae** (*Acacia*), p. 336		4(3).	Flowers 3–merous; plants weak annual herbs .... **Limnanthaceae**, p. 411
—	Shrubs or trees without thorns or pinnate leaves; fruit various but not a legume 11		—	Flowers 5–merous; plants various 5
11(10).	Stipular scars encircling the stem; perianth parts, stamens, and pistils spirally arranged .... **Magnoliaceae**, p. 418		5(4).	Plants evergreen shrubs .... **Celastraceae** (*Forsellesia*), p. 114
—	Stipular scars not encircling the stem; perianth parts and stamens in whorls; pistil usually 1 12		—	Plants various but, if shrubs, seldom if ever evergreen 6
12(11).	Filaments united into a tube around the pistil .... **Malvaceae**, p. 419		6(5).	Stamens inserted on a hypanthium; flowers perigynous .... **Rosaceae**, p. 519
—	Filaments not united into a tube, distinct, or in clusters and more or less united basally 13		—	Stamens inserted at base of ovary; flowers hypogynous .... **Ranunculaceae**, p. 503
13(12).	Inflorescence borne on a winglike bract; leaves oblique basally .... **Tiliaceae**, p. 610		7(1).	Plants usually with tendrils, trailing vines, monoecious .... **Cucurbitaceae**, p. 289
—	Inflorescence never as above; leaves various, but not as above .... **Rosaceae**, p. 519		—	Plants lacking tendrils, or if tendrils present the flowers perfect, usually not trailing (except in Passifloraceae) 8
14(8).	Sepals 2 (3 in *Argemone*, and the plants with milky juice) 15		8(7).	Styles 2–5, distinct to near the base 9
—	Sepals more than 2; plants with watery juice 16		—	Style 1 (or lacking), sometimes lobed or divided at the apex 24
15(14).	Sepals persistent; plants somewhat succulent .... **Portulacaceae**, p. 493		9(8).	Plants trees, shrubs, or vines 10
			—	Plants herbaceous 16

10(9).	Leaves scalelike .... **Tamaricaceae**, p. 610	—	Plants herbaceous (suffrutescent in some Pyrolaceae) 48
—	Leaves not scalelike, the blades typically well developed 11	30(29).	Flowers irregular 31
11(10).	Ovary inferior 12	—	Flowers regular 35
—	Ovary superior 13	31(30).	Petals 3, the lower 2 forming a keel .... **Polygalaceae**, p. 470
12(11).	Flowers in umbels; leaves palmately compound or palmately lobed .... **Araliaceae**, p. 50	—	Petals 4 or 5, the lower 2 sometimes forming a keel . 32
—	Flowers in racemes; leaves simple, variously veined and lobed .... **Saxifragaceae**, p. 557	32(31).	Flowers papilionaceous, the lower 2 petals often united and forming a keel; fruit a legume .... **Leguminosae**, p. 336
13(11).	Plants tendrillate vines; flower with a many-rayed, multicolored corona; ovary stipitate; styles 3 .... **Passifloraceae**, p. 453	—	Flowers not papilionaceous, the lower 2 petals not forming a keel; fruit a 3-valved, frequently 1-seeded capsule, or a spiny, 1-seeded pod 33
—	Plants shrubs or trees, lacking tendrils; flowers and ovaries various but not as above; styles 1 or 2 14	33(32).	Flowers yellow, borne in panicles in midsummer .. .. **Sapindaceae**, p. 556
14(13).	Leaves alternate; fruit a drupe ... **Anacardiaceae**, p. 46	—	Flowers variously colored or displayed, borne in springtime 34
—	Leaves opposite 15	34(33).	Plants low-growing shrubs; leaves simple; stamens usually 4 .... **Krameriaceae**, p. 327
15(14).	Plants mainly less than 0.5 m tall, to be sought in San Juan County; fruit a capsule .... **Frankeniaceae**, p. 306	—	Plants large shrubs or trees; leaves palmately compound; stamens 5–8 .... **Hippocastanaceae**, p. 315
—	Plants mainly 2–8 m tall or more; fruit a samaroid schizocarp .... **Aceraceae**, p. 41	35(30).	Leaves compound, with 2 or more leaflets 36
16(9).	Plants aquatic .... **Haloragaceae**, p. 314	—	Leaves simple, sometimes deeply divided or parted . 42
—	Plants terrestrial 17	36(35).	Leaves bipinnate; flowers purplish .. **Meliaceae**, p. 424
17(16).	Ovary inferior (partly so in some Saxifragaceae) 18	—	Leaves once pinnate; flowers variously colored 37
—	Ovary superior 19	37(36).	Leaves trifoliolate, alternate; fruit a samara with the wing continuous around it .... **Rutaceae**, p. 546
18(17).	Flowers in umbels; fruit a schizocarp .... **Umbelliferae**, p. 613	—	Leaves usually not trifoliolate; fruit various but, if samaroid, the wing not continuous around the seed and the leaves opposite 38
—	Flowers variously disposed, but not in umbels; fruit a many-seeded capsule or berry ... **Saxifragaceae**, p. 557	38(37).	Leaves opposite 39
19(17).	Leaves compound; leaflets 3 ..... **Oxalidaceae**, p. 449	—	Leaves alternate 40
—	Leaves simple 20	39(38).	Leaflets 3–9 or more; fruit a samara; plants usually trees; leaves deciduous .... **Oleaceae** (*Fraxinus*), p. 431
20(19).	Leaves opposite 21	—	Leaflets 2; fruit not a samara; plants shrubs; leaves persistent .... **Zygophyllaceae** (*Larrea*), p. 646
—	Leaves alternate 22	40(38).	Leaflets with few to several rounded teeth, these bearing a subcircular gland; trees malodorous; fruit of samaras .... **Simaroubaceae**, p. 604
21(20).	Sepals and petals 4–7; placentae 2–4, parietal .... **Frankeniaceae**, p. 306	—	Leaflets variously toothed to entire, but not bearing circular glands if toothed; trees or shrubs not especially malodorous; fruit of berries or legumes 41
—	Sepals and petals typically 4 or 5; placentae free-central or axile .... **Caryophyllaceae**, p. 101	41(40).	Leaves with leaflets spiny-toothed; sepals usually 6 in 2 whorls; petals usually 6 in 2 series; stamens 6 or 12 .... **Berberidaceae** (*Mahonia*), p. 55
22(21).	Sepals 2; plants succulent .... **Portulacaceae**, p. 493	—	Leaves not with spiny-toothed leaflets; sepals usually 5; petals 5 or fewer; stamens 5 or 10 .... **Leguminosae**, p. 336
—	Sepals more than 2; plants not succulent 23	42(35).	Plants subshrubs at most; fruit stipitate .... **Cruciferae** (*Stanleya*), p. 243
23(22).	Flowers irregular; capsule open at the top before maturity .... **Resedaceae**, p. 518	—	Plants trees or shrubs; fruit not stipitate (except in Rutaceae, which has entire, caducous leaves and blue flowers) 43
—	Flowers regular; capsules opening at maturity .... **Caryophyllaceae**, p. 101	43(42).	Flowers blue, 8–14 mm long; herbage glandular-punctate, strongly odoriferous; fruit stipitate .... **Rutaceae** (*Thamnosma*), p. 546
24(8).	Ovary inferior 25	—	Flowers variously colored, usually not blue, usually less than 8 mm long; herbage not glandular-punctate; fruit not stipitate 44
—	Ovary superior (sometimes enclosed by, but not adnate to, the floral tube) 29	44(43).	Plants with leaves modified as spines (foliage leaves in their axils), these sometimes 3-pronged; perianth conspicuous, usually 9– to 12-parted; fruit a berry; ovary 1-loculed ...... **Berberidaceae** (*Berberis*), p. 55
25(24).	Plants shrubs or subshrubs; leaves rough-hairy; petals hispid dorsally; known from Washington County .... **Loasaceae**, p. 413		
—	Plants variously woody or herbaceous; leaves not rough hairy; petals typically glabrous; distribution various 26		
26(25).	Plants herbaceous 27		
—	Plants woody shrubs 28		
27(26).	Flowers 4-merous .... **Onagraceae**, p. 434		
—	Flowers 5-merous .... **Saxifragaceae**, p. 557		
28(26).	Leaves opposite; flowers in cymes; stamens 4 .... **Cornaceae**, p. 241		
—	Leaves alternate; flowers solitary or in racemes; stamens 5 .... **Saxifragaceae**, p. 557		
29(24).	Plants trees or shrubs 30		

	Plants not spiny, or if so, not as above; perianth usually inconspicuous, usually 8– to 10–parted; fruit a capsule or a drupe; ovary 2– to 5–loculed 45
45(44).	Plants woody vines (lianas), typically with shaggy bark; leaves palmately veined or compound .... **Vitaceae**, p. 645
—	Plants not woody vines; leaves various 46
46(45).	Leaves leathery, evergreen; plants cultivated shrubs or small trees; fruit a globose, usually brightly colored berrylike drupe .... **Aquifoliaceae**, p. 50
—	Leaves thin, mainly deciduous; plants cultivated or indigenous; fruit variously shaped but seldom if ever brightly colored 47
47(46).	Stamens opposite the petals; ovary 2– or 3–loculed .... **Rhamnaceae**, p. 518
—	Stamens alternate with the petals or more numerous; ovary 2– to 5–loculed .... **Celastraceae**, p. 114
48(29).	Sepals 2 or 3 49
—	Sepals 4, 5, or more 53
49(48).	Plants succulent .... **Portulacaceae**, p. 493
—	Plants not succulent or not especially so 50
50(49).	Plants minute annual herbs with opposite leaves ... . **Elatinaceae**, p. 295
—	Plants of various duration; leaves alternate or basal and still essentially alternate 51
51(50).	Sepals caducous; stamens 6–12; leaves entire .... **Papaveraceae**, p. 450
—	Sepals persistent; stamens 6; leaves dissected 52
52(51).	Petals 4, in 2 unlike pairs; stamens diadelphous, in 2 sets of 3 .... **Fumariaceae**, p. 306
—	Petals 3, alike; stamens in 1 set **Limnanthaceae**, p. 411
53(48).	Flowers irregular 54
—	Flowers regular 58
54(53).	Petals typically 3; stamens commonly 8 .... **Polygalaceae**, p. 470
—	Petals mostly 5; stamens typically 10 or 5 55
55(54).	Flowers papilionaceous; fruit a legume .... **Leguminosae**, p. 336
—	Flowers not papilionaceous; fruit a capsule or separating into 3 indehiscent, 1–seeded carpels at maturity .. 56
56(55).	Plants typically evegreen, sometimes suffrutescent; anthers opening by apparently terminal pores .... **Pyrolaceae**, p. 501
—	Plants not evergreen or, if so, not otherwise as above . 57
57(56).	Leaves peltate; fruit separating into 3, indehiscent, 1–seeded segments at maturity . **Tropaeolaceae**, p. 611
—	Leaves not peltate; fruit a 1–loculed, 3–valved capsule .... **Violaceae**, p. 641
58(53).	Leaves compound or deeply pinnatifid 59
—	Leaves simple 62
59(58).	Leaves mostly basal, deeply pinnatifid .... **Geraniaceae**, p. 311
—	Leaves alternate or opposite 60
60(59).	Leaves opposite; leaflets 3–16 (but, if 3, these more or less spine-tipped) .. **Zygophyllaceae** (*Tribulus*), p. 646
—	Leaves alternate; leaflets usually 3 61
61(60).	Stamens 10; herbage with a sour taste .... **Oxalidaceae**, p. 449
—	Stamens 2–6; herbage with a sharp, biting flavor .. .. **Capparaceae**, p. 97
62(58).	Flowers imperfect; fruit an achene; plants with opposite leaves and stinging hairs or alternate and without stinging hairs .... **Urticaceae**, p. 637
—	Plants various, but not as above 63

63(62).	Plants minute annual herbs with opposite leaves, growing in mud .... **Elatinaceae**, p. 295
—	Plants various, but leaves alternate or mainly so 64
64(63).	Sepals and petals 4; stamens 6 (4 plus 2), 4, or rarely 2 .... **Cruciferae**, p. 243
—	Sepals and petals mostly 5; stamens 5–10 65
65(64).	Leaves with stipules; carpels tailed at maturity, separating from the base as 1–seeded, indehiscent segments .... **Geraniaceae**, p. 311
—	Leaves lacking stipules; carpels various but if as above, not long tailed 66
66(65).	Ovary 10–loculed, with 1 ovule per locule .... **Linaceae**, p. 412
—	Ovary 2– to 5–loculed 67
67(66).	Plants annual; leaves dissected; fruit 5–seeded, breaking into indehiscent nutlets ...... **Limnanthaceae**, p. 411
—	Plants perennial; leaves simple or scalelike; fruit a many-seeded capsule .... **Pyrolaceae**, p. 501

Key 4.
Corolla of united petals.

1.	Ovary inferior or partly so 2
—	Ovary superior 14
2(1)	Stamens more than 5; anthers opening by terminal pores .... **Ericaceae**, p. 296
—	Stamens 5 or fewer (if more as in *Adoxa*, the sepals 2–3); anthers not opening by terminal pores 3
3(2).	Stamens united by the anthers 4
—	Stamens separate 7
4(3).	Plants usually tendril bearing; flowers monoecious; fruit a pepo .... **Cucurbitaceae**, p. 289
—	Plants not tendril bearing; flowers usually perfect, at least some; fruit a capsule or an achene 5
5(4).	Flowers not in involucrate heads; stamens not adnate to the corolla .... **Campanulaceae**, p. 84
—	Flowers in involucrate heads; stamens adnate to the corolla 6
6(5).	Calyx cup-shaped, 4–lobed, persistent on the achenes; receptacular bracts awn-tipped .... **Dipsacaceae**, p. 294
—	Calyx consisting of a pappus of hairs, scales, or awns; receptacular bracts lacking or, if present, not as above .... **Compositae**, p. 131
7(3).	Basal leaves ternately compound, long-petioled 8
—	Basal leaves not ternate and long-petioled 9
8(7).	Plants aquatic or semiaquatic; leaflets 2–9 cm long or more .... **Menyanthaceae**, p. 424
—	Plants terrestrial; leaflets less than 2 cm long .... **Adoxaceae**, p. 43
9(7).	Leaves alternate 10
—	Leaves opposite or whorled 11
10(9).	Plants woody, trees or shrubs or, if herbaceous, the leaves doubly compound .... **Araliaceae**, p. 50
—	Plants herbaceous, the leaves simple .... **Campanulaceae**, p. 84
11(9).	Stamens 1–3; flowers irregular; plants frequently ill scented .... **Valerianaceae**, p. 638
—	Stamens 4 or 5; flowers regular or irregular 12
12(11).	Herbs with ovary 1–loculed; flowers in involucrate heads; fruit an achene .... **Dipsacaceae**, p. 289
—	Herbs or shrubs with ovary 2– to 5–loculed; flowers not in involucrate heads; fruit not an achene 13

13(12). Shrubs with broad leaves; ovary 3– to 5–loculed; leaves opposite or perfoliate, not whorled or with stipules .... **Caprifoliaceae**, p. 98
— Shrubs or more usually herbs, if shrubs, the flowers in globose heads; ovary 2–loculed; leaves opposite and having stipules, or whorled and lacking them .. .. **Rubiaceae**, p. 543
14(1). Stamens more than 5 15
— Stamens 5 or fewer 23
15(14). Corolla segments distinctly united, cup- or tube-shaped 16
— Corolla segments not distinctly united, usually united only near the base or only part of the segments united 18
16(15). Pistils more than 1; stamens twice as many as the petals .... **Crassulaceae**, p. 242
— Pistil 1 per flower; stamens the same number as the petals 17
17(16). Anther opening by a terminal pore; plants green ... . **Ericaceae**, p. 296
— Anther opening by a longitudinal slit; plants green or not .... **Pyrolaceae**, p. 501
18(15). Corolla irregular 19
— Corolla regular 21
19(18). Petals 5; fruit a legume .... **Leguminosae**, p. 336
— Petals 3 or 4; fruit not a legume 20
20(19). Leaves entire; petals 3 .... **Polygalaceae**, p. 470
— Leaves dissected; petals 4, in 2 unlike pairs .... **Fumariaceae**, p. 306
21(18). Flowers minute, in dense heads or spikes; stamens 10 to many; fruit a legume .... **Leguminosae**, p. 336
— Flowers not as above; fruit various but not a legume . 22
22(21). Filaments united well above the base; stamens many; leaves simple .... **Malvaceae**, p. 419
— Filaments united only at the base; stamens 10; leaves compound .... **Oxalidaceae**, p. 449
23(14). Plants parasitic, devoid of chlorophyll 24
— Plants usually not parasitic, possessing chlorophyll .. 25
24(23). Flowers regular, minute; stems slender, trailing, twining, yellow or orange vines ... **Cuscutaceae**, p. 291
— Flowers irregular; stems erect, not trailing or twining, yellow or purplish herbs .. **Orobanchaceae**, p. 448
25(23). Corolla irregular 26
— Corolla regular 31
26(25). Ovary with 1 ovule per locule, appearing 4–loculed, often 4–lobed; fruit consisting of 4 indehiscent, 1–seeded nutlets 27
— Ovary with more than 1 ovule per locule, usually neither 4–loculed nor 4–lobed; fruit a capsule 28
27(26). Ovary 4–lobed; style arising between lobes, cleft apically .... **Labiatae**, p. 328
— Ovary not or only slightly 4–lobed; style arising from apex of ovary, not cleft apically ... **Verbenaceae**, p. 639
28(26). Plants aquatic; leaves often dissected; corolla spurred; stamens 2 .... **Lentibulariaceae**, p. 411
— Plants usually terrestrial or, if aquatic, other than above 29
29(28). Plants trees or shrubs; seeds winged; capsule narrowly cylindrical to cigar shaped .. **Bignoniaceae**, p. 58
— Plants herbaceous or woody only at base; seeds not winged; capsules various, but not as above 30

30(29). Plants strongly viscid-pubescent; fruit a woody capsule with 2 recurved, thorn-like appendages, 6–15 cm long or more .... **Pedaliaceae**, p. 453
— Plants usually not viscid-pubescent; fruit not as above .... **Scrophulariaceae**, p. 567
31(25). Trees with blackish, reticulately patterned bark; corolla urceolate, typically 4–merous; fruit a several-loculed berry; calyx persistent on the fruit .... **Ebenaceae**, p. 294
— Trees, shrubs, or herbs; corolla various but typically 5–merous; fruit various, the calyx not persistent, or the plants herbaceous 32
32(31). Plants with milky juice (watery in some Apocynaceae); pistil of 2 carpels, these separate at base, united by stigmas and/or styles 33
— Plants usually without milky juice; pistil various, but never as above 34
33(32). Styles united; stamens appressed around the stigma, but free from it .... **Apocynaceae**, p. 47
— Styles distinct below; stamens adnate to the stylar column .... **Asclepiadaceae**, p. 51
34(32). Ovary 1–loculed, 1–ovuled; style and stigma 1; fruit hard, dry; apparent corolla actually a calyx subtended by an involucre .... **Nyctaginaceae**, p. 426
— Ovary and fruit various but not as above; corolla and calyx both present 35
35(34). Stamens as many as the corolla lobes and opposite them; flowers often (but not always) in umbels .... **Primulaceae**, p. 497
— Stamens as many as, or fewer than, the corolla lobes and alternate with them; flowers seldom if ever in umbels 36
36(35). Corolla small (2 mm long or less), scarious, veinless; capsule opening by a lid; leaves parallel veined or nearly so .... **Plantaginaceae**, p. 453
— Corolla various, but not as above; capsule usually not opening by a lid, or the fruit not a capsule; leaves usually distinctly net veined 37
37(36). Ovary 4–lobed, 4–loculed; fruit consisting of 1–4 nutlets at maturity .... **Boraginaceae**, p. 59
— Ovary not 4–lobed, 1– to 3–loculed; fruit a capsule or a berry 38
38(37). Style with 3 stigmatic lobes; ovary 3–loculed; fruit a 3–valved capsule .... **Polemoniaceae**, p. 455
— Style not with 3 stigmatic lobes; ovary 1– to 2–loculed; fruit not as above 39
39(38). Ovary 1–loculed 40
— Ovary usually with 2 or more locules 41
40(39). Leaves opposite or whorled, entire; style 1 or none; plants mostly glabrous; inflorescence not scorpioid .... **Gentianaceae**, p. 307
— Leaves usually alternate or, if opposite, not entire; styles 2 or single and 2–cleft apically; plants mostly hairy; inflorescence usually scorpioid .... **Hydrophyllaceae**, p. 316
41(39). Stems trailing or twining; leaves often hastately or sagittately lobed .... **Convolvulaceae**, p. 240
— Stems not trailing or twining; leaves seldom if ever hastately or sagittately lobed 42
42(41). Styles 2 or 2–branched apically; fruit a capsule; plants herbaceous (except Eriodictyon) .... **Hydrophyllaceae**, p. 316
— Style 1; stigma entire or merely lobed apically; fruit a capsule, drupe, or berry; plants herbs, shrubs, or small trees 43

43(42).	Stamens 2 44		—	Trees flowering as the leaves appear or after; flowers variously disposed 6
—	Stamens more than 2, or others represented by staminodes 46		5(4).	Petals lacking; sepals connate, the calyx merely lobed; leaves often deeply cut (more than half way to base); bark of upper branches silvery white; common in cultivation .... *A. saccharinum*
44(43).	Plants herbaceous **Scrophulariaceae** (*Veronica*), p. 567			
—	Plants woody 45			
45(44).	Leaves alternate .... **Ericaceae**, p. 296		—	Petals present; sepals distinct; leaves seldom incised to the middle; bark not especially silvery white; uncommon in cultivation .... *A. rubrum*
—	Leaves opposite .... **Oleaceae**, p. 431			
46(43).	Shrubs; leaves opposite or whorled .... **Buddlejaceae**, p. 84		6(4).	Leaves palmately 7– to 9–lobed 7
			—	Leaves palmately 3– to 5–lobed 8
—	Shrubs, herbs, or rarely trailing vines; leaves alternate 47		7(6).	Blades cut more than half way to the base, the lobes narrowed below the middle, long-acuminate ...... *A. palmatum*
47(46).	Hypanthium present, 4–angled ... **Lythraceae**, p. 417			
—	Hypanthium lacking 48		—	Blades cut to about 1/4 the distance to the base, the lobes not narrowed below, short-acuminate .... *A. circinatum*
48(47).	Plants coarse, hairy, biennial herbs; inflorescence a dense spike of yellow flowers .... **Scrophulariaceae** (*Verbascum*), p. 567			
			8(6).	Terminal lobe of leaf at least twice longer than broad .... *A. ginnala*
—	Plants shrubs, herbs, or trailing vines; inflorescence various but not a dense spike 49		—	Terminal lobe of leaf less than twice longer than broad 9
49(48).	Flowers 4–merous; plants shrubs or small trees, evergreen; leaves coriaceous .... **Aquifoliaceae**, p. 50		9(8).	Leaves mainly less than 8 cm wide; plants indigenous, rarely cultivated 10
			—	Leaves mainly 8–20 cm wide; plants cultivated 11
—	Flowers 5–merous; plants herbs or shrubs, typically deciduous; leaves thin, not especially coriaceous .. **Solanaceae**, p. 604		10(9).	Leaves glabrous, the sinuses narrow; sepals distinct, glabrous; petals commonly present .... *A. glabrum*
			—	Leaves pubescent beneath, with broadly rounded sinuses; sepals more or less connate, pubescent; petals lacking .... *A. grandidentatum*
			11(9).	Sinuses between principal lobes sharply angled; leaf lobes and teeth mostly blunt; inflorescence a raceme or panicle, much longer than broad .... *A. pseudoplatanus*
			—	Sinuses between principal lobes rounded; leaf lobes and teeth acuminate to acute 12
			12(11).	Samara wings widely diverging, essentially horizontal; plants with milky juice; leaf lobes attenuate-aristate; commonly cultivated .... *A. platanoides*
			—	Samara wings more or less pendulous; plants with watery juice; leaf lobes bluntly attenuate-acuminate apically; uncommonly cultivated .... *A. saccharum*

ACERACEAE A. L. Juss.
Maple Family

Trees or shrubs; leaves opposite, simple, or compound; flowers perfect, imperfect, or polygamous, regular, small, in lateral or terminal racemes, corymbs, or panicles; sepals 4–6, distinct or connate; petals 4–6, distinct or lacking; stamens 4–12, usually 8, the filaments distinct, arising from a disk or this lacking; pistil 1, the ovary superior, 2–loculed (rarely more); styles 2; fruit a flat, samaroid schizocarp; $x = 13$.

Acer L.

Leaves simple and usually palmately lobed and veined or pinnately to palmately compound; flowers inconspicuous; fruits with 2 wings, often falling as a unit. Note: The genus is represented by numerous cultivated and three native species, which also occur in cultivation. The plants persist following cultivation and occasionally become esablished. The vibrant display of colorful foliage in autumn by the members of this genus is outstanding.

Keller, A. C. 1942. *Acer glabrum* and its varieties. Amer. Midl. Nat. 27: 491–500.

1.	Leaves compound 2	
—	Leaves simple 3	
2(1).	Leaves pinnately 3– to 7–foliolate; plants dioecious, indigenous and cultivated .... *A. negundo*	
—	Leaves palmately 3–foliolate; plants polygamous, indigenous, seldom cultivated .... *A. glabrum*	
3(1).	Leaves with tips of lobes rounded; trees small, rounded, cultivated ornamentals .... *A. campestre*	
—	Leaves with lobe tips bluntly to sharply acute to acuminate or attenuate; trees or shrubs, cultivated or indigenous 4	
4(3).	Trees flowering before the leaves appear in late winter or early spring; flowers from buds of the previous year 5	

Acer campestre L. Hedge Maple. Small rounded trees mainly 3–6 m tall; bark corky, fissured; trunks mainly 2–4 dm thick; leaves 3–10 cm wide, 3– to 5–lobed, the lobes usually entire, rounded, green on both surfaces, puberulent along the veins, ciliate; flowers greenish, borne in corymbs; sepals green, narrowly oblong, 2.5–3 mm long, long hairy, colored like the petals and subequal to them, narrowly oblanceolate; samara wings almost horizontally spreading; $2n = 26$. Sparingly cultivated landscaping tree in Cache and Utah counties; widely grown in the U.S.; introduced from Eurasia; 11 (0).

Acer circinatum Pursh Vine Maple. Small trees, mainly 1–6 m tall, sometimes reclining; branchlets slender, glabrous or sparsely pilose; leaves 5–12 cm wide, 7– to 9–lobed, the sinuses extending 1/4–1/3 to the middle, glabrous or hairy along the veins above, green above, paler beneath; flowers few, borne in corymbs; sepals red purple, 4–6 mm long; petals shorter than the sepals, greenish or whitish; samaras glabrous, the wings almost horizontal; $n = 13$. Cultivated specimen plants; sparingly grown in Utah County; British Columbia to California; 2 (0).

Acer ginnala Maxim. Amur Maple. Small shrublike trees, mainly 2–6 m tall; branchlets slender, glabrous; leaf blades 3–12 cm long, 1.5–7.5 cm wide, 3–lobed, the terminal lobe more than twice longer than broad, doubly serrate, glabrous or sparing long hairy, green above, paler beneath; flowers corymbose; sepals greenish, 1.8–2.2 mm long, ovate-oval; petals white, subequal to the sepals; samaras sparingly long hairy and glandular, the wings strongly descending; 2n = 26. Frequently grown ornamental in Salt Lake and Utah counties; widely grown in the U.S.; introduced from Asia; 5 (0).

Acer glabrum Torr. Rocky Mountain Maple. [*A. tripartitum* Nutt; *A. glabrum* var. *tripartitum* (Nutt.) Pax; *A. neomexicanum* Greene; *A. glabrum* var. *neomexicanum* (Greene) Kearney & Peebles; A. glabrum ssp. neomexicanum (Greene) E. Murray]. Shrubs or small trees, mainly 2–8 m tall; herbage essentially glabrous; leaves palmately 3– to 5–lobed or 3–foliolate, mainly 2–8 cm wide, the lobes or leaflets rather sharply doubly serrate, acute apically; flowers in corymbose cymes, polygamous or dioecious; sepals greenish, 3–5 mm long; petals greenish, subequal to the sepals (rarely wanting?); samaras glabrous, the wings strongly descending. Pinyon-juniper, mountain brush, sagebrush, ponderosa pine, Douglas fir, lodgepole pine, and spruce-fir communities at 1675 to 3175 m in all Utah counties; Alaska south to California, Arizona, New Mexico, and Nebraska; 82 (xii). The phase of the species with small thickish leaves growing in Juab, Millard, and Beaver counties (at least) are assignable to **var. *diffusum* (Greene)** Smiley [*A. diffusum* Greene; *A. glabrum* ssp. *diffusum* (Greene) E. Murray]. The remainder probably best fit within **var. *glabrum*.**

Acer grandidentatum Nutt. Bigtooth Maple. [*A. saccharum* var. *grandidentatum* (Nutt.) Sudw.; *A. saccharum* ssp. *grandidentatum* (T. & G.) Desmarais; *A. saccharum* ssp. *grandidentatum* var. *trilobum* E. Murray, type from Ephraim Canyon]. Small trees, mainly 4–8 m tall; herbage more or less villous to puberulent at least on lower leaf surfaces; leaves palmately 3– to 5–lobed, mainly 2.5–10 (13) cm wide, lobed to near the middle, the lobes with few coarse teeth or lobes or almost entire; flowers in corymbose cymes, on pendulous elongate pedicels mainly 15–30 mm long, often long-hairy near the apex; sepals 3–5 mm long, greenish, broadly rounded, connate to near the middle or above; petals lacking; samaras more or less long-hairy, the wings descending to spreading; 2n = 26. Oak, oak-maple, sagebrush, Douglas fir, and white fir communities at 1280 to 2810 m in all Utah counties, except Daggett, Emery, and Wayne; Idaho and Wyoming, south to Nevada, Arizona, Mexico, and Oklahoma; 107 (vii). This plant is a principal component of the mountain brush community in Utah.

Acer negundo L. Boxelder. [*A. interior* Britt. in Britt. & Shafer; *A. negundo* var. *interior* (Britt.) Sarg.; *A. negundo* ssp. *interior* (Britt.) Love & Love; *A. kingii* Britt. in Britt. & Shafer, type from the Wasatch Mts.]. Trees mostly 4–12 m tall, dioecious; branchlets velvety hairy or glabrous; leaves pinnately 3– to 7–foliolate, or rarely twice compound; leaflets mostly 2–10 cm long and 12–25 mm wide, usually coarsely toothed or lobed, hairy to glabrous on one or both surfaces, especially along the veins; flowers on drooping, very long pedicels, the staminate in subumbellate clusters, the pistillate in racemes; sepals 4 or 5, very small, mostly 1–2 mm long; petals lacking; stamens long-exserted, lacking in pistillate flowers; ovary glabrous or hairy, lacking in staminate flowers; fruits usually glabrous, the wings descending- spreading; 2n = 26. Riparian and palustrine communities at 850 to 2440 m in all Utah counties (except Davis and Piute); widespread in the U.S. and southern Canada, south to Central America; 107 (xv). A few of our cultivated specimens, those with glaucous and glabrous branchlets, are probably referable to **var. *interior* (Britt.) Sarg.** The plants are common in cultivation but because of insect infestations and other problems they have the reputation of being "dirty trees."

Acer palmatum Thunb. Japanese maple. Small trees, mainly 3–7 m tall; branchlets slender, glabrous or nearly so; leaf blades 3–9 cm long, 2.5–10 cm wide, 5– to 9 (usually 7) -lobed, the lobes lanceolate tapering to the sinuses, serrate to doubly so, loosely and sparingly long hairy when young, finally glabrate or hairy in lower vein axils; flowers corymbose; sepals reddish, 2.2–2.6 mm long, cuneate- oblanceolate; petals drab purplish, shorter than the sepals; samaras sparingly long-hairy to glabrous, the wings almost horizontally spreading; n = 13. Cultivated ornamentals in Utah County; widely offered in nurseries, and grown through much of the U.S.; 7 (ii).

Acer platanoides L. Norway Maple. Trees, mainly 6–15 m tall; branchlets slender, glabrous; leaf blades mainly 3–15 cm long and 5–18 (22) cm wide, 5– to 7–lobed, the lobes often again lobed or toothed, each lobe or tooth attenuate-aristate, glabrous above, hairy in lower vein axils; flowers corymbose; sepals greenish, 5–6 mm long, oval-oblong, glabrous; petals yellowish green, surpassing the sepals, glabrous; staminal filaments glabrous; samaras sparingly hairy to glabrous, the wings almost horizontally spreading; 2n = 26. Cultivated shade and ornamental trees in Cache, Juab, Millard, Salt Lake, Tooele, Uintah, and Utah counties; widely grown in the U.S.; introduced from Eurasia; 48 (0). The specimens examined include numerous horticultural forms, including purple-leaved and variegated phases.

Acer pseudoplatanus L. Sycamore Maple. Trees, mainly 6–15 m tall; branchlets slender, glabrous; leaf blades mainly 4–13 cm long and 5–18 cm wide, usually 5–lobed, the lobes again bluntly serrate or doubly so, rather abruptly and bluntly acuminate apically, glabrous above, hairy along veins and sometimes overall below; flowers racemose or paniculate; sepals greenish, 1.4–2.5 mm long, ovate to lance-oblong, ciliate; petals greenish, linear, shorter or subequal to the sepals; staminal filaments hairy; fruit long hairy, becoming glabrate, the wings spreading- descending; 2n = 52. Commonly grown shade trees in Cache, Salt Lake, Sanpete, Tooele, and Utah counties; widely grown in the U.S.; introduced from Eurasia; 15 (i).

Acer rubrum L. Red Maple. Trees, mainly 5–15 m tall; branchlets slender, glabrous; leaf blades mainly 2.5–8 cm long and 2–9 cm wide, 3– to 5–lobed, the lobes again doubly serrate or shortly lobed, the apex rather abruptly acuminate, glabrous above, hairy in vein axils below; flowers umbellate from buds of previous season; sepals purplish, 3–3.5 mm long, oblanceolate; petals purplish, subequal to the petals; samaras glabrous, the wings descending to descending-spreading. Uncommonly grown shade and ornamental trees in Box Elder, Salt Lake, and Utah counties; indigenous in eastern U.S.; 10 (0).

Acer saccharinum L. Silver Maple. Trees, mainly 10–20 m tall; branchlets slender, glabrous; leaf blades mostly

6-14 cm long and 6-15 cm wide, usually 5-lobed, the lobes again doubly-serrate and long attenuate apically, the sinuses obtuse, glabrous or sparingly hairy on one or both sides; flowers corymbose, borne on long-hairy pedicels, from buds of the previous season; calyx campanulate, rather shallowly to deeply 5-lobed, 4-5 mm long; petals lacking; staminal filaments hairy; samaras hairy, the wings rather strongly pendulous. Shade trees, rather common, mainly in older plantings in probably all of lower elevation areas of Utah; indigenous in eastern U.S.; 22 (i). Several horticultural forms are represented. The silvery-white bark of upper branches are often diagnostic for our plants. The trees suffer from iron chlorosis in Utah and probably because of this affliction they are grown less often than formerly.

Acer saccharum L. Sugar Maple. Trees mainly 8-15 m tall; branchlets slender, glabrous; leaf blades mostly 4-18 cm long and 3.5-22 cm wide, usually 5-lobed, glabrous above, hairy in leaf axils below, the lobes entire or with few rounded to blunt teeth, the sinuses rounded, the lobes bluntly attenuate-acuminate apically; flowers corymbose on pedicels mainly 15-40 mm long, these glabrous or long-hairy apically; sepals 3-5 mm long, greenish, broadly rounded, carinate to near the middle or above; petals lacking; samaras glabrous, the wings descending-spreading; 2n = 26. Uncommonly grown shade tree in Cache and Utah counties; introduced from eastern U.S.; 6 (0). This is the sugar maple of commerce.

ADOXACEAE Trautv.
Moschatel Family

Small, perennial herbs from scaly rhizomes; stems erect, with a single pair of opposite, palmately veined and lobed leaves and 1-3 basal, long-petiolate, ternately compound leaves; flowers regular, perfect, borne in heads; sepals 2 or 3 (4); petals united, the corolla rotate, 4- to 5-lobed; stamens 8-10, distinct; anthers 1-loculed; filaments inserted in pairs in the sinuses of the corolla lobes; pistil 1, the ovary partly inferior, 4- or 5-loculed, each with a single ovule; styles 4 or 5, short; fruit a dry berry with 4 or 5 nutlets; x = 9.

Adoxa L.

Inconspicuous rhizomatous herbs; leaves ternately compound or palmately lobed, petiolate; flowers inconspicuous, yellowish green, borne in heads; fruit a dry berry.

Adoxa moschatellina L. Moschatel. Plants 0.6-1.8 dm tall, with a musky odor, from rhizomes with fleshy scales; basal leaves 4-16 cm long, the blades ternately compound, the 3 leaflets deeply 3-lobed, the lobes again lobed, green and glabrous; cauline leaves opposite, 0.8-4 cm long, the blades palmately lobed, the lobes again lobed or toothed; peduncles 2-5 cm long, glabrous; flowers sessile; sepals 1-1.5 mm long, green; petals 1-2 mm long, yellowish green; fruit 2-3 mm long; 2n = 36, 54. Woods and thickets at ca 2440 to 3050 m in the Abajo Mts., San Juan County and reported from the Uinta Mts.; widespread in North America; circumboreal; 1 (0).

AIZOACEAE Rudolphi
Carpetweed Family

Annual or perennial herbs; stems mostly prostrate or ascending, often succulent; leaves simple, opposite or whorled, entire, with or without stipules; flowers solitary or clustered in leaf axils, perfect, regular; calyx 4- to 5-lobed or -parted, the tube free or adnate to the ovary; petals lacking; stamens few to many; anthers small, 2-loculed, linear; ovary 1- to 5-loculed, inferior or superior; styles as many as locules; fruit a capsule; seeds numerous; x = 8, 9.

1. Leaves whorled; capsule loculicidally dehiscent; plants not succulent .... *Mollugo*
— Leaves opposite; capsule circumscissile 2
2(1). Stipules present; ovary 1- or 2-loculed; seeds several .... *Trianthema*
— Stipules lacking; ovary 3- to 5-loculed; seeds numerous .... *Sesuvium*

Mollugo L.

Annual, nonsucculent, much-branched herbs; leaves whorled; flowers small, perfect, without petals, with long filiform pedicels; calyx 5-parted, persistent, white inside; stamens 3-5, hypogynous; ovary 3-loculed, superior; capsule ovoid to ellipsoid, thin-walled, 3-valved, loculicidally dehiscent; seeds numerous.

Mollugo cerviana (L.) Ser. Carpetweed. [*Pharnaceum cervianum* L.]. Stems very slender, 4-10 (15) cm long, prostrate; leaves glaucous, linear, 5-10 in a whorl, the basal ones linear-spatulate, 3-15 mm long; flowers whorled; pedicels filiform; sepals ca 1.5 mm long; capsules subglobose; seeds 0.4-0.5 mm long; 2n = 18. Sandy sites in pinyon-juniper woodland at ca 1300 m in Kane County (Atwood 3101 BRY); Arizona to California; 1 (0).

Sesuvium L.

Annual or perennial succulent herbs with prostrate or ascending stems; leaves opposite, fleshy, without stipules; flowers perigynous, solitary in leaf axils, sessile or with short thick pedicels; calyx tube turbinate, adnate below on the ovary, 5-lobed, usually horned on the back near the apex; petals none; stamens 1 to many; ovary 2- to 5-loculed, with as many styles; capsule membranous, ovoid, circumscissile; seeds stalked, smooth or rarely rugose, minute, many in each cell.

Sesuvium verrucosum Raf. Seapurslane. [*S. sessile* authors, not Pers.]. Perennial, glabrous, freely branched herbs; stems 1-5 dm long, papillose; leaves broadly spatulate to linear, 1-4 cm long, the base narrowed and clasping, fleshy; flowers subsessile, 8-10 mm long; sepals 5-7 mm long, scarious-margined, short-horned near the apex; stamens numerous; capsule ca 5 mm long; seeds black, shining, ca 1 mm long. Low saline or alkaline areas and in riparian habitats in Grand, Millard, Salt Lake, San Juan, Uintah, Utah, and Wayne counties; California to Colorado, Missouri, Arkansas, Texas, and northern Mexico; 25 (ii).

Trianthema L.

Prostrate, annual herbs; leaves opposite, in unequal pairs; stipulate; flowers solitary, axillary; floral tube short; sepals 5; petals lacking; stamens 5-10, perigynous; ovary truncate, 1- or 2-loculed; styles 1 or 2; capsules shortly cylindric or turbinate, 1- to 5-seeded, tardily circumscissile, the upper part usually with 2 rounded marginal crests; seeds reniform, roughened.

Trianthema portulacastrum L. Horse Purslane. Annual succulent herbs, branching from the base, prostrate to decumbent, to 5 dm long or more; leaves broadly obovate to suborbicular or the smaller ones narrower, rounded to emarginate or apiculate, the blades mainly 1–4 cm long and 0.8–3 cm wide, with smaller ones on lateral branchlets; stipules scarious; flowers sessile and usually solitary in leaf axils; calyx lobes ovate-lanceolate to lanceolate, ca 2.5 mm long, pink purple within, with a dorsal appendage near the apex; capsules ca 4 mm long, the winged appendages at the apex prominent; seeds ca 2 mm wide. Waste places at ca 850 m in Washington County; widespread in southern U. S.; Central America; Old World; 1 (i).

AMARANTHACEAE A. L. Juss.

Amaranth Family

Annual herbs; leaves simple, entire, alternate or opposite; flowers perfect or imperfect, inconspicuous, with 3 dry, scarious, persistent, pungent bracts; calyx commonly of 5 persistent, usually scarious sepals; petals none; stamens as many as the sepals; pistil 1, the ovary superior, 1–loculed, usually with 2 or 3 stigmas; fruit a utricle, indehiscent or circumscissile; x = 6–13, 17+.

1. Leaves largely opposite; plants white stellate-hairy .... *Tidestromia*
— Leaves alternate; plants nearly or quite glabrous to villous or pilose 2

2(1). Pistillate flowers lacking a perianth, more or less concealed by broad, cordate, spine-tipped, scarious-margined bracts; plants rare in Kane County .. *Acanthochiton*
— Pistillate flowers with a perianth, not obscured by bracts; bracts slender; plants of various distribution and abundance 3

3(2). Staminal filaments distinct; ovary with 1 ovule; plants weedy and widespread .... *Amaranthus*
— Staminal filaments united at the base; ovary with 2 or more ovules; plants cultivated, rarely escaping ... *Celosia*

Acanthochiton Torr.

Glabrous annual dioecious herbs; stems striate, green and white, erect or decumbent-ascending, branched; leaves alternate; staminate flowers in clusters, borne in spikes, the perianth of 5 sepals; pistillate flowers with perianth obscured, subtended by cordate bracts, these finally spiny-margined.

Acanthochiton wrightii Torr. Hopiweed; Green Stripe. [*Amaranthus acanthochiton* (Torr.) Sauer]. Plants mainly 1–5 dm tall, erect, the branches decumbent or ascending, striate; leaves alternate, 1.5–8 cm long, 2–12 mm wide, linear to narrowly lanceolate, rounded apically, cuneate or linear to narrowly lanceolate, rounded apically, cuneate or attenuate basally, white-veined beneath, marginally wavy and crisped, with petioles 3–15 mm long; staminate flowers with bracts 2–3 mm long, the midribs prominent; sepals 5, of 2 lengths, the inner 2.5–3 mm long, the outer 3–4 mm long, apiculate; stamens 5, the anthers 4–loculed; pistillate flowers concealed by rigid, cordate bracts, these with usually 5 sepals each 4–5 mm long, with crenate margins; ovary ovoid, somewhat compressed, with 3 style branches; utricle ca 2 mm long, circumscissile; seeds erect, obovoid, 1–1.3 mm wide, lenticular, brown. Vanclevea community at ca 1340 m in Kane County (Atwood & Kaneko 3360, 3360a BRY); Arizona to Texas and Mexico; 2 (0).

Amaranthus L.

Glabrous annual monoecous or dioecious herbs; stems prostrate to erect; leaves alternate; flowers in dense terminal or axillary clusters, each subtended by 3 conspicuous green or red to purple bracts; sepals 2–5 (rarely 1), distinct; stamens 5 (1 or 3), distinct; anthers 4–loculed (apparently 2–loculed following dehiscence); ovary 1–loculed, with 2 or 3 stigmas; ovules 1; utricle 1–seeded, circumscissile to indehiscent; seeds erect, lenticular, lustrous.

1. Sepals of pistillate flowers broadened upward, the calyx more or less urceolate; plants dioecious ...... *A. palmeri*
— Sepals of pistillate flowers more or less attenuate apically, the calyx not urceolate; plants monoecious 2

2(1). Inflorescence of terminal and axillary spicate panicles .. 3
— Inflorescence of axillary spicate glomerules only 4

3(2). Pistillate sepals elliptic to narrowly lanceolate, the margins not overlapping; plants common weeds. *A. retroflexus*
— Pistillate sepals elliptic-obovate to broadly spatulate, the margins overlapping; plants uncommonly grown ornamentals, rarely escaping .... *A. hypochondriacus*

4(2). Leaves deeply and broadly emarginate apically; bracts and perianth inconspicuous; plants uncommon .... *A. lividus*
— Leaves rounded to retuse apically; bracts and perianth various; plants common or uncommon 5

5(4). Sepals 4 or 5; stems prostrate; seeds ca 1.5 mm wide .... *A. blitoides*
— Sepals 1–3; plants of various habit; seeds less than 1 mm wide 6

6(5). Sepals of pistillate flowers 2 or 3, but only 1 well developed; stems prostrate; plants uncommon .. *A. californicus*
— Sepals of pistillate flowers 3, about equal; stems ascending to erect; plants common and widespread .... *A. albus*

Amaranthus albus L. Pale Amaranth. Monoecious (rarely perfect) annuals, the main stem erect, the branches spreading or ascending, 1.5–10 dm tall; herbage pale green to whitish, glabrous, puberulent, or villous; leaves alternate, the petioles 3–40 mm long, the blades elliptic, oblong, spatulate, or obovate, mainly 1–7 cm long, cuneate basally, rounded to mucronate-cuspidate apically, the veins prominent; flowers in axillary clusters, usually shorter than the petioles; bracts green, rigid, 2–4 mm long, pungent, spreading; sepals 3, the staminate oblong, cuspidate, scarious, the pistillate oblong to linear, acute, 1–veined, often tinged with red; stamens 3; style branches 3; utricle circumscissile, rugose, exceeding the perianth; seed lenticular, 0.6–0.9 mm wide, dark brown, lustrous; 2n = 32. Weedy plants of open sites and cultivated land in probably all Utah counties; widespread in temperate North America; adventive from tropical America; 21 (iii).

Amaranthus blitoides Wats. Prostrate Pigweed. [*A. graecizans* authors, not L.]. Monoecious annuals; stems 1–6 dm long, prostrate, branched mainly from the base, glabrous or sparingly hairy; herbage pale green or whitish, sometimes tinged with red or purple; leaves alternate, the petioles 2–20 mm long, the blades obovate to oval, spatulate, or elliptic, mostly 8–40 mm long, cuneate to attenuate basally, rounded to acute apically, prominently veined; flowers in dense clusters, these usually shorter than the petioles; bracts oblong to lanceolate, subequal to the sepals, spinose; sepals 4 or 5, the staminate scarious, oblong, acute, the pistillate oblong, 2.5–3 mm long, 1-veined, green with white margins; stamens 3; style branches 3; utricle circumscissile; seeds 1.3–1.5 mm wide, black, shining or dull; 2n = 32. Weedy species of open sites and cultivated land at 790 to 2565 m in probably all Utah counties; widespread in temperate North America; Mexico; Europe; 41 (iii).

Amaranthus californicus (**Moq.**) **Wats.** California Amaranth. [*Mengea californica* Moq.]. Mostly monoecious annuals; stems 0.6–5 dm long, prostrate, branched from the base; herbage glabrous, whitish or tinged red; leaves alternate, the petioles 2–18 mm long, the blades 3–25 mm long, obovate to oblong, obtuse to rounded apically, cuneate to acute basally, the veins and margins white; flowers in axillary clusters; bracts lanceolate, subulate-aristate, subequal to the sepals; sepals in staminate flowers (2) 3, membranous, elliptic to lanceolate or oblong, mucronate, in pistillate flowers usually 1, inconspicuous; stamens 3; utricle irregularly dehiscent; seeds 0.6–0.9 mm wide, dark brown. Mud flats of Dry Lake, Cache County (Neese 12392 BRY); Alberta to Idaho, south to California and Texas; 1 (0).

Amaranthus hypochondriacus **L.** Grain Amaranth; Princes Feather. [*A. leucocarpus* Wats.; *A. caudatus* authors, not L.]. Annual, mostly monoecious herbs; stems erect, mainly 8–15 dm tall, glabrous, green or suffused with red; leaves alternate, the petioles subequal to the blades, these 2–10 cm long or more, elliptic to ovate, lanceolate, or oblong, acute apically, obtuse to acute basally; flowers in terminal panicles, with nodding or pendulous spikes, usually red; sepals of pistillate flowers with margins overlapping, obovate to spatulate; seeds ca 1 mm long, dull white to brown. Occasionally cultivated ornamental in lower elevation portions of Utah, rarely escaping; Arizona to Texas, south to Central America; Asia, Africa; 1 (0). This is one of the grain amaranths grown as potherbs and cereal crops in Central America and elsewhere. Another grain amaranth, **A. cruentus L.**, the purple amaranth, is also grown in Utah. It has erect purplish panicles.

Amaranthus lividus **L.** [*A. blitum* L.]. Monoecious annuals; stems mainly 5–9 dm tall, erect, glabrous, whitish or greenish; leaves alternate, the petioles 5–40 mm long, the blades rhombic to orbicular or ovate, maculate above, cuneate basally, emarginate to almost truncate apically; flowers in axillary clusters; bracteoles less than half as long as the perianth, ovate and wide at the base, acute; sepals 3 (5), oblong or linear to spatulate; utricle indehiscent; seeds brown, shining, ca 1 mm wide; 2n = 34. Moist site adjacent to Utah Lake, Utah County (Thorne 93 BRY); adventive from Europe; 1 (0).

Amaranthus palmeri **Wats.** Palmer Amaranth. Dioecious annuals; stems 6–10 dm tall, branched throughout, glabrous or villous; leaves alternate, the petioles slender and long, the blades 1–6 cm long, ovate to lanceolate, cuneate to rounded basally, acute or shortly acuminate apically; flowers borne in terminal and axillary clusters or panicles; bracts 4–6 mm long, much surpassing the sepals; staminate flowers with 5 stamens and 5 sepals; pistillate flowers with 5 recurved sepals; styles usually 2 (3); utricle circumscissile; seeds 1–1.3 mm wide, dark brown; n = 17. Open sites at ca 1400 m in Davis County; adventive from the southwestern U.S.; 1 (0). Specimens of **A. fimbriatus** (**Torr.**) **Benth** [*Serratia berlandieri* var. *fimbiata* Torr.] have recently been collected in Washington County. They differ from *A. palmeri* in having the calyx of female flowers fimbriate.

Amaranthus retroflexus **L.** Redroot Pigweed. Annual monoecious herbs; stems erect, 0.6–1.5 dm tall or more, puberulent to villous, white-striped or suffused with red; leaves alternate, long-petiolate, the blades mainly 1–8 cm long, usually hairy (at least on veins beneath), lanceolate to ovate, obtuse to acute basally, rounded to acute apically; flowers in dense terminal or axillary paniculate spikes; bracts ovate, subulate apically, at least the longest much surpassing the flowers; sepals of staminate flowers ovate to lanceolate, acute, scarious, those of pistillate flowers narrowly oblong, rounded to truncate apically, usually emarginate and often mucronate, scarious; stamens 5; style branches 3; utricle circumscissile, surpassed by the longest sepals; seeds ca 1 mm wide, dark brown or reddish; 2n = 34. Weedy species of gardens, other cultivated land, and open sites at 850 to 2440 m in probably all Utah counties; adventive from Central America, widespread in North America; 34 (v). Some specimens from Utah have been identified as *A. hybridus* L., a taxon that hybridizes with *A. retroflexus*. However, none of the specimens examined appear to be *A. hybridus*, which differs inter alia in having pistillate sepals shorter than the fruit to slightly longer and tapering to a terminal awn-tip. Some of our material shows variation encompasing both *A. retroflexus* and *A. hybridus* within the same specimen.

Celosia L.

Annual herbs; stems erect; leaves alternate; flowers perfect, borne in terminal or axillary spikes or in axillary clusters, commonly white, silvery, or colored, each one subtended by a bract and paired bractlets; sepals 5, scarious, striate; stamens 5, connate basally; ovary with 2 to several ovules; seeds 2 to several, lenticular, lustrous, often erect.

Celosia argentea **L.** Cockscomb. Stems mainly 3–10 dm tall, erect, glabrous; leaves 1.5–10 cm long, or more, lanceolate to ovate or linear, obtuse to acute basally, long-attenuate apically; panicle terminal or some axillary, mainly 2–20 cm long, varicolored; sepals 6–19 mm long, silvery or varicolored; style ca 3 mm long, exserted at maturity; seeds 2 to several, lustrous, 1–1.4 mm wide; 2n = 36, 72. Cultivated ornamentals, occasionally escaping in Utah; introduced from Central America; 3 (0). The f.*cristata* (**L.**) **Kuntze** [*C. cristata* L.] has cristate inflorescences.

Tidestromia Standley

Annual or perennial herbs; stems prostrate or ascending; herbage stellate-hairy; leaves opposite, petiolate; flowers perfect, clustered in leaf axils; bracts and bractlets scarious, pubescent; sepals 5, ovate-lanceolate, the bracteate subtending leaves finally hardened and more or less connate, forming an involucre; stamens 5, the filaments connate basally, forming a short cup, with or without intervening staminodia; anthers 2-loculed, apparently 1-loculed at dehiscence; ovary short, the style with a capitate, simple or 2-lobed stigma; ovule 1; utricle indehiscent.

1. Plants annual; staminodia short or lacking; sepals mainly 1.5–3 mm long .... *T. lanuginosa*
— Plants perennial; staminodia ca half as long as the stamens; sepals mainly 1–1.5 mm long .... *T. oblongifolia*

Tidestromia lanuginosa (**Nutt.**) **Standley** Espanta Vaqueros. [*Achyranthes lanuginosa* Nutt.]. Annual prostrate-spreading herbs to 10 dm wide; herbage gray green to white; leaves with petioles 0.5–2.5 cm long, the blades 5–40 mm long, ovate to orbicular or obovate, rounded to obtuse apically, cuneate to rounded basally, densely stellate hairy on both sides, finally glabrate; flower clusters axillary, subtended by involucrate leaves; perianth mainly 1.5–3 mm long, inconspicuous; seeds ca 0.5 mm wide. Warm, sand, and salt desert shrub communities at 915 to 1315 m in Kane, San Juan, and Washington counties; Nevada and California, east to South Dakota, and south to Mexico; 5 (i).

Tidestromia oblongifolia (Wats.) Standley [*Cladothrix oblongifolia* Wats.]. Perennial decumbent to ascending (or erect) herbs, sometimes woody at the base, and with thick woody taproots; stems 2–6 dm long; leaves shortpetioled, the blades 5–40 mm long, ovate to elliptic or orbicular, obtuse apically, rounded to acute basally, densely stellate on both sides; flower clusters axillary, the involucrate leaves finally hardened and connate basally; perianth mainly 1–1.5 mm long; staminodia present, ca half as long as the filaments; seeds ca 0.5 mm long. Creosote bush community at 850 to 1300 m in Washington County; Nevada, California, and east to Texas; 3 (0).

ANACARDIACEAE Lindl.

Cashew Family

Trees or shrubs; leaves alternate, simple, trifoliolate, or pinnately compound; flowers perfect or imperfect, regular, small, borne in terminal or axillary spikes or panicles; sepals 3–5, connate below, sometimes fused with the ovary; petals 3–5 or lacking, distinct; stamens usually 5 or 10; pistil 1, the ovary superior, 3–carpellate, unilocular, with a single functional ovule, the styles 2 or more; fruit a drupe; x = 7–16.

1. Leaves simple 2
— Leaves trifoliolate or pinnately compound 3
2(1). Leaves entire; panicles with plumose pedicels ... *Cotinus*
— Leaves toothed or lobed; panicles lacking plumose pedicels .... *Rhus*
3(1). Leaflets 3, mainly 5–11 cm long; fruit whitish to yellowish, glabrous .... *Toxicodendron*
— Leaflets 5–13 or more or, if 3, mainly 2–5 cm long 4
4(3). Leaf rachis very narrowly winged; petioles puberulent; leaflets 5–9, leathery, obtuse; street trees of Washington County .... *Pistacia*
— Leaf rachis not winged or, if so, broadly winged; leaflets not leathery, acuminate; shrubs of broad distribution .... *Rhus*

Cotinus Adanson

Shrubs or small trees, polygamous or dioecious; leaves alternate, the blades simple, entire, estipulate; flowers in large loose panicles, small, greenish or yellowish, the pedicels of sterile flowers elongating, clothed with long, spreading, multicellular hairs; sepals 5, short; petals 5, inconspicuous; stamens 5; drupe small.

Cotinus coggygria Scop. Smoke Tree. Plants mainly 2–3.5 m tall; branchlets brownish; leaves simple, entire, the blades 1.7–5 cm long, 1–3.5 cm wide, orbicular to broadly elliptic or oval, glabrous, prominently veined; panicles terminal, many-flowered; sepals green; petals yellowish green, 1.2–1.5 mm long; pedicels long-hairy, the crosswalls red to purplish; drupe seldom produced; n = 30. Cultivated ornamental, persisting, in towns and cities at lower elevations in Utah; introduced from Eurasia; 5 (0).

Pistacia L.

Small to medium dioecious trees, with a turpentine resin; leaves alternate, pinnately compound; flowers apetalous, small, borne in axillary panicles; staminate flowers with 1 or 2 sepals; stamens 3–5; pistillate flowers with 2–5 sepals; ovary 1–loculed; style 3–cleft; fruit a dry drupe.

Pistacia atlantica Desf. Pistacio. Trees, mainly 5–10 m tall; branchlets brownish; leaves odd-pinnately compound, with 5–9 lanceolate, thickish, obtuse leaflets; leaf rachis very narrowly winged; petioles puberulent; petals lacking; drupes reddish to blackish, glabrous, thinfleshed. Cultivated street trees and established in Washington County; introduced from Europe; 3 (i). It seems possible that these plants were introduced as rootstocks for the pistacio of commerce, *P. vera* L., which persists in cultivation and has escaped in the vicinity of Leeds, Washington County; 1 (i). A lone pistillate tree survives along a roadside south of Leeds. It was part of an orchard destroyed in construction of the highway. The pistacio is successfully grown at Littlefield, Arizona.

Rhus L.

Shrubs or small trees with milky or resinous juice, polygamous or dioecious; leaves alternate, simple, trifoliate, or odd-pinnately compound, estipulate; flowers small, borne in small axillary spikes or panicles or large terminal panicles; sepals 5; petals usually 5; stamens 5; ovary 1–loculed; styles 3; fruit a thinly fleshy to dry drupe.

Barkley, F. A. 1937. A monographic study of *Rhus* and its immediate allies in North and Central America, including the West Indies. Ann. Missouri Bot. Gard. 24: 265–498.
Barkley, F. A. 1940. *Schmaltzia*. Amer. Midl. Naturalist 24: 647–665.

1. Leaves simple or trifoliolate .... *R. aromatica*
— Leaves pinnately compound with 5 or more leaflets 2
2(1). Rachis of leaf winged; plants cultivated ...... *R. copallina*
— Rachis of leaf not winged; plants cultivated or indigenous 3
3(2). Branchlets glabrous and glaucous; plants indigenous and cultivated .... *R. glabra*
— Branchlets densely hairy, with long straight hairs and shorter glandular ones; plants cultivated ..... *R. typhina*

Rhus aromatica Ait. Skunkbush. Shrubs, mainly 0.5–2.5 m tall, spreading, forming small to large clumps or thickets; branchlets brown, becoming gray in age, densely puberulent; leaves simple and pinnately to subpalmately lobed or trifoliate, the leaflets 0.8–9.3 cm long, and 0.6–7 cm wide, lobed, ciliate, glabrous or puberulent on one or both sides; petioles puberulent; flowers in dense spikes or in short spicate panicles, mainly on short lateral branches, rarely terminal, yellowish, developing prior to the leaves, 2–3 mm long; sepals ca 1–1.5 mm long, often reddish; drupes red orange, 5–8 mm long. There are two indigenous phases of this species complex in Utah. They are ecologically correlated, but with intermediates along the ecotones. Examples of introduced phases, not herein identified to infraspecific category, are present in cultivation.

1. Leaves simple, merely pinnately to subpalmately lobed, often coriaceous; plants of xeric sites mainly in southeastern and southwestern Utah, rarely elsewhere .... *R. aromatica* var. *simplicifolia*
— Leaves trifoliolate, thin, seldom if at all coriaceous; plants of mesic sites, widespread ..... *R. aromatica* var. *trilobata*

Var. *simplicifolia* (Greene) Cronq. [*Rhus canadensis* var. *simplicifolia* Greene; *R. trilobata* var. *simplicifolia* (Greene) Barkley; *R. utahensis* Goodding, type from Diamond Valley; *Schmaltzia affinis* Greene, type from Kanab] Blackbrush, mixed desert shrub, and pinyon-juniper communities in xeric sites, often on rimrock, at 975 to 1925 m in Emery, Garfield, Grand, Kane, San Juan, Tooele (north of Ibapah), Washington, and Wayne counties; Arizona and Colorado; 40 (ix).

Var. *trilobata* (Nutt.) Gray [*R. trilobata* Nutt. ex T. & G.; *R. trilobata* var. *anisophylla* Jepson]. Stream banks and terraces, seep and spring margins, and on mesic slopes, mainly in riparian communities, at 885 to 2380 m in all Utah counties (except Rich and Summit); Alberta to Iowa, south to California, Mexico, and Texas; 142 (xxvii). The branchlets are supple and have been used by Indians prehistorically for tying and in basketry. The fruits are edible, but the seeds are large and the flesh thin.

***Rhus copallina* L.** Shining Sumac. Shrubs or small trees, mainly 2–4 m tall; branchlets villous-tomentose to glabrous; leaves 15–35 cm long; leaflets 7–11, sessile or the apical one petiolulate, lance-elliptic to lanceolate, glabrous or sparingly hairy, mainly 2.5–8 cm long, acute to acuminate; rachis winged; inflorescence to 12 cm long and about as broad; flowers greenish; fruit ca 4 mm long, with red glandular hairs. Cultivated ornamentals in lower elevation portions of Utah; introduced from the eastern U.S.; 2 (0).

***Rhus glabra* L.** Smooth Sumac. [*R. nitens* Greene, type from Provo; *R. cismontana* Greene]. Shrubs, mainly 1–3 m high; branchlets glabrous and glaucous; leaves mainly 10–30 cm long; leaflets 7–21 or more, lanceolate to lance-oblong, mainly 1.5–8 cm long, acuminate, serrate, the rachis not winged; inflorescence 7–15 cm long or more and 3–15 cm wide, puberulous; fruit red, viscid-hairy, ca 4 mm long; $2n = 30$. Desert shrub, riparian, juniper, and mountain brush communities at 1095 to 2290 m in Box Elder, Cache, Davis, Grand, Kane, Millard, Salt Lake, San Juan, Wasatch, and Washington counties; widespread in North America; 29 (i). The smooth sumac occurs in disjunct stands, often in dry sites, in widely scattered locations. It is a common species in ornamental plantings in Utah.

***Rhus typhina* L.** Velvet Sumac; Staghorn Sumac. Shrubs or small trees, mainly 2.5–5 m high; branchlets densely pubescent with mixed straight and glandular hairs; leaves mainly 15–35 cm long; leaflets 11–31, oblong-lanceolate, 5–10 cm long, acuminate, serrate or laciniate, the rachis not winged; flowers greenish, in dense terminal clusters 7–20 cm long; fruit red, hairy; $2n = 30$. Cultivated ornamental in lower elevation regions of the state; persisting; introduced from the eastern U.S.; 2 (0). We have both the dissected-leaved cultivar "laciniata" and the common phase with merely serrate leaflets.

***Toxicodendron* Miller**

Dioecious shrubs; leaves alternate, trifoliolate; flowers in axillary thyrsoid panicles or racemes; sepals 5, persistent, connate below; petals 5, veined, smaller in pistillate flowers; drupes subspherical, whitish to yellowish, glabrous.

Gillis, W. T. 1960. Taxonomic problems in poison ivy. Proc. Michigan Acad. Sci. 45: 27–34.
———. 1970. The systematics and ecology of poison-ivy and poison-oaks (*Toxicodendron*, Anacardiaceae). Rhodora 73: 72–159, 161–237, 370–443, 465–540.

***Toxicodendron rydbergii* (Small) Greene** Poison Ivy. [*Rhus rydbergii* Small ex Rydb.; *R. radicans* var. *rydbergii* (Small) Rehder; *R. toxicodendron* var. *rydbergii* (Small) Garrett; *T. radicans* var. *rydbergii* (Small) Erskine; *T. longipes* Greene, type from near Glenwood]. Shrubs, mainly 1–3 dm tall (rarely more); stems simple or sparingly branched, puberulent; leaves long-petiolate; leaflets 3 (rarely 4 or 5), the lateral leaflets often oblique, unequally lobed and smaller than the terminal; terminal leaflet 2–11 cm long, 1.5–10 cm wide, ovate to orbicular and toothed, acuminate, glabrous or puberulent mainly along the veins beneath, petiolulate; sepals ca 1 mm long, greenish; petals 2–3 mm long, yellowish, with dark veins; fruit subglobose, cream to yellow, 4–7 mm wide, glabrous; $2n = 30$. Riparian communities, mesic slopes, and in orchards, at 1125 to 2260 m in Box Elder, Cache, Daggett, Davis, Duchesne, Emery, Garfield, Grand, Juab, Kane, Millard, Salt Lake, San Juan, Summit, Uintah, Washington, and Weber counties; Alberta to Nova Scotia, south to Arizona, Texas, and Virginia; 18 (vii). This plant produces severe dermatitis in sensitive people; it should be avoided by everyone. The three large leaflets on leaves borne alternate on the stem of low shrubs and the waxy fruit are diagnostic.

APOCYNACEAE A. L. Juss.
Dogbane Family

Shrubs, trailing vines, or herbs, often with milky juice; stems erect or trailing; leaves opposite, alternate, or sometimes whorled, entire, estipulate; flowers regular, perfect, solitary and axillary, or cymose to paniculate; calyx 5–lobed, imbricate in bud, frequently with glandular appendages within; corolla 5–lobed, convolute and often twisted in bud, salverform to urceolate or campanulate, the tube often with appendages within; stamens 5, alternate with the corolla lobes, inserted on the throat or tube; anthers 4–loculed, often connivent around the stigma; pistil superior, the carpels 2, distinct or united apically, each 1–loculed; style simple or divided; fruit of 2 follicles; seeds commonly comose; $x = 8–12+$.

Woodson, R. E. 1938. Apocynaceae. N. Amer. Fl. 29: 103–192.

1. Leaves alternate .... *Amsonia*
— Leaves opposite or whorled 2
2(1). Herbs evergreen, trailing; corolla lavender blue to reddish purple; stamens inserted at summit of corolla tube; seeds not comose .... *Vinca*
— Herbs or shrubs, erect or ascending; corolla white, pink, or scarlet, seldom as above; stamens inserted at base of corolla tube; seeds comose 3

3(2). Corolla 5–8 mm long .... *Apocynum*
— Corolla 10–30 mm long or more 4
4(3). Plants low perennial herbs from fleshy rootstocks, indigenous in southeastern Utah .... *Cycladenia*
— Plants shrubs or small trees, cultivated ornamental .. .. *Nerium*

Amsonia Walter

Caulescent perennial herbs; stems erect or ascending; leaves alternate or verticillate; flowers many, bluish or pinkish, in terminal compound cymes; calyx 5–parted, the segments equal, acuminate; corolla salverform, the lobes spreading or reflexed; stamens inserted on corolla throat, included; style filiform; follicles terete, torulose, several-seeded; seeds cylindric or oblong, without a coma.

Woodson, R. 1928. Monograph of the Genus *Amsonia*. Ann. Missouri Bot. Gard. 15: 379–434.
McLaughlin, S. P. 1982. A revision of the southwestern species of *Amsonia*. Ann. Missouri Bot. Gard. 69: 336–350.

1. Follicles not markedly constricted between the seeds; corolla tube 6–10 mm long; leaves ovate, glabrous ... . *A. jonesii*
— Follicles markedly constricted between the seeds; corolla tube 7–12 mm long; leaves oblong-lanceolate to linear-lanceolate, glabrous or conspicuously pubescent .... *A. tomentosa*

Amsonia jonesii Woodson [*A. latifolia* Jones, type from Monroe Hot Springs]. Perennial herbs, glabrous, 1.5–5 dm tall; lower leaves ovate, 1.4–3 cm wide, the upper ones lanceolate, 0.3–1 cm wide; calyx lobes ovate to lanceolate, 1–4 mm long; corolla tube 6–10 mm long, broadest below the apex, slightly constricted at the opening, the lobes 4–8 mm long; follicles 1.5–9 cm long, not constricted between the seeds; seeds cylindrical, corky, 6–8 mm long, 2–2.5 mm wide. Blackbrush, sagebrush, mountain brush, pinyon- juniper and desert shrub communities at 1200 to 2150 m disjunctly in Duchesne, Emery, Garfield, Grand, San Juan, Sevier, Uintah, Washington, and Wayne counties; Arizona and Colorado; 30 (xiv).

Amsonia tomentosa Torr. & Frem. [*A. eastwoodiana* Rydb., type from near Moab; *A. brevifolia* Gray, type from Washington County]. Perennial herbs, glabrous or sparingly to densely tomentose, 2–6 dm tall; lower leaves lanceolate to linear-lanceolate, 3–12 mm wide, the upper leaves lanceolate to filiform, 1–5 mm wide; calyx lobes linear, 2–9 mm long; corolla tube 7–12 mm long, broadest at the apex, constricted at the orifice, the lobes 3–9 mm long; follicles constricted between the seeds, 2–8 cm long; seeds elliptic, corky, 8–21 mm long, 3–6 mm wide. Saltbush, blackbrush, ephedra and desert shrub communities at 950 to 1600 m in Emery, Garfield, Grand, Kane, San Juan, and Wayne counties; Arizona, New Mexico, Texas, and Mexico; 72 (xxxi). Our material belongs to **var. stenophylla** Kearney & Peebles. The typical variety occurs to the south of our area in Mohave County, Arizona and should be sought in Washington County. The glabrous phase of the species, long recognized as *A. eastwoodiana*, differs in no other perceptible way from tomentose plants with which it grows in much of the range in Utah. This plant simulates *Psoralidium junceum* and *Sphaeralcea janeae* in growth form and color, especially at maturity.

Apocynum L.

Perennial herbs, reproducing asexually from gemmiferous roots; stems branching; leaves opposite or rarely verticillate, not glandular, often mucronate; flowers small and pale, few to many on short pedicels in terminal or axillary cymes; calyx 5–parted nearly to the base, the lobes equal, scarcely imbricate; corolla campanulate to urceolate or cylindrical, the tube short with 5 small sagittate appendages at the base opposite the lobes, the limb 5–parted; stamens attached at corolla base; anthers connivent and agglutinated to the stigma, with enlarged narrowly 2–lobed connectives, the pollen grains maintained within persistent tetrads; ovary of separate carpels; ovules numerous; stigma sessile, mostly ovoid-fusiform; follicles 2, slender, terete; seeds numerous, truncate, comose. Note: At least some species of this genus are considered to be poisonous.

1. Corolla usually at least 5 mm long, pinkish, often more than twice the length of the calyx, the lobes spreading or reflexed; leaves mostly drooping to spreading 2
— Corolla less than 5 mm long, greenish-white to white, usually less than twice the length of the calyx lobes, erect or slightly spreading; leaves ascending 3
2(1). Corolla usually ca 3 times as long as the calyx; leaves drooping..................... *A. androsaemifolium*
— Corolla usually about twice as long as the calyx; leaves spreading .... *A. medium*
3(1). Leaves noticeably petiolate, narrowed at the base; follicles more than 12 cm long; coma of seeds 2–3 cm long .... *A. cannabinum*
— Leaves sessile or nearly so, cordate at base; follicles less than 12 cm long; coma of seeds 1–2 cm long .... *A. sibiricum*

Apocynum androsaemifolium L. Spreading Dogbane. Stems erect or ascending, 2–5 dm tall, glabrous, freely and dichotomously branched; leaves petiolate, drooping, ovate to oblong-lanceolate, 2–10 cm long, 1–6 cm wide, glabrous or rarely sparsely pilosulous above, sparsely pilosulous to densely tomentose beneath; calyx lobes ovate to ovate-lanceolate, 2–5 mm long, glabrous or rarely minutely pilosulous; corolla tubular to campanulate, 5–12 mm long, the lobes white, usually with pinkish veins, widely spreading or reflexed, glabrous externally; follicles 6–15 cm long, pendulous or erect at maturity.

1. Corolla campanulate, dilating apically; follicles pendulous .... *A. androsaemifolium* var. *androsaemifolium*
— Corolla urceolate, not distinctly dilating apically; follicles erect .... *A. androsaemifolium* var. *pumilum*

Var. androsaemifolium Oak, maple, populus, sagebrush, and spruce-fir communities at 1500 to 3350 m in Beaver, Box Elder, Cache, Daggett, Davis, Duchesne, Garfield, Grand, Iron, Juab, Millard, Salt Lake, San Juan, Sanpete, Summit, Tooele, Uintah, Utah, Wasatch, Washington, and Weber counties; Newfoundland to British Columbia, south to Georgia, Arizona, and California; 43 (iii).

Var. pumilum Gray Aspen, mountain brush, and coniferous forest communities at 2170 to 2650 m in Salt Lake, Utah, and Washington counties; Washington to Montana, south to California, and Wyoming; 5 (0).

Apocynum cannabinum L. Dogbane. [*A. pubescens* R. Br.; *A. suksdorfii* Greene; *A. cannabinum* var. *glaberri*-

mum A. DC. in DC.]. Stems erect or ascending, 3–9 dm tall, the branches opposite or subopposite, herbage glabrous or pubescent; leaves opposite or whorled, petiolate or the lowermost subsessile, ascending or somewhat spreading, ovate to lanceolate, 2–14 cm long, 1–7 cm wide, glabrous above, more or less densely pilosulous to tomentose beneath, or tomentose throughout; calyx lobes lanceolate or ovate-lanceolate, 3–4 mm long, glabrous; corolla cylindric to urceolate, 3–6 mm long, white to greenish, the lobes erect or somewhat spreading; follicles 12–20 cm long, glabrous, pendulous at maturity. Roadsides, fields, streambanks and disturbed sites mainly in riparian communities at 970 to 2350 m in all Utah counties, except Morgan, Rich, and Summit; widespread in the U.S. and Canada; 68 (vi). This species was used as a source of fiber for cordage of fine quality by prehistoric Indians in Utah.

Apocynum x medium Greene [*A. floribundum* Greene; *A. lividum* Greene]. Stems erect or ascending, 2–5 dm tall, freely and somewhat dichotomously branched, especially below, the branches ascending, alternate or opposite, glabrous to puberulent; leaves opposite, petiolate to subsessile, usually spreading, ovate to oblong-lanceolate, 6–10 cm long, 1.5–2 cm broad, occasionally somewhat cordate at the base, glabrous to somewhat puberulent above, more or less pilosulous to tomentose beneath; calyx segments ovate to lanceolate, 1.5–4 mm long, glabrous to sparsely pilosulous; corolla campanulate to broadly urceolate, 3–6 mm long, the lobes spreading, white or with pinkish veins; follicles 7–15 cm long, pendulous at maturity. Mixed deciduous woodlands at moderate to lower elevations in Box Elder, Cache, Davis, Garfield, Iron, Juab, Millard, Piute, Salt Lake, San Juan, and Weber counties; widespread in the U.S.; 15 (ii). This entity is intermediate between *A. androsaemifolium* and *A. cannabinum*. It tends to intergrade with both species and consists of hybrids and segregates involving the two species it resembles.

Apocynum sibiricum Jacq. Stems erect or ascending, 2–8 dm tall, glabrous, the branches opposite; leaves opposite or rarely whorled, sessile or subsessile, especially on the main stem, those of the upper branches often short-petiolate, ascending or somewhat spreading, oblong or oblong-lanceolate to oval, 1.5–14 cm long, 0.3–4.5 cm wide, obtuse to cordate at base, glabrous; inflorescence dense, bracts usually conspicuous and more or less subfoliaceous; calyx segments lanceolate, 2–4 mm long, glabrous; corolla urceolate to short-cylindric, about as long as broad, 3–5 mm long, white or greenish, the lobes erect or somewhat spreading; follicles 4–10 cm long, pendulous at maturity. Canyons, flats, streambanks and roadsides at 1385 to 1895 m in Box Elder, Emery, Juab, Uintah, and Utah counties; Oregon east to Newfoundland, south to California and New Mexico; 6 (0). Segregation of this entity from *A. cannabinum* is tenuous at best. Specimens vary in shape of leaf bases and degree of petiole development. Even those with cordate bases on lower and middle stem leaves are often shortly petiolate on uppermost leaves. Few plants are collected in fruit, and the value of the shorter fruit as a diagnostic feature is questionable. Possibly a better course would be to treat all of our material as *A. cannabinum*. I follow tradition in treating it otherwise.

Cycladenia Benth.

Perennial caulescent herbs; stems erect or spreading from a deeply set caudex; leaves opposite, broadly petiolate; flowers 2–8 on axillary peduncles; calyx 5-lobed, the lobes equal; corolla funnelform, with small appendages alternate with the lobes; stamens borne near base of tube; anthers connivent and sagittate; style filiform, with a conspicuous membraneous collar; ovary surrounded at base by 5 nectaries; follicles 2, separate, terete; seeds comose.

Cycladenia humilis Benth. Plants 11–36 cm tall, glabrous and glaucous, the lowermost leaves reduced to subamplexicaul bracts, enlarging and becoming green upwards; main foliage leaves 3.5–9.5 cm long, 2–6.5 cm wide, oval to orbicular or broadly obovate, tapering abruptly to the broad petiole, thickened, entire, the apex rounded to acute; pedicels 5–25 mm long; bracts linear-lanceolate, 3–9 mm long; calyx lobes 5–11 mm long, lance-linear, villous-pilose, somewhat accrescent in fruit; corolla rose purple, dimorphic, either broadly lobed, 23–28 mm long, and 19–31 mm wide or narrowly lobed, 18–21 mm long, and 13–19 mm wide, rose pink, more or less pilose; follicles 4.5–9.5 cm long; seeds brown, ca 7.5 mm long, the coma ca 20 mm long. Eriogonum-ephedra, mixed desert shrub, and juniper communities at 1340 to 1830 m, in gypsiferous, saline soils of the Cutler, Summerville, and Chinle formations, in Emery, Garfield, Grand, and Kane (?, taken by Siler in the 1870s, the location unknown) counties; California; 18 (vi). Our material belongs to var. *jonesii* (Eastw.) Welsh & Atwood [*C. jonesii* Eastw., type from San Rafael Swell]. The flowers are dimorphic within the few disjunct populations known for this entity in Utah. The similar var. *venusta* (Eastw.) Woodson is a montane plant of talus slopes and rocky places in coniferous forests of California. Our plant is evidently an obligate gypsophile of semibarren tracts on geological formations with poor water relationships.

Nerium L.

Shrubs or small trees; leaves whorled or rarely opposite, leathery; flowers showy, borne in terminal cymes; calyx with glands within near the base; corolla funnelform, with a cylindrical tube and campanulate throat, the 5 lobes broad, spreading, and twisted to the right; stamens attached to throat of corolla, included, the filaments very short, the anthers with long appendages apically and basally, connivent around the stigma and adhering to it; ovaries 2, with many ovules in each locule; style filiform; fruit of 2 elongated follicles; seeds comose.

Nerium oleander L. Oleander. Evergreen shrub to 5 m tall, nearly glabrous; leaves narrowly oblong-lanceolate, 1–3 dm long and to 3 cm wide, acuminate, tapering at base to a short petiole, dark green above, lustrous, paler beneath; inflorescence subcorymbose; flowers showy, scarlet to yellowish pink or white, freqently double; calyx 5-parted, the segments lanceolate to lance-ovate, 4–6 mm long; corolla glabrous externally, the tube 8–12 mm long, the 5 lobes obliquely obovate to oblong, 2–2.5 cm long, spreading; follicles 2, stout, 8–15 cm long; seeds compressed, densely comose. Cultivated and persisting below 1200 m in Washington County; introduced from the Mediterranean region and the Far East, now widely cultivated and naturalized in warmer areas of North America; 0 (0). This plant is deadly poisonous to animals and man.

Vinca L.

Erect or more often trailing perennial herbs; leaves opposite; flowers solitary in alternate leaf axils; calyx

5–lobed, the lobes nearly equal; corolla funnelform to salverform, the tube cylindrical, with the 5 lobes twisted to the left, hairy, thickened at the throat, not appendaged within; anthers not connivent, the connective produced apically into an appendage; ovary with 2 alternate nectaries; ovules numerous; follicles terete, slender; seeds naked.

1. Calyx lobes ciliate; leaves truncate or subcordate at the base .... *V. major*
— Calyx lobes glabrous; leaves narrowed at the base .... *V. minor*

Vinca major L. Trailing, slightly woody vines, rooting freely at the nodes; flowering stems and branches erect, 2–3 dm tall; leaves bright green, glabrous, 2–3 cm long, ovate, cordate at the base, obtuse to acute apically, with petioles 0.5–2 cm long, often ciliolate; pedicels slender, 3–5 cm long; calyx lobes subulate, ciliate, ca 1 cm long; corolla blue or violet, the tube 1.5–2 cm or more long. Cultivated and occasionally escaping below 1525 m in Utah and Washington counties; widely grown in the U.S., introduced from Europe; 3 (i).

Vinca minor L. Creeping or trailing evergreen herbs; leaves elliptic 1.5–6 cm long, 0.8–2.5 cm wide, obtuse to broadly acute apically, cuneate to acute basally, glabrous, shiny above; petioles 1–2 cm long; flowers solitary, with pedicels 1.5–3 cm long; calyx lobes lance-ovate, acute, to 3 mm long, glabrous; corolla funnelform, bright blue, rarely white, the tube 3–6 mm long, the throat conic to campanulate, 5–7 mm long; follicles slender, 2–7 cm long, rarely produced; n = 46. Cultivated and occasionally escaping below 2300 m in Cache, Salt Lake, Utah, and Washington counties; introduced from Europe, now widely grown in the U.S.; 5 (i).

AQUIFOLIACEAE Bartling
Holly Family

Evergreen shrubs or small trees: leaves alternate, simple, coriaceous, armed with spiny teeth; stipules minute, caducous; flowers usually imperfect, regular, small and inconspicuous, solitary or few in axillary cymes; sepals usually 4, more or less connate basally; petals usually 4, slightly connate basally; stamens or staminodes usually 4 (to 9), alternate with the petals; pistil 1, the ovary superior, 3– to many-loculed, the carpels as many as the locules; style 1 or lacking; fruit a globose, berrylike drupe with 2–8 bony 1–seeded divisions; x = 9, 10.

Ilex L.

Evergreen; leaves thick and shiny; flowers small, mostly in few-flowered axillary cymes; staminodia usually present in pistillate flowers, a rudimentary pistil present in most staminate flowers; fruit usually brightly colored. Note: Members of this family are known in Utah in cultivation only.

1. Flowers in axillary clusters on branches of the previous year .... *I. aquifolium*
— Flowers in solitary cymes on branches of the current year .... *I. opaca*

Ilex aquifolium L. English Holly. 2n = 40. Tall shrubs to small trees of ornamental plantings, rare in Utah; introduced from the Old World; 3 (0).

Ilex opaca Ait. American Holly. Low to moderate shrubs of ornamental plantings, occasionaly in Utah; introduced from the eastern U. S.; 3 (0).

ARALIACEAE A. L. Juss.
Ginseng Family

Herbs, climbing vines, or shrubs to small trees; leaves alternate, simple or compound (or 2– to 3–compound); flowers small, greenish or whitish, umbellate, the umbels simple or in paniculate clusters; calyx small, adnate to the ovary, the lobes minute or lacking; petals 5, typically distinct; stamens 5, alternate with the petals; ovary 1, inferior, 2– to 5–loculed, with 1 ovule per locule; styles 5 and separate or connate throughout; fruit a berry; x = 11, 12+.

1. Plants vinelike, with aerial holdfast roots, clinging to walls and other structures; leaves simple .... *Hedera*
— Plants herbs or shrubs, lacking aerial roots; leaves palmately or pinnately compound. 2
2(1). Leaves palmately compound, with 5 or 7 leaflets; plants shrubs .... *Acanthopanax*
— Leaves pinnately or ternately 1– to 3–compound, with many ultimate segments; plants herbs or small trees .... *Aralia*

Acanthopanax Miq.

Deciduous shrubs; stems armed with spines; leaves alternate, palmately compound; umbels simple or paniculate; flowers polygamous or perfect; calyx teeth minute or lacking; petals usually 5; stamens as many as the petals; ovary 2– to 5–loculed; styles free or connate; fruit globobose.

Acanthopanax sieboldianus Makino Shrubs; branchlets with nodal spines, grayish; petioles 3–10 cm long; leaflets 5, the terminal one oblanceolate to oblong-obovate, 3–7 cm long, 1–2.5 cm wide, toothed, glabrous; umbels solitary on short spurs, the peduncles 5–10 cm long; fruit globose, 6–7 mm long, black (seldom produced). Cultivated ornamental at lower elevations in Utah, persisting; introduced from China; 1 (i).

Aralia L.

Herbs or small trees; stems spiny or unarmed; leaves large, pinnately or ternately 1– to 3–compound; leaflets serrate; flowers in panicled umbels; calyx minute or lacking; stamens 5; petals 5; staminal filaments short; styles free above or throughout; fruit a berry.

1. Plants indigenous herbs, unarmed, known from Washington County .... *A. racemosa*
— Plants cultivated small trees; leaf rachis armed with prickles; known from ornamental plantings ... *A. spinosa*

Aralia racemosa L. American Spikenard. Stout perennial herbs, mainly 0.5–1.5 m tall; leaves 1–9 dm long or more and about as broad, the 3 primary divisions pinnately compound; leaflets ovate to orbicular, variable in shape and size on the same leaf, serrate to doubly serrate, acuminate apically, obliquely cordate basally; inflorescence a panicle with numerous umbels; fruit purplish black, 4–6 mm thick. Crevices in sandstone and on sandy detritis in the shaded defile of Zion Canyon, at ca 1220 m in Washington County; widespread in the northeastern U.S. and Canada, and southwestward to Arizona and New Mexico. Our material belongs to ssp. *bicrenata* (Woot. & Standl.) Welsh & Atwood; 3 (ii).

Aralia spinosa L. Hercules Club. Small trees, mainly 3–5 m tall; leaves commonly 5–10 dm long, 2- to 3-compound, the leaflets 4–10 cm long, acute to acuminate, serrate, acute to rounded basally; petioles and leaf rachis armed with stout prickles; inflorescence paniculate, with numerous umbels; fruit black (seldom produced). Cultivated botanical curiosity and ornamental of lower elevation cities and towns; introduced from the eastern U.S.; 1 (0).

Hedera L.

Evergreen woody climbers with adventitious holdfast roots along the stems; leaves simple, alternate, often with stellate hairs, estipulate; umbels paniculate; flowers perfect; calyx lobes 5, or lacking; petals 5; stamens 5; ovary usually 5–loculed; styles 5, connate; fruit berrylike.

Hedera helix L. English Ivy. Stems to 10 m long or more; leaves of various shapes and sizes, but usually 2–10 cm long, and often about as broad, ovate to elliptic, or lobed, the margins undulate to entire; flowers greenish yellow, 4–5 mm across; fruit globose, black, ca 6 mm wide; 2n = 48. Cultivated ornamental, growing on walls of houses and on utility poles, long persisting, in lower elevations of Utah; introduced from Europe; 2 (i).

ASCLEPIADACEAE R. Br.
Milkweed Family

Perennial herbs, vines, or shrubs with milky juice; leaves opposite, whorled, or sometimes alternate, without stipules; flowers perfect, regular, umbellate, 5–merous; calyx deeply lobed, the lobes mostly imbricate; corolla 5–lobed or -cleft, the lobes commonly valvate in bud, a 5–lobed crown (corona) usually present between the corolla and stamens and adnate to either or both; stamens 5, inserted on the corolla tube near its base, the filaments monadelphous or distinct, the anthers united and tipped with a scarious membrane inflexed on the summit of the stylar disk; pollen grains united into waxlike or granular pollinia; carpels 2, with superior ovaries and styles but united above by the peltate discoid stigma; fruit of 2 follicles; seeds many, usually with a long coma; x = 9–12.

1. Stems not twining .... *Asclepias*
— Stems twining 2
2(1). Corona lacking; corolla lobes hoodlike ...... *Cynanchum*
— Corona present; corolla rotate or open campanulate .. .. *Sarcostemma*

Asclepias L.

Annual or perennial herbs, with milky juice; stems prostrate to erect; leaves usually opposite, infrequently whorled or irregularly approximate; inflorescence terminal or lateral, umbelliformly cymose; calyx lobes 5, equal, divided nearly to the receptacle; corolla rotate, 5–lobed, the lobes reflexed, spreading or rarely erect, the gynostegium stipitate to subsessile, the corona of 5 hoods attached to the column and subtending the fused anthers, the hoods cucullate to clavate with various modifications, more or less stipitate to sessile and deeply saccate at the basal attachment to the column, usually bearing an internal horn or crest; anthers 2–locular, with more or less prominent corneous marginal wings enclosing the 5 stigmatic chambers and with membraneous apical appendages; pollinia paired and pendulous from the translator arms, flat and uniformly fertile, enclosing granular pollen with thin hyaline intine; stigma head peltate, more or less pentagonal; fruit a follicle containing many compressed comose or rarely naked seeds.

Woodson, R. E. 1954. The North American species of *Asclepias*. Annals Missouri Bot. Gard. 41: 1–211.

1. Hoods sessile, the basal attachment deeply saccate, the adnate horn often absent or reduced to an inconspicuous and isolated crest or terminal appendage ... 2
— Hoods more or less stipitate or substipitate, cucullate, the basal attachment not deeply saccate; horn commonly present 6
2(1). Hoods sharply deflexed from the anther head; leaves lanceolate .... *A. asperula*
— Hoods not deflexed from the anther head; leaves variously shaped 3
3(2). Hoods with 2 conspicuous external laminate basal appendages; flowers greenish to cream; leaves linear .... *A. rusbyi*
— Hoods without external basal appendages or merely keeled laterally; flowers variously colored, but if greenish to cream, the leaves not linear 4
4(3). Hoods closed or nearly so by the involute margins, deeply bifid dorsally; horn absent; sepals and petals greenish to cream; leaves broadly ovate to elliptic, more or less glaucous .... *A. cryptoceras*
— Hoods freely open above; horn on crest present; sepals and petals pink to purple; leaves ovate or linear 5
5(4). Leaves broadly ovate, 1–3 cm broad; plants densely tomentulose; follicles erect .... *A. ruthiae*
— Leaves linear, 1–2 mm broad; plants inconspicuously puberulent; follicles pendulous or weakly ascending .... *A. cutleri*
6(1). Anther head about as long as broad; column cylindric or conic; horns usually gradually tapered and arching over the anther head 7
— Anther head ca 3/4 as long as broad or somewhat shorter; column broadly obconic to essentially obsolete; horns usually abruptly beaked and sharply inflexed toward or over the anther head 11
7(6). Hoods with sharply incised marginal auricles; corolla orange or yellowish red; plants of southern Utah ... . *A. tuberosa*
— Hoods without sharply incised marginal auricles; plants variously distributed 8
8(7). Leaves opposite 9
— Leaves whorled or spirally approximate 10
9(8). Hoods subequal to the anther head; flowers bright pink; plants indigenous, growing in marshes or other wet sites .... *A. incarnata*
— Hoods 1/3–1/2 longer than the anther head; flowers bright red; plants cultivated .... *A. curassavica*
10(8). Inflorescence paired or clustered at the upper nodes; flowers grayish pink to white; leaves 10–25 mm broad .... *A. fascicularis*

—	Inflorescence usually solitary at the upper nodes; flowers white; leaves 1–5 mm broad . *A. subverticillata*
11(6).	Hoods gradually rounded to acuminate; horns fused to about the middle of the hood or less 12
—	Hoods truncate or very abruptly rounded apically .. 15
12(11).	Hoods usually broadly flattened dorsally, usually abruptly constricted to a short basal stipe; plants erect 13
—	Hoods narrowly keeled dorsally, sessile or subsessile; plants prostrate to ascending 14
13(12).	Hoods widely spreading, 3–4 times as long as the anther head, gradually acuminate .... *A. speciosa*
—	Hoods erect or only slightly spreading 2–3 times as long as the anther head, abruptly acute to acuminate .... *A. hallii*
14(12).	Hoods about twice as long as anther head; leaves lanceolate; plants inconspicuously puberulent .... *A. involucrata*
—	Hoods barely longer than anther head; leaves broadly ovate to ovate-lanceolate; plants conspicuously tomentulose .... *A. macrosperma*
15(11).	Inflorescence sessile or subsessile, the peduncles shorter than the subtending petioles or scarcely longer; leaves broadly oblong, tomentose, at least when young .... *A. latifolia*
—	Inflorescence pedunculate, peduncles much longer than the subtending petioles; leaves variously shaped 16
16(15).	Leaves narrowly lanceolate, cuneate, 0.6–2.5 cm broad; plants glabrous, occurring in eastern Utah .... *A. labriformis*
—	Leaves broadly lanceolate to obovate or broadly elliptic, 2.5–14 cm broad; plants pubescent, of various distribution 17
17(16).	Follicles erect on deflexed pedicels, smooth; sepals shorter than the corolla lobes; main leaves ovate, acute apically; plants of Washington County .. *A. erosa*
—	Follicles spreading to pendulous on spreading pedicels, with soft subulate processes; sepals subequal to corolla lobes; main leaves obovate to elliptic, truncate or rounded apically; endemic to Kane County .... *A. welshii*

Asclepias asperula (**Decne.**) **Woodson** Spider Milkweed. [*Acerates asperula* Decne. in A. DC; *A. decumbens* var. *erecta* Durand, type from vicinity of Great Salt Lake; *Asclepias capricornu* Woodson]. Low herbaceous perennials from stout rootstocks; stems usually clustered from the rootstock, simple, stout, ascending or decumbent, 2–4 (6) dm tall, minutely and roughly pilosulose; leaves 10–20 cm long, 1–3 cm broad, irregularly approximate, short petiolate, lanceolate to linear-lanceolate, acuminate, sparsely pilosulous; inflorescence terminal and solitary; peduncles 1–10 cm long, stout; flowers rather large; calyx lobes ovate to ovate-lanceolate, 4–5 mm long; corolla rotate, pale yellowish green, occasionally tinged with purple externally, the lobes 9–12 mm long; gynostegium sessile, the hoods broadly clavate-falciform, abruptly deflexed from the anther head, but with ascending bluntish apices, 8–10 mm long, greenish to purplish, the anther head depressed, ca 2 mm long and 5 mm broad; follicles erect on deflexed pedicels, fusiform, attenuate, 4–13 cm long, 1–25 mm broad, smooth. Warm desert shrub, sagebrush, pinyon-juniper, mountain brush and ponderosa pine communities at 1062 to 2700 m in Beaver, Box Elder, Cache, Daggett, Davis, Garfield, Grand, Iron, Juab, Kane, Millard, Salt Lake, San Juan, Tooele, Uintah, Utah, Washington and Wayne counties; Colorado, Texas to Idaho, Nevada, California and Mexico; 64 (iv). Our plants belong to **var. *asperula*.**

Asclepias cryptoceras **Wats.** Pallid Milkweed. Herbaceous perennials; stems decumbent, clustered, stout, somewhat flattened, simple, glabrous, 1–3 dm tall; leaves 4–9 cm long, 4–8 cm broad, opposite, short-petiolate, broadly oval to suborbicular, glaucous, glabrous; inflorescence terminal or lateral; peduncles lacking or to 7 cm long; flowers large; calyx lobes narrowly lanceolate, 6–7 mm long, glabrous; corolla reflexed-rotate, greenish yellow, the lobes 10–15 mm long; gynostegium sessile, pale rose, the hoods deeply saccate and decurrent upon the column, conspicuously 2–apiculate, 6–9 mm long, subequal to anther head, the horn very inconspicuous and incurved or absent, the anther head 3–3.5 mm long and 4–5 mm broad; follicles erect on erect pedicels, broadly fusiform, 4–7 cm long, 1.5–2.5 cm broad, smooth and glabrous. Blackbrush, saltbush, sagebrush, grassland, and pinyon-juniper communities at 1130 to 2000 m in Carbon, Daggett, Duchesne, Emery, Garfield, Grand, Juab, Kane, San Juan, Sanpete, Sevier, Uintah, and Wayne counties; California to Arizona, Colorado and Wyoming; 92 (iv). Our specimens belong to var. ***cryptoceras***.

Asclepias curassavica L. Tropics Milkweed. Herbaceous perennials; stems 3–8 dm tall, somewhat woody below, glabrous or puberulent above; leaves 5–15 cm long, 1.5–4 cm broad, opposite or whorled, short petiolate, oblong to lance-oblong, glabrous, inflorescence axillary and terminal; peduncles 2–8 cm long, slender; flowers rather large; calyx lobes lanceolate, 2–2.5 mm long; corolla rotate, red purple, ca 3 mm long, the horns broad, curved; follicles erect, fusiform, 4–10 cm long, acuminate, pubescent or glabrous; 2n = 22. A greenhouse and summer ornamental in Utah County; native to tropics; 3 (0).

Asclepias cutleri **Woodson** Cutler Milkweed. Small herbaceous perennials; stems slender, weak, simple or branching only from the base, 6–13 cm long, inconspicuously puberulent; leaves 3–8 cm long, 1–2 mm wide, irregularly approximate, sessile, filiform, minutely puberulent; inflorescence terminal or nearly so, usually solitary, sessile; flowers small; calyx lobes lanceolate, ca 3 mm long; corolla reflexed-rotate, pale greenish rose, the lobes ca 5 mm long; gynostegium subsessile, pale rose, the hoods shortly saccate, ca 1.5 mm long, truncate, but with prominent narrow marginal auricles, the horn tonguelike, slightly longer than the hood, the anther head ca 1 mm long and 1.5 mm broad; follicles pendulous to spreading on pendulous to weakly ascending pedicels, narrowly fusiform, 4–5 cm long, ca 8 mm broad, smooth, glabrous. Sand dunes and gravelly places in mixed desert shrub and pinyon-juniper communities at 1270 to 1420 m in Grand and San Juan counties; southeastern Utah and Arizona; a Colorado Plateau endemic; 7 (ii).

Asclepias erosa Torr. Desert Milkweed. [*A. leucophylla* Engelm. in Parry, type from near St. George]. Large herbaceous perennials; stems woody at the base, stout, simple, 10–20 dm tall, tomentose when young, glabrate with age; leaves 6–24 cm long, 2.5–11 cm broad, sessile, broadly ovate to elliptic-oblong, short acuminate apically, cordate basally, the margins minutely erose, white arachnoid-tomentulose when young, glabrate in age, pale green; inflorescences solitary or rarely paired, several at the upper nodes; peduncles 2–12 cm long, tomentulose; flowers large; calyx lobes lanceolate, 4–6 mm long; corolla pale yellowish green, the lobes 9–10 mm long; gynostegium short-stipitate, greenish white to cream, the column obconic, ca 2 mm long and 3 mm broad, the hoods broadly oval, obtuse apically, ca 4 mm long, the horn strongly adnate, incurved, subequal to the hood, the anther head truncately conic, ca 3 mm long and 4 mm broad; follicles erect on deflexed pedicels, broadly fusiform, 5–7 cm long, 2–3 cm broad, apiculate, apically smooth, glabrous; n = 11. Creosote bush, Joshua tree, and blackbrush communities below 1130 m in Washington County; Nevada to Arizona, California and Mexico; 4 (iii).

Asclepias fascicularis Decne. in DC. Mexican Milkweed. Herbaceous perennials from stout woody rootstocks; stems several, 4–10 dm tall, almost always with numerous microphyllous axillary branches, usually puberulent in decurrent lines from the nodes, or glabrous; leaves 3–12 cm long, 0.1–2.5 cm broad, opposite or whorled, oblong to linear-lanceolate, glabrous to minutely pilosulose beneath, short-petiolate; inflorescence paired or clustered at the upper nodes; peduncles slender 2–4 cm long; flowers small; calyx lobes narrowly triangular, 1–1.5 mm long; corolla reflexed-rotate, grayish pink or white, the lobes 3–4 mm long; gynostegium narrowly stipitate, grayish pink or white, the column cylindric, ca 1 mm long and broad, the hoods cucullate, oval, ca 1–1.5 mm long, the horn basal, narrowly acicular, ca half longer than the hood, gradually arching over the anther head, the anther head cylindric, ca 1.5 mm long and broad; follicles erect on erect pedicels, narrowly fusiform, ca 5–12 cm long, 0.7–1 cm broad, smooth, glabrous; n = 11. Sagebrush, mountain brush and pinyon-juniper communities at 1270 to 1600 m in Davis, Kane, and Utah counties; Idaho, Nevada, west to Washington, Oregon, and California; 6 (0).

Asclepias hallii Gray Hall Milkweed. Herbaceous perennials; stems mostly simple, 2–5 dm tall densely puberulent to glabrate; leaves 5–15 cm long 1.5–4 cm broad, irregularly approximate, ovate to broadly lanceolate, puberulent beneath, short-petiolate; inflorescences lateral and solitary at several of the upper nodes; peduncles 1–5 cm long; flowers rather large; calyx lobes lanceolate, ca 3 mm long; corolla reflexed-rotate, pale rose to purple, the lobes 6–8 mm long; gynostegium pale rose to cream, short-stipitate, the column obconic, ca 1 mm long and 1.5 mm broad, the hoods oblong-elliptic, acute, 4.5–5 mm long, the horn adnate about the middle, falciform, abruptly incurved, shorter than the hood, the anther head conic-truncate, ca 2 mm long and 3 mm broad; follicles erect on deflexed pedicels, broadly fusiform, short apiculate, 8–12 cm long, 1–3 cm broad, smooth, puberulent to glabrate. Rocky slopes in sagebrush, mountain brush, pinyon-juniper, ponderosa pine, and aspen communities at 1900 to 3070 m in Emery, Garfield, Grand, Kane, Morgan, Utah, and Washington counties and probably elsewhere in the state; Wyoming and Colorado to Nevada and Arizona; 5 (0). Most of the county records are reported by Woodson (1954).

Asclepias incarnata L. Swamp Milkweed. Herbaceous perennials from short rootstocks; stems stout, 4–15 dm tall, simple to much branched; leaves 7–18 cm long, 0.5–3 cm broad, linear-lanceolate, acute to acuminate apically, obtuse basally, short-petiolate; inflorescences commonly paired at the upper nodes, solitary below; peduncles 1.5–7 cm long; flowers small; calyx lobes linear-oblong, ca 1–1.5 mm long, corolla pink or rarely white, reflexed-rotate, the lobes 3–4 mm long; gynostegium pinkish, the column cylindric 1–1.5 mm long, 1 mm broad; hoods cucullate, rounded apically, ca 1.5 mm long, the horn acicular, incurved over the stigmatic head, slightly longer than the hood, the anther head ca 1.5 mm long; follicles erect on erect pedicels, fusiform, long attenuate, 7–9 cm long, 8–12 mm broad, smooth, glabrous to puberulent. In riparian or palustrine sites at 1470 to 2030 m in Box Elder, Cache, Davis, Juab, Salt Lake, Utah, and Weber counties; Canada to Florida and west to Utah and New Mexico; 19 (0).

Asclepias involucrata Engelm. ex Torr. Dwarf Milkweed. Low herbaceous perennials from woody subfusiform rootstalks; stems clustered, ascending or decumbent, slender, branching repeatedly, 3–25 cm long, minutely puberulent; leaves 1–12 cm long, 3–10 mm broad, irregularly approximate, sessile or nearly so, narrowly lanceolate, somewhat folded, inconspicuously pilosulose; inflorescence terminal or from the uppermost nodes, sessile; flowers rather small; calyx lobes ovate-lanceolate, 3–4 mm long; corolla reflexed-rotate, pale green or pinkish, the lobes 5–7 mm long; gynostegium short-stipitate, white with purple keels, the column obconic, 1–1.5 mm long and 1.5–2.5 mm broad, the hoods ovate, acute, 3–4 mm long, the horns adnate toward the base, falciform, incurved or ascending, subequal to hood, the anther head truncate-conic, ca 2 mm long and 3 mm broad; follicles erect on deflexed pedicels, stoutly fusiform, short apiculate, 4–7 cm long, 1.5–2 cm broad, nearly glabrous. Dry flats and washes with mixed desert shrub, sagebrush and pinyon-juniper communities at 1470 to 2000 m in San Juan County; Texas and Kansas to Colorado, Arizona and Mexico; 1 (0).

Asclepias labriformis Jones Jones Milkweed. Herbaceous perennials; stems erect, mostly simple, 2–5 dm tall, puberulent to glabrous; leaves 5–15 cm long, 0.6–2.5 cm broad, irregularly approximate, subsessile, lanceolate to linear-lanceolate, pale green, glabrous; inflorescences lateral and solitary at several of the upper nodes; peduncles stout, 0.4–3 cm long; flowers medium sized; calyx lobes lanceolate 4–5 mm long; corolla pale yellowish green, the lobes ca 8 mm long; gynostegium short-stipitate, cream, the column narrowly cylindrical, ca 1.5 mm long, 2 mm broad, the hoods subquadrate, truncate, ca 3–4 mm long, the horn half adnate, falciform, incurved, somewhat longer than hood, the anther head truncately conic, ca 2.5 mm long, 2 mm broad; follicles pendulous on spreading pedicels, ovoid to broadly fusiform, 4–7 cm long, 1.5–3 cm broad, smooth, glabrous. Sandy washes, canyons and flats with saltbush, Mormon tea, mixed desert shrub, and pinyon-juniper at 1533 to 2330 m in Duchesne, Emery, Garfield, San Juan, Uintah, and Wayne (type from Capitol Wash) counties; indigenous and endemic to Utah; 43 (x).

Asclepias latifolia (Torr.) Raf. Broadleaf Milkweed. [*A. obtusifolia* var. ? *latifolia* Torr.]. Herbaceous perennials; stems stout, usually simple, 2–6 dm tall, tomentulose when young, soon glabrate; leaves 4–16 cm long, 4–13 cm broad, opposite, very short-petiolate, broadly oval-obovate, retuse apically, glaucous; inflorescences lateral at several of the upper nodes; peduncles very short or subsessile; flowers rather large; calyx lobes ovate-lanceolate, ca 4 mm long; corolla reflexed-rotate, pale green, the lobes 11–12 mm long; gynostegium short-stipitate, greenish white, the column broadly obconic, ca 2 mm long and 3 mm broad, the hoods subquadrate, truncate or retuse, ca 4 mm long, the horn wholly adnate, very broadly falciform, sharply incurved, somewhat longer than the hood, the anther head ca 3 mm long, 4 mm broad, truncately conic; follicles erect on deflexed pedicels, broadly fusiform, apiculate, 6–8 cm long, 1.5–4 cm broad, smooth and nearly glabrous. Disturbed sites, with mixed desert shrub, sagebrush, juniper, and hanging garden communities at 1130 to 1660 m in Garfield, Grand, Kane and San Juan counties; Nebraska to Texas and California; 17 (ix).

Asclepias macrosperma **Eastw.** Eastwood Milkweed. [*A. involucrata* var. *tomentosa* Eastw., type from Courthouse Wash, Grand County]. Low herbaceous perennials from woody rootstocks; stems clustered, ascending or decumbent, branching, 6–25 cm long, densely tomentose; leaves 2–6 cm long, 1–2 cm broad, irregularly approximate, very short petiolate, ovate to ovate-lanceolate, densely tomentulose; inflorescence solitary and terminal, sessile; flowers small; calyx lobes ovate, ca 3 mm long; corolla reflexed-rotate, pale green, purplish without, the lobes 5–6 mm long; gynostegium very short stipitate, greenish white or cream, the column broadly obconic, ca 0.7 mm long and 1.3–1.5 mm broad, the hoods broadly ovate, obtuse, 2.5–3 mm long, the horn adnate toward the base, falciform, incurved, somewhat shorter than hood, the anther head ca 2 mm long and 3 mm broad; follicles erect on deflexed pedicels, broadly fusiform, short apiculate, 4–6 cm long, 1.5–2 cm broad, sparsely pilose to glabrous. Sandy sites in mixed desert shrub, blackbrush, sagebrush, and pinyon-juniper communities at 1130 to 2000 m in Emery, Garfield, Grand, Kane, San Juan, and Wayne counties; Arizona; 68 (xx).

Asclepias rusbyi **(Vail) Woodson** Rusby Milkweed. [*Acerates rusbyi* Vail; *Asclepias engelmanniana* var. *rusbyi* (Vail) Kearney]. Herbaceous perennials; stems erect, simple or branching from the caudex, 6–12 dm tall, glabrous; leaves 12–18 cm long, 1–4 (6) mm broad, linear, sessile or nearly so, laxly spreading or reflexed, glabrous; inflorescences lateral, from several of the upper nodes, subsessile to short pedunculate; flowers rather small; calyx lobes ovate-lanceolate to lanceolate, 3–4 mm long, reflexed, pilose; corolla reflexed-rotate, pale green, purplish without, the lobes ca 5 mm long; gynostegium sessile, the hoods deeply saccate, truncate, white to yellow, auriculate at the base, with well-developed winglike basal auricles, these rounded, entire, pale green; horn present as a small indistinct crest within the hood, the anther head conical, ca 2.5–3 mm long and 3 mm broad; follicles erect on deflexed pedicels, narrowly fusiform, attenuate, 8–12 cm long, ca 1.5 cm broad. Sagebrush, oak brush, pinyon-juniper, mountain brush, and ponderosa pine communities at 1270 to 2300 m in Grand, San Juan, and Washington counties; Nevada to Arizona; 7 (ii). Woodson reports a collection of *A. engelmanniana* from San Juan County, however the report of that Great Plains species requires substantiation.

Asclepias ruthiae **Maguire** Ruth Milkweed. Low, herbaceous perennials; stems ascending or decumbent, slender, simple or branched below ground, 6–9 cm long, minutely tomentulose; leaves 1–5 cm long, 1–3 cm broad, opposite, petiolate, broadly ovate, white-tomentulose, particularly the margins; inflorescences terminal, or occasionally lateral at the uppermost nodes, sessile; flowers small; calyx lobes ovate-lanceolate, ca 3 mm long; corolla reflexed-rotate, pale violet, the lobes ca 6 mm long; gynostegium subsessile, pale rose purple, the hoods saccate, truncate, ca 1 mm long, the horn tonguelike, subequal to the hood, the anther head truncate conic, ca 1 mm long and 1.5 mm broad; follicles erect on deflexed pedicels, broadly fusiform, apiculate, 3–4 cm long, ca 1.5 cm broad, smooth, glabrous. Mixed desert shrub, saltbush and pinyon-juniper communities at 1400 to 1800 m in Emery (type from Calf Spring Canyon), San Juan, Sevier, and Wayne counties; a Navajo Basin endemic; 43 (iii).

Asclepias speciosa **Torr.** Showy Milkweed. Herbaceous perennials; stems usually very stout, simple, 6–10 dm tall, usually densely white-tomentose; leaves 6–20 cm long, 3–14 cm broad, opposite, short petiolate, broadly ovate to ovate-lanceolate, obtuse apically, rounded to cordate basally, densely white-tomentose beneath, nearly glabrous above; inflorescences lateral and solitary at several of the upper nodes; peduncles stout, 1–10 cm long; flowers large and showy; calyx lobes lanceolate, 5–6 mm long; corolla purplish rose, the lobes 10–15 mm long; gynostegium pale rose or pinkish cream, subsessile, the column broadly obconic, ca 1 mm long and 3 mm broad, the hoods narrowly ovate-lanceolate, attenuate, widely spreading, 10–14 mm long, the horn adnate toward the base, falciform-acicular, sharply incurved, much shorter than the hoods, the anther head broadly truncate-conic, ca 3 mm long, and 4.5 mm wide; follicles erect on deflexed pedicels, fusiform, 9–12 cm long, 2–3 cm broad, densely spiny to smooth, white-tomentose; n = 11. Weed of fields, roadsides, riparian, and palustrine sites at 830 to 2550 m in nearly all counties throughout Utah; Manitoba to Minnesota, Texas and westward to British Columbia and California; 71 (vi).

Asclepias subverticillata **(Gray) Vail** Whorled Milkweed. [*A. verticillata* var. *subverticillata* Gray; *A. galioides* H.B.K.]. Herbaceous perennials from stout woody rootstocks; stems 1.5–7 (12) dm tall, mostly with sterile dwarf branchlets, occasionally simple, glabrous or with puberulent lines from the nodes; leaves 2–13 cm long, 1–3 (4) mm broad, whorled or opposite, linear, glabrous to pilosulose; inflorescences usually solitary at the upper nodes; peduncle slender, 1.5–3 cm long; corolla reflexed rotate, white, rarely with a greenish purple tinge, the lobes 3–5 mm long; gynostegium narrowly stipitate, white, the column cylindrical, ca 1 mm long, slightly narrower, the hoods cucullate, oval, ca 1.5 mm long, the horn basal, acicular, somewhat longer than the hoods, gradually arching over the anther head; anther head cylindric, ca 1.5 mm long and broad; follicles erect on erect pedicels, narrowly fusiform, ca 5–9 cm long, 6–8 mm broad, smooth, nearly glabrous. Roadsides and other disturbed sites in creosote bush, blackbrush, saltbush, sagebrush, rabbitbrush, pinyon-juniper and mountain brush communities at 830 to 2200 m in Emery, Garfield, Grand, Iron, Kane, Millard, San Juan, Sanpete, and Washington counties; Colorado, Texas, New Mexico, Arizona and Mexico; 32 (vi). Plants of this species are very poisonous to livestock.

Asclepias tuberosa **L.** Orange milkweed; Butterflyweed. Herbaceous perennials from deep woody rootstocks; stems stout, clustered, branching only at the inflorescence, 2-9 dm tall, hirsutulose or hispid; leaves 3-11 cm long, 0.3-3 cm broad, irregularly approximate, crowded, short petiolate, narrowly lanceolate to broadly oblanceolate, acuminate to rounded apically, cuneate to cordate basally; inflorescences terminal or nearly so, 1-several at the nodes; flowers moderately large; calyx lobes lance-trigonal, 2-3 mm long; corolla reflexed-rotate, orange to reddish or yellowish red, the lobes 7-8 mm long; gynostegium orange, rarely yellow, the column narrowly obconic, ca 2 mm long, 1.5 mm broad, the hoods cucullate, lanceolate, 4-5 mm long, the horn basal, narrowly acicular, slightly longer than the hoods, gradually arching over the anther head, the anther head cylindrical, ca 2 mm long and broad; follicles erect on deflexed pedicels, narrowly fusiform, 8-15 cm long 1-1.5 cm broad, smooth pilosulose. Sagebrush, mountain brush, pinyon-juniper and ponderosa pine communities at 1333 to 2330 m in Garfield, Kane, San Juan, and Washington counties; Minnesota, Michigan, to Colorado, Arizona, Texas, and Mexico; 13 (iv). Our plants belong to **ssp. terminalis** Woodson, which have leaves typically obtuse to truncate, varying to slightly cordate.

Asclepias welshii N. & P. Holmgren. Welsh Milkweed. Perennial herbs from extensive underground rootstocks; stems 2.5-10 dm tall, erect, stout, simple from the base; leaves 6-9 (15) cm long, 3-6 (8) cm broad, opposite, coriaceous, broadly elliptic to ovate or obovate, rounded to truncate and mucronate apically, rounded to cordate basally, short- petiolate; inflorescence lateral at upper nodes, white-lanate; peduncles 2-7 cm long; flowers rather large; calyx reflexed at base, linear 5-7.5 mm long; corolla 5-8 mm long, reflexed rotate, ovate, cream with a rose tinged middle; gynostegium cream to pale green, the column ca 0.5 mm long, 1.5-2.5 mm broad, pale green, the hood 2.5-3.2 mm long, broadly truncate, cream, the horn exserted from and attached at middle of the hood, falcately curved over the anther head, the anther head 1.2-1.5 mm long, 1.5-3.4 mm broad; follicles spreading to pendulous on spreading pedicels, broadly fusiform; 4-7.5 cm long, lanate to glabrate, bearing soft subulate processes. Coral Pink sand dunes (type locality) in sagebrush, juniper, and ponderosa pine communities at 1700 to 1900 m in Kane County; endemic; 6 (i).

Cynanchum L.

Shrubs or suffrutescent perennials; stems twining, slender; leaves opposite, linear or reduced; flowers small, in axillary umbels or small cymes; calyx 5-lobed, the lobes acute; corolla campanulate to urn-shaped, 5-lobed, crown lacking; pollen-masses solitary in each pollen sack, pendulous; follicles long acuminate, smooth, terete.

Cynanchum utahense (Engelm.) Woodson Swallowwort. [*Astephanus utahense* Engelm., type from near St. George]. Plants perennial from a branched woody caudex; stems slender, glabrous, 2-5 dm tall; leaves linear, acuminate, 2-3 cm long, spreading or reflexed; umbels short-pedunculate, 3- to 10-flowered, with a few subulate bracts; corolla dull yellow, ca 2 mm wide, the lobes ovate, somewhat hooded, puberulent within; anthers unappendaged apically; follicles fusiform, long-acuminate, 4-6 cm long; seeds rough granulate. Creosote bush, blackbrush, mesquite, and sagebrush communities below 1065 m in Washington County; Nevada, Arizona and California; 14 (vii).

Sarcostemma R. Br

Suffrutescent twining or trailing vines; leaves opposite; flowers umbellate or cymose; calyx deeply 5-lobed; corolla rotate to campanulate or salverform, 5-lobed; stamens 5, the filaments fused into a column, each filament bearing an inflated vesicular segment (corona-vesicle) just below the anther; anthers 2-celled, the membraneous dorsal appendage ovate to deltoid; pollinia solitary in each anther sac, pendulous; follicles fusiform to clavate.

Sarcostemma cynanchoides Decne. in DC. Climbing Milkweed. [*Funastrum heterophyllum* (Engelm.) Standley]. Stems twining or trailing, 1 m or more long, much branched, glabrous to puberulent; leaves to 6 cm long and 3.5 cm wide, broadly to narrowly ovate-lanceolate to triangular-lanceolate or linear-lanceolate, acute to acuminate apically, cordate to hastate or round-cuneate basally, sparsely puberulent on both surfaces, with one or more glands on the midrib near the base; inflorescence umbellate, to 20-flowered; peduncle slender to 60 cm long; bracts linear, minute; pedicels slender to 17 mm long; calyx lobes ovate to narrowly ovate, 2-3 mm long, pilosulose without, glabrous within; corolla rotate-campanulate, greenish white to purple or pinkish, the tube 1-2 mm long, the lobes ovate, acute to acuminate, 5-7 mm long, glabrous within, fimbriate-ciliate, the ring of the crown thin, revolute, not adnate to the base of the crown-vesicles, these 1.5 mm long; follicles fusiform, to 7 cm long, attenuate apically, puberulent. Creosote bush, yucca, desert shrub, and hanging garden communities up to 1333 m in Garfield, Kane, San Juan, and Washington counties; Texas to California and Mexico; 10 (vi).

BERBERIDACEAE A. L. Juss.
Barberry Family

Shrubs with yellow wood and inner bark; leaves alternate, simple or compound; flowers perfect, regular, borne in racemes, umbels, or corymbs, hypogynous; sepals and petals usually similar and in 2 sets of 3; stamens as many as the petals and opposite them; anthers opening by 2 hinged valves; pistil 1, the ovary superior, 1-loculed, 1-carpelled; style short or lacking; fruit a berry; x = 6, 7, 8, 10, 14.

1. Primary leaves modified as spines; foliage leaves simple, aggregated on axillary spurs .... *Berberis*
— Primary leaves pinnately compound, the leaflets spinose-toothed .... *Mahonia*

Berberis L.

Deciduous or evergreen shrubs; leaves alternate, the primary ones modified as simple or 3-pronged spines; foliage leaves clustered on axillary spurs, simple; flowers yellow, borne in racemes, subumbellate corymbs, or umbels; sepals 6, or 3 and subtended by 3 bractlets; petals 6, bearing 2 glands near the base; stamens 6, irritable; fruit a berry.

1. Leaves persistent, evergreen, leathery; plants cultivated 2
— Leaves deciduous, thin; plants cultivated or indigenous . 4
2(1). Leaves glaucous or white beneath, ca twice longer than broad or less .... *B. verruculosa*

— Leaves green on both sides, or pale beneath, ca 4–10 times longer than broad 3

3(2). Veins of leaves prominent; flowers mainly 6–10 or more, umbellate from bud scales .... *B. julianae*

— Veins of leaves not prominent; flowers mainly 2–6, corymbose from spur apices .... *B. sargentiana*

4(1). Racemes 10– to 15–flowered; leaves spinulose-serrate .... *B. vulgaris*

— Racemes 2– to 10–flowered; leaves entire or spinulose-serrate 5

5(4). Spines simple; leaves broadly obovate to spatulate, entire; plants cultivated, common .... *B. thunbergii*

— Spines 3– to 5–parted; leaves oblanceolate to elliptic, serrate or entire; plants indigenous in southeastern Utah .... *B. fendleri*

Berberis fendleri Gray Fendler Barberry. Shrubs, mainly 5–10 dm tall; branchlets purplish brown; spines 3 (5) -parted, 8–15 mm long; leaves 0.8–6 cm long, 4–14 mm wide, oblanceolate to spatulate, entire to spinulose-serrate; racemes terminating lateral branches, 5– to 10– flowered; pedicels 4–6 mm long; berries red, ovoid to ellipsoid, 5–7 mm long. Riparian, hanging garden, and pinyon-juniper communities at 1585 to 2290 m in Grand (?) and San Juan counties; Colorado and New Mexico; 4 (i). This is a handsome plant, with striking autumn foliage; it has horticultural potential.

Berberis julianae Schneider Evergreen shrubs, mainly 0.8–1.5 m tall; spines 3–parted, 12–40 mm long, stout, present almost throughout; leaves 1.8–7 cm long, 0.7–2 cm wide, oblanceolate to elliptic, leathery, spinose-serrate; umbels 6– to 15–flowered, sessile from nodal buds of the previous season; pedicels 7–15 mm long; berries black, with a waxy bloom, 6–8 mm long, the style apparent. Commonly cultivated ornamental in lower elevation communities in Utah; introduced from China; 7 (i).

Berberis sargentiana Schneider Evergreen shrubs, mainly 0.7–1.5 m tall; spines 3–parted, 6–22 mm long, slender; leaves 1.5–5 cm long, 3–8 mm wide, elliptic to narrowly lanceolate, leathery, revolute and more or less spinose-serrate; corymbs 2– to 6–flowered from apices of spur branches; pedicels 6–15 mm long; berries black, somewhat glaucous, 6–8 mm long, the stigma sessile. Commonly cultivated ornamentals of cities and towns at lower elevations in Utah; introduced from China; 3 (0).

Berberis thunbergii DC. Thunberg Barberry. Shrubs, mainly 0.8–1.5 m tall; leaves deciduous; spines simple, 5–15 mm long; leaves 0.6–3.5 cm long, 3–12 mm wide, spatulate to obovate, entire; inflorescence 1– to 3– flowered, from apices of spur branches; pedicels 2–8 mm long; berries bright red, ellipsoid, the stigma sessile; $2n = 28$. A very commonly cultivated ornamental in Utah; introduced from Japan; 9 (i). The purple-leaved cultivar is striking in autumn, when the branches are laden with scarlet berries.

Berberis verruculosa Hemsley & Wilson Evergreen shrubs, mainly 0.5–1 m tall; spines 3–parted, 6–15 mm long, rather robust; leaves 8–20 mm long, 5–10 mm wide, obovate to elliptic, coriaceous, green above, whitish glaucous beneath, subentire to spinose-serrate; flowers solitary; pedicels 4–12 mm long; fruit black, with a waxy bloom, 4–6 mm long. Cultivated ornamental in Utah; introduced from China; 2 (0).

Berberis vulgaris L. Common Barberry. Shrubs, mainly 0.8–1.5 m tall; spines 3–parted, 4–7 mm long; leaves 1.5–3.5 cm long, 5–16 mm wide, obovate to elliptic, spinulose-serrulate; flowers 10–15 or more in elongate racemes; pedicels 4–8 mm long; fruit scarlet to purplish, ovoid; $2n = 28$. Sparingly cultivated ornamental plants in Utah; introduced from Eurasia; 1 (0). This is the infamous alternate host of stem rust of cereals.

Mahonia Nutt.

Evergreen shrubs; leaves alternate, not modified as spines, none in axillary fascicles, pinnately compound; leaflets spinose-toothed, leathery; flowers yellow, borne in racemes or subcorymbose racemes; sepals 3 and closely subtended by 3 bracts or 6; petals 6; stamens 6; fruit a berry. Note: This genus is often considered by botanists (but not by horticulturalists) as a subgenus of *Berberis*. The modified primary leaves of *Berberis* are interpreted as mere reductions of the pinnately compound leaves of *Mahonia*, and not as fundamental generic features. And species of *Mahonia* can be crossed with those of *Berberis*, yielding yet another horticultural curiosity, i.e., *Mahoberberis*. However, the species of *Mahonia* are decidedly an evolutionary grouping, and since taxonomy should be practical as well as represent relationships it seems best to regard *Mahonia* as distinct from *Berberis*. Alternative names are presented for those who wish not to follow this treatment.

1. Shrubs mainly 8–30 dm tall or more; leaflets with 5–7 broadly triangular lobes or teeth; plants indigenous in southeastern Utah .... *M. fremontii*

— Shrubs mainly 1–10 dm tall; leaflets with 12–40 teeth; plants of various distribution 2

2(1). Leaflets mainly less than twice longer than broad, dull on both surfaces; plants commonly less than 3 dm tall, indigenous .... *M. repens*

— Leaflets mainly twice or more longer than broad, glossy on at least the upper surface; plants commonly over 5 dm tall .... *M. aquifolium*

Mahonia aquifolium (Pursh) Nutt. Shining Mahonia. [*Berberis aquifolium* Pursh]. Shrubs with erect stems, mainly 3–10 dm tall or more; leaves with 5–9 leaflets; leaflets 3–8 cm long, 2–6 cm wide, glossy above, glossy to dull beneath, with 12–40 spinose teeth; racemes several, 3–8 cm long; bractlets (outer sepals) 2–3 mm long; inner sepals 6–8 mm long, yellow; petals bilobed; staminal filaments commonly with 2 short teeth just below the anthers; berries blue, glaucous, 7–14 mm long; $2n = 28$, 56. Cultivated ornamentals in Utah; indigenous from British Columbia to Idaho and Oregon; 2 (i).

Mahonia fremontii (Torr.) Fedde Fremont Mahonia. [*Berberis fremontii* Torr.]. Shrubs with spreading branches, 15–30 dm tall and as broad; leaves with 3–9 leaflets; leaflets glaucous on both sides, 1–3 cm long, 0.6–3 cm broad, with 5–7 broad triangular spine-tipped lobes or teeth; bractlets (outer sepals) 3.5–4.5 mm long; inner sepals 5–8 mm long, yellow; petals broadly rounded; berries reddish or purplish, hollow, often open apically, the pericarp juicy and tart, 12–20 mm long. Warm, mixed, and salt desert shrub, pinyon-juniper, and mountain brush communities at 820 to 2380 m in Emery, Garfield, Grand, San Juan, Washington, and Wayne counties; Colorado, New Mexico, and Arizona; 71 (xii)

Mahonia repens (Lindl.) G. Don Oregon Grape; Creeping Mahonia. [*Berberis repens* Lindl.]. Shrubs, mainly 0.5–3 (5) dm tall; stems erect or more or less decumbent and often rooting (stoloniferous); leaflets 3–7 (9), 1.2–7 cm long, 1–5 cm wide, dull on both sides; bractlets (outer sepals) 2–3 mm long; inner sepals 5–8 mm long, yellow; petals bilobed; berries blue, glaucous, 6–10 mm long; 2n = 28. Riparian, sagebrush, pinyon-juniper, mountain brush, aspen, lodgepole pine, and mixed conifer communities at 1125 to 2980 m in all Utah counties; Washington and Alberta, south to California, Arizona, New Mexico, and Texas; 125 (x). This plant does well in cultivation, where it grows taller and simulates the closely related *M. aquifolium*, as do thick-leaved plants from Zion Canyon. Oregon grape is utilized in reclamation plantings.

BETULACEAE S. F. Gray

Birch Family

Monoecious, deciduous trees or shrubs; leaves alternate, simple, serrate to doubly serrate; stipules caducous; flowers imperfect; staminate flowers in spreading or pendulous catkins, subtended by scale bracts, with 2– to 4–parted perianth (or bracteoles) or the perianth lacking, and 2–many stamens; pistillate flowers in clusters, spikes, or scaly catkins, with perianth minute or lacking; ovary 2–loculed; styles 2; fruit a 1–loculed, 1–seeded nutlet or nut, with or without a foliaceous involucre; x = 8, 14.

1. Cultivated shrubs or small trees; scales of pistillate catkins deciduous, the involucre greatly enlarged, enclosing a single nutlet or nut 2
— Cultivated or indigenous shrubs or trees, or if as above (*Ostrya*) then indigenous in southeastern Utah 3

2(1). Involucres enclosing a nut 1–1.8 cm thick, lobed and the lobes laciniately fringed .... *Corylus*
— Involucres expanded, flat, subtending a nutlet to 5 mm thick, serrate .... *Carpinus*

3(1). Bracts of pistillate aments deciduous; nutlets wingless, enveloped by enlarged inflated involucres; plants of defiles, monolith bases, and alcoves in southeastern Utah .... *Ostrya*
— Bracts of pistillate aments persistent; nutlets often winged, lacking an involucre; plants of riparian areas, variously distributed 4

4(3). Pistillate aments 1–several, racemose, the bracts persistent, conelike; buds stalked; wood yellow orange.. *Alnus*
— Pistillate aments solitary; bracts deciduous with the nutlets; buds sessile; wood whitish .... *Betula*

Alnus Miller

Trees or shrubs; leaves alternate, simple, doubly serrate to lobed; flowers in catkins, borne with or before the leaves; staminate catkins 1–3, the bracts peltate, subtending 3–6 flowers, each flower with 2–4 stamens; pistillate catkins conelike, 1–several, the persistent woody bracts subtending 2 flowers; fruit a nutlet. Note: Potentially there are several cultivated species in Utah, including almost certainly *A. serrulata* (Ait.) Willd., but they are not treated herein because of taxonomic considerations.

Alnus incana (L.) Moench Thinleaf Alder. [*Betula alnus* var. *incana* L.; *A. tenuifolia* Nutt.; *A. incana* ssp. *tenuifolia* (Nutt.) Breitung]. Shrubs or small trees; stems commonly 1–4 (10) m tall, 0.2–2.5 dm thick; bark grayish to brownish; winter buds blunt; twigs puberulent and commonly glandular as well; leaf blades 2–9 (11) cm long, 1–6.5 (10) cm wide, ovate to elliptic, oblong, or oval, obtuse to rounded or less commonly acute or abruptly acuminate, obtuse to rounded or subcordate basally, doubly serrate with numerous teeth and shallowly lobed, the lower surface paler than the upper but not markedly so, hairy to glabrous above, usually hairy along the veins beneath; petioles 5–20 (30) mm long, puberulent; pistillate catkins arising from branches of the previous season, 3–9 per cluster, 9–15 mm long, 8–12 mm thick; wing of nutlet lacking; n = 14. Streambanks and terraces, and in seep and spring sites at 1250 to 2745 m in Box Elder, Cache, Daggett, Davis, Duchesne, Garfield, Morgan, Salt Lake, San Juan, Sevier, Summit, Tooele, Uintah, Wayne, and Weber counties; Alaska and Yukon, east to Nova Scotia, south to California, New Mexico, Arizona, and Pennsylvania; Eurasia; 37 (ii). Our material belongs to ssp. *rugosa* (**Duroi**) **R.T. Clausen** var. *occidentalis* (**Dippel**) **C.L. Hitchc.** [*A. occidentalis* Dippel]. The thinleaf alder is a portion of a huge circumboreal complex, with **ssp. *incana*** being the Old World portion and **var. *rugosa*** representing the eastern American plants.

Betula L.

Trees or shrubs; leaves alternate, simple, serrate to crenate, crenate-serrate, or doubly serrate; flowers in catkins, borne with or before the leaves; staminate catkins 1–4 per bud, pendulous or spreading in flower; pistillate catkins usually solitary, erect in flower; staminate flowers in clusters of 3, the stamens 2; bracts of pistillate catkins 3–lobed, deciduous; pistillate flowers 2–3 per bract; fruit a winged nutlet (samara).

Dugle, J. R. 1966. A taxonomic study of western Canadian species of the genus *Betula*. Canad. J. Bot. 44: 929–1007.

1. Plants low to moderately sized shrubs, 0.5–1.2 m tall; leaf blades 0.5–2.5 cm long, oval to orbicular, crenate to crenate-serrate, with usually 10 or fewer teeth per side .... *B. glandulosa*
— Plants tall shrubs or trees, mainly 4–8 m tall; leaf blades 1.5–8 cm long, ovate to deltoid, serrate to doubly serrate (or lobed), with usually 10–40 teeth per side 2

2(1). Bark shining, not exfoliating, brown to red brown or yellow brown; plants indigenous in riparian sites in Utah .... *B. occidentalis*
— Bark dull, white; plants cultivated in Utah 3

3(2). Bark exfoliating; leaves ovate, usually pubescent in vein axils beneath .... *B. papyrifera*
— Bark not exfoliating; leaves rhombic-ovate or laciniately lobed, glabrous beneath .... *B. pendula*

Betula glandulosa Michx. Glandular Birch; Swamp Birch. Shrubs, commonly 0.5–1.2 (2) m tall, with 1 to several main stems; bark not exfoliating, gray to brown or purplish, the lenticels not or seldom conspicuous; twigs puberulent, bearing yellowish crystalline resin glands; leaf blades 0.5–2.5 cm long, 0.5–2 cm broad, oval to orbicular, broadly elliptic, or obovate, or less commonly ovate, rounded to obtuse or rarely acute apically, cuneate to rounded basally, once crenate to crenate-serrate, mainly with 10 or fewer teeth per side, not hairy in lower vein axils, minutely hairy to glabrous on margins near the base; petioles 1–9 mm long, puberulent or glabrous;

pistillate catkins 7–20 mm long, 3–8 mm thick, the bracts commonly glabrous dorsally, ciliate; samara wings narrower than half the body width; 2n = 28. Meadows and streamsides in lodgepole pine and spruce-fir communities at (2135?) 2745 to 3355 m in Cache (?), Daggett, Duchesne, Summit, Utah, and Wasatch counties; Alaska and Yukon, east to Newfoundland, south to California, Colorado, and New York; Greenland; Asia; 15 (iv).

Betula occidentalis Hook. Water Birch. [*B. fontinalis* Sarg.]. Shrubs or small trees, commonly 3–6 m tall, with several trunks to 2.5 dm thick or more; bark not or scarcely exfoliating, reddish or yellowish brown to brown, shining, marked with pale horizontal lenticels; twigs pubescent to glabrous, bearing yellowish to reddish crystalline resin glands; leaf blades 1–5 cm long, 0.7–4 cm broad, ovate, acute, or abruptly acuminate apically, obtuse to rounded or less commonly cuneate to truncate basally, sharply and often doubly serrate, with mostly 15–25 teeth per side, not hairy in lower vein axils, minutely hairy to glabrous on margins near the base; petioles 5–15 mm long, glabrous or puberulent; pistillate catkins 15–40 mm long, 4–10 mm thick, the bracts puberulent and ciliate; samara wing subequal to width of the nutlet. Riparian communities and near seeps and springs at 1220 to 2685 m in all Utah counties; Alaska and Yukon, east to Mackenzie, south to California, Colorado, and South Dakota; 134 (xiv). This birch forms hybrids with *B. papyrifera*. Presumably the material constituting the type of *B. utahensis* Britt., from City Creek Canyon, was derived in that manner and has either persisted from a time when *B. papyrifera* was sympatric, or has been derived more recently through long-distance pollination.

Betula papyrifera Marshall Paper Birch. Trees, commonly 5–8 m tall or more, with 1 main trunk to 5 dm thick or more; bark exfoliating, creamy white, marked with pale elongate horizontal lenticels; twigs puberulent to glabrous, bearing yellowish to white crystalline resin glands, or these poorly developed or lacking; leaf blades 2.5–8 cm long, 1.3–7 cm broad, ovate to lance-ovate, acuminate to acute, rounded to truncate or subcordate basally, sharply and often doubly serrate and sometimes lobed as well, with usually 15–20 teeth per side, usually hairy in lower vein axils, hairy to glabrous on margins near the base and often over the upper surface; petioles 8–40 mm long, glabrous or puberulent; pistillate catkins 15–40 mm long, 5–15 mm thick; bracts glabrous or puberulent dorsally; samara wings broader than the body; 2n = 56, 70, 84. Cultivated ornamental in Utah; introduced from elsewhere in North America; 2 (0).

Betula pendula Roth European White Birch. Trees, commonly 5–10 m tall or more, with 1 main trunk to 5 dm thick or more; bark flaking off in layers, but not especially exfoliating, marked with elongate horizontal lenticels; twigs glabrous, with yellowish crystalline resin glands or these lacking, often pendulous; leaf blades 1.5–7 cm long, 1.2–5 cm wide, rhombic-ovate, acuminate apically, truncate to obtuse basally, sharply doubly serrate with 15–30 teeth per side or more, or laciniately lobed, not hairy in the vein axils below, hairy to glabrous on margins near the base; petioles 10–40 mm long; bracts puberulent or glabrous dorsally; pistillate catkins 20–50 mm long, 8–12 mm thick; samara wing broader than the body; 2n = 28. Cultivated ornamental in Utah; introduced from Eurasia; 7 (0).

Carpinus L.

Shrubs or small trees; leaves alternate, simple; staminate aments pendulous; bracts ovate, each with a single naked flower with several stamens; filaments short, divided apically and bearing 2 apically pilose half-anthers; pistillate aments slender, with ovate deciduous bracts, the flowers in pairs, with calyx much reduced; bracts accrescent in fruit; fruit a ribbed nutlet.

Carpinus caroliniana Walter Hornbeam. Plants mainly 3–8 m tall, with flattened trunk and smooth or gray striated bark (musclelike); leaves 2.5–12 cm long, ovate or lance-oblong to ovate or obovate, sharply doubly serrate, acuminate or acute; fruiting catkins 2–7 cm long, the bracts oblong-lanceolate, 1.2–3 cm long, often lobed at the base, entire or few-toothed. Cultivated ornamental plants in Utah; introduced from the eastern U. S.; 1 (0).

Corylus L.

Deciduous shrubs or small trees; leaves alternate, simple, doubly serrate; staminate catkins pendulous at anthesis, each scale subtending a pair of bractlets and a single naked flower with 4 stamens; filaments divided apically, each bearing 2 half-anthers; pistillate catkins small, ovoid, the flowers concealed by bud scales except for the elongate reddish stigmas, each flower subtended by a tiny bract and 2 bractlets, these greatly accrescent at maturity, finally enclosing the nut.

Corylus avellana L. European Filbert; Hazelnut. Shrubs, mainly 2–6 m tall; leaves 1.5–10 cm long, 1.2–8.5 cm broad, suborbicular to broadly obovate, abruptly acuminate apically, subcordate basally, hairy along the veins beneath, puberulent above; petioles 4–14 mm long, long-hairy; mature involucre densely hairy, subequal to or somewhat exceeding the nut, laciniately lobed; nuts 1–2 cm wide; 2n = 22. Cultivated nut plant, widely but not commonly grown in warmer portions of Utah; 6 (0). Other species, native elsewhere in North America are probably present.

Ostrya L.

Small trees; leaves alternate, simple, doubly serrate; staminate catkins pendulous, the scales abruptly acuminate, the flowers consisting of several stamens, each in a bract axil, the filaments short, each forked apically and the branches bearing pilose half-anthers; pistillate catkins slender, loosely flowered, the ovate hairy bracts caducous, subtending 2 flowers each enclosed within an ovoid pouch composed of united bract and bractlets; calyx minute; bracts accrescent and inflated in fruit, hoplike; fruit a compressed ovoid nutlet.

Ostrya knowltonii Cov. Western Hophornbeam. Small trees, mainly 2–6 m tall; trunks 3–18 cm thick; branchlets spreading-hairy and more or less stipitate-glandular, becoming glabrous; leaf blades 0.8–8 cm long, 0.8–5 cm wide, ovate to lance-ovate or elliptic, doubly serrate, acute apically, rounded to obtuse basally; fruiting aments 2–5 cm long, the individual sacs 10–25 mm long, greenish white to brownish. Bases of monoliths, defiles, and hanging gardens in sandstone areas at 1225 to 1710 m in Garfield, Grand, Kane, and San Juan counties; New Mexico, Arizona, and Texas; 17 (x).

BIGNONIACEAE A. L. Juss.

Catalpa Family

Shrubs or trees; leaves mostly opposite, simple or compound; flowers large and showy, perfect, irregular, in terminal panicles or spicate racemes; calyx short, bilabiate or unequally 4- to 5-toothed; corolla sympetalous, bilabiate, 5-lobed; stamens 2 or 4; pistil 1, the ovary superior, 2-carpelled; style 1; stigma bilobed; fruit an elongate 2-valved capsule; seeds many, large, flat, winged or comose; x = 20 (7).

1. Plants climbing or clambering woody vines; leaves compound .... *Campsis*
— Plants trees or shrubs; leaves simple 2
2(1). Leaves ovate, petioled; fertile stamens 2; plants cultivated, less commonly escaping and established .. *Catalpa*
— Leaves linear to lance-linear; fertile stamens 4; plants indigenous in Washington County, seldom cultivated
   .... *Chilopsis*

Campsis Lour.

Woody vines; leaves opposite, odd-pinnately compound, with toothed leaflets; flowers orange, in compact terminal panicles; calyx 5-lobed, suffused with red, leathery; corolla funnelform, expanded above the narrow tube, the 5 lobes spreading; stamens 4, in 2 pairs; ovary 2-loculed, surrounded at base by a large disk; fruit a fusiform capsule, the 2 valves separating from the septum; seeds numerous, compressed, with 2 large translucent wings.

Campsis radicans (**L.**) **Seemann** Trumpet-vine. [*Bignonia radicans* L.]. Stems mainly 1.5-10 m long, climbing by means of aerial rootlets or merely clambering over other plants or structures; leaves pinnately compound with 7-11 leaflets, these 2-8 cm long, oval to ovate or oblong, long-acuminate, serrate, dark green above, pale and hairy beneath (at least on the midrib); corolla tube orange, mainly 6-8 cm long; capsules commonly 7-16 cm long, sharply margined at maturity, tapering at both ends. Cultivated ornamental, persisting and spreading, at elevations below 1525 m in Grand, Utah, and Washington counties; widely cultivated in the U. S., indigenous in eastern U. S.; 4 (iii).

Catalpa Scop.

Trees; leaves opposite or sometimes whorled, long petioled, large, simple, entire or coarsely lobed, often with a pungent odor when crushed; flowers white, pinkish, or yellowish, in large showy terminal panicles; calyx splitting irregularly or 2-lipped; corolla campanulate, 2-lipped, with 2 smaller upper lobes and 3 larger lower lobes; fertile stamens 2, curved, the anther sacs divergent, included; style 2-lobed; fruit a cylindrical capsule, dehiscent by 2 valves at maturity; seeds small, compressed, with a tuft of white hairs at each end.

1. Leaves glabrous or soon glabrate beneath; corolla 1-2 cm broad, yellow, marked with orange and spotted with violet; capsules less than 8 mm thick .... *C. ovata*
— Leaves with persistent soft pubescence beneath; corolla 2-4 cm broad, white marked with yellow and brown purple; capsules mainly over 8 mm thick 2
2(1). Leaves long-acuminate apically; corolla ca 3-4 cm broad, the lower lobe notched .... *C. speciosa*
— Leaves abruptly short-acuminate; corolla 2-3 cm broad, the lower lobe not notched .... *C. bignonioides*

Catalpa bignonioides **Walter** Common Catalpa. Trees, 6-20 m tall, with rounded spreading canopy; leaves odoriferous, often whorled, the blades broadly cordate-ovate, 10-20 cm long, abruptly acuminate, sometimes with 2 small lateral lobes, pubescent on lower surface; panicles broadly pyramidal, 15-25 cm long, many-flowered; corollas 2-3 cm broad, white, with 2 yellow stripes inside and spotted with brown purple, the lower lobe not notched; capsules 12-37 cm long; n = 20. Cultivated ornamentals, escaping and established locally at or below 1750 m in Grand, Utah, and Wasatch counties; widely cultivated in the U. S.; 10 (ii).

Catalpa ovata **G. Don** Chinese Catalpa. Small trees or large shrubs; leaves abruptly acuminate, glabrous or soon glabrate beneath, often with sharp lateral lobes; corollas 1-2 cm broad, yellow, marked with orange and spotted with violet; capsules mainly less than 8 mm thick. Cultivated ornamental in Utah County; widely cultivated in the U. S., introduced from eastern Asia; 1 (0).

Catalpa speciosa **Warder** Showy Catalpa; Cigartree. Trees to 25 m tall or more; leaves nonodorous, cordate-ovate, 15-30 cm long, long acuminate, pubescent beneath; panicles rather few-flowered, ca 15 cm long and broader than long; flowers white, 3-4 cm broad, inconspicuously spotted on the throat, the lower lobe notched; capsules 20-50 cm long. Cultivated ornamentals, persisting and occasionally escaping, mainly below 1525 m in Davis, Grand, Utah, and Wayne counties; Illinois and Arkansas; widely cultivated in the U. S.; 14 (iii).

Chilopsis **D. Don**

Small trees or large shrubs to 10 m tall; leaves alternate or the lower opposite, simple, linear or lance-linear, entire; corolla white, often tinged, streaked, or spotted with purple; capsule cylindrical, 2-valved; seeds long-hairy.

Chilopsis linearis (**Cav.**) **Sweet** Desert Willow. [*Bignonia linearis* Cav.]. Branches laxly ascending or spreading; sterile branches glabrous or somewhat woolly, without terminal buds; leaves sessile or short-petiolate, 4-15 cm long, 0.3-1 cm wide, glabrous or often viscid; flowers fragrant; pedicels stout, pubescent, ca 2-5 mm long; calyx pubescent, bilabiate, 1-1.5 cm long; corolla broadly funnelform, 2-3.5 cm long; stamens 4, included; staminodia 1; capsule slender, mainly 15-30 cm long, ca 7 mm thick; seeds ca 6 mm long, with a fringe of hairs. Along stream courses below 1500 m in Washington County; Texas to Nevada, California, and Mexico; 20 (x). This plant has horticultural potential and phases with brightly colored flowers are being selected for introduction.

BORAGINACEAE A. L. Juss.
Borage Family

Plants herbaceous or shrubby; leaves simple, alternate, opposite, or whorled, entire and pubescent, hispid or setose; flowers perfect, regular, solitary or cymose; cymes glomerate-racemose or spicate, frequently unilateral and coiled (scorpioid), usually with bracts between, to one side of, or opposite the flowers; calyx usually 5-lobed or 5-parted, usually persistent, the lobes valvate; corolla 5-lobed, sometimes crested or appendaged in the throat; stamens 5, borne on the corolla tube alternate with the lobes; ovary superior, bicarpellate, usually 4-ovulate, entire or lobed, becoming tough or bony at maturity; fruit commonly breaking up into 4 single-seeded lobes (nutlets); style simple or 2–cleft, seated in the pericarp at the apex of the fruit or borne between the nutlets on the receptacle, or on an upward prolongation (gynobase); endosperm absent or scarce; embryo straight or curved; x = 4–12. Note: The classification of this family is based primarily upon the structure of the fruit. In many cases it is difficult to recognize the genus and almost impossible to obtain a precise identification of the species if the specimens lack mature fruiting structures.

1. Style 2–cleft; stigmas 2, distinct; flowers solitary or clustered in the stem forks .... *Tiquilia*
— Style simple; stigmas united 2
2(1). Style arising from the pericarp at the apex of the fruit, falling away with the nutlets; stigma annular-peltate, surmounted by a conical or cylindrical, simple or lobed appendage .... *Heliotropium*
— Style borne between the lobes of the nutlets, and attached to the receptacle or gynobase; stigma capitate, unappendaged 3
3(2). Nutlets with uncinate, glochidiate, or barbed prickles on the back, margins, or the apex 4
— Nutlets without hooked or barbed prickles 7
4(3). Nutlets subglobose, with dorsal surface rather uniformly covered with barbed prickles, no definite margins .... *Cynoglossum*
— Nutlets with a definite margin, the prickles confined to this (back may be muricate or tuberculate) 5
5(4). Nutlets stellately spreading, attached at the apical (radicle) end, armed with hooked appendages; plants small slender annuals .... *Pectocarya*
— Nutlets erect, incurved or weakly divergent, attached at or below the middle, i.e., toward the cotyledon end 6
6(5). Plants annual; pedicels erect or nearly so; styles surpassing the nutlets; subulate gynobase about as long as the nutlets .... *Lappula*
— Plants perennial or biennial; pedicels reflexed in fruit; styles usually shorter than the nutlets; pyramidal gynobase about half as long as the nutlets .... *Hackelia*
7(3). Corolla irregular, the upper lobes usually longer than the lower ones; stamens not all equal in length . *Echium*
— Corolla regular or nearly so 8
8(7). Calyx in fruit much enlarged, becoming conspicuously veiny, folded and flattened; stems procumbent, angled, with stiff retrorse bristles on the angles .... *Asperugo*
— Calyx in fruit little if any enlarged, not becoming veiny, folded, and flattened; stems various but not as above 9
9(8). Nutlet attachment surrounded by a swollen ring, leaving a distinct pit on the gynobase; plants of fields and waste places 10
— Nutlet attachment neither surrounded by a rim nor leaving a pit 11
10(9). Stamens appendaged dorsally, closely crowded around the style; corolla rotate .... *Borago*
— Stamens unappendaged, included within the tubular corolla .... *Anchusa*
11(9). Corolla normally blue (aberrant white-flowered plants occasionally are found), or reddish in the bud stage 12
— Corolla white, greenish white, yellow or orange 14

12(11). Nutlets with an oblique dorsal face encircled by an upturned flange or rim, this often irregularly toothed; plants depressed-pulvinate, seldom over 7 cm tall, of alpine areas in Utah .... *Eritrichium*
— Dorsal face of nutlet (if present) not encircled by an upturned flange or rim; plants not depressed-pulvinate, usually over 7 cm tall, most species growing below alpine areas in Utah 13
13(12). Corolla salverform, the lobes convolute in the bud; nutlets basally attached to a flat gynobase .... *Myosotis*
— Corolla with a tube and usually a campanulate throat, not salverform, the lobes imbricate in the bud; nutlets obliquely attached to a convex gynobase . *Mertensia*
14(11). Nutlets attached above the base along a usually open and generally basally forked central groove or slit, or by a triangular opening in the pericarp .... *Cryptantha*
— Nutlets lacking a distinct ventral groove or opening in the pericarp, this usually replaced by an elevated ventral keel 15
15(14). Plants perennial; nutlets attached by a broad, rounded, quite basal noncaruncular attachment, ovoid, smooth and shiny; corolla usually yellow or orange .... *Lithospermum*
— Plants annual; nutlets attached by a caruncular scar borne upon or at the basal end of the ventral keel, the attachment usually lateral or suprabasal, usually rough 16
16(15). Corolla white; cotyledons entire .... *Plagiobothrys*
— Corolla orange or yellow, the tube definitely longer than the calyx; cotyledons 2–lobed .... *Amsinckia*

Amsinckia Lehm.

Annual, pungent-bristly, herbaceous plants; stems erect or with spreading branches, leafy; leaves alternate, linear to ovate, usually veinless; racemes usually bractless; calyx cut to base into erect lanceolate or oblong lobes; corolla tubular or salverform, heterostyled, yellow or orange, tube cylindrical, glabrous, unappendaged; lobes spreading, rounded, imbricate; stamens included, affixed in the tube; filaments very short; anthers oblong; style obtuse, filiform, included; stigma capitate, emarginate; ovules 4; cotyledons 2–parted; nutlets 4, erect, angulate-ovoid, smooth or rough, unmargined, strongly keeled ventrally; gynobase pyramidal, ca half the height of the nutlet.

1. Corolla tube 20–nerved below attachment of stamens; calyx lobes unequal in width and reduced in numbers (2, 3, or 4) by fusion; nutlets tessellate .... *A. tessellata*
— Corolla tube 10–nerved below insertion of stamens; calyx lobes 5, distinct 2
2(1). Corolla orange yellow, 7–20 mm long, well exserted beyond the calyx; plants usually green; stems hirsute-bristly, but with few or no fine appressed hairs .... *A. intermedia*
— Corolla pale yellow, 4–7 mm long, little or not at all exserted beyond the calyx lobe; leaves pubescent with appressed or ascending hairs .... *A. menziesii*

Amsinckia intermedia Fisch. & Mey. Medium Fiddleneck. [*A. campestris* Greene; *A. velens* Macbr.; *A. intactilis* Macbr.; *A. arvensis* Suksd.]. Stems simple or much branched, erect to widely spreading, 3–9 dm tall, sparcely bristly otherwise usually glabrous except for a tomentose pubescence near the base of the spikes; basal and lower cauline leaves linear or linear-lanceolate to nearly ovate, usually clasping at base and acute at apex, thinly hirsute on both sides with spreading, often pustulate hairs; spikes short or usually elongating in fruit, usually leafy-bracted at base; calyx lobes linear attenuate, about half as long as the corolla, rufous-hispid on the back, densely white-hirsute on the margins; corolla orange yellow 8–10 mm long, the limb 3–6 mm wide; nutlets 2.5–3 mm long, incurved, grayish, narrowly keeled on the back and sharply rugose with the surface between papillate or muricate. Creosote bush, Joshua tree, and other warm desert shrub communities below 1200 m in Washington County; Washington, south to Baja California and eastward to Arizona, New Mexico, and west Texas; 15 (iii).

Amsickia menziesii (Lehm.) Nels. & Macbr. Menzies Fiddleneck. [*A. retrorsa* Suksd.; *A. parviflora* Heller; *A. rugosa* Rydb.; *A. helleri* Brand; *A. eatonii* Suksd., type from Utah]. Stems strictly erect, 3–8 dm tall, usually simple below the inflorescence, bristly-hirsute and often more or less cinereous with fine appressed hairs; leaves linear or the upper linear-lanceolate, hirsute on both sides with ascending or appressed hairs; inflorescence of 1 or few, strict, erect or ascending racemes, bractless; calyx lobes 5, distinct, 7–13 mm long, linear or linear-lanceolate; corolla light yellow, 5–7 mm long, the tube included or only slightly exserted beyond the calyx lobes; style 2.5–3 mm long; nutlets 2–3 mm long, broadly ovoid, densely tuberculate all over, with scattered larger tubercles intermixed, the latter on the central and lateral ridges when these are present. Mixed desert shrub, pinyon-juniper, sagebrush, mountain brush, and aspen-fir communities at 1400 to 2850 m in Beaver, Box Elder, Cache, Davis, Millard, Morgan, Rich, Salt Lake, Tooele, Utah, Wasatch, and Weber counties; Alaska, Yukon, south to California and Nevada; 20 (ii).

Amsinckia tessellata Gray Rough Fiddleneck. [*A. collina* Greene; *A. pustulata* Heller; *A. conica* Suksd.; *A. utahensis* Suksd., type from Salt Lake City]. Stems stout, branched throughout or sometimes simple below, 3–6 dm high, hispid with spreading bristles; leaves linear-lanceolate, 2–7 cm long, rather thinly hispid, the hairs pustulate at base, sessile except the narrowly oblanceolate basal ones; spikes elongating with age, often 5–12 cm long; calyx lobes (2) 3 or 4, when 4 with 1 broader and notched or 2–lobed at apex, when 3 a little broader and notched at apex, hispid and on the margins densely white-hirsute, 8–13 mm long; corolla orange, the tube 5–10 mm long, the limb 2.5–5 mm wide; nutlets 3–3.5 mm long, ovoid, the back low usually with a median line, densely tessellate or papillate, and often transversely rugose. Creosote bush, Joshua tree, mixed warm desert shrub, sagebrush, oak, and pinyon-juniper communities at 750 to 1900 m in Box Elder, Davis, Juab, Millard, Salt Lake, Sevier, Tooele, Utah, and Washington counties; Washington to Idaho, Arizona, and Baja California; 29 (iii).

Anchusa L.

Annual, biennial or perennial herbs with blue or purple flowers in panicled, scorpioid racemes; calyx divided into narrow lobes; corolla trumpet-shaped, the tube straight, the throat closed by scales, the limb with widely spreading lobes; stamens included; style slender; ovary 4–parted; nutlets 4, their attachment surrounded by an annular ring leaving a pit on the low gynobase.

1. Calyx lobes linear; flowers more than 12 mm wide .... *A. azurea*
— Calyx lobes triangular to lanceolate; flowers less than 8 mm wide .... *A. officinalis*

Anchusa azurea Miller Alkanet. Plants perennial, from a taproot; stems erect, branched from near the base, 3–10 dm tall or more, coarsely hirsute, the hairs often pustulate basally; basal leaves 6–20 cm long, oblong or lanceolate; calyx 8–12 mm long, the lobes linear, much longer than the tube; corolla ca 10–15 mm long, dark blue; nutlets oblong, erect; 2n = 32. Cultivated ornamental, escaping and persisting in Box Elder, Salt Lake, and Utah counties; introduced from the Mediterranean region of the Old World; 3 (i).

Anchusa officinalis L. Bugloss. Plants perennial, from a taproot; stems erect, branched from near the base, 3–10 dm high, coarsely hirsute, the hairs often pustulate at base; basal leaves 8–20 cm long, oblanceolate, the stem leaves lanceolate; calyx 5–8 mm long, the lobes lanceolate to narrowly triangular, about as long as the tube; corolla about 10 mm long, dark blue; nutlets 2–3 mm long, rugose or granulate, inserted by their bases on a flat gynobase; 2n = 16, 32. Disturbed hills, river bottoms, and ditch banks in sagebrush, cottonwood, maple, and oak communities at 1430 to 2500 m in Summit and Utah counties; introduced from Eurasia; 18 (ii).

Asperugo L.

Rough-hispid, annual, procumbent plants, with stiff bristly hairs; leaves alternate, or the upper sometimes opposite, entire; calyx campanulate, unequally 5–cleft, much enlarged and reticulate-veiny in fruit, lobes incised-dentate, the teeth often appearing as extra lobes in the sinuses; corolla tubular-campanulate, 5–lobed, 1 to 3 together on short, recurved pedicels in the upper leaf axils; stamens 5, inserted on the corolla tube, included; filaments very short; ovary 4–lobed; style short; stigma capitate; nutlets 4, ovoid, erect, granular-tuberculate, attached laterally above the middle to the elongate-conic receptacle, the scar not leaving a pit.

Asperugo procumbens L. Catchweed. Stems 2–6 dm long; diffusely branched, slender and procumbent or ascending, retrorsely short-hispid; leaves 1–4 cm long, obovate to oblanceolate, scabrous, obtuse to acutish at apex; fruiting calyx 8–15 mm wide; corolla small, 2–3 mm long, blue, purple or purplish red; nutlets obliquely ovoid, about 4 mm long, granulate-tuberculate; 2n = 24, 48. Roadsides, ditch banks, and other waste places at 1400 to 2450 m in Cache, Daggett, Davis, Salt Lake, Summit, Tooele, Utah, and Weber counties; native of Eurasia now widely scattered over the northern U. S.; 27 (ii).

Borago L.

Hirsute or hispid annual or biennial herbs with alternate, entire leaves and blue flowers in terminal leafy racemes; calyx deeply 5–cleft or 5–parted; corolla rotate, the tube very short, throat closed by scales, the limb 5-lobed, the lobes imbricated, acute; stamens 5, inserted on the corolla tube; filaments dilated below, narrowed above to a slender appendage; anthers linear, erect and connivent with a beaklike cone; ovary 4–divided; style filiform; nutlets 4, ovoid, erect, attached by their bases to the flat receptacle; scar of attachment large, concave.

Borago officinalis L. Common Borage. Stems erect, 5–8 dm tall, with ascending or spreading branches; leaves oblong to obovate, 5–11 cm long, rounded to acute at apex, the upper ones clasping, lower narrowed to a winged petiole; pedicels spreading or recurving, 2–5 cm long; calyx lobes linear-lanceolate, 7–10 mm long; corolla 15–20 mm broad, bright blue; anther-beak dark purple, ca 6–7 mm long; nutlets 4 mm long. Cultivated and escaping from gardens in Cache and Salt Lake counties; sparingly naturalized in the western U. S.; a native of Europe; 2 (0).

Cryptantha Lehm.

Annual or perennial, herbaceous or fruticulose; leaves opposite at base, or alternate throughout, firm, veinless; flowers white or rarely yellow, in bractless or bracted spikes or racemes; calyx divided to the base, the lobes erect or connivent, linear or oblong; mature calyx investing the nutlets and falling away entire, or the calyx persistent and the nutlets falling away separately; corolla with a short to somewhat elongate cylindrical tube with or without scales at the base of the tube, the throat with intruded appendages; corolla lobes imbricate, rounded spreading; style slender, short or long, included; stigma capitate; ovules 2–4; nutlets 1–4, erect, ovate to triangular, roughened or smooth, winged, margined or marginless, affixed laterally through a medial ventral and commonly basally forked groove to a usually columnar, subulate or pyramidal gynobase. Note: This is one of the most perplexing genera in the family. It is exclusively American, mostly in the western U. S., but common in the deserts of South America as well. Several species are reported to have some value as a forage for sheep.

1. Plants annual; stems slender (subgenus Krynitzkia) .. Key 1
— Plants biennial or perennial; stems coarse, usually with a basal tuft of leaves (subgenus Oreocarya) .... Key 2

Key 1.

Plants annual.

1. Nutlets with the margins decidedly winged or knifelike 2
— Nutlets with the margins rounded or angled, never with a marginal wing or knifelike edge 5
2(1). Pedicels usually evident, slender, 1–4 mm long; nutlets heteromorphic .... *C. racemosa*
— Pedicels obscure or none, less than 1 mm long 3
3(2). Nutlets heteromorphic, the odd nutlet abaxial .... *C. inaequata*
— Nutlets homomorphic, or if slightly heteromorphic the odd nutlet axial 4
4(3). Nutlets solitary or rarely 2; calyx obliquely conical at the base; corolla conspicuous; plants of Washington County .... *C. utahensis*
— Nutlets 4; calyx symmetrical; corolla inconspicuous; widespread species .... *C. pterocarya*
5(1). Nutlets all smooth 6
— Nutlets all rough or at least some of them so 10
6(5). Nutlets with an excentric groove; flowers in biserial naked spikes .... *C. affinis*
— Nutlets with a centrally placed groove 7
7(6). Nutlets broadly ovate; spikes usually geminate, the inflorescence projected above the leafy mass of the plant .... *C. torreyana*
— Nutlets oblong-ovate to narrowly-lanceolate 8
8(7). Style reaching 1/4–3/4 the height of the nutlets; calyx densely appressed hispid-villous, commonly lacking conspicuous spreading bristles .... *C. gracilis*
— Styles almost reaching the nutlet-tips or surpassing them 9
9(8). Margin of the nutlet acute, at least above the middle .... *C. watsonii*
— Margin of the nutlets rounded or obtuse ... *C. fendleri*
10(5). Nutlets not all alike, heteromorphic 11
— Nutlets all alike, homomorphic 16
11(10). Mature calyx strongly appressed to the flattened rachis, decidedly gibbous, persistent; plants limited to Washington County .... *C. dumetorum*
— Mature calyx somewhat spreading, not at all gibbous . 12
12(11). Odd nutlet abaxial, surpassed by the style 13
— Odd nutlet axial, surpassing the style 15
13(12). Spikes bracteate througout; calyx persistent .... *C. micrantha*
— Spikes naked or nearly so; calyx deciduous 14
14(13). Pedicels slender, 1–4 mm long; plants rare in Washington County .... *C. racemosa*
— Pedicels stout and obscure .... *C. angustifolia*
15(12). Odd nutlet spinular-muricate; consimilar nutlets with a deeply impressed scar; calyx lobes conspicuously thickened .... *C. crassisepala*
— Odd nutlet more or less granulate to nearly smooth, consimilar nutlets with a shallowly impressed scar; calyx lobes moderately thickened .... *C. kelseyana*
16(10). Calyx circumscissle .... *C. circumscissa*
— Calyx not circumscissle 17
17(16). Style surpassing the nutlets .... *C. micrantha*
— Style equal to or shorter than the nutlets 18
18(17). Nutlets and calyx recurved .... *C. recurvata*
— Nutlets and calyx straight 19
19(18). Nutlet usually solitary, abaxial; gynobase reaching 1/3 to 1/2 the height of the nutlet .... *C. decipiens*
— Nutlets usually 4 20
20(19). Nutlets decidedly ovate, with low inconspicuous rounded tuberculations .... *C. ambigua*
— Nutlets lanceolate to narrowly lanceolate, the tuberculations very prominent 21
21(20). Stems spreading hirsute; calyx densely and conspicuously long hairy; plants limited to Washington County .... *C. barbigera*
— Stems strigose or mostly so 22
22(21). Nutlets spinular-muricate; fruiting calyx 4–6 mm long; plants of Box Elder County .... *C. scoparia*
— Nutlets verrucose or verrucose-muricate; fruiting calyx 6–11 mm long; southern Utah ...... *C. nevadensis*

Key 2.

Plants biennial or perennial.

1. Corolla tube elongate, distinctly surpassing the calyx; flowers usually heterostyled 2
— Corolla tube short, scarcely if at all surpassing the calyx; flowers not heterostyled 17

2(1).	Nutlets smooth and shining 3		—	Scar of nutlets conspicuously open; style exceeding nutlets 3–8 mm; calyx 4.5–7 mm long in anthesis ... 15
—	Nutlets more or less roughened 8		15(14).	Scar of nutlets conspicuously open and surrounded by a definite elevated margin .... *C. flavoculata*
3(2).	Corolla yellow 4		—	Scar of nutlets slightly open and with only an inconspicuous elevated margin if any 16
—	Corolla white 5		16(15).	Leaves linear-spatulate; nutlets sharply and deeply rugose; corolla tube 5.5–7 mm long, the limb funnelform and with low, broad fornices .... *C. tenuis*
4(3).	Inflorescence an elongate, cylindrical thyrse; nutlets lanceolate with acute margins, usually only 1 maturing .... *C. flava*		—	Leaves obovate to broadly oblanceolate; nutlets with rounded ridges and tubercles; corolla tube 7–10 mm long; limb more spreading and with fornices long-papilose .... *C. wetherillii*
—	Inflorescence consisting of a large terminal cluster with 1 or more branches remote, at maturity frequently stalked, in much smaller lateral clusters; nutlets broadly ovate, with winged margins, all 4 usually maturing .... *C. confertiflora*		17(1).	Nutlets smooth on their dorsal surface 18
5(3).	Inflorescence capitate, 0.1–0.4 dm long; corolla limb 6–8 mm broad, the tube little surpassing the calyx; nutlets lanceolate .... *C. capitata*		—	Nutlets more or less roughened, muricate, rugose or tuberculate at least dorsally 19
—	Inflorescence elongate, 0.4–4.4 dm long; corolla limb 8–17 mm broad, the tube distinctly surpassing the calyx, except only slightly so in *C. barnebyi*; nutlets ovate 6		18(17).	Fruit depressed-globular, the nutlets not in contact by their margins; style exceeding mature fruit 1–3 mm; corolla tube 2.5–3 mm long .... *C. cinerea*
6(5).	Upper surface of the leaves glabrous .... *C. semiglabra*		—	Fruit conical, ovoid or lanceolate, the nutlets in contact by their margins or nearly so; style exceeding mature fruit 5–6 mm; corolla tube 5–7 mm long; plants of Uintah County .... *C. barnebyi*
—	Upper surface of the leaves pubescent 7		19(17).	Ventral surface of the nutlets smooth or nearly so ... 20
7(6).	Corolla limb 13–17 mm broad, the tube salverform; crests at base of tube absent; nutlets 3–3.5 mm long; plants of the north end of the San Rafael Swell, Emery County .... *C. johnstonii*		—	Ventral surface of nutlets roughened 23
—	Corolla limb 8–11 mm broad, the tube funnelform; crests at base of tube conspicuous; nutlets 3.5–4.5 mm long; plants of Uintah County .... *C. barnebyi*		20(19).	Nutlets bordered by a conspicuous wing; robust plants, 5–10 dm tall, with long, ebracteate spikes .... *C. setosissima*
8(2).	Nutlets uniformly muricate or papillose, or occasionally with some inconspicuous ridges 9		—	Nutlets never conspicuously winged, sometimes with an acute margin; plants shorter 21
—	Nutlets more or less rugose or tuberculate, or with a few inconspicuous murications 11		21(20).	Corolla tube 7–9 mm long; calyx 6–9 mm long in anthesis; plants of the Uinta Basin and San Rafael Swell .... *C. rollinsii*
9(8).	Leaves oblanceolate, silvery strigose, the hairs only inconspicuously pustulate; corolla 7–10 mm long, the fornices much longer than wide ..... *C. fulvocanescens*		—	Corolla tube 2–6 mm long; calyx 2.5–6 mm long in anthesis 22
—	Leaves spatulate-oblanceolate to spatulate, green, the surface exposed between the hairs (or the upper surface glabrous), evidently pustulate at least on the lower surface; corolla 10–15 mm long, the fornices low and broad 10		22(21).	Nutlets scarcely or not at all muricate between the rugae; strictly erect, conspicuously hispid perennials from northeastern Utah .... *C. stricta*
10(9).	Upper surface of the leaves glabrous ... *C. creutzfeldtii*		—	Nutlets distinctly muricate or tuberculate between the rugae and near the margins; plants of western Utah .... *C. rugulosa*
—	Upper surface of the leaves hairy .... *C. jonesiana*		23(19).	Nutlets conspicuously muricate or, in *C. humilis*, also with a few irregular low ridges 24
11(8).	Ventral surface of the nutlets smooth or nearly so; leaves strongly pustulate hairy on both surfaces; plants of the Uinta Basin and San Rafael Swell .... *C. rollinsii*		—	Nutlets not exclusively muricate, but rugose or tuberculate 26
—	Ventral surface of the nutlets distinctly roughened .. 12		24(23).	Pubescence of the leaves silky-strigose or strigillose but not subtomentose or tomentose; plants of the Uinta Basin .... *C. breviflora*
12(11).	Leaves conspicuously pustulate on both surfaces; corolla tube 12–16 mm long; calyx-lobes 7–10 mm long at anthesis .... *C. longiflora*		—	Pubescence of the leaves subtomentose or tomentose, also setose or strigose 25
—	Leaves sparsely if at all pustulate on the upper surface; corolla tube 6–12 mm long; calyx-lobes 3.5–7 mm long in anthesis 13		25(24).	Plants 0.3–1 dm tall; leaves 0.5–2.5 cm long; calyx 2–2.5 mm long in anthesis; corolla tube 1.8–2.2 mm long; plants of Millard County .... *C. compacta*
13(12).	Inflorescence subcapitate, less than 5 cm long; corolla tube 10–12 mm long; margins of nutlets not in contact; plants less than 1.5 dm tall, of Duchesne, Emery, and Uintah counties .... *C. paradoxa*		—	Plants 0.4–2.5 dm tall; leaves 2.5 cm long or longer; calyx 3–5 mm long in anthesis; corolla tube 3–5 mm long; widespread .... *C. humilis*
—	Inflorescence more elongate, 0.5–3 dm long; corolla tube 5–10 mm long; margins of nutlets in contact or nearly so; plants usually over 1.5 dm tall 14		26(23).	Scar of nutlets open some distance above the base .. 27
			—	Scar of nutlets closed or nearly so, without a conspicuous triangular opening near the base 34
14(13).	Scar of nutlets surrounded by an elevated margin but tightly closed; style exceeding nutlets 1–2 mm; calyx 3.5–4 mm long in anthesis .... *C. bakeri*		27(26).	Scar somewhat constricted some distance below the middle of the open portion 28
			—	Scar triangular and not constricted below the middle 29

28(27). Hairs with pustulate bases present on both leaf surfaces; elevated margin of nutlet scar very prominent .... C. mensana
— Hairs with pustulate bases present only on the lower leaf surface; elevated margin evident but not very prominent .... C. osterhoutii
29(27). Elevated margin evident around the nutlet scar 30
— Elevated margin not evident around the nutlet scar . 31
30(29). Cymules elongating and so the inflorescence broad; biennial or short-lived perennials; nutlets with an evident dorsal ridge; plants of Washington County .... C. virginensis
— Cymules short and the inflorescence narrow; long-lived caespitose perennials; nutlets with only a slight dorsal ridge, if any .... C. abata
31(29). Style exceeding mature fruit 1.6 mm or more; plants usually taller than 1.3 dm; widespread ..... C. humilis
— Style not exceeding the mature fruit by more than 0.5 mm; plants usually less than 1.3 dm tall 32
32(31). Corolla tube 2–2.6 mm long; nutlets 2.3–3 mm long; plants of the Pink Limestone member of the Wasatch Formation in Garfield County .... C. ochroleuca
— Corolla tube 3–4 mm long; nutlets 3–3.5 mm long . . 33
33(32). Ventral surface of nutlets deeply rugose and tuberculate; plants of southern Utah .... C. abata
— Ventral surface of nutlets indefinitely muricate; plants of Daggett and Rich counties ..... C. caespitosa
34(26). Upper surface of the leaves uniformly appressed strigose and without pustulate hairs 35
— Upper surface of the leaves with 2 distinct kinds of hairs, some pustulate at the base 37
35(34). Nutlets sharply rugose and tuberculate; scar tightly closed and surrounded by an elevated margin .... C. bakeri
— Nutlets not so sharply rugose or tuberculate; scar not surrounded by an elevated margin 36
36(35). Corolla tube 2–2.5 mm long; style exceeding mature fruit by 1 mm or less; plants of the Pink Limestone member of the Wasatch Formation, Garfield County .... C. ochroleuca
— Corolla tube 3.5 mm long or longer; style exceeding fruit by more than 1 mm; plants of northeastern Utah .... C. sericea
37(34). Mature calyx exceeding the nutlets by 2–4 mm; stem usually solitary; plants of Grand County ...... C. elata
— Mature calyx exceeding the nutlets by 4–8 mm 38
38(37). Nutlets muricate and rugulose; plants widespread in western Utah .... C. humilis
— Nutlets tuberculate, scarcely if at all rugulose; plants of Box Elder County and the Uinta Basin 39
39(38). Plants 1–2 dm tall; calyx 5–7 mm long in anthesis; corolla tube 3.5–5 mm long; style exceeding mature fruit by 2 mm; plant a Uinta Basin endemic .... C. grahamii
— Plants mostly over 2 dm tall; calyx 2–3 mm long in anthesis; corolla tube 2–2.5 mm long; style exceeding mature fruit by less than 1 mm; plants of Box Elder County .... C. interrupta

Cryptantha abata Johnston Low Cryptanth. [*Krynitzkia depressa* Jones, not *C. depressa* A. Nels.]. Plants perennial, arising from a strong, woody taproot, 5–18 cm high; stems many, 2–15 cm long, strigose and weakly setose; leaves oblanceolate to spatulate, obtuse, strigose, setose and subtomentose, the petioles ciliate; inflorescence narrow, 2–8 cm long; calyx segments lanceolate to ovate, 2.5–4 mm long in anthesis, in fruit becoming 5–8 mm long, setose; corolla white, the tube 3–4 mm long, with crests at base of tube conspicuous, the fornices yellow, rounded, papillose, ca 0.5 mm long; limb 7–8 mm wide; style exceeding mature fruit 0.5–1 mm; nutlets ovate, 3–3.5 mm long, 2–2.5 mm wide, usually all 4 maturing, margins in contact, obtuse to acute, dorsal surface carinate, tuberculate, muricate and sometimes with low inconspicuous ridges, ventral surface deeply and irregularly rugose; scar open, triangular, surrounded by a slightly elevated margin. Sagebrush, mountain brush, pinyon-juniper, ponderosa pine, and spruce communities at 2500 to 3000 m in Beaver, Garfield, Iron, Kane, Millard, Piute, Sevier, and Wayne counties; Arizona and western Nevada; 25 (vi).

Cryptantha affinis (Gray) Greene Ally Cryptanth. [*Krynitzkia affinis* Gray; *C. geminata* Greene; *C. confusa* Rydb.; *C. eastwoodiae* St. John]. Usually sparsely branched herb 1–2 (4) dm tall; branches commonly few and ascending but occasionally many from the base, hispid or short-hirsute throughout; leaves narrowly to broadly oblanceolate, 1–4 (5) cm long, 2.5–6 (8) mm broad, few, short-hirsute, usually minutely pustulate, obtuse or rounded at tip, lowest pair clearly opposite; spikes geminate or solitary, usually 2 to 8, becoming 15 mm long, slender, remotely flowered, commonly with a few large leafy bracts below; corolla inconspicuous, 1–2 mm long; limb ca 1.5 mm broad; fruiting calyx 2.5–4 mm long, usually about as broad as long, laterally compressed, ascending; pedicels 0.5–1 mm long; mature calyx lobes lanceolate, somewhat connivent, not greatly surpassing the nutlets, midrib weakly thickened on the abaxial lobe, sparsely hirsute, the margins appressed hispid; nutlets 4, homomorphous, smooth or very finely granulate, shiny, brownish to greenish, frequently mottled, 1.8–2.5 mm long, ovate, obliquely compressed, back low-convex, margins rounded; groove evidently excentric, closed, simple or shortly and unequally forked at the base; gynobase short, stout, ca 1/2 the height of the nutlets; style evidently surpassed by nutlets or rarely equaling them. Sagebrush, mountain brush, pinyon-juniper, and aspen communities at 2050 to 2800 m in Cache, Daggett, Juab, Salt Lake, Sevier, Uintah, and Utah counties; Washington and western Montana, south to Wyoming and California; 8 (i). The most obvious characteristics of this species are the obliquely compressed nutlets which result in the excentric position of the groove.

Cryptantha ambigua (Gray) Greene Wilkes Cryptanth. [*Eritrichium muriculatum* var. *ambiguum* Gray; *Krynitzkia ambigua* Gray; *C. polycarpa* Greene; *C. multicaulis* A. Nels.; *C. ambigua* var. *robustior* Brand]. Stems usually loosely branched from the base, ascending, 1–2.5 dm tall, hirsute and somewhat strigose; leaves linear to narrowly lanceolate, 2–3 (5) cm long, 1–4 (5) mm broad, obtuse to acutish, usually somewhat appressed hispid-hirsute, the hairs commonly pustulate at base; spikes often solitary, 5–15 cm long, bractless, or with the lowermost flowers bracteate, commonly not projecting clear of the leafy mass of the plant and not clearly differentiated from the leafy branches; corolla 1–2 mm broad, inconspicuous; fruiting calyces ovate-oblong, 4–7 mm long, crowded or distant, the tube rounded-obconic at base, lobes linear or linear-lanceolate more or less connivent, midrib thickened, tawny hirsute, margins strigose-hirsute; pedicels 0.5–0.9 mm long; nutlets 4, broadly ovoid, 1.6–2 mm long, granulate and coarsely tuberculate or rarely tending to be smooth toward the base, back low-convex, sides obtuse and rounded, groove closed or somewhat open at the always broadly forked base; gynobase narrow, 1–1.2 mm long, 2/3 the height of the nutlets; style reaching 4/5 to equaling the of nutlets. Dry slopes and ridges, sagebrush, mountain brush, pinyon-juniper, and aspen communities at 2100 to 2900 m in Beaver, Box Elder, Daggett, Duchesne, Grand, Juab, Sevier, Uintah, and Wasatch counties; Washington and Montana, south to northern Colorado, Nevada, and California; 8 (0).

Cryptantha angustifolia (**Torr.**) **Greene** Narrowleaf Cryptanth. [*Eritrichium angustifolium* Torr.; *Krynitzkia angustifolia* Gray]. Stems diffusely branched from the base, 0.5–3 dm tall, canescent, villous- hirsute, with light ashy gray hairs, the lowest branches decumbent or ascending; leaves narrowly linear, 1.5–4 cm long, 1–2 mm wide, hispid or strigose, somewhat pustulate; spikes usually geminate, ca 5 cm long, rather dense, bractless or with 1–2 bracts near the base; corolla minute, the tube 1–2 mm long, the limb 1–2.5 mm broad; fruiting calyces ovate-oblong, 3–4 mm long, stiffly ascending, strongly biseriate, slightly asymmetrical; pedicels less than 0.5 mm long; mature calyx-lobes linear-lanceolate, slightly connivent, hispid on the thickened midrib, short villous on the margins, abaxial lobe longest and most hirsute; nutlets usually 4, heteromorphous, ovate-oblong, brown or plumbeous with pale tuberculations or rarely muricatations, the back convex, the face flattish, the margins round or somewhat angular; odd nutlet next to the abaxial calyx lobe, a little larger and more persistent than the similarly colored and shaped consimilar nutlets, these ca 1 mm long; groove slightly open above, broadening at the base; gynobase columnar, equaled by consimilar nutlets but shorter than odd nutlet. Creosote, Joshua tree, and other warm desert communities below 930 m in Washington County; Nevada and Arizona to Baja California, Sonora and western Texas; 1 (i).

Cryptantha bakeri (**Greene**) **Payson** Baker Cryptanth. [*Oreocarya bakeri* Greene]. Biennial or short-lived perennials, 1–4 dm tall; stems 1–4 (6), 0.5–1.5 dm long, spreading setose-hirsute; leaves oblanceolate, obtuse, mostly basal, 3–6 (8) cm long, 0.5–1.2 (2) cm wide, strigose and spreading setose dorsally, pustulate, the ventral surface uniformly strigose and with few or no pustulate hairs; inflorescence narrow, 0.6–2.5 (3) dm long, setose-hirsute, the foliar bracts evident, slightly surpassing the individual cymes; calyx segments broadly lanceolate or ovoid, in anthesis 3.5–4 mm long, in fruit becoming 6–8 mm broad; nutlets ovate-lanceolate, 2.5–3 mm long, 1.5–2 mm wide, 3 to 4 usually maturing, margins obtuse, nearly in contact, the dorsal surface deeply and sharply rugose; scar closed, surrounded by a definitely elevated margin; style exceeding mature fruit 1–2 mm; n = 12. Sagebrush, oak brush, pinyon-juniper, mountain brush, ponderosa pine, aspen, and spruce fir communities at 1650 to 2700 m in Carbon, Duchesne, Garfield, Grand, Kane, San Juan, and Sevier counties; Colorado, New Mexico and Arizona; 35 (viii).

Cryptantha barbigera (**Gray**) **Greene** Bearded Cryptanth. [*Eritrichium barbigerum* Gray, type from St. George; *Krynitzkia barbigera* Gray; *K. mixta* Jones]. Stems erect, 1–4 dm high, solitary or several from the base, with branches strictly ascending or spreading, very bristly and sparsely if at all strigose, except in the inflorescence; leaves oblong to lance-linear, obtuse, 1–5 (7) cm long, 3–7 (13) mm broad, hirsute, more or less pustulate; spikes geminate or rarely solitary or ternate, naked, as much as 15 cm long; corolla inconspicuous, the limb 1–2 mm broad; fruiting calyx ascending, 5–10 mm long, narrowly ovoid-oblong to oblong-lanceolate, symmetrical, deciduous; pedicels 0.3–0.7 mm long, villous; mature calyx lobes lanceolate to linear-lanceolate, connivent above with recurving tips, the margin conspicuously long white-villous, midrib thickened and hirsute, abaxial lobe slightly the longest; nutlets 1–4 homomorphous, lance-ovoid, 1.5–2.5 mm long, strongly verrucose, usually brownish, the back convex, the edges obscurely angled or rounded, groove opened or closed but base gradually dilated to form a triangular areola; gynobase narrow, 2/3 to 3/4 the height of the nutlets; style reaching to or slightly beyond the nutlet tips. Joshua tree, blackbrush, and pinyon-juniper communities below 1500 m in Washington County; southern Nevada, California, Baja California, Arizona, New Mexico, Texas, and northern Sonora; 20 (vii).

Cryptantha barnebyi **Johnston** Barneby Cryptanth. Perennial, 1.5–3.5 dm tall; stems stout, erect, several, 0.8–1.2 dm long, conspicuously yellow-hispid; leaves oblanceolate, thick, acute, 5–9 cm long, 0.5–1.4 cm wide, coarsely appressed hispid-pustulate on both sides, and with some finer hairs beneath, the petioles conspicuously ciliate; inflorescence narrow, 1–1.5 dm long, densely yellowish hispid, foliar bracts evident to conspicuous; calyx segments lanceolate, in anthesis 5–7 mm long, in fruit becoming 8–13 mm long, yellowish hirsute; corolla white or light yellow, the tube 5–7 mm long, the crests at base of tube very conspicuous, the fornices yellow, emarginate, distinctly papillose, 0.5 mm long, the limb 8–11 mm wide; style exceeding mature fruit 5–6.5 mm; nutlets ovate, 3.5–4 mm long, 2.5–3 mm wide, all 4 usually maturing, the margins in contact, acute, smooth and glossy on both surfaces; scar closed, straight, and without an elevated margin. On white barren shale knolls of the Green River Formation in shadscale, rabbitbrush, sagebrush, and pinyon-juniper communities at 1850 to 2400 m in Uintah (type 30 mi S of Ouray) County; endemic; 45 (viii).

Cryptantha breviflora (**Osterh.**) **Payson** Short-flower Cryptanth. [*Oreocarya breviflora* Osterh., type N of Jensen]. Long-lived perennials, 1.6–3 dm tall; stems several, slender, 0.7–1.7 dm long, densely white setose at the base, strigose above; leaves oblanceolate to spatulate, 2.5–9 cm long, 0.4–1.4 cm wide, clustered at the ends of the branched caudices, the apices obtuse, the dorsal surface densely and uniformly silky-strigose with many very small pustules, ventral surface similar but with fewer pustules; inflorescence in flower narrow but becoming broad and open at maturity, 0.6–2.7 dm long, setose; calyx segments linear-lanceolate, 4.5–6 mm long in anthesis, in fruit becoming 7–9 mm long, setose; corolla white, 3.5–4.5 mm long, the crests at base of tube evident, the fornices yellow, rounded, ca 0.5 mm long, the limb 8–12 mm wide; nutlets lanceolate, 3.4–4 mm long, 2–2.5 mm wide, fewer than 4 nutlets maturing, the margins in contact, knifelike, the dorsal surface uniformly muricate or tuberculate, the ventral surface similar; scar open, narrowly triangular, margin not elevated; style exceeding mature fruit by 2 mm or less. Mostly heavy clay soils of the Morrison and Duchesne River formations in salt desert shrub, sagebrush, rabbitbrush, pinyon-juniper and mountian brush communities at 1600 to 2450 m in Duchesne, Uintah, and Wasatch counties; Colorado (?); a Uinta Basin endemic; 70 (vi).

Cryptantha caespitosa (A. Nels.) Payson Tufted Cryptanth. [*Oreocarya caespitosa* A. Nels.]. Densely caespitose or mat-forming perennials, 0.5–1.5 dm tall; stems 1 to many, arising from a much-branched woody caudex, 0.2–0.9 dm long, weakly setose and appressed strigose; leaves oblanceolate to spatulate, 1–3 cm long, 0.3–0.7 cm wide, pubescence of two kinds, strigose and appressed setose, becoming tomentulose toward the petiole; inflorescence narrow, 0.3–1 dm long; foliar bracts inconspicuous; calyx segments lanceolate, in anthesis 3–4 mm long, in fruit becoming 5–8 mm long, strigose and weakly setose, also somewhat tomentulose; corolla white, the tube 3–4 mm long, the crests at base of tube conspicuous, the fornices yellow, rounded, ca 0.5 mm long, the limb 4–7 mm wide, the margins acute, in contact, the dorsal surface with low rounded rugae, also tuberculate, and with numerous murications between the ridges, the ventral surface muricate; scar open, narrowly triangular, the margin not elevated; style equaling or 0.5 mm longer than mature fruit. Forb-grass, pinyon-juniper, mountain brush, limber pine, and spruce-fir forests at 1950 to 3120 m in Daggett and Rich counties; southern Wyoming; 5 (0).

Cryptantha capitata (Eastw.) Johnston Head Cryptanth. [*Oreocarya capitata* Eastw.]. Perennial, 1.5–2.7 dm tall; stems weak, 1–several, 12–24 cm long, appressed setose; leaves linear or very narrowlly oblanceolate, 3–8 cm long, 3–5 (8) mm wide, dorsal surface appressed setose-pustulate, ventral surface uniformly strigose and without pustules; inflorescence capitate, or with 1 or 2 glomerules below the terminal cluster, 1–4 cm long, spreading white setose; calyx segments linear-lanceolate, 7–9 mm long in anthesis, in fruit becoming 11–16 mm long, conspicuously setose-pustulate; corolla white, the tube 9–12 mm long, the crests at base of tube conspicuous, the fornices yellow, emarginate, ca 1 mm long, papillose, the limb 6–8 mm wide; nutlets lanceolate, 4–5 mm long, 2–3 mm wide, 2–4 usually maturing, the margins in contact, knifelike, both surfaces glossy-smooth; scar closed, straight, and without an elevated margin; style exceeding mature fruit 4–5 mm. Sandy to loamy soils in pinyon-juniper, sagebrush, and mountain brush communities at 1700 to 2520 m in Garfield, Kane, San Juan, and Wayne counties; northern Arizona; 6 (0).

Cryptantha cinerea (Torr.) Cronq. James Cryptanth. [*C. jamesii* (Torr.) Payson]. Perennials, 1–6 dm tall; stems 1 to many, 4–40 cm long, glabrous to conspicuously hirsute; leaves linear to broadly oblanceolate, obtuse to acute, 2–15 cm long, 0.2–1.5 cm wide, glabrous to hirsute, usually pustulate dorsally, the ventral surface lacking pustules or the pustules very inconspicuous; inflorescence open, the cymules usually elongating, tomentose to setose-hirsute; floral bracts inconspicuous to very conspicuous; calyx segments ovate-lanceolate, acute, in anthesis 3–4 mm long, in fruit 5–7 mm long, subtomentose to setose-hirsute; pedicels 1–3 mm long; corolla white, the tube 2.5–3 mm long, the crests at base of tube conspicuous, the fornices light yellow, emarginate, 0.5–1 mm long, the limb 5–8 mm broad; fruit oblate-ovoid, 1–4 nutlets maturing, ovate-lanceolate, 2–2.5 mm long, 1.5–2 mm wide, the acute margins not in contact, both surfaces smooth and glossy; scar straight, closed, extending from the base to near the apex; style exceeding mature fruit 1–3 mm.

1. Upper surface of the leaves glabrous, the petioles not ciliate-margined or the leaves tufted at the base ..... *C. cinerea* var. *pustulosa*
— Upper surface of the leaves pubescent; the petioles ciliate-margined; leaves tufted at the base 2
2(1). Stems simple; inflorescence capitate or nearly so; plants in our flora mostly limited to Navajo blow sand in Garfield, Kane, and Washington counties .... *C. cinerea* var. *arenicola*
— Stems branched from the base and above; inflorescence otherwise 3
3(2). Stems prostrate or weakly ascending, to 2 dm long; inflorescence conspicuously leafy-bracteate; plants of western Beaver, Iron, and Washington counties ..... *C. cinerea* var. *abortiva*
— Stems erect or ascending; inflorescence only weakly leafy bracteate 4
4(3). Stems freely branched at the base and upwards, leafy, often somewhat decumbent, sparsely appressed-hairy; cymules usually notably elongating .... *C. cinerea* var. *jamesii*
— Stems branched only near the base, pubescence mostly of dense cinereous and spreading setose hairs; leaves mostly in a dense basal tuft; cymules usually more congested .... *C. cinerea* var. *cinerea*

Var. abortiva (Greene) Cronq. [*Oreocarya abortiva* Greene; *C. jamesii* var. *abortiva* Payson]. Rocky or gravelly slopes and ridges in sagebrush, and pinyon-juniper communities at 1650 to 2600 m in Beaver, Iron, and Washington counties; Nevada and California; 3 (ii).

Var. arenicola Higgins & Welsh [*C. jamesii* var. *multicaulis* authors, not (Greene) Payson]. Usually on Navajo blow sands in sagebrush, pinyon-juniper, oak, mountain brush, and ponderosa pine communities at 1650 to 2600 m in Garfield, Kane, and Washington counties; Arizona; 20 (iii).

Var. cinerea [*Oreocarya cinerea* Greene; *C. jamesii* var. *cinerea* Payson; *C. jamesii* var. *setosa* (Jones) Johnston ex Tidestr.; *C. jamesii* var. *multicaulis* (Greene) Payson; *Krynitzkia multicaulis* var. *setosa* Jones, type from near Cove Fort, Millard County]. Sandy to clayey soils in sagebrush, pinyon-juniper, mountain brush, and ponderosa pine communities at 1650 to 2480 m in Beaver, Emery, Garfield, Iron, Kane, Piute, Sanpete, Sevier, Washington and Wayne counties; Arizona and Nevada; 60 (xii).

Var. jamesii (Payson) Cronq. [*Oreocarya disticha* Eastw., type from San Juan County; *C. jamesii* var. *disticha* Payson; *C. jamesii* Payson]. Mostly sandy or gravelly soils in mixed desert shrub, sagebrush, pinyon-juniper, and ponderosa pine communities at 1720 to 2550 m in Emery, Garfield, Kane, San Juan, and Wayne counties; Utah to Colorado, Arizona, New Mexico, South Dakota, Wyoming, Kansas, Oklahoma, Texas, and Mexico; 95 (xiii). At one time I thought var. *disticha* was distinct, but additional evidence indicates that var. *disticha* and var. *jamesii* are the same. Var. *jamesii* also introgresses with var. *pustulosa* especially in Garfield and Wayne counties.

Var. **pustulosa** (Rydb.) Higgins [*Oreocarya pustulosa* Rydb., type from Elk Mts., San Juan County; *C. jamesii* var. *pustulosa* (Rydb.) Harrington]. Sandy soils in mixed desert shrub, sagebrush, pinyon-juniper, and mountain brush communities at 1350 to 2300 m in Garfield, Kane, San Juan, and Wayne counties; Utah to southwestern Colorado, northwestern New Mexico and northeastern Arizona; 28(ii). Because var. *pustulosa* introgresses with var. *jamesii*, I cannot accept it at specific rank.

Cryptantha circumscissa **(H. & A.) Johnston** Opening Cryptanth. [*Lithospermum circumscissum* H. & A.; *C. depressa* A. Nels.]. Stems few to many from the base, strigose, more or less branched above, often forming a dense hemispherical mass 2–10 cm high, the outer ones often decumbent; leaves oblanceolate, 3–15 mm long, 1–2 mm broad, obtuse, the surface siliceous especially toward the pale base, strigose or short-hispid, obscurely pustulate; flowers in axils of foliaceous bracts in short somewhat indefinite racemelike clusters; corolla inconspicuous, 1–2 (3) mm broad; fruiting calyx 2.5–4 mm long, oblong-ovoid, united to near the middle, at maturity the upper half falling away by a circumscission just below the sinuses, basal part persistent, cupulate, appressed-hispid; mature calyx lobes narrowly linear-lanceolate, firm-herbaceous, scarcely ribbed, more or less hispid; pedicels obscure, ca 0.5 mm long; nutlets 4, homomorphous, or with the abaxial one slightly longer, smooth or inconspicuously muricate, triangular-ovoid or oblong-lanceolate, 1.2–1.7 mm long, back flattened, especially near the apex, margins angled, groove closed and forked at base; gynobase ca 2/3 height of nutlets; style equalling or barely exceeded by nutlets. Warm deseret shrub, blackbrush, salt desert shrub, sagebrush, pinyon-juniper, and mountain brush communities at 900 to 2500 m in Beaver, Box Elder, Duchesne, Emery, Garfield, Iron, Juab, Kane, Millard, Piute, San Juan, Sevier, Tooele, Uintah, and Washington counties; Washington to Idaho, Wyoming, Colorado, Arizona, and Baja California; South America; 67(vii).

Cryptantha compacta **Higgins** Mound Cryptanth. Densely caespitose perennials, 0.3–1 dm tall; stems numerous, arising from a woody root, 1–4 cm long, tomentose below, weakly strigose above; leaves oblanceolate to spatulate, obtuse, 5–15 (20) mm long, 2–4 mm wide, the dorsal surface with appressed setose-pustulate bristles, also densely strigose or subtomentose, the ventral surface similar but with fewer pustulate hairs, the petioles tomentose; inflorescence narrow, nearly capitate, 1–5 cm long; foliar bracts evident but not conspicuous; calyx segments lanceolate, 2–2.5 mm long in anthesis, in fruit becoming 3.5–4.5 mm long, densely white setose and tomentose; corolla white, the tube 1.8–2.2 mm long, the crests at base of tube evident, the fornices yellow, rounded, papillose, about 0.5 mm long, limb 4.5–5.5 mm wide; nutlets lance-ovate, acute, 2.5–3 mm long, 1.5–1.8 mm wide, only 1 or 2 maturing, the dorsal surface muricate or weakly tuberculate-rugulose, the ventral surface muricate, scar open, subulate to narrowly triangular, an elevated margin lacking; style equaling or shorter than mature fruit. Salt desert shrub and mixed desert shrub communities at 1900 to 2250 m in Beaver, Millard (type from 8 mi W of Desert Experimental Range), and Tooele counties; endemic; 6 (ii).

Cryptantha confertiflora **(Greene) Payson** Golden Cryptanth. [*Oreocarya confertiflora* Greene; *Krynitzkia leucophaea* var. *alata* Jones, type from Johnson, Kane County]. Perennial herbs, 1.7–4.3 dm tall; stems 1–7, slender, 1.5–2.5 dm long, tomentose at the base, strigose and setose upward; leaves linear to oblanceolate, 3–12 cm long, 2–16 mm wide, acute, the dorsal surface densely strigose and appressed setose with pustulate bases, the ventral surface uniformly strigose and with few or no pustules; inflorescence subcapitate, 3–20 cm long, strigose and with twisted setose hairs, bracts inconspicuous; calyx segments linear-lanceolate, in anthesis 6–8 mm long, in fruit becoming 10–14 mm long, strigose and spreading setose; corolla yellow, the tube 9–13 mm long, the fornices broad, emarginate, ca 1 mm long, the crests at base of tube evident or sometimes lacking, the limb 8–10 mm wide; nutlets triangular or ovate, 3.5–4 mm long, 2.5–3 mm wide, usually all 4 maturing, the margins narrowly winged, in contact, surfaces smooth and glossy; scar straight, closed, and lacking an elevated margin; plants distinctly heterostyled. Warm desert shrub, sagebrush, pinyon-juniper, mountain brush, ponderosa pine, and bristlecone pine communities at 900 to 2850 m in Emery, Garfield, Iron, Juab, Kane, Millard, San Juan, Sanpete, Sevier, Washington, and Wayne counties; Arizona, Nevada, and California; 110 (xii).

Cryptantha crassisepala **(T. & G.) Greene** Plains Cryptanth. [*Eritrichium crassisepalum* T. & G.]. Erect or widely spreading herb 5–15 cm high, stems normally numerous, loosely ascending, branched, hirsute to hispid; leaves oblanceolate, 2–3 (6) cm long, 3–4 (6) mm wide, rounded or obtuse, thickish, hirsute, pustulate, the upper scarcely reduced; spikes solitary or rarely geminate, naked or few bracted below, 5–8 (15) cm long, frequently produced from the lowest axils; corolla inconspicuous, 1–1.5 mm wide; fruiting calyces 6–7 (10) mm long, oblong-ovoid, somewhat asymmetrical, becoming distant below; mature calyx lobes linear-lanceolate, connivent above, midrib strongly thickened and indurate, hispid, the margins short-hispid; pedicels short, 0.5–1.2 mm long; nutlets 4 (1 or 2 rarely aborted), distinctly heteromorphous; odd nutlet next to the axial calyx lobe, persistent, 2–2.5 (3) mm long, brownish, ovoid, acute, finely granulate and spinular-muricate; consimilar nutlets readily deciduous, 1.2–1.5 (2) mm long, oblong-ovoid, thickish coarsely tuberculate, very obscurely if at all granulate, groove usually dilated and commonly excavated to form an areola occupying much of the ventral face of the nutlet; gynobase narrowly oblong, usually ca 2/3 height of consimilar nutlets; style equaling or a trifle exceeding the consimilar nutlets, surpassed by odd nutlet. Mixed desert shrub, blackbrush, sagebrush, and pinyon-juniper communities at 1200 to 1820 m in Carbon, Emery, Garfield, Grand, Kane, Millard, San Juan, Sevier, Washington, and Wayne counties; Colorado, Arizona, Texas, and Mexico; 94 (x). Our plants, which have a very narrow corolla limb, belong to var. **elachantha** Johnston.

Cryptantha creutzfeldtii **Welsh** Creutzfeldt-flower. Perennial herbs, 7–23 cm tall; stems many, arising from a multicipital caudex and stout black-barked taproot, the caudex branches 2–12 cm long clothed by marcescent leaf bases; leaves narrowly spatulate to oblanceolate, acute to obtuse, 2–8 cm long, 2–9 mm wide, coarsely appressed setose-pustulate (appearing ashy gray on leaves of previous years) below, the petioles long-setose; inflorescence an interrupted thyrse, with few to several clusters below the terminal subcapitate one; calyx segments lance-linear in anthesis, 6–8 mm long, in fruit 9–13 mm long, densely long-setose with yellowish, ascending bristles; corolla white, the tube 8–11 mm long, campanulate in the throat, the fornices low and broad, the crests at base of tube lacking, the limb 10–13 mm wide; nutlets lanceolate, 4–5 mm long, muricate; scar narrow, open, without an elevated margin. Mancos Shale in shadscale and matatriplex communities at 1600 to 1980 m in western Carbon and Emery (type from N of Emery) counties; endemic; 10 (0).

Cryptantha decipiens (Jones) **Heller** Beguiling Cryptanth. [*Krynitzkia decipiens* Jones]. Stems 1–4 dm high, slender, loosely branched, strigose and frequently short-hispid; leaves few, linear, obtuse, 1–3 cm long, 1–3 (4) mm wide, strigose and sometimes hispid, minutely pustulate; spikes slender, geminate or rarely solitary or ternate, bractless, usually becoming loosely flowered, 4–10 cm long; corolla inconspicuous in ours, 0.8–1 mm broad; fruiting calyx ovoid to ovoid-oblong, strictly ascending, asymmetrical, 2.5–7 (9) mm long, deciduous, sessile; mature calyx lobes lance-linear, decidedly connivent above with the tips frequently spreading or even recurving, midrib thickened and usually evidently hirsute, the margins strigose, abaxial lobe evidently the longest and most hirsute; nutlets 1 or rarely 2, next the abaxial calyx lobe, ovoid-lanceolate, 1.5–2.4 mm long, muriculate-granulate to tuberculate, usually brownish, the back convex, the sides rounded; groove open or closed but always dilated below to form a definite areola; gynobase 1/2–1/3 as high as nutlet; style much surpassed by nutlet, 1/2–2/3 the height of nutlet. Creosote, Joshua tree, blackbrush, and pinyon-juniper communities below 1500 m in Washington County; Nevada, Arizona, and California; 15 (viii).

Cryptantha dumetorum Greene Greene Cryptanth. [*Krynitzkia dumetorum* Greene]. Laxly branched closely strigose herb; stems at first erect but later commonly much elongated and sprawling or climbing among bushes; leaves lanceolate, thickish, 2–4 cm long, 2–4 (8) mm wide, sparsely appressed hirsute-villous, closely pustulate below and finely so above; spikes solitary or geminate, usually remotely flowered, 5–15 cm long, occasionally with foliaceous bracts toward base, the rachis brittle and tortuously flattened; corolla inconspicuous, ca 1 mm broad; fruiting calyx closely appressed to rachis, 2–3 mm long, very asymmetrical, not deciduous, the base very oblique and downwardly gibbous on axial side; mature calyx lobes connivent and subequal, the 3 abaxial lobes lanceolate, somewhat strigose, with the thickened midribs deflexed hirsute, the 2 axial lobes partly united, hirsute only on outer margins; nutlets 4, heteromorphous, granulate and muricate, odd nutlet persistent, axial, broadly lanceolate, 2–3 mm long, the base much developed and distorting the calyx, the groove open and broad; consimilar nutlets 1.5–2 mm long, deciduous, lanceolate, the groove closed or very narrow; gynobase narrow; style shortly surpassed by nutlets or reaching to their tips. Creosote bush, Joshua tree, and blackbrush communities below 1250 m in Washington County; Arizona, Nevada, and California; 2 (ii).

Cryptantha elata (**Eastw.**) **Payson** Tall Cryptanth. [*Oreocarya elata* Eastw.]. Biennial or short-lived perennials, 3–5 dm tall; stems 1–6, erect, stout, weakly setose with spreading white hairs, 9–15 cm long; leaves oblanceolate to spatulate, 2–5 cm long, 4–13 mm wide, the apices acute to obtuse, the blade abruptly tapering to the narrow petiole, the dorsal surface strigose and appressed setose, the ventral surface strigose, both surfaces pustulate; inflorescence spreading or open in age, 1.5–3.5 dm long, setose; foliar bracts inconspicuous; calyx segments lanceolate, in anthesis 3–4.5 mm long, in fruit becoming 7–8 mm long, hirsute; corolla white, the tube 3.5–5 mm long, the fornices yellow, rounded, papillose, ca 1 mm long, the crests at base of tube well developed, the limb 6–8 mm wide; nutlets lance-ovate, 4–4.5 mm long, 2–2.5 mm wide, usually all 4 maturing, the margins in contact, the dorsal surface densely tuberculate and somewhat rugulose, also covered with dense minute papillae, the ventral surface similar but the roughenings less prominent; scar closed, or narrowly open at the base, without an elevated margin; style exceeding mature fruit 0.5–2 mm. Mancos Shale in salt desert shrub communities at 1550 to 1700 m in Grand County; Colorado; 8 (ii).

Cryptantha fendleri (**Gray**) **Greene** [*Krynitzkia fendleri* Gray; *Eritrichium hispidum* var. *leiocarpum* Kuntze; *C. ramulosissima* A. Nels.; *C. wyomingensis* Gand.]. Stems erect, usually evident throughout and bearing lateral branches mostly above the middle, sometimes rather bushy-branched from near the base, 1–5 dm tall, more or less densely hispid and frequently appressedly so; leaves narrowly oblanceolate to linear, 2–5 cm long, 2–4 mm broad, appressed-hirsute, often pustulate on the lower surface; spikes solitary or geminate, 2–12 cm long, loosely flowered, bractless or rarely bracted below; corolla inconspicuous, ca 1 mm broad; fruiting calyces ovate-oblong, 4–5 (7) mm long, ascending, slightly asymmetrical, obscurely biserial; pedicels ca 0.5 mm long; mature calyx lobes linear to lance-linear, usually loosely connivent with the tips somewhat spreading, the midrib hirsute, thickened, the margins strigose; nutlets 4, homomorphous, or sometimes reduced to 1–3, smooth, somewhat shiny, lanceolate, acuminate, 1.5–2 mm long, convex on dorsal face, rounded or somewhat obtuse on the sides; groove closed above, but opening below into a definite deltoid areola at the base; gynobase subulate, twice the length of the style, at least 2/3 the height of the nutlets; style equaling or barely surpassing the nutlets. Sagebrush, pinyon-juniper, mountain brush, and ponderosa pine communities at 1850 to 2800 m in Daggett, Duchesne, Emery, Garfield, Iron, Kane, Sevier, Tooele, Uintah, Wasatch, Washington, and Wayne counties; Alberta and Saskatchewan south to Nebraska, New Mexico, Arizona to Oregon and Washington; 38 (viii).

Cryptantha flava (**A. Nels.**) **Payson** Yellow Cryptanth. [*Oreocarya flava* A. Nels.; *O. torva* A. Nels., type from Price]. Perennial, 1.3–4 dm tall; stems many from a multiple caudex, 8–26 cm long, densely setose, white-hairy at the base, becoming setose and strigose upward; leaves narrowly oblanceolate to nearly linear, acute, 2–9 cm long, 3–8 mm wide, the dorsal surface strigose and appressed setose with pustulate hairs, the ventral surface almost uniformly strigose and with the pustules less conspicuous; inflorescence narrow to somewhat open, 0.5–2.5 dm long, conspicuously yellow-setose, the floral bracts inconspicuous; calyx segments linear, in anthesis 8–10 mm long, in fruit becoming 9–12 mm long, densely setose, with yellowish hairs; pedicels 3–5 mm long in fruit; corolla yellow, the tube 9–12 mm long, the crests at the base of the tube absent or nearly so, the fornices yellow, truncate, emarginate, 1–1.5 mm long, the limb

8–10 mm wide; nutlets lanceolate, 3.4–4 mm long, 1.9–2.2 mm wide, 1 or 2 usually maturing, margins acute, in contact when more than 1 nutlet matures, both surfaces of nutlet smooth and glossy; scar straight, closed, elevated margin lacking; style exceeding mature fruit 3–7 mm (heterostyled); n = 12. Mixed desert shrub, salt desert shrub, sagebrush, Mormon tea, and pinyon-juniper communities at 1280 to 2600 m in Carbon, Daggett, Duchesne, Emery, Garfield, Grand, Kane, San Juan, Sevier, Uintah, and Wayne counties; Wyoming, Colorado, New Mexico, and Arizona; 250 (xxii).

Cryptantha flavoculata (A. Nels.) Payson Yellow-eye Cryptanth. [*Oreocarya flavoculata* A. Nels.]. Caespitose perennials, 10–37 cm tall; stems 1 to several, slender, 5–20 cm long, strigose and spreading setose with slender bristles; leaves linear-oblanceolate to spatulate, obtuse or sometimes acute, 3–11 cm long, 3–15 mm wide, densely strigose and weakly setose, the dorsal surface conspicuously pustulate, the ventral surface with a few pustules or sometimes silky-strigose; inflorescence narrow or sometimes slightly open and lax, 5–30 cm long; foliar bracts evident but not conspicuous; calyx segments in anthesis linear-lanceolate, 5–6 mm long, in fruit becoming 8–10 mm long and becoming broadly lanceolate to ovate; corolla white or pale-yellow, the tube 7–10 mm long, the crests at base of tube lacking, the fornices yellow, minutely papillose, 1–2 mm long, the limb 8–12 mm wide; nutlets lanceolate to lance-ovate, 2.5–3.5 mm long, 1.8–2 mm wide, usually all 4 maturing, with margins obtuse, in contact or slightly separated, the dorsal surface muricate, tuberculate, and with conspicuous ridges, sometimes nearly foveolate, the ventral surface tuberculate, rarely with ridges; scar open, constricted near the middle and surrounded by a highly elevated margin; style exceeding mature fruit 4–8 mm (heterostyled). Sagebrush, pinyon-juniper, mountain brush, and ponderosa pine communities at 1650 to 3160 m in all counties except Cache, Davis, Morgan, Piute, Rich, Salt Lake, and Weber counties; Wyoming and Colorado to New Mexico, Arizona, Nevada, and California; 315 (xxxv).

Cryptantha fulvocanescens (Wats.) Payson Yellow-hair Cryptanth. [*Krynitzkia echinoides* Jones, type from Paria Canyon; *C. fulvocanescens* var. *echinoides* (Jones) Higgins]. Densely caespitose perennials from a strongly lignified taproot, 8–30 cm tall; stems many from a multiple caudex, 5–13 cm long, white-hairy at the base, setose-hispid upward; leaves oblanceolate to spatulate, acute to obtuse, 1.5–7 cm long, 4–12 mm wide, uniformly strigose, with pustules mainly confined to the dorsal surface; inflorescence narrow or somewhat open at maturity, 3–19 cm long, white or yellowish setose; foliar bracts inconspicuous; calyx segments linear, 4–6 mm long in anthesis, in fruit becoming 9–13 mm long, densely white or yellowish setose; pedicels 2–10 mm long; corolla white, the tube 7–11 mm long, the crests at base of tube evident or lacking, the fornices yellow, emarginate or rounded, 0.7–1.3 mm long, the limb 7–9 mm broad, nutlets lance-ovate, 3.5–4.5 mm long, 2–3 mm broad, 1 or 2 usually maturing, margins acute to obtuse,, in contact when more than 1 nutlet matures, with both surfaces densely and uniformly muricate; scar open or nearly closed, an elevated margin lacking; style exceeding mature fruit 3–7 mm. Mancos Shale in salt desert shrub, blackbrush, sagebrush, and pinyon-juniper communities at 1500 to 2150 m in Carbon, Emery, Garfield, Grand, Kane, San Juan, and Wayne counties; Colorado, New Mexico and Arizona; 78 (ix).

Cryptantha gracilis Osterh. Slender Cryptanth. [*C. hillmannii* Nels. & Kennedy; *C. gracilis* var. *hillmannii* Munz & Johnston]. Stems slender, 10–20 cm high, erectly branched, usually solitary, densely spreading short-hirsute; leaves linear or narrowly oblanceolate, the lower 1.5–3 cm long, 1–3 mm broad, the upper usually much reduced, obtuse or rounded, ascendingly short-hirsute, minutely pustulate; spikes solitary or geminate, usually dense, 1–2 mm broad; fruiting calyx ovate, divaricate, 2–2.8 mm long, early deciduous, the base evidently conical, sessile; mature calyx lobes lanceolate, densely appressed tawny hispid-villous, with tips erect and midrib slightly thickened and inconspicuously short-hispid; nutlets 1 or rarely 2–3 and unequally developed, lanceolate, 1.5–2 mm long, 0.8–1 mm broad, smooth and shiny, acute, with back nearly flat and sides rounded at least near the apex; groove mostly opened to above the middle and scarcely forked below; style reaching to 2/3–3/4 height of nutlet. Mixed desert shrub, sagebrush, pinyon-juniper, and mountain brush communities at 1200 to 2800 m in Beaver, Box Elder, Carbon, Daggett, Emery, Garfield, Juab, Kane, Millard, Piute, Salt Lake, San Juan, Sanpete, Sevier, Uintah, Utah, and Washington counties; Idaho and Colorado to Arizona and California; 45 (vii).

Cryptantha grahamii Johnston Graham Cryptanth. Long-lived perennial from a thick woody taproot, 15–25 cm tall; stems several, 4–12 cm long, weakly spreading setose; leaves spatulate to oblanceolate, 2–4.5 cm long, 4–10 mm wide, conspicuously setose pustulate on both surfaces, with some finer pubescence beneath; inflorescence narrow, 0.4–1 dm long setose; foliar bracts evident but not conspicuous; calyx segments lanceolate, in anthesis 5–7 mm long, in fruit becoming 7–9 mm long, very setose; corolla white, the tube 3.5–5 mm long, constricted at the middlle, the crests at base of tube evident, the fornices yellow, emarginate, papillose, 0.5–1 mm long, the limb 11–15 mm wide; nutlets lanceolate, 3–3.8 mm long, 1.7–2 mm wide, 2–4 maturing, the margins in contact, acute, both surfaces with inconspicuous, small, low-rounded tubercles or some of these confluent into short irregular ridges; scar straight, open, narrowly linear, the margin not elevated; style coarse, exceeding mature fruit 1.8–2.2 mm. Green River Shale in mixed desert shrub, sagebrush, pinyon-junper, and mountain brush communities at 1520 to 2250 m in Duchesne and Uintah (type from mouth of Sand Wash) counties; endemic; 49 (ix).

Cryptantha humilis (Greene) Payson Dwarf Cryptanth. [*Oreocarya humilis* Greene; *O. shantzii* Tidestr., type from Great Salt Lake; *C. humilis* var. *shantzii* (Tidestr.) Higgins; *O. commixta* Macbr., type from Juab, Juab County; *C. humilis* var. *commixta* (Macbr.) Higgins; *C. nana* var. *ovina* Payson; *C. humilis* var. *ovina* (Payson) Higgins; *O. dolosa* Macbr., type from Logan]. Perennials, more or less densely caespitose, 5–30 cm tall; stems many, arising from the ends of the branched caudex, 2–15 cm long, strigose to spreading setose; leaves oblanceolate to spatulate, 1–6 cm long, 2–12 mm wide, strigose, setose or subtomentose, pustulate on both surfaces; inflorescence narrowly cylindrical to open and lax, 2–18 dm long, tomentose to conspicuously setose; calyx segments linear-lanceolate, in anthesis 2.5–4.5 mm long, in fruit becoming 6–13 mm long, setose or tomentose; corolla white, the tube 2.5–4.5 mm long, the crests at base of tube conspicuous to nearly obsolete; fornices yellow, more or less papillose, rounded, ca 0.5 mm long; limb 7–10 mm wide; nutlets lanceolate to ovate-lanceolate, 3–4.5 mm long, 1.8–3.2 mm wide, 1 to 4 of them maturing, the margins in contact, acute to somewhat obtuse, the dorsal surface muricate, tuberculate, or somewhat rugulose, the ventral surface indistinctly muricate or tuberculate; scar open, triangular, margin not elevated; style from shorter than to longer than the mature fruit by 2.5 mm. Salt desert shrub, mixed desert shrub, sagebrush, pinyon-juniper, and mountain brush communities at 1500 to 2540 m in all counties except Daggett, Rich, Summit, and Wasatch; Montana to Idaho, Colorado, Arizona, and California; 226 (xx). This is the most variable of all Utah cryptanthas. In a previous treatment of the Utah borages I recognized four varieties, but additional evidence does not support segregation of infraspecific taxa. Pubescence throughout the range is quite constant, however, some plants in the Uinta Basin are more hispid than otherwise. The style length varies from slightly shorter than to 2 mm longer than the mature nutlets. The nutlets can be densely and uniformly muricate to muricate-rugose. These characters exist in varying degree, but lack geographical correlation, except those plants in the Uinta Basin that tend to be shorter, more hispid, with short styles and quite rugulose nutlets. These might best be designated **C. humilis** var. **nana** (Eastw.) Higgins [*O. nana* Eastw.; *C. nana* (Eastw.) Payson].

***Cryptantha inaequata* Johnston** Unequal Cryptanth. Stems erect or ascending, 3–4 dm high, branched throughout or sometimes the basal branches elongated and simple or nearly so, hispid and strigose or hirsute toward the base; leaves linear to narrowly-oblanceolate, 2–4 cm long, acute, often becoming more or less convolute, more or less hispid, pustulate, especially beneath; spikes geminate or solitary, 4–12 cm long, at times few bracted below; corolla inconspicuous, the tube shorter than the calyx, the limb 1–2.5 mm broad; fruiting calyx ovoid-oblong, 2.5–3 mm long, ascending; pedicels less than 0.5 mm long; mature calyx lobes lanceolate, midrib thickened and hirsute, villous-ciliate on the margins; nutlets 4, heteromorphous, triangular-ovate, dark with small pale tuberculations, the margins acute, the groove closed above but below gradually enlarging into a shallow triangular areola; odd nutlet ca 1.7 mm long, somewhat persistent, slightly paler than the others, next to the abaxial calyx lobe; consimilar nutlets ca 1.3 mm long; gynobase equaling consimilar nutlets but surpassed by odd nutlet; style much surpassing the nutlets. Warm desert shrub, blackbrush, and mixed desert shrub communities below 1250 m in Kane, San Juan, and Washington counties; Utah to northern Arizona, and Nevada; 5 (i). This species is closely allied with *C. angustifolia* and has been placed in synonymy by Cronquist. However, *C. angustifolia* has smaller nutlets with the margins rounded and never knifelike or with a narrow winged-margin as in *C. inaequata*. *C. inaequata* is also a more robust plant. I find no basis on which to combine them.

***Cryptantha interrupta* (Greene) Payson** Basin Cryptanth. [*Oreocarya interrupta* Greene]. Long lived perennial, 1.7–6 dm tall; stems few to several, slender, 1–3.5 dm long, strigose and weakly setose with slender white hairs; leaves oblanceolate to spatulate, obtuse, 1.5–7 cm long, 4–12 mm wide, lower surface densely strigose and appressed setose-pustulate, the upper surface more finely strigose, the setose hairs less conspicuous and, pustules fewer; inflorescence narrow, interrupted, 1–2.5 dm long, densely setose, the cymes somewhat elongating at the top, the foliar bracts evident on lower part of stem; calyx segments lanceolate, 2–3 mm long in anthesis, in fruit becoming 5–8 mm long, setose; corolla white, the tube 2–2.5 mm long, the crests at base of tube conspicuous, the fornices light yellow, slightly emarginate, ca 0.5 mm long, the limb 5–6 mm wide; style exceeding mature fruit by less than 1 mm; nutlets lanceolate, 3.3–3.6 mm long, 1.7–2 mm wide, all four usually maturing, the margins in contact, acute and both surfaces tuberculate with scattered rounded tubercles, or sometimes nearly smooth; scar slightly open, linear, margin not elevated. On white tuffaceous slopes in rabbitbrush, salt desert shrub, sagebrush, and pinyon-juniper communities at 1720 to 2150 m in Box Elder County; Idaho, and Nevada; 6 (0).

***Cryptantha johnstonii* Higgins** Johnston Cryptanth. Caespitose perennial 1–2.5 dm tall; stems several, arising from the branched caudex, 6–13 cm long, very weakly strigose; leaves oblanceolate, the apices obtuse to acute, 2–6.5 cm long, 4–10 mm wide, the dorsal surface strigose with conspicuous pustulate hairs; inflorescence somewhat open, 5–20 cm long; floral bracts evident but not conspicuous, 1–2 cm long; calyx segments linear-lanceolate, in anthesis 5–6 mm long, in fruit becoming 8–10 mm long, strigose and spreading white-setose; pedicels 0.5–1 mm long; corolla white, the tube 12–15 mm long, flaring in the throat, the crests at base of tube lacking, the fornices yellow, 1–1.5 mm long, emarginate, papillose, the limb 13–17 mm broad; nutlets ovate, 3–3.5 mm long, 2.3–2.7 mm wide, usually all 4 maturing, the margins acute and knifelike, in contact, both surfaces smooth and glossy; scar straight, closed, an elevated margin lacking; style exceeding mature fruit 3–8 mm (heterostyled). Mixed desert shrub and pinyon-juniper communities at 1800 to 1950 m in Emery (type from Huntington road, ca 15 mi W. of US 50–6) County; endemic; 12 (iv).

***Cryptantha jonesiana* (Payson) Payson** Jones Cryptanth. [*Oreocarya jonesiana* Payson, type from San Rafael Swell]. Coarse caespitose perennials, 5–15 cm tall; stems many, arising from a thick-woody multiple caudex, 2–7 cm long, setose; leaves spatulate, 1–4 cm long, 4–13 mm wide, coarsely appressed setose-pustulate; leaf bases also setose with dense white hairs; inflorescence narrow, somewhat capitate, with 1–3 flowers in the axils of the bracts below the terminal cluster; calyx lobes lanceolate to nearly linear, in anthesis 5–7 mm long, in fruit becoming 7–10 mm long, densely setose, with ascending yellowish bristles; corolla white, the tube 10–15 mm long, campanulate in the throat, the fornices low and broad, papillose, the crests at base of tube lacking, the limb 9–13 mm wide; nutlets lanceolate, 3.5–4.5 mm long, densely and uniformly muricate, or with a few short low ridges; scar narrow, open, and without an elevated margin; style exceeding mature fruit 4–6 mm. Mixed desert shrub and pinyon-juniper communities at 1700 to 2280 m in Emery County; endemic; 25 (vi).

***Cryptantha kelseyana* Greene** Kelsey Cryptanth. [*C. pattersonii* authors, not (Gray) Greene]. Spreading or ascending hirsute herb 5–25 cm high; stems 1 to several, hirsute and also hispid-strigose; leaves linear or narrowly oblanceolate, 1.5–3 (4) cm long, 2–4 mm wide, rounded or obtuse, thickish, hirsute, pustulate, the upper ones scarcely reduced; spikes usually solitary, 4–9 cm long, naked or with a few bracts near the base; corolla minute, 1–2 mm broad; fruiting calyces 4–6 mm long, ovate-oblong, spreading, loose or dense, somewhat asymmetrical; pedicels short but definite, ca 0.8 mm long; mature calyx lobes linear, weakly connivent above, midrib thickened, hirsute to hispid, margins inconspicuously villous-strigose; nutlets 4, heteromorphous; odd nutlet next to the axial calyx lobe, broadly lance-ovoid, 2–2.6 mm long, smoothish or granulate or muriculate-granulate or rarely somewhat tuberculate, standing off slightly from the gynobase; consimilar nutlets lance- or oblong-ovoid, 1.8–2.3 mm long, coarsely tuberculate and usually granulate, darker than the odd nutlet, the sides rounded; groove narrow or closed, near base abruptly dilated to form a small triangular areola; gynobase subulate, a little longer than the style, 1/2 to 2/3 height of consimilar nutlets; style surpassed by odd nutlet and just surpassing or even exceeded by consimilar ones. Salt desert shrub, mixed desert shrub, sagebrush, pinyon-juniper, and mountain brush communities at 1580 to 2500 m in Carbon, Daggett, Duchesne, Grand, Millard, San Juan, Sevier, Tooele, Wasatch, and Weber counties; Saskatchewan and Montana south to Wyoming and Colorado; 40 (ii).

Cryptantha longiflora (A. Nels.) Payson Long-flower Cryptanth. [*Oreocarya longiflora* A. Nels.]. Short-lived perennial or possible biennials, 0.8–3 (5) dm tall; stems 1 to several, 5–10 cm long, setose and spreading hirsute; leaves spatulate, obovate or oblanceolate, 2–7 cm long, 5–15 mm wide, both surfaces strigose and strongly hirsute, pustulate; inflorescence broad and open, 7–25 cm long, setose; foliar bracts inconspicuous; calyx segments linear-lanceolate, in anthesis 7–10 mm long, in fruit becoming 10–16 mm long, setose; corolla white, the tube 12–14 mm long, lacking crests at base of tube, the fornices yellow, emarginate, broad, rounded, papillose, 0.5–1 mm long, the limb 9–11 mm wide; style exceeding mature fruit 4–9 mm (heterostyled); nutlets lanceolate-ovate 3–4 mm long, 2.2–2.6 mm wide, 2–4 maturing, both surfaces with tubercles and low rounded ridges; scar straight, closed or very narrowly open, with a slightly elevated margin. Salt desert shrub and pinyon-juniper communities at 1450 to 1830 m in Emery and Grand counties; Colorado; a Navajo Basin endemic; 15 (iii).

Cryptantha mensana (Jones) Payson Carbon Cryptanth. [*Krynitzkia mensana* Jones, type from Emery]. Short-lived perennials, 10–15 (20) cm tall; stems 1 to several, 5–12 cm long, setose-hirsute, with some finer strigose hairs beneath; leaves oblanceolate to spatulate, obtuse, 3–8 cm long, 5–14 mm wide, lower surface setose with pustulate hairs, also finely strigose, ventral surface strigose, less setose and with fewer pustules; inflorescence broad, open, 4–12 cm long, setose; foliar bracts well developed; calyx segments lanceolate, in anthesis 4–5 mm long, in fruit becoming 7–8 mm long, setose-hirsute; corolla white, the tube 3–4 mm long, the crests at base of tube lacking or nearly so, the fornices yellow, rounded, slightly papillose, ca 0.5 mm long, the limb 5–8 mm wide; nutlets ovoid, 3–3.5 mm long, 1.6–1.9 mm wide, the margins obtuse, not in contact, the dorsal surface rugose, tuberculate and somewhat muricate, the ventral surface conspicuously tuberculate; scar open, constricted at the middle and surrounded by a high elevated margin; style exceeding mature fruit 1.5–2 mm. Salt desert shrub, mixed desert shrub, sagebrush, pinyon-juniper, mountain brush, and ponderosa pine communities at 1750 to 2790 m in Carbon, Duchesne, Emery, Garfield, Grand, Sevier, Utah, and Wayne counties; endemic; 72 (viii).

Cryptantha micrantha (Torr.) **Johnston** Dye Cryptanth. [*Eritrichium micranthum* Torr.; *Eremocarya muricata* Rydb., type from southern Utah]. Slender, dichotomously branched herb 5–15 cm high, drying brownish; root and lower parts of the stem stained with a purple dye; leaves oblong-oblanceolate, 3–7 mm long, 0.8–1.4 mm wide, whitish strigose or short-hirsute, rounded at apex, uppermost scarcely reduced and extending through the inflorescence; spikes numerous, solitary or geminate, densely flowered and strongly unilateral, leafy-bracted throughout, 1–4 cm long; corolla inconspicuous to medium-sized, the limb 0.5–2.5 mm broad; fruiting calyx ovoid-oblong, 1.8–2.5 mm long, slightly asymmetrical, decidedly biseriate, the base broadly conical; pedicels 0.5–0.8 mm long; mature calyx lobes oblong-lanceolate, broad, erect, hirsute, the midrib not evidently thickened; nutlets 4, 1–1.3 mm long, smooth or tuberculate, with 1 sometimes a little longer and more persistent than the others, the groove extending full length of nutlet, narrow, scarcely broadened at base; gynobase subulate, nearly as long as the calyx, much surpassing the nutlets and bearing at its summit the sessile stigma. Creosote bush, warm desert shrub, blackbrush, saltbush, sagebrush, and pinyon-juniper communities below 1880 m in Garfield, Kane, Millard, San Juan, Tooele, and Washington counties; Texas and New Mexico to Oregon, California, and Baja California; 65 (xiv).

Cryptantha nevadensis Nels. & Kennedy Nevada Cryptanth. [*C. leptophylla* Rydb., type from St. George]. Stems slender, 1–5 dm high, 1 to several, erect or usually flexuous, appressed short-strigose, mostly laxly branched; leaves linear-oblanceolate to linear, acute or obtuse, 1–4 cm long, 1–5 (7) mm broad, not numerous, appressed short-hispid, more or less pustulate; spikes geminate or ternate, terminal, also scattered along the stem on short, slender, axillary branches, occasionally bracted toward the base, congested and somewhat glomerate or elongate and to 15 cm long; corolla inconspicuous, the limb 1–2 mm broad; fruiting calyx oblong-ovoid to lanceolate, 5–12 mm long, ascending, slightly asymmetrical; pedicels ca 0.5 mm long; mature calyx lobes lanceolate or linear, connivent above with the slender tips usually recurving, margins more or less villous and hispid on the somewhat thickened midrib; nutlets 4, homomorphous, verrucose or muriculate toward the tip, lanceolate to lance-ovoid, 2–2.9 mm long, back convex, the margins somewhat angled, groove open or closed, dilated below into a small areola; gynobase narrow; style reaching to or almost to the tips of the nutlets. Creosote bush, other warm desert shrub, sagebrush, and pinyon-juniper communities below 2130 m in Kane, San Juan, and Washington counties; Nevada and Arizona to California and Baja California; 16 (viii).

Cryptantha ochroleuca Higgins Yellowish Cryptanth. Low caespitose perennial, 2–13 cm tall; stems several, 1–4 cm long, strigose and weakly setose; leaves linear-oblanceolate to oblanceolate, the apices acute or sometimes obtuse, 1–2.5 cm long, 1–3 mm wide; basal leaves uniformly and densely strigose, sparsely setose, the petioles white-hairy; cauline leaves strigose and with some setose-pustulate bristles; inflorescence narrow, 2–7 cm long, weakly setose; calyx segments linear-lanceolate, 2–2.5 (3) mm long in anthesis, in fruit becoming 4–6 mm long, setose; corolla pale yellow, the tube 2–2.5 mm long, the crests at base of tube conspicuous, the fornices yellow, rounded, ca 0.3 mm long, the limb 4–5 mm wide; nutlets lanceolate, 2.5–3 mm long, 1.4–1.6 mm wide, usually only 1 maturing, the margin acute, the dorsal surface irregularly rugose with low rounded ridges, the ventral surface only slightly uneven; scar open, narrowly triangular, extending 3/4 the length of the nutlet and lacking an elevated margin; style scarcely surpassing the mature fruit. Pink Limestone member of the Wasatch Formation in pinyon-juniper, ponderosa pine, and bristlecone pine communities at 2300 to 2850 m in Garfield (type from Red Canyon) County; endemic; 9 (iii).

Cryptantha osterhoutii (**Payson**) **Payson** Osterhout Cryptanth. [*Oreocarya osterhoutii* Payson]. Densely caespitose perennials, 7–12 cm tall; stems slender, many, arising from the densely branched multiple caudex, 3–6 cm long, strigose and spreading setose; leaves spatulate to oblanceolate, obtuse, 1–3 cm long, 3–8 mm wide, the dorsal surface strigose and appressed setose, pustulate, the ventral strigose, not pustulate or the pustules inconspicuous; petioles ciliate; inflorescence open, 3–8 cm long, weakly white-setose; foliar bracts inconspicuous; calyx segments lanceolate, in anthesis 2.5–4 mm long, in fruit becoming 5–6.5 mm long, strigose and spreading white-setose; corolla white, the tube 2–3 mm long, the crest at base of tube usually evident but poorly developed, the fornices yellow, broad, emarginate, papillose, ca 0.5 mm long, the limb 5–7 mm wide; nutlets lanceolate, 2.7–3.2 mm long, 1.8–2.2 mm broad, usually less than 4 nutlets maturing, the obtuse margins not in contact, the dorsal surface carinate, sharply tuberculate and rugose, the ventral sharply tuberculate; scar open, constricted above the base, an elevated margin evident but not conspicuous; style exceeding mature fruit 0.2–0.7 mm. Blackbrush, mixed desert shrub, oak brush, salt bush, and pinyon-juniper communities at 1520 to 2000 m in Garfield, Grand, San Juan, and Wayne counties; Colorado; a Navajo Basin endemic; 29 (iv).

Cryptantha paradoxa (**A. Nels.**) **Payson** Paradox Cryptanth. [*Oreocarya paradoxa* A. Nels.]. Caespitose perennial, 4–12 cm tall; stems 1 to many, slender, 2–8 cm long, subtomentose near the base, weakly setose above; leaves oblanceolate to spatulate, usually folded, obtuse, 1.5–4 cm long, 2–4 mm wide, the dorsal surface with appressed setose-pustulate hairs, the ventral one uniformly strigose and without pustulate hairs; petioles ciliate; inflorescence subcapitate, 1–4 cm long, setose; foliar bracts inconspicuous; calyx segments linear-lanceolate in anthesis 5–6 mm long, in fruit becoming 6–8 mm long, weakly setose; corolla white, usually with a yellow tube 10–12 mm long, lacking crests at base of tube, the fornices yellow, broad, emarginate, papillose, ca 0.5 mm long, the limb 10–12 (15) mm wide; nutlets lanceolate, turgid, 2–3 mm long, 1.3–1.6 mm wide, all 4 usually maturing, the margins acute to obtuse, not in contact, the dorsal surface densely tuberculate and conspicuously rugose, the ventral surface tuberculate, also somewhat rugulose; scar open, constricted below the middle, the margin elevated; style exceeding mature fruit 4–9 mm. Salt desert shrub, mixed desert shrub, sagebrush, and pinyon-juniper communities at 1200 to 2150 m in Emery, Duchesne, and Uintah counties; Colorado and New Mexico; a Colorado Plateau endemic; 38 (v).

Cryptantha pterocarya (**Torr.**) **Greene** Wing-nut Cryptanth. Stems erect, branched throughout with ascending branches, 1–5 dm high, short-hirsute with either appressed or ascending slender strigose hairs; leaves linear or the reduced uppermost ones lanceolate or oblong, strigose, 1–2.5 (4) cm long, 1–3 (5) mm broad, obtuse, conspicuously pustulate below but usually finely so above; spikes geminate or rarely ternate or solitary, naked or bracted below, 2–6 (12) cm long, becoming loosely flowered; corolla inconspicuous, 0.5–1 (2) mm broad; fruiting calyces distinctly accrescent, (2) 3–5 mm long, symmetrical, ascending on short pedicels 0.5–1 mm long; mature calyx lobes ovate to lanceolate, somewhat connivent, thin, the margins more or less tawny, appressed hispid, the midrib slightly thickened and sparsely hispid; nutlets 4, homomorphous and all winged, or heteromorphous with axial nutlet wingless; body of nutlet broad or narrow, entire or crenate or lobed, the wing extending completely around the nutlet or only down the sides, groove open or closed (even in the same plant) and dilated below into an open excavated areola; gynobase slender, ca 2/3 height of nutlets; style subulate, slightly surpassing or somewhat surpassed by the wing-margin of the nutlets but always exceeding the body proper.

1. Nutlets heteromophous, the axial one wingless ......
   .... *C. pterocarya* var. *pterocarya*
— Nutlets homomorphous, all winged ....
   .... *C. pterocarya* var. *cycloptera*

Var. *cycloptera* (**Greene**) **Macbr.** [*Krynitzkia cycloptera* Greene]. Nutlets homomorphous, all winged, otherwise like the typical material. Creosote bush, blackbrush, sand sage, mixed desert shrub, and pinyon-juniper communities at 800 to 1330 m in Kane, San Juan, and Washington counties; California east to New Mexico and west Texas; 9 (ii).

Var. *pterocarya* [*Eritrichium pterocaryum* Torr.; *E. p.* var. *pectinatum* Gray, type from southern Utah]. Creosote bush, blackbrush, mixed desert shrub, salt desert shrub, sagebrush, and pinyon-juniper communities at 1200 to 2150 m in Beaver, Grand, Juab, Kane, Millard, San Juan, Sevier, Tooele, Uintah, and Washington counties; Washington and Idaho to New Mexico, Arizona, and Sonora Mexico; 116 (xv).

Cryptantha racemosa (**Wats.**) **Greene** Baja Cryptanth. [*Eritrichium racemosum* Wats.]. Long-lived annual, often decidedly suffruticose near the base, 1–10 dm high; stems single with numerous ascending branches or many and diffusely branched younger parts, green, strigose and commonly hirsute, the epidermis at length exfoliating leaving the older woody stems glabrous and brown; leaves oblanceolate, acute, hirsute, pustulate, the early ones 3–6 cm long, 6–12 mm broad; racemes apparently forked and paniculately disposed, minutely bracteate, 3–15 cm long; corolla very inconspicuous, the limb ca 1 mm wide; fruiting calyces oblong-ovoid, 2–4 mm long, slightly asymmetrical, ascending, tardily deciduous; pedicels usually well developed, 1–4 mm long, slender, frequently nodding; mature calyx lobes lance-linear, somewhat strigose, hirsute along the thickened midrib; nutlets 4, heteromorphous, triangular-ovate, the acute tips slightly outcurved; groove open or closed above but below broadening into a shallow broadly triangular areola; odd nutlet next to the abaxial calyx lobe, 1–2 mm long, somewhat persistent, finely muricate or tuberculate, light or dark, consimilar nutlets 0.8–1.5 mm long, acute, tending to be very narrowly wing-margined, dark with pallid tuberculations; gynobase subulate, 3/4 the length of odd nutlet and subequal to the consimilar nutlets; style much surpassing the nutlets. Creosote bush, Joshua tree, and other warm desert shrub communities below 950 m in Washington County; Nevada, Arizona, and southwestward to Baja California; 3 (0).

Cryptantha recurvata Cov. Recurved Cryptanth. Stems branched from the base, slender, ascending or decumbent at base, 1–3 dm high, strigose; root often dye-stained; basal leaves oblanceolate or spatulate, 1.5–2 cm long, the cauline ones remote, linear or lanceolate, 5–10 mm long, rounded or obtuse, rather finely appressed-hispid and minutely pustulate; spikes bractless, slender, loose, 2–10 cm long, solitary or geminate; corolla inconspicuous, ca 2 mm long, subtubular, not exserted, lobes short; fruiting calyx slender, asymmetrical, bent and recurved, 3–4 mm long, tardily deciduous, sessile; mature calyx lobes linear, the midrib somewhat thickened and hirsute, rarely merely strigose, the axial lobe longest, thickest, and most hirsute; ovules 2; nutlet 1, subpersistent, oblong-lanceolate, curved inwardly, dull brownish, granulate-muricate, next to the axial calyx lobe, the edges obtusish; groove somewhat oblique, narrow or closed, opening into a small basal areola; gynobase slender, ca 1/2 the length of the matured nutlet and subequal to the abortive ones; style commonly much surpassed by nutlet. Creosote bush, blackbrush, yucca, mixed desert shrub, sagebrush, and pinyon-juniper communities at 800 to 1500 m in Beaver, Box Elder, Emery, Garfield, Grand, Juab, Kane, Millard, San Juan, Sevier, Tooele, Uintah, and Washington counties; Oregon to Arizona, Nevada, and southern California; 35 (v).

Cryptantha rollinsii Johnston Rollins Cryptanth. Biennial herbs, 1–3.5 dm tall; stems 1 to several, 2–15 cm long, setose; leaves clustered at the base, gradually reduced upward, oblanceolate to spatulate, obtuse to acute, 2–5 cm long, 5–15 mm wide, setose and hispid, pustulate on both surfaces; inflorescence narrow to somewhat open at maturity, cylindrical or obovoid, racemes in dense glomerules, 3- to 6-flowered, hispid, 5–20 cm long; calyx segments linear, in anthesis 7–8 mm long, in fruit becoming 8–10 mm long, hispid; corolla white, campanulate, the tube 7–9 mm long, the crests at base of tube evident, the fornices yellow, papillose, ca 0.5–1 mm long, the limb 7–8 mm wide; plants slightly heterostyled; nutlets lanceolate, 3–4 mm long, 1–1.5 mm wide, obscurely rugulose and tuberculate on the dorsal surface, smooth ventrally; scar closed and without an elevated margin. Salt and mixed desert shrub, sagebrush, pinyon-juniper, and mountain brush communities at 1650 to 2250 m in Carbon, Duchesne, Emery, Uintah (type from 22 mi S of Ouray), and Wayne counties; Wyoming and Colorado; 89 (xiv).

Cryptantha rugulosa (**Payson**) **Payson** Tubercle Cryptanth. [*Oreocarya rugulosa* Payson, type from Fish Springs, Tooele County]. Biennial or short-lived perennial, 12–30 cm tall; stems slender, 1 to several, 8–16 cm long, spreading setose-hispid; leaves oblanceolate to spatulate, obtuse to acute, strigose and conspicuously setose-hispid, pustulate on both surfaces; inflorescence 2–20 cm long, hispid; foliar bracts inconspicuous; calyx segments linear-lanceolate, in anthesis 4–5 mm long, in fruit becoming 7–9 mm long, strigose and spreading hirsute; corolla white, the tube 3–4 mm long, with conspicuous crests at base of tube, the fornices rounded, distinctly papillose, ca 0.5 mm long, the limb 5–7 mm broad; nutlets lanceolate, 2.8–3.2 mm long, 1.3–1.7 mm wide, all 4 usually maturing, the margins in contact, acute, the dorsal surface with short low ridges, also somewhat tuberculate, the ventral surface smooth or nearly so; scar open, subulate, without an elevated margin; style exceeding mature fruit 1–1.5 mm. Salt desert shrub, rabbitbrush, horsebrush, mixed desert shrub, sagebrush, and pinyon-juniper communities at 1460 to 2620 m in Beaver, Box Elder, Juab, Millard, Piute, Sanpete, Sevier, and Tooele counties; eastern Nevada; a Great Basin endemic; 59 (viii).

Cryptantha scoparia A. Nels. Desert Cryptanth. Stems with several to many stiffly erect branches 10–35 cm tall, closely strigose and frequently also sparsely hispid; leaves linear to lance-linear, 2–4 cm long, obtuse, 1–3 mm broad, strictly ascending, strigose or appressed hirsute, finely pustulate; spikes stiff, bractless, solitary or geminate, 2–10 cm long; corolla inconspicuous, ca 1 mm wide, the tube subequal to the calyx; fruiting calyx ovate-oblong, 5–6 mm long, strictly ascending, subsessile, subsymmetrical, becoming rather obscurely biserial at maturity, 5–10 mm distant; mature calyx lobes linear-lanceolate, rather stiff, somewhat connivent above the tips, slightly spreading, midrib thickened and hirsute, the margins ascending-hispidulous, the axial lobe slightly longer; nutlets 4, homomorphous, lanceolate to broadly so, 1.8–2.2 mm long, antrorsely spinulose-muricate, especially near the apex, the margins and base rounded, the groove narrow and forked below where occasionally open to a small triangular areole; gynobase subulate, ca 3/4 heig' of nutlets; style reaching to tip of nutlets. Mixed desert shrub, sagebrush, and pinyon-juniper communities at 1450 to 1900 m in Box Elder and Weber counties; Washington, Oregon, Idaho, Wyoming, and Nevada; 2 (0).

Cryptantha semiglabra **Barneby** Pipe Springs Cryptanth. Erect perennials, 2–3 (4) dm tall; stems 1 to several, 9–18 (20) cm long, retrorsely strigose and weakly spreading setose; leaves oblanceolate, acute, 3–7 cm long, 3–6 cm wide, dorsal surface appressed setose-pustulate, the ventral glabrous, the old leaf bases long white-hairy; inflorescence narrow to somewhat open, 4–13 dm long; foliar bracts slightly surpassing the cymes, 1.5–2 cm long; calyx segments lanceolate, in anthesis 5–8 mm long, in fruit becoming 10–13 mm long, setose; pedicels 1–2 mm long; corolla white, the tube 10–12 mm long, the crests at base of tube conspicuous; fornices yellow, rounded, 1–1.2 mm long, obscurely papillose; limb 8–10 mm wide; nutlets ovate, 3.5–4 mm, 2–2.5 mm wide, usually all 4 maturing, the margins acute, in contact, both surfaces smooth and glossy; scar closed, elevated margin lacking, style surpassing the mature fruit 5–7 mm. Clay soils in mixed desert shrub, sagebrush, and pinyon-juniper communities at 1500 to 1730 m in Washington County; Arizona; 1 (0).

Cryptantha sericea (**Gray**) **Payson** Silky Cryptanth. [*Krynitzkia sericea* Gray]. Perennials, 1.5–4.3 (5) dm tall; stems 1 to several, branched from near the base, 5–12 (30) cm long, setose with spreading hairs; leaves oblanceolate to spatulate, obtuse, 2.5–10 (15) cm long, 0.5–2 cm wide, the dorsal surface strigose and weakly appressed to spreading setose, pustulate, the ventral silky-strigose, lacking pustules or these very inconspicuous; inflorescence narrow to somewhat open, 5–32 cm long, setose-hispid, foliar bracts 2–5 cm long; calyx segments lanceolate, 2.5–4 mm long in anthesis, in fruit becoming 6–8 mm long; pedicels 0.5–1 mm long; corolla white, the tube 2.5–3.5 mm long, the crests at base of tube conspicuous; fornices yellow, broad, depressed, 0.5–0.6 mm long; limb 7–9 mm wide; nutlets lanceolate, 2.5–3.5 mm long, 1.5–2 mm wide, usually all 4 maturing, the margins acute or narrowly winged, in contact, dorsal surface with low rounded tuberculations, also somewhat rugulose and muriculate, the ventral similar but the markings less evident; scar straight, closed and without an elevated margin; style exceeding mature fruit 0.5–1.3 mm. Salt desert shrub, sagebrush, pinyon-juniper, mountain brush, aspen, and spruce-fir communities at 1700 to 2950 m in Carbon, Daggett, Duchesne, Grand, Rich, Summit, and Uintah counties; Wyoming and Colorado; 125 (x).

Cryptantha setosissima (Gray) Payson Fish Lake Cryptanth. [*Eritrichium setosissimum* Gray, type from Fish Lake]. Biennial or short-lived perennial, 3–10 dm tall; stems usually 1–3, erect, 1.7–5 (10) dm long, hirsute; leaves clustered at the base, reduced upward, oblanceolate, the apices obtuse to acute, 3–13 cm long, 0.5–1.5 cm wide, setose and with some finer twisted pubescence beneath, with pustulate hairs numerous on both surfaces; inflorescence broad-topped due to the elongation of the scorpioid racemes, 1–5 dm long; calyx segments broadly lanceolate, 4–6 mm long in anthesis, in fruit becoming 9–11 mm long, setose, and strigose; corolla white, the tube 3–5 mm long, constricted above the ovary by the conspicuous ring of crests, the fornices yellow, emarginate, 0.5 mm long, the limb 7–9 mm broad; nutlets ovate, 5–6 mm long, 3.5–4.5 mm wide, papery, with a broad, winged margin, the dorsal surface muricate, and inconspicuously rugulose or tuberculate, the ventral smooth or nearly so; scar straight, narrow, slightly open, elevated margin lacking; style exceeding mature fruit 1–2 mm. Sagebrush, pinyon-juniper, oakbrush, mountain brush, aspen, and spruce-fir communities at 1950 to 3250 m in Garfield, Piute, Sevier, and Washington counties; Arizona and Nevada; 36 (xiii).

Cryptantha stricta (Osterh.) Payson Erect Cryptanth. [*Oreocarya stricta* Osterh.; *O. williamsii* A. Nels., type from Flaming Gorge, Daggett County]. Strict perennial, 10–39 cm tall; stems 1 to several, 4–20 cm long, strigose and conspicuously setose-hispid; leaves mostly basal, reduced upward, oblanceolate, acute, 2–7 cm long, 4–9 cm wide, retrorsely strigose and spreading setose-hispid, pustulate; inflorescence narrow, interrupted below the terminal cluster, 0.5–2 dm long, setose-hispid; foliar bracts conspicuous, especially near the base; calyx segments lanceolate, 4–6 mm long in anthesis, in fruit becoming 7–9 mm long, setose-hispid; corolla white, the tube 3–4 mm long, with conspicuous crests at base of tube, the fornices yellow, rounded, papillose, the limb 7–10 mm wide; nutlets lanceolate to elliptic, 3–3.5 mm long, 1.5–2 mm wide, usually all 4 maturing, the margins in contact, knifelike, with definite transverse ridges and also somewhat tuberculate or nearly smooth dorsally, ventrally smooth or nearly so; scar open, very narrowly linear, an elevated margin lacking; style exceeding mature fruit 1–1.5 mm. Sagebrush, pinyon-juniper, mountain brush, aspen, and spruce-fir communities at 1780 to 3100 m in Daggett, Summit, and Uintah counties; Wyoming and Colorado; 25 (iii).

Cryptantha tenuis (Eastw.) Payson Slender Cryptanth. [*Oreocarya tenuis* Eastw., type from Courthouse Wash, Grand County]. Caespitose perennials, 13–25 cm tall; stems slender, 1 to many, 8–12 cm long, strigose and weakly spreading setose; leaves linear-spatulate, mostly basal, obtuse, 2–5 cm long, 3–6 mm wide, dorsally strigose, weakly spreading setose, and conspicuously pustulate, ventrally uniformly strigose and without pustules; inflorescence narrow, interrupted, 6–14 cm long, weakly setose; foliar bracts inconspicuous; calyx segments linear-lanceolate, in anthesis 4.5–6 mm long, in fruit becoming 7–9 mm long, white-setose; corolla white, the tube 5.5–7 mm long, the crests at base of tube lacking or sometimes evident, the fornices yellow, broad, emarginate, papillose ca 0.5 mm long, the limb campanulate, 5–8 mm wide; nutlets lanceolate, 3–4 mm long, 1.8–2 mm wide, all 4 nutlets usually maturing, the margin acute to somewhat obtuse, nearly in contact, the dorsal surface carinate, sharply and deeply rugose, the ventral rugose; scar open, constricted above the base, and with an elevated margin; style exceeding mature fruit 3–4 mm. Blackbrush, salt desert shrub, mixed warm desert shrub, sagebrush, pinyon-juniper, and mountain brush communities at 1300 to 2280 m in Emery, Garfield, Grand, San Juan, and Wayne counties; endemic; 76 (xi).

Cryptantha torreyana (Gray) Greene Torrey Cryptanth. [*Krynitzkia torreyana* Gray]. Stems erect, 1–4 dm high, solitary or several with erect or more often spreading branches, finely strigose and sparsely hirsutulous; leaves oblanceolate to linear, strict or ascending, 2–5 (7) cm long, 3–6 (8) mm wide, obtuse or rounded, hispid, inconspicuously if at all pustulate; spikes commonly geminate, bractless, 4–8 (15) cm long, more or less projected from the leafy mass of the plant, very elongate and loosely flowered or congested and glomerate; corolla inconspicuous, ca 1 mm broad; fruiting calyx ovoid or oblong-ovoid, 2–7 mm long, ascending asymmetrical, base rounded or broadly conic; pedicels ca 0.5 mm long; mature calyx lobes lanceolate to linear-lanceolate, connivent above with tips usually spreading, midrib slightly thickened and hispid-hirsute, margins hispid-strigose; nutlets 4, occasionally 1 or more aborted, usually broadly ovate, 1.5–2.2 (2.5) mm long, 0.8–1.3 mm broad, smooth and polished, usually mottled, rarely finely granulate, back low and convex, the sides rounded or obtuse; groove broadly forked below and closed throughout; gynobase ca 1/2 height of nutlets, ca 1 mm long; style reaching to 2/3 height of nutlets or rarely even to their tips. Sagebrush, mountain brush, oak, maple, aspen, and spruce-fir communities at 1900 to 2930 m in Beaver, Cache, Carbon, Davis, Duchesne, Juab, Millard, Morgan, Rich, Salt Lake, Sanpete, Sevier, Utah, Wasatch, Washington, and Weber counties; Wyoming to California, north to Alaska; 35 (ii).

Cryptantha utahensis (Gray) Greene Utah Cryptanth. [*Krynitzkia utahensis* Gray, type from near St. George; *Eritrichium holopterum* var. *submolle* Gray, type from near St. George]. Plant usually with a main erect stem with a few scattered ascending or erect branches, 1–3 dm high, strigose or appressed short-hispid; leaves not numerous, strongly reduced above, linear to oblanceolate-linear, 1–5 (7) cm long, 1–4 mm wide, rounded at apex, commonly pustulate and short-hirsute especially beneath; spikes usually geminate, commonly 1–2.5 (5) cm long, dense, bractless; corolla rather conspicuous, 2–3 mm broad; fruiting calyx ovoid or ovoid-oblong, 2–3 (4) mm long, subsessile, spreading or somewhat recurved, deciduous, usually densely appressed hirsute and silky; mature calyx lobes lanceolate, connivent, midrib thick and usually brownish, and frequently bearing spreading or recurved hairs; corolla rotate, (2.5) 3–5 mm wide, white, with yellow or yellowish fornices more or less well developed; ovules 4, nutlets 1 or rarely 2, next to the abaxial calyx lobe, 1.7–2.5 mm long, 1–1.5 mm broad, pale, broadly lanceolate, granulate, muricate-papillate or rarely spinulose, back low-convex or flat, the margins sharply angled or with a very narrow knifelike margin; groove open, narrow, opening into a small areola below; gynobase subulate, ca 2/3 height of nutlet, not markedly differentiated from style; style usually slightly shorter than the nutlet. Gravelly to sandy washes in creosote, Joshua tree, blackbrush, warm desert shrub, and pinyon-juniper communities at or below 1200 m in Kane and Washington counties; Arizona, Nevada, and southern California; 12 (v).

Cryptantha virginensis (Jones) Payson Virgin Cryptanth. [*Krynitzkia glomerata* var. *virginensis* Jones, type from La Verkin]. Biennials, 1.5–3.5 (4) dm tall; stems 1 to several, arising from a stout taproot, 0.3–0.6 dm long, setose-hirsute; leaves oblanceolate to spatulate, obtuse, 3–10 (12) cm long, 0.5–1.5 cm wide, dorsal surface sparsely setose, pustulate, also with some fine tangled hair beneath, the ventral surface subtomentose and weakly appressed setose, with only a few pustulate hairs; inflorescence a broad thyrsus with the individual cymes much elongating, 0.5–3 dm long; foliar bracts conspicuous, 2–4 cm long; calyx segments linear-lanceolate, in anthesis 3–4 mm long, in fruit becoming 7–11 mm long, hirsute; corolla white, the tube 3–4 mm long, with conspicuous crests at base of tube, the fornices yellow, emarginate, papillose, ca 1 mm long, the limb 7–9 mm broad; nutlets ovate, 3.3–4.5 mm long, 2.4–2.6 mm wide, usually only 1–2 maturing, the margins in contact, acute, the dorsal surface with a distinct ridge, tuberculate and usually rugulose, ventrally very uneven with indeterminate rugae and tubercles; scar open and triangular with an elevated margin; style exceeding mature fruit 1–1.5 mm. Creosote, Joshua tree, mixed warm desert shrub, sagebrush, and pinyon-juniper communities below 1350 m in Washington County; Arizona, Nevada, and California; 23 (v).

Cryptantha watsoni (Gray) Greene Watson Cryptanth. [*Krynitzkia watsoni* Gray, type from the Wasatch Mts.]. Slender strictly branched hispid herb 1–3 dm high; stems solitary, sparsely or loosely branched, spreading short-hispid; leaves linear to oblanceolate, 1–4 (5) cm long, 1–4 (5) mm broad, obtuse or rounded, ascending hispid and merely pustulate; spikes solitary or in pairs, 1–4 (6) cm long, bractless or rarely bracted below; corolla inconspicuous, ca 1 mm broad; fruiting calyx ovoid to oblong-ovoid, 2–3.5 (4) mm long, subsessile, rounded at base, early deciduous, oldest ones becoming distant; mature calyx lobes lanceolate, the tips usually connivent, hirsute with ascending hairs, the midrib also with a few spreading bristles and scarcely thickened; nutlets 4, homomorphous or practically so, lanceolate, 1.5–2 mm long, about 0.8 mm broad, smooth, shiny or at times dulled by minute granulations, the back nearly flat, the margins definitely angled; groove closed or nearly so and forked at base; gynobase subulate, subequal to the nutlets; style equaling nutlets or a trifle surpassed by them. Sagebrush, pinyon-juniper, mountain brush, and ponderosa pine communities at 2200 to 2650 m in Beaver, Daggett, Duchesne, Juab, Salt Lake, Tooele, Uintah, Utah, and Wasatch counties; Washington and Montana, south to Colorado, Nevada, and California; 11 (i).

Cryptantha wetherillii (Eastw.) Payson [*Oreocarya wetherillii* Eastw., type from Courthouse Wash, Grand County; *Krynitzkia glomerata* var. *acuta* Jones, type from near Cisco, Grand County]. Biennial or short-lived perennials, 1–3.5 (4) dm tall; stems 1–6, 0.5–0.8 dm long, branched from the base with 1 stout and usually several low slender ascending stems; leaves clustered at the base, gradually reduced upward, spatulate to broadly oblanceolate, the apices obtuse to rounded, 2.5–5 cm long, 0.7–1.6 cm wide, strigose and appressed setose, the dorsal surface conspicuously pustulate, the ventral with few or no pustules; inflorescence becoming broad in age due to the elongation of the cymes, 0.6–3 dm long; calyx segments lanceolate, in anthesis 5–7 mm long, in fruit becoming 7–13 mm long, white-setose; corolla white, the tube 7–10 mm long, a crest at base of tube lacking; fornices light yellow, emarginate, papillose, ca 1 mm long; limb 6–13 mm wide; nutlets lanceolate, or ovate-lanceolate, 3.5–4 mm long, 2–2.5 mm wide, usually all 4 maturing, the margins acute, in contact, the dorsal surface distinctly tuberculate and often rugulose as well; scar open, linear, surrounded by a slightly elevated margin; style exceeding mature fruit 3–5 mm. Mixed salt desert shrub and pinyon-juniper communities at 1400 to 2100 m in Carbon, Emery, Garfield, Grand, and Wayne counties; endemic; 38 (vi).

Cynoglossum L.

Biennial or perennial or rarely annuals; leaves alternate, the basal ones long petioled; racemes elongating, usually without bracts, or rarely bracted at base; calyx cut to beyond the middle, somewhat accrescent, the lobes often spreading or reflexed in fruit; corolla cylindrical or funnelform, the tube short, the lobes broad, spreading, imbricate, throat with trapeziform oblong or subulate appendages; stamens included; filaments short; anthers oblong or elliptic; ovules 4; nutlets 4, equally divergent, depressed ovoid or orbicular, glochidiate, back flat or convex, frequently with an elevated margin, attached by a small or large medial to apical scar to a convex or pyramidal gynobase and frequently with a free subulate prolongation decurrent on the short entire style.

Cynoglossum officinale L. Houndstongue. Biennial, villous-tomentose throughout; stems stout, erect, leafy to the top, 4–5 dm high; lower leaves oblong to oblong-lanceolate, slender petiolate, 15–30 cm long, 2–7 cm wide; upper leaves lanceolate, acute or acuminate, sessile or the upper mostly clasping; racemes several to many simple or branched, sparingly bracted or bractless; much elongating in fruit; pedicels 5–12 mm long; calyx segments ovate-lanceolate, obtuse to acutish, 5–7 mm long in fruit, corolla reddish purple, the broad tube 3–5 mm long, the limb 6–8 mm broad; nutlets ascending on the pyramidal gynobases, ca 6 mm high, flattish on the upper surface and margined, splitting away from the gynobase at maturity but hanging attached to the subulate style; 2n = 24, 48. Sagebrush, pinyon-juniper, cottonwood, mountain brush, aspen, ponderosa pine, and spruce-fir communities at 1480 to 3000 m in Beaver, Box Elder, Cache, Carbon, Davis, Duchesne, Emery, Juab, Millard, Morgan, Rich, Salt Lake, Sanpete, Tooele, Utah, Wasatch, Washington, and Weber counties; native to Europe and Asia but widely distributed over North America; 136 (viii).

Echium L.

Biennial or possibly perennial, hispid, herbaceous plants; leaves alternate, entire; flowers blue to violet purple, in leafy bracted scorpioid, spikelike racemes; calyx 5-parted; corolla tubular-funnelform, irregular, usually 5-lobed, the throat not appendaged; stamens unequal, at least the longer ones exserted on long filaments; ovary 4-lobed, these separating in fruit; style 2-cleft at apex; nutlets erect, rugose, attached by their bases to a flat gynobase, the scar flat or somewhat concave, not leaving a pit.

Echium vulgare L. Stems 3–9 dm tall, erect, solitary or occasionally several, finely hispid-villous as well as shaggy coarsely hirsute; leaves appressed-hispid and the margin and midrib somewhat hirsute, with a strong midrib but very obscure or absent veins; lower leaves 8–16 cm long, oblanceolate, broadly stalked, forming a rosette that withers at anthesis; cauline leaves reduced upward, the middle ones linear- lanceolate, 3–9 cm long, contracted to a rounded sessile base; racemes short, lateral, disposed in a long narrow thyrse or open panicle; corolla bright blue, rarely rose or white, pubescent and sparsely setose, rather firm in texture, 10–15 mm long, the tube subequal to the calyx; stamens very unequal, the 2 pair slightly unequal but both surpassing the lower corolla lobe, the odd stamen included; nutlets ca 2 mm long, erect, rugose; 2n = 16, 32. Roadsides, fields, and waste places in mountain brush communities at 1820 to 1960 m in Echo Canyon, Summit County; widely distributed in North America; adventive from Europe; 3 (0).

Eritrichium Schrader

Low depressed cushionlike perennials, with the short stems densely clothed with small often imbricate leaves; flowers few in a racemelike cluster terminating the slender flowering stem; calyx lobes ascending, linear; corolla blue, funnelform, with a short tube; nutlets obliquely attached to the conical gynobase, smooth, the apex obliquely truncate, with a distinct, entire or toothed margin.

Eritrichium nanum (Vill.) Schrader Low Blue-eyes. [*Myosotis nana* Vill.]. Plants villous often silvery, forming a tuft ca 2–4 cm tall (not counting the flowering branches); leaves closely overlapping, 5–10 mm long, narrowly ovate to oblong or oblanceolate, 1.5–2 mm broad, acute or obtuse, pilose, with long white hairs, especially on the margins and tips; flower cluster compact when sessile among the leaves, or sometimes racemelike when borne on a leafy flowering branch up to 7 cm long; calyx lobes linear, 1.5–3 mm long; corolla tube equaling the calyx lobes, the limb variable in size (1) 4–5 (7) mm broad, bright blue (rarely white), the crests in the throat puberulent; nutlets smooth, with an entire margin to the truncated oblique portion, rarely with a few obscured teeth.

Alpine tundra and krumholtz communities at 3330 to 4350 m in Box Elder, Daggett, Duchesne, Summit, and Uintah counties; Alps of Europe, Asia, Alaska, Yukon, and in the Rocky Mountains from Montana to New Mexico and west to Oregon; 16 (i). Our plants belong to var. *elongatum* (**Rydb.**) **Cronq.** [*E. aretioides* var. *elongatum* Rydb.; *E. elongatum* var. *paysonii* Johnst., type from the Uinta Mts., Summit County].

Hackelia Opiz in Bercht.

Coarse biennial or perennial (rarely annual) herbs; leaves alternate, broad and veiny; flowers in naked or inconspicuous bracted racemes paniculately disposed; pedicels slender, recurving in fruit; calyx cut to the base into spreading ovate to oblong or lanceolate lobes; corolla white or blue, with a short or elongate tube, lobes rounded, imbricate; throat with trapeziform intruded appendages; stamens included, affixed at middle of tube; filaments slender, short; anthers oblong to elliptic; style slender, scarcely if at all surpassing the nutlets; stigma capitate; ovules 4; nutlets 4, erect, ovate, affixed ventrally to the pyramidal gynobase by a broad medial or submedial areola, the margin with subulate glochidiate appendages, these frequently confluent at the base, the back smooth or with glochidiate appendages.

1. Corolla white marked with blue .... *H. patens*
— Corolla blue, (rarely pinkish) with a white or yellow fornices 2

2(1). Cymes 1–3, mostly terminal; plants mostly alpine and subalpine, small, to 1.5 dm tall, of Juab County .... *H. ibapensis*
— Cymes more abundant; plants of lower elevations and over 3 dm tall 3

3(2). Stems several to many from a caudex; plants perennial; surface of the nutlet with (1) 3–8 intramarginal prickles .... *H. micrantha*
— Stems 1–3 (5) from the base; plants biennial or perennial; surface of the nutlet lacking intramarginal prickles or rarely with 1 or 2 .... *H. floribunda*

Hackelia floribunda (**Lehm.**) **Johnston** Showy Stickseed. [*Echinospermum floribundum* Lehm.]. Stem erect, stout, from a short-lived perennial root, 5–12 dm high, the rough pubescence deflexed, mixed with some spreading hairs; leaves oblanceolate to linear to oblong, appressed hirsutulous, the basal leaves petiolate, with spreading hairs, the stem leaves sessile above; racemes of the inflorescence many, rather strict, densely flowered; pedicels short, 5–7 mm long in fruit; corolla blue, 4–7 mm broad, appendages small, obscurely papillate, not closing the throat; nutlets 3–5 mm long, the face with a medial ridge, muriculate, hirsutulous, without short glochidiate prickles, the marginal spines much flattened at base, distinct or somewhat confluent, 4 to 6 on each side, mostly exceeding in width the face of the nutlet. Sagebrush, willow, mountain brush, aspen, lodgepole pine, Douglas fir, and spruce-fir communities at 1833 to 3470 m in Cache, Carbon, Daggett, Duchesne, Emery, Garfield, Grand, Iron, Juab, Piute, Rich, Salt Lake, Sanpete, Sevier, Summitt, Tooele, Uintah, Utah, Wasatch and Weber counties; British Columbia to Ontario, south to North Dakota, New Mexico, Arizona, and California; 86 (iii).

Hackelia ibapensis L. & J. Shultz Deep Creek Stickseed. Stems slender, clothed with the old leaf bases, up to 1.5 dm tall, strigose; basal leaves well developed, lance-elliptic, 1.5–8 (12) cm long, 6–11 (17) mm broad, the cauline reduced and narrowly lanceolate, thinly to densely strigose on both surfaces; inflorescence of a solitary terminal cyme, rarely with 2, laxly flowered; corolla blue; nutlets 2–3 mm long, the marginal prickles connate at base forming a distinct rim; intramarginal prickles several. Exposed granite outcrops in mountain brush and Douglas fir communities at 2500 m in Juab (type from Deep Creek Range) County; endemic; 3 (0).

Hackelia micrantha (Eastw.) J. Gentry Small-flower Stickseed. [*Lappula micrantha* Eastw.; *L. jessicae* McGregor]. Stems erect or ascending from a stout root, sparsely to rather densely villous-hirsute; basal leaves 8–15 cm long, the blades oblanceolate, 15–20 mm wide, narrowed to a winged petiole of about equal length; upper stem leaves sessile, lanceolate, acute, the reduced ones subtending the lower racemes often ovate-lanceolate; racemes several in an open panicle; pedicels slender, at length recurved, reflexed, 5–10 mm long; calyx lobes oblong to oblong-lanceolate, 2–3 mm long; corolla small, pale blue, 3.5–5 mm broad, the tube often whitish, 1.5–2 mm long, the lobes oblong-ovate, the crests yellowish, rounded, puberulent; nutlets 4–6 mm long, with ca 10 marginal prickles broadly dilated at base, distinct, often with a shorter one in between; dorsal face broadly ovate, usually flattened with a distinct median ridge, puberulent and in age more or less muriculate, usually with 1 or more short barbed prickles near the center; $n = 12$. Sagebrush, juniper, aspen, Douglas fir, and spruce-fir communities at 2540 to 3320 m in Box Elder, Cache, Carbon, Davis, Duchesne, Emery, Garfield, Juab, Piute, Rich, Salt Lake, Sanpete, Sevier, Summit, Utah, and Wasatch counties; British Columbia and Alberta, south to California, Nevada, and Colorado; 48 (ii). Immature specimens without fruit are difficult to distinguish from *H. floribunda*. However, in the field it is easily separated on the basis of perennial habit and numerous stems.

Hackelia patens (Nutt.) Johnston Pale Stickseed. [*Rochelia patens* Nutt.]. Stems 2 or 3, short, deflexed hirsute, with some hairs spreading, arising from a woody caudex covered with old leaf bases; basal leaves many, oblanceolate to lanceolate, the cauline leaves reduced upward, the pubescence rather dense, short, appressed; branches of the inflorescence 5- to 10-flowered; pedicels short, elongating somewhat in fruit; corolla whitish or light blue, usually quite large, the appendages somewhat broader than long, obscurely papillose; nutlets small, with marginal glochidiate prickles 3–5 on each side, a few short prickles interspersed, the longest prickles surpassing the body of the nutlet; face of nutlet with a faint median ridge, muricate and occasionally bearing 1 or 2 prickles; $n = 12$.

1. Hairs of the leaves and lower stem hirsute-spreading, with pustulate bases .... *H. patens* var. *harrisonii*
— Hairs of the leaves and lower stem strigose, and without pustulate bases .... *H. patens* var. *patens*

Var. **harrisonii** J. Gentry Sagebrush, mountain brush, pinyon-juniper, ponderosa pine, Douglas fir, and aspen communities at 1780 to 2800 m in Beaver, Utah, and Washington (type from Pine Valley) counties; endemic; 8 (ii). This variety intergrades with var. *patens* except in Washington County where it alone occurs.

Var. **patens** Sagebrush, pinyon-juniper, mountain brush, birch, aspen, and spruce-fir communities at 1500 to 3050 m in Beaver, Box Elder, Cache, Davis, Duchesne, Garfield, Iron, Kane, Juab, Millard, Morgan, Rich, Salt Lake, Sanpete, Summit, Tooele, Utah, Wasatch, and Weber counties; Montana to Wyoming, Idaho, and Nevada; 120 (iii).

Heliotropium L.

Annual or perennial, herbaceous or more or less shrubby plants; leaves small to large, sessile or petiolate; cymes unilateral and usually distinctly scorpioid, with or without bracts; corolla white, yellow or purple, variable in form, throat frequently pubescent inside; anthers included; filaments extremely short; style present or absent; stigma usually conic, mostly sterile, receptive only in a band around the base; fruit dry, at maturity breaking up into 4 single-seeded or 2 2–seeded nutlets; seeds with a thin endosperm.

1. Plant not succulent, hairy, never glaucous; fruit 2–lobed, each lobe splitting into 2 nutlets; stigma capped by a tuft of bristles; annuals .... *H. convolvulaceum*
— Plants very succulent, glabrous, usually glaucous; fruit not lobed; stigma discoid, naked; perennials ....
 .... *H. curassavicum*

Heliotropium convolvulaceum (Nutt.) Gray Showy Heliotrope. [*Euploca convolvulacea* Nutt.]. Annual, 1–4 dm tall; stems at first simple, but later developing elongate branches, that ascend or sprawl, strigose and cinereous; leaves numerous, strigose, pustulate, lanceolate to ovate, 10–14 mm long, 4–15 mm broad, acute apically, basally acute to rounded; petioles slender, 3–8 mm long, the midrib evident; flowers extra axillary, appearing to be borne along the elongating leafy branches; bracts foliose, numerous; calyx 5–lobed, at anthesis 4–6 mm long, becoming 6–9 mm long at maturity; lobes linear-lanceolate or linear, unequal; pedicels at anthesis 1–3 mm long, in fruit 3–5 mm long; corolla white with a yellow throat, fragrant, opening during the morning and evening, the limb widely funnelform, expanded, 15–22 mm broad, not lobed pentagonal, plicate in the bud with the sinus inflexed; tubular portion of corolla 8–10 mm long, strigose on the outside; anthers lanceolate, basifixed, 2–2.5 mm long; filaments ca 1 mm long; ovary globose, glabrous; style slender, 3–4 mm long; stigmatic head with a prominent stigmatic band 0.5 mm in diameter at the base, this surmounted by a truncate appendage bearing a cluster of hairs; fruit laterally compressed, hairy, 2–lobed, 3–4 mm long, 2–2.5 mm thick, 2–2.5 mm high; at maturity first dividing transversely and the lobes separating, forming halves, each with a broad flat commisural face; each half next dividing on the narrow longitudinal axis to form the asymmetrical single seeded nutlets. Sandy dunes in creosote, blackbrush, mesquite, mixed desert shrub, saltbush, rabbitbrush, and sagebrush communities at 950 to 1690 m in Emery, Garfield, Grand, Juab, Kane, Millard, San Juan, Tooele, Washington, and Wayne counties; Nebraska and Wyoming, south to northern Chihuahua and Texas, west to Arizona and Nevada; 43 (v).

Heliotropium curassavicum L. Salt Heliotrope. Annual or short-lived perennial, fleshy, glaucous and glabrous throughout; stems diffusely branched, 1–6 dm long; leaves succulent, varying from linear to obovate, but commonly spatulate, 1–4 cm long, obtuse, narrowed to a thick petiole; spike mostly in pairs, sometimes 3–5, often 6–12 cm long; calyx-segments ovate-lanceolate, acute, 2–3 mm long; corolla 3–5 mm long, white with a violet purple eye on the throat; stigma glabrous; stamens included, the anthers subsessile; fruit subglobose, at length separating into 4 nutlets; 2n = 26.

1. Corolla limb 5–16 mm wide, white or at most only purplish tinged or yellow at the throat; leaves 10–18 mm broad, spatulate to obovate .*H. curassavicum* var. *obovatum*

— Corolla limb 3–5 (7) mm wide, with a conspicuous purple eye and throat; leaves usually less than 9 mm wide, oblanceolate to slightly spatulate .... *H. curassavicum* var. *oculatum*

Var. *obovatum* DC. Saline seeps and roadsides in saltgrass, greasewood, saltbush, and other mixed salt desert shrub communities at 1125 to 1710 m in Box Elder, Cache, Garfield, Iron, Juab, Kane, Millard, Salt Lake, Uintah, and Utah counties; Washington, Oregon, and Nevada, east to Wyoming, Colorado, and New Mexico; 25 (ii).

Var. *oculatum* (Heller) Johnston [*Heliotropium oculatum* Heller]. Saline seeps, sidewalks, ditch banks in creosote, mesquite, saltbush, saltgrass, and other warm salt desert shrub communities below 1200 m in Kane? and Washington counties; Nevada, south to Baja California, and east to New Mexico and Texas; 18 (vi).

Lappula Gilib.

Annual herbs; leaves alternate, usually narrow, firm and veinless; flowers small, blue or white, on usually erect pedicels or rarely subsessile, in bracted racemes; calyx 5-parted into spreading lanceolate lobes; corolla with a rather short tube; lobes rounded, ascending imbricate; throat closed by intruded appendages; stamens affixed in the tube, included; filaments slender, short; anthers oblong, obtuse; style short, surmounting the subulate-columnar gynobase, commonly surpassing the mature nutlets; stigma subcapitate; ovules 4; nutlets 4, erect, smooth or verrucose, narrowly but firmly attached to the gynobase along the length of the well developed ventral keel, back angulate or margined by a single or double row of prickles which by confluence frequently form a winglike or cupulate border.

1. Nutlets with marginal prickles in at least 2 rows .... *L. squarrosa*

— Nutlets with the marginal prickles definitely in a single row .... *L. occidentalis*

Lappula occidentalis (Wats.) Greene Western Stickseed. [*Echinospermum redowskii* var. *occidentalis* Wats.; *L. texana* (Scheele) Britt.; *L. collina* Greene, type from Kingston; *E. collinum* (Greene) K. Schum.]. Annual, the stems simple or few branched at base and erect or sometimes diffuse, 15–35 cm tall, herbage more or less canescent with a strigose and villous pubescence; leaves narrowly linear to narrowly lanceolate or the lower narrowly oblanceolate, 1–3 cm long; flowers in the axils of the small foliaceous bracts forming open and at length elongated terminal racemes; pedicels 1–2 mm long; calyx-segments narrowly lanceolate, erect, or but little spreading in fruit, a little shorter than the corolla tube, corolla blue, 3–4 mm long, conspicuously crested on the throat; nutlets 2–2.5 mm long, bordered by a single row of barbed prickles, the prickles distinct at base or joined to form a cupulate margin, the dorsal area of nutlets distinctly tuberculate. Mixed warm desert shrub, sagebrush, pinyon-juniper, rabbitbrush, mountain brush, ponderosa pine, aspen, and spruce-fir communities at 830 to 3000 m in all Utah counties, except Davis and Morgan; throughout western North America; 240 (xiv). Our plants belong to var. *cupulata* (Gray) Higgins [*Echinospermum redowskii* var. *cupulatum* Gray].

Lappula squarrosa (Retz.) Dumort. European Stickseed. [*Myosotis squarrosa* Retz.; *L. echinata* Gilib.]. Annual, with erect, simple to freely branched stems 1.5–8 dm tall, villous-hirsute with upwardly more or less appressed hairs; lower leaves linear to linear-lanceolate or oblong, acute to obtuse, narrowed to a sessile base, closely ascending, 2–5 cm long, roughly pubescent like the stem, passing above into linear or lanceolate bracts of the usually numerous racemes; pedicels 1–3 mm long; calyx-lobes broadly linear, appressed bristly, in fruit spreading, 2.5–3 mm long; corolla bright blue, the limb 2–4 mm broad, the tube surpassing the calyx; nutlets 3–4 mm long, sharply verrucose or muricate dorsally, with 2 marginal rows of long slender bristles not confluent at the base, these sometimes irregularly distributed over the back; 2n = 48. Roadsides and disturbed sites in sagebrush, pinyon-juniper, mountain brush, aspen, sedge meadows, lodgepole pine, Douglas fir, and spruce-fir communities at 2050 to 2875 m in Cache, Daggett, Emery, Garfield, Grand, Morgan, Rich, Salt Lake, Sanpete, Sevier, Summit, Tooele, Uintah, and Utah counties; adventive from Eurasia, now widely distributed in Canada and northern parts of the U. S., especially the Rocky Mountains; 19 (iii).

Lithospermum L.

Annual or perennial, herbaceous or fruticose plants with alternate leaves; flowers white, yellow, or violet, in bracted racemes; calyx usually undivided; corolla tubular or salverform, the tube cylindrical, lobes spreading and imbricate, the throat with intruded appendages or with pubescent or glandular areas; stamens inserted in the tube, included; filaments short, anthers oblong, usually with apiculate connectives; style filiform; stigmas geminate; ovules 4; nutlets 4 or rarely fewer, erect, ovoid or angular, smooth or verrucose, inserted by a broad horizontal or slightly oblique basal areola; gynobase flat or very broadly pyramidal.

1. Flowers white; nutlets densely tuberculate and dull; plants annual .... *L. arvense*

— Flowers greenish to bright yellow; nutlets smooth; plants perennial 2

2(1). Corolla tube (in chasmogamous flowers) 15–35 mm long, the lobes erose .... *L. incisum*

— Corolla tube less than 13 mm long, the lobes entire or nearly so 3

3(2). Corolla greenish yellow to light yellow; flowers homostylic .... *L. ruderale*

— Corolla bright yellow or orange yellow; flowers heterostylic .... *L. multiflorum*

Lithospermum arvense L. Gromwell. Annual; stems erect, 2–7 dm tall, 1 to several, simple or sparsely branched above, hoary strigose; leaves 2–5 cm long, 2–8 mm broad, closely appressed-hispid, firm, veinless, with prominent midrib, pale beneath; basal leaves rosulate, oblanceolate, or spathulate; cauline leaves acute and lanceolate or linear; racemes bracteate, becoming loosely flowered; calyx parted into linear-subulate lobes, hispid, mature calyx with the erect or ascending lobes becoming 8–13 mm long, the very short tube oblique, pale and chartaceous; pedicels short and stout, ca 1 mm long; corolla white or yellowish or even purplish, 5–7 mm long, tubular-funnelform, the tube glabrous within and scarcely if at all surpassing the calyx; corolla throat gradually expanded, without protuberances, merely pubescent; corolla lobes ascending ovate, obtuse; nutlets brown, dull, roughened, tuberculate, or rough-wrinkled or pitted, ca 3 mm long, ovate; 2n = 14, 28, 42. Wet meadows, cottonwood, grassland, pinyon-juniper, mountain brush, aspen, and spruce-fir communities at 1420 to 3250 m in Cache, Davis, Salt Lake, Tooele, and Weber counties; native of Eurasia, now established as a weed over most of the U. S.; 9 (i).

Lithospermum incisum Lehm. Showy Stoneseed. Perennial plants from a thick woody root; stems 1–5 dm tall, usually several, erect or ascending, strigose to somewhat hirsute; leaves 10–50 mm long, linear to linear-oblong, strigose; inflorescence of terminal leafy racemes; calyx 6–10 mm long; corolla 10–30 mm long, yellow, the tube seldom over 2.5 mm wide when pressed, salverform, the limb 9–18 mm wide, the lobes fimbriate to toothed; nutlets 3–4 mm long, white and shining; cleistogamous flowers present, in fruit usually with recurved pedicels; flowers monomorphic. Sagebrush, pinyon-juniper, ephedra, mixed desert shrub, bitterbrush, mountain brush, ponderosa pine, and mountain mahogany communities at 1350 to 2870 m in all except Morgan, Piute, Rich, Summit, Tooele, and Wasatch counties; central U. S., Canada, and Mexico westward to British Columbia, Nevada, Arizona, New Mexico, and Texas; 136 (vi).

Lithospermum multiflorum Torr. Pretty Stoneseed. Perennial plants from a thick woody root containing a purple dye; stem 3–6 dm tall, more or less tufted, often virgately branched above, strigose-hispid; leaves 2–6 cm long, linear or linear-lanceolate, appressed-strigose above, hirsute beneath, becoming smaller and bractlike near the flowers, scarcely if at all longer than the calyx lobes and simulating them; flowers racemose, short-pediceled, often on several ascending corymblike branches; calyx lobes 4–6 mm long; corolla yellow or orange yellow, tubular funnelform, the tube 8–13 mm long, the lobes short, ca 2 mm long, rounded, not fimbriate; nutlets ca 3 mm long, white and shining; cleistogamous flowers lacking; flowers heterostyled and dimorphic. Sagebrush, pinyon-juniper, mountain brush, ponderosa pine, and aspen communities at 1850 to 2930 m in Beaver, Emery, Garfield, Iron, Kane, Piute, San Juan, Sevier, and Wayne counties; Colorado, south to Texas, Arizona, and northern Mexico; 29 (ii).

Lithospermum ruderale Dougl. ex Lehm. Contra Stoneseed. Stems usually several from a large root, erect or decumbent, rather stout, 2–5 dm high, simple or branched, hirsute and somewhat hispid to densely villous; leaves numerous, usually crowded above, mostly ascending or some reflexed, linear-lanceolate to lanceolate, 3–8 cm long, 2–12 mm wide, softly to rather harshly pubescent on both sides, scabrous on the margins, flowers in the axils of the upper leaves; pedicels stout, 1–3 mm long; calyx lobes in fruit subulate, 7–10 mm long; corolla pale, often greenish-yellow, 9–12 mm long, the tube broad, scarcely dilated at the throat, the lobes ca 3 mm long; nutlets broadly ovoid, 5–6 mm long, usually abruptly attenuate at the apex into a stout beak, whitish, smooth and highly polished. Sagebrush, pinyon-juniper, mountain brush, aspen, and Douglas fir communities at 1720 to 2800 m in all except Carbon, Garfield, Kane, Morgan, Piute, and Wayne counties; British Columbia and Alberta to California, Nevada, and Colorado; 80 (iv).

Mertensia Roth

Glabrous or pubescent caulescent perennial herbs with fleshy, fusiform, rhizomelike roots; leaves entire, linear to cordate, sessile or petiolate, alternate; stems 1 to many from each root, decumbent to erect, usually branched below the inflorescence, this a lax or congested ebracteate unilateral modified scorpioid cyme, or with the lowest flowers often single and subtended by leaves, often becoming panicled in age; calyx 5–parted, occasionally campanulate, the expanded limb shorter or longer than the tube, with or without fornices in the throat, blue, occasionally white or pink; filaments attached below the throat, the anthers exserted or included; style shorter or longer than the corolla, in some di- or trimorphic; stigma entire or slightly lobed; ovary 2–loculed, each locule 2–lobed; nutlets 4, attached laterally to the gynobase, usually rugose or pectinately rugose, coriaceous or smooth and shining, utriclelike.

1. Plants relatively tall and robust (4–15 dm tall when fully developed) with evident lateral veins in the cauline leaves; flowering in late spring and summer 2
— Plants smaller, seldom as much as 4 dm tall, usually without evident lateral veins in the cauline leaves, blooming as soon as snow and temperature permits 4

2(1). Leaves strigose at least on the upper surface; calyx lobes acute, often hairy on the back and ciliate-margined .. .. *M. franciscana*
— Leaves glabrous on both surfaces, or pustulate on the upper surface, but lacking hairs; calyx lobes ciliate but usually not hairy on the back 3

3(2). Calyx mostly 1.5–3(4) mm long at anthesis, cleft to the base or nearly so; limb of the corolla shorter than the tube, or subequal; leaves usually not acuminate. *M. ciliata*
— Calyx mostly longer than 5 mm and not cleft to the base; limb of the corolla longer than the tube; leaves usually acuminate .... *M. arizonica*

4(1). Style short, 1–2.5 mm long; anthers nearly sessile, attached in the corolla tube, not projecting beyond the throat .... *M. brevistyla*
— Style longer, 4 mm long or more; anthers with evident filaments, attached near the throat, the anther projecting beyond the throat 5

5(4). Corolla tube glabrous within, and usually much longer than the limb .... *M. oblongifolia*
— Corolla tube with a ring of hairs within, the tube shorter than to slightly longer than the limb 6

6(5). Upper surface of the leaves with loosely appressed hairs directed towards the margins of the leaves; stems 1–3 (5); plants of middle to lower elevations .... *M. fusiformis*

— Upper surface of the leaves glabrous or with strigose foreward pointing hairs; stems few to many; plants of middle and upper elevations .... *M. lanceolata*

Mertensia arizonica Greene Tall Bluebell. [*M. leonardii* Rydb., type from Mill Creek Canyon, Salt Lake County; *M. arizonica* var. *leonardii* (Rydb.) Johnston; *M. sampsonii* Tidestr., type from east of Ephraim; *M. toyabensis* var. *subnuda* Macbr.; *M. arizonica* var. *subnuda* (Macbr.) L. O. Williams]. Plants erect or ascending, 3–8 dm tall or more; stems 1 to several from each root stock; basal leaves narrowly to broadly ovate or oblong-lanceolate, 7.5–15 cm long, 2–6 cm broad, slightly decurrent on the petiole, this as long as the blade, glabrous but slightly papillate, the margin ciliate; lower cauline leaves spatulate to elliptical, usually petiolate, the petiole winged; upper cauline leaves usually sessile, elliptical to narrowly ovate, acute, 3–12 cm long, 1–5 cm broad, the base attenuate, acute apically; inflorescence of axillary peduncles with branches elongating in age; calyx 4–8 mm long, campanulate, glabrous on the back, hairy within, the lobes 1/2 or less the entire length of the calyx or cut nearly to the base, acute or obtuse, ciliate; pedicels 2–30 mm long, glabrous papillose or sometimes the papillae develping short hairs; corolla tube 6–9 mm long, with a definite ring of hairs at the base within; corolla limb 7–11 mm long, always longer than tube, moderately expanded; anthers 2.5–3.5 mm long, as long as or shorter and narrower than the filaments, these 3–4 mm long; fornices conspicuous, pubescent; style 10–15 mm long, usually shorter than the corolla; nutlets rugose, shorter than the calyx. Usually in moist canyons, streamsides, or moist meadows in pinyon-juniper, mountain brush, ponderosa pine, aspen, spruce-fir communities at 1920 to 3400 m in all except Daggett, Rich, Uintah, and Wayne counties; Wyoming, Colorado, and Utah; 136 (x).

Mertensia brevistyla Wats. Wasatch Bluebell. Plants with erect or ascending stems, 1–4 dm tall; 1 to many from each fusiform rootstalk, more or less pubescent; basal leaves broadly lanceolate to oblong, acute or obtuse, strigillose above, glabrous below, 5–13 cm long, 2–4 cm broad; petioles longer than the blade; cauline leaves obovate-oblong to narrowly elliptic, obtuse to acute, densely strigillose above, glabrous below, 2–6 cm long 0.5–3 cm broad; inflorescence congested at first, becoming paniculate in age; pedicels strigose, 1–14 mm long; calyx 2–5 mm long divided almost to the base, strigose, the lobes narrowly triangular to linear, acute, 1.5–4 mm long, 0.5–1 mm broad at the base; corolla tube 2–4 mm long, slightly shorter to a little longer than the calyx lobes, with or without a ring of scattered hair; corolla limb rotate, 4–6 mm long; anthers 1–1.3 mm long, longer than the filaments, inserted on the tube and not exceeding the throat; fornices more or less prominent; style shorter than the calyx lobes; nutlets rugose, 2–3.5 mm long. Sagebrush, mountain brush, aspen, and spruce-fir communities at 2000 to 3320 m in Box Elder, Cache, Davis, Duchesne, Juab, Salt Lake, San Juan (?), Sanpete, Summit (? type from the Wasatch Mts.), Tooele, Utah, Wasatch, and Weber counties; Wyoming and Colorado to Idaho and Nevada; 95 (i).

Mertensia ciliata (James) G. Don Mountain Bluebell. [*Pulmonaria ciliata* James]. Plants erect or ascending, 1–12 dm tall, usually with many stems from each rootstock; basal leaves variable, oblong to ovate, or lanceolate, subcordate, 4–15 cm long, 3–10 cm broad, ciliate on the margins, often papillate on the upper surface; petioles longer or shorter than the blades; cauline leaves lanceolate to ovate, acute, acuminate or obtuse at apex, attenuate to subcordate at the base, the lowermost short-petiolate, the uppermost sessile, ciliate on the margins, often papillate on the upper surface, often quite glaucous, thin in texture; pedicels 1–10 mm long, glabrous, papillose or rarely with a few hairs; inflorescence from the axils of leaves, the peduncles elongated in mature or well developed plants, in young plants the flowers aggregated at the top of the plant, each peduncle terminated in a modified ebracteate scorpioid cyme, or occasionally subumbellate; calyx lobes 1.5–3 mm long, glabrous on the back, ciliate to papillate on the margins, more or less strigose within, obtuse or rarely somewhat acute, divided almost or quite to the base, rarely enlarged in fruit; corolla tube 6–8 mm long, glabrous or with crisped hairs within; corolla limb 4–10 mm long, sometimes longer than the tube, moderately expanded; anthers 1–2.5 mm long, as long as or shorter and narrower than the expanded part of the filament; fornices prominent, glabrous, papillate or pubescent; style about as long as the corolla or exceeding it; nutlets rugose or mammilate. Mountain brush, aspen, Douglas fir, limber pine, ponderosa pine, lodgepole pine, spruce-fir, and alpine tundra communities at 2025 to 4050 m in Box Elder, Cache, Daggett, Davis, Duchesne, Grand, Salt Lake, San Juan, Summit, Utah, Uintah, and Wasatch counties; Montana to Colorado, New Mexico, and west to Oregon and California; 30 (v).

Mertensia franciscana Heller Flagstaff Bluebell. [*M. arizonica* var. *subnuda* L. O. Williams, type from Fish Lake, Sevier County]. Plants with erect or ascending stems, 1–10 (17) dm tall, usually several from each rootstock; basal leaves oblong-elliptic to elliptic, 6–20 cm long, 5–9 cm broad, the base subcordate to obtuse, the apex acuminte, acute or obtuse, short-strigillose above, glabrous or with spreading pubescence beneath; petioles longer or shorter than the blade; cauline leaves elliptical to narrowly ovate, 4–14 cm long, 1–5 cm broad, obtuse to acuminate, the lowermost petiolate, becoming sessile toward the inflorescence, strigillose on the upper surface, glabrous to densely pubescent with spreading hairs below; flowers of the inflorescence paniculately disposed in a bracteate modified scorpioid cyme, the branches of the inflorescence elongating in age; pedicels strigose, 1–20 mm long; calyx 2.5–5 mm long, divided almost to the base, the lobes linear to lanceolate, 1–2 mm wide at the base, acute, rarely obtuse, glabrous or pubescent on the back, strongly ciliate; corolla tube 5–9 mm long, glabrous or pubescent within; corolla limb 4–6 (9) mm long, subequal to or slightly shorter or longer than the corolla tube, moderately expanded; anthers 2.5–3 mm long, longer than the filaments; filaments 2–2.5 mm long, glabrous or with spreading hairs; fornices prominent, usually pubescent; style 9–20 mm long, usually shorter than the corolla, sometimes exceeding it; nutlets rugose and papilliferous. Moist canyons and streamsides or wet meadows in mountain brush, ponderosa pine, aspen, Douglas fir, willow, and spruce-fir communities at 2000 to 3520 m in Grand, San Juan, Sanpete, and Sevier counties; Colorado, New Mexico, and Arizona; 13 (0).

Mertensia fusiformis Greene Spindle Bluebell. Plants with erect stems or nearly so, 1–3 dm tall, glabrous or sparingly pubescent, 1 to few from each rootstock, this typically rather large and fusiform; basal leaves elliptic to oblong-ovate, 4–12 cm long, 1.5–3 cm broad, usually densely strigose above, glabrous below, petiole 7–12 cm long, cauline leaves linear-oblong to ovate-oblong, 1.5–10 cm long, 0.4–3 cm broad, sessile or the lowermost short-petiolate, more or less densely strigose above, glabrous below, usually quite obtuse, rarely somewhat acute; inflorescence usually congested, sometimes paniculate; pedicels 1–15 mm long, densely strigose, calyx 3–6 mm long, slightly accrescent, the lobes lanceolate to lance-ovate, 2–5 mm long, acute, ciliate, usually pubescent on the backs, occasionally nearly glabrous, not divided to the base; corolla tube 4–7 mm long, with a ring of crisp hairs within at the base; corolla limb 5–7 mm long, moderately expanded, usually subequal to or shorter than the limb, but sometimes longer; anthers 1.5–2.5 mm long; filaments 1–3 mm long; fornices present but usually not conspicuous, glabrous or nearly so; style usually surpassing the anthers, sometimes shorter; nutlets rugose, ca 3 mm long. Sagebrush, pinyon-juniper, mixed mountain brush, aspen, ponderosa pine, Douglas fir, and spruce communities at 2000 to 3500 m in Daggett, Duchesne, Grand, Iron, Kane, Salt Lake, San Juan, Sanpete, Summit, Uintah, Wasatch, and Washington counties; Colorado and Wyoming; 67 (ii).

Mertensia lanceolata (**Pursh**) **DC.** Lanceleaf Bluebell. [*Pulmonaria lanceolata* Pursh]. Stems 1 to many, 1–4.5 dm tall, erect or ascending, canescent to glabrous; leaves at base 1.5–14 cm long, 0.3–3.5 cm broad, glabrous to densely canescent on both surfaces, sessile or with the petioles longer than the blade; cauline leaves only moderately reduced toward the inflorescence, mostly sessile; inflorescence congested to loosely paniculate, especially in age; bracts only near the base; calyx 2–5 (8) mm long in fruit, divided to below the middle and mostly to near the base, the segments lanceolate to ovate-triangular, glabrous to strigose; pedicels 1–15 mm long, strigose to glabrous; corolla tube 3–7 mm long, with a ring of dense hairs near the base, the limb 3–9 mm broad, moderately expanded; fornices conspicuous, glabrous to pubescent; anthers 1–2 mm long, well exserted from the tube; style shorter or longer than the corolla tube; nutlets 2–3 mm long, rugose.

1. Leaves pubescent at least on one surface; plants usually occuring above 3000 m ...... *M. lanceolata* var. *lanceolata*
— Leaves glabrous on both surfaces; plants mostly well below 2800 m .... *M. lanceolata* var. *nivalis*

Var. *lanceolata* Fell fields, talus and open rocky slopes in mountain meadows, spruce-fir, krummholz, and alpine tundra communities at 3100 to 4110 m in Cache, Carbon, Daggett, Duchesne, Garfield, Grand, Salt Lake, San Juan, and Summit counties; Saskatchewan, Montana, and North Dakota, south through Idaho, Colorado, Wyoming, and Utah into northern New Mexico; 49 (i).

Var. *nivalis* (**Wats.**) **Higgins** [*M. paniculata* var. *nivalis* Wats., type from the Uinta Mts.; *M. lanceolata* var. *viridis* A. Nels.; *M. viridis* (A. Nels.) A. Nels.; ; *M. coriacea* var. *dilatata* A. Nels.; *M. viridis* var. *dilatata* (A. Nels.) L. O. Williams; *M. canescens* Rydb., not Kaulf.; *M. cana* Rydb.; *M. viridis* var. *cana* (Rydb.) L. O. Williams; *M. amoena* A. Nels.]. In open sites in sagebrush, pinyon-juniper, mountain mahogany, mountain shrub, and ponderosa pine communities at 1950 to 2800 m in Daggett and Uintah counties; Wyoming and Colorado; 15 (0). Attempts at further segregation of plants included within this variety are frought with difficulties; specimens separated by use of pubescence position on the leaves apparently do not represent taxa. Hence, a conservative approach is indicated.

Mertensia oblongifolia (**Nutt.**) **G. Don** Western Bluebell. [*Pulmonaria oblongifolia* Nutt.; *M. o.* var. *amoena* (A. Nels.) L. O. Williams; *M. o.* var. *nevadensis* (A. Nels.) L. O. Williams, *M. praecox* Smiley in Macbr., type from Logan]. Plants with erect or ascending stems, 1–3 dm tall, 1 to many from each elongated rootstock; blade of basal leaves 3–8 cm long, 0.5–2 cm broad, oblong or spatulate to narrowly oblong-ovate, usually obtuse, strigose above, glabrous below; petiole longer or shorter than the blade; cauline leaves sessile or the lowermost short-petiolate, linear-elliptical, 2–8 cm long, 3–15 mm broad, pubescent as on basal leaves; inflorescence congested, becoming panicled with age; pedicels strigose to essentially glabrous, 1–10 mm long; calyx 3–7 mm long, divided to within ca 1 mm of the base, the lobes linear to lanceolate-triangular, acute, ciliate, glabrous dorsally or rarely with a few hairs; corolla tube 5–12 mm long, usually quite glabrous within, occasionally with a few scattered hairs; corolla limb 4–7 mm broad; anthers 1.2–2 mm long, oblong and straight; filaments 2–4 mm long, usually longer and broader than the anthers; style exceeding the anthers; fornices prominent, glabrous or occasionally sparsely hairy; immature nutlets 3–4 mm long, rugose; $2n = 24$. Sagebrush, pinyon-juniper, mountain brush, aspen, and Douglas fir communities at 1600 to 3300 m in Box Elder, Cache, Juab, Millard, Morgan, Rich, Salt Lake, Summit, Tooele, Utah, and Weber counties; Washington and Montana, south to Idaho, Wyoming, Nevada, and California; 78 (ii).

Myosotis L.

Annual or perennial herbs; stems slender, usually erect; leaves alternate, entire; inflorescence racemose, bracted or bractless; calyx 5-parted, cut to beyond the middle into lanceolate or triangular lobes; corolla blue, white, or rarely pink, the tube short, salverform, the throat with prominent fornices; stamens adnate to the corolla tube, included or exserted; nutlets 4, small, ovoid, smooth and shiny, sharply margined, the attachment scar flat; gynobase short and depressed.

1. Calyx strigose; inflorescence terminal; corolla limb 5–10 mm wide, showy; plants known from Cache and Tooele counties .... *M. scorpioides*
— Calyx uncinate hairy; plants floriferous nearly to the base; corolla small, the limb 1–2 mm wide; plants known from Daggett County .... *M. micrantha*

Myosotis micrantha **Pallas** Annuals; stems to 2 dm tall, erect, simple or branched near the base, hirsute; leaves oblanceolate to oblong or elliptic, sessile or nearly so, 0.5–2 cm long, 3–7 mm wide, acute, plants floriferous to near the base, leafy-bracteate below, racemose above and without bracts; calyx in anthesis 3–5 mm long, the tube uncinate-hispid, the lobes strigose; pedicels spreading, 1–2 mm long; corolla blue, the tube short, the limb 1–2 mm wide; nutlets brown, distinctly surpassing the style. Our one record is from a ponderosa pine forest at 2150 m in Daggett County; introduced from Eurasia, now established in Canada and the northern U. S.; 1 (0).

Myosotis scorpioides **L.** Forget-me-not. Fibrous rooted perennial herbs; stems 2–6 dm tall, often branching at the base, commonly stoloniferous as well, strigose; leaves oblong to oblanceolate, sessile, 2–8 cm long, 7–15 (20) mm broad, obtuse; inflorescence terminal, the racemes usually in pairs, becoming loose and open; bracts lacking; calyx in anthesis 1.5–2.5 mm long, in fruit becoming 3–5 mm long, the segments triangular, short, strigose, equaling or shorter than the tube; pedicels in fruit spreading, 4–7 mm long; corolla blue with a yellow eye, the tube short, ca 2 mm long, the limb 5–8 (10) mm broad; nutlets angled, keeled on the inside, smooth; $2n = 63, 64, 66$. Moist places, our few records from along ditch banks and in sloughs at ca 1550 m in Cache and Tooele counties; introduced from Europe, now widely distributed in North America; 7 (0).

Pectocarya DC. ex Meissner

Low, often spreading annual herbs, with slender stems and narrowly linear leaves, canescent with appressed hairs; flowers scattered along the stems or branches, on short pedicels, solitary in the axils; calyx 5–parted, the lobes narrow, spreading or reflexed in fruit; corolla white, the tube shorter than the calyx, the lobes broadly oval, the throat nearly closed by prominent crests; stamens included; style very short; nutlets flattened, thin, widely divergent either radiately or in pairs, their margins with a row of hooked bristles, at least toward the apex.

1. Nutlets orbicular or nearly so, both the body and the very thin conspicuous wing beset with slender uncinate bristles .... *P. setosa*
— Nutlets oblong or linear, the body without uncinate bristles 2
2(1). Nutlets heteromorphic, 1 of each divergent pair wingless, or merely margined, the other with a broad somewhat incurved uncinate-toothed wing .... *P. heterocarpa*
— Nutlets not heteromorphic, all 4 wing-margined or toothed 3
3(2). Margin of nutlet conspicuous, the teeth confluent at base .... *P. platycarpa*
— Margin of the nutlet very narrow or wanting, the teeth nearly or quite distinct, subulate; nutlets strongly recurved .... *P. recurvata*

Pectocarya heterocarpa Johnston Unequal Combseed. [*P. penicillata* var. *heterocarpa* Johnston]. Diffusely branched from the base; stems slender, ascending or spreading, 3–15 cm long, strigose and canescent throughout; leaves narrowly linear, 1–3 cm long, 1–2 mm wide, the hairs on the basal ones often pustulate at base; corolla minute, its limb ca 1.5 mm broad; fruiting nutlets widely divergent, dissimilar, 2 narrower and with or without a narrow margin, and 2 prominently winged-margined, the wings pectinately bristly at the apex, irregular, few toothed and with or without scattered bristles on the sides. Creosote bush, Joshua tree, blackbrush, and other warm desert shrub communities at 730 to 1200 m in San Juan and Washington counties; Arizona, Nevada, California, Baja California, and Sonora Mexico; 16 (vii).

Pectocarya platycarpa Munz & Johnston Flattened Combseed. [*P. gracilis* var. *platycarpa* Munz & Johnston]. Stems slender, diffusely branched from the base, prostrate or widely ascending, 5–20 cm long, cinereous-strigillose throughout; leaves narrowly linear to linear-oblanceolate, 0.5–1.5 mm wide, 1–3.5 cm long; calyx lobes nearly as long as the nutlets; corolla 2 mm long; nutlets divergent in pairs, sometimes heteromorphous, linear-oblong or spatulate-oblong, 2.5–3 mm long, with a wide conspicuous stramineous margin bearing irregular uncinate-tipped teeth, the odd nutlet when differentiated, with more deeply dissected wing and with more pubescent body. Creosote bush, Joshua tree and other warm desert shrub communities at 730 to 900 m in Washington County; Arizona, Nevada, California and Baja; 6 (i).

Pectocarya recurvata Johnston Bent Combseed. Stems slender, simple below, with 2 to several erect or ascending branches above, or sometimes diffusely branched throughout and more spreading, 5–25 cm long; herbage cinereous-strigose; leaves narrowly linear, acute, 1–3.5 cm long, 0.5–2 mm broad; calyx lobes ca 2 mm long in fruit, acute; nutlets divergent in pairs, linear, strongly recurved, the wing divided to or almost to the body into prominent subulate straw colored uncinate bristles, the wing prolonged into a short scarious tip, the margin uncinate bristly. Creosote bush, Joshua tree, and other warm desert shrub communities at 730 to 900 m in Washington County; California, Nevada, Arizona, New Mexico, and Baja; 1(0). The species has been taken in Utah by Neese (12952 BRY).

Pectocarya setosa Gray Saucer Combseed. [*P. setosa* var. *aperta* Johnston; *P. setosa* var. *holoptera* Johnston; *Gruvelia setosa* (Gray) Rydb.]. Stems usually diffusely branched from the base, ascending, slender to rather stout, 5–20 cm tall, herbage rather thinly strigose and setose with spreading bristlelike hairs; leaves linear to linear-oblanceolate, 5–20 mm long; calyx lobes narrowly linear, 3–4 mm long in fruit, armed with 3–6 straight divergent bristles; nutlets divergent in pairs, broadly obovate to orbicular, two of them bordered with a thin scarious wing, the other two wingless, the body of the nutlets and usually the wing bearing slender uncinate bristles, the wing usually undulate and saucerlike. Desert almond, blackbrush, other warm and cool desert shrub, mountain brush, and pinyon-juniper communities at 800 to 1700 m in Washington County; Washington and Idaho, south to Nevada, Arizona, New Mexico, and Baja; 11 (iii).

Plagiobothrys Fisch. & Mey.

Slender, glabrate or mostly softly pubescent, annual or perennial herbs; leaves mostly linear or linear-lanceolate, alternate above and either opposite at base or forming a rosette; flowers in bractless or bracteate spikelike racemes, the racemes more or less scorpioid and usually elongated in fruit; corolla white, small, salveriform, with crests at the mouth of the throat; nutlets rugose, erect or incurved, attached at or below the middle of a depressed gynobase through a caruncular scar, this decurrent on the lower part of the ventral keel or situated at the lower end of the keel and sunken below its crest.

1. Leaves opposite, at least below; scar of nutlet lateral, oblique or basal 2
— Leaves alternate; scar of nutlet lateral, near middle of nutlet 3
2(1). Scar of nutlet nearly basal; calyx lobes becoming elongate and thickened, all tending to be directed toward the same side of the fruit; plants mostly prostrate .... *P. leptocladus*
— Scar lateral or basi-lateral; calyx lobes neither elongate or much thickened, semibasally disposed; plants prostrate to ascending or erect .... *P. scouleri*

3(2). Scar of nutlet elongate, extending along crest of the ventral keel; nutlets trigonous 4
— Scar round or nearly so, at or below end of ventral keel . 5
4(3). Corolla 4–10 mm broad; nutlets irregularly rugose; reported for Utah, but no specimens have been seen by me .... *P. kingii* (Wats.) Gray
— Corolla 1–2.5 mm broad; nutlets conspicuously tessellate .... *P. jonesii*
5(3). Nutlets 4, cruciform; calyx not circumscissile . *P. tenellus*
— Nutlets 1–4, not cruciform; calyx circumscissile .....
.... *P. arizonicus*

Plagiobothrys arizonicus (Gray) Greene Stainplant. [*Eritrichium canescens* var. *arizonicum* Gray]. Stems slender, several from the base, ascending, simple or few branched, 1–4 dm high, hirsute-hispid with spreading hairs and also rather sparingly villous pubescent; leaves hirsute-hispid with more or less appressed hairs, pustulate at base, without shorter puescence, the lower linear-lanceolate, 1.5–5 cm long, upper linear-oblong to lanceolate; roots, lower parts of stems and veins of the leaves, or sometimes the whole plant purplish and staining the pressing paper or mounting sheets; spikes at length elongated, remotely flowered and bractless or with a few foliaceous bracts; calyx ca 3 mm long, cleft to near the middle, the lobes narrow-attenuate, connivent, hirsute-hispid, the tube at length usually circumscissle near the base; corolla 2–2.5 mm broad; nutlets 1–4 commonly 2, ovoid and abruptly acute at apex, with median and lateral keels often tuberculate, and with connecting transverse rugae, the areolae between smooth or minutely papillate; scar median, seated in a sunken area at the base of the keel. Creosote bush, Joshua tree, blackbrush, other warm desert shrub, and pinyon-juniper communities at 830 to 1620 m in Davis and Washington counties; Nevada, Arizona, and California; 23 (iii).

Plagiobothrys jonesii Gray Jones Popcornflower. [*Sonnea jonesii* (Gray) Greene]. Stems erect, 1 to several from the base, divergently branched, 1–3 dm high, hispid with spreading bristly hairs pustulate at base and also retrorsely hairy; basal leaves linear or narrowly oblanceolate; cauline leaves mostly lanceolate with pubescence similar to the stem but thinner; racemes terminating the branches, mostly conspicuously leafy-bracted at base, 1.5–3 cm long, the lower leaves of the branches often bearing one or few axillary flowers; calyx lobes subulate-linear, 6–8 mm long, corolla 1–2 mm broad; nutlets 3 mm long, incurved and 4-angled by the dorsal and ventral keels and the 2 lateral ridges, abruptly pointed at apex, the keel and lateral angles tuberculate, the concave surface densely tessellate; scar narrow or medial-narrow merging into the keel above and with a diverging lateral ridge extending to either side. Creosote bush, Joshua tree, blackbrush, other warm desert shrub, and sagebrush communities at 830 to 1330 m in San Juan and Washington counties; California to Arizona and Nevada; 6 (i).

Plagiobothrys leptocladus (Greene) Johnston Hairy Popcornflower. [*Eritrichium californicum* var. *subglochidiatum* Gray; *Allocarya orthocarpa* Greene, type from Cache County; *A. versicolor* Brand]. Stems branched from the base, the branches prostrate, 1–3 dm long, straight, slender, and somewhat wiry, thinly strigose or glabrate, often floriferous nearly to the base; leaves narrowly linear, the lower 3–10 cm long, glabrous or nearly so above, thinly strigose beneath, the hairs mostly pustulate at base; racemes simple, becoming loosely flowered; mature calyx lobes usually accrescent, 3–8 mm long, ca 1 mm wide, connivent or sometimes spreading, more or less definitely curved toward one side; corolla 1–2 mm broad; nutlets narrowly to broadly lanceolate, acute; dorsal side keeled only above the middle, more or less obliquely or transversely rugose, smooth, granulate or penicillate-hairy; ventral side keeled down to the basal scar, this horizontal or slightly oblique, not surrounded by a ridge, but frequently with a downwardly directed dorsal flange. Wet, saline meadows and ditch banks, in greasewood, pickleweed, saltgrass, and other salt desert shrub commuties at 1420 to 2500 m in Box Elder, Cache, Davis, Salt Lake Counties; Oregon to the Dakotas, south to Baja; 15 (0).

Plagiobothrys scouleri (H. & A.) Johnston Scouler Popcornflower. [*Myosotis scouleri* H. & A.; *Allocarya scouleri* (H. & A.) Greene; *A. cusickii* Greene; *A. hispidula* Greene; *A. nitens* Greene; *A. cognata* Greene, type from Cache County; *P. scopulorum* Johnston; *P. nelsonii* Johnston; *P. nitens* (Greene) Johnston; *P. hispidulus* Johnston; *P. cusickii* Johnston; *P. cognatus* Johnston]. Plants more or less densely branched from the base; stems 5–25 cm tall, ascending or spreading, with stiff appressed hairs; leaves 1–8 cm long, linear to oblanceolate, strigose to somewhat setose; racemes rather lax, the bracts resembling the leaves; calyx 1–1.5 (2) mm long in flower, in fruit becoming 1.5–2.5 mm long; corolla ca 1.5 mm long, inconspicuous; nutlets 1.5–2 mm long, variously roughened, with or without setose projections. Sagebrush, mountain brush, pinyon-juniper, ponderosa pine, aspen, and spruce-fir communities at 1850 to 3500 m in Box Elder, Cache, Daggett, Davis, Duchesne, Emery, Garfield, Juab, Morgan, Rich, Sanpete, Sevier, Summit, Tooele, Uintah, Utah, Wasatch, Washington and Weber counties; Alaska to Manitoba, south to California, Nevada, and New Mexico; 60 (v). Our plant belongs to var. **penicillatus** (Greene) Cronq. [*Allocarya penicillata* Greene].

Plagiobothrys tenellus (Nutt.) Gray Slender Popcornflower. [*Myosotis tenella* Nutt.; *P. parvulus* Greene; *P. asper* Greene; *P. humifusus* Jones; *P. tenellus* var. *parvulus* subvar. *humifusus* (Jones) Brand]. Stems slender, erect, freely branched from the base or sometimes simple, 5–25 cm high, softly villous with spreading and reflexed hairs; leaves of the basal rosette oblong-lanceolate or oblanceolate, obtuse or acutish, villous, 1–2.5 cm long; cauline leaves distinct, the lower ones linear-oblong, the upper becoming lanceolate or ovate-lanceolate, gradually reduced in size; spikes elongated in age and loosely flowered, only the lowest flowers bracteate; calyx densely short-villous with whitish or more often rufous hairs, ca 3 mm long in age; corolla limb 2–3 mm broad; nutlets 1.5–2 mm long, thick cruciform, pale colored, sharply ridged dorsally and on the margins, the ridges commonly tuberculate. Sagebrush, mountain brush, pinyon-juniper, and ponderosa pine communities at 1525 to 2600 m in Salt Lake and Washington counties; British Columbia, south to California, Nevada, and Arizona; 3 (i).

Tiquilia Pers.

Herbaceous or suffruticose plants with slender, forking, usually prostrate or widely spreading stems; leaves small, entire, usually strongly veined, subsessile or petiolate; flowers small, typically white, commonly extra-axillary, along leafy twigs or at the forks of the branches, sometimes glomerate, opening in late afternoon; corolla with a short, cylindrical or ampliate tube and spreading lobes, the throat naked or sometimes appendaged; stamens 4–5, included, the filaments adnate to the corolla tube; style terminal on the ovary, short to long, bilobed; stigmas 2; ovary 2–loculed or sometimes 4–loculed by the septumlike placentae, entire or 4–lobed; fruit dry, pyramidal or hemispheric, divided into usually 4 1–seeded nutlets; nutlets more or less broadly united ventrally or joined to a central prolongation of the receptacle.

1. Fruit nearly globose, unlobed, breaking at maturity into quarter sections, each quarter forming a nutlet; leaves ovate to elliptic, white-tomentose, obscurely veined .... *T. canescens*

— Fruit deeply 4–lobed, the lobes joined only by their inner angle and each forming a nutlet; leaves not tomentose 2

2(1). Plants perennial; leaves not evidently nerved, lanceolate to linear, usually very pungently setose; petiole base expanded, indurate, usually villous; flowers solitary in the leaf axils; nutlets finely warty, ovate .... *T. hispidissima*

— Plants annual; leaves with evident impressed nerves, ovate or obovate to nearly orbicular; petiole base not expanded, or indurate or villous; flowers in dense clusters at the forks of the stem; nutlets smooth or granulate .... *T. nuttallii*

Tiquilia canescens (DC.) A. Richards. Hairy Tiquilia. [*Coldenia canescens* DC.; *Stegnocarpus canescens* (DC.) Torr.; *S. leiocarya* Torr.; *C. canescens* var. *subnuda* Johnston]. Plant suffrutescent, perennial, forming mats 2–6 dm in diameter; stems several to numerous, mostly prostrate; leaves numerous, white-tomentose; petiole slender, 2–7 mm long; leaf blade ovate to elliptic-lanceolate, obtuse to broadly acute at the ends, 7–10 (15) mm long, 2–7 (9) mm wide, thickish, the margins somewhat revolute, tomentose; flowers usually solitary in the axils of the leaves; calyx sessile, persistent, at anthesis 3–4 mm long, the lobes subequal to the corolla tube, unequal, finally lanceolate, usually long-attenuate, 4–7 mm long; corolla 5–6 (8) mm long, pink, rose, or rarely white, the lobes broad and rounded, 1.8–3 (4.5) mm wide, 1.5–2 (3.5) mm long, usually villous in the buds, the margins frequently erose; filaments all differing slightly in length and height of attachment; ovary 4–ovulate, at anthesis subglobose, ca 0.8 mm high, marked with 4 longitudinal grooves; fruit at maturity glabrous or hairy, ovoid-globose, 2.5–3 mm in diameter, 2–2.5 mm high, not lobed; nutlets bony, the back convex, densely and minutely tuberculate; style seated in the pericarp, persistent until the fruit fractures into 4 nutlets 1.5–2.5 mm long. Creosote bush, Joshua tree, and other warm desert shrub communities, on limestone, at ca 900 to 1200 m in Washington County; California, Nevada, Arizona, New Mexico, Texas, and Mexico; 5 (ii).

Tiquilia latior (**Johnston**) A. Richards. [*Coldenia hispidissima* var. *latior* Johnston; *Ehretia hispida* Torr.]. Plants prostrate, forming mats 2–6 dm in diameter; stems several to numerous, dicotomously branched, spreading from a woody taproot; leaves clustered, borne mostly on very short branchlets; petioles 1–2 mm long, broadest (1–1.5) mm at the base, becoming indurate, usually pallid, the margin hispid-cilliate and the back usually glabrous or nearly so; leaf blades linear, (rarely ovate or elliptic), revolute, 4–10 mm long, 0.6–1 mm wide, usually narrower than the indurate petiole-base; flowers borne among the leaves; calyx sessile, broadly and permanently attached in the leaf axil, at anthesis ca 3 mm long, lanceolate, united at the base, ciliate below the middle, frequently terminated with a stiff bristle; corolla usually pink, ca 7.5 mm long; corolla lobes rounded, spreading, ca 2 mm long and 3 mm broad; stamens 5, unequal in length and position within the corolla tube; style 1.5–2.2 mm long, somewhat flattened, the apex bilobed; nutlets ovoid, usually only 1 or 2 maturing, 1.1–1.4 mm long, 0.8–1 mm broad, papillate, the back convex and the ventral side rounded or somewhat flattened; scar open, 1.5–2 mm wide at base, nearly as long as the nutlet, surrounded by a narrow band of smooth nonpapillate pericarp, traversed down the center by a lineate ridge; n = 9. Salt and cool desert shrub and pinyon-juniper communities at 900 to 2100 m in Beaver, Emery, Garfield, Grand, Kane, Millard, San Juan, Washington, and Wayne counties; Arizona; a Colorado Plateau endemic; 41 (vi). *Tiquilia latior* is closely allied to *T. hispidissima* (T. & G.) A. Richards. of Texas, New Mexico, and Mexico, but has broader leaves, with densely ciliate petioles and larger papillate nutlets 1.5–2 mm long.

Tiquilia nuttallii (**Hook.**) A. Richards. [*Coldenia nuttallii* Hook.; *T. parviflora* Nutt. ex Hook.; *T. brevifolia* Nutt. ex Torr.]. Prostrate annual with slender, somewhat brittle, dichotomously branched stems forming mats 1–3.5 dm broad, finely strigose; leaves ovate to suborbicular, 4–8 mm long, narrowly revolute, often hispid on the margins, with 2–3 pairs of distinct veins dorsally, thinly strigose on the upper surface with somewhat stiff hairs, hirsute on the lower surface; petioles slender, usually as long or longer than the blade; flowers in compact clusters in the forks and at the ends of the branches; calyx lobes linear-subulate, 4–5 mm long, villous or setulose on the back, sparsely but conspicuously hispid on the margins; corolla pink or nearly white, little exceeding the calyx, the limb 2–2.5 mm broad, the tube with 5 triangular scales near the base; nutlets oblong-ovoid, smooth and shining. Salt desert shrub, creosote bush, mixed desert shrub, and pinyon-juniper communities at 900 to 1830 m in Beaver, Box Elder, Duchesne, Iron, Juab, Millard, Tooele, Uintah, and Washington counties; Washington to Idaho and Wyoming, south to California, and Arizona; 19 (i).

BUDDLEJACEAE Wilhelm

Logania Family

Shrubs or trees; leaves opposite, entire or crenate, stipulate; flowers regular, perfect, 4– or 5–merous, sympetalous, variously arranged; calyx deeply lobed or with separate sepals; corolla salverform to tubular or campanulate; stamens perigynous, as many as corolla lobes and alternate with them; pistil 1, the ovary 2–loculed, superior; style 1; fruit a capsule; x = 19.

Buddleja L.

Shrubs or small trees; leaves opposite, simple, entire to dentate; flowers mostly 4–merous; calyx campanulate;

corolla salverform or rotate-campanulate, the lobes ovate or rounded; anthers subsessile on throat or tube of corolla; capsule globose to ellipsoid, bivalved, septicidal; seeds many.

1. Leaves 5–25 cm long, lance-ovate to lanceolate; flowers lilac with an orange yellow center; plants cultivated .... B. davidii
— Leaves 1–3 cm long, linear, with revolute margins; flowers creamy yellow to purplish; plants indigenous in Washington County .... B. utahensis

Buddleja davidii Franchet Shrubs or small trees to 5 m tall; leaves 5–25 cm long, lance-ovate to lanceolate, acuminate, rounded to acute basally, shortly petiolate, serrate to doubly so, dark green above, densely white-tomentose below; flowers fragrant, lilac, with an orange yellow throat, in slender panicles 12–40 cm long; 2n = 76. Cultivated ornamental in Cache, Salt Lake, Utah, and Weber counties; widely grown in the U. S.; introduced from China; 6 (0).

Buddleja utahensis Cov. Low much-branched shrubs, 2–3 (5) dm tall, pubescent with densely lanate-tomentose stellate hairs; leaves 1–3 cm long, linear, with revolute margins, subsessile; axils often bearing fascicles of smaller leaves; flowers in glomerules, forming heads in an interrupted spike ca 10–15 mm thick; corolla creamy yellow to purplish, 4–5 mm long, the lobes ca 1 mm long, rounded, the tube tomentulose. Blackbrush, Joshua tree, and other warm desert shrub communities, mainly on limestone outcrops, at 800 to 1100 m in Washington (type from near St. George) County; Arizona, Nevada, and California; 8 (ii).

CACTACEAE A. L. Juss.
Cactus Family

Perennial succulent woody or herbaceous plants, with spiny, glochidiate, or rarely unarmed, globose, cylindric, columnar, or flattened stems; stems ribbed, smooth, or tuberculate; leaves lacking, or green, terete, and caducous (*Opuntia*); areoles axillary (regardless of apparent position), bearing wool, glochids, spines, branches, or flowers; perianth of numerous segments grading from sepals to petals, imbricate, the bases more or less united, inserted on a hypanthium; stamens numerous, variously inserted within the hypanthium tube; style 1; stigmatic lobes several; ovary inferior; fruit a dry or fleshy many-seeded berry; x = 11.

Benson, L. 1982. The cacti of the United States and Canada. Stanford University Press, California. 1044 pp.

1. Stems jointed, the joints flattened, clavate, or cylindric; areoles with glochids and/or spines (or spineless), subtended by caducous terete green leaves when young .... *Opuntia*
— Stems hemispheric or cylindroid, not jointed; areoles with hair or spines but no glochids 2
2(1). Flowers borne in axils of tubercles or at bases of grooves, removed from the spiniferous areoles; central spine hooked, dark purple; small hemispheric or cylindrical plants of Washington County .... *Mammillaria*
— Flowers borne variously, seldom as above; central spines hooked or straight but, if hooked, the plants not of Washington County 3
3(2). Stems with tubercles spirally arranged; tubercles distinctly grooved on upper side; flowers pink or yellow .... *Coryphantha*
— Stems ribbed; tubercles not grooved; flowers variously colored 4
4(3). Flowers borne laterally below the stem apex; hypanthium spiny .... *Echinocereus*
— Flowers terminal on the stems; hypanthium devoid of spines 5
5(4). Stems 20–50 cm in diameter or more, mainly 2–10 dm tall; upper axils and ovaries not woolly; plants of Washington County .... *Ferocactus*
— Stems usually much smaller, or if, as rarely, approaching the lower limits as described above, the ovaries and upper axils woolly 6
6(5). Stems mainly 12–25 cm in diameter; spines flattened, annular; ovaries and upper axils woolly; plants reported for Kane and Washington counties .... *Echinocactus*
— Stems mainly 3–10 cm in diameter; spines variously terete, subterete, or flattened, but not annular; ovaries and upper axils not or rarely woolly 7
7(6). Spines straight, purplish or reddish, 2–5 cm long or more; flowers rose pink; plants of the Beaver Dam Mountains, Washington County .... *Neolloydia*
— Spines hooked or some or all of them straight; flowers variously colored; plants not of western Washington County or, if so, the flowers yellow 8
8(7). Stems with spines all straight, depressed-hemispheric; flowers white to yellow or pale pinkish, mainly 1–2 cm long .... *Pediocactus*
— Stems with at least some spines hooked or, if straight, the flowers rose pink to violet or more than 2 cm long .... *Sclerocactus*

Coryphantha (Engelm.) Lem.

Plants depressed-hemispheric to hemispheric or shortly cylindric, solitary or colonial; tubercles separate; areoles circular; spines smooth; central spines 0 or 3–12 per areole, transitional to radials, straight, elliptic in cross-section; radial spines 12–40 per areole, straight, subterete; flowers axillary at tubercle base, at end of a felty persistent groove connected to the areole, borne near the summit of the stem, funnelform, the perianth pink purple to rose or yellow; fruit fleshy, green or red, indehiscent; seeds black or brown.

1. Flowers yellow; fruit red at maturity, globular; plants rare in Garfield and Kane counties .... *C. missouriensis*
— Flowers pink purple to rose; fruit green at maturity, ellipsoid; plants widely distributed .... *C. vivipara*

Coryphantha missouriensis (Sweet) Britt. & Rose [*Mammillaria missouriensis* Sweet]. Stems commonly solitary, depressed hemispheric, 2–5 cm tall, 3–8 cm wide; tubercles 6–9 mm long; areoles few; radial spines 10–19, spreading; flowers 3.8–5 cm wide and long; sepaloids greenish, the margins yellowish or whitish; petaloids yellow; filaments yellow; anthers yellow; style green, 12–25 mm long; fruit red, ca 1 cm thick; seeds black, 2–2.5 mm wide.; 2n = 44. Cool desert shrub, juniper, and ponderosa pine communities in Garfield (type from Hells Backbone) and Kane (lectotype from Buckskin Mts.) counties; Arizona. Our material belongs to var. *marstonii* (Clover) Benson [*C. marstonii* Clover]; 1 (0). The species is distributed from Montana east to North Dakota, south to Arizona, New Mexico, and Texas.

Coryphantha vivipara (Nutt.) Britt. & Rose [*Cactus viviparus* Nutt.; *Mammillaria vivipara* (Nutt.) Haw.]. Stems solitary or colonial, depressed hemispheric to short-cylindric, mainly 2–15 cm tall, 2–10 cm wide; tubercles 6–15 (23) mm long; areoles 1.5–3 mm wide; central spines 3–12, whitish basally, dark apically, mainly 12–20 mm long; radial spines 12–20, spreading, obscuring the stem; flowers 2.5–5 cm wide and long; sepaloids greenish, the margins variously colored; petaloids pink purple or rose; anthers yellow; fruit green, ellipsoid, 12–25 mm long; seeds brown, reticulate, 1.5–2 mm wide; n = 11. Like other of our depressed-hemispheric cacti, the plants of this species expand as they take up water in springtime. Following flowering the plants dry and shrink downward into the substrate surface. Plants conspicuous at flowering become difficult to observe when dormant. The juice of *C. vivipara* is apparently unique among our species in being non-mucilaginous. Three more or less distinctive varieties are present in Utah.

1. Central spines 4; flowers ca 3.8 cm wide; plants of Carbon and Uintah counties . . . *C. vivipara* var. *vivipara*
— Central spines 4–7; flowers wider or, if narrower, the plants not of northeastern Utah 2
2(1). Flowers ca 2.5–3 cm wide; radial spines 12–20, 9–12 mm long; petaloids yellowish, greenish, or pinkish; plants of Washington County . . . . . *C. vivipara* var. *deserti*
— Flowers mainly 3.8–5 cm wide; radial spines 20–30; petaloids pink purple to rose; plants rather broadly distributed . . . . *C. vivipara* var. *arizonica*

Var. *arizonica* (Engelm.) W. T. Marshall [*Mammillaria arizonica* Engelm.]. Desert shrub and pinyon-juniper communities at 1586 to 2440 m in Beaver, Garfield, Juab, Kane, Millard, Piute, San Juan, Sanpete, Sevier, Tooele, Washington, and Wayne counties; Nevada to Colorado, south to Arizona and New Mexico; 33 (xiii). This variety is locally common on limestone and dolomite outcrops and on gravels degraded from them. It is a beautiful plant when in flower, the violet flowers contrasting with the thatch of whitish spines.

Var. *deserti* (Engelm.) W. T. Marshall [*Mammillaria deserti* Engelm.; *M. chlorantha* Engelm., type from St. George]. The small yellowish to pinkish flowers are apparently diagnostic. Warm desert shrub communities at 760 to 980 m in Washington County; Arizona, Nevada, and California; 2 (i).

Var. *vivipara* Desert shrub and pinyon juniper communities in Carbon and Duchesne counties; Alberta to Manitoba, south to Oregon, New Mexico, and Texas; 1 (0).

Echinocactus Link & Otto

Stems solitary or few to many, subglobose to cylindric, woolly at the apex, few- to many-ribbed; areoles large; spines broad, flattened- triangular, with transverse annular rings; flowers borne subapically, yellow; floral tube bearing spiny persistent scales; ovary clothed with narrow scales having mats of wool; fruit densely white-woolly, dry at maturity; seeds black, shining.

Echinocactus polycephalus Engelm. & Bigel. Stems mostly 2–3 dm tall and 1–2 dm thick (or more); ribs 10–21; areoles 10–12 mm long; radial spines 8–10, 2.5–5 cm long, often reddish when young, subulate, triangular-flattened; central spines 3–5, stouter than the radial, annulate, curved but not hooked, 3–8 cm long; flowers 5–6 cm long, yellow; perianth segments narrowly oblong; fruits 15–25 mm long, dehiscing by a basal pore; seeds angular, black. Two varieties are potentially present in Utah.

1. Spines felty, at least when young; seeds papillate, dull or shining from the papillae; plants reported for Washington County . . . . *E. polycephalus* var. *polycephalus*
— Spines smooth or with scattered hairs; seeds smooth and shining; plants potentially of Kane County . . . . *E. polycephalus* var. *xeranthemoides*

Var. *polycephalus* Warm desert shrubland on the Beaver Dam slope, Washington County (reported by Meyer); Nevada, California, and Arizona; 0 (0).

Var. *xeranthemoides* Coult. Pinyon-juniper and desert shrub communities near Kanab, Kane County; Arizona; a Plateau endemic; 0 (0). This report is based on two collections by pioneer collectors, one by A. L. Siler (in "Kanab Mts.") in 1881 and the other by Dr. E. Palmer in 1877 (in "S. Utah"). Since it is probable that neither knew where the Utah-Arizona boundary was situated (it was surveyed in 1879), the collections noted could have been taken from nearby Arizona. However, the plants should be sought near Kanab.

Echinocereus Engelm. in Wisliz.

Stems erect or ascending, solitary or more usually colonial, cylindric or subcylindric; areole small; central spines 1–6; radial spines 5–12, acicular to subulate, flattened or subterete; flowers borne laterally, below the stem apex, the bud breaking through the epidermis above the areole, large and showy, pink purple to scarlet; stigmas green; fruit fleshy, spiny, not regularly dehiscent, the spine clusters deciduous as fruit matures.

1. Flowers pink purple to rose; stems solitary or few, often over 10 cm tall . . . . *E. engelmannii*
— Flowers scarlet; stems often 10 cm long or less, few to numerous in compact hemispheric clusters . . . . *E. triglochidiatus*

Echinocereus engelmannii (Parry) Lem. Engelmann Hedgehog-cactus. [*Cereus engelmannii* Parry]. Stems solitary or 2 to several (or rarely many) and loosely clustered, mainly 10–30 cm tall, 5–9 cm thick; ribs 10–13; tubercles not prominent; areoles small, subcircular; central spines 2–6, stout, more or less curved or twisted, 2–5 cm long; radial spines 6–12, 7–15 mm long, appressed and spreading; flowers 5–9 cm long, pink purple to rose; perianth segments oblong; fruit ovoid to oblong, green or turning red, the spine clusters deciduous; seeds black, globose, pitted, 1–1.5 mm long. This widely ranging southwestern species consists of a series of morphologically differing but intergrading segregates, which largely lack geographical integrity. Utah material is assigned to two of named segregates.

1. Lowermost central spines mainly 3.5–6 cm long, not markedly differing in color from the other spines; plants of canyons of the Colorado . . . . *E. engelmannii* var. *variegatus*
— Lowermost central spines mainly 2.8–4.5 cm long, often markedly differing in color from other spines; plants mainly not of the Colorado canyons . . . . *E. engelmannii* var. *chrysocentrus*

Var. *chrysocentrus* (Engelm. & Bigel.) Engelm. ex Rumpler [*Cereus engelmannii* var. *chrysocentrus* Engelm. & Bigel.; *E. engelmannii* var. *purpureus* L. Benson, type from near St. George]. Larrea, Joshua tree, shadscale, and mountain brush communities at 760 to 1865 m in Beaver, Juab, Kane (inter var. *variegatus*), Millard, and Washington counties; Nevada, California, and Arizona; 28 (iv). Maintenance of plants passing under the name of var. *purpureus* is only arbitrarily possible. They grade continuously in all known populations with var. *chrysocentrus*, and purple spined phases occur sporadically throughout the range of the species in Utah.

Var. *variegatus* (Engelm. & Bigel.) Engelm. ex Rumpler [*Cereus engelmannii* var. *variegatus* Engelm. & Bigel.]. Blackbrush, shadscale, and pinyon-juniper communities at 1125 to 1895 m in Garfield, Kane, and San Juan counties; Arizona; 6 (ii). Purported differences between this phase of the species complex and those noted for var. *chrysocentrus* are tenuous, and the two phases could be combined without doing serious injustice to their taxonomy.

***Echinocereus triglochidiatus* Engelm.** Claretcup. [*E. coccineus* Engelm.]. Stems few to several hundred in compact hemispheric clumps or mounds, mainly 8–15 cm long, 3–6 cm thick; ribs 9 or 10, the tubercles not prominent; areoles circular, bearing a white felty mat when young; central spines 1–3, 8–40 mm long or more, stout, straight or curved to twisted; radial spines 5–8, 4–35 mm long, not appressed, spreading; flowers 5–7.5 mm long, scarlet; perianth segments cuneate-obovate; fruit red at maturity, obovoid to cylindroid; seeds papillate, 1.5–2 mm long; n = 22. Two rather weakly separable varieties are present in Utah.

1. Central spines twisting or curved; flowers often over 4 cm wide; plants of Millard, Beaver and Washington counties .... *E. triglochidiatus* var. *mojavensis*
— Central spines straight; flowers often less than 4 cm wide; plants of broad distribution ....
.... *E. triglochidiatus* var. *melanacanthus*

Var. *melanacanthus* (Engelm.) L. Benson [*Cereus coccineus* var. *melanacanthus* Engelm.; *E. triglochidiatus* var. *inermis* (K. Schum.) Rowley; *E. phoeniceus* var. *inermis* K. Schum., neotype from San Juan County]. Blackbrush, ephedra, sagebrush, pinyon-juniper, mountain brush, and aspen communities at 975 to 2562 m in Beaver, Carbon, Daggett, Duchesne, Emery, Garfield, Grand, Iron, Juab, Kane, Millard, Piute, San Juan, Sanpete, Sevier, Tooele, Uintah, Utah, Washington, and Wayne counties; Nevada to Colorado, south to California, Arizona, New Mexico, Texas, and Mexico; 86 (xvi). There is a cline within the specimens from eastern Utah and western Colorado from densely spiny to no spines at all. The spineless plants have passed under the name of var. *inermis*, but they do not seem to represent a taxon.

Var. *mojavensis* (Engelm. & Bigel.) L. Benson [*Cereus mojavensis* Engelm. & Bigel.]. Mixed desert shrub, pinyon-juniper, and ponderosa pine communities at 1550 to 2257 m in Beaver, Millard, and Washington counties; 4 (i).

Ferocactus **Britt. & Rose**

Plants hemispheric to cylindric, massive; ribs thick, prominent, somewhat spirally arranged; spines coarse, the central ones flattened and transversely annulate, not hooked; areoles large, more or less woolly when young; flowers subterminal, yellow, funnelform; stamens numerous; ovary and floral tube scaly, not woolly; fruit oblong in outline, dry at maturity, dehiscent by a basal pore.

***Ferocactus acanthodes* (Lem.) Britt. & Rose** Barrel Cactus. [*Echinocactus acanthodes* Lem.]. Plants mainly 2–15 dm tall and 2–5 dm thick or more; ribs 20–30; areole 1–1.5 cm long, brown-woolly when young; spines pink, red, or yellow, the central ones 1–4, subulate, flattened, annulate, curved, 4–8 cm long or more; radial spines with mixed coarse and slender ones; flowers 4–6 cm long, the scales of the tube and ovary overlapping when young, ovate; perianth segments oblong to spatulate; filaments yellow; fruit 3–3.5 cm long; seeds 2–3 mm long, reticulate.; n = 11. Limestone and dolomite outcrops and gravels at 760 to 1375 m in Washington County; Nevada, California, and Arizona; 2 (i). Our material belongs to var. *lecontei* (Engelm.) Lindsay [*Echinocactus lecontei* Engelm.]. This is the largest cactus native to Utah; it is distinguished from *Echinocactus polycephalus* by the large size, glabrous ovaries, and merely short-woolly areoles.

Mammillaria **Haw.**

Subglobose to shortly cylindric plants, stems solitary or few; tubercles many, elongate, in spiral rows; areoles spiniferous; spines smooth, the central 1–4 straight or 1 or more hooked; flowers borne between tubercles, diurnal; fruit fleshy, red, lacking appendages, elongate.

***Mammillaria tetrancistra* Engelm.** [*Phellosperma tetrancistra* (Engelm.) Britt. & Rose]. Stems 4–10 cm tall or more, 4–6 cm wide; tubercles 4–10 mm long, more or less woolly in the axils when young; central spines 1–4, dark, 1 or more hooked, 10–15 mm long; radial spines 30–45, mostly whitish; flowers 25–30 mm long; sepaloids green with pink margins; petaloids rose to pink purple; fruit initially green and included below the spines, some finally scarlet and 12–20 mm long, often protruding beyond the spines. Warm desert shrub communities at 760 to 1300 m in Washington County; Nevada, California, Arizona; 6 (iii). The seeds evidently mature while the fruit is still green and included below the spines. Ultimately, at least some of the fruit ripens to scarlet and typically protrudes beyond the spines. Several stages of fruit development—from green (included) to red (included to exserted)—have been observed on a single plant.

Neolloydia **Britt. & Rose**

Subglobose to shortly cylindric plants, mostly solitary; ribbed and more or less tuberculate; areoles small; central spines 1 to several or lacking, straight (or curved), not hooked; radial spines 9–10; flowers borne subapically at the base of a woolly groove, purple or pink purple; fruit green, drying tan, dehiscing by a basal pore, the scales and their axils glabrous.

***Neolloydia johnsonii* (Parry) L. Benson** [*Echinocactus johnsonii* Parry in Engelm., type from near St. George; *Echinomastus johnsonii* (Parry) Baxter; *Ferocactus johnsonii* (Parry) Britt. & Rose]. Stems solitary, seldom branched, 8–20 cm tall, 5–10 cm thick, the ribs 17–21, obscured by interlocking spines; areoles with a short narrow woolly groove running to the axil; central spines pink to reddish or purplish, 3–4 cm long, terete; radial spines paler in color; flowers 5–6 cm long, purple or pink purple; fruit green, drying tan, oblong, 10–15 mm long, nearly naked, splitting dorsally; seeds ca 2.5 mm long, papillate; n = 11. Warm desert shrub community at 760 to 1250 m in Washington County; Nevada, Arizona, and California; 5 (ii).

Opuntia Miller

Stems jointed, the joints flattened, cylindric, or clavate; areoles with glochids (i.e., detachable barbed spinules), and commonly with 1 or more stout spines (less commonly spineless); spines naked or sheathed; leaves terete, fleshy, caducous; flowers borne in areoles of previous years growth, variously colored; floral tube cup-shaped; ovary with areoles; stamens numerous; stigmas short; fruit fleshy or dry, armed or unarmed; seeds with a bony aril, flattened.

1. Stem joints cylindric or clavate; spines with detachable epidermal sheaths, at least apically (subgenus Cylindropuntia) 2
— Stem joints flattened; spines not sheathed (subgenus Opuntia) 5
2(1). Stem joints clavate, 1 or few above ground, mainly 3–10 cm tall, arising from a tuberous subterranean joint; plants of Millard, Juab, and Tooele counties .... *O. pulchella*
— Stem joints cylindric, several to numerous above ground, mainly 3–20 dm tall, not arising from a tuberous joint; plants of various distribution 3
3(2). Joints mainly less than 2 cm thick; fruits fleshy at maturity; plants of rather broad distribution in Utah .... *O. whipplei*
— Joints mainly over 2.5 cm thick; fruits dry at maturity; plants mainly of Washington County 4
4(3). Ridge of tubercle on mature joints mainly 18–22 mm long, more than 3 times longer than broad; longer terminal joints mainly more than 15 cm long .... *O. acanthocarpa*
— Ridge of tubercle on mature joints mainly 10–15 mm long, only 1–2 times longer than broad; longer terminal joints mainly shorter than 15 cm long .... *O. echinocarpa*
5(1). Areoles with glochids only; spines not developed (except in hybrids with other taxa) or if present the glochids very numerous and 4–10 mm long; plants of Washington, Kane, and San Juan, and less commonly of Emery, Garfield and Wayne counties 6
— Areoles with glochids and spines, at least some, or if lacking (a condition probable in all species), of different distribution 8
6(5). Plants with pyriform, erect, usually bluish or purplish tinged joints, these erect, forming vase-shaped clusters; fruit dry at maturity .... *O. basilaris*
— Plants with ovate to elliptic, sprawling to decumbent, usually green joints, these often rooting in contact with the soil; fruit dry or not 7
7(6). Fruit fleshy; joints often wrinkled; plants sporadic in Utah except in Washington County, where locally abundant .... *O. macrorhiza*
— Fruit dry; joints seldom if ever wrinkled; plants mainly of Kane County .. spineless phase of *O. erinacea*
8(5). Fruit dry at maturity, finally tan, green or reddish when young; seeds mainly 4–8 mm long, rough and irregular in outline (key non-fruiting specimens both ways) 9
— Fruit fleshy at maturity, red or reddish purple to purple; seeds mainly 2.5–4.5 mm long, smooth and regular in outline 12
9(8). Largest joints 2–8 cm long, 1.5–3.5 cm wide, readily detachable (carried burlike by animals) .... *O. fragilis*
— Largest joints mainly 7–15 cm long or more, 4–12 cm broad or more, not readily detachable 10
10(9). Spines not especially flattened, even basally, terete or nearly so, or rarely lacking; plants transitional to the next .... *O. polyacantha*
— Spines at least somewhat flattened, at least basally, usually elliptic in cross-section 11
11(10). Spines less than 1 mm thick, more or less flexible; joints mainly 5–15 cm long and 3–10 cm wide; plants rather widespread, transitional to the next .. *O. erinacea*
— Spines over 1 mm thick (at least some), not especially flexible; joints often over 15 cm long and 10 cm wide; plants of the Glen Canyon vicinity (transitional to *O. phaeacantha*) .... *O. nicholii*
12(8). Spines terete to subterete, not flattened (except when hybridizing with *O. phaeacantha*), commonly 1–6 per areole 13
— Spines at least basally flattened, narrowly elliptic in cross-section, commonly 3 per areole 14
13(12). Spines gray or brownish, or more commonly lacking; plants usually prostrate, rooting along the joint in contact with the soil, not forming upright clumps; largest joints mainly less than 12 cm long; plants scattered in Utah .... *O. macrorhiza*
— Spines tan or variously colored; plants usually upright and with several joints standing above the ground; largest joints mainly more than 15 cm long; plants of San Juan and Washington counties .... *O. littoralis*
14(12). Joints subcircular in outline; spines all deflexed, yellow; plants of Washington County .... *O. chlorotica*
— Joints mainly obovate in outline; spines spreading in various directions, brown to tan or gray; plants of rather broad distribution .... *O. phaeacantha*

Opuntia acanthocarpa Engelm. & Bigel. Buckhorn Cholla. Shrubs, mainly 8–15 dm tall or more; trunk short; larger terminal joints mostly 12–50 cm long, 2–3 cm thick; tubercles decurrent along the joint, mostly 15–50 mm long and 4–6 mm wide; leaves caducous; areoles circular; spines 6–20 or more per areole, 1–4 cm long, variously colored, the sheaths straw colored; glochids minute; flowers 4–6 cm long; sepaloids greenish yellow to brownish; petaloids typically yellow orange to bronze; ovary spiny; anthers yellow, the filaments wine colored; fruit dry, tan or brown, spiny, 2–4 cm long; seeds 5–8 mm long, tan or whitish; 2n = 22. Creosote bush-Joshua tree, and other warm desert shrub communities at 760 to 1220 m in Washington County; Nevada, California, Arizona, and Mexico; 4 (i). Two varieties are reported from Utah by Benson (1982); **var. acanthocarpa**, with tubercular ridges 30–50 mm long and longer joints 25–50 cm long; and **var. coloradoensis** L. Benson, with tubercular ridges 15–22 mm long and larger joints 12–30 cm long. More specimens are necessary to determine the nature of the Utah materials.

Opuntia basilaris Engelm. & Bigel. Plants mainly 10–30 cm high and to 3 broad or more; joints typically blue, blue green or violet green, obpyriform, obovate, or spatulate, 5–30 cm long, 2.5–12 cm broad; areoles circular, 9–17 mm apart; spines lacking; glochids brown to tan; flowers 5–8 cm long; sepaloids greenish, edged with violet; petaloids violet; fruit 2.6–3.3 cm long, dry at maturity, green, becoming tan or gray; seeds ca 6–8 mm long, white or grayish; n = 11. Two more or less distinctive and geographically correlated varieties are present.

1. Joints obpyriform, seldom otherwise, suffused with violet or blue; glochids brown; plants of Washington and San Juan counties .... *O. basilaris* var. *basilaris*
— Joints mainly obovate to spatulate, suffused with blue or purple; glochids tan to yellowish; plants of Emery, Garfield, and Wayne counties ...... *O. basilaris* var. *heilii*

Var. ***basilaris*** Warm desert shrub community at 760 to 1770 m in San Juan (Cataract Canyon) and Washington counties; Nevada, California, Arizona, and Mexico; 9 (ii). The materials from Cataract Canyon differ in tenuous ways from the typical material in Washington County; they do not seem worthy of taxonomic recognition.

Var. ***heilii*** Welsh & Neese Salt desert shrub communities at 1460 to 1680 m in Emery, Garfield, and Wayne (type from SW of Hanksville) counties; endemic; 3 (0).

Opuntia chlorotica Engelm. & Bigel. Pancake Pricklypear. Shrubby plants, mainly 6–15 dm tall; trunk to 30 cm long; larger joints 15–20 cm long and about as broad, orbicular to suborbicular, blue green; areoles elliptic, ca 20 mm apart; spines present in all but basal areoles, yellowish, 1–6, all deflexed, straight or curved at the base, 2.5–4 cm long; glochids yellow; flowers 5–8 mm long; sepaloids and petaloids yellow, or suffused with red; ovary with glochids and some spinules; fruit fleshy, grayish, tinged with purple, lacking spines; seeds 2.2–3 mm long, tan, smooth. Creosote bush, blackbrush, and other warm desert shrub communities at 1400 m in Washington County (Beaver Dam Wash, Pine Valley Mts., and Zion Canyon); Nevada, California, Arizona, New Mexico, and Mexico; 3 (ii).

Opuntia echinocarpa Engelm. & Bigel. Pale Cholla. Shrubs, mainly 8–15 dm tall; trunk to 1/2 of plant height; larger terminal joints mainly less than 15 cm long (5–15), 2–4 cm thick; tubercles decurrent along the joint, mostly 6–15 mm long and 4–5 mm wide; leaves caducous; areoles circular; spines 3–12 per areole, 1–3 cm long, straw colored or silvery or yellow; sheathes colored like the spines; glochids minute; flowers 3–4.5 cm long; sepaloids and petaloids greenish-yellow, the outer sometimes suffused reddish; fruit dry, green, turning tan; n = 11. Creosote bush, Joshua tree, blackbrush, and shadscale communities at 760 to 1376 m in Beaver (?) and Washington counties; Nevada, California, Arizona, and Mexico; 3 (ii).

Opuntia erinacea Engelm. Plants mainly 10–30 cm tall and to 1 m in diameter or more; larger joints obovate to spatulate, glaucous, 5–19 cm long, 3–11 cm wide; areoles small, 4–18 mm apart; spines at all or most areoles or only in the upper ones (or lacking?), 4–9 per areole, deflexed, flexible, straight or irregularly curved, 0.5–10 cm long, less than 1 mm thick, at least some clearly flattened (at least basally); glochids yellow to tan or brown; flowers 4.5–7.5 cm long; sepaloids commonly greenish; petaloids yellow, bronze, pink, or violet; ovary usually spiny; fruit dry, tan to brown, spiny, 2.5–3 cm long, deciduous; seeds 4–6 mm long, whitish. Plants of this complex of morphologically differing forms intergrade freely among themselves, and they hybridize with the dry fruited *O. fragilis*, *O. nicholii*, and the varieties of *O. polyacantha*. Further, they hybridize with the fleshy fruited *O. phaeacantha*, *O. littoralis*, and likely with *O. macrorhiza*. Intergradation with *O. polyacantha* is sufficiently complete as to pose the question of whether maintainance of the proposed segregates within separate species is reasonable. I follow tradition in maintaining them thusly, because, if a case is made for combining these two species then a similiar case must be considered for union of all platyopuntias with which they intergrade into a single polymorphic species. The variants could then be recognized as belonging to numerous infraspecific taxa, approximately equal to the number of taxa recognized currently. Such a proposal would solve none of the basic problems resulting from intergradation of taxa, despite the convenience of having only one name at the specific level for all of the prickly pears. Four varieties are recognized.

1. Areoles with glochids only, lacking spines or essentially spineless; flowers ordinarily yellow; joints typically turgid; plants mainly of Kane County . *O. erinacea* var. *aurea*
— Areoles with glochids and spines; flowers rose to orange or yellow; plants of broad distribution 2

2(1). Spines lacking in much of the joint, mainly confined to the upper half or along the upper edge; plants widespread .... *O. erinacea* var. *utahensis*
— Spines present in much or all of the joint; plants of the southern half of Utah 3

3(2). Spines stiff, rigid, the longest mainly 1–4 cm long; plants widespread in southern Utah ....
.... *O. erinacea* var. *erinacea*
— Spines slender and more or less flexible, the longest 3–10 cm long; plants of Washington County ....
.... *O. erinacea* var. *ursina*

Var. ***aurea*** (Baxter) Welsh [*O. aurea* Baxter, type from Pipe Springs, Arizona; *O. basilaris* var. *aurea* (Baxter) W. T. Marshall. n = 33. Sagebrush, pinyon-juniper, and ponderosa pine communities at ca 1220 to 2075 m in Kane and Washington counties; Arizona; a Plateau endemic; 6 (iii). Intermediates occur between this taxon and adjacent phases of *O. erinacea* and *O. polyacantha*. Previous inclusion with *O. basilaris* was evidently erroneous. Lack of spines was the main criterion for inclusion within that entity, but the relationship clearly lies with *O. erinacea* var. *utahensis*.

Var. ***erinacea*** [*O. hystricina* sensu Utah authors]. Warm and mixed desert shrub communities at 885 to 1285 m in Beaver, Emery, Grand, Kane, Millard, San Juan, Washington, and Wayne counties; Nevada to Colorado, California, Arizona, and New Mexico; 20 (ix). It has been postulated (Benson 1982) that this phase of *O. erinacea* is one of the putative parents of *O. nicholii*, the other being *O. phaeacantha*. Along Glen Canyon there are many specimens that bridge this variety with *O. nicholii*.

Var. ***ursina*** (Weber) Parish in Jepson [*O. ursina* Weber in Boiss.; *O. rubrifolia* Engelm. ex Coult., type from St. George]. Warm desert shrub community at 760 to 900 m in Washington County; Nevada, California, and Arizona; 4 (ii). Our material shows evidence of mixing with var. *erinacea*.

Var. ***utahensis*** (Engelm.) L. Benson [*O. sphaerocarpa* var. *utahensis* Engelm.; *O. rhodantha* K. Schum.; *O. xanthostemma* K. Schum.; *O. erinacea* var. *xanthostemma* (K. Schum.) L. Benson]. Blackbrush, pinyon-juniper, sagebrush, mountain brush, ponderosa pine, and aspen communities at 1220 to 2810 m in most if not all Utah counties; Idaho to California, Arizona, New Mexico, and Wyoming; 24 (xiii). This variety is the counterpart of *O. polyacantha* var. *polyacantha*, with which it hybridizes wherever they meet.

Opuntia fragilis (**Nutt.**) **Haw.** Brittle Pricklypear. [*Cactus fragilis* Nutt.; *O. fragilis* var. *denudata* Wiegand & Backeberg; *O. brachyarthra* Engelm. & Bigel.; *O. fragilis* var. *brachyarthra* (Engelm. & Bigel.) Coult.]. Plants mat-forming, mainly 5–10 cm tall and to 5 dm wide; larger joints 2.5–7 cm long, 1–4 cm wide, obovate to ovate or orbicular to elliptic in outline, blue green, often to half as thick as wide or more, readily detached and transported by animals; leaves caducous; areoles 3–12 mm apart; spines in most areoles or only in the upper ones or sometimes lacking, 1–9 per areole, disoriented, 4–25 mm long or more, terete to somewhat flattened; glochids tan to brown; flowers 3.5–6 cm long; sepaloids greenish; petaloids yellowish, greenish, bronze, or violet; fruit dry, tan, spiny or spineless, 1.2–2.5 cm long; $n = 33$. This is a taxon of unusually great latitude of habitat types ranging from low elevation marshlands and riparian sites upwards to pinyon-juniper, ponderosa pine, sagebrush, mountain brush, and aspen communities at 1370 to 2565 m in Box Elder, Carbon, Duchesne, Emery, Garfield, Piute, Rich, Salt Lake, San Juan, Sanpete, Sevier, Uintah, Utah, Wayne, and Weber counties; British Columbia to Ontario, south to California, Nevada, Arizona, New Mexico, Texas, and Iowa; 13 (iii). Morphological amplitude of our specimens is greater than that reported for the species as a whole (Benson 1982), excluding hybrids presumably intermediate with both *O. erinacea* and *O. polyacantha*. Recognition of proposed infraspecific taxa seems moot.

Opuntia littoralis (**Engelm.**) **Britt. & Rose** [*Opuntia engelmannii* var. *littoralis* Engelm.]. Plants mainly 30–50 cm high and 0.5–1.2 m wide, erect or more or less sprawling; larger joints 10–18 cm long, 7–14 cm wide, obovate to orbicular, green or glaucous; areoles mainly 15–30 mm apart; spines in all, or only in the upper, areoles, 1–6 per areole, mainly 2–7 cm long, spreading to deflexed, straight or curved, terete to somewhat flattened; glochids yellowish to brown; flowers 5–7.5 cm long; sepaloids greenish; petaloids yellow, the bases sometimes violet or rose purple; fruit fleshy, reddish or purplish red, armed with glochids, 3–6 cm long; seeds 3–6 mm long, tan or gray. Pinyon-juniper community (?) in Washington and San Juan counties (reported by Benson 1982); Nevada, California, and Arizona; 1 (0). This plant can be mistaken for *O. phaeacantha*, with which it is at least partially sympatric. Our material is assigned to **var. *martiniana*** (**L. Benson**) **L. Benson** [*O. eriocentra* var. *martiniana* L. Benson].

Opuntia macrorhiza **Engelm.** Plains Pricklypear. [*O. utahensis* Purpus, type from the Pine Valley Mts. *O. compressa* Macbr.; *O. basilaris* var. *woodburyi* Earle, type from Fort Pierce Wash, Washington County, pro max. parte]. Plants mainly 7–15 cm high and 2–15 dm wide or more; larger joints 5–15 cm long, 5–7.5 cm wide, obovate to orbicular, glaucous; leaves caducous; spines mainly in upper areoles, 1–6 per areole, commonly deflexed, straight or slightly curved, 1.5–5 cm long, terete or somewhat flattened; glochids yellow to brown; flowers 5–6.5 cm long; sepaloids greenish or reddish; petaloids yellow or tinged reddish basally; ovary smooth at anthesis, with few areoles; fruit fleshy, purple or purplish, with some glochids, 2.5–5 cm long; seeds 4–5 mm long, tan or gray; $n = 22$. Mixed desert shrub, pinyon-juniper, and mountain brush communities in Garfield, Iron, Kane, Piute, Salt Lake, San Juan, and Washington counties, the reports mainly by Benson (1982); Idaho to Wisconsin, south to Mexico, Texas, and Louisiana; 14 (vii). This is mainly a Great Plains species, with extensions into Utah where some of the reports might represent recent introductions. However, the principal population in Utah is in eastern Washington and western Kane counties. The plants from there have long been mistaken for *O. basilaris*, but the stems sprawl and root along the points of contact. They form numerous hybrids with the similarly fleshy fruited *O. phaeacantha* var. *discata*. One of these hybrids is the basis of *O. basilaris* var. *woodburyi*. The similar ***O. humifusa*** (**Raf.**) **Raf.** is reported for Utah on the basis of a collection from Utah County (Mason 6570 US), which was taken from along a railroad right-of-way. It is distinguished by having green joints and 1 spine per areole.

Opuntia nicholii **L. Benson** Plants mainly 15–25 cm high and 0.5–2 m wide or more; larger joints 10–20 cm long and 5–12 cm wide, narrowly obovate to obovate, bluish green; areoles 10–20 mm apart; spines usually in all areoles, 4–7 per areole, deflexed, often twisted and curving, mainly 5–12 cm long, flattened, some more than 1 mm thick; glochids yellow or tan; flowers 6–7 (8) cm long; sepaloids green, edged with purple or yellow; petaloids violet or yellow; fruit dry, tan to brown, 2.5–3.5 cm long, more or less spiny; seeds 7–8 mm long, whitish. Salt desert shrub and warm desert shrub communities at 1200 to 1500 m in Garfield, Kane, and San Juan counties (where it grows along Glen Canyon); Arizona; a Plateau endemic; 3 (i). Evident intermediates between O. nicholii and both *O. erinacea* var. *erinacea* and *O. phaeacantha* are known. Benson (1982) postulates a hybrid origin for this entity, and states that it should probably best be treated at varietal rank, but with which species?

Opuntia phaeacantha **Engelm.** Berry Pricklypear. Plants 30–90 cm high and 3–15 dm wide or more; larger joints 10–40 cm long, 7–25 cm wide, obovate to suborbicular, bluish green; areoles mainly 20–34 mm apart; spines in most areoles or restricted to upper ones or along the margin, 1–9 per areole, or none, spreading or deflexed, 2–8 cm long, flattened at least basally in some; glochids brown, reddish, or tan; flowers 6–9 (10.5) cm long; sepaloids greenish, edged yellow or red; petaloids yellow, or suffused with red below; ovary spineless, but with glochids; fruit fleshy, purple to red purple, 3.5–8 cm long; seeds 4–5 mm long, tan to gray; $n = 22, 33$. Three intergrading varieties have been identified from Utah (Benson 1982); their recognition at taxonomic rank is questionable, at least as far as our specimens are concerned.

1. Larger joints 10–15 cm long and 7–10 cm wide; plants of Washington and San Juan counties .... *O. phaeacantha* var. *phaeacantha*
— Larger joints 12–40 cm long and 9–25 cm wide; plants of various distribution 2
2(1). Joints mainly 12–20 cm long ad 9–15 cm wide; plants rather broadly distributed in southern Utah .... *O. phaeacantha* var. *major*
— Joints mainly 20–40 cm long and 15–25 cm wide; plants of San Juan and Washington counties .... *O. phaeacantha* var. *discata*

Var. *discata* (**Griffiths**) **Benson & Walkington** [*O. discata* Griffiths *O. palmeri* Engelm., type from near St. George]. $n = 33$. Warm desert shrub, pinyon-juniper, and sagebrush communities at 905 to 1800 m in San Juan (White Canyon, associated with prehistoric Indian dwellings) and Washington counties; California to Texas and Mexico; 8 (v). Some plants presumed to belong to this variety from along Beaverdam Wash have very long flowers and ovaries. They do not differ otherwise from the remainder of the variety.

Var. major Engelm. 2n = 44, 66. Warm desert shrub, pinyon-juniper, and mountain brush communities in Garfield, Kane, San Juan, and Washington counties; Nevada to Kansas, south to California, Mexico, and Texas; 8 (iii). This is the common phase of the species in Utah; plants north of the southern tier of counties having fleshy fruit probably belong to *O. macrorhiza*, with which transitional forms are known.

Var. phaecantha This variety is reported by Benson (1982) from Washington County; Arizona, Colorado, New Mexico, Texas, and Mexico; 0 (0).

Opuntia polyacantha Haw. Central Pricklypear. [*O. barbata* K. Schum., type from the La Sal Mts.; *O. barbata* var. *gracillima* M. Brandegee ex Purpus, type from the the La Sal Mts.]. Plants mainly 5–20 cm high and 3–30 dm wide or more; larger joints 5–15 (20) cm long (rarely longer) and 4–12 cm wide, obovate to orbicular, bluish green, not readily detached; areoles mainly 5–15 mm apart; spines variously borne in some or all areoles or lacking, often 6–10 per areole, variously oriented (all erect or spreading or some or all deflexed), straight or curved, terete (or somewhat flattened); glochids yellowish or tan; flowers 5–8 cm long; sepaloids green, margined with yellow or red; petaloids yellow, bronze, or pink to violet; ovary with glochids and often spines; fruit dry, 2–4 cm long, spiny, tan or brownish, deciduous; seeds 3–6 mm long, tan to white. This species, along with *O. erinacea*, forms a plexus around which revolve all other species of subgenus Opuntia (the platyopuntias) in Utah. Members of this complex form hybrids with *O. fragilis*, *O. erinacea*, and *O. phaeacantha*. Those, in turn, are transitional to all other species. Diversity of form and joint size; spine length, number, size, cross-sectional shape, and color; flower size and color; and other features give indications of genetic variability, differential response to ecological situations, and of problems of interpretation. Three varieties from Utah are treated by Benson (1982); they are more or less sympatric and intergrading.

1. Spines slender, flexible and curving, often whitish; plants of Grand and San Juan counties . . . . *O. polyacantha* var. *trichophora*
— Spines slender to coarse, not flexible (penetrating the skin before collapsing), mostly straight, or sometimes lacking; plants of various distribution 2

2(1). Spines in lower areoles mainly less than 12 mm long, those of upper areoles mainly less than 4 cm long, or spines lacking . . . . *O. polyacantha* var. *polyacantha*
— Spines in lower areoles mainly over 12 mm long, those of upper areoles often over 4 cm long . . . . *O. polyacantha* var. *rufispina*

Var. polyacantha [*O. polyacantha* var. *watsonii* Coult., type presumably from Summit County; *O. juniperina* Britt. & Rose; *O. polyacantha* var. *juniperina* (Britt. & Rose) L. Benson]. 2n = 44. Salt Desert shrub, mixed desert shrub, pinyon-juniper, sagebrush, mountain brush, mixed conifer, and aspen communities at 1370 to 2565 m in probably all Utah counties; British Columbia to Saskatchewan, south to Nevada, New Mexico, and Oklahoma; 39 (v). This variety is transitional with the next.

Var. rufispina (Engelm. & Bigel.) L. Benson [*O. missouriensis* var. *rufispina* Engelm. & Bigel.]. 2n = 22, 66. Blackbrush, mixed desert shrub, and pinyon-juniper communities at 1370 to 2200 m in Carbon, Emery, Garfield, Grand, Kane, Millard, San Juan, Washington, and Wayne counties; Wyoming to Nevada and California, south to Arizona and Texas; 21 (iii). This assemblage seems not to represent plants with genetic affinities. Rather the specimens appear to be artificial aggregations of phenotypically similar individuals. The use of spine characteristics as diagnostic results in a dilemma; the assemblage thus defined should be allied genetically to be recognized at a taxonomic level, but the definition is faulty. Thus, those plants of the Colorado drainage system show evidence of derivation from hybridization with *O. erinacea*, the plants of the Great Basin do not.

Var. trichophora (Engelm. & Bigel.) Coult. [*O. missouriensis* var. *trichophora* Engelm. & Bigel.]. 2n = 22, 44. Desert shrub and pinyon-juniper communities at 1125 to 1250 m in Grand, Kane, and San Juan counties; Colorado, Arizona, and New Mexico, Oklahoma, and Texas; 6 (iv).

Opuntia pulchella Engelm. Sand Cholla. Plants mainly 3–10 cm tall and about as broad, arising from an areolate glochid-armed tuberous joint 2–7 cm thick or more; joints mainly 1–5 cm long and 0.6–1.5 (2) cm thick, clavate to cylindric, green or blue green; tubercles 5–9 mm long and 4–5 mm wide; areoles spineless or with the upper mainly spiniferous, 8–15 per areole, straight or curved, 1–3.5 cm long or more, strongly flattened, the epidermal sheath not at all or poorly developed; glochids yellow to brown; flowers 3–4.5 cm long; sepaloids green, margined with pink purple; petaloids purple to violet; fruit 2–3 cm long, fleshy, red, prominently areolate and spiny; seeds 3–4.5 mm long, whitish; 2n = 22. Salt and mixed desert shrub communities at 1430 to 1770 m in Juab, Millard, Tooele, and Washington (?) counties; Nevada; a Great Basin endemic; 8 (iv). This taxon was named three times from Utah, all of the names based on types taken from the Desert Experimental Range in western Millard County, i.e., *Micropuntia brachyropalia* Daston, *M. barkleyana* Daston, and *M. spectatissima* Daston. All are characteristic of the species as it occurs in Utah, and none are worthy of taxonomic consideration.

Opuntia whipplei Engelm. & Bigel. Whipple Cholla. Plants low shrubs or less commonly mat-forming, mainly 10–60 cm tall or more; larger joints 2–15 cm long and 1–2 (2.5) cm thick; tubercles 8–9 mm long, 3–5 mm wide; leaves caducous; spines 4–10 or more per areole, straight, mainly 0.6–3 cm long, subulate to flattened; glochids yellow to tan or whitish; flowers 2–4 cm long; sepaloids and petaloids yellowish or yellowish green; fruit fleshy, 2–3 cm long, yellow, glochidiate; seeds 2.5–3 mm long, tan; n = 11. Desert shrub-grass and pinyon-juniper communities at 1340 to 1895 m in Beaver, Grand (?), Iron, Kane, Millard, San Juan, and Washington counties; Nevada, Colorado, Arizona, and New Mexico; 19 (xi). Our material belongs to **var. *whipplei***. Specimens with short terminal joints have been regarded tentatively as var. *multigeniculata*, but they fit in a graded series with *O. whipplei* in a strict sense. Dwarf plants at the limits of their ecological tolerance seem to represent a "cactus-line," corresponding to the dwarf conifers at "timberline." This seems to be the case with *O. whipplei* at the northern margin of its distribution in southwestern Millard County. A taxon represents the sum of its characteristics, not merely those considered to be diagnostic. *O. multigeniculata* is evidently restricted to the Spring (Charleston) Mountains and vicinity in southern Nevada.

Pediocactus Britt. & Rose

Plants globose to depressed-hemispheric, solitary; tubercles spirally arranged; areoles woolly, at least when young, with spines various but not hooked; flowers subterminal, borne at one side of the areole at the tubercle apex, small; sepaloids shorter than the petaloids; fruit dry, green, becoming tan to brownish or yellowish, naked or scaly, dehiscing by a vertical slit; seeds black tuberculate.

Heil, K., B. Armstrong, and D. Schleser. 1981. A review of the genus *Pediocactus*. Cactus & Succ. J. (U.S.) 53: 17–39.

1. Central spines 1–7, 6–30 mm long or more 2
— Central spines lacking, the longest lateral spines mainly less than 6 mm long 3
2(1). Longest spines mainly 12–25 mm long; sepaloids long-fimbriate; plants known only from gypsiferous substrates in Washington and Kane counties .... *P. sileri*
— Longest spines mainly 6–12 mm long; sepaloids subentire or shortly fimbriate; plants broadly distributed, seldom as above .... *P. simpsoni*
3(1). Longest spines 4 mm long or less, white, or lacking, often obscured by a dense mat of persistent white hairs; flowers peach to pink; plants of western Wayne County .... *P. winkleri*
— Longest spines more than 4 mm long, pale yellowish, not obscured by hairs, the woolly hairs pale yellowish and caducous; flowers yellow to peach; plants of north-central Emery County .... *P. despainii*

Pediocactus despainii Welsh & Goodrich Despain Footcactus. Plants solitary or less commonly colonial, subglobose to depressed-hemispheric, 3.8–6 cm tall, 3–9.5 cm wide; tubercles 6–10 mm long, 5–11 mm wide; areoles elliptic, persistently white-woolly, the central spines lacking; radial spines 9–13, 1.7–6 mm long, pale yellowish; flowers 1.5–2.5 cm long, 1.8–2.5 cm wide; petaloids yellow bronze to bronze or pinkish; fruit green, drying reddish brown, smooth, obovoid, 9–11 mm long, 10–12 mm wide; seeds shiny black, papillate to ridged, 3–3.5 mm long. Open pinyon-juniper community on limestone gravels at ca 1830 m in Emery (type from San Rafael Swell) County; endemic; 5 (0).

Pediocactus sileri (Engelm.) L. Benson Gypsum Cactus. [*Echinocactus sileri* Engelm. in Coult.; *Utahia sileri* (Engelm.) Britt. & Rose]. Plants solitary (less commonly colonial), depressed-hemispheric to cylindroid, 5–25 cm high, 6–12 cm wide; tubercles 9–15 mm long, 6–11 mm wide; areoles circular, persistently white-woolly; central spines 3–7, 13–30 mm long, blackish brown when young, straight; radial spines 11–15, 11–21 mm long, white; flowers 18–22 mm long, 20–30 mm wide; petaloids yellow or yellowish with purple veins; sepaloids conspicuously fringed; fruit dry, greenish yellow, 1.2–1.5 cm long; seeds gray to black, 3.5–5 mm long; n = 11. Salt desert shrub community at ca 900 to 1590 m in Kane and Washington counties; Arizona; a Plateau endemic; 3 (i). The type locality for this remarkable species is Pipe Springs, Arizona, but those springs were thought by early collectors to be in Utah, hence the name *Utahia*, which commemorates an Arizona type.

Pediocactus simpsonii (Engelm.) Britt. & Rose Simpson Footcactus. [*Echinocactus simpsonii* Engelm.; *P. hermannii* W. T. Marshall, type from near Hatch; *E. simpsonii* var. *minor* Engelm.; *P. simpsonii* var. *minor* (Engelm.) Cockerell]. Stems solitary or colonial, depressed-hemispheric to subglobose, 2–15 cm high, 3–20 cm wide (or more); tubercles 5–25 mm long, 4–7 mm wide; areoles elliptic to subcircular; central spines (1–3) 4–10, mainly 5–25 mm long, brownish or blackish; radial spines mainly 10–25, white; flowers 1.2–3 cm long; petaloids whitish, pinkish, yellowish or greenish; sepaloids shortly fimbriate; fruit green, often turning reddish brown, with few scales, 6–11 mm long, 5–10 mm wide; seeds gray to black, tuberculate, 2–3 mm long. Shadscale, mixed desert shrub, pinyon-juniper, sagebrush, and Douglas fir communities at 1460 to 2830 m in Beaver, Box Elder, Carbon, Daggett, Duchesne, Emery, Garfield, Grand, Iron, Juab, Piute, San Juan, Sevier, Tooele, Uintah, Utah, Washington, and Wayne counties; Washington to Wyoming, south to Nevada, Arizona, and New Mexico; 41 (xiii). Segregation of this common species of cactus into varieties seems not to be practical or even possible for Utah materials. Ridgetops in some south-central mountain and plateaus support one to several plants per square foot, mainly flush with the ground surface.

Pediocactus winkleri Heil Winkler Footcactus. Plants solitary or sometimes colonial, 3.9–6.8 cm tall, 2.7–5 cm wide; tubercles 4–7 mm long, 5–7 mm wide; areoles elliptic, persistently white-woolly; central spines lacking; radial spines 8–14, 1.5–4 mm long, white; flowers 1.7–2.2 cm long, 1.7–3 cm wide; petaloids peach to pink; sepaloids like the petaloids, except the outer darker; fruit green, drying reddish brown, smooth, obovoid, 7–10 mm long, 8–11 mm wide; seeds shiny black, papillate to ribbed, 2.5–3 mm long. Salt desert shrub communities at 1460 to 1590 m in Wayne (type from east of Capitol Reef) County; endemic; 5 (0). This is a remarkable tiny plant of poor quality saline fine-textured substrates.

Sclerocactus Britt. & Rose

Plants subglobose, depressed-hemispheric, ovoid, obovoid, or cylindroid; ribs 8–17; tubercles coalescent; areoles circular to elliptic; central spines 0 or 1–10, usually 1 or more hooked, or all straight; radial spines 2–15, straight; flowers subterminal, borne on upper side of tubercle adjacent to the areole, the scar persisting; floral tube short; petaloids pink to violet, white, or yellow; fruit dry, green turning reddish to tan, naked or with scales, dehiscent by basal and horizontal or lateral and vertical slits; seeds black, papillate-reticulate. Note: The taxonomy of this genus is subject to interpretation because of the remarkable diversity of form present in each of the species complexes.

1   Plants depressed-hemispheric to subglobose; areoles retaining juvenile pubescent radial spines for some years, finally with 1 or 2 hooked central spines; plants of the Great Basin .... *S. pubispinus*
—   Plants variously shaped, but if as above and with areoles retaining juvenile radial spines for several years, the spines glabrous and plants not of the Great Basin 2

2(1). Flowers mostly 2–3 cm long and broad; yellowish, pink, or white with pale pink midrib dorsally; juvenile condition retained for several years; plants of Emery and Wayne counties .... S. wrightiae

— Flowers mostly 3.5–5 cm long and broad,rose pink to violet, white, or yellow; plants of broad distribution in eastern Utah .... S. whipplei

Sclerocactus pubispinus (Engelm.) L. Benson Basin Fishhook. [*Echinocactus pubispinus* Engelm.]. Plants solitary or sometimes colonial, depressed-hemispheric to ovoid, 1–10 cm high, 2–15 cm wide; ribs 6–13; tubercles more or less developed; areoles circular to elliptic; juvenile spines and often the others (in part) densely or sparingly white-pubescent, finally glabrate; central spines lacking or 1–5, the lower 1 often hooked, 1–3 cm long, the upper 1 flattened, 5–35 mm long, 0.7–2.2 mm wide; radial spines 5–12, spreading; flowers 2.5–3.5 cm long; sepaloids bronze to brownish; petaloids yellow, bronze, pink, or violet to rose purple; fruit dry, green or pink becoming tan to brownish, ellipsoid to obovoid, opening by vertical slits; seeds 2.8–3.4 mm long, papillate, black. This species was named simultaneously in 1863 as *Echinocactus pubispinus* and as *E. whipplei* var. *spinosior*. The type of the former was taken in Pleasant Valley, Juab County, Utah or in adjacent White Pine County, Nevada (the boundary bisects the valley), and that of the latter was taken in the Dugway or Thomas ranges in central northern Juab County. Both remained obscure for almost a century, with *S. pubispinus* being ignored and var. *spinosior* being placed with *S. whipplei* and interpreted as including what now belongs in var. *intermedius* of that species (sensu stricto, which it resembles in its broad upper central spines). The type of *S. pubispinus* is a juvenile plant lacking both flowers and fruit; that of var. *spinosior* consists of flowers and seeds. Modern interpretations are based on interpolations of presumed collection localities with known modern occurrences of these dwarf cacti. In a way these peculiar cacti share characteristics of *S. blainei* Welsh of southern Nevada (pubescent spines and the tendency to flattened upper central spine) and with *S. whipplei* (the flattened upper central spine). The smaller flowers and depressed growth form are diagnostic from both. There are two intergrading and partially sympatric varieties present.

1. Flowers rose to violet; widest upper central spines 1–2.2 mm wide .... *S. pubispinus* var. *spinosior*

— Flowers bronze to yellow; widest upper central spines 0.7–1 mm wide .... *S. pubispinus* var. *pubispinus*

Var. pubispinus Shadscale, sagebrush, winterfat, rabbitbrush, and pinyon-juniper communities on calcareous and dolomitic gravels and outcrops at 1800 to 1955 m in Beaver, Box Elder, Iron, Juab, Millard, and Tooele counties; Nevada; a Great Basin endemic; 13 (vi).

Var. spinosior (Engelm.) Welsh [*Echinocactus whipplei* var. *spinosior* Engelm.; *S. spinosior* (Engelm.) Woodruff & L. Benson]. Shadscale, rabbitbrush, sagebrush, and pinyon-juniper communities on calcareous and igneous gravels and clay silts at 1525 to 1985 m in Beaver, Juab, Millard, and Sevier counties; Arizona (?); 15 (x). The range of this variety is partially sympatric with var. *pubispinus* in western Beaver County and intermediates are known. The population from Sevier County differs in subtle ways (i.e., more like *S. whipplei*) from the remainder of the taxon, but does not seem to be worthy of taxonomic rank.

Sclerocactus whipplei (Engelm.) Britt. & Rose Whipple Fishhook. [*Echinocactus whipplei* Engelm. & Bigel.]. Plants solitary or in small colonies, depressed-hemispheric, obovoid, ovoid, or cylindric, 5–45 cm tall or more, 5–15 cm thick; ribs mainly 8–15, tuberculate; central spines 1–4, the lower 1 (sometimes 2–4) hooked or all straight, mainly 1–7.5 cm long, curved or some or all of them straight, the upper central spine (at least) usually pale and flattened to flat and ribbonlike, 1–5 cm long or more, 0.7–3.5 mm wide, erect; radial spines 7–12 or more; flowers 3.5–5 cm long; sepaloids greenish, margined with rose purple, pink, white, or yellow; petaloids pink, violet, white, or yellow; fruit dry, green, becoming tan, sparingly scaly, 1.2–2.5 cm long; seeds 2–3.4 mm long, black, papillate. Salt and mixed desert shrub, pinyon-juniper, sagebrush, and ponderosa pine communities at 1125 to 2440 m in Carbon, Duchesne, Emery, Garfield, Grand, Kane, San Juan, Sevier, Uintah, and Wayne counties; Colorado, New Mexico, Arizona, and Nevada; 134 (xxxi). Our materials belong to var. *roseus* (Clover) L. Benson [*S. havasupaiensis* var. *roseus* Clover; *S. intermedius* Peebles, type from Pipe Springs, Arizona; *S. whipplei* var. *intermedius* (Peebles) L. Benson; *S. parviflorus* var. *intermedius* (Peebles) Woodruff & Benson; *S. parviflorus* Clover & Jotter, type from Forbidding Canyon, San Juan County; *S. contortus* Heil, type from eastern Wayne County; *S. terrae-canyonae* Heil, type from Trachyte Wash, Garfield County; *E. glaucus* Purpus ex K. Schum.; *Sclerocactus glaucus* (K. Schum.) L. Benson; *S. whipplei* var. *glaucus* (Purpus) Welsh]. Spine characteristics are notoriously variable in this genus. Specimens of *Sclerocactus* with straight spines have long been known and they have been recognized as belonging to this genus. Their status has been open to question, because they differ in no other discernible way from the body of the species. Specimens with straight spines occur here and there through much of the range of *S. whipplei*, even though they are more common at the margin of the distribution in the Uinta Basin and in west-central Colorado. They are considered herein to be taxonomically neglegible. A peculiar phase of var. *roseus* from the Pariette Draw region of southeastern Duchesne County has a long juvenile stage, with the initial central spines very short (to 2 mm long) and hooked or straight. It does not seem to warrant taxonomic recognition. Some tiny juvenile plants have pubescent spines, but the juvenile stage is apparently arrested in most portions of this variety. The var. *roseus* is almost as variable as the species itself. It has been treated previously at specific rank (as three separate species) and as consisting of three varieties. With the degree of variability exhibited it is not surprising that so many divergent views should have developed; it is surprising that even more segregation was not attempted. *S. contortus* is a slender-spined phase in which the spines are contorted, a condition that seems to be unworthy of consideration from a taxonomic standpoint. *S. terrae-canyonae* appears to be more substantially based, with its long slender spines and yellow flowers. There is little correlation, however, between flower color and spine length. Long-spined phases are more common in the southeastern portion of Utah, but the flowers of the long-spined phases are mainly pink to violet. Spine length forms a cline from long in the southeast to short in the northwestern and northern portions of the range. The var. *intermedius* is more difficult to discount. In the extreme situations that variety has the uppermost central spine flattened and ribbonlike, commonly 1–3.5 mm wide at the base, and with var. *roseus* (or *S. parviflorus* per se) having the uppermost spine merely flattened and mainly 0.7–1 mm wide. There are as many intermediates as there are extremes, and until other diagnostic criteria are identified, it seems best to include all of the tremendous range of variation within an expanded var. *roseus* of *S. whipplei*.

Sclerocactus wrightiae L. Benson Wright Fishhook. Plants depressed-hemispheric to obovoid or short-cylindric, mainly 6–12 cm long and 4–8 cm thick; ribs mostly 8–13; tubercles more or less developed; areoles circular to elliptic; juvenile spines glabrous; central spines 1–4, the lower one often hooked on at least the upper tubercles, mostly 10–20 mm long, the uppermost 1–2.5 cm long, flattened, 0.8–1.5 mm wide; radial spines 8–11, spreading; flowers 2–3.5 cm long; sepaloids green or variously tinged with red or brown; petaloids yellowish to white or pink; fruit ellipsoid, 9–12 mm long; seeds black, tuberculate, 3–3.5 mm long. Salt desert shrub and shrubgrass to juniper communities at 1460 to 1865 m on Mancos Shale (Bluegate, Tununk, Emery, and Ferron members), Dakota, Morrison, Summerville, and Entrada formations in Emery and Wayne counties; endemic; 14 (iii). The small flowers and short spines are evidently diagnostic. Occasional intermediates with *S. whipplei* var. *roseus* occur in Emery County near the Sevier County line—at edaphic ecotones marking the boundary between shale and sandstone members of the Mancos Shale Formation.

CALLITRICHACEAE Link

Water-starwort Family

Slender, annual, monoecious, aquatic (terrestrial) herbs; leaves simple, opposite, entire, floating or submerged; flowers imperfect, solitary in leaf axils; calyx and corolla lacking; staminate flower of a single stamen; pistillate flowers of a single pistil, the ovary superior, 2–loculed (or 4–loculed by intrusion of false septae); styles 2, slender; fruit a schizocarp; x = 3, 5+.

Callitriche L.

Aquatic herbs, floating or submerged; flowers inconspicuous, much reduced; ovary superior; fruit a schizocarp.

Fassett, N. C. 1951. *Callitriche* in the New World. Rhodora 53: 137–55, 161–82, 185–94, 209–22.

1. Leaves uniformly linear-lanceolate to oblong, 3–13 mm long, the bases not connected by a ridge or wing . . . . *C. hermaphroditica*
— Leaves not uniformly linear-lanceolate, some or all of them broader or, if all linear, more than 13 mm long and bases connected by a ridge or wing 2
2(1). Fruit definitely obovoid, longer than broad, broadest about the middle, usually narrowly winged apically; plants common . . . . *C. verna*
— Fruit suborbicular in outline, about as broad as long, broadest near the middle, inconspicuously if at all winged apically . . . . *C. heterophylla*

Callitriche hermaphroditica L. Plants slender, mainly 5–30 cm long; leaves mainly 3–13 mm long, all linear-lanceolate to narrowly oblong, 1–veined and bidentate apically; flowers not subtended by bracts; fruits grayish, 1–2 mm long, narrowly winged apically, the pitlike depressions not in definite rows; 2n = 6. Reservoirs, stock ponds, lakes, and sluggish streams at 2150 to 3205 m in Cache, Daggett, Salt Lake, Sevier, Summit, Uintah, and Utah counties; Alaska east to Newfoundland, south to California, New Mexico, and New York; 7 (ii).

Callitriche hetrophylla Pursh Plants aquatic or growing on mud, mainly 5–25 cm long; leaves with small membraneous ridges connecting across the nodes, the emersed ones 1–nerved, 5–25 mm long, the vein not projecting into an apical notch, the floating and emergent ones broadly obovate to elliptic, mainly 1.5–5 mm wide (or more) and 3–nerved; floral bracts present; fruit more or less orbicular in outline, 0.6–1.2 mm long and about as broad, widest at or near the middle, wingless or rarely narrowly winged apically; 2n = 20. Beaver ponds, marshes, lakes, and reservoirs at 2285 to 3175 m in Daggett (?), Duchesne, Grand, Piute, Summit, Uintah, and Wasatch counties; widely distributed in North America; South America; 6 (i). Note: Reported for Utah (Flora Pacific N. W. 3:403. 1961) is *C. anceps* Fern. It differs in minor technical characteristics from *C. heterophylla*, i.e. the fruits are essentially orbicular, and the midnerve of the leaf projects beyond the leaf tip. The extent of plants with these features in Utah, if any, is unknown.

Callitriche verna L. Vernal Water-starwort. [*C. palustris* L.]. Plants slender, 5–30 cm long; leaves mainly 3–15 mm long, linear, 1–veined, and shallowly bidentate apically, the upper ones often enlarged and the terminal ones petioloid and narrowly obovate to spatulate, 3–veined, and often floating; flowers subtended by bracts; fruit tan to pale reddish brown, 1–2 mm long, more or less winged apically, the pitlike depressions arranged in rows; 2n = 20. Ponds, beaver ponds, sluggish streams, bogs, and muddy banks at 2400 to 3390 m in Cache, Daggett, Duchesne, Garfield, Rich, Salt Lake, Sevier, Summit, Uintah, Utah, and Washington counties; widespread in North America; Eurasia; 17 (i).

CAMPANULACEAE A. L. Juss.

Bellflower Family

Herbs, usually with milky juice; leaves simple, alternate, estipulate; flowers usually perfect; sepals 5, persistent; corolla 5–lobed, regular or irregular; stamens 5, distinct or connate; pistils 1; ovary inferior, 2– to 5–loculed; style 1; fruit a capsule, opening by pores, slits, or valves; x = 6–17. *Heterocodon rariflorum* Nutt. was discovered in Washington County too late for formal treatment.

1. Corolla regular; anthers and filaments distinct 2
— Corolla irregular, bilabiate; anthers distinct or connate . . 3
2(1). Flowers pedicellate or pedunculate, borne in racemes or solitary; plants perennial . . . . *Campanula*
— Flowers sessile, borne in spikes; plants annual . . *Triodanus*
3(1). Flowers minute, 2–3 mm broad; anthers alike, distinct; plants diminutive annuals with capillary branches . . . . *Nemacladus*
— Flowers larger, more than 5 mm broad; anthers united into a tube, with 2 shorter ones; plants annual or perennial, not with capillary branches 4

4(3). Flowers sessile, axillary; ovary linear, simulating a pedicel; capsule dehiscent by longitudinal slits ...... *Downingia*
— Flowers pedicellate; ovary obconic, opening by valves .. 5
5(4). Plants palustrine annuals; corolla 6–10 mm long ..... *Porterella*
— Plants perennial or annual; corolla 12–25 mm long or more ...... *Lobelia*

Campanula L.

Perennial herbs; leaves often dimorphic, the lower ones petiolate and with obovate to lanceolate or elliptic blades and the upper ones linear and sessile, or all petiolate and broader; flowers solitary or racemose, blue to violet or white, regular, campanulate; calyx 5–lobed; corolla 5–lobed; stigmas 3–5; ovary 3–loculed; fruit a capsule, opening subbasally or apically.

1. Cauline leaves lanceolate to lance-ovate, serrate, petiolate to stem middle or above; flowers several to numerous, borne in axils of foliose bracts; plants adventive .. *C. rapunculoides*
— Cauline leaves linear to narrowly oblong, entire, sessile or subsessile; flowers solitary or several; bracts much reduced; plants indigenous 2
2(1). Flowers solitary, 4–10 mm long; anthers 1.5–2.5 mm long; calyx lobes 3–4 mm long; plants of the Uinta and La Sal mts. ...... *C. uniflora*
— Flowers solitary or 2–several, 10–20 mm long; anthers 4–6 mm long; calyx lobes 5–10 mm long or more 3
3(2). Sepals 5–7 (8) mm long, entire; leaf bases and petioles glabrous or merely hispidulous; flowers nodding or erect; plants of the Uinta Mts. ...... *C. rotundifolia*
— Sepals 7–12 mm long or more, often callous-toothed; leaf bases and petioles ciliate (at least the lowermost); flowers erect; plants of central and southern Utah .... *C. parryi*

Campanula parryi Gray Parry Bellflower. Perennial herbs; stems 5–30 cm tall, pubescent at the base or glabrous throughout; lower leaves 1.5–6 cm long, spatulate to oblanceolate, entire or denticulate, ciliate on petiole at least, becoming sessile and linear upwards; flowers solitary (rarely 2), erect; sepals 7–12 mm long or more; corolla 10–20 mm long, campanulate, blue to purple or white; anthers 4–6.5 mm long; capsules opening by valves or pores near the summit. Ponderosa pine, aspen, and spruce-fir communities, often where wet, at 1920 to 3205 m in Beaver, Garfield, Iron, Kane, Piute, San Juan, Sanpete, Sevier, and Wayne counties; Colorado, Arizona, and New Mexico; 37 (iii).

Campanula rapunculoides L. Rover Bellflower. Perennial herbs from taproots and rootstocks; stems 2.5–14 dm tall, glabrous or pubescent; lower leaves long-petiolate, cordate-ovate; cauline leaves lance-ovate to lanceolate, the blades 2.5–10 cm long, 0.6–4.5 cm wide, short-petiolate, becoming sessile upwards, irregular serrate to doubly serrate; flowers blue, the corollas 15–30 mm long, the lobes 5–12 mm long, nodding, shortly pedicellate, secund; calyx lobes reflexed, 4–8 mm long; capsule opening basally; $2n = 68, 102$. Cultivated ornamental, persisting and escaping, at 1370 to 2505 m in Cache, Duchesne, Emery, Tooele, Uintah, and Utah counties; introduced from Eurasia; 7 (0).

Campanula rotundifolia L. Bluebells of Scotland. Perennial herbs from a slender rhizome or from a caudex, with 1 to several decumbent to ascending or erect stems 1–5 dm tall or more, glabrous and smooth or minutely scabrous in some; leaves mainly cauline, commonly 10–22, 0.7–8 cm long, 0.1–1.5 cm wide, more or less dimorphic, the lower ones often with broad, cordate-ovate, lanceolate, or elliptic blades and distinctive petioles, the upper ones reduced and linear to lance-linear to oblong, serrulate to entire, the margins scabrous; flowers solitary or 2–15; pedicels glabrous; calyx and ovary glabrous; calyx sinuses lacking appendages; corolla blue (rarely white), 15–30 mm long, the lobes ovate to broader, erect or ascending, much shorter than the tube; styles shorter than the corolla, but usually exceeding the tube; anthers 5–7 mm long; capsules obconic, nodding, opening at the base; $2n = 34, 66, 102$. Ponderosa pine, aspen, lodgepole pine, spruce-fir, and alpine tundra communities at 2285 to 3480 m in Daggett, Duchesne, Summit, Uintah, and Wasatch (?) counties; Alaska and Yukon, east to Newfoundland, south to California and Mexico; Eurasia; 38 (vi).

Campanula uniflora L. Arctic Harebell. Perennials, from elongate rhizomes; stems 1 to several, decumbent to erect, 3–20 (28) cm tall, glabrous, or more or less hairy in decurrent lines; leaves mainly cauline, usually 4–10, 0.5–4 (4.5) cm long, 1–5 (7) mm wide, elliptic to oblanceolate or linear, entire to crenulate, glabrous or nearly so, or the upper ones hairy, scabrous marginally; flowers solitary; pedicels glabrous or nearly so; calyx lobes lance-attenuate, entire; calyx and ovary hairy; corolla blue to purplish, 4–10 mm long, the lobes ovate, erect, about as long as the tube; styles shorter than the corolla, not exserted; anthers 1–2 mm long; capsules subcylindric, opening subapically; $2n = 34$. Alpine tundra, often in talus or rock stripes, at 3355 to 3965 m in Daggett, Duchesne, Grand, San Juan, Summit, and Uintah counties; Alaska and Yukon, south to Colorado; Eurasia; 8 (iii).

Downingia Torr.

Low, soft annual herbs; leaves sessile, simple, entire; flowers perfect, 5–merous, sessile in bract axils; sepals 5, more or less unequal; corolla bilabiate, blue, the lower lip maculate, the upper lobes smaller; stamens 5, unequal, the 2 smaller anthers each with a terminal tuft of bristles and often with a hornlike process also; capsule linear to fusiform, many times longer than thick, twisted (inverting the flowers), opening by slits.

Downingia laeta (Greene) Greene [*Bolelia laeta* Greene]. Plants 3–15 cm tall or more; leaves 3–20 mm long, 0.5–2 mm wide; corollas 4–6 mm long, pale blue or purplish, the lower lip marked with white or yellow, often with a purple stripe; filaments connate, the tube 1.5–2.5 mm long; anthers ciliate or glabrous; capsules 2–4 cm long, appearing like pedicels. Drying mud on reservoir and pond margins at 1525 to 1710 m in Cache, Rich, Salt Lake, and Utah counties; Montana and Saskatchewan, south to California, Nevada, and Wyoming; 8 (0).

Lobelia L.

Perennial or annual herbs; leaves sessile, alternate, simple, entire or toothed; flowers perfect, 5–merous, inverted at anthesis, pedicellate, borne in racemes; sepals 5, subequal; corolla irregular, bilabiate, blue or scarlet, the tube often split to the base on one side, the limb 5-lobed, the upper 3 lobes forming the apparent upper lip, the other 2 (on each side of the split) erect or recurved; anthers united into a tube around the style; ovary 2-loculed; fruit dehiscent near the apex.

1. Corollas scarlet, 22–40 mm long; plants of moist sites and hanging gardens in southern Utah .... *L. cardinalis*
— Corollas blue, violet, or white, 12–14 mm long; plants cultivated and escaping .... *L. erinus*

***Lobelia cardinalis* L.** Scarlet Lobelia; Cardinal-flower. [*L. splendens* L.]. Perennial herbs; stems simple, 3–10 dm tall or more, glabrous or pubescent, erect; leaves 3–15 cm long, 2–16 mm wide, narrowly elliptic to linear-lanceolate, irregularly denticulate to entire; racemes 7–36 cm long; pedicels 5–15 mm long; sepals 5–10 mm long; anthers bluish gray, 3–4 mm long, long-pilose on one side; capsule 4–5 mm long, 5–8 mm wide, 10-ribbed, papery. Drainages, seeps, springs, and hanging gardens at 900 to 1405 m in Kane, San Juan, and Washington counties; widely distributed in North America, Mexico, and Central America; 15 (vii). Our material belongs to ssp. *graminea* (**Lam.**) **McVaugh** [*L. graminea* Lam.]. This is a strikingly beautiful plant that flowers late in the summer and into the autumn.

***Lobelia erinus* L.** Annuals; stems diffusely branched, mainly 10–30 cm tall, glabrous or sparingly hairy, clump-forming; leaves 1–3 cm long, 2–12 mm wide, oblanceolate to elliptic, irregularly serrate to entire; racemes 2–5 cm long (mainly 1- to 5-flowered); pedicels 12–25 mm long, capillary; sepals 3–5 mm long; anthers bluish gray, 1–2 mm long, short- hairy on one side; capsules 4–5 mm long, 3–4 mm wide, ca 10-ribbed, papery. Cultivated ornamental "edging" plants, commonly grown and rarely escaping in low elevation communities in Utah; introduced from Africa; 3 (0).

Nemacladus Nutt.

Annuals with capillary, flexuous stems and branches; leaves basal, rosettiform, the cauline ones bracteate; flowers inverted, minute, borne in a panicle comprising most of plant height; sepals 5, triangular; corolla irregular, more or less bilabiate, the lower lip 3-lobed, the upper 2-lobed; staminal filaments connate, the anthers spreading, distinct; capsules bilocular, dehiscent by valves.

McVaugh, R. 1939. Some realignments of the genus *Nemacladus*. Amer. Midl. Naturalist 22: 521–550.

1. Stems purplish to dark brown; bracts mainly 2–7 mm long, linear to narrowly lanceolate; plants locally common in Washington County .... *N. glanduliferus*
— Stems silvery to gray; bracts mainly 0.5–2 mm long, lanceolate to lance-ovate; plants not definitely known from Utah, but possible in Beaver Dam Wash ....
.... *N. rubescens* Greene

***Nemacladus glanduliferus* Jepson** Diminutive puberulent or glabrous annuals; stems diffusely branched, 5–15 cm tall; pedicels spreading- ascending, 3–10 mm long, capillary; bracts 2–7 mm long; corolla 1.5–2.5 mm long, white, with lobes purple apically; sepals 0.8–1.2 mm long; anthers 0.2–0.3 mm long; capsule 1.5–3 mm long, half-inferior; n = 9. Warm desert shrub and mixed desert shrub communities at 750 to 1070 m in Washington County; Nevada, Arizona, California, and Mexico; 14 (0). Our material belongs to var. *orientalis* McVaugh.

Porterella Torr.

Annuals with thickish soft stems; leaves cauline, alternate, sessile; flowers inverted, solitary or few, pedicellate; sepals 5, subulate; corolla irregular, bilabiate, the lower lip 3-lobed, the upper 2-lobed; stamens connate; capsule bilocular, dehiscent by apical valves.

McVaugh, R. 1940. A revision of "Laurentia" and allied genera in North America. Bull. Torrey Bot. Club 67: 778–798.

***Porterella carnulosa* (H. & A.) Torr.** [*Lobelia carnulosa* H. & A.]. Plants erect or spreading; stems mainly 3–15 cm long, glabrous; leaves sessile, 4–20 mm long, 1–3 mm wide, linear-subulate to elliptic; bracts 4–15 mm long; sepals 2–6 (9) mm long; corolla showy, blue, with white or yellow center, 6–10 mm long, the upper 2 lobes erect; anthers bluish gray, 1.5–2.6 mm long; capsules 7–10 mm long, inferior. Pond and pool margins and streamsides at 2745 to 3205 m in Sevier and Summit counties; Oregon to Wyoming, south to California and Arizona; 5 (ii).

Triodanis Raf.

Annuals; stems simple or branched; leaves alternate, simple, transitional to foliose bracts; flowers solitary, axillary, sessile, the lowermost usually cleistogamous, at least some upper ones chasmogamous, with well-developed corollas; corollas regular; staminal filaments ciliate basally; capsules dehiscing by small valvelike openings.

***Triodanis perfoliata* (L.) Nieuwl.** Venus Looking-glass. [*Campanula perfoliata* L.; *Specularia perfoliata* (L.) DC.]. Plants erect, 1–6 dm tall; stems simple or with ascending branches; leaves 0.8–3 cm long, rounded-cordate, clasping, crenate, the lowermost shortly petiolate; calyx lobes 5–8 mm long, or smaller in cleistogamous flowers; corollas of upper flowers 8–13 mm long; capsules oblong to obovoid, 2- or 3-locular. Rare adventive weedy species of disturbed sites, known from Kane and Washington counties; throughout the U. S.; adventive from Europe; 2 (i).

CANNABACEAE Endl.
Hemp family

Plants herbaceous, with watery juice; leaves alternate opposite, palmately veined and lobed or divided to essentially compound; stipules persistent; flowers imperfect, the plants dioecious, regular, the staminate in open racemes or panicles, the pistillate in dense clusters; sepals 5, connate in pistillate flowers and enclosing the ovary; petals lacking; stamens 5; pistil 1, the ovary superior, 2-carpelled, the styles 2; fruit an achene; x = 8, 10.

1. Plants strong-smelling, stout, erect herbs; leaves palmately 5- to 9-parted .... *Cannabis*
— Plants rough-stemmed clambering vines; leaves coarsely 3- to 7-lobed .... *Humulus*

Cannabis L.

Plants dioecious or rarely some monoecous; leaves palmately lobed to parted and apparently compound, alternate or the lower opposite; flowers small, inconspicuous, the staminate in leafy panicles in upper axils; sepals 5, oblong; stamens 5; pistillate flowers in small clusters on leafy branches from upper axils, each flower subtended and enclosed by an acuminate bract, the calyx barely lobed, surrounding only the base of the ovary; stigmas 2, elongate; fruit a lenticular achene, enclosed within the accrescent bract; x = 10.

Small, E. and A. Cronquist. 1976. A practical and natural taxonomy for *Cannabis*. Taxon 5: 405–435.

Cannabis sativa L. Hemp, Marijuana, Hashish, Pot, Grass. Plants 6–20 dm tall or more, the stems simple or much branched; leaves long petioled, the blades 3- to 7-parted, the segments oblanceolate to elliptic, attenuate to acuminate apically, sharply serrate, mostly 4–12 cm long and 4–18 mm wide, scabrous and more or less glandular and pubescent; achenes mostly 3.5–4.5 mm long. Weedy species of roadsides and other disturbed sites in Cache, Grand, Salt Lake, Utah, Washington, and Weber counties; introduced from Eurasia; 8 (i). Cultivated historically in Utah for fiber produced from the stems, this is the commercial source of hemp; currently sporadic, or else grown illegally for its intoxicant properties. Utah materials are sufficiently rare as to give only hints as to the classification below the species level. It seems likely, however, that, at least historically, two phases have been grown in the state (for a complete review see Small and Cronquist l.c.). Ninteenth century plantings for hemp likely belonged to **ssp. *sativa*** , demonstrated to have only limited intoxicant properties. At least some of the recent introductions clearly belong to **ssp. *indica*** **(Lam.) Small & Cronq.** (*Cannabis indica* Lam.), which has demonstrated high intoxicant levels.

Humulus L.

Plants herbaceous, twining, perennial vines; stems scabrous; leaves opposite, broadly 3- to 5-lobed; flowers small, inconspicuous, the staminate in axillary panicles; sepals 5, distinct; stamens 5; pistillate flowers in short spikes, in pairs, with each pair subtended by a foliaceous bract; calyx membranous, unlobed, closely covering the ovary; stigmas 2, elongate; fruit an achene enclosed by the persistent calyx and accrescent bracts; x = 8.

Humulus americanus Nutt. American Hop. Plants twining, the stems to 20 dm long or longer; leaves ovate to orbicular in outline, deeply cordate basally, mostly 3–15 cm long and 2.8–16 cm wide, the lobes serrate to doubly so, attenuate to acuminate apically, rough-hairy above, glandular-dotted beneath; fruiting spikes usually 2–3.5 (4) cm long at maturity. Twining over shrubs and other vegetation at lower and middle elevations in Cache, Duchesne, Garfield, Grand, Millard, Piute, Salt Lake, Summit, Uintah, Utah, Wasatch, Washington, Wayne, and Weber counties, and probably throughout Utah; widespread in North America; 37 (vi). The hop of commerce, ***H. lupulus*** **L.**, or European hop, is grown in the U. S., where it has escaped and persists. Though not definitely known for Utah, the European hop might occur here. It can be distinguished by its unlobed leaves, or when lobed, the terminal lobe is less than twice longer than broad.

CAPPARACEAE A. L. Juss.
Caper Family

Annual herbs; leaves alternate, palmately 3- to 7-foliolate, rarely simple above; flowers in racemes, regular or irregular, perfect; sepals 4; petals 4, distinct; stamens 6 to many; pistil 1, the ovary superior, sessile or stipitate, 1-loculed, with 2 parietal placentae; pistil 1, the ovary superior, 2-carpellate; style 1, the stigma 2-lobed; fruit a 2-valved capsule; x = 8–17.

1. Stamens 8–16, unequal in length; herbage stipitate-glandular; petals white; staminal filaments purple . . . . *Polanisia*
— Stamens 6, equal in length; herbage glabrous or merely hairy, not glandular; petals yellow, yellow orange, or pink purple, rarely white; staminal filaments yellow or purple 2
2(1). Fruits about as broad as long or broader, few- to several-seeded; plants mainly 0.4–3 dm tall . . . . *Cleomella*
— Fruits several to many times longer than broad, many-seeded; plants 2–20 dm tall . . . . *Cleome*

Cleome L.

Annual herbs; leaves 3- to 7-foliolate; flowers in bracteate racemes, regular; sepals 4; petals 4, yellow or pink to pink purple (rarely white); stamens commonly 6; ovary stipitate, with a gland at the base; capsules linear to oblong, spreading or pendulous.

1. Petals pink or purplish to white; leaflets 3 . . . . . *C. serrulata*
— Petals yellow; leaflets 5–7 . . . . *C. lutea*

Cleome lutea **Hook.** Yellow Beeplant. Plants erect; stems mainly 3–15 dm tall, branched or simple; herbage glabrous or sparingly hairy; leaflets 3–5 (or 7), oblong to lanceolate or elliptic, 0.8–5 cm long, 2–10 mm wide, entire; racemes 2–10 cm long; pedicels 3–10 mm long; pods 1–3 (4) cm long, spreading, on stipes 6–12 mm long; 2n = 32. Warm, salt, and mixed desert shrub, pinyon-juniper, and ponderosa pine communities at 915 to 2380 m in all Utah counties, except Box Elder, Cache, Daggett, Davis, Morgan, Rich, Sanpete, and Summit; Washington to Nebraska, south to California, and New Mexico; 131 (xxiv).

Cleome serrulata **Pursh** Rocky Mountain Beeplant. Plants erect; stems mainly 3–20 dm tall or more, simple or branched above; herbage glabrous or sparingly hairy; leaflets 3, lanceolate, elliptic, or oblanceolate, 1.5–7 cm long, 4–14 mm wide, entire; racemes 5–25 cm long or more; pedicels 12–23 mm long; pods 2.5–7 cm long, descending to pendulous, on stipes 10–25 mm long; 2n = 32. Disturbed sites along roadsides, on stream traces and other pioneer situations at 900 to 2745 m in probably all Utah counties; Washington to Saskatchewan, south to California and Arizona; 92 (xvi). Two morphologically intergrading phases have been recognized in Utah: **var. *serrulata*** , with fruit 2–6 mm long and 2.5–6 cm wide, and seeds 3–4 mm long; and **var. *angusta*** **(Jones) Tidestr.** [*C. integrifolia* var. *angusta* Jones, type from Marysvale], with fruit 5–7 cm long and 3.5–5.5 mm wide, and seeds 4–4.5 mm long. The former is widespread and the latter is restricted to southern counties in Utah.

Cleomella DC.

Annual herbs; leaves alternate, 3-foliolate; racemes bracteate, many-flowered; flowers regular, sepals 4, tardily deciduous; petals 4, yellow, closed in bud; stamens 6, the anthers coiled when dry; capsules stipitate, obdeltoid, rhomboidal, deltoid, or ovoid, usually wider than long; seeds 2–12.

1. Leaflets 2–7 mm wide, ca 3.5 times longer than wide; petioles of lower leaves 6–27 mm long; plants of the Colorado Drainage system . . . . *C. palmeriana*
— Leaflets 1–3 mm wide, ca 5.5 times longer than wide; petioles of lower leaves 2–12 mm long; plants of the Great Basin . . . . *C. plocasperma*

Cleomella palmeriana Jones [*C. nana* Eastw., type from near Thompson Springs; *C. cornuta* Rydb., type from Cainville; *C. montrosae* Payson]. Plants glabrous; stems widely branching from the base, 6–30 cm tall; leaflets 9–20 mm long, 2–9 mm wide, elliptic to oblong or lance-oblong; petioles of lower leaves 6–27 mm long; bracts, except the lowermost, reduced to setae; pedicels 5–6 mm long; sepals entire, acute; petals 3–4 mm long, yellow or yellow orange, tipped with red in bud; filaments to ca twice as long as the petals; anthers 1.2–2 mm long; style ca 1 mm long; capsules obtuse apically, the base triangular-acute, 2–5 mm long, 5–9 mm wide, the valves acute; stipe 3–7 mm long, often recurved; seeds few to several. Salt desert shrub community, on Mancos Shale, Tropic Shale, and Morrison formations, at 1220 to 1830 m in Emery (type from Green River), Garfield, Grand, Kane, Uintah, and Wayne counties; Colorado; a Plateau endemic; 43 (iii). The material from Uintah County, disjunct considerably from the remainder of the species, has conspicuously horned fruit 8–9 mm wide. In specimens from the Navajo Basin the fruit is not especially horned and mostly 3–5 mm wide. The Uinta Basin specimens are designated as **var. *goodrichii*** Welsh.

Cleomella plocasperma Wats. Plants usually with several diffuse, strongly ascending branches; stems 15–40 cm tall; leaflets 7–18 mm long, 1–3 mm wide, linear to narrowly oblong; petioles 5–12 mm long; bracts unifoliolate and linear or the lowermost 3–foliolate; pedicels 9–15 mm long; petals 3–5 mm long; stamens 6–9 mm long; capsules ovoid, rhomboidal, or obovoid, 1.5–3 mm long, 3–6 mm wide, the valves hemispheric, deltoid, or conical; stipes 2–7 mm long; seeds several. Saline soils with saltgrass, greasewood and other halophytes at 1370 to 1680 m in Beaver, Iron, and Millard counties; Oregon and Idaho, south to California; 5 (ii).

Polanisia Raf.

Viscid-puberulent annuals with a rank odor; leaves petiolate, palmately 3–foliolate; racemes bracteate, many-flowered; sepals 4, free nearly to the base; petals 4, white, open in bud, irregular; stamens 6–20, of unequal length; anthers short; capsules elongate, erect, subsessile, glandular, the persistent valves dehiscing apically; seeds many.

Iltis, H. H. 1958. Studies in the Capparidaceae - IV. Brittonia 10: 33–58.

Polanisia dodecandra (L.) DC. Clammy-weed. [*Cleome docecandra* L.]. Plants 1.5–8 dm tall; petioles 0.5–5 cm long; leaflets 0.8–4.5 cm long, 3–18 mm wide, obovate to oblanceolate or elliptic; pedicels 10–21 mm long; sepals 4–5 mm long, purplish; petals white to cream, 8–12 mm long, emarginate, clawed; staminal filaments purple, long exserted; style 4–6 mm long; sepals subsessile, 3–5.5 cm long, 5–9 mm broad; 2n = 20. Warm, mixed, and salt desert shrub and juniper communities, often in wash bottoms, at 1065 to 1985 m in Box Elder, Carbon, Daggett, Duchesne, Salt Lake, Sevier, Tooele, Uintah, Utah, Washington, and Weber counties; British Columbia to North Dakota, south to California, Arizona, and Mexico; 27 (vi). Our material is assignable to **var. *trachysperma*** (T. & G.) Iltis [*P. trachysperma* T. & G.].

CAPRIFOLIACEAE A. L. Juss.
Honeysuckle Family

Subshrubs, shrubs, woody vines, or small trees; leaves opposite, (rarely some of them whorled, simple or compound, lacking stipules or rarely with them; flowers regular or irregular, perfect, arranged in cymes, or in axillary, pedunculate pairs, or in head- or spikelike clusters; calyx (3) 4– to 5–lobed; corolla of connate petals, typically 4– or 5–lobed; stamens 4 or 5; pistil 1; ovary inferior, 1– to 5–loculed; fruit a berry, drupe, or bristly achene; x = 8–12+.

1. Corolla rotate, regular or nearly so 2
— Corolla tubular, typically irregular 3
2(1). Leaves pinnately compound .... *Sambucus*
— Leaves simple, though often lobed .... *Viburnum*
3(1). Plants trailing, low subshrubs; flower pairs on long peduncles .... *Linnaea*
— Plants erect shrubs; flower pairs typically axillary and on moderately long peduncles, or sometimes clustered at stem ends 4
4(3). Stamens 5; locules all fertile, 2– to many-ovuled; flowers yellow, white, or pink, the lobes (or some of them) flaring; plants indigenous, cultivated or escaping and established .... *Lonicera*
— Stamens 4 or 5; locules 1–ovuled or infertile; flowers typically pink, the lobes flaring or not; plants indigenous or cultivated 5
5(4). Stamens 4, in 2 pairs; flowers pink, very showy, with flaring limb; pedicels, calyx, and fruit bristly with long, yellowish trichomes; plants cultivated and persisting .... *Kolkwitzia*
— Stamens 4 or 5, not paired; flowers pink to almost white, not very showy, the limb not especially flaring; pedicels, calyx, and fruit glabrous; plants indigenous, rarely cultivated .... *Symphoricarpos*

Kolkwitzia Grabner

Deciduous shrubs; branches with conspicuously exfoliating bark; leaves opposite, petiolate, serrate, lacking stipules; flowers showy, borne in axillary pairs forming terminal corymbs; calyx 5–lobed; corolla 5–lobed; campanulate; stamens 4; ovaries 2, each 3–loculed, but only 1 locule fertile and 1–ovuled; fruit a bristly achene.

Kolkwitzia amabilis Graebn. Beauty-bush. Shrubs to 3 m tall; leaves lanceolate, 2–8 cm long, 0.8–3 cm wide, acuminate apically, obtuse basally, entire or sparingly serrate, ciliate, sparingly yellowish hairy on the lower surface and sometimes also on the veins above; corymbs 3–8 cm long; flowers pink, the throat yellow, 17–22 mm long, with a short basal tube and a flaring limb, the upper pair of lobes erect, the lower 3 spreading; pedicels, sepals, and fruit bristly with long yellowish trichomes. Commonly cultivated ornamental, in lower elevation portions of Utah, long persisting but not escaping; introduced from China; 9 (i).

Linnaea L.

Subshrubs, the stems prostrate, rooting at the nodes; leaves opposite, simple; flowers borne in pairs on elongate, axillary peduncles; calyx 5–lobed; corolla 5–lobed, regular or nearly so, funnelform, regular or nearly so, pink or pinkish; stamens 4, in 2 pairs, attached near the corolla base; ovary 3-loculed, with only 1 functional; style elongate; fruit dry, 1-seeded.

Linnaea borealis L. Twin-flower. Stems mostly 1–12 dm long, trailing, moderately to sparingly hairy, becoming glabrous in age; flowering stems erect or ascending, commonly 1–15 cm long; leaves 0.5–2 cm long, 3–15 mm wide, elliptic to obovate or orbicular, or sometimes ovate, entire or few-toothed, rounded to obtuse apically, shortly petiolate, long-hairy to glabrous; peduncles axillary, 1.5–6 cm long, stipitate-glandular and often villous, bearing a pair of small bracts at the apex; flowers paired (rarely more); pedicels mostly 0.5–2 cm long, hairy like the peduncles; ovary glandular hairy; calyx 2–5 mm long; corolla 6–12 mm long, pink or pinkish, hairy within; fruit 1.5–3 mm long, hairy; $2n = 32$. Douglas fir, aspen, lodgepole pine, and spruce-fir communities at 1830 to 2930 m in Beaver, Daggett, Duchesne, Emery, Salt Lake, Sanpete, Summit, and Uintah counties; Alaska and Yukon, east to the Atlantic, south to California, Arizona, New Mexico, South Dakota, Indiana, and West Virginia; circumpolar; 12 (i). Our material is assignable to var. **longiflora** Torr. [*L. americana* Forbes; *L. borealis* var. *americana* (Forbes) Rehder; *L. borealis* ssp. *americana* [Forbes] Hulten; *L. longiflora* (Torr.) Howell; *L. borealis* ssp. *longiflora* (Torr.) Hulten].

Lonicera L.

Shrubs, the stems erect, or vines with twining stems; leaves opposite, simple; flowers borne in axillary, pedunculate pairs or in terminal head- or spikelike clusters; calyx 5-lobed; corolla 5-lobed, 2-lined, irregular, funnelform, swollen near the base, white, pink, orange, or yellowish; stamens 5, attached near the corolla base; ovary 2- or 3-loculed, each locule functional; style elongate; fruit a berry, several-seeded.

1. Plants trailing vines, cultivated .... *L. japonica*
— Plants erect shrubs, indigenous or cultivated 2

2(1). Flowers typically bright pink, fading yellow or brownish; branchlets hollow; plants cultivated, escaping, and established .... *L. tatarica*
— Flowers ochroleucous, whitish, or yellow; branchlets solid; plants cultivated or indigenous 3

3(2). Corollas yellow, often tinged purplish, glandular-hairy; flowers subtended by large foliose bracts; fruit black; plants indigenous .... *L. involucrata*
— Corollas pale yellowish to ochroleucous, glabrous; flowers not subtended by large bracts; fruit red; plants indigenous or cultivated 4

4(3). Leaves ciliate; petioles to 6 mm long; shrubs 1.5–2.5 m tall; flowers appearing before the leaves of the season; plants semi-evergreen, cultivated ..... *L. fragrantissima*
— Leaves typically not ciliate; petioles over 6 mm long; shrubs mainly less than 1.5 m tall; plants deciduous, indigenous .... *L. utahensis*

Lonicera fragrantissima Lindl. & Paxt. Fragrant Honeysuckle. Shrubs, mainly 1.5–2.5 m tall, semievergreen; leaves oval, 1.5–5 cm long or more, mainly 12–40 mm wide, glabrous except along the vines, ciliate, short-petiolate, apiculate, bicolored, green above, glaucous beneath; flowers creamy white, 10–12 mm long, the lobes widely flaring or reflexed, fragrant, borne in pairs on short peduncles, glabrous without; fruit red, seldom formed. Cultivated early-flowering ornamental with very fragrant flowers, long persisting in lower elevation portions of Utah; introduced from China; 4 (i).

Lonicera involucrata (Richards.) Banks ex Sprengel Black Twinberry. [*Xylosteum involucratum* Richards.; *L. ledebourii* Eschsch.]. Shrubs 0.5–2 m tall, the young stems glabrous, quadrangular; leaves 2.5–16 cm long, 1.5–7.5 cm broad, elliptic to oblanceolate or broadly lanceolate, entire, acuminate apically, acute to rounded basally, green and glabrous above or hairy along the midvein, pale and more or less hairy beneath, short-petiolate; peduncles 0.5–4 cm long; flowers borne in sessile pairs, subtended by broad, greenish or purplish black, foliose, glandular-hairy bracts commonly 1 cm long or more, these becoming black and fleshy at maturity; ovary glandular-hairy; calyx less than 1 mm long; corollas 12–18 mm long, yellow (sometimes tinged purplish), pubescent externally; fruit 8–12 mm long, black; $2n = 18$. Stream banks and other moist sites in riparian, aspen, Douglas fir, and spruce-fir communities at 1370 to 3235 m in Beaver, Cache, Carbon, Davis, Duchesne, Emery, Garfield, Grand, Iron, Kane, Piute, Rich, Salt Lake, San Juan, Sanpete, Sevier, Summit, Uintah, Wasatch, and Weber counties; Alaska east to Quebec, south to California, New Mexico, and Mexico; 75 (xiii).

Lonicera japonica Thunb. Japanese Honeysuckle. Plants clambering or twining woody vines with hollow stems, semievergreen, mainly 1–3 m long or more; leaves 2–8 cm long or more, ovate to oblong, pubescent beneath, becoming glabrous above, apiculate; flowers white, often suffused with purple without, aging yellow, borne in axillary pairs on lateral branchlets, often crowded apically, fragrant; peduncles short, bracteate; corolla mainly 30–40 mm long, pubescent, the tube slender, the limb 2-lipped; fruit black, seldom produced. Cultivated ornamental and reclamation plant in lower elevation portions of Utah, persisting; introduced from Asia; 3 (0).

Lonicera tatarica L. Tatarian Honeysuckle. Shrubs 1–3 m tall or more, the young stems glabrous, more or less terete; leaves 1.5–6 cm long, 0.5–3 cm broad, lanceolate to oblong, entire, attenuate to acuminate or acute apically, obtuse to rounded or subcordate basally, green and glabrous above, pale and glabrous beneath, short petiolate; peduncles 3–15 mm long; flowers borne in sessile pairs, subtended by narrow, scarious bracts; ovary glabrous; calyx less than 1 mm long; corollas 7–15 mm long, white to pink, glabrous; fruit 4–6 mm long, orange or red; $2n = 18$. Cultivated ornamental, escaping and established, in Carbon, Juab, Salt Lake, Summit, Uintah, and Utah counties; introduced from Eurasia; 10 (i).

Lonicera utahensis Wats. Utah Honeysuckle. Shrubs 1–2 m tall; leaves 1–8 cm long, 0.5–4 cm wide, elliptic to ovate or oblong, rounded to subcordate basally, rounded to obtuse apically, glabrous above, often pubescent beneath, short-petiolate; peduncles axillary, 5–35 mm long, with a pair of bracts apically; flowers paired; corolla ochroleucous or yellowish, 10–20 mm long, somewhat bilabiate, the lobes shorter than the tube, spurred basally; fruit 8–10 mm long, red. Mountain brush, ponderosa pine, aspen, and spruce-fir communities at 1500 to 3145 m in Cache, Carbon, Duchesne, Grand, Juab, Salt Lake (type from Cottonwood Canyon), San Juan, Sanpete, Sevier, Summit, Utah, and Wasatch counties; British Columbia and Alberta, south to California and Wyoming; 21 (i).

Sambucus L.

Shrubs, with coarse, erect or ascending stems; leaves opposite, pinnately compound, sometimes stipulate; flowers numerous, borne in terminal, compound, paniculate cymes; calyx 5-toothed; corolla 5-lobed, rotate, regular, white to cream; stamens 5, attached to the corolla tube; ovary 3- to 5-loculed, each locule functional; style very short or lacking; fruit a 3- to 5-seeded berry.

1. Cymes flat-topped, mainly 7–30 cm wide; flowers mainly cream colored; fruit black, with a glaucous, waxy bloom; plants typically below 2600 m elevation .... *S. caerulea*
— Cymes obconic to rounded, mainly less than 8 cm wide; flowers white to cream; fruit red (or rarely black?); plants typically above 2500 m elevation .... *S. racemosa*

Sambucus caerulea Raf. Blue Elderberry. [*S. glauca* Nutt. ex T. & G.]. Shrubs, mainly 2–4 dm tall, with soft branches, the pith brown, very thick; stipules linear, caducous or lacking; leaves typically 5- to 9-foliolate, the leaflets 3–15 cm long, 1–6 cm wide, lanceolate to lance-ovate or elliptic, often strongly oblique at the base, glabrous or hairy beneath; inflorescence pedunculate or sessile, corymbose (when alive), commonly 7–30 cm wide at anthesis, with mostly 4 or 5 rays, occasionally with subsidiary lateral inflorescences from the uppermost leaf axils; flowers cream to whitish, 4–7 mm wide, the corolla lobes longer than the tube; fruit 4–6 mm thick, subglobose, bluish black, glaucous and appearing powdery blue; 2n = 38. Riparian, sagebrush, mountain brush, pinyon-juniper, ponderosa pine, aspen, and spruce-fir communities at 1370 to 2810 m in probably all Utah counties; British Columbia to Montana, south to California, Arizona, and New Mexico; 130 (xxv). Two other elderberry species are known in cultivation in Utah. They are *S. canadensis* L. and *S. nigra* L. They resemble *S. caerulea* in general habit, but have black, not glaucous fruit. Additionally, *S. nigra* has mainly 5 leaflets per leaf, these again divided in some ornamental specimens.

Sambucus racemosa L. Red Elderberry. Shrubs, mainly 1–2 m tall, with subherbaceous pithy twigs; leaves 5–35 cm long, the leaflets (3) 5–7, 2.5–15 cm long, 1–6 cm broad, lanceolate to elliptical, oblanceolate, or oblong, acuminate apically, unequally obtuse to acute basally, serrate, petiolulate, green and glabrous or hairy only along the midvein beneath; flowers numerous, borne in terminal, paniculate cymes with a distinct central rachis extending beyond the lowermost branches, borne on leafy lateral shoots; calyx lobes less than 1 mm long; corollas white to cream, 4–6 mm wide, the lobes much longer than the tube; fruit 4–6 mm long, red (or black?); 2n = 36, 38. Aspen, lodgepole pine, and spruce-fir communities at 1860 to 3355 m in Beaver, Cache, Carbon, Duchesne, Emery, Garfield, Grand, Iron, Kane, Piute, Salt Lake, San Juan, Sanpete, Sevier, Summit, Tooele, Uintah, Utah, Wasatch, Washington, Wayne, and Weber counties; Alaska east to the Atlantic, south to California, Arizona, and New Mexico; 78 (xvii). Our material, or most of it, is assignable to var. *microbotrys* (Rydb.) Kearney & Peebles [*S. microbotrys* Rydb.], the typical red-fruited plants from the Rocky Mountains. However, the var. *melanocarpa* (Gray) McMinn [*S. melanocarpa* Gray], a black-fruited phase might be present in Utah.

Symphoricarpos Duhamel

Shrubs, with slender, erect, ascending, or less commonly prostrate stems; leaves opposite, simple; flowers solitary in leaf axils, or 2 to several borne in short, terminal or axillary spikes; calyx 5-lobed; corolla 5-lobed, more or less irregular, funnelform, white to pink; ovary 4-loculed, with only 2 functional; style elongate; fruit a 2-seeded berry.

Jones, G. N. 1940. Monograph of *Symphoricarpos*. J. Arnold Arb. 21: 201–253.
Wood, B. W. 1965. Revision of *Symphoricarpos* (Caprifoliaceae) in Utah. Proc. Utah Acad. 42: 203–213.

1. Corolla 10–18 mm long, salverform; plants typically of semi-arid sites .... *S. longiflorus*
— Corolla typically less than 10 mm long, or sometimes longer, campanulate to tubular-funnelform; plants typically of mesic sites or in cultivation 2
2(1). Flowers typically in terminal or axillary spikes; plants cultivated 3
— Flowers typically solitary in leaf axils, though sometimes in spikes; plants indigenous, rarely cultivated 4
3(2). Stamens included; berries white .... *S. albus*
— Stamens exserted; berries red .... *S. orbiculatus*
4(3). Corolla short-campanulate, about as broad as long (when pressed), the lobes subequal to the tube; staminal filaments longer than the anthers; plants of northeastern Utah .... *S. occidentalis*
— Corolla campanulate to tubular-funnelform, the lobes much shorter than the tube; staminal filaments shorter than the anthers; plants widespread and common .... *S. oreophilus*

Symphoricarpos albus (L.) Blake White Snowberry [*Viburnum album* L.; *S. racemosus* var. *laevigatus* Fern.]. Shrubs, mostly 0.5–1.5 m tall, the young stems glabrous, terete or nearly so; leaves 1.5–7 cm long, 0.8–4 cm broad, ovate to elliptic, entire or coarsely toothed, rounded to obtuse apically, rounded to obtuse basally, green and glabrous above, pale and sparingly hairy to glabrous beneath, short-petiolate; peduncles 3–30 mm long; flowers solitary, or more commonly 2 to several in axillary or terminal spikes, each closely subtended by a pair of minute bracts; ovary glabrous; corollas 5–7 mm long, white to pink, densely hairy within; fruit 6–12 mm long, white; 2n = 54, 72. Cultivated ornamental shrub, persisting in Salt Lake and Utah counties; introduced from elsewhere in North America; 3 (0).

Symphoricarpos longiflorus Gray Long-flower Snowberry. Shrubs, mostly 0.5–1 m tall, the young stems glabrous or pubescent, typically spreading at right-angles to the stem axis; leaves 6–15 (20) mm long, 2–7 (9) mm wide, oval to lanceolate or oblanceolate; flowers very fragrant, solitary or paired in leaf axils, or in small, terminal, few-flowered racemes; corollas 10–18 mm long, salverform, glabrous within and without, pale to deep pink, regular, the lobes much shorter than the tube; fruit white. Mixed desert shrub, pinyon-juniper, sagebrush, mountain brush, and ponderosa pine, grass, and hanging garden communities at 915 to 2900 m in Beaver, Daggett, Emery, Garfield, Grand, Iron, Juab, Kane, Millard, San Juan, Sevier, Tooele, Uintah, Washington, and Wayne counties; Oregon to Colorado, south to California, Arizona, New Mexico, Texas, and Mexico; 55 (xv).

Symphoricarpos occidentalis Hook. Wolfberry. Shrubs, mainly 0.4–1 m tall, spreading below ground, forming colonies, the young stems puberulent or glabrous; leaves 2–8 cm long, 1–5 cm wide, or larger on twigs of the season, elliptic to ovate, entire or coarsely toothed or lobed, glabrous above, puberulent beneath, at least on the main veins; flowers borne in axillary or terminal spikelike racemes; corollas 5–8 mm long, often wider than long (when pressed), hairy within, the lobes subequal to the tube; anthers shorter than the filaments; berries white. Riparian communities, mainly with cottonwood and willow species at 1525 to 1895 m in Daggett, Duchesne, and Uintah counties; British Columbia to Manitoba, south to Washington, Idaho, and Colorado; 9 (0).

Symphoricarpos orbiculatus Moench Coralberry. Shrubs, mainly 0.5–2 m tall; leaves 1–3.5 cm long, elliptical to oval, typically glabrous above, softly puberulent below; flowers borne in dense axillary or terminal spikelike racemes; corolla short-campanulate, pink, villous within, slightly ventricose on the lower sides, the lobes subequal to the tube; fruit coral red, ellipsoid. Cultivated ornamental in Utah County, persisting and escaping; introduced from the eastern U. S.; 1 (0).

Symphoricarpos oreophilus Gray Mountain Snowberry. Shrubs, mainly 0.5–1.5 m tall, the young stems puberulent to glabrous; leaves typically 1–5 cm long and 3–25 mm wide, elliptic to ovate, entire or few-toothed; flowers solitary or paired in leaf axils, and sometimes also in terminal, few-flowered racemes; corolla tubular-funnelform to campanulate, mostly 5–13 mm long, the lobes much shorter than the tube, hairy or glabrous; anthers longer than the filaments; fruit white. Sagebrush, ponderosa pine, Douglas fir, aspen, lodgepole pine, and spruce fir communities at 1370 to 3295 m in all Utah counties (except Davis and Morgan); British Columbia to Montana, south to California, Arizona, New Mexico, and Mexico; 219 (xxv). This common and widespread snowberry species is recognized as having three confluent and insubstantial varieties. The phase with decumbent, intricately branched stems from the western border counties is regarded as **var. parishii** (Rydb.) Cronq. [*S. parishii* Rydb.]. That with corollas 6–10 mm long and nutlets blunt to acutish at the base are assignable to **var. utahensis** (Rydb.) A. Nels. [*S. utahensis* Rydb., type from Logan; *S. vaccinoides* Rydb.; *S. rotundifolius* var. *vaccinoides* (Rydb.) A. Nels.; *S. tetonensis* A. Nels.], with most of the plants occurring in southern Utah. The final phase consists of plants with corollas mostly more than 10 mm long and with often acutish nutlets. It is the **var. oreophilus**.

Viburnum L.

Shrubs or small trees with simple, opposite, deciduous or evergreen leaves; flowers borne in corymbose clusters, radially symmetric, or some of the marginal flowers of an inflorescence enlarged and slightly irregular; calyx 5-lobed; corolla 5-lobed, rotate to campanulate; stamens 5; ovary 3-loculed, with 1 fertile locule and a single ovule; style short; stigma 3-lobed; fruit a drupe. Note: The genus is not well represented in herbaria, but the plants long persist following cultivation. They are included in the key but not described.

1. Leaves lobed, reddish in the fall; shrubs to 4 m tall (snowball bush) .... *V. opulus* L.
— Leaves not lobed 2
2(1). Leaves evergreen, entire or obscurely denticulate; petiole 12–25 mm long; shrubs to 3 m tall (wayfaring bush) .... *V. rhytidophyllum* Hemsl.
— Leaves typically deciduous; petiole more than 25 mm long; shrubs usually less than 2 m tall 3
3(2). Leaves 2.5–10 cm long; winter buds scaly .... *V. plicatum* Thunb.
— Leaves 10–20 cm long; winter buds naked .... *V. alnifolium* Marshall

CARYOPHYLLACEAE A. L. Juss.
Pink Family

Annual or perennial herbs; leaves opposite, entire, simple, with or without stipules; flowers usually perfect, regular; sepals 4 or 5, connate or persistent; petals 4 or 5 or lacking; stamens 4–10, distinct; anthers 2–loculed; pistil 1, the ovary superior, 1–loculed, with free-central placentation, or incompletely 3– to 5–loculed; styles 2–5 (rarely connate into 1); fruit a capsule or a utricle; $x = 5$–19.

1. Fruit a utricle, 1-seeded, indehiscent; petals lacking; flowers minute, whitish, greenish, or yellowish 2
— Fruit a capsule, many-seeded, dehiscent; petals usually present; flowers various 4
2(1). Plants pulvinate-caespitose perennials, usually montane or alpine .... *Paronychia*
— Plants annual, prostrate or ascending, not montane or alpine 3
3(2). Sepals united basally into a tube, the lobes white-hyaline, surpassing the ovary .... *Achyronychia*
— Sepals distinct or nearly so, the lobes green with hyaline margins, shorter than or subequal to the ovary .... *Herniaria*
4(1). Sepals connate into a tubular calyx; petals clawed 5
— Sepals distinct or nearly so; petals without claws 10
5(4). Ribs of calyx twice as many as the calyx teeth, ending both in apices and in sinuses 6
— Ribs of calyx as many as calyx teeth, the calyx 5-ribbed, 5-nerved, nerveless, or striate-nerved ... 7
6(5). Styles 3 (4); capsule opening by 6 (3, 4, or 8) teeth . *Silene*
— Styles 5 (4); capsules opening by 10 (8) teeth ... *Lychnis*
7(5). Calyx scarious between green midveins ending in teeth 8
— Calyx not at all scarious, green overall 9
8(7). Calyx urn-shaped, constricted apically; petals pink .... *Vaccaria*
— Calyx campanulate, not constricted apically; petals white or pink .... *Gypsophila*
9(7). Calyx closely subtended by and basally invested by long-aristate bractlets .... *Dianthus*
— Calyx not immediately subtended by bractlets, these, when present, not investing the base ...... *Saponaria*
10(4). Leaves with scarious or setaceous, persistent stipules 11
— Leaves lacking stipules 12
11(10). Plants erect-ascending; leaves rigid, with setaceous stipules .... *Loeflingia*
— Plants usually mat-forming; leaves not rigid, with scarious broad stipules .... *Spergularia*
12(10). Petals 2-cleft, -parted, or -lobed, or lacking (key both ways) 13
— Petals rounded, emarginate, irregularly toothed, or bilobed (see *Stellaria jamesiana* and *Arenaria fendleri*), or lacking 14

13(12).	Styles usually 5; capsule cylindrical, membranous, often 1 or 2 times longer than the calyx, dehiscent by 10 teeth .... *Cerastium*
—	Styles usually 3; capsule ovoid to short-cylindric, seldom equaling the calyx, dehiscent by 6 or 8 teeth .. .. *Stellaria*
14(12).	Plants annual, flowering at middle elevations in springtime; petals irregularly toothed ...... *Holosteum*
—	Plants perennial or annual; flowering in spring and summer at low to high elevations; petals not toothed 15
15(14).	Styles usually 3, or if more, opposite the sepals; valves of capsule alternate with sepals; plants common .... *Arenaria*
—	Styles usually 5, alternate with the sepals; valves of capsule opposite the sepals; plants uncommon .. *Sagina*

Achyronychia T. & G.

Glabrous annual herbs; leaves opposite, but unequal at each node, spatulate; stipules hyaline, conspicuous; flowers in axillary, cymose clusters; calyx 5–lobed, united at the base, the lobes white-scarious; petals lacking; stamens 10–15, with 1–5 fertile ones; styles 2–cleft; fruit a utricle, enclosed by the calyx; seeds minute, black.

Achyronychia cooperi T. & G. Plants prostrate, mainly 3–20 cm across; leaves spatulate, 3–18 mm long, 1.2–7 mm wide; flowers 2–3 mm long, the sepals green basally, with white-scarious lobes, these oval; seeds shining. Warm desert shrub, Washington (Atwood sn, missing at BRY) County; Arizona, Nevada, California, and Mexico; 1 (0).

Arenaria L.

Annual, biennial, or perennial herbs, with prostrate to erect stems; leaves opposite, linear to ovate, lacking stipules; flowers solitary or few to many, borne in open or contracted cymes; sepals 5, distinct or nearly so, 1– to 3–veined; petals 5, white, the blade entire or slightly emarginate but not bilobed (except in some *A. fendleri*), showy to inconspicuous or sometimes lacking; stamens usually 10, inserted at the base of a glandular disk; styles usually 3; ovary 1–loculed; capsule opening by 3 or 6 teeth.

Maguire, B. 1951. Studies in Caryophyllaceae - V. *Arenaria* in America north of Mexico. A conspectus. Amer. Midl. Naturalist 46: 493–511.
———. 1958. *Arenaria rossii* and some of its relatives in America. Rhodora 60: 710.

1.	Leaves ovate to lanceolate, oblanceolate, or broadly elliptic, acute to rounded apically 2
—	Leaves linear to oblong or subulate, acute, apiculate, or abruptly acute to blunt apically 5
2(1).	Leaves once to twice longer than broad, much shorter than the internodes; plants annual .... *A. serpyllifolia*
—	Leaves mainly 2–10 times longer than broad or more, from slightly shorter to much longer than the internodes; plants perennial 3
3(2).	Petals notched apically; leaves clasping .... *Stellaria jamesiana* (q.v.)
—	Petals not notched apically; leaves attenuate basally .. 4
4(3).	Stems from a branching caudex; inflorescence mainly terminal .... *A. lanuginosa*
—	Stems from a rhizome (seldom collected); inflorescences mainly from stem middle and sometimes also terminal .... *A. lateriflora*
5(1).	Inflorescence congested; leaves very slender, mainly basal, commonly 2–4 cm long .... *A. congesta*
—	Inflorescence more or less open; leaves slender and 1–4 or rarely to 11 cm long, or shorter and proportionately thicker 6
6(5).	Plants cushionlike, densely pulvinate-caespitose; inflorescence mainly 1–5 cm tall .... *A. hookeri*
—	Plants of various habit, but seldom both cushionlike and densely pulvinate-caespitose (see *A. obtusiloba*) .. 7
7(6).	Plants subshrubs, woody at the base; inflorescence comprising 1/4–1/3 of plant height ..... *A. macradenia*
—	Plants herbaceous, seldom if ever woody at the base; inflorescences various 8
8(7).	Leaves slender, mainly 1–4 (11) cm long; capsules dehiscent by 6 teeth; plants widespread .... *A. fendleri*
—	Leaves not especially slender, mainly less than 1 cm long; capsules dehiscent by 3 teeth; distribution various, but plants usually of alpine sites 9
9(8).	Plants soboliferous, with long trailing subterranean stems, forming mats; leaves 5–10 mm long . *A. nuttallii*
—	Plants not soboliferous, either annual, biennial, or short lived perennials from taproots, or caespitose perennials from a caudex; leaves often less than 5 mm long 10
10(9).	Plants caespitose perennials from a caudex; sepals obtuse .... *A. obtusiloba*
—	Plants delicate, clump-forming annuals, biennials, or short-lived perennials from taproots 11
11(10).	Stems capillary; sepals 2–3 mm long; plants diminutive annuals known from sandy soil in southwestern Utah .... *A. pusilla*
—	Stems not or seldom capillary; sepals often more than 3 mm long; plants biennial or short-lived perennials, of various distribution but seldom, if ever, as above... 12
12(11).	Sepals 1–veined; plants glabrous; inflorescence 1–flowered .... *A. rossii*
—	Sepals 3–veined; plants glandular-puberulent or glabrous; inflorescence 1– to several-flowered 13
13(12).	Leaves 3–veined; plants glandular-puberulent or sometimes glabrous .... *A. rubella*
—	Leaves 1–nerved; plants glabrous .... *A. filiorum*

Arenaria congesta Nutt. in T. & G. Head Sandwort. Perennial herbs from a woody caudex, mainly 0.6–3.3 dm tall, glabrous; basal leaves subulate to almost filiform, 0.8–6 (8) cm long, erect, flexible or stiff, apiculate to spinulose; cauline leaves of mainly 3–5 pairs, usually shorter than the basal and reduced upward; inflorescence capitate or subumbellate, many-flowered; bracts ovate-acuminate, hyaline, with a stout midrib extending to the acute-attenuatre tip; sepals 4–5.5 mm long, ovate, greenish, with broad scarious margins; petals 5–7.5 mm long, white; capsules subequal to or slightly longer than the sepals; seeds 1.5–2 mm long, flat, winged, brownish; 2n = 22. Two geographically correlated and morphologically differing varieties occur in Utah.

1.	Basal leaves mainly 0.8–2 cm long, rigid, spinulose, and often recurved; plants of the Great Basin. .... *A. congesta* var. *subcongesta*
—	Basal leaves mainly 1.5–6 (8) cm long, flexible, apiculate, usually straight; plants of the Wasatch Mts., Wasatch Plateau, and eastward. .... *A. congesta* var. *congesta*

Var. *congesta* [*A. congesta* var. *lithophila* Rydb., *A. lithophila* (Rydb.) Rydb.; *A. cephaloidea* Rydb.; *A. con*-

gesta var. *cephaloidea* (Rydb.) Maguire]. Pinyon-juniper, sagebrush, mountain brush, ponderosa pine, aspen, lodgepole pine, spruce-fir, and alpine tundra communities at 2135 to 3360 m in Cache, Daggett, Duchesne, Grand, Rich, San Juan, Summit, Uintah, Utah, and Wasatch counties; Washington and Montana, south to California, Nevada, and Colorado; 71 (x). Some specimens from high elevations in the Uinta Mts. are dwarf, but grade downward into the taller materials. They do not seem to warrant taxonomic recognition.

Var. *subcongesta* (Wats.) Wats. [*A. fendleri* var. *subcongesta* Wats.]. Sagebrush, white fir, and bristlecone pine communities at 2650 to 2780 m in Beaver, Millard, and Washington counties; California and Nevada; 3 (i). This material simulates *A. fendleri* in its vegetative features.

Arenaria fendleri Gray Fendler Sandwort. Perennial herbs from caudices and taproots, more or less caespitose; stems mainly 0.2–2.5 dm tall; basal leaves mainly 0.3–5 (11) cm long, pungent, straight, or recurved; cauline leaves mainly 2–6 pairs; inflorescence open, mostly (1) 2- to several-flowered, the flowers subsessile or on pedicels to 20 mm long, glandular or glabrous; sepals ovate to ovate-lanceolate, acute to obtuse, 3.5–6.5 mm long, greenish or suffused with purple, with broad scarious margins; petals 5–8 mm long, entire, emarginate, or bilobed, white to brown or pinkish red; capsule with 6 teeth, subequal to or much surpassing the sepals; seeds 1.5–2.5 (3) mm long, winged to essentially wingless; 2n = 22. The *fendleri* complex consists of a variable assemblage of more or less mat-forming, pungent-leaved, and open-cymose plants of broad distribution and broad ecological latitude. The Utah portion of the complex has been interpreted previously (Maguire 1951) as consisting of four species, with three of the species further subdivided into infraspecific taxa, for a total of nine taxa. That the taxa have long been regarded as similar is indicated by names combining various phases of the taxon in a general sense. The lack of consistent diagnostic criteria and considerations of morphological confluence indicate a conservative interpretation, a course followed herein. Four more or less distinctive and geographically correlated varieties are treated. The following tentative key will allow segregation of most specimens.

1. Sepals more or less hooded apically, thus apparently obtuse; plants usually definitely glaucous, known from Box Elder County .... *A. fendleri* var. *aculeata*
— Sepals not or seldom hooded apically, thus apparently acute or attenuate; plants green or glaucous 2
2(1). Sepals mainly 3.3–4.5 mm long, usually purplish medially; inflorescence usually glandular; plants of the Wasatch Mts. and high plateaus and westward .... *A. fendleri* var. *glabrescens*
— Sepals often more than 4.5 mm long, green or purplish medially; inflorescence glandular or glabrous; plants of eastern and southern Utah 3
3(2). Inflorescence glandular-pubescent (usually copiously so); stem with (4) 5 or more pairs of leaves, or the plants dwarf; distribution montane ..... *A. fendleri* var. *fendleri*
— Inflorescence glabrous or glandular; stem leaves mainly 2–4 pairs; plants usually of lower elevations .... *A. fendleri* var. *eastwoodiae*

Var. *aculeata* (Wats.) Welsh [*A. aculeata* Wats.]. Sagebrush and pinyon-juniper communities at 2135 to 2930 m in Box Elder County; Oregon to Montana, south to California and Nevada; 3 (0).

Var. *eastwoodiae* (**Rydb.**) Welsh [*A. eastwoodiae* Rydb.]. Shadscale, black sagebrush, big sagebrush, pinyon-juniper, ponderosa pine, and snowberry communities at 1130 to 2680 m in Carbon, Daggett, Duchesne, Emery, Garfield, Grand, Kane, San Juan, Uintah, Wasatch, Washington, and Wayne counties; Colorado, Arizona, and New Mexico; 103 (xvii). Flower color is usually creamy yellow or whitish, but many specimens have bronze, brownish, or red pink petals. The plants with colored flowers are often at the lowermost elevations, but flower color does not seem to be correlated with other features. They seem not to warrant taxonomic recognition. The antisepalous glands are conspicuous in this taxon.

Var. *fendleri* [*A. fendleri* var. *diffusa* T. C. Porter; *A. fendleri* var. *porteri* Rydb.; *A. fendleri* ssp. *brevifolia* Maguire (type from the La Sal Mts.) var. *brevicaulis* Maguire; *A. fendleri* var. *brevifolia* (Maguire) Maguire; *A. fendleri* var. *tweedyi* (Rydb.) Maguire; *A. tweedyi* Rydb.]. Mountain brush, sagebrush-snowberry, ponderosa pine, spruce-fir, and alpine tundra communities at 2895 to 3360 m in Garfield, Grand, and San Juan counties (Henry, La Sal, and Abajo mts.); Wyoming south to Arizona, New Mexico, and Texas; 13 (0). Utah materials have been treated (Maguire 1951) as belonging to varieties *brevifolia* and *tweedyi* (superceded by var. *brevicaulis* Maguire). Both have sepals in the size range of var. *glabrescens* (*A. kingii* sens. lat.), but have more attenuate sepals. However, besides the alpine dwarfs with short sepals, taller plants with long sepals are known (Neese et al 2192 BRY), as do specimens with very long leaves (Atwood 7061 BRY). These latter specimens would fall in still other of the named units recognized by Maguire (1951). They do not seem to represent taxa and a conservative treatment is indicated.

Var. *glabrescens* Wats. [*Stellaria kingii* Wats.; *A. kingii* (Wats.) Jones; *A. uintahensis* A. Nels.; *A. kingii* ssp. *uintahensis* (A. Nels.) Maguire; *A. aculeata* var. *uintahensis* (A. Nels.) Peck; *A. kingii* ssp. *plateauensis* Maguire, type from Cedar Breaks]. 2n = 44. Shadscale, horsebrush, pinyon-juniper, sagebrush, mountain brush, ponderosa pine, white fir-Douglas fir, aspen, bristlecone pine, spruce-fir, and grass-forb communities at 1555 to 3355 m in Beaver, Box Elder, Emery, Garfield, Iron, Juab, Kane, Millard, Piute, Rich, Sanpete, Sevier, Tooele, Utah, Washington, and Wayne counties; 118 (xv). Petals in many specimens are deeply bifid; in others they are rounded apically. Transitions are known, and there does not appear to be geographical or morphological correlation of bifid petals with other features. Some plants, apparently the *uintahensis* phase, tend to have rather obtusish sepals and pale midribs, but seem to lack geographical correlation.

Arenaria filiorum Maguire Beach Sandwort. Perennial (or biennial) herbs from taproots and weak caudices, cushionlike; stems mainly 1–4 cm tall; leaves mainly cauline, 2–5 pairs, 2–8 mm long, 1–veined, blunt; inflorescence glabrous, 1– to 3–flowered, the pedicels 1–8 mm long; sepals ovate-acuminate, green, 3–veined, the margins hyaline, 4–5.4 mm long; petals 3.5–4.5 mm long, entire, white; capsules 3–valved, shorter than the sepals; seeds 0.7–1 mm long, wingless. Spruce-fir, lake shore, and alpine tundra communities at 2745 to 3390 m in Duchesne, Iron, Juab, Kane (type from Navajo Lake), and Summit counties; endemic; 9 (i). This small cushion plant combines characteristics of both *A. rossii* and *A. rubella* (q.v.); the leaves are 1-veined and the plants glabrous as in the former, but the sepals are 3-veined and the inflorescence is mainly 2- or 3-flowered as in the latter. Some phases of *A. rossii* [e.g., var. *elegans* (C. & S.) Welsh] possess 3-veined sepals, but the flower number is seldom, if ever, more than one. Resolution of relationships of this taxon awaits revision of this complex in western North America.

Arenaria hookeri Nutt. in T. & G. Hooker Sandwort. Perennial, compactly pulvinate-caespitose herbs, forming dense cushions; stems mainly 1–5 cm tall, hirtellous or glabrous; basal leaves 1–6 cm long, pungent, spreading and rosettelike, shining, acicular; cauline leaves 1–3 pairs, hyaline-margined, acicular, pungent, 3–8 mm long; inflorescence of congested cymes, mainly 3- to several-flowered, the flowers sessile, subsessile, or on pedicels to 2 mm long; bracts inconspicuous, with broad hyaline margins; sepals 5–8 mm long, lance-acuminate, green, with broad hyaline margins, glabrous or scabrous; petals 4.5–8.5 mm long, about equaling the sepals, white, rounded to obtuse; capsules with 6 teeth, much shorter than the sepals; $2n = 44$. Pinyon-juniper, sagebrush, black sagebrush, and rabbitbrush communities at 1735 to 2930 m in Carbon, Daggett, Duchesne, Emery, Summit, Uintah, Utah, and Wayne counties; Nevada to Montana, south to South Dakota, Wyoming, Nebraska, and Colorado; 35 (iv). Our material belongs to **var. *desertorum*** Maguire (type from Uintah County). The Hooker sandwort is confined to the Colorado Plateau portion of Utah, except for a single record from Utah County (Brown s.n. 1927 BRY).

Arenaria lanuginosa (Michx.) Rohrb. [*Spergulastrum lanuginosum* Michx.] Perennial, loosely caespitose herbs from a caudex; stems mainly 5–16 cm tall, retrorsely scabrous; leaves mainly cauline, 3–6 pairs, 4–15 mm long, lanceolate to elliptic, (1) 2–4 mm wide, short-ciliate, acute, mucronate; inflorescence of open few- to several-flowered cymes; sepals lanceolate to ovate-lanceolate, attenuate-acuminate, 2.5–3.5 mm long, green, with hyaline margins; petals subequal to the sepals, white; capsules opening by 6 teeth; seeds 0.7–0.9 mm long, narrowly winged. Douglas fir-white fir woods at ca 2200 to 2640 m in Beaver, Piute, San Juan, and Washington counties; coastal eastern U. S. to Colorado and New Mexico; Mexico and South America; 8 (i). Utah material has been regarded at specific level as *A. saxosa* Gray [*A. lanuginosa* ssp. *saxosa* (Gray) Maguire]. The species has often been mistaken for a *Stellaria*, especially *S. longifolia*, and specimens should be sought in herbaria under that epithet. The presence of retrorse scaberulous pubescence in lines distinguishes *A. lanuginosa*.

Arenaria lateriflora L. Bluntleaf Sandwort. [*Moehringia lateriflora* (L.) Fenzl]. Perennial, rhizomatous or stoloniferous plants, the stems arising singly or in small clumps, the stolons or rhizomes lacking conspicuous persistent leaves; flowering stems 4–20 (25) cm tall, erect or ascending; basal leaves smaller than the middle and upper ones; cauline leaves 3–7 pairs, 3–26 mm long, 1–11 mm wide, the largest near the middle of the stems, lanceolate to elliptic or lance-oblong, pinnately 3- to 5-veined, acute to obtuse apically, minutely hairy below, especially along the midrib, ciliate, often with sterile axillary shoots; flowers solitary or 2–5 in terminal and/or lateral cymes; pedicels puberulent to glabrate or glabrous; sepals 2–3 mm long, erect or spreading, with white, scarious margins, obtuse apically; petals white, 3.5–8 mm long, obovate; capsules 2–4 mm long, 6-valved; seeds subreniform, 0.8–1.1 mm long, black, shining; $2n = 48$. Willow-wet meadow, aspen, lodgepole pine, and spruce-fir communities at 1830 to 2745 m in Cache, Daggett, Duchesne, Garfield, Salt Lake, and Weber counties; Alaska to Newfoundland, south to New Mexico, Missouri, and Rhode Island; Eurasia; 11 (0).

Arenaria macradenia Wats. Shrubby Sandwort. [*A. macradenia* ssp. *ferrisiae* Abrams]. Perennial subshrubs from woody bases; stems mainly 13–45 cm tall; leaves mainly cauline, 4–9 pairs, 0.7–4.5 (7) cm long, acicular or linear, pungent, often scabrous, straight or curved; inflorescence open, with 3 to many flowers, constituting ca 1/4–1/3 of plant height, the flowers on pedicels 3–35 mm long, glabrous or glandular; sepals ovate to lanceolate or elliptic, obtuse to acute, 3.4–6.5 mm long, green, the margins scarious; petals 6.5–9 mm long, entire, white; capsules with 6 teeth, surpassing the sepals; seeds 2–2.5 mm long, dark brown, spinulose-tuberculate, wingless. Creosote bush-Joshua tree, pinyon-juniper, mountain brush, ponderosa pine, white fir-Douglas fir, aspen, and alpine meadow communities at 915 to 2990 m in Beaver, Garfield, Juab, Kane, Millard, Piute, Sanpete, Sevier, Utah, Weber, and Washington counties; California, Nevada, and Arizona; 82 (xi). Our specimens have been regarded as belonging to ssp. *macradenia* and ssp. *ferrisiae*, but supposed differences in sepal length and woodiness are not apparent. Sepal size forms a continuum from 3.5–6.5 mm, and degree of woodiness is difficult to interpret. A conservative approach is indicated.

Arenaria macrophylla Hook. [*Moehringia macrophylla* Torr.]. Perennial, rhizomatous herbs, often mat-forming; stems decumbent to ascending or erect, mainly 5–15 cm tall; leaves in few to several pairs, lanceolate to elliptic or almost linear, acute, 1–7 cm long, 2–15 mm wide, scabrous; flowers solitary or 2–5 in terminal, pedunculate cymes; bracts minute; pedicels 8–30 mm long, scabrous marginally; petals subequal to or surpassing the sepals, white, rounded; capsules shorter than the calyx, 6-valved; seeds smooth. Plants of wooded areas and meadows, not definitely known from Utah, but to be expected; British Columbia to Labrador, south to California, New Mexico, and the Atlantic Coast; 0 (0).

Arenaria nuttallii Pax Nuttall Sandwort. [*A. nuttallii* var. *gracilipes* Jones, type from Piute County]. Perennial, mat-forming, soboliferous herbs from elongated caudex branches and taproot; herbage glandular-hairy; stems prostrate, the bases often buried, mainly 5–20 cm long; leaves cauline, mainly 3–5 pairs, often with fascicled axillary ones, 2–10 mm long, linear-subulate, 3-veined; inflorescence cymose, of solitary or 2–7 or more flowers, bracteate; pedicels 2–20 mm long; sepals lance-linear to ovate-lanceolate, 1- to 5-veined, 3.5–6 mm long, attenuate, usually purplish, with narrow hyaline margins; petals 3–7 mm long, white, entire; capsules shorter than the sepals, dehiscing by 3 valves; seeds ca 1.5 mm long, dark brown, papillate; $2n = 36$. Commonly in talus or on scree slopes at (2070) 2745 to 3660 m in Beaver, Duchesne, Juab, Piute, Rich, Salt Lake, Sanpete, and Utah counties; British Columbia to Alberta, south to California and Wyoming; 23 (iii). This is a distinctive species, especially adapted to shifting slopes, where it adjusts to downward creep by elongating from beneath the talus.

Arenaria obtusiloba (**Rydb.**) **Fern.** Rydberg Sandwort. [*A. obtusa* Torr., not All.; *Alsinopsis obtusiloba* Rydb.; *Minuartia obtusiloba* (Rydb.) House]. Perennial, clump-forming herbs, mainly 0.5–1.5 dm wide, arising from a weak to strong, spreading caudex, the caudex branches clothed with persistent leaves; flowering stems 1–8 cm tall, erect or ascending; basal leaves 3–8 mm long, commonly less than 1 mm wide, oblong to linear, 1–veined, abruptly obtuse apically, scabrous-ciliate almost or quite to the apex, or smooth; cauline leaves (0) 1–3 pairs, 2–8 mm long, commonly broader than the basal ones; flowers erect, solitary or in twos; pedicels glandular-hairy; sepals 2.5–4 (5) mm long, erect or ascending in flower, often purplish, the apex obtuse, 3–veined, glandular-hairy to glabrate; petals 4–6 (10) mm long, white, obovate; capsules 4–8 mm long, 3– or 4–valved, the seeds 0.8–1 mm long, smooth or nearly so; 2n = 26, 52, 78. Lodgepole pine, spruce-fir, and alpine tundra communities at 2895 to 3815 m in Beaver, Daggett, Duchesne, Garfield, Grand, Iron, Juab, Piute, Salt Lake, San Juan, and Uintah counties; Alaska to Labrador and Greenland, south to Oregon and New Mexico; 46 (x). This plant is often mistaken for *A. rubella* (q.v.), which has 3 veins in both sepals and leaves and has acute leaflets. The relationship of this taxon to *A. laricifolia* L. remains to be evaluated. Our plants have been assigned, in part, to *A. sajanensis* Willd. ex Schlecht., but that entity is not known from Utah.

Arenaria pusilla **Wats.** Dwarf Sandwort. [*Alsinopsis pusilla* (Wats.) Rydb.; *Minuartia pusilla* (Wats.) Mattf.]. Diminutive annuals; stems capillary, mainly 2–5 cm tall, simple or branched; leaves 2–5 mm long, 0.3–1 mm wide, lance-attenuate, more or less clasping basally, 1–veined, obtuse, lacking sterile axillary ones; flowers few to many, borne in open, leafy bracted cymes; sepals 2–3 mm long, 1–veined, lanceolate to lance-acuminate; petals shorter than the sepals or lacking; styles 3; capsule ovoid, subequal to the sepals, opening by 3 valves; seeds brown, less than 0.5 mm long, minutely ornamented. Ponderosa pine and mixed mountain brush community at 1525 to 1830 m in Washington County; Washington to Idaho, south to California, and Nevada; 3 (0).

Arenaria rossii **R. Br. ex Richards.** Ross Sandwort. Plants short-lived perennials, forming cushions or mats mainly 0.5–2 dm wide, arising from a taproot and caudex branches, the caudex branches prostrate to ascending, clothed with persistent leaves; flowering stems mainly 2–8 cm tall, erect; basal leaves 3–10 mm long, less than 1 mm wide, linear to lance-linear, 1–veined, obtuse apically, glabrous; cauline leaves (0) 2 or 3 pairs, 3–12 mm long, similar to the basal ones, often with short, sterile, axillary branches; flowers erect, solitary; pedicels glabrous; sepals 1.5–3.5 mm long, spreading to ascending in flower, more or less purplish throughout, acute apically; petals white, 2–3.5 mm long, oblanceolate (or lacking); capsules 1.5–2.5 mm long, 3–valved; seeds 0.6–1 mm long, reddish brown, minutely roughened. Alpine tundra at ca 3355 m in Sanpete County (Lewis 4760 BRY); Alaska and Yukon, south to Washington, Idaho, and Colorado; Asia, Greenland; 1 (0). This species is evidently rare in Utah, but should be expected in other locations.

Arenaria rubella (**Wahl.**) **J. E. Sm.** Reddish Sandwort. [*Alsine rubella* Wahl.; *Arenaria verna* var. *rubella* (Wahl.) Wats.; *Minuartia rubella* (Wahl.) Grabner; *Arenaria propinqua* Richards.; *Alsinopsis propinqua* (Richards.) Rydb.; *Arenaria verna* authors, not L.]. Short-lived perennial (or biennial) herbs, forming cushions or mats, 4–30 cm wide, arising from a taproot and branching caudex, the caudex branches prostrate to ascending or erect, clothed with persistent leaves; flowering stems mainly 2–10 cm tall, erect or ascending; basal leaves 3–10 (12) mm long, less than 1 mm wide, lance-linear, 3–veined, acute, glandular-puberulent to glabrate; cauline leaves 2–6 pairs, 1–14 mm long, similar to the basal ones but reduced upward, often with short, sterile, axillary branches; flowers erect, solitary, or more commonly 2–4; pedicels glandular-hairy; sepals (2) 3–4.5 mm long, spreading in flower, commonly more or less purplish throughout, acute to acuminate apically; petals white, 2.5–5 mm long, oblanceolate; capsules 3–5 mm long, 3– or 4–valved; seeds 0.4–0.6 mm long, reddish brown, minutely roughened; 2n = 24. Lodgepole pine, spruce-fir, forb-grass, sedge, and alpine tundra communities, often in gravel, at 2315 to 3660 m in Beaver, Daggett, Duchesne, Garfield, Juab, Kane, Piute, Salt Lake, Sanpete, Summit, Uintah, and Utah counties; Alaska east to Greenland, south to California, Nevada, New Mexico, and Gaspe; Eurasia; 26 (vii).

Arenaria serpyllifolia **L.** Thymeleaf Sandwort. Annual herbs from taproots; stems mainly 5–20 cm tall, retrorsely pubescent, ascending to erect, commonly branched; leaves much shorter than the internodes, 2–7 mm long, 1–4 mm wide, ovate to lanceolate, acute, sessile, or the lower ones short-petiolate; flowers few to several, borne in open cymes; pedicels slender, 2–10 mm long; sepals lanceolate, acute, 3–3.5 mm long, 3–veined; petals white, shorter than the sepals; capsules 3–4 mm long, 6–valved; seeds 0.5–0.6 mm long, grayish, tuberculate; 2n = 20, 44. Pinyon-juniper, mountain brush, and Douglas fir communities at 2195a to 2565 m in Cache and Juab counties; widely distributed in North America; adventive from Eurasia; 2 (0).

Cerastium **L.**

Annual, biennial, or perennial herbs, with prostrate to erect stems; leaves opposite, ovate to lanceolate, oblong, or linear, lacking stipules; flowers solitary and terminal or more commonly in terminal, open or congested cymes; sepals 5, distinct to the base, 1 (or 3–5) -veined; petals 5, white, showy to inconspicuous, the blade notched or bilobed, or sometimes lacking; stamens commonly 10, inserted at the base of the ovary; styles usually 5, opposite the sepals; ovary 1–loculed; capsule opening by 10 teeth.

Hulten, E. 1956. The *Cerastium alpinum* complex. A case of World-wide introgressive hybridization. Svensk. Bot. Tidskr. 50: 471–495.

1. Plants gray-tomentose; petals 12–18 mm long; cultivated ornamentals .... *C. tomentosum*
— Plants variously hairy, but seldom, if ever, gray-tomentose; petals less than 12 mm long; indigenous 2

2(1). Petals subequal to the sepals; plants annual, biennial, or short-lived, adventive perennials 3
— Petals longer than the sepals; plants indigenous perennials 4

3(2). Plants annual, the stems erect-ascending; petals sometimes exceeding the sepals .... *C. nutans*
— Plants biennial or perennial, rooting at lower nodes, the stems usually decumbent basally; petals about as long as the sepals .... *C. fontanum*

4(2). Leaves linear or lance-linear, oblong, or narrowly elliptic (the uppermost leaves sometimes broader); sterile branches arising in most leaf axils; bracts scarious-margined .... *C. arvense*

— Leaves lanceolate to elliptic, oblong, or oblanceolate; sterile branches lacking, or present only in lowermost axils; bracts not scarious-margined ..... *C. beeringianum*

Cerastium arvense L. Field Chickweed. Perennial herbs, forming loose to dense clumps 1–4 dm wide, arising from stems 6–30 cm long, decumbent, ascending, or erect, from glabrous to descending or spreading hairy on the middle internodes, becoming densely glandular upward; leaves (3) 10–25 mm long, 1–4 (7) mm wide, linear to lance-linear, oblong, or narrowly elliptic (or the uppermost sometimes lanceolate), 1-veined, more or less pubescent on both sides, ciliate, acute; most cauline leaves with sterile axillary shoots; flowers erect, commonly 3–9 mm long, glandular-hairy; bracts more or less scarious-margined; sepals 4–6.5 mm long, glandular, scarious-margined, the inner mostly broadly so; petals white, 8–12 mm long, deeply bilobed; capsules 6–10 mm long, cylindric; seeds 0.8–1.2 mm long, yellowish or brownish red; $2n = 36, 72, 90$. Pinyon-juniper, mountain brush, sagebrush, meadow, ponderosa pine, aspen, and lodgepole pine communities at 2375 to 3390 m in Beaver, Daggett, Duchesne, Emery, Garfield, Grand, Juab, San Juan, Sevier, Summit, Uintah, and Utah counties; widely distributed in North America; Eurasia; 24 (0). Apparent intermediates are known between this and the next following species.

Cerastium beeringianum C. & S. Bering Chickweed. [*C. variable* Goodding, type from Uintah County]. Perennial herbs, forming loose to dense mats of clumps, 4–40 cm wide or more, arising from taproots and more or less stoloniferous, prostrate stems; flowering stems 3–25 (35) cm tall, decumbent to erect, the lower internodes glabrous to moderately retrorsely hairy, becoming moderately to densely hairy and often glandular above; leaves 3–38 mm long, 2–8 mm wide, lanceolate to lance-oblong, or less commonly oblong to linear, 1-veined, more or less pubescent on both surfaces (or rarely glabrous except along the vein), ciliate, acute to obtuse or rounded; cauline leaves lacking sterile axillary shoots or bearing them only in the lowermost axils; flowers erect, commonly 3–6 in an open, more or less dichotomous cyme; pedicels 2–25 mm long, glandular-hairy or merely spreading hairy; uppermost bracts narrowly scarious-margined, the inner mostly broadly so; petals white, 6–12 mm long, deeply bilobed; capsules 8–12 mm long, cylindrical; seeds 0.7–1.1 mm long, yellowish to brownish; $2n = 72$. Silver sagebrush, aspen, lodgepole pine, spruce-fir, and alpine tundra communities at 2680 to 3965 m in Beaver, Daggett, Duchesne, Garfield, Grand, Iron, Juab, Piute, San Juan, Sevier, Summit, Uintah, Utah, and Wayne counties; Alaska east to the Mackenzie, south to California, Arizona, and Colorado; Asia; 56 (x).

Cerastium fontanum Baumg. Mouse-ear Chickweed. [*C. vulgatum* L. 1762, non 1755; *C. caespitosum* Gilib.]. Biennial or short-lived perennial herbs, forming clumps mostly 1–4 dm broad, arising from taproots and often from horizontal, stoloniferous branches; flowering stems 5–40 cm long, decumbent to erect, moderately or sparsely villous and often glandular as well; leaves 7–25 mm long, 2–10 mm wide, lanceolate to ovate, lance-oblong, or oblanceolate, 1-veined, coarsely pubescent on both surfaces, long-ciliate, obtuse to rounded or acute; stems lacking sterile, axillary shoots, but often with axillary flowering shoots; flowers erect or spreading, inconspicuous, commonly several to many in a congested to open cyme; pedicels mostly 3–10 mm long or more, spreading hairy; uppermost bracts distinctly scarious-margined; sepals 5–7 mm long, spreading hairy, scarious-margined; petals white, 5–7 mm long, bilobed; capsules 7–10 mm long, cylindrical; seeds 0.6–0.8 mm long, reddish brown; $2n = 140, 144$. Lawns, open sites, and often in riparian habitats, at 850 to 2746 m in Beaver, Box Elder, Daggett, Duchesne, Emery, Salt Lake, Summit, Tooele, Utah, and Washington counties; widely distributed in North America; adventive from Eurasia; 15 (iii).

Cerastium nutans Raf. Nodding Chickweed. [*C. nutans* var. *brachypodum* Engelm. ex Gray; *C. brachypodum* (Engelm.) Robins. ex Britt.]. Annual herbs, often branched from the base; stems erect-ascending, 5–20 cm tall or more, arising from taproots; herbage glandular-pubescent to pilose; leaves 5–25 mm long or more, spatulate to oblanceolate or linear-lanceolate, 1-veined, ciliate, acute to obtuse apically; cauline leaves lacking sterile axillary shoots; flowers nodding, commonly several in open, leafy-bracteate cymes; pedicels 3–20 mm long; sepals 3–5 mm long; petals subequal to or somewhat exceeding the sepals, or sometimes lacking; capsules cylindrical, much longer than the sepals; seeds brownish, 0.5–0.7 mm long; $2n = 35, 36$. Silver sagebrush and meadow communities at ca 2350 m in Box Elder, Daggett, Duchesne, Summit, and Uintah counties; widely distributed in North America; 5 (0).

Cerastium tomentosum L. Snow-in-summer; Dusty Miller. Perennial herbs, forming mats or clumps mainly 10–40 cm wide or more, arising from taproots and often from horizontal, substoloniferous branches; flowering stems mainly 15–30 cm tall; herbage gray-tomentose; leaves 12–35 mm long, 2–5 mm wide, lance-linear, rounded to obtuse or less commonly acute; stems not bearing sterile axillary shoots; flowers erect, showy, commonly several to many in a congested to open cyme; pedicels mainly 5–15 mm long, tomentose; bracts with distinctive, shining, scarious margins; sepals 6–7.5 mm long, tomentose, with broad, scarious margins; petals white, 12–15 mm long, bilobed; capsules longer than the calyx. Cultivated ornamental, persisting and rarely escaping in lower elevation cities and towns; introduced from Europe; 1 (0).

Dianthus L.

Annual, biennial, or perennial herbs with erect or ascending stems; leaves opposite, lance-linear to oblong or linear, lacking stipules; flowers solitary or 2–4, borne in branching cymes; sepals 5, united to near the apex, each several-veined; petals 5, pink to pink purple or white, showy, the blade spreading at right-angles to the claw, toothed; stamens 10, connate, adnate to the base of the petals; styles usually 2; ovary 1-loculed, stipitate; capsule opening by 4 or 5 valves.

1. Flowers solitary, or 2 to few in open clusters, pedicellate; episepalous bracts much shorter than the calyx .. .. *D. caryophyllus*

— Flowers clustered, few to many, sessile to short-pedicellate; episepalous bracts almost as long as the calyx or longer 2

2(1). Calyx and bracts villous-pilose; leaves linear . . *D. armeria*
— Calyx and bracts glabrous; leaves lanceolate to elliptic
.... *D. barbatus*

Dianthus armeria L. Grass Pink. Annual or biennial herbs; stems 1 to several, erect, mainly 2–6 dm tall, glabrous throughout, or puberulent below the nodes and in the inflorescence; leaves linear, erect, mainly 3–10 cm long and 2–5 mm wide, reduced upward; flowers several to many in a compact cyme; calyx tubular, 14–17 mm long, 20- to 25-veined, villous-pilose; episepalous bracts clasping, the base oval-oblong, the apex abruptly long-attenuate; petals 20–25 mm long, pink or red, the claws narrow, the blades obovate, dentate; stipe ca 1 mm long; capsule subequal to the calyx; 2n = 30. Cultivated ornamentals, escaping, and established in Rich and Summit counties; widely established in North America; introduced from Europe; 2 (0).

Dianthus barbatus L. Sweet William. Perennial herbs; stems erect (decumbent at the base), glabrous, 2–6 dm tall; leaves basal and cauline, mainly 2–9 cm long and 0.5–2 cm wide, lanceolate to oblanceolate or elliptic, the cauline ones in 4–10 pairs; flowers numerous, borne in compact cymes; calyx cylindric, 14–17 mm long, 35- to 40-veined, the teeth attenuate, glabrous; episepalous bracts broadly clasping basally, obovate, the apex abruptly long-attenuate, glabrous; petals narrowly clawed, the claws subequal to the calyx, the blade 6–12 mm long, rhombic-obovate, dentate, white to red or pink, or bicolored; carpophore 3–4 mm long; capsule subequal to the calyx; 2n = 30. Cultivated ornamental, escaping and persisting in Utah County; widely grown in North America; introduced from Eurasia; 4 (0).

Dianthus carophyllus L. Carnation. Perennial herbs (grown as annuals); stems 1 to several, erect, mainly 2–8 dm tall, glabrous and glaucous throughout or more or less puberulent; leaves linear to narrowly elliptic, 1.5–8 cm long, mainly 2–4 mm wide, reduced upward; flowers solitary or few in open cymes; calyx tubular 20–30 mm long, 30- to 50-veined, glabrous; episepalous bracts clasping, less than 1/4 as long as the tube, the apex abruptly short-attenuate; petals 30–50 mm long, the limb 10–15 mm wide or more, usually glabrous, dentate, pink, white, purple, or red; 2n = 30. Widely cultivated ornamental both in gardens and under glass in Utah; introduced from Eurasia; 3 (0). The closely allied maiden pink (**D. deltoides** L.) and rainbow pink (**D. chinensis** L.) are grown in Utah also. They differ from the carnation in the former being puberulent and in the latter having bearded petals.

Gypsophila L.

Perennial (or annual) herbs; leaves opposite, lacking stipules; flowers numerous, borne in open cymes, perfect or imperfect; bracts scarious; sepals 5, connate, with membranous sinuses; petals 5, white or pink to red, lacking appendages; stamens 10; styles 2; ovary 1-locuied; capsule ovoid, 4- to 6-valved; seeds reniform.

1. Inflorescence glabrous; petals white ...... *G. paniculata*
— Inflorescence glandular; petals pink to red ....
.... *G. scorzonerifolia*

Gypsophila paniculata L. Babysbreath. Perennial, glabrous herbs from a taproot; stems mainly 4–10 dm tall, usually freely branched and umbrelliform; leaves 1–5 cm long or more, lanceolate, 1-veined; flowers numerous on pedicels 2–8 mm long; calyx broadly campanulate, 1.5–2 mm long, with broad, white, membranous intervein areas below the sinuses, the veinal areas blackish; petals white, 2–3.5 mm long; capsules ovoid, 4-valved; seeds 1.5–2 mm long, blackish, cross-wrinkled and papillate. Established in ponderosa pine-mountain brush communities at ca 2090 m in Washington County; widely established in western U. S.; introduced from Eurasia; 2 (0).

Gypsophila scorzonerifolia Ser. in DC. Perennial herbs, glabrous below, glandular in inflorescence, from a caudex and taproot; stems mainly 3–10 dm tall, usually freely branched upward; leaves 1.5–8 cm long, 3–10 mm wide or more, oblong to lance-oblong or elliptic, 3- to 7-veined; flowers numerous, on pedicels 3–8 mm long; calyx obconic to subcylindric, 2.5–3.5 mm long, with moderately broad white membranous intervein areas below the sinuses, the veinal areas broad and green; petals pink to reddish, 4–5 mm long; capsules ovoid, 4-valved; seeds 1–1.5 mm long, blackish, ornamented; 2n = 34, 56, 68. Moist roadsides and pastures at 1300 to 1375 m in Salt Lake and Utah counties; introduced from Europe; 2 (0).

Herniaria L.

Annual or perennial herbs; stems prostrate, spreading; leaves opposite, entire; stipules scarious, minute; flowers very small, sessile in compact axillary clusters; sepals 4 or 5; petals lacking; stamens 2–5; style 2-cleft; fruit a utricle, 1-seeded.

Herniaria glabra L. Annuals or perennials with prostrate stems, these more or less retrorsely puberulent, mainly 3–15 cm long; leaves 3–7 mm long, 1–2 mm wide, narrowly oblanceolate to elliptic, glabrous; stipules 0.5–1 mm long; flowers green, ca 1 mm wide; sepals ca 1 mm long; utricle cylindroid, membranous, surpassing the sepals; seeds ca 0.5 mm long, black, shining. Cultivated ground cover, escaping in Salt Lake County (Arnow 5966 BRY); introduced from Europe; 1 (0).

Holosteum L.

Annual herbs; leaves basal and cauline, opposite, more or less connate-clasping; inflorescence umbellate, long-peduncled; flowers several to many, borne on long, reflexed pedicels; sepals 5, distinct; petals 5, white, irregularly toothed; styles 3; fruit a capsule; seeds reniform.

Holosteum umbellatum L. Plants erect or erect-ascending, simple or branched from the base, mainly 3–22 cm tall or more; basal leaves petiolate, oblanceolate to spatulate; cauline leaves 1–4 pairs, lanceolate to elliptic, 6–20 mm long or more, glandular-ciliate, obtuse to acute, entire; umbels 3– to 10-flowered, the pedicels of various lengths, erect in anthesis, soon deflexed; sepals 3.8–5.3 mm long, lanceolate, green or suffused with red, the margins hyaline; petals white or pinkish, subequal to or longer than the sepals; capsules cylindric, surpassing the sepals; seeds 0.7–1 mm long, brown, ornamented; 2n = 20. Open sites in mixed desert shrub, sagebrush, and mountain brush communities at 1340 to 1985 m in Box Elder, Cache, Davis, Millard, Salt Lake, Tooele, and Utah counties; widely established in North America; adventive from Europe; 14 (iii). Our material has glands mainly near the stem mid-length and is assigned to **var. umbellatum**.

Loeflingia L.

Annual, glandular-hairy, rigid herbs; leaves opposite, connate-clasping, subulate; flowers axillary, sessile or short-pedicellate, solitary or few; sepals 5, keeled, with rigid, long-attenuate, straight or recurved apices; petals 3–5 or none; stamens 3–5; style 1 or lacking; stigmas 3; capsule 3–valved; seeds numerous.

Loeflingia squarrosa Nutt. Spreading to erect, glandular-pubescent annuals, mainly 3–10 cm tall; leaves 4–6 mm long, subulate; sepals 3.5–4.5 mm long, rigid, toothed on each side; petals minute; capsules shorter than the calyx; seeds ca 0.4 mm long. Mountain brush community in Washington County (Atwood 5025 BRY); California, Arizona, and Mexico; 1 (0).

Lychnis L.

Perennial herbs with erect or ascending stems; leaves opposite or chiefly basal, lance-linear to oblong, lacking stipules; flowers solitary or 2–5 in congested or more or less open cymes; sepals 5, united to near the apex, each 2- to 4-veined; petals 5, white to pink or reddish, the blade spreading at right angles to the claw (when exserted from the calyx), retuse or 2- to 4-lobed; stamens 10, connate, adnate to the petal bases; styles 5 (4); ovary 1-loculed (or incompletely 5-loculed), stipitate; capsules opening by 5 (4) or 10 (8) teeth. Our species have been treated variously within *Melandrium* and *Gastrolychnis*, or as part of an expanded *Silene*, a seemingly practical option. I follow tradition in western American taxonomic works to avoid replacing well known names.

Baker, H. G. 1950. Dioecious *Melandrium* in western North America. Madrono 10: 218–221. 1950.
Maguire, B. 1950. A synopsis of the North American species of the subfamily Silenoidea. Rhodora 52: 233–245.

1. Petal claws exserted, the blades mostly 5–10 mm long, white; flowers imperfect, dimorphic, those of the pistillate ones much larger than those of the staminate . *L. alba*
— Petal claws mostly included, the blades less than 5 mm long, pinkish, purplish, or white; flowers perfect, all about the same size 2
2(1). Plants 3–15 (20) cm tall; flowers 1–3, nodding in bud; calyces inflated and turbinate in fruit .... *L. apetala*
— Plants mainly 20–60 cm tall; flowers 1–7, erect in bud; calyces inflated or not, cylindroid to turbinate in fruit .... *L. drummondii*

Lychnis alba Miller White Cockle. [*Melandrium album* (Miller) Garcke; *Silene alba* (Miller) Krause in Sturm]. Commonly dioecious perennial herbs; stems erect or ascending, mainly 4–10 dm tall, pubescent with multicellular hairs, these often glandular upwards; basal leaves petiolate, mainly 4–10 cm long, 0.5–2 cm wide; cauline leaves 4–10 pairs, often larger than the basal ones, becoming sessile and smaller upward; flowers more or less clustered, the clusters borne in open leafy-bracteate, paniculate cymes, dimorphic, with the pistillate ones much larger than the staminate; calyx subcylindric, the teeth triangular to lanceolate, the tube mainly 15–30 mm long, 10- to 20-veined, the staminate smaller and more obscurely veined than the pistillate, inflated in fruit; petals 5, white (or pink), 25–35 mm long, clawed, the blade 8–12 mm long, bilobed, appendaged at juncture of blade and claw; capsule ovoid-cylindric, 10–15 mm long, opening by 10 teeth; seeds reniform, 1.3–1.7 mm long, papillose; $2n = 22+xx$ or xy. Open sites, often in indigenous vegetation, at 1525 to 2930 m in Cache, Salt Lake, Utah, Wasatch, Wayne, and Weber counties; widespread in North America; adventive from Europe; 6 (0).

Lychnis apetala L. Nodding Lychnis. [*Melandrium apetalum* (L.) Fenzl; *Wahlbergella apetala* (L.) Fries; *Silene wahlbergella* Chaudhri, not *Silene apetala* Willd.]. Plants 2.3–18 cm tall, from a taproot and branching caudex, the stems erect or ascending, white-hairy below, becoming glandular-villous above with multicellular hairs, the cross-walls purplish; basal leaves 1–6 cm long, 2–5 mm wide, narrowly oblanceolate to elliptic, 1-veined, acute to obtuse apically, glabrous to puberulent on one or both sides, ciliate; cauline leaves 2 or 3 pairs, reduced upward, the upper ones commonly pubescent with multicellular hairs; flowers solitary or 2 or 3, nodding (or erect?) in bud, spreading to erect in fruit; pedicels villous with multicellular, glandular hairs, the cross walls purplish; calyx urn-shaped to campanulate, purplish veined and commonly more or less hairy like the pedicels, 10–19 mm long, 5–10 mm wide, the teeth 1.5–3 mm long, ciliate; petals pinkish to purplish or white, included or exserted 1–4 mm, bilobed; capsule erect, 10–18 mm long, opening by 5 or 6 (10 or 12) teeth; seeds 1–2 mm long, wing-margined or not; $2n = 24$. This species is circumpolar, with two more or less distinctive varieties present in Utah.

1. Basal leaves glabrous on both sides, ciliate; seeds evidently winged .... *L. apetala* var. *montana*
— Basal leaves retrosely puberulent and ciliate; seeds wingless or nearly so .... *L. apetala* var. *kingii*

Var. **kingii** (Wats.) Welsh [*L. kingii* Wats., type from Summit County; *Wahlbergella kingii* (Wats.) Rydb.]. Spruce-fir, sedge, and alpine tundra communities at 3290 to 3450 m in Daggett, Duchesne, Piute, San Juan, Summit, and Uintah counties; Colorado; 21 (iii). Pubescence and seed differences seem to be the main diagnostic features of this taxon. Attitude of flower buds and petal exsertion are difficult to apply as taxonomic tools. Even seed differences are only useable when mature fruit is collected, as rarely.

Var. **montana** (Wats.) C. L. Hitchc. [*L. montana* Wats; *L. apetala* ssp. *montana* (Wats.) Maguire]. Spruce-fir, meadow, and alpine tundra communities at 3355 to 3905 m in Duchesne, Grand, and San Juan counties; Montana, Wyoming, and Colorado; 3 (1).

Lychnis drummondii (Hook.) Wats. Drummond Catchfly. [*Silene drummondii* Wats; *Melandrium drummondii* (Hook.) Porsild; *Gastrolychnis drummondii* (Hook.) Love & Love; *L. striata* Rydb.; *L. drummondii* var. *striata* (Rydb.) Maguire]. Perennial herbs from a taproot and more or less branching caudex; stems mainly 15–50 dm tall, retrosely hairy below, usually becoming glandular-villous with multicellular hairs upward, the cross-walls purplish; basal leaves with blades 1–10 cm long, 3–15 mm wide, oblanceolate to elliptic or lanceolate, petiolate, glabrous or puberulent on both sides, ciliate; cauline leaves 2–5 pairs, becoming sessile and reduced upward; flowers 1 to several, in bracteate, racemose or paniculate cymes; calyx campanulate to turbinate, 10–15 mm long, expanding and investing the fruit, 10-veined; corolla white or tinged pink, included or the blade exserted; petals clawed, the blades 1–3 mm long, retuse or 2– to 4–lobed, appendaged; stipe short; capsule dehiscent by 5 (or 4) valves; seeds ca 0.7 mm long, brown, papillate. Sagebrush, pinyon-juniper, ponderosa pine, aspen, lodgepole pine, spruce-fir, and alpine tundra in Beaver, Box Elder, Cache, Carbon, Daggett, Duchesne, Emery, Garfield, Grand, Juab, Iron, Kane, Millard, Morgan, Piute, Rich, San Juan, Sanpete, Sevier, Summit, Tooele, and Uintah counties; Northwest Territories, south to Oregon, Nevada, Arizona, and Colorado; 49 (xi). This is a highly variable taxon, varying in stature, flower number, and glandularity. The variation seems to lack geological or morphological correlation.

Paronychia Miller

Perennial, pulvinate-caespitose herbs; leaves opposite, subulate or elliptic; stipules scarious; flowers sessile, solitary or few in cymes; calyx 5–lobed, the lobes awn-tipped; petals lacking; stamens 5, opposite the sepals; style 1, bifid; fruit a utricle, 1–seeded, enclosed in the calyx.

1. Leaves 0.8–1.2 mm wide, elliptic; plants mainly of high elevations in the Uinta Mts. .... *P. pulvinata*
— Leaves 0.4–0.6 mm wide, linear to subulate; plants of middle and lower elevations, variously distributed .... *P. sessiliflora*

Paronychia pulvinata Gray Plants cushionlike, mainly 6–15 cm wide; stems leafy throughout, usually lacking elongate internodes; leaves 1–6 mm long, elliptic, thickened, glabrous or scabrous-ciliate, yellow green, appearing veinless; stipules ovate, attenuate, surpassing the leaves in youth, scarious; flowers mostly solitary, sessile, borne among the leaves; bracts surpassing the flowers; sepals 2–2.5 mm long, ovate, cuspidate from below the tip; fruit ovoid, ca 1 mm long, excluding the style ca 0.5 mm long; seeds brown, ca 1 mm long; $2n = 32$. Spruce-fir and alpine tundra communities at 3295 to 3815 m in Daggett, Duchesne, Summit, and Uintah counties; Wyoming and Colorado; 13 (i). Our material has been assigned to var. *longiaristata* Chaudhri (type from Mt. Emmons).

Paronychia sessiliflora Nutt. Plants cushionlike, mainly 6–20 cm wide; stems lacking or with 1–3 elongate internodes; leaves 1–8.5 mm long, linear to subulate, scabrous, thickened, gray green, 1–veined; stipules narrowly lance-attenuate, suprassing or shorter than the leaves; flowers usually solitary; bracts subequal to the sepals; sepals 2.5–4 mm long, lanceolate, cuspidate-aristate from below the tip; fruit ovoid, ca 1 mm long, excluding the style ca 0.5 mm long; seed brown, ca 1 mm long; $2n = 64$. Shadscale, pinyon-juniper, sagebrush, mountain brush, and fringed sagebrush communities at 1675 to 2930 m in Beaver, Carbon, Daggett, Duchesne, Emery, Garfield, Grand, Millard, Piute, Rich, Sevier, Summit, Uintah, Wasatch, and Wayne counties; Alberta and Saskatchewan, south to New Mexico and Texas; 36 (vi).

Sagina L.

Annual, biennial, or perennial herbs with prostrate, ascending, or erect stems; leaves opposite, linear, basally connate, apiculate, lacking stipules; flowers solitary or 2 to several in cymes; sepals 4 or 5, distinct, obscurely 1– to 3–veined; petals 4 or 5, usually shorter than the sepals, white, the blades entire or slightly emarginate; stamens 4 or 5, alternate with the sepals; ovary 1–loculed; capsule opening by 4 or 5 valves, these opposite the sepals.

1. Sepals 5, usually erect in fruit; capsules sometimes reflexed; plants common in montane sites .... *S. saginoides*
— Sepals 4 (or 5) often spreading in fruit; capsules erect; plants uncommon in ruderal sites .... *S. procumbens*

Sagina procumbens L. Biennial or perennial (annual) herbs, usually glabrous; stems prostrate to ascending and often rooting at the nodes, mainly 2–15 cm long; rosette leaves mainly 8–25 mm long, the cauline 2–6 mm long, conspicuously sheathing, often with sterile axillary shoots; flowers solitary or 2 to several per stem; pedicels erect in fruit; sepals 4 (or 5), oval, rounded, 1.8–2.5 mm long, spreading in fruit; petals 0.5–1 mm long or lacking; capsules subequal to or surpassing the sepals, 4– or 5–valved; seeds ca 0.3 mm long; $2n = 22$. Weedy species of open disturbed sites at ca 1350 m in Salt Lake County (Arnow 5593 UT; BRY); widely scattered in North America; adventive from Eurasia; 1 (0)

Sagina saginoides (L.) Britt. [*Spergula saginoides* L.; *Sagina linnaei* Presl; *Alsinella saginoides* (L.) Greene]. Biennial to perennial (or annual?) herbs, forming compact to more or less diffuse clumps 3–15 (20) cm wide, arising from a taproot and branching caudex, the caudex branches prostrate to ascending; flowering stems 2–7 (14) cm long, decumbent to erect; basal leaves 4–25 mm long, usually less than 1 mm wide, linear, 1–veined, apiculate, glabrous; cauline leaves 2–5 pairs, often with clustered, secondary, axillary leaves; flowers erect, solitary and terminal or lateral, or 2 or 3 and the lower ones lateral; sepals 1.5–2 mm long, erect, greenish, the margins scarious, glabrous, obtuse to rounded; petals white, 1–1.5 mm long; capsules 2.5–3 mm long, 4– or 5–valved, sometimes recurved at maturity; seeds 0.2–0.4 mm long, reddish brown; $2n = 22$. Aspen and spruce-fir communities, often in sandy gravel along lake margins or stream banks, at 2560 to 3965 m in Box Elder, Cache, Daggett, Duchesne, Emery, Garfield, Grand, Juab, Piute, Rich, Salt Lake, Sanpete, Sevier, Summit, Tooele, Uintah, and Washington counties; Alaska to the Atlantic, south to California, Mexico, Michigan, and Quebec; circumboreal; 35 (ix).

Saponaria L.

Perennial herbs; stems erect, leafy; leaves opposite, palmately 3–veined; flowers numerous, in usually contracted cymes, perfect; calyx tubular, obscurely 15– to 25–veined; petals clawed, appendaged at juncture of blade and claw; stamens 10; styles 2; capsule opening by 4 teeth; seeds reniform.

Saponaria officinalis L. Bounding Bet; Soapwort. Robust perennial herbs, mainly 30–80 cm tall, forming clumps, glabrous or nearly so; leaves 3–12 cm long, 0.8–4 cm wide, lanceolate to elliptic or ovate-lanceolate, reduced and subsessile upward; flowers numerous, usually in compact cymes; calyx 15–25 mm long, the tube subcylindric or tapering toward the apex, the lobes ovate or triangular, acuminate; petals 3–4 cm long, white or pink, showy, the claw usually surpassing the calyx, the lobes spreading, obovate-cuneate, retuse; appendages 2, 1–2 mm long; seeds black, 1.6–1.8 mm long, pitted; $2n = 28$. Cultivated ornamental, escaping and established in Cache, Duchesne, Juab, Salt Lake, Uintah, Utah, and Washington counties; widely established in North America; introduced from Europe; 12 (0). Double flowered forms occur in cultivation. This plant is considered to be poisonous to livestock.

Silene L.

Annual or perennial herbs with prostrate, decumbent, ascending, or erect stems; leaves opposite, lanceolate to lance-oblong or linear, lacking stipules; flowers solitary or few to many in open to contracted cymes; sepals 5, united to near the apex, each 2 (or more) -veined; petals 5, white to pink, red, or purplish, the blade ascending or spreading, entire or several lobed apically, often appendaged at juncture of blade and claw; stamens 10, adnate to base of petals; styles commonly 3 (or less commonly 4 or 5); ovary 1 (or incompletely 3–5) -loculed, stipitate; capsule opening by 6 (rarely 8 or 10) teeth.

1. Plants pulvinate-caespitose, forming dense mats, acaulescent, of alpine sites; leaves linear to oblong, mainly 3–20 mm long, 1–2 mm wide .... *S. acaulis*
— Plants neither pulvinate-caespitose nor acaulescent, of various distribution; leaves various, but seldom if ever as above 2
2(1). Plants annual, from slender taproots 3
— Plants perennial, from a caudex and taproot 5
3(2). Plants densely pubescent throughout; reported from Cache and Weber counties .... *S. noctiflora*
— Plants glabrous or essentially so 4
4(3). Stems lacking glutinous bands; calyces cylindric-clavate, often pinkish; appendages 1–2.5 mm long; plants cultivated, escaping, and established ... *S. armeria*
— Stems with glutinous bands; calyx turbinate, green, with broad, scarious intervein areas; appendages less than 1 mm long; plants indigenous .... *S. antirrhina*
5(2). Plants soboliferous, growing in limestone on the Wasatch and Sevier-Paunsaugunt plateaus; petals bright pink the lobes 8–15 mm long .... *S. petersonii*
— Plants from a caudex, this not soboliferous, seldom with the above distribution; petals pink or white, the lobes mostly less than 7 mm long 6
6(5). Corollas white, usually less than 12 mm long; flowers solitary or in open leafy cymes .... *S. menziesii*
— Corollas white or pink, usually more than 12 mm long; flowers solitary or few to numerous, borne in open to contracted cymes, with much reduced bracts 7
7(6). Petal appendages commonly 4; petals equally 4–lobed; plants of northern Utah .... *S. oregona*
— Petal appendages commonly 2, or if 4, the blades bilobed or unequally 4–lobed; plants of various distribution 8
8(7). Calyx puberulent, lacking glands, papery, inflated in fruit; petal blades 4–7 mm long, bilobed or rarely with minute lateral teeth .... *S. douglasii*
— Calyx glandular, neither papery nor much inflated in fruit; petal blades various 9
9(8). Caudex branches with apparent, whitish internodes, usually several per plant; plants of Washington, Kane, and Beaver counties .... *S. verecunda*
— Caudex branches without apparent internodes, the stem bases often purplish, 1 or few per plant; plants of Piute, Garfield, San Juan, and Washington counties .... *S. scouleri*

Silene acaulis L. Moss Campion. Perennial, pulvinate-caespitose, dioecious or less commonly perfect plants forming dense mats or cushions 3–40 cm wide or more, arising from a taproot and branching caudex, the caudex branches prostrate to ascending or erect, clothed with persistent leaves; flowering stems lacking or to 5 cm tall, erect; basal leaves 5–35 mm long, 0.5–2 mm wide, linear to lance-linear, 1 (3) -veined, connate-sheathing basally, acute apically, ciliate with descending hairs, glabrous; cauline leaves lacking or sometimes with 1 (or 2) reduced pairs near the stem base; flowers erect, solitary, sessile or on pedicels 0.5–5 cm long; calyx cylindric to campanulate, 4.5–8.5 mm long (or longer), often purplish, glabrous, the teeth 1–2 mm long, ciliate; petals pink to pink purple or rarely white, 8–12 mm long, entire to emarginate, the appendages 0.5–1 mm long or lacking; capsules 3–locular, 4–10 mm long; seeds 0.8–1.2 mm long, light brown; 2n = 24. Spruce-fir and alpine tundra communities at 2680 to 3965 m in Beaver, Daggett, Duchesne, Grand, Juab, Piute, Salt Lake, San Juan, Summit, Uintah, and Utah counties; Alaska east to the Atlantic, south to Arizona, New Mexico, and New Hampshire; Eurasia; 40 (iii). Our plants are assignable to the weakly separable **var. subacaulescens (F. Williams) Fern. & St. John** [*S. acaulis* f. *subacaulescens* F. Williams; *S. acaulis* ssp. *subacaulescens* (F. Williams) Hulten].

Silene antirrhina L. [*S. antirrhina* var. *vacarifolia* Rydb.]. Annual herbs from yellowish taproots; stems erect or ascending, 10–80 cm tall, usually glandular in bands below the nodes, glabrous above, sometimes puberulent near the base; basal leaves spatulate to oblanceolate, 1–6 cm long, 2–15 mm wide or more, puberulent; cauline leaves 3–7 or more pairs, reduced, sessile, and glabrous upward; flowers solitary or more commonly few to many in a contracted to open cyme; calyx turbinate-campanulate, 4–10 mm long, 10–veined, glabrous; petals white or pink, the claw narrow, much longer than the obcordate blade; appendages 2, very short; capsules 3–locular; seeds brown, 0.5–0.7 mm wide, ornamented; $2n = 24$. Creosote bush, blackbrush, other warm desert shrub, pinyon-juniper, and mountain brush communities at 975 to 1740 m in Cache, Daggett, Garfield, Grand, Juab, Kane, Salt Lake, San Juan, Washington, and Weber counties; widely distributed in North America; 14 (vi).

Silene armeria L. Annual, glabrous or sparingly puberulent herbs; stems erect, 10–70 cm tall; leaves sessile and more or less clasping, the basal ones 1–5 cm long, 3–15 mm wide; cauline leaves in several to many pairs, oblong to lanceolate; flowers several to many in open to compact corymbose cymes; calyx narrowly cylindric to clavate, 13–17 mm long, 10–veined, glabrous, often suffused with pink; petals pink to lavender, the claw 6–8 mm long, the blade cuneate-obovate, 4–7 mm wide, rounded to emarginate, the appendages 2, 2–3 mm long; stipe 7–8 mm long, glabrous; capsule 3 (or 4) -loculed; seeds 0.5–0.7 mm long, rugose, brown; $2n = 24$. Cultivated ornamental, escaping and persisting in Salt Lake County (Neese 11134 BRY); widely established in the U. S.; introduced from Europe; 1 (0).

Silene douglasii Hook. Perennial plants, forming loose clumps, arising from a caudex and taproot, the caudex usually with apparent, often whitish, internodes; flowering stems 10–48 cm long or more, decumbent to ascending or erect; basal leaves smaller than the main cauline ones, often withered by flowering time; cauline leaves 2–8 cm long, 2–7 (12) mm broad, oblanceolate to narrowly lanceolate, 1 (several) -veined, slightly connate basally, acute apically, hairy above and beneath and rarely somewhat glandular; flowers erect or ascending, solitary or more commonly 2 to several, in open or contracted cymes; pedicels 3–35 mm long, retrorsely hairy;

calyx campanulate, (10) 12–15 mm long, greenish or purplish, with broad white interveins, becoming papery, pubescent to merely glabrous, the teeth 1.5–3 mm long, ciliate; petals white to greenish, pinkish, or purplish, 12–16 mm long, the blades bilobed, the appendages 1–3 mm long; capsules 1–loculed, 8–12 mm long (including stipe); seeds 1.1–1.4 mm long, reddish brown, roughened; 2n = 48. Sagebrush, mountain brush, white fir, mixed conifer, and grass-forb communities at 2040 to 3420 m in Beaver, Box Elder, Cache, Davis, Juab, Millard, Rich, Salt Lake, Utah, and Washington counties; British Columbia, south to California and Nevada; 44 (ii). A specimen from Box Elder County (Atwood & Goodrich 8333 BRY) has 4–lobed petal blades; it approaches *S. parryi* (Wats.) Hitchc. & Maguire in that feature.

Silene menziesii Hook. Menzies Campion. [*Anotites jonesii* Greene, type from American Fork Canyon]. Perennial, dioecious (or perfect?) plants forming loose clumps, arising from elongate rhizomes; flowering stems 5–30 cm long or more, sprawling to decumbent, ascending, or less commonly erect; basal leaves smaller than the main cauline ones and often withered by flowering time; cauline leaves 1–7 cm long, 3–22 mm wide, lanceolate to elliptic or oblong, 1 (several) -veined, not or only slightly connate basally, acuminate to attenuate or acute apically, glandular-pubescent above and beneath, especially along the veins, glandular-ciliate; flowers erect, solitary or more commonly 2 to several in open cymes; pedicels 4–30 mm long, glandular-hairy; calyx campanulate, 5–8 mm long, purplish or greenish, pubescent and glandular, the teeth 1.5–2.5 mm long, ciliate; petals white, 6–10 (12) mm long, the blades bilobed, the appendages 0.4 mm long or less; capsules 1–loculed, 5–8 mm long (including the stipe); seeds 0.7–0.9 mm long, black, shining; 2n = 24, 48. Riparian, meadow, mountain brush, mixed conifer, aspen, and spruce-fir communities, often in shade, at 1675 to 3175 m in Cache, Carbon, Duchesne, Emery, Garfield, Grand, Millard, Morgan, Piute, Rich, Salt Lake, Sanpete, Summit, Tooele, Uintah, Utah, Wasatch, and Weber counties; Alaska and Yukon, south to California and New Mexico; 49 (vii).

Silene noctiflora L. Night-flowering Catchfly. Annual, perfect or occasionally dioecious plants arising from taproots, the stems 2–8 dm tall, erect; leaves 2–10 cm long, 0.6–2.5 (4) cm broad, lanceolate to elliptic or the lower ones oblanceolate, several-veined, slightly connate basally, acute to attenuate or obtuse apically, hairy above and beneath, ciliate; flowers erect, solitary or few to several in open or contracted cymes; pedicels 0.3–1 cm long, glandular-hairy; calyx tubular, 12–25 mm long, greenish or pinkish, or scarious between the veins, glandular-hairy, the teeth 4–9 mm long, ciliate; petals white to pinkish, 20–35 mm long, the blades bilobed, the appendages 0.5–1.5 mm long; capsule 3–locular, 1.5–2.5 cm long (including the stipe); seeds 0.8–1.2 mm long, brown, roughened; 2n = 24. Reported from Cache and Weber counties; widely distributed in North America; introduced from Asia; 0 (0).

Silene oregana Wats. Oregon Campion. Perennial herbs from a taproot and branching, short-crowned caudex; stems 20–70 cm tall, ascending to erect, puberulent; cauline leaves 3–6 pairs, 2–8 cm long (or more), 3–12 mm wide, the main ones oblanceolate, acute to obtuse, petiolate, reduced and sessile or subsessile upward; inflorescence few-flowered, open or compact, stipitate-glandular; calyx narrowly campanulate, 10–16 mm long, glandular, the lobes 2–4 mm long; petals pink or white, the claws exserted, the 3 veins branched below junction with blade, the appendages 4 or 6, linear, 1–1.5 mm long; blades 3–5 mm long, deeply 4–lobed (the middle lobe sometimes again lobed); capsule 9–13 mm long (including the stipe); styles 3 or 4; seeds 1.2–1.8 mm wide, tuberculate; 2n = 24. Reported for Box Elder County (Hitchcock and Maguire 1947), but no specimens have been seen from Utah by me; Washington to Montana, south to California, Nevada, and Wyoming; 0 (0).

Silene petersonii Maguire Maguire Campion. Perennial, soboliferous herbs, from creeping, subrhizomatous caudex branches and taproot; stems 3–15 cm tall, erect or ascending, retrorsely puberulent and more or less glandular; leaves mainly cauline, 2–6 pairs, lanceolate to elliptic or oblanceolate, obtuse to acutish, hairy like the stems or glabrous above, 1–5 cm long, 2–10 mm wide, subsessile to sessile, 1–veined, reduced upward; flowers irregular, the upper petal reflexed, 1–5 in open cymes, nodding in bud and at anthesis; pedicels 5–25 mm long; calyx turbinate-campanulate, 13–19 mm long, glandular-villous with multicellular hairs, 10–veined, the veins broad, green or purplish, the lobes 3–6 mm long; petals 15–35 mm long, pink to purplish, the claws subequal to the calyx or shorter, the appendages very small or lacking, the blades 5–15 mm long and as broad, irregularly toothed, lobed, or 4–cleft; capsules included in the accrescent calyx; styles usually 3; seeds 2–2.5 mm wide, brown, tuberculate marginally. Ponderosa pine, aspen, and spruce-fir communities at 2135 to 3450 m in Garfield, Iron, Sanpete (type from Wasatch Plateau) and Sevier counties; endemic; 24 (iii). Specimens from Red Canyon, Garfield County (type locality) have been segregated as var. *minor* Hitchcock & Maguire. They were said to differ in lower stature [5–10 (12) cm vs 10–15 cm], shorter petals (1.5–2 cm vs 2–3 cm), and degree of lobing of the petal blade. Increased harshness of pubescence rounded out the list of supposed diagnostic features of the variety. The characteristics fail singly and in combination. There is a complete transition with the typical material. Short petals are found on short plants within the range of typical specimens and the reverse is true, also. Degree of lobing of petals varies on a single sheet of specimens collected at the same locality. Harshness of pubescence, whose interpretation is tenuous at best, fails as a diagnostic tool. Recognition of infraspecific taxa seems moot.

Silene scouleri Hook. Perennial herbs from a caudex and taproot; stems 20–80 cm tall, erect, puberulent, becoming glandular upward; leaves mainly cauline, 2–5 pair, 3–15 cm long, 4–15 mm wide, elliptic to oblanceolate or lance-linear, acute, puberulent, ciliate, reduced and becoming sessile upward, 1–veined; flowers few to numerous in an elongate, subverticillate (sometimes branched) cyme, often 2–8 per cluster; pedicels 0.5–3 cm long; calyx cylindric-campanulate, 10–18 mm long, glandular-puberulent, 10–veined, green with intervein areas scarious, the lobes 2–4 mm long; petals 10–20 mm long, purplish or whitish, the blades 3–6 mm long, usually equally 4–lobed, the appendages 2, 1–3 mm long; capsule included in the slightly accrescent calyx; styles 3; ovary 1–loculed; seeds reniform, 0.9–1.5 mm long, papillate; 2n = 48, 96. Ponderosa pine and pinyon-juniper communities at 2135 to 2440 m in Garfield, Iron, Piute, San Juan, and Washington counties; British Columbia to

Montana, south to California, Arizona, and New Mexico; 4 (i). Our material was designated provisionally as belonging to ssp. *hallii* (Wats.) Hitchc. & Maguire, but apparently does not differ in any substantial way from other southern specimens of the species regarded as ssp. *pringlei* (Wats.) Hitchc. & Maguire. Recognition of our material at infraspecific level seems moot, at least until many additional specimens are available for study.

Silene verecunda Wats. Perennial plants forming loose clumps, arising from a caudex and taproot, the caudex usually with apparent often whitish internodes; flowering stems 10–50 cm tall or more, decumbent to ascending or erect; basal leaves smaller than the main cauline ones, often withered by flowering time; cauline leaves mainly 2–10 cm long, 2–13 mm wide, narrowly oblanceolate to linear, 1–veined, slightly connate basally, acute to attenuate apically, scabrous or glabrous on one or both surfaces or more or less glandular; flowers erect or ascending, solitary or 2 or 3 in racemose array; pedicels 3–20 mm long, glandular; calyx tubular-campanulate, 10–14 mm long, the 10 veins green, the intervein areas scarious, usually glandular, the teeth 2–5 mm long, ciliate; petals white, greenish, or pink, 12–20 mm long, the blades bilobed, the appendages 0.5–1.5 mm long; capsules 1–loculed, 10–14 mm long (including stipe); seeds ca 1.5 mm long, brown, papillate. Pinyon-juniper and mountain brush community at ca 1250 m in Beaver, Kane, and Washington counties; California and Nevada; 2 (0). Our material has been assigned to **ssp. *andersonii*** (Clokey) Hitchc. & Maguire [*S. andersonii* Clokey]. This taxon simulates *S. douglasii* in habit, but has more slender flowers and differs in other technical ways.

Spergularia (Pers.) J. & C. Presl

Annual (biennial or perennial?) herbs with prostrate, decumbent, ascending, or erect stems; leaves opposite or clustered, linear, not connate basally, acute to apiculate; stipules membranous, connate-sheathing; flowers few to many, borne in open, leafy, terminal cymes; sepals 5, distinct, obscurely 1–veined; petals 5, white to pinkish, shorter than the sepals, the blades entire; stamens 2–10; styles 3; ovary 1–loculed; capsule opening by 3 valves.

1. Stamens 2–5; leaves, at least some, 10–30 mm long or more; plants of low elevations, saline, banks and shores .... *S. marina*
— Stamens 6–10; leaves mostly less than 10 mm long; plants of higher middle elevations in the Uinta Mts. .... *S. rubra*

Spergularia marina (L.) Griseb. [*Arenaria rubra* var. *marina* L.; *S. salina* J. & C. Presl; *Tissa salina* (J. & C. Presl) Britt.]. Annual, stipitate-glandular herbs; stems mostly 5–25 cm tall, ascending to erect, 0.6–2 mm thick; leaves opposite, seldom with sterile axillary ones, mostly 10–30 mm long, 0.6–1.2 mm wide, linear, shortly mucronate; stipules 2–4 mm long; flowers solitary in upper leaf axils, 2.5–5 mm long, glandular or glabrous; petals white to pink, shorter than the sepals; stamens 2–5; capsule surpassing the calyx; seeds 0.6–0.8 mm long, smooth to papillate, often winged; $2n = 18, 36$. Stream, pond, and lake margins, often on mud, at 1065 to 1375 m in Box Elder, Davis, Salt Lake, Uintah, Utah, and Washington counties; widespread in North America; adventive (?) from Europe; 9 (0).

Spergularia rubra (L.) J. & C. Presl [*Arenaria rubra* L.]. Plants annual (rarely biennial or perennial?), forming clumps or mats, arising from taproots; stems 5–30 cm long, prostrate to ascending; leaves 3–20 mm long, less than 1 mm wide, linear, mucronate, more or less glandular, usually with clustered secondary, axillary leaves; flowers few to many in leafy cymes; pedicels glandular-hairy; sepals 3–4.5 mm long, greenish, with scarious margins, obtuse apically, glandular-hairy; petals pinkish, 2–3.5 mm long; stamens 6–10; capsules 3–5 mm long, 3–valved; seeds 0.4–0.6 mm long, dark brown, not winged; $2n = 18, 36$. Mixed conifer, lodgepole pine, and meadow communities, especially on pond, stream, and lake shores and margins, but also on roadsides, at 2590 to 3050 m in Duchesne, Salt Lake, Summit, Uintah, Wasatch, and Weber counties; widespread in North America; adventive (?) from Europe; 10 (iii). This species is seldom weedy in Utah, growing instead as if it was indigenous. The introduced **S. media** (L.) Presl is reported for Davis County. It differs from *S. rubra* in having glabrous sepals and usually conspicuously winged seeds.

Stellaria L.

Annual or perennial herbs with prostrate to decumbent, ascending, or erect stems; leaves opposite (rarely whorled), linear-lanceolate to ovate, lacking stipules; flowers solitary in leaf axils, or few to many and borne in axillary or terminal cymes; sepals (4) 5, distinct, obscurely 1– to 3–veined; petals (4) 5, white, deeply to shallowly bilobed, or sometimes reduced or lacking; stamens usually 10, inserted at the base of the ovary; styles usually 3 (rarely 4 or 5); ovary 1–loculed; capsule opening by 6 (rarely by 8 or 10) teeth.

Fernald, M. L. 1940. *Stellaria calycantha*. Rhodora 42: 254–259.
Hulten, E. 1943. *Stellaria longipes* Goldie and its allies. Bot. Notiser 1943: 251–270.
Porsild, A. E. 1963. *Stellaria longipes* Goldie and its allies in North America. Nat. Mus. Canada Bull. 186: 1–35.

1. Lower leaves distinctly petiolate; pubescence of stems, pedicels, and petioles in longitudinal lines; plants weedy, occurring in open sites .... *S. media*
— Lower leaves sessile or short-petiolate; pubescence of stems, pedicels, and petioles uniform, or lacking, not in longidudinal lines; plants indigenous, broadly distributed 2
2(1). Herbage glandular-pubescent, at least in the inflorescence; petals longer than the sepals .... *S. jamesiana*
— Herbage glabrous or variously hairy, but seldom if ever glandular; petals shorter than to longer than the sepals or lacking 3
3(2). Plants subscapose, annual; stems filiform, not rooting at the lower nodes .... *S. nitens*
— Plants caulescent perennials, often rooting at lower nodes; stems various, but seldom filiform 4
4(3). Leaves about as broad as long; flowers solitary in leaf axils .... *S. obtusa*
— Leaves mainly 2– to 10–times longer than broad or more; flowers in terminal or axillary cymes, or some of them solitary in leaf axils 5
5(4). Leaves linear or narrowly oblong, minutely toothed marginally (use at least 30x); flowers few to many, borne in open, scarious-bracted cymes .... *S. longifolia*

— Leaves lance-linear to lance-attenuate, lance-ovate, or oblong, minutely toothed or entire marginally; flowers solitary or few to several in leafy or scarious-bracted cymes 6

6(5). Leaf margin minutely toothed (use a strong lens) and often ciliate with multicellular hairs; leaves commonly 3–9 mm wide; flowers axillary or in leafy bracted cymes .... *S. calycantha*

— Leaf margin entire (sometimes ciliate with multicellular hairs in *S. longipes*); leaves 1–3 mm wide, or if broader (as in *S. umbellata*), the pedicels often recurved and the petals lacking 7

7(6). Leaves lanceolate to elliptic, commonly less than 4 times longer than broad; flowers in terminal naked cymes; pedicels recurved .... *S. umbellata*

— Leaves lance-attenuate to lance-linear, commonly 5–10 times longer than broad or more; flowers solitary or 2–5 in open, scarious-bracted cymes; pedicels usually erect .... *S. longipes*

Stellaria calycantha (Ledeb.) **Bong.** [*Arenaria calycantha* Ledeb.]. Perennials from elongate rhizomes, forming moderate to large clumps; flowering stems 6–40 cm long or more, glabrous, scabrous, or sparsely short-villous; leaves 8–50 mm long, 2–9 mm wide, lanceolate to lance-attenuate, narrowly oblong or lance-linear, attenuate, acute, or acuminate, sessile or rarely short-petiolate, glabrous, minutely serrulate and usually long-ciliate marginally; flowers solitary and axillary, or often in open terminal cymes, the bracts scarious margined to green throughout; pedicels mostly 4–45 mm long, glabrous or scabrous; sepals 2–4.5 mm long, obscurely 1- to 3-veined, scarious margined, not ciliate; petals white, 1–4 mm long, or lacking; capsules straw colored to purplish, 3.5–6 mm long, opening by 6 teeth; seeds 0.6–1 mm long, reddish brown, smooth to slightly roughened; 2n = 26, 48, 52. Moist sites in Ponderosa pine, aspen, lodgepole pine, and spruce-fir communities at 2225 to 3175 m in Daggett, Duchesne, Garfield, Juab, Salt Lake, Sevier, Summit, and Wasatch counties; Alaska to the Atlantic, south to California, Wyoming, Michigan, and New York; Eurasia; 9 (i).

Stellaria jamesiana **Torr.** [*Arenaria jamesiana* (Torr.) Shinn.]. Perennial herbs, arising from rhizomes, these often with tuberous thickenings; stems mainly 1–5 dm long, usually decumbent-ascending; leaves spreading, lance-attenuate, 2–16 cm long, 2–20 mm wide; inflorescence open, glandular, the flowers solitary or more commonly 3 to several, borne on pedicels 4–45 mm long, sepals ovate-lanceolate, acute, 3.5–7 mm long, green with scarious margins, conspicuously glandular; petals mainly 6–10 mm long, white, bilobed; capsule shorter than the sepals, opening by 3 valves; seeds 2.5–3 mm long, dark brown, wrinkled; 2n = 26, 96. Mountain brush, sagebrush, ponderosa pine, Douglas fir-white fir, bristlecone pine, aspen, and spruce-fir communities at 1830 to 3235 m in probably all Utah counties; Washington to Idaho, south to California, Nevada, New Mexico, and Texas; 96 (xi). The maintenance of this taxon in *Stellaria* seems to be arbitrary. Placement within *Arenaria* by Shinners is not without merit, simulating as it does *A. lateriflora*. The bilobed petals probably have been derived secondarily at several places within the family (see *A. fendleri* var. *glabrescens*).

Stellaria longifolia **Torr.** Long-leaved Starwort. Perennials from elongate rhizomes forming loose clumps; flowering stems 10–35 cm long, glabrous or scabrous; leaves 8–40 mm long, 1–3 mm wide, linear to narrowly elliptic, attenuate to acute, sessile, glabrous or sparingly ciliate, minutely serrulate; flowers few to many in open cymes; bracts scarious throughout; pedicels commonly 3–35 mm long, glabrous or scabrous; sepals 2–3 mm long, obscurely 3–veined, scarious margined, not ciliate; petals white, 2–3.5 mm long, capsules 3–6 mm long, greenish or yellowish, opening by 6 teeth; seeds 0.7–0.8 mm long, brown, slightly roughened; 2n = 26. Meadows at ca 2135 m in Rich, Salt Lake, Sevier, Utah, and Wasatch counties; Alaska to Newfoundland, south to California, New Mexico, and South Carolina; 5 (0).

Stellaria longipes **Goldie** Long-stalked Starwort. Perennials from slender rhizomes, forming small to large clumps or mats; flowering stems 0.3–3.2 dm long, villous, glabrate, or glabrous; leaves 4–26 (40) mm long, 1–4 mm wide, lanceolate to lance-attenuate, sessile, sparingly villous, especially along the margin near the base or glabrous to glabrate, entire, smooth and shiny; flowers solitary or few to several in open cymes, the bracts scarious, scarious margined, or green throughout, ciliate or not; sepals 3–5 mm long, 3–veined, scarious margined, glabrous, usually not ciliate; petals white, 3–8 mm long; capsules 4–6 mm long, often purplish, opening by 6 teeth; seeds 0.6–0.9 mm long, brown, slightly roughened; 2n = 52, 78–104. The *longipes* complex has been treated as consisting of a series of morphologically variable entities, often recognized at specific level. Primary diagnostic criteria include pubescence or its lack, presence or absence of scarious bracts, and number of flowers. Our specimens fall into three categories; 1–flowered, glabrous, foliose-bracteate ones; several-flowered, pubescent, scarious-bracted ones; and several-flowered, glabrous, scarious-bracted ones. Most specimens that lack scarious bracts are 1–flowered, but the correlation might be moot; i.e., cymose inflorescences reduced to a single flower are subtended by foliose bracts. Only when secondary branches are developed do typical scarious or scarious-margined bracts develop. A conservative interpretation is indicated. Two names evidently apply to our material, but they might not represent taxa.

1. Flowers commonly more than 1 per stem, subtended by scarious-margined bracts .... *S. longipes* var. *longipes*

— Flowers commonly 1 per stem, not subtended by scarious bracts .... *S. longipes* var. *monantha*

Var. *longipes* [*S. stricta* Richards.; *S. longipes* var. *stricta* (Richards.) Rydb; *S. vestita* Greene; *S. longipes* var. *vestita* (Greene) Polunin]. Sagebrush, Douglas fir, ponderosa pine, aspen, lodgepole pine, forb-grass, spruce-fir, and alpine tundra communities at 2135 to 3480 m in Box Elder, Cache, Daggett, Duchesne, Emery, Garfield, Sevier, Summit, Uintah, and Wasatch counties; Alaska to Newfoundland, south to California, Arizona, New Mexico, Minnesota, and New York; Eurasia; 32 (iv). Both glabrous and long-haired phases occur together.

Var. *monantha* (**Hulten**) **Welsh** [*S. monantha* Hulten; *S. longipes* var. *altocaulis* (Hulten) C. L. Hitchc.]. Pinyon-juniper (?), lodgepole pine, spruce-fir, forb-grass, and alpine tundra at 2440 to 3660 m in Beaver, Duchesne, Garfield, Grand, Piute, Sevier, Summit, and Uintah counties; Alaska to Greenland, south to Washington, Idaho, and Colorado; 13 (ii). This phase seems to occur in more harsh sites than plants with open inflorescences. The morphology might represent ecological stress more than genetic differences.

Stellaria media (L.) Vill. Common Chickweed. [*Alsine media* L.]. Annual or biennial (overwintering) herbs from taproots, forming loose mats or clumps, the stems mostly 1–5 dm long, decumbent and rooting at the nodes, pubescent with multicellular hairs in longitudinal lines; leaves 10–50 (60) mm long, 5–25 (30) mm wide, the blades ovate to elliptic or broadly lanceolate, abruptly acuminate, glabrous above and beneath, ciliate, at least the lower ones with distinctive petioles to 2 mm long, the petioles pubescent; flowers axillary, or commonly few to several in short, leafy bracted cymes; pedicels 2–60 mm long, hairy like the stems; sepals 3–6 mm long, 3–veined, scarious-margined, ciliate basally, hairy dorsally; petals white, 2–6 mm long; capsules 4–8 mm long, straw colored or greenish, opening by 6 valves; seeds 0.8–1.2 mm long, uniformly warty; $2n = 40, 42, 44$. Weedy plants of open sites, usually associated with cultural activities, at 1280 to 2290 m in Box Elder, Cache, Millard, Salt Lake, Utah, and Weber counties; widely established in North America; adventive from Eurasia; 6 (0).

Stellaria nitens Nutt. Slender, erect annuals; stems 5–25 cm tall, subfiliform, the internodes much longer than the leaves, glabrous to sparingly villous; leaves mainly below middle of plant height; basal leaves petiolate, the petioles 2–8 mm long, the blades 1.5–6 (8) mm long, ovate to orbicular; cauline leaves becoming sessile and lance-attenuate to linear, 3–13 mm long, 0.5–2.5 mm wide; flowers few to many in open cymes; bracts scarious; sepals 3–5 mm long, lance-attenuate, acute, 3–veined; petals very short or lacking; capsules 2.2–3.5 mm long, straw colored, included in the calyx; seeds ca 0.5 mm long, minutely warty. Warm and mixed desert shrub communities at 915 to 1830 m in Cache, Salt Lake, Tooele, and Washington counties; British Columbia to Montana, south to California; Mexico; 4 (0). This is a diminutive, early flowering annual; it is easily overlooked and is often mistaken for *Holosteum umbellatum*, which differs inter alia in having reflexed pedicels.

Stellaria obtusa Engelm. Perennial, decumbent to ascending herbs forming mats; stems 2–10 cm long, glabrous or weakly and sparingly villous with moniliform hairs; leaves 2–12 mm long, 1.5–8 mm wide, ovate, lanceolate, or elliptic, long-ciliate basally, the lower ones short-petiolate, acute to obtuse; flowers solitary in leaf axils; pedicels 3–25 mm long, glabrous; sepals lance-ovate, 2–2.8 mm long, only slightly if at all scarious marginally, obtuse; petals lacking; stamens 8–10; capsule subglobose, subequal to the sepals; styles 3 or 4; seeds finely though distinctly ridged; $2n = 78$. Damp areas in sagebrush, aspen, spruce-fir, and grass-sedge communities at 1920 to 3390 m in Cache, Salt Lake, Sanpete, Summit, and Utah counties; British Columbia and Alberta, south to California and Colorado; 6 (0).

Stellaria umbellata Turcz. ex Kar. & Kir. Perennials from slender rhizomes, forming small to large clumps or mats; flowering stems 5–50 cm long, glabrous; leaves 5–30 mm long, 2–8 mm wide, lanceolate to oblong, or elliptic, acute, sessile or subsessile, glabrous, or ciliate near the base; flowers few to many in terminal and axillary cymes, the branches subumbellately arranged, ascending to spreading; pedicels 3–30 mm long, subtended by small, scarious bracts, often reflexed; sepals 2–3 mm long, ovate-lanceolate to lanceolate, scarious-margined, obscurely 3–veined, not ciliate; petals lacking; capsules 3–4.5 mm long, straw colored or greenish, opening by 6–8 teeth, surpassing the sepals; seeds 0.5–0.7 mm long, brownish, roughened; $2n = 26$. Moist sites in white fir, aspen, lodgepole pine, spruce-fir, and alpine tundra communities at 2315 to 3480 m in Beaver, Box Elder, Cache, Daggett, Duchesne, Emery, Garfield, Iron, Juab, Piute, Salt Lake, San Juan, Sanpete, Sevier, Summit, Uintah, and Wasatch counties; Alaska south to California and Colorado; Asia; 43 (x).

Vaccaria Medicus

Annual herbs with erect or ascending stems; leaves opposite, lanceolate to oblong, lacking stipules; flowers few to numerous in open cymes; sepals 5, united nearly to the apex; calyx tube 10–veined and sharply angled; petals 5, pinkish, the blade retuse; stamens 10; styles 2 (rarely 3); ovary 1–loculed; capsules opening by 4 teeth.

Vaccaria pyramidata Medicus [*Saponaria vaccaria* L.; *S. segetalis* Necker; *V. segetalis* (Necker) Garcke ex Asch.]. Plants annual from taproots, the stems 15–80 cm tall, glabrous; leaves 3–8 cm long, 5–25 mm wide, lanceolate to lance-ovate or oblong, 1– to 3–veined, attenuate to acute apically, connate-clasping and sessile or short-petiolate basally; flowers few to many in open cymes, erect to spreading; pedicels 5–60 mm long, glabrous; calyx constricted apically, 11–15 mm long in flower, the veins purplish or green, glabrous, inflated and strongly 5–angled in fruit, the teeth 1–2 mm long; petals pink, the blades 5–8 mm long, retuse; seeds 1.7–2.1 mm long, reddish black; $2n = 20, 24, 30, 60$. Open habitats and in sagebrush-bunchgrass and mountain brush communities at 1280 to 1525 m in Cache, Juab, Rich, Salt Lake, and Utah counties; widespread in the U. S.; introduced from Europe; 8 (0).

CELASTRACEAE R. Br. in Flinders

Stafftree Family

Shrubs; leaves alternate or opposite, simple, deciduous or persistent; stipules small, caducous, or lacking; flowers perfect, regular, inconspicuous; sepals 4 or 5, persistent; petals 4–5, distinct; stamens 4 or 5 or 8–10; pistil 1 (3), the ovary superior, 1 (2–5)-loculed; style 1, the stigmas 2– to 5–lobed; fruit a capsule or follicle; $x = 8–12$ or more.

1. Leaves opposite 2
— Leaves alternate 3
2(1). Plants low indigenous evergreen shrubs, flowering in late winter or early spring, usually as ground cover with conifers .... *Pachystima*
— Plants cultivated trees or shrubs, deciduous or evergreen .... *Euonymus*
2(1). Leaves deciduous, not leathery-thickened; herbage not scabrous; flowers axillary; plants rather broadly distributed .... *Forsellesia*
— Leaves persistent, leathery-thickened; herbage scabrous; flowers in panicles; plant of Washington County .... *Mortonia*

Euonymus L.

Shrubs or trees with opposite petioled leaves, deciduous or evergreen; stipules small or lacking; flowers cream to greenish or purplish, borne in axillary cymes, mostly 5- or 4-merous; sepals spreading or recurved; stamens inserted on a broad disk; pistil 1, the ovary 3- to 5-loculed, short; stigma 3- to 5-lobed; capsules loculicidally dehiscent, the seeds 1 or 2 per locule, enveloped by a red or white aril, this visible upon dehiscence of the fruit. Several species of this genus are known from cultivation in Utah. They are poorly represented in herbaria, and the extent of their distributions is not understood. The most common ones are keyed below, but descriptions are not provided.

1. Leaves persistent, leathery, green or varicolored .... (Japanese spindletree) *E. japonica* Thunb.
— Leaves deciduous, not especially thickened, typically green 2
2(1). Branches with conspicuous corky wings .... (Winged spindletree) *E. alata* (Thunb.) Sieb.
— Branches lacking wings 3
3(2). Leaves pubescent beneath; flowers purple .... (Wahoo; Burning bush) *E. atropurpurea* Jacq
— Leaves glabrous beneath (except sometimes along the midrib); flowers yellowish or greenish 4
4(3). Leaves abruptly long-acuminate; plants small trees .. .. (Bunge spindletree) *E. bungeana* Maxim.
— Leaves acute to abruptly short-acuminate; plants mainly shrubs 5
5(4). Capsules warty; aril red .... (Strawberry bush) *E. americana* L.
— Capsules smooth; aril white .... (European spindletree) *E. europaea* L.

Forsellesia Greene

Intricately branched deciduous shrubs; stems slender, greenish, angled, more or less thorny; leaves simple, entire; flowers usually axillary, pedicellate; sepals 5 (4 or 6), hyaline-margined; petals 5 (4 or 6), white, deciduous; stamens equal or unequal; pistils 1–3, distinct, the ovary, superior, 1-loculed; fruit a follicle. Recent treatments place this genus in the Crossosomataceae Engler in Engler & Pranth.

Ensign, M. 1942. A revision of the Celastraceous genus *Forsellesia* (*Glossopetalon*). Amer. Midl. Nat. 27: 501–511.

1. Stipules lacking or mostly less than 0.5 mm long, subulate; petals constricted below the apex; plants mainly of eastern Utah .... *F. meionandra*
— Stipules usually more than 0.5 mm long, often flattened and whitish; petals not subapically constricted; plants mainly of western Utah .... *F. nevadensis*

Forsellesia meionandra (**Koehne**) **Heller** Utah Greasebush. [*Glossopetalon meionandrum* Koehne]. Bluntly thorny shrub usually 2–8 dm tall; leaves 3–15 mm long, 1.5–4 mm wide, oblanceolate to elliptic, pubescent, acute; stipules subulate, 0.1–0.5 (0.7) mm long, not flattened; flowers axillary, inconspicuous; pedicels with scarious bracts at the base; sepals 2–3 mm long; petals oblanceolate, white, somewhat constricted subapicallly, 4–6 mm long, elliptic; follicles multinervate, concavo-convex, thin, 3.5–4.5 mm long and about as wide. Shadscale, sagebrush, mixed desert shrub and pinyon-juniper communities at 1495 to 2320 m in Carbon, Daggett, Duchesne, Emery, San Juan, Sanpete, Sevier, Uintah, and Wayne counties; Colorado; 36 (vii).

Forsellesia nevadensis (**Gray**) **Greene** Nevada Greasebush. [*Glossopetalon nevadense* Gray]. Thorny shrubs, mainly 1–6 (10) dm high; leaves 3–12 mm long, 1–4 mm wide, oblanceolate to obovate, acuminate to obtuse; stipules subulate to narrowly oblanceolate, whitish or purple, mainly 0.4–1 mm long; flowers axillary, inconspicuous; pedicels with scarious bracts or reduced leaves; sepals 1–3 mm long; petals oblanceolate, 4–7 mm long, not constricted subapically; follicles multinervate, concavo-convex, thin, 4–5 mm long and about as wide. Mixed desert shrub and pinyon-juniper communities, often on limestone outcrops, at 1370 to 1985 m in Beaver, Juab, Kane, Millard, Tooele, and Washington counties; Idaho, Nevada, California, and Arizona; 13 (0). The diagnostic criteria used to distinguish the two species are not always discrete, and there is justification for recognition of all of our materials as *F. nevadensis*. However, there does seem to be geographical correlation, and at least trends of differences.

Mortonia Gray

Yellowish, scabrous, rounded shrubs; leaves alternate, simple, leathery, subsessile; stipules minute, glandlike, caducous; flowers borne in terminal thyrsoid cymes; calyx campanulate, 10-ribbed, 5-lobed; petals 5, white; stamens 5; pistil 5-loculed; style 1; stigmas 5; ovules 2 per locule; seeds oblong.

Mortonia scabrella **Gray** Shrubs mainly 8–12 dm tall, hispidulous; leaves 8–15 mm long or more, oval to obovate, scabrellous, the margin thickened; petioles very short; inflorescence 2.5–8 cm long, glabrous. Creosote bush community, often on limestone, in lower reaches of Washington County; Nevada, Arizona, California, New Mexico, Texas, and Mexico; 4 (i). The material has been assigned to **var. utahensis** Cov. ex. Gray [*M. utahensis* (Cov.) A. Nels].

Pachystima Raf.

Shrubs; leaves opposite, evergreen, leathery, serrulate; stipules minute, caducous; flowers inconspicuous, perfect, solitary or in few-flowered, axillary cymes; sepals 4; petals 4; stamens borne on the margin of a disk; pistil 1, superior, 2-loculed; style short; stigma 2-lobed; fruit a loculicidal capsule [also *Paxistima*].

Pachystima myrsinites (**Pursh**) **Raf.** Mountain Lover. [*Ilex myrsinites* Pursh]. Spreading to prostrate low shrubs; stems mostly 3–8 dm long; branchlets corky, squarrish; leaves 8–33 (41) mm long, 2–15 mm wide, elliptic, oblong, oblanceolate, or ovate to obovate rounded to obtuse, truncate or marginate apically, glossy green above, pale beneath, short-petioled; peduncles 2–3 cm long; flowers 1–3; petals reddish brown, ovate; capsules 4–5 mm long. Pinyon-juniper, sagebrush, mountain brush, Douglas fir, ponderosa pine, lodgepole pine, aspen, tall forb, and spruce-fir communities at 1675 to 3209 m in probably all Utah counties; British Columbia south to California, Arizona, and New Mexico; 80 (ix).

CERATOPHYLLACEAE S. F. Gray

Hornwort Family

Herbaceous, submersed aquatics; leaves whorled, dichotomously dissected; flowers imperfect, solitary, sessile in leaf axils, each subtended by a calyxlike involucre of 8–14 bracts, lacking a perianth; stamens 12–16; pistil 1, the ovary superior, of 1 carpel; fruit an achene; x = 12 (mostly).

Ceratophyllum L.

The family consists of a single genus.

Ceratophyllum demersum L. Common Hornwort. Stems to 10 dm long or more, freely branched and forming tangled masses; leaves whorled 5–12 per node, each usually dichotomously 1 or 2 times divided, the segments linear and flat, antrorsely toothed; achenes 4–5 mm long, ellipsoidal, provided with 2 basal spines; $2n = 24, 28, 38, 48$. Ponds, reservoirs, lakes, and slowly flowing streams in Beaver, Box Elder, Cache, Davis, Juab, Salt Lake, Utah, and Wasatch counties; cosmopolitan; 8 (0). This plant simulates the algal genus *Chara*, which has whorled branches, not dichotomously branched whorled leaves. Because of this similarity it seems probable that the common hornwort has not been collected when it was assumed to be *Chara*, which is widespread and common.

CHENOPODIACEAE Vent.
Goosefoot Family

Herbs, subshrubs, or shrubs, often succulent or scurfy; leaves simple, alternate or opposite, estipulate; flowers inconspicuous, monoecious, dioecious, polygamous, or perfect; calyx persistent, 1- to 5-lobed, enclosing the fruit, or lacking in some pistillate flowers; corolla none; stamens opposite the calyx lobes and as many or fewer; pistil 1, the ovary superior, with 1–3 stigmas, 1–loculed and 1–ovuled; fruit a utricle; $x = 6–9$.

Welsh, S. L. 1984. Utah flora: Chenopodiaceae. Great Basin Naturalist 44: 183–209.

1.	Leaves scalelike; stems fleshy; plants of saline pans and other salty sites	2
—	Leaves well developed, not scalelike; stems not fleshy; plants of various habitats	3
2(1).	Leaves alternate; plants woody at the base, mainly 3–8 dm tall	*Allenrolfea*
—	Leaves opposite; plants herbaceous, mainly 0.5–3 dm tall	*Salicornia*
3(1).	Leaves opposite, united at the base; sepals strongly imbricate, scarcely united, chartaceous; plants rare in western Utah	*Nitrophila*
—	Leaves alternate or rarely some opposite; sepals slightly if at all imbricate, herbaceous	4
4(3).	Leaves or bracts of inflorescence tipped with a spine or a spinelike bristle	5
—	Leaves and floral bracts not bristle- or spine-tipped	6
5(4).	Leaves linear to subulate; bracts of inflorescence ovate-lanceolate, spine-tipped; fruiting sepals winged on the back; flowers not embedded in hair	*Salsola*
—	Leaves sausagelike, abruptly spine-tipped; bracts of inflorescence not different from the leaves; sepals ending in wings; flowers embedded in hair	*Halogeton*
6(4).	Leaves sub- or semicylindric to linear, usually fleshy	7
—	Leaves with flattened blades, not especially fleshy	9
7(6).	Shrubs, armed with thorny branchlets; staminate flowers in spikes, the pistillate flowers solitary and axillary	*Sarcobatus*
—	Shrubs or herbs, not armed; flowers perfect or both perfect and pistillate	8
8(7).	Herbage villous-tomentose; plants low subshrubs	*Kochia*
—	Herbage glabrous and glaucous, or puberulent; plants annual or perennial, or if subshrubs then tall	*Suaeda*
9(6).	Plants densely white-hairy with at least some dendritic hairs, these becoming golden brown in age; shrubs of broad distribution	*Ceratoides*
—	Plants variously hairy or glabrous, but not as above; shrubs or herbs of various distribution	10
10(9).	Flowers imperfect, the pistillate enclosed in 2 accrescent or connate bracts	11
—	Flowers perfect or some also pistillate, all with sepals and not enclosed by paired bracts	14
11(10).	Stigmas 4 or 5; plants green, cultivated potherbs	*Spinacia*
—	Stigmas 2; plants green, grayish, or yellow green, not or seldom cultivated	12
12(11).	Bracts dorsally compressed, variously tuberculate, smooth, or winged; pubescence of inflated hairs or none; plants shrubs or perennial or annual herbs; axillary rounded buds lacking	*Atriplex*
—	Bracts laterally compressed or 6- to 8-ribbed, lacking appendages; pubescence of simple or branched hairs; plants shrubby; axillary rounded buds present	13
13(12).	Shrubs with divaricate often thorny branches; bracts with margins thickened, spongy within; pubescence of branched hairs	*Grayia*
—	Shrubs with erect nonthorny branches; bracts with margins not spongy thickened, either obcompressed or dorsiventrally compressed and 6–ribbed; pubescence of scurfy or moniliform hairs	*Zuckia*
14(10).	Plants more or less tomentulous; calyx transversely winged in fruit	*Cycloloma*
—	Plants glabrous, scurfy, pilose, or otherwise pubescent, but not or seldom as above; calyx not transversely winged in fruit, except in *Kochia* (q.v.), and then not tomentulous	15
15(14).	Perianth developing conspicuous horizontal, scarious wings or armed with curved or uncinate spines	16
—	Perianth lobes rounded or keeled on the back, lacking wings or spines	17
16(15).	Perianth developing conspicuous, horizontal, scarious wings	*Kochia*
—	Perianth with lobes each armed with a curved or uncinate spine	*Bassia*
17(15).	Calyx becoming woody in age; ovary partly inferior; plants cultivated, rarely escaping	*Beta*
—	Calyx not woody; ovary superior; plants not or seldom cultivated	18
18(17).	Calyx lobes 5, largely concealing to exposing the fruit; stamens usually 5; herbage glabrous or scurfy	*Chenopodium*
—	Calyx lobes 1–3, the fruit largely exposed; stamens 1–3; plants not scurfy	19
19(18).	Leaves hastately lobed or if entire 4–12 mm long, the blades mainly 2–8 mm broad or more; calyx l-lobed; stamens 1; plants widespread, of many habitats	*Monolepis*
—	Leaves linear, 0.8–6 cm long, entire, 1–2 mm wide; calyx 1- to 3-lobed; stamens 1–3; plants of sandy, low elevation sites	*Corispermum*

Allenrolfea Kuntze

Succulent, glabrous subshrubs or shrubs; branches alternate, jointed; leaves reduced to fleshy scales, alternate; flowers perfect, sessile, borne spirally in 3's or 5's in axils of peltate bracts, arranged in cylindrical erect spikes; sepals reduced, 4- to 5-lobed; stamens 1 or 2, exserted; stigmas 2 or 3; utricle ovoid, flattened, the pericarp free, membranous.

Allenrolfea occidentalis (**Wats.**) **Kuntze** Iodine Bush; Pickleweed. [*Halostachys occidentalis* Wats., type from Raft River Valley]. Plants mainly 3–8 (12) dm tall, woody at least below, glabrous and more or less glaucous; stems constricted at the nodes, fleshy; leaves very short, obtusely triangular; spikes 6–30 mm long; calyx enclosing the fruit; seeds brown. Saline and alkaline pans, springs, and seeps, and in other saline sites, often with saltgrass, samphire, seepweed, and other halophytes, at 1380 to 1620 m in Beaver, Box Elder, Cache, Davis, Emery, Grand, Iron, Juab, Millard, Salt Lake, Tooele, Utah, Washington, Wayne, and Weber counties; Oregon and Idaho, south to California, Arizona, Texas, and Mexico; 25 (iv). The plants accomodate high soil salinity by including salt within the protoplast, giving the plant a salty flavor; hence the name "pickleweed."

Atriplex L.

Monoecious or dioecious herbs or shrubs, often with scurfy (mealy) collapsed hairs; leaves alternate or opposite; flowers small, inconspicuous, borne in axillary clusters, glomerules, or in spicate panicles; staminate flowers with 3- to 5-parted calyx, bractless, with 3–5 stamens; pistillate flowers without a perianth and the pistil naked or rarely with a perianth, commonly enclosed within a pair of foliaceous bracts, enlarged in fruit, variously thickened and appendaged; styles 2; utricle with the pericarp free; seeds flattened, mainly erect. The genus is complex both taxonomically and nomenclaturally. It consists of native and introduced herbs and shrubs. Indigenous shrubby species form hybrids with all or most of their constituent taxa, wherever they come in contact. The resulting plasticity allows these remarkable plants to occupy numerous habitats, but poses problems that preclude a "neat" taxonomic treatment. The following keys are tentative at best.

Hanson, C. A. 1962. Perennial *Atriplex* of Utah and the northern deserts. Unpublished M.S. thesis, Brigham Young University. 133 pp.

Stutz, H. C. and S. C. Sanderson. 1983. Evolutionary studies in *Atriplex*: Chromosome races of *A. confertifolia* (Shadscale). Amer. J. Bot. 70: 1536–1547.

Thorne, K. H. 1977. A revision of the herbaceous members of the genus *Atriplex* (Chenopodiaceae) for the state of Utah. Unpublished M.S. thesis, Brigham Young University. 78 pp.

1. Plants herbaceous annuals .... Key 1
— Plants woody, at least below, perennial .... Key 2

Key 1.

Plants herbaceous annuals.

1. Seeds of 2 types, black and brown; plants mainly introduced 2
— Seeds all alike, either black or brown; plants indigenous 5
2(1). Fruiting bracts orbicular or nearly so, the dorsal surfaces smooth, entire 3
— Fruiting bracts triangular to ovate or rhombic, the dorsal surfaces usually tubercled, denticulate to entire 4
3(2). Black seeds horizontal, enclosed in a membranous calyx; brown seeds vertical between very large bracts .... *A. hortensis*
— Black seeds vertical between small bracts; brown seeds vertical between large bracts ... *A. heterosperma*
4(2). Lowermost leaves ovate, dentate, sessile or short-petiolate .... *A. rosea*
— Lowermost leaves lanceolate, rhombic, triangular, or hastate, dentate or entire, petiolate .... *A. patula*
5(1). Plants dioecious, widely distributed in eastern Utah, rarely elsewhere .... *A. powellii*
— Plants monoecious, of various distribution 6
6(5). Fruiting bracts orbicular or nearly so, the margins denticulate to lacerate; plants of Washington County .... *A. elegans*
— Fruiting bracts various in outline, but, if orbicular, the margins entire and seldom if ever in Washington County 7
7(6). Lower leaves linear to lanceolate, sessile, not over 2.5 cm long .... *A. wolfii*
— Lower leaves rhombic, deltoid, or cordate, usually at least some petiolate 8
8(7). Fruiting bracts all stipitate, or some of them sessile to subsessile, when stipitate usually prominently tubercled; plants forming low rounded clumps on saline substrates in eastern Utah .... *A. saccaria*
— Fruiting bracts all sessile or subsessile, variously tubercled or, if stipitate, as in *A. graciliflora* and *A. pleiantha* the surfaces smooth; plants slender or clump-forming, variously distributed 9
9(8). Fruiting bracts truncate and with 3 minute teeth apically; upper leaves sessile .... *A. truncata*
— Fruiting bracts not truncate apically; upper leaves petiolate or subsessile 10
10(9). Fruiting bracts with margins dentate, foliaceous well below the apex, the surfaces sometimes with appendages .... *A. argentea*
— Fruiting bracts samaralike and orbicular or triangular-ovate, the surfaces smooth 11
11(10). Fruiting bracts samaralike and orbicular, enclosing a single flower; plants widespread in southeastern Utah .... *A. graciliflora*
— Fruiting bracts triangular-ovate, enclosing 2 to 6 flowers; plants known from southeastern San Juan County .... *A. pleiantha*

Key 2.

Plants woody, at least below, perennial.

1. Leaves dentate (at least some); plants of Washington County 2
— Leaves entire; plants of various distribution 3
2(1). Plants monoecious; leaves green, ovate, irregularly and shortly dentate; weedy species near St. George .... *A. semibaccata*
— Plants dioecious; leaves silvery white, orbicular, definitely and coarsely toothed; indigenous .... *A. hymenelytra*
3(1). Leaf blades subhastate; shrubs to 3 m tall; plants of Washington County 4
— Leaf blades attenuate to rounded basally; shrubs mainly less than 2 m tall; plants of various distribution 5
4(3). Branchlets terete .... *A. lentiformis*
— Branchlets angled .... *A. torreyi*
5(3). Bracts with 4 lateral wings or 4 rows of teeth; plants unarmed 6

	Bracts without lateral wings, merely tuberculate or smooth dorsally; plants sometimes thorny 8
6(5).	Leaves more than 8 mm wide; bract tip with or without lateral teeth; canyons of the Colorado .. *A. garrettii*
—	Leaves less than 8 mm wide; bract tip without lateral teeth; distribution various 7
7(6).	Bracts more than 9 mm wide, with tips not exceeding the wings; staminate flowers yellow; shrubs to 2 m tall, widely distributed .... *A. canescens*
—	Bracts less than 9 mm wide, with tips exceeding the wings; staminate flowers mostly brown; plants mainly 5–10 dm tall, of playas in western Utah ... *A. gardneri*
8(5).	Plants with thorny branches; bracts foliose, united only basally, the surfaces smooth; staminate flowers yellow; widely distributed .... *A. confertifolia*
—	Plants lacking thorny branches; bracts not foliose, at least 1/3 united and the surfaces with appendages; staminate flowers yellow or brown 9
9(8).	Leaves 2–4 mm wide; bracts with appendages on lower 1/3; staminate flowers in spikes; plants prostrate .... *A. corrugata*
—	Leaves often more than 4 mm wide; bracts with appendages various; staminate flowers mainly in panicles; plants not or seldom prostrate 10
10(9).	Leaves oblong-ovate to orbicular, more than 10 mm wide, the lowermost alternate; staminate glomerules very numerous; plants of San Juan County .. *A. obovata*
—	Leaves linear to oblong, mainly less than 10 mm wide, or if wider, the lowermost opposite; staminate glomerules merely numerous; plants of Great Basin and northern Colorado River drainage, less commonly in San Juan County .... *A. gardneri*

Atriplex argentea Nutt. Silver Orach. [*A. rydbergii* Standley, type from south of Moab]. Plants annual, monoecious; stems simple or freely branched; leaves petiolate or the upper ones subsessile, the blades 0.5–6 cm long, 0.4–5 cm wide, lance-ovate, lanceolate, deltoid, or cordate, runcinate to subhastate basally, obtuse to acute apically, entire or essentially so, scurfy (glabrous); staminate flowers with 5-parted calyx; fruiting bracts sessile or subsessile, 4–5 mm long, 4–10 mm wide, the margin foliaceous below the apex, dentate to laciniate, the face smooth, tubercled, or crested; seeds ca 2 mm wide, brown; $2n = 18$. Mat-atriplex, shadscale, and greasewood communities at 1125 to 1770 m in Duchesne, Emery, Kane, Rich, Salt Lake (?), San Juan, Sevier, Summit, and Uintah counties; widespread in western U. S. and Mexico; 20 (0).

Atriplex canescens (**Pursh**) **Nutt.** Four-wing Saltbush. [*Calligonum canescens* Pursh]. Dioecious or rarely monoecious shrubs, mainly 8–20 dm tall, not especially armed; leaves persistent, alternate, sessile or nearly so, 10–40 mm long, 2–8 mm wide, linear to oblanceolate, oblong, or obovate, entire, retuse to obtuse apically; staminate flowers yellow (rarely brown), in clusters 2–3 mm wide, in panicles; pistillate flowers borne in panicles 5–40 cm long; fruiting bracts 9–25 mm long and as wide, on pedicels 1–8 mm long, with 4 prominent wings extending the bract length, united throughout; surface of wings and body smooth or reticulate; wings dentate to entire; apex toothed; seeds 1.5–2.5 mm wide; $2n = 18$, 36, or higher. Sandy, commonly non-saline, sites in Joshua tree, blackbrush, greasewood, salt desert shrub, sagebrush, mountain brush, and pinyon-juniper communities at 670 to 2380 m in all Utah counties, except Cache, Morgan, Rich, Summit, and Wasatch; Washington to Alberta and South Dakota, south to Mexico and Texas; 199 (xxiv). This species forms hybrids with *A. confertifolia* and *A. gardneri* varieties (see var. *bonnevillensis*). Materials from the type locality of the species in South Dakota are low subherbaceous plants that differ from our shrubby tall material. In a strict sense the common phase of our plants seems best regarded at varietal status as **A. canescens var. *occidentalis* (Torr. & Frem.) Welsh & Stutz** [*Pterchiton occidentale* Torr. & Frem in Frem., type from the Great Salt Lake]. A second more or less distinctive phase occurs within the distribution of var. *occidentalis*, but is restricted ecologically and differs morphologically from that variety. These dune plants were noted by Hanson (1962) as having bracts to 25 mm wide and elongated internodes. To these features can be added the presence of adventitious roots at buried internodes and a diploid chromosome number of $2n = 18$. They are **var. *gigantea* Welsh & Stutz** and grow in interdune valleys of the Lynndyl sand dunes, Juab County (type locality), where they are encroached upon by the following dune. They survive being buried as the dune advances by producing adventitious roots along the stem and by continued growth above the encroaching sand. The species is an important browse plant for both wildlife and domestic livestock. It is used in reclamation projects and might be found established at sites beyond its usual range and habitat latitude.

Atriplex confertifolia (**Torr. & Frem.**) **Wats.** Shadscale [*Obione confertifolia* Torr. & Frem. in Frem., type from Weber County; *A. collina* Woot. & Standl.]. Dioecious spinescent shrubs, 3–8 dm tall; leaves persistent, alternate, with petioles 1–4 mm long; blades 9–25 mm long, 4–20 mm wide, orbicular to ovate, elliptic, or oval, entire, obtuse apically; staminate flowers yellow, in clusters 2–4 mm wide or in spikes to 1 cm long; inflorescence paniculate, 3–15 cm long; fruiting bracts sessile or subsessile, suborbicular to rhombic or elliptic, 4–12 mm long and wide, the surface smooth, lacking appendages; terminal teeth distinct, foliaceous, shorter than the bracts, entire or toothed below, spreading at maturity; seeds 1.5–2 mm broad; $2n = 18$, 36, 54, or higher. Gravelly to fine textured soils in greasewood, mat-atriplex, other salt desert shrub, sagebrush, and pinyon-juniper communities at 850 to 2140 m in all Utah counties, except Morgan, Rich, and Summit; Oregon east to North Dakota, south to California, Arizona, New Mexico, and Texas; 144 (xii). Shadscale forms hybrids with *A. canescens*, *A. garrettii*, *A. corrugata*, and *A. gardneri* varieties. This is a valuable browse plant for wildlife and livestock, especially sheep.

Atriplex corrugata **Wats.** Mat-saltbush; Mat-atriplex. Dioecious, low, spreading shrubs, mainly 3–15 cm tall and 3–15 dm broad; leaves persistent, sessile, opposite below, alternate above, 3–18 mm long, 1–6 mm wide, linear to linear-oblanceolate, or oblong, entire, obtuse apically; staminate flowers yellow to brownish, in clusters 3–6 mm wide, borne in spikes 1–8 cm long; pistillate flowers in leafy bracteate spikes 5–15 cm long; fruiting bracts sessile or subsessile, 3–5 mm long, 4–6 mm wide, densely tuberculate (or smooth), entire or undulate, rounded to acute apically; seeds ca 1.5 mm wide; $2n = 36$. Saline, usually fine-textured substrates derived from Mancos Shale, Tropic Shale, Morrison, Duchesne River, and other similar formations in mat-atriplex and Castle

Valley saltbush communities at 1220 to 2150 m in Carbon, Duchesne, Emery, Garfield, Grand, Kane, San Juan, Uintah, and Wayne counties; New Mexico and Colorado; 105 (viii). Mat-saltbush is known to form intermediates with both *A. confertifolia* and *A. gardneri* var. *cuneata*. This saltbush is a valuable browse plant on the sparsely vegetated clays and silts of eastern Utah, where it is often the only woody vegetation present.

Atriplex elegans (Moq.) D. Dietr. Wheelscale Orach. [*Obione elegans* Moq.]. Annual herbs, the stems erect to ascending or prostrate-decumbent, mainly 1–6 dm tall, scurfy to glabrate; leaves mostly alternate, subsessile or shortly petiolate, 5–30 mm long, 2–8 mm wide, elliptic to spatulate, oblanceolate, oblong, or obovate, the base cuneate, entire to denticulate, densely scurfy; flowers monoecious, in axillary clusters; staminate flowers with a 3–5 parted perianth; fruiting bracts shortly stalked, compressed, united except at the thin margin, orbicular, 2–4 mm wide, the margins dentate, the terminal teeth often prominent, the faces smooth; seeds 1–4 mm wide, brown. Disturbed sites at ca 885 m in Washington County; California, Nevada, Arizona, New Mexico, Texas, and Mexico; 3 (0).

Atriplex gardneri (Moq.) D. Dietr. Gardner Saltbush. [*Obione gardneri* Moq.; *A. nuttallii* Wats.]. Dioecious or monoecious shrubs or subshrubs, 1–10 dm tall, unarmed; leaves more or less persistent, alternate or opposite to subopposite, sessile to petiolate, linear to oblanceolate, obovate, spatulate, or orbicular, 5–55 mm long, 2–25 mm wide, entire (rarely dentate), retuse to obtuse or rounded apically; staminate flowers in spikes or panicles, 2–30 cm long, yellow or brown, in clusters 2–4 mm wide; pistillate flowers in spikes or panicles to 30 cm long; fruiting bracts 3–9 mm long, 2–9 mm wide, bearing tubercles or wings or the tubercles alligned in 4 rows or rarely smooth, the apex toothed and usually with 2 or more lateral teeth; seeds 1.5–2.5 mm wide, tan or brown. This is a widely distributed complex of intergrading genotypes of great plasticity. Plants of this complex occur commonly in fine-textured saline substrates in much of the western Great Plains and in the intermountain region. Diploids, triploids, tetraploids, and hexaploids (and higher polyploids, all multiples of the base number 9) are known within the complex, and hybrids are known between the constituents and the other woody species which they contact, i.e., *A. canescens*, *A. confertifolia* and *A. corrugata*. The treatment essentially follows the allignment of taxa suggested by Hanson (1962), with the exception that they are reduced to varietal status and var. *bonnevillensis* is placed within the *gardneri* phase and not with *A. canescens*. The use of the epithet *nuttallii* for this complex was reviewed by Hanson (1962), and other problems aside, the name *A. gardneri* clearly has priority over *A. nuttallii* and must be used according to stipulations of the International Code. Within Utah there are six morphologically intergrading entities that seem worthy of taxonomic recognition.

1. Fruiting bracts with 4 lateral wings or rows of tubercles; plants of valley bottoms and playas in Juab and Millard counties .... *A. gardneri* var. *bonnevillensis*
— Fruiting bracts lacking lateral wings, the tubercles, when present, often more or less aligned; plants of various distribution 2
2(1). Lower leaves opposite or subopposite; plants prostrate to ascending, in eastern Utah 3
— Lower leaves alternate; plants ascending to erect, of various distribution 4
3(2). Leaves mainly 10–25 mm wide, grayish green; bracts 5–9 mm wide, heavily tuberculate; plants of the Uinta and Navajo basins .... *A. gardneri* var. *cuneata*
— Leaves mainly 4–12 mm wide, green; bracts 2–5 mm wide, not tuberculate or the tubercles very short; plants of Daggett County .... *A. gardneri* var. *gardneri*
4(2). Staminate flowers mostly brown; fruiting bracts with apical teeth half-united, lacking lateral teeth; plants of Great Basin and Rich County .... *A. gardneri* var. *falcata*
— Staminate flowers mostly yellow; fruiting bracts with apical teeth free, subtended by lateral teeth; plants of various distribution 5
5(4). Leaves mainly 5–15 times longer than wide; fruiting inflorescences spicate; plants of Grand County .... *A. gardneri* var. *welshii*
— Leaves mainly less than 5 times longer than wide; fruiting inflorescences paniculate; plants of broad distribution .... *A. gardneri* var. *tridentata*

Var. bonnevillensis (C. A. Hanson) Welsh Bonneville Saltbush. [*A. bonnevillensis* C. A. Hanson, type from Pine Valley playa, Millard County]. Greasewood communities in valley bottoms and playas at 1500 to 1585 m in Juab and Millard counties; Nevada; 10 (i); $2n = 18$. The Bonneville saltbush is apparently a partially stabilized introgressant involving *A. gardneri* var. *falcata* and *A. canescens*. The habitat is intermediate between that of the parental taxa. There is evidence that the introgression is continuing in some populations at least.

Var. cuneata (A. Nels.) Welsh Castle Valley Saltbush. [*A. cuneata* A. Nels., type from Emery]. $2n = 18, 27, 36$, and higher. Saline fine textured substrates on Mancos Shale, and other formations of similar texture and salinity, in greasewood and mat-atriplex communities at 1220 to 2170 m in Carbon, Duchesne, Emery, Garfield, Grand, San Juan, and Uintah counties; Colorado and New Mexico; 166 (xxi). A series of at least partially stabilized introgressants between var. *cuneata* and var. *tridentata* in Carbon county form the basis of *A. cuneata* ssp. *introgressa* C. A. Hanson (type from Wellington). Possibly they warrant taxonomic recognition, but no nomenclatural combination is intended or implied herein.

Var. falcata (Jones) Welsh Jones Saltbush. [*A. nuttallii* var. *falcata* Jones; *A. falcata* (Jones) Standley]. Sagebrush, shadscale, and greasewood communities at 1310 to 1985 m in Box Elder, Iron, Juab, Millard, Rich, and Tooele counties; Washington to Montana, south to Nevada and Wyoming; 30 (i).

Var. gardneri Gardner Saltbush. Greasewood and sagebrush-saltbush communities at ca 1895 m in Daggett County; Wyoming, Colorado, and Montana; 6 (0).

Var. tridentata (Kuntze) Macbr. Basin Saltbush. [*A. tridentata* Kuntze, type from Corinne; *A. nuttallii* var. *utahensis* Jones, type from Salt Lake City]. $2n = 18, 36, 54$. Greasewood, shadscale, alkali saccaton, kochia, saltgrass, and sedge-rush communities at 1280 to 1985 m in Beaver, Box Elder, Cache, Carbon, Davis, Duchesne, Iron, Juab, Millard, Piute, Salt Lake, Sanpete, Sevier, Tooele, and Uintah counties; Colorado, Nevada, Idaho, and Wyoming; 95 (iv). Materials from eastern Utah have leaves narrower on the average than those from the Great Basin. The bracts are more heavily tuberculate also.

Var. *welshii* (C. A. Hanson) Welsh Welsh Saltbush. [*A. welshii* C. A. Hanson, type from south of Cisco]. 2n = 18. Mat-saltbush and Castle Valley saltbush communities at 1280 to 1315 m in Grand County; endemic; 10 (ii).

***Atriplex garrettii* Rydb.** Garrett Saltbush [*A. canescens* ssp. *garrettii* (Rydb.) H. & C.; *A. canescens* var. *garrettii* (Rydb.) L. Benson]. Dioecious (rarely monoecious) shrubs or subshrubs, mainly 2–6 dm tall, unarmed; leaves opposite or subopposite below, petiolate, the blades 8–55 mm long, 6–32 mm wide, ovate to obovate, elliptic, or orbicular, yellow green, sparingly scurfy, entire or repand-dentate, obtuse to cuneate basally, rounded to acute apically; staminate flowers brown to tan (rarely yellow), in clusters 2–4 mm wide on panicles 2–8 cm long; pistillate flowers in spikes or spicate panicles 4–30 cm long; fruiting bracts 6–10 mm long and wide, winged, the surface smooth, reticulate, or with flattened processes, toothed apically; seeds ca 2 mm wide, brown. 2n = 18. Shadscale, ephedra, eriogonum, blackbrush, and mixed shrub-grass communities on talus slopes of canyons of the Colorado at 1125 to 1895 m in Garfield, Grand (type from near Moab), Kane, San Juan, and Wayne counties; endemic; 41 (xii). This distinctive plant has been regarded as a portion of the variation within an expanded *A. canescens*, with which it shares the feature of 4-winged pistillate fruiting bracts, but it is possibly more closely allied with *A. confertifolia*, with which it hybridizes.

***Atriplex graciliflora* Jones** Blue Valley Orach. Monoecious annual herbs, mainly 1–3 dm tall, branching from the base; leaves alternate, petiolate, the blades 8–10 mm long and about as wide or wider, cordate-ovate to orbicular, cordate, or deltoid, truncate to cordate basally, rounded to obtuse or acute apically, entire; staminate flowers in panicles overtopping the foliage, the perianth 5-lobed; pistillate flowers axillary; fruiting bracts samara-like, 6–16 mm wide, stipitate, compressed, orbicular, oblong or cordate in outline, winged, the wings undulate or entire, the surfaces smooth; seeds ca 3 mm wide, dull white. Saltbush, seepweed, greasewood, rabbitbrush, and tamarix communities on saline, often salt encrusted, soils at 1125 to 1900 m in Carbon, Emery, Garfield, Kane, San Juan, eastern Sevier, and Wayne (type from Blue Valley) counties; Colorado; 18 (iii).

***Atriplex heterosperma* Bunge** Two-seed Orach. Monoecious annual herbs, mainly 1.5–14 dm tall, erect, branched from below the middle or above; leaves opposite or subopposite below, commonly alternate above and petiolate, the blades mainly 15–80 mm long and as wide or wider, hastately lobed, triangular, the base truncate to cordate or obtuse, acute apically; staminate flowers with 5 sepals; fruiting bracts 2–7 mm long, orbicular to suborbicular or ovate, entire, the surfaces smooth, dimorphic, the larger with a pale brown vertical seed 2–3 mm wide, the smaller with a shiny black vertical seed ca 1 mm wide; 2n = 36. Riparian and palustrine (less commonly ruderal) habitats in greasewood, saltgrass, cocklebur, tamarix, cottonwood, and rush-cattail communities at 1310 to 1985 m in Box Elder, Cache, Davis, Duchesne, Emery, Juab, Salt Lake, San Juan, Sanpete, Sevier, Summit, Uintah, and Weber counties; western North America; adventive from Eurasia; 29 (ii). This is a handsome vigorous annual that appears to be invading saline lowland and other disturbed areas throughout the state.

***Atriplex hortensis* L.** Garden Orach. Monoecious annual herbs, mainly 5–20 dm tall, erect, branching from the middle or above; leaves opposite or subopposite below, alternate above, petiolate, the blades commonly 1.5–13.5 cm long and 1–13 cm wide, ovate to lanceolate, not especially hastate, the base acute to cordate, acute to rounded apically; staminate flowers with 3–5 sepals; pistillate flowers dimorphic, the pistil vertical and enclosed in bracts or the pistil horizontal and enclosed in a 4- or 5-lobed calyx, both shortly pedicellate; fruiting bracts 8–19 mm long, orbicular to ovate, entire, the surfaces smooth, greenish or reddish; seeds dimorphic, either 2–4 mm wide and brown or ca 1 mm wide and black. Disturbed sites in riparian and ruderal habitats at 1310 to 2135 m in Cache, Duchesne, Salt Lake, Sanpete, Summit, Tooele, Uintah, Utah, and Weber counties; practically cosmopolitan, introduced from Eurasia; 12 (i). This plant is grown as a potherb, and is to be expected practically anywhere. It persists and escapes following cultivation.

***Atriplex hymenelytra* (Torr.) Wats.** Desert Holly. [*Obione hymenelytra* Torr.]. Dioecious shrubs, 3–10 dm tall or more, unarmed; leaves persistent, petiolate, alternate, the blades 10–40 mm long and as wide or wider, orbicular to reniform or oval, greenish to silvery white, permanently scurfy, prominently dentate; staminate flowers yellow to purple brown, in clusters 3–4 mm thick, borne in panicles to 3 cm long; pistillate bracts subsessile, 7–10 mm long and wide, orbicular to reniform, the margins entire to crenate; seeds ca 2 mm wide, brown; 2n = 18. Warm desert shrub community at ca 730 m in Washington County; Arizona, Nevada, California, and Mexico; 6 (ii). This is a handsome rounded shrub with silvery white foliage. It flowers very early in springtime.

***Atriplex lentiformis* (Torr.) Wats.** Big Saltbush. [*Obione lentiformis* Torr. in Sitgr.]. Dioecious or less commonly monoecious shrubs, mainly 10–25 dm tall, unarmed; branchlets terete; leaves persistent, alternate, petiolate, the blades 0.5–4 cm long, 0.3–3 cm wide, deltoid to rhombic, ovate, or oblong-elliptic, gray green, scurfy, entire to repand or subhastately lobed, rounded to obtuse apically; staminate flowers yellow, in clusters 1–2 mm wide, borne in panicles 0.5–5 dm long; fruiting bracts 3–4 mm long and wide, sessile, orbicular to oval, crenulate, rounded apically; seeds ca 1–1.5 mm wide, brown; n = 9. Drainages, stream and canal banks, and roadsides in warm desert shrub communities at 760 to 950 m in Washington County; Arizona, Nevada, California, and Mexico; 8 (0).

***Atriplex obovata* Moq.** New Mexico Saltbush. Dioecious shrubs, mainly 2–8 dm tall; leaves tardily deciduous, alternate, shortly petiolate, the blades 8–30 mm long, 6–20 mm wide, obovate to elliptic or orbicular, gray green, entire or rarely dentate, rounded to retuse or obtuse apically; staminate flowers yellow, in clusters 2–3 mm wide, borne in panicles 6–30 cm long; fruiting bracts 4–5 mm long, 5–9 mm wide, sessile, broadly cuneate, the surfaces smooth or rarely tubercled, the margins entire, the apical tooth subtended by 2–6 equal or smaller teeth; seeds 1–1.5 mm wide, brownish. Salt desert shrub and lower pinyon-juniper communities at ca 1525 to 1650 m in San Juan County; Arizona, New Mexico, and Mexico; 7 (ii).

***Atriplex patula* L.** Fat-hen Saltbush. Monoecious annual herbs, mainly 1.5–10 dm tall, prostrate-ascending or erect, simple or branched; leaves alternate or some or all opposite, petiolate, the blades mainly 1–12 cm long, 1–5 cm wide or more, ovate, deltoid, lance-ovate, or lance-linear, cordate to hastate, truncate, or acute to cuneate basally, rounded to obtuse or acute apically, entire to dentate or hastate, thin or thick, green, glabrous or scurfy; flowers in paniculate clusters; staminate flowers with 4 or 5 sepals; fruiting bracts sessile, subsessile, or rarely stipitate, 2–12 mm long, 3–9 mm wide, deltoid to ovate or rhombic, sometimes spongy-thickened, the margin entire or denticulate, the face smooth, roughened, or tuberculate; seeds vertical, dimorphic, either black and 1–2 mm wide or brown and 1–3 mm wide; n = 9, 18, 36. This species complex consists of plants with circumboreal representation. Our material apparently consists of introduced and indigenous portions of that complex, often treated at specific status or in various infraspecific categories. Treatment as a broad highly variable taxon consisting of two infraspecific taxa seems to best represent our material. The following key will serve to segregate most specimens.

1. Principal lower leaves triangular-hastate; leaf bases truncate, broadly cuneate, or subcordate; plants common .... *A. patula* var. *triangularis*
— Principal lower leaves various, but seldom hastate; leaf bases acute to cuneate; plants uncommon .... *A. patula* var. *patula*

Var. *patula* Sedge-reed, tamarix-Russian olive, and willow-cottonwood communities at 1370 to 1985 m in Rich, Sanpete, and Utah counties; widespread in North America; Eurasia; 3 (i). Materials tentatively assigned here are thin-leaved, but bracts vary in outline from rhombic to ovate or narrowly oblong and the surfaces from smooth to tubercled.

Var. *triangularis* (Willd.) K. Thorne & Welsh [*A. triangularis* Willd.; *A. hastata* authors, not L..; *Chenopodium subspicatum* Nutt.; *A. subspicata* (Nutt.) Rydb.; *A. carnosa* A. Nels.]. 2n = 18. Saltgrass, sedgerush, rush-cattail, and other palustrine and riparian habitats, usually in saline mucky soils at 850 to 1985 m in Cache, Davis, Duchesne, Emery, Millard, Salt Lake, San Juan, Sanpete, Uintah, and Washington counties; widely distributed in North America; Europe; 31 (v). Both thick and thin-leaved specimens are included in this variety. The problem of typification of *A. hastata* L. was reviewed by Taschereau (Canad. J. Bot. 50: 1585, 1972). That name evidently replaces the long established *A. calotheca* (Rafn) Fries, a plant not known from Utah. The next available epithet at specific rank is apparently *A. triangularis* Willd., herein treated at varietal level. It seems likely that another name might well supersede that name at varietal rank.

Atriplex pleiantha W. A. Weber Four-corners Orach. Monoecious, glabrous or sparingly scurfy annual herbs, mainly 0.5–1.5 dm tall, branching from the base; leaves alternate to subopposite, petiolate, the blades 5–15 mm long and about as wide, ovate to suborbicular, obtuse to acute apically, entire; staminate flowers in short terminal spikes; fruiting bracts 3–7 mm wide and about as long, triangular-ovate, short-stipitate, compressed, entire, enclosing 2–5 flowers, these with perianth well developed, consisting of 5 hyaline, sparsely ciliate scales 1–1.2 mm long; utricle black, smooth and shining, ca 1.5 mm long, falling at maturity. Salt desert shrub community, Morrison Formation, southeastern San Juan County; Colorado and New Mexico; a Navajo Basin endemic; 1 (i).

Atriplex powellii Wats. Powell Orach. Dioecious (sparingly monoecious) annual herbs; stems slender to stout, mainly 1–5 (7) dm tall, branching almost throughout; herbage pubescent with scurfy and arachnoid hairs; leaves alternate, petiolate, or the upper sessile, the blades 0.4–5 cm long, 0.2–3 cm wide, ovate to rhombic or elliptic, entire, rounded to cuneate basally, acute to obtuse apically, prominently 3–veined; staminate flowers with calyx 4– or 5–lobed; fruiting bracts sessile, 2–2.8 mm long, 1.5–3.2 mm wide, thick, united to the apex, ovate to oblong or broadly cuneate, truncate to cuspidate apically, the surfaces with thickened processes or rarely smooth; seeds ca 2 mm long, greenish; 2n = 18. Saline, usually fine textured substrates, in greasewood, rabbitbrush, shadscale, seepweed, mat-atriplex, juniper-pinyon, and blackbrush communities at 1220 to 1830 m in Carbon, Duchesne, Emery, Garfield, Grand, Iron, Kane, Millard, Sanpete, Sevier, Tooele, Uintah, Utah, and Wayne counties; Montana and South Dakota to Arizona and New Mexico; 54 (ix). This is the only annual atriplex in Utah that approaches being truely dioecious, but occasionally a few flowers of the opposite gender occur, resulting in sparingly monoecious individuals. The species in Utah is characteristic of the Colorado Drainage system. The few specimens from the Great Basin possibly represent recent introductions.

Atriplex rosea L. Tumbling Orach. Monoecious, coarse, annual herbs; stems simple or more commonly branching throughout, mainly 2–8 dm tall; herbage scurfy to glabrate; leaves alternate, petiolate, the blades mainly 1.2–7 cm long, 0.6–3.5 cm wide, ovate to lanceolate, acute to obtuse apically, irregularly dentate and often subhastately lobed; staminate flowers with 4 or 5 sepals; fruiting bracts sessile, 4–6 (8) mm long and as wide, ovate to rhomic, united to the middle, dentate, sharply tuberculate on the surfaces; seeds dimorphic, brown and 2–2.5 mm wide, or black and 1–2 mm wide; n = 9. Widely established weedy species of disturbed sites, often in riparian habitats or in barnyards or on animal bedgrounds, at 850 to 2560 m in all or nearly all Utah counties; widespread in North America; Eurasia and elsewhere; 47 (v).

Atriplex saccaria Wats. Stalked Orach. Low monoecious herbs, forming rounded clumps, mainly 0.5–2 (2.5) dm tall; stems usually branched from the base; herbage scurfy; leaves alternate or the lowermost subopposite, petiolate, the blades mainly 0.6–4 cm long, 0.4–3 cm wide, ovate to deltoid-ovate or oval, entire or in some the base subhastately lobed, truncate to subcordate or broadly cuneate basally, acute to rounded apically; staminate flowers with 5–parted perianth; fruiting bracts stipitate, or some subsessile to sessile, the faces smooth to coarsely tubercled or appendaged, mainly 4–6 mm long and as wide; seeds 2–3 mm wide, brownish to whitish; n = 9. Two rather continuously intergrading varieties are present.

1. Pedicels of fruiting bracts 3–8 mm long or more, the bracts all essentially alike; plants uncommon .... *A. saccaria* var. *caput-medusae*
— Pedicels of fruiting bracts mainly less than 3 mm long, the bracts of lower axils tending to be subsessile and less tubercled than the upper ones ... *A. saccaria* var. *saccaria*

Var. *caput-medusae* (Eastw.) Welsh Medusa-head Orach. [*A. caput-medusae* Eastw., type from Recapture Creek, San Juan County; *A. argentea* var. *caput-medusae* (Eastw.) Fosberg]. Greasewood, saltbush, and other salt-desert shrub communities at ca 1525 to 1600 m in Emery, San Juan (type from Recapture Creek), and Uintah counties; New Mexico and Arizona; 3 (i).

Var. *saccaria* [*A. cornuta* Jones, type from Green River]. Mat-atriplex, shadscale, greasewood, and pinyon-juniper communities at 1125 to 1830 m in Carbon, Daggett, Emery, Garfield, San Juan, eastern Sevier, Uintah, and Wayne counties; Wyoming to Arizona and New Mexico; 30 (vi).

Atriplex semibaccata R. Br. Australia Saltbush. Decumbent, monoecious subshrubs, mainly 5–30 cm high and to 1 m wide or more, unarmed; leaves alternate subsessile or shortly petiolate, mainly 0.8–3 cm long, 4–9 mm wide, obovate to oblong, remotely dentate, obtuse apically, attenuate basally, 1-veined; staminate flowers in clusters ca 1.5 mm wide; fruiting bracts sessile, 3–6 mm long and as wide, rhombic, united below, convex, red-fleshy at maturiy, obtuse to acute apically, strongly veined; seeds dimorphic, ca 1.5 mm long and black, and ca 2 mm long and brown; 2n = 18. Disturbed sites at ca 850 m in Washington County; southwestern U. S.; introduced from Australia; 5 (0).

Atriplex torreyi Wats. Torrey Saltbush. Dioecious shrubs, mostly 8–30 dm tall, forming broad clumps; branchlets angled, becoming bluntly thorny; leaves alternate, persistent, with petioles 1–4 mm long, the blades 0.5–3 cm long, 4–16 mm wide, ovate to deltoid, rhombic, oval, or lanceolate, entire (rarely toothed), obtuse apically, truncate to obtuse basally; staminate flowers yellow, in clusters ca 1 mm wide, in panicles mainly 1–3 dm long; fruiting bracts sessile, 2–3 mm long and wide, orbicular, crenate, rounded apically; seeds ca 1–1.4 mm wide, brown. Mesquite, creosote bush, shadscale and blackbrush communities at ca 800 to 900 m in Washington County; Nevada and California; 15 (ii).

Atriplex truncata (Torr.) Gray Wedge Orach. [*Obione truncata* Torr. in Wats.; *A. subdecumbens* Jones, type from Fish Lake, Sevier County]. Monoecious annual herbs; stems simple or more commonly branched throughout, mainly 3–8 dm tall; herbage scurfy, becoming glabrate; leaves alternate, petiolate below, sessile above, the blades mainly 4–30 mm long, 3–30 mm wide, ovate to deltoid or oval, acute to obtuse apically, entire or dentate, truncate or subhastate to rounded basally; staminate flowers with 4 or 5 sepals; fruiting bracts sessile, 2–3 mm long and as wide, broadly cuneate, truncate apically, with 3 (or more) teeth across the summit, the surfaces smooth (rarely tubercled); seeds 1–2 mm wide, brown; 2n = 18. Saline saltgrass-greasewood and other palustrine habitats at ca 850 to 1375 (2700) m in Box Elder, Cache, Carbon, Emery, Juab, Millard, Rich, Salt Lake, Sevier, Utah, Washington, and Wayne counties; British Columbia to Montana, south to California and New Mexico; 12 (0).

Atriplex wolfii Wats. Slender Orach. [*A. tenuissima* A. Nels., type from Gunnison]. Monoecious, slender, delicate, annual herbs; stems simple or more commonly branched throughout, mainly 0.7–3.5 dm tall; herbage scurfy; leaves alternate, sessile, mainly 0.4–2.5 cm long, 1–3 mm wide, linear to narrowly lanceolate; staminate flowers with 5 sepals; fruiting bracts sessile, 1.5–3 mm long, 1–2.5 mm wide, ovate to cuneate in outline, truncate to attenuate apically, the faces smooth or tuberculate; seeds 1–2 mm long, brown. Greasewood community at 1525 to 2135 m in Carbon, Duchesne, Emery, Garfield, Piute, Sanpete, Sevier, and Uintah counties; Wyoming and Colorado; 7 (ii).

Bassia All.

Annual herbs; leaves alternate, entire, sessile; herbage pilose or tomentose, at least in inflorescence; flowers perfect and pistillate, glomerate or solitary in leaf axils and in short axillary spikes, bracteate; calyx 5–lobed, depressed-globose, enclosing the fruit and usually prominently armed with a curved or hooked spine on the dorsal surface of each lobe; stamens 5, hypogynous; styles 1, with 2 (3) stigmas; fruit compressed; seed horizontal.

Bassia hyssopifolia (Pallas) Kuntze [*Salsola hyssopifolia* Pallas; *Echinopsilon hyssopifolius* (Pallas) Moq. in DC.; *Kochia hyssopifolia* (Pallas) Schrader]. Annual herbs, the main stem erect, the lower lateral ones often decumbent, 2–10 dm tall; herbage more or less lanate, especially in the inflorescence; leaves 4–40 mm long, 1–5 mm wide, linear to oblong or narrowly oblanceolate; flowers clustered in terminal or lateral spikes, or solitary in leaf axils; floral bracts reduced; pistillate and sterile flowers mixed with perfect ones; fruiting calyx ca 2 mm wide, each lobe with a stout curved to uncinate spine; pericarp membranous, planoconvex; 2n = 18. Commonly on saline substrates, often in riparian or palustrine habitats, in saltgrass, greasewood, horsebrush, shadscale, and cottonwood-tamarix communities at 850 to 2380 m in most, if not all, Utah counties; adventive from Eurasia; 34 (vii). This species forms apparent intergeneric hybrids with *Kochia scoparia*, from which *Bassia* differs in having spines on the sepals instead of horizontal flattened processes, inter alia.

Beta L.

Glabrous annual or biennial herbs; leaves alternate, petiolate, essentially entire; flowers perfect, solitary or borne in few-flowered cymes, these arranged in spicate terminal or axillary spikes; sepals 5; stamens 5; ovary partially inferior, connate with receptacle in fruit; stigmas 2 or 3; fruits adhering, fused by the swollen perianth and receptacle; seeds horizontal.

Beta vulgaris L. Stems mainly 4–10 dm tall or more; basal leaves well developed, long-petiolate, the blades mainly 5–25 cm long, 2–10 cm wide, undulate-crisped; inflorescence elongate, with lower bracts prominent; sepals incurved in fruit. Cultivated food plant, occasionally escaping but not persisting, mainly below 2135 m in much of Utah; introduced from Europe; 7 (0). This is the beet of commerce, including the red table beet and sugar beet. The latter was a major cash crop in Utah until the 1960's. Swiss chard, grown for use as a potherb, is a cultivar of this species.

Ceratoides Gagnebin

Monoecious tomentose shrubs; leaves alternate, entire; staminate flowers ebracteate, with calyx 4–lobed; stamens 4; pistillate flowers lacking a perianth, enclosed in 2 villous-pilose, partially connate bracteoles, the tips divergent and hornlike; styles 2, slender; pericarp thin, free from the seed [*Eurotia* Adanson; *Krascheninnikovia* Gueldenst.].

Howell, J. T. 1971. A new name for winterfat. Wasmann J. Biol. 29: 105.

Ceratoides lanata (Pursh) **J. T. Howell** Winterfat; White-sage. [*Diotis lanata* Pursh; *Eurotia lanata* (Pursh) Moq.]. Shrubs, woody for 0.2–8 dm above ground (or more), and with numerous annual branchlets mainly 0.5–3 (5) dm long; herbage stellate-hairy, commonly with longer straight hairs intermixed, the hairs white or becoming yellowish in age; leaves 1–4.5 cm long, 1–6.5 mm wide, linear to narrowly lanceolate, entire, revolute to almost flat, sessile above, short-petiolate below; flowers borne in dense axillary clusters or more or less spicate along branch tips; pistillate flowers 2–4 per axil; staminate flowers in spicate axillary clusters, the perianth segments 1.5–2 mm long; fruiting bracts 3–6 mm long, obscured by the long covering hair; 2n = 18. Shadscale, black sagebrush, bullgrass, sagebrush, and pinyon-juniper communities at 730 to 2840 m, known in all Utah counties except Davis, Morgan, Salt Lake, Summit, and Weber; Yukon to Saskatchewan, south to California, New Mexico, and Texas; 105 (ix). Three weakly differentiated morphological phases of winterfat are present in Utah. The common and most widely distributed phase (**var. *lanata***) is woody only at the base, has erect annual growth, and is moderately long-hairy as well as stellate. In Washington County and to a lesser extent in Kane County (mainly along Lake Powell) is a definitely woody plant with divaricate branches, which tend to persist as blunt thorns. The pubescence consists of a preponderance of stellate hairs, with few or none of the long slender ones being present. These plants are known as **var. *subspinosa* (Rydb.) J. T. Howell** [*Eurotia subspinosa* Rydb., type from Washington County]. A third phase is present in Grand and San Juan counties, where it grows in sandy parks surrounded by monoliths. The stems are woody for a distance of up to 8 dm or more, but the branching or current annual growth is erect as in var. *lanata* and the pubescence is intermediate between that of var. *subspinosa* and var. *lanata*. This latter plant is **var. *ruinina* Welsh**. This variety, named for Ruin Park (type locality), San Juan County, is a striking phase of the species, with individuals in the populations exceeding 12 dm in height. All phases are considered as valuable browse plants for livestock, especially for sheep.

Chenopodium L.

Annual herbs, glabrous, pubescent, glandular, or farinose (mealy); leaves alternate, flat, entire, toothed, or lobed; flowers perfect or some pistillate only, ebracteate, usually in cymes, variously arranged in spicate or paniculate inflorescences; calyx segments usually 4 or 5, persistent, flat or keeled, more or less covering the fruit, rarely becoming fleshy; stamens commonly 5; styles 2 (3); seeds lenticular, horizontal or vertical. Note: The genus is notoriously complex for several reasons. The floral features are greatly reduced and diagnostic characteristics are often based on either vegetative structures or on minutae of calyx, pericarp, and seed coat, which are often subject to interpretation and might be demonstrated ultimately as trivial. Nomenclature is tangled both within the native and introduced entities, leading to taxonomic treatments that do not satisfactorily circumscribe the taxa as represented by actual specimens. Further, there is variability within the diagnostic features leading to contradictory statements in taxonomic treatments; e.g., with regard to such characters as adherent versus non-adherent pericarps. Thus, the treatment presented below attempts to provide names for the taxa recognizeable in Utah based on examination of actual specimens. The entities seem to be real, but the names might be misapplied in some cases.

Bassett, I. J. and C. W. Crompton. 1982. The genus *Chenopodium* in Canada. Canad. J. Bot. 60: 586–610.
Wahl, H. A. 1954. A preliminary study of the genus *Chenopodium* in North America. Bartonia 27: 1–46.

1.	Plants with yellow glands or glandular hairs, not farinose, aromatic	2
—	Plants glabrous or farinose, eglandular, not aromatic	3
2(1).	Flowers solitary in small cymes, these spreading-recurved along the axis of an elongate panicle; plants common	*C. botrys*
—	Flowers in small sessile clusters, borne in bracteate panicles; plants rare	*C. ambrosioides*
3(1).	Seeds all, or at least some, vertical in the flowers (except sometimes in *C. glaucum*, keyed both ways)	4
—	Seeds usually all horizontal	6
4(3).	Leaves mainly 0.5–2 cm long, 2–7 mm wide, irregularly dentate, glaucous-farinose beneath	*C. glaucum*
—	Leaves mainly larger and often hastately lobed, green or reddish beneath	5
5(4).	Flowers in elongate axillary clusters, these forming erect or steeply ascending compact panicles; plants commonly palustrine	*C. rubrum*
—	Flowers in subglobose axillary clusters, these forming bracteate spikes; plants usually montane	*C. capitatum*
6(3).	Leaves 0.5–2 cm long, 2–7 mm wide, sinuate-dentate, glaucous-farinose beneath	*C. glaucum*
—	Leaves various, but not simultaneously as above	7
7(6).	Larger cauline leaves more or less cordate to truncate basally, often over 4 cm wide, glabrous; sepals not keeled dorsally; panicles large and open	*C. hybridum*
—	Larger cauline leaves with bases various, but seldom as above, and less than 4 cm wide, often farinose beneath; sepals usually keeled	8
8(7).	Larger cauline leaf blades mainly 3–5 or more times longer than wide, entire or with a pair of basal lobes	9
—	Larger cauline leaf blades mainly 1–3 times longer than broad, hastately lobed, toothed, or entire	10
9(8).	Leaves 1-veined, linear, entire; pericarp adherent to the seed	*C. leptophyllum*
—	Leaves 3-veined (at least near the base in larger ones), entire or with 2 basal lobes; pericarp not adherent to the seed	*C. dessicatum*
10(8).	Leaf blades hastately lobed (the lobe sometimes again lobed or toothed), or oval to elliptic and entire, or rarely with one or more teeth on the apical larger lobe; calyx lobes obscurely or narrowly membranous-margined; plants indigenous	11
—	Leaf blades sinuate dentate, ovate to lanceolate, or entire; calyx lobes with broad scarious margins; plants adventive	12
11(10).	Leaf blades not hastately lobed; pericarp usually adherent; plants usually of upper middle to higher elevations	*C. atrovirens*
—	Leaf blades often hastately lobed; pericarp not adherent; plants of wide altitudinal range	*C. fremontii*
12(10).	Fruits sharply angled on the margin; seed coat with minute rounded pits; plants uncommon in Washington and Sevier counties	*C. murale*

— Fruits rounded to obtuse on the margin; seed coat smooth to sculptured; plants common or uncommon, of various distribution .... *C. album*

Chenopodium album L. Lambsquarter; Pigweed. Erect annual herbs, the stems red-striate, 1–10 dm tall or more, simple or more commonly branched; herbage more or less farinose, at least when young; leaves petiolate, the blades 1–6.5 cm long, 0.5–5.6 cm wide, ovate to rhombic-ovate or lanceolate, sinuate-dentate and often subhastately lobed or the upper (rarely all) entire; flowers in dense glomerules, these spicate in upper axils; calyx with keeled lobes, enclosing the fruit; pericarp adherent; seeds horizontal, rounded marginally, smooth to sculptured, black, 1–1.5 mm wide; 2n = 18, 36, 54, 108. Two phases are present, which have been given taxonomic recognition.

1. Seeds sculptured, alveolate-reticulate or reticulate .... *C. album* var. *berlandieri*
— Seeds smooth or faintly striate .... *C. album* var. *album*

Var. album n = 26. Weedy plants of disturbed habitats at 850 to 2265 m, probably in all Utah counties; widespread in North America; adventive from Eurasia; 53 (v). This taxon has been confused with *C. fremontii*, q.v., but in those plants having mature fruits the adherent pericarps are diagnostic.

Var. berlandieri (Moq.) Mack. & Bush [*C. berlandieri* Moq.; *C. berlandieri* ssp. *zschackei* (Murray) Zabel; *C. zschackei* Murray]. n = 18. Weedy or pioneer plants of disturbed substrates in several plant communities at 1280 to 2585 m in Duchesne, Grand, Juab, Millard, San Juan, Sevier, and Utah counties; widespread in North America; 8 (v). The specimen from Utah county (Welsh 3798 BRY) has the fruit not adherent.

Chenopodium ambrosioides L. Mexican-tea. Aromatic annual herbs; stems erect or ascending, mainly 4–10 dm tall; herbage pubescent and with sessile glands; leaves short-petioled, the blades 2–10 cm long, 3–30 mm wide (or more?), usually lanceolate, dentate to laciniate; inflorescence paniculate, the cymes sessile on ultimate branches, usually bracteate; calyx united to the middle or above; seeds 0.5–0.8 mm wide; 2n = 32, 36, 48?, 64. Ruderal and garden weeds of Utah and Washington counties; widespread in tropical and temperate New World; adventive from Mexico; 2 (0).

Chenopodium atrovirens Rydb. Mountain Goosefoot. [*C. fremontii* var. *atrovirens* (Rydb.) Fosberg; *C. hians* Standley; *C. incognitum* Wahl]. Plants mainly 2–75 cm tall, the stems erect or steeply ascending, usually branched; herbage sparingly scurfy to glabrous; leaves petiolate, the blades 0.6–4 cm long, 2–23 mm wide, lanceolate to ovate, entire or obscurely hastately lobed, otherwise entire, obtuse basally; flowers clustered in leaf axils or in interrupted terminal spikes, the lower ones subtended by foliose bracts, becoming ebracteate upwards; perianth lobes free to below the middle, keeled dorsally; pericarp adherent or less commonly not adherent to the horizontal rugulose to smooth, obtusely margined seed; seeds ca 0.9–1.3 (1.5) mm wide; 2n = 18. Sagebrush, pinyon-juniper, mountain brush, ponderosa pine, Douglas fir, aspen-tall forb, and spruce-fir communities at 1705 to 3175 m in Beaver, Carbon, Daggett, Duchesne, Emery, Garfield, Grand, Rich, Salt Lake, Sanpete, Sevier, Summit, Utah, and Wasatch counties; British Columbia to Saskatchewan, south to California, Nevada, Colorado, and Iowa; 24 (ii). Relationship of this species probably lies closer to the *leptophyllum* end of the spectrum. The occasional specimens with free pericarps might indicate intermediacy with *C. fremontii*, however. There have been little agreement between previous authors as to whether the pericarp was adherent or not, but in the specimens examined from Utah, the pericarps are usually adherent.

Chenopodium botrys L. Jerusalem-oak. Aromatic annual herbs; stems commonly 1–5 dm tall, erect or ascending, usually branched; herbage glandular-villous; leaves petiolate, the blades sinuate-pinnatifid, the lobes again toothed or lobed, oblong to oval in outline; inflorescence an erect panicle of loosely spreading-recurved cymes, mainly shortly bracteate; sepals ca 1 mm long; seeds horizontal or vertical, 0.5–0.8 mm wide, dull, dark; 2n = 18, 36. Widespread ruderal weedy species, established locally in indigenous communities, especially in gravelly washes at 760 to 1985 m in Beaver, Box Elder, Cache, western Garfield, Iron, Juab, Millard, Salt Lake, Sanpete, Sevier, Tooele, Utah, and Washington counties; widely distributed in the U. S.; adventive from Eurasia; 22 (iii).

Chenopodium capitatum (L.) Asch. Strawberry-spinach. [*Blitum capitatum* L.; *C. overi* Aellen in Fedde?; *C. chenopodioides* authors, not (L.) Aellen]. Plants mainly 1–4 dm tall, the stems erect or decumbent-ascending, simple or more commonly branched from the base; herbage glabrous; leaves petiolate, the blades (1) 1.5–10 cm long, 1–10 cm wide, triangular-hastate to lanceolate, shallowly to deeply toothed or subentire, hastately lobed or the upper entire, acute to obtuse apically, often turning reddish; flowers clustered in axillary capitate spikes, the lower clusters subtended by foliose bracts, the upper ones ebracteate or with reduced bracts; perianth lobes free to below the middle, not mealy, dry and greenish or becoming fleshy and reddish, shorter than the fruit; pericarp adherent to the erect (or less commonly horizontal) seed; fruit ca 1 mm long; 2n = 18. Two varieties are present in Utah.

1. Flower clusters often over 5 mm wide, the calyx becoming red and fleshy at maturity; plants uncommon .... *C. capitatum* var. *capitatum*
— Flower clusters commonly less than 5 mm wide, the calyx not fleshy, though sometimes reddish at maturity; plants common .... *C. capitatum* var. *parvicapitatum*

Var. capitatum 2n = 18. Gravelly soil in lodgepole pine forest, north slope of Uinta Mts. (Welsh & Moore 6696 BRY), Summit County; Alaska to Quebec, south to California, New Mexico, and New England; 1 (i).

Var. parvicapitatum Welsh Mountain brush, ponderosa pine, aspen, and spruce-fir communities at 1860 to 3050 m in Beaver (type from the Tushar Mts.), Box Elder, Cache, Carbon, Daggett, Duchesne, Emery, Garfield, Grand, Iron, Juab, Kane, Millard, Morgan, Piute, Rich, Salt Lake, Sanpete, Summit, Tooele, Uintah, Utah, and Wasatch counties; British Columbia to Saskatchewan, south to California, Nevada, and Colorado; 60 (xiv). This common plant of montane habitats has been identified variously as *C. capitatum* or *C. rubrum* and, more recently, as *C. chenopodioides* (L.) Aellen. The latter plant is a portion of the flora of the Soviet Union and might represent nothing more than phases of *C. rubrum* sensu lato. Certainly the description provided for that entity (Flora USSR 6: 51. 1936) is not of our specimens, and neither is the habitat cited (i.e., "wet solonchaks"). The name given here is for the purpose of providing an unequivocal epithet for this montane western American phase of *C. capitatum*.

Chenopodium dessicatum A. Nels. Desert Goosefoot. [*C. pratericola* Rydb.; *C. petiolare* var. *leptophylloides* Murray; *C. leptophyllum* var. *dessicatum* (A. Nels.) Aellen; *C. leptophyllum* var. *oblongifolium* Wats.; *C. pratericola* var. *oblongifolium* (Wats.) Wahl]. Plants mainly 3–8 dm tall, the stems erect, simple or branched; herbage commonly more or less scurfy; leaves petiolate, the blades mostly 1.3–6 cm long and 2–10 (15) mm wide, linear to narrowly lanceolate or elliptic, entire or less commonly hastately lobed, cuneate basally; flowers clustered in terminal or axillary spicate panicles; perianth lobes free to below the middle, keeled dorsally; pericarp not adherent to the horizontal smooth to rugulose, obtusely margined seed; seeds 0.9–1.2 mm wide; 2n = 18. Shadscale, hopsage, rabbitbrush, tamarix-poplar, sagebrush, and pinyon-juniper communities at 850 to 1925 m in Duchesne, Emery, Garfield, Grand, Kane, Salt Lake, San Juan, Uintah, Utah, Washington, and Wayne counties; Yukon to Manitoba, south to California, New Mexico, and Nebraska; 26 (iii). Material assigned to *C. dessicatum* in a strict sense, as distinct from *C. pratericola* in a strict sense, has perianth lobes covering the mature fruit, as opposed to having the fruit not covered. The distinction is not great. Reduction of this taxon to *C. leptophyllum* (as var. *oblongifolium*) begs the question of a probably nearer relationship to *C. fremontii*, with which it shares non-adherent pericarp and broader, more veined leaves. Furthermore, apparent intermediates between *C. fremontii* and *C. dessicatum* exist. The entire complex is in need of monographic study.

Chenopodium fremontii Wats. Fremont Goosefoot. Plants mainly 1–8 (12) dm tall, the stems erect or ascending, usually branched; herbage more or less scurfy to glabrous; leaves petiolate, the blades 0.6–5 (6) cm long, and about as broad, less commonly 2–3 times longer than broad, triangular-ovate to ovate or lanceolate, commonly hastately lobed, the lobes often again lobed or toothed, otherwise entire or rarely with 1 or few teeth on the main apical lobe, broadly cuneate to subcordate basally; flowers clustered in large terminal and smaller lateral spikes, scurfy; perianth lobes free to below the middle, keeled dorsally; pericarp not adherent to the horizontal, smooth to rugulose, obtusely margined seed; seeds 0.9–1.2 mm wide; n = 9. This closely interrelated complex of forms involves the linear-leaved *C. leptophyllum* (q.v.) at one end of the spectrum and the broad-leaved phases of *C. fremontii* at the other end (and with both *C. atrovirens* and *C. dessicatum*, inter alia, between the extremes). The intervening plants have been regarded as species or some of them have been placed within expanded species concepts at both ends of the series. The course followed herein is a compromise between having one all inclusive species, with numerous varieties, and that of recognition of all of the named entities at specific level. The proposed treatment attempts to represent the major taxa as they occur in Utah; the synonymy might not be properly applied in all cases.

1. Plants mainly less than 25 cm tall, branching from the base, the curved ascending branches subequal to the main stem; leaves more or less white-farinose, at least beneath .... *C. fremontii* var. *incanum*

— Plants 0.5–8 dm tall, variously branched but, if as above, the lateral branches much shorter than the main stem; leaves white-farinose to glabrous ....
.... *C. fremontii* var. *fremontii*

Var. *fremontii* 2n = 18. Blackbrush (and other warm desert shrub), salt desert shrub, sagebrush, mountain brush, pinyon-juniper, aspen, and spruce-fir communities at 850 to 3050 m in Beaver, Carbon, Daggett, Duchesne, Emery, Garfield, Grand, Juab, Piute, Salt Lake, San Juan, Sevier, Summit, Tooele, Uintah, Utah, Washington, Wayne, and Weber counties; British Columbia to Manitoba, south to California and Mexico; 63 (xix). Specimens assigned here are not uniform with regard to leaf shape, plant height, and openness of the inflorescence. More work is indicated.

Var. *incanum* Wats. [*C. incanum* (Wats.) Heller; *C. watsonii* of authors, not A. Nels.?]. 2n = 18. Blackbrush, salt desert shrub, pinyon-juniper, mountain brush, and ponderosa pine communities at 850 to 2350 m in Carbon, Duchesne, Emery, Garfield, Juab, Kane, Piute, Salt Lake, Tooele, Uintah, Wayne, and Washington counties; Nevada to Nebraska, south to Texas and Mexico; 28 (iii). The variety is transitional to the next.

Chenopodium glaucum L. Oakleaf Goosefoot. [*C. salinum* Standley]. Plants mainly 3–30 cm long, the stems prostrate to ascending or erect, usually branched; herbage farinose, especially on lower leaf surfaces; leaves shortly petiolate, the blades 4–25 mm long, 2–10 mm wide, lanceolate to oblong or ovate, coarsely sinuate-dentate; flowers in clusters in numerous short, bracteate or ebracteate, axillary spikes and a terminal spicate panicle; perianth cleft almost to base, not enclosing the fruit; pericarp not adherent to the seed; seeds horizontal or vertical, 0.8–1.3 mm wide, smooth; 2n = 18, 36. Often in saline substrates on lake shores and stream banks, in sedge-rush, tamarix-sedge, rabbitbrush, pinyon-juniper, and aspen to spruce-fir communities at 1220 to 2745 m in Box Elder, Cache, Carbon, Duchesne, Garfield, Grand, Kane, Millard, Morgan, Rich, Salt Lake, Sanpete, Sevier, Tooele, Uintah, Utah, Wasatch, and Wayne counties; widespread in U.S. and Canada; Eurasia; 26 (xiv). Our material is assignable to **var. *salinum* (Standley) J. Boivin**, on the basis of its larger fruits (0.8–1.3 mm, not 0.6–0.9 mm).

Chenopodium hybridum L. Mapleleaf Goosefoot. [*C. gigantospermum* Aellen; *C. hybridum* var. *gigantospermum* (Aellen) Rouleau]. Plants mainly 2–10 dm tall, the stems erect, simple or branched; herbage glabrous, except in inflorescence; leaves alternate or the lower often opposite, long-petiolate, the blades commonly 1.7–10 cm long, 1.2–10 cm wide, ovate to deltoid ovate, sinuate-dentate to -lobate, with 2–4 teeth or lobes, cordate to truncate or obtuse basally; flowers in small cymes, these arranged in large terminal and smaller axillary panicles, more or less farinose and often sparingly glandular; perianth cleft nearly to the base, not strongly keeled dorsally; pericarp not or moderately adherent; seeds 1.2–1.9 mm wide, with obtuse margin, smooth or somewhat sculptured; 2n = 18, 36. Sagebrush, pinyon-juniper, mountain brush, ponderosa pine, and aspen communities, less commonly in riparian or palustrine habitats, at 1280 to 2135 m in Beaver, Cache, Daggett, Millard, Rich, Salt Lake, Sanpete, Summit, Tooele, Utah, and Weber counties; widely distributed in U. S. and Canada; Europe; 12 (iv). Our specimens have been identified as **var. *gigan-***

tospermum (Aellen) Rouleau, on the basis of the seed being less sculptured than in typical European material. Plants of this species formed huge stands in burned over sagebrush and aspen communities in Summit County in 1963.

Chenopodium leptophyllum (Moq.) Wats. Narrowleaf Goosefoot. [*C. album* var. *leptophyllum* Moq.]. Plants mainly 12–70 cm tall, erect or the branches ascending, simple or branched; leaves short-petiolate, the blades mainly 0.7–4 cm long, 1–5 (7) mm wide, linear to narrowly oblong or narrowly lanceolate, 1-veined, cuneate to acute basally, entire; flowers in loose to compact cymes aggregated into terminal or axillary spicate panicles; perianth lobes cleft to well below the middle, keeled dorsally; pericarp adherent; seeds horizontal, 0.9–1.1 mm wide, black, finely rugulose to smooth; $2n = 18$. Shadscale, greasewood, rabbitbrush, tamarix, sagebrush, fringed sagebrush, mountain brush, and aspen communities at 1125 to 2900 m in Cache, Carbon, Daggett, Duchesne, Emery, Garfield, Grand, Kane, Millard, Piute, Salt Lake, San Juan, Sevier, Summit, Tooele, Uintah, Washington, and Wayne counties; British Columbia to Saskatchewan, south to California and Mexico; 34 (xiii). The 1-veined, narrow leaf blades and adherent pericarps are apparently definitive for this plant.

Chenopodium murale L. Nettleleaf Goosefoot. Plants mainly 2–5 dm tall, the stems erect or with branches ascending; herbage glabrous or sparingly farinose, especially in inflorescences; leaves petiolate, the blades 1–5 (7) cm long and as broad or nearly so, ovate to oval or lanceolate, irregularly sinuate-dentate and some often subhastate, cuneate to subcordate basally; flowers sessile and solitary to clustered in axillary or terminal panicles, not much, if at all, surpassing the leaves; perianth lobes free to below the middle, keeled dorsally; pericarp adherent to the horizontal, rugulose to smooth, sharply margined seed; seeds 1–1.5 mm long; $2n = 18$. Ruderal weeds at 730 to 1620 m in Sevier, Wasatch, and Washington counties; widespread in U. S. and Canada; adventive from Eurasia; 5 (0).

Chenopodium rubrum L. Red Goosefoot. Plants mainly 0.5–10 dm tall, erect, simple or with steeply ascending branches; herbage glabrous or somewhat villous in inflorescence; leaves petiolate, the blades 0.7–9 cm long, 0.4–7 cm wide, trianglular to ovate, lanceolate, or elliptic, sinuate-dentate or -lobed, some often subhastate, fleshy and often suffused with red; flowers sessile, the clusters borne in simple or branched axillary and terminal spicate panicles; perianth lobes cleft to the middle or below, rounded or sometimes keeled dorsally; pericarp not adherent; seeds nearly always vertical, oval in outline, 0.7–1 mm long; $2n = 18, 36$. Saline moist substrates in palustrine and riparian habitats at 1280 to 2440 m in Box Elder, Cache, Daggett, Davis, Duchesne, Emery, Juab, Rich, Salt Lake, San Juan, Sevier, Uintah, Utah, and Weber counties; widespread in U. S. and Canada; Eurasia; 18 (i).

Corispermum L.

Annual herbs, often pubescent with stellate hairs; leaves alternate, sessile, entire, 1-veined; flowers perfect, solitary or clustered in bract axils, arranged in dense or lax spikes; perianth segments 1–3, minute, unequal, the posterior 1 largest, erect, 1-nerved, scarious; stamens 1–3 (5); stigmas 2, connate basally; achenes strongly flattened, plano-convex, indurate, the margin winged or acute.

Corispermum villosum Rydb. Bugseed; Tickseed. [*C. hyssopifolium* authors, not L.; *C. nitidum* authors, not Kit. in Schultes; *C. imbricatum* A. Nels.; *C. emarginatum* Rydb.; *C. marginale* Rydb.; *C. simplicissimum* Lunnell?]. Stems mainly 8–50 cm tall, commonly branched throughout, the lower branches often curved-ascending to ascending, usually reddish; herbage glabrous or sparsely to densely pubescent with soft, branched hairs; leaves alternate, sessile, semicylindric and more or less involute or subulate, mostly 0.8–6 cm long, 0.5–3 mm wide, apiculate; inflorescence slender and elongate or compact, the bracts broadly scarious-margined, with the lower ones narrower or broader than the fruit, the upper ones usually broader than the fruit; perianth mostly consisting of a single posterior erose segment (rarely with 2 additional small anterior-lateral segments); stamens 3–5; fruit 2.2–4.1 mm long, 1.6–2.6 mm wide, oval to suborbicular, smooth, glabrous, brownish; wings opaque, stramineus, 1/8–1/5 as wide as the body; $2n = 18$? Usually growing on sandy substrates, in ephedra, four-wing saltbush, rabbitbrush, scurfpea, and pinyon-juniper communities, at 1065 to 1955 m in Duchesne, Emery, Garfield, Grand, Juab, Kane, Millard, San Juan, Tooele, Uintah, and Wayne counties; Montana and North Dakota, south to Arizona, New Mexico, and Texas; 20 (iv). Our specimens have long been presumed to be introduced and conspecific with either *C. hyssopifolium* L. and/or *C. nitidum* Kit. in Schultes. While the details of those taxa are reported in contradictory fashion in Flora USSR (6: 146–149. 1936) versus the Flora Europaea (1: 99–100. 1964), it seems clear that our plants cannot be *C. hyssopifolium*, which is reported to have flat leaves. The description of our specimens more nearly fits that of *C. nitidum*, but that plant has filiform leaves, not semicylindric ones as in ours. Further, our plants grow in native plant communities and are not weedy; their distribution pattern is that of numerous other indigenous taxa. Plants of this genus were described historically from the edge of civilization and beyond in North America (Nuttall, Genera of North American Plants 1: 4. 1818; Hooker, Flora Boreale Americana 2: 126. 1838; Watson, Proc. Amer. Acad. 9: 123. 1874). While not definitive evidence of nativity these historical data are additive to the weight of evidence for the indigenous nature of the genus in North America. Because of these considerations I have, arbitrarily, chosen the earliest name available at specific level for our plants. Both elongate slender and short broad inflorescences are represented within our specimens, along with occasional intermediates. There is little correlatation of inflorescence type with other morphological features or geography.

Cycloloma Moq.

Annual herbs; stems commonly branched and forming rounded clumps; leaves alternate, sinuate-dentate; herbage more or less pubescent, eglandular; inflorescence a panicle of interrupted spikes; flowers sessile; perianth 5-lobed, the segments in fruit with a transverse wing on the back, the wings connate and completely encircling the fruit; stamens 5; achenes depressed; seeds horizontal.

Cycloloma atriplicifolia (Sprengel) Coult. Winged-pigweed. [*Salsola atriplicifolia* Sprengel]. Plants mainly 0.8–5 dm tall (rarely more), divaricately branched, the stems striate, loosely and sparingly tomentulous, becoming glabrate; leaves short-petioled to sessile, the blades 1–8 cm long, 2–15 mm wide, coarsely serrate-dentate, acute apically, cuneate basally; flowers perfect and pistillate; sepals 5, keeled, the perianth developing into a horizontal wing; wings white-hyaline, lobed or toothed, 4–5 mm in diameter, often red or purple at maturity; ovary tomentulose; styles 2 or 3; fruit enclosed in calyx; seed ca 1.5 mm wide, black, smooth; pericarp not adherent; 2n = 36. Sandy habitats in blackbrush, mixed desert shrub, and juniper communities at ca 1125 to 1465 m in Emery, Garfield, Grand, Kane, San Juan, Uintah, and Wayne counties; Manitoba and Indiana, south to Arizona and Texas; adventive in Europe; 11 (iii).

Grayia H. & A.

Monoecious shrubs or subshrubs; branches more or less thorny; axillary buds subglobose, prominent; leaves alternate, entire; herbage pubescent with simple and stellate hairs; flowers in terminal and axillary spicate panicles, imperfect; staminate flowers 2–5 in clusters in bract axils, not separately bracteolate, the perianth 4– or 5–lobed, subequal to the 4 or 5 stamens; pistillate flowers 1–several per bract, often some vestigial, each enclosed by 2 connate bracteoles, the more or less accrescent bracteoles obcompressed, the margins thickened and spongy within; fruits vertical; stigmas 2.

Grayia spinosa (Hook.) Moq. Hopsage. [*Chenopodium* (?) *spinosum* Hook.; *G. polygaloides* H. & A.]. Shrubs, mainly 5–12 (15) dm tall; branches gray brown; branchlets often spinose-persistent, the bark exfoliating in long strips, pubescent with scurfy and stellate hairs when young; leaves 5–30 mm long (or more), mainly 2–12 mm wide, spatulate to oblanceolate, entire, tapering to a short petiole; staminate flowers with usually 4–lobed perianth, enclosing the 4 stamens, 1.5–2 mm long; pistillate flowers in short spicate inflorescences, the subtending bracts reduced, enclosed by paired, accrescent, obcompressed bracts, orbicular or cordate, the wings thickened and spongy within, 6–15 mm wide, greenish, straw colored, or suffused with red; 2n = 36. Blackbrush, other warm desert shrub, shadscale, horsebrush, rabbitbrush, sagebrush, and pinyon-juniper communities at 760 to 2900 m in Beaver, Box Elder, Daggett, Davis, Duchesne, Garfield, Grand, Juab, Kane, Millard, Piute, San Juan, Sevier, Tooele, Uintah, Utah, Washington, and Weber counties; Washington to Montana, south to California, Arizona, and New Mexico; 91 (xvii). This is a valuable browse plant for livestock, especially for sheep. Locally it is called "applebush" because of its palatability. For consideration of relationship to *G. brandegei* Gray see discussion under *Zuckia*.

Halogeton C. A. Mey.

Annual herbs; leaves alternate, fleshy and sausagelike, bearing an apical slender spine; flowers perfect or partially pistillate, usually bracteolate; perianth of 5 segments, free nearly to the base, embedded in white hair, the segments gibbous, winged in fruit; stamens 2–5, connate basally into a glandular, hypogynous disk; stigmas 2; seeds vertical, laterally flattened, adherent to the pericarp.

Halogeton glomeratus (Bieb.) C. A. Mey. Halogeton. [*Anabasis glomerata* Bieb.]. Plants mainly 3–45 cm tall and as broad, glaucous, usually branched from the base, with curved-ascending branches; leaves mainly 3–15 mm long and 1–2 mm thick, terete, dilated and semiamplexicaul basally, obtuse and terminating in a deciduous slender spine ca 1–1.5 mm long, bearing a tuft of hairs and fasicled leaves in the axils; bracteoles ovate; perianth segments membranous, ovate to oblong, 1–veined, with lustrous, membranous, fanlike, veiny wings 2–3 mm long and 3–4 mm wide; stamens united into 2 clusters of 2 or 3, with 1 anther per cluster; fruit oval to obovate, 1.2–1.8 mm long, with an erect cusp on one or both sides; 2n = 18. Mainly in disturbed sites in cheatgrass, Russian thistle, salt grass, mixed desert shrub, salt desert shrub, and pinyon-juniper communities at 1220 to 1985 m in most, if not all, Utah counties; widely distributed in western U. S.; adventive from Eurasia; 47 (x). This plant was introduced into northern Nevada in the early 1930's (first collected in 1934), possibly for use in grazing experiments. It spread quickly into the desert lands of Nevada and western Utah, and subsequently into eastern Utah and other states. The plant is rich in oxalates and poses a serious threat to grazing animals, especially to sheep, which have suffered heavy death loses for several decades.

Kochia Roth

Annual herbs or subshrubs; leaves alternate (or some opposite), linear to narrowly lanceolate, in some fleshy and terete; flowers 1 to several, sessile in axils of foliose bracts, mostly perfect, 5–merous, the perianth lobes enclosing the fruit, keeled and horizontally winged; stamens mostly 5; stigmas 2 or 3; pericarp thin, free from the horizontal, smooth seed.

1. Plants subshrubs, introduced for reclamation . . . . K. prostrata
— Plants annual or perennial herbs, woody if at all only at the base 2
2(1). Plants annual, introduced weeds of disturbed habitats . . . . K. scoparia
— Plants perennial, woody at the base, indigenous in saline habitats . . . . K. americana

Kochia americana Wats. Gray Molly. [*K. americana* var. *vestita* Wats., type from shores of the Great Salt Lake; *K. vestita* (Wats.) Rydb.]. Plants mainly 5–30 cm tall, with erect branches from a woody base; herbage villous-pilose to glabrous; leaves 5–25 mm long, 1–2 mm wide, linear, semiterete and fleshy; flowers solitary or 2–5, sessile in axils of scarcely reduced leaves; inflorescence often more than half the branch length; perianth segments pubescent, at least apically, 1–1.5 mm long, hooded above, somewhat enlarged in fruit, ultimately keeled and with a membranous, striate wing to 2 mm long and 3 mm wide. Greasewood, seepweed, saltbush, saltgrass, matchweed, horsebrush, and pinyon-juniper communities at 1125 to 1985 m in Beaver, Box Elder, Cache, Davis, Emery, Garfield, Grand, Iron, Juab, Millard, Salt Lake, San Juan, Sanpete, Sevier, Tooele, Uintah, Utah, and Wayne counties; Oregon to Montana, south to California, Arizona, and New Mexico; 65 (v).

Kochia prostrata (L.) Schrader [*Salsola prostrata* L.]. Subshrubs, mainly 10–75 cm tall; stems erect or steeply ascending, more or less pubescent with short crinkly hairs often intermixed with longer villous ones; leaves 3–12 mm long, 0.3–0.7 mm wide, linear to filiform, flat; inflorescence spiciform to paniculate; flower clusters interrupted; perianth hairy dorsally, the appendages round, flat, and tuberculate or oblong and winglike (ca 1 mm long and 1.5 mm wide), often suffused with red; seeds oval or orbicular in outline, ca 2 mm wide, brown, smooth. This plant is being tried in reclamation plantings and is to be expected throughout Utah; introduced from Eurasia; 1 (0).

Kochia scoparia (L.) Schrader Summer-cypress. [*Chenopodium scoparium* L.]. Annual herbs, mainly 3–12 (15) dm tall, green, or suffused with red in autumn, simple or branched from the base, villous and often finely lanate to glabrous; leaves 0.8–4.5 (6) cm long, 1–4 mm wide, lanceolate to oblanceolate, elliptic or linear, usually 3- to 5-veined, glabrous or softly pilose below (and above) or glabrous above, generally ciliate, acute; inflorescence spicate, interrupted; fruiting perianth of perfect flowers glabrous dorsally, ciliate, mostly transversely keeled, tubercled or sometimes horizontally winged from middle of the keel; pistillate flowers often lacking a keel; seeds ovate in outline, 1.5–2 mm long; $2n = 18$. Disturbed roadsides, canal banks, field margins, and other waste places in salt marsh, sedge-rush, sagebrush, mountain brush, and pinyon-juniper communities at 850 to 1985 m in probably all Utah counties; widespread in North America; adventive from Eurasia; 38 (ii). The cultivated ornamental phase, with bright red or red orange foliage is **f. trichophylla** (Hort.) **Stapf ex Schinz & Thell.** [*K. trichophylla* Hort. ex Voss]. Some workers have suggested that our material belongs to the Russian species *K. iranica* (Hausskn. & Bornm.) Litv. [*Salsola iranica* Hausskn. & Bornm.], but materials examined fit neither key diagnostic nor other descriptive features of that taxon (see Flora USSR 6: 128–134. 1936].

Monolepis **Schrader**

Annual, polygamo-monoecious herbs; leaves simple, hastately lobed or entire, alternate, mealy to subglabrous, fleshy; flowers unisexual, inconspicuous, borne in axillary clusters; perianth consisting of 1 bractlike scale (rarely 2 or 3, or lacking), not enclosing the fruit; stigmas 2; pericarp reticulately patterned or warty, adherent to the erect seed.

1. Plants rare, dichotomously branched, the ultimate branches filiform; leaves small and inconspicuous, entire; flowers 5 or fewer per axillary cluster; perianth segments 1–3, linear .... *M. pusilla*

— Plants common, not dichotomously branched, the ultimate branches not filiform; leaves conspicuous, commonly at least some hastately lobed .... *M. nuttalliana*

Monolepis nuttalliana (Schultes) Greene Povertyweed. [*Blitum nuttallianum* Schultes in R. & S., based on *B. chenopodioides* Nutt.; *M. chenopodioides* (Nutt.) Moq. in DC.]. Plants mainly 4–30 cm tall, the stems prostrate or ascending to erect, simple or much branched from the base, mealy to subglabrous; leaves 5–50 mm long, the blades 1–15 mm wide, lanceolate to elliptic or oblong, with 1 pair of lateral lobes near the middle, reduced upwards and sometimes entire, the petiole 1–20 mm long; flowers borne in dense, sessile, axillary clusters; perianth segments 1–2 mm long, more or less acute apically; pericarp pitted, usually pale; fruit 0.9–1.5 mm broad; $2n = 18$. Pioneer plant of open sites in blackbrush, shadscale, mat saltbush, sagebrush, pinyon-juniper, mountain brush, ponderosa pine, Douglas fir-limber pine, aspen, and lodgepole pine communities at 850 to 3355 m in probably all Utah counties; Alaska and Yukon to California and New Mexico, east to Manitoba and Missouri; 70 (xiv). The correct name for this plant seems to be *M. chenopodioides*, there being no apparent obstacle to transfer of *Blitum chenopodioides* to *Monolepis*. However, I hesitate to suggest the change until further research corroborates such a necessity.

Monolepis pusilla Torr. ex Wats. Plants clump-forming, 3–20 cm tall and as broad or wider; leaves 4–12 mm long, oblong, entire, short-petioled; flowers 1–5, in sessile clusters; sepals spatulate, obtuse, 1–3; pericarp tuberculate; seeds dull, ca 0.5 mm wide. Salt desert shrub at ca 1500 m in Uintah County (Holmgren et al 1934 BRY); Oregon to Wyoming, south to California and Nevada; 1 (0).

Nitrophila **Wats.**

Perennial rhizomatous herbs from a definite caudex; leaves opposite, linear to oblong, fleshy, clasping, entire; flowers perfect, axillary, 2-bracteolate, solitary or in 3's; calyx chartaceous, with 5 sepals, these keeled, 1-veined; stamens 5, united basally into a perigynous disk; style filiform, with paired stigmas; fruit ovoid, beaked, included in connivent sepals; seeds vertical, lenticular.

Nitrophila occidentalis (Moq.) Wats. Niterwort. [*Banalia occidentalis* Moq. in DC.]. Plants glabrous, mainly 10–30 cm tall; stems with opposite branches, erect or ascending; leaves 0.7–2 cm long, linear, semicylindric, sessile, mucronate; flowers sessile; sepals 2–3 mm long, oblong, stramineus (pinkish when fresh); fruit brown; seed shining, black, ca 1 mm wide. Saline clay substrate in saltbush, pickleweed, alkali saccaton community in western Juab and Millard counties; Oregon to California, Nevada, and Mexico; 1 (i).

Salicornia **L.**

Annual or perennial herbs from taproots or rhizomes; leaves simple, scalelike, opposite, connate, glabrous; flowers perfect, borne sessile in opposite groups of 3, sunken in depressions of thickened, terminal spikes, subtended by scalelike bracts; perianth consisting of 4 connate segments free at the tip around a slitlike opening, enclosing the fruit; stigmas commonly 2; pericarp thin, free from the erect, retrorsely pubescent seed.

1. Plants annual, from slender taproots, the main stems mostly less than 3 mm thick; central flower much above the lateral ones .... *S. europaea*

— Plants perennial from thick rhizomes and fibrous roots, the main stem mostly more than 3 mm thick; central flower not much above the lateral ones .... *S. utahensis*

Salicornia europaea L. Annual Samphire. [*S. rubra* A. Nels.; *S. europaea* ssp. *rubra* (A. Nels.) Breitung; *S. trona* Lunell; *S. europaea* var. *trona* (Lunell) J. Boivin]. Plants annual, mainly 9–30 cm tall, from slender taproots; stems fleshy, erect or ascending, commonly branched, often reddish at maturity; leaves scalelike, often with a scarious margin; spikes 0.5–5 cm long, 2–3 mm thick, the joints 2–4 mm long; central flower much above the lateral ones; fruit dehiscent, the seeds falling separately; $2n = 18, 36$. Saline substrates in salt marsh, seepweed, poverty weed, alkali saccaton, and saltgrass communities at 1280 to 1465 m in Beaver, Box Elder, Cache, Davis, Juab,

Millard, Rich, Salt Lake, Tooele, Uintah, and Utah counties; widespread in North America; Eurasia; 19 (ii). Our specimens are supposed to differ from the more coastal ones in having slender spikes with joints about as thick as long, but no such correlation is apparent. Recognition of our material at infraspecific level seems moot.

Salicornia utahensis Tidestr. Utah Samphire. [*S. pacifica* var. *utahensis* (Tidestr.) Munz]. Plants perennial, mainly 9–30 cm tall, from coarse rhizomes and thick fibrous roots; stems fleshy, erect or ascending, not creeping and rooting at the nodes, turning gray or brown at maturity; leaves scalelike, with scarious margins; spikes 0.6–5 cm long, 3–4 mm thick the joints 3–5 mm long, breaking at the joints when mature, the central flower not especially above the others; fruit dehiscent, the seeds falling separately, sometimes adherent to the calyx. Saline substrates in salt marsh, pickleweed, and saltgrass communities at 1280 to 1405 m in Box Elder, Davis, Millard, Salt Lake, Tooele (type from shore of Great Salt Lake), Utah, and Weber counties; endemic ? (reported from California?); 11 (ii). Our material differs markedly from the similar coastal species assigned variously to either *S. pacifica* Standley or *S. virginica* L.

Salsola L.

Annual herbs; leaves alternate, entire, commonly spinulose; flowers perfect, 5–merous, solitary or clustered in axils of spiny bracts, each with 2 smaller bracteoles, borne in spicate inflorescences; fruiting perianth with winglike, mostly horizontally spreading ridges; stamens 5, usually inserted at the margin of a lobed disk; styles 2 or 3; fruit closely enveloped in the persistent calyx; seeds horizontal to oblique.

Beatley, J. C., 1973. Russian thistle (*Salsola*) species in western United States. J. Range Management 26: 225–226.

1. Leaves mainly 0.3–1.8 mm wide and 1.5–4 (6) cm long, slender; wings of fruiting calyx usually 1–2 mm long ... . S. *iberica*
— Leaves mainly 1–1.5 mm thick and 0.5–2 cm long, thick and rigid; wings of fruiting calyx mainly 3–4 mm long .. .. S. *paulsenii*

Salsola iberica Sennen & Pau Russian-thistle; Tumbleweed. [*S. pestifer* A. Nels.; *S. kali* authors, not L.?]. Plants with simple stems or freely branched, clump-forming, appearing taller than wide, 1–10 dm tall; stems with red purple, longitudinal striations, glabrous or pubescent; leaves mainly 1.5–4 (6) cm long, 0.3–0.8 mm wide, narrowly linear or filiform, spinulose apically, modified upward as spinescent bracts with expanded bases and scarious margins; perianth segments distinct, membranous in anthesis, in fruit becoming transversely winged, the wings mainly 1–2 mm long; seeds horizontal; $2n = 18$, 36. Weedy species of disturbed habitats at 760 to 2440 m in probably all Utah counties; widespread in North America; adventive from Asia; 62 (vi). This species forms intermediates with the next. It is unfortunate that so many specific epithets have been applied to this weedy introduction, and possibly even the above name is not unequivocally correct.

Salsola paulsenii Litv. Barbwire Russian-thistle. Plants commonly freely branched, clump-forming, appearing wider than tall, mainly 10–40 cm tall and usually broader; stems yellowish green, seldom purplish striate, pubescent or glabrous; leaves mainly 0.5–2 cm long, 1–2 mm thick, linear, rigid, spinose apically, modified upwards as spinose bracts with wide bases and scarious margins; perianth segments yellowish, membranous in anthesis, becoming transversely winged, the wings 2–4 mm long in fruit; seeds horizontal. Weedy species of disturbed habitats at 915 to 1830 m in Cache, Carbon, Garfield, Grand, Juab, Kane, Millard, Tooele, Washington, and Wayne counties; Arizona, Nevada, and California; 12 (iii). A third species, *S. collina* Pallas, might be present in Utah. It has long slender spicate inflorescences with appressed bracts and bracteoles, and in the lower portions the flowers are usually gall-like.

Sarcobatus Nees

Thorny shrubs; leaves mostly alternate, linear, fleshy, sessile; flowers imperfect, borne in axillary spikes, the staminate ones spirally arranged, ebracteate, and lacking a perianth; stamens 2 or 3, borne beneath stalked peltate scales; pistillate flowers sessile, 1 or 2, in the axils of scarcely reduced leaflike bracts, the pistil surrounded by a cuplike, shallowly lobed to subentire perianth, this accrescent and adherent to the fruit base, its upper portion flaring to form a broad, winglike border; seeds erect, flattened, orbicular.

Sarcobatus vermiculatus (Hook.) Torr. in Emory Greasewood. [*Batis* (?) *vermiculata* Hook.; *Fremontia vermicularis* (Hook.) Torr. & Frem.]. Shrubs, mainly 10–20 dm tall or more; branches rigid, spreading, often modified as thorns; leaves 0.3–4.5 cm long, 1–3 mm wide, semicylindric, linear; staminate spikes catkinlike, 1–4 cm long; pistillate flowers fewer than the staminate ones, the perianth ca 1 mm long; calyx wing 2–6 mm long; fruit 4–5 mm long, cup-shaped below the wing; seeds brown, ca. 2 mm long; $n = 18$. Greasewood, seepweed, saltbush, and other plant communities of saline substrates, at 1220 to 2170 m in most if not all Utah counties; Washington and Alberta to North Dakota, south to California, Arizona, New Mexico, and Texas; 72 (x). Our plants are referable to **var. vermiculatis**. This is an important browse species for cattle and sheep, even though potentially poisonous due to oxalate salts of sodium and potassium and oxalic acid.

Spinacia L.

Annual or biennial glabrous herbs; flowers borne in dense spicate inflorescences; pistillate flowers lacking a perianth, but with 2 (rarely 3 or 4) persistent bracteoles that enlarge, become connate, and harden in fruit; stigmas 4 or 5; seeds vertical.

Spinacia oleracea L. Spinach. Plants mainly 2–6 dm tall or more, erect; leaves mainly 3–15 cm long and 2–10 cm wide, ovate to triangular, often hastately lobed; bracteoles in fruit orbicular to obovate, often with a spreading spine at the apex. Cultivated potherb throughout Utah, mainly below 2135 m; introduced from Eurasia; 1 (0). This plant escapes occasionally, but does not persist.

Suaeda Forskal ex Scop.

Annual or perennial herbs or shrubs; leaves alternate, entire, terete or flattened, often succulent; flowers inconspicuous, mostly perfect, solitary or clustered in leaf axils, bracteate; calyx 5–lobed, fleshy, the lobes equal and unappendaged or unequal and some more or less corniculately appendaged; stamens 5, the filaments short; ovary subglobose or depressed; seeds horizontal or vertical.

1. Plants suffrutescent or definitely shrubby; leaves abruptly short-petiolate; calyx lobes equal, not appendaged, smooth dorsally; herbage glabrous or puberulent; seeds vertical or horizontal ...... *S. torreyana*
— Plants annual; leaves sessile; calyx lobes unequal in fruit, horned; herbage glabrous; seeds horizontal 2
2(1). Plants often over 3 dm tall, erect, not clump-forming, the branches stiffly erect-ascending; flowers mostly 3–7 per axil .... *S. calceoliformis*
— Plants mainly 0.5–3 dm tall, forming depressed rounded clumps, the branches spreading, more or less flexuous; flowers 1–3 per axil .... *S. occidentalis*

Suaeda calceoliformis (Hook.) Moq. Broom Seepweed. [*Chenopodium calceoliforme* Hook.; *S. depressa* authors, not (Pursh) Wats.; *S. depressa* var. *erecta* Wats.]. Plants glabrous, often glaucous, erect, simple or with erect-ascending branches and broomlike, 1–5 (8) dm tall; leaves mainly 1–4 cm long, 1–2 mm wide, linear or tapering from base to apex, semiterete, intergrading with floral bracts upwards; spikes slender; flowers sessile, mostly in clusters of 3–7; calyx lobes unequal, ca 1.5 mm long, at least some conspicuously horned; fruit horizontal; seeds smooth, 1–1.5 mm wide, dark brown; 2n = 54. Saline palustrine or riparian sites in saltgrass, greasewood, seepweed, alkali saccaton, and cattail-sedge communities at 850 to 2440 m in Box Elder, Cache, Davis, Duchesne, Emery, Grand, Millard, Rich, Salt Lake, Sanpete, Sevier, Uintah, Utah, Washington, and Weber counties; British Columbia to Saskatchewan, south to California, Arizona, New Mexico, and Texas; 35 (vi).

Suaeda occidentalis (Wats.) Wats. Western Seepweed. [*Schoberia occidentalis* Wats.]. Plants glabrous, often glaucous, forming depressed rounded clumps, not broomlike, mainly 0.5–3 dm tall, seldom simple; leaves mostly 0.5–2 (3) cm long, linear-oblong, semicylindric, intergrading with floral bracts upwards; spikes slender; flowers sessile, mostly 1–3 per cluster; calyx lobes unequal, ca 1.5 mm long, at least some conspicuously horned; fruit horizontal; seeds smooth 1–1.5 mm wide, dark brown. Saline palustrine or riparian habitats in greasewood, saltgrass, seepweed, and other such communities at 1280 to 2135 m in Beaver, Box Elder, Cache, Davis, Duchesne, Rich, Salt Lake, Sanpete, Sevier, and Utah counties; Washington to Wyoming, south to Nevada and Colorado; 12 (i).

Suaeda torreyana Wats. Torrey Seepweed. Plants glabrous or pubescent, sometimes glaucous, suffrutescent or definitely shrubby, 1–12 (15) dm tall or more, with slender ascending to spreading branches; leaves 0.5–3.5 cm long, 1–3 mm thick, subterete to flattened, abruptly short-petiolate, intergrading with floral bracts upwards; flowers 1–8 or more per axil; calyx lobes equal, ca 1.5–2 mm long, the lobes merely rounded dorsally, not horned or tuberculate; fruit horizontal or vertical; seeds 0.8–1.2 mm wide, black, shiny; n = 9, 18. Greasewood, seepweed, saltgrass, and other salt desert shrub communities, often in riparian or palustrine habitats at 1125 to 1955 m in Box Elder, Carbon, Cache, Davis, Duchesne, Emery, Garfield, Grand, Juab, Kane, Millard, Salt Lake, San Juan, Sevier, Tooele, Uintah, Utah, and Wayne counties; California, Nevada, Wyoming, Arizona, and Mexico; 79 (xiv). Materials cited with the above description belong to var. *torreyana* [*Chenopodina linearis* Torr., type from Great Salt Lake; *C. nigra* authors, not (Raf.) Macbr.] A second variety, sometimes treated at specific rank, is based on densely puberulent plants. It is var. **ramosissima** (Standley) Munz [*Dondia ramosissima* Standley; *S. intermedia* Wats., type from Sevier or Washington counties]. Saltgrass, shadscale, greasewood and other salt desert shrub communities at 850 to 1400 m in Juab, Millard, Salt Lake, Sevier, and Washington counties; Arizona and California; 11 (0). Phases of this variety from Washington County seem to have more steeply ascending and more elongate branches than those from elsewhere in Utah. Possibly all of the variants are taxonomically neglegible.

Zuckia Standley

Dioecious or less commonly monoecious shrubs or subshrubs; branches not thorny; axillary buds subglobose, prominent; leaves alternate, entire or more or less lobed; herbage more or less scurfy; staminate flowers 2–5 in clusters in bract axils, not separately bracteolate, the perianth 4– or 5–lobed, subequal to the 4 or 5 stamens; pistillate flowers 1 to several per bract, often some vestigial, each enclosed by 2 bracteoles, these dorsiventrally flattened and unequally 6–keeled or obcompressed and thin-margined, often subtended by a single filiform bractlet; fruits vertical or horizontal; stigmas 2.

Zuckia brandegei (Gray) Welsh & Stutz Siltbush. [*Grayia brandegei* Gray]. Plants mainly 1–5 dm tall; branching from a persistent woody base ca 0.5–2 dm tall, the annual stems erect or ascending; herbage more or less scurfy and less commonly with some moniliform hairs in the inflorescence; leaves subsessile or tapering to a short petiole, 13–80 mm long, 15–42 mm wide, linear or narrowly oblanceolate-spatulate to elliptic, ovate, obovate, or orbicular, entire or rarely hastately lobed; staminate flowers with a 4– or 5–lobed stramineous perianth, cleft to the middle or below, ca 1.5–1.8 mm long; pistillate bracts obcompressed or dorsiventrally compressed, with vertical or horizontal fruits respectively, but not exclusively, when mature either flattened and 4–8 mm wide or 6–keeled (with 4 small and 2 large keels) and 2–4 (5) mm wide; fruits included within the bracts; 2n = 36. It has long been recognized that *Grayia brandegei* was not a close congener of *G. spinosa*, type species of the genus *Grayia*, even though they shared features of bract compression and rounded axillary buds. They differ markedly in gross morphology, in vesture, and in nature of the bracts. The bracts of *G. spinosa* are thickened marginally, and filled internally with a spongy cellular matrix. Those of *brandegei* are thin margined, and do not possess a spongy cellular matrix. Some workers have suggested that *G. brandegei* should be removed from its long association with *G. spinosa* and placed within the *gardneri* phase of *Atriplex*. The differences between *Z. brandegei* and any of the *Atriplex* species are striking even though the plants in vegetative condition might be confused. Bract and bud differences in *Zuckia* are correlated, apparently, with C-3 type of photosynthesis and its attendent foliar morphology, while the shrubby atriplexes have the C-4 type of photosynthesis with its foliar morphology. Two more or less distinctive but intergrading phases are present.

1. Fruit dorsiventrally compressed; bracts 6–keeled; fruit mostly horizontal .... *Z. brandegei* var. *arizonica*

— Fruit obcompressed; bracts 2-winged, samaralike; fruit mostly vertical .... *Z. brandegei* var. *brandegei*

Var. *arizonica* (Standley) Welsh [*Z. arizonica* Standley]. Fine-textured or sandy, often saline and seliniferous substrates on Entrada, Morrison, and Duchesne River formations in Emery, Uintah, and Wayne counties; Arizona; 10 (i). Despite the distinctive bracts and horizontal fruits, the plants are well within the range of vegetative variation existing in plants of *Z. brandegei* in a strict sense. Staminate plants of this variety have not previously been discerned from among the general collections of *Z. brandegei*. Previous descriptions of *Zuckia* and its only species (*arizonica*) are conspicuous in lacking discussion of staminate features. Supposed differences in vesture between the taxa apparently do not exist, and other diagnostic morphology is unknown.

Var. *brandegei* Fine-textured, often saline and seleniferous substrates on the Duchesne R., Uinta, Kaiparowits, Summerville, Morrison, Chinle, and Moenkopi formations (and probably others) at 1280 to 2440 m in Daggett, Emery, Garfield, Grand, Kane, San Juan (? type from San Juan River), Sevier, Uintah, and Wayne counties; Colorado and Arizona; 66 (xii). Both broad and narrow-leaved phases of both of the varieties are known, and there is a tendency for the wideness of leaves to be geographically correlated (i.e., the broad-leaved populations from the Uinta Basin and those on "The Blues" of the Kaiparowits Formation northeast of Henrieville and north of Fourmile Bench), but the plants differ in no other discernible way. Probably these variants do not warrant taxonomic recognition.

COMPOSITAE Giseke

Sunflower Family

Annual, biennial, or perennial herbs, or shrubs; leaves alternate, opposite, or whorled, simple, pinnatifid, or compound; inflorescence of involucrate heads, these solitary or several in corymbose, racemose, paniculate, or cymose clusters; flowers few to numerous on a common receptacle, surrounded by green bracts forming a cup-shaped, cylindrical, or urn-shaped involucre enclosing the flowers in bud; heads entirely of tubular (disk) corollas, entirely of ligulate (ray) corollas, or with tubular corollas forming a central disk and an outer radiating row of ligulate corollas; receptacle flat, convex, conic, or cylindric, naked or bearing chaffy bracts, scales, or hairs; calyx lacking, or crowning the summit of the ovary and modified as a pappus of capillary bristles, scales, or awns; stamens alternate with corolla lobes; filaments free (rarely connate); the anthers united and forming a tube (rarely separate); ovary inferior, of 2 carpels, 1-loculed and with a single ovule; styles 1, 2-cleft, exserted through the anther tube; fruit an achene; $x = 2-19+$ [Asteraceae Dumort.]. Note: All involucral measurements are from dried pressed herbarium specimens. The width measurements are sometimes broader than in fresh material.

Welsh, S. L. 1983. Utah flora: Compositae (Asteraceae). Great Basin Naturalist 43: 179–357.

1. Corollas all raylike; plants usually with milky juice . Key 1
— Corollas not all raylike, some or all of them tubular; juice seldom if ever milky 2

2(1). Corollas all tubular; no ray flowers present, or the rays vestigial and minute .... Key 2
— Corollas not all tubular; ray flowers present 3
3(2). Pappus of capillary bristles, at least in part .... Key 3
— Pappus of awns or scales, or lacking 4
4(3). Pappus lacking .... Key 4
— Pappus present, of awns or scales .... Key 5

Key 1.

Corollas all raylike; plants usually with milky juice.

1. Pappus lacking 2
— Pappus present 3
2(1). Rays 10–20 mm long; plants glabrous, with leaves in a basal rosette .... *Atrichoseris*
— Rays 5–7 mm long; plants pubescent, with well-developed cauline leaves .... *Lapsana*
3(1). Pappus, at least in part, of plumose bristles 4
— Pappus of simple bristles, awns, or scales 7
4(3). Plants acaulescent, with merely bracteate stems ... . *Hypochaeris*
— Plants caulescent 5
5(4). Achenes not beaked, truncate at apex; involucres usually less than 15 mm long .... *Stephanomeria*
— Achenes tapering or beaked at apex; involucres usually more than 15 mm long 6
6(5). Leaves pinnatifid; corollas white or pinkish; involucre with an outer series of short bractlets; plants of southern Utah .... *Rafinesquia*
— Leaves not pinnatifid, entire; corollas yellow or purplish; involucre lacking short outer bractlets; plants widespread .... *Tragopogon*
7(3). Pappus of 1–3 series of unawned or awned scales 8
— Pappus of capillary bristles 9
8(7). Pappus of 2 or 3 series of unawned scales; corollas blue, closing by midmorning .... *Cichorium*
— Pappus scales in a single series, awned; corollas yellow, not closing by midmorning .... *Microseris*
9(7). Achenes more or less flattened; stems leafy; heads in panicles or in umbellate clusters 10
— Achenes not flattened; stems leafy or scapose; heads solitary or variously disposed 11
10(9). Involucres cylindric or ovoid-cylindric; achenes beaked; flowers yellow or blue .... *Lactuca*
— Involucres broadly campanulate to hemispheric; achenes not beaked; flowers yellow .... *Sonchus*
11(9). Corollas pink or purplish 12
— Corollas yellow or yellowish, or white or cream colored 14
12(11). Plants annual; heads mainly 5–7 mm long (from base of involucre to tip of pappus) .... *Prenanthella*
— Plants perennial; heads mainly 8–20 mm long or more 13
13(12). Plants with rigid spine-tipped branches . *Stephanomeria*
— Plants unarmed, the branches soft .... *Lygodesmia*
14(11). Leaves all basal; heads solitary on scapose peduncles 15
— Leaves not all basal, the stems leafy; heads not on scapose peduncles 17

15(14). Achenes not beaked, truncate; pappus bristles barbellate .... *Microseris*
— Achenes beaked or tapering to apex; pappus not of barbellate bristles 16

16(15). Achenes 10–ribbed or 10–nerved, not spinulose; involucral bracts usually imbricated in several series .... *Agoseris*
— Achenes 4– to 5–ribbed, spinulose, especially near apex; principal bracts in a single series, the outer much shorter .... *Taraxacum*

17(14). Achenes ridged or tuberculate between the angles; leaves either crustaceous-margined or peduncles stipitate-glandular; plants of southwestern Utah 18
— Achenes striate between the angles; leaves and peduncles otherwise (rarely glandular setose in some *Crepis* species); plants widely distributed 19

18(17). Plants depressed annuals with crustaceous-margined leaves, not stipitate-glandular; achenes abruptly beaked, transversely ridged between the ribs .... *Glyptopleura*
— Plants erect, lacking crustaceous-margined leaves, conspicuously stipitate-glandular above; achenes tapering to a beak, not transversely ridged .. *Calycoseris*

19(17). Pappus bristles early deciduous, more or less united below and falling together, only a few of the stout outer ones may be persistent .... *Malacothrix*
— Pappus bristles persistent or tardily deciduous, and then falling separately 20

20(19). Pappus tan to brown; involucral bracts not thickened .... *Hieracium*
— Pappus white or whitish; involucral bracts somewhat thickened at base or on midrib .... *Crepis*

Key 2.

Corollas all tubular; no ray flowers present.

1. Heads unisexual, the pistillate heads with 1–4 flowers enclosed in involucre; involucre burlike or nutlike, only style tips exserted 2
— Heads perfect or unisexual; involucre not burlike or nutlike 4

2(1). Involucral bracts of the staminate heads separate; fruiting involucres burlike, covered with hooked appendages .... *Xanthium*
— Involucral bracts of the staminate heads united; fruiting involucres various but, if burlike, lacking hooked appendages 3

3(2). Shrubs; fruiting involucre with several transverse, scarious wings; leaves or their lobes linear-filiform .... *Hymenoclea*
— Shrubs or herbs; fruiting involucre lacking transverse wings; leaves and their lobes not linear-filiform .... *Ambrosia*

4(1). Stamens not united by their anthers; flowers always unisexual, the pistillate corollas none or much reduced 5
— Stamens with united anthers or rarely not united in some species with perfect flowers, at least some flowers usually perfect 7

5(4). Achenes long-villous; leaves or their lobes linear-filiform .... *Oxytenia*
— Achenes not long-villous; leaves or their lobes not linear-filiform 6

6(5). Pistillate flowers subtended by large, chaffy scales simulating inner involucral bracts; achenes with pectinate or winged margins .... *Dicoria*
— Pistillate flowers subtended by chaffy scales or these lacking; achenes without pectinate or toothed wings .... *Iva*

7(4). Involucral bracts with translucent, usually yellow or orange dots .... *Porophyllum*
— Involucral bracts without distinct dots; pappus various, but not as above 8

8(7). Pappus of capillary bristles, at least in part, these smooth, scabrous, barbellate, or plumose 9
— Pappus lacking or, if present, not of capillary bristles 43

9(8). Leaves opposite or whorled, some or all cauline 10
— Leaves alternate, at least basally, or basal and actually alternate 14

10(9). Plants rushlike, xeric herbs of low elevations in Washington County .... *Bebbia*
— Plants not rushlike or xeric, of various habitats and distribution, but not as above 11

11(10). Corollas yellow; involucral bracts in 1 series or in 2 series, but all equal in length .... *Arnica*
— Corollas white, ochroleucous, flesh colored, blue, or purple; involucral bracts in 2 to several series 12

12(11). Pappus double—the outer series of short scales, the inner series of capillary bristles; shrubs with white bark .... *Hofmeistera*
— Pappus single, or the plants herbaceous; shrubs or herbs 13

13(12). Achenes 5–angled or 5–ribbed; involucral bracts subequal or in 2 series .... *Eupatorium*
— Achenes 10–angled or 10–ribbed; involucral bracts imbricated in several series of different lengths .... *Brickellia*

14(9). Leaves spinescent, usually with spiny teeth or lobes, rarely entire but then with spine-tipped apex, mostly thistlelike 15
— Leaves entire, denticulate or lobed, lacking spines, not thistlelike 19

15(14). Corollas of some or all flowers bilabiate; basal leaf axils woolly; leaves spinulose-dentate; flowers pink; arid sites in Kane and Washington counties .... *Perezia*
— Corollas not bilabiate; leaves not or seldom spinulose-dentate; basal leaf axils woolly; flowers pinkish white or cream; various distribution 16

16(15). Pappus of 2 series of awns, the outer long and naked, the inner short and hispidulous; flowers yellow .. *Cnicus*
— Pappus of plumose or barbellate capillary bristles; flowers not yellow 17

17(16). Pappus bristles plumose (rarely some otherwise); receptacle densely bristly .... *Cirsium*
— Pappus bristles merely barbellate 18

18(17). Receptacle densely bristly, not fleshy or honeycombed; heads nodding .... *Carduus*
— Receptacle not bristly or scarcely so, fleshy and honeycombed; heads not nodding .... *Onopordum*

19(14). Receptacle with dense bristles or narrow, chaffy scales between disk flowers 20
— Receptacle naked or at most short-hairy, never with dense bristles or scales 22

20(19). Involucral bracts with hooked spines; lower leaves large (resembling rhubarb), cordate at base ... *Arctium*
— Involucral bracts without spines, or spines not hooked; lower leaves not large and cordate at base .. 21

21(20). Receptacle chaffy except in center; plants small, woolly .... *Filago*
— Receptacle chaffy throughout; plants not small and woolly .... *Centaurea*

22(19). Heads unisexual; plants dioecious (staminate flowers may have styles but ovary does not develop) 23
— Heads with at least central flowers perfect 25

23(22). Plants shrubs or else woody at base, not tomentose; leaves sometimes toothed or lobed; involucral bracts not strongly scarious margined .... *Baccharis*
— Plants herbaceous, more or less tomentose; leaves entire; involucral bracts strongly scarious, at least along margins 24

24(23). Pappus bristles of pistillate flowers united at base and falling together; pappus bristles of staminate flowers usually club-shaped at apex; plants typically less than 30 cm tall; basal leaves commonly in a rosette; cauline leaves reduced and different in shape; leaves usually tomentose on both sides .... *Antennaria*
— Pappus of pistillate flowers separate at base and falling separately; pappus bristles of staminate flowers not club-shaped at apex; plants mostly over 30 cm tall; leaves all alike, usually green and glabrate above .... *Anaphalis*

25(22). Stems longitudinally brown-striate; involucral bracts imbricate, chartaceous, the inner with scarious margins and broadly rounded apices; shrubs with yellow flowers, of western Millard County .... *Lepidospartum*
— Stems striate or not; involucral bracts scarious, hyaline, or herbaceous but not as above; herbs, or shrubs with flowers and distribution various 26

26(25). Involucral bracts scarious or hyaline (only partly so in *Pluchea*) 27
— Involucral bracts herbaceous, at least in the center .. 29

27(26). Involucral bracts subscarious; corollas purplish; plants not tomentose, slender woody shrubs or annual to perennial herbs .... *Pluchea*
— Involucral bracts scarious; corollas rarely purplish; plants tomentose, prostrate to erect herbs 28

28(27). Plants perennial, subdioecious pistillate heads usually with a few, central, perfect flowers ...... *Anaphalis*
— Plants annual or perennial, not dioecious; heads all alike, the marginal flowers pistillate and central ones perfect .... *Gnaphalium*

29(26). Involucral bracts in a single series, a few very short ones may be present at the very base 30
— Involucral bracts of 2 or more series, these often of different lengths 33

30(29). Plants woody, shrubs; involucral bracts 4–6 per head .... *Tetradymia*
— Plants herbaceous; bracts more than 6 per head 31

31(30). Plants annual; heads with inner flowers perfect, the outer pistillate .... *Conyza*
— Plants perennial; heads with all flowers perfect 32

32(31). Style branches with a tuft of hairs near the truncate apex; involucral bracts in 1 series only (a few short bracts may be present) .... *Senecio*
— Style branches without a tuft of hairs near the truncate apex; involucral bracts actually in 2 or more series .... *Erigeron*

33(29). Pappus double, the outer series of short scales, the inner ones of capillary bristles; shrubs with white bark .... *Hofmeisteria*
— Pappus simple or else the plants herbaceous 34

34(33). Plants annual 35
— Plants perennial or woody shrubs 37

35(34). Plants low, depressed, scurfy pubescent herbs; leaves broadly ovate or roundish, entire or toothed .... *Psathyrotes*
— Plants not as above 36

36(35). Leaves all entire .... *Aster*
— Leaves toothed or lobed, at least the lower .... *Conyza*

37(34). Involucral bracts in more or less distinct vertical rows .... *Chrysothamnus*
— Involucral bracts not in vertical rows 38

38(37). Involucral bracts usually in one subequal series ... . *Erigeron*
— Involucral bracts imbricate, in 2 or more series 39

39(38). Involucral bracts not longitudinally striate; flowers commonly yellow 40
— Involucral bracts longitudinally striate; flowers commonly cream to off-white, or pink to purplish 41

40(39). Plants woody shrubs, evergreen; involucral bracts in 2 series, the outer subterete and subulate; plants of Washington County .... *Peucephyllum*
— Plants herbs or shrubs; involucral bracts in 2 or more series but, if 2, not terete and subulate ... *Haplopappus*

41(39). Flowers pink to purplish; plants of northwestern Utah .... *Eupatorium*
— Flowers cream to white; plants of various distribution 42

42(41). Pappus plumose; plants perennial herbs ...... *Kuhnia*
— Pappus scabrous or hispidulose; plants shrubs or herbs .... *Brickellia*

43(8). Receptacle with bristles or chaffy scales among the flowers 44
— Receptacle naked or merely short-hairy 52

44(43). Receptacle densely bristly .... *Centaurea*
— Receptacle with chaffy scales 45

45(44). Plants low woolly annuals; outer bracts boat-shaped and enclosing the achenes 46
— Plants various, but not low and woolly; outer bracts various but not usually enclosing the achenes 47

46(45). Stem leaves opposite; style lateral .... *Psilocarpus*
— Stem leaves alternate; style terminal .... *Stylocline*

47(45). Involucral bracts in 2 distinct sets—the outer herbaceous, the inner differing in shape and texture; leaves opposite, at least below, or alternate 48
— Involucral bracts not in 2 unlike sets; leaves alternate or basal 49

48(47). Leaves alternate throughout; outer involucral bracts ca 5, spreading, herbaceous, the inner (1–3 subtending pistillate flowers) larger and broader, becoming strongly accrescent and hooded in fruit .... *Dicoria*
— Leaves opposite, at least below; outer involucral bracts various, but not as above, not accrescent and hooded in fruit .... *Thelesperma*

49(47). Involucral bracts in 1 series, boat-shaped, each bract enclosing a marginal flower; rays short, yellow .. *Madia*
— Involucral bracts in 1 or more series, not boat-shaped and enclosing marginal flowers; rays lacking 50

50(49). Plants woody shrubs; mostly along the canyons of the Colorado and Green rivers .... *Encelia*
— Plants herbaceous; widely distributed 51

51(50). Receptacles high-conical, mostly over 3 cm long; stems leafy .... *Rudbeckia*
— Receptacles merely convex, much less than 3 cm long; leaves all basal .... *Enceliopsis*
52(43). Pappus none 53
— Pappus present 56
53(52). Leaves opposite, some cauline, somewhat connate at base; plants of Grand, San Juan, and Tooele counties .... *Flaveria*
— Leaves alternate or basal 54
54(53). Heads numerous, in spikes, racemes, or panicles; anthers with acute tips; receptacles flat; plants woody or herbaceous .... *Artemisia*
— Heads solitary on ends of stems, or sometimes corymbose or capitate; anthers with rounded tips; receptacles convex or conic; plants herbaceous, or woody only at base 55
55(54). Plants annual; heads solitary or paniculately arranged; leaves green and glabrous .... *Chamomilla*
— Plants perennial; heads corymbose or capitate; leaves usually silvery-canescent .... *Chrysanthemum*
56(52). Plants dioecious shrubs .... *Baccharis*
— Plants not dioecious herbs or shrubs 57
57(56). Pappus of 2–8 caducous awns; plants usually strongly glutinous .... *Grindelia*
— Pappus various, but not of 2–8 caducous awns 58
58(57). Leaves and involucre conspicuously punctate with translucent oil glands .... *Dyssodia*
— Leaves and involucre sometimes impressed-punctate, but without translucent oil glands 59
59(58). Pappus of 12 or more scale or bristlelike segments, these nearly or quite as long as achene 60
— Pappus of fewer than 12 scalelike segments or else much shorter than achene 61
60(59). Pappus of 12–16 linear, acuminate awns; involucres glutinous; leaves 3- to 5-nerved .... *Vanclevea*
— Pappus of ca 35 flattened, silvery scales and bristles of different widths; involucres not glutinous; leaves 1-nerved .... *Acamptopappus*
61(59). Achenes strongly compressed; pappus of 1 or 2 slender awns .... *Perityle*
— Achenes not compressed or, if so, the pappus not of 1 or 2 slender awns............................. 62
62(61). Pappus a crown with margins entire or of short scales united into a crown 63
— Pappus not as above 65
63(62). Plants annual; heads solitary or paniculately arranged; flowers all perfect; leaves green and glabrous .... *Chamomilla*
— Plants perennial; heads corymbose or capitate, rarely solitary, with some marginal flowers pistillate only; leaves mostly silvery-canescent 64
64(63). Plants 0.5–1 m tall; leaves doubly pinnately dissected, mainly 10–20 cm long .... *Tanacetum*
— Plants mainly less than 0.3 m tall; leaves entire, once pinnately dissected, ternate, merely toothed apically, or entire, mainly less than 10 cm long .... *Sphaeromeria*
65(62). Involucral bracts with a thin, scarious, white, yellow, or purplish margin and tip .... *Hymenopappus*
— Involucral bracts without a scarious, colored margin and tip 66

66(65). Plants scapose; leaves roundish, entire, or crenate .... *Chamaechaenactis*
— Plants leafy stemmed; leaves not roundish and entire or subentire 67
67(66). Pappus scales with a strong midrib; leaves lanceolate or linear, entire; southern Utah .... *Palafoxia*
— Pappus scales nerveless or essentially so; leaves, at least in part, toothed to pinnatifid; widely distributed .... *Chaenactis*

Key 3.

Corollas not all tubular; ray flowers present pappus of capillary bristles.

1. Rays white, pink, violet, or purple, not yellow 2
— Rays yellow or orange yellow 9
2(1). Pappus of numerous unequal bristles, alternating with shorter, lacerate scales; involucral bracts subequal; low winter annuals .... *Monoptilon*
— Pappus of numerous bristles; involucral bracts imbricate or subequal; plants various, but seldom low winter annuals 3
3(2). Pappus, at least of disk flowers, of several to many rigid bristles; achenes pubescent with 2–forked hairs or the hairs barbed at apex .... *Townsendia*
— Pappus, at least of disk flowers, of many capillary bristles, at least in part; achenes glabrous or pubescent with simple hairs 4
4(3). Rays very inconspicuous, shorter than the tube and scarcely if at all exceeding their pappus; central perfect flowers few; plants annual .... *Conyza*
— Rays usually conspicuous, longer than the tube and pappus; central perfect flowers several to many; plants annual, biennial, or perennial 5
5(4). Involucres subequal, rarely somewhat graduated; rays usually narrow; style tips very short, triangular, rounded, or obtuse .... *Erigeron*
— Involucres usually strongly graduated; rays comparatively broad; style tips ovate and acute to subulate, usually lanceolate 6
6(5). Plants perennial, rhizomatous, or annual, or, if from a caudex then ordinarily less than 10 cm tall (see also *Aster kingii*) 7
— Plants from a caudex or taproot 8
7(6). Low, white-rayed perennial herbs from spreading cordlike rootstocks, in arid sites; flowering in springtime .... *Leucelene*
— Low to tall, white- to pink- or purple-rayed annual or perennial herbs from rhizomes or fibrous roots (a caudex in *A. kingii*); mainly flowering in summer and autumn .... *Aster*
8(6). Plants herbaceous, from a taproot, biennial or perennial; heads usually several to numerous .... *Machaeranthera*
— Plants more or less woody, from a ligneous caudex; heads usually solitary and large (primary selenophytes) .... *Xylorhiza*
9(1). Leaves opposite, at least below 10
— Leaves alternate throughout 12
10(9). Plants subshrubs .... *Perityle*
— Plants herbaceous 11
11(10). Leaves with stiff marginal bristles; involucre and leaves with conspicuous oil glands; plants annual .. *Pectis*

	Leaves without stiff marginal bristles; involucre and leaves without oil glands; plants perennial ..... *Arnica*
12(9).	Plants 1–1.5 m tall, herbaceous; heads 3–5 cm wide; rays 1–2 cm long .... *Inula*
—	Plants various, usually less than 1 m tall or, if taller, woody; heads much smaller; rays seldom to 1 cm long 13
13(12).	Pappus of 2–8 stiff, caducous bristles; plants usually glutinous .... *Grindelia*
—	Pappus of numerous, usually soft, persistent bristles 14
14(13).	Pappus of ca 20 twisted, flattish bristles .... *Amphipappus*
—	Pappus of numerous, straight, capillary bristles 15
15(16).	Pappus double, the inner of numerous bristles, the outer sometimes scalelike 16
—	Pappus not double, of subequal capillary bristles only 17
16(15).	Leaves essentially filiform .... *Conyza*
—	Leaves not filiform, linear-oblong or broader .... *Heterotheca*
17(15).	Involucral bracts in distinct vertical ranks 18
—	Involucral bracts not in distinct vertical ranks 19
18(19).	Outer involucral bracts with loose herbaceous tips; erect stems perennial; plants shrubs; leaves deciduous .... *Chrysothamnus*
—	Outer involucral bracts without loose herbaceous tips; erect stems annual; plants herbaceous; leaves persistent .... *Petradoria*
19(17).	Involucral bracts in 1 series, frequently with some smaller bracts at base; style branches truncate apically .... *Senecio*
—	Involucral bracts neither in 1 series nor with smaller bracts at base; style branches without truncate tips 20
20(19).	Heads small, the involucres usually less than 6 mm high, usually very numerous and densely paniculate, rarely racemose or corymbose; plants rhizomatous, fibrous rooted .... *Solidago*
—	Heads medium to large, the involucres usually more than 6 mm high, neither very numerous nor densely paniculate; plants with taproots, occasionally also rhizomatous .... *Haplopappus*

Key 4.

Corollas not all tubular; ray flowers present; pappus lacking.

1.	Rays white, pink, or pink purple, sometimes yellow at base 2
—	Rays yellow, sometimes partly purplish or maroon ... 7
2(1).	Receptacle naked 3
—	Receptacle with chaffy scales 6
3(2).	Leaves all basal; plants scapose .... *Bellis*
—	Leaves not all basal, at least some cauline; plants caulescent 4
4(3).	Leaves palmately lobed, the blades about as broad as long; heads less than 6.5 mm high .... *Perityle*
—	Leaves pinnately lobed to serrate, the blades longer than broad; involucres more than 8 mm high, or otherwise differing from above 5
5(4).	Receptacle broad and flattish; involucral bracts with a dark brown submarginal line .... *Chrysanthemum*
—	Receptacle convex, conic, or hemispheric; involucral bracts without a dark brown submarginal line .... *Chamomilla*
6(2).	Heads small, numerous, in dense, flattish or rounded cymose panicles; plants perennial .... *Achillea*
—	Heads comparatively large, solitary or few; plants annual or perennial .... *Anthemis*
7(1).	Receptacles not chaffy 8
—	Receptacles chaffy, at least toward the margin 13
8(7).	Heads 1- or 2-flowered, in dense glomerate clusters, sessile in the forks of the stem, or terminal and leafy involucrate .... *Flaveria*
—	Heads several- to many-flowered, solitary on terminal peduncles 9
9(8).	Plants woolly 10
—	Plants not woolly 11
10(9).	Rays persistent, becoming papery .... *Baileya*
—	Rays not persistent .... *Eriophyllum*
11(9).	Involucre and leaves with translucent oil glands... *Pectis*
—	Involucre and leaves without translucent oil glands... 12
12(11).	Rays conspicuous; involucral bracts acuminate, without scarious margins .... *Bahia*
—	Rays minute; involucral bracts obtuse, with scarious margins .... *Tanacetum*
13(7).	Ray achenes partly or wholly enfolded by their involucral bracts; plants annual, glandular-viscid above .... *Madia*
—	Ray achenes not conspicuously enfolded by their involucral bracts or, if so, the plants perennial; plants perennial or, if annual, not glandular above 14
14(13).	Involucre distinctly double, the outer bracts herbaceous, the inner ones broader and united to about the middle .... *Thelesperma*
—	Involucre not double, the bracts distinct to the base .. 15
15(14).	Plants scapose perennials; leaves broad, silvery-pubescent, entire; heads very broad .... *Enceliopsis*
—	Plants leafy stemmed or subscapose; leaves various but not broad and silvery-pubescent or, if so, sagittate; heads broad or narrow 16
16(15).	Plants subscapose; leaves variously dissected or sagittate; heads broad .... *Balsamorhiza*
—	Plants with stems definitely leafy; leaves usually not dissected or sagittate 17
17(16).	Plants shrubby; achenes conspicuously ciliate on the margins, notched at the apex, very flat .... *Encelia*
—	Plants herbaceous; achenes not conspicuously ciliate on the margins 18
18(17).	Leaves doubly pinnately dissected; heads numerous in corymbose cymes .... *Achillea*
—	Leaves simple, entire or toothed to lobed; heads few to several 19
19(18).	Achenes 2-winged; disks 15–25 mm wide; leaves white-strigose beneath, green above .... *Verbesina*
—	Achenes not 2-winged; disks 6–15 mm wide; leaves green on both sides .... *Heliomeris*

Key 5.

Corollas not all tubular; ray flowers present; pappus of awns or scales.

1.	Receptacle chaffy 2
—	Receptacle not chaffy, either naked or bristly 17

2(1).	Pappus scales fimbriate; ray flowers 4 or 5, white, only slightly surpassing the disk; introduced weedy plants, to be expected in Utah .... *Galinsoga parviflora* Cav.	—	Plants biennial or perennial; heads 3–5 cm wide; receptacle hemispheric .... *Matricaria*

— Pappus scales or awns not fimbriate; ray flowers various in size and color; indigenous or introduced 3

21(18). Pappus of 1 plumose awn and a denticulate crown .... *Monoptilon*

— Pappus of 2 to several awns or scales 22

3(2). Receptacle bearing a row of chaffy scales between the ray flowers and the outer disk flowers, otherwise naked; pappus of 10–20 slender setiform scales .. *Layia*

22(21). Plants dwarf woolly annuals .... *Eriophyllum*

— Plants annual or perennial, not woolly .... *Townsendia*

— Receptacle chaffy throughout; pappus not of 10–20 slender scales 4

23(17). Receptacle densely bristly or hairy 24

— Receptacle naked 25

4(3). Ray achenes dorsiventrally compressed, the thickened margins attached to a contiguous pair of infertile disk flowers and the subtending bract, and falling as a unit; pulvinate herbs of eastern Utah and shrubs of southwestern Utah .... *Parthenium*

24(23). Heads very small; involucres less than 10 mm wide .... *Gutierrezia*

— Heads medium sized; involucres more than 10 mm wide *Gaillardia*

25(23). Pappus of 4 hyaline scales united at the base; rays reddish purple to yellow .... *Hulsea*

— Ray achenes various, but not as above; herbs or shrubs 5

— Pappus a crown, or of caducous or persistent awns or scales; rays mostly yellow 26

5(4). Pappus of awns only, without scales 6

26(25). Pappus a mere crown or of caducous awns 27

— Pappus, at least in part, of scales 10

— Pappus persistent, of awns or scales 29

6(5). Achenes flat and obcompressed; awns retrorsely hispid .... *Bidens*

27(26). Pappus of 2–8 caducous awns; plants glutinous .... *Grindelia*

— Achenes not obcompressed; awns not retrorsely hispid 7

— Pappus a short crown; plants seldom if ever glutinous 28

7(6). Achenes plump; pappus of 2 to several caducous awns .... *Helianthus*

28(27). Leaves entire, bristly margined basally .... *Pectis*

— Leaves 2– or 3–pinnate .... *Tanacetum*

— Achenes flat, very strongly compressed; pappus various 8

29(26). Pappus of 1 or 2 awns or scales (rarely more) with or without a crown .... *Perityle*

8(7). Plants scapose; heads large, solitary .... *Enceliopsis*

— Pappus of 4 to many awns or scales 30

— Plants leafy stemmed; heads medium sized, usually several 9

30(29). Pappus of about 20 slender, twisted awns; rays 1 or 2 small .... *Amphipappus*

9(8). Plants shrubby; achenes narrowly white-margined, the margin not continous between weak awns .. *Encelia*

— Pappus of 4–16 twisted or plane awns or scales; rays usually several 31

— Plants herbaceous annuals; achenes strongly whitemargined the margin continuous between stout awns .... *Geraea*

31(30). Pappus of several scales dissected nearly to base; dwarf woolly annuals .... *Syntrichopappus*

10(5). Achenes very flat, strongly compressed 11

— Pappus awns or scales not dissected or else plants perennial or woody 32

— Achenes not very flat, usually much thickened 13

11(10). Leaves once to twice pinnatifid .... *Anthemis*

32(31). Pappus of several more or less united scales; rays broad, papery, and persistent .... *Psilostrophe*

— Leaves not pinnatifid, entire or nearly so 12

— Pappus not of united scales; rays not papery and persistent (occasionally so in *Hymenoxys*) 33

12(11). Plants scapose .... *Enceliopsis*

— Plants leafy stemmed .... *Helianthella*

33(32). Leaves and involucre with conspicuous oil glands .... *Dyssodia*

13(10). Pappus caducous (of 2 awns and rarely some scales) .... *Helianthus*

— Leaves and involucre without conspicuous oil glands 34

— Pappus persistent 14

14(13). Inner involucral bracts united to middle into a cup .... *Thelesperma*

34(33). Achenes slender, elongate-clavate 35

— Achenes stouter, oblong or obovoid 36

— Inner involucral bracts not united into a cup 15

35(34). Plants woolly .... *Eriophyllum*

15(14). Receptacle merely convex; rays pistillate ..... *Wyethia*

— Plants merely strigose .... *Platyschkuhria*

— Receptacle conic or cylindric; rays neuter 16

36(34). Involucral bracts spreading or reflexed; receptacle convex to subglobose; leaves decurrent .... *Helenium*

16(15). Involucral bracts subequal, in 2 or 3 series .. *Rudbeckia*

— Involucral bracts unequal, in 2 series, the inner ones shorter .... *Ratibida*

— Involucral bracts appressed; receptacle almost flat; leaves not decurrent 37

17(1). Rays white or purple 18

— Rays yellow, sometimes marked with purple 23

37(36). Pappus of numerous scales; stems leafy; leaves linear or linear-spatulate, entire, 2.5 mm wide or less .... *Gutierrezia*

18(17). Pappus a short crown 19

— Pappus of awns or scales 21

— Pappus of about 5 scales; leaves lobed or, if entire, broader and mostly or entirely basal ...... *Hymenoxys*

19(18). Leaves entire or pinnately divided ... *Chrysanthemum*

— Leaves irregularly 2–3 times pinnately dissected ... 20

Acamptopappus Gray

20(19). Plants annual; heads 1–2.5 cm wide; receptacle conic, hollow .... *Chamomilla*

Shrubs with white bark; leaves alternate, entire, 1–nerved; heads yellow, discoid, subglobose, cymose at tips of branches; flowers all fertile; involucral bracts ca 4-seriate, strongly imbricate, broad, rounded, the tip greenish, the margin scarious, erose; receptacle convex, fimbrillate; style branches linear; achenes subturbinate, densely villous; pappus persistent, of ca 35 flattened silvery scales and bristles of different widths.

Acamptopappus sphaerocephalus (Harv. & Gray) Gray Goldenhead. [*Haplopappus sphaerocephalus* Harv. & Gray]. Low rounded shrubs to 1 m tall, much branched, glabrous throughout or scabrous along some leaf margins; leaves spatulate to almost linear, 4-28 mm long, 1-5 mm wide, obtuse to acute, mucronulate, thick, sessile; heads subglobose, 6-10 mm high; involucre 4-6 mm high; n = 9. Blackbrush, indigobush, and creosote bush communities at 850 to 1375 m in Kane, San Juan, and Washington counties, Arizona, Nevada, and California; 22 (iv).

Achillea L.

Perennial, rhizomatous, aromatic herbs, with watery juice; stems erect or ascending; leaves alternate, 1- to 3-pinnately dissected; heads several to many, borne in compact to open corymbose cymes; involucral bracts imbricate in several series, chaffy, the margins scarious and hyaline; receptacle chaffy; ray flowers present, usually 3-12, pistillate, fertile, yellow, white, pink, or pink purple; disk flowers mostly 10 or more, perfect, fertile; pappus none; style branches flattened; achenes compressed, callus-margined, glabrous, beakless.

1. Flowers yellow; leaves coarsely twice pinnately dissected; plants cultivated .... *A. filipendulina*
— Flowers white, pink, or pink purple; leaves finely 2-3 times dissected; plants indigenous or cultivated ....
.... *A. millefolium*

Achillea filipendulina Lam. Fernleaf Yarrow. Herbs, the stems erect, 8-12 dm tall or more, longitudinally furrowed and minutely glandular; leaves 4-35 cm long, doubly pinnatifid, the lateral lobes with one large lobe on the upper side; heads numerous, borne in hemispheric or flat- topped corymbose cymes; involucre 3-4 mm high, the bracts with pale scarous margins, villous; rays ca 5, to 1 mm long, yellow; disk flowers 30-40, yellow; achenes 1-2 mm long; 2n = 18, 36. Cultivated ornamental, Salt Lake and Utah counties, and to be expected elsewhere; introduced from Asia; 2 (0).

Achillea millefolium L. Milfoil Yarrow. Herbs, the rhizomes horizontal; stems ascending to erect, 0.5-10 dm tall, villous-tomentose, simple or branched above; leaves 2-26 cm long, reduced upwards, pinnately once to thrice dissected, the segments very slender; heads numerous, borne in hemispheric or flat-topped, corymbose cymes; involucres 4-6 mm high, the bracts dark to pale margined, villous to glabrate; rays usually ca 5, 2-3.5 mm long, white to pink or pink purple; disk flowers 10-20; achenes 1-2 mm long. Sagebrush, pinyon-juniper, cottonwood, juniper, rabbitbrush, ponderosa pine, mountain brush, aspen, Douglas fir, spruce-fir, and alpine tundra communities at 1070 to 3750 m in all Utah counties; widely distributed in North America; circumboreal; 133 (xv). Two very similar taxa are present in Utah; the common, indigenous **ssp. lanulosa (Nutt.) Piper** [*A. lanulosa* Nutt.; (n = 18)], and the introduced, cultivated, **ssp. millefolium** (n = 9, 27, 54). A trend is recognizable within ssp. *lanulosa*; the high elevation specimens tend to have dark involucral bracts, fewer heads, and lower stature. These alpine plants have been treated as var. *alpicola* (Rydb.) Garrett, but they intergrade completely with specimens attributable to var. *lanulosa*. Indeed, the two extremes can be found mounted on the same herbarium sheet, taken from the same locality.

Agoseris Raf.

Perennial scapose herbs with milky juice, from taproots; leaves all basal, entire to pinnately lobed or merely toothed; heads solitary on a naked scape; involucral bracts in 2 to several series, herbaceous, or the inner ones hyaline or nearly so; receptacle usually naked; corollas all raylike, perfect, yellow to orange, often drying pinkish or purplish; pappus of capillary bristles; style branches semicylindric; achenes angular or terete, prominently nerved, usually beaked.

1. Plants annual; achene beak 2-3 times as long as the body; rare in Utah .... *A. heterophylla*
— Plants perennial; achene beak less than half to 2 or more times as long as the body 2

2(1). Achene beak striate, mostly less than half as long as the body (longer in some var. *laciniata*); flowers yellow, often drying bluish to pinkish .... *A. glauca*
— Achene beaks scarcely striate, more than half to 2 or more times as long as the body 3

3(2). Flowers brownish orange to yellow orange, often drying purplish; achene beak less than twice as long as the body
.... *A. aurantiaca*
— Flowers yellow, often drying bluish or pinkish; achene beak more than twice as long as the body 4

4(3). Achene body with ribs swollen apically, forming a shoulder, tapering abruptly to the beak; involucral bracts in two sets .... *A. retrorsa*
— Achene body with ribs not swollen apically, rather gradually tapering to the beak; involucral bracts not especially in two sets .... *A. grandiflora*

Agoseris aurantiaca (Hook.) Greene Orange Agoseris. Plants 0.6-6.6 dm tall, from a simple or branched caudex; leaves 3.5-36 cm long, 0.5-3 cm broad, narrowly oblanceolate, entire to toothed or lobed, villous to glabrate; scapes villous-tomentose to nearly glabrous; involucres 10-27 mm long, 10-42 mm wide, the outer bracts villous to glabrate and ciliate, often purple-spotted; corollas brownish orange to yellow orange, often drying purplish; achene body 4-8 mm long, the slender beak not striate, from more than half as long to longer than the body; n = 18, 36. Two rather weak and intergrading phases are recognized at varietal rank.

1. Involucres with bracts subequal or nearly so, slender, tapering, some often over 20 mm long ....
.... *A. aurantiaca* var. *aurantiaca*
— Involucres with bracts definitely imbricate, broad, and rounded apically or abruptly tapering ....
.... *A. aurantiaca* var. *purpurea*

Var. aurantiaca [*Troximon aurantiacum* Hook.; *A. arizonica* Greene; *A. gracilens* (Gray) Kuntze; *A. longirostris* Greene, type from Fish Lake]. Sagebrush, mountain brush, juniper, pinyon-juniper, and alpine meadow communities at 1375 to 3355 m in Beaver, Box Elder, Carbon, Duchesne, Grand, Juab, Salt Lake, San

Juan, Sevier, Summit, Tooele, Uintah, Utah, Wasatch and Wayne counties; Alberta to British Columbia, south to California and New Mexico; 33 (v).

Var. purpurea (Gray) Cronq. [*Troximon aurantiacum* var. *purpureum* Gray; *A. purpurea* (Gray) Greene; *A. confinis* Greene, type from near Marysvale]. Mountain brush, aspen, aspen-fir, and spruce-fir communities at 1700 to 3425 m in Cache, Carbon, Emery, Garfield, Grand, Iron, Juab, Kane, Piute, Sanpete, Sevier, Summit, Tooele, Wasatch, and Washington counties; Montana to Arizona and New Mexico; 20 (vi).

***Agoseris glauca* (Pursh) Raf.** Pale Agoseris; Mountain Dandelion. Plants perennial, 0.2–6.4 dm tall, from a simple or branched caudex; leaves 2–26 cm long, 0.2–3 cm broad, narrowly oblanceolate to linear or spatulate to elliptic, entire or toothed to lobed, villous to glabrate; involucres 12–28 mm high, 0.8–4 cm wide, the outer bracts villous to glabrous, ciliate or not, sometimes purple-spotted; corollas yellow, often drying bluish to pinkish; achene body 4–10 mm long, the striate beak stout, to half as long as the body (slender and to as long as the body in some var. *laciniata*); n = 9, 18. Three intergrading and partially sympatric varieties are present in Utah.

1. Leaves laciniately toothed or lobed; plants of broad distribution, common .... *A. glauca* var. *laciniata*
— Leaves entire, rarely with a few teeth or lobes; plants variously distributed, locally common 2
2(1). Plants pubescent, at least below the heads, mainly of spruce-fir and alpine communities, sometimes lower .... *A. glauca* var. *dasycephala*
— Plants glabrous throughout, mainly of lower elevation wet meadows, but sometimes of high elevation meadows .... *A. glauca* var. *glauca*

Var. dasycephala (T. & G.) Jepson [*Troximon glaucum* var. *dasycephalum* T. & G.; *Ammogeton scorzoneraefolius* Schrader; *A. scorzoneraefolia* (Shrader) Greene; *T. pumilum* Nutt.; *A. pumila* (Nutt.) Rydb.; *A. glauca* var. *pumila* (Nutt.) Garrett; *A. villosa* Rydb.]. Sagebrush, mountain brush, aspen, spruce-fir, and alpine tundra communities at 1830 to 3385 m in Duchesne, Garfield, Iron, Kane, Millard, Piute, Salt Lake, San Juan, Sanpete, Sevier, Tooele, Uintah, Utah, and Washington counties; Alaska to Manitoba and south to Colorado. Plants of this variety pass by degree into each of the following; 29 (v).

Var. glauca [*Troximon glaucum* Pursh; *A. isomeris* Greene, type from the Uinta Mts.]. Meadows at 2325 to 3660 m in Box Elder, Cache, Davis, Duchesne, Iron, Juab, Kane, Salt Lake, Sanpete, Sevier, Summit, Uintah, Wasatch, Washington and Weber counties; British Columbia to Manitoba and south to California and Arizona; 22 (iv).

Var. laciniata (D. C. Eaton) Smiley [*Macrorhynchus glaucus* var. *laciniatus* D.C. Eaton; *Troximon parviflorum* Nutt; *A. parviflora* (Nutt.) D. Dietr.; *A. glauca* var. *parviflora* (Nutt.) Rydb.; *T. taracifolium* Nutt.; *A. taracifolia* (Nutt.) D. Dietr.; *A. taraxacoides* Greene, type from near Marysvale; *A. caudata* Greene, type from Salina Canyon; *A. agrestis* Osterh.; *A. glauca* var. *agrestis* (Osterh.) Q. Jones]. Sagebrush, mountain brush, juniper, pinyon-juniper, Douglas fir, aspen, and spruce-fir communities at 1300 to 3050 m in Beaver, Box Elder, Cache, Carbon, Daggett, Duchesne, Emery, Garfield, Grand, Juab, Kane, Millard, Piute, San Juan, Salt Lake, Sanpete, Sevier, Summit, Tooele, Uintah, Utah, Washington, Wayne, and Weber counties; Washington to Montana and south to Arizona; 84 (xii). The phase designated as var. *agrestis* blends completely in our area with that treated herein as var. *laciniata*.

***Agoseris grandiflora* (Nutt.) Greene** [*Stylopappus grandiflorus* Nutt.]. Plants perennial, 1.5–4.5 (7) dm tall, from a simple or branching caudex; leaves 8–25 cm long, 1–3 cm broad, narrowly oblanceolate, pinnatifid to subentire, villous to glabrate; involucres 15–38 mm long, 20–43 mm wide, the outer bracts villous-tomentose to glabrate, ciliate, often suffused with purple; corolla yellow, drying bluish to pinkish; achene body 4–7 mm long, the nerveless beak more than twice as long as the body; n = 9. Sagebrush and mountain brush communities at 1830 to 2135 m in Cache, Iron, Tooele, and Washington counties; British Columbia to California and Nevada; 4 (i).

***Agoseris heterophylla* (Nutt.) Greene** Annual Agoseris. [*Macrorhynchus heterophyllus* Nutt.]. Plants annual, 0.3–2.5 (4) dm tall, with 1 to several scapes from the base; leaves 1–20 (15) cm long, 3–15 mm wide, narrowly oblanceolate, toothed or pinnatifid to entire, all basal, or with some not strictly basal; involucres 5–20 mm long, 4–10 mm wide, sparingly villous with multicellular hairs, the cross-walls purplish; corolla yellow, sometimes turning pinkish on drying; achene body 2–5 mm long, prominently ribbed or winged, the beak 2–3 times as long as the body; n = 18. Sagebrush-grass and mountain brush communities at 1280 to 1435 m in Cache and Salt Lake counties; British Columbia to California and Arizona; 5 (0).

***Agoseris retrorsa* (Benth.) Greene** Retrorse Agoseris. [*Macrorhynchus retrorsus* Benth.]. Plants perennial, mainly 1–4 dm tall, with 1 to several scapes from the base; herbage woolly when young, becoming glabrate in age; leaves 8–20 cm long, 5–15 mm wide, oblanceolate to lance-elliptic, pinnately parted, the segments linear to lanceolate and retrorse, the terminal one very long; involucres 2.5–4 cm high, the phyllaries in 2 sets; corollas yellow, often drying pinkish; achene body 5–7 mm long, more or less truncate apically and abruptly narrowed into a slender beak about twice as long as the body. Mixed desert shrub and pinyon-juniper communities at ca 1525 to 1830 m in Juab and Washington counties; Washington to California and Nevada; 3 (0).

Ambrosia L.

Annual or perennial herbs or shrubs; leaves alternate or opposite, pinnately or palmately lobed, toothed, or dissected; heads unisexual, discoid; staminate heads in slender spicate, bractless racemes; involucre 5– to 12–lobed; receptacle flat, bearing flattened filiform-setose bracts; staminal filaments monadelphous, the anthers scarcely united; pistillate heads borne below the staminate ones, mostly axillary, their involucres closed, nutlike, armed with prickles arranged in one or more series; pistil naked, the corolla lacking; pappus lacking.

Payne, W. W. 1964. A reevaluation of the genus *Ambrosia*. J. Arnold Arboretum 45: 401–438.

1. Plants woody shrubs of southwestern Utah 2
— Plants annual or perennial herbs, of various distribution 3
2(1). Leaves mainly less than 15 mm long, pinnately lobed, the lobes again toothed or lobed, silvery-strigose overall .... *A. dumosa*

— Leaves mainly more than 20 mm long, merely toothed or lobed, the lobes not again toothed or lobed, bicolored, the upper surface green, the lower surface white-tomentose .... *A. eriocentra*

3(1). Leaves palmately lobed, the lobes serrate; plants tall coarse herbs .... *A. trifida*
— Leaves pinnatifid or pinnately lobed; plants slender herbs usually less than 5 dm tall 4

4(3). Leaves bicolored, the lower surface obscured by appressed white hairs; plants low rhizomatous perennials .... *A. tomentosa*
— Leaves various, but not definitely bicolored; plants from tap-roots or rhizomes but, if the latter, not as above 5

5(4). Plants perennial, rhizomatous; leaves opposite .... *A. psilostachya*
— Plants annual; leaves mainly alternate 6

6(5). Lower stems and leaves with pustular-based, stiff, multicellular hairs; plants often with lower lateral branches decumbent-ascending; burs with spines in more than 1 series .... *A. acanthicarpa*
— Lower stems lacking pustular-based hairs, all stems slender and curved ascending-appressed; burs with spines in one series .... *A. artemisiifolia*

Ambrosia acanthicarpa Hook. Bur Ragweed. [*Franseria acanthicarpa* (Hook.) Cov. Plants annual, 0.9–7.5 dm tall, often branching from the base, the lower branches commonly decumbent-ascending; pubescence of stiff multicellular hairs, the bases pustular; leaves mostly alternate, petiolate, the blades 0.9–4.5 cm long, 0.6–3.5 cm wide, bipinnatifid to pinnatifid; heads numerous in terminal or axillary racemes, staminate above, pistillate below; staminate heads short-pedunculate, not bracteate; pistillate solitary or clustered in upper axils, with 2–3 series of flattened, curved spines. Blackbrush, salt desert shrub, desert shrub, pinyon-juniper, and riparian communities, often in sandy substrates, at 850 to 2000 m, in Beaver, Box Elder, Cache, Carbon, Daggett, Davis, Duchesne, Emery, Garfield, Grand, Iron, Juab, Kane, Millard, San Juan, Sanpete, Tooele, Uintah, Utah, Washington, and Wayne counties; Washington to Saskatchewan, south to California, Arizona, and Texas; 60 (xi).

Ambrosia artemisiifolia L. Common Ragweed. [*A. elatior* L.]. Plants annual, mostly 3–9 dm tall, branching from above the middle; pubescence of lax multicellular hairs, the bases not pustular; leaves alternate, or the lower usually opposite, petiolate, the blades 2.5–8.5 cm long, 1.9–7.5 cm wide, 1- to 2-pinnatifid; heads numerous in terminal or axillary racemes, the staminate above, pistillate below, clustered or solitary, with 1 series of tuberculate spines; $2n = 36$. Moist disturbed sites at 1375 to 1680 m in Box Elder, Cache, Duchesne, Garfield, Juab, Millard, Salt Lake, San Juan, Utah, and Weber counties; widespread in North America; 11 (i).

Ambrosia dumosa (Gray) Payne Bur-sage. [*Franseria dumosa* Gray]. Shrubs, 2–6 dm tall, rounded, much-branched; branchlets white, subspinescent; pubescence dense, strigose; leaves alternate, petiolate, the blades 9–30 mm long, 5–15 mm wide, mostly bipinnatifid, uniformly hairy on both sides; staminate heads spicate, rather few; pistillate heads often scattered among the staminate; mature pistillate involucre armed with 20–35 lance-subulate spines; $n = 18$, 36, 54. Creosote bush, blackbrush, and Joshua tree communities at 670 to 1000 m in Washington County; Arizona, California, and Mexico; 28 (i).

Ambrosia eriocentra (Gray) Payne [*Franseria eriocentra* Gray]. Shrubs, 3–10 (12) dm tall, aromatic, the branchlets white, subspinescent, pubescence of white tomentum and coarse multicellular hairs; leaves alternate, subsessile, sinuately toothed to lobed or 1-pinnatifid, 8–40 (50) mm long, 2–20 mm wide; staminate heads more or less clustered; pistillate heads 1-flowered; pistillate involucre with 12–20 flattened, subulate spines. Creosote bush, blackbrush, and Joshua tree communities at 670 to 1000 m in Washington County; Arizona, Nevada, and California; 12 (ii).

Ambrosia psilostachya DC. Western Ragweed. [*A. coronopifolia* T. & G.]. Perennial herbs, mostly 3–6 dm tall, simple or branching above the middle; pubescence of harsh, spreading, multicellular, pustular-based hairs (at least in part); leaves opposite, at least below, petiolate to subsessile, the blades 4–10 cm long, 2.5–4.5 cm wide, mostly once pinnatifid; staminate heads in terminal or axillary spicate racemes; pistillate involucres merely tuberculate or quite unarmed; $2n = 18, 27, 36, 45, 54, 63, 72, 108$. Meadows, stream banks, and roadsides in sagebrush and other communities at 1300 to 2100 m in Cache, Davis, Juab, Millard, Salt Lake, Tooele, Utah, and Weber counties; Washington to Illinois, south to Arizona and Mexico; 20 (ii).

Ambrosia tomentosa Nutt. Low Ragweed. [*Franseria discolor* Nutt.; *F. tomentosa* (Nutt.) A. Nels., not *A. tomentosa* Gray]. Perennial rhizomatous herbs, mostly 1–3.5 dm tall, branching from above the base; pubescence of short, stiff, appressed hairs; leaves alternate, petiolate, the blades 2–15 cm long, 0.4–3.5 cm wide, 1- to 3-pinnatifid; staminate heads racemose; pistillate heads armed with 2 or 3 series of coarse spines. Meadows and stream banks at 1155 to 1525 m in Box Elder, Davis, Duchesne, Garfield, Grand, Uintah, and Weber counties; Wyoming and Colorado; 10 (iii).

Ambrosia trifida L. Giant Ragweed. Annual, robust herbs, 10–15 dm tall or more; pubescence spreading-hirsute to hispid, at least above; leaves opposite, petiolate, the blades palmately 3- to 5-lobed, or unlobed, mainly 5–20 cm long, 4–15 cm wide, scabrous on both surfaces, serrate; staminate involucres 3-nerved; pistillate involucres 5–10 mm long, bearing short spines at the tip; $n = 12$. Uncommon (introduced?) weedy plants of disturbed sites in Salt Lake County; widely distributed in North America; 1 (0).

Amphipappus T. & G.

Low shrubs; branches white-barked, divaricate; leaves alternate, entire, short-petiolate; heads small, radiate, few-flowered, clustered at tips of branches; involucre in ca 3 series, strongly imbricate, straw colored to greenish, the bracts broad, rounded; receptacle fimbrillate; ray flowers yellow, 1 or 2, small; disk flowers 3–6, perfect; ray achenes hairy, broadly oblanceolate, compressed, their pappus of more or less united bristles, awns, or scales; disk achenes undeveloped, glabrous or sparingly pilose, their pappus of twisted, hispidulous bristles or scales.

Porter, C. L. 1943. The genus *Amphipappus* Torr. & Gray. Amer. J. Bot. 30: 481–483.

Amphipappus fremontii T. & G. Chaff-bush. Shrubs 3–8 dm tall, the herbage scabrous-puberulent; leaves 5–12 mm long, 2–5 mm wide, oblanceolate to elliptic, cuneate basally, acute to obtuse and apiculate, green;

heads 4–6 mm high, the bracts greenish medially near the apex, the margins hyaline and more or less erose. Joshua tree and creosote bush communities at 700 to 900 m in Washington County; Nevada, Arizona, and California. Our material belongs to var. *spinosus* (A. Nels.) C.L. Porter [*Amphiachyris fremontii* var. *spinosus* A. Nels.; ssp. *spinosus* (A. Nels.) Keck]; 12 (iii).

Anaphalis DC.

Perennial, dioecious or polygamo-dioecious, rhizomatous herbs, with watery juice; stems ascending to erect, simple or branched above; leaves simple, alternate, entire; heads several to many, in hemispheric or flat-topped corymbose cymes; involucral bracts imbricate in several rows, chaffy, scarious, white, or with a dark triangular basal spot; receptacle naked; corollas of disk flowers only, imperfect, whitish, the pistillate heads sometimes bearing some central staminate flowers, the pistillate corollas tubular-filiform, the staminate corollas tubular-funnelform; pappus of capillary bristles; style branches somewhat flattened; achenes small, roughened, glabrous to sparingly hairy.

Anaphalis margaritacea (L.) Benth. & Hook. Pearly Everlasting. [*Gnaphalium margaritaceum* L.]. Plants 1.5–8 dm tall, the stems white villous-tomentose; leaves only gradually reduced upwards, 2.5–12 cm long, 0.5–2 cm wide, narrowly lanceolate to oblong, elliptic, or oblanceolate, sessile, entire, flat to slightly revolute, white-tomentose below, commonly less pubescent and greenish above; heads showy, the involucres 4–7 mm high, 5–10 mm broad, the bracts pearly-white, with a dark triangular base, glabrous; achenes ca 1 mm long; 2n = 28. Meadows, streambanks, and openings in ponderosa pine, lodgepole pine, boxelder, and aspen communities at 1150 to 2700 m in Box Elder, Cache, Daggett, Duchesne, Iron, Juab, Salt Lake, Summit, Uintah, Wasatch, Washington, and Weber counties; widely distributed in North America; Asia; 33 (vi).

Antennaria Gaertner

Perennial, dioecious herbs with stolons, caudices, or rhizomes, the juice watery; stems ascending to erect, usually simple; leaves simple, alternate and basal, the cauline generally reduced upward; heads solitary to many, borne in corymbose cymes; involucral bracts imbricate in several rows, scarious (at least marginally), often colored; receptacle naked; corollas of disk flowers only, imperfect, whitish or tawny; pistillate corollas tubular-filiform, the pappus of numerous capillary bristles; staminate corollas tubular-funnelform, the pappus of few clavate to barbellate, usually flattened bristles; style branches slightly flattened; achenes terete to slightly compressed, glabrous or papillose.

1.	Heads solitary; flowering stems usually less than 5 cm tall	*A. dimorpha*
—	Heads (1) 2 to many (see *A. rosulata*); flowering stems often more than 5 cm tall	2
2(1).	Upper leaf surface green; leaf blades broadly spatulate, rounded to obtuse	*A. neglecta*
—	Upper leaf surface not notably different from the lower; blades seldom both spatulate and rounded to obtuse	3
3(2).	Plants not forming mats, lacking leafy stolons, some caespitose from caudex or rhizome	4
—	Plants mat-forming, with leafy stolons	6
4(3).	Involucral bracts glabrous or nearly so, scarious near the base, white-opaque apically	*A. luzuloides*
—	Involucral bracts densely tomentose in the lower half, opaque to dark with pale scarious apices	5
5(4).	Involucral bracts blackish in aspect, the tips pale and scarious	*A. pulcherrima*
—	Involucral bracts opaque white, somewhat darkened at the middle	*A. anaphaloides*
6(3).	Terminal scarious portion of involucral bracts dirty brownish to blackish green on at least the middle and outer ones	7
—	Terminal scarious portion of involucral bracts white to pink, with a dark basal spot on some only	8
7(6).	Terminal scarious portion of involucral bracts blackish green; plants usually alpine in Uinta, Wasatch, and Tushar mts., and on the Markagunt Plateau	*A. alpina*
—	Terminal scarious portion of bracts merely discolored and pale brown, or the inner bracts whitish at the tips; plants usually of lower elevations	*A. umbrinella*
8(6).	Flowering stems less than 5 cm tall; heads 1 or 2; plants of Beaver, Garfield, Grand, Iron, Kane, and Wayne counties	*A. rosulata*
—	Flowering stems commonly more than 5 cm tall; heads usually 3 or more; plants of broad or other distribution	9
9(8).	Involucral bracts with a black spot between the tomentose greenish base and the opaque white-scarious apex	*A. corymbosa*
—	Involucral bracts lacking a conspicuous black spot	10
10(9).	Involucres mostly 4–7 mm high, often bright pink; pistillate corollas mostly 2–4.5 mm long	*A. microphylla*
—	Involucres mostly 7–11 mm high, seldom pink; pistillate corollas mostly 5–8 mm long	*A. parvifolia*

Antennaria alpina (L.) Gaertner Alpine Pussytoes. [*Gnaphalium alpinum* L.]. Plants caespitose from a caudex, mat-forming and stoloniferous, 2–13 cm tall; basal leaves 0.6–2.2 cm long, 2–6 mm wide, cuneate-oblanceolate to spatulate, acute to obtuse or rounded apically, grayish tomentose on both surfaces or greenish and subglabrous above on some leaves; heads 3–5, borne in subcapitate cymes; pistillate involucres 5–7 mm high, villous-tomentose below, the scarious tips of bracts uniformly blackish or brownish green, all rather blunt apically, often erose; staminate involucres mostly 4–5 mm high, the scarious tips of bracts often pale apically; achenes glabrous; 2n = 63. Lodgepole pine, spruce-fir, and alpine tundra communities at 3050 to 3550 m in Beaver, Cache, Daggett, Duchesne, Grand, Juab, Piute, Salt Lake, San Juan, Sevier, Uintah, and Utah counties; north to Alaska and east to Labrador; circumboreal; 27 (vii). Our material belongs to var. **media (Greene) Jepson** [*A. media* Greene; *A. austromontana* E. Nels., type from Tushar Mts.]. There is a tendency for some specimens to approach *A. parvifolia* in the Uinta Mts. and *A. umbrinella* elsewhere.

Antennaria anaphaloides Rydb. Pearly Pussytoes. Plants from a caudex, not mat-forming or stoloniferous, 1.5–3.5 (5) dm tall; basal leaves 2.5–19 cm long, 4–18 mm wide, narrowly oblanceolate to elliptic, tomentose on both surfaces; heads several to many in branching or compact cymes; pistillate involucres 5–8 mm high, villous-tomentose below, the scarious tips opaque white, all rounded or obtuse, often erose; staminate involucres 5–8 mm high, similar to the pistillate; achenes glabrous. Aspen, spruce-fir, sagebrush, and mountain brush communities at 2440 to 3325 m in Box Elder, Daggett, Summit, and Uintah counties; British Columbia to Montana and south to Nevada and Colorado; 7 (i).

Antennaria corymbosa E. Nels. Plains Pussytoes. [*A. nardina* Greene]. Plants caespitose, mat-forming and stoloniferous, 5–26 cm tall; basal leaves 0.6–3.7 cm long, 2–6 mm wide, narrowly oblanceolate to spatulate, acute to obtuse apically, gray to greenish and tomentose on both surfaces; heads commonly 3–8, in compact to branching cymes; pistillate and staminate involucres 4.5–6 mm high, the bracts green and tomentose basally, with a dark spot at the base of the white or sordid terminal portion; achenes puberulent; $2n = 28$. Alpine tundra, krummholz, spruce-fir, lodgepole pine, and willow-alder communities, often along stream banks and in wet meadows or bogs, at 2240 to 3355 m in Beaver, Duchesne, Garfield, Summit, and Uintah counties; Montana and Idaho to Colorado and California(?); 22 (iii). The main body of the species in Utah lies in the Uinta Mts., with outliers in the Stansbury and Tushar mts., and in the Markagunt Plateau.

Antennaria dimorpha (Nutt.) T. & G. Low Pussytoes. [*Gnaphalium dimorphum* Nutt.; *A. dimorpha* var. *macrocephala* D. C. Eaton, type from Salt Lake City]. Plants caespitose, mat-forming, rooting from short caudex branches, not truly stoloniferous, 1–5 (7) cm tall; basal leaves narrowly oblanceolate, 0.6–4 cm long, 1–14 mm wide, acute apically, grayish tomentose on both sides; heads solitary, terminal on short leafy stems; pistillate involucres (7) 10–18 mm long, the bracts strongly imbricated, slender, attenuate, green at base, suffused with brown above the base, the apical portions yellowish to brownish scarious; staminate involucres 6–9 mm long, tomentose at the base, brown above the base, the broad apical portion hyaline to scarious; achenes puberulent. Mat-saltbush, sagebrush, juniper, oak-serviceberry, ponderosa pine, and spruce-fir-lodgepole pine communities at 1430 to 3050 m in Beaver, Box Elder, Cache, Daggett, Duchesne, Iron, Juab, Kane, Millard, Salt Lake, Sanpete, Sevier, Summit, Tooele, Uintah, Utah, and Washington counties; British Columbia to Montana, south to California, Nevada, Colorado, and Nebraska; 37 (viii).

Antennaria luzuloides T. & G. Rush Pussytoes. [*A. oblanceolata* Rydb.]. Caespitose from a caudex, 1.1–5 (7) dm tall; basal leaves 2–5 (8) cm long, 2–8 mm wide, greenish, tomentose on both surfaces; heads numerous in a compact or more often branched corymbose inflorescence; pistillate and staminate involucres similar, 4–5 mm high, glabrous to the base, the bracts brownish scarious and more or less hyaline below, opaque whitish above; achenes puberulent. Openings in aspen-conifer and lodgepole pine-spruce communities at 2950 to 3050 m in Daggett, Duchesne, Summit, Uintah, and Wasatch counties; British Columbia to Montana, south to California, Nevada, and Colorado; 14 (i).

Antennaria microphylla Rydb. Rosy Pussytoes. [*A. rosea* Rydb.; *A. concinna* E. Nels.; *A. arida* A. Nels.]. Plants caespitose, stoloniferous and mat-forming, 0.4–3 (4) dm tall; basal leaves 0.5–3 cm long, 2–8 mm wide, oblanceolate to spatulate; heads 2–13 (or more), in congested to open cymes; pistillate involucres 4–7 mm high, the bracts tomentose below, greenish or scarious below the middle, often somewhat brownish below the scarious, whitish or pinkish, terminal portion; mainly known from pistillate individuals; achenes glabrous or sparingly hispidulous; $2n = 56, 126$. Sagebrush, juniper, ponderosa pine, Douglas fir, lodgepole pine, spruce-fir, and alpine meadow communities at 1830 to 3450 m in all Utah counties, except Davis, Morgan, and San Juan; Alaska to Ontario, south to California and New Mexico; 105 (xxiii).

Antennaria neglecta Greene Field Pussytoes. [*A. marginata* Greene]. Plants caespitose, stoloniferous and mat-forming, 5–15 (25) cm tall; basal leaves 1.8–3.5 cm long, 3–15 mm wide, spatulate, thinly tomentose to glabrous and green above, white-tomentose beneath; heads mainly 3–5, in compact cymes; pistillate involucres 6–11 mm high, the bracts tomentose on the greenish base, the apical scarious portion white or suffused with pink; staminate plants rare; achenes glabrous or minutely pubescent; $2n = 28$. Pinyon-juniper and shrub communities at 1525 to 1900 m in Emery, Grand, San Juan, Sanpete, Utah, and Washington counties; Alaska to Newfoundland, south to California, Arizona, and Virginia; 8 (i). Our few specimens are hardly adequate to represent this species clearly in Utah.

Antennaria parvifolia Nutt. Common Pussytoes. [*A. aprica* Greene; *A. obtusata* Greene, type from Uinta Mts., Uintah County]. Plants caespitose, stoloniferous and mat-forming, 3–15 cm tall; basal leaves 0.8–3.5 cm long, 3–8 mm wide, spatulate, obtuse to acute apically, tomentose on both sides; heads 2–6 or more; pistillate involucres 7–11 mm high, the bracts more or less tomentose on the greenish base, the scarious portion white, sordid, or pink; staminate plants rarely collected; achenes glabrous. Mountain brush, pinyon-juniper, sagebrush, ponderosa pine, aspen, lodgepole, and spruce-fir communities at 1650 to 3250 m in Beaver, Cache, Carbon, Daggett, Davis, Duchesne, Emery, Garfield, Juab, Kane, Millard, Piute, San Juan, Sanpete, Sevier, Summit, Tooele, Uintah, Utah, and Wayne counties; British Columbia to Manitoba, south to Arizona and New Mexico; 46 (iv).

Antennaria pulcherrima (Hook.) Greene Showy Pussytoes. [*A. carpathica* var. *pulcherrima* Hook.]. Plants from a caudex, not mat-forming or stoloniferous, 23–40 cm tall; basal leaves 4–19 cm long, 5–23 mm wide, narrowly to broadly oblanceolate to elliptic, tomentose on both surfaces; heads several to many in branching or compact cymes; pistillate and staminate involucres both 6.5–8 mm long, the bracts tomentose at the greenish base, the terminal scarious portion blackish to brownish or the apex whitish; achenes glabrous. Sedge-rush meadows, streamsides, and bogs at 2440 to 2800 m in Duchesne, Garfield, and Summit counties; Alaska to Newfoundland, south to Colorado; 6 (i).

Antennaria rosulata Rydb. Breaks Pussytoes. Plants caespitose, stoloniferous and mat-forming, 1–3 cm tall; basal leaves 0.5–1.1 cm long, 2–5 mm broad, spatulate, obtuse to rounded apically, tomentose on both surfaces; heads 1 or 2, terminating short erect branches; pistillate involucres 5–9 mm high, the outer bracts greenish and tomentose to the apex, the inner ones green at base, with scarious slender white tips; staminate involucres 4–5 mm high, the bracts densely tomentose at base, the broad scarious tips white opaque; achenes puberulent. Ponderosa pine, aspen, Douglas fir, limber pine, sagebrush, and spruce communities, and in alpine meadows, at 2600 to 3350 m in Beaver, Garfield, Grand, Iron, Kane, and Wayne counties; Colorado, New Mexico, and Arizona; 14 (i).

Antennaria umbrinella **Rydb.** Mountain Pussytoes. [*A. dioica* authors, not (L.) Gaertner]. Plants caespitose, mat-forming and stoloniferous, 2–14 mm tall; basal leaves 0.7–2 cm long, 2–15 mm wide, cuneate-oblanceolate to spatulate, acute to obtuse apically, tomentose on both sides; heads 2–6, borne in subcapitate cymes; pistillate involucres 5–8 mm long, the bracts greenish and tomentose at the base, the scarious tips dirty brownish to pale tan, or the innermost almost white, acute to rounded, usually erose; staminate plants rare in our region; achenes glabrous. Aspen communities and alpine meadows at 2745 to 3500 m in Duchesne, Garfield, Iron, Juab, and Summit counties; Alaska to Hudson Bay, south to California, Arizona, and Colorado; 12 (0). Specimens assigned here are more or less intermediate beween *A. alpina* and *A. microphylla*. Many more specimens are required to provide definitive information on this entity in Utah.

Anthemis L.

Annual or short-lived perennial aromatic herbs from taproots, the juice watery; stems erect, commonly branched; leaves alternate, 1–3 pinnately dissected; heads solitary on the uppermost branches; involucral bracts imbricated in several series, chaffy, the margins scarious or hyaline; receptacle hemispheric, chaffy at least near the middle; ray flowers present, white or yellow, usually 10 or more, sterile; disk flowers numerous, perfect, fertile; pappus none or a short crown; style branches flattened; achenes subterete or compressed, not callous-margined, glabrous, beakless.

1. Rays white; pappus lacking; disk commonly less than 10 mm broad .... *A. cotula*
— Rays yellow; pappus a short crown; disk commonly more than 12 mm broad .... *A. tinctoria*

Anthemis cotula **L.** Mayweed. Plants annual, 1–7.5 dm tall; stems simple or branched, ill-scented; leaves 1–6 cm long, twice pinnatifid, the ultimate segments lance-oblong, sparsely villous and glandular-dotted; heads borne solitary at the upper ends of the uppermost branches; ray flowers commonly 10–20, white, sterile, 5–10 mm long; disk flowers numerous; disk 4–10 (12) mm wide; receptacle chaffy only in the middle, the bracts narrowly subulate; achenes slightly flattened, glandular, the pappus lacking; 2n = 18. Introduced Old World weeds of fields, roadsides, revegetated woodlands, and other disturbed sites at 1280 to 1400 m in Cache, Duchesne, Morgan, Salt Lake, Sanpete, Utah, Wasatch, and Weber counties (likely elsewhere); widespread in North America; 7 (0).

Anthemis tinctoria **L.** Yellow Camomile. Plants short-lived perennials, 2.5–6 dm tall; stems simple or branched; leaves 1.5–7 cm long, 1– to 2– pinnatifid, the segments oblong in outline, merely toothed or lobed, villous-tomentose below, glabrous or glabrate above, sparsely glandular-dotted; heads borne solitary at ends of the uppermost branches; ray flowers 20–35, yellow, fertile, 7–14 mm long; disk flowers numerous; disk 12–15 mm wide or more; receptacle chaffy throughout, the bracts narrow and with yellow awn-tips; achenes compressed; pappus a short crown; 2n = 18. Old World cultivated ornamentals; widely planted and occasionally escaping (Salt Lake County, Garrett 8865 BRY); widespread in North America; 1 (0).

Arctium L.

Biennial, coarse herbs with watery juice, from a taproot; leaves rhubarblike, basal and alternate, entire or toothed; heads few to numerous in axillary or terminal corymbose or racemose inflorescences; flowers all tubular, perfect, the corollas pink to purplish; involucres urn-shaped, the bracts imbricate in many series, the tips slender and inwardly hooked; receptacle flat, densely bristly; achenes slightly compressed, more or less 3-angled, many-nerved, truncate apically; pappus of numerous, scaly, deciduous bristles.

1. Heads mainly 1.5–2.5 cm thick, arranged in racemelike axillary clusters, the terminal also racemelike .... *A. minus*
— Heads commonly over 2.5 cm thick, arranged in corymbose clusters, especially the terminal .... *A. lappa*

Arctium lappa **L.** Great Burdock. Plants 8–15 dm tall; basal leaves long-petiolate, the blades commonly 2–5 dm long, 1–3 dm broad, cordate-ovate, obtuse, thinly tomentose beneath, glabrous or nearly so above; inflorescence corymbosely disposed, the peduncles glandular or glandular-hairy; heads 2.5–4 cm broad, the involucre greenish stramineus, glabrous or glandular, often sparingly arachnoid-tomentose; 2n = 36. Cultivated for its edible roots, and persisting; introduced from Eurasia; 1 (0).

Arctium minus **(Hill) Bernh.** Burdock. Plants 5–15 dm tall; basal leaves long-petiolate, the blades commonly 1–3.5 (4) dm long, 1–3 dm wide, cordate-ovate, obtuse, thinly tomentose to glabrous beneath, glabrous above or nearly so; inflorescence racemosely disposed, the peduncles short or lacking; heads 1–2.5 cm thick (rarely more), the bracts glabrous or glandular to definitely arachnoid; 2n = 36. Introduced Old World weed of consequence in Cache, Juab, Millard, Piute, Sevier, Summit, Tooele, Uintah, Utah, Wasatch, Wayne, and Weber counties, and probably cosmopolitan; widespread in North America; Eurasia; 23 (i).

Arnica L.

Perennial herbs from rhizomes or caudices, the juice watery; stems erect, simple or branched above; leaves opposite or the uppermost alternate, simple, entire or toothed; heads solitary, or 3–9 (11) in corymbose clusters; involucral bracts subequal or evidently biseriate, herbaceous; receptacle naked, convex; ray flowers present, yellow or orange, several to many, fertile, or lacking (in *A. parryi*); disk flowers numerous, perfect, fertile; pappus of barbellate or subplumose capillary bristles; style branches flattened; achenes cylindrical, 5- to 10-nerved, pubescent to glabrate or glabrous, often glandular.

Maguire, B. 1943. A monograph of the genus *Arnica*. Brittonia 4: 386–510.

1. Heads discoid (rarely some with rays), the lateral (lower) ones spreading or reflexed, the uppermost one erect .... *A. parryi*
— Heads radiate, the lateral ones (if any) erect like the uppermost 2

2(1). Cauline leaves (4) 5–9 pairs; pappus brownish; heads often 5 or more per main stem 3

— Cauline leaves 1–4 (5) pairs; pappus white or brownish; heads mainly 1–4 per stem 4

3(2). Involucral bracts merely acute to abruptly rounded (rarely acuminate), bearing an apical or subapical tuft of hairs .... *A. chamissonis*

— Involucral bracts acuminate to attenuate, not especially more hairy at the apex .... *A. longifolia*

4(2). Leaves (at least the lower) cordate, ovate, or broadly ovate-lanceolate, often cordate, truncate, or obtuse basally, seldom cuneate 5

— Leaves narrowly lanceolate to lance-oblong or lance-olate, usually cuneate basally 7

5(4). Pappus brownish, subplumose; main cauline leaves obtuse to subcuneate basally .... *A. diversifolia*

— Pappus white, merely barbellate; main cauline leaves usually cordate, truncate, or obtuse basally 6

6(5). Blades of main cauline leaves much longer than the petiole, or sessile; achenes glabrous throughout, or at least near the base .... *A. latifolia*

— Blades of main cauline leaves subequal to or shorter than the petioles; achenes uniformly, though sometimes sparingly, hairy .... *A. cordifolia*

7(4). Pappus brownish, subplumose .... *A. mollis*

— Pappus usually white or tawny, merely barbellate 8

8(7). Heads turbinate-campanulate, commonly with 7–10 rays; lower cauline leaves sessile or nearly so . *A. rydbergii*

— Heads hemispheric, commonly with 10–20 rays; lower cauline leaves often petiolate 9

9(8). Old leaf bases bearing dense brown wool in the axils; disk corollas both spreading hairy and stipitate-glandular .... *A. fulgens*

— Old leaf bases lacking axillary tufts of hair, or with white hair only; disk corollas merely stipitate-glandular .... *A. sororia*

Arnica chamissonis Less. Chamisso Arnica. [*A. foliosa* Nutt.; *A. chamissonis* ssp. *foliosa* (Nutt.) Maguire; *A. foliosa* var. *incana* Gray]. Plants 1–6 (8) dm tall, the stems erect or ascending, simple or more commonly branched in the inflorescence, sparsely to densely villous with multicellular hairs and often glandular as well; basal leaves 3–11 (15) cm long, 3–16 (20) mm wide, lanceolate to oblong or oblanceolate, with 3–5 main veins, pilose to villous or tomentose, tapering to a slender petiole, entire to distinctly toothed, smaller than the cauline ones and often withered by flowering time; cauline leaves (4) 5–8 (9) pairs, lanceolate to lance-elliptic, the largest near the middle of stem or slightly below, the lower ones petiolate and with membranous connate-sheathing bases, the upper sessile, entire to distinctly toothed; heads (1) 3–9, the peduncle apex sparingly to densely villous with whitish hairs often intermixed with glands; involucres 9–15 mm high, the bracts lanceolate, obtuse, acute, or less commonly acuminate, sparsely to densely pilose, ciliate, the tips with a conspicuous tuft of whitish hairs; rays usually 10–16, yellow; achenes 4–6 mm long, hairy to glandular or glabrate; pappus brownish to straw colored, barbellate; 2n = 60. Stream banks, gravel bars and lake shores in aspen, willow, and spruce-fir communities at 2300 to 3350 m in Box Elder, Cache, Daggett, Duchesne, Emery, Garfield, Iron, Kane, Salt Lake, Sanpete, Sevier, Summit, Tooele, Uintah, Wasatch, Washington, and Wayne counties; Alaska to Hudson Bay, south to California and New Mexico; 29 (v). Maguire (1943) treated all Utah material as ssp. *foliosa* (Nutt.) Maguire. Cronquist (Univ. Washington Publ. Biol. 17(5): 45–54. 1955) cited var. *incana* (Gray) Hulten [ssp. *incana* (Gray) Maguire] from Utah. Our specimens are only arbitrarily separable into two phases, differing mainly in degree of pubescence.

Arnica cordifolia Hook. Heartleaf Arnica. Plants 1.5–4 dm tall, the stems erect or ascending, simple or branched above, sparsely villous with multicellular hairs and often glandular as well; basal leaves smaller than the cauline, often withered at anthesis; petioles of main leaves (at least) often longer than the blades; cauline leaves 2–4 (5) pairs, the blades 2–9 cm long (from sinus to apex), 1–9 cm wide, cordate-ovate to orbicular or reniform, or the uppermost lanceolate, the largest below the middle of the stem, the lower leaves petiolate, the upper ones sessile or subsessile, serrate-dentate to subentire; heads 1 (3), rarely more, the peduncle apex villous with whitish hairs often intermixed with glands; involucres 14–20 mm high, the bracts lanceolate to oblong, acuminate to acute, sparsely to densely pilose and often glandular-ciliate, the tip with a moderate tuft of hair; rays usually 10–15, yellow; achenes 4–5.5 mm long, uniformly hairy and often glandular; pappus white, barbellate; n = 19, 38+. Sagebrush, Douglas fir, white fir, lodgepole pine, ponderosa pine, aspen, and spruce-fir communities at 1525 to 3355 m in Beaver, Box Elder, Cache, Carbon, Daggett, Duchesne, Garfield, Iron, Juab, Piute, Salt Lake, San Juan, Sanpete, Sevier, Summit, Tooele, Uintah, Utah, Wasatch, Washington, Wayne, and Weber counties (likely universal); Alaska to Michigan, south to California, Arizona, New Mexico, and Nebraska; 102 (xii). The white pappus and cordate long-petiolate leaves are diagnostic for this species.

Arnica diversifolia Greene Varying Arnica. Plants 1.5–4.2 dm tall, the stems erect or ascending, simple or branched above, sparsely villous with multicellular hairs and often glandular, or almost glabrous; basal leaves smaller than the cauline and often withered by flowering time, borne on slender to broadly winged petioles shorter than or subequal to the blades; cauline leaves 2–4 (5) pairs, blades 2–8 cm long, 0.8–4 (6) cm wide, ovate or the uppermost lanceolate, the largest at the middle or below, becoming sessile to subsessile above, subentire or irregularly serrate; heads 1–3 or more, the peduncle apex sparsely to moderately villous with whitish hairs and often with glands; involucre 10–16 mm high; bracts lanceolate, acuminate to acute, sparsely to densely pilose and often glandular, ciliate, the tip lacking a tuft of hairs; rays usually 10–15, yellow; achenes 5–7 mm long, glabrous or sparsely and uniformly hairy; pappus brownish, subplumose; 2n = 54–57, 76. Streamsides, meadows, and scree slopes in spruce-fir and alpine tundra communities at 2560 to 3400 m in Duchesne, Grand, Piute, Salt Lake, San Juan, Sanpete, Summit, Uintah, and Utah counties; Alaska and Yukon, south to Oregon and Colorado; 19 (ii). This taxon is not well collected in Utah. The broad leaves and brownish subplumose pappus are diagnostic for these plants which might be regarded as consisting of a series of hybrid derivatives between *A. mollis* and *A. cordifolia*, *A. latifolia*, or *A. rydbergii*. More work is necessary.

Arnica fulgens Pursh Orange Arnica. [*A. pedunculata* Rydb.]. Plants 1.5–6 (7) dm tall, the stems erect, the basal leaf axils with tufts of long brown woolly hair, otherwise stipitate-glandular and often hairy as well; basal leaves smaller than the cauline, often withered at anthesis, with broadly winged petioles or subsessile; cauline leaves 2–4 pair, the blades oblanceolate to elliptic (often narrowly so), mostly 3–12 cm long, 0.6–4 cm wide, the largest ones near the base, becoming sessile upwards, subentire to entire; heads 1–3, the peduncle apex yellowish villous; involucre 10–15 (18) mm high, the bracts narrowly elliptic to lance-elliptic, attenuate to an obtuse or acute apex, villous, the tips pubescent within; rays mostly 10–20, yellow to yellow orange; achenes 4–5.5 mm long, densely hairy; pappus whitish to cream colored, barbellate; 2n = 38. Dry sagebrush and meadow communities at 2000 m in Daggett County; British Columbia to Saskatchewan, south to California, Nevada, and Colorado; 2 (0).

Arnica latifolia Bong. Broadleaf Arnica. [*A. gracilis* Rydb.; *A. jonesii* Rydb., type from Alta]. Plants 1–4 (6) dm tall, the stems erect or ascending, simple or branched above, sparsely villous with multicellular hairs and often glandular; basal leaves smaller than the cauline, usually withered by flowering time, the petioles (if any) usually shorter than the blades; cauline leaves 2–5 pairs, the blades 2–4.5 (7) cm broad, cordate-ovate to lanceolate, the largest ones at the middle or below, the lower ones with petioles shorter than the blades, the upper ones sessile or subsessile, serrate-dentate, less commonly entire or nearly so; heads 1–5 or rarely more, the peduncle apex sparsely to moderately villous with whitish or yellowish hairs and often glandular; involucres 9–17 mm high, the bracts lanceolate, acuminate to acute, sparsely pilose and often glandular, ciliate, lacking an apical tuft of hair; rays usually 8–12, yellow; achenes 5–8 mm long, glabrous or sparsely hairy, or glabrous in the lower portion; pappus white, barbellate; n = 19, ca 36. Lodgepole pine, spruce-fir, and alpine tundra communities at 2240 to 3400 m in Cache, Davis, Duchesne, Salt Lake, Summit, Uintah, Utah, and Wasatch counties; Alaska and Yukon to California and Colorado; 24 (v). Specimens available for study are variable. They occur in the Uinta and Wasatch mountains and on the Tavaputs Plateau. The var. *gracilis* (Rydb.) Cronq. was reported from Utah by Maguire (l.c., as *A. gracilis* Rydb.), but has not been seen by me. It differs from the bulk of our material in its small size (1–3 dm), more numerous heads (3–9), and narrow small involucre 9–13 mm high. The single collection cited by Maguire is from Salt Lake or Utah County. More material is necessary.

Arnica longifolia D. C. Eaton in Wats. Longleaf Arnica. [*A. caudata* Rydb., type from Big Cottonwood Canyon]. Plants 3–10.5 dm tall; stems erect or sprawling, tufted from caudex-like shortened rhizomes, simple or branched above, shortly villous to puberulent and often somewhat glandular-viscid; basal leaves lacking or soon withering, the cauline ones 5.5–20 cm long, 0.6–3 cm wide, lanceolate to elliptic, with 3–5 main veins, puberulent, all sessile, 5–7 pairs, the largest near the middle of the stem, the lower ones connate-sheathing, entire or nearly so; heads 1–9, the peduncle apex sparingly yellowish villous; involucres 6–13 mm high, the bracts lanceolate to lance-oblong, acute to acuminate, sparingly pilose and glandular, ciliate, the tips sparingly white-hairy; rays mainly 8–13, yellow; achenes 4.5–5.5 mm long, glabrate, or uniformly stipitate-glandular; pappus brownish to straw colored, barbellate; 2n = 56, 76. Snow flushes, talus, and stream banks in lodgepole pine, aspen, ponderosa pine, Douglas fir, white fir, and spruce-fir communities at 1890 to 3325 m in Box Elder, Cache, Duchesne, Emery, Garfield, Grand, Juab, Rich, Salt Lake, Sanpete, Summit (type from the Uinta Mts.),, Tooele, Utah, Wasatch, and Washington counties; Washington to Alberta, south to California, Nevada, and Colorado; 16 (i). Our material belongs to var. **longifolia**.

Arnica mollis Hook. Hairy Arnica. [*A. arachnoidea* Rydb., type from Big Cottonwood Canyon; *A. chamissonis* var. *longinodosa* A. Nels., type from near Marysvale; *A. ovata* Greene, type from Alta]. Plants 1.5–6.5 dm tall, the stems erect or ascending, loosely to compactly clump-forming, simple, or branched in inflorescence, puberulent to villous and glandular; basal leaves smaller than the cauline ones, often withered at anthesis, the cauline ones 4.5–18 cm long, 0.8–4 cm wide, oblanceolate to obovate, lanceolate or elliptic, the lower slenderly to broadly petiolate, becoming sessile upwards, 3–4 pairs, the largest below the middle, the lower connate-sheathing, entire to irregularly denticulate; heads 1–5 (7), the peduncle apex sparingly yellowish villous; involucres 10–17 mm high, the bracts lanceolate to lance-elliptic, acute to attenuate, sparingly to densely villous-pilose and more or less glandular, lacking a subapical tuft of hair; rays mainly 12–18, yellow; pappus brownish, subplumose; achenes pubescent to stipitate-glandular; 2n = 38, 57, 74, 76. Meadows, bogs, streambanks, seeps, talus slopes, and rock stripes in sagebrush, ponderosa pine, lodgepole pine, Douglas fir, white fir, aspen, spruce-fir, and alpine tundra communities at 1950 to 3550 m in Beaver, Box Elder, Cache, Daggett, Duchesne, Garfield, Grand, Juab, Kane, Piute, Rich, Salt Lake, San Juan, Sanpete, Summit, Tooele, Uintah, Utah, and Wasatch counties; British Columbia to California, Nevada, and Colorado; 50 (xii).

Arnica parryi Gray Rayless Arnica. Plants l.5–5 (6) dm tall, erect or ascending, from elongate rhizomes, simple or branched in inflorescence, villous and more or less glandular; basal leaves smaller than the cauline ones, often withered at anthesis, the cauline ones long-petioled below, becoming sessile upwards, the blades 2–9.5 cm long, 0.4–4 cm wide, lanceolate to ovate, the base obtuse to truncate or cuneate, 2–4 (5) pairs, the largest near the stem base, the lower connate-sheathing, entire to denticulate; heads 3–12, nodding in bud, the peduncle apex glandular-villous; involucres 10–16 mm high, the bracts narrowly lanceolate, acute to attenuate, glandular-villous, lacking a subapical tuft of hairs; rays lacking, or rarely present, yellow; pappus brownish, barbellate to subplumose; achenes glabrous to glandular or hairy; 2n = 76. Aspen and spruce-fir communities at 2415 to 3175 m in Cache, Carbon, Daggett (?), Garfield, Iron, Rich, Salt Lake, San Juan, Sanpete, Summit, and Wasatch counties; British Columbia and Alberta to California and Colorado; 12 (ii). Some specimens from Piute and San Juan counties have ray flowers well developed.

Arnica rydbergii Greene Rydberg Arnica. Plants 1–2.6 dm tall, erect or ascending, from elongate rhizomes, sparingly villous and shortly stipitate-glandular; basal leaves smaller than the cauline, sometimes bladeless, often withered at anthesis, the cauline ones short- to long-petioled below, becoming sessile upwards, the blades 2–5 cm long, 4–15 mm wide, lanceolate to elliptic, ovate, or obovate, the base obtuse to cuneate, 2 or 3 (4) pair, the largest at or near the middle of the stem, the lower connate-sheathing, entire or denticulate; heads 1–5, the peduncle apex yellowish villous, glandular; involucres 9–13 mm high, the bracts narrowly lanceolate, acute to attenuate, stipitate-glandular, ciliate, lacking a subapical tuft of hair; rays mainly 7–10, yellow; pappus white, barbellate; achenes shortly pilose; 2n = 76. Spruce fir and lodgepole pine forests in Duchesne, Summit, and Utah counties; British Columbia and Alberta to Oregon and Colorado; 12 (iii).

Arnica sororia Greene Meadow Arnica. Plants 1.5–6 dm tall, the stems erect, the basal axils lacking tufts of hair, otherwise more or less villous and glandular; basal leaves smaller than the cauline, often withered at anthesis, with winged to narrow petioles or subsessile; cauline leaves 2–4 pair, the blades lanceolate to elliptic, mostly 3–10 cm long, 0.5–2 cm wide, the largest ones near the base, becoming sessile upwards, mainly entire; heads 1–3, the peduncle apex sparingly villous; involucres 10–15 mm high, the bracts narrowly oblong-lanceolate, attenuate, villous, the tips more or less hairy within; rays mainly 9–15, yellow; achenes 4–6 mm long, densely short-hairy; pappus white, barbellate; 2n = 38. Meadows and foothills in sagebrush and aspen communities at 1675 to 2100 m in Cache, Rich, Sanpete, and Sevier counties; Alberta and British Columbia to Wyoming, Nevada, and California; 4 (0).

Artemisia L.

Annual, biennial, or perennial herbs, subshrubs, or shrubs from taproots, caudices, or rhizomes, the juice watery; stems decumbent to ascending or erect, simple or branched; leaves alternate or basal, entire or toothed, lobed, or divided; heads several to numerous, borne in spicate, racemose, or paniculate clusters; involucral bracts imbricate in several series, dry, at least the inner with scarious margins; receptacle naked or beset with long hairs, often glandular; corollas of disk flowers only (rarely with minute bilabiate ray flowers in *A. bigelovii*), perfect, or sometimes the central ones sterile, the marginal merely pistillate; marginal corollas tubular (or bilabiate), the central ones tubular-funnelform; pappus lacking, or a short crown; style branches flattened; achenes subterete or angular, glabrous.

Beetle, A. A. 1960. A study of sagebrush - The section Tridentate of *Artemisia*. Univ. Wyoming Agr. Expt. Sta. Bull. 368. 83 pp.
Keck, D. D. 1946. A revision of the *Artemisia vulgaris* complex in North America. Proc. Calif. Acad. 25: 421–468.

1.	Plants shrubs or subshrubs	2
—	Plants herbs	14
2(1).	Heads with both ray and disk flowers, the ray flowers 2-lipped; branchlets of inflorescence spreading to reflexed; plants mainly of rimrock areas in Colorado drainage, rarely in southern Great Basin	*A. bigelovii*
—	Heads discoid; branchlets of inflorescence variously disposed; plants seldom of rimrock, the distribution various	3
3(2).	Leaves 1- to 3-pinnately or ternately dissected, the segments linear	4
—	Leaves entire or toothed or, if lobed, the lobes oblong or broader or, if linear (see *A. filifolia*), tall shrubs of sandy areas at low elevations	7
4(3).	Plants silvery-canescent; receptacle hairy, growing commonly on windswept ridges, but not always so restricted	*A. frigida*
—	Plants green to gray green; receptacle glabrous or, if hairy, plants of low elevations	5
5(4).	Shrubs with spreading branches, spinescent, flowering in springtime	*A. spinescens*
—	Shrubs with erect or ascending branches, not spinescent, flowering in late summer and autumn	6
6(5).	Plants mainly less than 2 dm tall, indigenous on harsh substrates; leaves 0.3–1 cm long	*A. pygmaea*
—	Plants mainly over 5 dm tall, introduced; leaves mainly 2.5–6 cm long	*A. abrotanum*
7(3).	Leaves linear-filiform, less than 1 mm wide, entire, or 3-parted; tall plants of sandy low elevation sites	*A. filifolia*
—	Leaves broader, entire, or the segments broader than 1 mm wide; plants of various habitats and elevations	8
8(7).	Persistent leaves entire or with 1 or 2 teeth (the ephemeral ones often tridentate); heads borne in slender panicles; plants of high elevations	*A. cana*
—	Persistent and ephemeral leaves toothed or lobed at the apex; heads borne in slender spicate to broad panicles	9
9(8).	Plants usually less than 3 dm tall; leaves usually less than 1 cm long; foliage dull yellowish to lead gray or rarely silvery	*A. nova*
—	Plants mainly more than 3 dm tall; leaves usually more than 1 cm long (at least some); foliage silvery-canescent	10
10(9).	Leaves all, or many of them, deeply cleft into narrowly oblong lobes which may be further divided; flowers commonly 5–8 per head	*A. tripartita*
—	Leaves mainly merely toothed apically; flowers various	11
11(10).	Leaves coarsely and deeply 3-lobed, the lobes broad and rounded apically; inflorescence narrow, seldom over 1.5 cm wide; plants of Rich and Summit counties	*A. longiloba*
—	Leaves variously 3- to 5-toothed, seldom lobed; inflorescence various; plants variously distributed	12
12(11).	Inflorescence open, paniculate, commonly more than 2 cm wide; plants of broad distribution, our common sagebrush species	*A. tridentata*
—	Inflorescence narrow, spicate, commonly less than 1.5 cm wide; plants less broadly distributed	13
13(12).	Leaves mainly less than 1.5 cm long; heads small; plants often less than 4 dm tall, usually of middle elevations	*A. arbuscula*
—	Leaves mostly over 2 cm long; heads large; plants often over 4 dm tall, of high elevations	*A. spiciformis*
14(1).	Leaves all entire, or the lower ones toothed or lobed, glabrous and green above and beneath, or white-hairy on both surfaces (see also *A. carruthii* and *A. michauxiana*), usually much longer than broad	15
—	Leaves deeply incised, pinnatifid, or ternately divided, variously pubescent, various in length-width proportions	16
15(14).	Leaves green above and beneath; central flowers of heads with normal ovaries	*A. dracunculus*
—	Leaves white-hairy above and beneath or green above; central flowers of head with abortive ovaries	*A. ludoviciana*
16(14).	Plants annual or biennial from a taproot; leaves green, essentially glabrous; adventive	17
—	Plants perennial from a rhizome or caudex; leaves tomentose, strigose, or pilose	18

17(16). Inflorescence paniculate, loose and open; heads borne on short peduncles; involucres 1–2 mm high .... *A. annua*

— Inflorescence a spicate panicle, the branches appressed-ascending; heads sessile or nearly so; involucres more than 2 mm high .... *A. biennis*

18(16). Cauline leaves reduced upward, the largest leaves in a basal rosette, silvery villous to strigulose, scarcely tomentose and uniformly colored above and beneath; plants from caudices, only occasionally rhizomatous .. 19

— Cauline leaves not especially reduced upwards, seldom with a basal rosette, variously tomentose and often bicolored; plants often rhizomatous (except in *A. absinthium*) 22

19(18). Pubescence of leaves loosely villous to glabrous; corollas hairy, the receptacle glabrous; plants of high elevations in the Uinta and La Sal mts. 20

— Pubescence of leaves appressed strigose or villosulose; corollas glabrous or hairy but, if hairy, the receptacle long-villous; plants variously distributed 21

20(19). Involucres 3–4 mm high; plants of the La Sal Mts. .
.... *A. parryi*

— Involucres 4–5.3 mm high; plants of the Uinta Mts.
.... *A. norvegica*

21(19). Inflorescence a spicate raceme; receptacle and corolla long-villous; plants of high elevations ... *A. scopulorum*

— Inflorescence a slender panicle; receptacle and corollas glabrous; plants of low elevations, seldom of high elevations .... *A. campestris*

22(18). Receptacle beset with numerous long hairs between the flowers; leaves about equally hairy above as below; plants introduced, weedy, of low elevations ...
.... *A. absinthium*

— Receptacle naked; leaves more or less tomentose below, usually green or greenish above, or equally tomentose on both sides; plants indigenous, not weedy, of mid- to high elevations 23

23(22). Leaves entire or with entire lobes; plants of moderate elevations in central and southern Utah ... *A. carruthii*

— Leaves bipinnatifid, the lobes again toothed; plants of high elevations in the Uinta, Wasatch, and La Sal mts. .... *A. michauxiana*

Artemisia abrotanum L. Garden Sagebrush. Fragrant shrubs, mainly 5–10 dm tall; herbage greenish; leaves (2.5) 3–6 cm long or more, (1) 2 or 3 times pinnately dissected into linear-filiform seggments, glabrous or sparsely puberulent above, sparingly tomentulose beneath; inflorescence paniculate, leafy below, 15–40 cm long, 4–15 cm broad; flowers all fertile, the outer pistillate, the others perfect; receptacle glabrous; involucre mainly 2–2.5 mm high and 2.5–3 mm broad; bracts with scarious margins, canescent or somewhat tomentose to glabrous; ray flowers 5–15, the corolla ca 1.5 mm long; disk-flowers 10–20, the corolla campanulate, glabrous; achenes 4- or 5-angled, glabrous. Saltgrass meadow at a saline seep at 1200 m in Washington County (along the Motoqua Road); introduced as a garden plant in the U. S.; native to Eurasia; 1 (i).

Artemisia absinthium L. Absinthe. Perennial fragrant herbs from a rhizomatous caudex, 5–10 (12) dm tall, appressed sericeus; leaves bi- or tripinnatifid, the main lobes again lobed or toothed, silvery-sericeus on both surfaces, with very short tangled hairs, hardly tomentose, 1.5–5.5 cm long on flowering stems (2–10 cm long on sterile stems) the main ultimate segments mostly 2–4 mm wide, petiolate below, shortly petiolate and less commonly divided above; involucres 2–3 mm high, the bracts scarious over the greenish center, the margins brownish hyaline; flowers all fertile, the marginal ones pistillate; receptacles with numerous long slender hairs; achenes glabrous; $2n = 18$. Roadsides, streambanks, and abandoned fields in Garfield, Rich, and Utah counties; widely established in North America; adventive from Europe; 6 (ii).

Artemisia annua L. Sweet Wormwood. Annual fragrant herbs, mainly 0.3–1.5 (3) m tall; stems sparingly glandular; leaves 2- or 3-pinnatifid, the main lobes again lobed, green and minutely glandular on both surfaces, 1.5–8 (10) cm long, the main ultimate segments 0.5–3 mm wide, petiolate below, subsessile or shortly petiolate above; involucres 1.3–2 mm high; involucral bracts with green centers minutely glandular, the margins hyaline; receptacles naked; achenes glabrous; $2n = 18$. Introduced weedy species of disturbed sites in Sanpete and Washington counties; adventive from Eurasia, now widely naturalized in North America; 2 (0).

Artemisia arbuscula Nutt. Low Sagebrush. [*A. tridentata* ssp. *arbuscula* (Nutt.) H. & C.; *A. tridentata* var. *arbuscula* (Nutt.) McMinn]. Shrubs, commonly 2–4 (5) dm tall, the vegetative stems 1.5–10 cm long, the flowering stems erect, 8–30 cm long; leaves 0.4–1.6 cm long, shallowly 3- to 5-dentate to deeply lobed, cuneate basally, appressed canescent; inflorescence spicate, mostly less than 2 cm wide; involucres 4–6 mm long, campanulate; involucral bracts 4–8, canescent, the margins brownish scarious; flowers 4–9, all perfect; receptacle naked; achenes glabrous; $2n = 18, 36$. Pinyon-juniper, mountain brush, sagebrush, white fir, aspen, and spruce-fir communities at 1375 to 2550 m in Box Elder, Cache, Millard, Rich, Salt Lake, Summit, and Tooele counties; Washington to Montana, south to California and Nevada; 14 (0). *A. arbuscula*, or low sagebrush, has been confused with both *A. tridentata* and *A. nova*. It can be distinguished from the former by its narrow inflorescence, and from the latter by its canescent involucres. Beetle (l.c.) reports intermediates with *A. longiloba*, a taxon with broadly campanulate heads and bluntly lobed leaves.

Artemisia biennis Willd. Biennial Wormwood. Plants annual or biennial, with taproots, the stems 0.3–9 (10) dm tall or more, glabrous; basal leaves often withered by anthesis; cauline leaves well developed, 1.5–10 (15) cm long, once pinnately divided, the segments oblong to oblanceolate, again toothed, essentially glabrous, green; inflorescence spicate or in spicate panicles; heads numerous, crowded, sessile or subsessile, erect or nearly so; involucres 2–3 mm high, 2–4 mm broad, the bracts glabrous, greenish to yellowish, the margins hyaline; marginal flowers perfect, fertile, the corollas glabrous; receptacle and achenes glabrous; $n = 9$. Floodplains, lake beds and shores, mud flats, and pond margins at 1375 to 2900 m in Box Elder, Cache, Carbon, Duchesne, Garfield, Grand, Iron, Millard, Piute, Salt Lake, Sanpete, Tooele, Uintah, and Utah counties; widespread in North America, where presumeably indigenous in the western portion; Europe; 20 (v).

Artemisia bigelovii Gray Bigelow Sagebrush. Shrubs, commonly 2–7 (10) dm tall or more, the vegetative stems 1–3 dm long, the flowering stems erect, 3–4.5 dm tall; leaves 0.3–2.3 cm long, 1–7 mm wide, entire or shallowly 3-toothed, basally cuneate, appressed to loosely canescent-tomentose; inflorescence narrowly paniculate, mostly less than 4 cm wide, the branches often lax and with heads tending to be pendulous; involucres mainly 2.5–3.5 mm high, subcylindric to narrowly campanulate, the bracts 5–10, silvery canescent, with narrow scarious margins; flowers 3 or 4, imperfect or some perfect, the marginal pistillate (ray) flowers bilaterally symmetrical; receptacle naked; achenes glabrous; 2n = 18, 26, 35, 36. Rimrock areas in pinyon-juniper and mixed desert shrub communities at 975 to 2135 m in Duchesne, Emery, Garfield, Grand, Kane, Millard, San Juan, Sevier, Uintah, and Wayne counties; California and Nevada east to Colorado, New Mexico, and Texas; 45 (xvii).

Artemisia campestris L. Field Wormwood. Perennial herbs from a caudex and taproot, the stems (1.5) 2.5–7 dm tall (rarely taller), tomentose or glabrous; basal leaves well developed (often withered at anthesis), 2–12 cm long, 2- to 3-pinnatifid or ternate, the segments linear to narrowly oblong or spatulate, villous or pilose to glabrous on both sides; cauline leaves reduced upwards, once pinnatifid, ternate, or entire; inflorescence of narrow to lax panicles; heads numerous, shortly pedunculate on contracted to lax branchlets, finally pendulous; involucres 2.5–3.8 mm high, 2–2.3 mm wide, the bracts glabrous, greenish to yellowish, the margin hyaline; marginal flowers pistillate, fertile, the corollas glabrous; disk flowers sterile, the ovaries abortive; receptacle and achenes glabrous; 2n = 18, 36. Saltbush, greasewood, sagebrush, mountain brush, and pinyon-juniper communities, mainly in dunes and other sandy sites at 1250 to 2075 m in Duchesne, Emery, Garfield, Grand, Kane, San Juan, Sevier, Washington, and Wayne counties; Arizona, New Mexico, Colorado, Wyoming, and west to the Pacific; 25 (vii). Our material is assignable to ssp. *borealis* (Pallas) H. & C., in a broad sense, and belongs to var. *scouleriana* (Benth.) Cronq. [*A. pacifica* Nutt.; *A. campestris* ssp. *pacifica* (Nutt.) H. & C.; *A. forwoodii* authors, not Wats.; *A. caudata* authors, not Michx.] in a more narrow sense.

Artemisia cana Pursh Silver Sagebrush. Shrubs, commonly 2.5–12 (15) dm tall, the vegetative branches 1–3 (5) dm long; flowering stems erect, 1–3 dm tall; leaves 0.8–5.3 (7) cm long, linear to narrowly elliptic or oblong, entire, or some of them toothed or deeply lobed, usually acute basally, acute to obtuse apically, appressed tomentose; inflorescence narrowly spicate or glomerate-paniculate, mostly less than 5 cm wide, often conspicuously bracteate, the branches, when present, erect, the heads erect; involucres 3.3–6.1 mm high, 3.5–6 mm wide, campanulate; bracts numerous, the outer silvery canescent, with greenish median, the margins brownish scarious, rounded-erose; flowers 10–20, perfect; receptacle naked; achenes glabrous; 2n = 18, 36. Meadows and stream terraces, less commonly on moist slopes away from meadows and streams at 2270 to 3050 m in Cache, Carbon, Daggett, Duchesne, Emery, Garfield, Grand, Iron, Juab, Kane, Piute, Rich, Sanpete, Sevier, Summit, Uintah, Utah, Wasatch, and Washington counties; British Columbia to Saskatchewan, south to California, Nevada, and New Mexico. Our materials are assigned to var. *viscidula* Osterh. [*A. cana* ssp. *viscidula* (Osterh.) Beetle], which differs from typical var. *cana* in its smaller, narrower leaves and less canescent herbage. Silver sagebrush forms intermediates with both *A. tridentata* var. *vaseyana* and *A. spiciformis*, within whose altitudinal range it occurs, but whose habitats are ordinarily separate; 42 (viii).

Artemisia carruthii Wood ex Carruth Carruth Wormwood. [*A. wrightii* Gray; *A. vulgaris* ssp. *wrightii* (Gray) H. & C.]. Perennial herbs, with well-developed rhizomes, the stems 2–7 dm tall, sparingly to densely tomentose; basal leaves not well developed; cauline leaves various but usually pinnatifid with linear lobes, those of innovations and sometimes the primary ones at base of flowering stems entire or merely lobed, 0.6–3 cm long, the lobes 0.5–1.5 (2) mm wide, linear or narrowly oblong, tomentose on both sides, or less so above; inflorescence paniculate (narrowly so) or spicate; heads numerous, shortly pedunculate to sessile, erect; involucres 2.3–3 mm high, 2–2.5 mm wide, the bracts sparingly tomentose, pale greenish with hyaline margins; marginal flowers pistillate, fertile; central flowers perfect, fertile, the corollas glabrous; receptacle and achenes glabrous; 2n = 18. Sagebrush, mountain brush, aspen, and spruce-fir communities at 1890 to 3050 m in Beaver, Emery, Garfield, Iron, Kane, Piute, San Juan, Sevier, Utah, Washington, and Wayne counties; east to Kansas and south to Arizona, New Mexico, and Texas; 28 (ii). This taxon is allied to *A. ludoviciana*, and some specimens appear to be intermediate between them. There is justification for inclusion of *A. carruthii* within an expanded *A. ludoviciana*, but no formal proposal is intended or implied herein. The deeply pinnatisect main foliage leaves are thought to be diagnostic. The species has not been collected in sufficient numbers as to understand its distribution in any definitive manner.

Artemisia dracunculus L. Terragon. [*A. aromatica* A. Nels.; *A. dracunculoides* Pursh]. Shortly rhizomatous, perennial herbs, the stems (2) 5–12 (15) dm tall, glabrous (rarely tomentose?); leaves primarily cauline, entire or rarely a few of them cleft, 1.2–7.5 cm long, 1–6 mm wide, glabrous, green on both surfaces; inflorescence paniculate; heads numerous, short-pedunculate to subsessile, more or less pendulous; involucres 2–2.8 mm high, 2.2–3 mm wide, the bracts glabrous, greenish, with broad hyaline margins; marginal flowers pistillate, fertile; central flowers sterile, the ovaries abortive, the corolla glabrous (often glandular); receptacle and achenes glabrous; 2n = 18, 36, 54, 90. Rabbitbrush, sagebrush, skunkbush, wildrye, salt desert shrub, pinyon-juniper, ponderosa pine, aspen, spruce-fir, and hanging garden communities at 1220 to 3200 m in all Utah counties, except Carbon, Davis, Morgan, Summit, and Weber; Yukon southeast to Illinois and south to Mexico; 87 (xviii). Our material fits within the concept of var. *glauca* (Pallas) Besser in Hook. [*A. glauca* Pallas; *A. dracunculus* ssp. *glauca* (Pallas) H. & C.], which is probably not separable from var. *dracunculus* of the Old World.

Artemisia filifolia Torr. Sand Sagebrush; Old-man Sagebrush. Shrubs commonly 5–15 dm tall, the vegetative branches 1–3 dm long; flowering branches erect, 1.5–6 dm long; leaves 0.6–8 cm long, 0.3–1.5 mm wide, revolute (appearing terete) or somewhat flattened (and still revolute), entire or the lower ternate, appressed villous-tomentose; inflorescence paniculate, mostly more than 3 cm wide, conspicuously bracteate, the branches erect, the heads pendulous; involucres 1.6–2.2 mm long 1.5–2.2 mm wide, campanulate to subglobose; bracts 5–9, densely silvery canescent; flowers 3–9, the marginal ones pistillate, fertile, the central ones sterile; receptacle naked; achenes glabrous; 2n = 18. Sandy sites in blackbrush, creosote bush, ephedra, *Poliomintha*, *Eri-*

ogonum, rabbitbrush, and pinyon-juniper communities at 825 to 2290 m in Emery, Garfield, Grand, Iron, Kane, San Juan, Washington, and Wayne counties; Colorado and South Dakota, south to Arizona, Texas, and Mexico; 43 (vi).

Artemisia frigida **Willd.** Fringed Sagebrush; Prairie Sagewort. Shrubs 0.5–4.5 dm tall, white-tomentose to strigulose; flowering stems arising from short prostrate or ascending woody offsets; leaves of basal offsets much like the stem leaves, 0.5–1.5 (2.5) cm long, 2- to 3-ternately (or subpinnately) divided into linear segments mainly 0.3–0.8 mm wide, often with stipulelike divisions near the base, whitish pilose-tomentose throughout (fading brownish); inflorescence paniculate or less commonly borne sessile or on very short peduncles; involucres 2–3.5 mm high, 4–6 mm broad, the bracts pilose-tomentose, with brownish scarious margins; marginal flowers pistillate, fertile; central flowers perfect, fertile, the corolla glabrous (often glandular), yellow or tinged reddish; receptacle long-hairy; achenes glabrous; 2n = 18. Shadscale, sagebrush, pinyon-juniper, ponderosa pine, mountain brush, aspen, spruce, and alpine (often on windswept ridge crests) communities at 900 to 3480 m in Box Elder, Cache, Carbon, Daggett, Duchesne, Emery, Garfield, Grand, Juab, Kane, Millard, Piute, San Juan, Sanpete, Sevier, Summit, Uintah, Utah, Wasatch, Washington, and Wayne counties (likely elsewhere); Alaska to Quebec, south to Arizona and Kansas; Asia; 78 (x).

Artemisia longiloba **(Osterh.) Beetle** Longleaf Sagebrush. [*A. spiciformis* (?) *longiloba* Osterh.]. Shrubs, mainly 2–5 dm tall, appressed villous-tomentose; flowering stems 1–2 dm long; leaves 0.4–2 cm long, broadly cuneate, deeply 3-lobed, the lobes obtuse, appressed villous-tomentose; inflorescence spicate, the heads several, shortly pedunculate to sessile, erect; involucres 4–6 mm high, 3–5 mm wide, the 4–12 bracts villous-tomentose, green, with brownish scarious margins; marginal and central flowers perfect, fertile, the corolla glabrous (glandular), cream colored; receptacle and achenes glabrous; n = 9, 18. Sagebrush and grass communities at 1675 to 2440 m in Rich and Summit counties; Oregon to Montana, south to Nevada and Colorado; 4 (0). This entity is reported to grow in tight to heavy soils (Beetle 1960), and matures seed in July and August. The plants have large heads similar to those of *A. cana* and the low habit of *A. nova*. Possibly they would best be treated within an expanded *A. tridentata*, but no combination is proposed herein.

Artemisia ludoviciana **Nutt.** Louisiana Wormwood. Perennial rhizomatous herbs, the stems 2–10 dm tall (or more), white-tomentose or glabrate to glabrous; leaves mainly cauline, entire, lobed, or pinnately incised, white-tomentose below, green and glabrous or tomentose above (rarely glabrous throughout), 0.8–9 cm long, 0.1–1 (2) cm wide; inflorescence spicate to paniculate; heads numerous, shortly pedunculate to sessile, more or less pendulous; involucres 2.5–4.5 mm high, 3–7 mm wide (or more), the bracts tomentose to glabrous, with broad scarious margins; marginal flowers pistillate, fertile; central flowers perfect, fertile, the corolla glabrous, yellow; receptacle and achenes glabrous; 2n = 18, 36. This is a widespread species of many phases and habitats. In Utah there are five more or less distinctive varieties. Two of the varieties, *ludoviciana* and *incompta* are especially abundant, the remaining three less so. Not all specimens are readily separable into the named varieties, and the following key is arbitrary.

1. Inflorescence an open panicle, often more than 8 cm wide; plants of southern and southeastern Utah 2
— Inflorescence a spicate panicle, usually less than 6 cm wide; plants of various distribution 3
2(1). Leaves mainly less than 2.5 cm long, the margin often narrowly revolute .... *A. ludoviciana* var. *albula*
— Leaves mainly over 2 cm long, the margins not revolute .... *A. ludoviciana* var. *mexicana*
3(1). Leaves entire or less commonly some of them toothed or lobed .... *A. ludoviciana* var. *ludoviciana*
— Leaves more or less deeply parted or divided 4
4(3). Involucres 3.5–4.2 mm high, 4–7 mm wide .... *A. ludoviciana* var. *latiloba*
— Involucres 2.5–3.5 (3.8) mm high, 2.5–5 mm wide .. .. *A. ludoviciana* var. *incompta*

Var. *albula* (Wooton) Shinn. [*A. albula* Wooton, nomen novum pro *A. microcephala* Wooton]. This distinctive short-leaved variety has open inflorescences; it occurs in riparian areas with rabbitbrush, cottonwood, and copperweed at 880 to 1680 m in Emery, Garfield, Kane, San Juan, and Washington counties; Nevada and Colorado south to Mexico; 6 (iv).

Var. *incompta* (Nutt.) Cronq. [*A. incompta* Nutt.; *A. ludoviciana* ssp. *incompta* (Nutt.) Keck]. The deeply lobed or cleft leaves and compactly spicate inflorescence are diagnostic. The plants occur at moderate to high elevations (2135 to 3500 m) in aspen, spruce-fir, willow-wet meadow, and riparian communities in Cache, Carbon, Duchesne, Juab, Millard, Piute, Salt Lake, San Juan, Sanpete, Sevier, Summit, Tooele, Utah, and Wasatch counties; Oregon to Montana, south to California, Nevada, and Colorado; 85 (viii). This variety passes by degree into *A. michauxiana* at high elevations.

Var. *latiloba* Nutt. [*A. candicans* Rydb.; *A. ludoviciana* ssp. *candicans* (Rydb.) Keck]. This variety is similar to var. *incompta*, differing in larger (higher and wider) heads. Mountain brush and montane communities in Cache, Daggett, and Utah counties; Washington to Montana, south to California and Nevada; 3 (0). This entity is poorly understood in Utah.

Var. *ludoviciana* [*A. gnaphaloides* Nutt.; *A. ludoviciana* var. *gnaphaloides* (Nutt.) T. & G.; *A. purshianus* Besser in Hook.]. The typical variety is a plant with entire or cleft (rarely deeply cleft or parted) leaves and loose, but not open, inflorescences. Rabbitbrush, sagebrush-grass, mountain brush, pinyon-juniper, ponderosa pine, and hanging garden communities at 880 to 2750 m in all Utah counties, except for Morgan, Piute, and Summit (and likely there also); British Columbia to Ontario, south to California, Arizona, New Mexico, Texas, and Indiana; 86 (xvi).

Var. *mexicana* (Willd.) Fern. [*A. mexicana* Willd.; *A. ludoviciana* ssp. *mexicana* (Willd.) Keck]. This is the long-leaved plant with open inflorescences, which forms the counterpart of var. *albula*. It is a component of riparian, pinyon-juniper, ponderosa pine, and aspen communities at 750 to 2600 m in Garfield, Grand, Kane, San Juan, Washington, and Wayne counties; Colorado to Missouri, south to Mexico; 19 (viii).

Artemisia michauxiana **Besser** Michaux Wormwood. [*A. discolor* Dougl. ex Besser]. Perennial herbs, the stems 0.8–4 dm tall (rarely more), white tomentose to glabrate or glabrous; leaves mainly cauline, 0.5–4 (5) cm long and about as broad, bipinnately dissected, the secondary segments again toothed or lobed, acute, the uppermost seldom entire, commonly green above and tomentose beneath, but often green beneath also; inflorescence spicate; heads several to numerous, commonly pedicellate, erect or nodding; involucres 3.4–4.4 mm high, 3–6 mm wide, the bracts glabrous or sparingly tomentose, green, the broad margins brownish scarious and erose-ciliate; marginal flowers pistillate, fertile; central flowers pefect, fertile, the corolla glabrous (glandular), yellow; receptacle and achenes glabrous. Spruce-lodgepole pine and alpine tundra communities, often in boulder stripes and talus, at 2950 to 3500 m in Beaver, Duchesne, San Juan, Summit, Tooele, and Utah counties; British Columbia and Alberta south to Nevada and Wyoming; 11 (ii). Keck (1946) notes that A. michauxiana is connected through a series of intermediates with A. ludoviciana var. incompta in Nevada specimens. This is true for ours also. There appears to be some justification for treating A. michauxiana within an enlarged A. ludoviciana, but such a combination is not implied herein.

Artemisia norvegica Fries Spruce Wormwood. Perennial herbs, 2–4.1 dm tall, from a simple or branched caudex and stout taproot, the caudex branches short, clothed with persistent leaf bases, the flowering stems arising directly from the caudex, villous, often reddish; leaves of basal rosettes 2–19 cm long, bi- or tripinnatifid, the segments lance-attenuate to acute, villous on both surfaces; cauline leaves becoming smaller upwards, often with stipulelike divisions near the base; inflorescence racemose; heads several to numerous, finally nodding, the peduncles to 4.5 cm long; involucres 4–5.3 mm high, 6–11 mm wide, the bracts sparingly to densely villous-pilose, more or less green, the margins broadly dark brownish scarious; marginal flowers pistillate, fertile; central flowers perfect, fertile, the corollas long-hairy from near the base, cream colored; receptacle and achenes glabrous. Spruce-fir, lodgepole pine, and alpine tundra communities in Duchesne and Summit counties; Alaska east to Mackenzie, and south to California and Colorado; 4 (iii). Our material belongs to **var. piceetorum** Welsh & Goodrich.

Artemisia nova A. Nels. Black Sagebrush. [A. tridentata ssp. nova (A. Nels.) H. & C.; A. tridentata var. nova (A. Nels.) McMinn; A. arbuscula ssp. nova (A. Nels.) G. Ward]. Shrubs, 1–3 (5) dm tall, the main branches spreading, the vegetative stems 1–3 dm long (rarely more); flowering stems mainly 1.5–3 (4) dm long; leaves 0.3–2.1 cm long, shallowly to deeply 3- to 5-lobed or -toothed, the lobes or teeth rounded, often lead gray or gray green, cuneate basally, appressed canescent and often minutely punctate; inflorescence narrowly paniculate, seldom more than 3 cm wide; involucres 3.1–5.8 mm long, 1.4–3.4 mm wide, cylindric to narrowly campanulate; bracts 8–12, canescent to glabrous, green to yellowish, the margin hyaline; flowers 3–8, all perfect; receptacle glabrous; achenes glabrous; 2n = 18, 36. Horsebrush, greasewood, shadscale, ephedra, juniper, sagebrush, rabbitbrush, winterfat, pinyon-juniper, and mountain brush communities at 1400 to 2600 m in Beaver, Box Elder, Cache, Carbon, Daggett, Duchesne, Emery, Garfield, Grand, Iron, Juab, Kane, Millard, Piute, Rich, San Juan, Sanpete, Sevier, Summit, Tooele, Uintah, and Weber counties; Oregon to Montana, south to California, Arizona, and New Mexico; 57 (x). Black sagebrush forms intermediates with all other members of the section Tridentatae that it contacts. The intermediates form narrow bands along lines of contact, but generally the habitats are mutually exclusive. There is little justification for considering black sagebrush in an expanded A. tridentata unless one is willing to accept most of the remainder of the section as portions of that species also.

Artemisia parryi Gray Parry Wormwood. Perennial herbs, 0.8–2 (4) dm tall, from a simple or branched caudex and stout taproot, the caudex branches short, clothed with persistent leaf bases, the flowering stems arising directly from the caudex, sparingly and loosely villous to glabrous, often reddish; leaves of basal rosettes 2–4 (8) cm long, bipinnatifid, the segments oblong to lance-oblong, sparingly and loosely villous (to glabrous?); cauline leaves becoming smaller upwards; inflorescence racemose to subspicate; heads several to numerous, commonly nodding, the peduncles 1–5 mm long; involucres 3–4 mm long, 3–5 mm wide, the bracts sparingly villous to glabrate, with green to brownish middle and brownish scarious margins; marginal flowers pistillate, fertile; central flowers perfect, fertile, the corollas long-hairy (to glabrous?); receptacle and achenes glabrous. Alpine sites in the La Sal Mts. (Grand and San Juan counties); Colorado; 0 (0). The species is reported for Utah by Hall and Clements (l.c.), but no specimens have been seen from the state by me. Possibly it is only a phase of A. norvegica.

Artemisia pygmaea Gray Pygmy Sagebrush. Shrubs 0.5–2 dm tall, from superficial woody caudexlike branches and stout taproots, the vegetative stems to 0.5 dm long; flowering stems erect, to 2 dm tall; leaves 0.3–1 cm long, pinnately (or subbipinnately) 3- to 10-lobed, the lobes acute, yellow to gray green, sparingly villous to glabrous; inflorescence spicate or narrowly paniculate, less than 2 cm wide; involucres 5.2–6.3 mm high, 3–4.5 mm wide, cylindric or becoming campanulate upon drying; involucral bracts oblong, 15 or more, sparingly villous to glabrous, green, the margins stramineous hyaline; marginal flowers lacking; central flowers 3–5, perfect, fertile, the corollas cream colored, glandular; receptacle and achenes glabrous; 2n = 18, 36. Black sagebrush, rabbitbrush, shadscale, greasebush, juniper, pinyon-juniper, and ponderosa pine communities at 1600 to 2300 m in Beaver, Box Elder, Carbon, Duchesne, Emery, Garfield, Iron, Juab, Millard, Piute, Sevier, and Uintah counties; Arizona and Nevada; 32 (viii). This dwarf sagebrush occurs in peculiar edaphic situations on Green River Shale, in clay soils forming the matrix in igneous gravels, on calcareous gravels, and on dolomitic outcrops and gravels. It is often a component of communities that support rare plant species.

Artemisia scopulorum Gray Dwarf Sagewort. Perennial herbs, 0.5–3.7 dm tall, from a simple or branched caudex and stout taproot, the caudex branches short, clothed with persistent leaf bases, the flowering stems arising directly from the caudex, appressed pilose to loosely and sparingly villous, often reddish or purplish; leaves of basal rosettes 1.5–9 cm long, bipinnatifid, the segments oblong to elliptic, pubescent like the stems; inflorescence spicate to racemose; heads several to numerous, erect or nodding, the peduncles lacking, or to 2.3 cm long; involucres 3–5.2 mm high, 3–8 mm wide, the bracts villous, green to brownish in the middle, the margins brown scarious; marginal flowers pistillate, fertile; central flowers perfect, fertile, the corollas cream colored, long-hairy; receptacle copiously long-villous; achenes glabrous; 2n = 18. Talus slopes, moraines, and outwash plains and terraces in alpine tundra and meadows in spruce, lodgepole pine, and Douglas fir communities at 3050 to 4000 m in Boulder, Tushar, La Sal, and Uinta mts.; Beaver, Daggett, Duchesne, Garfield, Grand, Piute, San Juan, Summit, Uintah, and Wayne counties; Montana, Wyoming, Colorado, and New Mexico; 30 (xi). The hairy corollas and long-villous receptacles are diagnostic for this distinctive species.

Artemisia spiciformis Osterh. Osterhout Sagebrush. [*A. tridentata* ssp. *spiciformis* (Osterh.) Goodrich & McArthur]. Shrubs, mainly 5–8 dm tall, the vegetative stems 0.4–1 dm long, the flowering stems erect, 1.5–3.4 dm long; leaves 1.7–5.7 cm long, shallowly to deeply 3- to 5-lobed or -toothed, often widest below the teeth, the lobes acute to obtuse (or rounded) or lacking, gray green, long-cuneate basally, appressed villous-canescent; inflorescences narrowly paniculate, usually less than 4 cm wide; involucres 5–6.3 mm long, 3.5–7 mm wide, cylindric to campanulate; involucral bracts 8–12 or more, canescent to glabrate, green, with broad yellowish brown scarious margins; flowers 6–10 or more, all perfect; receptacle and achenes glabrous. Ridge margins and snowflushes in sagebrush-grass, snowberry, aspen, spruce-fir, and Douglas fir communities at 2680 to 3050 m in Cache, Duchesne, Emery, Juab, Sanpete, Sevier, Summit, Tooele, Utah, and Wasatch counties; Colorado and Wyoming; 16 (ii). The plants flower in July and August. This is the material that has long passed under the name of *A. rothrockii* Gray in Utah. Resemblance to that species appears to be superficial, with relationships running to both *A. cana* and *A. tridentata* var. *vaseyana*. Its habitat is intermediate between the high elevation, moderately xeric conditions, of var. *vaseyana* and the more mesic stream terrace and valley bottom of *A. cana*.

Artemisia spinescens D. C. Eaton in Wats. Budsage. Shrubs, flowering in springtime, the branches spreading and often prostrate, 0.5–3 dm long or more, the vegetative stems mainly 0.3–0.8 dm long, commonly surpassing the flowering stems; leaves 0.4–2 cm long, petiolate, the blade palmately 3- to 5-cleft, the main divisions again cleft, suborbicular in outline, villous; inflorescence of short leafy-bracted racemose or spicate branches, or of solitary heads, the rachis persistent as a thorn; involucres 2–3.5 mm high, 3.5–5 mm wide; involucral bracts 4–8, villous, green, with narrow hyaline margins; flowers 6–20 or more, the marginal ones pistillate, fertile, the central ones sterile; corollas copiously long-hairy; receptacle naked; achenes long-hairy; 2n = 18, 36. Silty, clayey, or gravelly, often saline, substrates in black sagebrush, shadscale, tetradymia, greasewood, blackbrush, juniper, and winterfat communities at 1200 to 1925 m in Beaver, Box Elder, Carbon, Duchesne, Emery, Garfield, Grand, Iron, Juab, Kane, Millard, Piute, San Juan, Sevier, Tooele, Uintah, and Utah counties; Oregon to Montana, south to California and New Mexico; 92 (vii). This low shrub is a principal browse plant for domestic livestock on the spring ranges of western and southern Utah.

Artemisia tridentata Nutt. Big or Common Sagebrush. Shrubs 4–20 (30) dm tall; branches spreading to erect, the vegetative branchlets 0.5–2 dm long; flowering stems mostly 1.5–4 dm long, usually much surpassing the vegetative ones; leaves 0.5–5 cm long, 3- to 5-toothed apically, or the upper ones entire, long-cuneate; inflorescence paniculate, 3–20 (15) cm wide; involucres 3–5 mm long, 2–4 mm wide, the bracts 10–20, green, canescent, the margins scarious; flowers 3–8, all perfect, the corollas cream colored, glandular; receptacle and achenes glabrous; 2n = 18, 36. Three more or less completely intergrading varieties are known from Utah; they tend to occupy distinctive habitats, but intermediates form wherever they meet. While it is not possible to segregate all specimens, the following key will prove useful to those who must manage the sagebrush lands of Utah and the west.

1. Vegetative stems short, standing at about the same height, the inflorescence rather uniformly overtopping them; plants of middle and higher elevations [ssp. *vaseyana* (Rydb.) Beetle] 2
— Vegetative stems short to long, the inflorescence not uniformly overtopping them; plants of low to moderate elevations [ssp. *tridentata*] 3

2(1). Flowers 4–6 per head; involucres less than 1.5 mm wide; inflorescence paniculate with numerous heads .... *A. tridentata* var. *pauciflora*
— Flowers 7–11 per head; involucres over 1.5 mm wide; inforescence narrow and spiciform, with relatively few heads .... *A. tridentata* var. *vaseyana*

3(1). Leaves mainly to 2 cm long or more, narrowly cuneate; plants of low to moderate elevations .... *A. tridentata* var. *tridentata*
— Leaves mainly less than 1.2 cm long, cuneate to cuneate-flabellate; plants mainly of moderate elevations, in drier sites ..... *A. tridentata* var. *wyomingensis*

Var. *pauciflora* Winward & Goodrich. Sagebrush, pinyon-juniper, mountain brush, and aspen communities at 1830 to 3050 m in all Utah counties; Washington to the Dakotas, south to California and Colorado; U. S.; 50 (v).

Var. *tridentata* Big Sagebrush. Sagebrush, juniper, pinyon-juniper, and rabbitbrush communities, in deep alluvial soils, at 1220 to 2410 m in most, if not all, Utah counties; Washington to Montana, south to California, Arizona, and New Mexico; 57 (xviii).

Var. *vaseyana* (Rydb.) J. Boivin Vasey Sagebrush. [*A. tridentata* ssp. *vaseyana* (Rydb.) Beetle, sens. str.]. Sagebrush, rabbitbrush, mountain brush, pinyon-juniper, aspen, Douglas fir, ponderosa pine, and spruce-fir communities at 2865 to 3050 m in Duchesne, Emery, and Sanpete counties; Oregon, Washington, and Idaho; 3 (i).

Var. *wyomingensis* (Beetle & A. Young) Welsh [*A. tridentata* ssp. *wyomingensis* Beetle & A. Young]. Wyoming Sagebrush. Shadscale, rabbitbrush, sagebrush, juniper, bitterbrush, and mountain mahogany communities at 1525 to 1980 m in Box Elder, Garfield, Emery, Millard, Rich, Tooele, and Uintah counties; Wyoming and Idaho to Colorado. This is the sagebrush of drier sites at middle elevations. Its distribution is poorly understood; likely it is widespread. Its recognition allows management considerations by professionals in the various state and federal agencies; 9 (0).

Artemisia tripartita Rydb. Threetip Sagebrush. Shrubs 2–20 dm tall, the branches erect, the vegetative ones 0.3–1.5 dm long, the flowering stems 0.6–3.5 dm long; leaves 1–4 cm long, deeply 3-cleft, the linear lobes 0.5–0.8 mm wide, canescent, the lobes sometimes again divided, or the upper ones entire; inflorescence paniculate, commonly 2–5 cm wide; involucres campanulate, 3–4 mm long, 1.5–4 mm wide; bracts many, imbricate, canescent and more or less green, the inner with broad brownish scarious margins; flowers 4–8, all perfect, the corollas stramineus to cream colored, more or less glandular; achenes and receptacle glabrous; n = 9. Sagebrush and mountain brush communities at ca 1525 to 1830 m in Box Elder and Cache counties; British Columbia to Montana, south to California and Colorado; 1 (0).

Aster L.

Annual or perennial herbs from rhizomes (suffrutescent in *A. spinosus*) or caudex, with watery juice; stems decumbent to ascending or erect, simple or branched; leaves alternate, simple, entire or toothed; heads solitary or few to several in corymbose clusters; involucral bracts strongly imbricate to subequal (or the outer surpassing the inner), herbaceous throughout, or with scarious margins near the base; receptacle flat or merely convex, naked; rays blue, purple, pink, or white, few to numerous, pistillate; disk flowers numerous, perfect, fertile, yellow or tinged reddish or purplish; pappus of capillary bristles; style branches flattened, oblong to lanceolate, mostly more than 0.5 mm long; achenes mostly several-nerved.

1. Plants suffrutescent, rushlike, armed with axillary or subaxillary thorns, from a deep-seated rhizome; known from Garfield, Kane, San Juan, and Wayne counties .... *A. spinosus*
— Plants herbaceous, annual or perennial, unarmed, from a taproot or rhizome; distribution various 2

2(1). Plants annual, from taproots 3
— Plants perennial, from rhizomes or subrhizomatous caudices, or from branching caudices 4

3(2). Involucral bracts definitely acute; rays wanting or nearly so, the pistillate corollas tubular, shorter than the style .... *A. brachyactis*
— Involucral bracts obtuse to obtusish; rays to 2 mm long, longer than the style .... *A. frondosus*

4(2). Plants with a well-developed caudex; involucral bracts reflexed, at least the outer; plants of rock crevices in the Wasatch and Canyon mts. ...... *A. kingii*
— Plants with caudex lacking or poorly developed, rhizomatous; involucral bracts not reflexed; plants of various habitats and localities 5

5(4). Leaves all erect-ascending, thickened, to about 4 mm wide; pappus double, the outer series of very short bristles; heads solitary; plants known from Box Elder County .... *A. scopulorum*
— Leaves various, seldom as above; pappus in one series, or rarely double; heads solitary to numerous; distribution various 6

6(5). Involucral bracts dry, chartaceous, with scarious tips (at least the innermost), with a distinctive mid-vein, not herbaceous (the outer sometimes so) 7
— Involucral bracts herbaceous at the tips or throughout, lacking a distinctive midvein 9

7(6). Involucral bracts (at least the outer) bluntly obtuse apically; herbage strongly glaucous; plants often of open calcareous sites .... *A. glaucodes*
— Involucral bracts acute; herbage green, not glaucous; plants of various habitats 8

8(7). Rays white (drying pinkish); main leaves often over 20 mm wide; plants 6–15 dm tall, of montane areas in central northern Utah .... *A. engelmannii*
— Rays purple or violet; leaves mainly less than 15 mm wide; plants 2–6 dm tall, of central northern and western Utah .... *A. perelegans*

9(6). Involucres and peduncles glandular 10
— Involucres and peduncles lacking glands or apparently so 14

10(9). Stems glabrous; leaves linear to linear-oblanceolate, 2–5 mm wide, 1.5–7 cm long; plants of saline or hot water seeps and springs .... *A. pauciflorus*
— Stems puberulent to villous or hirsute with multicellular hairs or glabrous but, if so, differing in other respects 11

11(10). Rays white; leaves glaucous; plants of central to south-central Utah .... *A. wasatchensis*
— Rays blue to purple, lavender, or violet 12

12(11). Cauline leaves not or only slightly clasping, 2–10 mm wide; involucres 5–8 mm high .... *A. campestris*
— Cauline leaves clasping the stem, mainly 15–40 mm wide; involucres 8–15 mm high 13

13(12). Stems mostly 4–10 dm tall or more; lower leaves sessile; rays typically more than 30 per head; plants introduced, escaping and established .. *A. novi-angliae*
— Stems mainly 2–4 dm tall; lower leaves petiolate; rays mainly less than 25; plants indigenous, in central northern Utah .... *A. integrifolius*

14(9). Pubescence occurring in decurrent lines below leaf bases, commonly not uniform below the heads, or only in the inflorescence; inflorescence often conic, mostly large and leafy .... *A. hesperius*
— Pubescence of stem uniform or, if in decurrent lines, uniform below the heads and confined to the inflorescence; inflorescence few- to many-flowered and not usually leafy to large and leafy (see also *A. eatonii*) .. 15

15(14). Rays white; involucral bracts strigulose dorsally (rarely glabrous), with spreading to squarrose minutely spinulose tips; heads numerous 16
— Rays pink to purple, or less commonly white; involucral bracts mucronate at the tip; heads few to numerous 18

16(15). Pappus purplish or brownish; involucres 6–9 mm tall, greenish throughout or commonly suffused with purple; plants rare in the Uinta Mts. .... *A. sibiricus*
— Pappus typically white; involucres less than 6.5 mm long; plants of various distribution 17

17(16). Rhizomes well developed, creeping; involucres 4.6–6.5 mm high, 7–9.5 mm wide (when pressed); plants of western Utah .... *A. falcatus*
— Rhizomes mainly poorly developed, or reduced and caudexlike; involucres 3.8–4.9 mm high, 4.5–6 mm wide; plants of eastern Utah .... *A. pansus*

18(15). Achenes glabrous or nearly so; herbage glabrous except for lines of puberulence in the inflorescence, tending to be glaucous; rare plants in southeastern Utah .... *A. laevis*
— Achenes pubescent, except in some *A. foliaceus*; herbage pubescent to almost glabrous, scarsely glaucous 19

19(18). Involucral bracts strongly imbricate, the outer ones at least obtuse or obtusish (sometimes acute), not foliaceous; pubescence below the heads harsh .. *A. chilensis*
— Involucral bracts not strongly imbricate or, if so, the bracts sharply acute, the outer ones acute or, if obtuse, foliaceous; pubescence below heads soft or minute 20

20(19). Inflorescence a long slender leafy panicle; heads numerous; stem pubescence short, uniform; leaves mostly more than 7 times longer than wide; rays usually pink to white .... *A. eatonii*

— Inflorescence an open or congested panicle; heads solitary to several; pubescence various; rays usually blue to violet 21

21(20). Involucral bracts slender, never foliaceous; leaves at mid-stem mostly less than 1 cm wide, mostly over 7 times longer than broad .... *A. occidentalis*

— Involucral bracts various, but some of them usually enlarged and foliaceous; leaves at mid-stem mostly less than 7 times longer than broad .... *A. foliaceus*

Aster brachyactis **Blake in Tidestr.** [*Tripolium angustum* Lindl. in Hook.; *A. angustus* (Lindl.) T. & G., not Nees; *Brachyactis angustus* (Lindl.) Britt. in Britt. & Br.]. Annual herbs, with taproots, glabrous throughout, except for leaf margins and involucral bracts; stems 0.9–5.3 (7) dm tall; leaves 1.3–8 (12) cm long, 1–7 (9) mm wide, linear to narrowly oblong, entire, the lower ones soon deciduous; heads few to numerous, in paniculate to spicate inflorescences; involucres 5.5–9.4 (11) mm high, 7–15 (17) mm wide, the bracts linear-oblong, acute to attenuate, herbaceous, subequal to somewhat imbricate, or some outer ones often surpassing the inner; marginal flowers pistillate, the corollas tubular filiform, lacking rays, much shorter than the styles; pappus abundant, white, longer than the corollas; $2n = 14$. Sandbars, terraces, stream banks, marshes and pond margins, often where saline, in tamarix, rush, rabbitbrush, and cottonwood communities at 1220 to 1525 m in Box Elder, Carbon, Davis, Duchesne, Emery, Garfield, Grand, Juab, Millard, Salt Lake, Sevier, Uintah, Utah, Wayne, and Weber counties; British Columbia to Minnesota, south to Washington and Colorado; 23 (i).

Aster campestris **Nutt.** Meadow Aster. Perennial rhizomatous herbs, glandular, at least in inflorescence; stems puberulent to glabrous, mainly 1–5 dm tall; leaves 2–8 cm long, 2–8 mm wide, linear to oblong, entire, sessile, sometimes clasping, the lower ones larger and more or less petiolate, or smaller, soon deciduous; heads solitary or several to many; involucres 5–8 mm high, glandular, the bracts subequal to definitely imbricate, acute or attenuate, with long herbaceous tips; rays 15–20, violet to purple, 6–12 mm long; $2n = 10$. Meadows at 1525 to 2475 m, reported for Utah (Univ. Washington Publ. Biol. 17(5): 77. 1955), but I have seen no specimens from the state.

Aster chilensis **Nees** Pacific Aster. [*A. halophilus* Greene, type from Becks Hot Springs, Salt Lake County; *A. leucopsis* Greene, type from Salt Lake City]. Perennial rhizomatous to subrhizomatous herbs, uniformly harshly strigose to strigulose, at least above; stems (0.8) 1.2–10.5 dm tall; leaves 0.6–16.5 cm long, 2–16 (20) mm wide, entire or nearly so, pubescent to glabrous, ciliate, the lower ones more or less petiolate, often deciduous at anthesis in taller plants, becoming smaller and sessile upwards, sometimes markedly reduced bracteate in inflorescence; inflorescence of 1 to many heads, narrow, corymbose, or open paniculate; involucres 5–8 mm high, 6–15 mm broad, the bracts imbricate, green-tipped (machaerantheroid), the chartaceous bases white to straw colored, the outer ones abruptly pointed but mucronate; rays commonly 15–40, purplish to violet (rarely white) or pink, 5–15 mm long; achenes pubescent; $n = 13, 18, 27, 36$. Alluvial fans, terraces, and slopes along stream and canal banks, in hanging gardens, rabbitbrush, sagebrush, grass-sedge, cottonwood-willow, ponderosa pine, juniper-pinyon, mountain brush, aspen, and spruce-fir communities at 850 to 3200 m in all Utah counties; Washington to Saskatchewan, south to California and New Mexico; 189 (xxviii). The Pacific aster is a generalized taxon with no clearly defined diagnostic features. It is separated from its near congeners by a group of intangible characteristics. Involucral bracts are definitely imbricate, with the greenish portion usually glabrous, and margins ciliate. The tips of outer bracts are often but not always obtuse, and the tip, even when abruptly contracted, is mucronate. These features, which I designate as "machaerantheroid," are shared to a greater or lesser extent with *A. eatonii*, *A. occidentalis*, and *A. foliaceus*. The harsh pubescence below the heads appears to be diagnostic, but is difficult to distinguish from the soft or merely puberulent vesture of closely related species. Not all specimens can be assigned with certainty to any of the taxa. There are two intergrading morphological phases of the Pacific aster, which are striking in their extremes, but which probably represent nothing more than developmental gradients. These are represented by plants with few flowers that lack distinctive reduced bracteate leaves in the inflorescence, and taller plants with more numerous heads and distinctively bracteate inflorescences. The inflorescences of the taller plants are mainly corymbiform, and not cylindroid as in *A. eatonii*. More work is indicated. Our material belongs to ssp. ***adscendens*** **(Lindl.) Cronq.** [*A. adscendens* Lindl. in Lindl.].

Aster eatonii **(Gray) Howell** Eaton Aster. [*A. foliaceus* var. *eatonii* Gray; *A. oregonus* authors, not (Nutt.) T. & G.]. Perennial rhizomatous to subrhizomatous herbs, uniformly puberulent, at least above (below the heads and sometimes on upper leaves), the stems (2.7) 6–10.5 dm tall, often reddish; leaves 0.8–15 cm long, 2–25 mm wide, entire or serrate, puberulent to glabrous, ciliate, the lowermost shortly petiolate, often deciduous in anthesis, becoming smaller and sessile upwards, linear to narrowly elliptic or lanceolate to oblanceolate; inflorescence of few to numerous heads, commonly open-cylindric to conic in form; involucres 4.5–8 (10) mm high, 6–10 mm wide, the bracts more or less subequal to indistinctly imbricate, green-tipped (but not especially machaerantheroid), the chartaceous bases white to straw colored, all or most of them mucronate; rays 20–40, commonly pink (sometimes white), 5–12 mm long; achenes pubescent; $n = 8, 9$. Gravel bars, stream terraces, meadows, canal banks, hanging gardens, and marshes at 1370 to 2325 m in Box Elder, Cache, Garfield, Grand, Iron, Juab, Kane, Salt Lake, Summit, Uintah, Utah, Wasatch, and Washington counties; British Columbia to Saskatchewan, south to California, Arizona, and New Mexico; 48 (ix). The pink or white rays, uniform upper stem puberulence and leaves many times longer than broad are diagnostic for most specimens. Reports of *A. junciformis* Rydb. for Utah appear to be based on slender phases of the Eaton aster with linear leaves and slender rhizomes.

Aster engelmannii **(D. C. Eaton) Gray** Engelmann Aster. [*A. elegans* var. *engelmannii* D. C. Eaton]. Perennial rhizomatous or subrhizomatous herbs, puberulent to sparingly villous with multicellular hairs, or somewhat glandular, the stems 2–15.2 dm tall, reddish at the base; leaves 2–13.5 cm long, 3–46 mm wide, elliptic to lanceolate, entire (or nearly so), sparingly puberulent to glabrous or sparsely villous, sessile, largest near midstem, the lowermost reduced to scales; inflorescence of 1 to numerous large heads, corymbose or conic; involucres 8–13 mm high, 11–25 mm wide, the bracts mainly strongly imbricate, with a definite midvein, commonly purplish (at least the inner), the outer sometimes green and more or less foliaceous, sometimes all greenish or straw colored to the tip, glabrous dorsally, ciliate; rays 8–23, white (drying pinkish), 12–25 mm long; achenes pubescent; n = 9. Mountain brush, juniper, Douglas fir, aspen, white fir, lodgepole pine, and spruce-fir communities at 1950 to 3200 m in Cache, Carbon, Davis, Duchesne, Juab, Rich, Salt Lake, Sanpete, Sevier, Summit, Tooele, Utah, and Wasatch counties; British Columbia and Alberta, south to Nevada and Colorado; 57 (vi).

Aster falcatus Lindl. [*A. multiflorus* var. *commutatus* T. & G.; *A. commutatus* (T. & G.) Gray]. Perennial rhizomatous herbs, villous or villous-hirsute with multicellular hairs, the stems 2.8–7.5 dm tall; leaves 1.2–6 (8) cm long, 2–8 mm wide, entire, antrorsely scaberulous on both surfaces (or glabrous), sessile, linear to narrowly oblong, often spinulose-mucronate, the lowermost often lacking at anthesis; inflorescences several- to many-headed, cylindroid; involucres 4.6–6.5 mm high, 7–9.5 mm wide, the bracts strongly to only somewhat imbricate, with a green tip, scaberulous to glabrous dorsally and ciliate; rays mainly 17–25, white (drying pale lavender in some), 6–8 mm long; achenes pubescent; n = 9. Oak, sagebrush, and ponderosa pine communities at 1525 to 2135 m in Kane, Utah, and Washington counties; Alaska to Minnesota, south to California, New Mexico, and Kansas; 6 (i). The species is closely allied to *A. pansus* (q.v.), which has smaller heads.

Aster foliaceus Lindl. in DC. Leafybract Aster. Perennial rhizomatous or subrhizomatous herbs, uniformly and shortly soft-villous below the heads, uniformly villous to glabrous below, or in lines below leaf bases, the stems 1.3–7 dm tall; leaves 1.8–16 cm long, 3–34 mm wide, entire or nearly so, strigose to glabrous, ciliate, the lower ones petiolate (often lacking at anthesis), becoming smaller and sessile (and more or less clasping) upwards; inflorescence of 1–19 (50) corymbosely arranged large and showy heads; involucres 6–12 mm high, 10–20 mm wide, the bracts imbricate to slightly so, foliaceous or slender, green with pale white to yellowish or brownish chartaceous bases (at least the inner), acute to obtuse or rounded, mucronate; rays mainly 15–50, pink to purple, blue, or violet, 9–16 (20) mm long; achenes hairy; n = 8, 25, 32. The leafybract aster is a portion of an assemblage that includes the concept of *A. subspicatus* Nees. Both *A. foliaceus* and *A. subspicatus* were described from coastal Alaska (Unalaska and Yakutat Bay respectively). Brownish bases of involucral bracts, commonly serrate leaves, and reddish pappus are supposedly diagnostic for *A. subspicatus*, which is not known from Utah, but some specimens of *A. foliaceus* have one of more of these features. In the Alaska Flora (Welsh 1974), I treated both species under the older name of *A. subspicatus*. Now, I follow tradition so as to avoid creation of synonyms should further study indicate a better course of action. Three more or less distinctive infraspecific taxa are present in Utah.

1. Involucral bracts foliaceous, 2–6 mm broad; plants uncommon .... *A. foliaceus* var. *canbyi*
— Involucral bracts not especially foliaceous, mainly less than 2 (2.5) mm wide; plants common to uncommon ... 2

2(1). Plants mainly 0.5–2.5 dm tall, decumbent or ascending; bracts often purple-margined or -tipped; known from high elevations, rare .... *A. foliaceus* var. *apricus*
— Plants often more than 2 dm tall, erect; bracts seldom as above; known from low to high elevations, common .. .. *A. foliaceus* var. *parryi*

Var. *apricus* Gray Meadows in spruce-fir forest at 3050 to 3660 m in Sanpete and Summit counties; British Columbia to Montana, south to California and Colorado; 3 (0).

Var. *canbyi* Gray Mountain brush, aspen, and spruce-fir communities at 1950 to 2900 m in Box Elder, Cache, Duchesne, Iron, Juab, Salt Lake, Summit, Utah, and Wasatch counties; Washington to Wyoming, south to California and New Mexico; 5 (0).

Var. *parryi* (D. C. Eaton) Gray [*A. adscendens* var. *parryi* D. C. Eaton; *A. foliaceus* var. *frondeus* Gray]. Meadows and openings in aspen, spruce, lodgepole pine, and Douglas fir communites at 1890 to 3265 m in Cache, Daggett, Duchesne, Garfield, Grand, Iron, Juab, Piute, Salt Lake, Sanpete, Summit, Tooele, Uintah, and Utah counties; Washington to Wyoming, south to California and New Mexico; 36 (viii). This is the phase of the leafy-bract aster which simulates *A. occidentalis* (q.v.), but which seldom has long peduncles, dark blue purple ray corollas, and much reduced upper stem leaves of that species.

Aster frondosus (Nutt.) T. & G. Leafy Aster. [*Tripolium frondosum* Nutt.; *Brachyactis frondosa* Gray]. Annual herbs from taproots; stems 0.2–3.6 cm tall; leaves 1–6 cm long, 2–12 mm wide, linear to oblong or oblanceolate, entire, the lower ones sometimes deciduous; heads few to numerous, in a narrow paniculate to spicate inflorescence; involucres 5–9 mm high, 6–13 mm wide, the bracts oblong to narrowly oblanceolate, obtuse or obtusish, herbaceous, subequal to moderately imbricate; marginal flowers pistillate, the rays developed, pink, to 2 mm long; pappus abundant, white, longer than the disk colollas. Lake shores, seep margins, wet meadows, and stream banks in salt-grass, tamarix, Russian olive, rabbitbrush, and greasewood communities at 1250 to 2270 m in Beaver, Duchesne, Garfield, Grand, Juab, Kane, Salt Lake, San Juan, Sanpete, Uintah, Utah, Washington, and Wayne counties; Washington to Wyoming, south to California and New Mexico; 26 (vi).

Aster glaucodes Blake Blueleaf Aster. Perennial rhizomatous herbs, glabrous and glaucous, or puberulent to glandular in the inflorescence; stems 1.1–7 dm tall; leaves 1.4–12.5 cm long, 4–25 mm wide, entire, lance-oblong to oblong or elliptic, glaucous, glabrous, sessile and clasping, the lower often lacking at anthesis, reduced upwards; heads few to numerous in corymbose inflorescences; involucres 6–9 mm tall, 7–9 mm wide, the bracts imbricate, dry, chartaceous throughout or sometimes some of them greenish, the midvein prominent, commonly suffused with pink or purple, mainly obtuse to less commonly acute apically; rays 10–20, white or pink, 11–17 mm long; n = 9. There are two varieties among our specimens.

1. Peduncles and/or involucres glandular-pubescent; plants of Washington and adjacent western Kane counties .... *A. glaucodes* var. *pulcher*
— Peduncles and involucres lacking glandular pubescence; plants widespread .... *A. glaucodes* var. *glaucodes*

Var. glaucodes This is the common phase of the species, often on calcareous substrates at higher elevations and in saline seeps at moderate to lower elevations in sagebrush, pinyon-juniper, mountain brush, ponderosa pine, ryegrass, spruce-fir, Douglas fir, lodgepole pine, and hanging garden communities at 1220 to 3050 m in Cache, Carbon, Daggett, Duchesne, Emery, Garfield, Grand, Kane, Salt Lake, San Juan, Sanpete, Sevier, Summit, Tooele, Uintah, Utah, Wasatch, Washington, and Wayne counties; Idaho and Wyoming, south to Arizona and Colorado; 56 (ix).

Var. pulcher (Blake) Kearney & Peebles [*A. glaucodes* ssp. *pulcher* Blake, type from Elk Ranch, Kane County]. Salt desert shrub, sagebrush, pinyon-juniper, and ponderosa pine communities at 825 to 2136 m in Washington and adjacent western Kane counties; Arizona; 6 (0).

Aster hesperius Gray Siskiyou Aster. [*A. laetivirens* Greene]. Perennial rhizomatous herbs, villous with multicellular hairs in decurrent lines from leaf bases, or less commonly almost glabrous and with decurrent lines below the heads; stems 3.6–9.5 (15) dm tall; leaves 3–17 (21) cm long, 5–27 mm wide, entire or serrate, glabrous or scabrous, ciliate, the lower ones commonly petiolate, often deciduous at anthesis, becoming smaller, sessile, and more or less clasping upwards, sometimes much reduced in inflorescence; heads few to numerous in open to narrow subcorymbose inflorescences; involucres 4.5–7 (8) mm high, 7–12 mm wide, the bracts imbricate to subequal, green-tipped, the chartaceous base white to straw colored, all acute and mucronate; rays commonly 20–50, pink to blue or white, 6–14 mm long; achenes hairy; n = 12, 32. Wet meadows, canal banks, and streamsides with sedges, rabbitbrush, willow, and other riparian communities at 850 to 2135 m in Box Elder, Cache, Duchesne, Garfield, Grand, Kane, Millard, Summit, Uintah, Utah, Wasatch, Washington, and Weber counties; Alberta to Saskatchewan, south to California, Arizona, New Mexico, and Missouri; 32 (vii). This plant occurs at lower elevations in Utah and has been confused with *A. foliaceus*, with which some plants share the subequal bracts. It has also been mistaken for *A. chilensis*, with which it is partially sympatric. The lack of uniformly disposed hair in the inflorescence appears to be diagnostic.

Aster integrifolius Nutt. Thickstem Aster. Perennial subrhizomatous herbs, glandular villous with multicellular hairs, at least above; stems 2.3–6.4 (7) dm tall; leaves 2.5–19 cm long, 8–50 mm wide, entire, oblanceolate to elliptic or lanceolate, glandular-villous, ciliate, the lower ones petiolate, becoming smaller, sessile and clasping upwards; heads few to several (numerous), large and showy, in elongate to subcorymbose clusters; involucres 8–13 (14) mm high, 12–23 mm wide, the bracts mainly subequal, green or suffused with purple, glandular dorsally, foliaceous or not; rays commonly 10–25, dark purple, 10–15 mm long. Meadows and moist woods in sedge-willow, sagebrush, Douglas fir, and spruce communities at 2275 to 3125 m in Cache, Emery, Piute, Rich, Salt Lake, Sanpete, Summit, and Wasatch counties; Washington and Montana, south to California and Colorado; 8 (0).

Aster kingii D. C. Eaton King Aster. [*Machaeranthera kingii* (D. C. Eaton) Cronq. & Keck]. Perennial herbs from a caudex and taproot, the caudex branches clothed with blackish or dark brown marcescent leaf bases, these scarious and ashy when young; stems 3–12 (15) cm long, more or less villous below, stipitate-glandular above; basal leaves 0.8–12 cm long, 3–22 mm wide, petiolate, the petiole bases expanded and scarious, the blades oblanceolate or spatulate, glabrous or glandular, or less commonly hispidulous or merely puberulent on one or both sides; heads 1–5, racemosely or corymbosely arranged; involucres 8–11 mm high, 10–16 mm wide; bracts glandular to shortly stipitate-glandular, herbaceous above the middle, scarious below, often suffused purplish, especially the inner, the tips of at least the outer reflexed; rays 15–27, white (often fading pale pink), 8–17 mm long, 1.5–2.8 mm wide; achenes ca 3.5 mm long. Douglas fir-white fir, mountain brush, and cottonwood communities at 1839 to 3050 m in Juab, Millard, Salt Lake (type from Cottonwood Canyon), and Utah counties; endemic; 21 (i). The southern populations have at least some toothed leaves and stems with longer stipitate-glandular hairs; they belong to var. **barnebyana** (Welsh & Goodrich) Welsh [*Machaeranthera kingii* var. *barnebyana* Welsh & Goodrich, type from Canyon Mts., Millard County].; 6 (0). Northern plants belong to var. *kingii*. Attempts to segregate genera within the Astereae are often frought with difficulties. This is especially true of that core of genera involving *Haplopappus*, *Machaeranthera*, *Xylorhiza*, and *Aster*. Cronquist and Keck (1957. Brittonia 9: 231–329) reconstituted the genus *Machaeranthera*, and included within that expanded generic definition those species treated elsewhere herein as *Machaeranthera* and *Xylorhiza*. Included within the series Integrifoliae of section Xylorhiza was *Aster kingii*. Watson (1978. Madrono 25: 205–210) has shown the chromosome number to be 2n=18 for *Aster kingii*, and he notes that its placement within *Machaeranthera* section Xylorhiza ". . is phenologically, ecologically, morphologically, and chromosomally anomalous . . ." The chromosome numbers reported for *Xylorhiza* are 2n=12 or 24; that of *Machaeranthera*, in a restricted sense, is 2n=8, 10, or 16; that of *Aster* is mainly 2n=18. The taproots and squarrose involucral bracts suggest an alliance with *Machaeranthera*, shorn of *Xylorhiza*, but the similarity seems superficial, especially in light of different chromosome numbers. Some asters in a strict sense, i.e., *A. alpigenus* Rydb., have a caudex, with the rhizome abbreviated. The logical conclusion of such an abbreviation is the caudex of *A. kingii*, and the squarrose bracts seem to have been secondarily derived, being present to a greater or lesser degree in other *Aster* species as well as in *Machaeranthera*. Hence, it seems best to treat this taxon within *Aster*.

Aster laevis L. Smooth Aster. Subrhizomatous perennial herbs, glabrous or nearly so; stems mainly 5–12 dm tall; leaves 0.8–14 cm long, 2–30 mm wide, entire or serrate, linear-subulate to lanceolate or elliptic, the lower ones petiolate, often lacking at anthesis, becoming smaller, sessile, and more or less clasping upwards; heads numerous, in corymbose inflorescences; involucres 5–8 mm high, 7–12 mm broad (when pressed), the bracts slender, green-tipped, the chartaceous bases straw colored to brownish or white, acute and mucronate; rays 15–30, blue or purple, 6–9 mm long; achenes glabrous; 2n = 24, 48. Riparian communities at ca 1400 m in Grand (and San Juan?) County; Yukon to Maine, south to Oregon, New Mexico, and Georgia; 1 (i). This plant is rare in collections from Utah, due presumeably to the paucity of late season collections from southeastern Utah.

Aster novae-angliae L. New England Aster. Rhizomatous to subrhizomatous perennial herbs, hirsute and more or less glandular, especially upward; stems 4–10 dm tall or more, very leafy; leaves 3–12 cm long, lanceolate to oblong, entire, sessile and cordate-clasping, becoming smaller and bracteate in the inflorescence; heads numerous, in corymbose inflorescences; involucres 8–11 mm high, hemispheric or campanulate, the bracts slender, green-tipped, whitish near the base, attenuated and spreading apically, glandular; rays typically more than 30 per head, rose to violet purple; achenes silky hairy. Cultivated ornamental in much of the lower elevation portions of Utah, now escaped and established locally in Juab, Salt Lake, and Uintah counties; introduced from the eastern U.S.; 3 (0).

Aster occidentalis (Nutt.) T. & G. Western Aster. [*Tripolium occidentale* Nutt.]. Rhizomatous or subrhizomatous perennial herbs, uniformly, softly, and often loosely villous (at least above); stems 0.9–8.5 dm tall; leaves 1–15 cm long, 1–20 mm wide, entire or toothed, glabrous or nearly so, ciliate, the lower ones petiolate, sometimes lacking at anthesis, rather abruptly smaller and finally sessile upwards; inflorescence mainly of 1–7 (rarely to 15), corymbosely arranged, large and showy heads; involucres 5–12 mm high, 7–20 mm wide, the bracts imbricate to subequal, slender, green, with pale yellowish to white or brownish chartaceous bases (at least the inner), mainly acute, mucronate; rays 20–50, blue to purple, 6–15 mm long; achenes hairy; 2n = 16, 27. Meadows and streamsides in lodgepole pine, cottonwood, willow, aspen, and spruce-fir communities at 2175 to 3175 m in Box Elder, Cache, Carbon, Daggett, Duchesne, Emery, Garfield, Grand, Iron, Kane, Piute, Sanpete, Sevier, Summit, Uintah, Wasatch, and Washington counties; Mackenzie to Colorado and California; 43 (ix). This species shares the features of soft loose pubescence and general aspect with the partially sympatric *A. foliaceus*. The very slender and abruptly reduced cauline leaves are diagnostic in most instances.

Aster pansus (**Blake**) **Cronq.** Elongate Aster. [*A. multiflorus* var. *pansus* Blake]. Subrhizomatous herbs, villous or villous-hirsute with multicellular hairs, the stems 3–12 (or more) dm tall; leaves 1–6 cm long, 2–8 mm wide, entire, antrorsely scaberulous on both surfaces, sessile, linear to narrowly oblong, often spinulose-mucronate, the lowermost commonly lacking at anthesis; inflorescence paniculate to secund-paniculate, narrow; involucres 3.8–4.9 mm high, 4.5–6 mm wide, the bracts strongly imbricate, green-tipped, scaberulous dorsally and ciliate; rays mainly 15–25, white, 3–8 mm long; achenes hairy; n = 5. Drainages, meadows, seeps, and hanging gardens at 1220 to 1890 m in Daggett, Grand, San Juan, and Uintah counties; British Columbia to Montana, south to Colorado and Nebraska; 6 (iv). This species forms the basis for inclusion in previous botanical works of the name *A. ericoides* L. in the Utah flora. It is closely allied to *A. falcatus*, but differs in the smaller heads, taller stature, and eastern distribution.

Aster pauciflorus Nutt. Alkali Aster. [*A. thermalis* Jones, type from Monroe Hot Springs]. Subrhizomatous perennial herbs, glabrous below, stipitate-glandular above and in inflorescence; stems 2–7.5 dm tall; leaves 1.1–12.5 cm long, 1–4 mm wide, entire, acicular to lance-linear or linear, glaucous, glabrous, all sessile or the lowermost petiolate, reduced upwards; heads few to several in corymbose inflorescences; involucres 4.3–7 mm long, 7–10 mm wide, the bracts imbricate to subequal, glandular dorsally, green throughout, narrow and acute; rays mainly 20–35, blue to purple, 5–12 mm long; achenes hairy; n = 9. Hot springs, stream terraces, and salt grass meadows, often in saline or alkaline substrates at 1300 to 2135 m in Box Elder, Duchesne, Emery, Juab, Kane, Millard, Salt Lake, Sanpete, Sevier, Summit, Uintah, and Utah counties; Saskatchewan to Nevada, Arizona, and Mexico. This distinctive glandular aster has been collected in full anthesis on 27 April growing in hot water at Monroe Hot Springs in Sevier County. It continues to flower into October; 21 (vi).

Aster perelegans Nels. & Macbr. Nuttall Aster. [*Eucephalus elegans* Nutt.; *A. elegans* (Nutt.) T. & G., not Willd.]. Subrhizomatous perennial herb, puberulent to glabrate (sometimes glandular); stems 3–7 dm tall; leaves 1.3–6.5 cm long, 3–14 mm wide, entire, oblong to oblong-lanceolate or elliptic, scabrous, firm, sessile, the lowermost reduced in size; heads 3–16, in corymbose inflorescences; involucres 7–10 mm high, 7–12 mm wide, the bracts chartaceous, imbricate, with prominent midvein, and acute to obtuse apex, the margins hyaline and ciliate, more or less puberulent dorsally; rays 5–16, dark purple, 7–13 mm long; achenes hairy. Sagebrush, mountain brush, Douglas fir, aspen, and limber pine communities at 1725 to 3050 m in Cache, Carbon, Duchesne, Juab, Millard, Rich, Salt Lake, Wasatch, and Weber counties; Oregon to Montana, south to Nevada; 21 (iv).

Aster scopulorum Gray Crag Aster. [*Chrysopsis alpina* Nutt., not *A. alpinus* L.]. Perennial subrhizomatous herbs with a woody caudex, villous on stems and peduncles; stems 4–12 cm tall; leaves 5–12 (15) mm long, 1–3 mm wide, overlapping, elliptic to oblong or linear, firm, scabrous or puberulent, often with some villous hairs above, spinulose-mucronate; heads solitary, pedunculate; involucre 7–11 mm high, 8–12 mm wide, the bracts imbricate, sparingly villous-hirsute and glandular, with a prominent midvein in the lower half, greenish, with chartaceous border and hyaline margins, acute; rays mainly 8–15, blue or purplish, 6–15 mm long; achenes hairy; n = 9, 18. Sagebrush community at 2440 to 2745 m in Box Elder County; Oregon to Montana, south to California and Nevada; 8 (0).

Aster sibiricus L. Siberian Aster. Plants with elongate rhizomes, the stems arising singly or few to several together, ascending to erect, 0.5–1.5 dm tall, sparsely to moderately villous with multicellular hairs, rarely glabrate; basal leaves smaller than the middle cauline ones, usually withered by flowering time; cauline leaves 2–8 (10) cm long, 0.5–3.5 cm broad, lanceolate to elliptic, oblong, or oblanceolate, sharply serrate to entire, sessile or the lower ones short-petiolate, not auriculate-clasping, acute to acuminate or less commonly obtuse apically, hairy above and beneath or glabrate to glabrous above; heads solitary or more commonly few in corymbose clusters, the peduncles villous, not glandular; involucres 2–15 mm high, not glandular; involucres 6–9 mm high, 10–18 mm broad, the bracts oblong to lance-oblong, hairy dorsally, not glandular, ciliate, in several subequal or distinctly imbricate series, greenish throughout or commonly suffused with purple; rays purple, mostly 8–15 mm long, 1.5–2.5 mm wide; pappus brown or suffused with purple; achenes several-nerved, hairy; n = 9, 18. Spruce community on limestone and quartzite contact zone at ca 3142 m in Summit County (Goodrich 16211 BRY); Alaska and Yukon, south to Oregon, Idaho, and Wyoming; Eurasia; 1 (0).

Aster spinosus Benth. Mexican Devilweed. Suffrutescent, rushlike plants from a deeply placed rhizome, glabrous; stems 6–12 (or more) dm tall, with axillary or supra-axillary thorns to 1.5 cm long; leaves 2–4 cm long, 2–5 mm wide, firm, entire to toothed, reduced above to scales; heads solitary at ends of branches, or some axillary; involucres 4–6 mm high, 6–8 mm wide, the bracts imbricate, slender, acute to acuminate, green, with prominent scarious margin; rays 15–30, white, very short; achenes glabrous; n = 9. Riparian communities below 1130 m in Garfield, Kane, San Juan, and Wayne counties; California to Texas, south to Central America; 9 (viii). The plant was originally collected at the mouth of Ticaboo Canyon, along the Colorado River in Glen Canyon (Lindsay 20, 1958 UT), at a site now inundated by Lake Powell. This is one of a series of extirpations related to construction of Glen Canyon Dam. Small populations of this plant were drowned by the high water of the 1983 flood at the mouth of Llewellyn Canyon and Wilson Creek, along Lake Powell. Fortunately, the plant is locally common in Cataract Canyon.

Aster wasatchensis (**Jones**) **Blake** Markagunt Aster. [*A. glaucus* var. *wasatchensis* Jones, type from near Marysvale]. Subrhizomatous perennial, glandular-puberulent; stems 3.5–6.5 dm tall; leaves 1.8–8.5 cm long, 6–24 mm wide, entire, lanceoate to oblong, or oblanceolate, glandular-puberulent to glabrous, firm, more or less glaucous, the lowermost often smaller and commonly lacking at anthesis; heads several to numerous, more or less corymbosely arranged; involucres 8–11.5 mm long, 10–20 mm wide, the bracts herbaceous throughout or the inner with scarious bases, glandular dorsally, abruptly acute to attenuate, apically; rays 15–25, white or pink, 10–20 mm long; achenes hairy. Pinyon-juniper, aspen, limber pine, and spruce-fir communities at 1890 to 3050 m in Garfield, Iron, Millard, and Piute counties; endemic; 14 (vi). This remarkable aster is unique in Utah in having foliaceous or subfoliaceous glandular involucral bracts and glaucous leaves.

Atrichoseris Gray

Annual scapose herbs, with milky juice, from taproots; leaves all basal, sinuate-dentate, often spotted; heads on slender peduncles, few to numerous, corymbosely arranged; involucre of about 12–15 subequal but biseriate, lance-linear scarious-margined bracts and some shorter outer bracts; receptacle naked; corollas all raylike, perfect, white; pappus lacking; achenes oblong, with corky-thickened ribs.

Atrichoseris platyphylla Gray Tobacco-weed; Gravel Ghost. Plants 3–10 dm tall (or more), from slender taproots; leaves 1.2–10.5 cm long, 0.5–6 cm wide, obovate to broadly spatulate, tapering abruptly to a broad petiole, sinuate-dentate, the teeth mucronate-cuspidate, glabrous, often mottled, more or less glaucous; involucres 6–8 mm high, 12–16 mm wide, the outer bracts ovate-lanceolate, hyaline, more or less scurfy, the inner ones lance-acuminate, with broad hyaline margins; corollas white, 8–20 mm long; achenes white, with corky ridges; n = 9. Joshua tree, ambrosia, yucca, cholla communities at 670 to 750 m in Washington County; California and Arizona; 11 (vi).

Baccharis L.

Dioecious shrubs; leaves alternate, entire or toothed; heads discoid, many-flowered, the corollas white, turbinate, borne in corymbose or paniculate clusters; involucres imbricate, the bracts chartaceous, whitish; pistillate heads with tubular-filiform obscurely toothed or truncate corollas, the pappus of copious capillary bristles; staminate heads of tubular 5–toothed corollas, the pappus (often scanty) of usually twisted clavellate scales; receptacle naked; style branches flattened; achenes subcylindric, 5– to 10–ribbed.

1. Plants low shrubs, mainly less than 5 dm tall; pappus reddish or purplish; known from San Juan County ... . *B. wrightii*
— Plants low to tall shrubs; pappus white or somewhat tawny but not reddish; known from various distributions 2
2(1). Branches fastigiate, deeply sulcate and more or less ridged, the leaves commonly deciduous at anthesis; achenes 10–ridged 3
— Branches not especially fastigiate, commonly spreading to ascending; leaves commonly persistent at flowering time; achenes 5– or 10–ribbed 4
3(2). Main leaves linear; pistillate pappus to 10 mm long or more in fruit; plant not definitely known for Utah, included previously on the basis of a specimen now presumed to be mislabeled .... *B. sarothroides* Gray
— Main leaves obovate-spatulate; pistillate pappus to 3 mm long in fruit .... *B. sergilloides*
4(2). Leaves long-cuneate basally, thickened, entire or few toothed towards apex; branches often subfastigiate, achenes 10–nerved; plants of Virgin and Colorado drainages 5
— Leaves not especially long-cuneate basally, commonly thin, entire, or toothed from below the middle; achenes 5–nerved 6
5(4) Staminate involucres 3.5–5.3 mm long, 3.7–4.8 mm wide; pistillate involucres 7.3–8.5 mm long; pappus 11–13 mm long; plants of Washington and Kane counties .... *B. emoryi*
— Staminate involucres 5.3–6 mm long, 5–10 mm wide; pistillate involucres 6–6.5 mm long; pappus 8–9.5 mm long; plants of Emery, Grand, Garfield, and San Juan counties .... *B. salicina*
6(4). Leaves mainly entire; panicles terminating short lateral branches; plants of Washington County ...... *B. viminea*
— Leaves usually serrate; panicles terminating main stems; plants of Washington and Kane counties .... *B. glutinosa*

Baccharis emoryi Gray in Torr. Emory Seepwillow. Shrubs, mainly 1–2 (3) m tall, the branches green to olive or brownish, ascending, subfastigiate, more or less glutinous; leaves 1.2–8.5 cm long, 3–20 mm wide, spatulate-oblanceolate to elliptic or linear, cuneate to a slender petiole, thick, entire or sparingly and irregularly toothed, obtuse to acute apically; heads numerous in a conic to pyramidal panicle; pistillate involucres 7.3–8.3 mm high, 4.5–7 mm wide, the bracts in several series, scarious, often glutinous, with thickened green or brown to reddish tips and hyaline margins; staminate involucres 3.7–5.3 mm high, 3.7–4.8 mm wide; pistillate corollas 4.5–5.5 mm long, the pappus 11–13 mm long; achenes 10–ribbed. Stream and canal banks and hanging gardens at 825 to 1220 m in Kane, San Juan, and Washington counties; Arizona, Texas, and California; Mexico; 19 (iv).

Baccharis glutinosa Pers. Sticky Seepwillow. Shrubs, mainly 1–3 m tall, the branches straw colored to brownish or greenish, ascending-spreading, not fastigiate, glutinous; leaves 1.2–12.5 (15) cm long, 4–18 mm wide, elliptic to narrowly lanceolate, acuminate to attenuate, cuneate to a short petiole, evenly serrate to entire; heads numerous in terminal cymose panicles (less commonly in lateral ones) with pistillate and staminate heads about the same size; involucres 3.5–4.5 mm high, 4–5.5 (7.5) mm wide; corollas 2.2–3 mm long, the pistillate pappus 3.5–4.5 mm long; involucral bracts in several series, chartaceous, greenish in the center, the margins scarious, not glutinous; achenes 5-ribbed; n = 9. Stream bars and banks, and in seeps, at 670 to 1130 m in Kane and Washington counties; Colorado and Nevada to Texas and California; South America; 6 (ii).

Baccharis salicina T. & G. Rio Grande Seepwillow. Shrubs, mainly 1.5–3 m tall, the branches green to brownish, subfastigiate, glutinous; leaves 1.4–8 cm long, 4–18 mm wide, elliptic to oblanceolate or linear, cuneate to a short petiole, thick or thin, entire or sparingly toothed or lobed mainly near the apex, acute to rounded apically; heads few to numerous in axillary and/or terminal panicles; pistillate involucres 6–6.5 mm high, 4–6 mm wide, the bracts in several series, scarious, often glutinous, with thickened greenish to reddish tips and hyaline margins; staminate involucres 5.3–6 mm high, 5–10 mm wide; pistillate corollas 2.5–3.5 mm long, the pappus 8–10 mm long; achenes 10-ribbed. Stream banks and hanging gardens at 1220 to 1525 m in Emery, Garfield, Grand, and San Juan counties; Colorado to Kansas, south to New Mexico and Texas; 8 (iv). Our material of *B. salicina* has long been mistaken for *B. emoryi*, to which it is allied. The shorter pistillate involucres and broader staminate involucres are diagnostic.

Baccharis sergilloides Gray Squaw Waterweed. Shrubs, mainly 0.3–2 m tall, the branches green to brown, fastigiate, glutinous, finally almost leafless; leaves 0.5–2.5 cm long, 1–10 mm wide, spatulate to obovate, entire or few-toothed, thick; heads numerous, borne in conic to pyramidal panicles; involucres 2.5–3.5 mm high, 2.5–3.5 mm wide, the bracts in several series; straw colored, or with thickened brownish centers; pappus 2.5–3 mm long; achenes 10-ribbed. Stream bars and banks at 670 to 825 m in Washington County; California and Arizona; 8 (iii).

Baccharis viminea DC. Mule-fat. Shrubs, mainly 2–3 m tall, the branches green to straw colored or brownish, spreading-ascending, not fastigiate, glutinous; leaves 0.8–9.5 cm long, 2–9 mm wide, elliptic to lance-elliptic or narrowly oblong, attenuate to acute, cuneate to a short petiole, entire to evenly serrate; heads few to many in terminal cymose panicles on short lateral branches, with pistillate and staminate heads about the same size; involucres 3–5.7 mm high, 6–9 mm wide; corollas 2.5–3.8 mm long; pistillate pappus 5–6 mm long; involucral bracts in several series, chartaceous, commonly with reddish centers, the margins scarious, not glutinous; achenes 5-ribbed. Stream bars and banks at 650 to 900 m in Washington County; California and Arizona; 10 (ii).

Baccharis wrightii Gray Low shrub or subshrub mainly 1–5 (7.5) dm tall, glabrous, typically branched from near the base; branchlets angled; leaves few, sessile, punctate, linear to narrowly lanceolate; lower leaves mainly 5–10 (25) mm long, oblanceolate to oblong, 1–3 mm broad or more, the upper ones usually subulate and with apices recurved, entire or serrulate, 1-veined; pistillate heads hemispheric, 9–12 mm long, the bracts lanceolate, acute to acuminate, green or brown veined, the margins scarious; receptacle flat, naked; corolla filiform, 3.7–4.7 mm long; pappus copious, multiseriate, antrorsely barbellate, mainly 10–15 mm long, pinkish or purplish (finally brownish); achene 3–5 mm long, glandular, 5- to 10-ribbed, transversely ridged; staminate involucre hemispheric, 8–9 mm long, the bracts linear to lanceolate, acute, the margins slightly serrate to entire, scarious, the corollas filiform, 4.6–5 mm long, the pappus subequal to the corolla, plumose-tiped, crisped; ovary abortive. Desert shrub communities at ca 1220 to 1375 m in San Juan County; Arizona to Oklahoma, south to Mexico; 1 (0). This species differs from all other *Baccharis* taxa in Utah by being a xerophyte.

Bahia Lag.

Biennial or short-lived perennial herbs with watery juice, arising from taproots; stems erect or ascending, puberulent; leaves alternate, once to twice ternately divided; heads few to numerous, in corymbose panicles; involucral bracts subequal, in 1 or 2 series, greenish; ray flowers present, yellow, pistillate, fertile; disk flowers perfect, fertile; pappus none; style branches flattened; achenes 4–angled, 12–curved.

Bahia dissecta (Gray) Britt. Cutleaf. [*Amauria* (?) *dissecta* Gray]. Biennial or short-lived perennial herbs, the stems 2–8 dm tall, minutely puberulent; leaves 1–10 cm long, the blade 1- to 3–ternately divided, oval to cordate in outline, strigulose; peduncles glandular hairy; involucres hemispheric, 3.4–6 mm high, 8–12 mm wide, the bracts more or less glandular hairy (or merely villous), greenish, abruptly contracted to a broadened apex; rays mainly 10–15, yellow, 4.5–9 mm long; achenes glabrous; n = 18. Sagebrush, pinyon-juniper, mountain brush, aspen, lodgepole pine, ponderosa pine, and spruce communities at 1700 to 2930 m in Beaver, Daggett, Duchesne, Emery, Garfield, Grand, Iron, Kane, Piute, Salt Lake, San Juan, Sevier, Summit, Uintah, Washington, and Wayne counties; Nevada to Wyoming, south to California, Arizona, and Mexico; 24 (iii). Those species treated elsewhere in this work as *Platyschkuhria* belong to *Bahia* in a broad sense, and are probably best treated in the latter genus, but their combination is not implied here.

Baileya Harv. & Gray

Annual, biennial, or perennial herbs from taproots, with watery juice; stem erect, white-tomentose; leaves alternate, once or twice pinnatifid to entire; heads solitary or few in cymose clusters; involucral bracts subequal, white-tomentose; receptacle naked; ray flowers persistent, yellow, pistillate, fertile; disk flowers perfect, fertile; pappus none; style branches short, truncate; achenes oblong or clavate, striate.

1. Ray flowers 7 or fewer; plants slender annuals with involucres less than 8 mm wide; reported from Washington County, but no specimens have been seen by me .... *B. pauciradiata* Harv. & Gray
— Ray flowers 20 or more; plants annual, biennial, or perennial, with involucres 10–26 mm wide 2
2(1). Rays 11–22 mm long; peduncles (4.5) 12–32 cm long in anthesis; involucres 5.7–7.5 mm high, 13–26 mm wide .... *B. multiradiata*

— Rays 8–10 mm long; peduncles 1–8 (11) cm long in anthesis; involucres 3–5.5 mm high, 10–16 mm wide
.... *B. pleniradiata*

Baileya multiradiata Harv. & Gray Desert Baileya. Biennial or short-lived perennial herbs; stems 1.9–5 (5.2) dm tall; herbage white-tomentose; leaves 0.8–10 cm long, the blade 1- to 2-pinnately lobed to entire, ovate-oval to linear; peduncles (4.5) 13–32 cm long in anthesis; involucres 5–7.5 mm high, 13–26 mm wide, the bracts slender, greenish; rays 25–40 or more, yellow, 11–22 mm long; achenes glabrous; n = 16, 17. Creosote bush, Joshua tree, burrobush, blackbrush, and sagebrush communities at 670 to 1320 m in western Kane and Washington counties; Nevada and California south to Mexico; 36 (iii).

Baileya pleniradiata Harv. & Gray Annual to short-lived perennial herbs; stems 0.8–5 dm tall; herbage white-tomentose; leaves 0.8–12 cm long, the blades 1- to 2-pinnately lobed to entire, obovate to linear; peduncles 1–8 (11) cm long in anthesis; involucres 3.5–5.5 mm high, 6–13 mm wide, the bracts slender, greenish; rays 18–58, yellow, 8–10 mm long; achenes glabrous. Creosote bush, blackbrush, shadscale, mesquite, sagebrush, and pinyon-juniper communities at 820 to 1100 m in Washington County; Nevada and California to Texas; Mexico; 29 (ii).

Balsamorhiza Nutt.

Perennial scapose or subscapose herbs from taproots, the juice watery; leaves mainly basal, simple and entire or variously pinnatifid, reduced and bractlike upwards; heads solitary, or few to several; involucral bracts in several series, imbricate or subequal, herbaceous; receptacle chaffy, convex, the bracts enclosing the achenes; ray flowers present, pistillate, fertile, usually yellow; disk flowers numerous, perfect, fertile, yellow; pappus none; style branches slender; achenes compressed. Note: The genus is notorious for the lack of genetic barriers to hybridization. Any two taxa can intergrade where they occur together.

1. Leaves sagittate, with entire margins .... *B. sagittata*
—   Leaves pinnatifid or variously cleft 2
2(1). Leaves mainly 3–6 dm long, with segments typically 5–12 cm long, these entire or few-lobed or -toothed .. .. *B. macrophylla*
—   Leaves mainly 1–3 dm long, with segments mostly 1–5 cm long, these entire or variously lobed or toothed 3
3(2). Involucral bracts abruptly tapering to a long-attenuate apex; stem leaves relatively well developed, pinnatifid or bipinnatifid; reported for northern Utah, but no specimens have been seen .... *B. hirsuta* Nutt.
—   Involucral bracts gradually tapering to an attenuate apex; stem leaves lacking or small and inconspicuous .... *B. hookeri*

Balsamorhiza hookeri Nutt. Hooker Balsamroot. Perennial scapose herbs from a thick taproot, mainly 0.9–4.5 (5.2) dm tall; leaves 6–30 cm long, (0.3) 1.5–11 cm wide, pinnatifid or bipinnatifid, the segments to 5.5 cm long; peduncles naked or with a few inconspicuous, linear, entire or pinnatifid bracts near the base; heads solitary; involucres 13–24 mm high, 21–47 mm wide, the bracts lance-linear, evenly tapering to the apex or somewhat enlarged at the base, long-ciliate, glandular to tomentose dorsally; rays mainly 10–16, yellow, 16–40 mm long; achenes glabrous. Phases of this taxon are known to form intermediates with *B. sagittata*, and presumeably with *B. macrophyllum*. Ours are separable into two modestly distinctive varieties.

1. Involucres densely villous-tomentose dorsally; plants of Daggett, Duchesne, and Uintah counties .... *B. hookeri* var. *neglecta*
—   Involucres glandular to glabrous dorsally; plants of broad distribution, occasionally of Daggett and Duchesne counties .... *B. hookeri* var. *hispidula*

Var. hispidula (W. Sharp) Cronq. [*B. hispidula* W. Sharp, type from Lake Point, Tooele County]. This is the common phase of the species in Utah, and it has been confused with *B. hirsuta* Nutt., with which it is compared in the key. Bunchgrass, sagebrush, mountain brush, juniper, pinyon-juniper, and salt desert shrub communities at 1240 to 2745 m in Beaver, Box Elder, Daggett, Duchesne, Juab, Salt Lake, Tooele, Uintah, Utah, Wasatch, and Washington counties; Nevada, Idaho, and Wyoming (?); 31 (ii).

Var. neglecta (W. Sharp) Cronq. [*B. hirsuta* var. *neglecta* W. Sharp]. Salt desert shrub, sagebrush, pinyon-juniper, and ponderosa pine communities at 1640 to 2625 m in Box Elder, Daggett, Duchesne, Tooele, and Uintah counties; Nevada, Idaho, and Wyoming (?); 18 (ii). Plants of this variety form hybrids with *B. sagittata*.

Balsamorhiza macrophylla Nutt. Cutleaf Balsamroot. Perennial scapose herbs from a thick taproot, mainly 3–7 dm tall; leaves 15–60 cm long, 3.7–25 cm wide, pinnatifid, the segments entire, few-toothed or -lobed, up to 12.5 cm long; peduncles sparingly long shaggy-villous, naked, or with one to few reduced leaves near the base; heads solitary; involucres 23–35 mm high, 30–60 mm wide, the bracts lance-linear, attenuate, long-ciliate, glandular and more or less long-villous dorsally; rays 9–14, yellow, 30–55 mm long; achenes glabrous. Mountain brush and sagebrush or bunchgrass communities at 1525 to 2290 m in Box Elder, Cache, Davis, Morgan, Salt Lake, Summit, Utah, and Weber counties; Idaho to Montana and Wyoming; 7 (0).

Balsamorhiza sagittata (Pursh) Nutt. Arrowleaf Balsamroot. [*Bupthalmium sagittatum* Pursh]. Perennial scapose herbs, from thick taproot, mainly 1.5–8 dm tall; leaves (including long slender petioles) 5–45 cm long, 1.5–15 cm wide, sagittate, entire, or the cauline ones from near the summit to near the middle of the subscapose stem and linear to elliptic; peduncles villous-tomentose; heads solitary (or with additional reduced ones); involucre 15–30 mm long, 20–50 mm wide, the bracts lance-linear, attenuate, villous-tomentose; rays 8–25, yellow, 25–60 mm long; achenes glabrous; n = 19. Sagebrush, mountain brush, pinyon-juniper, ponderosa pine, Douglas fir, aspen, and fir communities at 1340 to 3020 m in Beaver, Box Elder, Cache, Daggett, Davis, Garfield, Iron, Juab, Kane, Millard, San Juan, Salt Lake, Sanpete, Sevier, Summit, Tooele, Uintah, Utah, Washington, and Weber counties; British Columbia to Montana and South Dakota, south to California, Nevada, and Colorado; 43 (vii).

Bebbia Greene

Shrubs or subshrubs with rushlike, sparingly leafy, much-branched stems; leaves opposite below, often alternate above, entire or few toothed or lobed; heads solitary or few in loosely cymose clusters, discoid, yellow; involucres hemispheric, the bracts in ca 3 series, striate, shorter than the disk; receptacle, paleaceous, the bracts partially enfolding the achenes; corolla tube glandular, the limb hairy; achenes slender, pubescent with appressed hairs; pappus of 15–20 plumose bristles.

***Bebbia juncea* (Benth.) Greene** Sweetbush. [*Carphephorus juncea* Benth.]. Plants rounded, often much branched, mainly 6–10 dm tall, often rough-hairy; leaves mostly opposite, few, linear, entire or with a few linear lobes, mainly 2–7 cm long, 2–7 mm wide; involucres 4–8 mm high; bracts ovate-lanceolate to lanceolate, the outer typically harshly hairy with pustular-based hairs, the inner ones often becoming reddish or purplish; ray flowers lacking; disk flowers mainly 20–30, perfect, fertile, pale yellow; achenes trigonously compressed, with 2 sharp angles and a blunt one. Gravels of Beaver Dam Wash and adjacent area at ca 850 m in Washington County; California to Texas; Mexico; 2 (0).

Bellis L.

Scapose perennial herbs, with fibrous roots and short stolons, the juice watery; stems leafless simple; leaves all basal, simple, petiolate, toothed to entire; heads solitary; involucral bracts in 2 subequal series, herbaceous; receptacle conic to hemispheric, naked; rays white, pink, or purple, numerous, pistillate; disk flowers numerous, perfect, yellow; pappus lacking; style branches flattened; achenes flattened, usually 2-nerved, pubescent.

***Bellis perennis* L.** European Daisy. Plants 2–20 cm tall; leaves all basal, with short to long petioles, the blades 0.7–3 (4) cm long, 5–25 mm wide, obovate to oval or orbicular, dentate to entire, obtuse to rounded or emarginate apically, pubescent on both sides with coarse spreading hairs; scapes pubescent with ascending hairs; heads solitary; involucres 4–7 mm high, 9–15 mm wide, the bracts ovate to broadly lanceolate, rounded to obtuse apically, sparsely hairy dorsally, often suffused with purple, mostly 8–10 mm long, 1.5–2.5 mm wide; pappus lacking; achenes flattened; 2n = 18. Cultivated ornamental, escaping and persisting in lawns of lower valleys in Salt Lake and Utah counties; adventive from Europe; 4 (0).

Bidens L.

Annual herbs with fibrous roots, or rooting along the lower stem, the juice watery; stems decumbent to erect, commonly branched; leaves opposite, simple or pinnately compound; heads few to several in cymose inflorescences; involucral bracts in 2 series, the outer herbaceous, the inner somewhat petaloid and striate; receptacle flat or slightly convex, chaffy throughout, the chaff similar to the inner involucral bracts; ray flowers present, yellow, neutral or pistillate, or lacking; disk flowers numerous, perfect, fertile, yellow; pappus of (1) 2–4 awns or teeth, these retrorsely barbed, persistent; style branches flattened; achenes flattened, pubescent, usually 2- to 4-awned.

Sherff, E. E. 1937. The genus *Bidens*. Field Mus. Pub. Bot. 16: 1–709.

1. Leaves pinnately compound, with 3–5 leaflets, all petiolate .... *B. frondosa*
— Leaves simple, the middle and upper ones (at least) sessile or subsessile 2
2(1). Heads erect; disk corollas 4-lobed; anthers included; awns typically 3; plants local along the Wasatch Front .... *B. comosa*
— Heads typically nodding; disk corollas 5-lobed; anthers exserted; awns typically 4; plants of broad distribution .... *B. cernua*

***Bidens cernua* L.** Bur-marigold. Plants 1–13 dm tall, the stems sparingly spreading-hairy to glabrous; leaves simple, 1.5–15 cm long, 0.5–4 cm wide, narrowly lanceolate to lance-ovate, coarsely serrate to subentire, glabrous; heads nodding in age; outer involucral bracts 5–8, green, foliaceous, unequal, spreading or reflexed, the inner bracts erect, mostly 6–15 mm long; rays 6–8, yellow, or lacking; achenes mainly 5–7 mm long, tan, the 2–4 awns retrorsely barbed; 2n = 24. Wet meadows, bogs, stream banks, bars, and shores, at 1300 to 2380 m in Box Elder, Cache, Davis, Garfield, Juab, Kane, Piute, Salt Lake, San Juan, Sevier, Summit, Uintah, and Tooele counties; widely distributed in the Northern Hemisphere; 27 (iii).

***Bidens comosa* (Gray) Wiegand** [*B. connata* var. *comosa* Gray]. Plants 1–6 dm tall, the stems sparingly spreading-hairy to glabrous; leaves simple, 1.5–10 cm long, 0.4–2.8 cm wide, elliptic to lanceolate or oblanceolate, serrate to shallowly lobed, glabrous or sparingly hairy; heads erect in age; outer involucral bracts 7–12, green, foliaceous, unequal, spreading-ascending, the inner bracts erect, mostly 6–18 mm long; rays typically lacking; achenes mainly 5–7 mm long, brown, with usually 3 retrorsely barbed awns. Wet meadows, bogs, and other moist sites in Box Elder, Cache, Salt Lake, and Utah counties; widely distributed in the U.S.; 5 (0).

***Bidens frondosa* L.** Devil's Beggarticks. Plants 2–12 dm tall, the stems short-hairy to glabrous; leaves petiolate, pinnately compound with 3–5 leaflets, these 2–10 cm long, 0.5–3 cm wide, lanceolate, serrate; heads erect in age; outer involucral bracts 5–8, green, subfoliaceous, subequal, erect or spreading, the inner bracts erect, mostly 5–8 mm long; rays usually lacking; achenes 5–9 mm long, dark brown to black, the 2 awns barbed; 2n = 48. Marshes, pond and lake shores, bars, wet meadows, and irrigation canals at 1190 to 1650 m in Box Elder, Davis, Grand, Salt Lake, Uintah, Utah, Washington, and Weber counties; widespread in North America; 12 (ii). Note: The pan-boreal weed, ***Bidens tripartita* L.**, might occur in our area. It is distinguished from *B. frondosa* in its simple but trifid leaves, and from *B. cernua* in its petiolate trifid leaves. ***B. bigelovii* Gray** was collected once in Utah, from a ditch near Panguitch. It differs from *B. frondosa* in having the first main lobes of the leaves again divided, parted, or lobed, and linear achenes with 2–4 awns.

Brickellia Ell.

Perennial herbs, subshrubs or shrubs; leaves alternate or opposite, simple; heads campanulate or cylindric, cymose or paniculate, discoid; flowers all perfect, fertile; involucral bracts imbricate in several series, striate; receptacle almost flat, naked; style branches flattened, with long-papillate appendage; achenes 10-ribbed; pappus of barbellate, smooth, or subplumose bristles.

Robinson, B. L. 1917. A monograph of the genus *Brickellia*. Mem. Gray. Herb. 1: 1–151.

1. Leaves spinulose-serrate, or spinulose-tipped; low rounded shrubs of Washington, Kane, and San Juan counties .... *B. atractyloides*

— Leaves entire or toothed, not spinulose; herbs, sub-shrubs, or tall shrubs of various distribution 2

2(1). Plants herbaceous; heads reflexed, broadly campanulate; leaves sagittate- to cordate-ovate, longer than broad. .... *B. grandiflora*

— Plants, shrubs or subshrubs; heads narrowly cylindric or, if campanulate, erect; leaves ovate to linear or, if cordate, about as broad as long or broader 3

3(2). Leaves petiolate, the blades cordate-ovate to ovate or suborbicular, 1–5 cm broad .... *B. californica*

— Leaves sessile or subsessile, linear to narrowly lance-olate or, if broader, mainly less than 1 cm broad 4

4(3). Leaves linear to lanceolate or narrowly elliptic; shrubs 6–15 dm tall or more; flowers 3–5 per head .. *B. longifolia*

— Leaves ovate to oval or oblong to linear; shrubs or subshrubs less than 5 dm tall; flowers many per head ... 5

5(4). Leaves 5–10 times longer than broad or more, entire or nearly so, sessile; involucres 10–20 mm high ....
.... *B. oblongifolia*

— Leaves only somewhat longer than broad, often toothed or lobed, at least some evidently petiolate; involucres 8–12 mm high .... *B. microphylla*

Brickellia atractyloides Gray Spiny Brickellbush. Shrubs, much-branched, mostly 3–5 dm tall, the branchlets greenish to straw colored, soon gray; leaves alternate, short-petiolate, the blades 0.6–3.2 cm long, 0.3–2.2 cm wide, lance-ovate to ovate, obtuse to rounded basally, spinulose-serrate to entire, acuminate and spinulose-tipped, thick and prominently veined, glabrous or minutely glandular puberulent; heads solitary, terminating the branches; peduncles 1–5.2 cm long, glandular-puberulent; involucres 10–13.5 mm high, 8–16 mm wide, the outer bracts ovate-lanceolate, acuminate apically, many-veined, the inner narrower, glandular-puberulent dorsally; flowers 50–75 or more; achenes black, 3.8–4.2 mm long, hirtellous on the ribs; n = 9. Rock crevices and talus slopes, creosote bush, blackbrush, and indigo bush communities at 820 to 1130 m in Kane and San Juan (San Juan and Glen Canyon arms of Lake Powell), and Washington counties; Nevada and Arizona; 12 (ii). The type is from the Colorado River (Utah?), Palmer sn, 1870 (US).

Brickellia californica Gray California Brickellbush. [*Bulbostylis californica* T. & G.; *Coleosanthus californicus* (T. & G.) Kuntze]. Subshrubs, mainly 5–10 dm tall, the branchlets whitish to brownish; leaves alternate, petiolate, the blades 1.7–5.2 cm long, 1.3–4.5 cm wide, cordate-ovate to ovate or orbicular, truncate to cordate basally, crenate-serrate, rounded to obtuse apically, the veins not prominent, glandular-scabrous; heads clustered in a leafy-bracteate panicle, sessile or shortly pedunculate; involucres 5.5–8 mm high, 4–7 mm wide, the outer bracts very short, rounded apically, few-veined, the inner long and slender, often suffused with red or purple, glabrous; flowers 8–18; achenes straw colored, 2.5–3.5 mm long; n = 9. Canyons and rock outcrops at 825 to 2135 m in Beaver, Cache, Garfield, Juab, Kane, Millard, Salt Lake, San Juan, Sevier, Utah, and Washington counties; Colorado to California and south to Texas and Mexico; 15 (iii).

Brickellia grandiflora (Hook.) Nutt. Tasselflower. [*Eupatorium grandiflorum* Hook.; *Coleosanthes garrettii* A. Nels., type from City Creek Canyon, Salt Lake County]. Perennial herb, from a caudex and taproot, the stems green to straw colored, 2.5–9.5 dm tall; leaves alternate, petiolate, the blades sagittate to cordate-ovate, 1.5–9 (11) cm long, 0.6–6.5 cm wide, cordate to truncate basally, serrate to doubly so, attenuate to acuminate apically, the veins not prominent, minutely puberulent or hirtellous; heads several to numerous in a short corymbose panicle, commonly reflexed; involucres 7–12 mm high, 6–10 mm wide, the outer bracts lance-acuminate, the inner abruptly acuminate, puberulent dorsally; flowers mostly 20–40 (70); achenes brown to black, 3.5–4.5 mm long, hirtellous; n = 9. Pinyon-juniper, mountain brush, ponderosa pine, aspen, Douglas fir-white fir, spruce, and bristlecone pine communities at 1640 to 3200 m in Beaver, Cache, Daggett, Duchesne, Garfield, Iron, Juab, Kane, Millard, Piute, Salt Lake, San Juan, Sanpete, Tooele, Uintah, Utah, Wasatch, Washington, and Weber counties; Washington east to Missouri, south to Mexico; 34 (vii).

Brickellia longifolia Wats. Longleaf Brickellbush. [*Coleosanthus longifolia* (Wats.) Kuntze]. Shrubs, with erect stems and white to tan bark, mainly 10–15 dm tall; leaves alternate, sessile or subsessile, 1.2–13.5 cm long, 3–8 mm broad, lance-linear to lance-elliptic, obtuse to acute basally, attenuate apically, the veins not prominent, glabrous, glandular-resinous; heads numerous in panicles; involucres 3.4–6.2 mm high, 2.3–4 mm wide, the outer bracts ovate, acute, the inner, longer and slender, glabrous; flowers 3–5; achenes 1.8–2.4 mm long, brown, glabrous. Canyon bottoms, stream margins, seeps, and hanging gardens at 750 to 1590 m in Emery, Garfield, Grand, Kane, San Juan, Washington, and Wayne counties; California, Nevada, Arizona; 37 (xiii).

Brickellia microphylla (Nutt.) Gray Rough Brickellbush. [*Bulbostylis microphyllus* Nutt.]. Shrubs or subshrubs, with tan to whitish bark, mainly 2–7 dm tall; leaves alternate, shortly petiolate to subsessile or sessile, 3–14 (20) mm long, 1–9 (12) mm wide, ovate to suborbicular, toothed to entire, commonly glandular-villous or -hispidulose, the veins not especially prominent, rounded to acute apically; heads solitary or few at tips of branches, racemosely arranged in leafy-bracteate panicles; involucres 7–10.3 mm high, 4–8.5 mm wide, the outer bracts oval to ovate, with thickened glandular tips, the inner often lacking glands and more or less 3–lobed or 3–veined; flowers 8–18; achenes 3.5–4.3 mm long, blackish, hirtellous or glabrous; n = 9. Two distinctive phases, which have been treated at specific level, are present in Utah. There is justification for treating them at specific rank, but they are similar in vegetative features and general aspect.

1. Flowers 8–11 per head; involucres 7–10 mm long, 4–7.5 mm wide; plants of the Green, Colorado, and Virgin river systems .... *B. microphylla* var. *scabra*

— Flowers (12) 17–18 per head; involucres 8.5–10.3 mm long, 6.5–10 mm wide; plants of the Great Basin ......
.... *B. microphylla* var. *watsonii*

Var. *scabra* Gray [*B. scabra* (Gray) A. Nels.]. Blackbrush, rabbitbrush, sagebrush, shadscale, Grayia, greasewood, juniper, and pinyon-juniper communities mainly on sandstone outcrops at 885 to 2170 m in Daggett, Duchesne, Emery, Garfield, Grand, Kane, San Juan, Uintah, and Washington counties; Colorado, Nevada, Arizona. Our material is uniformly hispidulose-glandular along upper stems at least, and has 8–11 flowers per head; 35 (xi). Note: A peculiar specimen from San Juan County (Anderson A-6 BRY) has heads nearly all clustered at branch tips.

Var. **watsonii** (Robins.) Welsh [*B. watsonii* Robins., type from American Fork Canyon]. Sagebrush, shadscale, mountain brush, and juniper communities at 1525 to 2440 m in Juab, Millard, Salt Lake, Sevier, Tooele, and Utah counties; Nevada and California; 7 (ii). All modern floras distinguish *B. microphylla* by its heads "about 22–flowered". Our material fits well within the concept of *B. watsonii* Robins., which has heads "18–flowered". Specimens from the Great Basin of Utah are uniformly 18–flowered, except in depauperate heads which vary downwards to 12 flowers per head. Stems are villous to glandular-villous, with the type of *B. watsonii* Robins. (Watson 494 US) at the villous end of a cline.

Brickellia oblongifolia Nutt. Mohave Brickellbush. Subshrubs or subherbaceous, with green to tan branches, mainly 1–5.5 dm tall; leaves alternate, sessile or nearly so, 0.9–4 cm long, 1–11 (15) mm wide, elliptic to oblong, or lance-oblong, entire or essentially so, glandular-hispidulous, the veins not especially prominent, acute to attenuate or obtuse apically; heads solitary and terminating branches, or corymbosely arranged; involucres 10.8–15 mm long, 12–22 mm wide, the bracts all acute to acuminate, glabrous or glandular to glandular-puberulent; flowers (11) 26–40 (50); achenes 4.8–5.8 mm long, blackish, hispidulous; n = 9. Grayia, shadscale, rabbitbrush, blackbrush, desert almond, juniper, pinyon-juniper, and ponderosa pine communities at 1280 to 2500 m in Beaver, Box Elder, Duchesne, Emery, Garfield, Juab, Kane, Millard, San Juan, Sevier, Uintah, Utah, Wasatch, Wasatch, and Washington counties; British Columbia to Montana, south to California, Arizona, and New Mexico; 41 (x). Our material is assignable to **var. linifolia** (D. C. Eaton) Robins. [*B. linifolia* D. C. Eaton, type from Jordan Valley, American Fork], which is distinguished by its achenes being hispidulous, not glandular-hispidulous or glandular. The segregation is tenuous at best.

Calycoseris Gray

Annual subscapose or caulescent herbs, with milky juice, from taproots, beset with tacklike stipitate glands above; leaves mostly basal, pinnately parted; heads solitary or few on leafy-bracteate peduncles; involucral bracts in 2 series, herbaceous, the inner with hyaline margins; receptacle with capillary bristles; corollas all raylike, yellow or white tipped; achenes fusiform, 5– or 6–ribbed, tapering to a short beak, this produced apically into a low denticulate cup; pappus abundant, white, of barbellate capillary bristles falling attached.

1. Rays white, with pink or purple dots or streaks dorsally; stipitate glands pale .... *C. wrightii*
— Rays yellow; stipitate glands purple .... *C. parryi*

Calycoseris parryi Gray Purple Tackplant. Annual herbs, mainly 0.7–3 dm tall, the stems simple or with spreading branches; leaves basal and alternate along stem, pinnately parted, the lobes linear, reduced and entire above, glabrous except for a few tangled long hairs on lower surface; peduncles mainly 0.5–4 cm long, clad with tacklike, long-stipitate, purple or purplish black glands; involucres 11–15 mm high, 8–14 mm wide (when pressed), the bracts linear-subulate to lance-subulate, more or less stipitate-glandular, attenuate apically; rays yellow, 10–20 (25) mm long; pappus surpassing the achene; n = 7. Creosote bush and Joshua tree communities at 800 to 830 m in Washington County; California and Arizona; 2 (i).

Calycoseris wrightii Gray Pale Tackplant. Annual herbs, mainly 1.4–4 dm tall, the stems commonly with spreading branches; leaves basal and alternate along the stem, pinnately parted, the lobes linear, reduced and subentire upwards, glabrous except for a few tangled hairs on lower surface; peduncles mainly 0.3–5 cm long, clad with tacklike long-stipitate pale glands; involucres 12–17 mm long, 12–20 mm wide, the bracts linear-subulate to lance-subulate, more or less stipitate-glandular, attenuate apically, rays 10–25 mm long, white, with pink or purple markings dorsally; pappus shorter than achene; n = 7. Creosote bush and Joshua tree communities in Washington County; California, Nevada, Arizona; 6 (ii).

Carduus L.

Biennial or annual herbs with taproots, the juice watery; stems erect, simple or branched; leaves alternate simple, pinnatifid to bipinnatifid or merely pinnately lobed, often decurrent, spiny; heads solitary or few, borne in corymbose cymes; involucral bracts imbricated in several series, spine-tipped; receptacle hemispheric, densely bristly; disk flowers only present, perfect, red purple, with long slender lobes; pappus of barbellate bristles; style branches connate, shortly hairy at base of branches; achenes compressed.

Carduus nutans L. Nodding Thistle; Musk Thistle. Rank biennial or annual herbs, mostly 0.6–20 (25) dm tall; stems arachnoid-tomentose to glabrate; leaves alternate, decurrent, 3–40 cm long, 0.5–20 cm wide (or more), lance-linear to elliptic, glabrous, or tomentose along veins beneath; heads commonly solitary, nodding; involucres 20–30 mm long, 30–80 mm wide, the bracts 2–8 mm wide, ovate-lanceolate to lanceolate, glabrous or nearly so, spinose-tipped, at least the outermost reflexed near the middle, the midrib prominent; flowers red purple; achenes 3.5–4.5 mm long, smooth, marked with vertical lines, umbonate; n = 8, 20. Disturbed sites along roads and in fields and pastureland spreading into sagebrush, pinyon-juniper, and mountain brush communities, at 1340 to 2440 m in Cache, Daggett, Duchesne, Emery, Iron, Juab, Morgan, Salt Lake, Sanpete, and Utah counties, and probably universal; introduced Old World plants, now widely established in the United States; 15 (ii).

Centaurea L.

Annual, biennial, or perennial herbs with taproots or rhizomes, the juice watery; stems erect or ascending; leaves alternate, entire to pinnatifid; heads solitary, or few to numerous, discoid; involucral bracts imbricate in several series, spine-tipped or some of them enlarged and with scarious or hyaline erose to lacerate or pectinate appendages; receptacle bristly; flowers all tubular, perfect, or the marginal ones sterile and falsely subradiate; purple, blue, yellow, pink, or white; pappus of bristles, scales, or none; style branches more or less connate, with a thickened often hairy ring at the base; achenes obliquely or laterally attached to receptacle. Note: This is a large genus, mainly of the Mediterranean region of the Old World, but with some indigenous to North America,

Australia, and South America. All of ours are introduced, and the potential for other introductions in this remarkable genus is great. In Flora Europaea, our species are treated within three genera: *Amberboa* (Pers.) Less. (*C. moschata* L.), *Acroptilon* Cassini (*C. repens* L.) and *Centaurea* (for the others).

1. Involucral bracts definitely spine-tipped, at least some with spines 1–20 mm long 2
— Involucral bracts definitely not spine-tipped or, if shortly spinose as in *C. maculosa* and *C. scabiosa*, the heads 6–25 mm wide 5
2(1). Stem definitely winged, the leaf bases decurrent; pappus present (central flowers, at least) 3
— Stems angled, not winged; pappus none 4
3(2). Apical spine of involucral bract 5–9 mm long; plants arachnoid when young; flowers all with evident pappus .... *C. melitensis*
— Apical spine of involucral bract 11–20 mm long; plants persistently tomentose; flowers in center only with a pappus .... *C. solstitialis*
4(2). Apical spine of bracts 5–15 mm long or more .... *C. calcitrapa*
— Apical spine of bracts 1–4 mm long .... *C. virgata*
5(1). Leaves entire or merely toothed, not pinnatifid 6
— Leaves pinnatifid or deeply pinnately lobed 8
6(5). Leaves linear to lance-linear, entire or nearly so, less than 1 cm wide .... *C. cyanus*
— Leaves various but, if as above, the plants rhizomatous .. 7
7(6). Plants rhizomatous; leaves mainly 2–10 mm wide; pappus evident, 6–11 mm long .... *C. repens*
— Plants not rhizomatous; leaves 6–15 mm wide; pappus 2–5 mm long .... *C. jacea*
8(5). Leaves merely pinnately lobed; involucral bracts entire or nearly so .... *C. moschata*
— Leaves pinnately divided, the lobes linear to narrowly oblong; involucral bracts pectinately lobed 9
9(8). Involucres 15–25 mm wide; lobes of leaves often again toothed or lobed .... *C. scabiosa*
— Involucres mainly 6–10 mm wide; lobes of leaves usually entire .... *C. maculosa*

Centaurea calcitrapa L. Star-thistle. Biennial herbs, from taproots, the stems usually branched, 1–8 dm tall, arachnoid-villous to glabrate; leaves 0.5–4.5 cm long, pinnatifid, the lobes linear to oblong, attenuate, or the upper ones entire; heads few to numerous; involucres urn-shaped, 10–18 mm high, mainly 8–12 mm wide, the bracts weakly spinose-ciliate, with a stout apical spine mainly 5–30 mm long; flowers few, purple; pappus none; 2n = 20. Roadside weeds in Salt Lake and Utah counties; introduced from Eurasia; 2 (0).

Centaurea cyanus L. Bachelors Button; Cornflower. Annual or biennial herbs from taproots, the stem usually branched, mostly 1–8 (12) dm tall, arachnoid-tomentose; leaves 2–10 (13) cm long, 1–8 mm wide, entire or some with slender lobes, attenuate; heads few to numerous; involucres hemispheric, 10–16 mm high, 10–23 mm wide, the bracts with a tapering pectinate or fringed tip, often purplish suffused, the central apical tooth not especially spinose; flowers several, blue, purple, pink, or white, the marginal ones enlarged, irregular; pappus 2–3 mm long; 2n = 20, 24. Cultivated ornamental, now established in disturbed sites in Cache, Salt Lake, Tooele, Utah, Wasatch, and Washington counties; adventive from Europe; 6 (0).

Centaurea jacea L. Brownscale Centaurea. Perennial herbs from taproots, the stems simple or branched from the middle, mostly 5–12 dm tall, glabrous or somewhat arachnoid; leaves entire or toothed to shallowly lobed, the basal ovate to lanceolate, petiolate, becoming smaller upwards; heads few to numerous; involucre 12–18 mm high, 12–15 mm wide, ovoid, the bracts with orbicular appendages, scarious, brown, darker in middle, the outer denticulate to pectinate-lacerate, the inner less so and often bifid; flowers purple or white, the outer more or less radiate; pappus none or very short; 2n = 22, 44. Cultivated ornamental, now established in Salt Lake County; adventive from Europe; 0 (0). Note: the large headed *C. montana* L., is cultivated in Utah. It has wedge-shaped involucral bracts and decurrent large leaves.

Centaurea maculosa Lam. Spotted Centaurea. Biennial or short-lived perennial, the stems simple or commonly branched above the middle, mainly 3–10 (15) dm tall, tomentose and sparingly scabrous-puberulent; leaves 1–9 cm long, pinnatifid, the lobes linear to lanceolate or oblong, entire or variously toothed or lobed, reduced and bracteate in the inflorescence; heads few to many, hemispheric to vase-shaped; involucres 10–13 mm high, 10–13 mm wide, the bracts with short dark pectinate tip, the central tooth produced as a spine to 0.5 mm long; flowers pink or purplish, rarely white, the marginal ones radiate; pappus to 2 mm long, rarely lacking; 2n = 18, 36. Roadsides in Beaver, Cache, Juab, Tooele, and Uintah counties; adventive from Europe; 3 (i).

Centaurea melitensis L. Annual or biennial, the stems sparingly branched from middle or below, 1.5–8 dm tall, winged by decurrent leaf bases; basal and lower cauline leaves oblanceolate, toothed to lyrate-pinnatifid or sinuately lobed, reduced upwards, finally entire; heads solitary, terminating branches, or 2 or 3 in clusters; involucres 8–15 mm high, 8–12 mm wide, tapering apically, the middle and outer bracts spine-tipped, the spines 5–8 mm long; flowers yellow, all alike; pappus 1.5–3 mm long; 2n = 24. Adventive Old World species of disturbed sites in Salt Lake County (Without collector UT); 1 (0).

Centaurea moschata L. [*Amberboa moschata* (L.) DC.]. Annual herbs; simple or sparingly branched, mainly 3–7 dm long, sparingly tomentose; leaves 1–9.5 cm long, 1–3 cm wide, pinnatifid, the lowermost petiolate, becoming sessile upwards; heads solitary, on peduncles 8–15 cm long or more; involucres vase-shaped, 12–14 mm high, 18–22 mm wide, the bracts oval, with purplish margins, only the inner with broad, reflexed, entire appendage; flowers pink; pappus shorter to about equaling the achenes; 2n = 28, 32. Cultivated ornamental, escaping and persisting in Washington County; adventive from Asia; 1 (0).

Centaurea repens L. Russian Knapweed. [*C. picris* Pallas ex Willd.; *Acroptilon repens* (L.) DC.]. Perennial rhizomatous herbs, mostly 3–8 dm tall, arachnoid-tomentose to glabrate; leaves in a basal rosette and cauline, the basal leaves often withered by flowering time, the cauline mainly 1–6 cm long, 2–12 mm wide, entire or serrate; heads few to numerous, terminating branches; involucre 9–15 mm high, 5–12 mm wide, more or less urn-shaped, middle and outer bracts broad, glabrous, with broader rounded, subentire hyaline tips, the inner bracts narrow, tapering, and with plumose hairy tips; flowers pink to purplish, all alike; pappus bristle subplumose, 6–11 mm long; 2n = 36. Introduced Old World primary noxious weed, now widely established at 1220 to 2380 m in Cache, Daggett, Duchesne, Emery, Garfield, Grand, Kane, Salt Lake, San Juan, Sanpete, Tooele, Uintah, Utah, and Washington counties; widespread in North America; adventive from Eurasia; 28 (ii).

Centaurea scabiosa L. Perennial herbs, mostly 5–15 dm tall, scabrous-puberulent; leaves 4–20 cm long or more, the lowermost long-petiolate, once to twice pinnatisect, the segments linear to oblong, entire or dentate-serrate to lobed, the upper pinnately divided, sessile; heads few to several, terminating branches; involucres 13–20 mm high, 18–25 mm wide, ovoid-globose; bracts ovate, glabrous or arachnoid, the appendages triangular-ovate, brown or black, with pale brown teeth; flowers purple, alike or nearly so; pappus 4–5 mm long; 2n = 20, 40. Cultivated ornamental, persisting and escaping, in Salt Lake and Uintah counties; adventive from Europe: 2 (0).

Centaurea solstitialis L. Yellow Starthistle. Annual or biennial, grayish-tomentose, the stems 1–6 (10) dm tall, evidently winged; leaves mainly 1–12 (20) cm long, 0.1–3 (5) cm wide, the basal ones lyrate to pinnatifid, the cauline ones progressively smaller and entire upwards, linear to linear-subulate; heads few to numerous, terminating branches; involucres 8–15 mm high, 7–15 mm wide, urn-shaped, the middle and outer bracts with central apical spines 10–20 (30) mm long, the inner with a small hyaline appendage; flowers yellow, all alike; pappus of marginal flowers none, that of the central ones 3–5 mm long; 2n = 16. Roadsides and abandoned fields at 915 to 1900 m in Box Elder, Wasatch, Washington, and Weber counties; adventive from Europe; 4 (i).

Centaurea virgata Lam. Perennial, from a caudex, more or less grayish tomentose, the stems 4–9 dm tall, branched above; leaves mainly 0.5–15 cm long, 0.1–6 cm wide, the basal ones petiolate, once to twice pinnately divided, the lobes linear, these often again toothed or lobed; cauline leaves smaller, sessile, and lobed to entire; heads several to numerous, terminating short branches; involucre 7–10 mm high, 3–5 mm wide, the bracts pale or suffused with red or purple, with a slender apical spine 1–4 mm long; flowers pink; pappus about 1.5 mm long. Roadsides and other disturbed sites in Grand, Juab, and Utah counties at 1525 to 1830 m; adventive from Eurasia; 5 (i).

Chaenactis DC.

Annual, biennial, or perennial herbs, from taproots; leaves alternate or mainly basal, pinnately dissected to entire; heads solitary or few to several, borne in corymbose cymes, discoid, the flowers white, or cream to pink, all perfect, the marginal ones sometimes enlarged and raylike; involucral bracts in 1–3 series, herbaceous; receptacle flat, naked; pappus of 4–20 hyaline scales; style branches slightly compressed; achenes clavate, terete or more or less compressed.

Stockwell, P. 1940. A revision of the genus *Chaenactis*. Contr. Dudley Herb. 3: 89–168.

1. Plants perennial from a simple or branching caudex, 2–9 cm tall; stemless or with few short internodes, of high elevations .... *C. alpina*
— Plants annual or biennial, rarely perennial, the caudex seldom developed; stems mainly 10–30 cm tall or, if less, the plants definitely not perennial; distribution usually of middle and lower elevations 2

2(1). Basal rosette well developed; plants biennial or short-lived perennials; pappus scales 10–16 ...... *C. douglasii*
— Basal rosettes poorly if at all developed; plants annual; pappus scales 4 or 5 (rarely 8) 3

3(2). Lower and upper cauline leaves simple, the middle ones few-lobed; plants of Washington County .... *C. fremontii*
— Lower, middle, and upper leaves pinnately divided, or only the uppermost simple 4

4(3). Heads mostly 15–22 mm high; flowers pink, much surpassing the involucre; anthers included ... *C. macrantha*
— Heads mostly 8–10 mm high; flowers white or cream, only slightly surpassing the involucre; anthers exserted .. 5

5(4). Involucral bracts blunt or nearly acute apically; plants widely distributed .... *C. stevioides*
— Involucral bracts long-attenuate and bristle-tipped apically; plants of Washington and Millard counties .... *C. carphoclina*

Chaenactis alpina (Gray) Jones Alpine Dusty-maiden. [*C. douglasii* var *alpina* Gray, type from Alta]. Perennial, from a simple or branched, sometimes soboliferous caudex, 3.5–9 cm tall; stems with few contracted internodes, very short, or not developed; leaves 1.3–5 cm long, pinnately divided, the lobes again toothed or lobed, 1–7 mm long, gray tomentose to glabrate; heads solitary or sometimes 2, the peduncles tomentose or glandular, 0.5–6 cm long; involucres (7.5) 10–13 mm long, (8) 10–17 mm wide, the bracts often suffused with purple, glandular or tomentose; corolla purplish to white, glandular or sparingly tomentose; pappus of 10 oblong-spatulate rounded hyaline scales, in 2 series; achenes 6–8 mm long, hairy. Boulder stripes and talus in alpine tundra or upper montane communities at 2980 to 3965 m in Cache, Duchesne, Salt Lake, Summit, and Utah counties.; Oregon to Montana, California and Colorado; 10 (i). Our materials are separable into two more or less distinctive phases; a glandular phase, with distribution mainly in the Wasatch Mts., which is **var. alpina** [including *C. rubella* Greene; *C. alpina* var. *rubella* (Greene) Stockwell], and a tomentose phase, mainly from the Uinta Mts. which might be assignable to **var. leucopsis** (Greene) Cockerell [*C. leucopsis* Greene]. More work is necessary, including evaluation of the type specimen of var. *leucopsis*.

Chaenactis carphoclina Gray Annual, from a taproot, 6–28 (40) cm tall; stems well developed, more or less flexuous; leaves 0.8–5.6 cm long, mealy-puberulent, 1- to 2-pinnatifid, the segments linear-filiform, 1–20 mm long; heads few to numerous, on slender farinose to glandular peduncles 0.4–3 cm long; involucres 6–9 mm high, 6–15 mm wide, the bracts lance-attenuate into slender bristlelike tips, glandular; flowers white to cream; pappus of central flowers usually of 4 lance-acuminate scales, those of marginal flowers sometimes shorter; achenes 3.5–4.5 mm long, hairy; n = 8. Creosote bush and cool desert shrub communities at 850 to 1000 m in Millard and Washington counties; California, Nevada, Arizona; 13 (i).

Chaenactis douglasii (Hook.) H. & A. Douglas Dusty-maiden. [*Hymenopappus douglasii* Hook.; *C. achillefolia* H. & A.; *C. douglasii* var. *achilleifolia* (H. & A.) A. Nels.; *C. douglasii* var. *montana* Jones; *C. brachiata* Greene, type from Springdale; *C. brachiata* var. *stansburyi* Stockwell, type from Stansbury Island]. Biennial or short-lived perennial, from a taproot, seldom with a caudex, mainly 5–50 (60) cm tall, sparsely to densely tomentose; stems with few to many well-developed internodes; leaves 0.6–12 (15) cm long, 1–3 pinnatifid, the lobes 1–3 cm long, tomentose to glabrate; heads solitary or several in a corymbose cyme; involucre 7–16 mm high, 8–25 mm wide, the bracts glandular to glandular-tomentose, oblong to narrowly oblanceolate or linear, blunt apically; flowers white to pink; pappus of 10–16 scales in 2 series; achenes 6–8 mm long, hairy; 2n = 12, 37. Shadscale, sagebrush, pinyon-juniper, mountain brush, ponderosa pine, white fir, Douglas fir, aspen, and limber pine communities at 1340 to 3050 m in all Utah counties; British Columbia to Montana, south to California, Arizona, and Colorado; 132 (xx). It does not seem reasonable to attempt to segregate our materials into varieties. The variability apparently does not demonstrate geographic correlation.

Chaenactis fremontii Gray Fremont Dusty-maiden. Annual or winter annual, from a taproot, 10–25 (40) cm tall, glabrate or sparingly tomentose when young; leaves 0.6–6.5 cm long, the lower and upper simple, linear, the middle few-lobed, glabrous; heads solitary to several on tomentose to glabrate (glandular?); peduncles 1–5 cm long; involucres 8–10 mm high, 10–12 mm wide, glabrous or tomentose, attenuate but not caudate; flowers white to pinkish, the outer ones enlarged; pappus of central flowers of 4 scales; achenes hairy; n = 5. Creosote bush and Joshua tree communities at 670 to 885 m in Kane and Washington counties; Arizona, Nevada, California; 6 (i).

Chaenactis macrantha D. C. Eaton Showy Dusty-maiden. Annual or winter annual, from a taproot, mainly 6–25 cm tall, branching from the base or simple, floccose-tomentose to glabrate; leaves 0.5–5 cm long, 1– to 2-pinnatifid, the lobes to 1 cm long, broad, floccose to glabrate; heads solitary to several, on tomentose peduncles 0.5–5 cm long; involucres 12–17 mm high, 8–22 mm wide, the bracts oblong-lanceolate, rather abruptly short-acuminate, tomentose; corollas pink to white, all about alike; anthers included; pappus of 4 linear-oblong scales and 2–4 short outer ones or these lacking; achenes hairy; n = 6. Shadscale, pinyon-juniper, creosote bush, and blackbrush communities at 885 to 2135 m in Beaver, Box Elder, Juab, Kane, Millard, San Juan, Tooele, and Washington counties; California, Nevada, Arizona; 28 (v).

Chaenactis stevioides H. & A. Stevia Dusty-maiden. Annual or winter annual, from a taproot, mainly 4–42 cm tall, branching from the base or simple, more or less tomentose; leaves 0.3–10 cm long, 1–2 pinnatifid, the lobes to 2.5 cm long, linear to oblong, sometimes all or nearly all simple in depauperate specimens; heads solitary to several on glandular peduncles 0.3–3 cm long; involucres 6–11 mm high, 8–22 mm wide, the bracts oblong-lanceolate to linear, acute to shortly acuminate apically, glandular; corollas white to cream, the outer ones enlarged; pappus of 4 oblong-lanceolate scales; achenes hairy; n = 5. Creosote bush, blackbrush, matatriplex, shadscale, indigo bush, and juniper communities at 915 to 1891 m in Beaver, Box Elder, Carbon, Duchesne, Emery, Garfield, Grand, Juab, Kane, Millard, Piute, San Juan, Tooele, Uintah, Washington, Wayne, and Weber counties; Wyoming south to Nevada, west to California; 63 (vi).

Chamaechaenactis Rydb.

Perennial scapose herbs from a long-pilose caudex, clothed with marcescent leaf bases, and taproot, with watery juice; leaves all basal, petiolate, simple; heads solitary; involucres turbinate, the bracts subequal or the outer shorter; receptacle naked; rays none; disk flowers perfect, fertile, cream colored to pink; pappus of hyaline scales; style branches flattened, papillate; achenes 4–angled, hairy.

Chamaechaenactis scaposa (Eastw.) Rydb. Eastwood-plant. [*Chamaechaenactis scaposa* Eastw.; *C. s.* var. *parva* Preese & Turner, type from 32 mi S of Green River]. Plants 2–9 cm tall, the scapes long-villous; leaves petiolate, the blades 0.4–1.8 cm long, 3–13 (15) mm wide, lance-oblong to ovate, to oval or orbicular, obtuse to rounded apically, obtuse to truncate basally, villous beneath, strigose to strigulous or villous above; heads solitary; involucre 7–17 mm high, 10–23 mm wide, the bracts oblong or linear-oblong, the outer densely villous, green or suffused with red purple, the margin hyaline; corollas cream to pink; pappus scales oblanceolate-spatulate, rounded; achenes black, hirsute-pilose. Shadscale, galleta, pygmy sagebrush, mountain brush, pinyon-juniper, and ponderosa pine communities at 1580 to 2565 m in Carbon, Duchesne, Emery, Garfield, Grand, Kane, San Juan, Sevier, Uintah, and Wayne counties; Arizona and Colorado; 56 (v).

Chamomilla S. F. Gray

Annual herbs, aromatic in some; leaves alternate, bi- or tripinnatifid, with linear filiform ultimate segments; heads radiate or discoid, solitary or corymbose; involucral bracts greenish-chartaceous, the margins hyaline, in 2 or 3 series, subequal to imbricate; receptacle conic, hollow, naked; marginal flowers pistillate; rays (when present) white, the central disk flowers perfect and fertile, the style branches truncate, tufted-hairy apically; pappus a short crown of minute scales, or vestigial or lacking; achenes subcylindric, the ventral face with 3–5 narrow ribs, the dorsal face smooth and convex.

1. Heads radiate; disk corollas 5–lobed; involucre 11–25 mm in diameter .... C. rescuita
— Heads discoid; disk corollas 4–lobed; involucres 4–10 mm in diameter .... C. suaveolens

Chamomilla rescuita (L.) Rauschert Chamomile. [*Matricaria resutita* L.; *M. chamomilla* L.]. Annual herbs; stems 0.2–4 (6) dm tall, erect or ascending, branched above; herbage glabrous or puberulent; leaves 2–6 cm long; heads solitary or more commonly few to many and corymbosely arranged; involucres saucer-shaped, 3–4 mm high, 11–25 mm wide, the bracts subequal, the margins broadly hyaline, the midstripe greenish to brownish; rays 10–20, white, 4–10 mm long. Moist disturbed soils at low to moderate elevations in Cache, Salt Lake, Sanpete, and Wasatch counties; adventive from Europe; 2 (0).

Chamomilla suaveolens (Pursh) Rydb. Pineapple Weed. [*Santolina suaveolens* Pursh; *Matricaria matricarioides* (Less.) T. C. Porter]. Annual herbs; stems 0.4–4 dm tall, erect or ascending, branched from the base or simple; herbage glabrous or pubescent; leaves 1–5 (9) cm long; heads few to many, paniculately arranged; involucres saucer-shaped, 2–6 mm high, 4–10 mm wide, the bracts subequal to somewhat imbricate, the margins hyaline, the midstripe greenish; rays lacking; disk flowers 4–lobed; n = 9. Disturbed sites at 1310 to 2810 m in Box Elder, Cache, Carbon, Daggett, Davis, Duchesne, Rich,

Salt Lake, Sevier, Summit, Utah, Wasatch, and Weber counties; adventive from Europe; 26 (0).

Chrysanthemum L.

Perennial herbs from a rhizome or a caudex, with watery juice; stems erect or nearly so; leaves alternate, serrate to pinnatifid, heads solitary or few to numerous in open corymbose clusters; involucral bracts imbricate, in 2-4 series, greenish or straw colored, the margins brownish scarious; receptacle naked; ray flowers white, numerous, pistillate, fertile, or lacking; disk flowers numerous, perfect, fertile, yellow; pappus lacking or a short crown; style branches flattened; achenes several-nerved, beakless, glabrous.

1. Leaves finely serrate; heads usually numerous, small, commonly rayless .... *C. balsamita*
— Leaves coarsely serrate or pinnatifid; heads larger, fewer, commonly with rays 2
2(1). Heads solitary of few; involucres 7-10 mm high; rays 1-2 cm long; leaves serrate to more or less once pinnatifid .... *C. leucanthemum*
— Heads several to numerous; involucres 3-4.5 mm high; rays 2-6 mm long .... *C. parthenium*

Chrysanthemum balsamita L. Costmary. [*Balsamita major* Desf.]. Perennial herbs, from a caudex, commonly 5-10 (12) dm tall; stems strigose, at least above; leaves petiolate below, sessile or subsessile above, the blades 0.9-10 (15) cm long, 0.6-5 (8) cm wide, elliptic to oblanceolate, finely serrate, strigose; heads numerous, corymbose; involucres 3.7-4.6 mm high, 6-8 mm wide, the bracts oblong, sparingly strigose, the tip hyaline; ray flowers (when present) 4-6 mm long. Fields, roadsides, and cemetaries at 1370 to 2135 m Cache, Salt Lake, Sanpete, Summit, Tooele, Uintah, and Utah counties; escaped from cultivation, now widely established in the U. S.; 5 (i).

Chrysanthemum leucanthemum L. Oxeye-daisy. [*Leucanthemum vulgare* Lam.]. Perennial rhizomatous or subrhizomatous herbs, commonly 2-8 (10) dm tall; stems glabrous or nearly so, mainly simple; leaves petiolate below, becoming smaller and sessile above, the blades 0.8-5 cm long, oblanceolate to obovate or linear, serrate, crenate, or pinnately lobed, glabrous or villosulose; heads solitary; involucres 7-10 mm high, 15-23 mm wide, the bracts lance-ovate to oblong-linear, with brown margins, hyaline apically; rays mainly 15-30, white, 10-22 mm long; pappus none; n = 9, 18. Roadsides, fields, and other disturbed sites at 1525 to 2820 m in Salt Lake, Sanpete, Uintah, Utah, Wasatch, and Weber counties; widespread in North America; adventive from Eurasia; 6 (0).

Chrysanthemum parthenium (L.) Bernh. [*Matricaria parthenium* L.; *Leucanthemum parthenium* (L.) Gren. & Godron; *Pyrethrum parthenium* (L.) Sm.; *Tanacetum parthenum* (L.) Schultz-Bip.]. Perennial herbs with caudex and taproot; commonly 3-9 dm tall; stems glabrous, or puberulent above; leaves petiolate, becoming smaller, but still petiolate above, the blades 0.5-8 cm long, 0.6-4.5 (6) cm wide, pinnatifid or doubly so; heads several to numerous, the inflorescence corymbose; involucres 3-4.5 mm high, 7-10 mm wide, the bracts oblong, with a dark center, otherwise scarious except the tip hyaline; rays 10-20, white, 4-8 mm long; pappus a crown or none; 2n = 18. Cultivated ornamental, escaping and persisting at 1525 to 1950 m in Carbon, Salt Lake, Utah, and Weber counties; widely established in the United States; adventive from Europe; 5 (0).

Chrysothamnus Nutt.

Shrubs with white bark, or the surface obscured by a tomentum, this often glandular-resinous; leaves alternate, linear to oblong, or lanceolate, sessile, entire; heads white or yellow, discoid, narrow, in contracted to open paniculate inflorescences; flowers perfect, fertile; involucral bracts imbricate, more or less keeled, in 4 or 5 vertical or obscure ranks, chartaceous or coriaceous, or the tip herbaceous; receptacle naked; style branches flattened; achenes slender, flattened, angled, or terete, hairy or glabrous; pappus of numerous capillary bristles.

1. Flowers white; leaves terete; plants of western tier of counties (except Iron and Washington) .... *C. albidus*
— Flowers yellow; leaves various but, if terete, plants of Washington County or rarely elsewhere 2
2(1). Leaves terete, resinous punctate; stems more or less fastigiate; plants of Washington County ... *C. paniculatus*
— Leaves commonly more or less flattened, resinous-punctate or not; stems not especially fastigiate; plants of broad or other distribution 3
3(2). Stems obscured by a tomentum, this often impregnated with resinous-glandular material 4
— Stems glabrous or puberulent, the surface readily apparent 5
4(3). Involucral bracts long-attenuate, membranous; inflorescence more or less racemose .... *C. parryi*
— Involucral bracts obtuse to acute, rarely attenuate but, if so, chartaceous; inflorescence cymose .... *C. nauseosus*
5(3). Leaves lanceolate to lance-oblong, not contorted; shrubs mainly 6-20 dm tall; plants of the Green and Colorado river basins .... *C. linifolius*
— Leaves linear, oblong, or lanceolate but, if lanceolate, twisted and shrubs mainly less than 6 dm tall; distribution various 6
6(5). Achenes hairy 7
— Achenes lacking hairs, sometimes glandular or, if sparingly hairy, the involucre over 10 mm long 8
7(6). Involucral bracts acuminate-cuspidate; leaves 1-2 mm wide .... *C. greenei*
— Involucral bracts acute to obtuse; leaves various .... *C. viscidiflorus*
8(6). Flowers 10-12 mm long, surpassed by the pappus; plants of Emery, Wayne, and San Juan counties .... *C. pulchellus*
— Flowers 7-9 mm long, surpassing or subequal to the pappus; distribution various 9
9(8). Involucral bracts strongly ranked; involucres 9.2-13 mm long .... *C. depressus*
— Involucral bracts not strongly ranked; involucres 6.2-7.5 mm long .... *C. vaseyi*

Chrysothamnus albidus (Jones) Greene Alkali Rabbitbrush; White Rabbitbush. [*Bigelovia albida* Jones]. Shrubs, mainly 5-10 dm tall, more or less fastigiately branched, white-barked, glabrous, resinous-viscid, aromatic; leaves 0.5-3.5 cm long, terete, 0.5-1 mm thick, glandular-punctate, mucronate, crowded, often with axillary fascicles; heads clustered at branchlet apices; involucres 6.8-9 mm high, 3-7 mm wide, the bracts obscurely

4– to 5–ranked, the outer ones lance-ovate, thickened in lower half, abruptly subulate-attenuate, the inner oblong, acuminate to acute, the margin hyaline, glandular to tomentose; corollas white, 6–7.5 mm long; achenes 4–4.5 mm long, pilose and glandular; pappus abundant; 2n = 18. Local in salt grass, pickleweed, and alkali-saccaton communities at 1450 to 1650 m in Beaver, Box Elder, Juab, Millard, and Tooele counties; California, Nevada; 11 (v).

Chrysothamnus depressus Nutt. Dwarf Rabbitbrush. Low, spreading shrubs, the ascending to erect, subherbaceous stems 0.6–3 dm tall, white-barked, scabrous-puberulent or glandular-puberulent; leaves 0.4–2 cm long, 1–4 (5) mm wide, flat, narrowly lanceolate to oblanceolate or spatulate, flat, scabrous-puberulent, obtuse, rounded or sharply apiculate; heads clustered at branch apices; involucres 9.2–13 mm high, 4.5–7 mm wide, the bracts in 4 or 5 definite vertical ranks, keeled, lance-attenuate, the subulate tip soft, the outer more or less herbaceous (sometimes suffused with purple) and the inner with broad hyaline margins; corollas yellow, 7.5–9 mm long; achenes (5) 6–7 mm long, glabrous or sparingly stipitate-glandular; pappus off-white to brownish, abundant. Sagebrush, salt desert shrub, juniper, pinyon- juniper, mountain brush, ponderosa pine and alpine fir communities at 1550 to 2900 m in Carbon, Duchesne, Emery, Garfield, Iron, Juab, Kane, Millard, Piute, San Juan, Sanpete, Sevier, Summit, Uintah, Utah, Wasatch, Washington, and Wayne counties; Colorado, New Mexico, Arizona, and Nevada; 34 (iv).

Chrysothamnus greenei (Gray) Greene Greene Rabbitbrush. [*Bigelovia greenei* Gray]. Low, ascending to erect shrubs, with subherbaceous stems from a woody crown, mainly 1–3.5 dm tall, white-barked, glabrous; leaves 0.3–3.5 cm long, 0.8–1.2 mm wide, flat, linear, glabrous or scabrous-ciliate; heads numerous, corymbosely clustered at branch tips; involucres 5–7.1 mm high, 2.5–4 mm wide, the bracts obscurely ranked, the outer ones herbaceous-thickened near the tip, gradually acuminate-cuspidate, the inner ones abruptly narrowed, glabrous or more or less tomentose, narrowly if at all hyaline-margined; corollas yellow, 3.5–4.8 mm long; achenes 3.3–4 mm long, pilose. Rabbitbrush, black sagebrush, shadscale, winterfat, sagebrush, and pinyon-juniper communities at 1280 to 2745 m in Carbon, Duchesne, Emery, Garfield, Grand, Iron, Juab, Millard, Piute, Tooele, Uintah, Utah, and Wayne counties; Colorado, New Mexico, Arizona, and Nevada; 53 (vi). This entity forms intermediates with phases of *C. viscidiflorus*.

Chrysothamnus linifolius Greene Spreading Rabbitbrush. Tall shrubs, the branches erect-ascending, mainly 8–20 (35) dm tall, white-barked, glabrous; leaves 0.9–7.7 cm long, 1–9 mm wide, flat, plane (not contorted or rarely somewhat so), thick, oblong to elliptic or narrowly lanceolate, glabrous, scabrous-ciliate, attenuate to acute; heads numerous, corymbosely arranged at branch tips; involucres 4.3–7.2 mm long, 1.8–3 mm wide, the bracts indistinctly ranked, the outer distinctly herbaceous at tip, the inner often merely glandular thickened at midrib, all obtuse to rounded, glabrous; corollas yellow, 4.5–5.8 mm long; achenes 2.1–2.8 mm long, pilose; 2n = 18. Stream banks and terraces, irrigation canals, seeps and springs in riparian communities at 1130 to 2535 m in Carbon, Daggett, Duchesne, Emery, Garfield, Grand, Kane, San Juan, Sanpete, Sevier, Uintah, and Wayne counties; Montana to Arizona and New Mexico; 54 (xvii).

Chrysothamnus nauseosus (Pallas) Britt. Rubber Rabbitbrush. Low to tall shrubs, the branches erect-ascending, mainly 2–20 (30) dm tall, the bark obscured by a tomentum, this often resinous-glandular impregnated; leaves 0.6–7 (10) cm long, 0.5–5 (10) mm wide, 1– to 3–nerved, tomentose to glabrate or glabrous, subcylindric to flat, if the latter then commonly plane, linear to narrowly oblong, acute to apiculate apically; heads numerous, in terminal paniculate cymes; involucres (6) 6.5–11.5 (13) mm high, 1.5–7.2 mm wide, the bracts obscurely to definitely ranked, the outer ones sparingly tomentose to glabrous, the inner commonly glabrous, oblong, chartaceous to more or less herbaceous-thickened, obtuse to acute or shortly acuminate apically; corollas yellow or yellow-orange, 6–10.3 (12) mm long; achenes 2.5–5.5 mm long, glabrous or hairy; 2n = 18. The *nauseosus* complex in Utah is represented by a diverse assemblage of more or less geographically and ecologically segregated races, which are placed in some 14 varieties. The following arbitrary key will serve to identify most specimens.

1.	Shrubs usually 3 dm tall or lower, local endemics in Piute, Sanpete, Sevier, Carbon, Emery, Daggett, and Duchesne counties	2
—	Shrubs usually more than 3 dm tall, seldom lower, but then of different distribution	5
2(1).	Branchlets white-pannose; achenes pilose; plants known from Daggett and Uintah counties .... *C. nauseosus* var. *uintahensis*	
—	Branchlets various, but not otherwise as above; achenes glabrous; plants not from Uintah or Daggett counties	3
3(2).	Involucres glabrous, 8.5–9.5 mm high; plants of Emery, Carbon, Wasatch, and Duchesne counties .... *C. nauseosus* var. *psilocarpus*	
—	Involucres tomentose or glabrous, 10–12 (13.5) mm high; plants of Sanpete, Sevier, and Piute counties ...	4
4(3).	Involucres glabrous; corollas 7.8–9 mm long; plants local on Arapien shale in Sanpete and Sevier counties .... *C. nauseosus* var. *iridis*	
—	Involucres tomentose; corollas 10–12 mm high; plants local in Piute County .... *C. nauseosus*. var. *glareosus*	
5(1).	Achenes and ovaries glabrous	6
—	Achenes and ovaries pilose	9
6(5).	Flowers 5–8 mm long; involucres 7–8.5 (9) mm long, 1.5–3 mm wide (when pressed) .... *C. nauseosus* var. *leiospermus*	
—	Flowers 8.3–10 mm long; involucres 9–11 mm long, 3.7–7 mm wide (when pressed)	7
7(6).	Involucres subcylindric; plants of dunes and deep sands of western Utah and in the Uinta Basin .... *C. nauseosus* var. *turbinatus*	
—	Involucres tapering to the base; plants of south-central and southeastern Utah	8
8(7).	Achenes 5–5.5 mm long; plants low, commonly less than 5 dm tall, of San Juan and Emery counties .... *C. nauseosus* var. *bigelovii*	
—	Achenes 2.5–4 mm long; plants taller, commonly over 5 dm tall, known from Kane County .... *C. nauseosus* var. *nitidus*	

9(5). Involucres over 10 mm long; corollas 9.5–10.5 mm long 10
— Involucres 6.5–8.6 (9.5) mm long (to 11 mm long in var. *junceus*); corollas 5–8.6 (10) mm long 11

10(9). Involucres cylindric, the bracts neither strongly keeled nor ranked; plants of dune areas in western and northeastern Utah ... *C. nauseosus* var. *turbinatus*
— Involucres tapering, clavate, the bracts strongly keeled and aligned; plants of southeastern Utah ... . *C. nauseosus* var. *arenarius*

11(9). Leaves 3–5 (10) mm wide; plants of central to north-central Utah .... *C. nauseosus* var. *salicifolius*
— Leaves 0.5–3 mm wide; plants of various distribution 12

12(11). Corolla lobes commonly long-pilose (glabrate in age); leaves often deciduous by anthesis; plants of southeastern Utah .... *C. nauseosus* var. *junceus*
— Corolla lobes glabrous; leaves present or absent at anthesis; distribution various 13

13(12). Corolla lobes 0.4–0.9 mm long .... *C. nauseosus* var. *gnaphaloides*
— Corolla lobes 1–2 mm long 14

14(13). Leaves (1) 3– to 5–nerved, commonly 1–3 mm wide .... *C. nauseosus* var. *glabratus*
— Leaves 1–nerved, commonly 0.5–1.5 mm wide 15

15(14). Leaves and/or stems usually grayish or whitish tomentose or green, not especially yellow green; involucres more or less tomentose .... *C. nauseosus* var. *albicaulis*
— Leaves and/or stems usually yellowish green, the tomentum commonly resinous-matted; involucres glabrous .... *C. nauseosus* var. *consimilis*

Var. albicaulis (Nutt.) Rydb. [*C. speciosus* var. *albicaulis* Nutt.; *C. nauseosus* ssp. *albicaulis* (Nutt.) H. & C.]. Saltgrass, sagebrush, pinyon-juniper, and ponderosa pine communities at 1310 to 2290 m in Box Elder, Cache, Carbon, Iron, Juab, Kane, Millard, Morgan, Salt Lake, San Juan, Sanpete, Tooele, Uintah, Utah, Wasatch, and Weber counties; Oregon to Wyoming, south to California, Nevada, and New Mexico; 24 (i). This taxon forms intermediates with var. *glabratus*. In low elevation phases of saline substrates the stems are white-pannose.

Var. arenarius (L. C. Anderson) Welsh [*C. nauseosus* ssp. *arenarius* L. C. Anderson]. Sagebrush, juniper, and pinyon-juniper communities at 1675 to 1830 m in Grand, Kane, San Juan, and Washington counties; Arizona; 12 (v). This is a plant of deep sandy alluvium.

Var. bigelovii (Gray) Hall [*C. nauseosus* ssp. *bigelovii* (Gray) H. & C.; *Linosyris* (*Chrysothamnus*) *bigelovii* Gray]. Grayia and pinyon-juniper communities 1460 to 1950 m in Emery, Garfield, and San Juan counties; Arizona, Colorado, New Mexico; 6 (v). More collections of this entity are required.

Var. consimilis (Greene) Hall [*C. nauseosus* ssp. *consimilis* (Greene) H. & C.; *C. consimilis* Greene]. Saline meadows, riparian zones, and terraces in saltgrass-alkali saccaton, shadscale, sagebrush, rabbitbrush, mountain brush, pinyon-juniper, and ponderosa pine communities at 1280 to 3000 m in all Utah counties except Grand and San Juan; Oregon to Wyoming, south to California, Arizona and New Mexico; 100 (xxv). This is the common narrow-leaved phase with cone-shaped panicles. They occur frequently in saline moist sites, such as the travertine mounds at Monroe Hot Springs.

Var. glabratus (Gray) Cronq. [*Bigelovia graveolens* var. *glabrata* Gray; *C. nauseosus* ssp. *graveloens* (Gray) Piper; *C. nauseosus* var. *graveolens* (Gray) Hall]. Desert willow-baccharis, willow-cottonwood, greasewood- tamarix, sagebrush, shadscale, mountain brush, and ponderosa pine communities at 750 to 2475 m in Iron, Piute, Sanpete, Sevier, Summit, Utah, Wasatch, and Washington counties, and in all counties east of those; Idaho to North Dakota, south to Arizona, and New Mexico; 88 (xxiii).

Var. glareosus (Jones) Welsh [*Bigelovia glareosa* Jones, type from Marysvale; *C. nauseosus* ssp. *glareosa* (Jones) H. & C.]. The type specimen is lost, and the ultimate dispositon of this taxon is uncertain; it should be sought in the canyon north of Marysvale, on Tertiary igneous substrates; endemic; 0 (0).

Var. gnaphaloides (Greene) Hall [*C. speciosus* var. *gnaphaloides* Greene; *C. nauseosus* ssp. *hololeucus* (Gray) H. & C., in part; *C. zionis* A. Nels., type from 20 mi N of St. George]. Shadscale, pigmy sagebrush, rabbitbrush, sagebrush, and pinyon-juniper communities at 1070 to 2380 m; known in all Utah counties except Daggett, Kane, Morgan, Rich, Summit, and Wayne, and likely in them also; California, Nevada, and Arizona (?); 73 (vii). This taxon is a near ally of ssp. *hololeucus* (Gray) H. & C., and should that taxon be placed within a quadrinomial, then the var. *gnaphaloides* would be placed within it. However, no such combination is implied or proposed herein.

Var. iridis (L. C. Anderson) Welsh [*C. nauseosus* ssp. *iridis* L. C. Anderson, type from Rainbow Hills, Sevier County]. Rabbitbrush-sagebrush community on an incipient seep in Arapien shale at ca 1980 m in Sanpete and Sevier counties; endemic; 2 (i).

Var. junceus (Greene) Hall [*C. nauseosus* ssp. *junceus* (Greene) H. & C.; *Bigelovia juncea* Greene]. Blackbrush, shadscale, rabbitbrush, matchweed, and pinyon-juniper communities at 1220 to 1800 m in Emery, Garfield, Grand, Kane, San Juan and Wayne counties; Arizona; 18 (iv). The non-glandular, clear straw colored, long involucres with bracts usually aligned are distinctive of this variety.

Var. leiospermus (Gray) Hall [*Bigelovia leiosperma* Gray, type from St. George; *C. nauseosus* ssp. *leiospermus* (Gray) H. & C.; *B. leiosperma* var. *abbreviata* Jones, type from Clear Creek Canyon, Sevier County; *C. nauseosus* var. *abbreviata* (Jones) Welsh; *C. oliganthus* A. Nels, type from Zion Canyon]. Blackbrush, Grayia, shadscale, black sagebrush, Vanclevea, pinyon-juniper, and ponderosa pine communities at 1070 to 2745 m in Box Elder, Emery, Garfield, Grand, Juab, Kane, Millard, Piute, Sevier and Washington counties; Nevada, California; 15 (v). The materials from Emery and Grand counties have leaves which are very slender and subterete. The condition is presumeably derived from introgression with var. *bigelovii*.

Var. nitidus (L. C. Anderson) Welsh [*C. nauseosus* ssp. *nitidus* L. C. Anderson]. Vanclevea-ephedra community at about 1250 m in Garfield and Kane counties; Arizona; 1 (0). This variety has the general aspect of vars. *bigelovii* and *leiospermus*. It is a taller plant than either, and differs otherwise as set forth in the key.

Var. psilocarpus Blake [*C. nauseosus* ssp. *psilocarpus* (Blake) L. C. Anderson]. Sagebrush and salina wildrye communities at 1925 to 2290 m in Carbon, Duchesne,

Emery (type from Huntington Canyon), and Wasatch counties; endemic; 5 (0). These peculiar low shrubs occasionally produce taller intermediates with var. *glabratus* (q.v.).

Var. *salicifolius* (Rydb.) Hall [*C. salicifolius* Rydb., type from Strawberry Valley; *C. nauseosus* ssp. *salicifolius* (Rydb.) H. & C.]. Sagebrush, pinyon-juniper, mountain brush, and aspen communities at 1310 to 2870 m in Box Elder, Cache, Carbon, Duchesne, Emery, Juab, Salt Lake, Sanpete, Sevier, Summit, Tooele, Utah, and Wasatch counties; endemic; 19 (iii). This entity forms intermediates with var. *glabratus*, and might represent nothing more than a broad-leaved extension of that taxon.

Var. *turbinatus* (Jones) Blake [*Bigelovia turbinata* Jones, type from Canaan Ranch, Kane County; *C. nauseosus* ssp. *turbinatus* (Jones) H. & C.]. Rabbitbrush, saltbush, ephedra, juniper, and greasewood communities at 1370 to 1710 m in Beaver, Duchesne, Iron, Juab, Kane, Millard, and Uintah counties; Nevada (?); 48 (v). Both glabrous and pilose achenes occur in this distinctive taxon. It shares the feature of villous corolla lobes with the sand-loving var. *junceus* of the Navajo Basin. The Uintah Basin materials differ in the more keeled and attenuate involucres and flowers that are more exserted from the involucre.

Var. *uintahensis* (L. C. Anderson) Welsh stat. nov. [based on: *Chrysothamnus nauseosus* ssp. *uintahensis* L. C. Anderson Great Basin Naturalist 44: 416. 1984]. Rabbitbrush and juniper communities at 1675 to 2015 m in Daggett and Uintah counties; endemic; 12 (i). This is a peculiar dwarf white-pannose phase with heads clustered at branch ends, simulating *Tetradymia canescens* in habit and aspect. The plant appears to be a stabilized hybrid involving *C. nauseosus* and *C. parryi*.

Chrysothamnus paniculatus (Gray) Greene Mohave Rabbitbrush. [*Bigelovia paniculata* Gray]. Tall shrubs, the branches subfastigiate, mainly 6–20 dm tall, the bark green, becoming tan to gray in age, resinous-punctate; leaves 0.4–3 cm long, about 0.5 mm wide, linear-filiform, terete, mucronate apically; heads numerous, in usually conic panicles; involucres 4.8–6.5 mm high, 2–3 mm wide, the bracts indistinctly ranked, chartaceous-indurate, scarcely if at all glandular, thickened at midrib, obtuse, glandular; corollas yellow, 5.5–6 mm long; achenes 1.8–3.4 mm long, pilose; n = 9. Roadsides, stream banks, terraces, and slopes in creosote bush, Joshua tree, and baccharis communities at 670 to 1220 m in Washington County; Nevada, Arizona, California; 15 (iv). The plants begin to flower in October and continue into November.

Chrysothamnus parryi (Gray) Greene Parry Rabbitbrush. Low to moderate shrubs, the branches not especially fastigiate, mainly 2–6 dm tall, the bark pannose-tomentose or the tomentum glandular-resinous; leaves 0.6–6 (8) cm long, 1–2 mm wide, 1- to 3-nerved, green, viscid or sometimes tomentulose, flat, usually plane, linear to narrowly oblong; heads several to many, the inflorescences tending to be elongate and subracemose; involucres 9–14.5 mm high, 4–8 mm wide, the bracts obscurely to definitely ranked, puberulent to glabrous, the outer usually with elongate herbaceous tips, the inner chartaceous, with glandular-thickened midrib, abruptly to gradually acuminate-attenuate or attenuate; corollas yellow or creamy yellow, 8–10 mm long; achenes 3.3–7.5 mm long, pilose; 2n = 18. Plants of the *parryi* complex form hybrid derivatives with phases of *C. nauseosus*, and with other named segregates within the complex. Except for varieties *parryi* and *nevadensis*, only arbitrary segregation appears possible. Thus, the conservative treatment as outlined below seems to best reflect the nature of *C. parryi* in Utah.

1. Flowers usually more than 10 per head . . . . *C. parryi* var. *parryi*
— Flowers commonly 5–9 per head 2
2(1). Involucral bracts mainly 24–28; plants of southwestern Utah . . . . *C. parryi* var. *nevadensis*
— Involucral bracts mainly 12–22; plants of south central, central, and northeastern Utah . *C. parryi* var. *attenuatus*

Var. *attenuatus* (Jones) Kittell in Tidestr. & Kittell [*Bigelovia howardii* var. *attenuata* Jones, type from near Marysvale; *C. parryi* ssp. *attenuatus* (Jones) H. & C.; *C. affinis*. A. Nels.; *C. parryi* ssp. *affinis* (A. Nels.) L. C. Anderson; *Linosyris howardii* Parry in Gray; *C. parryi* ssp. *howardii* (Parry) H. & C.; *C. parryi* var. *howardii* (Parry) Kittell in Tidestr. & Kittell]. Meadows, sagebrush, juniper, pinyon-juniper, mountain brush, ponderosa pine, and aspen communities at 1740 to 2930 m in Beaver, Carbon, Daggett, Duchesne, Garfield, Grand, Iron, Kane, Piute, Sanpete, Sevier, Uintah, Utah, Wasatch, and Wayne counties; Wyoming and Nebraska, south to Arizona and New Mexico; 55 (xv). The *howardii* phase differs supposedly in the bracteate leaves overtopping the inflorescence and in the pale colored flowers; both characters fail as diagnostic features.

Var. *nevadensis* (Gray) Kittell in Tidestr. & Kittell [*Linosyris howardii* var. *nevadensis* Gray; *C. parryi* ssp. *nevadensis* (Gray) H. & C.]. Sagebrush, juniper, pinyon-juniper, mountain brush, and ponderosa pine communities at 1830 to 2565 m in Beaver, Iron, Millard, and Washington counties; Arizona; 10 (ii). The var. *nevadensis* differs only in degree from var. *attenuatus*, with which it is contiguous, if not partially sympatric, to the east. Should the two be combined, then the correct name will be var. *nevadensis*, since that name has priority in rank. Plants with leaves overtopping the inflorescence occur; technically they would key to the *howardii* phase of var. *attenuatus*.

Var. *parryi* [*Linosyris parryi* Gray]. Ponderosa pine and spruce-fir communities at 2075 to 2625 m in Beaver, Emery, Garfield, Kane, Millard, and Washington counties; Wyoming, Colorado, New Mexico, and Nevada; 9 (ii).

Chrysothamnus pulchellus (Gray) Greene Southwest Rabbitbrush. [*Bigelovia pulchella* Gray]. Low to moderately tall shrubs, the branches not fastigiate, mainly 5–10 dm tall, the bark white, becoming tan or brown in age, glabrous or puberulent above; leaves 0.4–3 cm long, 1–2 mm wide, linear to narrowly oblanceolate, glabrous or puberulent, flat or revolute, mucronate; heads few to many, in corymbose panicles; involucres 11.5–15 mm high, 4.5–6 mm wide, the bracts distinctly aligned, more or less herbaceous towards the apex, glandular, attenuate to sharply acute; corollas yellow, 9–10 (14) mm long; achenes 3.8–4.5 mm long, sparingly hirsute and glandular; 2n = 18. Shadscale, blackbrush, ephedra, pinyon-juniper, and ponderosa pine communities at 1370 to 2350 m in Emery, Wayne, and San Juan counties; Arizona to Kansas, south to Mexico; 4 (i). Our material belongs to **var. *baileyi* (Woot. & Standl.) Blake** [ssp. *baileyi* (Woot. & Standl.) H. & C.].

Chrysothamnus vaseyi (Gray) Greene Vasey Rabbitbrush. [*Bigelovia vaseyi* Gray]. Low shrubs, mainly 1–3 dm tall, the branches not especially fastigiate, the bark green, becoming whitish tan or finally gray in age, puberulent; leaves 0.3–3.7 cm long, 0.8–3 mm wide, linear to oblong or narrowly oblanceolate, glabrous or glandular, flat, plane, mucronate; heads numerous in compact terminal cymes; involucres 6.2–7.5 mm high, 3–6 mm wide, the bracts more or less aligned, commonly herbaceous or thickened near the apex, glandular, obtuse, the margins fimbriate-hyaline; corolla yellow, 4.8–7 mm long; achenes 2.6–4 mm long, glabrous. Meadows, sagebrush, rabbitbrush, juniper, mountain brush, and ponderosa pine communities at 1675 to 2900 m in Beaver, Carbon, Emery, Garfield, Juab, Kane, Iron, Piute, San Juan, Sanpete, Sevier, Utah, and Wasatch counties; Nevada, Wyoming, Colorado, New Mexico; 21 (ii).

Chrysothamnus viscidiflorus (Hook.) Nutt. Viscid Rabbitbrush. Low to moderate shrubs, mainly 2–10 dm tall, the branches fastigiate or not, the bark green to tan or white, finally gray in age, glabrous or puberulent; leaves 0.3–4.5(6) cm long, 0.5–4(10) mm wide, 1- to 5-nerved, linear to oblong, elliptic or oblanceolate, often twisted, mucronate; heads numerous, in compact to open terminal cymes; involucres 5–7.5 mm high, 2–4 mm wide, the bracts not well-aligned, commonly herbaceous or thickened near the apex (at least the outer), glandular or puberulent, obtuse, or abruptly acute, the margin narrow, hyaline; corollas yellow, 3.8–6 mm long; achenes 3–4 mm long, pilose; $2n = 18, 36$. The *viscidiflorus* complex is separable into two groups on the basis of pubescence of upper stems or the lack of pubescence. The segregation is not complete, because pubescence or its absence is not an absolute criterion. There is a cline in the amount of pubescence from abundant to little (or none), and the adoption of a position that one hair equals pubescence and, therefore one part of the complex and not the other, will lead to absurdity. Within the hairy phase of the complex are two more or less distinctive but largely sympatric varieties. The "glabrous" portion of the species is more difficult to separate into its constituent entities. Anderson (Great Basin Naturalist 40: 117–20, 1980) reviewed this portion of the complex; concluding that there are three taxa involved, i.e. ssp. *axillaris*, ssp. *viscidiflous* var. *viscidiflous*, and ssp. *viscidiflorus* var. *stenophyllus*. Only arbitrary separation of the three is possible, and segregation of the *axillaris* phase is problematical. In my view it is not practical to attempt recognition of more than two taxa, i.e. var. *stenophyllus* (including *axillaris*) and var. *viscidiflorus*. They are all recognized herein at varietal level, but probably would best fit within an expanded ssp. *viscidiflorus* as varieties (a course not intended or implied herein). The following key will allow for identification of most specimens.

1. Stems (at least above) and/or leaves puberulent to hispidulous 2
— Stems and leaves glabrous, or the leaves ciliate, or rarely with a few short hairs on stems or with glandular excrescences in the inflorescence 3
2(1). Leaves 0.5–2 mm wide; stems finely puberulent above .... *C. viscidiflorus* var. *puberulus*
— Leaves 2–5 mm wide; stems hispidulous-puberulent above .... *C. viscidiflorus* var. *lanceolatus*
3(1). Leaves 0.5–1.5 mm wide; plants mainly 2–3 dm tall .... *C. viscidiflorus* var. *stenophyllus*
— Leaves mainly 1–4 mm wide (or more); plants mainly 3–10 dm tall .... *C. viscidiflorus* var. *viscidiflorus*

Var. *lanceolatus* (Nutt.) Greene [*C. lanceolatus* Nutt.; *C. viscidiflorus* ssp. *lanceolatus* (Nutt.) H. & C.]. Sagebrush, pinyon-juniper, mountain brush, aspen, Douglas fir, lodgepole pine, spruce-fir, and alpine meadow communities at 1375 to 3200 m in all Utah counties except Kane and Washington, and likely there also; British Columbia to South Dakota, and south to California, Nevada, Arizona, and New Mexico; 112 (xii).

Var. *puberulus* (D. C. Eaton) Jepson [*Linosyris viscidiflora* var. *puberula* D. C. Eaton; *C. viscidiflorus* ssp. *puberulus* (D. C. Eaton) H. & C.; *C. marianus* Rydb., type from near Marysvale, Piute County]. Rabbitbrush, black sagebrush, shadscale, sagebrush, pinyon-juniper and ponderosa pine communities at 1460 to 2200 m in the western tier of counties, east to Carbon, Emery, Piute, Sevier, Utah, and Salt Lake counties; Oregon and Idaho south to California, Nevada, and Arizona; 44 (vii).

Var. *stenophyllus* (Gray) Hall [*Bigelovia douglasii* var. *stenophylla* Gray; *C. viscidiflorus* ssp. *stenophylla* (Gray) H. & C.; *C. axillaris* Keck; *C. viscidiflorus* ssp. *axillaris* (Keck) L. C. Anderson]. Ephedra, blackbrush, rabbitbrush, sagebrush, galleta, shadscale, and pinyon-juniper communities at 1280 to 2075 m in all Utah counties except Cache, Carbon, Davis, Dushesne, Morgan, Piute, Salt Lake, Sanpete, Sevier, Summit, Utah, Wasatch, and Weber; Oregon to Wyoming and south to California, Nevada, Arizona, and Colorado; 34 (vii).

Var. *viscidiflorus* [*Crinitaria viscidiflora* Hook.; *C. viscidiflorus* var. *pumilus* authors, not (Nutt.) Jepson (?); *B. douglasii* var. *spathulata* Jones, type from Fish Lake, Sevier County; *Linosyris serrulata* Torr. in Stansb., type from Salt Lake Valley]. Rabbitbrush, shadscale, sagebrush, pinyon-juniper, mountain brush, white fir, ponderosa pine, and aspen communities at 1460 to 2900 m in all or nearly all Utah counties; Washington to Nebraska, south to California, Nevada, Arizona, and Colorado; 100 (xx). The var. *viscidiflorus* forms intermediates with all other taxa in the species, and with *C. greenei* also.

Cichorium L.

Perennial herbs, with milky juice, from taproots; leaves alternate, toothed to pinnatifid; heads sessile or subsessile, numerous, borne in clusters at nodes of a spicate, simple, or branched inflorescence; involucral bracts biseriate, the outer shorter; corollas all raylike, perfect; pappus of 2 or 3 series of scales, sometimes minute; achenes angular or somewhat compressed, glabrous.

Cichorium intybus L. Chickory. Plants 3–10 dm tall or more, hirsute or glabrous; lower leaves petiolate, the blades 6–20 cm long, 1–5(7) cm wide, sinuate-dentate to runcinate-pinnatifid, becoming smaller and sessile upwards, some finally subentire; heads large and showy, 1–3 per node of inflorescence; flowers pure blue, rarely white; involucre 9–15 mm high, the outer bracts chartaceous at base, herbaceous apically; achenes 2–3 mm long; $2n = 16, 18, 20$. Roadsides and disturbed sites at 1340 to 2135 m in Cache, Duchesne, Iron, Kane, Salt Lake, Tooele, Utah, and Washington counties; widespread in North America; native of Eurasia; 8 (i). The herb *C. endiva* L. is grown in Utah; the extent is not known.

Cirsium Miller

Annual, biennial, or perennial, caulescent or acaulescent, spiny herbs from taproots, with caudices or rhizomes in some, the juice watery; leaves basal and cauline, alternate; heads solitary to several; involucral bracts in several series, subequal to imbricate, some or most of them spine-tipped; receptacle densely bristly; corollas all discoid, pink, purple, red, or creamy white, perfect or imperfect; pappus of plumose bristles (or those of the outermost flowers merely barbellate); style with a thickened minutely hairy ring below the nearly connate lobes; achenes glabrous, flattened or 4–angled, 4– to many-nerved. Note: This is a particularly complex genus taxonomically, with both introduced and indigenous species. The indigenous members are especially difficult, due in part to hybridization, mainly within species groups. The following treatment is tentative, but represents an attempt to categorize the variation present in Utah plants and to provide a legitimate name for each. Several taxa previously reported from the state are excluded, or they are treated within the constituent taxa. All involucral measurements are from pressed plants.

Moore, R. J. and C. Frankton. 1963a. Cytotaxonomic notes on some *Cirsium* species of the western United States. Canad. J. Bot. 41: 1553–1567.
_____. 1963b. A clarification of *Cirsium foliosum* and *Cirsium drummondii*. Canad. J. Bot. 42: 451–461.
_____. 1965. Cytotaxonomy of *Cirsium hookerianum* and related species. Canad. J. Bot. 43: 597–613.
_____. 1973. The *Cirsium arizonicum* complex of the southwestern United States. Canad. J. Bot. 52: 543–551.
Petrak, F. 1917. Die nordamerikanischen Arten der Gattung *Cirsium*. Beih. Bot. Centralbl. (Abt. 2), 35: 223–567.

1. Flowers mainly imperfect; heads unisexual; plants perennial, from rhizomes; introduced weed of consequence .... *C. arvense*
— Flowers perfect; plants biennial or perennial, seldom if ever with rhizomes 2

2(1). Leaves roughly hispid above, green; stems conspicuously winged decurrent; plants biennial, introduced .... *C. vulgare*
— Leaves villous, floccose, arachnoid, tomentose, or glabrous, white to gray or green; stems not winged-decurrent, except in some species; plants indigenous biennials or perennials 3

3(2). Basal rosettes to 10 dm across, the mature leaves commonly 10–30 cm wide, green, glabrate or glabrous on both sides; heads small, with long, tapering, recurved spines; plants of hanging gardens in southeastern Utah, rarely in drainages below them .... *C. rydbergii*
— Basal rosettes rarely to 5 dm across, the mature leaves usually less than 8 cm wide, floccose, tomentose, arachnoid, or glabrous on one or both sides; plants seldom of hanging gardens in southeastern Utah 4

4(3). Bracts, at least the innermost, conspicuously dilated (but not lacerate), or definitely tan to silvery in appearance, contrasting with the overall aspect of the bracts; plants commonly of meadows .... *C. scariosum*
— Bracts all spinose, or the innermost occasionally twisted to contorted at the tips, but not especially dilated or conspicuously different in color or texture from the overall aspect of bracts (see *C. centaureae*); plants of various habitats 5

5(4). Involucral bracts (at least the outer) pinnately spinose; plants green, with yellowish spines, of high elevations in the Wasatch, Tushar, and Uinta mts. . ... *C. eatonii*
— Involucral bracts not, or rarely, pinnately spinose (except in *C. clavatum*, *C. scopulorum*, and *C. ownbeyi* and the plants then of low elevations) 6

6(5). Heads 1.8–2.7 cm high, and about as wide; inner bracts with coarsely lacerate margins; plants of lower middle elevation meadows .... *C. centaureae*
— Heads 1.5–3 cm high, 1.5–4.5 (6) cm wide; inner bracts not lacerate; leaves thinly textured, finely to coarsely spined, definitely tomentose or glabrous; plants of various distribution 7

7(6). Herbage definitely white- to gray-tomentose (rarely green); involucres 1.5–2 cm high, 1.5–2.5 cm wide; plants of white shale outcrops in the Uinta Basin ... . *C. barnebyi*
— Herbage green, or white- to gray-tomentose; involucres mainly longer and broader but, if not, of different distribution 8

8(7). Stems definitely winged-decurrent; heads mainly 1.3–2 cm high, 1.2–3.2 cm wide; herbage white- to gray-tomentose; plants of Sanpete and Washington counties 9
— Stems not winged, or if so, the herbage green and glabrous or nearly so, or the heads commonly larger; plants of various distribution 10

9(8). Leaves of upper stem merely spinose-toothed, tapering from base to apex; plants of Washington County only .... *C. virginensis*
— Leaves of upper stem definitely lobed, the lobes spinose-toothed, with parallel sides from base to near apex; plants not of Washington County .. *C. subniveum*

10(8). Herbage glabrous or glabrate, green 11
— Herbage tomentose, floccose-tomentose, gray or white, or only the upper leaf surfaces green 16

11(10). Flowers bright red or carmine; corolla lobes 15–18 mm long; spines of middle involucral bracts 7–11 mm long or more; plants of San Juan County .. *C. rothrockii*
— Flowers pink, pink purple, or white; corolla lobes less than 15 mm long; spines of middle involucral bracts 1–6 mm long; plants of various distribution 12

12(11). Outer bracts not pinnately spinose; mainly low elevation plants, usually in gypsiferous soils, in the Navajo Basin .... *C. calcareum*
— Outer bracts more or less pinnately spinose; plants of the Navajo and Uinta basins 13

13(12). Stems strongly winged almost or quite the length of upper internodes; main upper leaves tripinnatifid; plants of lower elevations in northern Uintah and Daggett counties .... *C. ownbeyi*
— Stems not winged, or rarely some internodes with incipient winging; main upper leaves pinnatifid to bipinnatifid; plants of moderate to high elevations in the southern end of the Uinta Basin and southward 14

14(13). Involucral bracts ciliate with long yellowish or brownish multicellular hairs; spines of bracts 6–15 mm long or more; plants of the east Tavaputs Plateau and La Sal Mts. .... *C. scopulorum*
— Involucral bracts more or less ciliate with whitish hairs or a tomentum; spines of bracts mainly 3–7 mm long; plants from the Tavaputs Plateau and southwestward 15

15(14). Involucral bracts scabrous dorsally, at least the innermost; herbage not at all tomentose; plants of the Henry Mts. .... *C. calcareum*
— Involucral bracts not scabrous dorsally; herbage more or less tomentose; plants not of the Henry Mts. .... *C. clavatum*

16(10). Heads campanulate, mainly 3.5–6.5 cm wide at anthesis or, if narrower, the bracts commonly glandular-thickened dorsally 17

— Heads turbinate to subcylindric, mainly 2–3.5 cm wide at anthesis; involucral bracts seldom glandular-thickened dorsally 18

17(16). Involucral bracts appearing brown to gray brown, the spines arising from the body of the bract, not from spreading long-attenuate herbaceous terminal portions; bracts of inflorescence usually prominent; plants of broad distribution .... *C. undulatum*

— Involucral bracts appearing green or fresh green or at least herbaceous, the spines arising from the apex of spreading long-attenuate terminal portions; bracts of inflorescence much reduced; plants of various distribution .... *C. neomexicanum*

18(16). Corollas bright red or carmine; plants from Garfield and Iron counties southward .... *C. arizonicum*

— Corollas pale pink, pink, rose purple, or white; plants mainly from Garfield and Iron counties northward .. 19

19(18). Involucral bracts (at least the inner) tapering, wedge-shaped, definitely scabrous roughened on dorsal surface, often suffused with reddish or purplish .... *C. calcareum*

— Involucral bracts smooth dorsally, seldom only somewhat scabrous, not conspicuously tapering, and seldom conspicuously suffused with reddish or purplish .... *C. wheeleri*

Cirsium arizonicum (Gray) Petrak Arizona Thistle. [*Cnicus arizonicus* Gray]. Biennial or short-lived perennial herbs from a taproot, the caudex sometimes developed; leaves of basal rosettes 7–36 cm long, bipinnately lobed or parted, the lobes again lobed or toothed, the main spines 1–6 mm long, white to grayish tomentose below, more or less tomentose and greenish to green above; stems 4–7.5 dm tall, more or less floccose-tomentose; cauline leaves 3–35 cm long, 1–8 cm wide, with lobing and vesture similar to the basal, reduced and less deeply lobed upwards; involucres 22–30 mm high, 20–50 mm wide, subcylindric to turbinate, the bracts tomentose at margins, and over back, smooth and often shiny medially, rarely glandular-thickened, the apical portions, especially of the inner definitely scabrous; spines yellowish, 3–10 (15) mm long; corollas crimson to carmine, 25–34 mm long, the tube 8–13 mm long, throat 1.5–11 mm long, the lobes 10–19 mm long. Two more or less distinctive but intergrading phases are present.

1. Heads subcylindric to turbinate; spines 3–10 mm long; plants mainly of the Colorado drainage system (also in western Garfield, and in Iron counties) .... *C. arizonicum* var. *arizonicum*

— Heads turbinate to broadly so; spines 3–15 mm long or more; plants mainly of the Great Basin and Virgin drainages (also in eastern Iron and western Garfield counties) .... *C. arizonicum* var. *nidulum*

Var. *arizonicum* 2n = 30, 32, 34. Salt desert shrub, pinyon-juniper, ponderosa pine, spruce-fir, and hanging garden communities at 1220 to 3050 m in Garfield, Grand, Iron, Kane, Piute, San Juan, and Washington counties; Arizona; 26 (iv).

Var. *nidulum* (Jones) Welsh [*Cnicus nidulus* Jones, type from Paria, Garfield County]. 2n = 32, 34. Pinyon-juniper, mountain brush, aspen, ponderosa pine, Douglas fir, white fir, and spruce-fir communities at 1890 to 3200 m in Beaver, Garfield, Iron, Kane, Millard, Piute, San Juan, and Washington counties; Arizona, Nevada; 37 (iii). Relationships apparently lie with *C. rothrockii*, *C. calcareum*, and to a lesser extent, with *C. wheeleri*.

Cirsium arvense (L.) Scop. Creeping or Canada Thistle. [*Serratula arvensis* L.]. Perennial rhizomatous herbs, the stems mostly 5–10 dm tall, glabrous or sparingly tomentose; leaves 3–15 cm long, 1–6 cm broad, deeply pinnatifid or lobed to merely toothed, glabrous to tomentose above and beneath; heads several to many, mainly unisexual; involucres 10–20 (25) mm high, 10–25 mm wide, the bracts lance-ovate, at least the outer ones and often all of them spine-tipped, tomentose to glabrous; corollas pink purple to white; pappus of pistillate heads longer than the corollas, that of staminate heads shorter than the corollas; achenes 3–5 mm long; n = 17, 18. Roadsides, fields, and other disturbed sites, but also invading native plant communities, at 1280 to 2535 m, probably in all Utah counties; widespread in North America; adventive from Eurasia; 42 (iii). We have two phases of creeping thistle in Utah; the one with merely toothed (unlobed) leaves is **var. *mite*** Wimmer & Grab., and the common one with deeply lobed leaves is **var. *horridum*** Wimmer & Grab. This common weed and the bull thistle are our only two introduced thistles in the genus *Cirsium*, which makes up a huge assemblage in the Old World. We can expect more introductions.

Cirsium barnebyi Welsh & Neese Barneby Thistle. Perennial herbs from a caudex and taproot, the caudex clothed with black marcescent leaf bases; leaves of basal rosettes 11–25 cm long, bipinnately lobed or parted, the lobes again lobed or toothed, the main spines 3–5 mm long, whitish to grayish tomentose on both sides; stems 3–5 dm tall, whitish tomentose (rarely green); cauline leaves 2–30 cm long, 1–8 cm wide, with lobing and vesture similar to the basal, reduced and less deeply lobed upwards; involucres 15–22 mm high, 20–30 mm wide, turbinate, the bracts glabrate or sparingly arachnoid on margins, glutinous dorsal ridge inconspicuous, smooth medially, the apical portions of the inner often contorted, not scabrous dorsally; spines 2–7 mm long, flattened apically, more or less spreading; corollas bluish pink. Sagebrush, juniper, cryptantha, ephedra, wildrye, and rabbitbrush communities at 1525 to 2257 m in Uintah (type from near Ignacio) County; endemic; 7 (iii). The Barneby thistle is apparently related to the *undulatum* complex.

Cirsium calcareum (Jones) Woot. & Standl. Cainville Thistle. [*Cnicus calcareus* Jones, type from Cainville]. Perennial herbs from a caudex and taproot, the caudex with brownish black to castaneous marcescent leaf bases; leaves of basal rosettes 6–35 cm long, pinnatifid to bipinnatifid, glabrous and green or tomentose on one or both surfaces, the main spines 3–8 mm long; stems mainly 2–5 dm tall, glabrous or more or less floccose-tomentose, winged-decurrent or not; cauline leaves 3–28 cm long, 0.8–7 cm wide, bipinnatifid, with lobing and vesture like the basal, reduced upwards, the main spines 3–8 mm long; involucres 19–34 mm long, 15–45 mm wide, the bracts ovate-lanceolate to linear, more or less tomentose at the margins, smooth and often shiny medially, the dorsal ridge glandular-thickened or not, the apical portions of at least the inner scabrous; spines straw colored, 1.5–6 mm long; corollas pink to blue pink. The *calcareum* complex is a portion of the *arizonicum* group of thistles, and has long been misinterpreted. There are three more or less confluent varieties present in Utah. Specimens collected are few, especially in the critical southeastern portion of the state. More work is indicated.

1. Herbage permanently tomentose, the leaves grayish tomentose beneath .... *C. calcareum* var. *pulchellum*
— Herbage green, the leaves rarely sparingly tomentose along the midveins beneath 2
2(1). Leaves definitely decurrent, the stems winged 2–6 cm below leaf base; plants of San Juan and Wayne counties .... *C. calcareum* var. *calcareum*
— Leaves not or scarcely decurrent; plants of other distribution .... *C. calcareum* var. *bipinnatum*

Var. ***bipinnatum*** (Eastw.) Welsh [*Cnicus drummondii* var. *bipinnatum* Eastw.]. Aspen, Douglas fir, and riparian communities at 1130 to 3150 m in Garfield, Kane, and San Juan counties; Colorado, New Mexico, and Arizona; 8 (i).

Var. ***calcareum*** [*Cirsium pulchellum* var. *glabrescens* Petrak type from Elk Mts., San Juan County]. Riparian communities at 1460 to 2200 m in Carbon, San Juan, and Wayne counties; endemic (?); 4 (i). A peculiar plant with thin leaves that are glabrous on both sides and subentire is known from Cedar Canyon (Atwood and Higgins 5918 BRY). How it fits into the scheme of Utah thistles is not known, but the plant appears to be intermediate between this and some other thistles. The status of the Cainville thistle, as strictly interpreted, beyond Utah is unknown; it seems likely that it does not occur outside the state.

Var. ***pulchellum*** (Greene) Welsh [*Carduus pulchellus* Greene ex Rydb.]. Rabbitbrush, sagebrush, tamarix, pinyon-juniper, and aspen communities at 1340 to 2745 m in Carbon, Emery, Garfield, Grand, Kane, San Juan, Uintah, Utah, and Wayne counties; Colorado, New Mexico, Arizona; 41 (vii). Both winged and wingless stems are present within our material. There are plants from the San Rafael Swell with winged stems and they are similar to **C. ochrocentrum** Gray of New Mexico, but they appear to be transitional in every way with the wingless plants. And it seems probable that they are not conspecific with that plant as it occurs beyond Utah. Possibly they do warrant taxonomic recognition. Further collections are necessary.

Cirsium centaureae (Rydb.) K. Schum. Fringed Thistle. Perennial herbs from a simple caudex and taproot, the caudex with chestnut leaf bases; leaves of basal rosette 2–28 cm long, 1–8 cm wide, pinnatifid, the lobes often again toothed, tomentose below, thinly tomentose to glabrous above, the main spines 1–5 mm long; stems 3–12 dm tall, not succulent, arachnoid or glabrous; cauline leaves with lobing and vesture like the basal, the spines 3–8 mm long; involucres 18–27 mm high, and about as wide, the outer bracts lance-ovate, the inner with coarsely lacerate margins, usually dilated in the upper half, tomentose to glabrous on the margins, the dorsal ridge not well developed, the longest spines 2–5 mm long, straw colored; flowers white to pink or purple; n = 17, 18. Montane communities at 3355 m in San Juan County; Wyoming and Colorado; 2 (0).

Cirsium clavatum (Jones) Petrak Fish Lake Thistle. [*Cnicus clavatus* Jones, type from Fish Lake]. Perennial or biennial herbs from a taproot, and often with a caudex, the caudex clothed with marcescent chestnut brown leaf bases; leaves of basal rosettes 2.5–22 cm long, bipinnately parted to merely toothed, green on both sides or more or less tomentose below, the main spines 1–6 mm long; stems 3–10 dm tall, glabrous or thinly tomentose; cauline leaves 3–26 cm long, 0.5–7 cm wide, with lobing and vesture like the basal, reduced and less lobed above; involucres 18–23 (32) mm high, 22–30 (55) mm wide, the bracts more or less villous-tomentose on margins, the outer ones usually pinnately spiny, smooth medially, the dorsal ridge not especially glandular, apical portions of the inner ones often scabrous, sometimes slightly dilated-erose; spines yellowish, 3–8 (18) mm long; corollas white or less commonly pink. Sagebrush, meadow, aspen, Douglas fir, and spruce-fir communities at 2135 to 3200 m in Beaver, Carbon, Emery, Garfield, Grand, Kane, Piute, San Juan, Sanpete, Sevier, Uintah, and Wayne counties; endemic; 27 (viii). The Fish Lake thistle is apparently related to the allopatric *C. eatonii*. It is more or less transitional to *C. wheeleri*, and probably other taxa, especially those with scabrous inner bracts. Rarely some have decurrent leaf bases, and when the pinnately spinose bracts are poorly developed, this thistle approaches *C. calcareum*. Moore and Frankton (1965) proposed that *C. clavatum* was a hybrid between *C. eatonii* and *C. centaureae*. However, despite its possible origin from hybridization, the taxon seems to be organized on about the same basis as other thistles. Further, its distribution is distinct from that of the putative parents. There does not seem to be justification for recognition of this entity as a hybrid.

Cirsium eatonii (Gray) Robins. Eaton Thistle. [*Carduus eatonii* Gray]. Perennial herbs from a simple or rarely branched caudex and taproot, the caudex clothed with brownish black to brown marcescent leaf bases; leaves of basal rosette 4–20 cm long, more or less bipinnatifid, green and glabrous or nearly so on both sides, the main spines 1.5–4 mm long; stems 1.5–5 dm tall, glabrous or nearly so; cauline leaves 3–25 cm long, 0.6–5.5 cm wide, with lobing like the basal, reduced upwards; involucres 20–37 mm high, 25–50 mm wide, the bracts ovate-lanceolate to lance-linear, tomentose to long-villous marginally (rarely overall), the outer ones usually pinnately spiny, smooth to roughened medially, the dorsal ridge not developed, the apical portions of the inner ones sometimes contorted; spines 5–18 mm long, straw colored; corollas pink to white. Three more or less distinctive varieties are present.

1. Involucral bracts copiously gray to brown villous with multicellular hairs; corollas ocroleucous; plants of the Uinta Mts. from Lake Fork eastward .... *C. eatonii* var. *murdockii*
— Involucral bracts merely white-tomentose or rarely with short multicellular hairs; corollas mainly pink or rose; plants of western Uinta Mts., and elsewhere 2
2(1). Involucral bracts commonly suffused with dark purple; involucres not obscured by outer spinose bracts; plants of the Tushar Mts. .... *C. eatonii* var. *harrisonii*
— Involucral bracts green or variously purplish; involucres with copious pinnate spines, mainly obscuring the surface of inner bractlets; plants of western Uinta and Wasatch mountains and Great Basin ranges .... *C. eatonii* var. *eatonii*

Var. ***eatonii*** [*C. eriocephalum* var. *leiocephalum* D. C. Eaton; this is the basionym for *C. eatonii* in a strict sense, which was renamed by Gray in honor of D. C. Eaton who collected with Sereno Watson in 1869]. The lectotype came from the head of the Bear River, in Summit County (Watson 691, 1869 US), with syntypical material being taken under the same number in Cottonwood Canyon (now Salt Lake County]. Lodgepole pine and spruce communities upwards into alpine tundra at 2375 to 3420 m in Duchesne, Juab, Salt Lake, Summit, Tooele, and Weber counties; Nevada and Colorado; 31 (iv). Specimens from the Deep Creek Mts. have few lateral spines on the outer bracts, and approach *C. clavatum* in technical features. More material is needed to determine their status and relationships.

Var. *harrisonii* Welsh Talus slopes and alpine meadows at 2975 to 3450 m in Beaver and Piute counties; endemic; 6 (v). This low phase of the Eaton thistle stands geographically apart from the remainder of the species, isolated on the islandlike Tushar Mts.

Var. *murdockii* Welsh The plants grow in talus slopes and on rock stripes at 3230 to 3660 m in Daggett, Duchesne, and Uintah counties; endemic; 7 (iii). This variety has been regarded as constituting a portion of *C. tweedyi* (Rydb.) Petrak. That entity was reviewed by Moore and Frankton (1965) and was mapped to include northeastern Utah in its range. However, no specimens were cited from Utah. I have seen the type of *C. tweedyi* and other material within its range in northeastern Wyoming, and they differ in pubescence of involucral bracts being merely white-tomentose along the margins.

Cirsium neomexicanum Gray New Mexico Thistle. Biennial herbs from taproots; leaves of basal rosette 5–25 cm long (or more), pinnatifid, the lobes again toothed or lobed, white-tomentose below and less so above, the main spines 1–6 mm long; stems 6–15 dm tall, whitish-tomentose; cauline leaves 1.5–35 cm long, 0.5–7 cm wide, tomentose, appearing filmy greenish white, lobed like the basal ones, rather abruptly reduced upwards, finally minute spiny bracts; involucres 20–30 mm high, 40–65 mm wide, the bracts green or greenish, narrowly lanceolate, tomentose marginally (or overall), the outer ones often reflexed, the inner minutely serrulate-ciliate, long-attenuate apically, the spine a continuation of the attenuation, smooth medially, the glandular dorsal ridge more or less well developed, the apical portions of the inner often contorted; spines 1–9 mm long, yellowish; corollas creamy white.

1. Involucral bracts green throughout, the attenuate apex not differing in texture from the body of the bract .... *C. neomexicanum* var. *neomexicanum*

— Involucral bracts not green throughout, the attenuate apex differing in texture from the body of the bract .... *C. neomexicanum* var. *utahense*

Var. *neomexicanum* $2n = 30$. Creosote bush, Joshua tree, blackbrush, shadscale, sagebrush, and pinyon-juniper communities at 915 to 2050 m in Beaver, Garfield, Grand, Juab, Kane, Millard, San Juan, Tooele, and Washington counties; Nevada, Arizona, and New Mexico; 26 (vii). This is one of the most distinctive species of thistle in Utah. The tall slender stems, with one or few large heads with creamy white flowers, stand in candelabra form in the arid portions of western and southern Utah. Ghostlike stalks of previous years persist for a time, reminding one of the regime that allowed their growth.

Var. *utahense* (Petrak) Welsh [*C. utahense* Petrak, type from Silver Reef, Washington County]. $2n = 30, 32, 34$. Salt desert shrub, sagebrush, pinyon-juniper, and mountain brush communities at 1220 to 2300 m in Cache, Carbon, Emery, Millard, Rich, Salt Lake, Tooele, and Utah counties; Colorado (?); 24 (ii). This taxon has long been confused with *C. undulatum* with which it shares the grayish tomentum, large heads, and tall stature. They have been separated previously on the basis of glandular development of the dorsal ridge; a feature which is, unfortunately, not diagnostic. The long-attenuate bract apices from which the spines arise are apparently distinctive for this taxon. It is essentially intermediate between *undulatum* and *neomexicanum* in a strict sense. The type is from Silver Reef, Washington County, but the main area of distribution for this variety is apparently along the Wasatch Mountains in northern Utah. The more finely bracted plant occasional in northern Utah seems identical with **C. *davisii*** Cronq., from adjacent Idaho, but is recognizable mainly on intangible characters. Probably it belongs within the concept of var. *utahense*.

Cirsium ownbeyi Welsh Ownbey Thistle. Perennial herbs from caudex and taproot, the caudex with marcescent dark brown leaf bases; leaves of basal rosettes 5–13 cm long, 1.5–3 cm wide, tripinnatifid, green on both sides, sparingly tomentose along lower side of midrib; cauline leaves with vesture and lobing like the basal; stems 5–7 dm tall, winged-decurrent, sparingly tomentose; involucres 1.8–2.5 cm high, 1.5–2.5 cm wide, the outermost bracts more or less pinnately spinose, lance-attenuate, smooth medially, the dorsal ridge not well developed, not scabrous, sparingly tomentose along margins, the inner more or less contorted apically; spines 3–8 mm long; corollas rose pink. Juniper, sagebrush, and riparian communities at 1678 to 1891 m in Daggett and Uintah counties; Colorado; an eastern Uinta Mts. endemic; 2 (i). Relationships of the Ownbey thistle apparently belong with *C. eatonii*.

Cirsium rothrockii (Gray) Petrak Rothrock Thistle. [*Cnicus rothrockii* Gray; *Cn. rothrockii* var. *diffusus* Eastw., type from Willow Creek, San Juan County]. Perennial or biennial herbs from a caudex and taproot, the stems 5–8 dm tall, sparingly tomentose or glabrate to glabrous; cauline leaves 3.5–30 cm long, 2–9 cm wide, bipinnatifid, green and glabrous or nearly so on both sides, carried well to the inflorescence; involucres (19) 23–28 (34) mm long, 20–35 mm wide, the bracts lanceolate to lance-linear, more or less tomentose along the margins, smooth medially, the dorsal ridge not or only somewhat glandular, sometimes purplish apically, the apical portions of the inner definitely scabrous, the spines 7–17 mm long; corollas red to carmine; $2n = 30$. Mixed shrubs and ponderosa pine woods at 1830 to 2560 m in San Juan County; Arizona; 3 (0). This entity is poorly known in Utah; its relationship is with both *C. calcareum* and *C. arizonicum*. It is a green subglabrous plant with red flowers and long involucral spines.

Cirsium rydbergii Petrak Rydberg Thistle. [*Cirsium lactucinum* Rydb., type from Bluff]. Perennial herbs from a definite caudex and taproot, the caudex clothed with blackish brown leaf bases; leaves of basal rosette mainly 30–90 cm long, 15–40 cm wide, bipinnatifid, the lobes narrow to very broad, glabrous to glabrate on both surfaces, the main spines 2–11 mm long; stems 6–12 dm tall or more, glabrous; cauline leaves glabrous, less lobed and much reduced upwards; involucres 10–17 mm high (not measuring the reflexed outer bracts), 13–26 mm wide, the outer bracts lance-ovate, rather abruptly contracted into recurved spines 3–25 mm long, sparingly tomentose marginally; dorsal glandular ridge lacking, the inner attenuate, not scabrous; flowers pink; 2n = 34. Hanging gardens, or rarely in canyons below them, at 1125 to 1525 m in Garfield, Grand, Kane, San Juan, Wayne counties; Arizona (?).; 18 (v). Both *C. rydbergii* Petrak and *C. lactucinum* Rydberg are based on the same type collection from the hanging gardens near Bluff. The Rydberg thistle is a plant with huge basal rosettes, tall slender flowering stems, and small heads.

Cirsium scariosum Nutt. Meadow Thistle. [*Carduus lacerus* Rydb., type from near Midway; *Carduus olivescens* Rydb., type from the Aquarius Plateau; *Cnicus drummondii* var. *acaulescens* Gray; *C. foliosum* authors, not T. & G.; *C. drummondii* authors, not T. & G.]. Perennial herbs from a simple caudex and taproot, the caudex with chestnut leaf bases; leaves of basal rosette 2–28 cm long, 1–8 cm wide, merely spiny-toothed to bipinnatifid, the lobes often again toothed, tomentose to glabrate below, thinly tomentose to glabrous above, the main spines 1–5 mm long; stems lacking, or 1–12 dm tall or more, often succulent and edible, arachnoid to glabrous; cauline leaves (when stems present) bipinnatifid or merely pinnatifid, the spines 3–35 mm long, with vesture like the basal; involucre 22–35 mm high, 20–65 mm wide, the outer bracts lance-ovate, the inner progressively more lance-attenuate, smooth medially, the margins smooth to minutely scabrous, tomentose to glabrous on margins, the dorsal ridge not well developed, the longest spines mainly 2–5 mm long, straw colored, the inner with tips more or less contorted, dilated, or fimbriate, usually whitish or silvery; flowers white to pink or pink purple; 2n = 32, 34, 36. Our specimens fall into two rather distinctive varieties.

1. Heads 25–35 mm high, 35–80 mm wide; inner bracts slender, sometimes contorted, not especially dilated; plants mainly 6–12 dm tall .... *C. scariosum* var. *thorneae*

— Heads 22–30 mm high, 20–40 mm wide; inner bracts often dilated or contorted, sometimes fimbriate; plants 0–6 dm tall .... *C. scariosum* var. *scariosum*

Var. scariosum [*Cirsium acaule* var. *americanum* Gray]. This taxon, as here interpreted consists of an amazingly diverse assemblage which has passed under a series of names including those cited above; and, if it is demonstrated that *C. foliosum* (Hook.) DC. is actually conspecific, that name has priority. Saline seeps and salt marshes, streamsides, terraces, and other meadowlands at 1310 to 3175 m in Cache, Carbon, Duchesne, Emery, Garfield, Juab, Millard, Salt Lake, Sanpete, Sevier, Summit, Tooele, Utah, and Washington counties; British Columbia to Montana, south to California, Arizona, and Colorado; 43 (x). This phase of *C. scariosum* has passed under the names *C. acaulescens* (Gray) K. Schum., *C. coloradoense* (Rydb.) Cockerell; *C. tioganum* (Congdon) Petrak, *C. drummondii* T. & G., and *C. foliosum*. Nomenclature is still unclear, and more work is indicated. Our highly variable material is transitional from acaulescent to caulescent within populations, with stems, when present, fleshy and edible. This is our common thistle of meadowlands, and it is unfortunate that nomenclatural entanglements have not allowed selection of an unequivocal name. Reported for the state is **C. parryi** (Gray) Petrak (Harrington, Flora of Colorado, 1952), but I have seen no specimens of that entity from Utah. It would key to *C. scariosum* in the present work. It has densely arachnoid involucral bracts, with at least the innermost dilated-fringed at the tips; flowers are greenish yellow and the leaves are glabrate on both surfaces.

Var. thorneae Welsh Stream terraces and seeps or springs at 1650 to 2475 m in Beaver, Garfield, Iron, Kane, Millard, and Piute counties; endemic (?); 10 (vi). In addition to the features noted above, the cauline leaves are thick, with coarse veins, and spines 8–35 mm long.

Cirsium scopulorum (Greene) Cockerell in Daniels Carmine Thistle. [*Carduus scopulorum* Greene]. Perennial herbs from taproots; leaves of basal rosettes 3–28 cm long, 0.8–8 cm wide, with spines 2–6 mm long, unlobed to bipinnatifid, tomentose below, glabrate to glabrous and green above; stems mainly 3–7 dm tall, sparingly arachnoid, not winged-decurrent; cauline leaves mainly bipinnatifid, or the upper ones merely pinnatifid, green above, glabrous to sparingly tomentose below, rather gradually reduced upwards; heads in a compact subglobose terminal cluster; involucres 30–35 mm high, 30–55 mm wide, the bracts lance-attenuate, abundantly villous marginally, with long yellowish to brownish multicellular hairs, the outer ones usually pinnately spiny, the dorsal crest not glandular, smooth medially, the apical portions of the inner ones often contorted; spines 10–18 mm long, yellowish; corollas pale yellow to cream; 2n = 34, 37. Sagebrush, aspen, and spruce-fir communities at 2135 to 3000 m in Grand, San Juan (?), and Uintah counties; Colorado; 3 (0).

Cirsium subniveum Rydb. Perennial herbs from taproots; basal rosettes not seen; stems mainly 6–10 (13) dm tall, tomentose, winged-decurrent; cauline leaves 3–25 cm long or more, 1–6 cm wide, pinnatifid, tomentose on both sides, or less so above, the bases decurrent; involucres 17–25 mm high, 20–30 mm wide, the bracts ovate-lanceolate, smooth medially, the glandular dorsal ridge more or less developed, none scabrous, tomentose marginally; spines 3–5 mm long; corollas apparently white to cream; 2n = 34–36. Pinyon-juniper communty at 1890 m in Box Elder, Cache, Rich, and Sanpete counties; Oregon to Montana; 6 (0).

Cirsium undulatum (Nutt.) Sprengel Gray Thistle. [*Carduus undulatus* Nutt.]. Perennial herbs from a simple caudex and taproot, the caudex more or less clothed with persistent leaf bases; leaves of basal rosette mainly 7–25 cm long, 1.5–6 cm wide, merely toothed to bipinnatifid, tomentose on both sides, white-tomentose below, white to greenish above, the main spines 1–6 mm long; stems 2–10 (12) dm tall, tomentose; cauline leaves bipinnatifid or the upper ones merely pinnatifid, with vesture as in the basal ones, rather gradually reduced upwards; involucres (15) 20–30 mm high, 20–60 mm wide, the bracts brown or brownish, lance-ovate to lanceolate, tomentose on margins or overall, the dorsal ridge strongly glutinous to undeveloped, the spinose tips spreading, with yellowish spines mainly 2–5 (10) mm long, smooth medially, the apical portion of the innermost more or less contorted; corollas pink, pink purple, or creamy white.

1. Heads mainly less than 2.5 cm wide, even the largest, commonly (1) 3–10 or more per stem ....
   .... *C. undulatum* var. *tracyi*

— Heads mainly more than 2.5 cm wide, at least the largest, commonly 1–3 per stem .... *C. undulatum* var. *undulatum*

Var. tracyi (Rydb.) Welsh [*Carduus tracyi* Rydb.]. 2n = 24. Sagebrush, mountain brush, juniper, aspen, and

Douglas fir communities at 1525 to 2900 m in Duchesne, Emery, Grand, Juab, Summit, and Uintah counties; Colorado; 26 (iii). This variety grades with the type variety, and separation is at least partially arbitrary.

Var. undulatum n = 12, 13. [*C. undulatum* var. *albescens* D. C. Eaton, type from Stansbury Island, Tooele County]. Desert shrub, sagebrush, pinyon-juniper, mountain brush, ponderosa pine, and aspen communities at 1400 to 2600 m in Cache, Carbon, Daggett, Duchesne, Emery, Garfield, Juab, Grand, Rich, Salt Lake, San Juan, Sanpete, Sevier, Summit, Tooele, Uintah, Utah, Wasatch, and Weber counties; British Columbia to Minnesota, south to Arizona, New Mexico, and Missouri; 59 (vi).

Cirsium virginensis Welsh Virgin Thistle. Perennial (?) herbs from taproots; leaves of basal rosettes 6–35 cm long, 1–5 cm wide, unlobed, pubescent like the cauline ones, with spines 1–4 mm long; stems 6–15 dm tall, tomentose, winged by definitely decurrent leaf bases; cauline leaves 1.5–15 cm long or more, sinuate-dentate to pinnatifid, whitish tomentose on both sides, or greenish above, often reduced to spiny bracts upwards; involucres 13–20 mm tall, 12–32 mm wide, the bracts ovate-lanceolate to narrowly lanceolate, brownish to straw colored, or often suffused with purple, tomentose marginally (or overall), the outer not especially reflexed, the inner serrulate or entire, smooth medially, the glandular dorsal ridge more or less developed, the apical portions of the inner often contorted; spines 2–6 (8) mm long, yellowish; corollas pink to lavender (or white?). Saline seeps and stream terraces at 850 to 1270 m in Washington County; Arizona; 9 (i). The small heads and long decurrent leaf bases are diagnostic. The relationships of the Virgin thistle are unknown. It does not appear to be closely related to other species groups represented in our area.

Cirsium vulgare (Savi) Ten. Bull Thistle. [*Carduus vulgaris* Savi]. Biennial herbs from taproots; leaves of basal rosette mainly 5–25 cm long, 2–8 cm wide, merely doubly serrate-dentate to doubly pinnatifid, tomentose beneath, coarsely hispid above; stems mainly 3–12 (15) dm tall, spiny-winged by decurrent leaf bases; cauline leaves mainly bipinnatifid, with vesture as in the basal ones; involucres 28–40 mm high, 35–70 mm wide, the bracts narrowly lanceolate, with spreading spine-tips, tomentose marginally, the dorsal ridge not developed, the inner sometimes contorted apically; spines 1–4 mm long, yellowish; corollas rose purple; 2n = 68. Meadows, fields, roadsides, and other disturbed sites from 1340 to 2745 m in most, if not all, Utah counties; widespread in North America; 52 (i).

Cirsium wheeleri (Gray) Petrak Wheeler Thistle. [*Cnicus wheeleri* Gray]. Perennial or biennial herbs from a simple or branched caudex and taproot, the caudex clothed with persistent brown to dark brown leaf bases; leaves of basal rosettes mainly 7–20 cm long, 1–5 cm wide, once to twice pinnatifid, or merely toothed or spinose-serrate, grayish or whitish tomentose below, thinly so to glabrous and green above, the main spines 0.5–4 mm long; stems 2.5–7 dm tall; cauline leaves 2–25 (32) cm long, 0.5–5 (7) cm wide, with lobing and vesture similar to the basal, carried well to the inflorescence, though reduced above; involucres 20–27 mm high, 20–35 mm wide, the bracts lance-ovate to lance-linear, more or less tomentose along the margins, smooth medially, the dorsal ridge not or only somewhat glandular, sometimes purplish-tipped, the apical portions of at least the inner more or less scabrous; corollas pink to pink purple, or less commonly white; 2n = 22. Mountain brush, pinyon-juniper, white fir, aspen, and spruce-fir communities at (1980) 2165 to 3150 m in Beaver, Emery, Garfield, Iron, Juab, Kane, Millard, San Juan, Sanpete, Sevier, and Washington counties; Colorado, New Mexico, and Arizona. Our materials apparently intergrade with *C. undulatum*, *C. nidulum*, and possibly *C. scariosum*. The moderate sized heads, usually pink or pink purple flowers, low stature, essentially non-glandular bracts, and usually green upper leaf surface appear to be diagnostic. The phases from Cedar Canyon (Iron County), with merely spinose unlobed leaves, are striking, but probably not more than minor variants; 39 (iv).

Cnicus L.

Annual caulescent spiny herbs from taproots, the juice watery; leaves alternate; heads solitary, terminating branches; involucral bracts in several series, spine-tipped, the inner ones pinnately spiny; receptacle densely bristly; corollas all discoid, yellow, perfect; pappus in 2 series, the outer smooth, long, alternating with short sparingly pectinate ones; style with a ring of hairs at base of divergent branches; achenes terete, strongly ribbed, glabrous.

Cnicus benedictus L. Blessed Thistle. Plants 1–5 dm tall or more, branching from near the base; stems villous; leaves mainly 8–15 cm long, pinnatifid, more or less glandular and sparingly villous, the spines 0.5–1.5 mm long, the lower ones petiolate, becoming sessile above; involucres 3–4 cm high, closely subtended and obscured by the foliose bracteate upper leaves; corollas yellow; 2n = 22. Waste places and gardens at 885 m in Washington County; widespread in the U.S.; adventive from Europe; 2 (0).

Conyza Less.

Annual herbs from taproots, with watery juice; stems erect, commonly branched; leaves alternate, simple; heads numerous, in cylindric to conic panicles; involucral bracts more or less imbricate, herbaceous medially; receptacle flat or nearly so, naked; rays minute, white or purplish, scarcely surpassing the pappus; disk flowers seldom more than 20, perfect, fertile; pappus of capillary bristles; achenes 1- or 2-nerved or nerveless.

1. Leaves more or less lobed or pinnatifid, narrowed to a clasping base; plants rare in Washington County . . . . *C. coulteri*

— Leaves tyically entire, not narrowed to a clasping base; plants variously distributed 2

2(1). Involucre (and typically the stem) copiously hirsute; achenes more than 1 mm long; plants rare in Washington County . . . . *C. bonariensis*

— Involucre (and typically the stem) glabrous or nearly so; achenes less than 1 mm long . . . . *C. canadensis*

Conyza bonariensis (L.) Cronq. [*Erigeron bonariensis* L.]. Annuals, mainly 3–10 dm tall, simple or with erect leafy branches often overtopping the main stem, the herbage hirsute with spreading hairs and strigose, grayish green; leaves 1.5–6 (10) cm long, 2–8 mm wide, linear to oblanceolate, saliently toothed, often deciduous in lower stem region by anthesis; heads numerous, rather inconspicuous; involucres 4–5 mm high, at least some more than 10 mm wide when pressed, the bracts densely hirsute to hirsutulous, whitish on the inner face when reflexed; ray flowers in 2 or 3 rows, their corollas usually shorter than the pappus. Stream banks and irrigation canals in riparian plant communities at ca 850 m in Washington County; widespread in the warmer parts of North America; adventive from South America; 3 (i).

Conyza canadensis (L.) Cronq. Horseweed. [*Erigeron canadensis* L.]. Annuals, mainly 0.5–10 dm tall, glabrous or spreading-hairy; leaves 2–8 (10) cm long, 2–8 mm wide, linear to oblanceolate, ciliate-serrate, often deciduous by late anthesis; heads numerous, inconspicuous; involucres 2–3.5 (4) mm high, (2.5) 3–7 mm wide, the bracts lance-subulate, the midvein glandular-thickened, herbaceous medially, glabrous or strigose, green on the inner face when reflexed; rays white or purplish; $2n = 9$, 18, 54. Weedy species, often in riparian or other moist disturbed sites at 850 to 2135 m in all (?) Utah counties; widespread in North America; Europe; 30 (vi). Our material belongs to **var. glabrata (Gray) Cronq.** [*Erigeron canadensis* var. *glabrata* Gray].

Conyza coulteri Gray Annuals, mainly 2–10 dm tall, glandular-pubescent and villous or hirsute; stems leafy, erect, usually simple at the base, the inflorescence paniculate; leaves 2–6 cm long, 3–10 mm wide, oblong or narrowly obovate, with a few teeth or hallow lobes, rounded apically, rounded or usuallly ariculate and somewhat clasping basally; heads ca 4 mm high; ray flowers in 2 or 3 peripheral rows. Braided gravels in the creosote bush and Joshua tree community at ca 825 to 855 m in Beaver Dam Wash, Washington County; widely distributed in the Southwest; Mexico; 2 (i).

Crepis L.

Annual, biennial, or perennial caulescent or subacaulescent herbs, from taproots, with milky juice; leaves basal and cauline, alternate, pinnatifid to toothed or entire; heads few to numerous, in corymbose or paniculate clusters; involucral bracts in 1 or 2 series, herbaceous; receptacle naked; corollas all raylike, perfect, yellow or yellowish; pappus of numerous white capillary bristles; achenes terete or nearly so, 10– to 20–ribbed, often beaked.

1. Plants annual, adventive, of disturbed sites .. *C. capillaris*
— Plants perennial, indigenous, neither weedy nor of disturbed sites 2

2(1). Leaves and stems glabrous (or glandular-hispid only above); plants subacaulescent or subscapose 3
— Leaves and stems more or less tomentose or puberulent to setose or glandular hispid; plants caulescent 4

3(2). Plants less than 10 cm tall, soboliferous, of high elevations .... *C. nana*
— Plants mainly 15–40 cm tall, never soboliferous, of lower elevation meadows .... *C. runcinata*

4(2). Heads narrowly cylindric; involucral bracts 5–7 (8), the inner commonly glabrous; flowers mostly 5–10 .... *C. acuminata*
— Heads narrowly to broadly campanulate; involucral bracts 8–15, tomentose and often setose-hispid; flowers mostly 8–60 5

5(4). Leaf segments linear to narrowly lanceolate, entire or nearly so, the terminal lobe more than 5 cm long; achenes commonly green .... *C. atrabarba*

— Leaf segments narrowly lanceolate to triangular, some usually toothed, the terminal lobe less than 5 cm long; achenes mainly yellowish to brownish 6

6(5). Involucres more than twice longer than broad; leaves usually green, runcinate-pinnatifid .... *C. intermedia*
— Involucres less than twice longer than broad, or leaves not green or not runcinate-pinnatifid 7

7(6). Involucre and stems not or sparingly setose, but if setose, the setae gland-tipped .... *C. occidentalis*
— Involucre and/or stems conspicuously setose, the setae not glandular .... *C. modocensis*

Crepis acuminata Nutt. Mountain Hawksbeard. Perennial herbs, 2.5–8.5 dm tall, with 1 to several stems from a caudex, the caudex clothed with dark brown marcescent leaf bases; herbage more or less tomentose to glabrate; basal and lowermost cauline leaves 8–33 (40) cm long, 2–12 cm wide, petiolate, the blade elliptic to oblanceolate in outline, pinnatifid to runcinate-pinnatifid, the lobes triangular to narrowly subulate, sometimes toothed or lobed; heads mainly 20–75 or more, cylindric, 5– to 10–flowered; involucres (8) 9–13.5 (16) mm high, 3–7 mm wide, the inner ones 5–8, glabrous or sometimes shortly villous-tomentose, the outer bracts much shorter, commonly tomentose; corollas 10–18 mm long, yellow; achenes yellow to brown, narrowed above. Sagebrush, mountain brush, white fir, aspen, and spruce-fir communities at 1430 to 2900 m in most if not all Utah counties; Washington to Montana, south to California, Arizona, and New Mexico; 69 (viii).

Crepis atrabarba Heller Slender Hawksbeard. [*C. occidentalis* var. *gracilis* D. C. Eaton]. Perennial herbs, 2–4.5 dm tall, with 1 to several stems from a caudex, the caudex with dark brown to purplish marcescent leaf bases; herbage gray villous-tomentose to glabrate, basal and lowermost cauline leaves 6–22 cm long, 1.5–4 cm wide, petiolate, the blade lance-elliptic in outline, pinnatifid, the lobes linear or linear-subulate, the terminal lobe 5–9 cm long, entire; heads mainly 2–15, campanulate, commonly 10– to 40–flowered; involucres 9–15 mm long, 7–13 mm wide, the inner ones 8–10, usually grayish tomentulose and often with few glandless black setae; corollas 10–18 mm long, yellow; achenes usually greenish, attenuate at the apex. Sagebrush, ponderosa pine, Douglas fir, and white fir communities at 1890 to 2870 m in Cache, Daggett, Garfield, Salt Lake, Summit, Tooele, Uintah, and Utah counties; British Columbia and Alberta, south to Nevada and Colorado; 8 (i). The species is evidently uncommon in Utah. It is known to form apparent hybrids with *C. acuminata*, and probably with other taxa as well.

Crepis capillaris (L.) Wallr. Thread Hawksbeard. Annual or biennial herbs, the stems erect, simple or branched, mostly 1–6 dm tall, sparingly spreading-hairy; basal leaves 3–20 cm long, 0.5–3 cm broad, lanceolate to oblanceolate, denticulate to pinnatifid or bipinnatifid, glabrous or pubescent with stiff spreading hairs, especially along the lower midvein, petiolate; cauline leaves reduced upwards, sessile and auriculate-clasping; heads (1) several to numerous, mostly 20– to 60–flowered, borne in an open inflorescence; involucres 5–8 mm high, 5–14 mm wide, the inner bracts lance-attenuate, 8–16, tomentose, often glandular-hairy, glabrous within, the outer bracts lance-linear; achenes 2–5 mm long, pale brown to straw colored, not beaked; $2n = 6, 12, 24$.

Ruderal weed of Salt Lake County; widely scattered in North America; adventive from Europe; 1 (0).

Crepis intermedia Gray Gray Hawksbeard. [*C. barbigera* Leiberg, in part]. Perennial herbs, 2.5-7 dm tall, with 1-several stems from a caudex, the caudex clothed with pale to dark brown marcescent leaf bases; herbage more or less tomentose or villous; basal and lowermost cauline leaves 15-30 cm long, 2-10 cm wide, petiolate, the blade elliptic to oblanceolate in outline, pinnatifid to runcinate-pinnatifid, the lobes triangular to linear-subulate, sometimes toothed or lobed, the terminal lobe less than 5 cm long; heads mainly 10-60, campanulate, 7- to 16-flowered; involucres 11-20 mm high, 6-12 mm wide, the inner ones 7-12, tomentulose (rarely glabrate), sometimes setose with non-glandular setae, the outer bracts much shorter; corollas 13-20 mm long, yellow; achenes mainly yellowish or brownish, narrowed above. Sagebrush, pinyon-juniper, and mountain brush communities at 1525 to 2575 m in Beaver, Cache, Duchesne, Garfield, Grand, Iron, Kane, Salt Lake, Sevier, Tooele, Uintah, Utah, Wasatch, and Washington counties; Washington to Alberta, south to California, Nevada, and Colorado; 16 (iii). The *intermedia* assemblage consists of a series of apomictic intermediates involving *C. acuminata* as one of the parental types, and one or more of the other taxa (i.e., *occidentalis* or *modocensis*) to complete the complex. Included here is the concept of *C. barbigera* as it has been applied in Utah; it consists of a similar hybrid sequence of polyploid apomicts from outside our area.

Crepis modocensis Greene Modoc Hawksbeard. Perennial herbs, 1.5-3.7 dm tall, with 1-several stems from a caudex, the caudex clothed with pale to brown marcescent leaf bases (the stem base often yellow); herbage more or less tomentose; basal and lowermost cauline leaves 9-25 cm long, 2-5 cm wide, petiolate, the blade elliptic to oblanceolate in outline, bipinnatifid, the lobes linear to lance-subulate, again toothed or lobed, the terminal lobe less than 5 cm long; heads 1-9, 10- to 60-flowered; involucres 11-16 mm high, 11-23 mm wide, the inner bracts 10-15, tomentulose, commonly setose, the setae not glandular, the outer bracts much shorter; corollas 13-22 mm long, yellow; achenes greenish black to reddish brown, attenuate. Sagebrush, pinyon-juniper, and mountain brush communities at 1640 to 3175 m in Beaver, Box Elder, Cache, Daggett, Juab, Millard, Rich, Salt Lake, Sanpete, Sevier, Tooele, Uintah, and Utah counties; British Columbia to California, Nevada, and Colorado; 24 (0). The peculiar numerous slender lateral lobes of the deeply dissected or parted leaf blades are diagnostic.

Crepis nana Richards. Dwarf Hawksbeard. Perennial caespitose herbs, the stems much branched, often soboliferous, mostly 0.2-1.1 dm tall, contracted, usually obscured by the leaves, glabrous; basal leaves mainly 1-7.5 cm long, 0.2-1.8 cm wide, the blades spatulate to orbicular, elliptic, or ovate, glabrous, petiolate; cauline leaves similar to the basal, not clasping; heads few to numerous, mostly 4- to 12-flowered, borne in a compact cushionlike inflorescence; involucre 7-12 mm high, 3-6 mm wide, the inner bracts narrowly oblong, 8-12, greenish or blackish, glabrous, the outer much shorter; achenes brownish, ribbed, shortly beaked; 2n = 14. Alpine communities, mainly in talus, at 3050 to 3425 m in Beaver, Garfield, Juab, Piute, San Juan, and Utah or Salt Lake counties; Alaska to Labrador, south to California and Utah; 10 (iii).

Crepis occidentalis Nutt. Western Hawksbeard. Perennial herbs, 1-4 dm tall, with 1-several stems from a caudex, the caudex clothed with brown marcescent leaf bases (the stem base often yellow); herbage tomentose; basal leaves mainly 6-30 cm long, 1-5 cm wide, petiolate, the blade lanceolate to elliptic in outline, pinnatifid to bipinnatifid, the lobes triangular to oblong or linear-subulate, usually again toothed or lobed, the terminal lobe less than 5 cm long; heads 2-25, 12- to 30-flowered; involucres 10-20 mm high, 6-15 mm wide, the inner bracts (7) 8-13 (18), tomentose, the outer ones much shorter; corollas 10-22 mm long, yellow; achenes pale too dark brown, not much attenuate apically. There are three rather weak and arbitrarily recognizeable varieties of this species, with some geographical correlation, in Utah. Intermediates occur between the varieties and with other taxa as well.

1. Largest heads 12- to 14-flowered, with 8 or 9 involucral bracts; plants mainly of the Great Basin . . . . *C. occidentalis* var. *costata*
— Largest heads with more than 15 flowers, with 10-13 involucral bracts; plants of various distribution 2
2(1). Involucres with few glandular setae, or none; plants mainly 2-3 dm tall, of the Great Basin . . . . *C. occidentalis* var. *pumila*
— Involucres with few to numerous glandular setae; plants mainly 1-2 dm tall, of the Colorado drainage system, less commonly in the Great Basin . . . . *C. occidentalis* var. *occidentalis*

Var. costata Gray Sagebrush, pinyon-juniper, mountain brush, and aspen communities at 1525 to 2200 m in Box Elder, Cache, Juab, Millard, Salt Lake, Tooele (type from Stansbury Island), Utah, and Washington counties; British Columbia to California and Colorado; 23 (0).

Var. occidentalis Shadscale, rabbitbrush, sagebrush, pinyon-juniper, and ponderosa pine communiteis at 1280 to 2565 m in Beaver, Daggett, Duchesne, Garfield, Kane, Piute, San Juan, Sanpete, Sevier, and Washington counties; Oregon to Wyoming, south to California and New Mexico; 25 (v).

Var. pumila (Rydb.) Babc. & Stebbins [*Crepis pumila* Rydb.]. Sagebrush, pinyon-juniper, and mountain brush communities at 1700 to 2100 m in Millard and Tooele counties; Oregon to Montana, south to California and Nevada; 7 (0).

Crepis runcinata (James) T. & G. Meadow Hawksbeard. [*Hieracium runcinatum* James]. Perennial herbs, 1.5-5 (7) dm tall, with 1-several stems from a caudex, the short caudex clothed with brown marcescent leaf bases; herbage glabrous or hispid above (puberulent in some), not tomentose; basal leaves mainly 2-25 cm long, 1-6 (8) cm wide, petiolate or not, spatulate to oblanceolate, or the blades ovate to oval, oblong, or oblanceolate, more or less pinnatifid to lobed or entire, commonly glaucous; heads 1-30, with 20-50 flowers; involucres campanulate, 8-16 mm high, 6-15 mm wide or more, the inner bracts mainly 10-16, puberulent or hispid, the outer ones much shorter; corollas 9-18 mm long, yellow; achenes light to dark brown, attenuate, or shortly beaked; n = 11. Three distinctive varieties are present.

1. Involucres merely puberulent; plants mainly of saline meadows . . . . *C. runcinata* var. *glauca*
— Involucres hispid with black hairs (resembling species of *Hieracium*); plants of saline or non-saline sites 2

2(1). Basal leaves definitely petiolate, the blade 2–4 times longer than broad .... *C. runcinata* var. *hispidulosa*
— Basal leaves broadly winged-petiolate, the blade 4–8 times longer than broad ..... *C. runcinata* var. *runcinata*

Var. **glauca** (**Nutt.**) **Welsh** [*Crepidium glaucum* Nutt.]. Meadows, lake shores, seeps, and hot springs in salt grass, rush, alkali saccaton, and common reed communities at 1220 to 2200 m in Carbon, Daggett, Duchesne, Emery, Grand, Juab, Kane, Millard, Piute, San Juan, Sevier, Tooele, Uintah, Utah, and Wayne counties; Idaho to Saskatchewan, south to Arizona and New Mexico; 34 (xii). This variety has been collected in full flower on 27 April at Monroe Hot Springs.

Var. **hispidulosa** **Howell ex Rydb.** Sedge-willow and meadow communities at 1370 to 2535 m in Box Elder, Cache, Daggett, Duchesne, Garfield, Kane, Piute, Rich, Sanpete, Sevier, Summit, Uintah, Utah, and Wasatch counties; Washington to Montana, south to California and Colorado; 18 (ii). The meadows are seldom saline where this plant occurs.

Var. **runcinata** [*C. runcinata* var. *alpicola* Rydb.; *C. aculeolata* Greene, type from Utah]. Bogs in Salt Lake and Utah counties; Manitoba to Minnesota, south to Idaho and New Mexico; 1 (0). This variety is evidently uncommon in Utah.

Dicoria T. & G.

Annual herbs; leaves alternate or the lower ones opposite, simple, entire or toothed; heads unisexual or perfect, discoid; involucral bracts strongly dimorphic, the usually 5 outer ones small, herbaceous, the inner subtending the 1 or 2 pistillate flowers, subscarious, accrescent, much larger than the outer at maturity; chaff narrow, tardily deciduous; pistillate flowers without corolla; staminate flowers with funnelform corolla, the anthers distinct; achenes plano-convex, black, toothed to pectinately wing-margined; pappus lacking.

1. Foliose bracts of inflorescence orbicular to broadly ovate; plants of Washington County .... *D. canescens*
— Foliose bracts of inflorescence lance-ovate to lanceolate; plants of southeastern Utah .... *D. brandegei*

Dicoria brandegei **Gray** Brandegee Sandplant. [*D. paniculata* Eastw., type from between McElmo and Recapture, San Juan County; *D. wetherillii* Eastw., type from the San Juan River, a monstrous form]. Plants branched from the base upwards, 1.5–5.5 dm tall, the herbage white-pilosulose to strigose, the hairs multicellular; lower cauline leaves linear to lanceolate, more or less hastately lobed, toothed, or subentire, 1–7 cm long (including petiole), 0.2–1.5 cm wide; foliose bracts linear to oblong, lanceolate or ovate, rarely if ever orbicular, the blades 0.6–4 cm long; outer involucral bracts oblong, 1.5–3 cm long, the inner ones suborbicular, glandular-puberulent, accrescent in fruit; achenes 5–8 mm long, the winged margin toothed to pectinate, black like the body or pale. In dunes and other sandy sites, in wavy-leaf oak, eriogonum, amsonia, old-man sagebrush, rabbitbrush, ephedra, and vanclevea communities at 1130 to 1830 m in Emery, Garfield, Grand, Kane, San Juan, and Wayne counties; Arizona, New Mexico, and Colorado (?); 21 (v).

Dicoria canescens **Gray in Torr.** Gray Sandplant. Plants branched from base upwards, 2.5–9 dm high, the herbage white-pilosulose to strigose and glandular, the hairs multicellular; lower cauline leaves deltoid-lanceolate, dentate, 1–5 cm long; foliose bracts ovate to orbicular, the blades 0.6–1.5 cm long; outer involucral bracts oblong, 2–3 mm long, the inner ones suborbicular, glandular-puberulent, accrescent in fruit, to 10 mm long or more; achenes 5–6 mm long, the winged margin toothed to pectinate, black like the body, or pale; n = 18. Dunes and other sandy sites in blackbrush and creosote bush communities at 825 to 1000 m in Washington County; Arizona, Nevada, and California; 4 (i). Our material belongs to ssp. **clarkae** (**Kennedy**) **Keck** [*D. clarkae* Kennedy]..

Dyssodia Cav.

Annual or perennial herbs or subshrubs from taproots, the juice watery; herbage with conspicuous translucent oil glands; stems striate, numerous; leaves opposite or alternate, entire to pinnatisect; heads solitary at branch ends, or few to several in cymose clusters; involucral bracts in 2 series, distinct or united, and usually with a much shorter outer set; receptacle flat or convex, puberulent; ray flowers yellow, pistillate, fertile; disk flowers fertile; pappus of 10–15 bristle-tipped scales, or these dissected into 3 or more bristles; style branches with a short, conic appendage.

Strother J. L. 1969. Systematics of *Dyssodia* Cavanilles (Compositae: Tageteae). Univ. Calif. Publ. Bot. 48: 1–88.

1. Plants annual; leaves bipinnatisect; stems villosulous .... *D. papposa*
— Plants perennial, herbs or subshrubs; leaves simple or merely pinnatisect; stems hispidulous 2
2(1). Heads 1.2–2 cm tall or more; pappus of 10–15 scales, each dissected into numerous bristles .... *D. cooperi*
— Heads less than 1 cm tall; pappus not as above 3
3(2). Heads borne on elongate merely bracteate peduncles; leaves pinnately 5–lobed, shortly hispid; pappus scales tipped with usually a solitary bristle ..... *D. pentachaeta*
— Heads sessile or essentially so; leaves simple, entire or rarely irregularly few-lobed, glabrous or merely ciliate; pappus scales with 3–5 bristles .... *D. acerosa*

Dyssodia acerosa **DC.** Dogweed. [*Thymophylla acerosa* (DC.) Strother; *Aciphyllaea acerosa* (DC.) Gray; *Hymenatherum acerosum* (DC.) Gray]. Plants suffruticose, 10–25 cm tall, forming compact clumps, from taproots; herbage glabrous or villosulous; leaves opposite (or alternate above), simple or irregularly lobed, 3–18 mm long, 0.5–1 (2) mm wide, glandular, ciliate or glabrous; heads sessile or subsessile; involucres turbinate-cylindric, 5–7 mm high, 3–4 mm wide; involucral bracts ca 13, connate, each bract with conspicuous orange glands; ray flowers 7–8, lemon yellow; disk flowers 18–25, pale yellow; pappus of ca 20 scales, each dissected into 3–5 bristles; achenes dark brown, 3–3.5 mm long, strigose; n = 8, 13. Blackbrush communities at 1130 to 1350 m in Garfield, San Juan, and Washington counties; Arizona and New Mexico, south to Mexico; 5 (ii).

Dyssodia cooperi **Gray** Cooper Glandweed. [*Adenophyllum cooperi* (Gray) Strother; *Clomenocoma cooperi* (Gray) Rydb.]. Plants woody only at the base, 3–5 dm tall, forming rounded clumps, from taproots; herbage puberulous to hispidulous; leaves alternate, simple, lanceolate, spinulose-dentate, attenuate apically; heads pedunculate; involucres turbinate, 12–22 mm high,

15–30 mm wide when pressed; involucral bracts 20–30, strongly nerved, linear-lanceolate, with subulate-acuminate tips, each bract with a conspicuous subterminal orange gland; ray flowers 8–12, orange to yellow, often drying saffron, ca 1 cm long; disk flowers numerous, yellowish; pappus of 10–15 scales, each dissected into numerous bristles; achenes dark brown, ca 5 mm long; n = 13. Joshua tree and creosote bush communities at 760 to 1185 m in Washington County; California, Nevada, and Arizona; 4 (0).

Dyssodia papposa (Vent.) A. S. Hitchc. Pappose Glandweed. [*Tagetes papposa* Vent.]. Plants annual, 1.5–4 dm tall; herbage glabrous to sparingly puberulent; leaves opposite below, alternate above, 1.5–3 (5) cm long, pinnatisect into 11–15 lobes, these sometimes again lobed; heads shortly pedunculate to subsessile; involucres turbinate to campanulate, 6–10 mm high, and about as wide; involucral bracts 6–12, oblanceolate, with yellowish oil glands, connate only at the base; ray flowers 8 or fewer, yellow orange; disk flowers mainly 20–40, dull yellow; pappus of ca 20 scales, each dissected into 5–10 bristles; achenes black, 8–35 mm long; n = 13. Sandy roadsides at 1450 to 1500 m in Carbon, Duchesne, Sanpete, and Tooele counties; through much of the U. S. and Mexico; 4 (0).

Dyssodia pentachaeta (DC.) Robins. Scale Glandweed. [*Hymenatherum pentachaetum* DC.; *H. thurberi* Gray; *Thymophylla pentachaeta* (DC.) Small; *D. thurberi* (Gray) Robins.]. Plants suffruticose, 8–28 cm tall, forming rounded clumps, from taproots; leaves opposite, pinnately parted into 3–5 rigid linear lobes, 0.5–2 cm long, sparingly hirtellous; peduncles 1–8 cm long; involucres turbinate, 4.8–6 mm high, 5–10 mm wide; involucral bracts in 2 series, connate for much of their length, with distinctive yellow oil glands; ray flowers usually 13, bright yellow; disk flowers 50–70, dull yellow; pappus usually of 10 scales, these awnless or with 1–3 awns; achenes brown, 2.2–3 mm long, hispid to glabrous; n = 8, 13. Blackbrush, ephedra, shadscale, creosote bush, and Joshua tree communities at 700 to 1220 m in Garfield, Kane, San Juan, and Washington counties; Nevada and California to Texas and Mexico; South America; 30 (vi). Our material has been assigned to var. **belinidium (DC.) Strother** [*Hymenatherum belinidium* DC.; *Thymophylla pentachaeta* var. *belinidium* (DC.) Strother].

Encelia Adams

Shrubs; stems ascending to erect, grayish or whitish, the branchlets commonly pubescent; leaves alternate, simple, petiolate, entire or toothed; heads solitary or in cymose clusters, radiate or discoid; involucral bracts in 2 or 3 series; receptacle convex to flat, chaffy, the scales clasping the achenes and falling with them; ray flowers (when present) sterile, yellow; disk flowers perfect, yellow; pappus lacking (or of 2 awns); achenes flat, obovate, villous-ciliate and pubescent on the surfaces.

Blake, S. F. 1913. A revision of *Encelia* and some related genera. Proc. Amer. Acad. 49: 358–376.

1. Leaves white-tomentulose; peduncles glabrous; heads in branching cymes; plants rare in Washington County . . . . *E. farinosa*
— Leaves strigose to hispid, green; peduncles scabrous to strigose; heads solitary at branch ends; plants of Washington County and elsewhere . . . . *E. frutescens*

Encelia farinosa Gray Incienso. Plants mainly 3–10 dm tall, aromatic; leaves clustered at apex of current stems, 2–8 cm long, ovate, entire or toothed, silvery tomentose, petiolate; peduncles elongate, cymosely branched or simple; heads showy, the disk 1–1.5 cm wide; involucres 4–7 mm high, villous and glandular dotted; rays 8–12 mm long, orange yellow; achenes narrowly obovate; n = 18. Blackbrush community at 1280 m in Washington County; Nevada, Arizona, and California; Mexico; 1 (0).

Encelia frutescens (Gray) Gray Bush Encelia. [*Simsia frutescens* Gray]. Plants mainly 3–12 (15) dm tall; leaves scattered along current stems, the blades commonly 0.5–2.5 cm long, 0.3–2 dm wide, ovate to orbicular or lanceolate, entire or toothed, strigose to hispid with pustular-based hairs; heads showy or not, the disk 1–3 cm wide; involucres 6–10 mm high, strigose or glandular; rays lacking or 1–16 (or more), 2–12 mm long, yellow; achenes obovate; n = 18. Two distinctive varieties are present in Utah.

1. Herbage strigose, also with some pustular-based hairs; involucral bracts abruptly caudate-acuminate, strigose; plants of Washington County. . *E. frutescens* var. *virginensis*
— Herbage hispid with pustular-based hairs; involucral bracts gradually attenuate, more or less glandular (sometimes strigose); plants of the Colorado Basin . . . . *E. frutescens* var. *frutescens*

Var. **frutescens** [*E. frutescens* var. *resinosa* Jones in Blake]. Talus and slickrock in blackbrush and shadscale communities at 1130 to 1830 m in Emery, Grand, Kane, and San Juan counties; Arizona, California; 15 (vii). There is a cline of glandularity in leaves from definitely glandualr in the southern portion of the range to no glands at all in the northern material. Also, our plants vary from discord to radiate.

Var. **virginensis (A. Nels.) Blake** [*E. virginensis* A. Nels.]. Creosote bush, Joshua tree, and blackbrush communities at 760 to 1325 m in Washington County; Nevada, Arizona, and California; 23 (i).

Enceliopsis (Gray) A. Nels.

Perennial scapose or subscapose herbs, from tuberous roots or taproots and subterranean to superficial caudex; herbage pilosulose to velutinous; leaves all basal (rarely some reduced bracteate ones along the scape), the blades spatulate, lanceolate, oblanceolate, ovate, or orbicular; heads solitary; involucral bracts in 2 or 3 series, herbaceous throughout; receptacle flat to convex, chaffy, the scales clasping the achenes; rays yellow, sterile (but apparently pistillate), or lacking; disk flowers numerous, perfect, fertile, yellow; pappus of 2 awns and with or without small scales between, or none; achenes flattened, blackish.

1. Heads discoid; herbage pilose-hirsutulose; plants arising from a subterranean tuberous root . . . . *E. nutans*
— Heads radiate; herbage tomentulose; plants arising from a superficial caudex 2

2(1). Petioles broadly winged, mainly shorter than the blades; plants reported from the Virgin Narrows section of Washington County, but none have been seen by me [*Tithonia argophylla* D. C. Eaton, type from near St. George, or more likely from southern Nevada] . . . . *E. argophyllus* (D. C. Eaton) A. Nels.

— Petioles slender, not or only narrowly winged, mainly longer than the blades; plants commonly in eastern and west-central portions of Utah, but also in Washington County .... *E. nudicaulis*

Enceliopsis nudicaulis (Gray) A. Nels. Nakedstem. [*Encelia nudicaulis* Gray, type from Utah]. Scapose, caespitose perennials from a superficial, branching caudex, 10–43 (60) cm tall, the herbage tomentulose, silvery white; petioles 0.7–17 cm long, narrowly if at all winged; leaf blades 2–9 cm long, 1.3–10 cm wide, ovate to elliptic, orbicular or spatulate, cuneate to subcordate basally, obtuse to rounded apically; scapes often with a reduced foliose bract; involucres 1.3–2.2 cm high, 3–5.6 cm wide, the bracts ovate-lanceolate to lanceolate or linear-lanceolate, attenuate to acuminate; rays 13–21, yellow, 22–38 mm long; achenes 10–12 mm long, long silky-pilose, cuneate, black or dark brown; pappus commonly of 2 awns connected by a crown of short connate scales (or none); n = 17, 18. Commonly on gypsiferous semibarren knolls in blackbrush, rabbitbrush, ephedra, shadscale, grayia, and pinyon-juniper communities in Beaver, Box Elder, Emery, Garfield, Grand, Millard, Piute, San Juan, Sevier, Uintah, Washington, and Wayne counties; Idaho, Nevada, Arizona, and California; 70 (xiii). Specimens of *E. nudicaulis* from Washington County tend to have less well defined blades than for the species as a whole. However, they lack the broadly winged petioles of *E. argophylla* and are mostly in the size range of *E. nudicaulis*. The type of *Tithonia argophylla* D. C. Eaton (basionym of *E. argophylla*) was taken by Dr. Edward Palmer either near St. George (in 1870) or more likely in nearby southern Nevada. The specimen at US consists of basal leaves and a flower head, with the plant (Rep. Explor. 40th Parallel 5: 423. 1871) reputed to be 2 to 3 feet high (ca 60 to 90 cm), well within the size of the material of *E. argophylla* from southern Nevada but not exclusive of all material from the south end of the Beaver Dam Mts. in Utah.

Enceliopsis nutans (Eastw.) A. Nels. Noddinghead. [*Encelia nutans* Eastw.; *Verbesina scaposa* Jones, type from near Cisco, Grand County]. Scapose, discoid perennials, 10–25 cm tall, from a subterranean caudex (2–15 cm long) and tuberous root to 4 cm thick, the herbage strigose to pilosulose (antrorsely on the upper surface, retrorsely so below), green; petioles 2–6.5 cm long, often narrowly winged; leaf blades 2–7.5 cm long, 1.4–6 cm wide, ovate to orbicular or spatulate, cuneate basally, obtuse to rounded apically; scapes not bracteate; involucres 0.9–1.5 cm high, 2.5–4 cm wide, the bracts lance-attenuate; rays lacking; achenes 9–11 mm long, oblanceolate, long silky-pilose, brown; pappus lacking; n = 18. Mainly in finely textured soils in shadscale, budsage, galleta, and ephedra communities at 1310 to 1830 m in Carbon, Duchesne, Emery, Grand, Uintah, and Wayne counties; Colorado (a Colorado Basin endemic); 35 (iii).

Erigeron L.

Annual, biennial, or perennial herbs from caudices, rhizomes, stolons, or taproots, with watery juice; stems decumbent to ascending or erect, rarely prostrate; leaves alternate, simple, entire, toothed, or pinnatifid to palmatifid; heads solitary or few to numerous in corymbose or paniculate inflorescences; involucral bracts equal, or slightly to definitely imbricate, slender, herbaceous (or scarcely herbaceous) throughout; receptacle flat, naked; rays white, pink, purple, bluish, or yellow, pistillate, numerous or lacking; disk flowers numerous, yellow or tinged reddish; pappus of capillary bristles, sometimes with an outer series of short bristles or scales; style branches with lanceolate and acute or triangular and obtuse appendages; achenes flattened, 2 (rarely 4–14) –nerved. Note: This is a large and complex genus. The species, although mainly distinctive, are distinguished by minute features that can be interpreted variously. The genus is a near congener of both *Aster* and *Conyza*, and is not always separable from either.

Cronquist, A. C. 1947. Revision of the North American species of *Erigeron* north of Mexico. Brittonia 6: 121–302.

1. Plants with yellow ray flowers, known from Box Elder County .... *E. linearis*
— Plants with ray flowers pink, pink purple, blue, blue purple, or white, but not yellow, of various distribution . 2

2(1). Plants annual, biennial, or short-lived perennials from usually slender taproots, lacking rhizomes (except in some specimens of *E. proselyticus*) or woody caudices .... Key 1
— Plants definitely perennial, often from rhizomes or caudices 3

3(2). Plants silvery pubescent; achenes with 4 or more nerves; involucral bracts definitely imbricate ..... Key 2
— Plants green, or less commonly silvery pubescent; achenes with 2 nerves, or, if with more nerves; involucral bracts subequal 4

4(3). Involucres woolly-villous to spreading villous, or villous-hirsute with at least some long spreading multicellular hairs .... Key 3
— Involucres merely glandular, glabrous, puberulent, or with appressed simple or multicellular hairs, rarely with some spreading long hairs near the base .... Key 4

Key 1.

Plants annual, biennial, or short-lived perennials from slender taproots, lacking rhizomes or woody caudices.

1. Pistillate corollas very numerous, filiform, the rays short, erect, not exceeding the disk, or the inner ones tubular and lacking rays 2
— Pistillate corollas few to numerous (rarely lacking), the tube generally cylindric, the rays well developed and spreading, rarely reduced or absent 3

2(1). Cauline leaves narrowly lanceolate to oblong, or less commonly linear; rayless pistillate flowers present between the ray and disk flowers; inflorescence corymbose, the peduncles curved-ascending, or the heads solitary .... *E. acris*
— Cauline leaves linear to oblong; rayless pistillate flowers lacking; inflorescence racemose, the peduncles erect or nearly so, or the heads solitary . *E. lonchophyllus*

3(1). Pappus of ray and disk flowers unlike, that of the disk flowers composed of bristles and short outer setae, that of the ray flowers lacking bristles; plants tall adventive weedy species 4
— Pappus of ray and disk flowers alike, consisting of bristles, sometimes also with outer setae or scales; plants indigenous, low to tall 5

4(3). Foliage ample; plants mainly 6–12 (15) dm tall; pubescence of stem long and spreading (at least below); plants adventive, weedy .... *E. annuus*
— Foliage sparse; plants mainly 3–7 dm tall; pubescence various; plants to be sought in Utah . *E. strigosus* Muhl.

5(3).	Plants diffusely branched, annual; leaves linear to linear-oblong; hairs of stem short and incurved; pappus simple .... *E. bellidiastrum*	3(2).	Leaves pinnately lobed; plants of the La Sal Mts. ... . *E. mancus*
—	Plants various, but seldom as above or, if so, the pappus double 6	—	Leaves palmately lobed or divided; plants widespread .... *E. compositus*
6(5).	Disks mainly over 1 cm wide; stems commonly simple, with solitary or few heads, and broad cauline leaves .... *E. glabellus*	4(1).	Involucres long- and shaggy-villous, the hairs sometimes obscuring the bract surface from middle to base. .5
—	Disks mostly less than 1 cm wide; stems commonly branched, often with several to many heads 7	—	Involucres hirsute to shortly villous, or, if long and shaggy-villous, the hairs not obscuring the bract surface even in the lower portion 7
7(6).	Stems with hairs all spreading .... *E. divergens*	5(4).	Plants 4–7 dm tall or more, known from southeastern Utah .... *E. elatior*
—	Stems with hairs appressed or ascending, or glabrous . 8	—	Plants mainly 0.3–1 dm tall, the distribution various . 6
8(7).	Leaves entire; plants with sterile flagellate branches .... *E. flagellaris*	6(5).	Hairs of involucre with black or dark purple cross-walls; basal leaves rounded to retuse apically; plants of the La Sal Mts. .... *E. melanocephalus*
—	Leaves pinnately lobate or toothed or, if entire, the plants lacking sterile flagellate branches 9	—	Hairs of involucre with pale cross-walls or some with bright reddish purple to dark purple cross-walls; basal leaves acute to abruptly obtuse apically; plants of the Uinta, Deep Creek, Tushar, and La Sal mts. .. .. *E. simplex*
9(8).	Ray flowers commonly 40–80; plants psammophytes of eastern Washington and western Kane counties .. .. *E. religiosus*		
—	Ray flowers commonly 25–40; plants of various substrates in eastern Washington and Kane counties 10	7(4).	Cauline leaves ample, usually lanceolate or broader; plants tall, erect (more or less asterlike) 8
10(9).	Involucres 2.5 mm high or less; stems 1–11 cm tall; peduncles sparingly villous; plants known from seeps and moist sandstone in Zion National Park ..... *E. sionis*	—	Cauline leaves usually much reduced, subulate, linear, oblong oblanceolate or, if broader, the plants not tall or not erect 11
—	Involucres 2.5–3 mm high; stems 14–25 cm long; peduncles hirsute; plants from limestone and sandstone outcrops in eastern Iron and adjacent Kane counties .... *E. proselyticus*	8(7).	Hairs of involucre with black cross-walls near their bases; rays white; plants of the western Uinta and Wasatch mts. .... *E. coulteri*
		—	Hairs of involucre with pale cross-walls; rays white, pink, or purple; plants with abundance and distribution various 9

Key 2.

Plants perennial, silvery pubescent; achenes 4 (or more) –nerved; involucral bracts imbricate.

1.	Achenes glabrous, with 8–14 nerves; caudex clothed with marcescent leaf bases, the midribs evident in age; plants of higher elevations in southern Utah .... *E. canus*	9(8).	Plants with cauline leaves well developed and equably distributed, only gradually reduced upward, the middle ones as large as or larger than the lower ones .... *E. speciosus*
—	Achenes more or less hairy, with 3–8 nerves; caudex lacking marcescent leaf bases, or, if these present, the midribs not evident; plants of low to moderate elevations, more widely or otherwise distributed 2	—	Plants with cauline leaves rather abruptly reduced upward, those of the middle smaller than the lower ones 10
2(1).	Involucres villous-hirsute with multicellular spreading hairs, the bracts more or less glandular apically; achenes with 3–5 nerves .... *E. pulcherrimus*	10(9).	Involucres glandular or viscid towards the apex; stems curved at base .... *E. formosissimus*
—	Involucres more or less strigose with simple hairs; achenes 4- to 8-nerved 3	—	Involucres seldom if at all glandular or viscid; stems erect .... *E. glabellus*
3(2).	Basal leaves evident, tufted, persistent; heads one per stem; plants through much of Utah .... *E. argentatus*	11(7).	Ray flowers lacking .... *E. aphanactis*
—		—	Ray flowers present 12
—	Basal leaves mostly withered at anthesis, not forming a conspicuous tuft; plants mainly of southeastern Utah .... *E. utahensis*	12(11).	Plants subscapose, the bracteate leaves very small; caudex branches with persistent leaf bases 13
		—	Plants not subscapose, the leaves merely reduced upward; caudex branches with or without persistent leaf bases 16

Key 3.

Plants perennial; achenes mostly 2–nerved; bracts mostly subequal, villous with woolly or spreading multicellular hairs.

1.	Plants with pinnatifid or palmatifid or merely lobed leaves, low-spreading, more or less mat or clump forming, of high elevations 2	13(12).	Stems and involucres with long contorted villous hairs; plants of Box Elder and Daggett counties .... *E. nanus*
—	Plants with entire leaves or, if some of them lobed, otherwise differing; low to tall, of various elevations .. 4	—	Stems and involucres strigose, pilosulose, or hispidulous, the hairs appressed or ascending to spreading; distribution various 14
2(1).	Plants soboliferous, the caudex divided into elongate spreading branches; leaves merely toothed, or, if lobed, not as below .... *E. vagus*	14(13).	Leaves linear; herbage strigose; rays 7–11 mm long; plants mainly of lower elevations in the Navajo and Great basins .... *E. compactus*
—	Plants not soboliferous, the caudex branches short; leaves pinnately to palmately lobed or divided 3	—	Leaves narrowly oblanceolate to spatulate; herbage strigose to pilosulose or hispidulous; rays 4–8.2 mm long; plants of the Uinta Basin and Wasatch Plateau . 15
		15(14).	Involucres long-villous with spreading multicellular hairs; rays 6.8–8.2 mm long; plants of the Wasatch Plateau .... *E. carringtonae*

—	Involucres short-hispidulous; rays 4–6.5 mm long; plants of the Uinta Basin .... *E. untermannii*
16(12).	Caudex branches robust, 1–2.5 cm thick; plants of western Beaver and Washington counties .... *E. wahwahensis*
—	Caudex branches mainly less than 1.5 cm thick or, if thicker, plants of different distribution 17
17(16).	Stems spreading-hairy 18
—	Stems strigose, or with ascending hairs 21
18(17).	Leaves linear to linear-oblanceolate; plants without a prominent caudex, mainly of lower elevations .... *E. pumilus*
—	Leaves oblanceolate to spatulate; plants with prominent caudex, of low to middle or higher elevations .. 19
19(18).	Stems glandular, with sand grains adhering; plants of lower elevations in Emery and Wayne counties ... . *E. maguirei*
—	Stems lacking glands; plants of moderate and higher elevations 20
20(19).	Stems commonly purplish at the base; leaves thin; plants of broad or other distribution .... *E. eatonii*
—	Stems green throughout; leaves thickish; plants of the Uinta and Wasatch mts. .... *E. goodrichii*
21(17).	Caudex with spreading subrhizomatous branches, with numerous fibrous roots; stems and lower leaf bases purplish; plants sod-forming, of higher elevations .... *E. ursinus*
—	Caudex not subrhizomatous, seldom if ever with roots; stem and leaf bases not purple; plants of various elevations 22
22(21).	Stems decumbent, sharply bent from apex of caudex 23
—	Stems ascending to erect, not sharply bent from caudex apex 24
23(22).	Herbage glaucous, sparingly hairy; rays 15–22; basal leaf bases greatly expanded, long-ciliate; plants of eastern Washington County .... *E. canaani*
—	Herbage green, not especially glaucous, sparingly to moderately hairy; basal leaf bases not greatly expanded, short-ciliate; plants of broad distribution .. .. *E. eatonii*
24(22).	Cauline leaves moderately well developed, the basal ones linear-oblanceolate; involucres 9–12 mm wide; plants of lower elevations .... *E. engelmannii*
—	Cauline leaves much reduced, the basal ones spatulate; involucres less than 8 mm wide 25
25(24).	Basal leaves acute or acutish; rays blue to red-purple; pappus subequal to disk corollas; plants widely distributed .... *E. tener*
—	Basal leaves obtuse to rounded; rays white to pink; pappus shorter than disk corollas; plants of the Bear River Range, Cache County .... *E. cronquistii*

Key 4.

Plants perennial, green; achenes mostly 2-nerved; involucres mostly without long, spreading, multicellular hairs.

1.	Plants substoloniferous; leaves spatulate; involucres mainly less than 8 mm wide; plants of hanging gardens and moist canyon bottoms in San Juan County .... *E. kachinensis*
—	Plants not substoloniferous; leaves various; involucres mainly over 8 mm wide; plants not or seldom of hanging gardens, variously distributed 2
2(1).	Cauline leaves ample, usually lanceolate or broader; plants tall and erect (more or less asterlike) 3
—	Cauline leaves definitely reduced upward, mostly linear to oblanceolate, or broader in some low species; stems often spreading or decumbent 5
3(2).	Rays mainly 2–3 mm wide .... *E. peregrinus*
—	Rays 1–2 mm wide 4
4(3).	Cauline leaves glabrous or minutely glandular, not ciliate, subequal to or shorter than the internodes .... *E. superbus*
—	Cauline leaves ciliate or otherwise pubescent, sometimes also glandular, usually longer than the internodes .... *E. speciosus*
5(2).	Pubescence of the stem widely spreading or glandular-scabrous 6
—	Pubescence of the stem appressed, ascending, or lacking 9
6(5).	Involucre canescent with fine, white, appressed to ascending hairs, sometimes also glandular 7
—	Involucre glandular and more or less spreading-hairy or strigose 8
7(6).	Involucral bracts evidently thickened on the back; basal leaves typically obtuse to rounded apically; stems rarely purplish at the base .... *E. caespitosus*
—	Involucral bracts slightly if at all thickened on the back; basal leaves acute apically; stems often purplish at the base .... *E. corymbosus*
8(6).	Stems hirsute with short spreading hairs, conspicuously decumbent; involucres glandular and spreading hairy .... *E. jonesii*
—	Stems glandular-scabrous, ascending or erect; involucres glandular (rarely sparingly strigose) .... *E. nauseosus*
9(5).	Basal leaves broadly oblanceolate or usually broader, the blade well developed, usually abruptly contracted to the petiole 10
—	Basal leaves linear to oblanceolate or spatulate, tapering gradually to the petiole 12
10(9).	Rays purple; achenes 4– to 7-nerved; pappus simple .... *E. peregrinus*
—	Rays various; achenes 2–nerved (occasionaly more, but rays then pale and pappus double) 11
11(10).	Stems essentially scapose, the upper bracts linear; plants known from the Wasatch Mts. ...... *E. garrettii*
—	Stems subscapose, the upper bracts oblong; plants rather broadly distributed .... *E. leiomeris*
12(9).	Peduncles and involucres densely glandular, not hairy; stems glabrous or essentially so; plants of the Wasatch Mts. .... *E. arenarioides*
—	Peduncles not glandular, or, if so, the stem more or less hairy; involucres and distribution various 13
13(12).	Bases of basal leaves neither enlarged nor of different texture than the blades; blades linear or linear-filiform; plants known from Cache and Daggett counties .... *E. filifolius*
—	Bases of basal leaves somewhat enlarged, membranous or thickened, or otherwise different from above; blades not or only some linear 14
14(13).	Leaves glabrous or nearly so, the hairs, if present, short and appressed .... *E. leiomeris*
—	Leaves hairy, the hairs spreading or curved-ascending 15

15(14). Plants subscapose; cauline leaves reduced to acicular bracts; plants of the Uinta Basin and West Tavaputs Plateau .... *E. nematophyllus*

— Plants caulescent; cauline leaves well developed 16

16(15). Basal leaves linear to narrowly oblanceolate, less than 4 mm wide; cauline leaves mostly linear; inner pappus bristles mainly 5–8; plants of moist alcoves along Lake Powell .... *E. zothecinus*

— Basal leaves oblanceolate to spatulate, more than 4 mm wide; cauline leaves oblong to oblanceolate; inner pappus bristles mainly more than 8; plants not of moist alcoves 17

17(16). Stems decumbent-ascending, commonly curved at the base; basal leaves sheathing basally; heads mainly solitary .... *E. abajoensis*

— Stems erect or nearly so; basal leaves not especially sheathing; heads mainly 2–4 .... *E. awapensis*

Erigeron abajoensis Cronq. Abajo Daisy. Perennial herb, with a taproot and stout caudex, the caudex branches clothed with brown marcescent leaf bases; stems decumbent to spreading at the base, 5–20 cm long, strigose to strigulose, the hairs ascending; basal leaves oblanceolate, 2–7 cm long, 2–6 mm wide, more or less sheathing basally; cauline leaves several to many, oblong to lance-oblong, mostly 0.6–2.5 cm long, 1.5–4 mm wide; heads solitary, less commonly 2–4; involucres 4–5.2 mm high, 7–12 mm wide, the bracts subequal or slightly imbricate, somewhat thickened dorsally, greenish brown, strigose to strigulose, the hairs multicellular; rays about 40–60, pink purple to blue (or white), 3–8 mm long, 1–1.8 mm wide; pappus double, the inner of 12–20 bristles, the outer of setae or scales; achenes 2–nerved, hairy. Pinyon-juniper, ponderosa pine, and spruce-fir communities at 2135 to 3450 m in Garfield, Piute, San Juan (type from the Abajo Mts.), and Wayne counties; endemic; 4 (i).

Erigeron acris L. Bitter Fleabane. Short-lived perennial, with a slender taproot and poorly developed caudex; stems erect or decumbent at the base, 8–32 cm tall, spreading-hairy and more or less glandular; basal leaves spatulate-oblanceolate, 0.5–6.5 cm long, 2–10 mm wide, entire or sparingly toothed; cauline leaves several to many, oblong to narrowly oblanceolate, lanceolate, or linear, mostly 0.8–7 cm long, 1–8 mm wide; heads solitary, or more commonly few to numerous, on short to elongate peduncles; involucres 4.5–8 mm high, 9–17 mm wide, the bracts imbricate, not especially thickened, green or tinged pink apically in some, sparingly hairy with spreading to ascending stiff multicellular hairs and beset with short glandular processes; rays numerous, pink or white, erect, ca 2–4.5 mm long, the inner pistillate flowers eligulate, with corolla tubular; pappus of ca 25–35 slender barbellate white to reddish bristles, surpassing the disk corollas; achenes 2–nerved, sparingly hairy; n = 9. Lodgepole pine, spruce, and fir communities at 2800 to 3500 m in Duchesne, Summit, and Uintah counties; Alaska to Labrador, south to California, Colorado, Michigan, and Maine; circumboreal; 11 (ix). Varietal status of our few specimens is unclear. One of the specimens has few heads and essentially eglandular bracts, one is monocephalus and has glandular involucres, and the others are polycephalus and have glandular involucres. Names available are var. *asteroides* (Andrz.) DC. and var. *debilis* Gray, but there appear to be three rather poorly differentiated taxa involved. Decisions as to proper names must await further study.

Erigeron annuus (L.) Pers. Annual Fleabane. Plants annual, with slender taproots; stems erect, 6–12 (15) dm tall, sparingly to densely hirsute with long spreading hairs, becoming appressed upward; basal leaves commonly withered at anthesis, ovate to suborbicular, petiolate; cauline leaves numerous, lanceolate to oblong, mainly 1.5–8 cm long, 3–20 mm wide, serrate to entire; heads several to numerous, in a leafy inflorescence; involucres 7.5–12 mm wide, 3–5 mm high, the bracts subequal or the outer somewhat shorter, greenish to brownish, acuminate-attenuate, glandular and sparingly villous-hirsute with multicellular hairs; rays ca 80–125, white (rarely bluish), 4–10 mm long, 0.5–1 mm wide; pappus double; achenes 2–nerved, hairy; 2n = 54. Roadsides, fields, and other disturbed sites at 1370 to 1830 m in Utah and Wasatch counties; widespread in the U. S.; adventive from Europe; 8 (ii).

Erigeron aphanactis (Gray) Greene Hairy Daisy. [*E. concinnus* var. *aphanactis* Gray]. Perennials with definite branching caudex; stems decumbent to ascending or erect, 5–20 (30) cm tall, sparingly to copiously spreading-hirsute with multicellular hairs; basal leaves narrowly oblanceolate to spatulate, 0.5–8 cm long, 1–6 mm wide, petiolate; cauline leaves well developed or essentially lacking; heads solitary or several; involucres 7–15 mm wide, 3.5–6 mm high, finely to coarsely spreading-hirsute and sometimes also finely glandular; bracts subequal or somewhat imbricate, slender, acuminate, green or greenish brown, the midrib thickened; pistillate flowers present, tubular, eligulate, or sometimes with rays shorter than the disk; pappus double; achenes 2–nerved, sparsely hairy. This species is represented in Utah by two rather weak varieties.

1. Plants essentially scapose; corolla lobes sometimes becoming reddish or purplish ... *E. aphanactis* var. *congestus*

— Plants with leafy stems; corolla lobes commonly yellowish .... *E. aphanactis* var. *aphanactis*

Var. aphanactis Salt desert shrub, sagebrush, pinyon-juniper, and mountain brush communites at 1300 to 2700 m in Beaver, Garfield, Juab, Piute, Sanpete, Sevier, Washington, and Wayne counties; Oregon and Idaho south to California, Arizona, and Colorado; 21 (iii).

Var. congestus (Greene) Cronq. [*E. congestus* Greene]. Juniper-black sagebrush, sagebrush, and aspen communities at 1830 to 2600 m in Garfield and Sevier counties; California; 3 (0).

Erigeron arenarioides (D. C. Eaton) Gray Wasatch Daisy. [*E. stenophyllus* D. C. Eaton, not H. & A., type from Cottonwood Canyon, Salt Lake County; *Aster arenarioides* D. C. Eaton ex Gray]. Perennial herbs, with definite branching caudex, the caudex branches clothed with brownish marcescent leaf bases; stems ascending to erect, 6–25 (30) cm tall, slender, glabrous or glandular below the heads; leaves glabrous or sparingly strigose, the basal ones linear-filiform to linear-oblanceolate, 1.5–6 (8) cm long, 0.5–2 (4) mm wide, entire; heads solitary or 2 or 3 (rarely more); involucres 7–9 mm wide, 3.7–5 mm high, the bracts imbricate in several series, greenish brown, finely glandular, the tips often purplish; rays 10–25, blue, 4–8 mm long, 0.8–1.8 mm wide; pappus of ca 10–16 bristles, and with a few short setae; achenes 2–nerved, sparsely strigose. Crevices in limestone and quartzite outcrops, rarely in beach sand, at 1300 to 2440 m in Salt Lake, Tooele, Utah, and Weber counties; endemic; 8 (0).

Erigeron argentatus Gray Silver Daisy. [*Wyomingia argentata* (Gray) A. Nels.]. Perennial herbs, with definite branching caudex, the caudex branches more or less clothed with brown marcescent leaf bases, the midribs not especially persistent; stems erect, 9–28 (40) cm tall, finely strigose and silvery to gray-green; basal leaves tufted, spatulate to oblanceolate, 1.5–7 cm long, 1–4 (6) mm wide, petiolate, entire; cauline leaves reduced upward; heads solitary; involucres 10–18 mm wide, 5.5–9 mm high, the bracts strongly imbricate, silvery strigose with appressed antrorse hairs; rays ca 20–50, blue, lavender, or pink to white, 9–15 mm long, 1.6–2.8 mm wide; pappus double; achenes pilose; n = 9. Salt desert shrub, sagebrush, pinyon-juniper, and mountain brush communities at 1600 to 2440 m in Beaver, Box Elder, Emery, Garfield, Iron, Juab, Millard, Piute, Sanpete, Sevier, Tooele, and Utah counties; Nevada, California; 29 (iii).

Erigeron awapensis Welsh Awapa Daisy. Perennial herbs from a branching caudex, the caudex branches clothed with ragged brown marcescent leaf bases; stems erect or nearly so, 10–24 cm long, strigose, the hairs ascending; basal leaves 1.5–7 cm long, 2–8 mm wide, not especially sheathing; cauline leaves well developed, oblong to linear, mostly 1–4 cm long, 2–4 mm wide; heads 2–4, rarely solitary; involucres 3–9 mm wide, 3.7–4.5 mm high, the bracts more or less imbricate, thickened near the base dorsally, greenish, strigulose, the hairs multicellular; rays 35–45, pink purple to pink (or white?), 5–6 mm long, 0.9–1.8 mm wide; pappus apparently simple, of 15–20 slender bristles, and with a few inconspicuous shorter setae in some; achenes 2–nerved, hairy. Pinyon-juniper and sagebrush communities at 2135 to 2260 m in Garfield and Wayne (type from 26.5 mi S. of Antimony) counties; endemic; 2 (i).

Erigeron bellidiastrum Nutt. Pretty Daisy. Plants annual (or biennial), the stems 3.5–32 (50) cm tall, erect or ascending, often intricately branched; herbage strigulose with incurved multicellular hairs; leaves mainly cauline, 0.5–4 cm long, 1–3 (6) mm wide, linear to oblanceolate, entire (or sparingly toothed to pinnatifid), petiolate, becoming sessile upward; heads solitary to numerous; involucres 5–11 mm wide, 3–5 mm high, the bracts hirtellous with spreading curved multicellular hairs, thick, greenish, subequal, or the outermost shorter; rays ca 30–70, pink or white, 4–6 mm long, ca 1 mm wide; pappus of ca 15 deciduous bristles; achenes 2–nerved, hairy; n = 9. Vanclevea-ephedra, blackbrush, and pinyon-juniper communities at 1125 to 1830 m in Garfield, Grand, Kane, San Juan, and Washington counties; Wyoming and South Dakota to New Mexico and Texas; 23 (ii).

Erigeron caespitosus Nutt. Tufted Daisy. [*E. caespitosus* var. *laccoliticus* Jones, type from the Henry Mts.]. Perennial herbs with a branching caudex, the caudex branches clothed with brown or blackish marcescent leaf bases; stems decumbent at the base, 4–25 (30) cm tall, hirtellous with short spreading hairs (especially above); basal leaves oblanceolate to spatulate, generally rounded to obtuse apically, 1–9 cm long, 2–13 mm wide, 1– to 3–nerved; cauline leaves reduced upward; heads solitary or few to several; involucres 9–18 mm wide, 4–7 mm high, the bracts subequal to imbricate, thickened on the back, green, strigose to pilose with multicellular hairs spreading laterally from the midrib; rays ca 30–100, blue, pink, or white, 5–15 mm long, 1–2 mm wide; pappus double; achenes 2–nerved, hairy; 2n = 18. Sagebrush, pinyon-juniper, aspen, lodgepole pine, spruce, and tundra communities at 2135 to 3570 m in Daggett, Duchesne, Emery, Garfield, Salt Lake, Sevier, Summit, Uintah, Utah, Wasatch, and Wayne counties; Alaska and Yukon south to Arizona, New Mexico, and Nebraska; 40 (viii). This is a variable species with many phases in Utah, each differing in stature, nature of vesture, size of heads, and other features that fail singly and in combination as diagnostic criteria. The species interfaces with *E. nauseosus*, *E. abajoensis*, and probably with other taxa.

Erigeron canaani Welsh Canaan Daisy. Perennial herbs from a simple (or branched?) caudex, this clothed with brown marcescent leaf bases, the taproot prominent; stems 7–20 cm tall, decumbent to ascending, sometimes purplish at the base, sparingly pubescent with ascending hairs; leaves pubescent like the stem, the basal ones tufted, 1–nerved, 1.4–9 cm long, 0.7–1 mm wide, linear, involute, sharply acute, conspicuously expanded and long-ciliate basally; cauline leaves numerous, reduced upward; heads 1–3; involucres 9–13 mm wide, 5.3–6.5 mm high; bracts imbricate, conspicuously glandular and sparingly to moderately villous-pilose with multicellular hairs, green or variously suffused with purple; rays 15–22, white or pinkish, 3.5–5 mm long, 1.8–2.1 mm wide; pappus single, of ca 20 slender bristles; achenes 2–nerved, hairy. Ponderosa pine community at 1585 to 2075 m in Washington (type from Canaan Mt.) County; endemic; 5 (ii). The Canaan daisy is similar in general aspect to *E. eatonii* (q.v.). The involute linear glaucous leaves and few ray flowers appear to be diagnostic.

Erigeron canus Gray Hoary Daisy. Perennial herbs, with branching caudex, the caudex branches clothed with persistent leaf bases, the marcescent midribs prominent; stems erect or nearly so, 5–30 (35) cm tall, appressed strigose; basal leaves oblanceolate, mostly 1–6 cm long and 1–5 (7) mm wide, hairy like the stems; cauline leaves reduced upward; heads solitary (rarely up to 4); involucres 9–16 mm wide, 5–7 mm high, the bracts strigulose with ascending to spreading multicellular hairs and more or less glandular, imbricate; rays ca 30–40, blue or white, 7–12 mm long, 0.8–1.4 mm wide; pappus double; achenes ca 8– to 14–nerved. Gravelly substrates of the Pink Limestone member of the Wasatch Formation in ponderosa pine and sagebrush communities at 2300 to 2500 m in Garfield County; Wyoming and South Dakota to Arizona and New Mexico; 3 (i).

Erigeron carringtonae Welsh Carrington Daisy. Pulvinate perennial herbs with a pluricipital caudex, the branches clothed with conspicuous brown to straw colored or ashy marcescent leaf bases; leaves mainly basal, thickish, 0.6–3.5 cm long, 1–5 mm wide, spatulate to oblanceolate, strigose to pilosulose, obtuse to rounded apically; scapes 2.5–8 cm tall; heads solitary; involucres 9.8–15 mm wide, 5.8–7 mm high, the bracts imbricate, suffused with purple or green, the inner greenish with scarious margins, spreading-villous with long multicellular hairs; rays 18–30, pink to pink purple, 6.8–8.2 mm long, 1.4–2.3 mm wide; pappus double, the inner of 25–35 barbellate bristles, the outer of short setae; achenes 2–nerved, pilose. Meadows and escarpment margins, commonly on Flagstaff Limestone at 3050 to 3355 m in Emery and Sanpete (type from Wasatch Plateau) counties; endemic; 6 (i).

Erigeron compactus Blake Mound Daisy. [*E. pulvinatus* Rydb., type from the Deep Creek Mts.]. Perennial pulvinate herbs with a branching caudex, the caudex branches clothed with marcescent leaf bases; leaves mainly basal, 4–20 mm long, 0.6–1.4 mm wide, linear, finely strigose; scapes 2–10 cm tall; heads solitary; involucres 7–17 mm wide, 5–8.5 mm high, the bracts more or less imbricate, straw colored or greenish brown to green, hispidulous with short spreading hairs; rays mainly 15–50, white or pink, 7–11 mm long, 1.4–2.5 mm wide; pappus double; achenes 2–nerved. Two geographically segregated races are recognizeable as varieties.

1. Leaves yellowish green; involucral bracts appressed strigose; plants of the Great Basin .... E. compactus var. compactus
— Leaves grayish green; involucral bracts spreading-hispidulose; plants of the Colorado Drainage system ... . E. compactus var. consimilis

Var. compactus Pinyon-juniper community at 1830 to 2135 m in Beaver, Box Elder, Millard, and Tooele counties; Nevada and California; 4 (0).

Var. consimilis (Cronq.) Blake [E. consimilis Cronq.]. Salt desert shrub and pinyon-juniper communities in Daggett, Duchesene, Emery, and Wayne counties; Arizona; 11 (i). The general aspect of this variety is similar to that of E. pulcherrimus (q.v.), with which it is sympatric in much of its range; the 2–nerved achenes and low subscapose stems are diagnostic.

Erigeron compositus Pursh Fern-leaf Daisy. Perennial caespitose cushion plants, with a shortened much-branched caudex, the caudex branches densely clothed with brown marcescent leaf bases; herbage glandular and more or less spreading-hairy; leaves mainly basal, mostly 2– or 3–ternately lobed or dissected, 0.5–7 cm long; cauline leaves few and reduced upward, simple or ternate; stems subscapose, 2–20 (25) cm tall; heads solitary; involucres 8–20 mm wide, 5–10 mm high, the bracts glandular and spreading-hairy, commonly purplish at the tips; rays lacking, or developed and 20–60, blue, pink, or white, to 12 mm long and 2 mm wide; pappus simple; achenes 2–nerved, villous-hirsute; 2n = 36, 54. Sagebrush, rabbitbrush, aspen, aspen-fir, lodgepole pine, spruce-fir, and alpine tundra communities at 2375 to 3965 m in Beaver, Box Elder, Daggett, Duchesne, Emery, Garfield, Iron, Juab, Millard, Piute, Salt Lake, Sanpete, Sevier, Summit, Tooele, Uintah, Utah, Wasatch, Washington, Wayne, and Weber counties; Alaska to Greenland, south to California, Arizona, Colorado, South Dakota, and Quebec; 77 (xvi). This is an extremely variable apomictic species, with rare sexual individuals. Our material has been assigned to var. **glabratus** Macoun, which is separable from the type variety only problematically.

Erigeron corymbosus Nutt. Mountain Daisy. [E. nelsonii Greene]. Perennial herbs from a simple or branched caudex; stem short hirsute with spreaing hairs and sometimes with longer ones intermixed, purplish and often curved at the base, 1–5 dm tall; leaves pubescent like the stem, often ciliate with longer hairs; basal leaves 3–nerved, linear-oblanceolate, acute apically, tapering to a petiole, 1.5–15 cm long, 3–14 mm wide; cauline leaves reduced upward; heads 1–10 (16); involucre 7–13 mm wide, 5–7 mm high, villous-hirsute with ascending multicellular stiff hairs and more or less glandular; bracts only somewhat if at all thickened on the back, green to tan, with darker midrib and usually close tips; rays ca 35–65, blue or pink, 7–13 mm long, 1.2–2 mm wide; disc corollas (3) 4–5.3 mm long; pappus double, the inner or 20–30 firm bristles, the outer of evident setae or narrow scales; achenes 2–nerved, pubescent; 2n = 18. Sagebrush, rabbitbrush, grassland community at ca 2227 m in Rich County; British Columbia to Montana, south to Oregon, Idaho, and Wyoming; 1 (0).

Erigeron coulteri T. C. Porter in Port. & Coult. Coulter Daisy. Perennial herbs from a rhizome or caudex; stems more or less spreading-hairy, mainly 1–6 dm tall; basal and cauline leaves ample or the cauline ones somewhat reduced, entire or toothed, the largest 6–15 cm long, 1–2.5 cm wide, oblancolate to elliptic, lanceolate, oblong, or ovate; heads solitary or 2 or 3; involucres 10–15 mm wide, 6–10 mm high, the bracts densely white hirsute below with hairs having purplish black cross-walls, at least near the base, glandular to the tips; rays 40–80, ca 10–15 mm long, white to pink purple; pappus simple; achenes sparsely hairy; 2n = 18. Aspen and spruce-fir communities at 2235 to 2745 m in Salt Lake, Summit, and Utah counties; Oregon to Wyoming, south to California, Nevada, and New Mexico; 9 (0).

Erigeron cronquistii Maguire Cronquist Daisy. Perennial herb, with short caudex branches clothed with brown leaf bases; stems 1.5–7 cm long, sparingly strigose; basal leaves 0.5–4 cm long, spatulate to oblanceolate or elliptic, petiolate, sparingly strigose; cauline leaves few or wanting; heads solitary, sometimes 2; involucres 5–8 mm wide, 3–5 mm high, glandular and spreading-hirsute, the bracts imbricate, green, often suffused with purple; rays 10–25, white or pale pink, 5–6 mm long, 1.3–2.1 mm wide; pappus single, or with a few shorter outer ones; achenes 2–nerved, sparingly hairy. Limestone cliffs at 1750 to 2600 m in the Bear River Range (type from Logan Canyon), Cache County; endemic; 2 (0). This beautiful, tiny plant is a near congener of E. tener (q.v.).

Erigeron divergens T. & G. Spreading Daisy. [E. divaricatus Nutt., not Michx.; E. cinereus var. aridus Jones, type from Silver Reef, Washington County]. Annual, biennial, or short-lived perennial herbs from taproots; stems branched from the base and above, pubescent with spreading hairs, 0.5–5 (7) dm tall; basal leaves oblanceolate to spatulate, entire or sometimes lobed, mainly 1–7 cm long, 2–10 mm wide, spreading-hairy, petiolate, usually lacking at anthesis; cauline leaves reduced upward; heads several to numerous; involucres 7–11 mm wide, 4–5 mm high, finely glandular and hirsute with long, spreading hairs, the bracts green, attenuate; rays ca 75–150, blue, pink, or white, ca 5–10 mm long, 0.5–1.2 mm wide, sometimes scarcely developed; pappus double; achenes 2 (4) -nerved, sparsely hairy; n = 9, 18, 27, 36. Riparian, rabbitbrush, sagebrush, pinyon-juniper, mountain brush, ponderosa pine, and aspen-spruce communities at 975 to 2900 m in all Utah counties, except Box Elder, Carbon, Morgan, Rich, Sanpete, Summit, and Wayne; British Columbia to South Dakota, south to California, Arizona, and Mexico; 109 (xiii). Our materials have been segregated into two weak varieties differentiated as follows: var. **cinereus** Gray, with earliest flowers borne on long naked peduncles and plants later with long leafy stolons; and var. **divergens**, with earliest heads on leafy peduncles and plants not developing leafy stolons. The var. cinereus is evidently rare in Utah; 5 (ii).

Erigeron eatonii Gray Eaton Daisy. [E. eatonii f. molestus Cronq., type from the Stansbury Mountains]. Perennial herbs, from a short simple or branched caudex, this clothed with brown marcescent leaf bases, the taproot prominent; stems 5–38 cm tall, decumbent to ascending, usually purplish at the base, strigose or rarely more or less hirsute; leaves pubescent like the stem, the basal ones tufted, 1– (or more commonly) 3–nerved, acute, mainly 1.2–12 (15) cm long, 1–10 mm wide; cauline leaves numerous, reduced upward; heads 1–3 (7); involucres 8–15 mm wide, 5–8 mm high, the bracts imbricate, conspicuously glandular and more or less hirsute with spreading-ascending multicellular hairs, green or the tips purplish; rays ca 20–50, white to blue or pink, mainly 4–10 mm long, 1–2.5 mm wide; pappus single or with a few short outer setae; achenes 2 (3) -nerved; n = 9, 18. Sagebrush, mountain brush, pinyon-juniper, ponderosa pine, aspen, spruce-fir, and alpine tundra communities at 1890 to 3630 m in all Utah counties except for Box Elder and Morgan; Oregon to Wyoming, south to California, Arizona, and Colorado; 171 (xxv). The type is from the Uinta Mts., Summit County. This is a widespread and variable species, with variants differing in size, in head dimensions, and in nature of the pubescence. The hirsute phase from the Stansbury Mountains has been designated as f. *molestus* Cronq.

Erigeron elatior (Gray) Greene Tall Daisy. [*E. grandiflorus* var. *elatior* Gray]. Perennial herbs, from a short caudex (seldom collected); stems mainly 4–7 dm tall, often purplish below, leafy throughout, spreading-hairy and more or less glandular above; leaves mainly 2.2–10 cm long, 6–28 mm wide, the lowermost smaller than the middle ones and commonly withered at anthesis, ovate-lanceolate to lanceolate, entire, the lower petiolate, becoming sessile and somewhat clasping upward; heads 1–3 (6); involucres 12–20 mm wide, 7–11 mm high, the bracts densely woolly-villous with long, flattened, shiny, multicellular hairs, some of which may have reddish purple cross-walls, subequal, long- attenuate apically, the tips glandular, purple, and reflexed; rays ca 75–150, pink or pink purple (white), 12–20 mm long, 0.8–1.6 mm wide; pappus double; achenes 2–nerved, hairy; n = 9. Meadows and openings in mountain brush and spruce-fir communities at 2440 to 3050 m in the La Sal Mts. of Grand and San Juan counties; Colorado and Wyoming. This is a beautiful asterlike plant with equally leafy stems and densely villous involucres; 4 (0).

Erigeron engelmannii A. Nels. Engelmann Daisy. Perennial herbs, with short branching caudex, this clothed with straw colored to brown marcescent leaf bases; taproot definite; stems 3–24 (30) cm tall, decumbent to erect, strigose or the hairs ascending and multicellular; basal leaves 1–6 (10) cm long, 1.5–5 mm wide, linear-oblanceolate, the blades hairy like the stems, the basal margins long and coarsely ciliate; cauline leaves reduced but well distributed upward; heads 1–4; involucres 7–12 mm wide, 4–7 mm high, the bracts hirsute and more or less glandular, subequal, green, with brownish midrib and scarious apices; rays ca 35–100, white (rarely pink or blue), 5–12 mm long, 0.6–2 mm wide; pappus double; achenes 2–nerved, hairy. Salt desert shrub, sagebrush, rabbitbrush, and pinyon-juniper communities at 1370 to 2200 m in Box Elder, Cache, Daggett, Duchesne, Grand, Juab, Millard, Salt Lake, Sanpete, Sevier, Summit, Tooele, Uintah, and Utah counties; Oregon to Wyoming and Colorado; 28 (iii).

Erigeron filifolius Nutt. Thread-leaf Daisy. Perennial herbs, with branching woody caudex, the caudex branches clothed with brownish marcescent leaf bases; stems 10–30 (50) cm tall, more or less strigose; leaves 1–8 cm long, 0.3–3 mm wide, linear or filiform, strigose, the cauline ones distributed along the stem but smaller than the basal ones; heads 1 to several; involucres 5–15 mm wide, 4–6 mm high, the bracts villous to strigose and commonly glandular as well, subequal or somewhat imbricate, greenish; rays ca 15–75, blue to pink or white, 3–12 mm long, 1–2 mm wide; pappus single or with a few outer setae; achenes 2–nerved, more or less hairy; n = 9, 18. Sagebrush, aspen, and mixed conifer communities at ca 2380 m in Cache (C. P. Smith 1737 RM) and Daggett counties; British Columbia and Montana to California and Nevada; 1 (0).

Erigeron flagellaris Gray Trailing Daisy. Biennial or short-lived perennials, with a poorly developed caudex (if at all) and slender taproot; herbage strigose or with spreading hairs at stem base; stems 3–25 (40) cm tall, the fertile ones terminated by a solitary head, the sterile ones developed as leafy stolons; basal leaves 1–5 cm long, 1.5–8 mm wide, oblanceolate to spatulate; cauline leaves smaller upward, linear to oblanceolate; heads solitary; involucres 7–13 mm wide, 3.5–5 mm high, the bracts with appressed or spreading hairs, glandular, green to purplish; rays mostly ca 50–100, white, pink, or blue, 5–10 mm long, 0.8–1 mm wide; pappus double; achenes 2–nerved, hairy to almost glabrous; 2n = 27, 36, 45. Sagebrush, juniper, ponderosa pine, aspen, spruce-fir, and alpine meadow communities at 1980 to 3180 m in Beaver, Carbon, Daggett, Duchesne, Emery, Garfield, Grand, Iron, Kane, Millard, Piute, San Juan, Sanpete, Sevier, Summit, Uintah, Wasatch, Washington, and Wayne counties; British Columbia to Nevada, Arizona, and Texas; 63 (xi).

Erigeron formosissimus Greene Pretty Daisy. [*E. frucetorum* Rydb., type from the La Sal Mts.]. Perennial herbs, with a simple or sparingly branched subrhizomatous caudex; herbage variously hirsute, glandular, or glabrous, the stems more or less glandular above, mainly 1.5–3 (4.5) dm tall; basal leaves the largest, mainly 2–10 (15) cm long, 4–10 (15) mm wide, oblanceolate to spatulate; cauline leaves commonly much reduced upward, lanceolate to oblong or ovate; heads 1–6; involucres 10–20 mm wide, 5–8 mm high, the bracts subequal, linear, acuminate, glandular, and more or less hirsute; rays ca 75–150, 8–15 mm long, ca 1 mm wide, blue, pink, or white; pappus double; achenes 2–nerved, hairy; n = 9. Meadows in aspen and mountain brush communities at 2440 to 2840 m in Grand, Iron, Salt Lake, San Juan, and Sevier counties; Alberta south to Arizona and New Mexico; 5 (i). The species is poorly known in Utah (reports of the species in Iron County are from Cronquist 1947).

Erigeron garrettii A. Nels. Garrett Daisy. [*E. controversus* Greene, type from Alta].. Perennial subscapose herbs, with branching caudex, the caudex branches clothed with brown leaf bases; stems 3–23 cm tall, sparingly strigose; basal leaves 1.2–12 cm long, 3–13 mm wide, oblanceolate to spatulate, glabrous, sparingly ciliate; cauline leaves lacking or greatly reduced; heads solitary; involucres 8–17 mm wide, 5–8 mm high, the bracts finely strigose and obscurely glandular, moderately imbricate; rays ca 20–35, white to pink, 7–13 mm long, 1.4–2.7 mm wide; pappus double; achenes 2–nerved, hairy. Moist cliff faces and crevices at 2750 to 3570 m in Salt Lake (type from Big Cottonwood Canyon), Utah, and Wasatch counties; endemic; 17 (0).

Erigeron glabellus Nutt. Smooth Daisy. Perennial or biennial herbs with simple or branched caudices, the caudex, when present, clothed with brown to blackish leaf bases; herbage strigose to hirsute; stems 1–6.5 dm tall, erect or nearly so; basal and lower leaves mainly 3–15 cm long, 3–18 mm wide, oblanceolate, entire or toothed, petiolate; middle cauline leaves lanceolate to linear, reduced upward; heads 1–12 (15), borne on bracteate peduncles; involucres 10–20 mm wide, 5–9 mm high, the bracts subequal to slightly imbricate, acuminate, strigose to strigulose; rays ca 125–175, blue to pink, or white; pappus double; achenes 2–nerved, hairy; n = 9, 18. Meadows and stream sides at 1370 to 1770 m in Beaver, Cache, Daggett, Davis, Duchesne, Salt Lake, Uintah, Utah, and Wasatch counties; Alaska and Yukon, south to Colorado, South Dakota, and Wisconsin; 12 (0). This is a tall handsome daisy of lower elevations in Utah.

Erigeron goodrichii Welsh Goodrich Daisy. Perennial herbs from a stout taproot and caudex, the caudex branches with dark brown marcescent leaf bases; stems 3–10 cm tall, decumbent-ascending to erect, spreading-hairy; basal leaves 0.4–6 cm long, 1.2–5 mm wide, narrowly oblanceolate, the veins not apparent, pilosulose, obtuse apically; cauline leaves more or less developed, but much reduced upward; heads solitary; involucres 10.5–18 mm wide, 6.4–7.8 mm high; bracts imbricate, spreading villous-pilose with multicellular hairs, thickened basally, green or the apices suffused purplish, the inner with scarious margins, the attenuate apices more or less glandular and sometimes spreading; rays 40–65, pink purple to pink or white, 6.8–10.4 mm long, 1.5–2 mm wide; pappus apparently single, of 20–30 minutely barbellate bristles; achenes 2–nerved, pilose. Engelmann spruce krummholz and meadow communities, often on rock outcrops or talus at 3050 to 3400 m in Duchesne (type from the Uinta Mts.), Summit, Uintah, and Utah counties; endemic; 8 (0).

Erigeron jonesii Cronq. Jones Daisy. Perennial herbs, from a branching or simple caudex, the caudex branches clothed with blackish or dark brown marcescent leaf bases; herbage hirsute with short spreading hairs; stems mainly 10–25 cm tall, conspicuously decumbent and often purplish at the base; basal leaves 3–nerved, 1.5–8 cm long, 3–12 mm wide, oblanceolate to elliptic or spatulate, petiolate, entire or toothed; cauline leaves smaller than the basal; heads 1–4; involucres 9–15 mm wide, 5–7 mm high, the bracts glandular and spreading-hairy, slightly thickened dorsally, more or less imbricate, green, with tips often purplish; rays ca 25–50, blue, pink, or white, 4–8 mm long, 1.4–1.8 mm wide; pappus single or with a few short outer setae; achenes 2–nerved, hairy. Sagebrush, pinyon-juniper, mountain brush, and alpine meadow communities at 1890 to 3350 m in Juab, Tooele, and Washington counties; Nevada; 5 (iii). The Jones daisy simulates *E. eatonii* in habit and stature, but the definite spreading hairs of the herbage are apparently definitive in most instances. Possibly it would best be treated at some infraspecific rank within *E. eatonii*.

Erigeron kachinensis Welsh & Moore Kachina Daisy. Perennial herbs, from a short thick branching or simple caudex, the caudex branches clothed with brown marcescent leaf bases; herbage glabrous throughout; stems 6–18 cm tall, decumbent to erect; basal leaves 1.3–5 cm long, 2–13 mm wide, oblanceolate to obovate or spatulate, the blade tapering to the petiole, rounded or retuse apically, entire; cauline leaves 5–11, reduced upward; heads solitary or 2–4, the involucres 5–6 mm broad, 3.2–4 mm high, the bracts distinctly imbricate, some purplish at the tip, glabrous or minutely glandular; rays 10–15, white or pinkish, 3.5–5.5 mm long, 0.9–1.1 mm wide; pappus double; achenes 2–nerved, hairy. Seeps and hanging gardens at 1680 to 1890 m in White (type from Natural Bridges National Monument) and Dark Canyons and Elk Ridge, San Juan County, Utah, and Montrose County, Colorado; 3 (ii). This distinctive dwarf daisy is a Colorado Plateau endemic.

Erigeron leiomerus Gray Glaber Daisy. [*E. minusculus* Greene, type from Big Cottonwood Canyon, Salt Lake County]. Perennial herbs, from a branching caudex, the caudex branches clothed with brown marcescent leaf bases; herbage glabrous or merely strigose; stems 4–12 (15) cm tall, decumbent to erect; basal leaves 1.3–7 cm long, 2–11 (15) mm wide, oblanceolate to spatulate or obovate, rounded to retuse apically, enlarged and often purplish basally, glabrous or strigose to glabrate; cauline leaves reduced upward, usually several, becoming acutish; head solitary, the involucres 7–13 mm wide, 4–6 mm high, the bracts somewhat imbricate, purplish overall or at tips, finely glandular; rays ca 15–60, purplish to blue or white, 6–11 mm long, 1.5–2.5 mm wide; pappus double; achenes 2–nerved, short-hairy; 2n = 18. Talus slopes, boulder fields, and meadows in spruce and lodgepole pine and alpine tundra communities at 2950 to 3750 m in Beaver, Box Elder, Cache, Daggett, Duchesne, Juab, Piute, Salt Lake, Summit (type from the Uinta Mts.), Tooele, and Uintah counties; Nevada and Idaho to Wyoming, Colorado, and New Mexico; 27 (x).

Erigeron linearis (Hook.) Piper Yellow Daisy. [*Daucopappus linearis* Hook.]. Perennial herbs from a pluricipital caudex, the branches of the caudex clothed with broad clasping brownish marcescent leaf bases; herbage strigose; stems 5–20 cm tall; basal leaves 1–9 cm long, 0.5–3 mm wide, linear to linear-oblanceolate, acute, the bases enlarged, more or less sheathing, straw colored and strongly ciliate; cauline leaves reduced upward; heads solitary or 2 or 3; involucres 8–13 mm wide, 4–7 mm high, strigose-villous with multicellular hairs and more or less glandular; bracts subequal to somewhat imbricate, green or greenish to straw colored, attenuate, thickened dorsally; rays ca 20–45, yellow, 4–11 mm long, 1.3–2.5 mm wide; pappus double, the inner of 10–20 barbellate bristles, the outer of scales; achenes 2–nerved, short hairy; 2n = 18, 27. Sagebrush and juniper communities at 1675 to 2000+ m in Box Elder County; British Columbia, Washington, and Oregon, east to Idaho and Wyoming, and south to Nevada; 3 (0).

Erigeron lonchophyllus Hook. Longleaf Daisy. Short-lived perennial or biennial (?) herbs, with slender taproots and subfibrous roots from a poorly developed caudex; stems decumbent to erect, 5–55 (60) cm long, sparsely to densely spreading-hairy; basal leaves oblanceolate to spatulate, 1.2–11 (15) cm long, 2–12 mm wide; cauline leaves several to many, mostly 0.6–8 cm long, 2–6 mm wide; heads few to numerous, rarely solitary, borne on nearly erect peduncles; involucres 4–9 mm high, 7–17 mm wide, the bracts evidently imbricate, not especially thickened basally, greenish to brownish or yellowish, the tips commonly purplish, sparsely to moderately strigulose with multicellular hairs; rays numerous, white or pinkish, ca 2–4 mm long, lacking inner eligulate pistillate corollas; pappus of ca 20–30 slender barbellate white bristles, surpassing the disk corollas; achenes 2–nerved, sparsely hairy; n = 9. Marshes, stream banks, seeps, and wet meadows at 1370 to 2900 m in Beaver, Cache, Carbon, Daggett, Duchesne, Garfield, Grand, Juab, Piute, Rich, Salt Lake, Sanpete, Sevier, Summit, Tooele, Uintah, Utah, Wasatch, and Washington counties; Alaska and southern Yukon, south to California and New Mexico, and east to Quebec and South Dakota; 39 (vii).

Erigeron maguirei Cronq. Maguire Daisy. Perennial herbs, with a branching caudex, the caudex branches clothed with brown to straw colored marcescent leaf bases; herbage spreading hirsute; stems 7–18 cm high, decumbent to sprawling or erect; basal leaves 2–5 cm long, 3–8 mm wide, oblanceolate to spatulate, rounded apically; cauline leaves well developed, but somewhat reduced upward, becoming acutish; heads solitary or 2–5; involucres 5–6.5 mm high, 7–11 mm wide, the bracts imbricate, not much thickened, green or yellowish, the inner less pubescescent and with scarious purplish tips, all finely glandular also; rays 12–20, white or pinkish, ca 6–8 mm long, 1.1–2 mm wide; pappus of 13–25 slender barbellate sordid bristles, with a few shorter outer ones; achenes 2–nerved, hairy. Canyon bottoms in Wingate (?) and Navajo formations at 1640 to 1740 m in Emery (type from Calf Canyon) and Wayne counties; endemic; 5 (ii). For the past four decades the Maguire daisy was known officially from the type localtiy in the San Rafael Swell in Emery County. Now, other material has been discovered at BRY and relocated in the field, which is distinguishable only technically from specimens at the type locality. These latter specimens (from Capitol Reef, Wayne County) tend to have more heads per stem, have narrower ray corollas, and shorter disk corollas. All of these features may be the result of ecological responses, but the plants from Capitol Reef (type locality) are recognized as ***E. maguirei*** var. ***harrisonii*** Welsh.

Erigeron mancus Rydb. La Sal Daisy. [*E. pinnatisectus* (Gray) A. Nels. var. *insolens* Macbr. & Payson, type from the La Sal Mts.]. Pulvinate caespitose subscapose perennials from a usually branched caudex, the caudex clothed with dark brown to straw colored marcescent leaf bases; herbage more or less hirtellous and puberulent or minutely glandular; stems mainly 2–6 cm long, erect or ascending; basal leaves 1.2–4 cm long, 2–4 mm wide, pinnatifid, the lobes lanceolate, sometimes again lobed; cauline leaves much reduced; heads solitary; involucres 5–6.5 mm high, 7–12 mm wide, glandular, villous with multicellular hairs, the bracts subequal, somewhat thickened basally, the acuminate tips often purplish; ray flowers lacking; pappus simple or nearly so, of 20–30 bristles; achenes 2–nerved, hairy; n = 18. Alpine forb and grass-sedge communities at 3050 to 3660 m in the La Sal Mts. (type locality), astride the Grand - San Juan County line; endemic; 7 (0).

Erigeron melanocephalus (A. Nels.) A. Nels. Darkhead Daisy. [*E. uniflorus* var. *melanocephalus* A. Nels.]. Perennial herbs, from a simple or branched caudex, the caudex branches clothed with dark brown marcescent leaf bases; herbage more or less villous with multicellular hairs; stems commonly 5–12 cm tall, erect; basal leaves 0.8–6 cm long, oblanceolate to spatulate, rounded or retuse apically; cauline leaves much reduced upward; heads solitary; involucres 10–14 mm wide, 5–9 mm high, the bracts more or less densely villous with multicellular hairs, the cross-walls black or dark purple, equal, attenuate, green, with purplish tips or purplish throughout; rays 50–70, white or pink, 7–11 mm long, 1.2–2 mm wide; pappus single, of ca 20–25 bristles; achenes 2–nerved, sparsely hairy. Alpine meadows at 3355 to 3720 m in Grand and San Juan counties (La Sal Mts.); Wyoming, Colorado, and New Mexico; 4 (0). Specimens from the Uinta Mts., which have involucral hairs with purple cross-walls, have been assigned here previously, but they seem to represent nothing more than phases of *E. simplex* (q.v.).

Erigeron nanus Nutt. Dwarf Daisy. [*E. inamoenus* A. Nels.]. Perennial herbs, from a branching caudex, the caudex branches clothed with imbricate ashy to straw colored marcescent leaf bases; stems 3–8 cm high, villous with contorted multicellular hairs, subscapose; basal leaves linear-oblanceolate, 1.2–4 cm long, 1–2 mm wide, hirtellous to sparingly villous or glabrous, ciliate toward base with spreading long hairs, the bases conspicuously enlarged; heads solitary; involucres 7–13 mm wide, 5–8 mm high, long-villous with multicellular hairs and more or less finely glandular; bracts subequal, the midstripe brown to purplish, the margins green to scarious or purplish; rays 15–35, purplish, 5–10 mm long, 1.3–2.4 mm wide; pappus of 15–25 bristles and some outer setae; achenes 2–nerved, hirsute. Sagebrush and sagebrush-grass communities, often on windswept ridges, at 2135 to 3270 m in Box Elder and Daggett counties; Idaho and Wyoming; 5 (0).

Erigeron nauseosus (Jones) A. Nels. Marysvale Daisy. [*E. caespitosus* Nutt. var. *nauseosus* Jones, type from near Marysvale]. Perennial herbs, from a stout branching brittle caudex, the branches clothed with dark brown marcescent leaf bases, the taproot similarly colored; stems 6–25 cm tall, ascending to erect, glandular-scabrous; basal leaves 2.3–10 cm long, 2–15 mm wide, oblanceolate to spatulate, rounded apically, tapering to the petiole, commonly 3–nerved; cauline leaves well developed, only gradually reduced upward; heads solitary, rarely 2; involucres 8–17 mm wide, 5–8 mm high, finely glandular (rarely sparingly strigose as well); bracts imbricate, somewhat thickened, often purplish, attenuate; rays 30–60, white or purplish, 6–12 mm long, 1.3–2 mm wide; pappus double, the inner of 12–23 bristles, the outer of inconspicuous setae; achenes 2–nerved, hairy. Crevices in limestone, quartzite, and igneous outcrops, and in talus in pinyon-juniper, sagebrush, mountain brush, and Douglas fir-white fir communities at 1830 to 2900 m in Beaver, Garfield, Millard, Piute, and Sevier counties; White Pine County, Nevada; a Great Basin endemic; 24 (iii).

Erigeron nematophyllus Rydb. Needleleaf Daisy. Perennial herbs from a branching caudex, the caudex branches clothed with fibrous ashy to brown marcescent leaf bases; herbage strigose to subglabrous; stems 4–15 cm tall; basal leaves 1–8 cm long, 1–3 mm wide, linear to linear-oblanceolate, ciliate near the enlarged sheathing base; cauline leaves few and reduced, not especially exceeding the basal cluster; heads solitary; involucres 6–13 mm wide, 4–6.5 mm high; bracts more or less imbricate, moderately strigulose, green or brown, the inner often with scarious margins and purplish tips; rays 15–55, white (less commonly pink), 4–8 mm long, 1.2–2.3 mm wide; pappus of ca 15–25 bristles; achenes 2–nerved, shortly hairy. Sagebrush, mountain brush, and pinyon-juniper communities, often on Green River Shale, at 2280 to 2870 m in Carbon, Daggett, Duchesne, and Uintah counties; Wyoming and Colorado; 7 (i).

Erigeron peregrinus (Pursh) Greene Strange Daisy. [*E. callianthemus* Greene; *E. peregrinus* ssp. *callianthemus* (Greene) Cronq.; *E. regalis* Greene; *E. peregrinus* var. *eucallianthemus* Cronq.; *E. peregrinus* var. *scaposus* (T. & G.) Cronq.; *E. salsuginosus* var. *scaposus* T. & G.]. Perennial herbs, from a rhizome, the rhizome sometimes short, dark brown; stems 0.9–5.5 (7) dm tall, glabrous or sparingly to moderately villous below, often densely villous below the heads; basal leaves 2–16 (20) cm long, 0.8–3.2 (4.5) cm wide, oblanceolate to spatulate or obovate, tapering or abruptly contracted to the petiole, obtuse or rounded to acute apically, glabrous or rarely sparingly villous, ciliate; cauline leaves reduced upward, becoming sessile and more or less clasping; heads solitary, or 2–6; involucres 12–22 (25) mm wide, 6–9 (11) mm high; bracts subequal, reflexed at the attenuate apices, glandular and purplish throughout; rays ca 30–75, 8–17 (25) mm long, 1.8–4 mm wide, rose purple to white; pappus of ca 20–30 bristles, sometimes with a few outer setae; achenes 4– to 7–nerved, sparingly hairy; 2n = 18. Aspen, spruce-fir, lodgepole pine, and sedge communities at 2280 to 3570 m in Box Elder, Cache, Daggett, Duchesne, Garfield, Grand (?), Salt Lake, San Juan, Summit, Uintah, Wasatch, Weber, and Washington counties; Alaska south to California and New Mexico; 57 (x). Our materials were segregated by Cronquist (1947) into a dwarf alpine var. *scaposus* (T. & G.) Cronq. and a taller montane var. *eucallianthemus*. On the basis of the rather abundant materials at hand there does not seem to be any means of recognition of those taxa, except arbitrarily. Thus, all our specimens are herein considered as belonging to ssp. *callianthemus* (Greene) Cronq. var. *callianthemus*.

Erigeron proselyticus Nesom Professor Daisy. [*E. flagellaris* Gray var. *trilobatus* Maguire ex Cronq., type from Cedar Breaks]. Perennial herbs, from a subrhizomatous or substoloniferous caudex, the caudex branches with weakly persistent brown marcescent leaf bases; stems 14–25 cm tall, decumbent to ascending or erect, sparingly strigose; basal leaves 0.5–6.5 (7.5) cm long, 2–11 mm wide, oblanceolate to spatulate or linear, entire to pinnately few toothed or lobed, glabrous to sparingly strigose, acute to obtuse or rounded apically; cauline leaves gradually to abruptly reduced upward, entire or the lower few toothed; heads 3 to several; involucres 3.5–7 mm wide, 2.5–4.5 mm high, sparingly to moderately hirtellous; bracts subequal, brown, suffused with purple, or the inner greenish, with chartaceous margins; rays 22–46, white to purplish, 5.4–8.5 mm long, 1–1.4 mm wide; pappus double, the inner of 10–19 bristles, the outer of short setae; achenes 2– or 4–nerved, sparsely hairy; 2n = 18. Bristlecone pine, spruce-fir, and aspen communities on sandstone and marly limestone formations at 2440 to 3050 m in Iron and Kane counties; endemic; 10 (i).

Erigeron pulcherrimus Heller Basin Daisy. Perennial herbs, from a branching caudex, the caudex branches with exfoliating brownish bark, not especially clothed with persistent leaf bases; herbage silvery or grayish strigose; stems (5) 9–32 (35) cm tall, erect; basal leaves 0.8–7 cm long, 1–3 (5) mm wide, linear to linear-oblanceolate; cauline leaves reduced upward, but generally developed to stem middle or above; heads solitary; involucres 10–20 mm wide, 6–9 mm high, coarsely villous with spreading-ascending, multicellular hairs, obscurely glandular apically; bracts imbricate, greenish, the midrib often brown, the margins chartaceous, acuminate-attenuate, especially the inner; rays ca 25–60, white, pink, or violet, 8–15 mm long, 2–3.7 mm wide; pappus of ca 30–50 bristles, the outer series more or less developed; achenes (2) 3– to 5–nerved, densely hairy. Salt desert shrub and pinyon-juniper communities on saline and seleniferous clays, clay-silts, and gravelly pediments at 1310 to 2105 m in Carbon, Daggett, Duchesne, Emery, Grand, Uintah, and Wayne counties; Wyoming, Colorado, and New Mexico; 61 (xiv). Our materials have been treated as belonging to a wide-leaved (1.5–5 mm) var. *wyomingia* (Rydb.) Cronq. and a narrow-leaved (1–1.5 mm) var. *pulcherrima*. However, only arbitrary segregation appears to be possible, and it seems best not to attempt recognition of infraspecific taxa.

Erigeron pumilus Nutt. Vernal Daisy. [*E. pumilus* ssp. *concinnoides* var. *subglaber* Cronq., type from south of Monticello]. Perennial herbs, arising from a caudex, the branches clothed with ashy to brown marcescent leaf bases; herbage more or less hirsute with spreading hairs; stems 4–50 cm tall, leafy or subscapose; basal leaves 0.4–8 cm long, mostly 2–5 mm wide, linear-oblanceolate to oblanceolate; cauline leaves well-developed, somewhat reduced, or much reduced upward, or almost lacking; heads solitary or few to numerous; involucres 7–15 mm wide, 4–7 mm high, sparingly to densely spreading-villous with multicellular hairs; bracts subequal, acuminate to attenuate, green, with brownish midrib; rays mostly 50–100, white or pink to lavender, 6–15 mm long, 0.7–1.5 mm wide (or more); pappus double, the inner of 7–20 coarse bristles, the outer of evident bristles or scales; achenes 2–nerved, sparsely to moderately hairy; 2n = 18. Blackbrush, shadscale, sagebrush, pinyon-juniper, and mountain brush communities at 885 to 2960 m in all Utah counties; Washington to Saskatchewan, south to California, Arizona, New Mexico, and Kansas. Our highly variable material was segregated on technical characteristics by Cronquist (1947) into two subspecies, with some five varieties. The bulk of the Utah specimens belong to ssp. *concinnoides* Cronq., segregated in large measure from the much less common and more northern ssp. *intermedius* Cronq. by the fewer (7–15, not 13–20) inner pappus bristles and evidently puberulent (not glabrous or slightly puberulent) corolla tubes. The varieties *intermedius* (var. *euintermedius* Cronq.) and *gracilior* Cronq. of ssp. *intermedius* are only arbitrarily separable by stem thickness and head number. The weakly segregated varieties within ssp. *concinnoides*, **var. concinnoides** (var. *euconcinnoides* Cronq.), **var. subglaber Cronq.**, and **var. condensatus (D. C. Eaton) Cronq.**, differ in degree of development of cauline leaves and pubescence, with var. *concinnoides* having more equally leafy stems and var. *condensatus* tending to be subscapose. The var. *subglaber* has the stem only sparingly pubescent and the leaves sparsely to moderately pubescent. It seems best to treat our materal as belonging to two variable taxa; ssp. *intermedius* and ssp. *concinnoides*; 212 (xxvi).

Erigeron religiosus Cronq. Religious Daisy. Short-lived perennial (or biennial?) herbs from a slender taproot and poorly developed caudex; herbage more or less strigose and glandular below the heads; stems 6–35 cm tall, decumbent-ascending to erect; basal leaves 2–5.5 (7) cm long, 2–8 mm wide (or more), oblanceolate to spatulate, entire or some pinnately toothed or lobed; cauline leaves gradually reduced upward; heads 2 to numerous; involucres 5.5–7.5 mm wide, 2–3.5 mm high, sparingly to moderately villous and more or less glandular; bracts with brown midrib, somewhat thickened, scarious apically; rays 35–85, white or pinkish, 3.4–6.8 mm long, 0.5–1.4 mm wide; pappus double, the inner of 6–12 bristles, the outer of short setae; achenes 2-nerved, sparsely hairy. Ponderosa pine-oak and pinyon-juniper communities at 1525 to 1830 m in Kane and Washington (type from Clear Creek Canyon) counties; endemic; 20 (ix).

Erigeron simplex Greene Greene Daisy. Perennial herbs, from a simple or branched caudex, the caudex clothed with dark brown marcescent leaf bases; herbage more or less viscid-villous with multicellular hairs; stems commonly 2–15 (20) cm tall; basal leaves 0.8–6 (8) cm long, 2–10 (13) mm wide, oblanceolate to spatulate, obtuse to abruptly acute or mucronate apically; cauline leaves reduced; heads solitary; involucres 8–22 mm wide, 5–10 mm high, moderately to densely villous and somewhat viscid, the hairs with clear to reddish purple or purplish black cross-walls; bracts equal, suffused with purple or green, appressed or some reflexed; rays 50–125, blue purple to pink (or white), 7–11 mm long, 1.2–2.5 mm wide; pappus double, the inner of ca 10–15 barbellate bristles, the outer of conspicuous setae; achenes 2-nerved, sparsely hairy; 2n = 18. Lodgepole pine, Engelmann spruce, alpine fir, and alpine meadow and tundra communities at 3355 to 3660 m (in Deep Creek, Tushar, La Sal, and Uinta mts.) in Beaver, Daggett, Duchesne, Juab, Piute, San Juan, Summit, and Uintah counties; Oregon to Montana, south to Nevada, Arizona, and New Mexico; 30 (vi). Our variable materials include specimens with purplish black cross-walls of the multicellular hairs, especially on the involucres and below the heads. These have been placed with the similar and related *E. melanocephalus* (q.v.), but differ in shape of lower leaves and general aspect of the plants.

Erigeron sionis Cronq. Zion Daisy. Low perennial herbs, with short stoloniferous branches arising from a slender taproot; stems 1.5–13.5 cm long, decumbent to erect, glabrous or appressed pubescent; basal leaves 0.5–3.5 cm long, 2–10 mm wide, oblanceolate to obovate, entire or more commonly 3- to 5-lobed, glabrous or sparsely strigose; heads solitary or 2 to several; involucres 5–7 mm wide, 2–3 mm high, glandular and sparsely to moderately spreading-hairy; bracts suffused purplish or the inner green with chartaceous margins; rays 23–38, white, the midstripe below purplish, 3.9–6.1 mm long, 1–1.6 mm wide; pappus double, the inner of 7–13 bristles, the outer of slender setae; achenes 2-nerved, sparsely pubescent. Seeps and hanging gardens in ponderosa pine and riparian communities in Navajo and Wingate sandstones at 1350 to 1600 m in Zion National Park (type locality), Washington and Kane (?) counties; endemic; 9 (iii).

Erigeron speciosus (Lindl.) DC. Oregon Daisy. Rhizomatous perennial herbs with the caudex more or less developed; stems 1.5–9 cm tall, ascending to erect, spreading-hairy to subglabrous or glandular above; basal leaves often lacking at anthesis, the lowermost cauline ones oblanceolate to spatulate, petiolate, commonly 5–15 cm long, 4–20 mm wide; middle cauline leaves lanceolate to oval, oblanceolate, or elliptic, 2–11 cm long, 5–28 mm wide; upper leaves gradually to markedly reduced, lanceolate to obliquely ovate, ciliate, the surfaces glabrous, spreading-hairy, or glandular (or a combination); heads 1–15 (or more); involucres 11–22 mm wide, 5.5–9 mm high, glandular, with a few long hairs, or more or less spreading-hairy; bracts subequal, acuminate or attenuate, the tips more or less spreading, often suffused purplish; ray flowers ca 75–150, pink, pink purple or blue purple, or white, 7–18 mm long, 0.7–1 mm wide; pappus double, the inner of 20–30 bristles, the outer of more of less evident setae; achenes 2- to 4-nerved, hairy; 2n = 18, 36. The *speciosus* complex in Utah, as herein interpreted, consists of four variable, and more or less intergrading, largely sympatric infraspecific taxa. All have been treated previously at specific rank, or they have been treated within *E. speciosus*, in part. Cronquist (1947) discussed the problem of intermediacy in the complex but hesitated to combine the taxa because "such a treatment would distort the facts as well as being unwieldy." It is here contended that they are unwieldy apart; it seems therefore best to combine them as follows:

1. Leaves spreading-hairy on one or both surfaces; involucres spreading-hairy and more or less glandular .... *E. speciosus* var. *mollis*
— Leaves glabrous on both surfaces or minutely glandular, or with minute strigose hairs, rarely with a few spreading multicellular hairs 2
2(1). Leaves glandular on the surfaces (especially the upper ones), and also ciliate ...... *E. speciosus* var. *uintahensis*
— Leaves glabrous on both surfaces, ciliate 3
3(2). Involucral bracts merely glandular, rarely also somewhat spreading-hairy; upper leaves often ovate .... *E. speciosus* var. *macranthus*
— Involucral bracts glandular and commonly also spreading-hairy; upper leaves lance-attenuate .... *E. speciosus* var. *speciosus*

Var. macranthus (Nutt.) Cronq. [*E. grandiflorus* Nutt., not Hook.; *E. macranthus* Nutt.; *E. leiophyllus* Greene, type from Fort Douglas]. 2n = 18. Sagebrush, snowberry, aspen, spruce-fir, and alpine meadow communities at 1760 to 3420 m in Beaver, Box Elder, Cache, Carbon, Duchesne, Emery, Garfield, Grand, Juab, Kane, Millard, Piute, Rich, Salt Lake, Sanpete, Sevier, Tooele, Utah, Wasatch, Washington, and Weber counties; Washington and Alberta south to Nevada, Arizona, and New Mexico; 104 (xv). This is our most common phase, but it is only arbitrarily separable from var. *speciosus*, to which it is completely transitional.

Var. mollis (Gray) Welsh [*E. glabellus* var. *mollis* Gray]. Aspen, spruce-fir, and meadow communities at 2070 to 3050 m in Carbon, Duchesne, Garfield, Grand, Juab, Salt Lake, San Juan, Sanpete, Sevier, Uintah, Utah, and Wasatch counties; Montana to South Dakota, and south to New Mexico and Nebraska; 20 (i). This variety includes what has traditionally been called *E. subtrinervis* Rydb.

Var. speciosus [*Stenactis speciosa* Lindl.]. Mountain brush, sagebrush, ponderosa pine, aspen, spruce-fir, and alpine meadows at 2040 to 3300 m in Duchesne, Garfield, Grand, Iron, Juab, Piute, San Juan, Sanpete, Sevier, and

Utah counties; British Columbia and Montana, south to Nevada and New Mexico; 23 (iv).

Var. **uintahensis** (Cronq.) Welsh [*E. uintahensis* Cronq., type from Mill Creek, Summit County]. 2n = 36. Sagebrush, mountain brush, ponderosa pine, aspen, lodgepole pine, spruce-fir, and alpine meadow communities at 2070 to 3420 m in Beaver, Carbon, Daggett, Duchesne, Juab, Piute, Sanpete, Sevier, Summit, Uintah, Utah, and Wasatch counties; Wyoming; 39 (vi). The glandular condition of the leaves varies in amount and position, and the Uinta phase passes by degree into other taxa of the *speciosus* complex. Because of the intergradation it seems best that this most distinctive portion of the variation should be treated within an expanded *E. speciosus*.

***Erigeron superbus* Greene ex Rydb.** Splendid Daisy. Rhizomatous perennial herbs and with the caudex more or less developed, the perennating branches bearing brown marcescent leaf bases; herbage glabrous or glandular above and villous in some below the heads; stems 1–6 dm tall, erect; basal leaves smaller than the cauline and commonly present at anthesis, 3–15 cm long, 6–25 (33) mm wide, oblanceolate to obovate or spatulate, petiolate; middle cauline leaves somewhat smaller than the lower ones, oblong to elliptic or lanceolate, glandular (glabrous), the uppermost sessile and glandular, rarely some denticulate, not ciliate; heads 1–7; involucres 11–19 mm wide, 7–10 mm high; bracts subequal, glandular, sometimes with long spreading hairs near the base, acuminate, sometimes suffused purplish; rays 40–95, 1–2 mm wide, 12–20 mm long, rose purple or white; pappus double, the inner of 20–25 pinkish or tawny bristles, the outer of setae; achenes 2–nerved, hairy; 2n = 36. Aspen, Douglas fir, lodgepole pine, and spruce-fir communities at 2250 to 3050 m in Beaver, Carbon, Daggett, Duchesne, Garfield, Kane, Piute, San Juan, Summit, and Uintah counties; Wyoming south to Arizona and New Mexico; 18 (v).

***Erigeron tener* Gray** Thin Daisy. Perennial herbs, from a branching caudex, the slender branches with ashy to brownish marcescent leaf bases; herbage strigose; stems slender, decumbent, ascending, or erect, 3–15 cm tall; basal leaves 1–7.5 cm long, oblanceolate to elliptic, rhombic, or obovate, petiolate, acute to obtuse apically; cauline leaves much reduced; heads solitary or 2 or 3; involucres 6–10 (12) mm wide, 3.5–5 mm high, glandular and with spreading multicellular hairs; bracts imbricate, somewhat thickened, brownish, the inner membranous or somewhat scarious, sometimes suffused with purple; rays ca 15–40, purplish or white, 4–8 mm long, 1–1.7 mm wide; pappus double, the inner of 15–30 bristles, usually with slender outer setae; achenes 2–nerved, hairy to subglabrous. Sagebrush, mountain brush, pinyon-juniper, and white fir-Douglas fir communities, often on limestone outcrops at 1980 to 2900 m in Beaver, Juab, Millard, Rich, Sanpete, Summit, Tooele, and Utah counties; Oregon to Wyoming south to California and Nevada; 20 (i).

***Erigeron untermannii* Welsh & Goodrich** Untermann Daisy. Perennial pulvinate herbs with an intricately branched caudex, the caudex branches clothed with persistent leaf bases; leaves mainly basal, 0.8–3.3 cm long, 1–4 mm wide, narrowly oblanceolate to spatulate, pilosulose with ascending, often curved, hairs; scapes 2–6 cm tall; heads solitary; involucres 7–11 mm wide, 5–5.7 mm high, the bracts more or less imbricate, green, or the inner somewhat chartaceous, the margins hyaline, the tips suffused with purple (sometimes throughout), densely hispidulous with short spreading hairs; rays 14–26, white, 4–6.5 mm long, 1.5–2.1 mm wide; pappus apparently single, of ca 20 slender fragile bristles; achenes 2–nerved, pilose. Pinyon-juniper community on calcareous shales and sandstones of the Uinta and Green River formations at 2135 to 2380 m in Duchesne (type from Indian Canyon) and Uintah counties; endemic; 4 (0).

***Erigeron ursinus* D. C. Eaton** Bear Daisy. Perennial rhizomatous sod-forming herbs, the perennating organs arising from short superficial branches clothed with brown marcescent leaf bases; herbage subglabrous to strigose or variously ascending- or spreading-hairy; stems ascending, 5–25 (30) cm tall; basal leaves 1.2–12 cm long, 2–11 mm wide, oblanceolate to oblong, commonly acute or acutish apically, ciliate, the surfaces glabrous or variously hairy; cauline leaves reduced upward; heads solitary or 2 or 3; involucre 9–19 mm wide, 5–7 mm high, glandular and spreading-hairy with multicellular hairs; bracts subequal, green or suffused purplish at the usually reflexed tips; rays ca 30–100, pink or blue-purple, 6–15 mm long, 1–2 mm wide; pappus double, the inner of ca 10–20 bristles, the outer of setae or scales; achenes 2–nerved, hairy; n = 9. Sagebrush, aspen, lodgepole pine, and spruce-fir communities, often in forb-grass or forb-sedge meadows at 2440 to 3660 m in Beaver, Cache, Carbon, Daggett, Duchesne, Emery, Garfield, Grand, Iron, Juab, Kane, Piute, San Juan, Sanpete, Sevier, Summit (type from Bear River Canyon), Uintah, Utah, Wasatch, and Wayne counties; Idaho and Montana, south to Nevada and Arizona; 95 (x).

***Erigeron utahensis* Gray** Utah Daisy. Perennial herbs from a branching caudex, the branches with grayish marcescent leaf bases and usually densely clothed with white villous-pilose hairs; stems 10–60 cm tall, erect, appearing grayish or silvery due to strigose hairs; basal and lowermost cauline leaves 1.5–10 cm long, 1–6 mm wide, linear-oblanceolate, commonly withered or lacking at anthesis; cauline leaves gradually reduced upward; heads solitary or few to many; involucres 5–15 mm wide; 3–7 mm high, strigose and often glandular apically; bracts imbricate, brownish, the inner with scarious margins; rays ca 10–40, blue, pink, or white, 4–18 mm long, 1–2.7 mm wide; pappus double, the inner of ca 20–30 bristles, the outer of setae; achenes 4–nerved, more of less pilose; 2n = 18, 36. Two rather weakly separable varieties are present in Utah, as follows:

1. Stem bases not densely white-pilose; involucres mainly less than 8 mm wide; plants uncommon .... *E. utahensis* var. *sparsifolius*

— Stem bases densely white-pilose; involucres commonly more than 10 mm wide; plants common .... *E. utahensis* var. *utahensis*

Var. ***sparsifolius*** (Eastw.) Cronq. [*E. sparsifolius* Eastw. and *Wyomingia vivax* A. Nels, both types from San Juan County]. Sandstone outcrops in salt desert shrub and pinyon-juniper communities, often in shaded mesic areas, at 1220 to 1900 m in Emery, Garfield, Grand, Kane, and San Juan counties; Colorado and Arizona; 9 (iv).

Var. ***utahensis*** [*E. stenophyllus* var. *tetrapleuris* Gray, type from Kanab]. Creosote bush, blackbrush, warm desert shrub, pinyon-juniper, and Mountain brush communities at 900 to 2000 m in Emery, Garfield, Grand, Iron, Kane, San Juan, Washington, and Wayne counties; Colorado and Arizona; 75 (vii).

***Erigeron vagus* Payson** Payson Daisy. Caespitose perennial herbs, from a diffuse caudex, the branches commonly soboliferous; herbage moderately villous and glandular; leaves mainly basal, tufted at the apex of the caudex branches, 0.5–2.5 cm long, palmately 3-toothed or -lobed; heads solitary, subscapose; involucres 8–16 mm wide, 5–7.5 mm high, spreading-hairy and more or less glandular; bracts subequal, commonly suffused purplish at the attenuate apices; rays ca 25–35, white or pink, 4–7 mm long, 1–2 mm wide; pappus simple, of about 28 bristles; achenes 2-nerved, sparingly hairy. Ponderosa pine, western bristlecone pine, and sedge-forb communities at 2375 to 3660 m in Garfield, Grand (type from the La Sal Mts.), Iron, and San Juan counties; California east to Colorado; 18 (0).

***Erigeron wahwahensis* Welsh** Wah Wah Daisy. Perennial herbs, from a branched caudex, the caudex branches with conspicuous fibrous brown to ash colored marcescent leaf bases; stems 15–40 cm long, decumbent to ascending; basal leaves 3–18 cm long, 4–13 mm wide, linear-oblanceolate to oblanceolate or elliptic, 3-nerved, petiolate, appressed to spreading-hairy with curved hairs; cauline leaves reduced, sessile, and bracteate above; heads solitary or 2 or 3; involucres 13–17 mm wide, 6–7 mm high, spreading-villous with multicellular hairs, glandular apically; bracts imbricate, green, the tips reddish, thickened basally; rays 30–40, pink or white, 5.5–7 mm long, 1.7–2.2 mm wide; disk corollas 3.5–4.2 mm long, the tube ca 2 mm long, the lobes 0.4 mm long; pappus of 15–20 bristles, with inconspicuous outer setae; achenes 2-nerved, short-hairy. Sagebrush, oak-maple, and pinyon-juniper communities at 1670 to 2440 m in Beaver (type from Wah Wah Mts.) and Washington counties; endemic; 19 (iii). The Wah Wah daisy stands between the distributions of *E. jonesii* and *E. eatonii*, and it shares features of both. The specimens examined from Washington County have appressed strigose stems, and are highly variable. Those from the Wah Wah Mountains have spreading hairy stems. Additional work is indicated.

***Erigeron zothecinus* Welsh** Alcove Daisy. Perennial herb, with a taproot and well-developed caudex, the caudex branches clothed with brown to ashy leaf bases; stems erect or ascending a the base, 10–21 cm long, strigose to strigulose with stiff hairs, these ascending to spreading- ascending; basal leaves narrowly oblanceolate to linear, 1–4 mm wide, more or less sheathing basally; cauline leaves numerous, linear to linear-oblanceolate, 0.5–2.5 cm long, 1–2 mm wide; heads solitary or 2 or 3; involucre 4.5–5.2 mm high, 7–13 mm wide, the bracts slightly imbricated, somewhat thickened dorsally, green, strigose to strigulose or hirtellous with multicellular glandular hairs; rays ca 30–50, pink purple, 5–6 mm long, 0.8–1.2 mm wide; pappus double, the inner of 5–8 bristles, the outer of setae or scales; achenes 2-nerved, hairy. Saline seeps on vertical walls of alcoves at ca 1125 to 1160 m in eastern Kane County (type from along Lake Powell); endemic; 3 (iii).

Eriophyllum Lag.

Annual or perennial woolly herbs; leaves alternate, entire or toothed to lobed; heads solitary or corymbosely clustered, radiate; rays few, pistillate and fertile, yellow or white; involucres campanulate or hemispheric; bracts 1 (apparently 2) -seriate, firm, erect; receptacle flat to low-conic, naked; disk flowers perfect, fertile, the tube glandular or hairy; pappus of firm nerveless chaffy scales; style branches flattened; achenes 4-angled.

Constance, L. 1937. A systematic study of the genus *Eriophyllum*. Univ. California Publ. Bot. 18: 69–136.

1. Plants perennial .... *E. lanatum*
— Plants annual 2
2(1). Rays white; pappus of unequal scales .... *E. lanosum*
— Rays yellow; pappus of equal scales or reduced to a short crown .... *E. wallacei*

***Eriophyllum lanatum* (Pursh) Forbes** Pursh Woollyleaf. Perennial herbs, the herbage tomentose; stems erect or decumbent from a ligneus base, mainly 10–20 cm tall; leaves mainly 1–4 cm long, entire or 3– to 5-toothed or -lobed; heads solitary or corymbose on naked peduncles 3–10 cm long; involucres campanulate, 6–10 mm wide, 6–8 mm high; bracts 5–8 (10), carinate, distinct, the tips erect; rays 5–8 (10), yellow, 6–10 mm long, 2–5 mm wide; pappus of 8–10 variable scales; achenes 2.5–4 mm long, 4-angled, variously glabrous, hairy, or glandular; n = 8, 9, 10, 16, 24, 32. Sagebrush community (reported from Utah in the Pacific Northwest Flora); British Columbia to Montana, south to California, Nevada, Idaho, and Wyoming; 0 (0). Our material likely belongs to **var. *integrifolium* (Hook.) Smiley** [*Trichophyllum integrifolium* Hook.].

***Eriophyllum lanosum* (Gray) Gray** Gray Woollyleaf. [*Actinolepis lanosa* Gray]. Annual floccose-tomentose herbs; stems mainly 2–10 cm tall, simple and erect or branching from the base; leaves 0.5–1.8 cm long, 1–2 mm wide, linear to linear-oblanceolate, entire or essentially so; heads turbinate, solitary on naked peduncles 0.5–5 cm long; involucres 5–6.5 mm wide, 5–7 mm high; bracts 8–10, oblong, acute, distinct or nearly so; rays 5–10, white, 3–5 mm long, 2.5–3.5 mm wide; pappus of ca 5 slender hyaline awn-tipped scales; achenes 2.5–4.5 mm long, slender, sparsely strigulose; n = 4. Larrea, blackbrush, and Joshua tree communities at 700 to 900 m in Washington County; California, Nevada, and Arizona; 13 (i).

***Eriophyllum wallacei* (Gray) Gray** Wallace Woollyleaf. [*Bahia wallacei* Gray]. Annual tomentose herb; stems mainly 1–8 cm tall, simple or branched from the base; leaves 0.5–1.5 cm long, spatulate to obovate, entire or 3-lobed; heads solitary, turbinate-cylindric, on short peduncles; involucres 4–6 mm wide, 5–7 mm high; bracts 6–10, ovate, distinct; rays 5–10, yellow, 3–4 mm long, 2.5–3.5 mm wide; pappus of 6–10 scales or none; achenes ca 2 mm long, linear, hairy or glabrous; n = 5. Creoste bush, blackbrush, and Joshua tree communities at 700 to 1585 m in Washington County; California, Nevada, Arizona, and Mexico; 32 (iii).

Eupatorium L.

Perennial herbs; leaves alternate, opposite, or whorled, simple; heads discoid, the flowers all perfect and tubular; involucres cylindric to campanulate, the bracts striate, imbricate; receptacle naked, mainly flat; anthers obtuse and entire basally, or minutely sagittate; style branches with short stigmatic lines and an elongate papillate appendage; pappus of numerous capillary bristles; achenes 10-nerved.

1. Leaves alternate; plants of Box Elder and Tooele counties .... *E. occidentale*
— Leaves opposite or whorled; plants of various distribution 2
2(1). Leaves opposite; flowers white to cream .. *E. herbaceum*
— Leaves whorled; flowers purple or purplish .... *E. maculatum*

Eupatorium herbaceum (Gray) Greene White Thoroughwort. [*E. ageratifolium* var.? *herbaceum* Gray]. Perennial herbs from a woody caudex; stems 4–7 dm tall, branched above; herbage scabrous-puberulent; leaves mainly opposite, the blades 1.5–6 cm long, 0.5–4 cm wide, ovate, the bases cordate or truncate, coarsely crenate-serrate, acute; heads numerous, in dense corymbose clusters; involucres 3.5–5 mm wide, 3–4 mm high; bracts green, puberulent, subequal; corollas white; achenes black, 1.5–2 mm long; n = 17. Ponderosa pine and spruce-fir communities at 1585 to 2745 m in Grand, Kane, Piute, San Juan, and Washington counties; California Arizona, and Colorado; 5 (ii).

Eupatorium maculatum L. Joe-Pye Weed. Robust perennial herbs from short subrhizomatous caudices; stems mainly 6–15 dm tall, branching in the inflorescence; herbage puberulent and glandular-dotted; leaves in whorls of 3 or 4, mainly 6–25 cm long and 1.5–7 cm wide, lanceolate to lance-elliptic or lance-ovate, sharply serrate; heads numerous in corymbose clusters; involucres 3.5–5 mm wide, 6.5–9 mm high, the outer puberulent, the inner glabrous dorsally, often ciliate, purplish to straw colored; flowers purple; achenes ca 3 mm long, green to brown, glandular-dotted; 2n = 20. River and canal banks, wet meadows, bogs, and seeps at 1370 to 1865 m in Box Elder, Cache, Davis, Kane, Uintah, and Utah counties; British Columbia to Newfoundland, south to New Mexico, Illinois, and Michigan. Our material belongs to var. **bruneri** (Gray) Breitung [*E. bruneri* Gray]; 15 (i).

Eupatorium occidentale Hook. Western Eupatorium. Perennial herbs from a rhizome and with a branching caudex; stems 1.5–7 dm tall, often branched above; herbage scabrous-puberulent; leaves alternate, the blades mainly 1.5–6 cm long, 0.6–3 cm wide, deltoid or deltoid-ovate, serrate or subentire; heads numerous, in compact corymbose clusters; involucres 3–5 mm wide and as high; bracts subequal, puberulent, green or suffused with purple; flowers pink or purplish; achenes ca 3 mm long, brown, glandular-dotted. Rock crevices and talus (usually in quartzite) at 2135 to 2745 m in Box Elder and Tooele counties; Washington to Idaho, south to California and Nevada; 2 (0).

Filago L.

White-tomentose annual herbs; leaves entire, alternate; heads discoid, small, in capitate clusters; involucre reduced, the bracts resembling those of the receptacle; outer flowers pistillate, fertile, with tubular-filiform corolla, in several series, epappose and subtended by concave, partly enclosing bracts, the inner flowers bractless and with pappus of capillary bristles, the central flowers 2–5, apparently perfect, but often sterile, bractless and with capillary bristles; achenes subterete, nerveless.

Filago californica Nutt. Fluffweed. Annual herbs, the stems erect, simple or branched above, 0.5–3 dm tall; leaves 0.8–2 cm long, narrowly oblong to oblanceolate; heads ovoid, 3–4 mm high, subequal to involucrate leaves; bracts of outer pistillate flowers 8–10, tomentose, boat-shaped, the tips hyaline, the inner ones thinner and less hairy, the inner florets ca 12–20; inner achenes papillose; n = 14. Warm desert shrub at 915 to 1070 m in Washington County; Arizona and California; 1 (0).

Flaveria Juss.

Annual herbs; leaves opposite, sessile, more or less connate; heads several to numerous, in compact corymbose clusters; involucres cylindric; bracts carinate, striate, 2–5, subequal; receptacles naked; ray flowers pistillate, fertile, commonly 1 per head, yellowish, inconspicuous; disk flowers 2–5, perfect, fertile, yellowish; anthers not caudate at the base; pappus none; achenes 8- to 10-ribbed, glabrous.

Flaveria campestris J. R. Johnston Marshweed. Plants 12–85 cm tall, simple or branched, glabrous or hairy at the nodes; leaves 1–8 cm long, 0.4–1.5 cm wide, lance-oblong to linear, serrate to subentire, commonly 3-veined, glabrous; inflorescence leafy bracted; involucres 5–8 mm high, the longer inner bracts mostly 3, strongly keeled, glabrous; rays ca 1–2 mm long; achenes black, ca 3 mm long; n = 18. Sand bars, stream banks, and seeps at 1220 to 1680 m in Juab, Grand, San Juan, and Tooele counties; Colorado to Missouri, south to New Mexico and Texas; 8 (iii).

Gaillardia Foug.

Perennial (or biennial or annual) herbs; leaves alternate or mainly basal, entire or pinnatifid; heads radiate, the rays yellow, 3–lobed, neuter or sometimes pistillate and fertile; involucres 2- or 3-seriate, herbaceous, more or less spreading, reflexed in fruit; receptacle convex, with numerous setae; disk flowers perfect, fertile; anthers auricled at the base; pappus of 5–10 scarious, awned scales; achenes broadly obpyramidal, long-hairy.

1. Disk flowers purple or purplish 2
— Disk flowers yellow 3
2(1). Base of involucral bracts densely long-villous or the corolla lobes 5–11 mm long, or both; plants mainly montane in northeastern Utah .... *G. aristata*
— Base of involucral bracts not especially hairy, the corolla lobes mainly less than 5 mm long; plants of lower elevations in southeastern to southwestern Utah .... *G. pinnatifida*
3(1). Stems with well developed, pinnately dissected cauline leaves; plants of canyon bottoms of the Tavaputs Plateau .... *G. flava*
— Stems subscapose or, if the cauline leaves well developed, merely toothed or lobed, and plants mainly of other distribution 4
4(3). Pappus scales broadly oblong or oval, awnless or abruptly short awned; plants annual, of southern Utah .... *G. arizonica*
— Pappus scales oblong-lanceolate, awned; plants perennial, rarely flowering the first year 5
5(4). Leaves mainly basal, entire or rarely some of them toothed or lobed .... *G. parryi*
— Leaves cauline, toothed, lobed or entire .. *G. spathulata*

Gaillardia aristata Pursh Blanketflower. Perennial herbs from a slender taproot; stems 20–80 cm tall, commonly foliose to middle or above, less commonly with basal leaves only; leaves 1.5–16 cm long, 3–25 mm wide, oblong to oblanceolate or elliptic, entire or toothed to pinnatifid, puberulent and sparingly long-villous with multicellular hairs; heads solitary or few, long-peduncled; disk mainly 2–2.5 cm wide, purple; involucral bracts (and/or peduncle apex) commonly long-villous basally, green or suffused with purple, attenuate; rays 6–16, yellow, often purplish at the base, the lobes 5–12 mm long; setae of receptacle well developed; disk corollas densely woolly-villous, the hairs with reddish purple cross-walls, often obscuring the attenuate lobes; pappus of slender lance-attenuate scales, the caudate apex entire; achenes ca 1.5 mm long, rufous-pilose; n = 17, 34. Pinyon-juniper, ponderosa pine, aspen, lodgepole pine, and spruce-fir communities at 2135 to 2870 m in Daggett and Uintah counties; British Columbia to Saskatchewan, south to Oregon, Colorado, and South Dakota; 9 (ii). A specimen by Neese (5711 BRY) is only sparingly villous on the basal portion of the bracts. The species is known from cultivation in Utah and Emery counties; 3 (0).

Gaillardia arizonica Greene Arizona Blanketflower. Annual herbs from a slender taproot, mainly 15–40 cm tall; stems typically short, the leaves mostly near the plant base; leaves 1.9–10 cm long, 3–28 mm wide, oblanceolate to elliptic in outline, pinnatifid or less commonly subbipinnatifid, more or less glandular and sparingly to rather densely long-villous with multicellular hairs; heads solitary or few, long-peduncled; disk 1.5–2.5 cm wide, yellow; involucral bracts commonly long-villous near the base and on the margins, green or suffused with purple, attenuate; rays 5–16, yellow, the lobes 3–12 mm long; setae of receptacle well developed; disk corollas villous, the hairs with pale cross-walls, not obscuring the lobes; pappus of broadly blong or oval, awnless or awn-tipped, scales; achenes ca 1.5 mm long; n = 17. Warm desert shrub community at ca 900 m in Washington County; Nevada and Arizona; 1 (0).

Gaillardia flava Rydb. Yellowflower. Perennial herbs from a subrhizomatous woody caudex; stems 20–50 cm tall, foliose to the middle or above; leaves 2–5 cm long, 4–25 mm wide, pinnately incised, minutely puberulent and glandular-punctate; heads solitary, on peduncles to 25 cm long; disk 17–32 mm wide, yellow; involucral bracts sparingly to moderately villous, green, caudate-attenuate; rays 8–12, yellow, the lobes 3–5 mm long; setae of receptacle well developed, coarse and spinescent; disk corollas sparingly villous, the hairs with colorless cross-walls, the lobes acute; pappus scales oblong to oblanceolate, abruptly contracted to a barbellate appendage; achenes ca 1–1.5 mm long, yellowish pilose. Stream terraces and valley bottoms, commonly in cottonwood, willow, and tamarix communities at 1280 to 1650 m in Emery (type from Lower Crossing) and Grand counties; endemic; 6 (v). The plants are extremely resinous glandular, with a very bitter-flavored exudate.

Gaillardia parryi Greene Parry Blanketflower. [*G. acaulis* Gray, type from near St. George]. Perennial herbs from a woody caudex; stems 10–35 cm tall; foliose basally, less commonly with some leaves cauline; leaves 2.5–9 cm long, 8–25 mm wide, petiolate, the blades ovate to elliptic, sparingly puberulent, minutely glandular-punctate, entire or irregularly lobed, obtuse; heads solitary on scapose peduncles; disks 17–32 mm wide, yellow; involucral bracts sparingly villous, green, attenuate; rays ca 8–12, yellow, the lobes 3–5 mm long; setae of receptacle copious, surpassing achenes; disk corollas sparingly villous, hairs with translucent cross-walls, the lobes acutish; pappus scales lanceolate, rather abruptly contracted to a smooth bristle; achenes ca 1.5 mm long, yellowish pilose. Pinyon-juniper and ponderosa pine communities, often in disturbed sites, at 1525 to 1830 m in Garfield, Kane, and Washington counties; northern Arizona; 7 (i).

Gaillardia pinnatifida Torr. Hopi Blanketflower. [*G. mearnsii* Rydb.; *G. crassifolia* Nels. & Macbr., type from LaVerkin; *G. gracilis* A. Nels., type from Diamond Valley; and *G. straminea* A. Nels., type from LaVerkin]. Perennial (less commonly biennial or annual) herbs, the caudex seldom well developed; stems 8–55 cm tall, foliose to the middle, less commonly all leaves basal; leaves 1–7.5 cm long, 2–15 mm wide, petiolate; blades elliptic to oblanceolate or linear-oblong, puberulent and minutely glandular-punctate, pinnatifid to entire, acute to obtuse; heads solitary, on long peduncles; disks 15–35 mm wide, purple; involucral bracts moderately to sparingly villous, green or suffused purplish, caudate-attenuate; rays 7–12, yellow, the lobes 2–5 mm long; setae of receptacle spinescent; disk corollas sparingly villous, hairs with translucent or reddish cross-walls, the lobes acute; pappus scales oblanceolate, abruptly contracted to a scabrous awn; achenes ca 2 mm long, white-pilose. Blackbrush, shadscale, ephedra-Vanclevea, and pinyon-juniper communities at 915 to 1830 m in Carbon, Emery, Garfield, Grand, Kane, San Juan, Washington, and Wayne counties; Colorado and Arizona to Texas and Mexico; 75 (ix).

Gaillardia spathulata Gray Perennial herbs from a taproot and caudex; stems 6–35 cm tall, commonly foliose to middle or above; leaves 1–7.5 cm long, 0.4–2.3 cm wide, petiolate to sessile; blades oblanceolate to elliptic or ovate to oval, sparingly villous and glandular-punctate, entire or variously toothed or lobed, obtuse; heads solitary or few, on long peduncles; disks 18–33 mm wide, yellow; involucral bracts moderately to densely villous-pilose, green, lance-attenuate; rays 7–10, yellow, the lobes 2–4 mm long; setae of receptacle short, spinescent; disk corollas shortly villous on the obtuse lobes, the hairs with colorless cross-walls; pappus scales oblong-lanceolate, abruptly contracted to a scabrous awn; achenes ca 3.5 mm long, yellowish pilose. Salt desert shrub and shrub-grass communities at 1220 to 2320 m in Carbon, Emery, Garfield, Grand, and Wayne (type from Rabbit Valley) counties; endemic; 58 (xi).

Geraea T. & G.

Annual herbs; leaves alternate; heads radiate, showy, solitary or few in a corymbose panicle; involucres hemispheric, 2- or 3-seriate; bracts white-ciliate; receptacle convex, the bracts clasping the achenes; rays neuter, yellow; pappus of 2 awns, connected by a low whitish crown; disk achenes flat, cuneate, villous-ciliate, black.

Geraea canescens T. & G. Desert Sunflower. Annual herbs; stems 2–6 dm tall, simple or branched, white-hirsute, glandular; leaves 1–7 cm long, 0.8–4 cm wide, lanceolate to oblanceolate or ovate, acute to obtuse, entire or few-toothed, reduced upward; heads showy, borne on slender, often bracteate peduncles; involucres 10–25 mm wide, 7–12 mm high; bracts green, strongly ciliate, lance-acuminate; rays 10–21, yellow, 7–20 mm long; achenes 6–7 mm long; n = 18. Warm desert shrub communities at 700 to 900 m in Beaver Dam Wash, Washington County; Nevada, Arizona, and California; 6 (ii).

Glyptopleura D. C. Eaton

Low annual herbs with milky juice; leaves rosettiform, with a few-toothed, white crustaceous margin; heads many, short peduncled, the flowers all raylike, white or pale yellowish (or drying pinkish); involucres of 7–12 scarious-margined bracts subtended by a basal group of pinnatifid or toothed bractlets; pappus of capillary white bristles in several series, the outer falling separately; achenes oblong, 5-angled, each face with 2 rows of tubercles, abruptly beaked.

1. Ray flowers showy, long exserted, 1.5–2.5 cm long; plants of Washington County .... *G. setulosa*
— Ray flowers inconspicuous, only shortly exserted, mainly less than 10 mm long; plants broadly distributed .... *G. marginata*

Glyptopleura marginata D. C. Eaton Crustweed. Depressed annual herbs; stems 0.5–4 cm long; leaves crowded on the short stems, mainly 0.5–4 cm long, pinnatifid, the margins white-crustose, extended into irregular white processes; involucres 10–13 mm high, urceolate; bracts green, the margins hyaline; bractlets with white irregular branched processes, crustose at the apex; rays mainly 4–7 mm long, withered and pinkish on drying; achenes 4–5 mm long, tan, sculptured; n = 9. Desert shrub communities at 1240 to 1590 m in Beaver, Box Elder, Iron, Piute, San Juan, Sevier, and Uintah counties; Oregon and Nevada; 9 (i).

Glyptopleura setulosa Gray Low annual herbs; stems 1.5–6 cm long; leaves crowded on the short stems, mainly 0.3–5 cm long, pinnately lobed, the margins white-crustose into teeth; involucres 10–13 mm high, urceolate; bracts green or purplish tipped, the bracts with expanded apices bearing simple or coalescent processes; rays mainly 1.5–2.5 cm long, pale yellowish, showy; achenes 4–5 mm long, tan, sculptured. Larrea, blackbrush, and Joshua tree communities at 700 to 915 m in Washington (type from near St. George) County; Arizona, Nevada, and California; 6 (0).

Gnaphalium L.

Annual or perennial tomentose herbs; leaves alternate, entire; heads discoid, the flowers white, yellowish, or suffused with pink, borne in spikes, corymbs, or panicles; involucres campanulate to ovoid; bracts imbricate, scarious apically (at least); receptacle naked; outer flowers numerous, slender and pistillate, the few inner ones broader and perfect; style branches of inner flowers flattened, truncate, the stigmatic portion not sharply differentiated; anthers caudate; pappus of capillary bristles; achenes small, nerveless.

1. Heads small, the involucres 2–4 mm long; clusters of heads commonly surpassed or equaled by leafy bracts; plants mainly 4–20 cm tall 2
— Heads large, mostly 4–7 mm high; clusters of heads not or rarely surpassed by leafy bracts; plants often over 20 cm tall 4
2(1). Leaves spatulate to oblong, mainly 3–8 mm wide; plants loosely tomentose .... *G. palustre*
— Leaves linear to narrowly oblanceolate, mainly 1–3 mm wide; plants rather closely tomentose 3
3(2). Leafy bracts commonly less than 1.5 cm long, more loosely tomentose than the following ..... *G. exilifolium*
— Leafy bracts commonly more than 1.5 cm long, the tomentum appressed .... *G. uliginosum*
4(1). Herbage more or less glandular-hairy, at least the upper leaf surfaces .... *G. viscosum*
— Herbage more or less tomentose, not at all glandular ... 5
5(4). Pappus bristles ciliate near the base, basally united into small, easily fractured groups .... *G. luteo-album*
— Pappus bristles not ciliate, falling separately 6
6(5). Leaves strongly decurrent; bracts of involucre yellowish or fading yellowish; plants annual or biennial .. *G. chilense*
— Leaves not strongly decurrent or the plants perennial; bracts of involucre pearly white 7
7(6). Middle involucral bracts ca 3 times as long as broad, mostly acutish apically; leaves not strongly decurrent .... *G. wrightii*
— Middle involucral bracts ca twice as long as broad, mostly obtuse to rounded apically; leaves narrowly decurrent .... *G. microcephalum*

Gnaphalium chilense Sprengel Cottonbatting Cudweed. Annual or biennial herbs, the tomentose stems 15–40 cm tall or more; leaves 1.5–7 cm long, 2–8 mm wide, oblong to linear or the lowermost oblanceolate, decurrent, tomentose, reduced upward; heads numerous, in capitate clusters at stem apices; involucres 4–7 mm high, the bracts yellowish, tomentose only at the base; n = 14. Disturbed, often moist sites at 1370 to 1770 m in Daggett, Duchesne, Kane, Salt Lake, Utah, and Wasatch counties; British Columbia to Montana, south to California, Arizona, and Texas; 5 (0).

Gnaphalium exilifolium A. Nels. [*G. grayi* Nels. & Macbr.]. Annual herbs; stems 8–25 cm tall, simple or branching from the base, tomentum appressed or somewhat loose; leaves 0.4–4 cm long, 1–3 mm wide, linear to linear-oblanceolate; heads clustered, in capitate cymes or spicate, subtended by leafy bracts that surpass them; involucres ca 3 mm high; bracts with hyaline brownish tips, tomentose at the base; n = 7. Sedge-grass community, known in Utah from Piute, Wasatch, and Washington counties; Colorado, New Mexico, and Arizona; 3 (0). This plant simulates *G. uliginosum*, with which it has been synonymized by some workers. More material is necessary to provide a definitive solution as to its proper taxonomic position.

Gnaphalium luteo-album L. Annual herbs; stems mainly; 10–40 cm tall, simple or moderately branched, woolly; basal leaves oblanceolate, mainly 1–4.5 cm long and 4–10 mm wide; cauline leaves reduced upward; heads solitary or in cymose clusters; involucres 3.5–5 mm high; bracts hyaline, brownish stramineous, woolly only at the base, finally spreading; pappus bristlles mostly united at the base, falling in an easily fragmented ring. Weedy species known from ca 2135 m in Washington County; adventive from Europe; 1 (0).

Gnaphalium microcephalum Nutt. Perennial herbs; stems mainly 2–7 dm tall, usually several from a root crown; leaves mainly 3–10 cm long and 2–10 mm wide, broadly linear or oblanceolate, narrowly and often shortly decurrent at the base; heads typically numerous in cymose clusters, these commonly forming a broad, open inflorescence; involucre 4–7 mm high, the bracts numerous, imbricate, white to tan, woolly only at the base (if at all); pappus bristles numerous, falling separately. Open sites at ca 1830 m in Weber County; British Columbia to Montana, south to California and Colorado; 1 (0).

Gnaphalium palustre Nutt. Lowland Cudweed. Annual herbs; stems 3–20 (30) cm tall, simple or more commonly much branched, loosely tomentose; leaves 1–3.5 cm long, 2–6 (10) mm wide, oblong to oblanceolate; heads clustered in capitate terminal or axillary cymes, subtended by leafy bracts that equal or surpass them; involucres 3–4 mm high; bracts brown, usually with whitish tips, tomentose below; n = 7. Tamarix-willow, mountain brush, ponderosa pine, Douglas-fir, and sedge-grass communities, often on sand bars, lake shores, and pond margins, at 1370 to 2600 m in Box Elder, Cache, Garfield, Iron, Juab, Millard, Piute, Salt Lake, Sanpete, Sevier, Tooele, Uintah, Utah, Washington, Wayne, and Weber counties; British Columbia and Alberta, south to California and New Mexico; 51 (v).

Gnaphalium uliginosum L. Marsh Cudweed. Annual herbs; stems 3–15 (25) cm tall, simple or more commonly much branched, closely tomentose; leaves 1–5 cm long, 1–3 mm wide, linear to linear-oblanceolate; heads clustered in capitate terminal or axillary cymes, subtended by leafy bracts that much surpass them; involucres 3–4 mm high; bracts brown with pale tips, tomentose below; n = 7. Lake and pond margins and other disturbed sites at 1430 to 2830 m in Beaver, Daggett, Garfield, Iron, Sevier, Uintah, and Washington counties; widely distributed in North America; probably introduced from Europe; 8 (0).

Gnaphalium viscosum H.B.K. Viscid Cudweed. Annual or biennial herbs; stems 4–9 dm tall, simple or branched from the base, the stem conspicuously glandular-hairy, becoming tomentose upward; leaves numerous, oblanceolate to linear-oblong, evidently decurrent at the base, glandular-hairy on the upper surface, more or less woolly on the lower one, 4–10 cm long and 5–20 mm wide; inflorescence many-headed; involucre 4–7 mm high, the bracts numerous, yellowish to nearly white, woolly at the base, acute; pappus bristles distinct, falling separately. Mountain brush, juniper, ponderosa pine, and lodgepole pine communities at 2350 to 2715 m in Piute, Salt Lake, San Juan, Summit, and Uintah counties; British Columbia to Quebec, south to Mexico and Tennessee; 3 (0).

Gnaphalium wrightii Gray Wright Cudweed. Perennial herbs; stems 3–8 dm tall, branched in the inflorescence; leaves 1.5–7 cm long, lance-linear, the lower ones spatulate; panicle open, with capitate clusters of heads not subtended or surpassed by bracteate leaves; involucres 5–6 mm high; bracts pearly white, tomentose below. Ponderosa pine and live oak communities at 1585 to 1830 m in Garfield and Washington counties; California to Texas, south to Mexico; 10 (i).

Grindelia Willd.

Annual, biennial, or perennial herbs, sometimes woody at the base; leaves alternate, simple, more or less resinous-punctate, usually sessile, often clasping; heads radiate or discoid, the rays 10–45, pistillate, fertile, yellow; involucres imbricate, more or less resinous; bracts thickish, with pale appressed base and often squarrose or revolute herbaceous tips; receptacle naked, flattish; disk flowers fertile, yellow; style branches with slender hispidulous appendages; pappus of 2–8 stiff, often curved, deciduous awns; achenes compressed to angular, glabrous.

1. Heads discoid 2
— Heads radiate 3
2(1). Plants perennial; involucral bracts much thickened apically .... G. fastigiata
— Plants annual or biennial; involucral bracts only somewhat thickened .... G. aphanactis
3(1). Involucral bracts, at least middle and upper ones, with appressed or erect tips, these not revolute or thickened .... G. laciniata
— Involucral bracts spreading or recurved apically, often thickened apically 4
4(3). Rays mostly 12–25, rarely more; leaves entire or sharply toothed, not callous-serrulate; achenes usually with one or more knobs on the apical margin; plants perennial .... G. nana
— Rays mostly 25–40, rarely fewer; leaves regularly callous-serrulate to sharply toothed or entire; achenes mainly lacking apical knobs; plants biennial or perennial .... G. squarrosa

Grindelia aphanactis Rydb. Biennial herbs; stems 1.5–9 dm tall, uniformly leafy, glabrous; leaves mainly 2.5–7 cm long, 2–12 mm wide, oblong or oblanceolate, entire, crenulate-serrate or denticulate to pinnatifid, glabrous, the margin scabridulous; heads discoid, campanulate; involucres 7–20 mm high, 10–28 mm wide, resinous, mostly in 5 or 6 series, the upper portion loosely to moderately reflexed, glabrous; pappus awns 2 or 3; achenes 2.3–3 mm long, brown, mainly truncate apically; n = 12. Weedy species of disturbed sites in Kane and San Juan counties; Colorado, Arizona, and Texas; 1 (0).

Grindelia fastigiata Greene Erect Gumweed. Perennial herbs; stems 5–10 dm tall or taller, glabrous; leaves mainly 1.5–13 cm long, 10–18 mm wide, oblanceolate to lance-oblong, entire or denticulate to dentate or serrate, glabrous; heads discoid; involucres campanulate, 10–14 mm high, 9–17 mm broad; bracts in ca 6 series, only the upper third or fourth spreading, with revolute, thickened tips; pappus awns 2 or 3; achenes oblong, 3.5–5 mm long; n = 6. Sandy terraces and washes at 1125 to 1375 m in Emery, Grand, and San Juan counties; Colorado; a Plateau endemic; 12 (vi).

Grindelia laciniata Rydb. Perennial herbs; stems 2.5–4.5 dm tall, glabrous; leaves mainly 2–6 cm long, 3–1.5 mm broad, pinnatifid or the upper subentire or entire, narrowly oblanceolate to oblanceolate, glabrous; heads radiate; involucres 7–10 mm high and wide; bracts with upper 1/3–1/2 spreading, glabrous; pappus awns 3–5; achenes 2.5–3.5 mm long; n = 6. Sandy washes in San Juan County (type from San Juan County); Arizona; 1 (0).

Grindelia nana Nutt. Low Gumweed. [*G. brownii* Heller; *G. nana* f. *brownii* (Heller) Steyerm.]. Perennial herbs; stems 0.8–6.5 (8) dm tall, glabrous; leaves mainly 1.5–10 cm long, 5–30 mm wide, oblanceolate, scarcely clasping; heads radiate; involucres campanulate; bracts in 5–7 series, reflexed or revolute in the upper third to fifth; rays 11–28, yellow, 5–11 mm long; pappus awns 2; achenes 3.5–4 mm long; n = 12. Ruderal weed at ca 1585 to 1650 m in Cache County; Washington to Montana, south to California and Idaho; 4 (0).

Grindelia squarrosa (**Pursh**) **Dunal** Curly Gumweed. [*Donia squarrosa* Pursh]. Perennial or biennial herbs; stems 1–8 (10) dm tall, glabrous; leaves mostly 2–5 cm long, oblong, regularly callous toothed, sometimes sharply toothed or entire, the upper clasping; heads radiate, strongly resinous; bracts with the green tips strongly rolled back; rays 25–40, yellow, 7–15 mm long; pappus awns 2 or 3 (to 6); achenes 2.3–3 mm long; 2n = 12, 24. Two more or less distinctive varieties are present in Utah.

1. Main upper cauline leaves 2–4 times longer than broad, oblong-ovate to oblong ...... *G. squarrosa* var. *squarrosa*
— Main upper cauline leaves 5–8 times longer than broad, narrowly oblong to oblanceolate .... *G. squarrosa* var. *serrulata*

Var. *serrulata* (Rydb.) Steyerm. [*G. serrulata* Rydb.]. n = 6. Salt desert shrub, sagebrush, saline meadow, and mountain brush communities at 1310 to 1420 m in all Utah counties; Wyoming south to New Mexico and Arizona, and introduced widely elsewhere; 72 (vii).

Var. *squarrosa* [*G. serrulata* f. *depressa* Steyerm., type from west of Salt Lake City]. Waste places at 1300 to 2135 m in Duchesne, Juab, Salt Lake, Utah, and Wasatch counties; widespread mainly to the east of our area; 6 (0).

Gutierrezia Lag.

Perennial shrubs or subshrubs, glutinous, glabrous or hirtellous; leaves alternate, linear, often punctate; heads radiate, small, numerous; rays pistillate or neutral, yellow, or lacking; involucres cylindric to turbinate, the bracts imbricate, chartaceous; receptacles naked or bristly, convex; disk flowers few to many, yellow, perfect or sterile; pappus of 10–12 unequal scales; achenes obovoid or oblong, pubescent.

Lane, M. 1982. Generic limits of *Xanthocephalum*, *Gutierrezia*, *Amphiachris*, *Gymnosperma*, *Greenella*, and *Thurovia* (Compositae: Asteraceae). Systematic Botany 7: 405–417.
Solbrig, O. T. 1960. Cytotaxonomic and evolutionary studies in the North American species of *Gutierrezia* (Compositae). Contr. Gray Herb. 188: 1–63.

1. Heads cylindric, the ray and disk flowers 1 or 2 each .... *G. microcephala*
— Heads turbinate, with more than 4 flowers 2

2(1). Ray and disk flowers 3–8 each; involucres 2–3 mm thick; heads often clustered at ends of branches; plants widespread .... *G. sarothrae*
— Ray flowers 4–10, disk flowers 5–23; involucres 2–7 (9) mm thick; heads solitary or in pairs at branch ends; plants of restricted distribution 3

3(2). Disk flowers 5–12, 3.5–4.5 mm long; ray flowers 2–5 mm long; plants of Uintah County .... *G. pomariensis*
— Disk flowers 15–23, ca 3 mm long; ray flowers 5–7 (10) mm long; plants of eastern Millard County .... *G. petradoria*

Gutierrezia microcephala (**DC.**) **Gray** Thread Snakeweed. [*Brachyris microcephala* C.; *G. sarothrae* var. *microcephala* DC.) L. Benson; *Xanthocephalum microcephalum* (DC.) Shinn.]. Rounded shrub, 30–100 cm tall; stems slender, grayish to straw colored or green above, from a woody crown; leaves dimorphic, the cauline 2–5 cm long, 2–4 mm wide, linear or linear-lanceolate, and with shorter, narrower fasciculate axillary ones, often one or both lacking at anthesis; heads clustered at branch ends, sessile; involucre 3–4 mm long, 1–1.5 mm wide, cylindric; bracts fewer than 10, lanceolate, the tip greenish, slightly thickened; ray flowers 1 or 2, 3–4 mm long; disk flowers 1–3, 2–3 mm long; pappus of ca 8 scales; achenes of disk flowers abortive, those of ray flowers fertile, 2–3 mm long, hairy; n = 4, 8. Blackbrush, Vanclevea-ephedra, saltbush, purple sage, rabbitbrush, and pinyon-juniper communities at 850 to 1830 m in Emery, Garfield, Grand, Juab, Kane, Millard, San Juan, Utah, Washington, and Wayne counties; Nevada and California to Colorado, south to Texas and Mexico; 27 (viii).

Gutierrezia petradoria (**Welsh & Goodrich**) **Welsh** Goldenrod Snakeweed. [*Xanthocephalum petradoria* Welsh & Goodrich]. Perennial, suffrutescent; stems herbaceous except at the base, hirtellous, simple below the inflorescence, loosely caespitose, from a stout taproot and branching, mostly underground, woody caudex; leaves arranged singly along the stems, linear, 0.5–4.5 cm long, 1–3 (4) mm wide, reduced upward, secondary fascicled leaves in some lower axils; heads solitary or in pairs on bracteate peduncles, or some almost sessile; involucres 5–9 mm high, 3–7 mm wide (to 9 when pressed), campanulate, the bracts ca 20, in 3 (4) series, greenish, the tips thickened; ray flowers 4–10, 5–10 mm long, 1–4 mm wide, when fresh; disk flowers 15–23, ca 3 mm long; pappus scales ca 10–12; achenes 3–4 mm long, pubescent, abortive in disk flowers. Sagebrush, oakbrush, mountain mahogany, and white fir communuities at 1920 to 2590 m in eastern Millard County (Canyon and Pavant ranges); type from the Canyon Mountains; endemic; 9 (0).

Gutierrezia pomariensis (**Welsh**) **Welsh** Orchard Snakeweed. [*G. sarothrae* var. *pomariensis* Welsh; *Xanthocephalum sarothrae* var. *pomariensis* (Welsh) Welsh]. Orchard Snakeweed. Rounded subshrubs; stems 1.2–4.5 dm tall, several to many from a persistent woody base; leaves 1.5–5.2 cm long, 0.5–2.5 mm wide, linear, entire, glabrous or scabrous, glandular-punctate; heads in corymbose inflorescences, solitary or 2 or 3 clustered at stem ends; involucres 5–7.5 mm high, 2–5 mm broad, turbinate to cylindric; bracts broadly obtuse, with a greenish subapical spot, resin coated; ray flowers 5–9, the corollas 2–5 mm long; disk flowers 5–12, the corollas 3.5–4.5 mm long; pappus scales ca 5–8; achenes 1–2 mm long, hairy. Mixed desert shrub communities at 1460 to 2135 m in Duchesne and Uintah (type from Orchard Creek Draw, Dinosaur National Monument) counties; endemic; 17 (iv).

Gutierrezia sarothrae (**Pursh**) **Britt. & Rusby** Broom Snakeweed. [*Solidago sarothrae* Pursh; *Xanthocephalum sarothrae* (Pursh) Shinn.]. Rounded shrubs; stems 9–90 cm tall, profusely branched from the base, otherwise in the inflorescence, from a woody caudex and stout taproot; leaves dimorphic, the main cauline ones 2–7 cm long, 1–3 mm wide, linear to linear-lanceolate, the fascicled secondary ones in lower axils, entire, glabrous to tomentulose; heads in corymbose inflorescences, usually in clusters of 3–10 at branchlet ends, seldom solitary; involucres 3–4.5 mm high, 2–3.5 mm wide, turbinate; bracts narrow, acute, with green thickened tip; ray flowers 3–7, yellow, 2–5 mm long; disk flowers mostly 3–8, 2–3 mm long; pappus of 8–10 scales; achenes 1–2 mm long, hairy; n = 4. Warm desert shrub, sand sagebrush, live oak, sagebrush, rabbitbrush, mountain brush, and pinyon-juniper communities, often in disturbed sites, at 760 to 2440 m in probably all Utah counties; British

Columbia east to Saskatchewan and south to Mexico. Our variable material adjusts to disturbances and increases on grazed native rangelands; it is not considered to be palatable; 208 (xlv).

Haplopappus Cassini

Annual or perennial herbs, subshrubs, or shrubs, usually resinous or glandular; leaves alternate, entire or toothed to lobed; heads discoid or radiate, usually small to large, variously clustered or solitary; involucres cylindric to turbinate or campanulate, the bracts imbricate, not aligned; receptacle flat to convex, naked; rays yellow when present; disk flowers perfect, yellow; pappus of barbellate capillary bristles; achenes angled or striate to smooth.

Hall, H. M. 1928. The genus *Haplopappus*. Carnegie Institution of Washington. 391 pp.

1. Plants low, rounded, branched shrubs, or tall slender shrubs or subshrubs 2
— Plants annual or perennial herbs, branched or unbranched 11
2(1). Heads borne on stems 2.5–5 dm long; plants of saline sandy drainages or sandstone outcrops in southern Utah 3
— Heads borne on stems less than 2.5 dm long; plants of various substrates and distribution 4
3(2). Plants definitely shrubby, the mature branchlets ashy gray or white, of sandstone outcrops and canyons .. .. *H. scopulorum*
— Plants shrubby only at the base, the branchlets straw colored to greenish; plants of saline drainage bottoms and terraces .... *H. drummondii*
4(2). Stems of the season white-tomentose; involucres 10–13 mm long, the bracts only somewhat imbricate; plants commonly of high elevations .... *H. macronema*
— Stems glabrous, glandular, or hairy, not tomentose; involucres mainly less than 10 mm long, but, if longer, otherwise differing 5
5(4). Involucres campanulate, 8–12 mm long; heads showy, the rays 8–10 mm long; plants of lower elevations in Washington County .... *H. linearifolius*
— Involucres turbinate to cylindric or campanulate, commonly less than 8 mm long; heads not especially showy, the rays mainly 2–5 mm long, or lacking 6
6(5). Leaves densely glandular-punctate, linear; ray flowers present; plants known from Washington County .... *H. laricifolius*
— Leaves not glandular-punctate, narrowly to broadly oblanceolate or oblong; ray flowers lacking; plants more broadly or otherwise distributed 7
7(6). Heads 12–22 mm high, the bracts subequal, the outer herbaceous and the inner chartaceous and with broad hyaline margins; plants of limestone outcrops in the Paunsagunt and Markagunt plateaus, or mainly of igneous outcrops in the Pine Valley Mts. 8
— Heads 5.5–9.5 mm high, the bracts imbricate in several series, variously herbaceous or chartaceous, but seldom any with broad hyaline margins; plants of various substrates and distribution 9
8(7). Involucral bracts 1–nerved; achenes evenly though sparingly hairy; plants of the Pine Valley Mts., Washington County .... *H. crispus*
— Involucral bracts 3–nerved; achenes glabrous except for a few hairs apically; plants of the Paunsagunt and Markagunt plateaus .... *H. zionis*
9(7). Leaves densely stipitate-glandular, oblanceolate, acute, the margins not especially repand .. *H. watsonii*
— Leaves lacking stipitate glands, narrowly oblanceolate to oblong or, if oblanceolate, the margins repand-undulate 10
10(9). Leaves oblanceolate, 2–5 mm wide ...... *H. cervinus*
— Leaves narrowly oblanceolate to oblong, 0.5–2 mm wide .... *H. nanus*
11(1). Leaves strongly 3–nerved and veiny, thick and leathery; caudices thick, woody, branched 12
— Leaves not 3–nerved and veiny; caudices simple or, if branched, not woody 13
12(11). Bracts obtuse to rounded or less commonly acutish, strongly imbricate; plants of the Colorado drainage system .... *H. armerioides*
— Bracts acute to attenuate, subequal to strongly imbricate; plants widespread .... *H. acaulis*
13(11). Leaves with lobes or teeth spinulose tipped; involucral bracts spinulose tipped 14
— Leaves entire or toothed, but then not spinulose tipped; involucral bracts not spinulose tipped 15
14(13). Involucral bracts glabrous or glandular dorsally; leaves pinnatifid; plants perennial ...... *H. spinulosus*
— Involucral bracts strigose dorsally, also minutely ciliate; leaves lobed or merely toothed to entire; plants annual .... *H. gracilis*
15(13). Stems mainly 20–40 cm tall, loosely tomentose above; involucres 15–30 mm wide; plants evidently rare .... *H. croceus*
— Stems 5–20 cm tall or, if taller, not or seldom loosely tomentose, or the heads smaller 16
16(15). Heads racemosely or spicately arranged; stems erect or nearly so, not strongly bent at the base .... *H. racemosus*
— Heads solitary or corymbosely (rarely racemosely) arranged; stems strongly bent at the base 17
17(16). Involucres 12–15 mm high, 20–30 mm wide; plants not hairy in the leaf axils .... *H. clementis*
— Involucres 5–10 mm high, 10–20 mm wide or, if larger, the plants with hair tufts in basal leaf axils ... 18
18(16). Involucral bracts herbaceous throughout; achenes glabrous; plants rare, known only from the Tushar Mts. .... *H. apargoides*
— Involucral bracts herbaceous only apically; achenes hairy; plants locally common in saline meadows ... . *H. lanceolatus*

Haplopappus acaulis (**Nutt.**) **Gray** Stemless Goldenweed. [*Chrysopsis acaulis* Nutt.; *C. caespitosa* Nutt.]. Perennial caespitose herbs from a thick ligneous pluricipital caudex and stout taproot, the caudex branches clothed with brown to ashy marcescent leaf bases and leaves; herbage resinous, scabrous to glabrous; stems mainly 5–20 cm tall; basal leaves 0.3–6 cm long, 1.5–10 mm wide, rigid, narrowly to broadly oblanceolate, sharply mucronate, 1– to 3–nerved; cauline leaves few, developed or reduced upward; heads solitary (rarely 2); involucres hemispheric, 6–10 mm high, 8–20 mm wide; bracts in 3 series, more or less mucronate; rays 6–15, 8–10 mm long, 2–4 mm wide; pappus white to brownish; achenes silky-villous or glabrous; $2n = 36$. This is a variable taxon, with several morphological phases. Despite the tendency for some of the variations to be correlated geographically, it seems best to regard our materials as consisting of two mainly sympatric varieties.

1. Cauline leaves well developed, often the main foliage leaves; herbage merely resinous-glandular; plants of the Great Basin .... *H. acaulis* var. *glabratus*
— Cauline leaves usually much reduced, surpassed in size by the basal ones; herbage scaberulous or merely resinous-glandular; plants more widely distributed .... *H. acaulis* var. *acaulis*

Var. *acaulis* Sagebrush-grass, pinyon-juniper, mountain brush, ponderosa pine, western bristlecone, and spruce-fir communities at 1430 to 2685 m in Beaver, Box Elder, Cache, Daggett, Duchesne, Emery, Garfield, Juab, Kane, Millard, Sanpete, Sevier, Summit, Tooele, Uintah, and Utah counties; Oregon to Wyoming, south to California, Nevada, and Colorado; 60 (ix). There is a narrow-leaved glabrous phase of this taxon in the southern portion of Duchesne County, mainly on Green River Shale. Possibly it deserves recognition at some taxonomic rank.

Var. *glabratus* D. C. Eaton [*Stenotus falcatus* Rydb., type from Iron County; *H. falcatus* (Rydb.) Blake in Tidestr.; *S. latifolius* A. Nels., type from Utah County]. n = 9. Black sagebrush, wildrye, pinyon-juniper, mountain brush, and grass-shrub communities at 1525 to 2900 m in Beaver, Iron, Juab, Millard, Tooele, and Utah counties; Saskatchewan south and west to California and Nevada; 22 (iv).

Haplopappus apargoides Gray Perennial shortly caulescent herbs, 3–8 (15) cm tall, from a taproot and simple or branched caudex, this clothed with brown marcescent leaf bases; basal leaves mainly 2–6 cm long, 2–6 mm wide, lanceolate to narrowly oblanceolate; cauline leaves reduced upward, sessile, the margins scabrous or ciliate; herbage sparingly long-villous with multicellular hairs; heads solitary; involucres hemispheric, 8–12 mm high, 10–14 mm broad; bracts imbricate, lanceolate to oblong, acute, cuspidate, herbaceous almost or quite to the base, glabrous dorsally, the margins long-ciliate; ray flowers 15–40, yellow, 8–15 mm long; pappus tawny; achenes glabrous; n = 6. Alpine tundra community at 3355 m in Piute County (Tushar Mts.); California and Nevada; 1 (i). The specimen examined (Welsh and Thorne 12982 BRY) is tentatively assigned to this species, which is known otherwise only from the eastern Sierra Nevada and adjacent Nevada.

Haplopappus armerioides (Nutt.) Gray Thrifty Goldenweed. [*Stenotus armerioides* Nutt.]. Perennial caespitose herbs from a thick ligneous pluricipital caudex and stout taproot, the caudex branches clothed with brown to ashy marcescent leaf bases and leaves; herbage resinous-glandular, otherwise glabrous or with scabrous leaf margins; stems 0.5–20 cm tall; basal leaves 1.5–8 cm long, 1.5–10 mm wide, rigid, linear to oblanceolate, sharply mucronate, 1- to 3-nerved; cauline leaves few, reduced upward; heads solitary (rarely 2); involucres campanulate, 8–13 mm high, 10–18 mm wide; bracts in 3 or 4 series, imbricate, oblong to oval or obovate, obtuse, sometimes lobed below the apex, greenish near the apex, glabrous; rays 8–12, 10–12 mm long, yellow, 3–5 mm wide; pappus white; achenes silky-villous; n = 9. This distinctive species is represented in Utah by two phases, which are more or less morphologically distinctive and geographically correlated.

1. Stems mainly 3–8 cm tall; leaves linear to linear-oblanceolate, mainly 1–3 mm wide; plants of the Green River Formation, Duchesne and Uintah counties .... *H. armerioides* var. *gramineus*
— Stems usually over 8 cm tall; leaves oblanceolate, mainly 3–10 mm wide; plants widespread .... *H. armerioides* var. *armerioides*

Var. *armerioides* Blackbrush, black sagebrush, pigmy sagebrush, salt desert shrub, pinyon-juniper, mountain brush, and ponderosa pine communities at 1340 to 2120 m in Carbon, Daggett, Duchesne, Emery, Garfield, Grand, Kane, San Juan, Sevier, and Uintah counties; Montana to Arizona, east to New Mexico and Nebraska; 85 (xii).

Var. *gramineus* Welsh & F. G. Sm. Desert shrub and pinyon-juniper communities at ca 1585 to 1895 m in Duchesne and Uintah counties; endemic; 17 (0).

Haplopappus cervinus Wats. Antelope Goldenbush. Shrubs, 1–4 dm tall, much branched; branchlets grayish to straw colored; leaves 6–18 mm long, 2.2–6 mm wide, oblanceolate, straight or curved, entire or repand-undulate, attenuate basally, cuspidate apically, glabrous or resinous; heads few, cymose; peduncles 3–10 mm long; involucres 6.5–7.5 mm high, 5–8 mm wide; bracts imbricate in several series, the outer greenish ones narrowly acuminate with straight or spreading tips, the inner chartaceous ones narrowly oblong, acute or cuspidate, all glabrous but resinous; ray flowers 5–7, yellow, 2.5–4 mm long, ca 1 mm wide; disk flowers 5–11, glabrous or tube sparingly puberulent; pappus tawny; achenes strigose; n = 9. Black sagebrush, shadscale, pinyon-juniper, and mountain brush communities at 1670 to 2440 m in Millard and Sevier counties; Arizona; 5 (0). The type came from a place called Antelope Canyon (possibly in present-day western Millard County, Welsh 1982). More collections are needed.

Haplopappus clementis (Rydb.) Blake Clement Goldenweed. [*Pyrrocoma clementis* Rydb.; *P. subcaesia* Greene, type from Panguitch Lake; *P. lapathifolia* Greene, type from "Utah"; *P. chieranthifolia* Greene, type from W of Ephriam]. Perennial herbs from a simple caudex and stout taproot, the subrhizomatous caudex clothed with brown, often shredded marcescent leaf bases; stems 10–30 (40) cm tall, decumbent-ascending from an abruptly curved base, villous; basal leaves mostly 2–15 cm long, 4–17 mm wide, oblanceolate, glabrous or sparingly puberulent, entire or dentate, tapering to a petiole, acute; cauline leaves reduced upward, sessile and somewhat clasping; heads solitary (rarely 2 or 3); involucres broadly hemispheric, 8–16 mm high, 18–30 mm wide; bracts in several series, oblong to lanceolate, green throughout or the base chartaceous, villous; ray flowers 30–60, yellow or golden, 8–14 mm long; pappus tawny; achenes hairy; n = 18. Grass-sagebrush, spruce-fir, sedge-forb, and meadow communities at 2590 to 3390 m in Beaver, Daggett, Duchesne, Emery, Garfield, Iron, Kane, Piute, Sanpete, Summit, and Uintah counties; Wyoming to Colorado; 39 (vi). Specimens from Utah that have been determined as *H. integrifolius* Gray apparently fall here, including the type of *Pyrrocoma lapathifolia*, which was discussed by Hall (1920). Involucral bracts vary from herbaceous throughout to chartaceous at the base. More work is indicated. A report of *H. uniflorus* (Hook.) T. & G. (Hall 1920) from the Uinta Mts. probably belongs here.

Haplopappus crispus L. C. Anderson Pine Valley Goldenbush. Shrubs, much branched from the base, 3–5 dm tall (or more); branchlets covered with short-stalked glands; leaves 1.5–3 cm long, 3–8 mm wide, entire, green, spatulate to oblong-oblanceolate, acuminate, the margins undulate-crisped, glutinous with low glands, not crowded below the inflorescence; heads 1 or 2, or more commonly more per branch, loosely paniculate to congested and cymose; involucres campanulate, the heads 12.5–16 mm long, 5–9 mm wide; bracts in several series, finely glandular, the outermost green, leaflike; rays lacking; disk flowers 14–24, pale yellow; pappus tawny; achenes 6.5–8.5 mm long, sparsely but evenly hairy. Ponderosa pine, fir, manzanita, and aspen communities at (915?) 2471 to 3050 m in Washington (type from Pine Valley Mts.) and Millard (?) counties; endemic; 9 (0).

Haplopappus croceus Rydb. Perennial herbs, mainly 2–6 cm tall, from a simple caudex and stout taproot, the caudex clothed with fibrous marcescent leaf bases; basal leaves 8–20 cm long, 6–25 (40) mm wide, elliptic to oblanceolate, petiolate, entire or undulate, obtuse to acutish, glabrous or puberulent; cauline leaves reduced upward, sessile, more or less clasping; heads solitary (rarely more); involucres hemispheric, 12–18 mm high, 20–30 mm wide; bracts in several series, ovate to oblong or oblanceolate, herbaceous apically, chartaceous to leathery basally; ray flowers 25–70, burnt orange, 10–25 mm long; pappus brownish; achenes glabrous or pilose; n = 6, 12. Mountain brush community at ca 2470 m in San Juan (La Sal Mts.) and Washington (Kolob Reservoir) counties; Wyoming south to Arizona and New Mexico; 3 (0).

Haplopappus drummondii (T. & G.) Blake [*Linosyris drummondii* T. & G.]. Perennial subshrub, the stems subherbaceous, arising from a woody base, 25–75 cm tall, straw colored to tan, longitudinally striate, glabrous; leaves 1.5–7.5 cm long, 1–16 mm wide, entire or irregularly lobed, linear to spatulate, glabrous, resinous; heads few to numerous, borne in corymbose cymes, peduncled; involucres turbinate, 6–8 mm high, 4–7.2 mm wide; bracts in 4 or 5 series, lance-oblong, coriaceous, with a thick green or brownish subapical spot, acute, resinous; ray flowers lacking; pappus tawny; achenes silky. Saline riparian areas in greasewood, saltgrass, rabbitbrush, saltbush, and tamarix communities at 1050 to 1800 m in Emery, Garfield, Grand, Kane, and San Juan counties; Colorado, Arizona, New Mexico, and Texas; 25 (vii).

Haplopappus gracilis (**Nutt.**) Gray Slight Goldenweed. [*Dieteria gracilis* Nutt.]. Annual herbs, 3–25 (30) cm tall, commonly branched from near the base; leaves 4–25 mm long, 1–3 mm wide, linear to narrowly spatulate, spinulose-dentate to pinnatifid, white-strigose, progressively reduced and entire upward; heads solitary or few to several and corymbosely arranged; involucres 6–8.5 mm high, 8–12 mm wide; bracts in 5 or 6 series, linear-lanceolate, awn tipped, herbaceous medially, strigulose, not glandular; rays 15–30, yellow, 6–9 mm long; strigulose, not glandular; pappus tawny to white; achenes pilose; n = 2, 4. Creosote bush-matchweed, ponderosa pine, and spruce-fir communities at 850 to 960 m in Iron, Kane, and Washington counties; California to Colorado, south to Mexico; 10 (0).

Haplopappus lanceolatus (**Hook.**) T. & G. Meadow Goldenweed. [*Donia lanceolata* Hook.; *H. tenuicaulis* D. C. Eaton; *H. lanceolatus* var. *tenuicaulis* (D. C. Eaton) Gray; *Pyrrocoma subviscosa* Greene; *H. lanceolatus* ssp. *subviscosus* (Greene) Hall; *Donia uniflora* Hook.; *H. uniflorus* (Hook.) T. & G.]. Perennial herbs from a simple caudex and stout taproot, the caudex clothed with brown to ashy marcescent, often fibrous, leaf bases; stems decumbent-ascending, abruptly bent at the base, 5–68 cm long; basal leaves 3–16 cm long, 3–35 mm wide, elliptic-oblong or lanceolate, glabrous or tomentose, petiolate, entire or dentate to lobed, often densely tomentose in the axils; cauline leaves reduced upward, finally sessile and clasping; heads solitary or few to several, and subcorymbose or less commonly racemose; involucres hemispheric, 5–12 mm high, 10–18 mm wide; bracts imbricate in 3 or 4 series, with green tips, glabrous or tomentulose; ray flowers 10–45, yellow, 5–10 mm long; pappus tawny; achenes densely hairy; n = 12, 18. Saline meadows at 1300 to 2500 m in Beaver, Cache, Carbon, Duchesne, Emery, Garfield, Iron, Juab, Millard, Piute, Rich, Salt Lake, Sevier, Summit, Tooele, and Utah counties; Oregon to Saskatchewan, south to California, Nevada, Colorado, and Nebraska; 42 (xi). This is a highly variable taxon of saline meadows through much of our area. Heads vary from solitary to numerous, from solitary to corymbosely or racemosely arranged. Vesture is lacking or tomentose, or rarely glandular. Recognition of taxonomic categories within the variation appears to be only arbitrarily possible, and it seems best to treat our specimens conservatively.

Haplopappus laricifolius Gray Larchleaf Goldenbush. Rounded shrubs 3–8 dm tall; branchlets resinous, yellowish, becoming gray in age; leaves 5–18 mm long, 1–1.5 mm wide, thick, linear, resinous-punctate; heads few to several in compact cymes, shortly pedunculate; involucres campanulate, 3–5 mm high, 3–6 mm wide; bracts imbricate in ca 3 series, narrowly oblong, acute, yellowish or hyaline, glabrous or puberulent-ciliate; ray flowers 3–6, yellow, 4–5 mm long; disk flowers 9–16, glabrous or minutely pubescent; pappus tawny; achenes white hairy; n = 9. Saltgrass seep margin in warm desert shrub at 1220 m in Washington County; Arizona to Texas and Mexico; 6 (v).

Haplopappus linearifolius Gray Mohave Goldenbush. Shrubs, mainly 4–10 (12) dm tall; branchlets yellowish, resinous, becoming gray in age; leaves 6–28 mm long, 1–2.5 mm wide, thickish, linear to narrowly oblanceolate, resinous-punctate; heads few to many, solitary on naked peduncles mainly 2–7 cm long; involucres hemispheric, 8–10 mm high, 10–18 mm wide; bracts biseriate, lance-linear, acute or acuminate, herbaceous medially, sometimes minutely glandular; rays 12–18, yellow, 9–15 mm long, 4–5 mm wide; disk flowers numerous; pappus white; achenes densely hairy. Joshua tree, creosote bush, blackbrush, juniper, live oak, and sagebrush communities at 700 to 1375 m in Washington County; California, Nevada, Arizona, and Baja California; 30 (ii). Our material is assignable to **var. interior** (**Cov.**) Jones [*H. interior* Cov.; *H. linearifolius* ssp. *interior* (Cov.) Hall].

Haplopappus macronema Gray Cobwebby Goldenbush. [*Macronema discoideum* Nutt.]. Shrubs, mainly 1–5 dm tall; branchlets white-tomentose; leaves 8–32 mm long, 2–7 mm wide, oblanceolate to oblong, entire or more commonly undulate-crisped, acute to obtuse, mucronate, glandular-scabrous; heads solitary or 2 to several; involucres campanulate, 9–13 mm high, 6–12 mm wide; bracts subequal, the outer few herbaceous, oblong, the inner lance-acuminate, chartaceous, glandular-scabrous; ray flowers lacking; disk flowers 10–25; pappus tawny; achenes villous. Douglas fir, lodgepole pine, spruce-fir, and alpine tundra communities at 2135 to 3420 m in Beaver, Box Elder, Duchesne, Garfield, Iron, Juab, Piute, Salt Lake, Sanpete, Sevier, Tooele, Salt Lake, and Utah counties; Oregon to Wyoming, south to California, Nevada, and Colorado; 38 (vi).

Haplopappus nanus (Nutt.) **D. C. Eaton** Low Goldenbush. [*Ericameria nana* Nutt.]. Compact shrubs, mainly 1–3 (5) dm tall; branchlets yellowish, resinous, becoming gray in age; leaves 3–18 mm long, 0.5–2 mm broad, narrowly oblanceolate to linear, entire, acute, resinous but not punctate; heads solitary or few to several in compact cymes, sessile or shortly pedunculate; involucres narrowly turbinate, 5.5–8.5 mm high, 3–7 mm wide; bracts imbricate in 4 or 5 series, the outer often greenish medially, the inner chartaceous, with hyaline margins, glabrous; rays 1–7, yellow, 2–3 mm long; disk flowers 4–10; pappus tawny; achenes villous or glabrous; n = 9. Desert shrub, shrub-grass, and juniper or pinyon-juniper communities at 1310 to 2820 m in Beaver, Juab, Millard, Piute, Sevier, Tooele, and Washington counties; Oregon, California, Nevada, and Idaho; 18 (iii).

Haplopappus racemosus (Nutt.) **Torr.** Slender Goldenweed. [*Homopappus racemosus* Nutt.]. Perennial herbs, from a simple caudex and stout taproot, the caudex clothed with fibrous marcescent leaf bases; stems 20–60 (100) cm tall, erect, not abruptly bent at the base (in ours); basal leaves mainly 6–25 cm long, 5–30 mm wide, the blades elliptic to oblong or oblanceolate, petiolate, rigidly erect, entire or toothed, glabrous or puberulent; cauline leaves reduced, sessile, clasping; heads racemose, in panicles or spikes, shortly pedunclulate; involucres 8–12 mm high, 4–18 mm wide; bracts in 3 or 4 series, with green tips and coriaceous bases, abruptly pointed apically; rays 10–35, yellow, 5–12 mm long; pappus tawny; achenes hairy or glabrous. Saline meadows at 1370 to 1470 m in Millard and Utah counties; Oregon to Idaho, south to California and Nevada; 2 (i). Utah lies at the eastern margin of the range of this species complex, in which Hall (1928) recognized nine subspecies. Our material is hardly representative of the variation within the assemblage of forms which lie to the west of this region. One of our specimens (Welsh et al. 14514 BRY) belongs to the spiciform narrow-headed **var. *sessiliflorus* (Greene) Welsh** [*Pyrrocoma sessiliflora* Greene], and the other is a paniculiform large headed phase apparently nearest to **var. *prionophyllus* (Greene) Welsh** [*P. prionophylla* Greene]. Much more material is required to evaluate the nature of the specimens in Utah. Racemose phases of the closely related *H. lanceolatus* (q.v.) have been mistaken for *H. racemosus*. The erect or suberect stems and stiffly erect leaves appear to be diagnostic for our specimens of *H. racemosus*.

Haplopappus scopulorum (Jones) **Blake** Spindly Goldenbush. [*Bigelovia menziesii* var. *scopulorum* Jones, type from Zion Canyon; *H. scopulorum* var. *hirtellus* Blake, type from Cedar Canyon]. Shrubs, mainly 3–10 dm tall; branchlets green to straw colored or white, glabrous; leaves 0.7–7.8 cm long, 1–8 mm wide, narrowly lanceolate to oblong, entire, 3–nerved, glabrous, the margins scabrous, attenuate to a spinulose apex; heads few to many, borne in loose to subcompact cymes, peduncled; involucre narrowly campanulate, 6.5–9.5 mm high, 3–5.5 mm wide; bracts in 5 or 6 series, oblong, chartaceous and pale, or the tips greenish or often brownish, rounded-obtuse, glabrous, not resin coated; ray flowers lacking; disk flowers 10–20; pappus white; achenes whitepilose. Pinyon-juniper, mountain brush, and ponderosa pine communities at 1370 to 1830 m in Iron, Kane, San Juan, and Washington counties; Arizona; 10 (vii).

Haplopappus spinulosus (Pursh) **DC.** Spiny Goldenweed. [*Amellus spinulosus* Pursh]. Perennial herbs from a ligneus caudex; stems mainly 12–50 (60) cm tall, branching above the base; leaves 0.5–6 cm long, 1–10 mm wide, pinnatifid to bipinnatifid or the upper ones entire, or merely toothed, spinulose; heads solitary, or few in corymbose clusters; involucres 5–8 mm high, 8–12 mm wide; bracts in 4–6 series, linear, awn-tipped, herbaceous medially, glandular, not strigulose; rays 15–50, yellow, 8–10 mm long; pappus brownish; achenes pilose; n = 4, 5, 8. Desert shrub community at ca 1300 m in San Juan and Washington counties; Alberta to Minnesota, south to California, Arizona, New Mexico, Texas, and Mexico; 4 (ii).

Haplopappus watsonii **Gray** Watson Goldenbush. Shrubs, 1–4 dm tall; herbage stipitate-glandular; branchlets yellowish, becoming whitish to straw colored or grayish in age; leaves 4–28 mm long, 3–10 mm wide, oblanceolate to obovate or spatulate, entire or undulate, abruptly cuspidate-acuminnate apically; heads several to numerous (rarely some solitary) in loose cymes, the peduncles 1–7 mm long; involucres subcylindric to narrowly campanulate, 5.5–8 mm high, 3–7.6 mm wide; bracts in ca 5 series, the outer ones greenish, the inner chartaceous or greenish at the tips; rays 5–10, yellow, 4–6 mm long; disk flowers 5–15; pappus brownish; achenes hairy; n = 9. Rock outcrops (limestone, sandstone, or quartzite) in desert shrub, pinyon-juniper, mountain brush, and ponderosa pine communities at 1310 to 3440 m in Beaver, Box Elder, Cache, Davis, Juab, Millard, Salt Lake, Summit, Tooele, and Weber counties; Nevada and Utah; 33 (ii). Our material belongs to one of a vicarious pair of infraspecific taxa within the Great Basin known as **var. *rydbergii*** (Blake) Welsh [*H. rydbergii* Blake; *Macronema obovatum* Rydb., type from City Creek Canyon]. The var. *rydbergii* differs in having fewer disk flowers (5–15 not 15–25). Other supposedly diagnostic features (i.e., the green outer involucral bracts) fail, being present to a greater or lesser degree in both phases. The type variety has not been discovered in Utah, but should be expected in the western border region.

Haplopappus zionis **L. C. Anderson** Cedar Breaks Goldenbush. Shrubs, mainly 1–3 dm tall; herbage minutely and shortly stipitate-glandular; leaves 0.8–3.5 (4) cm long, 2–4.5 (7) mm wide, oblong to narrowly oblanceolate, 1–nerved, entire, abruptly mucronate; heads solitary or 2 or 3, in cymose clusters, peduncled; involucres cylindric-campanulate, 12–15 mm high, 6–12 mm wide; bracts subequal, herbaceous (outer) and greenish, the inner chartaceous or with a subapical green spot and broadly hyaline margins; rays lacking; disk flowers 8–21; pappus tawny; achenes glabrous below, strigose apically. Ponderosa pine and spruce-fir communities, commonly on limestone members of the Cedar Breaks (Wasatch) Formation, at 2440 to 3050 m in Garfield, Iron (type from Cedar Canyon), and Kane counties; endemic; 9 (i).

Helenium L.

Annual or perennial herbs; leaves alternate, glandular-punctate, decurrent or clasping; heads solitary or few to numerous in corymbose clusters, radiate, yellow; involucral bracts in 2 or 3 series, the bracts subequal or the inner shorter and narrower, herbaceous or essentially so, soon deflexed; receptacle naked, convex or conic; rays pistillate or neuter; disk flowers numerous, perfect; pappus of 5–10 scarious or hyaline scales; achenes truncately obpyramidal, 4- or 5-angled, with as many intermediate ribs.

1. Leaves sessile, clasping; stems not winged; plants of aspen communities and upward .... *H. hoopesii*
— Leaves decurrent; stems winged below the leaf bases; plants of riparian communities at lower elevations .... *H. autumnale*

Helenium autumnale L. Common Sneezeweed. Perennial herbs; stems mainly 1.5–10 (12) dm tall, puberulent and glandular, corymbosely branched above; leaves 1.5–15 cm long, 3–35 (40) mm wide, serrate to entire, glandular-punctate; heads 3 to many, the disk hemispheric to subglobose, yellow, 1–2 cm wide; rays 10–20, yellow, mainly 8–12 mm long, soon reflexed; pappus scales lance-ovate, with slender awn-tip as long as the body; achenes ca 1.5 mm long, hirsute and glandular; n = 16, 17, 18. Cattail-willow, tamarix-greasewood, sedgerush, and other riparian and palustrine communities at 1220 to 1830 m in Box Elder, Daggett, Emery, Rich, Uintah, and Utah counties; British Columbia to Quebec, south to Arizona, and Florida; 18 (iii). This species is poisonous to livestock.

Helenium hoopesii Gray Orange Sneezeweed. [*Heleniastrum hoopesii* (Gray) Kuntze; *Dugaldia hoopesii* (Gray) Rydb.]. Perennial herbs, mainly 2–8 (10) dm tall, with a subrhizomatous caudex and fibrous roots; herbage more or less villous-tomentose to glabrate; basal leaves 2–30 cm long, 0.5–5 cm wide, oblanceolate, tapering to a clasping base; cauline leaves reduced upward, oblanceolate to elliptic or lanceolate, entire; heads 2–11, in loose corymbs; disks hemispheric, 2–3.5 cm wide; involucres 5–8 mm high, the bracts lanceolate to elliptic; rays 13–21, yellow or yellow orange, 15–35 mm long, finally reflexed; pappus scales hyaline, lanceolate, attenuate; achenes 3–4 mm long, hairy; n = 15, 17. Sagebrush, mountain brush, aspen, and spruce-fir communities, often in openings or riparian zones, at 1830 to 3420 m in Beaver, Box Elder, Cache, Carbon, Duchesne, Garfield, Grand, Iron, Juab, Piute, Rich, San Juan, Sevier, Summit, Utah, Wasatch, and Washington counties; Oregon to Wyoming, south to California, Arizona, and New Mexico; 90 (xvi). This is a poisonous plant, causing spewing sickness in sheep.

Helianthella T. & G.

Perennial herbs; leaves simple, opposite or alternate, entire; heads radiate, solitary or few to several in loose subcorymbose clusters; bracts imbricate to subequal, more or less herbaceous; receptacle plano-convex, chaffy throughout, the persistent bracts clasping the achenes; disk flowers numerous, fertile, yellow, or purple; rays yellow; pappus of 2 slender awns and short scales; achenes strongly compressed at right-angles to involucral bracts.

Weber, W. A. 1952. The genus *Helianthella* (Compositae). Amer. Midl. Naturalist 48: 1–35.

1. Heads 3–12 or more, mainly less than 20 mm broad; rays 7–13 mm long, inconspicuous; disk flowers normally purple .... *H. microcephala*
— Heads solitary or 2 or 3, mainly over 20 mm broad; rays 15–30 mm long, showy; disk flowers yellow 2
2(1). Heads erect; involucral bracts lance-oblong, short-ciliate .... *H. uniflora*
— Heads nodding; involucral bracts oblong-ovate, long-ciliate with multicellular hairs .... *H. quinquenervis*

Helianthella microcephala (Gray) Gray Smallhead Sunflower. [*Encelia microcephala* Gray, type from Sierra Abajo, San Juan County]. Perennial herbs; stems 20–65 cm tall; herbage appressed hispidulous; basal leaves mainly 4–30 cm long, 0.5–3 cm wide, petiolate, the blades elliptic to lanceolate, scabrous and harshly ciliate, acute to obtuse; cauline leaves reduced upward; heads 3–12 or more; bracts imbricate in ca 3 series, oblong to lanceolate or oblanceolate, strigose and roughly ciliate and glandular; rays 8–10, yellow, 7–13 mm long; disk flowers commonly purple; achenes 7–8 mm long, long-pilose. Desert shrub, pinyon-juniper, ponderosa pine, mountain brush, and Douglas fir-limber pine communities at 1220 to 2745 m in Carbon, Duchesne, Emery, Garfield, Grand, Kane, San Juan, Sevier, Uintah, and Washington counties; Colorado and Arizona; 34 (x). In one plant from Navajo Mt. the disk flowers are apparently yellow. A single collection from west of Richfield (Welsh et al. 17487 BRY) is the only record examined for the Great Basin.

Helianthella quinquenervis (Hook.) Gray Fivenerve Sunflower. [*Helianthus quinquenervis* Hook.]. Perennial herbs; stems 5–15 dm tall, glabrous or villous above; basal leaves 0.3–40 cm long, 0.8–4 cm wide, petiolate, the blades elliptic to oblong or oblanceolate, entire, obtuse to acute; cauline leaves often enlarged to near stem middle then reduced, becoming subsessile or sessile, the largest (at least) prominently 5–nerved; heads nodding, solitary or 2 or 3; disk 2.5–4 cm wide; bracts ovate-lanceolate, acuminate, long-ciliate; rays 12–21, yellow, 15–35 mm long; achenes 8–10 mm long, pilose; n = 15. Sagebrush, aspen, ponderosa pine, and spruce-fir communities at 2115 to 3175 m in Carbon, Daggett, Duchesne, Emery, Garfield, Grand, San Juan, Sevier, Summit, Uintah, and Wasatch counties; Oregon to South Dakota, south to Nevada, Arizona, and New Mexico; 23 (0).

Helianthella uniflora (Nutt.) T. & G. Onehead Sunflower. [*Helianthus uniflorus* Nutt.; *Helianthella multicaulis* D. C. Eaton, type from Parleys Park, Summit County]. Perennial herbs from a branching caudex; stems mainly 3–10 dm tall, glabrous below or more or less spreading-hairy throughout; basal leaves 3–15 cm long, 0.6–5.5 cm wide, petiolate, the blades oblanceolate to elliptic or lanceolate, entire, obtuse to acute; cauline leaves often enlarged to near stem middle, then reduced, becoming sessile or subsessile, the largest prominently 3–nerved; heads erect, solitary or 2 or 3; disk 1.5–3 cm wide; bracts lance-linear, acuminate or obtuse, scabrous-puberulent, shortly ciliate; rays 13–17, yellow, 2–4.5 cm long; achenes 6–7 mm long, pilose. Sagebrush, pinyon-juniper, mountain brush, ponderosa pine, aspen, and spruce-fir communities at 1525 to 3175 m in Beaver, Box Elder, Cache, Carbon, Davis, Duchesne, Garfield, Grand, Iron, Millard, Morgan, Piute, Rich, Salt Lake, Sanpete, Sevier, Summit, Tooele, Uintah, Utah, Wasatch, and Weber counties; Alberta to Montana, south to Nevada and Colorado; 83 (v).

Helianthus L.

Annual or perennial herbs; leaves simple, opposite below, usually alternate above; heads radiate, showy, solitary or few in corymbs; involucral bracts imbricate or subequal, herbaceous; receptacle flat to convex, chaffy throughout, its bracts clasping the achenes; ray flowers conspicuous, yellow, neuter; disk flowers yellow or reddish, fertile; pappus usually of 2 main awns, scalelike at base, sometimes with additional scales present; achenes narrowly obovate in outline, 4-angled or obcompressed.

Blauer, A. C. 1965. *Helianthus* (Compositae) in Utah. Proc. Utah Academy 42: 240-251.
Heiser, C. B. Jr. 1947. Hybridization between sunflower species *Helianthus annuus* and *H. petiolaris*. Evolution 1: 249-262.

1. Plants perennial; disk flowers yellow; leaves mainly opposite, lanceolate to linear-lanceolate .... *H. nuttallii*
— Plants annual; disk flowers reddish brown to purplish; leaves mainly alternate; leaves lanceolate to ovate 2

2(1). Involucral bracts linear to narrowly lanceolate; pappus of numerous unequal scales .... *H. anomalus*
— Involucral bracts lanceolate to ovate; pappus commonly of 2 distinct awns 3

3(2). Involucral bracts ovate, rather abruptly narrowed to an acuminate tip, the central ones inconspicuously hairy; leaves often cordate and with serrate margins .... *H. annuus*
— Involucral bracts lanceolate, tapering to the tip, the central ones often white bearded apically; leaves seldom cordate, usually entire .... *H. petiolaris*

Helianthus annuus L. Common Sunflower. Annual herbs; stems commonly hispid and rough, 3-40 dm tall, simple or branched; leaves alternate above, (3) 5-40 cm long, 2-40 cm wide, lance-ovate to broadly ovate, acute to obtuse, serrate, truncate or cordate basally, hispid to hispidulous on both sides, petiolate; heads solitary or few; disks mainly 2-5 cm wide; involucral bracts lance-ovate to ovate, attenuate to caudate, hispid to hispidulous, ciliate; disk corolla lobes purplish red (rarely yellow); pappus of 2 awnlike ovate-lanceolate scales; achenes glabrous to strigose; n = 20, 34. Saltgrassmuhly grass, desert shrub, pinyon-juniper, and mountain brush communities, commonly where disturbed, at 1200 to 2440 m, probably in all Utah counties; widespread in the U. S., Canada, Mexico, and elsewhere; 70 (xv). Our common weedy sunflower is assignable to ssp. **lenticularis** (Dougl.) Cockerell [*H. lenticularis* Dougl.]; the cultivated largeheaded phase to var. **macrocarpus** (DC.) Cockerell [*H. macrocarpus* DC.].

Helianthus anomalus Blake Sand Sunflower. [*H. deserticola* Heiser, type from west of Hurricane]. Annual herbs; stems sparingly hispid to glabrate, 5-70 cm tall; leaves mainly alternate, petiolate, the blades 1.2-10 cm long, 0.4-4 cm wide, narrowly lanceolate to lance-ovate, yellowish green, acute, cuneate to obtuse basally, hispidulous to hispid on both sides; heads solitary or few, showy; disks mainly 12-24 mm wide; involucral bracts linear, commonly 10-25 mm long and 2-3 mm wide, hispid above, definitely hispid-ciliate, at least below, often some much surpassing the disk; disk corolla lobes purple; pappus of 2 large linear scales and numerous similar subequal scales; achenes 3.5-5.5 mm long, appressed pilose. Blackbrush, ephedra, purple-sage, Vanclevea, psorothamnus, and pinyon-juniper communities, commonly in dunes or other sandy sites, at 1150 to 1830 m in Emery, Garfield, Grand, Juab, Kane, Millard, San Juan, Tooele, Wayne (type from near Hanksville), and Washington counties; Arizona; 35 (xi). This is a ColoradoPlateau endemic, with an extension onto dunes of the eastern Great Basin.

Helianthus nuttallii T. & G. Nuttall Sunflower. [*H. bracteatus* E. E. Watson, type from Logan; *H. giganteus* var. *utahensis* D. C. Eaton, type from Wasatch Mountains; *H. utahensis* (D. C. Eaton) A. Nels.]. Perennial rhizomatous herbs with tuberous roots; stems 3-20 dm tall or more, glabrous or sparingly scabrous or hispid; leaves mainly opposite, shortly petiolate, the blades 4-16 cm long, 0.8-3 cm wide, narrowly lanceolate, acute to attenuate, entire or denticulate, cuneate basally, scabrous on both sides; heads solitary or few to many; disks mainly 12-28 mm wide; bracts lance-linear, 1.5-3 mm wide, subequal to, or surpassing, the disk, attenuate, appressed pubescent and more or less ciliate; disk corolla lobes yellow; pappus of 2 narrow awnlike scales; achenes 3-4 mm long, glabrous. Seeps, springs, wet meadows, and canal banks at 1280 to 2200 m in Cache, Carbon, Duchesne, Garfield, Juab, Rich, Salt Lake, Summit, Tooele, Uintah, Utah, Wasatch, Washington, and Weber counties; British Columbia to Saskatchewan, South to Nevada, Arizona, and New Mexico; 18 (0). Note: The perennial sunflower, *H. tuberosus* L., is grown for its edible roots in our area. It persists following cultivation and is difficult to eradicate. The leaves are broadly lanceolate to ovate.

Helianthus petiolaris Nutt. Prairie Sunflower. [*H. niveus* authors, not (Benth.) Brandegee]. Annual herbs; stems 0.5-12 dm tall, strigose to hispid or glabrous; leaves mainly alternate, petiolate, the blades 1-8 cm long, 4-25 (30) mm wide, lanceolate to ovate, acute to obtuse, entire or rarely serrate, cuneate to truncate basally, hispidulous to strigose; heads solitary or few; disk 10-25 mm wide; involucral bracts 2-5 mm wide, 7-15 mm long, lanceolate, acuminate to attenuate, hispidulous, usually short-ciliate; disk corolla lobes purplish; pappus of 2 lanceolate awnlike scales; achenes 3-4.5 mm long, hairy. Salt desert shrub, desert shrub, pinyon-juniper, and riparian communities, often where disturbed, at 1220 to 1920 m in Beaver, Duchesne, Emery, Garfield, Grand, Kane, Millard, San Juan, Sevier, Uintah, Wasatch, Washington, and Wayne counties; Alberta to Maine, south to California, Arizona, New Mexico, Texas, Louisiana, and South Carolina; 62 (xxii). Our material has been assigned to ssp. *fallax* Heiser. The material appears to be indigenous in the Colorado drainage system, but the rare specimens in the Great Basin seem to be adventive.

Heterotheca Cassini

Annual, biennial, or perennial herbs; leaves alternate, simple, entire; heads radiate; involucres campanulate to hemispheric; bracts numerous, narrow, imbricated in several series; receptacle convex, naked; rays yellow, pistillate and fertile; pappus of capillary bristles; disk flowers numerous, the pappus present and usually double, the inner of capillary bristles, the outer (when present) of short scales or bristles; achenes hairy.

Wagenknecht, B. L. 1960. Revision of *Heterotheca*, Section Heterotheca (Compositae). Rhodora 62: 61-76, 97-109.

1. Plants low, creeping, arising from subrhizomatous caudex branches; heads nodding, solitary or 2 or 3, known from sandy sites in Garfield, Kane, and Washington (?) counties .... *H. jonesii*
— Plants various in habit, the caudex branches, if present, not rhizomatous; heads few to numerous, seldom nodding, distribution various 2

2(1). Plants perennial, from a woody root crown, the stems numerous, forming rounded clumps, common . *H. villosa*

— Plants annual or biennial, the root crown herbaceous; stems solitary or few, not forming rounded clumps, rare   3

3(2). Upper leaves cordate-clasping basally; involucres glandular-puberulent and canescent .... *H. psammophila*

— Upper leaves tapering to a sessile base; involucres glandular, pubescent, not canescent .... *H. grandiflora*

***Heterotheca grandiflora* Nutt.** Telegraph Weed. Annual or biennial herbs; stems stout, branched above, 5–12 (20) dm tall, hirsute, glandular-pubescent; leaves 2–6 cm long, 0.8–2.5 cm wide (or more), ovate to elliptic, oblong, or oblanceolate, serrate, the lower petiolate and lobed at base; heads numerous; involucres 7–9 mm high; rays 25–35, 6–8 mm long, ca 1 mm wide, the tube hairy; disk flowers numerous, slender; pappus tawny; 2n = 18. Sandy roadside at ca 915 m in Washington County; California and Arizona; 4 (ii). Our material has the pappus merely tawny, not brick red as reported elsewhere for the species.

***Heterotheca jonesii* (Blake) Welsh & Atwood** Jones Goldenaster. [*Chrysopsis caespitosa* Jones, not Nutt.; *Chrysopsis jonesii* Blake]. Perennial caespitose herbs from a creeping subrhizomatous caudex; stems 4–8 cm tall, loosely villous; leaves 5–11 mm long, 1.5–4 mm wide, petiolate, the blades obovate to spatulate, pilose; heads solitary or 2 or 3; involucres 5–7.5 mm high, 6–10 mm wide, the bracts narrowly lance-oblong, strigose-pilose, the hyaline margins reddish; ray flowers 5–13, yellow, 4–6 mm long, 1.5–2.5 mm wide; pappus tawny; achenes 2–3 mm long, hairy. Ponderosa pine, manzanita, and Douglas fir communities, on sandstone or in sand, at 1580 to 2745 m in Garfield, Kane, and Washington (?) counties (the type presumably came from Springdale); endemic; 8 (iii).

***Heterotheca psammophila* Wagenkn.** Sand Goldenaster. [*H. subaxillaris*, authors]. Annual or biennial herbs; stems stout, 5–12 (20) dm tall, branching above, hispid hirsute and glandular upward; leaves mainly 1–7 cm long, ovate to lance-oblong, serrate to subentire the lower petiolate, the upper cordate-clasping; heads numerous; involucres 8–12 mm high, glandular and canescent; rays 20–30, yellow, mainly 3–7 mm long; pappus tawny; achenes 2.4–3.8 mm long; n = 9. Sandy roadside at ca 970 to 1350 m in Grand and Washington counties; Arizona to Texas, south to Mexico; 1 (i). The specimen from Grand County (Welsh & Moore 2745) is missing from BRY, the Washington County locality is reported by Meyer.

***Heterotheca villosa* (Pursh) Shinn.** Hairy Goldenaster. [*Amellus villosus* Pursh; *Chrysopsis villosa* (Pursh) Nutt. ex DC.]. Perennial herbs, from a ligneous root-crown and taproot; stems several to numerous, forming rounded clumps, mainly 1.5–5 dm tall; herbage hirsute to strigose and more or less glandular; leaves 0.5–5 cm long, 2–10 mm wide, oblanceolate to spatulate or elliptic, green or silvery to gray green, petiolate or subsessile; heads few to numerous, mainly corymbose; involucres 7–10 mm high, 7–12 mm wide; bracts lance-linear, green or chartaceous, the margins hyaline, sometimes reddish; rays 10–25, yellow, 6–10 mm long; pappus tawny; achenes 2–3 mm long, hairy; n = 9, 18. Our materials represent only a small portion of the vast array of variation within the *villosa* complex. Three infraspecific taxa are apparent among our specimens, but application of names is difficult. The following treatment is therefore tentative, with a definitive treatment awaiting monographic study.

1. Leaves green or gray green, the surface apparent through the spreading to subappressed hairs; plants widespread .... *H. villosa* var. *hispida*

— Leaves silvery or grayish, the surface seldom apparent through the usually appressed hairs; plants restricted .. 2

2(1). Stems mainly appressed hairy, or with some hairs ascending to spreading .... *H. villosa* var. *villosa*

— Stems with appressed and spreading contorted long hairs .... *H. villosa* var. *foliosa*

Var. *foliosa* (Nutt.) Harms [*Chrysopsis foliosa* Nutt.]. n = 18. Mountain brush and bunchgrass (*Elymus* et al.) communities at 1280 to 2135 m in Davis, Duchesne, Salt Lake, San Juan (?), Utah, and Weber counties; widespread in western U. S.; 9 (0).

Var. *hispida* (Hook.) Harms [*Diplopappus hispidus* Hook.; *Chrysopsis hispida* (Hook.) Nutt.; *C. villosa* var. *scabra* Eastw. (?), type from San Juan County; *C. viscida* var. *cinerascens* Blake, type from Beaver Canyon]. n = 9, 18. Shadscale-rabbitbrush, ephedra-lycium-dropseed, sagebrush-grass, pinyon-juniper, mountain brush, ponderosa pine-manzanita, and aspen communities at 1150 to 2745 m in all Utah counties; widely distributed in the western U. S.; 105 (xii).

Var. *villosa* Blackbrush, pinyon-juniper, and ponderosa pine communities at 1090 to 2785 m in Garfield, Kane, Washington, and Wayne counties; widespread in western U. S.; 8 (iii).

Hieracium L.

Perennial rhizomatous herbs with milky juice; leaves alternate or basal and still alternate, entire or toothed, simple; heads few to numerous, in corymbose clusters; flowers all raylike, yellow to orange or white; involucres cylindric to hemispheric; bracts more or less imbricate; receptacle naked; pappus usually of brownish capillary bristles; achenes terete or prismatic, more or less strongly ribbed.

1. Flowers white; stems long-setose basally if at all, the petioles and leaves long-setose on the lower midrib .... *H. albiflorum*

— Flowers yellow; stems long-setose throughout or only above; leaves variously setose or glabrous 2

2(1). Leaves long-hairy, mainly 10–20 cm long, the cauline ones well developed, reduced above .. *H. cynoglossoides*

— Leaves glabrous or short-hairy, 2–10 cm long, mainly basal, the stems merely bracteate 3

3(2). Involucres grayish tomentose and with a few long, white setae; plants of Washington County .... *H. fendleri*

— Involucres brownish to blackish tomentose and more or less conically setose with long hairs, the cross walls blackish; plants of the Uinta Mts. .... *H. gracile*

***Hieracium albiflorum* Hook.** White Hawkweed. Perennial herbs; stems erect, 15–75 cm tall, long-hairy at the base, becoming glabrous upward; basal leaves 2–12 (17) cm long, 0.8–4.5 cm wide, oblanceolate, petiolate, entire or remotely toothed, long-hairy on petioles and midvein, commonly long-ciliate; cauline leaves sessile, reduced upward; stellate hairs lacking; heads few to many, on slender peduncles, the inflorescence open; involucres 6–11 mm high, blackish green, glandular or sparingly long-hairy, or glabrous; flowers 12–35, white; pappus tawny; n = 9. Lodgepole pine, spruce, and spruce-fir communities at 1980 to 3420 m in Daggett, Duchesne, Sanpete, Summit, Uintah, and Utah counties; Alaska and Yukon to Saskatchewan, south to California, Nevada, and Colorado; 12 (ii).

Hieracium cynoglossoides Arv.-Touv. Houndstongue Hawkweed. [*H. griseum* Rydb.; *H. scouleri*, authors]. Perennial herbs; stems erect, 20–75 cm tall, pubescent with long coarse loose or spreading white setae that dry yellowish; basal leaves commonly withered at anthesis; lower cauline leaves 10–25 cm long, 1–3 cm wide, petiolate, the blades oblanceolate to elliptic, long-hairy; middle and upper leaves reduced, sessile; heads yellow, few to many, 15- to 40-flowered, corymbose, the inflorescence more or less open; involucres 7–12 mm high; bracts greenish, the margins chartaceous to hyaline, minutely stellate and stipitate-glandular; pappus tawny; n = 9. Grass-forb, aspen, and spruce-fir communities at 2000 to 2990 m in Box Elder, Duchesne, Salt Lake, Summit, Utah, Wasatch, and Weber counties; British Columbia and Alberta, south to Oregon and Wyoming; 18 (0).

Hieracium fendleri Schulz-Bip Fendler Hawkweed. Perennial herbs; stems erect, 20–50 cm tall, almost glabrous or with a few long, spreading white setae; basal leaves 5–12 cm long, 1.5–6 cm wide, oblanceolate to obovate, acute to rounded and apiculate, cuspidate-serrate, attenuate basally, sparingly hairy on both surfaces; stem leaves greatly reduced or lacking; heads 3–40 per plant; inner involucral bracts 10–12 mm long; achenes ca 5.5 mm long, tapered almost from the base to the summit; pappus sordid-white. Ponderosa pine, mountain brush, and cottonwood communities at ca 2250 m in Washington County; New Mexico, Arizona, Colorado, Texas, and Mexico; 4 (i).

Hieracium gracile Hook. Slender Hawkweed. [*H. utahense* Gand., type from Cache County]. Perennial herbs; stems erect, 8–40 cm tall, tomentulous to glabrous; basal leaves 2–10 cm long, 0.4–2 cm wide, petiolate; blades oblanceolate, entire or denticulate, stipitate-glandular to glabrous or less commonly with a few long blackish setae; cauline leaves much reduced; heads solitary or more commonly few to several; involucres greenish black, stellate hairy, and with long black setae; pappus tawny; n = 9. Lodgepole pine, spruce-fir, and grass-forb communities at 3050 to 3390 m in Daggett, Duchesne, Summit, Uintah, and Wasatch counties; Alaska and Yukon to Mackenzie, south to California and New Mexico; South America; 13 (v).

Hofmeisteria Walp.

Shrubs; leaves opposite below, alternate above, simple, petiolate; heads discoid, few to several in terminal corymbose clusters; involucre campanulate; bracts striate, narrow, imbricated; receptacle naked; disk flowers whitish; pappus of 10–12 scabrous bristles and other short scales; achenes 5-angled, callous-thickened.

Hofmeisteria pluriseta Gray Arrowleaf. Shrubs, low, rounded, and intricately branched, mostly 3–8 dm tall; branchlets green, glandular-puberulent, becoming white barked in age; leaves long petioled, the petioles 0.8–4 cm long, the blades hastately lobed to entire, 4–10 mm long, 2–4 mm wide; heads small; involucres 4–9 mm high; bracts 3-lined, acuminate; disk flowers whitish. Reported for Utah in Munz (Flora of California, p. 267); to be sought on rock outcrops at lower elevations in Washington County; Nevada, California, and Arizona; 0 (0).

Hulsea T. & G.

Perennial viscid-pubescent aromatic herbs; leaves alternate, simple; heads radiate; involucres hemispheric, the bracts subequal in 2 or 3 series, herbaceous, finally reflexed; receptacle convex, naked; ray flowers yellowish to purplish, pistillate, fertile; disk flowers perfect, fertile; pappus of 4 hyaline scales united at the base; achenes compressed, angled, villous.

Hulsea heterochroma Gray Plants from a stout taproot; herbage viscid-villous, scented, 3–10 (12) dm tall; basal leaves oblanceolate or spatulate, tapering to a broadly petioled base, dentate; cauline leaves mainly 3–10 cm long, 1.5–3.5 cm wide, sessile, and more or less clasping; heads conspicuous, in racemose or corymbose clusters; bracts with long-attenuate, often reddish tips, subequal to the disk; rays reddish to purple or yellowish, hairy and glandular; pappus scales unequal, lacerate; achenes 6–7 mm long; n = 19. Pinyon-juniper community at 2135 m in the Beaver Dam Mts., Washington County (Higgins 1410 BRY); Nevada and California; 4 (0).

Hymenoclea T. & G.

Xerophytic shrubs; leaves alternate, linear, usually entire; heads discoid, small, numerous, mostly glomerate-paniculate, with both sexes in each leaf axil, the staminate above the pistillate; staminate heads several-flowered; pistillate heads 1-flowered; involucre becoming indurated and beaked in fruit, the bracts persistent as scarious wings; pappus none.

Hymenoclea salsola T. & G. Burrobush. Shrubs, 6–12 (15) dm tall; branchlets green, becoming straw colored to gray in age; herbage yellow green, resinous, glabrous or scabrous; leaves 2–5 cm long, linear, entire; staminate heads 2–3 mm high, 2.5–5 mm wide, the bracts obtuse to rounded, ciliate on the hyaline margin; pistillate heads mainly 6–9 mm high at maturity, the middle and upper bracts with white, chartaceous, broadly rounded, erose margins, longitudinally veined. Blackbrush, creosote bush, and Joshua tree communities at 670 to 900 m in Washington County; Nevada, Arizona, and California; 19 (ii).

Hymenopappus L'Her.

Perennial herbs; leaves alternate or mainly basal (and still alternate), mainly pinnatifid; heads discoid, the flowers perfect; involucral bracts in 2 or 3 series, subequal, at least the inner with broad rounded scarious or hyaline margins; receptacle flat, naked or rarely chaffy; corollas yellow or white; anthers sagittate; pappus of several membranous scales; achenes 15- to 20-nerved, 4- or 5-angled.

Turner, B. L. 1956. A cytotaxonomic study of the genus *Hymenopappus* (Compositae). Rhodora 58: 163–308.

Hymenopappus filifolius Hook. Hyalineherb. Perennial subscapose herbs; stems 5–60 (100) cm tall, tomentose to glabrate; basal leaves 3–20 cm long, twice pinnately dissected, the ultimate divisions mainly 2–25 mm long, minutely punctate; cauline leaves lacking or several, much reduced upward; heads solitary or more commonly few to numerous, turbinate to campanulate, with

10–59 flowers or more, on peduncles 0.5–10 cm long or more; involucral bracts mainly 3–14 mm long; corollas yellow or white, 2–7 mm long; pappus of narrowly oblong scales; achenes 3–7 mm long, densely hairy; n = 17, 34. This is a polymorphic species, which consists of a series of geographic and/or edaphically correlated infraspecific taxa. Those taxa peripheral to the main body of the species in the Colorado Plateau province are the most distinctive. The following treatment differs from that of Turner (1956) and represents a more conservative approach.

1. Basal leaf axils sparingly tomentose or glabrous; stems scapose, or with 1 or 2 leaves; plants of high elevations
   .... *H. filifolius* var. *nudipes*
— Basal leaf axils prominently white-tomentose; stem leaves often more than 2; plants of middle and lower elevations 2

2(1). Corollas 2–3 mm long; flowers fewer than 30; plants of Daggett, Summit, and Uintah counties ....
   .... *H. filifolius* var. *luteus*
— Corollas 3–7 mm long or, if shorter, not of Daggett or Uintah counties; flowers in main heads often more than 30; leaves more coarsely dissected; plants of various distribution 3

3(2). Flowers white; achene hairs 0.5–1 mm long; plants of Washington County .... *H. filifolius* var. *eriopodus*
— Flowers yellow; achene hairs 1–2 mm long; plants more widely distributed 4

4(3). Leaves mainly basal; plants of the Great Basin ....
   .... *H. filifolius* var. *nanus*
— Leaves cauline and basal; plants of the Colorado drainage system .... *H. filifolius* var. *cinereus*

Var. *cinereus* (Rydb.) Johnston [*H. cinereus* Rydb.; *H. lugens* Greene; *H. filifolius* var. *lugens* (Greene) Jepson; *H. filifolius* var. *megacephalus* Turner, as to Utah materials; *H. pauciflorus* Johnston, type from near Bluff; *H. filifolius* var. *pauciflorus* (Johnston) Turner; *H. tomentosus* Rydb., type from St. George; *H. filifolius* var. *tomentosus* (Rydb.) Turner; *H. niveus* Rydb., type from Springdale]. Blackbrush, warm desert shrub, salt desert shrub, sand sagebrush-ephedra, pinyon-juniper, ponderosa pine, and sagebrush communities at 1065 to 2685 m in Carbon Daggett, Duchesne, Emery, Garfield, Grand, Kane, San Juan, Uintah, Washington, and Wayne counties; Colorado, California, Arizona, New Mexico, and Texas; 144 (xxix). The variety *cinereus*, as interpreted herein, includes three largely sympatric phases that were treated by Turner (1956) at varietal rank. Although there is a tendency for these phases to be geographically correlated, they are connected completely by series of intermediates, and they can be segregated only arbitrarily. Turner noted that herbarium specimens of var. *megacephalus* from eastern Utah, inter alia, carried a "hodgepodge of annotations: *H. lugens*, *cinereus*, *pauciflorus*, *tomentosus*, *eriopodus*, etc.," and further that "the possibility exists that the variety [*megacephalus*] here typified includes only the individuals from Clark County, Nevada, and vicinity, and that most of the remaining material to the east represents either a weakly defined separate variety or a common area of extensive hybridization and introgression among the several peripheral taxa mentioned above, . . ." A phase of the *cinereus* complex from Washington County has cauline leaves well developed and plant bases appearing bulbous due to a copious tomentum. These plants are apparently intermediate between var. *tomentosus* (Rydb.) Turner, in a narrow sense, and var. *eriopodus* (A. Nels.) Turner, which shares the feature of the "bulbous" bases.

Var. *eriopodus* (A. Nels.) Turner [*H. eriopodus* A. Nels., type from Diamond Valley]. Pinyon-juniper community at 1675 to 2135 m in Washington County; California to Nevada; 3 (0). This variety is evidently rare in Utah, and might best be treated within an expanded var. *cinereus* (q.v.).

Var. *luteus* (Nutt.) Turner [*H. luteus* Nutt.]. Salt desert shrub, mixed cool desert shrub, and pinyon-juniper communities at 1525 to 1830 m in Daggett, Summit, and Uintah counties; Wyoming and Colorado; 7 (ii). The small flowers and finely divided leaves appear to be diagnostic for this variety.

Var. *nanus* (Rydb.) Turner [*H. nanus* Rydb.]. Black sagebrush-rabbitbrush, pinyon-juniper, and ponderosa pine communities at 1490 to 2300 m in Beaver, Garfield, Iron, Juab, Millard, Piute, Sevier, Tooele, and Washington counties; Nevada, California, and Arizona; 26 (v).

Var. *nudipes* (Maguire) Turner. [*H. nudipes* var. *alpestris* Maguire, type from Cedar Breaks; *H. nudipes* Maguire, type from 15 mi N of Orderville; *H. filifolius* var. *alpestris* (Maguire) Turner]. Ponderosa pine, western bristlecone pine, sagebrush-grass, limber pine, aspen, and alpine tundra communities, commonly on limestone or thermally modified igneous outcrops, at 2445 to 3450 m, in Beaver, Carbon, Duchesne, Emery, Iron, Garfield, Kane, Millard, Piute, Sanpete, Sevier, Summit, Utah, and Washington counties; Wyoming; 46 (ix). This is the most distinctive of the varieties within *H. filifolius* in Utah.

Hymenoxys Cassini

Perennial or biennial herbs from a taproot and commonly with a pluricipital caudex; stems simple or branched; leaves basal or basal and cauline, simple and entire or pinnately to ternately divided; heads radiate, pedunculate; involucres hemispheric; bracts in 2 or 3 series, the outer distinct or connate basally, subequal or imbricate, herbaceous or cartilaginous; receptacle naked, hemispheric; ray flowers yellow, pistillate, fertile, prominently veined, 3-toothed; disk flowers perfect, fertile; pappus scales usually 5, hyaline, nerved or nerveless, the nerve often produced into an awn; achenes obpyramidal, more or less 5-angled, appressed hairy.

1. Leaves entire, essentially all basal 2
— Leaves pinnatifid or palmatifid, or some entire, the cauline ones well developed 5

2(1). Involucral bracts sparsely pubescent or glabrous apically, the margins thin and scarious or hyaline 3
— Involucral bracts moderately to densely villous-pilose or some rarely glabrous, the margins not at all or only narrowly scarious 4

3(2). Plants depressed pulvinate-caespitose, acaulescent; outer involucral bracts recurved, thickened and reddish apically; disks less than 10 mm wide .... *H. lapidicola*
— Plants merely caespitose, scapose; outer involucral bracts erect, not thickened and seldom reddish apically; disks over 10 mm wide .... *H. torreyana*

4(2). Plants pulvinate-caespitose; caudex branches clothed with a marcescent thatch of erect-ascending leaf bases; leaves mainly linear, cuspidate apically ..... *H. depressa*

— Plants seldom pulvinate-caespitose; caudex branches without a definite thatch of ascending or erect leaf bases; leaves various, sometimes cuspidate apically .. *H. acaulis*

5(1). Disks 18–30 mm wide or more; involucral bracts similar, distinct, in 2 or 3 indefinite subequal series; herbage villous-tomentose; plants of high elevations .... *H. grandiflora*

— Disks 7–22 mm wide; involucral bracts in 2 dissimilar series, the outer thickened and united at the base 6

6(5). Plants silvery-canescent; leaves entire or 3-cleft, the blades or segments 1.5–4 mm wide ...... *H. subintegra*

— Plants green or, if silvery-canescent, the leaves commonly 3- to 5-cleft 7

7(6). Plants apparently biennial, with an evident basal rosette and taproot; cauline leaves numerous, gradually reduced upward, ternate or palmatifid .... *H. cooperi*

— Plants perennial, from a taproot and caudex; cauline leaves rather well developed, palmatifid to entire 8

8(7). Stems merely glandular or glandular-scabrous; plants of low elevation saline meadows in western Utah .... *H. lemmonii*

— Stems more or less villous; plants of various habitats, but seldom of saline meadows and not of western Utah . 9

9(8). Stems few to several from a pluricipital caudex; leaf bases conspicuously long-villous below; leaf segments mainly 1–2.5 mm wide .... *H. richardsonii*

— Stems solitary or few from a simple or branched caudex; leaf bases glabrous or only somewhat hairy; leaf segments 2–6 mm wide .... *H. helenoides*

Hymenoxys acaulis (Pursh) Parker Stemless Woollybase. [*Gaillardia acaulis* Pursh]. Perennial caespitose herbs from a short multicipital caudex, the caudex branches clothed with short brownish or blackish marcescent leaf bases, 2–50 cm tall, villous to glabrous; leaves 1–6 cm long, 2–8 mm wide, all basal or some cauline, glandular-punctate or epunctate, linear to oblanceolate, entire; heads solitary (rarely 2); disk 7–20 mm broad; bracts distinct, in 2 or 3 subequal series, 4–9 mm high; rays 5–9, yellow, 6–15 mm long; pappus scales 2.5–4.5 mm long, acute or shortly awned; achenes 2.5–4.5 mm long. This is a complex entity, consisting of a series of morphological phases, which are more or less geographically or edaphically correlated.

1. Plants with 1–4 (or more) cauline leaves; stems simple or branched .... *H. acaulis* var. *ivesiana*

— Plants scapose, with cauline leaves lacking or rarely with 1; scapes unbranched 2

2(1). Leaves linear to linear-oblanceolate, conspicuously glandular-punctate, sparingly long-hairy to glabrous; plants of the Colorado drainage system .... *H. acaulis* var. *arizonica*

— Leaves narrowly to broadly oblanceolate, inconspicuously glandular-punctate, merely punctate, or epunctate, or plants of the Great Basin, densely pilose to villous or glabrous 3

3(2). Leaves epunctate or nearly so, glabrous or less commonly silky-hairy; plants of the Colorado rainage system .... *H. acaulis* var. *caespitosa*

— Leaves punctate, silky-hairy, or less commonly glabrous; plants of the Great Basin .. *H. acaulis* var. *acaulis*

Var. *acaulis* 2n = 70. Sagebrush, mixed desert shrub, pinyon-juniper, and bunchgrass communities, often on windswept ridges, at 1525 to 2990 m in Beaver, Box Elder, Juab, Millard, Sanpete, Sevier, Tooele, and Washington counties; Idaho east to Saskatchewan, south to Nevada, Colorado, and Texas; 42 (viii). Specimens from the Great Basin might not belong to var. *acaulis* in a strict sense, and perhaps should be regarded as a separate variety. The problem cannot be solved on the basis of Utah specimens alone.

Var. *arizonica* (Greene) Parker [*Tetraneuris arizonica* Greene]. 2n = 56. Salt and sandy desert shrub, pinyon-juniper, sagebrush, blue grama, aspen, Douglas fir, white fir, and ponderosa pine communities at 1220 to 3175 m in Carbon, Daggett, Duchesne, Emery, Grand, Kane, San Juan, and Uintah counties; Colorado and Arizona; 108 (xvi).

Var. *caespitosa* (A. Nels.) Parker [*Tetraneuris acaulis* var. *caespitosa* A. Nels.; *Tetraneuris epunctata* A. Nels., type from Dyer Mine]. 2n = 28. Shadscale-eriogonum, black sagebrush, sagebrush, pinyon-juniper, mountain brush, and alpine tundra, often on plateau margins and windswept ridges, at 1585 to 3510 m in Carbon, Daggett, Duchesne, Emery, Grand, Sanpete, Sevier, Summit, and Uintah counties; Wyoming south to New Mexico; 87 (viii).

Var. *ivesiana* (Greene) Parker [*Tetraneuris ivesiana* Greene; *H. ivesiana* (Greene) Parker; *H. argentea* (Gray) Parker; *Actinella argentea* Gray]. 2n = 28, 56. Sand sagebrush, ephedra, pinyon-juniper, and ponderosa pine communities at 1150 to 2505 m in Garfield, Grand, Kane, San Juan, and Wayne counties; Colorado, New Mexico, and Arizona; 66 (xii). This variety approaches phases of the partially sympatric var. *arizonica* in stature, and it is possible to confuse some specimens when cauline leaves are lacking and the stems are unbranched. The varieties *acaulis*, *arizonica*, and *caespitosa* are tetraploids, i. e. 2n=60, whereas var. *ivesiana* is diploid, i. e., 2n = 30. Because of this difference, Parker (1960. Leafl. W. Bot. 9: 93) elevated this taxon to specific rank.

Hymenoxys cooperi (Gray) Cockerell Cooper Hymenoxys. [*Actinella cooperi* Gray; *A. biennis* Gray, type from Washington County?; *H. acaulis* var. *cooperi* (Gray) Cockerell]. Biennial or short-lived perennial herbs; stems 16–60 (80) cm tall, leafy, simple below, branched in a corymbose inflorescence above, often reddish, scurfy villous, canescent; basal rosette leaves mainly 2–10 cm long, pinnately divided, the linear lobes often again divided, mainly 1–1.5 mm wide; stem leaves longer than the internodes; heads (1) 3–50; involucres 5–6 mm high, 10–24 mm wide, hemispheric; bracts thickened and united basally, more or less pubescent and glandular; rays 7–13, yellow, 6–15 mm long; pappus scales acuminate; achenes 2–3 mm long, densely pilose; n = 15. Sagebrush and pinyon-juniper communities at 975 to 2380 m in Garfield, Juab, Kane, and Washington counties; Nevada, California, and Arizona; 18 (i).

Hymenoxys depressa (T. & G.) Welsh & Reveal Low Hymenoxys. [*Actinella depressa* Gray]. Pulvinate-caespitose scapose perennial herbs from a multicipital caudex, the caudex branches clothed with conspicuous, commonly erect marcescent leaf bases (often forming a thatch), 1–4 cm tall; scapes villous; leaves 0.4–3 (4) cm long, 1–2 (4) mm wide, linear to oblanceolate, the outer sparingly if at all glandular-punctate, the inner definitely so, sparingly villous to glabrous, cuspidate; heads solitary; disk 6–10 mm wide; involucres 4–6 mm high; bracts in 2 or 3 subequal series, long villous, the margins nonscarious, the apices erect; rays 5–7, yellow, 3–6 mm long; pappus scales 2–3 mm long, long-acuminate; achenes 2–3 mm long. Ephedra, sagebrush, shadscale, and pinyon-juniper woodland at 1340 to 2170 m in Duchesne, Emery, and eastern Sevier counties; endemic? There is justification for inclusion of *H. depressa* within the *H. acaulis* complex, at some infraspecific rank. And the plants have been suggested as merely depauperate phases of that group. However, if they are ecologically controlled variations, they should be expected through much of the range of *H. acaulis*, but they are not. Dwarf forms of *H. acaulis*, especially of the var. *caespitosa*, have been mistaken for this species, but they are more hairy, have usually broader leaves, and lack glandular punctae. There is also a question of typification; the type of *H. depressus* was taken by John Charles Fremont, on his second expedition in the Rocky Mountains. Fremont evidently traversed the area occupied by *H. depressa* in 1844, and the material could have come from Duchesne County; 24 (iv).

Hymenoxys grandiflora (T. & G.) Parker Graylocks. [*Actinella grandiflora* T. & G.]. Perennial herbs from a taproot and usually simple caudex, this clothed with brown marcescent leaf bases; stems mainly 5–25 cm tall, 1 to several, simple or branched basally, densely villous; leaves basal and cauline, 2–10 cm long, 2– or 3–times ternately or palmately divided, the lobes linear, villous to glabrate; heads solitary; disk 1.5–3 cm wide or more; involucral bracts subequal, in 2 or 3 series, 8–14 (16) mm high, densely villous-tomentose; rays 15–50, yellow, 25–35 mm long; pappus scales 3.5–7 mm long, attenuate; achenes 3–5 mm long. Sedge-forb communities at or above timberline, often in talus or rock stripes, at 3050 to 3660 m in Duchesne, Grand, Salt Lake, San Juan, Summit, and Utah counties (Uinta, Wasatch, and La Sal mts.); Idaho to Montana, south to Colorado; 24 (ii). This is a strikingly beautiful yellow sunflower of alpine tundra in our mountains.

Hymenoxys helenioides (Rydb.) Cockerell [*Picradenia helenioides* Rydb.]. Perennial herbs from a simple or branched caudex, this clothed with broad brown marcescent leaf bases; stems mainly 25–45 cm tall, simple below, branched above, scurfy and more or less villous; leaves basal and cauline, mainly 5–15 cm long, entire or 2– to 5–lobed, the lobes mainly 3–8 mm wide, finely glandular-punctate, glabrous or puberulent; heads 3–13, in corymbose clusters; disks 10–21 mm wide; involucres 6.5–8 mm high, the outer bracts green, connate in the lower portion, more or less villous and glandular; rays 5–11, yellow, 8–19 mm long; pappus scales 2.5–3.5 mm long, acuminate; achenes 2.5–3 mm long. Mountain brush, sagebrush, and aspen communities, often in meadows, at 2440 to 2990 m in Emery, Garfield, Sanpete, and Sevier counties; Colorado and Arizona; 10 (i). This handsome plant has long remained obscure in Utah, partially due, no doubt, to its resemblance to *Helenium hoopesii* (q.v.), with which it occurs in the aspen communities of central and southern Utah.

Hymenoxys lapidicola Welsh & Neese Rock Hymenoxys. Pulvinate caespitose herbs from a multicipital caudex, this densely clothed with brown marcescent leaf bases, acaulescent; leaves all basal, 0.3–1.2 cm long, 0.8–2 mm wide, narrowly oblanceolate, the inner conspicuously glandular-punctate, the blades glabrous, the axils long-villous; heads solitary, immersed in the leaves; disks 5.5–9 mm wide; involucres 5–8 mm high; bracts distinct, in 2 or 3 subequal series, sparingly villous and suffused reddish, the margins scarious, the tips more or less squarrose-spreading and somewhat thickened; rays 5 or 6, yellow, 5–6 mm long; pappus scales lance-acuminate, 2.3–3 mm long; achenes 2–2.5 mm long, pilose. Pinyon-juniper and ponderosa pine-manzanita communities, often in rock crevices, at 1830 to 2476 m in Uintah (type from Blue Mt.) County; endemic; 4 (0).

Hymenoxys lemmonii (Greene) Cockerell Lemmon Hymenoxys. [*Picradenia lemmonii* Greene; *H. lemmonii* ssp. *greenei* Cockerell, type from Washington County (?)]. Perennial herbs from a taproot and short ligneus caudex, the caudex clothed with brown to straw colored or purplish clasping leaf bases; stems 20–60 cm tall, 1 to few, glabrous; leaves cauline and basal, 2–15 cm long, pinnately parted, the lobes linear, 2–3 mm wide, glabrous, glandular-punctate; cauline leaves longer than the internodes, the uppermost often unlobed; heads 5–12; involucres 4.5–7 mm high, hemispheric, 8–14 mm wide; outer bracts green, sparsely scurfy and glandular, thickened dorsally and connate below; rays 6–10, yellow, 6–13 mm long; pappus lance-attenuate, 1.6–2 mm long; achenes 2.5–3 mm long, pilose. Saline rabbitbrush-alkali sacaton meadows at 1660 m in Millard, Tooele, and Washington (?) counties; Nevada and California; 3 (i).

Hymenoxys richardsonii (Hook.) Cockerell Colorado Rubberweed. [*Picradenia richardsonii* Hook.; *H. r.* var. *utahensis* Cockerell, type from Emery]. Perennial caespitose herbs from a pluricipital ligneus caudex, the caudex branches clothed with a thatch of marcescent brown leaf bases, usually with villous leaf axils; stems few to numerous, 6–40 (50) cm tall, simple below, branched; leaves basal and cauline 2–12 (15) cm long, ternate or with 5–7 linear segments, or some entire, pubescent to glabrous; involucres hemispheric, 5–8 mm high, the outer bracts connate basally, thickened dorsally, green or chartaceous, more or less villous; rays 9–14, yellow, 8–20 mm long; pappus 2–4.5 mm long, acuminate; achenes 2.5–4 mm long, pilose; n = 15, 28. Salt desert shrub, cool desert shrub, pinyon-juniper, sagebrush, mountain brush, ponderosa pine, aspen, fir, and western bristlecone pine communities at 1460 to 2870 m. Our material falls into two varieties, a low plant with 1–5 large heads of Daggett and Uintah counties, belonging to var. *richardsonii*, and a taller plant with 3–20 smaller heads of Beaver, Carbon, Duchesne, Emery, Garfield, Kane, Millard, Piute, Sanpete, Sevier, Uintah, and Wayne counties, belonging to var. *florabunda* (Gray) Parker; Alberta and Saskatchewan to Arizona and Texas; 129 (xvii). The plants are considered poisonous to sheep, cattle, and goats.

Hymenoxys subintegra Cockerell Perennial (or biennial) herbs from a taproot; stems solitary or few, 10–60 cm tall, branched above; herbage silvery canescent; basal leaves often withered at flowering; cauline leaves numerous, 1.5–8 cm long, 2–4 mm wide, entire or 2– or 3–lobed; heads few to several; disks 9–12 mm wide; involucres 5–7 mm high; outer bracts connate basally, villous; rays ca 12–20, yellow, 5–10 mm long; pappus scales lance-acuminate, 2.8–3.2 mm long; achenes ca 3 mm long. Ponderosa pine, aspen, and spruce-fir communities in Sanpete (Maguire 20049 BRY) and Washington (reported by Meyer) counties; Arizona; 1 (0).

Hymenoxys torreyana (Nutt.) Parker [*Actinella torreyana* Nutt.]. Perennial caespitose scapose herbs from a stout pluricipital caudex, the caudex branches densely clothed with brown to straw colored or ashy leaf bases, 3–10 cm tall, villous; leaves 1–7.5 (9) cm long, 2–6 mm wide, all basal, glandular-punctate, narrowly oblanceolate, entire; heads solitary; disk 12–20 mm wide; involucres hemispheric, 5–10 mm high; bracts distinct, in 2 or 3 subequal series, less pubescent to glabrous apically, the margins scarious, not thickened apically, the tips erect, sometimes reddish; rays 10–16, yellow, 8–20 mm long; pappus scales ovate-acuminate, 2.8–3.5 mm long; achenes 2–3 mm long. Pinyon-juniper, sagebrush, and mountain brush communities at 1830 to 2200 m in Daggett and Uintah counties; Wyoming; 6 (0).

Hypochaeris L.

Perennial subscapose herbs from taproots, the juice milky; leaves primarily basal, simple, pinnately lobed to pinnatifid, the cauline leaves small and bractlike; heads solitary or few in a branching inflorescence; involucral bracts in several series, greenish black, the inner ones with hyaline margins; receptacle chaffy; corollas of ray flowers only, perfect, yellow or purplish on the dorsal surface; pappus of plumose capillary bristles; style branches semicylindrical; achenes several-nerved, subterete, minutely roughened, long beaked.

Hypochaeris radicata L. Cat's-ears. Plants 1.5–5 dm tall, the stems simple or branched above, glabrous or spreading-hairy below; basal leaves 3–16 (25) cm long, 0.5–3.5 (5) cm broad, oblanceolate, pinnately toothed or pinnatifid, sparsely to moderately spreading-hairy above and below, rounded to obtuse apically, tapering to a broad petiole basally; cauline leaves alternate, minute or lacking; heads solitary, or more commonly 2–5; peduncles glabrous; involucres 5–15 mm high, 7–20 mm wide; bracts glabrous or stiffly hairy along the midribs; corollas numerous, longer than the bracts; achenes 4–7 mm long, the beak mostly 2–3 mm long. Weedy species of disturbed soils in Davis and Salt Lake counties; widespread in North America; adventive from Europe; 2 (0).

Inula L.

Perennial tomentose herbs; leaves basal and cauline, alternate; heads radiate, large, hemispheric, few to numerous in cymose clusters; involucral bracts imbricate in several series; receptacle naked; ray flowers pistillate, yellow, 3-toothed; disk flowers perfect, fertile; anthers sagittate at the base; style branches of disk flowers linear; pappus of capillary bristles; achenes 4- or 5-ribbed.

Inula helenium L. Elecampane. Perennial herbs, mainly 6–20 dm tall, from thick roots; stems simple below; basal leaves 2–5 dm long, petiolate, the blades ovate to oblong, denticulate, rough-hairy above, velvety beneath; cauline leaves reduced upward, cordate-clasping, acute; heads large and showy; involucres 15–23 mm high, 30–50 mm wide; outer bracts foliaceous, ovate; ray flowers numerous, 18–30 mm long, narrow; achenes glabrous; $2n = 20$. Canal banks and moist meadows at 1370 to 1830 m in Sanpete and Utah counties; widespread in North America; adventive from Eurasia; 2 (i).

Iva L.

Annual or perennial herbs; leaves opposite, at least below; heads discoid, the pistillate flowers few, with corolla tubular or lacking; involucres campanulate; bracts subequal or imbricate in 1–3 series, sometimes with a short inner series subtending the achenes; receptacle chaffy, the receptacular bracts linear to spatulate; staminate flowers with abortive pistils, the styles undivided, the filaments monadelphous; anthers obtuse basally, almost distinct; pappus none; achenes compressed.

1. Leaves sessile or shortly petiolate, entire; plants rhizomatous, mainly less than 40 cm tall, of saline low-elevation sites .... *I. axillaris*
— Leaves petiolate, serrate; plants taprooted annuals, mainly much over 40 cm tall, ruderal weeds .. *I. xanthifolia*

Iva axillaris L. Poverty Weed. Perennial herbs from elongate rhizomes, scented; stems 15–50 (60) cm tall, branched from the base; herbage strigose to strigulose and more or less glandular; leaves opposite below, alternate above, 0.8–4.5 cm long, 4–15 mm wide, elliptic to obovate or lanceolate; heads numerous in terminal bracteate spicate clusters, nodding, 3–7 mm wide; bracts connate, shallowly 4- or 5-lobed; pistillate flowers 4–8, perfect; achenes 2–3 mm long, glandular. Commonly in saline riparian sites in warm desert shrub, salt desert shrub, pinyon-juniper, and aspen communities at 760 to 2440 m in all Utah counties; British Columbia to Manitoba, south to California, New Mexico, and Oklahoma; 60 (viii).

Iva xanthifolia Nutt. Marsh-elder. Coarse perennial herbs, mainly 4–25 dm tall, simple or branched, essentially glabrous below, glandular above; leaves opposite below, petiolate, the blades 4–20 cm long and about as wide, broadly ovate to lance-ovate, serrate and sometimes lobed, green above, canescent beneath; heads 2–4 mm thick, numerous, borne ebracteate in paniculate clusters; involucral bracts distinct, ovate; pistillate flowers 5; achenes sparsely pilose apically, ca 2 mm long.; $2n = 36$. Ruderal weeds of disturbed soils at 1370 to 2290 m in Beaver, Duchesne, Emery, Iron, Kane, Millard, Salt Lake, Sevier, Summit, Uintah, Utah, and Wayne counties; Alberta to Saskatchewan, south to Washington, Arizona, and New Mexico; widely distributed elsewhere; 22 (iii).

Kuhnia L.

Perennial herbs from a woody caudex and taproot; stems branched, erect or ascending; leaves alternate or some lower ones opposite; entire or lobed; heads discoid, several to numerous in paniculate clusters; involucres campanulate; involucral bracts in 4–7 series, the outer ones only graduated; receptacle naked; disk flowers perfect, fertile, whitish; style tips flattened, clavate; pappus of plumose bristles; achenes 10–ribbed.

Kuhnia chlorolepis Woot. & Standl. False Bonset. Perennial clump-forming herbs; stems 30–75 cm tall, much branched, minutely hairy; leaves 8–50 mm long (or more), 1–3 mm wide, entire or with a pair of basal lobes, linear; involucres 8–12 mm high; bracts linear to narrowly oblong, striate; corollas 6–7.5 mm long; achenes 4.8–5.2 mm long, dark brown, short-hairy; $n = 9$. Rabbitbrush community in intermittent stream courses at 1890 to 2045 m in Uintah County; Colorado to Arizona, New Mexico, Texas, and Mexico; 2 (i).

Lactuca L.

Annual, biennial, or perennial herbs; leaves alternate, entire or pinnatifid; flowers all raylike, yellow, blue, or white; heads paniculately arranged; involucres cylindrical; bracts imbricate in several series; receptacle flat, naked; pappus copious, of white or brownish capillary bristles; achenes oval, oblong, or linear in outline, compressed, ribbed on each face, short- to long-beaked.

1. Plants perennial, rhizomatous; rays long-exserted, blue .... L. tatarica
— Plants annual or biennial; rays not long-exserted, yellow (often fading blue) or blue to white (in L. biennis) 2
2(1). Achenes prominently 1-nerved on each side 3
— Achenes prominently several-nerved on each side 4
3(2). Involucres 10–15 mm high in fruit; pappus 5–7 mm long; achenes 4.5–6.5 mm long, including the beak .. .. L. canadensis
— Involucres 15–22 mm high in fruit; pappus 7–12 mm long; achenes 7–10 mm long .... L. ludoviciana
4(2). Involucres cylindrical at anthesis; flowers not fading blue; plants cultivated and occasionally escaping . L. sativa
— Involucres tapering to the apex at anthesis; flowers fading blue 5
5(4). Achenes with a long filiform beak as long as or longer than the body of the achene; pappus white .... L. serriola
— Achenes with a short beak much shorter than the body of the achene; pappus brownish .... L. biennis

Lactuca biennis (Moench) Fern. [*Sonchus biennis* Moench]. Annual or biennial, glabrous or hairy (on midvein of leaves) herbs; stems erect, mainly 6–20 (35) dm tall; leaves mainly 10–40 cm long, 4–20 mm wide, pinnatifid or merely toothed; heads 13- to 50-flowered, numerous, arranged in a narrow paniculate inflorescence; rays bluish to white; pappus brownish; achenes 4–5.5 mm long, prominently several nerved on each face, beakless or short beaked; $2n = 34$. Moist sites at ca 1800 m in Salt Lake County (Arnow 2561 BRY, UT); Alaska to Newfoundland, south to California, Colorado, and North Carolina; 1 (0).

Lactuca canadensis L. Annual or biennial, glabrous or hirsute herbs; stems erect, 3–25 dm tall; leaves mainly 10–35 cm long, 1–12 cm wide, entire to pinnatifid; heads mostly 13- to 22-flowered, arranged in open panicles; rays yellow (fading blue); pappus white; achenes black, obovate, transversely rugose and with 1 prominent longitudinal vein on each face, 4.5–6.5 mm long, including the beak from half as long to as long as the body; $2n = 34$. Weedy species of moist sites at ca 1155 m in Kane County (Atwood 4118 BRY); widespread in U. S.; 1 (0).

Lactuca ludoviciana (Nutt.) Riddell [*Sonchus ludovicianus* Nutt.]. Biennial or short-lived perennial herbs; stems 6–15 dm tall or more; leaves 10–35 cm long or more, mainly 1–10 (20) cm wide, commonly pinnatifid and weakly spinose-toothed, setose-hispid on the lower midrib, the uppermost auriculate-clasping; heads numerous in an open paniculate cluster, the peduncles bracteate; involucres 15–22 mm high in fruit; heads mostly 20- to 50-flowered, the flowers yellow or sometimes blue, fading blue; pappus white, 7–10 mm long at maturity; achenes flattened, blackish, with a longitudinal median nerve on each face, transversely rugulose, 4–5 mm long. Collected once in Salt Lake County (without collector, UT); widespread in the northwestern U. S.; 1 (0).

Lactuca sativa L. Lettuce. Annual herbs; stems erect, mostly 5–12 dm tall, glabrous; leaves mainly 10–30 cm long and as broad, undulate-crisped and serrate, glabrous; involucres 7–10 mm high; heads ca 15-flowered, the flowers yellow, not fading blue, numerous in a paniculate cluster; pappus white; achenes brownish, oblanceolate in outline, flattened, hispid apically, 3.5–4.5 mm long, with 5–7 longitudinal nerves on each face, the beak 2.5–3.5 mm long. Cultivated food plant in much of Utah; introduced from Europe; 2 (0).

Lactuca serriola L. Prickly Lettuce. [*L. scariola*, *scarriola*, orthographic variants]. Biennial or winter annual herbs; stems erect, 3–18 dm tall, hispid below or glabrous overall; leaves mainly 3–30 cm long, 1–10 cm wide, pinnatifid or pinnately lobed, or merely spinose-toothed, the blades vertically oriented (twisted at the base), setose-hispid on main veins beneath; involucres 7–15 mm high at maturity; heads mostly 6- to 12-flowered, the flowers yellow, fading blue, several to numerous in a paniculate cluster; pappus white; achenes brown, the body obovate to oblong in outline, flattened, hispid along margin apically, 3–4.5 mm long, with 5–8 longitudinal nerves on each face, the beak 3–7 mm long; $2n = 18$. Ruderal weeds at 850 to 2440 m, probably in all Utah counties; widely distributed in the U. S.; adventive from Europe; 37 (v). This species invades lower elevation range lands, where it is eaten by wildlife and livestock. It is reported to produce fertile hybrids with *L. sativa* (q.v.).

Lactuca tatarica (L.) C. A. Mey. Blue Lettuce. Perennial rhizomatous herbs; stems 2–12 dm tall, glabrous; leaves 4–20 cm long, 5–35 mm wide, linear to lanceolate or oblong, entire, toothed, lobed, or pinnatifid, short-petiolate below, sessile above; involucres 10–20 mm high; heads cylindric, 15- to 50-flowered, the flowers blue, numerous in an elongate paniculate cluster; pappus white; achenes black to pale, oblong-lanceolate in outline, flattened, 4–7 mm long, with several longitudinal nerves on each face, the beak ca 2 mm long; $2n = 16, 18$. Marshes, canal and stream banks, and roadsides at 1370 to 2440 m in Cache, Daggett, Duchesne, Garfield, Grand, Iron, Juab, Kane, Millard, Piute, Salt Lake, Sevier, Tooele, Uintah, Utah, Washington, and Weber counties; Alaska to Minnesota, south to California and Missouri; 39 (vii). Our material belongs to ssp. **pulchella (Pursh) Stebbins** [*Sonchus pulchellus* Pursh], the North American phase of a circumboreal species.

Lapsana L.

Annual herbs from taproots, the juice milky; leaves alternate, simple, subentire to toothed or lyrate-pinnatifid; heads numerous; involucral bracts in 2 series, the inner ones large and keeled, the outer minute, greenish; receptacle naked; corollas of ray flowers only, perfect, yellow; pappus none; style branches semicylindrical; achenes subterete, several-nerved, tapering at both ends, beakless.

Lapsana communis L. Nipplewort. Plants mostly 2.5–10 dm tall, the stems erect, simple or branched, pubescent with stipitate glands or glabrous; leaves mostly 3–10 cm long and 1.4–5 (7) cm wide, the blades subentire to toothed, or the lower ones lyrate-pinnatifid, sparsely hairy to glabrous above and below; heads numerous, the peduncles glabrous or nearly so; involucres 5–8 mm high, 3–9 mm broad, the bracts glabrous; flowers mostly 10–14; achenes 3–5 mm long; $2n = 14, 16$. Weedy species of disturbed sites in Salt Lake County (Arnow 4747, BRY; UT); widely established in North America; adventive from Eurasia; 1 (0).

Layia H. & A.

Annual herbs from taproots; leaves mainly alternate, subentire to toothed or pinnatifid; heads radiate, solitary or few to several, subcorymbose; ray and disk flowers both fertile; involucres campanulate to broadly hemispheric; bracts with thin margins abruptly dilated below, enclosing the ray achenes; receptacle plano-convex, chaffy marginally; ray flowers 8–24, white or yellow or with the tips white; pappus of numerous bristles, awns, or scales, the bristles often plumose below; ray achenes obcompressed, commonly glabrous and epappose; disk achenes pubescent and pappose.

1. Ray flowers yellow with a white tip; pappus setae merely scabrous; anthers black; plants rare in San Juan County .... *L. platyglossa*
— Ray flowers white; pappus setae plumose; anthers yellow; plants locally common, widespread .... *L. glandulosa*

Layia glandulosa (Hook.) H. & A. Tidytips. [*Blepharipappus glandulosus* Hook.]. Plants slender, the stems simple or branched, 0.8–3 dm tall or more, often reddish, with long spreading-ascending multicellular setae; leaves 0.8–6 cm long, 1.5–16 mm wide, often mainly basal, hispid, toothed to lobed, the cauline ones reduced upward and finally entire; heads solitary or 2 to numerous; involucres 6–9 mm high, 10–18 mm wide; bracts hispid and with some tacklike purplish black stipitate glands; rays white, 6–15 mm long; disk flowers numerous, yellow; ray achenes 3–4 mm long; disk achenes 3–6 mm long; pappus of 10–12 white flattened setose scales plumose to above the middle with straight capillary and tangled woolly hairs; n = 7, 8. Sagebrush-grass, grassland, and pinyon-juniper communities at 1370 to 1865 m in Daggett, Garfield, Juab, Kane, Millard, Salt Lake, Sanpete, Tooele, Utah, and Washington counties; British Columbia, south to Baja California and Arizona; 24 (i).

Layia platyglossa (Fisch. & Mey.) Gray [*Callichroa platyglossa* Fisch. & Mey.]. Plants slender, the stems erect, simple or branched, setose with long multicellular hairs, often reddish; leaves mainly 1–6 cm long, 2–7 mm wide, with long, slender, spreading multicellular hairs, the cauline leaves reduced upward and finally entire; heads solitary or few; involucres 6–12 mm high, 12–20 mm wide; bracts hairy like the leaves and with some tacklike purplish black stipitate glands; rays yellow with white tips, 6–18 mm long; disk flowers numerous; ray achenes 3–4 mm long; disk achenes 3–5 mm long; pappus of scabrous setae. Dunes at ca 1375 m in San Juan County (Harrison 2545 BRY); California. Our material apparently belongs to var. *breviseta* Gray [ssp. *campestris* Keck], and this is apparently the only known station for the species east of California. The collection was taken in 1927. The plants resemble those of *Gaillardia* in a general way, and our material has been filed for more than four decades in a folder of that genus.

Lepidospartum Gray

Shrubs; leaves alternate, linear, entire; heads several to numerous, in corymbose or racemose clusters; heads discoid, the flowers perfect, yellow; involucres subcylindric; bracts chartaceous, imbricate in several series, rounded apically (at least the inner); receptacle flat, naked; anthers sagittate; style branches flattened; pappus of copious capillary bristles; achenes oblanceolate in outline, long-pilose.

Lepidospartum latisquamum Wats. Nevada Broomshrub. Shrubs mainly 6–15 dm tall or more; branchlets with prominent longitudinal striae, the striae glandular, the intervening areas tomentose; leaves 0.5–3 cm long, linear, 0.5–1 mm wide, apiculate; heads 4– to 7– flowered; involucres 8–10 mm high, 3.5–6 mm wide; bracts chartaceous, tomentose, the outer apiculate, very short, the inner broadly rounded and more or less hyaline margined; achenes 4–5 mm long, long-pilose with copious white hairs 3–4 mm long. Rabbitbrush community along a wash at 1705 to 1740 m in Millard County (Pine Valley); Nevada and California; 7 (iii).

Leucelene Greene

Perennial rhizomatous herbs; leaves alternate, simple, entire, linear or subulate; heads radiate, solitary or few to many; involucres turbinate; bracts imbricate in several series, green, the margins scarious; ray flowers white or tinged pink, pistillate; disk flowers perfect, fertile, yellow; pappus of capillary bristles; achenes subcylindric or somewhat compressed.

Shinners, L. H. 1946. Revision of the genus *Leucelene* Greene. Wrightia 1: 82–89.

Leucelene ericoides (Torr.) Greene Rose-heath. [*Inula ? ericoides* Torr.; *L. arenosa* Heller; *Aster bellus* Blake; *A. leucelene* Blake; *A. hirtifolius* Blake*]*. Perennial herbs from a branching caudex and rhizome, simple or more commonly branched, 3–17 cm tall, strigose and more or less glandular; leaves 2–10 mm long, 1–2 (3) mm wide, linear to spatulate, becoming subulate upward; heads solitary or few to many; involucres 5–7 mm high, 5–12 mm wide; bracts in 3–5 series; rays 12–25, white to pink, 3–6 mm long; achenes appressed-hairy; n = 8, 16. Blackbrush, desert shrub, salt desert shrub, pinyon-juniper, and ponderosa pine communities at 1370 to 2595 m in Beaver, Carbon, Daggett, Duchesne, Emery, Garfield, Grand, Iron, Juab, Kane, Millard, Piute, Salt Lake, San Juan, Sanpete, Sevier, Tooele, Uintah, Utah, Washington, and Wayne counties; Nevada and California, east to Kansas, south to Arizona and Mexico; 145 (xvii).

Lygodesmia D. Don

Perennial or annual herbs with milky juice; leaves alternate or mainly basal and still alternate, entire or pinnatifid; heads solitary or few to many in corymbose or paniculate clusters; flowers all raylike, pink to pink purple or white; involucres cylindric; bracts mostly 5–9, with some more or less reduced outer ones; receptacle naked; pappus of numerous capillary bristles; achenes linear, subterete, prominently several nerved.

1. Rays 10–12 mm long, ca 4 mm wide; pappus 6–9 mm long .... *L. juncea*
— Rays 15–25 mm long, 6–10 mm wide; pappus 12–17 mm long 2
2(1). Flowers white (or pink?, and drying pinkish); stems ligneous, branching from the base, forming rounded clumps; leaves stiff, spreading; plants of Emery and Grand counties .... *L. entrada*
— Flowers pink or pink purple; stems various but, if branched from the base, the leaves either lax or the plants of different distribution .... *L. grandiflora*

Lygodesmia entrada Welsh & Goodrich Entrada Rushpink. Perennial herbs from a subterranean caudex, branching from the base, the branches ligneus and wiry, mainly 25–45 cm tall; leaves entire, linear or acicular, 5–30 (70) mm long; peduncles with numerous bracts, 12–20 cm long; involucral bracts hyaline-margined, the outer 5–10 mm long, fimbrillate, the inner ca 6, 16–18 mm long, puberulent at the apex; rays white, ca 3 cm long; pappus barbellate, sordid, 10–15 mm long; achene ribs glabrous. Juniper and mixed desert shrub communities at 1340 to 1465 m in Emery and Grand (type from near Courthouse Wash) counties; endemic; 3 (i). The status of this entity is unclear; certainly it is a portion of the *grandiflora* complex. Further work is indicated.

Lygodesmia grandiflora (Nutt.) T. & G. Showy Rush-pink. [*Erythremia grandiflora* Nutt.]. Perennial herbs from deeply placed elongate rhizomes; stems 0.6–5 dm tall, simple or branched from the base or above; leaves alternate, 0.5–10 cm long or more, 1–5 mm wide, attenuate, gradually to abruptly reduced upward; involucres cylindric, 18–25 mm high, densely hairy to glabrous (?), the outer mostly short and ovate to lance-ovate, the inner 5–9 equal, narrowly oblong; rays 5–10, pink, pink purple, or rarely white, mostly 2–4 cm long; pappus of numerous barbellate tawny bristles; achenes 12–18 mm long, ribbed, glabrous; x = 9. Our material consists of three taxa, which have been regarded at specific rank. Intermediates between the taxa suggest a more conservative approach.

1. Main involucral bracts 8 or 9; flowers 8–12; plants of east central and northeastern Utah .... L. *grandiflora* var. *grandiflora*
— Main involucral bracts 5 or 6; flowers 5–7 (10); plants of Southeastern and western Utah 2
2(1). Uppermost leaves reduced to linear scales mainly 3–10 mm long; achenes 13–19 mm long, smooth on the lower surface .... L. *grandiflora* var. *dianthopsis*
— Uppermost leaves not reduced to scales, mainly 20–40 mm long; achenes 10–13 mm long, rugose on the lower surface .... L. *grandiflora* var. *arizonica*

Var. arizonica (Tomb) Welsh [*L. arizonica* Tomb; *L. doloresensis* Tomb, at least for Utah specimens]. n = 9. Blackbrush-ephedra and Indian ricegrass-dropseed communities at 1125 to 1590 m in Kane and Wayne counties; Arizona; 7 (ii).

Var. dianthopsis (D. C. Eaton) Welsh [*L. juncea* var. *dianthopsis* D. C. Eaton in Wats., type from Great Salt Lake islands; *L. dianthopsis* (D. C. Eaton) Tomb]. n = 9. Sagebrush-grass, pinyon-juniper, and mountain brush communities at 1370 to 2440 m in Beaver, Cache, Kane, Millard, Salt Lake, Sevier, and Utah counties; Nevada; 22 (iii). Intermediate specimens transitional to var. *arizonica* occur in south central Utah.

Var. grandiflora [*L. grandiflora* var. *stricta* Maguire, type from south of Price]. n = 9. Shadscale, sagebrush, pinyon-juniper, mountain brush, ponderosa pine, and aspen-sagebrush communities at 1460 to 2750 m in Carbon, Daggett, Duchesne, Emery, Garfield, Grand, and Uintah counties; Wyoming south to New Mexico; 37 (vii). The var. *stricta* is a phase with stiffly erect leaves, but seems to represent only an ecological variant. Specimens of intermediate nature occur southward with both varieties *arizonica* and *dianthopsis*.

Lygodesmia juncea (Pursh) D. Don [*Prenanthes juncea* Pursh]. Perennial glabrous herbs from a deeply placed elongate root (rhizome?); stems mainly 1.5–6 dm tall, much branched; leaves stiff, entire, mainly 1–4 cm long, 1–4 mm wide, the upper ones reduced to subulate scales; heads few to several, mainly 5 (4–10) –flowered; flowers pink or less commonly white; involucres 9–16 mm high, with 4–8 main bracts and several shorter outer ones; pappus tawny; achenes ca 5–7 mm long, several nerved; n = 9. Our few specimens from sandy sites in mixed desert shrub and juniper communities at ca 1400 to 1590 m in Emery and Juab counties; British Columbia to Minnesota, south to Arizona and Arkansas; 3 (0). This is mainly a Great Plains species, with disjunct populations westward, often in sandy habitats.

Machaeranthera Nees

Annual, biennial, or perennial herbs from taproots; leaves alternate, entire or pinnatifid to toothed or lobed, spinulose apically and the teeth, when present, spinulose; heads solitary or 2 to numerous; involucral bracts in several series, herbaceous apically, chartaceous or coriaceous basally, mainly squarrose; rays pistillate and fertile, pink, lavender, pink purple, or white, or lacking; receptacle naked; anthers not caudate; pappus of capillary bristles; achenes narrowly oblong in outline.

Cronquist, A. and D. D. Keck. 1957. A reconstitution of the genus *Machaeranthera*. Brittonia 9: 231–239.

1. Heads discoid; leaves spinose-toothed .. *M. grindelioides*
— Heads radiate; leaves various, but not conspicuously spinose-toothed 2
2(1). Plants perennial, from a definite caudex, of montane sites, commonly on granite, limestone, or quartzite .. .. *Aster kingii* D.C. Eaton (q.v.)
— Plants biennial, winter annual, or annual (rarely short-lived perennial), the caudex not well developed; plants of various habitats and substrates 3
3(2). Leaves pinnately dissected; plants annual .... *M. tanacetifolia*
— Leaves merely toothed to entire; plants mainly biennial or short-lived perennial 4
4(3). Involucral bracts with green tip commonly equaling or longer than the chartaceous base, the long-tapering apices often reflexed .... *M. bigelovii*
— Involucral bracts with green tip much shorter than the hartaceous base, the reflexed to erect tips shortly attenuate to acute .... *M. canescens*

Machaeranthera bigelovii (Gray) Greene Bigelow Aster. [*Aster bigelovii* Gray; *M. mucronata* Greene, sensu Utah materials; *M. commixta* Greene, type from the Henry Mts.; *M. canescens* var. *commixta* (Greene) Welsh]. Short-lived perennial (biennial in some?) herbs from a taproot, a caudex not or only poorly developed; stems 11–35 cm long, puberulent below, becoming glandular to stipitate-glandular above; basal leaves often withered at anthesis; cauline leaves oblanceolate to linear or oblong, mainly 1–7.5 cm long, 1.5–8 mm wide, the surfaces glabrous and more or less glandular or stipitate-glandular, ciliate, entire to spinose-toothed; heads few to many in corymbose inflorescences; involucres 9–12 mm high, 12–23 mm wide; bracts lance-linear, attenuate apically, the green apex subequal to the coriaceous base, especially in the outer bracts, commonly spreading-reflexed, glandular and glandular-ciliate; rays 21–31, violet or pink purple, 10–15 mm long, 2–4.2 mm wide; pappus off-white; achenes glabrous or sparingly strigose, 2.5–4.2

mm long; n = 4. Mountain brush, aspen, spruce-fir, and alpine meadow communities at 2440 to 3355 m in Garfield, Iron, Kane, and Washington counties (Henry Mts., Markagunt Plateau, and Kolob Terrace); Colorado, New Mexico and Arizona; 18 (iii).

Machaeranthera canescens (Pursh) Gray Hoary Aster. [*Aster canescens* Pursh]. Biennial (winter annual) or short-lived perennial herbs from a taproot, a caudex seldom developed; stems 8–60 cm tall or more, variously glabrous, glandular, or puberulent; basal leaves withered or persistent at anthesis; cauline leaves linear to oblong or oblanceolate, 1–10 cm long, 1–22 mm wide, the surfaces glabrous, puberulent, or glandular, commonly ciliate, entire or toothed; heads few to many in paniculate or corymbose clusters; involucres 6–10 (12) mm high, 6–18 mm wide; bracts linear to oblong, attenuate to abruptly attenuate, the green apex commonly much shorter than the coriaceous base, sometimes spreading-reflexed, glandular and or puberulent; rays 15–25, pink to pink purple or white, 5–12 mm long, 1.5–2.5 mm wide; pappus off-white; achenes pilose, ca 2.5 mm long; n = 4. The *canescens* complex consists of a series of intergrading taxa, which, in the extreme, are distinctive and geographically or edaphically correlated. Many names have been applied to members of the complex, and specimens often bear annotations of several of the names involved. This is partially in recognition of the intermediate nature of the specimens and partially due to the quality of diagnostic criteria. It seems best to treat the materials from Utah as belonging to a single polymorphic species, consisting of four intergrading varieties.

1. Leaf surfaces glabrous, the upper leaves commonly glandular to stipitate-glandular; upper branches usually with numerous bracteate leaves; plants of southeastern Utah, rarely elsewhere .... *M. canescens* var. *aristata*
— Leaf surfaces puberulent, the upper leaves sometimes also glandular; upper branches lacking bracteate leaves or variously so; plants of broad distribution, but not of southeastern Utah 2

2(1). Upper branches with numerous bracteate leaves; rosette leaves abruptly and angularly lobed or toothed; plants biennial, of central and southwestern Utah ....
.... *M. canescens* var. *leucanthemifolia*
— Upper branches seldom especially bracteate; rosette leaves various; plants biennial or short-lived perennial, of various distribution 3

3(2). Involucral bracts 1–1.5 (2) mm broad, abruptly attenuate apically; plants often perennial, mainly of higher middle elevations .... *M. canescens* var. *latifolia*
— Involucral bracts 0.5–1 mm wide, rather gradually attenuate apically; plants often biennial, mainly of lower to middle elevations .... *M. canescens* var. *canescens*

Var. *aristata* (Eastw.) Turner comb. nov. [based on: *Aster canescens* var. *aristatus* Eastw., Proc. Calif. Acad. II. 6: 296. 1896, type from San Juan County; *A. c.* var. *vacans* (A. Nels.) Welsh; *M. pulverulenta* var. *vacans* A. Nels., type from San Juan County, Utah]. Salt desert shrub, mixed desert shrub, pinyon-juniper, and ponderosa pine communities at 1155 to 2380 m in Carbon, Emery, Garfield, Grand, Kane, Juab, San Juan, and Washington counties; Colorado, Arizona, and New Mexico; 61 (xix). This material has been treated as *M. linearis* Rydb., a glabrous-leaved phase of *M. canescens* whose type came from Yellowstone Park, Wyoming. Work of a monographic nature is necessary for the entire *canescens* complex. The earlier epithet at varietal rank, *aristatus*, was called to my attention by B. L. Turner, to whom the combination is gratefully attributed.

Var. *canescens* [*M. pulverulenta* (Nutt.) Greene]. Salt desert shrub, mixed desert shrub, pinyon-juniper, mountain brush, aspen-sagebrush, Douglas fir, and lodgepole pine communities at 1250 to 2900 m in Beaver, Carbon, Daggett, Duchesne, Emery, Garfield, Grand, Iron, Juab, Kane, Millard, Piute, Salt Lake, Sanpete, Sevier, Summit, Uintah, and Washington counties; British Columbia to Saskatchewan, south to California, Arizona, and Colorado; 102 (xiii). This is a variable complex of forms that differs in several morphological features, but further segregation seems unwarranted. I have been unable to distinguish *M. tephrodes* (Gray) Greene from among our rather large collection.

Var. *latifolia* (A. Nels.) Welsh comb. nov. [based on: *M. latifolia* A. Nels., Proc. Biol. Soc. Washington 20: 38. 1901, type from Big Cottonwood Canyon; *M. leptophylla* Rydb., type from Logan; *M. paniculata* A. Nels, type from Parleys Canyon; *M. rubricaulis* Rydb.; *Aster rubrotinctus* Blake]. n = 4. Mountain brush, aspen, Douglas fir, sagebrush, spruce-fir, and alpine meadow communities at 1705 to 3420 m in Beaver, Cache, Carbon, Duchesne, Emery, Garfield, Iron, Juab, Millard, Salt Lake, Sanpete, Sevier, Summit, Tooele, Uintah, Utah, and Wayne counties; Wyoming and Colorado; 82 (v).

Var. *leucanthemifolia* (Greene) Welsh [*Aster leucanthemifolius* Greene; *M. leucanthemifolia* (Greene) Greene]. n = 4. Blackbrush, mixed desert shrub, pinyon-juniper, mountain brush, and ponderosa pine communities at 915 to 2135 m in Beaver, Iron, Juab, Sanpete, Sevier, Utah, and Washington counties; Nevada and Arizona; 43 (xiii). This plant is mainly a xerophyte of sandy and silty habitats at lower elevations in the Great Basin and lower Virgin River drainage systems; it is transitional at higher elevations with the preceding varieties. Phases of var. *canescens* from northeastern Utah have been regarded as portions of this variety, but they seem not to fit the concept of var. *leucanthemifolia*, whose type is from Mineral County, Nevada.

Machaeranthera grindelioides (Nutt.) Shinn. Gumweed Aster. [*Eriocarpum grindelioides* Nutt.; *Haplopappus nuttallii* T. & G.]. Perennial herbs from a woody caudex and stout taproot, the caudex branches more or less clothed with marcescent leaf bases; stems 2–30 cm tall, pilosulose or spreading-hairy below, stipitate-glandular and/or hairy above; basal leaves withered or persistent at anthesis; cauline leaves oblanceolate to spatulate or oblong, mainly 0.5–4.5 cm long, 2–12 mm wide, serrate, the teeth with spinulose tips 1–3 mm long, the surfaces pilosulose and/or stipitate-glandular; heads solitary or few to many in corymbose clusters; involucres 6.5–9.5 mm high, 8–15 mm wide; bracts narrowly oblong, attenuate to an acute apex, the apical portion green or brown, the base chartaceous, erect, glandular; rays lacking; pappus off-white to brownish; achenes densely pilose, ca 3 mm long. Two distinctive varieties are present in Utah.

1. Plants dwarf, often monocephalous; leaves commonly clustered at stem bases; plants of semibarren habitats in the Great Basin .... *M. grindelioides* var. *depressa*
— Plants seldom dwarf, often with more than 1 head; leaves mainly cauline; plants seldom of the Great Basin ......
.... *M. grindelioides* var. *grindelioides*

Var. depressa (Maguire) Cronq. & Keck [*Haplopappus nuttallii* var. *depressa* Maguire, type from Millard County]. Mixed desert shrub, pinyon-juniper, and mountain brush communities at 1465 to 2320 m in Beaver, Juab, and Millard counties; Nevada, a Great Basin endemic; 24 (xiv).

Var. grindelioides Blackbrush, mixed desert shrub, sagebrush, pinyon-juniper, and mountain brush communities at 1340 to 3175 m in Carbon, Daggett, Duchesne, Emery, Garfield, Grand, Juab, Kane, Millard, Rich, San Juan, Sanpete, Sevier, Summit, Uintah, and Utah counties; Montana to Saskatchewan, south to Nevada, Arizona, and New Mexico; 98 (xiii). There is a tendency for leaves of plants from the Great Basin to be more glandular than for those in the main body of distribution in the Colorado drainage system.

***Machaeranthera tanacetifolia* (H.B.K.) Nees** Tansyleaf Aster. [*Aster tanacetifolius* H.B.K.]. Annual (winter annual) herbs; stems 8–50 cm tall, glandular-puberulent and more or less villous; leaves 1–6 cm long, 1- or 2-pinnatifid, the segments ending in spinulose bristles; heads 1 to many, in corymbose clusters; involucres 8–12 mm high, 10–18 mm wide, hemispheric; bracts lance-linear, attenuate, chartaceous basally, green apically, spreading, the reflexed tips glandular; rays 11–23 (36), pink purple or blue purple, 11–14 mm long; pappus off-white; achenes ca 2.5 mm long, pilose. Mixed desert shrub, salt desert shrub, and pinyon-juniper communities at 1125 to 1830 m in Emery, Garfield, Grand, Juab, Kane, San Juan, Sevier, Utah, Wasatch, Washington, and Wayne counties; 31 (vi). The plants are somewhat weedy, colonizing disturbed sandy and silty soils. The similar *M. parviflora* Gray [*Aster parvulus* Blake] is reported for Utah by various authors. It differs in having once-pinnatifid leaves, involucres 4–6 mm long, and rays 5–7 mm long. No material has been seen from Utah by me.

Madia Molina

Annual or biennial tar-scented herbs from taproots; leaves opposite below, alternate above, simple, entire; heads radiate, the rays pistillate, fertile, yellow, or inconspicuous; involucral bracts uniseriate, equal, enfolding the ray achenes; receptacle flat or convex, with a single series of bracts between the ray and disk flowers; disk flowers perfect; pappus none, a short crown, or a few scales; achenes finely striate, commonly incurved, compressed.

1. Heads turbinate-ovoid, 6–12 mm wide (when pressed); rays 4–7 mm long, showy .... *M. gracilis*
— Heads ellipsoid, 2–5 mm wide (when pressed); rays to 2.5 mm long, or lacking .... *M. glomerata*

***Madia glomerata* Hook.** Tarweed. Annual herbs; stems mainly 8–40 (60) cm tall; herbage strigose and with long setiform multicellular hairs on leaf bases and on stems above, and stipitate-glandular upward, malodorous; leaves 1.2–9 cm long, 1.5–7 mm wide, linear; heads in dense terminal clusters or sometimes open; involucres 5.5–9 mm high, 2–5 mm wide; rays inconspicuous, mostly 1.5–2.5 mm long, yellow or purplish; disk flowers 1–10; achenes 5-nerved, glabrous. Sagebrush, mountain brush, aspen, spruce-fir, grass-forb, and alpine meadow communities at 1830 to 3175 m in Cache, Carbon, Davis, Duchesne, Emery, Iron, Juab, Piute, Rich, Salt Lake, Sanpete, Sevier, Summit, Tooele, Uintah, Utah, Wasatch, Washington, and Weber counties; Alaska to Saskatchewan, south to California, Arizona, and Colorado; 38 (iv).

***Madia gracilis* (J. E. Sm.) Keck** [*Sclerocarpus gracilis* J. E. Sm. in Rees]. Annual herbs; stems mainly 10–60 (100) cm tall; herbage pilosulose, becoming hirsute with long multicellular hairs upward, stipitate-glandular with dark capitate glands on the peduncles and involucres; leaves 1–10 cm long, 2–7 (10) mm wide, linear to elliptic or oblong; heads several to many in an open corymbose cluster; involucres 6–11 mm high, 6–12 mm wide; rays conspicuous 5–13, yellow, 4–7 mm long; disk flowers 15–35; achenes often mottled. Opening in mountain brush community at ca 1925 m in Salt Lake County; British Columbia to Montana, south to California; 1 (0).

Malacothrix DC.

Annual (winter annual) or perennial herbs from taproots with milky juice; leaves alternate or mainly basal, mostly pinnatifid; heads of ray flowers only, long-peduncled, solitary or few to several and more or less corymbose; flowers yellow or white; involucres campanulate; bracts subequal in 2–4 series, with a few short outer ones; receptacle flat, setose or naked; rays 5–lobed; pappus of capillary bristles, these more or less united at the base and falling together or with some persistent; achenes columnar, glabrous, 10– to 15–ribbed, crowned or denticulate at the summit.

Williams, E. W. 1957. The genus *Malacothrix* (Compositae). Amer. Midl. Naturalist 58: 494–512.

1. Leaves merely dentate, elliptic to oblong or lanceolate, the cauline ones clasping basally; involucral bracts orbicular to ovate and with broad scarious margins .... *M. coulteri*
— Leaves pinnatifid or incised to pinnately lobed, the cauline ones not especially clasping; involucral bracts linear to narrowly lanceolate 2
2(1). Leaves linear-filiform or pinnately dissected into linear segments .... *M. glabrata*
— Leaves with triangular to oblong lobes or teeth, these sometimes attenuate but not linear 3
3(2). Involucres longer than broad (when pressed); persistent pappus setae 1 or 2; stems decidedly tapering upward; plants rare, in Washington County .... *M. clevelandii*
— Involucres broader than long (when pressed); persistent pappus setae 1–5 or lacking; stems various 4
4(3). Leaves with lateral lobes regularly toothed; involucres mainly less than 10 mm long; achenes 2–2.8 mm long; pappus bristles all deciduous .... *M. sonchoides*
— Leaves with lateral lobes irregularly toothed or lobed; involucres more than 10 mm long; achenes 3–4 mm long; pappus often with 1 or few persistent bristles ... . *M. torreyi*

***Malacothrix clevelandii* Gray** Cleveland Malacothrix. Annual herbs; stems mainly 10–40 cm tall, often branched, glabrous, commonly reddish; leaves basal and cauline, 1–10 cm long, 5–15 mm wide, oblanceolate to elliptic, pinnately lobed or merely toothed; heads few to many in a subcorymbose cluster; involucres campanulate, 6–7 mm high; main involucral bracts linear, glabrous, green, the tips often purple, the margins narrowly scarious; rays yellow or white, ca 2–3 mm long; pappus deciduous or with 1 or 2 persistent bristles; achenes ca 2 mm long, slender, longitudinally striate. Creosote bush, Joshua tree, pinyon-juniper and live oak communities at ca 1375 m in Washington County; California, Nevada, and Arizona; 3 (0).

Malacothrix coulteri Harv. & Gray in Gray Snakeshead Malacothrix. Annual (winter annual) herbs, stems mainly 10–50 cm tall, often branched, glabrous and straw colored to whitish tan; leaves basal and cauline 1.2–10 cm long, oblong to oblanceolate or lanceolate, the cauline ones clasping basally; heads few to several, corymbose; involucres hemispherical, 10–15 mm high; bracts imbricate, suborbicular to ovate, with broad scarious margins, the midline broad, purplish; rays yellow to off-white, 5–18 mm long; pappus with 1–4 persistent bristles; achenes 2–2.8 mm long, striate; n = 7. Warm desert shrub community at ca 950 m in Washington County (Galway sn BRY); Arizona and California; 2 (0).

Malacothrix glabrata (D. C. Eaton) Gray [*M. californica* var. *glabrata* D. C. Eaton]. Annual (winter-annual) or biennial herbs; stems mainly 10–60 cm tall, often branched from the base and above, glabrous; leaves basal and cauline, 0.5–15 cm long, pinnately lobed, glabrous or more or less villous, with rachis and lobes linear to linear-filiform, the cauline ones similar to the basal except reduced upward; head solitary or more commonly few to many and subcorymbosely arranged; involucres broadly campanulate 10–14 mm high, the main bracts linear to narrowly oblong, with narrow hyaline margins, glabrous, the outer bracts commonly more or less villous; rays yellow, 10–20 mm long; pappus with usually 2 persistent bristles; achenes 2–3 mm long, striate; n = 7. Joshua tree, blackbrush, Vanclevea-ephedra, and pinyon-juniper communities at 700 to 1525 m in Kane, Millard, San Juan, and Washington counties; Oregon to Idaho, south to California and Arizona; 22 (iv).

Malocothrix sonchoides (Nutt.) T. & G. [*Leptoseris sonchoides* Nutt.]. Annual or winter annual herbs; stems mainly 6–37 cm tall, often branched from the base and above, glabrous or with short yellowish glandular hairs in the inflorescence; leaves basal and cauline, 0.7–12 cm long, 1–28 mm wide, the basal ones at least pinnatifid and the lobes regularly toothed; heads solitary or more commonly few to many and subcorymbosely arranged; involucres campanulate 7.5–10.2 mm high, 6.5–12 (14) mm wide, the main bracts lance-oblong to linear, with narrowly hyaline margins, glabrous, the outer sometimes with yellowish stipitate glands; rays yellow, 7–12 mm long; pappus bristles all deciduous; achenes 2–2.8 mm long, striate; n = 7. Blackbrush, Krameria-psorothamnus, mixed desert shrub, sagebrush, and pinyon-juniper communities at 915 to 1856 m in Beaver, Duchesne, Emery, Garfield, Grand, Juab, Kane, Millard, San Juan, Tooele, Uintah, Washington, and Wayne counties; California and Nevada, east to Nebraska and New Mexico; 72 (vi).

Malacothrix torreyi Gray [*M. sonchoides* var. *torreyi* (Gray) E. W. Williams]. Annual or winter annual herbs; stems mainly 8–29 cm tall, often branched from the base and above, glabrous, or with yellowish stipitate glands in the inflorescence; leaves basal and cauline, 1.7–9.5 cm long, 5–27 mm wide, the basal ones at least pinnatifid, and the lobes irregularly toothed or lobed, often more or less white-villous; heads solitary, or more commonly few to several or many and subcorymbosely arranged; involucres broadly campanulate, 10.5–15 mm high, 12–21 mm wide, the main bracts lance-linear, with hyaline margins, glabrous or some with stipitate yellowish glands, the outer bracts often stipitate-glandular; rays yellow, 10–14 mm long; pappus all deciduous or with 1–5 persistent setae; achenes 3–4 mm long, striate; n = 7. Shadscale, greasewood, other salt desert shrub, and mixed desert shrub communities at 1460 to 1925 m in Beaver, Box Elder, Carbon, Duchesne, Emery, Garfield, Grand, Juab, Millard, Piute, Salt Lake, Sevier, Tooele (type from Great Salt Lake), and Uintah counties; Oregon to Wyoming, south to California and Arizona; 28 (i).

Matricaria L.

Biennial or perennial herbs; leaves alternate, 2- to 3-pinnatisect, with ultimate segments linear-filiform; heads radiate, few to many in corymbose clusters; involucres broadly campanulate, the bracts in several series, the margins scarious; receptacle hemispheric, solid, naked; rays pistillate, white; disk flowers 5–lobed, perfect, yellow; pappus a small crown; achenes laterally compressed, with 3 smooth ribs on the ventral surface and 1 or 2 (rarely more) resin glands at the apex of the dorsal face. Note: Tentatively I have chosen to follow authors of Flora Europaea (Vol. 4) in segregating *Chamomilla* (q.v.) from *Maticaria*. The genera are much alike and are separated mainly on technical characteristics that are discernible only when fruit is mature.

Matricaria maritima L. Biennial, or less commonly, perennial, essentially unscented herbs; stems 1–6 dm tall, glabrous or nearly so; leaves 1–8 cm long, the ultimate segments long and slender; heads several to many, the disk 8–15 mm wide; rays 10–25, white, 6–13 mm long; n = 18. Ruderal weed of moist sites at 1830 to 2135 m in Salt Lake, Sanpete, and Sevier counties; widespread in North America; adventive from Europe; 2 (i).

Microseris D. Don

Annual or perennial, scapose or caulescent taprooted herbs with milky juice; leaves alternate or all basal, entire or pinnatifid; heads many flowered, erect or nodding in bud; involucres cylindric to campanulate, the innermost bracts lance-attenuate, subequal, the outer ones shorter and imbricate; receptacle naked; corollas all raylike, showy, yellow to yellow orange (fading bluish); pappus of awn-tipped scales or of plumose capillary bristles; achenes columnar to fusiform, not or only short beaked, ca 10-ribbed.

1. Plants annual; pappus of 5 scales, each extended into a scabrous bristle apically .... *M. lindleyi*
— Plants perennial; pappus of numerous plumose capillary bristles arising from short scales .... *M. nutans*

Microseris lindleyi (DC.) Gray [*Calais lindleyi* DC.; *Uropappus linearifolius* Nutt.; *M. linearifolia* (Nutt.) Schultz-Bip.]. Annual herbs from slender taproots; herbage puberulent or glabrate; stems lacking or more or less developed, the scapose peduncles 10–25 cm high; leaves 6–15 (30) cm long, pinnately lobed to entire, linear to narrowly elliptic; heads many flowered, erect, the main bracts lance-attenuate, 15–30 mm long, subequal, the outer ones shorter and unequal; rays yellow (drying blue); pappus 10–20 mm long, silvery, deciduous, of 5 lance-linear scales, each terminating in a scabrous awn from a bifid apex; achenes dark brown, 9–13 mm long, tapering apically, scabrous on the ribs; n = 9. Blackbrush, creosote bush, and pinyon-juniper communities at 915 to 1375 m in Washington County; Washington to Idaho, south to Baja California and Arizona; 4 (i).

Microseris nutans (Geyer) Schultz-Bip. [*Scorzonella nutans* Geyer; *Ptelocalais macrolepis* Rydb., type from Salt Lake City]. Perennial herbs from tuberous roots; herbage glabrous or sparsely scurfy; stems more or less developed, the scapose peduncles mainly 12–40 (60) cm high, pinnately lobed to entire, linear to elliptic, lanceolate, or oblanceolate; heads solitary or 2–5, many flowered, nodding in bud, the main bracts 10–20 mm long, lance-attenuate, subequal, the outer bracts shorter and unequal; rays yellow (drying lavender or blue); pappus of numerous narrow scales, each with a plumose terminal bristle; n = 9. Sagebrush, pinyon-juniper, mountain brush, Douglas fir, and aspen communities at 1675 to 2745 m in Box Elder, Cache, Daggett, Davis, Juab, Millard, Rich, Salt Lake, Sanpete, Sevier, Summit, Uintah, Utah, and Weber counties; British Columbia to Montana, south to California, Nevada, and Colorado; 33 (ii).

Monoptilon T. & G. ex Gray

Annual herbs, branched from base, the herbage hispid; leaves alternate, spatulate, entire; heads radiate, solitary on branch tips, closely subtended by upper leaves; involucre campanulate, the bracts subequal, linear, herbaceous; receptacle flat, naked; ray flowers pistillate, white to pink; disk flowers perfect, fertile, yellow (purplish); pappus of a short scarious cup and 1 apically plumose bristle, or of numerous bristles alternating with short scales; achenes compressed, marginally nerved, pubescent.

1. Pappus of usually several nonplumose bristles alternating with scales; disk corollas sparsely if at all pilose; reported for Utah by Abrams and Ferris (Illustrated Flora of the Pacific States), but not seen by me .... *M. bellioides* (Gray) Hall
— Pappus consisting of minute scales and a single apically plumose bristle; disk corollas densely pilose below .... *M. bellidiforme*

Monoptilon bellidiforme T. & G. in Gray Depressed annual branching herbs; stems 1–5 cm high; leaves 4–10 mm long, 0.5–2.5 mm wide, narrowly oblanceolate; heads showy; involucres 4–5 mm high; bracts linear, attenuate or acuminate, hirsute, and minutely glandular; rays 12–20, ca 4–5 mm long, the tube pilose; pappus of 1 apically plumose bristle and several shorter scales, or the pappus rarely lacking; achenes ca 2 mm long; n = 8. Warm desert shrub at 700 to 900 m in Washington County; California, Nevada, and Arizona; 2 (0).

Onopordum L.

Biennial caulescent spiny herbs from taproots, the juice watery; leaves basal and cauline, alternate, winged-decurrent; heads solitary or few to several; involucral bracts in several series, imbricate, spine tipped; receptacle flat, fleshy, honeycombed, often with short bristles on the partitions, not densely bristly; corollas all discoid, reddish purple or pink, perfect; pappus bristles barbellate; achenes glabrous, subquadrangular, 4- or 5-ribbed.

Onopordum acanthium L. Scotch Thistle; Winged Thistle. Biennial herbs; stems mainly 5–15 (30) dm tall; leaves of basal rosettes 5–50 cm long or more and 2–15 cm wide, pinnately lobed and serrate-dentate, tomentose on both surfaces, but less so above, spinose; cauline leaves pinnatifid, tomentose to glabrate, strongly winged-decurrent along the stem length; involucres 25–35 mm high, 30–65 mm wide, the bracts lance-attenuate, with spreading spine tips, tomentose to glabrate marginally, the inner erect; spines 3–5 mm long, yellowish; corollas reddish purple to pink; 2n = 34. Ruderal weeds at low elevations in Millard, Salt Lake, Summit, Tooele, Utah, Wasatch, and Washington counties; adventive from Europe; 8 (ii). This handsome but troublesome thistle is spreading through the state, but less vigorously than the musk thistle, *Carduus nutans* (q.v.).

Oxytenia Nutt.

Perennial riparian herbs from a ligneus caudex; leaves alternate, pinnately divided or some entire, the segments linear-filiform, involute; heads discoid, numerous, in elongate paniculate inflorescences; marginal flowers 5, pistillate, inner flowers 10–30, staminate; flowers yellowish white; involucral bracts 5, orbicular, mucronate; receptacle chaffy, the chaffy bracts slender, with dilated villous tips; pappus lacking; achenes obovoid, densely villous, 1-ridged on each face.

Oxytenia acerosa Nutt. Copperweed. [*Iva acerosa* (Nutt.) R. Jackson]. Perennial herbs; stems erect, mainly 5–12 dm tall, broomlike in the inflorescence, striate; leaves 3–15 cm long, pinnately 3- to 7-lobed, or the upper ones simple; herbage strigulose; heads 3–4 mm wide, erect or ascending; involucral bracts herbaceous, strigulose; achenes 1.5–2 mm long, black, long villous-pilose. Saline riparian areas and near seeps and springs at 1220 to 2135 m in Carbon, Emery, Garfield, Grand, Kane, San Juan, and Washington counties; Colorado, New Mexico, Arizona, Nevada, and California; 25 (vii). Copperweed is poisonous to livestock.

Palafoxia Lag.

Annual herbs; leaves alternate, entire; heads discoid, few to several, corymbose or paniculate; involucres cylindric; bracts in 1 series, herbaceous; receptacle flat, naked; flowers white, all alike or the outer with unequal lobes; pappus scales 4–8, slender, unequal, with a strong nerve; achenes linear, quadrangular.

Palafoxia arida Turner & Morris Spanish Needle. [not *P. linearis* (Cav.) Lag., misapplied]. Annual herbs; herbage hispid with slender multicellular hairs, glandular upward; stems commonly branched above the base, 1–7 dm tall; leaves petiolate, the blades 1–7.5 cm long, 2–8 mm wide, linear-lanceolate, long-attenuate; involucres 12–18 mm high, glandular, and more or less hispid, 10- to 20-flowered, the cololllas white with pink exserted styles; pappus scales usually 4; achenes strigose; n = 12. Warm desert shrub community at 700 to 1000 m in Washington County; California to Arizona and Mexico; 5 (0).

Parthenium L.

Herbs or shrubs; leaves alternate, entire or lobed; heads solitary or few and more or less clustered, inconspicuously radiate; ray flowers 5, white, pistillate, fertile, persistent; disk flowers staminate; receptacle plano-convex, chaffy throughout; pappus of 2 or 3 awns or scales; ray achenes dorsiventrally compressed, rotund in outline, the margins thickened into riblike structures attached to the contiguous pair of infertile disk flowers and the subtending bract, the achene, the 2 attached flowers, and the bract falling as a unit.

Rollins, R. C. 1950. The guayle rubber plant and its relatives. Contr. Gray. Herb. 179: 1–73.

1. Plants shrubs, the internodes apparent; heads seldom solitary; known from Washington County .... *P. incanum*
— Plants pulvinate-caespitose herbs, the internodes not apparent; heads solitary; plants of eastern Utah .. *P. ligulatum*

Parthenium incanum H.B.K. Aromatic shrub, 4–10 dm tall, much branched, the branchlets loosely tomentose, becoming glabrate; leaves short-petioled, the blades 0.5–5 cm long, 0.4–1.5 cm wide, lobed, white-tomentose below, less so above; heads several to many, corymbose, 3–5 mm wide, outer involucral bracts oblong, acute, villous, the inner ones suborbicular, membranous; rays white, emarginate to incised, 1–2 mm long; pappus of 2 or more pubescent awns; achenes black, oblanceolate, 1.5–2 mm long, pubescent on the ventral surface; n = 18, 36. Limestone cliffs in creosote bush-blackbrush community at ca 1220 m in Washington County (Higgins 4102 BRY); Arizona to Texas, south to Mexico; 1 (0).

Parthenium ligulatum (Jones) Barneby [*P. alpinum* var. *ligulatum* Jones, type from Theodore (Duchesne); *Bolophyta ligulata* (Jones) W. A. Weber]. Pulvinate caespitose to merely caespitose acaulescent mound-forming herbs to ca 3 cm high, from a taproot and branched caudex, the caudex branches densely clothed with brownish marcescent leaf bases and often with ashy leaves of the previous year; leaves 3–20 mm long, 1.5–4 mm wide, spatulate to oblanceolate, strigose; heads solitary at branch ends, sessile, 5–7 mm high, 4.5–6 mm wide; outer bracts oblong, densely pubescent apically; pappus scales distinct or adnate to the corolla tube; rays white, 1–2 mm long, emarginate; achenes oblanceolate, densely hairy, 4–5 mm long, 2–3 mm wide. Barren or semibarren calciferous or gypsiferous outcrops of the Green River, Uinta, Ferron, and Carmel formations in salt desert shrub and pinyon-juniper communities at 1705 to 2135 m in Daggett, Duchesne, Emery, and Uintah counties; Colorado (a Colorado Plateau endemic); 42 (iv). This amazing plant is one of a series of edaphically restricted mound-formers in semibarren habitats on shales and clays of arid sites in Utah. It belongs to a closely related assemblage of two or three taxa within section Bolophytum, and has been regarded at specific status within the segregate genus *Bolophyta*. Its phylogenetic position was reviewed by Rollins (1950), and its status within *Parthenium* seems to represent best its generic affinities.

Pectis L.

Annual herbs; leaves opposite, entire, glandular-dotted; heads radiate, few to several in cymose clusters; involucres turbinate to subcylindric; bracts 3–12 in one series, expanded basally, enclosing the ray flowers, often with translucent glands; receptacle naked; ray flowers perfect, yellow; disk flowers few; anthers entire, obtuse at base; style branches short, hispidulous; pappus of short-plumose bristles on disk flowers, that of ray flowers a short crown of united scales; achenes terete.

1. Leaves dilated at base; pappus a short crown, with or without 1 or 2 awns .... *P. angustifolia*
— Leaves not dilated at base; pappus commonly of numerous plumose bristles .... *P. papposa*

Pectis angustifolia Torr. Annual herbs; stems dichotomously branched, 3–20 cm long, the gerbage greenish; leaves 10–50 mm long, 0.5–2 mm wide, with a few setae along the expanded base; heads on peduncles 0.2–1 cm long; involucres gibbous at the bawe, sparingly glandular; ray flowers yellow, 709, 4–6 mm long; achenes 4–5 mm long. Sandy sites in mixed desert shrub communities at 1300 to 1400 m in Kane and San Juan counties; Arizona to Nebraska, Texas, and Mexico; 8 (iii).

Pectis papposa Harv. & Gray in Gray Chinch-weed. Annual herbs; stems dichotomously branched, often decumbent-spreading, 5–20 (25) cm long, the herbage yellowish green; leaves 6–40 (60) mm long, 0.5–2 mm wide, with a few setae at the base, glabrous, bearing oval to elliptic large yellowish glands; heads on peduncles mainly 0.3–1 (2) cm long; involucres gibbous at the base, rounded dorsally, sparingly glandular like the leaves; ray flowers yellow, 7–9, ca 4–6 mm long; achenes 4–5 mm long, stipitate-glandular, n = 12. Sandy soil in warm and sandy desert shrub communities at 850 to 1300 m in Washington County; California to New Mexico, and south to Mexico; 8 (i).

Perezia Lag.

Perennial herbs from a caudex, this clothed with rusty woolly hairs; leaves alternate, simple, clasping; heads numerous in corymbose cymes, apparently discoid; involucres campanulate, strongly imbricate; flowers perfect, pink to pink purple, the corollas bilabiate, the outer lip 3-toothed, the inner lip recurved, 2-toothed; anthers appendaged; style branches flattened, truncate apically; pappus of white capillary bristles; achenes subterete, minutely glandular.

Perezia wrightii Gray [*Acourtia wrightii* (Gray) Reveal & King]. Perennial herbs; stems 4–6 (10) dm tall, often purplish at the base, the rusty hairs at stem base copious; leaves lance-oblong to ovate or lanceolate, spinulose-dentate, glandular-puberulent on both sides, the lower ones petiolate, becoming sessile and clasping upward; involucres 5–8 mm high and about as broad, the bracts graduated, the outer ones ovate, the inner lance-oblong, obtuse, green, the margins often purplish, ciliate; corollas pink to pink purple; achenes 4.8–5.2 mm long; n = 27. Warm desert shrub and juniper communities at 915 to 1525 m in Kane, San Juan, and Washington counties; Arizona to Texas, south to Mexico; 7 (i).

Perityle Benth.

Annual herbs or perennial subshrubs; leaves mostly opposite below, alternate above, simple, petiolate; heads few to numerous, corymbose, radiate or discoid; involucres hemispheric or turbinate, the bracts somewhat keeled, in 1 or 2 subequal series; receptacle flat, naked; ray flowers (when present) pistillate, white or yellow; disk flowers perfect; anthers subentire to auriculate at the base; style branches linear or subulate; achenes flattened; pappus of scales, or of 1 or 2 awnlike bristles, or lacking.

Powell, A. M. 1973. Taxonomy of *Perityle* section Laphamia (Compositae-Helenieae-Peritylinae). Sida 5: 61–128.
Powell, A. M. 1974. Taxonomy of *Perityle* section Perityle (Compositae-Peritylinae). Rhodora 76: 229–305.

1. Plants annual .... *P. emoryi*
— Plants subshrubs 2
2(1). Heads radiate; plants glandular-hispidulous, of the Great Basin .... *P. stansburyi*
— Heads discoid; plants villous or glandular-hispidulous, of the Colorado or Virgin drainages 3
3(2). Herbage short-villous and more or less glandular; pappus bristles 1; plants of Washington County .... *P. tenella*

— Herbage hispidulous; pappus of 3 (4) unequal bristles; plants of Grand County .... *P. specuicola*

Perityle emoryi **Torr. in Emory** Emory Rock-daisy. Annual herbs, mainly 2–5 dm tall; stems erect or spreading, commonly branched above, puberulent; leaves mostly alternate, petiolate, the bases 0.5–4 cm long, 0.6–3 (5) cm wide, ovate, cordate, or suborbicular, toothed, lobed, cleft, or divided, the lobes again toothed or lobed, hirsute to glandular-pubescent; heads radiate; involucres 5–6 mm high and usually broader; rays 8–12, white, 1.5–5 mm long; disk flowers numerous; pappus vestigial or a crown of scales and 1 slender bristle; achenes 2–3 mm long, the flattened faces nearly glabrous, the margin thickened and bearing short stiff hairs. Sand sagebrush community at lower elevations in Washington County (Tanner sn 1941 BRY); Nevada, California, Arizona, and Mexico; 1 (0).

Perityle specuicola **Welsh & Neese** Alcove Rock-daisy. Perennial suffruticose herbs, mainly 50–75 cm tall; stems sprawling or pendulous, much branched; herbage glandular-hispidulose; leaves mostly alternate, short-petiolate, the blades 3–6 mm long, 1.5–3 mm wide, ovate-elliptic, entire, hispidulous; heads few to many in a branching corymbose inflorescence; involucres 3.5–5 mm high, 5–6 mm wide; bracts 11–16, oblong to elliptic, keeled; ray flowers lacking; disk flowers numerous, ca 2.5 mm long, whitish (?); pappus of 3 unequal scabrous bristles and often with 1 apically flattened and sigmoid scale; achenes 3–3.8 mm long, the faces flattened, glabrous, the margin thickened and with short ascending hairs. Hanging garden communities at ca 1125 to 1220 m in Grand and San Juan counties; endemic; 5 (v).

Perityle stansburyi **(Gray) Macbr.** Stansbury Rock-daisy. [*Laphamia stansburyi* Gray, type from Stansbury Island]. Suffruticose perennials, clump-forming, 7–30 cm tall and as broad or more; herbage glandular-hirtellous; leaves mainly alternate, the blades 3–14 mm long, 1.5–12 mm wide, broadly ovate to deltoid or orbicular, typically few to several lobed; petioles 1–14 mm long; heads few to many in a branching corymbose inflorescence; involucres 5–6.5 mm high, 5–10 mm wide; bracts 16–21, lanceolate to oblanceolate, strongly keeled; ray flowers 10–14, yellow, 3–5.5 mm long; disk flowers numerous, yellow, 4–5 mm long; pappus of 1 stout bristle and a very short crown of hyaline scales; achenes 2.3–3.5 mm long, with thin callous margins, short-pubescent on margins and on faces; n = 17. Limestone, dolomite, and igneous ignimbrite (ash flow tuff) outcrops, in mixed desert shrub, pinyon-juniper, and mountain brush communities, at 1280 to 1895 m in Beaver, Juab, Millard, Salt Lake, Sanpete, Sevier, Tooele, and Utah counties; Nevada; a Great Basin endemic; 39 (v).

Perityle tenella **(Jones) Macbr.** Jones Rock-daisy. [*Laphamia palmeri* Gray, type from Beaver Dam, Arizona?, not *P. palmeri* Wats.; *L. palmeri* var. *tenella* Jones, type from Springdale]. Suffruticose perennials, clump-forming, 9–25 cm tall and as broad or more; herbage villous and glandular; leaves mainly alternate, the blades 4–13 mm long, 3–15 mm wide, deltoid-ovate, the base obtuse to truncate or cordate; petioles 1–8 mm long; heads solitary or few to many, corymbose; involucres 4–6.5 mm long, 5–10 mm wide; bracts 11–18, lance-elliptic, keeled; ray flowers absent; disk flowers numerous, yellow, 3–4 mm long; pappus of a single bristle; achenes 2.5–3 mm long, with thin callous margins, short-pubescent on margins and on faces. Joshua tree, creosote bush, blackbrush, warm desert shrub, pinyon-juniper, and ponderosa pine communities at 915 to 2135 m in Washington County; Arizona; a Mohave endemic; 16 (iv). Plants from the Beaver Dam Mts. have heads that average larger, but they seem not to differ otherwise from the typical materials taken near Zion National Park.

Petradoria Greene

Suffrutescent perennials from a taproot and woody caudex; stems herbaceous, leafy; leaves basal and cauline, alternate, entire, 3- to 5-veined, coriaceous; heads radiate (in ours), congested at branch ends in an open corymbose inflorescence; involucres cylindric; bracts in several series, in more or less vertical ranks; flowers 4–7, yellow, the corollas glabrous; pappus of brownish capillary bristles; achenes somewhat compressed, glabrous.

Anderson, L. C. 1964. Studies on *Petradoria* (Compositae); anatomy, cytology, taxonomy. Trans. Kansas Acad. Sci. 66: 632–684.

Petradoria pumila **(Nutt.) Greene** Rock Goldenrod. [*Chrysoma pumila* Nutt.]. Plants from a well-developed caudex, the caudex branches clothed with dark to ashy or tan marcescent leaf bases; leaves 1.5–12 cm long, 1–11 mm wide, oblanceolate to lanceolate, elliptic, or linear; cauline leaves reduced upward; heads numerous; involucres 5–9.5 mm high, 1.3–2.8 mm wide; involucral bracts 10–21, in 3–6 series, more or less keeeled; flowers 2–8, the rays 1–3, yellow, 4–9 mm long; achenes 4–5 mm long, glabrous, 6- to 9-nerved; 2n = 18. Shadscale, mixed desert shrub, pinyon-juniper, sagebrush, and ponderosa pine communities at 1525 to 3050 m in all (?) Utah counties; Idaho and Wyoming, south to Nevada, California, Arizona, and New Mexico; 100 (xv). Most of our specimens belong to the broad-leaved **var. *pumila***, but one specimen from Emery County (Harris 546 BRY) seems to be clearly allied to **var. *graminea*** **(Woot. & Standl.) Welsh** [*Petradoria graminea* Woot. & Standl.; *P. p.* ssp. *graminea* (Woot. & Standl.) L.C. Anderson). That taxon has been known previously only from Arizona.

Peucephyllum Gray

Aromatic evergreen shrubs; leaves alternate, linear, subterete, resinous-punctate, overlapping; heads solitary on branchlet tips, discoid; involucres campanulate, the bracts in 2 series, narrowly subulate, the outer ones subterete, the inner flattened; receptacle flat, naked; flowers perfect, fertile; corolla lobes short; achenes oblong-obconic, 10-striate, appressed-hairy; pappus of scabrous capillary bristles, some of them sometimes flattened and with hyaline margins.

Peucephyllum schottii **(Gray) Gray** Pigmy-cedar. [*Psathyrotes schottii* Gray]. Shrubs mainly 1–2 m tall, with a single stem, this much branched upward, rounded to flat-topped; leaves mainly 5–12 (20) mm long; heads subsessile; involucre 6–8 mm high, green and resin-coated; flowers yellow; achenes 3.5–4 mm long; pappus 3–4 mm long, tawny. Talus slopes in creosote bush community at ca 1037 m in the Beaver Dam Mts., Washington County; California, Nevada, and Arizona; 3 (i). This plant, more characteristic of the hotter deserts to the south, was discovered first in Utah by G. I. Baird, during April of 1986.

Platyschkuhria (Gray) Rydb.

Perennial herbs from a woody caudex and rootstock; leaves alternate, simple, coriaceous, often impressed-punctate; heads few to many in a cymose paniculate cluster, radiate, campanulate to hemispheric; involucral bracts subequal in 2 series; receptacle essentially flat, naked; rays pistillate, fertile, yellow; disk flowers numerous, perfect; anthers more or less sagittate basally; pappus of 8–16 scales with midribs sometimes produced apically; achenes narrowly obpyramidal and 4–sided, hairy or glabrous on the sides.

Ellison, W. L. 1971. Taxonomy of *Platyschkuhria* (Compositae). Brittonia 23: 269–279.

Platyschkuhria integrifolia (Gray) Rydb. [*Schkuhria integrifolia* Gray; *Bahia nudicaulis* Gray; *B. integrifolia* (Gray) Macbr.]. Perennial herbs; stems solitary or few to several, mainly 12–55 cm tall; herbage white-strigulose or stipitate-glandular, especially above; main leaves near the stem base, petiolate, the blades 1.5–9.5 cm long, 0.5–4 cm wide, ovate to lanceolate, elliptic, or oblanceolate; cauline leaves reduced upward, finally merely bracteate; heads (1) 2–10; rays 7–11, yellow, 6–14 mm long; achenes 5–8 mm long; n = 12, 24, 36. Three rather distinctive varieties are present in eastern Utah, as indicated below.

1. Stems leafy almost or quite to the apex; plants of southeastern San Juan County . *P. integrifolia* var. *oblongifolia*
— Stems leafy mainly below the middle; plants not known from San Juan County 2
2(1). Pubescence of upper stems merely white-strigulose; involucral bracts caudate-attenuate ....
.... *P. integrifolia* var. *ourolepis*
— Pubescence of upper stems stipitate-glandular; involucral bracts mainly obtuse to acute ....
.... *P. integrifolia* var. *desertorum*

Var. *desertorum* (Jones) Ellison [*Bahia desertorum* Jones, type from Cisco; *P. desertorum* (Jones) Rydb.]. n = 12, 36. Salt desert shrub, pinyon-juniper, and mountain brush communities, mainly in saline substrates, at 1280 to 2565 m in Carbon, Duchesne, Emery, Garfield, Grand, Sevier, Uintah, and Wayne counties; Colorado; a Colorado Plateau endemic. A report by Ellison (1971) of var. *integrifolia* (a Wyoming-Montana endemic) belongs here; 39 (xi). The var. *desertorum* is closely allied with var. *integrifolia*, as indicated by pubescence and bract shape similarities. This variety is transitional with var. *ourolepis*.

Var. *oblongifolia* (Gray) Ellison [*Schkuhria integrifolia* var. *oblongifolia* Gray; *Bahia oblongifolia* (Gray) Gray; *P. oblongifolia* (Gray) Rydb.]. n = 36. Desert shrub communities in San Juan County; Arizona, Colorado, New Mexico; 1 (i).

Var. *ourolepis* (Blake) Ellison [*Bahia ourolepis* Blake, type from Green River]. n = 12. Salt desert shrub and pinyon-juniper communities at 1280 to 1830 m in Duchesne, Emery, Grand, and Uintah counties; endemic; 24 (iii). The main body of this variety lies in Uintah County.

Pluchea Cassini

Shrubs or aromatic annual herbs; leaves alternate, simple, entire or crenate to serrulate, glabrous to glandular pubescent, or sericeus; heads discoid, few to numerous, aggregated in terminal cymose or corymbose clusters; involucres campanulate to cylindric; bracts imbricate in several series, herbaceous or scarious, the outer ones sericeus to glandular and sometimes ciliate; receptacle flat or concave, naked; outer flowers pistillate, numerous, their filiform corollas 3– or 4–toothed; central flowers perfect but the innermost sterile, their tubular corollas 5–toothed, that of the fertile flowers 3–lobed; anthers sagittate basally; pappus of outer flowers a single whorl of barbellate bristles or merely capillary bristles, those of inner flowers clavate apically, or all uniformly linear.

1. Plants woody shrubs; leaves narrowly lanceolate to lanceolate, densely sericeus .... *P. sericea*
— Plants herbaceous; leaves broadly elliptic to ovate-lanceolate not densely sericeus .... *P. camphorata*

Pluchea camphorata (L.) DC. Camphor-weed. [*Erigeron camphoratum* L.]. Annual or perennial herbs, 1.5–10 dm tall or more; stems glabrate below, puberulent upward, leafy to the summit; leaf blades elliptic to oblong-elliptic, 6–15 cm long, 1–7 cm wide, dentate-serrate to repand-serrate to subentire, sparsely resinous on both sides, puberulent beneath; heads in a paniculiform cyme; bracts with resin globules, the outermost sparingly puberulent and ciliate, the median and inner ones not pubescent; achenes usually less than 1 mm long, cylindric, 4– to 6–angled, brown to reddish brown, setose to hirtellous or glabrous. Reservoir bottom and stream margin at ca 850 m in Washington County; Arizona east to the southeastern U. S.; 4 (ii). Plants previously identified *P. camphorata* in Arizona were placed by R. K. Godfrey (Jour. Elisha Mitchell Sci. Soc. 68: 2: 238–272. 1952) into *P. purpurascens* (SW.) DC. Our material appears to more closely fit *P. camphorata*.

Pluchea sericea (Nutt.) Cov. Arrowweed. [*Polypappus sericeus* Nutt.; *Tessaria sericea* (Nutt.) Shinn.]. Shrubs with slender, erect, willowlike branches, mainly 0.8–3 m tall, sericeus throughout, longitudinally striate; leaves 0.8–4.5 cm long, 2–9 mm wide, elliptic to narrowly lanceolate or lanceolate, entire, sessile; heads more or less conspicuous; involucres 3.5–5 mm high, 4–7 mm wide; outer bracts ovate to ovate-lanceolate, abruptly acute, deciduous, often purplish; pistillate flowers purplish, numerous; perfect flowers purplish, fewer; achenes glabrous; pappus bristles of perfect flowers dilated apically. Riparian areas at 460 to 1220 m in Garfield, Kane, San Juan (?), and Washington counties; California to Texas, south to Mexico; 22 (iii). The genus *Pluchea*, in a broad sense, includes annual and perennial herbs and shrubs. *Tessaria*, when segregated from *Pluchea*, consists of the shrubby species that have dimorphic corollas and the inner perfect flowers with apically flared pappus bristles. The residue within *Pluchea* contains only herbs with uniformly 4–lobed or some 3–lobed corollas and pappus of uniform barbellate capillary bristles. I follow tradition in maintaining this species in *Pluchea*.

Porophyllum (Vaill.) Adanson

Suffruticose perennial; leaves alternate or opposite, simple, with at least some elliptic to oval oil glands; heads discoid, solitary, or few to several in corymbose clusters; involucres cylindric, the bracts usually 5, in subequal series, glandular like the leaves; receptacle naked; flowers perfect, fertile, purplish; anthers rounded basally; style branches slender, hirtellous, subulate; pappus of scabrous bristles; achenes slender, striate.

Porophyllum gracile Benth. Odora. Rounded bushy perennials from a woody base; stems much branched, 1.5–4 dm tall; herbage dark green or often purplish, glaucous, odoriferous; leaves 1–4 cm long, linear-filiform, entire; involucre subcylindric, 10–15 mm long; bracts 5, dark green, tinged purplish, oblong, the hyaline margin pink, gibbous basally, bearing conspicuous glands, especially apically; corollas purplish, white; pappus bristles pinkish; achenes 8–9 mm long, hispidulous; n = 24. Desert shrub communities in Washington County (Cottam 5522 UT); California to Arizona and Mexico; 1 (0).

Prenanthella Rydb.

Annual herbs, with milky juice; leaves basal and alternate, simple, pinnately lobed, toothed, or pinnatifid; heads small, few to numerous; involucres campanulate; bracts in 2 series, the inner subequal, 3 or 5, the outer much reduced, 1 or 2, herbaceous; flowers all raylike, 4 or 5; achenes 5-ribbed; pappus of white capillary bristles.

Prenanthella exigua (Gray) Rydb. [*Prenanthes exigua* Gray; *Lygodesmia exigua* (Gray) Gray]. Annual; stems branched from the base, forming rounded clumps, 7–24 (30) cm tall; inflorescence paniculate, comprising more than half the plant height; lower leaves 1–4 (6.54) cm long, 3–12 (20) mm wide, spatulate to oblanceolate, the rosette often withered at anthesis; cauline leaves reduced upward, finally bracteate scales; herbage sparingly stipitate-glandular; involucres 3–5.5 mm long, 1.2–3.5 mm wide; inner bracts oblong, herbaceous, apically constricted in bud; rays pink or white, 1.5–2 mm long; achenes 3–3.5 mm long, 5-ridged, scabrous; pappus of white capillary bristles. Blackbrush, creosote bush, other warm desert shrub, salt desert shrub, and pinyon-juniper communities at 850 to 1925 m in Beaver, Carbon, Emery, Garfield, Grand, Juab, Kane, Millard, San Juan, Tooele, Uintah, and Washington counties; California, Nevada, Colorado, Arizona, and New Mexico; 20 (ii).

Psathyrotes Gray

Annual or perennial (?) herbs; leaves alternate, petiolate, simple, entire or lobed to toothed; heads discoid, the flowers yellow or purplish in age; involucres campanulate; bracts biseriate, the outer often shorter or otherwise different; receptacle flat, naked; anthers minutely sagittate; style branches flattened; achenes hairy; pappus of capillary bristles.

1. Plants lanate-tomentose as well as scurfy; outer involucral bracts expanded apically, oblong-obovate; reported for Utah by Munz (A California Flora), but not seen by me .... *P. ramosissima* (Torr.) Gray
— Plants scurfy and less commonly somewhat tomentose; outer involucral bracts tapering apically, lanceolate 2

2(1). Leaves entire; herbage scurfy and with long-piliferous multicellular hairs .... *P. pilifera*
— Leaves toothed; herbage scurfy but not long-piliferous .... *P. annua*

Psathyrotes annua (Nutt.) Gray Mealy Rosettes. [*Bulbostylis annua* Nutt.]. Annual or winter annual herbs forming low rounded cushions, mainly 2–18 cm tall; leaves petiolate, the blades 5–17 mm long, 5–20 mm wide, orbicular to fan-shaped, dentate; herbage scurfy; heads few to numerous, corymbose; involucres 5.5–7.5 mm high, 5–8 mm wide; outer bracts lanceolate to oblong, more or less constricted above the middle, scurfy and ciliate; disk corollas 3.5–4.2 mm long, yellow, becoming purplish; achenes 2–2.5 mm long, pilose; n = 17. Warm desert shrub, salt desert shrub, and pinyon-juniper communities, commonly on limestone and dolomitic gravels, at 790 to 1740 m in Juab, Millard, Tooele, and Washington counties; Idaho south to California, Nevada, and Arizona; 29 (ix).

Psathyrotes pilifera Gray Annual or winter annual herbs forming hemispheric cushions, mainly 5–15 cm tall; leaves petiolate, the blades 5–15 mm long, 4–16 mm wide, obovate, ovate, or oval-elliptic, entire; herbage scurfy and piliferous with long multicellular hairs; heads few to many, corymbose; involucres 8.5–10 mm high, 4–5.5 mm wide; outer bracts lanceolate, seldom constricted above the middle, scurfy and with long-piliferous setae marginally; disk corollas 6–6.5 mm long, yellow, becoming purplish; achenes 3.8–4.8 mm long; n = 17. Warm desert shrub and salt desert shrub, commonly on gypsiferous substrates of the Moenkopi and Chinle formations, at 760 to 2260 m in Garfield, Grand, Kane, San Juan, and Washington counties; Arizona; endemic; 15 (vi).

Psilocarpus Nutt.

Low floccose-woolly annual herbs; leaves opposite, simple, entire; heads discoid, subglobose; involucre per se essentially lacking; receptacle chaffy, subglobose; pistillate flowers numerous, imbricate, each enclosed by and deciduous with its subtending bract, woolly, the sides meeting in the center, bearing below the rounded tip on inner side a scarious appendage; corollas filiform; pappus lacking; perfect flowers few, central, ebracteate, the corollas 4- or 5-toothed, epappose; anthers sagittate.

Psilocarpus brevissimus Nutt. Low white-woolly annuals; stems simple or with decumbent-prostrate branches mainly 1.5–20 cm long; leaves 5–15 mm long, 1–3 mm wide, spatulate to lanceolate, apiculate; heads solitary or clustered, long-woolly, ca 5–7 mm thick, subtending leaves about as long as the head or longer; pistillate flowers 20–34 or more, the enclosing bracts 2.5–3.2 mm long, woolly, the appendage horizontally produced to erect, ca 0.5 mm long; perfect flowers ca 6–10; achenes subcylindric, terete, 1.3–2 mm long; 2n = 28. Lake and reservoir beds at ca 1710 m in Cache and Salt Lake counties; Washington to Montana, south to California, Mexico, and South America; 6 (0).

Psilostrophe DC.

Perennial herbs or shrubs; leaves alternate, simple, entire or merely lobed; heads few to many, corymbose; involucres campanulate; bracts in 1 series, subequal; receptacle naked; ray flowers yellow, pistillate, becoming papery and persistent; disk flowers perfect, 5-lobed; anthers obtuse basally; style branches truncate; pappus of 4–6 hyaline scales; achenes obtusely angled, glabrous or hairy.

1. Plants shrubby; stems closely white-tomentose; of Washington County .... *P. cooperi*
— Plants herbaceous, from a definite caudex; stems glabrous or loosely tomentose; not known in Washington County 2

2(1). Stems loosely tomentose; involucres densely white villous-tomentose; plants of Grand County ..... *P. tagetina*
— Stems glabrous, or tomentose only at the base; involucres sparingly tomentose, greenish; plants of Wayne, Garfield, and Kane counties .... *P. sparsiflora*

Psilostrophe cooperi (Gray) Greene Whitestem Paperflower. [*Riddellia cooperi* Gray]. Shrubs; stems closely white-tomentose, mainly 30–60 cm tall; leaves 1.2–7 cm long, linear, entire, sparingly tomentose, finally glabrate; involucres tomentose, 6–8 mm high, 5–8 mm wide; rays 4–8, yellow, 8–20 mm long; pappus scales ca 2 mm long, acute; achenes whitish, 4.5–7 mm long; n = 16. Joshua tree, creosote bush, blackbrush, and pinyon-juniper communities at 915 to 2135 m in Washington County; Nevada, California, Arizona, and New Mexico; 15 (iv).

Psilostrophe sparsiflora (Gray) A. Nels. Greenstem Paperflower. [*Riddellia tagetina* var. *sparsiflora* Gray]. Perennial herbs from a caudex; stems 14–60 cm tall, densely to moderately pilose basally, sparingly villous-to-mentose upward; leaves 0.9–11.5 (14.5) cm long, 1–11 mm wide, spatulate to oblanceolate or linear, pubescent like the stems or glabrate; involucres 4.5–6 mm high, 4–6 mm wide; rays usually 3, yellow, 6–12 mm long; pappus scales 1.5–2.5 mm long, acutish; achenes yellowish, 2.5–3 mm long; n = 16. Salt desert shrub, pinyon-juniper, and sagebrush communities at 1430 to 2045 m in Garfield, Kane, and Wayne counties; Arizona, New Mexico, and Mexico; 42 (vi).

Psilostrophe tagetina (Nutt.) Greene Woolly Paperflower. [*Riddellia tagetina* Nutt.]. Perennial herbs from a caudex; stems 10–40 cm tall, densely white-woolly below, loosely tomentose upward; leaves 0.8–10 cm long, 2–15 mm wide, spatulate to oblanceolate, entire or lobed, pubescent like the stems; cauline leaves reduced upward; heads usually numerous in dense to loose corymbiform clusters; peduncles usually 5–20 (30) mm long; involucres loosely villous-tomentose, 5–7 mm high, ca 3–4 mm wide; rays 3–6, yellow, 6–12 (15) mm long; pappus scales ca 2 mm long, rounded; disk flowers 6–12; achenes whitish, ca 2–2.5 mm long. Sandy warm desert shrub community at ca 1285 m in Grand County; Arizona to Texas and Mexico; 4 (ii)

Rafinesquia Nutt.

Annual herbs with milky juice; stems fistulous; leaves alternate, pinnatifid; heads 2 to several, large, showy, with white or rose-tinged flowers; involucres essentially cylindric; bracts 7–15, in 2 series, the inner subequal, the outer ones much shorter, obtuse or subcordate basally; flowers all raylike; pappus white or tawny, of 8–15 slender long-plumose bristles; achenes terete, obscurely few ribbed, attenuate into a beak.

1. Rays mainly 5–8 mm long; achene beak as long as the body; plumose hairs of pappus straight ..... *R. californica*
— Rays mainly 12–18 mm long; achene beak shorter than the body; plumose hairs of pappus crinkled; plants of Washington County .... *R. neomexicana*

Rafinesquia californica Nutt. California Chicory. Annual (winter annual), mainly 2–7 (10) dm tall, simple or branched above; leaves mainly 5–15 cm long, oblong in outline, subentire to pinnatifid, the lower ones petioled, becoming auriculate-clasping and reduced upward; heads few; involucres 15–18 mm high; rays 5–8 mm long, white; 2n = 16. Uncommon in mixed warm desert shrub community at ca 1035 m in Washington County; Arizona, Nevada, and California; Mexico; 1 (i).

Rafinesquia neomexicana Gray Desert Chicory. Annual (winter annual) herbs; stems mainly 15–40 (50) cm tall, simple or branched, often growing up through shrubs; basal leaves 1.2–9 cm long, pinnatifid, often withered at anthesis; cauline leaves sessile and auriculate-clasping, reduced upward; involucre 15–25 mm high, 5–9 mm wide; main bracts lance-attenuate, the margins scarious, the outer ones more or less cordate basally; rays 12–18 mm long, white or suffused with pink, 5–toothed or –lobed apically; pappus bristles white, the bases flattened, plumose to near the apex; achenes 12–15 mm long, papillate-puberulent; n = 8. Joshua tree, creosote bush, blackbrush, and desert almond communities at 700 to 1070 m in Washington County; California to Texas and Mexico; 16 (ii).

Ratibida Raf.

Perennial herbs from a caudex and stout taproot; leaves alternate, pinnatifid; heads radiate, solitary or few and corymbose; rays neuter, commonly yellow (sometimes purple in part or throughout); involucre in 1 series, green; receptacle columnar, chaffy throughout, the bracts more or less clasping the achenes; disk flowers perfect, fertile; anthers sagittate; style branches flattened; achenes compressed at right angles to the involucral bracts, glabrous, the margins sometimes ciliate; pappus of an evident tooth and sometimes with a second one.

Ratibida columnifera (Nutt.) Woot. & Standl. Prairie Coneflower. [*Rudbeckia columnifera* Nutt.; *Ratibida columnaris* Pursh]. Perennial herbs; stems mainly 3–6 (12) dm tall, several, often branched above, strigose; leaves 2–9 cm long, pinnatifid, with the terminal division often the largest; heads borne on slender peduncles 6–18 cm long, the disk grayish in bud, purplish brown in flower, 1.5–3 cm long; rays 3–7, yellow (or purple), 1–3 (4.5) cm long, spreading or reflexed; pappus an evident awn-tooth on the inner angle of the achene, often also a shorter one on the outer edge; achenes ciliate and more or less winged on the inner edge; n = 14. Salt desert shrub and sagebrush communities at 1585 to 2565 m in Cache, Garfield, Grand, Millard, San Juan, and Washington counties; British Columbia to Minnesota, south to Arizona, Texas, and Mexico; 4 (i). Our material appears to be adventive from the main body of the species in the prairies and plains provinces to the east of Utah.

Rudbeckia L.

Perennial caulescent herbs; leaves alternate, serrate or pinnately to palmately lobed; heads radiate or discoid, the rays (when present) neuter, commonly yellow; involucral bracts in 2 or 3 series, mainly unequal, herbaceous, spreading or reflexed; receptacle conic or columnar, chaffy throughout, the bracts clasping the achenes; disk flowers fertile; anthers obtuse or sagittate basally; style branches flattened; pappus a crown or none; achenes quadrangular or flattened at right angles to the involucral bracts.

1. Heads radiate; disks 1–2 cm wide and about as long, little elongate in fruit; plants of San Juan County .....
   .... *R. laciniata*
— Heads discoid; disks 1.5–2.5 cm wide, mostly 2–5 cm long, elongating in fruit; plants not of San Juan County . 2

2(1). Leaves laciniately lobed; plants glabrous or merely scabrous- ciliate on leaf margins; known from Iron and Washington counties .... *R. montana*
— Leaves crenate-serrate, dentate, undulate, or entire, not lobed; plants evidently short-hairy to almost glabrous; known from mountains of central northern to south central Utah .... *R. occidentalis*

Rudbeckia laciniata L. Cutleaf Coneflower. Perennial herbs; stems erect from a coarse ligneus base, mainly 5–10 (20) dm tall, glabrous or scabrous-ciliate; leaves petiolate, the blades laciniate-pinnatifid to palmatifid, mainly 4–15 cm long and as broad; heads showy, the disk 1–2 cm wide and about as high; rays yellow, 6–16, deflexed-spreading, 2–5 cm long; pappus a short crown; n = 9. Moist meadows at 1890 to 2200 m in Grand and San Juan counties; Montana to Quebec, south to Arizona and Florida; 2 (0).

Rudbeckia montana Gray Montane Coneflower. [*R. occidentalis* var. *montana* (Gray) Perdue]. Perennial herbs; stems erect, from a short subrhizomatous caudex, 6–12 dm tall, glabrous; leaves petiolate, the blades laciniate-pinnatifid, mainly 4–20 cm long and about as broad; heads discoid, the disk 1.5–2.5 cm wide, 3–5 cm high; rays lacking; pappus an irregularly margined, almost toothed crown. Spruce-fir community at 2590 to 3050 m in Iron and Washington counties; Colorado; 2 (0). This material is probably best regarded as a variety of *R. occidentalis*.

Rudbeckia occidentalis Nutt. Western Coneflower. Perennial herbs; stems erect from a coarse ligneus rhizome, mainly 5–20 dm tall, glabrous or strigulose; leaves petiolate, the blades 5–20 cm long, 2.5–10 cm wide, ovate to ovate-lanceolate or lanceolate, attenuate to acuminate, entire, crenate-serrate, or dentate; heads discoid, the disks 1.5–2.5 cm wide, 3–6 cm long; rays lacking; pappus a short crown. Mountain brush, aspen, grass-tall forb, and spruce-fir communities at 2135 to 3175 m in Cache, Carbon, Davis, Duchesne, Emery, Piute, Salt Lake, Sanpete, Sevier, Summit, Tooele, Utah, Wasatch, and Weber counties; Washington to Montana, south to California and Wyoming; 42 (iii).

Senecio L.

Annual, biennial, or perennial herbs with rhizomes, caudices, or taproots, the juice watery; stems erect, ascending, or decumbent at the base; leaves alternate, simple, entire, toothed, or lobed to pinnatifid; heads solitary, or few to many in corymbose cymes; involucral bracts in 1 series, often with smaller bractlets at the base, green throughout or the margins scarious or hyaline, or variously colored; receptacle flat or convex, naked; ray flowers yellow or orange, or sometimes lacking; pappus of capillary bristles; style branches flattened; achenes terete, 5- to 10-nerved, glabrous or pubescent. Note: This genus consists of a series of species that intergrade freely when they are in contact with others of the group. Because of hybridization the species lines tend to be blurred, and it is not possible to place all specimens with confidence. Keys are, and have been, based on features that are subject to interpretation; the present one is not an exception, being tentative at best.

Barkley, T. M. 1978. *Senecio*. N. Amer. Flora II. 10: 50–139.

1. Leaves pinnatilobate with linear-filiform divisions or entire and linear-filiform; stems with leaves only gradually reduced upward, often more or less woody below 2
— Leaves variously lobed, toothed, or entire, but the segments and leaves not linear-filiform; stems with leaves various, seldom, if at all, woody at the base ... 3
2(1). Heads cylindric, subcylindric, or narrowly campanulate; main involucral bracts 8–13, the outer ones much reduced and inconspicuous; plants glabrous .... *S. spartioides*
— Heads campanulate to broadly campanulate; main involucral bracts 13–21, the outer ones conspicuous, or, if inconspicuous, the plants tomentose . *S. douglasii*
3(2). Heads nodding, especially in bud, or, if erect, with both distinctly black triangular tips on involucral bracts and cauline leaves prominently clasping 4
— Heads erect, even in bud; plants various but not as above 7
4(3). Heads discoid 5
— Heads radiate 6
5(4). Heads 8–12 mm high, 6–9 mm wide, narrowly campanulate, conspicuously pedunculate ...... *S. pudicus*
— Heads 12–20 mm high, 14–20 mm wide, broadly campanulate, short-pedunculate .... *S. bigelovii*
6(4). Heads erect, the involucral bracts black-tipped .... *S. crassulus*
— Heads nodding, the involucral bracts often suffused with purple throughout, but not especially black-tipped .... *S. amplectens*
7(3). Plants annual or winter annual, introduced weedy species .... *S. vulgaris*
— Plants perennial, indigenous species 8
8(7). Stems uniformly leafy to the inflorescence, or the leaves concentrated upward 9
— Stems few-leaved or the upper leaves definitely reduced in size and distribution 12
9(8). Stems 1–3 dm tall, more or less sprawling, arising from a subrhizomatous caudex .... *S. fremontii*
— Stems mostly 2–15 dm tall, erect or ascending, arising from a rhizome or a caudex 10
10(9). Plants mainly 2–4 dm tall; leaves pinnatifid to lobed or laciniate; involucral bracts with dark tips .... *S. eremophilus*
— Plants mainly 5–10 dm tall; leaves dentate to serrate; involucral bracts uniformly greenish or brownish ... 11
11(10). Leaf blades acute to obtuse basally, the teeth all about alike .... *S. serra*
— Leaf blades truncate to obtuse basally, or more or less hastately lobed, the lowermost teeth often the largest .... *S. triangularis*
12(8). Plants glaucous tall herbs, semiaquatic; leaves entire or denticulate, thick and leathery ..... *S. hydrophilus*
— Plants not or seldom glaucous, terrestrial (though sometimes growing where wet); leaves entire, toothed, or pinnatifid, not thick and leathery 13
13(12). Rays orange or orange red (see also *S. pauperculus*) .... *S. crocatus*
— Rays yellow or lacking 14
14(13). Heads discoid; plants tomentose, soboliferous .... *S. fendleri*
— Heads radiate (or rarely discoid in some individuals and the plants otherwise differing); plants tomentose, glabrate, or glabrous, not soboliferous (except in *S. werneriifolius*) 15
15(14). Leaves pinnatifid, at least the cauline ones, or the basal ones commonly rounded apically or oblanceolate to ovate or oval in outline 16
— Leaves serrate to entire, the basal ones variously shaped, but mainly acute to attenuate apically 20
16(15). Basal leaves distinctly pinnately divided, the lobes often again toothed .... *S. multilobatus*
— Basal leaves merely toothed to subentire 17

17(16).	Basal and lower cauline leaves entire to dentate, but not pinnatifid .... *S. hartianus*
—	Basal and lower cauline leaves toothed to pinnatifid . 18
18(17).	Middle and upper cauline leaves clasping with large auriculate bases .... *S. dimorphophyllus*
—	Middle and upper cauline leaves without a prominent clasping or auriculate base 19
19(18).	Basal leaves obovate to oblancoleater ovate, rounded apically, thickish; plants of dryish habitats .... *S. streptanthifolius*
—	Basal leaves oblanceolate to elliptic, obtuse, but usually pointed apically, thin; plants of wet meadows .... *S. pauperculus*
20(15).	Cauline leaves rounded and more or less clasping basally, long-attenuate apically, entire or denticulate .... *S. integerrimus*
—	Cauline leaves tapering to the base or with a few basal clasping lobes in some, usually not attenuate apically 21
21(20).	Stems subscapose, the cauline leaves none or few and bractlike; plants often soboliferous or with a branching rhizomatous caudex .... *S. werneriifolius*
—	Stems more or less leafy, the cauline leaves gradually reduced upward, but hardly bractlike; plants seldom as above 22
22(21).	Involucral bracts ca 8 or fewer; heads 5–6 mm wide, mainly 20–60 per inflorescence .... *S. atratus*
—	Involucral bracts mostly 13–21; heads 8–12 mm wide or more, fewer or larger than above 23
23(22).	Plants with glabrous achenes and more typically more than 30 cm tall, known from northwestern Box Elder County .... *S. foetidus*
—	Plants either with glabrous achenes and less than 30 cm tall or the achenes hirtellous or hispidulous 24
24(23).	Achenes glabrous; plants often less than 20 cm tall .. 25
—	Achenes hirtellous or hispidulous; plants often over 20 cm tall 26
25(24).	Main leaves regularly and evenly pinnatifid or pinnatisect; plants often with slender rhizomes .... *S. fendleri*
—	Main leaves entire to dentate, not as above; plants shortly rhizomatous .... *S. canus*
26(24)	Main leaves 10–15 cm long or more, entire or denticulate; plants 50–70 cm tall, of northern Utah .... *S. sphaerocephalus*
—	Main leaves 2–8 cm long, dentate, serrate, or subentire; plants mainly 15–35 cm tall, of southern Utah .... *S. neomexicanus*

Senecio amplectens Gray Alpine Groundsel. [*Ligularia amplectens* (Gray) W. A. Weber]. Perennial short-rhizomatous herbs; stems ascending to erect, mainly 8–30 cm tall; herbage glabrous or sparingly tomentose; main leaves middle and lower cauline, the lower ones broadly petiolate, more or less clasping the stem, the blades 3–10 cm long, 0.8–3 cm wide, dentate to shallowly lobed; cauline leaves becoming short-petiolate or sessile upward, finally bractlike; heads 1–5 (rarely more), conspicuously nodding, corymbose; involucres broadly hemispheric, 10–15 mm long and about as wide or wider; bracts mainly ca 21, usually brown, with scarious margins, glabrous; outer bracts to about half as long as the inner; rays 7–17, yellow, 10–25 mm long; pappus white; achenes glabrous. Spruce-fir and alpine tundra communities, often in talus or on ridge margins, at 3050 to 3570 m in Beaver, Grand, Piute, San Juan, Sanpete, and Utah counties (Wasatch, Tushar, and La Sal mts., and Wasatch Plateau); Montana, Wyoming, Colorado, and Nevada; 34 (v). Our material belongs to **var. holmii (Greene) Harrington** [*S. holmii* Greene; *Ligularia holmii* (Greene) W. A. Weber].

Senecio atratus Greene Black Groundsel. Perennial subrhizomatous herbs from a branching caudex; stems erect or ascending, 2–8 dm tall; herbage floccose-tomentose; basal and lower cauline leaves petiolate, mainly 8–30 cm long, 1–4 cm wide, the blade oblanceolate or oblong, conspicuously dentate to subentire; cauline leaves gradually reduced upward, becoming sessile or subsessile and finally bracteate; heads ca 15–60, in more or less compact corymbose clusters; involucres 6–8 mm high, 3–6 mm wide; main bracts 8 or fewer, greenish to brownish, the margins scarious, the tips black, tufted-hairy apically; rays 3–5, yellow, 4–8 mm long; pappus white; achenes glabrous; $2n = 40, 46, 90+$. Aspen, spruce-fir, mixed conifer, and tall forb communities at 2440 to 3335 m in Duchesne, Garfield, Iron, Salt Lake, San Juan, Sanpete, Summit, and Uintah counties; Colorado and New Mexico; 22 (iv).

Senecio bigelovii Gray in Torr. Bigelow Groundsel. [*Ligularia bigelovii* (Gray) W. A. Weber]. Perennial subrhizomatous herbs; stems erect, mainly 3–8 (10) dm tall; herbage floccose-tomentose to glabrate or glabrous; main leaves cauline, largest below, reduced gradually upward, petiolate below, sessile and clasping to auriculate above, mostly 7–15 cm long, 0.6–3 (5) cm wide, the blades oblanceolate to oblong or elliptic, subentire to serrate; heads 3–8, nodding, racemosely arranged; involucres 8–12 mm long, 12–25 mm wide; bracts mainly ca 21, usually brown, with scarious margins, the outer to half as long as the inner, all sparingly tomentose; ray flowers lacking; achenes glabrous. Mountain brush, ponderosa pine, aspen, and spruce-fir communities at 2745 to 3175 m in San Juan County; Wyoming south to New Mexico and Arizona; 4 (0). Our material has been assigned to **var. hallii Gray** [*Ligularia bigelovii* ssp. *hallii* (Gray) W. A. Weber].; the type variety is more southern.

Senecio canus Hook. Gray Goundsel. [*S. purshianus* Nutt.; *S. convallium* Greenman, type from Rabbit Valley]. Perennial short-rhizomatous herbs, often with a caudex; stems 8–30 cm tall (rarely more), erect or ascending; herbage woolly-tomentose; basal leaves petiolate, the blades 1–5 cm long, 3–30 mm wide, lanceolate to oblanceolate, elliptic or ovate, entire or denticulate, obtuse to rounded apically; cauline leaves reduced upward, the upper ones often clasping, finally bracteate, occasionally lobed in some introgressant forms; heads mainly 2–10, subumbellate or corymbose; involucres 3–8 mm long, 4–10 mm wide; main bracts 13–21, lance-attenuate, greenish or with brownish midstripe, glabrous or tomentose; outer bracts very short; rays 8–13, yellow, 5–10 mm long; achenes glabrous; $2n = 46, 69, 90, 132$. Pinyon-juniper, sagebrush, Douglas fir, aspen, spruce-fir, and alpine tundra communities, often in talus or on windswept ridges, at 2105 to 3815 m in Beaver, Box Elder, Cache, Carbon, Daggett, Duchesne, Garfield, Iron, Kane, Juab, Millard, Piute, San Juan, Sanpete, Sevier, Summit, Uintah, and Utah counties; British Columbia to Manitoba, south to California, Nevada, Colorado, and Kansas; 56 (xiii). This attractive grayish white species forms intermediates with *S. multilobatus*, *S. streptanthifolius*, and *S. werneriifolius*.

Senecio crassulus Gray Thick Groundsel. Perennial short-rhizomatous herbs, often with a caudex; stems 15–50 cm tall or more, erect; herbage glabrous; lower leaves broadly petiolate, the main ones 3–15 cm long, 0.6–3 (5) cm wide, lanceolate to elliptic or oblanceolate, dentate to entire; cauline leaves reduced upward, becoming sessile and clasping; heads solitary or 2–12, corymbose; involucres 8–13 mm high, 12–21 mm wide; main bracts 8–21, oblong to lance-oblong, greenish to brown, with scarious margins, the tips black and tufted-hairy; outer bracts to half as long as the inner or more; rays 8–13, yellow, 5–12 mm long; achenes glabrous. Aspen, lodgepole pine, and spruce-fir communities, often in forb-grass meadows, at 1830 to 3355 m in Box Elder, Cache, Carbon, Duchesne, Grand, Iron, Salt Lake, San Juan, Sanpete, Sevier, Summit, and Utah counties; Oregon to Montana, south to New Mexico; 34 (iv).

Senecio crocatus Rydb. Saffron Groundsel. Perennial subrhizomatous herbs, the caudex more or less developed; stems erect, mainly 2–8 dm tall; herbage glabrous or with minute hairs in the inflorescence; basal leaves with long slender petioles, the blades 1–8 cm long, 1–4 cm wide, ovate to oblong, lanceolate, or elliptic, subcordate to acute basally, often rounded apically, entire to crenate-dentate; cauline leaves reduced upward, becoming lobed or sublyrate, sessile and sometimes auriculate and/or clasping; heads mainly 3–30; involucres 4–8 mm long, 5–8 mm wide; main bracts 13–21, lance-oblong, green or suffused with red or purple; outer bracts very short; rays 6–13, orange or orange red; pappus white; achenes glabrous. Rush-grass, willow, aspen-forb, and lodgepole pine communities at 2195 to 2990 m in Cache, Duchesne, Rich, Summit, Utah, and Wasatch counties; Colorado; 23 (i). One specimen from Rich County (Thorne 1465 BRY) is apparently intermediate with *S. eremophilus*. The species is remarkably like the next.

Senecio dimorphophyllus Greene Different Groundsel. Perennial subrhizomatous herbs; stems erect, mainly 30–70 cm tall; herbage glabrous or essentially so; basal leaves with long slender petioles, the blades 1–7 cm long, 1–5 cm wide, oval to oblong or elliptic, subcordate to acute basally, commonly rounded apically; cauline leaves becoming sessile, lyrate-pinnatifid, and auriculate-clasping, the auricles often lobed; heads mainly 2–25, subumbellately to corymbosely arranged; involucres 5–8 mm high, 6–10 mm wide; main bracts 13–21, lance-attenuate, green, sometimes suffused reddish, the tips not black, tufted-hairy; outer bracts very short; rays 8–13, yellow, 5–8 mm long; pappus white; achenes glabrous. Two weakly discernible varieties are present in Utah.

1. Cauline leaves merely lobed to subentire; plants of the La Sal Mts. . . . . *S. dimorphophyllus* var. *intermedius*
— Cauline leaves sharply lobed; plants of Uinta Mts. and Wasatch Plateau . *S. dimorphophyllus* var. *dimorphophyllus*

Var. *dimorphophyllus* Aspen-tall forb and spruce-fir communities at 1860 to 3265 m in Duchesne, Emery, Sanpete, and Utah counties; Wyoming and Colorado; 9 (0). Utah materials approach *S. crocatus* in most morphological features, including the tall stature. If the flower color is discounted and the larger heads are not definitive, then the specimens could be considered as a portion of *S. crocatus*. Some specimens from Duchesne County appear to be transitional to *S. sphaerocephalus*.

Var. *intermedius* T. M. Barkley Wet meadows at 3050 to 3115 m in the La Sal Mts., Grand, San Juan (type from Geyser Pass), and Sanpete counties; endemic; 3 (0).

Senecio douglasii DC. Douglas Groundsel. Suffrutescent perennials; stems erect or ascending, mainly 3–8 (10) dm tall; herbage glabrous or tomentose; leaves simple and linear-filiform or pinnatifid into linear-filiform segments, 2–11 cm long, 0.8–3 mm wide; heads few to numerous, in paniculately branched subcorymbose cymes; involucres campanulate, mainly 5–10 mm long, 6–14 mm wide; main bracts 13–21, lance-oblong, green, with scarious margins, minutely tufted-hairy apically; the outer bracts short and inconspicuous or to half as long as the inner ones; rays 8–17, yellow, 10–18 mm long; pappus white; achenes hairy. Two infraspecific taxa, previously treated at specific rank with some justification, are present in Utah.

1. Herbage grayish or whitish tomentose; outer involucral bracts short and inconspicuous; plants rather broadly distributed . . . . *S. douglasii* var. *longilobus*
— Herbage green, glabrous or essentially so; outer involucral bracts to about half as long as the inner ones; plants of Washington County . . . . *S. douglasii* var. *monoensis*

Var. *longilobus* (Benth.) L. Benson [*S. longilobus* Benth.; *S. filifolius* var. *jamesii* T. & G., nom. illeg.]. n = 20. Warm desert shrub, salt desert shrub, sagebrush-rabbitbrush, saltgrass, and pinyon-juniper communities at 1095 to 2200 m in Beaver, Duchesne, Garfield, Iron, Kane, Millard, Piute, San Juan, Sevier, Washington, and Wayne counties; Arizona to Texas; 52 (x).

Var. *monoensis* (Greene) Jepson [*S. monoensis* Greene]. Creosote bush, blackbrush, other warm desert shrub, and pinyon-juniper communities at 760 to 1465 m in Washington County; California to Texas; 23 (vi).

Senecio eremophilus Richards. Desert Groundsel. Perennial subrhizomatous herbs; stems rather equably leafy, erect or ascending, mainly 2.5–9 dm tall; herbage glabrous or essentially so; lower leaves often deciduous or withered at anthesis; cauline leaves 2–15 cm long (or more), 0.4–5 (7) cm wide, oblanceolate to elliptic or lanceolate in outline, pinnatifid or pinnately lobed or toothed, the lower ones petiolate, becoming sessile upward; heads several to numerous, corymbose; involucres 5–8 mm high, 6–10 mm wide; main bracts 8–17, lance-oblong, brownish or greenish, with scarious margins, blackish tips, and hair-tufted apices; outer bracts very short; rays 7–10, yellow, 5–10 mm long; pappus white; achenes glabrous or puberulent along the ribs; n = 20. Grass-forb, ponderosa pine, aspen, lodgepole pine, spruce-fir, and alpine tundra communities, at 1615 to 3450 m in Beaver, Carbon, Duchesne, Emery, Garfield, Grand, Iron, Juab, Piute, Salt Lake, San Juan, Sanpete, Sevier, Summit, Tooele, Uintah, Utah, Wasatch, Washington, and Wayne counties; British Columbia and Mackenzie south to Arizona and New Mexico; 104 (xiv). Our material belongs to **var. *kingii* (Rydb.) Greenman** [*S. kingii* Rydb., type from Cottonwood Canyon]. This plant forms intermediates with *S. spartioides*.

Senecio fendleri Gray Fendler Groundsel. Perennial rhizomatous herbs, with a caudex more or less developed; stems mainly 5–30 cm tall, erect or ascending; herbage floccose-tomentose; basal leaves petiolate, the blades 1–6 cm long, 4–20 mm wide, pinnatifid or pinnately lobed; cauline leaves reduced upward, becoming sessile, finally bracteate; heads 3 to many, corymbose; involucres 4–6 mm high, 5–8 mm wide; main bracts ca 13, lance-attenuate, greenish, the margins scarious or hyaline, minutely hairy apically, more or less tomentose below; outer bracts very short; ray flowers lacking; pappus white; achenes glabrous; n = 23. Ridge tops on limestone barrens near Musinea Peak, at ca 2960 to 3295 m in Sanpete County, (Lewis 4274, 5516 BRY); Wyoming south to New Mexico; 2 (0). Our specimens approach S. *canus*, more or less.

Senecio foetidus Howell Foetid Senecio. Perennial herbs from a short erect root crown, mainly 3–10 dm tall; leaves thickish and somewhat succulent, generally dentate, the basal and lowermost cauline ones petiolate, the blades 6–25 cm long, 2–7 cm wide, elliptic to broadly oblanceolate; the middle and upper cauline leaves progressively reduced, finally sessile; heads numerous, the inflorescence congested; involucres 6–9 mm high, the bracts ca 13 (to 21), often black-tipped; rays few and to ca 8 mm long, or lacking; achenes glabrous. Wet meadows with false hellebore and hairgrass at ca 2680 m in Box Elder County (Goodrich 17824 BRY); British Columbia, Washington, Oregon, and Idaho; 1 (0).

Senecio fremontii T. & G. Fremont Groundsel. Perennial herbs, subrhizomatous or from a caudex and taproot; stems 0.6–4 dm tall; herbage glabrous; leaves cauline, 1–6 cm long, 0.5–2 cm wide, oblanceolate to obovate, shortly petiolate or sessile and somewhat clasping, dentate to subentire; heads 1–5; involucres 6–12 mm high, 7–12 mm wide; main bracts 8–17, lance-oblong or lance-attenuate, green or brown, the margins scarious, tufted hairy apically; outer bracts short and inconspicuous or to half as long as the inner ones; rays 7–10, yellow, 5–12 mm long; pappus white; achenes glabrous or hairy. Two rather weak varieties are present.

1. Involucres mostly 8–10 mm high; stems mostly less than 20 cm high .... S. *fremontii* var. *fremontii*

— Involucres 10–12 mm high; stems often over 30 cm tall .... S. *fremontii* var. *blitoides*

Var. *blitoides* (Greene) Cronq. [S. *blitoides* Greene]. 2n = 40. Alpine communities, often in talus or on rock outcrops, at 2745 to 3355 m in Salt Lake, Tooele, and Utah counties; Wyoming to Colorado; 9 (0).

Var. *fremontii* Spruce-lodgepole pine and alpine tundra communities at 3050 to 3965 m in Duchesne, Grand, San Juan, Salt Lake, Summit, Uintah, and Utah counties; British Columbia and Alberta, south to Oregon and Wyoming; 16 (v).

Senecio hartianus Heller Hart Groundsel. Perennial herbs from a subrhizomatous or stoloniferous caudex; stems erect, 2–5 dm tall; herbage floccose-tomentose, sometimes glabrate; basal leaves petiolate, the blades 1–5 cm long, 0.5–3 cm wide, oval to obovate or elliptic, serrate or crenate, rounded apically; cauline leaves reduced upward, subpinnatisect to entire; heads 3–12, corymbose; involucres 4–7 mm high and as broad; main bracts 13–21, lance-attenuate, greenish, the tips glabrous; rays ca 10–13, yellow, 5–8 mm long; achenes glabrous. Ponderosa pine community at ca 2290 m in Kane County (Atwood 7425 BRY); Arizona; 1 (0).

Senecio hydrophilus Nutt. Water Groundsel. Perennial subaquatic herbs from a caudex and fibrous roots; stems erect, mainly 4–10 dm tall; herbage glaucous, blue-green; basal and lower cauline leaves petiolate, the broad petioles with clasping bases, the blades 5–35 cm long or more, 1–10 cm wide, elliptic to oblanceolate, entire or denticulate, thick and leathery; cauline leaves reduced upward, becoming sessile, finally bracteate; heads numerous in a branching corymbose cluster; involucres 5–8 mm long, 4–76 mm wide; main bracts 8–13, oblong or lance-attenuate, yellowish, the tips often black, tufted-hairy; rays 3–5 or lacking, yellow, 3–8 mm long; pappus white; achenes glabrous. Stream banks, pond margins, and wet meadows at 1375 to 2745 m in Beaver, Box Elder, Cache, Carbon, Daggett, Duchesne, Garfield, Juab, Kane, Piute, Rich, Salt Lake, Sanpete, Sevier, Summit, Utah, and Wasatch counties; British Columbia, south to California and Colorado; 25 (iv).

Senecio integerrimus Nutt. Wet-the-bed; Gauge Plant. Perennial herbs with a short subrhizomatous caudex; stems mainly 1–6 (7) dm tall, erect; herbage arachnoid-villous or glabrate; basal and lower cauline leaves broadly petiolate, 3–20 cm long, 0.8–4 cm wide, lanceolate to elliptic or oblanceolate to oblong, entire or serrate to dentate, rounded to obtuse apically; cauline leaves reduced upward; heads few to many, in a corymbose to subumbellate cyme; involucres 6–12 mm high, 8–18 mm wide; main bracts 13–21, lance-attenuate, green, with scarious margins and black tips, the tips tufted-hairy; outer bracts very short; rays 8–13 (or lacking), yellow, 4–15 mm long; pappus white; achenes glabrous. Sagebrush, pinyon-juniper, forb-grass, mountain brush, ponderosa pine, aspen, and spruce-fir communities at 1460 to 3660 m in probably all Utah counties; British Columbia to Montana, south to California; 132 (xiv). Presumed hybrids with S. *dimorphophyllus* are known (Hansen sn 1976 BRY). This plant has been used by various livestock owners as a gauge to range condition. When it came into flower, the range was sufficiently developed for grazing to begin.

Senecio multilobatus T. & G. Uinta Groundsel. [S. *lapidum* Greenman, type from Silver Reef]. Perennial (or biennial?) herbs from a taproot; stems mainly 1–6 dm tall; herbage glabrous, glabrate, or tomentose throughout or only in axils of basal leaves; basal leaves 2–12 cm long, 0.3–3.5 cm wide, spatulate to obovate in outline, pinnatifid to lyrate-pinnatifid, the segments variously again toothed, petiolate; cauline leaves reduced upward, finally bracteate; heads few to many, corymbose or subumbellate; involucres 4–9 cm high, 4–10 mm wide; main bracts 13–21, lance-attenuate or oblong-attenuate, the margins scarious, the apices hair-tufted; rays 7–13, yellow, 4–10 mm long, or lacking; pappus white; achenes glabrous; 2n = 46. Blackbrush, desert shrub, pinyon-juniper, sagebrush, mountain brush, ponderosa pine, aspen, lodgepole pine, and spruce-fir communities at 915 to 3420 m in all Utah counties (type from the Uinta River); Idaho and Wyoming to California, Arizona, and New Mexico; 312 (xliv). This widespread and common species forms presumed hybrids with S. *streptanthifolius* and S. *neomexicanus*.

Senecio neomexicanus Gray Perennial (or biennial?) herbs from a taproot; stems 14–40 cm tall, erect; herbage tomentose; basal and lower cauline leaves petiolate, the blades 1–5 cm long, 0.6–2 cm wide, oblanceolate to obovate or oval, dentate, serrate or subentire, toothed to obtuse apically; cauline leaves reduced upward, toothed to lobed or entire, bracteate in inflorescence; heads few to many, corymbose or subumbellate; involucres 4–7 mm high, 5–12 mm wide; main bracts 13–21, lance-attenuate, green or brown, with scarious margins, not especially hairy apically; rays 8–13, yellow, 4–10 mm long; pappus white; achenes pubescent. Sagebrush, mountain brush, ponderosa pine, and aspen communities at 2105 to 3050 m in Garfield, Kane, San Juan, and Wayne counties; Colorado, New Mexico, and Arizona; 10 (0). Our materials are assigned to var. **mutabilis** (Greene) Barkley [*S. mutabilis* Greene]. Through this variety there is virtually a complete intergrading series into *S. werneriifolius*, *S. streptanthifolius*, and *S. multilobatus* (Barkley 1978).

Senecio pauperculus Michx. Balsam Groundsel. Perennial herbs from a subrhizomatous caudex; stems erect, mainly 2–4 dm tall; herbage glabrous or somewhat tomentose in axils of basal leaves; basal leaves petiolate, the blades mainly 2–6 cm long, 0.5–3 cm wide, oblanceolate to elliptic, obovate or ovate, crenate, dentate, or subentire; cuneate basally, toothed to obtuse apically; cauline leaves reduced upward, becoming sessile, pinnatifid, not especially auriculate, finally bracteate; heads few to many, corymbose or subumbellate; involucres 4–8 mm long, 5–9 mm wide; main bracts 13–21, lance-attenuate, often with scarious margins, the tips not especially tufted-hairy; outer bracts very short; rays 8–13, yellow or yellow orange, 4–10 mm long; pappus white; achenes glabrous or puberulent along the angles; n = 20, 22, 44. Lodgepole pine and spruce-fir communities, usually in moist meadows, at 2345 to 2745 m in Daggett, Garfield, Rich, Uintah, and Wasatch counties; Alaska to Labrador, south to Oregon and Georgia; 5 (0). Our material is intermediate to both *S. streptanthifolius* and *S. crocatus*.

Senecio pudicus Greene Butterweed. [*S. cernuus* Gray, not L.f.; *Ligularia pudica* (Greene) W. A. Weber]. Perennial herbs from a subrhizomatous caudex; stems 20–50 cm tall, erect; herbage glabrous; basal and lower cauline leaves petiolate, the blades 3–15 cm long, 0.5–3 cm wide, lanceolate to oblanceolate or narrowly elliptic, tapering basally, acute apically, entire or shallowly dentate; cauline leaves reduced upward, finally bracteate; heads few to many, nodding; involucres 5–9 mm long and as broad; main bracts 8–13, lance-oblong, green to brown, the margins scarious, tufted-hairy apically; outer bracts very short; ray flowers lacking; pappus white; achenes glabrous. Aspen, spruce-fir, and alpine tundra communities at 2650 to 3480 m in Carbon, Daggett, and Garfield counties; Colorado; 11 (i).

Senecio serra Hook. Saw Groundsel. Perennial herbs from a caudex, with coarse felt-covered roots; stems equally leafy, erect, 4–15 dm tall (or more), glabrous or sparingly tomentose; leaves 3–15 cm long, 0.4–4 cm wide, short-petiolate, the blades lanceolate to narrowly lanceolate or linear, dentate to subentire; heads several to numerous, corymbose; involucres 4–11 mm high, 2–10 mm wide; main bracts 8–13, lance-oblong, greenish to brownish, the margins scarious, black-tipped, hair tufted; outer bracts very short; rays 5–8, yellow, 3–10 mm long; pappus white; achenes glabrous or essentially so; n = 20. Two rather distinctive varieties are present.

1. Involucral bracts 4–6 mm long, 2–6 mm wide; disk flowers ca 12; plants of central and northern Utah ....
   .... *S. serra* var. *serra*
— Involucral bracts 6–8 mm long, 6–10 mm thick; disk flowers ca 20; plants of San Juan County ....
   .... *S. serra* var. *admirabilis*

Var. **admirabilis** (Greene) A. Nels. [*S. admirabilis* Greene]. Ponderosa pine and subalpine meadow communities at 1830 to 3185 m in San Juan County; Wyoming and Colorado; 2 (0).

Var. **serra** Sagebrush, mountain brush, aspen, forbgrass, lodgepole pine, and spruce-for communities at 1830 to 3035 m in Box Elder, Cache, Davis, Duchesne, Juab, Morgan, Rich, Salt Lake, Sanpete, Summit, Utah, Wasatch, and Weber counties; Washington to Montana, south to California and Nevada; 42 (vi).

Senecio spartioides T. & G. Broom Groundsel. Perennial herbs from a taproot; stems equally leafy, erect or ascending, 2–10 dm tall or more, often in clumps; herbage glabrous; leaves 2–10 cm long or more, linear, simple and entire or with linear lobes, mainly 1–3 mm wide (wider in some hybrid derivatives); heads several to many in branching corymbose cymes; involucres subcylindric to narrowly campanulate, 5–10 mm high, 4–8 mm wide; main bracts 8–13, lance-linear, green, the margins scarious, not tufted-hairy; outer bracts very short; rays 4–8, yellow, 7–12 mm long; pappus white; achenes white-hairy. Two intergrading varieties are present.

1. Leaves simple and unlobed, or, if lobed, lower cauline leaves often over 4 mm wide; plants widespread ....
   .... *S. spartioides* var. *spartioides*
— Leaves commonly with 4–6 lateral lobes, seldom if ever more than 2.5 mm wide; plants of southeastern Utah ....
   .... *S. spartioides* var. *multicapitatus*

Var. **multicapitatus** (Greenman) Welsh [*S. multicapitatus* Greenman in Rydb.]. Warm desert shrub and pinyon-juniper communities, often in saline riparian sites, at 1220 to 1895 m in Emery, Garfield, Grand, San Juan, and Wayne counties; Colorado, Arizona, New Mexico; and Texas; 11 (vii). Barkley (1978) hesitated to combine *S. multicapitatus* with *S. spartioides*, because of field distinctions. They are, however, much alike and evidently lack diagnostic criteria that will allow segregation of all specimens. Further, specimens intermediate between *S. spartioides* and *S. eremophilus* bear "multicapitatus" leaves. I follow a moderate course in maintaining this taxon at varietal level.

Var. **spartioides** [*S. incurvus* A. Nels., type from Zion National Park]. Warm desert shrub, pinyon-juniper, sagebrush, mountain bursh, and aspen communities, often in sand, at 1155 to 2870 m in Beaver, Duchesne, Emery, Garfield, Grand, Iron, Kane, Piute, San Juan, Sanpete, Sevier, Uintah, Washington, and Wayne counties; Wyoming to South Dakota, south to California and New Mexico; 75 (xxv). Intermediates are formed with *S. eremophilus*.

Senecio sphaerocephalus Greene Roundhead Groundsel. [*S. lugens* var. *hookeri* D. C. Eaton, type from Summit (?) County]. Perennial herbs from a short stout rhizome; stems erect or ascending, 3–8 dm tall; herbage tomentose; basal leaves petiolate, the blades 4–15 cm long, 1–3.5 cm wide, oblanceolate to elliptic, entire or denticulate, obtuse apically; cauline leaves reduced upward, becoming sessile, finally bracteate; heads few to many, corymbose; involucres 3–7 mm long, 6–12 mm wide; main bracts 13–21, oblong- to ovate-lanceolate, greenish or brownish, with scarious margins, the tips black, hair-tufted apically; outer bracts very short; rays

8–13, yellow, 4–10 mm long; pappus white; achenes hairy; 2n = 40. Lodgepole pine and spruce-fir communities, in meadows, at 2315 to 3205 m in Daggett, Duchesne, Summit, Uintah, and Wasatch counties; Oregon and Montana, south to Nevada and Wyoming; 10 (i).

Senecio streptanthifolius Greene Manyface Groundsel. [*S. aquariensis* Greenman, type from Aquarius Plateau; *S. jonesii* Rydb., type from Alta; *S. leonardii* Rydb., type from American Fork Canyon; *S. malmstenii* Blake in Tidestr., type from Little Podunk Creek, Kane County; *S. pammelii* Greene, type from Peterson, Morgan County; *S. platylobus* Rydb., type from the Wasatch Mts.; *S. rubricaulis* var. *aphanactis* Greenman, type from Logan; *S. wardii* Greene, type from Fish Lake Mt.]. Perennial herbs from a taproot and simple or branched and infrequently subrhizomatous caudex; stems erect, mainly 8–47 cm tall; herbage glabrous or rarely sparingly tomentose; leaves thickish; basal leaves petiolate, the blades 1–5 cm long, 0.3–3 cm wide, oblanceolate to obovate, suborbicular, elliptic, or ovate, crenate, dentate, or subentire, less commonly lobed; cauline leaves reduced upward, commonly some of them pinnatifid, finally bracteate; heads few to many, corymbose to subumbellate; involucres 4–8 mm high, 5–12 mm wide; main bracts 8–21, lance-oblong, green or brownish, the margins scarious, sparingly hair-tufted apically; outer bracts very short; rays 8–13, yellow, 5–8 mm long; pappus white; achenes glabrous; 2n = 23. Sagebrush, mountain brush, ponderosa pine, aspen, lodgepole pine, spruce-fir, and alpine tundra communities, often in meadows, at 1370 to 3415 m in all Utah counties, except Beaver and Iron; Yukon to Northwest Territories, south to California and New Mexico; 107 (vii). This species forms a plexus around which revolves such species as *S. pauperculus*, *S. multilobatus*, *S. neomexicanus*, and *S. canus*, as judged from morphological intermediates, which are presumed to be hybrids.

Senecio triangularis Hook. Arrowleaf Groundsel. Perennial herbs from a caudex and more or less well-developed rhizome; stems equally leafy, erect, 2.5–12 dm tall or more; herbage glabrous or sparingly tomentose; leaves petiolate, the blades mainly 3–15 cm long, 0.5–6 cm wide, lance-oblong to triangular, abruptly contracted or subhastate at the base, dentate to sinuate-denate or subentire, finally bracteate in the inflorescence; heads few to many, subcorymbose; involucres 6–12 mm high, 8–17 mm wide; main bracts 8–12, lance-attenuate, the margins often scarious, tufted-hairy apically; outer bracts very short; rays 5–9, yellow, 6–15 mm long; pappus white; achenes glabrous; n = 20, 40. Aspen-mountain brush, Douglas fir-white fir, lodgepole pine, and spruce-fir communities at 1765 to 3265 m in Cache, Daggett, Duchesne, Grand, Salt Lake, Summit, Uintah, Utah, Wasatch, and Weber counties; Alaska and Yukon, south to California and New Mexico; 25 (viii).

Senecio vulgaris L. Common Groundsel. Plants annual or biennial, with fibrous roots, 1–5.5 dm tall, the stems glabrous or sparingly villous; basal leaves smaller than the main cauline ones, often withered by anthesis; cauline leaves not much reduced upward, 2–10 cm long, 0.5–4.5 cm wide, irregularly pinnatifid, the lobes again toothed, glabrous or more or less villous, especially along the veins beneath, the lower ones petiolate, the upper ones becoming sessile and auriculate-clasping; heads few to many; involucres 5–8 mm high, 4–10 mm wide; outer bracts short and black tipped, the inner lance-linear, green, with scarious margins, black tipped; ray flowers lacking; pappus white; achenes hairy; 2n = 20, 40. Weedy species of disturbed sites in Cache, Salt Lake, and Utah counties; adventive from Europe; 6 (0).

Senecio werneriifolius (Gray) Gray Montane Groundsel. [*S. aureus* var. *werneriifolius* Gray]. Plants commonly rhizomatous or soboliferous herbs; stems erect or ascending, 3–18 cm tall; herbage tomentose, often glabrate or glabrous in age; basal leaves petiolate, the blades 0.6–3 cm long, 0.4–2 cm wide, oval to elliptic, obovate, or oblanceolate, thickish, sometimes revolute; cauline leaves few, commonly inconspicuous and bracteate; heads 1–4; involucres 4–10 mm long, 7–15 mm wide; main bracts 13–21, lance-oblong, green or suffused with purple, the margins scarious, hair tufted apically; outer bracts to half as long as the inner; rays 8–13, yellow, 4–10 mm long; pappus white; achenes glabrous; n = 23. Ponderosa pine, western bristlecone pine, aspen-conifer, and spruce-fir communities, often in semibarrens, at 2375 to 3600 m in Beaver, Duchesne, Garfield, Grand, Iron, Juab, Piute, Salt Lake, San Juan, Summit, and Utah counties; Idaho and Montana, south to California, Nevada, and Arizona; 28 (ii).

Solidago L.

Perennial herbs from a caudex or rhizome; leaves alternate, simple; heads numerous, radiate, yellow, borne in paniculate, racemose, or cymose clusters; involucres imbricate in several series or subequal, commonly chartaceous or with the tips green; receptacle flat, naked; ray flowers fertile; disk flowers perfect, fertile; anthers subentire basally; style branches with lanceolate appendages; pappus of capillary bristles; achenes few nerved, pubescent.

1. Heads in corymbs or flat-topped cymes; leaves punctate; plants of lower elevations riparian habitats [Euthamia] .... *S. occidentalis*
— Heads racemose or panicled; leaves not punctate; plants of various habitats 2

2(1). Stems glabrous 3
— Stems puberulent with short incurved hairs or villous with multicellular hairs 5

3(2). Involucres 2.5–4 mm long; plants definitely rhizomatous, of lower elevations .... *S. missouriensis*
— Involucres various; plants subrhizomatous, of high or low elevations 4

4(3). Plants with very large basal leaves, mainly 5–10 dm tall, known from saline seeps in western Utah and Washington County .... *S. spectabilis*
— Plants with basal leaves not especially large, mainly less than 5 dm tall, usually of high elevations ... *S. spathulata*

5(2). Stems villous with multicellular hairs; petioles long-ciliate .... *S. multiradiata*
— Stems puberulent with short incurved hairs; petioles scabrous or strigose marginally 6

6(5). Involucres 6–11 mm high, the outer bracts subfoliaceus .... *S. parryi*
— Involucres 2–5 mm high, the bracts not subfoliaceus ... 7

7(6). Leaves very numerous and much longer than the internodes, gradually attenuate or acuminate, not dimorphic, strongly 3–nerved .... *S. canadensis*

— Leaves not very numerous, often less than twice as long as the internodes, acute or rounded apically, often dimorphic, with lateral nerves obscure or moderately apparent 8

8(7). Leaves sparingly hairy to glabrous, the margins rough-hairy; plants widespread and common ..... *S. sparsiflora*

— Leaves cinereus-puberulent with disoriented hairs, the margins hairy like the surfaces; plants more restricted and less common .... *S. nana*

Solidago canadensis L. Goldenrod. [*S. altissima* L.; *S. lepida* DC.]. Perennial herbs from creeping rhizomes; stems 3–12 dm tall or more; herbage puberulent with short incurved hairs, or the stems glabrous below; basal leaves often deciduous or withered at anthesis; cauline leaves numerous and crowded, 2–10 cm long or more, 3–20 mm wide, lanceolate to lance-linear, or narrowly elliptic, tapering to a sessile base, 3–nerved, serrate to entire, attenuate to acuminate apically; inflorescence commonly (but not always) of recurved branches with secund heads; involucres 2–5 mm high and about as broad, the bracts lance- attenuate, scarious or greenish; rays 10–17, yellow, 1–3 mm long; n = 9, 20–24, 27. Riparian and other mesic sites at 350 to 2290 m in all Utah counties; widespread in North America; 87 (xvi). This plant serves as host for a peculiar red and black leaf beetle. A phase of the species is cultivated as an ornamental in Utah. Designation of varietal level in Utah seems academic. The species is transitional to *S. sparsiflora*.

Solidago missouriensis Nutt. Missouri Goldenrod. Perennial herbs from creeping rhizomes; stems 2–5 (9) dm tall; herbage glabrous or sparingly puberulent in inflorescence only; basal leaves oblanceolate, often withered at anthesis; main cauline leaves 2–13 cm long, 0.4–1.5 cm wide, oblanceolate to elliptic or linear, tapering to a sessile base, mainly 3–nerved, entire or essentially so, acute to obtuse apically; inflorescence compact, with ascending branches, somewhat or not at all secund; involucres mostly 3–5 mm high and as broad, the bracts lance-attenuate, greenish to scarious; rays 7–13, yellow, 2–3 mm long; 2n = 18, 36. Riparian communities at 1525 to 2475 m in Box Elder, Cache, Carbon, Daggett, Duchesne, Emery, Salt Lake, San Juan, Sanpete, Summit, Tooele, Uintah, and Wasatch counties; British Columbia to Ontario, south to Arizona, Texas, and Tennessee; 13 (ii).

Solidago multiradiata Ait. Low Goldenrod. [*S. ciliosa* Greene]. Perennial herbs from a rhizome or rhizomatous caudex; stems 5–45 cm tall; herbage villous with multicellular hairs, at least on upper stem and petiole bases; basal and lower cauline leaves 1.5–14 cm long, 5–24 mm wide, oblanceolate to spatulate or elliptic, tapering to a conspicuously ciliate petiole, obscurely 3–nerved, entire or serrate, rounded to obtuse apically; inflorescence loosely to densely corymbose; involucres 4–6 mm high, 5–7 mm wide, the bracts lance-oblong, green apically, with prominent midvein; rays ca 13, yellow, 4–5 mm long; 2n = 18, 27, 36. Aspen, lodgepole pine, spruce-fir, and alpine tundra communities at 2745 to 3660 m in Beaver, Cache, Carbon, Daggett, Duchesne, Garfield, Grand, Iron, Juab, Kane, Piute, Salt Lake, San Juan, Sanpete, Sevier, Summit, Uintah, and Utah counties; Alaska to Quebec, south to California and New Mexico; 84 (xvii). Our specimens belong to var. *scopulorum* Gray.

Solidago nana Nutt. Dwarf Goldenrod. [*S. radulina* Rydb., type from Cottonwood Canyon]. Perennial herbs from a rhizome or subrhizomatous caudex; stems 13–48 cm tall; herbage densely canescent with fine hairs of mixed orientation; basal and lower cauline leaves petiolate, 1.5–9 cm long, 0.7–2.3 cm wide, oblanceolate to spatulate, tapering to a petiole, weakly 3–nerved, entire or slightly toothed, rounded to obtuse apically; cauline leaves definitely reduced upward; inflorescence corymbose, seldom if at all secund; involucres 4–6 mm high and about as broad; rays 5–8, yellow, 3–4 mm long. Desert shrub upward to spruce-fir communities, mainly in riparian or wet meadow sites, at 1460 to 2745 m in Daggett, Duchesne, Emery, Kane, Salt Lake, Sevier, Summit, Uintah, Utah, and Wasatch counties; Idaho to Montana, south to Arizona and Colorado; 15 (i).

Solidago occidentalis (Nutt.) T. & G. Western Goldenrod. [*Euthamia occidentalis* Nutt.]. Perennial herbs from elongate rhizomes; stems erect, branched above, mainly 4–12 (20) dm tall; herbage essentially glabrous; leaves numerous, sessile, linear to lance-linear, 2–10 cm long, 1–10 mm wide; inflorescence usually large, leafy-bracted, broadly rounded; involucres 3.5–4.5 mm high and about as broad, the bracts narrowly oblong, greenish apically, the midnerve conspicuous; rays 15–30, yellow, 1.5–2.5 mm long; n = 9. Riparian habitats at 850 to 1650 m in Box Elder, Cache, Carbon, Duchesne, Emery, Garfield, Grand, Juab, Kane, Salt Lake, San Juan, Sevier, Uintah, Utah, Washington, and Weber counties; British Columbia and Alberta, south to California, New Mexico, and Nebraska; 42 (x). I follow tradition by including this taxon in *Solidago*; it might best be treated in *Euthamia*.

Solidago parryi (Gray) Greene Parry Goldenrod. [*Haplopappus parryi* Gray; *H. parryi* var. *minor* Gray, type from Alta]. Perennial rhizomatous herbs; stems erect or ascending, 8–50 cm tall; herbage scabrous to hispidulose; basal and lower cauline leaves petiolate, mainly 3–20 cm long, 0.9–3.8 cm wide, oblanceolate to elliptic, entire, obtuse to rounded apically; cauline leaves becoming sessile and smaller upward, more or less clasping; heads few to many in compact branched cymes; involucres 8–11 mm high, 7–14 mm wide; outer bracts ovate to ovate-lanceolate, green, ciliate, the bases often scarious; inner bracts narrower and with scarious or hyaline margins; rays 12–20–, yellow, 5–8 mm long; n = 9. Aspen, tall forb, lodgepole pine, spruce-fir, and alpine tundra communities at 2285 to 3570 m in Beaver, Carbon, Daggett, Duchesne, Emery, Garfield, Grand, Juab, Kane, Millard, Piute, Salt Lake, San Juan, Sanpete, Sevier, Summit, Tooele, Uintah, Utah, and Wasatch counties; Wyoming, New Mexico, Arizona; 50 (ix).

Solidago sparsiflora Gray Alcove Goldenrod. [*S. garrettii* Rydb., type from Big Cottonwood Canyon]. Perennial rhizomatous herbs; stems erect or ascending, mainly 15–50 dm tall; herbage puberulent (often sparingly so on leaf surfaces); leaves cauline or basal, oblanceolate to elliptic or spatulate, mainly 1–10 cm long, 2–25 mm wide, entire or less commonly some of them serrate, acute to attenuate or obtuse to rounded apically, often dimorphic, with the upper ones reduced in size; inflorescence a pyramidal to conic or cylindric cluster, compact or with branches curved and heads secund; involucres 4–6 mm high and about as broad; bracts oblong to subulate, chartaceous basally, green apically, the midvein conspicuous; rays 5–10 or more, yellow, 3–4 mm long; n = 9, 18. Pinyon-juniper, mountain brush, sagebrush, aspen, ponderosa pine, and spruce-fir communities at 1125 to 3050 m in all Utah counties (except Box Elder and

Morgan); Wyoming and South Dakota, south to Arizona and Nevada; 141 (xix). Our materials are far from uniform; in the hanging gardens of southeastern Utah they are transitional to *S. canadensis* (having more ray flowers), and at high elevations they are more or less intermediate with *S. spathulata*. Possible additional influence of *S. mollis* Bartling and/or *S. nemoralis* Ait. is indicated, although they are not known from the state currently.

Solidago spathulata DC. Coast Goldenrod. Perennial herbs from a subrhizomatous caudex; stems 5–30 cm tall (rarely more), erect or ascending; herbage glabrous or somewhat scabrous and often glutinous above; basal leaves oblanceolate to spatulate, 2–15 cm long, 8–30 mm wide, serrate to entire, obtuse to rounded apically; cauline leaves reduced upward, finally sessile and more or less clasping; inflorescence compact to elongate, narrow, the heads not secund; involucres 4–6 mm high and as broad or more; bracts oblong, scarious or greenish along the prominent midvein; rays 5–10, yellow, 2.5–4 mm long; n = 9. Aspen, spruce-fir, and alpine tundra communities at 2440 to 3510 m in Beaver (?), Daggett, Duchesne, Emery, Garfield, Grand, San Juan, Sevier, Summit, and Uintah counties; Alaska to Quebec, south to California, Arizona, and New Mexico; 21 (v). Two completely intergrading phases, regarded as varieties, are present in Utah; a tall montane phase known as **var. neomexicana (Gray) Cronq.** [*S. multiradiata* var. *neomexicana* Gray], and a dwarf alpine phase known as **var. nana (Gray) Cronq.** [*S. humilis* var. *nana* Gray; *S. decumbens* Greene].

Solidago spectabilis (D. C. Eaton) Gray Nevada Goldenrod. [*S. guiradonis* var. *spectabilis* D. C. Eaton]. Perennial herbs from a subrhizomatous caudex; stems 4–13 dm tall, erect or ascending; herbage glabrous or hispidulous above; basal leaves oblanceolate, petiolate, 9–20 cm long or more, 1–2 cm broad or more, the cauline leaves much reduced upward; inflorescence more or less paniculate, dense, often less than 10 cm long, the heads secund; involucres 3–4 mm high and as broad; bracts linear-oblong, obtusish, scarious or greenish along the midvien; involucres 4–4.5 mm high, rays 4–7 (11–15), yellow; disk flowers ca 8–10 (15–22); achenes puberulent. Saline seeps at ca 885 m in Millard and Washington counties; Oregon, California, and Nevada; 3 (ii). Our material differs from Nevada specimens in the more open inflorescence and fewer florets.

Sonchus L.

Annual or perennial herbs from taproots or deep-seated, rhizomelike roots, the juice milky; leaves chiefly cauline, alternate, simple, entire to lobed or pinnatifid; heads few to several; involucral bracts imbricate in several series, green or greenish (drying brownish), the inner ones with hyaline margins; receptacle naked; corollas of ray flowers only, yellow, perfect; pappus of capillary bristles; style branches semicylindrical; achenes compressed, several to many nerved, beakless, glabrous.

1. Plants perennial, spreading from rhizomelike roots; involucres more than 14 mm long in fruit 2
— Plants annual from taproots; involucres less than 14 mm long in fruit 3

2(1). Involucres and peduncles bearing coarse stipitate glands .... *S. arvensis*
— Involucres and peduncles glabrous or tomentose, not stipitate-glandular .... *S. uliginosus*

3(1). Leaves sharply and narrowly toothed, and sometimes lobed; achenes not transversely wrinkled, merely longitudinally nerved .... *S. asper*
— Leaves sharply and broadly toothed, or merely toothed and lyrate pinnatifid; achenes transversely wrinkled and longitudinally nerved .... *S. oleraceus*

Sonchus arvensis L. Field Sow-thistle. Plants perennial with deep-seated rhizomelike roots; stems 4–10 dm tall or more, pubescent with coarse stipitate glands, at least above, and often glabrous below; leaves 5–40 cm long, 0.8–10 cm broad, more or less pinnatifid, auriculate-clasping basally, acute to obtuse apically, prickly margined; heads few to several, the peduncles stipitate-glandular; involucres 14–20 mm high and 10–30 mm broad in fruit, the bracts lance-oblong to lance-linear, glandular like the peduncles; rays yellow, mostly 10–20 mm long; achenes transversely wrinkled; n = 18, 36, 54. Weedy species of disturbed soils at 1370 to 2135 m in Cache, Duchesne, Salt Lake, and Utah counties; widely distributed and considered as a "noxious" weed in North America; adventive from Europe; 10 (0).

Sonchus asper (L.) Hill Spiny Sow-thistle. [*S. oleraceus* var. *asper* L.; *S. a.* var. *glandulifera* Garrett, type from Utah?]. Plants annual from taproots; stems 3–10 dm tall, pubescent with coarse stipitate glands, at least above, often glabrous below (less commonly throughout); leaves 3–15 cm long, 1–5 cm broad, merely lobed or lobeless, auriculate-clasping basally, acute to acuminate or less commonly obtuse apically, the margins armed with slender sharp prickles; heads few to several, the peduncles stipitate-glandular or glabrous; involucres 9–14 mm long and 10–16 mm wide in fruit, the bracts lance-oblong to lance-linear, glabrous or with few stipitate glands; rays yellow, mostly 5–10 mm long; achenes 2–3 mm long, several nerved, not transversely wrinkled; 2n = 18, 36. Weed of disturbed sites at 760 to 2135 m in Box Elder, Cache, Duchesne, Garfield, Grand, Kane, Millard, Piute, Salt Lake, San Juan, Sevier, Tooele, Uintah, Utah, and Washington counties; widespread in North America; adventive from Europe; 27 (ii).

Sonchus oleraceus L. Common Sow-thistle. Plants annual from taproots, the stems 2–10 dm tall or more, glabrous throughout or sometimes with stipitate glands above; leaves 4–20 cm long, 0.6–10 cm broad, more or less lyrate-pinnatifid, auriculate-clasping basally, acute to obtuse apically, irregularly and broadly toothed, the teeth weakly prickly; heads few to several, the peduncles glabrous or stipitate-glandular; involucres 10–13 mm high and 8–20 mm broad in fruit, the bracts lance-linear to lance-oblong, glabrous or with a few stipitate glands; rays yellow, mostly 8–12 mm long; achenes 2–3 mm long, several nerved and transversely wrinkled; 2n = 18, 32, 36. Weeds of disturbed sites at 850 to 2135 m in Cache, Duchesne, Garfield, Salt Lake, Utah, and Washington counties; widely distributed in North America; adventive from Europe; 9 (i).

Sonchus uliginosus M. Bieb. Meadow Sow-thistle. Plants perennial from deeply seated rhizomelike roots; stems 4–10 dm tall or more; herbage glabrous or obscurely tomentose; leaves 5–40 cm long, 0.8–10 cm wide, pinnatifid, auriculate-clasping basally, acute to obtuse apically, prickly margined; heads few to several, the peduncles glabrous; involucres mainly 14–16 mm high and 10–20 mm broad in fruit; bracts lance-linear to -oblong, glabrous or tomentose; rays yellow, mostly 10–20 mm long; achenes 2–3.5 mm long, several nerved, transversely wrinkled; 2n = 18, 27, 36. Weeds of disturbed sites at 1220 to 2260 m in Daggett, Duchesne, Garfield, Grand, Juab, Salt Lake, Uintah, and Utah counties; widespread in North America; adventive from Europe; 22 (ii). Authors of Flora Europaea (Tutin et al. 1976) treat this entity as *S. arvensis* ssp. *uliginosus* (M. Bieb.) Nyman. Arnow et al. (Flora of the Central Wasatch Front, Utah) discount the usefulness of stipitate glands as diagnostic features, noting that glandular and eglandular plants occur together in the same populations, and that glands are not correlated with other features. On a statewide basis the plants act like legitimate taxa, and the eglandular plants do seem to have somewhat smaller heads.

Sphaeromeria Nutt.

Perennial herbs or subshrubs; leaves alternate or mainly basal, simple and entire or pinnatifid to palmatifid; heads discoid, few to several, corymbose to subcapitate; involucres hemispheric to campanulate; bracts in 2 or 3 series, imbricate to subequal; receptacle conic or concave, naked; outer flowers pistillate, fertile; disk flowers perfect, fertile; pappus lacking or a short crown; achenes usually 5- to 10-ribbed, glabrous or glandular.

Holmgren, A. H., L. M. Shultz, and T. K. Lowrey. 1976. *Sphaeromeria*, a genus closer to *Artemisia* than to *Tanacetum* (Asteraceae: Anthemidae). Brittonia 28: 255–262.

1. Heads capitately arranged on subscapose branches; plants pulvinate-caespitose, known from Garfield County. .... *S. capitata*
— Heads in paniculate or corymbose clusters on leafy branches; plants caulescent subshrubs, not of Garfield County 2
2(1). Leaves pinnatifid, at least some, tomentose; heads paniculate; plants of Washington County ...... *S. ruthiae*
— Leaves entire or pinnatifid, glabrous; heads corymbose; plants not of Washington County .... *S. diversifolia*

Sphaeromeria capitata Nutt. [*Tanacetum capitatum* (Nutt.) T. & G.]. Pulvinate-caespitose herbs; herbage canescent with malpighian hairs; stems subscapose, 2–12 (20) cm tall; leaves mainly basal, 4–10 mm long, 1- or 2-palmately lobed, the cauline entire and reduced upward; heads few to numerous in a compact headlike cluster; involucres 3–5 mm high, the broad bracts with hyaline margins; corollas 2.5–3 mm long. With western bristlecone pine on Cedar Breaks Limestone, at ca 2380 m in Garfield County; Montana and Wyoming; 1 (0).

Sphaeromeria diversifolia (D. C. Eaton) Rydb. [*Tanacetum diversifolium* D. C. Eaton, type from American Fork Canyon]. Subshrubs, mainly 1–4 dm tall; herbage glabrous; leaves simple, entire, or some of them pinnately lobed, 8–55 mm long, 0.5–5 mm wide, linear; heads several to many in compact to open corymbose clusters; involucres 3–4 mm high, the broad bracts with hyaline margins; corollas 2–2.5 mm long. Juniper, mountain brush, mixed conifer, and aspen communities upward to alpine tundra, often in rock crevices, at 1370 to 3205 m in Davis, Juab, Millard, Salt Lake, Tooele, and Utah counties; Nevada; 33 (i). This is a Great Basin endemic.

Sphaeromeria ruthiae Holmgren, Shultz, & Lowrey. Subshrubs, mainly 3–7 dm tall; herbage tomentose-canescent with malpighian hairs; leaves pinnately lobed or the upper ones entire, 1–9 cm long, 2–4 mm wide or more; heads several to many, paniculate; involucres 3–5 mm high, the broad bracts with hyaline margins; corollas 1.8–2 mm long, yellow. Crevices in Navajo Sandstone, ponderosa pine community, in Washington County (type from Zion Canyon); endemic; 5 (0).

Stephanomeria Nutt.

Annual, biennial, or perennial herbs with milky juice; leaves alternate, often pinnatifid; flowers all raylike, perfect, pink or white; involucres cylindric; main bracts few, subequal; outer bracts much shorter; receptacle naked; pappus of plumose bristles (barbellate in *S. spinosa*); achenes 5-angled or -ribbed.

1. Plants annual, from slender taproots .... *S. exigua*
— Plants perennial, a caudex often more or less developed 2
2(1). Plants spinescent; pappus barbellate .... *S. spinosa*
— Plants unarmed; pappus plumose 3
3(2). Involucres 12–15 mm high; heads with 10 or more flowers .... *S. parryi*
— Involucres 5–12 mm high (rarely higher); heads with 3–9 flowers 4
4(3). Main leaves runcinate-pinnatifid; plants commonly 1–2 dm tall .... *S. runcinata*
— Main leaves entire or pinnatifid, often deciduous at anthesis; plants commonly 2–8 dm tall 5
5(4). Stems very slender; leaves filiform to linear, entire or dentate; pappus bristles white (rarely brownish), plumose to the base .... *S. tenuifolia*
— Stems not very slender; leaves linear-subulate, often pinnatifid or lobed; pappus brownish, scabrous toward the base .... *S. pauciflora*

Stephanomeria exigua Nutt. Annual Wirelettuce. Annual or biennial (winter annual) herbs from slender taproots; herbage glabrous or puberulent; stems 5–60 cm tall, erect and commonly branched from the base upward, often fistulous; main leaves 1–6 cm long, pinnatifid to bipinnatifid, deciduous or withered by anthesis; cauline leaves soon reduced and bracteate upward; heads more or less corymbose, terminating bracteate branchlets; involucres 5–10 mm high, 3–4.5 mm wide; main bracts usually 3–5; rays pink or white, 3–5 mm long; pappus of white to off-white bristles plumose in the upper half; achenes 3–4 mm long, tuberculate; n = 8. Warm, mixed cool, and salt desert shrub, and pinyon-juniper communities, often in sand, at 850 to 2230 m in Beaver, Emery, Garfield, Grand, Iron, Kane, Millard, San Juan, Sevier, Tooele, Uintah, Utah, Washington, and Wayne counties; Oregon to Wyoming, south to California and New Mexico; 84 (x).

Stephanomeria parryi Gray Parry Wirelettuce. Perennial herbs; stems 1 to few, weak, branching, 8–25 cm tall; herbage glabrous; leaves 2–8 cm tall, runcinate-pinnatifid, thickish, the lobes weakly spinulose-tipped; heads terminating very short bracteate branches, 10- to 14-flowered; involucres 12–15 mm high; rays whitish, 15–20 mm long; pappus bristles tawny, scabrous at the base only; achenes 3–4 mm long, not rugose; n = 16. Blackbrush community at ca 1460 m in Kane County (Atwood & Allen 2822a BRY); California to Arizona; 1 (0).

Stephanomeria pauciflora (Torr.) A. Nels. in Coult. & Nels. Fewflower Wirelettuce. [*Prenanthes? pauciflora*

Torr.]. Perennial herbs (or somewhat woody below) from a caudex, branched from the base, mostly 30–60 cm tall; herbage glabrous; main leaves 2–7 cm long, runcinate-pinnatifid, the lobes weakly spinulose-toothed; heads terminating short to elongate branchlets, 3- to 5-flowered; involucres 8–10 high, 3–5 mm wide; main bracts 5; rays pink or white, mainly 4–7 mm long; pappus bristles brownish, plumose except at the base; achenes 3.5–7 mm long, striate, more or less wrinkled; n = 8. Warm, salt, and mixed desert shrub, and juniper communities, often in sandy soil, at 760 to 1525 m in Beaver, Garfield, Grand, Juab, Kane, Millard, San Juan (?), Tooele, and Washington counties; California to Kansas, south to Texas and Mexico; 26 (v).

Stephanomeria runcinata Nutt. Desert Wirelettuce. Perennial herbs from a caudex; stems branched from the base, mostly 8–25 (30) cm tall; herbage glabrous, scabrous, or sparingly villous; main leaves 2–7 cm long, runcinate-pinnatifid, the lobes merely cuspidate; heads terminating naked or sparingly bracteate branchlets, commonly 5-flowered; involucres 9–12 mm high, 3.5–7 mm wide; rays pink, mainly 8–12 mm long; pappus bristles white, plumose almost to the base; achenes 4–5 mm long, tuberculate; n = 8. Salt desert shrub and pinyon-juniper communities at 1250 to 2535 m in Daggett, Duchesne, Emery, Grand, Uintah, and Wayne counties; Montana to Nebraska and Colorado; 17 (i).

Stephanomeria spinosa (Nutt.) Tomb Thorn Wirelettuce. [*Lygodesmia spinosa* Nutt.]. Perennial herbs from a woody caudex, the caudex branches clothed with brownish marcescent leaf bases, the axils copiously villous-hairy; stems 11–52 cm tall, thorny; herbage glabrous upward or the branches puberulent; leaves linear 0.5–7 cm long, 1–3 mm wide, reduced to bracteate scales upward, often lacking at anthesis; heads terminal on short lateral branches or sessile, 3- to 5-flowered; involucres 5.7–10 mm high, 3–5 mm wide; main bracts oblong to lance-oblong, green or often suffused with purple; outer bracts proportionately broader; rays pink, 3–5 mm long; pappus bristles off-white, scabrous throughout; achenes 4–6.5 mm long, smooth; n = 8. Desert shrub, sagebrush-grass, pinyon-juniper, mixed conifer, and aspen communities, often in moist sites, at 1675 to 3050 m in Beaver, Box Elder, Emery, Garfield, Juab, Kane, Millard, Piute, Sevier, Tooele, Washington, and Wayne counties; British Columbia to Montana, south to California and Arizona; 41 (ii).

Stephanomeria tenuifolia (Torr.) Hall Slender Wirelettuce. [*Prenanthes? tenuifolia* Torr.]. Perennial herbs from a woody caudex; caudex branches lacking or with few marcescent leaf bases, not hairy; stems 25–100 cm tall or more; herbage glabrous or puberulent; leaves filiform to linear, 1–8 (11) cm long, 1–3 (8) mm wide, entire or dentate, much reduced upward; heads terminating elongate or short lateral bracteate branchlets, 5-flowered; involucres 8–11.2 (16) mm high, 3–5 mm wide; main bracts lance-oblong, green, puberulent or glabrous; outer bracts very short; rays 4–8 (10) mm long, pink; pappus bristles white, dull white, or less commonly brownish, plumose to the base; achenes 4–6 mm long, longitudinally ribbed, smooth. Two more or less distinctive phases are present, recognizable as varieties.

1. Involucre 10–16 mm high, the bracts attenuate; basal leaves bipinnatifid, at least some; plants of Uintah County .... *S. tenuifolia* var. *uintahensis*
— Involucres mainly 8–11.2 mm high, the bracts not especially attenuate; basal leaves seldom if ever bipinnatifid; plants of rather broad distribution
.... *S. tenuifolia* var. *tenuifolia*

Var. *tenuifolia* Desert shrub, hanging garden, pinyon-juniper, mountain brush, ponderosa pine, and white-fir communities, at 1155 to 2746 m in Beaver, Duchesne, Emery, Garfield, Grand, Iron, Kane, Piute, San Juan, Sevier, Uintah, Washington, and Wayne counties; British Columbia to Montana, south to California, Arizona, and Texas; 46 (xvi). The great sprawling plants of the canyonlands portion of Utah might be worthy of taxonomic consideration; sometimes they approach *S. pauciflora* in having tawny pappus bristles.

Var. *uintahensis* Goodrich & Welsh Ponderosa pine community at ca 2490 m in Uintah (type from Brownie Canyon) County; endemic; 2 (0).

Stylocline Nutt.

Woolly annual herbs; stems commonly branched; leaves alternate, simple, entire; heads discoid, leafy bracted; involucre per se lacking; outer receptacular bracts subtending and enclosing pistillate flowers; receptacle cylindric; pistillate flowers many, deciduous with the enclosing bract, the bract apex hyaline; corollas filiform; pappus none; perfect flowers (functionally staminate) few, surrounded by linear hyaline bracts; corollas tubular, the ovaries vestigial; pappus of 3–5 deciduous bristles; anthers sagittate basally; achenes ellipsoid, few nerved.

Stylocline micropoides Gray Desert Nest-straw. Annual woolly herbs; stems usually branched, 4–12 cm tall; leaves 4–12 mm long, 0.5–1.5 mm wide, acute; bracteate leaves 6–10 mm long, 1.5–2.5 mm wide, lanceolate; heads clustered at branch tips, densely woolly; pistillate flowers with bracts boat-shaped, densely long-woolly, hyaline margined; staminate flowers with pappus of 3–5 deciduous bristles; achenes ellipsoid, ca 1.5 mm long; n = 14. Blackbrush, bursage, and indigo bush communities at 915 to 1160 m in San Juan and Washington counties; California to New Mexico, south to Mexico; 3 (i).

Syntrichopappus Gray in Torr.

Annual herbs; stems simple or branched; leaves alternate (or some opposite below), simple, entire or lobed; heads radiate, many, terminating branchlets; involucres subcylindric, bracts few, in 1 series, partly enclosing ray achenes; receptacle flat, naked; ray flowers pistillate, fertile, yellow; disk flowers perfect, fertile, yellow; anthers obtuse at base; style branches flattened; pappus of barbellate bristles; achenes 5-angled.

Syntrichopappus fremontii Gray Annual herbs, 2–14 cm tall; herbage floccoso-tomentose; leaves 5–22 mm long, narrowly spatulate to spatulate, rounded to 3–lobed apically, cuneate basally; heads few to many; involucres 5–6 mm high, 3–4 mm wide; bracts 5, oblong, greenish, with scarious margins, abruptly acute apically; rays 5, yellow, 2–5 mm long; disk corollas numerous, yellow; pappus of white barbellate bristles falling together; n = 6. Joshua tree, creosote bush, blackbrush, sagebrush, and juniper communities at 760 to 1375 m in Washington County; California, Nevada, Arizona; 8 (i).

Tanacetum L.

Perennial herbs from a rhizome; leaves alternate, 2- to 3-pinnatifid; heads discoid, numerous, corymbose; flowers perfect; involucres hemispheric; bracts in 2 or 3 series, more or less imbricate, the margins scarious; receptacle low-convex, naked; anthers entire at the base; pappus a minute crown; achenes 5–angled, truncate.

Tanacetum vulgare L. Tansy. Aromatic, glabrous or sparingly tomentose perennials, 3–10 (15) dm tall; leaves 6–15 cm long, sessile or subsessile, the blades 2– to 3–pinnatifid; heads many, discoid, yellow; involucres ca 4–5 mm high and 6–10 mm broad; bracts lanceolate; marginal flowers 3–lobed; inner flowers 5–lobed; achenes glandular, 5–angled, ca 1 mm long; n = 9. Weedy species of disturbed soils at 1370 to 1985 m in Emery, Uintah, and Utah counties; widespread in the U.S.; adventive from Europe; 3 (0).

Taraxacum Weber

Perennial scapose herbs with milky juice, from taproots; leaves all basal, pinnatifid to subentire; heads solitary on a scape; involucral bracts in 2 series, herbaceous, the outer shorter, the inner often dilated or appendaged apically, usually with broad hyaline or scarious margins, at least basally; receptacle naked; corollas of ray flowers only, perfect, yellow; pappus of capillary bristles; style branches semicylindric; achenes angular or terete, prominently nerved or ribbed, usually spinulose or with ridges near the body apex, glabrous, beaked.

1. Inner involucral bracts commonly dilated or bearing appendages apically, over 10 cm long; plants indigenous, of high elevations .... *T. ceratophorum*
— Inner involucral bracts usually not dilated or with appendages apically; plants various 2

2(1). Outer bracts reflexed or spreading, the inner ones 12–18 mm long; achenes straw colored to olive drab or brownish; plants adventive .... *T. officinale*
— Outer bracts erect, the inner ones 6–10 mm long; achenes black to grayish; plants indigenous at high elevations .... *T. lyratum*

Taraxacum ceratophorum (Ledeb.) DC. Rough Dandelion. [*Leontodon ceratophorus* Ledeb.]. Plants mostly 4–10 cm tall, from a simple or branched caudex; leaves 4–8 cm long, 0.7–2 cm broad, subentire to toothed; scapes sparingly villous, moderately so below the head; involucres 12–17 mm high in flower, the outer bracts ovate to lanceolate, appressed or ascending, the inner ones lance-oblong, attenuate, the apex dilated or appendaged; rays yellow; achene bodies 3–7 mm long, straw colored to olive-drab or brownish, the beak usually 2–4 times longer than the body; pappus white; 2n = 16, 24, 32, 40, 80. Spruce krumnolz and sedge-forb meadows at 3230 to 3660 m in Daggett, Duchesne, Summit, and Uintah counties (Leidy Peak); Alaska to Yukon, east to the Atlantic, south to California, New Mexico, and Massachusetts; circumboreal; 6 (0).

Taraxacum lyratum (Ledeb.) DC. Alpine Dandelion. [*Leontodon lyratus* Ledeb.]. Plants mostly 2–8 cm tall, from a simple or branched caudex; leaves 1–6 cm long, 0.3–1 cm wide, pinnately lobed to pinnatifid or subentire; scapes glabrous or nearly so; involucres 6–10 mm high, the outer bracts lanceolate-ovate, appressed or ascending-spreading, the inner ones lance-oblong to oblong, scarcely or slightly dilated; rays yellow (fading bluish); achene bodies 3–6 mm long, black or grayish, the beak subequal to the body; pappus white. Alpine tundra and meadows in spruce-fir communities at 3325 to 3965 m in Daggett, Duchesne, San Juan, and Summit counties; Alaska and Yukon, south to Nevada, Arizona, and Colorado; Asia 7 (i).

Taraxacum officinale Weber ex Wiggers Common Dandelion. Plants mostly 3–60 cm tall, from a simple or branched caudex; leaves 5–40 cm long, 1–10 cm wide, pinnately lobed to pinnatifid, the terminal lobe broader than the lateral ones; scapes villous to subglabrous, often moderately to densely villous below the head; involucres 15–25 mm high in flower, the outer bracts lance-acuminate, reflexed, the inner ones lance-attenuate, not or scarcely dilated apically, rarely appendaged; rays yellow, or bluish externally; achene bodies 3–4 mm long, straw colored to olive drab, the beak usually 2–4 times longer than the body; pappus white; 2n = 24, 48. Ubiquitous brightly flowered weedy species at 885 to 3205 m throughout Utah; widespread in North America; adventive from Eurasia; 65 (xiii). This handsome plant is among the earliest of our spring flowers, and among the last to bloom in autumn.

Tetradymia DC.

Armed or unarmed shrubs; stems pannose-tomentose; leaves alternate, entire, foliaceous or modified as spines, with secondary leaves fasciculate in the axils; heads discoid, corymbose or racemose; involucres cylindric to turbinate or hemispheric; receptacle naked; bracts 4–6, equal or nearly so; flowers 4–8, yellow or cream; style branches truncate to rounded or conic apically; anthers sagittate basally; pappus of capillary bristles or barbellate scales; achenes striate. Note: At least some, and probably all, species are poisonous to livestock, especially to sheep, in which consumption results in a syndrome known as "big head."

Strother, J. L. 1974. Taxonomy of *Tetradymia* (Compositae: Senecioneae). Brittonia 26: 177–202.

1. Heads solitary or 2 or 3, axillary; primary leaves modified as spines 2
— Heads several to many in terminal corymbose clusters; primary leaves foliaceous or modified as spines 3

2(1). Spines commonly recurved, mainly 5–20 mm long, pannose-tomentose; achenes 6–8 mm long; plants widespread, not of Washington County ...... *T. spinosa*
— Spines straight, mainly 20–40 mm long, glabrescent; achenes 4–5 mm long; plants of Washington County .... *T. axillaris*

3(2). Primary leaves modified as persistent spreading, straight or recurved spines 5–25 mm long .... *T. nuttallii*
— Primary leaves not modified as persistent spines, if at all spinescent then appressed-ascending 4

4(3). Primary leaves linear-subulate, spinescent apically, appressed-ascending, tomentose; secondary leaves obtuse apically, glabrous or essentially so .... *T. glabrata*
— Primary leaves various but not spinescent, not contrasting in shape and pubescence with the secondary ones .... *T. canescens*

Tetradymia axillaris A. Nels. Longspine Horsebrush. Spiny shrubs, mainly 4–12 dm tall; branchlets evenly white-pannose; primary leaves modified as persistent spines 1–5 cm long, straight or becoming curved, tomentose at first, becoming glabrate; secondary leaves linear to spatulate, 2–12 mm long, essentially glabrous; heads solitary or 2 or 3, from nodes of the previous year; involucres 8–11 mm high; bracts 5, subequal, tomentose; flowers 5–7, pale yellow, the corollas 7.5–9 mm long; pappus of slender bristles; achenes 4.5–5.5 mm long; achenes pilose, the hairs 9–11 mm long; $2n = 60, 62$. Salt and warm desert shrub communities at 850 to 1375 m in Washington County; Nevada and California; 18 (ii). Our material belongs to var. **longispina** (**Jones**) **Strother** [*T. spinosa* var. *longispina* Jones, type from St. George].

Tetradymia canescens DC. Gray Horsebrush. [*T. linearis* Rydb., type from Iron Co.]. Unarmed shrubs, mainly 1–9 dm tall; branchlets white-pannose except for glabrate streaks below the primary leaves; primary leaves 0.5–4 cm long, 1–6 mm wide, lanceolate to oblanceolate or spatulate, tomentose; secondary leaves similar to the primary ones but shorter and narrower; heads few to several at branch tips; involucres 6–8 mm high or more; bracts 4, subequal, tomentose; flowers 4, yellow to cream, the corollas 7–11 mm long; pappus of white or tawny bristles; achenes 2.5–5 mm long, glabrous or hairy; $2n = 60, 62, 90, 120$. Sagebrush-grass, mountain brush, ponderosa pine, mixed conifer, and aspen communities at 1525 to 3150 m throughout Utah; British Columbia to Montana, south to California, Arizona, and New Mexico; 75 (viii).

Tetradymia glabrata T. & G. Littleleaf Horsebrush. Shrubs, mainly 3–12 dm tall; branchlets pannose except for glabrate or glabrous streaks below the primary leaves; primary leaves mainly 5–15 mm long, 0.8–1.4 mm wide, linear-subulate, spinose tipped, soon deciduous; secondary leaves linear to narrowly spatulate, glabrous or thinly tomentose; heads few to many on branch tips; involucres 7–10 mm high; bracts 4, subequal, tomentose to glabrous; flowers 4, yellow to cream, the corollas 9–10 mm long; pappus of white bristles; achenes 3–5 mm long, hirsute; $2n = 60, 120, 180$. Shadscale, greasewood, sagebrush, rabbitbrush, and juniper communities at 1370 to 2370 m in Box Elder, Emery, Juab, Millard, Sanpete, Sevier, Tooele, Uintah, and Wayne counties; Oregon and Idaho, south to California and Nevada; 44 (v).

Tetradymia nuttallii T. & G. Nuttall Horsebrush. Spinescent shrubs, 3–12 dm tall; branchlets white-pannose except for glabrescent streaks below the primary leaf bases; primary leaves modified as persistent straight or recurved spines 5–25 mm long, tomentose to glabrous; heads in terminal clusters of (2) 3–6; involucres 6–9 mm high; bracts 4, equal; flowers 4, yellow, the corollas 8–10 mm long; pappus of white or tawny bristles; achenes 4–6 mm long, hirsute; $2n = 60$. Shadscale, greasewood, sagebrush-rabbitbrush and pinyon-juniper communities at 1370 to 1830 m in Beaver, Box Elder, Daggett, Duchesne, Juab, Millard, Sanpete, Tooele, and Uintah counties; Wyoming and Nevada; 25 (i).

Tetradymia spinosa H. & A. Thorny Horsebrush. Spinescent shrubs, 3–12 dm tall; branchlets evenly pannose; primary leaves modified as spines, 5–20 mm long, tomentose, finally glabrate; secondary leaves linear to spatulate, glabrous or glabrescent; heads borne singly or in pairs, laterally, on stems of the previous season; involucres 8–12 mm high; bracts 4–6, subequal, tomentose; flowers 5–8, yellow, the corollas 6–10 mm long; pappus of slender bristles, white; achenes 6–8 mm long, hairy, the trichomes 9–12 mm long; $2n = 60$. Mixed desert shrub, shrub-grass, and pinyon-juniper communities at 1250 to 1925 m in Box Elder, Carbon, Daggett, Duchesne, Emery, Garfield, Grand, Iron, Juab, Millard, Salt Lake, Sevier, Tooele, Uintah, and Utah counties; Oregon to Montana and Wyoming, south to California, Nevada, and New Mexico; 15 (vi).

Thelesperma Less.

Perennial glabrous or sparingly puberulent herbs; leaves opposite, pinnately to palmately parted, or the upper ones entire; heads pedunculate, solitary or few per stem; involucres hemispheric to campanulate; bracts in 2 unlike series, the outer ones spreading and distinct, the inner ones connate to the middle and calyxlike; receptacle flat, chaffy with broad scarious scales; rays present (or lacking), neuter, yellow; disk flowers perfect, fertile; anthers not caudate basally; pappus of 2 retrorsely hispid awns, a crown, or lacking; achenes oblong to linear.

1. Plants 30–80 cm tall; rays normally lacking; pappus of 2 awns; known from San Juan and Washington counties
   .... *T. megapotamicum*
— Plants 3–35 cm tall; rays normally present; pappus a crown or none .... *T. subnudum*

Thelesperma megapotamicum (Sprengel) Kuntze Greenthread. [*Bidens megapotamica* Sprengel]. Perennial herbs from a caudex and stout root; stems 30–80 cm tall; leaves mainly 2–7 cm long, once or twice pinnatifid, the lobes linear, or the uppermost simple; outer bracts 4–6, oblong to ovate, obtuse, much shorter than the inner; inner bracts 6–12 mm high, connate to above the middle, the lobes with narrow scarious margins; rays lacking; disk flowers yellow (or brownish); pappus of 2 or 3 retrorsely hispid awns; outer achenes somewhat papillose dorsally; $n = 11, 17, 22, 44$. Desert shrub community at ca 915 to 1375 m in San Juan and Washington counties; Wyoming to Nebraska, south to Arizona, Texas, and Mexico; 5 (ii).

Thelesperma subnudum Gray Perennial herbs from a taproot and less commonly with a caudex and creeping rootstock; stems 3–35 cm tall, subscapose; leaves mainly at base of stem, 1.5–9 cm long, pinnately to subpalmately lobed or some or all of them entire; petioles often ciliate and blades more or less puberulent; involucres 6.3–14 mm high, 9–22 mm wide; outer bracts oblong to lanceolate, with narrow scarious margins, to half as long as the inner ones; inner bracts united to below the middle, conspicuously scarious-margined; rays present and bright yellow, 10–28 mm long and 6–18 mm wide, or lacking; disk flowers yellow; pappus a toothed crown or lacking; achenes glabrous or hairy apically, 3.5–4.5 mm long. This taxon is variable being radiate or discoid, in division of leaves, and in position of leaves along the stem. They occur mainly at elevations below 2135 m elevation. A dwarf alpine phase occurs above that elevation, and because of its small size, lack of rays, and apparent ecotypical differences these plants are herein designated at varietal level.

1. Plants mainly 3–7 cm tall; involucres 6.3–9 mm high, 9–14 mm wide; heads discoid .. *T. subnudum* var. *alpinum*
— Plants mainly 9–35 cm tall; involucres 8–14 mm high, 12–22 mm wide; heads commonly radiate ....
   .... *T. subnudum* var. *subnudum*

Var. alpinum Welsh Pinyon-juniper, mountain brush, and western bristlecone pine communities at ca 2745 m in Wayne (type from 3 mi N of Bicknell) County; endemic; 3 (0).

Var. *subnudum* Mixed desert shrub, salt desert shrub, and pinyon-juniper communities at 1065 to 2135 m in Carbon, Duchesne, Emery, Garfield, Grand, Iron (type from Red Creek), Kane, San Juan, Sanpete, Uintah, Washington, and Wayne counties; Colorado, Arizona, and New Mexico; 109 (xiv).

Townsendia Hook.

Annual, biennial, or perennial herbs, caulescent or acaulescent; leaves alternate, entire or rarely lobed or toothed; heads radiate, solitary or few, terminating branches, or sessile; receptacle convex, naked; involucres campanulate to hemispheric; bracts in 2–7 series; rays pistillate, fertile, the corollas white, pink, or yellow; disk flowers perfect, yellow; disk pappus of barbellate capillary bristles; ray pappus similar to that of the disk or shortened; achenes 2- or 3-ribbed, compressed, usually hairy.

Beaman, J. H. 1957. The systematics and evolution of *Townsendia* (Compositae). Contr. Gray Herb. 183: 1–151.
Reveal, J. L. 1970. A revision of the Utah species of *Townsendia* (Compositae). Great Basin Nat. 30: 23–52.

1. Plants caulescent, the internodes apparent, annual or biennial (short-lived perennial) 2
— Plants acaulescent, the internodes not elongating, perennial 5
2(1). Plants annual or winter annual; disk pappus shorter than disk-corollas; plants of southeastern Utah (Navajo Basin) .... *T. annua*
— Plants biennial or short-lived perennials; disk pappus subequal to or longer than the disk corollas 3
3(2). Achenial hairs unevenly branched; ray flowers usually dark pink purple dorsally; plants biennials of western Utah .... *T. florifer*
— Achenial hairs glochidiate; ray flowers variously colored, but, if dark pink purple dorsally, the plants perennial and of different distribution 4
4(3). Stems gray white, the pubescence dense; plants of broad distribution, perennial .... *T. incana*
— Stems thinly strigose, evident beneath the hairs; plants of the Uinta Basin, biennial .... *T. strigosa*
5(1). Involucral bracts linear to narrowly lanceolate, typically in 5–7 series 6
— Involucral bracts lanceolate to narrowly oblong, ovate, or elliptic, typically in 2–5 series 9
6(5). Involucral bracts hair tufted apically, linear, acuminate; plants of Carbon, Duchesne, and Daggett counties .... *T. hookeri*
— Involucral bracts not hair tufted apically, narrowly lanceolate, acute; plants variously distributed 7
7(6). Rays glandular dorsally; leaves canescent; plants of Duchesne and Uintah counties .... *T. mensana*
— Plants glabrous or sparingly pubescent dorsally; leaves greenish or grayish-canescent; plants not or seldom of Duchesne and Uintah counties 8
8(7). Disk pappus 3–6 mm long; leaves green, the midveins not conspicuous; plants of the Wasatch Plateau and Uinta Mts. .... *T. leptotes*
— Disk pappus 6–12 mm long; leaves grayish canescent, the midveins conspicuous; plants of Sevier, Iron, Wayne, and Garfield counties ...... *T. exscapa*
9(5). Rays yellow ventrally, densely glandular and often purplish dorsally; ray pappus 0.7–1 mm long; plants of Emery and eastern Sevier counties .... *T. aprica*
— Rays white or pink or bluish, or rarely yellow ventrally but, if yellow, the ray pappus 2–4.5 mm long and plants of other distribution 10
10(9). Plants villous with long, multicellular hairs, growing at high elevations in the Tushar Mts. ... *T. condensata*
— Plants not villous with long, multicellular hairs, of low to high elevations, but if high, not of the Tushar Mts. . 11
11(10). Plants green or greenish; flowers often bluish or purplish to pink, mainly of higher elevations in mountains and plateaus .... *T. montana*
— Plants grayish canescent or whitish; flowers seldom bluish or purplish, usually white to pink or yellowish ventrally; mainly of low to moderate elevations 12
12(11). Ray pappus 0.3–0.6 mm long; plants with densely white-hairy stems, mainly of eastern Utah ... *T. incana*
— Ray pappus 2–4.5 (6.2) mm long; plants not with densely white-hairy stems, mainly of western Utah . 13
13(12). Involucral bracts strigulose dorsally; ray pappus 3.1–6.2 mm long; plants evidently rare in western Utah .... *T. scapigera*
— Involucral bracts sparingly if at all strigulose; ray pappus 2–4.5 mm long; plants locally common in western Utah .... *T. jonesii*

Townsendia annua Beaman Caulescent annual or winter annual herbs, 2–18 cm tall; herbage strigose; leaves of basal rosettes soon withered or poorly developed; cauline leaves 5–28 mm long, 1–5 mm wide, oblanceolate to spatulate or linear, sparingly to moderately strigose, green or greenish; heads solitary or few; involucres 4.5–7 mm long, 6–14 mm wide; bracts in 2–4 series, green or suffused with purple, scarious, ciliate; rays 13–34, the corollas white or pink to lavender, 4–8 mm long, 1–2.3 mm wide, glabrous; disk corollas yellow, 2.2–3.5 mm long; achenes 1.9–2.6 mm long, pubescent with glochidiate hairs; ray pappus 0.2–0.8 mm long, that of disk flowers 1.8–3 mm long; n = 9. Sandy desert shrub and blackbrush communities at 1125 to 1590 m in Carbon, Emery, Garfield, Grand, Kane, and San Juan (type from 1.5 mi N of Bluff) counties; Colorado, Arizona, New Mexico, and Texas; 23 (v).

Townsendia aprica Welsh & Reveal Pulvinate-caespitose acaulescent perennial herbs from a caudex, 1.5–2.5 cm tall; leaves 7–13 (16) mm long, 1–3.5 mm wide, spatulate to oblanceolate, strigose; heads sessile, submersed in the leaves; involucres 4–8 mm high, 7–13 mm wide; bracts in 3–4 series, lanceolate, fimbriate, red-scarious, hyaline-ciliate, the outermost sparsely strigose; rays 13–21, the corollas yellow to golden ventrally, purplish dorsally and glandular, 4–7 mm long; disk corollas yellow, 3.7–4.5 mm long; achenes 2–2.5 mm long, 2-ribbed, the hairs glochidiate; ray pappus 0.7–1 mm long; pappus of disk flowers 4–5 mm long. Salt desert shrub and pinyon-juniper communities, commonly on clay or clay-silt exposures of the Mancos Shale (Blue Gate Member), at 1860 to 2440 m in Emery and adjacent Sevier (type from south of Fremont Junction) counties; endemic; 10 (ii). The yellow flowers and short pappus of ray flowers are diagnostic.

Townsendia condensata D. C. Eaton Pulvinate, acaulescent perennial herbs from a simple or branghed caudex, mainly 1–2 cm tall; leaves 6–15 mm long, 1–3 mm wide, spatulate, villous with multicellular hairs; heads sessile, submersed in the leaves; involucres 8–18 mm high, 10–40 mm wide; bracts in 3–5 series, lanceolate to linear, acuminate, the margins scarious and long-ciliate, pubescent with multicellular hairs, purplish to greenish dorsally; rays 22-60, the corollas white to pink or lavender, glandular on the dorsal surface, 8-16 mm long; disk corollas yellow, 4-6.5 mm long; achenes 3.2-4.5 mm long, 2-ribbed, moderately pubescent with simple or bifurcate hairs; ray and disk pappus ca 4-8 mm long. Alpine tundra at 3565 to 3630 m in the Tushar Mts., Piute County; disjunct in Wyoming and Montana; 2 (0). This distinctive dwarf plant was discovered in Utah by Alan Taye, authority on the flora of the Tushar Mts.

Townsendia exscapa (**Richards.**) **T. C. Porter** [*Aster? exscapa* Richards.]. Caespitose acaulescent perennial herbs from a simple or branched caudex, 2-3.5 cm high; leaves 0.6-5 cm long, 1-3.5 mm wide, oblanceolate to linear, acute and mucronate apically, strigose, with midvein apparent; involucres 10-18 mm high, 15-30 mm wide; bracts in 4-7 series, linear to narrowly lanceolate, ciliate on scarious margins, sparingly strigose to glabrous; ray flowers 21-40, the corollas white or pinkish, 8-15 mm long, 1.2-3 mm wide; disk corollas yellow; achenes 2- or 3-ribbed, pubescent with glochidiate hairs; ray pappus 4-8 mm long; disk pappus 6-12 mm long. Ponderosa pine, mountain sagebrush, and spruce-fir communities, often in meadows, at 2135 to 3295 m in Garfield, Iron, Sevier, and Wayne counties; British Columbia to Manitoba, south to Nevada, Arizona, Mexico, and Texas; 8 (ii).

Townsendia florifer (**Hook.**) **Gray** [*Erigeron? florifer* Hook.; *T. watsonii* Gray, type from Stansbury Island; *T. scapigera* var. *ambigua* Gray, type from Rabbit Valley; *T. florifer* var. *communis* Jones, type from Marysvale]. Caulescent winter annual or biennial herbs 3-20 cm tall; basal leaves 6-50 mm long, 3-12 mm wide, spatulate; cauline leaves narrowly oblanceolate to linear, 10-40 mm long, 1-5 mm wide, strigose, petiolate, grayish; heads solitary or few; involucres 6.5-13 mm high, 15-30 mm wide; bracts in 3 or 4 series, green or suffused with purple, scarious, ciliate; rays 13-34, the corollas white or pink ventrally, dark pink or lavender dorsally, 7-12 mm long, 1.2-3 mm wide, often glandular; disk corollas yellow, 3.3-6 mm long; achenes 3.3-4.5 mm long, pubescent with unequally forked hairs; ray pappus 1-6 mm long; disk pappus 3.5-7.5 mm long. Mixed desert shrub communities at 1280 to 1985 m in Beaver, Box Elder, Garfield, Juab, Millard, Sanpete, Sevier, Tooele, Utah, and Wayne counties; Washington to Idaho, Oregon, and Nevada; 56 (vii).

Townsendia hookeri **Beaman** Caespitose acaulescent perennial herbs from a simple or branched caudex, 2.5-3.5 cm high; leaves 10-40 mm long, 1-2.5 mm wide, linear to linear-oblanceolate, strigose; involucres 9-13 mm high, 9-14 mm wide; bracts in 5-7 series, linear to lance-linear, tufted-hairy apically, green or suffused with purple, strigose; rays 13-34, the corollas 6-9 mm long, 1-1.9 mm wide, white or pink ventrally, pinkish dorsally, glabrous; disk corollas yellow, 4.5-6 mm long; achenes 3.5-4.5 mm long, pubescent with glochidiate hairs; ray pappus 1-1.5 mm long; disk pappus 5.5-8.5 mm long. Sagebrush, sagebrush-grass, and mixed conifer communities at 2165 to 2716 m in Carbon, Daggett, Duchesne, and Uintah counties; Yukon to Saskatchewan, south to South Dakota and Colorado; 5 (0).

Townsendia incana **Nutt.** [*T. incana* var. *ambigua* Jones, type from Thompson]. Subcaulescent to acaulescent caespitose herbs, the caudex often branched; stems conspicuously white-strigose, mainly 2-6 cm high, forming clumps to 2 dm wide; leaves 5-40 mm long, 1-5 mm wide, spatulate to oblanceolate, strigose; heads solitary or few; involucres 7-11 mm high, 8-20 mm wide; bracts in 3 or 4 series, lanceolate, green, the margins scarious and ciliate, strigose; rays 13-34, the corollas white ventrally, pink to lavender dorsally, 6-10 mm long, 1.5-3 mm wide; achenes 2.5-4.5 mm long, pubescent with glochidiate hairs; ray pappus 0.3-0.6 mm long; disk pappus 4-7.5 mm long; n = 9, 27, 36. Blackbrush, salt desert shrub, mixed desert shrub, pinyon-juniper, and sagebrush communities at 1310 to 2290 m in Beaver, Carbon, Daggett, Duchesne, Emery, Garfield, Grand, Iron, Kane, Piute, San Juan, Sevier, Uintah, and Wayne counties; Wyoming to Nevada, Arizona, and New Mexico; 183 (xxiii). This is the common townsendia of the Colorado drainage system in Utah; its Great Basin counterpart is *T. jonesii*, from which it can be distinguished by the white strigose stems and shorter ray pappus.

Townsendia jonesii (**Beaman**) **Reveal** [*T. mensana* var. *jonesii* Beaman, type from Mammoth]. Subcaulescent to acaulescent caespitose herbs, the caudex commonly branched; stems not conspicuously white strigose, mainly 2-4 cm tall, forming clumps to 1 dm wide; leaves 10-40 mm long, 1-4 mm wide, oblanceolate to spatulate or almost linear, strigose; heads mostly solitary or involucres 9-12.5 mm high, 8-14 mm wide; bracts in 4 or 5 series, lanceolate, green or suffused purple, sparsely strigose; rays 13-21, the corollas white to pink, cream, or yellow ventrally, pink to red purple dorsally, glandular, 4-7 mm long; disk corollas yellow, ca 3.5 mm long; achenes 3-5.5 mm long; pubescent with glochidiate hairs; ray pappus 2-4.5 mm long; disk pappus 5-8 mm long. Two weak, but geographically and edaphically correlated, varieties are present.

1. Ray flowers yellow to lemon-yellow ventrally; plants of gypsiferous substates in Sevier and Piute counties . . . .
 . . . . *T. jonesii* var. *lutea*
— Ray flowers pink to white or cream ventrally; plants of various substrates, rather broadly distributed . . . .
 . . . . *T. jonesii* var. *jonesii*

Var. *jonesii* Sagebrush, shadscale, rabbitbrush, pinyon-juniper, and mountain brush communities at 1525 to 2745 m in Beaver, Juab, Millard, Sanpete, Sevier, and Washington counties; Nevada; 13 (ii). The type of *T. mensana* var. *jonesii* consists of strictly acaulescent plants with very slender leaves and smallish heads; it is unmatched in the specimens examined, and it is understandable why the taxon was placed initially with *T. mensana*.

Var. *lutea* **Welsh** Salt desert shrub and juniper communities at ca 1675 to 1830 m in Sevier and Piute counties (on Arapien shale and clays in volcanic rubble); endemic; 6 (i).

Townsendia leptotes (**Gray**) **Osterh.** [*T. sericea* var. *leptotes* Gray]. Perennial acaulescent herbs from a simple or more commonly branched caudex, 1-3 cm tall; herbage sparingly strigose, greene; leaves 0.6-4 cm long, 1.3-2.6 mm wide, linear to narrowly oblanceolate; involucres 5-10 mm high, 9-14 mm wide; bracts in 4-7 series, lanceolate to linear, the margins scarious, ciliate, often suffused purple; rays 13-34, the corollas white, cream, or pink ventrally, sometimes lavender dorsally, 6-10 mm long, 1.2-2 mm wide; disk corollas yellow, 3-5 mm long; achenes pubescent with glochidiate hairs; ray pappus 0.8–6.5 mm long; disk pappus like the ray pappus. Montane sagebrush and grass-forb communities, often on ridge crests and plateau margins at 2680 to 3145 m in Duchesne, Emery, Sanpete, and Summit counties (Uinta Mts. and Wasatch Plateau); Idaho and Montana, south to California, Nevada, and New Mexico; 7 (0).

Townsendia mensana Jones Perennial acaulescent herbs from a simple or more commonly branched caudex 1–2.5 cm high; herbage strigose; leaves 3–17 mm long, 0.6–1.3 mm wide, narrowly oblanceolate to linear; involucres 5–9 mm high, 7–10 mm wide; bracts in 4 or 5 series, lanceolate, the margin scarious and ciliate; rays 13–21, the corollas whitish, cream, or pinkish, glandular dorsally, 5–7.5 mm long, 0.9–1.4 mm wide; disk corollas yellow, 3.5–4.8 mm long; achenes pubescent with glochidiate hairs; ray pappus 2.5–4 mm long; disk pappus 5–6.5 mm long. Salt desert shrub, pinyon-juniper, and sagebrush communities, especially on barren and semibarren sites, at 1705 to 2715 m in Duchesne (type from near Duchesene, then Theodore) and Uintah counties; Colorado (?); a Uinta Basin endemic; 38 (v).

Townsendia montana Jones Perennial acaulescent or rarely subcaulescent herbs from a simple or branched caudex, sometimes with soboliferous rhizomatous branches, from a taproot, 2–6 cm high; herbage glabrate to strigose; leaves 5–40 mm long, 2–8 mm wide, spatulate, thickish; involucres 6–12 mm high, 8–20 mm wide; bracts in 3–6 series, oblong, obovate, oblanceolate or lanceolate, glabrous or sparingly strigose, the margins scarious; ciliate, often suffused with purple; rays 12–30, the corollas blue, pink, lavender, or white, 6–12 mm long, 1–3.5 mm wide; achenes 3.7–5.2 mm long, glabrous or sparingly pubescent with bifurcate or glochidiate hairs; ray and disk pappus alike, 3–5.5 mm long. Three more or less distinctive varieties are present.

1. Heads usually sessile; leaves mainly 1–3.5 mm wide, rather abruptly obtuse apically; plants of Garfield and Kane counties .... *T. montana* var. *minima*

— Heads usually at least shortly pedunclulate; leaves mainly broader (at least some), rounded to obtuse; plants not of Garfield or Kane counties 2

2(1). Leaves rounded apically, broadly spatulate; plants of calciferous outcrops in southern Duchesne, Wasatch, and Sanpete counties ...... *T. montana* var. *caelilinensis*

— Leaves obtuse to subacute apically; plants of various substrates in the Uinta and Wasatch mts. ....
.... *T. montana* var. *montana*

Var. *caelilinensis* Welsh Pinyon-juniper, spruce-fir, and limber pine communities on Flagstaff Limestone and Green River formations at 2135 to 3735 m in southern Duchesne, Wasatch, and Sanpete counties; endemic; 13 (i).

Var. *minima* (Eastw.) Beaman [*T. minima* Eastw., type from Bryce Canyon; *T. alpigena* var. *minima* (Eastw.) Dorn]. Ponderosa pine, western bristlecone, limber pine, and Douglas fir-white fir communities, on white and pink members of the Cedar Breaks Formation, at 2375 to 3115 m in Garfield and Kane counties; endemic; 14 (i).

Var. *montana* [*T. dejecta* A. Nels., type from Dyer Mine]. Spruce-fir and lodgepole pine communities at 3050 to 3510 m in Cache, Juab, Salt Lake (type from Alta), Summit, and Uintah counties; Idaho, Montana, and Wyoming; 3 (0).

Townsendia scapigera **D. C. Eaton in Wats.** Acaulescent or subacaulescent, caespitose, biennial or short-lived perennial herbs, mainly 2–7 cm tall, from a simple or branched caudex; herbage strigose to strigulose; leaves mainly 1–7 cm long, 2–9 mm wide, spatulate, obtuse or mucronate to emarginate; heads pedunculate to subsessile; involucres 7–14 mm high, 12–32 mm wide; bracts in 3 or 4 (5) series, lanceolate, acute, the margins lacerate-ciliate, scarious, strigose; rays 18–35, the corollas white to pink, glandular dorsally, 7–15 mm long; disk corollas yellow, 3.5–5.4 mm long; achenes 2–ribbed, pubescent with bifurcate hairs, 3.8–5.6 mm long; ray pappus 3.1–6.2 mm long; disk pappus 5–7.2 mm long. Sagebrush community to alpine tundra at ca 1525 to 3175 m in Millard County; California and Nevada; 1 (0).

Townsendia strigosa **Nutt.** [*T. incana* var. *prolixa* Jones, type from Duchesne Valley]. Caulescent biennial herbs; stems branched from the base and above, 3–15 cm long; herbage strigose to strigulose; basal leaves 1.2–4.5 cm long, 1.2–7 mm wide, oblanceolate to spatulate, more or less persistent; cauline leaves mostly smaller and narrower, often clustered below and overtopping the heads; involucres 5–10 mm high, 7–20 mm wide; bracts in 3 or 4 series, lance-ovate to lanceolate, the margins scarious, ciliate, strigose; rays 12–30, the corollas white to pink, sometimes darker dorsally, 5–14 mm long, 1.5–3 mm wide; disk corollas 3.3–5 mm long; achenes pubescent with glochidiate hairs; ray pappus 0.5–1.6 mm long; disk pappus 3.3–5 mm long; n = 9. Salt desert shrub, mixed desert shrub, and pinyon-juniper communities at 1460 to 1895 m in Daggett, Duchesne, and Uintah counties; Wyoming; 14 (ii).

Tragopogon L.

Biennial (annual or perennial) herbs from thickened taproots, the juice milky; leaves alternate, entire, clasping basally; heads solitary or few and corymbose; flowers all raylike, perfect, yellow or purple; involucres cylindric or campanulate; bracts uniseriate, equal; receptacle naked; pappus of plumose bristles united at the base; achenes 5- to 10-nerved, slender-beaked or the outer beakless.

Ownbey, M. 1950. Natural hybridization and amphiploidy in the genus *Tragopogon*. Amer. J. Bot. 37: 487–499.

1. Peduncles scarcely if at all inflated, even in fruit; achenes 15–25 mm long (including the beak); bracts subequal to the rays; plants rare in Utah .... *T. pratensis*

— Peduncles strongly inflated apically; achenes 25–36 mm long (including the beak); bracts usually longer than the rays; plants locally common 2

2(1). Rays purple; involucral bracts mainly 8 or 9 . *T. porrifolius*

— Rays yellow; involucral bracts usually 13 ...... *T. dubius*

Tragopogon dubius **Scop.** Biennial herbs; stems erect, 3–10 dm tall, simple or branched; leaves mainly 5–25 cm long, linear-subulate from an expanded base, floccose, becoming glabrate; peduncles enlarged and fistulous below the heads; involucres cylindic to campanulate; bracts commonly 13 (8 on later heads), 2.5–4 cm long in flower, 4–7 cm long in fruit; rays pale lemon yellow, shorter than the bracts; achenes 25–36 mm long; pappus whitish to tawny. Disturbed soils and in low quality range sites at 1370 to 3205 m in all Utah counties; widely distributed in the U. S.; adventive from Europe; 58 (vi).

Tragopogon porrifolius L. Oyster-plant; Salsify. Biennial herbs; stems erect, 3–10 dm tall, simple or branched above; leaves mainly 5–30 cm long, linear-subulate, the apex not recurved; peduncles enlarged and fistulose below the heads; involucres cylindric to campanulate; bracts commonly 8 (5–11), 2.5–4 cm long in flower, 4–7 cm long in fruit; rays purple, subequal to or shorter than the bracts; achenes 25–35 mm long, pappus brownish; n = 6. Cultivated plants, escaping and persisting on canal banks, in moist meadows, and along roadsides at 1370 to 2595 m in Cache, Carbon, Millard, Salt Lake, Sanpete, Summit, Washington, and Weber counties; widespread in much of the U. S.; introduced from Europe; 10 (0).

Tragopogon pratensis L. Biennial herbs; stems erect, 1.5–8 dm tall, simple or branched; leaves mainly 5–30 cm broad, tapering from a broadly expanded base to 2 cm wide, recurved apically; peduncles not especially enlarged in flower or in fruit; involucres campanulate; bracts commonly 8, 12–24 mm long in flower, 18–38 mm long in fruit; rays chrome-yellow, equaling or surpassing the bracts; achenes 15–25 mm long; pappus off-white; n = 6. Disturbed sites in Rich, Salt Lake, and Summit counties; widespread in the U. S.; adventive from Europe; 2 (0).

Vanclevea Greene

Shrubs; branchlets glutinous-resinous, green to tan, finally white to gray barked; leaves alternate, sessile, entire or serrate, falcately curved; heads discoid, yellow, solitary or cymose; involucres campanulate; bracts in 4 or 5 series, imbricate, glutinous; receptacle naked, resinous; styles long-exserted, the branches flattened, papillose; achenes clavate, 5-angled; pappus of 12–16 linear persistent slender scales.

Vanclevea stylosa (Eastw.) Greene Resinbush. [*Grindelia stylosa* Eastw., type from San Juan County]. Shrubs, mainly 5–12 dm tall; branchlets glutinous-resinous; bark tan to white or grayish black in age; leaves 0.6–3.5 cm long, 1–9 mm wide, narrowly lanceolate to oblong or elliptic, commonly entire, attenuate to a spinulose tip; heads solitary or more commonly few to many in corymbose or cymose clusters; involucres 8–10 mm high, 9–15 mm wide; bracts lanceolate to lance-attenuate, sometimes abruptly acuminate and recurved apically, resin coated; corollas yellow to cream, 6–7 mm long; achenes 4–5 mm long, compressed, glutinous and spreading hairy. Four-wing saltbush, ephedra, sand dropseed, Indian ricegrass, blackbrush, and juniper communities, in sand, at 1125 to 1620 m in Emery, Garfield, Grand, Kane, San Juan, and Wayne counties; Arizona (a Colorado Plateau endemic); 32 (viii). The genus is monotypic.

Verbesina L.

Annual (biennial or perennial?) herbs; leaves opposite, at least below, simple, toothed; heads radiate, showy; involucres biseriate, about equal, herbaceous; receptacle convex, chaffy, the bracts enfolding the achenes; rays yellow or yellow orange, pistillate; disk flowers perfect, fertile; anthers subentire basally; style branches with hispidulous appendages; pappus of 2 slender awns; achenes flattened, 2-winged.

Verbesina encelioides (Cav.) Benth. & Hook. Crownbeard. [*Ximenesia encelioides* Cav.]. Annual herbs; stems 4–10 dm tall, cinereous-strigose, often branched above; lowest leaves opposite, alternate upward, petiolate, often with stipulelike appendages at base; blades 1.2–10 cm long, 0.7–6 cm wide, ovate to lanceolate, acute to attenuate, irregularly toothed, strigose beneath, green and sparingly strigose above; involucres 7–12 mm high, 15–25 mm wide; bracts lance-ovate to lance-linear, herbaceous, strigose; rays 10–15, yellow or yellow-orange, 8–20 mm long; pappus of 2 short slender awns; achenes thickly 2 winged, pubescent; n = 17. Sagebrush, rabbitbrush, saltgrass, pinyon-juniper, and ponderosa pine communities, often in disturbed sites, at 1280 to 2260 m in Beaver, Garfield, Grand, Juab, Kane, San Juan, and Washington counties; Montana to California and Texas; 20 (v). Most of our material belongs to var. *exariculata* Robins. & Greenm. The bright flowers contrast sharply with the grayish strigose pubescence, resulting in a strikingly beautiful plant.

Viguiera H.B.K.

Annual or perennial herbs; leaves opposite (at least below), simple; heads radiate, solitary or cymose; involucres 2- or 3-seriate; rays yellow, neuter, pubescent dorsally; receptacles chaffy, the chaffy bracts clasping the achenes; disk flowers fertile; pappus none; achenes laterally compressed, 4-angled.

1. Plants perennial, widespread in montane habitats, less commonly in saline low elevation sites ..... *V. multiflora*
— Plants annual, restricted in low elevation saline habitats 2
2(1). Plants subscapose, with long naked peduncles; leaves ovate to lanceolate .... *V. soliceps*
— Plants caulescent, the peduncles bracteate or leafy; leaves linear 3
3(2). Leaves canescent with appressed hairs; plants of southern Utah .... *V. longifolia*
— Leaves hispidulous; plants of central and western Utah .... *V. ciliata*

Viguiera ciliata (Robins. & Greenm.) Blake Hairy Goldeneye. [*Heliomeris multiflora* var. *hispida* Gray; *H. hispida* (Gray) Cockerell; *Gymnolomia hispida* var. *ciliata* Robins. & Greenm., type from Utah]. Annual herbs; stems simple or variously branched, 10–70 cm tall, hispidulous; leaves 0.6–9 cm long, 1–3 mm wide, linear, hispid and hispid-ciliate, acute; heads solitary or 2–5 or more; disks 7–15 mm wide, the corollas yellow; rays ca 9–15, yellow, 6–13 mm long; involucral bracts 5.5–10 mm long, lance-attenuate, hispid and coarsely ciliate; pappus lacking; achenes ca 2.5 mm long, glabrous; n = 8. Saline marshes and meadows at ca 1300 to 1470 m in Millard, Salt Lake, and Utah counties; Arizona, New Mexico, and Mexico; 9 (ii).

Viguiera longifolia (Robins. & Greenm.) Blake [*Gymnolomia longifolia* Robins. & Greenm.; *Heliomeris longifolia* (Robins. & Greenm.) Cockerell]. Annual herbs; stems simple or variously branched, 14–60 cm tall, strigose; leaves 1–6 cm long, 1.2–7.5 mm wide, linear to oblong, strigose, rarely hispid-ciliate near the bases, acute; heads solitary, or 2 to numerous; disks 7–10 mm wide, the corollas yellow; rays ca 8–10, yellow, 6–12 mm long; involucral bracts lance-acuminate to -attenuate, strigose, not especially ciliate; pappus lacking; achenes 2–2.5 mm long, brown, glabrous; n = 8. Salt desert shrub and pinyon juniper communities at 1150 to 1525 m in Kane and Washington counties; Arizona to Texas and

Mexico; 7 (iii). Our material is assignable to var. *annua* (**Jones**) **Welsh** comb. nov. [based on: *Gymnolomia multiflora* var. *annua* Jones Proc. Calif. Acad. II. 5: 698. 1895, type from Utah?; *Heliomeris longifolia* var. *annua* (Jones) Yates; *V. annua* (Jones) Blake].

Viguiera multiflora (**Nutt.**) **Blake** Showy Goldeneye. [*Heliomeris multiflora* Nutt.]. Perennial herbs, from a woody taproot and pluricipital caudex; stems 2–10 (13) dm tall, strigose to scabrous-puberulent; leaves lanceolate to linear, mainly opposite, entire or serrate, 1–8 (10) cm long, 2–20 (25) mm wide, short-petiolate, plane or revolute, acute to obtuse; heads commonly 2 to several; disk 6–14 mm wide; involucral bracts linear or narrowly lanceolate, strigose; rays 10–14, yellow, 7–18 mm long; pappus lacking; achenes 1.2–1.8 mm long, brown, glabrous; n = 8, 9. Two weakly discernible varieties are included in our material.

1. Leaves commonly over 5 mm wide, plane; plants of mesic montane sites .... *V. multiflora* var. *multiflora*
— Leaves commonly less than 5 mm wide, the margins revolute; plants of arid plains and mountains ....
 .... *V. multiflora* var. *nevadensis*

Var. multiflora Sagebrush, juniper, cottonwood, pinyon-juniper, aspen, and spruce-fir communities, often in riparian sites, at 1340 to 2870 m in all Utah counties; Montana south to California, Arizona, and New Mexico; 137 (xvi).

Var. nevadensis (**A. Nels.**) **Blake** [*Gymonolmia nevadensis* A. Nels.; *Heliomeris multiflora* var. *nevadensis* (A. Nels.) Yates; *G. linearis* Rydb., type from near St. George]. Shadscale, mat-atriplex, pinyon-juniper, and mountain brush communities at 1370 to 2135 m in Grand, Juab, Uintah, and Washington counties; Nevada; 13 (i).

Viguiera soliceps Barneby Tropic Goldeneye. [*Heliomeris soliceps* (Barneby) Yates]. Annual herbs, 10–41 cm tall; stems branched below, terminating in subscapose, merely bracteate peduncles that overtop the foliage; leaves opposite below, the blades 15–38 mm long, 6–20 mm wide, ovate to lanceolate, strigose, 3-nerved, petiolate, obtuse to cuneate, becoming smaller upwards; peduncles 7–28 cm long; involucres biseriate, the bracts lance-acuminate, acute, 5–6 mm long, strigose; rays 10–12, yellow, 10–15 mm long; pappus lacking; achenes 2.8–3.3 mm long, blackish. Mat-saltbush community on Tropic Shale Formation at 1400 to 1470 m in Kane (type from Cottonwood Canyon) County; endemic. This is a striking species, forming masses of yellow blossoms in years of adequate rainfall; 5 (ii).

Wyethia Nutt.

Perennial herbs from thick taproots; stems erect or ascending; leaves alternate, simple; heads large, solitary or several, radiate; involucral bracts in 2–4 series, herbaceous or coriaceous; receptacle convex, chaffy, the bracts folded, persistent; rays yellow, pistillate, fertile; disk flowers perfect, yellow; pappus a crown of scales or lacking; achenes trigonal or 4-angled, glabrous or pubescent.

Weber, W. A. 1946. A taxonomic and cytological study of the genus *Wyethia*, family Compositae, with notes on the related genus *Balsamorhiza*. Amer. Midl. Naturalist 35: 400–452.

1. Leaves mainly cauline, the basal reduced or lacking, scabrous-roughened; plants of sandy desert shrublands
 .... *W. scabra*

— Leaves basal and cauline, the basal often larger than the cauline ones, smooth or, if rough-hairy, not of lower elevations 2

2(1). Herbage glabrous, resinous; upper leaves rounded and clasping basally .... *W. amplexicaulis*

— Herbage hirsute to glabrate; upper leaves petiolate ..
 .... *W. arizonica*

Wyethia amplexicaulis (**Nutt.**) **Nutt.** Mulesears. [*Espeletia amplexicaulis* Nutt.]. Perennial herbs; stems mostly 2.5–9 dm tall, glabrous; basal leaves 12–40 cm long, 2–15 cm wide, entire or dentate, petiolate, resinous; cauline leaves smaller, sessile, rounded and clasping basally; heads large, solitary or several; involucres hemispheric, 25–35 mm high, 25–50 mm wide; outer bracts foliaceous, subequal; rays 6–16, yellow, 2.5–4.5 cm long; pappus a crown, sometimes prolonged into filiform awns; achenes 8–10 mm long, glabrous; n = 19. Sagebrush, oak, pinyon-juniper, aspen-fir, and forb-grass communities at 1525 to 2745 m in Box Elder, Cache, Juab, Millard, Morgan, Salt Lake, Sanpete, Sevier, Summit, Tooele, Utah, Wasatch, Washington, and Weber counties; Washington to Montana, south to Nevada and Colorado; 38 (ii).

Wyethia arizonica Gray Perennial herbs; stems mainly 30–80 cm tall, spreading hairy, especially upward; basal leaves 15–40 cm long or more, 3–15 cm wide, petiolate, the blades oblanceolate to elliptic or lanceolate; cauline leaves smaller, attenuate basally to a short petiole; heads large, solitary or several; involucres hemispheric or campanulate, 20–30 mm high, 15–40 mm wide; outer bracts foliaceous, subequal; rays 6–16, yellow, 2.5–4 cm long; pappus a crown, sometimes prolonged into filiform awns; achenes 8–10 mm long, glabrous; n = 19. Pinyon-juniper, oak, and ponderosa pine communities at 1430 to 2440 m in Grand, Kane, San Juan, and Washington counties; Colorado, New Mexico, and Arizona; 9 (0).

Wyethia scabra Hook. Rough Mulesears. Robust, clump-forming perennial herbs; stems several to many, 1.5–6 dm high or more, scabrous and hispidulose; leaves mainly cauline, the lower ones rudimentary, 3–15 cm long, 3–17 mm wide, elliptic to oblong or linear, scabrous; heads solitary or few, terminating stems and branches; involucres hemispheric, 20–40 mm high, 20–55 mm wide; bracts lanceolate to linear, attenuate to caudate-attenuate; rays 10–23, yellow, 18–40 mm long; pappus a crown; achenes 6–8 mm long, glabrous. Three almost completely confluent varieties are present. Diagnostic features are based on the nature of surface and habit of the involucral bracts, which in the typical, common phase is almost sufficiently variable as to include the others. The following arbitrary key segregates specimens, if not taxa.

1. Involucral bracts long-attenuate from short dilated bases, ciliate with multicellular hairs, glabrous but with shiny resin droplets dorsally; plants of Kane County ..
 .... *W. scabra* var. *attenuata*

— Involucral bracts variable in shape, ciliate or not, scabrous to pubescent and more or less glandular dorsally, but seldom if ever with resin droplets 2

2(1). Involucral bracts closely imbricate, the outer recurved-spreading, pubescent with short fine hairs; plants of San Juan, Grand, and eastern Kane counties ....
 .... *W. scabra* var. *canescens*

— Involucral bracts various, scabrous to long-hairy dorsally; plants rather widely distributed ....
.... *W. scabra* var. *scabra*

Var. *attenuata* W. A. Weber Ponderosa pine, oak, and pinyon-juniper (less commonly in desert shrub) communities, in sand, at 1370 to 1985 m in Kane County (type from north of Kanab); Arizona; 13 (iii) This handsome plant is a botanical motif of the Coral Pink dunes area, and is also present on East Clark Bench.

Var. *canescens* W. A. Weber Warm desert shrub and mixed desert shrub communities at 1125 to 1680 m in Grand, Kane, and San Juan counties; Colorado, Arizona, and New Mexico; 4 (i). This is a variable entity transitional to the typical variety, especially in Grand and eastern Kane counties.

Var. *scabra* Blackbrush, Vanclevea-ephedra, other mixed desert shrub, pinyon-juniper, and ponderosa pine communities at 1220 to 2625 m in Carbon, Daggett, Duchesne, Emery, Garfield, Grand, Kane, and Uintah counties; Wyoming; 48 (vii).

Xanthium L.

Annual herbs with fleshy large cotyledons and a taproot; leaves alternate, petiolate, the blades broad, rough-hairy; heads unisexual, discoid, or the corolla lacking; staminate heads uppermost, many-flowered; involucral bracts in 1–3 series, separate; receptacle cylindric, chaffy; filaments monadelphous, the anthers separate; pistil vestigial, the styles unbranched; involucre of pistillate heads enclosing the 2 flowers, forming a 2-chambered bur armed with hooked prickles, the corolla lacking; achenes large, solitary in each chamber; pappus none.

***Xanthium strumarium* L.** Cocklebur. [*X. italicum* Moretti; *X. pensylvanicum* Wallr.]. Annual monoecious herbs; stem 1.5–10 dm tall or more, simple or branched, scabrous, often purple mottled; leaves petiolate, the blades mainly 2–12 cm long and about as broad, ovate to oval or orbicular, obtuse to cuneate or cordate basally, scabrous, dentate and often lobed; heads in few to many short axillary clusters; burs broadly cylindric to ovoid, 1–3.5 cm long, with 2 more or less incurved beaks apically, covered with stout hooked prickles; 2n = 36. Weedy species of cultivated and other disturbed lands, at 850 to 1925 m in much of Utah; adventive (?) from the eastern U. S.; 33 (iii). The seedlings are poisonous to livestock, and they produce dermatitis in some people. *X. spinosum* L. was recently discovered in Utah County. It has prominent 3-forked yellow spines 1–2 cm long.

Xylorhiza Nutt.

Subshrubs or suffrutescent perennial herbs; branchlets green to straw colored or whitish; leaves alternate, simple; heads solitary at branch ends; involucres campanulate to hemispheric; bracts imbricate in several series, herbaceous to largely scarious, erect; ray flowers pistillate, fertile, yellow; achenes somewhat compressed, hairy; pappus of tawny to whitish capillary bristles. Note: Members of this genus are all primary or secondary selenium indicators.

Cronquist, A. and D. D. Keck. 1957. A reconstitution of the genus *Machaeranthera*. Brittonia 9: 231–239.
Watson, T. J. 1977. The taxonomy of *Xylorhiza* (Asteraceae-Astereae). Brittonia 29: 199–216.

1. Leaves linear to linear-filiform, the margins entire and more or less involute; plants of Kane and Garfield counties .... *X. confertifolia*
— Leaves serrate to serrate-dentate, or, if entire, of other distribution (except *X. cronquistii*) 2

2(1). Leaves serrate to serrate-dentate (at least some); plants of south central and southwestern Utah canyons of the Colorado 3
— Leaves entire; plants of eastern Utah 4

3(2). Leaves only sparingly serrate, linear oblanceolate to elliptic; involucral bracts shortly attenuate, short-villous dorsally; plants of north central Kane County ...
.... *X. cronquistii*

— Leaves sharply serrate-dentate, narrowly oblanceolate, elliptic, oblong, or lanceolate; involucral bracts long-attenuate, glandular or villous-pilose dorsally; plants of canyons of the Colorado and southwestern Utah .....
.... *X. tortifolia*

4(2). Peduncles mainly less than 5 cm long; stems usually leafy to much above the middle .... *X. glabriuscula*
— Peduncles mainly more than 5 cm long; stems usually to the middle or below .... *X. venusta*

***Xylorhiza confertifolia* (Cronq.) T. J. Watson** Henrieville Woodyaster. [*Machaeranthera glabriuscula* var. *confertifolia* Cronq., type from NE of Henrieville]. Perennial herbs from a woody caudex and taproot, with rootstocks sometimes developed; stems 9–23 cm high, sparingly pilose to glabrate and sparingly to densely stipitate-glandular; leaves 1–4.5 cm long, 1–2.5 mm wide, linear, pilose to glabrate, commonly involute; peduncles 1.8–14 cm long; involucres 9–12 mm high, 12–18 mm wide; bracts lanceolate to lance-acuminate, pilose to glabrate and glandular; rays 4–12, white, 9–18 mm long, 2–4 mm wide; disk flowers yellow, the corollas 6–9 mm long; pappus of capillary bristles to 6.5 mm long; achenes 3.5–6 mm long, pubescent; 2n = 12. Salt desert shrub and pinyon-juniper communities at 1675 to 1985 m in Garfield and Kane counties; endemic; 6 (i).

***Xylorhiza cronquistii* Welsh & Atwood in Welsh** Cronquist Woodyaster. Subshrubs, forming rounded clumps, from a stout taproot; stems numerous, whitish, ca 30 cm tall, villous at the nodes, almost glabrous otherwise; leaves 2.5–5 cm long, 2.5–5 mm wide, linear-lanceolate, sparingly serrate-dentate to entire, sparsely villous, ciliate, the midrib prominent; heads solitary on branches; involucre 10–12.5 mm high, 13–19 mm wide; bracts oblanceolate to lance-attenuate, acute to acuminate, herbaceous above the middle, chartaceous below, short-villous and glandular dorsally; rays white, 14–16, 20–25 mm long; achenes compressed, villous; pappus of capillary bristles to 7.2 mm long. Pinyon-juniper community, on the Kaiparowits Formation, at 1890 to 2075 m in Kane (type from Horse Mt.) County; endemic; 1 (0).

***Xylorhiza glabriuscula* Nutt.** Smooth Woodyaster. Subshrubs or suffrutescent perennial herbs from a woody caudex and taproot; stems 7–37 cm tall, villous to glabrous; leaves 1–7.5 cm long, 1–9 mm wide, villous to glabrate, lanceolate to narrowly lanceolate or oblanceolate; heads solitary at branch ends; involucres 9–13 mm high, 15–27 mm wide; bracts lanceolate, attenuate to acute or acuminate, herbaceous above the middle, scarious below, villous to glabrous; rays 10–22, white to bluish or purplish, 11–20 mm long; achenes compressed, villous; pappus of capillary bristles to 5 mm long. Two allopatric varieties are present.

1. Leaves with attenuate bases; rays white; plants of Daggett County .... *X. glabriuscula* var. *glabriuscula*

— Leaves with truncate or rounded bases; rays bluish, purplish, or white; plants of San Juan County ....
.... *X. glabriuscula* var. *linearifolia*

Var. glabriuscula [*Aster glabriuscula* (Nutt.) T. & G.; *Machaeranthera glabriuscula* (Nutt.) Cronq. & Keck]. n = 12. Salt and mixed desert shrub communities at ca 1525 to 2135 m in Daggett County; Colorado, Montana, South Dakota, and Wyoming; 0 (0).

Var. linearifolia T. J. Watson n = 6, 12. Salt desert shrub community, mainly on Chinle and Moenkopi formations, in Grand (type from 6 mi NW of Moab) and San Juan counties; endemic; 3 (iii).

Xylorhiza tortifolia (T. & G.) **Greene** Hurtleaf Woodyaster. Subshrubs; stems 15–50 cm tall or more, villous or tomentose and more or less stipitate-glandular; leaves 1–10 cm long, 4–20 mm wide, lanceolate to elliptic or oblanceolate, villous to tomentose and glandular, spinulose-dentate; heads terminating branches; involucres mainly 12–20 mm high and 15–30 mm wide; bracts narrowly lance-attenuate to -acuminate, herbaceous above, scarious below; rays 17–60 or more, bluish or purplish to white, 10–33 mm long, 1.8–5.5 mm wide; pappus of capillary bristles to 9 mm long; achenes compressed, pilose. Two varieties are present.

1. Involucres merely glandular dorsally; plants of canyons of the Colorado .... *X. tortifolia* var. *imberbis*
— Involucres villous-pilose as well as glandular; plants of Washington County .... *X. tortifolia* var. *tortifolia*

Var. imberbis (Cronq.) T. J. Watson [*Machaeranthera tortifolia* var. *imberbis* Cronq., type from W of Moab]. n = 6. Blackbrush, pinyon-juniper and sagebrush communities at 1220 to 2290 m in Garfield, Grand, Kane, San Juan, and Wayne counties; Arizona (Colorado canyons endemic); 32 (viii).

Var. tortifolia [*Haplopappus tortifolius* T. & G.; *Aster abatus* Blake; *Machaeranthera tortifolia* (T. & G.) Cronq. & Keck; *Xylorhiza lanceolata* Rydb., type from St. George]. n = 6, 12. Blackbrush and other warm desert shrub communities at 760 to 1010 m in Washington County; Arizona, Nevada, and California; 10 (i).

Xylorhiza venusta (Jones) **Heller** Cisco Woodyaster. [*Aster venustus* Jones, type from Cisco]. Suffrutescent to herbaceous perennial herbs from a woody caudex and taproot; stems mainly 10–40 cm tall, glabrous to densely pilose; leaves 2.4–9 cm long, 2–17 mm wide, oblanceolate to spatulate, villous to glabrate, attenuate basally; heads terminating branches; peduncles 5–20 cm long; involucres 10–18 mm high, 18–50 mm wide; bracts lance-attenuate to caudate-acuminate, herbaceous above, scarious below; rays 12–36, white or bluish to purplish, 12–27 mm long; pappus bristles to 10 mm long; achenes sericeus; n = 6, 12. Salt desert shrub communities at 1250 to 1985 m in Carbon, Daggett, Emery, Garfield, Grand, San Juan, Uintah, and Wayne counties; Colorado (a Colorado Plateau endemic); 99 (xv).

CONVOLVULACEAE A. L. Juss.
Morning Glory Family

Annual or perennial herbs, vines, or shrubs; leaves simple or compound, alternate, entire or lobed, without stipules; flowers solitary or cymose, axillary or terminal, perfect; sepals 5, equal or unequal, separate or united near the base; corolla sympetalous, regular or nearly so, 5-angled to deeply 5-lobed; stamens 5, epipetalous; pistil 1, the ovary superior, 2 (1–5) –loculed; fruit a capsule, with 1 to several seeds; x = 7–15+.

1. Stigmas linear to oblong, more than twice as long as abroad 2
— Stigmas globose to reniform or flat-topped, as broad as long or broader 4
2(1). Corolla nearly rotate when fully expanded, 5–18 mm wide .... *Evolvulus*
— Corolla funnelform to campanulate, 16–80 mm wide ... 3
3(2). Calyx not enclosed by large bracts; stigmas linear, flattened .... *Convolvulus*
— Calyx enclosed by 2 large bracts; stigmas oblong, slightly or not at all flattened .... *Calystegia*
4(1). Corollas imbricate in bud, white; styles 2, entire; stigmas capitate .... *Cressa*
— Corollas plicate-convolute in bud; style 1; stigma 1, globose or essentially so .... *Ipomoea*

Calystegia R. Br.

Prostrate or twining perennial herbs; leaves petiolate, glabrous, entire or lobed, sagittate to hastate basally; flowers axillary, usually solitary, bracts 2, mostly large and leaflike; sepals 5, enclosed by bracts; corolla campanulate to funnelform, white or pink, with 5 stripes on the outside; stigmas 2–lobed, lobes mostly oblong or elliptic; ovary 1–loculed; fruit a capsule.

1. Calyx enclosed or closely subtended by a pair of large sepaloid bracts; plants introduced .... *C. sepium*
— Calyx subtended by more remote bracts, these not especially sepaloid; plants indigenous .... *C. longipes*

Calystegia longipes (Wats.) **Brummitt** [*Convolvulus longipes* Wats.]. Perennial herbs; stems woody at the base, branched, erect to ascending, glabrous, slender, 3–10 dm long; leaves remote, linear to lance-hastate, 1–3 (5) cm long, the upper gradually reduced to linear bracts; peduncles slender, 1- or 2-flowered, 5–18 cm long; bracts lance-linear, adjacent to or remote from the calyx; sepals broadly oval, rounded and mucronate apically, unequal, 6–10 mm long; corolla white to cream, often with lavender veins, 2.5–3.5 cm long; stigmas linear-oblong; seeds dark. Mixed desert shrub communities below 1300 m in Washington County; Nevada, Arizona, and California; 3 (ii).

Calystegia sepium (L.) **R. Br.** [*Convolvulus sepium* L.]. Perennial herbs from elongate rootstocks; stems trailing and twining, to 2 m long; leaves long-petiolate, cordate at base, acuminate apically; flowers axillary, solitary or paired; peduncle short at anthesis, later elongating; floral bracts 2, elliptic-ovate, laterally overlapping, 12–25 mm long, ca twice longer than the calyx and enclosing it; corolla funnelform, 5-angled, 4–8 cm long, white; ovary 1–loculed; style simple; capsule 2- to 4–seeded; n = 11, 12. Moist sites along streams and lakes at 1270 to 1590 m in Cache, Salt Lake, Uintah, Utah, and Weber counties; adventive from Europe, now widely naturalized in the U. S.; 10 (0).

Convolvulus L.

Annual or perennial herbs; stems ascending to trailing or twining; leaves petiolate to subsessile, entire or lobed;

flowers axillary, solitary or in loose or congested cymes, both pedunculate and pedicellate; calyx with small slender bracts or none; corolla funnelform, 5–angled or shallowly lobed, white, pink, or suffused with purple; ovary 2–loculed; style simple; stigmas 2, linear, more or less flattened; capsule 2– to 4–seeded.

1. Calyx 3–5 mm long, inconspicuously pubescent or glabrate; perennial from deeply set creeping rootstocks, forming extensive beds; plants broadly distributed .... *C. arvensis*

— Calyx 6–12 mm long, densely pubescent; perennial from a taproot, not creeping; plants of Washington County .... *C. equitans*

***Convolvulus arvensis* L.** Bindweed. Deeply rooted perennial with trailing or twining stems to 1 m or more long, glabrous or somewhat hairy; leaves variable in form, but usually oblong-elliptic to deltoid-ovate, mostly 1.5–3.5 cm long, the petioles slender; flowers solitary, on peduncles subequal to the leaves; pedicels shorter than the peduncles; sepals elliptic-orbicular, obtuse, ca 3 mm long; corolla open funnelform, white or occasionally pink or lavender pink on the outside with broad vertical stripes, 1.5–2 cm long; n = 12, 24, 25. This is a pestiferous weed of roadsides, railroads, fields (especially so in dry farming grain areas), gardens, and waste places at 930 to 2800 m in probably all Utah counties; adventive from Eurasia, of cosmopolitan distribution; 72 (iii).

***Convolvulus equitans* Benth.** [*C. incanus* authors, not Vahl.; *C. hermannioides* Gray]. Stems radiating from a root crown, prostrate and clambering or rarely twining, densely pubescent; leaf blades 1–7 cm long, 0.2–4 cm wide, ovate-elliptic, triangular-lanceolate, or narrowly oblong, the hastate bases entire or variously lobed or toothed, the hairs loosely appressed; flowers solitary or 2 or 3; peduncles mainly 2–8 cm long; pedicels shorter than the peduncles; sepals more or less auricled basally; corolla funnelform, 5–angled, 1.5–3 cm long, pink or white, the center often red. Joshua tree and creosote bush communities at ca 1342 m in the Beaver Dam Mts., Washington County (S L & E R Welsh 23038 BRY); Colorado and Kansas, south to Mexico and Texas; 2 (i).

Cressa L.

Low, densely appressed-pubescent perennials with deep vertical or obliquely branching rhizomes, forming colonies; stems erect or decumbent, freely branched; leaves sessile, small, entire; flowers axillary, solitary, short-pedicellate; sepals 5, united at base, laterally overlapping, elliptic; corolla white, funnelform, 5–lobed, surpassing the calyx; stamens exserted; styles 2; stigma capitate; capsule 1– to 4–seeded.

***Cressa truxillensis* H.B.K.** [*C. erecta* Rydb., type from Becks Hot Springs, Salt Lake County]. Low, tufted, gray woolly-villous herbs, 1–2 dm tall; leaves oblong-ovate, sessile, 5–10 mm long; flowers solitary, axillary, small; sepals 4–5 mm long, canescent; corolla ca 6 mm long, the ovate-lanceolate spreading or reflexed lobes ca 2 mm long and 1–1.5 mm wide; ovary and capsule pubescent. Saline or alkaline habitats below 1600 m in Box Elder, Davis, Garfield, Juab, Millard, Salt Lake, Tooele, Utah, Washington, and Weber counties; California to Oregon, Arizona, and Mexico; 30 (ii).

Evolvulus L.

Low perennials with erect to ascending stems; leaves sessile or subsessile, entire; flowers axillary, solitary or few in cymes; sepals 5; corolla rotate to shallowly funnelform, 5–angled or lobed, blue to lavender or white; styles 2, each with 2 branches; stigmas filiform; ovary 1– or 2–loculed; capsule 1– to 4–seeded.

***Evolvulus nuttallianus* Schultes in R. & S.** [*E. pilosus* Nutt.]. Stems 5–25 cm long, usually many from a shortly creeping root crown, densely pilose; leaf blades lanceolate to elliptic-oblong, 6–20 mm long, 1–8 mm wide, exceeding the flowers; peduncles absent; pedicels shorter than the calyx; corolla lavender to nearly white. Rocky or sandy sites in mixed desert shrub communities at 1430 to 1900 m in Kane, San Juan, Tooele, and Wayne counties; Arizona to Montana, south to Texas; 12 (ii).

Ipomoea L.

Annual or perennial herbs or vines; stems usually trailing, creeping, or twining; leaves sessile to petiolate, simple or palmately compound, entire or toothed to lobed; flowers axillary or terminal, solitary to numerous; sepals 5, commonly laterally overlapping; corolla 5–angled or shallowly 5–lobed, salverform to funnelform, or campanulate, showy; variously colored; stamens included or exserted; ovary 1– to 3–loculed; style simple; stigma globose or with 2 or 3–lobes; capsule 1– to several-seeded.

1. Pedicels and peduncles with reflexed or spreading hairs .... *I. purpurea*

— Pedicels and peduncles glabrous or nearly so ..... *I. batatas*

***Ipomoea batatas* (L.) Lam.** Sweet Potato, Yam. [*Convolvulus batatas* L.]. Glabrous, trailing or twining herbs, from tuberous roots; leaf blades variable, deltoid-ovate in outline, mostly indented at base, entire to deeply 3– to 5–lobed, 4–10 cm long and wide; peduncles 1– to several-flowered; sepals oblong-lanceolate, acuminate; corolla pink to purplish, 4–7 cm long; 2n = 90. Sparingly cultivated plant in lower elevation regions of Utah; introduced from tropical regions; 2 (i).

***Ipomoea purpurea* (L.) Roth** Morning-glory. [*Convolvulus purpureus* L.]. Annual twining, hairy. herbs; leaves broadly cordate-ovate, 7–12 cm long, entire, short-acuminate, pubescent; peduncles bearing 1–5 flowers; sepals lanceolate to oblong, acute, 12–16 mm long, pubescent; corolla funnelform, purple to blue, pink, or white, 5–6 mm long; 2n = 30, 32. Cultivated ornamental, persisting, escaping, and established along fence rows, city dumps, and waste places mainly below 1500 m in Salt Lake, Utah, and Washington counties; widely grown in the U. S.; 3 (i).

CORNACEAE Dumort.
Dogwood Family

Shrubs; leaves opposite, entire, simple; inflorescences cymose or capitate, often involucrate; flowers regular, perfect; sepals 4 or 5, small or lacking; petals 4 or 5, distinct; stamens 4 or 5 and alternate to the petals; pistil 1, the ovary inferior, mostly 2–loculed and with 1 ovule per locule; style 1; fruit a drupe, with 1 or 2 seeds; x = 8–13, 19.

Cornus L.

Shrubs with stems red to purplish or yellowish; leaves ovate to elliptic or obovate; flowers in corymbose cymes, small; sepals 4, very small, greenish to white and sometimes with a purple tip; stamens 4; fruit white. Note: The yellow-flowered *C. mas* L. (Cornelian cherry) is sparingly cultivated in Utah. It flowers very early, typically in March and April. And, the showy dogwood, *C. florida* L. is likewise present in both white land lavender flowered phases.

Cornus sericea L. Red-osier Dogwood; Kinnikinik; American Dogwood. *Thelycrania sericea* (L.) Dandy; *C. stolonifera* Michx.; *C. sericea* ssp. *stolonifera* (Michx.) Fosberg f. *stolonifera* (Michx.) Fosberg; *Suida stolonifera* (Michx.) Rydb.]. Clump-forming shrubs, mainly 1.5–4 m tall, and often as broad, with branches red to purplish or yellowish, subglabrous to strigulose, the older stems grayish green and mostly glabrous; leaves 1–12 cm long, lanceolate to ovate or elliptic, acute to acuminate, green above, pale beneath, sparingly strigulose; flowers numerous, in cymes, strigose; sepals ca 0.5 mm long; petals white to cream, 2–3.5 mm long; styles 1.5–3 mm long; drupes white, 7–9 mm thick, sparsely strigulose; 2n = 22. Streambanks and other moist sites at 1370 to 3050 m in all Utah counties; widespread in North America; 130 (ii). Our plants belong to var. *sericea*. Under stipulations of the Sydney version of International Code of Botanical Nomenclature there seems to be no alternative to the use of *C. sericea* L. for our distinctive plant, which has gone for so long under the name *C. stolonifera*. Some of our material, especially specimens from the Pine Valley Mts., approaches var. *occidentalis* T. & G. in being densely villous-pilose in the inflorescence.

CRASSULACEAE A. DC. in Lam. & DC.

Stonecrop Family

Plants perennial or annual, succulent herbs; leaves alternate or opposite, simple, fleshy, entire or toothed; flowers perfect or imperfect, regular, borne in cymose clusters or solitary in leaf axils; sepals 4 or 5, distinct or basally connate; petals 4 or 5, distinct; stamens 4, 8, or 10; pistils 4 or 5, distinct or connate at the base, superior, each 1-carpelled, the styles 1 per pistil; fruit a follicle; x = 4–22+.

1. Plants diminutive annuals; flowers solitary, axillary; stamens usually 4 .... *Tillaea*
— Plants small to moderate perennials; flowers borne in terminal, cymose clusters; stamens 8 or 10 2

2(1). Flowering stems lateral, arising from axils of leaves of the basal rosette; petals brick red .... *Dudleya*
— Flowering stems terminal, the leaves borne at the base of the stem; petals typically yellow or whitish .... *Sedum*

Dudleya Britt. & Rose

Perennial herbs; leaves mainly in basal rosettes, fleshy, flattened; flowering stems axillary, with cauline leaves much reduced; flowers in terminal paniculate or cymose clusters; calyx deeply 5-lobed, with erect lance-linear segments; corolla of 5 petals united above the base; stamens 10, borne on the corolla tube; carpels 5, more or less united below; seeds numerous.

Dudleya pulverulenta (Nutt.) Britt. & Rose Liveforever. [*Echeveria pulverulenta* Nutt. in T. & G.]. Plants mainly 2–4 dm tall, typically orange-red and transitional to purplish upward; herbage coated with a white mealy powder throughout; caudex 4–9 cm thick; rosette leaves several, 8–25 cm long, 4–10 cm wide, obovate-spatulate, spreading; cauline leaves many, broadly ovate, cordate-clasping, acute; inflorescence branches 1–4 dm long, each with 10–30 flowers; pedicels slender, 5–30 mm long; calyx segments lanceolate, acute, 4–8 mm long; petals red, 12–18 mm long, connate nearly to the middle; carpels more or less distinct, 7–8 mm long. Crevices in limestone at ca 1070 m in the Beaver Dam Mts., Washington County; Arizona and California; 3 (i). Our materials belong to **var. arizonica (Rose) Welsh** stat. nov. [based on *Dudleya arizonica* Rose Addisonia 8: 35. 1923; *Echeveria arizonica* (Rose) Kearney & Peebles; *D. pulverulenta* ssp. *arizonica* (Rose) Clokey]. This plant was discovered in the Beaver Dam Mts. by Larry C. Higgins in the spring of 1986. The Utah materials represent a disjunction from the species distribution in Arizona.

Sedum L.

Perennial herbs from taproots and caudices, the caudex branches often subrhizomatous or substoloniferous; stems decumbent to ascending or erect; leaves alternate or opposite, subterete to flattened, entire or toothed; flowers showy, borne in subcapitate to paniculate cymes; sepals usually 5, persistent; petals usually 5, yellow, greenish yellow, pink, or dark purple; pistils usually 5; follicles erect.

1. Flowers white, 2.5–3 mm long; calyx ca 1 mm long; plants cultivated and escaping .... *S. album*
— Flowers yellow, cream, pink, or purple, usually 5–10 mm long or more; calyx 1–6 mm long; plants indigenous 2

2(1). Flowers yellow or cream; leaves subterete 3
— Flowers pink to purple or white; leaves flattened 4

3(2). Leaves opposite or subopposite; petals connate at the base .... *S. debile*
— Leaves alternate; petals distinct .... *S. lanceolatum*

4(3). Flowers pink, borne in elongate, capitate clusters; plants of the Uinta Mts. and high central plateaus .... *S. rhodanthum*
— Flowers dark purple, borne in capitate cymes as broad as long or broader; plants of the Raft River and Deep Creek mts. .... *S. rosea*

Sedum album L. Plants caespitose, the caudex branches prostrate, more or less rhizomatous or stoloniferous; flowering stems erect or ascending from decumbent bases, mostly 8–20 cm tall; leaves alternate, 2–15 mm long, subterete, sessile; flowers numerous in open, corymbiform cymes; calyx 5-lobed, ca 1 mm long; corolla white 2.5–3.5 mm long; stamens subequal to the corolla lobes. Cultivated ornamental, escaping and persisting in Utah County; introduced from Eurasia; 2 (0).

Sedum debile Wats. Opposite Stonecrop. Plants caespitose, the rhizomatous to stoloniferous caudex branches prostrate; flowering stems erect from a decumbent base, mainly 4–12 cm tall; leaves opposite (or nearly so), subterete, fleshy, sessile, easily deciduous, 3–8 mm long; flowers few to many in loose, terminal cymes; calyx usually 5-lobed, 2.5–3.5 mm long; corolla yellow to cream, the petals 5–8 mm long, connate basally; stamens 10, shorter than the petals; pistils surpassing the petals; follicles connate basally; 2n = 14–18. Mountain brush, Douglas fir-white fir, ponderosa pine, aspen, and alpine tundra communities, often on rock outcrops and talus, at 1575 to 3115 m in Box Elder, Cache, Davis, Duchesne, Juab, Millard, Piute, Salt Lake, Sanpete, Summit, Tooele, Utah, Wasatch, Washington, and Weber counties; Oregon and Idaho, south to Nevada and Wyoming; 43 (iv).

Sedum lanceolatum Torr. Common Stonecrop. [*S. stenopetalum*, authors, not Pursh, misapplied; *S. meehanii* Gray (?), type from City Creek Canyon]. Plants caespitose, the subrhizomatous or stoloniferous caudex branches prostrate; flowering stems erect from decumbent bases, mostly 4–25 cm tall; leaves alternate, subterete, sessile, easily deciduous, 3–20 mm long; flowers few to many in loose to compact, terminal cymes; calyx usually 5-lobed, 2.5–4 mm long; corolla yellow, the petals 5–8 mm long, distinct; stamens usually 10, subequal to the petals; follicles connate at the base; n = 8. Sagebrush, pinyon-juniper, mountain brush, aspen, lodgepole pine, and spruce-fir communities at 1765 to 3660 m in all Utah counties; Alaska and Yukon, south to California, Arizona, New Mexico, and Nebraska; 90 (xi).

Sedum rhodanthum Gray Pink Stonecrop. Perennial herbs from a taproot and simple or branched caudex, this clothed with marcescent, brown leaf bases; stems erect or more or less decumbent at the base, mostly 5–35 (40) cm tall; leaves alternate, flat, sessile, 6–30 mm long, 2–6 mm wide, entire (or somewhat toothed), persistent; flowers numerous, borne in compact, terminal, usually elongate, cymose inflorescences; calyx 5-lobed, 4–6 mm long, usually tinged pink; corolla of 5 distinct petals, 8–12 mm long, pink to rose or white; stamens shorter than the petals; follicles connate at the base; 2n = 14. Spruce-fir and alpine tundra communities, often in damp meadows and talus, at 2745 to 3600 m in Beaver, Daggett, Duchesne, Iron, Garfield, Sevier, Summit, and Uintah counties; Montana to Arizona and Colorado; 26 (viii).

Sedum rosea (L.) Scop. Roseroot. [*Rhodiola rosea* L.]. Perennial herbs from a taproot and vertical caudex, this clothed with persistent, scalelike leaves; stems erect or ascending, usually simple, 3–30 (45) cm tall; leaves alternate, flat, ovate to elliptic or oblanceolate, reduced and scalelike below, becoming larger upward, 4–40 mm long, 2–15 mm wide, toothed to entire; sepals 1.2–2 mm long, persistent, distinct, dark purple; petals distinct, 1–3 mm long, dark purple; stamens longer than the petals, the filaments purple or yellow, lacking in pistillate flowers; follicles erect, distinct; 2n = 22, 33, 36. Talus slopes in alpine tundra at ca 2925 to 3050 m in Box Elder (Raft River Mts.) and Juab (Deep Creek Mts.) counties; Alaska and Yukon, east to the Atlantic, south to California, Nevada, Colorado, and Maine; 2 (0). Our material belongs to **var. integrifolium** (Raf.) Berger (*Rhodiola integrifolia* Raf.; *S. integrifolium* (Raf.) Hulten].

Tillaea L.

Diminutive, annual herbs from taproots; stems prostrate, ascending, or erect; leaves opposite, flattened, entire; flowers minute, solitary in leaf axils; sepals usually 4, persistent; petals usually 4, whitish; pistils commonly 4; follicles erect.

Tillaea aquatica L. Pygmyweed. Stems typically prostrate and more or less stoloniferous, 1–8 cm long; leaves linear to narrowly oblanceolate, 1–6 mm long, 0.2–1 mm wide, flattened, entire, connate-sheathing basally; sepals 0.5–1 mm long, connate to near the middle; petals 1–2 mm long, whitish, membranous; stamens shorter than the petals; follicles distinct; 2n = 42. Shallow water in a wet meadow at ca 2100 m in Uintah County (Neese 13963 BRY); widespread in North America; circumboreal; 1 (0).

CRUCIFERAE A. L. Juss.
Mustard Family

Annual to perennial herbs or subshrubs, often with pungent watery juice; leaves alternate or basal and still alternate, simple to compound, estipulate; flowers perfect, regular or nearly so, hypogynous, borne in racemes, spikes, or corymbs; sepals 4, greenish to colored, the outer 2 often somewhat gibbous at the base; petals 4, rarely lacking, yellow, white, or pink to blue or purple, commonly clawed, the blade spreading in the form of a cross (hence Cruciferae); stamens 6, with the outer 2 shorter than the other 4 (tetradynamous), rarely 4 or 2; nectar glands commonly 4; ovary superior, 2-loculed or, less commonly, only 1-loculed, usually with a thin partition (replum) between the 2 marginal placentae from which, when mature, the valves usually separate; stigma rather small, entire to shallowly 2-lobed; fruit sessile or stipitate, typically dehiscent or, if indehiscent, constricted between the seeds (torulose) and sometimes breaking transversely, several times longer than broad (silique) or 1 to 3 times as long as broad (silicle), the sides flattened or compressed parallel with the replum or, less commonly, flattened contrary to the replum; seeds (1) 2 to several in a single row per locule (uniseriate) or more or less distinctly arranged in 2 rows (biseriate), smooth to striate or pitted and plump to flattened or even wing-margined, exalbuminous; x = 5–12+ [Brassicaceae Burnett].

Welsh, S. L. and J. R. Reveal. 1977. Utah flora: Brassicaceae (Cruciferae). Great Basin Naturalist 37: 279–365.

1. Plants with cauline leaves both sessile and auriculate (at least some), or auriculate and petiolate (in some *Barbarea*, *Nasturtium*, and *Rorippa*) 2
— Plants without cauline leaves or with cauline leaves not both auriculate and sessile, either petioled or merely sessile 4

2(1). Petals yellow; plants glabrous or with simple hairs, rarely with some malpighian hairs in *Thelypodiopsis* .... Key 1
— Petals white, pink, lavender, chestnut, or purple, but not yellow (cream colored in *Camelina* and *Arabis*); plants glabrous or variously pubescent 3

3(2). Plants glabrous or with simple trichomes only ..... Key 2
— Plants pubescent with at least some malpighian, branched, or stellate hairs .... Key 3

4(1). Petals yellow, sometimes fading white or pinkish to purplish 5
— Petals white, pink, lavender, purple, or chestnut, but not yellow 6

5(4). Plants glabrous or with simple hairs only .... Key 4
— Plants pubescent with malpighian, branched, or stellate hairs .... Key 5

6(4). Plants glabrous or with simple hairs only .... Key 6
— Plants pubescent with at least some malpighian, branched, or stellate hairs .... Key 7

Key 1.

Cauline leaves auriculate; petals yellow.

1. Uppermost cauline leaves falsely perfoliate-clasping; basal leaves finely dissected, usually lacking at anthesis; seeds 2, one in each locule; silicles about as broad as long (*L. perfoliatum*) .... *Lepidium*
— Uppermost cauline leaves various, but not perfoliate-clasping, more than twice as long as broad; basal leaves pinnatifid to dentate or subentire; seeds several to many, or reduced to 1 in *Isatis*; fruit more than twice as long as broad 2

2(1). Cauline leaves lyrate-pinnatifid or falsely petiolate above the auriculate base 3
— Cauline leaves entire, ovate to oblong or lanceolate, undulate or rarely toothed, sessile 4

3(2). Styles (0.5) 1–2 mm long or more, abruptly contracted at the stigma; fruit 15 mm long or more, many times longer than broad; plants not rhizomatous ..... *Barbarea*
— Styles 0.5–0.8 (1.3) mm long or, if longer, the plants rhizomatous, tapering to the stigma; fruit less than 10 mm long (rarely longer), 1 to several times longer than broad .... *Rorippa*

4(2). Cauline leaves hastately lobed; pedicels deflexed in fruit; silicles cuneate, winged-flattened, 1-seeded, 1-loculed .... *Isatis*
— Cauline leaves clasping-auriculate; pedicels spreading-ascending to ascending in fruit; siliques linear, not or only somewhat flattened, not winged, many-seeded, 2-loculed 5

5(4). Cauline leaves broadly rounded to truncate apically, strictly entire, elliptic or oblong to lance-oblong .... *Conringia*
— Cauline leaves rounded to acute apically, more or less dentate to entire, tapering from base to apex 6

6(5). Plants biennial; weeds of cultivated or disturbed sites; basal leaves lyrate pinnatifid (*B. campestris*) .... *Brassica*
— Plants perennial or biennial, rarely annual; indigenous plants of clay soils; basal leaves dentate to entire 7

7(6). Stamens exserted; siliques long-stipitate, the stipe 1–2 cm long or more (*S. viridiflora*) .... *Stanleya*
— Stamens included; siliques sessile or only short-stipitate .... *Thelypodiopsis*

Key 2.

Cauline leaves auriculate; petals white, pink, lavender, or chestnut; herbage glabrous or with simple trichomes only.

1. Leaves pinnately compound or pinnatifid; plants aquatic, glabrous (or nearly so); flowers white .. *Nasturtium*
— Leaves simple, entire or merely toothed; plants terrestrial; flowers white, pink, lavender, or chestnut 2

2(1). Uppermost cauline leaves commonly rounded to emarginate or truncate apically, the lower ones dentate apically; flowers chestnut to brown purple or purple; siliques 3–5 (6) mm broad .... *Streptanthus*
— Uppermost cauline leaves attenuate to acute apically, the lower ones various but not apically dentate only; flowers white, pink, or lavender 3

3(2). Plants annual or winter annual; pedicels recurved in fruit; fruit indehiscent, winged-flattened, 1-seeded, 1-loculed; restricted to Washington County .... *Thysanocarpus*
— Plants annual, biennial, or perennial; pedicels spreading-ascending to erect, rarely descending in some *Thlaspi*; fruit dehiscent, or indehiscent in *Cardaria*, winged-flattened to subterete, 2 or more seeded; distribution various 4

4(3). Limb of petal 4–6 (7) mm long; sepals mostly 4–7 mm long; siliques often more than 5 cm long .. *Thelypodiopsis*
— Limb of petal 2–3 mm long or less; sepals mostly 2–4 (7) mm long; fruit less than 5 cm long, except in *Arabis* 5

5(4). Fruit 10–30 times longer than broad or more, linear or narrowly oblong in outline 6
— Fruit 1–4 times longer than broad or less, clavate to obcordate, ovate, or cordate-reniform in outline 8

6(5). Pedicels mostly 8–12 mm long, erect or nearly so; plants of middle altitudes in mountains (*A. hirsuta*) .... *Arabis*
— Pedicels 2–7 mm long, spreading-ascending to descending; plants of low to moderate elevation 7

7(6). Plants 10 dm tall or more, biennial, the basal leaves often withered at anthesis; cauline leaves numerous, somewhat hastately and acutely auriculed; known only from central northern Utah (*T. rollinsii*) ... *Thelypodium*
— Plants 4 dm tall or less, annual or winter annual, the basal leaves not withered at anthesis; cauline leaves few to several, cordate-auricled to merely sessile; known only from Washington County (*C. cooperi*) .. *Caulanthus*

8(5). Seeds and ovules 3 to many per fruit, usually 2 or more in each locule; fruit conspicuously winged or more than twice as long as broad .... *Thlaspi*
— Seeds and ovules 2 per fruit, 1 in each locule; fruit not or only somewhat winged 9

9(8). Fruit broader than long, indehiscent; upper cauline leaves ovate; racemes 2–5 cm long in fruit, numerous .... *Cardaria*
— Fruit longer than broad, dehiscent; upper cauline leaves lanceolate; racemes 5–10 cm long or more in fruit, few (*L. campestre*) .... *Lepidium*

Key 3.

Cauline leaves auriculate; petals white, pink, or lavender (except in *Camelina* and *Arabis*); herbage with malpighian, branched, or stellate hairs.

1. Plants indigenous biennials or perennials of broad distribution in native plant communities; fruits siliques, several to many times longer than broad; flowers pink, lavender, or white (cream in *Arabis glabra*) 2
— Plants adventive annuals or winter annual of disturbed or cultivated places; fruits silicles, less than 3 times longer than broad; flowers usually white 3

2(1). Plants pubescent with mixed simple, forked, and branched hairs, known from north-central and northeastern Utah; siliques subquadrangular ..... *Halimolobos*
— Plants variously pubescent but not as above or of that distribution; siliques distinctly flattened .... *Arabis*

3(1). Plants flowering in early springtime; siliques triangular-obcordate, compressed .... *Capsella*
— Plants flowering in late springtime and summer; siliques obovoid, terete or nearly so .... *Camelina*

Key 4.

Cauline leaves sessile or petiolate, not auriculate or, if hastately lobed, petiolate; flowers yellow; herbage glabrous or with simple hairs only.

1. Cauline leaves both hastately lobed and petiolate, the leaf blades triangular-ovate or lanceolate, entire; plants mostly 8 dm tall or more, of middle elevations often in dense vegetation .... *Chlorocrambe*
— Cauline leaves pinnatifid to entire, the leaf-blades not hastately lobed; plants low to tall but, if tall, seldom if ever of middle elevations in dense vegetation 2

2(1). Leaves all simple and entire or sparingly toothed; plants perennial from a caudex 3
— Leaves (at least some) pinnatifid or definitely and regularly toothed (or, if all entire, then plant not arising from a caudex); plants annual, biennial, or perennial, sometimes with a distinct caudex 5
3(2). Plants low, less than 1 dm tall; cauline leaves lacking; silicles lanceolate in outline, 1–4 times longer than broad (*D. densifolia*) .... *Draba*
— Plants 1–5 dm tall or more; cauline leaves present; siliques linear, many times longer than broad 4
4(3). Stamens long-exserted from the flower; pedicels spreading; siliques stipitate, the stipes 10–25 mm long or more .... *Stanleya*
— Stamens not exserted beyond the flower; pedicels ascending to suberect; siliques sessile or subsessile, or the stipe less than 1 mm long .... *Schoencrambe*
5(2). Plants growing in mud, along beaches, or in or near streams; petals usually less than 3 mm long; siliques usually less than 12 mm long .... *Rorippa*
— Plants seldom if ever in perennially moist sites; petals more than 4 mm long or siliques more than 12 mm long, or both 6
6(5). Leaves glaucous, thickened; perennial herbs from an often woody caudex; stamens long-exserted; siliques stipitate, the stipes 10–25 mm long .... *Stanleya*
— Leaves not glaucous or, if so, not especially thickened; annual, biennial, or perennial herbs without a distinct caudex (except *Schoencrambe*); stamens not exserted; siliques not stipitate 7
7(6). Plants perennial, rhizomatous, indigenous, of lower elevations in native plant communities ..... *Schoencrambe*
— Plants annual or biennial, adventive, of disturbed or cultivated places 8
8(7). Fruits dehiscent to the apex; pedicels spreading-ascending to ascending or, if erect (as in *S. officinalis*), the petals only 3–4 mm long .... *Sisymbrium*
— Fruits with long, indehiscent sterile apices; pedicels ascending or, if appressed (as in *B. nigra*), the petals 5–8 mm long or more 9
9(8). Plants mostly 2–5 dm tall; leaves mainly basal; fruiting racemes longer than the leafy stems; seeds biseriate .... *Diplotaxis*
— Plants mostly 4–10 dm tall or more; leaves basal and cauline; fruiting racemes shorter than the leafy stems; seeds uniseriate .... *Brassica*

Key 5.

Cauline leaves sessile, petiolate, or lacking; petals yellow; herbage pubescent with malpighian, branched, or stellate hairs.

1. Leaves once to twice or rarely thrice pinnately dissected or compound; plants annual or winter annual .... *Descurainia*
— Leaves simple and entire or merely toothed or lobed; plants annual, biennial, or perennial 2
2(1). Cauline leaves lacking, all leaves basal .... *Draba*
— Cauline leaves present, at least one, the basal leaves present or absent 3
3(2). Plants pubescent with appressed, Y-shaped and/or malpighian hairs; siliques many times longer than broad .... *Erysimum*
— Plants pubescent with branched, dendritic, or appressed stellate hairs; silicles from about as broad as long to ca 5 times longer than broad 4
4(3). Pubescence of branched or dendritic hairs, rarely with some stellate; silicles lance-ovate to oblong or elliptic, usually more than twice as long as broad .... *Draba*
— Pubescence of appressed stellate hairs only (except for *Alyssum saxatile*); silicles orbicular in outline or subglobose to bladdery inflated, not over twice as long as broad 5
5(4). Silicles compressed, lens-shaped; plants annual .. *Alyssum*
— Silicles either subglobose or greatly inflated and terete or didymous; plants annual, biennial, or perennial 6
6(5). Basal leaf blades often more than 20 mm wide; silicles (excluding style) 8–10 mm long or more, didymous and bladdery inflated, cordate at the base .... *Physaria*
— Basal leaf blades rarely to 20 mm wide; silicles (excluding style) 3–6 mm long, not didymous, seldom bladdery inflated, not cordate at the base .... *Lesquerella*

Key 6.

Cauline leaves sessile, petiolate, or lacking; petals white, pink, lavender, purple, or chestnut; herbage glabrous or with simple hairs only.

1. Herbage stipitate-glandular, or glabrous and the plants scapose and with large flowers; plants either alpine perennials or winter annuals of low elevations . 2
— Herbage glabrous or with simple hairs; plants of various habits, habitats, and duration 3
2(1). Plants scapose alpine perennials; leaves basal; petal blades 8–10 mm long or more; siliques flattened, dehiscent .... *Parrya*
— Plants winter annuals; leaves cauline and basal; petal blades 2–4 mm long; siliques terete, indehiscent .. .. *Chorispora*
3(1). Blades of cauline leaves hastately lobed, otherwise entire, triangular-ovate to triangular-lanceolate; plants 8–10 dm tall or more; siliques stipitate, the stipe 2–7 mm long .... *Chlorocrambe*
— Blades of cauline leaves, if any, not both hastately lobed and entire; plants low to tall; fruit subsessile or sessile on the pedicel (except in *Lunaria*) 4
4(3). Plants dwarf scapose perennials; leaves often less than 10 mm long; scapes often less than 10 cm tall .. .. *Draba*
— Plants caulescent, at least some leaves cauline; leaves mostly more than 10 mm long; stems usually over 10 cm tall 5
5(4). Plants slender annuals, 1–2.5 dm tall; leaves entire or merely serrate (at least the cauline ones); fruits obovate in outline and several-seeded or, if linear, less than 15 mm long 6
— Plants annual, biennial, or perennial, 1.5–10 dm tall or more; leaves variously toothed, lobed, or entire; fruits obovate in outline and 2-seeded, or lance-ovoid, or, if linear, mostly more than 15 mm long 7
6(5). Petals 1.2 mm long or less; silicles obovate, compressed, 2–3 mm long .... *Hutchinsia*
— Petals 2–2.9 mm long; siliques linear, terete, 9–14 mm long .... *Arabidopsis*
7(5). Basal leaves long-petiolate, the blades oblong to lanceolate, mostly 10–30 (50) cm long or more; cauline leaves lance-oblong, irregularly crenate-serrate; petals white; silicles less than twice longer than broad, abortive; plants cultivated but escaping and often persistent .... *Armoracia*

— Basal leaves sessile, poorly developed or, if long-petiolate, the blades seldom as above, usually less than 10 cm long; cauline leaves not both lance-oblong and irregularly crenate-serrate; petals pink, white, or chestnut purple; fruit various; plants indigenous or cultivated 8

8(7). Petal blades 8–12 mm long, pink to lavender; silicles more than 20 mm wide; style 6–8 mm long; leaves cordate-ovate, irregularly toothed .... *Lunaria*

— Petal blades usually less than 8 mm long; fruit much less than 10 mm wide; style less than 3 mm long; leaves various but not cordate-ovate (except in *Cardamine cordifolia*) 9

9(8). Seeds and ovules 2, 1 in each locule; silicles obcordate, orbicular, or elliptic, usually about as long as broad .... *Lepidium*

— Seeds and ovules more than 2, usually more than 2 in each locule; silicles lance-oblong or siliques linear and 3 to many times longer than broad 10

10(9). Plants annual, with thickened tuberous roots; flowers white or pink, the petal limb more than 5 mm long; siliques terete, indehiscent .... *Raphanus*

— Plants perennial, biennial, or, if annual, not from thickened roots; flowers variously colored but, if white or pink, the petal limb less than 5 mm long; fruit terete or flattened, dehiscent 11

11(10). Cauline leaves ovate, cordate-ovate, or pinnately compound; flowers white; plants of moist sites .... *Cardamine*

— Cauline leaves pinnatifid or entire to subentire and linear to elliptic, cordate-deltoid, or deltoid, but not as above; flowers white or brown purple, lavender or with purple veins; plants of various habitats 12

12(11). Flowers minute, less than 1.2 mm long; silicles lance-oblong, less than 10 mm long; leaves lyrate-pinnatifid (*R. tenerrima*) .... *Rorippa*

— Flowers 2–6 mm long or more; siliques linear to oblong, much more than 10 mm long; leaves pinnatifid to toothed or entire 13

13(12). Flowers subsessile, the pedicels less than 2 (4) mm long; leaves, at least some, definitely pinnatifid; petals mostly brown purple .... *Caulanthus*

— Flowers pedicellate, the pedicels usually over 2 mm long, or the leaves entire; petals variously colored, but, if brown purple, the pedicels definitely longer than 2 mm 14

14(13). Basal leaves long-petiolate, the blades reniform; cauline leaves becoming short-petiolate upward, the blades cordate-deltoid to deltoid, irregularly toothed or lobed; plants with odor of garlic .... *Alliaria*

— Basal leaves variously petiolate to sessile, the blades various but seldom if ever as above; cauline leaves often sessile throughout, usually entire; plants not smelling of garlic 15

15(14). Plants annual; pedicels curved-descending; siliques compressed, the sutures parallel, not torulose .... *Streptanthella*

— Plants biennial or perennial; pedicels spreading to spreading-ascending (declined in *Thelypodium wrightii*); siliques with parallel sides or else torulose 16

16(15). Petals strongly purple-veined; leaves linear or oblong; plants from a caudex, known from Emery and Uintah counties .... *Schoencrambe*

— Petals not or seldom purple veined; leaves various; plants with or without a caudex, of various distribution 17

17(16). Plants erect biennials; basal rosette prominent, but usually withered at anthesis; pedicels spreading or declined; siliques torulose; flowers numerous to very numerous .... *Thelypodium*

— Plants ascending to erect perennials; basal rosette not especially prominent; pedicels spreading-ascending; siliques with parallel sides; flowers several to many (*A. nuttallii*) .... *Arabis*

] Key 7.

Plants with cauline leaves sessile, petiolate, or lacking; petals white, pink, lavender, or purple; herbage pubescent with malpighian, branched, or stellate hairs.

1. Leaves pinnately lobed; herbage cinereous pubescent; plants of high elevations .... *Smelowskia*

— Leaves entire to serrate or sinuate-dentate; herbage green or glaucous (cinereous in *Dithyrea*); plants of low to high elevations 2

2(1). Plants scapose, the leaves basal, mostly less than 1 dm tall .... *Draba*

— Plants with at least some leaves cauline, often more than 1 dm tall 3

3(2). Petal blades more than 6 mm long; sepals definitely saccate, 6–8 mm long; cauline leaves sinuate-dentate; plants 5–10 dm tall or more, cultivated and frequently escaping .... *Hesperis*

— Petal blades less than 6 mm long; sepals not especially saccate, 1–6 mm long; cauline leaves entire or sinuate-dentate; plants (0.5) 1–4 dm tall, indigenous or adventive, not cultivated (except *Lobularia*) 4

4(3). Style (1.5) 2–3 mm long; ovaries appressed stellate pubescent, about twice as long as broad, dehiscent; seeds several .... *Berteroa*

— Style 0.2–1 (1.5) mm long; ovaries variously pubescent but, if appressed stellate, much broader than long, dehiscent or indehiscent; seeds 1 to many 5

5(4). Pedicels 5–21 mm long, spreading; silicles with stellate or dendritic pubescence, more than twice as broad as long, 2–seeded .... *Dithyrea*

— Pedicels less than 8 mm long or not spreading, or, if spreading, the fruit not with stellate or dendritic hairs and broader than long; seeds 1, 2, or more 6

6(5). Petals minute, less than 2 mm long; seeds 1 or 2; fruit indehiscent; plants annual or winter annual 7

— Petals mostly 2–10 mm long or more; seeds 2 to many; plants annual, biennial, or perennial 8

7(6). Ovary and fruit with dendritic pubescence; fruit ovoid, with a curved stylar beak up to 1 mm long, 2–seeded .... *Euclidium*

— Ovary and fruit with simple hairs; fruit orbicular, compressed, lacking a stylar beak, 1–seeded; plant only recently known from Washington County .... *Athysanus pusillus* (Hook.) Greene

8(6). Pedicels less than 2 mm long, appressed or ascending; leaves sinuate-dentate; siliques ascending, 33–63 mm long, pubescent with dendritic or branched hairs, the stigma oblique .... *Malcolmia*

— Pedicels 2–10 mm long, descending to spreading or ascending; fruit various, sometimes pubescent with dendritic or branched hairs, the stigma terminal 9

9(8). Pedicels descending; fruit pendulous, 13–65 mm long; plants perennial .... *Arabis*

— Pedicels spreading to ascending; fruit erect or spreading to ascending, 2–14 mm long; plants annual or perennial 10

10(9). Leaves mainly cauline, strigose with malpighian hairs; fruit about as broad as long; seeds 2; plants cultivated and often escaping .... *Lobularia*

— Leaves mainly basal, with stalked dendritic or stellate hairs; seeds usually more than 2; plants indigenous or adventive 11

11(10). Fruit linear to narrowly oblong, terete or subterete, less than 1 mm broad; plants slender annuals .... *Arabidopsis*

— Fruit lance-ovate to lanceolate or oblong in outline, flattened, 1.5–3 mm broad; plants annual or perennial .... *Draba*

Alliaria Adanson

Plants glabrous or pubescent with simple hairs, biennial or annual, with odor of garlic; stems erect, simple or branched from the base; leaves alternate, the blades reniform to deltoid, long-petiolate below, becoming short-petiolate to subsessile upward; flowers borne in racemes; sepals 4, deciduous; petals white, spatulate, short-clawed; stamens 6; style short; stigma capitate; fruit a sessile silique, linear, 4–angled, the valves 3–veined.

Alliaria officinalis Andrz. Garlic Mustard. [*Erysimum alliaria* L.; *Sisymbrium alliaria* (L.) Scop.; *Alliaria alliaria* (L.) Britt.]. Herbs from a taproot, almost glabrous or sparsely pubescent with simple hairs; stems erect, 3–10 dm tall; leaves petiolate, the basal often withered by anthesis, the blades reniform; cauline leaves with cordate-deltoid to deltoid blades 1.5–6 cm wide, coarsely and irregularly toothed to sinuate-dentate; racemes simple (sometimes branched), often bracteate basally; pedicels 2–6 mm long; petals white, 5–6 mm long, much surpassing the sepals; siliques spreading, arched upward, 4–6 cm long, more or less quadrangular, somewhat torulose; seeds ca 3 mm long, black; 2n = 16, 18, or ca 21. Disturbed sites in Park City, Summit County; sporadic in North America; adventive from Eurasia; 2 (0).

Alyssum L.

Stellate-pubescent annuals or perennials from taproots; leaves alternate, simple, entire, tapering to base, not auriculate; flowers in racemes, the pedicels spreading-ascending to ascending or erect, not subtended by bracts; sepals 4, deciduous or persistent; petals 4, yellow (often fading cream or white), emarginate; stamens 6, at least the 2 shorter filaments with a whitish process near base; style slender; stigma capitate; fruit a sessile silicle, less than twice longer than broad, broadly elliptic to oval in outline, compressed parallel to the septum; valves veinless; seeds 1 or 2 per locule.

1. Cauline leaves, at least some, more than 4 cm long; flowers bright yellow, fading cream; petal blades obcordate, about as broad as long; plants perennial, cultivated and occasionally escaping .... *A. saxatile*

— Cauline leaves less than 4 cm long; flowers pale yellow (fading white); petal blades cuneate, much longer than broad; plants adventive annuals of arid sites 2

2(1). Silicles glabrous or essentially so, orbicular in outline; style 0.5–0.8 mm long, persistent .... *A. desertorum*

— Silicles stellate-pubescent, orbicular to elliptic in outline; style various 3

3(2). Silicles 4–5 mm broad; style 0.8–1.2 mm long; pubescence of coarse spreading-ascending stellate hairs .... *A. minus*

— Silicles 2.8–4 mm broad; style 0.3–0.6 mm long; pubescence of delicate appressed-stellate hairs 4

4(3). Silicles emarginate at the style, about as broad as long; hairs of fruit minute, not or seldom overlapping .... *A. alyssioides*

— Silicles truncate at the style, longer than broad; hairs of fruit ample, usually overlapping .... *A. szowitsianum*

Alyssum alyssioides (L.) L. Alyssum. [*Clypeola alyssioides* L.]. Annual, the stems ascending to erect, simple or branched from near base, (3) 6–25 cm tall; leaves spatulate to oblanceolate, 5–20 (37) mm long, 2–6 mm broad, stellate and green on both surfaces; pedicels spreading-ascending, 1.5–4 (5) mm long; sepals 2.1–2.8 (3) mm long, green, stellate-pubescent, persistent; petals 3–4.2 mm long, yellow, fading white, cuneate, surpassing sepals; silicles 3–4 mm long and almost as broad; valves minutely stellate-pubescent, emarginate at the style; style persistent, mostly 0.3–0.6 mm long; n = 12, 16. Bunchgrass, sagebrush, pinyon-juniper, and mountain brush communities at 1370 to 2440 m in Box Elder, Cache, Millard, Salt Lake, Tooele, Utah, Wasatch, and Weber counties; widespread in North America; adventive from Europe; 27 (0).

Alyssum desertorum Stapf Desert Alyssum. Annual, the stems decumbent to ascending or erect, simple or branched from near the base, (3) 4–20 cm tall; leaves spatulate to oblanceolate or almost linear, 4–22 mm long, 2–3 mm wide, stellate and green on both surfaces; pedicels spreading-ascending, 1.8–2.5 mm long; sepals 1.2–1.9 mm long, often suffused with red, stellate-pubescent; petals 2.3–2.8 mm long, yellow, fading white, cuneate, surpassing the sepals; silicles 2.9–3.8 mm long and about as broad, the valves glabrous or rarely with scattered stellate hairs, emarginate at the style, this persistent, 0.5–0.8 mm long; 2n = 16, 32. Mixed desert shrub, sagebrush, bunchgrass, pinyon-juniper, ponderosa pine, and aspen communities at 1280 to 2200 m in Beaver, Box Elder, Carbon, Davis, Juab, Millard, Morgan, Salt Lake, Sanpete, Tooele, Utah, and Weber counties; Washington, Oregon, Idaho, and Montana; adventive from Asia Minor and adjacent southeastern Europe; 24 (v).

Alyssum minus (L.) Roth [*Clypeola minor* L.]. Annual, the stems ascending to erect, simple or branched from near the base, 0.9–2.5 dm tall; leaves spatulate to oblanceolate, 4–25 mm long, 2–7 mm broad, coarsely stellate and green on both surfaces; pedicels spreading-ascending, 2.8–4.9 mm long; sepals (1.5) 1.7–2.5 mm long, often suffused with red, stellate-pubescent, caducous; petals 2.7–3.6 mm long, yellow, fading white, surpassing the sepals; silicles (3.5) 4–5.2 mm long and about as broad or broader, the valves coarsely stellate with ascending rays, emarginate at the style, this persistent, 0.8–1.2 mm long; 2n = 16, 48. Sagebrush, mountain brush, and pinyon-juniper communities at 1525 to 1895 m in Grand, Salt Lake, and Uintah counties; California and Colorado; adventive from the Old World; 5 (0).

Alyssum saxatile L. Sweet Alyssum. Annual, the stems sprawling- decumbent to ascending or erect, branched from the base and above, 20–35 cm tall; leaves oblanceolate to elliptic or oblong, 90–135 mm long, 2–17 mm wide, minutely stellate and with long, forked, and simple hairs, and green on both surfaces; pedicels spreading-ascending, 3.5–8 mm long; sepals 1.7–3 mm long, green or cream, loosely stellate to glabrate, caducous; petals 4–5 mm long, yellow, fading cream, much surpassing the sepals; silicles 3.5–4.5 mm long or more and almost as broad; valves glabrous, truncate to rounded at style, this persistent, 0.5–1 mm long; 2n = 16. Cultivated ornamental, occasionally escaping; introduced from southern Europe; 4 (0).

Alyssum szowitsianum Fisch. & Mey. Szowits Alyssum. Annual, the stems decumbent-ascending to erect, usually branched from near the base, 5–15 cm tall; leaves oblanceolate, 8–25 mm long, 1–5 mm broad, stellate and green or suffused with red on both surfaces; pedicels ascending to erect, 2.7–4.2 mm long; sepals 0.9–1.2 (2) mm long, often suffused with red, stellate-pubescent, caducous; petals 1.7–2 mm long, yellow, fading white, only slightly surpassing the sepals; silicles 4–5 (6) mm long, longer than broad; valves densely stellate, truncate at style, this persistent, 0.5–0.6 mm long; 2n = 16. Bunchgrass, sagebrush, and mountain brush communities at 1340 to 1922 m in Millard and Salt Lake counties; adventive from Europe; 7 (0).

Arabidopsis (DC.) Schur

Annual, glabrous, or pubescent with simple or branched hairs, from taproots; leaves alternate or basal (and still alternate), simple, entire, or remotely serrate, tapering to the base, not auriculate; flowers in racemes; pedicels spreading-ascending, not subtended by bracts; sepals 4, deciduous; petals 4, white, not emarginate; stamens 6, at least the 2 shorter filaments subtended by a semicircular gland; style very short, tapering, the stigma not enlarged; silique sessile, several times longer than broad, subterete; valves with conspicuous mid-nerve; seeds several to many in each locule.

Arabidopsis thaliana (L.) Schur Mouse-ear Cress. (*Arabis thaliana* L.). Slender annuals; stems erect or nearly so, usually branched throughout, 8–30 (40) cm tall; leaves mainly basal, spatulate to oblong, 3–30 (50) mm long, 2–8 mm broad, remotely toothed to subentire, pubescent with simple or 2– to 4 (5) –rayed hairs, green; cauline leaves much smaller than the basal ones, lance-oblong, sessile or nearly so, 5–20 mm long, 2–5 mm wide; pedicels very slender, 2.5–10 mm long; sepals 1.2–1.7 mm long, green to cream or reddish tinged, sparingly long-hairy; petals 2–2.9 mm long, white, spatulate, surpassing the sepals; siliques 9–14 mm long, 0.5–0.8 mm broad; valves glabrous; style 0.2–0.3 mm long; n = 5, 7. Bunchgrass, sagebrush, and mountain brush communities at 1310 to 2135 m in Davis, Morgan, Salt Lake, Tooele, and Weber counties; widespread in temperate North America; adventive from Europe; 8 (0).

Arabis L.

Biennial or perennial, glabrous or pubescent with simple, branched, or stellate hairs; leaves alternate and basal, simple, entire, dentate, serrate, or sinuate, tapering to the base or the cauline blades sessile and usually auriculate; flowers in racemes; pedicels erect, ascending, spreading-ascending, spreading, descending, or reflexed, not subtended by bracts; sepals 4, deciduous; petals 4, white, pink, lavender, or purple (cream in *A. glabra*); stamens 6, at least the 2 shorter filaments subtended by glands; style prominent to lacking; stigma entire to lobed; fruit a sessile or substipitate silique many times longer than broad, laterally flattened; valves usually with a midnerve; seeds numerous.

Rollins, R. C. 1941. Monographic study of *Arabis* in western America. Rhodora 43: 289–325, 348–411, 425–481.

1.	Cauline leaves usually attenuate to rounded basally, either petiolate or sessile but not auriculate (rarely so in *A. demissa*) 2
—	Cauline leaves auriculate, at least some 6
2(1).	Petals 9–18.5 mm long; petal limb divaricate; ovary and silique densely pubescent; herbage pubescent with minute dendritic hairs .... *A. pulchra*
—	Petals (4.5) 5–8 (9) mm long; petal limb ascending to erect; ovary and silique glabrous or sparingly hairy only; herbage glabrous or with simple or branched hairs 3
3(2).	Pedicels divaricate to descending; siliques descending to pendulous 4
—	Pedicels spreading-ascending to erect; siliques ascending to erect 5
4(3).	Tufts of basal leaves mostly 3–4 cm long; siliques typically hairy; plants of the northern Wasatch Mts. and in Box Elder County .... *A. lasiocarpa*
—	Tufts of basal leaves mostly less than 3 cm long; siliques glabrous; plants not of the northern Wasatch, seldom of Box Elder County .... *A. demissa*
5(3).	Pedicels and siliques erect; siliques 3–6 cm long; plants widespread in northern and central Utah ... . *A. hirsuta*
—	Pedicels and siliques merely ascending; siliques 1–3 cm long; plants known from Cache and Wasatch Counties .... *A. nuttallii*
6(1).	Lower leaves and/or stems pubescent exclusively with malpighian hairs (at least some); flowers white to pink; pedicels and siliques erect; plants widespread .... *A. drummondii*
—	Lower leaves and/or stems glabrous or variously pubescent but not exclusively of malpighian hairs or, if so, otherwise various; flowers white, pink, lavender, or purple (cream in *A. glabra*); pedicels and siliques variously disposed but sometimes erect 7
7(6).	Basal leaves more or less hirsute and usually ciliate with long simple or forked hairs, not both hirsute and with dendritic hairs on the blade surfaces 8
—	Basal leaves more or less densely pubescent with dendritic (rarely malpighian) hairs, rarely also with a few long simple or forked hairs along the leaf bases .. 12
8(7).	Leaves of basal tuft linear, thinly pubescent with dendritic and simple or forked hairs; plants known from Garfield County .... *A. schistacea*
—	Leaves of basal tuft broader than linear, variously pubescent; plants of various distribution 9
9(8).	Flowers cream to white or pinkish; pedicels and siliques ascending to erect 10
—	Flowers pink to lavender or purple, rarely white; pedicels and siliques spreading to descending 11
10(8).	Stigma expanded, 0.8–1.1 mm broad, much wider than the style base; outer sepals not gibbous at the base; petals cream to rarely pinkish; siliques not strongly compressed; plants from Sevier County northward .... *A. glabra*
—	Stigma not obviously expanded, 0.3–0.6 mm broad, not much wider than the style base; outer sepals gibbous at the base; petals white to pink; siliques definitely compressed; plants of broader distribution .... *A. hirsuta*
11(9).	Stems solitary or few, 2.5–6 dm tall; cauline leaves 1–4 cm long; seeds biseriate .... *A. holboellii*

— Stems several to many, 1–3 dm tall; cauline leaves 0.5–1 cm long; seeds uniseriate .... A. demissa

12(7). Fruiting pedicels (but not necessarily the siliques) ascending to erect; siliques erect to spreading or even curved pendulous 13

— Fruiting pedicels spreading to descending, pendulous, or appressed downward along the axis of the raceme 18

13(12). Leaves, stems, and pedicels (and sometimes also the siliques) hoary with minute soft hairs; plants of central western Utah .... A. shockleyi

— Leaves variously pubescent; stems and pedicels glabrous or puberulent, or the stems hairy near the base only; plants of various distributions 14

14(13). Pedicels erect; siliques erect or steeply ascending; plants 0.5–2.5 dm tall, of alpine sites at high elevations .... A. lyallii

— Pedicels spreading-ascending to ascending; siliques spreading to more or less pendulous; plants (2) 2.5–5 dm tall or more, of moderate to low elevations 15

15(14). Plants forming broad mats or carpets, the caudex branches numerous, persistent for several years, with successive branches clothed with marcescent leaf bases; plants of northeastern Uintah County .... A. vivariensis

— Plants seldom if ever mat-forming, the caudex branches typically lacking or short and not with successive branches clothed with marcescent leaf bases; plants of various distribution 16

16(15). Stems usually solitary, 3–10 dm tall (usually pubescent with malpighian hairs); cauline leaves 10 or more, well developed, closely positioned, and often overlapping .... A. holboellii

— Stems usually more than one, 2.5–5 dm tall; cauline leaves usually fewer than 8, poorly developed, commonly widely spaced and only the lowermost overlapping 17

17(16). Stems arising from between the basal rosette and a tuft of ascending leaves; basal leaves mostly 2.5–5 cm long or more .... A. selbyi

— Stems arising from the basal rosette, a secondary tuft of leaves lacking or poorly developed; basal leaves mostly 0.5–2.5 cm long .... A. microphylla

18(12). Lowermost leaves entire, poorly developed, smaller than the main cauline ones; cauline leaves linear to narrowly oblong; siliques more or less finely pubescent .... A. pulchra

— Lowermost leaves in a rosette, more or less well developed, usually oblanceolate to spatulate and broader than the main cauline ones, often toothed; siliques not or rarely pubescent (except in A. puberula) 19

19(18). Stems usually 3 or more, arising between basal rosette and an ascending-erect tuft of leaves .... A. perennans

— Stems usually solitary, rarely 3 or more but then branches not arising from between a basal rosette and a tuft of ascending leaves; lowermost leaves entire or toothed 20

20(19). Leaves, stems, pedicels, and fruit densely pubescent with minute hairs; plants of western and northwestern Utah .... A. puberula

— Leaves, stems, and pedicels variously pubescent but seldom if ever all of them hairy at once 21

21(20). Stems usually numerous, less than 2 dm tall; flowers bright pink to lavender or purple; plants of high elevations in Wasatch and Uinta mts. .... A. lemmonii

— Stems usually solitary or, if more, over 3 dm tall; flowers pale pink to lavender or white; plants of lower to moderate elevations, not of alpine sites 22

22(21). Petals 8–12 mm long or more; pedicels and siliques merely spreading; plants evidently uncommon in northern Utah .... A. sparsiflora

— Petals 4–9 mm long; pedicels and siliques spreading-descending to reflexed .... A. holboellii

Arabis demissa Greene [*Boechera demissa* (Greene) W. A. Weber; *A. demissa* var. *russeola* Rollins, type from 18 mi N of Vernal; *A. rugocarpa* Osterh.; *A. aprica* Osterh. ex A. Nels. in Coult. & Nels.; *A. pendulina* Greene; *B. pendulina* (Greene) W. A. Weber; *A. setulosa* Greene, type from near Marysvale, Piute County; *A. diehlii* Jones, type from Mt. Belknap, Beaver County; *A. nevadensis* Tidestr.]. Perennial, the stems 0.6–3.5 (4) dm tall, (1) several to many from a simple or branched caudex, arising from between the basal rosette and a usually well-developed tuft of ascending leaves, hirsute with simple hairs below or glabrous throughout; basal leaves 0.8–4 cm long, 1–10 mm wide, spatulate to narrowly oblanceolate, entire, ciliate and hairy on one or both surfaces with long simple or forked hairs or totally glabrous; cauline leaves much shorter than the internodes, few, 0.3–1.2 cm long, 0.1–0.6 cm broad, oblong to lanceolate, entire, sessile and auriculate or typically not auriculate (but sometimes slightly clasping), hairy or the upper glabrous; pedicels 3–6 (10) mm long, arched downward, glabrous; sepals 2.5–3.6 (4) mm long, usually purplish; petals 4.5–6.8 mm long, pink to lavender or white, spatulate, erect or ascending; siliques (10) 13–40 mm long, 1.5–2.3 (3) mm wide, pendulous, the valves glabrous, nerved below the middle; style obsolete or very short, the stigma not enlarged; seeds biseriate or uniseriate. Pinyon-juniper, sagebrush, grassland, mountain brush, ponderosa pine, Douglas fir, white fir, and aspen communities at 1670 to 2930 m elevation in Beaver, Box Elder, Daggett, Duchesne, Emery, Garfield, Kane, Millard, Piute, Sevier, Tooele, Uintah, Wasatch, Washington, and Wayne counties; Nevada east to Colorado and Wyoming; 55 (viii). There does not appear to be any basis for segregation of *A. pendulina* or any of the proposed infraspecific taxa in this species. The taxon is more broadly distributed in Utah than first thought, and all diagnostic features used to segregate *A. demissa* from *A. pendulina*, and the infraspecific taxa proposed in each, fail singly and together. And, there is no apparent geographic correlation.

Arabis drummondii Gray [*Turritis stricta* Graham, not *A. stricta* Hudson; *Streptanthus angustifolius* Nutt. ex T. & G., not *A. angustifolia* Lam.; *Boechera drummondii* (Gray) Love & Love; *A. connexa* Greene; *A. oxyphylla* Greene; *A. albertina* Greene; *A. philonipha* A. Nels. ex Rydb.; *T. drummondii* (Gray) Lunell; *A. drummondii* var. *connexa* (Greene) Fern.; *A. drummondii* var. *oxyphylla* (Greene) Hopkins]. Biennial or short-lived perennial, the stems (0.8) 1.2–9 (9.7) dm tall, solitary or 2 to several from a simple or branched caudex, not arising from between basal rosette and tuft of ascending leaves, usually glabrous throughout or strigose with malpighian hairs at base only; basal leaves (1.1) 1.5–7 (8) cm long, 0.2–1.2 cm wide, oblanceolate, entire, subglabrous or pubescent with malpighian hairs; cauline leaves usually longer than the internodes, numerous, (1) 1.5–6 cm long, 0.2–1.5 cm wide, oblong to lanceolate, usually entire, sessile and auriculate, usually glabrous; pedicels 7–15 mm long in fruit, erect, glabrous; sepals (3) 3.3–5.7 mm long, glabrous; petals 6.5–10.5 mm long, white to pink, spatulate, ascending to erect; siliques (27) 35–95 (110) mm long, (1.2) 1.5–2 (3) mm wide, erect; valves glabrous, nerved to middle or above; style short or obsolete; seeds biseriate; n = 7. Sagebrush, aspen, lodgepole pine, and spruce-fir communities at 2315 to 3815 m in probably all Utah counties; Alaska and Yukon east to the Atlantic, south to California, Arizona, and New Mexico; 136 (xxiii).

Arabis glabra (L.) Bernh. Tower Mustard. Biennial or rarely perennial; stems 3–8 (10) dm tall, solitary or 2 or 3, from a taproot, not arising from between the basal rosette and a tuft of ascending leaves, usually hirsute with simple or forked hairs, rarely with appressed dendritic hairs at base only; basal leaves 3–10 (15) cm long, 0.8–3.5 (5) cm wide, oblong-oblanceolate to spatulate, sinuate-dentate to entire, more or less ciliate with forked or dendritic hairs, the surfaces (especially the veins) hirsute with simple or forked hairs; cauline leaves usually longer than internodes (at least below), numerous, 1.5–9 (12) cm long, 0.4–2.3 (3.5) cm wide, lanceolate, denticulate to entire, sessile and auriculate, glabrous at least above; pedicels 4–12 mm long in fruit, erect, often appressed, glabrous; sepals 3.3–4.5 mm long, glabrous, often tinged reddish purple; petals 4.7–6 (7) mm long, cream or rarely pinkish, narrowly spatulate, ascending to erect; siliques 40–90 mm long, 1–1.5 mm wide, strictly erect; valves glabrous, nerved to the middle or above; style ca 0.5 mm long; stigma 0.8–1.1 mm broad; seeds more or less biseriate; n = 6. Sagebrush, pinyon-juniper, mountain brush, white fir, aspen, and spruce-fir communities at 1525 to 3205 m in Box Elder, Cache, Daggett, Duchesne, Iron, Juab, Millard, Morgan, Salt Lake, San Juan, Sevier, Summit, Tooele, Uintah, Utah, and Weber counties; widespread in North America; Europe; 28 (vi). Two varieties have been recognized in Utah; **var. *furcatipilis*** Hopkins (type from Logan Canyon), with hairs at stem base appressed and several-branched, and **var. *glabra*** [*Turritis glabra* L.; *A. perfoliata* Lam.; *T. macrocarpa* Nutt. ex T. & G.; *A. macrocarpa* (Nutt.) Torr.], with stem base hirsute with simple or merely forked hairs.

Arabis hirsuta (L.) Scop. Hairy Rockcress. [*Turritis hirsuta* L.; *T. ovata* Pursh; *A. h.* var. *ovata* (Pursh) T. & G.?; *A. pycnocarpa* Hopkins; *A. h.* var. *pycnocarpa* (Hopkins) Rollins; *A. h.* var. *laevis* Tuzon, type from American Fork Canyon]. Biennial or perennial; stems (0.9) 1.5–6 (7) dm tall, solitary or more commonly 2–6 or more, from a simple or branched caudex, not arising from between the basal rosette and a tuft of ascending leaves, hirsute with simple or forked hairs, at the base at least; basal leaves (0.8) 1.5–6 (8) cm long, 0.3–1.8 (3) cm wide, elliptic to oblong or oblanceolate, entire or more or less dentate, ciliate with simple or forked hairs, the surfaces glabrous or more or less hirsute with simple or forked hairs; cauline leaves usually longer than the internodes at least below, numerous, (0.6) 1–5 cm long, 0.2–2 cm wide, oblong to lanceolate, toothed to entire, sessile and auriculate, rarely merely sessile, hirsute or glabrous; pedicels 3–18 mm long in fruit, erect, appressed, glabrous; sepals 2.2–4 (4.5) mm long, glabrous or sparingly hairy, seldom tinged reddish purple; petals 3.2–7 (9) mm long, white or pink, oblong to spatulate, ascending to erect; siliques 30–55 (60) mm long, 1–1.5 (2) mm wide; valves glabrous, erect, nerved to above the middle; style 0.3–1 mm long; stigma 0.3–0.7 mm broad; seeds uniseriate; n = 16. Sagebrush, pinyon-juniper, mountain brush, ponderosa pine, aspen, and Douglas fir communities at 1645 to 3205 m in Beaver, Box Elder, Cache, Carbon, Daggett, Duchesne, Emery, Garfield, Grand, Juab, Kane, Millard, Piute, Rich, Salt Lake, Sevier, Summit, Uintah, Utah, and Wasatch counties; Alaska and Yukon east to the Atlantic and south to California, Arizona, and New Mexico; circumboreal; 34 (ix). Two varieties have been proposed but they are completely transitional in Utah and it seems best to regard all of our material as belonging to **var. *glabrata*** T. & G. [*Turritis spathulata* Nutt. ex T. & G.; *A. rupestris* Nutt. ex T. & G.; *A. pycnocarpa* var. *glabra* (T. & G.) Hopkins].

Arabis holboellii Hornem. Holboell Rockcress. [*Boechera holboellii* (Hornem.) Love & Love; *Turritis brachycarpa* T. & G.; *A. drummondii* var. *brachycarpa* (T. & G.) Gray; *A. confinis* var. *brachycarpa* (T. & G.) Wats. & Coult.; *A. brachycarpa* (T. & G.) Britt., not Rupr.; *A. confinis* Wats.; *A. divaricarpa* A. Nels.; *A. stokesiae* Rydb., type from Parleys Canyon]. Biennial or perennial; stems (1) 2–11.5 dm tall, solitary or less commonly 2–6, from a simple or branching caudex, not arising from between the basal rosette and a tuft of ascending leaves, pubescent with appressed or spreading hairs, at least at the base; basal leaves (1) 1.5–5.5 (7) cm long, 0.2–8 mm wide, elliptic to oblanceolate, entire to dentate, with dendritic hairs on margins and usually on the surfaces, rarely with some simple or forked hairs near petiole base, or exclusively hairy with simple or forked trichomes; cauline leaves usually longer than internodes, at least below, numerous, (0.5) 1.2–6.5 cm long, 0.2–0.9 cm wide, oblong to lanceolate, entire or some toothed, sessile and auriculate, pubescent to glabrous; pedicels 4–23 mm long in fruit, reflexed to loosely descending or spreading-ascending, glabrous or pubescent; sepals 2.3–5 mm long, pubescent or glabrous, often tinged reddish; petals 4–9 (10) mm long, pink to lavender or white, spatulate, erect or ascending; siliques (20) 25–70 (85) mm long, 1–2.5 mm wide, reflexed to loosely pendulous or spreading, the valves glabrous, nerved to below or above middle; style obsolete or very short; stigma not much enlarged; seeds uni- or biseriate; 2n = 22. Included here, and possibly keying mostly to var. *pinetorum*, are the specimens previously regarded in Utah as *A. confinis* Wats. [*A. divaricarpa* A. Nels.]. The specimens keyed arbitrarily to that name are scattered throughout the state, but seem not to represent a taxon; rather, they are an assemblage arrived at arbitrarily. Three more or less intergrading varieties are recognized.

1. Basal leaves more or less hirsute and usually ciliate with long, simple or forked hairs, not both hirsute and with dendritic hairs on the blade surfaces .... *A. holboellii* var. *fendleri*
— Basal leaves more or less densely pubescent with dendritic hairs, rarely also with a few long, simple or forked hairs along the leaf bases 2
2(1). Pedicels gently curved downward; pods pendulous, often somewhat curved inward . *A. holboellii* var. *pinetorum*
— Pedicels abruptly curved at the base, deflexed; siliques strictly reflexed to descending and often straight .... *A. holboellii* var. *secunda*

Var. *fendleri* Wats. in Gray [*Boechera fendleri* (Wats.) W. A. Weber; *A. spatifolia* Rydb.; *A. fendleri* var. *spatifolia* (Rydb.) Rollins]. Pinyon-juniper, ponderosa pine, mountain brush, aspen, Douglas fir, and spruce-fir communities at (1220) 2010 to 3145 m in Beaver, Carbon, Duchesne, Emery, Garfield, Iron, Kane, Millard, Piute, San Juan, Sevier, Wasatch, and Washington counties; Wyoming and Colorado, west to Nevada, south to Texas and Mexico; 26 (viii). This taxon has been traditionally treated at specific rank, but morphological intermediates occur between it and the following varieties. And, general aspect of this variety agrees with those of *A. holboellii* generally.

Var. *pinetorum* (Tidestr.) Rollins [*A. pinetorum* Tidestr.; *A. lignifera* A. Nels.]. Sagebrush, pinyon-juniper, mountain brush, cottonwood, ponderosa pine, aspen, lodgepole pine, and spruce-fir communities at 1280 to 2810 m in probably all Utah counties; Saskatchewan to British Columbia south to California, Colorado, and Nebraska; 171 (xvii). A dwarf phase, mainly 0.8–2.5 dm tall, occurs in the northern tier of counties and has been designated **var. *pendulocarpa* (A. Nels.) Rollins** [*A. pendulocarpa* A. Nels.]. I can find no morphological criteria that distinguish *A. lignifera* from *A. holboellii* var. *pinetorum*, at least in Utah specimens.

Var. *secunda* (Howell) Jepson [*A. retrofracta* Graham; *Boechera retrofracta* (Graham) Love & Love; *Turritis retrofracta* (Graham) Hook.; *Streptanthus virgatus* Nutt. ex T. & G.; *A. secunda* Howell; *A. arcuata* var. *secunda* (Howell) Robins.; *A. holboellii* var. *retrofracta* (Graham) Rydb.; *A. rhodantha* Greene; *A. exilis* A. Nels.; *A. tenuis* Greene; *A. lignipes* A. Nels.; *A. consanguinea* Greene; *A. kockii* Blank., not Jordan; *A. sparsiflora* var. *secunda* (Howell) Piper; *A. caduca* A. Nels.; *A. macdougalii* Rydb.]. n = 7, 8. Sagebrush, pinyon-juniper, mountain brush, ponderosa pine, aspen, and lodgepole pine communities at 1495 to 2870 m in all Utah counties, except Grand, Iron, Kane, and Wayne; Quebec to Alaska and British Columbia, south to California and Colorado; 46 (ix). This entity is transitional, at least in part, with var. *pinetorum*.

***Arabis lasiocarpa* Rollins** Wasatch Rockcress. Perennial, the stems 1–2.5 dm tall, few to several from a branching, typically elongate caudex, arising at the base of tufts of ascending to erect leaves, pubescent with minute dendritic or stellate hairs; basal leaves mainly 1–4 cm long, 1.5–5 mm wide, spatulate to oblong, usually entire, pubescent overall with minute dendritic or stellate hairs; cauline leaves shorter than the internodes, few, 0.8–1.6 cm long, 1.5–5 mm wide, elliptic to oblong or lanceolate, sessile, not auriculate, pubescent like the lower leaves; pedicels 4–9 mm long, curved descending, pubescent; sepals 3.5–4.5 mm long, erect, pubescent; petals 7–8 mm long, erect, spatulate, purple; siliques mainly 3–5 cm long, 1.5–2 mm wide, nerved near the base, pubescent with forked or stellate hairs, or rarely glabrous. Sagebrush, mountain brush, aspen, and spruce-fir communities at ca 1830 to 2900 m in Box Elder, Cache, Rich (type from 6 mi W of Garden City), and Salt Lake counties; endemic; 5 (0).

***Arabis lemmonii* Wats.** Lemmon Rockcress. [*A. canescens* var. *latifolia* Wats.; *A. latifolia* (Wats.) Piper; *A. bracteolata* Greene; *A. egglestonii* Rydb.]. Perennial, the stems 0.5–2 dm tall, several to many from a branching typically subterranean caudex, often with tufts of ascending leaves, pubescent to glabrous; basal leaves 0.8–2 (2.5) cm long, 0.2–0.5 cm wide, spatulate to oblanceolate, usually entire, marginally pubescent with dendritic hairs, rarely with some hairs simple or forked on the petiole bases, surfaces densely pubescent with dendritic hairs; cauline leaves shorter or longer than the internodes, few, 4–15 mm long, 1–3 mm wide, elliptic-oblong to lanceolate, entire, sessile and auriculate (at least some), pubescent to glabrous; pedicels 2–6 mm long in fruit, ascending to spreading, glabrous or pubescent; sepals 2–3.5 mm long, glabrous or pubescent, often purplish, petals 4–5.5 (6) mm long, pink to lavender, spatulate, erect to ascending; siliques mostly 20–50 mm long, (1.5) 2–2.5 (3) mm wide, ascending, spreading or somewhat pendulous; valves glabrous, nerved to middle; seeds uniseriate. Alpine tundra, often in talus or rock stripes, at 2440 to 3965 m elevation, in Box Elder, Cache, Duchesne, Garfield, Juab, Summit, and Salt Lake counties; Alaska and Yukon south to California, Colorado, and Montana; 9 (0). Our plant is **var. *lemmonii*.**

***Arabis lyallii* Wats.** Lyall Rockcress. [*A. drummondii* var. *alpina* Wats.; *A. oreophila* Rydb., type from Salt Lake County; *A. armerifolia* Greene; *A. densa* Greene; *A. multiceps* Greene; *A. drummondii* var. *lyallii* (Wats.) Jepson; *A. drummondii* var. *oreophila* (Rydb.) Hopkins]. Perennial; stems 3–25 cm tall, few to many from a branching, often subterranean, caudex, often with tufts of ascending leaves, glabrous; basal leaves 5–25 mm long, 2–6 mm wide, oblanceolate to spatulate, entire, glabrous or the margin and surfaces sparingly pubescent with dendritic hairs; cauline leaves shorter to longer than the internodes, few, 6–20 mm long, 2–6 mm wide, ovate to lanceolate or oblong, entire, sessile and at least some auriculate, usually glabrous; pedicels 4–10 (13) mm long, erect-ascending, glabrous; sepals 3.3–4.5 mm long, glabrous, not or only slightly gibbous, often purplish; petals 7–9 (10) mm long, bright pink to lavender, spatulate, erect or ascending; silique 20–60 mm long, 2–3 mm wide, erect-ascending; valves glabrous, nerved to the middle; style short, stigma not or slightly enlarged; seeds uniseriate or more or less biseriate. Alpine tundra, krummholz, glacial moraines and among subalpine conifers, at 3050 to 3815 m elevation in Daggett, Duchesne, Salt Lake, Uintah, Utah, Wasatch, and Weber counties; Yukon and British Columbia south to California, Nevada, and western Wyoming; 13 (ii). Our material is var. *lyallii*.

***Arabis microphylla* Nutt. ex T. & G.** Small-leaf Rockcress. Perennial; stems 0.5–3 (5) dm tall, several to many from a branching, often subterranean or elongate caudex, with tufts of ascending leaves lacking or poorly developed, or produced on innovations, glabrous throughout or hairy below with simple or forked hairs; basal leaves 7–25 mm long, 2–5 mm wide, oblanceolate to spatulate, entire or rarely toothed, the margins with dendritic or merely sessile and forked hairs, densely pubescent with minute dendritic hairs; cauline leaves often shorter than the internodes, few, 6–20 mm long, 1–6 mm wide, oblong to lance-linear, entire or some toothed, sessile and auriculate, glabrous or the lowermost pubescent; pedicels 4–13 (15) mm long, ascending to spreading-ascending, glabrous or pubescent; sepals 2.5–4 mm long, glabrous or pubescent, often purplish; petals 5–8 mm long, pink to lavender, spatulate, ascending to erect; siliques (20) 25–60 mm long, 1.2–2 mm wide, erect to spreading, nerved at the base only; style developed or obsolete; stigma not expanded; seeds uniseriate; n = 7, 14. Sagebrush, mountain brush, pinyon-juniper, aspen, white-fir and Douglas fir communities at 1370 to 2990 m Box Elder, Cache, Carbon, Daggett, Davis, Duchesne, Garfield, Juab, Millard, Piute, Rich, Salt Lake, Sanpete, Sevier, Uintah, Utah, and Weber Counties; British Columbia to Montana, south to Oregon, Nevada, and western Wyoming; 28 (iii). Our material has been segregated into two varieties; **var. macounii (Wats.) Rollins** [*A. macounii* Wats.; *A. densicaulis* A. Nels.], with stems 2.5–5 dm tall and with both pedicels and siliques spreading, and **var. microphylla** [*A. tenuicola* Greene], with stems less than 2 dm tall and with pedicels divaricate but siliques erect. They are totally confluent and only arbitrarily separable.

***Arabis nuttallii* Robins. in Gray** Nuttall Rockcress. [*A. spathulata* Nutt. ex T. & G., not DC.; *Erysimum nuttallii* (Robins.) Kuntze; *A. bridgeri* Jones; *A. macella* Piper]. Perennial; stems 0.9–3 dm tall, several to many from a branching caudex, tufts of ascending leaves lacking, glabrous throughout or hirsute below with simple or forked hairs; basal leaves 1–4 cm long, 4–12 mm wide, oblanceolate, usually entire, ciliate and hairy on lower surface (at least) with long simple and often some forked hairs; cauline leaves shorter than the internodes, few, 5–20 mm long, 1–7 mm wide, elliptic to oblong, lanceolate or oblanceolate, entire, sessile or subsessile, not auriculate, hairy or the upper glabrous; pedicels 5–20 mm long, spreading-ascending, glabrous; sepals 3–4 mm long, glabrous or sparingly hirsute, usually green or cream; petals (5) 6–8 mm long, white or lavender, spatulate, more or less spreading; siliques (8) 12–20 mm long, 1–1.5 mm wide, erect to spreading; valves glabrous, nerveless or faintly nerved; style ca 1 mm long; stigma not especially expanded; seeds uniseriate. Meadows, typically in sagebrush and aspen communities at 1370 to 2960 m in Cache and Wasatch counties; Alberta south to Wyoming; 5 (0).

***Arabis perennans* Wats.** Common Rockcress. [*Boechera perennans* (Wats.) W. A. Weber; *A. arcuata* var. *perennans* (Wats.) Jones; *A. gracilenta* Greene; *A. eremophila* Greene; *A. recondita* Greene; *A. angulata* Greene ex Woot. & Standl.; *A. falactoria* Rollins, type from south of Lynn, Box Elder County]. Perennial; stems 0.9–5.5 (6) dm tall, (1) several to many from simple or branching herbaceous to woody caudex, typically arising from between basal rosette and a tuft of ascending leaves; bassal leaves pubescent with dendritic hairs on margin and on surfaces, rarely with simple or forked hairs along petiole base; cauline leaves longer than the internodes at least below, several, 0.7–4 cm long, 2–8 mm wide, oblong to lanceolate, entire to toothed, hairy or the upper glabrous; pedicels 4–24 mm long, spreading to arched downward, glabrous or pubescent; sepals 3–4 (4.5) mm long, often purplish, usually dendritic hairy; petals 5–7 (9) mm long, pink to lavender, spatulate, erect or spreading; siliques (20) 27–55 (60) mm long, 1.2–2 mm wide, spreading to pendulous; valves glabrous, nerveless or nerved at the base; style obsolete or very short; stigma not enlarged; seeds uniseriate. Warm and salt desert shrub, pinyon-juniper, sagebrush, mountain brush, ponderosa pine, and aspen communities at 915 to 3050 m in all Utah counties, except Weber, Cache, Rich, and Summit; Colorado and New Mexico to Nevada, California, and Baja California; 180 (xxxviii). Specimens of the wide-ranging *A. perennans* are transitional to other species contacted, especially *A. microphylla* and *A. holboellii*. The phase of the species known as *A. falactoria* is evidently intermediate between *A. perennans* and *A. demissa*.

***Arabis puberula* Nutt. ex T. & G.** Puberulent Rockcress. [*A. beckwithii* Wats.; *Erysimum puberulum* (Nutt.) Kuntze; *A. subpinnatifida* var. *beckwithii* (Wats.) Jepson; *A. arida* Greene; *A. lignipes* var. *impar* A. Nels.; *A. sabulosa* Jones; *A. sabulosa* var. *frigida* Jones; *A. sabulosa* var. *colorata* Jones; *A. subpinnatifida* var. *impar* (A. Nels.) Rollins]. Perennial or infrequently biennial; stems (7) 10–30 cm tall or more, solitary or few from a simple caudex, not arising from between a basal rosette and a tuft of ascending leaves, pubescent throughout with dendritic hairs, rarely glabrous above; basal leaves 1–2.5 (3) cm long, 2–6 mm wide, oblanceolate, entire or toothed, pubescent with minute dendritic hairs; cauline leaves usually longer than internodes, several to many, 1–3 cm long, 2–8 mm wide, toothed to entire, sessile and at least some auriculate, hairy like the basal leaves; pedicels 2–7 mm long, arched downward, pubescent; sepals 3.5–6 mm long, dendritic hairy; petals 7–11 mm long, pink to white, spatulate, erect or ascending; siliques (25) 30–50 (60) mm long, 2–3 mm wide, pendulous to reflexed; valves copiously hairy to glabrate, nerved to below the middle; style obsolete; stigma not enlarged; seeds uniseriate. Sagebrush and mountain brush communities at ca 1555 to 1830 m in Box Elder and Juab (?), counties; Washington and Idaho south to California and Nevada; 1 (0). A closely related species, *A. cobrensis* Jones, is to be sought in extreme northern Utah. The species occurs just north of the state line in Uinta County, Wyoming, and in northeastern Elko County, Nevada, but as yet it is not known from Utah. *A. cobrensis* is a well-defined perennial with a much-branched caudex, the basal leaves are only 1–3 mm wide, the few cauline leaves are remote and only 1–3 mm wide, and the petals are 4–6 mm long. It is to be sought in Daggett and Box Elder counties.

***Arabis pulchra* Jones** Pretty Rockcress. Perennial; stems 1.5–6 dm tall, solitary or several from a branching herbaceous to woody caudex, not arising from between the basal rosette and a tuft of ascending leaves, pubescent with dendritic hairs throughout or glabrous above; basal leaves in poorly developed rosettes, 1–6 cm long, 2–6 mm wide, narrowly oblanceolate to spatulate, entire, densely hairy with minute hairs; cauline leaves shorter or longer than the internodes, many, 1.2–6 cm long, 1–5 mm wide, entire, sessile and mostly not auriculate, hairy like the basal leaves; pedicels 5–18 mm long, recurved to pendulous in fruit, pubescent to glabrate; sepals 5–8.2 mm long, more or less gibbous, dendritic-pubescent, often purplish; petals 9–18.5 mm long, pale pink to white or pink to lavender or purple, spatulate to obovate-spatulate, spreading or ascending to erect; siliques 35–55 (65) mm long, 1.8–2.5 (3) mm wide, pendulous to reflexed; valves copiously hairy to glabrate, nerved to middle or above; style obsolete or very short, not expanded; seeds biseriate; n = 21. Two distinctive varieties are known in Utah.

1. Flowers lavender to purple, 9–11 mm long; petals ascending to erect .... *A. pulchra* var. *munciensis*
— Flowers pale pink or white, less commonly lavender, 10–18.5 mm long; petals spreading . *A. pulchra* var. *pallens*

Var. munciensis Jones Joshua tree, creosote bush, blackbrush, pinyon-juniper, and sagebrush communities at 790 to 1770 m in Beaver, Juab, Iron, Kane, Millard, Tooele, and Washington counties; Nevada and California; 15 (iv).

Var. *pallens* Jones [*A. formosa* Greene]. Desert shrublands and pinyon-juniper community at 1130 to 2075 m in Carbon, Daggett, Duchesne, Emery, Garfield, Grand (type from Westwater), Kane, San Juan, Uintah, and Wayne counties; Colorado, New Mexico, and Arizona; 83 (xiii). A phase of var. *pallens* from Duchesne County with spreading siliques has been designated as var. *duchesnensis* Rollins (type from 3.8 mi E of Duchesne). In my view, it is taxonomically neglegible.

Arabis schistacea **Rollins** Schist Rockcress. Perennial, the stems 1–2 dm tall, usually several from a branching caudex, arising between a basal rosette and an ascending tuft of leaves, glabrous; basal leaves 1–3.5 cm long, 1–2 mm wide, linear to narrowly spatulate, entire, sparingly hairy with dendritic or simple hairs on the margin and surfaces, the petiole bases with simple or forked hairs; cauline leaves shorter than the internodes, few to several, 5–15 mm long, 1–2.5 mm wide, narrowly oblong to linear, entire, glabrous, auriculate; pedicels 3–7 mm long in fruit, curved descending; sepals 2.5–3.5 mm long, often purplish, glabrous; petals 4–6 mm long, whitish, spatulate, erect or ascending; siliques 15–35 mm long, 1–1.8 mm wide, pendulous, the valves glabrous, nerved at the base or nerveless; style very short, the stigma not expanded. Sagebrush community at ca 2440 m in Garfield County; Nevada; a Great Basin endemic; 1 (0).

Arabis selbyi **Rydb.** Selby Rockcress. [*Boechera selbyi* (Rydb.) W. A. Weber]. Perennial, the stems 1.5–4 (5) dm tall, usually several from a simple or branching caudex, arising from between basal rosette and an ascending tuft of leaves, pubescent below with dendritic or forked hairs; basal leaves 1.5–4 (6) cm long, 0.2–1 cm wide, oblanceolate to spatulate, usually entire, densely to sparsely pubescent with dendritic hairs on the margin and surfaces, sometimes with simple or branched hairs along the petiole base; cauline leaves usually shorter than internodes, few to several, 0.3–2.5 (3) cm long, 1–4 mm wide, narrowly oblong to lanceolate, entire, hairy or glabrous; pedicels 4–15 mm long in fruit, spreading-ascending, glabrous or pubescent; sepals 2.6–3.5 (4) mm long, often purplish, usually dendritic hairy; petals 5–7 (8) mm long, pink to lavender, spatulate, erect or ascending; siliques 30–55 (60) mm long, 1.2–1.8 mm wide, spreading-ascending to more or less pendulous; valves glabrous, nerved at base or nerveless; style obsolete or very short; stigma not enlarged; seeds uniseriate. Sagebrush, pinyon-juniper, mountain brush, ponderosa pine, and Douglas fir communities mostly at 1500 to 2625 m elevation in Carbon, Daggett, Duchesne, Emery, Garfield, Grand, Kane, San Juan, Sevier, Uintah, Utah, and Wayne counties; western Colorado and northwestern New Mexico; 40 (vii). Only the slightly more ascending pedicels and usually entire or subentire leaves serve to distinguish this entity from *A. perennans*. Possibly, it would be treated better at infraspecific level within that species.

Arabis shockleyi **Munz** Shockley Rockcress. [*A. inyoensis* Rollins?]. Perennial or infrequently biennial; stems 1.2–3.5 (4) dm tall, solitary or 2–4 from a simple or branching caudex, not arising from between a basal rosette and a tuft of ascending leaves, densely pubescent throughout with minute dendritic hairs, often somewhat less densely so above; basal leaves 1.6–3 (3.5) cm long, 5–9 mm wide, oblanceolate to spatulate, entire, densely pubescent throughout with dendritic hairs only; cauline leaves much longer than internodes, numerous, 1.2–3 cm long, 5–12 mm wide, ovate-lanceolate to lance-attenuate, entire or nearly so, sessile and auriculate, hairy as on the basal leaves; pedicels 12–17 mm long in fruit, ascending, pubescent; sepals 5.5–7 mm long, often reddish, dendritic hairy; petals 7.5–10 mm long, pink to lavender, spatulate, erect or ascending; siliques 42–65 mm long, 1.2–1.8 mm wide ascending to spreading; valves glabrous, nerved to about middle; style up to 0.8 mm long; stigma not much expanded; seeds biseriate. Desert shrublands and pinyon-juniper communities mostly 1430 to 1900 m in Juab, Millard, and Tooele counties; Nevada and California; 16 (iv).

Arabis sparsiflora **Nutt. ex T. & G.** Sickle Rockcress. Perennial; stems 2.3–10 dm tall, solitary or less commonly 2 or more from a simple or branching caudex, not arising from between the basal rosette and a tuft of ascending leaves, pubescent with appressed or spreading hairs at least below; basal leaves 3–7 (9) cm long, 3–6 (9) mm wide, oblanceolate, entire or dentate, pubescent with coarse dendritic hairs usually on both surfaces; cauline leaves longer than the internodes, numerous, 1.5–6 cm long or more, 3–7 mm wide, oblong to lanceolate, entire or the lower toothed, sessile and auriculate, hairy like the basal leaves or wholly glabrous; pedicels 5–15 mm long, ascending-spreading, glabrous or pubescent; sepals 4.7–6 mm long, dendritic hairy; petals 8–12 mm long, pink to lavender or purple, spatulate, ascending to erect; siliques 6–10 (12) cm long, 1.2–2 mm wide, ascending to curved-descending; style obsolete or nearly so; stigma not expanded; seeds uniseriate; n = 21, 22. Sagebrush, mountain brush, pinyon-juniper, ponderosa pine, and aspen communities at 1675 to 2685 m in Box Elder, Cache, Davis, Duchesne, Emery, Rich, Salt Lake, Sanpete, Uintah, Utah, and Weber counties; British Columbia and Alberta south to California and Wyoming; 13 (iii). Plants with entire leaves and spreading-ascending pedicels have been designated as **var. sparsiflora** [*A. peramoena* Greene; *A. sparsiflora* var. *peramoena* (Greene) Rollins], and those with dentate leaves and spreading pedicels are known as **var. subvillosa (Wats.) Rollins** [*A. arcuata* var. *subvillosa* Wats.; *A. perelegans* A. Nels. in Coult. & Nels.].

Arabis vivariensis **Welsh** Park Rockcress. Perennial, forming mats or carpets to 1 m wide or more, the caudex branches bearing marcescent leaf bases, the branches of several seasons evident back from the branch ends, horizontally spreading to decumbent, finally erect and bearing flowering stems of the season or terminating in leafy rosettes, the flowering stems mainly 8–32 cm tall, puberulent with minute dendritic trichomes or glabrous above; basal leaves and those of the innovations 0.7–3 cm long, 1.2–4 mm wide, oblanceolate to elliptic, the blade tapering to a long, slender petiole, green to gray, pubescent overall with minute dendritic hairs, acute; cauline leaves 3–13 mm long, 1–2.5 mm wide, oblong to lanceolate or lance-subulate, puberulent to glabrous, much reduced upward; pedicels ascending to erect, 5–15 mm long in fruit, glabrous or minutely puberulent; sepals 2.5–4.5 mm long, the outer pair gibbous at the base, the inner ones less so, commonly purplish, glabrous to puberulent; petals 7–9 mm long, tapering to a basal claw, purplish; siliques 3–7 cm long, 1–1.5 mm wide, glabrous, nerved at the base, erect-ascending, typically curved or contorted, the style to 0.5 mm long; seeds uniseriate, ca 1.2 mm long, narrowly winged apically. Limestone and sandstone outcrops in mixed desert shrub and pinyon-juniper communities at 1525 to 1830 m in Uintah (type from Jones Hole) County; endemic (?); 5 (iv). This plant is closely similar to *A. fernaldiana* Rollins from Nevada, differing from the type variety of that species in smaller flowers and from the species generally in the shorter style.

Armoracia Gaertner

Plants glabrous perennials from tuberous-thickened taproots; leaves alternate, simple, crenately toothed or lobed, petiolate to subsessile, not auriculate; flowers racemose, the pedicels ascending, not subtended by bracts; sepals 4, deciduous; petals 4, white, not emarginate; stamens 6, at least the 2 short stamens subtended by glands; style short, the capitate stigma hemispheric; fruit a silicle, bilocular, about as long as broad, obovoid-ellipsoid, the valves with an inconspicuous midnerve; seeds apparently never developing.

Armoracia rusticana Gaertner Horse-radish. [*Cochlearia armoracia* L.; *Nasturtium armoracia* (L.) Fries; *Rorippa armoracia* (L.) A. S. Hitchc.; *Radicula armoracia* (L.) Gray; *A. armoracia* (L.) Cockerell]. Perennial; stems 6–10 dm tall or more, few to many arising from summit of root crown; basal leaves oblong to oblong-lanceolate, the blades 15–50 cm long, 10–15 cm wide, base cordate to rounded; petioles 0.6–4 dm long or more; cauline leaves reduced and only short-petiolate to subsessile upwards, lanceolate to elliptic and crenately toothed to lobed, glabrous; pedicels mostly 8–11 mm long, ascending in fruit; sepals 2–2.5 mm long, caducous, greenish, glabrous; petals 4.2–4.5 (5) mm long, white, obovate-spatulate, surpassing the sepals; silicles 3–6 mm long and about as wide; valves glabrous; style 0.2–0.3 mm long, hemispheric stigma to 0.5 mm broad or more; n = 32. Cultivated and occasionally escaping and established at 1370 to 1985 m in Cache, Salt Lake, Summit, Utah, and Weber counties, and most likely elsewhere in agricultural regions of the state; widespread in North America; introduced from Eurasia; 2 (0).

Barbarea R. Br.

Plants glabrous to sparsely hirsute biennials or rarely annuals from taproots; leaves alternate, lyrate-pinnatifid to pinnately compound, the cauline ones auriculate-clasping and often falsely petiolate above a clasping base; flowers ebracteate racemes, the pedicels erect; sepals 4, deciduous; petals 4, yellow, truncate to rounded apically; stamens 6, the filaments lacking glandular processes; style stout, abruptly contracted to capitate stigma; fruit a silique, many times longer than broad, linear, only slightly compressed, more or less contracted between the seeds; valves 1-nerved; seeds numerous, uniseriate.

1. Style 2–3 mm long, long-beaked .... *B. vulgaris*
— Styles 0.5–2 mm long, short-beaked .... *B. orthoceras*

Barbarea orthoceras Ledeb. Wintercress. [*B. americana* Rydb.; *Campe orthoceras* (Ledeb.) Heller; *B. orthoceras* var. *dolichocarpa* Fern.]. Stems erect, 1.5–10 dm tall, glabrous; basal leaves lyrate-pinnatifid to pinnately compound, rarely reduced to the terminal lobe, mostly (1.5) 4–15 (20) cm long, 1–2.5 (4) cm wide, glabrous or petiole and lower lobes sparsely hirsute; cauline leaves reduced upwards, auriculate-clasping; pedicels 2–4 mm long, glabrous, ascending; sepals 2.5–3.5 mm long, yellowish, glabrous; petals 4–5.5 mm long, yellow, spatulate to oblanceolate, ascending-spreading; siliques (15) 20–50 mm long, 1.5–2.5 mm wide, erect or ascending; valves glabrous, prominently nerved to apex; style beaklike, (0.5) 1–2 mm long, abruptly contracted to the stigma; seeds uniseriate, pitted; n = 8. Moist meadows and riparian habitats in pinyon-juniper, sagebrush, mountain brush, ponderosa pine, aspen, lodgepole pine, and spruce-fir communities at 1350 to 3235 m in Box Elder, Cache, Daggett, Davis, Duchesne, Juab, Rich, Salt Lake, Sevier, Summit, Uintah, Utah, Wasatch, Washington, and Weber counties, and to be expected almost throughout the state; Alaska and Yukon east to the Atlantic and south to California, Nevada, and Colorado; Eurasia; 24 (ii).

Barbarea vulgaris R. Br. [*Erysimum barbarea* L.]. Stems erect, 1.5–10 dm tall, glabrous; basal leaves generally lobed, mostly 2–15 cm long and 1–3 cm wide, glabrous; cauline leaves reduced upward, auriculate clasping; pedicels 2–5 mm long, glabrous, ascending; sepals 3–4 mm long, yellowish, glabrous; petals 6–8 mm long, yellow, spatulate, to oblanceolate, ascending-spreading; siliques 10–30 mm long, 1–2 mm thick, erect or ascending, the valves glabrous, prominently nerved; style definitely beaklike, 2–3 mm long; n = 8. Riparian or other moist sites in shadscale, sagebrush, sedge-rush, and aspen communities at 1350 to 2290 m in Davis, Juab, Salt Lake, Summit, Utah, Wasatch, and Weber counties; adventive from Eurasia; 8 (i).

Berteroa DC.

Stellate-pubescent annuals or infrequently winter annuals from taproots; leaves alternate and basal, simple, entire, reduced upwardly and sessile, not auriculate; flowers in ebracteate racemes, the pedicels erect-ascending; sepals 4, deciduous; petals 4, white, deeply emarginate and often bilobed; stamens 6, the filaments lacking glandular processes; style long, slender, the stigma capitate; fruit a silicle, 1–3 times longer than broad, compressed parallel to the septum; valves 1-nerved or nerveless; seeds several.

Berteroa incana (L.) DC. [*Alyssum incanum* L.] Stems erect, 3–10 dm tall or more, appressed stellate-hairy; basal leaves oblanceolate, 3–5 cm long, entire, petiolate, usually withered at anthesis; cauline leaves reduced upwards, sessile or short-petiolate below, stellate-hairy; pedicels erect or ascending, 4–10 mm long, stellate and sometimes more or less hirsute; sepals 2–3 mm long, greenish to whitish, stellate-hairy; petals 4–6 mm long, white, deeply bilobed; silicles 5–7 mm long, 2–3 mm wide, moderately inflated, stellate hairy; style (1.5) 2–3 mm long, persistent; n = 8. Roadsides and other disturbed places in Daggett and San Juan counties; widespread in North America; Europe; 2 (0).

Brassica L.

Glabrous or hirsute annuals from taproots; leaves alternate and basal, variously lobed to entire, basal ones often lyrate-pinnatifid, reduced upwardly and petiolate to sessile or auriculate; flowers in ebracteate racemes, the pedicels erect or ascending; sepals 4, deciduous; petals 4, yellow; stamens 6, the filaments lacking glandular processes; style slender to thick, mostly well developed; stigma capitate; fruit a silique, several to many times longer than broad, linear, terete or nearly so, often more or less constricted; valves 1- to 3-nerved, the apical portion produced into a stout 1 to 3 (5) -nerved beak; seeds several to many, uniseriate. Note: Several cultivated members of this genus are present in our region in addition to the weedy adventives distinguished below. They are: *Brassica caulorapa* Pasquale (kohlrabi); *B. napobrassica* Miller (rutabaga); *B. oleracea* L. var. *botrytis* L. (cauliflower), var. *capitata* L. (cabbage), var. *gemifera* Zenker (brussel sprout), and var. *italica* Plenck (broccoli); and *B. rapa* L. (turnip).

1. Cauline leaves sessile, auriculate-clasping, glaucous and entire or nearly so .... *B. campestris*
— Cauline leaves petiolate and not auriculate or, if rarely so, falsely petiolate above the clasping base 2
2(1). Valves of fruit, and often the pedicels and raceme rachis, hirsute with course, spreading hairs; plants cultivated and escaping .... *B. hirta*
— Valves of fruit, pedicels, and raceme rachis glabrous; plants adventive weeds 3
3(2). Silique with a flattened, 2-edged or angular beak, the valves and beak strongly 3 (5) -nerved .... *B. kaber*
— Silique with a cylindrical or rarely slender-conic beak, the valves and beak with 1 nerve (rarely 2 additional delicate ones) 4
4(3). Pedicels 2–6 mm long; siliques 1–2.5 cm long, 1–1.8 mm wide, ascending-appressed .... *B. nigra*
— Pedicels mostly (5) 10–15 mm long; siliques 2–4 cm long, 2–3 mm wide, ascending to erect but not appressed .... *B. juncea*

***Brassica campestris* L.** Rape. [*B. rapa* authors, not L.; *B. napus* authors, not L.]. Stems erect, glabrous or with very few hairs, 2.5–10 dm long or more, simple or branched; basal leaves lyrate-pinnatifid, 5–18 cm long, the terminal lobe mostly 2–5 cm wide, crenate-dentate; lower cauline leaves similar to the basal ones, reduced upwards, becoming auriculate-clasping and dentate to entire; pedicels 7–20 mm long, slender, ascending, glabrous; sepals 4.5–6 mm long, yellowish to greenish; petals 6–10 mm long, yellow; siliques 30–70 mm long, (1.5) 2.5–3.5 mm thick, the beak 8–15 mm long, 1–nerved; valves conspicuously 1–nerved and with 2 more or less delicate lateral nerves, glabrous; n = 10, 11. Cultivated fields, roadsides, fence rows, saline saltgrass meadows, sedge-rush, and other plant communities at 975 to 3020 m in Box Elder, Cache, Davis, Duchesne, Rich, Sanpete, Summit, Utah, Washington, and Weber counties; widespread in temperate regions of the world; adventive from Europe; 11 (ii).

***Brassica hirta* Moench** White Mustard. [*Sinapsis alba* L.; *B. alba* (L.) Rabenh., not Gilib.] Plants erect, pubescent with coarse, descending hairs at least below, the stems 2–10 dm tall, usually branched; basal leaves lyrate-pinnatifid, mostly 5–15 cm long, the terminal lobe 3–10 cm wide, obscurely crenate-dentate; cauline leaves reduced upwards, usually all petiolate, becoming merely lobed, not auriculate; pedicels 5–10 mm long, slender or stout, spreading, often hirsute; sepals 4–5 mm long, yellowish, glabrous; petals 7–10 mm long, yellow; siliques (20) 30–50 mm long, 3–4.5 mm wide, the beak 8–16 mm long, 3 (5) –nerved, the valves conspicuously 3–nerved, hirsute; n = 12. Cultivated white mustard of commerce and rarely escaping, but potentially a noxious weed of cultivated land and to be expected in agricultural regions of the state; widespread in the western U. S.; adventive or introduced from Europe; 1 (i).

***Brassica juncea* (L.) Czernj.** Indian Mustard. [*Sinapsis juncea* L.]. Erect, glabrous or hirsute, the stems 3–10 dm tall or more, usually branched; basal leaves lyrate-pinnatifid, 8–25 cm long, terminal lobe 5–15 cm wide, crenate to dentate or lobed; cauline leaves reduced upwards, short-petiolate to sessile, not auriculate-clasping; pedicels 8–17 mm long, slender to stout, ascending, glabrous; sepals 4–6 mm long, yellowish, glabrous; petals (5.5) 7–12 mm long, yellow; siliques 20–50 mm long, 2–3 mm wide, the beak 6–12 mm long, 1–veined; valves 1 (or lightly 3) –nerved, glabrous; n = 18. Weedy species of disturbed soils at 825 to 2595 m in Box Elder, Duchesne, San Juan, and Washington counties, but not yet common in the state; introduced from Asia; 4 (0).

***Brassica kaber* (DC.) L. Wheeler** Charlock. [*Sinapsis arvensis* L.; *B. arvensis* Rabenh., not L.; *S. kaber* DC.] Erect, pubescent with coarse spreading hairs at least below; stems 3–10 dm tall or more, simple or branched; basal leaves lyrate-pinnatifid to merely dentate, 5–20 cm long, 3–10 cm wide; cauline leaves reduced upwards, short-petiolate or sessile, not auriculate-clasping or, if apparently so, falsely petiolate or leaves sinuate-dentate; pedicels 2–6 mm long, ascending, stout, glabrous; sepals 4–5 mm long, yellowish, glabrous; petals 8–14 mm long, yellow; siliques 30–50 mm long, 2–3 mm thick, the beak 7–15 mm long, 3–veined; valves 3 (5) –nerved, glabrous; n = 9. Roadsides, fields, and ditch banks at 825 to 1985 m in Box Elder, Cache, Juab, Salt Lake, Sevier, Summit, Utah, and Washington counties; widespread in temperate portions of the world; adventive from Europe; 9 (iii).

***Brassica nigra* (L.) Koch in Roehl.** Black Mustard. [*Sinapsis nigra* L.]. Plants erect, glabrous or more usually sparsely to densely hirsute-hispid at least near the base; stems 3–12 dm tall or more, usually branched; basal leaves lyrate-pinnatifid to lobed or serrate-dentate, 5–25 cm long, 2–15 cm wide; cauline leaves reduced upwards, short-petiolate to sessile, not auriculate; pedicels 2–6 mm long, erect, stout, glabrous; sepals 3–4 mm long, yellowish, glabrous; petals (5) 7–12 (15) mm long, yellow; siliques 10–25 mm long, 1–2 mm wide, the beak 1–5 mm long, 1–veined; valves with 1 mid-nerve and 2 faint lateral ones, glabrous; n = 8, 16. Roadsides, fields, and other disturbed places at 1280 to 2260 m in Beaver, Box Elder, Cache, Carbon, Davis, Duchesne, Garfield, Juab, Rich, Salt Lake, San Juan, Sanpete, Utah, Wasatch, Washington, Weber, and perhaps all Utah counties; widespread in North America; adventive from Europe; 30 (iii).

Camelina Crantz

Pubescent with forked or stellate hairs, annual, from taproots; leaves alternate, simple, entire, auriculate-clasping basally; flowers racemose in ebracteate racemes, the pedicels ascending; sepals 4, deciduous; petals 4, pale yellowish; stamens 6, the filaments lacking glandular processes; style slender, the stigma capitate; fruit a silicle less than twice longer than broad, obovoid, somewhat compressed parallel to the septum; valve 1–nerved; seeds several per locule, biseriate.

***Camelina microcarpa* Andrz. ex DC.** Falseflax. Stems erect, (0.8) 1.5–8 dm tall or more, hirsute to subappressed with simple and forked to stellate hairs at least near base; leaves mainly cauline, the basal mostly 1–7 cm long, entire or obscurely toothed, usually withered by late anthesis; cauline leaves reduced upward, at least the upper ones auriculate; pedicels spreading-ascending, (6) 8–18 mm long, glabrous; sepals 2–2.7 mm long, often reddish, more or less villous; petals 3–4 (5) mm long, white or nearly so, rounded apically; silicle 5–6.5 mm long, 3–4 mm wide, moderately inflated, glabrous; style 1–2 (2.5) mm long, persistent; n = 20. Roadsides, foothills, gardens, and other disturbed moist to dry sites at 1280 to 2745 m in probably all Utah counties; widespread in North America; adventive from Asia; 61 (iv).

Capsella Medicus

Stellate-pubescent and often with coarse simple hairs also, annual from taproots; leaves alternate or basal, simple, dentate or variously toothed or lobed to entire, the cauline ones auriculate-clasping; flowers in ebracteate racemes the pedicels spreading-ascending; sepals 4, deciduous; petals 4, white; stamens 6, the filaments lacking glandular processes; style short, the stigma capitate; fruit a silicle, less than twice longer than broad, cuneate-obcordate in outline, compressed at right angles to the septum; valves reticulately veined, strongly keeled; seeds many per locule.

Capsella bursa-pastoris (L.) Medicus Shepherds Purse. [*Thlaspi bursa-pastoris* L.; *Bursa pastoris* Weber in Wiggers; *B. bursa-pastoris* (L.) Britt.]. Stems erect, 1–5 dm tall, stellate-pubescent and more or less hirsute; basal leaves oblanceolate in outline, 2.5–16 (20) cm long, 0.5–2.8 (4) cm wide, lyrate-pinnatifid to merely toothed or subentire; cauline leaves much reduced upwards, sessile and auriculate; sepals 1.2–2.5 mm long, often reddish, pubescent or glabrous; petals 2–4 mm long, white to pinkish, apex rounded; silicles 4.5–8 mm long, 3–5 (6) mm wide, cuneate-obcordate, glabrous; style 0.3–0.6 (1) mm long, persistent; n = 16, 20. Disturbed sites at 850 to 2930 m probably in all Utah counties; widespread in North America; introduced from Europe; 55 (v). We may have, in our material from Utah, two other species from Europe which are infrequently recognized; flowers pinkish (*C. rubella* Reuter) and a white flowered form with the style 0.5–1 mm long (*C. thracica* Velen.).

Cardamine L.

Annual, biennial, or perennial from taproots or rhizomes, glabrous or with simple hairs; leaves alternate, sometimes with basal rosettes, simple to pinnately compound, petiolate, not auriculate; flowers racemose or rarely subcorymbose, the pedicels spreading- ascending to ascending, not subtended by bracts; sepals 4, deciduous; petals 4, white to pinkish; stamens 6, the filaments lacking glandular processes; style stout; stigma capitate; fruit a silique, several to many times longer than broad, slightly compressed parallel to septum; valves obscurely 1 (3) –nerved or nerveless; seeds several to many, uniseriate.

Schulz, O. E. 1903. Monographie der Gattung *Cardamine*. Bot. Jahrb. Syst. 32:280–623.

1. Leaves all simple, cordate-ovate to orbicular; petals 7–12 mm long; plants of stream and seep margins at middle and higher elevations .... *C. cordifolia*
— Leaves pinnately compound, at least the lower ones; petals 2–7 mm long; plants of spring and seep margins, occasionally elsewhere, mainly at middle to low elevations 2

2(1). Leaflets usually 3–5, the terminal leaflet, at least, more than 10 mm wide; upper leaves simple, with broadly ovate blades; petals 3–7 mm long; siliques 1–2 mm wide
.... *C. breweri*
— Leaflets usually 6–11 (rarely 3–5), the terminal leaflet usually less than 10 mm wide; upper leaves compound; petals 2–3 mm long; siliques 0.7–1 mm wide 3

3(2). Cauline leaves mostly 2–4 cm long, with narrow, non-decurrent lateral segments 1–3 mm wide, linear to linear-spatulate or narrowly oblong; plants rare ......
.... *C. parviflora*
— Cauline leaves mostly longer than 4 cm, or the lateral leaflets broader than 3 mm; plants locally common 4

4(3). Lateral leaflets, at least of the lower cauline leaves, obovate to oblanceolate; siliques usually more than 1 mm broad .... *C. oligosperma*
— Lateral leaflets of cauline leaves linear to oblong or oblanceolate; siliques usually less than 1 mm broad ..
.... *C. pensylvanica*

Cardamine breweri Wats. Brewer Bittercress. [*C. vallicola* Greene]. Perennial, rhizomatous, erect or descending; stems (2) 2.5–5 (6) dm tall, glabrous or pubescent with simple hairs near the base; leaves mostly cauline, pinnately compound with 3–5 (rarely more in ours) leaflets, or basal and upper ones simple, mostly 1–7 cm long, the lateral leaflets mostly 10–25 mm long and 0.4–1.2 mm wide, the terminal segment 12–35 mm long and 13–30 mm wide, subentire to sharply toothed, ovate to orbicular, glabrous or sparsely hirsute; pedicels 4–10 mm long or more, glabrous, ascending; sepals 1.5–2.5 mm long, whitish, glabrous or sparingly simple-hairy; petals 3–7 mm long, white, rarely pinkish, spatulate-obovate, spreading; siliques 17–30 mm long, 1–1.8 mm wide, erect or ascending; valves glabrous, obscurely 1 (3) –nerved; style 0.5–2 mm long, tapering to stigma; seeds 1–1.5 mm long, smooth; 2n = 24. Stream sides and seep margins at 1370 to 2440 m in Tooele, Salt Lake, Summit, Wasatch, Utah, and Weber counties; Alaska and British Columbia south to California, Nevada, and Colorado; 6 (i). Our plant is var. *breweri*.

Cardamine cordifolia Gray Heartleaf Bittercress. [*C. cordifolia* var. *pubescens* Gray, type from Thousand Lake Mt.; *C. infausta* Greene; *C. uintahensis* F. Hermann, type from Mt. Elizabeth Ridge, Summit County]. Perennial, rhizomatous, erect or ascending; stems (1.5) 2–8 dm tall, glabrous to more or less densely pubescent with simple hairs near the base; leaves mostly cauline, all simple, the blade mostly (1.5) 2–6 (8) cm long, 1.3–5 (7) cm wide, cordate-ovate or broader, usually sinuate-crenate, glabrous or rarely pubescent; pedicels mostly 10–20 mm long, glabrous or hairy, ascending-spreading; sepals 3–5 mm long, greenish, glabrous or sparingly hairy; petals 7–12 mm long, white, obovate-spatulate, spreading; siliques 20–35 mm long, (1) 1.5–3 mm wide, ascending to erect; valves glabrous, obscurely 1-nerved; style 0.5–2 mm long, or more; seeds 1.5 mm long or more, smooth. Stream sides and seeps in aspen, spruce-fir, and alpine tundra communities at 1675 to 3600 m in Beaver, Box Elder, Cache, Carbon, Davis, Duchesne, Garfield, Grand, Iron, Juab, Piute, Salt Lake, San Juan, Sanpete, Sevier, Summit, Tooele, Utah, Wasatch, Washington,

Wayne, and Weber counties; British Columbia to Wyoming south to California, Nevada, and New Mexico; 101 (xvii). Our plants belong to var. *cordifolia*.

Cardamine oligosperma Nutt. ex T. & G. Nuttall Bittercress. Annual or biennial (short-lived perennial?), with taproots, but sometimes rooting from the lower nodes; stems (4) 10–45 cm tall, pubescent with spreading hairs to glabrate or glabrous; basal leaves with (1–3) 7–11 oval to ovate leaflets, the lateral leaflets 3–20 mm long, 2–15 cm wide, entire or lobed, the terminal leaflet 3–23 mm long, 4–16 mm broad, reniform and entire to cuneate-ovate or orbicular and lobed, glabrous or pubescent; cauline leaves reduced upward; pedicels 4–15 mm long, spreading- ascending to suberect, glabrous; sepals 1.5–2 mm long, glabrous, greenish or purplish; petals 3–5 mm long, white; siliques 15–30 mm long, 1.2–2 mm wide, glabrous, erect, the style 0.4–1 mm long. Weedy plants of gardens and other cultivated lands and in native plant communities at 1370 to 3390 m in Cache, Salt Lake, Summit, and Utah counties; Alaska and Yukon, south to California; Asia; 9 (0). This taxon is sometimes transitional to *C. pensylvanica*, and both are portions of the *C. hirsuta* L. complex. None of our material exactly matches *C. hirsuta* as it occurs in the eastern U. S.

Cardamine parviflora L. Smallflower Bittercress. Annual or biennial from a taproot, the stem erect, usually solitary, 10–30 cm tall, glabrous; basal leaves with 3–5 pairs of oblong to cuneate-obovate, lateral leaflets 2–4 (5) mm long, and 2–3 mm wide, these entire or slightly lobed, the terminal leaflet broadly cordate to orbicular, 3–8 (10) cm long and 5–10 (12) mm wide; cauline leaves reduced upward, mostly 2–4 cm long, with 3–6 pairs of lateral leaflets, the segments similar to the basal ones, only the lateral ones slightly narrower, linear to linear-spatulate or narrowly oblong, entire or toothed, 5–12 mm long and 1–3 mm wide, not decurrent; pedicels 3–7 mm long, spreading-ascending, glabrous; sepals 1.2–1.6 mm long, greenish, glabrous; petals 2–3 mm long, white, oblanceolate, spreading; siliques 12–30 mm long, 0.5–1 mm wide, erect or nearly so, glabrous; style 0.3–0.6 mm long; seeds 1–1.5 mm long, smooth. Moist sites at ca 2440 m in Duchesne County; widespread and common in the eastern U. S.; Europe; 1 (0).

Cardamine pensylvanica Muhl. ex Willd. Muhlenberg Bittercress. [*C. flexuosa* ssp. *pensylvanica* (Muhl.) Schulz; *C. hirsuta* var. *pensylvanica* (Muhl.) Graf]. Annual or biennial from a taproot; stems erect, usually solitary, 1.5–3.5 dm tall, glabrous or pubescent; basal leaves with 7–11 pairs of oval to lanceolate or oblanceolate leaflets, the lateral leaflets 3–15 mm long and 2–12 mm wide, entire or lobed, the terminal leaflet orbicular to cuneate-oblanceolate, 4–20 mm long and 3–15 mm wide; cauline leaves reduced upwards, mostly 4–8 cm long, with 3–5 pairs of lateral leaflets, these broadly oblong to oval, the terminal one cuneate-obovate, entire or toothed, (0.5) 1–3 cm long and 0.5–2 cm wide, decurrent; pedicels 3–10 mm long, spreading-ascending, glabrous; sepals 1.2–1.8 mm long, pinkish, glabrous; petals 2–3 mm long, white, oblanceolate, spreading; siliques 15–30 mm long, 0.7–1 mm wide, erect, glabrous; style 0.4–0.8 mm long; seeds 1–1.5 mm long, smooth; 2n = 32. Stream sides and other moist areas at 1555 to 2475 m in Box Elder, Cache, Duchesne, Salt Lake, Summit, and Uintah counties; widespread in North America; 9 (i).

Cardaria Desv.

Pubescent rhizomatous perennials; leaves alternate, sinuate-dentate, auriculate-clasping; flowers in ebracteate clustered (paniculate) racemes; pedicels spreading-ascending; sepals 4, caducous; petals 4, white, the apex rounded; stamens 6, the filaments lacking glandular processes; style slender, prominent; stigma capitate; fruit a silicle, usually broader than long, compressed at right angles to septum, indehiscent or tardily so; valves reticulately veined; seeds 1 (rarely 2) per locule.

Rollins, R. C. 1940. On two weedy crucifers. Rhodora 42: 302–306.

1. Silicles obcordate in outline, glabrous; plants widespread and common .... *C. draba*
— Silicles orbicular in outline, pubescent or glabrous; plants uncommon 2
2(1). Silicles puberulent, 1.5–2.5 mm long and about as wide .... *C. pubescens*
— Silicles glabrous, 3–4 mm long and about as wide .... *C. chalepense*

Cardaria chalepensis (L.) Hand.-Mazz. [*Lepidium chalepense* L., *L. draba* ssp. *chalepense* (L.) Thell., *C. draba* ssp. *chalepensis* (L.) Schulz]. Plants decumbent to ascending or erect, 2–6 dm tall, glabrous; leaves elliptic to oblong or lanceolate, 0.8–10 cm long, 3–20 mm wide, sinuate-dentate to entire, the upper sessile and auriculate; pedicels 2–8 mm long in fruit, spreading-ascending, glabrous; sepals 1.2–1.8 mm long, greenish to whitish, glabrous; petals 2–3 mm long, white, spatulate, spreading; silicles (excluding the style) 3–4 mm long, 3–4 mm wide, erect, glabrous; style 0.7–1 mm long; seeds 1 or 2. Moist soil, at 1280 to 1891 m in disturbed sites in Daggett, Davis, Salt Lake, San Juan, and Weber counties; adventive from Europe; 5 (i).

Cardaria draba (L.) Desv. Whitetop. [*Lepidium draba* L.; *Cochlearia draba* (L.) L.; *Physolepidium repens* Schrenk ex Fisch. & Mey.; *L. repens* (Schrenk) Boiss.; *C. repens* (Schrenk) Jarm.]. Plants decumbent to ascending or erect; stems (1.2) 1.5–6 dm tall, puberulent to hirtellous with usually descending simple hairs; leaves elliptic to oblong, ovate, or oblanceolate, 0.9–9.8 cm long, 0.6–3.5 cm wide, sinuate-dentate to irregularly toothed, the lower ones petiolate, the upper sessile and auriculate, puberulent to hirtellous with usually retrorse simple hairs; pedicels 5–12 mm long in fruit, spreading-ascending, glabrous or puberulent; sepals 1.2–2 mm long, greenish, usually glabrous; petals 2–3.5 (4) mm long, white, broadly spatulate, spreading; silicles (excluding the style) 2–3.8 mm long, 3.5–5.7 mm wide, erect, glabrous; style 0.6–1.2 mm long; seeds 1–2 mm long; n = 31, 32. Cultivated and waste places at 1280 to 2684 m in Beaver, Cache, Davis, Duchesne, Iron, Juab, Millard, Piute, Salt Lake, Sanpete, Sevier, Tooele, Utah, Washington, and Weber counties; widespread in the U. S. and Canada; adventive from Europe; 55 (ii).

Cardaria pubescens (C. A. Mey.) Jarm. Hairy Whitetop. [*Hymenophysa pubescens* C. A. Mey. in Ledeb.]. Plants ascending to erect; stems 1.5–4 dm tall, puberulent to hirtellous with usually descending simple hairs; leaves elliptic to oblong or oblanceolate, 0.6–6 cm long, 3–15 mm wide or longer, irregularly sinuate-dentate, the lower petiolate, the upper sessile and auriculate, puberulent to hirtellous with usually simple hairs; pedicels 6–10 mm long in fruit, ascending, hairy; sepals 1.8–2 mm long, greenish, hairy; petals 3.5–4 mm long, white, broadly spatulate, spreading; silicles (excluding the style) 1.5–2.5 mm long and about as wide, erect, puberulent; style 0.7–1.2 mm long. Agricultural lands and disturbed places at 1645 to 2260 m in Box Elder, Daggett, Rich, Salt Lake, and Summit counties; widespread in the U. S. and Canada; adventive from Asia; 9 (ii). Our plants are **var. elongata Rollins.** As here defined, the genus *Cardaria* includes *Hymenophysa*. Except for an occasional publication, this seems to be strictly an American concept as most workers in Europe and Asia prefer to distinguish between the two.

Caulanthus Wats.

Annual to perennial, from taproots, glabrous or with simple or malpighian hairs; leaves alternate or mostly basal, simple, lyrate-pinnatifid, pinnatifid, toothed or subentire, petiolate or sessile and auriculate; flowers in ebracteate racemes; sepals 4, deciduous; petals 4, white, yellow, or chestnut brown to purple; stamens 6, the filaments lacking glandular processes; style obsolete or slender and conspicuous; stigma capitate and sometimes distinctly bilobed; fruit a sessile or subsessile silique, many times longer than broad, terete or more or less compressed; valves 1 (3) –nerved; seeds several to many uniseriate.

Payson, E. B. 1922. A monographic study of *Thelypodium* and its immediate allies. Ann. Missouri Bot. Gard. 9: 233–324.
Rollins, R. C. 1971. Protogyny in the Cruciferae and notes on *Arabis* and *Caulanthus*. Contr. Gray Herb. 201: 3–10.

1. Cauline leaves sessile and auriculate at the base; plants of southwestern Utah .... *C. cooperi*
— Cauline leaves petiolate or sessile but not auriculate; plants of broad distribution, sometimes of southwestern Utah 2
2(1). Stems usually conspicuously inflated, glabrous or nearly so; plants perennial .... *C. crassicaulis*
— Stems not inflated; hispid (at least below); plants annual or biennial 3
3(2). Pedicels very short, 1–2 mm long, soon recurved; siliques up to 4 cm long, descending; plants known only from Washington County .... *C. lasiophyllus*
— Pedicels 3–7 mm long, spreading-ascending; siliques 4.5–13.5 cm long, ascending to curved-pendulous; plants known from Beaver, Millard, and Tooele counties .... *C. pilosus*

***Caulanthus cooperi* (Wats.) Payson** [*Thelypodium cooperi* Wats.; *Guillenia cooperi* (Wats.) Greene]. Annual, erect or sprawling; stems not inflated, 1–7.5 dm tall, glabrous and often glaucous or sparsely pubescent with simple or malpighian hairs; leaves mainly cauline, the lower ones 1–7 cm long, 0.4–2 cm wide or more, obscurely sinuate-dentate, glabrous, the cauline ones reduced upwardly, mostly 1–7.5 cm long and 2–12 mm wide, mostly entire with at least the uppermost auriculate; pedicels 1–4 mm long, soon recurved, glabrous; sepals 5–6.5 (7) mm long, green or reddish, glabrous; petals 6–9 mm long, yellowish, suffused with purple, narrowly spatulate, ascending; anthers 1–1.3 mm long; siliques 20–45 mm long, 2–3 mm wide, sessile, descending, compressed, glabrous; style 1–2.5 mm long; stigma not expanded, shortly bilobed; n = 14. Creosote bush and Joshua tree communities at 760 to 855 m in Washington County; Arizona, Nevada, and California; 4 (ii).

***Caulanthus crassicaulis* (Torr.) Wats.** Spindlestem. Perennial, erect; stems usually strongly inflated, (2) 3.2–9.7 (10.8) dm tall, glabrous and glaucous; leaves mainly basal, the lower ones 3–12 (17) cm long, 0.3–3 cm wide, lyrate-pinnatifid to entire, glabrous, the cauline ones much reduced upwardly, linear to narrowly oblanceolate, petiolate and not auriculate; pedicels 1–4 mm long, stout, ascending, glabrous or more commonly hirsute at least apically; sepals (7) 9–13 mm long, brown to brown purple, narrowly spatulate, spreading-ascending; anthers 3.9–6.3 mm long; siliques 7–14 cm long, 1.5–2 mm wide, sessile, ascending to erect, glabrous; style obsolete; stigma more or less expanded, the lobes up to 0.8 mm long; n = 12.

1. Sepals more or less hirsute; plants of broad distribution .... *C. crassicaulis* var. *crassicaulis*
— Sepals glabrous or with a few hairs; plants of various distributions 2
2(1). Stigmas deeply divided; plants common .... *C. crassicaulis* var. *glaber*
— Stigmas shallowly lobed; plants evidently rare .... *C. crassicaulis* var. *major*

Var. *crassicaulis* [*Streptanthus crassicaulis* Torr., type from Great Salt Lake; *C. senilis* Heller]. Salt desert shrub, blackbrush, sagebrush, pinyon-juniper, and ponderosa pine communities at 1525 to 2870 m in Box Elder, Carbon, Daggett, Duchesne, Emery, Garfield, Juab, Kane, Sanpete, Sevier, and Uintah counties; Idaho, Nevada, Colorado, Arizona, and California; 53 (viii).

Var. *glaber* Jones [*C. glaber* (Jones) Rydb.]. Sagebrush, pinyon-juniper, ponderosa pine, white fir, aspen, and spruce-fir communities at 1310 to 3050 m in in Beaver, Duchesne, Emery, Garfield, Kane (type from Sink Valley), Millard, Piute, San Juan, and Washington counties; southern Nevada; 36 (xiii). This taxon is questionably distinct from the former variety and the following, differing from the first in being glabrous and from the second in the less deeply lobed stigma. The plants are mostly allopatric with var. *crassicaulis*, but sympatric with var. *major*.

Var. *major* Jones [*C. major* (Jones) Payson; *C. procerus* authors, not Wats.]. Sagebrush, ponderosa pine, aspen, and fir communities at 2500 to 3175 m in the Henry Mts. (type locality), Garfield County; Nevada and California; 3 (0).

***Caulanthus lasiophyllus* (H. & A.) Payson** [*Turritis lasiophylla* H. & A.; *Thelypodium lasiophyllum* (H. & A.) Greene; *Sisymbrium lasiophyllum* (H. & A.) M. Brandegee; *Guillenia lasiophylla* (H. & A.) Greene]. Annual; stems not inflated, 1–8 (12) dm tall, more or less hirsute with simple or rarely forked hairs; leaves mainly cauline, these 0.7–15 cm long, 0.1–5.5 cm wide, irregularly pinnatifid, petiolate and not auriculate; pedicels 1–2 mm long, deflexed in fruit, glabrous or sparingly hirsute; sepals 2–3 mm long, often purplish, glabrous; petals 3–5 mm long, yellowish, oblong-spatulate, not constricted at juncture of blade and claw, ascending-spreading; anthers 0.6–1 (1.5) mm long; siliques 25–45 (60) mm long, 0.8–1.1 mm wide, sessile, terete, reflexed-descending, glabrous; style 0.8–1.3 mm long, the stigma small, obscurely lobed. Creosote bush, Joshua tree, and blackbrush communities at 790 to 1040 m in Washington County; California, Nevada, and Arizona; 9 (iii). Our material has been designated as **var. *utahensis* (Rydb.)**

Payson [*Thelypodium utahense* Rydb., type from St. George; *T. lasiophyllum* var. *utahense* (Rydb.) Jepson].

Caulanthus pilosus Wats. [*Streptanthus pilosus* (Wats.) Jepson]. Biennial or infrequently annual; stems not inflated, mostly 4–10 dm tall or more, hirsute with simple hairs at least below; leaves mostly basal, these 3–15 cm long, 0.5–3.5 cm wide, irregularly pinnatifid, hirsute, the cauline ones only slightly reduced and shorter upwardly, petiolate and not auriculate; pedicels 4–9 mm long, spreading-ascending, glabrous or nearly so; sepals 5–7 (9) mm long, often purplish, glabrous or hairy; petals 7.5–9 (10) mm long, white, suffused with purple or pink, spatulate-lanceolate, constricted at juncture of blade and claw, ascending-spreading; anthers 2–4 mm long; siliques 7–11.5 cm long, 0.8–1 (1.5) mm wide; style short, the stigma bilobed. Shadscale-winter fat and other cool desert shrub and grass communities at 1370 to 1770 m in Beaver, Millard, and Tooele counties; Oregon and Idaho south to California and Nevada; 13 (i).

Chlorocrambe Rydb.

Glabrous perennials from a stout caudex; leaves alternate and mainly cauline, simple, more or less hastate, entire or sinuately lobed, petiolate, not both sessile and auriculate; flowers in ebracteate racemes; sepals 4, deciduous; petals 4, white; stamens 6, the filaments lacking glandular processes; style obsolete or up to 0.5 mm long; stigma small, entire; fruit a stipitate silique, many times longer than broad, subterete; valves 1 (3–5) –nerved; seeds uniseriate.

Chlorocrambe hastatus (Wats.) Rydb. [*Caulanthus hastatus* Wats., type from Wasatch Mts.]. Stems erect, 6–18 dm tall, usually simple, glabrous and glaucous; leaves with slender petioles 1–16 cm long, the blades hastate to ovate or lanceolate, 3–13.5 cm long, 1–8.5 cm wide, more or less hastate, entire or sinuate-lobed; pedicels spreading to reflexed, 5–10 mm long, glabrous; sepals usually surpassing the petals, their tips coiled, yellowish green, glabrous; petals 4.5–8 mm long, white, mostly 4–6 mm long, the blade constricted at juncture with claw, ascending-spreading; siliques 4–10.5 cm long, 1.8–2.5 mm wide, spreading to curved descending, glabrous; stipe 2–7 mm long; style to 0.5 mm long; stigma not lobed. Mountain brush, aspen, white fir, Douglas fir, and spruce-fir communities at 1800 to 3050 m in in Cache, Davis, Morgan, Salt Lake, Tooele, Utah, Wasatch, and Weber counties; Wallowa Mts., Oregon; 21 (i).

Chorispora R. Br.

Annuals from taproots, stipitate-glandular, sometimes also hirsute; leaves alternate and basal, simple, sinuate-dentate to pinnatifid or entire, the cauline ones petiolate to sessile but not auriculate; flowers in ebracteate racemes; sepals 4, deciduous; petals 4, pink to lavender; stamens 6, the filaments lacking glandular processes; style apical on a slender sterile beak, the stigma minute, bilobed; fruit a silique, many times longer than broad, terete, indehiscent, breaking at maturity into 1–seeded segments; valves 1 (3 or more) –nerved; seeds uniseriate.

Chorispora tenella (Pallas) DC. Musk-mustard. [*Raphanus tenellus* Pallas; *Chorispermum tenellum* (Pallas) R. Br. ex Ait.]. Stems decumbent-ascending to erect, 0.2–4.5 dm tall, simple or branched from the base, stipitate-glandular and often simple hirsute at least at the base; leaves mainly cauline, 0.5–8.5 cm long, 0.1–2.8 cm wide, sinuate-dentate, pinnatifid, or entire, petiolate or sessile but not auriculate; pedicels spreading-ascending, 2–6 mm long, stipitate-glandular and often sparingly hirsute; sepals 4.4–6.7 mm long, reddish or purplish, stipitate-glandular; petals 9–12.5 mm long, pink to lavender, rounded apically, spreading; siliques 3–4.5 cm long, curved-ascending, stipitate-glandular, the beak 8–22 mm long; style obsolete or very short, the stigma minute; n = 7. Roadsides, fields, and other disturbed sites at 1370 to 2320 m probably throughout the state; Washington and Idaho south to California, Arizona and Colorado; adventive from Asia; 55 (x). The musky odor is evident even when driving along roads where the plant is abundant as a weed.

Conringia Adanson

Glabrous and glaucous annuals or biennials, from taproots; leaves alternate and basal (and still alternate), simple, entire, tapering to the base or cauline, sessile, and auriculate-clasping; flowers in ebracteate racemes, the pedicels ascending to curved-erect; sepals 4, deciduous; petals 4, yellow or cream; stamens 6, the filaments lacking glandular appendages; style stout, the stigma lobed; fruit a sessile, slender silique many times longer than broad, quadrangular, the valves 1– to 3–nerved; seeds numerous, uniseriate.

Conringia orientalis (L.) **Dumort.** Hares-ear Mustard. [*Brassica orientalis* L.]. Annual or winter annual; stems 1.9–5 (7) dm tall, solitary or 2 or 3 from the base, glabrous; basal leaves 3–6 (9) cm long, 1.2–3 cm wide, entire, glabrous; cauline leaves several, 1.7–12 cm long, 0.8–5.7 cm wide, ovate to oblong to elliptic, shorter to longer than internodes, glabrous, entire; pedicels 5–14 mm long, ascending to curved-erect, glabrous; sepals 4.3–6 (8) mm long, glabrous, often reddish tinged, acute; petals 6.2–10 (12) mm long, yellow to cream, spatulate, spreading-ascending; siliques 7–10 (13) cm long, 1.5–2 mm thick, erect; valves glabrous; style up to 1 mm long; stigma small; seed numerous; n = 7. Disturbed sites in meadow, pinyon-juniper, sagebrush, and mountain brush communities at 1370 to 2170 m in Box Elder, Cache, Juab, Kane, Rich, Salt Lake, San Juan, Sanpete, Utah, and Washington counties; widely distributed in North America; adventive from Europe; 14 (i).

Descurainia Webb

Stellate-pubescent, stipitate-glandular, or glabrate annuals or biennials from slender to stout taproots; leaves basal and cauline, alternate, 1–3 times pinnately compound or pinnatifid, not auriculate basally, flowers in ebracteate racemes; sepals 4, deciduous; petals 4, yellow to cream; stamens 6, the filaments lacking glandular processes; style short or obsolete, stigma capitate; fruit a silique more than (3) 5 times longer than broad, linear to oblong or clavate, terete or nearly so; valves 1–nerved, glabrous; seeds several to many, uniseriate or biseriate.

Detling, L. E. 1939. *Descurainia* in North America. Amer. Midl. Naturalist 22: 481–520.

Schulz, O. E. 1924. *Descurainia*. Pflanzenr. IV. 105 (Heft 86): 481–520.

1. Upper leaves bi- or tripinnate; siliques narrowly linear, mostly about 20 (10–30) mm long; seeds usually more than 20, uniseriate; replum 2– to 3–nerved; tall to low plants of low elevations .... *D. sophia*

— Upper leaves once-pinnate; siliques clavate, elliptic, or, if linear, less than 20–seeded and less than 15 mm long; replum nerveless or 1–nerved 2

2(1). Siliques clavate or linear to elliptic, rounded to pointed above; seeds often in 2 rows or at least partially so; replum usually nerveless; plants of middle to lower elevations .... *D. pinnata*

— Siliques linear or elliptic, usually pointed above; seeds in one row; replum 1–nerved; plants of middle and higher elevations 3

3(2). Siliques 7–14 mm long, linear or less commonly ellipsoid; pedicels appressed-erect or ascending; seeds mostly 4–10 per locule .... *D. richardsonii*

— Siliques 3.3–7.2 mm long, ellipsoid; pedicels ascending to spreading; seeds mostly 1–3 per locule ....
.... *D. californica*

Descurainia californica (Gray) **Schulz** California Tansymustard. [*Smelowskia californica* Gray]. Annual or winter annual to biennial; stems 4–13 dm tall or more, simple or more commonly profusely branching from most of upper leaf and bract axils, minutely dendritic-stellate pubescent to almost or quite glabrous below, glabrous above; leaves basal and cauline, 2–7 cm long, the lower once-pinnately compound with 2–4 pairs of entire to incised pinnae, the upper reduced, once-pinnate or pinnatifid; pedicels 3–7 mm long, spreading to spreading-ascending, glabrous; sepals spreading, 0.8–1.3 mm long, yellow or greenish, glabrous; petals 1.1–1.6 mm long, yellow; siliques 3.3–7.2 mm long, 0.9–1.3 mm wide, erect or ascending, not appressed; style 0.3–0.7 mm long; seeds uniseriate, 1–3 per locule; n = 14. Juniper, sagebrush, Douglas fir, white fir, aspen, and spruce-fir communities at 1645 to 3480 m in probably all Utah counties; Wyoming, Colorado, and New Mexico westward to California; 87 (xx). This taxon is a mirror-image congener of *D. richardsonii* (q.v.), especially of those taxa of that species with ascending pedicels.

Descurainia pinnata (Walter) **Britt.** Pinnate Tansymustard. [*Erysimum pinnatum* Walter; *Sisymbrium pinnatum* (Walter) Greene; *Sophia pinnata* (Walter) Howell]. Annual or winter annual; stems 1–10 dm tall, stellate-pubescent and sometimes also stipitate-glandular at least below, simple or highly branched above; leaves basal and cauline, 2–10 cm long, once to twice pinnatifid, the segments linear to oblong, often toothed, the upper reduced and usually once pinnatifid; pedicels 3–23 mm long spreading, stellate-pubescent to glabrous; sepals 1–2.2 mm long, yellowish to greenish or violet, stellate-pubescent to glabrous; petals 1.5–3 mm long, cream to yellow; siliques 3–15 (rarely 20) mm long, 1–2 mm wide, clavate to oblong or linear, very rarely ellipsoid; style up to 0.3 mm long; seeds 1–20 per locule, biseriate in part; n = 7, 14, 21. This polymorphic species occurs in all Utah counties. The morphology is sufficiently variable as to invite segregation of infraspecific taxa, and the work of Detling (l.c.) is classic and instructive. In spite of his careful work, I have synonymized some of the segregates recognized by Detling, because of nearly identical overall nature of the plants, otherwise set apart by minute and inconsistent diagnostic characters. The proposed segregates intergrade endlessly, and there is no apparent geographical or ecological correlation of the diagnostic features. The following key gives the impression of such a correlation, in part, but the characters are applied arbitrarily. Possibly trends are indicated, and for that reason, I retain the key to varieties.

1. Pedicels (17) 18–24 mm long; siliques mostly 12–18 mm long; petals yellow, over 2 mm long; terminal leaflet of uppermost leaves linear, entire, and more than 2 cm long; plants of Grand, San Juan, and Uintah counties
.... *D. pinnata* var. *paysonii*

— Pedicels usually less than 15 mm long; siliques often less than 12 mm long (up to 20 mm in var. *filipes*); terminal leaflet various but not or seldom as above; petals yellow to cream, shorter or longer than 2 mm in length; distribution various 2

2(1). Stems moderately to densely stipitate-glandular 3

— Stems stellate-hairy or glabrous, not at all, or seldom stipitate-glandular 4

3(2). Flowers with calyx usually rose colored; corolla 1.5–2 (2.2) mm long; siliques usually 10 mm long or less ....
.... *D. pinnata* var. *osmiarum*

— Flowers with calyx yellowish; corolla 2–3 mm long; siliques often more than 10 mm long ....
.... *D. pinnata* var. *filipes*

4(2). Siliques usually shorter than pedicels, mostly 10–21 mm long, not or only somewhat clavate in outline ....
.... *D. pinnata* var. *osmiarum*

— Siliques usually subequal to or longer (rarely shorter) than pedicels, mostly 3–12 mm long, more or less clavate or elliptic in outline 5

5(4). Siliques 3–8 mm long, borne on pedicels 4–12 mm long; plants of warm deserts in Washington County
.... *D. pinnata* var. *glabra*

— Siliques 8–10 mm long, or, if only 4–8 mm long, not of Washington County; pedicels various 6

6(5). Pedicels 4–6 (8) mm long; flowers 1–1.5 mm long ....
.... *D. pinnata* var. *nelsonii*

— Pedicels (6) 8–12 mm long; flowers 2–3 mm long ....
.... *D. pinnata* var. *intermedia*

Var. filipes (Gray) **Peck** [*Sisymbrium incisum* var. *filipes* Gray; *Sisymbrium longipedicellata* Fourn.; *Sophia filipes* (Gray) Heller; *Sisymbrium gracilis* Rydb.; *D. rydbergii* var. *eglandulosa* Schulz; *D. longipedicellata* (Fourn.) Schulz; *D. longipedicellata* var. *glandulosa* Schulz; *Sisymbrium glandifera* Osterh.; *Sisymbrium longipedicellata* var. *glandulosa* (Schulz) St. John; *D. pinnata* ssp. *filipes* (Gray) Detl.]. Creosote bush, blackbrush, pinyon-juniper, and ponderosa pine communities at 975 to 2300 m in Beaver, Davis, Emery, Grand, Juab, Kane, Millard, Salt Lake, San Juan, Sanpete, Sevier, Tooele, Uintah, Utah, Washington, and Wayne counties; British Columbia and Alberta south to California, Arizona, and Colorado; 44 (viii).

Var. glabra (Woot. & Standl.) **Shinn.** [*Sophia glabra* Woot. & Standl.; *D. pinnata* ssp. *glabra* (Woot. & Standl.) Detl.]. Creosote bush, Joshua tree, blackbrush, and pinyon-juniper communities at 750 to 1220 m elevation in Washington County; Arizona and New Mexico west to California and northern Mexico; 7 (iv). Plants similar to those designated within this species occur elsewhere in Utah, but are arbitrarily assigned to other taxa, especially to var. *nelsonii* or var. *intermedia*.

Var. intermedia (Rydb.) **C. L. Hitchc.** [*Sophia intermedia* Rydb.; *D. intermedia* (Rydb.) Daniels; *Sisymbrium intermedium* (Rydb.) Garrett; *D. pinnata* ssp. *intermedia* (Rydb.) Detl.]. Mixed desert shrub, pinyon-juniper, sagebrush, and upward to montane communities at 1125 to 2745 m elevation in Garfield, Kane, Millard, Uintah, and Wayne counties; British Columbia and Alberta south to California, Nevada, and Colorado; 15 (iv).

Var. nelsonii (Rydb.) Peck [*Sophia nelsonii* Rydb.; *D. brachycarpa* var. *nelsonii* (Rydb.) Schulz]. Shadscale, other salt desert shrub, blackbrush, other warm desert shrub, pinyon-juniper, mountain brush, and aspen communities at 1160 to 3250 m elevation in Carbon, Duchesne, Garfield, Juab, Kane, Millard, San Juan, Sevier, Tooele, Uintah, and Wasatch counties; Washington east to Montana, south to Nevada, and Colorado; 20 (iv).

Var. osmiarum (Cockerell) Shinn. [*Sophia andrenarum* var. *osmiarum* Cockerell; *S. halictorum* Cockerell; *S. andrenarum* Cockerell; *Sisymbrium halictorum* (Cockerell) K. Schum.; *D. halictorum* (Cockerell) Schulz; *D. halictorum* var. *andenarum* (Cockerell) Schulz; *D. halictorum* var. *osmiarum* (Cockerell) Schulz; *D. andrenarum* (Cockerell) Cory; *D. pinnata* ssp. *halictorum* (Cockerell) Detl.]. Shadscale, other salt desert shrub, blackbrush, other warm desert shrub, cool desert shrub, pinyon-juniper, mountain brush, ponderosa pine, aspen, and spruce-fir communities at 850 to 2930 m in probably all Utah counties; Oregon to Wyoming south to Mexico, Oklahoma, and Arkansas; 120 (xxiv). As interpreted here, var. *osmiarum* is expanded to include the concept of *Sophia halictorum* Cockerell, which differs from var. *osmiarum* only in lacking glandular hairs. The plants are identical otherwise. Too, the smaller flowers and pinkish calyces of var. *osmiarum* are transitional with the larger flowers with yellowish calyces of var. *filipes*.

Var. paysonii (Detl.) Welsh & Reveal [*D. pinnata* ssp. *paysonii* Detl.]. Uncommon in pinyon-juniper and sagebrush zones at lower elevations in Grand, San Juan, and Uintah counties; Wyoming and Colorado south into Arizona; 4 (0). Only the longer pedicels keep this out of var. *osmiarum* in a broad sense.

Descurainia richardsonii (Sweet) Schulz Richardson Tansy-mustard. [*Sisymbrium canescens* Richards., not *D. canescens* Nutt., *S. richardsonii* Sweet; *S. canescens* var. *major* Hook.; *Sophia richardsoniana* Rydb.]. Annual or winter annual to biennial; stems (1.5) 3–12 dm tall or more, simple or more commonly profusely branching from upper leaf and bract axils, minutely dendritic-stellate pubescent and sometimes also minutely stipitate-glandular to almost or quite glabrous below, glabrous or glandular above; leaves basal and cauline, 1.5–8 cm long, the lower 1- to 2-pinnatifid with 2–4 pairs of toothed or lobed or subentire pinnae, the upper reduced and only once-pinnatifid; pedicels (2.5) 3–9 mm long, spreading to erect, pubescent or glabrous; sepals spreading, 0.9–1.7 mm long, yellow or greenish, glabrous or hairy; petals 1.3–2.8 mm long, yellow; siliques 7–14 mm long, 0.8–1.2 mm wide, ascending or erect, often appressed; style 0.2–0.6 mm long; seeds uniseriate, (1–3) 4–10 per locule; n = 7, 14, 21. Three weakly defined variants can be at least arbitrarily distinguished as in the following key:

1. Pedicels and siliques erect, more or less appressed to raceme axis . . . . *D. richardsonii* var. *brevipes*
— Pedicels ascending, siliques erect or ascending, neither one appressed to raceme axis 2

2(1). Plants not glandular . . . . *D. richardsonii* var. *sonnei*
— Plants stipitate-glandular . . . . *D. richardsonii* var. *viscosa*

Var. brevipes (Nutt.) Welsh & Reveal [*Sisymbrium canescens* var. *brevipes* Nutt. ex T. & G.; *Sophia procera* Greene; *Sisymbrium procerum* (Greene) K. Schum.; *Sophia brevipes* (Nutt.) Rydb.; *D. richardsonii* var. *macrosperma* Schulz, type from Alta, Salt Lake County; *D. richardsonii* ssp. *procera* (Greene) Detl.; *D. richardsonii* var. *procera* (Greene) Breitung]. Aspen, lodgepole pine, spruce-fir, and alpine communities at 2105 to 3480 m elevation in Cache, Duchesne, Garfield, Iron, Piute, Salt Lake, Summit, and Wasatch counties; Idaho and Montana south to New Mexico; 30 (x).

Var. sonnei (Robins.) C. L. Hitchc. [*Sisymbrium incisum* Engelm. in Gray; *D. incisa* (Engelm.) Britt.; *S. incisum* var. *sonnei* Robins.; *Sophia sonnei* (Robins.) Greene; *Sophia incisa* (Engelm.) Greene; *S. leptophylla* Rydb.; *S. serrata* Greene; *S. purpurascens* Rydb; *Sisymbrium leptophyllum* (Rydb.) Nels. & Macbr.; *D. serrata* (Greene) Schulz; *D. incisa* var. *leptophylla* (Rydb.) Schulz; *D. richardsonii* ssp. *incisa* (Engelm.) Detl.]. Sagebrush, aspen, and alpine communities at 1890 to 3185 m elevation in Cache, Duchesne, Garfield, Grand, Iron, Salt Lake, Sevier, Uintah, Utah, and Wasatch counties; Idaho and Montana south to Mexico; 26 (i). This phase of *D. richardsonii* closely simulates *D. californica* (q.v.).

Var. viscosa (Rydb.) Peck [*Sophia viscosa* Rydb.; *Sisymbrium viscosum* (Rydb.) Blank.; *D. rydbergii* Schulz; *D. richardsonii* ssp. *viscosa* (Rydb.) Detl.]. n = 7. Alder, aspen, and spruce-fir communities at 2450 to 3050 m in Duchesne and Wasatch counties; British Columbia and Alberta south to California, Arizona, and New Mexico; 5 (0). The var. *richardsonii* is known to occur just north and east of Utah in southwestern Wyoming and northwestern Colorado. It differs from var. *brevipes* in being canescent as opposed to moderately pubescent to nearly glabrous.

Descurainia sophia (L.) Webb ex Prantl [*Sisymbrium sophia* L.; *Sophia sophia* (L.) Britt.; *S. parviflora* Standley]. Annual or infrequently winter annual; stems 1.7–8.5 (10) dm tall or more, simple or more commonly branched above, softly dendritic-hairy or with mixed simple and dendritic trichomes at least below; leaves basal and cauline, 1–12 cm long, the lower 2–3 times pinnately compound to pinnatifid, with 2–6 pairs of pinnatifid pinnae, the upper ones smaller and usually twice pinnately compound or pinnatifid; pedicels 4–17 mm long, ascending, puberulent or glabrous; sepals erect, 2–3.1 mm long, yellowish, glabrous or hairy; petals 2.2–3 mm long, cream; siliques (10) 12–27 (30) mm long, 0.8–1.2 mm wide, ascending-erect; style 0.1–0.3 mm long; seeds uniseriate, mostly 10–25 mm long; n = 10, 14, 28. Roadsides, corrals, agricultural lands, and other disturbed sites at 700 to 2450 m in all Utah counties; widespread in North America; adventive from Europe; 90 (xiv).

Diplotaxis DC.

Glabrous or simple-hirsute annuals or biennials, from taproots; leaves alternate, mostly basal, pinnatifid to irregularly toothed, reduced upwards and petiolate to merely sessile but not auriculate; pedicels ascending or ascending-spreading, ebracteate; sepals 4, deciduous; petals 4, yellow or sometimes fading rose; stamens 6, the filaments lacking glandular processes; style stout, well developed; stigma capitate; fruit a silique, many times longer than broad, linear, somewhat flattened parallel to partition, valves 1- to 3-nerved; beak not nerved; seeds numerous, biseriate.

Diplotaxis muralis (L.) DC. [*Sisymbrium murale* L.]. Erect or ascending, pubescent with coarse, descending hairs at least below; stems 0.7–5 dm tall, usually branched; basal leaves lyrate-pinnatifid to irregularly lobed, mostly 2.5–9.5 cm long, 0.5–3.5 cm wide; cauline leaves usually much reduced upwards, all petiolate, not auriculate; pedicels 6–23 mm long or more, slender, ascending to spreading-ascending, often hirsute; sepals 3–4.5 mm long, purplish tinged, glabrous or hirsute; petals 4.5–7.5 mm long, yellow or sometimes fading rose; siliques 17–33 mm long, 1.5–2.8 mm wide; style 1.5–2.5 mm long; valves glabrous, lightly 1- to 3-nerved; $2n = 42$. Fields and disturbed sites in Cache, Salt Lake, and Utah counties; scattered in North America; adventive from Europe; 3 (0).

Dithyrea Harv.

Stellate- or dendritic-hairy annual or winter annual from a taproot; leaves alternate, simple, subentire to toothed, lobed or pinnatifid, petiolate to sessile but not auriculate; flowers in ebracteate racemes; sepals 4, deciduous; petals 4, white; stamens 6, the filaments lacking glandular processes; style broad and stout; stigma enlarged-capitate; fruit a silicle, more than twice longer than broad, spectaclelike, compressed at right angles to the replum; valves reticulately veined; seeds 1 per locule [*Dimorphocarpa* Rollins].

Dithyrea wislizenii Engelm. in Wisliz. Spectacle-pod. [*Dimorphocarpa wislizenii* (Engelm.) Rollins]. Erect or ascending, pubescent with soft stellate or dendritic hairs; stems 0.7–5 dm tall, simple or branched; basal leaves often withered by anthesis; cauline leaves 1.2–9.5 cm long, 2–25 mm wide, pinnatifid, sinuately dentate, irregularly lobed, or entire, moderately to densely hairy like stems, reduced upward; pedicels 5–21 mm long in fruit, spreading, dendritic-hairy; sepals 3–5.5 mm long, greenish, yellowish, or purplish, usually dendritic-hairy; petals 4.8–8 mm long, white, the claws occasionally lavender, the blades orbicular to spatulate, 2.5–5.5 mm wide; silicles 4–6.5 mm long (from apex of the short stipe), 9–14 mm wide; valves pubescent; seeds 1 per locule; $2n = 18$. Sandy dunes and other sandy sites in warm or mixed desert shrub, pinyon-juniper, and ponderosa pine communities at 820 to 2135 m in Emery, Garfield, Grand, Kane, San Juan, Washington, and Wayne counties; Nevada, Colorado, Arizona, New Mexico, and Mexico; 87 (x). Segregation of *D. wislizenii* and related species within *Dimorphocarpa* is based on dissimilarities of about the same order of magnitude as occur within other genera in this family, where generic lines are notoriously indistinct.

Draba L.

Plants with stellate, dendritic, forked, or simple hairs, or glabrate, annual, biennial, or perennial, from taproots and often with well-developed caudices; leaves all basal or some cauline, alternate, simple, entire or toothed, tapering to the base or rounded, not auriculate; flowers in ebracteate racemes; sepals 4, deciduous; petals 4, white, yellow or cream, rounded to bifid apically; stamens 6, the filaments lacking glandular processes; style obsolete to prominent and slender; stigma obscurely bilobed; fruit a silicle or infrequently a short silique, mostly 1–10 times longer than broad, oval to ovate, lanceolate, or linear, compressed parallel to the septum, plane or twisted, straight or curved; valves obscurely 1-nerved or nerveless; seeds biseriate, usually numerous.

Hartman, R. L., et al. 1975. Biosystematics of *Draba cuneifolia* and *D. platycarpa* (Cruciferae) with emphasis on volatile and flavonoid constituents. Brittonia 27: 317–327.

Hitchcock, C. L. 1941. A revision of the Drabas of western North America. Univ. Washington Publ. Biol. 11: 1–132.

Payson, E. B. 1917. The perennial scapose Drabas of North America. Amer. J. Bot. 4: 253–267.

1. Plants scapose; leaves all basal, mostly depressed-caespitose perennials (except in *D. verna*, which has deeply bilobed petals) .... Key 1
— Plants with 1 to many cauline leaves in addition to basal ones, annual (and the petals not bilobed), biennial, or perennial. .... Key 2

Key 1.

Plants scapose.

1. Plants annual, flowering in spring; style scarcely if at all evident (up to 0.1 mm long) .... *D. verna*
— Plants perennial, flowering in spring and in summer; style 0.2–2.5 mm long 2

2(1). Petals white 3
— Petals yellow 7

3(2). Plants pubescent throughout with doubly pectinate hairs; plants of Daggett and Uintah counties .... *D. oligosperma*
— Plants variously hairy, but either stellate or with simple or forked hairs only, seldom or rarely pubescent throughout; plants known from Raft River, Uinta, and Wasatch mts. and plateaus of southern Utah. 4

4(3). Leaves with simple hairs only; petals 2–3 mm long; style up to 0.2 mm long .... *D. fladnizensis*
— Leaves variously pubescent; petals mostly 3–5 mm long; style 0.2–1.5 mm long 5

5(4). Leaves cinereus with appressed hairs, sometimes stellate; plants of various distribution .. *D. lonchocarpa*
— Leaves glabrous above and below, or pubescent with simple or forked hairs and similarly ciliate; plants of the Raft River Mts. and plateaus of southern Utah ... 6

6(5). Styles mainly 0.5 mm long or shorter; seeds 6–12 per silicle; plants of mountains and plateaus of southern Utah .... *D. subalpina*
— Styles 0.5–1.5 mm long; seeds 1 or 2 per silicle; plants of the Grouse Creek Mts. .... *D. douglasii*

7(2). Leaves glabrous or with unforked hairs only, often merely ciliate 8
— Leaves pubescent, at least some of the hairs forked, stellate, or doubly pectinate 10

8(7). Styles 1–2.5 mm long; petals 5–6 mm long; plants of Box Elder, Cache, and Weber counties ... *D. maguirei*
— Styles usually less than 1 mm long; petals less than 5 mm long; distribution various 9

9(8). Styles not over 0.2 mm long; plants biennial or short-lived perennial; leaves not densely imbricated; petals 1.5–3 mm long .... *D. crassifolia*
— Styles 0.2–1 mm long; plants perennial, densely caespitose, with closely imbricated leaves; petals often more than 3 mm long .... *D. densifolia*

10(7). Lower side of leaves (at least) with appressed, pectinately branched hairs .... *D. oligosperma*
— Lower side of leaves glabrous or pubescent but, if so, with merely stellate or forked hairs, not pectinate ... 11

11(10). Leaves almost glabrous, ciliate with forked or dendritic hairs; plants of Box Elder and Weber counties .... *D. maguirei*

	Leaves more or less pubescent on one or both surfaces	12
12(11).	Stems glabrous, at least above	13
—	Stems pubescent throughout	14
13(12).	Basal leaves 10–35 mm long, 3–10 mm wide; silicles 7–14 mm long, 2–4 mm wide; plants mainly of western Kane and eastern Washington counties	*D. asprella*
—	Basal leaves 6–12 mm long, 9–14 mm wide; silicles 4–10 mm long, 1.5–3 mm wide; plants of Box Elder, Cache, and Weber counties	*D. maguirei*
14(12).	Leaves densely cinereus pubescent, individual hairs almost indistinguishable; plants known from Uinta Mts.	*D. ventosa*
—	Leaves not densely cinereus pubescent, individual hairs apparent; plants of Beaver and Piute counties	*D. sobolifera*

Key 2.

Plants with 1 or more cauline leaves.

1.	Plants annual; style obsolete or rarely up to 0.2 mm long	2
—	Plants biennial or perennial; style mostly 0.2–1 mm long or more	7
2(1).	Upper portion of stem, including pedicels, pubescent	3
—	Upper portion of stem glabrous	4
3(2).	Flowers white; plants widely distributed at low elevations	*D. cuneifolia*
—	Flowers yellow; plants of montane regions in northern Utah	*D. rectifructa*
4(2).	Upper leaf surfaces usually glabrous; cauline leaves 1 or 2, rarely lacking	*D. crassifolia*
—	Upper leaf surfaces usually pubescent; cauline leaves 1–5 or more	5
5(4).	Petals white; leaves entire or nearly so; silicles less than 2 mm wide; plants of low elevations	*D. reptans*
—	Petals yellow, sometimes fading whitish; silicles at least 2 mm wide or the plants of montane places	6
6(5).	Pedicels usually at least 1.5 times longer than the silicles; plants usually of low elevations	*D. nemorosa*
—	Pedicels rarely up to 1.5 times longer than the silicles; plants usually of montane places	*D. stenoloba*
7(1).	Petals white; plants of high elevations	8
—	Petals yellow or cream, sometimes fading whitish; plants of low to moderate or high elevations	10
8(7).	Styles obsolete or nearly so, less than 0.2 mm long	*D. fladnizensis*
—	Styles 0.2–0.8 mm long	9
9(8).	Silicles glabrous or merely ciliate; cauline leaves 1 or 2	*D. lonchocarpa*
—	Silicles pubescent on valves; cauline leaves 1–10	*D. lanceolata*
10(7).	Cauline leaves solitary; plants of low elevations, restricted to Kane and Washington counties (disjunct in western Juab County)	*D. asprella*
—	Cauline leaves usually 2–20 or, if solitary, not of the above counties	11
11(10).	Plants with a definite caudex, the branches densely clothed with threadlike grayish leaf bases; plants of the Deep Creek Mts.	*D. kassii*
—	Plants with caudex not well developed, or if so, seldom if ever as above, and not or seldom of the Deep Creek Mts.	12
12(11).	Styles obsolete or nearly so, up to 0.2 mm long	*D. stenoloba*
—	Styles 0.2–2.5 mm long	13
13(12).	Petals 2.8–3.8 mm long; pubescence of leaves of stiff, stalked, 2– to 5–rayed hairs; silicles pubescent with simple or stalked and forked hairs	*D. brachystylis*
—	Petals (3.8) 4–8 mm long; pubescence of leaves various; silicles glabrous or pubescent with stellate to simple hairs	14
14(13).	Leaf surfaces glabrous, ciliate or not; silicles glabrous	*D. crassa*
—	Leaf surfaces with stalked 2– to 4–rayed, forked, or simple hairs, margins ciliate or not; silicles glabrous or hairy	15
15(14).	Leaves solitary, subtending lowermost flower or branch of inflorescence; plants of Box Elder and Cache counties	*D. incerta*
—	Leaves 2–20; plants of various distribution	16
16(15).	Leaves bright green to somewhat grayish; cauline leaves often denticulate; silicles plane or slightly contorted; plants of Grand and San Juan counties	*D. spectabilis*
—	Leaves grayish green; cauline leaves entire; silicles usually contorted; plants of broad distribution	*D. aurea*

Draba asprella Greene Zion Draba. Perennial, caespitose; stems 0.5–1.4 (2) dm tall, arising from a branching caudex, hirsute with mixed simple and forked to dendritic hairs; leaves all basal, rarely with 1 cauline, 1–3.5 cm long, 0.3–1 cm wide, oblanceolate to spatulate, entire or obscurely denticulate, green, the surfaces more or less pubescent with usually stalked and 4–rayed hairs; racemes simple, 10– to 30–flowered, elongating in fruit; pedicels 3–15 mm long, ascending, glabrous; sepals 1.8–2.5 mm long, greenish, stellate-hairy; petals 3.4–5 (6) mm long, yellow to yellow orange, obovate-spatulate, rounded; silicles 7–14 mm long, 1.5–4 mm wide, lance-elliptic, glabrous or ciliate; style 0.8–1.1 mm long; seeds 12–20. Mixed warm desert shrub, mountain brush, and ponderosa pine-manzanita communities at 1050 to 2600 m elevation in western Juab, western Kane, and eastern Washington counties; Arizona; 12 (ii). Our material belongs to **var. zionensis** (C. L. Hitchc.) Welsh & Reveal [*D. zionensis* C. L. Hitchc., type from Zion Canyon]. The var. *asprella* is restricted to Arizona.

Draba aurea Vahl in Hornem. Golden Draba. [*D. luteola* Greene; *D. surculifera* A. Nels.; *D. aureiformis* Rydb.; *D. uber* A. Nels.; *D. mccallae* Rydb.; *D. decumbens* Rydb.; *D. aurea* var. *luteola* (Greene) Schulz; *D. aurea* var. *aureiformis* (Rydb.) Schulz; *D. aurea* var. *decumbens* (Rydb.) Schulz; *D. aureiformis* var. *leiocarpa* Payson & St. John]. Perennial, not caespitose; stems 0.7–4 (5) dm tall, simple or few from a branching caudex, pubescent throughout with simple hairs often intermixed with forked ones; basal leaves 0.8–4 cm long, 2–13 mm wide, oblanceolate, entire or serrulate, green or grayish, the surfaces pubescent with stalked 4–rayed hairs; cauline leaves mostly 3–20, lanceolate to ovate or oblanceolate, 0.5–3 cm long, 3–12 mm wide, entire or less commonly denticulate, pubescent like the basal ones; racemes simple or branched, several- to many-flowered, much elongating in fruit; pedicels 3–15 (20) mm long, ascending to erect, pubescent; sepals 2–3.5 mm long, greenish, pubescent; petals (3.5) 4–6 mm long, yellow, spatulate to obovate, rounded to emarginate; fruit 8–17 mm long, plane or more commonly contorted, 2–4 mm wide, ovate-lanceolate to lanceolate or elliptic, pubescent with simple, branched, or stellate hairs; style (0.3) 0.8–1.3 (1.5) mm long; seeds 20–50; n = 32. Sagebrush-grass, aspen, Douglas fir, lodgepole pine, spruce-fir, and alpine tundra at 2300 to 3815 m in Beaver, Carbon, Daggett, Duchesne, Garfield, Iron, Juab, Piute, Salt Lake, San Juan, Sevier, Summit, Uintah, and Utah counties; Alaska and Yukon east to Labrador and Greenland, south to Arizona and New Mexico; 63 (xii). This is a highly variable taxon in our region.

Draba brachystylis Rydb. Wasatch Draba. Biennial (annual) or short-lived perennial, not caespitose; stems (1) 1.3–2.5 (3) dm tall, usually branched, pubescent throughout with simple, branched, or stellate hairs; basal leaves 1–5.3 cm long, 3–15 mm wide, oblanceolate, entire or denticulate, the surfaces pubescent with stalked 2- to 5-rayed hairs; cauline leaves (1) 2–8, ovate to obovate, elliptic, or lanceolate, 6–28 mm long, 2–11 mm wide, denticulate to entire, pubescent like the basal ones; racemes simple or branched, several- to many-flowered, elongating in fruit; pedicels 1–10 mm long, spreading to spreading-ascending, pubescent; sepals 2–2.7 mm long, yellowish, pubescent; petals 2.8–3.8 mm long, yellow, spatulate, rounded to emarginate; fruit (7) 10–15 mm long, 2–3.5 mm wide, plane, obliquely oblong-elliptic, pubescent with simple and branched hairs; style 0.4–0.8 mm long; seeds 20–30. Aspen and white fir-Douglas fir communities at 1675 to 2990 m in Duchesne, Salt Lake, and Utah counties; Spring Mt., Clark County, Nevada; 5 (i). This is a poorly known and rarely collected plant with affinities to both *D. aurea* and *D. rectifructa*. The type is from the Wasatch Mts.

Draba crassa Rydb. Thick Draba. [*D. chrysantha* var. *crassa* (Rydb.) Schulz]. Plants not caespitose, the stems 8–18 cm tall, arising from a thickened caudex clothed with numerous marcescent leaf-bases, moderately hairy with simple or branched hairs, at least below; basal leaves 1.5–8 cm long, 3–12 mm wide, elliptic to oblanceolate, entire or obscurely toothed, glabrous or merely ciliate; cauline leaves 2–8, ovate to elliptic or obovate, 0.5–2 cm long, 2–8 mm wide, entire or nearly so, usually glabrous; racemes few- to many-flowered, elongating in fruit; pedicels 3–10 mm long, ascending, softly villous; sepals 2.3–3 mm long, greenish, suffused with purple, pubescent; petals 4–8 mm long, yellow, obovate, rounded; silicles 8–15 mm long, 2.5–4 (5) mm wide, plane or contorted, ovate to lanceolate or elliptic, glabrous; style 0.8–1.2 mm long; seeds 14–26. Alpine tundra, typically in talus or rock stripes, at 3355 to 3965 m in the Uinta Mts., Duchesne and Summit counties; Montana, Wyoming, and Colorado; 2 (0).

Draba crassifolia Graham Hairy Draba. [*D. parryi* Rydb.; *D. crassifolia* var. *parryi* (Rydb.) Schulz]. Biennial (annual) or short-lived perennial, not caespitose, the stems 2–12 (20) cm tall, arising from a rosulate tuft of leaves, usually glabrous except for a few hairs near base; basal leaves 3–15 (23) mm long, 1–3 mm wide, narrowly spatulate, entire, surfaces usually glabrous, sometimes ciliate; cauline leaves lacking or 1 or 2, very small; racemes 2- to several-flowered, congested or elongating in fruit; pedicels 2–10 mm long, curved-ascending, glabrous; sepals 1–1.4 mm long, greenish, glabrous; petals 1.7–2.5 mm long, yellow but rarely fading white, elliptic-spatulate, emarginate; silicles 5–10 mm long, 1.5–2.5 mm wide, plane, glabrous, lance-elliptic; style to 0.5 mm long; seeds 10–60; n = 20. Spruce-fir, lodgepole pine, and alpine tundra, often in talus or in rock stripes, at 3050 to 3800 m in Daggett, Duchesne, Garfield, Grand, Iron, Juab, Piute, Salt Lake, San Juan, Sanpete, Sevier, Summit, Uintah, and Wayne counties; widespread from Alaska and Yukon east to Greenland and south to California, Arizona, and Colorado; Europe; 13 (iii). This entity apparently intergrades with some phases of *D. stenoloba* in Utah, from which it is difficult, if not impossible, to segregate all specimens.

Draba cuneifolia Nutt. ex T. & G. Wedgeleaf. Annual, not caespitose, 1–15 (20) cm tall, the very short leafy stems arising from a taproot, simple or branched, pubescent with dendritic hairs almost throughout; basal leaves 0.5–4 cm long, 2–27 mm wide, suborbicular to oblanceolate or cuneate-spatulate, dentate to entire, the surfaces hirsute with stalked 2- to 4-rayed hairs, sometimes intermixed with simple hairs; cauline leaves few to several, usually much reduced, pubescent like the basal ones; racemes 3- to many-flowered, congested or elongating in fruit; pedicels 1–7 mm long, spreading to ascending, dendritic-hairy; sepals 1.5–2.5 mm long, greenish, pubescent; petals 3–4.5 (5) mm long, occasionally small or even lacking in cleistogamous flowers, white, spatulate, rounded to emarginate; silicles 4–13 (15) mm long, 1.8–3.8 (5) mm wide, plane, strigose, oblong-elliptic; style up to 0.2 mm long; seeds 20 or more. Warm desert shrublands upward to mountain brush, pinyon-juniper, and ponderosa pine communities, at 750 to 2535 m in Beaver, Box Elder, Emery, Grand, Iron, Juab, Kane, Millard, Salt Lake, San Juan, Sanpete, Sevier, Tooele, Uintah, Utah, and Washington counties; Washington and Idaho south to Mexico, Texas, and Arkansas; 72 (ix). Two varieties are present. The bulk of our material has racemes seldom half as long as the plant height and is assignable to **var. *cuneifolia*** [*D. helleri* Small; *D. ammophila* Heller; *D. cuneifolia* var. *helleri* (Small) Schulz; *D. cuneifolia* var. *leiocarpa* Schulz]. Some of the specimens from Washington County have the racemes elongated to half or more the plant height and are included within **var. *platycarpa*** (T. & G.) Wats. [*D. platycarpa* T. & G.; *D. viperensis* St. John].

Draba densifolia Nutt. ex T. & G. Rockcress Draba. Perennial, pulvinate-caespitose and matted, scapose, arising from compacted caudex branches clothed with marcescent leaf-bases; scapes 5–15 cm tall, glabrous to pubescent throughout; leaves 2–9 mm long, mostly 1–3 mm wide, oblong to oblanceolate, the surfaces glabrous or with few-forked or dendritic hairs beneath, more or less ciliate with stiff, coarse, simple or forked hairs; racemes 2- to 10 (or more) -flowered, not elongated in fruit; pedicels 0.5–2 mm long, ascending, glabrous; sepals 2–3 mm long, greenish, glabrous or pubescent; petals 2–6 mm long, yellow, obovate, truncate to emarginate; silicles 2–7 mm long, 2–3.5 mm wide, ovate to elliptic, glabrous; styles 0.2–1 mm long; seeds 2–12. Alpine tundra, often in rock stripes, talus, or meadows, less commonly in spruce-fir communities at 3140 to 3815 m in the Wasatch and Uinta mts. in Daggett, Duchesne, Juab, Salt Lake, Summit, Uintah, and Utah counties;

British Columbia and Montana south to California, Nevada, and Wyoming; 16 (0). Two weakly defined, sympatric variants are recognized.

1. Styles 0.2–0.5 mm long; plants glabrous except for cilia of the leaves .... *D. densifolia* var. *apiculata*
— Styles 0.5–1 mm long; plants sometimes hairy on the lower leaf surfaces and on the scape ....
.... *D. densifolia* var. *densifolia*

Var. *apiculata* (C. L. Hitchc.) Welsh [*D. apiculata* C. L. Hitchc., type from Uinta Mts., Summit County]. *D. apiculata* var. *daviesiae* C. L. Hitchc.; *D. densifolia* var. *daviesiae* (C. L. Hitchc.) Welsh & Reveal]. This is a poorly differentiated phase of alpine sites in Duchesne, Salt Lake, Summit (type from LaMotte Peak), and Uintah counties; Wyoming and Montana.

Var. *densifolia* [*D. glacialis* var. *pectinata* Wats.; *D. mulfordae* Payson; *D. nelsonii* Macbr. & Payson; *D. globosa* Payson; *D. sphaerula* Macbr. & Payson; *D. pectinata* (Wats.) Rydb.; *D. caeruleomontana* Payson & St. John; *D. caeruleomontana* var. *piperi* Payson & St. John; *D. densifolia* f. *nelsonii* (Macbr. & Payson) Schulz; *D. globosa* var. *sphaerula* (Macbr. & Payson) Schulz]. This is the common phase of the species in alpine sites of Daggett, Duchesne, Salt Lake, Summit, and Uintah counties; distribution of the species.

Draba douglasii Gray Douglas Draba. Tufted perennial from a stout taproot; leaves basal, thick, leathery, oblanceolate, the midrib prominent, 5–12 mm long, 1–2 mm broad, ciliate with simple or forked hairs and often pubescent with unbranched or forked hairs; scapes 1–3 cm tall, softly pubescent with simple or forked hairs; pedicels hairy like the scapes, subequal to the fruits; sepals 2–2.5 mm long, glabrous or sparingly pubescent; petals 4–5 mm long, white; silicles ovoid, the valves thick and leathery, little flattened, mainly 3–7 mm long, pubescent with short, retrorsely pubescent hairs; style 0.5–1.5 mm long, slender; ovules 4, the seeds 1 or 2, ca 2 mm long. Windswept ridge crests at ca 2440 m in Box Elder County (Goodrich & Atwood 17127); Washington to California and Nevada; 1 (0). Our material belongs to var. *douglasii*.

Draba fladnizensis Wulfen in Jacq. Arctic Draba. [*D. pattersonii* Schulz; *D. pattersonii* var. *hirticaulis* Schulz; *D. pattersonii* var. *dasycarpa* Schulz]. Perennial, not caespitose, the stems 2–9 dm tall, glabrous or pubescent at least near the base with simple or forked hairs; basal leaves 3–10 mm long, 1–2 mm wide, oblanceolate, the surfaces glabrous or moderately hairy with simple or 2-forked hairs, ciliate; cauline leaves 1 or 2, greatly reduced; racemes 3– to several-flowered; pedicels 1–3 mm long, ascending to spreading, glabrous; sepals 1.2–1.8 mm long, greenish, glabrous; petals 1.8–2.5 mm long, white (rarely pink), spatulate, rounded to retuse; silicles 3–6 mm long, 1.2–2 mm wide, oblong-ovate, glabrous; style essentially lacking; seeds 10–20; n = 8. Alpine tundra in Uinta and La Sal mts. above 3100 m in Daggett (?), Grand, San Juan, and Uintah (?) counties; Alaska to Mackenzie south to Colorado; 4 (0). Our material is difficult to interpret from the dwarf alpine specimens of *D. stenoloba* (q.v.), but apparently the white petals are diagnostic.

Draba incerta Payson [*D. laevicapsula* Payson; *D. incerta* var. *laevicapsula* (Payson) Payson & St. John]. Perennial, caespitose but loosely so, 2–15 (20) cm tall; stems pubescent with stellate or dendritic and sometimes simple hairs; basal leaves 5–18 (25) mm long, 1–3 mm wide, narrowly oblanceolate, surfaces with at least some doubly pectinate hairs, often intermixed with other types of pubescence, ciliate with simple to pectinately branched hairs; cauline leaves 1 or lacking; racemes 5– to many-flowered, elongating in fruit; pedicels 2–12 mm long, ascending, usually hairy; sepals 2.5–3.5 mm long, greenish or suffused with purple, pubescent; petals 4–5.5 mm long, yellow but fading cream, cuneate-obovate, broadly emarginate; silicles 6–10 mm long, 2.5–3.5 mm wide, ovate to lanceolate, plane, pubescent or glabrous; seeds 8–14. Ridge crests and talus at ca 2980 to 3050 m in the Raft River and northern Wasatch mts., Box Elder and Cache counties; Alaska and Yukon south to Washington and Wyoming; 2 (0).

Draba kassii Welsh Kass Rockcress. Perennial, caespitose, from a definite, branching, subligneus caudex, this clothed with persistent, filiform, threadlike, leaf bases; stems 2–13 cm tall, glabrous or sparingly hirsute with mixed simple and forked to dendritic hairs; leaves all basal, rarely with 1 cauline, 1.8–4.8 cm long, 2–6 mm wide, narrowly oblanceolate to spatulate, entire or obscurely and sparingly denticulate, green, the surfaces glabrous or pubescent with forked hairs, sparingly ciliate with simple or forked hairs; racemes simple, 2– to 9-flowered, elongating in fruit; pedicels 2–10 (15) mm long, ascending, glabrous; sepals 1.5–2.4 mm long, greenish, sparingly hairy, with simple or forked hairs; petals 4.6–5.9 mm long, yellow, obovate-spatulate, rounded; silicles 3–10 (14) mm long, 0.8–2.5 mm wide, elliptic to oblong, glabrous; style 1–2 mm long; seeds 2–14. Pinyon-juniper, white fir, and mountain brush communities, mainly in crevices in granite, at 2135 to 2440 m in the Deep Creek Mts., Tooele County; endemic; 3 (0).

Draba lanceolata Royle [*D. cana* Rydb.; *D. valida* Gooding, type from Uintah County]. Perennial, loosely caespitose; caudex simple or branched, the stems 0.5–3.5 dm tall, pubescent throughout with soft many-branched hairs; basal leaves 0.5–4 cm long, 1–4 mm wide, oblanceolate, entire, pubescence of overlapping, stellate or branched hairs; cauline leaves several, commonly toothed; racemes several- to many-flowered, sometimes with solitary flowers in upper leaf axils; pedicels 2–9 mm long, erect, usually appressed to rachis, pubescent; sepals 1.5–2 mm long, sparsely pilose; petals 2.2–4 mm long, white, cuneate-obovate, more or less emarginate; silicles 5–12 mm long, 1.5–2.5 mm wide, narrowly lanceolate to oblong, plane or contorted, softly pubescent, rarely glabrous; style 0.2–0.8 mm long; seeds 20 or more; 2n = 32. Spruce-fir and alpine tundra communities, often in talus or rock stripes, at 3200 to 3910 m in Daggett, Duchesne, Juab, Piute, Salt Lake, Summit, Uintah, and Utah counties; Alaska and Yukon, south to Nevada and Colorado; 12 (0).

Draba lonchocarpa Rydb. Perennial, loosely to densely caespitose; caudex usually branched, scapose or rarely with one cauline leaf, the scape 1–12 cm tall, glabrous or pubescent with soft many-branched hairs; leaves 5–15 mm long, 1–4 mm wide, pubescent with usually overlapping or stellate (or rarely some simple) hairs, ciliate with stellate (or with some simple) hairs; racemes 3– to 12-flowered, contracted or elongating in fruit; pedicels 1–6 (11) mm long, ascending to erect, glabrous; sepals 1.5–2 mm long, glabrous or pubescent; petals 2.5–4 mm long, white; silicles 5–14 mm long, 1–2 mm wide, linear to lance-linear or oblong, plane or twisted, glabrous or pubescent; style 0.2–0.5 mm long; seeds 8–30; n = 8. Spruce-fir and alpine tundra, often in talus or rock stripes, at 3050 to 3910 m in Box Elder, Daggett, Duchesne, Grand, Juab, Piute, Salt Lake, San Juan, Summit, Uintah, and Utah counties; Alaska and Yukon south to Oregon and Colorado; 22 (ii). Our material has been treated as portions of an expanded *D. nivalis* Lilj., from which it differs in technical features of pubescence and silicle characteristics. Two sympatric variants have been designated among the Utah material.

1. Silicles mostly less than 7 mm long, elliptic to linear .... *D. lonchocarpa* var. *exigua*
— Silicles mostly more than 10 mm long, linear to narrowly elliptic .... *D. lonchocarpa* var. *lonchocarpa*

Var. exigua Schulz [*D. nivalis* var. *exigua* (Schulz) C. L. Hitchc.]. Alpine tundra in Uinta and La Sal mts., Duchesne, Grand, and Summit counties; Wyoming and Colorado.
Var. lonchocarpa [*D. nivalis* var. *elongata* Wats.; *D. lonchocarpa* var. *dasycarpa* Schulz; *D. lonchocarpa* var. *vestita* Schulz; *D. lonchocarpa* var. *semitonsa* St. John]. Alpine sites from 3050 to 3900 m elevation in the Wasatch, Uinta, and La Sal mts. of Cache, Duchesne, Grand, Salt Lake, San Juan, and Summit counties; Alaska south to Oregon and Wyoming.

***Draba maguirei* C. L. Hitchc.** Maguire Draba. [*D. maguirei* var. *burkei* C. L. Hitchc., type from Wellsville Mts., Box Elder County]. Perennial, caespitose, with substoloniferous branches, scapose; scapes 2–20 cm tall, glabrous or with a few forked hairs near base; leaves 3–15 mm long, 1–4 mm wide, oblong-oblanceolate to obovate-oblanceolate, the surfaces glabrous or nearly so, ciliate with simple, forked, or 4-rayed, shortly stalked hairs; racemes few- to several-flowered, elongating in fruit; pedicels 2–10 (15) mm long, ascending, glabrous; sepals 2–3 mm long, yellowish, glabrous; petals 4.5–6 mm long, yellowish, spatulate, rounded; fruit 4–9 mm long, 2–3.5 mm wide, ovate to lanceolate, oblique, glabrous or scaberulous; style 1–1.5 mm long; seeds 2–8. Talus slopes and rocky outcrops in Douglas fir and mixed conifer communities at 1830 to 2930 m in Box Elder, Cache (type from Mt. Naomi), and Weber counties; endemic; 5 (0).

***Draba nemorosa* L.** Woods Draba. [*D. dictyota* Greene]. Annual, from a slender taproot; stems simple or branched, 0.5–2.5 dm tall, pubescent with mixed forked and stellate hairs, or less commonly with some simple ones, or even glabrate; leaves 3–30 mm long, 2–8 mm wide, oblanceolate to lanceolate, ovate or oblong, entire or toothed, pubescent with branched or simple hairs; racemes few- to many-flowered, much elongating in fruit; pedicels 5–26 mm long, spreading-ascending, glabrous; sepals 1–1.5 mm long, green to yellowish or suffused purple, pilose to glabrous; petals 1.2–4 mm long, yellow to white; silicles 4–10 mm long, 1.5–3 mm wide, oblong to oblanceolate or elliptic, plane, glabrous; style obsolete; seeds 25 or more; n = 8. Bunchgrass, sagebrush, mountain brush, ponderosa pine, and white fir communities, mainly in exposed, disturbed sites, at 1280 to 2475 m elevation in Box Elder, Cache, Carbon, Daggett, Davis, Duchesne, Grand, Juab, Millard, Morgan, Salt Lake, Summit, Uintah, Utah, and Weber counties; common throughout much of North America and Eurasia; 33 (iii).

***Draba oligosperma* Hook.** Doublecomb Draba. Perennial, caespitose, the caudex much branched, scapose, the scapes 1–10 cm tall, pubescent throughout with pectinate or stellate hairs or glabrous, at least below; leaves 3–12 mm long, 1–2 mm wide, linear to spatulate or oblong, surfaces (one or both) pubescent with sessile, appressed, doubly pectinate hairs, commonly ciliate with at least some pectinately branched hairs; racemes 2- to 15-flowered, only moderately elongating in fruit; pedicels 1–10 mm long, ascending, glabrous or pubescent; sepals 1.5–2.5 mm long, yellowish, pubescent; petals 3–5 mm long, yellow or white, obovate, rounded to emarginate; silicles 3–8 mm long, 2–4 mm wide, ovate to oval or oblong, plane, glabrous to pubescent; style 0.1–1.2 mm long; seeds 2–10. Two rather weakly separable but geographically correlated varieties are present.

1. Petals yellow; pubescence of silicles, when present, of simple or forked hairs only; plants widespread ....
   .... *D. oligosperma* var. *oligosperma*
— Petals evidently white; pubescence of silicles at least in part of doubly pectinate hairs; plants of Daggett and Uintah counties .... *D. oligosperma* var. *juniperina*

Var. juniperina (Dorn) Welsh [*D. juniperina* Dorn; *D. oligosperma* var. *pectinipila* authors, not *D. pectinipila* Rollins]. Pinyon-juniper, sagebrush, mountain brush, ponderosa pine, and Douglas fir communities at 1920 to 3205 m in Daggett and Uintah counties; Wyoming; a Bridger Basin—Uinta Mt. endemic; 17 (v).

Var. oligosperma [*D. oligosperma* var. *andina* (Nutt.) ex T. & G.; *D. andina* (Nutt.) A. Nels.; *D. saximontana* A. Nels.; *D. oligosperma* var. *microcarpa* Blank.; *D. oligosperma* var. *saximontana* (A. Nels.) Schulz; *D. oligosperma* var. *leiocarpa* Schulz]. Sagebrush, white fir, spruce-fir, and alpine tundra communities at 2135 to 3815 m in Box Elder, Cache, Daggett, Duchesne, Juab, Rich, Salt Lake, Sanpete, Summit, and Utah counties; British Columbia and Alberta south to California, Nevada, and Colorado; 25 (iii).

***Draba rectifructa* C. L. Hitchc.** Mountain Draba. [*D. montana* Wats., not Bergeret]. Annual, from a slender taproot, the stems simple or branched, 1–22 cm tall, pubescent throughout, although sparsely so above, in some with branched and/or simple hairs; leaves 5–30 mm long, 1–7 mm wide, lanceolate to oblanceolate, entire, pubescent with branched and simple hairs intermixed; racemes several- to many-flowered, much elongating in fruit; pedicels 2–6 mm long, spreading to curved-ascending, pubescent; sepals 1.2–1.7 mm long, greenish, pubescent; petals 2–4 mm long, yellow, narrowly spatulate, usually emarginate; silicles (4) 6–10 mm long, 2–2.5 mm wide, obliquely oblong, pubescent; style obsolete; seeds 40 or more. Sagebrush, ponderosa pine, aspen, lodgepole pine, spruce-fir, and alpine meadow communities at 2135 to 3205 m in Carbon, Duchesne, Garfield, Grand, Juab, Piute, Salt Lake, Salt Lake, Sevier, Summit, and Wasatch counties; Colorado, New Mexico, and Arizona; 21 (ii).

***Draba reptans* (Lam.) Fern.** Dwarf Draba. Annual, from a slender taproot, 2–10 (20) cm tall, with the very short leafy stem simple or sometimes branched, pubescent with simple to stellate hairs below, glabrous above; basal leaves 3–18 mm long, 1–10 mm wide, spatulate to ovate or obovate, usually entire, the surfaces pubescent with branched or forked hairs or upper surface with simple hairs only; cauline leaves few, usually reduced, pubescent like basal ones, or upper surface with simple hairs; racemes (1) several- to many-flowered, compact in fruit; pedicels 1–6 mm long, ascending, glabrous; sepals 1.5–2.5 mm long, greenish or yellowish, usually pubescent; petals 3–5 mm long, white, ovate, rounded, sometimes smaller or lacking in cleistogamous flowers; silicles 5–20 mm long, 1–2 mm wide, oblong, nearly erect, pubescent or glabrous; style up to 0.15 mm long; seeds 15 or more. Shadscale, other salt desert shrub, bunchgrass, sagebrush, mountain brush, and pinyon-juniper communities at 1280 to 2200 m in Beaver, Box Elder, Cache, Daggett, Grand, Millard, Salt Lake, San Juan, Tooele, Uintah, Utah, and Washington counties; widespread in the U. S.; 29 (ii). Two completely confluent and sympatric varieties have been recognized; those with leaves pubescent primarily with simple hairs are **var. reptans** [*Arabis reptans* Lam.; *D. caroliniana* Walter; *D. micrantha* Nutt. ex T. & G.; *D. coloradoensis* Rydb.; *D. reptans* var. *micrantha* (Nutt.) Fern.], and those with leaf hairs predominantly forked or branched are **var. stellifera (Schulz) C. L. Hitchc.** [*D. caroliniana* f. *stellifera* Schulz; *D. caroliniana* ssp. *stellifera* (Schulz) Payson & St. John]. They do not seem to represent taxa. *Draba reptans* is similar to and not always distinct from *D. cuneifolia* (q.v.).

***Draba sobolifera* Rydb.** Creeping Draba. [*D. uncinalis* Rydb., type from Tate Mine, Tushar Mts., Piute County]. *D. sobolifera* var. *uncinalis* (Rydb.) Schulz; *D. ramulosa* Rollins, type from Mt. Belknap]. Perennial, caespitose, the caudex with elongate, soboliferous branches, scapose or with some leaves; scapes 1–8 cm tall, pubescent with intermixed stellate, branched, and simple hairs; leaves 5–20 mm long, 2–5 mm wide, obovate to oblanceolate, pubescent with stalked stellate or 4-rayed hairs at least beneath, and usually ciliate with simple hairs at base; racemes (2) 5- to 20-flowered, compact to elongating in fruit; pedicels 3–8 mm long, stellate or with branched hairs; sepals 1.8–2.5 (3) mm long; petals 4–5 mm long, yellow, obovate; silicles 3–8 mm long, 2.5–4 mm wide, ovate to elliptic, pubescent to glabrous; style 0.4–1 mm long; seeds 4–12. Alpine tundra and spruce-fir communities at 3050 to 3695 m in the Tushar Mts., Beaver and Piute counties; endemic; 14 (vi). The types of both *D. sobolifera* and *D. uncinalis* were collected by M. E. Jones in the Delano Peak (Tate Mine) area west of Marysvale, Piute County. It is remarkable that three names are based on specimens from the Tushar Mts. The species is allied to *D. ventosa* Gray, q.v.

***Draba spectabilis* Greene** Splendid Draba. [*D. spectabilis* var. *glabrescens* Schulz, type from the La Sal Mts.]. Perennial, not caespitose, the caudex branched; stems mostly 1–4 dm tall, usually simple, pubescent with simple or forked hairs; basal leaves (0.5) 1–4 cm long, 2–10 mm wide, obovate to spatulate, subentire to denticulate, green, surfaces with subsessile 4-rayed or forked hairs, infrequently with the upper surface glabrous; cauline leaves mostly 3–15, ovate to lanceolate, subentire to sharply toothed, 5–20 mm long, 2–15 mm wide, pubescent like basal ones; racemes several- to many-flowered, much elongating in fruit; pedicels 5–15 mm long or more, ascending, glabrous or hairy; sepals 2–3.5 mm long, yellowish, pubescent; petals 4.5–7 mm long, yellow but fading white, elliptic, rounded; silicles 5–14 mm long, 2–3 mm wide, lanceolate to ovate, plane, glabrous or pubescent; style 0.8–2.5 mm long; seeds 10–20. Mountain brush, white fir, ponderosa pine, spruce-fir, and alpine tundra communities at 3050 to 3660 m in the La Sal and Abajo mts., Grand and San Juan counties; Colorado, New Mexico, Arizona; 13 (0). Our material is assignable to **var. spectabilis**.

***Draba stenoloba* Ledeb.** Alaska Draba. [*D. nemorosa* var. *stenoloba* (Ledeb.) Jones; *D. nitida* Greene; *D. deflexa* Greene; *D. nitida* var. *nana* Schulz; *D. nitida* var. *praelonga* Schulz]. Plants (annual?) biennial or short-lived perennial, from a taproot, the caudex more or less developed; stems 3–25 cm tall, glabrous or sometimes hirsute below; basal leaves 0.3–4 cm long, 2–8 mm wide, entire to denticulate, pubescent with simple to forked or branched hairs, or with one or both surfaces glabrous on some leaves; cauline leaves (0) 1–8, ovate to lanceolate or elliptic, entire or denticulate, similar to the basal ones in all respects; racemes several- to many-flowered, elongating in fruit; pedicels 1–14 mm long, ascending, glabrous; sepals 1–2.2 mm long, greenish, glabrous; petals 2–4.5 mm long, yellow to cream or fading white, spatulate, rounded to emarginate; fruit 6–18 mm long, 1.5–2.2 mm wide, linear to oblong or elliptic, usually glabrous; style up to 0.2 mm long; seeds 16 or more; n = 20. Aspen, lodgepole pine, spruce-fir, grass-forb, and alpine tundra communities at 1830 to 3510 m in Beaver, Box Elder, Cache, Carbon, Daggett, Duchesne, Emery, Garfield, Iron, Juab, Piute, Rich, Salt Lake, Sanpete, Sevier, Summit, Tooele, Uintah, Utah, Wasatch, Washington, and Wayne counties; Alaska and Yukon south to California, Nevada, and Colorado; 77 (xv).

***Draba subalpina* Goodman & Hitchc.** Breaks Draba. [*D. oriebata* authors, not Macbr. & Payson]. Perennial, caespitose, the caudex simple or branched, clothed with marcescent leaves, scapose; scapes 1–12 cm tall, rarely with one cauline leaf, glabrous throughout or pubescent with simple or forked hairs at least near base; leaves 3–18 mm long, 1–4 mm wide, oblong to spatulate, surfaces glabrous or sparingly hirsute, ciliate with coarse simple or less commonly forked or branched hairs; racemes few- to many-flowered, only moderately elongating in fruit; pedicels (1) 2–10 mm long, usually purplish, glabrous; petals 4–5 mm long, white, cuneate-spatulate, emarginate; silicles 3–8 mm long, 2–4 mm wide, ovate to elliptic, plane or more or less contorted, glabrous; style 0.6–1 mm long; seeds 6–12. Ponderosa pine, pinyon-juniper, Douglas fir, bristlecone pine, and spruce-fir communities, on the Pink and White Limestone members of the Wasatch Formation, from 2130 to 3295 m in Garfield, Iron (type from Cedar Breaks), Kane, and Millard counties; endemic; 42 (vi).

***Draba ventosa* Gray** Tundra Draba. Perennial, caespitose, the caudex usually branched, more or less clothed with marcescent leaves, scapose; scapes 2–6 cm tall, pubescent with simple and forked or sometimes stellate hairs; leaves 5–12 mm long, 2–4 mm wide, elliptic to lanceolate, surfaces pubescent with simple, forked, or branched to stellate hairs; racemes 3- to many-flowered, little elongating in fruit; pedicels mostly 4–8 mm long, ascending, densely pilose to stellate; sepals 2–2.5 mm long, greenish or yellowish, pilose; petals 4–5 mm long, yellow, obovate; silicles 5–8 mm long, 3.5–5.5 mm wide, oval to ovate, plane, densely hairy; style 0.6–1.2 mm long; seeds 10–16. Alpine tundra and talus at 3050 to 3800 m in the Uinta Mts., Duchesne and Summit counties; Wyoming; 2 (0). Utah plants are assignable to var. *ventosa*.

Draba verna L. Spring Draba. Diminutive annual from a slender taproot, scapose; scapes 2–5 (12) cm tall, glabrous throughout or pubescent near the base only; leaves 3–10 (20) mm long, 0.8–3 mm wide, spatulate to oblanceolate, entire or toothed, pubescent with branched hairs; racemes few- to many-flowered, elongating in fruit; pedicels 2–12 mm long, ascending, glabrous; sepals 0.6–1.1 mm long, greenish or tinged purplish, glabrous or pubescent; petals 1.5–2.5 mm long, white, deeply bilobed; silicles 2–10 mm long, 1–4 mm wide, elliptic to obovate, glabrous; style up to 0.2 mm long; n = 7, 12, 15, 16, 17, 18, 19, 20, 26, 27, 28, 29, 30, 31, 32, 47. Dry open, commonly disturbed, sites at 1375 to 2440 m in Cache, Davis, Duchesne, Juab, Rich, Salt Lake, Utah, Wasatch, and Weber counties; widespread in North America; Asia; 15 (0). Our material cannot be segregated into the two variants reported for North America.

Erysimum L.

Plants pubescent with 2– or 3 (4) –rayed hairs, annual to perennial, from taproots; leaves alternate or basal and still alternate, simple, entire to toothed, not auriculate; flowers in ebracteate racemes; sepals 4, deciduous; petals 4, yellow to orange or burnt-orange to purple; stamens 6, the filaments lacking glandular processes; style prominent, short to elongate; stigma bilobed; fruit a silique, many times longer than broad, compressed parallel to the partition or subterete; valves 1– to several-nerved; seeds uniseriate, many per locule.

Rossbach, G. B. 1958. The genus *Erysimum* (Cruciferae) in North America north of Mexico—A key to species and varieties. Madrono 14: 261–267.

1. Petals (10) 12–20 mm long or more; style mostly 1.5–3 mm long; plants biennial or perennial, widespread in indigenous plant communities .... *E. asperum*
— Petals 3.5–11 mm long; style usually less than 1.5 mm long (longer in *E. repandum*); plants annual or biennial to short-lived perennial, restricted weedy sites or less commonly in indigenous plant communities 2
2(1). Petals 3.5–5 mm long; siliques 12–27 mm long; plants uncommon, annual weeds of moist sites .... *E. cheiranthoides*
— Petals mostly 5–11 mm long; siliques (15) 25–100 mm long; plants common, annual, biennial or perennial, of various habitats 3
3(2). Pedicels usually more than 5 mm long, more slender than fruit; siliques ascending to erect, less than 50 mm long; plants indigenous biennial or short-lived perennial .... *E. inconspicuum*
— Pedicels 2–5 mm long, almost or quite as thick as fruit; siliques spreading to curved-ascending, at least some often more than 50 mm long; plants adventive, annual weeds .... *E. repandum*

Erysimum asperum (Nutt.) DC. Wallflower; Pretty Wallflower. [*Cheiranthus asper* Nutt.; *Cheirinia brachycarpa* Rydb., type from the Abajo Mts.; *Cheiranthus capitatus* Dougl. ex Hook.; *E. elatum* Nutt. ex T. & G.; *E. capitatum* (Dougl.) Greene; *C. elatus* (Nutt.) Greene; *C. asperrimus* Greene; *C. argillosus* Rydb.; *C. bakeri* Greene; *Cheirinia elata* (Nutt.) Rydb.; *E. asperrimum* (Greene) Rydb.; *E. oblanceolatum* Rydb.; *E. bakeri* (Greene) Rydb.; *E. aridum* A. Nels.; *E. capitatum* var. *argillosum* (Greene) R. J. Davis; *E. asperum* var. *purshii* Durand, type from near Great Salt Lake; *E. asperum* var. *amoenum* (Greene) Reveal; *Chieranthus nivalis* var. *amoenus* Greene; *Cheirinia amoena* (Greene) Rydb.; *E. wheeleri* Rothr.; *Cheirinia wheeleri* (Rothr.) Rydb.]. Biennial or short-lived perennial, with simple or less commonly branched caudex; stems 1.2–8.5 (10) dm tall or more; basal leaves 2–10 (12) cm long, 2–14 mm wide, sublinear to elliptic or spatulate, entire or denticulate, grayish to green, pubescent with malpighian or Y-shaped appressed hairs; cauline leaves 1.1–10.4 cm long, 1–15 mm wide, much reduced to little if at all reduced upward, variously shaped, entire or toothed, pubescent like basal ones; racemes much elongating in fruit; pedicels 3–17 mm long, spreading-ascending to ascending, usually more slender than the fruit; sepals 7.5–14 mm long, yellowish or purplish; petals 12–28 mm long, yellow to yellow orange or burnt-orange; siliques (1.7) 2–11.5 cm long, 1–2.5 mm thick, subquadrangular to somewhat flattened, ascending to erect or less commonly spreading-ascending; style 1–4 (5) mm long; seeds wingless or winged only near tip, 1.5–2.3 mm long; n = 18. Warm desert shrub, cool desert shrub, mountain brush, pinyon-juniper, ponderosa pine, aspen, Douglas fir, and spruce-fir to alpine tundra communities, at 760 to 3800 m throughout Utah; Yukon Territory south to California and Arizona, and eastward to Oklahoma, Kansas, and Minnesota; 295 (xxxvii). This pretty wallflower demonstrates greater ecological latitude than any other Utah plant. Our tremendously variable material does not lend itself to infraspecific segregation.

Erysimum cheiranthoides L. Treacle. [*Cheirinia cheiranthoides* (L.) Link; *Cheiranthus cheiranthoides* (L.) Heller]. Annual, with stems simple or branched, 2–12 dm tall; leaves 2–8 cm long, 2–15 mm wide, linear to oblong, lanceolate, or oblanceolate, entire or denticulate, green, pubescent with malpighian or Y-shaped hairs; racemes much elongating in fruit; pedicels 4–15 mm long, spreading-ascending, very slender; sepals 2–3 mm long, yellowish or greenish; petals 3–6 mm long, pale yellow; siliques 12–27 mm long, ca 1 mm wide, subterete, ascending to erect; style 0.8–1 mm long; seeds 1–1.2 mm long, not winged; n = 8. Moist places, in meadows and along roadsides, mainly in sagebrush, grass, aspen, and spruce-fir communities, at 1430 to 2625 m in Cache, Davis, Duchesne, Juab, Rich, Salt Lake, Utah, Wasatch, and Weber counties; widespread in North America; adventive from Eurasia; 8 (0).

Erysimum inconspicuum (Wats.) MacMillan Lesser Wallflower. [*E. asperum* var. *inconspicuum* Wats.; *E. parviflorum* Nutt. ex T. & G., not Pers.; *E. syrticolum* Sheldon; *Cheiranthus inconspicuus* (Wats.) Greene; *Cheirinia inconspicua* (Wats.) Rydb.; *C. syrticola* (Sheldon) Rydb.]. Biennial or short-lived perennial with usually unbranched caudex; stems mostly 2–10 dm tall, usually simple; leaves 1.5–8 cm long, 2–8 cm wide, linear to oblong, elliptic, lanceolate or oblanceolate, pubescent with malpighian or Y-shaped hairs; racemes elongating in fruit; pedicels 3–8 mm long, ascending; sepals 4–7 mm long, greenish or purplish; petals (6) 7–10 (11) mm long, pale to bright yellow; siliques 15–50 mm long, 1–2 mm thick, quadrangular, erect or ascending; style 1–1.5 mm long; seeds ca 1.5 mm long, not winged; n = 8. Sagebrush, aspen, and spruce-fir communities at 1370 to 2745 in Duchesne, Garfield, Sevier, Summit, Tooele, and, Wasatch counties; Alaska and Yukon south to Oregon, and Colorado, and east to central Canada and the north-central U. S.; 20 (0). This species resembles *E. asperum* in general habit, but is more restricted in its distribution in Utah.

Erysimum repandum L. Spreading Wallflower. [*Cheirinia repanda* (L.) Link]. Annual; stems 0.8–3 (5) dm tall, simple to much-branched; leaves 0.7–11 (15) cm long, 1–8 (12) mm wide, oblong, oblanceolate to linear, pubescent with malpighian and Y-shaped hairs; pedicels 2–5 mm long, spreading, almost or quite as thick as the fruit; sepals 3.5–5.8 mm long, yellowish or greenish; petals 5.2–8 mm long, yellow; siliques 26–85 mm long, 0.9–1.5 mm thick, quadrangular, spreading to ascending, rarely descending; style 1–3 mm long; seeds ca 1 mm long, not winged; n = 8. Disturbed sites, mainly in bunchgrass, sagebrush, and pinyon-juniper communities, at 1310 to 2075 m in Beaver, Cache, Davis, Grand, Juab, Millard, Salt Lake, Sanpete, Sevier, Summit, Utah, and Washington counties; widespread in North America; adventive from Europe; 32 (v).

Euclidium R. Br.

Plants pubescent with forked hairs, annual, from taproots; leaves alternate, simple, entire to remotely serrulate, petiolate to subsessile, not auriculate; flowers solitary and axillary or subaxillary, or borne in elongate racemes; pedicels sometimes subtended by bracts; sepals 4, caducous; petals 4, white, minute; stamens 6; style very short atop beak of fruit, the stigma bilobed; fruit a silicle, bilocular, tardily dehiscent; seed 1 per locule.

Euclidium syriacum (L.) R. Br. Syria-weed. [*Anastatica syriaca* L.]. Plants 0.4–5 dm tall, simple to much-branched, pubescent with forked hairs; leaves mainly cauline, 0.7–6.5 cm long, 2–17 mm wide, oblanceolate to elliptic or lanceolate, pubescent; pedicels 0.5–1 mm long, ascending to erect, pubescent; sepals 0.7–0.9 mm long, tinged with purple; petals 0.8–1.1 mm long, white, spatulate, erect; silicles (including beak) 2.8–4 mm long, the body 1.2–2 mm wide, pubescent with simple or forked hairs, beak 1.2–1.5 mm long; style short or obsolete; n = 7. Roadsides, vacant lots, and dry foothills in Box Elder, Cache, Juab, Millard, Salt Lake, Sanpete, Summit, and Utah counties; Idaho, Washington; adventive from Europe; 14 (iii).

Halimolobos Tausch

Plants pubescent with simple, forked, and branched hairs, (annual) biennial or perennial, arising from a taproot; leaves alternate or basal and still alternate, simple, dentate to subentire, tapering basally or the cauline sessile and auriculate; flowers in ebracteate racemes; pedicels ascending; sepals 4; petals 4, white; stamens 6; style prominent; stigma small, entire; fruit a sessile, slender silique, many times longer than broad, terete to quadrangular, the valves strongly 1–nerved; seeds biseriate, numerous.

Rollins, R. C. 1943. Generic revisions in the Cruciferae: *Halimolobos*. Contr. Dudley Herb. 2: 241–265.
Rollins, R. C. 1952. A note on *Halimolobos*. Rhodora 54: 161–163.

Halimolobos virgata (Nutt.) Schulz Strictweed. [*Sismybrium virgatum* Nutt. ex T. & G.; *Hesperis virgata* (Nutt.) Kuntze; *Stenophragma virgatum* (Nutt.) Greene; *Arabis brebneriana* A. Nels.; *Pilosella virgata* (Nutt.) Rydb.; *P. stenocarpa* Rydb.; *Arabidopsis virgata* (Nutt.) Rydb.; *Arabidopsis stenocarpa* (Rydb.) Rydb.]. Biennial but occasionally flowering the first year; stems 1–3.5 dm tall, simple or branched, pubescent with mixed simple, forked, and branched hairs at least below; basal leaves 3–6 cm long, 5–18 mm wide, oblanceolate to lanceolate, denticulate to dentate, rarely entire; cauline leaves several, reduced upwardly, at least the uppermost both sessile and auriculate; pedicels 7–11 mm long, ascending, glabrous or puberulent; sepals 2.5–3 mm long, greenish, pubescent; petals 4–4.5 mm long, white, the veins often suffused with pink or purple, usually erect; siliques 15–40 mm long, mostly 1–1.5 mm wide, subquadrangular, erect; valves glabrous, strongly nerved; style 0.2–0.5 mm long; seeds irregularly biserrate. Sagebrush, meadows, and aspen communities at 2135 to 2745 m in Daggett, Millard, Wasatch, and Weber (?) counties; Colorado, Nevada, Wyoming, Idaho, and northward to Alberta and Saskatchewan; 4 (0). The number of generic segregates listed above are an indication of the difficulties involved in placing this unusual species in a genus. It simulates an *Arabis*, from which it is outwardly separable only with difficulty, but on technical grounds the plant is probably more closely allied to *Sisymybrium* in a broad sense. Our Utah material, and that from southwestern Wyoming, differs slightly from that found elsewhere and may deserve varietal recognition.

Hesperis L.

Plants pubescent with simple and forked hairs, perennial, from taproots; leaves alternate, simple, mainly cauline, petiolate to subsessile, not auriculate; flowers in ebracteate racemes; sepals 4, deciduous; petals 4, pink to lavender or white; stamens 6, the filaments lacking glandular processes; style obsolete; stigma bilobed, massive; fruit a silique, many times longer than broad, subterete, tardily dehiscent; valves 1 (3) –nerved; seeds uniseriate.

Hesperis matronalis L. Dame's Violet; Sweet Rocket. Herbs with one to several stems, these simple or branched, mostly 5–12 dm tall or more, pubescent with mixed forked and simple hairs; leaves 2–15 (20) cm long, 0.6–3.5 (4) cm wide, ovate-lanceolate to elliptic or lanceolate, sinuate-dentate to serrate, pubescent; pedicels 8–21 mm long, ascending to spreading, pubescent; sepals 6.5–8 mm long, often suffused with red or purple, pubescent; petals 15–25 mm long, pink to lavender or white, obovate, spreading; siliques 3–10 cm long, 1–2 mm wide, subterete, erect or ascending, puberulent; style obsolete; stigma deeply bilobed; seeds numerous in each locule; 2n = 7, 12, 14. Cultivated ornamental, persisting and escaping, now widely established in cultivated lands especially along irrigation canals at 1370 to 1895 m in Cache, Uintah, Utah, and Summit counties; introduced from Europe; 4 (i).

Hutchinsia R. Br.

Glabrous, annual or winter annual, from taproots; leaves alternate, simple, not auriculate; flowers borne in ebracteate racemes; sepals 4, deciduous; petals 4, white; stamens 6, the filaments lacking glandular processes; style obsolete or very short; stigma capitate, entire; fruit a silicle, only somewhat longer than broad, strongly compressed at right angles to septum; valves reticulately veined; seeds several.

Hutchinsia procumbens (L.) Desv. Slenderweed. [*Lepidium procumbens* L.; *Hymenolobus divaricatus* Nutt. ex T. & G.; *Hymenolobus erectus* Nutt. ex T. & G.]. Slender, diminutive annuals; stems glabrous, simple or more commonly branched, 0.5–3 dm long, erect or prostrate; leaves basal and cauline, 0.5–3 cm long, 1–13 mm wide, ovate to lanceolate, oblanceolate, or nearly linear, entire to pinnatifid, petiolate to subsessile; pedicels 3–8 mm long, ascending to spreading-ascending, glabrous; sepals 0.7–1.1 mm long, greenish or purplish, glabrous; petals 0.8–1.3 mm long, white, spatulate, rounded to retuse; silicles 2.4–4.2 mm long, 1.5–2 mm wide, elliptic to obovate, truncate, rounded, or somewhat emarginate, glabrous; style to 0.2 mm long; seeds several per locule; n = 6, 12. Moist to dry sites, hanging gardens, drainage banks, sidewalks, roadsides, seeps, playas, and peat bogs from 850 to 2135 m in Beaver, Box Elder, Cache, Daggett, Davis, Garfield, Kane, Millard, Salt Lake, San Juan, Sanpete, Tooele, Uintah, Utah, Washington, and Weber counties; widely distributed in North America; Eurasia; 30 (iii).

Isatis L.

Plants pubescent with long simple hairs at least below, biennial or short-lived perennial, from strong taproots; leaves simple, alternate, the basal petiolate, the cauline hastately-auriculate; flowers racemose or paniculate; pedicels not subtended by bracts; sepals 4, deciduous; petals 4, yellow; stamens 6, the filaments lacking glandular processes; style obsolete, bilobed; stigma sessile; fruit a silicle, samaroid, indehiscent, flattened at right angles to the plane of the partition (which is lacking), more or less reticulate; seed solitary.

Isatis tinctoria L. Dyer's Woad. Stems erect, 3.5–10 dm tall or more, glabrous throughout or hirsute with long simple hairs at base; basal leaves 3.5–15 (18) cm long, 0.8–4 cm wide, oblanceolate to elliptic, subentire to crenulate, ciliate to pilose with simple hairs; cauline leaves gradually reduced upwards, lanceolate to elliptic, entire, hastately-auriculate, glabrous or pilose on veins beneath and often ciliate; pedicels 4.5–9 mm long, reflexed, glabrous; sepals 1.8–2.3 mm long, yellowish, glabrous; petals 3–4.2 mm long, yellow, spatulate, rounded; silicles mostly 10–18 mm long, 4–7 mm wide, cuneate-oblong to oblanceolate, more or less truncate-rounded apically, glabrous; stigmas sessile; n = 13, 14. Roadsides, abandoned fields, and dry foothills in bunchgrass, sagebrush, and mountain brush communities, at 1310 to 2135 m in Box Elder, Cache, Daggett, Davis, Duchesne, Juab, Millard, Morgan, Salt Lake, Sanpete, Summit, Uintah, Utah, and Weber counties; widely established in the U.S.; adventive from Europe; 26 (ii). This plant is the source of a blue dye and was widely cultivated in the recent past. It seems to be spreading outwardly from its initial points of establishment in Box Elder County, where it has been established as a weed since at least 1947.

Lepidium L.

Plants glabrous or with simple hairs, annual, biennial or perennial, from taproots; leaves alternate or basal and still alternate, simple, entire or variously toothed or bi- or tripinnatifid, petiolate or sessile, auriculate in some; flowers in ebracteate racemes; sepals 4, caducous or persistent; petals 4, yellow or white, infrequently lacking; stamens 6, rarely 2 or 4, the filaments lacking glandular processes; style obsolete or well developed; stigma capitate; fruit a silicle, usually less than twice longer than broad, compressed at right angles to septum, dehiscent; valves more or less reticulately veined; seeds 1 per locule. (Note: Measurement of silicle length includes the style.)

Hitchcock, C. L. 1936. The genus *Lepidium* in the United States. Madrono 3: 265–320.

1.	Cauline leaves perfoliate-clasping or auriculate	2
—	Cauline leaves petiolate to sessile but not auriculate or clasping	3
2(1).	Petals 2–2.5 mm long, white; cauline leaves lanceolate to lance-oblong, merely auriculate .. *L. campestre*	
—	Petals 1–2 mm long, yellow; cauline leaves oval to ovate, pseudoperfoliate .... *L. perfoliatum*	
3(1).	Plants arising from thickened, well-developed caudices; leaves all entire; stems seldom more than 2.5 dm tall; plants rare, south-central to northern Utah ..	4
—	Plants from root crowns, or from simple or branched caudices, these seldom thickened; leaves entire, toothed or pinnatifid; stems short to long; plants widespread	5
4(3).	Leaves linear; petals more than 3 mm long; plants of xeric shales, known only from Duchesne County .... *L. barnebyanum*	
—	Leaves oblanceolate to elliptic; petals less than 3 mm long; plants of moist meadows from Beaver County northward but not of Duchesne County .... *L. integrifolium*	
5(3).	Plants 0.8–1.5 m tall or more; leaves entire or merely crenate-serrate; silicles 1.5–2 mm long; known mainly from riparian and palustine habitats .... *L. latifolium*	
—	Plants mainly less than 0.8 m tall, or if taller, the leaves typically pinnately lobed to pinnatifid; silicles mainly more than 2 mm long; plants of various habitats, sometimes riparian or palustrine	6
6(5).	Styles 0.3–1 mm long; plants perennial or biennial ...	7
—	Styles obsolete or only up to 0.3 mm long; plants annual, rarely if ever biennial	9
7(6).	Plants pulvinate-caespitose, 1–5 cm tall, from a branching caudex clothed with marcescent leaf bases; leaves grayish puberulent, the basal ones 3- to 5-lobed; plants known from the San Francisco Mts. .... *L. ostleri*	
—	Plants various, seldom if ever pulvinate-caespitose, mainly over 5 cm tall, from a taproot or branching caudex, if the latter, not especially clothed with leaf bases; leaves green to grayish puberulent, the basal ones often more than 5-lobed; plants of various distribution	8
8(7).	Stems woody well above the base; petals mostly 3 mm long or more; silicles 4.5–7.5 mm long, obovate; plants of Washington County .... *L. fremontii*	
—	Stems slightly if at all woody above the base; petals mostly less than 3 mm long (but sometimes longer); silicles 2–4 mm long, ovate; plants almost cosmopolitan in Utah .... *L. montanum*	
9(6).	Sepals persistent, enclosing the mature silicle, these plainly reticulate; pedicels narrowly wing-margined; plants evidently rare .... *L. strictum*	
—	Sepals deciduous at or shortly following anthesis; silicles plainly to slightly reticulate; pedicels terete to strongly flattened; plants at least locally common ...	10

10(9). Fruits emarginate and with prominent, acute, and divergent toothlike apices .... *L. dictyotum*
— Fruits emarginate but lobes on each side of sinus neither acute nor divergent 11

11(10). Pedicels much flattened, about twice broader than thick 12
— Pedicels not strongly flattened, not twice as broad as thick 13

12(11). Plants short-hirsute to hispid hairy; silicles pubescent .... *L. lasiocarpum*
— Plants softly hairy; silicles glabrous or pubescent .. .. *L. densiflorum*

13(11). Petals lacking or, if present, generally shorter than the sepals; silicles oblong-obovate to obovate .... *L. densiflorum*
— Petals present and usually surpassing the sepals; silicles elliptic-rotund to orbicular .... *L. virginicum*

Lepidium barnebyanum Reveal Ridgecress. [*L. montanum* ssp. *demissum* C. L. Hitchc. (type from Indian Canyon), not *L. demissum* C. L. Hitchc.]. Perennial, densely pulvinate-caespitose, from a thickened branched caudex, this clothed with marcescent leaf bases; stems erect, 7–12 cm tall, subglabrous or minutely recurved scaberulous, simple or branched from the upper axils; basal leaves 1–4 cm long, 1–2 (3) mm wide, somewhat flattened and minutely wing-margined, scaberulous; sepals 2–2.3 mm long, greenish, glabrous or puberulent; petals 3.5–4.2 mm long, white, obovate, spreading; stamens 6; silicles 3–6.2 mm long, 3–4 mm wide, lanceolate to elliptic, glabrous, plane, wingless; style 0.5–1.2 mm long. White shale outcrops, Uinta Formation, in a pinyon-juniper community at 1890 to 1985 m, growing with mound-forming species, mainly on ridge crests, Indian Canyon, Duchesne County; endemic; 10 (vii).

Lepidium campestre (L.) R. Br. Fieldcress. [*Thlaspi campestre* L.]. Annual, lacking a caudex; stems 1.5–6 dm tall, hirtellous throughout with simple hairs; basal leaves 3–12 cm long, 8–15 mm wide, oblanceolate, entire or variously lobed; cauline leaves numerous, reduced upwards, becoming sessile and auriculate, usually denticulate; pedicels mostly 4–7 mm long, spreading, slightly flattened, hirtellous; sepals 1.3–2 mm long, greenish or varously tinged, hirtellous; petals 1.7–2.5 mm long, white, spatulate, ascending; stamens 6; silicles 5–6 mm long, oblong-ovate, glabrous or puberulent, concave, wingless, slightly emarginate; style 0.2–0.6 mm long; 2n = 16, 32. Roadsides and other disturbed sites from 1370 to 2410 m in Cache, Salt Lake, San Juan, Sanpete, Summit, Uintah, and Utah counties; widely established in North America; Asia; 11 (0).

Lepidium densiflorum Schrader Densecress. Annual, lacking a caudex; stems 0.3–5 dm tall or more, densely finely hairy to puberulent throughout; basal leaves 1.2–11 cm long, 3–22 mm wide, oblanceolate, entire and more commonly pinnately lobed; cauline leaves several to many, reduced upwards, petiolate to sessile, not auriculate, lobed, toothed, or entire; pedicels 1.5–3 mm long, spreading to ascending, subterete to conspicuously flattened, glabrous to puberulent; sepals 0.6–1.1 mm long, often purplish, glabrous; petals 0.7–1 mm long, white, narrowly oblong, sometimes lacking; stamen 1; silicles 2.5–3.5 mm long and about as wide, elliptic to oval or obovoid, glabrous or rarely pubescent, shallowly notched apically, the teeth rounded; style lacking; n = 16. Our material is separable into three more or less distinctive variants.

1. Silicles pubescent; pedicels usually somewhat flattened, especially on lower side; plants evidently rare .... *L. densiflorum* var. *pubicarpum*
— Silicles glabrous; pedicels definitely flattened or else subterete; plants common 2

2(1). Pedicels subterete; silicles averaging less than 3 mm long; plants commonly of waste places .... *L. densiflorum* var. *densiflorum*
— Pedicels definitely flattened, especially on upper side; silicles averaging at least 3 mm long; plants often growing in indigenous communities .... *L. densiflorum* var. *ramosum*

Var. **densiflorum** [*L. neglectum* Thell.; *L. bourgeauanum* Thell.; *L. densiflorum* var. *bourgeauanum* (Thell.) C. L. Hitchc.; *L. densiflorum* var. *macrocarpum* Mulligan]. Warm desert shrub, salt desert shrub, salt grass, sagebrush, pinyon-juniper, ponderosa pine, aspen, and spruce-fir communities at 850 to 3020 m in most, if not all, Utah counties; widespread in North America; 64 (viii).

Var. **pubicarpum** (A. Nels.) Thell. [*L. pubicarpum* A. Nels.]. Uncommon in sage brush and mountain brush communities at 1370 to 1925 m in Box Elder, Salt Lake, Sanpete, Tooele, and Utah counties; Washington to Montana, south to California; 5 (0).

Var. **ramosum** (A. Nels.) Thell. [*L. ramosum* A. Nels.; *L. densiflorum* var. *pubicaule* Thell.]. Warm and salt desert shrub, sagebrush, pinyon-juniper, and ponderosa pine communities at 1129 to 2320 m in Daggett, Duchesne, Emery, Garfield, Grand, Kane, San Juan, Sanpete, Uintah, and Utah counties; Wyoming south to New Mexico, west to California; 33 (viii).

Lepidium dictyotum Gray [*L. dictyotum* var. *macrocarpum* Thell.]. Annual, lacking a caudex; stems 2–15 cm tall, pubescent; basal leaves usually pinnatifid with 2–5 pairs of linear lobes; cauline leaves mostly entire, linear, reduced upwards; pedicels 1.5–3.5 mm long, flattened, spreading to reflexed; sepals 0.7–1 mm long, greenish, pubescent; petals 1–1.2 mm long, white, oblong, or usually lacking; stamens typically 4; silicles 3.2–4.5 mm long, 2–2.5 mm wide, ovate, glabrous or hirtellous, notched apically, teeth prolonged and usually divergent, acute; style lacking. Evidently rare in Utah, known from greasewood communities at ca 4220 m in Salt Lake and Weber counties; Washington to Idaho, south to California; 2 (0).

Lepidium fremontii Wats. Subshrubs, definitely woody above base; stems mostly 3–8 dm tall, glabrous and glaucous; leaves cauline, mostly 1.5–5 (8) cm long, pinnatifid into linear-oblong acute lobes up to 2 mm wide, becoming simple above, not auriculate; bracts subtending branches of the inflorescence often more than 2 cm long, linear; pedicels 3–8 mm long, ascending to spreading or curved-ascending, glabrous; sepals 1.5–2.5 mm long, green or more or less hyaline, glabrous; petals 3–4.5 mm long, white, obovate; stamens typically 6; silicles 4.5–7.5 mm long, 5.2–6.5 mm wide, obovate, glabrous, slightly notched apically; style 0.4–1.3 mm long; n = 32. Blackbrush, creosote brush, and other warm desert shrub communities from 700 to 1525 m in Washington County; Arizona, Nevada, and California; 33 (ii). The long, straight lobes of leaves, typically not sinuately lobed, are more or less diagnostic from *L. montanum* var. *jonesii*, with which this species is partially sympatric in Washington County.

Lepidium integrifolium Nutt. [*L. utahense* Jones, type from Milford; *L. zionis* A. Nels., type from Richfield; *L. montanum* var. *integrifolium* (Nutt.) C. L. Hitchc.]. Perennial, caespitose, from a thickened, usually branched caudex, more or less clothed with marcescent leaf bases; stems decumbent-ascending, 1.5–2.5 dm tall, minutely puberulent, simple or branched from upper axils; basal leaves 3–8.5 cm long, 6–25 mm wide, elliptic to oblanceolate, entire; cauline leaves mostly 1–4 cm long, 2–12 mm wide, gradually reduced upward, glabrous or nearly so; pedicels 5–8.5 mm long, spreading to ascending, puberulent; sepals 1.5–2 mm long, greenish, glabrous or pubescent; petals 2.7–3.1 mm long, white, obovate; stamens typically 6; silicles 3–4.2 mm long, 1.7–2.7 mm wide, ovate to lance-ovate, glabrous, plane, wingless; style 0.4–0.7 mm long. Saline or cool wet meadows at 1280 to 1925 m in Beaver, Rich, Sanpete, Sevier, and Uintah counties; Wyoming; 6 (i). The populations in central Utah are known only from historic collections. This distinctive entity has been collected only rarely, possibly because the habitat type has been exploited as marginal pastureland in Utah and Wyoming, and probably because its habitat is now occupied by the similar *Cardaria draba*, which is bypassed by most collectors.

Lepidium lasiocarpum Nutt. ex T. & G. Hispidcress. Annual, lacking a caudex; stems 2–30 cm tall or more, hispidulous to distinctly hispid with simple hairs; basal leaves mostly 1–8.5 cm long, 3–9 mm wide, oblanceolate, pinnatifid or merely lobed or toothed; cauline leaves few to several, reduced upwards, petiolate to sessile, not auriculate; pedicels 1.5–5 mm long, spreading to ascending, conspicuously flattened, usually puberulent to hispidulous; sepals 1–1.2 mm long, greenish or reddish to purplish, glabrous; petals 1–1.3 mm long, white, narrowly oblong, sometimes lacking; stamens 2 or 4; silicles 3.4–4.5 mm long, 2.7–4 mm wide, elliptic to rotund or obovate, hispidulous, notched apically, the teeth sometimes elongate-winged, rounded; style lacking or nearly so. Plants with somewhat attenuate apical teeth on the silicles approach *L. dictyotum*, and possibly the account for some of the reports of that entity from Utah. Two scarcely distinguishable varieties are known from Utah.

1. Pedicels glabrous or merely puberulent on lower side; stems, at most, hispidulous . *L. lasiocarpum* var. *georginum*
— Pedicels hispidulous on lower side; stems definitely hispid .... *L. lasiocarpum* var. *lasiocarpum*

Var. ***georginum*** (Rydb.) C. L. Hitchc. [*L. georginum* Rydb., type from southern Utah; *L. lasiocarpum* ssp. *georginum* (Rydb.) Thell.]. Warm and salt desert shrublands and lower pinyon-juniper woodlands at 1155 to 1985 m in Beaver, Emery, Garfield, Grand, Juab, Kane, San Juan, Tooele, Uintah, and Washington counties; Arizona and California; 18 (iv).

Var. ***lasiocarpum*** [*L. palmeri* Wats.; *L. lasiophyllum* Brandegee; *L. lasiocarpum* ssp. *palmeri* (Wats.) Thell.; *L. lasiocarpum* ssp. *lasiophyllum* (Brandegee) Thell.]. Warm, salt, and cool desert shrublands and pinyon-juniper communities at 820 to 1770 m in Garfield, Kane, Millard, Tooele, Utah, and Washington counties; Arizona and California; 30 (ix).

Lepidium latifolium L. Perennial herbs; stems 8–15 dm tall or more, glabrous; leaves basal and cauline, 1–15 (30) cm long, (0.4) 1–8 cm wide, the basal ones petiolate, orbicular to obovate, elliptic, or lanceolate, typically lacking at anthesis, the cauline becoming sessile upward, not auriculate, entire or irregularly crenate-serrate; bracts subtending branches of inflorescence 0.5–2 cm long; pedicels 1.5–4 mm long, spreading to ascending, glabrous; sepals ca 1 mm long, often tinged with blue or purple, glabrous or puberulent; petals ca 1.5 mm long, white, obovate; stamens typically 6; silicles 1.5–2 mm long, ca 1.5 mm wide, ovate, glabrous; style ca 0.2 mm long; 2n = 24, 48. Riparian and palustrine habitats, or less commonly in dryish barrow pits and roadsides at 1250 to 2410 m in Daggett, Grand, Iron, Rich, Salt Lake, Uintah, Utah, Wasatch, Washington, and Weber counties; widespread in the U. S.; introduced from Eurasia; 26 (iv). This species was introduced into Utah sometime prior to 1970 and is now spreading rapidly over the state.

Lepidium montanum Nutt. Perennial or less commonly biennial herbs or rarely suffrutescent, seldom definitely woody above base; stems 0.3–12 dm tall or more, glabrous or variously puberulent to hirtellous; leaves basal or basal and cauline, 0.5–12.5 cm long, 3–25 mm wide, variously shaped, entire to pinnatifid, often at least uppermost cauline ones simple and entire; bracts subtending branches of inflorescence seldom more than 1 cm long; pedicels 3–8 mm long, spreading to ascending, straight, or more or less sigmoid-curved, glabrous or puberulent; sepals 1.2–1.7 mm long, green or variously tinged, glabrous or puberulent; petals 2.5–3.5 mm long, white, obovate to spatulate: stamens typically 6; silicles 2.8–4.1 mm long, 2.1–2.5 mm wide, ovate to elliptic, glabrous or rarely puberulent; style 0.3–0.9 mm long; n = 16. There is an amazing amount of variation within *L. montanum* as interpreted herein. The more stable phases have been treated previously either as species or as varieties. There is justificatiion for recognizing all of these entities at the specific level, but the convenience of having an inclusive species with several variants seems to best represent the situation within Utah.

1. Plants mostly 6–12 dm tall, of eastern to southeastern Utah; cauline leaves simple, toothed or entire, often more than 4 mm wide; flowering from midsummer to autumn .... *L. montanum* var. *spathulatum*
— Plants usually less than 6 dm tall, of various distribution; cauline leaves various but, if simple and entire, usually less than 4 mm wide; flowering in springtime and midsummer (at higher elevations) 2

2(1). Basal leaves well developed, usually over 5 cm long and 0.6 cm wide; cauline leaves few, reduced upwards; plants usually montane 3
— Basal leaves poorly if at all developed (or rarely so), usually less than 5 cm long and 0.6 cm wide; cauline leaves numerous to few, gradually reduced upwards; plants seldom montane 4

3(2). Stems puberulent almost or quite throughout .... *L. montanum* var. *alpinum*
— Stems glabrous almost or quite throughout .... *L. montanum* var. *heterophyllum*

4(2). Stems glabrous or essentially so (rarely puberulent); plants somewhat woody at the base 5
— Stems puberulent almost or quite throughout; plants herbaceous or somewhat woody at the base 6

5(4). Basal leaves divided, not entire; plants (0.7) 1–6 (10) dm high, widespread in eastern, southeastern, and southern Utah .... *L. montanum* var. *jonesii*

— Basal leaves entire or some divided; plants 2–5 (7) cm high, rare and local, Aquarius and Paunsagunt plateaus, Garfield County .... *L. montanum* var. *neeseae*

6(4). Plants perennial, restricted to Kane County; caudex woody; silicles puberulent; stems 0.5–1 dm tall .... *L. montanum* var. *stellae*

— Plants biennial or perennial, widespread in western Utah; caudex lacking or only slightly woody; silicles glabrous; stems often over 1 dm tall .... *L. montanum* var. *montanum*

Var. *alpinum* Wats. [*L. scopulorum* Jones]. Sagebrush and spruce-fir communities at 1525 to 3050 m in Box Elder and Salt Lake (type from Cottonwood Canyon) counties; endemic; 4 (0).

Var. *heterophyllum* (Wats.) C. L. Hitchc. [*L. integrifolium* var. *heterophyllum* Wats.; *L. heterophyllum* (Wats.) Jones]. Mountain brush, aspen, and white fir communities at 1830 to 2320 m in Iron, Millard, Piute, and Sevier counties; endemic; 4 (i).

Var. *jonesii* (Rydb.) C. L. Hitchc. [*L. jonesii* Rydb., type from St. George; *L. crandallii* Rydb.; *L. tortum* L. O. Williams; *L. alyssioides* var. *jonesii* (Rydb.) Thell.; *L. alyssioides* var. *stenocarpum* Thell.]. n = 16. Shadscale, other salt desert shrub, warm desert shrub, cool desert shrub, sagebrush, and pinyon-juniper communities at 850 to 2135 m in Carbon, Daggett, Duchesne, Emery, Garfield, Grand, Kane, San Juan, Uintah, Utah, Washington, and Wayne counties; Arizona and Nevada, east to Colorado; 195 (xxix).

Var. *montanum* [*L. corymbosum* H. & A.; *L. utaviense* Regal; *L. brachybotryum* Rydb., type from Juab, Juab County; *L. philonitrum* Nels. & Macbr.; *L. albiflorum* Nels. & Macbr.; *L. montanum* var. *stenocarpum* Thell.; *L. scopulorum* f. *nanum* Thell.; *L. montanum* var. *canescens* (Thell.) C. L. Hitchc.]. Shadscale, other salt desert shrub, mixed desert shrub, sagebrush communities at 1340 to 2135 m in in Beaver, Box Elder, Cache (?), Duchesne, Iron, Juab, Millard, Piute, Rich, Sevier, Summit, Tooele, Utah, and Washington counties; Idaho and Wyoming south to Arizona, Nevada, and California; 68 (xv). The phase distinguished as var. *canescens* by Hitchcock passes into var. *montanum* through a series of morphological intermediates; thus it is placed here in synonymy with var. *montanum*. Nevertheless, the material is hardly uniform, with some individuals having strict branches and only sparingly puberulent vesture. Possibly additional segregation is advisable.

Var. *neeseae* Welsh & Reveal Ponderosa pine and spruce-fir communities mainly on the Pink and White Limestone members of the Wasatch Formation, but also on the Navajo Sandstone, at ca 2250 m in Garfield County; endemic; 4 (i).

Var. *spathulatum* (Robins.) C. L. Hitchc. [*L. scopulorum* var. *spathulatum* Robins.; *L. spathulatum* Vasey ex Robins., not Phil.; *Thelypodium crenatum* Greene; *L. crenatum* (Greene) Rydb.; *L. vaseyanum* Thell.; *L. montanum* var. *eastwoodiae* sensu Utah material; *L. montanum* var. *alyssioides* sensu Utah material]. Riparian and pinyon-juniper communities at 1125 to 2260 m in Garfield, Grand, Kane, San Juan, and Uintah counties; Colorado; 38 (xvii). This is the tall phase of the species that flowers in midsummer and autumn in southeastern Utah, it is sufficiently distinct from the other varieties of *L. montanum* to be recognized at the specific level. If that is done, the correct name at the specific level would be *L. crenatum*.

Var. *stellae* Welsh & Reveal White silty barrens of the Winsor Formation, with mound-forming plants, in a sagebrush-juniper community at ca 1700 m elevation in Kane (type from SE of Cannonville) County; endemic; 3 (i).

***Lepidium ostleri* Welsh & Goodrich** Ostler Pepperplant. Perennial, pulvinate-caespitose herbs from a branching caudex, this clothed with marcescent gray leaf bases; stems erect, 1–3.5 (5) cm tall, hirsute; leaves 4–15 mm long, linear and entire or the basal ones 3- to 5-lobate, hirsute, the cauline lacking or few; racemes ca 1 cm long in flower and 1–2 cm long in fruit, 5- to 35-flowered; pedicels 2–3 mm long in fruit; sepals 1.3–2 mm long, obtuse, hyaline, roughly pilose, often purplish; petals 2–3 mm long, white or suffused with purple; silicle 2.5–3 mm long, 2.3–2.5 mm wide, broadly ovate, to 1 mm thick, the sinus 2 mm deep; style 0.4–0.7 mm long. Limestone outcrops typically in pinyon-juniper community at ca 1765 to 2075 m at Frisco (type locality), San Francisco Mts., Beaver County; endemic; 7 (iii). This is a dwarf plant of crevices in rocky outcrops.

***Lepidium perfoliatum* L.** Peppergrass. Annual, lacking a caudex; stems 0.7–4 (6) dm tall, hirtellous below, glabrous and glaucous above; basal leaves 0.8–15 cm long, 2–28 mm wide, bi- or tripinnatifid into narrow segments; cauline leaves with lowermost like the basal ones, these transitional upward into entire and falsely perfoliate-clasping ones; pedicels 3–6.5 mm long, spreading-ascending, glabrous, subterete; sepals 0.7–1.2 mm long, often purplish, glabrous or pubescent; petals 0.8–1.5 mm long, yellow, narrowly spatulate, ascending; stamens usually 6; silicles 3.5–4.3 mm long, 3.2–3.6 mm wide, oval to elliptic, glabrous, plane, wingless or essentially so, slightly emarginate; style 0.1–0.3 mm long; n = 8. Ruderal weed, often in disturbed sites, in creosote bush, blackbrush, other warm desert shrub, shadscale, greasewood, galleta, pinyon-juniper, and sagebrush communities at 850 to 2440 m in Beaver, Box Elder, Cache, Carbon, Daggett, Davis, Garfield, Grand, Iron, Juab, Kane, Millard, Piute, Rich, Salt Lake, San Juan, Sanpete, Tooele, Uintah, Utah, Washington, and Weber counties; widespread in the U. S.; adventive from Europe; 66 (iii).

***Lepidium strictum* (Wats.) Rattan** Erectcress. [*L. oxycarpum* var. *strictum* Wats.; *L. reticulatum* Howell; *L. pubescens* authors, not Desv.]. Annual, lacking a caudex; stems 0.5–2 dm long, prostrate to rarely erect, pubescent; basal leaves mostly 3–7 cm long, 1–2 cm wide, bipinnatifid; cauline leaves less divided, uppermost sometimes entire; pedicels mostly 2–3 mm long, ascending, flattened, and narrowly wing-margined; sepals 1–1.5 mm long, persistent, purplish, pubescent; petals minute or lacking; stamens 2; silicles 2.2–3.5 mm long, 2–3 mm broad, oval to oblong-ovate, planely reticulate, slightly winged, concave, notched apically; style lacking. Apparently rare in Utah and possibly not a part of the continuing flora; cited by Hitchcock (1936) from Castle Gate, Carbon County, collected by Grant in 1900; 0 (0).

***Lepidium virginicum* L.** Virginiacress. Annual, lacking a caudex; stems 1.5–7 dm tall, pubescent throughout or glabrous above; basal leaves 1.3–15 cm long, 0.3–3.5 (5) cm wide, coarsely toothed to lobed; cauline leaves reduced upwards, uppermost usually entire; pedicels 2–6 mm long, spreading, terete or nearly so, pubescent or glabrous; sepals 0.6–1 mm long; petals 0.7–1.5 (3) mm long, white, spatulate, rarely lacking; silicles 2.5–4 mm long, 2.2–3.5 mm broad, elliptic to orbicular, usually glabrous, plane, notched apically; style lacking; n = 16. Joshua tree, creosote bush, other warm desert shrub, mixed desert shrub, pinyon-juniper, ponderosa pine, white fir, and spruce-fir communities at 915 to 2655 m in Beaver, Cache, Davis, Duchesne, Morgan, Piute, Salt Lake, San Juan, Sevier, Summit, Uintah, Utah, Wasatch, Washington, and Weber counties; widely distributed in North America; 29 (i). Our material belongs to **var. pubescens** (Greene) Thell. [*L. intermedium* var. *pubescens* Greene; *L. hirsutum* Rydb.].

Lesquerella Wats.

Stellate-pubescent, annual (biennial) or perennial herbs arising from taproots; leaves basal and cauline, alternate, simple, tapering to base or merely sessile, not auriculate; flowers racemose; pedicels straight, sigmoid, reflexed, spreading or ascending, not subtended by bracts; sepals 4, deciduous; petals 4, yellow, rounded; stamens 6, the filaments lacking glandular processes; style slender; stigma capitate; fruit a sessile to substipitate silicle, less than twice longer than broad, varying in outline, compressed parallel with or contrary to septum; valves veinless; seeds 2–10 per locule, biseriate.

Maguire, B., and A. H. Holmgren. 1951. Botany of the intermountain region - II. *Lesquerella*. Madroño 11: 172–184.
Payson, E. B. 1921. A monograph of the genus *Lesquerella*. Ann. Missouri Bot. Gard. 8: 103–236.
Rollins, R. C., and E. A. Shaw. 1973. The genus *Lesquerella* in North America. Cambridge, Mass.: Harvard University Press. 288 p.

1. Lowermost leaves narrow, mostly 1–5 mm wide, blade and petiole indistinct, or, if so, as in *L. rectipes*, the plants tall and of low elevations, tufted at stem base; basal and cauline leaves alike in general shape . . . . Key 1
— Lowermost leaves often over 5 mm wide, the blade distinct from petiole, forming a rosette; basal leaves clearly of a different shape from cauline leaves . . . . Key 2

Key 1.

Basal and cauline leaves alike.

1. Silicles and ovaries glabrous 2
— Silicles and ovaries stellate pubescent 4
2(1). Stems mostly 0.6–1.8 dm tall; silicle bodies 4 mm long or more; plants of San Juan County . . . . *L. fendleri*
— Stems 1–5 cm tall; silicle bodies usually less than 4 mm long; plants of south-central Utah 3
3(2). Plants densely pulvinate-caespitose, cushionlike, with usually numerous caudex branches; restricted to white shale outcrops southeast of Cannonville, Kane County . . . . *L. tumulosa*
— Plants caespitose in small compact to loose clumps, with one to several caudex branches; restricted to limestones on the Paunsaugunt Plateau . . . . *L. rubricundula*
4(1). Plants definitely pulvinate-caespitose, usually less than 1 dm tall; style mostly 1–2 mm long; known from Washington and Kane counties . . . . *L. arizonica*
— Plants various but, if pulvinate-caespitose and less than 1 dm tall, the style usually 2–4 mm long or more, or distribution otherwise 5
5(4). Plants usually less than 1 dm tall; basal leaves linear to narrowly spatulate; usually of middle to high elevations . 6
— Plants usually 1.5–2 dm tall or more; basal leaves spatulate to oblanceolate; usually of middle to lower elevations 7

6(5). Pedicels usually strongly S-shaped; inner basal leaves usually flat . . . . *L. alpina*
— Pedicels usually straight or only slightly curved; inner basal leaves usually involute . . . . *L. intermedia*
7(5). Pedicels generally recurved or arched in fruit, or less commonly almost straight; blades of basal leaves rarely more than 4 mm wide . . . . *L. ludoviciana*
— Pedicels more or less S-shaped; blades of basal leaves often more than 4 mm wide . . . . *L. rectipes*

Key 2.

Basal and cauline leaves different.

1. Plants slender annnuals; stems erect or ascending, mostly 1–4 dm tall; known from Washington County . . . . *L. tenella*
— Plants caespitose perennials; stems decumbent to erect, mostly less than 1 (1.5) dm tall; distribution otherwise . . 2
2(1). Plants with caudex definitely branched, known from low to moderate elevations of Duchesne, Emery, and Uintah counties . . . . *L. subumbellata*
— Plants with caudex simple or rarely branched; distribution various, but usually of higher elevation 3
3(2). Silicles ellipsoid to subglobose or obovoid to obdeltoid, compressed contrary to the septum 4
— Silicles of various shape but, if as above, compressed parallel to the septum 5
4(3). Racemes loose to elongated in fruit, often secund; silicles sparingly pubescent with hairs 0.2–0.3 mm broad; plants of the Wasatch and Uinta mts. . . . . *L. utahensis*
— Racemes short and dense in fruit, not at all secund; silicles moderately pubescent with hairs 0.2–0.4 mm broad; plants of the Wasatch Plateau southward . . . . *L. wardii*
5(3). Silicles ovoid, compressed at apex or margins; plants from Beaver County northward . . . . *L. occidentalis*
— Silicles various but not compressed at the apex or margins; plants of various distribution 6
6(5). Basal leaves angular, the blades usually deltoid or hastate, narrowed abruptly to the petiole; plants of Rich and possibly Summit counties . . . . *L. prostrata*
— Basal leaves not angular, the blades elliptic to obovate, narrowed gradually to the petiole; plants of various distribution 7
7(6). Silicles either definitely compressed apically or truncate to emarginate apically; plants from Utah County southward 8
— Silicles more or less acute apically; plants from Utah County northward 9
8(7). Silicles obcordate to obdeltoid, sparingly pubescent; largest basal leaves usually less than 1 cm wide; plants of Carbon, Emery, Sanpete, and Utah counties . . . . *L. hemiphysaria*
— Silicles ellipsoid, densely pubescent; largest basal leaves usually more than 1 cm wide; plants of western Utah . . . . *L. kingii*
9(7). Pedicels loosely S-shaped; silicles sparsely pubescent with appressed or erect hairs; caudex branches lacking conspicuous leaf scars and bases; plants of Cache and Rich counties . . . . *L. multiceps*
— Pedicels straight or curved; silicles densely pubescent with erect hairs; caudex branches with conspicuous scars and/or leaf bases; plants of Davis, Salt Lake, Utah, and Wasatch counties . . . . *L. garrettii*

Lesquerella alpina (Nutt.) Wats. Alpine Bladderpod. Perennial, arising from simple or branched caudices, these often clothed with marcescent leaf-bases; herbage pubescent with stellate, 4- to 7-branched hairs; stems 1–10 cm tall, rarely more, erect, simple; basal leaves mostly 0.5–5 cm long, 1–4 mm wide, linear to very narrowly spatulate, gradually narrowed to base; cauline leaves 5–20 mm long(or more), 1–3 cm wide; pedicels 4–10 mm long or more, straight, curved, or S-shaped; sepals 3.5–6 (7) mm long; petals 4–8 mm long, yellow, spatulate; silicles (excluding style) mostly 3–5 mm long, sessile or subsessile, ovoid, more or less compressed apically; valves pubescent externally and sometimes internally also; style mostly 2–4 mm long; ovules 2–6 per locule; n = 5. Two varieties which lack definite diagnostic criteria are represented among our materials.

1. Leaves spatulate, at least some; perhaps not distinct from the next .... *L. alpina* var. *alpina*
— Leaves uniformly narrow, linear to linear-spatulate .... *L. alpina* var. *parvula*

Var. alpina [*Vesicaria alpina* Nutt. ex T. & G.; *L. spatulata* Rydb.; *L. curvipes* A. Nels.; *L. alpina* var. *spatulata* (Rydb.) Payson; *L. condensata* var. *laevis* Payson; *L. alpina* var. *laevis* (Payson) C. L. Hitchc.]. Sagebrush and juniper communities, often on ridgetops, at 2135 to 3050 m elevation in Daggett, Duchesne, Sanpete, and Uintah counties; widespread northward to Alberta and Saskatchewan; 8 (0).

Var. parvula (Greene) Welsh & Reveal [*L. parvula* Greene; *L. alpina* ssp. *parvula* (Greene) Rollins & Shaw]. Pinyon-juniper and Douglas fir communities at 1830 to 2960 m in Daggett and Wasatch counties; Colorado and Wyoming; 13 (ii). The var. *condensata* (A. Nels.) C. L. Hitchc. is to be sought in extreme northern Daggett County; it is currently known from just north of the Utah line in Uinta County, Wyoming. That variety consists of small, tufted plants with the stems barely exserted beyond the leaves.

Lesquerella arizonica Wats. Arizona Bladderpod. [*L. arizonica* var. *nudicaulis* Payson]. Perennial, more or less pulvinate-caespitose, with few to several (or many) caudex branches, these often with marcescent leaf-bases; herbage densely pubescent with stellate hairs; stems 2–10 cm tall, rarely more, erect, simple; basal leaves 0.5–3.5 cm long, 1–4 mm wide, oblanceolate to spatulate, gradually narrowed basally; cauline leaves 0.5–2.5 cm long or more. 1–3 mm wide; pedicels 3–10 mm long or more, straight or curved, ascending; sepals 4–6.5 mm long; petals 5.5–7 mm long, yellow, suborbicular; silicles (excluding style) 3–5 mm long, sessile or substipitate, ovoid to ellipsoid, rounded and compressed apically; valves pubescent externally; style 0.5–2 (4) mm long; ovules 2–5 (8) per locule; n = 5, 10. Mountain brush, sagebrush, and pinyon-juniper communities at 1280 to 2750 m in Kane and Washington counties; Arizona; 6 (0).

Lesquerella fendleri (Gray) Wats. Fendler Bladderpod. [*Vesicaria fendleri* Gray; *V. stenophylla* Gray; *L. folicacea* Greene; *L. stenophylla* (Gray) Rydb.; *L. praecox* Woot. & Standl.]. Perennial, caespitose, the caudex simple or few- to several-branched, the branches naked or with few marcescent leaf-bases; herbage usually densely pubescent with stellate hairs; stems 5–25 cm tall, rarely more, erect to decumbent, simple; basal leaves 1–4 cm long, 1–6 cm wide, elliptic to oblanceolate, gradually tapering to the base; cauline leaves 0.4–2 (2.5) cm long, mostly 1–5 mm wide; pedicels 7–15 (20) mm long, straight or curved to S-shaped, spreading to suberect; sepals 4.5–8 mm long; petals 6–12 mm long, yellow, obovate; silicles (excluding style) 4.5–8 mm long, sessile or subsessile, ellipsoid to ovoid, rounded apically, the valves glabrous; style 2–6 mm long; ovules 6–16 per locule; n = 6. Shadscale, blackbrush, and juniper communities at ca 1700 m in San Juan County; Colorado and Arizona east to Texas and south to Mexico; 7 (i).

Lesquerella garretti Payson Garrett Bladderpod. Perennial, caespitose, the caudex simple or few- to several-branched, the branches clothed with marcescent leaf-bases and scars; herbage pubescent with stellate hairs; stems 5–15 cm tall, decumbent-spreading to suberect, simple; basal leaves 1–3.5 (4) cm long, 2–7 mm wide, elliptic to obovate, differentiated into blade and petiole; cauline leaves 3–13 mm long 1–4 mm wide; pedicels 3.5–7 mm long, straight or curved, spreading-ascending; sepals 4.5–6 mm long; petals 6–9 mm long, yellow, spatulate to obovate; silicles (excluding style) 3.5–4 mm long, short-stipitate, subglobose or obovoid, the valves pubescent externally, glabrous within; style 4–7 mm long; ovules 2–4 per locule. Spruce-fir and alpine tundra, often in talus or on rock outcrops, at 3050 to 3660 m in Davis, Salt Lake (type from Big Cottonwood Canyon), Utah, and Wasatch counties; endemic; 6 (0).

Lesquerella hemiphysaria Maguire Skyline Bladderpod. Perennial, caespitose, the caudex simple or few-branched, the branches more or less clothed with marcescent leaf-bases; herbage pubescent with stellate hairs; stems 3–10 (15) cm tall, decumbent-spreading or rarely erect; basal leaves 0.5–3.5 (5.5) cm long, 4–10 (15) mm wide, obovate to elliptic, diferentiated into blade and petiole; cauline leaves 3–15 mm long, 2–5 mm wide; pedicels 3–7 mm long, spreading, ascending, or recurved, S-shaped or curved; sepals 4–7 mm long; petals 5–7 mm long, yellow, oblanceolate; silicles (excluding style) 3–7 mm long, sessile or substipitate, obcordate in outline, the valves more or less pubescent externally, glabrous within; style 3–6 mm long; seeds 4–8 per locule. Two scarcely differentiated varieties are known:

1. Silicles glabrous or nearly so; plants of the West Tavaputs Plateau .... *L. hemiphysaria* var. *lucens*
— Silicles uniformly pubescent throughout; plants of the Wasatch Plateau and southwestern rim of the Uinta Basin .... *L. hemiphysaria* var. *hemiphysaria*

Var. hemiphysaria Sagebrush, aspen, Douglas fir, and meadow communities at 2135 to 3355 m in Duchesne, Emery, Sanpete (type from Wasatch Plateau), Utah, and Wasatch counties; endemic; 31 (v).

Var. lucens Welsh & Reveal Sagebrush and spruce-fir communities at 2700 to 2800 m on the West Tavaputs Plateau (type from Range Creek), Carbon County; endemic; 4 (0).

Lesquerella intermedia (Wats.) Heller Watson Bladderpod. [*L. alpina* var. *intermedia* Wats.]. Perennial, caespitose, the caudex simple or few-branched, the branches more or less clothed with marcescent leaf-bases; herbage pubescent with stellate hairs; stems 2–15 (25) cm tall, erect or ascending, simple; basal leaves 1–5 cm long, 1–2 mm wide, linear or only slightly expanded apically, the inner usually involute, tapering gradually to base; cauline leaves 0.8–3.5 cm long, 1–3 mm wide; pedicels 4–12 mm long, spreading to ascending, straight or curved; sepals 2.8–6.5 mm long; petals 5.5–10.5 mm long, yellow, spatulate; silicles (excluding style) 4–7 mm long, sessile or substipitate, ellipsoid or ovoid, acute and somewhat flattened apically, pubescent externally, glabrous or pubescent within; style (2) 3–4.5 mm long; seeds (4) 6–8 (10) per locule; n = 18. Sagebrush, pinyon-juniper, ponderosa pine, aspen, and bristle-cone pine communities at 1525 to 2840 m in in Beaver, Duchesne, Emery, Garfield, Iron, Kane, Piute, Sanpete, Sevier, Washington, and Wayne counties; Arizona, New Mexico; 52 (ix). This taxon approaches *L. alpina* on the one hand and *L. rectipes* on the other. Definitive features to separate alpine phases of *L. intermedia* from *L. alpina* are subject to interpretation. Only arbitrary separation seems possible.

***Lesquerella kingii* Wats.** King Bladderpod. Perennial, caespitose, the caudex usually simple, densely clothed with marcescescent leaf-bases; herbage pubescent with stellate hairs; stems 3–18 dm tall, decumbent to suberect, simple; basal leaves 1.5–4.5 (6) cm long, 4–20 mm wide, the blade spatulate to oval, obovate or ovate, sharply differentiated from petiole; cauline leaves 0.5–2 cm long, 1–7 mm wide; pedicels 4–10 mm long or more, curved to straight or S-shaped, ascending to descending; sepals 4–8 mm long; petals 5.5–12 mm long, yellow, spatulate; siliques (excluding style) 3.5–9 mm long, sessile to substipitate, ellipsoid, compressed apically; valves pubescent externally, pubescent to glabrous within; style 2–5 mm long; seeds 4–8 per locule; n = 5. Shadscale, rabbitbrush, sagebrush, pinyon-juniper, and ponderosa pine communities at 1370 to 3450 m in Beaver, Emery, Garfield, Iron, Juab, Kane, Millard, Piute, Sevier, Tooele, and Washington counties; Oregon, California, and Nevada; 26 (viii). Utah material, which is not always distinguishable from *L. wardii*, belongs to **var. *parvifolia* (Maguire & Holmgren) Welsh & Reveal** [*L. occidentalis* var. *parvifolia* Maguire & Holmgren; *L. latifolia* A. Nels.; *L. barnebyi* Maguire; *L. kingii* ssp. *latifolia* (A. Nels.) Rollins & Shaw]. The var. *cordiformis* (Rollins) Maguire & Holmgren is to be sought in the deserts of extreme western Utah.

***Lesquerella ludoviciana* (Nutt.) Wats.** Silver Bladderpod. [*Myagrum argenteum* Pursh; *Alyssum ludovicianum* Nutt.; *L. argentea* (Pursh) MacMillan, not *L. argentea* (Schauer) Wats.]. Perennial, loosely caespitose, the caudex simple to many-branched, the branches clothed with marcescent leaf-bases; herbage pubescent with stellate hairs; stems 0.7–3 (4) dm tall, ascending to erect or the outer decumbent, simple; basal leaves 1–9 cm long, (1) 2–8 mm wide, spatulate to oblanceolate or appearing linear when folded; cauline leaves 1–6 cm long, 1–6 mm wide; pedicels 5–15 mm long or more, straight or curved, ascending to recurved; sepals 4–7 mm long; petals 5–10 mm long, yellow, oblanceolate to obovate; silicles (excluding style) 3–6 mm long, sessile or nearly so, subglobose or obovoid, the valves pubescent externally and usually pubescent within; style 3–5 mm long; seeds 2–8 per locule; n = 5, 15. Pinyon-juniper, sagebrush, and upward to spruce-fir communities at 1405 to 2750 m in Beaver, Carbon, Daggett, Duchesne, Emery, Garfield, Kane, Uintah, and Wayne counties; Minnesota and Illinois, west to Montana, and south to Nevada, Colorado, and Kansas; 53 (vii).

***Lesquerella multiceps* Maguire** Wasatch Bladderpod. Perennial, caespitose, the caudex simple or several- to many-branched, more or less clothed with marcescent leaf-bases; herbage pubescent with stellate hairs; stems 3–20 cm long, rarely more, prostrate to erect; basal leaves 1–6 cm long, 4–15 (20) mm wide, blades elliptic to obovate, differentiated from long slender petioles; cauline leaves 0.4–2 cm long, 1–5 mm wide; pedicels 3–10 mm long, S-shaped; sepals 4.3–7 mm long; petals 5–10 mm long, yellow, spatulate; silicles (excluding style) 3–5.5 mm long, sessile or nearly so, ovoid; valves pubescent externally, glabrous within; style 3–6.5 mm long; ovules 2–4 per locule. Mountain brush and aspen communities at 1830 to 2960 m elevation in Cache (type from Tony Grove Lake) and Rich counties; Idaho, Wyoming; 6 (0).

***Lesquerella occidentalis* Wats.** Western Bladderpod. Perennial, caespitose, the caudex simple or few-branched, more or less clothed with marcescent leaf-bases; herbage pubescent with stellate hairs; stems 2–15 cm long, rarely more, prostrate, decumbent, ascending or erect, usually simple; basal leaves 0.5–8 cm long, 3–15 mm wide, the blades spatulate to oval or obovate, differentiated from petioles; cauline leaves 4–15 mm long, 1–8 mm wide; pedicels 3–10 mm long or more, sigmoid or straight, ascending; sepals 4–7 mm long; petals 6–10 mm long, yellow, spatulate; silicles (excluding style) 4–6 mm long, ellipsoid, usually compressed apically, the valves pubescent externally and also within; style 2–4 mm long; seeds 2–8 per locule. Sagebrush, pinyon-juniper, aspen, Douglas fir, and alpine meadow communities, often in talus or on rock outcrops, at 2010 to 2355 in Beaver, Box Elder, Davis, Juab, Millard, Tooele, and Weber counties; Idaho and Nevada to California and Oregon; 24 (0). Our material belongs to **var. *cinerascens* Maguire & Holmgren** [*L. occidentalis* ssp. *cinerascens* (Maguire & Holmgren) Rollins & Shaw; *L. goodrichii* Rollins, type from Millard County]. I can see no way to segregate our materials into additional subordinate taxa.

***Lesquerella prostrata* A. Nels.** Rich Bladderpod. Perennial, caespitose, the caudex simple or few-branched, the branches more or less clothed with marcescent leaf-bases; herbage pubescent with stellate hairs; stems 3–15 cm tall, decumbent to ascending, simple; basal leaves 1–5 cm long, 5–10 (15) mm wide, the blades deltate to hastate, more or less angular, differentiated from slender petioles; cauline leaves 4–15 mm long, 1–5 cm wide; pedicels 4–10 mm long, curved to straight or S-shaped, spreading; sepals 5–6 mm long; petals 5–9 mm long, yellow, spatulate; silicles (excluding style) 4–7 mm long, sessile or nearly so, ovoid to ellipsoid, not flattened apically, the valves pubescent externally, glabrous or sparsely hairy within; style 3–6 mm long; seeds 2–4 per locule. Sagebrush, grass, and juniper communities at 1830 to 2135 m in Rich County; Wyoming and Idaho; 4 (i).

***Lesquerella rectipes* Woot. & Standl.** Colorado Bladderpod. Perennial, loosely caespitose, the caudex simple or few-branched, more or less clothed by marcescent leaf-bases; stems 0.4–4 dm tall or more, decumbent to ascending or erect, simple or branched; basal leaves 1–8 (10) cm long, 3–12 mm wide, oblanceolate to elliptic, entire or toothed, tapering gradually to base; cauline leaves 0.6–3 cm long, 2–10 mm wide; pedicels 4–10 mm long or more, straight or S-shaped, spreading to ascending or recurved; sepals 4–8 mm long; petals 6–10 mm long or more, yellow, spatulate to obovate; fruit (excluding style) 4–7 mm long, substipitate to subsessile, ovoid to ellipsoid, pubescent externally and glabrous or pubescent within; style 2–7 mm long; seeds 4–8 per locule; 2n = 18. Blackbrush, other warm desert shrub, pinyon-juniper, and ponderosa pine communities from 1125 to 2565 m in Carbon, Duchesne, Emery, Garfield, Grand, Kane, San Juan, Uintah, Washington and Wayne counties; Arizona, Colorado and New Mexico; 63 (xviii). This is the most common of the tall species of *Lesquerella* in southeastern Utah.

Lesquerella rubicundula **Rollins** Breaks Bladderpod. [*L. hitchcockii* ssp. *rubicundula* (Rollins) Maguire & Holmgren]. Perennial, loosely caespitose, not matted, the caudex simple or with few to several branches, clothed with marcescent leaf bases; herbage pubescent with stellate hairs; stems 1–5 cm tall, erect, simple; leaves mainly basal, 3–12 mm long, 1–2 mm wide, not differentiated into blade and petiole; pedicels 1–6 mm long, straight or curved, ascending; sepals 3.2–4.5 mm long; petals 4.5–7 mm long, yellow, spatulate; silicles (excluding style) 3–5 mm long, sessile to substipitate, ovoid, valves glabrous externally and within; style 2.8–3.5 mm long; ovules 2–4 per locule. Ponderosa pine, bristlecone pine, and spruce-fir communities on the Pink and White Limestone members of the Wasatch Formation, at 2350 to 3355 m in Garfield (type locality at Red Canyon), Iron, and Kane counties; endemic; 22 (i).

Lesquerella subumbellata **Rollins** Bladderpod. Perennial, caespitose, the caudex several-branched, clothed with marcescent leaf bases and often with leaves as well; herbage pubescent with stellate hairs; stems 1–8 (10) cm tall, ascending to erect, simple; basal leaves 0.8–3 cm long, 2–6 (10) mm wide, at least outer and usually most of them obovate-spatulate, gradually tapering to petiole; cauline leaves 0.3–2 cm long, 1–4 mm wide; pedicels 2–5 mm long, straight or curved, ascending; sepals 3–4.5 mm long; petals 5–6 mm long, yellow, spatulate; silicles (excluding style) 3.5–5 mm long, subsessile, ellipsoid, the valves pubescent externally; style 1.5–2.5 mm long; seeds 2–6 per locule. Shadscale, pinyon-juniper, and sagebrush communities at 1645 to 2440 m in Duchesne, Emery, and Uintah (type locality 18 miles north of Vernal) counties; Colorado, Wyoming; 52 (v). Rollins and Shaw (1973) place this taxon within the range of variation of *Lesquerella alpina*. That disposition ignores the basic continuity of *L. subumbellata*, and likewise denies the apparent relationship of *L. alpina* in quite another direction, i.e., with *L. intermedia*. The uniformly broad blades of basal leaves, although tenuous as diagnostic features, are on the order of other such characters in this genus, with its critically defined species.

Lesquerella tenella **A. Nels.** Slender Bladderpod. [*L. gordonii* var. *sessilis* Wats., type from St. George; *L. palmeri* authors, not Wats.]. Annual, the herbage pubescent with stellate hairs; stems 1–5 dm tall or more, spreading-decumbent to erect, simple or branched; basal leaves 1.5–6 cm long, 2–15 mm wide, the blades elliptic, sometimes toothed, differentiated from the petiole; cauline leaves 0.4–4 cm long, 2–10 cm wide; pedicels 4–12 (18) mm long or more, S-shaped, spreading to recurved; sepals 3.5–7 mm long; petals 5–10 mm long, yellow to orange, orbicular to obovate; silicles (excluding style) 3.5–5 mm long, sessile to substipitate, globose to obovoid, the valves pubescent externally and internally; style 2–4.5 mm long; ovules 2–6 per locule; 2n = 10. Blackbrush, creosote brush, and Joshua tree communities at 790 to 1130 m in Washington County; Nevada and California south to Mexico; 12 (ii).

Lesquerella tumulosa **(Barneby) Reveal** Kodachrome Bladderpod. [*L. hitchcockii* ssp. *tumulosa* Barneby, type from southeast of Cannonville]. Perennial, pulvinate-caespitose, and mound-forming, densely matted, the caudex many-branched, clothed with marcescent leaves and leaf bases; herbage pubescent with stellate hairs; stems 1–4 cm tall, erect, simple; leaves mainly basal, 2–10 (15) mm long, (0.7) 1–1.5 (2) mm wide, not differentiated into blade and petiole; pedicels 2–5 mm long, straight or S-shaped, spreading to ascending; sepals 2.8–4 mm long; petals 5–7 mm long, yellow, spatulate; silicles (excluding style) 2.7–3.8 mm long, substipitate, ovoid, the valves glabrous externally and internally; style 1.9–2.7 mm long; ovules 2–4 per locule. White, bare shale knolls (Winsor Member of the Carmel Formation) among scattered juniper in a *Bouteloua* grassland, Kane County; endemic; 8 (ii). This curious species grows with several other mound-forming plants at the type locality.

Lesquerella utahensis **Rydb.** Utah Bladderpod. Perennial, caespitose, the caudex usually simple, more or less clothed with persistent leaf-bases; herbage pubescent with stellate hairs; stems 2–15 cm tall or more, decumbent to ascending or erect; basal leaves 1.2–6.5 cm long, 1–5 mm wide; pedicels 3–10 mm long, sigmoid or curved, ascending; sepals 4.5–5.7 (7) mm long; petals 6.5–9 mm long, yellow, spatulate; siliques (excluding style) 3.5–4.5 (6) mm long, substipitate to sessile, globose-ellipsoid, compressed contrary to partition, the valves pubescent externally, glabrous internally; style 4–5.5 (6.5) mm long; seeds 3–6 per locule. Sagebrush, meadows, and spruce-fir communities at 2255 to 3355 m in Daggett, Duchesne, Salt Lake, Summit, Uintah, Utah (the type is from American Fork Canyon), and Wasatch counties; endemic; 26 (i).

Lesquerella wardii **Wats.** Ward Bladderpod. Perennial, the caudex usually simple, clothed with marcescent leaf-bases; herbage pubescent with stellate hairs; stems 2–20 cm tall, rarely more, decumbent to ascending or erect, simple; basal leaves 1–4.5 (6) cm long, 3–15 (20) mm wide, the blades deltoid to orbicular or elliptic, differentiated from slender petiole; pedicels 2–7 mm long, straight, curved, or S-shaped, ascending; sepals 4–6.3 (8) mm long; petals (5.5) 6–9 (11) mm long, yellow, spatulate; silicles (3.5) 4–6.8 (8) mm long, substipitate to sessile, ellipsoid or ovoid, not compressed apically, pubescent externally and usually so internally; style (1) 2.2–4 (7) mm long; seeds 2–8 per locule. Pinyon-juniper, mountain brush, sagebrush, meadow, and spruce-fir communities at 1890 to 3355 m in Beaver, Garfield (type area is Aquarius Plateau), Iron, Kane, Piute, Sanpete, and Wayne counties; endemic; 55 (xiv). The Ward bladderpod is most similar to the partially sympatric *L. kingii* var. *parvifolia* from which it is distinguished inter alia in the silicles being rounded, not compressed, apically.

Lobularia Desv.

Annual or rarely biennial herbs, pubescent with malpighian hairs, arising from taproots; leaves alternate, simple, entire, petiolate or sessile, not auriculate; flowers in ebracteate racemes; sepals 4, deciduous; petals 4, white, pink, or lavender; stamens 6, the filaments lacking glandular processes; style short, the stigma capitate; fruit a silicle, about as broad as long, compressed parallel to septum, dehiscent; valves 1-nerved; seeds 1 per locule.

Lobularia maritima **(L.) Desv.** Sweet Alyssum. [*Clypeola maritima* L.; *Alyssum maritimum* (L.) Lam.]. Annual, the stems spreading-decumbent or ascending to erect, usually much branched, 0.8–3 dm long or more; leaves linear-oblanceolate, 0.8–3 (4) cm long, 1–4 mm wide, strigose on both surfaces; pedicels spreading-ascending, 5–9 mm long; sepals 1–1.5 (2) mm long, green or purplish, pubescent; petals 2.5–4 mm long, white, pink or lavender, the blades obovate; silicles 2.5–3.5 mm long and about as broad; valves sparingly strigose; style 0.3–0.6 mm long. Ornamental border plant with sweetly scented flowers, escaping and occasionally persistent; introduced from Europe; 7 (0).

Lunaria L.

Annual or biennial herbs, pubescent with simple or with some forked hairs, arising from taproots; leaves usually alternate and basal, simple, dentate to lobed, petiolate, not auriculate; flowers in ebracteate racemes; sepals 4, deciduous; petals 4, pink to lilac, blue, or purplish; stamens 6, the filaments lacking glandular processes; style elongate, persistent; stigma capitate, bilobed; fruit a stipitate silicle, about as broad as long, compressed parallel to septum, dehiscent; valves reticulately veined; seeds 3–5 per locule.

Lunaria annua L. Honesty; Moonwort; Satin-flower. Stems one to several, simple or branched, mostly 4–10 dm tall or more, pubescent with simple and rarely some forked, hairs; leaves 2.5–10 (15) cm long, 1.5–8 cm wide, rarely more, ovate to lance-ovate or cordate, dentate to lobed, pubescent with simple or rarely forked hairs; pedicels 10–22 mm long, spreading-ascending, pubescent; sepals 6–8 mm long, greenish or variously suffused, sparsely pubescent; petals 14–20 mm long, pink to lilac or blue to purplish, obovate; silicles 32–45 mm long, 25–35 mm wide, borne on stipes 7–12 mm long, much flattened, the valves glabrous; style 6–8 mm long; 2n = 30. Ornamental, cultivated for the showy flowers and distinctive fruiting inflorescences, occasionally escaping and persisting in Cache, Salt Lake, and Utah counties; introduced from Europe; 8 (i).

Malcolmia (L.) R. Br.

Annuals, pubescent with forked or 3-rayed hairs, from taproots; leaves alternate and basal, simple, sinuate-dentate, the cauline ones petiolate to sessile, not auriculate; flowers racemose, pedicels seldom subtended by bracts; sepals 4, tardily deciduous; petals 4, pink to lavender; stamens 6, the filaments lacking glandular processes; style tapering; stigma oblique; fruit a silique, many times longer than broad, subterete, dehiscent; valves nerveless; seeds uniseriate.

Malcolmia africana R. Br. in Ait. African Mustard. Stems decumbent-ascending to erect, 3–40 cm tall, simple or branched almost throughout, pubescent with forked or 3-rayed hairs; leaves mainly basal to mainly cauline, 1.2–9 cm long, 0.3–2.3 cm wide, oblanceolate to elliptic, sinuate-dentate, petiolate to sessile, not auriculate; pedicels spreading, 1–2 mm long, pubescent; sepals 3.7–5.2 mm long, often reddish or purplish, pubescent; petals 6.2–9.5 mm long, pink to lavender, rounded apically, spreading; siliques 33–66 mm long, straight, pubescent; style to 1 mm long; stigma oblique; 2n = 28. Roadsides, foothills, and other disturbed sites in many plant communities at 850 to 2685 m in most if not all Utah counties; widespread throughout the Great Basin and adjacent areas of the western U. S.; adventive from Africa; 83 (viii). This mustard lacks the distinctive musky odor of the look-alike *Chorispora tenella* q.v.

Nasturtium R. Br.

Glabrous perennials from subrhizomatous stolons; leaves alternate, simple or some pinnately compound, petiolate and auriculate; flowers in ebracteate racemes; sepals 4, deciduous; petals 4, white; stamens 6, the filaments lacking glandular processes; style stout, well developed, the stigma capitate, bilobed; fruit a silique, several times longer than broad, oblong, somewhat compressed parallel to septum, the valves 1-nerved; seeds several to many, biseriate.

Nasturtium officinale R. Br. in Ait. Water Cress. [*Sisymbrium nasturtium-aquaticum* L.; *Rorippa nasturtium-aquaticum* (L.) Schinz & Thell.]. Plants submersed or emergent, with succulent stems (0.3) 1–8 (10) dm long or more, glabrous; leaves 1–10 cm long, with terminal lobe usually the largest, falsely petiolate and narrowly auriculate-clasping basally; pedicels 5–13 (20) mm long, spreading to spreading-ascending, glabrous or pubescent; sepals 2–3 mm long, green or with white tips, glabrous; petals 3–4.7 mm long, white, rarely with purplish veins, oblanceolate; siliques 10–18 (25) mm long, 1.8–2.4 mm wide; style 0.7–1.2 mm long; n = 7, 14, 16. Seeps, springs, and sluggish to swift streams, usually in flowing water at 850 to 2745 m in all Utah counties; widely established in North America; introduced from Europe; 72 (vii). The plant is eaten as a salad green, but poses health problems due to contamination with various infective organisms.

Parrya R. Br.

Perennial, glabrous, herbs, pubescent with stipitate-glandular hairs, arising from taproots and with simple to branched caudices; leaves mainly basal, simple, dentate, tapering basally, not auriculate; flowers in ebracteate racemes; sepals 4, deciduous; petals 4, pink to lavender; stamens 6, the filaments lacking glandular processes; style stout, persistent, the stigma bilobed; fruit a silique, usually several times longer than broad, oblong, constricted between seeds, strongly compressed parallel to the septum; valves 1-nerved; seeds 1 to several per locule.

Parrya rydbergii Botsch. Uinta Parrya. [*P. platycarpa* Rydb., not Hook. f. & Thomas]. Rosulate low perennials, the caudex clothed with marcescent leaf bases; stems scapose, 7–12 cm tall, 2–8 cm to base of raceme, the herbage stipitate-glandular; leaves 3–10 cm long, 6–20 mm wide, oblanceolate to elliptic; racemes 3- to 10-flowered; pedicels 4–20 mm long, stout, steeply ascending; sepals 7.2–9.3 mm long, purplish, stipitate-glandular or glabrous; petals 16–23 mm long, pink to lavender, cuneate-spatulate, emarginate; siliques 25–47 mm long, 3–3.5 mm wide, straight or curved, the midnerve prominent, glabrous or stipitate-glandular; style 0.3–0.6 mm long; stigma deeply bilobed; seeds 1–4 per locule. Alpine tundra, usually in rock stripes and talus, at 3325 to 3965 m in the Uinta Mts., Daggett, Duchesne, Summit (type from Uinta Mts.), and Uintah counties; endemic; 13 (iii).

Physaria (Nutt.) Gray

Stellate-pubescent perennials, arising from taproots; leaves mainly basal, the cauline ones much reduced, alternate, simple, tapering basally or merely sessile, not auriculate; flowers racemose; pedicels straight, curved, sigmoid, spreading or descending, not subtended by bracts; sepals 4, deciduous; petals 4, yellow, rarely purple, rounded; stamens 6, the filaments lacking glandular processes; style slender, the stigma capitate; fruit a sessile, bladdery-inflated silicle, often broader than long, varying in outline, compressed (if at all) contrary to septum, the valves veinless; seeds 2–6 per locule, biseriate. The diagnostic features used to segregate entities of this genus are mainly based on the shape of the mature silicle and the length of the style. These features are subject to variation within rather broad limits, and not all specimens will fit neatly into the taxa in the following key. Immature specimens are particularly difficult to place, and fruiting specimens are more easily distinguished than flowering ones.

Mulligan, G. A. 1967. Cytotaxonomy of *Physaria acutifolia*, *P. chambersii*, and *P. newberryi* (Cruciferae). Canad. J. Bot. 45: 1887–1898.
Payson, E. B. 1918. Notes on certain Cruciferae, *Physaria*. Ann. Missouri Bot. Gard. 5: 143–147.
Rollins, R. C. 1939. The cruciferous genus *Physaria*. Ann. Missouri Bot. Gard. 5: 143–147.
Waite, S. B. 1973. A taxonomic revision of *Physaria* (Cruciferae) in Utah. Great Basin Nat. 33: 31–36.

1. Styles 2–3 (4) mm long; valves of silicles sharply angled at maturity; plants of Garfield, Grand, Kane, San Juan, and Washington counties .... *P. newberryi*
— Styles mostly 4–8 mm long or more; valves of silicles obtusely angled to rounded at maturity; distribution various 2
2(1). Basal leaves sinuate-dentate to lobed, sometimes entire; cauline leaves often toothed; plants of the Uinta Basin .... *P. floribunda*
— Basal leaves entire or less commonly sinuate-dentate; cauline leaves entire; plants of broad distribution 3
3(2). Sinuses of silicles indented above and below, when immature equally rounded to apex and base; valves rounded, not angled at maturity; plants mainly of the Colorado River drainage system .... *P. acutifolia*
— Sinuses of silicles only slightly indented to rounded below, deeply indented above, when immature tapering to base and obcordate in outline; valves obtusely angled at maturity; plants mainly of Great Basin and valleys of Virgin River drainage system ... *P. chambersii*

Physaria acutifolia Rydb. Rydberg Twinpod. [*P. australis* sensu Rollins]. Perennial, caespitose, the caudex usually simple, clothed with marcescent leaf bases; stems 0.3–2 (2.5) dm long, decumbent to ascending or erect, simple; basal leaves 1.7–9.5 cm long, 0.6–4.5 cm wide, the blade orbicular to ovate or obovate, less commonly lance-elliptic, entire, angular, or toothed to repand; cauline leaves greatly reduced; pedicels mostly 7–16 mm long, ascending to descending; sepals 5.2–8.7 mm long; petals 7.5–12.5 mm long, yellow or purple, spatulate; silicles bladdery-inflated, sinuses indented above and somewhat less so below, the valves 7–15 mm long in fruit, the surface rounded in cross-section, papery to membranous; style 3.8–7 mm long; n = 5. This taxon might best be regarded at infraspecific level, along with *P. chambersii*, in an expanded *P. newberryi*. The taxa differ in about the same magnitude as do varieties in other genera in this family. Three weakly differentiated and intergrading morphological subunits are recognized at varietal level.

1. Leaves of basal rosette, repand to sinuately lobed; plants of the rim of the Uinta Basin, especially along the south and southwest; flowers yellow or purple externally .... *P. acutifolia* var. *purpurea*
— Leaves of basal rosette entire or rarely somewhat sinuately lobed; plants of various distribution; flowers typically yellow 2
2(1). Plants often with a branching caudex, the branches commonly slender; leaves of basal rosette mostly less than 12 (15) mm wide; plants of high elevations in northwestern Duchesne and adjacent Wasatch counties ... . *P. acutifolia* var. *stylosa*
— Plants seldom with a branching caudex, the branches, when present, typically thickened; leaves of basal rosette typically more than 12 mm wide; plants widepread .... *P. acutifolia* var. *acutifolia*

Var. acutifolia Salt desert shrub, mixed desert shrub, sagebrush-grass, pinyon-juniper, mountain brush, ponderosa pine, aspen, and spruce-fir communities at 1130 to 2745 m in Carbon, Daggett, Duchesne, Emery, Garfield, Grand, Kane, Morgan, Rich, San Juan, Sevier, Summit, Uintah, and Wayne counties; Wyoming and Colorado; 173 (xxxii).

Var. purpurea Welsh & Reveal [*P. repandum* Rollins, type from Duchesne County]. Pinyon-juniper, sagebrush, ponderosa pine, Douglas fir, and limber pine communities at 2135 to 2870 m in Carbon, Duchesne, Grand (type from north of Thompson), Uintah, and Wasatch counties; endemic; 14 (v).

Var. stylosa (Rollins) Welsh [*P. stylosa* Rollins]. Spruce-fir communities and alpine tundra, mainly in talus slopes, at 2955 to 3450 m in northwestern Duchesne and adjacent Wasatch counties; endemic; 3 (0).

Physaria chambersii Rollins Perennial, caespitose, the caudex usually simple, clothed with marcescent leaf bases; stems 0.3–2.5 dm long, decumbent to ascending or erect, simple; basal leaves 1–7.5 (13) cm long, 0.3–3.5 cm wide, the blade orbicular to ovate, obovate, elliptic or oblanceolate, entire or repand, angular, or toothed; cauline leaves greatly reduced; pedicels mostly 4–18 mm long, ascending to descending; sepals 6–7.2 mm long; petals 8–13 mm long, yellow, spatulate; silicles bladderyinflated, the upper sinus deeply indented and with a broad, V-shaped sinus, the lower shallow or lacking; valves 11–17 mm long in mature fruit, the surface obtusely angled at edges, papery to membranous; style 4–8 mm long; n = 4, 8, 12. This entity is more or less confluent and partially sympatric with *P. acutifolia* eastward and with *P. newberryi* southward. Three weakly definable, morphologically intergrading and possibly taxonomically insignificant, varieties are recognized.

1. Plants with branching soboliferous caudex branches (less commonly some single), growing on the Pink and White Limestone members of the Wasatch Formation .... *P. chambersii* var. *sobolifera*
— Plants with simple caudices, rarely some branching but seldom if ever especially soboliferous, growing in various substrates, sometimes as above 2
2(1). Trichomes with peltate rays more or less confluent, disklike with rays projecting as teeth; plants of western Garfield and Kane counties, transitional to the next .. .. *P. chambersii* var. *membranacea*
— Trichomes with peltate rays more or less distinct, not disklike and with rays projecting as teeth; plants widely distributed .... *P. chambersii* var. *chambersii*

Var. chambersii Warm desert shrub, sagebrush, pinyon-juniper, mountain brush, ponderosa pine, bristlecone pine, and spruce-fir communities at 820 to 3420 m in Beaver, Boxelder, Cache, Carbon, Emery, Garfield, Iron, Juab, Kane, Millard, Piute, Rich, Sanpete, Sevier, Summit, Tooele, Washington, and Wayne counties; Arizona and Utah west to California and Oregon; 188 (xxviii).

Var. membranacea Rollins [*P. lepidota* Rollins, type from west of Mt. Carmel]. Pinyon-juniper, salt desert shrub, mountain brush, ponderosa pine, and aspen communities at 1525 to 2440 m in Kane and Garfield (type from Red Canyon) counties; endemic; 18 (ii). Plants with the peculiar, gear-shaped trichomes grade by degree into the typical variety with which they are sympatric.

Var. sobolifera Welsh Ponderosa pine, bristlecone pine, and Douglas fir communities at 2135 to 2900 m in western Garfield County (type locality); endemic; 3 (0).

Physaria floribunda Rydb. Grand Junction Twinpod. [*P. grahamii* Morton, type from Chandler Canyon, Uintah County]. Perennial, caespitose, the caudex simple, clothed with marcescent leaf bases; stems 5–15 (20) cm long, ascending to erect, simple; basal leaves 2–15 cm long, 0.8–5 cm wide, blade oblanceolate to spatulate, toothed to pinnatifid; cauline leaves often toothed, much reduced; pedicels 5–20 mm long, ascending; sepals 4.5–7 mm long; petals 6.5–10 mm long, yellow, spatulate; fruit strongly inflated, cordate, 4–9 mm long, 3–10 mm wide; style 3.5–8 mm long. Mixed desert shrub, known from southern Uintah County; Colorado and New Mexico; 1 (0). I can find no reason for recognition of *P. grahamii* as distinct from *P. floribunda*.

Physaria newberryi Gray Newberry Twinpod. [*Coulterina newberryi* (Gray) Kuntze; *P. didymocarpa* var. *newberryi* (Gray) Jones; *P. newberryi* var. *racemosa* Rollins, type from 13.5 mi N of St. George]. Perennial, caespitose, the caudex simple, clothed with marcescent leaf bases; stems 4–22 cm long, decumbent to ascending or erect; basal leaves 2–7.5 cm long, 0.8–4 cm wide, the blade obovate to orbicular, ovate, or spatulate, angled, indistinctly toothed or more commonly entire; cauline leaves much reduced, entire; pedicels 5–17 mm long, ascending to descending; sepals 6–8.5 mm long; petals 6.5–12 mm long, yellow, spatulate; silicles bladdery-inflated, the upper sinus deeply indented, the lower one shallow or lacking; valves 8–11 mm long or more in fruit, sharply angled at the margins, papery; style 2–3.5 mm long; n = 8. Warm desert shrub, galleta-shadscale, other salt desert shrub, pinyon-juniper, and ponderosa pine communities at 885 to 2350 m in Garfield, Grand, Kane, San Juan, and Washington counties; Arizona and New Mexico; 20 (iv). The phase of the species known as var. *racemosa* is completely transitional with typical materials, and appears to be taxonomically negligable. Where this species contacts either *P. chambersii* or *P. acutifolia* there are transitional specimens. Style length forms a continuum from one entity to another, and errors in determination are possible, especially in material that lack mature fruits. A better course, perhaps would be to regard *P. acutifolia* and *P. chambersii* as infraspecific taxa within *P. newberryi*, a course not intended or implied herein. A quadrinomial hierarchy of categories would seem to best represent the taxa of this complex in Utah.

Raphanus L.

Pubescent annuals with simple hairs, from tuberous taproots; leaves alternate and basal, simple, lyrate-pinnatifid, cauline ones petiolate to subsessile, not auriculate; flowers in ebracteate racemes; sepals 4, deciduous; petals 4, white or pink to lavender; stamens 6, filaments lacking glandular processes; style apical on a tapering sterile beak, the stigma minute, bilobed; fruit a silique, many times longer than broad, terete, indehiscent, breaking irregularly at maturity into segments; valves several-grooved; seeds uniseriate.

Raphanus sativus L. Radish. Plants erect, the stems mostly 4–10 dm tall, simple or more commonly branched, more or less hispid with simple hairs; leaves basal and cauline, mostly 2–18 cm long, 0.5–6.5 cm wide, lyrate-pinnatifid, hispid, the cauline ones reduced upwardly; pedicels 10–20 mm long or more, spreading to ascending; sepals 7–9 mm long, green to reddish or purplish, glabrous; petals 12–20 mm long, white to pink or lavender; siliques 30–60 mm long, 5–10 mm wide, the beak 10–25 mm long; n = 9. Cultivated for the edible roots, rarely escaping but usually not persisting; introduced from Europe; 1 (0). The yellow-flowered species, *R. raphanistrum* L., is an occasional weed that is to be expected in the state.

Rorippa Scop.

Plants glabrous to hirsute with simple hairs, annual, biennial or perennial, from taproots or rhizomes; leaves alternate and basal, lyrate-pinnatifid, sinuately toothed, lobed, or uppermost subentire, petiolate to sessile, auriculate in some; flowers in ebracteate racemes; sepals 4, deciduous; petals 4, yellow, fading pinkish in some, truncate to rounded apically; stamens 6, the filaments lacking glandular processes; style prominent to almost lacking, the stigma capitate; fruit a sessile silique or silicle from one to several times longer than broad, terete or somewhat flattened; seeds several to numerous, mostly biseriate.

Stuckey, R. L. 1972. Taxonomy and distribution of the genus *Rorippa* (Cruciferae) in North America. Sida 4: 279–430.

1. Plants perennial, with creeping rhizomes; petals 2.5–5 mm long 2
— Plants annual or biennial, rarely perennial, from taproots; petals 0.7–2.5 (3) mm long 4
2(1). Plants 2.5–6 dm tall; siliques 8–10 (15) mm long, fertile .... *R. sylvestris*
— Plants of various height but, if as above, the fruit globose and usually seedless 3
3(2). Plants 3–9 dm tall; sepals 1–2 mm long; fruit globose, usually seedless; introduced weeds of cultivated lands .... *R. austriaca*
— Plants 1–3 dm tall; sepals 2–4 mm long; fruit ovoid to oblong, bearing seeds; indigenous plants of saline valley bottoms .... *R. sinuata*
4(1). Siliques globose, 1–1.3 times longer than broad; partition circular in outline .... *R. sphaerocarpa*
— Siliques cylindroid or tapering at the apex, 2–5 times longer than wide or more; partition oblong to triangular in outline 5
5(4). Plants prostrate to decumbent; stems 2 dm long or less; siliques tapering to apex, minutely papillose .... *R. tenerrima*
— Plants prostrate to decumbent or erect; stems often over 2 dm long; siliques cylindroid or tapering to apex, smooth 6

6(5). Plants erect; stems mostly 3–10 dm long; pedicels usually as long as the fruit, mostly 4–12 mm long .... R. islandica
— Plants prostrate, decumbent, ascending, or erect, mostly 2–5 dm tall; pedicels usually shorter than the fruit, mostly 2–4 mm long .... R. curvipes

Rorippa austriaca (Crantz) Besser Austrian Fieldcress. [*Nasturtium austriacum* Crantz; *Radicula austriaca* (Crantz) Small]. Perennial, from thickened rhizomes, the stems 3–9 dm tall, erect, slender, finely puberulent; leaves 3–6 cm long, oblong to oblong-ovate, unequally serrate, glabrous, narrowed to a petiolelike auriculate base; racemes 7–12 cm long, in terminal panicles; pedicels 4–10 mm long in fruit, spreading-ascending; sepals 1–2 mm long; petals 3–15 mm long, yellow; silicles subglobose, 1.5–3 mm long; 2n = 16. Occasional to rare weed of cultivated places, reported for south-central Utah by Stuckey (1972); 0 (0).

Rorippa curvipes Greene Annual or short-lived perennial, arising from taproots; stems (1) 2–5 dm long, prostrate to decumbent, ascending or erect, glabrous; leaves (2) 4–8 (12) cm long, 3–15 mm wide, oblong, obovate, spatulate or oblanceolate, entire, crenate, irregularly serrate, or the lower ones pinnatifid, petiolate to sessile, often auriculate; racemes 4–15 cm long, in axillary and terminal racemes; pedicels mostly 2–5 mm long in fruit, spreading to ascending or descending; sepals 0.8–1.7 mm long; petals 0.5–2.8 mm long, yellow, rarely fading pinkish; silicles 1.8–8.7 mm long, 0.6–2.3 mm wide; style 0.2–1.3 mm long; 2n = 16. Three morphologically intergrading phases are recognized. They might not represent taxa.

1. Petals 1.2–2.8 mm long, longer than the sepals; siliques usually acute; plants prostrate to decumbent .... R. curvipes var. alpina
— Petals mostly 0.5–1.5 mm long, about as long as or shorter than the sepals; siliques acute or obtuse; plants prostrate to decumbent or erect 2

2(1). Siliques 1.4–5 mm long, acute to somewhat obtuse; petals 0.5–1 (1.3) mm long ..... R. curvipes var. curvipes
— Siliques 3.5–8 mm long or more, obtuse apically; petals 1–1.5 mm long .... R. curvipes var. integra

Var. alpina (Wats.) Stuckey [*Nasturtium obtusum* var. *alpinum* Wats., type from the Uinta Mts.; *R. obtusa* var. *alpina* (Wats.) Britt.; *R. alpina* (Wats.) Rydb.; *Radicula alpina* (Wats.) Greene]. Bogs and seeps, mainly in sagebrush and mountain brush communities upward to alpine tundra at 1370 to 3390 m in Cache, Daggett, Duchesne, Emery, Garfield, Summit, Uintah, and Utah counties; Idaho, Wyoming, and Colorado; 16 (ii).

Var. curvipes [*Cardamine palustris* var. *jonesii* Kuntze, type from City Creek Canyon; *Rorippa underwoodii* Rydb.; *Radicula curvipes* (Greene) Greene; *Radicula underwoodii* (Rydb.) Heller; *Rorippa obtusa* authors, not (Nutt.) Britt.; *Rorippa curvisiliqua* authors, not (Hook.) Bessey ex Britt.]. Ponds, streams, and seep margins in riparian, mountain brush, aspen, lodgepole pine, and spruce-fir communities, at 1370 to 3355 m in Beaver, Daggett, Duchesne, Garfield, Grand, Iron, Kane, Millard, Piute, Salt Lake, San Juan, Sanpete, Sevier, Uintah, Utah, Wasatch, Washington, and Wayne counties; Alberta and Saskatchewan east to Wisconsin and south to California, Arizona, New Mexico and Kansas; 50 (xi).

Var. integra (Rydb.) Stuckey [*R. integra* Rydb., type from Wasatch Mts.; *Radicula integra* (Rydb.) Heller; *Rorippa obtusa* var. *integra* (Rydb.) Marie-Victorin]. Stream banks and beaches, typically in spruce-fir communities, but sometimes much below them, at 2070 to 3390 m elevation in Box Elder, Daggett, Duchesne, Emery, Piute, Summit, Uintah, Utah, and Wasatch (the probable type area) counties; Alberta south to California and east to Montana and Wyoming; 16 (ii).

Rorippa islandica (Oeder) Borbas [*Sisymbrium islandicum* Oeder]. Annual or biennial, arising from taproots, the stems 3–10 dm tall, erect, glabrous or pubescent; leaves mostly 2–10 cm long, 0.6–2.5 cm wide, more or less pinnatifid, or cauline ones merely toothed, petiolate to sessile and more or less auriculate; racemes 3–10 cm long or more, both axillary and terminal; pedicels mostly 4–10 mm long in fruit, ascending to spreading or descending; sepals 1.2–2.5 mm long; petals 0.8–3.5 mm long, yellow, fading pinkish or purplish; siliques 3–8 mm long, rarely more, 2–3 mm wide; valves glabrous; style 0.2–1.2 mm long. Presented herein is a conservative approach to the species, maintaining *R. palustris* (L.) Besser [*Sisymbrium amphibium* var. *palustre* L.; *Sisymbrium palustre* (L.) Pollich; *Radicula palustris* (L.) Moench; *Nasturium palustre* (L.) DC.] within the broad definition of *R. islandica*, thereby differing from the recent treatment of the species complex in North America by Stuckey (1972). Within the North American propulation, Stuckey recognizes several subspecies and varieties, denoting a bewildering array of morphological and geographically reinforced infraspecific variation. In Utah, we have only 2 of the 11 entities recognized.

1. Leaves glabrous beneath; stems glabrous or pubescent only at the base .... R. islandica var. glabra
— Leaves hairy beneath; stems pubescent usually to the apex .... R. islandica var. hispida

Var. glabra (Schulz) Welsh & Reveal [*Nasturtium palustre* var. *glabrum* Schulz; *Radicula glabra* (Schulz) Britt.; *R. palustris* ssp. *glabra* (Schulz) Stuckey]. Palustrine and riparian sites, often in mud, at 1280 to 3175 m in Box Elder, Cache, Duchesne, Garfield, Rich, Sevier, and Utah counties; Alberta south to New Mexico; Cuba (the type area); 13 (0).

Var. hispida (Desv.) Butters & Abbe [*Brachylobus hispidus* Desv.; *Sisymbrium hispidum* (Desv.) Poir.; *Nasturtium hispidum* (Desv.) DC.; *N. palustre* var. *hispidum* (Desv.) Gray; *Radicula hispida* (Desv.) Britt.; *R. palustris* var. *hispida* (Desv.) Robins.; *Rorippa palustris* ssp. *hispida* (Desv.) Jonsel]. Palustrine and riparian sites, often in mud, from 1280 to 2600 m in Cache, Davis, Duchesne, Grand, Salt Lake, San Juan, Sevier, Summit, Uintah, Utah, and Weber counties; widespread in North America; 20 (i).

Rorippa sinuata (Nutt.) A. S. Hitchc. [*Nasturtium sinuatum* Nutt. ex T. & G.; *Radicula sinuata* (Nutt.) Greene]. Perennial, arising from a slender rhizome; stems 0.8–5 dm long, often more or less decumbent, glabrous to minutely hairy; leaves (1) 2–8 cm long, oblong in outline, pinnatifid, petiolate to sessile and somewhat auriculate; racemes 3–15 (25) cm long, in axillary and terminal racemes; pedicels 5–11 mm long in fruit, ascending to recurved; sepals 2.7–4.5 mm long; petals 3.5–5.5 mm long, yellow, fading light yellow; siliques 5–12 mm long, ca 1.5 mm wide, glabrous to roughened, narrowed to a style, this 1–2.5 mm long. Palustrine, riparian, and other moist, often saline, sites at 1125 to 1435 m in Uintah and San Juan counties; widespread in the western states; 6 (i).

Rorippa sphaerocarpa (Gray) Britt. [*Nasturtium sphaerocarpum* Gray; *N. obtusum* var. *sphaerocarpum* (Gray) Wats. ex Allen; *Radicula sphaerocarpa* (Gray) Greene; *R. obtusa* var. *sphaerocarpa* (Gray) Robins.; *Rorippa obtusa* var. *sphaerocarpa* (Gray) Cory]. Annual, arising from taproots; stems 1–4 dm long, decumbent to erect, sparingly hirsute below; leaves 3–10 cm long, 0.3–1.5 (3) cm wide, entire, crenate, serrate or pinnatifid; racemes 2–10 cm long, in axillary and terminal racemes; pedicels 1.5–4.2 mm long, ascending to recurved, often secund; sepals 0.7–1.3 mm long; petals 0.6–1.2 mm long, yellow, fading pinkish; fruit subglobose, 1–2.5 mm long and as wide or 1–1.3 times longer than broad; style 0.4–0.7 mm long. Moist sites at 1370 to 3050 m in Garfield and Utah counties where evidently rare; Arizona and New Mexico northward to Wyoming; 3 (0). This species approaches both *R. curvipes* and *R. islandica* var. *hispida*. Despite the apparent convergences, *R. sphaerocarpa* appears to exist as a functional entity and not just as an extreme morphological phase of either of the two taxa it approaches.

Rorippa sylvestris (L.) Besser [*Sisymbrium sylvestre* L.]. Rhizomatous perennial; stems 2.5–6 dm tall, usually branched above, glabrous or hirsute below; leaves 4–10 cm long, ovate or oblong to oblanceolate or obovate, lobed to pinnatifid, the divisions narrow, incised or sharply toothed to entire, petiolate; pedicels spreading to ascending, typically 6–12 mm long; sepals 2–3 mm long, the outer pair not saccate; petals 3–4 mm long; siliques linear, 8–10 (15) mm long, usually not curved; style ca 1 mm long; n = 16, 20, 24. Mud of a lake shore at 1590 m in Wasatch County; widely distributed in the U. S.; introduced from Europe; 1 (0).

Rorippa tenerrima Greene [*Radicula tenerrima* (Greene) Greene]. Annual, arising from taproots, the stems 0.4–2 dm long, decumbent to prostrate, glabrous; leaves 1–5 (8) cm long, 5–15 mm wide, lyrate-pinnatifid to subentire; racemes 2–10 cm long, in axillary and terminal racemes; pedicels 1–3 (4) mm long, ascending to spreading; sepals 0.7–1.2 mm long; petals 0.6–0.8 mm long, yellow; siliques 3–8 mm long, 0.8–2 mm wide, tapering to apex; style 0.3–1 mm long. Palustrine and riparian sites at 1403 to 2990 m in Cache, Daggett, Emery, Garfield, Juab, Kane, Iron, Millard, Salt Lake, Sanpete, Sevier, and Washington counties; Washington eastward to North Dakota and southward to Mexico; 23 (iv).

Schoencrambe Greene

Glabrous, perennial, arising from a branching caudex or creeping rhizome; leaves alternate and some basal or all cauline, some pinnatifid or all entire, petiolate to sessile, not auriculate; flowers in ebracteate racemes; sepals 4, deciduous; petals 4, yellow or white with purple veins, or lavender; stamens 6, the filaments lacking glandular processes; style almost lacking, the stigma expanded and bilobed or unexpanded and not bilobed; silique many times longer than broad, terete or slightly flattened; valves 1-nerved; seeds uniseriate.

Rollins, R. C. 1938. *Glaucocarpum*, a new genus in the Cruciferae. Madroño 4: 232–235.

1. Flowers yellow; plants from rhizomes and/or caudices .. 2
— Flowers lavender or white with purplish veins; plants from caudices 3
2(1). Plants clump-forming, arising from superficial caudices; known only from Uintah County .... *S. suffrutescens*
— Plants not clump-forming, arising from rhizomes and subterranean caudices; plants widespread .... *S. linifolia*
3(2). Cauline leaves linear to narrowly oblong; plants of Uintah County .... *S. argillacea*
— Cauline leaves elliptic to oblanceolate; plants of Emery County .... *S. barnebyi*

Schoencrambe argillacea (Welsh & Atwood) Rollins [*Theylpodiopsis argillacea* Welsh & Atwood, type from Big Pack Mt., Uintah County]. Perennial, glabrous and glaucous herbs; stems 13–30 cm tall, simple or branched, arising from a branching, ligneus caudex; leaves 9–35 mm long, 0.8–2 mm wide, sessile throughout, the cauline not auriculate, linear, somewhat thickened, acute or rounded; racemes 5- to 22-flowered; pedicels 7–18 mm long, ascending; sepals 4.2–6.5 mm long, violet, with hyaline margins; petals 7.8–10.9 mm long, 2.5–3.2 mm wide, white or lilac, with conspicuous purple veins, the blade not distinguished from the claw; anthers 1.7–2.5 mm long; siliques 18–25 mm long, 1–1.2 mm wide, sessile, terete, ascending to erect, typically curved; style 0.5–1 mm long, obconic; stigma bilobed. Shadscale, Indian ricegrass, pygmy sagebrush, and other mixed desert shrub communities on the lower Uinta and upper Green River Shale formations at 1525 to 1720 m in Uintah County; endemic; 5 (0).

Schoencrambe barnebyi (Welsh & Atwood) Rollins [*Thelypodiopsis barnebyi* Welsh & Atwood, type from San Rafael Swell, Emery County]. Perennial, hirtellous to glabrous and glaucous herbs; stems 10–25 cm tall, simple or branched, arising from a branching, ligneus caudex; leaves 13–51 mm long, 4–24 mm wide, petiolate, the petioles 0.4–10 mm long, the blades oblong to elliptic or oblanceolate, entire or incipiently toothed, not auriculate, rounded to obtuse apically; racemes 2- to 8-flowered (or more); pedicels 4–13 mm long in flower to 27 mm long in fruit, ascending to spreading; sepals 6–8 mm long, green to violet, with hyaline margins; petals 9.5–12 mm long, 2.5–3.5 mm wide, white or lavender, conspicuously purplish veined, the claw narrower than the blade; anthers 2.5–3 mm long; siliques 34–65 mm long, 1–2 mm wide, subsessile, terete, ascending to spreading; style 1–1.2 mm long, obconic; stigma bilobed. Mixed shadscale, eriogonum, and ephedra communities on the Chinle Formation at ca 1705 to 1740 m in Emery County; endemic; 7 (ii). This beautiful plant is a mirror-image congener of *S. argillacea* but differs in being larger in all its parts. The problem of generic disposition was discussed with the protologue of *Thelypodiopsis barnebyi* (Brittonia 33: 301. 1981), where it was noted that the "ultimate disposition of these unique species awaits reevaluation of generic relationships among primitive cruciferous genera, . . ." Placement within *Schoencrambe* by Rollins satisfies that requirement in part, at least, but leads to inclusion of *Glaucocarpum suffrutescens* in this genus also.

Schoencrambe linifolia (Nutt.) Greene [*Nasturtium linifolium* Nutt.; *N. pumilum* Nutt., not J. St. Hil.; *Sisymbrium linifolium* (Nutt.) Nutt ex T. & G.; *Sisymbrium pygmaeum* Nutt. ex T. & G.; *Erysimum glaberrimum* H.

& A.; *Schoencrambe pygmaea* (Nutt.) Greene; *S. pinnata* Greene, type from Garfield County?; *S. decumbens* Rydb.; *S. linifolia* var. *pinnata* (Greene) A. Nels. in Coult. & Nels.; *Sisymbrium decumbens* (Rydb.) Blank.; *S. linifolium* var. *pinnatum* (Greene) Schulz; *S. linifolium* var. *decumbens* (Rydb.) Schulz]. Perennial, glabrous and usually glaucous, arising from a simple or more usually branched caudex, this from a creeping, deeply placed rhizome; stems (1.5) 2–9 dm tall, erect or less commonly ascending to decumbent, simple or branched; leaves 1.3–9.3 cm long, 1–25 mm wide, entire to deeply pinnatifid, the basal ones often deciduous by midanthesis; pedicels 2–9 (10) mm long in fruit, ascending to spreading; sepals 4.3–6 mm long, yellowish; petals 7.5–11 mm long, yellow, spreading; siliques 25–65 (75) mm long, 0.8–1.2 mm wide, terete, sessile, erect or ascending; style 0.3–0.6 mm long; stigma bilobed; 2n = 14. Salt desert shrub, sagebrush, pinyon-juniper, mountain brush, and aspen communities at 1155 to 2690 m in all Utah counties, except Cache, Davis, Salt Lake, San Juan, Washington, and Weber; British Columbia east to Montana and south to Nevada and New Mexico; 131 (xix). Despite the widely sporadic distribution in the Great Basin, this is essentially a plant of the Colorado Basin in Utah. The indecision as to the generic position of this plant is indicated by the numerous generic names associated with it. The plants simulate both *Caulanthus* and *Sisymbrium* in a broad sense but do not show apparent relationship with the introduced annual or biennial weedy species of *Sisymbrium*. It seems best to place this plant within *Schoencrambe*, even though the principle diagnostic features involve vegetative instead of floral characteristics.

Schoencrambe suffrutescens (Rollins) Welsh & Chatterley [*Thelypodium suffrutescens* Rollins in Graham, type from Big Pack Mt., Uintah County; *Glaucocarpum suffrutescens* (Rollins) Rollins]. Perennial clump-forming herbs; stems herbaceous 1–25 dm tall, glabrous, arising from a stout taproot and branching subligneous caudex; leaves (7) 10–25 mm long, 3–10 mm wide, elliptic to lanceolate or oblanceolate, shortly petiolate to subsessile, not auriculate; pedicels (2.5) 3–12 mm long, curved-ascending to erect, glabrous; sepals 4–6 mm long, yellowish or greenish; petals 9–11 mm long, yellow, spatulate, ascending to spreading; siliques 10–20 mm long, 1.2–3 mm wide, flattened, erect, glabrous, strongly 1-veined; style (0.5) 1–2 mm long; stigma capitate, entire; seeds 4–8 per locule. Shadscale, pygmy sagebrush, mountain mahogany, juniper, and other mixed desert shrub communities at 1645 to 1830 m on calcareous shale of the Green River Formation, west of Willow Creek in the vicinity of Big Pack Mt., Uintah County and along the Wrinkles Road west of the Green River, Duchesne County; endemic; 15 (iv). In general aspect, involving nature of the caudex, habit, and arrangement of branches and leaves, this plant is essentially identical with *S. argillacea* and *S. barnebyi*. *S. suffrutescens* differs from the *S. argillacea* and *S. barnebyi* in about the same manner as they differ from *S. linifolia*. And, it seems probable that all will be differently placed in the future, possibly in an enlarged *Thelypodium* or *Caulanthus*. The feature of veination of the silique is reminiscent of that of *Streptanthus cordatus*.

Sisymbrium L.

Glabrous or hirsute annuals or rarely biennials, arising from taproots; leaves alternate and basal, variously lobed to entire, the lower ones usually pinnatifid, reduced upwards, petiolate to sessile, not auriculate; flowers racemose; pedicels spreading to erect, not subtended by bracts; sepals 4, deciduous; petals 4, yellow; stamens 6, the filaments lacking glandular processes; style almost lacking, the stigma bilobed; fruit a sessile silique many times longer than broad, linear to tapering, terete, the valves usually 3-nerved; seeds several to many, uniseriate.

1. Leaves strongly dimorphic, the lower lyrate-pinnatifid, the uppermost with linear-filiform lobes; pedicels ascending; siliques 50–90 mm long, spreading-ascending to erect, not appressed; pedicels 6.3–8.5 mm long . . . . *S. altissimum*
— Leaves not dimorphic, the upper and lower lobed about the same; pedicels ascending to erect; siliques 10–45 mm long, erect or ascending, appressed or not; petals 2–4 mm long 2

2(1). Siliques and pedicels ascending to spreading, not appressed to the rachis; siliques (20) 25–45 mm long . *S. irio*
— Siliques and pedicels appressed-erect, siliques 10–15 mm long . . . . *S. officinale*

Sisymbrium altissimum L. Jim Hill Mustard; Tumbling Mustard. [*Norta altissima* Britt.]. Annual; stems 2.5–10 dm tall or more, sparingly to densely hirsute near the base, usually glabrous above; leaves 1–20 cm long or more, petiolate, the lower ones pinnatifid or merely lobed, becoming pinnatifid into linear-filiform segments upwardly; pedicels 4–10 mm long, stout, almost or quite as thick as the silique, ascending to spreading; sepals 3.6–5 mm long, often yellowish; petals 6.3–8.5 mm long, yellow, fading white, obovate to spatulate, spreading; siliques 50–90 mm long, 1–1.5 mm wide, terete, spreading to ascending; valves evidently 3-nerved; 2n = 14. Disturbed sites, but often in indigenous plant communities, at 820 to 2410 m in all Utah counties; widespread in North America; adventive from Europe; 104 (ix).

Sisymbrium irio L. [*Norta irio* (L.) Britt.]. Annual; stems 2–8 dm tall, erect, glabrous; leaves 1.5–10 (20) cm long, runcinate-pinnatifid, reduced upwardly; pedicels 6–10 mm long, slender, ascending; sepals 2–2.5 (3) mm long, greenish or yellowish; petals 3–4 mm long, yellow, oblanceolate, spreading-ascending; siliques (20) 25–45 mm long, 0.8–1 mm wide, ascending; 2n = 14, 28. Weed of disturbed sites at 820 to 1680 m in San Juan, Utah, and Washington counties; California to Arizona; adventive from Europe; 8 (iii).

Sisymbrium officinale (L.) Scop. Hedge Mustard. [*Erysimum officinale* L.]. Annual; stems 2.5–8 dm tall or more, hispid-hirsute throughout; leaves 1.5–20 cm long or more, lyrate-pinnatifid to pinnatifid, not especially dimorphic, the upper ones merely reduced; pedicels 2–3 mm long, stout, erect, the tip about as thick as the silique; sepals 1.5–2.2 mm long, green or yellow; petals 3–4 mm long, yellow, fading white, narrowly oblanceolate, ascending; siliques 8–15 mm long, appressed-erect, tapering to a beaklike tip; valves 3-nerved; 2n = 14. Uncommon weedy plant of disturbed sites in Utah and Salt Lake counties; widespread in North America; adventive from Europe; 2 (0).

Smelowskia C. A. Mey.

Pubescent pulvinately caespitose perennials, arising from a branching caudex, the pubescence with branched and often with some simple hairs; leaves alternate or chiefly basal, usually pinnatifid, petiolate to sessile, not auriculate; flowers subcorymbose to racemose; pedicels not subtended by bracts; sepals 4, deciduous; petals 4, white rarely purplish; stamens 6, the filaments lacking glandular processes; style prominent; stigma expanded; fruit a silique, 3 to several times longer than broad, subterete or compressed; valves 1-nerved; seeds few to several per locule.

Drury, W. H. and R. C. Rollins. 1952. The North American representatives of *Smelowskia*. Rhodora 54: 85–119.

Smelowskia calycina **C. A. Mey.** Perennial, caespitose, with short to elongate caudex branches, these clothed with marcescent leaf bases; stems 4–20 cm tall, pubescent with short, branched and long, simple or branched hairs, or rarely glabrous; basal leaves 0.5–5 cm long, 4–16 mm wide, pinnately divided; cauline leaves reduced upwardly; pedicels mostly 3–8 mm long in fruit, ascending to spreading-ascending, pubescent; sepals 2–3.2 mm long, often tinged purplish, pubescent; petals 3–8 mm long, white to cream or tinged pink to lavender, ovate, spreading; siliques 5–9 mm long, slightly flattened parallel to the septum or terete; style 0.2–1 mm long; seeds 6–10; 2n = 12. Alpine tundra, mainly in talus and rock stripes, at 2900 to 3965 m in Beaver, Daggett, Duchesne, Grand, Iron, Piute, Salt Lake, San Juan, Sevier, Summit, Uintah, and Utah counties; Alaska Yukon and Northwest Territories south to Nevada and Colorado; 66 (xii). Our material is **var. *americana* (Regel & Herder) Drury & Rollins** [*Hutchinsia calycina* var. *americana* Regel & Herder; *S. americana* (Regel & Herder) Rydb.; *S. lineariloba* Rydb.; *S. lineariloba* f. *virescens* Schulz].

Stanleya **L.**

Perennial, glabrous to simple pubescent herbs or subshrubs arising from taproots or caudices; leaves alternate and basal, pinnatifid to entire, petiolate to sessile, auriculate in some; flowers showy, racemose; pedicels not subtended by bracts; sepals 4, deciduous; petals 4, yellow to greenish yellow; stamens 6, the filaments lacking glandular processes, long-exserted; style lacking to prominent, the stigma small; fruit a long-stipitate silique, linear, terete to compressed; valves with 1 or more nerves; seeds numerous, uniseriate.

Rollins, R. C. 1939. The cruciferaceous genus *Stanleya*. Lloydia 2: 109–127.

1. Middle and upper cauline leaves sessile, auriculate; leaves all entire; petal-claw glabrous on both surfaces .... *S. viridiflora*
— Middle and upper cauline leaves petiolate to subsessile, not auriculate; leaves various; petal-claw pubescent on the inner or outer surfaces 2
2(1). Leaves entire or merely toothed, the basal ones ovate to elliptic, cauline ones ovate to lanceolate; plants of east-central and northeastern Utah .... *S. integrifolia*
— Leaves, at least some, pinnatifid, only the upper ones entire and usually lanceolate to lance-linear; distribution various 3
3(2). Plants woody at base, the caudex well developed; blades of petals mostly 1–3 mm wide; our most common and widespread species of the genus .... *S. pinnata*
— Plants herbaceous, the caudex not developed; blades of petals 3–6 mm wide or more; an uncommon to rare plant of southeastern Utah, if it occurs here at all .... *S. albescens* Jones

Stanleya integrifolia **James** [*S. pinnatifida* var. *integrifolia* (James) Robins.; *S. glauca* var. *latifolia* Cockerell; *S. pinnata* var. *integrifolia* (James) Rollins]. Perennial, the caudex well-developed and more or less woody; stems 2.5–9 dm tall or more, sparingly hairy, glaucous, simple or branched; leaves 0.8–12.3 cm long, 0.3–3.5 cm wide, elliptic to ovate or lanceolate, all entire or lower ones merely toothed, puberulent, glaucous; sepals mostly 9–19 mm long, yellowish, reflexed, glabrous; petals 10.5–15.5 mm long, the blade 1.4–2.6 mm wide, the claw hairy within; siliques 33–75 mm long, 1.2–1.8 mm wide, subterete; stipe 10–25 mm long. Shadscale, other salt and mixed desert shrub and pinyon-juniper communities at 1460 to 2745 m in Carbon, Daggett, Duchesne, Emery, Grand, Uintah, and Wayne counties; Wyoming east to Kansas; 28 (iii). Our material is distinctive and about equivalent in diagnostic features as exist among the other species of *Stanleya* in Utah.

Stanleya pinnata **(Pursh) Britt.** Prince's Plume. [*Cleome pinnata* Pursh; *S. pinnatifida* Nutt.; *S. heterophylla* Nutt. ex T. & G.; *S. fruticosa* Nutt.; *S. arcuata* Rydb.; *S. canescens* Rydb.; *S. glauca* Rydb.]. Perennial, the caudex well developed and slightly to very woody; stems (2.5) 3.5–12 (15) dm tall or more, glabrous to pilose, glaucous, simple or branched; leaves mostly 5–18 cm long, 2–5 cm wide or more, lanceolate to elliptic in outline, pinnatifid or the upper usually entire and narrowly lanceolate to elliptic, not auriculate, glabrous or sparsely pilose; pedicels 4–13 mm long, spreading; sepals 11–22 mm long, yellowish, reflexed, glabrous; petals 11–17 mm long, the blade 1–3.8 mm wide, the claw hairy within; siliques (30) 35–70 (80) mm long, 1.2–2 mm wide, subterete to flattened; stipe 12–24 mm long, puberulent at base; n = 14. Seleniferous soils derived from shales, mudstones, and siltstones in many geological formations, in salt desert shrub, mixed desert shrub, sagebrush, pinyon-juniper, and mountain brush communities at 915 to 2290 m in all Utah counties except Cache, Davis, Morgan, Salt Lake, Summit, Wasatch, and Weber; Idaho east to North Dakota south to California, Arizona, and New Mexico; 161 (xxvi). This plant is a primary indicator of the poisonous element selenium, the growth in any given place provides evidence of the presence of that element in the soil. Our material belongs to **var. *pinnata***, but the **var. *gibberosa* Rollins**, known from Uinta County, Wyoming, is to be sought in Summit County, Utah. Its leaves are all bipinnate. The **var. *inyoensis* (Munz & Roos) Reveal** may be encountered in Washington County; it differs from var. *pinnata* in being a subshrub.

Stanleya viridiflora **Nutt. ex T. & G.** Desert Plume. [*S. collina* Jones]. Perennial, the caudex simple or branched, somewhat woody at base; stems 3–12 dm tall or more, glabrous and glaucous, simple or branched; leaves 7–30 cm long, the basal ones 1.8–8.7 cm wide, petiolate, the blades entire or somewhat runcinate-pinnatifid, the upper entire, sessile and auriculate, glabrous; pedicels mostly 6–12 mm long, spreading; sepals 11–20 mm long, yellow, reflexed, glabrous; petals 10–19 mm long, the blade 0.8–1.5 (3) mm wide, not much wider than the glabrous claw; siliques 35–70 mm long, subterete; stipe 14–25 mm long; 2n = 28. Gypsiferous or clay substrates in sagebrush, wildrye, pinyon-juniper, and mountain brush communities at 1525 to 2745 m in Duchesne, Emery, Garfield, Salt Lake (?), Summit, Uintah, and Wayne counties; Idaho and Wyoming to Nevada and Oregon; 18 (iii).

Streptanthella Rydb.

Glabrous, annual or winter annual, arising from taproots; leaves alternate, entire to shallowly dentate, tapering to the base, not auriculate; flowers racemose; pedicels spreading-recurved to recurved, not subtended by bracts; sepals 4, deciduous; petals 4, white or purplish; stamens 6, the filaments lacking glandular processes; style obsolete, the small capitate stigma borne atop a beaklike extension of the fruit, this a subsessile, slender silique many times longer than broad, strongly compressed; valves 1-nerved, dehiscent at the base only; seeds several to many per locule, uniseriate.

Streptanthella longirostris (Wats.) Rydb. [*Arabis longirostris* Wats., type from Stansbury Island; *Streptanthus longirostris* (Wats.) Wats.; *Thelypodium longirostris* (Wats.) Jepson]. Annual, the stems erect or asccending, usually branched throughout, 1–5 dm tall; leaves 1.5–8.5 cm long, 1–12 mm wide, oblanceolate to elliptic or lance-linear, sinuate-dentate to entire, the lower usually deciduous by anthesis, reduced upwardly; pedicels 1.5–6 mm long curved in fruit; sepals 2–4.8 mm long, greenish or purplish, with scarious margins; petals 5–8 mm long, white with purplish veins; siliques 30–60 mm long, 1.5–2 mm wide, reflexed-descending; valves narrowed apically into an indehiscent beak 3–7 mm long; style obsolete; 2n = 28. Sandy, clayey, or gravelly soils in warm desert shrub, mixed desert shrub, pinyon-juniper, and low elevation grassland communities from 750 to 2350 m in Beaver, Box Elder, Carbon, Daggett, Duchesne, Emery, Garfield, Grand, Juab, Kane, Millard, San Juan, Sevier, Tooele, Uintah, Washington, and Wayne counties; Washington east to Wyoming and southward to Baja California and New Mexico; 137 (xxiii).

Streptanthus Nutt.

Glabrous, perennial, arising from taproots, rarely rhizomatous; leaves alternate and basal, simple, dentate to entire, the basal ones petiolate, the upper becoming sessile, entire, and auriculate; flowers in ebracteate racemes; sepals 4, deciduous; petals 4, chestnut to brown purple or purple; stamens 6, often in 3 pairs as regards length and position, the filaments lacking glandular processes; style conspicuous, expanded; stigma bilobed; fruit a short-stipitate silique many times longer than broad, much flattened; valves with one main nerve and more or less reticulate lateral nerves; seeds several to many per locule, uniseriate.

Streptanthus cordatus Nutt. ex T. & G. [*Euklisia cordata* (Nutt.) Rydb.; *Cartiera cordata* (Nutt.) Greene; *S. crassifolius* Greene]. Perennial, the stems not inflated, 1.8–5.7 (8) dm tall, glabrous and glaucous; basal leaves 1.5–8 (15) cm long, 0.5–2 (5) cm wide, spatulate to obovoid or oblanceolate, variously dentate, often sharply ciliate at least basally; cauline leaves becoming sessile and auriculate, ovate to oval or lanceolate; pedicels 4–9 mm long, ascending, glabrous; sepals (5) 7–10.5 mm long, usually purplish, glabrous except apically; petals 10–14.5 mm long, purple to chestnut, the broad claw not constricted at the juncture of blade, ascending to recurved; siliques 50–85 mm long, 3–5.8 mm wide, ascending to erect, glabrous; style 1–3 mm long, expanded upwards, the stigma bilobed; n = 14. Shadscale, greasewood, horsebrush, sagebrush, pinyon-juniper, mountain brush, and Douglas fir-white fir communities at 1065 to 2960 m in all Utah counties; Oregon to Wyoming south to California, Arizona, and New Mexico; 165 (xxix). A specimen collected near Kanosh, Millard County, by Pickford (130–OGDF) is definitely rhizomatous.

Thelypodiopsis Rydb.

Glabrous or pubescent with simple hairs, annual or biennial to perennial, arising from taproots; leaves alternate or alternate and basal, simple, oblong-oblanceolate to lanceolate or ovate, toothed to entire, petiolate or sessile and auriculate or merely sessile; flowers in racemes; pedicels ascending or spreading-ascending, not subtended by bracts except in some lowermost flowers; sepals 4, deciduous; petals 4, white, pink, lavender, or yellow; stamens 6, the filaments lacking processes; style expanded upwards; stigma bilobed or not; fruit a sessile to stipitate silique, many times longer than broad, subterete to terete; valves 1 (3–5) -nerved; seeds uniseriate. Note: The species included herein have been treated as members of *Streptanthus*, *Thelypodium*, an expanded *Sisymbrium*, and even *Caulanthus*. The generic problem has been reviewed by Al-Shebaz (1973; see *Thelypodium*), but the solution adopted by that author, and followed in part here, is not altogether acceptable for our species. *Thelypodium sagittatum* mirrors both *Thelypodiopsis elegans* and *T. ambigua* and is again placed in this genus also.

1. Petals yellow 2
— Petals pink, lavender, or white 3

2(1). Pedicels, sepals, and siliques glabrous; petal blade not especially constricted at juncture with petaloid claw .. .. *T. aurea*

— Pedicels, sepals, and siliques sparingly villous; petal blade conspicuously constricted at juncture with broadened membranous claw .... *T. divaricata*

3(1). Style expanded, the stigma bilobed; plants of eastern Utah, and in Kane County 4
— Style not expanded or the stigma not bilobed; plants of the Great Basin or of montane sites 5

4(3). Pedicels, sepals, and siliques glabrous; basal leaves often over 6 times longer than broad; plants known from Kane County .... *T. ambigua*

— Pedicels, sepals, and siliques sparingly villous; basal leaves usually less than 5 times longer than broad; plants of eastern Utah .... *T. elegans*

5(3). Stigma not bilobed, the style not expanded; plants of moist sites from low to upper middle elevations in northern Utah .... *T. sagittata*

— Stigma bilobed or the style expanded (or both); siliques conspicuously torulose, often contorted; plants of xeric sites in western Utah .... *T. vermicularis*

Thelypodiopsis ambigua (Wats.) Al-Shebaz [*Thelypodium ambiguum* Wats.]. Biennial or short-lived perennial; stems not inflated, 2–10 (12.5) dm tall, glabrous throughout; leaves basal and cauline, the basal ones 3–15 (20) cm long, 0.6–2.8 cm wide, sinuate-dentate to entire, the cauline reduced upwards, becoming entire and lance-ovate to lanceolate, 0.5–6.6 cm long, 0.3–1.8 cm wide, sessile and auriculate; pedicels 4–12 cm long, curved-ascending to spreading-ascending, glabrous; sepals 5–7 mm long, often pinkish hyaline, glabrous, erect; petals 9–12 mm long, the blade 3–4 mm wide, pink to lavender or white, the claw broad, slightly constricted at juncture of blade; anthers 2.5–3.5 mm long; siliques 55–75 mm long, 1.2–2 mm wide, short-stipitate, terete, ascending to erect; style 1–2.5 mm long, expanded upwards; stigmatic lobes developed. Pinyon-juniper community from 1530 to 1830 m in Kane (Eastwood & Howell 9300 US; Welsh & Atwood 9706 BRY) and Washington (Palmer 27, southern Utah, 1877 US) counties; Mohave County, Arizona; 3 (i).

Thelypodiopsis aurea (Eastw.) Rydb. [*Thelypodium aureum* Eastw.; *Sisymbrium aureum* (Eastw.) Payson]. Biennial or rarely short-lived perennial; stems not inflated, 1–7 dm tall, glabrous throughout or somewhat pubescent below with flattened, flexuous hairs; leaves basal and cauline, the basal ones 2.5–6.5 cm long, 5–13 mm wide, irregularly toothed to subentire, the cauline reduced upwardly, becoming entire and lanceolate to ovate-lanceolate, 1.3–6.5 cm long, 4–23 mm wide, sessile and auriculate; pedicels 5–9 mm long, spreading-ascending, glabrous; sepals 5–9.3 mm long, yellowish, glabrous, ascending; petals 6.9–12.5 mm long, the blade not expanded, 1.5–3.5 mm wide, yellowish, the claw very broad, the ascending limb not especially constricted at juncture with claw; anthers 2.7–3.7 mm long; siliques 40–70 mm long, 1–1.5 mm wide, stipitate, the stipe 2–7 mm long, terete, ascending to erect, glabrous; style 1–3 mm long, expanded upwards; stigmatic lobes well developed; n = 11. Clay to sandy soils in semidesert shrub and pinyon-juniper communities at 1310 to 2135 m in San Juan County; Colorado and New Mexico; 11 (ii). This is a beautiful plant in full flower, with its bright golden flowers.

Thelypodiopsis divaricata (Rollins) Welsh & Reveal [*Caulanthus divaricatus* Rollins, type from 10 mi E of Hite]. Biennial or winter annual, erect, simple; stems not inflated, 2–11 dm tall, the herbage more or less pubescent with long, tangled hairs, at least below; leaves basal and cauline, the basal ones 1.5–10 cm long, 4–28 mm wide, irregularly toothed or rarely pinnatifid, subentire or dentate at apex only, the cauline reduced upwardly, becoming entire and ovate-lanceolate, 0.7–9 cm long, 4–28 mm wide, sessile and auriculate; pedicels 5–21 mm long, spreading-ascending, sparingly villous to glabrate; sepals (3.5) 4.8–7 mm long, yellowish, sparingly villous; petals 7–10 mm long, yellowish, the claw very broad, the ascending limb constricted at juncture with the claw; anthers 2.5–3.8 mm long; siliques 40–80 (90) mm long, 1–1.5 mm wide, sessile or subsessile, terete, ascending to erect, sparingly villous when young, becoming glabrous with age; style 1–2 mm long, with the stigma reduced or expanded and lobes not or only moderately well developed; n = 11. Fine textured substrates in shadscale, blackbrush, and pinyon-juniper communities at 1220 to 2075 m in Carbon, Emery, Garfield, Grand, San Juan, and Wayne counties; endemic; 48 (ix). The tall branched stems persist for a while, standing ghostlike above the barren substrates of southeastern Utah.

Thelypodiopsis elegans (Jones) Rydb. [*Thelypodium elegans* Jones, type from Westwater, Grand County; *Sisymbrium elegans* (Jones) Payson; *Streptanthus wyomingensis* A. Nels.; *Thelypodiopsis wyomingensis* (A. Nels.) Rydb.]. Biennial, or rarely winter annual, or a short-lived perennial; stems 1.2–9.5 dm tall, almost or quite glabrous to densely pubescent below with flattened, flexuous hairs; leaves basal and cauline, the basal ones 1–6.5 cm long, 3–15 mm wide, sinuate-dentate to irregularly toothed or entire, the cauline reduced upwardly, becoming entire and lanceolate to ovate-lanceolate or oblong, 0.3–8.5 (12) cm long, 3–20 mm wide, sessile and auriculate; pedicels 5–18 mm long, curved-ascending to spreading-ascending, glabrous or sparingly villous, erect; sepals 4–8 mm long, usually purplish, glabrous; petals 11–14.5 mm long, the blade 3–6 mm wide, pink to lavender or white with pinkish veins, the claw very broad, not constricted at juncture with blade; anthers 2.6–3.2 mm long; siliques 45–75 mm long, 1.2–1.8 mm wide, sessile or subsessile, terete, ascending to erect, glabrous or sparingly villous; style 1.8–3 mm long, expanded upwards; stigmatic lobes well developed. Fine to coarse textured substates in salt desert shrub, sagebrush, pinyon-juniper, and mountain brush communities at 1280 to 2075 m in Duchesne, Grand, and Uintah counties; Wyoming and Colorado; 65 (xiii).

Thelypodiopsis sagittata (Nutt.) Schulz Biennial or short-lived perennial; stems 2–10 dm tall, rarely more, simple or more usually branched, glabrous or hirsute with simple hairs below; basal leaves 2–25 cm long, 1–5 cm wide, ovate to oblanceolate or oblong, entire; cauline leaves 0.7–10 (14) cm long, 2–28 mm wide, ovate to lanceolate or lance-oblong, entire, glabrous or puberulent especially on the veins, auriculate; pedicels 2.5–15 (20) mm long, spreading to spreading-ascending; sepals 2.5–10 mm long, green to purplish, glabrous; petals 5–15 mm long, 1–4 mm wide, white to lavender or purple; anthers 1.5–5 mm long; siliques 10–65 mm long, 0.5–1.2 mm wide, stipitate, the stipes 0.3–2 mm long; style 0.5–3 mm long, more or less cylindric; stigma not bilobed. This species simulates *T. elegans* and *T. ambigua* in some features and care should be taken in distinguishing them. The species is divisible into two distinctive varieties in Utah.

1. Petals linear to oblanceolate, 1.5 mm wide or less; pedicels 2.5–9 mm long; plants of southwestern Utah .. .. *T. sagittata* var. *ovalifolia*

— Petals oblanceolate to spatulate, 1–4 mm wide; pedicels 5–15 mm long or more; distribution various .... *T. sagittata* var. *sagittata*

Var. ovalifolia (Rydb.) Welsh [*Thelypodium ovalifolium* Rydb., type from Panguitch Lake; *T. palmeri* Rydb., type from southern Utah; *T. sagittatum* ssp. *ovalifolium* (Rydb.) Al-Shebaz; *T. sagittatum* var. *ovalifolium* (Rydb.) Welsh & Reveal]. n = 13. Saline meadows at 1585 to 2565 m in Garfield, Iron, Juab, and Kane counties; Nevada; a Great Basin endemic; 5 (0).

Var. sagittata [*Streptanthus sagittatus* Nutt; *Pachypodium sagittatum* (Nutt.) Nutt. ex T. & G.; *Thelypodium sagittatum* (Nutt.) Endl. in Walp.; *T. nuttallii* Wats.; *T. amplifolium* Greene; *T. torulosum* Heller; *T. macropetalum* Rydb., type from Davis County; *Thelypodiopsis nuttallii* (Wats.) Schulz; *T. torulosa* (Heller) Schulz]. Clay or silty, often saline, soils in shadscale, greasewood, saltgrass, sedge grass, aspen, and sagebrush communities at 1280 to 2690 m in Box Elder, Cache, Davis, Duchesne, Emery, Juab, Millard, Salt Lake, Summit, Tooele, Uintah, Utah, Wasatch, Wayne, and Weber, counties; Washington and Montana south to Nevada, Utah, and Colorado; 26 (i).

Thelypodiopsis vermicularis (Welsh & Reveal) Rollins [*Thelypodium sagittatum* var. *vermicularis* Welsh & Reveal, type from 4 mi SE of Sigurd]. Biennial herbs; stems 8–60 cm tall, rarely more, simple or more usually branched, glabrous or hirsutulous with simple hairs below; basal leaves 2–15 cm long, 1–4 cm wide, ovate to oblanceolate or oblong, entire or toothed; cauline leaves 0.7–9 cm long, 0.2–2.3 cm wide, ovate to lanceolate or lance-oblong, entire or less commonly toothed, glabrous, auriculate; pedicels 6–10 mm long, spreading to spreading-ascending; sepals 3–7 mm long, white to green or yellowish, glabrous; petals 7–12 mm long, 2–4 mm wide, white; anthers 2–3 mm long; siliques 22–39 mm long, 1–2 mm wide, stipitate, the stipes 0.3–2 mm long; style often expanded, 0.5–3 mm long, the stigma obscurely 2-lobed. Salt desert shrub communities at 1310 to 2200 m in Box Elder, Juab, Millard, Sanpete, Sevier, Tooele, and Utah counties; Nevada; a Great Basin endemic; 28 (v). Expansion of styles has been regarded as a fundamental feature in segregation of *Thelypodiopsis* from *Thelypodium*. This species has the style only obscurely expanded and was therefore placed within *Thelypodium* initially. Failure of that feature as diagnostic has resulted in realignment of this taxon and of *T. sagittata*.

Thelypodium Endl.

Glabrous or pubescent with simple hairs, annuals, biennials, or short-lived perennials, arising from taproots; leaves alternate and basal, simple, lyrate-pinnatifid, toothed or entire, petiolate or sessile, auriculate in some; flowers racemose, pedicels not subtended by bracts; sepals 4, deciduous; petals 4, white, pink, lavender, or purple; stamens 6, the filaments lacking glandular processes; style slender to stout, cylindric to somewhat expanded upwards, the stigma small, entire or somewhat bilobed; fruit a stipitate silique many times longer than broad, terete or somewhat flattened; valves 1 (3) –nerved; seeds uniseriate. Note: *Pennellia micrantha* (Gray) Nieuwl. [*Thelypodium micranthum* (Gray) Wats.; *Heterothrix micrantha* (Gray) Rydb.] has been repeatedly reported from Utah, but no specimens are definitely known from the state. It differs from the species of *Thelypodium* reported here in having stellately pubescent leaves.

Al-Shebaz, I. A. 1973. The biosystematics of the genus *Thelypodium* (Cruciferae). Contr. Gray Herb. 204:3–148.
Payson, E. B. 1922. A monographic study of *Thelypodium* and its immediate allies. Ann. Missouri Bot. Gard. 9:233–324.

1. Cauline leaves definitely auriculate (see also *Thelypodiopsis*) 2
—   Cauline leaves petiolate or sessile but not auriculate 3
2(1). Inflorescence compact, not elongating in fruit, seldom more than 6 cm long; plants strict, known from south-central to southern and southeastern Utah .... *T. rollinsii*
—   Inflorescence soon elongating, 5–20 cm long or more; plants ascending to erect, of greasewood bottoms in western Utah .... *T. flexuosum*
3(1). Plants robust biennials, mostly 5–15 dm tall or more; racemes borne in corymbose panicles .... *T. integrifolium*
—   Plants various but, if (as rarely) as above, the racemes seldom if ever borne in corymbose panicles 4
4(3). Racemes mostly 1–8 dm long or more; petals 8 mm long or more, differentiated into a blade and claw; plants known only from Box Elder County ...... *T. millefolium*
—   Racemes mostly less than 1 dm long; petals 4–8 mm long, hardly differentiated into a blade and claw; plants not of Box Elder County 5
5(4). Sepals ascending to spreading; petals 1.5–3.5 mm wide; plants of eastern Utah .... *T. laxiflorum*
—   Sepals spreading to reflexed; petals 1–2 mm wide; plants of the Henry Mts., Garfield County .... *T. wrightii*

Thelypodium flexuosum Robins. in Gray Perennial from a woody taproot and caudex, this simple or branched, clothed with persistent leaf bases; stems few to several, decumbent to erect, 1.5–6 dm tall; basal leaves petiolate, 2–12 cm long or more, 0.5–3 cm wide, lanceolate to oblong or oblanceolate, cuneate to attenuate at the base; cauline leaves linear to linear-lanceolate, or lanceolate, sagittate to clasping basally, glabrous, entire, inflorescence corymbose, elongating in fruit; sepals 3–4.5 mm long, green with white margins; petals lavender to white, 6–10 mm long, 1.5–3.5 mm wide; anthers 1–2.5 mm long; pedicels 3–10 (15) mm long, spreading to spreading-ascending; siliques 1.1–4.2 mm long, terete, torulose, straight or curved, erect to ascending; style 0.3–3 mm long; stigma entire; stipes 0.2–1 mm long; seeds uniseriate. Fine textured substrates in valley bottoms in greasewood communities at 1525 to 1650 m in Beaver and Tooele counties; Nevada, Idaho, Oregon, and California; 2 (0).

Thelypodium integrifolium (Nutt.) Endl. Biennial, the stems mostly 5–30 dm tall, simple or branched, glabrous, glaucous; basal leaves 5–30 cm long, 1.5–10 cm wide or more, spatulate to lanceolate, ovate or obovate, entire to denticulate; cauline leaves 2–12 (15) cm long, 2–20 mm wide, oblanceolate to elliptic or lanceolate, sessile, not auriculate; pedicels 3–10 mm long, variously spreading, strongly to moderately flattened at base; sepals 3–7 mm long, often purplish, glabrous; petals 5–10 mm long, mostly 1–2 mm wide, white to lavender or purple; anthers 1–3 mm long; siliques 10–45 mm long, 1–2 mm wide, stipitate, the stipes up to 2 mm long; style slender, 0.5–3 mm long; stigma entire; n = 13. Four poorly defined and arbitrarily separable varieties are present.

1. Mature fruiting pedicels whitish, stout; flowers white; plants of southwestern Utah .. *T. integrifolium* var. *affine*
—   Mature fruiting pedicels not whitish or only seldom so, slender or stout; flowers pink to lavender or white; plants not of southwestern Utah 2
2(1). Mature fruiting pedicels not or only somewhat flattened at base; plants mostly of south-central to north-central Utah .... *T. integrifolium* var. *integrifolium*
—   Mature fruiting pedicels strongly flattened at base; plants of eastern and southeastern or northwestern Utah 3
3(2). Racemes mostly 1–8 cm long in fruit; stipes often less than 1 mm long; plants of northwestern Utah ....
    .... *T. integrifolium* var. *complanatum*
—   Racemes mostly more than 8 cm long in fruit; stipes often more than 1 mm long; plants of eastern and southeastern Utah .... *T. integrifolium* var. *gracilipes*

Var. affine (Greene) Welsh & Reveal [*Thelypodium affine* Greene; *T. rhomboideum* Greene; *Pleurophragma rhomboideum* (Greene) Schulz; *T. integrifolium* ssp. *affine* (Greene) Al-Shebaz]. Saline seeps, springs, irrigation canals, and stream sides from 850 to 1100 m elevation in Iron, Kane, Millard, and Washington counties; Nevada and California; 12 (iii).

Var. complanatum (Al-Shebaz) Welsh & Reveal [*T. integrifolium* ssp. *complanatum* Al-Shebaz]. Saline seeps and along streams at 1280 to 2075 m in Beaver, Box Elder, Juab, Millard, and Tooele counties; Nevada, California and Oregon; 12 (ii). This taxon is completely transitional to var. *affine*, and might not be worthy of taxonomic consideration.

Var. **gracilipes** Robins. [*T. gracilipes* (Robins.) Rydb.; *Pleurophragma gracilipes* (Robins.) Rydb.; *P. platypodium* Rydb., the type is from Moab; *T. rhomboideum* var. *gracilipes* (Robins.) Payson; *T. integrifolium* ssp. *gracilipes* (Robins.) Al-Shebaz]. Canyon bottoms, terraces, and hanging gardens at 1125 to 2440 m in Carbon, Daggett, Duchesne, Garfield, Grand, Kane, San Juan, Uintah, and Wayne counties; Arizona, Colorado, and New Mexico; 31 (xv). This variety is completely transitional to var. *integrifolium*.

Var. **integrifolium** [*Pachypodium integrifolium* Nutt. ex T. & G.; *Pleurophragma integrifolium* (Nutt.) Rydb.; *T. lilacinum* Greene; *Pleurophragma lilacinum* (Greene) Rydb.; *T. lilacinum* var. *subumbellatum* Payson]. n = 13. Marshes, seeps, stream sides, and other moist sites in Cache, Garfield, Iron, Piute, Rich, Salt Lake, Sevier, and Utah counties; Washington east to the Dakotas and south to Oregon, Colorado, and Nebraska; 16 (iii).

Thelypodium laxiflorum Al-Shebaz [*T. wrightii* var. *tenellum* Jones, type from Slate Canyon, Utah County; *Stanleyella wrightii* var. *tenella* (Jones) Payson]. Biennial; stems 1.5–15 dm tall or more, simple, glabrous throughout or hispid near base; basal leaves 3–15 cm long, 1–6 cm wide, pinnatifid to toothed; cauline leaves 3–9 cm long, 2–20 mm wide, subentire to entire, not auriculate; pedicels 4–13 mm long, spreading to descending, somewhat flattened at base; sepals 2.5–6 mm long, white to lavender, glabrous, ascending to spreading; petals 4–8 mm long, 1.5–3.5 mm wide, white to lavender; anthers 1–3 mm long; siliques 20–75 mm long, 0.3–1.2 mm wide, stipitate, the stipes 0.2–2 mm long; style 0.5–2 mm long. Pinyon-juniper, mountain brush, ponderosa pine, aspen, Douglas fir-white fir, and spruce-fir communities at 1155 to 3050 m in Beaver, Carbon, Duchesne, Grand, Iron, Kane, San Juan, Utah, and Washington counties; Nevada to Colorado; 15 (vii). This species is closely related to *T. wrightii*.

Thelypodium milleflorum A. Nels. [*T. laciniatum* var. *milleflorum* (A. Nels.) Payson]. Biennial, the stems mostly 4–12 dm tall, almost always hollow, simple or branched, glabrous and glaucous; basal leaves 6–15 cm long, 1–7 cm wide, oblong to lanceolate or ovate, toothed or pinnatifid; cauline leaves similar to the basal ones only gradually reduced upwardly, petiolate, not auriculate; pedicels 2–6 mm long, curved-ascending, slightly flattened at base; sepals 4–9 mm long, creamy white, glabrous; petals 8–15 mm long, 1–2 mm wide, white; anthers 2.5–5 mm long; siliques 25–85 mm long, 0.8–1.8 mm wide, stipitate, the stipes 0.5–5 mm long; style stout, 0.5–2 mm long; stigma entire; n = 13. Dry sites in sagebrush, juniper, and salt desert shrub communities at ca 1370 m in Box Elder County; British Columbia south to California and east to Idaho; 1 (0).

Thelypodium rollinsii Al-Shebaz Biennial, the stems (4) 5–16 (20) dm tall, simple or branched, glabrous and glaucous; basal leaves 1.3–7 cm long, 4–18 mm wide, spatulate to oblanceolate, toothed to subentire; cauline leaves 1–6 cm long, 1–8 mm wide, linear to narrowly lanceolate or oblong, erect, not flattened; sepals 4–7 mm long, lavender to purplish, glabrous; petals 6–10 mm long, 1.2–3 mm wide, lavender to purplish; anthers 2–4 mm long; siliques 20–60 mm long, 0.7–1 mm wide, stipi-tate, the stipe 0.5–6 mm long; style 0.5–2 mm long; stigma entire; n = 13. Saline substrates in greasewood, saltgrass, and other salt tolerant commuities at 1495 to 1650 m in Beaver, Carbon, Juab (type from 12 mi N of Scipio), Millard, Piute, Sanpete, and Sevier counties; endemic; 5 (iii).

Thelypodium wrightii Gray [*Stanleyella wrightii* (Gray) Rydb.]. Biennial, the stems 1.5–10 dm tall or more, simple or more commonly branched, glabrous throughout or hispid near the base; basal leaves 3–15 cm long, 1–6 cm wide, pinnatifid to lyrate-pinnatifid; cauline leaves 3–9 cm long, 2–20 mm wide, pinnatifid to sinuate-dentate, rarely subentire, not auriculate; pedicels 4–13 mm long, spreading to descending, somewhat flattened at base; sepals 2.5–6 mm long, white to lavender, glabrous, spreading to reflexed; petals 4–8 mm long, 1–2 mm wide, white to lavender; anthers 1–3 mm long; siliques 20–75 mm long, 1–1.5 mm wide, stipitate, the stipes 0.2–2 mm long; style 0.5–1 mm long. Known in Utah only from a Jones collection from the Henry Mts., Garfield County; Colorado and Arizona east to Texas and northern Mexico; 0 (0). The Utah material, if it is properly identified, is far removed from the remainder of the species.

Thlaspi L.

Glabrous, annual or perennial, arising from taproots; leaves cauline or cauline and basal alternate, simple, entire to dentate or lobed; flowers in ebracteate racemes; sepals 4, deciduous; petals 4, white, sometimes pinkish or lavender; stamens 6, the filaments lacking glandular processes; fruit a sessile silicle, compressed contrary to septum, often more or less wing-margined; style obsolete or slender and conspicuous; stigma capitate; seeds 2 to several per locule, uniseriate.

Holmgren, P. K. 1971. A biosystematic study of North American *Thlaspi montanum* and its allies. Mem. New York Bot. Gard. 21:1–106.
Payson, E. B. 1926. The genus *Thlaspi* in North America. Univ. Wyoming Publ. Bot. 1:145–186.

1. Plants annual; style obsolete or to 0.2 mm long; fruit orbicular in outline, conspicuously winged ..... *T. arvense*
— Plants perennial; style 0.3–2.5 mm long or more; fruit oblanceolate to obcordate or obovate in outline, narrowly wing-margined or not winged at all .... *T. montanum*

Thlaspi arvense L. Annual, the stems mostly 1–7 dm tall; basal leaves usually deciduous by anthesis; cauline leaves 1–8 cm long, 2–25 mm wide, elliptic to lanceolate or oblanceolate, sinuate-dentate to pinnatifid or subentire, the uppermost sessile and auriculate; pedicels 5–12 mm long or more, spreading to curved-ascending; sepals 1.5–2.5 mm long, green with whitish margins; petals 3–4.5 mm long, white; silicles 10–17 mm long, 7–12 mm wide, strongly compressed, wing-margined all around; style almost obsolete.; n = 7. Weedy species of roadsides, meadows, fields, and other disturbed places, almost ubiquitous, at 1310 to 2745 m in all Utah counties except Carbon, Emery, Garfield, Kane, Piute, San Juan, and Wayne; widespread in North America; adventive from Europe; 23 (i).

Thlaspi montanum L. [*T. alpestre* of authors, not L.; *T. cochleariforme* DC.; *T. nuttallii* Rydb.; *T. alpestre* var. *glaucum* A. Nels.; *T. glaucum* (A. Nels.) A. Nels.; *T. coloradoense* Rydb.; *T. purpurascens* Rydb.; *T. alpestre* var. *purpurascens* (Rydb.) Ostenf.; *T. glaucum* var. *pedunculatum* Payson; *T. glaucum* var. *hesperium* Payson;

T. hesperium (Payson) G. Jones; *T. fendleri* var. *coloradoense* (Rydb.) Maguire; *T. fendleri* var. *tenuipes* Maguire, type from Sanpete County; *T. fendleri* var. *hesperium* (Payson) C. L. Hitchc.; *T. prolixum* A. Nels., type from near Marysvale]. Perennial, with simple or branched caudex; stems 0.2–4 dm tall; basal leaves 0.9–3.5 cm long or more, 2–10 (15) mm wide, oval to oblong or spatulate-oblanceolate; cauline leaves 0.5–2.5 cm long, 1–12 mm wide, sessile and auriculate; pedicels 2–15 mm long, spreading or spreading-ascending; sepals 1.5–3.5 mm long, greenish to purplish; petals 3.5–7.5 mm long, white, pinkish or lavender, spatulate; silicles 3–8 mm long, rarely more, 1.5–5 mm wide, obovate to obcordate, winged or not; style 0.5–2.5 mm long; n = 7, 14. Sagebrush-grass, meadow, mountain brush, aspen, lodgepole pine, and spruce-fir communities at 1280 to 3815 m in all Utah counties; Washington east to Montana and south to California, Arizona, New Mexico, and Texas; Eurasia; 143 (xxi). Our material belongs to var. **montanum**. Some specimens from the southern part of the state approach the more southern var. *fendleri* (Gray) P. K. Holmgren, which has styles longer than 2.5 mm in length, petals 6–13 mm long, and silicles 7–16 mm long and 4–9 mm wide.

Thysanocarpus Hook.

Glabrous or pubescent with simple hairs, annual, arising from taproots; leaves cauline, alternate, simple, oblong-lanceolate to linear, toothed to entire; flowers in ebracteate racemes; pedicels curved; sepals 4, deciduous; petals 4, white or tinged pink or purple; stamens 6, the filaments lacking glandular processes; fruit a strongly compressed, unilocular, 1-seeded, indehiscent silicle, the margin winged.

Thysanocarpus curvipes Hook. [*T. trichocarpus* Rydb., type from Silver Reef, Washington County]. Annual, the stems 1–5 dm tall or more, simple or branched, sometimes hirsute below, otherwise glabrous; leaves 1.2–5 cm long, 1–10 mm wide, lance-oblong to elliptic or linear, sinuate-dentate to entire, transitional upwards to smaller sessile and auriculate blades; pedicels 3–7 mm long, recurved in fruit; sepals 1–1.5 mm long, often purplish; petals 1.5–2 mm long, white or tinged purplish or pinkish; silicles 4.5–8 mm long, 3–4 mm wide, ovate to obovate, often plano-convex, glabrous or pubescent, the wing crenate and sometimes perforate; style 0.4–0.6 mm long. Warm desert shrub communities from 670 to 1500 m in Washington County; British Columbia south to California and eastward to Arizona; 13 (iv). Our material belongs to var. **eradiatus** Jepson.

CUCURBITACEAE A. L. Juss.

Gourd Family

Trailing or climbing annual or perennial herbs typically with tendrils; leaves alternate, broad, usually simple but often deeply cut; tendrils simple or branched; flowers imperfect, the plants monoecious or sometimes dioecious; sepals 5; petals 5, connate or almost distinct; stamens 5, with 2 pairs often united, mostly connate by the anthers; ovary inferior; carpels 3; fruit a pepo or bladdery pod; x = 7–14.

1. Fruit fleshy, with a hard or firm rind, indehiscent, often edible 2
— Fruit berrylike, podlike, or bladdery, dry or fleshy, mostly inedible 5
2(1). Corolla bellshaped and distinctly united, 5-lobed to the middle or more .... *Cucurbita*
— Corolla variously shaped, the petals almost distinct or deeply parted 3
3(2). Tendrils unbranched .... *Cucumis*
— Tendrils branched 4
4(3). Leaves not lobed; flowers white, opening in the evening .... *Lagenaria*
— Leaves pinnatifid; flowers yellow, opening in the morning .... *Citrullus*
5(1). Staminal filaments connate; ovules erect .... *Echinocystis*
— Staminal filaments distinct; ovules various 6
6(5). Tendrils present; plants dioecious; fruit globular . *Bryonia*
— Tendrils lacking; plants monoecious; fruit various .... *Ecballium*

Bryonia L.

Perennial, monoecious (or dioecious) herbs with tuberous roots; stems with simple tenrils; leaves angled or lobed; flowers small, the staminate in peduncled racemes or cymes, the pistillate solitary or in short axillary clusters; calyx short-campanulate, 5-lobed; corolla rotate or campanulate, 5-parted; pistillate flowers with 3–5 more or less obsolete stamens; staminate flowers with 3 stamens, the anthers distinct; style slender, 3-parted; stigmas simple or 2-lobed; fruit a spherical berry.

Bryonia alba L. Vines glabrous to hairy, mostly 1–2 m long; leaf blades 3–10 cm long, cordate to triangular-ovate, 5- to 7-lobed; staminate flowers 7–15 in racemes with peduncles to 20 cm long; pistillate flowers corymbiform on shorter peduncles in lower nodes; calyx lobes 1.5–2 mm long; corolla 4–5 mm long, bluish yellow or yellowish white with green nerves; fruit black, globose, 7–8 mm thick; 2n = 20. Riparian areas and along fence rows or clambering on other plants, escaped from cultivation and established from 1470 to 1870 m in Box Elder, Cache, Summit, Utah, and Wasatch counties; introduced from Europe; 7 (0).

Citrullus Schrader

Monoecious annual herbs; stems procumbent or climbing; tendrils simple or branched; leaves pinnatifid, with the divisions again lobed; flowers solitary, yellow; corolla rotate, deeply parted; staminal filaments and anthers free; female flowers with 3 small staminodes; stigmas 3; fruit globular or oblong.

Citrullus lanatus (Thunb.) Mansfeld Watermelon. [*Momordica lanata* Thunb.; *C. vulgaris* Schrader]. Pubescent, sprawling herbs; leaves mainly 5–29 cm long, pinnately deeply divided; tendrils branched or less commonly simple; flowers solitary; corolla to ca 4 cm wide, rotate, the 5 lobes obovate, obtuse; fruit round to oblong, to 6 dm long or more, glabrous, green or variously striped, or whitish or yellowish, the flesh red, yellow, or greenish; seed compressed, black, white, or reddish; 2n = 22. Widely grown and escaping along roadsides and sandbars of the Colorado River in Grand County, and elsewhere; introduced from Africa; 2 (i).

Cucumis L.

Annual or perennial, mostly monoecious, procumbent herbs; stems pubescent, climbing by unbranched tendrils; leaves entire or somewhat lobed; flowers yellow, the male usually in clusters, the pistillate solitary; corolla rotate or campanulate, deeply 5-parted; anthers free; stigmas 3–5; fruit globular to elongate, glabrous, pubescent, or echinate.

1. Fruit prickly at maturity, usually several times longer than broad .... C. sativus
— Fruit netted-ribbed and rough at maturity, usually more or less pubescent, only slightly longer than broad . C. melo

Cucumis melo L. Cantelope; Muskmelon. Stems pubescent, striate or angled; leaves orbicular-ovate to subreniform, mainly 4–13 cm wide, angled but not distinctly lobed, pubescent and scabrous; flowers ca 25 mm wide; corolla lobes obtuse; fruit mostly globular or ellipsoid, more or less furrowed, with yellow or greenish pulp; $2n = 24, 48$. Cultivated fruit plant, sometimes escaping, especially in warmer portions of the state; introduced from tropical Africa and Asia; 1 (0).

Cucumis sativus L. Cucumber. Stems pubescent, strongly angled; leaves triangular-ovate, angled or somewhat 3-lobed, mainly 6–14 cm long, the lobes acute; flowers ca 25–30 mm wide; corolla lobes acute; fruit cylindroid, usually several times longer than broad, prickly, green or pale whitish green, the pulp white to greenish; $2n = 14, 24$. Cultivated fruit plant, occasionally escaping in Salt Lake and Utah counties; introduced from Asia; 3 (i).

Cucurbita L.

Annual or perennial monoecious herbs; stems scabrous, long trailing or climbing with branched tendrils (except in some cultivars); leaves simple, commonly lobed, coarse and often scabrous; flowers solitary, large, the staminate ones long-peduncled; corolla yellow, campanulate, lobed to near the middle or less; stamens 3, the filaments free, the anthers united; pistillate flowers with small staminodes; ovary 1-loculed; stigmas 3–5, each 2-lobed; fruit smooth or roughened, fleshy, mostly with a hard rind.

1. Plants perennial, from a greatly thickened root; fruit 5–9 cm thick, globose or nearly so; striped or mottled with pale and dark green; indigenous in southern Utah . 2
— Plants annual, from taproots; fruit various; grown for food and as ornamental plants 3

2(1). Leaf blades triangular-ovate, not especially lobed or cleft; 1–2.5 dm long; plants common in Washington County, less common elsewhere .... C. foetidissima
— Leaf blades palmately 5-lobed or cleft, less than 1 dm long; plants rare in Washinton County .... C. palmata

3(1). Plants rough-hairy; leaves strongly lobed and triangular apically, dentate or serrate; peduncle expanded near the fruit, not flared; seeds thick-margined; buds pointed preeceding anthesis; calyx lobes of staminate flowers foliaceous .... C. pepo
— Plants soft, variously pilose; leaves shallowly if at all lobed, round or triangular apically, mostly not or only obscurely serrate; buds various; calyx lobes of staminate flowers various 4

4(3). Seeds thin, with a hyaline ragged edge that often wears away; leaves broad-ovate to somewhat triangular; buds pointed in evening prior to anthesis; peduncle flared near fruit base; calyx lobes of staminate flowers often foliaceous .... C. moschata

— Seeds plump, the margins obtuse and more or less elevated; leaves circular to reniform; buds not pointed in evening prior to anthesis; peduncle short and spongy, nearly cylindric, not flared near the fruit; calyx of staminate flowers not foliaceous .... C. maxima

Cucurbita foetidissima H.B.K. Stinking Gourd; Stinking Cucumber. Plants robust, from a large, fusiform root to 10 kilograms or more; stems several from a root crown, to 6 m long or more; leaves coarse and thick, triangular-ovate, angulate or shallowly lobed, broadly rounded to cordate basally, acute to acuminate apically, 1–3 dm long or more, ill-scented, grayish green, scabrous; corolla 6–10 dm long or more, funnelform; fruit globose, green, with pale stripes, yellowish when ripe, to 8 cm thick. Wash bottoms, roadsides, and farmsteads in the creosote bush, Joshua tree, blackbrush, rabbitbrush, live oak, and pinyon-juniper communities at 830 to 1900 m in San Juan and Washington counties; Nevada and California, east to Missouri and south to Texas; 10 (vi). The fruit of this gourd was used by prehistoric peoples as a storage and drinking vessel. It occurred previously along Glen Canyon astride the Kane—San Juan county line, but is not known to have survived the filling of Lake Powell. It persists in the Needles section of Canyonlands National Park, above The Jump in Salt Canyon.

Cucurbita maxima Duchesne in Lam. Fall Squash; Winter Squash; Pumpkin. Stems short or long-trailing, the pubescence soft to the touch; leaves not rigid, orbicular or nearly so, not lobed, the base deeply cordate, the margins shallowly serrate; corolla with broad, usually reflexed lobes; peduncle short and cylindrical or enlarged at middle, often spongy, not expanded at attachment; fruit glaucous or bluish, but variously colored and shaped; seeds plump, the margin obtuse. Cultivated, but seldom escaping, in much of Utah below 1830 m (less commonly above that elevation under glass); introduced from Central America; 1 (i). The Hubbard and turban squashes belong to this species.

Cucurbita moschata Duchesne ex Poir. Cushaw; Butternut; Winter Crookneck. Stems long-trailing, soft to the touch; leaves limp, velvety, broadly ovate to nearly orbicular, not lobed; corolla with widely spreading crinkly lobes, the tube broad at the base and usually not bulging above; calyx lobes often long and expanded in fruit; seeds thin, the margin hyaline when fresh. Commonly and widely cultivated food plants in Utah; introduced from Central America; 3 (0).

Cucurbita palmata Wats. California Gourd. [C. californica Torr. ex Wats.]. Stems 5–15 dm long; leaves grayish with whitish streaks along veins on the upper surface, palmately 5-lobed, 3–9 cm long and about as broad, the lobes triangular to deltoid-lanceolate; corolla yellowish, 4–6 cm long; fruit globose 6–9 cm in diameter, green with bands of greenish white not well defined; seeds white, narrow-ovate, 10–14 mm long. Dry sandy flats in creosote bush and blackbrush communities at 1050 m and lower in Washington County; Arizona and California; 4 (ii).

Cucurbita pepo L. Field Pumpkin. Stems with stiff, translucent hairs; leaves stiff and rigid, erect, triangular or ovate-triangular, mainly 10–35 cm long, pointed, usually lobed and apiculate, the margins irregularly sharp-serrate; corolla with erect or spreading pointed lobes; peduncle strongly angled, expanded near the fruit; fruit large, orange, furrowed; $n = 10$. Widely grown food and ornamental plants in Utah, seldom escaping; introduced from Central America; 3 (0). The shape and form of the fruit are highly variable. The cultivated varieties hybridize freely, and seed from them will often yield several kinds of fruit, most of them inferior to the cultivars from which they were derived. The var. *melopepo* Alef. includes the bush squashes, such as summer crookneck, warty, and zucchini squashes.

Ecballuim A. Rich.

Trailing monoecious perennials from a tuberous root, lacking tendrils; flowers yellow, pedunculate, the pistillate solitary and often in the same leaf axil with the staminate raceme; corolla rotate or campanulate, deeply 5-parted, the lobes acute; anthers free; style very short; stigmas 3, bilobed; fruit much elongated.

Ecballium elaterium (L.) A. Rich. in Bory Squirting Cucumber. [*Momordica elaterium* L.]. Stems hairy, mainly 1.5–6 dm long; leaves long petiolate, the blades 4–10 cm long, cordate to triangular, entire, denticulate, or shallowly lobed, undulate, rather fleshy; corolla of staminate flowers 28–30 mm long; fruit 4–5 cm long, ca 2.5 cm thick, greenish, rough hairy, finally detaching from the peduncle and squirting brownish seeds from the basal aperature. Cultivated, botanical curiosity; introduced from Europe; 1 (0).

Echinocystis T. & G.

Annual, climbing or clambering, monoecious plants with branched tendrils; leaves lobed or angled; flowers white or whitish, small, usually numerous, the pistillate solitary, the staminate in racemes or panicles; corolla rotate, 5- or 6-lobed; filaments connate, the anthers connate or distinct; ovary usually 2-loculed; ovules erect, 2 per locule; fruit fleshy or dry.

Echinocystis lobata (Michx.) T. & G. Wild Cucumber. [*Sicyos lobata* Michx.]. Stems slender, angled, glabrous or hairy at the nodes, mainly 2–7 m long, with branched tendrils; leaves 7–12.5 cm long, cordate-ovate, shallowly 3- to 7-lobed, the lobes apiculate, the margins entire or sparsely serrate; pistillate flowers solitary, the staminate few to numerous; fruit ovoid, 3–5 cm long, ca 2.5 cm thick, bladdery or inflated, somewhat beaked with weak glabrous prickles to ca 6 mm long, bursting irregularly near the apex; 2n = 32. Cultivated ornamental, escaping and established along fence rows, canal banks, and roadsides at 1380 to 2130 m in Salt Lake, Uintah, Utah, and Weber counties; widespread in the U. S.; 6 (ii).

Lagenaria Ser.

Annual, monoecious or rarely dioecious vines with branched tendrils; flowers solitary, white, showy, soon withering, vespertine, 5-merous, the staminate on very long peduncles; anthers weakly connate; fruit various, of many sizes and shapes, the exocarp hard and durable.

Lagenaria siceraria (Molina) Standley Bottle-gourd. [*Cucurbita siceraria* Molina]. Musky-scented, soft, viscid-pubescent plants; leaves ovate to ovate-reniform, cordate basally, 15–30 cm wide, usually not lobed, the margins apiculate-dentate; flowers white, 5–10 cm wide, the staminate on long peduncles that commonly surpass the leaves; anthers only slightly coherent; fruit to 3 dm long or more, varing from globular to bottle-shaped, dumbell-shaped, or clublike, crookneck, or coiled; 2n = 22. Cultivated ornamental, grown for the fruit used as containers and decoration; probably of Old World origin, but grown in the New World since antiquity; 1 (i).

CUSCUTACEAE Dumort.
Dodder Family
Contributed by Gary I. Baird

Rootless, twining, parasitic herbs, attached to a wide variety of herbaceous and woody hosts by means of discoid haustoria, mostly annual, achlorophyllous; stems glabrous, usually much branched and tangled, white to yellow to orange; leaves alternate, reduced to minute scales or lacking; inflorescence cymose, 1- to many-flowered; flowers perfect, 4- or 5-merous, white to yellowish, sessile or pedicelled, infrequently with subtending bracts; calyx united, rarely separate, persistent, usually glabrous, often pellucid; corolla tubular to campanulate to urceolate, usually whitish, glabrous, often persistent around the developing capsule, commonly with a ring of fringed, united scales attached to the inside of the corolla tube and alternate with the corolla lobes (infrastaminal scales); stamens alternate with and mostly shorter than the corolla lobes, inserted on the corolla tube below or at the sinuses; filaments mostly subulate, usually equaling the anthers; pistil 1, superior, 2-carpeled and 2-loculed, usually enlarging and distending or rupturing the perianth; styles 2, distinct, often unequal; stigmas capitate, occasionally linear; fruit a capsule, indehiscent or some basally circumscissle; seeds 1–4; embryo acotyledoneous; x = 17.

Cuscuta L.

The single genus has the characteristics of the family.

Yuncker, T. G. 1965. *Cuscuta*. North American Flora. Series II, part 4: 1–40.

1. Stigmas linear; capsule basally circumscissle; inflorescence a compact glomerule .... *C. approximata*
— Stigmas capitate; capsule usually indehiscent; inflorescence compact to open but not a glomerule 2
2(1). Infrastaminal scales lacking .... *C. occidentalis*
— Infrastaminal scales present 3
3(2). Sepals with a single, large projection; corolla very papillate; plants of Millard County .... *C. warneri*
— Sepals and corolla various but not as above; plants common or uncommon but not limited to Millard County 4
4(3). Flowers mostly 4–merous .... *C. cephalanthi*
— Flowers rarely other than 5–merous 5
5(4). Flowers closely subtended by 1–3 sepaloid bracts; sepals distinct to the base .... *C. cuspidata*
— Flowers mostly ebracteate; sepals united, at least partly so at the base 6
6(5). Capsule ovoid, either 1–seeded and small or seeds mostly 4 and the capsule greatly enlarged 7
— Capsule globose, mostly 2- to 4-seeded, neither very small nor greatly enlarging 9
7(6). Capsule large, ca 5 mm long and mostly 4–seeded; plants of montane areas .... *C. megalocarpa*
— Capsule small, 1–3 mm long, 1–seeded; plants of desert and saline areas 8

8(7). Flowers 1–2 mm long, sessile, solitary or few in a group; sepal and petal lobes ovate-orbicular .... *C. denticulata*

— Flowers 2–3 mm long, pedicelled, in cymose clusters; sepal and petal lobes lanceolate .... *C. salina*

9(6). Capsules circumscissle at the base; sepals carinate; plants of southern Utah .... *C. applanata*

— Capsules indehiscent; sepals sometimes nerved but not carinate; plants mostly widespread 10

10(9). Corolla often nerved; capsule slightly crest-thickened; flowers sometimes bracteated ...... *C. indecora*

— Corolla not papillose-nerved; capsule mostly flattened to rounded on top, sometimes with thickened cells but not crested; flowers ebracteate 11

11(10). Capsule papillate-granular on upper half; corolla remaining about the expanded capsule; plants disjunct in the Uintah Basin .... *C. glabrior*

— Capsule smooth to tuberculate-thickened apically; corolla remaining at the base of the expanded capsule; plants common throughout eastern and southern Utah .... *C. pentagona*

Cuscuta applanata Engelm. Wing Dodder. [*C. alata* Brandegee]. Stems medium, 0.3 mm in diameter (when dry); inflorescence a few-flowered, compact cluster; flowers 5-merous, 1–2 mm long, subsessile, often basally thickened; calyx united, the lobes 1 mm long, broadly triangular, obtuse to broadly acute, not or just overlapping, pellucid, irregularly carinate at the midrib and sinuses (or merely thickened), the carina extending down onto the very short pedicel, this equaling or exceeding the corolla tube; corolla campanulate, the lobes 1 mm long, broadly lanceolate, acute, the margins irregular, erect to spreading, slightly thickened at the midribs, somewhat saccate between the calyx lobes; stamens shorter than the corolla lobes; filaments slender, equaling or longer than the anthers; anthers ovate, 0.5 mm long; infrastaminal scales reaching the filaments, with the long fringe exserted; ovary globose; styles slender-subulate, 1–1.5 mm long; stigmas capitate; capsule flattened-globose, angled by the developing seeds, the apical depression evident, surrounded by the distended corolla, easily circumscissle by a definite or indefinite basal line, the septae remaining with the base; seeds 4, oval to wedge-shaped. On a variety of hosts; known in Utah from a single collection in Washington County (B. Albee 3374 UT); Arizona and New Mexico to Mexico; 1 (0).

Cuscuta approximata Babington Slender Dodder. [*C. planiflora* var. *approximata* Engelm.; *C. cupulata* Engelm.; *C. urceolata* Kuntze; *C. approximata* ssp. *approximata* Feinbrun; *C. gracilis* Rydb.; *C. anthemi* A. Nels.]. Stems medium, to 0.3 mm in diameter (when dry), whitish to yellow (orange); inflorescence a few- to many-flowered glomerule; flowers sessile, 2.5–3 (3.5) mm long; calyx united, the lobes 1.5–2 mm long, usually as broad as long, pellucid, slightly overlapping, the apex slightly attenuate and thickened, often recurved; corolla campanulate, equal to or slightly exceeding the calyx, the lobes 1 mm long, obtusish, whitish, entire; stamens shorter than the corolla lobes; filaments subulate, equaling the anthers; anthers oval, 0.3 mm long, often with a purplish stripe; infrastaminal scales inserted, just reaching the filaments, apically fringed with short processes; ovary globose; styles 0.5 mm long, slender; stigmas linear, 0.5 mm long; capsule globose, the apical depression evident or not, surrounded by the distended corolla, basally circumscissle along a definite line; seeds usually 4, oblong. On a variety of herbaceous plants, especially cultivated crops and particularly alfalfa in Cache, Grand, Salt Lake, Uintah, and Utah counties; introduced from the Mediterranean region of Europe; 21 (0).

Cuscuta cephalanthi Engelm. Slenderflower Dodder. [*C. tenuiflora* Engelm.; *Epithymum cephalanthi* (Engelm.) Nieuwl. & Lunell]. Stems medium, 0.3 mm in diameter (when dry), whitish to yellow; inflorescence a few- to many-flowered cluster; flowers mostly 4–merous, sessile, ca 2 mm long; calyx united, the lobes 1–1.5 mm long, overlapping, shorter than the corolla tube, pellucid, sometimes glanular; corolla tubular to campanulate, distended by the capsule and often persisting about the top, the lobes ovate, obtuse, 1–1.5 mm long, erect to spreading, entire; stamens equaling the corolla lobes, the filaments subulate and stout, equaling the anthers, these oval to round; infrastaminal scales inserted, shorter than or subequal to the filaments, shortly fringed; ovary globose; styles 0.5–1.5 mm long; stigmas capitate; capsule globose, apically depressed, not thickened around the aperture, indehiscent, often surrounded by the withered corolla; seeds usually 2, oval. Known from a variety of hosts, in Salt Lake County; Washington to California, east to the Atlantic and Gulf states; 1 (0).

Cuscuta cuspidata Engelm. Toothed Dodder. [*Grammica cuspidata* (Englem.) Hadac & Chrtek]. Stems medium, 0.3 mm in diameter (when dry), yellowish (greenish); inflorescense a loose, many-flowered, paniculate cyme; flowers 5–merous, 2.5–4 (5) mm long, pedicelled, often closely subtended by 1–3 sepaloid bracts; calyx separate, the sepals 1.5–2 mm long, ovate to orbicular, the whole calyx becoming somewhat spherical, acutish and often cuspidate-mucronate, the margin serrulate to entire, strongly overlapping, midrib often nerved with large, thickened, linear cells; corolla tubular, the lobes 1.5 mm long, ovate to triangular, acute and sometimes cuspidate, the margin sometimes serrulate, midrib nerved as in calyx, spreading to reflexed; stamens shorter than corolla lobes; filaments subulate, shorter than the anthers, these oval to oblong, 0.6 mm long; infrastaminal scales inserted, not reaching the filaments, with a short irregular fringe; ovary globose; styles 1.5 mm long; stigmas capitate; capsule globose, the apical depression small, surrounded by the persistent perianth, indehiscent; seeds usually 4, obovate. On a variety of hosts, in Utah and Weber counties; North Dakota to Indiana, south to Texas and Louisiana; 2 (0).

Cuscuta denticulata Engelm. Smalltooth Dodder. [*Grammica denticulata* (Engelm.) W. A. Weber]. Stems slender, 1.5–2 mm diameter (when dry); inflorescence a single flower or sometimes a loose, few-flowered cluster; flowers 5–merous, 1.5–2 mm long, white, subsessile, often with a small bract(s); calyx united, at least near the base, the lobes 1.2–1.7 mm long, strongly overlapping, orbicular, the whole becoming spherical, denticulate to erose; corolla campanulate, becoming urceolate, the lobes to 1 mm long, oval to oblong, obtuse to rounded, spreading to less often erect, overlapping, white; stamens shorter than the corolla lobes; filaments subulate, equaling the anthers, these oval, 0.3 mm long; infrastaminal scales inserted, reaching the filaments, with an irregular, denticulate fringe; ovary ovoid to conical; styles 0.5 mm long; stigmas small, capitate; capsule ovoid to conical, the apical depression not evident, surrounded by the persistent perianth, indehiscent; mostly 1-seeded, the seed globose-ovoid. This is a desert species found growing on several genera of desert shrubs, often germinating in place, in Box Elder, Duchesne, Emery, Grand, Juab, Kane, Tooele, Uintah, and Washington (type from Rio Virgin) counties; western U. S.; 12 (i).

Cuscuta glabrior (Engelm.) Yuncker Smooth Dodder. [*C. verrucosa* var. *glabrior* Engelm.; *C. verrucosa* Engelm., not Sweet; *C. arvensis* var. *verrucosa* (Engelm.) Engelm.; *C. pentagona* var. *verrucosa* (Engelm.) Yuncker; *C. glabrior* f. *pedicellata* Yuncker]. Stems medium, 0.3 mm in diameter (when dry); inflorescence a loose cymose cluster; flowers 5-merous, 2-3 mm long, pedicellate; calyx united, the lobes 1 mm long, triangular to obtuse, slightly overlapping, pellucid with chartaceous margins, equaling or surpassing the corolla tube; corolla campanulate, the lobes 1 mm long, triangular to broadly lanceolate, acute to slightly apiculate, entire, spreading to reflexed with inflexed tips; stamens shorter than the corolla lobes; filaments subulate, slightly shorter than the anthers, these oval, ca 0.5 mm long; infrastaminal scales inserted, just reaching the filaments, with a short fringe; ovary globose to obconic; styles ca 0.7 mm long; stigmas capitate; capsule globose to obconic, membraneous, finely papillose-granular on the upper half, slightly thickened about the evident apical depression, the perianth persistent around the expanded capsule; seeds mostly 2, oval. On a variety of hosts, potentially, but known in Utah from near Ouray, Uintah County; New Mexico to Louisiana and southward into Mexico; 1 (0). This taxon is closely allied to *C. pentagona* Engelm. and very similar to it in habit.

Cuscuta indecora Choisy Plain Dodder. [*C. decora* Engelm.; *C. decora* var. *indecora* (Choisy) Engelm.; *C. decora* var. *subnuda* Engelm.; *C. indecora* var. *subnuda* (Engelm.) Yuncker; *Epithymum indecorum* (Choisy) Nieuwl. & Lunell; *Grammica indecora* (Choisy) W. A. Weber; *C. verrucosa* var. *hispidula* Engelm.; *C. hispidula* (Engelm.) Engelm.; *C. indecora* var. *hispidula* Yuncker; *C. neuropetala* Engelm.; *C. neuropetala* var. *minor* Engelm.; *C. neuropetala* var. *littoralis* Engelm.; *C. indecora* var. *neuropetala* (Engelm.) A. S. Hitchc.; *Grammica indecora* ssp. *neuropetala* (Engelm.) W. A. Weber; *C. pulcherrima* Scheele; *C. decora* var. *pulcherrima* (Scheele) Engelm.; *C. indecora* var. *bifida* Yuncker]. Stems medium to coarse, 0.3-0.5 mm in diameter (when dry), yellowish; inflorescence a many-flowered, loose paniculate cyme; flowers 5-merous, pedicellate, 3-4 mm long, commonly granular, mostly ebracteate; calyx united, the lobes 1.5-3 mm long, triangular-ovate, acute, entire to uneven, sometimes with a nerved midrib, rarely, if ever, overlapping; corolla campanulate, the lobes 1-1.5 mm long, triangular-ovate, erect to spreading, the tip often inflexed, acute, entire to granulate, often with a nerved midrib; stamens shorter than the corolla lobes; filaments slender, equal to the anthers, these oblong to triangular, 0.7 mm long; infrastaminal scales inserted or slightly exserted, abundantly fringed; ovary globose to slightly ovoid; styles 0.5-1.5 mm long; stigmas capitate; capsules globose, the apical depression evident, often slightly thickened apically, indehiscent; seeds mostly 4, oval. Known from several hosts in Davis, Millard, Salt Lake, Uintah, and Utah counties; widespread throughout the U. S. and tropical America; 24 (iv).

Cuscuta megalocarpa Rydb. Largefruit Dodder. [*C. umbosa* Hook.; *C. gronovii* var. *curta* Engelm.; *C. curta* (Engelm.) Rydb.]. Stems coarse to robust, 0.5 mm in diameter (when dry), pale yellow; inflorescence a compact, few-flowered, paniculate cluster; flowers 5-merous, on short, stout pedicels, often with pellucid cells, 2-3 mm long; calyx united, much shorter than the corolla tube, the lobes 1-1.5 mm long, ovate, obtuse, entire, slightly overlapping; corolla campanulate, the lobes 0.7-0.9 mm long, obtuse to rounded, spreading to reflexed, entire; stamens shorter than or equaling the corolla lobes; filaments subulate, equaling the anthers; anthers oval, 0.6 mm long; infrastaminal scales inserted, not reaching the filaments, slightly fringed; ovary ovoid, the apex much thickened; styles short, 0.3-0.5 mm long; stigmas capitate; capsule large, 3-6 mm long, indehiscent; seeds 2-4, large, mostly over 2.0 mm long, roundish. On woody plants at higher elevations in Salt Lake and Utah counties; Saskatchewan to Manitoba, south to New Mexico; 7 (0).

Cuscuta occidentalis Millsp. Western Dodder. [*C. californica* var. *breviflora* Engelm.; *Grammica occidentalis* (Millsp.) Hadac & Chrtek]. Stems medium, to 0.3 mm in diameter (when dry), yellowish; inflorescence a small, compact to loose cluster; flowers 5-merous, 2-3 mm long, sessile or subsessile; calyx united, equaling the corolla tube or exceeding it, the lobes 2 mm long, narrowly ovate, acuminate, becoming thickened at the base, entire, not overlapping; corolla campanulate-globose, sometimes saccate between the calyx lobes, these lanceolate, 1.5-2 mm long, acuminate, erect to spreading, entire; stamens shorter than corolla lobes; filaments subulate, mostly longer than the anthers; anthers oval, 0.3-0.5 mm long; infrastaminal scales lacking; ovary globose; styles 0.5-1 (1.5) mm long; stigmas capitate; capsules globose, the apical depression present, membraneous, surrounded by the persistent corolla, with the lobes often spread outward; seeds 4, oval. On several hosts in Emery, Morgan, Salt Lake, and Summit counties; Pacific Coast states inland to Idaho, Wyoming and Colorado; 17 (0). A closely related species, *C. californica* Choisy, may be present in Utah. It is distinguished from *C. occidentalis* by its longer, linear anthers (0.7-1 mm long), longer, exserted styles (1.5-2 mm), and by being pedicelled.

Cuscuta pentagona Engelm. Field Dodder. [*C. arvensis* Beyrich ex Engelm. in Gray; *C. arvensis* var. *pentagona* (Engelm.) Engelm.; *Epithymum arvense* Nieuwl. & Lunell; *Grammica pentagona* (Engelm.) W. A. Weber; *C. pentagona* var. *calycina* Engelm.; *C. arvensis* var. *calycina* (Engelm.) Beyrich ex Engelm. in Gray; *C. campestris* Yuncker; *Grammica campestris* (Yuncker) Hadac & Chrtek]. Stems medium, 0.4 mm diameter (when dry), yellowish; inflorescence a loose to crowded cluster; flowers 5-merous, 1-2 (3) mm long, pedicelled, the pedicels mostly stout, often slightly tuberculate throughout; calyx united, the lobes ca 0.9 mm long, as broad as long, ovate, entire, pellucid, overlapping, equaling the corolla tube; corolla campanulate, the lobes ca 1 mm long, triangular to broadly lanceolate, acute, entire, spreading, the tip often inflexed; stamens shorter than corolla lobes; filaments subulate, equaling the anthers; anthers oval, to 0.5 mm long; infrastaminal scales equaling the corolla tube, with the long fringe exserted; ovary globose to obconic, often visible and exposed at anthesis; styles 0.5-1 mm long; stigmas capitate; capsules globose to obconic, membraneous, often turbiculate around the evident apical depression, indehiscent, the perianth persistent at the base of the capsule; seeds 2-4, ovate to oval.

Widespread species known from several hosts in Carbon, Kane, Salt Lake, San Juan, Sanpete, Uintah, Utah, Washington and Wayne counties; throughout the U. S. and tropical America; 48 (ii).

Cuscuta salina Engelm. in Gray Salt Dodder. [*C. california* var. *squamigera* Engelm.; *C. squamigera* (Engelm.) Piper; *C. salina* var. *squamigera* (Engelm.) Yuncker; *Grammica salina* (Engelm.) Taylor & Macbryde; *C. veatchii* var. *apoda* Yuncker; *C. nevadensis* Johnston; *C. salina* var. *apoda* (Yuncker) Yuncker]. Stems slender, orangish; inflorescence a few-flowered, umbellate-cymose cluster; flowers 5-merous, 2-4 mm long, short pedicelled; calyx united, the lobes 1-2 mm long, lanceolate to narrowly ovate, acute to acuminate, entire, slightly overlapping at the base, equaling the corolla tube; corolla tubular-campanulate, the lobes 1-2 mm long, erect to spreading, entire, occasionally granular; stamens shorter than the corolla lobes; filaments subulate, equaling the anthers; anthers oval to oblong, 0.5 mm long; infrastaminal scales inserted, with a short fringe; ovary ovoid; styles 0.4-0.8 mm long; stigmas capitate; capsule ovoid, slightly attenuate, apical depression wanting, surrounded by the withered corolla, indehiscent; seed 1, globose-ovoid. On a variety of hosts but especially on members of Chenopodeaceae in Davis, Salt Lake, Tooele, Washington, and Weber counties; British Columbia to Baja and east to Arizona; 11 (0).

Cuscuta warneri Yuncker Warner Dodder. Stems slender, yellow; inflorescence a few- to many-flowered cluster; flowers 5-merous, 2.0 mm long, short pedicelled, very papillous to turbiculate; calyx united, the lobes 0.8-1 mm long, triangular-ovate, shorter than the corolla tube, each lobe with a thickened, acute, conelike projection, 0.5-0.7 mm long; corolla tubular-campanulate, the lobes 1-1.2 mm long, triangular-ovate, acute, erect to inflexed, prominently papillose; stamens shorter than corolla lobes; filament subulate, equaling the anthers; anthers ovoid; infrastaminal scale inserted, sparingly toothed at the truncate apex, not or just reaching the filaments; ovary globose; styles slender, 0.5-0.7 mm long; stigmas capitate; capsule globose, with a prominent, shallowly lobed collar around the apical depression, smooth, membraneous, finely striate, surrounded by the persistent perianth; seeds usually 2, oval. Known only from the type collection near Fillmore, Millard County; endemic; 1 (0). At the type locality, the Warner dodder was growing historically in a weedy area between a road and an alfalfa field on *Phyla cuneifolia*. The host plant is common in the area but several attempts at finding the Warner dodder have been unsuccessful.

DIPSACACEAE A. L. Juss.

Teasel Family

Plants herbaceous; stems stout, prickly; leaves opposite or whorled, simple to pinnately dissected; flowers epigynous, perfect, aggregated into dense heads with involucrate bracts; calyx cup-shaped, 4-toothed; corolla 4-lobed, irregular, the lobes imbricate; stamens 4, distinct, or seldom 2 or 3, attached near the throat of the corolla tube, alternate with the lobes; pistil 1, the ovary inferior, 1-loculed; style 1, slender, with an entire or 2-lobed stigma; fruit an achene, crowned by the persistent calyx; $x = 5, 10$.

Dipsacus L.

Coarse biennial or perennial herbs; stems erect, stout, prickly; leaves opposite, sessile, usually connate; heads ovoid to cylindrical, dense; involucral bracts elongate, linear, spine-tipped, the bracts of the receptacle ovate to lanceolate, acuminate or awn-tipped; calyx minute, cupulate, 4-lobed and 4-angled, persistent on the achene; corolla 4-lobed, elongate; stamens 4.

Dipsacus sylvestris Hudson Teasel. [*D. fullonum* L. ssp. *sylvestris* Clapham in Clapham et al.]. Stems stout, 8-20 dm tall, striate, with prickles on the angles; leaves 20-60 cm long, sessile, the upper ones perfoliate, lanceolate to oblong-lanceolate, crenate-serrate or the upper entire, prickly, especially on the veins; heads 5-9 cm long, ovoid or somewhat cylindrical; involucral bracts linear, prickly on the margin, curving upward and as long as or longer than the head; bractlets of the receptacle with a stiff, straight awn; calyx silky, ca 1 mm long; corolla 8-14 mm long, the tube whitish, with pale purplish lobes ca 1 mm long; fruit ca 5 mm long; $2n = 16, 18$. Moist roadsides, ditch blanks, marshes, and waste places at 1430 to 2660 m in Box Elder, Cache, Daggett, Davis, Salt Lake, Utah, and Wasatch counties; widespread in the U. S.; native to Europe; 9 (i). This plant is used in dried floral arrangements.

EBENACEAE Gurke in Engler & Prantl

Ebony Family

Trees with alternate, entire, deciduous leaves; flowers regular, imperfect, axillary, solitary or in cymes; calyx 4- or 5-lobed, persistent and accrescent in fruit; corolla 4- or 5-lobed, urceolate to campanulate; stamens 3 to many, inserted at corolla base; pistil 1, the ovary superior, usually many-loculed; ovules 1 or 2 per locule; fruit a several-locular berry; $x = 15$.

Diospyros L.

Trees with blackish, reticulately patterned bark, dioecious or essentially so; flowers axillary; corollas dimorphic, those of the pistillate flowers much larger than those of the staminate ones.

Diospyros virginiana L. Common Persimmon. Trees, mainly 5-12 m tall, the blackish wood very hard; leaves petiolate, the blades ovate-oblong to elliptic, 2.5-10 cm long or more, 1.5-7 cm wide, abruptly acuminate apically, rounded to cordate basally; peduncles very short; corolla of pistillate flowers pale yellow, campanulate-urceolate, 1-1.5 cm long, those of the staminate flowers much smaller; styles 4, bilobed at apex; ovary 8-loculed; fruit mainly 1-4 cm thick, astringent when unripe. Cultivated ornamental and fruit tree, escaping and established in Washington County; indigenous in the eastern U. S.; 4 (i).

ELAEAGNACEAE A. L. Juss.

Oleaster Family

Shrubs or trees with stellate or peltate, scalelike trichomes; leaves simple, alternate or opposite, entire; flowers perfect or imperfect, regular, borne in axillary clusters; perianth 4-lobed, in a single whorl from the apex of the hypanthium; stamens 4-8; pistil 1, the ovary superior (though often appearing inferior), 1-loculed; styles and stigmas 1; fruit a dry, indehiscent achene enveloped by the fleshy, persistent hypanthium, hence drupelike; x = 6, 10, 11, 13.

1. Leaves alternate; flowers perfect or polygamous; stamens 4 .... *Elaeagnus*
— Leaves opposite; flowers perfect (or rarely imperfect); stamens 8 .... *Shepherdia*

Elaeagnus L.

Shrubs or trees with ascending, alternate branches; leaves alternate; flowers axillary, perfect or polygamous; hypanthium investing the ovary and produced upward as a tube usually longer than the spreading sepals, deciduous above the ovary, not glandular-thickened at the lobe apex; stamens 4; ovary wall bony.

1. Branchlets and leaves with silvery peltate scales only; leaves lance-oblong to linear-oblong or elliptic; cultivated, escaping, and established .... *E. angustifolia*
— Branchlets and leaves with both silvery and brown peltate scales; leaves ovate to obovate, oblong, or elliptic; indigenous in Daggett and Summit counties ...... *E. commutata*

Elaeagnus angustifolia L. Russian Olive; Oleaster. Trees of small to moderate stature and rapid growth, mainly 5-12 m tall, with trunks 1-5 dm thick; stems often armed with coarse thorns; leaves 2-9 cm long, 5-38 mm wide, lanceolate to elliptic, silvery with peltate scales on both surfaces but less so above, hence the leaves often bicolored; flowers deathly fragrant, 8-12 mm long, silvery, the lobes yellow ventrally and somewhat stellate-hairy; stamens 4; fruit olivelike, silvery or finally brown and glabrate, 1-2 cm long; 2n = 12, 28. Cultivated shade tree, escaped and established in probably all Utah counties; widespread in the U.S.; introduced from Europe; 55 (iv). This plant is naturalized along drainages and in moist meadows over vast areas of the state.

Elaeagnus commutata Bernh. Silverberry. [*E. argentea* Pursh, not Moench]. Shrubs, mostly 1-2 m tall, the branchlets with brownish, peltate scales, unarmed; leaves 1.3-7 cm long, 5-30 mm wide, elliptic to oblanceolate or lanceolate, acute to obtuse or rounded apically, acute to obtuse basally, silvery on both sides and with some brown scales intermixed; flowers 1-4 per axil, 10-15 mm long, the lobes yellowish; stamens 4; fruit 6-10 mm long, the hypanthium base mealy, silvery; 2n = 28. Riparian zone with willow and poplar at ca 1830 to 2440 m in Daggett and Summit counties; Alaska and Yukon, east to Quebec, south to Idaho, Wyoming, and Minnesota; 5 (i).

Shepherdia Nutt.

Shrubs with spreading to ascending, opposite branches; leaves opposite, green or greenish above, pale beneath, sparsely to densely brownish or silvery with peltate scales or stellate hairs; flowers axillary, perfect (or imperfect); perianth short-tubular, investing the ovary in pistillate flowers, not deciduous above the ovary; staminate flowers with 8 stamens alternating with glandular thickenings at the base of the perianth lobes; ovary wall somewhat woody, enclosed by the juicy hypanthium; fruit drupelike.

1. Leaves evergreen, oval to orbicular or ovate, rounded to obtuse at the base; fruit subglobose, the surface obscured by silvery scales; plants of southern Utah .... *S. rotundifolia*
— Leaves deciduous, variously shaped, but usually much longer than broad; fruit red, the surface sparingly if at all scaly 2
2(1). Plants armed with stout thorns, 1-4 m tall; leaves cuneate at the base; fruit red, edible; plants of riparian habitats .... *S. argentea*
— Plants unarmed, commonly 0.5-2 m tall; leaves usually rounded at the base; fruit yellowish red, unpalatable .... *S. canadensis*

Shepherdia argentea (Pursh) Nutt. Silver Buffaloberry. [*Hippophae argentae* Pursh; *Lepargyraea argentea* (Pursh) Nutt.]. Shrubs with spreading to ascending, opposite, frequently thorny branches; 2-4 m tall, the branchlets covered with silvery peltate scales; leaves petiolate, the blades 0.5-6 cm long, mainly 3-14 mm wide, oblong, elliptic, or lanceolate to oblanceolate, cuneate at the base, rounded to obtuse apically; petioles 2-12 mm long; flowers subsessile, clustered at the nodes, precocious, 2.5-4 mm long, the lobes yellowish; fruit 4-7 mm long, red, tart, edible; 2n = 26. Stream banks and terraces, often where moist or wet, at 1400 to 2290 m in Beaver, Daggett, Duchesne, Emery, Garfield, Grand, Kane, Millard, Piute, San Juan, Sevier, Uintah, Utah, Tooele, and Wayne counties; British Columbia to Saskatchewan, south to California, Nevada, New Mexico, and North Dakota; 36 (vi).

Shepherdia canadensis (L.) Nutt. Soapberry. [*Hippophae canadensis* L.; *Lepargyraea canadensis* (L.) Greene]. Shrubs, mostly 0.5-2 m tall, the branchlets clothed with brown peltate scales; leaves 0.5-8 cm long, 3-30 (48) mm wide, ovate to lanceolate, rounded apically and basally, green above, pale or brownish beneath, sparingly to rather densely brownish scaly; flowers 1 to several per axil, 2-3 mm long, the lobes stramineous brownish; fruit 4-7 mm long, juicy, red, bitter and soapy; 2n = 22. White fir, Douglas fir, aspen, bristlecone pine, lodgepole pine, and spruce-fir communities at 2040 to 3205 m in Box Elder, Cache, Carbon, Daggett, Duchesne, Emery, Garfield, Grand, Juab, Kane, Piute, Salt Lake, Sanpete, Sevier, Summit, Uintah, and Utah counties; Alaska to the Atlantic, south to Oregon, Nevada, Colorado, and New York; 48 (iv). The ill-flavored fruit is sometimes eaten, whipped with cream and with sugar added, as a frothy "ice cream."

Shepherdia rotundifolia Parry Roundleaf Buffaloberry. [*Lepargyraea rotundifolia* (Parry) Greene]. Shrubs mainly 1-2 m tall, but 1-4 m wide, the branchlets stellate-hairy with white or yellowish trichomes; leaves 0.5-4.2 cm long, 5-38 mm wide, ovate, oval, orbicular, lanceolate, rounded apically, the base rounded to obtuse, silvery green above, pale beneath, clothed with trichomes; flowers 1 to few per axil, 3.5-5 mm long, the lobes yellowish, opening flat; fruit 5-8 mm long, the surface obscured by stellate hairs. Blackbrush, ephedra, rabbitbrush, shadscale, pinyon-juniper, and ponderosa pine communities at 1220 to 2565 m in Emery, Garfield, Kane (type from valley of the Virgin); San Juan, Sevier (eastern border), Washington, and Wayne counties; Arizona (Colorado Plateau endemic; 70 (ix). This is a beautiful shrub. It festoons slopes with silvery clumps.

ELATINACEAE Dumort.
Waterwort Family

Annual herbs; leaves simple, opposite, with paired stipules between them; flowers small, regular, perfect, axillary; sepals 2–5; petals 2–5, distinct; stamens 2–5 or 4–10; pistil 1, the ovary superior, 2- to 5-loculed, with axile placentation; style 1; fruit a septicidal capsule, the seeds several to many; x = 6, 9.

1. Flowers 5–merous; herbage glandular-pubescent; sepals scarious margined .... Bergia
— Flowers 2– to 4–merous; herbage glabrous; sepals not scarious margined .... Elatine

Bergia L.

Annual, prostrate or ascending glandular-pubescent herbs; flowers solitary or fascicled, pedicelled, 5-merous; sepals cuspidate, the midrib thickened, scarious margined; petals oblong; stamens 5 or 10; capsules included in the calyx.

Bergia texana (Hook.) Seub. [*Merimea texana* Hook.]. Stems reclining or ascending, radiating from a taproot, mainly 5–40 cm long; leaves petiolate, lanceolate to oblanceolate or elliptic, 5–23 mm long, 3–7 mm wide, glandular-serrate; sepals 2.5–4 mm long; petals white; seeds brown and shining, ca 0.3 mm long, obscurely quadrate-reticulate, curved. Drying mudflats along streams and on pond margins at ca 1400 to 1435 m in Millard and Uintah counties; Washington to California, east to Mississippi; 5 (i).

Elatine L.

Aquatic or amphibious (terrestrial), dwarf annuals; stems erect or prostrate; herbage glabrous; leaves opposite, sessile or petioled; flowers 1 or 2 per node; petals membranous; styles or stigmas 2–4; capsule membranous, septicidally dehiscent. Note: The genus is represented poorly in our herbaria, being collected only rarely. Therefore, all decisions presented below are tentative.

1. Capsules 4–loculed; seeds strongly curved, J- or U-shaped; stamens 8 .... *E. californica*
— Capsules 3–loculed; seeds gently curved, not as above; stamens usually 3 .... *E. triandra*

Elatine californica Gray. California Waterwort. Plants prostrate, mat-forming; leaves subsessile to short-petiolate, 4–12 mm long, obovate to oblanceolate; flowers pedicellate, the pedicels elongating at maturity, finally 1–2 times longer than the fruit; sepals 4, equal to subequal, oblong, united below and enlarging with the fruit; petals 4, obovate; stamens 8; capsule 4–carpelled; seeds strongly bent, rounded at one end, truncate at the other with a subapiculate base. Mud or shallow water of ponds and lakes at ca 1708 m in Cache County (Neese 12386 BRY); Washington to Montana, south to California and Arizona; 1 (0).

Elatine triandra Schkuhr Three-lobed waterwort. [*E. gracilis* Mason?; *E. rubella* Rydb.?]. Plants prostrate, mat-forming; leaves subsessile or short-petiolate, 3–15 mm long, linear to narrowly oblanceolate to spatulate or oblong; flowers sesile; sepals 3, subequal; petals 3, orbicular; stamens usually 3; capsule 3–carpelled; seeds merely curved, with mainly 15–25 pits per longitudinal row, rounded at both ends. Pond and lake margins at 2440 to 2700 m in Salt Lake and Utah counties; Washington and Alberta south to Arizona, New Mexico, Texas, and Mexico; Eurasia; 6 (0). Examination of scanning electron photomicrographs (BRY) of all specimens at BRY and UT did not show any convincing evidence for separation of our materials into more than the one species. The similar *E. brachysperma* Gray might occur in Utah. It differs in having seeds with 9–15 pits per longitudinal row and leaves rounded (not emarginate) apically.

ERICACEAE A. L. Juss.
Heath Family

Shrubs or subshrubs; leaves simple, alternate, sometimes leathery or persistent; flowers perfect, regular, axillary, in terminal clusters, or solitary; sepals mostly 4 or 5, distinct or more or less connate; petals mostly 4 or 5, connate or distinct, the corolla rotate to funnelform or urn-shaped; stamens as many as the corolla lobes and alternate with them or twice as many, the anthers dehiscent by terminal pores or by longitudal slits; pistil 1, the ovary superior or inferior, usually with 4–10 carpels and locules; style 1, the stigma capitate or lobed; fruit a capsule or a berry; x = 8–23.

Welsh, S. L. 1980. Ericaceae. pp. 40–44. In: Utah flora: Miscellaneous families. Great Basin Naturalist 40: 38–58.

1. Ovary inferior or apparently so 2
— Ovary superior 3
2(1). Plants prostrate shrublets, rooting along the stems; ovary superior but surrounded by the fleshy calyx when ripe and apparently inferior .... *Gaultheria*
— Plants erect or ascending, rooting only at the base; ovary inferior .... *Vaccinium*
3(1). Flowers borne in terminal corymbs, white, the segments of the corolla much longer than the short tube; leaves punctate below with yellow-glandular dots . *Ledum*
— Flowers solitary and axillary or in axillary racemes, rarely terminal, pink to lavender, the segments of the corolla much shorter than the tube; leaves lacking glandular punctae 4
4(3). Corolla broadly saucer-shaped or rotate, not constricted at the apex .... *Kalmia*
— Corolla campanulate to urn-shaped, often more or less constricted at the throat 5
5(4). Corolla campanulate; anthers lacking appendages; fruit a capsule embedded in a fleshy calyx .... *Gaultheria*
— Corolla urn-shaped; anthers 2–awned; fruit a berry .. .. *Arctostaphylos*

Arctostaphylos Adams

Evergreen prostrate to ascending or erect shrubs, often with purplish to orange brown, smooth bark; leaves alternate, simple entire, leathery-thickened; flowers in terminal panicles or racemes, perfect, regular; sepals usually 5; petals usually 5, united almost to the tips; corolla urn-shaped; stamens usually 10, included; anthers opening by falsely terminal pores, each with 2 hornlike appendages; ovary superior, usually 5–loculed; fruit fleshy, berrylike, 1– to several-seeded.

Adams, J. E. 1940. A systematic study of the genus *Arctostaphylos*. J. Elisha Mitchell Soc. 56: 1–62.
Eastwood, A. 1934. A revision of *Arctostaphylos* with keys and descriptions. Leaft. West. Bot. 1: 105–127.

1. Plants with creeping-prostrate stems; leaves obovate-spatulate, commonly less than 1.5 cm long ... *A. uva-ursi*

— Plants with stems ascending to erect; leaves mostly ovate to lanceolate or elliptic, often more than 2 cm long 2

2(1). Calyx and pedicels puberulent with spreading glandular hairs; twigs and leaves puberulent throughout with spreading hairs; plants of Washington County . *A. pringlei*

— Calyx glabrous or nearly so; twigs and leaves puberulent or sessile to sparingly stipitate-glandular or almost or quite glabrous; plants of various distribution 3

3(2). Twigs and axis of inflorescence white-puberulent, not glandular; plants of Washington and Kane counties .. .. *A. pungens*

— Twigs and axis of inflorescence glandular to glandular-puberulent; plants widely distributed .... *A. patula*

Arctostaphylos patula Greene Greenleaf Manzanita. (*Uva-ursi patula* (Greene) Abrams; *A. pungens* var. *platyphylla* Gray; *A. platyphylla* (Gray) Kuntze; *A. obtusifolia* Piper; *A. patula* var. *incarnata* Jepson; *A. pinetorum* Rollins; *A. parryana* var. *pinetorum* (Rollins) Weisl. & B. Schreiber). Rounded shrubs with gnarled stems to 15 cm long or more, the bark smooth, cinnamon to reddish brown or purplish in color; branchlets glandular-puberulent and sometimes with long-spreading hairs as well; leaf blades (0.8) 1.8–4.7 cm long, (0.6) 1.5–4 cm wide, ovate to elliptic, lanceolate, or orbicular, obtuse to acute apically, rounded to truncate basally, glabrous or glandular, yellow green; petioles pubescent like the twigs; inflorescence paniculate, the axis and bracts glandular-puberulent and sometimes with some long hairs; pedicels glabrous; sepals glabrous; corolla pink to white, 5–8 mm long; ovary glabrous; fruit 8–11 mm thick, depressed-globose, glabrous, white to brown, with nutlets separable or not; n = 13. Usually associated with ponderosa pine at 1370 to 2830 m in Beaver, Daggett, Duchesne, Emery, Garfield, Iron, Juab, Kane, Millard, San Juan, Sanpete, Sevier, Summit, Tooele, Uintah, Utah, Wasatch, and Washington counties; Colorado, Nevada, Oregon, Arizona, and California; 115 (xvii). The names *A. patula* and *A. platyphylla* both date as species from the same year, 1891; the question of which has priority is difficult to ascertain.

Arctostaphylos pringlei Parry Pink-bracted Manzanita. Rounded, erect shrubs to 20 dm tall or more, the bark smooth, dull red brown; branchlets densely glandular-hairy with long-spreading hairs; leaf blades (1.2) 1.8–4.2 cm long, (4) 8–20 mm wide, elliptic to lance-elliptic or lanceolate, obtuse to acute apically, truncate to rounded or obtuse basally, glandular-pubescent, gray green; petioles pubescent like the twigs; inflorescence paniculate or racemose, the axis and bracts glandular-hairy; corolla pink, 6.5–8.5 mm long; ovary glandular-hairy; fruit 6–10 mm thick, ovoid, glandular-hairy, red, with nutlets inseparable. Oak-juniper, manzanita, garrya, and mixed warm desert shrub communities at 1840 to 2750 m in Washington County; Arizona, California and Baja California; 5 (i).

Arctostaphylos pungens H.B.K. Mexican Manzanita. Erect or ascending, rounded shrubs to 20 dm tall or more, the bark smooth, red brown; branchlets canescent with a dense pubescence; leaf blades 1.6–4.7 (6) cm long, 0.5–3.2 cm wide, ovate to elliptic or oblong, rounded to acute apically, acute to rounded basally, puberulent on one or both sides, bright green; petioles pubescent like the twigs; inflorescence paniculate, the axis and bracts canescent; pedicels glabrous; sepals glabrous; corolla pink to white, 5.5–8.5 mm long; ovary glabrous; fruit 5–8 mm thick, depressed-globose, glabrous, brownish red, with nutlets separable or not; n = 13. Pinyon, juniper, live oak communites at 1005 to 1895 m in Washington and Kane counties; California, Arizona, New Mexico, Texas; Mexico; 35 (viii).

Arctostaphylos uva-ursi (L.) Sprengel Kinnikinnick; Bearberry; Sandberry. [*Arbutus uva-ursi* L., *Uva ursi procumbens* Moench; *Mairania uva-ursi* (L.) Desv.; *U. buxifolia* S. F. Gray; *Arctostaphylos officinalis* Wimmer & Grab.; *A. procumbens* Patze, Mey. & Elkan; *U. uva-ursi* (L.) Britt. in Britt. & Br.; *Arctostaphylos media* Greene; *A. uva-ursi* var. *coactilis* Fern. & Macbr.; *A. uva-ursi* var. *adenotricha* Fern. & Macbr.]. Prostrate shrub with stoloniferous rooting stems, mat-forming, the branches ascending, the internodes usually apparent, puberulent and sometimes glandular, the bark exfoliating exposing dull brown under bark; leaf blades (0.6) 1–2.7 (3) cm long, 3–12 mm wide, oblanceolate to spatulate, rounded apically, cuneate to acute basally, glabrous or puberulent, especially on the margins, green; inflorescence racemose, the axis and bracts glandular; pedicels glabrous; corolla pink to white, 4–5.2 mm long; ovary glabrous; fruit 6–11 mm thick, globose, bright red, with separable nutlets; n = 13, 26. Ground layer in coniferous forests at 2140 to 3510 m in Daggett, Duchesne, Garfield, Salt Lake, San Juan, Sevier, Summit, Uintah, and Wasatch counties; Alaska and Yukon east to the Atlantic, south to California, New Mexico, Illinois, and Georgia; Eurasia; 22 (iii). This forms occasional hybrids with *A. patula*.

Gaultheria L.

Prostrate shrubs, the branchlets rooting; leaves alternate, thin, serrulate; flowers axillary, solitary, perfect, regular; calyx 5–lobed, united, enlarging and becoming fleshy at maturity; corolla campanulate, the lobes shorter than the tube; stamens usually 10, included, the filaments flattened, tapering to the apex; anthers opening by terminal pores, not awned; ovary superior, usually 5–loculed; fruit a loculicidally dehiscent capsule enclosed by the fleshy expanded calyx.

Gaultheria humifusa (Graham) Rydb. Alpine Wintergreen. [*Vaccinium humifusum* Graham; *G. myrsinites* Hook.]. Prostrate, scarcely woody plants with creeping, rooting stems to 2 dm long, glabrous or puberulent; leaves 6–15 mm long, 4–13 mm wide, oval to ovate or elliptic, rounded to obtuse apically and basally, serrulate; flowers solitary, axillary; calyx glabrous; corolla 3–4 mm long, campanulate, pink; fruit 5–7 mm thick, subglobose, red. Ground layer in coniferous forests and margins at 2900 to 3355 m in Duchesne and Summit counties; Colorado to California, north to Alberta and British Columbia; 5 (i).

Kalmia L.

Low shrubs with puberulent branches; leaves opposite, evergreen, leathery, decurrent, entire, revolute, glaucous beneath; flowers in terminal leafy-bracted corymbs or solitary, perfect, regular; calyx 5–lobed, the segments almost distinct; corolla bowl-shaped, the lobes shorter than the tube, the tube with 10 pouches in which the anthers are enclosed in bud; stamens usually 10, the filaments flattened, hairy below; anthers opening throughout, unawned; ovary superior, 5–loculed; fruit a septicidally dehiscent capsule.

Kalmia microphylla (Hook.) Heller Bog Laurel. [*K. glauca* var. *microphylla* Hook.; *K. polifolia* var. *microphylla* (Hook.) Rehder). Erect slender shrubs, 7–15 cm tall; leaves 6–18 (30) mm long, 2–8 (12) mm wide, lance-oblong to elliptic, revolute, shining and green above, grayish beneath; corymbs mostly 2- to 6-flowered, the pedicels 1–3 cm long; sepals glabrous, ciliate; corollas 11–14 mm broad, pink; capsules 4–6 mm broad; 2n = 24. Spruce-fir and lodgepole pine communities, typically in meadows and along streams and lake margins at 2900 to 3750 m in Daggett, Duchesne, Summit, and Uintah counties; Alaska and Yukon south to California and Colorado; 19 (iv).

Ledum L.

Erect or spreading shrubs with glandular-puberulent branchlets; leaves alternate, evergreen, leathery, entire, revolute, pale below; flowers in terminal corymbs, perfect, regular; calyx small, the segments almost distinct; corolla rotate, the 5 petals distinct or nearly so; stamens usually 5–10, the filaments almost filiform, usually hairy below; anthers opening by terminal pores, unawned; ovary superior, 5-loculed; fruit a septicidally 5-valved capsule, opening at the base. Note: At least some species of this genus are poisonous to livestock.

Ledum glandulosum Nutt. Trapper's Tea. Plants mostly 5–15 dm tall, the branchlets puberulent and glandular dotted; leaves 1.1–3.4 (4) cm long, 4–14 (18) mm wide, elliptic to oblong, rounded to acute apically and basally, green above, pale to grayish beneath, glandular, the margin more or less revolute; flowers white, the segments to 5 mm long or more; pedicels commonly 1–2.5 cm long, puberulent near the base; capsules 3–6 mm long, puberulent and glandular. Meadows, stream banks, and bogs in open forest at 2600 to 3295 m in Duchesne, Salt Lake, Summit, and Uintah counties; British Columbia east to Montana and south to California, Nevada, and Wyoming; 13 (0).

Vaccinium L.

Decumbent-ascending to erect shrubs; leaves alternate, deciduous or more or less evergreen, entire or serrulate, flat, green or pale beneath; flowers solitary, axillary, or in terminal clusters, perfect, regular; calyx 4- to 6-lobed, united at the base; corolla urn-shaped or campanulate, the 4–6 lobes shorter than the tube; stamens 8–12, the filaments usually glabrous; anthers opening by pores at the ends of tubular beaks, usually 2-awned; ovary inferior, usually 4-locular; fruit a several-seeded berry.

Camp, W. H. 1942. A survey of the American species of *Vaccinium*, subgenus Euvaccinium. Brittonia 4: 205–247.

1. Branches bright green and angled; plants often less than 3 dm tall 2
— Branches neither bright green nor angled, or sometimes irregularly angled when dry; plants often more than 3 dm tall 3

2(1). Fruit red; grooves of branches usually glabrous; leaves often less than 12 mm long .... *V. scoparium*
— Fruit blue black or black; grooves of branches usually puberulent; leaves often over 12 mm long ... *V. myrtillus*

3(1). Flowers in clusters of 2–4, or solitary; leaves entire; calyx deeply lobed, the lobes persistent in fruit .... *V. occidentale*

— Flowers solitary in leaf axils; leaves more or less serrate; calyx shallowly lobed, the lobes deciduous in fruit 4

4(3). Plants mostly 1–3 dm tall; leaves serrate above the middle and unconspicuously below the middle, mainly 1–3 (4) cm long, oblanceolate to obovate . *V. caespitosum*
— Plants mostly 4–7 dm tall or more; leaves serrate to the base or nearly so, commonly 2–6 cm long, elliptic to ovate .... *V. membranaceum*

Vaccinium caespitosum Michx. Dwarf Huckleberry. Plants 1–3 dm tall; twigs brownish, somewhat angled, puberulent or glabrous; leaves 0.7–4 cm long, 3–20 mm wide, oblanceolate to obovate, obtuse or less commonly acute to rounded apically, usually cuneate basally, serrulate from tip to below the middle; flower solitary, axillary, whitish to pink, the corollas 5–6 mm long, twice as long as thick; calyx obscurely lobed, the lobes decidous in fruit; berries blue glaucous, subglobose, 5–8 mm broad, edible and good. Spruce-fir and alpine tundra communities at 2225 to 3420 m in Daggett, Duchesne, Emery, Garfield, Grand, San Juan, Summit, and Uintah counties; Alaska and Yukon east to Newfoundland and New Hampshire, south to California and Colorado; 17 (ii). Materials from Utah have previously passed under the names *V. membranaceum* Dougl. (see below) and *V. globulare* Rydb. The latter is not known for the state.

Vaccinium membranaceum Dougl. Mountain Huckleberry. Shrubs mostly 3–7 dm tall or more; twigs brownish, glabrous or puberulent; leaves 1.8–7 cm long, 1–3.4 cm wide, elliptic or less commonly ovate to obovate, acute to obtuse apically, acute to rounded basally, serrate almost throughout; flowers solitary, axillary, yellowish pink, the corollas ca 6 mm long, ca 1/3 longer than broad; calyx obscurely lobed, the lobes deciduous, berries purple, not glaucous, 7–9 mm broad, edible and good. Ponderosa pine, Douglas fir, and spruce-fir communities at 2500 to 3145 m in Cache, Carbon, Duchesne, San Juan, Summit, and Wasatch counties; British Columbia southward to California, Idaho, and Monatana; 11 (0).

Vaccinium myrtillus L. Dwarf Billberry. [*V. oreophilum* Rydb., in part, type from the Uinta Mts.]. Plants mostly 0.5–3 dm tall; twigs seldom numerous and broomlike, green sharply angled, puberulent; leaves 1.1–3.9 cm long, 6–16 mm wide, ovate to lanceolate or elliptic, acute to obtuse apically, obtuse to rounded basally, serrulate almost or quite from base to apex; flowers solitary, axillary, pink, the corollas 4–5 cm long; calyx shallowly lobed; berry usually bluish, 5–8 mm broad; 2n = 24. Ponderosa pine, aspen, sagebrush, lodgepole pine, and spruce-fir communities at 2285 to 3355 m in Daggett, Duchesne, San Juan, Summit, and Uintah counties; British Columbia and Alberta south to Arizona and New Mexico; Eurasia; 16 (i). The dwarf billberry is a near congener of the very common *V. scoparium*, but can be distinguished by the larger size of its leaves and flowers and by the puberulent stems.

Vaccinium occidentale Gray Western Huckleberry. Plants mostly 2–6 dm tall, the twigs round, usually glabrous; leaves 6–21 mm long, 4–12 mm wide, oblanceolate, rounded to obtuse apically, acute basally, entire; flowers 2–4, or less commonly solitary in the axils, pinkish, the corollas 3.5–6 mm long; calyx definitely lobed, the lobes persistent in fruit; berries blue, glaucous, 4–6 mm thick. Lodgepole pine and spruce-fir communities at 2805 to 3510 m in in Duchesne, Morgan, Summit, and

Wasatch counties; British Columbia south to California and Idaho; 16 (viii).

Vaccinium scoparium Leiberg Grouseberry. [*V. myrtillus* var. *microphyllum* Hook.; *V. microphyllum* (Hook.) Rydb., not Rein.; *V. erythrococcum* Rydb.]. Plants mostly 1–2.5 dm tall, the twigs numerous, broomlike, sharply angled, usually glabrous; leaves 6–13 mm long, 3–7 mm wide, ovate, obtuse to acute apically, rounded to obtuse basally, serrulate throughout; flowers solitary, axillary, pinkish, the corollas 2.5–3.5 mm long; calyx very shallowly lobed; berry bright red, drying red purple, 4–6 mm thick. Lodgepole pine and spruce-fir communities at 2500 to 3508 m in Carbon, Daggett, Duchesne, Salt Lake, Summit, Uintah, and Wasatch counties; British Columbia and Alberta south to California and Colrado; 32 (iii).

EUPHORBIACEAE A. L. Juss.
Spurge Family

Annual or perennial herbs or woody plants, the juice milky or watery; leaves simple, alternate or opposite (or whorled); flowers imperfect; sepals 4–6 or lacking (a calyxlike involucre, a cyathium, present in *Euphorbia*); petals 4–6, distinct, or commonly lacking; stamens 1 to many; pistil 1, the ovary superior, 3–loculed; styles 3 or 3–lobed; ovules 1 or 2 per locule; fruit a capsule; x = 6–14+.

Cronquist, A. 1983. Euphorbiaceae. Unpublished mss. Intermountain Flora. New York Botanical Garden. 57 p.

1. Plants with milky juice; flowers borne in cyathia . . . . *Euphorbia*
— Plants with watery (though sometimes colored) juice; flowers with a calyx 2

2(1). Plants with persistent, opposite, linear, foliaceous cotyledons; plants annual, of sandy habitats in Kane and Washington counties . . . . *Reverchonia*
— Plants with alternate leaves, the cotyledons not persistent; plants of various duration, habitats, and distribution 3

3(2). Leaves palmately lobed, peltate, mostly over 5 cm wide; plants cultivated . . . . *Ricinus*
— Leaves pinnate, toothed or entire, mostly less than 2 cm wide; plants indigenous 4

4(3). Leaves long-petiolate, pubescent with appressed stellate hairs; plants of sandy sites in western and southern Utah . . . . *Croton*
— Leaves short-petiolate, pubescent with long, spreading, branched or simple hairs; plants mainly of limestone outcrops or gravels in western and southwestern Utah 5

5(4). Plants forming low, rounded clumps; pubescence of long, branched, stellate hairs; leaves not sharply serrate . . . . *Eremocarpus*
— Plants not especially clump-forming; pubescence mainly of long, simple, stinging hairs; leaves sharply serrate . . . . *Tragia*

Croton L.

Annual or perennial, appressed-stellate herbs; leaves alternate, petioled, simple, estipulate; flowers imperfect, the inflorescence racemose; calyx mostly 5–lobed; petals lacking; nectary disk present, the lobes or glands alternate with the petals; receptacle usually hairy; stamens mostly 5 or more; pistillate flowers typically below the staminate; ovary typically 3–loculed; styles as many as the locules; capsule usually 3–lobed, 3– (or 1) –seeded; seeds carunculate, smooth, shiny.

Ferguson, A. M. 1901. Crotons of the United States. Ann. Rep. Missouri Bot. Gard. 12: 33–73.

1. Plants green, sparsely stellate-hairy, annual . . . . *C. texensis*
— Plants gray or gray green, densely stellate-hairy, perennial . . . . *C. californicus*

Croton californicus Muell.-Arg. Mohave Croton. [*C. longipes* Jones, type from near Leeds; *C. californicus* (var.) *longipes* (Jones) A. Ferg.]. Dioecious grayish or whitish perennials from taproots; herbage with appressed stellate hairs; stems simple or branched, mostly 2–10 dm tall; leaf blades 1–5 cm long, 5–25 mm wide, lance-oblong to elliptic or ovate, greenish above entire, obtuse to rounded apically; petiole 0.5–4 cm long; racemes mostly 1–3 cm long; calyx of staminate flowers 5–lobed, the lobes 1.5–2 mm long; petals none in flowers of either sex; stamens mostly 10–15, subequal to the calyx lobes; ovary 3–loculed; styles 1–2.5 mm long, palmately cleft or bifid; capsules 3–6 mm long; seeds 3, the caruncle large. Sand dunes and other sandy sites with sand sagebrush at 820 to 1220 m in Washington County; California, Nevada, and Arizona; 24 (v).

Croton texensis (Klotzsch) Muell.-Arg. Doveweed. [*Hendecandra texensis* Klotzsch]. Dioecious, scented, yellow green annuals from taproots, with sparse, appressed stellate hairs, commonly branched, mostly 1–6 dm tall; leaf blades mostly 1.5–6.5 cm long, 2–28 mm broad, lanceolate, to elliptic or oblong, entire, acute to acutish apically; petioles 0.5–1.5 cm long; cymes mostly 1–2 cm long; staminate calyx deeply 5–lobed, the lobes 1–1.5 mm long; petals lacking; stamens mostly 8–12, subequal to the calyx lobes; ovary 3–loculed; styles 1–2 mm long, 2 or more times bifid; capsules 4–6 mm long; seeds 1–3, 3.5–4 mm long; caruncle large. Sand dunes and other sandy sites at 1125 to 1680 in Juab, Kane, Millard, and San Juan counties; Arizona to Texas and Mexico, and northeast to South Dakota; 17 (v).

Eremocarpus Benth.

Low, gray, scented, monoecious, annual herbs from taproots, pubescent with elongate stellate and stinging hairs; leaves alternate, entire, 3–veined; staminate flowers in terminal cymes, with 5– or 6–lobed calyx, lacking petals, with 6 or 7 stamens, the receptacle hairy; pistillate flowers, 1–3, lacking a perianth; ovary with 4 or 5 small glands at the base, 1–loculed, the style not divided; capsule 2–valved, 1–sided.

Eremocarpus setigerus (Hook.) Benth. Turkeymullein. [*Croton setigerus* Hook.]. Plants dichotomously branched from the base, forming cushionlike clumps 3–20 cm tall; leaves 1.8–6 cm long, ovate to suborbicular; petioles about as long as the blades; staminate flowers pedicillate, the calyx ca 2 mm long, surpassed by the stamens; pistil pubescent; capsules ca 3–5 mm long; seeds ellipsoid, somewhat ridged. Creosote bush community at ca 824 m in Washington County; Washington and Idaho, south to California and Arizona; 3 (ii).

Euphorbia L.

Annual or perennial herbs; leaves simple, alternate, opposite, or whorled; flowers monoecious, borne in in volucres called cyathia, these simulating a single flower and often with petaloid appendages on marginal glands; pistillate flowers solitary in center of the cyathium, pedicelled, lacking a calyx and corolla, the ovary 3–loculed; styles 3, usually bifid; staminate flowers without a calyx or corolla, in 5 fascicles, 1 to several per fascicle, the fascicles opposite the lobes of the involucre, each pedicel of 1 stamen often with a minute bract at base; capsules 3–loculed, 3–seeded, usually nodding. Note: The poinsettia of commerce, **Euphorbia pulcherrima Willd. ex Klotsch**, is widely grown under glass in Utah.

1.	Upper bractate leaves white-margined or cyathium with conspicuous white petaloid appendages, the plants then perennial	2
—	Upper bractate leaves green or red and the cyathium not with conspicuous white petaloid appendages or, if so, the plants mostly annual	3
2(1).	Plants prostrate or ascending perennials; leaves less than 1 cm long; indigenous in Iron, Kane, and Washington counties	*E. albomarginata*
—	Plants erect annuals; leaves mainly 1.5–4 cm long; adventive or cultivated broadly	*E. marginata*
3(2).	Leaves all linear, mainly 4–10 times longer than broad; plants annual, the branches spreading umbrellalike above the ground or ascending-spreading	4
—	Leaves not all linear, or if so, the plants perennial and the branches various	5
4(3).	Capsules less than 1.7 mm long; stems usually less than 1 mm thick; seeds with transverse ridges; plants known from Washngton County	*E. revoluta*
—	Capsules more than 1.8 mm long; stems usually at least 1 mm thick; seeds smooth; plants of rather broad distribution	*E. parryi*
5(3).	Leaves dimorphic, the lower obovate to elliptic, the upper ones linear; glands of the involucre 5; cyathia terminal, borne in forks of the stem	*E. nephradenia*
—	Leaves usually all about alike; glands of involucre 1 or usually 4, the cyathia variously disposed	6
6(5).	Uppermost leaves red or red spotted; foliage leaves often more or less lobed and violin-shaped; plants adventive in St. George	*E. cyathiflora*
—	Uppermost leaves green like the others; foliage leaves various, but not as above; plants indigenous or adventive	7
7(6).	Leaves irregularly toothed or lobed; cyathium with a solitary gland; weedy species of gardens	*E. dentata*
—	Leaves entire or essentially so; cyathia usually with 4 glands; plants weedy or not	8
8(7).	Leaves symmetrical at the base, mainly alternate; inflorescence terminal, an umbel subtended by whorled leafy bracts; plants usually branched above, erect	9
—	Leaves asymmetrical at the base, opposite; inflorescence of small clusters along the stems; plants branching from the base, often prostrate	14
9(8).	Plants annual, from slender taproots	10
—	Plants perennial, from caudices	11
10(9).	Lower cauline leaves with distinctive filiform petioles; glands of cyathia horned at each end	*E. peplus*
—	Lower cauline leaves tapering to a broad, short, petiolar base; glands of cyathia entire	*E. spathulata*
11(9).	Leaves thick, apparently spirally arranged, prominently mucronate; plants cultivated and escaping	*E. myrsinites*
—	Leaves not especially thickened or spirally arranged, not mucronate; plants indigenous or adventive	12
12(11).	Leaves orbicular to ovate or broadly elliptic; plants from a caudex and taproot, indigenous	*E. brachycera*
—	Leaves linear to narrowly oblong; plants from rootstock or rhizomes, sod-forming	13
13(12).	Main leaves over 3 cm long and more than 3 mm wide; plants 3–7 dm tall	*E. esula*
—	Main leaves less than 2.5 cm long and 3 mm wide; plants 1–3 (4) dm tall	*E. cyperissias*
14(8).	Plants with a pluricipital caudex, perennial	*E. fendleri*
—	Plants with slender taproots, annual	15
15(14).	Glands of cyathia lacking appendages	16
—	Glands of cyathia appendaged, this more or less petaloid	17
16(15).	Leaves mainly 1.5–5 mm long, 1–3 mm wide, subsessile, or shortly petiolate; seeds 0.8–1.3 mm long	*E. micromera*
—	Leaves mainly 5–12 mm long and 4–7 mm wide, petiolate; seeds 1.3–1.8 mm long	*E. ocellata*
17(15).	Herbage, ovary, and capsules evidently hairy	18
—	Herbage, capsules, and fruits glabrous, but sometimes sparingly hairy when young, or the herbage hairy, but ovaries and capsules glabrous	20
18(17).	Glandular appendages deeply cleft into 3–5 attenuate segments; leaves mostly 2–7 mm long, entire; plants indigenous in Washington County	*E. setiloba*
—	Glandular appendages entire or toothed or lobed, but not with slender segments; leaves mostly more than 7 mm long, more or less serrulate; plants adventive	19
19(18).	Seeds with low, rounded, transverse ridges not whitened at the summit; capsules appressed hairy; styles divided less than half way to the base	*E. maculata*
—	Seeds with narrow, sharp, transverse ridges whitened on the summit; capsules spreading-hairy; styles divided almost to the base	*E. prostrata*
20(17).	Stems erect or erect-ascending, the larger leaves mainly 1.5–3 cm long	*E. hyssopifolia*
—	Stems prostrate, decumbent, or ascending, the larger leaves seldom over 1 cm long	21
21(20).	Seeds with 3 or 4 transverse ridges	*E. glyptosperma*
—	Seeds not transversely ridged	*E. serpyllifolia*

Euphorbia albomarginata T. & G. Rattlesnake-weed. Perennial herbs from a caudex, glabrous; stems 3–25 cm long, prostrate; leaves opposite, 3–8 mm long, orbicular to ovate or oblong, the margin thin, hyaline, and entire; stipules united, forming a white-membranous scale with fringed or entire margin; cyathia solitary, axillary, 1.2–2 mm long, campanulate; glands transversely oblong, concave, brown to red brown, 0.5–0.8 mm long, the petaloid appendages white, conspicuous, entire or 3– to 5–lobed; staminate flowers 15–30; capsules 1.6–2.3 mm long; glabrous, ovoid, sharply angled; seeds rounded to quadrangular, cylindroid, 1.2–1.7 mm long, white, the surfaces smooth; n = 18, 24. Creosote bush, Joshua tree, burrobush, sand sagebrush, shadscale, and pinyon-juniper communities at 700 to 1830 m in Iron, Kane, and Washington counties; California, Nevada, Arizona, New Mexico, and Texas; 23 (vii).

Euphorbia brachycera Engelm. in Emory. Shorthorn Spurge. [*E. montanum* var. *robusta* Engelm.; *E. robusta*

(Engelm.) Small ex Britt. & Br.; *Tithymalus robusta* (Engelm.) Small ex Rydb.; *E. r.* var. *interioris* J. S. B. Norton, type from Wasatch Mts.]. Perennial herbs from a stout caudex, glabrous and glaucous or velvety hairy; stems 10–45 cm tall, erect or steeply ascending, usually numerous; leaves alternate (the cauline at least), typically 7–28 mm long and 3–25 mm wide, orbicular to spatulate or oblong, entire, the lower ones often mere bracts, the upper ones subverticillate below the umbellate inflorescence; branches of inflorescences dichotomous, with foliose, opposite bracts subtending each dichotomy; cyathia 2–3 mm long, obconic; glands 4, yellowish, transversely oblong, 1–2 mm wide, toothed marginally, the 2 lateral teeth somewhat enlarged; staminate flowers several; fruit 3–5 mm long, glabrous or pubescent; seeds 2.5–3.5 mm long, the surface pitted to almost smooth, whitish. Hanging garden, pinyon-juniper, mountain brush, sagebrush, and aspen communities at 1190 to 3205 m in Beaver, Daggett, Duchesne, Emery, Garfield, Grand, Iron, Juab, Morgan, Piute, Salt Lake, San Juan, Sanpete, Sevier, Summit, Uintah, Utah, Wasatch, Washington and Wayne counties; Montana and South Dakota, south to California, Nevada, Arizona, New Mexico, and Texas; 88 (xx).

Euphorbia cyathophora Murray Annual Poinsettia. [*E. heterophylla* authors, not L.]. Annual herbs from a taproot, glabrous or essentially so; stems mostly 2–5 (7) dm tall, erect, simple or branched; leaves opposite below, becoming alternate above, petiolate, the blades often oblong in outline, but typically lobed and more or less violin-shaped, green above and pale beneath, or the uppermost red; stipules minute and glandlike; cyathia irregularly clustered, these subtended by reddish, foliose bracts, urceolate-campanulate, 2–2.5 mm long, toothed; gland usually solitary, deeply cupped, shallowly bilabiate, the opening narrowly oblong, green, sessile, not appendaged; staminate flowers many per cyathium; pistillate flowers long-exserted and reflexed at maturity; styles 3, bifid; capsule with 3 rounded lobes, 3–4 mm long; seed ovoid to subglobose, 2.9–3.1 mm long, dark brown, tuberculate; caruncle small. Along irrigation canals at ca 900 m in Washington County; widely distributed in North America; 2 (ii).

Euphorbia cyparissias L. Cypress Spurge. [*Tithymalus cyparissias* (L.) Hill]. Perennial herbs from a caudex and elongate, horizontal rhizome, glabrous and sometimes glaucous; stems mainly 10–40 cm tall, erect or steeply ascending, few to several; leaves alternate (at least the cauline ones), typically 5–25 mm long and 1.2–3 mm wide, linear to narrowly spatulate, entire, the lower one bracteate, those on branches below inflorescence very slender; inflorescence umbellate, but with some flowers from axils below the umbels, the rays ultimately dichotomous and with paired, broad, yellowish green bracts; cyathia 2–3 mm long, campanulate; glands 4, yellowish, transversely oblong, 1–1.5 mm wide, horned on each side; staminate flowers few; fruit ca 3 mm long, glandular-roughened; seeds 1.5–2 mm long, smooth; 2n = 20, 36, 40. Weedy species of disturbed sites at lower elevations, and cultivated ornamental, in Cache, Salt Lake, and Utah counties; widely naturalized in U. S.; introduced from Europe; 6 (0).

Euphorbia dentata Michx. Toothed Spurge. [*Poinsettia dentata* (Michx.) Klotzsch & Garcke]. Annual herbs from taproots, puberulent to scabrous; stems erect, typically 2–7 dm tall, simple or branched; leaves mainly opposite, with petioles 0.5–2.5 cm long, the blades 1–9 cm long, 0.5–2.5 cm wide, lanceolate to ovate or orbicular, serrate to serrate-dentate; stipules reduced to glands or lacking; bracteate leaves subtending the inflorescence usually congested and enlarged; cyathia 2–3 mm long; glands solitary, with white, fimbriate lobes; fruit 3–3.5 mm high, glabrous; seeds 2.2–2.8 mm long, dark brown, tuberculate; n = 28. Weedy species of gardens and other disturbed sites at 1310 to 1465 m in Cache, Salt Lake, Tooele, and Utah counties; indigenous from Wyoming to Wisconsin, south to Arizona and Mexico; widely established elsewhere; 13 (0).

Euphorbia esula L. Leafy Spurge. [*Tithymalus esula* (L.) Hill]. Perennial herbs from a caudex and elongate, horizontal rhizome, glabrous; stems mainly 3–7 dm tall, erect or steeply ascending, few to several; leaves alternate (at least the cauline ones), typically 25–60 mm long and 2.5–7 mm wide, linear to narrowly oblong, entire, the lower ones often deciduous at anthesis, those on branches below the inflorescence not very slender; inflorescence umbellate, but with some flowers from axils below the umbels, the rays ultimately dichotomous and with paired, broad, greenish or yellowish bracts; cyathia 2–3 mm high; glands 4, yellowish, campanulate, transversely oblong, prominently horned on both edges; staminate flowers few; fruit 3–3.5 mm long, granular or tubercled; seeds 2–2.5 mm long, smooth; 2n = 16, 60, 64. Weedy species of open habitats at 1400 to 2900 m in Box Elder, Cache, Salt Lake, Sanpete, Sevier, Uintah, Utah, Wasatch, and Weber counties; widely established in the U. S.; introduced from Eurasia; 13 (0).

Euphorbia fendleri T. & G. Fendler Euphorb. Perennial herbs from a caudex, glabrous, often suffused with purple; stems 5–20 cm long, prostrate; leaves opposite, 3–11 mm long, ovate to oval or lance-ovate to elliptic, the margins thick; stipules separate, linear-subulate, entire or forked, glabrous; cyathia solitary, axillary, 1.3–2 mm long, campanulate; glands transversely elliptic, concave, reddish, ca 1 mm long, the appendages white, crenate, about as wide as the gland, inconspicuous; staminate flowers 15–35; capsules typically 2–2.5 mm long, glabrous, rounded; seeds quadrangular, 2–2.2 mm long, white, the facets smooth or wrinkled. Creosote bush, blackbrush, salt desert shrub, mixed desert shrub, pinyon-juniper and ponderosa pine communitites at 975 to 2320 m in Beaver, Carbon, Daggett, Duchesne, Emery, Garfield, Grand, Iron, Juab, Kane, Millard, San Juan, Uintah, Washington, and Wayne counties; Wyoming and South Dakota, south and southwest to California, Nevada, Arizona, New Mexico, and Texas; 196 (xxx). Should this species be confused with *E. albomarginata* the separate stipules are diagnostic.

Euphorbia glyptosperma Engelm. in Emory Ridge-seeded Spurge. Annual herbs from a slender taproot, glabrous; stems 3–20 cm long, prostrate or ascending; leaves opposite, 3–12 (15) mm long, entire or toothed apically, oblong, short-petiolate; stipules subulate, alternate, laciniate or dissected; cyathia solitary, axillary, 0.4–1 mm long, obconic; glands transversely oblong-elliptic, 0.2–0.4 mm long, with white appendages as wide as or wider than the glands, these more or less crenate; staminate flowers commonly 4 (1–5); capsules 1.5–1.7 mm long, glabrous, rather sharply angled; seeds quadrangular, 1–1.3 mm long, white to tan, the surfaces with 3 or

4 transverse ridges; n = 11. Weedy species of open places such as gardens and roadsides, but also in blackbrush, mixed desert shrub, pinyon-juniper and ponderosa pine communities at 850 to 2105 m in Box Elder, Cache, Carbon, Daggett, Davis, Duchesne, Garfield, Kane, Millard, Salt Lake, San Juan, Tooele, Uintah, Utah, Washington, and Wayne counties; widespread in the U. S.; 41 (xi).

Euphorbia hyssopifolia L. Annual herbs from slender taproots, glabrous or with long, spreading, soft hairs; stems 10–35 cm long, erect or ascending; leaves opposite 3–35 mm long, 4–10 mm wide, oblong to elliptic, serrulate throughout, short-petiolate; stipules separate or connate, more or less lacerate; cyathia solitary, terminal and axillary, glabrous, campanulate to obconic, 1–1.5 mm high; glands transversely elliptic, 0.3–0.4 mm long, greenish, with a white, uniform, entire appendage; staminate flowers mostly 4–15; capsule glabrous, 1.6–2.1 mm long, rounded; seeds quadrangular, 1–1.4 mm long, the faces irregularly ridged, depressed, or pitted, dark brown. Weedy species in Washington County; widely distributed in warm temperate U. S.; 6 (iv).

Euphorbia maculata L. Spotted Spurge. [*E. supina* Raf.]. Annual herbs from slender taproots, villous to puberulent; stem 3–40 cm long, prostrate; leaves opposite, mostly 4–15 mm long, 2–5 mm wide, oblong to elliptic or ovate, serrulate to subentire, sometimes with a purple blotch medially; stipules distinct, subulate, often lacerate; cyathia few to several, congested, on short lateral branches, obconic, 0.5–1 mm tall; glands transversely elongate, ca 0.2 mm long, with a narrow, white, crenulate appendage; staminate flowers 2–5; capsule strigose, sharply angled, ca 1.4–1.5 mm long; seeds quadrangular, ca 1 mm long, the surfaces smooth or with low transverse ridges; n = 28. Weedy species of open sites at 850 to 1405 m in Cache, Salt Lake, Utah, Washington, and Weber counties; widespread in temperate North America; 13 (vi).

Euphorbia marginata Pursh Snow on the Mountain. Annual herbs from slender taproots, glabrous to long-hairy; stems mostly 20–100 cm tall, erect; leaves alternate or subopposite above, mostly 1.5–10 cm long, 1–3 cm wide, oblong to ovate or elliptic, entire; inflorescence cymose, the bracteate leaves whorled or opposite, lanceolate, entire, white margined or entirely white; cyathia subcylindric to campanulate, mostly 3–4 mm long, hairy, the 5 glands oblong, depressed, with white appendages 2–3 mm long; capsules 5–8 mm wide, serrate, usually pubescent; seeds ca 4 mm long; 2n = 56. Gardens, canal banks, and other open sites at 950 to 1405 m in Salt Lake, Utah, and Washington counties; indigenous in central U. S., now widely grown as an ornamental and escaping or established; 2 (ii). This and other cultivated euphorbs result in human poisonings in the U. S.

Euphorbia micromera Boiss. [*E. podagrica* Johnst.]. Annual herbs from slender taproots; glabrous; stems 7–25 cm long, prostrate; leaves opposite, 2–7 mm long 1–4 mm wide, entire, short-petiolate to subsessile; stipules usually distinct, subulate and more or less laciniate or pectinately ciliate; cyathia campanulate, 0.5–0.9 mm tall; glands discoid, pink or red purple, lacking appendages or these mere rudiments; staminate flowers 2–5; capsules glabrous, 1.2–1.4 mm long, rounded; seeds quadrangular, whitish, 0.8–1.3 mm long, the surface smooth or faintly wrinkled. Weedy species of open sites near habitations, and in creosote bush, blackbrush, and mixed desert shrub communities at 850 to 1740 m in Garfield, San Juan, Wayne, and Washington counties; California, Nevada, New Mexico, Texas, and Mexico; 8 (v).

Euphorbia myrsinites L. Blue Spurge. Coarse, perennial, clumpforming herbs, glabrous, glaucous; stems mainly 6–45 cm long, sprawling- ascending; leaves alternate but spirally arranged, rosetttelike at the stem apex, 0.8–3 cm long, 0.6–2.5 cm wide, oblanceolate to obovate, subsessile, cuneate basally, apiculate; stipules lacking; inflorescence umbellate, the rays dichotomously branched, each dichotomy subtended by paired orbicular, green bracts; cyathia campanulate, 4–5 mm long, the glands transversely crescent-shaped, produced at each end as lobed horns; capsules glabrous, 5–7 mm high; seeds 3–4.5 mm long, sculptured to almost smooth; 2n = 20. Cultivated ornamental, escaping, persisting and established in Cache, Salt Lake, Summit, Uintah, and Utah counties; sporadic in western U. S.; 8 (i).

Euphorbia nephradenia Barneby Utah Spurge. Annual herbs from slender taproots, strigulose; stems erect or ascending, branched; leaves opposite, dimorphic, the lower ones lanceolate to oblanceolate, obovate, or elliptic, 10–33 mm long, 4–9 mm wide, the upper linear to linear-lanceolate, 10–40 mm long, 1–2.5 mm wide, short-petiolate; stipules lacking; cyathia solitary in stem forks or leaf axils, 1–1.5 mm high, strigulose, campanulate; glands greenish, transversely elliptic, 0.8–1 mm wide, with yellow green, entire appendages ca 0.2–0.4 mm long; capsules 2.8–3.2 mm long, rounded; seeds ovoid, 2.1–2.6 mm long, grayish mottled, smooth or punctate, mat-saltbush, blackbrush, ephedra, and mixed sandy desert shrub communities at 1155 to 1465 m in Emery, Garfield, Kane (type from Cottonwood Canyon), and Wayne counties; Colorado (a Plateau endemic); 10 (iv).

Euphorbia ocellata Dur. & Hilg. Eyed Spurge. Annual herbs from slender taproots, glabrous; stems prostrate, mainly 8–20 cm long; leaves opposite, dimorphic, those on the main stems obliquely ovate-lanceolate, acute, the blades 5–10 mm long and 2–7 mm wide, those of congested lateral branches much smaller, all short-petiolate; stipules mostly distinct, subulate or wider, entire or fimbriate; cyathia solitary, solitary, turbinate to campanulate, 1.5–2 mm long; glands not appendaged, discoid to ellipsoid, ca 0.6 mm wide; staminate flowers 40–60; capsules glabrous, 2–2.5 mm high, obtuse; seeds ovoid, 1.3–1.6 mm long, smooth or nearly so, whitish or brownish. Sand dunes and other sandy sites in salt and sand desert shrub communities at ca 1730 to 1465 m in Millard, Tooele, and Utah counties; Idaho, Nevada, and California; 8 (iii). Our plants are assignable to **var. arenicola (Parish) Jepson** [*E. arenicola* Parish].

Euphorbia parryi Engelm. Parry Spurge. Annual herbs from slender taproots, glabrous, spreading umbrel-lalike to prostrate; leaves opposite, mainly 10–35 mm long and 1–3 mm wide, linear, entire; stipules distinct, linear, often fimbriate; cyathia clustered, mainly 1–13, campanulate, 1.2–1.7 mm high; glands transversely oval, ca 0.4 mm long, the appendage ca 0.2–0.4 mm wide, yellowish, entire; capsules 1.5–2 mm long, the lobes obtuse; seeds ovoid, 3–angled, 1.2–1.8 mm long, roughened to smooth, grayish mottled to whitish. Sand dunes and other sandy sites in blackbrush, sand sagebrush, ephedra, purple-sage, and pinyon-juniper communities at 820 to 1865 m in Emery, Garfield, Grand, Kane, San Juan, Washington (type from near St. Georege), and Wayne counties; Nevada, California, Arizona, New Mexico, Texas, and Mexico; 84 (xxvi).

Euphorbia peplus L. Petty Spurge. Annual herbs from slender taproots, glabrous; stems erect, mainly 10–30 cm tall, simple or branched from near the base; cauline leaves alternate, mainly 3–25 mm long and 3–15 mm wide, obovate to oval, entire, short-petiolate; bracteate leaves ovate; inflorescence umbellate, the rays dichotomously branched, each dichotomy subtended by paired orbicular bracts; cyathia ca 1–1.5 mm high, campanulate; glands transversely elongate, yellow, bearing a hornlike projection at each end; capsules ca 2 mm long, each lobe with a dorsal keel; seeds oblong, 1.2–1.5 mm long, subtriangular, the outer faces each with 3 or 4 pits, the 2 inner ones with a longitudinal groove; $2n = 16$. Weeds of gardens and other open sites at 1310 to 1405 m in Cache, Carbon, Salt Lake, and Utah counties; widely established in North America; adventive from Eurasia; 8 (0).

Euphorbia prostrata Ait. Prostrate Spurge. Annual herbs from slender taproots, villous; stems 10–35 cm long, prostrate; leaves opposite, mostly 3–11 mm long and 1.5–4.5 mm wide, oblong to elliptic, entire or serrulate, not blotched; stipules usually distinct, subulate, often lacerate; cyathia few to several, congested on short lateral branches, obconic, 0.6–0.9 mm tall; glands transversely oval to oblong, 0.2–0.3 mm long, with a white appendage broader than the gland; staminate flowers 4; capsules glabrous or thinly pilose on the faces, long-hairy on the angles, 1–1.4 mm long; seeds quadrangular, ca 1 mm long, the surfaces with transverse ridges; $2n = 18, 20$. Cracks in sidewalks and other disturbed sites at ca 1400 m in Salt Lake, Utah, and Washington counties; widely established in the U. S.; introduced from eastern U. S.; 5 (iii).

Euphorbia revoluta Engelm. Revolute Spurge. Annual herbs from slender taproots, glabrous; stems 3–23 cm long, erect, with branches spreading-ascending, usually less than 1 mm thick; leaves opposite, mostly 3–25 mm long and 1–1.2 mm wide, entire, revolute; stipules distinct, entire, linear-subulate; cyathia solitary in the forks or appearing cymose and clustered apically, 0.7–0.9 mm long, campanulate; glands 4, discoid, to 0.3 mm long, the appendages white to purple, or obsolete, usually ovate, about equaling the gland; staminate flowers 3–8; capsules 1.3–1.7 mm long, ovoid, glabrous; seeds sharply angled, 1.2–1.5 mm long, transversely ridged. Creosote bush and Joshua tree community at ca 1220 m in Iron and Washington counties; Arizona to Texas and Colorado, south to Mexico; 3 (i).

Euphorbia serpyllifolia Pers. Thyme-leaved Spurge. Annual herbs from slender taproots, glabrous; stems prostrate, 5–30 cm long; leaves opposite, 3–12 mm long and 1.5–4 mm wide, oblong to obovate, entire or serrulate towards the apex; stipules distinct, subulate, often fimbrillate; cyathia solitary, 1.2–1.5 mm long, campanulate; glands transversely oblong, 0.2–0.5 mm long, with narrow white appendages with crenate margins; staminate flowers 5–10 or more; capsules sharply angled, 1.5–1.9 mm long; seeds sharply quadrangular, 1–1.4 mm long, the surfaces smooth or somewhat pitted, white to brownish; $n = 11$. Gardens, roadside, and other open habitats in blackbrush, sagebrush, pinyon-juniper and mountain brush communities at 1155 to 2745 m in Beaver, Box Elder, Cache, Duchesne, Emery, Garfield, Juab, Millard, Salt Lake, Sanpete, Sevier, Tooele, Uintah, Utah, and Washington counties; widely distributed in North America; 29 (vii).

Euphorbia setiloba Engelm. in Torr. Fringed Spurge. [*Chamaesyce setiloba* (Engelm.) Millsp. ex Parish]. Annual herbs from slender taproots, spreading hairy throughout; stems prostrate, mainly 10–20 cm long; leaves opposite, mostly 2–7 mm long and 2–5 mm wide, elliptic to ovate-oblong or oblanceolate, entire; stipules obsolete; cyathia clustered, ca 1 mm long, turbinate, constricted above; glands 4, each with a deeply cleft white or pinkish petaloid appendage ca 0.3–0.7 mm long, the segments slender and attenuate; staminate flowers 3–7; capsules evidently 3-lobed, mostly 1–1.2 mm thick; seeds strongly angled, mainly 0.8–1 mm long, transversely wrinkled-rugose, brownish. Creosote bush, Joshua tree, and other warm desert shrub communities at 1000 to 1375 m in Washington County; California and Nevada east to Texas and south to Mexico; 6 (v).

Euphorbia spathulata Lam. Prairie Spurge. Annual herbs from slender taproots, glabrous; stems 8–30 (45) cm tall, simple or branched; leaves alternate, 6–30 mm long, 4–20 mm wide, obovate to spatulate, serrulate (subentire), subsessile; stipules lacking; inflorescence umbellate, the rays dichotomously branched, each dichotomy subtended by ovate-oblong paired green bracts; cyathia ca 1 mm tall, campanulate; glands small, yellow, transversely oval; capsules 2–3 mm long, warty; seeds 1.2–1.5 mm long, brown, reticulate to ellipsoid. Pinyon-juniper and ponderosa pine communities at ca 1525 to 1830 m in Salt Lake, San Juan, and Washington counties; widespread in western U. S.; Mexico and South America; 5 (i).

Reverchonia Gray

Glabrous, monoecious, annual herbs from taproots; leaves alternate (the cotyledons foliaceous and opposite), entire, stipulate; flowers borne in compact, bracteolate cymes in leaf axils on lateral branches; pistillate flowers central, the staminate lateral; staminate flowers with 4-lobed calyx, the stamens arising from the margin of a 4-lobed disk, the 2 stamens opposite the outer lobes of the calyx; petals lacking; pistillate flowers with 6-lobed calyx; petals lacking; pistil of 3 united carpels; stigma 3, bilobed; capsules 3-locular, the seeds 2 per locule, trigonous, grooved along the ventral margin.

Webster, G. L. 1963. The genus *Reverchonia* (Euphorbiaceae). Rhodora 65:193–207.

Reverchonia arenaria Gray Sandspurge. Plants 4–30 cm tall, erect; stems with spreading, curved-ascending branches from the base; leaves mainly 15–40 mm long, linear-oblong to narrowly elliptic, subsessile or short petiolate; stipules membranous, lance-acuminate, often red purple; sepals ca 1.5–2.5 mm long, constricted above the middle, the staminate one more or less inflexed; capsule pedicellate, this recurved, subglobose, 7–10 mm wide; seeds 4.5–6.5 mm long, trigonous, dark reddish brown, smooth dorsally, papillate on radial surfaces. Sand dunes with Vanclevea, yucca, dropseed, scurfpea, sophora, sagebrush, ponderosa pine, and pinyon-juniper communities at 1220 to 1895 m in Kane and Washington counties; Arizona, New Mexico, Texas, Oklahoma, and New Mexico; 14 (vi).

Ricinus L.

Monoecious coarse annuals, glabrous, glaucous; leaves alternate, petiolate, palmately lobed; flowers racemose or paniculate, the pistillate above the staminate; calyx 3- to 5-lobed; petals lacking; disk none; stamens many, the filaments branched; ovary 3-loculed and 3-ovuled; styles 3, bifid and plumose; capsules with spinelike projections; seeds smooth, variously marked and colored.

Ricinus communis L. Castor-bean. Ornamental annual herbs, mainly 10–20 dm tall; leaves palmately 5- to 11-lobed, 5–15(25) cm wide or more, the lobes serrate or doubly so; seeds oblong-ellipsoid, flattened, 10–14 mm long, lustrous, variously red- or purple-mottled; $2n = 10, 20$. Cultivated border plant in lower elevation portions of Utah, seldom escaping but not persisting; widely grown in the U. S.; introduced from the Old World; 1 (0). In warmer areas the plant is a shrub, but is grown in Utah as a frost sensitive annual. The plants are considered to be poisonous to humans, especially the attractive beetlelike seeds.

Tragia L.

Monoecious perennial herbs with watery juice and stinging hairs; leaves alternate, stipulate, petiolate, simple, serrate; flowers borne in bracteate racemes, these opposite the leaves or terminal, the staminate ones above, the pistillate few and placed below, each with a bract; petals lacking; staminate calyx 3- to 5-lobed; stamens commonly 3–5, the filaments short; pistillate calyx 3- to 8-lobed; style 3-lobed; capsule 3-loculed, bristly, 3-seeded.

Tragia ramosa Torr. [*T. stylaris* Muell.-Arg.]. Perennial herbs from a woody root and caudex; herbage pubescent with coarse stinging hairs and the stem puberulent as well; stems erect or ascending, mainly 10–30 cm tall; leaves 10–30 mm long, 2–15 mm wide, lance-ovate, ovate, or oblong-lanceolate, coarsely and sharply serrate; stipules triangular-attenuate, often purplish; staminate flowers 2–4, the calyx ca 1 mm long; pistillate flowers solitary, the calyx 1.5–2 mm long; capsule 6–8 mm wide; seeds globose, brown, smooth, ca 2 mm thick. Ledges and coarse talus in mixed desert shrub, shrub-grass, and juniper communities at ca 915 to 1710 m in Beaver and Washington counties; California and Nevada, east to Kansas and Missouri; 5 (iii).

FAGACEAE Dumort.
Beech Family

Deciduous trees or shrubs; leaves alternate, lobed, dentate, or less commonly entire; flowers monoecious, usually axillary on young shoots; staminate flowers with a 4- to 7-lobed perianth, borne in slender catkins or in peduncled heads; pistillate flowers solitary or in threes, with a 4- to 8-lobed perianth, borne sessile, subsessile, or in pedunculate, subcapitate catkins; pistil 1, the ovary inferior, usually 3-loculed and with 3 styles, a single ovule maturing; fruit a nut; $x = 11–13$. Note: The American chestnut, *Castanea dentata* (Marshall) Borkh., has been grown successfully in Utah.

1. Staminate flowers in pendulous, long-peduncled heads; pistillate flowers usually in pairs; winter buds long, slender, sharp-pointed; leaves 2-ranked, with prominent lateral veins .... *Fagus*
— Staminate flowers in slender catkins; pistillate flowers solitary; winter buds seldom, if ever, long and slender; leaves variously arranged, but not as above .... *Quercus*

Fagus L.

Trees; bark smooth, ash-gray or purplish; leaves simple, strongly straight-veined; flowers appearing with the leaves, the staminate ones yellowish, in pendulous heads; stamens 8–16, with slender filaments; pistillate flowers usually paired at peduncle apex, invested by bractlets, the inner 4 forming a connate involucre; calyx lobes 6; ovary 3-loculed, with 2 ovules per locule; nutlets 2 per coriaceous involucre.

1. Leaves with 9–14 pairs of veins, serrate .... *F. grandifolia*
— Leaves with 5–9 pairs of veins, denticulate to undulate .... *F. sylvatica*

Fagus grandifolia Ehrh. American Beech. Trees to 25 m tall or more and with trunk diameter to 1 m or more; bark smooth, ash-gray; branches horizontally spreading; leaves petiolate, the blades 2–13 cm long, 1–7 cm wide, ovate-ellitpic to oblong-elliptic, cuneate to rounded or subcordate basally, strongly 9- to 14-veined; staminate flowers yellow, with 5- to 7-cleft calyx; nutlets usually 2, thin-shelled, edible. Cultivated shade and ornamental tree, long persisting at lower elevation portions of Utah; indigenous in eastern U. S.; 3 (0).

Fagus sylvatica L. European Beech. Trees to 15 m tall or more and with a trunk diameter of 6 dm or more; bark smooth, gray to purplish; branches variously oriented; leaves petiolate, the blades 2–9 cm long, 1.5–5 cm wide, elliptic to obovate, cuneate to obtuse basally, strongly 5- to 9-veined; staminate flowers yellowish; nutlets usually 2, edible; $2n = 24$. Cultivated ornamental trees, long persisting in low elevation portions of Utah; introduced from Europe; 12 (0). Numerous cultivars are present, including pendulous and purple-leaved forms.

Quercus L.

Trees or shrubs; wood hard, ring-porous, with prominent rays; pith stellate in cross-section; leaves alternate, lobed, toothed, or entire; staminate flowers in usually pendulous, naked catkins; bracts caducous; calyx with 2–8 lobes; stamens 3–12; pistillate flowers with a subtrilocular, 6-ovuled ovary; stigma 3-lobed, enclosed by a scaly involucre, this hardened and cuplike, surrounding the base of the nut or acorn.

Tucker, J. M. 1961. Studies in the *Quercus undulata* complex. I. A preliminary statement. Amer. J. Bot. 48: 202–208.
_____. 1961. Studies in the *Quercus undulata* complex. II. The contribution of *Quercus turbinella*. Amer. J. Bot. 48: 329–339.
_____. 1970. Studies in the *Quercus undulata* complex. IV. The contribution of *Quercus havardii*. Amer. J. Bot. 57: 71–84.
_____. 1971. Studies in the *Quercus undulata* complex. V. The type of *Quercus undulata*. Amer. J. Bot. 58: 329–341.
Welsh, S. L. 1986. *Quercus* (Fagaceae) in the Utah flora. Great Basin Nat. 46: 107–111.

1. Plants shrubby, seldom arborescent; leaves with lobes spinescent or sharply angled, or with regularly 3–6 pairs of lateral lobes, or seldom entire; indigenous or sometimes cultivated 2
— Plants becoming trees; leaves variously lobed, seldom or sometimes as above, in part; cultivated 5

2(1). Leaves evergreen, the lobes of teeth spinescent, or seldom entire; plants of Washington, and, less commonly, of Kane and San Juan counties, hybridizing with the following .... *Q. turbinella*
— Leaves deciduous (persistent in some hybrids), the lobes variously angled or rounded, but seldom, if ever, truly spinescent; plants of broad or other distribution ... 3

3(2). Leaf lobes typically 1–2 times longer than the width of the leaf axis, rounded to obtuse or less commonly acute and often bilobed apically; plants broadly distributed .... *Q. gambelii*

— Leaf lobes seldom if ever as long as the width of the leaf axis, acute to acuminate apically 4

4(3). Plants deciduous, colonial in sandy sites, typically with branch ends protruding above the substrate 1–10 dm; acorns mostly over 15 mm long and about as broad; hybridizing with the former along canyons .... *Q. havardii*

— Plants semi-evergreen, mainly 10–30 dm tall or more, or if deciduous, then typically hairy above and densely so beneath, forming clones within and adjacent to stands of *Q. gambelii*, and occuring as scattered individuals where *Q. gambelii* and *Q. turbinella* or *Q. havardii* coexist; acorns typically less than 15 mm long and less than 10 mm wide, if formed at all .... *Q. pauciloba* and *Q. eastwoodiae* (hybrids)

5(1). Leaves or lobes bristle-tipped .... *Q. rubra*

— Leaves or lobes rounded to obtuse, or sometimes merely mucronate 6

6(5). Leaves shallowly lobed to sinuate-dentate; petiole very short; bark exfoliating; acorns pedunculate ... *Q. bicolor*

— Leaves mostly deeply lobed; petioles typically more than 12 mm long; bark various; acorns sessile or pedunculate (in *Q. robur*) 7

7(6). Leaves glabrous; acorns on long peduncles .... *Q. robur*

— Leaves pubescent or glaucous, at least beneath; acorns sessile or subsessile 8

8(7). Leaves with 2 or 3 pairs of lobes, the one sinus extending nearly to the midrib; acorn cup often with long, fringing scales; plants commonly grown .. *Q. macrocarpa*

— Leaves with 3–5 pairs of lobes, the sinuses half-way or more to the midrib; acorn cup not distinctly fringed; plants uncommonly grown .... *Q. alba*

Quercus alba L. White Oak. Trees to 25 m tall or more; bark pale, flaky; branchlets greenish to reddish brown; leaves white hairy when young, finally glabrate and glaucous beneath, 3–16 cm long or more, obovate to oblong-obovate in outline, the sinuses obliquely descending about half way or more to the midrib, the lobes 4–10, oblong, entire or again apically lobed; involucral cup sessile or subsessile, ca 1/3 as long as the acorn, bowl shaped, clothed with imbricate, puberulent scales; acorn 2–3 cm long. Cultivated shade and specimen trees at scattered sites in lower elevation portions of Utah; introduced from the eastern U. S.; 4 (0). This species is thought to be an eastern vicariad of *Q. gambelii*, q.v.

Quercus bicolor Willd. in Muhl. Swamp-white Oak. Trees to 25 m tall or more; bark gray, flaky; branchlets grayish to brown; leaves white tomentose beneath, puberulent to glabrate above, 4–22 cm long or more, obovate to oblanceolate in outline, the sinuses extending mostly much less than half way to the midrib, the lobes 10–20, rounded to triangular; involucral cups borne on peduncles 2–10 cm long, ca 1/3 as long as the acorn, clothed with woody scales (the upper ones awn-pointed); acorns 2–3 cm long. This is a rather commonly grown shade tree in lower elevation portions of Utah; introduced from the eastern U.S.; 5 (0).

Quercus eastwoodiae Rydb. (hybrid) Plants assignable to this hybrid are similar to those described as *Q. x pauciloba*, but they tend to be more densely hairy, to have hairs persistent on the upper surface, even in age, and to be less distinctly green, even if bicolored. They represent the hybrids and presumed introgressants involving *Q. havardii* and *Q. gambelii* as parental types. The plants are present in blackbrush, other warm desert shrub, mountain brush, pinyon-juniper, and Douglas-fir communities at 1130 to 1830 m in Garfield, Grand, Kane, and San Juan (type from Butler Wash) counties; Arizona; 13 (ii).

Quercus gambelii Nutt. Gambel Oak. [*Q. stellata* var. *utahensis* A. DC., type from west of Salt Lake City (?); *Q. utahensis* (A. DC.) Rydb.]. Clonal, deciduous shrubs or small trees, or less commonly trees to 10 m tall and with a trunk diameter to 6 dm or more thick, spreading by rhizomes; leaves densely grayish or yellowish stellate hairy on both surfaces when young, in age stellate hairy and paler beneath but glabrate and green and subglossy above, 2.4–17 cm long, 1–11 cm wide, obovate to elliptic in outline, the sinuses obliquely descending about 1/4–3/4 to the midrib, the lateral lobes (0) 2–10, oblong to lance-oblong, entire or notched apically and sometimes again laterally; staminate catkins 3.5–5 cm long; involucral cup 3–10 mm long, 10–17 mm wide, short-pedunculate to subsessile, ca 1/4–1/2 the length of the acorn, clothed with imbricate, densely hairy scales; acorns 8–18 mm long, 7–15 mm thick. Mountain brush, sagebrush, pinyon-juniper, ponderosa pine, and aspen communities at 1125 to 2745 m in all Utah counties, except Daggett and Rich (?); Wyoming, Colorado, New Mexico, Arizona, Nevada, Texas, and Mexico; 206 (xxi). The plants extend below the usual elevational range for the species along the canyons in the Colorado drainage system. Gambel oak is central to a series of problematical taxa, belonging in a broad sense to the *Q. undulata* Torr. complex, in which every degree of consanguinity is recognized. See *Q. x pauciloba* and *Q. x eastwoodiae* in this treatment for discussion of hybrids. Plants with acorns 27–35 mm long and cups 20–25 mm wide from the east side of Lake Powell in San Juan County are regarded as **var. bonina** Welsh (type from Goodhope Bay).

Quercus havardii Rydb. Shinnery Oak. [*Q. undulata* authors, not Torr.]. Clonal, deciduous, sand-binding shrubs, or, less commonly, small trees to 2 m or more; leaves densely grayish to yellowish stellate-hairy on both sides when young, less densely so in age, but only slightly, if at all, paler beneath than above, even in age, 1.5–5.5 cm long, 0.9–3.3 cm wide, oblanceolate to elliptic in outline, with usually 6–10 toothlike lateral lobes, these typically apiculate-acuminate and sometimes further notched or toothed apically; catkins 1–2.5 cm long; involucral cups 7–10 mm long, 14–18 mm wide, subsessile, ca 1/4–1/3 the length of the acorn, clothed with imbricate, densely hairy scales; acorns 15–23 mm long, 14–18 mm thick. Blackbrush, ephedra, Vanclevea, purple sage, and pinyon-juniper communities, usually in sand, at 1125 to 2135 m in Emery, Garfield, Grand, Kane, San Juan, and Wayne counties; Arizona, New Mexico, Oklahoma, and Texas; 54 (x). The shinnery oak, as it occurs in Utah and adjacent Arizona, is more or less influenced by intergradation with the partially sympatric *Q. gambelii* and *Q. turbinella* (Tucker 1970). Intermediates between both of those parental types and *Q. havardii* are known. However, in the sandy footslope of the San Rafael Swell in Emery and Wayne counties and adjacent portions of the Navajo Basin of Utah and Arizona, the species is more or less stable and tends to be habitat specific. Our material is regarded as **var. tuckeri** Welsh (type from E of Kens Lake, San Juan County).

Quercus macrocarpa **Michx.** Bur Oak. Trees to 25 m tall or more and with trunks to 1 m in diameter or more; twigs often corky winged; bark rough, blackish gray; leaves deciduous, 5–20 cm long, 3–15 cm wide, obovate in outline, rather deeply incised with 2–4 sinuses per side, at least one pair of sinuses reaching almost to the midrib, the lobes oblong or clavate, usually bilobed apically, finally rounded, the lower surfaces villous and gray-puberulent; petioles 5–25 mm long; staminate catkins 3–4 cm long; pistillate catkins short-pedunculate, few-flowered; acorns maturing in 1 year, the cups solitary, on peduncles 1–2 cm long; involucral cups 1.5–2.5 cm long or more, 2.5–6 cm wide, the margins fringed with attenuate upper scales; acorns 2.5–5 cm long, 1.8–4 cm wide. This is a rather commonly cultivated shade tree in Utah; introduced from the eastern U. S.; 14 (0).

Quercus pauciloba **Rydb.** (hybrid) Clonal, deciduous to semievergreen shrubs or small trees mainly 2–4 m tall and with trunks 4–15 cm in diameter; leaves stellate-hairy on both surfaces when young, becoming sparingly so to glabrate on one or both sides in age, bicolored (more or less), typically green to dark green above and paler beneath, 2–10 cm long, 1–7 cm wide, usually with (0) 4–8 toothlike lateral lobes, these typically apiculate and sometimes apiculate-acuminate, rarely some of them again notched or toothed; staminate catkins 3–4 cm long; pistillate catkins, mature involucral cups, and mature acorns not present in specimens examined. Sagebrush, mountain brush, pinyon-juniper, and ponderosa pine communities at 1220 to 2045 m in Beaver, Iron, Juab, Kane, Millard, Salt Lake, Tooele, Utah, Washington, and Weber counties; Colorado, Arizona, and Nevada; 11 (0). Specimens designated as *Q.* x *pauciloba* consist of an aggregation of hybrids and presumed introgressants involving *Q. gambelii* and *Q. turbinella* as parental types.

Quercus robur **L.** English Oak. Trees to 25 m tall or more and with trunks to 1 m in diameter; bark grayish, flaky; leaves deciduous, 4–15 cm long, 2–11 cm wide, obovate to oblanceolate in outline, with 4–7 pairs of lobes, the sinuses descending 1/4–3/4 the distance to the midvein, sometimes bilobed apically, otherwise rounded; petioles 3–5 mm long; staminate catkins 3–5 cm long; pistillate catkins few-flowered, on peduncles 2–7 cm long; cups 9–12 mm long, 15–18 mm wide, ca 1/4 as long as the acorn; acorns 2.5–3.5 cm long; 2n = 24. This species is rather commonly grown in Utah; introduced from Europe; 9 (0).

Quercus rubra **L.** Red Oak. [*Q. borealis* Michx. f.]. Trees to 25 m tall or more and with trunks to 1 m in diameter; leaves deciduous, 6–20 cm long, 4–15 cm wide, oblanceolate to obovate in outline, with 3–5 pairs of lateral lobes, the sinuses descending 1/4–3/4 of the distance to the midvein, the lobes often again toothed or lobed, aristate-attenuate; petiole 1.5–5 cm long; staminate catkins 5–10 cm long; pistillate catkins on peduncles to ca 0.5 cm long; cup 4–7 mm long, 18–25 mm wide, ca 1/4 as long as the acorn; acorns 2–3 cm long. Cultivated shade tree in Utah; introduced from the eastern U. S.; 14 (0).

Quercus turbinella **Greene** Turbinella Live-oak. Clump-forming (clonal?) evergreen shrubs or, less commonly, small trees, mainly 1–4 m tall and with stem diameters to 2 dm; leaves yellowish stellate-hairy on both surfaces when young, finally glabrate and glaucous above, not especially bicolored, typically 1.3–4 cm long, 0.7–2.4 cm wide, lanceolate to oblong or suborbicular in outline, with 2–6 pairs of lateral, spine-tipped teeth or entire; staminate catkins 1–3 cm long; involucral cup 6–8 mm long, 10–14 mm wide, ca 1/4 as long as the acorn; acorns 12–24 mm long, 7–10 mm thick. Chaparral (oak, manzanita, ceanothus), pinyon-juniper, and riparian communities at 820 to 1710 m in Kane, San Juan, and Washington counties; Nevada and Arizona; 39 (iv).

FRANKENIACEAE S. F. Gray

Alkali-heath Family

Annual herbs or shrubs; leaves simple, entire, opposite, decussate; stipules lacking, but bases of leaves of the pair connected by a ciliate line; flowers regular, solitary or cymose; sepals 4–7, the tube of the calyx more or less angled, persistent in fruit; petals 4–7, distinct; stamens 4–7; pistil 1, the ovary superior, 1-loculed, with 2–4 parietal placentae; styles 2–4, partly united; fruit a capsule; x = 10, 15.

Frankenia L

A single genus with characteristics of the family.

1. Plants annual, with very slender stems, known from Salt Lake County .... *F. pulverulenta*
— Plants woody, low shrubs, not definitely known from Utah, but to be expected in eastern San Juan County .... *F. jamesii* Torr. in Gray

Frankenia pulverulenta **L.** Annual herbs, the stems mainly 3–30 cm long, procumbent and often mat-forming or rarely erect, puberulent; leaves mainly 1–5 mm long, 0.5–4 mm wide, obovate to oblong or spatulate, usually flat, glabrous or sparingly puberulent above, densely puberulent beneath; flowers dense, secund, in short, terminal or axillary spikes; calyx 2.5–4 mm long; petals 3.5–5 mm long, oblong to obovate, pale to deep violet; 2n = 20. Saline substrates at ca 1280 m in Salt Lake County; introduced from Europe; 4 (0).

FUMARIACEAE DC.

Fumitory Family

Annual, biennial, or perennial herbs with watery juice; leaves alternate or basal; flowers perfect, irregular; sepals 2, bractlike; petals 4, the 2 outer ones spreading at the apex and one or both saccate or spurred at the base, the 2 inner ones united over the stigma; stamens 6, diadelphous, 3 per set; pistil 1, the ovary superior, 1-loculed; style 1, the stigma bilobed; fruit a capsule; x = 6, 7, 8+.

1. Flowers 2-spurred or bigibbous at the base, heart-shaped, solitary .... *Dicentra*
— Flowers 1-spurred, not heart-shaped, several to numerous, borne in racemes or panicles 2

2(1). Flowers purplish, mainly 5–10 mm long; fruit 1-seeded, globose, indehiscent .... *Fumaria*
— Flowers yellow, white, or purplish, mostly 12–25 mm long; fruit few- to many-seeded, dehiscent ..... *Corydalis*

Corydalis Medicus

Annual, biennial, or perennial herbs from taproots; leaves basal and cauline, once or twice compound; flowers in terminal or axillary racemes or panicles, showy; sepals 2, scalelike, caducous; petals 4, the outer pair dissimilar, only 1 petal spurred or saccate at the base; capsules slender or thickened.

Ownbey, G. B. 1947. Monograph of the North American species of *Corydalis*. Ann. Missouri Bot. Gard. 34: 197–258.

1. Stems stout, usually over 50 cm tall; corolla white, rose, or purple, 17–25 mm long, the spur longer than the body; capsule reflexed .... *C. caseyana*
— Stems slender, less than 50 cm tall; corolla yellow, 12–18 mm long; capsules erect-ascending .... *C. aurea*

Corydalis aurea **Willd.** Golden Corydalis. [*C. engelmannii* var. *exaltata* Fedde, type from the La Sal Mts]. Annual or biennial (or perennial), glaucous herbs from taproots, 6–40 cm long; stems simple or with few to several decumbent-ascending branches; leaves 1–4 times pinnately compound, the ultimate segments linear to oblong; racemes few- to many-flowered, the pedicels longer than the spur; sepals 1–3 mm long, yellowish or whitish; corolla yellow, 1–18 mm long, the spur 3–67 mm long; capsules 1.5–3 cm long, torulose, usually curved-ascending. Warm desert shrub, cool desert shrub, pinyon-juniper, sagebursh, mountain brush, aspen, spruce-fir, and alpine tundra communities at 800 to 3355 m in probably all Utah counties; Alaska and Yukon east to the Atlantic, south to Mexico; 125 (xiv). This is a beautiful ubiquitous plant.

Corydalis caseyana **Gray** [*Capnodes brachycarpum* Rydb., type from Alta]. Perennial herbs from tuberous roots, mainly 1–10 dm tall, fistulous, simple or branched above; herbage glabrous, glaucous; leaves cauline, 2–3 times pinnately compound or pinnatifid, the ultimate segments elliptic to lanceolate or oblong; flowers numerous, borne in racemes or panicles, the pedicels shorter than the spurs; sepals ovate, 2–4 mm long, whitish; corolla 17–25 mm long, white or pinkish, purple-tipped, the spur 7–13 mm long; capsule 0.8–1.5 cm long, ovoid to ellipsoid, usually reflexed. Mountain brush, aspen, spruce-fir, and alpine meadow communities at ca 1890 to 3050 m in Salt Lake, Utah, Wasatch, and Weber counties; Oregon to California and Colorado; 11 (0). Utah materials belong to ssp. *brachycarpa* (**Rydb.**) **Ownbey.**

Dicentra **Bernh.**

Perennial, scapose, glaucous herbs from fascicled, tuberous roots; leaves basal, once or twice ternately compound; flowers solitary, pinkish; sepals 2, deciduous; outer petals saccate, the tips recurved; inner petals enlarged and connate apically, often sagittate above the claw, fruit an elongate capsule with several seeds.

Dicentra uniflora **Kellogg** Steershead. Low perennial herbs from tuberous roots; scales and leaves thin and pale below ground level; leaves long-petiolate, subequal to the scapes; scapes 4–10 cm tall, bearing 1 or 2 bracts below the solitary flower; sepals ca 4 mm long; corolla pinkish or white, cordate basally, the outer petals widely spreading; inner petals erect, often purple-tipped; capsule ovoid-ellipsoid, 10–13 mm long. Mountain brush and aspen-forb communities at 1525 to 2745 m in Box Elder, Cache, Millard, Salt Lake, Summit, Tooele, Utah, and Wasatch counties; Washington to Wyoming, south to California; 10 (0). Bleeding-heart, ***D. spectabilis*** (**L.**) **Lem.**, is a commonly grown ornamental in Utah.

Fumaria **L.**

Annual, caulescent, glabrous and glaucous annuals from taproots; leaves ternately 2 or 3 times compound into narrow segments; flowers small, borne in racemes; sepals 2, small, closely appressed; petals gibbous basally, elongate and more or less connivent, the outer pair dilated apically, the inner pair coherent at the apex; fruit 1–seeded, indehiscent, globose.

Fumaria officinalis **L.** Erect or sprawling annual herbs, mainly 1.5–6 dm tall; leaves petiolate, with filiform petiolules and very slender ultimate segments; flowers few to numerous in axillary racemes; petals reddish purple, the tips dark purple, 5–10 mm long, the spur 2–4 mm long; fruit ca 2.5 mm thick; 2n = 28, 32. Weedy species of disturbed sites in sagebrush, mixed desert shrub, and mountain brush communities at 1310 to 1740 m in Cache, Millard, Salt Lake, and Utah counties; sporadic in the U. S.; adventive from Europe; 11 (i).

GARRYACEAE Lindl.

Silk-tassel Family

Evergreen shrubs or small trees with somewhat quadrangular branchlets; leaves simple, opposite, leathery; flowers dioecious and apetalous, borne in pendulous catkins, the staminate borne in 3's in the axil of each of the decussate, connate bracts, these with 4–lobed calyx (the lobes often connate apically); pistillate flowers solitary in bract axils, with calyx lacking or 2–lobed; stamens 4, alternate with the sepals; pistil 1, the ovary inferior, 1–loculed; styles 2, persistent; fruit a 2–seeded berry; seeds dark purple to black; x = 11.

Garrya **Dougl.**

Evergreen shrubs; branchlets 4–angled; leaves shortly petiolate; flowers unisexual; pistillate flowers almost sessile, solitary in bract axils; fruit a bitter berry.

Garrya flavescens **Wats.** Shrubs 1.5–3.5 m tall, grayish; twigs and leaves cinereous-pubescent; leaves ovate to broadly elliptic, 2–3.5 cm wide, 3.5–6 cm long, the petioles 3–10 mm long; aments 3–5 cm long, densely silky; fruit broadly ovoid, 6–8 mm long, silky. Ponderosa pine, mountain brush, and pinyon-juniper communities at 1300 to 2500 m in Washington County; Nevada and Arizona; 9 (iv). The plants flower at a season when collectors are not usually active, which results in herbarium specimens being mostly in fruit. The flavor of the plant is bad. Flowering occurs in (December) January and February.

GENTIANACEAE A. L. Juss.

Gentian Family

Annual, biennial, or perennial herbs with bitter juice; stems prostrate to erect, glabrous; leaves mostly opposite or whorled, the cauline ones sessile, often somewhat connate and clasping, the basal ones often petiolate, simple, entire, lacking stipules; flowers perfect, regular, solitary, or in cymose clusters; calyx 2– to 5–lobed; corolla 4– or 5–lobed, the lobes often convolute in bud; stamens 4 or 5, alternate with the corolla lobes, adnate to the corolla tube; pistil 1, the ovary superior, 1–loculed, with 2 parietal placentae; style 1 or lacking; stigma usually 2–lobed; fruit a 2–valved capsule; x = 5–13.

Allred, K. 1976. The plant family Gentianiaceae in Utah. Great Basin Naturalist 36: 483–495.
Holmgren, N. H. 1984. Gentianaceae. pp. 4–23. In: Intermountain Flora. Vol. 4. New York Botanical Garden, Bronx, New York.

1. Corolla rotate, with 1 or 2 fringed glands on the upper surface of each petal .... *Swertia*
— Corolla tubular-campanulate, funnelform, or salverform, without fringed glands 2
2(1). Corolla pink to rose; anthers becoming spirally twisted in age .... *Centaurium*
— Corolla usually blue, yellowish, or whitish, but not pink; anthers not spirally twisted in age 3
3(2). Corolla plicate, folded in the sinuses; calyx tube with an intermembrane rim .... *Gentiana*
— Corolla not plicate in the sinuses; calyx tube without an intermembrane rim 4
4(3). Corolla 5.5–16 (24) mm long, the lobes entire; seeds nearly smooth .... *Gentianella*
— Corolla (16) 28–70 mm long, the lobes fringed; seeds papillose .... *Gentinopsis*

Centaurium Hill

Plants annual, the stems erect, glabrous, sparingly branched; leaves opposite, sessile or clasping; flowers white to pink or rose, in terminal cymes; calyx 4– or 5–lobed, the segments linear; corolla salverform, with 4 or 5 lobes; stamens inserted on the corolla tube near the throat, alternate with the lobes, the anthers exserted and twisting at maturity; ovary 1–loculed, the style filiform and caducous; fruit a fusiform or oblong-ovoid capsule.

1. Corolla lobes 7–12 mm long, only somewhat shorter than the tube; anthers linear; basal leaves spatulate ....
.... *C. calycosum*
— Corolla lobes 3–7 mm long, half as long as the tube; anthers oblong; basal leaves oblong to linear .. *C. exaltatum*

Centaurium calycosum (Buckley) Fern. Buckley Centaury. [*Erythraea calycosa* Buckley; *E. calycosa* var. *arizonica* Gray; *C. calycosa* var. *arizonica* (Gray) Tidestr.]. Stems erect, angled, 12–50 cm tall, simple or branched above; leaves 2–5 cm long, 4–12 mm wide, oblong to elliptic or lance-linear, or the basal ones spatulate; inflorescence racemose; pedicels longer than the calyx; calyx lobes 8–12 mm long; corolla tube 10–15 mm long, the lobes 7–12 mm long, oblong to elliptic or obovate, fresh pink; capsules 8–14 mm long, cylindrical; n = 40. Saline seeps and other moist sites, usually growing with saltgrass, shadscale, greasewood, pickleweed, and alkali saccaton, in the riparian, blackbrush, pinyon-juniper and mountain brush communities at 930 to 2270 m in Garfield, San Juan, and Washington counties; California to Texas, south to Mexico; 12 (viii). This is a beautiful plant, when in full flower, but it is not often collected because of the lateness of flowering (mainly in September and October).

Centaurium exaltatum (Griseb.) Wight ex Piper Great Basin Centaury. [*Cicendia exaltata* Griseb. in Hook.; *Centaurium namophilum* var. *nevadense* Broome; *Erythraea nuttallii* Wats.; *C. nuttallii* (Wats.) Heller]. Stems erect, glabrous, sparsely branched above, 4–48 cm tall; leaves linear to lanceolate, acute, 1–5 cm long, 2–10 mm wide; flowers in terminal, compound cymes; pedicels much longer than the calyx; calyx lobes 6–10 mm long; corolla tube 6–12 mm long, the lobes 3–7 mm long, lanceolate to oblong-elliptic, acute to obtuse, whitish to rose pink; capsule 7–15 mm long, cylindric to fusiform; n = 17. Hanging gardens, seeps, and moist riparian sites, often where saline, usually in riparian or palustrine communities at 930 to 2075 m in Beaver, Box Elder, Cache, Daggett, Duchesne, Garfield, Grand, Kane, Juab, Millard, Salt Lake, San Juan, Uintah, Utah, Washington, and Wayne counties; British Columbia to Idaho, south to Baja California and Colorado; 38 (xi).

Gentiana L.

Annual or perennial herbs; stems usually erect, simple or sparingly branched; leaves opposite, sessile, or sheathing basally; flowers, solitary or in axillary or terminal cymes; calyx 4– or 5–lobed, the tube well developed, often forming an intercalycine membrane between the lobes; corolla plicate, cylindrical, tubular to funnelform or campanulate, the lobes 4 or 5, shorter than the tube, often with teeth or folds in the sinuses, blue, purple, white, or yellow; stamens inserted on the corolla at or below the middle; style short or lacking; stigmas 2; capsule sessile or stipitate, 2–loculed, septicidal.

1. Capsules stipitate; corolla 10–20 mm long, terminal; calyx 7–12 mm long; plants diminutive annuals or biennials .... *G. prostrata*
— Capsules sessile; corolla 22–50 mm long; calyx 10–30 mm long; plants perennial 2
2(1). Leaves mostly basal, linear to oblanceolate, the cauline ones greatly reduced; corolla whitish or pale yellowish, with dark stripes .... *G. algida*
— Leaves all cauline, the basal ones much reduced; corolla blue to gentian violet 3
3(2). Flowers solitary, terminal (rarely 2 or 3); margins of calyx and leaves glabrous .... *G. calycosa*
— Flowers usually more than one; margins of calyx and leaves puberulent 4
4(3). Flowers terminal or subterminal; corolla 3–4.5 cm long; calyx tube 1–1.8 cm long; floral bracts ovate to ovate-lanceolate, often somewhat scarious .... *G. parryi*
— Flowers several to many, some axillary and well below the stem apex; corolla 2.3–3 (4) cm long; calyx tube 0.4–1 cm long; floral bracts lanceolate to linear . *G. affinis*

Gentiana affinis Griseb. Rocky Mountain Gentian. [*G. forwoodii* Gray]. Perennial herbs; stems 1–5.7 dm tall, usually several, mostly erect; leaves 1.5–3.5 cm long, 6–18 mm wide, all cauline, numerous, ovate, oblong, or broadly lanceolate, puberulent on the margins and in lines below the leaf bases; inflorescence cymose, the flowers several to many; bracts lanceolate; calyx 4–18 mm long, funnelform to campanulate, the tube 4–10 mm long; corolla 2.3–3 (4) cm long, funnelform, strongly plicate, blue to purple or violet, the lobes 3–6 mm long; capsules subequal to the calyx, dehiscing apically. Meadows and streamsides in ponderosa pine, lodgepole pine, aspen, and spruce-fir communities at (1340) 2540 to 3850 m in all Utah counties, except Beaver, Box Elder, Davis, Millard, Morgan, Wayne, and Weber; British Columbia to Montana, south to California and New Mexico; 55 (ix).

Gentiana algida Pallas Arctic Gentian. [*G. romanzovii* Ledeb. ex Bunge]. Plants perennial, the stems 0.5–2 dm tall, erect or ascending, simple; leaves mainly basal, connate-sheathing, the cauline leaves commonly in 2-3 pairs below the bracts (uppermost leaves), 2-12 cm long, 2-10 mm broad, linear to narrowly oblong; flowers solitary or in terminal cymes; calyx 5-toothed, 15-23 mm long, the teeth unequal, longer or shorter than the tube; corolla funnelform, cream to yellowish green, mottled or striped with purple, 3.5-5.3 cm long, the folds yellowish; pistil stipitate; capsule oblong-elliptic in outline; 2n = 24, 26. Alpine meadows and slopes with spruce-fir and alpine tundra communities at 3000 to 3965 m in Daggett, Duchesne, San Juan, Summit, and Uintah counties; Alaska and Yukon, south to New Mexico; 14 (ii).

Gentiana calycosa Griseb. Explorer Gentian. [*G. calycosa* ssp. *asepala* Maguire, type from Mt. Agassiz]. Plants perennial, the stems 0.3-2 (3) dm tall, decumbent to ascending or erect; leaves cauline and basal, soon withering, broadly ovate to elliptic-ovate, 1-2 (2.5) cm long, 5-15 (20) mm wide, glabrous; flowers usually solitary (rarely 2 or 3), terminal; calyx 10-18 mm long, funnelform, the lobes 2-8 mm long, ovate to lanceolate, or almost lacking; corolla 30-40 mm long, tubular-funnelform to campanulate, deep blue, usually streaked with green, the lobes 5-10 mm long, rounded, the sinus plaits lacerate; capsule sessile, to 24 mm long; n = 13. Meadows in willow, aspen, and spruce-fir communities at 2700 to 3900 m in Daggett, Duchesne, Garfield, Summit, Uintah, Wasatch, and Washington counties; British Columbia and Alberta, south to California, Arizona, and Wyoming; 44 (ix).

Gentiana parryi Engelm. Parry Gentian. [*G. bracteosa* Greene]. Plants perennial, the stems 0.6-3 (4.5) dm tall; leaves oval to ovate or lanceolate, all cauline, 13-32 (40) mm long, 8-21 mm wide, puberulent on the margins and in decurrent lines on the stem; inflorescence cymose, with (1) 2-4 flowers, compact; bracts ovate to oval; calyx 10-18 (23) mm long, campanulate, the lobes 1-8 mm long, lanceolate; corolla 30-40 (50) mm long, tubular-funnelform, blue and streaked with green, the lobes 3-8 mm long, broadly rounded to mucronate apically; capsule ca 20 mm long. Meadows and slopes in mountain brush and spruce-fir communities at 1900 to 3850 m in Garfield, Grand, Iron, Piute, San Juan, and Wasatch counties; Wyoming and Colorado to New Mexico and Arizona; 8 (iii).

Gentiana prostrata Haenke ex Jacq. Moss Gentian. Plants annual or biennial, the stems 2-15 (25) cm long, prostrate or ascending, branched from the base, arising from a slender taproot; leaves chiefly cauline, connate-sheathing, 2-15 mm long (including the sheath), 1-5 mm wide; flowers terminal, solitary; calyx with 4 or 5 lobes, 8-16 mm long, the teeth shorter than the tube; corolla tubular, blue (or the tube whitish), 11-25 mm long; pistil stipitate; capsule lance-ovoid, long-stipitate; 2n = 32-36. Moist sites in sedge-willow, lodgepole pine, and spruce-fir communities at 2400 to 3850 m in Daggett, Duchesne, Emery, Rich, San Juan, Sevier, Summit, and Uintah counties; Alaska and Yukon, south to California, Arizona, and Colorado; 20 (vii).

Gentianella Moench

Annual or biennial herbs from taproots; stems erect, simple or branched, glabrous; leaves opposite, sessile; inflorescence cymose, terminal and/or axillary; flowers showy; calyx lobes free or united at the base, the sinuses without an inner membrane; corolla tubular to campanulate, blue or lilac, the lobes entire, without plaits in the sinuses; stamens 4 or 5, inserted on the corolla tube; nectary glands at the base of the tube; pistil sessile, the ovary 1-loculed; stigmas 2, persistent; capsule dehiscing apically; seeds numerous, smooth.

Gillett, J. M. 1957. A revision of the North American species of *Gentianella* Moench. Ann. Missouri Bot. Gard. 44: 195-269.

1. Flowers solitary on a naked peduncle 2-10 cm long; plants less than 13 cm tall .... *G. tenella*
— Flowers several to many in short-pedunculate clusters; plants often more than 13 cm tall 2
2(1). Calyx lobes very unequal, the outer two large and foliaceous, 7-20 mm long, enclosing the smaller more slender ones .... *G. heterosepala*
— Calyx lobes equal or nearly so 3
3(2). Plants caespitose, 2-10 cm tall, much branched; corolla yellowish, whitish, or suffused with pale blue ....
.... *G. tortuosa*
— Plants erect, mainly 10-50 cm tall, often simple; corolla blue, blue purple, or sometimes yellow or whitish ...
.... *G. amarella*

Gentianella amarella (L.) Borner Northern Gentian. [*Gentiana amarella* L.; *G. plebeia* Cham. ex Bunge; *G. strictiflora* A. Nels.; *G. acuta* Michx.; *Gentianella amarella* ssp. *acuta* (Michx.) Hulten]. Annual or biennial, the stems 8-50 cm tall, erect or ascending, simple or more commonly branched, from a taproot; basal leaves oblanceolate to elliptic; cauline leaves ovate to lanceolate, 1-6 cm long, 3-15 mm wide, not connate; flowers several to many in cymes, or axillary, the pedicels shorter than the next lower internode; calyx 5-7 mm long, usually 5-lobed, the lobes longer than the tube; corolla tubular, 9-15 mm long, pink purple, bluish purple, pinkish, blue, or yellowish green (or whitish), the lobes fringed within (at least some); pistil sessile; capsule cylindrical. Meadows and stream banks in aspen, grass-forb, lodgepole pine, spruce-fir, and alpine tundra communities at 2400 to 4050 m in Beaver, Cache, Carbon, Daggett (?), Duchesne, Emery, Garfield, Iron, Piute, Salt Lake, Sanpete, Sevier, Summit, Uintah, Utah, and Wasatch counties; Alaska to the Atlantic, south to California, Arizona, New Mexico, North Dakota, and Vermont; circumboreal; 51 (xv). This taxon is closely allied to *G. heterosepala*, q. v.

Gentianella heterosepala (Engelm.) Holub Engelmann Gentian. [*Gentiana heterosepala* Engelm.; *Gentianella amarella* ssp. *heterosepala* (Engelm.) J. M. Gillett]. Plants annual or biennial, the stems mainly 10-45 cm tall, erect or ascending, simple or more or less branched; basal leaves 10-35 mm long, oblanceolate to spatulate; cauline leaves 2-5.5 cm long, oblanceolate to elliptic; inflorescence cymose, borne in axils of the 4 to 6 pairs of upper leaves; pedicels slender, 2-7 cm long; calyx lobes dimorphic, the outer 2 large and foliaceous, 8-15 mm long, the inner ones linear-lanceolate, mostly hidden by the outer, one of the small sepals almost free, the other two variously connate; corolla 10-20 mm long, tubular-campanulate, bluish, whitish, or yellowish, the lanceolate lobes 4-7 mm long, the lobes fringed within (at least some); pistil sessile; capsule cylindrical. Sagebrush, mountain brush, aspen, lodgepole pine, grass-forb, and spruce-fir communities at 1980 to 3500 m in Cache, Carbon, Daggett, Duchesne, Emery, Garfield, Grand, Iron, Kane, Piute,

Salt Lake, San Juan, Sanpete, Summit, Uintah, Utah, Wasatch, and Washington counties; Arizona, Colorado, and New Mexico; 85 (xiii).

Gentianella tenella (Rottb.) Borner Lapland Gentian. [*Gentiana tenella* Rottb.]. Annual, the stems 5–20 cm tall, ascending to erect, simple or more commonly branched, from taproots; basal leaves elliptic to obovate; cauline leaves lanceolate to elliptic, 4–10 mm long, 1–4 mm wide, not connate (except for the lower ones); flowers solitary, terminal or axillary, the pedicels longer than the next lower internode; calyx 4–10 mm long, the lobes distinct, dimorphic, the outer lobes foliaceous, ovate to lanceolate, the inner ones lanceolate, slightly shorter than the outer; corolla 7–16 mm long, blue to white, tubular-campanulate, the lobes 2.5–6 mm long, oblong-ovate, each lobe bearing 2 blunt, fimbriate scales; 2n = 10. Spruce-fir and alpine tundra communities at 2400 to 3800 m in Duchesne, Juab, Garfield, Summit, and Uintah counties; Alaska to Greenland, south to California, Nevada, Arizona, and New Mexico; Eurasia; 10 (i).

Gentianella tortuosa (Jones) J. M. Gillett Jones Gentian. [*Gentiana tortuosa* Jones, type from Panguitch Lake]. Clump-forming annual, mainly 2–10 cm tall; stems much branched, spreading to erect; basal leaves elliptic to spatulate, rounded or obtuse apically, 5–20 mm long, 2–6 mm wide; cauline leaves elliptic, oblong, or lanceolate, 5–35 mm long; flowers solitary in the leaf axils or on short axillary branches; pedicels 3–15 mm long; calyx 5–11 mm long, the tube 1–2 mm long, the lobes linear to oblanceolate, subequal, the margins hyaline; corolla yellowish, white, or blue, 7–16 mm long, funnelform to campanulate, the lobes 3–6 mm long, lanceolate, subequal to the tube, each lobe bearing 2 fringed scales within. Sagebrush, grass-forb, ponderosa pine, limber and bristlecone pine, and spruce-fir communities, usually on Tertiary limestone formations, at 2000 to 3400 m in Emery, Garfield, Iron, Kane, Sanpete, Sevier, Utah, and Wasatch counties; Nevada; 28 (vi).

Gentianopsis Ma

Annual, biennial, or perennial herbs; stems erect, often branched, glabrous; leaves opposite, sessile; inflorescence terminal or axillary; flowers solitary or cymose, 4–merous; calyx usually 4–angled, the lobes with thin, hyaline margins, the two inner ones triangular to ovate, the outer ones linear to lanceolate; corolla large and showy, tubular-funnelform to campanulate, the lobes fimbriate marginally, without folded plaits in the sinuses or fringed scales at the base of the lobe, blue or rarely whitish; stamens epipetalous; pistil stipitate, the ovary 1–loculed; stigmas 2, large; capsule 2–valved, septicidal; seeds numerous, papillate.

Iltis, H. H. 1965. The genus Gentianopsis (Gentianaceae): transfers and phytogeographic comments. Sida 2: 129–154.

1. Flowers borne on long slender pedicels 2–16 cm long; plants annual or biennial from taproots .... *G. detonsa*
— Flowers sessile or short-pedicellate; plants perennial from rhizomes .... *G. barbellata*

Gentianopsis barbellata (Engelm.) Iltis Barbellate Gentian. [*Gentiana barbellata* Engelm.; *Gentianella barbellata* (Engelm.) J. M. Gillett]. Rhizomatous, acaulescent or short-caulescent perennials, mainly 5–15 cm tall; stems glabrous; leaves mostly basal, oblanceolate, rounded or obtuse apically, the base connate-clasping, 2–8 cm long, 3–12 mm wide; flowers solitary and terminal, sessile or short-pedicellate; calyx funnelform, 11–25 mm long, the lobes triangular to lanceolate, equal, the sinuses acute, spanned by a membrane bearing 1–6 minute processes; corolla deep blue, narrowly funnelform, 24–45 mm long, the lobes oblong, 15–25 mm long, the lateral margins conspicuously fimbriate, the apex erose-dentate. Spruce-fir and alpine tundra communities, typically on limestone and often in meadows, at 2500 to 3750 m in Duchesne, Grand, Sanpete, and Summit counties; Wyoming south to Arizona and New Mexico; 9 (0).

Gentianopsis detonsa (Rottb.) Ma Meadow Gentian. [*Gentiana detonsa* Rottb.; *Gentianella detonsa* (Rottb.) G. Don; *Gentiana thermalis* Kuntze; *Gentiana elegans* A. Nels.; *Gentianella detonsa* ssp. *elegans* (A. Nels.) J. M. Gillett]. Annual or biennial, mainly 0.5–5 (9) dm tall, from taproots; stems branched from the base or above, rarely simple, glabrous; basal leaves forming a rosette or reduced to 1 pair, obovate-elliptic to spatulate, 2–6.5 cm long, 1–15 mm wide; cauline leaves linear to spatulate; flowers solitary and terminal, on pedicels 1–30 cm long; calyx broadly funnelform, the tube 9–14 mm long, the lobes ovate-triangular, acute, 8–20 mm long; corolla gentian violet, dark blue, or clear blue, narrowly funnelform, 25–60 mm long, the lobes 10–25 mm long, the lateral margins ciliate to fimbriate or toothed, the apex erose to dentate. Meadows in sagebrush, mountain brush, aspen, and spruce-fir communities at 1700 to 3550 m in Carbon, Daggett, Emery, Garfield, Iron, Juab, San Juan, Sanpete, Sevier, Summit, Uintah (?), Utah, and Wasatch counties; Alaska to Newfoundland, south to Mexico; circumboreal; 50 (viii). Our plants belong to var. *elegans* (A. Nels.) N. Holmgren.

Swertia L.

Glabrous perennial herbs; leaves (alternate) opposite or whorled, rarely alternate; flowers 4– or 5–merous, solitary or more often several to numerous in terminal and axillary cymes; calyx lobed nearly to the base; corolla rotate to shallowly campanulate, with short to long, often fringed scalelike processes borne ventrally at the base, the tube short, each lobe with 1 or 2 marginally fringed nectary glands near the middle or below; stamens 4 or 5, attached near the base of the corolla tube, the filaments free or expanded and connate near the base; ovary ovoid, the style one or lacking; capsule 1–loculed; seeds numerous, compound, often wing-margined.

St. John, H. 1941. Revision of the genus *Swertia* (Gentianaceae) of the Americas and the reduction of *Frasera*. Amer. Midl. Naturalist 26: 1–29.

1. Corolla normally dark blue, the lobes 4 or 5, each with a pair of sparsely fringed nectar pits at the base; stem leaves alternate, mostly of higher elevations . *S. perennis*
— Corolla usually greenish white or yellowish green, often purple spotted, the lobes usually 4, each with 1 or 2 large, copiously fringed glands and pits; stem leaves opposite or whorled; plants of middle and lower elevations 2

2(1). Leaves whorled, not white-margined, puberulent; glands 2 per corolla lobe .... *S. radiata*
— Leaves opposite or whorled, usually white-margined, glabrous; glands 1 per corolla lobe 3

3(2). Leaves linear; inflorescence a simple, narrow thyrse; plants perennial, known from western Millard County .... *S. gypsicola*

— Leaves oblong to oblanceolate, mostly 3–18 mm wide; inflorescence broadly paniculate; plants biennial, widespread 4

4(3). Leaves opposite; glands lobed at the base, sagittate; plants 7–12 dm tall or more, growing in sandy sites in southeastern Utah .... *S. utahensis*

— Leaves in whorls of 4; glands 2–lobed at apex; plants mainly 1–6 dm tall, growing in various substrates in the southern half of Utah .... *S. albomarginata*

Swertia albomarginata (Wats.) Kuntze White-margined Swertia. [*Frasera albomarginata* Wats., type from near St. George]. Glabrous perennial, from a well-developed, yellow taproot; stems solitary or few, normally much branched above the base, somewhat glaucous, 2–6 dm tall; leaves in whorls of 4, the bases connate-sheathing, the basal narrowly oblanceolate, mainly 4–11 cm long and 5–9 mm wide, white-margined, the cauline ones reduced and linear; flowers in broad paniculate cymes; calyx 2–6 mm long, the lobes narrowly lanceolate to nearly ovate; corolla white or greenish white, marked with green dots, the lobes 7–11 mm long; gland solitary on each lobe, long and slender, bilobed at the tip, fringed with short, soft, flat, white hairs; coronal scales absent or few; capsule conic, attenuate, 10–17 mm long. Warm desert shrub, sagebrush, pinyon-juniper, and mountain brush communities at 1470 to 2430 m in Beaver, Emery, Garfield, Iron, Kane, Millard, San Juan, Washington, and Wayne counties; Colorado, New Mexico, Arizona, Nevada, and California; 48 (xv).

Swertia gypsicola Barneby White River Swertia. [*Frasera gypsicola* (Barneby) D. Post]. Perennial, glabrous herbs from a branched caudex, and stout yellow root; stems erect, 10–35 cm tall; leaves linear, opposite, connate-clasping, the basal ones crowded, 3.5–8 cm long, the cauline 1.5–3.5 cm long; inflorescence a racemose thyrse of opposite 1- to 3-flowered cymes; calyx 3–4 mm long, the lobes broadly lanceolate, the margins whitescarious; corolla cream with purple spots, the lobes lanceolate 5–8 mm long; glands on each lobe small and circular, the edge with long white fimbriae all around; subapical scales ca 2 mm long, lacerate; capsule 10–12 mm long, oblong. Greasewood and shadscale dominated valley bottoms at ca 1525 m in western Millard County; Nevada; a Great Basin endemic; 1 (0).

Swertia perennis L. Felwort. [*S. fritillaria* Rydb., type from Big Cottonwood Canyon]. Perennial, the stems 1–6 dm tall; basal leaves obovate or oblong-elliptic, petiolate, 4–20 (25) cm long, 1–4.5 cm wide; cauline leaves shortpetiolate to sessile, reduced in size upward, clasping but not connate; sepals essentially distinct, 5–12 mm long, narrowly oblanceolate, acuminate; corolla rotate, 10–15 mm long, dark bluish purple to whitish, often spotted; pistil sessile; capsule lance-ovoid; 2n = 28. Meadows, stream banks, and other moist sites in mountain brush, aspen, lodgepole pine, spruce-fir, and alpine tundra communities at 2375 to 3850 m in Daggett, Duchesne, Emery, Salt Lake, Sanpete, Sevier, Summit, Uintah, and Wasatch counties; Alaska to California; Eurasia; 47 (viii).

Swertia radiata (Kellogg) Kuntze Elkweed. [*Tessaranthium radiatum* Kellogg; *Frasera speciosa* Dougl. ex Griseb. in Hook., not *Swertia speciosa* Wallich ex G. Don]. Glabrous to puberulent perennials from a thick taproot and caudex, the caudex transversely marked with leaf scars and at the summit with marcescent leaf bases; stems simple, erect, stout, 5–20 dm tall, yellowish to bluish green; leaves mostly whorled, slightly sheathing, puberulent to nearly glabrous, the basal ones 2–5 dm long, spatulate or elliptic-oblong to oblanceolate, the cauline ones reduced upward, oblanceolate to lanceolate; inflorescence an open or congested thyrse of whorled cymes; calyx lanceolate, 10–22 mm long; corolla pale green to whitish with purple dots, the lobes 10–25 mm long, obovate to oblanceolate; glands 2 on each lobe, elliptic, the margins long-fimbriate all around; subapical scales 7–9 mm long, deeply lacerate; capsule oblong, 20–25 mm long; 2n = 78. Sagebrush, mountain brush, ponderosa pine, Douglas fir, lodgepole pine, aspen, and spruce-fir communities at 1890 to 3500 m in probably all Utah counties; Washington to South Dakota, south to California, Nevada, Arizona, New Mexico, Texas, and Mexico; 83 (xvi).

Swertia utahensis (Jones) St. John Utah Swertia. [*Frasera paniculata* Torr. in Whipple, not *Swertia paniculata* Wallich; *Frasera utahensis* Jones, type from the Buckskin Mountains, Kane Co., Utah or Coconino Co., Arizona]. Perennial (biennial?) herbs from a thick, yellow taproot and caudex, the caudex with transverse leaf scars and at the summit clothed with persistent leaf bases; stems branched above, 7–15 dm tall; leaves opposite, sheathing, white-margined, often folded and curved, the basal ones 7–20 cm long, 7–20 mm wide, lance-linear to oblanceolate, the cauline ones greatly reduced upward, lanceolate, sessile; inflorescence a broad, pyramidical cluster 15–50 cm or more broad; calyx 3–6 mm long, lanceolate to ovate, acuminate, with white scarious margins; corolla white to greenish yellow, with dark green dots, the lobes 6–10 mm long; gland 1 per lobe, broad, oblong, lobed at the base, fringed with long fimbriae all around; subapical scales lacking; capsule elliptic-ovoid, attenuate, 10–16 mm long. Usually in sandy habitats in salt desert shrub, warm desert shrub, and pinyon-juniper communities at 1220 to 1895 m in Emery, Garfield, Grand, Kane, San Juan, and Wayne counties; New Mexico, Arizona, and Nevada (?); 29 (viii).

GERANIACEAE A. L. Juss.
Geranium Family

Annual or perennial herbs; leaves opposite or alternate, stipulate, simple or compound; inflorescence umbellate; flowers perfect, mostly regular; sepals and petals 5, distinct; stamens 5 or 10, the filaments more or less united at the base; pistil 1, the ovary superior, usually 5–loculed; style 1, with 5 stigmatic lobes; fruit dry, with 1 seed per locule, the valves separating from the base and coiling at maturity; x = 7–14.

1. Flowers with a spur adnate to the pedicel; plants cultivated .... *Pelargonium*

— Flowers lacking a spur; plants rarely if ever cultivated .. 2

2(1). Leaves palmately lobed or divided; stamens usually all with anthers .... *Geranium*

— Leaves pinnately lobed or dissected; stamens with anthers 5 .... *Erodium*

Erodium L'Her.

Annual herbs; leaves basal and cauline, pinnately lobed or parted, opposite; pedicels commonly recurved in fruit;

flowers 5–merous; fertile stamens 5, alternating with 5 scalelike staminodia; style column elongate, the styles bearded inside, spirally coiled at maturity; carpel bodies spindle-shaped, indehiscent.

1. Leaves with confluent blades, these merely 3–lobed and pinnately veined; plants of Washington County .... *E. texanum*
— Leaves pinnately dissected, the blade usually bipinnatifid; plants widely distributed .... *E. cicutarium*

Erodium cicutarium (L.) L'Her. Storksbill. [*Geranium cicutarium* L.]. Stems erect or finally prostrate to decumbent, 0.5–8 dm long (or more), strigulose as well as glandular; leaves 1–12 cm long, doubly pinnately dissected or parted; stipules lanceolate; peduncles 1–15 cm long; pedicels 1–10, typically 6–18 mm long; sepals 3–6 mm long, mucronate and bristle-tipped; petals pink to lilac, 5–7 mm long, spotted; stylar beak 2–4 cm long; carpel bodies 4–5 mm long, stiffly pilose; 2n = 36, 38, 40. Widely distributed herbs of open sites in numerous plant communities at 820 to 2475 m in all or nearly all Utah counties; widespread in western U. S.; adventive from Europe; 70 (vii). This plant was noted by Fremont in 1844, and it is potentially indigenous. Certainly it now appears like a native.

Erodium texanum Gray Stems decumbent-ascending, mainly 0.6–4 dm long, puberulent (not glandular); leaves 2–6 cm long, the blades 0.8–3 cm long, ovate to suborbicular, cordate basally, 3–lobed; stipules lanceolate to triangular; peduncles 1–3 cm long; pedicels 1–3, mainly 3–10 mm long; sepals 5–8 mm long, purple-veined, apiculate, but not bristle-tipped; style column 4–6 cm long; style bodies 6–8 mm long, ascending hairy with stiff straight hairs. Creosote bush, Joshua tree, and shadscale communities at ca 760 to 915 m in Washington County; Nevada, California, Arizona, New Mexico, and Texas; 4 (0).

Geranuim L.

Perennial, annual, or biennial herbs; leaves alternate or opposite, or chiefly basal, palmately lobed or divided; flowers often large, showy, borne solitary or in umbels on axillary peduncles; sepals 5; petals 5, soon deciduous; stamens 10, usually all anther-bearing; styles much longer than the ovary, curved or coiling in fruit. Note: Diagnostic criteria have been based previously on such features as kind of pubescence and its position on plant parts, but those features do not appear to be of primary diagnostic value. Thus, keys are difficult to compose that will allow accurate identification of all specimens. Because of these factors the following key is tentative at best.

1. Petals 5–10 mm long; plants annual or biennial from slender taproots 2
— Petals mainly 10–26 mm long; plants perennial from caudices or rhizomes 4
2(1). Fertile stamens 5; sepals not apiculate; seeds smooth .... *G. pusillum*
— Fertile stamens 10; sepals apiculate; seeds pitted 3
3(2). Beak of stylar column less than 3 mm long, the stigmas less than 1 mm long; pedicels subequal to the calyx at maturity .... *G. carolinianum*
— Beak of stylar column 4–5 mm long, the stigmas ca 1 mm long; pedicels much surpassing the calyx at maturity .... *G. bicknellii*

4(1). Petals white, the veins sometimes pink or purple, or rarely suffused with pale lavender; style branches 2–5 mm long; plants common and widespread .... *G. richardsonii*
— Petals pink, rose purple, pink purple, white or shades of lavender and violet; style branches 4.5–7 mm long; plants various 5
5(4). Blades of lower leaves usually more than 7 cm broad and 4 cm long from sinus to apex; plants 2–10 dm tall, widespread in northern Utah, less common elsewhere .... *G. viscosissimum*
— Blades of lower leaves usually less than 7 cm broad and less than 4 cm long from sinus to apex; plants 1–4 (7) dm tall, mainly of the southern 1/2 of Utah ... *G. caespitosum*

Geranium bicknellii Britt. Bicknell Cranesbill. Annual or biennial, the stems 1.5–5 dm tall, erect or decumbent, spreading hairy and often glandular, from a slender taproot; leaf blades cordate to rounded in outline, 1–3.5 cm long from sinus to apex, the segments not distinct; flowers usually 2 per peduncle; fruiting pedicels usually 2–3 times longer than the calyx; sepals 4–8 mm long, aristate; petals pink, 3.5–7 mm long; stylar column 14–20 mm long including the beak, pubescent, the beak 3–6 mm long, pubescent; style branches 1–1.5 mm long. Shady moist roadside at 1830 to 2440 m in Cache, Daggett, Salt Lake, and Utah counties; Alaska and Yukon east to Newfoundland, south to Washington, Michigan, and New York; 3 (0).

Geranium caespitosum James Small-leaf Geranium. [*G. marginale* Rydb., type from Aquarius Plateau]. Perennial, few- to many-stemmed from a definite caudex, mostly 15–70 cm tall; herbage variously retrorsely, spreading, or appressed pubescent and often glandular as well; blades of basal leaves palmately 5–lobed, the median lobe mainly 0.8–3.6 cm long, 1.5–7 cm wide, this cuneate basally and usually with 3 abruptly acuminate to obtuse teeth apically, the petioles 3–20 (28) cm long; fruiting pedicels much longer than the calyx; flowers in pairs or not; sepals 6.7–11 mm long, apiculate; petals white, pink, lavender, or pink purple, 6.7–16.5 mm long; stylar column 10–18 mm long, including the beak, this 3–5 mm long, pubescent or glabrous; style branches 4.5–5.5 mm long. Sagebrush, mountain brush, pinyon-juniper, ponderosa pine, aspen, and spruce-fir communities, often where moist and shaded, at 1830 to 3465 m in Beaver, Carbon, Duchesne, Garfield, Grand, Iron, Millard, Piute, San Juan, Sanpete, Sevier, Tooele, Washington, and Wayne counties; Colorado, New Mexico, and Arizona; 59 (vii). The small-leaf geranium is typical of the problems of interpretation of perennial geraniums generally; i.e., they tend to merge morphologically (hybridize?) when the taxa meet, and there are few, if any, definitive diagnostic features (including proportion of the petal surface bearing hairs and pubescence type and position elsewhere on the plants). These problems have led to contradictory and often unsatisfactory treatments of our perennial geranium species. Interpretation of the *caespitosum* complex is further complicated by its tangled nomenclatural history, beginning at its inception. The nomenclatural problem was reviewed in part by Osterhout (Bull. Torrey Club 50: 81–84. 1923), who noted the presence of a possibly earlier name for this species. That name, if determined to be properly published, has priority, evidently, over the epithet *caespitosum*. Aside from that problem, the original description of *G. caespitosum* is brief and does not characterize the species. Further, no type specimen is known. Typification is possible only through interpolation from published notes of the collector (Dr. James on the Long Expedition). This was not established until the concept of the species as it occurs in New Mexico had been misapplied under the name *G. caespitosum* by later authors. The confusion is still evident in recently published works. The complex has been regarded as including three or four taxa treated at specific or infraspecific rank. The main diagnostic feature used in segregation of those taxa includes kind and position of pubescence. Separation of the largely sympatric Utah materials into three groups (that do not seem to represent taxa) is possible only arbitrarily, and it does not seem advisable under the circumstances to treat them at taxonomic rank. Eglandular material in Utah has been regarded as *G. marginale* Rydb. (more or less equivalent with *G. atropurpureum* Heller as far as being eglandular, but probably not the same taxonomically), but passes by degree into the other two possibilities. Plants with glandular pedicels, lower stem internode, and lower petioles are *G. caespitosum* var. *parryi* (Englem.) W. A. Weber [*G. fremontii* var. *parryi* Engelm.]. Plants of this type are not always separable from dwarf plants of *G. richardsonii*, with which they are partially sympatric. Those with glandular hairs on pedicels and calyces but not on the lower internode or lower petioles are assignable to *G. fremontii* Torr., in a strict sense, as near as I can determine from descriptions and contradictory contemporary treatments. Adding to the confusion, the name *G. fremontii* has been misapplied to the similar but larger *G. viscosissimum* (q.v.), common in northern Utah. Nomenclatural combinations are neither intended nor implied herein.

Geranium carolinianum L. Carolina Cranesbill. Annual or biennial, the stems 1.5–5 dm tall, usually erect, retrorsely hairy and often glandular, from a slender taproot; leaf blades cordate to ovate in outline, 1–4 cm long from sinus to apex, the segments not distinct; flowers 2 or more per peduncle, these very short; fruiting pedicels less than 1.5 times longer than the calyx; sepals 5–8 mm long, apiculate; petals pink, 5–7 mm long; stylar column 10–15 mm long (including beak), pubescent, the beak 1–2 mm long; style branches 1 mm long or less. Open sites at lower middle elevations in Davis, Salt Lake, and Weber counties; widely occurring in North America; 3 (0).

Geranium pusillum L. Slender Cranesbill. Plants annual or biennial, the stems 1–5 dm tall, prostrate to erect, pubescent with appressed or spreading hairs and often glandular, from a slender taproot; leaf blades cordate to orbicular in outline, 1.5–4 (7) cm wide, the segments not distinct; flowers mostly 2 per peduncle, these mainly 2–4 times longer than the calyx; sepals 2.5–4 mm long, not apiculate; petals red violet to lavender, mostly 3–4 mm long; stylar column 6–9 mm long, the beak very short; style branches ca 0.5–0.8 mm long; 2n = 26. Open sites at ca 1370 to 1525 m in Cache, Salt Lake, and Utah counties; widely established in the U. S.; adventive from Europe; 4 (0).

Geranium richardsonii Fisch. & Trautv. Richardson Geranium. Perennial herbs, the stems usually 1 per caudex branch, 1–10 (13) dm tall, erect, branched above, retrorsely hairy, glandular, or glabrous below, mostly glandular-hairy above, from a caudexlike rhizome; leaf blades mainly 1–8 cm long from sinus to apex, 1.5–17 cm wide, the segments not distinct, the median one cuneate basally and 3– to 7–lobed or toothed apically; flowers usually 2 per peduncle, this usually 2–4 (or more) times longer than the calyx; sepals 6–11 mm long, apiculate, with stipitate-glandular hairs; petals white to pale lavender, the veins often reddish or purplish, 12–22 mm long, hairy on the inner surface; style column stipitate-glandular, 16–28 mm long, including the beak, this 3–4.5 mm long; style branches 2–5 mm long; 2n = 26. Mountain brush, pinyon-juniper, ponderosa pine, aspen, lodgepole pine and spruce-fir communities at 1735 to 3235 m in probably all Utah counties; Yukon south to California, Arizona, and New Mexico; 182 (xvii). This is the most widely distributed of the native geraniums in Utah. It is sympatric with both *G. viscosissimum* and *G. caespitosum* and is not always reliably separable from either of them. In most cases the white flowers and smaller floral parts distinguishes it from *G. viscosossimum*, but its separation from *G. caespitosum*, especially from those phases with white flowers, is more tenuous.

Geranium viscosissimum Fisch. & Mey. Sticky Geranium. Perennial herbs, the stems often 2 per caudex branch, 2–12 dm tall, erect, branching above, retrorsely puberulent to glandular below, retrorsely hairy to stipitate-glandular above, from a caudexlike rhizome; leaf blades mostly 2–8 cm long from sinus to apex and 3–17 cm wide, the segments not distinct, the median one cuneate basally and 3– to 15–toothed or -lobed apical flowers usually 2 per peduncle, this mainly 2–6 (or more) times longer than the calyx; sepals 7–12 mm long, apiculate, pubescent with glandular or eglandular hairs; petals pink to lavender or purple, 12–26 mm long, the lines usually purplish, hairy on the inner surface; stylar column stipitate-glandular, 20–35 mm long including the beak, this 3–5 mm long; style branches 4.5–7 mm long. Sagebrush, mountain brush, aspen, tall forb, and spruce-fir communities at 1765 to 3145 m in Beaver, Box Elder, Cache, Carbon, Daggett, Davis, Duchesne, Juab, Rich, Salt Lake, San Juan, Sanpete, Summit, Uintah, Utah, Wasatch, and Weber counties; British Columbia to Saskatchewan, south to California, Nevada, and Colorado; 126 (ix). Our specimens have been regarded as **var. nervosum** (Rydb.) C. L. Hitchc. [*G. nervosum* Rydb.], but the original description of that taxon more closely describes *G. richardsonii*. Despite this problem, the relationships of this most beautiful of our geraniums apparently lies to the north. It is replaced south of Sanpete County (with outliers in Beaver and San Juan counties) by phases of *G. caespitosum*, which contain within a broad concept the Fremont geranium, *G. fremontii* Torr. (q.v.), a plant similar in major respects with *G. viscosissimum*. A similar series of glandular versus nonglandular plants occurs within *G. viscosissimum* as is present in *G. caespitosum*, but they have not been considered at taxonomic rank.

Pelargonuim L'Her.

Plants herbaceous, perennial (under glass) herbs; leaves alternate or opposite, usually palmately veined and lobed; flowers small to large, showy, borne in umbels on axillary peduncles; sepals 5 produced into a spur adnate to the pedicel; petals 5, the upper 2 differing in size from the lower 3; stamens 10, usually only 5 with anthers; stylar column much longer than the ovary, usually coiling in fruit.

Pelargonium hortorum Bailey Common Geranium. Stems mostly 2–5 dm long, somewhat succulent, fragrant, pubescent; leaves alternate (rarely opposite), the blades 2.5–10 cm broad, orbicular to reniform, shallowly lobed and crenate; flowers several to many in involucrate umbels; petals pink, salmon, red, white, or variously spotted, 10–25 mm long; 2n = 18, 24, 30, 34. This is the common geranium widely grown as a border plant and as a household and ornamental plant throughout temperate and warm temperate regions of the earth; 4 (0).

GUTTIFERAE A. L. Juss.
St. Johnswort Family

Herbs; leaves opposite, simple, entire, glandular-punctate; stipules lacking; flowers perfect, regular, cymose; sepals 5 (4); petals distinct, 5 (4), yellow; stamens numerous, often in 3–5 clusters, with the filaments united below; pistil 1, the ovary superior, 1- to 5-loculed; styles 3–5, distinct or united; fruit a capsule with 3–5 parietal or axile placentae; x = 7, 8, 9, 10. [Clusiaceae Lindl.; Hypericaceae A. L. Juss.].

Hypericum L.

Perennial glabrous herbs; leaves opposite, simple, sessile, entire, glandular-punctate; flowers perfect, regular; ovary 3-loculed or with 3 placentae; capsule dehiscent, many-seeded.

1. Stems procumbent, matforming, rooting at the lower nodes, 3–25 cm long; leaves 3–11 mm long, without black glandular punctae; flowers few, cymose, solitary, or occasionally axillary; petals salmon colored, 2–4 mm long .... *H. anagalloides*
— Stems erect, from creeping rootstocks, 20–70 cm long; leaves 10–35 mm long, with black glandular punctae; flowers several to numerous; petals yellow, 7–14 mm long .... *H. formosum*

Hypericum anagalloides C. & S. Plants small, slender stemmed annuals; stems 5–25 cm long, the branches ascending; leaves 5- to 7-veined, lacking glandular punctae marginally, elliptic to orbicular; cymes 1- to 3-flowered or solitary in leaf axils; sepals elliptic, unequal; petals 3–5 mm long, salmon colored; stamens 15–21; capsule to 3 mm long. Wet meadows and bogs in spruce-fir, lodgepole pine, and aspen communities at 2800 to 3400 m in Duchesne and Summit counties; British Columbia to Montana, south to Arizona and California; 5 (i).

Hypericum formosum H.B.K. [*H. scouleri* Hook.]. Erect herbaceous perennials, with horizontal rootstocks; herbage glandular-punctate; stems 20–70 cm tall, simple below, occasionally branched above; leaves 1–3.5 cm long, oval or elliptic, or sometimes oblong-ovate, obtuse to acuminate; flowers few to many in branching cymes; sepals 2–5 mm long, ovate, glandular punctate; petals 6–15 mm long; stamens numerous; capsule ca 8 mm long; n = 8. Stream banks and meadows in sagebrush, mountain brush, lodgepole pine, pinyon, ponderosa pine, and aspen communities at 1370 to 3090 m in Box Elder, Cache, Daggett, Duchesne, Emery, Garfield, Grand, Juab, Kane, Millard, Morgan, Salt Lake, San Juan, Summit, Tooele, Uintah, Utah, Wasatch, and Washington counties; Washington to Wyoming, south to California, Arizona, and New Mexico; 31 (vii).

HALORAGACEAE R. Br. in Flinders
Water-milfoil Family

Aquatic or semiaquatic, herbaceous perennials; leaves whorled or alternate, finely dissected; flowers regular, small, unisexual or perfect, solitary in the upper leaf axils or more often in terminal spikes, 3- or 4-merous; sepals obsolete or 4 and persistent in fruit; petals 2–4, small, or sometimes wanting; stamens 4 or 8, the filaments slender, distinct; anthers large; pistil 1, the ovary inferior, 1- to 4-loculed, with (1) 2–4 distinct styles; fruit indehiscent, nutlike or drupelike; x = 7.

Myriophyllum L.

Stems slender, floating and submersed or growing on mud; leaves whorled or alternate, finely pinnately divided; flowers unisexual, mostly in terminal spikes; calyx with 4 short lobes or lacking; petals 2–4 when present; stamens 8 or sometimes 4; anthers large; fruit bony, splitting into 2–4 (1–seeded) nutlets.

1. Pistillate and staminate flora bracts strongly lobed or dissected, mostly much longer than the flower and fruit .... *M. verticillatum*
— Pistillate floral bracts shortly toothed; staminate bracts entire, all floral bracts equaling the flowers or shorter (the pistillate sometimes longer) .... *M. exalbescens*

Myriophyllum exalbescens Fern. [*M. spicatum* ssp. *exalbescens* (Fern.) Hulten; *M. spicatum* var. *exalbescens* (Fern.) Jepson]. Stems 3–8 dm long, loosely erect; leaves whorled, 1–3 cm long, with segments in 4–12 pair; inflorescences submersed or sometimes emergent, the spikes 3–10 cm long, upper flowers staminate, the lower pistillate; bracts whorled, entire or the lower ones deeply lobed, shorter than or exceeding the flowers; stamens 8. fruit 2–3 mm long. Ponds, lakes, reservoirs, or in mud at 1370 to 2745 m in Cache, Daggett, Duchesne, Garfield, Millard, Rich, Sanpete, Sevier, Uintah, Utah, Wasatch, and Washington counties; widespread in North America; 31 (ii).

Myriophyllum verticillatum L. Stems 3–10 dm long; leaves 1–4 cm long, with 9–15 pairs of segments, whorled; flowers whorled, in an interrupted spike, 4–12 cm long; bracts usually much longer than the flowers and pinnatifid; staminate flowers with petals, these yellow green, deciduous; pistillate flowers lacking petals, but with persistent sepals; fruit 2–3 mm long; 2n = 28. Our few collections from Salem Pond and the Millrace in Provo, Utah County; widespread in North America; circumboreal; 3 (0).

HAMAMILIDACEAE R. Br. in Abel
Witch-hazel Family

Trees; leaves alternate, simple; stipules deciduous; flowers in heads or clustered, typically monoecious; flowers regular, bracteate, usually sessile; calyx tube present, the lobes present or absent; petals typically lacking; stamens numerous, in globular masses; pistil 1, the ovary 2-carpelled; styles 2; fruit a 2-beaked, 2-loculed woody capsule, usually with 1 or 2 seeds per locule; x = 8, 12, 15, 16.

Liquidambar L.

Monoecious trees with corky branches; leaves palmately 3- to 7-lobed, fragrant when bruised; flowers apetalous; staminate flowers intermixed with small scales in globose heads disposed in terminal racemes; pistillate flowers in pedunculate globose heads, consisting of coherent 2-loculed, 2-beaked pistils subtended by scales; fruiting head globose, woody.

Liquidambar stryaciflua L. Sweetgum. Trees to 20 m tall or more; bark grayish brown, furrowed, commonly with corky ridges on the branchlets, often exuding gum (that is said to be pleasant to chew); leaves palmately lobed, mainly 4–10 cm long and about as wide, brilliantly colored in autumn, truncate to subcordate at the base; fruiting head woody, subglobose, mainly 2–3 cm in diameter, borne on peduncles mainly 2–5 cm long; capsules woody, each producing 1 or 2 seeds and many abortive seeds resembling sawdust. Commonly cultivated ornamental trees of great beauty, especially in autumn, in much of lower elevation Utah; Indiana to New York, south to Texas and Florida; Mexico and Central America; 11 (0).

HIPPOCASTANACEAE DC.
Horsechestnut Family

Contributed by Kaye Hugie Thorne

Perennial shrubs or trees with large winter buds covered by resinous scales; leaves opposite, palmate, estipulate; flowers irregular; calyx 5-lobed or free; petals 4–5, distinct, an irregular disk between petals and stamens; stamens 4–9, distinct; pistil 1, the ovary superior, 3-carpellate (rarely 1 or 2 by abortion), each locule with 2 ovules; style and stigma simple; fruit a leathery capsule opening by 3 valves; seed solitary; x = 20.

Aesculus L.

Deciduous trees or shrubs; leaves opposite, palmate; inflorescence an erect terminal panicle; flowers perfect, showy, white, pink, red, yellow, or yellow green, sometimes spotted; sepals 5-lobed; petals 4 or 5, clawed, equal or unequal; stamens 5–8; ovary superior, 3-loculed; fruit a leathery capsule, globose, smooth, prickled or spiny; seed large lustrous dark brown with a white hilum.

1. Flowers scarlet to flesh pink; leaflets cuneate-obovate, to 16 cm long; fruit smooth, sometimes with a few small prickles .... *A. x carnea*
— Flowers white, yellow, or greenish-yellow, sometimes spotted red or yellow; leaflets obovate, elliptic, to 32 cm long; fruit prickled or spiny 2
2(1). Flowers white, spotted red or yellow; leaflets obovate to 32 cm; panicle to 33 cm long; fruit spiny .... *A. hippocastanum*
— Flowers yellow or greenish yellow 3
3(2). Flowers yellow, to 3 cm long; petals unequal; stamens included; fruit smooth .... *A. octandra*
— Flowers greenish yellow to yellow; petals equal; stamens exserted; fruit with prickles .... *A. glabra*

Aesculus x carnea Hayne in Guimpel, Otto & Hayne. Red Horsechestnut. Trees to 25 m tall; leaflets usually 5, subsessile, cuneate-obovate, 7–16 cm long, dark green above, lighter green beneath, twice serrate; inflorescence to 20 cm long; flowers scarlet, deep red, or flesh-colored, the claws shorter than or equal to the sepals; petals glandular, villous on the margins; stamens exserted; fruit subglobose, 3–4 cm across, smooth or occasionally with a few small prickles. Cultivated ornamental in lower elevation portions of Salt Lake and Utah counties; sporadic in cultivation elsewhere in the U. S.; 10 (0). This is a hybrid between *A. hippocastanum* and *A. pavia*. Cultivars include "Briotii," with scarlet flowers and "rosea" with pink flowers.

Aesculus glabra Willd. Ohio Buckeye. Trees to 10 m high; leaflets 5, subsessile, elliptic-obovate, cuneate, acuminate, 8–12 (14) cm long, serrulate, green above, lighter green and pubescent beneath when young; inflorescence to 15 cm long; flowers yellow to greenish yellow, 2–3 cm long, the claws slightly longer than the sepals; petals equal; stamens exserted; fruit obovoid, with prickles. Cultivated ornamental in Cache, Salt Lake, and Utah counties; widely grown in the the U. S.; indigenous to the eastern U. S.; 15 (0). Variants include var. *arguta* (Buckley) Robins. [*A. arguta* Buckley] the Texas buckeye, which has leaflets ovate-elliptic and attenuate at both ends.

Aesculus hippocastanum L. Common Horsechestnut. Tree to 25 m tall; leaflets 5–7, subsessile, cuneate-obovate, acuminate, twice serrate 10–24 cm long, dark green above, rusty tomentose beneath when young; inflorescence to 30 cm long; flowers white, usually red spotted or tinged, the claws equal to or shorter than the sepals; stamens exserted; fruit subglobose spiny, 3–6 cm across; 2n = 40. Widely cultivated as a street tree and in landscape plantings in Cache, Salt Lake, Utah, and Weber counties; widely grown in the U. S.; introduced from Europe; 17 (0). Cultivars include "alba", flowers pure white; "Baumannii," flowers white double; "rosea" flowers pink; and "rubicundula" flowers red.

Aesculus octandra Marshall Sweet Buckeye. Trees to 30 m long; leaflets 5, elliptic, or narrow elliptic to obovate, cuneate, acuminate, twice serrate, green above, light to yellow green beneath pubescent when young; inflorescence to 15 cm long; flowers yellow, ca 3 cm long; petals distinctly unequal, the claws longer than the sepals; stamens usually included; fruit subglobose, 5–6 cm across, smooth. Cultivated shade and ornamental trees in Cache, Salt Lake, Utah, and Weber counties; native to the American midwest; 9 (0).

HIPPURIDACEAE Link
Marestail Family

Aquatic or amphibious herbs from elongate rhizomes; stems wholly or partly submersed; leaves whorled, entire, linear, sessile; flowers inconspicuous, perfect or sometimes unisexual, in the upper leaf axils; calyx reduced to a rim at the top of the inferior ovary; petals none; stamen 1; pistil 1, 1-carpellate; fruit drupelike; x = 16.

Hippuris L.

Stems simple, mostly submersed, erect; leaves in whorls of 6 or more; flowers axillary, mostly perfect; sepals very small; stamen on top of the ovary, this 1-loculed, 1-ovuled; style filiform, placed in a groove in the anther; fruit 1-seeded, indehiscent.

Hippuris vulgaris L. Common Marestail. Stems erect, typically 1–4 dm tall (at least as regards the aerial portion, the submersed portion often much longer), 1.5–5 mm thick; leaves narrow, entire, spreading, simple, in whorls of (6) 8–12, with the internodes progressively shorter upward, the middle stem leaves often the longest, mostly 6–30 (50) mm long, 1–2 mm wide, in many whorls; flowers mostly perfect; staminal filament persistent, 0.5–1 mm long; fruit black, 1.5–2.5 mm long; 2n = 32. Streams, lakes, and ponds at 1370 to 2810 m in Cache, Duchesne, Garfield, Grand, Kane, Millard, Rich, Sanpete, Sevier, Summit, Uintah, Utah, Wasatch, and Wayne counties; widespread in North America; circumboreal and in South America; 28 (i).

HYDROPHYLLACEAE R. Br.
Waterleaf Family

Perennial, biennial, or annual herbs, or shrubs; leaves simple or pinnatifid; flowers prefect, regular, 5–merous, commonly in cymes, these mostly scorpioid; calyx lobes 5, similar or dissimilar, sometimes accrescent in fruit; corolla lobes 5; stamens 5, exserted or included; pistil 1, 2 carpellate, the ovary superior, 1–loculed or more or less completely 2–loculed; styles 2, or if 1 then 2–cleft; fruit a longitudinally dehiscent capsule; seeds 1 to many; x = 5–13.

Atwood, N. D. 1976. The Hydrophyllace of Utah. Great Basin Nat. 36: 1–55.

1. Plants aromatic shrubs; leaves leathery, evergreen; known from Washington County .... *Eriodictyon*
— Plants annual or perennial herbs; leaves not leathery or evergreen 2
2(1). Calyx lobes dimorphic, the 3 outer ones conspicuously enlarged, cordate and veiny in fruit, the 2 inner ones linear; known from Washington County .... *Tricardia*
— Calyx lobes subequal, or if somewhat unequal, not as above 3
3(1). Plants acaulescent; flowers solitary at ends of elongate, naked peduncles .... *Hesperochiron*
— Plants mostly caulescent; flowers in scorpioid cymes or solitary in leaf axils 4
4(3). Ovary unilocular 5
— Ovary more or less bilocular 7
5(4). Plants perennial; stamens exserted .... *Hydrophyllum*
— Plants annual; stamens included 6
6(5). Herbage glabrate; stems sharply angled and armed with minute, reflexed prickles; seeds usually 1 .... *Nemophila*
— Herbage viscid and scented; stems not as above; seeds 7–15 .... *Eucrypta*
7(4). Stamens unequally inserted on the corolla tube; flowers axillary, solitary in small dense leafy clusters ...... *Nama*
— Stamens equally inserted on the corolla tube; flowers mostly in cymes 8
8(7). Corolla pale yellow, marcescent; flowers long-pedicellate, the pedicels and flowers ca 1 cm long, pendulous .... *Emmenanthe*
— Corolla blue, purple, or white, or if yellow, less than 1 cm long .... *Phacelia*

Emmenanthe Benth.

Plants annual, villous and glandular; leaves alternate, sessile or nearly so, oblong, pinnatifid; inflorescence with pendulous flowers on filiform pedicels; corolla pale yellow, marcescent; sepals shorter than the calyx; capsules unilocular, oblong, compressed; seeds numerous, reticulate, compressed.

***Emmenanthe penduliflora* Benth.** Whispering-bells. Annual, 1.9–6.4 dm tall; stems erect, hirsute and glandular-viscid; leaves alternate, pinnatifid; inflorescence of scorpioid terminal cymes, hirsute; sepals ovate to lanceolate, 6–10 mm long, hirsute; corolla light yellow, campanulate, 8–12 mm long, 6–10 mm broad, pendulous; stamens included, subequal, usually inserted at the base of the corolla tube; style included, 2–cleft at the apex; ovules numerous, pendulous; seeds ca 15, dark brown, 1.4–2.5 mm long; 2n = 36. Mixed shrub and pinyon-juniper communities, or in burned areas, at ca 1220 to 2380 m in Washington County; Arizona, Nevada, and California; 6 (i).

Eriodictyon Benth.

Aromatic, evergreen shrubs from underground rootstocks; leaves dark green, resinous above and tomentose beneath; inflorescence with numerous scorpioid cymes in terminal panicles; corolla white to purple, deeply lobed, funnelform; sepals subequal, deeply divided; stamens included; style divided to the base; capsules 4–valved, cartilaginous; seeds 2–6, ridged longitudinally, flattened.

***Eriodictyon angustifolium* Nutt.** Yerba-santa. Evergreen, glabrous and glutinous shrubs, 5–20 dm tall; leaves alternate, linear, 4–8 cm long, 2–5 cm broad, revolute, thick, entire to toothed; inflorescence of terminal, branched, scorpioid cymes; calyx-lobes subequal, linear, 3 mm long; corolla white, deciduous, narrowly campanulate, 5–6 mm long; stamens included; style divided to base; seeds 6 or fewer. Warm desert shrub, juniper, pinyon-juniper, and mountain brush communities in open, sandy or gravelly soils at 730 to 2135 m in Washington County; Nevada and Arizona; 36 (iii).

Eucrypta Nutt.

Annual, fragile, glandular herb; leaves pinnatifid, the lower petioled, upper auriculate-clasping; corolla white to blue or yellowish, campanulate; calyx divided at least half its length, shorter than the corolla; stamens included; style divided only at the apex; capsules unilocular, ovoid, enclosed by the enlarged calyx; seeds dimorphic, brown to black.

***Eucrypta micrantha* (Torr.) Heller** Desert Eucrypta. [*Phacelia micrantha* Torr.; *P. pinetorum* Jones, type from Deep Creek Mts.; *Ellisia micrantha* (Torr.) Brand; *Nyctelea pinetorum* (Jones) Tides.]. Annual, 0.5–2.5 dm tall; stems weak, diffuse; leaves pinnatifid, the lower leaves petiolate, the upper auriculate clasping; inflorescence of terminal or axillary cymes; pedicels filiform; calyx-lobes oblong to spatulate, 1.5–2 mm long, stipitate-glandular; corolla campanulate, purplish, blue or white, the tube yellow, 2–4 mm long and broad; stamens included; style included, bifid at apex; mature capsule unilocular; seeds 5–15, dimorphic, brown to black; 2n = 12, 18, 24. Shaded or protected areas in the desert or canyons from the creosote bush community upward to the pinyon-juniper zone at 760 to 1375 m in Juab, Kane, and Washington counties; Texas, west to California; 17 (v).

Hesperochiron Wats.

Acaulescent perennials from a thick root; leaves simple, petioled, in a basal rosette; flowers long-petioled, solitary in the leaf axils; corolla white to blue, funnelform to rotate; sepals unequal, ciliate; stamens included, inserted on the tube of the corolla; filaments basally dilated; style bifid at the apex; capsule unilocular; seeds numerous, dark brown, pitted.

1. Corolla rotate or bowl-shaped, densely long-hairy within, the lobes mostly longer than the tube; flowers few, 1–5 (7); plants usually in the foothills or mountains in moist areas, rarely in alkaline or saline sites; plants not known from Summit County .... *H. pumilus*
— Corolla funnelform or campanulate, subglabrous or short-hairy within, the lobes equaling or usually shorter than the tube; flowers more numerous, rarely as few as 5; plants mostly of alkaline or saline soils at lower elevations, reported from Summit County .... *H. californicus* (Benth.) Wats.

Hesperochiron pumilus (Griseb.) **Porter** [*Villarsia pumila* Griseb.; *Capnorea pumila* (Dougl.) Greene; *C. watsoniana* Greene]. Dwarf, acaulescent perennials, 2–12 cm tall, short-villous; leaves in a basal rosette, petiolate, simple, linear to oblanceolate, 1.5–5 cm long; peduncles solitary in the axils, erect; sepals linear to oblong, usually unequal, 3–8 mm long; corolla rotate white to light blue, 0.5–1.6 cm long, densely hairy within; stamens included, inserted unequally on the corolla tube, the filaments dilated below; style included, 2-cleft at apex; mature capsule unilocular; seeds numerous, dark brown, alveolate; $2n = 16$. Moist ground along streams and in meadows, in mountain brush and aspen communities and in desert springs, mostly between 1675 and 2595 m in Box Elder, Emery, Juab, Millard, Rich, Salt Lake, Sanpete, Summit, Uintah, Utah, Washington, and Weber counties; Washington to Montana, south to California and Arizona; 11 (ii).

Hydrophyllum L.

Perennial herbs; stems erect, succulent, from horizontal rhizomes, these bearing fleshy fibrous or tuberous roots; leaves pinnately compound, mostly basal, oblong, ovate or oval in outline; petioles slightly dilated and clasping at the base, ciliate; inflorescence composed of 1 to several globose or lax cymes, short pubescent or strigose and hispid; calyx divided nearly to the base, the lobes linear, oblong or lanceolate; corolla campanulate purplish to blue, white or violet; stamens 5, exserted; style 1, exserted 5–10 mm, cleft 1–2 mm; stigmas capitate; ovules attached to the front of the 2 large pariental placentae; seeds 1–3, brown, subglobose, reticulate.

1. Flowers in dense capitate clusters; peduncles shorter than the petioles of the subtending leaves; anthers short-oblong, 0.6–1 mm long .... *H. capitatum*
— Flowers in open clusters; peduncles longer than the petioles of the subtending leaves; anthers linear to oblong, 1–2 mm long 2
2(1). Leaflets acuminate, with 8–12 acuminate teeth; cymes lax in fruit .... *H. fendleri*
— Leaflets obtuse to abruptly acute, with 3–6 obtuse to acute teeth, cymes compact in fruit ...... *H. occidentalis*

Hydrophyllum capitatum Dougl. ex Benth. Plants 1–5 dm high, from short rhizomes, these bearing a fascicle of tuberous roots; stems short; leaves pinnately compound, ovate to oval in outline, strigose, the blade 2.5–10 cm long, 2–13 cm wide, the primary divisions 5–7, obovate to oblong or lanceolate, the lobes and divisions acute, obtuse or mucronate; inflorescence of 1 to several globose cymes, the peduncles 1–5 cm long, shorter than the subtending leaves, mostly recurved in fruit; pedicels 2–5 mm long; sepals obtuse or abruptly acute, 3–4 mm long, 1.5 mm broad or less, ciliate and strigose; corolla 5–9 mm long, purplish, blue or white; stamens exserted 5 mm; style exserted 5–10 mm; seeds normally 2, light brown, 2–3 mm in diameter; $2n = 18$.

1. Cymes lax (at least in fruit); pedicels 7–19 mm long, reflexed in fruit; plants low (2.5 dm tall or less) and more or less acaulescent .... *H. capitatum* var. *alpinum*
— Cymes capitate even in fruit; pedicels 2.5–5 mm long, not reflexed in fruit; plants usually taller (1–5 dm high) and caulescent .... *H. capitatum* var. *capitatum*

Var. *alpinum* Wats. [*H. alpinum* Greene ex Brand]. Sagebrush, pinyon-juniper, aspen, spruce-fir, and grassy meadow communities at 2010 to 2870 m in Deep Creek, Goose Creek, and Raft River mts. in Box Elder, Juab, and Tooele counties; Oregon and Idaho, south to California and Nevada; 8 (i).

Var. *capitatum* [*H. densiflorum* Nutt.; *H. pumilum* Geyer ex Hook.; *H. capitatum* var. *pumilum* (Geyer) Hook.; *H. capitatum* var. *pumilum* subvar. *densum* Brand; *H. capitatum* var. *alpinum* Wats. subvar. *laxum* Brand]. Mountain brush, aspen, sagebrush, and mixed conifer communities at 1340 to 2745 m in Beaver, Box Elder, Carbon, Davis, Duchesne, Emery, Salt Lake, Sanpete, Sevier, Summit, Utah, Wasatch, and Weber counties; Washington to Idaho, south to Colordo; 62 (iii).

Hydrophyllum fendleri (Gray) **Heller** [*H. occidentale* var. *fendleri* Gray; *H. albifrons* var. *fendleri* (Gray) Brand]. Perennial, 2–9 dm tall from short rhizomes, these bearing fleshy fibrous roots; stems erect, retrorse-hispid; leaves pinnately compound, ovate or oval in outline, strigose, the blade 6–30 cm long, with 9–13 primary divisions, these ovate to lanceolate, acuminate, coarsely serrate or incised; inflorescence of 1 to several lax cymes; peduncles 3–17 cm long, often branched, mostly longer than the subtending leaves (at least in fruit); pedicels 2–7 mm long; sepals linear to lanceolate (in fruit), 4–6 mm long, 1–2 mm broad, ciliate and strigose, often hispid dorsally, corolla 6–10 mm long, white to violet; stamens exserted 4–6 mm; style exserted 5–7 mm; seeds 1–3, light brown, 2.5–3 mm in diameter; $2n = 18$. Moist ground in mountain brush and ponderosa pine communities at ca 2684 m in the Abajo Mts., San Juan County; British Columbia to Idaho, south to Oregon and Colorado; 1 (i).

Hydrophyllum occidentale (Wats.) **Gray** Western Waterleaf. [*H. macrophyllum* var. *occidentale* Wats.; *H. occidentale* var. *watsonii* Gray, type from the Wasatch Mts.; *H. watsonii* (Gray) Rydb.]. Perennial, 1–6 dm tall, rhizome bearing fleshy fibrous roots; stems erect, short pubescent to more or less retrorse-hispid; leaves pinnately compound, oblong in outline, strigose, the blade 5–28 cm long, with 7–15 primary divisions, these broadly oblong to ovate, obtuse, incised or lobed; inflorescence with one to several globose cymes, peduncles 5–27 cm long, usually exceeding the subtending leaves; pedicels 2–5 mm long; sepals narrowly lanceolate, 3.5–5 mm long, 1–2 mm wide, strigulose dorsally to hispid ciliate on the margins; corolla 7–10 mm long, white to violet; stamens exserted 4–6 mm; style exserted 5–8 mm; seeds 1–2, brown, 2.5–3.1 mm in diameter; $2n = 18$. Mountain brush, sagebrush, pinyon-juniper, aspen, and spruce-fir communities at 1525 to 3050 m in Beaver, Iron, Juab, Millard, Salt Lake, Sanpete, Sevier, Tooele, Utah, and Washington counties; Oregon south to California, east to Arizona; 49 (v).

Nama L.

Low, branching annuals; leaves alternate, entire, hirsutulous to hispid, retrorse to erect; inflorescence of terminal nonscorpioid cymes; sepals subequal, linear to lanceolate; corolla purple, or lavender, deciduous, tubular to funnelform; stamens included, borne unequally on the corolla tube; style included, divided to the base or 2-lobed at the apex; mature capsule falsely bilocular by intrusion of the placentae; ovules numerous; seeds numerous, brown, mostly reticulate.

1. Style shallowly 2-lobed at the apex; corolla tubular, 3–5 mm long .... *N. densum*
— Style divided to the base; corolla mostly 8–15 mm long or, if shorter, the shorter stem hairs retrorse (corolla 4–7 mm long in *N. retrorsum*) 2

2(1). Leaves mostly in clusters at the ends of the branches and in a basal rosette; herbage hirsutulous or pilose .... *N. demissum*
— Leaves well distributed along the stem; herbage hirsute or hispid 3

3(2). Stems erect, fastigiate; shorter stem hairs retrorse; corolla 4–7 mm long .... *N. retrorsum*
— Stems more or less spreading; stem hairs ascending; corolla 7–15 mm long .... *N. hispidum*

Nama demissum Gray [*Conanthus demissus* (Gray) Heller; *N. demissum* var. *deserti* Brand]. Diffusely branched annuals, 2–14 cm tall, hirsute; leaves entire, 1–3.2 cm long, 1–5 mm wide; flowers solitary to several, subsessile; sepals linear to lanceolate, 5–8 mm long; corolla broadly funnelform, 8–16 mm long, 6–11 mm wide; style divided to base, 3–5 mm long; seeds mostly 10–15, 0.5 mm long, dark brown, pitted and reticulate; $2n = 14$. Desert shrub communities in open gravelly or sandy sites, at 670 to 1560 m in Beaver, Millard, Tooele, and Washington counties; Arizona, Nevada, and Mexico; 21 (ii).

Nama densum Lemmon Compact Nama. [*Nama demissa* Wats.; *Conanthus parviflorus* Greenman]. Dichotomously branched annuals, the branches prostrate, or under harsh conditions simple and depauperate, spreading hirsute, 1–10 cm across; leaves oblanceolate, 0.4–3 cm long, 1–5 mm wide, entire; flowers sessile and solitary in the upper leaf axils; calyx-lobes linear, 3.8–7 mm long, 1–3 mm wide; corolla funnelform, white to lavender, 4–7.2 mm long, the tube long, the lobes short; style 0.5–2 mm long, 2-lobed at the apex; capsule 2–3.5 mm long; seeds ca 15, 0.5–0.9 mm long, dark brown to blackish, pitted and reticulate; $2n = 14, 28$. Sandy soils or sometimes loose gravel in juniper, sagebrush, and desert shrub communities at 1430 to 1865 m in Box Elder, Duchesne, Grand, Juab, Kane, Millard, San Juan, Tooele, Uintah, and Washington counties; Idaho and Washington, south to California; 21 (iii). Our plants belong to var. *parviflorum* (Greenman) C. L. Hitchc. [*Conanthus parviflorus* Greenman].

Nama hispidum Gray [*Marilaunidium tenue* Small; *M. hispidum* (Gray) Kuntze; *Conanthus hispidus* (Gray) Heller]. Leafy, branched annuals, 1–3 dm tall; stems more or less spreading, hispid; leaves 1–7 cm long, 2–6 mm wide, entire, revolute, finely glandular and with somewhat appressed, nonglandular hairs, oblanceolate to spatulate; flowers solitary to several in terminal cymes; sepals linear to lanceolate, 5–8 mm long; corolla purple, broadly funnelform, 8–14 mm long, 7–8 mm wide; style 2–5 mm long, cleft to the base; seeds numerous, 0.5 mm long, yellowish brown, reticulate; $2n = 14$. Sandy shore and drainage bottoms adjacent to Lake Powell, at ca 1220 m, in Kane County; California to Texas; 4 (iv).

Nama retrorsum J. T. Howell Leafy branched annuals, 1–3 dm tall; stems erect, fastigiate, hirsute, the shorter stem hairs retrorse; leaves 1.5–5 cm long, 2–5 mm wide, entire; flowers sessile and solitary in the upper leaf axils; sepals linear, 5 mm long, hirsute; corolla purple, funnelform, 4–7 mm long; seeds 0.6–0.8 mm long. Mixed desert shrub, mostly in sand, at 1460 to 2135 m in Garfield, Grand, Kane, and San Juan counties; northeastern Arizona; 14 (v).

Nemophila Nutt. ex Barton

Delicate annuals, 0.5–3 dm tall; stems sharply angled (obscurely so in *N. parviflora*), glabrous, except for small, reflexed prickles; leaves alternate or opposite, 0.7–2.9 cm long, 1.5–4 cm wide, pinnately divided, sparsely hispid; flowers solitary in the upper leaf axils, pedicellate; calyx divided nearly to the base, the lobes linear to lanceolate, 3 mm long, 1–2 mm wide, with reflexed auricles; corolla narrowly campanulate, purplish or white, 1.5–2.9 mm wide; stamens included, equally inserted on the corolla; style 0.5–1 mm long; seeds usually 1–4, globose, 2–3 mm long, brick-red, pitted.

1. Leaves all alternate; seeds usually 1; calyx 3 mm long; style cleft only at the apex; capsule shorter than the strongly accrescent calyx; plants widespread . *N. breviflora*
— Leaves all opposite; seeds mostly 2–4; calyx 1–3 mm long; style cleft ca 1/2 its length; capsule exceeding the calyx; presently known from Cache and Weber counties .... *N. parviflora*

Nemophila breviflora Gray [*Viticella breviflora* (Gray) Macbr.; *N. petrophila* L. O. Williams]. Stems weak, 0.5–2 dm long, sharply angled; leaves alternate, pinnately divided, 0.7–3 cm long, 1.5–3.9 cm broad, sparsely hispid; sepals broadly campanulate, linear to lanceolate, 3 mm long, 1–2 mm broad; corolla narrowly campanulate, white or purplish, 1.5–2.9 mm broad, shorter than the calyx; style 0.5–1 mm long, cleft at the apex; mature capsule exceeded by the accrescent calyx; seed usually 1, globose, 2–4 mm in diameter, deeply pitted in rows; cucullus reduced, persistent; $2n = 18$. Mountain brush, sagebrush, and aspen communities at 1460 to 2715 m in Box Elder, Cache, Carbon, Daggett, Juab, Salt Lake, Summit (type from Parleys Park), Utah, Wasatch, and Weber counties; California east to Colorado, north to Montana and southern British Columbia; 22 (i). This inconspicuous plant is typically in the understory of aspen and maples in moist, rich soil.

Nemophila parviflora Dougl. ex Benth. [*N. inconspicua* Henderson; *N. explicata* A. Nels.; *N. parviflora* var. *typica* Brand subvar. *inconspicua* Brand; *Viticella parviflora* var. *austinae* (Eastw.) Macbr.]. Stems obscurely angled, 0.5–3 dm long; leaves all opposite, 1–1.5 cm long, 2–2.4 mm broad, sparsely hispid; sepals lanceolate, the auricles 0.2–0.4 mm long; corolla campanulate, white or bluish, 1.5–3 mm broad, barely exceeding the calyx; style 0.6–0.8 mm long, cleft ca 1/2 its length; mature capsule exceeding the calyx; seeds mostly 2–4, ovoid, 2–2.5 mm long, shallowly pitted; cucullus deciduous; 2n = 18. Mountain brush community at ca 1525 m in Cache and Weber counties; British Columbia and Idaho, south to California; 1 (0). Our plants are referable to var. **austinae** (Eastw.) Brand [*N. austinae* Eastw.].

Phacelia A. L. Juss.

Herbaceous, annual, biennial or perennials, mostly pubescent and with glandular hairs; leaves mostly alternate, the lower sometimes opposite, entire to pinnately compound; flowers few to numerous in variously disposed scorpioid cymes, or apparently in lax racemes; calyx divided nearly to the base; corolla tubular to broadly campanulate, blue, purplish, violet or white, mostly deciduous, a few species with a tardily deciduous corolla; stamens included or exserted, equally inserted at the base of the corolla tube, with a pair of scales attached to the base of the corolla and filaments; style included or exserted, bifid, mostly pubescent; capsule unilocular (or nearly bilocular by intrusion of the placentae); seeds 1 to numerous, variously roughened, boat-shaped, terete, angled or flattened.

1. Stamens exserted from the corolla for more than 2 mm . 2
— Stamens included in the corolla or exserted up to 2 mm . 3
2(1). Seeds excavated on the ventral surface, on 1 or both sides of a prominent ridge .... Key 1
— Seeds not excavated ventrally .... Key 2
3(1). Ovules 4; seeds 2–4 in each capsule .... Key 3
— Ovules mostly 10 or more; seeds mostly 6 or more (rarely 4 in *P. demissa*) .... Key 4

Key 1.

Stamens exserted from the corolla;
Seeds excavated on the ventral surface.

1. Corolla small, 4 mm long or less, white, blue, or lavender, the lobes erose; plants of southern Utah .... *P. alba*
— Corolla more than 4 mm long, white or variously colored 2
2(1). Corolla tubular, pale 3
— Corolla campanulate, purple, lavender, or white (appearing tubular in some pressed specimens) 5
3(2). Seeds 3.5–4 mm long; cauline leaves sessile or nearly so, auriculate; plants of Emery, Wayne, and Washington counties .... *P. rafaelensis*
— Seeds shorter than 3 mm, black 4
4(3). Inflorescence thyrsoid; stems solitary or branched from near the base; plants of Washington and Iron counties .... *P. palmeri*
— Inflorescence open; stems branched throughout, especially at base; plants of Kane and San Juan counties .... *P. constancei*
5(2). Leaves pinnately compound to doubly so, finely dissected 6
— Leaves simple or, if compound, not finely so, the divisions over 5 mm wide 7
6(5). Mature seeds 2.4 mm long, excavated only along 1 side of the ridge; leaves lacking stipitate glands; plants of Utah County .... *P. argillacea*

— Mature seeds over 2.4 mm long, excavated on both sides of the ridge; leaves strigose and stipitate-glandular; plants of Uintah County .... *P. glandulosa*
7(5). Corolla distinctly bicolored, the tube yellow or white and the lobes blue 8
— Corolla not distinctly bicolored, blue, purple, or white 10
8(7). Cauline leaves sessile, auriculate; plants robust, 0.8–5.8 dm tall, endemic in Sanpete and Sevier counties .... *P. utahensis*
— Cauline leaves distinctyly petiolate; plants not especially robust, less than 2.7 dm tall 9
9(8). Stems branched at the base; leaves simple, strigose and glandular; corolla tube white; seeds corrugated on the margins and on the ridge, the dorsal surface smooth; plants of San Juan and Grand counties .... *P. howelliana*
— Stems simple or branched above; leaves essentially glabrous; corolla tube yellowish; seeds mostly lacking corrugations, the dorsal surface deeply pitted; plants to be expected in southeastern Utah .... *P. splendens* Eastw.
10(7). Corolla lavender; seeds lacking ventral corrugations; plants of Kane and San Juan counties ... *P. integrifolia*
— Corolla blue or purple; seeds corrugated ventrally .. 11
11(10). Mature seeds corrugated only on the ridge; pubescence of the stems densely hispid, glandular above; plants of Washington County ...... *P. ambigua*
— Mature seeds with the margins and ridge corrugated; pubescence of stems mostly glandular, sometimes finely so 12
12(11). Mature seeds light to dark brown; anthers yellow; corolla deep blue, broadly campanulate; stems mostly branched throughout; plants broadly distributed .... *P. crenulata*
— Mature seeds reddish; anthers yellow or blue; corolla pale blue, the lobes not widely spreading; stems solitary, or branching at the base; plants of eastern Kane and Garfield counties .... *P. mammallarensis*

Key 2.

Stamens exserted from the corolla;
seeds not excavated ventrally.

1. Leaves pinnately compound, the segments variously toothed 2
— Leaves entire to shallowly lobed or pinnate, the segments entire 3
2(1). Plants perennial; basal leaves mainly entire (sometimes with 1 or 2 lateral lobes); corolla white to lavender ... . *P. hastata*
— Plants biennial or short-lived perennial; basal leaves pinnately dissected, with 1–4 pairs of lateral lobes; corolla white to cream .... *P. heterophylla*
3(1). Plants annual, of Washington County; stems growing up through shrubs; seeds typically 4; corolla broadly campanulate deciduous .... *P. vallis-mortae*
— Plants biennial or perennial, of broad distribution; stems stout; seeds 8–18; corolla pelviform, marcescent . 4
4(3). Corolla hairy within; stem bases with marcescent leaves; inflorescence virgate .... *P. sericea*
— Corolla glabrous within; stem bases not with marcescent leaves; inflorescence compact, shortly if at all virgate .... *P. franklinii*

Key 3.

Stamens included in the corolla; ovules 4.

1. Leaves pinnately compound; plants long bristly hispid, especially in the inflorescence 2
— Leaves simple to crenate or lobed, seldom compound; plants with short, spreading hairs or more often with viscid, stipitate-glandular hairs and often scented 3

2(1). Corolla broadly campanulate, 6–12 mm long, 8–10 mm wide; pedicels short and stout, 1–1.5 mm long; calyx 5–8 mm long, about twice as long as the capsule and subequal to the corolla; plants growing up through shrubs .... *P. vallis-mortae*
— Corolla tubular-campanulate, 4–7 mm long; pedicels more slender and averaging 2 mm long; calyx 7–10 mm long, 3–4 times longer than the capsule and longer than the corolla .... *P. cryptantha*

3(1). Leaves simple or with 1 or 2 small lobes near the base; seeds not deeply excavated, less than 2 mm long; plants non-odorous, with spreading hairs and some glands above .... *P. austromontana*
— Leaves regularly lobed or crenate; seeds deeply excavated on both sides of a prominent ventral ridge, 2.6–3.4 mm long; plants scented, stipitate-glandular throughout and with some spreading hairs 4

4(3). Corolla 3–4 mm long, campanulate, the lobes mauve to blue (drying white in fruit); sepals narrowly oblanceolate, 2.5–4 mm long, ca 1 mm wide; plants brittle, breaking easily when picked; capsule 2.5–3.5 mm long; seeds 2–2.5 mm long .... *P. coerulea*
— Corolla 5–6 mm long, rotate to campanulate, pale violet or white; sepals oblanceolate to spatulate, 3–6 mm long, 1.5–3.3 mm wide; plants not brittle; capsule 3.3–3.7 mm long; seeds 2.7–3.4 mm long .... *P. anelsonii*

Key 4.

Stamens included in the corolla; ovules 10 or more.

1. Flowers 4–merous (less commonly 5–merous); corolla white, ca 1.5 mm long, persistent and surrounding the lower 1/2 of the capsule, this 1.5–3 mm long; plants of salt flats .... *P. tetramera*
— Flowers 5–merous and not with the above combination of characters 2

2(1). Flowers all yellow or tinged with lavender in age, marcescent; plants of saline areas .... *P. lutea*
— Flowers blue, lavender, violet, purple, or white (the tube yellow in some species), mostly deciduous 3

3(2). Leaves orbicular to reniform; pedicels in fruit surpassing the calyx; corollas white; plants of limestone cliffs in Washington County .... *P. laxiflora*
— Leaves not orbicular to reniform; pedicels various; corollas typically not white; plants of various substrates and sometimes of Washington County 4

4(3). Seeds transversely corrugated, numerous (10 or more) 5
— Seeds terete or angled and mostly foveolate or reticulate, but not excavated ventrally 7

5(4). Corolla 7–17 mm long, more than twice the calyx length; plants of Washington County ..... *P. fremontii*
— Corolla 2–4.5 mm long, shorter than or equaling the calyx 6

6(5). Stems with black, capitate glands, at least above; calyx lobes spatulate; plants disjunct in widely scattered localities, rare .... *P. affinis*
— Stems lacking capitate glands; calyx lobes linear to oblanceolate; plants widespread .... *P. ivesiana*

7(4). Corolla campanulate to rotate or pelviform 8
— Corolla tubular or tubular-campanulate 9

8(7). Style branches 2–3 mm long; stems 0.3–1.5 dm tall; corolla 4–6 mm long, campanulate, leaves mostly entire .... *P. curvipes*
— Style branches 4.5–8 mm long; stems 1–5 dm tall; corolla 6–9 mm long, open-campanulate; leaves entire or with 1–4 lobes .... *P. linearis*

9(7). Ovules 8–16 per ovary 10
— Ovules 20 or more per ovary 11

10(9). Leaves oblong to elliptic; style and branches 1.5 mm long; filaments glabrous; flowers in dense sessile clusters; plants of Kane and Washington counties .... *P. cephalotes*
— Leaves broadly ovate to orbicular; style including branches 1.5–4 mm long; filaments sparsely hairy; flowers in racemes, these 1–4 m long; plants more broadly distributed .... *P. demissa*

11(9). Corolla 8–14 mm long; style including branches .5–5 mm long; plants of Garfield, Kane, and Washington counties .... *P. pulchella*
— Corolla 8 mm long or less; style, including branches, 3 mm long or less; plants of various distribution 12

12(11). Stem pubescence finely glandular-puberulent 13
— Stem pubescence glandular-villous or glandular-hirsutulous, or if glandular-puberulent, the leaves dentate to crenate 14

13(12). Staminal filaments glabrous; style, including branches, 2.5–3 mm long; corolla 3–4.5 mm long; plants of San Juan and Wayne counties .... *P. indecora*
— Staminal filaments sparsely hairy below; style, including branches, 1.5–2 mm long; corolla 5–6 mm long; plants to be expected in western Utah .... *P. parishii* Gray

14(12). Leaves coarsely toothed; seeds 60–100, 0.5 mm long .... *P. rotundifolia*
— Leaves entire to repand, crenate, or dentate; seeds 60 or fewer (if 60 then less than 0.5 mm long) 15

15(14). Leaves dentate to crenate; style, including branches, 2–3 mm long; capsule 4–6 mm long; seeds 40–50, 1–1.3 mm long; plants to be expected in Washington County .... *P. peirsoniana*
— Leaves entire; style, including branches, 1.5 mm long; capsule 2.5–4 mm long; seeds 22–37 or 60 per capsule, 1 mm long or shorter 16

16(15). Corolla tubular, marcescent; seeds ca 60, 0.3–0.4 mm long, reticulate; plants to be expected in Washington County .... *P. saxicola* Gray
— Corolla tubular-campanulate, deciduous; seeds 22–36 per capsule, 0.6–1 mm long, pitted; plants of western Utah and Uintah County .... *P. incana*

Phacelia affinis Gray Annual, 0.5–3 dm tall; stems densely pubescent and stipitate-glandular, at least above, the glands mostly black; leaves mainly in the lower half, pinnately divided, oblong to lanceolate, 1–5 cm long, petiolate, the petioles to 4 cm long; inflorescence of elongate, scorpioid cymes, the flowers subsessile, or the pedicel to 2 mm long on the lowest flowers; sepals oblanceolate to spatulate, 3–4 mm long in flower, 6–10 mm long in fruit; corolla pale lavender or white with a pale yellowish tube, narrowly campanulate, 3–5 mm long; stamens included, unequal; filaments glabrous; style included, 1–2 mm long; capsule 4–5 mm long, shorter than the tube; ovules 13–40; seeds ovate to oblong, brown, reticulate and transversely corrugated, 0.7–1.2 mm long; 2n = 22, 24. Blackbrush, creosote bush, mountain brush, and pinyon-juniper communities at 915 to 1800 m in Kane, Millard, and Washington counties; California to Arizona, south to Mexico; 8 (v).

Phacelia alba Rydb. [*P. neomexicana* var. *alba* (Rydb.) Brand; *P. neomexicana* var. *coulteri* (Greenman) Brand subvar. *foliosissima* Brand; *P. glandulosa* ssp. *euglandulosa* Brand var. *elatior* Brand, in part]. Annual, 0.5–7 dm tall; stems simple to branched, erect or ascending, leafy, hirsute to setose and stipitate-glandular; leaves pinnatifid to subbipinnatifid, 2–10 cm long, 2–8 cm broad, short-hairy and somewhat glandular, the lowermost long-petiolate, the upper sessile or subsessile; inflorescence of dense, terminal, compound scorpioid cymes, the lower ones often simple, densely glandular and puberulent to hirsute, the cymes 1–2 cm long in flower, to 8 cm long in fruit, the flowers subsessile, or the pedicels to 1 mm long; sepals linear to oblanceolate, 3.5–4 mm long; corolla white to lavender or pale blue, tubular-campanulate, 3–4 mm long and wide, the lobes pubescent, erose-denticulate; stamens definitely exserted 2–4 mm; style exserted, divided to below the middle; capsule 3–3.3 mm long, ovoid to subglobose, puberulent and glandular; ovules 4; seeds 4, or 1 or 2, light to dark brown, 2.4–3 mm long, uniformly alveolate and cymbiform, the ventral surface shallowly excavated on both sides of the ridge, lacking corrugations, the margins thick and entire. Sagebrush, pinyon-juniper, mountain brush, ponderosa pine, aspen, and spruce-fir communities at 1370 to 2930 m in Garfield, Kane, Iron, Piute, Sevier, Washington, and Wayne counties; Wyoming, south to Arizona and New Mexico; 29 (viii).

Phacelia ambigua Jones [*P. crenulata* var. *ambigua* (Jones) Macbr.]. Annual, 1.5–4 dm tall; stems erect, usually branched from the base, spreading-hispid, puberulent, and somewhat stipitate-glandular; leaves irregularly lobed to subpinnate, 0.5–13 cm long, 0.5–4.5 cm wide, long-petiolate to nearly sessile above, strigose to hispid and somewhat glandular; inflorescence of terminal, scorpioid cymes, these elongating to 12 cm in fruit; sepals elliptic to oblanceolate 2.7–5.1 mm long, 1–1.3 mm wide, puberulent, hispid, and stipitate-glandular; corolla purple or dull lavender, 4–10 mm long and broad, pubescent; stamens and style exserted 4–10 mm; style bifid, pubescent below; capsule 3–3.5 mm long, puberulent and glandular; ovules 4; seeds 4, ovate, reddish to brown, 2.5–3.3 mm long, 1.3–1.8 mm wide, alveolate, cymbiform, the ventral surface excavated on both sides of the ridge, this corrugated on one side. Joshua tree, creosote bush, and other warm desert shrub communities at 820 to 1525 m in Washington County; Nevada, Arizona, and California; 33 (ii).

Phacelia anelsonii Macbr. Erect, scented annual, 1.5–5 dm tall, usually simple, with brownish, stipitate-glandular hairs and some nonglandular ones; leafy throughout; leaves short-petiolate below, becoming sessile above, 1.5–8 cm long, pinnately lobed; inflorescence of simple or compound scorpioid cymes, usually terminal, but sometimes on leafy lateral branches, the individual cymes 1–5 cm long, setose and stipitate-glandular; sepals oblanceolate to spatulate, 3–6 mm long, surpassing the capsule; corolla rotate to campanulate, lavender to white, 6 mm long and wide; stamens included, the anthers yellow; style cleft to below the middle, included, 3.5–4.8 mm long, shorter than the stamens, glandular and puberulent near the base; capsule 3.5–3.7 mm long, glandular-spotted throughout and pilose on the upper half; ovules 4; seeds oblong, pale brown, the margins entire, the ventral surface strongly alveolate, divided by a prominent ridge, this corrugated along one side, the dorsal surface alveolate, 2.7–3.4 mm long, 1–1.3 mm wide. Creosote bush and Joshua tree communities at 760 to 1220 m in Washington County; Nevada and California; 9 (ii).

Phacelia argillacea Atwood Clay Phacelia. Winter annual, 1–3.6 dm tall, simple to much branched; stems finely pubescent, leafy; leaves deeply pinnatifid, 0.8–5 cm long, 5–25 mm wide, strigose, petiolate; inflorescence of compound, scorpioid cymes, stipitate-glandular and with nonglandular hairs; sepals oblanceolate to spatulate, 2–3.8 mm long, 1 mm wide; corolla blue to violet, campanulate, 4–5.5 mm long and wide, the lobes pubescent; stamens and style exserted 6–7 mm; style bifid 2/3 its length; capsule subglobose, 3.2–3.3 mm long, glandular; mature seeds 4, ovate to elliptic, 2.4 mm long, 1.1 mm wide, pitted, the ridge curved and more or less excavated along one side, brown. Steep hillsides in a sparse juniper-pinyon and mountain brush community, on Green River Shale, at ca 2015 m in Spanish Fork Canyon (type locality), Utah County; endemic; 7 (i).

Phacelia austromontana J. T. Howell Annual, 0.5–2 dm tall, widely branched from the base, hirsute and glandular; leaves entire or few-toothed, 1–3 cm long, linear, lanceolate, or oblong; inflorescence of few- to many-flowered cymes; sepals unequal, linear to oblanceolate, hirsute and glandular; corolla lavender or pale blue, open campanulate, 3–5 mm long; stamens subequal to the corolla; style included, pubescent at the base, 3–4 mm long; capsule 3–3.4 mm long, ovate, puberulent, beaked; seeds 2–4, coarsely pitted, pale brown, 1.4–1.8 mm long; 2n = 18. Ponderosa pine and mountain brush communities at ca 2135 m in Washington County (Oak Grove, Cottam 8831 UT); Nevada and California; 1 (0).

Phacelia cephalotes Gray Annual, 0.5–1.3 dm tall; stems low, almost prostrate, glandular-villous; leaves entire, oblong, ovate to elliptic, petiolate, hirsutulous and glandular, the lower petioles longer than the blade, 0.5–1.8 cm long; inflorescence of compact, sessile or subsessile scorpioid cymes, these terminal on the branches of the stems; sepals oblanceolate, 3–4 mm long in flower, to 8 mm long in fruit; corolla lavender, inconspicuous, 3–5 mm long, the tube yellowish; stamens and style included; capsule ovate, 3–4 mm long, shorter than the calyx; seeds 8–12, oblong, angular, the angles denticulate, pitted, 1–1.5 mm long; n = 11. Salt desert shrub and juniper communities, mainly on the Chinle Formation, at 1065 to 1525 m in Kane, San Juan, and Washington (type locality) counties; Arizona; a Plateau endemic; 9 (v).

Phacelia coerulea Greene Annual, 0.5–6 dm tall; stems erect, simple or branched throughout, sparsely to densely stipitate-glandular and mostly with some stiff nonglandular hairs as well; leaves deeply sinuate to pinnatifid, oblong to ovate, strigose and glandular, the margins crenate, petiolate; inflorescence terminal on the main stem and on secondary branches; sepals oblanceolate to spatulate, 2.5–4 mm long; corolla pale mauve to blue (drying white), 3-4 mm long, campanulate, the tube yellowish; stamens mostly included, though sometimes exserted 1-2 mm; anthers yellow, the filaments bluish; style subequal to the stamens, deeply cleft; capsule globose, 2.5-3.5 mm long; ovules 4; mature seeds 4, dark brown, 2-2.5 mm long, the ventral surface pitted and divided by a prominent ridge, this corrugated along one side, the margins usually corrugated and the dorsal surface pitted, with 0.3-0.4 mm of the margin slightly elevated and smoother than the pitted center. Creosote bush and other warm desert shrub communities at 760 to 915 m in Washington County; Nevada and California to Texas and south to Mexico; 1 (0).

Phacelia constancei Atwood Erect biennial herb, 1.5-4.3 dm tall; stems coarse, 1 to several, reddish, hirsutulous to hirsute and finely glandular; leaves basal and cauline, linear to lanceolate, revolute, the lower short-petiolate or sessile and larger than the upper, the upper sessile; inflorescence of terminal cymes; sepals elliptic to oblanceolate, 3-4 mm long; corolla white to pale lavender, 5-6 mm long; stamens exserted 3-4 mm; seeds 4, black, 2.5-2.8 mm long, 1-1.2 mm wide, elliptic, the margins corrugated, the ventral surface finely pitted, excavated and divided by a prominent ridge, this corrugated along one side, the dorsal surface finely pitted; n = 11. Salt desert shrub and juniper communities at 1370 to 3050 m in Garfield, Kane, and San Juan Counties; Arizona; 21 (xv).

Phacelia crenulata Torr. Annual, 0.3-8.3 dm tall; stems simple or more commonly branched, stipitate-glandular with short or long trichomes, and usually with some nonglandular hairs intermixed; leaves oblong-elliptic to suborbicular, subentire to deeply lobed, the lower lobes sometimes almost distinct, the lower ones the largest, petiolate, the upper mostly reduced and finally sessile; inflorescence scorpioid, terminal on stems and branches; sepals elliptic to oblanceolate, 3-10 mm long; corolla blue violet to purple, campanulate, 4-7 mm long; stamens exserted 3-11 mm; style exserted, bifid 1/2 its length, glandular-pubescent below; capsule globose to subglobose, 2.6-4.1 mm long, 2.8-3.2 mm wide, puberulent to glandular; seeds 4, elliptic to oblong, brown, 2.8-4 mm long, 1.3-3.2 mm wide, pitted, the ventral surface corrugated on the margins and one side of the ridge; 2n = 22. Three varieties are recognized.

1. Leaves linear; plants of eastern Kane County ....
   .... *P. crenulata* var. *angustifolia*
— Leaves variously shaped, but seldom if ever all linear ... 2
2(1). Stems greenish yellow, with uniformly and finely glandular and more or less appressed non glandular hairs; plants widespread in Utah ... *P. crenulata* var. *corrugata*
— Stems usually reddish, pubescent with long to short stipitate-glandular hairs; plants of south-central Utah
   .... *P. crenulata* var. *crenulata*

Var. *angustifolia* Atwood Salt desert shrub community at 1220 to 1830 m in eastern Kane County; endemic; 13 (vii).

Var. *corrugata* (A. Nels.) Brand [*P. corrugata* A. Nels.]. Salt and cool desert shrub, pinyon-juniper, and mountain brush communities at 1065 to 2595 m in Beaver, Carbon, Duchesne, Emery, Garfield, Grand, Juab, Kane, Millard, San Juan, Tooele, Uintah, and Wayne counties; Nevada, Colorado, Arizona, and New Mexico; 167 (xli). This variety intergrades with var. *crenulata*, but is generally distinctive. It is replaced in Washington County by *P. ambigua* Jones.

Var. *crenulata* [*P. orbicularis* Rydb., type from the Henry Mts.]. Salt and cool desert shrub and pinyon-juniper communities in Beaver, Emery, Garfield, Piute, and Sevier counties; California, Arizona, and Nevada; 28 (viii).

Phacelia cryptantha Greene Annual, 1-4 dm tall; stems simple or branched, glandular-puberulent and hispid, at least above; leaves pinnately compound, reduced above, the leaflets toothed; inflorescence of terminal cymes, to 5 cm in fruit; sepals 4-7 mm long in flower, to 10 mm long in fruit, linear-oblanceolate, with spreading, hispid hairs and densely puberulent to glandular; corolla lavender, tubular-campanulate, 4-7 mm long; pedicels slender and evident, especially in fruit, 2-4 mm long, densely retrorsely pubescent; stamens included, 4-7 mm long; style 3-4 mm long, pubescent below; capsule globose, 2-2.5 mm long, densely pubescent with short and long hairs; 2n = 22. Joshua tree, creosote bush, other warm desert shrub, chaparral, and pinyon-juniper communities at 760 to 1220 m in Washington County; Arizona, Nevada, and California; 4 (i).

Phacelia curvipes Torr. ex Wats. Annual, diffusely branched from the base, the branches spreading to ascending, hirsute, hirsutulous, and sometimes shortly glandular above, 3-15 cm tall; leaves mostly basal, oblong to oblong-ovate, mostly entire, occasionally with 1-3 lobes, the lower the larger and definitely petiolate, the upper smaller and with short petioles; cymes laxly few- to many-flowered, pedicellate, the pedicels short in flower, to 1.5 cm long in fruit, spreading or deflexed; sepals linear to linear-oblanceolate, 3-10 mm long, hirsute, somewhat unequal; corolla lavender to pale bluish, the throat yellowish, open campanulate, 3.5-6 mm long, deciduous; stamens and style included, 2-5 mm long, sparingly hairy; capsules 4-5 mm long, ovoid, flattened; seeds 6-17, oblong, pitted, 0.8-1.2 mm long. Warm desert shrub, pinyon-juniper, and mountain brush communities at 1350 to 2170 m in Washington County; Nevada, California, and Arizona; 8 (v).

Phacelia demissa Gray Brittle Phacelia. Annual, branched from the base (except in depauperate plants), mostly broader than high, 0.3-2 dm tall; stems brittle and succulent when fresh, densely glandular villous, usually reddish; leaves mostly cauline, broadly ovate to orbicular, 1-2.6 cm long, entire to undulate, the lower ones long-petiolate, the upper short-petiolate; inflorescence of terminal or axillary, sessile scorpioid cymes; sepals oblong to oblanceolate, 5-7 mm long in fruit; corolla lavender to purple with a yellow tube, easily deciduous, 3.5-9 mm long, tubular-campanulate when fully open, sessile to short-pedicellate; stamens included; style included, hairy, shortly bifid; capsule oblong, 2.5-4 mm long, finely pubescent; ovules 4-20; seeds ovate to elliptic, 1.1-1.5 mm long, brown to reddish, pitted-reticulate; 2n = 24. Two varieties are known.

1. Plants densely glandular-villous to glandular-puberulent throughout, averaging 7-10 (20) cm tall, diffusely branched; corolla tube pale yellow; sepals equaling or somewhat surpassing the capsule, but not twice as long; seeds mostly 10-16, 1.1-1.3 mm long ....
   .... *P. demissa* var. *demissa*

— Plant finely but not densely glandular-puberlent throughout, the leaves mostly glabrous except on the margins, plants mostly less than 5 (10) cm tall and not diffusely branched; corolla tube bright yellow, often more showy than the otherwise colored lobes; sepals twice as long as the capsule; seeds 4–12, 1.2–1.5 mm long .... *P. demissa* var. *minor*

Var. demissa [*P. demissa* var. *heterotricha* J. T. Howell, type from Piute County; *P. nudicaulis* Eastw., type from Moab]. Salt desert shrub, mainly on Mancos Shale, Tropic Shale, and igneous gravels at 1220 to 2200 m in Carbon, Emery, Grand, Kane, Piute, Sevier, and Wayne counties; emdemic; 53 (xviii).

Var. minor Atwood Salt desert shrub community on Morrison, Duchesne River, and Mancos Shale formations at 1430 to 1895 m in Daggett, Duchesne, and Uintah counties; endemic; 19 (iv).

Phacelia franklinii (R. Br.) Gray [*Eutoca franklinii* R. Br.]. Annual or biennial, 1.9–4.4 dm tall, hirsute-puberulent to spreading-hirsute and viscid; stems solitary, or with several subsidiary ones; leaves both basal and cauline, pinnatifid or subpinnatifid, 1.5–7 cm long, petiolate, especially the lower; inflorescence terminal on the stems and branches, scattered along the plant, but concentrated upward; corolla purplish, broadly campanulate, 6–9 mm long and somewhat wider, pubescent externally; stamens exserted ca 2 mm; style exserted, cleft to just above the middle; ovules 26–46; seeds ca 17, 1–1.3 mm long, dark brown to blackish, pitted-reticulate; 2n = 22. Sagebrush, aspen, and spruce-fir communities at 2315 to 2715 m in Wasatch counties; Alaska and Yukon, south to Idaho and Wyoming; 10 (ii).

Phacelia fremontii Torr. in Ives Annual, 0.4–2.6 dm tall; stems 1 to several from the base, nearly erect to ascending, puberulent to hirsutulous, retrorsely hairy below, somewhat glandular above; leaves pinnately divided, mostly basal, petiolate, the blade oblong to elliptic, 2–6 cm long; flowers in terminal cymes, the flowers pedicellate, the pedicel 1–4 mm long; sepals linear to oblanceolate, 4–6 mm long in flower and to 9 mm in fruit, longer than the capsule; corolla lobes lavender, violet to blue (rarely white), the tube bright yellow, tubular-funnelform to funnelform-campanulate, 6–16 mm long; stamens included, unequal, 4–6 mm long; style included, 3.5–5.5 mm long, pubescent to near the middle; ovules ca 12–40, oblong to obovate, ca 1 mm long, angular, brown, corrugated and alveolate-reticulate; 2n = 26. Joshua tree, creosote bush, chaparral, and pinyon-juniper communities at 700 to 1375 m in Washington County; Arizona, Nevada, and California; 34 (vi).

Phacelia glandulosa Nutt. [*Eutoca glandulosa* Hook.]. Biennial herbs, 0.6–3.6 dm tall; stems simple or branched, erect, stout, densely stipitate-glandular and spreading hirsute; leaves lanceolate to oblong in outline, pinnatifid, 1–7 (10) cm long, 0.5–3 cm wide, glandular and densely hirsute, the lower petiolate, the upper subsessile; inflorescence of congested terminal cymes, these stipitate-glandular and hirsute, the cymes to 10 cm in fruit; sepals elliptical to oblanceolate, 3–4 mm long, 1.2–1.4 mm wide; corolla blue to purplish, 5–7 mm long and somewhat broader, the lobes pubescent and more or less crenate; stamens and style exserted 5–9 mm, the style bifid 3/4 its length, the lower 1/4 pubescent; capsule subglobose, 3.5–4 mm long, 3.2–3.3 mm wide, glandular and hirsute; mature seeds elliptic to oblong, brown, 2.4–3.3 mm long, 1.1–1.4 mm wide, pitted, the ventral surface excavated on both sides of the ridge. Salt and mixed desert shrub and pinyon-juniper communities on the Green River Formation at 1525 to 1925 m in Grand (?) and Uintah counties; Idaho and Montana, south to Colorado; 4 (0).

Phacelia hastata Dougl. ex Lehm. [*P. leucophylla* Torr. in Frem., *P. hastata* var. *leucophylla* (Torr.) Cronq.; *P. alpina* Rydb.; *P. heterophylla* var. *alpina* (Rydb.) A. Nels. in Coult. & Nels.]. Perennial, 2–4.5 dm tall; stems 1 to several, erect to ascending, strigose and hispid; leaves entire or sometimes with 1–2 lateral pairs of lobes near the base, lanceolate to ovate or oblanceolate, in a well-developed basal rosette; inflorescence of scorpioid cymes, simple or branched; sepals elliptic to linear-lanceolate, 3–6 mm long, strigose and hirsute to hispid; corolla campanulate, white to lavender or purple, 4–6 mm long; stamens exserted, filaments hairy at the middle; style exserted, pubescent at the base, bifid ca 1/3 the length; capsule ovoid, ca 3 mm long; seeds 1–2, oblong, 2–2.6 mm long, brown; n = 11. Sagebrush, mountain brush, ponderosa pine, Douglas fir, aspen, spruce-fir, and grass-forb communities at 1340 to 3510 m in all Utah counties; British Columbia and Alberta to California, Colorado, and Nebraska; 115 (ix). Segregation of our material as var. *alpina* (Rydb.) Cronq. can be applied only arbitrarily, and it seems best to treat all Utah specimens as portions of a polymorphic **var. hastata**.

Phacelia heterophylla Pursh Biennial or perennial, 2–6 dm tall; stems simple, sometimes branched, erect, stout, strigose to hispid, not silvery but sometimes also glandular; leaves basal and cauline, the lower petiolate, entire or with 1–4 pairs of lobes; inflorescence somewhat virgate, typically branched, densely pilose to spreading-hispid; sepals lanceolate to oblong, 3–6 mm long, unequal; corolla dirty white to purplish, campanulate, 4–7 mm long; capsule ovoid, 2–3 mm long, pubescent; seeds 1 or 2, 2–2.5 mm long, brown; n = 11. Sagebrush, mountain brush, aspen, and spruce-fir communities at 1370 to 3295 m in all Utah counties except, Davis, Daggett, Morgan, Rich, and Wayne; Washington to Montana south to Mexico; 72 (vi).

Phacelia howelliana Atwood Annual, 0.9–2.3 dm tall; stems mostly branched and leafy at the base, glandular and hirsute; leaves oblong to oval, 2–6 cm long, 1–2.5 cm wide, irregularly crenate to lobed, strigose and slightly glandular, the petiole to 5 cm long; inflorescence of branched scorpioid cymes; pedicels to 2 mm long; sepals linear to narrowly oblanceolate, 3.5–4.5 mm long, 1–1.2 mm wide, glandular and hirsute; corolla 5–6 mm long, 6–7 mm wide, rotate to funnelform, the lobes pale violet to blue, the tube white; stamens and style exserted 3–4 mm, the style shorter than the stamens, bifid 3/4 its length, the lower 1/4 pubescent; capsule oblong to subglobose, glandular and hirsutulous, especially near the apex; seeds 4, brown, 3.2–4 mm long, 1.4–1.8 mm wide, elliptic, the margins corrugated, involute to flattened, the ventral surface pitted, excavated and divided by a prominent ridge, this sometimes curved to one side and barely corrugated, the dorsal surface reddish brown, smooth and surrounded by a lighter margin. Salt and warm desert shrub and pinyon-juniper communities at 1125 to 1525 m in Grand, San Juan (type from Bluff), and Wayne counties; Arizona; a Plateau endemic; 10 (v).

Phacelia incana Brand Annual, 5–15 cm tall; stems simple to more often branching at the base or throughout, glandular-villous, the hairs spreading; leaves mostly cauline, elliptic to ovate, entire, 1–1.4 cm long, 3–8 (13) mm wide, petiolate; flowers in terminal, elongate cymes, laxly few-flowered; sepals linear to spatulate or oblanceolate, 3.5–6 mm long; corolla white to pale lavender, small, 3.5–4.5 mm long, tubular-campanulate; stamens included, the filaments hairy at base; style included, 1–1.7 mm long, shortly bifid, pubescent; capsule oblong, 3–4 mm long, shorter than the calyx; seeds 12–26, angular, 0.8–0.9 mm long, brown. Salt and cool desert shrub, sagebrush, and pinyon-juniper communities at 1400 to 2320 m in Beaver, Juab, Kane, Millard, Sanpete, Tooele (type from Dugway), and Uintah counties; Idaho, Nevada, Wyoming, and Colorado; 13 (i).

Phacelia indecora J. T. Howell Bluff Phacelia. Annual, 3–14 cm tall; stems erect to spreading, branched, glandular; leaves elliptic to oblong, 4–26 mm long, hirsutulous and glandular; sepals oblanceolate, 3–5 mm long; corolla narrowly campanulate, pale blue, 3–4 mm long, the lobes pubescent, the tube pale yellow and streaked with blue lines; capsule elliptic, 3–4 mm long; seeds ca 40. Salt desert shrub community at ca 1370 m in San Juan (type from Bluff) County; endemic; 1 (0). Some have contended that this is only a depauperate specimen of *P. lemmonii* (q.v), but the characters outlined by Howell, a student of the genus, seem more important than reduction to synonymy would indicate. The continued recognition here points attention to the ellusive Bluff plant and future workers can search for it.

Phacelia integrifolia Torr. Annual or winter annual, 1–6 dm tall; stems simple or somewhat branched, densely stipitate-glandular and spreading-hirsute; leaves simple, crenate to somewhat cleft, oblong to ovate or lanceolate, strigose, finely glandular and with spreading, setose hairs, the lower leaves the largest, petiolate, the upper smaller, short-petiolate to sessile; inflorescence terminal, elongate; sepals oblanceolate to elliptic, 3–4.5 mm long in flower, 4.4–6.5 mm long in fruit; corolla lavender to bluish, 5–6.5 mm long and broad, tubular-campanulate; stamens and style exserted 5–6 mm, the style bifid 2/3–3/4 its length, puberulent below; capsule ovoid to globose, 3.2–5.3 mm long, pubescent; mature seeds 4, oblong to elliptic, dark brown to black, 3.1–4.5 mm long, 1.7–2.2 mm wide, the transverse ridges on the dorsal surface quite distinct, the ventral surface lacking corrugations, the ridge often curved to one side; n = 22. Blackbrush, other warm desert shrub and shrub-grass, and mountain brush communities at 1190 to 1710 m in Kane and San Juan counties; Arizona to Texas and Mexico; 15 (vii).

Phacelia ivesiana Torr. in Ives Annual, 0.4–2.7 dm tall, hirsutulous and mostly finely glandular (rarely dark or black); stems prostrate or ascending, branched; leaves pinnately divided or lobed, oblong to lanceolate, 1–5 cm long, basal and cauline; inflorescence of laxly flowered terminal scorpioid cymes; sepals oblong to oblanceolate, 2–4 mm long in flower, elongating to 7 mm in fruit; corolla white, inconspicuous, 2–4 mm long, the tube yellowish; stamens included, the filaments glabrous; style included, divided 1/4 the length, glabrous; capsule oblong, 3–4.5 mm long, pubescent, especially apically; seeds 8–15, brown, 1–1.5 mm long, corrugated transversely and alveolate; n = 22. Two varieties are treated.

1. Style 2–3 mm long; leaves subpinnatifid; staminal filaments sparingly hairy; corolla usually 4–6.5 mm long, the lobes lavender; plants of Box Elder County .... *P. ivesiana* var. *glandulifera*
— Style 0.7–2 mm long; leaves merely pinnatifid, rarely subpinnatifid; staminal filaments glabrous; corolla mostly 2.5–4 mm long, the lobes white; plants widespread ... . *P. ivesiana* var. *ivesiana*

Var. *glandulifera* A. Nels. [*P. glandulifera* (A. Nels.) Piper; *P. ivesiana* f. *glandulifera* (A. Nels.) Brand]. Sagebrush and juniper communities, often of tuffaceous outcrops, in Box Elder County; Washington to Wyoming, south to California and Nevada; 1 (i).

Var. *ivesiana* Creosote bush, other warm desert shrub, cool desert shrub, sagebrush, and pinyon-juniper communities at 760 to 2410 m in all Utah counties, except Box Elder, Cache, Davis, Morgan, Salt Lake, Summit, Utah, and Wasatch; California to Wyoming, south to New Mexico and Arizona; 96 (xviii).

Phacelia laxiflora J. T. Howell [*P. perityloides* var. *laxiflora* (J. T. Howell) Cronq.]. Perennial, with branched, often trailing stems, the stems 4–20 cm long, glandular-puberulent; leaves orbicular-reniform, 1–3 cm long, 0.5–3.5 cm wide, glandular-hirsutulous, dentate to shallowly lobed, cordate basally, petiolate, the petiole 1–4 cm long; flowers 5–merous on laxly flowered racemes, the pedicels 1–7 mm long; sepals 2–4 mm long in flower, elongating to 8 mm long in fruit; corolla tubular-campanulate, 6–8 mm long, white; stamens included, somewhat unequal; style 4 mm long, glabrous except at the ovary base; ovules numerous; mature seeds not developed. North exposures on limestone cliffs in crevices and at bases of cliffs in shaded areas along the Virgin River Gorge in Washington County at 965 m; adjacent Mohave County, Arizona; 2 (ii).

Phacelia lemmonii Gray Annual, 0.5–2.5 dm tall; stems erect, glandular-pubescent, little branched; leaves cauline or basal, petiolate below to sessile above, elliptic to ovate, entire to toothed; cymes terminal on the main stems and branches; flowers short-pedicellate, the pedicels stout, ca 1 mm long in flower and 3–5 mm long in fruit; sepals linear-oblong to spatulate, 2.5–4 mm wide; anthers included; style hardly divided, 1.5–3 mm long; capsule ellipsoid-cylindric; seeds ca 35–60, ovoid to elliptic, 0.4–0.5 mm long, coarsely pitted; n = 44, 48. Pinyon-juniper community (an understory species) or in moist sites, at 1830 to 2105 m in Beaver and Kane counties; Arizona, Nevada, and California; 3 (i).

Phacelia linearis (Pursh) Holz. [*Eutoca heterophylla* Torr., type from Valley of the Great Salt Lake]. Annual, 1–5 dm tall; stems erect, simple or branched, densely pubescent, the hairs minute; leaves linear to lanceolate, 2–7 cm long, entire or with 1–4 lobes, the cauline short-petiolate below to sessile above; inflorescence of cymes from the leaf axils, mostly above the stem middle, the cymes many-flowered and crowded; sepals linear to oblanceolate, 3–6 mm long in flower, to 12 mm long in fruit, united basally, with spreading stiff hairs marginally; corolla bluish to purple or white, broadly campanulate, showy, 8–10 mm long, 8–17 mm wide; stamens and styles included or exserted 1–2 mm, the filaments hairy; style bifid 1/3 the length, pubescent; capsule ovoid to oblong, 5–7 mm long; seeds mostly 6–15, oval to oblong, 1.4–1.6 mm long, dark brown to black, pitted; 2n = 22. Sagebrush, mountain brush, and pinyon-juniper communities at 1370 to 2440 m in Beaver, Box Elder, Cache, Davis, Juab, Kane, Millard, Salt Lake, Sanpete, Sevier, Summit, Tooele, Utah, Wasatch, and Washington counties; British Columbia and Alberta, south to California, Nevada, and Wyoming; 52 (i).

Phacelia lutea (H. & A.) J. T. Howell [*Eutoca lutea* H. & A.; *Miltitzia lutea* (H. & A.) DC.; *Emmenanthe lutea* (H. & A.) Gray; *E. foliosa* Jones, ;type from Deep Creek, Tooele (?) County]. *M. foliosa* (Jones) Brand; *P. foliosa* (Jones) Phil.; *E. salina* A. Nels; *M. salina* (A. Nels.) Rydb.; *P. salina* (A. Nels.) J. T. Howell]. Annual from a slender taproot, 2–10 cm tall; stems 1 to several, hirsutulous to finely glandular; leaves oblong, elliptic to oblanceolate, 0.5–3 cm long, entire to pinnately lobed; inflorescence of compact to laxly flowered cymes; sepals linear to oblanceolate, 3–7 mm long, hirsutulous; corolla yellow (sometimes tinged with lavender), 3–5 mm long, tubular-campanulate; stamens and style included, the filaments glabrous; style hairy below; capsule elliptic, oval, or oblong, pubescent, 3–6 mm long; seeds 7–20, 1–2 mm long, corrugated, pitted, brown; 2n = 24. Salt and mixed desert shrub communities, often in saline substrates, at 1370 to 1680 m in Sanpete and Tooele counties; Oregon to Montana, south to California, Arizona, and Colorado; 2 (0). Our plants belong to var. **scopulina** (A. Nels.) Cronq. [*Emmenanthe scopulina* A. Nels.; *M. lutea* var. *scopulina* (A. Nels) Brand; *M. scopulina* (A. Nels.) Rydb.; *P. scopulina* (A. Nels.) J. T. Howell].

Phacelia mammillarensis Atwood Nipple Phacelia. Annual, 0.9–5 dm tall; stems erect, simple or branched below, stout, yellowish or green, densely stipitate-glandular; leaves simple, oblong to lanceolate, irregularly crenate to dentate, 1–7 cm long, 0.5–3 cm wide, stipitate-glandular, spreading-setose and strigulous; lower leaves petiolate to sessile upward; inflorescence of terminal or lateral, compact, scorpioid cymes, these stipitate-glandular, puberulent, and spreading-hispid; sepals elliptic to oblanceolate, 4–6 mm long, 1–2 mm wide; corolla tubular-funnelform, the lobes pale blue to white, 5–8 mm long; stamens and style exserted 5–10 mm, the anthers lavender, the style divided ca 1/2 the length, the lower 1/4 pubescent; capsule subglobose, 4–5 mm long, pubescent; seeds 4, 3 mm long, 1.5 mm wide, brown, pitted dorsally, the ventral surface pitted, excavated and divided by a prominent ridge, one side of the ridge and the margins corrugated. Salt and mixed desert shrub communities at 1220 to 1830 m in eastern Kane (type locality) and Garfield counties; endemic; 18 (xiii).

Phacelia palmeri Torr. ex Wats. Palmer Phacelia. [*P. foetida* Gooding, type from Diamond Valley; *P. palmeri* var. *foetida* (Gooding) Brand; *P. integrifolia* var. *palmeri* (Torr.) Gray]. Robust biennial, 2–9 dm tall, viscid, mephitic; stems usually solitary or branched from the base, erect, wandlike, densely stipitate-glandular, hirsute and pilose (hairs heteromorphic); leaves basal (largest) and with well-developed cauline leaves, oblong to lanceolate, irregularly sinuate, crenate, or pinnatifid, short-petiolate; inflorescence dense, with many short, scorpioid cymes along the stems; sepals oblong to spatulate, 4–5 mm long, 1–1.8 mm wide, glandular and with spreading hirsute hairs; corolla tubular, pale lavender or white, 5–7 mm long; stamens and style exserted, the style bifid 2/3 the length, pubescent below; ovules 4; seeds 4, elliptic, black, 2.5 mm long, 1.5 mm wide, excavated on both sides of the prominent ridge, this corrugated on one side, pitted, the margins and furrows or grooves corrugated (in part), the dorsal surface longitudinally pitted and transversely ridged; 2n = 22. Salt desert shrub community, mainly on volcanic debris and the Moenkopi Formation, at 790 to 1285 m in Washington County (type locality); Nevada and Arizona; a Virgin-Colorado endemic; 19 (viii).

Phacelia peirsoniana J. T. Howell Annual, 1–2.5 dm tall; stems simple or more often branched, glandular-villous; leaves simple, mostly cauline, the blade crenate to dentate, 1–5 cm long, cordate to truncate, petiolate; inflorescence of laxly flowered, terminal cymes; sepals linear-oblanceoolate to spatulate, 3–4 mm long in flower, strongly accrescent and to 8 mm long in fruit; corolla tubular-campanulate, lavender or violet, 4–6 mm long, 3–4 mm wide; stamens and style included, the filaments hairy below; style 2–3 mm long; capsule oblong, 4–6 mm long; seeds ca 4, ovoid or angular, 1–1.3 mm long, pitted-reticulate; 2n = 24. Sagebrush and pinyon-juniper community at ca 1370 m in upper Beaverdam Wash, Washington County; Nevada and California; 1 (0).

Phacelia pulchella Gray [*P. pulchella* f. *luteola* Brand]. Annual, branched from the base, 0.5–2 dm tall; stems spreading to erect, leafy, succulent and brittle, finely glandular; leaves cauline, entire to toothed, oblong to orbicular, 1–3 cm long, 1–2 cm wide; inflorescence terminal, laxly flowered below to compact apically; sepals oblanceolate to spatulate, 3–5 mm long in flower, to 8 mm long in fruit; corolla violet to purple, the tube yellow, campanulate, 6–15 mm long and as broad, deciduous; stamens included; style included, bifid apically, 3.5–5 mm long; capsule oblong to ovoid, 3–5 mm long, hirsutulous above; seeds 28–50, oblong to elliptic, 0.5–1 mm long, brown, pitted; 2n = 24. Two varieties are present.

1. Leaves mostly entire or repand; corolla lobes purple; seeds mostly less than 35 per capsule; plants of western Kane and Washington counties . *P. pulchella* var. *pulchella*

— Leaves commonly crenate to dentate; corolla lobes of paler color, more violet; seeds mainly over 35 per capsule; plants of eastern Kane and Garfield counties ....
.... *P. pulchella* var. *sabulonum*

Var. *pulchella* Creosote brush, blackbrush, other warm desert shrub, and pinyon-juniper communities at 850 to 1830 m in Kane and Washington (type locality) counties; Arizona; 21 (viii). Some plants in Washington county are intermediate to var. *gooddingii* (Brand) J. T. Howell.

Var. *sabulonum* J. T. Howell Mixed desert shrub communities at 1155 to 1375 m in eastern Garfield and Kane (type locality) counties; endemic; 12 (viii).

Phacelia rafaelensis Atwood Erect biennial herbs, 0.8–5.4 dm tall; stems stout, simple or sometimes branched at the base, olive-green to brownish, glandular and hirsute; basal leaves petiolate, dentate, crenate, or pinnatifid, 2–7 cm long, 5–15 mm wide, strigose to hirsute, the cauline ones sessile, undulate to crenate or dentate, oblong-lanceolate, 1–10 cm long, 0.5–3.5 cm wide, strigose to hirsute and sparsely stipitate-glandular; inflorescence mainly terminal, paniculate, some axillary; flowers almost sessile; sepals oblanceolate to spatulate, 3–4 mm long in flower, 5–6 mm long in fruit, 1–1.7 mm wide, glandular and hirsute; corolla tubular, pale and grooved, with the lobes somewhat spreading, 5–6 mm long; stamens and style exserted 3–5 mm, the anthers dull; style bifid 3/4 its length, the lower half pubescent; capsule subglobose, 4–5 mm long, stipitate-glandular and hirsute; mature seeds 4, elliptic to oblong, 3.5–4 mm long, 1.5–2 mm wide, the ventral surface alveolate, paler than the dorsal surface, excavated on both sides of a prominent ridge, this sometimes corrugated along one side, the margins usually entire, the dorsal surface brown and less deeply pitted, the surface often smoothish. Salt desert shrub and pinyon-juniper communities, mainly on the Carmel, Chinle, Summerville, and Moenkopi (Sinbad member) formations at 1065 to 1830 m in Emery, Wayne (type locality), and Washington counties; Arizona; a Plateau endemic; 38 (xxiv). The Washington county plants are allopatric with *P. palmeri*.

Phacelia rotundifolia Torr. ex Wats. Limestone Phacelia. Annual, 0.3–2 dm tall; stems simple or more commonly branched from below, stipitate-glandular and spreading hirsute; leaves rotund, mostly if not all cauline, coarsely toothed, 4–21 mm long and broad, petiolate, the petiole to 4 cm long, the lower ones longer than the blade; inflorescence of terminal or axillary cymes; sepals linear, 3–6 mm long, 1 mm wide; corolla pale lavender or white, tubular, inconspicuous; stamens included, the filaments glabrous; style included, bifid; capsule oblong, shorter than the calyx, 4 mm long; ovules numerous, sometimes more than 100; seeds 0.4–0.5 mm long, brown, pitted; $2n = 24$. Mixed desert shrub communities at 850 to 1525 m in Kane and Washington (probable type locality) counties; Arizona, Nevada, and California; 14 (iii).

Phacelia sericea (Graham) Gray [*Eutoca sericea* Graham]. Stout, erect perennial, 1–5 dm tall; stems 1 to several, the pubescence silky, the hairs appressed to spreading; leaves pinnatifid, the segments also cleft or entire, the lower well developed, reduced upward; inflorescence a dense terminal thyrse, usually elongate, the cymes many, compact, short; corolla dark purple or blue, persistent, campanulate, 5–7 mm long and as broad, pubescent; stamens long-exserted, the filaments hairy at the base; style long-exserted, bifid 1/2 the length; capsule ovoid, 4–6.5 mm long, pubescent; ovules 20–40; seeds 10–20, oblong, 1–2 mm long, brown to black, reticulate and grooved longitudinally; $2n = 22$. Sagebrush, aspen, and spruce-fir communities at 2285 to 3540 m in Beaver, Cache, Carbon, Daggett, Duchesne, Emery, Garfield, Grand, Juab, Piute, Rich, San Juan, Sanpete, Sevier, Summit, Tooele, Uintah, and Utah counties; Alaska south to Nevada and Colorado; 80 (v). Our plants belong to var. *ciliosa* Rydb, but var. *sericea* has been reported from the La Sal Mts.

Phacelia tetramera J. T. Howell Diminutive annual, freely branched, prostrate or forming mats, spreading-hirsutulous; leaves oblanceolate, petiolate, the blade 0.5–2 cm long, mostly entire or some leaves toothed; cymes numerous, terminal, or in upper leaf axils, few-flowered, the flowers 4–merous (rarely some 5–merous), short-pedicellate to sessile; corolla white, persistent, 1–1.5 mm long, 1/2 as long as the sepals; stamens included; style shorter than the stamens, 0.3–0.4 mm long, slightly lobed; seeds 6–10, 0.7–1 mm long, pitted, reticulate, and corrugated; $2n = 22$. Samphire community, where definitely saline, at ca 1310 m in Weber County; Nevada, Oregon, and California; 1 (0).

Phacelia utahensis Voss Stout, erect biennials, 0.8–5.8 dm tall; stems mostly solitary, brownish to yellowish, densely stipitate-glandular, finely pubescent and with some spreading, non-glandular hairs; leaves well developed, both basal and cauline, crowded, short-petiolate below, but most leaves sessile, especially upward, the upper also auriculate, the largest to 13 cm long, crenate to pinnately lobed; inflorescence of many, densely flowered cymes, elongated on the upper portion of stems and branches; sepals linear-oblanceolate, 3–4 mm long, ca 1 mm wide; corolla showy, bluish lavender, the tube yellowish, broadly campanulate, 4–5 mm long, ca 6 mm wide; stamens exserted 9–10 mm; filaments violet, the anthers yellow; style exserted ca 10 mm, bifid 3–4 the length, setose and glandular below; capsule globose, 3.5–4.1 mm long, 2.6–3.5 mm wide, glandular and setose; mature seeds 4, elliptical, dark (reddish), 3–3.5 mm long, the dorsal surface faintly pitted, the ventral surface excavated on both sides of the ridge, often paler than the dorsal surface, pitted with the markings in excavations longer (transversely) than those of the ridge or margins, the ridge sometimes faintly corrugated on one side. Salt desert shrub community on the Arapien Shale Formation at 1675 to 1740 m in Sanpete (type locality) and Sevier counties, and probably on the Mancos Shale Formation in Carbon County; endemic; 17 (vi).

Phacelia vallis-mortae Voss Annual, 2–5 dm tall; stems diffusely branched, weak, supported by shrubs, spreading-hispid and glandular above; leaves pinnately divided, evenly distributed from base upward, somewhat reduced above, to 6 cm long, spreading-hispid; cymes terminal on the main stem and branches, densely flowered, the pubescence dense and coarse, of hispid and glandular hairs; sepals linear-oblanceolate, 4–6 mm long in flower, to 13 mm in fruit, hispid, several times longer than the capsule; corolla lavender to violet, broadly campanulate, 7–12 mm long and broad; stamens barely included to slightly exserted (1–2 mm); style bifid to the middle; capsule globose, 3–4 mm long; seeds mostly 4, ovoid, 2.5–3 mm long, brown, pitted; $2n = 22$. Joshua tree, creosote bush, and other warm desert shrub, chaparral, and pinyon-juniper communities at 790 to 1220 m in Washington County; Arizona, Nevada, and California; 17 (vii).

Tricardia Torr. ex Wats.

Perennial from a taproot; stems branched from the base; leaves mostly basal, entire; cauline leaves alternate; flowers in short, naked, terminal cymes; calyx cleft almost to the base, the sepals heteromorphic, the three outer lobes cordate and enlarged, veiny in fruit, the 2 inner ones narrow; corolla broadly urceolate-campanulate, white and purple; stamens equally inserted on the corolla tube, included, the filaments unequal and with a pair of appendages; style included, bifid; ovules 8; seeds 4–8.

Tricardia watsonii Torr. ex Wats. Stems 1 to several from a stout caudex, erect to ascending, 0.5–4 dm long, loosely villous when young, finally glabrate; leaves mostly basal, elliptic to elliptic-ovate, 2–9 cm long, petiolate, the cauline ones smaller and with shorter petioles, ovate to lanceolate; inflorescence of terminal cymes; sepals very unequal, the outer 4–9 mm long in flower, 10–25 mm long in fruit; corolla white with a purple ring around the throat and onto the lobes, 5–8 mm long; capsule compressed, 7–10 mm long; seeds 4–8, brown, 3–5 mm long, oblong, pitted; $2n = 16$. Joshua tree, creosote bush, other warm and cool desert shrub communities at 730 to 1100 m in Washington County; Arizona, Nevada, and California; 8 (iii).

JUGLANDACEAE A. Richard ex Kunth

Walnut Family

Trees with alternate pinnately compound leaves; flowers monoecious, the staminate in long drooping catkins, the pistillate solitary or few in a cluster; staminate calyx 2- to 5-lobed, the stamens 3-40; pistillate flower bracteate and usually with 2 bracteoles, the calyx 4-lobed and with 4 small petals; pistil 1, the ovary inferior, 1- or incompletely 2- to 4-loculed; styles 2; fruit drupaceous; x = 16.

1. Pith transversely partitioned; husk of fruit indehiscent; leaflets at midleaf the largest .... *Juglans*
— Pith not paritioned; husk of fruit splitting and deciduous; leaflets at leaf apex the largest .... *Carya*

Carya L.

Trees; pith continuous; leaves odd-pinnate, the terminal 3 the largest; staminate catkins borne in peduncled groups of 3; staminate calyx 2- or 3-lobed, subtended by a bract; stamens 3-10; pistillate flowers solitary or few to several, each subtended by a 4-lobed perianthlike involucre; fruit with dehiscent exocarp.

Carya illinoensis (Wangenh.) K. Koch Pecan. [*Juglans illinoensis* Wangenh.]. Trees with spreading branches and rounded crowns; bark deeply furrowed; leaflets 11-17, oblong-lanceolate, the lateral ones falcate, the terminal on a petiolule ca 2-4 mm long; fruits in spikes of 3-10, ellipsoid or cylindric; nuts 2-5 cm long, brown, smooth, short-pointed, the kernel edible. Commonly grown shade and nut tree in Washington County and less common elsewhere in Utah; introduced from the eastern U. S.; 2 (i). The shagbark hickory, **Carya ovata (Miller) K. Koch**, has been successfully grown in Utah, but is not known in contemporaneous plantings.

Juglans L.

Trees; pith chambered; leaves pinnate, the largest leaflets at about midleaf; staminate flowers in pendulous sessile catkins; stamens 8-40, closely subtended by a bract; bracteoles 2; sepals usually 4; pistillate flowers solitary or several in clusters, each with a bract and 2 bracteoles; sepals 4; fruit with an indehiscent exocarp.

1. Leaflets entire or nearly so, usually 5-9; bark whitish, often smooth .... *J. regia*
— Leaflets serrate or crenate-serrate, usually 9-19 (23); bark blackish, often furrowed 2
2(1). Pith dark brown; bark with smooth ridges; fruit ovoid to ellipsoid, usually longer than broad .... *J. cinerea*
— Pith light brown; bark with rough ridges; fruit subglobose, little if any longer than broad 3
3(2). Fruit more than 35 mm in diameter; leaflets 11-23; trees commonly grown and widespread .... *J. nigra*
— Fruit less than 35 mm in diameter; leaflets mainly 9-13; trees uncommon in Washington County .... *J. major*

Juglans cinerea L. Butternut. Trees to 20 m or more; bark grayish brown to black, with smooth ridges; pith dark brown; leaflets 11-19, lanceolate to elliptic, long-acuminate; fruit ellipsoid-oblong, 4-7 cm long, tapering to the apex; endocarp ellipsoid to ovoid, longer than thick. Sparingly grown shade and nut tree in Utah; introduced from the eastern U. S.; 4 (0).

Juglans major (Torr.) Heller Arizona Walnut. [*J. rupestris* var. *major* Torr.]. Trees to 15 m tall or more; bark grayish to brown, furrowed on older stems; pith light brown; leaflets usually 9-13, rarely more, lanceolate, acuminate apically, coarsely dentate-serrate; fruit subglobose, 2.5-3.5 cm thick; endocarp about as long as broad, with a small edible kernel. Cultivated shade and ornamental tree in Washington County; introduced from the southwestern U. S.; 4 (ii). This species is known from a grove along Beaver Dam Wash, where it was probably initially planted, but is now established.

Juglans nigra L. Black Walnut. Trees to 25 m or more; bark black, the ridges rough; leaflets 11-23, lanceolate to oblong-lanceolate, serrate; fruit subglobose, mostly 4-8 cm thick, abruptly pointed; endocarp subglobose, about as long as thick; 2n = 32. Widely and commonly grown shade and nut tree, persistent and escaping in much of lower elevation Utah; introduced from the eastern U. S.; 13 (i). The wood is considered to be the premier cabinet wood in North America. It finishes easily and well, and is in great demand.

Juglans regia L. English Walnut. Trees to 15 m or more; bark whitish, smooth or furrowed in age; leaflets 5-9, ellipsoid to oblong-lanceolate, entire; fruit subglobose, mostly 4-9 cm thick; endocarp subglobose, with prominent ridges margining the sutures, the shell moderately thin and the edible kernel large; 2n = 32. Rather widely grown nut tree in Utah; persisting but not known to escape; introduced from Eurasia; 10 (0).

KRAMERIACEAE Dumort.

Ratany Family

Shrubs with divaricate branches; herbage grayish pubescent; leaves alternate, simple, entire, estipulate; flowers perfect, irregular, solitary, axillary; pedicels usually with 2 opposite foliaceous bracts; sepals 4 or 5, unequal; petals 5, the upper 3 long-clawed, distinct or partially connate and often purplish in color, the 2 others broad, thick, sessile, usually greenish and glandlike; stamens 4, free or adnate to claw of upper petal, the anthers dehiscent by pores; pistil 1, the ovary superior, 1-loculed; style 1; ovules 2; fruit an indehiscent pod, armed with prickles; x = 6.

Krameria L.

A single genus with characteristics of the family.

1. Branchlets modified as thickened thorns 0.8-1.2 mm in diameter at base; spines of fruit barbed at apex only ... . *K. grayi*
— Branchlets not modified as thorns or, if so, less than 0.6 mm in diameter; spines of fruit with barbs scattered or, rarely, barbless .... *K. parviflora*

Krameria grayi Rose & Painter White Ratany. Shrubs, branched, 2.5-6 dm tall, 1-3 mm wide; leaves 5-7 (25) mm long, 1-3 mm wide, lance-ovate to lanceolate, elliptic or oblong, more or less spinulose-tipped, tomentose on both surfaces; pedicels not glandular-pubescent; upper petals 2.5-3.5 mm long, 0.3-0.5 mm wide, yellowish with a purplish tip; sepals 4.5-6.5 mm long, villous-pilose dorsally, pilose to glabrate within, purplish; prickles of the fruit 2-6 mm long at maturity, each with a whorl of barbs at the apex; pods subglobose, 6-10 mm in diameter, hirsute over the surface and on the bases of the prickles. Blackbrush and creosote communities at 670 to 1170 m in western Washington County; California, Nevada, Arizona, New Mexico, Texas, and Mexico; 4 (i).

Krameria parvifolia Benth. Range Ratany. [*K. glandulosa* Rose & Painter; *K. parvifolia* var. *glandulosa* (Rose & Painter) Macbr.; *K. imparata* (Britt.) Macbr.]. Shrubs, intricately branched, 2–6 dm tall and as wide; leaves 3–15 mm long, 0.3–1 mm wide, linear to oblong, callous- to spinulose-tipped, tomentose on both surfaces; pedicels glandular or not; upper petals 2.5–2.8 mm long, retrorsely barbed along the rachis; pods subglobose, 5–9 mm in diameter, pilose-hirsute on the surface. Joshua tree, blackbrush, creosote bush, and bursage communities, 750 to 1070 m in Washington County; California, Nevada, Arizona, New Mexico, Texas, and Mexico; 21 (i). The materials demonstrate variation in glandular condition of pedicels, sepals, and bracts. The variation seems to be haphazard, with little or no correlation with other features or ecology. Hence, included herein as synonyms are those names involved with recognition of glandular and nonglandular phases.

LABIATAE A. L. Juss.

Mint Family

Often aromatic, annual or perennial herbs or shrubs, ordinarily with square or 4–angled stems; leaves simple, opposite or rarely whorled; flowers perfect, mostly irregular, borne in various types of cymose inflorescence; calyx of 5 united sepals, regular or irregular, usually 5–lobed, or the lobes obscure; corolla of 5 united petals, usually bilabiate, 5–lobed, or apparently 4–lobed by fusion of the upper 2 lobes; stamens 4, in 2 unequal pairs, or only 2 by abortion; pistil 1, the ovary superior, 2–carpelled, falsely 4–loculed and 4–lobed; style 1, usually bifid apically; fruit a schizocarp, breaking at maturity into 4, 1–seeded, nutlets; x = 5–11+. [Lamiaceae Lindl].

1. Stamens with functional anthers 2 2
— Stamens with functional anthers 4 6
2(1). Shrubs or herbs of western and southwestern Utah, or widely cultivated; inflorescence a capitate broadly bracteate cyme, or with one or more verticillasters below the terminal one; connective between the anther sacs elongate, only 1 branch usually bearing an anther sac .... *Salvia*
— Shrubs or herbs of various distribution but, if as above, the anther sacs not widely separated by a connective, the flowers variously arranged 3
3(2). Corolla regular or nearly so; plants of palustrine habitats .... *Lycopus*
— Corolla distinctly irregular, 2–lipped; plants not or rarely of palustrine habitats 4
4(3). Plants shrubs; herbage with a feltlike tomentum; plants of sandy sites in southeastern Utah . *Poliomintha*
— Plants herbaceous, annual to perennial, glabrous or variously hairy, but not with a feltlike tomentum; plants not of deep sand, distribution various 5
5(4). Flowers in dense subglobose clusters, these terminal and solitary or several and axillary, forming interrupted verticillasters; calyx teeth about equal. . *Monarda*
— Flowers seldom more than 6, in axillary clusters, but these not dense or subglobose, the clusters usually many; calyx more or less 2–lipped .... *Hedeoma*
6(1). Calyx teeth 10, hooked at apex; stems densely white-woolly .... *Marrubium*
— Calyx teeth 5 or fewer, not hooked apically; stems usually not white-woolly 7

7(6). Calyx 2–lipped, the lips entire 8
— Calyx not 2–lipped, or if so the lips lobed or toothed . 9
8(7). Plants shrubs; calyx inflated in fruit, not crested or gibbous on the back .... *Salazaria*
— Plants herbs; calyx with a gibbous or helmetlike crest on the back .... *Scutellaria*
9(7). Corolla seemingly 1–lipped, or 2–lipped with the upper lip cleft to the base; ovary merely 4–lobed, not cleft to the base, the nutlets laterally attached 10
— Corolla either evidently bilabiate or nearly regular; ovary cleft to the base, the nutlets basally attached . . 11
10(9). Lobes of the upper lip of corolla well separated, displaced onto the lateral margins of the lower lip, the central lobe of lower lip much larger than other 4 lobes .... *Teucrium*
— Lobes of the upper lip of corolla adjacent, only slightly displaced, appearing as three minute lobes .... *Ajuga*
11(9). Corolla regular or nearly so, obscurely 2–lipped 12
— Corolla definitely 2–lipped 13
12(11). Anther sacs parallel; flowers either in terminal spikes or in dense axillary clusters .... *Mentha*
— Anther sacs divergent; flowers in terminal globose clusters .... *Monardella*
13(11). Calyx enlarging into a flaring, veiny funnel ca 2.5 cm wide; plants glabrous .... *Molucella*
— Calyx not especially enlarging, not as above; plants pubescent or glabrous 14
14(13). Bracts spinose-toothed to entire; upper calyx tooth ovate, twice as broad as the others .... *Dracocephalum*
— Bracts, if present, not spinose-toothed; calyx regular, slightly irregular, or if 2–lipped, not as above 15
15(14). Plants low perennials or subshrubs with creeping stems rooting at the nodes; calyx bilabiate; stamens exserted .... *Thymus*
— Plants mostly erect annuals or perennials, usually not rooting at the nodes or the calyx and stamens otherwise 16
16(15). Anther sacs parallel or nearly so 17
— Anther sacs widely divergent or placed end to end . . 19
17(16). Leaves palmately 3– to 5–cleft .... *Leonurus*
— Leaves merely toothed 18
18(17). Leaves ovate to ovate-deltoid, definitely petiolate; flowers many, densely crowded in continuous or interrupted spikes; upper (inner) pair of stamens longer than the lower .... *Agastache*
— Leaves lanceolate to oblong-lanceolate, sessile; flowers in rather loose spikelike racemes, but not very numerous; upper (inner) pair of stamens shorter than the lower .... *Physostegia*
19(16). Flowers, at least some, in clusters in axils of upper leaves, or in a single terminal leafy-bracted cluster . . 20
— Flowers in dense or somewhat interrupted spikes, these leafy-bracteate, if at all, only at the base 22
20(19). Flowers subtended by short, setaeous, conspicuously hirsute-ciliate bracts; stems erect .... *Satureja*
— Flowers without such bracts; stems decumbent 21
21(20). Upper pair of stamens longer than the lower; calyx teeth ca one-third as long as the tube; plants perennial .... *Glecoma*
— Upper pair of stamens shorter than the lower; calyx teeth ca one-half as long as the tube; plants annual or biennial .... *Lamium*

22(19). Calyx definitely 2-lipped; flowers in dense uninterrupted spikes, the bracts ovate to reniform, not at all leaflike .... *Prunella*
— Calyx teeth equal or nearly so; spikes rather interrupted, the bracts rather leaflike 23
23(22). Calyx 5- to 10-veined; lower stamens longer than the upper .... *Stachys*
— Calyx 15-veined; upper stamens longer than the lower .... *Nepeta*

Agastache Clayton

Perennial herbs; stems leafy and tall, arising from creeping rootstocks; leaves ovate or deltoid-ovate, crenate-serrate, petiolate; flowers in dense sessile clusters, these spicate or paniculate; calyx tubular or campanulate, 5-toothed, the teeth subequal or the upper somewhat longer, often whitish or colored other than green; corolla purplish, rose, or whitish, 2-lipped, the upper lip erect, concave, 2-lobed, the lower somewhat spreading and 3-cleft; stamens 4, the anther sacs parallel or nearly so.

1. Upper lip of corolla prominently thrust foreward, 2-2.5 mm long; stamens all parallel, both pair exserted similarly from the corolla tube; plants not definitely known for Utah, but to be sought in Grand and San Juan counties .... *A. pallidiflora* (Heller) Rydb.
— Upper lip of corolla not prominently thrust foreward, 1-1.5 mm long; lower pair of stamens ascending under upper lip, the upper pair thrust down and crossing the lower .... *A. urticifolia*

Agastache urticifolia (Benth.) Kuntze Horse-nettle. [*Lophanthus urticifolius* Benth.]. Stems 6-15 (20) dm tall, glabrous or nearly so; leaves 3.5-8 cm long, mostly ovate or deltoid-ovate, usually glabrous above, the lower surface glabrous to pubescent, green but paler than the upper; inflorescence verticellate, 4-15 cm long, 2-3 cm thick, spikelike or paniculate; calyx 4-10 mm long, green or rose, at least on the teeth and summit of the tube, this 4-7 mm long, the teeth 2.5-5 mm long, deltoid-lanate; corolla white to rose or violet, the tube ca 8-13 mm long, the upper lip 1-1.5 mm long, not prominently thrust foreward; lower stamens ascending under the upper lip, the upper stamens thrust down and crossing the lower; n = 9. Sagebrush, mountain brush, ponderosa pine, aspen, and spruce-fir communities at 1650 to 3300 m in all Utah counties, except Daggett, Emery, Grand, Morgan, Summit, Uintah, and Wayne; British Columbia, south to California and Colorado; 106 (xiii).

Ajuga L.

Annual or perennial herbs; stems decumbent and often stoloniferous; leaves coarsely toothed or incised, rarely entire; inflorescence terminal, spikelike and consisting of 2 to many whorls, dense, or interrupted below; calyx not bilabiate; corollas blue, white, or rose, withering and often persisting in fruit, the upper lip obsolete and reduced to 3 minute toothlike lobes; stamens 4.

Ajuga reptans L. Carpet-bugle. Low perennial herbs, spreading by leafy stolons, matforming; flowering stems to 3 dm tall, nearly glabrous; leaves obovate to oblong-spatulate, 2-5 cm long, the larger ones commonly sinuate-dentate, the lower ones tapering to a broad petiole, the upper ones sessile; flower sessile, in whorls of 2-6 in the axils of bracteate leaves, aggregated into a terminal leafy spike; calyx 4-6 mm long, sparsely villous, the 5 lobes ovate-triangular, somewhat shorter than the tube; corolla blue to purplish, the upper lip very short, bilobed, the lower lip elongate and dilated, the lateral lobes shorter and narrower than the broad, bilobed midlobe; stamens 4, unequal, included; n = 16, 32. Cultivated ornamental ground cover, persisting and occasionally escaping, mainly below ca 1600 m, in Cache, Davis, Salt Lake, and Weber counties; introduced from Europe, widely cultivated in the U. S.; 4 (0).

Dracocephalum L.

Annual, biennial, or short-lived perennial herbs; stems erect; leaves petiolate, sharply serrate or the lower incised; flowers blue to rose pink, crowded in dense terminal or axillary clusters, with conspicuous spinulose-pectinate bracts; calyx campanulate-tubular, 5-toothed, the upper tooth enlarged; corolla tube enlarging above, 2-lipped, the upper lip spreading and 3-lobed, the middle lobe larger and sometimes notched apically; stamens 4, paired, the upper pair longer, the anthers with 2 divergent sacs.

1. Calyx 6-8 mm long, the teeth 1-2 mm long; bracts entire or nearly so, narrow .... *D. thymiflorum*
— Calyx 9-15 mm long, the teeth 4-6 mm long; bracts aristately toothed, broad .... *D. parviflorum*

Dracocephalum parviflorum Nutt. [*Moldavica parviflora* (Nutt.) Britt.]. Stems 1-6 dm tall, more or less hairy; leaves lanceolate to oblong or occasionally ovate, 1-6 cm long, coarsely serrate, the upper with spinulose teeth, more or less puberulent, at least below; bracts ovate or oblong, pectinate with awn-pointed teeth; calyx 9-15 mm long; corolla bluish to pale rose, scarcely longer than the calyx; n = 7. Common in sagebrush, pinyon-juniper, mountain brush, ponderosa pine, aspen-tall forb, and lodgepole pine communities at 1470 to 3300 m in Cache, Daggett, Duchesne, Garfield, Grand, Iron, Juab, Kane, Millard, Piute, Salt Lake, San Juan, Sanpete, Sevier, Summit, Tooele, Uintah, Utah, Wasatch, Washington, and Weber counties; New England to Alaska, south to Arizona and New Mexico; 55 (vii).

Dracocephalum thymiflorum L. [*Moldavica thymiflora* (L.) Rydb.]. Stems 1-5 dm tall, pubescent with minute incurved hairs; leaves lanceolate to oblong or the lower ovate, 1-4 cm long, entire to serrate, the upper nearly entire, or with a few minute teeth, more or less puberulent; bracts ovate to lanceolate, entire or with a few teeth; calyx 6-8 mm long; corolla bluish, slightly exceeding the calyx, pubescent; n = 7. Pinyon-juniper and sagebrush community at ca 2670 m in Sevier County (Welsh et al. 13762 BRY); sporadic in North America; adventive from Eurasia; 1 (i).

Glecoma L.

Low creeping perennial herbs; stems leafy; leaves long-petiolate, orbicular or reniform, crenate; flowers rather large, blue or purple, in small axillary clusters; calyx tubular, somewhat unequally 5-toothed; corolla tube exserted, enlarged above, 2-lipped, the upper lip erect, 2-lobed or emarginate; stamens 4, paired, the upper pair longer; anther sacs divergent.

Glecoma hederacea L. [*Nepeta glecoma* Benth.; *N. hederacea* (L.) Trev.]. Fibrous-rooted perennial from slender stolons and rhizomes; stems lax, 1-4 dm long, retrorsely scabrous to glabrous or nearly so, pilose at the nodes; leaves rotund-cordate to reniform, glabrous to hirsute, the margins crenate, 1–3 cm long; flowers shortly pedicellate; calyx narrow, tubular, 5–6 mm long, hirtellous, scabrous, the upper teeth the longer; corolla blue violet, purple-maculate, (9) 13–23 mm long, those with reduced anthers the shorter; $2n = 18, 36$. Cultivated, escaping, and established in moist disturbed habitats at ca 1400 to 1590 in Box Elder, Salt Lake, and Utah counties; native of Eurasia, now well established in the U. S.; adventive from Eurasia; 8 (0).

Hedeoma Pers.

Aromatic annual or perennial herbs; stems erect or ascending; leaves small, entire or toothed, sessile or petiolate; flowers in small cymules or solitary in axils of upper leaves; calyx tubular, 13–ribbed, gibbous at the base, pubescent, 5–toothed; corolla blue or purple, 2–lipped, the upper lip erect, entire or 2–lobed, the lower 3–lobed and spreading; stamens 4, usually only 1 pair fertile.

1. Calyx teeth connivent at maturity, the aperature closed
   .... *H. drummondii*
— Calyx teeth spreading, the calyx bilabiate .... *H. nana*

Hedeoma drummondii Benth. Mock Pennyroyal. Annual or perennial; stems few to numerous, ascending to decumbent, 0.6–6 dm tall, densely pubescent with retrorse tightly curled hairs; leaves linear to elliptic- oblong or ovate, 5–14 mm long, 1–4 mm wide, entire or obscurely crenate, rounded to attenuate at the base, glabrate above, densely pubescent below and on the margins; axillary cymes crowded or well spaced along the inflorescence; flowers 1–5 per cyme; calyx 5–7 mm long, conspicuously saccate for ca 2/3 its length, the tube above the distended region constricted and tapering to the neck, completely or almost closing the orifice following anthesis, the teeth convergent; corolla blue or whitish in the throat, 7–15 mm long, the upper lip ligulate, straight, concave and subgaleate, ca 1.5 mm long, the lower lip spreading, 2.5–5 mm wide; $2n = 34, 36$. Mixed desert shrub, sagebrush, mountain brush, and pinyon-juniper communities at 1130 to 2600 m in Beaver, Duchesne (?), Garfield, Grand, Iron, Kane, Millard, San Juan, Sevier, Tooele, and Washington counties; North Dakota and Montana, south to Texas, Arizona, and Northern Mexico; 35 (x).

Hedeoma nana (Torr.) Briq. [*H. dentata* var. *nana* Torr.]. Annual or perennial; stems few to very numerous, ascending to erect, 3–20 cm tall, cinereus puberulent with retrorse hairs; leaves ovate to elliptic, 4–10 mm long, 1.5–5 mm wide, entire, obtuse to acute at the base, puberulent on one or both sides, not especially ciliate; axillary cymes well spaced along the inflorescence; flowers 1–4 per cyme; calyx 4–5 mm long, tapering from the opening to the base, the teeth more or less spreading, the upper 3 triangular-subulate, the lower pair linear-acicular and longer; corolla 6–8 mm long, pale purple, the upper lip concave, ca 1 mm long, the lower lip spreading, blotched white and purple lined; $2n = 36$. Hanging gardens and warm desert shrub communities at 915 to 1190 m in Kane, San Juan, and Washington counties; Arizona, Nevada, and California; 4 (ii).

Lamium L.

Annual or perennial herbs; stems leafy, branching; leaves mostly toothed, orbicular to ovate; flowers verticillate in axillary or terminal clusters, purplish red; calyx tubular-campanulate, usually 5–nerved, 5–toothed, the teeth sharp-pointed, equal or the upper longer; corolla 2–lipped, the tube somewhat longer than the calyx, the upper lip erect, concave and usually entire, more or less hairy, lower lip spreading, 3–lobed, the middle lobe enlarged and notched, the lateral lobes small; stamens 4, the upper pair shorter; anthers divergent, with 2 sacs, ascending under upper lip.

1. Upper leaves sessile or clasping .... *L. amplexicaule*
— Upper leaves petiolate .... *L. purpureum*

Lamium amplexicaule L. Dead-nettle. Weedy annuals; stems branched from the base, more or less decumbent, 1–4 dm tall, sparsely pubescent; leaves broadly ovate to rounded, truncate, or cordate basally, coarsely crenate, the lower petioled, the upper sessile, 1–2.5 cm wide; flowers in axillary and terminal clusters; calyx pubescent, 4–5 mm long, the teeth erect; corolla purplish red, 12–16 mm long, the tube very slender, the upper lip pubescent; $2n = 18$. Lawns, fields, roadsides, and other disturbed sites at 850 to 1700 m in Cache, Davis, Salt Lake, Utah, Washington, and Weber counties; adventive from Europe, now widely established in much of the U. S.; 14 (i).

Lamium purpureum L. Red Henbit. Weedy annuals; stems commonly branched below, decumbent to ascending, 1–4 dm tall; leaves all petiolate, the upper ones only shortly so, the blades broadly ovate, usually acutish apically, cordate basally; corolla lilac, 10–16 mm long, the tube rather stout, the lateral lobes reduced to short teeth; $2n = 18$. Ditch banks, roadsides, and other disturbed sites at ca 1600 m in Cache, Morgan, and Salt Lake counties; adventive from Europe; now widely established in the U. S.; 7 (0).

Leonurus L.

Perennial herbs; stems erect, leafy; leaves petiolate, the blades 3– to 5–cleft or -parted; flowers pink, purple, or white, borne in dense axillary clusters near the stem apex; calyx tubular- campanulate, 5–veined, with 5, nearly equal, triangular-aristate teeth; corolla strongly 2–lipped, the upper lip erect and slightly concave, entire, the lower lip spreading, 3–lobed, the central lobe larger, truncate or emarginate; stamens 4, paired, the upper pair shorter, ascending under the upper lip; anthers with 2 parallel sacs.

Leonurus cardiaca L. Motherwort. Perennial herbs with erect, strigose stems 3–6 dm tall or more; leaves palmately 3– to 5–cleft, the lobes entire or toothed, the lower leaves broadly ovate, 2–10 cm long and about as broad, the upper mostly narrower; calyx 6–8 mm long, the teeth spreading or reflexed, subequal; corolla 6–10 mm long, pale purple, rose, or whitish, densely pubescent without, especially on back of upper lip; $2n = 18$. Ruderal weed of disturbed sites at lower elevations, but widely established in the mountain brush, mixed conifer, and aspen zones at 1000 to 2250 m in Beaver, Cache, Davis, Duchesne, Juab, Millard, Salt Lake, Summit, Tooele, Uintah, Utah, Wasatch, and Washington counties; adventive from Asia, now widely established in the U. S.; 32 (ii). This plant was formerly cultivated as a home remedy.

Lycopus L.

Perennial, mostly stoloniferous herbs resembling *Mentha*, but not fragrant; leaves mostly toothed or pinnatifid;

floral bracts similar to the leaves and much longer than the dense axillary whorls of small, mostly white flowers; calyx campanulate to ovoid, 4- to 5-toothed, naked in the throat; corolla more or less campanulate; stamens 2, distant, the upper pair of sterile rudiments or wanting.

1. Leaves pinnatifid, at least the lower ones; calyx teeth awn-tipped; corolla equaling or slightly surpassing the calyx; plants lacking stolons .... *L. americanus*
— Leaves serrate or the lower rarely incised; calyx teeth acuminate; corolla slightly exceeding the calyx; plants stoloniferous .... *L. asper*

Lycopus americanus Muhl. ex Barton American Bugleweed. [*L. sinuatus* L.]. Stems erect, slender, 2–9 dm tall, glabrous or sparingly appressed pubescent with dark hairs; leaves petiolate, to ca 10 cm long and 3 cm wide, the lower primary ones incised or pinnatifid, lanceolate to narrowly ovate or oval, the upper ones lanceolate, sinuate to sharply toothed; bracts short; calyx teeth with long subulate tips; corolla white, the tube scarcely longer than the calyx teeth, the filaments exserted; nutlets 1–1.5 mm long, 0.6–1 mm wide, with entire or barely undulate angles; n = 11. Marshes and other wet sites in palustrine and riparian habitats at 1350 to 2550 m in Box Elder, Cache, Duchesne, Grand, Salt Lake, Uintah, and Utah counties; Newfoundland to British Columbia, south to Florida, Texas, New Mexico, and California; 14 (i).

Lycopus asper Greene Rough Bugleweed. [*L. lucidus* sensu American authors, not Turcz. ex Benth.]. Erect perennials with stolons and elongate turions; stems 2–8 dm tall, simple or rarely branched above; leaves lanceolate or oblong-lanceolate to elliptic, 2–8 cm long, 1–3 cm wide, acute or acuminate apically, tapering to the subsessile base, sharply and evenly serrate; calyx 2–3 mm long, the lobes lance-subulate, acuminate, longer than the tube, hispidulous marginally; corolla white, slightly longer than the calyx; rudimentary stamens slender; nutlets 1.6–2.1 mm long, 1.4–1.8 mm broad, truncate apically; n = 11. Marshes and other wet sites at 1400 to 1800 m in Box Elder, Cache, Davis, Duchesne, Grand, Salt Lake, Uintah, and Utah counties; Alaska to Saskatchewan, south to California, Texas, and Missouri; 31 (i).

Marrubium L.

Perennial herbs; stems several, erect, simple or branched, woolly, with bitter juice; leaves petiolate, toothed, rugose; flowers small, white, in dense axillary clusters; calyx cylindric, 5- to 10-veined, regularly 5- or 10-toothed, the teeth acute or aristate, spreading or recurved, nearly equal; corolla 2-lipped, the upper lip erect, entire or emarginate, the lower lip spreading, 3-cleft, its middle lobe often emarginate; stamens 4, included, didynamous, the upper pair shorter; anthers 2-loculed; style shortly cleft; ovary deeply 4-lobed, smooth.

Marrubium vulgare L. Common Horehound. Stems erect, stout, white-woolly, especially below, 2–10 dm tall; leaves 1–5 cm long, oval to nearly orbicular, crenate-dentate, rugose veined, obtuse to rounded apically, narrowed to a subcordate base, woolly beneath, canescent above, the petioles 1–2 cm long; flowers in dense axillary clusters, whitish; calyx teeth usually 10, subulate, recurved; 2n = 34. Disturbed sites throughout the state, where especially common on sheep bedgrounds, road sides, and trails, at 850 to 2800 m; adventive from Eurasia, now widely established in North America; 61 (x).

Mentha L.

Aromatic perennial herbs from rhizomes; stems erect, often diffusely branched; leaves opposite, punctate, toothed, usually petiolate; flowers in dense axillary clusters or in terminal spikes; calyx campanulate to cylindric, 10-veined, regular or slightly 2-lipped, 5-toothed; corolla funnelform or campanulate, 2-lipped, the tube shorter than the calyx, the upper lip entire or emarginate, the lower lip 3-lobed; stamens 4, equal, included or exserted; nutlets ovoid, smooth.

1. Flowers in dense axillary whorls; plants indigenous in moist habitats .... *M. arvensis*
— Flowers mainly in terminal spikes; plants cultivated, persisting, and escaping 2
2(1). Leaves sessile or subsessile; spikes narrow, mostly interrupted .... *M. spicata*
— Leaves petiolate; spikes thick, dense 3
3(2). Calyx hirsute; leaves lanceolate, acute ...... *M. piperita*
— Calyx glabrous; leaves ovate, obtuse .... *M. citrata*

Mentha arvensis L. Field Mint. [*M. canadensis* L.; *M. borealis* Michx.]. Stems 1.5–8 dm tall, ascending to erect, simple or branched; leaves short-petiolate, the blades elliptic to lanceolate, ovate, or oval, 1–8 cm long, 0.7–3 cm wide, once to twice serrate, acuminate, acute, obtuse or rounded apically, usually glandular-punctate; calyx 2–3 mm long, pubescent and often glandular, the triangular teeth shorter than the tube; corolla purplish to pinkish or white, 4–6 mm long; 2n = 24, 36, 72, 90. Moist places, especially along stream and ditch banks, at 750 to 3000 m in all Utah counties; circumboreal, widely distributed in North America; 114 (xxv). Our plants belong to var. **glabrata** (Benth.) Fern. [*M. canadensis* var. *glabrata* Benth.].

Mentha citrata Ehrh. Bergamot Mint; Lemmon Mint. Stems 3–6 dm tall, from leafy-bracted stolons, usually purplish; leaves with the odor of lemon when crushed, petiolate, the blades mainly 2–6 cm long, 1.5–4 cm wide, ovate to oval, obtuse or the upper acute, rounded or subcordate basally, thin, rather shallowly serrate; flower whorls in thick dense terminal spikes, sometimes also some in the upper leaf axils; spikes mostly 2–2.5 cm long; calyx teeth subulate, glabrous, shorter than the tube; corolla rose, about twice as long as the calyx; 2n = 36, 48. Cultivated spice plant, escaping and established along irrigation canals in Washington County; introduced from Europe; sparingly established in the western U.S.; 2 (0). This plant is sometimes regarded as *M. piperita* var. *citrata* (Ehrh.) Jacq.

Mentha piperita L. Peppermint. Stems mainly 3–8 dm tall, often from leafy-bracteate stolons; leaves petiolate, the blades lance-ovate to elliptic, 1.5–6 cm long, 1–3 cm wide, serrate, glandular-punctate; calyx 2.5–4 mm long, the tube glandular, the lance-acuminate teeth hispid, shorter than the tube; corolla pink purple to white, 4–5 mm long. Stream and ditch banks, bottom lands, and moist roadsides at 1200 to 1800 m in Salt Lake, Tooele, Utah, and Washington counties; cultivated spice plant, escaping and established; widespread in North America; introduced from Europe; 9 (i). Peppermint is considered to have been derived by hybridization between the European species *M. aquatica* L. and *M. spicata* L.

Mentha spicata L. Spearmint. Stems mainly 3–10 dm tall or more, usually branched, often from leafy-bracteate stolons; leaves sessile or subsessile, the blades 2–8 cm long, 0.6–2.5 cm broad, lanceolate to lance-oblong, elliptic, or ovate, serrate, glandular-punctate; calyx 1.5–2 mm long, the tube glandular, the lance-acuminate teeth hispid, subequal to the tube; corolla pink to white, 3–4 mm long. Cultivated aromatic herbs, escaping, and now established along stream and canal banks and in other moist sites at 100 to 1900 m in Box Elder, Cache, Emery, Millard, Salt Lake, Tooele, Utah, Wasatch, Washington and Weber counties; introduced from Europe, now widely distributed in North America; 22 (i).

Molucella L.

Annual herbs, glabrous; leaves coarsely toothed, rounded, petiolate; flowers several, in axils of upper leaves; calyx with lobes wholly united into a flaring funnel shaped structure; corolla white or pinkish, the upper lip concave or galeate, including the stamens; stamens 4, paired; nutlets truncate at the apex.

Molucella laevis L. Stems erect, mainly 2–8 dm tall, simple or branched; leaves 2–5 cm long, suborbicular, coarsely toothed, the teeth broadly rounded; petioles subequal to the blade; flowers fragrant, several in the axils of upper leaves, subtended by several slender spreading or reflexed spines; calyx campanulate, the lobes united into a broadly dilated, funnelform, membraneous, reticulately veined structure 2–3.5 cm long, subtended by several spines; corolla included, white, tipped with pink, 2-lipped, the upper lip arched and including the stamens. Creosote bush to pinyon-juniper communities at 960 to 1850 m in Washington County; cultivated ornamental, introduced from Asia, escaping in the southwestern U. S.; 10 (iii).

Monarda L.

Erect or ascending herbs or shrubs; leaves odoriferous, ovate to lanceolate or linear, entire or toothed; flowers in a few verticils closely subtended by bracts to form head-like clusters, these in interrupted spikes and/or terminating the branches; calyx tubular, 13- to 15-veined, usually hairy in the throat; corolla elongated, strongly 2-lipped, with a slightly expanded throat; lips linear or oblong, somewhat equal, the upper lip erect and arched, entire or slightly notched, the lower lip spreading and 3-lobed apically, the lateral lobes ovate and obtuse, the middle lobe narrower and emarginate; stamens 2, elongated, ascending.

1. Heads solitary and terminal, rarely 2; upper corolla lip somewhat arcuate, but usually erect and straight; stamens exserted .... *M. fistulosa*
— Heads 2 or more, forming an interrupted spike; upper corolla lip sicklelike; stamens usually not exserted .... *M. pectinata*

Monarda fistulosa L. Beebalm. Stems 3–10 dm tall, rarely branched, retrorsely pubescent above or spreading hairy, or glabrous below or throughout; leaves 3.5–9 cm long, 1–3.5 cm broad, ovate to ovate-lanceolate or lanceolate, serrate to entire, softly hirtellous to glabrate; petioles 2–10 mm long; glomerules 15–25 mm broad, the outer bracts subfoliar, frequently tinged pinkish, reflexed; calyx 7–12 mm long, sometimes tinged pinkish, puberulent, the teeth acuminate, ca 1 mm long, the throat hirsute within; corolla lavender to rose purple, 2.5–3.8 cm long, pubescent, the tube 1.5–2.5 cm long, enlarged upwards; n = 17, 18. Mountain brush, pinyon-juniper, and ponderosa pine communities at 2050 to 2600 m in San Juan County; British Columbia to Manitoba, south to Arizona, Texas, and Georgia; 3 (0). Our indigenous plants belong to **var. *menthifolia* (Graham) Fern.** [*M. menthifolia* Graham]. The species is known from cultivation in Utah County; 1 (i).

Monarda pectinata Nutt. Plains Beebalm. Stems mainly 2–3 dm tall, commonly branched, retrorsely hairy; leaves 2–5 cm long, 6–12 mm wide, oblong to oblong-lanceolate, glabrous or sparingly puberulent, remotely serrate to subentire; petioles 5–15 mm long; glomerules 1.5–2.5 cm broad; foliar bracts nearly or quite glabrous, straight and flat, divergent from the base, greenish, 2–7 mm wide, acuminate to a spinose slender bristle, the margins regularly pectinate-ciliate; calyx tube 6–8 mm long, puberulent or glabrous, the throat hirsute, the slender teeth 2–3 mm long, often colored; corolla pink or nearly white, the tube 8–14 mm long, the lips subequal and usually shorter than the tube; n = 18. Pinyon-juniper, ponderosa pine, and sagebrush communities at 1400 to 1700 m in Kane, San Juan, and Washington counties; Colorado and Nebraska, south to Arizona, New Mexico, and Texas; 6 (ii). The plants are strongly and pervasively scented.

Monardella Benth.

Aromatic perennial herbs; stems usually branched; leaves elliptic or oblong, entire or denticulate; flowers borne in a terminal head subtended by broad, often colored, bracts; calyx tubular, nearly equally 5-toothed, the tube usually 15-veined, glabrous in the throat, the upper lip erect, 2-cleft, the lower 3-parted, all lobes linear to narrowly oblong; corolla white with purple spots or rose colored, the tube longer than the calyx, the lobes oblong-linear, subequal; stamens 4, exserted; nutlets smooth, oblong.

Monardella odoratissima Benth. [*Madronella oblongifolia* Rydb., type from Mt. Nebo; *Madronella sessilifolia* Rydb., type from Washington County; *M. parvifolia* Greene]. Plants clumpforming; stems mostly woody below, ascending to erect, puberulent above, 2–4 dm tall; leaves 2–3 cm long, narrowly to broadly lanceolate or oblong, tapering to a subsessile base, green but puberulent; bracts broadly ovate to orbicular, subequal to the calyces, rounded or obtuse apically, membraneous and rose-purple to whitish, pubescent on veins and ciliate; calyx 5–8 mm long, 13-veined; corolla pale purple to rose, ca 15 mm long, the tube surpassing the calyx, the lobes very slender. Sagebrush, mountain brush, pinyon-juniper, ponderosa pine, aspen, tall forb, and spruce-fir communities at 2170 to 3700 m in all or nearly all Utah counties; Washington east to Wyoming, south to California and New Mexico; 148 (xii).

Nepeta L.

Aromatic perennial herbs; stems leafy and branched; leaves truncate or subcordate basally, rather coarsely toothed; flowers cymose, white to purplish mottled, borne in interrupted spikes; calyx tubular or campanulate, the tube somewhat constricted above, the 5 teeth deltoid, subulate, subequal in length, but the lower 3 basally joined; corolla tube longer than the calyx,

2-lipped, the upper lip erect, notched, the lower lip spreading and 3-lobed, the middle lobe larger; stamens 4, paired, the upper pair longer, exserted; anthers with 2 divergent sacs; nutlets smooth.

Nepeta cataria L. Catnip. Stems erect, 3–10 dm tall, with ascending branches, canescent-tomentose; leaves ovate to oblong, 2–9 cm long, 0.8–6 cm wide, usually cordate basally, coarsely crenate-serrate, with petioles 1–4 cm long; calyx urceolate, very pubescent, ca 6 mm long, the teeth subulate; corolla whitish, spotted with purple, 7–12 mm long; 2n = 18, 30, 34, 36. Usually in moist sites in sagebrush, mountain brush, and pinyon-juniper communities at 1400 to 2300 m in Box Elder, Cache, Davis, Kane, Millard, Piute, Salt Lake, Sanpete, Tooele, Uintah, Utah, Wasatch, Washington, and Weber counties; adventive from Europe, now widely distributed in North America; 54 (iii).

Physostegia Benth.

Perennial herbs; stems erect, leafy; leaves sessile, toothed; flowers purple to rose pink, in terminal, showy, spikelike racemes; calyx campanulate-tubular, more or less inflated, the 5 teeth subequal, short; corolla with the tube dilated upward, 2-lipped, the upper erect and slightly concave, entire, the lower lip spreading and 3-lobed, the middle lobe emarginate; stamens 4, paired, the uppermost pair shorter, ascending under the upper lip, the anther sacs parallel; nutlets smooth.

1. Flowers usually less than 15 mm long; leaves thin in texture .... *P. parviflora*
— Flowers mainly 20–25 mm long or more; leaves firm in texture .... *P. virginiana*

Physostegia parviflora Nutt. ex Gray Obedient Plant. [*Dracocephalum nuttallii* Britt.]. Stems arising from vertical rhizomes, simple, or branched above, mainly 2–10 dm tall; leaves 3–10 cm long, 5–20 mm wide, the lower gradually reduced and deciduous, linear-oblong to elliptic-oblong, sessile, serrate or subentire, those near the inflorescence lanceolate or lance-ovate; flowers subsessile in closely flowered racemes 2–8 cm long; calyx 4–6 mm long, finely glandular-puberulent, shortly and unevenly toothed; corolla lavender-purple, 12–16 mm long, the lips 3–5 mm long. Stream terraces and other palustrine and riparian sites in Cache and Rich counties; British Columbia to Saskatchewan, south to Oregon, Idaho, and Nebraska; 4 (0).

Physostegia virginiana (L.) Benth. [*Dracocephalum virginianum* L.]. Stems simple or branched above, 2–7 dm tall or more; leaves 3–10 cm long, 5–15 mm wide, gradually reduced upwards, linear-elliptic, sessile, sharply serrate, the uppermost becoming nearly linear; flowers subsessile in closely flowered racemes 2–15 cm long; calyx 5–8 mm long, finely glandular-puberulent, the teeth 1.5–2.5 mm long, unequal; corolla lavender-purple, 20–25 mm long or more, the lips 5–10 mm long. Cultivated ornamental, persisting, and to be expected in moist disturbed sites in lower elevations portions of Utah County; eastern U. S.; 2 (0).

Poliomintha Gray

Hoary canescent shrubs; leaves entire; flowers pink, purple, or white, clustered in the leafy upper axils; calyx tubular, 13- to 15-veined, the tube shaggy-pilose, the teeth equal or nearly so, the throat strongly constricted; corolla 2-lipped, the upper lip erect and emarginate, the lower lip 3-cleft, more or less constricted in the tube with coarse ascending hairs; fertile stamens 2, ascending against the upper lip; nutlets smooth, oblong.

Poliomintha incana (Torr.) Gray Purple Sage. [*Hedeoma incana* Torr.]. Shrubs, mainly 3–10 dm tall; leaves sessile, 1–3 cm long, linear to oblong, densely white-tomentose with simple hairs, veinless, the upper floral bracts usually shorter than the flowers; calyx 15-veined, oblong, or cylindric, densely white villous, 6–7 mm long, the subulate teeth conspicuous; corolla 10–14 mm long, lavender to whitish, with purple dots on the lower lip, the tube pilose at the summit. Sandy sites, often on stabilized dunes with ephedra, blackbrush, rabbitbrush, other mixed desert shrub, and pinyon-juniper communities at 1300 to 2000 m in Emery, Garfield, Grand, Kane, San Juan, and Wayne counties; California and Nevada to Texas; 104 (xiii).

Prunella L.

Perennial herbs from tap or fibrous roots and often with rhizomes, pubescent with multicellular hairs; leaves petiolate, crenate to entire; flowers short-pedicellate, borne in clusters aggregated into dense spikelike inflorescences, the bracts entire, ciliate; calyx 2-lipped, the upper lip shallowly 3-toothed, the lower lip deeply cleft into 2 narrow teeth; corolla bilabiate, the upper lip entire or nearly so, the lower lip 3-lobed; stamens with anthers 4, the filaments notched near the apex, the anthers glabrous.

Prunella vulgaris L. Heal-all. Stems 0.6–5 dm long, ascending to erect or simple; leaf blades lance-ovate to oblong or elliptic, 2–9 cm long, 0.7–4 cm broad, minutely puberulent to glabrous; spikes 1–2 cm broad, 2–8 cm long; calyx 6–10 mm long, sparsely villous, purplish, the lower teeth subequal to the tube; corolla pink purple to pink or white, 12–18 mm long, glabrous. Wet meadows, streamsides, and other moist sites in ponderosa pine, cottonwood, lodgepole pine, aspen, and spruce communities at 1370 to 3400 m in Duchesne, Emery, Garfield, Iron, Kane, Morgan, Rich, Salt Lake, Summit, Uintah, Utah, Wasatch, and Weber counties; widespread in temperate North America; Eurasia; 23 (iv). Our materials have been assigned to **var. lanceolata** (**Barton**) **Fern.** [*P. pensylvanica* var. *lanceolata* Barton].

Salazaria Torr.

Divaricately branched shrubs; stems canescent, spinescent; leaves small, entire or rarely toothed, shortly petiolate; flowers in loose spikelike racemes; calyx equally 2-lobed, the lips entire, becoming inflated and globular in fruit; corolla purple, 2-lipped, the upper lip arched, the lower with recurved sides and small lateral lobes; anthers ciliate, the upper pair of stamens 2-loculed, the lower pair 1-loculed; nutlets tuberculate.

Salazaria mexicana Torr. Bladder Sage; Paperbag Bush. Rounded shrubs, mainly 3–10 dm tall, the soft twigs pubescent; leaves green, 1–1.5 cm long, glabrous or minutely hispidulous, oblong or broadly lanceolate; racemes 5–10 cm long; calyx 8 mm long in anthesis, in fruit becoming inflated papery and 1–2 cm long; corolla 15–18 mm long, pubescent externally, the throat pale, the lips dark purple; nutlets olive brown, peltate on raised gynobases, tessellately tubercled. Joshua tree, blackbrush, mixed desert shrub, and pinyon-juniper communities at 940 to 1400 m in Washington County; California to Texas and Mexico; 24 (vii).

Salvia L.

Herbs or shrubs, usually strongly aromatic; leaves opposite, sometimes mostly basal, entire or toothed; flowers in terminal bracteate spikelike racemes; calyx tubular to campanulate or ovoid, 2–lipped, the upper lip entire or 3–toothed, the lower with 2 lobes; corolla tubular, strongly 2–lipped, the upper lip concave, entire or 2–notched, the lower lip longer than the upper, 3–toothed and spreading; stamens 2, the connective between the anther sacs well developed and usually longer than the filament, bearing 1 sac terminal on 1 branch, the other lacking a sac and deflexed; style 2–cleft apically, the ovary deeply 4–parted; nutlets smooth.

1. Flowers 4–6 cm long, bright red; cultivated herbs ... . S. splendens
— Flowers less than 2.5 cm long, not bright red; cultivated or indigenous herbs or shrubs 2
2(1). Plants shrubs or subshrubs 3
— Plants herbaceous 4
3(2). Leaves white-woolly beneath; cultivated garden sage . ... S. officinalis
— Leaves not white-woolly; indigenous shrubs of south-central and western Utah .... S. dorrii
4(2). Corolla pale yellow; herbage floccose woolly, eventually partly glabrate .... S. aethiopsis L.
— Corolla blue to white or marked with yellow; herbage coarsely hairy to subglabrate 5
5(4). Leaves dissected, mainly basal; lower anther sac fertile; plants annual, indigenous in Washington County .... S. columbariae
— Leaves not dissected, not mainly basal; anther sacs various; plants annual or perennial, variously distributed ... 6
6(5). Leaves ovate, coarsely dentate, the lower ones 6–25 cm long .... S. scalarea
— Leaves oblong to lanceolate or linear, entire or denticulate, mainly less than 10 cm long 7
7(6). Corolla 15 mm long or more, the tube evidently exserted; calyx puberulent to tomentose ...... S. azurea
— Corolla less than 15 mm long, the tube scarcely exserted; calyx only minutely hairy on the veins .... S. reflexa

Salvia azurea Lam. Perennial herbs; stems erect, 3–16 dm tall, glabrous or pubescent with reflexed hairs; leaves 5–10 cm long, lanceolate to oblong or linear, obtuse to acute, tapering to a very short petiole, serrate to entire; inflorescence elongate, spikelike, composed of numerous verticels, these dense above, remote below; bracts subulate; calyx oblong-campanulate, 6–9 mm long, slightly 2-lipped, the upper lip very broad and obtuse, the lower lobes narrower and all similar; corolla deep blue to nearly white, 15–20 mm long; anterior part of connective deflexed, ca 4 mm long, linear. Cultivated ornamental, escaping and persisting in disturbed sites at 1400 to 1800 m in Box Elder, Cache, Salt Lake, and Utah counties; South Carolina to Florida, west to Nebraska, Colorado, Texas, and Mexico; 5 (0).

Salvia columbariae Benth. Chia. Annual; stems branching and leafy at the base, 1–5 dm tall, cinereous with short retrorse hairs; basal leaves 3–12 cm long, the petioles subequal to the blades, oblong-ovate, pinnatifid or bipinnatifid into toothed or incised divisions; flowers in capitate verticels 1–3 cm wide, terminating the stems and branches; bracts suborbicular, awn-tipped, green or often suffused with purple; calyx ca 1 cm long, purplish, arcuate, the middle spinose tooth of the upper lip with small lateral lobes and a large emarginate middle lobe; corolla purplish or white; stamens with upper arm of connective linear and lacking an anther, the lower branch deflexed and with a fertile anther. Creosote bush, Joshua tree, blackbrush, and pinyon-juniper communities at 800 to 1200 m in Washington County; California, Nevada, Arizona, and Mexico; 16 (v).

Salvia dorrii (Kellogg) Abrams Dorr Sage. [*Audibertia dorrii* Kellogg; *Audibertiella argentea* Rydb.; *S. pilosa* Merriam; *S. carnosa* Dougl.; *S. carnosa* var. *pilosa* Jepson]. Low rounded shrubs, mainly 3–8 dm tall, densely scurfy-canescent, punctate-glandular; leaves 10–25 mm long, 7–15 mm wide, obovate to spatulate or oval, rounded or emarginate apically, abruptly narrowed to a slender petiole, this 5–8 mm long; inflorescence of (1) 2–5 verticels 1.5–2.5 cm wide; bracts oblong-elliptic to round, purplish or greenish, conspicuously ciliate, pilose or glabrous on the back; calyx ca 6 mm long, the lower lip deeply 2–toothed, the upper entire; corolla blue, 10–12 mm long, the upper lip erect and 2–3 mm long, the lower 3–lobed with the middle lobe erose; stamens conspicuously exserted, the upper ones short and sterile. Creosote bush, Joshua tree, blackbrush, shadscale, sagebrush, mountain brush, and pinyon-juniper communities at 830 to 2600 m in Beaver, Box Elder, Iron, Juab, Kane, Millard, Tooele, and Washington counties; Idaho, Nevada, Arizona, and California; 60 (xiv). Our materials are assignable to **ssp. dorrii**, but some of the plants with nearly glabrous bracts from Washington County approach **ssp. argentea (Rydb.) Munz.**

Salvia officinalis L. Garden Sage. Subshrubs, mainly 2–7 dm tall, more or less white woolly; leaves oblong, 2–5 cm long, petiolate, entire or crenate, acute or obtuse, the surface rugose; inflorescence racemose, the racemes mostly simple, with few interrupted verticels; bracts ovate to ovate-lanceolate, acuminate; calyx campanulate, membranaceous, purplish tinged, somewhat 2-lipped, the 2 lower teeth longer than the 3 upper ones, all subulate-acuminate; corolla 12–20 mm long, purple, blue, or white, with a hairy ring inside; $2n = 14$. Cultivated spice plant, rarely escaping, but long persisting at 1300 to 1600 m in Salt Lake and Utah counties; introduced from Europe; 5 (0).

Salvia reflexa Hornem. [*S. lanceolata* Willd.]. Annuals; stems simple to much branched, 2–7 dm tall, pubescent or glabrous; leaves 2.5–5 cm long, 4–12 mm wide, lanceolate to linear, obtuse to acute apically, tapering to a slender petiole to 15 mm long, irregularly serrate to subentire; flowers 1–3 per whorl, the winged pedicels 2–3 mm long, borne in a slender interrupted spike; bracts lanceolate, deciduous, 2–5 mm long; calyx 4–8 mm long, deeply 2-lipped, minutely hairy on the veins, cleft ca one-half to the base, the upper lip entire; corolla blue to whitish, only slightly exserted, the tube 4–4.5 mm long, the erect upper lip 2.5–3 mm long, the lower lip rounded, 4.5–5 mm wide; $2n = 20$. Desert shrub, sagebrush, pinyon-juniper, and riparian sites, often in disturbed lands, at 1300 to 1800 m in Beaver, Grand, Salt Lake, Uintah, Utah, and Wayne counties; Wisconsin to Montana, south to Arkansas, Texas, Arizona, and Mexico; 8 (0).

Salvia sclarea L. Coarse biennials; stems erect, 5–15 dm tall, spreading hairy, with the hairs gland-tipped above; lowermost leaves long-petiolate, the blades 7–20 cm long, ovate to ovate-oblong, rugose, with toothed margins, subcordate basally; cauline leaves reduced upwards; inflorescence freely branched, with the verticels of flowers scattered along the branches; bracts conspicuous, 1–3 cm long, broadly ovate, caudate-acuminate, purplish tinged; calyx 2–lipped, glandular, the upper lip well developed, aristate, lateral teeth 1.5–3 mm long, well separated from the shorter central tooth; corolla blue to white, or marked with yellow, 15–30 mm long, the upper lip strongly arched, more or less galeate, longer than the tube and lower lip; stamens exserted; 2n = 22. Disturbed sites at 1400 to 2000 m in Cache, Salt Lake, and Wasatch counties; widespread in the western U. S.; adventive from Europe; 3 (0).

Salvia splendens Sellow Annual herbs; stems erect, mainly 2–9 dm tall, commonly branched, glabrous or the upper parts villous with colored hairs; leaves 5–8 cm long, ovate, petiolate, acuminate, dentate; inflorescence terminating the branches, racemose or paniculate, the verticels 2– to 6–flowered; bracts ovate, acuminate, 12–16 mm long, reddish; calyx campanulate, mainly 2–3 cm long, with 3 broadly ovate acute teeth, scarlet; corolla 4–5 cm long, scarlet the tube yellowish, glabrous to villous, with a hairy ring inside; n = 8, 22. Cultivated ornamental in greenhouses and gardens at lower elevations in Utah; introduced from Brazil; 2 (0).

Satureja L.

Perennial herbs; stems leafy, erect, hirsute; leaves oval to narrowly ovate, subentire; flowers white to purple, usually rose colored, in dense cymules, these terminal and headlike or forming an interrupted spike; calyx tubular, 2–lipped, the upper with 3 short teeth; corolla 2–lipped, the upper lip mostly erect, somewhat concave, entire or emarginate, the lower lip spreading and 3–lobed; stamens 4, paired, the upper slightly shorter; anthers with 2 divergent sacs; nutlets smooth.

Satureja vulgaris (**L.**) **Fritsch** [*Clinopodium vulgare* L.]. Stems 1–5 dm tall, from rhizomes, simple, more or less hairy; leaves 1–4 cm long, 6–15 mm wide, ovate, entire or minutely denticulate; flowers in dense axillary and terminal clusters; calyx 8–9 mm long, villous-hirsute, the lobes subulate-aristate; corolla 8–12 mm long, purple or white, pubescent externally; 2n = 20, 40. Sagebrush, mountain brush, and ponderosa pine communities at ca 2300 m in Washington County; Manitoba to Nova Scotia, south to Arizona, New Mexico, and North Carolina; Eurasia; 3 (ii).

Scutellaria L.

Perennial herbs, commonly rhizomatous, not aromatic; leaves opposite, entire or toothed; flowers 1–3 in leaf axils or in bracteate racemes or spikes; calyx campanulate in flower, splitting to the base at maturity, strongly 2–lipped, the lips entire, the upper longer lip usually falling away, in fruit with an appendage or scale; corolla blue, violet, or white, well exserted, 2–lipped, dilated above the throat, the upper lip arched, entire or emarginate, the lower lip spreading or deflexed, with a large central lobe and two small lateral ones; stamens 4, in 2 pairs, ascending under the upper lip, ciliate or bearded; nutlets variously marked.

1. Leaves crenate, thin, mostly 3–9 cm long; corolla dull blue; plants common .... S. *galericulata*
— Leaves entire or subentire, firm, mostly less than 3 cm long; corolla blue; plants uncommon to rare 2

2(1). Plants 0.5–1 dm tall; stems retrorsely pubescent . S. *nana*
— Plants usually 1–3 dm tall; stems with hairs spreading or ascending .... S. *antirrhinoides*

Scutellaria antirrhinoides **Benth.** Skullcap. Perennial herbs; stems 1–3 dm tall or more, puberulent with ascending incurved hairs, and more or less glandular puberulent; leaves 1–2 (3) cm long, the blades oblong-ovate to oblong-elliptic, entire; petioles 5–20 mm long; flowers solitary in axils of upper leaves; corolla violet blue, the tube and galea 12–22 mm long, the throat closed. Mountain brush community at 1500 to 2000 m in Morgan and Weber counties; Washington to Idaho, south to California; 2 (0).

Scutellaria galericulata **L.** Perennial herbs; stems 1.5–6 dm tall, commonly pubescent on angles with short curly hairs, sometimes also glandular pubescent; leaves 3–6 cm long or more, oblong-ovate to lanceolate, truncate to subcordate basally, acute, crenate-serrate, glabrous above or nearly so, paler and pubescent or glabrous beneath; petioles 1–3 mm long; flowers solitary in axils of upper leaves; corolla blue, 14–22 mm long; n = 8, 16. Moist meadows and streamside communities at 1800 to 2700 m in Beaver, Box Elder, Cache, Davis, Garfield, Piute, Salt Lake, Sevier, Uintah, Utah, Wasatch, and Weber counties; Alaska to Newfoundland, south to California, Arizona, Wisconsin, and California; 21 (iv).

Scutellaria nana **Gray** Perennial herbs; stems 5–10 cm tall, branched near the base, densely and finely retrorsely puberulent; leaves 1–2 cm long, 3–7 mm wide, oblanceolate to oblong-ovate, or elliptic, entire; flowers solitary in leaf axils; corolla blue, with a pale tube, 14–22 mm long. Sagebrush and pinyon-juniper communities at 1800 to 2300 m in Iron and Washington counties; Oregon and Idaho, south to California and Nevada; 4 (iii). Our plants are assignable to **var. *sapphirina* Barneby**.

Stachys L.

Perennial herbs from rhizomes; stems erect, leafy; leaves oblong, sessile or nearly so, toothed; flowers few to many, borne in whorls, in dense or interrupted terminal spikes, or also in upper leaf axils; calyx turbinate or campanulate, 5– to 10–veined, enlarged somewhat at maturity, the 5 teeth nearly equal, more or less spinulose apically; corolla white, pallid, or rose purple, 2–lipped, the upper lip erect, entire or notched, lower lip spreading, 3– or rarely 2–lobed; stamens 4, the upper pair shorter, ascending under the upper lip; nutlets ovoid or oblong.

1. Lower leaves with petioles ca 6 mm long or longer; pubescence of cobwebby, soft-woolly hairs 2
— Lower leaves sessile or with petioles to 4 mm long; pubescence pilose or canescent, not soft woolly; plants of various distribution 3

2(1). Leaves definitely bicolored; spikes ultimately much interrupted; plants of Washington County ...... S. *albens*
— Leaves not much bicolored, about as pale above as beneath; inflorescence compact; plants cultivated and escaping .... S. *byzantina*

3(1). Leaves 15–40 mm wide, toothed, clothed with spreading, pilose hairs; plants widespread .... S. *palustris*
— Leaves 7–12 mm wide, entire or very finely toothed, clothed with silvery strigose-canescent hairs; plants of Kane County .... S. *rothrockii*

Stachys albens Gray Stems stout, 3–10 dm tall, simple or branched, densely cobwebby; leaves 3–12 cm long, narrowly to broadly ovate-oblong, cordate basally, villous-tomentose (especially beneath), crenate-serrate, the petioles 0.6–5 cm long; spikes 1–2 dm long, more or less interrupted at maturity; calyx 5–7 mm long, woolly, the deltoid-ovate teeth cuspidate; corolla white or pinkish, with purple veins, the tube 6–8 mm long, the upper lip 3.5–5.5 mm long, the lower 5–8 mm long. Riparian, creosote bush, pinyon-juniper, mountain brush, and aspen communities at 800 to 2100 m in Washington county; Nevada and California; 10 (v).

Stachys byzantina C. **Koch.** Woolly Betony; Lambsears. [*S. olympica* authors, not Koch; *S. lanata* Jacq, not Crantz]. Stems 2.5–8 dm tall, simple or branched at the base; leaves 4–9 (10) cm long, 1.5–3.5 cm wide, the petioles 0.5–3 cm long, the blades lanceolate to ovate or elliptic; entire to finely toothed; white wooly on both sides; spikes 3–15 cm long, compact or the lower verticel separated; calyx 8–12 mm long, the tube ca 3–4 mm long; corolla pink to purple, densely woolly. Cultivated oramental, escaping and persisting, in sagebrush community at 2257 m in Summit County; introduced from Asia; 2(0).

Stachys palustris L. Marsh Betony. [*S. asperrima* Rydb. Type from Jordan, Salt Lake County]. Stems 1.5–8 dm tall, spreading hirsute or villous, especially above; leaves 4–8 cm long, ovate-oblong, villous or pilose with spreading hairs, crenate-serrate; calyx 5–9 mm long, hispidulose, the deltoid teeth cuspidate; corolla pale rose, veined with deeper red, the tube subequal to the calyx, (8) 10–15 mm long, the lower lip 6–8 mm long, villous on the back; n = 32. Riparian or palustrine communities at 1500 to 2800 m in Cache, Daggett, Duchesne, Garfield, Iron, Kane, Piute, Rich, Salt Lake, Uintah, Utah, Wasatch, and Weber counties; widely distributed in North America; circumboreal; 21 (iii). Our plants have been assigned to **var. pilosa** (**Nutt.**) **Fern.** [*S. pilosa* Nutt.].

Stachys rothrockii Gray Stems 1–3.5 dm tall, simple or branched from the summit of a caudex, strigose-canescent or tomentose; leaves 2–6 cm long, 7–12 mm wide, sessile or subsessile, oblong or narrowly elliptic-lanceolate, entire or finely toothed; spikes 8–15 cm long; calyx 6–8 mm long, the deltoid lobes shortly cuspidate; corolla whitish, with purple lines, 10–13 mm long, the lower lip 3–4 mm long, the upper lip 4–5 mm long, woolly villous externally. Tropic Shale outcrops in pinyon-juniper and salt desert shrub communities at 1615 to 2200 m in Kane County; New Mexico and Arizona; 4 (ii).

Teucrium L.

Perennial herbs; leaves simple, serrate; flowers terminal, in slender spikes; calyx saccate, toothed or deeply 5-lobed; corolla pinkish, bluish, or nearly white, the upper lip very short, deeply notched, the lower lip conspicuous and spreading, with small lateral lobes; stamens 4, paired; nutlets roughened.

Teucrium canadense L. American Germander. Stems 3–8 dm tall, erect, branching mostly above, if at all, spreading hairy throughout, more or less glandular upward; leaves 4–9 cm long, ovate, oblong, oval, or lanceolate, serrate, villous, at least beneath; calyx 5–7 mm long, villous to tomentose, the teeth unequal, shorter than the tube; corolla 7–15 mm long, rose to purplish, or sometimes cream colored; n = 16. Riparian and palustrine habitats at 1400 to 2100 m in Cache and Utah counties; widespread in North America and Mexico; 2 (0). Our plants have been assigned to **var. occidentale** (**Gray**) **McClintock & Epling** [*T. occidentale* Gray]. They are characterized by having glandular hairs, especially in the inflorescence.

Thymus L.

Shrubs or subshrubs; stems decumbent to erect; leaves small, entire; flowers verticillate in axils of upper leaves; calyx 2-lipped, 10- to 13-veined, villous at the throat within, the upper lip 3-toothed, the lower one more deeply cleft into 2 narrow lobes; corolla 2-lipped, the upper lip nearly flat, the lower 3-lobed and spreading; stamens 4, mostly exserted, subequal or the lower pair the longer.

Thymus serpyllum L. Thyme. Stems 10–25 cm tall, the slender branches somewhat woody, with dense whitish pubescence; leaves 3–10 mm long, petiolate, elliptic to lanceolate, ovate, or obovate; inflorescence dense and often headlike or racemose and interrupted, the whorls rather loosely several- to many-flowered; calyx ca 3 mm long, hairy in the throat, the teeth of upper lip lanceolate, of the lower lip subulate and ciliate; corolla lilac or puplish, ca 4 mm long. Cultivated ornamental, persisting, but not established, at lower elevation regions of Salt Lake and Utah counties; introduced from Europe; 4 (0). Another species, **Thymus vulgaris L.**, with the leaves sessile rather than petiolate is to be expected in cultivation in Utah.

LEGUMINOSAE A. L. Juss.

Legume Family

Herbs, shrubs, or trees; leaves alternate, pinnately or palmately compound, or simple, stipulate; flowers perfect, irregular or regular, usually borne in racemes; calyx 5-lobed; petals 5 (a banner, 2 wings, and 2 keels) or fewer, less commonly reduced to 1 (banner), or lacking; stamens 10 or 5, or numerous, diadelphous, monadelphous, or distinct; pistil 1, the ovary superior, 1- or 2-loculed, 1-carpelled, the style and stigma 1; fruit (pod) a legume or loment, sessile, subsessile, stipitate, or with a gynophore, dehiscent or indehiscent; x = 5–14. [Fabaceae Lindl.].

Barneby, R. 1984. Fabales of the Intermountain Flora. Unpublished mss.
Isely, D. 1973. Leguminosae of the United States: I. Mimosoideae. Mem. New York Bot. Gard. 25(1): 1–52.
———. 1975. Leguminosae of the United States: II. Caesalpinoideae. Mem. N. Y. Bot. Gard. 25(2): 1–228.
Welsh, S. L. 1963. Legumes of Utah I. Preliminary report. Proc. Utah Acad. 40: 200–201.
———. 1964. Legumes of Utah II. Conspectus of the genera. Proc. Utah Acad. 41: 84–86.
———. 1978. Utah flora: Fabaceae (Leguminosae). Great Basin Naturalist 38: 225–367.

1. Flowers regular, in dense heads or compact spicate racemes; stamens 5 or numerous (Mimosoideae) ... Key 1

— Flowers irregular (only slightly so in some); stamens 10 or fewer 2

2(1). Corolla not papilionaceous, sometimes nearly regular, the upper petal enclosed by the others; stamens 10 or fewer, commonly distinct (Caesalpinoideae) ...... Key 2

— Corolla papilionaceous, the upper petal (banner) enclosing the wing and keel petals in bud, much reduced in *Amorpha*, or lacking in *Dalea*, *Psorothamnus*, and *Parryella*; stamens 10 or 5 (Papilionoideae) 3

3(2). Plants woody; trees, shrubs, or woody vines ...... Key 3

— Plants herbaceous perennials or annuals 4

4(3). Leaves even-pinnate .... Key 4

— Leaves odd-pinnate or simple 5

5(4). Leaflets 3 only .... Key 5

— Leaflets 5 or more, or the leaves simple .... Key 6

Key 1.

Flowers regular; stamens 5 or numerous (Mimosoidea).

1. Plants herbaceous; flowers whitish, in compact heads; stamens shortly exserted .... *Desmanthus*

— Plants woody, either trees or shrubs; flowers variously colored, in racemes or spikes or, if in heads, the stamens long-exserted 2

2(1). Trees, unarmed, cultivated; flowers in umbellate heads; stamens long-exserted, the filaments commonly 20–30 mm long .... *Albizia*

— Trees or shrubs, armed, indigenous; flowers in spicate racemes; stamens included or shortly exserted, the filaments less than 5 mm long 3

3(2). Spines recurved; pods flat, 10–20 mm broad, brown at maturity .... *Acacia*

— Spines straight; pods spirally coiled or if flattened, less than 10 mm broad and yellowish to tan at maturity .... *Prosopis*

Key 2.

Corolla not papilionaceous (Caesalpinoideae).

1. Leaves simple, the blades rotund-ovate; flowers pink, appearing before the leaves .... *Cercis*

— Leaves once or twice compound; flowers yellow, white, or greenish, appearing after the leaves 2

2(1). Shrubs or herbs; flowers with yellow petals, the stamens exserted or not; plants indigenous in eastern Utah or cultivated and naturalized in Washington County .... *Caesalpinia*

— Trees; flowers with yellow, white, or greenish yellow petals, the stamens included or not much exserted; distribution broad 3

3(2). Leaves subsessile, the pinnae in 1 or 2 pairs and with numerous leaflets 2–4 mm long and 0.5–1 mm wide; flowers bright yellow .... *Parkinsonia*

— Leaves typically petiolate, the pinnae various, but leaflets not both numerous and 2–4 mm long; flowers white or greenish yellow 4

4(3). Leaves once to twice pinnate; branches often armed; flowers greenish yellow, borne in spicate racemes; pods long and straplike .... *Gleditsia*

— Leaves bipinnate; branches unarmed; flower white, long-stalked, in open panicles; pods thick .. *Gymnocladus*

Key 3.

Trees, shrubs, or woody vines (Papilionoideae).

1. Leaves even-pinnate, the rachis produced apically as a bristle; flowers yellow .... *Caragana*

— Leaves simple or odd-pinnate; flowers variously colored 2

2(1). Leaves simple or the lower ones 3-foliolate; plants shrubs; flowers yellow, solitary or borne in erect racemes 3

— Leaves compound; plants varying in one or more ways from above 4

3(2). Calyx split above, hence 1-lipped, with 5 minute teeth; flowers borne in erect racemes; plants known from Washington County .... *Spartium*

— Calyx bilabiate, the upper lip 2-lobed, the lower 3-lobed; flowers mostly solitary, axillary; plants known from Weber County .... *Cytissus*

4(2). Plants twining woody vines; flowers large and showy, borne in terminal, pendulous racemes .... *Wisteria*

— Plants trees or shrubs; flowers various, usually borne in axillary, erect or pendulous racemes 5

5(4). Leaflets 3; flowers yellow, borne in pendulous racemes .... *Laburnum*

— Leaflets 5 or more; flower white, pink, indigo, or yellow, borne in erect or spreading racemes 6

6(5). Herbage glandular-punctate; indigenous shrubs with petals indigo or lacking 7

— Herbage not glandular-punctate; cultivated or indigenous shrubs or trees; petals white, pink, or yellow or, if indigo (as in *Amorpha*), the corolla reduced to a single petal (the banner) 8

7(6). Petals lacking; leaflets linear; plants of Grand and San Juan counties .... *Parryella*

— Petals present; leaflets broad; plants of southern and southeastern Utah .... *Psorothamnus*

8(6). Petal 1, the banner only present, indigo; sparingly cultivated shrubs .... *Amorpha*

— Petals 5, white, pink, or yellow; shrubs or trees, sparingly to commonly cultivated 9

9(8). Plants shrubs; pods bladdery-inflated; flowers yellow; ornamental and roadside plants .... *Colutea*

— Plants trees or shrubs; pods flat or terete; flowers white or pink 10

10(9). Branches armed with stipular spines or internodal hispid processes; staminal filaments diadelphous; petals pink or white .... *Robinia*

— Branches unarmed; staminal filaments distinct; petals white 11

11(10). Leaf bases hollow, covering superposed buds; pods flat, not constricted between the seeds ..... *Cladrastis*

— Leaf bases solid, not covering buds; pods terete, constricted between the seeds .... *Sophora*

Key 4.

Leaves even-pinnate (Papilionoideae).

1. Flowers yellow; fruit ripening below ground, tardily dehiscent, constricted between the seeds .... *Arachis*

— Flowers white, pink, red, lavender, or cream; fruit borne above ground, not constricted between the seeds 2

2(1). Style strongly dilated; sepals foliaceous; plants cultivated .... *Pisum*
— Style not strongly dilated; sepals not foliaceous; plants indigenous or cultivated 3
3(2). Style bearded down one side; wings of corolla essentially free from the keel .... *Lathyrus*
— Style bearded in a tuft or ring at apex; wings of corolla adherent to the keel .... *Vicia*

Key 5.

Leaflets three (Papilionoideae).

1. Leaves palmate, the terminal leaflet neither stalked nor jointed 2
— Leaves pinnate, the terminal leaflet stalked or jointed 4
2(1). Flowers golden yellow, the banner orbicular, large; legumes narrowly oblong, erect or ascending; staminal filaments distinct .... *Thermopsis*
— Flowers ochroleucous to white or pink to pink purple, the banner not orbicular, moderate to small in size; staminal filaments diadelphous 3
3(2). Leaflets usually toothed; flowers mostly in heads, commonly pink or white .... *Trifolium*
— Leaflets entire; flowers not in heads, commonly ochroleucous or pink .... *Astragalus*
4(1). Herbage glandular-punctate; indigenous plants with usually linear to oblanceolate leaflets 5
— Herbage not glandular-punctate; indigenous or cultivated plants with spatulate to obovate or oblanceolate to ovate leaflets 6
5(4). Plants caulescent, with 5 or more developed internodes; pods not included in the calyx at maturity .... *Psoralidium*
— Plants acaulescent or short-caulescent, usually with fewer than 5 developed internodes; pods included in the calyx at maturity .... *Pediomelum*
6(4). Leaflets entire 7
— Leaflets toothed (except in some *Trifolium* species) .. 9
7(6). Flowers in umbels, loosely capitate, or solitary in leaf axils, (white) yellow or suffused with orange .... *Lotus*
— Flowers in interrupted racemes or panicles, purplish . 8
8(7). Leaflets stipellate; pods several-seeded, several to many times longer than broad .... *Phaseolus*
— Leaflets lacking stipels; pods 1-seeded, only somewhat longer than broad .... *Lespedeza*
9(6). Flowers usually in heads; corolla persistent, investing the fruit; fruit straight .... *Trifolium*
— Flowers usually in racemes; corolla not persistent; fruit straight or curved to coiled 10
10(9). Leaflets toothed along the distal 1/2 or more; racemes elongate, several times longer than broad ... *Melilotus*
— Leaflets toothed along the distal 1/3 only (except in some *Trigonella*); racemes compact or loose, seldom more than twice longer than broad 11
11(10). Fruit straight or falcately curved, prominently veined on the valves; flowers yellow; terminal leaflet with an apical spinose cusp, rarely as much as twice longer than broad; plants rare .... *Trigonella*
— Fruit coiled or curved, veined or not; flowers pink, lavender, whitish, or yellow; terminal leaflet seldom strongly cuspidate apically, usually more than twice longer than broad; plants common .... *Medicago*

Key 6.

Leaflets (4) 5 or more, or leaves simple (Papilionoideae).

1. Leaves palmately compound, with usually 5–11 leaflets, long-petiolate 2
— Leaves pinnately compound or, if rarely palmately compound (as in some *Lotus* species), sessile or with only 4 leaflets 4
2(1). Herbage not glandular-dotted; leaflets usually 7–11, variously shaped; stamens monadelphous; pods several-seeded .... *Lupinus*
— Herbage glandular-dotted; leaflets usually 5, broadly obovate-spatulate; stamens usually diadelphous; pods 1-seeded 3
3(2). Plants caulescent, usually with 5 or more developed internodes; pods not included within the calyx at maturity .... *Psoralidium*
— Plants subacaulescent to short-caulescent, usually with fewer than 5 developed internodes; pods included within the calyx at maturity .... *Pediomelum*
4(1). Herbage glandular-dotted 5
— Herbage not glandular-dotted 6
5(4). Racemes spicate; legumes 1-seeded, not bearing appendages; stamens 5; petals (except banner) inserted on staminal tube .... *Dalea*
— Racemes not spicate; legumes several-seeded, bearing hooked appendages; stamens 10; petals not inserted on staminal tube .... *Glycyrrhiza*
6(4). Terminal leaflet of lower leaves several times larger than the lateral; inflorescence a many-flowered head closely subtended by foliose involucral bracts; flowers yellow; introduced, rare .... *Anthyllis*
— Terminal leaflet not much larger than the lateral; inflorescence a raceme or an umbel, lacking foliose bracts (except in *Lotus*); flower color various 7
7(6). Margin of leaflets toothed; corolla persistent, investing the fruit .... *Trifolium*
— Margin of leaflets entire; corolla usually deciduous ... 8
8(7). Flowers in umbels, loosely capitate, or solitary in leaf axils; petals yellow, often suffused with orange, or pink 9
— Flowers in racemes or cymes; petals usually not yellow 10
9(8). Leaflets 3–5; flowers yellow .... *Lotus*
— Leaflets 9–23; flowers pink to pink purple ... *Coronilla*
10(8). Keel petals much longer than the wings; fruit a flattened loment 11
— Keel petals subequal to the wings or shorter; fruit a legume (a terete loment in *Sophora*) 12
11(10). Fruit 4- to several-seeded, not spiny (except in *H. boreale* var. *gremiale*); plants indigenous .. *Hedysarum*
— Fruit 1- to 2-seeded, more or less spiny-toothed; plants adventive, cultivated and escaping . *Onobrychis*
12(10). Stipules spiny; flowers dirty whitish .... *Peteria*
— Stipules various, but not spiny; flowers seldom if ever dirty whitish 13
13(12). Staminal filaments distinct; fruit a terete to somewhat flattened loment; plants with blue or white flowers, usually of sandy sites .... *Sophora*
— Staminal filaments diadelphous or monadelphous; fruit a legume; plants from a caudex and/or taproot, rarely rhizomatous; habitats various 14

14(13). Keel with a porrect beak; ventral suture of legume forming a partial or complete partition; plants usually acaulescent .... *Oxytropis*

— Keel beakless, or the beak diverging from the floral axis; ventral suture usually not produced internally, the dorsal usually produced in bilocular fruits; plants usually caulescent 15

15(14). Stamens monadelphous; flowers blue .... *Galega*

— Stamens diadelphous; flowers pink purple, pink, lavender, ochroleucous, red, white, or variously suffused, but not blue 16

16(15). Flowers red orange when fresh; plants adventive .... *Sphaerophysa*

— Flowers pink, pink purple, lavender, or white to ochroleucous; plants indigenous, or rarely adventive .... *Astragalus*

Acacia Miller

Armed trees; leaves alternate, often clustered on short axillary shoots, bipinnate, petiolate, the pinnae bearing several leaflets; internodal spines curved; stipules small and deciduous; flowers numerous, borne in elongate spikes; calyx 5-lobed; corolla regular, 5-lobed, inconspicuous; stamens numerous, included, distinct; ovary substipitate; pods flattened, indehiscent.

Acacia greggii Gray Catclaw Acacia. Small trees to 4 m tall, the branches armed with curved internodal spines; leaves to ca 4 cm long, with 2 pairs of pinnae, each with 4-6 pairs of obovate to oblong leaflets 3-6 mm long, puberulent on both surfaces; petioles 2-5 mm long, bearing a solitary gland between the lower pair of pinnae; spikes mostly 3-6 cm long (including peduncles); flowers fragrant, 2-2.5 mm long; petals greenish, like the sepals; legumes flattened, oblong, usually curved, 5-10 cm long, 10-20 mm wide, constricted between the seeds; seeds 5-7 mm broad, nearly circular. Warm desert shrub, drainage-terrace vegetation, at ca 870 m in Washington County; Nevada, California, Arizona, New Mexico, Texas, and Mexico; 10 (iii). Our material belongs to var. *arizonica* Isely.

Albizia Durazz.

Unarmed trees; leaves alternate, not clustered, bipinnate, petiolate, the several pairs of pinnae each with numerous oblique leaflets; stipules small and caducous; flowers several to many, in umbellate heads; calyx tubular, 5-lobed; corolla united, funnelform, the 5 lobes shorter than the tube; stamens numerous, united into a tube basally, long-exserted; pods flattened, dehiscent.

Albizia julibrissin Durazz. Silk-tree; Mimosa. Small tree to 3 m tall or more and as broad or broader; leaves to 25 cm long or more (including petiole), with 5-10 (15) pairs of pinnae, each with 12-25 (30) pairs of leaflets 7-15 mm long, puberulent, if at all, on rachis and leaflet margins; petioles 3-6 cm long, each with a single large flattened gland; calyx 3-3.5 mm long; corolla 7.5-9.5 mm long, cream to greenish; staminal filaments exserted 20-30 mm, brightly rose pink to reddish in color; pods 12-20 cm long, 15-25 mm wide, oblong, flattened, membranous; $2n = 26$. Cultivated ornamental at lower elevations in much of Utah, but frost sensitive; introduced from Asia; 7 (i).

Amorpha L.

Cultivated shrubs; leaves alternate, odd-pinnate, the leaflets marked with dots, usually with stipels; flowers purple, borne in terminal spicate racemes; calyx 5-toothed, persistent; banner present (wings and keel lacking), wrapped around the stamens and style; stamens 10, monadelphous at the base only, otherwise distinct; pods 1- to 2-seeded, tardily dehiscent.

1. Plants usually less than 1 m tall; petioles short, usually shorter than width of lowest leaflets .... *A. canescens*
— Plants usually more than 1 m tall; petioles elongate, longer than the width of the lowest leaflets .... *A. fruticosa*

Amorpha canescens Pursh Lead Plant. Subshrub, the erect branches, mostly 4-10 dm tall, the herbage densely white-villous; leaves subsessile, 2.5-12 cm long, with 15-51 leaflets, these elliptic to lance-elliptic or oblong, green above, white-hairy beneath; racemes clustered, paniculate; calyx tube white-villous; pods white-villous, the style almost as long as the body. Cultivated ornamental in some communities in northern Utah (Reimschussel s.n., BRY); introduced from the Great Plains; indigenous from Canada south to Texas and New Mexico; 2 (0).

Amorpha fruticosa L. False Indigo; Bastard Indigo. Shrub to 3 m tall or more, the herbage sparingly pubescent to glabrate; leaves long-petioled, with 13-35 leaflets, these elliptic to oblong, green on both sides, the lower only somewhat paler and strigulose; racemes clustered, paniculate; calyx tube glabrous; pods glabrous, the style much shorter than the body; $2n = 40$. Cultivated ornamental and botanical curiosity in northern Utah; introduced from eastern U. S.; indigenous in much of eastern North America and southwestward to Arizona; 5 (0).

Anthyllis L.

Cultivated herbaceous perennial; leaves odd-pinnate; stipules small, adnate to the petiole, the lowermost somewhat sheathing; flowers many, borne in pedunculate heads or headlike clusters; calyx tubular, 5-lobed; corolla papilionaceous; stamens 10, monadelphous; pods invested by the accrescent calyx, 1- or few-seeded.

Anthyllis vulneraria L. Kidney Vetch, Woundwort. Stems arising from a caudex, 8-30 cm tall, decumbent to erect; leaves 2-7 cm long, odd-pinnate, with usually 5-9 leaflets, the terminal leaflet of lowermost leaves much larger than the lateral ones; peduncles 5-16 cm long, usually with a foliose bract below the inforescence; heads 1 to few, each closely subtended by foliose bracts; flowers 10-15 mm long, sessile, yellow (or suffused with red); calyx pilose, much inflated at maturity; $2n = 12$. Introduced forage and reseeding plant, known from Sanpete County, but to be expected elsewhere; indigenous to Eurasia; 1 (0).

Arachis L.

Cultivated annual herbs; leaves even-pinnate, lacking tentrils; stipules prominent, long-attenuate, adnate to the petiole and almost sheathing the stem; flowers yellow, papilionaceous, few or solitary in the axils, sessile, hypanthium elongating and pushing the developing ovary underground; stamens diadelphous, usually 9 and 1; pods maturing underground, indehiscent, constricted between the seeds.

Arachis hypogaea L. Peanut. Stems from a taproot, mostly 20–50 cm tall; leaves mostly 4–15 cm long, even-pinnate, with 2 pairs of leaflets, 2.2–6 cm long, 0.8–2.5 cm wide, entire; stipules 20–35 mm long; flowers yellow, soon withering, usually only the lowermost producing fruits. Sparingly cultivated plants, rarely escaping but not persisting, in Utah; introduced from Brazil (?); 2 (i).

Astragalus L.

Plants annual or perennial, caulescent or acaulescent, from a taproot, a caudex commonly developed, rarely with a rhizome; leaves alternate, odd-pinnate, trifoliolate, or simple; stipules adnate to the petiole base, sometimes connate-sheathing around the stem; flowers papilionaceous, in axillary racemes, each subtended by a single bract; bracteoles 1 or 2 or lacking, attached at base of calyx or on pedicel; calyx 5-toothed; petals 5, pink, lavender, pink purple, orcholeucous, or white, or variously suffused, the keel shorter than the wings, rounded to attenuate apically; stamens diadelphous; ovary enclosed in the staminal sheath, the style glabrous; pods variable in size, shape, and dehiscence, unilocular to bilocular, sessile, subsessile, or stipitate (or with a gynophore). Note: This is a large and complex genus, certainly the largest genus of flowering plants in Utah, and because of this, the keys to species are constructed as to reflect political geographic subdivisions of the state. This makes it possible to identify unknown plants without the effort of struggling through a single interminably long key. Keys and descriptions are based on some 5000 specimens from Utah, a quarter of them collected by the author.

Barneby, R. C. 1964. Atlas of North American *Astragalus*. Mem. N. Y. Bot. Gard. 13: 1–199.
Jones, M. E. 1923. Revision of North American species of *Astragalus*. By the Author, Salt Lake City, Utah. 288 pp.
Rydberg, P. A. 1929. Astragalanae. North American Flora 24: 251–462.

1. Plants of northwestern Utah (i.e., Box Elder, Cache, Davis, Juab, Morgan, Salt Lake, Tooele, Utah, and Weber counties) .... Key 1
— Plants not from northwestern Utah 2
2(1). Plants of south-central and southwestern Utah (i.e., Beaver, Iron, Millard, Piute, Sanpete and Sevier counties) .... Key 2
— Plants not from south-central and southwestern Utah ... 3
3(2). Plants of Washington County .... Key 3
— Plants not of Washington County 4
4(3). Plants of northeastern Utah (i.e., Daggett, Duchesne, Rich, Summit, Uintah, and Wasatch counties) .... Key 4
— Plants of east-central and southeastern Utah 5
5(4). Plants of Carbon, Emery, and Wayne counties .... Key 5
— Plants of Garfield, Grand, Kane, and San Juan counties . 6
6(5). Plants of Grand and San Juan counties .... Key 6
— Plants of Garfield and Kane counties .... Key 7

Key 1.

Plants of Box Elder, Cache, Davis, Juab, Morgan, Salt Lake, Tooele, and Weber counties.

1. Plants prostrate; leaves to 1 cm long, with spinulose-tipped leaflets all decurrent; usually of high elevations .... *A. kentrophyta*
— Plants of various habit; leaves often more than 1 cm long, not or rarely spinose, the lateral ones usually jointed to the rachis; distribution various 2

2(1). Leaves all simple; plants strictly acaulescent; dwarf, mat-forming plants of high elevations in Cache County .... *A. spatulatus*
— Leaves with (3) 5–25 leaflets or more, or only the uppermost simple; plants caulescent or acaulescent, of low to high elevations 3
3(2). Plants rushlike or sprawling, with slender leaflets, or the uppermost leaves often simple, the terminal leaflet confluent with the rachis 4
— Plants not rushlike, the leaflets various but usually not slender and with the uppermost leaves simple, the terminal leaflet jointed to the rachis 8
4(3). Flowers 15–19 mm long; stems arising from a superficial woody caudex 5
— Flowers less than 10 mm long; stems arising from a slender, subrhizomatous, subterranean caudex 6
5(4). Flowers ochroleucous; calyx brown; legumes long-stipitate, pendulous .... *A. lonchocarpus*
— Flowers bicolored, pink purple with white wing-tips; calyx purplish; legumes sessile, erect .... *A. toanus*
6(4). Pods bladdery-inflated; pubescence of malpighian (dolabriform) hairs; stems usually sprawling .... *A. ceramicus*
— Pods oblong, not inflated; pubescence of basifixed hairs; stems erect 7
7(6). Caudex superficial; leaflets and leaf rachis commonly expanded into flat, grasslike blades; ovules 10–16; pods mostly 10–17 mm long, 3–3.7 mm broad; plants of moist meadows and streambanks, rare .... *A. diversifolius*
— Caudex subterranean; leaflets and leaf rachis usually very narrow; ovules more than 17; pods 25–45 mm long, 2.5–3.3 mm broad; plants of dry hillsides, common .... *A. convallarius*
8(3). Calyx tube less than 4 mm long (longer in some *A. filipes*), campanulate or short-cylindric 9
— Calyx tube more than 4 mm long (as short as 3.6 mm in some *A. anserinus*), cylindric to long-cylindric ... 19
9(8). Plants acaulescent, with a distinctive thatch of persistent leaf bases; pods bladdery inflated; dwarf, at high elevations in western ranges and in the south-central plateaus, rare .... *A. platytropis*
— Plants short- to long-caulescent, not with marcescent leaf bases; pods various; of low to high elevations ... 10
10(9). Plants annual, from slender taproots, usually of low elevation arid sites 11
— Plants perennial, with well-developed caudices; commonly montane 12
11(10). Pods bladdery-inflated, unilocular; flowers 3 or more per raceme; plants of sandy sites .... *A. geyeri*
— Pods curved-oblong, bilocular or nearly so; flowers often 1 or 2 per raceme; plants of various soils ..... *A. nuttallianus*
12(10). Stipules all distinct, not connate-sheathing around the stem 13
— Stipules connate-sheathing around the stem, at least the lowermost 14
13(12). Pods bladdery-inflated, bilocular; leaflets oblanceolate or broader; flowers pale; plants not growing up through shrubs .... *A. lentiginosus*
— Pods oblong, not inflated, unilocular; leaflets narrowly oblong; flowers dirty purplish; plants growing up through shrubs .... *A. pinonis*
14(12). Pubescence malpighian; pods falcately curved, bilocular .... *A. falcatus*

—	Pubescence basifixed; pods straight, unilocular 15
15(14).	Flowers usually some shade of pink or lavender, sometimes as below; pods stipitate or sessile, neither strongly flattened nor bisulcate 16
—	Flowers ochroleucous; pods stipitate, either laterally flattened or bisulcate; inflorescence several times longer than broad 17
16(15).	Keel with a prominent upturned beak; pods and ovaries sessile, laterally compressed; plants common .... *A. miser*
—	Keel merely rounded apically; pods and ovaries stipitate, 3-angled; plants rare .... *A. alpinus*
17(15).	Leaflets linear to oblong; plants mainly 4–9 dm tall; pods long-stipitate, the stipes 6–9 mm long; plants of northwestern Box Elder County .... *A. filipes*
—	Leaflets oblong to lanceolate or oblanceolate; plants mainly less than 4 dm tall; pods with stipes less than 4 mm long; plants of broad or other distribution 18
18(17).	Stipules turning black on drying; flowers usually 15 or fewer; pods strongly laterally compressed .. *A. tenellus*
—	Stipules not turning black on drying; flowers usually many more than 15; pods bisulcate ...... *A. bisulcatus*
19(8).	Plants acaulescent or subcaulescent, the internodes seldom apparent, the stems then prostrate; herbage usually grayish hairy 20
—	Plants caulescent, the internodes not obscured by leaf bases and stipules (subacaulescent in *A. megacarpus*); stems usually erect or ascending (reclining in *A. chamaemeniscus*); herbage commonly green (grayish hairy in *A. anserinus*) 28
20(19).	Wing tips bilobed; plants with malpighian hairs; flowers commonly 12 mm long or less .... *A. calycosus*
—	Wing tips entire; pubescence basifixed; flowers usually 15 mm long or more 21
21(20).	Plants strictly acaulescent, clothed below with a persistent thatch of leaf bases and stipules 22
—	Plants either not strictly acaulescent or the thatch not, or only poorly, developed 23
22(21).	Flowers ochroleucous, the keel purple-tipped; pods thinly long-pilose, the valves apparent through the pubescence .... *A. eurekensis*
—	Flowers pink purple throughout; pods densely woolly-villous, the valves obscured by hairs .... *A. newberryi*
23(21).	Flowers 9–11.2 mm long, pink purple; pods stramineous, the valves apparent through the thin villous vesture; plants of northwestern Box Elder County .... *A. anserinus*
—	Flowers 14–20 mm long or more, variously colored; pods various; plants of broad or other distribution ... 24
24(23).	Leaves very densely hirsute-tomentose, with the longer hairs straight and spirally twisted; pods bilocular, densely woolly-hairy .... *A. mollissimus*
—	Leaves variously pubescent, but if densely tomentose, the hairs all fine, sinuous, and cottony, none straight and spirally twisted 25
25(24).	Pubescence of leaves, and commonly of entire plant, softly villous-tomentose, consisting of fine, cottony, contorted or entangled hairs; pods both villous-tomentose and hirsute 26
—	Pubescence of leaves various but not of extremely fine entangled hairs; pods merely strigose, or both villous-hirsute and tomentose 27
26(25).	Leaflets mostly obovate and obtuse; flowers bright pink purple .... *A. utahensis*
—	Leaflets various but where the range of this and the preceding overlap (in Box Elder County), either elliptic or the petals whitish .... *A. purshii*
27(25).	Pods hirsute and tomentose, the valves obscured by the long hairs; plants uncommon in northwestern Utah .... *A. piutensis*
—	Pods strigillose to strigose, the valves not obscured by the short hairs; plants common in northwestern Utah .... *A. argophyllus*
28(19).	Plants subcaulescent; inforescences with 1–4 pink purple flowers, soon surpassed by the leaves; pods unilocular, 25–60 mm long, bladdery-inflated .... *A. megacarpus*
—	Plants caulescent; inflorescences with typically 5 to many ochroleucous to pink purple flowers, often surpassing the leaves; pods uni- or bilocular, often less than 25 mm long, or flowers not pink purple, bladdery-inflated or not 29
29(28).	Stems arising from slender rhizomelike caudex branches; flowers subsessile in headlike racemes, erect or ascending; pods erect, long-pilose, less than 12 mm long .... *A. agrestis*
—	Stems arising from a woody caudex; flowers variously arranged but, if headlike, commonly spreading; pods various but, if erect, not long-pilose and often over 12 mm long 30
30(29).	Pubescence of herbage consisting largely or entirely of malpighian hairs; pods erect, oblong-cylindric, fully bilocular; flowers ochroleucous; plants flowering in June and July .... *A. canadensis*
—	Pubescence of herbage consisting of basifixed hairs; pods not as above; flowers variously colored; plants flowering in springtime 31
31(30).	Stipules connate-sheathing, at least some 32
—	Stipules all distinct 34
32(31).	Stems and leaves long-hairy; plants with nodding white flowers; pods pendulous, stipitate, the body narrowly oblong, straight, glabrous, bilocular .... *A. drummondii*
—	Stems and leaves merely strigose; flowers ochroleucous; pods differing 33
33(32).	Pods and ovaries glabrous, the body more than 12 mm long when mature, curved, trigonous, bilocular; plants of foothills and mountains, not with odor of selenium .... *A. scopulorum*
—	Pods and ovaries usually strigose, the body often less than 12 mm long when mature, straight, bisulcate, unilocular; plants of low elevations in clay soils .... *A. bisulcatus*
34(31).	Flowers small, the banner 12 mm long or less 35
—	Flowers larger, the banner 12.5–28 mm long 37
35(34).	Plants cushionlike, grayish hairy; pods stramineus, the valves easily discerned through the thin villous vesture; known from northwestern Box Elder County .... *A. anserinus*
—	Plants of various habitat, but not both cushionlike and grayish hairy; pods of various shape and vesture, but not villous; distribution various 36
36(35).	Pods narrowly lanceolate to lance-elliptic in outline, never inflated, semibilocular; plants of western Box Elder County .... *A. iodanthus*
—	Pods greatly inflated, bilocular; plants of various distribution .... *A. lentiginosus*

37(34). Flowers ochroleucous, or less commonly pink purple; pods either bladdery-inflated or leathery and dorsiventrally compressed, borne on a stipelike gynophore (a stalk of receptacular origin), usually jointed to the pod 38

— Flowers pink purple or bicolored, not or seldom ochroleucous; pods bladery inflated or not, sessile or nearly so (short-stipitate in *A. chamaemeniscus*) 39

38(37). Pods bladdery-inflated, 1-loculed, commonly more than 30 mm long, the stipe (gynophore) more than 2 mm long in flower .... *A. oophorus*

— Pods leathery, subunilocular, never bladdery, dorsiventrally compressed, 15–30 mm long, the stipe (gynophore) 2 mm long or less in flower .. *A. beckwithii*

39(37). Flowers reddish violet to pink purple; pods triquetrous, bilocular, the valves leathery, cross-reticulate; plants known from Beaver and Iron counties .. .. *A. chamaemeniscus*

— Flowers pink purple or bicolored; pods various, but not both triquetrous and bilocular or leathery and cross-reticulate; plants widespread 40

40(39). Flowers usually bicolored, borne in compact racemes; pods oblong in outline, leathery, unilocular .... *A. cibarius*

— Flowers usually pink purple, borne in short to elongate racemes; pods bladdery-inflated, membranous to papery, bilocular .... *A. lentiginosus*

Key 2.

Plants of Beaver, Iron, Millard, Piute, Sanpete, and Sevier counties.

1. Plants mat-forming to erect, with leaflets all spinulose-tipped and decurrent along the rachis .... *A. kentrophyta*

— Plants various but not with both spinulose tips on leaflets and the leaflets all decurrent 2

2(1). Leaves simple, the blades oval to orbicular .... *A. asclepiadoides*

— Leaves plurifoliolate, or if some simple, not as above . 3

3(2). Plants rushlike or sprawling, the terminal leaflet confluent with the rachis and the upper leaves often simple 4

— Plants various, but seldom rushlike, the terminal leaflet jointed to the rachis and all leaves commonly plurifoliolate 9

4(3). Flowers less than 10 mm long; stems arising from a slender subterranean, subrhizomatous caudex 5

— Flowers 12–20 mm long; stems arising from a superficial caudex 6

5(4). Pods bladdery-inflated; pubescence of malpighian hairs; stems usually sprawling .... *A. ceramicus*

— Pods narrowly oblong, not inflated; pubescence basifixed; stems erect .... *A. convallarius*

6(4). Flowers ochroleucous or very pale fresh pink 7

— Flowers pink purple or bicolored 8

7(6). Racemes commonly shorter than the subtending leaves; calyx not brown; pods sessile, curved; plants in rounded clumps, not especially rushlike .... *A. tetrapterus*

— Racemes much longer than subtending leaves; calyx brown; pods long-stipitate, pendulous, straight; plants rushlike .... *A. lonchocarpus*

8(6). Uppermost leaves often simple; flowers bicolored; pods sessile, erect; plants of western Millard County .... *A. toanus*

— Uppermost leaves usually with tiny leaflets; flowers uniformly pink purple; pods stipitate, pendulous; plants of Sevier County .... *A. coltonii*

9(3). Plants annual, usually growing in sand; pods bladdery-inflated .... *A. geyeri*

— Plants perennial, of various soils; pods various 10

10(9). Pubescence of herbage consisting largely or entirely of malpighian hairs 11

— Pubescence of herbage consisting of simple basifixed hairs 15

11(10). Plants acaulescent; stipules all distinct 12

— Plants caulescent; stipules connate, at least those at the lowermost nodes 13

12(11). Wing tips deeply cleft apically; flowers usually bicolored, or varicolored in populations; pods bilocular, oblong .... *A. calycosus*

— Wing tips entire; flowers ochroleucous suffused with purple; pods unilocular, obliquely ovoid; plants of eastern Sevier County .... *A. consobrinus*

13(11). Flowers yellowish; pods erect; plants erect, with odor of selenium .... *A. flavus*

— Flowers pink purple or ochroleucous; pods ascending; plants spreading, not with odor of selenium 14

14(13). Racemes 1– to 3–flowered; flowers pink purple; leaflets 7–11; plants rare in Sanpete (?) County .... *A. sesquiflorus*

— Racemes 7– to many-flowered; flowers ochroleucous or suffused with dull purple; leaflets 11–17; plants of Beaver and Iron counties .... *A. humistratus*

15(10). Flowers small, the banner 12 mm long or less 16

— Flowers larger, the banner 12.5–20 mm long or more 30

16(15). Stipules all distinct, not even the lowermost connate 17

— Stipules connate into a sheath, at least at the lowermost nodes 21

17(16). Flowers 5.5–8.5 mm long, bicolored; pods narrowly oblong, 3–angled, stipitate, the stipe 1.4–2 mm long .... *A. straturensis*

— Flowers over 8.5 mm long or, if shorter, not bicolored; pods either bladdery-inflated or sessile, or both 18

18(17). Flowers dull purplish; pods oblong in outline; caudex usually subterranean; plants growing up through sagebrush .... *A. pinonis*

— Flowers ochroleucous to pink purple; pods bladderyinflated; caudex superficial 19

19(18). Flowers very small, ochroleucous, the banner 5.2–8 mm long; pods diaphanous, unilocular; plants of the Sevier Valley and less commonly around seeps and springs in western valleys .... *A. wardii*

— Flowers larger, or purplish, or the pods not especially diaphanous, or else bilocular; plants of various distribution 20

20(19). Pods with a stipe 1–2.5 mm long, unilocular; flowers usually pink-purple; plants of plateaus in Wayne, Garfield, and Piute counties .... *A. serpens*

— Pods sessile, bilocular; flowers ochroleucous tinged purplish; plants mainly of western Beaver and Iron counties .... *A. lentiginosus*

21(16). Stipules turning black on drying; flowers ochroleucous; pods stipitate, the body strongly laterally flattened .... *A. tenellus*

—	Stipules not turning black on drying; flowers variously colored; pods stipitate or sessile, the body not as above 22
22(21).	Plants subcaulescent, the internodes obscured by stipule and leaf bases 23
—	Plants short- to long-caulescent, the internodes apparent though sometimes short 25
23(22).	Flowers ochroleucous; plants of lake shores and ridges in eastern Iron County .... *A. limocharis*
—	Flowers pink purple, with white wing tips; plants of ridge tops in western Beaver and Sanpete counties .. 24
24(23).	Leaflets glabrous above; pods unilocular; plants of Sanpete County .... *A. montii*
—	Leaflets strigose above; pods semibilocular; plants of western Beaver County .... *A. platytropis*
25(22).	Plants erect or ascending; pods stipitate, pendulous . 26
—	Plants prostrate to decumbent or erect; pods short-stipitate to subsessile or sessile, usually not pendulous 27
26(25).	Flowers deflexed, numerous; plants with odor of selenium .... *A. bisulcatus*
—	Flowers ascending to spreading; plants lacking odor of selenium .... *A. australis*
27(25).	Plants high montane or alpine dwarfs with bladdery-inflated pods, known from Tertiary igneous gravels in Iron, Garfield, Piute, and Sevier counties . *A. perianus*
—	Plants of high to low elevations, spreading, the pods oblong or merely turgid 28
28(27).	Flowers 4.5–6 mm long; peduncles capillary; pods sausagelike; plants of Iron and Piute counties .... *A. brandegei*
—	Flowers 5.9–11 mm long; peduncles not capillary; pods and distribution various 29
29(28).	Plants erect; keel with an elongate erect beak; pods oblong, not at all inflated .... *A. miser*
—	Plants spreading-decumbent, keel tip merely rounded; pods inflated .... *A. subcinereus*
30(15).	Plants acaulescent or subcaulescent; herbage grayish or silvery pubescent 31
—	Plants caulescent (except subcaulescent in some *A. megacarpus*); herbage usually green 42
31(30).	Plants strictly acaulescent; leaflets 3–11, or flowers 1–6 per raceme, or both; thatch of persistent leaf bases and stipules often obscuring the caudex branches 32
—	Plants not strictly acaulescent or, if so, the leaflets more than 11, or flowers more than 8, or both; thatch of persistent leaf bases and stipules poorly developed or lacking 36
32(31).	Herbage pubescent with malpighian hairs (even though the attachment just above the base); flowers ochroleucous, tinged with purple, the keel purple-tipped; plants local in south-central Utah 33
—	Herbage pubescent with basifixed hairs; flowers ochroleucous or pink purple, the keel purple-tipped; plants more widely distributed 34
33(32).	Leaflets (1) 3–5, ovate to obovate, obtuse to emarginate; plants of Sevier County .... *A. loanus*
—	Leaflets 5–9, lanceolate to obovate, typically acute; plants not known from Sevier County .... *A. welshii*
34(32).	Pods merely strigulose, the hairs not over 1 mm long; flowers and fruit included in the short foliage; plants of central Millard County .... *A. uncialis*

—	Pods long-hirsute, with stiff hairs mainly 2.5–4.5 mm long; flowers and fruit variously included or surpassing the foliage; plants of various or other distribution 35
35(34).	Petals commonly ochroleucous, sometimes faintly suffused with purple; valves of pod scarcely obscured by curly hairs .... *A. eurekensis*
—	Petals pink purple; valves of pod obscured by contorted underhairs .... *A. newberryi*
36(31).	Herbage pubescent with malpighian hairs; pods strigose to strigillose; plants of Iron County .... *A. amphioxys*
—	Herbage pubescent with basifixed hairs; pods variously pubescent; plants of broad or other distribution 37
37(36).	Leaves very densely hirsute with the longer hairs spirally twisted; pods bilocular, densely long-hairy .... *A. mollissimus*
—	Leaves variably pubescent, if densely tomentose, the hairs all extremely fine, sinuous, and cottony, none straight and spirally twisted; pods unilocular, variously hairy 38
38(37).	Pubescence of leaves (and commonly of entire plant) softly villous-tomentose, composed of extremely fine, cottony, or contorted and entangled hairs; pods both villous-tomentose and hirsute .... *A. utahensis*
—	Pubescence of leaves various, composed either of straight, appressed of narrowly ascending hairs, or of spreading-incurved and sometimes sinuous and contorted hairs; pods strigulose, villous, or hirsute 39
39(38).	Pods strigulose to strigulose, the valves not obscured by the pubescence .... *A. argophyllus*
—	Pods hirsute or tomentose or both, the valves usually obscured by the pubescence 40
40(39).	Valves of pod hirsute with lustrous hairs, not obscured by the hairs; plants of sandy sites at low elevations in Millard County .... *A. callithrix*
—	Valves of pods shaggy-hirsute and tomentose, almost or quite concealed by the hairs; plants of various distribution 41
41(40).	Leaf pubescence appressed or nearly so; petals not very strongly graduated, the banner 17–21 and the keel 15–19 mm long; pods 20–35 mm long; ovules 27–36; plants widespread .... *A. piutensis*
—	Leaf pubescence mostly ascending; petals strongly graduated, the banner 18–22.5 and the keel 12–13.5 mm long; pods to 12 mm long; ovules 14–16; plants local in Utah County .... *A. desereticus*
42(30).	Plants subcaulescent; inflorescences with 1–4 pink purple flowers, soon surpassed by the leaves; pods unilocular, 25–70 mm long, bladdery-inflated .... *A. megacarpus*
—	Plants caulescent; inflorescences with 5 to many ochroleucous to pink purple flowers, often surpassing the leaves; pods uni- or bilocular, often less than 25 mm long, bladdery-inflated or not 43
43(42).	Stems arising from slender, rhizomelike caudex branches; flowers subsessile in headlike racemes, erect or ascending; pods erect, long-pilose, less than 12 mm long .... *A. agrestis*
—	Stems arising from a woody caudex; flowers variously arranged but, if in headlike racemes, usually spreading; pods various but, if erect, not long-pilose and often over 12 mm long 44
44(43).	Stipules connate into a sheath, at least at the lowermost nodes .... (Note: go to couplet 29, Key 1)

—	Stipules all distinct 45
45(44).	Plants odoriferous selenophytes of clay soils; flowers ochroleucous; pods leathery-woody, cylindric to ovoid, spreading to ascending 46
—	Plants not with odor of selenium, of various soils; flowers ochroleucous, pink purple, or bicolored; pods inflated, often bladdery and membranous but, if leathery, of different shape 47
46(45).	Calyx ochroleucous or whitish, as pale as the petals; pods cylindric, steeply ascending; plants of eastern Sevier County .... *A. pattersonii*
—	Calyx of somewhat different hue from the petals; pods normally spreading-ascending and ellipsoid or broadly cylindric; plants of Beaver, Iron, Millard, and western Sevier counties .... *A. praelongus*
47(45).	Pods and ovaries long-hairy, at maturity plumply ovoid; flowers ochroleucous, steeply ascending; plants introduced in Sanpete County, but to be expected elsewhere .... *A. cicer*
—	Pods and ovaries strigose to glabrous, at maturity variously shaped; flowers ochroleucous, pink purple, or bicolored, seldom steeply ascending; plants indigenous and widespread 48
48(47).	Flowers ochroleucous 49
—	Flowers pink purple or bicolored 50
49(48).	Pods bladdery-inflated, unilocular, usually more than 30 mm long at maturity; stipe more than 2 mm long, subequal to the calyx tube .... *A. oophorus*
—	Pods leathery, subunilocular, never bladdery, dorsiventrally compressed, 15–30 mm long; stipe 2 mm long or less in flower .... *A. beckwithii*
50(48).	Flowers bicolored, the wing tips white, borne in subcapitate racemes; pods oblong, leathery, sessile .... *A. cibarius*
—	Flowers pink purple, borne in open racemes; pods various, but not as above 51
51(50).	Pods and ovaries bilocular, sessile or nearly so; plants widespread and common .... *A. lentiginosus*
—	Pods and ovaries subunilocular or unilocular, shortly to moderately stipitate; plants of rather restricted range in Beaver, Iron, and western Millard counties 52
52(51).	Pods bladdery-inflated, unilocular, usually more than 30 mm long; stipe subequal to the calyx tube, this more than 7 mm long .... *A. oophorus*
—	Pods leathery, subunilocular, never bladdery, dorsiventrally compressed 15–30 mm long; stipe shorter than the calyx tube, this less than 7 mm long .... *A. beckwithii*

Key 3.

Plants of Washington County.

1.	Plants rushlike or sprawling, the terminal leaflet confluent with the rachis and the upper leaves often simple 2
—	Plants various, but seldom rushlike, the terminal leaflet jointed to the rachis and all leaves plurifoliolate 5
2(1).	Flowers more than 12 mm long, the petals ochroleucous or suffused with pale pink; pods curved, dorsiventrally compressed .... *A. tetrapterus*
—	Flowers less than 12 mm long, the petals variously colored; pods neither curved nor dorsiventrally compresed 3
3(2).	Pods and ovaries stipitate, at maturity bladdery-inflated; pubescence malpighian; stems from elongated rhizomelike caudex branches .... *A. ceramicus*
—	Pods and ovaries sessile or subsessile, oblong in outline, not inflated; pubescence basifixed; stems from a superficial to deep-seated caudex 4
4(3).	Stipules connate into a sheath, at least at lowermost nodes; plants of northwestern Washington County .... *A. convallarius*
—	Stipules all distinct; plants of eastern Washington County .... *A. lancearius*
5(1).	Plants slender, diminutive annuals with tiny flowers and curved bilocular pods .... *A. nuttallianus*
—	Plants perennial and otherwise commonly differing from above 6
6(5).	Pubescence of herbage consisting largely or entirely of malpighian hairs 7
—	Pubescence of herbage consisting of basifixed hairs (incipiently malpighian in some *A. piutensis*) 11
7(6).	Flowers pink purple, more than 12 mm long; pods 20–30 mm long or more .... *A. amphioxys*
—	Flowers variously colored but, if pink-purple, usually less than 12 mm long or pods shorter than 12 mm long 8
8(7).	Plants acaulescent; wing petals deeply cleft apically; pods bilocular .... *A. calycosus*
—	Plants with well developed stems; wing petals entire apically; pods uni- or bilocular 9
9(8).	Flowers ochroleucous, nodding at anthesis; pods erect, bilocular, oblong-cylindric ...... *A. canadensis*
—	Flowers ochroleucous to pink purple, ascending at anthesis; pods erect to spreading, variously shaped, unilocular 10
10(9).	Stems prostrate-spreading; pods curved, not bisulcate ventrally; plants not with odor of selenium .... *A. humistratus*
—	Stems erect or ascending; pods straight, bisulcate ventrally; plants with odor of selenium ...... *A. flavus*
11(6).	Flowers small, the banner 12 mm long or less 12
—	Flowers larger, the banner 12.5–25 mm long or more 15
12(11).	Flowers 6.5–8.5 mm long, bicolored; pods narrowly oblong, 3-angled, stipitate, the stipe 1.4–2 mm long .... *A. straturensis*
—	Flowers over 8.5 mm long or, if shorter, not bicolored; pods bladdery-inflated, sessile or subsessile ... 13
13(12).	Plants subacaulescent, usually less than 5 cm tall; flowers 1–5; rare in sandy sites in eastern Washington County .... *A. striatiflorus*
—	Plants caulescent, usually over 20 cm tall; flowers numerous; plants of various habitats and distribution 14
14(13).	Pods bilocular, diaphanous; lowermost stipules distinct .... *A. lentiginosus*
—	Pods unilocular, opaque and usually mottled; lowermost stipules shortly connate-sheathing .... *A. subcinereus*
15(11).	Plants strictly acaulescent or subacaulescent; herbage grayish, silvery hairy, thinly pilose and green 16
—	Plants caulescent; herbage usually green 23
16(15).	Plants strictly acaulescent 17
—	Plants subacaulescent 18
17(16).	Caudex branches obscured by a thatch of persistent leaf bases; herbage grayish or silvery hairy; leaflets 11 or fewer .... *A. newberryi*

— Caudex simple, not obscured by persistent leaf bases; herbage green; leaflets often more than 11, at least on some leaves .... *A. holmgreniorum*

18(16). Leaflets more than 21 on at least some mature leaves; pods strigose or strigulose .... *A. tephrodes*

— Leaflets fewer than 21 on all leaves or the pods densely villous 19

19(18). Leaves very densely hirsute with the longer hairs spirally twisted; pods bilocular, densely long-hairy .... *A. mollissimus*

— Leaves variably pubescent, if densely tomentose, the hairs all extremely fine, sinuous, and cottony, none straight and spirally twisted; pods unilocular, variously hairy 20

20(19). Pods densely long-hairy 21

— Pods merely strigose or strigulose 22

21(20). Leaflets oval to orbicular, rounded apically, white cottony-hairy; calyx tube more than 4 mm wide (when pressed) .... *A. utahensis*

— Leaflets elliptic to obovate, obtuse or acute to rounded apically, silvery strigose; calyx tube less than 4 mm wide .... *A. piutensis*

22(20). Leaflets mostly acute; pods brightly mottled; plants of rocky ledges and talus of sandstone canyons and escarpments .... *A. zionis*

— Leaflets obtuse to acute; pods not mottled; plants mostly of humus in mountain brush and upwards .. .. *A. argophyllus*

23(15). Stipules connate-sheathing, at least at the lowermost nodes 24

— Stipules all distinct 25

24(23). Petals ochroleucous, the keel purple-tipped; pods pendulous, the body oblong-cylindric, bisulcate; plants with odor of selenium .... *A. bisulcatus*

— Petals pink purple or ochroleucous; pods erect, the body ovoid, not bisulcate .... *A. ampullarius*

25(23). Plants odoriferous selenophytes, usually of clay soils; flowers ochroleucous or pink purple with white wing-tips; pods oblong-cylindric, ascending or spreading . 26

— Plants not odoriferous selenophytes, of various soils; flowers variously colored; pods various 27

26(25). Flowers bicolored; calyx suffused dark purple; pods erect-ascending .... *A. preussii*

— Flowers ochroleucous; calyx greenish; pods ascending-spreading .... *A. praelongus*

27(25). Pods and ovaries stipitate, the stipe at maturity subequal to or surpassing the calyx 28

— Pods and ovaries sessile or nearly so 29

28(27). Stems decumbent to ascending; pods bladdery-inflated, unilocular, the body usually over 25 mm long; flowers ochroleucous or pink purple ...... *A. oophorus*

— Stems typically erect; pods not or only somewhat inflated, bilocular, the body usually less than 20 mm long; flowers ochroleucous or pink purple .... *A. eremiticus*

29(27). Pods bladdery-inflated, ovoid or merely curved-oblong, bilocular, sessile; flowers variously colored, but not usually bicolored .... *A. lentiginosus*

— Pods not bladdery, oblong, usually curved, bi- or unilocular; flowers commonly bicolored 30

30(29). Flowers borne in subcapitate racemes; pods dorsiventrally compressed, unilocular .... *A. cibarius*

— Flowers borne in elongate racemes; pods laterally or trigonously compressed, bilocular .... *A. ensiformis*

Key 4.

Plants of Daggett, Duchesne, Rich, Summit, Uintah, and Wasatch counties.

1. Plants matforming or erect, with leaflets all spinulose-tipped and decurrent along the rachis .... *A. kentrophyta*

— Plants various, but not with the leaflets both spinulose-tipped and decurrent on the rachis 2

2(1). Leaves simple, oval to orbicular; plants with odor of selenium .... *A. asclepiadoides*

— Leaves various but, if simple, linear to linear-oblanceolate; plants with or without odor of selenium 3

3(2). Leaves simple (rarely with leaflets on some leaves), the blades grasslike; plants acaulescent 4

— Leaves plurifoliolate, or rarely trifoliolate, or only the uppermost simple; plants various but, if acaulescent, never as above 5

4(3). Leaves not over 8 cm long; racemes with flowers usually fewer than 8, less than 3.5 cm long in fruit .... *A. spatulatus*

— Leaves usually more than 8 cm long; racemes with flowers usually more than 8, more than 4.5 cm long in fruit .... *A. chloodes*

5(3). Plants rushlike or sprawling, the terminal leaflet confluent with the rachis, and some of the upper leaves often simple (see also *A. duchesnensis*) 6

— Plants various, but seldom rushlike, the terminal leaflet jointed to the rachis and all leaves plurifoliolate (except in some *A. detritalis*, *A. aretioides* and *A. gilviflorus*) 10

6(5). Flowers 10 mm long or less; stems arising from a slender subterranean subrhizomatous caudex 7

— Flowers 15-20 mm long or more; stems arising from superficial caudices 8

7(6). Pods bladdery-inflated; pubescence of malpighian hairs; stems usually sprawling .... *A. ceramicus*

— Pods narrowly oblong, not inflated; pubescence of basifixed hairs; stems commonly erect .... *A. convallarius*

8(6). Flowers pink purple; pods leathery-woody, laterally compressed; plants selenophytes of clays and silts, restricted to Uintah County .... *A. saurinus*

— Flowers ochroleucous to yellow; pods various, but not as above; plants selenophytes or not, from Uintah or Daggett counties 9

9(8). Pods stipitate, pendulous; plants not with odor of selenium, usually 3.5 dm tall or more, known only from Uintah County .... *A. hamiltonii*

— Pods sessile or subsessile, ascending; plants seleniferous, less than 3.5 dm tall, known only from Daggett County .... *A. nelsonianus*

10(5). Plants annual, usually growing in sand; pods bladdery-inflated .... *A. geyeri*

— Plants perennial, of various soils; pods various 11

11(10). Pubescence of herbage consisting largely or entirely of malpighian hairs 12

— Pubescence of herbage consisting of basifixed hairs . 18

12(11). Plants acaulescent or subacaulescent, the herbage grayish or whitish pubescent 13

— Plants caulescent, the herbage commonly green 16

13(12). Flowers ochroleucous, borne sessile among the leaves; leaves trifoliolate; plants of Rich and Summit counties .... A. gilviflorus

— Flowers pink purple or pale and suffused with purple, pedunculate; leaves with 5 or more leaflets on at least some leaves (trifoliolate in A. aretioides) plants of Daggett, Duchesne, and Uintah counties 14

14(13). Leaves with leaflets 3, silvery strigose; flowers 6–8 mm long; plants local in Daggett County .. A. aretioides

— Leaves simple or with 3–17 leaflets, strigose but not silvery; flowers 13–20 mm long or more; plants more widely distributed 15

15(14). Plants strictly acaulescent; leaflets narrowly oblong, spinulose-tipped; pods linear-oblong ..... A. detritalis

— Plants subacaulescent; leaflets obovate, rounded to obtuse apically; pods ellipsoid .... A. chamaeleuce

16(12). Plants selenophytes, usually of clay soils; flowers yellowish; pods unilocular .... A. flavus

— Plants not selenophytes, of various soils; flowers ochroleucous or pink purple; pods bilocular 17

17(16). Flowers pink purple, erect or steeply ascending at anthesis; stems from a caudex; plants known from Daggett County .... A. adsurgens

— Flowers ochroleucous, nodding at anthesis; stems from creeping rhizomes .... A. canadensis

18(11). Flowers small, the banner 12 mm long or less 19

— Flowers larger, the banner 12.5–25 mm long or more 29

19(18). Stipules all distinct, not even the lowermost connate 20

— Stipules connate-sheathing, at least at the lowest nodes 24

20(19). Plants dwarf, arising from a deeply seated caudex and elongate rhizomelike branches; flowers ochroleucous; pods bladdery-inflated, strigose, unilocular; plants of Green River Shale, Uintah, Wasatch, and Utah counties .... A. lutosus

— Plants differing from above, the caudex superficial; flowers pink purple or ochroleucous; pods various but, if bladdery-inflated, bilocular or spreading-hairy 21

21(20). Flowers ochroleucous, suffused faintly with purple; pods bladdery-inflated, strigose to glabrous; plants of Wasatch and Summit counties .... A. lentiginosus

— Flowers pink purple or bicolored; pods not inflated or, if so, spreading-hairy; plants of Summit, Duchesne, and Uintah counties 22

22(21). Pods strigulose with black and white hairs; plants of willow communities at ca 2730 m on the north slope of Uinta Mts. .... A. robbinsii

— Pods strigose to pilose with usually whitish hairs; plants of low elevations in the Uinta Basin 23

23(22). Leaflets linear to narrowly oblong; calyx teeth broadly triangular, 1 mm long or less; pods pendulous, straight, strigose .... A. duchesnensis

— Leaflets oblanceolate to obovate; calyx teeth narrowly subulate, more than 2 mm long; pods sessile, inflated, spreading, curved, long-pilose ..... A. pubentissimus

24(19). Plants dwarf, less than 6 cm tall; caudex branches with a persistent thatch of marcescent stipules; flowers minute, 6.5 mm long or less; pods bladdery-inflated; known from Rich County .... A. jejunus

— Plants taller; caudex branches not as above; flowers commonly larger; pods not bladdery-inflated; distribution various 25

25(24). Stipules turning black on drying; flowers ochroleucous; pods short-stipitate, strongly laterally flattened .... A. tenellus

— Stipules not turning black on drying; flowers pink purple or ochroleucous; pods various but, if as above, the flowers pink-purple 26

26(25). Plants odoriferous selenophytes; flowers pink purple or ochroleucous, the keel purple-tipped, numerous, nodding at anthesis; pods bisulcate ventrally .... A. bisulcatus

— Plants not smelling of selenium; flowers pink purple or ochroleucous; pods not bisulcate ventrally 27

27(26). Flowers ochroleucous; pods long-stipitate more than 20 mm long at maturity, laterally compressed but not flattened; plants known from Wasatch and Duchesne counties .... A. australis

— Flowers pink purple or, if stipitate, less than 15 mm long and flattened laterally 28

28(27). Pods and ovaries sessile or substipitate, the body flattened; plants known from Duchesne County ... . A. wingatanus

— Pods and ovaries sessile or substipitate, the body various, but not strongly flattened; distribution various .... A. miser

29(18). Plants acaulescent or subacaulescent; herbage grayish or silvery hairy 30

— Plants caulescent (subacaulescent in some A. megacarpus); herbage usually green 34

30(29). Leaves very densely hirsute-tomentose, with the longer hairs straight and spirally twisted; pods bilocular, densely woolly-hairy .... A. mollissimus

— Leaves variously pubescent but, if densely tomentose, the hairs all fine, sinuous, and cottony, none straight and spirally twisted 31

31(30). Plants strictly acaulescent or subacaulescent; flowers pink purple on ascending-erect peduncles; pods only moderately pilose-hirsute; known only from central Uintah County .... A. equisolensis

— Plants subacaulescent to shortly caulescent; flowers ochroleucous or pink purple but, if the latter, the peduncles often reclining (at least in fruit); pods strigose, villosulose, or densely woolly-hairy; distribution various, but usually not of central Uintah County (except for some A. purshii) 32

32(31). Pubescence of leaves various but not of extremely fine entangled hairs; pods merely strigose .... A. argophyllus

— Pubescence of leaves softly villous-tomentose, consisting of fine, cottony, contorted or entangled hairs; pods both villous-tomentose and hirsute 33

33(32). Leaflets mostly obovate and obtuse; flowers bright pink purple .... A. utahensis

— Leaflets either elliptic and acute or petals ochroleucous .... A. purshii

34(29). Plants subacaulescent; flowers pink purple, 1–5 per raceme; pods bladdery-inflated, unilocular, commonly over 30 mm long .... A. megacarpus

— Plants differing from above in one or more ways from above 35

35(34). Plants odoriferous selenophytes, usually of clay soils at lower elevations; flowers nodding at anthesis ... 36

— Plants not with odor of selenium, usually of loamy soils at moderate elevations 37

36(35). Pods and ovaries strigose, at maturity ventrally bisulcate, mostly less than 15 mm long .... A. bisulcatus

— Pods and ovaries glabrous, at maturity trigonous, not bisulcate, mostly over 15 mm long ...... A. racemosus

37(35). Stems arising from slender, rhizomelike caudex branches; flowers subsessile in headlike racemes, erect or ascending; pods erect, long-pilose, less than 12 mm long .... A. agrestis

— Stems arising from a caudex; flowers variously arranged but, if in headlike racemes, spreading; pods not erect, usually more than 15 mm long 38

38(37). Flowers in compact, headlike racemes; petals bicolored; pods oblong, dorsiventrally compressed .... A. cibarius

— Flowers in elongate racemes or, if somewhat shortened, the petals pink purple or the pods bladdery-inflated 39

39(38). Flowers ochroleucous; pods stipitate, pendulous .. .. A. scopulorum

— Flowers pink purple; pods sessile, spreading .... A. lentiginosus

Key 5.

Plants of Carbon, Emery, and Wayne counties.

1. Plants mat-forming or erect, with leaflets all spinulose-tipped and decurrent along the rachis .... A. kentrophyta

— Plants various, but not with both leaflets spinulose-tipped and decurrent along the rachis 2

2(1). Leaves simple, oval to orbicular; pods erect, inflated, stipitate .... A. asclepiadoides

— Leaves usually plurifoliolate, at least some (simple in A. spatulatus); pods various, but not as above 3

3(2). Leaves simple; plants pulvinate-caespitose, of Emery County .... A. spatulatus

— Leaves plurifoliolate, at least some; plants of various distribution 4

4(3). Plants rushlike or sprawling, the terminal leaflet confluent with the rachis, some of the uppermost leaves sometimes simple 5

— Plants various, but not or seldom rushlike, the terminal leaflet jointed to the rachis and all leaves plurifoliolate (rarely the lowermost simple) 14

5(4). Pubescence of herbage consisting largely or entirely of malpighian hairs 6

— Pubescence of herbage consisting of basifixed hairs .. 7

6(5). Flowers 10 mm long or less; stems arising singly from elongate, rhizomelike caudex branches; pods bladdery-inflated, stipitate; plants not with odor of selenium .... A. ceramicus

— Flowers 15 mm long or more; stems arising several together from a subterranean caudex; pods laterally compressed, sessile; plants with odor of selenium .. .. A. woodruffii

7(5). Flowers 10 mm long or less 8

— Flowers 11–25 mm long or more 10

8(7). Flower numerous; calyx densely pilose; pods short, spreading-ascending .... A. moencoppensis

— Flowers 12 or fewer; calyx merely strigose; pods narrowly oblong, declined to pendulous 9

9(8). Pods and ovaries stipitate; flowers bright pink purple; plants of sandy washes, known only from the Waterpocket Fold .... A. harrisonii

— Pods and ovaries sessile; flowers dull purplish; plants of different distribution .... A. convallarius

10(7). Flowers ochroleucous; calyx brown; pods stipitate, pendulous; plants often over 50 cm tall .... A. lonchocarpus

— Flowers pink to pink purple; calyx often cyaneus, not brown; pods various, usually sessile or, if stipitate, less than 40 cm tall 11

11(10). Plants tall selenophytes; flowers 20 mm long or more; plants of clay or silt in San Rafael Swell .. A. rafaelensis

— Plants not with odor of selenium; flowers 17 mm long or less; plants of various soils and distribution 12

12(11). Pods and ovaries sessile or nearly so; petals pale pink or tinged with purplish; plants of San Rafael Swell and southward .... A. episcopus

— Pods and ovaries stipitate, at maturity the stipe 3–6 mm long or more; petals mostly bright pink purple; plants usually west or east of the San Rafael Swell proper 13

13(12). Caudex superficial; flowers spreading to declined at anthesis; plants mostly west of the San Rafael Swell proper .... A. coltonii

— Caudex subterranean; flowers ascending at anthesis; plants east of the San Rafael Swell proper .... A. nidularius

14(4). Plants definitely annual; flowers usually less than 8 mm long; pods bladdery-inflated and unilocular, or curved-oblong and bilocular 15

— Plants perennial, though sometimes flowering the first year, but the flowers then mostly over 8 mm long; pods various 16

15(14). Pods bladdery-inflated unilocular; stems and peduncles often 1 mm wide or more .... A. geyeri

— Pods curved-oblong, bilocular; stems and peduncles filiform, mostly less than 1 mm thick ... A. nuttallianus

16(14). Herbage pubescent largely or entirely with malpighian hairs 17

— Herbage pubescent with basifixed hairs 25

17(16). Plants caulescent; pods short, ascending; stipules connate, at least the lowermost 18

— Plants acaulescent or subacaulescent; pods various; stipules all distinct 19

18(17). Plants odoriferous selenophytes, of low elevations; pods unilocular .... A. flavus

— Plants not smelling of selenium, of moderate to high elevations; pods bilocular .... A. adsurgens

19(17). Wing tips deeply cleft apically; flowers usually bicolored or varicolored in populations; pods bilocular, oblong .... A. calycosus

— Wing tips entire; flowers various in color; pods unilocular, variously shaped 20

20(19). Leaflets (1) 3–5 (more in A. welshii); plants strictly acaulescent, the caudex branches obscured by a thatch of persistent leaf bases; pods spreading-hairy 21

— Leaflets mostly more than 5, at least on some leaves; plants various but, if strictly acaulescent, the thatch poorly or not developed; pods strigose 22

21(20). Leaflets 1–3 on most leaves; flowers pink purple; pods with hairs to 2 mm long; plants rather widespread in the region .... A. musiniensis

— Leaflets 5–11 on most leaves; flowers ochroleucous or tinged purplish; pods with hair 2–2.5 mm long; plants of western Wayne and Garfield counties .... A. welshii

22(20). Flowers ochroleucous, tinged purplish, 11–12.5 mm long; pods laterally compressed only near the apex; plants of western Emery and Wayne counties .... *A. consobrinus*
— Flowers pink purple or bicolored, mostly 15–25 mm long; pods laterally compressed throughout or dorsiventrally so 23

23(22). Pods laterally compressed, straight, persistent; flowers often pale or dull purplish; plants common in the vicinity of the San Rafael swell .... *A. cymboides*
— Pods dorsiventrally compressed, usually curved, deciduous; flowers pink purple or bicolored; plants common to uncommon in the region 24

24(23). Flower usually bicolored; walls of pod at least 1 mm thick, the endocarp and exocarp separated by a thick mesocarp; plants rare in the region .... *A. chamaeleuce*
— Flowers usually bright pink purple; walls of pod less than 1 mm thick, becoming leathery when ripe; plants common .... *A. amphioxys*

25(16). Flowers small, the banner 12 mm long or less 26
— Flowers larger, the banner 12.5–15 mm long or more 39

26(25). Stipules all distinct, not even the lowermost connate 27
— Stipules connate-sheathing at least at the lowermost nodes 32

27(26). Flowers 5 mm long or less, 2–5, borne on linear-filiform peduncles; plants sprawling, usually of volcanic gravels .... *A. brandegei*
— Flowers commonly 7 mm long or more, usually more than 5, borne on substantial poduncles; plants ascending to erect, never really sprawling, seldom if ever of volcanic soils (except *A. serpens*) 28

28(27). Plants subacaulescent, the stems shorter than the inflorescence; stipules prominent ...... *A. desperatus*
— Plants caulescent, the stems longer than the inforescence; stipules hardly prominent 29

29(28). Pods strongly mottled, bladdery-inflated, merely strigose; plants known from western Wayne County .... *A. serpens*
— Pods not or slightly mottled, inflated but hardly bladdery, villous with spreading hairs; plants mostly along the sandy eastern portion of the region 30

30(29). Racemes with 10 or more flowers; plants rare in the region .... *A. pubentissimus*
— Racemes with 10 or fewer flowers; plants rare to locally common 31

31(30). Pods 5–8 mm in diameter, curved-oblong; ovules 10–19; plants rare in the region .... *A. sabulonum*
— Pods 8–11 mm in diameter, ovoid-ellipsoid; ovules 20–28; plants locally common .... *A. pardalinus*

32(26). Stipules turning black on drying; flowers ochroleucous; pods short-stipitate, strongly laterally flattened; plants of the high plateau sections .... *A. tenellus*
— Stipules not turning black on drying; flowers various; pods various but, if as above, the flowers pink purple or plants of low elevations 33

33(32). Stems shorter than the inflorescences; flowers pink purple, numerous; pods erect, sessile .... *A. moencoppensis*
— Stems longer than the inflorescences or, if shorter, the flowers 10 or fewer per raceme; pods seldom erect 34

34(33). Plants odoriferous selenophytes; flowers ochroleucous, numerous, nodding at anthesis; pods bisulcate ventrally .... *A. bisulcatus*

— Plants not smelling of selenium; flowers pink purple or sometimes ochroleucous, ascending at anthesis; pods not bisulcate ventrally 35

35(34). Pods and ovaries stipitate, the body flattened or triquetrous 36
— Pods and ovaries sessile or subsessile, the body various but not strongly flattened 37

36(35). Pods not strongly compressed; wing petals bilobed; plants of high elevation in Emery County . *A. australis*
— Pods strongly compressed; wing petals entire; plants of lower elevations in Carbon and Emery counties .... *A. wingatanus*

37(35). Plants with elongate clambering stems, known from moist meadows in western Wayne, Piute, and Sevier counties .... *A. bodinii*
— Plants with short to elongate, erect to sprawling stems, of different distribution or habitat 38

38(37). Plants with sprawling stems, commonly of pinyon-juniper and desert shrublands; pods mostly 15 mm long or less; known from lower elevations Carbon and Emery counties .... *A. flexuosus*
— Plants with upright stems, usually of aspen or spruce-fir woodlands; pods laterally compressed, usually 15 mm long or more .... *A. miser*

39(25). Plants acaulescent or subacaulescent; herbage grayish or silvery hairy 40
— Plants caulescent (subacaulescent in *A. megacarpus*); herbage usually green 43

40(39). Herbage merely strigose; flowers 13–15 mm long; stipules prominent; pods hirsute .... *A. barnebyi*
— Herbage tomentose to strigose; flowers 15–25 mm long; stipules not prominent; pods variously pubescent 41

41(40). Leaves very densely hirsute-tomentose, with the longer hairs straight and spirally twisted; pods bilocular, densely woolly-hairy .... *A. mollissimus*
— Leaves variously pubescent but, if densely tomentose, the hairs all fine, sinuous, and cottony, none straight and spirally twisted 42

42(41). Pubescence of leaves softly villous-tomentose, consisting of fine, cottony, contorted or entangled hairs; pods both villous-tomentose and hirsute .. *A. utahensis*
— Pubescence of leaves strigose, not of fine entangled hairs; pods strigose .... *A. argophyllus*

43(39). Plants subacaulescent; flowers pink purple, 1–5 per raceme; pods bladdery-inflated, unilocular, often more than 30 mm long .... *A. megacarpus*
— Plants differing in one or more ways from above 44

44(43). Plants odoriferous selenophytes, usually of clay soils at low elevations; flowers ochroleucous and nodding or the calyx purple and flowers ascending 45
— Plants not smelling of selenium, of various soil types and elevations; flowers not with either of the combinations noted above 48

45(44). Flowers ochroleucous, nodding at anthesis; calyx whitish, cream, or greenish; pods woody 46
— Flowers pink purple or bicolored, ascending at anthesis; calyx purple; pods leathery 47

46(45). Calyx and petals concolorous both whitish to cream; pods cylindric, sessile, steeply ascending; plants of Carbon County .... *A. pattersonii*
— Calyx and petals discolorous, the calyx often greenish, the petals ochroleucous; pods spreading ascending or, if steeply ascending, stipitate .... *A. praelongus*

47(45). Pods horizontally spreading or declined, borne on ascending to reclining peduncles; stems seldom more than 10 cm long; plants rare .... *A. eastwoodae*
— Pods erect or steeply ascending, borne on erect peduncles; stems mostly much more than 10 cm long; plants common .... *A. preussii*

48(44). Stems arising from slender, rhizomelike caudex branches; flowers subsessile in headlike racemes, erect or ascending; pods erect, long-pilose, less than 12 mm long .... *A. agrestis*
— Stems arising from a caudex; flowers variously arranged, but not or rarely headlike, spreading at anthesis; pods not as above 49

49(48). Flowers pink to pink purple; pods sessile, bilocular .... *A. lentiginosus*
— Flowers ochroleucous; pods stipitate, uni- or bilocular 50

50(49). Stipules connate-sheathing at least some; pods bilocular, triquetrous, 3–5 mm wide .... *A. scopulorum*
— Stipules all distinct; pods unilocular, bladdery-inflated, 10–30 mm wide .... *A. oophorus*

Key 6.

Plants of Grand and San Juan counties.

1. Leaves simple, oval to orbicular .... *A. asclepiadoides*
— Leaves usually plurifoliolate (simple in *A. spatulatus*) 2

2(1). Leaves simple; plants pulvinate-caespitose, of Grand County .... *A. spatulatus*
— Leaves usually plurifoliolate; plant seldom as above, of various distribution 3

3(2). Plants rushlike or sprawling, the terminal leaflet of at least uppermost leaves confluent with the rachis, some of the uppermost leaves often simple 4
— Plants various but seldom rushlike, the terminal leaflet jointed to the rachis in all leaves, and all leaves plurifoliolate (rarely the lowermost simple) 8

4(3). Pubescence of herbage consisting of malpighian hairs; pods bladdery-inflated .... *A. ceramicus*
— Pubescence of herbage consisting of basifixed hairs; pods not bladdery-inflated 5

5(4). Flowers 10 mm long or less 6
— Flowers 12 mm long or more 7

6(5). Stems shorter than the inflorescence; flowers numerous, pink purple; pods spreading-ascending .... *A. moencoppensis*
— Stems longer than the inflorescences; flowers usually 15 or fewer, dull pink-purple; pods spreading-pendulous .... *A. convallarius*

7(5). Flowers ochroleucous; calyx brown; plants commonly 50 cm tall or more .... *A. lonchocarpus*
— Flowers pink purple; calyx greenish or blackish; plants commonly less than 45 cm tall, local in White Canyon, San Juan County .... *A. nidularius*

8(3). Plants definitely annual (see also *A. sabulonum*); flowers usually less than 8 mm long; pods bladdery-inflated and unilocular, or curved-oblong and bilocular 9
— Plants perennial, though sometimes flowering the first year, but the flowers then mostly more than 8 mm long; pods various 10

9(8). Pods bladdery-inflated, unilocular; stems and peduncles often 1 mm thick or more .... *A. geyeri*

— Pods curved-oblong, bilocular or nearly so; stems and peduncles filiform, mostly less than 1 mm thick ... . *A. nuttallianus*

10(8). Herbage pubescent largely or entirely with malpighian hairs 11
— Herbage pubescent with basifixed hairs 19

11(10). Plants caulescent selenophytes; pods erect, less than 12 mm long .... *A. flavus*
— Plants acaulescent or subcaulescent, not smelling of selenium; pods various 12

12(11). Stems diffuse and prostrate, sometimes matted; racemes 1- to 3-flowered .... *A. sesquiflorus*
— Stems various, but not as above; racemes commonly with more than 3 flowers 13

13(12). Plants strictly acaulescent; flowers less than 12 mm long, or leaflets 5 or fewer or, if more the pods strongly bicarinate 14
— Plants subcaulescent (see also *A. piscator*); flowers various; leaflets 5 or more on at least some leaves ... 16

14(13). Flowers 12 mm long or less; wing petals bilobed apically; pods bilocular .... *A. calycosus*
— Flowers 18–25 mm long; wing petals entire apically; pods unilocular 15

15(14). Leaflets 5–11; pods laterally compressed, strongly bicarinate at maturity; plants of the Canyonlands of Utah .... *A. piscator*
— Leaflets (1) 3–5; pods dorsiventrally compressed, not bicarinate at maturity; plants seldom in the canyonlands proper .... *A. musiniensis*

16(13). Pods narrowly oblong to oblong-ellipsoid, straight, laterally compressed when ripe; flowers typically pale .... *A. cymboides*
— Pods obliquely ovoid to ellipsoid, mostly curved, if straight then dorsally compressed; flowers pale or bright pink purple 17

17(16). Walls of pod at least 1 mm thick, the exocarp and endocarp separated by a thick mesocarp; petals mostly bicolored; plants flowering earlier than the following, when sympatric (as sometimes with *A. amphioxys*) .... *A. chamaeleuce*
— Walls of pod much less than 1 mm thick, becoming leathery when ripe; petals pink purple or bicolored . 18

18(17). Pods persistent or tardily deciduous, mostly lance-ovoid in outline; plants rare in San Juan County ... . *A. missouriensis*
— Pods readily deciduous, ellipsoid in outline; plants common .... *A. amphioxys*

19(10). Flowers small, the banner 12 mm long or less 20
— Flowers larger, the banner 12.5–25 mm long or more 33

20(19). Stipules all distinct, not even the lowermost connate 21
— Stipules connate-sheathing at least at the lowermost node 26

21(20). Plants subcaulescent, the stems shorter than the inflorescences; stipules prominent 22
— Plants caulescent, the stems longer than the inflorescences; stipules inconspicuous 23

22(21). Pods spreading hairy, unilocular, dorsally compressed; plants widespread .... *A. desperatus*
— Pods strigose, bilocular, laterally compressed; plants of the Cedar Mesa Sandstone, San Juan County ... . *A. monumentalis*

23(21). Stems arising from elongate rhizomelike caudex branches; pods narrowly oblong, pendulous; plants of San Juan County .... *A. cronquistii*

— Stems from a superficial caudex; pods ovoid to ellipsoid, mostly spreading; plants of various distribution 24

24(23). Petals whitish, with pink veins; pods ovoid-acuminate, strongly beaked, strigose; plants of Colorado River Canyon east from Moab .... *A. wetherillii*

— Petals pink purple or less commonly ochroleucous; pods ellipsoid to ovoid-ellipsoid, spreading hairy; plants of various distribution 25

25(24). Flowers 8.8–11.7 mm long, commonly more than 10 per raceme; pods mostly more than 8 mm in diameter; plants of the Tavaputs escarpment .... *A. pubentissimus*

— Flowers 5.8–8.2 mm long, commonly fewer than 10 per raceme; pods less than 8 mm in diameter; plants of San Juan County .... *A. sabulonum*

26(20). Stipules turning black on drying; flowers ochroleucous; pods short-stipitate; strongly laterally flattened .... *A. tenellus*

— Stipules not blackening on drying; flowers various; pods various but, if as above, the flowers pink purple 27

27(26). Stems shorter than the inflorescences; flowers pink purple, numerous; pods erect, sessile .... *A. moencoppensis*

— Stems longer than the inflorescences or, if shorter, the flowers 10 or fewer per raceme; pods seldom erect 28

28(27). Plants odoriferous selenophytes; flowers ochroleucous, numerous, nodding at anthesis; pods bisulcate ventrally .... *A. bisulcatus*

— Plants not smelling of selenium; flowers pink purple or sometimes ochroleucous, ascending at anthesis; pods not bisulcate ventrally 29

29(28). Pods sessile or subsessile, the body oblong 30

— Pods stipitate or the body bladdery-inflated 31

30(29). Plants with sprawling or slender and erect stems; pods mostly less than 18 mm long, subterete; mostly of mountain brush and pinyon-juniper communities .... *A. flexuosus*

— Plants with erect stems; pods laterally compressed usually 19 mm long or more; mostly in aspen or spruce-fir communities .... *A. miser*

31(29). Flowers 8 mm long or less; pods strongly laterally flattened; plants mostly in pinyon-juniper woodlands .... *A. wingatanus*

— Flowers 8.5–11.5 mm long; pods various, but not as above; plants of various habitats 32

32(31). Herbage fresh green; racemes usually with fewer than 10 flowers; pods narrowly oblong, not inflated; plants of the Tavaputs Plateau, Grand County .... *A. alpinus*

— Herbage often cinereous; racemes with more than 10 flowers; pods bladdery-inflated; plants of lower elevations in San Juan County .... *A. fucatus*

33(19). Plants acaulescent or subacaulescent; herbage often grayish or silvery hairy 34

— Plants caulescent; herbage often green 37

34(33). Flowers 15 mm long or less; pods merely strigose, linear-oblong; plants of sandstone formations in San Juan County .... *A. cottamii*

— Flowers more than 15 mm long; pods variously pubescent, ovoid-ellipsoid to ellipsoid; plants of various habitats and distributions 35

35(34). Leaflets mostly 17 or more per leaf, densely hirsute-tomentose, with longer hairs straight and spirally twisted; pods bilocular, densely woolly-hairy .... *A. mollissimus*

— Leaflets mostly 15 or fewer, mostly strigose; pods unilocular 36

36(35). Leaflets mostly acute; pods brightly mottled; plants of rocky ledges and talus of sandstone canyons and escarpments .... *A. zionis*

— Leaflets obtuse to acute; pods not mottled; plants mostly of humus in mountain brush and upwards .. .. *A. argophyllus*

37(33). Plants odoriferous selenophytes, usually of clay soils 38

— Plants not smelling of selenium, usually not of clay soil 44

38(37). Stipules connate into a bidentate sheath, at least at the lowermost nodes; flowers pink purple, declined at anthesis; pods stipitate, pendulous, bisulcate ... . *A. bisulcatus*

— Stipules all distinct, even at the lowermost nodes; flowers variously colored but, if declined at anthesis, ochroleucous; pods sessile or stipitate, ascending to spreading, not bisulcate 39

39(38). Calyx tube purple; petals pink purple or bicolored, rarely whitish; pods leathery-inflated 40

— Calyx tube green to whitish, not purplish; petals ochroleucous to whitish 41

40(39). Pods horizontally spreading or declined, borne on ascending to reclining peduncles; stems seldom more than 10 cm long; plants rare .... *A. eastwoodae*

— Pods erect or steeply ascending, borne on erect peduncles; stems mostly much longer than 10 cm; plants locally common .... *A. preussii*

41(39). Flowers declined at anthesis; pods ascending to erect, 9 mm in diameter or less, leathery-woody in texture 42

— Flowers spreading to ascending; pods spreading to declined, usually over 10 mm in diameter, leathery in texture 43

42(41). Calyx and petals concolorous, both whitish to cream-colored; pods cylindric, sessile, steeply ascending; plants not definitely known from the region but nearby in Colorado .... *A. pattersonii*

— Calyx and petals discolorous, the calyx greenish, the petals ochroleucous; pods spreading-ascending or, if steeply ascending, stipitate; plants common in the region .... *A. praelongus*

43(41). Flowers 17–18 mm long, the petals whitish; calyx tube 5.5–6.3 mm long; plants of Paradox and Morrison formations, La Sal Mts. .... *A. iselyi*

— Flowers 28–31 mm long, the petals ochroleucous; calyx tube 11.5–14 mm long; plants of the Mancos and Morrison formations north of the Colorado River .... *A. sabulosus*

44(37). Stems arising from slender, rhizomelike caudex branches; flowers subsessile in headlike racemes, erect or ascending; pods erect, long-pilose, less than 12 mm long .... *A. agrestis*

— Stems arising from a caudex; flowers variously arranged, but seldom headlike and ascending to spreading; pods more than 12 mm long, spreading to pendulous 45

45(44). Pods and ovaries sessile, spreading, either bladdery-inflated or curved-oblong to straight, bilocular .... *A. lentiginosus*

— Pods and ovaries stipitate, descending to pendulous, not much inflated or curved, uni- or bilocular 46

46(45). Flowers ochroleucous; pods bilocular; plants fresh green .... *A. scopulorum*

— Flowers pink purple; pods unilocular; plants more or less cinereus .... *A. coltonii*

Key 7.

Plants of Garfield and Kane counties.

1. Plants mat-forming or erect, with leaflets all spinulose-tipped and decurrent along the rachis .... *A. kentrophyta*

— Plants various, but not with leaflets both spinulose-tipped and decurrent along the rachis 2

2(1). Leaves simple, oval to orbicular .... *A. asclepiadoides*

— Leaves plurifoliolate, at least some 3

3(2). Plants annual (see also *A. subulonum*); flowers commonly less than 8 mm long; pods curved-oblong and bilocular or bladdery-inflated and ovoid 4

— Plants perennial, though sometimes flowering the first year, and then differing in one or more respects from above 6

4(3). Pods bladdery-inflated, unilouclar; plants of eastern Garfield County .... *A. geyeri*

— Pods curved oblong, bilocular; plants of Kane and Garfield counties 5

5(4). Keel tips rounded; leaflets truncate-retuse on all leaves; pods deciduous, dehiscing at both ends after falling .... *A. emoryanus*

— Keel tips pointed; leaflets not truncate-retuse on all leaves; pods persistent, dehiscent at tip only .... *A. nuttallianus*

6(3). Plants with terminal leaflets confluent with the rachis, the uppermost leaves sometimes simple 7

— Plants with terminal leaflets jointed to the rachis, the leaves all plurifoliolate (or the lowermost simple) ... 16

7(6). Pubescence of herbage consisting largely or entirely of malpighian hairs 8

— Pubescence of herbage consisting of basifixed hairs .. 9

8(7). Flowers 10 mm long or less; stems arising singly from elongate, rhizomelike caudex branches; pods bladdery-inflated, stipitate .... *A. ceramicus*

— Flowers 15 mm long or more; stem arising several together from a subterranean caudex; pods laterally compressed sessile .... *A. woodruffii*

9(7). Flowers ochroleucous, or whitish tinged with pink, 15 mm long or more 10

— Flowers pink purple, bicolored, or pale, 12 mm long or less 11

10(9). Inflorescences usually shorter than the subtending leaves; calyx not brown; pods sessile, curved; plants clump-forming not slender and tall, of western Kane County .... *A. tetrapterus*

— Inflorescence much longer than the subtending leaves; calyx brown; pods long-stipitate, pendulous, straight; plants tall and slender, widespread .... *A. lonchocarpus*

11(9). Stems shorter than the inforescences; flowers numerous; calyx densely pilose; pods short, spreading-ascending .... *A. moencoppensis*

— Stems shorter or longer than the inforescences; flowers few to numerous; calyx merely strigose; pods various but not as above 12

12(11). Flowers dull pink purple, usually 10 mm long or less; pods narrowly oblong, sessile, the body 4 mm wide or less .... *A. convallarius*

— Flowers pale to bright pink purple, usually over 10 mm long, or pods more than 5 mm wide 13

13(12). Pods and ovaries sessile; petals pale; plants west and south of the Henry Mts. 14

— Pods and ovaries stipitate, at maturity the stipe 3–6 mm long or more; plants west or east of those mountains 15

14(13). Calyx short-cylindric, the tube longer than broad, suffused purplish or very pale, white-strigose; ovules 16–26 .... *A. episcopus*

— Calyx campanulate, the tube about as long as broad, not purplish, black-strigose; ovules 8–14 . *A. lancearius*

15(13). Caudex superficial; flowers spreading or declined at anthesis; plants west of the Henry Mts. ..... *A. coltonii*

— Caudex subterranean; flowers ascending at anthesis; plants east of the Henry Mts. .... *A. nidularius*

16(6). Herbage pubescent, largely or entirely, with malpighian hairs 17

— Herbage pubescent with basifixed hairs 25

17(16). Plants caulescent; pods short, ascending 18

— Plants acaulescent or subacaulescent; pods various .. 20

18(17). Plants erect or descending; flowers mostly more than 12 mm long; odoriferous selenophytes ...... *A. flavus*

— Plants prostrate-spreading; flowers less than 12 mm long; not smelling of selenium 19

19(18). Racemes 1- to 3-flowered; flowers pink purple; leaflets 7–11; plants of sandstone escarpments and ledges, mostly at lower elevations ..... *A. sesquiflorus*

— Racemes 7- to many-flowered; flowers dull purplish; leaflets 11–17; plants mostly of higher plateaus .... *A. humistratus*

20(17). Plants subacaulescent, with one or more internodes usually apparent; caudex branches only rarely with persistent leaf bases 21

— Plants strictly acaulescent, the caudex branches usually with a persistent thatch of leaf bases 22

21(20). Flowers usually bicolored; walls of pods at least 1 mm thick, the exocarp and endocarp separated by a thick mesocarp; plants rare in the region, flowering earlier than the next .... *A. chamaeleuce*

— Flowers usually bright pink purple; walls of pod less than 1 mm thick, becoming leathery when ripe; plants common .... *A. amphioxys*

22(20). Flowers mostly less than 12.5 mm long; pods less than 17 mm long 23

— Flowers more than 15 mm long; pods more than 20 mm long 24

23(22). Flowers pink to pink purple, or bicolored, or ochroleucous, the wing petals emarginate apically; pods linear-oblong, bilocular .... *A. calycosus*

— Flowers ochroleucous, the wing petals entire apically; pods ovoid-ellipsoid, unilocular .. *A. consobrinus*

24(22). Leaflets 1–3 on most leaves; flowers pink purple; pods with hairs to 2 mm long; plants mostly east of the high plateaus .... *A. musiniensis*

— Leaflets 5–11 on most leaves; flowers ochroleucous or tinged purplish; pods with hairs 2–2.5 mm long; plants of western Garfield County .... *A. welshii*

25(16). Flowers small, the banner 12 mm long or less 26
— Flowers larger, the banner 12.5–25 mm long or more
   40

26(25). Flowers 6 mm long or less; pods unilocular, borne on slender peduncles, resupinate; stems sprawling; plants usually of volcanic gravels .... A. brandegei
— Flowers over 6 mm long, and the plants differing in other ways, usually not of volcanic gravels (or seldom so) 27

27(26). Plants with stipules all distinct, not even the lowermost connate 28
— Plants with stipules connate-sheathing, at least the lowermost 33

28(27). Plants subacaulescent; stipules prominent; stems shorter than the inflorescences 29
— Plants caulescent; stipules not conspicuous; stems longer than the inflorescences 30

29(28). Pods spreading-hairy unilocular, dorsally compressed; plants widespread along canyons of the Colorado .... A. desperatus
— Pods strigose, bilocular, laterally compressed; plants of Cedar Mesa Sandstone, eastern Garfield County
   .... A. monumentalis

30(28). Pods bladdery-inflated, diaphanous, strigose to glabrous 31
— Pods only moderately inflated, opaque, spreading-hairy 32

31(30). Flowers ochroleucous; pods sessile, glabrous, or nearly so, usually not mottled .... A. wardii
— Flowers pink purple; pods strigose, short-stipitate, strongly mottled .... A. serpens

32(30). Pods 5–8 mm in diameter, curved-oblong; ovules 10–19; plants of eastern Kane County ....
   .... A. sabulonum
— Pods 8–11 mm in diameter, ovoid-ellipsoid; ovules 20–28; plants of eastern Garfield County ....
   .... A. pardalinus

33(27). Plants subacaulescent, the stems shorter than inflorescences; pods bladdery-inflated; growing in sand, on beaches, or volcanic gravels 34
— Plants caulescent, the stems shorter or longer than the inflorescences, when shorter, the pods not inflated; growing in various soil types and habitats 36

34(33). Flowers ochroleucous or pink; plants of beaches in western Kane County or of the Table Cliff Plateau vicinity .... A. limnocharis
— Flowers pink purple to whitish; plants of sandy sites at low elevations or of volcanic gravels at high elevations 35

35(34). Flowers 8–9 mm long, the banner not strongly veined; plants of volcanic gravels at high elevations in western Garfield County .... A. perianus
— Flowers 10–12 mm long, the banner strongly veined; plants of sandy sites at low elevations in Kane County
   .... A. striatiflorus

36(33). Stipules turning black on drying; flowers ochroleucous; pods short-stipitate, strongly laterally flattened
   .... A. tenellus
— Stipules not blackening on drying; flowers various; pods various, but not as above 37

37(36). Stems shorter than the inflorescences; flowers pink purple, numerous; calyx densely villous; pods spreading-ascending, sessile .... A. moencoppensis

— Stems longer than the inflorescences or, if shorter, the flowers 10 or fewer, or the pods not spreading-ascending; calyx merely strigose 38

38(37). Pods oblong, not inflated; plants of aspen and spruce-fir communities at higher elevations .... A. miser
— Pods inflated to bladdery-inflated; plants commonly of ponderosa pine, pinyon-juniper, or desert shrub communities at moderate to lower elevations 39

39(38). Calyces and stems silvery canescent, the hairs appressed; pod bladdery-inflated, more than 12 mm in diameter; ovules 21–32; plants of sandy sites at low elevations in eastern Garfield County ...... A. fucatus
— Calyces and stems not silvery canescent, the hairs spreading; pods moderately inflated, less than 13 mm in diameter; ovules 10–20; plants of moderate elevations in Kane and Garfield counties .... A. subcinereus

40(25). Plants acaulescent or subacaulescent; herbage often grayish or silvery-hairy 41
— Plants caulescent; herbage often green 48

41(40). Flowers 15 mm long or less, pink purple; pods curved, 10–12 mm long, densely spreading-hairy; plants rare and local, Garfield County .... A. barnebyi
— Flowers more than 15 mm long; pods curved or straight, more than 12 mm long, variously hairy; plants of various distribution 42

42(41). Plants strictly acaulescent; caudex branches with a thatch of persistent stipules 43
— Plants subacaulescent; caudex branches with thatch poorly developed or lacking 45

43(42). Flowers ochroleucous, 15–23 mm long; pods lance-ovoid, strigose; plants known only from the Henry Mts. .... A. henrimontanensis
— Flowers ochroleucous or pink purple, 18–25 mm long or more; pods spreading-hairy or villous; plants of various distribution but not as above 44

44(43). Flowers ochroleucous, sometimes suffused with purple; valves of pod scarcely obscured by the curly hairs
   .... A. eurekensis
— Flowers pink purple; valves of pod obscured by contorted under hairs .... A. newberryi

45(42). Leaflets mostly 17 or more per leaf, densely hirsute-tomentose, with longer straight hairs spirally twisted; pod bilocular, densely woolly hairy ..... A. mollissimus
— Leaflets mostly fewer than 17 per leaf, variously hairy, but not with longer straight hairs spirally twisted; pod unilocular, variously hairy 46

46(45). Leaflets oval to obovate, rounded apically; pods woolly hairy .... A. utahensis
— Leaflets oblanceolate to elliptic, obtuse to acute apically; pods merely strigose 47

47(46). Leaflets mostly acute; pods brightly mottled; plants of rocky ledges and talus of sandstone canyons and escarpments, mostly at lower elevations .... A. zionis
— Leaflets obtuse to acute; pods not mottled; plants of loamy soils in mountain brush and upward ....
   .... A. argophyllus

48(40). Plants odoriferous selenophytes, usually of clay soils
   49
— Plants not smelling of selenium, usually not of clay soil 52

49(48). Calyx tube purple; petals pink purple or bicolored; flowers erect-ascending at anthesis; pods leathery-inflated, ascending .... A. preussii

	Calyx tube greenish or whitish, not purplish; petals ochroleucous, or keel tip purple, descending at anthesis; pods various 50
50(49).	Stipules connate into a bidentate sheath, at least at the lowermost nodes; pods stipitate, pendulous, bisulcate .... A. bisulcatus
—	Stipules all distinct, even at the lowermost nodes, pods sessile or shortly stipitate, ascending to spreading, not bisulcate 51
51(50).	Calyx and petals concolorous, both whitish to cream colored; pods cylindric, sessile, steeply ascending; plants rare in the Henry Mts. and Capitol Reef .... A. pattersonii
—	Calyx and petals discolorous, the calyx greenish, the petals ochroleucous; pods broadly ellipsoid, sessile or shortly stipitate, spreading-ascending; plants common .... A. praelongus
52(48).	Plants subcaulescent; flowers pink purple, 1–5 per raceme; pods bladdery-inflated, unilocular, often more than 30 mm long .... A. megacarpus
—	Plants differing in one or more respects from above . 53
53(52).	Stems arising from slender, rhizomelike caudex branches; flowers subsessile in headlike racemes, erect or ascending; pods erect, long-pilose, less than 12 mm long .... A. agrestis
—	Stems arising from a superficial or subterranean caudex; flowers variously arranged, but seldom headlike and ascending to spreading; pods more than 17 mm long, spreading to pendulous 54
54(53).	Stipules connate-sheathing, at least at the lowermost nodes; caudex subterranean 55
—	Stipules all distinct, even at the lowermost nodes; caudex superficial to subterranean 56
55(54).	Pods and ovaries long-stipitate; plants known from clay soils of the Chinle formation ...... A. ampullarius
—	Pods and ovaries sessile or subsessile; plants known from sandy alluvium .... A. hallii
56(54).	Pods and ovaries stipitate, the stipe at maturity subequal to or surpassing the calyx 57
—	Pods and ovaries sessile or nearly so 58
57(56).	Stems decumbent to ascending; pods bladdery-inflated, unilocular, the body mostly more than 25 mm long; flowers ochroleucous; plants commonly of higher middle elevations .... A. oophorus
—	Stems erect; pods not or only somewhat inflated, bilocular, the body usually less than 20 mm long; flowers ochroleucous or pink purple; plants of lower elevations .... A. eremiticus
58(56).	Flowers borne in subcapitate racemes; pods curved-oblong, dorsiventrally compressed, unilocular .... A. cibarius
—	Flowers usually in elongate racemes; pods ovoid to curved-oblong, bilocular, compression variable 59
59(58).	Plants from a subterranean caudex, usually less than 25 mm tall; pods curved-oblong, laterally or trigonously compressed; known from the Kaiparowits, Henry Mts., and Circle Cliffs regions, at moderate elevations .... A. malacoides
—	Plants from a superficial caudex, often more than 25 cm tall; pods curved-oblong and dorsiventrally compressed, or bladdery-inflated and ovoid; widespread .... A. lentiginosus

Astragalus adsurgens Pallas Standing Milkvetch. [*A. striatus* Nutt. ex T. & G.; *A. nitidus* Dougl. ex Hook.; *A. nitidus* var. *robustior* (Hook.) Jones]. Perennial, caulescent, 15–45 cm tall, from a branching caudex; pubescence malpighian; stems decumbent-ascending to erect; stipules 6–10 mm long, at least the lowermost connate-sheathing; leaves 6–12 cm long; leaflets 15–23, 13–28 mm long, 3–9 mm wide, oblong to elliptic, acute or obtuse; peduncles 6–16 cm long; racemes 16- to 50-flowered, the flowers ascending at anthesis, the axis 1.5–13 cm long in fruit; bracts 2–5 mm long; pedicels to 0.8 mm long; bracteoles 0; calyx 5.8–10.5 mm long, the tube 4.4–7 mm long, short-cylindric, strigulose, the teeth 1.4–4.2 mm long, subulate; flowers 13–16 mm long, pink purple; pods erect, sessile, ovoid-oblong, 7–12 mm long, 2.3–3.8 mm thick, bilocular; ovules 9–16; 2n = 32. Juniper-sagebrush communities at 2000 to 3175 m in Carbon, Daggett, Emery, Sanpete, and Sevier counties; Alaska east to Manitoba and south to Washington, Idaho, Colorado, Nebraska, and Iowa; Eurasia; 18 (iii). Our material belongs to ssp. *robustior* (Hook.) Welsh [*A. adsurgens* var. *robustior* Hook.].

Astragalus agrestis Dougl. ex G. Don Field Milkvetch. [*A. dasyglottis* Fisch., ex DC., not Pallas; *A. goniatus* Nutt. ex T. & G.; *A. hypoglottis* var. *polyspermus* T. & G.; *A. agrestis* var. *polyspermus* (T. & G.) Jones]. Perennial, caulescent, 9–43 cm tall, from a subterranean caudex and long rhizomelike caudex branches; pubescence basifixed; stems erect to decumbent-clambering; stipules 4–11 mm long, at least the lowermost connate-sheathing; leaves 2–10 cm long; leaflets 13–23, 4–18 mm long, 2–5 mm wide, narrowly elliptic to lance-oblong, obtuse to retuse or acute, strigulose above and below; peduncles 1.5–15 cm long; racemes subcapitate, 5- to 15-flowered, the flowers ascending-erect at anthesis, the axis 0.5–2.5 mm long in fruit; bracts 3–7 mm long; pedicels 0.5–1.5 mm long; bracteoles 0; calyx 7–12.5 mm long, the tube 5–7.8 mm long, cylindric, villous, the teeth 2.5–5.5 mm long, linear; flowers 17–24 mm long, pink purple, ochroleucous, or almost white; pods short-stipitate, the stipe 0.3–1 mm long, the body 7–10 mm long, 2.8–4.5 mm thick, unilocular, oblong-ellipsoid, silky-villous; ovules 14–26; 2n = 16. Meadows and openings in sagebrush and aspen at 1850 to 3050 m in Box Elder, Cache, Carbon, Daggett, Duchesne, Emery, Garfield, Juab, Kane, Rich, San Juan, Sanpete, Sevier, Summit, Tooele, Uintah, Utah, Wasatch, and Wayne counties; Yukon east to Ontario and south to California, Nevada, New Mexico, Nebraska, and Iowa; 77 (xvi).

Astragalus alpinus L. Alpine Milkvetch. Perennial, caulescent, 2–30 cm tall, from a subterranean caudex and rhizomatous caudex branches; pubescence basifixed; stems decumbent to ascending; stipules 1.5–8 mm long, at least the lowermost connate-sheathing; leaves 3–15 cm long; leaflets 15–26, 6–20 mm long, 2–10 mm wide, ovate to elliptic or oblong, retuse to rounded, strigulose above and below; peduncles 3–15 cm long; racemes 5- to 17-flowered, the flowers erect to declined at anthesis, the axis 0.5–5 cm long in fruit; bracts 1–2.5 mm long; pedicels 0.5–2.3 mm long; bracteoles 0; calyx 3.2–6.3 mm long, the tube 2–3.5 mm long, campanulate, strigulose, the teeth 1–3.2 mm long; flowers 9–12 mm long, pink purple; pods pendulous, stipitate, the stipe 2–5 mm long, the body oblong-lanceolate in outline, 10–17 mm long, 1.5–4 mm thick, strigulose, semibilocular; ovules 5–11; 2n = 16, 32. Aspen and coniferous woods at 2430 to 2730 m in Box Elder (Raft River), Grand (on Tavaputs

Plateau), and Salt Lake (at Brighton) counties; Alaska to Nova Scotia and south to Oregon, Nevada, New Mexico, Wisconsin and Vermont; Eurasia; 3 (0).

Astragalus amphioxys Gray Crescent Milkvetch. Perennial (rarely flowering the first year), subacaulescent to shortly caulescent, 2–35 cm tall, from a weak caudex; pubescence malpighian; stems lacking or up to 20 cm long, the internodes often concealed by stipules; stipules 2–13 mm long, all distinct or the lowermost sometimes connate-sheathing; leaves 2–13 mm long; leaflets (1) 5–21, 3–20 mm long, 1–9 mm wide, elliptic to obovate or oblanceolate, obtuse to acute, strigose on both sides; peduncles 2–15 (20) cm long; racemes 2- to 13-flowered, the flowers ascending at anthesis, the axis 1–6.5 cm long in fruit; bracts 2.5–8 mm long; pedicels 0.6–2.5 mm long; bracteoles 0–2; calyx 6.3–14.2 mm long, the tube 5.2 mm long cylindric, strigose, usually purplish, the teeth 1.1–4.5 mm long, subulate; flowers 16.5–31 mm long, pink purple, rarely white; pods ascending, sessile, 1.5–5 cm long, 5–12 mm thick, usually curved, mostly dorsiventrally compressed, unilocular, strigose; ovules 42–70; 2n = 22. Two rather feeble and sympatric varieties are present.

1. Banner twice as long as the calyx or less; keel 14.3–19 mm long .... *A. amphioxys* var. *amphioxys*
— Banner more than twice as long as the calyx; keel 17–25 mm long .... *A. amphioxys* var. *vespertinus*

Var. amphioxys [*A. shortianus* var. (?) *minor* Gray; *Xylophacos amphioxys* (Gray) Rydb.; *X. aragalloides* Rydb. type from St. George; *X. melanocalyx* Rydb., type from the Beaver Dam Mts.; *A. amphioxys* var. *melanocalyx* (Rydb.) Tidestr.; *A. marcus-jonesii* Munz]. Creosote bush, Joshua tree, blackbrush, indigo bush, salt desert shrub, pinyon-juniper, and less commonly in mountain brush communities at 670 to 2000 m in Emery, Garfield, Grand, Iron, Kane, San Juan, Washington, and Wayne counties; Nevada, Arizona, and New Mexico; 148 (xxxix).

Var. vespertinus (Sheldon) Jones [*A. vespertinus* Sheldon]. Desert shrub, sagebrush, pinyon-juniper, and rarely in mountain brush communities at 670 to 1530 m in Emery, Garfield, Grand, Kane, San Juan, Washington, and Wayne counties; Colorado, Arizona, Nevada, and New Mexico; 144 (xlv).

Astragalus ampullarius Wats. Gumbo Milkvetch. [*Tragacantha ampullaria* (Wats.) Kuntze; *Phaca ampullaria* (Wats.) Rydb.]. Perennial, shortly caulescent, 2–28 cm tall, from a deep subterranean caudex; pubescence basifixed; stems prostrate-ascending, radiating; stipules 2–6 mm long, at least the lowermost connate-sheathing; leaves 3–14 cm long; leaflets 7–15 (19), 4–15 mm long, 2–12 mm wide, obovate, rounded to emarginate, strigose on both sides or glabrous above; peduncles 0.5–9.5 cm long; racemes 5- to 30-flowered, the flowers ascending at anthesis, the axis 1.2–13 cm long in fruit; bracts 1.5–3 mm long; pedicels 1–3 mm long; bracteoles 2; calyx 4.8–7.5 mm long, the tube 4.2–6 mm long, short-cylindric, black strigose, the teeth 0.5–1.5 mm long, broadly triangular; flowers 13.5–22 mm long, pink purple with white wing tips, or ochroleucous; pods ascending-erect, stipitate, the stipe 9–19 mm long, the body ovoid to subglobose, inflated, 12–20 mm long, 8–11 mm thick, subunilocular, glabrous or nearly so; ovules ca. 12. Clay soils of the Chinle and Tropic (?) Shale formations at 970 to 1650 m in Kane (west of the Cockscomb, the type from near Kanab) and Washington counties; Mohave and Coconino counties, Arizona; 13 (iii). The gumbo milkvetch is one of the most unusual of the vast array of species in Utah. Its propensity for the variegated, saline shales of the Chinle Formation near Kanab has long been known. The hypogeous caudex and short, prostrate-ascending stems, which persist in rosette form with marcescent stems and pods of previous years circular-reclining, bleached and skeleton-like, are quite unlike any of the taxa within the region. Phases of *A. eremiticus*, q.v., with inflated ovoid-oblong pods resemble *A. ampullarius*, but the caudex is typically superficial, the stipules are all distinct, and the stems commonly are erect (except in *A. eremiticus* var. *ampullarioides*, q.v.).

Astragalus anserinus Atwood, Goodrich, & Welsh Goose Creek Milkvetch. Dwarf, tufted or matted, shortly caulescent, perennial herbs from a slender taproot; stems 3–11 cm long, decumbent-spreading; herbage villous-tomentose; stipules all free; leaves 1–4 cm long; leaflets 5–15, 3.2–6.5 mm long, obovate; peduncles 1.1–2.4 cm long; racemes with 3–7 flowers, the axis 1–5 mm long, little if at all elongating in fruit; bracteoles lacking; bracts ca 2 mm long, lance-subulate; pedicels 0.6–1.2 mm long; calyx tube 3.6–4.8 mm long, the teeth 1.1–1.8 mm long, subulate; flowers 9–11.2 mm long, pink purple; pods sessile, 9–12 mm long, 5–7 mm wide, deciduous from within the calyx, dorsiventrally compressed, falcately curved, conspicuously trigonous-beaked, thinly villous; ovules 16–20. Sagebrush, rabbitbrush, and juniper communities, on white tuffaceous outcrops at 1525 to 1590 m in northwestern Box Elder (type from Goose Creek) County; Nevada and Idaho; 6 (0).

Astragalus aretioides (Jones) Barneby Cushion Orophaca. [*A. sericoleucus* var. *aretioides* Jones]. Perennial, pulvinate-caespitose, from a branching caudex; pubescence malpighian; stems almost entirely concealed by stipules; stipules 3.5–7 mm long, connate-sheathing, hyaline, glabrous dorsally or nearly so; leaves 0.6–2 cm long, palmately trifoliolate, the leaflets 3–7.5 mm long, 1.2–1.8 mm wide, spatulate to elliptic, acute, silvery strigose on both sides; peduncles 7–15 mm long; racemes 2- or 3-flowered, the flowers ascending, the axis very short; bracts 2–3 mm long; pedicels 1–1.5 mm long; bracteoles 0; calyx 3.3–4.2 mm long, the tube 2.1–2.3 mm long, campanulate, densely long-strigose, the teeth 1.2–2 mm long; flowers 6–8 mm long, pink purple (rarely white); pods ascending, sessile, 4–5 mm long, 1.2–2 mm thick, densely hairy, unilocular. White tufaceous outcrops of the Browns Park Formation at ca 1769 m in Daggett County; Montana and Wyoming; 3 (0).

Astragalus argophyllus Nutt. ex T. & G. Meadow Milkvetch. Perennial, acaulescent to subacaulescent, 1.5–12 cm tall, arising from a superficial caudex; pubescence basifixed; stems obsolete or to 10 cm long, prostrate; stipules 2–10 mm long, all distinct; leaves 1.5–12 cm long; leaflets 7–21, 2–12 mm long, 1–6 mm wide, elliptic, oblanceolate or obovate, acute to obtuse, pilose above and below; peduncles to 9 cm long; racemes 1- to 6-flowered, the flowers ascending at anthesis, the axis little elongating in fruit; bracts 1.8–6.5 mm long; pedicels 1.2–3.8 mm long, bracteoles 0–2; calyx 9–16.8 mm long, the tube 6.5–11.8 mm long, cylindric, pilose with mixed black and white hairs, the teeth 1.5–6 mm long, linear; flowers 15–25 mm long, pink purple; pods ascending, sessile, 1.5–3.7 mm long, 5–13 mm thick, ovoid-acuminate, unilocular, strigose or rarely villous, the valves not obscured; ovules 25–43. Three rather distinctive varieties are present.

1. Flowers 15–17.5 mm long, the keel 12–15.2 mm long; pods curved-ellipsoid, densely silky strigose, 3–4 times longer than broad ..... *A. argophyllus* var. *panguicensis*
— Flowers 18–25 mm long, the keel 15.9–21 mm long; pods various, variably pubescent, mostly 2–3 times longer than broad 2
2(1). Petals bright pink purple, the flowers 22–26 mm long; plants of meadows, streambanks, and lake shores .... *A. argophyllus* var. *argophyllus*
— Petals dull pink purple or pale, the flowers 18–22.5 mm long; plants of sagebrush and mountain brush communities .... *A. argophyllus* var. *martinii*

Var. argophyllus [*Xylophacos argophyllus* (Nutt.) Rydb.; *A. uintensis* Jones, in part (lectotype from Salt Lake Valley); *X. uintensis* (Jones) Rydb.]. Meadows, stream banks, and lake shores at 1400 to 1970 m in Box Elder, Cache, Carbon, Salt Lake, San Juan, Sanpete, Summit, Utah, Wasatch, and Weber counties; Nevada, Idaho, and Wyoming; 22 (iii).

Var. martinii Jones [*A. argophyllus* var. *cnicensis* Jones, type from Thistle; *A. argophyllus* var. *pephragmenoides* Barneby]. Sagebrush, mountain brush, aspen, and spruce-fir communities at 1700 to 3030 m in Beaver, Cache, Carbon, Daggett, Duchesne, Emery, Grand, Iron, Juab, Millard, Piute, Rich, Salt Lake, Sanpete, Sevier, Summit, Tooele, Uintah, Utah, Wasatch, and Wayne counties; Idaho, Wyoming, and Colorado; 127 (xxi).

Var. panguicensis (Jones) Jones [*A. chamaeleuce* var. *panguicensis* Jones, type from Panguitch Lake; *A. panguicensis* (Jones) Jones; *Batidophaca sabinarum* Rydb., type from Cedar Canyon; *A. sabinarum* (Rydb.) Barneby]. Ponderosa pine, aspen-Douglas fir-limber pine, white fir, sagebrush, and pinyon-juniper communities at 2130 to 2900 m in Garfield, Iron, Kane and Washington counties; Arizona; 59 (xiii).

Astragalus asclepiadoides Jones Milkweed Milkvetch. [*Jonesiella asclepiadoides* (Jones) Rydb.]. Perennial, caulescens, 7–62 cm tall, arising from a usually superficial caudex; pubescence basifixed; stems glabrous, erect; stipules 2–15 mm long, all distinct; leaves simple, 1.5–6.5 cm long, 1–5.5 cm wide, ovate, orbicular, or cordate, obtuse to rounded or retuse, glabrous; peduncles 0.5–4.5 cm long; racemes 2– to 12–flowered, the flowers ascending, the axis 0.4–2.5 cm long in fruit; bracts 1–5 mm long; pedicels 1–5 mm long; bracteoles usually 2; calyx 10–17 mm long, the tube 8.3–13 mm long, cylindric, strigose with black hairs, the teeth 1.5–3.8 mm long, linear to subulate; flowers 17–27 mm long, suffused purple or almost ochroleucous; pods erect-ascending, stipitate, the stipe 10–21 mm long, the body ovoid or ovoid-ellipsoid, inflated, 25–35 mm long, 11–16 mm thick, unilocular, glabrous; ovules ca 40; 2n = 24. Saline desert shrub vegetative types on Mancos Shale, Tropic Shale, Carmel, Moenkopi, Arapien, and Duchesne River formations at 1250 to 2320 m in Carbon, Duchesne, Emery, Garfield, Grand (type from Cisco), Sanpete, Sevier, Uintah, and Wayne counties; Colorado; 89 (xxix). This singular selenophyte was first collected by M. E. Jones (1923) "... on sand-bars along the Price River..." in September of 1888, but that habitat is apparently unique, for it is commonly distributed on the clay soils of the Mancos Shale in that vicinity.

Astragalus australis Fisch. Subarctic Milkvetch. [*A. aboriginorum* Richards., for a complete list of synonyms, see Barneby 1964]. Perennial, caulescent, 6–20 cm tall, from a superficial caudex; pubescence basifixed; stems pubescent, erect; stipules 2–7 cm long, at least the lowermost connate-sheathing; leaves 2–7 cm long, sessile; leaflets 7–15, 3–22 mm long, 1–7 mm wide, elliptic, acute, villous to glabrate on both sides; peduncles 2–8.5 cm long; racemes 2– to 30–flowered, compact and ascending at anthesis, the axis 1–10 cm long in fruit; bracts 1.2–5 mm long; pedicels 0.8–3.5 mm long; bracteoles 0; calyx 3.7–6.4 mm long, the tube 2.1–3.9 mm long, campanulate, villous, the teeth 1–3 mm long, subulate; flowers 7.5–12.6 mm long, ochroleucous or suffused with pink, the wing petals bilobed apically; pods pendulous, stipitate, the stipe 2.5–6 mm long, the body obliquely and narrowly elliptic in outline, 13–27 mm long, 3–6 mm wide, semibilocular, glabrous; ovules 8–16; 2n = 16, 32. Sagebrush-grass, spruce-fir, and alpine tundra communities at 1750 to 3630 m in Beaver, Duchesne, Emery, Piute, Sanpete, and Wasatch counties; Alaska east to Gaspe, and south to Oregon, Nevada, Colorado, and South Dakota; Eurasia; 8 (ii). The distribution in Utah is sporadic.

Astragalus barnebyi Welsh & Atwood Barneby Milkvetch. [*A. desperatus* var. *conspectus* Barneby]. Perennial, acaulescent or subacaulescent, 1.5–5 cm tall, from a branching caudex; pubescence basifixed; stems 0–5 cm long, mostly obscured by stipules, these 2–7 mm long, at least some connate-sheathing; leaves 1.5–5 cm long; leaflets 7–17, 3–9 mm long, 0.9–3.2 mm wide, elliptic to oblanceolate, acute to obtuse, strigose on both sides; peduncles 0.5–5.2 cm long; racemes 2– to 8–flowered, the flowers ascending at anthesis, 0.5–2.5 cm long in fruit; bracts 2–4 mm long; pedicels 0.5–1.5 mm long; calyx 6.1–7.7 (8.4) mm long, the tube 5.2–6.5 mm long; short-cylindric, pilose with mixed black and white hairs, the teeth 0.9–1.7 mm long, subulate; flowers 12.2–15 mm long, pink purple or bicolored; pods declined, sessile or short-stipitate, ovoid-ellipsoid, curved, 12–19 mm long, 5–6 mm thick, subunilocular, long silky-pilose; ovules ca 20. Pinyon-juniper woods and mixed desert shrublands on platy shales of the Carmel or on sandstones of Jurassic and Cretaceous ages at 1430 to 1830 m in eastern Garfield and Wayne counties; Navajo County, Arizona; 12 (vi). The Barneby milkvetch is a near congener of *A. desperatus*, q.v., from which it differs in the larger size of flowers and parts and in the usually more compact habit of growth.

Astragalus beckwithii T. & G. Beckwith Milkvetch. Perennial, caulescent, 5–40 (70) cm tall, from a branching caudex; pubescence basifixed; stems decumbent to ascending or erect; stipules 2–10 mm long, all distinct; leaves 2–15 cm long; leaflets (7) 11–27, 3–17 (25) mm long, 2–9.6 mm wide, orbicular to obovate, obtuse to retuse, glabrous to glabrate on both sides; peduncles 3–15 cm long; racemes 7– to 16–flowered, the flowers ascending at anthesis, the axis 1–7 cm long in fruit; bracts 1–7 mm long; pedicels 1–3.5 mm long; bracteoles 2; calyx 7–9.5 mm long, the tube 3.5–5.7 mm long, short-cylindric, sparingly strigose to glabrous, the teeth 2.5–4.4 mm long, subulate; flowers 14.5–21 mm long, ochroleucous to whitish or pinkpurple; pods ascending to declined, stipitate, the stipe (gynophore) 1.5–5 mm long, the body obliquely ellipsoid, leathery, unilocular; ovules 18–41; 2n = 22. Two distinctive varieties are known in Utah.

1. Flowers pink purple or bicolored; plants of western Beaver, Juab, and Millard counties ....
   .... *A. beckwithii* var. *purpureus*
— Flowers ochroleucous, concolorous; plants of central to northwestern Utah .... *A. beckwithii* var. *beckwithii*

Var. **beckwithii** [*Tragacantha beckwithii* (T. & G.) Kuntze; *Phaca beckwithii* (T. & G.) Piper; *Phacomene beckwithii* (T. & G.) Rydb.]. Juniper-pinyon, sagebrush, bunch-grass, and mountain brush communities at 1330 to 2300 m in Beaver, Carbon, Box Elder, Davis, Millard, Morgan, Piute, Salt Lake, Sevier, Tooele (type from W of Lone Rock); Utah, and Weber counties; Nevada; 90 (xii). Plants of var. *beckwithii* are indistinguishable in anthesis from those of *A. oophorus* var. *caulescens*, with which they are sympatric in part of their range (q.v.).

Var. **purpureus** Jones [*A. artemisiarum* Jones; *Phaca artemisiarum* (Jones) Rydb.; *Phacomene artemisiarum* (Jones) Rydb.]. Pinyon-juniper and cool-desert shrublands at 1370 to 2200 m in western Beaver, Juab (type from Deep Creek Mts.), Millard, and Tooele counties; Nevada; 34 (vi).

Astragalus bisulcatus (Hook.) Gray Two-grooved Milkvetch. Perennial, caulescent, 15–70 cm tall, from a caudex; pubescence basifixed; stems erect; stipules 2.5–10 mm long, at least the lowermost connate-sheathing; leaves 3–13.5 cm long; leaflets (7) 15–35, 5–33 mm long, 1.5–11 mm wide, lance-oblong to oblong, elliptic or oblanceolate 2–16 cm long; racemes 25– to 80–flowered, the flowers declined at anthesis, the axis 3–25 cm long in fruit; bracts 2.5–6 mm long; pedicels 1–3.5 mm long; bracteoles 0–2; calyx 3.5–9.6 mm long, gibbous-saccate, the tube 2.8–5.7 mm long, obliquely campanulate, sparingly strigose, the teeth 1–3 mm long, subulate; flowers 8–18 mm long, ochroleucous, whitish, or pink purple, the keel-tip usually purple; pods pendulous, stipitate, the stipe about equaling the calyx tube, the body ellipsoid, 6.5–20 mm long, 2–4.5 mm thick, dorsiventrally compressed, bisulcate, unilocular, strigose or glabrous; ovules 5–15; 2n = 24. The two-grooved milkvetch is an ill-scented primary indicator of selenium in most areas, but non-scented populations are known. Three distinctive varieties are present in Utah.

1. Flowers bright pink purple; calyx purple; plants of Daggett, Uintah and Grand counties ....
   .... *A. bisulcatus* var. *bisulcatus*
— Flowers ochroleucous or whitish, the keel often purple-tipped; calyx ochroleucous, whitish, or greenish; distribution various 2

2(1). Flowers 8–11 mm long; body of pod 6.5–9.5 mm long, prominently reticulate; plants usually montane in Utah
   .... *A. bisulcatus* var. *haydenianus*
— Flowers 10–18 mm long; body of pod 10–17 mm long, smooth or faintly reticulate; plants usually not montane
   .... *A. bisulcatus* var. *major*

Var. **bisulcatus** [*Phaca bisulcata* Hook.; *Tragacantha bisulcata* (Hook.) Kuntze; *Diholcos bisulcatus* (Hook.) Rydb.]. River terraces at 1530 to 2200 m in Daggett, Grand, and Uintah counties; Alberta east to Manitoba and south to Idaho, New Mexico, and Kansas; 20 (v).

Var. **haydenianus** (Gray) Barneby [*A. haydenianus* Gray ex Brandegee; *Tragacantha haydeniana* (Gray) Kuntze; *Diholcos haydenianus* (Gray) Rydb.]. Sagebrush-mountain brush communities at 2130 to 3265 m in Carbon, Duchesne, Grand, San Juan, Sanpete, Sevier, Uintah, and Utah counties; Wyoming, Colorado, New Mexico, and Arizona; 20 (v).

Var. **major** (Jones) Welsh [*A. haydenianus* var. *major* Jones, type from Johnson, Kane County; *A. scobatinatulus* Sheldon; *Diholcos scobatinatulus* (Sheldon) Rydb.]. Pinyon-juniper, sagebrush, mountain brush, and salt desert shrub at 1530 to 2440 m elevation in Beaver, Garfield, Juab, Kane, Rich, Sanpete, Sevier, and Washington counties; Wyoming, Colorado, and Arizona; 30 (viii). This taxon was included within an expanded var. *bisulcatus* by Barneby (1964), who considered it as being taxonomically negligible. However, var. *major* differs in about the same manner and degree as do other varieties in *Astragalus*. It is the dominant type within Utah. A population of var. *major* from north of Glendale in western Kane County lacks the characteristic odor of selenium commonly associated with all varieties of *A. bisulcatus*.

Astragalus bodinii Sheldon Bodin Milkvetch. [*Phaca bodinii* (Sheldon) Rydb.; *A. debilis* sensu authors, not Gray (?); *A. yukonis* Jones; *P. yukonis* (Jones) Rydb.; *A. bodinii* var. *yukonis* (Jones) J. Boivin]. Perennial, caulescent, 15–80 cm long, from a superficial or buried caudex; pubescence basifixed; stems straggling on other plants; stipules 1–7 mm long all connate-clasping; leaves 2.5–8.5 cm long; leaflets 7–17, 2–12 mm long, 1–7 mm wide, oblanceolate to obovate, ovate or elliptic, rounded to emarginate or acute, strigose beneath, glabrous above; peduncles 1.5–10 cm long; racemes 3– to 15–flowered, the flowers ascending at anthesis, the axis 0.5–9 cm long in fruit; bracts 0.5–2.5 mm long; pedicels 0.7–2.2 mm long; bracteoles 0; calyx 4.5–5.2 mm long, the tube 3.4–3.8 mm long, campanulate, black-strigose, the teeth 1.5–1.8 mm long, subulate; flowers 9.5–10.2 mm long, pink purple; pods ascending to spreading, stipitate, the stipe (gynophore) to 1 mm long, the body ellipsoid, 5.5–10 mm long, 3–4.5 mm thick, somewhat trigonous, unilocular, strigose; ovules 2–10; 2n = 24, 32. Wet meadows at 2000 to 2300 m in Rabbit Valley near Loa, Wayne County and along Otter Creek in Piute and Sevier counties; Alaska east to Newfoundland and south to Alberta and Manitoba, and from Wyoming, Nebraska, and Colorado; 11 (iii). Of the great number of species present in Utah, only a handful are true mesophytes. Included in that category are *A. agrestis*, *A. argophyllus* var. *argophyllus*, *A. bodinii*, *A. canadensis*, and *A. diversifolius*. They grow almost exclusively in meadows or other sites that are moist through much of the year. The habitat has been exploited to an extent almost unknown in more arid situations and is seldom explored botanically because of the exploitation, fencing, and private ownership. Despite that exploitation, *A. bodinii* has persisted from its initial discovery by Lester F. Ward (602, US, BRY) in Rabbit Valley, Utah on 18 August, 1875. Possibly *A. bodinii* will yet be found in other meadowlands of the state.

Astragalus brandegei T. C. Porter in Port. & Coult. Brandegee Milkvetch. Perennial, though sometimes flowering as an annual, caulescent, 5–35 (40) cm long, from a branching caudex; pubescence basifixed; stems prostrate-spreading, very slender; stipules 1.5–5 mm long, at least the lowermost usually connate-clasping; leaves 2–11.5 cm long; leaflets 5–15, 6–27 mm long, 0.5–2.6 mm wide, linear-filiform to narrowly oblong, acute to obtuse, strigose beneath, glabrous above; peduncles 2.5–14 cm long, very slender; racemes 1– to

5-flowered, the flowers ascending at anthesis, the axis 0.5–6 cm long in fruit; bracts 1–2 mm long; pedicels 1.2–4 mm long; bracteoles 2; calyx 2.7–4 mm long, the tube 1.8–2.5 mm long, campanulate, black-strigose, the teeth 0.9–2 mm long, subulate; flowers 4.5–6 mm long, ochroleucous or tinged violet; pods pendulous to ascending, sessile or subsessile, the body obovoid to oblong-ellipsoid, 10–18 mm long, 3.5–5 mm thick, slightly dorsiventrally compressed, semibilocular, strigose; n = 11. Volcanic gravels in mixed shrublands or pinyon-juniper at 1650 to 2430 m in Carbon, Emery, Garfield, Iron, Piute, and Wayne counties; Colorado, New Mexico, and Arizona; 10 (v). This is a cryptic plant that is seldom collected, because of its inconspicuous, tiny flowers, slender peduncles, and slender prostrate stems; 10 (v).

***Astragalus callithrix* Barneby** Callaway Milkvetch. Perennial, subacaulescent, 2–11 cm tall, from a caudex; pubescence basifixed; stems lacking or to 10 cm long, prostrate, the internodes often concealed by stipules; stipules 2–5 mm long, all distinct, leaves 2–11 cm long; leaflets 9–21, 2–13 mm long, 1.5–10 mm wide, obovate, suborbicular, or lanceolate, obtuse to truncate or emarginate, villous on both sides; peduncles 2–8 (12) cm long; racemes 5- to 15-flowered, the flowers ascending at anthesis, the axis 0.5–6 cm long in fruit; bracts 3–7.5 mm long; pedicels 1–1.5 mm long; bracteoles 0–2; calyx 6.8–13.3 mm long, the tube 5.5–10.8 mm long, cylindric, villous-pilose, purplish, the teeth 1–3.2 mm long; flowers bright pink purple, 16–26 mm long; pods ascending-spreading, sessile, oblong-ovoid, 10–20 mm long, 5–7.5 mm thick, dorsiventrally compressed, curved, long hairy (the valves not obscured), unilocular; ovules 24–34. Sandy flats and dunes in mixed desert shrublands and juniper communities at 1550 to 1710 m in western Millard County; Nye County, Nevada; a Great Basin endemic; 11 (vi).

***Astragalus calycosus* Torr. ex Wats.** Torrey Milkvetch. Perennial low acaulescent herbs, 1–12 cm tall, from a branching caudex; pubescence malpighian; stems lacking or to 2 cm long, the internodes concealed by stipules, these 1.5–6 mm long, all distinct; leaves 1–8 (12) cm long; leaflets 3–13, 2–19 mm long, 1–7 mm wide, obovate, oblanceolate, or elliptic, obtuse to acute, silvery strigose on both sides; peduncles 0.5–10 cm long, rarely longer; racemes 1- to 8-flowered, the flowers ascending to spreading at anthesis, the axis 0.2–2.5 cm long in fruit; bracts 0.5–2 mm long; pedicels 0.7–3 mm long; bracteoles lacking or minute; calyx 5–8.5 mm long, the tube 4–6.7 mm long, campanulate to short-cylindric, strigose, the teeth 1–4.2 mm long, subulate; flowers 10–16.5 mm long, varicolored, ochroleucous to shades of pink and purple, with white or pale wing-tips, the wings bilobed apically; pods ascending, sessile, narrowly oblong, usually curved, 8–25 mm long, 3–4.5 mm thick, laterally compressed, bilocular, strigose; 2n = 22. Rather widespread and distinctive, the plants of Torrey milkvetch are separable into three varieties, two rare and one common.

1. Leaves with 5–13 leaflets along a rachis usually more than 1 cm long; scapes erect-ascending, usually over 7 cm long; raceme axis usually over 2 cm long; plants of Washington County .... *A. calycosus* var. *scaposus*
— Leaves with 3–13 leaflets along a rachis usually less than 1 cm long; scapes ascending or decumbent, 1–7 cm long; raceme axis less than 2 cm long; plants of various distribution 2

2(1). Leaflets 7–13, mostly 2–6 mm long; plants alpine in extreme west-central Utah, rare ....
    .... *A. calycosus* var. *mancus*
— Leaflets 3–7, mostly 5–19 mm long; plants of lower elevations, widespread, except in the northeastern third of Utah .... *A. calycosus* var. *calycosus*

Var. *calycosus* [*Tragacantha calycosa* (Torr.) Kuntze; *Hamosa calycosa* (Torr.) Rydb.; *A. brevicaulis* A. Nels.]. Mixed desert shrublands, pinyon-juniper, and ponderosa pine woods at 1430 to 2730 m in Beaver, Box Elder, Emery, Garfield, Iron, Juab, Kane, Millard, Piute, San Juan, Sanpete, Sevier, Tooele, Utah, Washington, and Wayne counties; Wyoming, Idaho, Nevada, California, and Arizona; 155 (xxv).

Var. *mancus* (Rydb.) Barneby [*Hamosa manca* Rydb; *A. mancus* (Rydb.) Wheeler]. Ridgetops at 2650 to 3660 m in Deep Creek Mts., Juab County, Nevada; 6 (i).

Var. *scaposus* (Gray) Jones [*A. scaposus* Gray; *Hamosa scaposa* (Gray) Rydb.; *A. candicans* Greene]. Ridgetops at ca 2730 m in Beaverdam Mts., Washington County; Nevada, Arizona, Colorado, and New Mexico; 1 (0).

***Astragalus canadensis* L.** Canada Milkvetch. Perennial, caulescent, 1.5–12 dm tall, erect or ascending; stipules 3–12 mm long or more, at least the lowermost connate-sheathing; leaves 4–30 cm long; leaflets 13–35, 10–52 mm long, 6–16 mm wide, lanceolate, lance-oblong, or elliptic, obtuse to emarginate, strigose on both sides, or glabrous above; peduncles 4–22 cm long; racemes many-flowered, the flowers spreading-declined at anthesis, the axis 2.5–16 cm long in fruit; bracts 1.5–10 mm long; pedicels 0.5–3.5 mm long; bracteoles 0–2; calyx 4.6–10.5 mm long, the tube 4–8.5 mm long, short-cylindric, strigose, the teeth 1.2–4.4 mm long, subulate or triangular; flowers 13.5–17.5 mm long, ochroleucous; pods erect, sessile or subsessile, cylindroid, 10–20 mm long, 2.9–5.2 mm thick, bilocular, strigose or glabrous; ovules 16–28; 2n = 16. Two varieties are known from Utah. These are peculiar among our many low-elevation species in flowering in midsummer.

1. Pods and ovaries glabrous, terete at maturity, not sulcate dorsally; calyx teeth 2.5–4.4 mm long; plants mostly 4–12 dm tall .... *A. canadensis* var. *canadensis*
— Pods and ovaries pubescent, sulcate dorsally at maturity; calyx teeth mostly 1–2.5 mm long; plants 1–5 dm tall ..
    .... *A. canadensis* var. *brevidens*

Var. *brevidens* (Gand.) Barneby [*A. mortonii* f. *brevidens* Gand.]. Meadows, stream banks, lake shores and hillsides at 1830 to 2300 m in Box Elder, Cache, Daggett, Rich, and Summit counties; Montana, Idaho, Oregon, California, Nevada, Wyoming, and Colorado; 10 (v).

Var. *canadensis* [*A. carolinianus* L.]. Stream terraces and lake shores at 1370 to 1600 m in Cache, Piute, Salt Lake, Utah, and Wasatch counties; British Columbia east to Ontario and south to Washington, New Mexico, Texas, Louisiana, Alabama, and South Carolina; 6 (i).

***Astragalus ceramicus* Sheldon** Painted Milkvetch. [*Phaca picta* Gray; *A. pictus* (Gray) Gray, not Boiss. & Gaill.; *A. pictus* var. *foliolosus* Gray; *A. pictus* var. *angustus* Jones, type from Montezuma Canyon, San Juan County; *A. pictus* var. *magnus* Jones, type from Silver Reef, Washington County; *Tragacantha picta* (Gray) Kuntze; *A. angustus* var. *pictus* (Gray) Jones; *A. angustus* var. *ceramicus* (Sheldon) Jones; *A. ceramicus* var. *jonesii*

Sheldon; *A. angustus* (Jones) Jones]. Perennial, caulescent, 3–40 cm tall, from elongate rhizomelike caudex branches and deeply buried caudex; pubescence malpighian; stems sprawling to erect; stipules 1.5–9 mm long, at least some connate-sheathing; leaves 2–17 cm long; leaflets 3–13 or only 1, the terminal continuous with the rachis, 3–30 mm long, 0.5–3 mm wide, filiform to narrowly oblong, obtuse to retuse or acute; peduncles 0.7–7.5 cm long; racemes 2– to 15–flowered (rarely more), the flowers ascending to declined at anthesis, the axis 1–12 cm long in fruit; bracts 1–2.5 mm long; pedicels 0.7–3.1 cmm long; bracteoles 0; calyx 3.1–4.2 mm long, the tube 2.1–3.3 mm long, campanulate, strigose, the teeth 1–2.4 mm long, subulate; flowers 6.3–9.5 mm long, dull purplish to pink or rarely whitish; pods pendulous, stipitate, the stipe 1–3.3 mm long, the body bladdery-inflated, ellipsoid to glabrous, unilocular; ovules 12–29; $2n = 22$. Sandy soils in pinyon-juniper, sagebrush, stream bank, grassland, and mixed desert shrub communities at 1270 to 2360 m in Beaver, Emery, Garfield, Grand, Juab, Kane, Millard, San Juan, Tooele, Uintah, Washington and Wayne counties; Idaho to Montana and North Dakota, south to Arizona, New Mexico, and Oklahoma; 94 (xix). Our highly variable materials belong to var. *ceramicus*. Specimens from the Uinta Basin have the long leaves typical of the Great Plains phase.

Astragalus chamaeleuce Gray in Ives Cicada Milkvetch. [*Phaca pygmaea* Nutt.; *Tragacantha pygmaea* (Nutt.) Kuntze; *A. cicadae* var. *laccoliticus* Jones, type from Henry Mts.; *A. pygmaeus* var. *laccoliticus* (Jones) Jones]. Perennial, acaulescent to subacaulaescent, 2–10 cm tall, from a caudex; pubescence malpighian; stems lacking or to 6 cm long and prostrate the internodes mostly obscured by stipules; stipules 2–7 mm long, all distinct; leaves 2–10 cm long; leaflets 5–17, 4–15 mm long, 2–10 mm wide, obovate to oblanceolate, obtuse to truncate or emarginate, strigose on both sides; peduncles 1–8 cm long; racemes 2– to 11–flowered, the flowers spreading-ascending, the axis 0.9–2 cm long in fruit; bracts 2–5 mm long; pedicels 1–3.5 mm long, short-cylindric, strigose, the teeth 1.5–2.9 mm long; flowers ochroleucous or tinged purplish to pink purple; pods ascending, sessile, oblong-ovoid or ellipsoid, 2–4 cm long, 7–16 mm thick, the fleshy valves ca 3 mm thick, shrinking in ripening, the papery exocarp ultimately lustrous, separating from the veins beneath and appearing quite smooth, mottled, strigose, unilocular; ovules 37–46; $2n = 22$. Juniper-pinyon, sagebrush, mixed desert shrub, and grassland communities at 1530 to 2130 m in Daggett, Duchesne, Emery, Garfield, Grand, and Uintah counties; Colorado and Wyoming; 30 (xi). The cicada milkvetch is the only subacaulescent species of *Astragalus* of its type with malpighian hairs known from the Uinta Basin. To the south of there it is rare, and mingles with both *A. amphioxys* and *A. cymboides*. Mature fruit is necessary for positive identification of *A. chamaeleuce* from those entities. It flowers earlier than either of those species where they are sympatric, and typically bears mature fruit when the others are in full anthesis. The phase from west of the Henry Mts., with pods oval instead of oblong in outline are within the circumscriptions of *A. cicadae* var. *laccoliticus* Jones and probably deserve taxonomic recognition. The plants are intermediate between *A. consobrinus* and *A. chamaeleuce*.

Astragalus chamaemeniscus Barneby Ground-crescent Milkvetch. Perennial, caulescent, from a woody shallowly subterranean caudex, the stems 3–18 cm long, reclining and radiating; pubescence basifixed; stipules 3–7 mm long, the lower ones almost or quite connate-sheathing; leaves 2–8 cm long; leaflets 9–10, 3–11 mm long, 1.5–8 mm wide, obovate to oblanceolate, obtuse or emarginate, loosely pilose below, glabrous above; peduncles 1.5–5 cm long; racemes loosely 3– to 10–flowered, the flowers ascending, the axis not much elongating, 0.5–2 cm long in fruit; bracts 2.5–4.5 mm long; pedicels 1.5–3 mm long; bracteoles 2; calyx 5.8–10 mm long, the tube campanulate-cylindric, 4.7–7 mm long, the teeth 1.1–3 mm long, subulate; flowers 14–19 mm long, reddish violet to pink purple; pods ascending to declined, curved, 2–4.5 cm long, 5–8 mm thick, with a stipe 1–2.5 mm long, triquetrous, sulcate dorsally, the valves stramineus, leathery, prominently reticulate, thinly villosulous, bilocular. Rabbitchush, ricegrass, and galleta communities at 1525 to 1620 m in Iron and Beaver counties; eastern Nevada; a Great Basin endemic; 2 (0).

Astragalus chloodes Barneby Grass Milkvetch. Perennial, acaulescent or subacaulescent, 5–24 cm tall, from a branching caudex; pubescence malpighian; stems obscured by stipules; stipules 2–8 mm long, all usually connate-sheathing; leaves simple 1–13 (17) cm long, 1–3 mm wide, flat or involute, strigose on both sides; peduncles 2–9 cm long; racemes loosely 7– to 23–flowered, the flowers ascending at anthesis, the axis 4.5–24 cm long in fruit; bracts 2–4.5 mm long; pedicels 1–2.5 mm long; bracteoles 0; calyx 4.5–8.5 mm long, the tube 2–3 mm long, campanulate, strigose, the teeth 2.5–5.2 mm long, rigid-spreading; flowers 6.2–8.2 mm long, pink purple; pods erect or ascending, sessile, obliquely lanceolate or oblong in outline, curved, 8–12 mm long, 1.7–3 mm wide, glabrous or strigose, unilocular; ovules 4–8. Entrada, Navajo, Frontier, Dakota, and other sandstone hogbacks and cuestas in pinyon-juniper and mixed desert shrub communities at 1450 to 1700 m in Uintah (type from 6 mi SE of Jensen) County; endemic; 36 (ix). The grass milkvetch simulates a grass, not only in its narrow leaves, but also in the flowers in bud, which resemble grass spikelets.

Astragalus cibarius Sheldon Browse Milkvetch. [*A. webberi* var. *cibarius* (Sheldon) Jones; *Xylophacos cibarius* (Sheldon) Rydb.; *A. arietinus* Jones, type from Cedar City]. Perennial, caulescent, 6–30 cm tall, from a branching caudex; pubescence basifixed; stems decumbent to ascending; stipules 3–8 mm long, all distinct; leaves 3.5–10 cm long; leaflets 11–19, 4–17 mm long, 2–13 mm wide, obovate, oblong, or oblanceolate, obtuse or retuse, strigose beneath, glabrous above; peduncles 3–8 cm long; racemes 4– to 14–flowered, subcapitate at early anthesis, the flowers spreading-ascending, the axis 0.5–2.7 cm long in fruit; bracts 2–4 mm long; pedicels 1–2.5 mm long; bracteoles 0–2; calyx 6.4–9.2 mm long, the tube 5–7 mm long, cylindric, strigose, the teeth 1.4–2.5 mm long; flowers 15–19 mm long, pink purple with white wing-tips or whitish to ochroleucous and variously tinged; pods ascending, subsessile, ellipsoid to oblong, 17–32 mm long, 7–10 mm thick, curved to almost straight, strigose, unilocular, woody or stiffly leathery; ovules 27–32; $2n = 22$. Mountain brush, sagebrush, juniper-pinyon, and mixed desert shrub communities at 1630 to 2430 m in Beaver, Box Elder, Cache, Davis, Garfield, Iron, Juab, Millard, Morgan, Rich, Salt Lake, Sanpete, Sevier, Summit, Tooele, Utah (type from Utah Valley), Wasatch, Washington, and Weber counties; Nevada,

Idaho, Montana, Wyoming, and Colorado; 151 (xix). This species is allied to *A . ensiformis* and *A . malacoides*, but is easily distinguished from both of them on geographical and morphological grounds.

Astragalus cicer L. Chickpea Milkvetch. [*Cystium cicer* (L.) Steven]. Perennial, caulescent. 20–70 cm, tall or more, from a branching caudex; pubescence basifixed; stems prostrate to ascending; stipules 2–8 mm long, at least the lowermost connate-sheathing; leaves 4–21 cm long; leaflets 17–27 (31), 5–40 mm long, 2–14 mm wide, lance-elliptic to oblong, acute to obtuse, strigose on both sides or glabrous above; peduncles 4–12 cm long; racemes densely 10- to 30-flowered, the flowers ascending at anthesis, the axis 2–7 cm long in fruit; bracts 2–6.5 mm long; pedicels 0.3–1.5 mm long; bracteoles 0; calyx 6–9 mm long, the tube 5–6 mm long, short-cylindric, strigulose, the teeth 1.6–3 mm long; flowers 12.5–16.5 mm long, ochroleucous; pods ascending, or by crowding, spreading, subsessile, the body ovoid or subglobose, strongly inflated, 6–14 mm long, 6–10 mm thick, pilose, bilocular; 2n = 64. Introduced forage plant, escaping and persisting in pinyon-juniper, sagebrush, mountain brush, and aspen communities at 1770 to 2930 m in Box Elder, Garfield, Iron, Sanpete, and Washington counties; indigenous to Europe; 10 (0).

Astragalus coltonii Jones Colton Milkvetch. Perennial, caulescent, 10–75 cm tall, from a branching caudex; pubescence basifixed; stems erect or ascending; stipules 1–7 mm long, all distinct; leaves 2–10 cm long; leaflets 3–19, or the uppermost leaves simple, 3–20 mm long, 0.3–3 mm wide, linear, narrowly oblong or ovate, strigose on both sides; peduncles 4–30 cm long; racemes loosely 5- to 30-flowered, the flowers spreading-declined at anthesis, the axis 3–20 cm long in fruit; bracts 0.5–3.2 mm long; pedicels 0.8–2.5 mm long; bracteoles 0; calyx 4.5–8 mm long, the tube 4–6.7 mm long, cylindric, strigose, purplish, the teeth 0.6–2.3 mm long, broadly subulate; flowers 12–19 mm long, pink purple; pods pendulous, stipitate, the stipe 4–11 mm long, the body oblong to oblanceolate in outline, 19–35 mm long, 3–6 mm wide, strongly laterally flattened, glabrous, unilocular; ovules 14–20. Two allopatric and distinctive varieties occur in Utah.

1. Leaves all odd-pinnate, with 9–19 leaflets, the terminal one jointed on all leaves; plants of Grand and San Juan counties .... *A . coltonii* var. *moabensis*

— Leaves odd-pinnate, or the uppermost simple, with 3–9 leaflets, the terminal one continous with the rachis; plants not from Grand or San Juan counties ....
.... *A . coltonii* var. *coltonii*

Var. coltonii [*Homalobus coltonii* (Jones) Rydb.; *A. coltonii* var. *aphyllus* Jones, type from Richfield]. Bunchgrass, salt desert shrub, pinyon-juniper, and mountain brush communities at 1470 to 2300 m in Carbon (type from Castle Gate), Emery, Garfield, Kane, Millard, Sevier, and Wayne counties; endemic; 61 (xii).

Var. moabensis Jones (*A . coltoni*i var. *foliosus* Jones ex Eastw.; *Homalobus canovirens* Rydb., type from La Sal Mts.; *A . canovirens* (Rydb.) Barneby]. Pinyon-juniper and mountain brush communities at 1400 to 2300 m in Grand and San Juan (type from Monticello) counties; Wyoming, Colorado, New Mexico, and Arizona; 31 (xii).

Astragalus consobrinus (Barneby) Welsh Bicknell Milkvetch. [*A . castaneiformis* var. *consobrinus* Barneby, type from near Bicknell]. Perennial, sometimes flowering the first year, acaulescent, 1–5 cm tall, the caudex branches obscured by persistent leaf bases and stipules; pubescence malpighian; stems essentially lacking; stipules 3–7 mm long, all distinct; leaves 1–5 cm long; leaflets 3–11, 2–8 mm long, 1.5–4 mm wide, obovate to oblanceolate or orbicular, rounded to obtuse or acute, strigose on both surfaces; peduncles 0.4–3 cm long; racemes 2- to 10-flowered, the flowers ascending at anthesis, the axis to 1 cm long in fruit; bracts 1.5–3.5 mm long; pedicels 1–2 mm long; bracteoles 0; calyx 5.5–8.9 mm long, the tube 4.1–6.8 mm long, cylindric, strigose, the teeth 1.4–1.7 mm long, subulate; flowers 10–15.5 mm long, ochroleucous suffused with purple; pods ascending, sessile, obliquely ovoid or lance-ovoid, 11–19 mm long, 3–8 mm thick, strigose, unilocular; ovules 18–33. Sagebrush-grassland and pinyon-juniper communities at 1830 to 2200 m in Emery, Garfield, Piute, Sevier, and Wayne counties; endemic; 11 (vi).

Astragalus convallarius Greene Lesser Rushy Milkvetch. Perennial, caulescent, 10–60 cm tall, from a subterranean caudex; pubescence basifixed; stems erect or ascending, stipules 2–7 mm long, at least the lowermost connate-sheathing; leaves 2–11 cm long; leaflets, when present, 3–13, the uppermost leaves reduced to the rachis, mostly 5–30 mm long, 0.5–4 mm wide, linear to oblong or oblanceolate; peduncles 1–14 cm long; racemes 3- to 25-flowered, the flowers ascending to declined at anthesis, the axis 2–20 cm long in fruit; bracts 0.5–2.3 mm long; pedicels 1–8 mm long; bracteoles 0–2; calyx 4–6.3 mm long, the tube 3.4–5.4 mm long, triangular-subulate; flowers 6.4–12 mm long, ochroleucous or variously tinged or veined with purple; pods pendulous to spreading, sessile, linear to narrowly oblong, straight, 13–50 mm long, 2.3–4 mm thick, laterally compressed, strigose, unilocular; ovules 11–26. This is possibly the most widespread of the milkvetch species within Utah. Two allopatric varieties are present.

1. Pods 25–50 mm long, 2.3–3.3 mm wide; plants widespread, but not in western Iron and Washington counties
.... *A . convallarius* var. *convallarius*

— Pods 13–25 mm long, 3.4–4 mm wide; plants of western Iron and Washington counties ....
.... *A . convallarius* var. *finitimus*

Var. convallarius [*Homalobus campestris* Nutt. ex T. & G.; *Tragacantha campestris* (Nutt.) Kuntze; *A . serotinus* var. *campestris* (Nutt.) Jones; *Phaca convallaria* (Nutt.) Greene; *Homalobus junceus* Nutt. ex T. & G.; *A . junceus* (Nutt.) Gray, not Ledeb. ex Sprengel; *T . juncea* (Nutt) Kuntze; *A . diversifolius* var. *junceus* (Nutt.) Jones; *A . diversifolius* var. *roborum* Jones; *A . junciformis* A. Nels; *Homalobus junciformis* (A. Nels.) Rydb.; *A . junceus* var. *attenuatus* Jones, type from Price]. Mixed desert shrub, sagebrush, pinyon-juniper, mountain brush, ponderosa pine, and aspen communities at 1400 to 2900 m in all except Grand, Iron, and Washington counties; Idaho and Montana, south to Nevada and Colorado; 206 (xxvii).

Var. finitimus Barneby Pinyon-juniper and sagebrush communities at 1700 to 2270 m in western Iron and Washington (type from 3 mi S of Enterprise) counties; Lincoln County, Nevada; 14 (iii).

Astragalus cottamii Welsh Cottam Milkvetch. [*Astragalus monumentalis* var. *cottamii* (Welsh) Isely]. Perennial, sometimes flowering the first year, acaulescent or subacaulescent, 1.2–8 cm tall, from a branching caudex; pubescence basifixed; stems lacking or 0.5–6 cm long, the internodes mostly obscured by stipules, these 2–6 mm long, all distinct; leaves 1.2–8 cm long; leaflets (5) 9–19 (21), 2–9 mm long, 1–4.2 mm wide, elliptic to oval or oblanceolate, acute to obtuse, strigose on both sides or glabrous above; peduncles 0.7–7 cm long; racemes 3- to 9-flowered, the flowers ascending at anthesis, the axis 0.5–2 cm long in fruit; bracteoles 0–2; calyx 6.2–8 mm long, the tube 4.8–6.7 mm long, cylindric, strigulose, purplish, the teeth 1.2–2 mm long, subulate; flowers 11–17 mm long, pink purple or bicolored; pods spreading-descending, sessile, curved, oblong to oblong-lanceolate in outline, triquetrous, the dorsal suture sulcate, bilocular, strigose, usually purple-blotched. Rimrock and ledges of Cedar Mesa, Kayenta, and Entrada sandstone formations and in the sandy canyons cut from them, in pinyon-juniper and blackbrush communities on Entrada, Navajo, Cedar Mesa, and White Rim formations at 1310 to 1895 m in San Juan (type from east of Clay Hills Divide) County; Arizona and New Mexico; A Colorado Plateau endemic; 14 (ix).

Astragalus cronquistii Barneby Cronquist Milkvetch. Perennial, caulescent, 1.5–4 dm tall, from a subterranean caudex; pubescence basifixed; stems decumbent-ascending; stipules 2–6 mm long, all distinct; leaves 1.5–4.5 cm long; leaflets 7–15, 6–23 mm long, 1.5–4 mm wide, oblong to narrowly elliptic, retuse to truncate, strigose beneath, glabrate above; peduncles 2–6.5 cm long; racemes 6- to 20-flowered, the flowers declined at anthesis, the axis 0.5–8.5 cm long in fruit; bracts 0.6–1.2 mm long; pedicels 1.5–2.5 mm long; bracteoles 1; calyx 3.8–5.3 mm long, the tube 3.3–4 mm long, campanulate, strigose, the teeth 0.5–1.3 mm long, triangular; flowers 8–9 mm long, dull pink purple; pods declined-pendulous, sessile or subsessile, the body narrowly elliptic in outline, 13–30 mm long, 3–4.8 mm wide, trigonous, grooved dorsally, strigose, semibilocular. Blackbrush and salt desert shrub on Cutler and Morrison formations at 1430 to 1525 m in San Juan (type from W of Comb Reef) County; Montezuma County, Colorado; a Plateau endemic; 11 (iii).

Astragalus cymboides Jones Canoe Milkvetch. [*Xylophacos cymboides* (Jones) Rydb.; *A. amphioxys* var. *cymbellus* Jones, type from San Rafael Swell]. Perennial, acaulescent or subacaulescent, 2.5–8 cm tall, from a simple or branched caudex; pubescence malpighian; stems lacking or 0.5–3 cm long, the internodes mostly obscured by stipules; stipules 3–8 mm long, all distinct; leaves 2.5–8 cm long; leaflets (1–3) 5–18, 3–13 mm long, 2–9 mm wide, obovate, elliptic or oblanceolate, obtuse to acute, pubescent on both surfaces; peduncles 2–8 cm long; racemes 3- to 9-flowered, the flowers ascending at anthesis, the axis 0.5–2 cm long in fruit; bracts 1–4 mm long; pedicels 0.7–2.5 mm long; bracteoles 0; calyx 7.6–10.2 mm long, the tube 5.9–8 mm long, cylindric, strigose, the teeth 1–2.3 mm long, subulate; flowers 15–18.5 mm long, ochroleucous or suffused purplish, or pink purple; pods ascending, sessile, oblong to oblong-elliptic in outline, straight, 17–30 mm long, 6–9.5 mm wide, laterally compressed, the valves papery or cellular-spongy, the exocarp in time exfoliating, strigose, unilocular; ovules 39–57; 2n = 24 (?). Salt desert shrub and pinyon-juniper communities at 1600 to 2300 m in Carbon, Emery (type from Huntington), and Grand counties; endemic; 59 (x). The small-flowered phases of canoe milkvetch approach *A. consobrinus* and the large-flowered phases are difficult to distinguish, in anthesis, from *A. amphioxys*. Further work is indicated.

Astragalus desereticus Barneby Deseret Milkvetch. Perennial, acaulescent or subacaulescent, 4–15 cm tall, from a caudex; pubescence basifixed; stems to 6 cm long, the internodes more or less obscured by stipules, these 3.5–7 mm long, all distinct; leaves 4–12 cm long; leaflets 11–17, 2–14 mm long, 1.5–8 mm wide, elliptic to obovate, short-acuminate to acute, strigulose-villosulous on both sides; peduncles 2–5.5 cm long; racemes 5- to 10-flowered, the flowers ascending at anthesis, the axis 0.5–2 cm long in fruit; bracts 3–6 mm long; pedicels 2–3 mm long; bracteoles 0–2; calyx 8.4–12 mm long, the tube 6.2–7.5 mm long, cylindric, villous, the teeth 2–4 mm long, subulate; flowers 18–22.5 mm long, whitish suffused with pale purple, the keel purple-tipped; pods ascending, sessile or substipitate, ovoid-ellipsoid, curved, 10–20 mm long, 5–10 mm thick, densely hirsute with lustrous hairs; ovules 14–16. Sagebrush-juniper community at 1830 m in Utah (type from near Birdseye) County; endemic; 7 (i). This species has long remained obscure, but occurs within a short distance from a road that has been traversed by botanists for almost a century since its initial discovery. It was rediscovered by Elizabeth Neese in May of 1981, after a hiatus of some six decades. The plants were initially thought to be closely allied to *A. argophyllus*, but the relationship is probably nearer to *A. purshii*.

Astragalus desperatus Jones Rimrock Milkvetch. [*Batidophaca desperata* (Jones) Rydb.; *Tium desperatum* (Jones) Rydb.]. Perennial, acaulescent or subacaulescent, 1–12 cm tall, from a branching caudex; pubescence basifixed; stems to 8 cm long, the internodes often obscured by stipules; stipules 1.5–7 mm long, at least some connate-sheathing; leaves 1–12 cm long; leaflets 7–17, 2–13 mm long, 1–5 mm wide, elliptic to oblanceolate or obovate, acute to obtuse, strigose on both sides or glabrate above; peduncles 0.5–13 cm long; racemes 3- to 18-flowered, the flowers ascending to declined at anthesis, the axis 0.4–13 cm long in fruit; bracts 1.5–5 mm long; pedicels 0.5–1.4 mm long; bracteoles 0–2; calyx 3.5–6 mm long, the tube 2.5–4 mm long, campanulate, strigose-pilose, the teeth 0.8–2.6 mm long, subulate; flowers 6–9 mm long, pink purple or bicolored; pods declined to deflexed, sessile or short-stipitate, the stipe (gynophore) to 1.2 mm long, the body obliquely ovoid to lance-ellipsoid, curved, 6–19 mm long, 3–6 mm thick, hirsute with lustrous hairs, unilocular; ovules 16–28; 2n = 24. Two more or less distinctive but partially sympatric varieties are present.

1. Pods lunately curved, at maturity mainly 11–19 mm long; plants commonly shortly caulescent; racemes typically with 6–28 flowers; widespread . . . . *A. desperatus* var. *desperatus*
— Pods straight or nearly so, at maturity 6–11 mm long; plants commonly acaulescent; racemes typically with 3–6 flowers; northern and western San Rafael Swell . . . . *A. desperatus* var. *petrophilus*

Var. *desperatus* Mixed desert shrub and pinyon-juniper communities, often on rimrock, at 1130 to 1900 m in Emery, Garfield, Grand (type from near Cisco), Kane, San Juan, and Wayne counties; Colorado and Arizona; a Plateau endemic; 107 (xxxiv).

Var. petrophilus Jones [*Batidophaca petrophila* (Jones) Rydb.]. Pinyon-juniper and mixed desert shrub communities at 1430 to 2135 m in Emery (type from San Rafael Swell) County; endemic; 22 (v).

Astragalus detritalis Jones Debris Milkvetch. [*Homalobus detritalis* (Jones) Rydb.; *A. spectabilis* C. L. Porter, not Schischkin, the type from southwest of Vernal]. Perennial, acaulescent, 0.5–8 cm tall, from a branching caudex; pubescence malpighian; stems essentially lacking; stipules 3–10 mm long, at least some (usually all) connate-sheathing; leaves 0.5–8 cm long, simple or with 3–7 leaflets, 3–30 mm long, and 0.5–2.5 mm wide, narrowly oblanceolate to linear, spinulose-tipped, strigose on both sides; peduncles 1–9 cm long; racemes 2- to 8-flowered, the flowers ascending at anthesis, the axis 0.9–3.8 cm long; bracts 2.5–7 mm long; pedicels 0.5–2.5 mm long; bracteoles 0–2; calyx 5–9.6 mm long, the tube 3.1–5.4 mm long, campanulate, strigose, the teeth 1.6–4.7 mm long, subulate; flowers 13–30 mm long, pink purple; pods erect to straight or curved, 15–38 mm long, 2–3.5 mm wide, laterally compressed, mottled, strigose, unilocular; ovules 15–24. Pinyon-juniper and mixed desert shrub communities at 1650 to 1950 m in Duchesne (type from southwest of Duchesne) and Uintah counties; Rio Blanco County, Colorado; a Uinta Basin endemic; 38 (x).

Astragalus diversifolius Gray Mesic Milkvetch. [*Homalobus orthocarpus* Nutt. ex T. & G., not *A. orthocarpus* Boiss.; *A. campestris* var. *diversifolius* (Gray) Macbr.; *A. junceus* var. *orthocarpus* (Nutt.) Jones; *A. junceus* var. *diversifolius* (Gray) Jones; *A. convallarius* var. *diversifolius* (Gray) Tidestr.; *A. ibapensis* Jones, type from Deep Creek Mts.; *Atelphragma ibapense* (Jones) Rydb.]. Perennial, caulescent, 20–50 cm tall, from a subterranean to superficial caudex and stout taproot; pubescence basifixed; stems prostrate to ascending; stipules 1–3 mm long, the lowest connate-sheathing; leaves 1.5–7 cm long; leaflets 1–7, 4–47 (67) mm long, narrowly elliptic, linear, oblanceolate, or lanceolate, the uppermost often simple, strigose on both sides; peduncles 2–15 cm long; racemes 3- to 8-flowered, the flowers ascending at anthesis, the axis 0.5–3 mm long in fruit; bracts 0.7–2.5 mm long; pedicels 1.8–4 mm long; bracteoles 0–2; calyx 4.4–6.7 mm long, the tube 3.2–4.7 mm long, campanulate, strigose, the teeth 1–2 mm long, subulate; flowers 8–23.5 mm long, greenish white, often tinged with purple; pods ascending to declined, sessile or substipitate, the body narrowly oblong, 10–17 mm long, 2.7–4 mm wide, strongly laterally compressed, strigose, unilocular; ovules 10–16. Moist, often saline meadows at 1340 to 1700 m in Juab and Tooele counties; Idaho; 6 (ii). The mesic milkvetch was rediscovered at Juab by Sherel Goodrich in 1980; the next previous collection was taken there in 1902.

Astragalus drummondii Dougl. ex Hook. Drummond Milkvetch. [*Tragacantha drummondii* (Dougl.) Kuntze; *Tium drummondii* (Dougl.) Rydb.]. Perennial, caulescent, 25–60 cm tall, from a branching, subterranean caudex; pubescence basifixed; stems erect or ascending; stipules 3–12 mm long, at least some connate-sheathing; leaves 4–13 mm long; leaflets 17–33, 4–33 mm long, 2–12 mm wide, oblong to oblanceolate or obovate, obtuse to truncate or emarginate, villous-pilose beneath, usually glabrous above; peduncles 4–12 cm long; racemes 14- to 30-flowered, the flowers spreading-declined at anthesis, the axis 3–22 cm long in fruit; bracts 2–5 mm long; pedicels 1.5–5 mm long; bracteoles 0–2; calyx 7–12.5 mm long, the tube 4.7–8 mm long, short-cylindric, sparingly villous, the teeth 1.7–4.5 mm long, subulate; flowers 18–25 mm long, whitish to ochroleucous, the keel purple-tipped; pods pendulous, stipitate, the stipe 5–11 mm long, the body narrowly oblong to oblanceolate in outline, 17–32 mm long, 3.5–5.5 mm thick, trigonous, sulcate dorsally, glabrous, bilocular; ovules 14–30; 2n = 22. Pinyon-juniper, ponderosa pine, and mountain brush communities at 1530 to 2130 m in Beaver, San Juan, Sevier, and Utah counties; Alberta and Sasketchewan, south to New Mexico; 22 (vi). Recent collections from near Natural Bridges National Monument in San Juan County appear to be waifs.

Astragalus duchesnensis Jones Duchesne Milkvetch. [*Lonchophaca duchesnensis* (Jones) Rydb.]. Perennial, caulescent, 15–40 cm tall, from a branching caudex; pubescence basifixed; stems straggling to ascending or erect; stipules 3–8 mm long, all distinct; leaves 2–10 cm long; leaflets 5–15, 3–20 mm long, 0.5–3 mm wide, linear to oblong or narrowly oblanceolate, obtuse to retuse, strigose on both sides or glabrate above, the uppermost leaflet sometimes continuous with the rachis; peduncles 3–10.5 cm long, racemes 6- to 22-flowered, the flowers ascending at anthesis, the axis 2.5–12 cm long in fruit; bracts 0.7–2 mm long; pedicels 0.8–2.2 mm long; bracteoles 0; calyx 3.6–5.5 mm long, the tube 3.1–4.3 mm long, campanulate, usually dull purple, the teeth 0.4–1 mm long, triangular; flowers 8.5–12.5 mm long, pink purple with white wing-tips; pods declined, sessile, oblong to narrowly oblanceolate in outline, 20–35 mm long, 3.3–5 mm thick, dorsiventrally compressed in the lower half, becoming laterally compressed in the distal portion, strigose, unilocular; ovules 21–31. Sandy or gravelly to fine-textured clay soils in mixed desert shrub and pinyon-juniper communities at 1430 to 1830 m in Daggett, Duchesne (type from Theodore to Myton), and Uintah counties; Rio Blanco County, Colorado; a Uinta Basin endemic; 73 (xviii).

Astragalus eastwoodae Jones Eastwood Milkvetch. [*A. preussii* var. *sulcatus* Jones (type from SE of Kane Spring); *Phaca eastwoodae* (Jones) Rydb.; *A. preussii* var. *eastwoodae* (Jones) Jones]. Perennial, short caulescent, 8–20 cm tall, from a branching caudex; pubescence lacking except on the calyx; basifixed; stems 2–14 cm long, decumbent to ascending; stipules 2–6.5 mm long, all distinct; leaves 3–13 cm long; leaflets 13–25, 1–15 mm long, 1–5 mm broad, elliptic to lance-elliptic, oblanceolate or obovate, obtuse to truncate-emarginate, glabrous; peduncles 2–10.5 cm long; bracts 1.5–4.5 mm long; pedicels 1.5–3.5 mm long; bracteoles 2; calyx 10–12.2 mm long, the tube 8–9.5 mm long, cylindric, purple, sparsely black-strigose, the teeth 1.3–2.7 mm long, subulate; flowers 18–22 mm long, pink purple; pods spreading to declined, stipitate, the stipe 1.5–4.5 mm long, the inflated body oblong-ellipsoid, 14–26 mm long, 7–14.5 mm thick, the valves papery and straw colored, unilocular, glabrous; ovules 20–38; 2n = 24, 26. Mixed desert shrub and pinyon-juniper communities at 1330 to 1830 m in seliniferous, often fine-textured soils, in Emery, Garfield, Grand, and San Juan counties; Colorado; 10 (ii). The Eastwood milkvetch is closely allied to *A. preussii*. It differs mainly in the shorter stems, smaller leaflets, and spreading-descending, thin-textured pods.

Astragalus emoryanus (Rydb.) Cory Emory Milkvetch. [*Hamosa emoryana* Rydb.]. Annual or winter annual, caulescent, 4–45 cm tall, from a slender taproot; pubescence basifixed; stems prostrate; stipules 1.5–3.6 mm long, all distinct; leaves 1–4.5 cm long; leaflets 11–19, 2–10 mm long, 1–6 mm wide, oval-obovate to obcordate or oblanceolate, obtuse to retuse on all leaves, sparingly strigose on both sides or glabrate above; peduncles 2–6 cm long; racemes 2- to 10-flowered, the flowers spreading at anthesis, the axis 0.3–2.5 cm long in fruit; bracts 0.5–2 mm long; bracteoles 0; calyx 3.6–6 mm long, the tube 1.9–3.5 mm long, campanulate, strigose, the teeth 1.3–2.5 mm long; flowers 7.3–11 mm long, pink purple; pods declined to ascending, sessile, narrowly oblong, curved, 0.8–2.2 cm long, 2.2–4.3 mm wide, trigonous, glabrous, bilocular; ovules 10–16. Pinyon-juniper and sagebrush communities at 1700 m along the strike valley at the plunging east margin of the East Kaibab monocline, Kane County; Arizona, New Mexico, Texas, and Mexico; 2 (i). This is a poorly known entity in Utah. It is very similar in most salient features with *A. nuttallianus*, from which it can be distinguished by the deciduous, straw colored pods, merely strigose calyx teeth, retuse-obtuse leaflets on all leaves, and rounded keel-tip. This latter feature is shared with *A. nuttallianus* var. *micranthiformis*, but that taxon is known to occur in Kane County only east of the Cockscomb.

Astragalus ensiformis Jones Pagumpa Milkvetch. [*A. ursinus* Jones, not Gray; *Hamosa ensiformis* (Jones) Rydb; *A. ensiformis* var. *gracilior* Barneby (type from S of Veyo); *A. minthorniae* var. *gracilior* (Barneby) Barneby]. Perennial, caulescent, 8–45 cm tall, from a superficial to subterranean caudex; pubescence basifixed; stems decumbent to erect; stipules 4–10 mm long, all distinct; leaves 4–15.5 cm long; leaflets (5) 11–23, 6–24 mm long, 1.5–13.5 mm wide, ovate to oblong, obovate, or oblanceolate, obtuse to retuse, strigose (sometimes sparingly so) beneath, strigose to glabrous above; peduncles 2.5–13 cm long; racemes 12- to 30-flowered, the flowers ascending to declined at anthesis, the axis 3.5–13.5 cm long in fruit; bracts 2–6 mm long; pedicels 1–3.5 mm long; bracteoles 0–2; calyx 5.2–7.8 mm long, the tube 4.5–6.5 mm long, short-cylindric, pilosulose with black hairs, the teeth 1.2–2.5 mm long, subulate; flowers 13–17 mm long, purplish to pink purple, the wing-tips pale to white; pods ascending to descending, sessile or substipitate, the body narrowly oblong, curved, 15–30 mm long, 4–6 mm thick, subterete (compressed laterally when pressed), bilocular, strigose to strigulose; ovules 24–36. Pinyon-juniper, sagebrush, salt desert shrub, and blackbrush communities at 1230 to 2350 m in Washington County; Mohave County, Arizona; 21 (vi).

Astragalus episcopus Wats. Bishop Milkvetch. [*Tragacantha episcopa* (Wats.) Kuntze; *Homalobus episcopus* (Wats) Rydb.; *A. kaibensis* Jones; *Lonchophaca kaibensis* (Jones) Rydb.]. Perennial, caulescent, rushlike, 20–45 cm tall, arising from a subterranean caudex; pubescence basifixed; stems erect or ascending; stipules 2–13 mm long, all distinct; leaves 2–10 cm long, most of them reduced to the rachis, some with leaflets 3–13 in number, these 1–15 mm long and 0.5–2 mm wide, linear to elliptic or oblong, acute to obtuse or emarginate, strigose on both sides; peduncles 6–23 cm long; racemes very loosely 6- to 30-flowered, the flowers ascending at anthesis, the axis 3–30 cm long in fruit; bracts 1.3–3 mm long; pedicels 1.5–3.5 mm long; bracteoles 0–2; calyx 4.1–7 mm long, the tube 3.4–5.2 (6) mm long, short cylindric, always much longer than broad, suffused with purple or very pale, white-strigose, the teeth 0.6–2.2 mm long, triangular to subulate; flowers 10–15.5 mm long, pale pink or whitish to pink purple; pods pendulous, sessile or subsessile, the body oblong to lance-elliptic in outline, slightly curved to straight, 14–32 mm long, 4–8 mm wide, laterally compressed, glabrous to strigose, straw colored, or suffused with purple or mottled; unilocular; ovules 16–26; n = 11. Mixed desert shrub and pinyon-juniper communities, often in clay or silty soils, at 1270 to 1700 m in Emery, Garfield, Kane (type locality?), San Juan, and Wayne counties; Arizona and New Mexico; 56 (xiv). This taxon is closely allied to *A. lancearius*, and a case can be made for its inclusion within that entity at infraspecific level.

Astragalus equisolensis Neese & Welsh Horseshoe Milkvetch. Perennial, acaulescent or subacaulescent, 5–15 cm tall, from a branching caudex; pubescence basifixed; stipules 2–5 cm long; leaves 1.5–9 cm long; leaflets 5–17, 3–12 mm long, 1.5–5 mm wide, elliptic, oblanceolate, or obovate, acute to obtuse, strigose on both sides; peduncles erect, 2–9 cm long; racemes 4- to 13-flowered, the flowers ascending or spreading at anthesis, the axis 1.5–8 cm long in fruit; bracts 2–4.5 mm long; pedicels 0.5–2 mm long; bracteoles 0; calyx 6–8.5 mm long, the tube 4.5–6 mm long, cylindric, strigose, the teeth 1.2–2.5 mm long, subulate; flowers 12–16 mm long, purplish; pods declined to deflexed, sessile or stipitate, obliquely ovoid or lance-ellipsoid, lunately curved, dorsiventrally compressed, constricted distally at the beak, laterally compressed, incurved, 10–14 mm long, 3.5–6.5 mm wide, hirsute, unilocular, the valves thickly papery; ovules 20. Sagebrush, shadscale, horsebrush, and other mixed desert shrub communities on the Duchesne River Formation at 1460 to 1580 m in Uintah (type from E of Horseshoe Bend) County; endemic; 11 (ix).

Astragalus eremiticus Sheldon Hermit Milkvetch. [*A. arrectus* var. *eremiticus* (Sheldon) Jones]. Perennial, caulescent, 20–45 cm tall, from a branching commonly superficial caudex; pubescence basifixed; stems erect or ascending; stipules 3–11 mm long, 1.5–12 mm wide, ovate to oblong, elliptic or narrowly oblong, obtuse to retuse, strigose beneath, glabrate to glabrous above; peduncles 2.5–15 cm long; racemes 10- to 26-flowered, the flowers ascending to declined at anthesis, the axis 5–17 cm long in fruit; bracts 1.5–4 mm long; pedicels 0.7–3.5 mm long; bracteoles 0–2; calyx 4.5–8.7 mm long, the tube 4–6.8 mm long, short-cylindric, strigose the teeth 0.7–2 mm long, triangular to subulate; flowers 12–18 mm long, ochroleucous, pink purple, or merely tinged purplish; pods erect, stipitate, the stipe 6–15 mm long, the body oblong to ellipsoid or obliquely ellipsoid, 12–27 mm long, 3.5–8 mm thick, trigonous, glabrous, bilocular; ovules 17–32; 2n = 24. Juniper-pinyon, live oak, and sagebrush communities at 1130 to 1830 m in Iron, Kane (west of Cockscomb), and Washington (type from Beaver Dam Mts.) counties; Arizona, Nevada, and Idaho; 18 (iii). Plants from west of the Gunlock intersection at Shem, with inflated and proportionately shorter pod bodies, decumbent stems, and subterranean caudices, simulate the rare *A. ampullarius*, which occupies similar clay soils also on the Chinle Formation. The Shem plants are designated as **var. *ampullarioides*** Welsh.

Astragalus eurekensis Jones Eureka Milkvetch. [*Xylophaca eurekensis* (Jones) Rydb.; *X. medius* Rydb., type from Lake Point, Tooele County]. Perennial, acaulescent, 2-15 cm tall, the caudex branches obscured by persistent leaf bases; pubescence basifixed; stipules 3-11 mm long, all distinct; leaves 2-15 cm long; leaflets (3) 5-19, 3-35 mm long, 2-8 mm wide, elliptic to oblong, acute, gray or silvery strigose on both surfaces; peduncles 1-13 cm long, pilose; racemes 3- to 7-flowered, the flowers ascending at anthesis, the axis 0.2-2 cm long in fruit; bracts 4-8 mm long; pedicels 1.2-3 mm long; calyx 10.5-15.5 mm long, the tube 8.5-10.5 mm long, cylindric, pilose-villous, the teeth 2.5-5.7 mm long, subulate; flowers 22-27 mm long, ochroleucous, faintly to strongly suffused with purple or rarely pink purple; pods ascending, sessile, obliquely lance-ovoid, 15-40 mm long, 5-9 mm thick, villous-hirsute, unilocular; ovules 26-36. Sagebrush, pinyon-juniper, and mountain brush communities at 1370 to 2135 m in Beaver, Garfield, Iron, Juab (type from Eureka),, Millard, Sanpete, Sevier, Tooele, Utah, and Wasatch counties; endemic; 83 (xv). When, as rarely, the flower color is bright pink purple this entity, lacking fruit is exceedingly difficult to distinguish from *A. newberryi* (q.v.).

Astragalus falcatus Lam. Russian Sickle Milkvetch. Perennial, caulescent, 40-90 cm tall, from a branching caudex; pubescence malpighian; stems ascending to erect, forming large clumps; stipules 2-12 mm long, at least some connate-sheathing; leaves 5-22 cm long; leaflets 19-37, 6-35 mm long, 1.5-10 mm wide, oblong to elliptic or oblanceolate, acute to apiculate, strigose below, glabrous above, green on both sides; peduncles 6-17 cm long; racemes 20- to 70-flowered, the flowers declined at anthesis, the axis 3-20 cm long in fruit; bracts 2-5 mm long; pedicels 32-65 mm long, recurved in fruit; bracteoles 2; calyx 3.6-4.7 mm long, the tube 3-3.5 mm long, campanulate, strigose, the teeth 0.5-1.2 mm long, triangular; flowers 9-11 mm long, greenish white, sometimes suffused with purple; pods decurved, subsessile, curved-oblong, 13-23 mm long, 2.5-4.5 mm wide, triangular, strigose, bilocular; ovules 12-14; n = 8. Introduced soil stabilization plant, established in Duchesne and Juab counties; southeastern Europe; 3 (iii). This is a robust perennial, capable of surviving in harsh, fine-textured soils in the mountain brush and pinyon-juniper communities. The plant is also suspected of being poisonous to livestock.

Astragalus filipes Torr. Perennial, caulescent, 2-9 dm tall; pubescence basifixed; stems erect or ascending, forming large clumps; stipules 2-5 mm long, at least the lowermost connate-sheathing; leaves 2.5-12 cm long; leaflets (5) 9-23, 3-25 (30) mm long, 0.5-5 mm wide, linear to narrowly elliptic or oblong, obtuse to truncate, retuse or subacute, green and glabrous or strigose on one or both sides; peduncles 4.5-22 cm long; racemes loosely (4) 6- to 30-flowered, the flowers spreading to declined, ochroleucous; bracts 1-3.5 mm long; pedicels 1.5-6 mm long; bracteoles typically lacking; calyx 4-7.7 mm long, the tube 3.3-6.4 mm long, subcylindric to campanulate, strigulose with black and white hairs, the teeth 0.5-1.5 mm long, triangular; flowers 10-14.3 mm long; pod pendulous, stipitate, the stipe 6-16 mm long, the body 17-30 mm long, 3.5-6.5 mm wide, oblong to elliptic in outline, the sutures commonly parallel, strongly laterally compressed, the valves papery, glabrous to strigulose; ovules 11-22; n = 11, 12. Juniper and sagebrush communities at 1645 to 1710 m in Box Elder County; Washington and Idaho south to California and Nevada; 3 (0).

Astragalus flavus Nutt. ex T. & G. Yellow Milkvetch. Perennial, caulescent, 5-30 cm tall, from a branching caudex; pubescence malpigian; stems decumbent to ascending or erect; stipules 2-10 mm long, all connate-sheathing; leaves 3-15 (18) cm long; leaflets (5) 9-21, 3-31 mm long, 0.5-6 mm wide, linear, narrowly oblong, or oblanceolate to ovate, obtuse to acute, silvery strigose (greenish) on both sides or glabrate to glabrous above; peduncles 3-23 cm long, racemes 6- to 30-flowered, the flowers ascending at anthesis, the axis 2-12 cm long in fruit; bracts 1.5-5 mm long; pedicels 0.7-1.2 mm long; bracteoles 0; calyx 5.5-9.5 mm long, the tube 3-5.2 mm long, campanulate, strigose to pilose, the teeth 2-6 mm long, yellow to ochroleucous, whitish, lilac, or pink purple; pods erect, sessile, oblong, 7-13 mm long, 3.5-5.5 mm thick, straight, dorsiventrally compressed, strigose, unilocular; ovules 6-17; 2n = 24, 26. Three more or less distinctive but intergrading varieties, all primary selenophytes, are present in Utah.

1. Calyx shaggy long-villous, the teeth 3-6 mm long, equaling or longer than the tube; flowers pink purple; plants of Emery, Grand, Wayne, and Garfield counties
 .... *A. flavus* var. *argillosus*
— Calyx strigose to short-villous, the teeth 2-3 (4) mm long, shorter than the tube; flowers yellow to white or tinged purplish, rarely pink purple; plants of various distribution 2

2(1). Flowers whitish or yellowish, sometimes tinged with purple, rarely pink-purple; keel 6-8 mm long; plants from central and southern counties ....
 .... *A. flavus* var. *candicans*
— Flowers cream to yellow; keel 8-10 mm long; plants from east-central to northeastern Utah ....
 .... *A. flavus* var. *flavus*

Var. *argillosus* (Jones) Barneby Clay Milkvetch. [*A. argillosus* Jones, type from Green River, Emery County; *Cnemidophacos argillosus* (Jones) Rydb.]. Mancos shale, Summerville, Cedar Mountain and Morrison formations, on saline clays and silts with salt desert shrubs at 1230 to 1600 m in Emery (type from Green river), Garfield, Grand, and Wayne counties; endemic; 39 (vii).

Var. *candicans* Gray St. George Milkvetch. [*A. confertiflorus* Gray, type from near St. George; *Cnemidophacos confertiflorus* (Gray) Rydb.]. Mancos and Tropic shales, Moenkopi, Chinle, Morrison, and Kaiparowits formations and other saline clays and silts at 900 to 2130 m in Garfield, Kane, San Juan, Sevier (type from near Richfield), and Washington counties; Arizona and Nevada; 51 (x). The var. *candicans* passes into var. *flavus* in central and eastern Utah.

Var. *flavus* Yellow Milkvetch. [*Cnemidophacos flavus* (Nutt.) Rydb.; *Tragacantha flaviflora* Kuntze; *A. confertiflorus* var. *flaviflorus* (Kuntze) Jones]. Mancos Shale, Chinle, Moenkopi, Duchesne River, Uinta, and other formations composed of saline silts and clays in salt desert shrub and pinyon-juniper communities at 1230 to 1730 m in Carbon, Daggett, Duchesne, Emery, Grand, San Juan, Sanpete, Sevier, and Uintah counties; Arizona, New Mexico, Colorado, and Wyoming; 75 (xxii). Sheep poisoning attributable to var. *flavus* is known from the lower elevation portions of the Uinta Basin.

Astragalus flexuosus (Hook.) G. Don Prairie Milkvetch. Perennial, caulescent, 1–6 dm tall, from a branching caudex; pubescence basifixed; stems decumbent or ascending; stipules 1–7 mm long, at least the lowermost connate- sheathing; leaves 1.5–9 cm long; leaflets (5) 9–25, 1–8 mm wide, linear or oblong to oblanceolate or obovate, obtuse to truncate or retuse, strigose to glabrate beneath, usually glabrous above; peduncles 1.5–14 cm long; racemes 7– to 26–flowered, the flowers spreading at anthesis 0.6–4.5 mm long; pedicels 0.7–3.5 mm long; bracteoles 0–2; calyx 3.3–5.8 mm long, the tube 2.4–4.3 mm long, campanulate, strigose, the teeth 0.5–1.7 mm long, subulate; flowers 7–11 mm long, pink purple to dull purplish; pods descending to spreading, sessile or short-stipitate, the stipe 0.5–1.3 mm long, the body oblong to oblanceolate or elliptic in outline, 8–24 mm long, 2.7–4.8 mm thick, subterete or variously somewhat flattened, strigose to glabrous, unilocular; ovules 12–25; 2n = 22. Two rather distinctive varietes are present, both confined to eastern Utah.

1. Calyx tube 2.4–2.7 mm long; pods sessile or nearly so; plants spreading decumbent, in pinyon-juniper and mixed desert shrublands .... *A. flexuosus* var. *diehlii*
— Calyx tube 2.7–4.3 mm long; pods subsessile to shortly stipitate; plants of pinyon-juniper and mountain brush communities .... *A. flexuosus* var. *flexuosus*

Var. *diehlii* (Jones) Barneby [*A. diehlii* Jones, type from Farnham, Carbon County; *Pisophaca diehlii* (Jones) Rydb.]. Salt desert shrub and pinyon-juniper communities at 1370 to 1670 m in Carbon, Emery, Grand, and San Juan counties; Colorado; 13 (vii).

Var. *flexuosus* [*Phaca flexuosa* Hook.; *Homalobus flexuosus* (Hook.) Rydb.; *Pisophaca flexuosa* (Hook.) Rydb.]. Pinyon-juniper and mountain brush communities at 1685 to 2135 m in Grand and San Juan counties; British Columbia east to Ontario and south to New Mexico, Nebraska, and Minnesota; 2 (0).

Astragalus fucatus Barneby Hopi Milkvetch. [*A. subcinereus* sensu Jones, not Gray, q.v.]. Perennial, caulescent, 7–45 cm tall, from a subterranean to superficial caudex; pubescence basifixed; stems ascending to erect or sprawling; stipules 1–5.5 mm long, the lowest connate-sheathing; leaves 2–12.5 cm long; leaflets 9–17, 3–20 (25) mm long, 0.5–4 mm wide, obtuse to retuse, strigose beneath, glabrous above; peduncles 1–6.5 cm long; racemes 9– to 22–flowered, the flowers ascending to declined at anthesis, the axis 2–11.5 cm long in fruit; bracts 0.8–2 mm long; pedicels 0.7–3.5 mm long; bracteoles 0; calyx 3.3–5.4 mm long, the tube 2.3–3.3 mm long, campanulate, strigose, the teeth 0.8–2.2 mm long; flowers 6.4–8.7 mm long, pink purple; pods spreading to declined, sessile, bladdery-inflated, ovoid, ellipsoid, or subglobose, 17–32 mm long, 12–20 mm wide (when pressed), mottled, strigose, unilocular; ovules 21–32. Mixed sandy desert shrub communities at 1330 to 1830 m in Emery, Garfield, San Juan, and Wayne counties; New Mexico and Arizona; 27 (ix).

Astragalus geyeri Gray Geyer Milkvetch. [*Phaca annua* Geyer, not *A. annuus* DC.]. Annual (or rarely biennial), caulescent, 6–27 cm long, from a slender taproot; pubescence basifixed; stems prostrate to ascending or erect; stipules 1.5–4 mm long, all distinct; leaves 2–10.5 cm long; leaflets 3–13, 3–17 mm long, 1–5.2 mm wide, linear to oblong or narrowly elliptic, obtuse to retuse, strigose beneath and strigose to glabrous above; peduncles 0.6–1.5 cm long; racemes 2– to 8–flowered, the flowers ascending at anthesis, the axis 0.8–1.5 cm long in fruit; bracts 0.7–2 mm long; pedicels 0.6–1.5 mm long; bracteoles 0–1; calyx campanulate, strigose, the teeth 0.6–1.5 mm long; flowers 5–7.6 mm long, pale, suffused with purple, or pink purple; pods spreading to declined, bladdery-inflated, obliquely ovoid, 15–24 mm long, 6–12 mm wide (when pressed), strigose, unilocular; ovules 7–18. Sandy soil in mixed desert shrub communities at 1370 to 1830 m in Beaver, Daggett, Duchesne, Emery, Garfield, Grand, Iron, Juab, Millard, Salt Lake, Tooele, Uintah, and Wayne counties; Oregon east to Wyoming, south to California, Nevada, and Colorado; 54 (xi). Plants grown from seed in a greenhouse produced mature fruit and seed in 58 days.

Astragalus gilviflorus Sheldon Plains Orophaca. [*A. triphyllus* Pursh, not Pallas; *Tragacantha triphylla* (Pursh) Kuntze; *Orophaca triphylla* (Pursh) Britt.; *Phaca caespitosa* Nutt., not *A. argophyllus* Nutt. ex T. & G.]. Perennial, acaulescent, 1.5–13 cm tall, from a branching caudex; pubescence malpighian; stems entirely obscured by stipules; stipules 6–18 mm long, connate-sheathing; leaves 1–13 cm long, palmately trifoliolate, the leaflets 7–20 mm long, 2–7 mm wide, spatulate to elliptic, acute to obtuse, silvery strigose on both sides; peduncles obsolete; racemes capitate, 1– to 3–flowered, the flowers erect, the axis very short in fruit; bracts 4.5–7.6 mm long, tridentate; pedicels 0–1.6 mm long; bracteoles 0; calyx 9.3–18 mm long, the tube 10–14 mm long, cylindric, loosely villous, the teeth 1.6–4 mm long; flowers 17–28 mm long, whitish to ochroleucous; pods erect, sessile, ovoid-ellipsoid, 6–10 mm long, 2.5–5 mm thick, densely hairy, unilocular; 2n = 24. Sagebrush community at ca 2130 m in Rich and Summit counties; Alberta east to Ontario and south to Wyoming and Nebraska; 7 (ii).

Astragalus hallii Gray Hall Milkvetch. Perennial, caulescent, 12–50 cm tall, from a subterranean caudex; pubescence basifixed; stems decumbent to ascending or erect; stipules 1–7 mm long, at least the lowermost connate-sheathing; leaves 2–9 cm long; leaflets 11–23, 2–11 mm long, 1–7 mm wide, obovate to oblanceolate or elliptic, retuse to truncate or obtuse, strigulose beneath, sparingly hairy or glabrous above; peduncles 3–9.5 cm long; racemes 9– to 25–flowered, the flowers spreading-declined at anthesis, the axis 1–7 cm long in fruit; bracts 1.5–5 mm long; pedicels 1.2–4 mm long; bracteoles 0–2; calyx 6–7 mm long, the tube 5.2–6.2 mm long, triangular; flowers 12.8–15 mm long, pink purple; pods spreading to declined, short-stipitate, the stipe 1.5–3.5 mm long, the inflated body cylindroid to obliquely ovoid-ellipsoid, 19–27 mm long, 8–12 mm thick, strigulose, unilocular; ovules 20–37. Pinyon-juniper and mountain brush communities at 1600 to 2130 m in Garfield (?) and Kane counties; Arizona and New Mexico; 7 (i). Our plants belong to **var. *fallax* (Wats.) Barneby** [*A. fallax* Wats.; *A. famelicus* Sheldon, not *A. fallax* Fisch.; *A. gracilentus* var. *fallax* (Wats.) Jones; *Pisophaca famelica* (Sheldon) Rydb.]. They seem not to differ from those of the Flagstaff Plateau in north-central Arizona.

Astragalus hamiltonii C. L. Porter Hamilton Milkvetch. [*A. lonchocarpus* var. *hamiltonii* (C. L. Porter) Isely]. Perennial, caulescent, 25–60 cm long, from a shallowly subterranean caudex; pubescence basifixed; stems erect; stipules 1.5–9.5 mm long, all distinct or rarely some shortly connate-sheathing; leaves 3–8 cm long, the uppermost (and sometimes the lowermost) simple, the others with leaflets 3–7, 10–47 mm long, 2–7 mm wide, elliptic to narrowly oblanceolate, obtuse to retuse, strigose on both sides, the terminal leaflet continuous with the rachis; peduncles 2.5–15.5 cm long; racemes 7- to 30-flowered, the flowers spreading-declined at anthesis, the axis 2–11 cm long in fruit; bracts 1–2.5 mm long; pedicels 1.2–3 mm long; bracteoles 0–2; calyx 8.2–11 mm long, light brown, the tube 6.5–9.2 mm long, cylindric, gibbous, strigose, the teeth 1.7–2.6 mm long, subulate; flowers 20–24 mm long, ochroleucous, concolorous; pods pendulous, stipitate, the stipe 9–12 mm long, the body ellipsoid, 25–35 mm long, 4–7.5 mm thick, dorsiventrally compressed, strigose, unilocular; ovules 16–22. Duchesne River and Wasatch formations at 1600 to 1770 m in the juniper-pinyon community of western Uintah (type from 5 mi SW of Vernal) County; endemic; 25 (v). This is a mirrored-image congener of *A. lonchocarpus* (q.v.), which is known in the Uinta Basin only from eastern Uintah County, adjacent to Colorado.

Astragalus harrisonii Barneby Harrison Milkvetch. Perennial, caulescent, rushlike, 40–70 cm tall, from a subterranean caudex; pubescence basifixed; stems diffusely interbranched, in clumps to 1 m wide or more; stipules 1–5 mm long. all distinct; leaves 1.5–6.5 cm long, the uppermost simple, with the terminal leaflet expanded and confluent with the rachis, the others with leaflets 3–9, 2–11 mm long, 0.5–1.5 mm wide, linear-elliptic, acute, strigose on both sides; peduncles 6–19 cm long; racemes loosely 3- to 15-flowered, the flowers ascending at anthesis, the axis 5–40 cm long in fruit; bracts 0.5–1.1 mm long; pedicels 1.5–5.5 mm long; calyx 2.7–4.6 mm long, the tube 1.5–3.7 mm long, campanulate, strigose, the teeth 0.5–1.9 mm long, triangular; flowers 9–10.5 mm long, pink purple; pods pendulous, stipitate, the stipe 3–4 mm long, the body narrowly ellipsoid, straight or curved, 17–28 mm long, dorsiventrally compressed, strigose to glabrous, unilocular; ovules 10–12. Pinyon-juniper community at 1650 m in Capitol Reef (type locality) and Waterpocket Fold, Wayne County; endemic; 13 (vii). The Harrison milkvetch is a near congener of *A. nidularius* (q.v.).

Astragalus henrimontanensis Welsh Dana Milkvetch. [*Astragalus stocksii* Welsh, not Benth. ex Bunge]. Perennial, strictly acaulescent, 4–15 cm tall, from a branching caudex, the branches clothed with coarse, persistent leaf bases; pubescence basifixed; stipules 3–8 mm long, all distinct; leaves 2.7–12.5 cm long; leaflets 7–17, 3–13 mm long, 1.5–6 mm wide, elliptic to oblanceolate, mucronate, acute to obtuse or truncate, strigose on both sides; peduncles 1.1–8 cm long; racemes 2- to 11-flowered, the flowers ascending at anthesis, the axis 0.8–2.2 cm long in fruit; bracts 1.8–5.5 mm long; pedicels 1.3–2.5 mm long; bracteoles 0–2; calyx 10.2–15 mm long, the tube 8.2–11.5 mm long, cylindric, strigulose, the teeth 1.9–3.5 mm long, subulate; flowers 15–23 mm long, ochroleucous, the wings and keel purple-tipped; pods ascending, sessile, lance-ovoid, slightly incurved, 22–35 mm long, 5–11 mm thick, somewhat dorsiventrally compressed, strigose, unilocular. Ponderosa pine, pinyon-juniper, and sagebrush communities at ca 2430 m in the Henry Mts.(type locality) and Aquarius Plateau, Garfield County; endemic; 17 (v). The Dana milkvetch combines features of *A. argophyllus* var. *panguicensis*, which sometimes has persistent leaf bases but a tendency to caulescence; however, the Dana milkvetch has the habit and flower color of *A. eurekensis*. Specimens identified previously by other workers as *A. henrimontanensis* from northern Arizona are *A. argophyllus* var. *panguicensis*.

Astragalus holmgreniorum Barneby Holmgren Milkvetch. Perennial, strictly acaulescent, 4–12 cm tall, arising from a thickened root crown and taproot; pubescence basifixed; stipules papery, 3–8 mm long, all distinct; leaves 4–21.5 cm long; leaflets 5–23, 6–16 mm long, 3.5–12 mm wide, broadly obovate to obcordate, emarginate, pilose below, green and glabrous above; peduncles 2–9 cm long, procumbent; racemes 4– to 16-flowered, the flowers ascending to spreading at anthesis, the axis 0.4–6.5 cm long in fruit; bracts 4–8 mm long; pedicels 1–2 mm long; calyx 10.5–12.5 mm long, the tube 8–9.5 mm long, cylindric, thinly white pilose, the teeth 2–3.5 mm long, subulate; flowers 18–23.5 mm long, pink purple; pods reclining, 2.5–5.5 cm long, 6–9 mm thick, curved-elliptic, trigonously compressed, keeled on the ventral suture, openly sulcate dorsally, the valves coriaceous, glabrous, more or less reticulate, bilocular; ovules 30–34. Warm desert shrub at ca 820 to 850 m in Washington County; Mohave County Arizona; a Virgin-Mohave endemic; 3 (0). This plant was first taken by Melvin Ogden southeast of St. George on 19 May 1941 (BRY).

Astragalus humistratus Gray Goundcover Milkvetch. Perennial, caulescent, 7–80 cm long, radiating from a caudex; pubescence malpighian; stems prostrate; stipules 2–9 mm long, 0.5–6 mm wide, elliptic to oblong or oblanceolate, acute, strigose on both sides or glabrate above; peduncles 2–9 cm long; racemes 3- to 20-flowered, the flowers ascending at anthesis, the axis 1–12 cm long in fruit; bracts 1.5–7 mm long; pedicels 0.4–2.2 mm long; bracteoles 0–2; calyx 4.5–8.8 mm long, the tube 2.4–4.1 mm long, campanulate, strigose, the teeth 1.4–5 mm long; flowers 7–11.5 mm long, greenish to ochroleucous, often suffused or veined purplish, or pink purple; pods ascending to spreading, obliquely ovoid or oblong-ellipsoid, 6–14 mm long, 3.5–5.7 mm wide, variously compressed, strigose, unilocular; n = 12. Mountain brush, cool desert shrub, pinyon, juniper, ponderosa pine, and aspen communities at 1600 to 2745 m in Beaver, Garfield, Iron, Kane, and Washington counties; Nevada, Arizona, and New Mexico; 49 (xvi). Our material belongs to **var. *humivagans*** (**Rydb.**) **Barneby** [*Batidophaca humivagans* Rydb.].

Astragalus iodanthus Wats. Humboldt River Milkvetch. [*Tragacantha iodantha* (Wats.) Kuntze; *Xylophacos iodantha* (Wats) Rydb.; *A. iodanthus* var. *diaphanoides* Barneby]. Perennial, caulescent, 8–35 cm long, from a branching caudex; pubescence basifixed; stems prostrate to decumbent; stipules 2–6 mm long, all distinct; leaves 2–8 cm long; leaflets 7–21, 3–18 mm long, 2–12 mm wide, obovate to oblong or oblanceolate to elliptic, truncate to retuse, obtuse, or mucronate, sparingly strigose to glabrous on both sides; peduncles 1–4.5 cm long, shorter than the leaf; racemes 7- to 17-flowered, the flowers ascending to spreading at anthesis, the axis 0.5–4.5 cm long in fruit; bracts 1–3 mm long; pedicels 0.3–2 mm long; bracteoles 2; calyx 5–8 mm long, the tube 3.4–5 mm long, short-cylindric, strigose, the teeth 1.3–3 mm long; flowers 12–15.5 mm long, pink purple to pale; pods ascending to declined, sessile, the body curved through a half-circle, 20–40 mm long, 5–8 mm thick, dorsiventrally or trigonously compressed, obliquely lanceolate, unilocular to semibilocular, strigose; ovules 14–30; 2n = 22. Juniper-sagebrush community at ca 1649 m in Box Elder County; Oregon, California, and Nevada; 1 (0).

Astragalus iselyi Welsh Isely Milkvetch. Perennial, caulescent, 8–25 cm tall, from a branching caudex; pubescence basifixed; stems ascending to erect; stipules 3–9 mm long, all distinct; leaves 3.2–22 cm long; leaflets (1) 3–13, 7–23 (45) mm long, 3–19 (40) mm wide, elliptic to rhombic, acute to mucronate, sparsely strigose to glabrate on both sides; peduncles 1.4–10 cm long, racemes 7– to 20–flowered, the flowers spreading at anthesis, the axis 1–3 cm long in fruit; bracts 2–4.5 mm long; pedicels 0.8–2.5 mm long; bracteoles 0; calyx 6.7–10 mm long, the tube 5.5–6.3 mm long, cylindric, strigulose, the teeth 1.8–3.1 mm long, subulate; flowers 17–18 mm long, ochroleucous, concolorous; pods spreading-declined, sessile or subsessile, inflated, subcylindric, 25–38 mm long, 10–15 mm thick, strigose, leathery, unilocular; ovules 38–44. Morrison and Paradox formations in pinyon-juniper and salt desert shrub communities at 1530 to 1830 m on the western foothills of the La Sal Mts., Grand and San Juan (type from Brumley Ridge) counties; endemic; 20 (vii). This early flowering (March - April) selenophyte is allied to *A. sabulosus* (q.v.).

Astragalus jejunus Wats. Starvling Milkvetch. [*Tragacantha jejuna* (Wats.) Kuntze; *Phaca jejuna* (Wats.) Rydb.]. Perennial, acaulescent, 1–5 cm tall, from a much-branched caudex; pubescence basifixed; stems obsolete, obscured by stipules and leaf bases; stipules 1.5–3 mm long, all connate-sheathing; leaves 1–4 cm long; leaflets 9–15, 1–5 mm long, 0.5–1 mm wide, linear to narrowly elliptic, obtuse to acute, involute, strigose on both sides; peduncles 0.5–3.5 cm long; racemes 3– to 7–flowered, the flowers spreading at anthesis, the axis 0.2–1 cm long in fruit; bracts 1–1.5 mm long; pedicels 1–2.5 mm long, very slender; bracteoles 0; calyx 2.3–3 mm long, the tube 1.5–2 mm long, campanulate, strigose, the teeth 0.5–1 mm long, subulate; flowers 5–6.5 mm long, pink purple; pods spreading, sessile, bladdery-inflated, subglobose, 10–17 mm long, 7–11 mm thick, mottled, strigose, unilocular; ovules 10–14. Sagebrush and sagebrush-juniper communities, often on windswept ridgetops at 1830 to 2300 m in Rich County; southwestern Wyoming and east-central Nevada; 10 (i). The type is from the Bear River Valley. This tiny plant has as its nearest congener *A. limnocharis* (q.v.), a plant of higher elevations in central southern Utah.

Astragalus kentrophyta Gray Kentrophyta. Perennial, caulescent, mat-forming to erect, 15–45 cm long, from a caudex and stolonlike, creeping stems; pubescence basifixed or malpighian; stems prostrate to erect, compact to elongate; stipules 1.5–5 mm long, at least some connate-sheathing; leaves 0.4–2.6 cm long; leaflets 3–9, 3–13 mm long, 0.5–1.5 mm wide, linear to narrowly elliptic or lanceolate, all continuous with the rachis and spinulose-tipped, strigose on both sides; peduncles to 1.5 cm long; racemes 1– to 3–flowered, the flowers declined at anthesis, the axis to 0.5 mm long in fruit; bracts 0.8–3.5 mm long; pedicels 0.5–2 mm long; bracteoles 0; calyx 2.4–8.3 mm long, campanulate, strigose, the teeth 1.5–5 mm long, subulate; flowers 4.5–10 mm long, pink-purple or whitish, ochroleucous or purple tinged; pods declined or spreading, sessile, elliptic to oblong or lance-acuminate in outline, usually curved, 4–10 mm long, 1.3–4 mm wide, strigose, unilocular; ovules 2–8; 2n = 24. Four rather distinctive varieties occur in Utah.

1. Pubescence entirely of basifixed hairs; plants prostrate, of barrens at high elevations ....
 .... *A. kentrophyta* var. *implexus*
— Pubescence mostly of malpighian hairs; plants prostrate to erect, of low to high elevations 2

2(1). Calyx 6–8.3 mm long, the teeth 3.4–5 mm long; pods 7–10 mm long; plants prostrate, of sandy sites in canyons of the Colorado ....
 .... *A. kentrophyta* var. *coloradoensis*
— Calyx 3.4–5.5 mm long, the teeth 1.5–2.4 mm long; pods 4–7 mm long; plants erect or prostrate, usually of clay or silty soils at low to higher elevations 3

3(2). Plants prostrate; pods 3–4.5 mm long, beakless or nearly so; known from Daggett County ....
 .... *A. kentrophyta* var. *jessiae*
— Plants erect or prostrate; pods 4–7 mm long, beaked; known from Duchesne County southward ....
 .... *A. kentrophyta* var. *elatus*

Var. *coloradoensis* Jones Canyon Kentrophyta. [*A. montanus* var. *coloradoensis* (Jones) Jones; *Kentrophyta coloradoensis* (Jones) Rydb.]. Wash bottoms and rimrock in mixed desert shrub communities at 970 to 1630 m in Emery, Garfield, Kane, and Wayne counties; Coconino County, Arizona; a Colorado Plateau endemic; 19 (viii).

Var. *elatus* Wats. Tall Kentrophyta. [*A. viridis* var. *impensus* Sheldon; *A. kentrophyta* var. *impensus* (Sheldon) Jones; *A. impensus* (Sheldon) Woot. & Standl; *A. montanus* var. *impensus* (Sheldon) Jones; *A. tegetarius* var. *elatus* (Wats.) Barneby]. Mixed desert and salt desert shrub, juniper-pinyon, ponderosa pine, and pine-spruce communities, often in flood plains, at 1530 to 2600 m in Beaver, Duchesne, Emery, Garfield, Grand, Iron, Juab, Kane, Millard, Piute, Sevier, Uintah, and Wayne counties; California, Nevada, Arizona, Colorado, and New Mexico; 79 (xxviii). Both erect and prostrate phases are known, and the presence of malphigian hairs is diagnostic.

Var. *implexus* (Canby) Barneby Mountain Kentrophyta. [*A. tegetarius* var. *implexus* Canby; *A. tegetarius* Wats.; *Tragacantha tegetaria* (Wats.) Kuntze; *Homalobus tegetarius* (Wats.) Rydb.; *A. montanus* var. *tegetarius* (Wats.) Jones; *A. tegetarius* var. *rotundus* Jones, type from Loa; *A. kentrophyta* var. *rotundus* (Jones) Jones]. Ridgetops and breaks, commonly in barrens, at 2130 to 3500 m in Box Elder, Cache, Garfield, Iron, Juab, Kane, Salt Lake, Sanpete, Sevier, Summit, Tooele, and Utah counties; Oregon to Montana, south to California, Nevada, and New Mexico; 58 (ix).

Var. *jessiae* (Peck) Barneby Jessie Kentrophyta. [*A. jessiae* Peck]. White volcanic ash "barrens" at 1770 m in Daggett County at 1740 to 1800 m in Daggett County; Oregon, Idaho, and Wyoming; 6 (ii).

Astragalus lancearius Gray Lancer Milkvetch. [*Homalobus lancearius* (Gray) Rydb.; *A. episcopus* var. *lancearius* (Gray) Isely]. Perennial, caulescent, rushlike, 15–55 cm tall, arising from a subterranean caudex; pubescence basifixed; stems erect or ascending; stipules 2–7 mm long, all distinct; leaves 1.5–10.5 cm long, some or most of them reduced to the rachis, some or most of them with leaflets 3–7 in number, 2–14 mm long, 0.5–1.5 mm wide, linear to oblong, acute to obtuse, strigose on both sides; peduncles 4–23 cm long; racemes very loosely 6- to 25-flowered, the flowers ascending at anthesis, the axis 3–19 (26) cm long in fruit; bracts 1–1.5 mm long; pedicels 1–3 mm long; bracteoles 0–2; calyx 3.5–5.2 mm long, the tube 2.8–4.2 mm long, campanulate, about as broad as long, campanulate, black-strigose, the teeth 0.5–1.2 mm long, triangular; flowers 8.8–11.5 mm long, pink purple or merely tinged purplish; pods deflexed, sessile or subsessile, the body lance-oblong to oblong or lance-elliptic in outline, slightly curved or straight, 20–35 mm long, 5–9 mm wide, laterally compressed, glabrous to strigose, brown to straw colored, unilocular; ovules 8–14. Mixed desert shrub and pinyon-juniper communities at 1270 to 1730 m in Kane and Washington counties; Coconino and Mohave counties, Arizona; a Colorado Plateau endemic; 12 (iii). This is a near ally of *A. episcopus* (q.v.), differing diagnostically mainly in the proportionally shorter calyces.

Astragalus lentiginosus Dougl. ex Hook. Freckled Milkvetch. Perennial, caulescent, mostly 1.5–6 dm tall, from a caudex; pubescence basifixed; stipules 1.5–7 mm long or more, all distinct; leaves 2.4–15 cm long; leaflets 9–23, 2–23 mm long, 1–13 mm wide, elliptic to ovate or lanceolate, obtuse to rounded, emarginate, or acute, pubescent to glabrous on one or both sides; peduncles 1–14 cm long, sometimes more; racemes (5) 11- to 30-flowered, the flowers ascending to declined at anthesis, the axis 1–18 cm long in fruit; pedicels 1–4 mm long; bracts 1.5–6 mm long; bracteoles 0–2; calyx 3.5–11.6 mm long, the tube 3–9 mm long, cylindric to short-cylindric, strigose, the teeth 0.6–2.5 mm long, subulate or triangular; flowers 8.4–22 mm long, pink purple, ochroleucous, whitish, or variously suffused with pink or purple; pods ascending to declined, sessile or with a gynophore, variable in outline, either inflated and ovoid (12–26 mm long, 5–20 mm thick) or not inflated and oblong in outline (15–25 mm long, 3–7.5 mm thick), strigose or glabrous, mottled or not, leathery to membranous, bilocular; ovules 16–38; n = 11. Numerous varieties are known to occur in Utah.

1. Flowers small, the keel 5.5–8.5 mm long 2
— Flowers larger, the keel 9–16 mm long or more 4
2(1). Raceme axis much elongating in fruit, 4–16 cm long or more; plants of Washington County .... *A. lentiginosus* var. *fremontii*
— Raceme axis little elongating in fruit, not over 4 cm long; plants not of Washington County 3
3(2). Pods opaque, stiffly papery; plants rare in west-central Utah .... *A. lentiginosus* var. *scorpionis*
— Pods transparent to translucent, thinly papery; plants locally common and rather widespread in western to central and northern counties.. *A. lentiginosus* var. *salinus*
4(1). Pods stipitate, the stipes 0.5–1.5 mm long; flowers 20–23 mm long; plants of greasewood communities in western Utah .... *A. lentiginosus* var. *pohlii*
— Pods sessile or essentially so; flowers mainly less than 20 mm long 5
5(4). Raceme axis much elongating in fruit, usually more than 8 cm long 6
— Raceme axis not much elongating in fruit, seldom as much as 7 cm long in fruit........................ 8

6(5). Flowers ochroleucous or faintly suffused with purple; pods diaphanous, glabrous; plants of Washington County .... *A. lentiginosus* var. *vitreus*
— Flowers ordinarily pink purple to pink; pods usually opaque or strigose or both; distrigution various 7
7(6). Pods bladdery-inflated; herbage cinereus, the stems canescent; plants known from the western slope of the Beaverdam Mts., Washington County .... *A. lentiginosus* var. *stramineus*
— Pods not or scarcely inflated; herbage greenish, the stems not or seldom canescent; plants of various distribution .... *A. lentiginosus* var. *palans*
8(5). Pods thinly papery, diaphanous; plants of eastern Kane County .... *A. lentiginosus* var. *wahweapensis*
— Pods stiffly papery to leathery, opaque or nearly so; plants variously distributed 9
9(8). Leaflets 9–17; peduncles 1–5 cm long, shorter than the subtending leaves; flowers pale pink to whitish, or less commonly pink purple; plants of Rich and Daggett counties .... *A. lentiginosus* var. *platyphyllidius*
— Leaflets 15–23; peduncles 2.5–8 cm long, mostly longer than the subtending leaves; plants variously distributed .... *A. lentiginosus* var. *araneosus*

Var. *araneosus* (Sheldon) Barneby Cobweb Milkvetch. [*Astragalus araneosus* Sheldon, type from Frisco; *A. palans* var. *araneosus* (Sheldon) Jones; *Cystium araneosum* (Sheldon) Rydb.; *A. lentiginosus* var. *chartaceous* Jones, type from Ephraim]. Sagebrush, mixed desert shrub, pinyon-juniper, and less commonly in mountain brush communities at 1270 to 2430 m in Box Elder, Beaver, Garfield, Iron, Juab, Kane, Millard, Piute, Sanpete, Sevier, Tooele, and Wayne counties; Nevada; 216 (lix). This variety is the common phase of the freckled milkvetch in south-central to western Utah. It approaches var. *diphysus* (Gray) Jones in its technical features, but specimens from Utah assigned previously to that taxon belong to other varieties. The problems of inclusion of this variety within an expanded var. *diphysus* involve negation of the morphological trends within var. *araneosus* proper. Material from the Henry Mts., with a slightly lax inflorescence belongs to var. *wahweapensis* (q.v.).

Var. *fremontii* (Gray) Wats. Fremont Milkvetch. [*A. fremontii* Gray ex Torr.; *A. coulteri* var. *fremontii* (Gray) Jones; *Cystium fremontii* (Gray) Rydb.]. Braided stream gravels and slopes in creosote bush, Joshua tree, and pinyon-juniper communities in western Washington County; California and Nevada; 12 (ii).

Var. *palans* (Jones) Jones Straggling Milkvetch. [*A. palans* Jones, type from Montezuma Canyon; *Tium palans* (Jones) Rydb.; *A. ursinus* Gray, type from Bear Valley, Utah; *A. lentiginosus* var. *ursinus* (Gray) Barneby; *A. bryantii* Barneby]. Salt desert shrub, blackbrush, juniper, pinyon-juniper, and mixed desert shrub communities at 1130 to 1900 m in Carbon, Emery, Grand, Kane, San Juan, Washington, and Wayne counties; Colorado and Arizona; 117 (xxxiii). The straggling milkvetch is the common phase of *A. lentiginosus* in the canyons of the Colorado River. It is distinctive in having oblong, usually curved pods, seldom more than 7 mm thick, pale pink purple flowers, and elongate inflorescences. The type locality of the long sought *A. ursinus* Gray remains in question. Specimens taken recently in the south end of the Beaverdam Mts., Washington

County, approach the type materials in general conformation and suggest that the specimens on which *A. ursinus* was based came from Washington (not Iron) County. The name *A. ursinus* is here taken as a synonym of *A. lentiginosus* var. *palans*. Specimens cited previously as *A. bryantii* do not differ in any remarkable way from var. *palans*.

Var. *platyphyllidius* (Rydb.) Barneby Broad-leaved Milkvetch. [*Cystium platyphyllidium* Rydb.]. Sagebrush, pinyon-juniper, and mountain brush communities at 1700 to 2135 m in Daggett and Rich counties; Oregon and California, east to Wyoming and Colorado; 4 (i). The peduncles and commonly the fruiting racemes are surpassed by the subtending leaves. This feature, coupled with leaflets that average fewer per leaf seem to distinguish var. *platyphyllidius* from var. *araneosus*. Flower color varies from whitish, with keel tip purple, to pink purple throughout.

Var. *pohlii* Welsh & Barneby Pohl Milkvetch. Greasewood communities at 1330 to 1650 m in Rush and Skull valleys, Tooele (type from 4.5 mi N of Vernon) County; endemic; 7 (iii).

Var. *salinus* (Howell) Barneby Salt Milkvetch. [*A. salinus* Howell; *Cystium salinum* (Howell) Rydb.]. Pinyon-juniper, mountain brush, and sagebrush communities at 1770 to 2430 m in Beaver, Box Elder, Iron, Juab, Millard, Morgan, Sanpete, Summit, and Wasatch counties; California, Oregon, Nevada, Idaho, and Wyoming; 33 (viii). The tiny flowers and diaphanous pods are diagnostic.

Var. *scorpionis* Jones Scorpion Milkvetch. [*Cystium scorpionis* (Jones) Rydb.]. Sagebrush community upward to timberline at 2130 to 3350 m in western Juab County; Nevada; 1 (0). This rare small-flowered plant differs from var. *salinus* mainly in its opaque leathery pods.

Var. *stramineus* (Rydb.) Barneby Straw Milkvetch. [*Cystium stramineum* Rydb., type from southern Utah]. Mixed warm desert shrub community at 900 to 1000 m on the western slope of the Beaver Dam Mts., Washington County; Arizona and Nevada. This is the only member of the species having moderate sized flowers known to occur in Washington County; it is a Beaver Dam-Virgin endemic; 6 (0).

Var. *vitreus* Barneby Glass Milkvetch. Creosote bush and mixed warm desert shrub communities at 1830 to 1430 m in central and western Kane and eastern Washington (type from 5 mi W of Leeds) counties; Mohave and Coconino counties, Arizona; 21 (iii). In its typical phase, north of Washington, the flowers are very pale, but become bright pink purple eastward. The large, glassy pods are distinctive.

Var. *wahweapensis* Welsh Pinyon-juniper and mixed desert shrub communities at 1860 to 2135 m in Garfield (Henry Mts.) and Kane (Four Mile Bench, type locality) counties; endemic; 28 (viii). This plant is characterized by its diaphanous pods and moderately elongate racemes. It is more or less intermediate between varieties *araneosus* and *vitreus*.

Astragalus limnocharis **Barneby** Navajo Lake Milkvetch. Perennial, subacaulescent, 1–5 cm tall, arising from a branching caudex; pubescence basifixed; stems prostrate to erect; stipules 2–4 mm long, all connate-sheathing; leaves 1.5–7 cm long; leaflets (5) 7–13, 2–9 mm long, 1–2 mm wide, lanceolate to elliptic or oblong, obtuse, strigose beneath, long-ciliate on the involute margin, glabrous above; peduncles 2–5 cm long, reclining in fruit; racemes 2- to 10–flowered, the flowers spreading to declined at anthesis, the axis 0.2–0.5 cm long in fruit; bracts 1–3 mm long; pedicels 0.8–1.5 mm long; bracteoles 0; calyx 2.8–3.6 mm long, the tube 2–2.5 mm long, campanulate, strigose, the teeth 0.7–1.6 mm long, subulate; flowers 6.2–7.5 mm long, ochroleucous or pink purple; pods spreading, sessile, ovoid, bladdery-inflated, 9–18 mm long, 7–13 mm thick, mottled, strigose, unilocular; ovules 10–12. Lake shores and limestone breaks on Pink and White Limestone members of the Wasatch Formation, at 2670 to 3400 m in Garfield, Iron, and Kane (type from Navajo Lake), counties; endemic; 14 (v). This "lake beauty" has a principal locality on the beach at Navajo Lake, where it is evidently dispersed by wave action. Plants occur along the terraces below the high water line as the lake recedes. Evidently, the nearest allies of the Navajo Lake milkvetch are *A. jejunus* (q.v.), known in Utah only from Rich County, and *A. montii* of the Wasatch Plateau. A phase with pink purple flowers is present on the Table Cliff Plateau and vicinity in Garfield County. These latter plants are soboliferous, possibly an ecological response to the steep, unstable limestone slopes that serve as habitat; they are designated as **var. *tabulaeus* Welsh**. Specimens from Navajo Lake and vicinity have ochroleucous flowers and typically lack sobols. They are **var. *limnocharis***.

Astragalus loanus **Barneby** Loa Milkvetch. [*A. newberryi* var. *wardianus* Barneby, type from east of Glenwood]. Perennial, acaulescent, 3–6 cm tall, the caudex branches clothed with a thatch of persistent leaf bases; pubescence incipiently malpighian; stipules 5–10 mm long, all distinct; leaves 2–8 cm long; leaflets (1) 3–5, 6–25 mm long, 5–20 mm wide, obovate to rhombic, elliptic, or oval, obtuse, densely strigose on both sides; peduncles 1–3 cm long; racemes 2- to 7–flowered, the flowers erect-ascending at anthesis, the axis 0.2–0.8 cm long in fruit; bracts 2.5–6 mm long; pedicels 1.2–4 mm long; bracteoles 0–2; calyx 12–17 mm long, the tube 10–14 mm long, cylindric, loosely strigulose, the teeth 1.8–5.5 mm long, linear-subulate; flowers 20–28 mm long, ochroleucous or greenish white, often tinged faintly purplish, the keel-tip purple; pods ascending, sessile, inflated, ovoid, 17–30 mm long, 13–23 mm wide (when pressed), the valves red purple to straw colored, hirsute with lustrous long hairs; unilocular; ovules 28–38. Sagebrush and pinyon-juniper communities, exclusively on igneous gravels at 1920 to 2075 m in central Sevier County; endemic; 8 (iv). Careful examination of the type of *A. loanus* has demonstrated that two taxa were involved within the initial concept of *A. loanus*. The discordant material has been segregated as a separate species, *A. welshii* Barneby (q.v.).

Astragalus lonchocarpus **Torr.** Great Rushy Milkvetch. [*Phaca macrocarpa* Gray; *Tragacantha lonchocarpa* (Torr.) Kuntze; *Homalobus macrocarpus* (Gray) Rydb.; *Lonchophaca macrocarpa* (Gray) Rydb.; *A. macer* A. Nels.; *L. macra* (A. Nels.) Rydb.]. Perennial, caulescent, 30–84 cm tall, from a shallowly subterranean caudex; pubescence basifixed; stems erect, often in dense clumps; stipules 1–9 mm long, all distinct; leaves 2–13 cm long, the uppermost and sometimes all simple, the lower with leaflets 3–9, 2–36 mm long, 0.5–4 mm wide, linear to narrowly oblanceolate, obtuse to acute, strigose on both sides or glabrous above; peduncles 6–24 cm long;

racemes loosely 7- to 40-flowered, the flowers spreading-declined at anthesis, the axis 3.5–45 cm long in fruit; bracts 0.8–2.5 mm long; pedicels 1.3–4.5 mm long; bracteoles 0; calyx 5.8–10.3 mm long, usually brown, the tube 5–8 mm long, cylindric, gibbous, strigose, the teeth 0.6–2 mm long, subulate; flowers 13–20 mm long, ochroleucous to almost white, concolorous; pods pendulous, stipitate, the stipe 3–15 mm long, the body elliptic to oblong in outline, 22–50 mm long, 3.3–6.2 mm wide, dorsiventrally compressed, the valves often brownish, strigose, unilocular; ovules 12–26; 2n = 22. Salt desert shrub, blackbrush, and pinyon-juniper communities at 1170 to 2530 m in Beaver, Box Elder, Carbon, Emery, Garfield, Grand, Iron, Juab, Kane, Millard, Piute, San Juan, Sevier, Tooele, and Wayne counties; Colorado, New Mexico, Arizona, and Nevada; 135 (xxx). This species is allied with *A. hamiltonii* (q.v.).

Astragalus lutosus Jones Dragon Milkvetch. Perennial, short-caulescent, 2–10 tall, from a subterranean caudex; pubescence basifixed; stems prostrate to ascending, radiating; stipules 2–5 mm long, all distinct or some shortly connate-sheathing; leaves 1–5.5 cm long; leaflets 11–27, 1–12 mm long, 1–5 mm wide, obovate to elliptic or oblong, often folded, obtuse to retuse, gray-strigulose on both sides or glabrous above; peduncles 0.5–4 cm long; racemes 4- to 10-flowered, the flowers ascending-spreading at anthesis, the axis 0.3–1 cm long in fruit; bracts 1.5–2.5 mm long; pedicels 1.2–3 mm long; bracteoles 0; calyx 4.8–10.4 mm long, the tube 3.6–7.4 mm long, short-cylindric, strigulose, the teeth 1.2–3 mm long, subulate; flowers 9–17 mm long, white to ochroleucous, the keel-tip purplish; pods spreading to ascending, stipitate, the stipe (gynophore) 1–4.5 mm long, the body ovoid-ellipsoid, bladdery-inflated, 15–37 mm long, 8–23 mm thick, strigose, unilocular. Barrens, often with other mound-forming species in mixed desert shrub, pinyon-juniper, mountain brush, and limber pine-Douglas fir communities, on outcrops of Green River Shale at 1645 to 2870 m in Uintah (type from White River), Utah, and Wasatch counties; Rio Blanco County, Colorado; 33 (i). The Dragon milkvetch long remained obscure, but exploration during the past decade has demonstrated a rather broad distribution for this plant of peculiar shale strata.

Astragalus malacoides Barneby Kaiparowits Milkvetch. Perennial, short-caulescent, 7–26 cm tall, from a slightly subterranean caudex; pubescence basifixed; stems decumbent or prostrate to ascending; stipules 2–7 mm long, all distinct; leaves 4.5–14 cm long; leaflets 15–29, 3–25 mm long, 2–12 mm wide, obovate to oblong or elliptic, obtuse to emarginate, hirtellous beneath, glabrous above; peduncles 4–12 mm long; racemes 10- to 24-flowered, the flowers ascending-spreading at anthesis, the axis 1.2–10 cm long in fruit; bracts 3–5 mm long; pedicels 1–2.5 mm long; bracteoles 0–2; calyx 10–15 mm long, the tube 7–12 mm long, cylindric, hirsutulous, the teeth 2–6.2 mm long, linear-subulate; flowers 16–22 mm long, pink purple; pods declined to ascending, short-stipitate, the stipe 2–3 mm long, the body oblong, curved, 25–40 mm long, 5–8 mm wide, laterally compressed, hirsutulous, bilocular; ovules 24–30. Straight Cliffs, Wahweap, Kaiparowits, Ferron, and Moenkopi formations (inter alia), usually in clay or silty soils, in juniper-pinyon and mixed desert shrub communities at 1600 to 2400 m in Garfield (Circle Cliffs, Tarantula Mesa, and Henry Mts.) and Kane (Kaiparowits vicinity, type locality) counties; endemic; 28 (vi).

Astragalus megacarpus (Nutt.) Gray Great Bladdery Milkvetch. [*Phaca megacarpa* Nutt. ex T. & G.; *Tragacantha megacarpa* (Nutt.) Kuntze; *A. megacarpus* var. *prodigus* Sheldon]. Perennial, short-caulescent or subacaulescent, 3–15 (17) cm tall, arising from a caudex; pubescence basifixed; stems 1–5 cm long, in dense, leafy clumps, the internodes mostly obscured by stipules, these 2–7 mm long and all distinct; leaves 2–15 (17) cm long; leaflets 7–19, 3–19 mm long, 2–12 mm wide, obovate, ovate, elliptic, or suborbicular, obtuse to retuse and mucronate, strigose to glabrous beneath, glabrous above; peduncles 0.5–6 cm long, much shorter than the leaves; racemes 1- to 7-flowered, the flowers ascending in anthesis, the axis 0.2–2.5 cm long in fruit; bracts 2–5 mm long; pedicels 3.5–8 mm long; bracteoles 0–2; calyx 7–10.2 mm long, the tube 5.2–8.5 mm long, cylindric, strigose, the teeth 1.8–3.5 mm long; flowers 15–23 mm long, pink purple; pods ascending to descending, stipitate, the stipe (gynophore) 2–4 mm long, the body bladdery- inflated, ellipsoid, 35–70 (88) mm long, 15–45 mm wide (when pressed), commonly mottled, strigose, unilocular; ovules 38–54. Commonly on clay soils, these often saline or calciferous, in salt desert shrub, sagebrush, oakbrush, pinyon-juniper, and ponderosa pine communites at 1650 to 3050 m in Carbon, Duchesne, Emery, Garfield, Iron (type of var. *parryi* from Cedar Canyon), Juab, Kane, Millard, Sanpete, and Sevier counties; Nevada, Wyoming, and Colorado; 56 (xii). The great bladdery milkvetch is among our most distinctive species of *Astragalus*. A plant with its large pods declined on the ground around the base is a surprising sight. Our plants belong to **var. parryi** Gray ex Wats., sens. str., but this variety differs mainly in flower color (though our specimens seem to have smaller calyces also) from **var. megacarpus**, sens. str.

Astragalus miser Dougl. ex Hook. Weedy Milkvetch. Perennial, caulescent or short-caulescent, 3–35 cm tall, from a branching caudex; pubescence basifixed; stems decumbent to erect, 1–25 cm long or more; stipules 1.5–7 mm long, at least some connate-sheathing; leaves 1.5–20 cm long or more; leaflets 3–21, 3–35 mm long, 0.5–7 mm wide, linear to oblong, elliptic, or oval, acute to obtuse or emarginate, strigose beneath, glabrous or glabrate above; peduncles 2–14 cm long; racemes 3- to 20-flowered, the flowers spreading-declined at anthesis, the axis 1–10 cm long in fruit; calyx 2.1–5.2 mm long, the tube 1.9–2.9 mm long, campanulate, strigose, the teeth 0.5–2.5 mm long, subulate; flowers 5.9–10 mm long, pink purple, ochroleucous, or whitish, often suffused or veined with purple; keel with an upturned, usually purple beak; pods declined-pendulous, sessile or nearly so, narrowly oblong or oblanceolate in outline, 12–25 mm long, 2–4 mm wide, strigose, unilocular; ovules 8–19; 2n = 22. The weedy milkvetch is a widespread, poisonous plant of middle and upper elevations in much of Utah. Both cattle and sheep are poisoned by this plant. Two rather tenuous varieties are known, one common the other evidently rare.

1. Leaflets 3–11; flowers 6–8 mm long; pods narrowly oblong in outline; plants rare, scattered in northern Utah . . . . *A. miser* var. *tenuifolius*

— Leaflets mostly 11–21; flowers 8–11 mm long; pods oblanceolate in outline; plants common in Utah . . . . *A. miser* var. *oblongifolius*

Var. **oblongifolius** (Rydb.) Cronq. Rydberg Weedy Milkvetch. [*Homalobus oblongifolius* Rydb.; *A. hylophilus* var. *oblongifolius* (Rydb.) Macbr.; *A. decumbens* var. *oblongifolius* (Rydb.) Cronq.; *Homalobus humilis* Rydb., type from the Tushar Mts.; *A. carltonii* Macbr., not *A. humilus* M. Bieb.]. In aspen, mixed aspen-conifer, and coniferous woodlands, and in sagebrush, mountain brush, and alpine meadow communities at 1830 to 3570 m in all Utah counties except Davis, Morgan, San Juan, and Washington; Nevada, Arizona, Wyoming, and Colorado; 211 (xxxiv).

Var. **tenuifolius** (Nutt.) Barneby Garrett's Weedy Milkvetch. [*Homalobus tenuifolius* Nutt.; *H. paucijugus* Rydb.; *A. garrettii* Macbr., not *A. paucijugus* Schrenk]. Sagebrush upward to mountain summits at 1970 to 3050 m in Rich and Salt Lake (type from Big Cottonwood Canyon) counties; Idaho, Nevada, and Wyoming; 4 (i).

Astragalus missouriensis Nutt. Missouri Milkvetch. Perennial, subacaulescent, 3–12 cm tall, from a caudex; pubescence malpighian; stems prostrate, to 8 cm long, the internodes often concealed by stipules; stipules 2–9 mm long, all distinct; leaves 1.5–12 cm long; leaflets 5–15, 3–15 mm long, 1–8 mm wide, elliptic to obovate, acute to mucronate or obtuse, strigose on both sides; peduncles 1.5–8 cm long; racemes 3– to 12–flowered, the flowers ascending at anthesis, the axis 0.4–3 cm long, in fruit; bracts 2.5–8 mm long; pedicels 1–3.5 mm long; bracteoles 0–1; calyx 8.5–13 mm long, the tube 7–10 mm long, cylindric, strigose, the teeth 1.5–3 mm long, subulate; flowers 15–22 mm long, pink purple, rarely white; pods ascending to descending, sessile, ellipsoid, 15–25 mm long, 7–9 mm thick, curved, dorsiventrally compressed, strigose, unilocular; ovules 35–55. Pinyon-juniper and sagebrush communities at 1600 to 2430 m in San Juan County; Alberta to Manitoba and Minnesota, south to New Mexico and Texas; 2 (0). Our material belongs to var. **amphibolus** Barneby.

Astragalus moencoppensis Jones Moenkopi Milkvetch. [*Cnemidophacos moencoppensis* (Jones) Rydb.]. Perennial, caulescent, 9–60 cm tall, from a branching caudex; pubescence basifixed; stems erect or ascending, commonly shorter than the longest peduncles; stipules 1.5–7 mm long, at least some connate-sheathing; leaves 4–16.5 cm long; leaflets 5–15, 2–23 mm long, 0.3–2 mm wide, filiform to linear or narrowly elliptic, acute to obtuse, the terminal often continuous with the rachis, strigose below, glabrous on the involute upper surface; peduncles 4–25 cm long; racemes 6– to 34–flowered, the flowers ascending at anthesis, the axis 3–25 cm long in fruit; bracts 1.5–3.5 mm long; pedicels 0.3–2 mm long; bracteoles 0–1; calyx 5–7.5 mm long, the tube 3–4 mm long, campanulate, white-pilose, the teeth 1.8–3.5 mm long, lance-subulate; flowers 8–11 mm long, pink purple; pods ascending-spreading, sessile, ovoid to ellipsoid, 6–7 mm long, 2.3–3.4 mm thick, strigulose, unilocular. Salt desert shrub, mixed desert shrub, and pinyon-juniper communities, usually in silty or clay soils, at 1330 to 2130 m in Emery, Garfield, Grand, Kane, San Juan, and Wayne counties; Arizona; a Colorado Plateau endemic; 63 (xviii). The Moenkopi milkvetch is a primary selenium indicator of distinctive mien and odor.

Astragalus mollissimus Torr. Woolly Locoweed. Perennial, acaulescent, 6–45 cm tall, from a caudex; pubescence basifixed; stems mostly obscured by stipules, these 4–13 mm long, all distinct; leaves 2–28 cm long; leaflets 15–35, 2–18 mm long, 1–14 mmm wide, obovate to suborbicular or elliptic, obtuse to retuse or acute, densely woolly-tomentose on both sides; peduncles 2.5–24 cm long; racemes 7– to 20–flowered, the flowers ascending at anthesis, the axis 1.5–18 cm long in fruit; bracts 2.5–8 mm long; pedicels 0.5–3 mm long; bracteoles 0–1; calyx 11–15.5 mm long, the tube 7.7–13 mm long, cylindric, villous, the teeth 2–4.2 mm long, subulate; flowers 18–25 mm long, pink purple; pods descending, sessile, ovoid, 11–23 mm long, 6–11 mm thick, curved, densely villous-tomentose, bilocular; ovules 28–38; 2n = 22. Salt desert shrub, mixed desert shrub, grassland, and pinyon-juniper communities at 1130 to 2330 m in Beaver, Carbon, Daggett, Duchesne, Emery, Garfield, Grand, Iron, Kane, Millard, Piute, San Juan, Sanpete, Sevier, Tooele, Uintah, Washington, and Wayne counties; Wyoming and Nebraska, south to Nevada, Texas, and Mexico; 193 (xlv). Our material belongs to var. **thompsonae** (Wats.) Barneby [*A. thompsonae* Wats. type from Kanab; *Tragacantha thompsonae* (Wats.) Kuntze; *A. bigelovii* var. *thompsonae* (Wats.) Jones; *A. syrticolus* Sheldon]. The woolly locoweed is one of the earliest of the *Astragalus* species to flower in Utah. It has been taken in flower as early as February in southern counties. The species as a whole is recognized as being poisonous. The plants are seldom abundant, except along roadsides, and this might account for the lack of reports of poisoning from this entity in Utah.

Astragalus montii Welsh Heliotrope Milkvetch. [*A. limnocharis* var. *montii* (Welsh) Isely]. Perennial, subacaulescent, 1–5 cm tall, arising from a branching caudex; pubescence basifixed; stems ascending to erect; stipules 2–4 mm long, all connate-sheathing; leaves 1.3–4.8 cm long; leaflets 5–13, 2–8 mm long, 1–2 mm wide, lanceolate to oblong or elliptic, strigose beneath, not ciliate on the involute margin, glabrous above; peduncles 0.8–4.5 cm long, reclining in fruit; racemes 2– to 8–flowered, the flowers ascending to spreading in anthesis, the axis 0.2–0.5 cm long in fruit; bracts 1–3 mm long; pedicels 0.8–1.5 mm long; bracteoles 0; calyx 3.3–4 mm long, the tube 2.2–2.5 mm long, campanulate, strigose, the teeth 0.6–1.5 mm long, triangular-subulate; flowers 7.2–8 mm long, pink purple, the wing-tips white; pods spreading, sessile, ovoid, bladdery-inflated, 11–18 mm long, 8–12 mm thick, mottled, strigose, unilocular; ovules 10. The Heliotrope milkvetch is known only from Flagstaff Limestone at ca 3350 m on the Wasatch Plateau, San Pete (type from Heliotrope Mt.) County; endemic; 6 (i). The species is a near congener of *A. limnocharis* from the Markagunt and Table Cliff plateaus and vicinities, differing inter alia in the pink purple flowers with contrasting white wing-tips, which are larger, and in the merely strigose (not long-ciliate) leaflet margins. The phase of *A. limnocharis* (var. *tabulaeus*) with pink purple flowers does not have the contrasting white wing tips and is strongly soboliferous.

Astragalus monumentalis Barneby Monument Milkvetch. Perennial, acaulescent or subacaulescent, 3–18 cm tall, from a branching caudex; pubescence basifixed or shortly malpighian; stems 1–6 cm long, ascending, the internodes commonly concealed by stipules, these 2–4 mm long, all distinct; leaves 1.5–8 (11) cm long; leaflets (5) 9–17 (21), 2–9 mm long, 1–4 mm broad, oval to obovate, elliptic, or oblanceolate, strigulose beneath, glabrous or glabrate above; peduncles 1–12 cm long;

racemes 3- to 9-flowered, the flowers ascending at anthesis, the axis 0.5-7 cm long in fruit; bracts 1.5-5 mm long; pedicels 0.8-2.2 mm long; bracteoles 0; calyx 3.6-4.5 mm long, the tube 3.1-3.5 mm long, campanulate, strigose, purplish, the teeth 0.5-1 mm long; flowers 8-9 mm long, pink purple; pods ascending, sessile or nearly so, narrowly oblong to lanceolate in outline, 12-21 mm long, 2.3-3 mm wide, straight or curved, triangular in cross-section, the dorsal suture sulcate, strigose, bilocular, often mottled. Rimrock and other slickrock sites in mixed desert shrub and pinyon-juniper communities at 1230 to 1870 m in Garfield and San Juan (type from White Canyon) counties; endemic; 27 (xviii). This is a mirror-image congener of *A. cottamii* (q.v.), differing in its smaller floral parts and overall flower size and in the incipient malpighian pubescence.

Astragalus musiniensis Jones Ferron Milkvetch. [*Xylophacos musiniensis* (Jones) Rydb.; *A. musiniensis* var. *newberryoides* Jones, type from the San Rafael Swell]. Perennial, acaulescent, 3-13 cm tall, from a caudex, the branches often clothed with a thatch of persistent leaf bases; pubescence basifixed or incipiently malpighian; stipules 3.5-10 mm long, all distinct; leaves 1.5-13 cm long; leaflets (1) 3-5, 5-35 mm long, 2-11 mm wide, elliptic to lanceolate, acute to obtuse, strigose on both surfaces; peduncles 0.5-7 cm long; racemes 1- to 6-flowered, the flowers erect-ascending at anthesis, the axis 0.2-1.4 cm long in fruit; bracts 1.6-4 mm long; pedicels 1.2-4 mm long; bracteoles 0; calyx 12-15.5 mm long, the tube 9.5-12.7 mm long, cylindric, strigose-pilose, the teeth 1.5-4 mm long; flowers 20-28 mm long, pink purple; pods ascending, sessile, obliquely ovoid, 15-36 mm long, 8-17 mm thick, dorsiventrally compressed, hirsutulous to villous-hirsute, unilocular. Salt desert shrub, mixed desert shrub, and pinyon-juniper communities at 1430 to 2130 m in Carbon, Emery (type from 2 miles south of Ferron), Garfield, Grand, Kane, and Wayne counties; endemic; 55 (xvii). The species was named by Marcus E. Jones for a peak on the Wasatch Plateau many miles remote from the type locality.

Astragalus nelsonianus Barneby Nelson Milkvetch. [*A. pectinatus* var. *platyphyllus* Jones]. Perennial, caulescent, 10-30 cm tall, from a subterranean caudex; pubescence basifixed; stems decumbent to ascending or erect; stipules 4-13 mm long, at least some connate-sheathing; leaves 2.5-9 cm long; leaflets 3-13, 10-45 mm long, 2-5 mm wide, narrowly oblong, obtuse to apiculate, strigose on both surfaces, the terminal leaflet continuous with the rachis; peduncles 3-12 cm long; racemes 6- to 20-flowered, the flowers ascending at anthesis, the axis 2-12 cm long in fruit; bracts 2.5-7 mm long; pedicels 1.5-4 mm long; bracteoles 1; calyx 10-14.5 mm long, the tube 7-10.2 mm long, cylindric, strigose, the teeth 2-4.5 mm long; flowers 10-14 mm long, white, concolorous; pods deflexed, sessile, oblong-ellipsoid, 13-33 mm long, 6-12 mm wide, finally laterally compressed, glabrous or pubescent, unilocular; ovules 20-28. Saline soils in desert shrub communities at 2070 to 2100 m in Daggett County; Wyoming; 4 (i). The Nelson milkvetch differs from *A. pectinatus* Dougl. ex G. Don in much the same manner that *A. hamiltonii* differs from *A. lonchocarpus*. The leaves and leaflets of both *A. nelsonii* and *A. hamiltonii* are much broader than those of their counterparts.

Astragalus newberryi Gray Newberry Milkvetch. [*Xylophacos newberryi* (Gray) Rydb.; *A. newberryi* var. *castoreus* Jones, type from Beaverdam Mts.]. Perennial, acaulescent, 2-12 cm tall, from a caudex, the branches commonly clothed with a thatch of persistent leaf bases; pubescence basifixed; stipules 3-11 mm long, all distinct; leaves 1.5-14 cm long; leaflets 3-15, 3-20 mm long, 2-14 mm wide, obovate to elliptic, oblanceolate, or orbicular, acute to obtuse or retuse, villous-tomentulose on both surfaces; peduncles 0.5-11 cm long; racemes 2- to 8-flowered, ascending in flower, the axis 0.2-2.7 cm long in fruit; bracts 3.5-10 mm long; pedicels 1.4-5 mm long; bracteoles 0-2; calyx 11.5-20 mm long, the tube 8-17 mm long, cylindric, villous, the teeth 1.9-6 mm long, subulate; flowers 20-32 mm long, pink purple; pods ovoid, curved, 18-23 mm long, 7-12 mm thick, densely villous-tomentose, unilocular; ovules 20-40; n = 11. Salt desert shrub, mixed desert shrub, sagebrush, mountain brush, and pinyon-juniper communities at 830 to 2300 m in Beaver, Box Elder, Garfield, Iron, Juab, Kane, Millard, Piute, Sevier, Tooele, and Washington counties; Oregon, California, Nevada, Arizona, and New Mexico; 116 (xxxiv). Dwarf material from the vicinity of Escalante and Kanab east to Glen Canyon belongs to **var. newberryi**, the remainder of our specimens are assigneable to **var. castoreus** Jones.

Astragalus nidularius Barneby Birds-nest Milkvetch. Perennial, caulescent, 15-51 cm tall, from a subterranean caudex; pubescence basifixed; stems ascending to erect, often branched; stipules 1-6 mm long, at least some connate-sheathing; leaves 1.5-7 cm long; leaflets 5-11, 2-20 mm long, 1.3-2 mm wide, linear to oblong, obtuse to emarginate or acute, the terminal leaflet of upper leaves sometimes continous with the rachis, pubescent on both sides or glabrous above; peduncles 4-16 cm long; racemes (3) 8- to 33-flowered, the flowers ascending to declined at anthesis, the axis 1.5-28 cm long in fruit; bracts 1.2-2.2 mm long; pedicels 1.2-3 mm long; bracteoles 0-2; calyx 3.8-7 mm long, the tube 3.3-5.5 mm long, campanulate, strigose, the teeth 1.5-2.2 mm long, subulate; flowers 10-15 mm long, pink purple; pods pendulous, stipitate, the stipe 3.5-6 mm long, the body narrowly oblong, 20-32 mm long, 3.5-4.5 mm thick, dorsiventrally compressed, strigose, unilocular; ovules 20-24. Pinyon-juniper and mixed desert shrub communities at 1370 to 1900 m in Garfield, San Juan (type from White Canyon) and Wayne counties; endemic; 20 (xii). The birds-nest milkvetch is a close ally of *A. harrisonii* (q.v.), from which it is separated geographically and in technical morphological features.

Astragalus nuttallianus A. DC. Small-flowered Milkvetch. Annual or winter-annual, caulescent, 2-35 cm long, from a taproot; pubescence basifixed; stems prostrate to decumbent or erect; stipules 1-5 mm long, all distinct; leaves 1-8.5 cm long; leaflets 7-19, 2-14 mm long, 1-6 mm wide, obovate to elliptic or oblong, acute to rounded or retuse to emarginate, strigose to glabrous on both sides; peduncles 1-8 cm long; racemes 1- to 7-flowered, the flowers ascending to declined at anthesis, the axis 0.2-2 cm long in fruit; bracts 0.5-2.5 mm long, pedicels 0.4-1.6 mm long; bracteoles 0; calyx 3.2-5.4 mm long, the tube 1.9-2.8 mm long, campanulate, strigose, the teeth 1-2.2 mm long, subulate; flowers 4.1-7.6 mm long, pink purple to whitish or faintly tinged with purple; pods ascending to declined, sessile or subsessile, narrowly oblong, curved, 12-20 mm long, 1.9-3.3 mm wide, glabrous to strigose, bilocular or nearly so; ovules 12–18; n = 11, 12. There are two allopatric varieties in Utah.

1. Leaflets 9–15; axis of raceme 0.5–2 cm long in fruit; keel-tip blunt; plants of eastern Kane County and eastward .... A. *nuttallianus* var. *micranthiformis*
— Leaflets 7–11; axis of raceme 0.1–1 cm long in fruit; keel-tip acute to subacute, usually beaked; plants from Washington and Tooele counties ....
.... A. *nuttallianus* var. *imperfectus*

Var. *imperfectus* (Rydb.) Barneby [*Hamosa imperfecta* Rydb.]. Creosote bush, warm desert shrub, and cool desert shrub communities at 670 to 1470 m in Washington and disjunctly in Tooele (Stansbury Island) counties; Nevada, California, and Arizona; 38 (viii).

Var. *micranthiformis* Barneby Blackbrush, mixed desert shrub, and pinyon-juniper communities at 1130 to 1670 m in Emery, Garfield, Grand, Kane, and San Juan counties; Arizona, Colorado, and New Mexico; 26 (xvi).

Astragalus oophorus Wats. Egg Milkvetch. Perennial, caulescent, 15–30 cm tall, from a caudex; pubescence basifixed; stems decumbent to ascending, radiating from the caudex; stipules 1.5–7 mm long, all distinct; leaves 3–21 cm long; leaflets 9–25, 3–20 mm long, 2–11 mm wide, oval to obovate or orbicular, obtuse to retuse or mucronate, glabrous on both surfaces, often ciliate; peduncles 4–13 cm long; racemes 4- to 13-flowered, the flowers spreading at anthesis, the axis 1–8 cm long in fruit; bracts 1.5–5 mm long; pedicels 2–6 mm long; bracteoles 0–1; calyx 6–12 mm long, the tube 4–8.5 mm long, cylindric or short-cylindric, glabrous or sparingly strigose, the teeth 2–5 mm long, subulate; flowers 17–24 mm long; ochroleucous, concolorous, or pink purple and with white wing-tips; pods spreading to pendulous, stipitate, the stipe (gynophore) 3.5–12 mm long, the body bladdery-inflated, ellipsoid, 25–55 mm long, 10–30 mm wide (when pressed), glabrous, unilocular, often mottled; ovules 28–54; 2n = 24. Two distinctive varieties are present in Utah.

1. Flowers pink purple; calyx tube 7.8–9 mm long; stipe 10–12 mm long; plants of western Iron and Beaver counties .... A. *oophorus* var. *lonchocalyx*
— Flowers ochroleucous; calyx tube 4–7 mm long; stipe 3.5–8.5 mm long; plants widespread ....
.... A. *oophorus* var. *caulescens*

Var. *caulescens* (Jones) Jones Pallid Egg Milkvetch. [A. *megacarpus* var. *caulescens* Jones, type from Loa Pass; A. *artipes* Gray; *Phaca artipes* (Gray) Rydb.; A. *oophorus* var. *artipes* (Gray) Jones]. Sagebrush, pinyon-juniper, and mountain brush communities at 1370 to 2430 m in Beaver, Garfield, Grand, Iron, Juab, Kane, Millard, Piute, Sanpete, Sevier, Tooele, Utah, and Washington counties; Colorado, Arizona, and Nevada; 99 (xix).

Var. *lonchocalyx* Barneby Pink Egg Milkvetch. Pinyon-juniper, sagebrush, and mixed desert shrub communities at 1770 to 2300 m in western Iron and Beaver counties; Lincoln County, Nevada; a Great Basin endemic; 7 (iv).

Astragalus pardalinus (Rydb.) Barneby Panther Milkvetch. [*Phaca pardalina* Rydb., type from Cedar Mt., Emery County]. Perennial (short-lived) or functionally annual, caulescent, 8–30 (35) cm tall, often clump-forming and about as broad as tall; pubescence basifixed; stems decument to ascending, forming rounded clumps; stipules 2–5 mm long, all distinct or some shortly connate-sheathing; leaves 3–7 cm long; leaflets 11–17, 3–20 mm long, 1–5 mm wide, oblong to oblanceolate or obovate, truncate to retuse, mucronate, or acute, strigulose on both sides; peduncles 1–4 cm long; racemes 3- to 8-flowered, the flowers ascending at anthesis, the axis 1–4 cm long in fruit; bracts 1–3.5 mm long; pedicels 0.8–3.6 mm long; bracteoles 0–2; calyx 4.8–6.6 mm long, the tube 2.3–2.8 mm long, campanulate, villous, the teeth 2.3–2.8 mm long, subulate; flowers 6.3–8.2 mm long, pink purple (soon fading yellowish); pods declined, sessile or nearly so, obliquely ovoid-ellipsoid, inflated, 13–21 mm long, 8–11 mm thick, straight or only slightly curved, villosulous with spreading-curved hairs, unilocular; ovules 20–28. Mixed desert shrub and pinyon-juniper communities, usually in sandy soils, at 1270 to 1600 m in Emery (type from Cedar Mt.), Garfield, and Wayne counties; endemic; 17 (vii). The panther milkvetch shares several salient features with the closely related but largely allopatric taxa A. *pubentissimus* and A. *sabulonum* (q.v). From both of these, A. *pardalinus* differs in its straight or nearly straight pods, with 20–28 ovules (not 9–19). The main area of distribution of the panther milkvetch is on the sandy eastern foot of the San Rafael Swell, which breaks abruptly to the canyons of the Green and Colorado rivers.

Astragalus pattersonii Gray ex Brand. Patterson Milkvetch. [*Tragacantha pattersonii* (Gray) Kuntze; *Phacopsis pattersonii* (Gray) Rydb.; *Rydbergiella pattersonii* (Gray) Fedde; *Jonesiella pattersonii* (Gray) Rydb.]. Perennial, caulescent, 20–45 (50) cm tall, from a branching caudex; stems decumbent to ascending or erect; stipules 3–8 mm long, all usually distinct; leaves 5–13 cm long; leaflets 7–15 or more, 6–38 mm long, 3–16 mm wide, elliptic to lanceolate, oblanceolate, or obovate, obtuse to acute, retuse or mucronate, strigose to glabrous on both sides; peduncles 3–18 cm long; racemes 6- to 24-flowered, the flowers declined-nodding at anthesis, the axis 2–15 cm long in fruit; bracts 2–8 mm long; pedicels 1–4.5 mm long; bracteoles 2; calyx 8.8–14.2 mm long, the tube 6–8.8 mm long, cylindric, gibbous, pale tan or whitish, colored like the petals, thinly strigulose, the teeth 2.3–6.5 mm long, subulate; flowers 14–22 mm long, white, concolorous or the keel tip faintly purplish; pods erect, sessile, cylindric to ellipsoid or ovoid, 17–35 mm long, 6–10 mm thick, glabrous or puberulent, unilocular; ovules 22–38; 2n = 24. Pinyon-juniper and mixed desert shrub communities at 1470 to 2300 m in Carbon, Emery, Garfield, Sevier, Uintah, and Wayne counties; Colorado; 30 (xii). This handsome milkvetch occurs in a crescent below the coal measures from east of Price to south of Emery, on foothills and along washes in the Blue Gate member of the Mancos Shale. The Arizona materials are known from the valley of Kanab Creek, and this plant might also occur on seleniferous muds and silts in adjacent Kane County. The plant is similar in many ways to A. *praelongus* (q.v.), from which it differs in the whitish concolorous calyces and petals, spreading calyx teeth, and erect pods.

Astragalus perianus Barneby Rydberg Milkvetch. Perennial, short-caulescent, 1–6 cm tall, from a shallowly subterranean, branching caudex; pubescence basifixed; stems prostrate, 3–12 cm long; stipules 1–2.5 mm long, at least some connate-sheathing; leaves 1–3 cm long; leaflets 7–19, 1–5 mm long, 1–3 mm wide, oval to obovate, retuse, strigulose on both sides or glabrous above; peduncles 0.3–2.2 cm long, racemes 2- to 6-flowered, the flowers spreading at anthesis, the axis 0.2–0.8 cm long in fruit; bracts 0.8–1.2 mm long; pedicels 1.4–2.5 mm long; bracteoles 0; calyx 3.5–4.2 mm long, the tube 2.3–3 mm long, campanulate, pilosulous, purplish, the teeth 1–1.4 mm long, subulate; flowers 6.8–8.5 mm long, whitish, faintly suffused with pink or purple; pods ascending to declined, sessile, bladdery-inflated, ovoid, 10–23 mm long, 8–14 mm wide (when pressed), strigose, purple-mottled, unilocular; ovules 18–20. Tertiary igneous gravels, often on barrens in alpine or montane sites in Beaver, Iron, Garfield, Kane (?), and Piute (type from west of Marysvale) counties; endemic; 30 (vi). The Rydberg milkvetch most closely resembles *A. serpens* (q.v.).

Astragalus pinonis Jones Pinyon Milkvetch. [*Pisophaca pinonis* (Jones) Rydb.]. Perennial, caulescent, 10–55 cm tall, from a shallowly subterranean caudex; pubescence basifixed; stems erect or reclining, commonly growing through sagebrush canopy; stipules 1.5–5 mm long, all distinct; leaves 2–11 cm long; leaflets 9–19, 2–18 mm long, 1–4 mm wide, linear to oblong, obtuse to retuse, strigose on both surfaces; peduncles 1.5–8 cm long; racemes 5- to 19-flowered, the flowers ascending-spreading at anthesis, the axis 2–7 cm long in fruit; bracts 1–2 mm long; pedicels 1–3 mm long; bracteoles 0–1; calyx 4.3–5.6 mm long, the tube 2.3–3.8 mm long, campanulate, strigose, the teeth 1–2 mm long, subulate; flowers 8.2–10.3 mm long, greenish to ochroleucous, suffused with purple; pods spreading-declined, subsessile, oblong-ellipsoid, straight to slightly curved, 20–35 mm long, 5.5–8.5 mm thick, terete or nearly so, strigose, unilocular; ovules 32–42. With black sagebrush in juniper and pinyon-juniper woodlands at 1580 to 2275 m in Beaver and Juab counties; Nevada; a Great Basin endemic. The pinyon milkvetch is obscure and apparently rare in Utah; 5 (iii). The type was collected at Frisco, Beaver County, where M. E. Jones found the plant in 1880 (on June 22). It grows up through low plants of black sagebrush and is difficult to see. The plants resemble those of *A. convallarius*, but the thicker pods, jointed terminal leaflets, and superficial caudex of *A. pinonis* are diagnostic.

Astragalus piscator Barneby & Welsh Fisher Milkvetch. Acaulescent or subacaulescent herbs from a vertical taproot, perennial of short duration flowering the first year, the leaves and scapiform peduncles arising from the root-crown at soil level, this more or less clothed with a persistent thatch of leaf bases, strigose throughout with appressed dolabriform hairs, the leaflets yellow green above, gray beneath; stipules ovate-acuminate 3–9 mm, usually closely imbricated, strigose dorsally, persistent; leaves (3) 4–10 (16) cm long; leaflets of most leaves 5–11 (13), elliptic or lance-elliptic, acute or subobtuse, (5) 7–17 (32) long, 2–4 (6) mm wide, those of some early leaves only 1–3 and rhombic ovate; peduncles (1) 2–6 (9) cm, ascending at anthesis, procumbent in fruit, the pods humistrate; racemes shortly loosely 3- to 10-flowered, the axis becoming 4–15 (20) mm in fruit; calyx 11–14.5 mm, either black- or partly white-strigose, the cylindric tube 8.5–11 mm long, 3–4 mm wide, the linear-subulate teeth 2–3.5 (4) mm long; corolla of *A. amphioxys*, the banner 18–24 mm long, the obtuse keel 16–18 mm long; ovary strigulose, the ovules ca 40; pod ascending, sessile, deciduous from the receptacle, in profile lance-elliptic, shallowly lunate-incurved, obtuse at the base, acuminate distally, 24–40 long, 8–15 mm wide, somewhat laterally compressed but the valves dilated near the middle into an obtuse longitudinal ridge, both sutures becoming sharply prominulous at maturity, the moderately fleshy, densely strigose, purplish-mottled valves becoming stiffly chartaceous or subcoriaceous (but not pithy) and ca 0.5 mm thick when dry, dehiscent after falling through the gaping beak. Sandy soils of valley benches and in gullied foothills, on Moenkopi, Cutler, and White Rim formations, 1550 to 1750 m, known only from the lower Grand River Valley in southeastern Grand, northern San Juan (type from Salt Creek) and extreme eastern Wayne counties; Colorado; endemic; 11 (vi).

Astragalus piutensis Barneby & Mabberley Sevier Milkvetch. [*Xylophacos marianus* Rydb., type from Marysvale; *A. marianus* (Rydb.) Barneby]. Perennial, acaulescent or subacaulescent, 3–10 cm tall, from a branching caudex; pubescence basifixed (or incipiently malpighian; stems 0–6 cm long, the internodes mostly concealed by stipules, these 2–7 mm long, all distinct; leaves 1.2–8.5 mm long; leaflets 7–17, 3–11 mm long, 1–4 mm broad, obovate to oblanceolate, obtuse to emarginate or acuminate to acute, strigose on both sides; peduncles 1–8 cm long; racemes 2- to 10-flowered, the flowers ascending at anthesis, the axis 0.2–3.5 cm long in fruit; bracts 2–4.5 mm long; pedicels 0.8–3 mm long; bracteoles 0–2; calyx 10–13.2 mm long, the tube 7.3–9.4 mm long, cylindric, pilosulous, the teeth 2–3.5 mm long, subulate; flowers 17–24 mm long, pink purple, often pale; pods spreading-ascending, sessile or nearly so, 10–23 mm long, 7–12 mm thick, ovoid-acuminate, densely shaggy-hirsute, unilocular; ovules 27–36. Oak-sagebrush, mixed warm and cool desert shrub, pinyon-juniper, and aspen-white fir communities at 900 to 2430 m in Beaver, Garfield, Iron, Juab, Millard, Piute, Sevier, Tooele, Washington, and Wayne counties; Nevada; 71 (xxv). The Sevier milkvetch resembles *A. argophyllus* var. *martinii*, but the shaggy-hirsute pods and flowers that average smaller are diagnostic.

Astragalus platytropis Gray Broad-keeled Milkvetch. [*Tragacantha platytropis* (Gray) Kuntze; *Phaca platytropis* (Gray) Rydb.; *Cystium platytrope* (Gray) Rydb.]. Perennial, acaulescent or nearly so, 2–7 cm tall, from a branching caudex, the branches often with a thatch of persistent leaf bases; pubescence basifixed; stems 0–2 cm long, prostrate-ascending or erect; stipules 1.5–5 mm long, at least some connate-sheathing; leaves 1–7 cm long; leaflets 5–15, 2–11 mm long, 1.5–7 mm wide, elliptic to obovate, oblong, or oval, acute to obtuse or retuse, silvery strigose on both sides; peduncles 1.5–6.5 cm long; racemes 2- to 9-flowered, the flowers ascending at anthesis, the axis 2–6 mm long in fruit; bracts 0.6–2 mm long; pedicels 0.7–1.9 mm long; bracteoles 0–2; calyx 3–5.4 mm long, the tube 2–3.4 mm long, campanulate, strigose, the teeth 0.2–2.1 mm long, subulate; flowers 7–9.5 mm long, pink purple; pods ascending, sessile, bladdery-inflated, ovoid to subglobose, 15–33 mm long, 10–22 mm wide (when pressed), purple-mottled, strigulose, semibilocular; ovules 26–34. Ridge tops in shrub communities at 2400 to 3500 m in western Beaver, Kane, Millard, and Tooele counties; Nevada and California; 4 (i).

Astragalus praelongus Sheldon Stinking Milkvetch. Perennial, caulescent, 1–9 dm tall, from a branching caudex; pubescence basifixed; stems erect or ascending, forming clumps; stipules 2.5–9 mm long, all distinct; leaves 3–22 cm long; leaflets 7–33, 3–50 mm long, 2–24 mm wide, obovate, elliptic, oblong, lanceolate, or oblanceolate, obtuse or retuse to acute, sparingly strigose beneath, glabrous above; peduncles 4–26 cm long; racemes 10- to 33-flowered, the flowers deflexed at anthesis, the axis 3–16 cm long in fruit; bracts 1–7 mm long; pedicels 1–7 mm long; bracteoles 2; calyx 5.8–14 mm long, the tube 4.4–7.5 mm long, cylindric, gibbous, glabrous, or thinly strigose, green or yellowish, usually differently colored than the petals, the teeth 0.3–6 mm long, subulate; flowers 15–24 mm long, ochroleucous, the keel often faintly purplish tipped; pods erect to declined, sessile, subsessile, or stipitate, inflated, ellipsoid, ovoid, obovoid, or subglobose, 18–42 mm long, 5–25 mm thick, usually straight, glabrous or puberulent, subunilocular, leathery-woody; ovules 40–75; n = 12. The stinking milkvetch is represented in Utah by three more or less distinctive varieties.

1. Pods long-stipitate, the stipe 4–8 mm long; plants of San Juan County .... A. praelongus var. lonchopus
— Pods sessile or short-stipitate, the stipe, when present, less than 3 mm long; plants not of San Juan County 2
2(1). Pods narrowly elliptic to oblong, 6–10 mm thick; plants of Beaver, Carbon, Iron, Millard, and Sevier counties eastward to east-central Utah .. A. praelongus var. ellisiae
— Pods broadly oblong to elliptic, 10–15 mm thick; plants of Garfield, Kane, Washington, and Wayne counties, and disjunctly in Sevier County ....
.... A. praelongus var. praelongus

Var. ellisiae (Rydb.) Barneby in Turner [*Jonesiella ellisiae* Rydb.]. Clay soil, commonly on Mancos Shale, Moenkopi, and Chinle formations, but also on alluvial substrates containing selenium, in warm and salt desert shrub and pinyon-juniper communities, at 1330 to 1970 m in Beaver, Carbon, Emery, Grand, Iron, Millard, Sevier, and Wayne counties; Colorado, New Mexico, and Texas; 81 (xxii). The Ellis stinking milkvetch passes by degree into var. *praelongus*. It seems clear however, that most of the Great Basin populations are more closely allied with those of the San Rafael Swell region than they are with those of Washington County. The relationship appears to have resulted from migration of plants of the *ellisiae* type through passes in the Wasatch Plateau rather than by a north-south exchange of typical phases.

Var. lonchopus Barneby Seleniferous fine-textured substrates in blackbrush, mixed desert shrub, and pinyon-juniper communities at 1300 to 1925 m in southern San Juan County; Arizona; a Colorado Plateau endemic; 13 (iv).

Var. praelongus [*A. procerus* Gray, not Boiss. & Hausskn.; *A. pattersonii* var. *praelongus* (Sheldon) Jones; *Phacopsis praelongus* (Sheldon) Rydb.; *Rydbergiella praelonga* (Sheldon) Fedde & Sydow; *Jonesiella praelonga* (Sheldon) Rydb.]. Clay and silt of the Mancos Shale, Tropic Shale, Moenkopi, and Chinle formations, and other seleniferous soils, in salt desert shrub and pinyon-juniper communities at 770 to 2530 m in Garfield, Kane, Sevier, Washington, and Wayne counties; Nevada, Arizona, and New Mexico; 105 (xxxii). The type variety of stinking milkvetch demonstrates much variation. The most extreme phase occurs in Washington and portions of Kane counties. It has fistulous tall stems and greatly thickened pods. Still another from Washington County has pinkish lavender flowers. Most of these plants are primary selenophytes, however, a small population from near the mouth of Zion Canyon lacked the characteristic odor of selenium and presumably was not seleniferous.

Astragalus preussii Gray Preuss Milkvetch. Perennial or annual, caulescent, mostly 12–45 cm long, from a woody caudex; pubescence basifixed; stems erect or ascending, forming clumps; stipules 2–7 mm long, all distinct; leaves 3.5–13 cm long; leaflets 7–25, 6–28 mm long, 1–6 mm wide, obovate or obcordate to oblong, narrowly elliptic, lanceolate, or linear, emarginate to rounded, obtuse, or acute, glabrous; peduncles 2–15 cm long; racemes 3– to 22–flowered, the flowers ascending, the axis 1–20 cm long in fruit; bracts 1.5–4 mm long; pedicels 1–5.5 mm long; bracteoles 2; calyx 6.4–12.3 mm long the tube 5.1–9.7 mm long, cylindric, thinly strigose, purple, the teeth 1.3–2.6 mm long, subulate; flowers 14–24 mm long, pink purple, bicolored, or white; pods erect to ascending, stipitate or subsessile, the stipe 2–7 mm long, the body oblong-ellipsoid, inflated, 12–34 mm long, 6–13 mm thick, glabrous or puberulent, stiffly papery to leathery, unilocular; ovules 20–44; 2n = 24. Three rather distinctive varieties are present in Utah.

1. Plants annual; leaflets 5–11, 7–12 mm wide; flowers white or merely tinged purplish .. A. preussii var. cutleri
— Plants short-lived perennial; leaflets mostly 7–25, 1–6 mm wide; flowers purple 2
2(1). Pods sessile or nearly so; racemes 4–20 cm long in fruit; plants of south-western Washington County ....
.... A. preussii var. laxiflorus
— Pods stipitate; racemes 1–9 cm long in fruit; plants of eastern Kane County and northeastward ....
.... A. preussii var. preussii

Var. cutleri Barneby Warm desert shrub community at ca 1160 m, at Copper Canyon, San Juan (type from Monument Valley) County; endemic; 2 (0).

Var. laxiflorus Gray [*A. preussii* var. *laxispicatus* Sheldon; *Phaca laxiflora* (Gray) Rydb.]. Creosote bush community at 2200 to 2500 m in the Beaver Dam Mountains, Washington County; Nevada. This variety is not known from Utah in modern collections, but contemporary specimens have been seen from Nevada; 1 (0).

Var. preussii [*Phaca preussii* (Gray) Rydb.; *Tragacantha preussii* (Gray) Kuntze; *Rydbergiella preussii* (Gray) Rydb.; *A. preussii* var. *latus* Jones, type from Green River; *A. preussii* var. *sulcatus* Jones (in part), type from Westwater; *A. preussii* var. *arctus* Sheldon; *Rydbergiella arcta* (Sheldon) Rydb.; *A. arctus* (Sheldon) Tidestr.; *Jonesiella arcta* (Sheldon) Rydb.]. Blackbrush, mixed desert shrub communities on seleniferous clays and silts at 1170 to 1570 m in Emery, Garfield, Grand, Kane, San Juan, and Wayne counties; Arizona, Nevada, and California; 87 (xxvi). Our materials belong to var. *latus* Jones, sens. str., which differ from the more western var. *preussii* sens., str., in having a larger number of leaflets (17–25, not 11–15). Both (in sens. lat.) can be treated conservatively as belonging to a polymorphic var. *preussii*. The species is named after Charles Preuss, topographer with John Charles Fremont on his second expedition in 1843–44.

Astragalus pubentissimus T. & G. Green River Milkvetch. [*A. multicaulis* Nutt. ex T. & G., not Ledeb.; *Tragacantha pubentissima* (T. & G.) Kuntze; *Phaca pubentissima* (T. & G.) Rydb.]. Perennial, or functionally annual, caulescent, 9–25 cm tall, from a rather weak caudex; pubescence basifixed; stems decumbent to ascending, radiating from the caudex; stipules 1–4.5 mm long, all distinct; leaves 2–8 cm long; leaflets 5–15, 2–14 mm long, 1.5–5 mm wide, oblong to ovate, obovate, or elliptic, obtuse to retuse, mucronate, or acute, villosulous on both side; peduncles 1–3.5 cm long; racemes 3- to 12-flowered, the flowers spreading at anthesis, the axis 0.4–3.5 cm long in fruit; bracts 1–3 mm long; pedicels 0.5–2 mm long; bracteoles 0; calyx 4.8–6.3 mm long, the tube 2.8–4.2 mm long, campanulate, villosulous, the teeth 1.8–2.8 mm long, subulate; flowers 8.8–12.2 mm long, pink purple or ochroleucous and suffused with purple; pods spreading-declined, sessile, inflated, obliquely lance-ellipsoid, 12–20 mm long, 4–8 mm thick, shaggy-pilose, unilocular; ovules 9–18.

1. Plants ascending to erect; flowers monochrome in populations .... *A. pubentissimus* var. *pubentissimus*
— Plants spreading-decumbent; flowers polychrome in populations .... *A. pubentissimus* var. *peabodianus*

Var. *peabodianus* (Jones) Welsh [*A. peabodianus* Jones, type from Thompson Springs, Grand County]. The Peabody phase of the Green River milkvetch occurs in entrenched channels cut into the escarpments draining the south and west flanks of the Tavaputs Plateaus, in pinyon-juniper and mixed desert shrub communities at 1300 to 1770 m in Emery and Grand (type from Thompson's Springs) counties; endemic; 10 (ix).

Var. *pubentissimus* The Green River milkvetch occurs in pinyon-juniper and mixed desert shrub communities at 1525 to 1925 m in Duchesne and Uintah counties; Wyoming and Colorado; 37 (iv). Reports of livestock poisoning, mainly of sheep, are attributable to this species. In years when it is abundant, the plants constitute the principle forb in much of the lower middle reaches of the Uinta Basin, reminiscent of seeded alfalfa fields among the open juniper-pinyon woodlands.

Astragalus purshii Dougl. ex Hook. Pursh Milkvetch. Perennial, acaulescent or subacaulescent, 4.5–13 cm tall, from thatch-clothed caudex branches; pubescence basifixed; stems 0–2 cm long; leaflets 5–17, 2–14 mm long, 1–7 mm wide, elliptic to oblanceolate, acute, densely villous on both sides; peduncles 1.5–10.5 cm long; racemes 1- to 7-flowered, the flowers ascending at anthesis, the axis 0.3–2 cm long in fruit; bracts 2–7 mm long; pedicels 1–5 mm long; bracteoles 0; calyx 12–16 mm long, the tube 8.5–12.5 mm long, cylindric, villous-pilose, the teeth 2.2–6 mm long, subulate; flowers 19–26 mm long, whitish to ochroleucous or pink purple; pods ascending, sessile, obliquely ovoid, usually curved, 13–26 mm long, 5–11 mm wide, shaggy villous, unilocular; ovules 20–34; 2n = 22. Two varieties are present in Utah.

1. Flowers whitish or ochroleucous, discolorous, the keel purple-tipped; plants of broad distribution in northernmost counties .... *A. purshii* var. *purshii*
— Flowers pink purple, concolorous; plants rare in Box Elder and Rich counties .... *A. purshii* var. *glareosus*

Var. *glareosus* (Dougl.) Barneby [*A. glareosus* Dougl. ex Hook.; *Tragacantha glareosa* (Dougl.) Kuntze; *A. inflexus* var. *glareosus* (Dougl.) Jones; *Phaca glareosa* (Dougl.) Piper; *Xylophacos glareosus* (Dougl.) Rydb.]. Sagebrush community at 1530 to 1830 m in Box Elder and Rich counties; British Columbia, south to Nevada; 2 (0). This variety simulates *A. utahensis* (q.v.) both in having pink purple flowers and in habit. Leaflet shape and flower structure are important diagnostic features.

Var. *purshii* [*Tragacantha purshii* (Dougl.) Kuntze; *Phaca purshii* (Dougl.) Piper; *Xylophacos purshii* (Dougl.) Rydb.; *A. purshii* var. *interior* Jones]. Sagebrush, desert shrub, and pinyon-juniper communities at 1530 to 2270 m in Box Elder, Daggett, Duchesne, Rich, Summit, and Uintah counties; Washington to Alberta and Saskatchewan, south to California, Nevada, Colorado, and South Dakota; 59 (vii).

Astragalus racemosus Pursh Alkali Milkvetch. Perennial, caulescent, 20–53 cm tall, from a branching caudex; pubescence basifixed; stems erect or ascending, forming clumps; stipules 3–8 mm long, at least some connate-sheathing; leaves 4–15 cm long; leaflets 9–19, 3–35 mm long, 1.5–13 mm wide, lance-elliptic to linear-lanceolate, acute to acuminate, glabrous or glabrate on both surfaces, sometimes ciliate; peduncles 3–13 mm long; racemes 9- to 45-flowered, the flowers nodding at anthesis, the axis 5–20 cm long in fruit; bracts 1.5–7.5 mm long; pedicels 2–8 mm long; bracteoles 0–2; calyx 8.6–12 mm long, the tube 5–6 mm long, short-cylindric, gibbous, glabrous or sparingly strigose, the teeth 3.3–6 mm long, subulate; flowers 14–17 mm long, ochroleucous, concolorous, or with the keel purple-tipped; pods pendulous, stipitate, the stipe 4–6 mm long, the oblong-ellipsoid body 10–21 mm long, 4–8 mm wide, triquetrous, glabrous, unilocular; ovules 12–22; 2n = 24. Uinta and Duchesne River formations, rarely otherwise on saline clay and silty soils, in salt desert shrub and pinyon-juniper communities at 1570 to 1970 m in in Duchesne and Uintah counties; Wyoming; 48 (vi). Our plants belong to var. **treleasei** C. L. Porter. They are primary selenium indicators. Consumption by cattle of this and other selenophytes results in "alkali disease," or "blind staggers." Plants from along Hill Creek in Uintah County have concolorous flowers; those from elsewhere are discolorous.

Astragalus rafaelensis Jones San Rafael Milkvetch. [*Cnemidophacos rafaelensis* (Jones) Rydb.]. Perennial, caulescent, rushlike, 32–65 cm tall, from a branching caudex; pubescence basifixed; stems erect or ascending, arranged in clumps; stipules 1–5 mm long, at least some connate-sheathing; leaves 2.5–14.8 cm long, all simple or some with leaflets 3–5, the lateral ones 3–20 mm long, 0.5–1.5 mm wide, linear to oblong, acute, glabrate beneath, glabrous above, the terminal leaflet longer and confluent with the rachis; peduncles 11–27 cm long; racemes loosely 5–flowered, the flowers ascending to declined in anthesis, the axis 2–5 cm long in fruit; bracts 1.2–3.5 mm long; pedicels 2–5.5 mm long; bracteoles 2; calyx 6–9.6 mm long, the tube 5.2–7.5 mm long, short-cylindric, thinly strigose to glabrate, the teeth (0.8) 1.1–2.1 mm long, triangular; flowers 19–26 mm long, pale pinkpurple or bicolored; pods deflexed, sessile, oblong-elliptic in outline, 12–25 mm long, 5–7 mm wide, laterally compressed, glabrous, leathery-woody, unilocular; ovules 18–20; 2n = 22. Seliniferous clays and silts of the Buckhorn Conglomerate, Morrison, Summerville, Chinle, and Moenkopi formations at 1370 to 1625 m in salt desert shrub community in Emery and Grand Counties; near Gateway, Colorado; a Navajo Basin endemic; 19 (vii). There are two main populations in Emery County; one at the east base of Cedar Mountain (type locality) and the other in the vicinity of Window-blind Butte in the San Rafael Swell proper. This is a handsome selenophyte, with close affinities to *A. toanus* and *A. saurinus* in Utah and *A. linifolius* in Colorado, especially the latter. It has lesser affinities to *A. woodruffii* (q.v.).

Astragalus robbinsii (Oakes) Gray Robbins Milkvetch. [*Phaca robbinsii* Oakes]. Perennial, caulescent, 10–40 (60) cm long, from a branching caudex; pubescence basifixed; stems erect to ascending; stipules 1.5-6 mm long, at least some of the lower ones usually connate-sheathing; leaves 2–12 cm long; leaflets (7) 9–15, 7–32 mm long, 4–8 mm broad, lanceolate to oblong, obtuse or emarginate, strigulose beneath, glabrous or glabrate above; peduncles 3.5–19 cm long; racemes 5- to 25–flowered, the flowers declined at anthesis, the axis 3–18 cm long in fruit; bracts 1–3.5 mm long; pedicels 0.5–3 mm long; bracteoles usually 0; calyx 4–6.8 mm long, the tube 3.2–4.5 mm long, campanulate, strigulose, the teeth 0.7–2.3 mm long, subulate; flowers 7.2–11.5 mm long, purple or whitish; pods pendulous, stipitate, the stipe 1.5–6.5 mm long, the body ellipsoid, 13–25 mm long, 3.5–5.5 mm thick, obtusely trigonous, flattened dorsally, the valves loosely strigulose with black and white hairs, the septum 0.2–1.5 mm wide; ovules 6–10. With willows at 2730 m on the north slope of Uinta Mountains, Summit County;.Alaska east to Newfoundland, south to Oregon, Nevada, Colorado, and Vermont; 2 (0). A specimen taken in 1914 was previously identified tentatively as *A. eucosmus* B. Robins. A second collection, by Sherel Goodrich in 1984 (BRY) at the site of the earlier discovery removes doubt as to the determination.

Astragalus sabulonum Gray Gravel Milkvetch. [*Phaca sabulonum* (Gray) Rydb.; *A. virgineus* Sheldon]. Annual, winter annual, or biennial, caulescent, 4–30 cm tall, radiating from a taproot; pubescence basifixed; stems decumbent to ascending, rarely erect; stipules 1–4 mm long, all distinct; leaves 1.5–7 cm long; leaflets 9–15, 2–13 mm long, 1–5 mm wide, oblanceolate to oblong or obovate, retuse to truncate or obtuse, loosely villous on both surfaces or glabrate above; peduncles 0.5–4 cm long; racemes 2- to 7-flowered, the flowers ascending-spreading at anthesis, the axis 0.3–2 cm long in fruit; bracts 1–2.5 mm long; pedicels 0.8–2 mm long; bracteoles 0; calyx 3.8–5.5 mm long, the tube 1.8–2.5 mm long, campanulate, hirsutulous, the teeth 1.8–3 mm long, subulate; flowers 6.2–8 mm long, pink purple or less commonly ochroleucous and tinged with purple; pods spreading-declined, sessile, obliquely ovoid, inflated, 9–17 mm long, 5–8 mm thick, curved, white-hirsutulous, unilocular; ovules 10–19; $2n = 22, 24$. Mixed desert shrub, salt desert shrub, and lower pinyon-juniper communities at 1130 to 1700 m in Emery, Garfield, Grand, Kane, San Juan, and Wayne counties; California, Nevada, Arizona, and New Mexico; 36 (xv).

Astragalus sabulosus Jones Cisco Milkvetch. [*Jonesiella sabulosa* (Jones) Rydb.]. Perennial, caulescent, 13–38 cm tall, from a woody caudex; pubescence basifixed; stems decumbent to ascending or erect, forming clumps; stipules 4–9 mm long, all distinct; leaves 3–10.5 cm long; leaflets 5–11, 6–35 (50) mm long, 3–17 mm wide, rhombic-oval to obovate or elliptic, mucronate, strigose to glabrous on both sides; peduncles 3.5–7 cm long; racemes 4- to 10–flowered, the flowers ascending-spreading at anthesis, the axis 0.5–2 cm long in fruit; bracts 2–6 mm long; pedicels 2–5 mm long; bracteoles 0–2; calyx 15–17.5 mm long, the tube 11.5–14 mm long, cylindric, strigulose, the teeth 3–4 mm long, subulate; flowers 27–34 mm long, whitish to ochroleucous, fading yellowish; pods spreading to declined, subsessile, inflated, cylindroid, 20–48 mm long, 10–15 mm thick, stiffly papery to leathery, strigose, unilocular; ovules 55–59; $2n = 26$. Salt desert shrub at 1300 to 1600 m on Mancos Shale and Morrison formations in the Grand River Valley, Grand (type from Cisco) County; endemic; 16 (ix). The Cisco milkvetch is a primary selenium indicator, with close affinities to *A. iselyi* (q.v.), from which it can be distinguished by its larger yellowish flowers. The Cisco milkvetch is also allied with *A. praelongus*, which has much smaller flowers and pods. The flowers of *A. sabulosus* are the largest within *Astragalus* in Utah, and possibly elsewhere (though they are not the longest).

Astragalus saurinus Barneby Dinosaur Milkvetch. Perennial, caulescent, rushlike, 25–45 cm tall, from a shallowly subterranean caudex; pubescence basifixed; stems erect or ascending, arranged in clumps; stipules 1.2–7 mm long, usually at least some shortly connate-sheathing; leaves 2.5–9 cm long, the uppermost usually simple, the others with leaflets 3–9, mostly 10–28 mm long, 0.5–2 mm wide, linear to linear-elliptic, obtuse to acute, strigose on both surfaces, the terminal leaflet confluent with the rachis; peduncles 7–21 cm long; racemes 3- to 15–flowered, the flowers ascending-spreading at anthesis, the axis 0.5–6 cm long in fruit; bracts 1.2–2 mm long; pedicels 1.5–3 mm long; bracteoles 1–2; calyx 6.4–9.6 mm long, the tube 5.6–6.7 mm long, cylindric, the teeth 0.9–2.9 mm long, triangular-subulate; flowers 18–22 mm long, bicolored, pink purple, with white wingtips, rarely all white; pods deflexed, sessile or nearly so, narrowly oblong in outline, straight or curved, 15–35 mm long, 4.4–6 mm wide, laterally compressed, strigose to glabrate, unilocular; ovules 19–29; $2n = 22$. Duchesne River, Morrison, Carmel, Chinle, and Moenkopi formations in salt desert shrub and pinyon-juniper communities at 1430 to 1700 m in Uintah (type from 6 mi N of Jensen) County; endemic; 42 (xiii). This is a remarkably beautiful plant on the seleniferous substrates in Dinosaur National Monument.

Astragalus scopulorum T. C. Porter Rocky Mountain Milkvetch. [*Tragacantha scopulorum* (T. C. Porter) Kuntze; *Tium scopulorum* (T. C. Porter) Rydb.]. Perennial, caulescent, 15–48 cm tall, from a shallowly subterranean caudex; pubescence basifixed; stems decumbent to ascending, radiating from the caudex; stipules 3–9 mm long, at least some connate-sheathing; leaves 1.5–8.5 cm long; leaflets 15–29, 2–18 mm long, 1–8 mm wide, oblong to elliptic or oblanceolate, some narrowly so, acute to obtuse or mucronate, thinly strigose to glabrous beneath, glabrous above, thinly ciliate; peduncles 2–14 cm long; racemes 4- to 22–flowered, the flowers declined to nodding at anthesis, the axis 1–7 cm long in fruit; bracts 1.5–7 mm long; pedicels 1–4 mm long; bracteoles 0–2; calyx 9–11.5 mm long, the tube 6.8–8.5 mm long, subcylindric, strigulose, the teeth 1.5–3.5 mm long; flowers 18–24 mm long, ochroeucous, concolorous or the keel faintly purplish; pods pendulous, stipitate, the stipe 4–9 mm long, the body oblong, straight or curved, 18–35 mm long, 3–6.5 mm wide, triquetrous, glabrous, bilocular; $2n = 22$. Mountain brush, sagebrush, ponderosa pine, pinyon-juniper, and aspen-white fir communities at 1670 to 2430 m in Carbon, Grand, eastern Juab, Millard, San Juan, Sanpete, Sevier, Summit, Utah, and Wasatch counties; Colorado, Arizona, and New Mexico; 41 (xii). The previous report of *A. adanus* A. Nels. for Utah belongs here.

Astragalus serpens Jones Plateau Milkvetch. [*Phaca serpens* (Jones) Rydb.]. Perennial, caulescent, 3.5–23 cm long, radiating from a weak caudex; pubescence basifixed; stems ascending to erect; stipules 1.5–3.5 mm long, all distinct; leaves 1.5–4.5 cm long; leaflets 9–15, 2–9 mm long, 1–4 mm wide, obovate to oblanceolate or elliptic, obtuse to emarginate, strigose-pilosulous beneath and above, or glabrate above; peduncles 0.7–2.5 cm long; racemes shortly 2- to 9-flowered, the flowers spreading at anthesis, the axis 0.2–0.8 cm long in fruit; bracts 1–1.5 mm long; pedicels 1–1.8 mm long; bracteoles 0; calyx 4.2–5 mm long, the tube 2.7–3.5 mm long, campanulate, white strigulose (some black hairs sometimes present), the teeth 1.5–2 mm long, subulate; flowers 6.6–8.6 mm long, purplish to pink purple or whitish; pods ascending to declined, stipitate, the stipe (gynophore) 0.7–1.5 cm long, the body bladdery-inflated, ovoid or ellipsoid, 13–29 mm long, 7–17 mm wide (when pressed), red mottled, strigose, unilocular. Sagebrush, pinyon-juniper, aspen, and aspen-fir communities at 2070 to 2780 m in eastern Piute, southeastern Sevier, and western Wayne (type from Loa Pass) counties; endemic; 11 (v). The closely similar *A. perianus* (q.v.) is distinguished by its prostrate habit, sessile or substipitate pods, subterranean root crown, and connate stipules. In some years *A. serpens* is abundant in the Loa Pass vicinity, where it occupies interspaces in low sagebrush on igneous gravel.

Astragalus sesquiflorus Wats. Sandstone Milkvetch. [*Tragacantha sesquiflora* (Wats.) Kuntze; *Batidophaca sesquiflora* (Wats.) Rydb.; *A. sesquiflorus* var. *brevipes* Barneby]. Perennial, caulescent, prostrate, often matforming, 0.5–5 cm tall, radiating from a branching caudex; pubescence malpighian; stems 5–28 cm long or more; stipules 1.5–4.5 mm long, all connate-sheathing; leaves 1–4 cm long; leaflets 7–13, 1.5–10 mm long, 0.6–2 mm wide, elliptic to obovate, acute to obtuse, strigose on both sides (commonly involute); peduncles 0.8–4.5 cm long; racemes 1- to 4-flowered, the flowers ascending at anthesis, the axis very short in fruit; bracts 1.2–3 mm long; pedicels 0.7–4 mm long; bracteoles 0–2; calyx 3.7–5.5 mm long, the tube 1.5–2.8 mm long, campanulate, strigulose, the teeth 1.9–3 mm long, subulate; flowers 6–8 mm long, pinkpurple; pods spreading-ascending, sessile or subsessile, the body obliquely oblong in outline, 8–10 mm long, 3–4 mm wide, trigonously compressed, strigulose, unilocular; ovules 7–10. Sandy sites, often on sandstone in sagebrush, mixed desert shrub, pinyon-juniper, and ponderosa pine or aspen communities at 1470 to 2930 m in Garfield, Kane (type from near Kanab), San Juan, Wayne, and disjunctly (accidentally?) in Sanpete counties; Colorado and Arizona; a Mohave-San Juan endemic; 28 (vii). The sandstone milkvetch is more closely allied to *A. humistratus* than to other species of milkvetch in Utah.

Astragalus spatulatus Sheldon Draba Milkvetch. [*Homalobus caespitosus* Nutt.; *A. caespitosus* (Nutt.) Gray, not Pallas; *Tragacantha caespitosa* (Nutt.) Kuntze; *A. simplicifolius* var. *spatulatus* (Sheldon) Jones; *H. canescens* Nutt. ex T. & G.; *H. brachycarpus* Nutt. ex T. & G.; *A. simplex* Tidestr., not *A. brachycarpus* Bieb.; *A. spatulatus* var. *simplex* Tidestr.; *H. uniflorus* Rydb.; *A. spatulatus* var. *uniflorus* (Rydb.) Barneby]. Perennial, acaulescent, 1.5–9 (12) cm tall, from a branching caudex; pubescence malpighian; stems obscured by marcescent leaf bases and stipules, tufted, sometimes matforming; stipules 2–7 mm long, all connate-sheathing; leaves all or mostly simple (reduced to phyllodia), 0.8–10 cm long, only rarely with leaflets 3–5 on some leaves and with the terminal one confluent with the rachis, oblanceolate to linear, acute, mucronate, or spinulose, strigose on both surfaces; peduncles 0.4–9 cm long; racemes 1- to 11-flowered, the flowers ascending at anthesis, the axis 0.2–3 cm long in fruit; bracts 0.6–3 mm long; calyx 2.6–5 mm long, the tube 1.9–3.4 mm long, campanulate, strigose, the teeth 0.5–2.5 mm long, subulate; flowers 5.7–9.5 mm long, pink purplish to ochroleucous or whitish (in populations); pods erect, sessile, lanceolate to lance-oblong in outline, 4–13 mm long, 1.5–3 mm wide, stright or slightly curved, strigose to rarely glabrous, unilocular; ovules 4–12; $2n = 24$. Pinyon-juniper, sagebrush, and mountain brush communities, often on exposed ridges, at 1530 to 2630 m in Cache, Carbon, Daggett, Duchesne, Emery, Grand, Rich, Summit, Uintah, and Wasatch counties; Alberta and Saskatchewan, south to Idaho, Colorado, and Nebraska; 106 (xix). The draba milkvetch is most nearly allied to *A. detritalis* and *A. chloodes*, among our species.

Astragalus straturensis Jones Silver Reef Milkvetch. [*Atelophragma straturense* (Jones) Rydb.; *Hamosa atratiformis* Rydb., type from Washington County; *Tium atratiforme* (Rydb.) Rydb.]. Perennial, caulescent, 13–36 cm tall, from a superficial to shallowly subterranean caudex; pubescence basifixed; stems decumbent to ascending or erect, forming clumps; stipules 1–2.5 mm long, all distinct; leaves 3.5–9.5 cm long; leaflets 9–19, 3–13 mm long, 1–5 mm wide, oblong to linear or oval, obtuse to retuse, strigose beneath, glabrous above; peduncles 1.5–7.5 cm long; racemes 9- to 25-flowered, the flowers ascending to declined at anthesis, the axis 1.5–14 cm long in fruit; bracts 0.8–1.5 mm long; pedicels 0.8–1.5 mm long; bracteoles 0; calyx 3.5–4.2 mm long, the tube 2.5–3.5 mm long, campanulate, strigose, the teeth 0.7–1 mm long, triangular; flowers 6.5–8.5 mm long, pink purple with white wing-tips; pods pendulous, stipitate, the stipe 1.4–2 mm long, the body oblong in outline, curved or straight, 10–15 mm long, 2.2–3 mm wide, strigose, bilocular; ovules 10–13; $n = 11$. Sagebrush, pinyon-juniper, and mountain brush communities at 1530 to 2130 m in Beaver, Iron, southeastern Millard, and Washington (type from Silver Reef) counties; Arizona and Nevada; 15 (iv). The Silver Reef milkvetch has no close relatives in the milkvetch flora of Utah. In flower size and general habit, it resembles *A. wingatanus* of southeastern Utah, but the resemblance is only superficial.

Astragalus striatiflorus Jones Escarpment Milkvetch. Perennial, subacaulescent to short-caulescent, 1.5–6 cm tall, radiating from a branching, usually subterranean caudex; pubescence basifixed; stems 0–5 cm long, only the tips produced above the sand; stipules 2–4 mm long, all connate-sheathing; leaves 1–4 cm long; leaflets 5–13, 1–7 mm long, 0.8–2.5 mm wide, ovate to obovate or oblanceolate, obtuse, mucronate, or emarginate, pilosulous; peduncles 1–3 cm long; racemes 2- to 5-flowered, the flowers ascending at anthesis, the axis 0.2–1 cm long; pedicels 1–1.5 mm long; bracteoles 0; calyx 5.5–7 mm long, the tube 3–4 mm long, campanulate, hirsutulous, the teeth 1.8–3 mm long, subulate; flowers 9–12 mm long, pinkpurple or whitish and commonly suffused with purple, the keel-tip purple, long-attenuated, with the capitate stigma protruding; pods spreading, sessile, the body bladdery-inflated, ellipsoid, 12–18 mm long, 8–15 mm wide (when pressed), mottled, spreading-hairy, bilocular. Interdune valleys, sandy depressions on ledges, and bars and terraces in stream channels at 1530 to 1900 m in Kane (from the Paria River westward) and eastern Washington (type from Springdale) counties; Coconino County; Arizona; a Plateau endemic; 14 (vi). The escarpment milkvetch is a singular plant, resembling *A. perianus* more closely than any other in Utah. The attenuated keel-tip, protruding stigma, and inflated, bilocular pods are both unusual and diagnostic features.

Astragalus subcinereus Gray Silver Milkvetch. [*Phaca subcinerea* (Gray) Rydb.; *A. sileranus* Jones, type from Sink Valley (near Alton); *Phaca silerana* (Jones) Rydb.; *A. sileranus* var. *caraicus* Jones, type from Elk Head Ranch on the upper Virgin River]. Perennial, caulescent, 14–90 cm long, radiating from a subterranean, branching caudex; pubescence basifixed; stems prostrate to weakly ascending; stipules 1.5–6.5 mm long, at least some connate-sheathing; leaves 1.5–8.5 cm long; leaflets 9–23, 2–16 mm long, 1–8.5 (10) mm wide, oblong to oblanceolate or obovate, obtuse, emarginate, or retuse, villosulous on both surfaces or glabrate above; peduncles 1.5–10 cm long; racemes 5- to 37-flowered, the flowers ascending to declined at anthesis, the axis 1–7 cm long in fruit; bracts 1–3 mm long; pedicels 0.5–2.5 mm long; bracteoles 0–1; calyx 3.4–6.3 mm long, the tube 2.3–3.6 mm long, campanulate, villosulous, the teeth 0.9–2.9 mm long, subulate; flowers 6–11 mm long, ochroleucous and commonly suffused with purple; pods spreading to declined, subsessile, inflated, ovoid-ellipsoid to ellipsoid, 12–27 mm long, (3.5) 6–13 mm wide (when pressed), subterete to dorsiventrally compressed, thinly villosulous, mottled; ovules 10–20. Much of the material from Kane, Garfield, and Washington counties differs from the typical plants in Mohave County, Arizona in being more leafy (the leaflets 4–10 mm broad), in having longer stems (3–7 dm long), and in having more firmly walled pods 15–28 mm long, and 6–10 (13) mm thick. These Utah plants belong, sens. str., to var. *cariacus* Jones, but the features are weak and overlapping at best. Two varieties are treated:

1. Mature pods elliptic-oblong to oblong, 3.5'–5.5 mm wide; flowers 8.5–11 mm long; stems 40–90 cm long; plants of igneous gravels in eastern Sevier and western Emery counties . . . . *A. subcinereus* var. *basalticus*

— Mature pods ovoid-ellipsoid, 6–13 mm wide; flowers 6–9 mm long; stems mostly 14–70 cm long; plants commonly of sedimentary gravels, sometimes from igneous substrates, but not in Emery or Sevier counties . . . . *A. subcinereus* var. *subcinereus*

Var. *basalticus* Welsh Pinyon-juniper and ponderosa pine communities at 1380 to 2430 m in western Emery and eastern Sevier (type from 10 mi S of Fremont Junction) counties; endemic; 7 (v). Specimens of var. *basalticus* grow sympatrically with *A. flexuosus* var. *diehlii*. When material of the latter variety is robust, it approaches var. *basalticus* in habit, but not in pod and flower size. Indicated, however, is a close alignment between the two taxa, and var. *basalticus* could as well be treated within an expanded *A. flexuosus* or var. *diehlii* within *A. subcinereus*.

Var. *subcinereus* Ponderosa pine, pinyon-juniper, and sagebrush communities at 1670 to 2270 m in Garfield, Iron, Kane, and Washington counties; Lincoln County, Nevada; Mohave and Coconino counties, Arizona; 18 (ii). The type is from southern Utah or northern Arizona.

Astragalus tenellus Pursh Pulse Milkvetch. [*Ervum multiflorum* Pursh; *Homalobus multiflorus* (Pursh) Nutt. ex T. & G.; *A. multiflorus* (Pursh) Gray; *Tragacantha multiflora* (Pursh) Kuntze; *H. tenellus* (Pursh) Britt. in Britt. & Br.; *H. strigulosus* Rydb.; *A. tenellus* f. *strigulosus* (Rydb.) Macbr.; *A. tenellus* var. *strigulosus* (Rydb.) F. Hermann]. Perennial, caulescent, 10–52 cm tall, from a branching caudex; pubescence basifixed; stems erect or ascending, or less commonly decumbent; forming clumps; stipules 1.5–7 mm long, turning black in drying, at least some connate-sheathing; leaves 2–9 cm long; leaflets 11–21, 3–24 mm long, 0.4–6 mm wide, narrowly oblong to elliptic, linear, oblanceolate, or obovate, acute to obtuse, mucronate, or emarginate, thinly strigose beneath, glabrous above; peduncles 0.2–4 cm long, often paired; racemes (1) 3- to 23-flowered, the flowers ascending at anthesis, the axis 0.5–11 cm long in fruit; bracts 0.5–2.7 mm long; pedicels 0.7–3.2 mm long; bracteoles 0–2; calyx 2.6–5.2 mm long, the tube 2–2.7 mm long, campanulate, strigose, the teeth 0.7–2.5 mm long; flowers 6–9 mm long, white to ochroleucous; pods pendulous, stipitate, the stipe 0.6–5.5 mm long, the body elliptic to oblong in outline, straight or curved, 7–16 mm long, 2.5–4.5 mm wide, laterally flattened, glabrous or less commonly strigose, unilocular; ovules 3–9; 2n = 16, 24. Mountain brush, sagebrush, pinyon-juniper, ponderosa pine, aspen-fir, and spruce-fir communities at 1670 to 2900 m in Box Elder, Cache, Carbon, Daggett, Duchesne, Emery, Garfield, Grand, Iron, Millard, Piute, Sanpete, Sevier, Summit, Uintah, Utah, Wasatch, and Wayne counties; Yukon east to northern Manitoba, south to Nevada, New Mexico, Nebraska, and Minnesota; 147 (xxx).

Astragalus tephrodes Gray Ashen Milkvetch. Perennial, acaulescent to short-caulescent, 5–15 cm tall, from a branching, typically spreading caudex; pubescence basifixed; stems 0–15 cm long, the internodes often obscured by stipules; stipules 2–11 mm long, all distinct or rarely some shortly connate-sheathing; leaves 4–16 cm long; leaflets 11–31, 3–17 mm long, 2–11 mm broad, obovate to oblanceolate, elliptic, or orbicular, obtuse, acute, or emarginate, strigulose to pilosulous on both sides or glabrous above; peduncles 4–15 cm long; racemes 10- to 25-flowered, the flowers ascending at anthesis, the axis 1.5–8 cm long in fruit; bracts 1.5–9 mm long; pedicels 0.6–3.4 mm long; bracteoles 0; calyx 8.8–12.7 mm long, the tube 7.1–10 mm long, cylindric, pilosulous, the teeth 1.7–2.8 mm long, subulate; flowers 15–24 mm long, pink purple; pods ascending, ellipsoid to lance-ellipsoid, 17–30 mm long, 6–10 mm thick, strigulose to pilosulous, unilocular; ovules 24–35; 2n = 22. Mountain brush community at 2000 to 2500 m in Washington County; California, Arizona, and New Mexico; 6 (i). Our material belongs to the highly variable **var. *brachylobus*** (Gray) Barney [*A. shortianus* var. *brachylobus* Gray; *Xylophacos brachylobus* (Gray) Rydb.; *A. curtilobus* Tidestr., not *A. brachylobus* DC.].

Astragalus tetrapterus Gray Four-wing Milkvetch. [*Pterophacos tetrapterus* (Gray) Rydb.]. Perennial, caulescent, 10–35 cm tall, from a subterranean caudex; pubescence basifixed; stems erect or ascending or finally decumbent; stipules 2–5.5 mm long, all distinct; leaves 1.5–8.5 cm long; leaflets 9–23, 1–33 mm long, 0.3–3.2

mm wide, linear, narrowly oblong, or elliptic, obtuse to acute, strigose to glabrous on both sides, at least some terminal leaflets confluent with the rachis in the uppermost leaves; peduncles 1–6.5 cm long; racemes 6– to 15–flowered, the flowers ascending at anthesis, the axis 1–4 cm long in fruit; bracts 1.5–3.5 mm long; pedicels 1.4–4.3 mm long; bracteoles 0–2; calyx 5.5–8.7 mm long, the tube 4.7–7 mm long, cylindric, strigose, the teeth 0.8–2.8 mm long, subulate; flowers 15–19 mm long, white to yellowish and tinged faintly with pink, the keel faintly purple-tipped; pods pendulous, sessile, obliquely oblong in outline, curved or coiled, 20–40 mm long, 6–10 mm wide, succulent at first, ultimately (by collapse of fleshy walls), sharply 4–angled, glabrous or strigose, unilocular; ovules 28–38; 2n = 22. Pinyon-juniper and sagebrush communities at 1030 to 2130 m in Beaver, Iron, Kane, and Washington (type from 25 miles north of St. George) counties; Oregon, Nevada, and Arizona; 20 (iv). The four-wing milkvetch has been reported to produce locoism in livestock, but it is seldom sufficiently abundant to produce serious, large-scale losses.

Astragalus toanus Jones Toano Milkvetch. [*Cnemidophacos toanus* (Jones) Rydb.]. Perennial, caulescent, rushlike, 15–50 cm tall, from shallowly subterranean to superficial caudex; pubescence basifixed; stems erect or ascending, in clumps; stipules 1.5–6.5 mm long, at least some connate-sheathing; leaves 2–10 cm long, the uppermost, rarely all, simple, or with the terminal leaflet confluent with the rachis, the lower ones with leaflets 3–9, 3–30 mm long, 1.4–2.5 mm wide, linear-filiform to oblong, obtuse to acute, strigose or glabrous on both sides; peduncles 6–25 cm long; racemes 7– to 35–flowered, the flowers ascending at anthesis, the axis 3–30 cm long in fruit; bracts 1–3 mm long; pedicels 0.8–3.5 mm long; bracteoles 0–2; calyx 4.6–8 mm long, the tube 4.1–6.4 mm long, short-cylindric, strigose, the teeth 0.5–2 mm long, subulate; flowers 15–20 mm long, pink purple, with wing tips white; pods erect, sessile, oblong in outline, 13–25 mm long, 3.7–5.5 mm wide, slightly compressed laterally, glabrous to strigose, unilocular; ovules 14–26; 2n = 22. Seleniferous clay soils in salt desert shrub communities at 1530 to 1770 m in Box Elder and Millard counties; Nevada, Oregon, and Idaho; 17 (i). This species has affinities with both *A. saurinus* of the Uinta Basin and *A. rafaelensis* of the San Rafael Swell.

Astragalus uncialis Barneby Currant Milkvetch. Perennial, acaulescent, 1.5–7.5 cm tall, arising from a branching caudex, forming tufts; stems obscured by stipules, these 3–6.5 mm long, all distinct; pubescence basifixed; leaves 1.5–7.5 cm long; leaflets 3–5, 5–17 mm long, 2–6 mm wide, oblanceolate, elliptic, or narrowly obovate, acute to almost obtuse; peduncles 0.5–3.5 mm long, ascending at anthesis, prostrate in fruit; racemes 1– to 3–flowered, the flowers almost erect, the axis very short in fruit; bracts 1.5–3.5 mm long; pedicels 1.5–3 mm long; bracteoles 0–1; calyx 12–17 mm long, cylindric, strigulose with white and sometimes with some black hairs, the tube 10.2–13 mm long, the teeth 1.8–3.5 mm long, subulate; flowers 24.5–32 mm long, pink purple or purple; pods ascending, obliquely lance-ellipsoid, 2–3 cm long, 8–12 mm thick, dorsiventrally compressed, the valves thinly fleshy, strigulose, becoming leathery, unilocular; ovules 38–54. Shadscale and budsage communities at 1400 to 1620 m in Millard County; Nye County, Nevada; a Great Basin endemic; 11 (ii). Long known only from the vicinity of Currant, Nevada, the principal population of this distinctive species is apparently in Utah, north of Sevier (dry) Lake, Millard County, where it grows in an extensive shadscale-budsage community on ancient terraces of Lake Bonneville. The species is closely allied to *A. newberryi*.

Astragalus utahensis (Torr.) T. & G. Utah milkvetch. [*Phaca mollissima* var. *utahensis* Torr. in Stansbury, type from Stansbury Island; *Tragacantha utahensis* (Torr.) Kuntze; *Xylophacos utahensis* (Torr.) Rydb.]. Perennial, subacaulescent, mostly 2–12 cm tall, radiating from a branching caudex and more or less matforming; pubescence basifixed; stems 0–10 cm long, the internodes usually concealed by stipules; stipules 3–10 mm long, all distinct; leaves 1.5–12 cm long; leaflets 9–19, 2–15 mm long, 2–12 mm wide, obovate or suborbicular to ovate, obtuse to emarginate, rarely acute, densely villous-tomentose on both surfaces; peduncles 1–14 cm long; racemes 2– to 8–flowered, the flowers ascending at anthesis, the axis 0.4–2.6 cm long in fruit; bracts 4–9 mm long; pedicels 2–4.3 mm long; bracteoles 0–2; calyx 12–14 mm long, the tube 8.5–13 mm long, cylindric, villous-tomentose, the teeth 2–4.5 mm long, lance-subulate; flowers 23–31 mm long, pink purple (rarely white); pods ascending, sessile or stipitate, the stipe (gynophore) 1–2.5 mm long, the body 17–30 mm long, 5.5–7.5 mm wide, obscured by long shaggy-villous hairs, unilocular; ovules 22–31; 2n = 22. Sagebrush, pinyon-juniper, mountain brush, and grassland communities at 1250 to 2130 m in all Utah counties except Daggett, Emery, Grand, Iron, Kane, Rich, San Juan, and Uintah; Idaho and Nevada; 228 (xxi). The Utah milkvetch is known locally as "ladyslipper," because of a fancied resemblance of the large flowers to softly cushiony house-slippers. The plants are abundant along the Wasatch front, where they flower in April and May, much to the delight of beginning students in taxonomy, each of whom feels compelled to collect at least one plant and deposit it in a herbarium.

Astragalus wardii Gray Ward Milkvetch. [*Phaca wardii* (Gray) Rydb.]. Perennial, caulescent, 9–50 cm tall, from a branching, superficial (rarely subterranean) caudex; pubescence basifixed; stems decumbent to erect; stipules 1–3 mm long, all distinct; leaves 3–10 cm long; leaflets 17–23, 3–11 mm long, 0.8–6 mm wide, oblanceolate to elliptic or narrowly oblong to linear, obtuse to retuse or emarginate, strigose only on the midrib beneath and on the margins; peduncles 1.5–5 cm long; racemes loosely 5– to 15–flowered, the flowers ascending to declined at anthesis, the axis 1–5 cm long in fruit; bracts 1–2.2 mm long; pedicels 1–3.4 mm long; bracteoles 0–2; calyx 2.9–4.6 mm long, the tube 1.7–2.3 mm long, campanulate, strigose with black, less commonly black mixed with white, hairs, the teeth 1–2.4 mm long, subulate; flowers 5–8 mm long, whitish or ochroleucous; pods pendulous to spreading, sessile or on a gynophore about as broad as long, bladdery-inflated, 15–28 mm long, 9–17 mm wide (when pressed), mottled or not, glabrous, unilocular; ovules 12–17. Sagebrush, cottonwood, pinyon-juniper, ponderosa pine, spruce-fir, and less commonly in grassland and salt desert shrub communities at 1530 to 2730 m in Beaver, Garfield, Kane, Piute, and Sevier counties; endemic; 32 (ix). The Ward milkvetch (type from the Aquarius Plateau) is similar in its inflated, unilocular pods, distinct stipules, and small flowers with the closely contiguous, if not sympatric, *A. serpens*, from which it can be distinguished by the more numerous leaflets, predominately black strigose calyces, and glabrous pods. The presence of an incipient gynophore in *A. wardii* strengthens the similarity between these distinctive species.

Astragalus welshii **Barneby** Welsh Milkvetch. [*A. loanus* Barneby, in part, but not as to type]. Perennial, acaulescent, 4–20 cm tall, the caudex branches clothed with a thatch of persistent leaf bases; pubescence incipiently malpighian; stipules 5–12 mm long, all distinct; leaves 3–20 cm long; leaflets 5–11, 6–25 mm long, 2–8 mm wide, lanceolate to elliptic or obovate, typically acute, densely strigose on both sides; peduncles 1.5–11 cm long, typically recurved at maturity; racemes 5- to 7-flowered, the flowers erect-ascending at anthesis, the axis 0.2–0.8 cm long in fruit; bracts 2.5–6 mm long; pedicels 1.2–4 mm long; bracteoles 0–2; calyx 11–14 mm long, the tube 8.5–11 mm long, cylindric, loosely strigulose, the teeth 1–3.5 mm long, linear-subulate; flowers 20–23 mm long, ochroleucous or greenish white, often tinged faintly purplish, the keel-tip purple; pods ascending (humistrate), sessile, inflated, ovoid, 17–23 mm long, 10–15 mm wide (when pressed), conspicuously beaked, the valves reddish to straw colored, hirsute with lustrous long hairs; unilocular; ovules 24–38. Sagebrush, pinyon-juniper and sagebrush-aspen communities, exclusively on igneous gravels at 2135 to 2810 m in western Garfield, eastern Iron, Piute (type from west of Loa Pass), and Wayne counties; endemic; 14 (v). The Welsh milkvetch has long masqueraded within the concept of *A. loanus* (q.v.), from which it differs in the more numerous, acute leaflets, smaller flowers and pods, and more robust size.

Astragalus wetherillii **Jones** Wetherill Milkvetch. [*Phaca wetherillii* (Jones) Rydb.]. Perennial, caulescent, 4–26 cm tall, from a branching caudex; pubescence basifixed; stems decumbent to ascending, forming clumps; stipules 1.2–3.5 mm long, all distinct; leaves 2–10 cm long; leaflets 7–15, 3–14 mm long, 2–9 mm wide, obovate to oval, obtuse to emarginate or mucronate, thinly strigose beneath, glabrous above; peduncles 1.5–4.5 cm long; racemes 2- to 9-flowered, the flowers ascending to declined at anthesis, the axis 0.3–2.3 mm long in fruit; bracts 1–2.5 mm long; pedicels 1–2.5 mm long; bracteoles 0; calyx 4.6–6.2 mm long, the tube 2.8–3.8 mm long, campanulate, strigose, the teeth 1.8–2.4 mm long, subulate; flowers 7.5–11 mm long, whitish or tinged lavender; pods spreading to declined or shortly stipitate, the stipe (gynophore) 2–2.5 mm long, the body inflated, ovoid-ellipsoid, slightly curved, 14–22 mm long, strigulose, unilocular; ovules 9–13. Mountain brush and pinyon-juniper communities at 1430 to 1770 m in Grand County (collected by Eastwood according to Jones 1923); central western Colorado; 0 (0). The only Utah collection was taken from the canyon of the Colorado River east of Moab. It has not been recollected in many years, but likely it persists along the canyon slopes on the shaded side. It is a peculiar species, not quite like any other in Utah. The pods of early produced flowers are often mature and brown while progressively younger pods and flowers continue to be produced as the plants elongate. The species is locally common in the sandstone canyons east of Grand Junction, Colorado.

Astragalus wingatanus **Wats.** Fort Wingate Milkvetch. [*Homalobus wingatanus* (Wats.) Rydb.; *A. dodgeanus* Jones, type from Thompson's Springs; *A. wingatanus* var. *dodgeanus* (Jones) Jones]. Perennial, caulescent, 15–45 (60) cm tall, from a subterranean caudex; pubescence basifixed; stems spreading-ascending, forming diffuse clumps; stipules 1.5–5 mm long, at least some connate-sheathing; leaves 1.5–6.5 cm long; leaflets 7–15 (17), 3–18 mm long, 0.4–3.6 mm wide, linear to narrowly oblong, elliptic, or oblanceolate, acute, obtuse, or retuse, strigose to glabrous beneath, glabrous above, often ciliate; peduncles 2–14 cm long; racemes very loosely 7- to 35-flowered, the flowers ascending at anthesis, the axis 3–8 cm long in fruit; bracts 0.5–2 mm long; pedicels 0.8–3 mm long; bracteoles 0–2; calyx 2.5–3.7 mm long, the tube 1.5–2.6 mm long, campanulate, strigose, the teeth 0.4–1.4 mm long, triangular-subulate; flowers 5.5–8 mm long, pink purple, the wing-tips white or pale; pods deflexed, subsessile or short-stipitate, the stipe to 1.7 mm long, the body elliptic to oblong in outline, straight or slightly curved, 9–15 mm long, 3–4.5 mm wide, compressed, glabrous, unilocular; ovules 4–8; n = 11. Pinyon-juniper, mixed desert shrub, salt desert shrub, and less commonly in mountain brush communities at 1530 to 2130 m in Carbon, Duchesne, Emery, Garfield, Grand, and San Juan counties; Colorado, New Mexico, and Arizona; 43 (xv). This species simulates *A. tenellus* (q.v.) in its small flowers and laterally flattened pods. The buried caudex and elongating fruiting racemes are diagnostic for *A. wingatanus*, and indicate relationships elsewhere (i.e., with scytocarpous taxa such as *A. flexuosus* var. *diehlii*).

Astragalus woodruffii **Jones** Woodruff Milkvetch. [*Homalobus woodruffii* (Jones) Rydb.]. Perennial, caulescent, rushlike, 25–55 (65) cm tall, from a deeply subterranean caudex; pubescence basifixed or incipiently malpighian; stems erect or ascending in broomlike clumps; stipules 10–25 mm long, at least some usually connate-sheathing; leaves 1.5–6.5 (8) cm long, at lest the upper ones simple, the others with decurrent leaflets 2–9, 2–17 mm long, 0.5–2 mm wide, acute to obtuse, silvery strigose on both surfaces, the terminal leaflet decurrent also; peduncles 3.5–16 cm long; racemes 8- to 45-flowered, the flowers ascending at anthesis, the axis 2–25 cm long in fruit; bracts 2.5–7 mm long; pedicels 1–5 mm long; bracteoles 0–2; calyx 7.2–10.9 mm long, the tube 4.2–6.6 mm long, short-cylindric, pilosulous, the teeth 2.7–6 mm long, lance-subulate; flowers 12–19 mm long, pink to red purple with pale or white wing tips; pods erect, sessile, oblong in outline, straight to slightly curved, 14–20 mm long, 2.5–4.8 mm wide, laterally compressed, strigose, unilocular. Seleniferous, sandy or sandy-silts with sandloving desert shrubs, mainly on the Entrada Formation, at 1330 to 1670 m in southeastern Emery (type from the San Rafael Swell), eastern Garfield and Wayne counties; endemic; 32 (vii). This handsome milkvetch has affinities with *A. rafaelensis*, *A. saurinus*, and *A. toanus*, but has been placed in its own section by Barneby (1964: 436). The species was described by Jones (1923: 78) as being "the most beautiful species of the genus when the whole mass is ablaze with pink-purple bloom." The plant is a primary selenium indicator.

Astragalus zionis **Jones** Zion Milkvetch. [*Xylophacos zionis* (Jones) Rydb.]. Perennial, subacaulescent or short-caulescent, 3–23 cm tall, from a branching caudex, this sometimes clothed with a persistent thatch of leaf bases; pubescence basifixed; stems 0–11 cm long, prostrate to ascending, the internodes often concealed by stipules, these 1.5–5.5 mm long, all distinct or some shortly connate-sheathing; leaves mostly 2–15 cm long; leaflets 13–25, 2–16 mm long, 1–6 mm wide, elliptic or ovate, acute or less commonly obtuse, silvery villous on both sides; peduncles 0.5–15 cm long; racemes 1- to 11-flowered, the flowers ascending at anthesis, the axis 0.3–6 cm long in fruit, bracts 2–5 mm long; pedicels 1–3 mm long; bracteoles 0–2; calyx 8.3–18 mm long, the tube 6.5–12.7 mm long, cylindric, villous, the teeth 1.5–5.7 mm long, subulate; flowers 18–26 mm long, pink purple or sometimes pale; pods ascending, sessile, obliquely ovoid-oblong in outline, 15–30 mm long, 5.5–9 mm wide, usually curved, dorsiventrally compressed, strigose or villosulous, brightly mottled, usually unilocular; ovules 24–30; 2n = 22. On sandstone and in sandy and gravelly soils in mixed desert shrub, mountain brush, and riparian communities at 1340 to 2430 m in Garfield, Kane, San Juan, and Washington (type from Springdale) counties; Arizona; a Mohave-San Juan endemic; 46 (xiv).

Caesalpinia L.

Unarmed shrubs or perennial herbs, glandular-punctate in part; leaves alternate, bipinnate, petiolate, the pinnae bearing several leaflets; stipules small, persistent; flowers several to many, borne in racemes, the axis of the raceme densely glandular; calyx 5-lobed; corolla irregular; petals 5, conspicuous, the uppermost not enclosing the others in bud; stamens 10, long-exserted, brightly colored and showy, distinct; pistils sessile; pods flattened, dehiscent; x = 12.

1. Plants herbaceous, arising from prostrate rhizomes; leaves odd-pinnate, the pinnae 3–9; pedicels 2–5 mm long; flowers 6–12 mm long; known from eastern Utah . . . . *C. repens*
— Plants woody, 1–2.5 m tall; leaves even-pinnate, the pinnae 12–20 or more; flowers ca 2 cm long, with stamens 6–9 cm long; known from cultivation and established in Washington County . . . . *C. gilliesii*

Caesalpinia gilliesii (Wallich) Dietr. Poinciana; Bird-of-paradise. [*Poinciana gilliesii* Wallich ex Hook.; *Erythrostemon gilliesii* (Wallich) Klotzsch]. Shrubs, commonly (1) 1.5–2.5 m tall; leaves 15–28 cm long, with 8–12 pairs of pinnae, each with 7–11 pairs of elliptic to oblong leaflets 3–8 mm long, glabrous; petioles 2–4 cm long; flowers very showy; sepals distinct, 15–20 mm long, stipitate glandular and puberulent; petals yellow with orange markings, 20–35 mm long; staminal filaments red, 6–9 cm long; pods ascending, sessile, oblong in outline, 5.5–12 cm long, 14–20 mm wide, glandular-dotted, dehiscent. Cultivated ornamental of startling beauty, escaping and established at low elevation portions of Washington County; Texas to California; 7 (iii).

Caesalpinia repens Eastw. Creeping Rush-pea. [*Hoffmanseggia repens* (Eastw.) Cockerell; *Moparia repens* (Eastw.) Britt. & Rose]. Subacaulescent or shortly caulescent, 5–12.5 cm tall (above ground), from a deeply subterranean caudex, the branches below ground 3–15 cm long, pale; pubescence basifixed; leaves 2.5–9.5 cm long; pinnae 3–7 (9); leaflets 4–14, 3–12 mm long, 1–6 mm wide, asymmetrically obovate-elliptic to oblong, crowded, entire, villosulous; peduncles 1.2–6 cm long; racemes 7- to 26-flowered, the flowers spreading at anthesis, the axis 3–8 cm long in fruit; bracts 3–7 mm long, caducous; bracteoles 0; pedicels 2–7 mm long; calyx 8–10.5 mm long, the tube 2–4.5 mm long, campanulate, retrorsely villosulous, the teeth 6–8.5 mm long, oblong-lanceolate, villosulous; flowers opening flat or nearly so, the petals yellow, 10–12 mm long, red-spotted near the base, the whole fading pink orange; pods pendulous, oblong, 20–50 mm long, 10–20 mm wide, membranous, pilosulous. Sandy deserts with ephedra, Indian ricegrass, and other arenophilus plants at 1430 to 1670 m in Emery, Garfield, Grand, San Juan, and Wayne counties; Colorado; a Colorado Plateau endemic; 41 (vi). This is a striking plant in the light of early morning along the sandy stretches near Hanksville.

Caragana Fabr.

Shrubs; leaves alternate, even-pinnate, the rachis extended as a bristle or spine; stipules small and deciduous or persistent as spines; flowers solitary, yellow, showy; calyx campanulate or turbinate, obscurely to conspicuously 5-toothed; corolla papilionaceous; stamens 10, diadelphous; ovary sessile; pods subcylindric, linear-oblong, straight, glabrous, elastically dehiscent, the valves coiling upon dehiscence.

Caragana arborescens Lam. Pea-tree. [*Robinia caragana* L.; *C. sibirica* Medicus; *C. inermis* Moench; *C. caragana* (L.) Karsten; *Aspalathus caragana* (L.) Kuntze]. Shrubs to 4 m tall or more; leaves 4–10 cm long; leaflets 8–12, 12–25 mm long, 5–15 mm wide, lance-oblong to elliptic or oval, cuspidate, villous above and below, becoming glabrate in age; stipules slender, occasionally persisting as spines; bracts reduced to rudiments at juncture of peduncle and pedicel; flowers 17–23 mm long, borne singly on peduncles 12–35 mm long, few to several from each bud; pedicels 5–15 mm long; calyx turbinate, pubescent, the tube 4.5–7.5 mm long, the teeth small or obsolete, the margin villous; pods ascending to declined, sessile, linear-oblong, straight, 35–55 mm long, 4–7 mm thick, glabrous, the valves drying brown; 2n = 16. Cultivated ornamental and erosion control plant in Kane, Millard, Salt Lake, Sanpete, Sevier, Summit, Uintah, and Utah counties; introduced from Siberia; 15 (ii).

Cercis L.

Small trees or shrubs; leaves alternate, simple, palmately veined, cordate-ovate or orbicular; stipules deciduous; flowers clustered, appearing before the leaves from spurs on old branches or cauliflorus, pink, showy; calyx turbinate-campanulate, shallowly 5-lobed; corolla irregular, the keel larger than the banner; stamens 10, distinct; ovary subsessile; pods short-stipitate, laterally flattened, the ventral suture somewhat winged, indehiscent or tardily dehiscent; x = 7.

1. Leaves ovate- or cordate-acuminate; flowers 9–12 mm long; plants cultivated, introduced . . . . *C. canadensis*
— Leaves cordate-reniform or broader than long, rounded or emarginate, not acuminate apically; flowers 12–14 mm long or more . . . . *C. occidentalis*

Cercis canadensis L. American Redbud; Judas Tree. Shrub or small tree, mostly 1.5–3 m tall, rarely more; leaves commonly ovate-cordate, truncate to cordate basally, acuminate to acute apically, 3.5–10 cm long, longer than broad to somewhat broader than long, glabrate to puberulent beneath; flowers appearing before the leaves, clustered on spurs, cauliflorus; pedicels 6–10 mm long; calyx asymmetric, 2.5–3.5 mm long, 4.3–6.3 mm wide; corolla pink (pink purple), 9–12 mm long; keel petals 3.8–5 mm wide; pods pendulous-spreading, laterally flattened, 40–80 mm long, 8–18 mm wide, winged, the wings 0.8–2.1 mm wide, glabrous; n = 7. Cultivated ornamental at 1370 to 1570 m in Salt Lake and Utah counties, and undoubtedly elsewhere; introduced from the eastern U. S., where indigenous from Nebraska south to Texas, east to the Atlantic; 17 (ii). Our materials seem to belong to var. *canadensis*.

Cercis occidentalis Torr. ex Gray Western Redbud. Small tree or less commonly a shrub, mostly 1.5–3.5 m tall, rarely more; leaves cordate-reniform, cordate basally, rounded to emarginate apically, 2–7 cm long, commonly broader than long, glabrous or puberulent along vein axils beneath; flowers appearing before the leaves, clustered on spurs, cauliflorus; pedicels 8–12 mm long; calyx asymmetric, 3–4.5 mm long, 5.5–8 mm wide; corolla pink to pink purple, 12–15 mm long; keel petals 5.5–8 mm wide; pods short-stipitate, 4–10 cm long, 13–20 mm wide, winged, the wings 1.5–2.5 mm wide, glabrous. Indigenous, or rarely cultivated, at 770 to 1230 m in sandstone canyons and alcoves, often in hanging gardens, in Garfield, Kane, Salt Lake (in cultivation), San Juan, and Washington counties; California, Nevada, and Arizona; 41 (xii). Variation in *Cercis* has been summarized by Isely (1975). Our material was recognized by him as being distinctive as a population when compared to typical California specimens. Plants from Utah, Arizona, and Nevada clearly belong to var. **orbiculata** (**Greene**) **Tidestr. in Tidestr. & Kittell** [*C. orbiculata* Greene, type from Diamond Valley, Washington County].

Cladrastis Raf.

Unarmed trees; leaves alternate, odd-pinnately compound, petiolate; stipules lacking; axillary buds covered by the leaf base; flowers in panicles, white, showy; calyx turbinate-campanulate, 5-toothed, stamens 10, distinct or nearly so, included; ovary stipitate; pods pendulous with the panicles, oblong in outline, laterally flattened, often irregularly constricted (by abortion of the seeds), dehiscent.

Cladrastis kentukea (**Dum.-Cours.**) **Rudd** Yellowwood. [*Sophora kentukea* Dum.-Cours.; *Virgilia lutea* Michx.; *C. lutea* (Michx.) K. Koch; *C. tinctoria* Raf.]. Tree to 10 m tall, rarely more; bark smooth; wood yellow; leaves 15–32 cm long; leaflets 7–9 (11), 4–11 cm long, 2.3–7 cm wide, elliptic to ovate or obovate, acute to obtuse basally, acuminate to acute or obtuse apically, borne on petiolules, glabrous; panicles 20–40 cm long; flowers numerous, white, 15–25 mm long; calyx 8.5–9.5 mm long, villous, the rounded lobes 1.2–1.8 mm long, densely villous; pods stipitate, the stipe 4–6 mm long, usually shorter than the calyx, the body 55–80 mm long, 8–10 mm wide, flattened, glabrous; 2n = 28. Cultivated ornamental tree of lower elevations in Cache, Salt Lake, and Utah counties; introduced from the eastern U. S.; 8 (0). This is a beautiful tree, worthy of being cultivated widely.

Colutea L.

Unarmed shrubs; leaves alternate, odd-pinnately compound, petiolate; stipules more or less persistent, papery; flowers in racemes, yellow, showy; calyx turbinate-campanulate, 5-toothed; petals papilionaceous; stamens diadelphous, 9 and 1; ovary stipitate, the style hirsute along the ventral margin; pods pendulous, bladdery-inflated, strigulose, the incurved style persistent.

Colutea arborescens L. Bladder-senna. Shrubs, 1.2–3.5 m tall, sometimes dying to the ground; herbage strigose; leaves 6–12 cm long; leaflets 9–13, 14–23 (30) mm long, 7.5–12.5 mm wide, obovate to oblanceolate or elliptic, rounded to emarginate, mucronate, strigose beneath, glabrous above, the veins apparent; racemes 4– to 11-flowered; pedicels 3.5–9 mm long; bracts 1.2–4.2 mm long; bracteoles 2; calyx 4.5–7.5 mm long, the tube 4.2–5.2 mm long, the teeth 1.2–2.2 mm long; flowers 16–22 mm long, yellow, the banner abruptly reflexed, the wings strongly angled upward, surpassed by the keel; pods stipitate, the stipe 6–10 mm long, the body bladdery-inflated, ellipsoid, 45–60 mm long (or more), 16–24 mm wide (when pressed), usually mottled, indehiscent; 2n = 16. Cultivated, mainly as an ornamental, in Salt Lake and Utah counties, but recently it has been used in erosion control plantings, especially on clay roadsides in Juab and Sanpete counties at elevations of 1670 to 2500 m; introduced from Europe and Africa; 8 (i).

Coronilla L.

Perennial herbs, caulescent, from a taproot and caudex; leaves alternate, pinnately compound; stipules herbaceous, becoming chartaceous; flowers papilionaceous, in axillary, pedunculate umbels; bracts minute, scarious; calyx 5-toothed; petals 5, pink or white, the keel long-attenuate; stamens 10, diadelphous; ovary enclosed in the staminal sheath, the style glabrous; fruit a loment, 4-angled.

Coronilla varia L. Crown-vetch. Perennial, 3–10.5 dm tall, with spreading stems, glabrous to scabrous; stipules small; leaflets 9–23, 5–21 mm long, 1.5–11 mm wide, oblong to obovate or oblanceolate, acute to retuse apically; peduncles 3–11 cm long; bracts minute; flowers 11–20, 9–12 mm long; calyx 1.5–2 mm long, the teeth much shorter than the tube; loments 35–50 mm long; 2n = 24, 48. Cultivated erosion control plant, escaping and persisting in Salt Lake, Sanpete, Sevier, Utah, and Washington counties; introduced from Europe: 5 (ii).

Cytisus L.

Deciduous or evergreen shrubs; leaves alternate, trifoliolate or simple; stipules minute, thickened; flowers papilionaceous, axillary, solitary or sometimes 2 or 3; bracts minute; calyx bilabiate, the upper lip 2-lobed, the lower 3-lobed; petals 5, yellow, the keel not especially alternate; stamens 10, monadelphous, with 4 longer than the other, the style curved, broadened near the concave top; pods flattened, several-seeded.

Cytisus scoparius (L.) Link Scots Broom. [*Spartium scoparium* L.]. Shrubs to 2 m tall or more, the branchlets green, strongly angled; leaves trifoliolate, becoming simple upward; flowers solitary, rarely 2 or 3, 16–22 mm long, yellow or tinged purple; pods villous along the margin; 2n = 46, 48. Escaped cultivated plants in Weber County; widespread in the Pacific coastal states; introduced from Europe; 9 (0).

Dalea Lucanus emend. Barneby

Perennial herbs, unarmed; leaves alternate, odd-pinnate with 3 or more leaflets, glandular-dotted; stipules linear to subulate; peduncles opposite the leaves; flowers in dense spikes; calyx campanulate, 5- to 10-ribbed and 5-lobed; corolla papilionaceous, the banner petal attached near the rim of the floral cup, the outer four variously inserted or all near the rim of the staminal tube; stamens 5, monadelphous; petals white, yellowish, pink, or pink purple to indigo; pods 1 to 2-seeded, indehiscent, included in the calyx or slightly exceeding it; x = 7 or 8. Note: Included herein are the herbaceous species traditionally placed in *Petalostemon*, and excluded are those shrubby species now regarded as *Psorothamnus* (q.v.).

Wemple, D. K. 1970. Revision of the genus *Petalostemon* (Leguminosae). Iowa State J. Sci. 45: 1–102.

1. Stems prostrate; spike slender, commonly less than 8 mm thick; leaflets 5–17; plants of Kane, San Juan, and Washington counties .... *D. lanata*
— Stems decumbent to ascending or erect; spikes thicker, commonly 10 mm or more; leaflets 7 (9) or fewer; distribution various 2

2(1). Herbage pilose, pilosulous, or hirsute with lustrous hairs .... *D. flavescens*
— Herbage glabrous 3

3(2). Petals pink to pink purple; calyx oblique, the teeth long-pilose .... *D. searlsiae*
— Petals white (fading cream); calyx symmetric, the teeth ciliate .... *D. oligophylla*

Dalea flavescens (Wats.) Welsh Kanab Prairie-clover. [*Petalostemon flavescens* Wats., type from near Kanab; *Kunistera flavescens* (Wats.) Kuntze]. Stems 20–52 cm tall, from a superficial to subterranean caudex, glabrous to strigulose or pilose; stipules 1–4 mm long, lance-subulate to linear, persistent; leaves 1.5–4.7 cm long; leaflets 3–7, 5–20 mm long, 1–9 mm wide, folded or flat, oblong to oblanceolate or elliptic to linear, lustrous strigulose or pilose on both sides, glandular beneath; terminal leaflet petiolulate on a continuation of the rachis or subsessile; peduncles 1.5–20 cm long, glabrous to sparingly or densely pilose- or villous-hirsute; spikes 1.5–9 (14) cm long, 10–18 mm wide (when pressed), the rachis spreading-hairy; bracts 4–8 mm long, lance-aristate to lance-subulate or lanceolate, villous to pilose and sometimes glandular; calyx 4–7 mm long, the tube 10-ribbed, the tube not translucent, the teeth 1.5–4 mm long; flowers 6.2–11 mm long, the petals white, fading cream; pistils 7.5–13 mm long, the style 6–9.5 mm long; pods villous. Two rather distinctive varieties are recognized:

1. Spikes dense, the longest seldom more than 5 cm long; calyx teeth 1.5–2.5 mm long; pistils 7.5–10.3 mm long; styles 6–8 mm long; plants widespread ....
.... *D. flavescens* var. *flavescens*
— Spikes rather lax, the longest 5–14 cm long; calyx teeth 2.7–4 mm long; pistils 11.5–13 mm long; styles 8.5–9.5 mm long; plants of central Glen Canyon ....
.... *D. flavescens* var. *epica*

Var. epica (Welsh) Welsh & Chatterley Hole-in-the-Rock Prairie-clover. [*D. epica* Welsh, type from east of Halls Crossing]. Sandstone bedrock and sand in blackbrush and mixed desert shrub communities at 1430 to 1530 m in eastern Garfield and southwestern San Juan counties; endemic; 6 (v).

Var. flavescens Grasslands, mixed desert shrub, blackbrush, and pinyon-juniper communities, commonly in sandy soils, at 970 to 1870 m in Emery, Garfield, Grand, Kane, San Juan, and Wayne counties; Arizona, a Navajo Basin endemic; 46 (xxiii). The Kanab prairie-clover is remarkably variable in several features, e.g., in number and position of glands, in vesture of the peduncle, and in width of the spike. In one specimen (Holmgren & Goddard 9990, frag. BRY) from the bottom of Glen Canyon near the mouth of Bridge (Forbidding) Canyon (now beneath Lake Powell), the calyx teeth are glandular. This seems to represent an extension of the glandular condition from bracts, where they do or don't exist, to the calyx teeth, where they tend not to exist. Spike width varies with maturity of flowers and fruit, but there is a tendency, especially along the canyons of the Colorado River, towards very broad flowering spikes (12–16 mm), and these might warrant taxonomic recognition when more material becomes available. Plants from East Clark Bench and eastward to Last Chance Canyon in Kane County, mainly occupy stabilized dunes dominated by *Vanclevea*, *Ephedra*, and *Oryzopsis* mixtures, and have moderately to densely hairy peduncles which vary from 3–10 cm in length at anthesis. Most of the remainder of the plants in Utah have glabrous to sparingly hairy peduncles, which are 9–20 cm long at anthesis. Later, it might become prudent to treat these variants at some taxonomic level.

Dalea lanata Sprengel Woolly Dalea. [*Parosela lanata* (Sprengel) Britt.; *D. terminalis* (Jones) Heller]. Stems prostrate, 15–60 cm long or more, from a subterranean to superficial caudex, pilosulous to glabrous; stipules 1–2 mm long, subulate, more or less persistent; leaves 0.9–3 cm long; leaflets 5–17, 1.5–10 mm long, 1–5.5 mm wide, obovate to cuneate, truncate to emarginate, commonly folded, pilosulous or glabrous; peduncles 0.5–2.5 cm long; spikes 1.8–7.5 cm long, lax in mid- to late-anthesis and in fruit; bracts ovate-acuminate, pilosulous and with one or more large glands; calyx 3.2–4.6 mm long, the tube 2.1–2.7 mm long, glabrous, the teeth 0.9–1.9 mm long, pilosulous dorsally and ciliate; flowers 6–7 mm long, the petals indigo to rose pink; pistils 5–6 mm long, the style 3.5–4.5 mm long; pods villous to glabrous; 2n = 14. Stabilized dunes and other sandy sites in salt desert shrub and blackbrush communities at 970 to 1525 m in Kane, San Juan, and Washington counties; Arizona, Colorado, and Kansas, south to Mexico; 10 (ii). Our material belongs to **var. terminalis (Jones) Barneby**. This variety differs from **var. lanata** inter alia in the glabrous, shining, membranous calyx tube. Some plants from San Juan County (Harrison 12194 et al. BRY) are glabrous throughout, but seem not to differ otherwise.

Dalea oligophylla (Torr.) Shinn. Western Prairie-clover. [*Petalostemon gracilis* var. *oligophyllus* Torr. in Emory; *P. gracilis* Gray, not *D. gracilis* Kunth; *Kuhnistera occidentalis* Gray ex Heller; *P. occidentale* (Gray) Fern.; *K. candida* var. *occidentalis* (Gray) Rydb.; *K. oligophylla* (Torr.) Heller; *P. oligophyllus* (Torr.) Rydb.; *P. truncatus* Rydb.; *P. sonorae* Rydb.; *P. candidus* var. *oligophyllus* (Torr.) F. Hermann; *D. candida* var. *oligophylla* (Torr.) Shinn.]. Stems decumbent to erect, 4–9 dm tall, from a superficial caudex, glabrous; stipules 1–4.5 mm long, fragile, often coiled; leaves 1.5–5.2 cm long; leaflets 4–9, 5–27 mm long, 1–7 mm wide, oblanceolate to elliptic or oblong, truncate to emarginate, commonly folded, glabrous, glandular beneath; peduncles 1.5–15 cm long, glabrous; spikes 1–6.8 cm long, 8–12 mm thick, the rachis commonly glabrous; bracts lance-acuminate, caducous; calyx strongly 10-ribbed, usually pubescent between the ribs, the tube 2.3-3 mm long, the lobes 1-1.3 mm long; pistils 9-11 mm long, the style 8-10 mm long; pods glabrous or sparingly hairy apically. Sandy drainages and crevices in rimrock in mixed desert shrub, blackbrush, pinyon-juniper, and hanging garden communities at 1070 to 1830 m in Emery, Garfield, Grand, Kane, San Juan, and Wayne counties; Alberta to Saskatchewan, south to Arizona, New Mexico, Texas, Mexico, and Iowa; 45 (ix).

Dalea searlsiae (Gray) **Barneby** Searls Prairie-clover. [*Petalostemon searlsiae* Gray; *Kuhnistera searlsiae* (Gray) Kuntze]. Stems decumbent to erect, 23-65 cm tall, from a superficial caudex, glabrous; stipules 1-3 mm long, deciduous; leaves 1.4-5.2 cm long; leaflets 5-11, 3-18 mm long, 0.8-6 mm wide, oblanceolate to elliptic or almost linear, truncate to emarginate or acute, commonly involute or folded, glabrous, glandular beneath; peduncles 3.5-22 cm long, glabrous; spikes 1.6-9 (13) cm long, 8-12 mm wide, the rachis glabrous to hairy; bracts obovate to lanceolate, acuminate-aristate, deciduous; calyx 10-ribbed, the tube glabrous to moderately pilose, 2.4-2.9 mm long, the lobes 1-1.4 mm long, usually long-pilose; pistils 5-7 mm long, the style 4-5 mm long; pods villous. Sagebrush, pinyon-juniper, warm-desert shrub, cool desert shrub, or less commonly in spruce-fir communities, at 1230 to 2800 m in Box Elder, Garfield, Iron, Juab, Kane, Tooele, and Washington counties; Arizona, Nevada, and California; 46 (ii). This species is allied to *D. ornata* (Dougl.) Barneby, whose range is contiguous to the northwest.

Desmanthus Willd.

Perennial herbs, unarmed; leaves bipinnate; flowers whitish, numerous, borne in axillary, long-pedunculate heads; calyx united, campanulate, 5-toothed; petals 5, distinct or connate only near the base; stamens 5, long-exserted; fruit obliquely oblong, glabrous, borne in compact subglobose clusters; seeds few.

Desmanthus illinoensis (**Michx.**) **Macmillan** Illinois Mimosa. [*Mimosa illinoensis* Michx.]. Plants erect, mainly 8-12 dm tall or more, strongly angled; stipules slender, 6-10 mm long; leaves bipinnate, 3-10 cm long, with 6-12 pairs of pinnae, 2-4 cm long, these each with 20-30 pairs of leaflets 2-5 mm long; peduncles 2-6 cm long, ascending or curved- ascending; petals ca 2 mm long; pods strongly curved or twisted together in a dense capitate cluster, 10-25 mm long, 4-7 mm wide; seeds ca 2.5-5 mm long. Cultivated botanical curiosity, escaping and established, especially along irrigation canals in Washington County; indigenous from the northeastern U. S., south to Colorado, New Mexico, and Florida; 5 (i).

Galega L.

Perennial, caulescent, from a caudex and stout taproot; leaves alternate, odd-pinnate; stipules sagittate, distinct; flowers papilionaceous, borne in axillary racemes, each subtended by a bract; bracteoles 0; calyx 5-toothed; petals 5, purplish blue, the keel subequal to the wings, attenuate apically; stamens monadelphous; ovary enclosed in the staminal sheath, the style glabrous; pods sessile, narrowly oblong in outline, more or less constricted between the seeds, unilocular.

Galega officinalis L. Goatsrue; Professor-weed. Perennial, caulescent, 5-10 dm tall, from a branching caudex; pubescence basifixed; stems erect or ascending; stipules 7-16 mm long, sagittate, the basal lobes again once or twice lobed; leaves 3-22 cm long; leaflets 9-15, 7-52 mm long, 3-17 mm wide, lanceolate to elliptic, cuspidate-aristate apically, glabrous above, thinly pilose along veins beneath; peduncles 3-9 cm long; racemes 20- to 36-flowered, the flowers declined at anthesis, the axis 2.5-7 cm long in fruit; bracts 3.5-6 mm long, some commonly semisagittate; calyx 4-5.5 mm long, the tube 2-2.5 mm long, campanulate, glabrous, the teeth 2-3.2 mm long, subulate; flowers 9.5-12.5 mm long, blue purple (fading cream); pods ascending, subcylindric, 18-38 mm long, 2-3 mm thick, longitudinally nerved, glabrous, unilocular; $2n = 16$. Introduced, established weedy plant, known in Utah only from Cache County, where it has grown continuously since at least 1909; adventive from Europe; 10 (0).

Gleditsia L.

Trees, often armed with simple or branched thorns; leaves alternate, deciduous, pinnately once to twice compound (with both kinds of leaves often on the same branch, or some intermediate); stipules minute, caducous; leaflets 14-36 on once-pinnate leaves and on the 3-5 pinnae; flowers in spikelike, axillary racemes, polygamous, almost regular; sepals equal or nearly so; petals 3-5, very narrow, yellowish, the uppermost internal in bud; stamens 3-10, distinct, the anthers in pistillate flowers abortive; pods flattened, straplike, indehiscent.

Gleditsia triacanthos L. Honey Locust. Trees to 20 m tall or more; bark smooth; leaflets 12-35 cm long, the pinnae 3-5 (8) when bipinnate, the leaflets 14-36 on once pinnately compound leaves and on pinnae, 10-42 mm long, 3-14 mm wide, lanceolate to oblong, crenate, obtuse to cuspidate, glabrous above, puberulent along veins beneath; racemes many-flowered, 3-7 cm long, short-pedunculate or subsessile; sepals separate; petals 4-5 mm long, greenish; pods sessile, oblong in outline, laterally flattened, 7-35 cm long, 15-30 mm wide, curved, indehiscent, the seeds imbedded in tissue; $2n = 28$. Cultivated shade tree, rarely escaping, in Beaver, Cache, Davis, Grand, Millard, Salt Lake, Utah, Wayne, Washington, and Weber counties; 29 (v).

Glycyrrhiza L.

Perennial, caulescent, from stout, sweet roots; leaves alternate, odd-pinnate, glandular-punctate; stipules subulate, distinct; flowers papilionaceous, in axillary racemes, each subtended by a lanceolate, deciduous bract; bracteoles 0; calyx 5-toothed; petals 5, white to cream, the keel shorter than the wings; stamens 10, diadelphous; ovary enclosed in the staminal sheath, the style glabrous; pods sessile, elliptic to oblong in outline, burlike, armed with uncinate appendages, or smooth, indehiscent, few-seeded; $x = 8$.

Glycyrrhiza lepidota **Pursh** Licorice. Plants 4-12 dm tall, from a deep-seated root; stipules 2-7 mm long, subulate; leaves 8-19 cm long; leaflets 13-19, 8-53 mm long, 3-15 mm wide, lanceolate to oblong, mucronate, glabrous above, glandular-dotted and puberulent beneath; peduncles often paired, 3-8 cm long; racemes 20- to 50-flowered, the flowers ascending at anthesis, the rachis 2.5-9 cm long in fruit; bracts 5-8 mm long, caducous; calyx 4.8-6.9 mm long, the tube 2.5-4.9 mm long, campanulate to short-cylindric, stipitate-glandular, the teeth 1.5–3.6 mm long; flowers 9.1–13 mm long, white to cream; pods spreading, laterally compressed, oblong, 13–20 mm long, the body 5–7 mm wide, beset with hooked prickles, simulating cockleburs. Terraces, streamsides, seeps, and other semimoist sites in streamside, creosote bush, greasewood, mixed desert shrub, pinyon-juniper, and cottonwood-willow communities at 670 to 2470 m in all Utah counties, except Carbon, Davis, Duchesne, Morgan, and Tooele; widespread in the U. S., except for the southeast; 78 (xviii). The licorice of commerce, **G. glabra** L., is reported from ditches and meadows in Cache County. It differs from *G. lepidota* mainly in the smooth pods, lacking uncinate appendages.

Gymnocladus L.

Unarmed trees; leaves alternate, deciduous, bipinnately compound; stipules small, deciduous; leaflets 9–15 on each of the 3–7 pinnae, or the lowermost pinnae represented by leaflets; flowers in terminal panicles, dioecious or polygamous, regular or nearly so; calyx 5-lobed; petals 5, distinct; pods broad-oblong, hard, thick, flattened, pulpy.

Gymnocladus dioica (L.) K. Koch Kentucky Coffeetree. [*Guilandina dioica* L.]. Tree to 20 m or more; leaves 20–60 (90) cm long; pinnae 3–7 pair, each with 9–15 leaflets, these 17–70 mm long, 7–30 mm wide, ovate to lanceolate, entire, acuminate, pilose (especially along the veins) beneath, glabrous to glabrate above; panicles many-flowered; pistillate panicles to 25 cm long, the staminate much smaller and more dense; flowers 10–13 mm long, greenish to bluish white; fruit 10–20 cm long, 3–5 cm broad, persistent, tardily dehiscent, the large seeds imbedded in greenish, initially gelatinous tissue. Cultivated ornamental and botanical curiosity in Salt Lake, Utah, and Weber counties; introduced from the eastern U. S.; 8 (i).

Hedysarum L.

Perennial herbs, caulescent, from a caudex and taproot; leaves alternate, odd-pinnate; stipules adnate to the petiole base, at least the lowermost connate-sheathing; flowers papilionaceous, in axillary racemes, each subtended by a bract; bracteoles 2; calyx 5-toothed; petals 5, red purple to pink or pink purple, the keel much longer than the wings, abruptly bow-shaped; stamens 10, diadelphous; ovary enclosed in the staminal sheath, the style glabrous; loments with 2–8 segments, prominently reticulate.

Northstrom, T. E. 1974. The genus *Hedysarum* in North America. Ph.D. Dissertation. Brigham Young University. 187 pp.
Northstrom, T. E. and S. L. Welsh. 1970. Revision of the *Hedysarum boreale* complex. Great Basin Naturalist 30: 109–130.
Rollins, R. C. 1940. Studies in the genus *Hedysarum* in North America. Rhodora 42: 217–238.

1. Leaflets thin, the veins readily apparent; fruit segments winged; calyx lobes unequal, shorter than the tube; plants of eastern and northern Utah .... *H. occidentale*
— Leaflets thick, the veins not apparent; fruit segments not winged; calyx lobes subequal, longer than the tube; plants widespread .... *H. boreale*

Hedysarum boreale Nutt. Northern Sweetvetch. Perennial, caulescent, 17–70 cm tall, from branching, subterranean to superficial caudex; pubescence basifixed; stems decumbent to erect; stipules 2–10 mm long, at least some connate-sheathing; leaves 3–12 cm long; leaflets 5–15, 7–35 mm long, 2–19 mm wide, oblong to elliptic, lance-oblong, or ovate (rarely linear), strigose on both sides or glabrate to glabrous above; peduncles 2.8–15 cm long; racemes 5– to 45–flowered, the flowers ascending at anthesis, the axis 5–28.5 cm long in fruit; bracts 2–4 mm long; pedicels 0.8–4.5 mm long; bracteoles 2; calyx 4.5–8 mm long, the tube 2.5–3.5 mm long, campanulate, strigose, the teeth 2–6 mm long, subulate; flowers 10–19 mm long, red purple to pink or pink purple, less commonly white; loments stipitate, pendulous to spreading, with 2–8 segments, not winged, prominently reticulate. Our material belongs to ssp. *boreale* and is separable into two varieties.

1. Segments of loment bearing spines on lateral surfaces; plants of Uintah County, transitional to the next ......
.... *H. boreale* var. *gremiale*
— Segments of the loment lacking spines; plants widespread
.... *H. boreale* var. *boreale*

Var. boreale [*H. carnulosum* Greene; *H. pabulare* A. Nels.; *H. pabulare* var. *rivulare* L. O. Williams; *H. mackenziei* var. *pabulare* (A. Nels.) Kearney & Peebles; *H. utahense* Rydb., type from Salt Lake City; *H. boreale* var. *utahense* (Rydb.) Rollins; *H. canescens* Nutt. ex T. & G.; *H. cinerascens* (Rydb.) Rollins; *H. boreale* var. *obovatum* Rollins]. Mixed desert shrub, pinyon-juniper, mountain brush, ponderosa pine, and aspen communities at 1175 to 2500 m in all Utah counties, except Beaver, Iron, Morgan, Piute, and Summit; Alberta east to Manitoba, south to Nevada, Arizona, New Mexico, and Texas; 217 (xxxi). The use of leaf pubescence, or lack thereof, to segregate the var. *boreale* into further taxa is a function in frustration, leading to two essentially sympatric phases that might reflect ecology more than genetics.

Var. gremiale (Rollins) Northstrom & Welsh [*H. gremiale* Rollins, type from north of Vernal]. Pinyon-juniper and mountain brush communities at 1470 to 1670 m in Uintah County; endemic; 13 (0). The spines on loment segments vary from few (or none) to numerous, indicating a complete transition with var. *boreale*. Incipient spines are present in specimens of var. *boreale* from outside of Utah, but nowhere are the spines so well or so consistently developed as in Uintah County.

Hedysarum occidentale Greene Western Sweetvetch. [*H. lancifolium* Rydb.; *H. marginatum* Greene; *H. uintahense* A. Nels.]. Perennial, caulescent, 30–90 cm tall, from a branching, superficial caudex; pubescence basifixed; stems ascending to erect; stipules 10–17 mm long; leaves 8–20 cm long; leaflets 11–19, 9–37 mm long, 4–16 mm wide, ovate to lance-ovate or elliptic, apiculate to emarginate, strigose on both sides or glabrous above; peduncles 3.7–15 cm long; racemes 10– to 50–flowered, the flowers spreading to declined at anthesis, the axis 6–14 cm long in fruit; bracts 2–8 mm long; bracteoles 2; calyx 3.5–11 mm long, the tube 2.3–6 mm long, campanulate, glabrous to strigose, the teeth 0.5–2 mm long, triangular; flowers 16–23 mm long, pink to red-purple; loments stipitate, pendulous, with 1–5 segments, winged. Mountain brush, sagebrush, and lower sprucefir-aspen communities at 1770 to 2430 m in Carbon, Duchesne, Emery, Rich, and Summit counties; British Columbia south to Washington, Montana, Idaho, Wyoming, and Colorado; 17 (vii). Utah plants have been collected in flower and fruit in mid- to late summer. Materials from western Duchesne, Carbon, and Emery counties differ from the body of the species in leaflet features and are separable as **var. *canone* Welsh** (type from Soldier Creek, Carbon County); 17 (vii). Specimens from Rich, Summit, and northern Duchesne counties belong to **var. *occidentale*.**

Laburnum Medicus

Trees, unarmed; leaves alternate, palmately trifoliolate; stipules lacking; flowers in terminal, pendulous racemes, perfect; calyx 2–lipped, the upper lip 2–toothed, the lower lip with 3 coalescent teeth; petals all distinct; stamens 10, monadelphous; ovary stipitate; pods pendulous with the raceme, narrowly oblong, laterally compressed, more or less constricted between the few to several seeds.

***Laburnum anagyroides* Medicus** Golden-chain; Beantree. [*L. vulgare* Bercht. & Presl; *Cytissus laburnum* L.]. Slender trees to 6 m tall; leaves (including petioles 1–7 cm long) 2.5–15 cm long; leaflets 3, palmate, 1.4–7.5 cm long, 0.7–3.5 cm wide, lanceolate to elliptic or oblanceolate, acute to obtuse or rounded, strigulose to glabrate beneath, glabrous above, often ciliate; peduncles 1.2–3.8 cm long; racemes (7) 15– to 50–flowered, the flowers inverted and spreading at anthesis, the axis 15–30 cm long in fruit; bracts lacking; bracteoles 0; pedicels 8–14 mm long; calyx oblique, 4.5–5.5 mm long, the tube 3–4 mm long, glabrous, the lobes ca 1.5 mm long, tufted hairy; flowers 14–17 mm long, yellow; pods stipitate, the stipe 2–5 mm long, the body oblong in outline, 30–50 mm long, 5–8 mm wide, strigose, tardily dehiscent; 2n = 50. Cultivated ornamental tree of great beauty, in cities and towns in much of lower elevation Utah; introduced from southern Europe; 7 (0).

Lathyrus L.

Annual or perennial herbs, clambering, trailing, or climbing; leaves alternate, even-pinnately compound, the rachis terminating in a bristle or prehensile tendril; stipules herbaceous, semihastate or semisagittate; leaflets 2–12, very variable; flowers in axillary racemes, papilionaceous; calyx 5–toothed, obliquely campanulate; petals 5, white or cream to pink, purplish, or otherwise (in cultivated types), the wings not adnate to the keel, but fitted together in a groove; stamens 10, diadelphous; style laterally compressed, bearded along the ventral (upper) edge; pods oblong, several-seeded, the valves coiling upon dehiscence.

Hitchcock, C. L. 1952. A revision of the North American species of *Lathyrus*. Univ. Washington Publ. Biol. 15: 1–104.
Welsh, S. L. 1965. Legumes of Utah III: *Lathyrus* L. Proc. Utah Acad. Sci. 42: 214–221.

1. Leaflets 2; stem winged; plants introduced, annual or perennial 2
— Leaflets 4 or more; stems angled but not winged; plants indigenous, perennial 4
2(1). Plants annual, pubescent; flowers 25–30 mm long ... . *L. odoratus*
— Plants perennial, glabrous; flowers 12–25 mm long 3
3(2). Leaflets narrowly lanceolate to elliptic; flowers 15–18 mm long .... *L. sylvestris*
— Leaflets lance-elliptic to oblong or ovate; flowers 20–25 mm long .... *L. latifolius*

4(1). Keel conspicuously shorter than the wings; calyx glabrous or the teeth merely ciliate, the lower tooth usually longer than the tube; stipules large, foliaceous; petals pink-purple, rarely white .... *L. pauciflorus*
— Keel commonly subequal to the wings; calyx often hairy, the lower tooth shorter than the tube; stipules not foliaceous; flowers pink purple, pale lavender, pinkish violet, cream, or white 5
5(4). Flowers 12–22 mm long; petals pale lavender tinged to pinkish violet, cream, or white, often polychrome in populations; plants common at middle elevations, especially in aspen, flowering in summer ...... *L. lanszwertii*
— Flowers 15–30 mm long; petals bright pink to blue purple; plants widespread at lower elevations, flowering mainly in springtime 6
6(5). Plants villous; calyx 5–8 mm long, the tube 3.5–5.5 mm long; pods not stipitate; plants common .. *L. brachycalyx*
— Plants glabrous or sparingly villous; calyx 8–12 mm long, the tube 5–7 mm long; pods stipitate, the stipe 4–6 mm long; plants uncommon in southeastern Utah .... *L. eucosmus*

***Lathyrus brachycalyx* Rydb.** Rydberg Sweetpea. Perennial, clambering herbs, decumbent to erect, 10–50 cm long, the herbage villous or sometimes glabrous above (exceptionally throughout); stipules 6–15 mm long, semisagittate; leaves 2–9 cm long (excluding tendrils); leaflets 6–12, 5–50 (70) mm long, 2–10 mm wide, linear to elliptic, oblong, lanceolate, or oblanceolate; tendrils simple or branched; peduncles 4–10 cm long; racemes 2– to 5–flowered, the flowers spreading at anthesis; calyx tube 3.5–7 mm long, campanulate, the teeth 1.5–4 mm long, triangular to triangular-lanceolate; flowers 15–25 mm long, pink to pink purple (white); pods 30–45 mm long, 5–8 mm wide; 2n = 14. There are two more or less distinctive and allopatric varieties in Utah.

1. Plants typically villous; leaflets commonly 10–25 (30) mm long; banner not deeply cordate apically, the blade longer than broad; Great Basin .. *L. brachycalyx* var. *brachycalyx*
— Plants typically glabrous or sparingly villous; leaflets commonly 25–70 mm long; banner often deeply cordate apically, the blade as broad as long or broader; plants mainly of the southern tier of counties *L. brachycalyx* var. *zionis*

Var. *brachycalyx* Mixed desert shrub, pinyon-juniper, and mountain brush communities at 1575 to 2600 m in Beaver, Box Elder, Juab, Millard, Salt Lake (type from City Creek Canyon), Sanpete, Tooele, and Utah counties; Nevada; a Great Basin endemic; 65 (x).

Var. *zionis* (C. L. Hitchc.) Welsh Zion Sweetpea. [*L. zionis* C. L. Hitchc., type from 10 mi E of Zion National Park]. Sandy soils in pinyon-juniper, mixed desert shrub, and riparian communities at 1200 to 2500 m in Garfield, Grand, Kane, San Juan, and Washington counties; Arizona, Colorado, and New Mexico; 53 (xi).

***Lathyrus eucosmus* Butters & St. John** Seemly Sweetpea. [*L. brachycalyx* ssp. *eucosmus* (Butters & St. John) Welsh; *L. brachycalyx* var. *eucosmus* (Butters & St. John) Welsh]. Plants 1.5–4 (6) dm tall; stipules 8–20 mm long, semisagittate; leaves 3–10 (13) cm long; leaflets 6–10, mainly 3–6.5 cm long, 4–12 (16) mm wide, narrowly elliptic to lance-elliptic, coriaceous; peduncles 3–12 cm long; racemes 2– to 5–flowered; calyx 8–12 mm long, the tube 5–7 mm long, prominently veined, the teeth 2–5 mm long; flowers 20–30 mm long; pods stipitate, the stipe 4–6 mm long, the body 3–4.5 cm long, 7–10 mm wide; 2n = 14. Clay soils in washes in salt desert shrub communities at 1450 to 1700 m in Emery, Grand, and San Juan counties; Colorado, New Mexico, and Arizona; Mexico; 5 (0). It is with some hesitance that I reverse a stand of more than a decade of reduction of this entity within *L. brachycalyx*, but the floral features seem overwhelming, and the potential exists for nearer relationships elsewhere.

Lathyrus lanszwertii Kellogg Lanszwert Sweetpea. Plants clambering, decumbent to erect, 20–60 cm tall, the herbage glabrous to villous; stipules 7–20 mm long, semisagittate; leaves 2–14 cm long (excluding tendrils); leaflets 4–12, 7–75 mm long, 3–18 (26) mm wide, elliptic to lanceolate, oblanceolate, or oval; tendrils short and simple to more commonly branched and prehensile; peduncles 2–8.5 cm long; racemes 2– to 5–flowered, the flowers spreading at anthesis; calyx tube 3.5–6 mm long, campanulate, the lower lateral teeth 1.8–4.2 mm long, triangular to lanceolate; flowers 12–22 mm long, pink purple to white or cream and commonly suffused or veined with pink or purple; pods 30–60 mm long, 3–7 mm wide; 2n = 14, 28. Three rather poorly defined and apparently intergrading varieties are present in Utah.

1. Tendrils reduced to a simple filiform stalk, rarely coiled; leaflets commonly 6 only; plants of southeastern Utah .... *L. lanszwertii* var. *arizonica*
— Tendrils commonly branched and/or coiled; leaflets often more than 6; plants widespread 2

2(1). Flowers white, less commonly suffused or veined with pink or purple, mostly 15–22 mm long; plants more abundant southward in Utah .... *L. lanszwertii* var. *laetivirens*
— Flowers pink purple or suffused with purple or almost white, commonly 12–17 mm long; plants more abundant northward in Utah ... *L. lanszwertii* var. *lanszwertii*

Var. *arizonicus* (Britt.) Welsh [*L. arizonicus* Britt.; *L. leucanthus* Rydb.]. Mountain brush, ponderosa pine, aspen, and alpine communities at 2440 to 3355 m in Grand and San Juan counties; Colorado, Arizona, and New Mexico; 10 (0).

Var. *laetivirens* (Greene) Welsh [*L. laetivirens* Greene ex Rydb.; *L. leucanthus* var. *laetivirens* (Greene) C. L. Hitchc.]. Riparian, mountain brush, aspen, coniferous forest and other montane communities at 1830 to 3130 m in Beaver, Carbon, Emery, Garfield, Grand, Millard, Salt Lake, San Juan, Sanpete, Sevier, Utah, and Washington counties; Colorado and Arizona; 53 (xiii). Mostly the members of this variety are distinctive by their broader leaflets and larger flowers, but they are transitional in every way with the following.

Var. *lanszwertii* [*L. coriaceus* White, type from the Wasatch Mts.]. Aspen, Douglas fir, spruce-fir, and less commonly in mountain brush communities at 1650 to 2440 m in Davis, Duchesne, Garfield, Rich, Tooele, Salt Lake, Sanpete, Sevier, Summit, Utah, and Wasatch counties; Washington, Oregon, California, Idaho, and Nevada; 101 (xxv). Plants with truly purple or pink purple flowers are the exception in Utah. Flower color changes subtly from population to population and even within populations. Recognition of more than one entity from among the variation would be arbitrary at best, and would serve no useful purpose.

Lathyrus latifolius L. Perennial Sweetpea. Perennial, climbing or clambering vines, 8–20 dm tall, the stems broadly winged, glabrous; stipules 9–40 mm long, semihastate to semisagittate; leaves 6–12 cm long (excluding tendrils); leaflets 2, 35–80 (150) mm long, 5–23 (50) mm wide, lance-elliptic to oblong or ovate; tendrils branched, coiled; peduncles 7–15 cm long; racemes 5– to 15–flowered, the flowers spreading at anthesis; calyx tube 5.8–6.2 mm long, campanulate, the lower lateral teeth 3–6 mm long, lanceolate; flowers 20–25 mm long, pink purple, pink or white; pods 6–8 cm long, 7–10 mm wide, glabrous; 2n = 14. Cultivated, escaping and now established, mainly along canal banks in Carbon, Grand, Millard, Uintah, and Utah counties; introduced from Europe; 10 (ii).

Lathyrus odoratus L. Sweetpea. Annual, climbing vines, 8–30 dm tall, the stems broadly winged, pubescent; stipules 10–30 mm long, semihastate; leaves 6–15 cm long (excluding tendrils); leaflets 2, 25–85 mm long, 8–40 mm wide, elliptic to ovate or oblanceolate; tendrils well developed, prehensile; peduncles 3–28 cm long; racemes 2– to 5–flowered, the flowers spreading at anthesis; calyx tubes 5.5–7.5 mm long, campanulate, spreading-hairy, the lower lateral teeth 5–8 mm long, lanceolate; flowers 25–37 mm long, varicolored; pods 3–6 cm long, 5–8 mm wide, pubescent; 2n = 14. Cultivated ornamental of greenhouse and outside plantings, growing best in cool middle elevations; indtroduced from Europe; 4 (ii).

Lathyrus pauciflorus Fern. Utah Sweetpea. Perennial, 2–10 dm tall or more, climbing vines, glabrous; stems merely angled; stipules 8–32 mm long, the larger ones, at least, foliose and toothed; leaves 2–12.5 cm long (excluding tendrils); leaflets 8–12, 14–50 mm long, 8–32 mm wide, ovate to elliptic; tendrils well developed, prehensile; peduncles 3.5–24 cm long; racemes 3– to 10–flowered, the flowers spreading at anthesis; calyx tube 5–7.3 mm long, obliquely campanulate, more or less gibbous, the lower lateral teeth 2.5–7 mm long, often curved and spreading; flowers 13–27 mm long, pink to pink purple, with keel usually pale or white; pods 4–7.5 cm long, 7–11 mm wide; 2n = 14. Sagebrush, mountain brush, aspen, lodgepole pine, mixed conifer, and meadow communities at 1370 to 2900 m in Box Elder, Cache, Davis, Millard, Morgan, Salt Lake, Sanpete, Sevier, Summit, Tooele, Utah, Wasatch, and Weber counties; Oregon, Idaho, Colorado, and California; 92 (xiii). Our material belongs to **var. *utahensis*** (Jones) Peck [*L. utahensis* Jones, type from Sevier County; *L. bradfieldianus* A. Nels.]. Apparent hybrids are known between this and both varieties *laetivirens* and *lanszwertii* of *L. lanszwertii*.

Lathyrus sylvestris L. Scots Sweetpea. Perennial, 6–20 dm tall, clambering, the stems broadly winged, glabrous; stipules 20–34 mm long, semisagittate; leaves 3–12 (15) cm long (excluding tendrils); leaflets 2, 3–12 cm long, 5–40 mm wide, linear-lanceolate to lanceolate or elliptic; peduncles 8–22 cm long; racemes 4– to 9–flowered, the flowers ascending-spreading at anthesis; calyx tube 4–5 mm long, campanulate, the lower lateral teeth 1.7–4 mm long, lanceolate; flowers 15–18 mm long, red or red purple; pods 4–6 cm long, 5–8 mm wide, glabrous; 2n = 14. Cultivated ornamental, persisting, escaping, and established in Utah and Sevier counties; introduced from Europe; 2 (0).

Lespedeza Michx.

Perennial herbs (or woody at the base) from a caudex; leaves alternate, pinnately trifoliolate; stipules inconspicuous; flowers papilionaceous, in axillary racemes or subpaniculate, each subtended by a bract; bracteoles 2, attached at base of calyx; calyx 5–toothed; petals 5, pink purple; stamens 10, diadelphous; ovary 1-ovuled, the style incurved and beardless, the stigma small and terminal; pod short, partially included in the calyx.

Lespedeza thunbergii (DC.) Nakai. Thunberg Bushclover. [*Desmodium thunbergii* DC.]. Perennial, 8–10 dm tall or more and 15–20 dm wide, clumpforming; pubescence basifixed; stems striate; petioles 3–8 mm long; leaflets 3, lacking stipules, commonly 2–5 cm long, elliptic to oblong or ovate, glabrous above, strigose beneath; peduncles 0.5–1.2 cm long; racemes many-flowered; flowers 10–12 mm long (or more?), pink purple; bracteoles lanceolate; pods obovate to oblong, to 10 mm long; $n = 11$. Cultivated ornamental plant in Utah County; introduced from China; 1 (i).

Lotus L.

Plants annual or perennial herbs or suffrutescent, caulescent, from a taproot and caudex; leaves alternate, pinnately (or appearing palmately) compound; stipules foliaceous, scarious, or glandlike; flowers papilionaceous, in axillary pedunculate umbels or solitary; bracts leaf-like; calyx 5-toothed; petals 5, yellow or white, sometimes suffused with red, the keel long-attenuate; stamens 10, diadelphous; ovary enclosed in the staminal sheath, the style glabrous; pods flattened or subterete, straight, 1- to several-seeded, dehiscent; $x = 6, 7, 10$.

Ottley, A. M. 1944. The American Loti with special consideration of a proposed new section, Simpeteria. Brittonia 5: 81–123.

1. Plants annual, prostrate to ascending; flowers sessile in leaf axils; known from Washington County 2
— Plants perennial, sometimes flowering the first year, prostrate to erect 3
2(1). Plants subglabrous or merely strigose; calyx teeth subequal to the tube; pods 10–15 mm long ... *L. denticulatus*
— Plants villous; calyx teeth much longer than the tube; pods 5–10 mm long .... *L. humistratus*
3(1). Flowers sessile, solitary in leaf axils, or on peduncles to 2.6 cm long; plants indigenous in San Juan County ... . *L. wrightii*
— Flowers pedunculate, solitary, or 2 to several; plants variously distributed, seldom or not of San Juan County 4
4(3). Plants suffruticose, rigid, commonly erect, internodes greatly exceeding leaf length; bracts of inflorescence 1 or 0; plants of Washington County .... *L. rigidus*
— Plants herbaceous, prostrate to ascending or erect; internode and leaf length various; bracts of inflorescence 1 or more; distribution various 5
5(4). Stipules leaflet-like; plants introduced, cultivated and escaping 6
— Stipules reduced to glands; plants indigenous 7
6(5). Flowers 3 or 4, each 7–9 mm long; leaflets of main leaves lance-linear to narrowly elliptic, acute ... *L. tenuis*
— Flowers 5–12, each 8–12 mm long; leaflets of main leaves obovate, rounded .... *L. corniculatus*
7(5). Leaves sessile, the leaflets strictly palmate, usually drying a lead-green color; plants widely distributed in central to southwestern Utah .... *L. utahensis*
— Leaves short-petiolate, the rachis elongate and at least 1 leaflet pinnately disposed; plants of western Kane, Iron, and and Washington counties .... *L. plebeius*

Lotus corniculatus L. Birds-foot Trefoil. Perennial, 1–5 dm long, with ascending or procumbent stems, glabrous or strigose; stipules foliar, almost or quite as large as the leaflets; leaflets 3, 5–15 mm long, 2–8 mm wide, obovate, rounded apically; peduncles 0.5–7.5 cm long; bracts 1- to 3-foliolate; flowers (1 or 2) mostly 5–12, 8–12 mm long, yellow; calyx 3–4 mm long, the teeth subequal to the tube; pods linear, 20–35 mm long, 2–3.5 mm thick, subterete, straight, glabrous; $2n = 12, 24, 32$. Cultivated forage plant of moist pastures, persisting in Cache, Millard, Utah, and Washington counties; introduced from Europe; 5 (i).

Lotus denticulatus (Drew) Greene Mohave Trefoil. [*Hosackia denticulata* Drew; *Anisolotus denticulatus* (Drew) Heller; *L. subpinnatus* authors, not Lag.?]. Annual, 3–30 cm long, with prostrate to ascending stems, glabrous to strigose; stipules reduced to glands; leaves pinnate, the rachis flattened; leaflets 3–5, 1.5–15 mm long, 0.8–7 (10) mm wide, obovate, obtuse to truncate apically; peduncles lacking, the flowers solitary, axillary, 4.5–7 mm long, white or suffused with yellow; calyx 2.8–5 mm long, the teeth subequal to the tube; pods 10–15 mm long, 2.8–3.1 mm wide, compressed, sparingly strigose; $2n = 12$. Salt and warm desert shrub communities at 870 to 1100 m in Washington County; Nevada and California; 9 (i).

Lotus humistratus Greene Low Trefoil. Annual, 6–30 cm long, with prostrate to ascending stems; stipules reduced to glands; leaves pinnate, the rachis flattened; leaflets 3–5, 3–15 mm long, 1–8 mm wide, obovate to oblanceolate, obtuse apically; peduncles lacking, the flowers solitary, axillary, 4–6 mm long, yellow or tinged red; calyx 3–4 mm long, the teeth much longer than the tube; pods 5–10 mm long, 2.5–3.5 mm wide, laterally compressed, strigulose-villous. Sandy or gravelly sites in creosote bush and other warm desert shrub communities at 670 to 950 m in Iron and Washington counties; Arizona, New Mexico, and California; Mexico; 9 (i).

Lotus plebeius (Brandegee) Barneby Long-bracted Trefoil. [*Hosackia plebeia* Brandegee; *L. neomexicanus* Greene; *H. rigida* var. *nummularia* Jones, type from Rockville; *Anisolotus nummularius* (Jones) Woot. & Standl.; *L. numularius* (Jones) Tidestr., not Reichenb. ex Steudel; *L. nummulus* Dayton; *L. longebracteatus* Rydb., type from near St. George; *A. longebracteatus* (Rydb.) Rydb.; *L. oroboides* var. *nanus* Isely]. Perennial, 7–38 cm long, with prostrate to decumbent stems radiating from a herbaceous, superficial caudex; stipules reduced to glands; leaves petiolate, pinnate, with at least 1 leaf commonly placed along the short rachis; leaflets 3 or 4, 2–22 mm long, 1–8 mm wide, oblanceolate to elliptic or oval (on lowermost leaves), obtuse to acute; peduncles 0.5–6.5 cm long; bracts 1- to 3-foliolate; flowers 1 or 2, 12–17 mm long, yellow, suffused with red; calyx 5–8 mm long, the tube 2.8–5.1 mm long, strigose, the teeth 2.2–8.9 mm long, shorter than the tube; pods narrowly oblong, 15–28 mm long, 3–4 mm wide, straight, strigose. Sandy and gravelly sites in desert shrub, riparian, and pinyon-juniper communities at 670 to 1600 m in Iron, Kane (west of the Paria River), and Washington counties; Nevada and Arizona; 50 (viii). The long-bracted trefoil, as it occurs in Utah, was considered by Ottley (1944: 109–113) to consist of hybrids between *L. rigidus* and *L. utahensis*. The plants differ markedly from either of the putative parents, and have features not shared by them. The stems of *L. plebeius* are prostrate-decumbent or procumbent from a herbaceous caudex, differing in this regard from both of the presumed parental species.

Lotus rigidus (Benth.) Greene Bush Trefoil. [*Hosackia rigida* Benth.; *Anisolotus rigidus* Rydb.]. Perennial, 25–70 cm tall, with erect or ascending stems commonly woody at the base, from a ligneous or subligneous, superficial caudex; stipules reduced to glands; leaves petiolate, the rachis flattened, pinnate, with at least 1 leaflet along the rachis; leaflets 3–5, 2–20 mm long, 0.8–4 mm wide, oblanceolate to oblong, obtuse to emarginate; peduncles 2.3–14 cm long; bracts 1- to 3-foliolate; flowers 1–3, 12–23 mm long, yellow suffused with red; calyx 6.2–10 mm long, the tube 4.2–6.5 mm long, cylindro-campanulate, strigose, the teeth 2–4 mm long, shorter than the tube; pods narrowly oblong, 32–45 mm long, 3.7–4.2 mm wide, straight, glabrous. Sandstone outcrops and sandy or gravelly to clay banks and terraces at 750 to 1070 m in Washington County; Nevada, Arizona, California, and Mexico; 15 (iv). The plants form rounded clumps with very brittle stems, making difficult the task of representing it well on herbarium mounts.

Lotus tenuis Kit. in Willd. Slender Trefoil. [*L. corniculatus* var. *tenuifolius* L.; *L. tenuifolius* (L.) Reichenb.]. Perennial, 2–6 dm tall, with weak, ascending stems, glabrous; stipules foliar, almost or quite as large as the leaflets; leaflets 3, 5–15 mm long, 2–4 mm wide, lanceolate to narrowly oblanceolate or lance-linear, acute apically; peduncles 2–7.5 cm long; bracts 1- to 3-foliolate; flowers 2–4, 7–9 mm long, yellow (often drying blue); calyx 4–5 mm long, the teeth subequal to the tube; pods linear, 25–30 mm long, 2–3 mm thick, subterete, glabrous; 2n = 12. Cultivated forage plant of moist meadows, persisting in Utah and Washington counties; introduced from Europe; 2 (0).

Lotus utahensis Ottley Utah Trefoil. Perennial, 15–43 cm tall, with erect-ascending stems from a shallowly subterranean, subligneus caudex; stipules reduced to glands; leaves sessile, palmate; leaflets 3–5, 2–23 mm long, 1.5–7 mm wide, spatulate to oblanceolate or oblong, obtuse to acute; peduncles 1.4–7.5 cm long; bracts 1- to 3-foliolate; flowers 1–4, 12–16 mm long, yellow, suffused with red; calyx 4.5–8.7 mm long, the tube 3.3–4.5 mm long, cylindro-campanulate, strigose, the teeth 2.1–4.2 mm long, shorter than the tube; pods narrowly oblong, 22–35 mm long, 2.5–3.5 mm wide, minutely strigulose to glabrate, shining straight. Sagebrush, pinyon-juniper, mountain brush, aspen, and spurce-fir communities at 1470 to 2730 m in Beaver, Garfield, Iron, Kane (type from W of Carmel), Millard, Piute, Sevier, Utah, Washington, and Wayne counties; Nevada and Arizona; 110 (xxvi).

Lotus wrightii (Gray) Greene Wright Trefoil. [*Hosackia wrightii* Gray; *Anisolotus wrightii* (Gray) Rydb.]. Perennial, 12–60 cm tall, with erect-ascending stems from a commonly superficial caudex; stipules reduced to glands; leaves petiolate (sometimes shortly so), palmate; leaflets 3–5, 2–22 mm long, 1–5 mm wide, spatulate to oblanceolate, oblong, linear, obtuse to acute; peduncles 0–2.6 cm long; bracts 1- to 5-foliolate; flowers commonly solitary, rarely 2, 14–18 mm long, yellow suffused with red; calyx 7.5–9.2 mm long, longer than or subequal to the tube; pods narrowly oblong, 25–34 mm long, 2–2.6 mm wide, strigulose to villosulous, straight. Ponderosa pine woods at 1830 to 2130 m in San Juan County; Colorado, Arizona, and New Mexico; 4 (0).

Lupinus L.

Plants annual or perennial herbs; leaves alternate, palmately compound; stipules slender, persistent; flowers borne in terminal racemes, perfect; calyx bilabiate, the lips entire or toothed, commonly with bracteoles; petals usually blue or blue purple, less commonly whitish, yellow, or reddish, the banner variously reflexed, glabrous or variously hairy dorsally, the wings mostly glabrous, the keel glabrous or ciliate on upper (less commonly lower) edges; stamens 10, monadelphous, with 5 long filaments alternating with 5 short ones; pods laterally compressed, 2- to several-seeded. Note: The genus is notoriously difficult because of lack of clear diagnostic features. Taxa tend to grade morphologically into each other, probably due to hybridization. The basic chromosome number is x = 12, but most of ours are polyploids, and numerous aneuploids are known. Wide ranging perennial taxa tend to intergrade with all others they contact. Because of these problems and the likelihood of cleistogamy in some taxa it is not possible to assign all specimens to described entities. The following summary treatment is tentative.

Cox, B. J. 1970. A monograph of the *Lupinus ornatus* complex. M. S. Thesis, University of Missouri, 133 pp.
Dunn, D. B. 1956. Leguminosae of Nevada, Par. II. *Lupinus*. Contr. Flora of Nevada 39: 1–64.
———. 1957. A revision of the *Lupinus arbustus* complex of the Laxiflori. Madrono 14: 54–73.
———. 1969. *Lupinus pusillus* and its relationship. Amer. Midl. Naturalist 62: 500–510.
———. 1964. *Lupinus*. pp. 140–43. In: Welsh, S. L., et al. Common Utah Plants. Brigham Young University Press, Provo, Utah. 312 pp.
Fleak, L. S. and D. B. Dunn. 1971. Nomenclature of the *Lupinus sericeus* complex (Papilionaceae). Trans. Missouri Acad. 5: 85–88.
Hess, L. W. 1969. The biosystematics of the *Lupinus argenteus* complex and allies. Ph.D. dissertation, University of Missouri. 230 pp.

1. Plants annual, the cotyledons commonly persistent .. .. Key 1
— Plants perennial, the cotyledons not present at flowering .... Key 2

Key 1.

Plants annual.

1. Leaflets long-pilose on both surfaces; cotyledons petiolate; ovules 4–6; plants of Washington County .... L. concinnus
— Leaflets variously hairy beneath, glabrous or glabrate above; cotyledons sessile; ovules 2–6; plants of various distribution 2

2(1). Flowers borne in subcapitate racemes, the rachis commonly 2 cm long or shorter 3
— Flowers borne in elongate racemes, the rachis commonly 2 cm long or longer 4

3(2). Plants subcaulescent or acaulescent, the internodes seldom to 1 cm long; upper calyx lip 2 mm long or less, entire .... L. brevicaulis
— Plants caulescent, with several developed internodes, at least some more than 2 cm long; upper calyx lip 3–6 mm long, bilobed .... L. kingii

4(2). Plants caulescent; cotyledons petiolate; keel petals ciliate, at least on lower margin near the base; ovules 4–6; plants of Washington County .... L. sparsiflorus
— Plants acaulescent, subcaulescent or short caulescent; cotyledons sessile; keel petals ciliate on upper edges toward the apex; ovules 2; plants of various distribution . 5

5(4). Peduncles 1.5–7.5 cm long; pods not constricted between the seeds; plants subcaulescent to short caulescent, known from Washington County ... L. flavoculatus
— Peduncles 0–3.5 cm long; pods constricted between the seeds; plants mostly short caulescent, widespread in Utah .... L. pusillus

Key 2.

Plants perennial.

1. Leaflets glabrous above 2
— Leaflets pubescent above 4
2(1). Banner reflexed at or below the midpoint, glabrous or pubescent distally along the crest; plants of Washington County .... *L. latifolius*
— Banner reflexed above (beyond) the midpoint, hairy if at all beneath the upper calyx lip or, if as above, plants not of Washington County 3
3(2). Leaves mainly basal, the petioles 8–13 cm long or more, coarsely hirsute; stems from a rhizome; plants of Grand and San Juan counties .... *L. polyphyllus*
— Leaves mainly well distributed along the stem, the petioles commonly less than 8 cm long, strigose or silvery hairy; stems from a caudex; plants of various distribution .... *L. argenteus*
4(1). Banner glabrous dorsally 5
— Banner pubescent on the back, at least beneath upper lip of calyx 7
5(4). Flowers 10–13.5 mm long; calyx saccate-gibbous or shortly spurred; leaves mainly basal; plants of Duchesne, Garfield, Kane, and Uintah counties .... *L. polyphyllus*
— Flowers 7–13 mm long; calyx not especially gibbous; leaves distributed along the stem, or all leaves basal; plants variously distributed 6
6(5). Plants acaulescent or short-caulescent; leaves essentially all basal .... *L. lepidus*
— Plants caulescent; leaves distributed along the stem .... *L. argenteus*
7(4). Calyx with a gibbous-saccate spur at base of upper lip; wings pubescent, or keel ciliate below the claws (rarely glabrous) 8
— Calyx at most gibbbous at base of upper lip; wings and lower edge of keel glabrous 9
8(7). Wings not ciliate; flowers yellow or varicolored in populations; plants of far western Utah .... *L. arbustus*
— Wings or keel (or both) ciliate below the claws (rarely glabrous); flowers commonly blue purple; plants widespread .... *L. caudatus*
9(7). Leaves mainly basal; plants commonly less than 5 dm tall, known from Juab, Millard, Tooele, and Utah counties .... *L. polyphyllus*
— Leaves mainly well distributed along the stems; plants often over 5 dm tall or of different distribution or both 10
10(9). Banner reflexed at or below the midpoint, strigose to thinly strigose near the tip or rarely hairy along the crest .... *L. sericeus*
— Banner reflexed beyond the midpoint, strigose on the back beneath the calyx lobe or over much of the back 11
11(10). Stems velvety or woolly hairy; plants commonly of meadows and stream terraces in Duchesne, Salt Lake, Summit, Uintah, Utah and Wasatch counties .... *L. leucophyllus*
— Stems with appressed, ascending, spreading, or retrorse hairs, but not velvety or woolly; distribution and habitat various .... *L. argenteus*

Lupinus arbustus Dougl. ex Lindl. Spur Lupine. Perennial, 26–60 cm tall, from a woody caudex; pubescence of stems minutely strigulose; leaves scattered along the stem, but with much shorter petioles upward; petioles 2.5–16 cm long; leaflets 7–13, 24–50 mm long, 3–10 mm wide, oblanceolate, pilose on both surfaces; peduncles 2–5 cm long; racemes 12– to 46–flowered, the axis 2.5–18 cm long in anthesis, 4–23 cm long in fruit; flowers 10–14 mm long, yellow or white or blue purple or all of these in populations or in the same raceme; pedicels 1.5–6 mm long; calyx with a gibbous-saccate spur at base of upper lip, the spur 1.5–2.5 mm long; banner with a central white, yellowish, or brownish spot, pubescent dorsally, reflexed near the midpoint; wings ciliate along the upper edge near the apex; keel ciliate along the upper margin; ovules 5 or 6. Sagebrush and pinyon-juniper communities at 2135 to 2440 m in the Deep Creek Mountains, Juab and Tooele counties; Washington to Montana, south to California and Nevada; 11 (ii). Our material is assigned to **var. calcaratus (Kellogg) Welsh**, a taxon of questionable significance among the range of variation in *L. arbustus* [*L. calcaratus* Kellogg; *L. laxiflorus* var. *calcaratus* (Kellogg) Dunn]. This is perhaps our most beautiful lupine species.

Lupinus argenteus Pursh Silvery Lupine. Plants perennial, 18–90 cm tall, from a superficial caudex, puberulent to strigose on stems and petioles; leaves mainly cauline; petioles 1.5–8 cm long; leaflets 6–9, 7–95 mm long, 2–22 mm wide, oblanceolate to spatulate or almost linear, flat or folded, strigulose to strigose on both surfaces or almost or quite glabrous above; peduncles 1.5–14.5 cm long; racemes 15– to 92–flowered, 5–24 cm long in anthesis, the axis 6–29 cm long in fruit; flowers (5–7) 8.5–16 mm long, blue purple, blue, white, or rarely other hues; pedicels 1.5–6 mm long; calyx gibbous or rounded at base of upper lip; banner with a central yellow or white spot, pubescent or glabrous dorsally, reflexed above the midpoint, the wings and keel glabrous or variously sparingly ciliate; ovules 3–6. The silvery lupine is represented in Utah by several more or less distinctive but intergrading varieties. Furthermore, at least some of the phases grade into other taxa, especially into *L. caudatus*, but also into *L. sericeus*. Silvery lupine, along with those species, constitutes the most common and most widespread of the perennial lupines in the state. Possibly all of them are poisonous to livestock due to the presence of alkaloids. The large proportion of the specimens encountered can be segregated by use of the following, admittedly arbitrary key.

1. Leaflets more or less evenly pubescent above, generally folded, the upper surface thus obscured; plants generally of middle and lower elevations 2
— Leaflets glabrous above, or with hairs scattered or merely with a few near the margin, commonly flat and green, but sometimes folded; plants generally of middle and upper elevations 4
2(1). Stems with spreading or retrorse hairs; flowers mainly 7–10 mm long; plants of Washington County .... *L. argenteus* var. *palmeri*
— Stems with appressed or ascending hairs; flower size various; plants of various distribution 3
3(2). Flowers 12–14 mm long; plants of sandy washes in Emery, Garfield, Grand, and San Juan counties .... *L. argenteus* var. *moabensis*
— Flowers 8–12 mm long; plants widespread in the western U. S. .... *L. argenteus* var. *argenteus*

4(1). Flowers 5–7 (7.5) mm long; pedicels typically more than 2.5 mm long; plants uncommon in the northern Wasatch .... *L. argenteus* var. *parviflorus*
— Flowers typically larger or, if almost as small, the pedicels less than 2.5 mm long and plants of different distribution 5

5(4). Flowers mainly 6–8 mm long, with pedicels less than 3.5 mm long; plants of Grand and San Juan counties .. .. *L. argenteus* var. *fulvomaculatus*
— Flowers mainly 8–11 mm long and with pedicels more than 3.5 mm long; plants of broader distribution .... *L. argenteus* var. *rubricaulis*

Var. argenteus [*L. tenellus* Dougl. ex G. Don; *L. argenteus* ssp. *argenteus* var. *tenellus* (Dougl.) Dunn; *L. alpestris* as to concept, but not as to type; *L. lucidulus* Rydb.; *L. laxus* Rydb.; *L. decumbens* var. *argentatus* Rydb.; *L. macounii* Rydb.; *L. stenophyllus* Rydb.; *L. hillii* authors, not Greene; *L. pulcher* Eastw., type from south of Cedar City; *L. garrettianus* C. P. Sm., type from 1 mi west of Duchesne; *L. spathulatus* var. *boreus* C. P. Sm.]. Creosote bush, shadscale, ephedra, grass, sagebrush, pinyon-juniper, mountain brush, ponderosa pine, aspen, Douglas fir, and less commonly spruce-fir communities at 1065 to 3450 m in all Utah counties (except San Juan); Alberta and Saskatchewan, south to Oregon, California, Nevada, New Mexico, and the Dakotas; 185 (li).

Var. fulvomaculatus (Payson) Barneby [*L. fulvomaculatus* Payson; *L. parviflorus* var. *fulvomaculatus* (Payson) Harmon]. Mountain brush, aspen, and spruce-fir communities at 2590 to 3050 m in Grand and San Juan counties; Colorado; 5 (0).

Var. moabensis Welsh Blackbrush and mixed desert shrub communities at 1280 to 1895 m in Emery, Grand, San Juan, and Wayne counties; Mesa County, Colorado; a Navajo Basin endemic; 23 (v).

Var. palmeri (Wats.) Barneby [*L. palmeri* Wats.]. Sagebrush and pinyon-juniper communities at ca 1525 to 2135 m in Iron and Washington counties; New Mexico, Arizona, Nevada, and California; 3 (i).

Var. parviflorus (Nutt.) C. L. Hitchc. [*L. parviflorus* Nutt. ex H. & A.; *L. argenteus* ssp. *parviflorus* (Nutt.) Phillips]. Aspen, lodgepole pine, and spruce-fir communities at 2270 to 2730 m in Box Elder, Cache, Duchesne, Salt Lake, and Summit counties; Montana and Idaho to Wyoming and South Dakota; 7 (0).

Var. rubricaulis (Greene) Welsh [*L. rubricaulis* Greene; *L. caudatus* var. *rubricaulis* (Greene) C. P. Sm.; *L. argenteus* ssp. *rubricaulis* (Greene) Hess & Dunn; *L. argenteus* var. *boreus* (C. P. Sm.) Welsh, at least as to concept; *L. spathulatus* Rydb., type from Wasatch Mts.; *L. argenteus* ssp. *spathulatus* (Rydb.) Hess & Dunn; *L. alpestris* A. Nels., as to type, but not as to concept; *L. maculatus* Rydb., type from P. V. (Pleasant Valley) Junction]. Aspen, meadow, mixed conifer, riparian, and spruce-fir communities at 2470 to 3370 m in Beaver, Carbon, Daggett, Davis, Duchesne, Emery, Garfield, Iron, Juab, Piute, Salt Lake, Sanpete, Sevier, Summit, Tooele, Uintah, Utah, Wasatch, Wayne, and Weber counties; Idaho, Montana, and South Dakota, south to Nevada and New Mexico; 129 (xxix). Two potential varieties are separable only arbitrarily, one with single inflorescences per stem (var. *rubricaulis*), and the other with more than the terminal inflorescence developing (*argentatus* phase). The specimens are wholly confluent geographically and seem unworthy of taxonomic segregation. No combinations are intended or implied however. It is with some reluctance that I synonymize *L. maculatus*. The materials previously assigned there, especially that from the east side of Mt. Timpanogos, are indeed singular, some even with an incipient spur and shaggy pubescence, but they fit within the whole as herein interpreted.

Lupinus brevicaulis Wats. Shortstem lupine. Annual, 4–11 cm tall from a taproot; cotyledons sessile; stems 0–2 cm long, when developed at all usually obscured by the leaf bases; leaves in a basal tuft; petioles 0.8–6.5 cm long; leaflets 3–9, 5–18 mm long, 1.5–9 mm wide, oblanceolate, flat or folded, pilose beneath, glabrous above (except marginally in some); peduncles 0.6–6.5 cm long; racemes 4– to 12–flowered, 1–2.5 cm long in anthesis, the axis 1.5–3 cm long in fruit; flowers 5.2–7 mm long, blue purple or white; pedicels 0.3–0.8 mm long; calyx tapering to the pedicel, the upper lip very short; banner with a central yellow spot, glabrous dorsally, reflexed near the midpoint; ovules 2 or 3. Creosote bush, blackbrush, salt desert shrub, pinyon-juniper, and sagebrush communities at 830 to 1970 m in Beaver, Carbon, Daggett, Emery, Garfield, Grand, Iron, Millard, Salt Lake, San Juan, Uintah, Utah, and Washington counties; Oregon to Colorado, south to Arizona and New Mexico; the short-stemmed lupine apparently forms hybrids with *L. flavoculatus*; 30 (iv).

Lupinus caudatus Kellogg Spurred Lupine. Perennial, 21–80 cm tall, from a woody caudex; leaves mainly cauline; petioles 1–12 cm long, commonly 2–8 cm long; leaflets 5–9, 10–45 (60) mm long, 2–14 cm wide, oblanceolate to elliptic or narrowly oblanceolate, pilose on both surfaces or glabrate above; peduncles 1–6.5 cm long; racemes 10– to 57–flowered, 3–16 cm long in anthesis, the axis 4.5–17 cm long in fruit; flowers 8–12.5 (13.5) mm long, blue purple or less commonly white; pedicels 1–3 (5) mm long; calyx with a gibbous-saccate spur 0.2–1.5 (2) mm long at the base of the upper lip; banner pubescent dorsally, reflexed at or beyond the midpoint; wings commonly ciliate above and near the claws, the keel commonly ciliate above and near the claws; ovules 3–5. Three rather weak varieties are known from Utah. They are separable only arbitrarily, but seem to represent at least trends within the variation.

1. Banner reflexed at the midpoint; leaflets rather broadly oblanceolate; plants rare in Kane County .... *L. caudatus* var. *cutleri*
— Banner reflexed beyond the midpoint; leaflets only rarely broadly oblanceolate; plants of various distribution 2

2(1). Leaflets bicolored, green to yellow green above, dull green to grayish beneath; plants uncommon in southern Utah .... *L. caudatus* var. *argophyllus*
— Leaflets more uniformly colored, either green or gray on both surfaces or only somewhat bicolored; plants common in much of Utah ..... *L. caudatus* var. *utahensis*

Var. argophyllus (Gray) Welsh [*L. decumbens* var. *argophyllus* Gray; *L. caudatus* ssp. *argophyllus* (Gray) Phillips; *L. argophyllus* (Gray) Cockerell; *L. laxiflorus* var. *argophyllus* (Gray) Jones; *L. argenteus* var. *argophyllus* (Gray) Wats.; *L. helleri* Greene; *L. aduncus* Greene]. Pinyon-juniper, mountain brush, ponderosa pine, and grassland communities at 1570 to 2430 m in Beaver, Garfield, Kane, and San Juan counties; Wyoming to New Mexico; 9 (iii).

Var. cutleri (Eastw.) Welsh [*L. cutleri* Eastw.; *L. caudatus* ssp. *cutleri* (Eastw.) Hess]. Pinyon-juniper woodland at 1570 m along the Cockscomb in Kane County; Arizona, Colorado, and New Mexico; 1 (i).

Var. utahensis (Wats.) Welsh [*L. holosericeus* var. *utahensis* Wats., type from Wasatch Mts.; *L. argentinus* Rydb.; *L. leucophyllus* var. *lupinus* Rydb., type from Bears Ears, Elk Mt.; *L. utahensis* Mold.]. Sagebrush, pinyon-juniper, mountain brush, ponderosa pine, aspen, mixed conifer and grassland communities at 1470 to 2730 m in Beaver, Box Elder, Cache, Davis, Duchesne, Garfield, Iron, Juab, Kane, Millard, Morgan, Piute, Rich, Salt Lake, San Juan, Tooele, Uintah, Utah, Wasatch, Wayne, and Weber counties; Oregon, Idaho, and Wyoming, south to California, Nevada, Arizona, and Colorado; 104 (xiv).

Lupinus concinnus Agardh Elegant Lupine. Annual, caulescent, 6.5–20 cm tall, from a taproot; cotyledons petiolate; stems with apparent internodes; leaves mainly cauline, the lowermost the smallest; petioles 0.6–8 cm long; leaflets 5–8, 4–24 mm long, 1.5–7.5 mm wide, obovate to oblanceolate, pilose on both sides; peduncles 0.4–2 cm long; racemes 6- to 17–flowered, 2–4 cm long in anthesis, the axis 4–7 cm long in fruit; flowers 7–8.8 mm long, blue purple (rarely white); pedicels 0.5–1.2 mm long; calyx tapering to the pedicel; upper lip well developed; banner with a central yellow spot, glabrous dorsally, reflexed at or below the midpoint; ovules 2–4. Warm desert shrublands at 770 to 1070 m in Washington County; New Mexico, Arizona, and Nevada; 16 (iv). Our material belongs to **var. orcuttii** C. P. Sm. [*L. micensis* Jones; *L. orcuttii* Wats.; *L. concinnus* ssp. *orcuttii* (Wats.) Dunn].

Lupinus flavoculatus Heller Yellow-eye Lupine. [*L. rubens* var. *flavoculatus* (Heller) C. P. Sm.; *L. odoratus* var. *flavoculatus* (Heller) Jepson]. Annual, 3–11 cm tall from a taproot; cotyledons sessile; stems 0–2 cm long, often with at least 1 apparent internode; leaves mainly basal; petioles 0.5–8.5 cm long; leaflets 6–10, 5–20 mm long, 2–9 mm wide, obovate to oblanceolate, flat or folded, pilose to glabrate beneath, glabrous above, long ciliate; peduncles 1.5–7.5 cm long; racemes 9- to 16–flowered, 1.2–3 cm long in anthesis, the axis 2–4 cm long in fruit; flowers 5.1–7.2 mm long, bluepurple or less commonly white; pedicels 0.5–2 mm long; calyx tapering to the pedicel, the upper lip short; banner with a central yellow spot, glabrous dorsally, reflexed near the midpoint; ovules 2–4. Warm desert shrub and pinyon-juniper communities at 770 to 2330 m in Washington County; Nevada and California; 12 (0).

Lupinus kingii Wats. King Lupine. [*L. sileri* Wats., type from southern Utah; *L. capitatus* Greene]. Annual, 6–24 cm tall, from a taproot; cotyledons sessile; stems with 2 or more apparent internodes; leaves mainly cauline; petioles 0.8–3.5 cm long; leaflets 5–7, 7–23 mm long, 2–6.5 mm wide, oblanceolate, flat or folded, long-pilose on both surfaces or glabrous medially above or overall; peduncles 0–9.5 cm long; racemes 5- to 12–flowered, 1–2.3 cm long at anthesis, the axis 1.8–3.7 cm long in fruit; flowers 7.8–9.2 mm long, blue purple or less commonly pallid; calyx somewhat gibbous at base of upper lip; banner with a central yellow spot, glabrous dorsally, reflexed below the midpoint; ovules usually 2. Two varieties are present.

1. Racemes sessile, the peduncles obsolete, contained within the foliage .... *L. kingii* var. *argillaceus*
— Racemes pedunculate, the peduncles 1–9.5 cm long, commonly produced beyond the foliage ....
.... *L. kingii* var. *kingii*

Var. argillaceus (Woot. & Standl.) C. P. Sm. [*L. argillaceus* Woot. & Standl.]. Pinyon-juniper community at 2400 m in the Henry Mts., Garfield County; New Mexico; 1 (0).

Var. kingii Sagebrush, pinyon-juniper, mountain brush, and ponderosa pine communities at 1830 to 2440 m in Garfield, Iron, Kane, Piute, San Juan, Sevier, Wasatch (type from Heber Valley), and Washington counties; Nevada, Arizona, New Mexico, and Colorado; 45 (vii).

Lupinus latifolius Agardh Broad-leaved Lupine. Perennial, 3–12 dm tall, from a branching caudex; pubescence appressed strigose or almost lacking; leaves mainly cauline; petioles 4–20 cm long; leaflets 5–11, 25–90 mm long, 4–20 mm wide, oblong to elliptic or oblanceolate, flat, glabrous above, thinly appressed-strigose beneath; racemes 10- to 35–flowered, 8–25 cm long at anthesis, 10–40 cm long in fruit; flowers 10–14 mm long, blue to pinkish, fading brown; pedicels 6–12 mm long; calyx not gibbous at the base; banner with a central yellow spot, glabrous dorsally, reflexed below the midpoint; ovules 7–10. Oakbrush and streamside communities at 1230 m in Washington (Zion National Park) County; California, Oregon, Washington. Our materials appear to belong to **var. columbianus** (Heller) C. P. Sm. [*L. columbianus* Heller; *L. leucanthus* Rydb., type from Springdale; *L. latifolius* ssp. *leucanthus* (Rydb.) Kenney & Dunn]; 2 (0).

Lupinus lepidus Dougl. Stemless Lupine. Perennial, 2.5–40 cm tall, caespitose from a branching caudex, acaulescent or subacaulescent to short-caulescent; leaves mainly basal; petioles 1–10 cm long; leaflets 5–9, 3–40 mm long, 1.5–9 mm wide, oblanceolate to elliptic, flat or folded, mucronate, pilose on both surfaces, or less so above; peduncles 0–3 cm long; racemes 12– to 85–flowered, 1–23 cm long at anthesis, the axis 2–30 cm long in fruit; flowers 7–14 mm long, blue purple or white; pedicels 1–4 mm long; calyx tapering to the petiole or only somewhat gibbous at base of upper lip; banner with a central yellow spot, glabrous dorsally, reflexed at or below the midpoint; ovules 2–4. Two markedly distinctive varieties are present in Utah.

1. Plants low, 3–12 cm tall; racemes sessile, surpassed by the foliage; distribution montane . *L. lepidus* var. *utahensis*
— Plants 15–40 cm tall; racemes typically short-pedunculate, much surpassing the foliage; plants of lower elevations in western Iron and Washington counties ....
.... *L. lepidus* var. *aridus*

Var. aridus (Dougl.) Jepson Rolled Lupine. [*L. aridus* Dougl. ex Lindl.; *L. volutans* Greene]. Sagebrush and pinyon-juniper communities at 1670 to 1850 m in southwestern Iron and northwestern Washington counties; Washington, Oregon, and Nevada; 8 (i).

Var. utahensis (Wats.) C. L. Hitchc. [*L. aridus* var. *utahensis* Wats., type from Parleys Park, Summit County; *L. watsonii* Heller, type from Parleys Park; *L. caespitosus* var. *utahensis* (Wats.) B. Cox; *L. caespitosus* Nutt. in T. & G.; *L. lepidus* ssp. *caespitosus* (Nutt.)

Detl.]. Meadows, open deciduous woodland, mixed conifer, and sagebrush communities at 2130 to 3350 m Beaver, Cache, Carbon, Daggett, Emery, Garfield, Grand, Iron, Kane, Millard, Piute, Rich, Salt Lake, Sanpete, Sevier, Uintah, Utah, Wasatch, Washington and Wayne counties; Idaho and Montana, south to to California, Nevada, and Colorado; 68 (xxii). It is with some reluctance that I accept the conservative treatment presented herein instead of the more liberal earlier presentation (Welsh 1978). Consideration is here given to the broader problems of interpretation of *Lupinus* as it exists in the western states. The concepts included herein support ideology elsewhere within the Leguminosae. Historical accidents of previous approaches have been largely overcome by David Dunn and his students. Hopefully, the present treatment represents a further step in reducing the problems of this amazing group of plants.

Lupinus leucophyllus Dougl. ex Lindl. White-leaved Lupine. [*L. eatonanus* C. P. Sm., type from Hailstone, Wasatch County]. Perennial, 40–90 cm tall or more, from a woody caudex; pubescence of stems dense, appressed to spreading; leaves mainly cauline; petioles 1.2–16 cm long; leaflets 7–10, 9–70 mm long, 3–13 mm wide, oblanceolate, flat or folded, villous-pilose on both surfaces; peduncles 2.5–5.5 cm long; racemes 22- to 70-flowered, 7–18 cm long in anthesis, the axis 8–19 cm long in fruit; pedicels 2.5–5 mm long; flowers 10–15 mm long, blue purple or less commonly pallid; calyx gibbous to saccate-gibbous at base of upper lip; banner with a central yellow spot, densely hairy dorsally, reflexed above the midpoint; ovules 4–6. Sagebrush, grass, and riparian communities at 1450 to 2300 m in Duchesne, Morgan, Summit, Uintah, and Wasatch counties; Washington south to Nevada and east to Wyoming; 13 (iv). This lupine forms apparent hybrids with *L. caudatus* and *L. sericeus*.

Lupinus polyphyllus Lindl. Showy Lupine. Plants perennial, 13–70 cm tall, from a subterranean to superficial caudex; pubescence of stems spreading- hirsute to hirsute-pilose or appressed-ascending; leaves mainly basal or basal and cauline; petioles of lowermost leaves 3.5–30 cm long; leaflets 8–13, 12–75 mm long, 1.2–7 mm wide, oblanceolate to obovate, plane or folded, pilose beneath, glabrous above or pilose on both sides; peduncles 3.5–13 cm long; racemes 13- to 68-flowered, 6–25 cm long in anthesis, 8–36 cm long in fruit; flowers 10–16 mm long, blue purple or rarely white; pedicels 3–9 mm long; calyx gibbous or saccate-spurred at base of upper lip; banner with a central yellow or white spot, glabrous dorsally, reflexed near or above the midpoint; wings glabrous; keel sparingly ciliate near the apex; ovules 3–7. Three distinctive varieties are present in Utah.

1. Stems from a subterranean caudex; plants of San Juan and Grand counties ..... *L. polyphyllus* var. *ammophilus*
— Stems from a superficial caudex; plants of Uintah and Kane counties or from western Utah 2

2(1). Leaflets silky-strigose on both surfaces; petioles of basal leaves only somewhat (if at all) longer than the cauline ones; plants of Duchesne, Uintah, Garfield, and Kane counties .... *L. polyphyllus* var. *humicola*

— Leaflets glabrous above or essentially so; petioles of basal leaves markedly longer than the cauline ones; plants of Juab, Millard, Tooele, and Utah counties ... . *L. polyphyllus* var. *prunophilus*

Var. ammophilus (**Greene**) **Barneby** Sand Lupine. [*L. ammophilus* Greene]. Sagebrush, pinyon-juniper, mountain brush, ponderosa pine, and aspen-Douglas fir woods, commonly in sandy soils, at 1830 to 2730 m in Grand and San Juan counties; Colorado and New Mexico; 12 (iv).

Var. humicola (**A. Nels.**) **Barneby** [*L. humicola* A. Nels.; *L. wyethii* Wats.]. Wash bottoms and terraces or rimrock in pinyon-juniper and riparian communities at 1470 to 1900 m in Duchesne, Garfield, Kane, and Uintah Counties; British Columbia and Alberta south to California, Nevada, and Colorado; 10 (iii).

Var. prunophilus (**Jones**) **Phillips** Robinson Lupine. [*L. prunophilus* Jones, type from Robinson, Juab County; *L. wyethii* var. *prunophilus* (Jones) C. P. Sm.; *L. arcticus* var. *prunophilus* (Jones) C. P. Sm.; *L. tooelensis* C. P. Sm., type from Johnson Canyon, Deep Creek Mts.]. Sagebrush, pinyon-juniper, and mountain brush communities at 1530 to 2130 m in Juab, Millard, Tooele, and Utah counties; Washington to Nevada, east to Montana, Wyoming, and Colorado; 21 (iv).

Lupinus pusillus Pursh Dwarf Lupine. Annual, 3–24 cm tall, from a taproot; cotyledons sessile; pubescence of stems and petioles spreading long-hairy; leaves mainly cauline; petioles 1–9 cm long; leaflets 3–9 (14), 11–48 mm long, 2–10 mm wide, oblanceolate, flat or folded, long-pilose beneath, glabrous above; peduncles 0.5–3.5 cm long; racemes 4- to 38-flowered, 1–17 cm long in anthesis, the axis 4–21 cm long in fruit; flowers 8.5–12 mm long, blue or variously pink or white; pedicels 1–3.5 mm long; calyx tapering to the pedicel; banner with a central yellow spot, glabrous dorsally, reflexed near the midpoint; ovules 2; pods constricted between the seeds. Blackbrush, mixed desert shrub, pinyon-juniper, and mountain brush communities, commonly in sand, at 800 to 1970 m in Beaver, Daggett, Duchesne, Emery, Garfield, Grand, Kane, Juab, Millard, San Juan, Tooele, Uintah, and Wayne counties; Washington to Alberta amd Saskatchewan south to Oregon, Nevada, Arizona, New Mexico, and Kansas; 177 (xxix). Three more or less completely intergrading varieties have been recognized from among our specimens. Plants with short peduncles, seldom more than 1 cm long have been designated as **var. intermontanus** (**Heller**) **C. P. Sm.** Those with elongate peduncles and glabrous calyx tube and pedicels are **var. rubens** (**Rydb.**) **Welsh** [*L. rubens* Rydb., type from St. George], and those with pilose calyx tube and pedicels are .**var. pusillus**. They do seem to represent trends within the variation, and there is some geographical basis for them. However, there is poor geographic correlation and sufficient intergradation as to make segregation of the materials moot.

Lupinus sericeus Pursh Silky Lupine. Perennial, 3–12 dm tall, from a branching caudex; pubescence of stems short-villous to pilose or strigose, sometimes spreading; petioles 1.2–9 cm long; leaflets 5–9, 7–78 mm long, 2–15 mm wide, oblanceolate, commonly flat (at least some), pilose to puberulent on both surfaces or glabrous to glabrate above; peduncles 1.3–9 (12) cm long; racemes 14- to 70-flowered, 6–28 cm long in anthesis, the axis 8–37 cm long in fruit; flowers 10–16 mm long, blue, blue purple, pale, or white; pedicels 2–7 mm long; calyx more or less gibbous at the base of the upper lip; banner with yellow or brown eyespot, strigose along the dorsal crest or more widely; ovules 5–7; $2n = 48$. Widely distributed in Utah, and constituting with *L. argenteus* and *L. caudatus* one of three important species complexes in Utah. They occupy most of the range of perennial lupines in the state. Four more or less completely confluent varieties are recognized.

1. Stems with spreading to retrorse hairs; flowers pale or blue; plants of Washington County ....
.... *L. sericeus* var. *jonesii*
— Stems with appressed or ascending hairs, the hairs seldom if ever spreading to retrorse 2

2(1). Flowers white in populations, the eyespot and veins commonly dark brown or fading dark brown; plants mostly from Sevier County southward ....
.... *L. sericeus* var. *barbiger*
— Flowers commonly blue or blue purple in populations, intergrading with the preceding when in contact, the eyespot yellow or brown 3

3(2). Leaflets sparingly pubescent to glabrous above; plants mostly of central southern Utah; intergrading with the following .... *L. sericeus* var. *marianus*
— Leaflets uniformly puberulent to pilose above; plants of broad distribution .... *L. sericeus* var. *sericeus*

Var. barbiger (Wats.) Welsh Sink Lupine. [*L. barbiger* Wats., type from Kane County]. Sagebrush and mountain brush communities at 1930 to 2730 m in Garfield, Kane, Piute, Sevier, Washington and Wayne counties; endemic; 40 (xvii).

Var. jonesii (Rydb.) Welsh Jones Lupine. [*L. jonesii* Rydb., type from Silver Reef, Washington County]. Pinyon-juniper and oak-ponderosa pine communities at ca 2000 m in southwestern Kane and Washington Counties; endemic; 8(i).

Var. marianus (Rydb.) Welsh Marysvale Lupine. [*L. marianus* Rydb., type from Bullion Creek, Piute County; *L. sericeus* ssp. *marianus* (Rydb.) Fleak & Dunn]. Sagebrush, pinyon-juniper, mountain brush, ponderosa pine, aspen, and spruce-fir communities at 1770 to 2870 m in eastern Beaver, Garfield, Piute (type from Bullion Creek), and Sevier counties; endemic; 15 (vi).

Var. sericeus [*L. aegra-ovium* C. P. Sm., type from Salina Canyon; *L. huffmanii* C. P. Sm., type from Salina Canyon; *L. sericeus* ssp. *huffmanii* (C. P. Sm.) Fleak & Dunn; *L. larsonanus* C. P. Sm., type from Salina Canyon; *L. puroviridis* C. P. Sm., type from Salina Canyon; *L. quercus-jugi* C. P. Sm., type from Salina Canyon vicinity; *L. rickeri* C. P. Sm., type from Salina Canyon; *L. egglestonianus* C. P. Sm.; *L. salinensis* C. P. Sm., type from Salina Canyon; *L. flexuosus* Lindl.; *L. sericeus* var. *flexuosus* (Lindl.) C. P. Sm.]. Sagebrush, mountain brush, pinyon-juniper, ponderosa pine, aspen, spruce-fir, and alpine meadow communities at 1770 to 3130 m in Beaver, Cache, Carbon, Davis, Duchesne, Emery, Garfield, Grand, Juab, Kane, Millard, Piute, Salt Lake, Sanpete, Sevier, Summit, Tooele, Uintah, Utah, Wasatch, Wayne, and Weber counties; British Columbia east to Alberta, south to California, Arizona, and New Mexico; 109 (xxxvii).

Lupinus sparsiflorus Benth. Mohave Lupine. Annual, 9–45 cm tall, from a taproot; cotyledons sessile; pubescence of stems appressed or ascending, of two lengths; leaves mainly cauline; petioles 1.4–8 cm long; leaflets 5–9, 5–35 mm long, 1–4 mm wide, elliptic-oblong to oblanceolate, flat or folded, long-pilose on both surfaces, less densely hairy above; peduncles 1.3–5 cm long; racemes 8– to 46–flowered, 4–23 cm long in anthesis, the axis 6–29 cm long in fruit; flowers 8.5–11 mm long, blue purple or rarely white; pedicels 2–5 mm long; calyx gibbous at base of upper lip; banner with a central yellow spot, glabrous dorsally, reflexed beyond the midpoint; ovules 4–6. Joshua tree, creosote bush, and mixed warm desert shrub communities, often on riparian gravels, at 670 to 1100 m in Washington County; Arizona, Nevada, California, and Mexico; 23 (vii).

Medicago L.

Annual or perennial herbs, caulescent from a taproot or caudex; leaves alternate, pinnately trifoliolate, the leaflets serrate in the distal half or less; stipules herbaceous, often toothed; flowers papilionaceous, borne in axillary, pedunculate racemes or heads; bracts subulate; calyx 5–toothed; petals 5, yellow, white, blue, pink, lavender, or purple; stamens 10, diadelphous; ovary enfolded by the staminal sheath, the style subulate, irritable; pods curved to spirally coiled, 1– to several-seeded, indehiscent, reticulate or spiny.

1. Flowers 2–3 mm long; inflorescence less than 10 mm long in anthesis; pods coiled through a single spiral, 1–seeded, unarmed; plants annual, prostrate to decumbent or rarely erect .... *M. lupulina*
— Flowers 4–10 mm long; inflorescence longer than 10 mm (including flower length), or pods differing from above; plants various 2

2(1). Flowers 4–5 mm long, yellow, 2–5 per raceme; racemes less than 10 mm long; pods armed with prickles, several-seeded; plants annual .... *M. polymorpha*
— Flowers 6–10 mm long, yellow or blue, lavender, pink, purple, or white, 6 to many on at least some racemes; racemes longer than 10 mm; pods unarmed, several-seeded; plants typically perennial 3

3(2). Flowers yellow (sometimes tinged violet); pods merely curved; plants uncommon .... *M. falcata*
— Flowers blue, pink, lavender, purple, or white; pods spirally coiled; plants common .... *M. sativa*

Medicago falcata L. Yellow Alfalfa. Perennial (rarely functionally annual), 4–10 dm tall or more, the stems erect or ascending, strigulose; stipules 4–12 mm long, persistent, conspicuously veined; leaves short-petiolate, the leaflets linear, oblong, oblanceolate, or elliptic, 6–20 mm long, 1–6 (10) mm wide, few-toothed, tridentate, or merely apiculate apically, strigulose beneath; peduncles subequal to the subtending leaves or longer; racemes 6– to 20–flowered, mostly 10–20 mm long; flowers 6–9 mm long, yellow, sometimes suffused with violet; calyx campanulate, the tube 1–2 mm long, the teeth 1.5–3 mm long, lance-subulate; pods 6–10 mm long, merely curved, unarmed, several-seeded; $2n = 16, 32$. Sparingly cultivated forage plant, escaping and persisting; introduced from the Old World; 3 (ii). This plant forms hybrids with *M. sativa*, and is sometimes considered a phase of that species.

Medicago lupulina L. Black Medick; Hop Clover. Annual, the stems prostrate to decumbent or sometimes erect, 10–40 cm long; stipules entire or nearly so, 3–6 mm long, persistent; leaves short-petiolate, the leaflets cuneate to obcordate, 4–15 mm long, 2–12 mm wide, toothed in the apical 1/3 (rarely more), pubescent to glabrous; peduncles mostly equaling or surpassing the subtending leaves; racemes 6– to 25–flowered, less than 25 mm long in fruit; flowers 2–3 mm long, yellow; calyx campanulate, about 1 mm long; pods spiral through ca 1 coil, unarmed, 1-seeded; 2n = 16, 28, 32. Introduced weedy species of lawns, fields, and other open sites in practically all of Utah; introduced from Europe; 49 (v).

Medicago polymorpha L. Bur Clover. [*M. hispida* Gaertn.]. Annual, the stems prostrate to erect, 1-4 dm long; stipules deeply divided into long teeth, mostly 3-7 mm long; leaves short-petiolate, the leaflets cuneate to obovate or obcordate, 10-25 mm long, 6-18 mm wide, toothed into the apical 1/3 or more, pubescent to glabrous; peduncles mostly shorter than the subtending leaves; racemes 2- to 5-flowered, less than 10 mm long; flowers 4-5 mm long, yellow; calyx campanulate, the tube 1-1.5 mm long, the lance-subulate teeth 1-2 mm long; pods spirally coiled, armed with spines, several-seeded; 2n = 14, 16. Rare weedy species of cultivated land in Utah, but to be expected almost anywhere; introduced from Europe; 1 (0).

Medicago sativa L. Alfalfa; Lucern. Perennial, or functionally annual, the stems 4-12 dm long or more, ascending to erect, finally sprawling, strigulose; stipules entire or toothed, 4-12 mm long, persistent; leaves short-petiolate, the leaflets elliptic to oblanceolate, 8-40 mm long, 2-15 mm wide, apically few-toothed, pubescent; peduncles often surpassing the subtending leaves; racemes 6- to 25-flowered, 10-35 mm long or more; flowers 6-10 mm long, blue, lavender, pink, purple, or white; calyx campanulate to short-cylindric, the tube 1.5-2.5 mm long, the lance-subulate teeth 2-4 mm long; pods spirally coiled, unarmed, several-seeded; 2n = 16, 32, 64. Introduced forage plant, escaping and persisting, now almost or quite cosmopolitan at moderate and lower elevations in Utah; introduced from Europe. This is possibly the most important single forage species in Utah. It has been grown since antiquity in the Old World.

Melilotus L.

Plants annual or biennial herbs, caulescent, from a stout taproot; leaves alternate, pinnately trifoliolate, the leaflets dentate-serrate in the distal half or more; stipules herbaceous, distinct, subulate, entire or hastately lobed; flowers papilionaceous, borne in axillary, pedunculate racemes; bracts subulate; calyx 5-toothed; petals 5, white or yellow, the keel obtuse; stamens 10, diadelphous; ovary enfolded by the staminal sheath, the style subulate, not irritable; pods straight, ovoid, reticulately veined or cross-ribbed, unarmed, glabrous, 1- to 2-seeded, indehiscent; x = 8.

1. Flowers 2-3 mm long, yellow; pedicels less than 1 mm long; plants of Washington County .... *M. indicus*
— Flowers 3-7 mm long, white or yellow; pedicels 1-2 mm long; plants widespread 2
2(1). Flowers white; pods reticulately veined .... *M. albus*
— Flowers yellow; pods cross-ribbed .... *M. officinalis*

Melilotus albus Desr. ex Lam. White Sweet-clover. Annual or biennial, the stems commonly 5-15 dm tall or more, erect, strigulose; stipules entire or hastately lobed, mostly 5-10 mm long, persistent; leaves short-petiolate, the leaflets obovate to elliptic or oblanceolate, 8-35 mm long, 1-15 mm wide, pubescent or glabrous; peduncles commonly surpassing the subtending leaves; racemes 38- to 115-flowered, 2.8-12.5 cm long or more; flowers 4-5.5 mm long, white; calyx campanulate, the tube 1.2-1.8 mm long, the teeth 1-1.5 mm long, acuminate; pods 2.5-6 mm long, reticulately veined, 1- to 2-seeded; 2n = 16. Introduced forage plant, now widely established at 1065 to 2135 m in Beaver, Box Elder, Cache, Daggett, Emery, Garfield, Grand, Juab, Kane, Millard, Salt Lake, San Juan, Sanpete, Sevier, Tooele, Uintah, Utah, Washington, and Wayne counties; introduced from Europe; 66 (xii). This plant is an excellent source of honey for domestic bees. The plants contain coumarin, and are responsible for bloat and subsequent death in livestock.

Melilotus indicus (L.) All. India Sweet-clover. [*Trifolium melilotus-indica* L.]. Annual or winter annual, the stems 2-8 dm tall, erect, glabrous or strigose; stipules lance-subulate, 4-6 mm long, persistent; leaves short-petiolate, the leaflets obovate to oblanceolate or elliptic, 7-30 mm long, 3-14 mm wide, glabrous; peduncles shorter to longer than the subtending leaves; racemes 21- to 43-flowered, 8-30 mm long; flowers 2-3 mm long, yellow (fading white); calyx campanulate, the tube 0.6-0.8 mm long, the teeth 0.5-0.7 mm long, triangular; pods 1.5-2 mm long, reticulately veined, 1-seeded; 2n = 16. Adventive weedy species at 820 to 915 m in Washington County; introduced from Eurasia; 4 (0).

Melilotus officinalis (L.) Pallas Yellow Sweet-clover. [*Trifolium melilotus-officinalis* L.]. Annual, winter annual, or biennial, the stems 5-15 dm tall or more, erect, strigulose; stipules entire or with 1-3 basal teeth, 3-10 mm long, persistent; leaves shortly petiolate, the leaflets cuneate to elliptic or oblanceolate, 8-38 mm long, 3-16 mm wide, pubescent or glabrous; peduncles shorter to longer than the subtending leaves; racemes 20- to 65-flowered, 1.8-11 (14) cm long; flowers 4.5-6 (7) mm long, yellow, fading cream; calyx campanulate, the tube 1-1.8 mm long, the teeth 1-1.5 mm long, acuminate; pods 3-5 mm long, cross-ribbed, 1- or 2-seeded; 2n = 16. Common ruderal weed of almost cosmopolitan distribution at 1220 to 2440 m in probably all Utah counties; introduced from Europe; 83 (xiii).

Onobrychis Adanson

Perennial herbs, caulescent, from a caudex and taproot; leaves alternate, odd-pinnate; stipules adnate to the petiole base, the lowermost amplexicaul but not connate; flowers papilionaceous, in axillary racemes, each subtended by a bract; bracteoles 2; calyx 5-toothed; petals 5, red purple to lavender or pink, the keel much longer than the wings, abruptly bow shaped; stamens 10, essentially diadelphous; ovary enclosed in the staminal sheath, the style glabrous; loment reduced to 1 segment, this armed with prickles.

Onobrychis viciifolia Scop. Sainfoin; Holy Clover. [*Hedysarum onobrychis* L.; *O. onobrychis* (L.) Rydb.; *O. sativa* Lam.]. Perennial, caulescent, 20-45 cm tall, from a branching, superficial caudex; stems ascending to erect; stipules 3-12 mm long, all more or less amplexicaul; leaves 3-12 mm long; leaflets 11-21 (27), 8-25 mm long, 2-7 mm wide, oblong to elliptic or oblanceolate, pilose mainly along veins beneath, glabrous above; peduncles 8-19 cm long; racemes 14- to 39 (50) -flowered, the flowers ascending-spreading at anthesis, the axis 4-14 cm long in fruit; bracts 2.5-4.5 mm long; pedicels 0.2-1.5 mm long; bracteoles 1; calyx 5.5-6.5 mm long, the tube 2.3-3 mm long, campanulate, the teeth 2.2-4 mm long, subulate; flowers 10-13 mm long, red purple, lavender, or pink; loment sessile, ascending, armed with prickles; 2n = 28. Introduced forage plant, escaping and persisting in Juab, Millard, Rich, Salt Lake, Sanpete, Summit, and Utah counties; native to Europe; 10 (i). The genus is closely allied to *Hedysarum* (q.v.).

Oxytropis DC.

Perennial, caulescent, or acaulescent herbs, from a taproot and caudex; leaves alternate or basal, odd-pinnate; stipules adnate to the petiole base, often connate-sheathing; flowers papilionaceous, scapose or borne in axillary racemes, each subtended by a single bract; bracteoles 0 (rarely 2); calyx 5-toothed; petals 5, pink, pink purple, or white, the keel shorter than the wings, the keel-tip produced into a porrect beak; stamens 10, diadelphous; ovary enfolded in the staminal sheath, the style glabrous; pods sessile or stipitate, straight, erect, ascending, or spreading-declined, 1- or 2-loculed, or partially 2-loculed by intrusion of the ventral (upper) suture, dehiscent apically or throughout; $x = 8$.

Barneby, R. C. 1952. A revision of the North American species of *Oxytropis* DC. Proc. California Acad. Sci. IV. 27: 177–312.

1. Plants shortly caulescent; stipules only somewhat adnate to the petioles; flowers 5–9 mm long; pods spreading-declined .... *O. deflexa*
— Plants acaulescent, scapose; stipules adnate to the petioles through half their length or more; flowers commonly more than 9 mm long; pods erect or ascending 2

2(1). Bracts, calyx teeth, pods, and sometimes other plant parts glandular-viscid; flowers either pink-purple or cream .... *O. viscida*
— Bracts, calyx teeth, pods, and other plant parts pilose to villosulous, but not at all glandular-viscid 3

3(2). Racemes 1- to 5-flowered, subcapitate 4
— Racemes 6- to many-flowered, the axis elongate, only rarely subcapitate 7

4(3). Calyx swollen at full anthesis, becoming inflated and finally enclosing the pod; pods rarely longer than 10 mm; plants of Daggett and Duchesne counties .... *O. multiceps*
— Calyx campanulate, not turgid, not becoming inflated or enclosing the fruit, at length rupturing along one side; pods often over 10 mm long 5

5(4). Pods oblong-ellipsoid, not inflated, leathery in texture; flowers 1–4; plants of mountain summits in north central to southern Utah .... *O. parryi*
— Pods ovoid, inflated, papery in texture; flower number various; plants distributed variously 6

6(5). Flowers 6–12 mm long; leaflets 7–17; plants of limestone, gravel, and basalt in southern Utah .... *O. oreophila*
— Flowers (11) 14–17 mm long; leaflets 3–7; plants of shale and limestone in eastern, central, and southern Utah .... *O. jonesii*

7(3). Plants dwarf, seldom over 10 cm tall; flowers 6–12 mm long; growing in limestone, gravel, and basalt in southern Utah .... *O. oreophila*
— Plants mainly over 10 cm tall; flowers 14–27 mm long; growing on various substrates and with distribution various but seldom as above 8

8(7). Petals bright pink purple 9
— Petals white or ochroleucous to yellowish 10

9(8). Pubescence of basifixed hairs; calyx somewhat swollen and accrescent; plants of Daggett County ... *O. besseyi*
— Pubescence of malpighian hairs; calyx not swollen or enlarged; plants of wide distribution .... *O. lambertii*

10(8). Wing petals dilated apically, at least 5 mm wide; plants broadly distributed .... *O. sericea*
— Wing petals not especially dilated, less than 5 mm wide; plants known from the north slope of the Uinta Mts. .... *O. campestris*

Oxytropis besseyi (**Rydb.**) **Blank.** Bessey Locoweed. [*Aragallus besseyi* Rydb.]. Caespitose, acaulescent, 8–28 cm tall; pubescence basifixed, silky pilose to subtomentose; stipules pilose-tomentose; leaves 6–20 cm long; leaflets 11–13 (25), 6–25 mm long, 2–6 mm wide, lanceolate to lance-oblong or elliptic, silky-pilose; scapes 6–19 cm long, subtomentose; racemes 8– to 22–flowered, the flowers spreading-ascending at maturity, the axis 2.5–8 cm long in fruit; bracts silky-pilose; flowers 18–25 mm long, brilliant pink purple; calyx 10–13 mm long, the tube swollen at anthesis, 7.5–9.5 mm long, the teeth 2.5–4 mm long, lance-subulate; pods sessile or nearly so, strongly inflated, the body ovoid, 5–8 mm thick, semibilocular, densely villous. Two varieties occur in Utah.

1. Scapes much surpassing even the longer leaves; leaflets mainly 5–9; pods 4–6 mm thick .... *O. besseyi* var. *ventosa*
— Scapes only somewhat surpassing the leaves; leaflets mainly 9–15 or more; pods 5–8 mm thick .... *O. besseyi* var. *obnapiformis*

Var. *obnapiformis* (C. L. Porter) Welsh [*O. obnapiformis* C. L. Porter; *O. nana* var. *obnapiformis* (C. L. Porter) Isely]. Pinyon-juniper community at 1765 to 2105 m in Daggett County; Wyoming and Colorado; 10 (i).
Var. *ventosa* (Greene) Barneby [*Aragallus ventosus* Greene; *O. nana* var. *ventosa* (Greene) Isely]. Sagebrush-grass and pinyon-juniper communities at 1750 to 1922 m in Daggett County; Wyoming; 4 (i).

Oxytropis campestris (**L.**) **DC.** Yellow Locoweed. [*Astragalus campestris* L.]. Caespitose, acaulescent, 4–18 cm tall; pubescence basifixed, mainly pilose; stipules glabrous or sparingly pilose; leaves 1.5–12 cm long; leaflets 7–17, 3–23 mm long, 1–6 mm wide, oblong to lanceolate or obovate; scapes 2.5–15 cm long, pilose to glabrate; flowers 14–20 mm long, ochroleucous; calyx 7–10 mm long, the tube cylindric, the teeth 1.5–3 mm long, triangular-subulate; pods 8–18 mm long, erect, sessile or subsessile, pilose, partially bilocular by intrusion of the ventral suture. Meadows and open woods at 2270 to 2600 m in Summit County (Goodmans Ranch, Bear River, E. & L. Payson 4868 POM); Oregon, Washington, Idaho, Montana, and Colorado. The range as given is for var. *cusickii* (**Greenman**) **Barneby** [*O. cusickii* Greenman]. The species proper is circumboreal.

Oxytropis deflexa (**Pallas**) **DC.** Stemmed Oxytrope. [*Astragalus deflexus* Pallas]. Shortly caulescent to subacaulescent, (5) 7–48 cm tall; pubescence basifixed, villous-pilose; leaves 2–20 cm long; leaflets 23–41, 3–24 mm long, 1–7 mm wide, lance-oblong to lanceolate, pilose on both surfaces, quite sessile; peduncles 3.5–32 cm long, villous-pilose; racemes 3– to 25–flowered, the flowers ascending to declined at anthesis, the axis 3.5–10 cm long in fruit; bracts pilose; flowers 5–10.5 mm long, whitish, lilac, or blue purple; calyx 4–8 mm long, the tube 2–3.5 (4.2) mm long, campanulate, the teeth 1.5–5 mm long, lance-subulate; pods spreading-declined, subsessile to shortly stipitate, the body oblong to ellipsoid, 8–18 mm long, 3–4.5 mm wide, subunilocular, pilosulous. Two rather distinctive varieties occur in Utah.

1. Plants subacaulescent; flowers 9–10.5 mm long, blue-purple; racemes 15–22 mm broad at anthesis; high elevations in the Uinta Mts. . . . . *O. deflexa* var. *deflexa*

— Plants commonly short caulescent; flowers 5–9 mm long, whitish, lilac, or blue purple; racemes commonly less than 17 mm wide at anthesis; moderate elevations, widespread . . . . *O. deflexa* var. *sericea*

Var. *deflexa* Alpine meadows at 2750 to 3350 m in Daggett, Duchesne, and Summit counties; Colorado and north to the Yukon; Asia; 11 (0). The North American material here assigned to var. *deflexa* might well represent a distinct taxon.

Var. *sericea* T. & G. Moist meadows, streambanks, and gravel bars in aspen, mixed conifer, and sagebrush communities at 2430 to 3050 m in Emery, Garfield, Grand, Iron, Sanpete, Sevier, Summit, Uintah, and Wayne counties; Alaska and Yukon south to California, New Mexico, and North Dakota; 21 (vi).

Oxytropis jonesii Barneby Jones Oxytrope. Acaulescent, densely pulvinate-caespitose; pubescence basifixed, villous-pilose; leaves 0.7–3 cm long; leaflets 1–7, 2–11 mm long, 1–3.5 mm wide, lanceolate to lance-oblong, pilose to strigose on both surfaces, quite sessile; scapes 0–3.5 cm long, villous-pilose; racemes 1- to 5-flowered, the flowers ascending at fruit; bracts pilose; flowers (11) 14–16.5 mm long, pink purple; calyx 7.5–9.5 mm long, the tube shortly cylindric, 4.5–7 mm long, the teeth 1.5–3 mm long, triangular to subulate; pods sessile or subsessile, erect, bladdery-inflated, 14–25 mm long, 8–13 mm wide (when pressed), semibilocular, villous-pilose. Ponderosa pine, western bristlecone pine, and mixed desert shrub communities on Flagstaff Limestone, Pink Limestone member of Wasatch Formation, and on Green River Shale, at 1930 to 2430 m in Emery, Garfield (type from Red Canyon), Grand, Iron, Sanpete, and Uintah counties; endemic; 27 (vi). The Jones oxytrope is probably better treated as a variety of *O. oreophila*, but no such combination is intended or implied herein.

Oxytropis lambertii Pursh Lambert Locoweed. [*O. lambertii* var. *bigelovii* Gray; *Aragallus bigelovii* (Gray) Greene; *Astragalus lambertii* var. *bigelovii* (Gray) Tidestr.; *Aragallus metcalfei* Greene; *A. knowltonii* Greene; *A. patens* Rydb.; *O. patens* (Rydb.) A. Nels.; *O. bilocularis* A. Nels.]. Caespitose, acaulescent, (10) 14–50 cm tall; pubescence malpighian, strigose; stipules pilose; leaves 3–24 cm long; leaflets 7–13, 7–45 mm long, 2–8 mm wide, lanceolate to oblong or linear; scapes 4–28 cm long, strigose; racemes 8- to 40-flowered, the flowers ascending at anthesis, the axis 3–23 cm long in fruit; bracts strigose; flowers 17–25 mm long, pink purple; calyx 6.5–10 mm long, the cylindric tube 4.5–7.5 mm long, the teeth 1.5–4.5 mm long, subulate; pods sessile or shortly stipitate, erect or ascending at maturity, cylindroid to lance-acuminate in outline, the body 15–27 mm long, 2.5–6 mm thick, bilocular, strigose to strigulose. Mixed desert shrub, pinyon-juniper, sagebrush, and grass communities at 1230 to 2730 m in Carbon, Duchesne, Emery, Garfield, Kane, San Juan, Sanpete, Sevier, Wasatch, and Wayne counties; Saskatchewan and Manitoba to Arizona and New Mexico; 75 (xv). Our material belongs to **var. *bigelovii*** Gray. In places where *O. lambertii* contacts *O. sericea*, hybrid swarms of great variability and beauty occur. No such extensive hybrid populations are known for Utah, possibly because the points of contact between the species are few or non-existent.

Oxytropis multiceps T. & G. Rocky Mountain Oxytrope. [*Spiesia multiceps* (T. & G) Kuntze; *Aragallus multiceps* (T. & G.) Heller; *Astragalus bisontum* Tidestr.; *O. multiceps* var. *minor* Gray; *Aragallus multiceps* var. *minor* (Gray) A. Nels.; *A. minor* (Gray) Cockerell ex Daniels; *O. minor* (Gray) Cockerell; *Astragalus bisontum* var. *minor* (Gray) Tidestr.]. Caespitosae, acaulescent, pulvinate; pubescence basifixed, silky pilose; stipules amplexicaul but distinct, silky-pilose; leaves 1–5 cm long; leaflets 5–9, 3–13 mm long, 1–4 mm wide, lanceolate to elliptic, oblong, or oblanceolate, silky-pilose; scapes 1–4 cm long, long-villous; racemes 1- to 4-flowered, the flowers ascending-spreading at anthesis, the axis to 0.5–1 cm long in fruit; bracts thinly pilose; flowers 17–24 mm long, bright pink purple; calyx 7–13 (20) mm long, the tube swollen at anthesis, becoming bladdery-inflated and investing the fruit at maturity, 5.5–10 mm long (8–18 mm in fruit), the teeth 2–3 mm long, triangular-subulate; pods included within the swollen calyx, stipitate, the stipe 0.5–1.5 mm long, the ovoid-ellipsoid body 6–10 mm long, 3–5 mm wide, subunilocular, short-villous. Pinyon-juniper and ponderosa pine communities at 1830 to 2270 m in Daggett and Duchesne counties; Wyoming, Colorado, and Nebraska; 5 (0).

Oxytropis oreophila Gray Mountain Oxytrope. Caespitose, acaulescent, often densely pulvinate, 1–14 (23) cm tall; pubescence basifixed, silky-pilose; stipules silky-pilose; leaves 0.5–8.5 cm long; leaflets (5) 7–17, 1–15 mm long, 0.5–4 mm wide, lanceolate, elliptic, oval, or ovate, pilose; scapes 0–12 (21) cm long, pilose to hirsute; racemes (1) 2- to 12-flowered, the flowers ascending-spreading at anthesis, the axis to 1 cm long in fruit; bracts pilose; flowers 6–12.5 mm long, pink to pink purple or white; calyx 4.5–8 mm long, the tube 3.2–5.5 mm long, campanulate to short-cylindric, the teeth 1.3–2.5 mm long, subulate; pods sessile, bladdery-inflated (7) 9–17 mm long, 5–14 mm wide (when pressed), subunilocular, hirsutulous to villous. Two varieties are present.

1. Plants pulvinate caespitose; leaves 0.5–2 cm long, the leaflets 7–11; pods 10–12 mm long, usually less than 6 mm wide; flowers mostly less than 10 mm long . . . . *O. oreophila* var. *juniperina*

— Plants densely to loosely caespitose; leaves often more than 2 cm long, the leaflets 9–15 (17); pods 9–17 mm long, more than 6 mm wide; flowers commmonly 10–12.5 mm long . . . . *O. oreophila* var. *oreophila*

Var. *juniperina* Welsh Pinyon-juniper, mountain brush, fringed sagebrush, ponderosa pine, and bristlecone pine communities at 1891 to 2565 m in Beaver, Garfield, Iron, Piute, Sevier, and Wayne (type from 1 mi E of Bicknell) counties; Nevada; 25 (vii). This is the compact, xeric phase of the species, that often shares similar habitats with *O. jonesii* (q.v.).

Var. *oreophila* [*Spiesia oreophila* (Gray) Kuntze; *Aragallus oreophilus* (Gray) A. Nels.; *Astragalus oreophilus* (Gray) Tidestr.; *A. munzii* Wheeler]. Alpine tundra, ridge tops, meadows, and spruce-fir communities at 2255 to 3785 m in Beaver, Garfield, Iron, Kane, Piute, Washington, and Wayne counties; California, Nevada, Arizona; 68 (xiv).

Oxytropis parryi Gray Parry Oxytrope. [*Spiesia parryi* (Gray) Kuntze; *Aragallus parryi* (Gray) Greene; *Astragalus parryanus* Tidestr., not *A. parryi* Gray]. Caespitose, acaulescent, 2–11 cm tall; pubescence basifixed, silky-pilose; leaves 1.5–7 cm long, leaflets 7–17, 2–9 (12) mm long, 0.8–3 mm wide, oblong to elliptic or lanceolate, pilose; scapes 1.2–8 (10) cm long, pilose; racemes 1– to 3 (4) -flowered, the flowers erect or ascending, the axis 0.5–1 cm long in fruit; bracts pilose; flowers 7.5–12 mm long, pink purple; calyx 5–8 mm long, the tube 3–5.5 mm long, campanulate to short-cylindric, the teeth 1.5–2.5 mm long, triangular-subulate; pods erect, sessile, oblong to lance-ovoid or ovoid, 13–22 mm long, 4–8 mm thick, bilocular or nearly so, pilosulous. Alpine tundra, ridge tops, and meadows, at 2700 to 3785 m in Beaver, Carbon, Duchesne, Emery, Garfield, Grand, Juab, Piute, San Juan, Summit, Tooele, Wasatch and Wayne counties; Idaho, Wyoming, California, Colorado, and New Mexico; 38 (viii).

Oxytropis sericea Nutt. in T. & G. Silky or White Locoweed. [*O. lambertii* var. *sericea* (Nutt.) Gray; *Spiesia lambertii* var. *sericea* (Nutt.) Rydb.; *Aragallus lambertii* var. *sericeus* (Nutt.) A. Nels.; *Aragallus majusculus* Greene, type from Mt. Ellen, Henry Mts.]. Caespitose, acaulescent, 13–32 cm tall, pubescence basifixed, silky-pilose; stipules pilose to subtomentose; leaves 3.5–21 cm long; leaflets 9–23, 4–32 (40) mm long, 1.5–10 mm wide, lanceolate to oblong, elliptic, or ovate, pilose; scapes 7–26 cm long, pilose; racemes 6– to 27–flowered, the flowers ascending to spreading, the axis 1.5–12 cm long in fruit; bracts pilose; flowers 15–26 mm long, white or tinged with purple; calyx 8–12 mm long, the tube cylindric, the teeth triangular to subulate; pods erect, sessile, the body subcylindric to ovoid-oblong, 10–25 mm long, 4–7.5 mm thick, bilocular or nearly so, strigose or pilosulous. Sagebrush, pinyon-juniper, and grassland (rarely mixed desert shrub) communities at 1670 to 3350 m in Box Elder, Carbon, Daggett, Duchesne, Emery, Garfield, Kane, Piute, Summit, and Wayne counties; Yukon south to Nevada, New Mexico, and Oklahoma. Our material belongs to **var. *sericea*;** 58 (x).

Oxytropis viscida Nutt. Viscid Locoweed. [*O. campestris* var. *viscida* (Nutt.) Wats.; *Spiesia viscida* (Nutt.) Tidestr.]. Caespitose, acaulescent, 4–18 cm tall; pubescence basifixed, spreading-hairy; stipules glabrous dorsally, commonly prominently glandular; leaves 2–17 cm long; leaflets 19–39 or more, 1.5–20 mm long, 1–5 mm wide, oblong to lanceolate or elliptic, glabrate to glabrous on both sides, sometimes glandular; scapes 2–12.5 cm long, spreading-hairy; racemes 3– to 20-flowered, the flowers spreading-ascending, the axis 2–7 cm long in fruit; bracts glandular; flowers 11–19 mm long, whitish, lilac, or pink purple; calyx 5–11 mm long, the shortly cylindric tube 4–7 mm long the teeth 1.5–3.5 mm long, trianglar-subulate, commonly glandular; pods erect, sessile, ovoid to subcylindric, 8–20 mm long, 4–5 mm thick, bilocular, glandular. Pinyon-juniper, grassland, and spruce-fir communities at 2430 to 3355 m in Emery, Salt Lake, Sanpete, Sevier, Utah, and Wayne counties; Alaska east to Gaspe and south to California, Nevada, Colorado, and Minnesota; 16 (iv). Our material belongs to **var. *viscida*.** This is a portion of a circumboreal complex with great variation. Utah materials demonstrate two of the extreme phases of that variety. Naming of the phases within the species would lead to an endless entaglement. Thus, the whitish flowered plants from Emery, Sanpete, and Sevier counties are considered as a minor element in this evidently polygenetic and polymorphic series.

Parkinsonia L.

Trees, armed; leaves alternate, pinnately twice-compound, subsessile; leaflets numerous; flowers in short racemes or some solitary in leaf axils, only slightly irregular; sepals 5, distinct; petals 5, subequal, the uppermost enclosed by the others in bud; stamens 10, distinct, hairy at the base; ovary several-ovuled, shortly stipitate.

Parkinsonia aculeata L. Retama. Small, open trees, mainly 5–8 m tall; branchlets green, armed with needle-sharp, somewhat recurved spines representing modified leaf rachises; leaves with 1 or 2 pair of pinnae, these mainly 1–4 (5) dm long, each with a flattened rachis; leaflets very numerous, 2–4 mm long and 0.5–1 mm wide; flowers ca 1 cm wide, yellow; legumes constricted between the seeds, 3–10 cm long. Cultivated ornamental of great beauty, escaping and persisting in Washington County; widespread in the southern U. S.; 1 (0).

Parryella T. & G.

Unarmed shrubs; leaves alternate, pinnate, with numerous leaflets, glandular-dotted; stipules subulate, caducous; peduncles opposite the leaves; flowers in loose spicate racemes or panicles; calyx turbinate-campanulate, 10-ribbed near the base, 5-lobed; petals lacking; stamens 10, the filaments distinct, inserted on the hypanthium; pods 1-seeded (2-ovuled), indehiscent, obliquely ovoid, glandular-dotted, exserted from the calyx.

Rydberg, P. A. 1919. *Parryella*. N. Amer. Fl. 24: 25–26.

Parryella filifolia T. & G. ex Gray Narrow-leaf Dunebroom. Shrubs to 15 dm tall or more, often partially buried in sand, the branchlets strigose and with mammiform, glandular protuberances; stipules 1–2 mm long, chestnut brown, fragile; leaves 3.5–13 cm long; leaflets 8–40, 10–21 mm long, linear, all involute, strigose to glabrate and glandular on the visible surface; peduncles 0–1 cm long; racemes 4–10 cm long, the main ones often branched near the base, forming panicles, 35– to 90-flowered, the axis not much elongating in fruit; bracts reduced to gland-tipped vestiges; calyx 2.5–3.1 mm long, 10-ribbed near the base, the tube opaque, the teeth 0.2–0.5 mm long, ciliate; flowers 5–6.5 mm long in anthesis, the stamens long-exserted; pods short-stipitate, the body 5–6.5 mm long, 2.5–3 mm wide, the pilose style base persistent, glabrous, glandular-punctate; n = 10. Stabilized dune sands at 1470 to 1570 m in Grand and San Juan counties; New Mexico and Arizona; 10 (vi). This plant shows promise in reclamation of sandy sites.

Pediomelum Rydb.

Perennial herbs, unarmed, from deep-seated, tuberous roots; leaves alternate, palmately 3– to 7-foliolate, glandular-dotted; stipules triangular to subulate; flowers borne in axillary racemes or spikelike racemes; calyx subcylindric, gibbous at the base on the upper side, 5-lobed, the lowest lobe the longest; stamens 10, diadelphous (rarely all connate); petals blue to purple, lavender, or white; pods 1-seeded, irregularly dehiscent, usually included within the calyx. Note: The name is from the

Greek words for prairie and apple, a literal translation of the French "Pomme de Prairie," a reference to the edible, tuberous roots of most species. The species have been traditionally assigned within an expanded *Psoralea* L., a genus now restricted to South African plants (see also *Psoralidium*).

Rydberg, P. A. 1919. *Pediomelum*. N. Amer. Fl. 24(1): 17–24.
Ockendon, D. J. 1965. A taxonomic study of *Psoralea* subgenus Pediomelum (Leguminosae). Southwestern Nat. 10: 81–124.

1. Leaflets conspicuously bicolored, cinereus beneath, bright green and glabrous to glabrate above (except along some veins); plants known only from southern Kane County and adjacent Arizona .... *P. epipsilum*
— Leaflets not bicolored, or if green above, the surface strigose to pilose also 2
2(1). Plants definitely caulescent; leaflets green on both sides; flowers mainly less than 12 mm long; plants of Grand and San Juan County .... *P. aromaticum*
— Plants acaulescent or short-caulescent; leaflets bicolored; flowers mostly more than 12 mm long; plants variously distributed 3
3(1). Leaflets commonly 3, strongly white strigose along the veins above; plants of the Paunsaugunt and Paria regions .... *P. pariense*
— Leaflets commonly 5, uniformly strigose or pilose above, or thinly strigose along the veins; plants variously distributed 4
4(3). Calyx lobes very unequal, the lower one much enlarged; seeds reticulate; plant known from near the Utah-Arizona boundary in Arizona ....
 .... *P. castoreum* (Wats.) Rydb.
— Calyx lobes subequal; seeds smooth; plants variously distributed 5
5(4). Petioles and peduncles pubescent with appressed or ascending hairs; plants widespread in eastern Utah ..
 .... *P. megalanthum*
— Petioles and peduncles retrorsely hairy; plants of Washington County .... *P. mephiticum*

Pediomelum aromaticum (Payson) Welsh Paradox Breadroot. [*Psoralea aromatica* Payson; *P. rafaelensis* Jones, type from the La Sal Mts.; *P. rafaelensis* var. *magna* Jones, type from the San Rafael Swell (?)]. Caulescent, 8–15 (20) cm tall, from slender, subterranean caudex branches arising from deep-seated tuberous roots; stems with 2–4 (5) elongated internodes, strigose to strigulose; leaves mainly 5-foliolate; petioles 1.2–8 cm long, pubescent like the stems; leaflets 6–26 mm long, 3–16 mm wide, cuneate-obovate, gray green, strigulose, and punctate beneath, yellow green, punctate, and strigose overall or only along the veins above; stipules scarious, 2–9 mm long; peduncles 0.2–0.5 cm long; racemes 3- to 7-flowered, 1–2 cm long; pedicels 1–2.5 mm long; bracts lanceolate, 4–7 mm long; flowers 10–13 mm long, the banner, wings, and keel more or less suffused with purple; calyx 10–11 mm long, the tube 3–4 mm long, strongly gibbous, the lower tooth 5–7 mm long, about twice as broad as the lower lateral ones; pods to 15 mm long. Pinyon-juniper and mixed desert shrub communities at ca 1530 m in Emery (?), Grand, and San Juan counties; Paradox Basin, Colorado; a Navajo Basin endemic; 10 (vi). Material from San Juan County has smaller flowers than noted for the species generally and is designated as **var. tuhyi** Welsh.

Pediomelum epipsilum (Barneby) Welsh Kane Breadroot. [*Psoralea epipsila* Barneby, type from 17 mi E of Kanab]. Subacaulescent or short-caulescent, 3.5–10.5 cm tall, from slender, subterranean caudex branches arising from deep-seated tuberous roots; stems with 2 or 3 elongated internodes, or none, strigose to ascending hairy; leaves mainly 5-foliolate; petioles 0.8–5 cm long, pubescent like the stems; leaflets 5, 8–25 mm long, 3–15 mm wide, obovate, grayish, strigulose, and punctate beneath, bright yellow green, punctate and glabrous to thinly strigose above (especially along the veins); peduncles to 5 cm long; racemes 2–4 cm long; pedicels ca 3 mm long; bracts broadly lanceolate, 10–13 mm long; flowers 11–14 mm long, the banner, wings, and keel pale violet; calyx 11–14 mm long, the tube 5–6 mm long, strongly gibbous, the lower tooth ca 8 mm long, the upper ones shorter and narrower; pods 1-seeded. Pinyon-juniper woodland at ca 1670 m on Chinle and Moenkopi formations in Kane County, and in adjacent Arizona; 4 (0).

Pediomelum megalanthum (Woot. & Standl.) Rydb. Large-flowered Breadroot. [*Psoralea megalantha* Woot. & Standl.]. Subacaulescent to caulescent, 4–25 cm tall, from slender, subterranean caudex branches arising from deep-seated tuberous roots; stems with 0–5 elongated internodes, incurved-strigose or with a few (rarely most) hairs spreading-ascending; leaves mainly 5 (8) -foliolate; petioles 1.2–9.5 cm long, pubescent like the stems or more commonly spreading hairy; leaflets 9–34 mm long, 4–23 mm wide, cuneate-obovate to subrhombic, gray green, strigulose, and punctate beneath, yellow green, punctate, and strigose overall above; stipules scarious 2–15 mm long; peduncles 1.4–5 cm long; racemes mainly 6- to 24-flowered, 2–5 cm long; pedicels 1.5–5 mm long; bracts commonly bidentate, lance-ovate, 3–12 mm long; flowers 12.5–21 mm long, the banner, wings, and keel commonly purple or suffused with purple; calyx 12.5–18.5 mm long, the tube 5.5–8 (9) mm long, only somewhat broader than the lateral ones; pods included in the calyx. Mixed desert shrub, juniper-pinyon woodland, and blackbrush communities at 1430 to 1830 m in Duchesne, Grand, San Juan, Uintah, and Wayne counties; Colorado and New Mexico; 54 (xvii).

Pediomelum mephiticum (Wats.) Rydb. Skunk Breadroot. [*Psoralea mephitica* Wats., type from near St. George; *Pediomelum retrorsum* Rydb; *Psoralea retrorsa* (Rydb.) Tidestr. in Tidestr. & Kittell; *P. mephitica* var. *retrorsa* (Rydb.) Kearney & Peebles]. Aaculescent to subacaulescent, 4.5–15 cm tall, from slender, subterranean caudex branches arising from deep-seated tuberous roots; stems lacking or with very short internodes above ground, retrorsely hairy; leaves mainly 5-foliolate; petioles 3.2–12 cm long, retrorsely hairy; leaflets 11–35 mm long, 4–28 mm wide, obovate to broadly so, gray green, punctate beneath, green to yellow green, strigose overall (more densely along veins), and punctate above; stipules scarious, 4–12 mm long; peduncles 2–6 cm long; racemes mainly 12- to 35-flowered, 1.5–4.5 cm long; pedicels 1.5–3.5 mm long; bracts mainly not bidentate, elliptic-obovate, 5–12 mm long; flowers 10.5–12.8 mm long, the banner whitish or yellowish, the wings and keel purple or suffused with purple; calyx 9–12.5 mm long, the tube 3.5–4.5 mm long, strongly gibbous at the base, the lower tooth 4.5–9 mm long, ca twice as broad as the others; pods included in the calyx. Blackbrush and pinyon-juniper communities at 1470 to 1700 m in Washington County; Arizona, Nevada, and California; 17 (iii). This species was described by Sereno Watson simultaneously with *Psoralea castorea* (Proc. Amer. Acad. 14: 291. 1879). The type locality of both of these taxa is cited in the original descriptions as "near Beaver City, S. Utah." The types were both taken by Dr. E. Palmer (No. 96 for *P. castorea*, and No. 97 for *P. mephitica*). Palmer (Amer. Nat. 12: 601. 1878) in describing the use of these plants by Indians (as food) noted that *P. castorea* Wats. "grows in exposed sandy localities between Beaver Dams, Arizona, and St. Thomas, Nevada" [sic]. Of *P. mephitica*, Palmer (l.c.) notes "it is abundant on the lower places between the hills southeast of St. George, Southern Utah, and the Pah-Utes resort there to collect its roots" [sic]. Thus, the type localities can be inferred as being in northwestern Arizona or southeastern Nevada for *P. castorea*, and in the hills to the southeast of St. George for *P. mephitica*. There is no evidence to indicate that *P. castorea* was ever collected in Utah, but it is known from a site ca 1.5 km south of the boundary along Beaverdam Wash.

Pediomelum pariense (Welsh & Atwood) Welsh Paria Breadroot. [*Psoralea pariensis* Welsh & Atwood, type from Bryce Canyon]. Acaulescent or subacaulescent, 2–9 cm tall, from slender, subterranean caudex branches arising from deep-seated tuberous roots; stems lacking or with very short internodes above ground, strigose; leaves mainly 3-foliolate; petioles 1.3–7 cm long, strigose; leaflets 9–25 mm long, 7–22 mm wide, obovate or orbicular, cuneate, rounded to truncate or emarginate apically, gray green, glandular, and strigose beneath, yellow green, glandular, and strongly strigose along veins above; stipules scarious, 4–10 mm long; peduncles 0.5–2.8 cm long, the hairs appressed to ascending; racemes mainly 6- to 15-flowered, 1–2.5 cm long; pedicels 1–3.8 mm long; bracts abruptly acuminate, 4–8 mm long; flowers 8.8–12.5 cm long, the banner cream to ochroleucous, the wings and keel purple or suffused with purple; calyx 8.6–11.4 mm long, the tube 3.3–4.6 mm long, more or less gibbous at the base, the lower calyx teeth 5.3–6.8 mm long, almost twice broader than the others; pods included in the calyx, ca 9 mm long. Ponderosa pine and juniper pinyon woods at 1700 to 2430 m in Garfield (type from Bryce Canyon) and Kane counties; endemic; 8 (ii).

Peteria Gray

Perennial, caulescent, from a subterranean caudex, arising from deep-seated tuberous roots; leaves alternate, odd-pinnate; stipules only slightly adnate to the petiole base, spinescent; flowers papilionaceous, borne in terminal racemes, each subtended by a bract, or the lowermost flower sometimes axillary; bracteoles 0; calyx 5-toothed; petals 5, whitish to ochroleucous, rarely pinkish; stamens 10, diadelphous; style hairy at apex; pods narrowly oblong, straight, few- to several-seeded, laterally compressed, dehiscent, the valves coiling upon dehiscence.

Peteria thompsonae Wats. Thompson Peteria. [*P. nevadensis* Tidestr.]. Perennial, 11–48 cm tall; pubescence of basifixed, flattened, appressed hairs; stems erect to sprawling; stipular spines 1.8–6.5 mm long; leaves 4–18 cm long; leaflets 9–27, 6–17 mm long, 3–11 mm wide, elliptic to ovate, oblong, or oval, strigose on both sides or glabrate above; peduncles 0–4 cm long, racemes 5- to 44-flowered, the flowers ascending at anthesis, the axis 6–18 cm long in fruit; bracts 5–9 mm long, stipitate-glandular; pedicels 1–5 mm long, strigulose and stipitate-glandular, the teeth 3.8–9 mm long, lance-subulate; flowers 18–22 mm long, whitish to ochroleucous, rarely pinkish; pods descending, stipitate, the stipe to 11 mm long, the body 48–55 (70) mm long, 5–7 mm wide, glabrous, more or less constricted between the seeds. Pinyon-juniper and mixed desert shrub communities at 1230 to 1870 m in Emery, Grand, Juab, Kane, San Juan, and Washington counties; Arizona, Nevada, and Idaho; 27 (x). The type specimen was collected by Mrs. Ellen Thompson, sister of John Wesley Powell, at Kanab, where she lived in 1872. This is a singular and striking plant.

Phaseolus L.

Annual herbs, from a taproot; leaves alternate, pinnately trifoliolate; stipules herbaceous, distinct; flowers papilionaceous, axillary or in axillary racemes, each subtended by a bract; bracteoles 2, attached at base of calyx; calyx 5-toothed; petals 5, pink to purplish or white; stamens 10, diadelphous; ovary few- to several-ovuled, the style twisted or coiled in the keel, bearded toward the apex, the stigma oblique; pods linear to oblong, laterally flattened to subterete, the valves coiling upon dehiscence.

Phaseolus vulgaris L. Kidney Bean. Annual, 3–20 dm tall or more, clump-forming or twining and vinelike; pubescence basifixed, villosulous or finally glabrate; petioles 4–20 cm long; leaflets 3, stipellate, commonly 4–10 cm long and 2–8 cm wide, ovate to lanceolate, pilosulous to villosulous on both sides, mainly along the veins; peduncles 0–2 cm long; racemes few-flowered; flowers 12–16 mm long; bracteoles ovate-lanceolate, prominently veined; pods slender, subterete to flattened, most 6–15 cm long. This is the table bean of commerce. It is cultivated widely, escaping commonly, but hardly persisting; native of the New World (?); 2 (i). Additional species are present in the cultivated flora of Utah, but the extent is unknown, and they are here excluded. Among them are the scarlet runner bean (*P. coccineus* L.) and the lima bean (*P. limensis* Macfady.).

Pisum L.

Plants herbaceous, annual, from a taproot; stipules prominent, larger than the leaflets, semisagittate to ovate or reniform; stems clambering, not winged; leaves alternate, even-pinnate, the terminal extension of the rachis forming prehensile tendrils; racemes axillary, pedunculate, 1- to few-flowered; stamens 10, diadelphous; style laterally compressed, bearded along the ventral edge; pods 1-loculed, oblong in outline, several-seeded, the valves coiling upon dehiscence.

Pisum sativum L. Garden Pea. Plants 2–20 dm long or more, sprawling or clambering, the stems merely angled, glabrous; stipules foliaceous, larger than the leaflets; leaflets 4–6, elliptic to ovate or oblong-lanceolate, 9–60 mm long, 6–40 mm wide, glabrous; tendrils with 1–3 pairs of lateral branches, prehensile; flowers 1–3 per raceme, white, red, or bicolored, 18–30 mm long; calyx 12–18 mm long, the teeth longer than the tube; pods commonly 4–10 cm long, glabrous; $2n = 14, 15, 28, 56, 84, 98, 112, 224$. Cultivated pea of commerce, widely grown, escaping but not persisting; introduced from Eurasia; 4 (i). Several horticultural forms are grown, most of them with white flowers.

Prosopis L.

Armed shrubs or small trees; leaves alternate, bipinnate, with an obscure gland between the lower pair of pinnae; leaflets several to numerous; stipules small or modified as spines; flowers perfect, borne in spikelike racemes, yellowish to ocroleucous; calyx 5-toothed; corolla regular, the 5 petals distinct or nearly so; stamens 10, distinct, exserted, the anthers terminally glandular; ovary pubescent, sessile or stipitate, the stigma concave; pods indehiscent, narrowly oblong and more or less constricted between the seeds or spirally coiled.

Johnston, M. C. 1962. The North American Mesquites, *Prosopis* Sect. Algarobia (Leguminosae). Brittonia 4: 72–90.
Isely, D. 1973. Prosopis. Mem. New York Bot. Gard. 25(1): 116–122.

1. Leaflets 5–8 pairs, mostly less than 10 mm long; spines stipular, paired; pods coiled spirally into a woody cylinder .... *P. pubescens*
— Leaflets 10–18 pairs or more, often exceeding 10 mm in length; spines axillary, solitary or paired; pods narrowly oblong, laterally compressed, straight or nearly so .... *P. glandulosa*

Prosopis glandulosa Torr. Honey Mesquite. [*Algarobia glandulosa* (Torr.) T. & G.; *P. juliflora* var. *glandulosa* (Torr.) Cockerell; *Neltuma glandulosa* (Torr) Britt. & Rose; *P. odorata* Torr. & Frem. in Frem., as to flowering portions]. Armed shrubs or broad-crowned trees, commonly 3–5 m tall and as broad or more; leaves petiolate, the rachis produced as a spine; pinnae 2; leaflets 7–17 pairs, oblong to narrowly oblong, 7–38 mm long, 1–4 mm wide, glabrous on both sides, sometimes ciliate; spines nodal but not stipular, single or paired, 3–35 mm long or more; stipules inconspicuous, subulate; flowers in yellowish to greenish spikes, clustered from spur branches, ascending to declined; ovary pilose; pods stipitate, the stipe 5–8 mm long, the body narrowly oblong to linear, often curved, subterete to somewhat flattened, mostly 10 to 20 cm long, a bony endocarp around each seed; $2n = 26, 52, 56, 112$. Terraces and bars, or in creosote bush communities at 670 to 1170 m in Washington County; California, Nevada, Arizona, New Mexico, Kansas, Oklahoma, Texas, and southward; 19 (i). Our material belongs to var. ***torreyana*** (L. Benson) M. C. Johnston [*P. juliflora* var. *torreyana* L. Benson].

Prosopis pubescens Benth. Screwbean. [*P. odorata* Torr. & Frem. in Frem., as to fruit; *Strombocarpa pubescens* (Benth.) Gray; *S. odorata* (Torr. & Frem.) Gray]. Armed shrubs or small trees with slender branches, commonly 2.5–3.5 m tall; leaves shortly petiolate, the rachis produced as a spine; pinnae 2 (rarely 4); leaflets 5–8 pairs, elliptic to oblong, puberulent to glabrate; spines paired at nodes, apparently stipular, mostly 5–25 mm long; flowers in clusters or solitary, yellowish spikes; ovary villous; pods tightly coiled into a woody cylinder, 30–50 mm long, 4–6 mm thick; $2n = 56$. Benchlands, slopes, and terraces along drainages at 730 to 900 m in Washington County; California, Nevada, Arizona, New Mexico, Texas, and Mexico; 10 (i).

Psoralidium Rydb.

Perennial herbs, unarmed, from rhizomes; leaves alternate, pedately 3-foliolate, glandular-dotted; stipules triangular to subulate; flowers in axillary, pedunculate, interrupted spikelike racemes or racemes; calyx campanulate, the tube short, 5-lobed, the lowermost lobe typically longer than the others; corolla papilionaceous; stamens 10, diadelphous; petals bluish or purplish to lavender, or less commonly white; pods 1–seeded, indehiscent, usually not included in the calyx.

Rydberg, P. A. 1919. *Psoralidium*. N. Amer. Fl. 24: 12–17.

1. Peduncles typically 15–40 cm long or more; leaves few, mostly deciduous by anthesis, the leaflets sharply acuminate; plants of southeastern Utah ...... *P. junceum*
— Peduncles mainly shorter than 15 cm; leaves numerous, persisting through the season, the leaflets obtuse to rounded or cuspidate; distribution various 2
2(1). Plants commonly 4–10 dm tall, at least some leaves 4- to 5-foliolate; flowers mainly indigo; plants of Garfield, Iron, Kane, and Washington counties ..... *P. tenuiflorum*
— Plants often less than 4 dm tall, all leaves commonly 3-foliolate; flowers white to purple; plants widespread in Utah .... *P. lanceolatum*

Psoralidium junceum (Eastw.) Rydb. Rush Scurfpea. [*Psoralea juncea* Eastw., type from Epsom Creek, San Juan County]. Caulescent, 4.8–9 dm tall, from a rhizome; stems with 5 or more elongated internodes, strigose; leaves commonly 3-foliolate, often deciduous by anthesis; petioles 1.4–7 cm long, pubescent like the stems; leaflets 3, 19–44 mm long, 3–7 mm wide, oblanceolate to elliptic, acuminate apically, strigose and glandular on both surfaces; stipules acuminate, strigose; peduncles (8) 11–48 cm long; racemes 7- to 20-flowered, 5–11 cm long; bracts lance-acuminate, glabrous dorsally, 1.5–2.5 mm long, deciduous; flowers 4.2–5.8 mm long, the petals indigo; calyx 2.7–3.4 mm long, the tube 1.3–2.5 mm long, campanulate, not especially gibbous, the lower tooth 0.7–1.4 mm long, longer than the others; pods 1–seeded, densely silky-villous. Stabilized dunes and other sandy sites in Garfield, Kane and San Juan counties; Coconino County, Arizona; a Glen Canyon-San Juan endemic; 23 (vi).

Psoralidium lanceolatum (Pursh) Rydb. Dune Scurfpea. Caulescent, 1.5–6.8 dm tall, from a rhizome, clump-forming, the stems with 5 or more elongated internodes, glabrous or strigose; leaves persistent at flowering; petioles 0.8–3 cm long, strigose to glabrate; leaflets 3, 14–50 mm long, 0.5–9 mm wide, oblanceolate below, becoming linear upward, obtuse to acute or cuspidate, sparingly strigose on both sides or glabrous above, yellow green; stipules lance-attenuate, 3–16 mm long; peduncles 3.3–24 cm long; racemes 5- to 41-flowered, 1.2–17 cm long; bracts ovate to elliptic or lanceolate, glabrous dorsally, 1.3–2.8 mm long, persistent; flowers 4.8–6.3 mm long, blue, white, or bicolored; calyx 1.6–2.8 mm long, the tube 1.3–2 mm long, campanulate, not especially gibbous, the lower tooth 0.2–0.8 mm long, not much larger than the others; pods 1–seeded, conspicuously glandular. Three varieties are recognized.

1. Racemes lax, 2.5–17 cm long at anthesis, the flower nodes widely separated, often with 2 or 3 flowers per node; blue flower color predominating; plants of southeastern Utah .... *P. lanceolatum* var. *stenophyllum*
— Racemes compact to moderately lax, 1.2–5.5 cm long at anthesis, the flower nodes seldom widely spaced or mainly with more than 3 flowers per node; blue and white flowers variably abundant; plants of various distribution 2

2(1). Racemes (1.2) 2.7–5.5 cm long, with (17) 27–41 flowers; plants of the Great Basin ....
.... *P. lanceolatum* var. *stenostachys*
— Racemes 1–2.5 (2.8) cm long, with 5–24 flowers; plants of the Colorado Basin ... *P. lanceolatum* var. *lanceolatum*

Var. *lanceolatum* Plains Scurfpea. [*Psoralea lanceolata* Pursh; *P. elliptica* Pursh; *Lotodes ellipticum* (Pursh) Kuntze; *P. laxiflora* Nutt. ex T. & G.; *P. micrantha* Gray ex Torr.; *P. micranthum* (Gray) Rydb.]. Sand dunes and other sandy sites at 1230 to 1770 m in Daggett, Garfield, Grand, Kane, Uintah, Washington, and Wayne counties; Washington and Saskatchewan, south to California, Arizona, New Mexico, and Oklahoma; 48 (ix).

Var. *stenophyllum* (**Rydb.**) **Welsh** [*Psoralea stenophylla* Rydb., type from Wilson Mesa, Grand County; *Psoralidium stenophyllum* (Rydb.) Rydb.; *Psoralea lanceolata* var. *stenophyllum* (Rydb.) Toft & Welsh]. Sandy sites at 1270 to 1830 m in Emery, Garfield,, Grand, Kane, San Juan, and Wayne counties; endemic; 38 (xxii).

Var. *stenostachys* (**Rydb.**) **Welsh** [*Psoralea stenostachys* Rydb., type from Government Well, Tooele County; *Psoralidium stenostachys* (Rydb.) Rydb.; *Psoralea lanceolata* var. *stenostachys* (Rydb.) Toft & Welsh]. Dunes and other sandy sites at 1340 to 1590 m in Juab, Millard, Salt Lake, Tooele (type from Government Well) and Weber counties; endemic; 33 (ii). Specimens from Salt Lake and Weber counties are placed here tentatively because of the nature of the existing materials. The plants have not been collected in those counties in recent times, and they might well be extinct. The question of their varietal status might therefore be moot.

Psoralidium tenuiflorum (**Pursh**) **Rydb.** Prairie Scurfpea. [*Psoralea tenuiflora* Pursh; *P. obtusiloba* T. & G.; *Psoralidium bigelovii* Rydb.; *Psoralea bigelovii* (Rydb.) Tidestr.; *P. tenuiflora* var. *bigelovii* (Rydb.) Macbr.; *P. floribunda* T. & G.; *Psoralidium floribundum* (Nutt.) Rydb.]. Caulescent, 4–10 dm tall or more, forming large clumps; stems with 8 or more elongated internodes, strigose; leaves 3– to 5–foliolate, persistent at flowering; petioles 0.1–2.2 cm long, strigose to strigulose; leaflets 3–5, 6–40 mm long, 2–16 mm wide, oblanceolate throughout, mainly rounded to a cuspidate apex, gray green and strigose beneath, yellow green and glabrous or pubescent only along the veins above; stipules scarious, 2–13 mm long, strigose; peduncles 0.5–7 cm long or lacking, sometimes bracteate or some flowers axillary; racemes mainly 7– to 21–flowered, 1–5.9 cm long; bracts ovate-acuminate, glabrous dorsally, 1.5–2.5 mm long, persistent; flowers 4.5–6 mm long, the petals indigo; calyx 2.5–3.2 mm long, the tube 1.5–2 mm long, campanulate, not gibbous, the lower tooth 0.8–1.7 mm long, noticeably larger than the others; pods 1–seeded, glabrous, conspicuously glandular. Pinyon-juniper, sagebrush, and mountain brush communities at 1700 to 2200 m in Garfield, Iron, Kane, and Washington counties; North Dakota and Montana, south to Arizona, Texas, and Mexico; 17 (iv). The segregation of varieties from among our materials seems unwarranted.

Psorothamnus Rydb.

Shrubs, typically armed; leaves alternate, odd-pinnate, with 5 or more leaflets, glandular-dotted; stipules subulate or vestigial; peduncles opposite the leaves; flowers in lax racemes; calyx campanulate, 10–ribbed, 5–lobed; corolla papilionaceous, the petals all inserted on the hypanthium; stamens 9 or 10, monadelphous; petals mainly indigo; pods 1– to 2–seeded, indehiscent, exserted from the calyx; x = 10. Note: Traditional treatments have placed the species of *Psorothamnus* within an expanded *Dalea* (q.v.).

Rydberg, P. A. 1919. *Psorodendron* Rydb., and *Psorothamnus* Rydb. N. Amer. Fl. 24: 41–48.

1. Branchlets densely reflexed-puberulent, bearing conspicuous yellow to red orange resinous glands 2
— Branchlets merely appressed-strigose, lacking glands or only obscurely glandular 3
2(1). Calyx lobes ovate to oval, obtuse to rounded, or the lowermost ovate-acute .... *P. thompsonae*
— Calyx lobes all lance-attenuate .... *P. polydenius*
3(1). Calyx villous with contorted, spreading hairs, the lateral teeth linear-lanceolate, quite as long as the tube; plants rare in Kane County .... *P. arborescens*
— Calyx strigose with appressed hairs or glabrate, the lateral teeth lance-attenuate, mostly shorter than the tube; plants locally common in Garfield, Kane, San Juan, and Washington counties .... *P. fremontii*

Psorothamnus arborescens (**Torr.**) **Barneby** Beauty Indigo-bush. [*Dalea arborescens* Torr. in Gray; *Parosela arborescens* (Torr.) Heller; *Psorodendron arborescens* (Torr.) Rydb.; *Dalea amoena* Wats.; *Parosela amoena* (Wats.) Vail; *Psorodendron amoenum* (Wats.) Rydb.]. Armed shrubs, 4–10 dm tall or more; branchlets strigose to strigulose, sparingly glandular; stipules 1–2 mm long, subulate; leaves 1.4–3.8 cm long; leaflets 7–15, 1–10 (12) mm long, 0.7–1.5 mm wide, glandular beneath, strigose on both sides, linear to narrowly oblong, obtuse to rounded, the uppermost lateral leaflet often confluent with the terminal; peduncles 0.2–0.6 cm long or the lowermost flower axillary; racemes 11– to 21–flowered, 1.8–4.5 cm long, the rachis pilosulous; bracts 1.5–3.5 mm long, lance-aristate, villosulous; calyx 8–10 mm long, the tube 3.8–4.8 mm long, definitely 10–ribbed, villous, the teeth 3.6–5.2 mm long, villous, not markedly differing in width; flowers 8.1–10.6 mm long, indigo; pods conspicuously glandular-dotted, pilosulous. Mixed desert shrub at 1230 to 1530 m in Kane County; Arizona and Nevada; 2 (i). Our material belongs to var. ***pubescens*** (**Parish**) **Barneby** [*Parosela johnsonii* var. *pubescens* Parish; *Psorodendron pubescens* (Parish) Rydb.; *D. fremontii* var. *pubescens* (Parish) L. Benson; *D. amoena* var. *pubescens* (Parish) Peebles].

Psorothamnus fremontii (**Torr.**) **Barneby** Fremont Indigo-bush. [*Dalea fremontii* Torr. ex Gray; *Parosela fremontii* (Torr.) Vail; *Psorodendron fremontii* (Torr.) Rydb.; *Dalea johnsonii* Wats., type from near St. George; *Parosela johnsonii* (Wats.) Vail; *Psorodendron johnsonii* (Wats.) Rydb.; *Parosela johnsonii* var. *minutiflora* Parish; *D. fremontii* var. *minutifolia* (Parish) L. Benson]. Armed shrubs, 5–15 dm tall or more; branchlets strigose, sparingly if at all glandular; stipules 0.3–1.5 mm long, fragile; leaves 1.8–6.5 cm long; leaflets (1) 3–9, 3–14 mm long, 0.8–6 mm wide, glandular beneath, strigose on both sides, linear to oblong or elliptic, obtuse to rounded, the uppermost often confluent with the rachis; peduncles 0–2.2 cm long, or some flowers axillary; racemes 7– to 41–flowered, 3.5–14.3 cm long, the rachis strigose to strigulose; bracts 0.8–1.8 mm long, lanceolate, glabrous or else foliose and with 1–4 leaflets; calyx 4.5–6.5 mm long, the tube 2.5–3.3 mm long, obscurely if at all 10–ribbed, appressed strigose to glabrate, the teeth 1.8–3.2 mm long, strigose, the upper markedly wider than the others; flowers 8.5–12 mm long, indigo; pods glandular-dotted; 2n = 20. Blackbrush, creosote bush, mixed desert shrub, and (less commonly) pinyon-juniper communities at 730 to 2270 m in Garfield, Kane, San Juan, and Washington counties; Nevada, Arizona, and California; 69 (xi). The plants are strikingly beautiful, contrasting indigo flowers against grayish foliage.

Psorothamnus polydenius (Torr.) Rydb. Glandular Indigo-bush. Armed shrubs, 1.5–8 dm tall or more; branchlets velvety with retrorse short hairs, conspicuously glandular with yellow or red orange resinous glands; stipules vestigial or to 0.5 mm long; leaves 0.4–2.8 cm long; leaflets 5–13, 1.2–6.5 mm long, 1–5 mm wide, oval to obovate, or obcordate, glandular beneath, strigose on both sides, obtuse, rounded, or emarginate, the uppermost often confluent to each other and to the rachis; peduncles 0.8–2.7 cm long; racemes mainly 6– to 18–flowered, 2–5 cm long, the rachis retrorsely hairy; bracts 1–2.5 mm long, lanceolate, pilose; calyx 10–ribbed, the teeth 3.5–4.5 mm long, villous, lance-subulate, all about alike; flowers 7.5–9 mm long, indigo; pods glandular-dotted; n = 10. Two distinctive, geographically correlated varieties are present.

1. Branches strongly divaricate; leaflets flat, commonly less than 4 mm long; plants of Washington County . . . .
. . . . *D. polydenius* var. *polydenius*

— Branches merely ascending, or rarely some divaricate; leaflets curved, at least some over 4 mm long; plants of eastern Emery County . . . . *D. polydenius* var. *jonesii*

Var. **jonesii** Barneby [*Dalea nummularia* Jones, type from Green River]. Salt desert shrub community on Mancos Shale Formation (Blue Gate and Tununk members) and less commonly elsewhere at ca 1470 m in eastern Emery County; endemic; 9 (i). The shrubs are typically less than 5 dm tall.

Var. **polydenius** [*Dalea polydenia* Torr. ex Wats.; *Parosela polydenia* (Torr.) Heller]. Creosote bush and Joshua tree communities on pedimental gravels and limestone at ca 1220 to 1345 m in Beaver Dam Mts., Washington County; Nevada and California; 6 (iii). The single colony known to me in Utah consists of shrubs ca 10–12 dm tall in the Limestone Knolls section of the Beaver Dam Mts., west of Castle Cliff.

Psorothamnus thompsonae (Vail) Welsh & Atwood Thompson Indigo-bush. [*Parosela thompsonae* Vail]. Armed shrubs, 2.5–8 dm tall or more; branchlets velvety with retrorse short hairs, conspicuously glandular with yellow to orange red mammiform resinous glands; stipules vestigial or to 0.8– mm long; leaves 0.7–5 cm long; leaflets 7–17, 1–10 mm long, 0.6–4 mmm wide, linear to oblong, oval, or obcordate, glandular and strigose to glabrate beneath, strigose above, the uppermost jointed to the rachis; peduncles 0.4–1.8 cm long, rarely obsolete and some flowers axillary; racemes mainly 8– to 25–flowered, 2–9 cm long, the rachis retrorsely hairy; bracts 0.5–1.5 mm long, lanceolate, glabrous or hairy, soon deciduous; calyx 3.7–5 mm long, the tube 2.1–3.2 mm long, conspicuously 10–ribed, glabrous or villous, ovate to oval, obtuse or only the lowermost acute; flowers 7.8–10.8 mm long, indigo or purple pink; pods glandular-dotted. Two distinctive varieties are known.

1. Calyx tube villous; leaflets linear; plants of San Juan County . . . . *P. thompsonae* var. *whitingii*

— Calyx tube glabrous; leaflets oblong to oval or obcordate; plants of San Juan County and elsewhere . . . .
. . . . *P. thompsonae* var. *thompsonae*

Var. **thompsonae** [*Parosela thompsonae* Vail; *Dalea thompsonae* (Vail) L. O. Williams]. Mixed and salt desert shrub communities at 1170 to 2270 m in Emery, Garfield, Kane (?, the type supposedly from near Kanab but more likely from along the Canyons of the Colorado east of there), San Juan, and Wayne counties; endemic; 25 (v). There are no modern collections of this taxon from Kane County.

Var. **whitingii (Kearney & Peebles) Barneby** [*Dalea whitingii* Kearney & Peebles]. Mixed desert shrub community at 1230 to 1530 m in Monument Valley and west to Navajo Mt., San Juan County; Arizona; a Navajo Basin endemic; 4 (0).

Robinia L.

Shrubs or trees, often armed; leaves alternate, odd-pinnate, the leaflets petiolulate and stipulate; stipules setaceous and caducous or persistent as spines; flowers in axillary racemes, white, pinkish, or pink, very showy, often aromatic; calyx campanulate to turbinate, the 5 teeth triangular to triangular-acuminate; corolla papilionaceous; stamens 10, diadelphous; ovary subsessile or sessile; pods oblong, straight, laterally flattened, tardily dehiscent; x = 10, 11.

1. Uppermost pair of calyx teeth connate almost to the apex, forming an emarginate lip; branchlets and peduncles lacking hispid processes . . . . *R. pseudoacacia*

— Uppermost pair of calyx teeth free for ca two-thirds of their length; branchlets or peduncles hispid 2

2(1). Branchlets and peduncles both glandular-hispid; shrubs, cultivated . . . . *R. hispida*

— Branchlets lacking processes; peduncles glandular-hispid; trees, cultivated or indigenous in Washington County . . . . *R. neomexicana*

Robinia hispida L. Rose-acacia. Shrubs, arising from root sprouts, spreading, commonly 1–2 m tall, except when grafted; branchlets both hispid and glandular-hispid; leaves 9.5–27 cm long; leaflets 7–13, 11–60 mm long, 7–40 mm wide, ovate or lance-ovate, elliptic, or oblong, obtuse, cuspidate, sparingly villous beneath, glabrous above; petioles hispid near the base; stipules obsolete; peduncles 1–4 cm long, hispid; racemes 3– to 10–flowered, 3–8 cm long; flowers 18–30 mm long, rose pink; calyx turbinate-campanulate, the tube 4.5–6 mm long, the teeth 3.5–7 mm long, triangular-acuminate; pods hispid, seldom produced. Cultivated ornamental in Utah County; indigenous from Virginia and Kentucky to Georgia and Alabama; 10 (i). Evidence indicates that rose-acacia is a sterile triploid, at least in part. Lack of fruit in plants from Utah seems to support this contention.

Robinia neomexicana Gray New Mexico Locust. [*R. neomexicana* var. *luxurians* Dieck; *R. luxurians* (Dieck) Schneider ex Silva Tarouca & Schneider; *R. breviloba* Rydb.; *R. subvelutina* Rydb.]. Small trees or shrubs, mainly 1–8 m tall; branchlets villosulous, rarely glandular; leaves 8–20 (28) cm long; leaflets 9–19, 12–40 mm long, 7–20 mm wide, lance-oblong to oblong, obtuse, cuspidate, sparingly pubescent above and below, finally glabrate on one or both sides; petioles villosulous near the base; stipules 3–10 mm long or more, often spiny; peduncles 1–4 cm long, glandular-pubescent to hispid throughout; racemes 3- to 14-flowered, 3–8 cm long; flowers 16–24 mm long, pink; calyx campanulate, the tube 5 mm long, the teeth 3–5 mm long, triangular-acuminate; pods 40–80 m long, glandular-pubescent to hispid or glabrous. Talus slopes and terraces in Zion Canyon, Beaverdam Wash, and near Enterprise Reservoir in Washington County; Colorado, Nevada, Arizona, New Mexico, Texas, and Mexico; 5 (i). The plant is also known from cultivation in Cache, Salt Lake, and Utah counties; 4 (i). The pink-flowered tree locusts of our region are all more or less involved through hybridization with the New Mexico locust and the black locust.

Robinia pseudoacacia L. Black Locust. Trees to 25 m tall or more; branchlets puberulent to villosulous, lacking glands; leaves (6) 8.5–26 cm long; leaflets (5) 11–25, 11–60 mm long, 6–30 (38) mm wide, lance-oblong to oblong, obtuse to retuse or cuspidate, puberulent to glabrate on both sides; petioles villosulous to pilose near the base; stipules minute or represented by spines; peduncles 1.4–4.3 cm long, puberulent to villosulous; racemes 11- to 27-flowered, 4–13 cm long; flowers 12–20 (23) mm long, white (fading cream), or pale paink; calyx broadly campanulate to turbinate, the tube 3.5–5.5 mm long, the teeth 1.5–2 mm long, the upper pair connate, the sinus shallow; pods 4–12 cm long, glabrous; 2n = 22. Cultivated ornamental, street, and shade trees, long persisting, escaping, and established in Cache, Carbon, Grand, Kane, Millard, Salt Lake, Sanpete, Utah, Washington, and Wayne counties; indigenous in the eastern U. S.; 34 (x). This is a handsome shade tree with very hard wood. The flowers yield nectar and the wood has been used for fence posts.

Sophora L.

Trees or herbs, unarmed; leaves alternate, odd-pinnately compound; stipules obsolete or herbaceous; flowers in terminal racemes or panicles, perfect, papilionaceous; calyx 5-toothed; petals all distinct; stamens 10, distinct or essentially so; ovary stipitate; pods spreading to pendulous, subterete, constricted between the seeds, indehiscent or tardily so; x = 9.

1. Plants trees, cultivated; flowers borne in panicles in midsummer .... S. japonica
— Plants herbaceous, indigenous; flowers in racemes, opening in springtime 2

2(1). Flowers blue purple to blue; leaflets more than 5 times longer than broad, silvery hairy .... S. stenophylla
— Flowers white to cream; leaflets less than 5 times longer than broad, gray green .... S. nuttalliana

Sophora japonica L. Japanese Pagoda-tree. Trees to 12 m tall, the bark green and smooth for some years; branchlets sparingly villosulous to glabrate; leaf bases expanded, obscuring the axillary buds; leaves 12–23 cm long; leaflets 7–17, 14–60 mm long, 8–29 mm wide, lance-oblong to lanceolate, acute and apiculate, strigose beneath, glabrate to glabrous above; stipules obsolete; panicles many-flowered, 15–40 cm long; flowers 11–14 mm long, white to cream; calyx broadly campanulate 5.5–6.5 mm long, the tube 4.5–5.1 mm long, the teeth 0.8–1.2 mm long, broadly triangular; pods mainly 5–10 cm long, fleshy, constricted between the seeds; 2n = 26, 28. Cultivated ornamental, widely planted in lower elevation communities in Utah; introduced from China; 10 (i).

Sophora nuttalliana Turner Silky Sophora. [*S. sericea* Nutt.]. Perennial, caulescent, 12–27 (30) cm tall, from a rhizome; pubescence essentially basifixed; stems ascending to erect; stipules 1.5–4 mm long, distinct, caducous; leaves 2.5–6 (7) cm long; leaflets 13–23, 2–11 mm long, 0.8–6 mm wide, oblong-obovate to obovate, rounded to retuse, often folded, strigose beneath, glabrous above; peduncles 0.8–3 cm long; racemes mainly 6- to 52-flowered, 3–9 cm long at anthesis, the axis 3–13 cm long in fruit; bracts 3–7 mm long; pedicels 2–4 mm long; bracteoles 1; calyx 8–10.5 mm long, gibbous, the tube 6.4–8.5 mm long, obliquely short-cylindric, the teeth 2–2.5 mm long, triangular; flowers 14–19 mm long, white to cream (fading cream); pods erect-ascending, stipitate, the stipe 6–12 mm long, the body 12–40 mm long, 3–4.5 mm thick, constricted tightly between the usually 1–3 seeds, strigose. River bottoms and roadsides at 1070 to 1670 m in San Juan and Washington counties; Wyoming and South Dakota, south to Arizona, New Mexico, and Texas; 6 (0).

Sophora stenophylla Gray Silvery Sophora. Perennial, caulescent, 13–41 cm tall (above ground), from deeply seated rhizomes; pubescence basifixed; stems ascending to erect; stipules 3–12 mm long, distinct, caducous, or obsolete; leaves 1.7–5.6 mm wide, linear to narrowly oblong, acute to attenuate, silvery pilosulous, the pubescence fading yellowish in time; peduncles 1.7–5 cm long; racemes mainly 12- to 39-flowered, 5–17 cm long in anthesis, 6–23.5 cm long in fruit; bracts 3–7 mm long; pedicels 1–8 mm long; bracteoles 0; calyx 6.5–10.8 mm long, gibbous, the tube 4.8–7.2 mm long, obliquely short-cylindric, the teeth 1.7–3.6 mm long, ovate-triangular; flowers 15–27 mm long, blue purple to blue; pods spreading declined, stipitate, the stipe 8–16 mm long, the body 15–60 mm long, 6–8 mm wide, strongly constricted between the usually 1–5 seeds, strigose. Sand dunes and other sandy sites, mainly in mixed desert shrub communities, at 900 to 2270 m in Emery, Garfield (at Red Canyon, with ponderosa pine), Grand, Kane, San Juan, Uintah, Washington, and Wayne counties; New Mexico and Arizona; 48 (viii).

Spartium L.

Shrubs, unarmed; leaves alternate to subopposite, simple; stipules obsolete; flowers in terminal, erect racemes, papilionaceous, perfect; calyx 1-lipped, the 5 teeth marginal, all on the lower side of the calyx, the three central ones approximate, the lateral ones somewhat removed; petals yellow, the keel pubescent along the lower edge and with a porrect beak; stamens 10, monadelphous; ovary sessile; pods spreading-ascending, laterally compressed, many-seeded.

Spartium junceum L. Spanish Broom. [*Genista juncea* (L.) Lam.]. Shrubs, 15–25 dm tall; leaves simple, 8–27 mm long, 0.5–4 mm wide, linear to narrowly oblanceolate, acute to obtuse, strigose on both sides; peduncles 3–22 cm long (from last leaf to first flower); racemes 3- to 16-flowered, 3–16 cm long; bracts minute, caducous; bracteoles 0; pedicels 1–4 mm long; calyx oblique, 7–8 mm long, glabrous, the teeth minute; flowers 21–26 mm long, yellow; pods sessile, 5–8 cm long, 5–7 mm wide, strigose, dehiscent; 2n = 52, 54. Cultivated ornamental in Washington County; introduced from southern Europe; 4 (ii).

Sphaerophysa DC.

Perennial, caulescent, from rhizomes; leaves alternate, odd-pinnate; stipules adnate to the petiole base, all distinct; flowers papilionaceous, borne in axillary racemes, each subtended by a single bract; bracteoles 1; calyx 5-toothed; petals 5, dull red, drying lavender to brown; stamens 10, diadelphous; ovary enfolded by the staminal sheath; style glabrous except for a tuft of hair below the stigma; pods stipitate, bladdery-inflated, subunilocular.

***Sphaerophysa salsula* (Pallas) DC.** [*Phaca salsula* Pallas; *Swainsona salsula* (Pallas) Taub. in Engler & Prantl]. Perennial, caulescent, 4-7 dm tall, from a deeply placed rhizome; pubescence basifixed; stipules 1-4 mm long, all distinct; leaves 3-10 cm long; leaflets 15-25, 3-18 mm long, 1-7 mm wide, oblong-obovate to elliptic, retuse to obtuse and apiculate, strigose beneath, glabrous above; peduncles 2.5-9 cm long; racemes 5- to 17-flowered, the flowers ascending in anthesis, finally nodding, the axis 2.5-9 cm long in fruit; bracts 1-2 mm long; pedicels 2.5-8 mm long; bracteoles 2; calyx 5-6 mm long, the tube 3.8-4.6 mm long, campanulate, the teeth 1.2-2 mm long, triangular; flowers 12-14 mm long, dull red, fading lavender to brown; pods ascending to declined, stipitate, the stipe 4-7 mm long, the body bladdery-inflated, ovoid, 13-24 mm long, 9-20 mm wide (when pressed), unilocular, strigulose. Introduced weedy species known from the Uinta Basin; widespread in the western U. S.; adventive from Asia; 1 (0). This plant resembles an *Astragalus* species, and has been mistaken twice as belonging to previously undescribed and unnamed indigenous species (*A. violaceus* St. John; *A. iochrous* Barneby). Generic concepts revolving around this genus are unresolved, and it seems likely that the plants will ultimately be placed in some earlier named genus.

Thermopsis R. Br.

Perennial herbs, caulescent, from rhizomes; leaves alternate, palmately trifoliolate; stipules foliaceous; flowers papilionaceous, borne in terminal racemes; bracts herbaceous, persistent; calyx 5-toothed; petals 5, yellow or suffused with purple, the keel rounded; stamens 10, distinct; ovary stipitate, the style glabrous; pods narrowly oblong, flattened, many-seeded; x = 9.

1. Pods straight, not especially lomentlike, erect or ascending; plants mostly 20-70 cm tall or more, common through most of Utah .... *T. montana*
— Pods curved, lomentlike, spreading to recurved; plants mostly 14-40 cm tall, known from Daggett and Uintah counties .... *T. rhombifolia*

***Thermopsis montana* Nutt. in T. & G.** Golden Pea; Yellow Pea. Caulescent, 2-7.5 (10) dm tall, the stems erect, pilosulous to glabrate; stipules foliar, lanceolate to ovate, 13-60 mm long, 3-30 mm wide; petioles 0.8-3.7 cm long; leaflets 3, 21-92 mm long, 5-36 mm wide, elliptic to lanceolate or oblanceolate, acute to rounded, pilosulous beneath, glabrous to glabrate above; peduncles 2.2-13 cm long; bracts 5-11 mm long; racemes mainly 2- to 23-flowered, 6-25 cm long in anthesis, the axis 9-28 cm long in fruit; flowers 20-26 mm long, yellow; calyx 10.2-12.3 mm long, the tube 7-8.3 mm long, obliquely campanulate, the teeth 3.1-4 mm long, ovate-triangular; pods erect or ascending, stipitate, the stipe 2.5-6 mm long, the body 40-54 mm long, 5-7 mm wide, pilose, stramineus or turning black; 2n = 18. Moist sites along streams, in meadows, around seeps and springs at 1250 to 3420 m in Box Elder, Daggett, Duchesne, Garfield, Juab, Kane, Millard, Morgan, Piute, San Juan, Sanpete, Sevier, Summit, Uintah, Utah, Wasatch, Washington, Wayne, and Weber counties; British Columbia to Montana, south to California, Arizona, and New Mexico; 126 (xvii). The abundant materials present in Utah herbaria demonstrate a wide range of variation in leaflet and stipule size and shape. There appears to be no basis for segregation of subordinate taxa, even those from the southern tier of counties known previously as *T. pinetorum* Greene or *T. ovata* (Robins.) Rydb. [*T. montana* ssp. *ovata* Robins ex Piper; *T. montana* var. *ovata* (Robins.) St. John; *T. rhombifolia* var. *ovata* (Robins.) Isely]. The leaves of plants from Washington County average larger, but that is hardly a basis for segregation. When included within an expanded *T. rhombifolia* our material is regarded as *T. rhombifolia* var. *montana* (Nutt.) Isely.

***Thermopsis rhombifolia* Nutt. ex Richards.** Caulescent, 15-40 cm tall, the stems erect, glabrate; stipules foliose, ovate to lanceolate, 6-30 mm long, 2-22 mm wide; petioles 0.3-2.5 cm long; leaflets 3, 15-47 mm long, 7-25 mm wide, obovate to oblanceolate, obtuse to rounded, glabrous on both sides; peduncles 0.5-5.8 cm long; bracts simple to foliose; racemes mainly 4- to 30-flowered, 2-10 cm long in anthesis, the axis 2-12 cm long in fruit; flowers 18-22 mm long; calyx 7.5-10 mm long the tube 4.5-6 mm long, the teeth 3-4 mm long, triangular-ovate; pods divaricate, finally recurved, lomentlike, stipitate, the stipe 1.5-4 mm long, the body 25-70 mm long, 5-7 mm wide, pilose to glabrate; 2n = 18. Sandy and clay soils, mainly where moist, in mixed desert shrub and pinyon-juniper communities at 1500 to 1800 m in Daggett and Uintah counties; Alberta and Saskatchewan south to Colorado and Nebraska; 21 (ii).

Trifolium L.

Perennial or short-lived perennial or annual, caulescent or acaulescent, from taproot and caudex, rhizome, or stolon; leaves alternate, palmately to pinnately 3-foliolate, or rarely 4- to 7-foliolate, commonly serrate throughout, rarely entire; stipules membraneous to foliaceous, often connate; flowers papilionaceous, borne in terminal or axillary, pedunculate to sessile, subcapitate heads or racemes; calyx 5-toothed; petals 5, pink, white, or red purple, withering and persistent, finally investing the pod; stamens 10, diadelphous; pods usually shorter than the calyx, indehiscent, 1- to several-seeded.

Gillett, J. M. 1965. Taxonomy of *Trifolium*: Five American species of section Lupinaster (Leguminosae). Brittonia 17: 107-112.
———. 1969. Taxonomy of *Trifolium* (Leguminosae). II. The *T. longipes* complex in North America. Canad. J. Bot. 47: 93-113.
———. 1971. Taxonomy of *Trifolium* (Leguminosae). III. *T. eriocephalum*. Canad. J. Bot. 49: 395-405.
———. 1972. Taxonomy of *Trifolium* (Leguminosae). IV. The American species of section Lupinaster (Adanson) Seringe. Canad J. Bot. 50: 1975-2007.
Hermann, F. J. 1953. A botanical synopsis of the cultivated clovers. USDA Agricultural Monograph 22. Washington, D. C.
McDermott, L. F. 1910. An illustrated key to the North American species of *Trifolium*. San Francisco. 325 pp.
Martin, J. S. 1946. Notes on *Trifolium eriocephalum* Nutt. Madroño 8: 152-157.

1. Plants acaulescent or subacaulescent, mainly 1.5–10 cm tall 2
— Plants caulescent, with 1 or more elongated internodes (see also *T. parryi*), mainly 10–60 cm tall .. 7
2(1). Heads 1- to 4-flowered, essentially sessile, the flowers 15–23 mm long; plants of high elevations in the Uinta and La Sal mts. .... *T. nanum*
— Heads several-flowered, sessile or pedunculate, the flowers either shorter or the heads pedunculate; plants of various distribution 3
3(2). Plants densely pulvinate-caespitose, matted 4
— Plants loosely caespitose, not especially mat-forming . 5
4(3). Calyces densely woolly-villous; herbage silvery-hairy; plants of central Beaver and western Millard counties .... *T. andersonii*
— Calyces strigose to strigulose; herbage green, merely strigose; plants of Daggett, Millard, Rich, San Juan, and Summit counties .... *T. andinum*
5(3). Leaflets toothed; calyces villosulous to pilosulous; flowers less than 9 mm long; plants of moderate elevations, broadly distributed .... *T. gymnocarpon*
— Leaflets entire or, if toothed, the calyces glabrous; flowers more than 9 mm long; plants of high elevations 6
6(5). Leaflets entire; calyces strigose; plants of Uinta, La Sal, Abajo, and Henry mts. .... *T. dasyphyllum*
— Leaflets toothed; calyces glabrous; plants of Uinta, La Sal, and Abajo mts. .... *T. parryi*
7(1). Plants stoloniferous, prostrate and rooting at the nodes; flowers white; calyx not bladdery-inflated; introduced, widespread .... *T. repens*
— Plants not stoloniferous (except in some *T. fragiferum*), usually erect or ascending, not or seldom rooting at the nodes; flowers mainly pink to red purple, though sometimes white 8
8(7). Calyx soon bladdery-inflated and enclosing the corolla; plants introduced .... *T. fragiferum*
— Calyx not accrescent or only slightly so, never enclosing the corolla; plants various 9
9(8). Heads sessile or nearly so, commonly immediately subtended by a trifoliolate bract; plants cultivated, escaping, and persisting .... *T. pratense*
— Heads with well-developed peduncles, not immediately subtended by foliose bracts; plants indigenous or cultivated 10
10(9). Peduncles bent at the apex, the heads reflexed or appearing turned to one side 11
— Peduncles straight, the heads erect 13
11(10). Calyx tube and teeth villous; ovary with hairs near the apex along the ventral side; plants of lower middle elevations in Juab, Piute, Sevier, and Beaver (?) counties .... *T. eriocephalum*
— Calyx tube and teeth sparingly villosulous to glabrous; ovary glabrous or scaly, but not hairy as above; plants of various distribution 12
12(11). Heads definitely longer than broad; calyx teeth shorter than the tube; basal leaves prominent, long-petioled, plants of lower to middle elevations in Beaver, Iron, and Washington counties . *T. macilentum*
— Heads about as broad as long; calyx teeth subequal to the tube; basal leaves more shortly petioled than the subbasal ones; plants of higher elevations, of broad distribution but not as above .... *T. kingii*
13(10). Heads subtended by a spinose-toothed involucre ... 14

— Heads lacking an involucre; plants from a caudex, variously distributed, common or uncommon 16
14(13). Plants perennial from a caudex or from rhizomes, known from Juab, Kane, and Tooele counties .... *T. wormskjoldii*
— Plants annual from a taproot, though often more or less rhizomatous also; distribution various 15
15(14). Lower calyx teeth usually bi- or trifid; involucre shallowly rotund-lobate, the lobes erose-dentate, the teeth less than 2 mm long; plant report apparently due to mislabeled specimen from Nevada .... *T. cyathiferum* Lindl.
— Lower calyx teeth entire; involucre irregularly lobed and lacerate about half its length, the teeth often more than 2 mm long; plants of northern Utah .... *T. variegatum*
16(13). Flowers mainly 7–9 mm long; heads axillary, from the uppermost nodes; plants cultivated, escaping and persisting .... *T. hybridum*
— Flowers 10–16 mm long; heads terminal, solitary; plants indigenous 17
17(16). Calyx glabrous; flowers mainly 14–16 mm long; plants rare in Sevier and Piute (?) counties .... *T. beckwethii*
— Calyx villosulous to pilosulous; flowers mainly 10–12 mm long; plants widespread in Utah .... *T. longipes*

Trifolium andersonii Gray Frisco Clover. Acaulescent, densely pulvinate-caespitose, mat-forming, 0.8–3 cm tall, from woody caudex branches and a thick taproot, the stems obscured by imbricated stipules and persistent leaf bases; petioles 0.3–1.1 cm long; leaflets 3, 3–8 mm long, 1.5–3.4 mm wide, oblanceolate to obovate, entire or toothed near the apex and the teeth more less obscured by pubescence, villosulous on both sides with hairs 0.3–0.7 mm long, commonly folded, mucronate; stipules scarious, pilose, 5–9 mm long; heads lacking an involucre, borne subsessile or on peduncles 0.2–0.6 cm long, these densely pilose; flowers 4–9, the banner reddish purple, the keel and wings pale, 8–9 mm long, on pedicels ca 0.5 mm long (obscured by pilose hairs); calyx 8.5–9.5 mm long, the tube 3–4 mm long, campanulate, obscured by villous hairs to 1.5 mm long, the teeth 5–7 mm long, subulate, villous; pods enclosed within the marcescent petals and calyx; $2n = 16$. Volcanic gravels and limestone in pinyon-juniper woods at ca 2100 to 2230 m in Beaver (San Francisco Mts.) and western Millard (Tunnel Springs Range) counties; California and Nevada. Our material, as described above, belongs to **var. friscanum** Welsh (type from Frisco, Beaver County); 8 (ii).

Trifolium andinum Nutt. in T. & G. Andean Clover. Acaulescent, densely pulvinate-caespitose, mat-forming, 0.3–5 cm tall, from woody caudex branches and a thick taproot, the stems obscured by imbricated stipules and persistent leaf bases; petioles 0–2.3 cm long; leaflets 3, 2–12.5 mm long, 1.2–5 mm wide, oblanceolate to obovate, toothed in the apical third, strigose to strigulose on both sides, commonly folded, abruptly acuminate; stipules scarious, yellowish, glabrous except toward the tip, 3–9 mm long; heads closely subtended by reduced leaflike bracts, these with or without a trifoliolate bract, sessile or appearing pedunculate by elongation of internodes on floriferous branches, the internodes glabrous; flowers 7–15, in 2 closely associated heads, the banner violet purple, the keel and wings ochroleucous,

9–12 mm long, on pedicels 0.5–1 mm long, these glabrous to strigose; calyx 5–7 (8.5) mm long, the tube 2.2–3.6 (4) mm long, sparingly pilose to glabrous, the teeth 2–3.8 (4.5) mm long, lance-subulate, pilose; pods 4–5 mm long, 2.3–2.7 mm wide. Ponderosa pine, sagebrush, or mixed shrub communities at 1970 to 2730 m in Daggett, Millard (Canyon Range), Rich, San Juan (Navajo Mt.), and Summit counties, Wyoming, Nevada, and Arizona; 17 (i). The tripartite distribution of this clover is without precedence among the legumes of Utah. The plants from Navajo Mountain differ in features of flower, mainly in calyx teeth much shorter than the tube; possibly they are of taxonomic significance.

Trifolium beckwethii Brewer ex Wats. Beckwith Clover. Caulescent, 12–40 cm tall, from rhizomes and thick roots; petioles of lower leaves 4–15 cm long, becoming shorter upward; leaflets 3, 12–45 (50) mm long, 7–20 mm wide, elliptic to lance-elliptic or oblong, toothed from near the base, glabrous, flat, emarginate to apiculate; stipules prominent, herbaceous to scarious, 12–28 mm long; heads not subtended by an involucre, ca as broad as long, 25–35 mm wide, terminal or axillary, on peduncles 5–26 cm long, these glabrous; flowers numerous, the petals pink, fading brown, 12–16.5 mm long; calyx 6.5–8 mm long, the tube 2.5–3.1 mm long, gibbous, glabrous, 5-veined, the teeth 3.5–4.9 mm long, lance-subulate, glabrous; pods 4.8–5.2 mm long, 1.9–2.1 mm wide; 2n = 48. Meadows at 2070 to 2200 m in Piute (?) and Sevier counties; California, Oregon, Nevada, Idaho, Montana, and South Dakota; 3 (ii). Utah materials have flowers and parts larger than for the species as a whole. At the known locality in Utah, the plants grow with *T. eriocephalum*.

Trifolium dasyphyllum T. & G. Uinta Clover. Acaulescent, loosely mat-forming, 2–14 cm tall, from a caudex and thick taproot, the stems obscured by imbricated stipules and leaf bases; petioles 0.3–4 cm long; leaflets 3, 3–28 mm long, 1–5 mm wide, oblanceolate, entire, strigose beneath, strigose to glabrate above, flat or folded, sharply apiculate; stiples scarious, glabrous or strigose, 5–17 mm long; heads closely subtended by a long-lobed involucre, 11–18 mm wide, terminal, sessile or on peduncles 0.5–10 cm long, these strigose; flowers 6–16, erect, the banner violet to ochroleucous, the keel and wings purple, or all purple, 11–13 mm long; pedicels 0.5–1.5 mm long; calyx 5.7–9.1 mm long, the tube 2.5–2.9 mm long, strigose to glabrate, the teeth 1.9–6.4 mm long, subulate, strigose to glabrate; pods 4–6 mm long. Alpine meadows at 2730 to 3800 m in Daggett, Garfield, Grand, San Juan, Summit, and Uintah counties; Montana, Wyoming, Colorado, and New Mexico; 28 (0). Our materials belong to var. **uintense** (**Rydb.**) Welsh [*T. uintense* Rydb., type from Uinta Mts., Summit County; *T. dasyphyllum* f. *uintense* (Rydb.) McDermott; *T. dasyphyllum* ssp. *uintense* (Rydb.) J. M. Gillett].

Trifolium eriocephalum Nutt. in T. & G. Woolly Clover. Caulescent, 12–36 (42) cm tall, from a caudex and tuberous to slender taproot; petioles 0.8–10 cm long; leaflets 3, 5–53 mm long, 5–12 mm wide, oblong to oblong-lanceolate or elliptic, toothed from near the base, thinly pilose to pilosulous on both sides, flat, obtuse to acute, apiculate; stipules foliaceous, glabrous or pilosulous, 15–55 mm long; heads lacking an involucre, 20–30 mm long, 20–27 mm wide, terminal, on peduncles 4–11 cm long, bent apically, thinly villosulous; flowers numerous, curved at base and reflexed, the petals purple to red purple, rarely white, fading brown, 12–13.5 mm long; pedicels essentially obsolete; calyx 6.5–8 mm long, the tube 2.5–3.3 mm long, villosulous, the teeth 3.5–5.1 mm long, villosulous to glabrate, subulate; pods 2.5–7 mm long; 2n = 16. Meadows at 1530 to 2270 m in Beaver (?), Juab, Piute, and Sevier counties; Washington, Idaho, Montana, California, and Nevada; 6 (ii). Our few specimens are assignable to var. **villiferum** (**House**) **Martin** [*T. villiferum* House, type from southern Utah. *T. eriocephalum* f. *villiferum* (House) McDermott; *T. eriocephalum* ssp. *villiferum* (House) J. M. Gillett]. The type (US) of *T. villiferum* is labeled as "S. Utah. Dr. E. Palmer 91, 1877." Gillett (1971) cites the type locality as Beaver City, Beaver Co., Utah." The problem of the type locality was discussed by Welsh (1978 p. 356).

Trifolium fragiferum L. Strawberry Clover. Caulescent, 5–30 cm long, rhizomatous and sometimes stoloniferous, decumbent to ascending; petioles 0.5–13 cm long, obovate, toothed from near the base, truncate to retuse and apiculate, flat, glabrous to glabrate on both sides; stipules scarious 8–20 mm long; heads involucrate, many-flowered, subglobose, 10–22 mm wide, on peduncles 2–17 cm long, these glabrous; flowers 4–6 mm long, purplish, finally included within the accrescent calyx; calyx finally bladdery-inflated, pilose, reticulately veined around translucent lacunae; 2n = 16. Meadows, roadsides, and other disturbed sites practically throughout Utah; introduced from Europe; 21 (iii).

Trifolium gymnocarpon Nutt. in T. & G. Nuttall Clover; Dwarf Clover. [*T. subcaulescens* Gray in Ives; *T. gymnocarpon* var. *subcaulescens* (Gray) A. Nels.; *T. nemorale* Greene; *T. plummerae* Wats; *T. gymnocarpon* f. *plummerae* (Wats.) McDermott; *T. gynmocarpon* var. *plummerae* (Wats.) Martin; *T. gymnocarpon* ssp. *plummerae* (Wats.) J. M. Gillett]. Acaulescent to short caulescent, 4–16 cm tall, from a caudex and taproot, the stems mainly obscured by imbricated stipules and leaf bases; petioles 1–9 cm long; leaflets 3–5, 6–23 mm long, 2–10 mm wide, elliptic to oblong, ovate or obovate, toothed from near the base, pilosulous beneath, glabrous to pilosulous above, flat or folded, rounded to acute and apiculate; stipules scarious to herbaceous, 6–23 mm long; heads without an involucre, hemispheric, commonly 12–18 mm wide, terminal and axillary, on peduncles 1–6.5 cm long, these strigulose, erect to bent apically; flowers 6–15, the lower ones reflexed in age, the petals pink to lavender or purple, 8.5–11 mm long; pedicels 0.5–1 mm long; calyx 4.4–7 mm long, the tube 2.2–3.3 mm long, strigose, the teeth 1.8–3.7 mm long, subulate, strigose; pods 4–8 mm long, 3–4.5 mm wide. Mixed desert shrub, sagebrush, pinyon-juniper, mountain brush, ponderosa pine, Douglas fir, and spruce-fir communities at 1530 to 2900 m in Beaver, Cache, Carbon, Daggett, Duchesne, Emery, Garfield, Iron, Kane, Millard, Piute, Rich, San Juan, Sanpete, Sevier, Summit, Uintah, Utah, and Wayne counties; Oregon, California, Nevada, Colorado, Arizona, Idaho, and Wyoming; 55 (xv). Attempts at recognition of infraspecific taxa appear to be futile.

Trifolium hybridum L. Alsike Clover. Caulescent, 15–70 cm tall, erect or ascending (rarely decumbent), from a caudex and taproot; petioles 0–16 cm long; leaflets 3, 5–38 mm long, 3–28 mm wide, oval to lance-elliptic, ovate or obovate, flat, toothed from near the base, glabrous on both sides, obtuse to retuse and apiculate; stipules herbaceous, 5–25 mm long; heads without an involucre, 12–25 mm wide, terminal and axillary on peduncles 1.5–13 cm long, these glabrous or glabrate, erect; flowers many, the lower reflexed in age, the petals white to pink or reddish, fading red brown, 5–9 mm long; calyx 2.7–4 mm long, the tube 1.2–1.6 mm long, glabrous, scarious, the teeth 1–2.5 mm long, subulate, glabrous; pods 1- to 3-seeded; 2n = 16, 32. Cultivated, short-lived, forage plant, escaping and persisting (?) in much of Utah; introduced from Europe; 27 (iii).

Trifolium kingii Wats. King Clover. Caulescent, (2) 7–40 cm tall, erect or ascending, from a caudex and taproot; petioles 0.8–15 cm long, the longest near the base but not basal; leaflets 3, 5–80 mm long, 4–26 mm wide, elliptic to lance-elliptic or lanceolate, flat, toothed from near the base, glabrous on both sides, obtuse to acute or attenuate and apiculate; stipules foliaceous, 8–30 mm long; heads nodding to suberect, without an involucre, 15–32 mm long, 15–30 mm wide (when pressed), terminal, on peduncles 3–13 cm long, these glabrous; flowers many, reflexed, the petals violet to purplish (rarely white), 12–16 mm long; calyx 5.3–9 mm long, glabrous, the tube 2.1–3.5 mm long, the teeth 1.8–6 mm long, subulate; pods 1- to 3-seeded; 2n = 16. Meadows and open woods at 2270 to 3700 m in Beaver, Emery, Garfield, Grand, Piute, San Juan, Sanpete, Sevier, Summit (type from Parleys Park), and Utah counties; endemic; 37 (v). With its nodding heads and reflexed flowers, the King clover simulates *T. eriocephalum* in all salient aspects. The main differential feature involves the pubescence of herbage and calyces.

Trifolium longipes Nutt. in T. & G. Rydberg Clover. Caulescent (rarely acaulescent), 5–31 (37) cm tall, erect or ascending from a branching caudex and stout to slender taproot; petioles 1.2–10 cm long; leaflets 3, 5–47 (57) mm long, 3–18 mm wide, narrowly oblong to elliptic, oblanceolate, or obovate, flat, toothed from near the base, pilosulous beneath, glabrous above, acute to obtuse and apiculate; stipules foliaceous, 8–40 mm long; heads erect, without an involucre, 17–31 mm long, 15–33 mm wide (when pressed), terminal on peduncles 0.5–17 cm long, these strigulose; flowers many, finally reflexed, the petals whitish to pink or purple; 11–13 mm long; calyx 4.5–7.8 mm long, the tube 1.6–2.5 mm long, scarious, pilose distally, the teeth 2.9–5.8 mm long, pilose, subulate; pods 1- to 4-seeded; 2n = 16, 32, 48. This is the common clover in the mountains of Utah. There are two rather distinctive and largely allopatric varities.

1. Leaflets of main leaves commonly more than 5 times longer than broad; roots slender, not much enlarged; plants of Uinta, Deep Creek, La Sal, and Abajo mts. .... *T. longipes* var. *reflexum*

— Leaflets of main leaves commonly less than 4 times longer than broad; roots tuberous-thickened; plants mainly of southwestern Utah .... *T. longipes* var. *brachypus*

Var. *brachypus* Wats. [*T. longipes* var. *pygmaeum* sensu Gray in Ives, the name used descriptively; *T. longipes* ssp. *pygmaeum* J. M. Gillett; *T. brachypus* (Wats.) Blank.; *T. rusbyi* Greene; *T. confusum* Rydb., type from southern Utah, i.e., Sheep Range, Cedar City; *T. oreganum* f. *rusbyi* (Greene) McDermott; *T. oreganum* f. *brachypus* (Wats.) McDermott; *T. longipes* var. *rusbyi* (Greene) Harrington]. Alpine meadows, open woods, stream banks, and grasslands at 1830 to 3500 m in Beaver, Emery, Garfield, Iron (type from near Cedar City), Kane, Piute, San Juan, Sevier, and Washington counties; Colorado, New Mexico, and Arizona; 57 (xiii).

Var. *reflexum* A. Nels. [*T. longipes* ssp. *reflexum* (A. Nels.) J. M. Gillett; *T. rydbergii* Greene; *T. oreganum* var. *rydbergii* (Greene) McDermott]. Meadows, stream banks, woods, and willow communities at 1870 to 3050 m in Cache, Daggett, Duchesne, Grand, Juab (Deep Creek Mts.), Salt Lake, San Juan, Summit, Uintah, Utah, and Wasatch counties; Wyoming, Colorado, and New Mexico; 35 (iv).

Trifolium macilentum Greene Lean Clover. [*T. kingii* ssp. *macilentum* (Greene) J. M. Gillett]. Caulescent, 12–35 cm tall, ascending to decumbent, from a caudex and taproot; petioles 2–16 cm long, the longest definitely basal; leaflets 3, 14–45 mm long, 4–25 mm wide, those of basal leaves broadly ovate to lance-ovate, those of cauline leaves narrowly lanceolate, flat, toothed from near the base, glabrous on both sides, retuse to obtuse, acute to attenuate and apiculate; stipules dimorphic, the lower ones scarious and very long, the cauline ones herbaceous and small; heads nodding to suberect, without an involucre, 21–40 mm long, 14–35 mm wide (when pressed), terminal on peduncles 4.5–15 cm long, these glabrous; flowers many, reflexed, violet to purplish, 13.5–17 mm long; calyx 4–5.7 mm long, pilose, the tube 2.2–3 mm long, the teeth 1.5–3.3 mm long, subulate; pods 1- to 3-seeded. Mountain brush, pinyon-juniper, and ponderosa pine communities at 1370 to 2270 m in western Beaver, Iron, and Washington (type from southern Utah) counties; Nevada and California; 20 (v). The lean clover was subordinated by Gillett (1972) within *T. kingii*, on the basis of specimens from Nevada that were mistaken for typical *T. kingii*. The lean clover is a plant of dryish hillsides at lower elevations, differing further in its strongly dimorphic leaves, flowers that average larger, and pilose (though sparingly) calyces that average shorter. Thus, it differs from *T. kingii* in about the same order of magnitude as does *T. eriocephalum*. Our material belongs to var. *macilentum*.

Trifolium nanum Torr. Dwarf Clover; Tundra Clover. Acaulescent, pulvinate-caespitose, 2–4 cm tall, from a caudex and taproot, the stems obscured by imbricated stipules and leaf bases; petioles 0.3–2 mm long; leaflets 3, 3–11 mm long, 1–5 mm wide, oblanceolate to obovate, toothed to entire, glabrous or with some hairs on the lower surface, folded or flat, acute to mucronate; stipules scarious to herbaceous; heads 1- to 4-flowered, with an involucre of distinct to connate bracts, terminal, sessile or on peduncles 0.3–4 cm long, these glabrate to glabrous; flowers 15–23 mm long, pale purplish (fading dark violet), erect; pedicels 1–2 mm long; calyx 5–7 mm long, the tube 3.5–4 mm long, scarious, glabrous, the teeth 2.2–2.8 mm long, triangular-subulate, glabrous; pods 1- to 4-seeded; 2n = 16. Alpine meadows at 2900 to 3730 m in Daggett, Duchesne (?), Grand, San Juan, Summit, and Uintah counties; Montana, Wyoming, Colorado, and New Mexico; 11 (0). The dwarf clover is one of our most distinctive, yet poorly collected clovers.

Trifolium parryi Gray Parry Clover. Acaulescent or short caulescent and with 1 elongate internode, 4–25 cm tall, from a caudex and taproot; petioles 0.6–13 cm long; leaflets 3, 5–43 mm long, 1.5–13 (16) mm wide, oblanceolate or obovate to elliptic or oblong, flat, toothed from near the base, glabrous on both sides, acute to obtuse and apiculate; stipules scarious to herbaceous, 6–18 mm long; heads 5- to 20–flowered, subtended by involucral bracts, terminal, on peduncles 1.8–22 cm long, these glabrous or sparingly hairy near the apex; flowers 12–17 mm long, the petals pale to dark pink purple (fading dark violet), erect; pedicels 0–1 mm long; calyx 4–7.1 mm long, the tube 2–3.9 mm long, scarious, glabrous, the teeth 2–3.2 mm long, lance-subulate; pods 1- to 4–seeded; 2n = 16, 32. Alpine meadows, openings in spruce and other coniferous woods, and on talus slopes at 2730 to 3800 m in Daggett, Duchesne, San Juan, Summit, Uintah, and Wasatch counties; Montana, Wyoming, and Colorado; 41 (v). Our materials are segregated as **var. *montanense* (Rydb.) Welsh** [*T. montanense* Rydb; *T. parryi* ssp. *montanense* (Rydb.) J. M. Gillett; *T. inequale* Rydb., type from Bear River Canyon, Uinta Mts.]., but it is doubtfully distinct from from var. *parryi*.

***Trifolium pratense* L.** Red Clover. Caulescent, short-lived perennial, 18–60 cm tall or more, from a taproot, erect or ascending; petioles 0.8–19 cm long; leaflets 3, 11–54 mm long, 8–28 mm wide, elliptic to lanceolate, ovate, or obovate, flat, toothed from near the base (the teeth inconspicuous), long-pilose beneath, glabrous above, obtuse to retuse; stipules scarious to subherbaceous, 11–24 mm long; heads closely subtended by one or more foliose bracts, these often 3–foliolate, sessile, or spreading hairy peduncles to 3 cm long, many-flowered, 22–36 mm long, 20–34 mm wide, axillary, erect; flowers 13–20 mm long, deep red; calyx 7.5–9.7 mm long, the tube 3.2–4.1 mm long, strigose, scarious, the teeth 4.3–5.6 mm long, subulate, pilose; pods 2–seeded; 2n = 14, 28, 48. Cultivated forage plant, escaping but seldom persisting in probably all Utah counties; introduced from Europe; 33 (iv).

***Trifolium repens* L.** White Clover. Caulescent, 8–35 cm tall, the stems stoloniferous, creeping and rooting at the nodes, the petioles and peduncles often arising at right-angles to the stem axis, radiating from a root crown; petioles 1.8–24 cm long; leaflets 3, 5–22 mm long, 4–18 mm wide, obcordate or obovate to oval or elliptic, flat, toothed from near the base, glabrous on both sides, truncate to emarginate; stipules scarious, 3–10 mm long; heads without an involucre, many-flowered, 10–32 mm long, 15–30 mm wide, axillary, on peduncles 6–33 cm long, these glabrous or sparingly pilose, erect; flowers 5–9 (10) mm long, white or pinkish, fading brown, the lower reflexed in age; calyx 3.2–5.4 mm long, the tube 2.2–2.7 mm long, scarious, glabrous, the teeth 1–2.7 mm long, subulate, glabrous; pedicels 1–6.4 mm long; pods 1- to 3–seeded; 2n = 16, 22, 30, 32. Cultivated forage and pasture plant now established through most of Utah; introduced from Europe; 53 (vi).

***Trifolium variegatum* Nutt. in T. & G.** Variegated Clover. Annual, caulescent, with prostrate to erect stems 1–4 (6) dm long; petioles 0.5–4.5 cm long; leaflets 3, 3–27 mm long, 1.5–10 mm wide, obcordate to obovate or oblanceolate, flat, sharply toothed from near the base, glabrous on both sides; stipules herbaceous, ovate, laciniately toothed; heads involucrate, the involucre flaring, saucer-shaped, lobed and lacerate, 3- to 40–flowered, 6–20 mm broad, axillary on peduncles 0.8–6.5 cm long, these glabrous; flowers 5–20 mm long, purplish, often white-tipped, fading brown, ascending to erect; calyx 5- to 20–nerved, the teeth subulate-setaceous, glabrous; pedicels very short; pods 1- to 20–seeded; 2n = 16. Stream banks and roadsides, at 1300 to 1800 m in Box Elder, Cache, Salt Lake, and Weber counties; British Columbia and Montana, south to California and Nevada; 4 (0).

***Trifolium wormskjoldii* Lehm.** Wormskjold Clover. [*T. involucratum* Ortega, not Lam.; *T. willdenovii* Sprengel; *T. fimbriatum* Lindl.; *T. involucratum* var. *fimbriatum* (Lindl.) McDermott; *T. willdenovii* var. *fimbriatum* (Lindl.) Ewan; *T. spinulosum* Dougl. ex Hook.; *T. heterodon* T. & G.]. Caulescent, 12–35 cm tall or more, from a taproot, or sometimes rooting at decumbent stem bases, erect or ascending; petioles 1.2–4 cm long; leaflets 3, 6–30 mm long, 3–14 mm wide, oblanceolate to elliptic or obovate, flat, toothed from near the base, glabrous throughout; stipules herbaceous, 8–15 mm long, many-toothed; heads subtended by an involucre, 20–30 mm wide, many flowered, axillary, erect; flowers 10–18 mm long, reddish to purple; calyx 7–9 mm long, the tube 2.9–3.7 mm long, glabrous, the teeth 4.1–5.3 mm long, subulate, glabrous; pods 1- to 4–seeded; 2n = 16, 32. Meadows at lower middle elevations in Juab, Kane, and Tooele counties; British Columbia and Idaho south to California, New Mexico, and Mexico; 3 (0).

Trigonella L.

Annual or short-lived perennial, caulescent from a taproot; leaves alternate, pinnately 3–foliolate, serrate in the apical 1/3–1/2; stipules herbaceous, distinct, toothed or entire; flowers papilionaceous, borne in axillary racemes or subumbellately disposed; calyx 5–toothed; petals 5, yellow; stamens 10, diadelphous; pods laterally compressed, much surpassing the calyx, several-seeded.

***Trigonella corniculata* (L.) L.** [*Trifolium* (*Melilotus*) *corniculata* L.]. Annual, caulescent, 12–26 cm tall; stipules 3–6 mm long, laciniately toothed; petioles 0.5–4.5 cm long; leaflets 3, the terminal on a short continuation of the rachis, 7–20 mm long, 2.5–16 mm wide, obovate to obcordate, toothed in the apical half, sparingly pilose (mainly along veins beneath), glabrous above, apiculate; peduncles 0.8–2.8 cm long, sparingly pilose; racemes 5- to 16–flowered, 0.8–1.4 cm long; pedicels 1–2 mm long; flowers 5.5–6.5 mm long, yellow; calyx 2.8–3.2 mm long, the tube 1.4–1.8 mm long, sparingly pilose, the teeth 1.2–1.4 mm long, subulate; pods sessile, 12–15 mm long, 2–2.5 mm wide, curved, reticulately veined, the veins leaving the margins at about right-angles, glabrous; 2n = 16. Introduced revegetation plant, not known to persist, but to be expected; known from Sanpete County; introduced from Europe; 2 (0).

Vicia L.

Annual or perennial herbs, clambering, trailing, or climbing; leaves alternate, even-pinnately compound, the rachis terminating in a usually prehensile tendril; stipules herbaceous, entire to semisagittate; leaflets 4–12 or more, very variable; flowers solitary, axillary, or in axillary racemes, papilionaceous; calyx 5–toothed, obliquely campanulate to short-cylindric; petals 5, pink to white, the wings adnate to the keel; stamens 10, diadelphous; style filiform, bearded around the circumference below the stigma; pods oblong, 2- to several-seeded, the valves coiling upon dehiscence.

Hermann, F. J. 1960. Vetches of the United States—native, naturalized, and cultivated. USDA Agr. Handb. 168: 1–84.

1. Flowers 15 or more in dense, secund racemes; introduced plants of cultivated lands and other sites . *V. villosa*
— Flowers 10 or fewer, in secund racemes or otherwise; introduced or indigenous plants 2
2(1). Flowers 5–8 mm long; plants very slender, indigenous in southern Utah .... *V. ludoviciana*
— Flowers 12–25 mm long or more; plants not very slender, indigenous and widespread or cultivated 3
3(2). Flowers 3–10 in pedunculate racemes; plants indigenous, widespread .... *V. americana*
— Flowers 1–3, sessile or very shortly pedunculate in leaf axils; plants uncommon, cultivated .... *V. faba*

Vicia americana Muhl. ex Willd. American Vetch. Perennial, 1.2–12.7 dm tall, the stems glabrous or pubescent; stipules 3–10 mm long, semisagittate, deeply toothed in the lower portion; leaves (excluding tendrils) 2–3 cm long; leaflets 8–16, 3–44 mm long, 1–19 mm wide, linear, elliptic, oblong, ovate, lanceolate, oblanceolate, or obovate, glabrous to pubescent, acute to truncate, rounded, or retuse and apiculate, less commonly toothed apically; tendrils branched or simple; peduncles 1.8–6.7 cm long; racemes 3- to 7 (10)-flowered, the flowers spreading at anthesis; calyx 6.2–8.4 mm long, the tube 4.8–6.5 mm long, the lowermost tooth 0.7–1.9 (2.5) mm long, triangular; flowers 13–22 (25) mm long, pink to pink purple; pods stipitate, the stipe 2.5–4.5 mm long, the body 23–35 mm long, 6–8 mm wide, glabrous; $2n = 14$. This widespread, indigenous vetch is extremely variable with regard to leaflet length-width ratios and shape. Thickness of leaflets and pubescence also varies considerably. Several subordinate taxa have been based on these variations, but continued recognition seems possible only when diagnostic criteria are arbitrarily applied, and even then with great difficulty, especially in plants with dimorphic leaves. More importantly, much of the variation seems to be ecologically influenced, and further recognition of most of the types seems unwarranted. Therefore, it seems best to treat only two infraspecific taxa, the one widespread and common, the other restricted and rare.

1. Leaflets 20–40 mm long, 2–5 mm wide, coriaceous, pubescent with short curved hairs; lateral veins prominent, leaving the midrib at a narrow angle; plants of Daggett and Uintah counties .... *V. americana* var. *minor*
— Leaflets 3–44 mm long, 1–19 mm wide, thin or glabrous, or both; lateral veins prominent or not, leaving the midrib at a wide angle; plants not of Daggett or Uintah counties .... *V. americana* var. *americana*

Var. americana [*V. oregana* Nutt. in T. & G.; *V. americana* var. *oregana* (Nutt.) A. Nels. in Coult. & Nels.; *V. americana* ssp. *oregana* (Nutt.) Abrams; *V. truncata* Nutt. in T. & G.; *V. americana* var. *truncata* (Nutt.) Brewer in Brewer & Wats.]. Sagebrush, pinyon-juniper, mountain brush, ponderosa pine, aspen, and spruce-fir communities at 1270 to 3570 m in all Utah counties, except Box Elder, Daggett, Grand, and Uintah; British Columbia east to Ontario and south to Mexico, Arizona, New Mexico, Kansas, Missouri, and Virginia; 221 (xlii).

Var. minor Hook. [*V. sylvatica* Nutt.; *V. sparsifolia* Nutt. ex T. & G.; *Lathyrus linearis* Nutt. ex T. & G.; *V. linearis* (Nutt.) Greene; *L. dissitifolius* Nutt. ex T. & G.; *V. dissitifolius* (Nutt.) Rydb.; *V. americana* var. *angustifolia* Nees in Neuwied; *V. caespitosa* A. Nels.; *V. trifida* Dietr.]. Pinyon-juniper and mixed desert shrub communities at 1670 to 1830 m in Daggett and Uintah counties; Alberta to North Dakota, south to Colorado and Texas; 9 (i).

Vicia faba L. Broadbean. Annual, 4–10 dm tall or more, the stems glabrous; stipules 4–15 mm long, semisagittate, not or only somewhat toothed; leaves (excluding tendrils) 6–14 cm long; leaflets 2–6, 3–7.5 (10) cm long, 1.3–4 cm wide, ovate-lanceolate to elliptic or obovate, glabrous or with short, curved hairs on veins and margin, obtuse to acute and apiculate; tendrils unbranched; peduncles very short or obsolete; racemes 1- to 4-flowered, the flowers erect-ascending; calyx 12–16 mm long, the shortly cylindric tube 7.5–9.5 mm long, the lowermost tooth 3.5–6.5 mm long, lance-subulate; flowers 25–31 mm long, white, spotted with maroon or blackish violet; pods subsessile to substipitate, the body 7–13 cm long or more 1–3 cm wide, glabrous; $2n = 12, 14, 16$. Sparingly cultivated vegetable plant, mainly for the large, edible seeds, known from Wasatch County; introduced from Asia; 1 (i).

Vicia ludoviciana Nutt. Louisiana Vetch. [*V. exigua* Nutt. ex T. & G.; *V. thurberi* Wats.; *V. producta* Rydb.]. Annual or winter annual, 3–8.5 dm tall, the stems glabrous or puberulent; stipules semihastate to linear-oblong, 1–4 mm long; leaves (excluding tendrils) 1.8–5.5 cm long; leaflets 6–10, 7–28 mm long, 0.6–4 mm wide, linear to oblong or oblanceolate, pilosulous to glabrous, obtuse to acute and mucronate; tendrils branched and prehensile; peduncles 0.4–4.8 cm long; racemes 1 (2)-flowered, the flowers ascending to spreading at anthesis; calyx 2.2–3.3 mm long, the campanulate tube 1.4–1.9 mm long, the lowermost tooth 0.8–1.4 mm long, lance-subulate; flowers 6.3–7.4 mm long, lavender; pods stipitate, the stipe 0.8–1.4 mm long, the body 16–28 mm long, 5–6.2 mm wide, glabrous. Blackbrush, creosote bush, mixed desert shrub, and pinyon-juniper communities at 900 to 1730 m in San Juan, Washington, and eastern Wayne counties; Oregon, California, Nevada, Arizona, New Mexico, Colorado, and Texas; 8 (i).

Vicia villosa Roth Hairy Vetch. Annual or biennial, 5–20 dm tall, the stems spreading-hairy; stipules toothed or entire, 5–15 mm long; leaves (excluding tendrils) 2.3–8 cm long; leaflets 10–18, 8–30 mm long, 1–6 mm wide, linear to oblong or narrowly lanceolate, long-pilose or hirsute on both sides, acute to obtuse and apiculate; peduncles 1.8–7.5 cm long; racemes mainly 15- to 25-flowered, the flowers declined at anthesis; calyx 7–7.8 mm long, the gibbous tube 3.8–4.7 mm long, the teeth 3.1–4.3 mm long, subulate, pilose; flowers 15–17 mm long, pink purple or reddish violet; pods 20–30 mm long, 7–10 mm wide, glabrous; $2n = 14$. Weedy introduction in cultivated lands and other disturbed sites, often along fence-rows in Salt Lake, Sanpete, Utah, Washington, and Weber counties; adventive from Europe; 9 (0).

Wisteria Nutt.

Twining woody vines, unarmed; leaves alternate, odd-pinnate with 7 or more leaflets; stipules ovate to lance-linear, caducous; peduncles opposite the leaves; flowers in elongate, pendulous racemes; calyx turbinate-campanulate, 2-lipped, the upper lip shortly 2-toothed, the lower 3-toothed; corolla papilionaceous, the banner strongly reflexed; stamens 10, diadelphous; petals white, pink, lavender, purple, or blue; pods stipitate, laterally flattened, constricted between the seeds.

1. Longest pedicels 12–28 mm long or more; calyx lobes very short; pods silky to velvety hairy ...... *W. floribunda*
— Longest pedicels 6–11 mm long; calyx lobes almost as long as the tube; pods glabrous .... *W. macrostachya*

Wisteria floribunda (Willd.) DC. Japanese Wisteria. [*Glycine floribunda* Willd.; *Kraunia floribunda* (Willd.) Taub.; *W. macrobotrys* Sieb. & Zucc.]. Twining vines to several meters long; leaves petiolate, 9–35 cm long; leaflets 9–15, 1.3–9 cm long, 9–30 mm wide, ovate to lance-oblong, long-attenuate, pilose to glabrate or glabrous on both sides; peduncles 0.5–5 cm long; racemes pendulous, many-flowered, 25–50 cm long; flowers 15–20 mm long, white, pink, purple, or lavender; calyx tube strigulose, 3.5–4.3 mm long, the upper teeth almost completely connate, the lowermost tooth 1.3–2.1 mm long, triangular; pods 10–18 cm long, 1.5–1.8 cm wide, tapering to the base, velvety hairy in lines. Cultivated ornamental of great beauty in both grafted and free growing forms in Utah and Washington counties; introduced from Japan; 4 (iii).

Wisteria macrostachya Nutt. Nuttall Wisteria. Freely twining vines to several meters long; leaves 15–40 cm long; leaflets 9–13, 2.8–9 cm long, 9–33 mm wide, ovate to lance-oblong, lance-acuminate, pilose along the veins beneath at maturity, pilose to glabrate or glabrous otherwise; peduncles 1–5 cm long; racemes pendulous, many-flowered, 22–55 cm long; flowers 18–22 mm long, lilac purple; calyx tube pilosulous, 4.5–5.5 mm long, the upper teeth cleft, the lowermost tooth 5.5–6.5 mm long, ovate-acuminate; pods 7–12 cm long, 1.1–1.5 cm wide, oblong, glabrous. Ornamental of charm and beauty in Utah, Washington, and Wayne counties; introduced from the southeastern U. S.; 5 (iii).

LENTIBULARIACEAE
Rich. in Poileau & Turpin
Bladderwort Family

Herbaceous, aquatic perennials; leaves alternate or basal, typically submersed and finely divided, bearing insectiverous bladders; flowers perfect, irregular, few to several in scapose racemes; calyx 2–lobed; corolla of 5 united petals, bilabiate, the lower lip saccate or spurred at the base; stamens 2, sometimes with 2 staminodia also; pistil 1, the ovary superior, unilocular, bicarpellate; style 1 or lacking, the stigma 2–lobed; fruit a capsule; x = 6, 8, 9, 11, 21.

Utricularia L.

Perennial herbs without roots or apparently so; leaves all submersed and finely dissected, or sometimes emergent and with leaves somewhat reduced, alternate, bearing bladderlike floats that trap small animals; flowers showy; calyx with 2 entire lobes; corolla with the upper lip entire or obscurely 2–lobed, the lower entire or 3–lobed, produced into a basal spur.

1. Leaf segments terete; bladders not on separate branches; leaves commonly 1–4 cm long or more; flowers 12–18 mm long .... *U. vulgaris*
— Leaf segments flat; bladders borne on separate branches or the flowers less than 10 mm long; leaves 0.5–1.5 cm long 2

2(1). Bladders borne on separate branches; flowers 8–12 mm long................................. *U. intermedia*
— Bladders borne on the leaves; flowers 5–8 mm long .. .. *U. minor*

Utricularia intermedia Hayne ex Schrader Flatleaf Bladderwort. Plants of shallow water, the stems growing on the bottom; leaves alternate, mostly 0.5–1.5 cm long, dissected into flat, linear segments; bladders 2–4 mm long, borne on separate, leafless branches; winter buds 5–7 mm long; scapes 5–20 cm long, emergent; flowers 2–5, yellow, 8–12 mm long, the spur about as long as the lower lip; fruiting pedicels erect. Ponds and lakes in Cache County; widely distributed in North America; Eurasia; 1 (0).

Utricularia minor L. Lesser Bladderwort. Plants of shallow water, growing along the bottom or floating; leaves alternate, mostly 4–10 mm long, divided into flat, linear segments; bladders 1–2 mm long, borne on the leaves; winter buds 2–5 mm long; scapes 5–12 (15) cm long, emergent; flowers 2–9, yellow, 5–8 mm long, the spur short and saccate or lacking; fruiting pedicels recurved. Ponds and lakes in Cache, Uintah, and Utah counties; widely distributed in North America; circumboreal; 4 (0).

Utricularia vulgaris L. Common bladderwort. Plants of shallow to deep water, rarely terrestrial, the stems coarse, usually floating; leaves alternate, 1–4 cm long, divided into numerous terete, filiform segments; bladders 1–3 mm wide, borne on the leaves; winter buds 8–15 mm long; scapes mostly 8–25 cm long, emergent; flowers 5–15 (rarely more), yellow, 12–18 mm long, the spur well developed, curved, shorter than the lower lip; fruiting pedicels recurved; n = 21, 22. Ponds, lakes, sluggish streams, or merely wet mud at 1340 to 3050 m in Cache, Duchesne, Garfield, and Millard counties; widespread in North America; circumboreal; 12 (i). This and the other bladderwort species in Utah are poorly represented in collections, usually present only as vegetative fragments taken with other aquatic vegetation. In 1983 Scipio Lake in eastern Millard County turned golden yellow with the flowers of common bladderwort. The entire lake surface was clogged with *Utricularia* and other floating aquatics.

LIMNANTHACEAE R. Br.
Meadowfoam Family

Low, weak, glabrous, annual herbs; leaves alternate, pinnately dissected, estipulate; flowers perfect, regular, solitary, axillary, mostly 3–merous; sepals distinct or essentially so, valvate, persistent; petals convolute, distinct; stamens 6–10, distinct, glandular at the base; pistil 1, the ovary 3–lobed, the carpels almost separate, connected by a common gynobasic style; ovules 1 per locule; fruit separating into 1–seeded segments (nutlets); x = 5.

Floerkea Willd.

This genus has the characters as noted for the family.
Floerkea proserpinacoides Willd. False Mermaid. Plants weak, decumbent-ascending or erect; stems 3–30 cm long, simple or branched; leaves pinnately dissected (compound), with mostly 3–15 segments, these typically 3–15 mm long and 1–4 mm wide, elliptic to oblanceolate; pedicels axillary, solitary, 1–4 cm long; sepals green,

foliaceous, ovate to lance-ovate, acute to acuminate, 2–6 mm long; petals white, 1–2 mm long, with small pyramidal tubercles; n = 5. Mountain brush, sagebrush, aspen, and spruce-fir communities, often in spring, pond, and stream margins and in other moist situations, at 1890 to 2625 m in Box Elder, Cache, Duchesne, Emery, Morgan, Salt Lake, Summit, Utah (?), and Wasatch counties; British Columbia to Nova Scotia, south to California, New Mexico, and Georgia; 14 (0).

LINACEAE S. F. Gray
Flax Family

Annual or perennial herbs; leaves alternate simple; flowers perfect, regular; sepals 5, persistent; petals 5, distinct, blue, yellow, or orange, soon falling; stamens 5, the filaments united at the base, (sometimes with 5 stammodia as well); pistil 1, the ovary superior, of 5 united carpels; styles distinct or united below; capsule 10-loculed and 10-seeded; x = 6–11+.

Linum L.

A single genus with characters of the family.

Rogers, C. M. 1968. Yellow-flowered species of *Linum* in Central America and western North America. Brittonia 209: 107–135.
———. 1979. A new combination in *Linum*. Phytologia 41:447–448.
Mosquin, T. 1971. Biosystematic studies in the North American species of *Linum*, section Adenolinum (Linaceae). Canad. J., Bot. 49: 1379–1388.

1. Petals blue, or rarely white; stigma elongate, or at least longer than wide; staminodes alternating with stamens .... *L. perenne*
— Petals yellow or orange; stigmas capitate; staminodes lacking 2

2(1). Styles distinct; sepals not or scarcely aristate, the outer ones entire or with a few glandular teeth; fruit separating into ten 1-seeded segments .... *L. kingii*
— Styles united nearly to the apex; sepals spinose-aristate, the outer with numerous glandular teeth; fruit separating into five 2-seeded segments 3

3(2). Plants short-lived perennials (rarely annual); sepals persistent in fruit; secondary partitions of carpels incomplete, lacerate-fringed .... *L. subteres*
— Plants annual; sepals deciduous; secondary partitions complete 4

4(3). Stems densely hirtellous with spreading hairs; petals mostly orange to salmon, the bases reddish . *L. puberulum*
— Stems glabrous or scaberulous; petals yellow to yellow orange 5

5(4). Styles 2–4 mm long; stems scaberulous below, usually simple below the middle .... *L. australe*
— Styles 4.5–7 mm long; stems glabrous, usually branched from near the base .... *L. aristatum*

Linum aristatum Engelm. Broom-flax. Annual, glabrous herbs from taproots; stems 8–45 cm tall, usually branched from near the base; leaves mainly 3–15 mm long, appressed-ascending, 0.5–1 mm wide; stipular glands present; flowers terminal, the peduncles mostly 5–35 mm long; sepals 5–9 mm long, lanceolate, acuminate to a slender awn-tip, glandular ciliate, hardly deciduous; petals yellow to yellow orange, caducous, mostly 7–12 mm long; anthers 0.8–1 mm long; staminodes lacking; style 4–7 mm long, cleft at the apex; fruit ellipsoid, 3.5–4.5 mm high, opening at maturity into five 2-seeded segments, the secondary partitions hyaline, almost complete; n = 15. Blackbrush, ephedra, sand sagebrush, Vanclevea, sagebrush, pinyon-juniper, and ponderosa pine communities, in sand dunes and other sandy sites, at 1125 to 1985 m in Emery, Garfield, Grand, Kane, San Juan, and Wayne counties; Colorado, Arizona, New Mexico, Texas, and Mexico; 30 (iii).

Linum australe Heller Small Yellow-flax. Annual, scabrous herbs from taproots; stems 10–40 cm tall, usually branched from above the base; leaves mainly 6–20 mm long, 0.5–1.5 mm wide, appressed-ascending or ascending; stipular glands absent or present; flowers terminal, the peduncles mainly 3–15 mm long; sepals 4.5–7 mm long, lanceolate, alternate to a short awn-tip, glandular ciliate, deciduous; petals yellow or yellow orange, not reddish at the base, mostly 5–9 mm long; anthers 0.4–0.8 mm long; style 2.5–4 mm long, subequal to the stamens, cleft at the tip, the stigmas capitate; fruit 3.5–4.5 mm long, ovoid, splitting at maturity into five 2-seeded segments, the secondary partitions hyaline, almost complete; n = 15. Ponderosa pine, sagebrush, or pinyon-juniper communities at ca 2350 to 2745 m in Garfield, Kane, San Juan, and Sevier counties; Alberta south to Nevada, Arizona, and northern Mexico; 6 (0).

Linum kingii Wats. [*L. kingii* var. *sedoides* T. C. Porter in Hayden, type from the Uinta Mts.; *L. kingii* var. *pinetorum* Jones, type from above Tropic]. Perennial, glabrous, glaucous herb from a caudex and stout taproot; stems 5–35 cm long, usually branching from the base; leaves 2–20 (25) mm long, 1–3 mm wide, ascending to spreading-ascending, linear to oblong; flowers in cymose clusters; sepals 2.5–3.5 mm long, ovate to oblong, acute or acutish, all or only the inner glandular-ciliate; petals yellow, 5–12 mm long, tardily deciduous; anthers 1.5–2.5 mm long; staminodia lacking; styles distinct, 4–7 mm long; fruit ovoid, 2.5–4 mm long, incompletely 10-loculed, opening from the top, with both primary and secondary partition long-ciliate. Salt desert shrub (rarely), juniper, ponderosa pine, Douglas fir, sagebrush, lodgepole pine, aspen, and spruce-fir communities, mainly in calcareous substrates, at 1400 to 3400 m in Box Elder, Cache, Carbon, Daggett, Duchesne, Emery, Garfield, Iron, Juab, Kane, Millard, Salt Lake, Sanpete, Sevier, Summit (type from Uinta Mts.), Uintah, Utah, Wasatch, Washington, and Weber counties; Nevada, Idaho, Colorado, and Wyoming; 45 (viii).

Linum perenne L. Blue flax. Perennial, glabrous, glaucous herbs from a caudex and stout taproot; stems mostly 15–80 cm tall, usually simple; leaves 4–30 mm long, 0.5–4 mm wide, ascending to spreading-ascending, linear to oblong, the tips subulate; flowers in cymose clusters; sepals 3.5–6 mm long, elliptic to broadly so, entire; petals blue with a whitish or yellowish base, or white, 12–23 mm long; staminodia subulate, small, ca 1 mm long; styles distinct; fruit ovoid to globose, 6–8 mm long, opening from the top, with both primary and secondary sutures fringed-ciliate; 2n = 18. Sagebrush, pinyon-juniper, mountain brush, mixed desert shrub, Salina wildrye, aspen, and spruce-fir communities at 1220 to 3355 m in all Utah counties; Alaska to Mackenzie, south to Mexico; Asia; 123 (xi). The North American phase of the *perenne* complex differs from the Eurasian counterpart primarily in homostylic flowers. Possibly our material is best treated as **ssp. *lewisii* (Pursh) Hulten** [*L. lewisii*

Pursh; *L. perenne* var. *lewisii* (Pursh) Eaton & Wright]. The cultivated flax **L. usitatissimum** L. occurs rarely as an escape. It is distinguished from *L. perenne*, by being annual and by having linear ciliate sepals and styles connected below the middle.

Linum puberulum (Engelm.) Heller [*L. rigidum* var. *puberulum* Engelm.]. Annual, hirtellous herbs from a taproot; stems 4–20 cm tall, usually branched from the base; leaves 3–15 mm long, 0.5–1 mm wide, appressed-ascending to ascending, linear subulate; stipular glands often present; flowers terminal on bracteate peduncles mostly 0.5–2.5 cm long; sepals 4–7 mm long, lanceolate, glandular ciliate; petals orange to salmon, with a dark base, 8–15 mm long; staminodia lacking; styles 3.5–6 mm long, cleft apically; fruit ovoid, 3–4 mm long, opening at maturity into five 2–seeded segments, the secondary partitions hyaline, almost complete; n = 15. Salt desert shrub and pinyon-juniper communities at 1125 to 2625 m in Duchesne, Emery, Garfield, San Juan, Sevier, Wasatch, and Wayne counties; Wyoming and Nebraska, south to California, Arizona, Texas, and Mexico; 16 (iii).

Linum subteres (Trel.) Winkler Utah Yellow-flax. [*L. aristatum* var. *subteres* Trel., type from Willow Creek, San Juan County]. Perennial (or annual), glabrous, glaucous herbs from taproots; stems mainly 10–45 cm tall, usually branched above; leaves 4–20 mm long, 0.5–2 mm wide, oblong to linear-subulate; stipular glands lacking; flowers terminal on slender sparingly bracteate peduncles mainly 0.5–6 cm long; sepals 4–7 mm long, lanceolate, acuminate to alternate, long; anthers 1–2 mm long; staminodes lacking; style 4.5–9 mm long, cleft apically; fruit ovoid, 3.5–4.5 mm high, opening at maturity into five 2–seeded segments, the secondary partitions prominently lacerate-fringed, incomplete; n = 15. Mixed desert shrub and pinyon-juniper communities, often in sand, at 1280 to 2075 m in Duchesne, Emery, Garfield, Kane, Millard, San Juan, Uintah, Washington, and Wayne counties; Nevada and Arizona; 42 (xiv).

LOASACEAE Dumort.
Stickleaf Family
Contributed by
L. C. Higgins (*Eucnide*; *Petalonyx*)
K. H. Thorne (*Mentzelia*)

Annual, perennial, or shrubby plants with barbed or sometimes stinging hairs; leaves opposite or alternate, entire or lobed, estipulate; flowers regular, bisexual, solitary or cymose; floral tube adnate to the ovary; sepals 4–5, persistent; petals 4–5, distinct or united, sometimes with petaloid staminodia alternate with the petals; stamens 5 to many; filaments narrow to petaloid, free or fused in fascicles; ovary inferior, 1– to 3–loculed; style 1, usually persistent; ovules 1 to many; fruit capsular, with 1 to many seeds; x = 7–15+.

1. Stamens 5; style entire; seed solitary; plants shrubby .... *Petalonyx*
— Stamens many; style often cleft; seeds few to many; plants herbs or subshrubs 2
2(1). Stamens inserted on the petals; placentae 5; petals cream; plants suffruticose, of Washington County . *Eucnide*
— Stamens not inserted on the petals; placentae 3; petals yellow to cream; plants typically herbaceous, widely distributed .... *Mentzelia*

Eucnide Zucc.

Suffruticose; herbage armed with short, barbed and stinging hairs; leaves alternate, petiolate, toothed or lobed; flowers mostly in terminal, bracteate cymes; sepals 5, persistent; petals 5, deciduous, united at the base, yellow or yellowish; stamens numerous, the filaments connate in a broad band below and adnate to the petals; style 5–cleft, the lobes often twisted; ovary 1–loculed and with 5 prominent placentae; ovules numerous; capsule obovoid, opening by 5 valves; seeds numerous, minute, longitudinally striate.

Eucnide urens (Gray) Parry Rocknettle. [*Mentzelia urens* Gray]. Clump-forming, 3–8 dm tall, the branches spreading or decumbent, straw colored, villous, also hispid; leaves ovate, coarsely toothed, 2–6 cm long and about as broad; petioles short or lacking; flowers corymbose; pedicels stout, 6–12 mm long; sepals lance-oblong, 20–25 mm long; petals pale yellow or cream, obovate, 2.5–4 cm long, the mucronate tip hispid; stamens 10–15 mm long, numerous; capsule obovoid, 10–12 mm long. Creosote bush and Joshua tree communities at ca 700 to 915 m in Washington County (type from near St. George); Arizona, Nevada, California, and Mexico; 7 (iv).

Mentzelia L.

Annual, biennial, or perennial, herbaceous or suffrutescent from a taproot and with a caudex in some; stems simple or branched, glabrous or hairy, the outer bark white, shining, exfoliating; leaves alternate, simple, entire, crenate, dentate, sinuate, lobed, or pinnatifid; hairs glochidiate, hispid, or pustulate; inflorescence terminal, of solitary or few- to many-flowered cymes, usually subtended by bracts; calyx lobes distinct; petals 5 or 10, distinct; stamens numerous, filamentous or some petaloid, some cuspidate apically; ovary unilocular; placentae parietal; ovules horizontal or pendulous; fruit a capsule; seeds flat, winged or not, or irregularly angled or prismatic.

1. Plants annual; seeds pendulous in the capsule, irregularly angled or prismatic 2
— Plants perennial, biennial, or rarely annual; seeds flat, winged or not, horizontal in the capsule 7
2(1). Seeds prismatic, grooved on the angles ... *M. dispersa*
— Seeds irregularly angled 3
3(2). Leaves entire, rarely slightly dentate; seeds 1.8 mm long; plants of Uintah and Grand counties .... *M. thompsonii*
— Leaves pinnatifid, lobed, or occasionally entire; seeds less than 1.2 mm long 4
4(3). Capsules erect, straight or rarely some mature ones curved about 90 degrees 5
— Capsules, at least the mature ones, sigmoid or C-shaped 6
5(4). Bracts lanceolate, linear, pinnatifid, lobed, or entire, without a whitish spot at the base dorsally; petals yellow, to 8 mm long; plants common and widespread .... *M. albicaulis*
— Bracts oblanceolate, trilobed, angled, or entire, with a whitish spot at the base dorsally; petals yellow, to 6 (8) mm long; plants uncommon and restricted .... *M. montana*
6(4). Bracts lanceolate, entire or remotely lobed; petals to 10 mm long; plants of Washington County .... *M. californica*

—	Bracts ovate, entire or remotely lobed or toothed; petals to 5 mm long; seed less than 1 mm long; plants of Kane, Tooele, and Washington counties .. *M. obscura*
7(1).	Filaments petaloid, distinctly cuspidate apically; plants of Washington County .... *M. tricuspis*
—	Filaments petaloid or filamentous but, if the former, not cuspidate apically; plants of various distribution .. 8
8(7).	Petals 5 9
—	Petals 10, the inner 5 somewhat narrower and shorter 14
9(8).	Petals 5–8 cm long; outer series of stamens filamentous; plants mainly 3–10 dm tall .... *M. laevicaulis*
—	Petals less than 5 cm long; outer series of stamens petaloid 10
10(9).	Petals longer than 2 cm; stems branching above; leaves broadly lanceolate, linear, or oblong, entire or coarsely dentate, mainly 10–25 cm long; plants of moist habitats .... *M. rusbyi*
—	Petals less than 2 cm long; stems, leaves, and habitats various 11
11(10).	Outer petals pubescent dorsally; leaves ovate, oblanceolate, or broadly lanceolate, crenulate or dentate; plants of Grand County .... *M. marginata*
—	Outer petals glabrous; leaves and distribution various 12
12(11).	Leaves entire or crenulate; plants of Sanpete and Sevier counties .... *M. argillosa*
—	Leaves dentate, pinnately lobed, or pinnatifid; plants not of Sanpete or Sevier counties 13
13(12).	Leaves ovate, broadly elliptic or lanceolate, dentate; outer petaloid stamens spatulate; plants of Grand County .... *M. schultziorum*
—	Leaves pinnatifid or trilobed; outer petaloid stamens broadly spatulate; plants of Carbon, Duchesne, Emery, Piute, and Uintah counties ..... *M. multicaulis*
14(8).	Outer petals pubescent dorsally; stems usually reddish beneath exfoliating bark; plants of southeastern Utah .... *M. cronquistii*
—	Outer petals glabrous or rarely sparsely ciliate apically; stem stramineus beneath exfoliating bark 15
15(14).	Leaves narrowly oblong, coarsely dentate to shortly lobed; petals sometimes ciliate apically; seeds narrowly winged; plants biennial .... *M. pumila*
—	Leaves oblong, lanceolate, or ovate, dentate; petals and seeds various; plants annual, biennial, or perennial 16
16(15).	Plants annual or apparently so, divaricately branching above; leaves ovate, dentate; seeds to 4 mm long, broadly winged .... *M. pterosperma*
—	Plants biennial or perennial, variously branched or simple; leaves various; seeds usually less than 3.5 mm long, winged or not 17
17(16).	Leaves lanceolate, oblong, dentate, broadly lobed to coarsely pinnatifid, the sinuses broad; outer stamens petaloid; seeds ca 3.5 mm long, winged .. *M. multiflora*
—	Leaves various in outline, pinnatifid almost to the midrib, the lobes narrow; outer stamens rarely petaloid; seeds brownish gray, 2.5–3.5 mm long, winged or almost wingless 18
18(17).	Plants branched above; seed almost wingless; plants to be sought in eastern Utah .... *M. laciniata* (Rydb.) Darlington
—	Plants branched from the base; seeds winged .... *M. pumila*

Mentzelia albicaulis Dougl. ex Hook. [*M. albicaulis* var. *gracilis* (Rydb.) Darlington]. Annual; stems 1–4 dm tall, simple or branched, white or greenish white, decumbent to erect, usually glabrous; leaves sessile, the lower ones oblanceolate, lanceolate, or linear, 3–10 (15) cm long, entire, lobed, or pinnatifid, the upper ones lanceolate or linear, the base sometimes clasping, lobed or pinnatifid; bracts linear or lanceolate, entire, angled, lobed, or pinnatifid, without a whitish spot at the base dorsally; inflorescence terminal; flowers 1–3, the 2 lateral ones pedunculate, subtended by bracts; calyx lobes erect, 2–4 mm long; petals 5, obovate, 2–8 (9) mm long; capsule cylindric, tapering to the narrow base, 10–17 (23) mm long, 2–3 (4) mm wide, erect at maturity, straight or curved; seeds irregularly angled, white or tan, tuberculate; n = 36. Warm, mixed, and salt desert shrub, sagebrush, grassland, pinyon-juniper, and ponderosa pine communities, typically on sandy or gravelly soils, at 1065 to 2420 m in all Utah counties, except Morgan, Rich, Summit, and Wasatch; British Columbia south to California, Texas, and Mexico; 153 (iii).

Mentzelia argillosa Darlington Arapien Stickleaf. Perennial, 2–3 dm tall; stems glabrous, white, branched from the base; leaves oblanceolate to elliptic, entire or crenulate, 1–4.1 cm long, 0.6–1 cm wide; bracts linear, entire; inflorescence 1- or 2-flowered; calyx lobes 5–7.5 mm long; petals 5, obovate, 5–20 mm long; outer stamens somewhat dilated; calyx turbinate to campanulate, 8–10 mm long, ca 6 mm wide; seeds ca 1.8 mm long, scarcely winged; n = 11. Salt desert shrub and pinyon-juniper communities on the Arapien Shale Formation, at 1705 to 1895 m in Sanpete and Sevier (type from Vermillion) counties; Garfield County, Colorado; 12 (i).

Mentzelia californica Thompson & Roberts Annual, 1–4 dm tall; stems branched or simple, white, glabrous, decumbent to erect; lower leaves lanceolate to narrowly oblong, entire or pinnatifid, 3–7 cm long, 5–15 mm wide; middle leaves lanceolate, pinnatifid or lobed to entire; upper leaves and bracts lanceolate, remotely lobed or entire; inflorescence 1- to 3-flowered, the 2 lateral flowers pedunculed, subtended by bracts; calyx lobes 2–6 mm long, erect; petals obovate, 3–10 mm long, yellow, spotted orange basally; capsules 2–3.5 cm long, 1–3 mm wide, sigmoid or C-shaped at maturity; seed irregularly angled, whitish or tan, tuberculate; n = 27. Creosote bush, Joshua tree, pinyon-juniper, and sagebrush communities at ca 840 m in Washington County; Washington to Colorado, south to California and Texas; 3 (0).

Mentzelia cronquistii Thompson & Prigge Cronquist Stickleaf. Perennial, the stems erect, simple, or branched from the base, 1–4 dm tall, reddish beneath the exfoliating white bark; lower leaves lanceolate to oblanceolate, mainly 5–14 cm long, 0.8–3.7 cm wide, dentate or with rounded lobes; upper leaves narrowly lanceolate, dentate or lobed; bracts linear, entire; calyx lobes 7–10 mm long; petals 10, narrowly obovate, 12–15 mm long, the outer ones pubescent, the inner shorter and narrower; outer stamens petaloid, to 2.5 mm wide; capsule 8–10 (15) mm long, campanulate; seeds ovate, ca 2.5 mm long, shortly winged, tan to dark gray, tessellate; n = 10. Sagebrush, rabbitbrush, salt desert shrub, and pinyon-juniper communities, in various substrates, at 975 to 2075 m in Emery, Garfield, Grand, Kane, San Juan (type from 10 mi E of Hite), Sevier, and Wayne counties; Colorado, Arizona, and New Mexico; 63 (i). This species is similar to *M.*

multiflora, with which it is sympatric, but can be distinguished by the pubescent petals, reddish under bark, and narrowly winged seed.

Mentzelia dispersa Wats. [*M. dispersa* var. *obtusa* Jepson; *M. albicaulis* var. *integrifolia* Wats., type from Antelope Island]. Annual, 1–3 dm tall; stems erect, simple or branched, white to grayish green, usually glabrous below; leaves sessile, the lower ones oblanceolate, lanceolate or linear, 2–8.3 cm long, 9–20 mm wide, entire, remotely lobed or dentate; upper leaves lanceolate, entire, sometimes angled or toothed; inflorescence terminal, 1– to 3–flowered, the 2 lateral ones pedunculate, subtended by bracts; calyx lobes 1–2 mm long, erect; petals 5, obovate, 1–3 (6) mm long; capsules narrowly cylindric, 10–15 (21) mm long, 1–3 mm wide, erect at maturity or slightly curving; seeds prismatic, grooved on 3 angles, muricate, gray brown; n = 9. Mountain brush, sagebrush, and aspen communities at 1555 to 2870 m in Box Elder, Cache, Carbon, Davis, Juab, Millard, Rich, Sanpete, Summit, Tooele, Utah, Wasatch, and Weber counties; Oregon to North Dakota, south to California, Nevada, and New Mexico; 29 (iii).

Mentzelia laevicaulis (Dougl.) T. & G. [*M. acuminata* Tidestr.; *M. ornata* Torr., not T. & G.]. Perennial; stems stout, coarse, 3–10 dm tall, branched or simple, glabrous below, hirsute above; basal leaves 2–15 (20) cm long, oblanceolate to lanceolate, dentate, coarsely dentate, or dentate lacerate; upper leaves lanceolate to lance-ovate, dentate or dentate-lacerate, acuminate; bracts linear to lanceolate, entire or remotely dentate, long-acuminate; inflorescence terminal; calyx lobes 2–4.2 cm long; petals 5, lanceolate, 4.5–8.2 cm long; stamens filiform, or the outer slightly dilated; capsule broadly turbinate, obtuse basally, 3.5–4 (5.3) cm long, usually striate; seeds 4.1 mm long, white, broadly winged; n = 11. Salt desert shrub, mountain brush, sagebrush, rabbitbrush, and pinyon-juniper communities at 1370 to 2440 m in Beaver, Box Elder, Davis, Duchesne, Garfield, Iron, Juab, Millard, Rich, Sanpete, Sevier, Summit, Utah, and Weber counties; Washington to Montana, south to California, Nevada, and Wyoming; 74 (ii).

Mentzelia marginata (Osterh.) **Thompson & Prigge** [*Nuttallia marginata* Osterh.]. Biennial or short-lived perennial; stems erect, branched from the base, 1–2.5 dm tall, white, hirsute; rosette leaves oblanceolate, crenate or sinuate; lower cauline leaves oblanceolate to broadly ovate, crenate or somewhat lobed, mainly 2–6 cm long, the upper ones ovate, crenate, sometimes clasping; bracts linear, entire; calyx lobes 5–8 mm long; petals 5, ovate, 10–13 mm long, pubescent dorsally; outer stamens dilated, to 3 mm wide, pubescent at the base; capsules campanulate, 8–12 mm long; seeds ovate, ca 2 mm long, narrowly winged, tan to dark gray; n = 10. Salt desert shrub, on fine-textured substrates, at ca 1525 m in Grand County; Mesa and Ouray counties, Colorado; a Plateau endemic; 2 (0).

Mentzelia montana (A. Davidson) **A. Davidson** [*Acrolasia montana* A. Davidson]. Annual; stems erect, sparingly branched, 2–4 dm tall; basal leaves lanceolate, oblong, or ovate-lanceolate, entire, lobed, or pinnatifid, the upper ones similar but smaller, usually entire or lobed apically; bracts lanceolate to lance-ovate, lobed, angled, or entire with a whitish spot basally on the dorsal side; inflorescence of 1–3 flowers, the 2 lateral ones pedunculate and subtended by bracts; calyx lobes 1–3 (4) mm long; petals 5, obovate, 2–8 mm long, sometimes with an orange spot basally; capsule cylindric, tapering to a narrow base, 5–15 mm long, 2–3 mm thick, erect, rarely slightly curving; seed irregularly angled, tuberculate; n = 18. Desert shrub, sagebrush, mountain brush, pinyon-juniper, and ponderosa pine communities at 1340 to 1710 m in Millard, Salt Lake, San Juan, Tooele, and Uintah counties; Oregon to Colorado, south to California, Texas, and Mexico; 12 (0). This taxon is difficult to separate from *M. albicaulis* unless the specimens have all taxonomic features.

Mentzelia multicaulis (Osterh.) **Goodman** [*Touterea multicaulis* Osterh.; *M. pumila* var. *multicaulis* (Osterh.) A. Nels.]. Perennial, from a branching woody caudex; stems diffusely branched, glabrous; leaves oblong or lanceolate in outline, deeply pinnatifid, the lobes narrow, obtuse, 3– or more-lobed, the apical lobe elongate; upper leaves and bracts entire, linear; inflorescence of 1–3 flowers; calyx lobes 10–12 mm long; petals 5, lanceolate, 7–16 mm long; outer petaloid stamens spatulate, broader than the petals but shorter; capsule 10–15 mm long, more or less campanulate; seeds 1.5–3.4 mm long, tan or whitish, winged or essentially wingless; n = 11. Two varieties are present.

1. Plants 2–3 dm tall, suffrutescent; leaves pinnatifid almost to the midrib, the lobes narrow, obtuse apically; seed scarcely winged; plants of the Uinta Basin . . . .
   . . . . *M. multicaulis* var. *multicaulis*
— Plants mainly 2–5 dm tall, suffrutescent; leaves trilobed, the lateral lobes near the blade base, the apical lobe elongate; seeds winged 3–3.5 mm long; plants of eastern Carbon and Emery counties . . . . *M. multicaulis* var. *librina*

Var. *librina* Thorne & F. G. Sm. Sagebrush, rabbitbrush, and pinyon-juniper communities at ca 1890 m on Mancos Shale and Price River formations, in Carbon and Emery (type from mouth of Horse Canyon) counties; endemic; 4 (0).

Var. *multicaulis* Sagebrush, rabbitbrush, and pinyon-juniper communities at 1525 to 2135 m on Moenkopi, Green River, Uinta, and Duchesne River formations, in Carbon, Duchesne, Piute, and Uintah counties; Rio Blanco County; Colorado; a Plateau endemic; 37 (iii).

Mentzelia multiflora (Nutt.) **Gray** Perennial, from a stout taproot; stems erect, branched from the base, rarely simple, 4–8 dm tall, glabrous below, hirsute above, tan beneath exfoliating bark; lower leaves lanceolate to oblanceolate, 2.5–12 cm long, pinnatifid, the lobes and sinuses rounded or the lobes acute, or dentate, the middle and upper leaves with similar, pinnatifid lobes but smaller; bracts linear, sometimes with a few remote lobes or revolute; inflorescence corymbose, of 1–3 flowers; calyx lobes ca 10 mm long; petals 10, oblong or oblanceolate, 9–17 (20) mm long, 6–8 mm wide, the inner 5 somewhat shorter and narrower; outer stamens petaloid, to 2 mm wide or merely dilated; capsule urceolate to campanulate, 1.5–2.1 cm long, ca 10 mm wide; seed to 3.5 mm long, broadly winged, whitish or tan; n = 9. Two varieties are present.

1. Plants biennial or short-lived perennial; leaves remotely dentate or shortly lobed, or some of them entire; known from Washington County . . . . . . *M. multiflora* var. *integra*
— Plants perennial; leaves pinnatifid to lobed, rarely if ever some of them entire; known from southeastern Utah . . .
   . . . . *M. multiflora* var. *multiflora*

Var. integra Jones [*M. integra* (Jones.) Tidestr.; *Nuttallia lobata* Rydb., type from near St. George, not *M. lobata* Walp.]. Warm desert shrub, chaparral, salt desert shrub, and pinyon-juniper communities at 700 to 2135 m in Beaver, Iron, Kane, and Washington (type from near Rockville) counties; Arizona and Nevada; 33 (iv). This plant has passed under practically every name recognized in the Utah mentzelias. It is the Virgin River vicariad of var. *multiflora*.

Var. multiflora [*Bartonia multiflora* Nutt.; *M. pumila* Kuntze, not (Nutt.) T. & G.; *M. pumila* var. *multiflora* (Nutt.) Urban & Gilg; *M. speciosa* Osterh.]. Salt desert shrub, warm desert shrub, sagebrush communities at 1250 to 1830 m in Emery, Garfield, Grand, Kane, Piute, San Juan, Sevier, and Wayne counties; Wyoming south to New Mexico and Texas, west to California; Mexico; 49 (0).

Mentzelia obscura Thompson & Roberts Annual; stems erect, branching from the base, 1–3 dm tall; basal leaves lanceolate, usually pinnatifid or acutely lobed; upper leaves ovate-lanceolate, entire or with a few, acute lobes; bracts ovate, entire, lacking a white spot at the base on the dorsal side; calyx lobes 2–3 mm long; petals 5, ovate, 3–5 (6) mm long, yellow with an orange spot basally; capsule cylindric, tapering to the base, sigmoid or c-shaped at maturity; seed irregularly angled, the margins rounded, the surface papillose, 0.8–1 mm long, white or tan; n = 18. Disturbed sites, along roads or in wash bottoms in creosote bush, Joshua tree, salt desert shrub, sagebrush, and pinyon-juniper communities at 915 to 1745 m in Kane, Tooele, and Washington counties; Arizona, Nevada, and California; Mexico; 6 (0).

Mentzelia pterosperma Eastw. Annual or perennial, flowering the first year; stems white, hirsute, divaricately branching above or below; rosette leaves oblanceolate, petiolate, crenate or with short, rounded lobes 3–7.4 cm long; lower cauline leaves lanceolate, crenulate or dentate; bracts linear, entire; calyx lobes 6–11 mm long; petals 10, obovate to lanceolate, 9–15 (20) mm long, 4–8 mm wide; outer stamens petaloid, 5–8 mm wide; capsule obconic, 9–14 (20) mm long, 8–10 mm wide; seeds ovate, to 4.1 mm long, broadly winged, white, granular; n = 11. Creosote bush, shadscale, sagebrush, and pinyon-juniper communities at 700 to 1525 m in Duchesne, Emery, Grand, Iron, Kane, Millard, San Juan (type from Willow Creek), Sevier, Tooele, Uintah, Washington, and Wayne counties; California, Nevada, Arizona, and Colorado; 21 (ii). In Washington County the stems of *M. pterosperma* are typically divaricately branched above the middle, the leaves are larger, broader, and dentately margined, but in northern areas the plants are compact, branching from the base, the flowers and leaves somewhat smaller, and these are crenate to sinuate marginally.

Mentzelia pumila (Nutt.) T. & G. [*Bartonia pumila* Nutt.]. Biennial or short-lived perennial from a stout taproot; stems whitish, simple or branched and more or less flexuous, 2–6 dm tall; rosette leaves lanceolate, oblanceolate or oblong, 8.5–10 (20) cm long, petiolate, sinuate-dentate, shortly lobed or pinnatifid, the apex acute or obtuse; lower cauline leaves lanceolate or oblong, shortly petiolate, 1.9–8.5 cm long, sinuate-dentate to shallowly pinnatifid, the sinuses broad, or deeply pinnatifid, the sinuses narrow; middle and upper leaves sessile, lanceolate, irregularly dentate or lobed, the sinuses broad, more or less clasping, or the leaves pinnatifid to laciniate, the sinuses narrow and the base not clasping; bracts linear, entire; inflorescence of 1–3 flowers; calyx lobes 4–8 (10) mm long; petals 10, lanceolate or obovate, sometimes ciliate at the tip, 9–15 mm long; outer stamens petaloid; capsule urceolate or cylindric, tapering to a narrow base, sometimes striate 15–20 mm long; seeds 2.5–3.5 mm long, narrowly or broadly winged, tan or gray; n = 11. Two varieties are treated.

1. Leaves deeply pinnatifid, the lobes obtuse or acute; upper leaves oblong, cut almost to midrib; seeds gray, winged, 3–3.5 mm long .... *M. pumila* var. *lagarosa*
— Leaves sinuate-dentate, shortly lobed, or merely pinnatifid, cut less than half way to midrib; upper leaves lanceolate to ovate-lanceolate, the base apparently clasping, sinuate, dentate, or lobed; seed tan, narrowly winged, 2.5–3 mm long .... *M. pumila* var. *pumila*

Var. lagarosa Thorne Mixed desert shrub, sagebrush, grass, pinyon-juniper, and bristle cone pine communities on limestone, clay, shale, and igneous gravels at 1765 to 2745 m in Duchesne, Garfield, Kane, Piute, Sevier, Uintah (type from near Watson), and Wayne counties; Lincoln County; Nevada; 9 (i).

Var. pumila Sagebrush, grass, scattered juniper, and mountain brush communities at 1675 to 2045 m in Daggett and Duchesne counties; Colorado and Wyoming; 5 (0).

Mentzelia rusbyi Woot. [*M. pumila* var. *rusbyi* (Woot.) Urban & Gilg; *M. nuda* var. *rusbyi* (Woot.) Harrington]. Biennial or short-lived perennial from a stout taproot; stems usually branched above or simple, whitish, hirsute, 2–5 dm tall; rosette leaves oblanceolate or oblong, entire, sinuate or remotely dentate, 5–20 (25) cm long; lower leaves lanceolate or oblong, coarsely dentate or dentate, 7–15 cm long, ca 10–15 mm wide, petiolate; middle and upper leaves sessile, ovate-lanceolate, 3–7 cm long, ca 2.1 cm wide, coarsely dentate; bract on lower 1/3 of capsule lanceolate, entire, dentate, or laciniate; inflorescence congested; calyx lobes 5.4–8 cm long; petals 5, creamy yellow, obovate to oblanceolate, 1.5–3 cm long, 4–8 mm wide; outer stamens petaloid, to 1.4 cm wide; capsules cylindrical, 15–30 mm long, 10–13 mm wide; seeds ovate, to 4 mm long, broadly winged, whitish tan, truncately tuberculate; n = 10. Stream banks, meadows, and other moist sites in sagebrush, rabbitbrush, mountain brush, and pinyon-juniper communities at 1830 to 2565 m in Carbon, Daggett, Duchesne, Emery, Garfield, Grand, Sanpete, Uintah, Utah, and Wasatch counties; Colorado, New Mexico, Arizona, and Wyoming; 16 (0).

Mentzelia shultziorum Prigge Perennial, suffruticose, from a woody caudex; stems diffusely branched, 3–5 dm tall, greenish white, striate, more or less flexuous, hispid above; lower leaves ovate-lanceolate to broadly elliptic, dentate, sinuate, obtuse apically, shortly petiolate, 10–31 mm long, 7–20 mm wide; upper leaves ovate, elliptic, dentate, obtuse to acute apically, 7–21 mm long, 5–14 mm wide; bracts linear to oblanceolate, entire or occasionally denticulate; inflorescence terminal; calyx lobes 6–10 mm long; petals 5, obovate, 9–14 mm long; outer stamens petaloid, spatulate, shorter than the petals; capsule 5–8 (10) mm long, urceolate, rounded basally; seeds oval, 3–4 mm long, 2–2.5 mm wide, pale brown. Shadscale, eriogonum, ephedra community on the Cutler Formation (and Paradox?) along Onion Creek (type locality) Grand County; endemic; 4 (0).

Mentzelia thompsonii Glad Annual; stems erect, 1–2 dm tall, simple or moderately branched, somewhat compact; lower leaves lanceolate or ovate-lanceolate, to 7 cm long, entire or occasionally with a few lobes, tapering to a narrow petiole; upper leaves lance-ovate, smaller, usually entire, sessile; bracts oblanceolate, entire; inflorescence crowded, terminal, sometimes obscured by leaves or bracts; calyx lobes 2–3 mm long; petals 5, obovate, 1–4 mm long, retuse; capsules erect, 12–20 mm long, to 4 mm wide, cylindric, scarcely tapering basally; seed irregularly angled, truncate-tuberculate, ca 1.5–1.8 mm long. Salt desert shrub communities at 1495 to 1895 m mostly on Mancos Shale in Grand and Uintah counties; Colorado; a Plateau endemic; 7 (0).

Mentzelia tricuspis Gray Annual; stems erect, 1–3 dm tall, branched from the base or simple, greenish white; rosette leaves lanceolate, crenate or coarsely dentate, 6.2–9 cm long; lower cauline leaves lanceolate or oblanceolate, dentate or coarsely dentate, short-petiolate, the upper ones similar but smaller; bracts lanceolate, deeply dentate or laciniate; calyx lobes 1–1.5 cm long; petals 5, cream colored, broadly obovate, mucronate, 0.8–3.5 cm long; outer stamens dilated, tricusped apically, the 2 lateral cusps usually longer than the anther-bearing, central cusp; capsules cylindric, tapering to the base, 1–18 mm long, reflexed at maturity; seeds ovate, constricted at the middle, tuberculate; n = 10. Creosote bush, Joshua tree, and other warm desert shrub communities at ca 730 m in Washington County; Nevada, Arizona, and California; 1 (0).

Petalonyx Gray

Shrubs or subshrubs; leaves alternate, petiolate or sessile, entire, toothed or crenate; inflorescence spicate or racemose, bracteate or ebracteate; calyx tube adnate to the ovary, the lobes free; petals cream to white, clawed, glabrous ventrally, hispid dorsally; stamens 5, free; style filamentous; stigma unbranched; fruit an irregularly dehiscent capsule; seed 1, oblong-ovoid, shining, brownish.

Davis, W. S. and H. J. Thompson. 1967. A revision of *Petalonyx*. Madrono 19: 1–18. 1967.

1. Subshrubs, predominantly herbaceous; leaf margins coarsely few-toothed; inflorescence a bracteate raceme; petals less than 11 mm long .... *P. nitidus*
— Shrubs, predominantly woody; leaf margins entire to crenate; inflorescence a naked, terminal raceme; petals more than 11 mm long .... *P. parryi*

Petalonyx nitidus Wats. [*P. thurberi* var. *nitidus* (Wats.) Jones]. Suffruticose perennial; stems 1.5–4.5 dm tall; leaves ovate to broadly ovate, 1.5–4.5 cm long, 1–3 cm wide, serrate to coarsely few-toothed, the surfaces muricate-scabrous; petiole 1–5 mm long; inflorescence a dense, 10- to 30-flowered raceme; bracts ovate, scabrous, short-ciliate, 6–8 mm long; sepals linear, 1.4–3 mm long; petals cream, 5–11 mm long; claws linear, 4–7 mm long, 0.3–0.4 mm wide; stamens exserted, 7–14 mm long; capsule 1.3–3 mm long; n = 23. Warm desert shrub at ca 850 m in Washington County; Arizona, Nevada, and California; 1 (0).

Petalonyx parryi Gray [*P. nitidus* ssp. *parryi* (Gray) Urban & Gilg]. Shrubby, 8–15 dm high; leaves oblong-ovate to broadly elliptic, 1.5–4 cm long, 1.2–3 cm wide, entire or crenate, the surface muricate-scabrous; petiole 0.5–2.5 mm long; inflorescence a naked spicate raceme, congested apically, to 65-flowered; bracts lanceolate to lance-ovate, scabrous, 6–14 mm long; sepals linear, 2.5–4 mm long; petals cream, 10–15 mm long; claws linear, 6.3–10 mm long; stamens exserted, 11–17 mm long; capsules 2–4 mm long; n = 23. Warm desert shrub at ca 780 to 915 m in Washington (type from St. George) County; Nevada, Arizona, and California; 6 (iv).

LYTHRACEAE J. St. Hil.
Loosestrife Family

Herbs, shrubs, or trees; leaves opposite or alternate, simple, entire; flowers perfect, regular or occasionally irregular, solitary or clustered; calyx 4- to 6-toothed, often with many accessory ones in the sinuses, the tube free from the ovary but closely investing it; petals 4 or 5 (7), borne on the throat of the hypanthium; stamens 4 to many, borne on the hypanthium margin; pistil 1, the ovary superior, 2- to 6-loculed, becoming 1-loculed; style 1; stigma capitate; fruit a capsule; x = 5–11. Note: The beautiful shrub, *Lagerstroemia indica* L. or crepe myrtle, is grown as an ornamental in Washington County.

1. Hypanthium much longer than broad, strongly nerved; petals 5–7, usually showy .... *Lythrum*
— Hypanthium about as long as broad, not strongly nerved; petals usually 4, not showy 2
2(1). Hypanthium lacking intersepalar appendages; petals lacking .... *Didiplis*
— Hypanthium with intersepalar appendages; petals 4, often deciduous .... *Ammannia*

Ammannia L.

Glabrous, annual herbs; stems commonly 4-angled; leaves opposite, sessile, entire; flowers small, borne in axillary clusters; sepals 4, the hypanthium campanulate or globose, 4-angled, with small, hornlike appendages in each sinus; petals 4, caducous; stamens 4–8; ovary 2- to 4-loculed; capsule opening irregularly.

Ammannia coccinea Rottb. Plants ascending to erect, the lower branches often decumbent-ascending, 1–5 dm tall; leaves 1.5–8 cm long, 2–8 mm wide, linear to oblong or lanceolate to oblanceolate, acute apically, cordate-clasping basally; cymes closely (1) 2- to 5-flowered, nearly sessile; calyx 3–5 mm long and 3–5 mm wide in fruit, the lobes short; corolla pink to purple, 1–2 mm long, the petals obovate; capsule ca 4 mm long; n = 33. Mud flats, canal banks, marshes, and flood plains at 1220 to 1435 m in Cache, Emery, Millard, and Uintah counties; widely distributed in the U. S.; South America; 9 (i). Utah plants seem to belong to ssp. *robusta* (Heer & Regel) Koehne [*A. robusta* Heer & Regel].

Didiplis Raf.

Glabrous, annual, amphibious herbs; leaves opposite, narrow, sessile; flowers sessile, minute, solitary in leaf axils, lacking bracteoles; hypanthium campanulate; sepals 4, appearing as triangular lobes; intersepalar appendages lacking; petals none; stamens 2 or 4, included; ovary imperfectly 2-locular; style short or obsolete; disk lacking; fruit globose, indehiscent; seeds many.

Didiplis diandra (**Nutt.**) **Wood** [*Peplis? diandra* Nutt. ex DC.; *Didiplis linearis* Raf.] Plants submersed or growing on mud; stems 1–4 dm long, usually branched; leaves numerous, dimorphic, the submersed ones linear and 1–2.5 cm long, sessile and truncate basally, the emersed ones shorter, narrowly elliptic, and tapering at the base; flowers ca 1 mm long at anthesis, enlarging to 2 mm long in fruit; sepals subequal to the hypanthium; fruit 2 mm thick; seeds many, ca 0.7 mm long, oblanceolate, planoconvex. Marshy places (reported by Tidestrom 1925) from Fish Lake, Sevier County; Minnesota to Ohio, south to Texas and Florida; 0 (0).

Lythrum L.

Annual or perennial herbs; stems mostly 4–angled; leaves opposite, alternate, or whorled, entire; flowers usually solitary in the upper axils; hypanthium cylindrical, 8– to 12–ribbed, 4– to 7–toothed, with an equal number of alternating teeth in the sinuses; petals 4–7, attached to the hypanthium margin; stamens 4–14, borne on the hypanthium margin; capsules cylindrical, included in the calyx tube, 2–loculed, regular, dehiscent.

1. Leaves mostly alternate, the main ones 1.5–5 cm long; stamens 4–7; plants indigenous in Washington County .... *L. californicum*
— Leaves opposite, the main ones 5–10 cm long; stamens 10–14; plants adventive in central Utah ...... *L. salicaria*

Lythrum californicum T. & G. Perennial herbs from creeping rhizomes; stems 4–10 (13) dm tall, pale green and glabrous, erect, much branched; leaves alternate, 1.5–5 cm long, 1–8 mm wide, linear to oblong or the lowermost lanceolate to oblanceolate, glabrous; calyx tube cylindrical, glabrous, 5–7 mm long, the teeth subulate; petals 4–7, obovate, pink purple, 4–6 mm long; capsule cylindroid to club shaped. Saline seeps and canal banks in creosote bush, blackbrush, and pinyon-juniper communities at 900 to 1340 m in Washington County; California to Kansas and Texas, south to Mexico; 7 (iv).

Lythrum salicaria L. Loosestrife. Perennial herbs from creeping rhizomes; stems 7–30 dm tall, green, with coarse hairs, erect, much branched; leaves opposite or whorled, 2–10 cm long, 4–20 mm wide, lanceolate to rounded or cordate at the base, acute to attenuate apically, hairy; calyx tube 5–8 mm long, cylindrical, coarsely hairy, the appendage of the calyx twice or more longer than the teeth; petals rose purple, 5–10 mm long; stamens 10–14; capsule cylindroid to club shaped; $2n = 30, 60$. Cultivated ornamental, escaping and established in moist sites at 1390 to 2070 m in Cache, Davis, Emery, Salt Lake, Utah, and Weber counties; introduced from Europe; 8 (i).

MAGNOLIACEAE A. L. Juss.

Magnolia Family

Deciduous or evergreen trees or shrubs; leaves alternate, simple, entire of lobed, stipulate, the stipules enclosing the buds, deciduous or caducous, and leaving a circular scar; flowers regular, perfect solitary, terminal and axillary, large and showy, the floral parts spirally arranged; sepals often 3, the petals 6 to many; stamens numerous, separate, hypogynous, the anthers 2–loculed; pistils several to many, each 1–loculed and 1–carpelled; style 1, the stigma 1; fruit a follicle or samara.

1. Leaves lobed, truncate or broadly retuse at the apex; flowers borne after the leaves .... *Liriodendron*
— Leaves entire, acute, or acuminate; flowers borne before or after the leaves .... *Magnolia*

Liriodendron L.

Trees, the leaves large and 4–lobed; flowers large, not brightly colored; sepals 3, soon reflexed; petals 6, ascending to erect, forming a tuliplike corolla; anthers extrose; pistils many, en masse becoming conelike, the individual samaras eventually deciduous.

Liriodendron tulipfera L. Tulip Tree, Yellow Poplar. Deciduous, cultivated trees to 40 m tall or more, the trunks to 10 dm in diameter or more; leaves long-petioled, the blades 6–15 cm long and almost as wide; flowers solitary, terminal; sepals green; petals 3.7–6 cm long, yellow green, with usual basal orange spot within; samaras narrow, 3–4 cm long; $2n = 38$. Occasional shade tree in more moderate, low elevation portions of Utah; introduced from the eastern U. S.; 8 (0).

Magnolia L.

Trees or shrubs; leaves large, entire; flowers large, conspicuous or inconspicuous; sepals 3, colored like the petals; petals 6–12, erect or spreading; anthers introrse; pistils many, en masse becoming conelike, the individual follicles finally dehiscent.

1. Plants shrubs or small trees, deciduous; flowers showy, cream to pink or suffused with rose or lavender, borne before the leaves appear .... *M. soulangeana*
— Plants moderate to large trees, deciduous or evergreen; flowers greenish and inconspicuous or, if showy, white in color and trees evergreen 2
2(1). Plants evergreen, the leaves dark green, leathery; flowers white .... *M. grandiflora*
— Plants deciduous, the leaves not both dark green and leathery; flowers greenish yellow .... *M. acuminata*

Magnolia acuminata L. Cucumber-tree. Deciduous trees to 30 m tall or more; leaves short-petioled, the blades 8–25 (3) cm long and 4–15 cm wide; flowers solitary, terminal; perianth greenish yellow, 5–8 cm long. Cultivated shade tree, uncommon, hardy in the major cities of the state; introduced from the eastern U. S.; 2 (0).

Magnolia grandiflora L. Bull Bay. Evergreen tree to 30 m tall; leaves evergreen, short-petioled, the blades mostly 8–20 cm long and 3–8 cm wide; flowers solitary, terminal; perianth white, mostly 8–12 cm long; $2n = 114$. Cultivated ornamental, uncommon, not hardy except in favorable sites in moderate to warm portions in Utah; introduced from southeastern U. S.; 1 (0).

Magnolia soulangeana Soul.-Bod. Showy Magnolia. Shrubs or small trees to about 4 m tall; leaves deciduous, short-petioled, the blades mostly 8–14 cm long and 3.5–10 cm wide; flowers solitary, terminal; perianth cream to pink or suffused with rose or lavender, 6–12 cm long or more; $2n = 95, 123, 133, 152$. Cultivated ornamental, occasional in more moderate climatic areas of Utah; a hybrid of *M. denudata* Descr. and *M. liliflora* Desr., both native of China; 3 (0).

MALVACEAE A. L. Juss.

Mallow Family

Herbs or less commonly shrubs, usually pubescent with branched or stellate hairs, annual, biennial, of perennial with mucilaginous juice; leaves alternate, simple, mostly palmately veined, stipulate; flowers perfect (or imperfect), regular, solitary or in thyrsoid cymes or more or less racemose or paniculate, sometimes with an involucel of sepaloid bractlets; sepals 5, more or less persistent; petals 5, separate, adnate to the staminal sheath; stamens numerous, united by the filaments (monadelphous); ovary superior, 3- to many-loculed; fruit a capsule or schizocarp; x = 6–17+, 20+.

Welsh, S. L. 1980. Utah flora: Malvaceae. Great Basin Naturalist 40: 27–37.

1. Involucel lacking 2
— Involucel of 1 or more bractlets or, if lacking (as some *Sphaeralcea* specimens), the flowers orange (grenadine) 3

2(1). Petals white, pink, or lavender; plants of moist sites, usually at middle and higher elevations .... *Sidalcea*
— Petals yellow or pink to red; plants of cultivated lands or of arid sites, usually at middle and lower elevations .. .. *Abutilon*

3(1). Petals orange or rarely purplish pink; indigenous perennial herbs of arid habitats at middle and lower elevations .... *Sphaeralcea*
— Petals variously colored, but not orange; indigenous or adventive perennial, biennial, or annual plants of various distribution 4

4(3). Flowers rose pink or rarely white; plants indigenous, 7–15 dm tall, perennial, of middle and higher elevations .... *Iliamna*
— Flowers white, pink, rose, yellow, or other hues; plants differing in one or more ways from above 5

5(4). Flowers mostly 6–10 cm broad, opening flat; plants tall adventive or cultivated biennials .... *Althaea*
— Flowers less than 6 cm broad or, if broader, the plants shrubby 6

6(5). Style branches 5, elongate; fruit a capsule; plants low annuals or shrubs .... *Hibiscus*
— Style branches more than 5, short; fruit a schizocarp; plants annual or biennial 7

7(6). Style branches filiform, with elongate stigmatic lines; plants annual or biennial .... *Malva*
— Style branches with capitate or truncate stigmas 8

8(7). Petals yellow, or orange to pink or red; plants annual with awned carpels, or subshrubs with unawned carpels .... *Abutilon*
— Petals yellowish white to lavender or whitish; carpels few to many, not awned; plants spreading annuals or herbaceous perennials 9

9(8). Petals yellow white; leaves reniform-orbicular, merely crenate-serrate .... *Malvella*
— Petals lavender or whitish; leaves palmately cleft, with rounded lobes .... *Malvastrum*

Abutilon Miller

Plants herbaceous, annual or perennial, with stellate or simple hairs; leaves alternate, petioled, cordate at base, not or only obscurely lobed; flowers solitary and axillary or in leafy panicles; involucel lacking; calyx 5–cleft; corolla yellow to orange pink or red; fruit truncate-cylindric or subglobose, the carpels smooth sided, dehiscent nearly to the base; ovules 2 or more per carpel.

1. Plants perennial, with slender spreading or trailing branchlets; carpels 5, lacking awn-beaks; plants rare, known only from Washington County .... *A. parvulum*
— Plants annual, with robust erect stems; carpels usually more than 10, each with a long divergent awn; plants uncommon, in agriculture regions .... *A. theophrasti*

Abutilon parvulum Gray Perennial, the stems slender and spreading or trailing, the caudex woody, grayish tomentose with minute stellate hairs, the branchlets pilose; leaves 0.5–5 cm long, ovate, cordate basally, dentate and sometimes obscurely 3–lobed; peduncles slender, axillary, 1–flowered, longer than the leaves; calyx lobes ovate-acuminate, reflexed in fruit; petals orange pink to red or sometimes yellowish, 4–6 mm long; carpels 5, somewhat tomentose, to 8 mm long; n = 7. Known in Utah only from Veyo, Washington County (Meyer 4111); Colorado to California and south to Texas and Mexico, 1 (0).

Abutilon theophrasti Medicus Velvet-leaf. Annual, the stems robust, erect, velvety and cinereous with short, soft hairs; leaves 3–10 cm long (from sinus to apex) and as broad or broader, orbicular-ovate, cordate at the base, abruptly acuminate at the apex, velvety pubescent; peduncles shorter than the leaves, 1– or few-flowered; calyx lobes broadly ovate-acuminate; petals yellow, to ca 6 mm long; carpels 10 or more, each with a long divergent awn; 2n = 42. Adventive weedy species of disturbed or cultivated areas, occasional in Davis, Sanpete, Tooele, Utah, and Washington counties; widespread in North America; native to Europe; 5 (i).

Althaea L.

Plants herbaceous, biennial, with coarse stellate hairs; leaves alternate, petiolate, cordate at the base, lobed; flowers solitary or in racemes; involucel of 6–9 bractlets, connate at the base; calyx 5–cleft; corolla of various colors; fruit flattened wheellike, invested by the calyx, the numerous carpels separating at maturity.

Althaea rosea (L.) Cav. Hollyhock. [*Alcea rosea* L.]. Coarse biennials to 20 dm tall or more, the stems erect, stellate-hairy; leaves (3–) 5– to 7–lobed, mostly 3–15 cm long (from sinus to apex) and often much broader; flowers shortly pedicellate, 6–12 cm wide or more, variously colored, often rose to pink or lavender or sometimes white, usually with a dark center; calyx lobes triangular, investing the fruit at maturity, the involucel calyxlike; carpels numerous, stellate along the margins, and reticulate on the sides, 5–7 mm long; 2n = 42. Cultivated ornamental, persisting and escaping, to be expected in all counties in Utah; widespread in North America; introduced from China; 17 (i).

Hibiscus L.

Plants herbaceous or woody, annual or perennial, with stellate or simple hairs; leaves alternate, petiolate, obtuse to truncate or cordate basally, lobed to incised; flowers axillary, solitary; involucel of 5–10 distinct bractlets; calyx 5–cleft, more or less accrescent in fruit; fruit a loculicidal capsule, the carpels 5; seeds several in each locule.

1. Plants annual; calyx strongly veined; petals cream colored, with a purple center .... *H. trionum*
— Plants shrubs; calyx herbaceous, not distinctly veined; petals variously colored, but usually rose pink to lavender .... *H. syriacus*

Hibiscus syriacus L. Althaea; Rose-of-Sharon. Shrubs, 20–40 dm tall or more, glabrous or softly stellate-hairy; leaves 2.5–8 cm long, 1.5–6 cm wide, triangular-ovate to rhombic, strongly 3–ribbed, commonly 3–lobed; flowers axillary, 4–7.5 cm wide; bractlets usually 5, linear, subequal to the calyx, glabrous to obscurely hairy; corolla variously colored and often double; fruit oblong-ovoid, to 25 mm long; n = 20, 40. Cultivated ornamental, rarely persisting; widely cultivated in North America; introduced from eastern Asia; 4 (i).

Hibiscus trionum L. Flower-of-an-Hour. Annual, commonly 1.5–5 dm tall, the lower branches often prostrate, coarsely hispid-stellate to glabrate; leaves 3–lobed or more commonly 3– to 5–parted, the main lobes cuneate basally, the middle lobe the largest; flowers solitary, axillary, mostly 3–6 cm wide; bractlets usually 10, linear, often coarsely hispid, much shorter than the fruiting calyx; corolla cream colored to yellowish, with a purple center, closing in shade; 2n = 28, 56. Weedy species of cultivated land at lower elevations in Box Elder, Cache, Davis, Salt Lake, Sanpete, Utah, and Washington counties; widespread in North America; adventive from central Africa; 8 (i).

Iliamna Greene

Plants herbaceous, perennial, sparingly and minutely stellate-hairy; leaves alternate, petiolate, cordate to truncate narrow, the margin lobed; flowers in thyrsoid panicles; involucel of 3 narrow, persistent bractlets; calyx 5–cleft; fruit a loculicidal capsule, the carpels many; seeds usually 3 in each locule.

Wiggins I. L. 1936. A resurrection and revision of the genus *Iliamna* Greene. Contr. Dudley Herb. 1: 213–229.

Iliamna rivularis (**Dougl.**) **Greene** Wild Hollyhock. [*Malva rivularis* Dougl. ex. Hook.; *Sphaeralcea rivularis* (Dougl.) Torr. ex. Gray; *Phymosia rivularis* (Dougl.) Rydb.]. Perennial, the stems few to many from a woody caudex, mostly 7–15 dm tall, minutely stellate-puberulent, green; leaves 3– to 7–lobed, cordate to truncate basally, 2.5–15 cm long (from petiole apex to tip), 2–16 cm broad, the lobes triangular, crenate-serrate, finely stellate; pedicels mostly less than 1 cm long; bractlets linear-lanceolate, shorter than the calyx; calyx lobes 3–5 mm long (to 8 mm long in fruit); petals rose pink (rarely white), 20–37 mm long; carpels 6–10 mm long in fruit, hispid and stellate; n = 33. Along streams and other mesic sites in mountain brush, ponderosa pine, aspen, and spruce-fir communities, 1430 to 2900 m in Cache, Daggett, Davis, Duchesne, Iron, Juab, Kane, Millard, Morgan, Piute, Salt Lake, Sanpete, Sevier, Summit, Tooele, Utah, Wasatch, and Weber counties; Colorado, Idaho, Nevada, and Washington; 59 (vii).

Malva L.

Herbaceous, annual, biennial, or perennial, from taproots, the pubescence simple to branched or stellate; leaves alternate, petiolate, usually more or less cordate basally, commonly lobed; flowers in axillary clusters (sometimes solitary) or in subterminal panicles; involucel of 3 narrow to broad persistent bractlets; calyx 5–cleft; fruit a schizocarp, the carpels mostly 10–15.

1. Petals commonly 1.5–2 cm long, very showy; bractlets of involucel ovate to oblong .... *M. sylvestris*
— Petals usually less that 1 cm long; bractlets of involucel linear to narrowly lanceolate 2
2(1). Stems prostrate spreading from the caudex; leaves obscurely lobed; plant a common weedy species .... *M. neglecta*
— Stems erect; leaves definately lobed; plant cultivated, rarely escaping .... *M. verticillata*

Malva neglecta Wallr. Cheeses; Mallow. Annual or biennial, the stems prostrate-spreading, commonly 1–6 dm long, stellate hairy; leaf blades reniform-orbicular, 0.6–3 cm long (from sinus to apex) or more, and much broader, crenate and not at all to only shallowly 5– to 7–lobed, the petioles to 20 cm long or more; flowers clustered (or solitary) in the axils; bractlets linear; calyx (3) 4–6 mm long at anthesis, the lobes acuminate; petals white to pink or lilac, about twice as long as the sepals; carpels hairy, rounded on the back; 2n = 42. Weeds of disturbed sites and cultivated land, in much of Utah (specimens known from Cache, Garfield, Grand, Iron, Kane, Millard, Rich, Salt Lake, San Juan, Summit, Uintah, Utah, and Washington counties); widespread in North America; adventive from Eurasia; 22 (ii). Note: Two other species *M. parviflora* L. and *M. rotundifolia* L., might be present in Utah. They are similar to *M. neglecta*, but have petals subequal to the sepals. *Malva parviflora* has glabrous petal claws, whereas in *M. rotundifolia* the claws are bearded.

Malva sylvestris L. High Mallow. Biennial, the stems ascending, mostly 3–10 dm tall, rough hairy to glabrate; leaf blades 3–8 cm long or more and often broader, orbicular to cordate or reniform, crenate and with 5–7 lobes, the petioles to 10 cm long or more; flowers clustered in the leaf axils; bractlets ovate to elliptic; calyx 5–7 mm long at anthesis, the lobes short and broad; petals 15–20 mm long, rose purple; carpels glabrous or nearly so, sharp edged; 2n = 42. Cultivated ornamental, rarely escaping; widespread in North America; adventive from Europe; 3 (0).

Malva verticillata L. Curled Mallow. Annual, the stems erect, mostly to 10 dm tall or more, sparingly stellate-hairy; leaf blades mostly 1.5–7 cm long and as broad or broader, orbicular to reniform, undulate-crisped and distinctly 5– to 7–lobed, long-petioled; flowers solitary or clustered, subsessile or some pedicelled; bractlets linear to narrowly lanceolate; calyx 3.5–5 mm long, the lobes acuminate; petals white, only somewhat surpassing the sepals; carpels glabrous, the edges rounded; 2n = 84 (ca 76, ca 126). Cultivated ornamental, rarely escaping (Washington County, Galaway in 1934 BRY); widely scattered in the U. S.; adventive from the Old World; 1 (0). Our material belongs to **var. crispa** L.

Malvastrum Gray

Plants herbaceous, annuals, stellate-hairy; leaves alternate, petiolate, the blades subcordate to truncate basally, palmately lobed; flowers solitary in the axils or in the terminal bracteate clusters; involucel of usually 3 slender bractlets; calyx 5–cleft, the lobes long-acuminate; carpels 10–15; fruit a schizocarp.

Malvastrum exile Gray [*Malveopsis exile* (Gray) Kuntze; *Eremalche exile* (Gray) Greene; *Sphaeralcea exile* (Gray) Jepson]. Annual, the stems spreading-decumbent to prostrate, branching from near the base, 0.3–4 dm long, rather sparingly stellate-hairy; leaf blades suborbicular, 8–32 mm wide, palmately 3- to 5-cleft, with rounded or cuspidate teeth; petioles 1–5 cm long; bractlets narrowly lanceolate to sublinear; calyx 3–5 mm long; petals whitish to pinkish or lavender, only somewhat surpassing the sepals; carpels transversely wrinkled. Blackbrush and creosote brush communities at 850 to 1200 m in Washington counties; Arizona and southern California; 6 (0).

Malvella Jaub. & Spach

Herbaceous, perennial, decumbent from spreading rhizomes, densely stellate-canescent; leaves alternate, petiolate, crenate-serrate; flowers solitary in leaf axils, the calyx bearing 1–3 linear, deciduous bractlets; calyx 5-lobed; petals 5, stellate-pubescent externally where exposed in bud; carpels 5–10, 1-seeded; fruit a schizocarp.

Fryxell, P. A. The North American Malvellas (Malvaceae). Southwestern Naturalist 19: 97–103.

Malvella leprosa (**Ortega**) **Krap.** Alkali Mallow; Dollar Weed. [*Malva leprosa* Ortega; *Sida hederacea* (Dougl.) Torr.; *Malva hederacea* Dougl.; *M. californica* Presl; *Disella hederacea* (Dougl.) Greene]. Perennial, the stems from elongate rhizomes, decumbent to prostrate, the surface obscured by overlapping stellate hairs, 1–4 dm long; leaf blades reniform to orbicular, often oblique, dentate, obscurely if at all lobed, the petioles 0.3–2.5 (3) cm long; bractlets sublinear; calyx 5–7 mm long; petals cream or yellowish (fading orange), 10–12 mm long; carpels reticulate on the sides. Saline meadows and seeps, at lower elevations in Cache, Emery, Millard, Salt Lake, San Juan, Tooele, Uintah, Utah, Washington, and Weber counties; Washington south to California, Texas, and Mexico; South America; 19 (iv).

Sidalcea Gray

Herbaceous, perennial from taproots or short rhizomes, usually stellate and somewhat hirsute; leaves alternate, petiolate, often dimorphic, the lowermost merely palmately lobed, the upper ones commonly cleft and with linear lobes; flowers borne in semispicate racemes, dimorphic, those of plants with perfect flowers the largest; involucel lacking; calyx 5-cleft; carpels 5–10, 1-seeded, tardily separating.

Hitchcock, C. L. 1957. A study of the perennial species of *Sidalcea*. Univ. Wash. Publ. Biol. 18: 1–79.
Roush, E. M. F. 1931. A monograph of the genus *Sidalcea*. Ann. Missouri Bot. Gard. 18: 117–244.

1. Petals white or merely pinkish tinged, often drying yellow; anthers bluish pink; plants rhizomatous; stems hirsute below .... S. *candida*
— Petals pink to lavender; anthers usually yellow to white; plants rhizomatous or not; stems hirsute to glabrous or tomentose below 2
2(1). Plants from rather fleshy taproots; stems commonly hirsute below; calyx hirsute with pustulose hairs (at least in part) .... S. *neomexicana*
— Plants often rhizomatous; stems stellate to glabrous below; calyx seldom with pustulose hairs .... S. *oregana*

Sidalcea candida Gray Plants from slender rhizomes, the stems 4–10 dm tall, glabrous to hirsute with simple hairs below, more or less stellate above; leaf blades 6–20 cm wide, the basal ones shallowly 5- to 7-lobed and coarsely crenate, the upper ones divided into 3–5 entire segments; calyx 7–10 mm long, variously stellate-hairy and glandular puberulent; petals white to pinkish, often drying yellow, 12–20 mm long; carpels ca 3 mm long. Stream banks, lake shores, and seeps, mainly in sagebrush, mountain brush, and aspen communities, at 1370 to 2870 m in Beaver, Cache, Daggett, Duchesne, Garfield, Grand, Kane, Iron, Millard, Morgan, Piute, Salt Lake, San Juan, Sanpete, Sevier, Summit, Uintah, Utah, and Wasatch counties; Wyoming and Colorado west to Nevada and south to New Mexico; 47 (vi). Plants with the calyx lobes rather uniformly hairy have been designated as **var. *candida***, while those with the base of the calyx more hairy than the lobes have been regarded as **var. *glabrata* C. L. Hitchc.**

Sidalcea neomexicana Gray Plants from thickened tap or fascicled roots, the stems 2–9 (10) dm tall, hirsute below (or rarely glabrous) with simple or bifurcate hairs; leaf blades 1.5–11 cm wide, the basal ones crenate to shallowly 5- to 7-lobed, the cauline ones divided usually into 5 laciniate to entire segments; calyx 5–10 mm long, usually with some simple pustulose hairs interspersed with the stellate ones; petals rose pink (fading blue purple), 11–19 mm long; carpels 2–3 mm long. Wet meadows, stream banks and seeps at 1370 to 2150 m in Box Elder, Cache, Davis, Garfield, Iron, Juab, Piute, Rich, Salt Lake, San Juan, Sanpete, Sevier, Summit, Tooele, Utah, Wasatch, and Washington counties; Oregon, Idaho, and Wyoming, south to California, Arizona, and Mexico; 40 (iv). Our materials have been treated as belonging to two or more or less confluent and at least partially sympatric varieties. The **var. *neomexicana*** is distinguished by having hairs of the lower stem nearly all simple, the calyx coarsely hirsute (lacking appressed stellate hairs), and upper stems usually glabrous. In **var. *crenulata*** [*S. crenulata* A. Nels., type from Juab, Utah; *S. neomexicana* ssp. *crenulata* (A. Nels.) C. L. Hitchc.] the hairs of the lower stem are often forked, the calyx is pubescent with appressed stellate hairs in addition to coarse ones, and the upper stems are often stellate hairy.

Sidalcea oregana (**Nutt.**) **Gray** [*Sida oregana* Nutt. ex T. & G.; *Sida nervata* A. Nels.]. Plants from taproot, lacking or rarely with rhizomes, the stems 3–11 dm tall or more, glabrous or usually appressed-stellate hairy below, appressed-stellate above; leaf blades 2.5–17 cm wide, the basal ones shallowly 5- to 7-lobed and coarsely crenate, the cauline ones deeply lobed, with 3–7 coarsely toothed to entire lobes; calyx 3.5–9 mm long, variously stellate-hairy and sometimes bristly; petals 7–23 mm long, pale pink to rose pink (fading blue purple); carpels 2.5–3 mm long. Meadows, stream banks, and open woods, at 1680 to 2750 m in Cache, Davis, Juab, Morgan, Rich, Salt Lake, Sanpete, Sevier, Summit, Utah, Wasatch, and Weber counties; Washington and Idaho south to California, Nevada, and Utah. Our materials belong to **var. *oregana***; 52 (ii).

Sphaeralcea J. St. Hil.

Herbaceous, perennial, from taproots or rhizomes, glabrescent to canescent with stellate hairs; leaves alternate, petiolate, sometimes dimorphic, the lowermost merely toothed or palmately lobed (rarely entire), the upper ones cleft to entire; flowers borne in racemose to thyrsoid cymes; involucel of 3 or fewer filiform bractlets; calyx 5-cleft; carpels 8–20, the seeds 1 or 2 per carpel; fruit a schizocarp, the mature fruit segments divided into a basal indehiscent, reticulate portion and an apical dehiscent portion.

Jefferies, J. A. M. 1972. A revision of the genus *Sphaeralcea* (Malvaceae) for the state of Utah. Unpublished thesis. Brigham Young University. 92 pp.

1. Inflorescence racemose, rarely with more than one flower per node or, if more as in S. *caespitosa*, the plants restricted to Millard County 2
— Inflorescence thyrsoid to thyrsoid-glomerate, with usually more than one flower per node; distribution various 6

2(1). Leaf blades only slightly, if at all, 3- to 5-lobed, the margins irregularly crenate-dentate; hairs with rays radiating in more than a single plane; plants seldom more than 1.5 dm tall, known only from western Beaver and Millard counties .... S. *caespitosa*
— Leaf blades distinctly 2- to 5-lobed, -parted, or -divided; hairs of rays radiating in a single plane (except S. *coccinea*); plants often 1.5 dm tall or more, of different distribution 3

3(2). Leaves trifoliolate, the leaflets linear to narrowly oblanceolate and entire, or the upper ones simple and entire; plants of southeastern Utah 4
— Leaves various but, if trifoliolate, the leaflets oblanceolate and entire to toothed or, if the uppermost simple, then toothed or lobed; distribution various 5

4(3). Leaves green; rays of hairs steeply ascending; plants simulating *Psoralidium junceum* .... S. *janeae*
— Leaves gray; rays of hairs spreading; plants not psoraloid .... S. *leptophylla*

5(3). Lowermost leaves simple and entire or trifoliolate, or some broadly toothed or lobed; involucel present; rays of hairs steeply ascending .... S. *psoraloides*
— Lowermost leaves usually 3- to 5-lobed, the lobes usually toothed or again lobed; involucel present or lacking (caducous); rays of hairs usually radiating in several planes; plants of broad distribution .. S. *coccinea*

6(1). Plants only sparingly pubescent, the herbage bright green 7
— Plants moderately to densely pubescent, the herbage yellowish, whitish, or grayish 9

7(6). Leaves 3- to 5-parted or -divided, the lobes with narrow, regularly pinnatifid margins, the teeth nearly at right angles to the vein; carpels often with transparent lacunae, 4–6 mm high; plants rare, of southern Utah only .... S. *rusbyi*
— Leaves variously lobed, divided, or parted, the lobes with broader margins irregularly toothed or lobed, but not as above; carpels with opaque lacunae, 3–4.5 mm high 8

8(7). Leaves slightly lobed, the margins unevenly toothed or, in some, deeply parted or divided with the margin coarsely and irregularly lobed, the base subcordate to cuneate; plants of northern Utah .... S. *munroana*
— Leaves 3- to 5-parted or -divided, the margins regularly cleft, lobed, or toothed, the base subcordate to deeply cordate; plants mostly of southern Utah .....
.... S. *grossulariifolia*

9(6). Inflorescence loosely thyrsoid (appearing paniculate), leafy; flowers not numerous at each node; peduncles generally elongate; calyx surpassing the fruit; carpels with reticulae extending onto back of carpel; plants of southwestern Utah .... S. *ambigua*
— Inflorescence contracted thyrsoid-glomerate; flowers often numerous at each node, not especially leafy; calyx often shorter than the fruit; carpels with reticulae confined to lateral face of carpel; plants of various distribution 10

10(9). Leaves 3- to 5-cleft, -parted, or -divided, carpels with well-defined recticulae on less than half of carpel face; plants of all but the northeastern one-fourth of Utah, transitional to the following .... S. *grossulariifolia*
— Leaves shallowly 3- to 5-lobed; carpels with well-defined to nearly obscure reticulae on the lower 1/3 of the carpel; plants mainly of eastern and southern Utah, scattered elsewhere .... S. *parvifolia*

Sphaeralcea ambigua Gray Stems arising from a woody caudex, several to numerous, 3–10 dm tall, whitish to yellowish canescent; leaf blades 1–6 cm long (from sinus to apex), 0.8–5 cm wide, thickish, usually rugose, with veins prominent beneath, ovoid, deltoid, or nearly orbicular, the base cordate to deeply cordate, obscurely to definitely 3- to 5-lobed, the lobes crenate; inflorescence an open panicle, sometimes narrowly thyrsoid; pedicels usually shorter than calyx; calyx uniformly pubescent to glabrate, 6–20 mm long at anthesis, the lobes lanceolate to acuminate; petals 15–22 mm long, orange to orange pink (fading pinkish); carpels 12–16 mm high, the indehiscent portion comprising about 1/3 of the carpel, prominently reticulate; n = 5 (2n = 37). Creosote bush, blackbrush, and mixed warm desert shrub communities at 700 to 1750 m in western Kane and Washington counties; Nevada, Arizona, and California; Mexico; 15 (iv). Our materials belong to var. *ambigua*.

Sphaeralcea caespitosa Jones Jones Globemallow. Stems solitary or more commonly few to several from the summit of a branching woody caudex, 0.2–2.5 dm tall, whitish to grayish canescent; leaf blades 1.2–5.5 cm long, 1.2–6 cm wide, thickish, not rugose, veins apparent but not especially prominent, ovate to deltoid or orbicular, the base truncate to obtuse, obscurely if at all lobed, crenate to crenate-dentate; inflorescence thyrsoid, the flowers tightly clustered or solitary; pedicels shorter than the calyx; calyx uniformly stellate, the rays of hairs not radiating in a single plane, the lobes lance-acuminate; petals 15–21 mm long, orange; carpels 12–14, 4–6 mm high, the indehiscent protion forming slightly more than 1/3 of the carpel, reticulate on the sides. Mixed desert shrub communities (shadscale, rabbitbrush, winterfat), mainly on the Sevy Dolomite Formation and on calcareous gravels, at 1370 to 1955 m in Beaver (type from Wah Wah) and Millard counties; endemic; 25 (viii). Material from Nevada identified as S. *caespitosa* passes by degree into S. *ambigua*, and possibly is not equivalent to that in Utah.

Sphaeralcea coccinea (Nutt.) Rydb. Common Globemallow. [*Malva coccinea* Nutt.; *Cristaria coccinea* (Nutt.) DC.; *Malvastrum coccineum* (Nutt.) Gray; *Sida dissecta* Nutt.; *M. coccineum* var. *dissectum* (Nutt.) Gray; *M. dissectum* (Nutt.) Cockerell; *Sphaeralcea dissecta* (Nutt.) Rydb.; S. *coccinea* ssp. *dissecta* (Nutt.) Kearney; S. *coccinea* var. *dissecta* (Nutt.) Kearney; *M. c.* var. *elatum* Baker; *M. elatum* (Baker) A. Nels.; S. *elata* (Baker)

Rybd.; *S. c.* var. *elata* (Baker) Kearney; *Malvastrum cockerellii* A. Nels.; *M. micranthum* Woot. & Standl.]. Stems solitary or few to many from the apex of a woody caudex or less commonly from creeping rhizomes, 0.6–4.2 dm tall, white to yellowish canescent; leaf blades 1.1–3.7 cm long, 1.2–5.2 cm wide, usually wider than long, ovate to cordate-ovate in outline, the base often cordate, usually 3- to 5-lobed, with main divisions cleft almost or quite to the base, the lobes usually again toothed or lobed; inflorescence racemose, sometimes paniculate, rarely thyrsoid; pedicels shorter than calyx; calyx uniformly stellate, the rays of hairs not radiating in a single plane, the lobes lance-acuminate; petals 8–15 mm long, orange; carpels 8–14, 2–3 mm high, the indehiscent part forming 2/3 or more of the carpel, reticulate on the sides and on the back. Blackbrush, salt desert shrub, mixed desert shrub, juniper-pinyon, mountain brush, and ponderosa pine communities at 920 to 2750 m, in all counties (except Morgan); Saskatchewan and Alberta south to Arizona, New Mexico, and Texas; 214 (xix). Our materials have been recognized as belonging to vars. *dissecta* and *elata*, but the segregation of these entities appears to have been wholly arbitrary, with intermediates more numerous than the supposed taxa.

Sphaeralcea grossulariifolia (H. & A.) **Rydb.** Gooseberry-leaved Globemallow. [*Sida grossulariifolia* H. & A.; *Malvastrum grossulariifolium* (H. & A.) Gray; *Sphaeralcea pedata* Torr. in Gray; *S. g.* ssp. *pedata* (Torr.) Kearney; *S. g.* var. *pedata* (Torr.) Kearney]. Stems few to many from a woody caudex, 1–10 dm tall or more, whitish, to yellowish canescent to subglabrous and green; leaf blades 1.3–5 cm long, 1.3–5 cm wide, usually longer than wide, ovate to cordate-ovate in outline, the base cordate to truncate or obtuse, usually 3- to 5-lobed, the main division usually cleft or parted to irregularly toothed; inflorescence thyrsoid, with usually more than 1 flower per node; pedicels shorter than to much longer than the calyx; calyx uniformly stellate, the rays of hairs not radiating in a single plane, the lobes ovate to lance-acuminate; petals 8–20 mm long, orange or rarely rose pink; carpels 10–14, 2.5–4.5 mm high, the indehiscent portion forming from 2/5–3/5 of the carpel, reticulate on the sides. A phase with green herbage and thin leaves occurs along Glen Canyon. It is distinctive where it grows alone, but is transitional to the typical material when in contact. The plants are identified as follows:

1. Herbage bright green; leaves thin-textured, often narrowly lobed; plants of eastern Kane and Garfield, and western San Juan counties .. *S. grossulariifolia* var. *moorei*
— Herbage gray green to whitish canescent; leaves thick textured; plants widespread ....
.... *S. grossulariifolia* var. *grossulariifolia*

Var. grossulariifolia Salt, warm, and cool desert shrub, pinyon-juniper, and ponderosa pine communities at ca 790 to 2290 m in Beaver, Box Elder, Carbon, Emery, Garfield, Juab, Kane, Millard, Piute, Salt Lake, San Juan, Sanpete, Tooele, Uintah, Utah, Washington, and Wayne counties; Washington, Oregon, California, Nevada, and Arizona; 159 (xv). This is the common and widely distributed phase of the species in Utah. It forms intermediates with *S. coccinea*, *S. parvifolia*, and with the more northern *S. munroana*. The report by Kearney (l. c., p. 90) of *S. digitata* (Greene) Rydb. apparently belongs here.

Var. moorei Welsh Moore Globemallow. Warm desert shrub and hanging garden communities at 1125 to 1375 m in Garfield, Kane, and San Juan counties; endemic; 16 (vii). This is the low elevation phase of sandy tracts along Glen Canyon (type locality) and the San Juan River. Some plants with uncut leaves are included here also. The variety is transitional to the typical material, but not moreso than for taxa as a whole in this genus and the plants could reasonably be treated at specific rank.

Sphaeralcea janeae (Welsh) **Welsh** Jane Globemallow. [*S. leptophylla* var. *janeae* Welsh, type from White Rim, San Juan County]. Stems few to many from a woody caudex, 3–9 dm tall, yellow green throughout; leaf blades 12–30 mm long, digitately 3-lobed, the lobes entire, linear to linear-oblanceolate, 0.8–3 mm wide, or the upper leaves simple and linear; inflorescence racemose, elongate, usually with one flower per node; pedicels from much shorter to longer than the calyx; calyx uniformly stellate, the rays of hairs radiating in a single plain, the lobes lance-attenuate; petals 8–12 mm long, orange; carpels 7–10, 3–3.5 mm high, the indehiscent portion forming 3/4–4/5 the carpel, coarsely reticulate, rough but not ridged or tuberculate on the back. Warm and salt desert shrub on the White Rim and Organ Rock members of the Cutler formation at 1220 to 1405 m in Grand, San Juan, and Wayne counties; endemic; 5 (iv). This plant simulates *S. leptophylla* in general habit during early anthesis, but at maturity they are dissimilar in both color and stature. The plants become much taller than the allopatric *S. leptophylla*, simulating the growth form and color of *Psoralidium junceum* and *Amsonia tomentosa*.

Sphaeralcea leptophylla (Gray) **Rydb.** [*Malvastrum leptophyllum* Gray]. Stems few to many from woody a caudex, 2–5.5 dm tall, grayish canescent throughout; leaf blades 10–32 mm long, digitately 3-lobed, the lobes entire, linear to oblanceolate, 1–4 mm wide, or the upper leaves simple and linear; inlorescence racemose, elongate, usually with one flower per node; pedicels from much shorter to longer than the calyx; calyx uniformly stellate, the rays of hairs radiating in a single plain, the lobes lance-attenuate; petals 8–12 mm long, orange; carpels 7–9, 3–3.5 mm high, the indehiscent portion forming 2/3–3/4 of the carpel, coarsely reticulate, ridged or tuberculate on the back. Blackbrush and mixed semidesert shrub communities at 1200 to 1520 m, in Emery, Garfield, Grand, San Juan, and Wayne counties; New Mexico, Arizona, Texas, and Mexico; 11 (iv).

Sphaeralcea munroana (**Dougl.**) **Spach in Gray** Munroe globemallow. [*Malva munroana* Dougl. in Lindl.; *Nuttallia munroana* (Dougl.) Nutt.; *Malvastrum munroanum* (Dougl.) Gray; *S. subrhomboidea* Rydb., type from Midway, Wasatch County; *S. m.* ssp. *subrhomboidea* (Rydb.) Kearney; *S. m.* var. *subrhomboidea* (Rydb.) Kearney]. Stems several to many from branching caudex, 1.8–7 dm tall or more, yellowish green to somehat grayish canescent, the foilage usually bright green; leaf blades 1–6 cm long, 0.8–6 cm wide, ovate to orbicular or rhombic in outline, the base truncate to obtuse or subcuneate, usually 3- to 5-lobed, the sinuses shallow to very deep, the main divisions merely toothed or the lateral ones incised; inflorescence narrowly thyrsoid, usually with more than 1 flower per node; pedicels usually much shorter than the calyx; calyx uniformly stellate, the rays of hairs not radiating in a single plane, the lobes deltoid-ovate to ovate; petals 8–15 mm long, orange;

carpels 10–13, 2.5–3 mm high, the indehiscent portion forming about half the carpel, reticulate on the sides. Mixed desert shrub, or more commonly, in sagebrush and mountain brush communities at 1370 to 2440 m, in Box Elder, Cache, Duchesne, Morgan, Rich, Salt Lake, Summit, Tooele, Utah, Wasatch, and Weber counties; Montana, Idaho, Washington, Wyoming, Nevada, and California; 35 (i). This entity is much like both *S. parvifolia* and *S. grossulariifolia*, but is typically montane, whereas those entities are mainly of lower elevations. The green color of herbage is diagnostic of *S. munroana* from both, except for the allopatric var. *moorei* of *S. grossulariifolia*.

Sphaeralcea parvifolia A. Nels. Nelson Globemallow. [*S. marginata* York, ex Rydb., *S. arizonica* Heller, ex Rydb.]. Stems few to many from branching woody caudex, 1.5–5.5 cm long, 1.2–5.2 cm wide, ovate to orbicular, reniform, or cordate-ovate, the base cordate to truncate or obtuse, usually shallowly 3- to 5-lobed, the sinuses usually shallow, the lobes crenate-dentate; inflorescence commonly narrowly thyrsoid, typically with more than 1 flower per node; pedicels usually shorter than the calyx; calyx uniformly stellate, the rays of hairs not in a single plane, the lobes lance-ovate to deltoid; petals 7–15 mm long, orange; carpels 10–12, 3–4 mm high, the indehiscent part forming from 1/4–1/3 of the carpel, faintly reticulate on the sides. Blackbrush, other warm desert shrub, salt desert shrub, sagebrush, pinyon-juniper, and mountain brush communities at 820 to 2700 m in Daggett, Duchesne, Emery, Garfield, Grand, Iron, Juab, Kane, Millard, Piute, San Juan, Sevier, Tooele, Uintah, Washington, and Wayne counties; Nevada, Arizona, New Mexico, and California; 187 (xxxi). *S. parvifolia* has been compared by Kearney (l.c.) with *S. ambiqua*, which it resembles. The relationship of *S. parvifolia* in Utah seems to lie with the largely sympatric *S. grossulariifolia*, but it is probably transitional with both *S. ambigua* and *S. rusbyi*.

Sphaeralcea psoraloides Welsh Psoralea Globemallow. Stems few to many from branching caudex, 1.4–3.5 dm tall or more, sparsely yellowish canescent, the foilage yellow green; leaf blades 1.3–6 cm long, 0.4–3.8 cm wide, oblanceolate to cuneate-ovate in outline, cuneate to obtuse or rounded basally, trifoliolate or simple to 3–lobed below, deeply 3–5-cleft above, the lobes entire to few toothed or lobed, usually more than 5 mm wide; inflorescence racemose, the flowers solitary in the upper axils; calyx uniformly stellate, the rays of hairs radiating in a single plane, the lobes lance-acuminate; petals 8–15 mm long, orange, carpels 10, 3–4 mm high, the indehiscent part forming 2/3–3/4 of the carpel, reticulate on the sides, the reticulum extending on to the margins of the back. Zuckia-ephedra community on saline and gypsiferous Entrada siltstone at 1220 to 1830 m in southwestern Emery and adjacent Wayne (type from Salt Wash) counties; endemic; 6 (ii).

Sphaeralcea rusbyi Gray Stems few to many from a caudex, or rarely subrhizomatous, mostly 2–6.5 (8.5) dm tall, yellowish green to somewhat grayish canescent; leaf blades 1.3–4.5 cm long, 1.2–4 cm wide, ovate to orbicular in outline, the base truncate-obtuse to prominently cordate, parted to divided or merely cleft, the lobes again toothed (the teeth spreading at nearly right angles); inflorescence thyrsoid to paniculate, with more than 1 flower per node; pedicels usually shorter (to much longer) than the calyx; bractlets often dark red; calyx uniformly stellate (more densely so than on the herbage), the rays of hairs not radiating in a single plain, the lobes ovate to lance-ovate; petals 9–18 mm long, orange; carpels 10–12, 4–6 mm high, the indehiscent part forming from 1/4–3/5 of the carpel, finely reticulate on the sides. Blackbrush, creosote brush, and mixed warm desert shrub communities at 820 to 1220 m in Washington County; Arizona; 5 (i). *S. rusbyi* forms apparent intermediates with phases of *S. grossulariifolia* and *S. parvifolia*.

MELIACEAE A. L. Juss.
Mahogany Family

Trees with hard, scented wood; leaves usually alternate, pinnate, estipulate; flowers perfect, regular, often in cymose panicles; sepals 4–6, connate at the base; petals 4–6, distinct; stamens 8–10, commonly monadelphous; pistil 1, the ovary superior, 2- to 5-loculed, with 2 ovules per locule; style 1 or lacking; fruit a drupaceous berry; seeds often winged; x = 10–14+.

Melia L.

Deciduous trees; leaves large, 1- to 3-times pinnately compound; flowers white to purple, in large axillary many-flowered panicles; fruit yellow, persistent.

Melia azedarach L. Chinaberry. Trees to 15 m tall, with a broad rounded crown; leaves usually bipinnate; leaflets numerous, ovate to lance-elliptic, acuminate, to ca 6 cm long and 3 cm wide, crenate-dentate; flowers fragrant, on pedicels to 3 mm long; sepals 1–2 mm long, elliptic; petals oblanceolate to spatulate, white to pale lavender, ca 1 cm long; staminal tube purplish, cylindrical, 8–10 mm long, with numerous irregular teeth at the orifice and 10–12 included anthers; fruit yellow, subglobose, fleshy, 8–15 mm thick, bitter, a single seed in each locule; 2n = 28. Cultivated, escaping, and persisting (established ?) at 850 to 915 m in Washington County; widely grown in warm portions of the U. S.; introduced from Asia; 8 (iii).

MENYANTHACEAE Dumort.
Buckbean Family

Plants aquatic or semiaquatic, glabrous herbs with thick rhizomes; leaves alternate, long-petiolate, trifoliolate, the petioles sheathing basally; flowers perfect, regular, arranged in paniculate cymes or racemes; sepals 5, adnate to the lower portion of the ovary; petals 5, united, the corolla rotate or funnelform; stamens 5, alternate with the corolla lobes; pistil 1, the ovary partially inferior, 1-loculed; styles long or wanting, the stigma simple or 2-lobed; fruit a septicidal capsule, tardily dehiscent to indehiscent; seeds smooth and shining; x = 9, 17.

Menyanthes L.

Plants scapose, the leaves all basal; calyx 5-lobed, the lobes distinct; corolla lobes spreading, each with numerous scaly hairs on the inner surface; capsules 1-loculed.

Menyanthes trifoliata L. Buckbean. Rhizomes covered with membranous leaf bases; petioles 5–30 cm long, the sheathing stipules subequal in length to the free petioles;

leaflets elliptic to oblanceolate or obovate, 2–9 (12) cm long, 0.9–5 cm broad; scapes 5–30 cm long; sepals 2–5 mm long, oblong, spreading, the adnate calyx tube short-conic; corolla white to pink, funnelform, the tube 5–8 mm long, the lobes 5–8 mm long, usually purplish tinged apically; capsules ellipsoid, almost superior. Lake margins and bogs, mainly at 2895 to 3145 m in Duchesne, Summit, Uintah, and Wasatch counties, but also at ca 1680 m in Kane (Hidden Lake near Orderville) County; Alaska and Yukon, south to California and Colorado, and east to the Atlantic; circumboreal; 10 (0).

MORACEAE Link

Mulberry Family

Deciduous trees or shrubs with milky juice; leaves alternate, simple, pinnately or palmately veined, entire, serrate, or lobed, stipulate, the stipules small and distinct or each pair forming a cap over the bud and leaving a scar around the stem; flowers imperfect, minute, regular, borne in cymes or much modified inflorescences; sepals usually 4 (2 in *Ficus*) stamens 4, the filaments distinct; pistil 1, the ovary typically superior, 1–loculed, the styles and stigmas 2 (1 in *Maclura*); fruit a multiple (*Morus*, *Maclura*) or a syconium (*Ficus*); x = 7 to many.

1. Fruit a fleshy hollow receptacle with flowers borne inside (syconium); leaves palmately veined and lobed; cultivated plants in Washington County, and under glass elsewhere .... *Ficus*

— Fruit a multiple (formed of several flowers and a common axis); leaves various 2

2(1). Leaves crenate-serrate, palmately veined and often palmate lobed as well; flowers, both sterile and fertile, borne in catkinlike spikes; fruit seldom more than 1 cm thick .... *Morus*

— Leaves entire, pinnately veined, not lobed; flowers borne in dissimilar inflorescences, the sterile in racemes, the fertile in globular heads; fruit globular, more than 5 cm thick .... *Maclura*

Ficus L.

Trees or large shrubs; leaves alternate, simple, palmately veined and lobed, the stipules forming a circular scar around the stem; flowers minute, numerous, borne inside a hollow receptacle which ripens to form a syconium; staminate perianth 2– to 6–parted, with 1 or 2 stamens; pistillate perianth reduced or lacking; receptacles perfect or imperfect; fruits of individual flowers of achenes.

Ficus carica L. Common Fig. Deciduous trees to 8 m tall, rarely more, often sprawling in age; leaves prominently veined, thick, to 25 cm long or more and to 20 cm broad, 3– to 5–lobed, the lobes undulate-serrulate; fruits obovoid, mostly 2.4–4.5 cm long and 2–3 cm thick; 2n = 26. Cultivated fruit plant in Washington (and formerly at Hite, Garfield) County, frost sensitive elsewhere except under glass; introduced from the Mediterranean region of the Old World; 6 (iii). This is the fig of commerce. Plants in Washington County bear heavily, the fruit being of excellent quality, produced in two or more successive crops per year.

Malcura Nutt.

Dioecious trees with hard yellow wood; leaves entire, the stipules minute, the scar not encircling the stem; staminate flowers numerous in loose, peduncled, axillary heads or umbels, the calyx 4–parted and with 4 stamens; pistillate flowers coherent in dense, globose, axillary heads, the calyx 4–lobed, the single filiform style very long; fruit a globose multiple.

Malcura pomifera (**Raf.**) **Schneider** Osage Orange. [*Toxylon pomiferum* Raf.]. Trees to 10 m tall, rarely more; stems usually armed with stout thorns 1–2 cm long; leaves petiolate, the blades 5–10 cm long and 1.8–6.5 cm wide, ovate, entire, rounded to obtuse basally, attenuate to acuminate apically; clusters of staminate flowers 2.5–3.5 cm across; heads of pistillate flowers 2–2.5 cm across; multiple fruit mostly 8–14 cm thick. Cultivated ornamental and botanical curiosity of low elevation regions in Utah, long persisting; introduced from the eastern U. S.; 5 (i). The wood of this tree is very strong, and has served as a source of bows for American Indians and others.

Morus L.

Dioecious trees; leaves palmately veined, serrate to dentate, sometimes lobed; stipules lanceolate, the scar not encircling the stem; flowers monoecious or dioecious, those of both sexes borne in stalked, axillary, catkinlike clusters; calyx 4–parted; stamens 4; styles 2, deeply parted; fruit a multiple.

1. Leaves glabrous above and beneath or pubescent beneath only along main veins and/or in vein axils; our common mulberry .... *M. alba*

— Leaves pubescent over much of the lower surface, scabrous above; rarely cultivated .... *M. nigra*

Morus alba L. White Mulberry. Cultivated ornamental and shade tree to 10 m tall or more; leaves obliquely ovate and crenate-serrate or irregular lobed, mostly 3.5–14 cm long and 2.5–10 cm wide, truncate to subcordate basally, acute to acuminate apically, glabrous above and below except along veins and in vein axils; fruit 1–2 cm long and 0.6–1 cm thick, white, pink, red purple, or nearly black; 2n = 14, 28, 42. Persisting and occasionally escaping in most of Utah at lower elevations; introduced from China; widespread in North America; 33 (vi). This plant was introduced into southern Utah to provide food for silkworms in an attempt to develop a silk industry. The fruit is edible, but is consumed mainly by birds. Reports from Utah of red mulberry, *M. rubra* L., belong here. Red mulberry is easily recognized by the densely hairy lower and scabrous upper leaf surfaces. So-called fruitless phases are known.

Morus nigra L. Black Mulberry. Small trees to about 10 m; leaves cordate-ovate, crenate-serrate, seldom lobed, 5–20 cm long, 3–15 cm, wide, cordate basally, obtuse to acuminate apically, scabrous above and hairy over veins and at least some intervein areas below; fruit 1–2.5 cm long and to 1 cm thick, purple to black; 2n = 28, 42+. Sparingly cultivated ornamental, mainly in warm regions of Washington County; widely cultivated in temperate regions of the earth for its fruit; introduced from Asia; 5 (0).

NYCTAGINACEAE A. L. Juss.
Four O'Clock Family

Annual or perennial herbs, often with stems swollen at the nodes; leaves simple, mostly opposite; flowers perfect, regular, the perianth reduced to a petaloid connate calyx, this subtended by sepaloid bracts; stamens 1 to many; pistil 1, 1–carpellate, the ovary superior, enclosed by the persistent perianth base (the anthocarp), 1–loculed, 1–ovuled; style 1; fruit indehiscent; seed erect; x = 10, 13, 17, 29, 33+.

1. Perianth mainly 1–4 cm long, tubular-funnelform, the tube mostly 1–3 mm wide; flowers in heads 2
— Perianth mostly 0.5–1 cm long and campanulate, or if longer and funnelform, the tube more than 3 mm wide (compare *Selinocarpus*); flowers not in heads 3
2(1). Perianth 5–lobed; receptacle rounded to conic, lacking pedicels; wings of fruit, if any, thickish and opaque, not continous around the fruit; plants perennial or annual . . . . *Abronia*
— Perianth 4– or 5–lobed; receptacle flat, bearing pedicels mainly 1–2 mm long; wings of fruit thin, translucent, continous around the fruit; plants annual . . . . *Tripterocalyx*
3(1). Plants prostrate, the stems radiating from the root crown; perianth pink, showy . . . . *Allionia*
— Plants with stems ascending to erect; perianth whitish, greenish, or pink 4
4(3). Plants annual from a slender taproot; perianth less than 5 mm long . . . . *Boerhaavia*
— Plants perennial from a caudex; perianth more than 1 cm long 5
5(4). Fruits winged, the wings conspicuous, 2–3 mm wide; perianth greenish, 3–4 cm long, the tube less than 5 mm wide . . . . *Selinocarpus*
— Fruits wingless; perianth white to pink or pink purple, 1.5–6 cm long, the tube more than 5 mm wide . . *Mirabilis*

Abronia Juss.

Annual or perennial herbs from a taproot or caudex; leaves opposite, unequal, the blades entire or lobed; inflorescence capitate, the 5–10 bracts distinct; receptacle rounded to conic, lacking pedicels; perianth 2–parted, the upper part corollalike and deciduous, the lower part forming the anthocarp; flowers 5–merous; perianth lobes flaring; anthocarps winged or merely ribbed; wings, if present, usually 2–5, opaque, not surrounding the anthocarp, the veins not prominent; achenes free.

Galloway, L. A. 1975. Systematics of the North American desert species of *Abronia* and *Tripterocalyx* (Nyctaginaceae). Brittonia 24: 328–347.

1. Perianth bright pink purple (magenta); plants annual, densely viscid-villous, of Washington County . . *A. villosa*
— Perianth white or pale pinkish to rose; plants perennial, variously hairy or villous 2
2(1). Plants subscapose, the leaves essentially all basal; caudex commonly with a thatch of marcescent leaf bases . . . . *A. nana*
— Plants definitely caulescent, the peduncles evidently axillary; caudex seldom with a thatch of marcescent leaf bases 3
3(2). Bracts subtending heads mainly 1.5–5 mm wide; fruits, at least some, usually winged, the firm thin wings wider than the body near the apex; plants of Cache County . . . . *A. mellifera*
— Bracts subtending inflorescence mainly 4–10 mm wide or more; fruits wingless or winged, but if so, seldom wider than the body; plants of various distribution 4
4(3). Plants essentially glabrous; stems suffused with purple (at least basally); involucral bracts orbicular to obovate; anthocarps glabrous, wingless, finally beakless; known from clayey and silty substrates in Grand and Uintah counties . . . . *A. argillosa*
— Plants variably pubescent and glandular (sand grains often adhering), usually with moniliform hairs in the inflorescence; stems green, pinkish, or uncommonly suffused with purple; known mainly from sandy substrates, widespread in Utah . . . . *A. fragrans*

Abronia argillosa **Welsh & Goodrich** Clay-verbena. Plants perennial from a branching caudex; stems slender, (6) 15–30 (38) cm tall, glabrous, suffused purplish throughout or at the base only; leaf blades 5–35 (55) mm long, 3–38 mm wide, elliptic, ovate, obovate, or suborbicular, entire, glabrous, glaucous, thick; peduncles 1–8 (12) cm long, glabrous or rarely puberulent; bracts 7–15 mm long, 6–15 mm wide, oval to orbicular, scarious, white or suffused with purple, glabrous dorsally, ciliate; flowers 15–20 per head; perianth tube 10–15 mm long, greenish, glabrous or rarely puberulent, the limb ca 6 mm wide, white; receptacle shortly conic; anthocarps wingless, the beak short or lacking, slightly angled or not, mainly 7–9 mm long, 3–4 mm wide, scarious, glabrous, rugose, oblanceolate in outline. Mat-saltbush, other salt desert shrub, and pinyon-juniper communities, on Mancos Shale and Green River formations, at 1310 to 1830 m in Grand (type from 6 mi S of Cisco) and Uintah counties, Utah and Mesa County, Colorado; 11 (vi). This is a beautiful plant, with the glaucous, glabrous, and thickened foliage blending with the sere substrate of gray to whitish shaley clays that support it, and with the large usually pure white bracts and flowers standing ghostlike above.

Abronia fragrans **Nutt. ex Hook.** Fragrant Sand-verbena. [*A. fragrans* var. *pterocarpa* Jones, type from St. George; *A. bakeri* Greene; *A. elliptica* A. Nels.; *A. fragrans* var. *elliptica* (A. Nels.) Jones; *A. fallax* Heimerl ex Rydb., type from Salt Lake City; *A. glabra* Rydb.; *A. pumila* Rydb., type from Emery; *A. salsa* Rydb., type from Salt Lake City; *A. turbinata* var. *marginata* Eastw., type from Bartons range, San Juan County]. Plants perennial from a taproot, the caudex sometimes developed; stems thickish to slender, 1.8–8 (10) dm tall, glabrous to densely glandular-hairy (sand grains adhering) below or throughout, green to straw colored to reddish tinged, rarely suffused purplish; leaf blades 8–90 mm long, 3–30 mm wide, lanceolate to ovate or lance-linear, entire, glabrous or hairy; peduncles 2–18 cm long, puberulent to villous or glandular; bracts 5–25 mm long, 2–12 mm wide, ovate to lanceolate, obovate, or elliptic, puberulent to glandular-villous; flowers 25–80 per head; perianth tube 10–25 mm long, greenish to rose or purplish, puberulent to glandular-villous, the limb 5–10 mm wide, usually white; anthocarps 5–12 mm long, 2.5–8.5 mm wide, obdeltoid to obliquely so in outline, often polymorphic in a head, 1– to 5–winged or -angled, puberulent to subglabrous. Creosote bush, blackbrush, Vancleveaephedra, shadscale, mixed desert shrub, pinyon-juniper, sagebrush, and ponderosa pine communities, in various substrates but often in sand, at 750 to 2500 m in all Utah counties, except Morgan and Rich; Montana to South Dakota, south to Nevada, Arizona, New Mexico, Texas, and Mexico; 229 (xxxi). Our material has been regarded as belonging to at least two entities, either at specific or varietal rank, i.e., as *A. fragrans* and *A. elliptica*, or as *A. fragrans* var. *fragrans* and *A. fragrans* var. *elliptica*. Main diagnostic features used to segregate Utah specimens into two taxa involve the nature of the anthocarp, with var. *elliptica* being 2-winged, with a groove between the wings, and with var. *fragrans* being 2- to 5-winged, but not especially grooved. The variation in structure of the anthocarp in Utah includes both types, often within the same head, with the latter one being most common in southeastern Utah, but the separation seems to be arbitrary and is not correlated with other features.

Abronia mellifera Dougl. ex Hook. Honey Sand-verbena. [*A. lanceolata* Rydb.]. Plants perennial from a taproot, the caudex sometimes developed; stems 2–8 dm tall, usually green (or reddish at the nodes), glabrous to glandular puberulent; bracts 5–12 mm long, 1–5 mm wide, lanceolate to ovate, elliptic, or obovate, glabrous to glandular; flowers 25–60 per head; perianth tube 15–25 mm long, glabrous to glandular, usually greenish, the limb 7–12 mm wide; anthocarps 6–10 mm long, 4–10 mm wide, beaked, puberulent to glandular, winged (rarely wingless), obdeltoid in outline. Cool desert shrub community at ca 1450 m in Cache County (fide Galloway 1975); Washington, Oregon, Idaho, and Wyoming; 0 (0). The similarity of this plant to *A. fragrans*, sens. lat., is striking. It differs only in minor technical characteristics, especially in the narrow bracts, and could be expanded to include *A. fragrans* without serious injustice to its taxonomy. I follow tradition in maintaining them separately. Occasional specimens from within the range of *A. fragrans* have slender bracts, even in spring flowering plants. However, this is especially true for plants flowering a second time late in the season.

Abronia nana Wats. Low Sand-verbena. Plants 5–16 (20) cm tall, perennial from a taproot and usually with a well-developed, branched caudex, this typically clothed with marcescent leaf bases; stems without developed internodes, or rarely short-caulescent; herbage more or less glandular-puberulent to glandular-villous; leaf blades 6–30 (45) mm long, 3–15 mm wide, elliptic to oval, oblong, or ovate, entire; peduncles 3–12 cm long; bracts 6–10 mm long, 3–7 mm wide, elliptic to obovate or ovate; flowers 7–25 per head; perianth tube 10–18 mm long, rose to purplish or white, the limb 5–8 mm wide; anthocarps 7–9 mm long, 5–8 mm wide, obcordate to obdeltoid, 5-winged. Warm, salt, and mixed desert shrub and pinyon-juniper communities at 1585 to 2170 m in Beaver (type from near Beaver), Emery, Garfield, Juab, Kane, Millard, Sevier, and Washington counties; Colorado, Nevada, Arizona, and California; 40 (viii). Most of our specimens are strictly acaulescent and have best developed leaves mainly less than twice as long as broad. These represent var. **nana**. However, in the San Rafael Swell, Emery County, Utah, there occur caulescent plants with main leaves 2–4 times longer than broad. These are designated as var. **harrisii** Welsh.

Abronia villosa Wats. Sticky Sand-verbena. Annual herbs from a taproot; stems decumbent to prostrate or ascending, mainly 0.8–5 dm long; herbage glandular puberulent to viscid-villous (sand grains adhering); leaf blades 8–50 mm long, 4–45 mm wide, elliptic to ovate, oblong, deltoid, suborbicular, or rhombic, succulent, entire or sinuate; peduncles 2–10 cm long; bracts 3–11 mm long, 1–3 mm wide, lanceolate to narrowly ovate; flowers 15–35; perianth tube 13–20 mm long, the limb 6–13 mm wide, pink purple (magenta) or rarely white; anthocarps 5–8 mm long, 4–13 mm wide, often pitted. Creosote bush and other warm desert shrub communities at 750 to 900 m in Washington County; Nevada, Arizona, California, and Mexico; 3 (i).

Allionia L.

Annual or perennial herbs; stems prostrate, the branches more or less dichotomous; herbage glandular-villous; leaves opposite, very unequal in the pair, ovate to oblong; flowers sessile, perfect, borne in peduncled axillary clusters of 3, each subtended by a bract that encloses the fruit; perianth campanulate, 4- or 5-lobed; stamens 4–7, exserted; stigma capitate; fruit leathery, flattened, the inner face with 2 rows of stipitate glands, the margins inrolled, toothed.

Allionia incarnata L. Trailing Four-O'Clock; Windmills. Perennial or winter annual herbs; stems prostrate, 1–8 dm long or more, radiating from the root crown; leaf blades 8–30 (55) mm long, 3–23 (43) mm wide, ovate to elliptic, entire or undulate to roundly lobed or toothed; peduncles 3–25 mm long; bracts ovate, 4–9 mm long; perianth pink-purple to magenta, rarely white, 6–15 mm long; fruits 3–4 mm long, the inner side 3-veined, the margins usually toothed and incurved. Warm, salt, and mixed desert shrub communities at 800 to 1830 m in Garfield, Kane, San Juan, Washington, and Wayne counties; Colorado, Nevada, California, and Texas, south to Mexico; South America; 32 (iv). The showy bright flowers contrast strongly with the dull, even drab, herbage.

Boerhaavia L.

Annual herbs; stems usually branched from the base; herbage usually glandular-viscid; leaves opposite, petiolate, usually unequal in the pair; flowers perfect, small, variously arranged; bracts small, distinct; perianth funnelform to campanulate, shallowly 5-lobed; stamens 1–5, the filaments connate below; ovary stipitate; style slender, the stigma peltate; anthocarp obovoid or obpyramidal, nerved or angled.

Boerhaavia spicata Choisy Spiderling. [*B. spicata* var. *torreyana* Wats.; *B. torreyana* (Wats.) Standley]. Plants erect or procumbent, 2–6 dm long; stems slender, puberulent to glandular-hirtellous or viscid; leaf blades 1–4.5 cm long, 0.4–2.5 cm wide, lanceolate to ovate, obtuse to rounded or acute apically, entire to sinuate, green above, paler beneath; inflorescence racemose to paniculate, much branched, the branches capillary, glabrous or sparingly glandular; flowers with pedicels 0.5–2 mm long, spaced along the axis; bracts ovate, scarious, 0.5–0.8 mm long, soon deciduous; perianth ca 1 mm long, pink to white; stamens 1 or 2; anthocarp 2.5–3 mm long, clavate, rounded apically, 5-ridged, the furrows open and with v-shaped indentations into the ridges. Sandy soil in pinyon-juniper community at ca 1650 m in Kane County (Atwood & Kaneko 3389 BRY); Arizona, New Mexico, and Texas; 2 (i).

Mirabilis L.

Perennial herbs, often from tuberous roots; stems erect to sprawling, simple or branched from the base or throughout; leaves opposite, subequal in the pair; involucres axillary or terminal or both, 1- to 10-flowered, of 5 distinct bracts or these connate from below to above the middle, sometimes accrescent in fruit; perianth variously colored, the tube constricted above the ovary, the limb campanulate to salverform or funnelform; stamens 3–6, the filaments filiform, united into a cup below; anthocarp 5-angled or -ribbed.

Pilz, G. E. 1978. Systematics of *Mirabilis* subgenus Quamoclidion (Nyctaginaceae). Madrono 25: 113–176.

1. Perianths varicolored, 4–6 cm long; plants cultivated . . . . M. jalapa L.
— Perianths not especially varicolored in populations, mainly shorter than 4 cm long (except in *M. multiflora*); plants indigenous 2

2(1). Leaves linear to linear-lanceolate, mainly 10–20 times (or more) longer than broad 3
— Leaves lanceolate, ovate, cordate, or orbicular, mainly less than 5 times longer than broad 4

3(2). Stems, involucres, perianths, and anthocarps glabrous or sparingly strigose; plants reported for Utah, but no specimens have been seen . . . M. glabra (Wats.) Standley
— Stems, involucres, perianths, and anthocarps sparingly to copiously strigose to pilose and often more or less glandular; plants widespread and locally common in much of Utah . . . . M. linearis

4(2). Leaves lanceolate to ovate (rarely broader), cuneate to obtuse; plants of western and southwestern Utah . . . . M. pumila
— Leaves ovate to cordate or orbicular, truncate to cordate basally; plants variously distributed 5

5(4). Involucres papery, conspicuously veined, much enlarged in fruit; plants rare in Utah . . . . M. nyctaginea
— Involucres herbaceous, not conspicuously veined or expanding in fruit 6

6(5). Perianths 0.8–1.2 cm long, white (rarely pinkish); plants of Washington County . . . . M. bigelovii
— Perianths variously shorter or longer than above, but if shorter, not white and the plants not of Washington County 7

7(6). Plants with sprawling, slender stems; perianth 7–9 mm long; uncommon plants of mesic sites . . . M. oxybaphoides
— Plants with thickish, decumbent-ascending stems; perianths 15–60 mm long 8

8(7). Limb of perianth campanulate; perianths mainly 15–20 mm long; plants of low quality substrates in the Uinta Basin, central western Utah, and Kane County . M. alipes
— Limb of perianth rotate; perianth mainly 30–60 mm long; plants often with pinyon and juniper in the southern third of the state . . . . M. multiflora

Mirabilis alipes (Wats.) Pilz Watson Four-O'Clock. [*Hermidium alipes* Wats; *H. alipes* var. *pallidum* C. L. Porter, type from Uintah County]. Plants mainly 15–35 (40) cm tall, clump-forming; stems arising from a caudex, this atop a tuberous root, decumbent-ascending, constricted below ground level, mainly 3–6 mm thick, tapering upwards; herbage pale green, glaucous, glabrous or sparingly puberulent; leaves petiolate to subsessile, the blades mainly 2.5–7 (9) cm long and 1.5–6.5 cm wide, suborbicular to ovate or lanceolate, entire, rounded to obtuse or acute apically; peduncles 3–10 mm long, erect; bracts of involucre distinct or united to the middle, mainly 15–30 mm long; involucres 6- to 9-flowered, the flowers on pedicels to 1 mm long, borne on midvein of the bract; perianth campanulate, 15–20 mm long, pink purple, pink, or white, glabrous; stamens subequal to the perianth, the filaments pubescent below; anthocarp ellipsoidal, 5.5–7 mm long, obscurely 10-ribbed, mucilaginous when wet. Mat-saltbush, shadscale, other salt desert shrub, mixed desert shrub, and pinyon-juniper communities at 1250 to 1955 m in Duchesne, Juab, Kane, Millard, Tooele, and Uintah counties; Colorado, Nevada, and California; 41 (v). This is an attractive thick-leaved plant, in which the bracts are often suffused with purple and are almost as colorful as the perianth. Its distribution is tripartite; i.e., Uinta Basin, on Duchesne River and Uinta formations; Kane County, on Tropic Shale Formation, and Great Basin, on various formations.

Mirabilis bigelovii Gray Bigelow Four-O'Clock. Plants erect or ascending from a woody caudex, clump-forming, mainly 2–5 dm tall; stems slender, mainly 2–3.5 mm thick; herbage glabrous or scaberulous to puberulent or viscid-villous; leaves short-petiolate, the blades 0.8–3.8 cm long, 0.5–3.4 cm wide, orbicular to broadly ovate, entire, rounded to obtuse apically, truncate to subcordate basally; involucres on peduncles 5–6 mm long, the bracts connate to above the middle, mainly 4–5 mm long; perianth 8–12 mm long, white to pale pink or suffused with purple; fruit ovoid-ellipsoid, ca 3 mm long, smooth or rugulose. Creosote-bush, Joshua tree, and blackbrush communities at 820 to 1340 m in Washington County; Arizona, Nevada, and California; 13 (i). Our plants have been assigned to **var. retrorsa (Heller) Munz** [*M. retrorsa* Heller; *Hesperonia glutinosa* (A. Nels.) Standley; *M. glutinosa* A. Nels., type from near St. George, not Kuntze; *M. limosa* A. Nels.].

Mirabilis linearis (Pursh) Heimerl Narrowleaf Umbrellawort. [*Allionia linearis* Pursh; *Oxybaphus linearis* (Pursh) Robins.; *M. glabra* authors, not (Wats.) Standley?]. Plants erect, from a woody caudex, clump-forming, mainly 2–10 dm tall; stems glabrous or puberulent below, viscid-puberulent to viscid-villous above, usually glaucous; leaves sessile or short-petiolate, linear to linear-lanceolate (rarely lanceolate), mostly 2–10 cm long, 1–10 (13) mm wide, entire to sparingly toothed, green on both sides or glaucous below; inflorescence paniculate (a branching cyme); involucres of connate bracts, these ca 4 mm long at anthesis, 6–10 mm long in fruit, viscid-villous, usually 3-flowered; perianth 8–12 mm long, white to pink or red purple, the limb deeply lobed; anthocarps 4–5 mm long, obovoid, 5-angled, pubescent, the ridges cut with triangular notches. Two rather weak varieties are present in Utah.

1. Flowers mainly red purple; leaves linear-lanceolate, more or less petiolate; plants uncommon . . . . M. linearis var. decipiens
— Flowers mainly white to pink; leaves linear, sessile; plants common and widespread . . . . M. linearis var. linearis

Var. decipiens (Standley) Welsh [*Allionia decipiens* Standley]. Pinyon-juniper, ponderosa pine, riparian, aspen, and spruce-fir communities at 1890 to 3050 m in Garfield, Piute, and San Juan counties; Colorado, Arizona, and New Mexico; 11 (v).

Var. linearis [*Oxybaphus angustifolius* var. *viscidus* Eastw., type from Butler Spring, San Juan County; *O. glaber* Wats., type from Kanab; *Mirabilis glabra* (Wats.) Standley]. Salt Desert shrub, mixed desert shrub, riparian, hanging garden, pinyon-juniper, sagebrush, mountain brush, and aspen communities at 1125 to 2535 m in Beaver, Daggett, Duchesne, Emery, Garfield, Grand, Juab, Kane, Millard, Salt Lake, San Juan, Sevier, Tooele, Uintah, Utah, Washington, and Wayne counties; Montana to South Dakota, south to Arizona, New Mexico, Texas, Missouri, and Mexico; 90 (xxv).

Mirabilis multiflora (Torr.) Gray in Torr. Large Four-O'Clock. [*Oxybaphus multiflorus* Torr.; *Allionia multiflora* (Torr.) D. C. Eaton; *Quamoclidion multiflorum* (Torr.) Torr. ex Gray]. Plants mainly 3–8 dm tall, from a caudex; stems decumbent to ascending, clump-forming,, mainly 4–10 mm thick, tapering upwards; herbage green, puberulent to glabrous; leaves short-petiolate, the blades mainly 2–10 cm long and about as broad, orbicular to ovate, entire, truncate to subcordate basally, rounded to obtuse or acute apically; peduncles 0.5–7 cm long; involucres canpanulate, 20–35 mm long, connate to above the middle, 5-lobed, 6-flowered, the flowers on pedicels to 2 mm long, on midvein of the subtending bract; perianth 4–6 cm long, funnelform, pink-purple (magenta), the tube colored like the limb or green; stamens exserted, 5–10 or more, the filaments glabrous to pubescent; anthocarps ellipsoidal, 6–11 mm long, smooth or with 10 tan longitudinal ribs alternating with 10 dark brown ones. Creosote bush, blackbrush, and pinyon-juniper communities at 760 to 2380 m in Carbon, Garfield, Grand, Kane, San Juan, Washington, and Wayne counties; Nevada, Colorado, New Mexico, Arizona, California, Texas, and Mexico; 44 (ix). Three difficultly discernable and partially sympatric varieties have been recognized from Utah. The varieties *multiflora* and *glandulosa* (Standley) Macbr. occur in southeastern Utah, and the var. *pubescens* Wats. is present in Washington County. Diagnostic features include fruit characteristics (mucilaginous or not, and ribbed versus smooth) and shape of bract lobes. The latter character grades continuously and the former is not possible to apply in most cases due to lack of fruiting specimens. Recognition of infraspecific taxa therefore seems moot.

Mirabilis nyctaginea (Michx.) Macmillan Wild Four-O'Clock. [*Allionia nyctaginea* Michx.; *Oxybaphus nyctagineus* (Michx.) Sweet]. Plants erect, mainly 3–10 dm tall, from a caudex; stems branched, glabrous or somewhat hairy, ridged and often purplish; leaves petiolate to sessile, the blades 3–12 cm long, ovate-lanceolate to cordate, entire, cordate to truncate basally, acute to attenuate apically; flowers 3–5 per involucre, these borne on umbellate peduncles in forked terminal clusters, the involucres 5–6 mm long in anthesis, 10–15 mm long in fruit, 5-lobed, scarious and veiny in fruit; perianth 5-lobed, pink to purple or reddish; stamens 3–5, exserted; anthocarps 4–6 mm long, cylindroic to obovoid, rugose or warty, 5-ribbed; $2n = 58$. Ruderal habitat at ca 1525 m in Utah County (Arnow and Albee 4103 BRY); Montana to Wisconsin, south to Colorado, Texas, Mexico, and Alabama; adventive elsewhere in the U. S.; 1 (0). The specimen examined is likely introduced, but the possibility of persistence and of reintroduction is great.

Mirabilis oxybaphoides (Gray) Gray in Torr. Spreading Four-O'Clock. [*Quamoclidion oxybaphoides* Gray; *Allioniella oxybaphoides* (Gray) Rydb.]. Plants prostrate, decumbent, or ascending (often clambering), mainly 3–8 dm long, forming extensive mats or clumps; herbage viscid-puberulent to glabrous, the raphide bundles apparent within the tissues; stems swollen at the nodes; leaves petiolate, the blades mainly 1.5–7 cm long and 1–6 cm wide, cordate-ovate to deltoid, entire, cordate to subcordate basally, alternate to acuminate apically; inflorescence of axillary or terminal cymes; involucres 4–6 mm long, deeply 5-cleft; perianth 5–9 mm long, pink to pink purple or white; anthocarps 2.5–3 mm long, obovoid-ellipsoid, rugose or smooth. Riparian, mountain brush, and pinyon-juniper communities at 1645 to 2500 m in Beaver, Garfield, Iron, Piute, San Juan, and Wayne counties; Colorado, Arizona, New Mexico, Texas, and Mexico; 6 (iii).

Mirabilis pumila (Standley) Standley Standley Four-O'Clock. [*Allionia pumila* Standley; *Oxybaphus pumilus* (Standley) Standley]. Plants erect or ascending, mainly 2–5 dm long, clumpforming; herbage more or less glandular-hirtellous or viscid-villous, with raphide clusters apparent in the tissues; stems not much swollen at the nodes; leaves petiolate to subsessile, the blades mainly 1.8–6 cm long, 0.6–2.5 cm wide, ovate to lanceolate, sinuate to entire, cuneate to obtuse basally, attenuate to a rounded apex; inflorescence axillary or in terminal paniculate cymes; involucres 3–4 mm long at anthesis, 7–10 mm long in fruit, 5-lobed; perianth 8–11 mm long, pink; stamens exserted; anthocarps 4.5–5.2 mm long, ellipsoid above the constriction, 5-ribbed, pubescent. Desert shrub communities at ca 1615 to 1770 m in Beaver, Millard, and Washington counties; Nevada, California, Arizona, and New Mexico; 6 (i). Dwarf plants, lacking viscid-villous pubescence in the inflorescence, are known from Washington County (Atwood 9386 and Higgins 15921, both BRY).

Selinocarpus Gray

Perennial herbs; stems erect or ascending, dichotomously branched; leaves opposite, unequal in the pair, the blades thick; flowers perfect, appearing axillary, sessile or short-pedicellate, often cleistogamous, each subtended by 2 or 3 narrow bracts; perianth tubular-funnelform; stamens 3–5, the filaments filiform, adherent to the perianth tube; ovary oblong, with filiform style and peltate stigma; anthocarp compressed, broadly 3- to 5-winged, the wings striate but not veined.

Selinocarpus diffusus Gray. Moonpod. Plants 0.8–3 dm tall, clump-forming, from a caudex and woody rootcrown; herbage with short, appressed, white, inflated hairs and often more or less glandular; leaves petiolate, the blades 0.8–2.5 cm long and about as broad, orbicular to oval or ovate, truncate to obtuse basally, rounded to obtuse apically; bracts linear-subulate, 3–6 mm long; perianth greenish, 3–4 cm long, caducous, slender, the limb 10–15 mm wide, greenish yellow; anthocarps 6–7 mm long the wings 2–3 mm long. Creosote bush and other warm desert shrub communities at 850 to 1000 m in San Juan and Washington counties; California, Nevada, Arizona, New Mexico, and Texas; 11 (i). Our material is regarded as ssp. **nevadensis** Standley [*S. nevadensis* (Standley) Fowler & Turner]. It differs in features of leaves, and the perianth is said to be more slender than those of the more eastern ssp. *diffusus*. Our specimens are largely, if not entirely, cleistogamous and typically lack a developed perianth.

***Tripterocalyx* (Torr.) Hook.**

Annual, more or less succulent herbs from a taproot; stems simple or branched, often reddish; herbage usually viscid to glandular hairy, sometimes glabrous; leaves opposite, subequal or unequal in the pair; inflorescence capitate, subtended by distinct bracts, or these connate near the base; receptacles flat or nearly so, bearing pedicel-like projections; perianth 2–parted, the lower portion persistent (forming the anthocarp), the upper part deciduous; flowers perfect; perianth 4– or 5–lobed; stamens 3–5; stigmas linear; anthocarps surrounded completely by 2–4 thin, membranous, conspicuously veined wings.

1. Flowers 2 cm long or less, the limb of perianth 3–6 mm wide .... *T. micranthus*
— Flowers more than 2 cm long, the limb of perianth 8–11 mm wide or more .... *T. carneus*

***Tripterocalyx carneus* (Greene) Galloway** Wooton Sandverbena. [*Abronia carnea* Greene]. Plants more or less succulent, erect or procumbent to ascending; stems mainly 1–7 dm long, variously pubescent and glandular (sand grains adhering) or glabrous; petioles 1–6 cm long; leaf blades 1–8 cm long, 0.6–5 cm wide, ovate to elliptic, oblong, or ovate-lanceolate, entire to undulate, truncate to cuneate basally, obtuse to acute apically; peduncles 1–11 cm long; bracts 5–8, 5–14 mm long, linear-lanceolate to lance-attenuate; flowers 10–25 per head; perianth tube 12–30 mm long, pink to pinkish red, the limb 8–11 mm wide, white to pink; anthocarps 2– to 4–winged, 13–27 mm long. Blackbrush, ephedra-Vanclevea, sand sagebrush, and indigo bush communities at 1125 to 1375 m in Garfield, Kane, and San Juan counties; Arizona, Colorado, New Mexico, Texas, and Mexico; 17 (iii). Our material is regarded as **var. *wootonii* (Standley) Galloway** [*T. wootonii* Standley]. This is a remarkably beautiful species, when in full flower.

***Tripterocalyx micranthus* (Torr.) Hook.** Sandpuffs. [*Abronia micrantha* Torr. in Frem.; *A. micrantha* var. *pedunculata* Jones, type from St. George; *A. pedunculata* (Jones) Rydb.; *T. pedunculatus* (Jones) Standley]. Plants not especially succulent, decumbent to ascending or erect; stems mainly 1–6 dm long, viscid (with sand adhering) or glabrous to glandular-puberulent; petioles 1–4 cm long; leaf blades 1–6 cm long, 0.4–2.5 cm wide, lanceolate to lance-ovate, oblong, or elliptic, entire to sinuate or undulate, obtuse basally; bracts 3–9 mm long; peduncles 1.5–10 cm long; flowers 5–15 per head; perianth tube 6–13 mm long, green to pink, the limb 3–6 mm wide; anthocarps 2– to 4–winged, 10–20 mm long. Creosote bush, blackbrush, sandy desert shrub, and mixed desert shrub, usually in sand, at 820 to 1770 m in Daggett, Duchesne, Emery, Garfield, Grand, Juab, Kane, Millard, San Juan, Tooele, Uintah, Washington, and Wayne counties; Montana and North Dakota, south to Nevada, California, Arizona, New Mexico, and Kansas; 52 (ix).

NYMPHAEACEAE Salisb.
Waterlily Family

Perennial aquatic plants from submersed rhizomes; leaves alternate, simple, the blades floating on the surface or rarely emergent, cordate; flowers perfect, regular, solitary, long-pediceled; perianth segments distinct; sepals mostly 3–12, green or petaloid; petals usually many, showy (scalelike in *Nuphar*); stamens numerous; pistil 1; ovary superior or partly inferior, several- to many-loculed; fruit a leathery berry; x = 12–29+.

1. Flowers yellow; sepals yellow, 5–12, more conspicuous than the petals; plants of high elevation lakes and ponds .... *Nuphar*
— Flowers usually pink; sepals greenish, 4, the petals conspicuous; plants of low elevation ponds .... *Nymphaea*

Nuphar J. E. Sm.

Herbs with thick rhizomes; leaves alternate (appearing spirally arranged), arising from the rhizome, the blade laterally attached; flowers long-pedicellate, solitary, arising from the rhizomes, yellowish; sepals 5–12, yellowish to greenish or reddish, petaloid; petals 10–20, small and inconspicuous; stamens numerous, the filaments flattened; pistil 1, the ovary with several carpels and several locules; stigma broad, forming a circular disk; fruit a capsule, many-seeded, opening irregularly.

Beal, E. O. 1956. Taxonomic revision of the genus *Nuphar*, Sm. of North America and Europe. J. Elisha Mitchell Sci. Soc. 72(2): 317–346.

***Nuphar polysepalum* Engelm.** Yellow Pondlily; Spatterdock. [*N. luteum* ssp. *polysepalum* (Engelm.) Beal]. Coarse aquatic herbs; leaves with petioles to 10 dm long or more (length depending on water depth), the blades floating to emergent, or rarely submersed, ovate, sagittately lobed basally, 8–25 cm long from sinus to apex, 10–23 cm wide, leathery; sepals 5–12, yellow or tinged with green or red, 3–6 cm long, 2–5 cm wide; petals yellowish to purple, subequal to the stamens; fruit ovoid, mostly 4–6 cm long. Ponds and lakes, especially at a depth of ca 8–14 dm, at 2745 to 3355 m in Duchesne, Summit, and Uintah counties; Alaska to Mackenzie, south to California, Colorado, and South Dakota; 11 (ii). Position of plants in lakes and ponds is evidently determined by water depth, the rhizomes surviving in water of a certain depth only, forming a ringlike zone between the margin and the center of the body of water. Thus, the plants are seldom available to collectors, except by wading into icy water by the most determined of botanists.

Nymphaea L.

Herbs with thick rhizomes; leaves alternate, appearing spirally arranged, arising from the rhizome, the blade laterally attached; flowers long-pedunculate, solitary, arising from the rhizome; sepals usually 4, greenish; petals numerous, pink, white, or variously colored, conspicuous; stamens numerous, transitional with the petals, the filaments flattened; pistil 1, the ovary with several to many carpels and locules; stigmas several to many, broad, petaloid; fruit a capsule, many-seeded, tardily dehiscent.

***Nymphaea odorata* Ait.** Fragrant Waterlily. Slender herbs; leaves with petioles to 12 dm long or more, depending on water depth, the blades floating, orbicular, sagittately lobed basally, mainly 10–20 cm long from sinus to apex, 15–25 mm wide, leathery; sepals 4 or 5, 3–7 cm long; petals numerous, pink or white, subequal to the sepals, longer than the stamens; capsules 2–4 cm long. Ponds and springs at ca 850 to 1710 m in Kane, Utah, and Washington (?) counties; eastern U. S.; probably introduced in Utah and now established; 3 (i). The plants from Kane County have fresh pink flowers; those from Utah County are startlingly white.

OLEACEAE Hoffsgg. & Link
Olive Family

Trees or shrubs; leaves opposite (or rarely alternate), simple or pinnately compound, stipulate; flowers perfect or imperfect, borne in axillary or terminal racemose, paniculate, or thyrsoid inflorescences; calyx commonly 4-lobed or absent; corolla usually of 4 united or distinct petals, or lacking; stamens 2, distinct; pistil 1, the ovary superior, 2-carpelled and 2-loculed; style 1, or lacking, the stigmas 1 or 2; fruit a berry, drupe, loculicidal capsule, circumscissile capsule, or samara; x = 10, 11, 13, 14, 23, 24.

1. Leaves pinnately compound; fruit a samara ..... *Fraxinus*
— Leaves simple, or rarely compound; fruit various 2
2(1). Leaves ovate to orbicular, crenate-serrate; fruit a samara; plants indigenous .... *Fraxinus*
— Leaves various, but seldom ovate to orbicular and crenate-serrate; fruit a drupe, capsule, or berry; plants cultivated or indigenous 3
3(2). Shrubs with yellow flowers appearing before the leaves; plants cultivated .... *Forsythia*
— Shrubs, subshrubs, or trees with flowers variously colored but, if yellow, not as above, and appearing with or after the leaves (before in *Forestiera*) 4
4(3). Corolla none or rudimentary, the flowers often unisexual; fruit a drupe; shrubs of stream banks in southeastern Utah .... *Forestiera*
— Corolla well developed, the flowers perfect; fruit a berry or capsule 5
5(4). Corolla yellow; fruit a membranous, circumscissile capsule; plants indigenous subshrubs of southern Utah .. .. *Menodora*
— Corolla commonly lavender to red, purple, white, or cream; fruit a loculicidal capsule or a berry; plants cultivated shrubs or trees 6
6(5). Flower clusters usually less than 6 cm long; flowers white to cream; fruit a berry .... *Ligustrum*
— Flower clusters usually 6-30 cm long or more; flowers lavender to red, purple, lilac, white, or cream .... *Syringa*

Forestiera Poir.

Sprawling indigenous shrubs; leaves opposite, simple, serrate to entire; flowers inconspicuous, polygamo-dioecious, borne sessile or in cymes, appearing before the leaves; calyx minute, unequally 5- to 6-cleft, or lacking; corolla lacking, or rarely with 2 or 3 petals; stamens 2 or 4; ovary 2-loculed, with 2 ovules per locule; style slender; stigma 1; fruit a drupe.

Forestiera pubescens Nutt. Desert Olive. [*F. neomexicana* Gray; *Adelia neomexicana* (Gray) Kuntze; *A. parvifolia* Cov.]. Shrubs to 2 m tall or more; leaves (0.8) 1.5-5.5 cm long, (0.3) 0.5-2 cm wide, oblanceolate to elliptic, entire to serrulate; staminate flowers sessile; pistillate flowers pedicellate; drupe 5-7 (8) mm long, ellipsoid, blue black. Sandy terraces along the Colorado and San Juan rivers and tributaries at 1130 to 1800 m in Emery, Garfield, Grand, Kane, and San Juan counties; California east to Oklahoma and Texas, and south to Chihuahua; 8 (iii). The fruit is eaten by fox and coyotes, and the purple-stained, stone-laden fecal pellets are to be found far from the rivers. Long known as *F. neomexicana*, our materials form a portion of a complex whose definition is included within *F. pubescens*, and that name has priority.

Forsythia Vahl

Cultivated shrubs; leaves opposite, simple or some compound, entire to serrate; flowers perfect, showy, borne in axillary clusters of 3-5, or solitary, appearing before the leaves; calyx 4-lobed; corolla 4-lobed, campanulate; stamens 2, inserted at corolla base; ovary 2-loculed, with several ovules per locule; fruit a loculicidal capsule, with many winged seeds (ours seldom fruiting).

Forsythia suspensa (Thunb.) **Vahl** Golden-bell. [*Syringa suspensa* Thunb.]. Shrub to 2 m tall or more; branchlets somewhat 4-angled; leaves 6-10 cm long, ovate to lanceolate, acute apically, cuneate to rounded basally, usually serrate; flowers to 25 mm long, golden yellow; fruit lance-ovoid, to 15 mm long, seldom developing. Cultivated ornamental, common, persisting but not spreading at lower elevations throughout Utah; widespread; introduced from China. Numerous horticultural varieties are present; 6 (0).

Fraxinus L.

Deciduous, cultivated and/or indigenous trees or shrubs; winter buds often prominent, gray to brown or black; leaves opposite, pinnately compound (simple in *F. anomala*)); flowers perfect or unisexual, inconspicuous, borne in panicles; calyx 4-lobed or lacking; corolla lacking or of 2 or more usually distinct petals; stamens commonly 2; ovary 2-loculed; styles 1; stigmas 1 or 2; fruit a samara.

1. Leaves normally simple, sometimes with 1 or 2 leaflets below the terminal one; indigenous shrubs or small trees of eastern and southern Utah .... *F. anomala*
— Leaves normally pinnately compound with 5-9 or more leaflets; trees, either indigenous or cultivated 2
2(1). Branchlets, petioles, and axis of panicle commonly spreading hairy, seldom glabrous; leaflets usually 5 or fewer; trees, indigenous in southwestern Utah, cultivated elsewhere .... *F. velutina*
— Branchlets, petioles, and axis of panicle variously hairy or glabrous, but seldom spreading hairy; leaflets usually 7 or more; trees, cultivated and sometimes escaping 3
3(2). Flowers appearing after leaves formed; corolla present .... *F. ornus*
— Flowers appearing before leaves formed; corolla lacking 4
4(3). Fruit with calyx persisting as a campanulate cap; anthers oblong; leaflets usually 5-7 5
— Fruit with calyx early deciduous or lacking (except *F. quadrangulata*); anthers often cordate; leaflets usually 9-11 or more 6
5(4). Petiolules of middle and lower mature leaflets wingless nearly their entire length; winter buds black; leaf scars horseshoe-shaped; wing of fruit terminal, not or only slightly decurrent .... *F. americana*

— Petiolules of middle and lower mature leaflets winged nearly to the base; winter buds brown; leaf scars semicircular or shield-shaped; wing of fruit decurrent to below the middle .... *F. pennsylvanica*

6(4). Branchlets 4-sided, 4-angled; bark broken into plates; flowers with a minute, deciduous calyx ....
.... *F. quadrangulata*

— Branchlets terete, not or only slightly 4-angled; bark smooth or irregularly roughened; flowers with calyx lacking 7

7(6). Leaflets glabrous or somewhat hairy along veins beneath; commonly cultivated tree .... *F. excelsior*

— Leaflets definitely pubescent beneath, especially along the veins, the long reddish hairs extend onto and along the leaf rachis; uncommon to rarely cultivated trees ..
.... *F. nigra*

Fraxinus americana L. White Ash. Moderate to large trees; branchlets terete, green to brown, glabrous; winter buds black; leaflets usually 7 (5–9), 6–15 cm long, petiolulate, ovate to lanceolate, acuminate apically, cuneate to rounded basally, entire to serrate, glaucus beneath and usually glabrous; anthers oblong, apiculate; calyx persistent; corolla lacking; samaras (20) 25–35 mm long, 4–7 mm wide, the wing not decurrent along the terete base. Shade tree of lower elevations in Utah; introduced from eastern North America; 5 (0).

Fraxinus anomala Torr. ex Wats. Singleleaf Ash. Shrub or small tree, commonly 2.5–4 m tall, usually with many stems; branchlets 4-angled; leaves glabrous, ovate, crenate-serrate to subentire, 1.5–6.5 cm long, 1–6 cm wide, acute to obtuse or subcordate basally, acute to rounded or emarginate apically, sometimes 2- or 3-foliolate or transitional to simple; flowers usually perfect; anthers oblong; calyx campanulate, persistent; petals lacking; samaras winged almost to the base, 12–27 mm long, 5–11 mm wide. Mixed desert shrub, mainly on rimrock or along drainages, and in pinyon-juniper woodland at 900 to 2625 m in Emery, Garfield, Grand, Iron, Kane, San Juan (? type from Labrynth Canyon), Uintah, Washington, and Wayne counties; Colorado, New Mexico, Arizona, and California; 80 (xv).

Fraxinus excelsior L. European Ash. Moderate to large trees; branchlets terete, glabrous; winter buds black; leaflets 7–11, 5–12 cm long, sessile, ovate to oblong or laneolate, acuminate apically, cuneate basally, serrate, green beneath, glabrous except along midrib, the hairs sometimes extending to the rachis; flowers polygamous; anthers ovoid; calyx lacking; corolla lacking; samaras 25–35 (40) mm long, 5–11 mm wide, the blade decurrent almost or quite to the base of the flattenend body; $2n = 46$. Shade trees of habitations and streets at lower elevations throughout Utah; introduced from Europe; 11(0).

Fraxinus nigra Marshall Black Ash. Moderate trees; branchlets terete, glabrous; winter buds black; leaflets 7–11, mostly 6–12 cm long, sessile, lanceolate to oblong, obtuse to rounded basally, long-acuminate apically, serrate, green and glabrous except reddish-hairy along veins, the pubescence extending along the leaf rachis; flowers dioecious; anthers oblong; calyx lacking; corolla lacking; samaras mostly 25–35 mm long and 6–10 mm broad, the blade decurrent to the base of the flattened body; $2n = 46$. Sparingly cultivated shade tree at lower elevations in at least the major population centers; introduced from eastern North America; 4 (0).

Fraxinus ornus L. Flowering Ash. Small to moderate trees; branchlets terete; winter buds gray to brownish; leaflets usually 7 (7–11), mostly 2.5–7 cm long, petiolulate, lance-ovate to obovate (terminal one), rounded to obtuse basally, acuminate apically, crenate-serrate, glabrous except along midrib; flowers perfect; calyx present, persistent, with 4 triangular-acuminate, spreading lobes; petals present, linear; samaras 20–25 mm long, 3–6 mm wide, the blade terminal on the terete base. Rarely cultivated shade and ornamental tree of lower elevations in Utah; introduced from Europe; 2 (0).

Fraxinus pennsylvanica Marshall Red Ash. Moderate trees; branchlets terete, pubescent to glabrous, sometimes glandular; winter buds olive to brown; leaflets usually 7 (5–9), 6–15 cm long, petiolulate, lanceolate to lance-oblong, acuminate apically, acute to obtuse or rounded basally, serrate to entire, green and glabrous or hairy (especially along the veins) beneath; anthers oblong, apiculate; calyx campanulate, persistent; corolla lacking; samaras 27–40 (50) mm long, the blade decurrent to the middle of the terete body or below; $2n = 46$. Common shade tree of lower elevations throughout Utah, persisting and escaping in Box Elder, Cache, Davis, Iron, Juab, Millard, Salt Lake, Utah, and Washington counties; introduced from eastern North America; 28 (0). The escaped plants have become established along streams and on lake margins at lower elevations. Much of our material has glabrous branchlets and petioles and has been designated as **F. pennsylvanica** var. *lanceolata* (**Borkh.**) **Sarg.** (*F. lanceolata* Borkh.). This phase is known as green ash.

Fraxinus quadrangulata Michx. Blue Ash. Small to moderate trees; branchlets sharply 4-angled, glabrous; winter buds black; leaflets 7–11, mostly 5–12 cm long, petiolulate, lanceolate, acute to rounded basally, acute to acuminate apically, serrate, glabrous except along the midrib or rarely hairy over the lower surface; flowers perfect; calyx minute, caducous; corolla lacking; anthers cordate-oblong, blunt; samaras 20–40 (50) mm long, the blade decurrent to the base of the flattened body; $2n = 46$. Sparingly cultivated shade tree at lower elevations in Utah; introduced from eastern North America; 1 (0).

Fraxinus velutina Torr. Velvet Ash. (*F. pennsylvanica* ssp. *velutina* (Torr.) G. N. Miller). Moderate trees; branchlets terete, densely spreading hairy to merely sparingly so or glabrous; winter buds brown; leaflets 3–5 (or leaves simple), lanceolate to ovate, elliptic, or orbicular, petiolulate, cuneate to acute basally, acuminate to rounded apically, serrate, glabrous or hairy over the lower surface; flowers imperfect; calyx campanulate, persistent; corolla lacking; anthers oblong, apiculate; samaras 16–34 mm long, 4–6 mm wide, the blade decurrent about half way along the terete body. Indigenous tree of stream courses and flood plains in Washington and Iron counties, and cultivated there and elsewhere in Utah; Arizona and New Mexico; 25 (ii). The plants with coriaceous leaflets has been treated as var. *coriacea* (**Wats.**) **Rehder** (*F. coriacea* Wats.), but seem not to be worthy of taxonomic recognition, at least in Utah. Note: The shrubby *Fraxinus dipetala* H. & A. is reported for Utah in Kearney & Peebles, 1961. Flora of Arizona, supplement p. 1063. The related **F. cuspidata** Torr. is known from adjacent Mohave and Coconino counties, Arizona, and might occur in Utah. Both species have corollas present; the former has two petals and the latter has four.

Ligustrum L.

Shrubs; leaves opposite, simple, entire; flowers perfect, white, showy through small, borne in terminal panicles, appearing after the leaves; calyx 4-toothed; corolla 4-lobed, funnelform; stamens 2, inserted on the corolla tube; ovary 2-loculed, 1- or 2-seeded; fruit a berry.

Ligustrum vulgare L. Common Privet. Deciduous or semievergreen shrub to 3 m tall or more, with puberulent to glabrate branchlets; leaves 2–6 cm long, 0.8–2 cm wide, oblong to elliptic or ovate-lanceolate, glabrous; panicle dense, 3–6 cm long; corolla tube shorter than the lobes, white; anthers exserted; fruit 6–8 mm long, black, ovoid to subglobose; $2n = 46$. Cultivated hedge plant throughout Utah at lower elevations, persisting and established in Utah County; introduced from Europe; 3 (0).

Menodora Humb. & Bonpl.

Subshrubs; leaves alternate or the lowermost opposite, simple, sessile or nearly so; flowers perfect, arranged in cymes; calyx 5- to 15-lobed; corolla yellow, subrotate, 5- to 6-lobed; stamens 2, inserted on the corolla tube; ovary 2-loculed, with 2–4 ovules per locule; style slender, the stigma capitate; fruit a circumscissile capsule.

Steyermark, J. A. 1932. Revision of the genus *Menodora*. Ann. Missouri Bot. Gard. 19: 87–176.

1. Plants armed with spines; corolla lobes white, shorter than the tube; known from southwestern Washington County .... *M. spinescens*
— Plants unarmed; corolla lobes yellow, longer than the tube; known from a broader distribution .... *M. scabra*

Menodora scabra Gray Plants erect or ascending, commonly 2–3.5 dm tall, woody at the base only; leaves 0.5–2.9 cm long, 2–5 mm wide, narrowly elliptic to oblong or lanceolate, glabrous or scaberulous; calyx minutely puberulent, the lobes linear; corolla bright yellow, subrotate, the lobes 5–9 mm long; capsule 8–12 mm thick, membranous; seeds 4–5 mm long; $2n = 22$. Pinyon-juniper community at 1525 to 1830 m in Garfield, Kane, and Washington counties; California, Arizona, New Mexico, Texas, and Mexico; 3 (i).

Menodora spinescens Gray Low rounded shrubs 1.5–8 dm tall, the branches spreading, ultimately modified as thorns; leaves 4–12 mm long, 2–4 mm wide, linear to oblong or spatulate, entire, appressed puberulent; calyx lobes linear-subulate; corolla white, tinged brownish-purple, 9–15 mm long, funnelform, the lobes 3.5–4.5 mm long; capsule 7–9 mm thick; seeds 5–6 mm long. Mixed warm desert shrub at ca 670 to 765 m on the a lacustrine member of the Muddy Creek Formation, in southwestern Washington County; California, Nevada, and Arizona; 2 (0).

Syringa L.

Shrubs or small trees; leaves opposite, simple, petiolate; flowers perfect, in terminal or lateral panicles; calyx campanulate, 4-toothed to nearly truncate, persistent; corolla tublar, the limb 4-lobed and rotate or nearly so; stamens 2, inserted on the corolla tube; ovary 2-loculed, each locule with usually 2 ovules; style with a 2-lobed stigma; fruit a loculicidal capsule.

1. Flowers cream to whitish, borne in large panicles; corolla tube 1–2.2 mm long, only half as long as the calyx; fragrance musky, not that of lilac; plants flowering in summer, often treelike 2
— Flowers lilac, violet, purplish white, or white; corolla tube mostly 6–12 mm long or more, several times longer than the calyx; fragrance usually of lilac; plants commonly shrubs, flowering in spring or summer 3
2(1). Leaves ovate, rounded or subcordate basally, the veins prominent on the lower surface .... *S. amurensis*
— Leaves lanceolate to elliptic or ovate-lanceolate, obtuse to cuneate basally, the veins not prominent .... *S. pekinensis*
3(1). Panicles from terminal buds; leaves of current season borne on branch with panicle; plants flowering in summer .... *S. villosa*
— Panicles from lateral (or terminal) buds, the terminal buds often lacking; leaves of current season not borne on the branch with panicle; plants flowering in springtime . 4
4(3). Leaves ovate to cordate, the base subcordate to obtuse; our most common species .... *S. vulgaris*
— Leaves lanceolate to elliptic or ovate, obtuse to cuneate basally; common to uncommon 5
5(4). Leaves mostly less than 4 cm long, some often irregularly lobed; individual panicles short, mostly 7 cm long or less .... *S. persica*
— Leaves often over 4 cm long, entire; individual panicles usually 8–12 cm long .... *S. x chinensis*

Syringa amurensis Rupr. Amur Lilac. Shrubs or small trees to 5 m tall or more; leaf blades 3.5–13 cm long, 1.3–8 cm wide, ovate, rounded to obtuse or short acuminate basally, acuminate apically, the lower surface hairy to glabrous, the veins prominent; petioles mostly 1–2 cm long; panicles 10–15 cm long, the clusters of panicles usually much longer; flowers cream to white; stamens exserted. Sparingly cultivated ornamental of lower elevations in Utah; introduced from Japan; flowering in summer; 4 (0).

Syringa x chinensis Willd. Chinese Lilac. Shrub to 4 m tall or more, with spreading and often arching branches; leaves 2.5–8 cm long, 1.5–4 (5) cm wide, ovate-lanceolate, obtuse to cuneate basally, acuminate apically, glabrous, the veins not prominent; petioles 0.5–1.5 cm long; panicles mostly 8–12 cm long, the clusters of panicles much longer; flowers purple, lilac, or otherwise; stamens included. Commonly cultivated ornamental almost throughout Utah; introduced from the Old World; 3 (0). This plant is evidently of hybrid origin, having resulted from a cross between *S. persica* and *S. vulgaris* q.v.; flowering in springtime.

Syringa pekinensis Rupr. Peking Lilac. Shrub or small tree to 5 m tall or more, with spreading branches; leaves 5–12 cm long, 2–4 (6) cm wide, lanceolate to ovate, cuneate basally, acuminate apically, glabrous, the veins not prominent; petioles 1.5–3 cm long; panicles mostly 8–15 cm long, the clusters of panicles to 30 cm long or more; flowers cream to yellow white; stamens exserted. Uncommon, cultivated ornamental in northern Utah, but to be expected elsewhere; introduced from China; flowering in early summer; 1 (0).

Syringa persica L. Persian Lilac. Shrub to 2 m tall, with upright to arching branches; leaves 1.5–6 cm long, 0.6–3 cm wide, lanceolate to elliptic, sometimes lobed, cuneate to obtuse basally, acute to acuminate apically, glabrous, the veins not prominent; petioles 0.5–1 cm long; panicles mostly 3–7 cm long; flowers usually lilac but purple phases are known; stamens included. Uncommonly cultivated ornamental, especially in northern Utah; introduced from Asia Minor; flowering in springtime; 5 (0).

Syringa villosa **Vahl** Shrub to 3 m tall, rarely more, with erect branches; leaves 4–15 cm long (or more), 2.5–9 cm wide, ovate to elliptic, acute basally, abruptly acuminate apically, spreading hairy below, especially along the prominent veins; petioles 0.8–2 cm long; panicles mostly 10–18 cm long; flowers pink lilac to white; stamens included. Sparingly but widely planted ornamental, mainly in northern Utah; introduced from China; flowering in summer; 4 (0).

Syringa vulgaris **L.** Common Lilac. Shrubs to 4 m tall or more, the branches usually erect; leaves 3–12 cm long, 1.5–8 cm wide, ovate to cordate, cordate to rounded, truncate or obtuse basally, acute to acuminate apically, glabrous; petioles 0.8–3 cm long; panicles mostly 10–20 cm long; flowers lilac or white, seldom purple; stamens included; 2n = 44, 46. Abundantly cultivated ornamental, long persisting in most of Utah; introduced from Europe; flowering in springtime; 7 (0). Many horticultural forms are known.

ONAGRACEAE A. L. Juss.

Evening Primrose Family

Caulescent or acaulescent herbs, or rarely woody plants; leaves alternate, opposite, or basal; flowers perfect; hypanthium adnate to the ovary and usually prolonged beyond; sepals 4 or 2; petals distinct, 4 or 2, inserted near or at the summit of the hypanthium; stamens as many or twice as many as the petals; ovary inferior, usually 4–loculed; styles 1; stigma capitate, 4–lobed, or discoid; fruit a capsule, nut, or berry; x = 6–18.

1. Sepals, petals, and stamens 2; leaves opposite; fruit indehiscent, about as broad as long, usually bearing hooked hairs .... *Circaea*
— Sepals and petals 4; stamens usually 8; fruit various, but not usually bearing hooked hairs 2
2(1). Seeds with a tuft of hairs at one end 3
— Seeds naked or pubescent overall 4
3(2). Hypanthium less than 1 cm long, not scaly within; flowers pink, lavender, white, or yellowish .... *Epilobium*
— Hypanthium 1.5–3 cm long, with a row of 8 scales within; flowers scarlet .... *Zauschneria*
4(2). Fruit indehiscent; flowers pink to red orange or white, often irregular .... *Gaura*
— Fruit dehiscent, capsular; flowers usually white or yellow, regular or nearly so 5
5(4). Ovary 2–loculed; hypanthium not prolonged beyond the apex of the ovary; flowers minute, the petals 1.5 mm long or less .... *Gayophytum*
— Ovary 4–loculed; hypanthium prolonged beyond the ovary; flowers larger, the petals mostly more than 2 mm long 6
6(5). Sepals erect; petals small or lacking .... *Boisduvalia*
— Sepals reflexed, or the apices united and turned to one side at anthesis; petals small to large 7
7(6). Anthers attached near one end; petals pink to lavender or white .... *Clarkia*
— Anthers versatile or if attached at one end, the petals not white; petals white, yellow, or lavender 8
8(7). Stigmas peltate; plants caulescent from a ligneous caudex .... *Calylophus*
— Stigmas 4–cleft or hemispheric to subglobose; plants of various habit 9
9(8). Stigmas 4–cleft; pollen hanging in long strands .... *Oenothera*
— Stigmas hemispheric to subglobose, entire or merely 4–lobed; pollen not as above .... *Camissonia*

Boisduvalia Spach

Annual, caulescent herbs; leaves opposite below, alternate above; flowers axillary, solitary; hypanthium shortly prolonged beyond the ovary with hairs below the summit; sepals 4, erect, deciduous with the hypanthium in fruit; petals 4, deeply emarginate; stamens 8; stigmas 4–lobed or capitate; ovary 4–loculed; fruit a capsule; seeds lacking a tuft of hair.

1. Sepals 2–4 mm long; petals 4–8 mm long; capsule with 1 row of seeds per locule .... *B. densifolia*
— Sepals ca 2 mm long; petals 2–4 mm long; capsule with 2 rows of seeds per locule .... *B. glabella*

Boisduvalia densiflora (**Lindl.**) **Wats.** [*Oenothera densiflora* Lindl.]. Stems erect or ascending, simple or branched, commonly 1–6 dm tall; herbage pilose with hairs mainly 0.5–2.5 mm long and sometimes glandular as well; leaves sessile, 1.5–5 cm long, 3–10 mm wide, lanceolate to lance-ovate, serrulate to entire; flowers subtended by broad, leafy bracts; hypanthium 1–4 mm long; sepals 2–4 mm long; petals lavender to pink or white, 4–8 mm long; stigma obscurely 4–lobed; capsules 5–10 mm long, dehiscent to the base; seeds 1 row per locule, 1.2–1.8 mm long. Marshy areas at ca 1310 to 1435 m in Cache and Weber counties; British Columbia to Montana, south to California and Idaho; 1 (0).

Boisduvalia glabella (**Nutt.**) **Walp.** [*Oenothera glabella* Nutt. ex T. & G.]. Stems simple and erect or with decumbent branches from the base, mainly 1–3 dm long; herbage puberulent to sparingly villous with crinkly hairs; leaves sessile, 1–2.2 cm long, 2–7 mm wide, lanceolate to ovate-lanceolate, serrulate; flowers subtended by broad, leafy bracts; sepals 1–2 mm long; petals purplish, 1–4 mm long; capsules 5–9 mm long; seeds in 2 rows per locule, 1–1.3 mm long. Shores of lakes and ponds at 1525 to 2200 m in Cache, Salt Lake, Uintah, and Utah counties; British Columbia to the Dakotas, south to California, Nevada, and Arizona; 3 (0).

Calylophus Spach

Perennial, caulescent herbs (or somewhat suffrutescent); leaves alternate; flowers axillary, opening at most times of day; hypanthium prolonged much beyond the ovary, deciduous; sepals 4, reflexed in anthesis; petals 4; stamens 8, the anthers versatile; stigma peltate, discoid to shallowly lobed; capsules sessile; seeds in 2 rows per locule.

Calylophus lavandulifolius (**T. & G.**) **Raven** [*Oenothera lavandulifolia* T. & G.; *Galpinsia lavandulifolia* (T. & G.) Small]. Plants perennial from a taproot and ligneous, pluricipital caudex; stems decumbent to erect, mainly 2–22 cm long; herbage strigulose to villous; leaves 0.6–3.5 cm long, 1–6 mm wide, linear to oblong, elliptic, or oblanceolate, cuneate basally, obtuse to acute apically; flowers axillary, solitary, the buds erect, expanded below the sepals; hypanthium 2.5–8 cm long above the ovary; sepals 10–23 mm long; anthers 5–10 mm long; capsules 1–2 cm long; seeds 1.5–2.5 mm long; n = 7. Blackbrush, salt and mixed desert shrub, sagebrush, pinyon-juniper, and ponderosa pine communities at 1370 to 2750 m in Beaver, Carbon, Daggett, Duchesne, Emery, Garfield, Grand, Juab, Kane, Millard, San Juan, Sanpete, Uintah, Washington, and Wayne counties; Nevada and Arizona, east to Texas, and north to Wyoming and South Dakota; 125 (xvi).

Camissonia Link

Annual or perennial herbs; leaves alternate or basal; flowers axillary, sessile or pedicellate, often disposed in spikes or racemes, opening in day or evening; hypanthium prolonged or short, deciduous from the fruit; sepals 4, reflexed at anthesis; petals 4, inconspicuous to showy, yellow, white, or lavender, fading orange to purple; stamens 8 (4 in *C. exilis*), the anthers versatile or basifixed; stigma hemispheric to globular, sometimes shallowly 4-lobed; capsules loculicidal; seeds in 1 or 2 rows per locule, lacking a tuft of hairs

Munz, P. A. 1928. Studies in Onagraceae. II. Revision of North American species of Sphaerostigma, genus *Oenothera*. Bot. Gaz. 85: 233–270.
Raven, P. H. 1962. The systematics of *Oenothera* subgenus Chylismia. Univ. California Publ. Bot. 34: 1–122.
———. 1969. A revision of the genus *Camissonia* (Onagraceae). Contr. U.S. Natl. Herb. 376: 161–396.

1.	Plants acaulescent perennials; flowers yellow, the ovary produced into a sterile tip mainly 1–8 cm long below the hypanthium proper	2
—	Plants caulescent annuals; flowers yellow, white or lavender, the ovary not produced below the hypanthium	3
2(1).	Leaves entire or lobed only near the base; herbage glabrous or essentially so .... *C. subacaulis*	
—	Leaves (or most of them) irregularly pinnatifid throughout; herbage short-hairy to almost glabrous .... *C. breviflora*	
3(1).	Capsules sessile or nearly so, straight, curved, or contorted; seeds in 1 row per locule	4
—	Capsules pedicellate (shortly so in some *C. brevipes*), straight to curved; seeds mainly in 2 rows per locule	11
4(3).	Flowers yellow, opening in morning	5
—	Flowers white, opening in evening	7
5(4).	Capsules 0.5–1 cm long, thickened near the base; leaves congested at the stem ends .... *C. andina*	
—	Capsules mainly more than 1.5 cm long, not thickened at the base; leaves not clustered at stem ends	6
6(5).	Leaves mainly basal or nearly so; herbage spreading-hairy; plants of Beaver, Iron, and Washington counties .... *C. pusilla*	
—	Leaves cauline, usually lacking near the base; herbage glabrous to strigose; plants of rather broad distribution .... *C. parvula*	
7(4).	Capsule cylindric, not tapering toward the tip	8
—	Capsule thickened basally, tapering toward the tip	9
8(7).	Petals 4–8 mm long; style exserted 4–8 mm beyond the hypanthium, the stigma surpassing the petals; anthers 1.5–2 mm long; plants of Washington County .... *C. refracta*	
—	Petals 1.5–3 mm long; styles exserted 1–2 mm beyond the hypanthium, the stigma not surpassing the petals; anthers 0.5–1 mm long; plants of Kane, Tooele, and Washington counties . *C. chamaenerioides*	
9(7).	Petals 3.5–8 mm long; style exserted 2.5–8 mm beyond the hypanthium; plants of the Great Basin and Washington and eastern Kane counties .... *C. boothii*	
—	Petals 1–3 mm long; style exserted 0.5–2.5 mm beyond the hypanthium; plants of various distribution	10
10(9).	Leaves entire, mainly basal; plants rather broadly distributed .... *C. minor*	
—	Leaves toothed, mainly cauline; plants of Washington County .... *C. gouldii*	
11(3).	Stamens 4; plants reported from the Cockscomb vicinity in eastern Kane County, but the report has not been verified .... *C. exilis* (Raven) Raven	
—	Stamens 8; plants of various distribution in Utah	12
12(11).	Petals white, with the base yellow; seeds with paired, marginal wings .... *C. pterosperma*	
—	Petals yellow or lavender, or if white, the base not yellow; seeds not winged	13
13(12).	Petals lavender, both at anthesis and upon drying; plants from eastern Kane County .... *C. atwoodii*	
—	Petals yellowish or white at anthesis, sometimes fading lavender; plants of various distribution	14
14(13).	Capsules 4 cm long or more; petals 10–14 mm long; plants of Kane and Washington counties .. *C. brevipes*	
—	Fruit mainly 0.2–3.5 cm long, but if to 4 cm long, the petals usually shorter; plants of various distribution	15
15(14).	Capsules 2–10 mm long; pedicels filiform, 2–10 mm long; herbage pilose with long, slender, white hairs; plants of Washington County .... *C. parryi*	
—	Capsules 10–35 mm long; pedicels filiform or not, of various lengths; herbage glabrous or variously hairy; plants of various distribution	16
16(15).	Plants mostly 2–6 dm tall or more; capsules 1.2–2 mm wide	17
—	Plants often less than 2.5 dm tall; capsules 2–3.5 mm wide	18
17(16).	Petals 5–10 mm long; anthers 1.5–4 mm long; plants of (Kane?) and Washington County ...... *C. multijuga*	
—	Petals 1.5–5 mm long; anthers 0.5–1.5 mm long; plants of broad distribution .... *C. walkeri*	
18(16).	Petals 2–5 mm long; anthers 0.5–2.5 mm long .... *C. scapoidea*	
—	Petals 7–12 mm long; anthers 2.5–3 mm long	19
19(18).	Staminal filaments of 2 lengths, the longer ones ca 1.5–2 mm longer than the others; petals bright yellow; plants of eastern Utah .... *C. eastwoodiae*	
—	Staminal filaments subequal; petals white; plants of western Utah .... *C. clavaeformis*	

***Camissonia andina* (Nutt.) Raven** [*Oenothera andina* Nutt. ex T. & G.]. Annual herbs from taproots; stems simple or more commonly branched from the base, erect or the branches curved-ascending, mainly 1–8 cm tall and 2–15 cm wide, with leaves clustered near the stem ends; herbage rather densely strigulose-puberulent; leaves 0.6–3 cm long, 1–3 mm wide, linear to linear-oblanceolate, cuneate to a slender petiole, entire; flowers solitary, axillary in most leaf axils, minute (self pollinated or cleistogamous); hypanthium prolonged 0.8–2 mm beyond the ovary; sepals 1–2.5 mm long, reflexed at anthesis; petals yellow, 1–2.5 mm long; stamens dimorphic, or one set sometimes obsolete; anthers 0.3–0.5 mm long; style exserted 0.7–2 mm beyond the hypanthium; capsule sessile, 6–11 mm long, tapering from the thick base; seeds in 1 row per locule, 0.7–1.3 mm long, brown, shining; n = 14, 21. Sagebrush communities at ca 1675 to 2290 m in Beaver, Cache, and Wasatch counties; British Columbia to Alberta, south to California, Nevada, and Wyoming; 0 (0). Reports of distribution are taken from Raven (1969).

Camissonia atwoodii Cronq. [*C. megalantha* authors, not (Munz) Raven]. Robust annual (or winter annual) herbs from taproot; stems 0.5–1.5 m tall and 2–8 dm broad, usually branched, and with leaves basally disposed or near branch bases; herbage stipitate-glandular; leaves petiolate, the blades 1.2–7.5 cm long, 0.8–5 cm wide, ovate to lanceolate or suborbicular, subcordate to truncate basally, serrulate to serrate-dentate; flowers numerous in panicles, open pollinated, showy; hypanthium 0.6–1 mm long; sepals 5–7 mm long, reflexed in anthesis; petals lavender with purple spots, 8–14 mm long; stamens dimorphic, the longer ones surpassing the shorter ones 3–4 mm, the filaments copiously pilose near the base; anthers 1.5–2 mm long; capsules 11–20 mm long, the seeds in 2 rows per locule, 1.5–1.8 mm long. Salt desert shrub community at ca 1525 m in eastern Kane County (Smokey Mt.); endemic; 2 (0).

Camissonia boothi (Dougl.) Raven [*Oenothera boothii* Dougl.]. Annual herbs from taproots; stems simple or branched from near the base, erect or the branches curved-ascending, mainly 4–30 cm tall and 6–45 cm wide, the leaves clustered near the base or scattered along the stem; herbage glabrous, strigulose, pilose, or stipitate-glandular; leaves 0.8–14 cm long, 2.5–26 mm wide, the blades elliptic to lanceolate or ovate, entire or irregularly toothed, tapering to a petiole; flowers numerous, but solitary in bract axils, clustered at stem apices, opening in evening; hypanthium prolonged 2–8 mm beyond the ovary; sepals 3–6.7 mm long, reflexed separately or some clinging at the apices; petals white, 4.2–7.3 mm long, rounded apically; stamens subequal; anthers 1.3–2.5 mm long; style exserted 3.7–7.5 mm beyond hypanthium; capsule sessile, 8–25 mm long, curved, tapering from the thickened base; seeds in 1 row per locule, 1–1.5 mm long, pale; n = 7. Two distinctive varieties are present in our material.

1. Sepals 3–4.5 mm long; style exserted 3.7–4.2 mm beyond the hypanthium; herbage strigulose to pilose or stipitate- glandular; plants of the Great Basin in Utah .. .. *C. boothii* var. *villosa*

— Sepals 4.7–6.7 mm long; style exserted 4.5–7.5 mm beyond the hypanthium; herbage glabrous or puberulent; plants of Kane, and Washington counties .... *C. boothii* var. *condensata*

Var. *condensata* (Munz) Cronq. [*Oenothera decorticans* var. *condensata* Munz; *C. boothii* ssp. *condensata* (Munz) Raven]. Creosote bush, bursage, Joshua tree, shadscale, and other warm or mixed desert shrub communities at 750 to 1220 m in Kane and Washington counties; Arizona, Nevada, California, and Mexico; 19 (vii).

Var. *villosa* (Wats.) Cronq. [*Oenothera alyssoides* var. *villosa* Wats., type from Great Salt Lake; *Sphaerostigma utahense* Small, type from Milford; *Oenothera gauriflora* var. *hitchcockii* H. Levl., type from Simpson's Park (probably Tooele County); *C. boothii* ssp. *alyssoides* (H. & A.) Raven]. Shadscale, galleta, hopsage, rabbitbrush, desert almond, sagebrush, and pinyon-juniper communities at 1280 to 1925 m in Beaver, Box Elder, Cache, Davis, Iron, Juab, Millard, Salt Lake, Sanpete, Tooele, and Utah counties; Oregon and Nevada; 39 (vi). Pubescence varies from merely crisp-puberulent to villous-pilose and more or less glandular.

Camissonia breviflora (T. & G.) Raven [*Oenothera breviflora* T. & G; *Taraxia breviflora* (T. & G.) Nutt. ex Small]. Perennial, acaulescent herbs from taproots and a simple or branched caudex; herbage crisp-puberulent to hirtellous or almost glabrous; leaves all basal, 2–14 cm long, 3–20 mm wide, most or all of them lobed to irregularly pinnatifid almost throughout; flowers opening in morning (self pollinated?), sessile in leaf axils, numerous, clustered; ovary apex prolonged into a slender tube 5–15 mm long below the hypanthium proper—this 15–25 mm long; sepals 2.8–6.5 mm long, reflexed at anthesis or sometimes some adherent at the apex; petals 5–8.5 mm long, yellow, fading cream; stamens dimorphic; anthers 0.7–1.4 mm long; style exserted 1.5–5 mm beyond hypanthium; capsules sessile, ovoid to lance-ovoid, 8–20 mm long, 4–5 mm thick, beaked; seeds 1.2–1.5 mm long, tan; n = 7. Aspen, sagebrush, and spruce-fir communities, mainly in damp meadows and on streambanks and lake shores, at 1585 to 3325 m in Beaver, Cache, Carbon, Garfield, Iron, Kane, Piute, Rich, Sanpete, Sevier, Summit, Utah, and Wasatch counties; British Columbia and Alberta to California and Nevada; 27 (v).

Camissonia brevipes (Gray) Raven [*Oenothera* 'brevipes Gray; *Chylismia brevipes* (Gray) Small; *Oe. brevipes* var. *pallidula* Munz; *C. brevipes* ssp. *pallidula* (Munz) Raven]. Annual herbs from taproots; stems simple or branched above, mainly 15–48 (60) cm tall, with leaves mainly near the base; herbage variably spreading-hairy to crisp-puberulent, glabrous, or more or less glandular; leaves petiolate, 1.5–18 cm long, the blades consisting of a terminal, ovate to lanceolate, serrate-dentate lobe 1–7.5 (10) cm long and 0.7–4 (6) cm wide, and with (0) 2 to numerous, much smaller, irregular, lateral lobes along the rachis; flowers few to numerous in bracteate racemes or panicles, open-pollinated, showy; hypanthium prolonged 3–8 mm beyond the ovary; sepals 5–14 mm long, glabrous, puberulent, or with long, spreading hairs; petals yellow, often red-spotted, 8–20 mm long; stamens 8, subequal; anthers 2.5–7 mm long; style exserted 7–15 mm beyond hypanthium; fruits pedicellate, subcylindric, mostly 4–5 cm long, the pedicel 3–12 mm long; seeds in 2 rows per locule, 1–1.5 mm long, not winged; n = 7. Creosote bush, Joshua tree, and other warm desert shrub communities at 750 to 1100 m in Kane and Washington counties; Arizona, Nevada, and California; 21 (vi). The specimen from eastern Kane County (Welsh & Neese 21002 BRY) was initially identified as *C. eastwoodiae* (q.v.), but is immediately distinctive because of its copious long, slender hairs.

Camissonia chamaeneroides (Gray) Raven [*Oenothera chamaeneroides* Gray; *Sphaerostigma chamaeneroides* (Gray) Small]. Annual herbs from taproots; stems simple or branched from near the base, 12–40 cm tall, with the largest leaves near the base; herbage sparingly stipitate-glandular to crisp-puberulent; leaves petiolate, the blades entire or irregularly toothed, mainly 0.8–6 cm long and 1–17 mm wide, elliptic to lanceolate, tapering to a short or elongate petiole; flowers few to numerous, borne in terminal racemes or panicles, inconspicuous; hypanthium prolonged 1.5–2.3 mm beyond the ovary; sepals 1.5–2.5 mm long, puberulent, spreading-reflexed; petals white, fading pinkish, 1.8–3 mm long; stamens 8, subequal; anthers 0.5–1.1 mm long; style exserted 1–2 mm beyond the hypanthium; capsules sessile, cylindric, mostly 3–6.5 cm long, ca 1 mm thick, curved; seeds in 1 row per locule, ca 0.9–1 mm long; $n = 7$. Creosote bush, bursage, Joshua tree, other warm desert shrub, sagebrush, and pinyon-juniper communities at 820 to 1375 m in Kane, Tooele, and Washington counties; California east to Texas, south to Mexico; 10 (ii).

Camissonia claviformis (**Torr. & Frem.**) **Raven** [*Oenothera claviformis* Torr. & Frem. in Frem.; *Oe. scapoidea* var. *claviformis* (Torr. & Frem.) Wats.; *Chylismia claviformis* (Torr. & Frem.) Heller]. Annual herbs from taproots; stems simple or branching from near the base, mainly 7–40 cm tall, with leaves basal or near the base; herbage scabrid-puberulent, puberulent, or sessile-glandular; leaves petiolate, the blades consisting of a terminal, ovate to lanceolate or elliptic, serrate to serrate-dentate lobe 1–5.5 cm long and 3–28 mm wide and (0) few to numerous, irregular, lateral, much smaller segments; flowers few to numerous in nodding, terminal racemes, conspicuous; hypanthium 5–7 mm long; sepals 4–6 mm long, reflexed-spreading, some often adherent apically; petals white, fading cream or pink, 5–7 mm long; stamens subequal; anthers 2.5–5 mm long; fruit with pedicels 4–16 mm long, the capsules subcylindric to subclavate, 12–25 mm long; seeds in 2 rows per locule, 1–1.5 mm long, not winged; $n = 7$. Two rather distinctive varieties occur in Utah.

1. Lateral leaflets lacking or few and small; leaves in a definite basal rosette; plants of Beaver and Millard counties .... *C. claviformis* var. *purpureus*
— Lateral leaflets usually present and conspicuous; leaves cauline in the lower 1/4–1/2 of the stem; plants of Washington County .... *C. claviformis* var. *claviformis*

Var. *claviformis* Salt desert shrub and creosote bush-Joshua tree communities at 750 to 1220 m in Washington County; Arizona, Nevada, and California; 5 (iii).

Var. *purpurascens* (**Wats.**) **Cronq.** [*Oenothera scapoidea* var. *purpurascens* Wats.; *C. claviformis* ssp. *integrior* (Raven) Raven; *Oe. claviformis* ssp. *integrior* Raven]. Shadscale, ricegrass, winterfat, galleta, and rabbitbrush communities at 1555 to 1710 m in Beaver and Millard counties; Nevada and California; 11 (i).

Camissonia eastwoodiae (**Munz**) **Raven** [*Oenothera scapoidea* var. *eastwoodiae* Munz; *Oenothera eastwoodiae* (Munz) Raven]. Annual herbs from taproots; stems simple or branched from near the base, mainly 6–40 cm tall, with leaves in the lower quarter; herbage glabrous or less commonly puberulent or glandular; leaves petiolate, the blade simple, entire or irregularly toothed, 0.8–7 cm long, 0.6–4 cm wide; flowers numerous in terminal, nodding racemes, conspicuous; hypanthium 3.5–9 mm long; sepals 4–9 mm long, spreading-reflexed, some often adherent apically; petals yellow, spotted red, fading cream, 5.5–15 mm long; stamens dimorphic, the longer ones 1.5–2 mm longer than the others; pedicels slender, 4–20 mm long or more; capsules 1.8–4 cm long, subcylindric, usually curved; seeds in 2 rows per locule, 1.2–1.7 mm long, tan; $n = 7$. Mat-saltbush, shadscale, blackbrush, and juniper communities at 1190 to 1800 m in Carbon, Emery, Garfield, Grand, Kane, San Juan, and Wayne counties; Colorado and Arizona (at Lee's Ferry); a Colorado Plateau endemic; 51 (vii). Plants of this species occur mainly on the Bluegate and Tununk members of the Mancos Shale Formation and on the Tropic Shale (Tununk equivalent?), but are also present on other strata in the vicinity.

Camissonia gouldii **Raven** [*Oenothera gouldii* (Raven) Welsh & Atwood]. Annual, caulescent herbs from taproots; stems simple or branched from near the base, mainly 6–20 cm tall; herbage glandular-hairy throughout; leaves petiolate, the blade mostly 1–1.5 cm long and 0.5–1 cm wide, elliptic to almost rhombic, usually denticulate; flowers (self pollinated) few to many in terminal, nodding spikes, inconspicuous; hypanthium 1.5–3 mm long; sepals 1–1.5 mm long; petals white, fading pink, 1.5–3 mm long; stamens subequal; anthers 0.4–0.9 mm long; style exserted 1.5–2.5 mm beyond the hypanthium; capsules sessile, 8–12 mm long; seeds in 1 row per locule, ca 1 mm long, smooth. Volcanic ash at 1068 m, with *Phacelia palmeri*, in Washington (type from N of St. George) County; Arizona; 2 (0). this plant is allied to *C. boothii* var. *villosa* (q.v.).

Camissonia minor (**A. Nels.**) **Raven** [*Oenothera alyssoides* var. *minutiflora* Wats., type from Stansbury Island; *Sphaerostigma minor* A. Nels.; *Oe. minor* (A. Nels.) Munz]. Annual, caulescent herbs from taproots; stems 2–28 cm tall, usually branched from near the base, 4–20 cm broad, the largest leaves at the base; herbage strigulose to crisp-puberulent, often more or less glandular upward; leaves 0.5–4.8 cm long, 0.2–1.5 cm wide, lanceolate to elliptic or oblanceolate, petiolate; flowers few to many, borne singly in axils of foliose bracts from near the stem base and upward; hypanthium 0.5–1.9 mm long; sepals 0.8–1.8 mm long, reflexed; petals white, drying pink, 0.8–1.5 mm long; stamens subequal, the anthers 0.4–0.8 mm long; style exserted 0.5–1.5 mm beyond the hypanthium; capsule sessile, contorted, 8–25 mm long, subcylindric to tapering; seeds 1.1–1.2 mm long, gray; $n = 7$. Sagebrush, winterfat, rabbitbrush, shadscale, greasewood, and mixed grass communities at 1430 to 1985 m in Beaver, Box Elder, Carbon, Emery, Juab, Millard, Salt Lake, Sanpete, Sevier, Tooele,, Uintah, and Utah counties; Washington to Wyoming, south to California and Nevada; 11 (ii).

Camissonia multijuga (**Wats.**) **Raven** [*Oenothera multijuga* Wats., type from near Kanab?]. Annual, subscapose herbs from taproots and basal rosettes; stems mainly 1–12 dm tall, nearly leafless above the rosette; herbage villous-pilose to puberulent, becoming glabrous or glandular upward; leaves 4–30 cm long, the blades oblanceolate in outline, sometimes purple-dotted, commonly pinnate-pinnatifid into few to numerous lateral segments (or these lacking in depauperate plants), the terminal segments 1–6 cm long, 0.8–4 cm wide, petiolate; flowers numerous, borne in minutely bracteate panicles; hypanthium 1–3 mm long; sepals 3–8 mm long, spreading to deflexed at anthesis, the tips often adherent; petals yellow to cream, fading cream or purplish, 4–14 mm long; stamens dimorphic, the shorter ones surpassed by 1–1.5 mm; anthers 2–4 mm long; style surpassing the hypanthium by 3.5–8 mm; capsules pedicellate, 10–50 mm long, 1.3–2 mm thick, straight to gently curved; pedicels 7–20 mm long; n = 7. Creosote bush, bursage, Joshua tree, shadscale, mesquite, blackbrush, sagebrush, live oak-cliffrose, and juniper communities at 730 to 1285 m in Kane (?) and Washington counties; Arizona and Nevada; 25 (ii).

Camissonia parryi (Wats.) Raven [*Oenothera parryi* Wats, type from near St. George; *Oe. tenuissima* Jones, type from Rockville]. Caulescent, annual herbs from taproots; stems mainly 5–40 cm tall, leafy mainly below the middle; herbage sparingly to densely pilose with straight or contorted hairs mainly 0.5–2.3 mm long; leaves petiolate, the blades simple (or less commonly with minute lateral segments also), 0.8–5 cm long, 0.6–3.5 cm wide, ovate to elliptic or oblanceolate to lanceolate; flowers numerous in terminal and lateral, bracteate panicles; hypanthiumn 0.5–2 mm long; sepals 1.5–4 mm long, reflexed or spreading; petals bright yellow, often red-spotted, fading yellow or rose; stamens dimorphic, the longer ones surpassing the others by 0.5–1 mm or more; anthers 0.9–1.2 mm long; style surpassing the hypanthium by 3–4 mm; capsules 4–10 mm long, subcylindric, straight or curved; seeds in 1 row per locule, 0.7–1.2 mm long; n = 7. Creosote bush, shadscale, mesquite, bursage, and other warm-desert shrub communities at 730 to 1100 m in Washington County; Arizona; a Virgin-Mohave endemic; 16 (i).

Camissonia parvula (Nutt.) Raven [*Oenothera parvula* Nutt. ex T. & G.]. Diminutive, caulescent, annual herbs from taproots; stems 1–18 cm tall, simple or branched from near the base, the lower branches often subopposite, leafy mainly above the base; herbage strigose to sparingly hirtellous or subglabrous; leaves linear, the lowermost subopposite, mainly 5–30 mm long, 0.4–1.2 mm wide, involute, essentially entire; flowers few to many, borne singly in bract or leaf axils from above the middle of the stem; hypanthium 0.8–1.5 mm long; sepals 1–2.5 mm long; petals yellow, 2.5–4.3 mm long; stamens dimorphic, the longer ones somewhat surpassing the others; style exserted ca 1–2 mm beyond the hypanthium; capsules 1.5–3 cm long, 1 mm thick or less; seeds in 1 row per locule, 0.7–0.9 mm long, brown; n = 7. Shadscale, rabbitbrush, sagebrush, pinyon-juniper, and lodgepole pine communities at 1280 to 2745 m in Beaver, Duchesne, Garfield, Kane, Tooele, Uintah, and Washington counties; Washington to Wyoming, south to California, Nevada, and Colorado; 26 (iii).

Camissonia pterosperma (Wats.) Raven [*Oenothera pterosperma* Wats.]. Annual, caulescent herbs from taproots; stems erect, typically branched from near the base, 1–15 cm tall; herbage hirsute-hirtellous to subglabrous, becoming glandular above; leaves scattered along the stem, the lower ones opposite or nearly so, short-petiolate, 3–15 mm long, 1–5 mm wide, obovate to elliptic or lanceolate, entire; flowers opening during the day, few to many, solitary in leaf axils; hypanthium 1–4 mm long; sepals 1.5–2.5 mm long, reflexed at anthesis; petals 1.5–3 mm long; stamens 8, the longer ones surpassing the others by 0.5–0.7 mm; style surpassing the hypanthium by 2–3 mm; anthers 0.3–0.4 mm long; capsules pedicellate, 12–18 mm long, curved-ascending; pedicels 4–9 mm long; seeds in 1 row per locule, 1–1.5 mm long, with a pair of broad, thick wings marginally; n = 7. Mixed shrub and pinyon-juniper communities at 1220 to 1830 m in Kane, Millard, Tooele, Uintah, and Washington counties; Arizona, Nevada, and California; 5 (0).

Camissonia pusilla Raven Annual, caulescent herbs from a taproot; stems 2–15 (22) cm tall, simple or typically branched from the base, with leaves mainly clustered near the base; herbage spreading-hairy below, becoming glandular to almost glabrous above; leaves 1–3 cm long, 0.4–1.8 mm wide, linear, entire or irregularly serrate-serrulate; flowers few to many, sessile or short-pedicellate in axils of bracteate leaves; hypanthium 0.8–1.5 mm long; sepals 1.2–2.5 mm long, separately reflexed at anthesis; petals yellow, often fading orange to lavender, 1.5–3.5 mm long, commonly red-spotted near the base; stamens 8, dimorphic, the longer ones surpassing the others by ca 1 mm; anthers 0.3–0.5 mm long; style exserted 1–2 mm beyond the hypanthium; capsules subsessile or with pedicels to 2 (4) mm long, 1.8–4 cm long, 0.6–0.9 mm thick; seeds in 1 row per locule, 0.7–0.8 mm long, shining, brown; n = 7. Mixed desert shrub, sagebrush, and mountain brush communities at ca 1220 to 1830 m in Beaver, Iron, and Washington counties; Washington and Idaho, south to California and Nevada; 3 (0).

Camissonia refracta (Wats.) Raven [*Oenothera refracta* Wats.]. Annual, caulescent herbs from taproots; stems typically 6–30 (45) cm tall, soon branched from near the base and 6–30 cm broad or more, the branches curved-ascending; herbage more or less strigose to stipitate-glandular or glabrous; leaves basally disposed initially, finally mainly along the stem, the lower ones commonly withered and fallen by anthesis, mostly 1–6 cm long, 1–8 mm wide, narrowly elliptic to lance-linear, weakly and irregularly denticulate, the basal ones petiolate; flowers several to numerous, aggregated in terminal, nodding, bracteate spikes or panicles, sessile or nearly so (the slender ovary simulating a pedicel), opening in afternoon (?); hypanthium 4–6.5 mm long; sepals 3–6 mm long, reflexed to spreading at anthesis; petals white, fading cream to yellow or pinkish, 3.5–8 mm long; stamens 8, subequal; anthers 1.5–2.5 mm long; style exserted 3–8 mm beyond the hypanthium; capsules sessile or nearly so, 2–5 cm long, 0.7–1 mm thick, straight or contorted, ascending to reflexed; seeds in 1 row per locule, 1–1.5 mm long; n = 7. Creosote bush and other warm desert shrub communities at 670 to 950 m in Washington County; Nevada, California, and Arizona; 4 (ii).

Camissonia scapoidea (T. & G.) Raven [*Oenothera scapoidea* T. & G.; *Chylisma scapoidea* var. *seorsa* A. Nels.]. Annual herbs from taproots; stems mainly 3–30 (45) cm tall, simple or branched from near the base, with main foliage leaves in lower portion; herbage glabrous, strigose, puberulent, or sparingly glandular; leaves petiolate, the blades simple (or less commonly with few to many, small, lateral segments), 1–6 cm long, 4–35 mm wide, lanceolate, ovate, or elliptic, entire to denticulate; flowers several to many, aggregated in terminal, minutely bracteate, nodding racemes; hypanthium 1–4.5 mm long; sepals 1.2–5 mm long, reflexed at anthesis; petals yellow, often red-spotted near the base, 2–5.5 mm long; stamens 8, dimorphic, the longer usually surpassing the shorter by 1–2 mm; anthers 0.5–2.5 mm long; pedicels ascending to deflexed, 5–20 mm long; capsules 10–30 (40) mm long, 1.8–3 mm thick; seeds in 2 rows per locule, 1–2 mm long, not winged; n = 7, 14. Two rather distinctive varieties are present in Utah.

1. Leaves with several prominent lateral segments, borne in a loose basal rosette; plants mainly of southwestern Utah
.... *C. scapoidea* var. *utahensis*
— Leaves usually lacking lateral segments but, if so, these small and inconspicuous; plants of broad distribution, but evidently missing from southwestern Utah ....
.... *C. scapoidea* var. *scapoidea*

Var. scapoidea [*Oenothera scapoidea* ssp. *brachycarpa* Raven; *C. scapoidea* ssp. *brachycarpa* (Raven) Raven]. Mixed salt and cool desert shrub and pinyon-juniper communities at 1310 to 1985 m in Box Elder, Carbon, Daggett, Duchesne, Emery, Garfield, Grand, Juab, Kane, Millard, Sanpete, Sevier, Tooele, Uintah, and Wayne counties; Oregon to Wyoming, south to Nevada, Arizona, and New Mexico; 73 (ix).

Var. utahensis (Raven) Welsh [*Oenothera scapoidea* ssp. *utahensis* Raven, type from Black Rock; *C. scapoidea* ssp. *utahensis* (Raven) Raven Brittonia 16: 282. 1964]. Mixed desert shrub communities at ca 1525 to 1944 m in Beaver, Juab, Millard, Salt Lake, and Sevier counties; Nevada, Oregon, and Idaho; 7 (iii). This phase of the species has been mistaken for various other species, especially *C. multijuga*, from which it differs in the type of pubescence, thicker capsules, and usually longer hypanthium.

Camissonia subacaulis (Pursh) Raven [*Jussiea subacaulis* Pursh; *Taraxia subacaulis* (Pursh) Rydb.; *Oenothera subacaulis* (Pursh) Garrett; *Oe. heterantha* Nutt. var. *taraxifolia* Wats; *Oe. subacaulis* var. *taraxifolia* (Wats.) Jepson]. Perennial, acaulescent herbs from thick taproots and a simple (or branched) caudex; herbage glabrous or the leaves merely puberulent; leaves all basal, 2–16(20) cm long, 3–35(50) mm wide, the blade entire or irregularly and sparingly lobed near the base; flowers opening in morning, sessile in leaf axils, numerous, clustered; ovary apex prolonged into a slender tube 1.5–7 cm long below the hypanthium proper—this 1.5–3 mm long; sepals 6–15 mm long, reflexed at anthesis, separate; petals 7–15 mm long, yellow, fading yellowish or cream; stamens dimorphic; anthers 1–2.5 mm long; style exserted 2–7 mm beyond the hypanthium; capsules sessile, ovoid to lance-ovoid, 10–25 mm long, 5–7 mm thick, beaked; seeds 1.3–1.9 mm long; n = 7. Sagebrush, aspen, and grass-sedge communities at 2070 to 2850 m in Cache, Carbon (?), Duchesne, Emery, Salt Lake, Sanpete, Summit, Utah, and Wasatch counties; Washington to Montana, south to California, Nevada, and Colorado; 21 (i).

Camissonia walkeri (A. Nels.) Raven [*Oenothera brevipes* var. *parviflora* Wats., type from near St. George, not *Oe. parviflora* L.; *Chylisma walkeri* A. Nels.; *Oe. walkeri* (A. Nels.) Raven; *Oe. scapoidea* var. *tortilis* Jepson; *C. walkeri* ssp. *tortilis* (Jepson) Raven; *Oe. multijuga* var. *orientalis* Munz, type from near Moab]. Annual or winter annual, subscapose herbs from taproots and a basal rosette; stems 0.8–7.5 dm tall, nearly leafless above the rosette; herbage typically spreading hairy, at least below, the hairs, when present, mainly 0.5–1.5 mm long, and/or more or less glandular puberulent; leaves 2–22 cm long, the blades oblanceolate to elliptic or ovate in outline, often purple-dotted, commonly pinnate-pinnatifid into few to numerous lateral segments, the terminal segment 1–5 cm long, 1–3 cm wide, petiolate; flowers numerous, borne in minutely bracteate panicles; hypanthium 0.5–2 mm long; sepals 1.2–4 (5) mm long, reflexed at anthesis; petals 1.2–5 (6) mm long, yellow, fading pinkish to lavender; stamens 8, dimorphic, the longer ones surpassing the others by ca 1 mm; anthers 0.5–1.5 mm long; style exceeding the hypanthium by 1–3 mm; capsules pedicellate, 12–30 (40) mm long, mostly 1.2–1.8 mm thick, straight to gently curved; pedicels 4–20 mm long; n = 7, 14. Blackbrush, rabbitbrush, matchweed, shadscale and pinyon-juniper communities at 1125 to 1770 m in Beaver, Box Elder, Emery, Garfield, Grand, Juab, Kane, Millard, San Juan, Tooele, Uintah, Washington, and Wayne counties; Colorado, Nevada, Arizona, and California; 48 (xii). Recognition of infraspecific taxa seems not to be warranted.

Circaea L.

Perennial herbs from rhizomes, stolons, or tubers; leaves opposite, petioled; flowers irregular, inconspicuous, borne in terminal racemes or panicles; hypanthium prolonged somewhat beyond the ovary; sepals 2, reflexed; petals 2, emarginate; stamens 2, alternate with the petals; ovary 1- or 2-loculed, the ovules 1 per locule; fruit indehiscent, pubescent with hooked or straight hairs.

Munz, P. A. 1965. *Circaea*. N. Amer. Fl. II. 5: 24–25.

Circaea alpina L. Enchanter's Nightshade. Plants mostly 0.5–5.5 dm tall; stems simple or branched; leaves petiolate, the blades 1–7 (9.5) cm long, 0.8–5.7 cm wide, ovate to cordate or suborbicular, the base cordate to truncate or rounded basally, acute to acuminate apically, irregularly serrate-dentate to subentire; racemes few to many-flowered, the pedicels subtended by minute bracts; sepals 1–2 mm long, white to pinkish, or greenish at the apex; petals 1–2 mm long, white to pinkish, notched to near the middle; fruit obovoid, 1-loculed, 1–2 mm long, clothed with usually uncinate hairs; 2n = 22, 24. Mountain brush, white fir-Douglas fir, and aspen communities at 1890 to 2350 m in Box Elder, Cache, Davis, Duchesne (?), Juab, Salt Lake, Tooele, Uintah, Utah, and Weber counties; widespread in North America; 18 (ii). Our material belongs to **var. pacifica** (Asch. & Mag.) Jones [*C. pacifica* Asch. & Mag.].

Clarkia Pursh

Annual herbs; leaves simple, alternate or the lower ones opposite; flowers few to several, borne singly in axils of more or less foliose bracts; hypanthium shortly prolonged beyond the ovary, deciduous after anthesis; sepals 4, spreading or reflexed at anthesis; petals 4, usually brightly colored; stamens 8 (or 4); anthers basifixed; stigma 4-lobed; capsule elongate.

Clarkia rhomboidea Dougl. Broad-leaved Clarkia. Plants mainly 1.5–5 (10) dm tall, simple or branched above; herbage puberulent to subglabrous; leaves petiolate, the blade 1.2–6 cm long, 0.4–3.2 cm wide, ovate, ovate-lanceolate, or elliptic, obtuse to acute apically, acute basally; flowers short-pedicellate, borne in nodding, bracteate, terminal, spicate racemes; hypanthium 1–3 mm long; sepals 5–9 mm long, lanceolate, green or purplish; petals 6–12 mm long, pink purple (or white); stamens 8, 1–4 mm long, often curved at anthesis; fruit 1–3 cm long, 4-angled, straight or curved. Mountain brush communities at 1585 to 2475 m in Box Elder, Cache, Davis, Salt Lake, Utah, Wasatch, and Weber counties; Washington and Idaho, south to California and Arizona; 11 (0).

Epilobium L.

Perennial, or less commonly, annual herbs from rhizomes, stolons, bulblike offsets (turions), and taproots or fibrous roots; stems decumbent to ascending or erect; leaves all opposite, the upper ones alternate, or all alternate (or rarely whorled), simple, entire to toothed; flowers perfect, regular or nearly so, solitary in leaf axils or in terminal racemes; hypanthium short or lacking; sepals 4; petals 4, entire or emarginate; stamens 8; stigma capitate, cylindric, or 4-lobed; fruit a loculicidal, 4-carpellate, 4-loculed, elongate capsule; seeds with a tuft of hair. Note: This genus is considered to be taxonomically complex, partly due to hybridization within the section Epilobium, and partly due to its circumboreal distribution, wherein different interpretations are the result of poor typification of ancient collections and poor descriptions that tend to be inclusive, but not exclusive.

1. Plants annual from taproots, growing in dryish sites .... *E. brachycarpum*
— Plants perennial from rhizomes, stolons, turions, or fibrous roots 2

2(1). Petals (8) 10–30 mm long, rounded apically, spreading; hypanthium not extending beyond the ovary (the calyx cleft to the apex of the ovary); stigma 4-lobed (Chamaenerion, the fireweeds) 3
— Petals 2–10 mm long, emarginate apically; hypanthium produced beyond the ovary (Epilobium, the willowherbs) 4

3(2). Plants erect; style pubescent at the base; floral bracts much reduced; herbage glabrous or nearly so; inflorescence with usually more than 15 flowers; plants widespread .... *E. angustifolium*
— Plants decumbent; styles glabrous; floral bracts foliose; herbage usually minutely pubescent; inflorescence with usually fewer than 10 flowers; plants of Salt Lake and Summit counties .... *E. latifolium*

4(2). Stigma 4-lobed; plants of arid sites .... *E. nevadense*
— Stigma not 4-lobed; plants often of moist sites 5

5(4). Plants often grayish-hairy, at least in the inflorescence, producing slender stolons ending in turions; leaves linear to narrowly lanceolate 6
— Plants glabrous to glandular, but if hairy, not grayish, with or without stolons and/or turions; leaves typically lanceolate or broader, rarely some of them linear . 7

6(5). Upper leaf surface glabrous or nearly so; plants not definitely known from Utah, but to be expected in the Uinta Mts .... *E. palustre* L.
— Upper leaf surface finely pubescent .. *E. leptophyllum*

7(5). Plants producing bulblike offsets (turions) at the base of the stem, the fleshy, overlapping scales often persistent at the base of the current stem 8
— Plants arising from short to elongate rhizomes or sobols, or neither, not producing turions 10

8(7). Petals 5–10 mm long; stems coarse, simple or with erect-ascending branches from near the apex .... *E. glandulosum*
— Petals mostly 2–5 mm long; stems typically slender, simple or branching from near the base 9

9(8). Leaves mostly sessile and clasping; fruit sessile or short-pedicellate, the pedicels to 5 (8) mm long .... *E. saxinmontanum*

— Leaves sessile or petiolate, but the blades not clasping; fruit with pedicels 5–10 mm long or more .... *E. halleanum*

10(7). Plants often more than 3 dm tall, from leafy, overwintering basal rosettes, not with sobols or rhizomes; leaves often denticulate; seeds minutely striate-roughened (use 10–30 x) .... *E. ciliatum*
— Plants mostly less than 3 dm tall, either soboliferous or subrhizomatous, or the stems borne at the summit of a root-crown; seeds variously roughened but typically not striate 11

11(10). Stems glabrous and glaucous, at least below the inflorescence .... *E. glaberrimum*
— Stems pubescent in lines or generally 12

12(11). Leaves sessile, steeply ascending, at least the upper oblong to linear; plants not definitely known from Utah, but to be sought in the Uinta Mts .... *E. oregonense* Hausskn.
— Leaves various, sessile or petiolate, spreading-ascending or spreading, lanceolate to ovate 13

13(12). Plants typically 2–4 dm tall, the stems straight; capsules typically 4–7 cm long .... *E. hornemannii*
— Plants typically 0.3–2 dm tall, the stems often S-shaped; capsules typically 1.5–4 cm long .... *E. alpinum*

Epilobium alpinum L. Alpine Willowherb. Stems often S-shaped, 1–20 (25) cm long, arising from short rhizomes or sobols, lacking turions, puberulent in lines or sometimes generally, or somewhat glandular upward; leaves sessile or short-petiolate, 4–30 mm long, ovate to obovate, lanceolate, oblanceolate, or elliptic, entire or obscurely toothed; flowers few to several in terminal, bracteate, nodding to erect racemes; pedicels 2–10 mm long; hypanthium very short; sepals 2.2–4 mm long, commonly purplish; petals 3.5–6 mm long; capsules 1.5–3.2 cm long, 0.8–2 mm wide; seeds 0.7–2 mm long, the coma white; 2n = 36. The approach to the *alpinum* complex by C. L. Hitchcock in his summary review of the plants in the Pacific Northwest, in which *anagallidifolium*, *clavatum*, *hornemannii*, and *lactiflorum* are regarded as varieties of an inclusive *E. alpinum*, has considerable merit. The subunits are closely allied, a fact noted in my summary revision of the Alaskan flora, where Hitchcock was followed. The present treatment represents a compromise between the views of Hitchcock and traditionalists, who still recognize the four taxa at specific rank. The complex, as treated herein consists of two species, *E. alpinum* and *E. hornemmannii*, each with two varieties.

1. Capsules linear, mainly 0.8–1.2 mm wide; inflorescence nodding in bud; leaves lanceolate to elliptic or oblong; seeds smooth, ca 1 mm long ..... *E. alpinum* var. *alpinum*
— Capsules clavate, mainly 1.5–2 mm wide; inflorescence often erect in bud; seeds roughened, ca 1.5–2 mm long; leaves ovate, more or less serrulate .... *E. alpinum* var. *clavatum*

Var. *alpinum* [*E. anagallidifolium* Lam.]. Lodgepole pine, spruce-fir, grass-sedge, and alpine tundra communities at 2055 to 3390 m in Box Elder, Duchesne, Juab, and Summit counties; widespread in North America; circumboreal; 8 (ii). The original description of *E. alpinum* by Linnaeus is sufficiently broad as to include not only this portion of the complex but others as well. I follow recent tradition in restricting the concept of this species to include what has been called *E. anagallidifolium* Lam. Stipulations of the International Code of Botanical Nomenclature allow for typification of a name, even where the concept is based on discordant elements, when at least some portion is included within the concept. There does not seem to be any obstacle to using the name to include the traditionally accepted portion of the species complex.

Var. clavatum (Trel.) C. L. Hitchc. [*E. clavatum* Trel.]. Lodgepole pine, spruce-fir, and alpine tundra communities at 2680 to 3355 m in Box Elder, Cache, Duchesne, Salt Lake, Sanpete, Summit, Uintah, Utah, and Wasatch counties; Alaska to Northwest Territories, south to California and Nevada; Eurasia; 20 (0).

***Epilobium angustifolium* L.** Fireweed. Plants mostly (3) 5–20 dm tall, arising from rhizomelike roots that bear buds freely; stems often purplish, at least above, basally decumbent to erect, usually simple (less commonly branched), glabrous below, commonly puberulent above; leaves alternate, lanceolate to elliptic (or linear), 5–20 cm long, 0.5–4 cm broad, entire or nearly so, sessile or subsessile, glabrous or pubescent only along the lower midvein; sepals 7.5–17 mm long, puberulent; petals 8–20 mm long, pink purple, pink, or rarely white; styles longer than the stamens, pubescent near the base; stigma 4–lobed; capsules 8–9 cm long, pubescent; seeds 1–1.5 mm long, the hair white to dingy; n = 36. Sagebrush, mountain brush, Douglas fir, aspen, lodgepole pine, and spruce-fir communities at 1525 to 3620 m in all Utah counites, except Carbon, Daggett, Emery, Kane, Morgan, and Salt Lake; widely distributed in North America; 77 (ix). Most of our material belongs to **var. canescens** Wood [*Chamaenerion angustifolium* var. *abbreviatum* Lunell; *E. angustifolium* var. *abbreviatum* (Lunell) Munz; *C. angustifolium* var. *platyphyllum* Daniels; *E. angustifolium* var. *platyphyllum* (Daniels) Fern.; *E. angustifolium* ssp. *circumvagum* Mosquin].

***Epilobium brachycarpum* Presl** Autumn Willowherb. [*E. paniculatum* Nutt. ex T. & G.; *E. tracyi* Rydb., type from Ogden]. Annual herbs from taproots; stems mostly (2) 3–7 (10) dm tall, simple to much branched; herbage glabrous to glandular-puberulent, the bark often exfoliating; leaves mostly alternate, 0.8–7 cm long, 1–9 mm wide, linear to lance-linear or elliptic, often folded, entire or toothed, spinulose to glandular-subulate, the axils often with clusters of secondary leaves; flowers few to numerous in terminal, bracteate racemes or panicles; pedicels filiform, 5–20 mm long, the bracts often adnate to the petiole base; hypanthium 1–6 mm long; sepals 1–4 mm long; petals white or pink, 3–8 mm long; stigma shortly 4–lobed; capsules 1.3–3 cm long, 1–2.2 mm wide; seeds 1.5–2.5 mm long, the coma caducous; n = 18. Sagebrush, pinyon-juniper, mountain brush, ponderosa pine, Douglas fir, and aspen communities at 1525 to 2745 m in all Utah counties, except Box Elder, Carbon, Emery, Grand, San Juan, Wasatch, and Wayne; British Columbia to the Dakotas, south to California, New Mexico, and Mexico; 81 (vii).

***Epilobium ciliatum* Raf.** Northern Willowherb. [*E. watsonii* Barbey in Brewer & Wats.; *E. ciliatum* ssp. *watsonii* (Barbey) Hoch & Raven; *E. americanum* Hausskn.; *E. adenocaulon* ssp. *americanum* (Hausskn.) Love & Love; *E. adenocaulon* Hausskn.; *E. glandulosum* var. *adenocaulon* (Hausskn.) Fern. Perennial (or annual?) herbs from a shallowly subterranean cluster of spreading roots, lacking rhizomes (but rooting along the stem when buried) and turions; stems mainly 2–12 dm tall, erect, simple or branched, pubescent in lines below the leaf bases, or puberulent almost overall, sometimes grayish canescent, often glandular upward; leaves opposite and some alternate, lanceolate or lance-ovate, serrulate to subentire; flowers numerous, borne in terminal, leafy-bracteate racemes or panicles; hypanthium 0.5–2 mm long; sepals 2–6 mm long, sometimes suffused with red; petals 2–5 mm long, white or pink; capsules 4–10 cm long, 0.9–1.8 mm wide; pedicels 2–20 mm long or more; seeds 0.7–1.2 mm long, longitudinally striate; n = 18, 36+. Willow-cottonwood, hanging garden, sagebrush, pinyon, juniper, white fir-Douglas fir, aspen, lodgepole pine, spruce-fir, and grass-sedge communities at 885 to 3050 m in all Utah counties; widespread in North America; Mexico; Central and South America; 206 (xxxi). This is the most common species of willowherb in Utah. It is almost ubiquitous along streams, lake margins, irrigation canals, and near seeps and springs. It hybridizes with *E. glaberrimum*, *E. glandulosum*, *E. halleanum*, *E. hornemannii*, *E. leptophyllum*, and *E. saximontanum*, inter alia.

***Epilobium glaberrimum* Barbey in Brewer & Wats.** Barbey Willowherb. Perennial herbs from sobols or short wiry rhizomes, lacking turions; stems 1–6 dm tall; herbage glabrous and glaucous throughout or only sparingly glandular-puberulent in the inflorescence; leaves mostly opposite, 1–7 cm long, lanceolate to ovate, subsessile, clasping, denticulate to entire; flowers few to many in terminal, bracteate racemes; hypanthium 1–2 mm long; sepals 3–7 mm long; petals 3–10 mm long, pink to pink purple or white; capsules 3–7 cm long; pedicels 5–25 mm long; seeds ca 1 mm long, longitudinally finely striate, the coma caducous; n = 18. Alpine or subalpine talus, ridges, and stream courses at 2195 to 2745 m in Salt Lake County; northwestern U. S.; 1 (0).

***Epilobium glandulosum* Lehm.** Glandular Willowherb. [*E. ciliatum* ssp. *glandulosum* (Lehm.) Hoch & Raven; *E. brevistylum* Barbey in Brewer & Wats.]. Perennial herbs from taproots or fibrous roots or from short to elongate rhizomes; stems mainly 3–10 dm tall, sometimes purplish, erect, simple or branched above, glabrous below, often glandular above, the pubescence commonly in lines below the leaf bases; leaves mostly opposite, 2.5–10 cm long, 0.5–4 cm wide, narrowly lanceolate to ovate-lanceolate, rounded to subcordate basally, acuminate to acute (the apex usually blunt) apically, serrulate, glabrous or pubescent, sessile and sometimes clasping to short-petiolate; flowers numerous in terminal, bracteate racemes or panicles; sepals 3–5 mm long, glandular-puberulent; petals 3–10 mm long, pink to purplish; capsules 3–7 cm long, usually glandular puberulent; seeds 1–1.8 mm long, raggedly roughened in parallel lines, the coma dingy or white. Streamsides and other moist sites at 1495 to 2440 m in Juab, Kane, Rich, Salt Lake, Tooele, Uintah, and Utah counties; widely distributed in North America; Asia; 8 (0). This entity is perplexingly difficult to distinguish from *E. ciliatum*, but the more conspicuously roughened and mostly larger seeds, and flowers that average larger, seem diagnostic, even when turions (only rarely collected) are not present. The basal part of the stem is often submersed in water or mud and supports numerous adventitious roots along the submersed portion. The species hybridizes with *E. ciliatum*, *E. hornemannii*, and *E. saximontanum*, inter alia.

Epilobium halleanum Hausskn. Plants perennial, the stems typically decumbent basally, otherwise erect and usually straight, arising from turions and short, rhizomatous sobols, mainly 1–3 (5) dm tall, puberulent in lines to subglabrous or glandular upward; leaves mainly opposite, sessile or short-petiolate, 5–40 mm long, 2–10 (15) mm wide, lanceolate to elliptic, oblong, or sublinear, mostly erect; flowers solitary or few to several in bracteate, terminal racemes, these more or less nodding; hypanthium very short; sepals 1.5–3 mm long; petals 3–5 mm long, white, pink, or lavender; pedicels 5–35 mm long; capsules 2–5 cm long; seeds 1–1.5 mm long, minutely roughened, the coma deciduous. Aspen, lodgepole pine, spruce-fir, and sedge-grass communities, often where wet, at 2530 to 3355 m in Beaver, Box Elder, Cache, Daggett, Duchesne, Emery, Grand, Iron, Juab, Morgan, Salt Lake, Sanpete, Sevier, Summit, Tooele, Uintah, Utah, Wasatch, and Weber counties; British Columbia to Saskatchewan, south to California, Nevada, Arizona, Colorado, and South Dakota; 22 (iii).

Epilobium hornemannii Reichenb. Hornemann Willowherb. Stems decumbent basally, otherwise straight or merely curved, arising from rhizomatous sobols, lacking turions, mainly 3–30 cm tall, puberulent in lines or sometimes generally, or somewhat glandular upward; leaves sessile to short-petiolate, 8–40 mm long, ovate to lanceolate or elliptic, entire or serrulate; flowers few to many in terminal, bracteate, nodding to erect racemes; pedicels 3–42 mm long or more; hypanthium very short; sepals 2.2–4 mm long, often purplish; petals 3–8 mm long, pink, pink purple, or white; capsules 2–6 cm long, 1–2 mm wide; seeds 0.9–2 mm long, the coma white to dingy, deciduous; $2n = 36$. Two more or less distinctive varieties are present.

1. Petals white (less commonly pinkish), 2–4 (5) mm long; seeds smooth .... *E. hornemannii* var. *lactiflorum*
— Petals pink to pink purple (rarely white), 5–8 mm long; seeds more or less roughened ....
.... *E. hornemannii* var. *hornemannii*

Var. *hornemannii* [*E. nutans* Hornem., not F. W. Schmidt; *E. alpinum* var. *nutans* (Hornem.) Lehm. in Hook.]. Sagebrush, aspen, spruce-fir, lodgepole pine, and sedge-forb communities, mainly in wet areas, at 2500 to 3890 m in Beaver, Cache, Box Elder, Duchesne, Emery, Garfield, Juab, Piute, Salt Lake, San Juan, Sanpete, Sevier, Summit, Tooele, Wasatch, and Weber counties; widespread in North America; circumboreal; 42 (vi).

Var. *lactiflorum* (Hausskn.) D. Love [*E. lactiflorum* Hausskn.; *E. alpinum* var. *lactiflorum* (Hausskn.) C. L. Hitchc.]. White fir-Douglas fir, aspen, spruce-fir, and sedge-forb communities, mainly in wet areas, at 1890 to 3175 m in Duchesne, Juab, Millard, Summit, Uintah, Utah, and Wasatch counties; widespread in North America; circumboreal; 22 (iii).

Epilobium latifolium L. Dwarf Fireweed. Plant perennial, the stems mostly 1–4 (7) dm long, arising from a caudex, lacking rhizomes, occasionally purplish, decumbent to ascending, glabrous below, puberulent above; leaves opposite (or whorled below), usually alternate above, lanceolate to elliptic or ovate, 1.5–8 cm long, 0.5–3 cm broad, entire or denticulate, sessile or nearly so, glaucous, usually pubescent; sepals 10–18 mm long, puberulent; petals 15–30 mm long, bright pink to pink purple or rarely white; style shorter than the stamens, glabrous, the stigma 4–lobed; capsules 3–9 (10) cm long, glabrate to pubescent, usually purplish; seeds (1) 1.5–2 mm long, smooth, the hair dingy; $2n = 36, 54, 72$. Lodgepole pine, aspen, and spruce-fir communities, usually on gravel bars along streams, at ca 2285 to 2745 m in Salt Lake and Summit counties; widely distributed in North America; circumboreal; 4 (ii).

Epilobium leptophyllum Raf. Slender Willowherb. [*E. oliganthum* var. *gracile* Farw.]. Perennial herbs, the stems erect or bent at the base, arising from rhizomes or stolons, these terminated in late summer by a turion, mainly 4.5–8 (10) dm tall, puberulent overall or glabrous below, the inflorescence sometimes somewhat glandular; leaves mainly opposite, sessile or nearly so, 1–7 cm long, 1–6 mm wide, linear to lance-linear, often with fascicled, axillary leaves, puberulent overall, the lower ones often withered at anthesis; flowers numerous in terminal, bracteate racemes or panicles; hypanthium short; sepals 2–3 mm long; petals 4–6 mm long, white, pink, or lavender; pedicels 3–18 mm long; capsules 3–7 cm long, 1–2 mm thick; seeds 1.3–1.5 mm long, smooth; coma more or less persistent. Cattail-bullrush and sedge-spikerush communities at 1420 to 1830 m in Cache, Duchesne, and Uintah counties; British Columbia to Newfoundland, south to California, Nevada, Kansas, and North Carolina; 3 (0). Differences between this species and *E. palustre* are slight, and it might be best regarded at infraspecific rank within that species. Such a combination is neither intended nor implied herein.

Epilobium nevadense Munz Nevada Willowherb. Perennial, suffrutescent plants from persistent woody branches and a stout taproot; stems 15–40 cm tall, erect or curved-ascending, puberulent to glabrous or glandular upward; leaves mostly alternate, or the lower ones opposite, sessile or short-petiolate, the blades 4–20 (30) mm long, 1–6 mm wide, oblanceolate to elliptic, folded, entire or denticulate, often with fascicled, axillary leaves, puberulent to glabrous, the lower ones often withered at anthesis; flowers few to several in terminal, bracteate, racemes or panicles; hypanthium 2–4.5 mm long; sepals 2–4 mm long, purplish; petals 5–7.5 mm long, pink purple; stigma with 4 broad lobes; pedicels 1–2 mm long; capsules 8–15 mm long; seeds 2–2.5 mm long, the coma deciduous. Mountain brush community at 2135 to 2685 m in Millard and Washington counties; Nevada; a Basin and Range endemic; 5 (0).

Epilobium saximontanum Hausskn. Rocky Mt. Willowherb. [*E. palmeri* Rydb., type from S. Utah]. Perennial herbs from taproots and turions; stems mainly 1–5 dm tall, puberulent to glandular-puberulent upward, or subglabrous; leaves mostly opposite, sessile or short-petiolate and often clasping, the blades 8–55 mm long, 3–16 (25) mm wide, lanceolate to elliptic or lance-ovate, entire or denticulate; flowers few to many in terminal, bracteate racemes; hypanthium short; sepals 1–3 mm long; petals 2–4.5 mm long, white (rarely pink); capsules 2–6 cm long, erect (often appressed to the stem), sessile or the pedicel 1–8 mm long; seeds 1–1.5 mm long, minutely roughened, longitudinally striate; coma deciduous. Sagebrush, mountain brush, willow-cottonwood, white fir, lodgepole pine, and spruce-fir communities at 2075 to 3420 m in Beaver, Box Elder, Cache, Carbon, Daggett, Davis, Duchesne, Emery, Garfield, Grand, Iron, Juab, Millard, Piute, San Juan, Sanpete, Sevier, Summit,

Tooele, Uintah, Wasatch, and Washington counties; Oregon to Alberta and Montana, south to California and New Mexico; disjunctly in eastern Canada; 57 (ix). This species forms hybrids with *E. glandulosum*, *E. halleanum*, *E. ciliatum*, and *E. hornemannii*.

Gaura L.

Annual, biennial, or perennial herbs (sometimes suffrutescent); leaves mainly cauline, attenuate; flowers numerous, borne in terminal spikes or racemes; hypanthium tubular-funnelform, deciduous at maturity; sepals 4, reflexed at anthesis; petals 4, somewhat disposed to one side (appearing irregular); stamens 8, with a scale at each filament base; ovary 4–loculed, each locule with 1 or 2 ovules; stigma 4–lobed; fruit hard, indehiscent.

1. Plants mainly 1.5–3.5 (4.5) dm tall, perennial, clump-forming; flowers modestly showy; known from Cache, Daggett, San Juan, and Washington counties .. *G. coccinea*
— Plants mainly 5–15 dm tall or more, annual or biennial, not clump-forming; flowers tiny, inconspicuous; broadly distributed .... *G. parviflora*

Gaura coccinea Pursh Scarlet Gaura. Perennial herbs from a caudex and spreading rootstock; stems clustered, mostly 1.5–4.5 dm tall, strigose to villous or subglabrous; leaves 0.5–5 cm long, 1–7 mm wide, lanceolate to elliptic or oblong to linear, entire or few-toothed; flowers few to numerous in spikes or panicles; bracts 1–5 mm long, more or less persistent; hypanthium 3–11 mm long; sepals 4.5–9 mm long; petals 3–7 mm long, pink to salmon, becoming red orange to maroon in age; anthers 3–5 mm long; fruit 4–angled, 3–8 mm long; seeds usually 3 or 4; n = 14. Creosote bush, Joshua tree, blackbrush, shadscale, juniper, mountain brush, and sagebrush communities at 885 to 1740 m in Cache, Daggett, San Juan, and Washington counties; Alberta to Manitoba, south to California, Arizona, New Mexico, Texas, and Mexico; 24 (iv).

Gaura parviflora Dougl. ex Lehm. in Hook. Willow Gaura. Annual or biennial, tall, plain herbs; stems mainly 5–15 dm tall or more, spreading-hairy and glandular; leaves 2–10 cm long, 2–42 mm wide, elliptic to lanceolate, oblanceolate or oblong, the lower ones often withered at anthesis, the upper ones greatly reduced; flowers numerous in elongate spikes or panicles; bracts 1–5 mm long, caducous; hypanthium 1.5–5 mm long; sepals 2–3.5 mm long; petals 1.5–3 mm long, whitish or pinkish, radially disposed; anthers 0.5–1 mm long; fruit 5–10 mm long, 4–angled; seeds 3 or 4. Moist riparian, riverine, or palustrine communities at 885 to 1955 m in Cache, Duchesne, Emery, Garfield, Grand, Juab, Kane, Millard, Salt Lake, San Juan, Tooele, Uintah, Utah, Washington, Wayne, and Weber counties; Washington to Indiana, south to California, Arizona, Texas, and Louisiana; 33 (iv).

Gayophytum A. L. Juss.

Annual herbs from taproots; stems erect, simple or branched, the cortex often exfoliating; leaves alternate (or the lower subopposite); flowers few to numerous in leafy-bracteate racemes, spikes, or panicles or solitary in leaf axils, opening in morning; sepals 4, reflexed individually or in pairs at anthesis; hypanthium very short; petals 4, white, variously marked with yellow, becoming pink to red in age; stamens 8, dimorphic; anthers versatile; stigma entire, subglobose or hemispheric; ovary 2–loculed; capsules 4–valved, subsessile or pedicellate. Note: This is a difficult genus, due in part to the existence of diploids, tetraploids, hybrids, and a complacent to overlapping morphology. Often the taxa are poorly marked, and the following arbitrary key is tentative at best.

Cronquist, A. 1984. *Gayophytum*. Intermountain Flora Project, unpublished manuscript.
Lewis, H. and J. Szweykowski. 1964. The genus *Gayophytum* (Onagraceae). Brittonia 16: 343–391.

1. Flowers and fruits sessile or the pedicels less than 2 mm long; fruit erect 2
— Flowers and fruits evidently pedicellate, the pedicels more than 2 mm long; fruits erect to deflexed 4
2(1). Capsules strongly dorsiventrally compressed, much broader than thick; dorsal and ventral valves of capsule remaining attached to the septum at maturity, the lateral ones free; plants rare .... *G. humile*
— Capsules quadrangular (and subterete) or somewhat dorsiventrally compressed, seldom much broader than thick; valves of capsule all free from the septum; plants uncommon to common 3
3(2). Plants branched only in the lower half; secondary branches few or none; capsules somewhat broader than thick .... *G. racemosum*
— Plants branched throughout or at least in the upper half; capsules about as broad as thick .... *G. decipiens*
4(1). Petals typically 3–6 mm long, the flowers modestly showy .... *G. diffusum*
— Petals 0.5–3 mm long, the flowers inconspicuous 5
5(4). Capsules 3–6 mm long; pedicels often spreading to reflexed; petals 0.5–1.5 mm long; seeds in 2 rows per locule .... *G. ramosissimum*
— Capsules 6–12 mm long; pedicels variously oriented; seeds in 1 row per locule 6
6(5). Lateral branches separated by 2 or more nodes; first flower borne in the lowermost 1–3 (5) nodes.. *G. decipiens*
— Lateral branches borne at adjacent nodes or separated by only 1 node; first flower borne commonly 5 or more nodes above the base .... *G. lasiospermum*

Gayophytum decipiens Lewis & Szweykowski Plants erect, typically branched, usually with 2 or more nodes between the branches, 5–50 cm tall; stems glabrous or with spreading or appressed hairs; leaves 4–20 (30) mm long, 1–2.5 (4) mm wide, linear or essentially so, sessile or with petioles to 5 mm long; flowers subsessile or the pedicels 0.5–2 mm long, the first one borne 1–3 (5) nodes above the base; petals 1–1.8 mm long; anthers ca 0.2 mm long, usually connivent with the stigma; capsules 6–15 mm long, 0.6–0.8 (1) mm thick, constricted between the seeds, opening by 4 subequal valves free from the septum; pedicels 0.3–3 (5) mm long; seeds 8–25, in 1 row per locule, 0.8–1.8 mm long, dark brown. Sagebrush, mountain brush, pinyon-juniper, and spruce-fir communities at 2255 to 2810 m in Beaver, Cache, Duchesne, Garfield, Salt Lake, San Juan, Sevier, Summit, Uintah, Wasatch, and Washington counties; Washington to Montana, south to California, Arizona, and Colorado; 12 (0).

Gayophytum diffusum T. & G. Plants erect, usually branched, mainly 1–5 dm tall; stems glabrous to strigose, puberulent, or hirtellous; leaves 8–50 mm long, 1–4 mm wide, linear; flowers showy (for the genus), pedicellate, the lowest ones commonly 5 nodes above the base; sepals 2.5–5 mm long; petals 3–6 mm long; capsules 5–12 mm long, 0.8–1 mm thick, constricted between the seeds; pedicels 3–11 mm long; seeds 3–18, in 1 row per locule, 1–1.6 mm long; 2n = 28. Sagebrush and mountain brush communities reported in Cache, Juab, Summit, and Utah counties; Washington to Wyoming, south to California and Nevada; 0 (0).

Gayophytum humile **A. L. Juss.** [*G. nuttallii* T. & G.]. Plants erect, typically branched in the lower half, mainly 5–30 cm tall; stems glabrous or merely glandular; leaves 10–25 mm long, 1–3 mm wide, linear to oblong; flowers sessile or nearly so, the first one borne 1–3 nodes from the base; petals 0.5–1.5 mm long; anthers 0.2–0.3 mm long, connivent with the stigma; capsules 8–17 mm long, 1–2 mm broad, much-flattened, the dorsal and ventral valves twice as wide as the lateral ones and remaining attached to the septum; pedicels less than 0.5 mm long; seeds 24–50, in 1 row per locule, 0.7–1.1 mm long. Reported for Cache County by Albee (unpublished map 1985) and Wasatch County by Cronquist (1984); Washington to Montana, south to California and Wyoming; 0 (0).

Gayophytum lasiospermum **Greene** [*G. nuttallii* authors, not Hook.; *G. intermedium* Rydb.; *G. diffusum* ssp. *parviflorum* Lewis & Szweykowski]. Plants erect, typically branched, usually at consecutive nodes, mainly 5–50 cm tall; stems glabrous to strigulose, puberulent, or hirtellous; leaves 4–50 mm long, 0.8–4 mm wide, linear lanceolate to linear or oblong; flowers pedicellate, the first borne 5–10 nodes above the base; sepals 0.8–2 mm long; petals 1.5–2.5 mm long; anthers minute; capsules 6–13 mm long, 0.8–1 mm thick, constricted between the seeds, opening by 4 subequal valves free from the septum; pedicels 2–10 mm long, ascending to deflexed; seeds 6–16 in 1 row per locule, 1–1.6 mm long. Sagebrush, mountain brush, ponderosa pine, aspen, and spruce-fir communities at 1765 to 3175 m in all Utah counties, except Box Elder, Iron, Juab, Millard, San Juan, and Wayne; British Columbia to Montana and South Dakota, south to California, Arizona, and New Mexico; Mexico; 48 (xi).

Gayophytum racemosum **T. & G.** Plants erect, usually branched in the lower half, unbranched above, mainly 5–30 (4) cm tall; stems glabrous or puberulent to hirtellous; leaves 6–25 mm long, 0.8–3 mm wide, linear to narrowly elliptic; flowers subsessile, usually the first borne 1–3 nodes above the base; petals 1.3–1.8 mm long; anthers 0.2–0.3 mm long, connivent with the stigma; capsules 10–15 mm long, 0.8–1.4 mm wide, flattened, splitting into 4 valves free from the septum, the dorsal and ventral valves somewhat broader than the others; pedicels 0.4–2 mm long; seeds 10–34, with 1 row in each locule, 0.8–1 mm long. Sagebrush, pinyon-juniper, ponderosa pine, Douglas fir, aspen, and lodgepole pine communities at 1830 to 3172 m in Beaver, Cache, Daggett, Duchesne, Juab, Millard, Salt Lake, Summit, Uintah, and Utah counties; Washington to Montana, south to California, Nevada, and Colorado; 16 (0).

Gayophytum ramosissimum **T. & G.** Plants erect, branched usually at every other node; stems glabrous or puberulent; leaves 6–40 mm long, 1–5 mm wide, linear or essentially so; flowers pedicellate, the first ones borne 5–10 or more nodes above the base; petals 0.7–1.5 mm long; anthers conivent with the stigma; capsules 3–6 (9) mm long, ca 1 mm thick, the 4 subequal valves free from the septum; pedicels (3) 5–12 mm long, ascending to deflexed; seeds 10–30, usually in 2 rows per locule, 1–1.5 mm long. Pinyon-juniper, sagebrush, mountain brush, ponderosa pine, aspen, and lodgepole pine communities at 1735 to 2870 m in probably all Utah counties; Washington to Montana, south to California, Nevada, Arizona, and New Mexico; 76 (iii).

Oenothera L.

Annual, biennial, or more commonly perennial herbs; leaves alternate or basal; flowers typically sessile in leaf or bract axils, fragrant, opening in evening (pollinated by moths); hypanthium much surpassing the ovary, deciduous in fruit; sepals 4, reflexed at anthesis; petals 4, white or yellow, fading pink, lavender, orange, bronze, or purple to brownish; stamens 8, dimorphic or subequal, the anthers versatile; stigma 4-lobed; capsules loculidial; seeds lacking a coma.

Cronquist, A. 1984. *Oenothera*. Intermountain Flora Project. Unpublished manuscript.

1.	Plants acaulescent or essentially so	2
—	Plants caulescent	6
2(1).	Petals white when fresh, fading pink to lavender or purple	3
—	Petals yellow when fresh, fading yellow to bronze or pinkish	4
3(2).	Plants annual; petals less than 2.5 cm long; flowers not fragrant; known from San Juan and Washington counties	*Oe. cavernae*
—	Plants perennial; petals mostly more than 2.5 cm long; flowers fragrant; widespread	*Oe. caespitosa*
4(2).	Plants annual (or winter annual), occuring in Washington (also in Emery and Kane?) County; capsules ribbed, but hardly winged	*Oe. primiveris*
—	Plants perennial, seldom of Washington County; capsules winged	5
5(4).	Leaves pubescent, subentire; capsules broadly winged throughout; petals 3–7 cm long	*Oe. howardii*
—	Leaves glabrous or essentially so; capsules winged apically or basally, not throughout; petals 1.5–5 cm long	*Oe. flava*
6(1).	Flowers yellow when fresh, fading orange to bronze or purplish	7
—	Flowers white or pink when fresh, fading pink to lavender	9
7(6).	Petals 1–2 cm long; sepals 1–2 cm long; style exserted from hypanthium ca 1.5 cm	*Oe. biennis*
—	Petals 2.5–5 cm long; sepals 2–4.5 cm long; styles exserted from the hypanthium 2–4 cm	8
8(7).	Hypanthium 6–12 cm long; plants of southeastern and southern Utah	*Oe. longissima*
—	Hypanthium 2.5–5 cm long; plants variously distributed	*Oe. elata*
9(6).	Plants annual, biennial, or perennial, from taproots, not or seldom sod-forming, seldom from creeping rootstocks	10
—	Plants perennial from creeping rootstocks, sod-forming	12
10(9).	Leaves strongly dimorphic, the lower ones subentire to merely toothed or lobed, the cauline ones deeply pinnatifid; capsules erect to steeply ascending, with 2 rows of seeds per locule; plants of southeastern Utah	*Oe. albicaulis*

— Leaves not strongly dimorphic, all entire or all pinnatifid; capsules variously oriented, with 1 row of seeds per locule; plants of other distribution 11

11(10). Leaves commonly subentire; plants of Beaver, Iron (?), and Washington counties .... *Oe. deltoides*

— Leaves lobed to pinnatifid; plants of the Uinta Basin .... *Oe. pallida*

12(9). Petals bright pink when fresh and when dried; plants cultivated and escaping .... *Oe. speciosa*

— Petals white when fresh, fading pink to lavender when dried; plants indigenous 13

13(12). Herbage conspicuously pubescent; petals typically 2.5–3.5 cm long; plants of Beaver, Iron, Millard, and Washington counties .... *Oe. californica*

— Herbage not especially pubescent; petals various; plants of broad or other distribution 14

14(13). Petals 1–1.5 (2) cm long; leaves deeply pinnatifid, the lobes and rachis mainly 2.5 mm wide or less; throat of hypanthium with copious, long, white hairs; plants mostly of upper elevations .... *Oe. coronopifolia*

— Petals typically 1.5–3 cm long; leaves subentire to variously lobed, the lobes and rachis commonly over 3 mm wide; throat of hypanthium glabrous or essentially so; plants of upper to lower elevations . *Oe. pallida*

Oenothera albicaulis Pursh Whitestem Evening-primrose. [*Anogra albicaulis* (Pursh) Britt.]. Annual or winter annual herbs from taproots, mainly 5–25 (30) cm tall; stems simple or branched from the base, not exfoliating; herbage puberulent and sometimes with longer hairs intermixed (especially on the sepals); basal leaves in a rosette, 1–10 cm long, 3–25 mm wide, entire or more or less lobed, but evidently different than the cauline ones, those deeply pinnatifid, both segments and rachis narrow; flowers nodding in bud; hypanthium 1.5–3 cm long; sepals 1.5–2.5 cm long; petals white, fading pink, typically 2–3.5 cm long; anthers 6–10 mm long; style exserted 1.5–2.2 cm beyond the hypanthium; capsules erect-ascending, 2–4 cm long, 3–4 mm thick; seeds ellipsoid, ca 1 mm long. Blackbrush, spiny hopsage, shadscale, galleta-ricegrass, and pinyon-juniper communities at 1250 to 1830 m in Garfield, Grand, Kane, San Juan, and Wayne counties; Montana and South Dakota, south to Arizona, Colorado, Texas, and Mexico; 31 (iv).

Oenothera biennis L. Biennial Evening-primrose. [*Oe. villosa* Thunb.]. Biennial (or annual to perennial) herbs, mainly 3–12 (15) dm tall or more; stems simple or sparingly branched, strigose to hirsute and usually with longer hairs intermixed; basal and lower cauline leaves petiolate, becoming sessile or subsessile upward, mainly 3–20 cm long, 0.6–3.5 cm wide, lanceolate to oblong or oblanceolate, acute to attenuate, entire to undulate or serrate-dentate; flowers few to numerous in terminal, bracteate spikes, self-pollinated; hypanthium 2–4 cm long; sepals 1–2 cm long; petals 1–2 cm long, yellow, often whitish or orange in age; anthers 4–8 cm long; style protruding 0.8–1.5 cm beyond hypanthium; capsules erect-ascending, 1.8–4 cm long, tapering; seeds numerous, in 2 rows per locule, ca 1–1.5 mm long; 2n = 14. Moist, disturbed sites at 1280 to 2350 m in Box Elder, Cache, Daggett, Duchesne, Rich, San Juan, Summit, Uintah, Utah, and Wasatch counties; widespread in the U. S. and Canada; 22 (i).

Oenothera caespitosa Nutt. Morning-lily. Perennial herbs from taproots and simple or branched, sometimes spreading caudices or rootstocks, acaulescent or rarely with stems to 5 dm long; herbage typically puberulent to villous (especially along leaf margins), glandular, or glabrous; leaves typically in a basal rosette, less commonly along a stem (up to 60 cm long) with elongate internodes, mostly 1.5–20 (30) cm long, 0.5–4 cm wide, petiolate, the blades entire, toothed, lobed, or pinnatifid; flowers solitary in leaf axils, sessile or pedicellate, erect or drooping in bud, opening in evening (pollinated by moths), fragrant; hypanthium 3–14 cm long or more, with nectary glands at the base within; sepals 1.5–5.5 cm long, reflexed at anthesis; petals white, fading white to pink or purplish, 2–6 cm long; anthers 6–15 mm long or more; capsules erect-ascending, 2–5 mm long, woody, tuberculate; seeds numerous, in 2 rows per locule, 2.5–3.5 mm long; 2n = 14. This is a most beautiful species, which occurs throughout Utah, except at highest elevations. The morphology is tremendously variable, and some of that variability is geographically or edaphically correlated. Segregation of the units of variability into taxonomic groups is, however, based on morphologically transitional features. The confluence is sufficiently great that only arbitrary separation is possible in the following key, but at least most of the determinations seem to represent taxa and not mere mechanical separation of similar specimens. The following key will serve to segregate most specimens into the proposed infraspecific taxa.

1. Hypanthium mostly (6) 7–14 cm long or more; petals 3–6 cm long, often white to pink in age 2

— Hypanthium mostly 3–8 cm long; petals 2–4 cm long, often fading pink to purple 3

2(1). Leaves merely toothed, at least in large part .... *Oe. caespitosa* var. *macroglottis*

— Leaves more or less pinnatifid; plants common over much of Utah .... *Oe. caespitosa* var. *marginata*

3(1). Plants glabrous or nearly so, or the leaves puberulent marginally with hairs mainly less than 0.5 mm long .. .. *Oe. caespitosa* var. *caespitosa*

— Plants definitely hairy, with hairs mainly 0.5–1 mm long, at least some 4

4(3). Plants more or less villous throughout, tending to be colonial, forming dense mats, confluent with the next, mainly of the western half of Utah .... *Oe. caespitosa* var. *crinita*

— Plants pubescent mainly along leaf margins and on the flowers, seldom colonial or, if so, not forming dense mats, mainly of eastern Utah .... *Oe. caespitosa* var. *navajoensis*

Var. caespitosa [*Oe. montana* Nutt. in T. & G.; *Oe. marginata* var. *purpurea* Wats.; *Pachylophus canescens* Piper]. Cheatgrass, bunchgrass, sagebrush, pinyon-juniper, and mountain brush communities at 1370 to 2015 m in Cache, Daggett, Duchesne, Emery, Juab, Morgan, Sanpete, Sevier, Uintah, and Utah counties; Washington to Saskatchewan, south to Oregon, Nevada, Colorado, and Nebraska; 16 (ii). This taxon is transitional to varieties *marginata* and *navajoensis*.

Var. crinita (Rydb.) Munz [*Pachylophus crinitus* Rydb., type from Rabbit Valley; *Oe. caespitosa* var. *jonesii* Munz, type from Fish Springs]. 2n = 14. Shadscale, horsebrush, pinyon-juniper, mountain brush, sagebrush, ponderosa pine, and bristlecone pine communities at 1400 to 2595 m in Beaver, Box Elder, Davis, Garfield,

Iron, Juab, Kane, Millard, Piute, Sevier, Utah, and Wayne counties; Nevada and California; 44 (vi). The type of this variety was taken at its easternmost known locality, immediately adjacent to the known distribution of var. *navajoensis*. Should the type be interpreted as representing that taxon, the varietal names would have to be realigned. Specimens from the Red Canyon vicinity and from similar fine-textured substrates in adjacent Kane County are very densely hairy, and often compactly colonial. They might constitute still another taxon of about equal validity as those recognized herein.

Var. *macroglottis* (Rydb.) Cronq. [*Pachylophus macroglottis* Rydb.; *Oe. caespitosa* ssp. *macroglottis* (Rydb.) W. L. Wagner, Stockhouse, & Klein]. Igneous substrates in pinyon-juniper and mountain brush communities in the La Sal and Abajo mountains of Grand and San Juan counties; Wyoming south to New Mexico; 2 (0).

Var. *marginata* (Nutt.) Munz [*Oe. marginata* Nutt. ex H. & A.]. Galleta, sagebrush, mountain brush, pinyon-juniper, ponderosa pine, and aspen communities at 1310 to 2900 m in Beaver, Box Elder, Cache, Daggett, Davis, Duchesne, Emery, Garfield, Grand, Juab, Kane, Millard, Salt Lake, San Juan, Sanpete, Sevier, Summit, Tooele, Uintah, Wasatch, Washington, Wayne, and Weber counties; Washington to Wyoming, south to California, Nevada, Arizona, New Mexico, Texas, and Mexico; 94 (iv). This taxon forms the center with which all other phases are involved; it forms intermediates with all of them.

Var. *navajoensis* (W. L. Wagner, Stockhouse, & W. Klein) Cronq. [*Oe. caespitosa* ssp. *navajoensis* W. L. Wagner, Stockhouse, & W. Klein, type from Paria River, Kane Co., Utah]. Shadscale, greasewood, blackbrush, galleta, rabbitbrush, ephedra, sagebrush, pinyon-juniper, and mountain brush communities at 1125 to 2380 m in Carbon, Daggett, Duchesne, Emery, Garfield, Grand, Kane, San Juan, Uintah, and Wayne counties; Colorado, New Mexico, and Arizona; 70 (xvi). Two of the other recognized segregates of *Oe. caespitosa* grow at or near the type locality of var. *navajoensis*; i.e., *crinita* and *marginata*. Distinctive plants of var. *navajoensis* occur mostly to the east of the type locality along the canyons of the Colorado. Plants placed within this variety in the Uinta Basin are often more or less morphologically intermediate to varieties *marginata* and *caespitosa*.

Oenothera californica (Wats.) Wats. Watson Evening-primrose. [*Oe. albicaulis* var. *californica* Wats.; *Oe. pallida* var. *californica* (Wats.) Jepson; *Oe. californica* ssp. *avita* W. Klein, type from near Leeds; *Oe. avita* (W. Klein) W. Klein]. Perennial herbs from spreading rootstocks, acaulescent to strongly caulescent, 5–50 cm tall, simple or branched from near the base; epidermis often exfoliating; herbage puberulent and often with longer spreading hairs intermixed (especially on the calyx); basal rosette developed or not, often withered in caulescent forms; leaves petiolate or sessile upwards, the blades 1.5–12 cm long, 0.4–2 cm wide, irregularly toothed to subentire or pinnatifid; flowers sessile in middle to upper leaf axils, nodding in bud; hypanthium 2–4.5 cm long; sepals 12–25 mm long, reflexed (often in pairs) at anthesis; petals white, drying pink to lavender or white, 2–4 cm long; anthers 7–10 mm long; style extending 1.5–2.5 cm beyond the hypanthium; capsules spreading to spreading-ascending, sessile, 2.5–5 cm long, 2–4 mm thick; seeds in 1 row per locule, 1.5–2.2 mm long. Creosote bush-Joshua tree, blackbrush, other warm desert shrub, mountain brush, sagebrush, pinyon-juniper and ponderosa pine communities at 820 to 2500 m in Beaver, Iron, Millard, and Washington counties; Nevada, California, Arizona, and Mexico; 28 (v).

Oenothera cavernae Munz Munz Evening-primrose. Annual herbs from taproots, acaulescent or essentially so; herbage glandular-pubescent, or with some longer hairs intermixed; leaves all basal, mostly 2–15 cm long, 0.8–2.5 cm wide, oblanceolate in outline, lyrate-pinnatifid, the terminal segment large, usually with numerous, much smaller, lateral segments; flowers sessile, opening in evening, not fragrant; hypanthium prolonged 2–4 cm beyond the capsule; sepals 5–12 mm long, often reflexed in pairs; petals white, fading white or pinkish, 0.6–2.5 cm long; anthers 2–6 mm long; capsules lance-ovoid or ellipsoid, 1.5–3 cm long, tuberculate; seeds in 2 rows per locule, 2.5–3 mm long. Warm desert shrub communities in San Juan and Washington counties (Cronquist 1984); Arizona and Nevada; 0 (0).

Oenothera coronopifolia T. & G. Rootstock Evening-primrose. Perennial herbs from a spreading rootstock, mainly 1–6 dm tall; stems simple or branched at ground level, the epidermis not exfoliating; herbage strigose to puberulent and often with some longer hairs intermixed; leaves mainly cauline, 1–7 cm long, 0.8–1.5 cm wide, deeply pinnatifid and with narrow segments or the lower ones merely toothed or lobed; fascicled axillary leaves often present; flowers solitary in upper axils, nodding in bud; hypanthium 1–2.5 cm long, the throat with long white hairs; sepals 1–2.2 cm long, separately reflexed at anthesis; petals white, fading pink or white, 1–2.2 cm long; anthers 4–7 mm long; style exserted 6–18 mm beyond the hypanthium; fruits erect or erect-ascending, 1–2.5 cm long, 3–5 mm thick; seeds in 2 rows per locule, ca 2 mm long. Sagebrush, mountain brush, pinyon-juniper, ponderosa pine, Douglas fir, lodgepole pine and aspen communities at 2135 to 2810 m in Duchesne, Garfield, Grand, Juab, Millard, Piute, Rich, San Juan, Sevier, Summit, Tooele, Uintah, Wasatch, and Wayne counties; Idaho to South Dakota, south to Arizona and New Mexico; 42 (ii).

Oenothera deltoides Torr. & Frem. Annual Evening-primrose. Annual herbs from taproots, caulescent to subacaulescent; stems 0.5–3.5 dm tall, simple or more commonly branched from the base, the epidermis typically exfoliating, glabrous or short-hairy; leaves mainly cauline, puberulent to glabrous, 2–15 cm long, 3–25 mm wide, the blades oblanceolate to elliptic or oblong, subentire, toothed, or lyrate-pinnatifid, long-petioled below, becoming subsessile upward; flowers sessile in middle to upper leaf axils, tending to nod in bud; hypanthium 2–3.5 cm long; sepals 1–2.5 cm long, reflexed separately at anthesis; petals white, fading white (cream) or pink, 1.5–4 cm long; anthers 5–10 mm long; style exserted 12–25 mm beyond the hypanthium; capsules spreading to spreading-ascending, often curved, sessile, 2–5 cm long, 2–5 mm thick; seeds in 1 row per locule; 1.5–2 mm long. Two varieties are present in southwestern Utah.

1. Leaves green or yellow green, the blades subentire to serrulate-denticulate; plants commonly of low elevation sandy sites .... *Oe. deltoides* var. *decumbens*

— Leaves gray greeen, the blades, or some of them, more or less lyrate-pinnatifid; plants mainly of shadscale and pinyon-juniper communites at higher elevations ......
.... *Oe. deltoides* var. *deltoides*

Var. *decumbens* (Parry) Munz [*Oe. albicaulis* var. (?) *decumbens* Parry, type from near St. George; *Oe. ambigua* Wats., type from near St. George; *Oe. deltoides* var. *ambigua* (Wats.) Munz; *Oe. deltoides* ssp. *ambigua* (Wats.) W. Klein]. Creosote bush and blackbrush communities, mainly in sand, at 760 to 915 m in Washington County; Nevada and Arizona; 4 (0).

Var. *deltoides* [*Anogra deltoides* (Torr. & Frem.) Small]. Shadscale and pinyon-juniper communities at ca 1370 to 1560 m in Beaver, Iron (?), and Washington counties; Nevada, Arizona, California, and Mexico; 3 (0).

Oenothera elata H.B.K. Hooker Evening-primrose. [*Oe. hookeri* T. & G.; *Oe. hookeri* var. *angustifolia* Gates, type from S of Thistle, Utah County]. Biennial (rarely perennial) herbs from taproots; stems mainly 3–15 dm tall, simple or branched; herbage hirsute to puberulent or with long hairs intermixed; leaves mainly cauline, the lower ones petiolate, becoming sessile or subsessile upward, 3–30 cm long, 0.5–3.8 cm wide; flowers sessile or subsessile in upper leaf axils, erect or nodding in bud; hypanthium 2.5–5 cm long; sepals (2) 2.5–4 cm long; petals 2.5–4.5 cm long, yellow, fading yellow to orange or purple; anthers 7–12 mm long; style surpassing the hypanthium by 2–3.5 cm; capsules 2–4.5 cm long, erect, 4–6 mm thick; seeds in 2 rows per locule, ca 1.5 mm long. Typically in riparian, riverine, or palustrine sites in sagebrush, rabbitbrush, pinyon-juniper, ponderosa pine, lodgepole pine, aspen, and spruce-fir communities at 1370 to 2840 m in most Utah counties; Washington to Montana, south to Panama; 69 (iv). The flowers are strikingly beautiful, but the plants tend to be coarse and open.

Oenothera flava (A. Nels.) Garrett Yellow Evening-primrose. Perennial acaulescent herbs from a taproot and simple or branched caudex, sometimes spreading by rootstocks; leaves basal, 5–30 cm long, 0.3–6 cm wide, subentire or more commonly toothed or pinnatifid, puberulent to subglabrous; flowers solitary in leaf axils; hypanthium 3–10 cm long or more; sepals 1–4 cm long, reflexed as a unit and with red margins; petals 1–5 cm long, yellow, fading red orange to bronze or pink; capsules 1–3 cm long, evidently winged above the middle, the wings 1–5 mm wide; seeds in 2 rows per locule, 1–2.6 mm long. Two rather distinctive varieties are present.

1. Flowers large and showy, the petals 2.8–5 cm long; plants of Daggett and Uintah counties .. *Oe. flava* var. *acutissima*
— Flowers smaller, but still more or less showy, the petals 1–3 cm long; plants widespread .... *Oe. flava* var. *flava*

Var. *acutissima* (W. L. Wagner) Welsh [*Oe. acutissima* W. L. Wagner, type from Flaming Gorge vicinity]. n = 7. Sagebrush, grass-forb, and ponderosa pine communities, often where temporarily moist in spring and early summer, at 1830 to 2380 m in Daggett and Uintah counties; Colorado; 9 (i).

Var. *flava* [*Lavauxia flava* A. Nels.; *Oe. triloba* var. *ecristata* Jones, type from Kanab]. Palustrine, riparian, and moist meadow habitats in sagebrush, ponderosa pine, aspen, lodgepole pine, and spruce-fir communities at 1250 to 3355 m in Beaver, Cache, Daggett, Duchesne, Emery, Garfield, Grand, Iron, Juab, Millard, Piute, San Juan, Sanpete, Sevier, Summit, Wasatch, Washington, and Wayne counties; Washington to Saskatchewan, south to California, Arizona, New Mexico, and Texas; 66 (ix).

Oenothera howardii (A. Nels.) W. L. Wagner Bronze Evening-primrose. [*Lavauxia howardii* A. Nels., type from Dugway, Utah; *Oe. brachycarpa* (Gray) Britt., misapplied]. Perennial, acaulescent herbs from a taproot and simple or branched caudex; leaves 2–20 cm long, 0.6–3 cm wide, the blades oblanceolate to elliptic, entire or repandly toothed, petiolate; herbage puberulent to glabrous; flowers solitary in leaf axils, large and showy; hypanthium 5–10 cm long or more; sepals 3–7 cm long, usually reflexed as a unit; petals 3–7 cm long, yellow, fading bronze, yellow, maroon, or red; capsules 3–3.5 cm long, 1–2 cm thick, broadly winged; seeds in 1 row per locule (or 2 rowed near the base), mostly 2–5 mm long. Shadscale, other salt-desert shrub, sagebrush, pinyon-juniper, and ponderosa pine communities at 1525 to 2410 m in Beaver, Daggett, Duchesne, Emery, Garfield, Juab, Millard, San Juan, Sevier, Tooele, Uintah, Utah, Washington, and Wayne counties; Nevada, Colorado, and Kansas; 43 (vi).

Oenothera longissima Rydb. Bridges Evening-primrose. Perennial (or biennial) herbs from taproots; stems (2) 4–12 dm tall or more, simple or branched, puberulent and often with longer hairs intermixed; leaves basal and cauline, mostly 2–25 (35) cm long, 0.3–4.5 cm wide, lanceolate to oblanceolate, oblong, or elliptic, entire to undulate-serrate, petiolate or the upper sessile; flowers solitary, sessile, in upper leaf axils; hypanthium 6–12 cm long; sepals 1.4–4.6 (5.5) cm long, reflexed separately, in pairs, or all together at anthesis; petals yellow, fading yellow to purple or orange, 2.5–5 cm long; anthers 9–18 mm long; style protruding 2–5.5 cm beyond the hypanthium; capsules erect, 2–4.5 cm long, 4–7 mm thick; seeds in 2 rows per locule, ca 1–1.5 mm long. Riparian and hanging garden communities at 790 to 1925 m in Garfield, Grand, Kane, San Juan (type from Natural Bridges National Monument), Washington, and Wayne counties; Colorado, Arizona, Nevada, and California; 29 (ix).

Oenothera pallida Lindl. Pale Evening-primrose. [*Anogra pallida* (Lindl.) Britt.; *Oe. pallida* var. *leptophylla* T. & G.; *Oe. albicaulis* var. *runcinata* Engelm.; *Oe. pallida* var. *latifolia* Rydb.; *Oe. trichocalyx* Nutt. in T. & G.; *Oe. pallida* ssp. *trichocalyx* (Nutt.) Munz & W. Klein]. Perennial or annual to biennial herbs from rootstocks or taproots, a caudex often more or less developed; stems mainly 1–7 dm tall, usually with exfoliating epidermis, glabrous or puberulent; leaves 1–8 cm long, 1–20 mm wide, oblanceolate, elliptic, oblong, linear, lanceolate, or ovate, entire, toothed, lobed, or pinnatifid, glabrous to puberulent or villosulose; flowers sessile, solitary in leaf axils, nodding in bud; hypanthium 1.5–3.5 cm long, the mouth lacking hairs within; sepals 7–25 mm long, typically reflexed together or in pairs, puberulent or with spreading, long hairs intermixed; petals white, fading pink to cream or lavender, 1–3 cm long; style exserted 1–2.2 cm beyond the hypanthium; capsules spreading, straight, curved, or S-shaped, 1.5–5 cm long, 1.5–3 mm thick; seeds in 1 row per locule, 1.4–2.2 mm long. Shadscale-greasewood, mixed warm desert shrub, blackbrush, sagebrush-rabbitbrush, pinyon-juniper, mountain brush, and ponderosa pine communities at 1155 to 2505 m in probably all Utah counties; Washington to South Dakota, south to Nevada, Arizona, New Mexico, and Texas; 227 (xxxiv). Our material has received varying treatments in the past several decades. It has been regarded as belonging to *Oe. pallida* and *Oe. trichocalyx*, or when these were combined, as an expanded *Oe. pallida* with two subspecies. The subspecies *pallida* has been further proposed for segregation into three varieties. Careful examination of our abundant material has not demonstrated consistent diagnostic criteria for segregation of any of the proposed infraspecific categories; either the criteria occur haphazardly, resulting in sympatry of the proposed entities, or the plants cannot be separated except arbitrarily. Some of the plants in the Uinta Basin simulate *Oe. albicaulis* in being annual and having a distinctive basal rosette, but others from the Basin are perennial from spreading rootstocks. Indeed, the phalanx of specimens from the Uinta Basin contains almost every variation found within the specimens from elsewhere in Utah. There does not seem to be any practical way to recognize the annual plants (as ssp. *trichocalyx*), which sometimes have long, spreading hairs on the calyx, without regarding the remainder of the material within the Basin as ssp. *pallida*. Specimens from elsewhere are annual and have long hairy calyces also.

Oenothera primiveris Gray Early Evening-primrose. [*Oe. johnsonii* Parry, type from near St. George]. Annual or winter annual herbs from taproots, acaulescent or essentially so; herbage with spreading, long, white hairs produced above a shorter layer of hairs; leaves 2–15 cm long (or more), 0.4–4 cm wide, oblanceolate in outline, petiolate, the blades entire to deeply once or twice pinnatifid; flowers solitary, sessile in leaf axils, erect in bud; hypanthium 4–8 cm long; sepals 20–32 mm long, reflexed separately, in pairs, or wholly at anthesis; petals yellow, fading yellow or pink to purplish, 2–4 cm long; anthers 7–10 mm long; style exserted 1.5–2.5 cm beyond the hypanthium; capsules 2–4.5 cm long; seeds in 2 rows per locule, 2.5–3 mm long; 2n = 14. Joshua tree, creosote bush, blackbrush, and pinyon-juniper communities at 750 to 1500 m in Emery (?), Kane (?), and Washington counties; New Mexico and Texas to California and Mexico; 18 (v). Our material has been designated as **var. bufonis** (**Jones**) Cronq. [*Oe. bufonis* Jones]. This species has been collected in full flower as early as 29 January.

Oenothera speciosa Nutt. Nuttall Evening-primrose. Perennial caulescent herbs from spreading rootstocks; stems mainly 2–5 dm tall, erect or sprawling; herbage strigose to puberulent; leaves 0.9–6 cm long, 2–12 mm wide, lanceolate to elliptic or oblanceolate; flowers solitary in axils of upper bracteate leaves, nodding in bud; hypanthium 10–15 mm long; sepals 15–20 mm long, reflexed in pairs or wholly at anthesis; petals 2.5–4 cm long, bright pink, fading pink; anthers 6–8 mm long; style extending 15–20 mm beyond the hypanthium; capsules 1–2 cm long. Cultivated ornamental, persisting and escaping in Utah and Washington counties; indigenous from Kansas to Missouri and Texas; 2 (0).

Zauschneria Presl

Perennial herbs from a caudex, often woody at the base; leaves cauline, opposite or alternate above; flowers scarlet, showy, somewhat zygomorphic, pollinated by hummingbirds; hypanthium produced much beyond the ovary; sepals 4, erect or nearly so; petals 4, notched apically; stamens 8; style exserted; stigma 4–lobed; fruit a capsule; seeds with a coma.

Zauschneria latifolia (**Hook.**) **Greene** Hummingbird Flower. [*Z. californica* var. *latifolia* Hook.]. Plants clump-forming, 1.5–4 (6) dm tall; herbage glandular-puberulent to stipitate-glandular and commonly with some longer white hairs intermixed, or subglabrous; leaves 0.8–4 cm long, 2–15 mm wide, lanceolate to elliptic or ovate, serrate, cuspidate apically; flowers sessile or subsessile in terminal, bracteate, spicate racemes; hypanthium 15–28 mm long, scarlet; sepals 6–12 mm long, scarlet; petals 10–15 mm long, scarlet; capsules 18–25 mm long, puberulent; seeds many, ca 1.5 mm long; 2n = 30. Rock outcrops and talus slopes in juniper, mountain brush, ponderosa pine, white fir, Douglas fir, and aspen communities at 1675 to 3070 m in Box Elder, Cache, Davis, Juab, Kane, Millard, Morgan, Salt Lake, San Juan, Sanpete, Sevier, Tooele, Utah, Wasatch, Washington, and Weber counties; Idaho and Wyoming to California, Nevada, and Arizona; 46 (v). Our material belongs to **var. garrettii** (**A. Nels.**) **Hilend** [*Z. garrettii* A. Nels, type from Big Cottonwood Canyon, Salt Lake County; *Epilobium canum* ssp. *garrettii* (A. Nels.) Raven]. I follow tradition in maintaining the genus *Zauschneria* apart from the closely allied *Epilobium*, to which it is related through *E. nevadense*, of section Cordylophorum. The bright red, hummingbird-pollinated, slightly zygomorphic flowers provide the basis for this decision.

OROBANCHACEAE Vent.

Broomrape Family

Herbaceous perennials, lacking chlorophyll, and parasitic on the roots of other green plants; stems fleshy; leaves reduced and scalelike, alternate; flowers mostly perfect, rarely dioecious, irregular, in spikelike or racemose clusters; sepals united, the 4 or 5 lobes subequal; petals united, irregular, the 5 lobes oblique, persistent; stamens 4, didynamous, included in the corolla tube and adnate to it; pistil 1, the ovary 1–loculed, with 4 parietal placenta; style 1; stigma peltate; fruit a 2–valved capsule; seeds many; x = 12, 18–21.

Orobanche L.

Plants commonly parasitic on species of *Artemisia*, glandular-pubescent throughout; stems bracteate, purplish, brownish, or yellowish white; flowers solitary or clustered in terminal spikes; calyx equally 5–lobed; corolla purplish, reddish, or yellowish, the tube slightly curved, the limb bilabiate; capsule 1–loculed, with 4 placentae.

1. Flowers without bracts, solitary or several on a stem, borne on long slender pedicels 2
— Flowers with 2 (or 3) bracts, sessile or pedicellate; inflorescence a spike, corymb, or panicle 3

2(1). Stems slender, very short, bearing 1 or sometimes 2 or 3 flowers; pedicels many times longer than the stem; calyx lobes awl-shaped, with a long tip, longer than the tube .... *O. uniflora*
— Stems stout, each bearing 3–12 pedicels, these subequal to the stem or shorter; calyx lobes triangular, shorter than the tube .... *O. fasciculata*

3(1). Corolla lobes rounded, obtuse, the upper ones erect; corolla tube much paler than the lobes; styles persistent .... *O. multiflora*
— Corolla lobes triangular, acute, the upper ones erect or reflexed; corolla tube not markedly paler than the lobes; style deciduous or persistent 4

4(3). Inflorescence compactly corymbose; anthers woolly; calyx lobes 9–16 mm long; plants to 10 cm tall ....
.... *O. corymbosa*

— Inflorescence an elongate spike or raceme, the axis much elongating in age; anthers woolly to glabrous; calyx lobes 5–20 (12) mm long; plants 10–40 cm tall ..
.... *O. ludoviciana*

Orobanche corymbosa (Rydb.) Ferris [*Myzorhiza corymbosa* Rydb.]. Fleshy yellowish or purplish herbs; stems mostly below ground level, 5–13 cm tall, glabrous to glandular-puberulent; inflorescence corymbose, dense, broad, 2.5–6 cm long; pedicels 4–15 (30) mm long; bract 1, with 1 or 2 smaller bractlets above; calyx 12–20 mm long, the tube 2–7 mm long, the lobes 9–16 mm long, linear; corolla 18–25 mm long, pale purplish, streaked with pink lines and mottled with yellow near the base of the lower lobes, these 3–6 mm long, the lower ones spreading; anthers woolly. Shadscale, greasewood, sagebrush, pinyon-juniper, mountain brush, and mixed conifer communities at 1200 to 2670 m in Beaver, Duchesne, Garfield, Juab, Millard, Salt Lake, Sevier, Tooele, Uintah, Utah, and Washington counties; Washington to Wyoming, south to California and Nevada; 26 (vi).

Orobanche fasciculata Nutt. Cluster Cancerroot. Fleshy purplish or yellowish herbs; stems forked, mostly below ground level, extending above ground for 5–10 cm, glandular pubescent; inflorescence of 3–12 erect, crowded, naked, axillary pedicels to 12 cm long; bractlets near the calyx lacking; calyx 7–10 mm long, campanulate, the lobes 3–5 mm long, triangular, shorter than the tube; corolla purple to pinkish or yellow, the tube curved, 15–30 mm long, the lobes rounded, spreading, 2–5 mm long; anthers glabrous to woolly. Warm desert shrub, saltbush, greasewood, sagebrush, pinyon-juniper, mountain brush, aspen, and fir communities at 1230 to 3260 m in all Utah counties; Yukon Territory, south to California and Mexico, east to Michigan and Texas; 85 (xxv). The material with yellow flowers has been designated as **var. lutea** (Parry) Achey (*Phelipaea lutea* Parry), but its distribution is apparently random, indicating a minor, reoccurring genetic trait.

Orobanche ludoviciana Nutt. Fleshy yellowish or purplish herbs; stems mostly simple, 10–30 (40) cm tall, viscid pubescent to nearly glabrous below; inflorescence spicate or racemose, elongating in age, to 20 cm long; pedicels 0–15 mm long; bractlets adjacent to the calyx 1 or 2, calyx 7–15 mm long, the lobes lance-linear, somewhat unequal in length, 5–12 mm long; corolla dull purple, the throat pale, 17–28 mm long, the lobes triangular-lanceolate, acute or rounded, 3–7 mm long; anthers glabrous to woolly pubescent. Records indicate that this plant parasitizes species of *Artemisia*, *Quercus*, *Gutierrezia*, and probably others. Two rather weak varieties occur in Utah.

1. Corolla lobes rounded or obtuse; calyx lobes 9–12 mm long .... *O. ludoviciana* var. *araneosa*

— Corolla lobes acute; calyx lobes 5–8 mm long ....
.... *O. ludoviciana* var. *cooperi*

Var. arenosa (Suksd.) Cronq. [*O. multiflora* var. *araneosa* (Suksd.) Munz]. Creosote bush, bursage, saltbush, blackbrush, sagebrush, pinyon-juniper, and mountain brush communities at 760 to 2800 m in Emery, Garfield, Grand, Kane, San Juan, Uintah, Utah, and Washington counties; Colorado to Washington, California, and Arizona; 20 (vii).

Var. cooperi (Gray) Beck [*Aphyllon cooperi* Gray; *O. cooperi* (Gray) Heller]. Creosote bush, bursage, blackbrush, Mormon tea, and sagebrush communities below 1330 m in Kane, San Juan, and Washington counties; Nevada to Texas, south to Sonora and Baja; 3 (i).

Orobanche multiflora Nutt. Plants commonly stout, mainly 1–3 dm tall, viscid-pubescent, often grayish, sometimes branched; cauline scales 5–12 mm long; inflorescence dense, corymbosoid or spicate, often branched, purplish; flowers sessile or short-pedicellate; bracteoles adnate or adjacent to calyx; calyx 8–17 mm long, the lobes linear to lance-attenuate; corolla 15–35 mm long, the lips 5–12 mm long, pale purple to yellow, the upper lip erect, the tube pale purple to yellow or whitish, the lobes rounded; anthers woolly or glabrous; stigma peltate; capsule subequal to the calyx. Sandy sites in blackbrush and riparian communities at 1125 to 1345 m in Grand and San Juan counties; Washington to Wyoming, south to California, Texas, and Mexico; 2 (0). The two specimens tentatively assigned here have the technical features of the species. They have been mistaken for the similar *O. ludoviciana*. More work is indicated.

Orobanche uniflora L. One-flower Cancerroot. Small pale, yellowish herbs; stems mostly below ground level, very short and scaly, 0.5–5 cm long, finely glandular puberulent; pedicels 1–3 per stem, mostly 3–10 cm long, much longer than the stem, bractless; calyx 6–12 mm long, the lobes narrowly triangular-lanceolate, 4–9 mm long; corolla curved, 20–25 mm long, whitish, with 2 yellow, bearded folds in the throat, and sometimes with pale purplish lines, the lobes rounded, 2–7 mm long; anthers glabrous to woolly. Sagebrush, pinyon-juniper, mountain brush, ponderosa pine, aspen, and spruce-fir communities at 1870 to 3200 m in Beaver, Cache, Davis, Duchesne, Iron, Millard, Rich, Salt Lake, San Juan, Summit, Tooele, Uintah, and Utah counties; Alaska to Newfoundland, south to California and Florida; 8 (i).

OXALIDACEAE R. Br. in Tuckey

Woodsorrel Family

Herbs with sour juice; leaves palmately 3-foliolate, alternate or basal; flowers in cymose or umbellate inflorescence, or solitary on axillary peduncles; flowers perfect, regular; sepals 5; petals 5; stamens 10, united at the base; pistil 1, the ovary superior, 5-loculed, with 5 styles; fruit a capsule; $x = 5–12$.

Oxalis L.

There is a single genus in Utah with characteristics noted for the family.

1. Stipules present; fruiting pedicels deflexed; hairs straight, simple .... *O. corniculata*

— Stipules lacking; fruiting pedicels spreading to erect; hairs contorted, moniliform .... *O. dillenii*

Oxalis corniculata L. Creeping Woodsorrel. Perennial herbs from taproots, caulescent, often rooting at the nodes; stems mainly 0.5–5 dm long, procumbent to ascending or erect, pubescent with stiff, straight, appressed to spreading simple hairs; stipules present; leaflets obcordate, 0.5–2 cm long; flowers 1–7, umbellate, on axillary peduncles; sepals 5, elliptic; petals 5, yellow, 4–8 mm long; capsules erect on pedicels that deflex at maturity; seeds reddish brown with transverse ridges. Ornamental and weedy species of gardens, lawns, and greenhouses at 850 to 1405 m in Salt Lake, Utah, and Washington counties; widespread in the U. S.; native to Europe; 5 (0).

Oxalis dillenii Jacq. [*O. stricta* authors, not L.] Erect Woodsorrel. Perennial herbs from fibrous roots, caulescent, not rooting at the nodes; stems mainly 1–3 dm long, erect or eventually decumbent, pubescent with ascending incurved hairs and with some long, contorted, moniliform ones; leaflets obcordate, 0.4–2 cm long; flowers 2–7, umbellate on axillary peduncles; sepals 5, oblong; petals 5, yellow, 5–10 mm long; capsules erect, the pedicels not deflexed at maturity; seeds reddish brown, with transverse ridges. Cultivated ornamental and weedy species at ca 1300 to 1405 m in Cache and Utah counties; widespread in Eastern U. S.; 3 (0).

PAEONIACEAE Rudolphi
Peony Family

Robust perennial herbs; leaves alternate or mostly basal, large, ternately compound, the leaflets again lobed, divided, or toothed, estipulate; flowers large, showy, regular; sepals 5, herbaceous, persistent; petals 5, purplish to red purple; stamens numerous, arising from a fleshy disk; pistils usually 5; fruit a many-seeded follicle; x = 5.

Paeonia L.

There exists a single genus with characteristics of the family.

Paeonia brownii Dougl. ex Hook. Peony. Plants glabrous and glaucous, clump-forming; stems erect or decumbent at the base, 2–6 cm tall; leaves cauline, the ultimate segments 3–10 mm wide; flowers solitary on short leafy, terminal peduncles; sepals unequal, leathery, green or purplish, 1–2 cm long; petals 5, purplish to red purple, deciduous; follicles 3–5 cm long, spreading. Sagebrush and mountain brush communities at ca 1525 to 2135 m in Box Elder County; Washington to Wyoming and Nevada; 2 (0). Many ornamental cultivars are grown in Utah, where they persist for many years. Mainly they are derivatives of **P. lactiflora** Pallas. The genus *Paeonia* is sometimes placed within the Ranunculaceae.

PAPAVERACEAE A. L. Juss.
Poppy Family

Annual or perennial herbs, usually with milky juice; leaves alternate, opposite, or basal, entire, lobed, or divided; flowers solitary or few to several in cymes or racemes; sepals 2 or 3, caducous; petals 4–12, white, yellow, rose, purple, or other colors; stamens few to numerous, the filaments distinct; pistil 1, the ovary superior, unilocular or many loculed by intrusion of false septae; stigmas usually sessile, of the same number as the carpels; x = 5, 6, 7, 8–11, or 19.

1. Leaves opposite, or appearing basal, entire; plants scapose, of Washington County .... *Platystemon*
— Leaves alternate, variously lobed or toothed; plants caulescent or scapose, of various distribution 2
2(1). Herbage, sepals, and capsules armed with spines; leaves cauline; plants coarse and thistlelike; sepals 3 ..
.... *Argemone*
— Herbage pubescent to glabrous, not armed with spines; leaves cauline or basal; plants not thistlelike; sepals usually 2 3
3(2). Leaves mainly 3-toothed or -lobed apically, the lobes each with a slender spinulose tip, long hairy; flowers white, showy; plants of gypsiferous substrates in Washington County .... *Arctomecon*
— Leaves variously lobed or toothed but not as above, sometimes hispid-hairy, but the plants then otherwise; distribution various 4
4(3). Leaves multifid into linear or oblong segments; plants indigenous mainly in Washington County, cultivated elsewhere .... *Eschscholzia*
— Leaves broadly lobed; flowers variously colored; distribution various but, if indigenous, of the Uinta Mts 5
5(4). Leaves broadly pinnately lobed; flowers yellow, less than 1 cm long, arranged in umbels; plants adventive .
.... *Chelodonium*
— Leaves simple, toothed or variously pinnately lobed; flowers variously colored, often over 1 cm long; plants cultivated, adventive, or indigenous 6
6(5). Capsule linear-attenuate; petals brick red; leaves cauline; plants adventive .... *Roemeria*
— Capsule broader than long or about as broad as long; leaves cauline or basal; plants adventive or indigenous
.... *Papaver*

Arctomecon Torr. & Frem.

Perennial herbs from a caudex and long taproot; leaves cauline and basal, long-hirsute, mostly toothed apically and the teeth spinulose-tipped; flowers large, showy, solitary, or borne in leafy-bracteate cymes, nodding in bud; sepals 2 or 3; petals 4–6; stamens many; styles united, short, the stigmas connate, 3–6; capsule ovoid, 3- to 6-valved; seeds several.

Arctomecon humilis Cov. Low Bearclaw-poppy. Plants forming rounded clumps, 10–25 cm tall, from a branching caudex, this clothed with brown to ash colored marcescent leaf bases; leaves 0.5–8 cm long and 4–16 mm wide, cuneate to obovate, usually 3- or 4-toothed or lobed apically, glaucous and hirtellous as well as sparing long pilose; peduncles mainly 2–9 cm long; sepals 8–12 mm long, glabrous; petals 4–6, white, oval to obovate, mainly 2–4 cm long; capsules obovoid, about as broad as long; seeds shiny black, 2.5–3 mm long, ca 1.2 mm thick, conspicuously arilate. Moenkopi Formation in mixed warm desert shrub communities at ca 790 to 915 m in Washington (type from near St. George) County; endemic (?); 15 (ii). This edaphically restricted endemic belongs to a genus consisting of only two other species; *A. californica* Torr. & Frem., of southern Nevada, and *A. merriamii* Cov., of southern Nevada and adjacent California. All are rare, or moderately so, and thus are subject to endangerment by industrial development and by other impacts of mankind. As irreplacable portions of our natural heritage they should be regarded as national prizes, as jewels of great price, and protected for future generations, whose advocacy this generation must represent.

Argemone L.

Perennial, robust, thistlelike herbs with yellow or orange latex; stems glaucous, 1 or few, erect or ascending, branched or simple, prickly; leaves sessile, alternate, cauline, oblanceolate to obovate in outline, pinnately lobed or pinnatifid, prickly; sepals caducous, 3 (or more), horned apically and prickly; petals large, white, usually 6; stamens numerous, the filaments filiform; anthers linear, coiled at maturity; stigma 3- to 7-lobed; capsules 3- to 7-carpellate, unilocular, ellipsoid to ovoid, glabrous or spinescent; seeds numerous, subglobose.

Ownbey, G. B. 1958. Monograph of the genus *Argemone* for North America and the West Indies. Mem. Torrey Bot. Club 21(1): 1-159.

1. Leaves green, armed with stout even-sized spines, mainly on the veins; sepal horns unarmed; capsules 25-30 mm long, sparingly spiny; plants of eastern Utah .... *A. corymbosa*
— Leaves glaucous, armed with spines on veins and on intervein areas, the spines not even-sized; sepal horns spiny; capsules 35-50 mm long, often armed with numerous spines of different sizes; plants of western and southern Utah .... *A. munita*

Argemone corymbosa Greene San Rafael Pricklypoppy. Plants 2-6 dm tall, moderately branched, green (or pinkish); leaves glaucous or green beneath, green above, 3-16 cm long, oblanceolate, lobed, the sinuses extending one-half or more to the midrib, armed with stout prickles on veins beneath, smooth or sparingly spiny above; sepal horns sometimes flattened; flowers 4-8 cm wide; petals white, the outer ones as broad as long, the inner ones much broader than long; stamens subequal to or shorter than the ovary; stigmas green or purplish; capsules lance-ovoid, 25-35 mm long, armed with stout, even-sized spreading spines; seeds bluish black, ca 2 mm long. Chinle, Morrison, Entrada, and Mancos formations in mixed salt and sandy desert shrub communities at 1155 to 1560 m in Emery, Garfield, Grand, Kane, San Juan, and Wayne counties; Arizona and California; 15 (iv). The specimens from the Colorado Plateau belong to ssp. *arenicola* G. B. **Ownbey** (type from 37.9 mi SW of Green River, Emery County].

Argemone munita Dur. & Hilg. Armed Prickly-poppy. Plants 4-10 dm tall, moderately branched, often purplish; leaves commonly glaucous on both surfaces, typically 3-15 cm long, oblanceolate to obovate, lobed, the lobes commonly broad and rounded in outline, the sinuses shallow or extending more than half way to the midrib, armed with uneven-sized prickles on veins and commonly on intervening areas; sepal horns usually very prickly; flowers 7-12 cm wide; petals white, obovate to cuneate; stamens subequal to the ovary; stigmas purple; capsules ovoid to elliptic, 30-55 mm long, armed with straight or incurved spines, these interspersed with smaller spines or prickles; seeds ca 2.2-2.6 mm long. Creosote bush-Joshua tree, blackbrush, mixed desert shrub, and pinyon-juniper communities at 760 to 2290 m in Beaver, Box Elder, Cache, Davis, Garfield, Juab, Iron, Kane, Millard, Piute, Salt Lake, Sanpete, Sevier, Tooele, Utah, Washington, and Weber counties; Nevada, California, and Arizona, New Mexico, and northern Mexico; 54 (v). Our specimens belong to ssp. *rotundata* (Rydb.) G. B. **Ownbey** [*A. rotundata* Rydb.]. The plants and exudates dry black, giving a drab appearance to the herbage, but the flowers with purple stigmas and yellow stamens are remarkably beautiful, the petals with the appearance of dainty handkerchiefs.

Chelodonuim L.

Biennial herbs with saffron colored juice; leaves alternate, pinnatifid, broadly lobed, the lobes again lobed or crenately toothed; flowers few, borne in umbels; sepals 2, deciduous; petals 4, yellow; stamens numerous, the filaments long and the anthers short; ovary narrowly cylindric, glabrous; stigma 2-lobed; capsules siliquelike, torulose.

Chelodonium majus L. Celandine. Plants mainly 2-8 dm tall; leaves mostly cauline, pinnatifid into 3-9 segments, these obovate to oblong or orbicular in outline and 2-8 cm broad; umbels few-flowered; sepals glabrous; petals 6-10 mm long; anthers 1-1.5 mm long; capsules 3-6 cm long, 1-2 mm wide. Cultivated ornamental, sometimes escaping and persisting in low elevation portions of Utah; sporadic in the U. S.; introduced from Eurasia; 1 (0).

Eschscholzia Cham. in Nees

Annual or winter annual herbs, the juice watery; leaves alternate or mostly basal, glabrous, ternately dissected; flowers yellow or orange (or varicolored), peduncled; receptacle dilated, forming a funnel at the pistil base; sepals 2, united into a cap, this pushed away by the expanding petals; petals 4, rarely more; stamens 12 to many, the filaments short; anthers linear; ovary cylindric, 1-loculed, with 2 placentae; stigma 4- to 6-lobed; capsule 2-valved. Note: The alternative spelling is "*Eschscholtzia*."

Fedde, F. 1909. *Eschscholtzia* Cham. Das Pflanzenreich IV. Fam. 104. Heft 40: 144-202.

1. Receptacle with 2 rims, the inner erect and hyaline, the outer widely spreading (to 1 cm across); plants known in cultivation only, rarely spreading .... *E. californica*
— Receptacle with only an erect hyaline rim, of if with an outer rim, this not well developed; plants indigenous to southwestern Utah 2
2(1). Leaves all in a basal tuft, the plants scapose .... *E. glyptosperma*
— Leaves cauline and basal 3
3(2). Petals 3-8 mm long; capsules 3-6 cm long .... *E. minutiflora*
— Petals 20-35 mm long; capsules 5-11 cm long .... *E. mexicana*

Eschscholzia californica Cham. California Poppy. Annual (or perennial?) herbs; stems branched, glaucous, mainly 2-5 dm tall, erect or spreading; leaves ternately dissected, the blades 2-8 cm long; basal leaves long petiolate; cauline leaves usually well developed; peduncles 3-15 dm long; receptacles with 2 rims, the inner hyaline, erect, the outer spreading and 2-4 mm wide; sheathing sepals 1-4 cm long; petals orange to yellow, 2-67 cm long; capsules 3-10 cm long; 2n = 12. Cultivated ornamental in lower elevation portions of Utah, occasionally persisting; Washington to California; 2 (0). The species is sometimes included in reclamation seeding mixtures, and might be expected anywhere. Plants identified as belonging to this species from among our indigenous representatives almost certainly belong to the closely allied *E. mexicana* (q.v.).

Eschscholzia glyptosperma Greene Scapose California-poppy. Annual herbs, glaucous, glabrous, scapose, 0.6–3 dm tall; leaves forming a basal tuft, the blades 1–3 cm long, finely ternately dissected; buds erect; receptacle with a narrow erect hyaline rim, lacking an outer one; sheathing sepals ovoid to lance-ovoid, 8–12 mm long, acuminate to acute; petals yellow, 10–25 mm long; capsules 4–7 cm long. Creosote bush, Joshua tree, blackbrush, and other warm desert shrub communities at 760 to 980 m in Iron and Washington counties; California, Nevada, and Arizona; 11 (ii).

Eschscholzia mexicana Greene Annual herbs, mainly 1–3.5 dm tall, glabrous and glaucous, caulescent, or less commonly subscapose; leaves basal and cauline (at least some), the basal ones at least twice as long as the cauline; blades 1.5–5 cm long, ternately dissected; receptacle with a narrow erect hyaline rim and a short flaring outer one; sheathing sepals 0.8–1.3 cm long, lance-ovoid, abruptly apiculate; petals yellow to orange or bronze, 15–35 mm long; capsules 5–11 cm long. Creosote bush-Joshua tree community at ca 1000 to 1250 m in Washington County; Arizona, New Mexico, Texas, and Mexico; 4 (ii). This is the most showy of our indigenous species, and evidently it is closely allied to *E. californica*. The plant is uncommon in Washington County, but stains hillsides bright yellow in years when moisture condiditions are adequate.

Eschscholzia minutiflora Wats. [*E. ludens* Greene, type from St. George]. Annual herbs, mainly 0.5–4.5 dm tall, glabrous, glaucous, and caulescent; leaves basal and cauline, the basal ones at least twice longer than the cauline ones; blades mostly 1–3 cm long, receptacle obconic, the outer rim erect, with a hyaline inner margin; sheathing sepals 4–6 mm long, short-apiculate; petals yellow, 3–8 mm long; capsules 3–6 cm long. Creosote bush-Joshua tree, blackbrush, and mixed desert shrub communities at 760 to 1590 m in Millard and Washington counties; Arizona, Nevada, California, and Mexico; 19 (iii).

Papaver L.

Annual, biennial, or perennial herbs with hispid hairy, glabrate, or glabrous herbage; leaves all basal, or cauline and alternate, merely lobed to once or twice pinnatifid; flowers solitary on scapes or on axillary peduncles; nodding in bud, erect or spreading in flower and fruit; capsules oblong to elliptic or cup-shaped, deshicent by a ring of small valves near margin of stigmatic disk; stigmas radiating from the apex.

1. Plants dwarf, scapose perennials; leaves all basal; petals yellow; indigenous in the Uinta Mts. ...... *P. radicatum*
— Plants with cauline leaves, annual or perennial; petals not yellow; adventive or cultivated species, not of the Uinta Mts. 2
2(1). Stem leaves with cordate-clasping bases 3
— Stem leaves petiolate or sessile, but not cordate-clasping 4
3(2). Leaves pinnately lobed .... *P. glaucum*
— Leaves coarsely doubly serrate and undulate-crisped . ... *P. somniferum*
4(2). Stems and peduncles with long, soft, spreading hairs . ... *P. rhoeas*
— Stems and peduncles with coarse, appressed hairs ... . *P. orientale*

Papaver glaucum Boiss. & Hausskn. Tulip Poppy. Plants mainly 2–5 dm tall; herbage glabrous or sparingly appressed hairy, glaucous; cauline leaves cordate-clasping basally, mainly 2–10 cm long, pinnatifid, with ascending lobes again toothed or lobed; flowers 6–10 cm wide; petals scarlet, spotted basally; capsules ca 20 mm long. Widely cultivated ornamentals, escaping and persisting, but hardly established in Juab County (Neese & Goodrich 10410 BRY); introduced from western Asia; 1 (0).

Papaver orientale L. Oriental Poppy. Plants mainly 6–10 dm tall; herbage coarsely hairy, the lower stems and leaves with spreading hairs, the upper stems and peduncles with appressed hairs; cauline leaves petiolate, reduced and becoming sessile upward, mostly 7–45 cm long, pinnatifid, with spreading, serrate, setiferous, oblong to linear-lanceolate lobes; flowers 8–16 cm long; petals orange to scarlet, spotted basally; capsules 20–25 mm long; 2n = 28, 42. Widely cultivated ornamentals, long persisting in Cache, Sanpete, Tooele, and Utah counties; introduced from Eurasia; 4 (0).

Papaver radicatum Rottb. Plants scapose, caespitose, mostly 4–11 cm tall, from a branching caudex, this clothed with marcescent leaf bases; leaves numerous, in a basal tuft, 1–3 cm long, pinnately lobed, the blades 0.8–2.3 cm long, coarsely strigulose; scapes 3–10.5 cm long, with spreading-ascending, blackish hairs; sepals 8–10 mm long, copiously black hairy; petals yellow, 10–13 mm long, more or less persistent; capsule obconic to obovoid, 11–13 mm long, 5–7 mm wide, beset with stiff, black, ascending bristles; 2n = 42, 56. Rockstripes in alpine tundra communites at 3385 to 3905 m in Duchesne, and Summit counties; widespread in arctic portions of North America, extending south to Colorado; circumboreal; 5 (i). Our material has been variously interpreted. It belongs to a complex of problematical phases in need of monographic study on a worldwide basis. A similar dwarf plant from Montana was named at specific rank by Rydberg, and ours seem to be similar or identical and can be known as **var. *pygmaeum*** (**Rydb.**) Welsh [*P. pygmaeum* Rydb.; *P. kluanense* D. Love].

Papaver rhoeas L. Corn Poppy. Plants mainly 3–7 dm tall; herbage pubescent with spreading soft hairs; cauline leaves petiolate, or becoming sessile upward, mostly 2–8 cm long, pinnately lobed, the spreading-ascending lobes again sharply toothed; flowers mostly 4–10 cm wide; petals usually scarlet, spotted basally; capsule glabrous, subglobose, 10–15 mm long. Widely cultivated ornamentals, escaping and persisting in Cache, Utah, and Washington counties; introduced from Eurasia; 2 (0).

Papaver somniferum L. Opium Poppy. Plants commonly 2–10 dm tall; herbage glaucous and glabrous or with some spreading hairs upward; leaves simple, sessile, cordate-clasping, mostly 4–14 cm long, coarsely doubly serrate and more or less undulate-crisped; flowers 6–10 cm wide; petals white to purple; capsule glabrous, subglobose, 2–5 cm long. Previously widely cultivated ornamental and source of commercial poppy seed, now uncommon but still occasionally grown; introduced from Eurasia; 1 (0). This plant is the source of opium and its derivative, morphine, the "principium somniferum."

Platystemon Benth.

Annual, subscapose herbs; leaves opposite or mainly basal, entire; sepals 3, deciduous; petals 6, white or yellowish, deciduous; stamens many, the filaments flattened; anthers linear; ovary of 6 or more united carpels, these separate in fruit, each with several seeds, breaking at maturity into 1-seeded, indehiscent segments; seeds quadrangular, sculptured.

Platystemon californicum Benth. [*P. remotus* Greene, type from near St. George; *P. rigidulus* Greene, type from near St. George; *P. terminii* Fedde, type from Diamond Valley]. Plants mainly 1-2.5 dm tall, many-stemmed, pubescent with long, spreading, soft hairs; leaves mainly basal or near the stem base, opposite, 1.2-5 cm long, lance-linear, sessile or subsessile; flowers solitary, on peduncles mainly 10-20 cm long; sepals 6-10 mm long, villous-pilose with long hairs; petals cream (fading yellow), 8-12 mm long (or more); carpels 10-20 mm long, beaked. Creosote bush, blackbrush, and mixed desert shrub communities at 850 to 1680 m in Washington County; Nevada, California, and Arizona; 4 (0).

Roemeria Medicus

Annual herbs; leaves alternate, petiolate, once to twice pinnatifid; sepals 2, separate, deciduous; petals 4, red (fading violet); stamens many, the filaments black; anthers yellow; ovary with 2-4 united carpels; capsules ellipsoid to attenuate, many times longer than broad.

Roemeria refracta DC. Plants mainly 3-8 dm tall, branching from the base, pubescent with long, spreading, soft hairs; leaves 1-15 cm long, petiolate, pinnatifid or doubly so, the axis sometimes broadly foliar, the ultimate lobes linear, ascending, and again toothed or lobed; flowers solitary, on peduncles 3-25 cm long or more; sepals 15-25 mm long, sparingly long-hairy; petals red (fading violet), with a dark basal spot, 3-4 cm long; capsules 4.5-7 cm long, 3-5 mm thick, glabrous; 2n = 14. Weedy species of fields, roadsides, and canal banks in Box Elder and Cache counties; sparingly established in the U. S.; adventive from Eurasia; 2 (0).

PASSIFLORACEAE A. L. Juss.
Passion-flower Family

Herbaceous or woody vines, climbing by simple tendrils; leaves alternate, simple, stipulate; flowers regular, solitary or fascicled; sepals 5; petals 5; corona present and conspicuous in several radiating series; stamens 5, elevated on an androgynophore; pistil 1, the ovary superior, 3-loculed; styles 3, the stigmas capitate; fruit a unilocular berry; seeds completely enclosed in a fleshy aril; x = 6, 9-11.

Passiflora L.

A single genus, with characters of the family.
Passiflora caerulea L. Blue Passionflower. Vines to 4 m long or more; stems somewhat angled and more or less grooved; leaves deeply 5- to 9-lobed, the lobes entire or merely undulate; stipules ovate, foliose; flowers 5-10 cm wide; sepals and petals white or pinkish, subequal in length; corona filamentous, in 4 rings, all purple at the base, white at the middle, and blue apically; fruit ovoid-ellipsoid, ca 5 cm long, yellow. Cultivated ornamental vine of great beauty, long persisting, in lower elevation portions of Utah (and under glass); introduced from South America; 4 (0). This plant and **P. incarnata** L. (the maypop) have long been grown in Utah under glass, and artificial hybrids have been produced. *P. incarnata* has serrate leaf lobes.

PEDALIACEAE R. Br.
Sesame Family

Coarse herbs with glandular or eglandular trichomes; leaves opposite or the upper sometimes alternate, simple, entire to variously lobed; flowers perfect, racemose or solitary; calyx lobes usually 5, distinct or mostly united; corolla irregular, sympetalous, slightly spurred or saccate at the base, the oblique limb with 5 imbricate lobes; stamens 4, paired, the fifth rudimentary, or stamens 2, with 3 rudimentary ones; pistil 1, the ovary superior, 2-loculed; fruit with a soft exocarp when young, at maturity the hard endocarp splitting apically into 2 lateral, curved beaks; x = 8, 13, 14, 15, 18.

Proboscidea Schmidel

Herbs; leaves opposite or alternate; flowers racemose, terminal, but appearing lateral; bracts small and deciduous or lacking; calyx unequally 5-lobed, cleft to near the base, membranous; corolla 5-lobed, slightly bilabiate, the limb oblique; stamens 4, paired, the fifth a staminodium; ovary 2-loculed; fruit a drupaceous capsule with fleshy exocarp and woody, reticulate, dehiscent endocarp, which gradually tapers and splits into 2 long, stout, incurved beaks.

Proboscidea parviflora (Woot.) Woot. & Standl. [*Martynia parviflora* Woot.]. Annuals, from a stout taproot; stems somewhat fistulose, much branched and ascending or spreading, to 6 dm tall and 2 m broad, viscid-pubescent; leaves cordate, entire to shallowly lobed or undulate, the mature ones 15-25 cm long and almost as broad; petioles subequal to the blades; racemes 2- to 10-flowered; calyx 9-12 mm long at anthesis, slightly accrescent and deciduous at maturity; corolla dirty yellow to pinkish or purplish, with a dark purple spot on the upper side, 24-35 mm long, the limb 15-35 mm broad; fruit at maturity strongly incurved, 15-35 cm long, crested along the concave side at the broadened basal portion. Fields, roadsides, and other open sites or in desert washes in the creosote bush and blackbrush communities at 750 to 1200 m in Washington and Utah counties; Nevada, Arizona, and Mexico; 10 (iii). The young capsules are edible, and very agreeable as pickles, and the seeds are oily and of good flavor. Another species, **P. louisianica** (Miller) Thell. [*Martynia louisiana* Miller], may eventually be found in the southern counties. The racemes are up to 20-flowered, the corolla is 35-40 mm long, and the calyx to ca 20 mm long.

PLANTAGINACEAE A. L. Juss.
Plantain Family

Annual or perennial, acaulescent or short-stemmed herbs; leaves all basal or nearly so; flowers sympetalous, small, perfect or imperfect, regular, borne in bracteate spikes or in heads; sepals 4; corolla scarious, 4-lobed; stamens 4, alternate with the corolla lobes, or only 2; pistil

1, the ovary superior, with 1–4 locules, the carpels 2; style 1; fruit a circumscissile capsule; x = 4–12+.

Plantago L.

Plants from taproots; leaves simple, entire or variously lobed; flowers several to many, inconspicuous, each subtended by a bract; calyx sometimes irregular; corolla scarious or membranous, persistent; stamens included or exserted; fruit included in the calyx or exserted.

1. Leaves linear to filiform, rarely more than 1 cm wide; plants annual 2
— Leaves lanceolate or broader, rarely less than 1 cm broad; plants perennial or biennial 4
2(1). Stamens 2; corolla lobes (in some flowers) erect and closing over the capsule; flowers more or less dioecious or polygamous; leaves linear-filiform, ca 1 mm wide .. .. P. elongata
— Stamens 4; corolla lobes spreading or reflexed; flowers all perfect; leaves linear, over 1 mm wide 3
3(2). Hairs on upper part of scape spreading at right-angles; bracts broadly ovate, with broad scarious margins .... P. insularis
— Hairs on upper part of scape usually closely ascending or appressed; bracts subulate or narrowly lanceolate, not at all or indistinctly scarious margined .... P. patagonica
4(1). Leaf blades broadly ovate, abruptly contracted to the petiole; seeds 6–20 per capsule; spikes mostly more than 6 cm long .... P. major
— Leaf blades lanceolate, oblanceolate, lance-oblong, or elliptic, gradually tapering into the petiole; seeds 2–4 per capsule; spikes various, but often less than 6 cm long 5
5(4). Outer pair of sepals (those adjacent to the bract) connate, appearing as a solitary, 2-veined, apically notched or entire sepal; capsule circumscissile near the middle; plants adventive, weedy; spikes compact .... P. lanceolata
— Outer pair of sepals distinct; plants indigenous, not weedy; spikes loose to compact 6
6(5). Crown of plants conspicuously woolly-hairy at base of petioles; spikes usually over 5 cm long; leaves thick .. .. P. eriopoda
— Crown of plants not woolly; spikes usually less than 5 cm long; leaves rather thin .... P. tweedyi

Plantago elongata Pursh [*P. myosuroides* Rydb., type from Salt Lake County]. Plants annual; leaves linear-filiform, 3–8 cm long, ca 1 mm wide, cinereous puberulent; scapes several to many, 3–10 cm long, exceeding the leaves; spikes loosely flowered, 2–10 cm long; bracts triangular-ovate, subequal to the calyx; flowers dioecious or polygamous; sepals 1.5–2 mm long; corolla lobes 0.5–1 mm long, on some flowers becoming erect and closing over the capsule; stamens 2; capsule ovoid, rounded apically, 1.5–3.5 mm long; seeds ca 4, nearly flat on both sides. Moist to dry usually saline sites in salt desert shrub, saltgrass, and pickleweed communities at 1280 to 1600 m in Cache, Salt Lake, Uintah, and Utah counties; Alberta and Montana, south to Oregon, Texas and Florida; 5 (0).

Plantago eriopoda Torr. Plants perennial, with reddish yellow wool on the root crown; leaves thickish, oblanceolate to narrowly elliptic-ovate, 3- to 9-veined or -ribbed, 6–20 cm long, entire to repand-dentate; scapes 10–40 cm long; spikes lax below, dense above, 4–12 cm long, somewhat pilose; bracts subequal to the calyx, broadly ovate, with a broadly rounded keel; sepals round-elliptic, scarcely keeled, ca 2 mm long; corolla lobes 1–2 mm long, spreading at maturity; stamens 4; capsule ovoid-cylindroid, the apex truncate, ca 3 mm long, circumscissile below the middle; seeds 2–4, reddish brown, shining, elliptic, flat, ca 2 mm long; 2n = 24. Meadows and riparian habitats in rush, sagebrush, and rabbitbrush communities at 1700 to 2500 m in Box Elder, Cache, Daggett, Duchesne, Garfield, Kane, Rich, Salt Lake, Sanpete, Utah, and Wasatch counties; Alaska and Yukon, east to Mackenzie, south to Mexico, and in eastern North America; 12 (i).

Plantago insularis Eastw. [*P. fastigiata* Morris]. Low annuals; stems short, with even shorter branches, villous and tomentose throughout; leaves 5–10 cm long, 3–8 mm wide, linear-lanceolate, acute, mostly entire; scapes axillary, erect or ascending, to 18 cm long, pilose or tomentose; spikes densely many-flowered, erect, shortly cylindrical, to 2 cm long and 8 mm thick, villous to tomentose; bracts ovate to oblong, scarious with green or brown rigid glabrous to villous midrib, about equaling the calyx; calyx lobes ovate to obovate, the midribs green or brown; corolla lobes ovate, apiculate, concave, ca 2 mm long and 1.5 mm wide, reflexed-spreading, with a brown spot at base of each; capsules twice as long as the calyx, oval, rounded apically, ca 4 mm long and 2 mm thick; seeds 2, brown, narrowly oblong, very finely pitted; 2n = 8. Creosote bush, Joshua tree, and blackbrush communities at 730 to 1130 m in Washington County; Nevada and California to Texas, south to Mexico; 17 (iii).

Plantago lanceolata L. Plants perennial, not woolly at the base, 1.5–5 dm tall; leaves elliptic to narrowly lanceolate or oblanceolate, 5–30 cm long, 0.3–4 cm broad, somewhat expanded but not membranous at the base, 3- to several-veined, entire to obscurely denticulate; inflorecence dense, 1–8 cm long; bracts ovate, the apex sometimes acuminate and slightly surpassing the flowers, dorsally pubescent to glabrous, ciliate; scapes strigose; corolla lobes 2–2.5 mm long, spreading; stamens 4; capsule 2–4 mm long, circumcissile below the middle, the seeds usually 2; 2n = 12, 24. Widely distributed weedy plants in numerous vegetative types from 800 to 3170 m in most if not all Utah counties; adventive from Eurasia, now widely naturalized in the U. S.; 27 (ii).

Plantago major L. Broadleaf Plantain. [*P. major* var. *pachyphylla* Pilger, type from Salt Lake City]. Plants perennial, not woolly at the base, mostly 1–5 dm tall; leaves ovate to lanceolate or broadly elliptic, acute to cordate basally, short- to long-petiolate, the blades 3–15 cm long, 2–12 cm broad, expanded and often somewhat membranous basally, 5- to several-veined, denticulate to entire; inflorescence dense to lax (especially below), 3–25 cm long, essentially glabrous; bracts ovate, shorter than the flowers, glabrous, not ciliate; corolla lobes spreading to reflexed, ca 1 mm long; stamens 4; capsules 3–4 mm long, circumcissile below the middle, the seeds several to many; 2n = 12, 24, 36. Widespread weedy species of lawns, fields, and other disturbed sites at 790 to 2700 m in most if not all Utah counties; adventive from Europe, now widely established in North America; 42 (vi).

Plantago patagonica Jacq. [*P. purshii* R. & S.]. Annuals, usually woolly villous throughout; leaves linear to narrowly oblanceolate, acute to acuminate, the apex commonly callous-tipped, 1–20 cm long, 2–15 mm wide; scapes shorter to longer than the leaves; spikes dense, cylindrical, 1–10 cm long, 5–8 mm thick; bracts linear to linear-subulate, shorter or longer than the flowers; sepals narrowly oblong-ovate; petals ca 2 mm long, suborbicular to ovate, spreading; stamens 4; capsule ca 3.5 mm long; seeds 2; 2n = 20. Joshua tree, blackbrush, creosote bush, shadscale, sagebrush, pinyon-juniper, and mountain brush communities at 850 to 2150 in Beaver, Cache, Carbon, Daggett, Duchesne, Emery, Garfield, Grand, Kane, Rich, Salt Lake, San Juan, Sevier, Uintah, Washington, and Weber counties; British Columbia to Saskatchewan, south to California and Texas; 110 (xviii). Our plants have been placed within three weak varieties., i.e., **var. gnaphaloides (Nutt.) Gray**, with bracts in lower part of spike shorter than to slightly longer than the calyces, **var. spinulosa (Decne.) Gray**, with long conspicuous bracts, and **var. breviscapa (Shinn.) Shinn.**, with scapes shorter than the spikes. Perhaps none of them are worthy of taxonomic recognition in our area.

Plantago tweedyi Gray Plants perennial, not woolly at the base, mostly glabrous or somewhat pubescent on the scapes; leaves 3–7 cm long, rather thin, lanceolate to spatulate, tapering gradually to the petiolar base, 3- to 5-veined, entire or nearly so; scapes 10–20 cm long; spikes 1–10 cm long, dense to lax; bracts subequal to or shorter than the calyx; sepals 1.5–2.5 mm long; corolla lobes ca 1 mm long; spreading; stamens 4; capsule narrowly ovoid, circumscissile below the middle, long exserted from the calyx; seeds 2–4; 2n = 12. Meadows, mainly in aspen, fir, and spruce-fir communities at 2100 to 3870 m in Beaver, Cache, Carbon, Duchesne, Emery, Iron, Piute, Salt Lake, San Juan, Sanpete, Sevier, Summit, Utah, and Wasatch counties; Alberta and Idaho to Colorado; 48 (i).

PLATANACEAE Dumort.

Sycamore Family

Moneoecious trees with exfoliating, platy bark; leaves alternate, deciduous, palmately 3- to 7-lobed, the petioles hollow basally and enclosing the axillary buds; flowers numerous, in globose, unisexual heads, minute; sepals 3–8, scalelike; petals 3–6, minute, in staminate flowers scarious pointed; stamens 3–8; pistils 3–8, each 1-loculed, surrounded by hairs and staminodia; style terminal; ovules 1 (or 2); fruit an obconic, 1-seeded nutlet; x = 7.

1. Leaves with the terminal lobe twice longer than broad; fruiting heads usually 3 or 4 .... *P. orientalis*
— Leaves with terminal lobe as broad as long or broader; fruiting heads 1 or 2 .... *P. occidentalis*

Platanus occidentalis L. American Sycamore. Trees of large stature and rapid growth, mainly 10–25 m tall and with trunk diameter 5–10 dm or more; bark platy, exfoliating, thus varicolored white or yellowish green and brown to blackish; leaves with petioles 1.5–9 cm long, the blades 5–27 mm long, from petiole attachment to apex and 6–32 cm wide, the terminal lobe as broad or broader than long, tomentose on one or both sides when young, finally glabrous, or pubescent only along the veins beneath; fruiting heads 1 or 2 per peduncle. Commonly grown shade and street trees in Utah mainly below 1830 m elevation; introduced from the eastern U. S.; 22 (0).

Our plants lack the snow white exfoliating bark of trees of flood plains in the eastern states, but it is not uncommon for tree species grown in Utah to exhibit bark and stature differences from those in their native habitats. A part of the variation might represent products of hybridization between *P. occidentalis* and *P. orientalis*, known as *P.* x *acerifolia* Willd., which has been distinguished on mainly intangible diagnostic features. I have been unable to distinguish the hybrids from among the specimens available for study. This species suffers in seasons of abnormally high precipitation from a blight that kills all of the leaves, but the trees recover and grow new leaves late in the season.

Platanus orientalis L. Oriental Planetree. Trees of large stature, mainly 10–25 m tall and with trunk diameter of 5–10 dm or more; bark platy, grayish to greenish white in a mosaic pattern; leaves with petioles mainly 3–11 mm long, the blades typically 5–15 cm long, from petiole attachment to apex, 6–20 cm wide, the terminal lobe about twice as long as broad, tomentose on one or both surfaces, becoming glabrous or pubescent only along the veins beneath; fruiting heads (2) 3–6. Uncommonly grown shade and street tree in lower elevation portions of Utah; introduced from Eurasia; 3 (0).

POLEMONIACEAE A. L. Juss.

Phlox Family

Annual, biennial, or perennial herbs, or some suffrutescent; leaves simple alternate or opposite, entire or pinnatifid to palmatifid or compound; flowers perfect, solitary or variously arranged; calyx 5-lobed; corolla usually of 5 united petals; stamens usually 5, inserted on the corolla tube, alternate with the corolla lobes; pistil 1, the ovary superior, 3-loculed; style 1; stigmas 3-lobed; fruit a capsule; x = 6–9.

1. Foliage leaves represented only by persistent, connate, entire cotyledons and a whorl of entire bracts subtending the inflorescence; plants diminutive annuals .... *Gymnosteris*
— Foliage leaves proper present, alternate or opposite; plants annual, biennial, or perennial 2
2(1). Leaves mostly opposite, at least below 3
— Leaves mostly alternate, even below 8
3(2). Leaves entire, acicular or flat; staminal filaments unequally inserted on the tube 4
— Leaves palmatifid or pinnatifid; staminal filaments variously inserted on the tube 5
4(3). Plants perennial; seeds not mucilaginous when wet .... *Phlox*
— Plants annual; seeds mucilaginous when wet .... *Microsteris*
5(3). Leaves petiolate, pinnatifid; plants delicate annuals with pungent leaf lobes and headlike flower clusters .... *Navarretia*
— Leaves sessile, palmatifid; plants annual or perennial . 6
6(5). Plants annual; leaves flexible; seeds mucilaginous when wet .... *Linanthus*
— Plants perennial; leaves stiff or flexible; seeds not mucilaginous when wet 7
7(6). Leaves stiff, pungent; flowers opening in evening .. .. *Leptodactylon*

—	Leaves flexible, not pungent; flowers opening any time .... *Linanthastrum*
8(2).	Leaves palmatifid, sessile, pungent .... *Leptodactylon*
—	Leaves various, but not both palmatifid and sessile ... 9
9(8).	Leaves simple, with definite broad blades, petiolate or tapering to a sessile base .... *Collomia*
—	Leaves simple, pinnatifid or more or less compound .. 10
10(9).	Leaves pinnate-pinnatifid, with definite broad leaflets in the lower portion .... *Polemonium*
—	Leaves pinnatifid, bipinnatifid, or palmatifid, lacking definite broad leaflets 11
11(10).	Calyx lobes unequal; leaves pinnatifid, the rachis narrow .... *Eriastrum*
—	Calyx lobes subequal; leaves pinnatifid, bipinnatifid, or palmatifid, the rachis often broad 12
12(11).	Leaves with bristle-tipped teeth or lobes; plants annual .... *Langloisia*
—	Leaves with spinulose-apiculate or acute to rounded teeth or lobes; plants annual, biennial, or perennial .... *Gilia*

Collomia Nutt.

Annual or perennial herbs; stems simple or branched; leaves alternate (except the lower ones), linear to obovate, entire or pinnatifid; flowers in cymes, capitate, or axillary; calyx tube green or chartaceous-membranous below the sinuses, campanulate to obconic, the sinuses in age with a projecting fold; corolla funnelform, salmon, lavender, pink, or white, the lobes spatulate to lanceolate; stamens equally or unequally inserted on the corolla tube; capsule ellipsoid to obovoid, each locule 1- or 2-seeded, these more or less mucilaginous when wet.

1.	Plants perennial from a soboliferous caudex, of alpine sites .... *C. debilis*
—	Plants annual from taproot, from lower elevations to alpine habitats 2
2(1).	Flowers solitary (or paired) in leaf axils; leaves linear; plants uncommon .... *C. tenella*
—	Flowers in capitate, terminal clusters; leaves lance-elliptic to elliptic, or oblanceolate; plants not uncommon 3
3(2).	Corolla mostly more than 15 mm long, the tube much longer than the calyx, salmon .... *C. grandiflora*
—	Corolla mostly 5-15 mm long, the tube slightly surpassing the calyx, pink to lavender, bluish, or white 4
4(3).	Stamens equally inserted, the filaments of unequal lengths; plants typically branched from near the base; leaves typically less than 5 mm wide; plants rare .... *C. tinctoria*
—	Stamens unequally inserted, the filaments subequal; plants typically unbranched; leaves mainly more than 5 mm wide; plants common .... *C. linearis*

Collomia debilis (Wats.) Greene [*Gilia debilis* Wats., type from Salt Lake County]. Perennial soboliferous herbs from a subterranean root crown and taproot, the caudex branches elongate, terminating in innovations and flowering stems; flowering stems 2.5-12 cm tall (above ground level); herbage pubescent with moniliform, glandular hairs; leaves alternate, mainly 3-40 mm long, enlarging above the stem base, with the largest displayed below the flowers, the blades obovate to oblancolate, entire to toothed or pinnatifid; flowers solitary or few in compact terminal clusters, sessile or pedicellate; calyx 8-11 mm long, the teeth unequal, subequal to or shorter than the tube; corolla 15-35 mm long, lavender, pink, blue, white, or cream, the lobes 5-7 mm long; stamens exserted; style exserted; capsule 4-5 mm long. Alpine sites, usually in talus, at 2560 to 3660 m in Juab, Salt Lake, Utah, and Wasatch counties; Washington to Montana, south to California, Nevada, and Wyoming; 25 (0). The elongate caudex branches adjust to creep in talus slopes.

Collomia grandiflora Dougl. ex Lindl. Plants annual, erect, simple or branched; herbage glabrous, puberulent, or glandular; stems mainly 1-5 dm tall or more; leaves alternate (except sometimes at stem base), 1.5-6.5 cm long, 3-13 mm wide, oblanceolate to elliptic, lanceolate, or almost linear, entire, transitional upward into lance-ovate bracts; flowers sessile or short-stipitate, in dense terminal or axillary heads; calyx 7-10 mm long, with lanceolate or triangular lobes; corolla 15-30 mm long, salmon or white, funnelform to salverform; stamens unequally inserted, of unequal length; stigmas included; capsule ca 5 mm long. Pinyon-juniper, sagebrush, mountain brush, aspen, and spruce-fir communities at 1430 to 2440 m in Beaver, Cache, Davis, Juab, Millard, Piute, Salt Lake, Sanpete, Sevier, Tooele, Utah, Wasatch, Washington, and Weber counties; British Columbia to Montana, south to California, Arizona, and Colorado; 25 (iii).

Collomia linearis Nutt. Plants annual, erect, simple or branched; herbage puberulent to glandular-villous; stems mainly 0.6-5 dm tall or more; leaves alternate (except at the lowermost nodes), 1.2-6 cm long, 3-10 mm wide, elliptic to lanceolate or oblanceolate, entire, transitional upward into lanceolate to lance-ovate bracts; flowers sessile in dense terminal or axillary heads; calyx 4-7 mm long, finally papery, with triangular lobes; corolla 8-15 mm long, pink to purplish, narrowly funnelform; stamens unequally inserted; stigma included; capsule 5-6 mm long; $2n = 16$. Sagebrush, mountain brush, aspen, spruce-fir, and grass-forb communities at 1495 to 3265 m in probably all Utah counties; British Columbia to Ontario, south to California, Arizona, New Mexico, Nebraska, and Wisconsin; 94 (ix).

Collomia tenella Gray Plants annual, erect, usually branched; herbage glandular-puberulent; stems mainly 0.4-1 dm tall, the branches very slender; leaves alternate (or the lower ones opposite), 1-4 (5) cm long, 1-5 mm wide, linear, entire; flowers solitary (or paired) in leaf axils almost to the stem base; calyx 3-4 mm long, with triangular lobes; corolla 4-6 mm long, lavender to pink or white, narrowly funnelform; stamens unequally inserted; stigma included; capsules subequal to the calyx. Sagebrush, mountain brush, aspen, and spruce-fir communities at 1980 to 2625 m in Cache, Juab, Millard, Salt Lake, Summit (type from Parleys Park) and Weber counties; Washington to Wyoming and Nevada; 7 (i).

Collomia tinctoria Gray Plants annual, forming low clumps, usually branched from near the base; herbage glandular-villous or -pilose; stems mainly 5-15 cm tall, the branches slender; leaves alternate (or the lower ones opposite), 1-4 (5) cm long, 1-5 mm wide, linear, entire; flowers mainly in clusters of 2 or 3 (5), short-pedicellate, the clusters closely subtended by foliose bracts that surpass them; calyx 3-4 mm long in anthesis, to 9 mm long in fruit, the lobes subulate; corolla 8-14 mm long, pinkish or lavender, narrowly funnelform; stamens equally inserted, the filaments of differing lengths; capsules much shorter than the calyx. Wet sedge-grass meadow with scattered wilows at 2690 m in Sevier County (Fish Lake); Washington and Idaho, south to California and Nevada; 1 (0). Our single collection (Thorne et al. 4171 BRY) seems to be slightly more robust than specimens from Nevada and California.

Eriastrum Woot. & Standl.

Annual herbs; stems simple or more commonly branched; herbage puberulent to arachnoid or villous; leaves alternate (rarely some opposite), entire and linear or pinnatifid; flowers sessile, borne in bracteate heads; calyx 5-lobed, the lobes subulate and pungent, the sinuses with a hyaline membrane; corollas funnelform, blue to pink or purplish to white; stamens exserted or included; capsules ovoid; seeds several per locule, usually mucilaginous when wet.

1. Corollas showy, usually bright blue lavender, 13–19 mm long (including the lobes), somewhat irregular; staminal filaments unequal, the longer ones 4–8 mm long .... *E. eremicum*
— Corollas not especially showy, pale blue lavender to white, 6–13 mm long (including lobes), regular; staminal filaments subequal, less than 3 mm long 2
2(1). Plants erect or ascending, usually taller than broad; anthers 0.7–1.3 mm long .... *E. sparsiflorum*
— Plants prostrate-ascending, usually broader than tall; anthers 0.4–0.6 mm long .... *E. diffusum*

Eriastrum diffusum (Gray) Mason [*Gilia filifolia* var. *diffusa* Gray]. Plants annual, diffusely branched from near the base, usually broader than tall, the stems 3–15 cm long, pilose to glabrous; leaves 5–23 mm long, simple, linear, or pinnatifid into 3–5 narrow lobes; flowers 3–20 in capitate clusters, lanate; bracts 3- to 7-lobed; calyx 5–7 mm long, the lobes unequal, lance-acicular, spinulose-tipped; corolla pale blue to white, 7–9 mm long, almost regular; stamens 1–2 mm long, the anthers 0.3–0.4 mm long; capsules 3–4 mm long; locules with 2 or 3 seeds. Creosote bush, blackbrush, Vanclevea-ephedra, shadscale, and juniper communities at 800 to 1620 m in Kane, Millard, and Washington counties; California and Nevada to Texas and Mexico; 11 (0).

Eriastrum eremicum (Jepson) Jepson [*Huegelia eremica* Jepson]. Plants annual, diffusely branched from near the base, often taller than broad, the stems 2–25 cm long, floccose to glabrous; leaves 0.6–4 cm long, simple and entire or pinnatifid into 3–5 or more lobes; flowers 2–10 in capitate, lanate clusters; bracts 3- to 7-lobed; calyx 5–7 mm long, the lobes subequal, lance-acicular, spinulose-tipped, glandular; corolla 12–15 mm long, more or less irregular, blue violet, showy; stamens unequal, the longest exserted, 4–8 mm long, the anthers ca 1 mm long; capsules 4–6 mm long; locules several-seeded. Creosote bush, Joshua tree, blackbrush, other warm-desert shrub, and pinyon-juniper communities at 850 to 1680 m in Washington County; California, Arizona, and Nevada; 16 (i).

Eriastrum sparsiflorum (Eastw.) Mason [*Gilia sparsiflora* Eastw.]. Annual herbs, simple or much branched from near the base, usually taller than broad; stems 4–25 cm tall or more, commonly floccose; leaves 0.4–3 cm long, linear, or with a pair of lobes at the base; flowers 2–5 in capitate, lanate clusters; bracts simple or 3- to 5-lobed; calyx 4–5 mm long, the lobes subequal, lance-acicular, spinulose-tipped; corolla 7–13 mm long, pale blue to white or pinkish; stamens 2–3 mm long, the anthers 0.7–1.3 mm long; capsules 3–5 mm long; seeds 2–4 per locule. Creosote bush, blackbrush, hopsage, ephedra, and pinyon-juniper communities at 850 to 2000 m in Garfield, San Juan, Tooele, and Washington counties; Washington to Idaho, south to California, Nevada, and Arizona; 11 (0).

Gilia Ruiz & Pavon

Annual, biennial, or perennial herbs from taproots; leaves alternate, entire or pinnatifid to palmatifid, often borne in a basal rosette and more or less well developed along the stem; flowers solitary and axillary, paniculate, thyrsoid, or capitate; calyx lobes 5, subequal, the ribs green, the interveinal areas usually membranous; corolla 5-lobed, funnelform to salverform, regular or nearly so, scarlet, lavender, blue, violet, pink, yellow, or white; stamens more or less equally inserted on the corolla tube, the filaments subequal to unequal in length; capsules dehiscent apically; seeds 1 to many per locule.

Constance, L. and R. Rollins. 1936. A revision of *Gilia congesta* and its allies. Amer. J. bot. 23: 433–440.
Grant, V. 1950. Genetic and taxonomic studies in *Gilia*. I. *Gilia capitata*. El Aliso 2: 239–316.
———. 1954. Genetic and taxonomic studies in *Gilia*. VI. Interspecific relationships in the leafy-stemmed Gilias. El Aliso 3: 35–49.
———. 1956. A synopsis of *Ipomopsis*. El Aliso 3: 351–362.
Wherry, E. T. 1946. The *Gilia aggregata* group. Bull. Torrey Club 73: 194–202.

1. Inflorescences capitate, thyrsoid, or spicate 2
— Inflorescences of solitary axillary flowers, or more commonly in open paniculate clusters 10
2(1). Inflorescence thyrsoid, usually more than half the plant height; plants of Carbon, Duchesne, Emery, and Uintah counties .... *G. stenothrysa*
— Inflorescence capitate, spicate, or thyrsoid, but in any case less than half the plant height; plants of various distribution 3
3(2). Cauline leaves apically 3-lobed or -toothed, at least some more or less palmatifid; plants perennial, with a definite caudex, occuring in the Tushar Mts. and at Cedar Breaks .... *G. tridactyla*
— Cauline leaves various, but if as above, otherwise differing; plants of various duration and distribution .. 4
4(3). Inflorescences spicate; cauline leaves 3–5 cm long; plants of Daggett County .... *G. spicata*
— Inflorescence capitate, or with branches terminating in capitate clusters; cauline leaves usually much less than 3 cm long; plants of various distribution 5
5(4). Inflorescences not especially involucrate; plants perennial; locules 1-seeded 6
— Inflorescences usually definitely involucrate; plants annual; locules with more than 1 seed (except in some *G. gunnisonii*) 7
6(5). Flowers cream (fading yellow), the corolla tube much exserted from the calyx; calyx 5.5–8.5 mm long; plants often woody at the base; staminal filaments subequal to the anthers in length; plants of eastern Utah .... *G. roseata*
— Flowers white, sometimes fading yellowish, the corolla tube included in the calyx; calyx 3–5 mm long; plants seldom woody at the base; staminal filaments much longer than the anthers; plants widely distributed .... *G. congesta*

7(5).	Leaves entire or nearly all entire 8	—	Corollas mostly 4–9 mm long; capsules 1.5–3 mm long; plants reported for northern Utah .. *G. capillaris*
—	Leaves pinnatifid to subpalmatifid 9	21(18).	Basal and lower cauline leaves and stem base villous to pilose; plants of Washington County 22
8(7).	Leaves and bracts with definite flat blades, tapering to the base; plants of western Utah .... *G. depressa*	—	Basal and lower cauline leaves and stem base glabrous or glandular (or rarely arachnoid-tomentose); distribution various 23
—	Leaves and bracts linear or acicular, not tapering to the base; plants of southeastern Utah ... *G. gunnisonii*	22(21).	Corolla tube much longer than the calyx .... *G. scopulorum*
9(7).	Leaves trifid or pinnately 5-cleft; corolla lobes mostly 2–3.5 mm long; plants of eastern Utah .... *G. pumila*	—	Corolla tube included in the calyx .... *G. stellata*
—	Leaves sharply pinnatifid, usually with 5–9 teeth or lobes; corolla lobes 1–1.8 mm long; plants widespread .... *G. polycladon*	23(21).	Upper stem with purplish black, peglike glands; corolla limb as wide as the corolla length; plants of Kane County .... *G. flavocincta*
10(1).	Plants perennial from a branching caudex; flowers red; plants from western Wayne County . *G. caespitosa*	—	Upper stem variously glandular or glabrous, seldom as above (in *G. inconspicua*); corolla limb narrower than the corolla length; plants more widely distributed 24
—	Plants annual or biennial, a caudex not developed; flowers variously colored; distribution various 11	24(23).	Lower leaves spatulate-oblanceolate, toothed or lobed, the lobes not exceeding twice the width of the rachis; corolla 2–6.5 mm long, the lobes tridentate or cuspidate-acuminate apically .... *G. leptomeria*
11(10).	Overall corolla length 1.5–4.5 cm or more 12		
—	Overall corolla length 0.2–1 cm 15		
12(11).	Basal leaves obovate-spatulate, merely dentate; corollas usually carmine; plants of southeastern Utah .... *G. subnuda*	—	Lower leaves variously shaped, pinnatifid to bipinnatifid, but if as above, the corolla more than 7 mm long and the lobes commonly rounded 25
—	Basal leaves variously shaped, definitely pinnatifid; flowers variously colored; distribution various 13	25(24).	Leaves pinnatifid to bipinnatifid and with a broad to narrow rachis; stem base and lower leaves often arachnoid-tomentose; corolla tube included in the calyx .... *G. inconspicua*
13(12).	Basal leaves with a rachis 2–4 mm wide; flowers lavender to blue purple, funnelform, 1.5–2 cm long; plants of San Juan County .... *G. haydenii*		
—	Basal leaves with the rachis commonly less than 2 mm wide; flowers scarlet, salmon, pink, white, or blue; distribution various 14	—	Leaves usually bipinnatifid, the rachis much narrower than the length of the divisions; corolla tube surpassing the calyx .... *G. hutchinsifolia*
14(13).	Flowers usually white to blue; lateral lobes of cauline leaves mostly 1–2 cm long; plants of lower elevations in southeastern Utah .... *G. longiflora*		
—	Flowers scarlet, salmon, pink, or white; lateral lobes of cauline leaves usually less than 1 cm long; distribution almost universal .... *G. aggregata*		
15(11).	Leaves about as broad as long, dentate, the teeth spinulose; plants uncommon, of southeastern and southern Utah .... *G. latifolia*		
—	Leaves usually much longer than broad, entire to dentate and sometimes with spinulose teeth; distribution various 16		
16(15).	Corolla yellow; leaves linear-filiform, mainly cauline; plants of Washington County .... *G. filiformis*		
—	Corolla lavender, pink, blue, or purple (rarely white); leaves mainly basal or mainly cauline; plants of various distribution 17		
17(16).	Stamens exserted, subequal to or longer than the corolla lobes; flowers usually blue; plants of eastern to southcentral Utah .... *G. pinnatifida*		
—	Stamens included, much surpassed by the corolla lobes; flowers variously colored; distribution various . . 18		
18(17).	Leaves entire or nearly so (sometimes prominently lobed in *G. gilioides*), all or mainly cauline, the lowermost sometimes opposite 19		
—	Leaves toothed, pinnatifid, or palmatifid, mainly in a basal rosette 21		
19(18).	Leaves definitely lobed or strongly toothed, at least some; plants of southwestern Utah .... *G. gilioides*		
—	Leaves entire; plants of south-central to northern Utah 20		
20(19).	Corollas 1.5–3 mm long; capsules 1.5–2 mm long; plants along the central highlands from south-central to northern Utah .... *G. tenerrima*		

Gilia aggregata (Pursh) Sprengel Scarlet Gilia. [*Cantua aggregata* Pursh; *Ipomopsis aggregata* (Pursh) V. Grant; *G. a.* ssp. *a.* var. attenuata f. utahensis Brand, type from Alta]. Biennial (or perennial?), mephitic herbs; stems arising from a basal rosette and taproot, 1–10 dm tall, stipitate-glandular or with crinkly white hairs; basal leaves mainly 2–8 cm long, pinnatifid to sub-bipinnatifid, the rachis and segments narrow, the segments mainly 1–12 mm long; cauline leaves reduced upward; flowers short-pedicellate, thyrsoid to more or less paniculate, this more or less bracteate; corolla 1.5–5 cm long, the lobes spreading, 5–12 mm long, scarlet, pink, salmon, bluish, or white and sometimes maculate; staminal filaments subequally inserted on the corolla tube, the anthers included or exserted; seeds 1 to several per locule; capsules 4–8 mm long; seeds 2–4 mm long, becoming mucilaginous when wet; n = 7. Warm desert shrub, sagebrush, pinyon-juniper, mountain brush, ponderosa pine, aspen, spruce-fir, and alpine meadow communities at 820 to 3295 m in all Utah counties; British Columbia to Montana, south to California, Arizona, New Mexico, Texas, and Mexico; 297 (xxxviii). This strikingly beautiful species, with the odor of a skunk, is practically ubiquitous in Utah, but is collected even where uncommon because it is so easily seen. There are four more or less completely intergrading phases among our specimens that have been recognized at various taxonomic ranks. Their recognition at varietal rank is tenuous, but some of the phases are remarkably distinct in the extreme. Because of the apparent differences, the plants are traditionally segregated at varietal level. They consist of: 1) high elevation plants with multicolored flowers having the corolla tube mainly less than 3 mm wide apically, known as **var. macrosiphon**

Kearney & Peebles [*G. tenuituba* Rydb., type from near Beaver]; 2) ubiquitous, scarlet-flowered plants with corolla tube more than 3 mm wide apically and calyx lobes usually longer than the tube in anthesis, known as **var. *aggregata***; 3) scarlet-flowered plants with calyx teeth subequal to or slightly longer than the broad tube and exserted anthers from Emery, Garfield, Grand, and San Juan counties, known as **var. *maculata* Jones**; and 4) scarlet, short-flowered plants with calyx teeth subequal to the broad tube or slightly longer and with stamens included from Iron, Kane, and Washington counties, known as **var. *arizonica* (Greene) Fosberg** [*Callisteris arizonica* Greene; *G. arizonica* (Greene) Rydb. The varieties *macrosiphon* and *arizonica* are most easily distinguished from the type variety, but transitional specimens are present.

***Gilia caespitosa* Gray** Rabbit Valley Gilia. Pulvinate-caespitose herbs from a taproot and branching caudex, the caudex clothed with persistent leaf bases and terminated by rosettes of leaves; herbage stipitate-glandular, commonly with adherent sand grains; flowering stems 3–8.5 cm tall; basal leaves 3–20 mm long, 1–3 mm wide, oblanceolate to linear, entire or few-lobed, spinulose-apiculate; cauline leaves with few, reduced leaves; flowers solitary or few to several in cymes; calyx 4–6 mm long, the lobes subequal to the tube; corolla scarlet (sometimes fading maroon or blue purple); salverform, the tube 9–17 mm long, the lobes 4–6 mm long; stamens included, the filaments subequally inserted; capsule 4–5 mm long, few-seeded; seeds not mucilaginous when wet; 2n = 14. Pinyon-juniper-cercocarpus communities at 1735 to 2595 m on Carmel and Navajo formations in Wayne County (type from Rabbit Valley); endemic; 6 (ii). This beautiful low plant was collected first in 1875 by L. F. Ward of the Powell Survey, but remained obscure for almost seven decades until it was rediscovered by Ripley and Barneby in 1947.

***Gilia capillaris* Kellogg** Annual, mephitic herbs, with stipitate-glandular stems and branches, these mainly 0.5–3 dm tall; leaves 0.5–3.5 cm long or more, 1–4 mm wide, linear to narrowly oblong, alternate or the lower ones opposite, all cauline and reduced upward; flowers terminal and solitary or somewhat cymose; calyx 2.3–5 mm long, the lobes often unequal, about as long as the tube or shorter; corolla lavender to bluish or white, 4–9 mm long, the lobes short; the filaments equal; capsule 3–4 mm long; seeds 0.8–1.5 mm long, mucilaginous when wet. Reported for northern Utah in the Intermountain Flora (Cronquist 1984), but not seen by me; Washington to Idaho, south to California and Nevada; 0 (0).

***Gilia congesta* Hook.** [*Ipomopsis congesta* (Hook.) V. Grant]. Perennial herbs or subshrubs from taproots, the caudex more or less developed, superficial, below, or above ground; stems 0.5–8 dm tall, simple or branched; herbage with white, villous tomentum or varying to glabrous; leaves basal and/or cauline, entire (and linear) to pinnatifid or palmatifid, 0.5–4 cm long or more; flowers numerous in dense, terminal, more or less bracteate, cymose heads; calyx 3–5 mm long, the lobes spinulose-tipped, usually shorter than the tube; corolla white (or purplish), the tube 3–5 mm long, subequal to the calyx or slightly longer; corolla lobes 1.5–3 mm long; stamens attached to the sinuses, the filaments longer than the anthers, these 0.4–0.8 mm long, produced above the corolla tube; ovules 1 or 2 per locule; capsule 2.5–4 mm long, usually with 1 seed per locule; seeds 1.6–2.5 mm long, mucilaginous when wet. Four rather clearly definable but partially confluent varieties are present in Utah.

1. Leaves glabrous (rarely some tomentose), usually green 2
— Leaves arachnoid-tomentose, usually appearing whitish or grayish 3
2(1). Leaves entire (rarely some trifid); plants 3–12 cm tall .... *G. congesta* var. *crebrifolia*
— Leaves commonly trifid (or some pinnatifid to entire); plants mainly 10–30 cm tall .. *G. congesta* var. *palmifrons*
3(1). Plants mainly 1.5–8 dm tall, usually woody at the base; leaves commonly entire; plants of southern Utah .... *G. congesta* var. *frutescens*
— Plants mainly 0.3–2 dm tall, seldom woody at the base; leaves palmatifid to pinnatifid (or some entire); plants widespread .... *G. congesta* var. *congesta*

Var. *congesta* [*G. congesta* var. *paniculata* Jones, type from Huntington; *G. burleyana* A. Nels.; *G. congesta* var. *burleyana* (A. Nels.) Const. & Rollins]. Shadscale, greasewood, sagebrush, and pinyon-juniper communities at 1525 to 2290 m in Beaver, Box Elder, Carbon, Daggett, Duchesne, Emery, Iron, Juab, Millard, Piute (?), Sanpete, Sevier, Tooele, Uintah, Washington, and Wayne counties; Oregon to Wyoming, south to California, Nevada, and Colorado; 105 (xxii).

Var. *crebrifolia* (Nutt.) Gray [*G. crebrifolia* Nutt.]. Sagebrush, pinyon-juniper, mountain brush, white fir and bristlecone pine communities at 1830 to 2780 m in Beaver, Millard, Rich, San Juan, and Summit counties; Montana and Wyoming south to New Mexico; 13 (iii). The disjunct distribution of this variety is unique for infraspecific taxa in this species. Perhaps this indicates arbitrary assignment of similar but independently derived populations. Independent origin of Beaver and Millard counties material is suggested by the presence of some specimens transitional to var. *congesta*.

Var. *frutescens* (Rydb.) Cronq. [*G. frutescens* Rydb., type from Springdale, Washington County; *Ipomopsis frutescens* (Rydb.) V. Grant]. Sandy desert shrub and grassland, pinyon-juniper, hanging garden, mountain brush, pinyon-juniper, and ponderosa pine communities at 1155 to 2290 m in Garfield, Kane, San Juan, and Washington counties; Colorado and Arizona; 24 (ix). This is the most distinctive phase of the *congesta* complex. Plants are often woody for 1–2 dm above ground level.

Var. *palmifrons* (Brand) Cronq. [*G. congesta* ssp. *palmifrons* Brand]. Shadscale, greasewood, rabbitbrush, sagebrush, and pinyon-juniper communities at 1495 to 1680 m in Beaver, Juab, Millard, and Tooele counties; Oregon, Idaho, and Nevada; 9 (0). Identification of our specimens with this variety is tentative. More work is indicated.

***Gilia depressa* Jones** [*Ipomopsis depressa* (Jones) V. Grant]. Annual, mephitic herbs, the herbage puberulent to villous with crinkly multicellular hairs, these often glandular; stems branching from the base (or simple), 2.5–12 cm long; leaves mostly cauline, 0.5–1.8 cm long, 1.5–6 mm wide, elliptic to lanceolate or linear, entire or irregularly toothed; flowers few, in leafy-bracteate cymes or some axillary and solitary; calyx 3.5–6 mm long, the lobes spine-tipped, subequal to or longer than the tube; corolla white, 5–8 mm long, almost regular, the lobes

1–1.8 mm long; stamens inserted near the top of the corolla tube; anthers exserted; capsule 3–4 mm long, shorter than the calyx; seeds several per locule, 1.2–1.4 mm long, mucilaginous when wet. Shadscale, winterfat, galleta, and Indian ricegrass communities at 1340 to 1710 m in Beaver, Juab, and Millard (type from Deseret) counties; Nevada and California; 6 (0).

Gilia filiformis Parry ex Gray Yellow Gilia. Annual, slender herbs, 0.8–1.5 dm tall, branched from below the middle, stipitate-glandular to glabrous; leaves cauline, 0.6–3 cm long, 0.5–1 mm wide, linear to filiform; flowers solitary or few in open cymes, the pedicels capillary; calyx 2.5–4 mm long, the lobes lance-acuminate, usually longer than the tube; corolla yellow, 3.5–7 mm long, campanulate, lobed almost to the base; staminal filaments inserted near the base of the tube; capsules 2.5–3 mm long; seeds several to many per locule, ca 0.8 mm long, mucilaginous when wet. Creosote bush and Joshua tree communities at 820 to 980 m in Washington County (type from St. George); Arizona, Nevada, and California; 3 (0).

Gilia flavocincta A. Nels. Annual or winter annual herbs from taproots and basal rosettes; stems mainly 1–3 dm tall, arachnoid-tomentose below and stipitate-glandular with purplish-black, peglike glands upward; basal leaves 0.8–6 cm long, pinnatifid to bipinnatifid, the rachis narrower than the length of the longest lobes; inflorescences compactly to openly cymose, inconspicuously bracteate; calyx 3–5 mm long, the trianuglar lobes shorter than the tube; corolla white to pale lavender, with yellow tube, 7–9 mm long, the limb 7–9 mm wide; stamens equally inserted; capsules equaling or surpassing the calyx; seeds 1 per locule, ca 3.5 mm long. Salt desert shrub at ca 1340 m in Kane County (Atwood 4547 BRY); Arizona and New Mexico; 1 (0). Our specimen has been identified as belonging to ssp. **australis** (A. & V. Grant) Day & V. Grant [*G. opthalmoides* ssp. *australis* A. & V. Grant].

Gilia gilioides (Benth.) Greene [*Colomia gilioides* Benth.]. Mephitic caulescent annual; stems ascending to almost erect, mostly 0.5–3 dm tall, glandular-hairy; leaves mainly or all cauline, entire or some or most of them with 1 or more irregularly disposed lateral lobes, mainly 1–3 cm long, 1–3 mm wide exclusive of the lobes; flowers terminating branches, but often appearing lateral because of the branching pattern; pedicels short to elongate, often paired; calyx 3–4.5 mm long at anthesis, with rather narrow intercostal membranes, the slender lobes subequal to or longer than the tube; corolla blue to violet, 5–8mm long, funnelform; staminal filaments subequally inserted on the corolla tube; anthers included or essentially so, surpassed by the corolla lobes; capsule 2.5–3.5 mm long; seeds ellipsoid, 1.5–2.5 mm long, usually 1 per locule, becoming mucilaginous when wet; $2n = 18$. Pinyon-juniper and sagebrush communities at ca 1745 m in northwestern Washington County; Arizona, Nevada, California, and Oregon; 1 (0). The species is known from Utah by a collection taken by L. C. Higgins in 1985 (15737 BRY).

Gilia gunnisonii T. & G. Gunnison Gilia. [*Ipomopsis gunnisonii* (T. & G.) V. Grant]. Annual (or winter annual) herbs; herbage puberulent to stipitate-glandular or glabrous; stems simple or branched from near the base, 6–25 cm long; leaves cauline, 0.5–4 cm long, 0.5–1.2 mm wide, linear, entire, or with 2–4 lateral lobes; flowers few to many in terminal, cymose, more or less bracteate clusters; calyx 3.5–4.5 mm long, the lance-subulate teeth spinescent, shorter than the tube, often ruptured in fruit; corolla white to lavender, 6–10 mm long; stamens equally inserted, unequal, exserted; capsule 2.5–3.5 mm long; seeds 1 or 2 per locule, 1.5–2.5 mm long. Sandy desert shrub, shadscale, sagebrush, purple sage, and pinyon-juniper communities at 1125 to 1830 m in Emery, Garfield, Grand, Kane, and San Juan counties; Arizona and New Mexico; 72 (xi). The type was collected at Green River by F. Creutzfeldt of the ill-fated Gunnison expedition in 1853.

Gilia haydenii Gray Hayden Gilia. Biennial (or winter annual) herbs from taproots and basal rosettes; stems mainly 1.5–4.5 dm tall; herbage glandular to stipitate-glandular or coarsely villous (especially on the leaves); basal leaves pinnatifid, the rachis broad, but usually narrower than the length of the longest lateral lobes; inflorescence paniculately cymose, inconspicuously bracteate; calyx 3.5–5.5 mm long, the narrowly triangular lobes shorter than the tube; corolla rose purple to pink lavender, drying blue purple, the tube 13–16 mm long, the limb 10–12 mm wide; stamens equally inserted, the anthers slightly exserted; capsules subequal to the calyx; seeds several per locule. Blackbrush, matchweed, and shadscale communities at ca 1370 to 1525 m in southeastern San Juan County; Colorado and Mexico; 4 (0).

Gilia hutchinsifolia Rydb. [*G. arenaria* ssp. *leptantha* var. *rubella* Brand, type from St. George; *G. leptomeria* ssp. *rubella* (Brand) Mason & Grant]. Annual (or winter annual) herbs from taproots and basal rosettes; stems mainly 0.7–3.5 dm tall; herbage stipitate-glandular and sometimes with nonglandular hairs as well; basal leaves 1–11 cm long, doubly pinnatifid, the rachis narrower than the length of the primary divisions; cauline leaves reduced upward; inflorescence paniculately cymose, inconspicuously bracteate; calyx 3–3.8 mm long, the oblong to triangular lobes subequal to or shorter than the tube; corolla 7–13 mm long, lavender to white, the lobes rounded to obtuse, the limb 5–10 mm wide, the tube white to yellowish; stamens attached at the corolla sinuses, the anthers surpassed by the corolla lobes; capsules 4–5.5 mm long, subequal to the calyx; locules several-seeded, the seeds 0.6–0.9 mm long, not mucilaginous when wet. Blackbrush, other warm desert shrub, shadscale, winterfat, horsebrush, ephedra, and Indian ricegrass-galleta grass communities at 915 to 1740 m in Garfield, Juab, Kane, Millard, San Juan, Tooele (Carrington Island), and Washington counties; Arizona, Nevada, and California; 26 (iv).

Gilia inconspicua (J. E. Sm.) Sweet [*Ipomopsis inconspicua* J. E. Sm.; *G. sinuata* Dougl. ex Benth; *G. inconspicua* var. *variegata* Brand, type from Kanab; *G. opthalmoides* Brand; *G. brecciarum* Jones; *G. straminea* Rydb., type from St. George; *G. clokeyi* Mason; *G. ochroleuca* ssp. *transmontana* Mason & Grant]. Annual (or winter annual) herbs from taproots and basal rosettes; stems mainly 0.5–4.5 dm tall; herbage typically arachnoid-tomentose below, becoming stipitate-glandular and often with peglike, purplish glands above; basal leaves 0.3–8 cm long, pinnatifid to bipinnatifid or merely lobed, the rachis narrower to broader than the length of the lateral segments; cauline leaves much reduced upward; inflorescence spicately to paniculately cymose; calyx 3–6.1 mm long, the triangular-attenuate lobes subequal to or shorter than the tube; corolla 4–12 mm long, funnelform, blue to lavender or white, the limb mostly 4–7 mm wide, the lobes rounded to obtuse apically; stamens inserted at the sinuses of the corolla, subequal to or longer than the anthers, these ca 0.5 mm long; capsules 3–6 mm long; seeds 1.4–2 mm long, mucilaginous when wet. Creosote bush, Joshua tree, blackbrush, other warm desert shrub, shadscale, hopsage, pinyon-juniper, sagebrush, and mountain brush communities at 760 to 2290 m in Beaver, Box Elder, Carbon, Daggett, Duchesne, Emery, Garfield, Juab, Kane, Millard, Piute, San Juan, Sanpete, Sevier, Tooele, Uintah, Utah, Washington, and Wayne counties; Washington to Wyoming, south to California, Nevada, Arizona, and New Mexico; 185 (xviii). Plants with merely lobed to once-pinnatifid leaves belong to **var. sinuata** (Hook.) Gray; those with bipinnatifid leaves are **var. inconspicuua**. It is unfortunate that this distinctive species should have borne so many epithets, but the large number of names is indicative of the variation within this and the other annual, small-flowered gilias.

Gilia latifolia Wats. Spiny Gilia. Annual herbs from taproots, the basal rosette not well developed; stems 0.5–4.6 dm tall, usually branched from near the base; herbage stipitate-glandular; leaves petiolate, the blades 0.5–4.5 cm long, 4–30 mm wide, ovate to oval or elliptic, coarsely dentate, the teeth spinulose-tipped; flowers numerous in paniculate cymes; calyx 2.9–4.5 mm long, the subulate teeth spinulose apically; corolla 6–11 mm long, salverform to funnelform, pink to purplish; stamens inserted in the corolla tube; capsule shorter to longer than the calyx; seeds several per locule, 0.5–0.9 mm long, yellow brown, not mucilaginous when wet. Mixed warm and cool desert shrub communities at 1000 to 1590 m in Emery, Garfield, Kane, San Juan, Washington (type from near St. George), and Wayne counties; California, Arizona, and Nevada; 13 (vi). The specimens from the San Rafael Swell, Cataract Canyon, and Warm Creek appear to have somewhat smaller capsules, more confluent leaves, and ultimately more intricately branched inflorescences than do those from the body of the species to the west. Possibly they warrant taxonomic recognition, but no single feature or combination of features seem sufficient to segregate all specimens.

Gilia leptomeria Gray Annual, ill-scented herbs from taproots and basal rosettes; stems 0.4–2.2 (3) dm tall; herbage hirtellous to stipitate-glandular; basal leaves 0.4–8.5 cm long, 1.5–18 mm wide, oblanceolate to spatulate, dentate to pinnately lobed the rachis broader than the length of the lateral teeth or lobes; cauline leaves much reduced upward; flowers few to numerous in paniculate cymes; calyx 1.5–3.5 mm long, the teeth triangular-oblong, shorter than the tube; corolla 2–6.5 mm long, salverform, pink to lavender or white, the lobes shorter than the tube, mucronate or tridentate; filaments subequal, inserted at the sinuses; anthers ca 0.5 mm long; seeds several per locule, 0.5–0.9 mm long, not mucilaginous when wet. Blackbrush, other warm desert shrub, greasewood, shadscale, sagebrush, mountain brush, pinyon-juniper and ponderosa pine communities at 820 to 2350 m in Beaver, Box Elder, Carbon, Daggett, Duchesne, Emery, Garfield, Grand, Iron, Juab, Kane, Millard, Piute, Salt Lake, San Juan, Sanpete (?), Sevier, Tooele, Uintah, Utah, Washington (type from near St. George), and Wayne counties; Washington to Wyoming, south to California, Nevada, Arizona, and New Mexico; 98 (xiv). Two varieties are recognized; **var. micromeria** (Gray) Cronq. [*G. micromeria* Gray; *G. leptomeria* ssp. *micromeria* (Gray) Mason & Grant], with calyx less than 2.5 and corolla less than 3.5 mm long; and **var. leptomeria** [*G. triodon* Eastw., type from San Juan County; *G. leptomeria* var. *tridentata* Jones, type from near Emery], with longer calyx and corollas.

Gilia longiflora (Torr.) D. Don [*Cantua longiflora* Torr.; *Ipomopsis longiflora* (Torr.) V. Grant]. Annual (winter annual) or biennial, inodorous (?) herbs; stems arising from a taproot, the basal rosette poorly developed and usually withered at anthesis, 1.5–6 dm tall; herbage sparingly white villous (leaves), glabrous, or minutely glandular (on calyces especially); cauline leaves 1–5 cm long, pinnatilobate, the rachis and segments linear, somewhat reduced upward; flowers on short to very long pedicels, borne in more or less flat-topped, paniculate, cymes, these more or less bracteate; calyx 4.5–8 mm long, the lance-attenuate lobes subequal to the tube or shorter; corollas 2.5–5.5 cm long, the lobes spreading, 5–12 mm long, pale blue to white, salverform; staminal filaments subequally inserted on the corolla tube, the anthers included or exserted; capsules 7–14 mm long; seeds 3–4 mm long, becoming mucilaginous when wet. Sandy desert shrub, warm desert shrub, sagebrush, and pinyon-juniper communities at 975 to 2135 m in Garfield, Grand, Kane, San Juan, Washington, and Wayne counties; South Dakota to Texas, west to Colorado and Arizona; 28 (ix).

Gilia pinnatifida Nutt. [*G. calcarea* Jones; *G. mcvickerae* Jones, type from near Marysvale]. Annual (winter annual) or biennial to short-lived perennial herbs from taproots and basal rosettes; stems 1–6 dm tall; herbage glandular to stipitate-glandular or subglabrous; basal leaves 1–7 cm long, 5–15 mm wide, the rachis narrow to broad, but mostly narrower than the length of the lateral segments; inflorescence paniculately cymose, more or less bracteate; calyx 2.5–5.2 mm long, the lobes shorter than the tube; corolla blue to blue purple, the tube 3–9 mm long, the lobes 3–5 mm long, spreading; stamens inserted below the sinuses, long-exserted; capsules subequal to or longer than the calyx; seeds 2 or more per locule, mainly 1.2–1.5 mm long, not mucilaginous when wet. Mixed desert shrub and grass, sagebrush, pinyon-juniper, and mixed conifer communities at 1585 to 2900 m in Daggett, Emery, Garfield, Iron, Kane, Piute, Sevier, and Uintah counties; Wyoming and Nebraska to New Mexico; 23 (v).

Gilia polycladon Torr. [*Ipomopsis polycladon* (Torr.) V. Grant]. Annual or winter annual herbs; herbage more or less villous with crinkly, multicellualar hairs and commonly more or less glandular; stems branched from near the base (or simple), 2.5–24 cm long; leaves basal and cauline, the basal ones 0.6–3 cm long, pinnatifid, the cauline ones few, subequal to the bracts; flowers few to several in leafy-bracteate, terminal cymes, the bracts pinnately toothed or lobed; calyx 4–6 mm long, the thick-end, ovate-acuminate lobes spinulose-tipped, shorter than the tube; corolla white, 3–6 mm long, the lobes 1–1.8 mm long; stamens inserted near the sinuses of the corolla; anthers barely exserted; capsules 3–4.5 mm long; seeds several per locule, 1.5–2.5 mm long, mucilaginous when wet. Creosote bush, Joshua tree, blackbrush, other warm desert shrub, shadscale, galleta, horsebrush, sagebrush, and pinyon-juniper communities at 730 to 1895 m in Beaver, Carbon, Duchesne, Emery, Garfield, Grand,

Iron (?) Juab, Kane, Millard, San Juan, Sevier, Tooele, Uintah, Washington, and Wayne counties; California, Nevada, Idaho, Colorado, Arizona, New Mexico, and Texas; 71 (vi).

Gilia pumila Nutt. [*Ipomopsis pumila* (Nutt.) V. Grant]. Annual or winter annual herbs; herbage more or less villous with crinkly, multicellular hairs and sometimes finely glandular; stems branching near the base or above, or simple, 2–23 cm long; leaves mainly cauline, the basal rosettes only weakly developed, pinnately or subpalmately 3- to 7-lobed, or the lower ones entire, mainly 0.6–3 cm long; flowers few to several in leafy-bracteate, terminal cymes, or with some solitary in leaf axils, the bracts pinnatilobate or palmatifid like the leaves; calyx 3.5–5.2 mm long, the lance-acuminate lobes spinulose (or bristle) -tipped, subequal to or shorter than the tube; corolla tube 7–9 mm long, mucilaginous when wet. Shadscale-greasewood, other salt desert shrub, sagebrush, and pinyon-juniper communities at 1370 to 1955 m in Carbon, Daggett, Duchesne, Emery, Garfield, Grand, Kane, Sevier, Uintah, and Wayne (?) counties; Colorado, Arizona, New Mexico, and Texas; 33 (v).

Gilia roseata Rydb. [*Ipomopsis roseata* (Rydb.) V. Grant; *G. congesta* var. *nuda* Eastw., type from San Juan County]. Perennial herbs or subshrubs from taproots, the caudex typically well developed, usually above ground; stems 5–35 cm tall, branched; herbage sparingly villous with crinkly, multicellular hairs to subglabrous or minutely to stipitate glandular; leaves mainly cauline, entire (and linear) or more commonly pinnatifid and with 2–4 pairs of lobes, 0.5–4 cm long or more; flowers few to numerous in dense, more or less bracteate, cymose heads; calyx 5.5–8.5 mm long, the ovate-acuminate lobes spinulose apically; corolla tube 7–9 mm long, well-exserted from the calyx, the lobes 2.5–4 mm long; stamens inserted at the sinuses, the filaments about as long as the anthers, produced beyond the corolla tube; capsules 3–5 mm long, with 1 seed per locule, the seeds 3–4 mm long, mucilaginous when wet. Blackbrush, warm, salt, and sand desert shrub, pinyon-juniper, and sagebrush communities at 1190 to 2290 m in Duchesne, Emery, Garfield, Grand, San Juan, and Uintah counties; Colorado; a Colorado Plateau endemic; 57 (viii). Gross appearance of this plant places it with *G. congesta*, but it is probably more closely allied to *G. spicata* (q.v.).

Gilia scopulorum Jones [*G. scopulorum* var. *deformis* Brand, type from near St. George]. Annual, mephitic to inodorous herbs from taproots, the basal rosettes developed or not; stems 1.5–4 dm tall; herbage glandular-villous, with long multicellular, often crinkly hairs, or merely stipitate-glandular upwards; leaves basal or borne in the lower 1/4–1/2 of the plant, 1–6 (10) cm long, 5–22 (35) mm wide, irregularly pinnately lobed to sub-bipinnatifid, the upper ones reduced, bracteate; flowers numerous, borne in open, paniculate, more or less bracteate cymes; calyx 3.5–5.8 mm long, the lance-subulate lobes usually shorter than the tube; corollas 10–16 mm long, the tube much longer than the calyx, salverform to funnelform, the lobes acute to obtuse; filaments subequal, inserted below the sinuses; anthers ca 0.5–0.6 mm long; seeds few per locule, 1.4–2 mm long, mucilaginous when wet. Creosote bush, Joshua tree, blackbrush, and mountain brush communities at 730 to 1100 m in Washington County (type from near St. George); Arizona, Nevada, and California; 17 (iv). Irregularly lobed leaves are almost a hallmark of this species

Gilia spicata Nutt. [*Ipomopsis spicata* (Nutt.) V. Grant]. Perennial herbs from a taproot and simple or branched caudex, the caudex clothed with persistent leaf bases; stems 0.5–4 dm tall, erect, simple; herbage villous-tomentose, becoming glandular upward; leaves basal and cauline, 1.5–7 cm long, 1–2 mm wide, linear and entire or pinnatilobate; flowers numerous, borne in a compact, cylindroid, thyrsoid-spicate inflorescence, this often more than half the plant height; calyx 4.5–6 mm long, the triangular lobes spinulose-tipped, usually shorter than the tube; corolla whitish to yellowish brown, the tube 5–7 mm long, equaling or somewhat longer than the calyx, the lobes spreading, 2–4 mm long; staminal filaments attached at the sinuses, very short, the anthers 0.6–1 mm long; capsules subequal to the calyx, each locule 1-seeded; seeds 2.5–3 mm long, mucilaginous when wet. Juniper and sagebrush communities, often on barrens, at 1750 to 1890 m in Daggett County; Idaho and Montana, south to New Mexico; 6 (i). Our material belongs to var. *spicata*. There is a specimen at BRY (Cora Shoop 43, 16 May 1930) that bears the locality data of "near Ogden, Utah." The supposition that the specimen is correctly labeled is supported by a previous report for the species from Weber or Morgan County by the Hayden expedition in 1871.

Gilia stellata Heller Annual herbs from taproots and basal rosettes; stems 1.5–4 dm tall; herbage white-villous (especially on leaves and lower stem), often stipitate-glandular or intermixed glandular and villous trichomes upward; basal leaves 1.5–8.5 cm long, 3–30 mm wide, once or more commonly twice pinnatifid; cauline leaves reduced upward; flowers several to many, borne in paniculate, more or less bracteate cymes; calyx 2.5–5.5 mm long, the lobes triangular-attenuate, shorter than the tube, bearing purple, capitate glands on multicellular hairs; corolla 5–8 mm long, the tube included in the calyx, salverform or funnelform, lavender to white; staminal filaments inserted at the sinuses, the filaments about as long as the anthers, those 0.2–0.4 mm long; capsules equaling or surpassing the calyx; seeds 1.2–1.6 mm long. Creosote bush, Joshua tree, and other warm desert shrub communities at 730 to 915 m in Washington County; Arizona, Nevada, and California; 5 (ii).

Gilia stenothyrsa Gray Biennial (or perennial?) herbs from taproots and thick basal rosettes; stems 1–5 (6) dm tall; herbage minutely glandular to subglabrous below, becoming stipitate-glandular above; leaves 1–5 (6) cm long, 3–12 mm wide, pinnately toothed and with the blade broader than the length of the teeth or lobes, or pinnatifid and with the segments much longer than the rachis width; cauline leaves much reduced above; inflorescence thyrsoid-paniculate, cylindroid to obconic, comprising half or more of the plant height, more or less bracteate; calyx 3.5–6 mm long, the triangular lobes usually shorter than the tube, bristle-tipped; corolla white to lavender, blue, or purple, the tube 6–10 mm long, the lobes spreading, 3–5 mm long; staminal filaments inserted in the corolla tube, long-exserted; capsules surpassing the calyx; seeds several per locule, 1.5–2 mm long, not mucilaginous when wet. Shadscale, greasewood, other salt desert shrub, black sagebrush, and pinyon-juniper communities, often on barrens, at 1555 to 2840 m in Duchesne (type locality), Emery, and Uintah counties; endemic; 65 (x). This plant was collected first by John Charles Fremont on his second expedition in 1844.

Gilia subnuda Torr. ex Gray [*G. superba* Eastw., type from Hatche's Wash, San Juan County]. Biennial or perennial, more or less mephitic herbs from taproots and basal rosettes, a caudex more or less developed, simple or branched, clothed with marcescent leaf bases; stems 1.3–5 dm tall; herbage sparingly to densely glandular-viscid, typically with adhering sand grains, becoming stipitate-glandular above, or with the leaves more or less puberulent; basal leaves 2–9 cm long, 0.6–2.5 cm wide, spatulate to oblanceolate or obovate, merely toothed or pinnately lobed to entire, the blade much broader than the length of the teeth or lobes; inflorescence paniculately cymose, inconspicuously bracteate, the flowers typically clustered near branch ends; calyx 5–7.5 mm long, the triangular to lance-attenuate lobes longer to shorter than the tube; corolla carmine to vermillion, the tube 11–19 mm long, the limb 12–18 mm wide; stamens equally inserted, the anthers included; capsules subequal to the calyx; seeds few per locule, not mucilaginous when wet. Warm desert shrub, sagebrush, and pinyon-juniper communities in Emery, Garfield, Kane, San Juan, and Wayne counties; Colorado, New Mexico, and Arizona; 62 (xviii).

Gilia tenerrima Gray Annual herbs from taproots; stems 5–15 (20) cm tall; herbage stipitate-glandular, the capitate glands purplish black, or the leaves subglabrous; leaves mainly 3–20 mm long, 0.4–4 mm wide, linear to narrowly oblong, cauline and alternate or the basal in a weakly developed rosette, reduced in the inflorescence; flowers numerous in diffusely branched, paniculate cymose, the inflorescences often comprising more than half the plant height; calyx 1–1.8 mm long, the triangular lobes shorter than or subequal to the tube; corolla white to lavender, campanulate, 1.5–3.5 mm long, the lobes subequal to the tube or shorter; filaments inserted below the sinuses; capsules subequal to the calyx; seeds 0.8–1.2 mm long, mucilaginous when wet. Mountain brush, sagebrush, juniper, and ponderosa pine communities at 2070 to 2595 m in Cache, Salt Lake, Sevier, and Summit (type from valley of the Bear River) counties; Oregon to Montana, south to Nevada and Wyoming; 4 (i).

Gilia tridactyla Rydb. [*G. spicata* var. *tridactyla* (Rydb.) Const. & Rollins]. Perennial herbs from a taproot and usually branched caudex, the caudex clothed with brown, marcescent leaf bases; stems 5–12 (15) cm tall, erect, simple or branched above; herbage villous-tomentose, becoming villous-pilose upward; basal leaves rosettelike at the caudex apices, linear and entire or trifid, 0.3–2.5 cm long, 1–5 mm wide; flowers numerous, borne in compact, inconspicuously bracteate, terminal or lateral, headlike cymes, these usually much less than half the plant height; calyx 5–6.2 mm long, the triangular-acuminate teeth shorter than the tube; corolla whitish or yellowish (fading brown), the tube 6–8 mm long, commonly surpassing the calyx, the lobes spreading, 2–4 mm long; staminal filaments inserted in the corolla tube, the anthers 0.8–1 mm long; capsules subequal to the calyx or shorter; seeds 2.2–4 mm long, mucilaginous when wet. Spruce-fir, grass-forb, and sedge-forb communities, often on talus, at 3050 to 3450 m in Iron and Piute (type from Brigham Peak) counties; endemic; 9 (iii).

Gymnosteris Greene

Diminutive, annual herbs from slender taproots and with persistent connate-clasping cotyledons; stems naked except for the few-flowered, terminal, bracteate heads, the bracts connate basally; calyx 5-lobed, urn-shaped, scarious basally, the lobes green or suffused with red; corolla salverform to funnelform, persistent; stamens sessile, inserted in throat of corolla; capsules many-seeded; seeds mucilaginous when wet.

Gymnosteris parvula (**Rydb.**) **Heller** [*Gilia parvula* Rydb.]. Plants 1–5 cm tall; cotyledons 1–3 mm long, often purplish; bracts 4–13 mm long, 1–5 mm wide, lance-attenuate to ovate, often purplish; calyx 3–5 mm long, the triangular-acuminate teeth bristle-tipped, usually shorter than the tube; corollas 5–7 mm long, white to yellowish, persistent, the tube expanded and calyptralike on the fruit; capsule 3.5–4.5 mm long. Sagebrush-grass, ponderosa pine, aspen, and lodgepole pine communities at 2255 to 2745 m in Sevier and Uintah counties; Oregon to Wyoming, south to California, Nevada, and Colorado; 6 (ii).

Langloisia Greene

Low, annual (or short-lived perennial) herbs from taproots; leaves alternate, the basal rosette more or less developed, pinnately toothed to lobed, or pinnately bristly in the lower portion and with bristle-tipped teeth or lobes near the apex; flowers few in terminal, bracteate, cymose heads, or some axillary; bracts foliose, with teeth and bristles like the leaves; calyx 5-lobed, the green lobes terminating in long, white bristles or spines, the tube scarious between the ribs; corolla tubular-funnelform, regular or irregular; stamens 5, inserted in the corolla throat; capsules 2- to many-seeded; seeds mucilaginous when wet.

1. Leaves pinnately toothed almost to the base, the teeth each with a single apical bristle; corollas irregular ..... *L. schottii*
— Leaves pinnately bristly in the lower portion with paired or trifid spines or bristles, toothed mainly above the middle; corolla almost regular .... *L. setosissima*

Langloisia schottii (**Torr.**) **Greene** [*Navarretia schottii* Torr. in Emory]. Plants simple or typically branched from the base and with the lower branches ultimately prostrate, 2–11 cm long; herbage villous with long, crinkly, multicellular hair; leaves 0.5–3 cm long, mostly 2–6 mm wide, narrowly oblanceolate to linear, pinnately toothed, with each tooth bearing a single, apical bristle; bracts cuneate-oblanceolate, mainly 1–2.2 cm long, similar to the leaves; calyx 4–5.5 mm long to base of bristles, these 2–2.5 mm long, the lobes exclusive of spines subequal to the tube; corolla 8–12 mm long, white, yellowish, or pink, the upper lip 3-lobed, purple maculate; stamens long-exserted; capsules 3–4 mm long; seeds 0.6–1.2 mm long. Creosote bush, bursage, and Joshua tree communities at 730 to 980 m in Washington County; Arizona, Nevada, and California; 5 (i).

Langloisia setosissima (**T. & G.**) **Greene** [*Navarretia setosissima* T. & G. ex Torr. in Ives, type possibly from Washington County; *L. setosissima* var. *campyclados* Brand, type from near St. George]. Plants annual or rarely perennial (?), simple or typically branched from near the base, the lower branches commonly prostrate, 1–14 cm long; herbage more or less villous to subglabrous; leaves 0.6–3.5 mm long, linear and entire or most of them cuneate and with 3 apical to subapical teeth, but some also linear to oblanceolate and with 1 or 2 pairs of lateral teeth or lobes, the segments with a single spine apically, the margin below armed with usually paired or 3-pronged spines; bracts similar to the leaves; calyx 5–6 mm long, to the base of the spines, these 3.5–5.5 mm long, the lobes, exclusive of spines, subequal to the tube; corolla 12–17 mm long, violet to lavender, lined and spotted with purple; stamens exserted; capsules 5–6 mm long; seeds ca 1 mm long. Creosote bush, bursage, Joshua tree, blackbrush, other warm desert shrub, and pinyon-juniper communities at 730 to 1315 m in Kane and Washington counties; Arizona, Nevada, and California; 20 (0). The type material was taken by John Charles Fremont during his expedition of 1844 (possibly on May 11) at about the time when his companion Tabeau was killed by Indians. This is possibly the earliest collection from southern Utah.

Leptodactylon H. & A.

Perennial herbs or subshrubs, sprawling to compact; leaves opposite or alternate, palmatifid or subpinnatifid into linear, pungent lobes, often with axillary, tufted leaves; flowers solitary and axillary or in terminal cymes; calyx lobes entire, pungent, the sinuses membranous; corolla funnelform to salverform, white to cream (rarely pink to purplish); stamens inserted in the corolla tube or throat, included; filaments subequal to the anthers; capsules subcylindric, the locules several-seeded.

1. Leaves mainly alternate; plants loosely open; flowers 5–merous .... L. pungens
— Leaves opposite; plants pulvinate-caespitose to clump-forming 2
2(1). Corolla lobes 10–14 mm long; flowers ordinarily 6–merous; plants clump-forming; ovary 3– or 4–carpelled .... L. watsonii
— Corolla lobes 3–6 mm long; flowers ordinarily 4–merous; plants pulvinate-caespitose (phloxlike); ovary usually 3–carpelled .... L. caespitosum

Leptodactylon caespitosum Nutt. Pulvinate-caespitose perennial herbs, mat-forming, mainly 10–50 cm wide; stems obscured by overlapping, 3– to 5–cleft, spinulose, opposite leaves; herbage glabrous to glandular or the leaves irregularly ciliate; flowers 4–merous; calyx 6–8.2 mm long, the lobes unequal, much shorter than the sparingly villous tube; corolla white or cream, 12–20 mm long, the lobes 3–6 mm long; ovary 2–loculed; stigmas 2 (rarely 3). Sagebrush, pinyon-juniper, ponderosa pine, and bristlecone pine communities, often in barrens, at 1675 to 2350 m in Daggett (Browns Park), Emery (Buckhorn and Salt Washes), Garfield (Red Canyon-Bryce Canyon-Johns Valley), and Juab (House Range) counties; Wyoming, Colorado, Nebraska, and Nevada; 12 (i). Specimens of this taxon are often misidentified as *Phlox* species. The ropelike nature of the caudex branches, with marcescent, grayish leaves of previous seasons, and the usually 4–merous corollas are diagnostic. The disjunct distribution of this plant is related, at least in part, to the peculiar edaphic conditions of the substrates occupied.

Leptodactylon pungens (Torr.) Nutt. [*Cantua pungens* Torr.; *L. brevifolium* Rydb., type from Iron County]. More or less open shrubs or subshrubs, mainly 1–5 dm tall, more or less aromatic; herbage puberulent to glandular-puberulent or -villous to glabrous; stems obscured or the internodes apparent; leaves alternate to subopposite (or the lower opposite), 3– to 9–cleft, spinulose; flowers 5–merous; calyx 7–11.5 mm long, the lobes more or less unequal, commonly shorter than the tube; corolla 12–25 mm long, the lobes 5–15 mm long, cream to yellowish, often suffused with purple externally; capsule 4–5.5 mm long, usually 3–loculed; stigmas 3; $2n = 18$. Shadscale, sagebrush, mountain brush, pinyon-juniper, and ponderosa pine communities at 1400 to 2745 m in all Utah counties, except Daggett, Davis, Grand, Morgan, Salt Lake, Sanpete, Summit, Utah, Wasatch, and Weber; British Columbia to Montana, south to California, Arizona, New Mexico, and Nebraska; 97 (viii).

Leptodactylon watsonii (Gray) Rydb. [*Gilia watsonii* Gray, type from Salt Lake County; *G. floribunda* var. *arida* Jones, type from Wayne County]. Plants cushion-like (depressed hemispheric) to open and sprawling (or pendulous) subshrubs; herbage puberulent to stipitate-glandular or glabrous; leaves opposite, 3– to 9–cleft, spinulose; flowers 6–merous; calyx 7–12 mm long, the lobes unequal, shorter than the tube; corolla 15–28 mm long, the lobes 7–15 mm long, white to cream; capsules 4–5 mm long, usually 4–loculed; stigmas 4. Blackbrush, sagebrush, pinyon-juniper, mountain brush, and mixed conifer communities (often in crevices) at 1220 to 3050 m in Beaver, Box Elder, Cache, Emery, Garfield, Grand, Kane, Salt Lake, San Juan, Sanpete, Tooele, and Utah counties; Idaho to Wyoming, south to Nevada and Colorado; 32 (iv).

Linanthastrum Ewan

Perennial subshrubs, erect to ascending; leaves opposite, palmatifid or trifid into linear, subulate-tipped lobes, often with tufted, axillary leaves; flowers borne in bracteate, terminal, headlike cymes, 5–merous; calyx lobes entire, subulate, the sinuses narrowly membranous; corolla salverform, white to cream; stamens inserted in the throat, slightly exserted; capsules 3–loculed, the locules few-seeded.

Patterson, R. 1977. A revision of *Linanthus* sect. Siphonella (Polemoniaceae). Madroño 24: 36–48.

Linanthastrum nuttallii (Gray) Ewan [*Gilia nuttallii* Gray; *Linanthus nuttallii* (Gray) Greene ex Milliken; *Leptodactylon nuttallii* (Gray) Rydb.]. Plants 8–25 (30) cm tall, clump-forming, compact to open; herbage puberulent (stems) or glabrous (otherwise); leaves opposite, 5– to 9–cleft, the lobes linear, subulate to minutely spinulose apically; flowers subsessile; calyx 6–9.5 mm long, the lobes lance-linear to subulate, 3–veined, ciliate with slender hairs; corolla tube 6–10 mm long, white, the lobes 5–8 mm long, spreading; capsules 4–5.5 mm long; $2n = 18$. Sagebrush, pinyon-juniper, ponderosa pine, mountain brush, bristlecone pine, aspen, and spruce-fir communities at 1370 to 3265 m in Cache, Garfield, Kane, Salt Lake, Uintah, Utah, and Washington counties; Washington to Montana, south to California, Arizona, New Mexico, and Mexico; 34 (iv).

Linanthus Benth.

Annual or winter annual herbs from taproots; leaves opposite (the upper sometimes alternate), palmatifid, the segments linear; flowers solitary in leaf axils or paniculately to capitately cymose at branch ends; calyx 5–lobed, usually membranous below the sinuses; corolla campanulate to funnelform or salverform; stamens subequally inserted in the corolla tube or throat; capsules 3– or

4-loculed, ellipsoid to cylindric; seeds 1 to many per locule.

1. Flowers typically long-pedicellate; distribution various . 2
— Flowers sessile or subsessile; known from Kane and Washington counties 4
2(1). Corollas bright yellow, 7–15 mm long; plants rare, known from Washington County .... *L. aureus*
— Corollas white or pink to purplish, 1.5–6 mm long; plants uncommon to common, widely distributed 3
3(2). Corollas 1.5–2.5 mm long, less than 1.5 times longer than the calyx, glabrous within .... *L. harknessii*
— Corollas 2.5–5 mm long, ca 1.5 times longer than the calyx, hairy or glabrous within .... *L. septentrionalis*
4(1). Inflorescence few- to several-flowered, capitate; plants caespitose, forming low clumps or mats ..... *L. demissus*
— Inflorescence of 1 or few flowers, not capitate; plants neither clump- nor mat-forming 5
5(4). Corolla 18–25 mm long or more, the lobes 5–10 mm wide .... *L. dichotomus*
— Corolla 8–15 mm long or more, the lobes less than 5 mm wide .... *L. bigelovii*

Linanthus aureus (**Nutt.**) **Greene** [*Gilia aurea* Nutt.]. Plants 5–15 cm tall; stems simple or branched from near the base; herbage subglabrous or puberulent and often glandular; leaves much shorter than the internodes, 3- to 7-cleft, the lobes linear to oblong, 2–6 mm long; flowers on pedicels mainly 3–15 mm long; calyx 5–7 mm long, the triangular-ovate lobes shorter than the tube, glandular-hirtellous, ciliate; corolla 6–14 mm long, yellow, funnelform, the tube included in the calyx; stamens glabrous, inserted in the throat of the corolla; capsules included; n = 9. Ponderosa pine community at ca 1650 m in Washington County; Arizona, New Mexico, Nevada, California, and Mexico; 1 (0). This is a strikingly beautiful plant. Our single specimen is without collector.

Linanthus bigelovii (**Gray**) **Greene** [*Gilia bigelovii* Gray]. Plants 6–40 cm tall; stems simple or variously branched; herbage sparingly villous to glabrous or glandular; leaves longer to much shorter than the internodes, entire or 2–3 cleft, mainly 0.8–4 cm long; flowers solitary in axils of dichotomous cymes; calyx 8–14 mm long, the thickened, subulate, spinulose-tipped lobes much shorter than the tube, often curved; corolla 10–15 mm long, surpassing the calyx by 2–5 mm, white to cream or suffused with purple in part; stamens glabrous, inserted in the corolla tube; capsules included; n = 9. Creosote bush, bursage, Joshua tree, and pinyon-juniper communities at 800 to 1650 m in Kane and Washington counties; Arizona, Nevada, and California; 7 (i).

Linanthus demissus (**Gray**) **Greene** [*Gilia demissa* Gray; *G. dactylophyllum* Torr. in Ives, nomen confusum?; *L. dactylophyllum* (Torr.) Rydb.]. Prostrate to ascending or erect, clump- or mat-forming plants, 2–10 cm long; stems simple or more commonly branched from the base; herbage more or less villous with crinkly, multicellular hairs and often glandular; leaves shorter to longer than the internodes, simple or 3- to 5-cleft, the lobes 2–10 mm long, acicular, minutely spinulose-tipped; calyx 3–6.5 mm long, the unequal lobes lance-attenuate, spinulose-tipped, much longer than the tube; corolla white, often streaked with purple, 6.5–8.5 mm long, campanulate, the lobes 2.5–3.5 mm wide; stamens inserted at the base of the throat, glabrous, included; capsule included in the calyx; n = 9. Creosote bush, bursage, Joshua tree, and blackbrush communities at 730 to 1130 m in Washington County; Arizona, Nevada, and California; 13 (ii).

Linanthus dichotomus **Benth.** Evening Snow. Plants erect, simple or dichotomously branched, 3–15 cm tall; herbage glabrous or sparingly villous-ciliate on calyx lobes; leaves opposite, simple or 3- to 7-parted, the lobes (or leaves) mainly 10–20 mm long, linear, apiculate; flowers short-pedicellate in axils of dichotomous, cymose branches, or terminal; calyx 8–14 mm long, markedly bicolored, the ribs green, alternating with broad hyaline membranes, these sometimes produced as a low protruberance at the sinuses, the lance-oblong lobes much shorter than the tube, apiculate; corolla 15–30 mm long, funnelform, white or marked with purple, the lobes 5–10 mm broad, spreading; capsule included in the calyx. Ponderosa pine-mountain brush community at ca 1800 m in Washington County (Franklin and Baird 468 BRY); Arizona, Nevada, and California; 1 (0).

Linanthus harknessii (**Curran**) **Greene** [*Gilia harknessii* Curran]. Plants erect or ascending, simple or branched, mainly 4–25 cm long, glabrous or puberulent; leaves opposite or the upper alternate, simple or palmatifid with 3–7 linear lobes mainly 4–12 mm long; flowers on pedicels mainly 3–15 mm long, solitary or in open cymes; calyx 2.5–3.8 mm long, the lance-attenuate lobes somewhat shorter than the tube, apiculate; corolla 1.5–3.5 mm long, white to blue, funnelform, subequal to the calyx or only somewhat longer; capsule included in the calyx, the locules 1–seeded; seeds mucilaginous when wet. Open sites in mountain brush, sagebrush, and aspen communities at 2070 to 2440 m in Cache, Morgan, Rich, Salt Lake, and Wasatch counties; Oregon to Idaho, south to California and Nevada; 5 (0).

Linanthus septentrionalis **Mason** Plants erect or ascending, simple or branched, mainly 4–25 cm long, glabrous or puberulent; leaves opposite, simple, or most of them palmately 5- to 7-cleft, the segments mainly 3–20 mm long, linear; calyx 3–4.5 mm long, the lance-attenuate lobes shorter than the tube; corolla 2.5–4 mm long, from slightly surpassing to 1.5 times longer than the calyx; capsule included in the calyx, the locules 1 (or 2) -seeded; seeds becoming mucilaginous when wet. Sagebrush, aspen, and sprude-fir communities at 2135 to 2900 m in Cache, Carbon, Duchesne, Emery, Morgan, Rich, Salt Lake, Sanpete, Summit, Uintah, Utah, and Wasatch counties; British Columbia and Alberta south to California, Nevada, and Colorado; 42 (iii).

Microsteris Greene

Simple or branched low annual herbs; leaves opposite below, the upper alternate, lanceolate, entire; flowers pedicellate, often in axillary pairs; calyx 5–lobed, with broad hyaline membranes between the ribs, the lobes subequal; corolla salverform; stamens unequally inserted on the corolla tube, included; ovary 3–loculed; stigmas 3–lobed; capsule bursting the calyx in age; locules 1–seeded; seeds mucilaginous when wet.

Microsteris gracilis (**Hook.**) **Greene** [*Gilia gracilis* Hook.; *Phlox gracilis* (Hook.) Greene; *G. gracilis* ssp. *spirillifera* var. *nana* Brand, type from Fillmore]. Plants mainly 1–20 cm tall, branched above the base or simple; herbage more or less villous with multicellular hairs and often glandular above; leaves 5–26 (30) mm long, 2–7 mm wide, subsessile to short-petiolate, lanceolate to linear or obovate; cotyledons often persistent, short-petiolate, oval or obovate; calyx 4.5–8 mm long, the lobes usually shorter than the tube; corolla 6–12 mm long, surpassing the calyx, the limb pink to lavender or white; capsule shorter than the calyx, the limb pink to lavender or white; capsule shorter than the calyx; seeds ca 3 mm long. Sagebrush, pinyon-juniper, mountain brush, and ponderosa pine communities at 1220 to 2505 m in Beaver, Box Elder, Cache, Dagget, Davis, Duchesne, Iron, Juab, Millard, Morgan, Salt Lake, San Juan, Sanpete, Sevier, Summit, Tooele, Uintah, Utah, Washington, and Weber counties; British Columbia to Montana, south to California, Arizona, and New Mexico; Mexico and South America; 68 (viii). Our specimens belong to var. **humilior** (**Hook.**) **Cronq.** [*Collomia gracilis* var. *humilior* Hook.].

Navarretia Ruiz & Pavon

Annual herbs; stems erect, simple or branched; leaves alternate, entire or pinnatilobate, bracteate upward, acerose or spine-tipped; flowers borne in dense, bracteate, cymose heads; calyx 5 (or 4) -lobed, the lobes much longer than the tube, entire or toothed; corolla funnelform or salverform; stamens variously inserted in the throat or at the sinuses; stigmas 2 or 3; capsule 1– to 3-loculed; seeds solitary or few to many per locule.

Crampton, B. 1954. Morphological and ecological considerations in classification of *Navarretia* (Polemoniaceae). Madrono 12: 225–288.

1. Corollas yellow; plants glandular; stigmas typically 3 .... N. breweri
— Corollas white to bluish or lavender; plants villous, puberulent, or almost glabrous; stigmas typically 2 .... N. intertexta

Navarretia breweri (**Gray**) **Greene** [*Gilia breweri* Gray]. Plants erect, 1–12 cm tall; herbage glandular-puberulent; leaves 0.4–2 (3) cm long, entire or with 1–4 pairs of lateral, linear, pungent lobes; bracts mainly 4–12 mm long, pinnatilobate, the pinnae often again 3–lobed; calyx 6–10 mm long, the lobes unequal, much longer than the scarious, hyaline tube, acerose; corolla 5–7 mm long, yellow, funnelform; stigma 3–lobed; capsule few-seeded, dehiscent from the base. Pinyon-juniper, mountain brush, ponderosa pine, aspen, and grass-forb communities at 2070 to 3250 m in Cache, Duchesne, Morgan, Rich, Salt Lake, San Juan, Sevier, Summit, Uintah, Utah, Wasatch, and Washington counties; Washington south to California, Arizona, and Colorado; 30 (iv).

Navarretia intertexta (**Benth.**) **Hook.** [*Aegochloa intertexta* Benth.; *N. propinqua* Suksd.]. Plants simple or branched, mainly 3–15 cm tall; herbage more or less villous (retrorse below the heads); leaves simple or pinnatifid-bipinnatifid, mainly 0.8–2.5 cm long, the lobes acicular-linear, pungent; bracts like the pinnatifid leaves; calyx 8–10 mm long, the lobes simple or pinnatifid, much longer than the scarious base; corolla white to blue or lavender, 5–9 mm long, funnelform; stigmas 2–lobed; capsule 1-loculed, several-seeded. Sagebrush, mountain brush, aspen, and grass-forb communities at 1710 to 2290 m in Cache, Morgan, Summit, Wasatch, Washington, and Weber counties; Washington and Saskatchewan, south to California, Arizona, and Colorado; 8 (0).

Phlox L.

Perennial herbs; stems erect or prostrate to decumbent or ascending; leaves opposite, simple, entire; flowers in corymbiform or paniculate cymes or solitary, showy; calyx 5-lobed and -ribed, typically with scarious margins and scarious, intercostal membranes between the ribs; corolla salverform, the throat constricted; stamens unequally inserted; ovary 3–lobuled; seeds 1 to several per locule, not mucilaginous when wet.

1. Plants 4–10 dm tall or more; leaves 1–3 cm wide; cultivated ornamentals .... *P. paniculata*
— Plants usually less than 3 dm tall; leaves less than 0.5 cm wide; cultivated or more commonly indigenous .. 2

2(1). Stems sprawling, forming mats mostly 5–10 dm broad or more; flowers usually more than 1 in terminal, bracteate cymes; plants cultivated 3
— Stems erect or more or less cushion or mat-forming, the mats (when formed) mainly less than 3 dm broad; flowers solitary at stem ends or more than 1 in terminal, bracteate cymes; plants indigenous 4

3(2). Corolla lobes deeply notched .... *P. subulata*
— Corolla lobes rounded to retuse or erose .... *P. nivalis*

4(2). Plants more or less open, erect to sprawling, the internodes apparent; flowers usually more than 1 in terminal, bracteate cymes (except in *P. austromontana*) 5
— Plants pulvinate-caespitose to mat-forming or cushionlike, often compact; internodes apparent or not; flowers solitary at stem ends 7

5(4). Leaves 3–4 mm wide, oblong-elliptic to lanceolate; plants of San Juan County .... *P. cluteana*
— Leaves mainly 1–2.5 mm wide, linear to acicular; distribution various 6

6(5). Calyces glabrous or merely villous apically, villous on lobes ventrally; flowers solitary at stem tips; plants of Kane and Washington counties ..... *P. austromontana*
— Calyces villous-glandular externally and on the lobes ventrally; flowers commonly more than 1 in terminal, bracteate cymes; plants broadly distributed .... *P. longifolia*

7(4). Plants markedly glandular (at least the calyces) 8
— Plants not glandular (or with sessile, granular glands) 9

8(7). Corolla tube coarsely glandular hairy, with long, multicellular hairs; intercostal membranes carinate .... *P. gladiformis*
— Corolla tube glabrous or with sessile glands; intercostal membranes not carinate .... *P. pulvinata*

9(7). Plants more or less softly woolly-tomentose, the tomentum white 10
— Plants variously hairy to glabrous, but not woolly or tomentose 11

10(9). Plants compactly pulvinate-caespitose, forming rounded cushions or mounds; calyx 3–4.5 mm long; plants of central western Utah .... *P. muscoides*
— Plants mat-forming, or if mound or cushionlike, not compactly pulvinate-caespitose; calyx 5–7.5 mm long; plants broadly distributed .... *P. hoodii*

11(9). Leaves mainly 2–5 mm long; plants forming rounded cushions and mounds, typically pulvinate-caespitose .... *P. tumulosa*

— Leaves mainly 6–35 mm long or more; plants matforming or cushionlike, but not especially pulvinate-caespitose 12

12(11). Calyx pubescent with short, stiff, spreading hairs .. .. *P. griseola*

— Calyx glabrous or pubescent with crinkly or contorted hairs 13

13(12). Flowers pedicellate, the pedicels 3–10 mm long; calyx not carinate on the intercostal membrane; plants of northern Utah .... *P. multiflora*

— Flowers sessile or subsessile; calyx carinate on the intercostal membrane; plants of the southern 2/3 of Utah .... *P. austromontana*

Phlox austromontana Cov. Desert Phlox. Plants caespitose, cushion or matlike, from a pluricipital caudex and taproot, mainly 0.5–3 dm wide; herbage pilose-puberulent to subglabrous or the calyx glabrous to villous externally; leaves opposite, mainly 5–20 mm long, simple, linear-subulate; flowers solitary, sessile or subsessile at branch tips; calyx urceolate to campanulate, glabrous to villous, the intercostal membranes carinate, the lobes villous internally; corolla white, blue, pink, lavender, or yellowish, the tube 8–15 mm long; styles 2–9 mm long. This is a complex assemblage of variants, some sufficiently distinctive and with geographic correlation as to warrant taxonomic recognition. The morphology is, however, wholly confluent. Trends within the variation are recognized at varietal level.

1. Plants more or less open, the internodes typically apparent; plants of western Kane and much of Washington counties 2

— Plants variously open to compact; distribution various .. 3

2(1). Calyx usually glabrous, the leaves (or some of them) 20–35 mm long; corollas usually bright pink; morphology transitional to the next . *P. austromontana* var. *jonesii*

— Calyx usually at least moderately villous, the leaves typically 10–22 mm long; corolla commonly white ... . *P. austromontana* var. *prostrata*

3(2). Flowers yellowish (fading lemon yellow); leaves 10–25 mm long; calyx campanulate .... *P. austromontana* var. *lutescens*

— Flowers white, pink, or lavender (sometimes fading cream); leaves mostly less than 15 mm long; calyx turbinate to subcylindric .... *P. austromontana* var. *austromontana*

Var. austromontana [*P. densa* Brand, type from Frisco]. Mixed desert shrub, salt desert shrub, pinyon-juniper, sagebrush, mountain brush, and ponderosa pine communities at 1525 to 3050 m in Beaver, Carbon, Duchesne, Emery, Garfield, Iron, Juab, Kane, Millard, Piute, Sanpete, Sevier, Tooele, Uintah, Washington (type from Beaver Dam Mts.), and Wayne counties; Nevada, California, and Arizona; 159 (xxiii).

Var. jonesii (Wherry) Welsh [*P. jonesii* Wherry, type from Zion Canyon]. Ponderosa pine, pinyon-juniper, and mountain brush communities at 1435 to 2600 m in Kane and Washington counties; endemic; 12 (ii). This variety forms intermediates with both var. *prostrata* and var. *austromontana*. It is partially sympatric with both.

Var. lutescens Welsh Blackbrush, ash, and squawbush community at ca 1220 m in eastern Garfield (type along the margin of Cataract Canyon) County; endemic; 1 (i). The plants are suffrutescent, rounded, cushions growing in crevices in rimrock of Cedar Mesa Sandstone. Because of the peculiar flower color and growth habit, these plants were initially mistaken for the superficially similar *Leptodactylon watsonii*.

Var. prostrata E. Nels. Mountain brush and pinyon-juniper communities at 1220 to 2135 m in Washington (type from Silver Reef) County; endemic (?); 14 (0).

Phlox cluteana A. Nels. Navajo Mt. Phlox. Plants with stems single or more or less clumped from subterranean, pluricipital, subrhizomatous caudices, mainly 4–12 (20) cm tall; herbage glandular-puberulent to -villous or merely villous to puberulent; leaves elliptic to linear, oblong, oblanceolate, or lanceolate, 0.8–3 (5) cm long, 2–5 mm wide; flowers on pedicels 3–15 mm long, borne singly or 2 to several in terminal cymes; calyx 7–9 mm long, the intercostal membranes not carinate; corolla tube 14–18 mm long, the lobes 7–10 mm long, pink to lavender or white; stamens included or slightly exserted; style 9–14 mm long. Ponderosa pine community at ca 2285 to 3050 m in San Juan (type from Navajo Mt.) County; Arizona; 4 (0).

Phlox gladiformis (Jones) E. Nels. Cedar Canyon Phlox. [*Phlox longifolia* var. *gladiformis*, type from Cedar Canyon, Iron County; *P. caesia* Eastw., type from Bryce Canyon]. Plants caespitose, forming cushions or mounds to ca 3 dm wide, from a taproot and pluricipital caudex; herbage glandular-hairy throughout, often with some adhering sand grains; leaves 8–20.(25) mm long, 1–2.5 mm wide, rigid, lance-linear to linear-subulate, spinulose-tipped; flowers subsessile or on pedicels to 3 mm long, borne at stem tips; calyx 3–9 mm long, the subulate lobes spinulose, the intercostal membrane carinate towards the base; corolla tube 8–14 mm long, glandular-villous, the lobes white to pink or lavender, 5–9 mm long; stamens included; style 2–6 mm long. Pinyon-juniper, ponderosa pine, and bristlecone pine communities at 1980 to 2535 m in Garfield, Iron, and Washington counties; Nevada; 20 (ii).

Phlox griseola Wherry Grayleaf Phlox. Plants caespitose, mat-forming, from taproots and pluricipital caudices, the caudex branches usually short and more or less tightly clustered, commonly 5–12 cm wide; herbage hirsute or hirtellous, with stiff, spreading trichomes, not glandular; leaves 4–12 mm long, lance- to linear-subulate, spinulose-tipped; flowers solitary, sessile, borne apically on branches; calyx 7–10 mm long, hirsute like the leaves, the lobes linear-subulate, sharply spinulose, intercostal membranes not carinate; corolla tube 9–14 mm long, the lobes pink to lavender or white, 5–6 mm long; anthers included; styles 1.5–3 mm long. Pinyon-juniper and mountain brush communities at 1525 to 2000 m in Beaver, Iron, and Washington counties; Nevada (Great Basin endemic); 7 (ii). This plant is apparently allied closely with *P. tumulosa*, which is intermediate between *P. griseola* and *P. muscoides*. A specimen tentatively assigned here (Higgins 14407 BRY) has glandular calyces, a feature not known in this species. The specimen was previously identified as *P. pulvinata* but is quite unlike that species. Ultimate disposition awaits additional collections.

Phlox hoodii Richards. Carpet Phlox. Plants cushion or matlike from a taproot and pluricipital, commonly prostrate caudex, mainly 5–30 cm wide; herbage more or less arachnoid-woolly with white hairs; leaves typically 2–12 mm long, 0.5–1 mm long, linear, spinulose-tipped; flowers solitary and sessile at stem ends; calyx 5–7.5 mm long, more or less tomentose or villous with white, multicellular hairs, the lobes linear-subulate, spinulose, intercostal membranes not carinate; corolla tube 6–15 mm long, the lobes 4–8 mm long, white, pink, lavender, or blue; stamens included; style 2–5 mm long; 2n = 28. Shadscale, greasewood, horsebrush, budsage, pinyon-juniper, ponderosa pine, and less commonly alpine parkland communities at 1460 to 3265 m in virtually all Utah counties, except Beaver, Iron, Kane, and Washington; Alaska and Yukon, south to California, Nevada, Arizona, and Colorado; 122 (xix). Our material belongs to var. **canescens** (T. & G.) Peck [*P. canescens* T. & G., type from S of Great Salt Lake].

Phlox longifolia **Nutt.** Longleaf Phlox. [*P. speciosa* var. *stansburyi* Torr. in Emory; *P. longifolia* var. *stansburyi* (Torr.) Gray, *P. longifolia* var. *stansburyi* f. *brevifolia* Gray; *P. longifolia* var. *brevifolia* (Gray) Gray; *P. grahamii* Wherry, type from Minnie Maud Creek, Duchesne County, teratological with smut]. Plants with stems solitary or more or less clumped from superficial to subterranean, pluricipital, subrhizomatous caudices, mainly 3–40 cm tall, often woody below; herbage more or less glandular-villous (especially in inflorescence); leaves 1–8 cm long or more, 1–6 mm wide, lance-subulate, linear-subulate, lance-ovate, or elliptic; flowers in leafy-bracteate, terminal cymes, subsessile or on pedicels 3–20 mm long (or more); calyx 7–13 mm long, the intercostal membranes carinate; corolla tube 10–30 mm long, mostly 1–3 times longer than the calyx, the lobes 7–15 mm long, pink to lavender or white; stamens included or somewhat exserted; style 6–20 mm long; 2n = 28. Salt and mixed cool desert shrub, shrub-grass, sagebrush, pinyon-juniper, mountain brush, ponderosa pine, aspen, and spruce-fir communities at 1340 to 2930 m in all Utah counties; British Columbia to Montana, south to California, Arizona, and New Mexico; 261 (xxxv). This species consists of a complex aggregation of forms. Some of the phases are distinctive, when viewed in the extreme, but they are connected in continuous, intergrading series. Recognition of infraspecific taxa seems to be moot.

Phlox multiflora **A. Nels.** Plants cushion or matlike from a taproot and pluricipital caudex, mainly 8–25 cm wide; herbage glabrous or glandular-scaberulous; leaves 8–30 mm long, 1–2 mm wide, linear-subulate, spinulose-apiculate; flowers solitary at branch ends, subsessile or on pedicels 3–10 mm long; calyx 11–15 mm long, glabrous, the lobes lance-subulate, spinulose, the intercostal membranes not carinate; corolla tube 10–15 mm long, the lobes white to pink or bluish, 6–11 mm long; stamens included or slightly exserted; styles 5–8 mm long; 2n = 14. Sagebrush, ponderosa pine, aspen, Douglas fir, and lodgepole pine communities at 2200 to 3265 m in Box Elder, Daggett, Rich, Summit, and Uintah counties; Idaho and Montana, south to Nevada, and Colorado; 19 (i).

Phlox muscoides **Nutt.** Moss Phlox. Plants densely pulvinate-caespitose, cushion or moundlike from a taproot and pluricipital, spreading to ascending caudex, mainly 3–10 cm wide; herbage arachnoid-woolly with tangled white hairs; leaves 2–5 mm long, oblong- to lance-subulate, 0.5–1 mm wide; flowers solitary and sessile at stem ends; calyx 3–4.5 mm long, tomentose, the lobes lance-subulate to narrowly triangular, spinulose, the intercostal membranes not carinate, obscured by the tomentum; corolla tube 5–10 mm long, the lobes white to pink, blue, or lavender, 3–5 mm long; stamens included; style 2–5 mm long; 2n = 14. Black sagebrush, budsage, shadscale, and pinyon-juniper communities at 1400 to 2105 m in Beaver, Juab, and Millard counties; Oregon to Montana, south to Nevada, Colorado, and Nebraska; 21 (vi). The morphology of this taxon lies between those of *P. hoodii* and *P. tumulosa*, at least in part. Suggested is the possibility that *P. muscoides*, as present in Utah, might have arisen independently from those in other places, the similar morphologies having been arrived at more than once. More work is indicated.

Phlox nivalis **Lodd.** Plants forming broad, carpetlike mats or cushions from a taproot, the caudex branches ultimately prostrate-spreading, the flowering stems erect-ascending, mainly 5–10 dm broad or more; herbage hirtellous to villous with multicellular hairs; leaves mainly 10–20 mm long, 1–2.5 mm wide, linear to elliptic or narrowly lanceolate, ciliate; flowers few to several in terminal, bracteate cymes, the pedicels 3–25 mm long or more; calyx 7–9 mm long, the lobes lance-attenuate, spinulose, the intercostal membranes not carinate, spreading-hairy; corolla tube 12–16 mm long, the lobes purple, pink, or white, rounded to retuse or erose apically, 7–9 mm long; stamens included; style 4–6 mm long. Cultivated ornamental of gardens, long-persisting in lower elevation portions of Utah; introduced from the eastern U. S.; 1 (0). The material examined is markedly similar to *P. subulata*, another beautiful ornamental species grown in our region.

Phlox paniculata **L.** Garden Phlox. Plants erect, 6–10 dm tall or more; herbage puberulent to subglabrous; leaves 1.2–12 cm long, 0.8–5 cm wide, lanceolate to elliptic or ovate, subsessile or short-petiolate, acuminate apically; flowers numerous, borne in terminal, compound, often closely aggregated, corymbiform cymes, subsessile or short-pedicellate; calyx 6–9 mm long, the lance-linear lobes not spinulose, glabrous or puberulent; corolla tube 20–28 mm long, the lobes 8–12 mm long, pink, lavender, blue, or white; stamens included; 2n = 14. Cultivated ornamentals, long persisting, in lower elevation portions of Utah; introduced from the eastern U. S.; 3 (i).

Phlox pulvinata **(Wherry) Cronq.** [*P. caespitosa* ssp. *pulvinata* Wherry; *P. caespitosa* authors, misapplied]. Plants cushion or matlike from taproots and multicipital caudices, these often horizontally spreading, mainly 8–25 cm wide; herbage more or less glandular, especially on the calyx; leaves 2–12 mm long, 0.6–1.5 mm wide, linear to oblong or lance-oblong, spinulose-tipped; flowers solitary, sessile at branch tip; calyx 4.6–9 mm long, the lobes lance-subulate, spinulose-tipped, the intercostal membrane not carinate; corolla tube 8–13 mm long, glabrous, the lobes 4–8 mm long, white or less commonly blue to pink; stamens included; style 2–6 mm long; 2n = 14. Sagebrush, aspen, lodgepole pine, grass-forb, and alpine tundra communities at 2315 to 3450 m in Beaver, Cache, Daggett, Duchesne, Emery, Garfield, Iron, Juab, Piute, Salt Lake, Sanpete, Summit, Tooele, and Wayne counties; Oregon to Montana, south to California, Nevada, and New Mexico; 66 (xi).

Phlox subulata **L.** Plants forming broad, carpetlike mats or cushions from a taproot and intricately branched, spreading caudex, mainly 5–10 dm broad or more; herbage hirtellous to villous with multicellular hairs;

leaves 6–20 mm long, 1–1.5 mm wide, linear to elliptic, rather abruptly apiculate, ciliate near the base; flowers few to several in terminal, bracteate cymes, the pedicels 3–15 mm long; calyx 6–8 mm long, the lobes lance-attenuate, spinulose, the intercostal membranes not carinate; corolla tube 10–13 mm long, the lobes lavender to pink or white, emarginate apically, 7–10 mm long; stamens included or slightly exserted; styles 5–8 mm long. Cultivated ornamentals, long persisting, in lower elevation portions of Utah; introduced from the eastern U. S.; 1 (i).

Phlox tumulosa Wherry Plants densely pulvinate-caespitose, cushion or moundlike herbs from a taproot and compactly pluricipital caudex, mainly 3–15 cm wide; herbage glabrous or more or less pubescent with spreading, multicellular hairs; leaves 1–5 mm long, 0.5–1.3 mm wide, oblong-ovate to oblong, or almost linear, spinulose-tipped; flowers solitary and sessile at branch ends; calyx 4–5 mm long, villous-pilose, the lobes lance-subulate, spinulose, the intercostal membranes not carinate, not obscured by hairs; corolla tube 8–11 mm long, the lobes 3–5 mm long, white to pink or lavender; stamens not exserted; style 1.5–3.5 mm long; 2n = 14, 28. Black and big sagebrush and pinyon-juniper communities at 1675 to 2075 m in Beaver and Iron counties; Nevada (Great Basin endemic); 8 (i). This entity is intermediate in its morphology between *P. muscoides* and *P. griseola*. It has the short leaves of the former and intermediate pubescence, but some specimens are transitional.

Polemonium L.

Annual or perennial, commonly mephitic herbs from taproots, rhizomes, or caudices; stems erect or ascending, simple or branched; leaves pinnate-pinnatifid, alternate; flowers in terminal or axillary, often corymbose cymes; calyx herbaceous, campanulate, 5-lobed; corolla 5-lobed, rotate-campanulate to funnelform, the tube and throat often indistinct, mostly blue violet to lavender or white; stamens inserted on the tube, exserted or included; capsule 3-locular, each locule with 1 to many seeds.

1. Corolla ochroleucous to pale yellow; leaflets rather remote; plants known from a single early collection from the Tushar Mts., Piute County .... *P. brandegei* (Gray) Greene
— Corolla blue, royal blue, purple, or white, but not otherwise as above; plants known from modern collections . 2
2(1). Corolla tubular-funnelform; plants subscapose, from definite caudices; leaflets pseudoverticillate, crowded .... *P. viscosum*
— Corolla rotate-campanulate; plants caulescent, from caudices or rhizomes; leaflets not pseudoverticillate, not especially crowded or overlapping 3
3(2). Plants annual; corolla 2–6 mm long; plants of lower elevations .... *P. micranthum*
— Plants perennial; corolla 10–15 mm long; montane or alpine 4
4(3). Stems decumbent basally, from a horizontal rhizome; plants commonly of wet meadows and streamsides ... . *P. caeruleum*
— Stems erect basally and throughout, from vertical caudices; plants of mesic, but scarcely hydrophytic communities 5
5(4). Plants mostly 0.6–2.5 dm tall, compactly clump-forming .... *P. pulcherrimum*
— Plants mostly 4–12 dm tall, loosely clump-forming .. .. *P. foliossimum*

Polemonium caeruleum L. Blue Jacobsladder. [*P. occidentale* Greene; *P. caeruleum* var. *occidentale* (Greene) St. John; *P. occidentale* ssp. *amygdalinum* Wherry]. Mephitic, perennial herbs from horizontal rhizomes, the stems decumbent basally, otherwise erect, mainly 4–10 dm tall; herbage glandular-villous to glabrate; leaves 3–25 cm long, the basal and lower ones long-petiolate, with 9–27 leaflets, these lanceolate to elliptic, 5–40 mm long, 1–12 mm wide, the terminal 3 more or less confluent; reduced upward; flowers in narrow, thyrsoid inflorescences; calyx 5–8 mm long at anthesis, the lobes subequal to or shorter than the tube; stamens about equaling the corolla lobes; corollas 10–15 mm long and about as broad, blue purple or white; seeds not mucilaginous when wet; 2n = 18. Moist to wet sites in meadows and along streams at 1340 to 3085 m in Cache, Carbon, Daggett, Duchesne, Emery, Garfield, Iron, Kane, Salt Lake, Sanpete, Sevier, Summit, Utah, and Wasatch counties; Alaska and Yukon, south to California and Colorado; 56 (iv). Our material belongs to var. **pterospermum** Benth. in A. DC.; at subspecific level it is ssp. *amygdalinum* (Wherry) Munz.

Polemonium foliosissimum Gray Leafy Jacobsladder. Plants perennial from a subrhizomatous, vertical caudex; stems erect, mainly 4–12 dm tall, leafy throughout, lacking a basal tuft of leaves; herbage sparingly to densely villous to puberulent and often glandular upward; leaves mainly 3–15 cm long; leaflets (3) 5–25, lanceolate to elliptic or narrowly oblong, 8–50 mm long, 2–12 mm wide, the terminal the largest, this and the first lateral leaflets commonly confluent; inflorescence in terminal and axillary corymbiform cymes; flowers sessile to pedicellate, more or less compactly clustered; calyx 4.5–9 mm long at anthesis, the triangular-attenuate lobes shorter than the tube to much longer; corolla 10–18 mm long, and as wide or wider, white, blue violet, or cream, open-campanulate; stamens subequal to the corolla lobes; seeds mucilaginous when wet. Two rather distinctive and geographically correlated varieties are present.

1. Flowers white; plants robust, mainly 8–12 dm tall, from the northern Wasatch Plateau and northward .... *P. foliosissimum* var. *alpinum*
— Flowers blue purple (rarely white); plants less robust, commonly 5–9 dm tall, from the northern Wasatch Plateau southward and eastward .... *P. foliosissimum* var. *foliosissimum*

Var. **alpinum** Brand [*P. albiflorum* Eastw., type from Scofield; *P. foliosissimum* ssp. *albiflorum* (Eastw.) Brand]. Sagebrush, mountain brush, aspen, and spruce-fir communities at 1830 to 3050 m in Cache, Carbon, Davis, Duchesne, Emery, Juab, Morgan, Salt Lake (type from Alta), Summit, Tooele, Utah, Wasatch, and Weber counties; Nevada and Wyoming; 62 (vi).

Var. **foliosissimum** Mountain brush, ponderosa pine, aspen, tall forb, and spruce-fir communities at 1950 to 3295 m in Emery, Grand, Juab, San Juan, Sanpete, Sevier, and Uintah counties; Colorado, Arizona, and New Mexico; 64 (iv).

Polemonium micranthum Benth. Annual herbs from taproots; stem mainly 3–25 cm long, erect or the branches prostrate-ascending; herbage glandular-villous to glabrate; leaves alternate, the basal rosette more or less developed, 1–6 cm long; leaflets 5–15, elliptic to lanceolate or ovate, 1.5–10 mm long, 1–4 mm wide; flowers solitary, terminal; corolla 2–6 mm long, white, the lobes subequal to the tube; stamens included; seeds mucilaginous when wet; 2n = 18. Cheatgrass, filaree, and other annual dominated communities at 1310 to 1590 m in Cache, Davis, Salt Lake, and Utah counties; British Columbia to Montana, south to California and Nevada; South America; 7 (0).

Polemonium pulcherrimum Hook. Pretty Jacobsladder. Perennial herbs from a vertical, subrhizomatous caudex; stems mainly 5–25 cm tall, often branched above; herbage glandular-villous; leaves 3–20 cm long, the lower ones more or less tufted and with broadly sheathing bases; leaflets 7–25, lanceolate to oblanceolate or elliptic to ovate, 2–20 mm long, 1–9 mm wide, the three terminal ones confluent; flowers few to many in congested, corymbiform cymes; calyx 4–7 mm long at anthesis, the triangular lobes subequal to or longer than the tube; corolla rotate-campanulate, 8–14 mm long and about as wide, blue to blue violet (or white); stamens subequal to the lobes; seeds not mucilaginous when wet; 2n = 18. Ponderosa pine, aspen, spruce-fir, and grass-forb communities at 2315 to 3510 m in Beaver, Cache, Carbon, Daggett, Duchesne, Emery, Garfield, Grand, Iron, Juab, Piute, Salt Lake, San Juan, Sanpete, Sevier, Summit, Tooele, Uintah, Utah, Washington, and Wayne counties; Alaska and Yukon, south to California, Arizona, and New Mexico; 62 (xii). Our specimens belong to var. *delicatum* (Rydb.) Cronq. [*P. delicatum* Rydb.].

Polemonium viscosum Nutt. Viscid Jacobsladder. Strongly mephitic, perennial herbs from a definite caudex, this typically clothed with persistent leaf bases; stems mainly 5–25 cm tall, subscapose, the cauline leaves unequally spaced and often much reduced; herbage glandular-villous; basal leaves 3–15 cm long, strongly sheathing basally; leaflets very numerous, 1–7 mm long, 0.5–3 mm wide, most or all of them cleft to the base or nearly so, thus pseudoverticillate, the rachis more or less winged; flowers several to many in dense, subcapitate cymes; calyx 7–12 mm long at anthesis, the lobes shorter than the tube; corolla funnelform, 15–30 mm long, longer than wide, dark blue purple; stamens commonly shorter than the corolla; seeds not mucilaginous when wet; 2n = 18. Spruce-fir and alpine tundra communities, often on talus, at 3050 to 3815 m in Beaver, Box Elder, Daggett, Duchesne, Garfield, Grand, Juab, Piute, Salt Lake, San Juan, Sanpete, Summit, Utah, Uintah, and Wayne counties; Washington to Alberta, south to Nevada, Arizona, and New Mexico; 50 (vii).

POLYGALACEAE R. Br. in Flinders
Milkwort Family

Herbs, subshrubs or shrubs; leaves simple, mostly alternate (or whorled); stipules lacking or mere glands; flowers perfect, irregular, borne in racemose, spicate, or paniculate clusters; sepals 5, distinct, the 2 inner ones larger and often petaloid; petals 3 (rarely 5), the upper 2 usually free, the lower one forming a keel; stamens commonly 8 in 2 series, the filaments united, forming a sheath; anthers opening by subterminal pores; pistil 1, the ovary superior, usually of 1 or 2 carpels; style 1; fruit usually capsular; x = 5–11+.

Polygala L.

There is a single genus with characteristics of the family.

1. Plants annual herbs, with whorled cauline leaves, evidently rare .... *P. verticillata*
— Plants, shrubs or subshrubs, with alternate leaves 2
2(1). Flowers pink purple, with a yellow keel, 8–14 mm long; plants suffrutescent .... *P. subspinosa*
— Flowers cream to whitish or yellowish green, 3–5 mm long; plants shrubs .... *P. acanthoclada*

Polygala acanthoclada Gray Thorny Milkwort. Thorny shrubs with intricately branched stems, mainy 2–12 dm tall; herbage densely hairy or subglabrous, or glabrous in part; leaves 3–25 mm long, 1–5 mm wide, narrowly oblanceolate-spatulate to elliptic; racemes few-flowered, the pedicels subtended by a short bract and by a pair of bracteoles, both of these deciduous; flowers 3–5 mm long, cream to yellowish green, the upper petals often purple-tipped, the keel greenish; fruit broadly elliptic, sessile, 3–6 mm long; 2n = 18, 36. Blackbrush, shadscale, horsebrush, rabbitbrush, and pinyon-juniper communities at 1150 to 2135 m in Box Elder, Emery, Juab, Millard, San Juan, Tooele, and Washington counties; Nevada, California; and Arizona; 20 (v). Two weakly separable varieties are present in Utah, both based on types from along the San Juan River in Utah. Plants designated as var. *acanthoclada* have pedicels usually spreading hairy and shorter than the flowers. It is known in Utah only in San Juan County. Those known as var. *intricata* Eastw. [*P. intermontana* Wendt] have usually glabrous pedicels a little longer than the flowers. They occur in the range noted for the species as a whole.

Polygala subspinosa Wats. Cushion Milkwort. Subshrubs, forming rounded clumps mainly 5–20 cm high, thorny; herbage hirtellous to subglabrous; leaves 8–30 mm long, 2–10 mm wide, oblanceolate to spatulate or elliptic; racemes several-flowered; pedicels 4–10 mm long; flowers 8–14 mm long, pink to pink purple, the keel yellow; fruit suborbicular, 4–8 mm long, 2n = 18, 36. Shadscale, sagebrush, pinyon-juniper, mountain brush, and ponderosa pine at 1310 to 2290 m in Emery, Garfield, Grand, Juab, Iron, Millard, San Juan, Sanpete, Sevier, Tooele, Utah, Washington, and Wayne counties; Colorado, New Mexico, Arizona, Nevada, and California; 82 (xii). Report of the species from Vernal is based on a mislabeled specimen.

Polygala verticillata L. Annual herb; stems simple or branched above, 0.8–4 dm tall; herbage glabrous; leaves mostly 10–20 mm long, 1–3 mm wide, linear to oblong, whorled; peduncles elongate; racemes spicate, many-flowered; flowers white, greenish or pinkish, 1–1.5 mm long, the perianth deciduous; capsules ca 1.5 mm long. Meadow at ca 2074 m in Uintah County; (Hutchings 285, 1932 BRY) widely distributed in eastern U. S. and Canada; 1 (0).

POLYGONACEAE A. L. Juss.
Buckwheat Family

Annual or perennial herbs, subshrubs, shrubs, or twining vines; leaves simple, alternate, opposite, or whorled; stipules forming a sheath (ocrea) or absent; flowers perfect or polygamo-dioecious, regular; perianth 2- to 6-parted or -cleft, not readily identifiable as sepals and petals;

stamens 2–9; pistil 1, the ovary superior, 1–loculed, 1–ovuled; styles 2 or 3; fruit an achene; x = 7–13.

Welsh, S. L. 1984. Utah flora: Polygonaceae. Great Basin Nat. 44: 519–557.

1. Sheathing stipules lacking; flowers subtended by a campanulate, obconic, or cylindric involucre, or by a folded, 2–toothed bract 2
— Sheathing stipules present; flowers not subtended by an involucre or with a folded, 2–toothed bract 5
2(1). Flowers solitary, subtended by a single, folded, 2–toothed bract, this accrescent and prominently veined in fruit; plants slender, weak, broad-leaved annuals, known from Washington County ...... *Pterostegia*
— Flowers solitary (in *Chorizanthe*) or several, arising from a campanulate, obconic, or cylindric involucre; plants various, but not as above 3
3(2). Involucres with lobes or teeth not spiny; bracts unarmed; plants annual, perennial, or shrubby .... *Eriogonum*
— Involucres and bracts armed with spines; plants annual . 4
4(3). Involucres with 2 or more flowers, the lobes tipped with straight spines or bristles; main bracts of inflorescence connate-perfoliate, disklike .... *Oxytheca*
— Involucres usually with 1 flower, the lobes tipped with hooked or straight spines; bracts not both perfoliate and disklike .... *Chorizanthe*
5(1). Leaves all basal, the blades reniform; sepals 4; styles 2; plants of high elevations .... *Oxyria*
— Leaves cauline or basal but, if basal, the blades not reniform and plants not or seldom of high elevations; sepals 5 or 6; styles 2 or 3; plants variously distributed .. 6
6(5). Sepals 5 (rarely 4), all similar and erect in fruit ...... *Polygonum*
— Sepals 6, in 2 sets, the inner ones erect and winged in fruit, or the wings from the achenes, the outer sepals reflexed and often smaller 7
7(6). Stipular sheathes large and prominent; stamens 8–10; leaf blades ovate to orbicular; plants cultivated and persisting .... *Rheum*
— Stipular sheathes not prominent, evanescent; stamens 6; leaf blades narrower; plants indigenous or adventive, not cultivated .... *Rumex*

Chorizanthe R. Br. ex Benth.

Annual herbs; stems more or less dichotomously branched or simple; leaves basal or cauline and alternate, entire, the upper ones often reduced to opposite or whorled bracts; inflorescence cymose or capitate; involucres sessile, cylindric to urn-shaped or funnelform, mostly 1–flowered, 3– to 6–angled or -ribbed and 3– to 6–toothed or -cleft, the teeth spreading, armed with straight or recurved awns; flowers pedicellate or subsessile; bractlets lacking; perianth 6–parted or -cleft; stamens 3–9; styles 3; achenes glabrous, 3–angled.

1. Foliar bracts 3–lobed; involucres with 3, broad, horizontally spreading, saccate horns at the base . *C. thurberi*
— Foliar bracts entire; involucres not horned at base 2
2(1). Involucres 6–ribbed, the 6 teeth sparingly recurved apically, less than 2 mm long; stems very brittle and soon falling apart; foliage leaves reduced to subulate bracts .... *C. brevicornu*
— Involucres 3–angled, the 3 teeth straight, more than 5 mm long; stems not brittle (the plants persisting and burlike); stems with some foliar leaves like the basal ones .... *C. rigida*

Chorizanthe brevicornu Torr. Short Spine-flower. Plants erect or ascending, mainly 5–28 cm tall; stems usually several from the base, strigulose, breaking at the nodes when dry; leaves mostly basal, 1–6 cm long, 2–8 mm wide, narrowly oblanceolate, reduced to opposite bracts upward; involucres solitary in axils of branches, subcylindric, conspicuously 6–ridged, straight or curved, ca 4 mm long, the lobes with recurved spinose teeth; flowers 3–4 mm long, glabrous, the perianth lobes whitish, subequal; stamens 3; achenes ca 2 mm long. Creosote bush, blackbrush, and other warm and salt desert shrub communities at 760 to 1220 m in Grand, Kane, San Juan, and Washington counties; Nevada, Arizona, and California; 14 (iv).

Chorizanthe rigida (Torr.) T. & G. Rigid Spine-flower. [*Acanthogonum rigidum* Torr.]. Stems simple, obscured by bracts and leaf bases; main leaves orbicular to obovate, woolly beneath, green and sparingly tomentose above; secondary leaves bracteate, lanceolate to subulate, spine-tipped, indurate and thorny at maturity; inflorescence dense, with involucres clustered in bract axils; tube of involucre ca 2 mm long, 3–angled, with 3 broad, spreading, unequal, straight, spine-tipped lobes 4–12 mm long; perianth yellowish, almost included; stamens 9; achenes ovoid, prominently beaked, ca 2 mm long. Creosote bush, Joshua tree, and other warm desert shrub communities at 760 to 1130 m in Washington County; Arizona, Nevada, California, and Mexico; 8 (i).

Chorizanthe thurberi (Gray) Wats. Thurber Spine-flower. [*Centrostegia thurberi* Gray]. Plants erect, usually simple from the basal rosette and typically dichotomously branched upward, 4–16 cm tall; basal leaves 4–30 mm long, 3–6 mm wide, spatulate, subglabrous; foliar bracts 3–lobed, spine-tipped, 2–4 mm long; involucres solitary, borne in branch axils, 4–6 mm long, 5–toothed, the teeth armed with straight spines, 3–angled and with 3 saccate, spinose horns near the base; perianth included, pubescent; stamens 6 or 9; achenes ca 1.5 mm long. Creosote bush, blackbrush, mountain brush, and pinyon-juniper communities at 850 to 1700 m in Garfield (?), Kane, and Washington counties; Arizona, Nevada, and California; 12 (i). The report for Garfield County is based on a specimen collected "4 miles south of Boulder (probably Nevada), in creosote bush," and might be mislabeled.

Eriogonum Michx.

Annual or perennial herbs, subshrubs, or shrubs; leaves basal or cauline and alternate, or with scalelike to foliaceous alternate or whorled bracts, entire, estipulate; flowers perfect or imperfect, borne in campanulate, obconic, or cylindric involucres; involucres 4– to 10–lobed or -toothed, or less often in 2 whorls or 3 more or less distinct bracts, awnless, few- to many-flowered, sessile or stipitate; perianth petaloid, 6–segmented, in 2 series; flowers pedicellate, subsessile, or the base attenuated and stipelike; stamens 9, the filaments filiform; ovary 1–loculed, with 3 styles and capitate stigmas; achenes 3–angled or -winged. Note: This is a dual genus, consisting of annual species distinguished by minute diagnostic characteristics, and of perennial herbs, subshrubs, and shrubs that are connected through series of intermediates that defy segregation and construction of keys based on characters similar to those used in the annual species. Taxonomic problems are not easily resolved, and the approach presented below is only tentative.

Reveal, J. L. 1967. Notes on Eriogonum - V. A revision of the *Eriogonum corymbosum* complex. Great Basin Nat. 27: 183–220
_____. 1969. A revision of the genus *Eriogonum* (Polygonaceae). Unpublished Ph.D. dissertation. Brigham Young University. 546 p.
_____. 1973. *Eriogonum* (Polygonaceae) of Utah. Phytologia 25: 169–217.
Reveal J. L., and B. J. Ertter. 1976. Reestablishment of *Stenogonum* Nutt. (Polygonaceae). Great Basin Nat. 36: 272–280.
Hess, W. J. and J. L. Reveal. 1976. A revision of *Eriogonum* (Polygonaceae), Subgenus Pterogonum. Great Basin Nat. 36: 281–333.

1. Plants annual (except in some *E. inflatum*, q.v.), from slender taproots .... Key 1
— Plants perennial herbs, subshrubs, or shrubs 2
2(1). Plants definitely shrubby, the stems developed above ground level and with 1 to several elongated internodes .... Key 2
— Plants acaulescent or, if caulescent, the stems prostrate at ground level or the internodes very short and obscured by a tomentum 3
3(2). Flowers with attenuated, stipelike bases, yellow to reddish yellow or cream; plants often with prostrate-spreading stems .... Key 3
— Flowers not with stipelike bases, variously colored; plants seldom with prostrate-spreading stems 4
4(3). Plants pulvinate-caespitose, mound-forming; inflorescences mainly 0.5–5 cm tall; leaves less than 1 cm long .... Key 4
— Plants not simultaneously pulvinate-caespitose, mound-forming, less than 5 cm tall, and with leaves less than 1 cm long .... Key 5

Key 1.

Plants annual (except in some *E. inflatum*).

1. Involucres angled to strongly ribbed, usually tightly appressed vertically to the stems and always sessile .. 2
— Involucres smooth, not ribbed or angled, usually stipitate or, if sessile, not vertically appressed to the stems 9
2(1). Leaves puberulent to villous beneath, but not tomentose 3
— Leaves tomentose, at least beneath 5
3(2). Cauline leaves more or less bracteate, the blades not well developed, soft-hairy; involucres 4-lobed; flowers white, suffused with red, glabrous or sometimes hispidulous, 1–1.8 mm long; plants of southwestern Utah .... *E. puberulum*
— Cauline leaves with well-developed blades, variously hairy; involucres 5-lobed; flowers variously colored, 1.5–2 mm long; distribution various 4
4(3). Outer perianth segments broadly obovoid, hooded, markedly ciliate, otherwise glabrous; plants of Kane County .... *E. darrovii*
— Outer perianth segments oblong to ovate, not hooded or markedly ciliate, otherwise hispidulous and more or less glandular; plants of central and eastern Utah .... *E. divaricatum*
5(2). Foliage leaves cauline and basal; plants of Kane and Washington (?) counties .... *E. polycladon*
— Foliage leaves all basal (some seldom cauline and the plants of other distribution); plants variously distributed 6
6(5). Stems tomentose to floccose-tomentose 7
— Stems glabrous 8
7(6). Flowers yellow to red, the outer perianth segments broadly obovate .... *E. nidularium*
— Flowers white, the outer segments narrowly obovate .... *E. palmerianum*

8(6). Involucres 3–5 mm long; achenes ca 2 mm long; plants of Kane and Washington counties ....
.... *E. davidsonii*
— Involucres 1–2.5 mm long; achenes ca 1 mm long; plants of Beaver County .... *E. baileyi*
9(1). Leaves glabrous or variously pubescent but not tomentose or lanate, even on the lower blade surface
.... 10
— Leaves tomentose to lanate on the lower blade surface, at least 15
10(9). Involucres in 2 whorls, each whorl 3-lobed 11
— Involucre consisting of a single whorl, this usually 4- or 5-lobed 12
11(10). Foliage leaves all basal; peduncles abruptly bent above the middle .... *E. flexum*
— Foliage leaves both cauline and basal; peduncles straight or gently curved .... *E. salsuginosum*
12(10). Stems usually strongly inflated; plants of broad distribution, annual or perennial .... *E. inflatum*
— Stems not inflated (except in some *E. trichopes*, q.v.); plants annual, of various distribution 13
13(12). Flowers glabrous, white, fading yellowish; plants of eastern Utah .... *E. gordonii*
— Flowers hairy, yellowish or reddish; plants of western and southwestern Utah 14
14(13). Branches of inflorescence with stipitate, usually purplish glands; flowers densely villous; involucres 5-lobed; plants of western Utah ...... *E. howellianum*
— Branches of inflorescence not glandular; involucres 4-lobed; plants of Washington County .... *E. trichopes*
15(9). Foliage leaves both cauline and basal 16
— Foliage leaves all basal (except in some *E. cernuum*, q.v.) 17
16(15). Flowers glabrous, yellow, the outer perianth segments cordate-ovate; leaves linear to narrowly oblanceolate .... *E. pharnaceoides*
— Flowers minutely glandular-puberulent, white to yellowish or pink, the outer segments oval; leaves obovate to lanceolate .... *E. maculatum*
17(15). Involucres minute, 0.3–1 mm long 18
— Involucres 1–3 mm long 20
18(17). Flowers yellow, soon suffused with red, glabrous; plants of southeastern Utah .... *E. wetherillii*
— Flowers white to pink or yellow, hairy or glabrous; plants of southwestern Utah 19
19(18). Flowers yellow; outer perianth lobes with saccate-dilated bases, pubescent .... *E. thomasii*
— Flowers white to pink; outer perianth lobes not swollen at the base, pubescent or glabrous ....
.... *E. subreniforme*
20(17). Branches of inflorescence or stipes with dark, stipitate glands; plants of Washington County ....
.... *E. brachypodum*
— Branches of inflorescence or stipes glabrous (or tomentulose) or, if stipes glandular (as in *E. nutans*), plants not of Washington County 21
21(20). Involucres and flowers puberulent and more or less glandular; plants of Washington County ... *E. pusillum*
— Involucres and flowers glabrous (or tomentulose); plants variously distributed 22
22(21). Branches of inflorescence more or less tomentose (at least when young) and glandular; leaf margins conspicuously undulate-crisped; plants of eastern Utah
.... *E. scabrellum*

	Branches of inflorescence glabrous; leaf margins not especially undulate; plants variously distributed 23
23(22).	Outer perianth segments merely truncate to obtuse basally; involucres usually stipitate 24
—	Outer perianth segments cordate at the base; involucres usually sessile 25
24(23).	Stipes glabrous; outer perianth segments violin-shaped, constricted below the middle, the margins undulate, more or less saccate below; plants common and widespread .... *E. cernuum*
—	Stipes stipitate-glandular; outer segments obovate, not constricted below the middle, the margins not especially saccate; plants uncommon .... *E. nutans*
25(23).	Involucres erect on branches of inflorescence; plants of Iron and Washington counties .... *E. insigne*
—	Involucres deflexed on branches of inflorescence; plants of various distribution 26
26(25).	Involucres broadly campanulate, broader than long; flowers yellow to reddish yellow; plants widespread .... *E. hookeri*
—	Involucres obconic, somewhat longer than broad; flowers white to pink; plants of western and southern Utah .... *E. deflexum*

Key 2.

Plants definitely shrubby.

1.	Flowers pubescent, white to pink; leaves fascicled in at least some axils; plants of Washington County ... . *E. fasciculatum*	
—	Flowers glabrous, variously colored; leaves fascicled or not; plants variously distributed 2	
2(1).	Stems angled or ribbed and more or less grooved, or conspicuously flexuous; plants of Washington County 3	
—	Stems rounded or terete and not especially, if at all, flexuous; plants variously distributed 4	
3(2).	Stems both flexuous and grooved, usually glabrous; involucres 0.7–1.5 mm long .... *E. heermannii*	
—	Stems flexuous, almost terete, tomentose; involucres 2–2.5 mm long .... *E. plumatella*	
4(2).	Plants completely glabrous; plants of Emery and Wayne counties .... *E. corymbosum*	
—	Plants pubescent or tomentose, at least in part, variously distributed 5	
5(4).	Leaves oval to oblong or elliptic, the blades mostly less than 3 times longer than broad 6	
—	Leaves linear to narrowly oblong or narrowly elliptic, the blades mostly 5–10 times longer than broad 10	
6(5).	Flowers not much exserted from the involucres; leaves typically densely tomentose on both surfaces; plants mainly of the Great Basin and western Kane and eastern Washington counties .... *E. nummulare*	
—	Flowers conspicuously exserted from the involucres; leaves often only thinly tomentose above; plants of broad or other distribution 7	
7(6).	Inflorescences ca 1/2 as long as plant height, subequal to the leafy portion of current annual growth; involucres racemosely arranged; plants of Washington County .... *E. wrightii*	
—	Inflorescences usually much less than 1/4 the plant height or, if longer, as in some *E. thompsonae*, the involucres not racemosely arranged; plants variously distributed 8	

8(7).	Inflorescence mostly 1/3–1/2 the plant height; leaves mostly near the base of the annual shoots; plants of Kane and Washington counties .... *E. thompsonae*
—	Inflorescence mostly less than 1/4 the plant height; leaves mostly cauline, except in some *E. corymbosum*; plants of broad distribution 9
9(8).	Leaf apices acute, the blades mostly elliptic and more or less revolute, mainly less than 8 mm wide; plants widespread .... *E. microthecum*
—	Leaf apices typically rounded, the blades orbicular to oblong or ovate to obovate, seldom if at all revoute, mainly more than 8 mm wide .... *E. corymbosum*
10(5).	Leaves flat, slightly if at all revolute 11
—	Leaves revolute, the lower surface largely obscured by revolute margins 13
11(10).	Plants semishrubby only; stems of current growth dying to plant base .... *E. brevicaule*
—	Plants definitely shrubby; stems of current growth not dying to plant base each year 12
12(11).	Leaves with margins at least somewhat revolute; plants broadly distributed .... *E. corymbosum*
—	Leaves mainly flat; plants of northern Uintah County .... *E. lonchophyllum*
13(10).	Inflorescences mainly 8–20 cm long or more; involucres racemose; plants of sandy tracts in the Navajo Basin .... *E. leptocladon*
—	Inflorescences mainly 1–6 cm high; involucres cymose; plants variously distributed, but usually not in deep sand 14
14(13).	Plants low, mat-forming; flowers bicolored, pink and white .... *E. bicolor*
—	Plants low to tall, but not mat-forming; flowers white, pink, reddish, or yellow 15
15(14).	Flowers yellow; plants of clay and silt substrates in eastern Grand County .... *E. contortum*
—	Flowers white, pink, or reddish; plants variously distributed 16
16(15).	Leaf axils (at least some) with fascicled leaves; plants of San Juan County .... *E. clavellatum*
—	Leaf axils seldom with fascicled leaves (if ever); plants of various distribution 17
17(16).	Leaves linear, rather tightly revolute, mainly 2–6 cm long, inflorescence dense, glabrous, green; plants of San Juan County .... *E. leptophyllum*
—	Leaves elliptic to linear, flat to revolute but, if revolute, mainly less than 2 cm long and the inflorescence open and tomentose; plants variously distributed ... 18
18(17).	Leaves mainly 3–7 cm long; involucres clustered on inflorescence branch tips; plants of Duchesne County and sometimes elsewhere .... *E. corymbosum*
—	Leaves mainly 0.8–3 cm long; involucres not clustered; plants widespread .... *E. microthecum*

Key 3.

Flowers with attenuated, stipelike bases.

1.	Stems with whorled, foliose bracts near the middle; plants of northern Utah .... *E. heracleoides*
—	Stems lacking whorled bracteate leaves, or these closely subtending the plants inflorescences; variously distributed 2
2(1).	Flowers glabrous .... *E. umbellatum*
—	Flowers hairy 3

3(2). Stems with whorled bracteate leaves subtending the umbellate inflorescence; involucres to 6 mm wide or more .... *E. jamesii*

— Stems lacking whorled bracteate leaves subtending the capitate inflorescence; involucres to 3 mm wide .... *E. caespitosum*

Key 4.

Plants pulvinate caespitose, mound-forming; inflorescences less than 5 cm tall and leaves less than 1 cm long.

1. Leaf blades oval, almost or quite as broad as long .... *E. ovalifolium*

— Leaf blades longer than broad 2

2(1). Scapes, if present, glabrous; plants of central and south central Utah .... *E. panguicense*

— Scapes, if present, tomentose; plants of various distribution 3

3(2). Flowers glabrous, 2–3 mm long; plants of San Francisco Range .... *E. soredium*

— Flowers hairy, mainly 2–4 mm long; distribution various 4

4(3). Ovaries and achenes pubescent; flowers white or yellow; plants of lower elevations in Great Basin and in eastern Utah .... *E. shockleyi*

— Ovaries and achenes glabrous; flowers yellow, white, or pink; plants variously distributed 5

5(4). Heads 10–15 mm wide, usually evidently pedunculate and definitely bracteate; plants mainly of the Great Basin .... *E. villiflorum*

— Heads less than 10 mm wide, usually sessile and not evidently bracteate; plants of eastern and south-central Utah 6

6(5). Flowers white to rose, 3.5–4 mm long; involucres 6- to 8-lobed; plants of eastern Utah .... *E. tumulosum*

— Flowers yellow 1.8–2.5 mm long; involucre 4-lobed; plants of Garfield County .... *E. aretioides*

Key 5.

Plants herbaceous perennials with leaves more than 1 cm long and with stems or scapes more than 5 cm tall.

1. Caudex branches or root crown 1–2.5 cm thick, clothed with persistent leaf bases, these with persistent, coarse, villous-pilose hairs; plants wandlike, mainly 3–12 dm tall .... *E. alatum*

— Caudex branches or root crown less than 1 cm thick, or if thicker, not villous-pilose 2

2(1). Inflorescence racemose or paniculate, the involucres spaced along elongate, erect branches ... *E. racemosum*

— Inflorescence cymose, the involucres clustered on short, spreading branches 3

3(2). Leaf blades all oval to orbicular and about as broad as long; inflorescence branching or capitate but, if the latter, the inflorescence mostly 15–30 mm wide 4

— Leaf blades, at least some, much longer than broad or, if as above, the inflorescence capitate and 5–14 mm wide 5

4(3). Involucres capitate; flowers white, pink, or yellow; plants widespread .... *E. ovalifolium*

— Involucres borne in open cymes; flowers white or pink; plants of central to western Utah .... *E. batemanii*

5(3). Plants strictly acaulescent above caudex branches 6

— Plants short-caulescent, the internodes apparent, though short and obscured by dense tomentosum or, if acaulescent, the inflorescence branched 7

6(5). Scapes glabrous; flowers white; plants mainly of southern highlands .... *E. panguicense*

— Scapes tomentose; flowers white, pink, cream, or yellow .... *E. brevicaule*

7(5). Scapes or peduncles pubescent or, if glabrous (as in some *E. spathulatum*), the plants of Beaver County .. 8

— Scapes or peduncles glabrous; plants variously distributed 9

8(7). Involucres capitate; plants of northern Utah .... *E. brevicaule*

— Involucres in branching cymes or subcapitate; plants of central and western Utah .... *E. spathulatum*

9(7). Flowers yellow; involucres not in capitate clusters; leaves linear to lanceolate or oblanceolate . *E. brevicaule*

— Flowers white or pink; leaves oblong to elliptic or ovate-lanceolate 10

10(9). Leaves broadly elliptic to ovate-lanceolate, the blades usually less than 3 times longer than broad *E. batemanii*

— Leaves narrowly elliptic, commonly 5–8 times longer than broad .... *E. lonchophyllum*

Eriogonum alatum Torr. in Sitgr. Winged Buckwheat. [*E. triste* Wats., type from Kane County; *E. alatum* ssp. *triste* (Wats.) Stokes]. Perennial herbs, mainly 3–12 dm tall, from a taproot and thick rootcrown, this 1–3 cm thick or more and clothed with persistent, coarsely villous-pilose leaf bases, the pith chambered; leaves mainly 3–12 (20) cm long, 3–15 mm wide, narrowly oblanceolate to lanceolate, strigose on one or both surfaces; cauline leaves reduced upwards; inflorescence cymose-paniculate; stipes erect, 3–20 mm long; involucres obconic to campanulate, 2–4.5 mm long, pilosulose to glabrous, 5-lobed; perianth yellowish to greenish, 1.5–2.8 mm long, the segments oblong, united to near the middle; achenes 5–9 mm long, 3–6 mm wide, glabrous, 3-winged the entire length; n = 20. Sagebrush, mixed desert shrub, pinyon-juniper, and mountain brush communities at 1155 to 2685 m in Carbon, Daggett, Duchesne, Emery, Garfield, Grand, Iron, Kane, San Juan, Sevier, Uintah, Wasatch, Washington, and Wayne counties; Wyoming to Nebraska, south to Arizona, New Mexico, Texas, and Mexico; 66 (x).

Eriogonum aretioides Barneby Widtsoe Buckwheat. Pulvinate-caespitose, mound-forming, herbaceous perennials from a pluricipital caudex and woody taproot, the caudex branches mainly 20–50, the taproot clothed with shreddy castaneous to blackish bark; leaves 1–3.5 mm long, 0.8–1.2 mm wide, oblanceolate in outline, revolute, the lower surface obscured, white-pilose, sessile; inflorescence of solitary, sessile involucres, not borne above the rosettes, these campanulate, 2.8–3.2 mm long, 2–4 mm wide, villous, 4-lobed; flowers yellow, 2–2.2 mm long, pilose, the segments lance-ovoid; achenes brown, ca 2 mm long, glabrous. Bristlecone pine, ponderosa pine, Douglas fir, and Rocky Mountain juniper communities, on Pink Limestone Member of the Wasatch Formation, at 2255 to 2655 m in Garfield (type from near Widtsoe) County; endemic; 5 (0).

Eriogonum baileyi Wats. Bailey Buckwheat. Annual herbs, mainly 10–30 cm tall; leaves all basal; blades orbicular or obovate, mainly 5–20 mm long and about as broad, tomentose on one or both sides; petioles 5–30 mm long; inflorescences much-branched, spreading; involucres sessile, subcylindric, 1.5–2.5 mm long, 5-lobed, glabrous, vertically appressed; flowers white to pink, 1.5–2 mm long, glabrous, the outer segments oblong to obovate, slightly constricted near the middle, the inner segments narrower; achenes brown, ca 1 mm long. Sagebrush-rabbitbrush and mountain mahogany communities at 1830 to 2200 m in Beaver County; Oregon and Idaho, south to Nevada and California; 2 (ii).

Eriogonum batemanii Jones Bateman Buckwheat. Perennial herbs, mainly 10–45 cm tall; leaves all basal; blades 1–3.5 cm long, 5–16 mm wide, oval, orbicular, elliptic, or lance-oblong, tomentose on one or both surfaces, flat marginally, obtuse to rounded apically; petioles 8–25 mm long; inflorescences usually glabrous, open, cymose-paniculate, the branches spreading-ascending; involucres sessile, clustered or solitary, narrowly campanulate or obconic, 2–4 mm long, with 5, hyaline, rounded lobes; flowers white, 1.5–2.8 mm long, glabrous, the outer segments obovate, the inner ones slightly narrower; achenes brown, 2.5–3 mm long; n = 20. Three rather weak but geographically correlated varieties occur in Utah.

1. Leaf blades mainly 2–3 times longer than broad; plants of eastern Utah .... *E. batemanii* var. *batemanii*

— Leaf blades about as broad as long; plants of western and central Utah 2

2(1). Involucres capitate, mainly 2–5, terminating long naked branches; plants of western Beaver and Millard counties .... *E. batemanii* var. *eremicum*

— Involucres cymose, mainly 1–5 in branching terminal cymes on rather short branches; plants of Piute and Sevier counties .... *E. batemanii* var. *ostlundii*

Var. *batemanii* Mixed desert shrub and pinyon-juniper communities at 1615 to 2515 m in Carbon (type from Price Valley), Duchesne, Emery, Garfield, and Uintah counties; Colorado; a Plateau endemic; 41 (viii). Plants with both capitate and cymose involucres in the branched inflorescences and very short broad leaf blades occur in this entity. Thus, the variation is similar to that represented in the two following varieties, which are distinguished on features not exclusive with them. A specimen from Horn Mt., Emery County (Foster 8257 BRY) simulates *E. lonchophyllum* var. *lonchophyllum*, and suggests a relationship between *E. batemanii* and that taxon.

Var. *eremicum* (Reveal) Welsh Hermit Buckwheat. [*E. eremicum* Reveal, type from SE of Garrison, Millard County]. Shadscale, desert shrub, and juniper communities at 1555 to 1925 m in Beaver and Millard (type from southeast of Garrison) counties; endemic; 12 (viii). Substrates include limestone and dolomite. Specimens herein assigned to *E. spathulatum* (q.v.), but having glabrous inflorescences, appear to be intermediate towards this variety.

Var. *ostlundii* (Jones) Welsh Elsinore Buckwheat. [*E. ostlundii* Jones, type from near Elsinore, Sevier County]. *E. spathuliforme* Rydb., type from Piute County]. Shadscale, mixed desert shrub, juniper, and ponderosa pine communities, often on igneous gravels, at 1675 to 1985 m in Piute and Sevier (type from near Elsinore) counties; endemic; 27 (iii).

Eriogonum bicolor Jones Pretty Buckwheat. [*E. microthecum* ssp. *bicolor* (Jones) Stokes]. Mound-forming shrubs, mainly 2–8 cm tall, the horizontal spreading stems mainly 5–20 cm long; leaves caulescent, mostly 5–15 mm long, 1–3 mm wide, clavate, the lower surface more or less obscured by revolute margins, tomentose; current stems white-tomentose; inflorescence umbellate-cymose, on peduncles 3–15 mm long; involucres obconic to broadly campanulate, 2–4 mm long, tomentose to glabrous, with 5 acutish to rounded lobes; flowers white to pink or rose, the midveins often pink to red purple, 2.2–4 mm long, glabrous, the outer segments obovate, the inner ones oblanceolate to elliptic; achenes brown, 3–3.5 mm long. Shadscale, mat-atriplex, other salt and mixed desert shrub, and pinyon-juniper communities at 1340 to 1985 m in Carbon, Emery, Garfield, Grand (type from Thompsons Springs), San Juan, Sevier, and Wayne counties; endemic; 54 (vi).

Eriogonum brachypodum T. & G. Parry Buckwheat. [*E. parryi* Gray, type from St. George; *E. deflexum* ssp. *parryi* (Gray) Stokes; *E. deflexum* var. *brachypodum* (T. & G.) Munz; *E. deflexum* ssp. *brachypodum* (T. & G.) Stokes]. Annual herbs, mainly 5–30 cm tall; leaves all basal, the blades 0.8–4 cm long and about as wide or wider, orbicular to reniform, white-tomentose, at least beneath; inflorescences umbellate, the branches glandular; involucres on stipes 3–15 mm long, usually deflexed, glandular; involucres 1–2.5 mm long, obconic to campanulate, usually glandular, with 5 triangular-acute teeth; flowers white or suffused with red, 1.5–2.8 mm long, glabrous, the outer segments ovate-cordate, the inner ones oblanceolate; achenes brown, 1.5–2 mm long; 2n = 40. Creosote bush, other warm desert shrub, and shadscale communities at 760 to 1550 m in Washington County; California, Nevada, and Arizona; 16 (ii).

Eriogonum brevicaule Nutt. Shortstem Buckwheat. Plants perennial; stems of the year dying to the base, mainly 3–35 cm tall, glabrous or tomentose; leaves all basal or some with obvious short stems, the short internodes obscured by a tomentum, 0.3–10 cm long, 1–9 mm wide, tomentose on one or both surfaces, flat to revolute, entire or undulate, linear to elliptic, oblanceolate, or lanceolate; petioles 1–40 mm long; inflorescences cymose, capitate, or cymose-umbellate; involucres solitary or clustered, obconic to campanulate, 1.5–4.5 mm long, tomentose to glabrous, with 5 acute lobes; flowers yellow to cream, white, or suffused with pink, glabrous, the segments ovate to oblong, lanceolate, oval, or obovate; achenes 1.5–3.5 mm long, brown. The *brevicaule* complex typifies the problematical nature of interpretation of perennial members of the genus. Floral morphology is sufficiently reduced and uniform as to lack definitive diagnostic criteria in most instances. Inflorescence structure is only somewhat more useful, but is often variable within a population, ranging from capitate to branched. Flower color is useful in a general sense only, often varying from white to yellow or even pink within a population. Pubescence appears, at first, to be of substantial value, but the use of this criterion fails also. The attempt here is to bring together those members of the group as they occur in Utah, meanwhile acknowledging the problems of recognition of all specimens within a constituent entity. Further, an indication of intermediacy, whether phenotypic, due to ecological response, or genotypic, due to hybridization, is presented. Phenotypic variation in response to different, often subtle, environmental conditions are apparently great. However, some part of the variation is due to hybridization of phases of this complex among themselves and with phases of the *E. corymbosum*, *E. lonchophyllum*, *E. microthecum*, and possibly other complexes. The following treatment should be regarded as tentative at best.

1. Inflorescences branched from well below the middle of the plant height; plants of the southern Uinta Basin .. .. *E. brevicaule* var. *ephedroides*
— Inflorescence capitate and unbranched or branched from above the middle of the plant height (seldom below in some var. *viridulum* q.v.) 2

2(1). Leaves revolute, the lower surface completely obscured by the margin; plants of the northern and western Uinta Basin .... *E. brevicaule* var. *viridulum*
— Leaves revolute or flat, the lower surface readily apparent or, if not, the plants of other distribution 3

3(2). Leaves both flat and scapes monocephalous; plants of northern Utah 4
— Leaves revolute or, if flat, the plants usually with a branching inflorescence 5

4(3). Plants strictly acaulescent; plants of western Box Elder County .... *E. brevicaule* var. *desertorum*
— Plants short-caulescent, the internodes obscured by a white tomentum; plants of Cache, Morgan, and Rich counties .... *E. brevicaule* var. *loganum*

5(3). Plants with a definitely woody caudex, this clothed with black, marcescent leaf bases; leaves usually undulately partially revolute; typically growing in crevices or ridge crests .... *E. brevicaule* var. *nanum*
— Plants with subligneous caudex, this only sometimes with blackish marcescent leaf bases; leaves various, but sometimes as above; of various habitats 6

6(5). Stems glabrous, the inflorescences branching in the upper 1/3–1/4; plants transitional with the following .... *E. brevicaule* var. *brevicaule*
— Stems tomentose or, if glabrous, the inflorescence branching in the upper 1/4 7

7(6). Flowers white, suffused with pink, or yellow, borne in capitate or branched inflorescences; plants of Minnie Maud Creek and Mt. Bartles vicinity .... *E. brevicaule* var. *promiscuum*
— Flowers usually yellow, borne in capitate or branched inflorescences; plants broadly distributed .... *E. brevicaule* var. *laxifolium*

Var. *brevicaule* [*E. campanulatum* Nutt.; *E. confertiflorum* var. *stansburyi* Benth. in DC., type from Salt Lake; *E. brevicaule* var. *aureum* Benth. in DC.; *E. nudicaule* ssp. *garrettii* Stokes, type from near Echo Reservoir; *E. nudicaule* ssp. *parleyense* Stokes, type from Parleys Canyon]. Sagebrush, juniper, mountain brush, pinyon-juniper, aspen, and spruce-fir communities at 1460 to 2745 m in Daggett, Davis, Salt Lake, Summit, and Utah counties; Idaho, Wyoming, and Colorado; 60 (viii). This variety, as interpreted here, includes var. *wasatchense* (Jones) Reveal [*E. wasatchense* Jones, type from American Fork Canyon], a narrow-leaved phase completely transitional with more typical var. *brevicaule* northward. The narrow-leaved phase is also transitional with var. *laxifolium* (q.v.) southward, and both varieties *brevicaule* and *laxifolium* intergrade upwards with the aggregation of forms treated herein as var. *nanum* (q.v.).

Hybrids between var. *brevicaule* and *E. corymbosum* are known from Wyoming.

Var. *desertorum* (Maguire) Welsh Desert Buckwheat. [*E. chrysocephalum* ssp. *desertorum* Maguire in Maguire & Holmgren; *E. desertorum* (Maguire) R. J. Davis]. Sagebrush, bitterbrush, and juniper communities at ca 1585 to 2440 m in Box Elder County; Nevada; 3 (0). This variety simulates the capitate phase of var. *laxifolium*, differing conspicuously only in the flat leaf blades.

Var. *ephedroides* (Reveal) Welsh Ephedra Buckwheat. [*E. ephedroides* Reveal, type from 10 mi S of Bonanza, Uintah County]. Shadscale, thistle, mixed desert shrub and open pinyon-juniper communities, on Green River Formation, at 1525 to 2075 m in Uintah (type from south of Bonanza) County; endemic; 28 (iii). This most distinctive phase of the *brevicaule* complex forms apparent hybrids with *E. corymbosum* in the eastern part of its range.

Var. *laxifolium* (T. & G.) Reveal Varying Buckwheat. [*E. kingii* var. *laxifolium* T. & G, type from Wasatch Mts.; *E. chrysocephalum* Gray; *E. chrysocephalum* var. *angustum* Jones, type from Johnsons Pass, Tooele County; *E. nudicaule* ssp. *angustum* (Jones) Stokes; *E. brevicaule* var. *pumilum* Stokes ex Jones, type from between Colton and Kyune, Carbon County; *E. nudicaule* ssp. *pumilum* (Stokes) Stokes; *E. tenellum* ssp. *cottamii* Stokes, type from Utah County; *E. brevicaule* var. *cottamii* (Stokes) Reveal; *E. medium* Rydb, type from Mt. Nebo]. Mountain brush, sagebrush, pinyon-juniper, ponderosa pine, and aspen communities at 1645 to 3390 m in Duchesne, Emery, Juab, Millard, Salt Lake, Sanpete, Sevier, Tooele, and Utah counties; endemic; 64 (xv). This variety consists of plants with both capitate and open inflorescences, slender to broad leaves, revolute to flat leaves, usually tomentose (but sometimes glabrous) inflorescences, and other diversity. The var. *cottamii* is based on the densely tomentose plants of western ranges, but these are transitional completely at higher elevations with other phases of var. *laxifolium*, and show affinity with *E. spathulatum* (q.v.) downward. Apparent hybrids occur with *E. lonchophyllum* (see var. *promiscuum*).

Var. *loganum* (A. Nels.) Welsh Logan Buckwheat. [*E. loganum* A. Nels., type from Logan; *E. chrysocephalum* ssp. *loganum* (A. Nels.) Stokes]. Sagebrush-bunchgrass communities at 1460 to 2045 m in Cache (type from Logan), Morgan, and Rich counties; endemic; 8 (0). This material differs only superficially from var. *nanum*, a higher elevation phase with similar well-developed woody base.

Var. *nanum* (Reveal) Welsh Dwarf Buckwheat. [*E. nanum* Reveal, type from Willard Peak; *E. grayi* Reveal, type from Lake Blanche, Salt Lake County]. Sagebrush, mountain brush, spruce-fir, and alpine tundra communities, in crevices in limestone or quartzite outcrops, or on windswept ridges or in talus slopes at 2010 to 3510 m in Box Elder, Cache, Juab, Millard, Salt Lake, Tooele, Utah, and Weber counties; endemic; 35 (i). This assemblage consists of crevice plants and other dwarf, high elevation phases that apparently do not have genetic integrity. Their recognition at any taxonomic rank is, therefore, problematical, and they are treated here for convenience only.

Var. *promiscuum* Welsh Green River Shale outcrops in sagebrush, aspen, and spruce-fir communities at 2285 to 3060 m in Carbon (type from Mt. Bartles) County; endemic; 6 (ii). The Mt. Bartles buckwheat is similar in some respects with var. *nanum*, but seems to have a separate origin. The plants appear to have arisen through hybridization of portions of *E. brevicaule* var. *laxifolium* with *E. corymbosum* var. *hylophilum*, and with a possible infusion of *E. lonchophyllum* var. *lonchophyllum*. Flowers are predominantly white suffused with pink, and in the upper elevational reaches have capitate inflorescences. Downward the inflorescences are branched and the plants are transitional to *E. corymbosum*. Yellow flowered individuals give evidence of contribution from *E. brevicaule* var. *laxifolium*; 12 (iii).

Var. *viridulum* (Reveal) Welsh Duchesne Buckwheat. [*E. viridulum* Reveal, type from 8 mi E of Duchesne]. Pinyon-juniper, shadscale, and mixed desert shrub communities at 1555 to 2135 m in Duchesne and Uintah counties; endemic; 58 (v). Hybrids between var. *viridulum* and *E. corymbosum*, which simulate *E. corymbosum* var. *aureum*, are locally common in Duchesne and eastern Utah counties. They have been described as both *E. corymbosum* var. *albogilvum* Reveal (type from Indian Canyon) and *E.* x *duchesnense* Reveal (type from Indian Canyon); 18 (iii). The var. *viridulum* is closely allied to var. *ephedroides*, standing about midway between that entity and var. *brevicaule*. There is an admitted close affinity with var. *laxifolium* westward. Some plants from near Split Mt., Uintah County, are apparently transitional with *E. lonchophyllum* var. *saurinum* and *E. microthecum*.

Eriogonum caespitosum Nutt. Mat Buckwheat. Plants perennial, mat-forming, mainly 1–4 dm across, the vegetative stems persistent, with branches woody and usually clothed with gray to black leaves and bases; flowering stems scapose, arising from rosettelike branches, mainly 0.5–10 cm long or lacking; leaves 2–12 mm long, 1.5–5 mm wide, spatulate to oblanceolate, elliptic or oval, tomentose, flat or essentially so, short-petiolate; inflorescence capitate, not subtended by bracts; involucres campanulate, with the tubes 2–3.5 mm long and 3–5 mm wide, the lobes oblong, 2–3.5 mm long; flowers yellow or suffused with red, 2.5–10 mm long including the stipitate base, pilose to villous, the segments oblanceolate; achenes 3.5–5 mm long; n = 20. Sagebrush, pinyon-juniper, and mountain brush communities at 1525 to 2290 m in Beaver, Box Elder, Iron, Juab, Millard, Rich, Summit, and Washington counties; Oregon to Montana, south to California, Nevada, and Colorado; 21 (iv).

Eriogonum cernuum Nutt. Nodding Buckwheat. [*E. cernuum* var. *tenue* T. & G., type from Weber Valley; *E. cernuum* var. *umbraticum* Eastw., type from McElmo Creek, San Juan County]. Plants annual, becoming umbelliform, mainly 5–45 cm tall; leaves all basal or cauline up to 10 cm above the base, the blades 3–25 mm long and about as wide, ovate to oval or orbicular, tomentose on one or both sides; petioles 3–40 mm long; inflorescence glabrous, open, the branches spreading or ascending; involucres usually stalked (except in var. *vimineum*), often deflexed, obconic to campanulate, 1–2 mm long, glabrous, the 5 teeth acute; flowers white, 1–2.5 mm long, glabrous, the outer segments constricted below the middle, the margins undulate, often more or less saccate basally, the inner ones obovate; achenes 1.5–2 mm long. Shadscale, other salt desert shrub, sagebrush, pinyon-juniper, mountain brush, ponderosa pine, aspen, and spruce-fir communities at 1220 to 2810 m in all Utah counties, except Cache, Morgan, and Summit; Canada south to California, Arizona, and New Mexico; 142 (xxvii). A phase with sessile involucres and somewhat larger flowers occurs in Beaver and Millard counties; i.e., **var. *viminale*** (Stokes) Reveal in Munz [*E. cernuum* ssp. *viminale* Stokes]; 5 (0). A few plants from sandy sites in Kane County have inflorescences more paniculiform than usual and more uniformly short-stipitate involucres. Possibly they are of taxonomic significance.

Eriogonum clavellatum Small Comb Wash Buckwheat. Shrubs, mainly 7–20 cm tall, clump-forming; leaves 3–15 mm long, 0.5–2 mm wide, narrowly oblanceolate to oblong, white-tomentose beneath, less densely so above, revolute, often with fascicled secondary ones in at least some axils; petioles very short; inflorescence cymose-umbellate, mainly 1–2.5 cm wide, glabrous; involucres on stipes mainly 1–4 mm long, glabrous, obconic to campanulate, 3.5–4.5 mm long, with 5-acutish teeth; flowers white or suffused with pink, glabrous, 3–3.5 mm long, the outer segments obovate to broadly spatulate, the inner ones narrower; achenes 3–3.5 mm long. Shadscale and blackbrush communities at ca 1325 to 1680 m in San Juan (type from Bartons range) County; Colorado; a Plateau endemic. 10 (iv). This taxon might be best treated as variety of *E. fasciculatum*, with which it is closely allied. However, no such combination is proposed herein.

Eriogonum contortum Small ex Rydb. Grand Buckwheat. [*E. effusum* ssp. *contortum* (Small) Stokes]. Shrubs mainly 5–20 cm tall, clump-forming; leaves 5–20 mm long, 1–2 mm wide, linear to narrowly oblanceolate, revolute, tomentose on one or both sides; petioles very short; inflorescence cymose to cymose-umbellate, the involucres not clustered, tomentose to glabrous, involucres 1–2.5 mm long, obconic to campanulate, glabrous, the 5 teeth acutish; flowers yellow, 1.5–2.5 mm long, glabrous, the segments oblong to obovate; achenes 2–2.5 mm long. Shadscale and other salt desert shrub communities at ca 1280 to 1525 m in Grand County; Colorado; 8 (ii). This low shrub is allied to the *brevicaule* complex.

Eriogonum corymbosum Benth. Corymb Buckwheat. Low to tall shrubs or subshrubs, 0.7–12 dm tall, clump-forming (seldom mat-forming); leaves 0.7–9 cm long, lanceolate to elliptic, orbicular, oblanceolate, spatulate, or linear, tomentose on one or both sides or glabrous, the margins flat to revolute; petioles 2–18 mm long; inflorescences cymose, the branches ascending to spreading or divaricate, glabrous or tomentose; involucres 1.5–4 mm long, obconic or campanulate, glabrous or tomentose, with 5 or 6 acute teeth; flowers white, suffused with pink or red, or yellow, 1.5–4.5 mm long, glabrous, the segments obovate to lanceolate or spatulate; achenes 2–3 mm long; n = 20. This is a huge and complex species group, involving numerous morphological variants, some of which are edaphically and geographically correlated. Diagnostic criteria are few, and are often based on vegetative characteristics that form continuous clines. The species is pivotal to *E. thompsonae*, *E. lonchophyllum*, *E. leptocladon*, and *E. brevicaule*, forming hybrids with all of them. At the margins of ecological tolerance the species undergoes reduction of internode length and concurrent elongation of the inflorescence. Yellow flowers are apparently derived, at least in part, from hybridization with other species having yellow flowers (see *E.* x *duchenense*, under *E. brevicaule*, and both *E. thompsonae* and *E. leptocladon*). The following treatment is preliminary, but allows recognition of the more important phases of the complex. There are other forms, possibly ecotypes or incipient ecotypes, that might be worthy of recognition, but those must await more definitive work.

1. Internodes of annual growth short, the inflorescence usually much longer than the vegetative branch 2
— Internodes of annual growth elongate, the inflorescence subequal to or shorter than the vegetative branch 4
2(1). Inflorescence tomentose; plants of the Sevier River drainage, Sink Valley, and Thousand Lake Mt. .... E. corymbosum var. revealianum
— Inflorescence glabrous; plants of various distribution ... 3
3(2). Leaves crenately revolute; plants of the Henry Mts. ... . E. corymbosum var. cronquistii
— Leaves flat or essentially so, the margins not especially crenate or revolute; plants of San Juan County .... E. corymbosum var. humivagans
4(1). Flowers yellow or pale yellow 5
— Flowers white or variously suffused with pink or red ... 6
5(4). Leaves glabrous on both surfaces; inflorescences glabrous; plants of southeastern Emery and eastern Wayne counties .... E. corymbosum var. smithii
— Leaves tomentose on one or both surfaces; inflorescences glabrous or tomentose; plants of different distribution .... E. corymbosum var. aureum
6(4). Leaf blades as long as broad or nearly so; plants forming clumps mainly 6–20 dm broad; inflorescence intricately and divaricately branched; plants typically of rimrock along the canyons of the Colorado River .... E. corymbosum var. orbiculatum
— Leaf blades much longer than broad; plants mainly less than 6 dm wide; inflorescence with branches not especially divaricate or sometimes so, but then of different substrates and distrubution 7
7(6). Leaves mainly 3–9 cm long, more or less revolute, but not especially crenate-revolute; plants of southeastern Duchesne County ...... E. corymbosum var. hylophilum
— Leaves mainly 0.5–4.5 cm long, usually crenate-revolute, less commonly flat; plants widespread ..../. E. corymbosum var. corymbosum

Var. aureum (Jones) Reveal Golden Buckwheat. [*E. aureum* Jones, type from near St. George; *E. aureum* var. *glutinosum* Jones; *E. corymbosum* var. *glutinosum* (Jones) Jones; *E. fruticosum* var. *glutinosum* (Jones) A. Nels.; *E. fruticosum* A. Nels.; *E. crispum* L. O. Williams, type from Cedar Canyon, Iron County]. Salt and mixed desert shrub and pinyon-juniper communties at 1065 to 2565 m in Emery, Garfield, Kane, Washington, and Wayne counties; Arizona; 27 (v). It is doubtful whether the yellow-flowered material constitutes "a taxon" in the usual sense. The assemblage is held together by the feature of flower color alone, a character hardly viewed as reliable in some portions of the genus, and the plants are almost as variable as those of var. *corymbosum*, with which they are largely sympatric. Similar yellow flowered plants from the Uinta Basin result from hybridization of *E. brevicaule* with *E. corymbosum*. Specimens from Washington County are transitional into *E. thompsonae*.

Var. corymbosum [*E. corymbosum* var. *divaricatum* T. & G., type from Green River; *E. corymbosum* ssp. *divaricatum* (T. & G.) Stokes; *E. divergens* Small; *E. effusum* ssp. *corymbosum* (Benth.) Stokes; *E. effusum* var. *durum* Stokes, type from Sunnyside; *E. corymbosum*

var. *erectum* Reveal & Brotherson, type from west of Duchesne; *E. corymbosum* var. *velutinum* Reveal; *E. lancefolium* Reveal & Brotherson, type from east of Wellington; *E. corymbosum* var. *davidsei* Reveal, type from Wellington]. Shadscale, other salt desert shrub, sagebrush, mixed desert shrub, and pinyon-juniper communities at 1400 to 2440 m, often on fine-textured or sandy soils, in Carbon, Daggett, Duchesne, Emery, Garfield, Grand, Kane, San Juan, Sevier, Uintah, Wasatch, and Wayne counties; Colorado and Arizona; 178 (xxxv). This variety is pivotal between *E. brevicaule* and *E. lonchophyllum* and other taxa.

Var. cronquistii (Reveal) Welsh Cronquist Buckwheat. [*E. cronquistii* Reveal, type from the Henry Mts.].. Pinyon, holodiscus, rabbitbrush, and rock-spiraea communities at ca 2680 to 2715 m in the Henry Mts., Garfield County; endemic; 3 (0). A closely similar plant is known from Thousand Lake Mt., but is tomentose throughout, except for the flowers, and is here assigned to var. *revealianum*.

Var. humivagans (Reveal) Welsh San Juan Buckwheat. [*E. humivagans* Reveal, type from 13.5 mi E of Monticello]. Woody aster, rabbitbrush, and pinyon-juniper communities at 1675 to 2105 m in San Juan (type from east of Monticello) County; endemic; 3 (i).

Var. hylophilum (Reveal & Brotherson) Welsh Gate Canyon Buckwheat. [*E. hylophilum* Reveal & Brotherson]. Juniper and pinyon-juniper communities at 2040 to 2535 m in Duchesne (type from Gate Canyon) County; endemic; 6 (0). Materials included within this variety are intermediate between *E. brevicaule* var. *promiscuum* and *E. corymbosum* var. *corymbosum*, especially that phase called *E. lancifolium* (q.v.). The variety is also influenced more or less by *E. brevicaule* var. *laxifolium*.

Var. orbiculatum (Stokes) Reveal & Brotherson Rimrock Buckwheat. [*E. effusum* ssp. *orbiculatum* Stokes, type from Green River, Emery County]. Eriogonum, mixed desert shrub, hanging garden, and pinyon-juniper communities, often on sandstone, at 1125 to 2200 m, in Emery, Garfield, Grand, Kane, San Juan, and Wayne counties; Arizona and New Mexico; 49 (xix). Materials designated as var. *velutinum* Reveal are transitional between var. *orbiculatum* and var. *corymbosum*, at least in Utah specimens.

Var. revealianum (Welsh) Reveal Reveal Buckwheat. [*E. revealianum* Welsh, type from south of Antimony]. Sagebrush, pinyon-juniper, and bristlecone pine communities at 2135 to 2745 m in igneous gravels or clay-silts in Garfield, Kane, Piute, and Wayne counties; endemic; 13 (v). A specimen from the south end of Thousand Lake Mt. (Atwood & Thompson 7645 BRY) is like var. *cronquistii* in habit, but has broader involucres and pubescence of var. *revealinanum*. Specimens from Kane County indicate a possible relationship with *E. thompsonae*.

Var. smithii (Reveal) Welsh Flat Top Buckwheat. [*E. smithii* Reveal, type from The Big Flat Top, Emery County]. Purple-sage, matchweed, ephedra-Indian ricegrass, and rabbitbrush communities, on the Entrada Formation and on stabilized dunes, at ca 1585 to 1710 m in Emery and Wayne counties; endemic; 12 (ii). This is the most striking phase within the *corymbosum* complex. Its origin is problematical, but the possibility of hybridization can not be discounted. Putative hybrids between var. *corymbosum* and *E. leptocladon* (q.v.) suggest such a possibility.

Eriogonum darrovii Hook. Darrow Buckwheat. Annuals, mainly 3–15 cm tall, usually branched from near the base; leaves mainly cauline, the blades 4–15 mm long, 3–13 mm wide, puberulent to villous-pilose on both sides, obovate to ovate, elliptic or orbicular; inflorescences axillary; involucres sessile, campanulate, 2–2.5 mm long, pilose, with 5 lance-ovate lobes; flowers yellow or pink, 1–2.5 mm long, hairy near the base, the outer segments broadly obovate, hooded, and conspicuously ciliate, the inner ones narrower; achenes ca 1 mm long. Pinyon-juniper community at ca 1860 m in Kane County; Arizona and Nevada; 1 (0).

Eriogonum davidsonii Greene Davidson Buckwheat. [*E. baileyi* var. *davidsonii* (Greene) Jones; *E. molestum* var. *davidsonii* (Greene) Jepson; *E. juncinellum* Gand.; *E. vimineum* ssp. *juncinellum* (Gand.) Stokes]. Annuals, 6–40 cm tall; leaves all basal (rarely some above the base), the blades 6–20 mm long and as wide or wider, orbicular, white-tomentose beneath and above or glabrate above; petioles 3–20 mm long; inflorescences glabrous, the branches erect-ascending; involucres sessile or terminal, narrowly obconic, 2.5–5 mm long, glabrous, the 5 teeth acutish; flowers white to pink, 1.5–2 mm long, glabrous, the segments obovate to oblong; achenes ca 2 mm long; n = 20. Creosote bush, Joshua tree, mixed warm desert shrub, and pinyon-juniper communities at 795 to 1680 m in Kane and Washington counties; California, Nevada, Arizona, and Mexico; 7 (0).

Eriogonum deflexum Torr. in Ives Skeletonweed Buckwheat. [*E. deflexum* var. *nevadense* Reveal]. Annuals, 5–40 (50) cm tall; leaves all basal; blades 6–30 (40) mm long and as wide or wider, orbicular to subreniform, rounded or cordate basally, tomentose on one or both sides; petioles 0.3–7 cm long; inflorescences usually spreading and umbrellalike, glabrous; involucres stipitate to subsessile or sessile, glabrous, deflexed, obconic to somewhat campanulate, mainly 1.5–2 mm long, the 5 teeth acutish; flowers white, sometimes pinkish, 1–2 mm long, glabrous, the outer segments cordate, the inner ones narrower; achenes 1.5–2 mm long; 2n = 40. Creosote bush, Joshua tree, blackbrush, other warm desert shrub, shadscale, and juniper communities at 760 to 1985 m in Garfield, Juab, Kane, Millard, San Juan, Washington, and Wayne counties; Nevada, Arizona, California and Mexico; 39 (xiii). This species is a close ally of *E. hookeri*, *E. brachypodum*, and *E. insigne*, all of which have been included previously within an expanded *E. deflexum*. Some specimens from Washington County have strict branches like *E. insigne*, but are otherwise *E. deflexum*. I follow recent tradition in treating the taxa as separate species. The var. *nevadensis*, in Utah at least, lacks both geographical and morphological continuity. Our material belongs to var. *deflexum*.

Eriogonum divaricatum Hook. Spreading Buckwheat. Annuals, prostrate to decumbent-ascending, the stems 5–22 cm long, dichotomously branched; leaves cauline and basal, the blades 3–30 mm long, 3–20 mm wide, oval to orbicular, puberulent with crinkly hairs; involucres sessile, borne in axils of bracteate leaves on spreading-decurved branches, obconic, 1–2 mm long, pilose, 5–lobed; flowers yellowish or suffused with red, 1–2 mm long, puberulent and glandular, the segments oblong to lanceolate; achenes 1.5–2 mm long. Shadscale, mixed desert shrub, and pinyon-juniper communities at 1155 to 2015 m in Emery, Garfield, Millard, San Juan, Sevier, Uintah, and Wayne counties; Wyoming, Colorado, New Mexico, and Arizona; 12 (i).

Eriogonum fasciculatum Benth. Mojave Buckwheat. Shrubs, mainly 2–8 dm tall, clump-forming; leaves cauline, often with some fascicled ones in lower axils, 4–18 mm long, 1–5 mm wide, usually more or less revolute, linear to narrowly oblong or oblanceolate, more or less tomentose on one or both sides; inflorescences long-peduncled, divarcately branched or subcapitate, tomentulose; involucres obconic to campanulate, 2–3.5 mm long, the 5 obtusish lobes with hyaline margins; flowers white to pink, 2–3 mm long, villous-pilose, the segments obovate; achenes 2–2.5 mm long. Warm desert shrub communities at 730 to 1495 m in Washington County; Nevada, California, Arizona, and Mexico; 25 (i). Our material has been assigned to var. **polifolium** (Benth.) T. & G. [*E. polifolium* Benth. in DC.]. A specimen with provenience of Emery County (Cottam 5224A BRY) is extant, but might be mislabeled.

Eriogonum flexum Jones Bent Buckwheat. [*E. flexum* var. *ferronis* Jones, type from near Ferron; *Stenogonum flexum* (Jones) Reveal & J. T. Howell]. Annuals, 4–35 cm tall; leaves all basal (rarely some whorled at nodes of inflorescence); blades 3–28 mm long and about as wide, orbicular to oval, truncate to subcordate basally, puberulent to glabrous and sometimes glandular on one or both surfaces; petioles 3–40 mm long; involucres stipitate, the filiform stipes commonly abruptly bent below the involucre and often glandular below, campanulate, in 2 whorls, each 3–lobed; flowers yellow, 1.5–4 mm long, puberulent, the segments lanceolate; achenes 2–2.5 mm long; n = 20. Shadscale, mat-saltbush, blackbrush, and pinyon-juniper communities, often on fine-textured substrates, at 1430 to 1865 m in Carbon, Emery, Garfield, Kane, San Juan, Uintah, and Wayne counties; Colorado and Arizona; 34 (iii). Although regarded by some workers as belonging, with *E. salsuginosum*, in the segregate genus, *Stenogonum*, because of their peculiar involucres, both species appear to be more nearly allied to species within *Eriogonum* proper than they are to each other.

Eriogonum gordonii Benth. in DC. Gordon Buckwheat. Annuals, mainly 8–60 cm tall; leaves all basal; blades 9–55 mm long, oval to suborbicular, obtuse to truncate or cordate basally, green above, paler beneath, softly spreading-hairy; petioles 0.5–10 cm long or more; inflorescences spreading-ascending, glabrous or hairy; involucres on stipes mainly 3–20 mm long, obconic-campanulate, 0.6–1.3 mm long, glabrous, with 5 obtusish teeth; flowers white, 1–2.5 mm long, glabrous, the segments obovate to oblong or oblanceolate; achenes 1.8–2.5 mm long. Salt desert shrub, shadscale, and juniper or pinyon-juniper communities, on fine-textured saline soils, at 1110 to 2015 m in Carbon, Daggett, Duchesne, Emery, Garfield, Grand, Kane, and Uintah counties; Wyoming to Nebraska, south to Arizona, and New Mexico; 49 (vi).

Eriogonum heermannii Dur. & Hilg. Heermann Buckwheat. Shrubs, mainly 1–6 dm tall, clump-forming, with intricately and divaricately branched inflorescences appearing cushionlike; leaves mainly 3–17 mm long, 2–5 mm wide, the blades elliptic to spatulate, tomentose on one or both sides, more or less revolute; petioles 3–10 mm long; inflorescence cymose, the branches angled or ribbed and sulcate between the ribs; involucres sessile, glabrous, campanulate, 0.6–1.5 mm long, with 5 rounded teeth; flowers white (yellowish?), 1.5–3 mm long, glabrous, the outer segments obovate, the inner ones narrower; achenes 2–2.5 mm long. Blackbrush, mixed desert shrub, mountain brush, and pinyon-juniper communities (often on rock outcrops) at ca 1220 to 2135 m in Washington County; Nevada, Arizona, and California; 7 (i). All material from Utah examined by me belongs to var. *sulcatum* (Wats.) Munz & Reveal (*E. sulcatum* Wats., type from near St. George]. The var. *subracemosum* (Stokes) Reveal [*E. howellii* var. *subracemosum* Stokes] is present in the region also. It differs in having stems less angled and involucres more racemosely arranged.

Eriogonum heracleoides Nutt. Whorled Buckwheat. [*E. heracleoides* var. *utahense* Gand., type from Cache County]. Perennial, mat-forming, mainly 2–6 dm across; vegetative stems persistent, the branches woody and more or less clothed with persistent, gray to brown or blackish leaves and bases; flowering stems with whorled leaves near the middle, arising from rosettelike bases, mainly 1.5–5 dm tall; leaves 2–7 cm long, 3–15 mm wide, the blades elliptic to oblong or oblanceolate, tomentose on one or both sides, entire, flat or essentially so; petioles 3–30 mm long; inflorescences umbellate or twice umbellate, rarely capitate, tomentose; involucres sessile or on stipes to 40 mm long, obconic to campanulate, 4–10 cm long, the lobes subequal to the tube or longer; flowers white or cream (or yellow), 4–9 mm long, including the stipitate base, the segments spatulate to elliptic or oblong; achenes 2–5 mm long. Sagebrush, mountain brush, juniper, pinyon-juniper, Douglas fir, and aspen communities at 1310 to 3050 m in Box Elder, Cache, Daggett, Davis, Duchesne, Juab, Millard, Salt Lake, Sanpete, Summit, Tooele, Uintah, Utah, and Wasatch counties; Canada, south to California, Nevada, and Wyoming; 83 (ii). This plant forms putative hybrids with phases of *E. umbellatum*. A specimen with features of *E. heracleoides*, but with yellow flowers (Neese 14148 BRY) might indicate hybridization.

Eriogonum hookeri Wats. Watson Buckwheat. [*E. deflexum* ssp. *hookeri* (Wats.) Stokes; *E. deflexum* ssp. *hookeri* var. *gilvum* Stokes, type from American Fork Canyon]. Annuals, mainly 8–60 mm tall; leaves all basal; blades mainly 10–50 mm long and as broad or broader, orbicular to reniform, tomentose on both sides, obtuse to cordate basally, flat to undulate; inflorescences glabrous, umbrellalike; involucres sessile, deflexed, campanulate to hemispheric, 1–2 mm long, glabrous; flowers yellow, soon suffused with pink to dark red, 1.5–2.7 mm long, glabrous, the outer segments cordate, the inner ones narrower; achenes 2–2.5 mm long; 2n = 40. Mixed desert shrub, sagebrush, pinyon-juniper, aspen, and spruce-fir communities at 1135 to 3050 m in Beaver, Box Elder, Carbon, Duchesne, Emery, Garfield, Iron, Juab, Millard, Piute, San Juan, Sevier, Tooele, Uintah, Utah (type from American Fork Canyon), and Wayne counties; Wyoming, Colorado, Arizona, Nevada, and California; 70 (xviii).

Eriogonum howellianum Reveal Howell Buckwheat. Annual, 5–30 cm tall, simple or branched from the base; leaves all basal; blades 6–25 (30) mm long and about as wide, oval to suborbicular, pubescent with long, soft, spreading hairs on at least the lower surface, obtuse to subcordate basally; petioles 3–40 mm long; inflorescences divaricately branched, the branches with scattered, stipitate, dark glands; involucres with filiform stipes 3–20 mm long or more, obconic to campanulate, glabrous, 1.3–2 mm long, usually 5-toothed; flowers yellowish or reddish, 1–2 mm long, the segments lanceolate, mostly obscured by spreading-villous hairs; achenes 1.5–2 mm long. Desert shrub, desert almond, and shadscale communities at 1460 to 1740 m in Juab, Millard (type from southeast of Garrison) and Tooele counties; Nevada; a Great Basin endemic; 8 (i). This taxon is allied to *E. inflatum* and *E. flexum*.

Eriogonum inflatum T. & G. Bottlebush; Bottlestopper; Desert trumpet. Annual or perennial herbs, mainly 8–100 cm tall; leaves all basal; blades 4–30 mm long and about as wide or wider, orbicular to oblong or reniform, hirtellous on one or both sides, obtuse to cordate basally, entire to undulate-crisped; petioles 0.5–6 cm long; peduncles and usually the primary and secondary rays of inflorescence inflated, rarely not; inflorescence umbellate-cymose; involucres borne on glabrous, capillary to filiform stipes 5–45 mm long or more, obconic, 0.7–1.5 mm long, glabrous, the 5 lobes acutish; flowers yellow or reddish, 1–2.5 mm long, densely strigose, the segments lanceolate to ovate; achenes 2–2.5 mm long; n = 16. Warm, mixed, and salt desert shrub and pinyon-juniper communities at 760 to 1955 m in Carbon, Duchesne, Emery, Garfield, Grand, Kane, San Juan, Uintah, Washington, and Wayne counties; California, Nevada, Colorado, Arizona, New Mexico, and Mexico; 90 (xvii). Annuals within this species have been regarded as var. *fusiforme* (Small) Reveal [*E. fusiforme* Small], and perennials as var. *inflatum*. The former occurs at the margins of the range of the latter, but is also sympatric. The segregation appears to be moot, owing to the flowering of specimens of both phases during the initial year.

Eriogonum insigne Wats. Unique Buckwheat. [*E. deflexum* var. *insigne* (Wats.) Jones; *E. deflexum* ssp. *insigne* (Wats.) Stokes; *E. exaltatum* Jones; *E. deflexum* ssp. *exaltatum* (Jones) Stokes]. Annuals, mainly 8–100 cm tall; leaves all basal; blades 8–50 mm long (or more) and as wide or wider, orbicular to reniform, obtuse to cordate basally, tomentose on one or both sides; petioles 0.6–10 cm long; peduncles simple or branched from the base, inflorescences open cymose, the branches glabrous, erect to spreading; involucres sessile or with stipes to 6 mm long, obconic to campanulate, 2–3 mm long, glabrous, the 5 teeth obtusish; flowers white or suffused with pink, 1.5–2 mm long, glabrous, the outer segments cordate to oblong-cordate, the inner ones narrower; achenes 2–2.5 mm long; 2n = 40. Creosote bush, other warm desert shrub, and mixed desert shrub communities at 730 to 1170 m in Iron (type from Red Creek) and Washington counties; California, Nevada, and Arizona; 6 (0).

Eriogonum jamesii Benth. in DC. James Buckwheat. Mat-forming perennials, mainly 1–6 dm wide; vegetative stems persistent, the branches woody, usually clothed with persistent, ashy to dark brown leaf bases; flowering stems subscapose, arising from rosettelike branches, mainly 6–30 cm long; leaves 1–9 cm long, 4–20 mm wide, the blades elliptic to obovate or ovate, tomentose on one or both sides, entire or undulate, flat or essentially so; petioles 0.5–6 cm long; inflorescences capitate or once or twice umbellate, tomentose, with foliose bracts at the nodes; involucres sessile, campanulate, 3–14 mm long, tomentose, the 5–8 teeth obtusish, erect to spreading; flowers yellow, 4–11 mm long, including the stipitate base, the segments spatulate to obovate; achenes 4–5 mm long. Sagebrush, mountain brush, pinyon-juniper, and ponderosa pine communities at 1585 to 2685 m in Carbon, Duchesne, Emery, Kane, San Juan, Sevier, Utah, Washington, and Wayne counties; Wyoming to Kansas, south to Arizona, New Mexico, Texas, and Mexico; 40 (viii). This is a remarkably beautiful species, with its bright sulfur yellow flowers. The species varies from population to population and specimens from Utah have been regarded as belonging to two varieties, although more segregation seems possible. Dwarf plants from Washington and adjacent Kane counties have pilose hairs over the tomentum on the upper leaf surfaces; they belong to var. *rupicola* Reveal (type from Zion National Park). The remainder of the Utah specimens are included within var. *flavescens* Wats., but that taxon consists of variants of about equal rank to var. *rupicola*. Specimens from San Juan County have capitate inflorescences, and material from western Emery County has huge involucres.

Eriogonum leptocladon T. & G. Sand Buckwheat. Shrubs, mainly 2–10 dm tall or more, clump-forming; leaves often deciduous at anthesis, mainly 10–45 mm long, 2–10 mm wide, linear to narrowly lanceolate or oblanceolate, more or less revolute to flat, tomentose on one or both sides; petioles 1–6 mm long; inflorescences tomentose or glabrous, much longer than the vegetative stems; involucres cymose-racemose, sessile or nearly so, obconic to campanulate, 1.5–3 mm long, glabrous or tomentose, the 5 teeth acute to rounded; flowers yellow, yellowish, or white and often suffused with pink, 2–3.5 mm long, glabrous, the segments obovate; achenes 2.5–3.5 mm long; n = 20. Three rather weak varieties are present.

1. Flowers yellow; plants of the central Canyonlands vicinity .... *E. leptocladon* var. *leptocladon*
— Flowers white; plants sometimes distributed as above, or otherwise 2
2(1). Branches of inflorescence yellowish green, glabrous or rarely tomentose; plants of Garfield and Kane counties .... *E. leptocladon* var. *papiliunculum*
— Branches of inflorescence green to gray green, tomentose or glabrous; plants of broad or other distribution .... *E. leptocladon* var. *ramosissimum*

Var. *leptocladon* [*E. microthecum* var. *leptocladon* (T. & G.) T. & G.; *E. effusum* ssp. *leptocladon* (T. & G.) Stokes; *E. effusum* ssp. *pallidum* var. *shandsii* Stokes, type from Indian Creek, San Juan County]. Purple-sage, ephedra, sand sagebrush, blackbrush, saltbush, and pinyon-juniper communities, usually in sand or on stabilized dunes, at 1340 to 1895 m in Emery, Garfield, Grand (type from Green River), San Juan, Sevier, and Wayne counties; endemic; 39 (x). This phase forms putative hybrids with *E. corymbosum* var. *corymbosum* (Neese 6829 - 6833 BRY). The apparent backcrosses to *corymbosum* have broad leaves and yellowish flowers or are broad leaved and have white or pinkish flowers. The latter plants simulate var. *ramosissimum* and suggest at least one possible origin for that entity.

Var. *papiliunculum* Reveal Little-butterfly Buckwheat. Ephedra-Vanclevea, sand sagebrush, other sand desert shrub, and juniper communities at 1400 to 1830 m in Garfield, Kane, San Juan, and Wayne counties; Arizona (?); 10 (ii). These plants have broader leaves than in var. *leptocladon* and yellowish green inflorescences.

They are intermediate in most respects between var. *ramosissimum* and *E. corymbosum*, with possibly both var. *corymbosum* and var. *aureum* as contributors. Specimens transitional to both var. *ramosissimum* and *E. corymbosum* var. *aureum* are known.

Var. *ramosissimum* (Eastw.) Reveal Eastwood Buckwheat. [*E. ramosissimum* Eastw., type from near Butler Wash, San Juan County]. Vanclevea, yucca, purple-sage, sand sagebrush, blackbrush, and juniper communities at 1310 to 1770 m in Garfield, Kane, San Juan, and Wayne counties; Arizona, Colorado, and New Mexico; 14 (ii). This plant appears to be closely allied to *E. wrightii*, q.v.

Eriogonum leptophyllum (Torr.) Woot. & Standl. Slenderleaf Buckwheat. [*E. effusum* var. *leptophyllum* Torr. in Sitgr.; *E. microthecum* var. *leptophyllum* (Torr.) T. & G.]. Shrubs, mainly 1.5–5 (6) dm tall; vegetative branches with leaves separated by short internodes; leaves (1.5) 2–6 cm long, 1–3 mm wide, linear, tomentose on the almost obscured lower surfaces, tightly revolute; petioles 0.4–1 mm long; peduncles and inflorescences glabrous, cymose, dense and tufted with numerous glabrous branches; involucres obonic, 2–3 mm long, glabrous, 5–lobed; flowers white to pink, 2–4 mm long, glabrous, the segments subequal; achenes 3.5–4 mm long. Desert shrub community at ca 1700 m in San Juan County (Heil & Porter sn BRY); Colorado, Arizona, and New Mexico; 1 (0).

Eriogonum lonchophyllum T. & G. Longleaf Buckwheat. Subshrubs or shrubs, mainly 8–80 cm tall; vegetative branches with leaves all at base of current growth or with leaves separated by elongated internodes; leaves mainly 2–11 cm long, 2–12 mm wide, linear to elliptic, lanceolate, or oblanceolate, tomentose on one or both sides, margins entire to crenate, plane to revolute; petioles 3–20 mm long; peduncles and inflorescences glabrous or tomentose, cymose-corymbose to cymose-capitate; involucres usually sessile, obconic to campanulate, 2–4 mm long, glabrous, 5–lobed; flowers white, cream, or suffused with pink, 2–4 mm long, glabrous, the segments subequal; achenes 2.5–3.5 mm long. As is typical of other species complexes in the perennial versus shrubby species in *Eriogonum*, the *E. lonchophyllum* phases demonstrate genetic compatibility with members of other complexes. And, these likewise tend to precipitate out more or less uniform phases on distinctive soils or geologic substrates. Problems of interpretation of the distinctive groupings, their orgins, and relationships are not made easier by the linear systems of classification and nomenclature usual in plant taxonomy. Instead of taxa (both ecotypes and microspecies) being related by descent from a common ancestor, they might have resulted from a reticulate relationship involving two or more parental taxa. There are two more or less distinctive taxa in Utah that fall within the circumscription of *E. lonchophyllum*, as described above. In species of genera in other families these would be regarded as belonging to the same taxon, in a broad sense, but here they might have had separate origins. The following treatment is, therefore, tentative.

1. Plants acaulescent or essentially so, the internodes of vegetative stems very short, growing on ridge crests along the Tavaputs divide and elsewhere ....
 .... *E. lonchophyllum* var. *lonchophyllum*

— Plants definitely caulescent, the internodes of vegetative stems readily apparent, growing on Mowry Shale and closely adjacent strata in northern Uintah County . . . . .
.... *E. lonchophyllum* var. *saurinum*

Var. lonchophyllum [*E. intermontanum* Reveal, type from the Roan Cliffs, Grand County]. Sagebrush, mountain brush, and Douglas fir communities, mainly on Green River and other calcareous formations at 2285 to 2745 m in Emery, Grand, and Uintah counties; Colorado and New Mexico; 11 (ii). This variety forms intermediates with *E. corymbosum* down slope in Uintah County (the Rainbow phase); 21 (0). The apparent hybrids are transitional from one extreme to the other, with individuals simulating not only var. *saurinum* but also the *E. lancifolium* and *E. corymbosum* var. *davidsei* phases of *E. corymbosum* var. *corymbosum* (q.v.). The similarity of this taxon to both *E. batemanii* var. *batemanii* and *E. spathulatum* is great. It also approaches *E. brevicaule* through the var. *promiscuum*.

Var. saurinum (Reveal) Welsh Dinosaur or Mowry Buckwheat. [*E. saurinum* Reveal, type from 10 mi E of Vernal]. Eriogonum, juniper, serviceberry, pinyon-juniper, and ponderosa pine, mainly on Wasatch, Mowry, Curtis, Entrada, Carmel, and Moenkopi formations, at 1585 to 1895 m in northern Uintah County; endemic; 33 (vi). Much of var. *saurinum* grows on the siliceous, acidic Mowry Shale Formation. That material, though variable, is the most uniform phase of the variety. Evidence exists that even the Mowry Shale phase is partially, at least, a product of introgression with *E. corymbosum*. On other formations adjacent to the Mowry Shale the plants vary from the type; e.g., in the Steinaker Reservoir area (Curtis, Entrada, and Carmel formations) the inflorescences are suggestive of those of *E. brevicaule* var. *viridulum* on the one hand and *E. microthecum* on the other; in the Asphalt Ridge (Wasatch Formation) vicinity the plants bear features of *E. corymbosum*; and, in the Bourdette Draw vicinity (Moenkopi Formation), south of Blue Mountain, the plants again share features of *E. brevicaule*, in a broad sense. Though trends exist that indicate direct relationship with *E. lonchophyllum*, this variety might represent mainly recombinants of various *E. brevicaule* and *E. corymbosum* introgressants. More work is indicated.

Eriogonum maculatum Heller Spotted Buckwheat. [*E. angulosum* var. *maculatum* (Heller) Jepson; *E. angulosum* ssp. *maculatum* (Heller) Stokes]. Annuals, mainly 8–37 cm tall or more; leaves basal and cauline (foliose bracteate); basal leaf blades 5–25 mm long, 3–15 mm wide, oval to obovate or elliptic, tomentose on one or both sides; petioles 3–15 mm long; bracteate leaves reduced and becoming sessile upwards; inflorescences tri- or dichotomous, tomentose; involucres on filiform stipes 5–30 mm long or more, broadly campanulate, 1–2.5 mm long, glandular-puberulent, with 5 broad teeth; flowers white to yellowish or pink, 1.5–2.8 mm long, glandular-puberulent, the outer segments ovate and cupulate, shorter than the slender inner ones; achenes 1–1.5 mm long. Creosote bush, Joshua tree, blackbrush, pinyon-juniper, live oak, and mixed desert shrub communities, at 730 to 1830 m in Box Elder, Juab, Millard, Tooele, and Washington counties; Washington and Idaho, south to California, Nevada, and Arizona; 25 (iv).

Eriogonum microthecum Nutt. Slender Buckwheat. Shrubs, mainly 4–100 cm tall, clump-forming; leaves 4–35 mm long, 1–7 mm wide, elliptic to linear or oblanceolate, tomentose on one or both sides, the margins flat or revolute; petioles 1–5 mm long; inflorescences cymose, the branches ascending to spreading, glabrous or tomentose; involucres sessile to short-stipitate, obconic, 2–3.5 mm long, tomentose or glabrous, with 5 obtusish to rounded teeth; flowers white or suffused with pink, 2–3.2 mm long, glabrous, the segments obovate; achenes 2–3 mm long. Salt and mixed desert shrub, sagebrush, pinyon-juniper, pondorosa pine, mountain brush, and white fir communities at 1125 to 2900 m in all Utah counties except Davis, Duchesne, Morgan, Salt Lake, Sanpete, Wasatch, and Weber; Washington to Montana, south to California, Nevada, Arizona, and New Mexico; 166 (xxxvi). There are two intergrading phases of this species in Utah, distinguished only by leaves being flat or revolute. The former have been designated as **var. laxiflorum** Hook. [*E. tenellum* var. *grandiflorum* Gand., type from Utah], and the latter as **var. foliosum** (T. & G.) Reveal [*E. effusum* var. *foliosum* T. & G.; *E. simpsonii* Benth. in DC.; *E. friscanum* Jones, type from Frisco; *E. nelsonii* L. O. Williams, type from Geyser Basin, San Juan County]. Specimens that are intermediate between *E. microthecum* and *E. brevicaule* are known (Neese 14531 a - c BRY), and likewise with *E. lonchophyllum* var. *saurinum* (Neese 8495 BRY). Despite its tendency to form intermediates with other taxa the slender buckwheat is not known to hybridize with *E. corymbosum*, with which it is typically contrasted in keys.

Eriogonum nidularium Cov. Birdnest Buckwheat. Annuals, mainly 5–20 cm tall, usually with erect-ascending branches from near the base; leaves all basal, 3–20 mm long and as wide, orbicular, tomentose on one or both sides; petioles 4–30 mm long; inflorescences densely branched, tomentose; involucres sessile, obconic, 0.6–1 mm long, appressed-erect, 5–toothed; flowers yellowish or reddish, 1.5–3 mm long, glabrous, the outer segments broadly obovate to flabellate, the inner ones narrower; achenes ca 1 mm long. Mixed desert shrub at ca 1065 to 1220 m in Washington County; Oregon to Idaho, south to California and Arizona; 3 (0).

Eriogonum nummulare Jones Coin Buckwheat. Shrubs or subshrubs, sprawling to erect, mainly 1–8 dm tall, clump-forming; leaves 4–30 mm long, 4–17 mm wide, orbicular to elliptic, lanceolate, or obovate, tomentose on both surfaces, plane or undulate; petioles 1–15 mm long; inflorescences cymose or cymose-racemose, tomentose or glabrous, the branches erect-ascending or spreading; involucres sessile or on stipes 1–2 mm long, obconic, 1.5–3.5 mm long, tomentose or glabrous, 5–toothed; flowers white or suffused with pink, 1.5–3 mm long, the segments obovate to oblong; achenes 1.5–3.5 mm long; n = 40. Two varieties occur in Utah.

1. Inflorescences glabrous; involucres narrowly obconic, glabrous; plants uncommon ....
.... *E. nummulare* var. *ammophilum*
— Inflorescences tomentose; involucres broadly obconic, tomentose; plants locally common ....
.... *E. nummulare* var. *nummulare*

Var. ammophilum (Reveal) Welsh Ibex Buckwheat. [*E. ammophilum* Reveal, type from Ibex Warm Point, Millard County]. Shadscale, horsebrush, winterfat, rabbitbrush, ephedra, and pinyon-juniper communities at 1460 to 1830 m in Millard (type from Ibex Warm Point)

County; endemic; 8 (v). These plants are intermediate between *E. nummulare*, in a strict sense, and *E. batemanii* var. *eremicum*. They share the caulescent habit of the former with the glabrous inflorescences and involucres of the latter. The distribution is intermediate between the two.

Var. *nummulare* [*E. kearneyi* Tidestr., type from w of Tooele; *E. dudleyanum* Stokes, type from Skull Valley]. Fourwing saltbush, rabbitbrush, sagebrush, salt desert shrub, and juniper communities at 1095 to 1985 m in Juab, Kane, Millard, Tooele (type from Dutch Mountain), and Washington counties; California, Nevada, and Arizona; 30 (v). Specimens from sandy areas of eastern Tooele County south to Kane and Washington counties (i.e., *E. kearneyi* sens. str.) have leaves proportionally longer than broad, but the variation is continuous westward with more typical material.

Eriogonum nutans T. & G. Dugway Buckwheat. [*E. deflexum* ssp. *ultrum* Stokes, type from Sevier Valley; *E. rubiflorum* Jones, type from Dugway, Tooele County]. Annuals, mainly 5–30 cm tall; leaves all basal; blades 5–25 mm long and as wide or wider, orbicular to reniform, obtuse to cordate basally, tomentose on one or both sides; petioles 5–28 mm long; inflorescences more or less trichotomously branched, glabrous or more or less stipitate-glandular; involucres with slender stipes mainly 3–12 mm long, finally decurved, broadly campanulate, 2–3 mm long, more or less glandular, the 5 teeth with hyaline margins; flowers white or suffused with pink or red, glabrous, 2–3 mm long, the outer segments oblong-obovate, the inner ones narrower; achenes 1.5–2 mm long. Shadscale and sagebrush communities at ca 1525 to 1830 m in Beaver, Carbon, Millard, Sanpete, Sevier, and Tooele counties; Oregon and Nevada; 6 (ii).

Eriogonum ovalifolium Nutt. Cushion Buckwheat. Pulvinate-caespitose, often mound-forming perennials, mainly 0.5–4 dm across; vegetative branches clothed with persistent, ashy to black leaf bases, terminated by rosettes of leaves; fertile stems scapose, 1–30 cm tall; leaf blades 2–6 cm long, 1–15 mm wide, tomentose on both surfaces, orbicular to elliptic, oblanceolate, or spatulate; petioles 1–50 mm long or more; inflorescences capitate, tomentose; involucres solitary or few to several, obconic to campanulate, 2–5.6 mm long, tomentose, with 5 teeth; flowers white, cream, yellow, or suffused with pink, red, or purple, 3–7 mm long, glabrous, the outer segments oval to orbicular, the inner ones narrower; achenes 2–3 mm long; n = 20. Shadscale, bullgrass, winterfat, Grayia, sagebrush, pinyon-juniper, fringed sagebrush, and alpine meadow communities at 1370 to 3420 m in all Utah counties, except Davis, Morgan, Summit, and Wasatch; Canada, south to California, Arizona, and New Mexico; 202 (xxv). This species has been treated as having three varieties in Utah; var. *ovalifolium* [*E. ovalifolium* var. *utahense* Gand., type from Cache County?], with white or whitish flowers that ultimately turn pink, red, or purple; var. *multiscapum* Gand., with yellow flowers; and var. *nivale* (Canby) Jones [*E. nivale* Canby], a dwarf, small-flowered plant of high elevations. The segregation has not proven to be more than arbitrary, with diagnostic features segregating specimens, not taxa.

Eriogonum palmerianum Reveal in Munz Palmer Buckwheat. [*E. plumatella* var. *palmeri* T. & G.; *E. baileyi* var. *tomentosum* Wats.]. Annuals, mainly 6–25 (30) cm tall; leaves all basal (rarely some cauline); blades 4–23 mm long and as wide or wider, orbicular to subreniform, obtuse to cordate basally, tomentose on one or both sides; petioles 3–40 mm long; inflorescences branched from near the base, tomentose, the branches often divaricate; involucres sessile, appressed, obconic, 1.2–2 mm long, tomentose, with 5 acute teeth; flowers white or pink, 1.5–2.4 mm long, glabrous, the outer segments broadly oblanceolate or obovate, the inner ones narrower; rower; achenes 1.5–2 mm long; n = 20. Blackbrush, shadscale, cheatgrass, rabbitbrush, desert almond, sagebrush, and pinyon-juniper communities at 1155 to 1985 m in Beaver, Box Elder, Garfield, Grand, Iron, Juab, Kane, Millard, San Juan, Sevier, Tooele, and Washington counties; Nevada to Colorado, California, Arizona, and New Mexico; 44 (x).

Eriogonum panguicense (Jones) Reveal Panguitch Buckwheat. [*E. pauciflorum* var. *panguicense* Jones, type from Panguitch; *E. spathulatum* var. *panguicense* (Jones) Stokes; *E. chrysocephalum* var. *alpestre* Stokes, type from Cedar Breaks; *E. panguicense* var. *alpestre* (Stokes) Reveal]. Pulvinate to caespitose perennial herbs, mainly 5–20 cm across; vegetative stems abbreviated, more or less clothed with ashy to black leaf bases and terminated by clustered leaves; flowering stems scapose, 2–30 cm long, glabrous; leaves 4–70 mm long, 2–8 (10) mm wide, linear to elliptic, oblanceolate, lanceolate, ovate, or obovate, obtuse to cuneate basally, plane or somewhat revolute; petioles 1–12 mm long; inflorescences glabrous, capitate or rarely branched; involucres sessile, several, obconic to campanulate, 2–3.7 mm long, the 5 teeth acute to obtuse; flowers white, often suffused with red, 2–3 mm long, glabrous, the segments oblong to lance-oblong; achenes 3–4 mm long. Pinyon-juniper, sagebrush, ponderosa pine, pygmy sagebrush, bristlecone pine, and spruce-fir communities, usually on limestone, at 1675 to 3355 m in Garfield, Iron, Kane, Millard, Sanpete, Sevier, and Washington counties; endemic; 48 (xi). This attractive buckwheat is closely allied to both *E. batemanii* and *E. spathulatum*, with whom it is partially sympatric. The species differs from both, however, in the usually unbranched inflorescences and smaller stature. It consists of a series of more or less disjunct populations growing on peculiar calcareous strata. Each population differs in subtle ways from all others, and if one is chosen for varietal status, the remainder require similar recognition. The overall status, as a mosiac of variation, seems to dictate against recognition of infraspecific categories.

Eriogonum pharnaceoides Torr. in Sitgr. Wirestem Buckwheat. Annuals, mainly 6–30 cm long; leaves basal and cauline (foliose bracteate); blades 8–35 mm long, 1–6.5 mm wide, linear to narrowly oblanceolate, tomentose on one or both sides; petioles 1–5 mm long or lacking; inflorescences cymose, tomentulose; involucres on filiform stipes mostly 8–50 mm long, these often curved, campanulate, usually pilose, 3–4 mm long, with 5 oblong teeth; flowers yellow, 2–3 mm long, glabrous, the outer segments cordate and more or less cupulate, the inner ones narrower and surpassing the outer; achenes 1.5–2 mm long. Pinyon-juniper, oakbrush, and ponderosa pine communities at ca 1830 to 2640 m in Iron and Washington counties; Nevada, Arizona, and New Mexico; 5 (0). Our material belongs to **var. *cervinum*** Reveal (type from Pine Valley Mts.).

Eriogonum plumatella Dur. & Hilg. Obscure Buckwheat. [*E. palmeri* Wats., type from Washington County]. Low, rounded shrubs, mainly 3–6 dm tall;

stems leafy in the lower portion, white-tomentose, branched above; leaves 8–15 mm long, oblanceolate to elliptic, oblong, or oblong-lanceolate, more or less revolute, acute, tomentose, with short, slender petioles; branches of inflorescence divaricate, green; involucres borne on widely spreading branches, these terminating in branches with shortened internodes, the involucres solitary or crowded, glabrous, sessile, obconic to subcylindric, ca 2.5 mm long; perianth segments glabrous, white, 1.5–2.5 mm long, the outer segments ovate, the inner narrower; filaments pilose below; achenes narrow, subangular. Creosote bush community at ca 1035 m in the Beaverdam Mts., Washington County; Arizona, Nevada, and California; 3 (i).

Eriogonum polycladon Benth. in DC. Leafy Buckwheat. [*E. vimineum* ssp. *polycladon* (Benth.) Stokes]. Annuals, mainly 15–60 cm tall, the leafy stems erect; leaves basal and cauline; blades 6–18 mm long, 4–13 mm wide, obovate to elliptic, ovate, or suborbicular, tomentose on one or both sides; petioles 2–15 mm long; inflorescences tomentose, the branches erect-ascending; involucres sessile, appressed-erect, 1.5–2.5 mm long, glabrous or tomentose, with 5 obtuse teeth; flowers white or suffused with pink, 1.5–2.5 mm long, glabrous, the outer segments broadly obovate, the inner somewhat narrower; achenes 1–1.5 mm long. Sagebrush and pinyon-juniper communities at ca 1675 to 1830 m in Kane County; Arizona, New Mexico, Texas, and Mexico; 6 (ii).

Eriogonum puberulum Wats. Red Creek Buckwheat. Annuals, mainly 4–30 cm tall; leaves basal and cauline (leafy bracteate); blades 2–15 mm long and about as wide, obovate to orbicular, puberulent to pilosulose on one or both sides; petioles 1–15 mm long; inflorescences puberulent, more or less dichotomously branched; involucres obconic, 0.6–1.5 mm long, mainly obscured by cupulate, long-lobed, nodal bracts, with 5 obtuse lobes; flowers white or suffused with red, 1.5–2.2 mm long, glabrous or scabrous, the segments oblong, sometimes somewhat cordate basally; achenes ca 1 mm long. Blackbrush, pinyon-juniper, mountain brush, and ponderosa pine communities at 1050 to 2745 m in Beaver, Iron (type from Red Creek), Millard, and Washington counties; Nevada; 7 (0).

Eriogonum pusillum T. & G. Low Buckwheat. [*E. reniforme* ssp. *pusillum* (T. & G.) Stokes]. Annuals, 5–30 cm tall; leaves all basal; blades 3–20 mm long and about as wide, obovate to oval, tomentose on one or both sides; petioles 6–30 mm long; inflorescences more or less trichotomous, glabrous or the bracts glandular; involucres on slender, glabrous stipes 3–40 mm long, campanulate, 1–1.7 mm long, glandular-puberulent, the 5 lobes acute to obtuse; flowers yellow, 2–2.5 mm long, glandular-scaberulous, the segments oblong; achenes ca 1 mm long. Creosote bush and Joshua tree communities at ca 760 m in Washington County; Oregon and Idaho, south to California and Arizona; 2 (i).

Eriogonum racemosum Nutt. Redroot Buckwheat. Perennial, scapose or subscapose herbs, 1.6–10 dm tall, from a simple or branched caudex; leaves all basal or some foliose-bracteate ones at nodes of inflorescence; blades 1–10 cm long, 6–38 mm wide, elliptic, oblong, oval, or ovate, tomentose on one or both sides, obtuse to truncate or cordate basally; petioles 0.6–10 cm long or more; inflorescences often swollen below the nodes, simple or branched, the branches erect-ascending, tomentose or glabrous; involucres sessile, racemosely arranged, obconic to campanulate, 2–6 mm long, tomentose or glabrous, with 5 acute teeth; flowers white or suffused with pink, rose, or scarlet, 2.5–5.5 mm long, glabrous, the segments oblong or oblanceolate; achenes 3–4.5 mm long; n = 18. Two varieties occur in Utah.

1. Flowering stems usually definitely swollen below the first branches of the inflorescence and often upward as well, glabrous or sometimes tomentose; plants of Kane and Washington counties . . . . *E. racemosum* var. *zionis*
— Flowering stems not at all or only occasionally somewhat swollen, tomentose or occasionally glabrous; plants widespread . . . . *E. racemosum* var. *racemosum*

Var. racemosum Sagebrush, pinyon-juniper, mountain brush, ponderosa pine, aspen, and spruce-fir communities at 1525 to 2745 m in Beaver, Cache, Davis, Duchesne, Emery, Garfield, Grand, Iron, Juab, Kane, Millard, Piute, Salt Lake, San Juan, Sanpete, Sevier, Summit, Tooele, Utah, Washington, and Wayne counties; Nevada, Colorado, Arizona, and New Mexico; 108 (xiv).

Var. zionis (J. T. Howell) Welsh Zion Buckwheat. [*E. zionis* J. T. Howell, type from Zion National Park]. Mountain brush, juniper-manzanita, and ponderosa pine communities at 1340 to 1830 m in Kane and Washington counties; Arizona; 9 (v). Specimens are known that grade morphologically with var. *racemosum*; i.e., plants with glabrous stems are essentially non-fistulose and some with fistulose stems are tomentose throughout. The phase with scarlet flowers from nearby in Arizona are very similar to specimens of var. *racemosum* with deep rose colored flowers. The variety might ultimately be discovered in Utah. It is regarded as **E. racemosum var. coccineum** (J. T. Howell) Welsh [*E. zionis* var. *coccineum* J. T. Howell].

Eriogonum salsuginosum (Nutt.) Hook. Smooth Buckwheat. [*Stenogonum salsuginosum* Nutt.]. Annuals, mainly 3–26 cm tall, clump-forming, 3–40 cm wide; leaves basal and cauline (foliose bracteate); blades 2–20 mm long, 2–12 mm wide, spatulate to oblanceolate, obovate, or linear, tapering to broad petioles 2–20 mm long or sessile, glabrous on both sides; inflorescence more or less dichotomous, glabrous or minutely glandular; involucres sessile or on stipes to 4 cm long, these curved-ascending, broadly campanulate, in 2 whorls, each 3-lobed; flowers yellow, 1.5–3 mm long, puberulent, the segments lanceolate; achenes 2–2.5 mm long. Shadscale, mat-atriplex, and pinyon-juniper communities at 1370 to 2760 m in Carbon, Daggett, Duchesne, Emery, Garfield, San Juan, and Uintah counties; Wyoming, Colorado, Nevada, Arizona, and New Mexico; 48 (viii).

Eriogonum scabrellum Reveal Westwater Buckwheat. Annuals, mainly 20–60 cm tall; leaves all basal, usually persistent at anthesis and beyond; blades 1–6 cm long and about as wide, orbicular to suborbicular, cordate basally, the margin strongly undulate-crisped, tomentose on one or both sides; petioles 8–50 mm long; inflorescences spreading-ascending to umbrellalike, tomentose and glandular; involucres sessile, erect or deurved, on usually decurved branchlets, obconic, 1.5–2.5 mm long, with 5 acute teeth; flowers white or suffused with pink or red, 1.5–2.2 mm long, the outer segments obovate, the inner ones narrower; achenes 1.8–2.2 mm long. Salt desert shrub communities at ca 1220 to 1740 m in Garfield,

Grand (type from Westwater), Kane, and San Juan counties; Colorado and New Mexico; 7 (i).

Eriogonum shockleyi Wats. Shockley Buckwheat. [*E. pulvinatum* Small, type from Milford; *E. longilobum* Jones, type from near Price]. Pulvinate-caespitose, scapose, mound-forming perennials, mainly 2–5 cm tall, 5–40 cm across or more, from a woody, pluricipital caudex, the branches clothed with marcescent leaf bases and terminated by rosettes; leaf blades 2–12 mm long, 1–6 mm wide, obovate, oblanceolate, elliptic, or spatulate, tomentose on one or both sides; petioles 1–10 mm long, or lacking; inflorescences capitate; involucres sessile, campanulate, 2–6 mm long, tomentose, with 5 (or more) ovate to lanceolate lobes; flowers white, cream, yellow, or suffused with red, 2.5–4.5 mm long, pilose, the segments oblong to obovate; achenes 2.5–3 mm long. Blackbrush, shadscale, mixed desert shrub, sagebrush, and pinyon-juniper communities, often on fine-textured substrates, at 1280 to 1955 m in Beaver, Box Elder, Carbon, Daggett, Duchesne, Emery, Garfield, Grand, Iron, Juab, Kane, Millard, San Juan, Sevier, Tooele, Uintah, and Wayne counties; Idaho to Colorado, south to California, Arizona, and New Mexico; 107 (xv). Specimens from Utah have been treated in two varieties; i.e., var. *longilobum* (Jones) Reveal, with larger, more deeply cut involucres, occupying eastern Utah, and var. *shockleyi*, with shorter, less deeply cut involucres, occupying western Utah. Some of the plants from eastern Utah do have large involucres, but many do not. A large number of plants from western Utah have yellow flowers, but only very few from eastern Utah bear yellow flowers. A conservative interpretation is indicated.

Eriogonum soredium Reveal Frisco Buckwheat. Densely matted, pulvinate-caespitose, scapose, mound-forming perennials, mainly 2–4 cm tall, 10–50 cm across, from a pluricipital caudex, the branches clothed with marcescent leaf bases and terminated by rosettes; leaves 2–5 mm long, 0.7–2 mm wide, elliptic to oblong, white-tomentose on both surfaces, revolute; petioles 0.6–3 mm long; inflorescences capitate, tomentose; involucres sessile, obconic, 1.5–2.5 mm long, obscured by a dense tomentum, with 4 or 5 teeth; flowers white or suffused with pink, 2–3 mm long, glabrous, the outer segments obovate, the inner ones narrower; achenes 2–2.5 mm long. Sagebrush and juniper communities, on white limestone outcrops, at 2010 to 2230 m in Beaver (type from Frisco) County; endemic; 5 (ii).

Eriogonum spathulatum Gray Sevier Buckwheat. Perennial herbs, 10–40 cm tall, from a branching caudex; leaves subbasal, at least some internodes apparent, but obscured by a dense tomentum; blades 1–8 cm long, 3–15 mm wide, obovate to spatulate, elliptic, or linear, usually 1.5–5 times longer than wide or more, tomentose on one or both sides, acute to cuneate basally; petioles 3–30 mm long; inflorescences tomentose or glabrous, more or less trichotomous, the branches ascending; involucres sessile, clustered at branch ends, obconic, 2–4 mm long, tomentose or glabrous, with 5 acute teeth; flowers white or yellow, 2–3.5 mm long, glabrous, the segments oblong; achenes 2–3.5 mm long. Two more or less geographically correlated varieties are present.

1. Flowers yellow; leaf blades mainly less than twice as long as broad .... *E. spathulatum* var. *natum*
— Flowers white, or rarely yellow; leaf blades usually more than twice longer than broad ....
.... *E. spathulatum* var. *spathulatum*

Var. **natum** (Reveal) Welsh Son Buckwheat. [*E. natum* Reveal, type from 43 mi SW of Delta]. Shadscale community on ancient marly playa remnants at 1440 to 1500 m in Millard County; endemic; 10 (ii).

Var. **spathulatum** [*E. nudicaule* ssp. *ochroflorum* Stokes, type from Clear Creek Canyon, Sevier County]. Greasewood, shadscale, rabbitbrush, ephedra, and pinyon-juniper communities at 1405 to 2135 m in Beaver, Millard, Sanpete, Sevier (type from Sevier River Valley), and Wayne counties; endemic; 47 (xiv). Both this and var. *natum* show affinities with *E. brevicaule* var. *laxifolium* (q.v.), especially through the densely hairy, low elevation *cottamii* phase, whose distribution is immediately adjacent to the north. The relationship is also through the *laxifolium* phase proper northeastward in Sanpete County. Plants with glabrous inflorescences and involucres from the vicinity of Frisco and the Shauntie Hills in Beaver County have about the same integrity as does var. *ammophilum* of the *E. nummulare* complex. Probably they have similar hybrid origin, with one parent represented by *E. batemanii* var. *eremicum*, but the other putative parent is different. These glabrous plants are similar to phases of *E. panguicense*, but the inflorescences are consistently branched.

Eriogonum subreniforme Wats. Stokes Buckwheat. [*E. filicaule* Stokes, type from Springdale]. Annuals, mainly 5–40 cm tall; leaves all basal; blades 4–30 mm long and about as broad or broader, orbicular to reniform, tomentose on one or both sides, truncate to cordate basally; petioles 6–60 mm long; inflorescence more or less trichotomous, glabrous, the branches ascending to spreading; involucres on filiform stipes mostly 3–25 mm long, glabrous, obconic, mostly 0.5–1 mm long, with 5 acute teeth; flowers white to rose, 1–2 mm long, glabrous or distinctly puberulent, the segments elliptic to lance-elliptic or spatulate; achenes 1.5–2 mm long. Creosote bush, shadscale, eriogonum, sagebrush, and pinyon-juniper communities at 850 to 1985 m in Garfield, Kane, and Washington (type from St. George) counties; Arizona and New Mexico; 17 (ii). Specimens from Garfield and Kane counties have glabrous flowers.

Eriogonum thomasii Torr. Thomas Buckwheat. [*E. minutiflorum* Wats.]. Annuals, mainly 5–30 cm tall; leaves all basal; blades 4–20 mm long and about as wide, orbicular to subreniform, tomentose on one or both sides, obtuse to subcordate basally; petioles 3–30 mm long; inflorescences more or less polychotomous, glabrous, the branches spreading to ascending; involucres on stipes mainly 3–30 mm long, glabrous, obconic to campanulate, 0.6–1.2 mm long, the 5 teeth obtuse; flowers yellow, 0.8–2 mm long, hispidulous near the base, the outer segments becoming saccate at maturity, the inner ones narrow and not saccate; achenes ca 1 mm long; n = 20. Creosote bush community at ca 850 to 915 m in Washington County; California, Nevada, Arizona, and Mexico; 6 (i).

Eriogonum thompsonae Wats. Ellen Buckwheat. [*E. corymbosum* var. *matthewsiae* Reveal, type from Springdale]. Perennial subshrubs or shrubs, mainly 2–8 dm tall, clump-forming; leaves subbasal or definitely cauline; blades 10–60 mm long, 8–28 mm wide, oblong to elliptic, lanceolate or ovate, tomentose on one or both sides, the margins entire, flat or undulate and sometimes crisped; petioles 1.5–10 cm long; inflorescences more or less trichotomous, glabrous or less commonly tomentose, the branches spreading to ascending; involucres sessile, narrowly obconic, 2.5–3.8 mm long, glabrous or tomentose, the teeth rounded and more or less hyaline; flowers yellow or white, 2.5–4 mm long, glabrous, the segments oblong or obovate; achenes 2–3 mm long. Blackbrush, salt desert shrub, and pinyon-juniper communities, mainly on Chinle and Moenkopi formations, at 1125 to 1830 m in Kane and Washington counties; Arizona, a Mohave Strip endemic; 32 (iii). The *thompsonae* complex consists of a series of morphological subunits, each more or less distinctive, but only arbitrarily separable. They are based on application of the 2^n formula, where "n" equals the number of characters contrasted; i.e., yellow or white flowers with subscapose or caulescent habit. Plants with yellow flowers and subscapose habit are var. *thompsonae* (type from near Kanab); those with white flowers and subscapose habit are var. *albiflorum* Reveal [type from W of Virgin; *E. corymbosum* var. *matthewsiae*, in part]; those with yellow flowers and caulescent habit are *E. corymbosum* var. *aureum*, in part (Shivwits phase); and those with white flowers and caulescent habit are *E. corymbosum* var. *matthewsiae* (Springdale phase), at least in part. Some plants of the Shivwits phase often have yellow flowers, but they display all shades from sulphur yellow to cream and white, and the yellow flowers seem to have been secondarily derived from *E. corymbosum* var. *aureum* (*E. aureum* Jones, in a strict sense). Occasional specimens of the varicolored materials have loosely tomentose inflorescences and the involucres are shortly obconic as in var. *aureum*. In other specimens of the Shivwits phase the narrowly obconic involucres are essentially like those of var. *albiflorum*. The recognition of any of these phases at taxonomic rank is problematical because of intermediates connecting most if not all of them.

Eriogonum trichopes Torr. Slender-stipe Buckwheat. [*E. trichopodum* Torr. in DC.; *E. trichopodum* var. *minus* Benth. in DC.]. Annuals, 8–45 cm tall; leaves all basal; blades mainly 5–30 mm long, 4–25 mm wide, oval to orbicular, hirtellous on one or both sides, obtuse to cordate basally, entire to undulate-crisped; petioles 3–40 mm long or more; peduncles and primary rays of inflorescence inflated or not; inflorescence polychotomous; involucres borne on capillary stipes 3–18 mm long, obconic to campanulate, 0.4–1 mm long, glabrous, 4-lobed; flowers yellowish, 1–2 mm long, strigulose, the segments lance-ovate; achenes 1.5–2 mm long; n = 16. Warm desert shrub communities at 760 to 980 m in Washington County; Nevada and California to New Mexico, south to Mexico; 10 (i). This species simulates the annual phase of *E. inflatum* in having inflated stems in some plants. The usually more numerous branches from the lowest node of the inflorescence, and flowers and involucres that average smaller, are diagnostic.

Eriogonum tumulosum (Barneby) Reveal Woodside Buckwheat. [*E. villiflorum* var. *tumulosum* Barneby, type from SW of Woodside]. Pulvinate-caespitose, mound-forming, herbaceous perennials from a pluricipital caudex and woody taproot, the caudex branches clothed with persistent leaves and bases, the roots with shaggy castaneous to blackish bark; leaves 3–7 mm long, 0.7–1.5 mm wide, oblanceolate to elliptic, tomentose to pilose on both surfaces, revolute; petioles very short; scapes to ca 1 cm long or lacking; inflorescences capitate; involucres campanulate, 2–4 mm long, villous, 7- to 10-lobed; flowers white or suffused with pink, 3–4 mm long, pilose, the segments oblong to oblanceolate; achenes ca 2 mm long. Mixed desert shrub and pinyon-juniper communities at 1525 to 2170 m in Duchesne, Emery, and Uintah counties; Colorado; a Colorado Plateau endemic; 16 (ii).

Eriogonum umbellatum Torr. Sulfur Buckwheat. Perennial herbs or subshrubs, mat-forming, mainly 1–10 dm across; vegetative stems persistent, the branches woody and usually more or less clothed with persistent ashy, castaneous, or blackish leaves and bases; flowering stems scapose, arising from rosettelike stem apices, mainly 10–60 cm tall; leaf blades 4–30 mm long, 2–20 mm wide, ovate to oval, elliptic, lanceolate, or oblanceolate, tomentose or glabrous on one or both sides, flat or nearly so; petioles 2–15 mm long; inflorescence umbellate (or compound) or capitate, often immediately subtended by foliose bracts; involucres terminating rays or sessile, obconic to campanulate, the tube 2–6 mm long and 1.5–10 mm wide, the lobes 1–6 mm long; flowers creamy white to yellow and often suffused with red or purple, 2.5–10 mm long (including the stipitate base), the segments spatulate to ovate; achenes 2–5 mm long; n = 40. This species is a portion of a huge assemblage occupying much of the western U. S. There are four more or less geographically correlated varieties present.

1. Flowers creamy white .... *E. umbellatum* var. *majus*
— Flowers yellow 2
2(1). Inflorescences of compound umbels, at least some; plants mainly of middle to lower elevations in the southern 2/3 of Utah .... *E. umbellatum* var. *subaridum*
— Inflorescences merely umbellate or capitate 3
3(2). Inflorescences capitate or rarely some branched; leaves glabrous on both sides; plants of high elevations .....
.... *E. umbellatum* var. *porteri*
— Inflorescences umbellate; leaves variously pubescent, sometimes as above; plants of moderate to high elevations .... *E. umbellatum* var. *umbellatum*

Var. *majus* Hook. Cream Buckwheat. [*E. subalpinum* Greene; *E. umbellatum* var. *subalpinum* (Greene) Jones; *E. umbellatum* ssp. *subalpinum* (Greene) Stokes; *E. heracleoides* var. *subalpinum* (Greene) R. J. Davis; *E. umbellatum* ssp. *majus* (Hook.) Piper; *E. aridum* Greene; *E. umbellatum* ssp. *aridum* (Greene) Stokes; *E. umbellatum* var. *aridum* (Greene) C. L. Hitchc.; *E. umbellatum* var. *dicrocephalum* Gand.; *E. umbellatum* var. *desereticum* Reveal, type from Mt. Timpanogos]. Sagebrush, mountain brush, pinyon-juniper, Douglas fir-white fir, aspen, lodgepole pine, and spruce-fir communities at 1495 to 3420 m in Beaver, Box Elder, Cache, Carbon, Daggett, Davis, Duchesne, Garfield, Juab, Millard, Rich, Salt Lake, Sanpete, Sevier, Summit, Tooele, Wayne, and Weber counties; Canada, south to California and Nevada; 73 (vi). This plant forms apparent hybrids (Neese 14620 A-E BRY) with *E. heracleoides*. It is also identical, except for flower color, with var. *umbellatum*, and has a similar sequence of pubescence forms.

Var. *porteri* (Small) Stokes Porter Buckwheat. [*E. porteri* Small, type from Bear River Canyon, Summit County]. Ponderosa pine, aspen, spruce-fir, lodgepole pine, and alpine meadow and talus communities at 2500 to 3700 m in Beaver, Duchesne, Iron, Sanpete, Sevier, Summit, and Uintah counties; Nevada and Colorado; 41 (xi).

Var. *subaridum* Stokes Arid Buckwheat. [*E. umbellatum* ssp. *subaridum* (Stokes) Munz; *E. biumbellatum* Rydb., type from Fish Lake; *E. ferrissii* A. Nels.; *E. umbellatum* ssp. *ferrissii* (A. Nels.) Stokes]. Sagebrush, mountain brush, pinyon-juniper, and Douglas fir communities at 1370 to 2745 m in Beaver, Emery, Garfield, Iron, Juab, Kane, Millard, San Juan, Sanpete, Sevier, Summit, Tooele, Washington, and Wayne counties; Colorado, Arizona, Nevada, and California; 79 (xvi). Occasional specimens share features, especially simple inflorescences and pubescence phases, with other varieties of the species.

Var. *umbellatum* [*E. luteum* Small ex Rydb.; *E. rydbergii* Greene; *E. cupreum* Gand.; *E. glaberrimum* var. *aureum* Gand.; *E. umbellatum* var. *aureum* (Gand.) Reveal; *E. neglectum* Greene; *E. azaleastrum* Greene; *E. umbelliferum* Small; *E. umbellatum* var. *umbelliferum* (Small) Stokes; *E. marginale* Gand.; *E. umbellatum* var. *intectum* A. Nels; *E. umbellatum* var. *glabratum* Stokes, type from Huntington Canyon]. Sagebrush, mountain brush, pinyon-juniper, ponderosa pine, white fir, aspen, spruce-fir, and alpine meadow communities at 1765 to 3450 m in Beaver, Box Elder, Carbon, Daggett, Duchesne, Emery, Grand, Juab, Millard, Piute, Salt Lake, San Juan, Sanpete, Sevier, Summit, Tooele, Uintah, Utah, Wasatch, Washington, and Wayne counties; Washington to Montana, south to California, Nevada, and Colorado; 111 (xv).

Eriogonum villiflorum Gray Gray Buckwheat. Pulvinate-caespitose, mound-forming, herbaceous perennials from a pluricipital caudex and woody taproot, the caudex branches clothed with persistent ashy to castaneous or blackish leaf bases and with shaggy blackish bark; leaves 4–15 mm long, 0.7–2 mm wide, oblanceolate to elliptic, villous-pilose on both sides, more or less revolute; petioles very short; scapes mainly 1–5 cm long; inflorescences subcapitate to shortly umbellate; involucres sessile or short-stipitate, campanulate, 3–5 mm long, villous-pilose, with 6–10 lobes; flowers white or suffused with pink, 3–4 mm long, pilose, the segments oblong; achenes 2–3 mm long. Sagebrush, pygmy sagebrush, mixed desert shrub, and pinyon-juniper communities at 1555 to 2350 m in Beaver, Iron, Juab, Millard, and Sanpete counties; Nevada; a Great Basin endemic; 17 (iii).

Eriogonum wetherillii Eastw. Wetherill Buckwheat. [*E. sessile* Stokes ex Jones; *E. filiforme* L. O. Williams, type from near Hanksville]. Annuals, 5–30 cm high, ultimately forming cushionlike, intricately branched clumps, mainly 8–40 cm wide; leaves all basal; blades 4–40 mm long and about as wide, orbicular to oval, tomentose on one or both sides, obtuse to subcordate basally; petioles 5–50 mm long; inflorescences intricately branched, glabrous, ultimately gray to red purple; involucres on filiform stipes, mainly 3–16 mm long or sessile, obconic, glabrous, 0.5–1 mm long, with 4 teeth; involucres yellow, soon suffused with red, 0.6–1.5 mm long, glabrous, the segments elliptic to obovate; achenes 0.6–1 mm long. Blackbrush, shadscale, mixed desert shrub, and pinyon-juniper communities (and often along roadsides) at 1125 to 2135 m in Emery, Garfield, Grand, Kane, San Juan (type from along the San Juan River), Sevier, and Wayne counties; Colorado, New Mexico, and Arizona; 68 (xii).

Eriogonum wrightii Torr. in DC. Wright Buckwheat. Shrubs, mainly 2–5 dm tall; leaves caulescent, mainly 5–25 mm long and 3–10 mm wide, elliptic to oblanceolate, tomentose on both sides, plane or more or less revolute; petioles 1–6 mm long; inflorescence erect-ascending, tomentose, more or less racemose; involucres sessile, obconic, tomentose, 2–4 mm long, with 5 teeth; flowers white or suffused with pink, 3–4 mm long, glabrous, the segments obovate; achenes 2–3 mm long. Pinyon-juniper and mountain brush communities at ca 1190 m in Washington County (upper Beaverdam and Manganese washes); California to Texas, south to Mexico; 4 (ii).

Oxyria Hill

Perennial, subrhizomatous herbs, from long taproots; leaves simple, alternate or mostly basal; stipules sheathing; flowers numerous, borne in panicles, not subtended by an involucre; perianth of 4 sepaloid segments, glabrous; stamens 6; pistil 2-carpelled, the ovary 1-loculed, 1-ovuled; styles 2, short, the stigmas fringed; fruit a flattened, wing-margined achene.

Oxyria digyna (L.) Hill Mountain Sorrel. [*Rumex digynus* L.]. Plants mainly 2–35 cm tall, the herbage often reddish tinged; stems usually simple, the juice acrid; leaves mostly basal; petioles 1–15 cm long; blades 5–50 mm long and as wide or wider, reniform to orbicular, cordate basally; panicles 2–20 cm long; perianth 1–2.5 mm long, the 2 segments at achene edges more slender than those on the flat sides; achenes flattened, 3–6 mm broad, prominently winged; $2n = 14, 42$. Lodgepole pine, spruce-fir, and alpine meadow communities, often in talus, at 2560 to 3965 m in Beaver, Box Elder, Cache, Daggett, Duchesne, Grand, Juab, Piute, Salt Lake, San Juan, Sanpete, Summit, Uintah, Utah, Wasatch, and Weber counties; Alaska and Yukon, east to Labrador, south to California, Arizona, and New Mexico; circumboreal; 38 (ix).

Oxytheca Nutt.

Annuals; stems dichotomously branched; leaves basal; bracts connate, in 3's, foliaceous; involucres few-flowered, stipitate, more or less campanulate, 4-lobed, the lobes awn-tipped; flowers pedicellate; perianth 6-parted; stamens 9; achenes ovoid.

Oxytheca perfoliata T. & G. [*Eriogonum perfoliatum* (T. & G.) Stokes]. Plants 6–20 cm tall or more, erect or spreading-ascending; leaves basal and cauline (leafy bracteate), the basal ones 1–4 cm long, spatulate to oblanceolate, sparingly hirsute to glabrous, ciliate; inflorescences short-pedunculate, then dichotomous or trichotomous, with each node bearing a connate-perfoliate, foliaceous, 3-lobed bract ca 1–2 cm wide, the lobes spinose-tipped; internodes of inflorescence more or less stipitate-glandular; involucres solitary, obconic, 3–6 mm long, including spines, 4-lobed, each lobe spinose-tipped; flowers several, cream to whitish, ca 1.5 mm long, coarsely strigose, the segments lanceolate; achenes ca 2 mm long; $n = 20$. Warm desert shrub communities at ca 950 m in Washington County; Arizona, Nevada, and California; 6 (i).

Polygonum L.

Plants annual, biennial, or perennial herbs from taproots or rhizomes; leaves alternate, cauline or basal; stipules sheathing; flowers solitary or clustered in leaf axils or in axillary or terminal spikelike racemes or panicles, not subtended by a regular involucre; perianth of 5 petaloid (or sepaloid) segments; stamens 8 (5 and 3) or lacking; pistils usually 3-carpelled, the ovary 1-loculed, 1-ovuled; styles 2 or 3, often very short; achenes lens-shaped or 3-angled.

1.	Leaves with subcordate, cordate, or hastate bases; flowers in axillary racemes or panicles; plants cultivated ornamentals, escaping and persisting, or occurring as weeds	2
—	Leaves various but not cordate or hastate basally; flowers variously arranged but not as above; plants indigenous or adventive, weedy or not	4
2(1).	Stems not twining; leaves broadly obovate, obtuse to subcordate basally; plants clump-forming, cultivated ornamentals, escaping and persisting	P. cuspidatum
—	Stems twining; leaves cordate to hastate; plants sprawling or twining on other plants or structures	3
3(2).	Plants perennial; flowers showy, whitish; fruit broadly winged; cultivated ornamentals, escaping and persisting	P. aubertii
—	Plants annual; flowers not showy, greenish; fruit not winged; adventive weedy species	P. convolvulus
4(1).	Stems erect, from an expanded to somewhat bulbous caudex; leaves mostly basal; flowers in terminal spicate racemes; plants mostly of higher elevations	5
—	Stems of various habit, but not from a caudex or, if so, the plants otherwise different from above; flowers axillary or in axillary and terminal spikelike racemes or panicles	6
5(4).	Racemes slender, mainly 4–6 mm thick, the lower flowers at least replaced by bulblets	P. viviparum
—	Racemes mainly 10–25 mm thick, the flowers not replaced by bulblets	P. bistortoides
6(4).	Leaves not jointed at the base; flowers in terminal and (or) axillary spikes or racemes	7
—	Leaves with a hingelike joint at the point of attachment of leaf base with sheath; flowers in small, axillary clusters or solitary	11
7(6).	Inflorescences all terminal, usually solitary; plants perennial, aquatic or semiaquatic to terrestrial; flowers bright pink	P. amphibium
—	Inflorescences not all terminal, at least some axillary; plants mostly annual, seldom aquatic (but sometimes so); flowers pink, green, or white	8
8(7).	Stipular sheaths lacking marginal bristles (or merely short-ciliate); veins of the outer pair of perianth segments branched and recurved at the tip	P. lapathifolium
—	Stipular sheaths usually with well-developed marginal bristles; veins of the outer pair of perianth segments not branched and recurved at the tip	9
9(8).	Plants perennial from rhizomes, growing in or near water; spikes slender, mostly less than 5 mm broad, often paired; not definitely known from Utah, but to be expected	P. hydropiperoides Michx.
—	Plants annual from taproots, growing in moist sites, but not aquatic; spikes slender to thick, not or seldom paired	10
10(9).	Mature perianth glandular-punctate, greenish to white (or pinkish); spikes slender, arching, interrupted near the base	P. hydropiper
—	Mature perianth not glandular-punctate, pink to purplish; spikes dense, erect or nearly so, not or rarely interrupted	P. persicaria
11(6).	Flowers in terminal, leafy-bracteate spikes; plants mainly less than 10 cm tall	P. kelloggii
—	Flowers in axillary clusters or solitary, or in terminal spikes with bracts much reduced; plant height various	12
12(11).	Leaves ovate to broadly elliptic, scarcely reduced upward; plants mainly less than 10 cm tall	P. minimum
—	Leaves linear to narrowly elliptic, lanceolate, or oblanceolate, more or less reduced upward; plant height various	13
13(12).	Flowers borne in elongate, spikelike racemes; leaves much reduced and bractlike upwards; plants usually erect and with branches erect-ascending	P. ramosissimum
—	Flowers borne in axils of foliage leaves, these sometimes reduced but not especially bracteate upward; plants of various habit	14
14(13).	Plants mainly prostrate; leaves mostly flat and with prominent lateral veins, often deciduous in fruit	P. aviculare
—	Plants mainly erect or ascending; leaves flat to revolute, the veins inconspicuous, usually persistent	P. douglasii

Polygonum amphibium L. Water Smartweed. [*P. coccineum* Muhl. in Willd.]. Perennial aquatic or terrestrial, rhizomatous or stoloniferous herbs, the herbage coarsely strigose to glabrous or stipitate-glandular; stems prostrate (often floating) or erect; leaf blades mainly 3–18 cm long, 1–6 cm wide, lanceolate to oblong or elliptic, acute to alternate or rounded apically, obtuse to truncate basally; petioles 0.5–7 cm long; stipules cylindric, 0.5–3 cm long, glabrous to coarsely strigose; panicles 1 or 2, spikelike, 1–8 cm long, the peduncles glabrous, glandular, or strigose and also more or less glandular; pedicels 1–2 mm long; flowers bright pink, 4–5 mm long, the segments oblong, subequal; stamens 8, exserted; style 2–4 mm long; achenes lenticular 2–3 mm long, brown, shining or dull; $2n = 88$. Springs, streams, ponds, lakes, reservoirs, and irrigation canals at 1340 to 2865 m in Beaver, Box Elder, Cache, Daggett, Duchesne, Garfield, Millard, Piute, Rich, Salt Lake, Sanpete, Sevier, Summit, Uintah, Utah, Wasatch, Wayne, and Weber counties; widely distributed in North America; cosmopolitan (except Australia?); 47 (v). Traditional separation of this taxon into two species on the basis of pubescence and panicle differences is not supported by the cline of variation connecting the distinctive extremes.

Polygonum aubertii L. Henry Silver Lace-vine. Perennial, twining herbs; stems mainly 2–7 m long or more; herbage glabrous or scabrous; ocrea soon deciduous, the margin not ciliate; leaf blades 1–8 cm long and 1–6 cm wide, cordate-ovate, cordate basally, attenuate to acuminate apically; petioles 0.5–5 cm long; panicles open, axillary or terminal, 5–15 cm long or more; flowers usually white, 7–10 mm long, including the attenuate winged base, fragrant; fruit lenticular (?), seldom formed. Cultivated ornamental, escaping and persisting in Utah County; introduced from China; 2 (0).

Polygonum aviculare L. Knotweed; Chivalry-grass; Dishwater-grass. Annuals, prostrate to ascending or erect, the stems striate, terete or angled, mostly 1–10 dm long; leaves usually not crowded, 5–40 mm long and 2–10 mm wide, oblong to elliptic or oblanceolate, smaller on the branchlets than on the main stem, acute to obtuse or rounded, the blade sessile or short-petiolate above the basal joint; stipules shredded, 3–6 mm long; flowers 1–5, axillary; pedicels included or shortly exserted; perianth 2–3 mm long, united ca 1/3 the length, 5-lobed, the lobes greenish with white or pink edges, the outer lobes only slightly broader than the inner; styles 3; achenes 3-angled, brown; 2n = 22, 40, 60. Weedy species of open sites at 760 to 3085 m in probably all Utah counties; widespread in most continents; 59 (vii). The plants tolerate trampling and similar abuse that forces other plants to yield way to this vigorous species.

Polygonum bistortoides Pursh American Bistort. [*P. bistorta* var. *oblongifolium* Meissner in DC.; *P. bistorta* var. *linearifolium* Wats.]. Perennials, erect, from thickened bulblike bases and rhizomes, the stems mainly 1–8 dm tall; basal leaves well developed, mainly 5–30 cm long, the blades 2–20 cm long and 0.3–3.5 cm wide, lanceolate to elliptic or linear, attenuate to obtuse or rounded apically, cuneate to obtuse basally; petioles usually well developed, not jointed; stipules mainly 1.5–8 cm long, sometimes flaring apically; cauline leaf blades reduced upward; flowers numerous, borne in terminal spikelike racemes, 1–7 cm long; perianth 4–6 mm long; connate only near the base, white or sometimes pinkish, the segments subequal; stamens 8, exserted; styles 3, exserted; achenes brown, shining, ca 4 mm long. Aspen, lodgepole pine, and spruce-fir communities, usually in moist meadows, at 2070 to 3510 m in all Utah counties, except Davis and Salt Lake; British Columbia to Montana, south to California, Arizona, and New Mexico; 68 (x). This species differs in degree only from *P. bistorta* L. of the Old World and Alaska-Yukon-Mackenzie. The synonyms indicate the views of some previous workers in this genus. Additional work might indicate a more conservative view than that followed here.

Polygonum convolvulus L. Black Bindweed. Annuals, erect (when young) or soon prostrate or twining, the stems 1–15 dm long or more; leaves with long petioles not jointed basally, the blades 1–8 cm long (from sinus to apex), 0.7–5 cm wide, sagittate-ovate, acuminate; stipules 2–5 mm long, shredded and soon deciduous; flowers few to many, borne in axillary or terminal racemes; perianth 4–4.5 mm long, greenish, 5-lobed, the outer lobes keeled; styles 3-cleft; achenes 3-angled, black, usually shining; 2n = 40. Weedy species of gardens, fields, and other open habitats at 850 to 1680 m in Beaver, Cache, Juab, Salt Lake, Sevier, Utah, and Washington counties; widespread in North America; adventive from Europe; 11 (0).

Polygonum cuspidatum Sieb. & Zucc. Fleece-flower. [*P. zuccarinii* Small]. Perennial, dioecious, erect or ascending herbs, mainly 8–15 dm tall; leaves petiolate, the blades mostly 5–15 cm long and 3–10 (12) cm wide, ovate, cuneate to truncate or subcordate basally, abruptly acuminate apically; stipules 4–8 mm long, soon deciduous; flowers 4–5 mm long or more, including the winged, stipelike base, cream to greenish, functionally imperfect, enlarging in fruit; styles 3; achenes 3-angled, black, smooth, shining, ca 3 mm long; 2n = 88. Cultivated ornamentals, escaping and persisting, at 1220 to 1830 m in Duchesne, Salt Lake, and Utah counties; widely grown in the U. S.; introduced from Asia; 5 (0).

Polygonum douglasii Greene Douglas Knotweed. Annuals, mainly 3–45 cm tall or more, erect or ascending; leaves 6–50 mm long, 1–8 mm wide, linear to oblong, lanceolate or oblanceolate, gradually reduced upward, jointed at the base; stipules lacerate, 3–12 mm long; flowers axillary, usually 1–4 per node, the pedicels erect or reflexed, 1–4 mm long; perianth 2–4.3 mm long, the segments green with white or pink to reddish margins, or white to pink overall, united only near the base; achenes 3-angled, black, smooth and shining, 2.5–3.5 mm long. Two rather well-defined but largely sympatric varieties are present in Utah.

1. Flowers deflexed, stipitate above a joint at pedicel apex, the stipe 0.1–0.2 mm long and persistent on the flower base .... *P. douglasii* var. *douglasii*
— Flowers erect, not stipitate, the base sessile on the joint, dehiscing without a peglike stipe at the base .... *P. douglasii* var. *johnstonii*

Var. **douglasii** Sagebrush, mountain brush, pinyon-juniper, ponderosa pine, Douglas fir-white fir, aspen, lodgepole pine, and spruce-fir communities at 1705 to 3145 m in Cache, Carbon, Daggett, Duchesne, Garfield, Grand, Juab, Kane, Millard, Salt Lake, Sanpete, Sevier, Summit, Tooele, Uintah, Utah, Wasatch, and Weber counties; widely distributed in North America; 56 (v).

Var. **johnstonii** Munz Sawatch Knotweed. [*P. sawatchense* Small; *P. utahense* Brenckle & Cottam, type from 6 mi N of Escalante]. Pinyon-junper, mountain brush, sagebrush, and spruce-fir communities at 1675 to 2625 m in Beaver, Carbon, Daggett, Duchesne, Garfield, Grand, Iron, Juab, Kane, Millard, Piute, Rich, San Juan, Sevier, Summit, Tooele, Uintah, Utah, and Washington counties; Washington to North Dakota, south to California, Arizona, and Colorado; 44 (viii). A phase with flowers almost completely white or pink, that tend to open wide (apparently *P. utahense*, sens. str.), occurs in sandy soils in the ponderosa pine and adjacent plant communities in eastern Washington and western Kane and Garfield counties. Possibly these plants are worthy of taxonomic recognition. More work is indicated, but similar plants occur elsewhere within the range of var. *johnstonii*.

Polygonum hydropiper L. Water-pepper. Plants annual (sometimes perennial?), the stems occasionally rooting at the nodes, mainly 3–8 dm tall; leaves with short petioles or else subsessile, not jointed at the base, the blades 3–10 cm long, 0.5–3 cm broad, lanceolate to elliptic, acute to acuminate apically, acute to cuneate basally, sparsely strigose to glabrous, ciliate; stipules 8–15 mm long, not shredded, strigose to glabrous, ciliate with long bristles; flowers several to many, borne in terminal and usually also in lateral, spikelike, interrupted racemes 2–8 cm long; perianth 2.5–4 mm long, glandular-dotted, united ca 1/3 the length, usually 4-lobed, the lobes greenish with white or pink margins; styles 2 or 3, distinct; achenes lens shaped or 3-angled, brown; 2n = 20, 22, 24. Irrigation ditches, roadsides, and bottomlands at ca 1340 to 1375 m in Cache, Salt Lake, and Utah counties; widespread in North America; adventive from Europe; 3 (0). The herbage has a peppery flavor.

Polygonum kelloggii Greene Kellogg Knotweed. Annuals, erect or ascending, 1–9 cm tall, the stems angled, simple or branched; leaves 3–20 (25) mm long, 0.5–2 mm wide, usually crowded and bracteate upwards (surpassing the flowers), and sometimes white margined, sessile or nearly so, jointed at the base; stipules lacerate, 2–7 mm long; pedicels mostly included; perianth 1.5–2.5 mm long, connate in lower 1/3, the 5 lobes subequal or the outer ones largest, green with white or pink margins; stamens 8, the 5 outer ones with linear filaments and usually abortive anthers; stigmas 3; achenes 3–angled, 1.5–2 mm long, yellow to brownish, shining and smooth or brown and dull. Mountain brush, sagebrush, ponderosa pine, meadows, lodgepole pine, aspen, and spruce-fir communities at 1830 to 3235 m in Cache, Daggett, Duchesne, Emery, Garfield, Morgan, Rich, Salt Lake, San Juan, Sevier, Summit, Uintah, Wasatch, and Washington counties; British Columbia to Montana, south to California, Arizona, and Colorado; 29 (iii).

Polygonum lapathifolium L. Willow-weed. [*P. nodosum* Pers.; *P. scabrum* Moench; *P. pensylvanicum* authors, not L.]. Plants annual, erect or prostrate (rarely rooting at the nodes), 1–9 dm long; leaves petiolate to subsessile, not jointed at the base; blades 2–20 cm long, 0.6–7 cm wide, lanceolate to oblong or elliptic, acuminate to acute (abruptly rounded) apically, acute to cuneate basally, glabrous or pubescent, ciliate or glabrous marginally; stipules 5–20 mm long, not shredded, glabrous to pubescent, sparsely short-ciliate to glabrous apically; flowers several to many, borne in spikelike racemes, often aggregated in panicles, the peduncles often stipitate (or sessile) -glandular; perianth 2–3 mm long, not (or sometimes) glandular-dotted, united only near the base, 4– to 5–lobed, the lobes greenish, white, or pink, finally strongly veined, the veins branched apically and the ends recurved; styles 2 or 3; achenes lens-shaped or 3–angled, brown, lustrous; $2n = 22$. Bogs, marshes, sand bars, stream and river margins at 790 to 2135 m in Box Elder, Cache, Daggett, Davis, Garfield, Grand, Millard, Piute, Rich, Salt Lake, San Juan, Uintah, Utah, Washington, Wayne, and Weber counties; widely scattered in North America; adventive (or indigenous in part?) from Eurasia; 43 (vi).

Polygonum minimum Wats. Broadleaf Knotweed. Annuals, ascending to erect, the stems not conspicuously striate, terete or triangular, 5–10 (25) cm long; leaves crowded only near the stem tips, 5–15 mm long, 2–8 mm wide, elliptic, ovate, or obovate, somewhat smaller above, acute to mucronate apically, acute basally, the blades sessile at the basal joint; stipules shredded, 2–4 mm long; flowers 1–4 axillary; pedicels included; perianth 1.5–2 mm long, united ca 1/3 the length, 5–lobed, the lobes greenish with white or pink edges, subequal; stigmas 3; achenes 3–angled, black, lustrous. Spruce-fir and alpine communities, often in rockstripes or talus, at ca 2745 to 3390 m in Cache, Salt Lake, and Summit (type from the Uinta Mts.) counties; Alaska south to California, Nevada, and Colorado; 3 (0).

Polygonum persicaria L. Ladysthumb. Annuals, erect to ascending, mainly 1.5–10 dm tall; leaves petiolate to subsessile, not jointed at the base; blades 1.5–15 cm long, 0.4–4 cm wide, lanceolate to elliptic or oblong, acuminate to attenuate apically, acute to cuneate basally, with a purplish spot near the center, usually glabrous, ciliate; stipules 5–15 mm long, not shredded, usually pubescent, long-ciliate apically; flowers several to numerous, borne in terminal and usually axillary racemes; perianth 1.5–3 mm long, not glandular-dotted, united only near the base, 5–lobed, the lobes pinkish or whitish, not strongly veined and with vein ends not recurved; styles 2 or 3; achenes lens shaped or 3–angled, black, lustrous; $2n = 22, 40, 44$. Fence lines, canal banks, marshes, pond margins, fields, gardens, and pastures at 915 to 2135 m in Box Elder, Cache, Duchesne, Garfield, Rich, Salt Lake, Sevier, Uintah, Utah, Wasatch, Washington, and Weber counties; widespread in North America; Eurasia; 24 (0).

Polygonum ramosissimum Michx. Bushy Knotweed. Annuals, ascending or erect, the stems striate and somewhat angled, 1–10 dm tall; leaves not crowded, 10–50 mm long, 2–6 mm wide, linear-oblong to lance-elliptic, usually acute, gradually reduced upward, short-petiolate above the joint; stipules shredded, 5–10 mm long; pedicels exserted; perianth 2.4–4.4 mm long, united ca 1/3 the length, 5–lobed, green or with pink, white, or yellow margins, the outer ones broader than the inner; stigmas 3; achenes 3–angled, brown to black, lustrous. Open sites and (mainly) in saline meadows at 1340 to 1770 m in Box Elder, Cache, Davis, Duchesne, Juab, Millard, Salt Lake, Sevier, Uintah, Utah, Wayne, and Weber counties; widespread in North America; Europe; 23 (0). The closely allied, but hardly differentiated and possibly identical, *P. argyrocoleon* Steudel, has been identified from Utah. The material grades continously with *P. ramosissimum*, and the older name is applied. That taxon might be valid beyond Utah.

Polygonum viviparum L. Alpine Bistort. Perennials, erect from short, expanded bases; stems 7–40 (55) cm tall; basal leaves well developed, 3–25 cm long, the blades 1.5–13 cm long, 3–25 mm wide, oblong to elliptic, lanceolate, or oval, attenuate to acute apically, cuneate to subcordate basally; petioles well developed, not jointed; cauline leaves reduced upward; stipules 1–6 cm long, not shredded, often flaring and brownish apically, the upper ones seldom bladeless; flowers several to numerous, borne in terminal, spikelike racemes 2–12 cm long, at least the lower (sometimes all) replaced by bulblets; perianth 2–3.5 mm long, the lobes connate only near the base, 5–lobed, greenish with white (cream) to pink margins, subequal; stamens often vestigial; styles 3, exserted; achenes 3–angled, brownish, lustrous, seldom developing; $2n = 66, 80, 88, 98+$. Sedge-grass meadows and alder-birch-willow streamside habitats, mainly in lodgepole pine and spruce-fir communities, at 2470 to 3570 m in Cache, Daggett, Duchesne, Emery, Garfield, Iron, Salt Lake, San Juan, Sevier, Summit, and Uintah counties; Alaska east to Newfoundland, south to Oregon, Nevada, New Mexico, Minnesota, and Maine; 21 (vi).

Pterostegia Fisch. & Mey.

Annuals, with diffuse, dichotomous, slender stems; leaves opposite, the blades broadly elliptic to ovate, entire or 2–lobed; bracts foliaceous, each subtending a solitary flower, 2–lobed, enlarging and becoming scarious and reticulate in fruit, enclosing the achene, with 2 pouches on the back; calyx 6–lobed, the lobes narrowly lanceolate; stamens 3 or 6; style 3–cleft; achenes 3–angled, glabrous.

Pterostegia drymarioides Fisch. & Mey. Plants prostrate-spreading, mainly 10–30 cm long; leaves petiolate, the blades 3–20 mm long, 3–25 mm wide, entire or bilobed, sparingly long-hairy; bracts 1–3 mm long in fruit, the margin subentire to lacerate; perianth ca 1 mm long, the segments white (or reddish); achenes ca 1 mm long. Rock crevices at ca 850 m in Washington County; Oregon to California, Nevada, and Arizona; 4 (i).

Rheum L.

Perennials from stout tuberous roots; leaves alternate, cauline or basal; stipules sheathing, prominent; flowers numberous, borne in terminal or axillary panicles, not subtended by regular involucres; perianth of 6 petaloid segments, open and spreading; stamens mostly 9 (rarely 6); pistils 3–carpellate, the ovary 1–loculed, 1 ovuled; styles 3; fruit a strongly winged achene.

Rheum rhubarbarum L. Rhubarb; Pie-plant. [*R. rhaponticum* L., misapplied]. Plants erect, the stems striate, fistulose, 3–20 dm tall or more, glabrous or nearly so; leaves mostly basal, the edible petioles thick; blades cordate-ovate to orbicular or reniform, entire but undulate; stipules long-sheathing; flowers numerous in branching panicles; perianth greenish white; achenes 6–12 mm long, winged. Cultivated fruit substitute and extender in all Utah counties; persisting but not escaping; introduced from Eurasia; 3 (0). This plant is used alone and in mixtures with other fruits to make delicious pies, jams, jellies, and beverages.

Rumex L.

Annual, biennial, or perennial herbs from stout taproots or rhizomes; leaves alternate, basal or mostly cauline, gradually reduced upward; stipules sheathing; flowers borne in panicles, not subtended by a regular involucre; perianth of 6 (rarely 4), petaloid or sepaloid segments, the inner 3 segments enlarging in fruit and forming the "wings" or "valves" enclosing the fruit, the midveins of the valves sometimes thickened and forming grainlike tuberosities on the segments; stamens usually 6; pistil 3–carpelled, the ovary 1–loculed, 1–ovuled; styles 3; fruit a 3–angled achene.

Rechinger, K. H. Jr. 1937. The North American species of *Rumex*. Field Museum Publ. Bot. 17:1–151.

1.	Flowers mostly or entirely imperfect; plants usually dioecious; leaves hastate or elliptic to oblanceolate	2
—	Flowers all or mostly perfect; leaves various	3
2(1).	Leaves all or some of them hastate; plants rhizomatous, sod-forming, weedy	*R. acetosella*
—	Leaves elliptic, tapering at both ends; plants from thick taproots, not sod-forming, not weedy	*R. paucifolius*
3(1).	Plants rhizomatous, the rhizomes black, spreading; valves of fruit mainly 1–20 mm wide	*R. venosus*
—	Plants from taproots (sometimes tuberous); valves of fruit less than 10 mm wide, or if wider (as in *R. hymenosepalus*, from deeply set tuberous roots	4
4(3).	Plants from deeply set tuberous roots; valves of fruit usually 10–20 mm wide when mature; habitats in sand dunes and other sandy sites	*R. hymenosepalus*
—	Plants from a superficial taproot; valves of fruit less than 10 mm wide at maturity; habitats various, but seldom if ever as above	5
5(4).	Valves toothed along the margins, the teeth at least 1 mm long at maturity	6
—	Valves entire or toothed but, if toothed, the teeth less than 1 mm long	9
6(5).	Tuberosities lacking or forming on only 1 or 2 of the valves; basal leaves mainly 5–10 cm wide or more	*R. obtusifolius*
—	Tuberosities usually forming on all valves; leaves mostly less than 4 cm wide (wider in *R. occidentalis*)	7
7(6).	Plants perennial; inflorescences paniculate, not especially verticillate in lower nodes	*R. stenophyllus*
—	Plants annual; inflorescences of verticillate panicles, the verticels apparent in lower nodes and sometimes throughout	8
8(7).	Valves 4–6 mm long at maturity; teeth subulate; tuberosities more than 0.5 mm wide; leaves not papillose	*R. dentatus*
—	Valves 2–3 mm long; teeth bristlelike; tuberosities less than 0.5 mm wide; leaves papillose, at least some	*R. maritimus*
9(5).	Stems with axillary branches at some or all nodes below the inflorescence, usually decumbent-ascending	*R. salicifolius*
—	Stems seldom with axillary branches below the inflorescence (except in some *R. occidentalis*, erect or essentially so	10
10(9).	Valves without tuberosities, even in fruit	*R. occidentalis*
—	Valves with tuberosities on 1 or more of them	11
11(10).	Valves cordate, 5–9 mm long; basal and lower leaves typically rounded to truncate or cordate, the margins not especially crisped	*R. patentia*
—	Valves triangular-ovate, mostly 3–5 mm long; basal and lower leaves rounded to acute basally, the margins strongly crisped	*R. crispus*

Rumex acetosella L. Sheep Sorrel. Perennial, dioecious, erect herbs from slender rhizomes; stems 1–6 dm tall, usually unbranched below the inflorescence; basal leaves long-petiolate; cauline leaves becoming short-petiolate to subsessile; blades 1–8 cm long, 2–25 mm wide, oblong to ovate, linear, lanceolate, or elliptic, hastately lobed basally, attenuate, acute or obtuse apically; flowers numerous, imperfect, borne in leafless panicles, often purplish tinged; fruiting pedicels jointed at flower base; perianth segments 0.5–1.8 mm long in flower, the outer ones not relexed, the inner ones enlarging and investing the achene, 1–2 mm long, ovate, entire, lacking tuberosities; achenes 1–2 mm long, yellowish brown, lustrous, sometimes adherent to the valves; $2n = 14, 28, 38, 42+$. Roadsides, meadows, and other open sites at 1370 to 2745 m in Beaver, Cache, Carbon, Davis, Duchesne, Emery, Grand, Piute, Salt Lake, Sanpete, Summit, Uintah, Washington, and Weber counties; widespread in North America; adventive from Eurasia; 25 (vi).

Rumex crispus L. Curled Dock. Perennial erect herbs from taproots; stems 3–10 dm tall or more; basal leaves long-petiolate; blades 8–40 cm long, 1.2–6 cm wide, oblong-lanceolate to elliptic, acute to rounded basally, acuminate to acute apically, undulate-crisped (the margin appearing irregularly lobed due to numerous overlapping folds in pressed specimens; cauline leaves somewhat smaller upward, short-petiolate; flowers numerous, perfect, borne in panicles with large leafy bracts to midlength or above, usually greenish; fruiting pedicels jointed above the base; perianth 1.5–2 mm long, the outer segments not reflexed; inner segments much enlarged in fruit, 3–5 mm long, cordate to deltoid or ovate, denticulate to entire, usually each (sometimes only 1 or 2) bearing a reticulately patterned tuberosity almost half as long as the segment; achenes 2–3 mm long, brown, lustrous; $2n = 60$. Weedy plants of open sites at 760 to 2440 m in probably all Utah counties; widespread in North America; adventive from Eurasia; 72 (v).

Rumex dentatus L. Annual or biennial herbs, erect, from tap or fibrous roots; stems mainly 2–7 dm tall; leaves cauline or essentially so, the lower ones long-petiolate; blades 1–6 cm long, oblong, rounded to subcorate basally, rounded to acute apically; flowers mainly perfect, borne in verticillate panicles; pedicels thickened apically, jointed below midlength; valves in fruit triangular, 4–6 mm long, toothed marginally, the teeth 1.5–2 mm long, usually all with a pronounced tuberosity; 2n = 40. Moist, open sites at ca 1340 m in Salt Lake County (Arnow 5263 UT); adventive from Asia; 1 (0).

Rumex hymenosepalus Torr. Canaigre. Perennial herbs, from deeply seated, tuberous roots; stems mainly 2–10 dm tall; lower leaves long-petiolate; blades mainly 8–25 cm long, 2–12 cm wide, elliptic to lanceolate or oblanceolate, cuneate basally, acute to acuminate apically, more or less fleshy; cauline leaves reduced and short- petiolate upward; stipular sheaths 1–4 cm long; panicles compact, 10–35 (40) cm long, usually pinkish; pedicels 4–12 mm long, jointed near the middle; perianth 2–4 mm long at anthesis, the valves 8–18 mm long in fruit, cordate-ovate to suborbicular, reticulate, rounded apically; 2n = 40. Blackbrush, Vanclevea, ephedra, and other sandy desert shrub communities at 760 to 1680 m in Daggett, Garfield, Grand, Kane, San Juan, Uintah, and Washington counties; California, Nevada, Arizona, New Mexico, Texas, Colorado, and Wyoming; Mexico; 49 (vi).

Rumex maritimus L. Golden Dock. [*R. maritimus* var. *athrix* St. John, type from Vermillion]. Annual (or biennial?) herbs, erect from taproots; stems 0.5–8 dm tall; basal leaves usually reduced; cauline leaves well developed, but reduced in size upward, short-petiolate; blades 2–15 cm long, 1–4 cm wide, oblong to lanceolate, rounded to surcordate or acute basally, acute to acuminate or obtuse apically, undulate to plane; flowers numerous, borne in compact axillary clusters, the inflorescence leafy throughout or nearly so, often half the total plant height, greenish; pedicels jointed near or at the base; perianth 1–2 mm long in flower, the outer ones not reflexed; inner segments 3–7 mm long (including the acuminate apex) in fruit, ovate, with 2–4 slender teeth per segment, each tooth 1.5–5 mm long, the valves each usually with a well-developed tuberosity ca 1/2 as long as the segment; achenes 1.5–2 mm long, brown, lustrous; 2n = 40. Lake shores, stream margins, pond and seep margins, and other moist sites at 1220 to 2565 m in Beaver, Box Elder, Cache, Carbon, Daggett, Davis, Duchesne, Emery, Garfield, Juab, Kane, Piute, Rich, Salt Lake, Sanpete, Sevier, Summit, Uintah, and Wayne counties; widespread in North and South America; Europe; 40 (ii). Our specimens belong to var. *fuegineus* **(Phil.) Dusen** [*R. fuegineus* Phil.; *R. maritimus* ssp. *fuegineus* (Phil.) Hulten].

Rumex obtusifolius L. Bitter Dock. Perennial, erect herbs from taproots; stems 4–12 dm tall (or more), usually unbranched below the inflorescence; basal leaves long-petioled; blades 10–40 cm long, 4–15 cm wide, ovate to oblong or lanceolate, cordate to truncate basally, obtuse to acute or acuminate apically, undulate; cauline leaves like the basal ones, somewhat smaller and with shorter petioles upward; flowers numerous, perfect, borne in panicles with leafy bracts to the middle or above, usually greenish; perianth segments 2–3 mm long, the outer ones not reflexed; inner segments 3.5–5 mm long in fruit, ovate, with 4–6 teeth per segment, each tooth 0.5–2 mm long, at least some valves with a prominent tuberosity; achenes 1.5–2 mm long, brown, lustrous; 2n = 20, 40. Ruderal weeds, mainly on canal and stream banks, at 1370 to 2290 m Cache, Davis, Duchesne, Salt Lake, Tooele, and Utah counties; widespread in North America; adventive from Eurasia; 13 (i).

Rumex occidentalis Wats. Western Dock. [*R. subalpinus* Jones, type from near Marysvale]. Perennial, erect herbs from taproots; stems 5–20 dm tall, usually unbranched below the inflorescence, often reddish tinged; basal leaves long-petioled; blades 0.6–4 dm long, 3–15 cm wide, oblong to ovate or oblong-lanceolate, cordate to truncate or obtuse basally, rounded to obtuse or acute apically, usually more or less undulate-crisped; cauline leaves reduced upward; flowers numerous, perfect, borne in panicles with leafy bracts only near the base, greenish; fruiting pedicels obscurely jointed near or below the middle; perianth segments 2–4 mm long, the outer ones not reflexed, the inner ones 4–10 mm long in fruit, ovate to oval (mostly longer than broad), denticulate to entire, lacking tuberosities; achenes 3–4 mm long, brown, lustrous; 2n = 160. Meadows, aspen, and spruce-fir communities at 1830 to 3175 m in Beaver, Daggett, Duchesne, Garfield, Salt Lake, Sanpete, Sevier, Summit, Uintah, and Wasatch counties; Alaska to Quebec, south to California, Nevada, New Mexico, and South Dakota; 13 (0).

Rumex patentia L. Perennial, erect herbs from a taproot; stems mainly 6–15 dm tall, unbranched below the inflorescence; basal leaves long-petiolate; blades mainly 10–30 cm long and 6–15 cm wide, ovate-oblong to lanceolate or oblong, subcordate to truncate or acute basally, acute to acuminate apically; panicles dense, 2–5 dm long, leafy bracteate to the middle; pedicels jointed at or below the middle; flowers perfect, the outer segments 1.5–2 mm long, finally reflexed, the inner ones 5–9 mm long in fruit, ovate to suborbicular and cordate basally, entire to denticulate, 1 valve (only) with a tuberosity; achenes 3–3.5 mm long; 2n = 60. Weedy species of open sites at 1340 to 2440 m in Cache, Davis, Salt Lake, and Utah counties; widely distributed in North America; introduced from Eurasia; 6 (i). This species is not clearly differentiated from *R. occidentalis*, q.v., and evidently forms intermediates with both *R. crispus* and *R. obtusifolius*.

Rumex paucifolius Nutt. Alpine Sorrel. Perennial, dioecious herbs from a taproot and thick root-crown; stems mainly 1–7 dm tall, unbranched below the inflorescence; basal leaves well developed, petiolate; blades 2–13 cm long, elliptic, acute to attenuate at both ends, much reduced upward; inflorescence essentially ebracteate, often as mauch as half the plant height; flowers imperfect, commonly red; pedicels jointed near the middle; outer perianth segments not reflexed; valves 3–4 mm long, cordate to suborbicular, lacking tuberosities; achenes smooth, ca 1.5 mm long; 2n = 14, 28. Meadows in aspen and spruce-fir communities at 2095 to 3050 m in Beaver, Cache, Davis, Rich, Salt Lake, Summit, and Wasatch counties; British Columbia and Alberta, south to California and Colorado; 23 (i).

Rumex salicifolius Weinm. Beach Dock. Perennial, decumbent to ascending (or erect) herbs from taproots, mainly 2–6 dm tall, branching from the lower nodes; leaves mostly cauline, short-petiolate, not much reduced upward; blades 3–20 cm long, 3–30 mm wide, narrowly lanceolate to oblong or linear, acute to rounded basally, acute apically, plane to undulate, not crisped; flowers numerous, perfect, borne in panicles, these more or less leafy-bracteate, usually greenish; fruiting pedicels jointed near the base; perianth segments 1–2 mm long, the outer ones not reflexed, the inner 2–4 mm long in fruit, ovate to deltoid, entire to denticulate, with tuberosities on all valves or lacking on all valves; achenes 1.5–2.5 mm long, brown, lustrous. Salt grass, salt desert shrub, sagebrush, pinyon-juniper, mountain brush, aspen-tall forb, Douglas fir, and spruce-fir communities at 1340 to 3205 m in most if not all Utah counties; Alaska to Quebec, south to California, Texas, and New York; 79 (viii). Our material has been treated within two varieties; var. *montigenitus* Jepson, with tuberosities lacking on the valves, and var. *mexicanus* (Meissner) C. L. Hitchc. [R. m*exicanus* Meissner; R. *utahensis* Rech. f., type from Kyune, Carbon County] with tuberosities on the valves. Transitional specimens connect the varieties, which are not geographically correlated. Both of the varieties are regarded as phases within ssp. *triangulivalvis* Danser.

Rumex stenophyllus Ledeb. Perennial, erect herbs from taproots, mainly 3–9 dm tall; leaves basal and cauline, petiolate; blades 4–20 cm long, 1–5 cm wide, lanceolate to lance-oblong or elliptic, obtuse to acute basally, acute to attenuate apically; panicles loose to dense, mainly 2–4 dm long; pedicels jointed below the middle; outer perianth segments 1–2 mm long, the valves with tuberosities; achenes 2–2.5 mm long, lustrous. Palustrine, riparian, and lacustrine habitats at ca 1400 to 1590 m in Duchesne and Uintah counties; Wyoming; adventive from Eurasia (?); 5 (0). These plants are more or less intermediate between *R. obtusifolius* and *R. crispus*, neither of which is known from the locality where this species occurs.

Rumex venosus Pursh Perennial herbs from creeping rhizomes; stems erect, 1–5 dm tall, usually branched; stipules conspicuous, 1–5 cm long; leaves cauline, the lowermost lacking blades; blades mostly 2–14 cm long, 1–6 cm wide, ovate to elliptic or oblong, leathery, obtuse to acute basally; flowers numerous, in more or less leafy bracteate panicles; pedicels jointed near the middle; perianth segments 4–5 mm long, the valves 15–35 mm long, usually suffused with red, orbicular to subreniform, cordate basally, rounded apically, reticulate, lacking tuberosities; achenes 5–6 mm long, smooth; 2n = 40. Sand dunes and other sandy habitats at 1370 to 2230 m in Cache, Davis, Emery, Grand, Juab, Kane, Millard, Salt Lake, Tooele, Uintah, Utah, Washington, and Weber counties; British Columbia to Saskatchewan, south to California, New Mexico, and Nebraska; 15 (iii).

PORTULACEAE A. L. Juss.
Purslane Family

Annual or perennial, often succulent herbs; leaves simple, entire, alternate, opposite, or mostly basal; flowers perfect, regular or nearly so; sepals 2, rarely more; petals mostly 4 or 5, or in *Lewisia* more numerous; stamens few to many, opposite the petals; style 2- to 8-cleft or -divided, the branches stigmatic on the inner side; pistil 1, the ovary superior or partly inferior, 1-loculed, dehiscing by 2 or 3 valves or circumscissile; styles 1–3 or more; seeds 1 to many; x = 4–42+.

1. Capsule circumscissile 2
— Capsule 2- or 3-valved, dehiscing from the apex 3
2(1). Calyx 2-lobed, the tube adherent to the ovary; capsule circumscissile near the middle, the calyx deciduous with the top of the capsule; stems leafy .... *Portulaca*
— Calyx of distinct sepals, free from the ovary; capsule circumscissile near the base and splitting longitudinally upward; leaves mostly basal .... *Lewisia*
3(1). Sepals deciduous; plants perennial .... *Talinum*
— Sepals persistent; plants annual or perennial 4
4(3). Stigmas and valves of capsule 2; plants annual ....
.... *Calyptridium*
— Stigmas and valves of capsule 3; plants annual or perennial 5
5(4). Plants perennial, with large corms or tuberous roots; cauline leaves in one pair; ovules 6 .... *Claytonia*
— Plants annual or, if perennial, with stolons terminating in bulblets; cauline leaves in several pairs or, if only 1 pair, connate-perfoliate; ovules 3 .... *Montia*

Calyptridium Nutt. in T. & G.

Annual or perennial herbs; leaves alternate or basal, spatulate; inflorescence of scorpioid spikes or the spikes umbellate, capitate, or paniculate; flowers perfect; sepals 2, persistent, scarious or scarious-margined; petals 2 or 4, withering and enclosing the style or summit of the capsule (hence *Calyptridium*); stamens 1–3; style simple, short- or long-filiform, with 2 stigmas; capsule membranous, 2-valved, with few to many seeds.

Hinton, W. F. 1975. Systematics of the *Calyptridium umbellatum* complex (Portulaceae). Brittonia 25: 197–208.

1. Inflorescence in headlike clusters; style long-filiform; petals in age twisting about the style; stamens exserted; sepals wholly scarious; plants alpine ..... *C. umbellatum*
— Inflorescence not headlike; style short; petals ultimately enclosing the capsule; stamens included; sepals merely with scarious margins; plants seldom if ever alpine 2
2(1). Capsule 2–4 times as long as the sepals; sepals 1–2 mm long .... *C. monandrum*
— Capsule less than twice as long as the sepals and often obscured by them; sepals 2–3.5 mm long ...... *C. parryi*

Calyptridium monandrum Nutt. in T. & G. Prostrate to ascending annual herbs; stems branched from the base, 5–18 cm long; leaves mostly in a basal rosette, 2–6 cm long, linear-spatulate, the cauline ones scattered, somewhat reduced; flowers subsessile in short spikes aggregated in a terminal panicle, the branchlets scorpioid, secund; sepals ovate to deltoid, scarious-margined, 1–2 mm long; petals commonly 3, white, ovate, ca 1 mm long; capsule compressed, linear-oblong, curved, many veined, 4–8 mm long; seeds 5–10, black, shining. Creosote bush, blackbrush, Joshua tree, and pinyon-juniper communities at 800 to 1750 m in Washington County; Nevada, Arizona, California, and Mexico; 5 (ii).

Calyptridium parryi Gray Prostrate to ascending annual herbs; stems branched from the base, 5–20 cm long; leaves mostly 1–6 cm long, 3–10 mm wide, oblanceolate, the cauline ones somewhat reduced; flowers subsessile or the lower ones short-pedicellate; sepals 2–3.5 mm long; petals usually 4, shorter than the sepals; capsule 1–2 times longer than the sepals or wholly obscured by them; seeds dull, muricate, 0.6–0.7 mm long. Ponderosa pine community at ca 2625 m in Sevier County (Albee 4850 BRY); Nevada and California; 1 (0).

***Calyptridium umbellatum* (Torr.) Greene** [*Spraguea umbellata* Torr.]. Glabrous, annual or perennial herbs; stems 5–25 cm long; leaves mainly basal, 1.5–7 cm long, the blades spatulate to obovate, petiolate; cauline leaves reduced or lacking; inflorescence umbellate-cymose, appearing capitate; flowers pink or white, closely crowded; bracts scarious; sepals pink or white, orbicular-reniform, 4.5–8 mm long, scarious, except for the greenish center, accrescent; petals 4, 3–6 mm long, oblong or ovate; stamens usually 3, mostly exserted; pistil 5–6 mm long, exserted, the stigma 2-lobed; capsule ovate, 3–4 mm long, 2- to 10-seeded; 2n = 22. Mainly in spruce-fir and alpine tundra communities at 2775 to 3355 m in Beaver, Box Elder, Davis, Duchesne, Piute, and Summit counties; British Columbia to Montana, south to California, Nevada, and Wyoming; 7 (i). Our material is referable to var. *caudicifera* Gray, which has a caudex and glomerate-capitate inflorescence.

Claytonia L.

Glabrous, perennial herbs from deep-seated corms or fleshy roots; basal leaves 1 to many; cauline leaves 2, opposite or nearly so; inflorescence terminal, racemose; bracts present only at raceme base; flowers perfect, regular; sepals 2, herbaceous, persistent; petals usually 5, rose to white or yellowish; stamens 5, adnate to the petals; styles 3, united at the base; capsule ovate, 3-valved, the valve margins involute in age; seeds 2–6, dark, shining.

1. Plants with fleshy, purplish red roots; basal leaves numerous, the petioles winged; petals 5–10 mm long, white to pink, clawed; plants of high elevations ...... *C. megarrhiza*
— Plants from globose to ellipsoid corms; basal leaves few or none; petals 7–12 mm long, white to pink; plants of various elevations .... *C. lanceolata*

***Claytonia lanceolata* Pursh** Spring-beauty. [*C. multicaulis* A. Nels.; *C. multiscapa* Rydb.; *C. rosea* Rydb.]. Prevernal perennials from ellipsoid to globose corms 1–2 cm in diameter; stems solitary or more commonly 2 to several from each corm, mostly 5–20 cm tall (above ground); basal leaves present or lacking, 5–10 cm long, petioled, oblanceolate to oblong or ovate; cauline leaves 2–6 cm long or more, linear to oblong-lanceolate, sessile, with 1–5 veins; sepals 4–7 mm long, ovate to oval; petals 7–12 mm long, white to pink, rounded to emarginate apically; capsule subequal to or shorter than the sepals; seeds ca 2 mm long; n = 8, 12, 16. Pinyon-juniper, mountain brush, aspen, fir, lodgepole pine, and spruce-fir communities at 1585 to 3400 m in nearly all counties in Utah; British Columbia and Alberta south to California, Arizona, and New Mexico; 78 (viii). This is a beautiful plant, with its fresh pink to white flowers and secund reflexed pedicels. Its flowering marks the upward advance of springtime from foothills to mountain summits.

***Claytonia megarhiza* (Gray) Parry in Wats.** [*C. arctica* var. *megarhiza* Gray]. Perennial herbs from thick, purplish red taproots and a thick, often monocephalous caudex, this marked by horizontal leaf scars; stems mostly 5–15 cm tall, several per caudex crown; basal leaves 1–15 cm long, the blades oblanceolate to nearly orbicular; cauline leaves 1–3 cm long, opposite or alternate, linear to spatulate; inflorescence subequal to the leaves or shorter; sepals 4–7 mm long, ovate, acute; petals 5–10 mm long, white to pink, clawed; capsule 4–6 mm long; seeds 2–2.4 mm long. Crevices and talus slopes in alpine tundra at 3080 to 4120 m in Duchesne, Grand, San Juan, Summit, and Uintah counties; Washington to Alberta, south to Nevada and Colorado; 10 (0).

Lewisia Pursh

Perennial, somewhat fleshy herbs from fleshy roots or corms; stems scapose, erect; leaves mostly basal, narrow, linear to oblanceolate; flowers solitary or borne in panicles, usually showy, the petals soon withering; sepals 2–9, persistent; petals 4–18; stamens 5 to many; styles 3–8; ovary superior; capsule circumcissile at the base, then splitting upward; seeds usually many.

1. Sepals 4–9, petaloid; bracts 3–7; pedicels jointed to the peduncles .... *L. rediviva*
— Sepals 2; bracts 1 or 2; pedicels not jointed to the peduncles 2
2(1). Basal leaves lacking or solitary, but with 2–5 linear cauline leaves subtending the inflorescence; plants from subglobose corms, known from Salt Lake and Summit counties .... *L. triphylla*
— Basal leaves numerous; plants from tuberous roots, of various distribution 3
3(2). Floral bracts similar to sepals and subtending them; sepals entire; plants mainly of Washington County; flowers large and showy .... *L. brachycalyx*
— Floral bracts dissimilar to the sepals and separated from them; sepals denticulate; plants of broad distribution .... *L. pygmaea*

***Lewisia brachycalyx* Engelm. ex Gray** [*Oreobroma brachycalyx* (Engelm.) Howell; *L. brachycarpa* Wats.]. Caudex short, thick; stems many, 2–6 cm long; leaves many, fleshy, 3–6 cm long, oblanceolate; flowers usually solitary, sessile, subtended by sepaloid bracts; sepals 6–8 mm long, ovate, acute, entire; petals 12–18 (25) mm long, white; capsule 8–9 mm long; seeds many, ca 1.5 mm long, black, shining. Meadows and wash bottoms in mountain brush, pinyon-juniper, and ponderosa pine communities at 1530 to 2600 m in Washington County; New Mexico, Arizona, Nevada, and California; 9 (vi).

***Lewisia pygmaea* (Gray) Robins.** [*Talinum pygmaeum* Gray; *Calandrina pygmaea* (Gray) Gray; *C. grayi* Britt.; *Claytonia grayana* Kuntze; *Oreobroma pygmaea* (Gray) Howell; *Calandrina nevadensis* Gray; *L. pygmaea* var. *nevadensis* (Gray) Fosberg]. Acaulescent to subcaulescent perennials from thickened roots and usually monocephalous caudices, the caudex marked with leaf scars and more or less persistent leaf bases; stems several to many, 2–6 cm long, partly buried; leaves basal, 2–8 cm long, 1–6 mm wide, numerous, linear to narrowly oblanceolate; scapes 1.5–5 cm long, bracteate at or below the middle; bracts hyaline, lanceolate, 6–10 mm long, opposite; sepals 2, ovate to orbicular, acute to rounded or truncate, entire to dentate apically with glandular teeth; petals 6–8, pink or white, 8–16 mm long; stamens 5–8; stigmas 3–5; capsule 4–6 mm long; seeds 1–1.3 mm long. Sagebrush, mountain brush, aspen, lodgepole pine, spruce-fir, and alpine tundra at 2285 to 3920 m in Beaver, Box Elder, Cache, Daggett, Duchesne, Garfield, Grand, Iron, Juab, Millard, Piute, San Juan, Sanpete, Sevier,

Summit, Tooele, Uintah, Utah, Wasatch, Washington, and Wayne counties; Washington to Montana, south to California, Arizona, and New Mexico; 73 (xiv). There is tremendous variation in plant height, leaf shape and size, and in other features, but it does not seem practical to segregate infraspecific taxa.

Lewisia rediviva Pursh Bitterroot. Plants with thick, fleshy roots and short caudices; stems many, 1–3 cm long, 1-flowered; leaves densely clustered at the caudex crown, 2–5 cm long, linear to clavate, obtuse, fleshy; bracts 5–8, whorled, scarious; pedicels easily disjointed in age; sepals 4–8, rose to white, 1.5–2.5 cm long; petals 12–18, 2–2.5 cm long, rose or sometimes white; stamens 35–50; styles 5–8; capsule 5–6 mm long; seeds dark, shining, 2–2.5 mm long; 2n = 26. Sagebrush, pinyon-juniper, mountain brush, Douglas fir, and aspen communities at 1460 to 3150 m in Beaver, Box Elder, Daggett, Davis, Iron, Juab, Millard, Piute, Salt Lake, San Juan, Sanpete, Tooele, Utah, and Washington counties; British Columbia to Montana, south to California, Nevada, and Colorado; 24 (iv). The roots were eaten in springtime by Indians in the western U. S. The plant bears its scientific name because the specimens taken by Lewis (of the Lewis and Clark expedition) continued to grow at the herbarium of the Philadelphia Academy of Sciences.

Lewisia triphylla (Wats.) Robins. in Gray [*Claytonia triphylla* Wats.]. Plants from deep seated, subglobose corms 5–8 mm thick; stems 1–5, slender, 3–10 cm long; basal leaves solitary or lacking; cauline leaves 2–5, whorled, linear, 1–5 cm long, 1–2.5 mm wide; inflorescence subumbellate or paniculate, bracteate, 2- to 15-flowered; sepals 2, oval, entire, 2.5–4 mm long; petals 5–8, white or pink, 4–5 mm long; seeds black, shining, ca 1 mm long. Meadows in sedge-forb and spruce-fir communities at 2745 to 3235 m in Cache, Duchesne, Morgan, Rich, Salt Lake, and Summit counties; Washington to Montana, south to California, Nevada, and Colorado; 6 (0). This is an obscure, poorly collected entity that should be expected beyond the range of the few specimens examined.

Montia L.

Annual or perennial, somewhat succulent herbs from fibrous roots, or with rhizomes or stolons; cauline leaves 2 and opposite or 2 to several and opposite or alternate; basal leaves well developed and petiolate or lacking; flowers showy or inconspicuous, usually several in axillary or terminal racemes; sepals 2, persistent; petals usually 5 (2–6), distinct or basally united; stamens 2–5, opposite the petals and usually adnate to them at the base; styles deeply 3-cleft; capsules 3-valved; seeds 1–3, smooth or tuberculate, often shining.

1. Cauline leaves alternate, linear; plants perennial .... *M. linearis*
— Cauline leaves opposite, not linear; plants annual or perennial 2
2(1). Stems with 2 to several pairs of leaves; plants rhizomatous perennials .... *M. chamissoi*
— Stems with a single pair of leaves; plants annual or perennial, but if the latter, not strongly rhizomatous ... 3
3(2). Plants perennial, the cauline leaves not connate-perfoliate; petals 9–12 mm long .... *M. cordifolia*
— Plants annual, with a single pair of connate-perfoliate leaves; petals 2–7 mm long .... *M. perfoliata*

Montia chamissoi (Ledeb.) Robins. & Fern. Toad-lily. [*Claytonia chamissoi* Ledeb.]. Plants perennial, 5–25 cm tall, from slender rhizomes and stolons (which produce bulbletlike offsets); flowering stems often branched; basal leaves reduced or lacking; cauline leaves opposite, 4 to several, 1–5 cm long, 2–18 mm wide, oblanceolate to elliptic, tapering to the petiole; inflorescence erect, racemose, terminal and axillary; flowers few to several, sometimes replaced by bulblets; pedicels 8–30 mm long in flower, nodding in bud, a single bract at the base of the lowest pedicel; sepals 2, 2–3 mm long; petals 5, white (or pinkish), 5–8 mm long; stamens commonly 5; capsules 2–3 mm long; seeds 1–3; n = 11. Moist meadows, stream banks, seeps, and springs at 1890 to 3520 m in Box Elder, Cache, Davis, Emery, Garfield, Kane, Sevier, Summit, Utah, Wasatch, Washington, and Wayne counties; Alaska to Manitoba, south to California, New Mexico, and Iowa; 13 (i).

Montia cordifolia (Wats.) Pax & K. Hoffm. in Engler & Prantl [*Claytonia cordifolia* Wats.]. Perennial herbs with short rhizomes; stems 1 to few, 1–3 (5) dm tall; basal leaves round-reniform to ovate, the blades 2–5 cm long and wide, the petioles 5–15 cm long; stem leaves 2, opposite, sessile, 1.5–3 cm long; racemes bractless, 4- to 9-flowered; pedicels 1–2 cm long; sepals suborbicular, 3–4.5 mm long; petals white, obovate, 6–12 mm long; capsule ca 4 mm long; seeds black, roundish, shining; 2n = 10. Moist sites with willow, birch, aspen, and fir at 1650 to 2330 m in Cache, Salt Lake, and Weber counties; British Columbia to Montana, south to California; 4 (0).

Montia linearis (Dougl.) Greene [*Claytonia linearis* Dougl. ex Hook.]. Annual, erect, branching herbs; stems slender, 5–20 cm tall; leaves alternate, linear, 2–3.5 cm long, the petioles with enlarged, scarious bases; racemes terminal, secund, lax, 2- to 10-flowered; pedicels 6–15 mm long, spreading or recurved; sepals rounded, ca 4 mm long, white-margined; petals unequal, obovate, white, ca 5 mm long; capsule ovoid, ca 4 mm long; seeds 3, lens-shaped, muricate marginally, 1.5–2 mm long. Meadows in sagebrush, mountain brush, mixed conifer, and riparian communities at ca 2000 m in Morgan and Weber counties; British Columbia to Montana, south to California and Idaho; 2 (0).

Montia perfoliata (Donn) Howell Miners-lettuce. [*Claytonia perfoliata* Donn]. Annual herbs from slender taproots; flowering stems few to several from the root crown, 5–20 cm tall; basal leaves 2–15 cm long, somewhat sheathing basally, the blades linear to spatulate or broader; cauline leaves 2, opposite, usually connate-perfoliate and forming a discoid structure 0.8–3 cm broad; inflorescence racemose, terminal and sometimes also axillary; flowers mostly 3–8; pedicels 2–10 mm long or longer, nodding, only the lowermost subtended by a bract; sepals 2, 1.5–3 mm long, rounded; petals 5, white or pinkish, 3–5 mm long; stamens 5; capsule 2–4 mm long; seeds usually 3, black, lustrous; 2n = 12, 24, 32, 36, 48, 60. Creosote bush, blackbrush, sagebrush, pinyon-juniper, mountain brush, riparian, and spruce-fir communities at 800 to 3300 m in Beaver, Box Elder, Cache, Davis, Duchesne, Juab, Millard, Morgan, Salt Lake, Sanpete, Summit, Tooele, Utah, Washington, and Weber counties; British Columbia to Montana, south to California and Nevada; 56 (xiv). Plants with linear basal leaves have been segregated as **var. utahensis** (Rydb.) Munz [*Limnia utahensis* Rydb., type from St. George]. There appears to be some geographic correlation with the linear leaf type, i.e., the plants occur in Washington County. The broad-leaved material is referable to **var. perfoliata**, and is widely distributed, including Washington County.

Portulaca L.

Annual or perennial, succulent herbs; leaves opposite or alternate, flat or terete; stipules scarious, reduced to hair tufts, or lacking; flowers perfect, solitary or crowded at the stem and branch ends, opening only in full light; sepals 2, united below and adnate to the ovary; petals 5 or rarely many, inserted on the calyx; stamens 8 to many; style deeply 3- to 9-parted; ovary partly to wholly inferior; capsule 1-loculed, circumscissile, many-seeded; seeds round-reniform.

1. Leaves terete, linear or narrowly oblanceolate; plants with long hairs in the inflorescence and leaf axils 2
— Leaves broad and flat, thick and fleshy; plants glabrous . 3
2(1). Plants cultivated; flowers variously colored; leaf axils with long, thin hairs, but not woolly *P. grandiflora* Hook.
— Plants indigenous; flowers yellow, orange, or bronze; axils of leaves with long, white, woolly hairs .... *P. parvula*
3(1). Seeds conspicuously and sharply granulate or echinate, 0.9–1 mm wide; leaves often retuse or emarginate apically; style lobes mostly 3 or 4; plants not definitely known from Utah, but to be expected in the southern counties .... *P. retusa* Engelm.
— Seeds minutely rounded tuberculate, less than 0.8 mm wide; leaves rounded or truncate apically; style lobes 5 or 6; plants widespread .... *P. oleracea*

Portulaca oleracea L. Purslane; Pusley; Mother-of-millions. Glabrous, fleshy, prostrate annuals; stems 5–50 cm long or more; leaves alternate, 6–30 mm long, 2–13 mm wide, the blades obovate-cuneate to spatulate, rounded to truncate apically; flowers clustered or solitary, sessile, the hairs subtending them inconspicuous or lacking; sepals broadly ovate to orbicular, 2.8–4.5 mm long, 2.3–3.8 mm wide, keeled, acutish; petals yellowish, 3–4.6 mm long, 1.8–3 mm wide; stamens 6–10; style lobes 5 or 6; capsules 6–9 mm long, circumscissile; seeds black, 0.4–0.6 mm long, minutely punctate; $2n = 36, 54$. Weedy species of gardens and fields, but widespread in disturbed sites in indigenous plant communities at 830 to 2440 m in all Utah counties; widely distributed in North America; cosmopolitan; 17 (vii). This is a persistent weed of gardens, yielding abundant work for gardeners, who harvest it in large quantities, only to find that it continues to flower and mature seed when pulled and left to dry. The plant is edible, and serves as feed for poultry, for those hapless gardeners who have both poultry and purslane.

Portulaca parvula Gray Dwarf Purslane. Plants with stems prostrate to ascending, annual; stems mainly 5–15 cm long, usually branched from the base; leaves 4–13 mm long, 1–2 mm thick, succulent, somewhat compressed; leaf axils with tufts of white hair 3–7 mm long; inflorescence apical, capitate, with 2–10 flowers; bracts subtending inflorescence 3–8 mm long; sepals ca 2–2.5 mm long; petals yellow, orange, or bronze, 2–2.5 mm long; capsule 1.5–2 mm thick, the persistent basal part campanulate, short-stipitate; seeds ca 0.5 mm wide, black, stellate-tubercular. Sandy sites in mixed warm desert shrub and pinyon-juniper communities at 1220 to 1895 m in Kane and San Juan counties; Arizona to Oklahoma and Texas, south to Mexico; 5 (iii). Previous reports of *P. mundula* Gray belong here.

Talinum Adanson

Perennial herbs, often with thickened roots; stems often short; leaves fleshy, alternate to subopposite or mainly basal, entire, flat or terete; flowers often showy, borne on long or short peduncled cymes, or sometimes only 1 to several in leaf axils; sepals 2, deciduous; petals 5 or more, soon withering; stamens few to numerous; filaments filiform; ovary superior; styles 3, more or less connate; capsule 1-loculed, 3-valved; seeds flattened.

1. Flowers solitary or few in axillary cymes; caudex branches spreading, forming mats of rosettes; leaves 4–9 mm long .... *T. brevifolium*
— Flowers in terminal cymes; caudex branches erect-ascending, not forming mats; leaves mainly 15–50 mm long 2
2(1). Stamens 4–8; stems and peduncles erect or ascending, often more than 6 cm tall .... *T. parviflorum*
— Stamens 10; stems and peduncles more or less spreading, less than 6 cm tall .... *T. thompsonii*

Talinum brevifolium Torr. Mat-forming, fleshy herbs, with spreading caudex or superficial branches and stout roots; vegetative and flowering stems short, seldom to 5 cm long; leaves arranged in rosettelike clusters or the internodes very short, subterete, narrowly spatulate to clavate or linear, 4–9 mm long, 1–1.8 mm wide, obtuse apically; flowers solitary in the axils of upper leaves, borne on pedicels 3–3.5 mm long; sepals 3–3.7 mm long, 3.8–4.4 mm wide, oval to orbicular; petals mostly 8–10 mm long, 4–5 mm wide, obovate, rose or white; stamens ca 20–25; style as long as the ovary, 3-cleft; capsule subglobose to ovoid or ellipsoid, 3.5–3.7 mm long; seeds almost smooth. Sandstone depressions and crevices in Navajo and Cedar Mesa formations (and probably others) at 1555 to 2130 m in Emery, Garfield, San Juan, Washington, and Wayne counties; Arizona to Texas; 15 (viii). Flowers in most materials examined in the field are rose colored and 10–12 mm across, but plants from near Fish Canyon on Cedar Mesa bear white flowers, some of them dimorphic. The small flowers average about 5 mm broad and the large ones about 10 mm broad. A plant from the Comb Reef, west of Bluff, had pink flowers 25 mm broad. More work is required to evaluate the importance of this diversity.

Talinum parviflorum Nutt. Low, clump-forming, perennial herbs, mainly 5–19 cm tall, with fleshy roots; stems short, much surpassed by the peduncle and inflorescence; leaves 15–50 mm long, 0.8–2.5 mm wide, linear, terete or nearly so, broadened at the base; inflorescence cymose, bracteate at the forks; peduncles slender, 3–15 cm long; pedicels slender, 1–4 mm long; sepals 2.7–4 mm long, 1.5–3.2 mm wide, ovate to oval-ovate, deciduous; petals 5.5–7 mm long, 2.2–2.6 mm wide, pink to pink purple, obovate or elliptic; stamens 4–8, the anthers oblong; style longer than the stamens; stigma subcapitate; capsule 3.5–4.5 mm long, ellipsoid; seeds smooth, 0.8–0.9 mm wide. Usually on rock outcrops (sandstone or igneous) in black sagebrush, pinyon-juniper, and ponderosa pine communities at 1525 to 2685 m in Emery, Garfield, Iron, and Kane counties; North Dakota and Minnesota, south to Texas and west to Arizona; 7 (ii).

Talinum thompsonii **Atwood & Welsh** [*T. validulum* authors, not Greene]. Low, clump-forming, perennial herbs to 10 cm wide, from a fusiform or cylindric, reddish, tuberous root and short rootcrown bearing branches of the season; stems spreading, rosettelike; leaves 8–32 mm long, fleshy, cylindroid, mainly 2–3 mm thick (when pressed), with an auriculate, clasping base; flowers (1) 3–6 in open cymes, ca 1 cm across; petals pink; sepals 4.3–4.8 mm long, ovate, reticulately veined, greenish or brownish, with scarious margins, abruptly acuminate apically, tardily deciduous; stamens 10; capsules 6–6.5 mm long, 3.2–3.8 mm wide, keeled along the sutures apically; seeds grayish-black, 1.2–1.3 mm long. Silicious conglomeratic gravels in pinyon-juniper and ponderosa pine communities at ca 2290 m in Emery (type from Cedar Mt.) County; endemic; 3 (ii).

PRIMULACEAE Vent.
Primrose Family

Annual or perennial herbs; leaves simple, alternate, opposite, or whorled; flowers perfect, regular, variously arranged, terminal or axillary; sepals commonly 5, more or less united; petals 5, united or lacking (in *Glaux*); stamens 5, opposite the corolla lobes, sometimes alternating with 5 staminodia; pistil 1, the ovary superior to partly inferior, 1–loculed, 5–carpelled; style 1, the stigma capitate; fruit a capsule; x = 5, 8–15, 17, 19, 22.

1. Leaves all basal; flowers borne in umbels 2
— Leaves cauline; flowers axillary, solitary or few to many, or in terminal racemes or panicles 4
2(1). Corolla lobes distinctly reflexed; flowers nodding . . . . *Dodecatheon*
— Corolla lobes spreading to erect or, if reflexed, the flowers ca 3 mm long; flowers erect or spreading 3
3(2). Flowers white, mostly less than 6 mm long or broad; corolla tube usually shorter than the calyx . . . . *Androsace*
— Flowers pink to rose or red purple, rarely white, commonly more than 6 mm long and broad; corolla tube longer than the calyx . . . . *Primula*
4(1). Corolla lacking; plants often of saline meadows . . . . *Glaux*
— Corolla present; plants of various habitats 5
5(4). Inflorescence terminal, of numerous flowers in racemes or panicles . . . . *Samolus*
— Inflorescence lateral, the flowers solitary or in dense axillary clusters 6
6(5). Plants indigenous, perennial; flowers sulfur yellow, borne singly, in pairs, or in dense axillary clusters . . . . *Lysimachia*
— Plants adventive, annual; flowers salmon, borne singly on axillary pedicels . . . . *Anagallis*

Anagallis L.

Annual herbs from taproots; stems branched from the base; leaves cauline, opposite or whorled; flowers borne singly in leaf axils, pedicellate; calyx deeply 5–cleft; corolla 5–lobed, rotate; capsule circumscissile.

Anagallis arvensis **L.** Scarlet Pimpernel; Poor-mans Weatherglass. Stems prostrate or ascending, mainly 10–25 (40) cm long, glabrous; leaves 5–15 mm long, ovate, entire, sessile, more or less clasping basally; pedicels 10–30 mm long, filiform, usually recurved in fruit; sepals 2–4 mm long, narrowly lanceolate, acuminate, almost distinct; corolla 5–8 mm broad, salmon, ciliolate; staminal filaments hairy; capsule globose, 3–5 mm wide. Weedy species of cultivated sites in Salt Lake County; widely distributed in the western U. S.; adventive from Europe; 1 (0).

Androsace L.

Annual or perennial, scapose herbs; leaves in basal rosettes, simple, usually persistent; herbage pubescent with stellate, forked, or simple hairs, or glabrous; flowers in umbels of 2–20 or more, rarely solitary, borne on peduncles that arise from the basal rosette; calyx 5–lobed, the tube nearly equaling the lobes; corolla tubular, contracted and sometimes with 5 crests at the throat, 5–lobed; stamens 5, the filaments adnate to the corolla; capsules subglobose.

Robbins, G. T. 1944. North American species of *Androsace*. Amer. Midl. Naturalist 32: 137–163.

1. Plants mat-forming perennials; herbage with long, moniliform hairs; corollas 5–6 mm wide; known from the La Sal Mts. . . . . *A. chamaejasme*
— Plants annual or biennial (?); herbage glabrous or puberulent; corollas 2–3 mm wide; distribution various . . . 2
2(1). Leaf blades abruptly narrowed to a petiole; calyx not keeled, the lobes 3–veined; corolla lobes reflexed . . . . *A. filiformis*
— Leaf blades tapering to the base; calyx keeled, the lobes 1–veined; corolla lobes spreading to erect 3
3(2). Bracts subtending umbel commonly more than 3 times longer than broad, attenuate apically; calyx lobes not incurved; plants common . . . . *A. septentrionalis*
— Bracts subtending umbel less than 3 times longer than broad, abruptly acute apically; calyx lobes ultimately incurved; plants uncommon . . . . *A. occidentalis*

Androsace chamaejasme **Host** Plants mat-forming perennials, with a caudex and prostrate stems, each with a terminal rosette, the flowering stems 1.2–5 cm tall; leaves several to many, 2–8 mm long, glabrous above, pilose beneath, ciliate with moniliform hairs; peduncles 1 per rosette, pubescent with moniliform hairs; flowers 2 to several, on pedicels 1–3 mm long, the umbels subtended by few to several saccate bracts; calyx 2–3 mm long, with teeth subequal to or shorter than the tube; corolla white to cream (fading pinkish), with a yellow center, the tube subequal to the calyx, the lobes 1.5–3 mm long; capsules 2–3 mm long. Alpine tundra at 3050 to 3845 m in Grand (?) and San Juan counties; Alaska to Mackenzie, south to Colorado; Eurasia; 5 (0). Our material is assignable to **var. carinata** (**Torr.**) **Knuth** [*A. carinata* Torr.; *A. chamaejasme* ssp. *carinata* (Torr.) Hulten]. The few specimens available for study do not appear to differ in any real way from those of interior Alaska, and they might best be treated within **ssp. lehmanniana** (**Sprengel**) **Hulten** [*A. lehmanniana* Sprengel], but nomenclatural combination is not intended or implied herein.

Androsace filiformis **Retz.** [*A. capillaris* Greene]. Annual herbs; leaves in a basal rosette, 6–20 (30) mm long, the blades ovate to deltoid, toothed, abruptly narrowed to broad petioles; scapes few to numerous, 0–8 (12) cm tall; glabrous or sparingly glandular or puberulent above; bracts several, 1–4 mm long; pedicels capillary, 1–4 cm long, glabrous or puberulent; calyx hemispheric, 1.4–2.1

mm long, the lobes triangular, 3-veined; corolla tube subequal to the sepals, the lobes 0.8-1.2 mm long; capsules surpassing the calyx, translucent; seeds brownish, 0.2-0.3 mm long. Meadows at 2135 to 2595 m in Cache, Duchesne, Emery, and Salt Lake counties; Washington to Montana, south to Oregon and Colorado; 5 (0).

Androsace occidentalis Pursh [*A. simplex* Rydb.]. Annual herbs; leaves in a basal rosette, 6-20 (30) mm long, lanceolate to oblanceolate or spatulate, not differentiated into blade and petiole, entire or toothed, puberulent with simple hairs; scapes 1 to many, 2-10 cm tall, puberulent with forked hairs; bracts lanceolate to elliptic, pubescent with forked hairs; umbels 3- to 10-flowered; pedicels slender, 3-20 (30) mm long; calyx turbinate-campanulate, 3.8-5 mm long, puberulent, keeled below each lobe, the lobes subequal to the tube; corolla white, included in the calyx; capsule globose, subequal to the calyx tube, opaque; seeds brown, ca 1 mm long. Mountain brush and pinyon-juniper communities at ca 1280 to 2135 m in Beaver, Davis, Kane, Salt Lake, and Utah counties; British Columbia to California and New Mexico; 5 (0).

Androsace septentrionalis L. Annual (or biennial?) herbs; leaves in a dense basal rosette, 5-60 mm long, oblanceolate to spatulate, usually with 5 or more apical teeth, sparsely to densely puberulent with simple or forked hairs, not differentiated into blade and petiole; scapes 1 to many, 1-20 cm tall or more, puberulent; umbels with several to many flowers; bracts lanceolate to oblong or lance-linear; pedicels 3-40 mm long or more; calyx turbinate-campanulate, 2.5-4 mm long, keeled below the lobes, those shorter than the tube; corolla white, subequal to the calyx, the lobes 0.5-1 mm long; capsules 2-4 mm long; 2n = 20. Pinyon-juniper, sagebrush, mountain brush, ponderosa pine, Douglas fir-white fir, aspen, lodgepole pine, grass-forb, spruce-fir, and alpine tundra communities at 1950 to 2965 m in most, if not all, Utah counties; Alaska to Labrador, south to California, Arizona, and New Mexico; circumboreal; 175 (xxiii).

Dodecatheon L.

Perennial, subrhizomatous, scapose herbs; leaves all basal, petiolate, the blades broad; herbage glabrous or glandular-pubescent; flowers (1) 2 to many, borne in terminal umbels, nodding at anthesis, erect in fruit; calyx cup-shaped, 4- or 5-lobed; corolla 4- or 5-lobed, the lobes much longer than the tube, reflexed; stamens 4 or 5, the filaments short, distinct or connate, adnate to the corolla tube; anthers connivent around the style; anther sacs with a prominent connective that is dilated at the base; capsules elongate.

1. Corolla and sepal lobes 4; stigma prominently dilated
   .... *D. alpinum*
— Corolla and sepal lobes 5; stigma various 2

2(1). Leaves abruptly contracted to petioles, the blades ovate to subcordate, sinuate-dentate; plants rare . *D. dentatum*
— Leaves tapering to the petioles, the blades mostly spatulate to oblanceolate or elliptic, entire or variously toothed 3

3(2). Staminal filaments mostly 1-3.5 mm long, united into a yellow or black tube; stigmas not especially enlarged; our most common species .... *D. pulchellum*
— Staminal filaments less than 1 mm long, distinct; stigmas noticeably enlarged; plants rare in western Utah
   .... *D. redolens*

Dodecatheon alpinum (Gray) Greene Alpine Shootingstar. [*D. meadia* var. *alpinum* Gray; *D. jeffreyi* var. *alpinum* (Gray) Gray; *D. alpinum* ssp. *majus* H. J. Thompson]. Plants 8-38 cm tall, the roots yellowish; leaves 1.5-17 cm long, 3-17 mm wide, spatulate to oblanceolate, entire or irregularly sinuate; scapes 6-33 cm long, glabrous or sparingly glandular-pubescent; umbels 2- to 6-flowered, the pedicels 2-45 mm long; calyx tube 3-4 mm long, the teeth 2.5-3.5 mm long; corolla magenta to lavender, 8-20 mm long; staminal filaments 0.5-1 mm long, distinct or narrowly connate, black; anthers 5-8.5 mm long, purplish black, the connective black, roughened; stigma conspicuously enlarged; capsules 5-8 mm long, oblong-ovoid. Alpine meadows in lodgepole pine, spruce-fir and tundra communities at 2195 to 3510 m in Beaver, Duchesne, Garfield, Iron, Juab, Salt Lake, Summit, Utah, Wasatch, and Washington counties; Oregon, Nevada, and California; 25 (iii).

Dodecatheon dentatum Hook. Plants 12-40 cm tall, the roots yellowish; leaves 2-15 cm long, 8-40 (60) mm wide, the blades ovate to elliptic or subcordate, abruptly contracted to a broad petiole, sinuate-dentate; scapes 10-35 cm long, glabrous or sparsely glandular-pubescent; umbels 2- to 12-flowered, the pedicels 0.8-7 cm long; calyx tube 1.5-2.5 mm long, the lobes 2-3 mm long, triangular; corolla creamy white (fading lavender), the lobes 12-20 mm long, the tube yellowish; staminal filaments less than 1 mm long, distinct, purple; anthers 6-7 mm long, deep red (or purple?); stigma not conspicuously enlarged; capsule 6-10 mm long, cylindric-ovoid. Crevices at ca 2375 to 2440 m in Salt Lake County; British Columbia to Oregon and Idaho; 2 (0).

Dodecatheon pulchellum (Raf.) Merr. Pretty Shooting-star. Plants (6) 12-65 cm tall, the roots pale; leaves 3-30 (35) cm long, 0.4-6 (7.5) cm wide, oblanceolate to elliptic, tapering to the petiole, usually entire; scapes 7-50 cm long; umbels (1) 3- to 20-flowered; pedicels 1-7 cm long; calyx tube 2-3.5 mm long, the lobes 2.5-6 mm long; corolla magenta to lavender (or white), 9-21 mm long; staminal filaments 0.5-3.5 mm long, united into a distinctive yellow or black tube; anthers 3-8 mm long, the connective maroon to black or yellow, smooth; stigma not conspicuously enlarged; capsules 7-17 mm long, cylindric to ovoid. This is a highly variable species, treated herein as separable into two intergrading phases at taxonomic rank.

1. Leaves mainly 10-35 cm long and 3-7.5 cm wide; petals mostly over 12 mm long; plants of hanging gardens and seeps, mainly below 1220 m in Kane and Washington counties .... *D. pulchellum* var. *zionense*

— Leaves mainly less than 10 cm long and less than 3 cm wide; petals typically less than 12 mm long; plants typically of moist sites above 1220 m, widely distributed ...
   .... *D. pulchellum* var. *pulchellum*

Var. pulchellum [*Exinia pulchella* Raf.; *D. meadia* var. *pauciflorum* Dur.; *D. pauciflorum* (Dur.) Greene; *D. radicatum* Greene]. n = 22, 44. Moist sites in meadows, seeps (sometimes markedly saline), and springs from 1370 to 3355 m in all Utah counties (except perhaps Grand and San Juan); Alaska and Yukon, south to Mexico, and disjunctly in the eastern U. S.; 85 (xi). Plants with black staminal tube have been designated as var. *monanthum* (Greene) C. L. Hitchc. [*D. pauciflorum* var. *monanthum* Greene], but there does not appear to be geographical correlation. The plants with yellow staminal tube belong to var. *pulchellum*. This plant often stains huge areas in wet meadows with the pink of its masses of flowers. Specimens from low elevation saline seeps and springs in the western counties have thick coriaceous leaves and tend to have very short pedicels. Perhaps they will prove to be worthy of taxonomic recognition.

Var. *zionense* (Eastw.) Welsh [*D. zionense* Eastw., type from Zion Canyon]. Seeps and hanging gardens at 1125 to 1285 m in Kane, San Juan (?), and Washington counties; endemic; 14 (v). This is a showy phase of the species simulating, but unmatched by, races from elsewhere within the huge distribution of the complex. The plants tend to maintain their characteristics even where growing in open sunny sites such as occasionally in seeps along Lake Powell; the features are thus not merely ecologically induced.

Dodecatheon redolens (Hall) **H. J. Thompson** [*D. jeffreyi* var. *redolens* Hall]. Plants 20–60 cm tall, the roots pale; leaves 8–40 cm long, 2–6 cm wide, oblanceolate, entire, tapering to a broad petiole; scapes 25–60 cm tall; umbel 5- to 10-flowered; pedicels 1–5 cm long or more; calyx tube 3–4 mm long, the lobes 5–8 mm long, lanceolate; corolla magenta to lavender, 15–25 mm long; staminal filaments less than 1 mm long, distinct, black; anthers 7–11 mm long, the connective rough, maroon to black; stigmas conspicuously enlarged; capsules 8–14 mm long. Streamsides at ca 2867 m in the Deep Creek Mts., Juab County; Nevada and California; 1 (0). This plant is questionably segregated from *D. jeffreyi* Van Houtte.

Glaux L.

Succulent, perennial herbs from short rhizomes and fibrous or tuberous roots; leaves opposite below, subopposite to alternate above, entire, sessile; herbage glabrous; flowers solitary, sessile or subsessile in the axils near the middle of the stem; calyx cupshaped, the 5 petaloid lobes equaling or surpassing the tube; corolla lacking; stamens 5, alternate with the calyx lobes; capsule subglobose, few-seeded.

Glaux maritima L. Sea Milkwort. Plants 3–25 (30) cm tall; leaves 3–20 (25) mm long, oval to narrowly oblong, jointed to the stem; calyx 3–5 mm long, the lobes white or pinkish; stamens subequal to the calyx lobes, inserted at the base of the ovary; capsules 2–3 mm long; 2n = 30. Seeps, springs, streambanks, and moist meadows, often where saline, at 1280 to 2200 m in Box Elder, Cache, Daggett, Duchesne, Garfield, Grand, Juab, Millard, Piute, Rich, Salt Lake, Sanpete, Sevier, Tooele, Utah, and Wayne counties; widely distributed in North America; circumboreal; 43 (v).

Lysimachia L.

Perennial, rhizomatous herbs; leaves opposite or whorled, large, minutely spotted with red, sessile or petiolate; flowers solitary or 2 to many, borne in middle to upper axils; calyx of usually 5 more or less united sepals; corolla usually of 5 petals more or less united at the base; stamens usually 5, attached to the ovary base; capsules subglobose.

Ray, J. D. Jr. 1956. The genus *Lysimachia* in the New World. Illinois Biol. Monographs 24(3–4): 1–160.

1. Flowers small, numerous, borne in dense, pedunculate racemes; leaves sessile .... *L. thyrsiflora*
— Flowers large, solitary or in pairs; leaves petiolate . *L. ciliata*

Lysimachia ciliata L. Fringed Loosestrife. [*Steironema ciliatum* (L.) Raf.]. Stems erect, 2.5–10 (13) dm tall, simple or branched above, glabrous; leaves cauline, the lower withered by anthesis, the petioles 3–20 mm long, coarsely ciliate, the blades 3.5–15 cm long, 0.9–6.5 cm wide, ovate to lanceolate, rounded to subcordate basally, obtuse to acuminate-attenuate apically, ciliate, flowers solitary or in pairs; pedicels 1–8 cm long; calyx 3.5–8 mm long, 5-lobed, the lobes lance-acuminate; corolla bright yellow (fading whitish), rotate, the lobes 7–12 mm long, glandular within; stamens 5, alternating with 5 narrow staminodia; capsule shorter than the calyx; seeds 10–20; 2n = 34, 92, 96, 98, 100, 108, 112. Stream and river banks and bogs at 1340 to 1405 m in Cache, Davis, Salt Lake, Utah, and Weber counties; Canada south to Oregon, New Mexico, and Georgia; 6 (i).

Lysimachia thyrsiflora L. Tufted Loosestrife. Plants 2–8 dm tall; stems erect, simple; leaves scalelike below, enlarged above, but the lower often withered at anthesis, 3–16 cm long, 0.5–6 cm wide, oblong to lanceolate, elliptic, or oblanceolate, sessile; flowers in dense, pedunculate racemes; pedicels 1–4 mm long; calyx glandular-spotted, 5 (7) -lobed, the lobes 1.8–3.5 mm long; corolla yellow, spotted with purple, 5 (7) -lobed, the lobes 3–7 mm long; ovary dark, glandular; style 4–6 mm long; capsule ca 2.5 mm wide, few-seeded; 2n = 40, 42. Marshes at 1340 to 1405 m in Cache, Utah, and Weber counties; widespread in North America; Eurasia; 3 (0).

Primula L.

Perennial, scapose herbs; leaves all basal, simple, neither conspicuously imbricate nor persistent; herbage glabrous, glandular, or mealy; flowers 1 to many in involucrate umbels; calyx tubular, 5-lobed; corolla tubular, not conspicuously contracted, the crests absent or reduced, 5-lobed, the lobes usually emarginate; stamens 5, the filaments adnate to the corolla; capsule cylindric to ovoid.

1. Involucres, calyces, and leaves (at least beneath) conspicuously white-mealy; bracts swollen at the base 2
— Involucres, calyces, and leaves (even beneath) not (or only somewhat) mealy, either glabrous or glandular; bracts not swollen at the base 3
2(1). Corollas 6–10 mm wide; calyx teeth parallel, the margins touching; plants of calcareous bogs at middle elevations .... *P. incana*
— Corollas 10–16 mm wide; calyx teeth diverging, the margins well separated; plants of hanging gardens at lower elevations .... *P. specuicola*
3(1). Calyx conspicuously glandular, 7–16 mm long; leaves mainly 15–25 cm long and 1.5–4 cm wide; plants of high elevation, montane sites .... *P. parryi*
— Calyx glabrous or sparingly mealy, 5.5–12 mm long; leaves 1.5–11 cm long and 0.4–2.2 cm wide; plants of limestone crevices and slopes at middle elevations 4
4(3). Calyx 5.5–7.5 mm long (to 9 mm long in fruit); calyx lobes 4–5 mm wide, obovate; plants of Logan Canyon .... *P. maguirei*
— Calyx 8–10 mm long (to 12 mm long in fruit); corolla lobes 4–12 mm wide, broadly obovate; plants of the House Range .... *P. domensis*

Primula domensis Kass & Welsh House Range Primrose. Plants 7–15 cm tall; leaves 2–8 (11) cm long, 5–22 mm wide, oblanceolate to spatulate, dentate or subentire, tapering to a broad petiole, green and more or less glandular on both sides; bracts usually 3, 1.5–10 mm long, lanceolate, not swollen at the base, glabrous or mealy; peduncle apex glabrous or somewhat mealy; umbels 1- to 5-flowered, the pedicels 5–22 mm long; calyx 8–12 mm long, mealy or glabrous, the teeth shorter than the tube; corolla rose to lavender, the tube surpassing the calyx, but less than twice as long, the limb 12–25 mm wide, the lobes shorter than the tube, 4–12 mm wide; capsule to 8 mm long, not surpassing the calyx. Crevices in limestone at 2590 to 2745 m in the House Range (type locality), Millard County; endemic; 3 (0). This primrose is nearer geographically to the *P. nevadensis* N. Holmgren, of eastern Nevada, than it is to *P. maguirei* of Logan Canyon. In *P. nevadensis* the cuneate, markedly dentate leaves commonly overtopping the inflorescence are diagnostic. Although grossly similar to *P. maguirei*, the House Range primrose has much larger calyces, on the average and in the extreme, more consistently toothed leaves, and corolla tubes shorter in proportion to the calyx length. All belong to the *P. cusickiana* Gray complex, and might be best treated within that entity at infraspecific rank; a proposal beyond the scope of this work, and neither intended nor implied herein.

Primula incana **Jones** Silvery Primrose. Plants 5–40 cm tall; leaves 1.5–8 cm long, 0.5–1.7 cm wide, oblanceolate to spatulate, shallowly denticulate, tapering to broad petioles, usually white-mealy below, green above; bracts lanceolate to linear, 3–10 mm long, swollen at the base, more or less mealy; peduncle apex often white-mealy; umbels 2- to 15-flowered, the pedicels 2–15 (25) mm long; calyx 5.5–10 mm long, mealy, the teeth shorter than the tube, erect, the margins touching or slightly diverging; corolla lilac, the tube surpassing the calyx, the limb 6–10 mm wide, the lobes shorter than the tube; capsule only slightly exceeding the calyx; seeds roughened. Wet meadows and calcareous, often quaking, bogs at 2012 to 2655 m in Daggett and Garfield (type from the vicinity of Tropic) counties; Alaska and Yukon to Hudson Bay, south to Colorado; 4 (0).

Primula maguirei **L. O. Williams** Maguire Primrose. Plants 4–15 cm tall; leaves 1.5–7 cm long, 4–15 (25) mm wide, oblanceolate to spatulate, subentire or irregularly toothed, tapering to slender petioles, green and more or less glandular on both sides; bracts 1–7 mm long, lanceolate to elliptic, not swollen at the base, glabrous or mealy; peduncle apex glabrous or sparingly mealy; umbels 1- to 3-flowered, the pedicels 2–15 mm long; calyx 5.5–7.5 mm long, mealy or glabrous, the teeth shorter than the tube, erect, the margins diverging; corolla rose to lavender, the tube surpassing the calyx (ca twice as long), the limb 10–15 (25) mm wide, the lobes shorter than the tube, 4–5 mm wide; capsule surpassing the calyx. Limestone crevices and slopes at ca 1550 to 1680 m in Cache (type from Logan Canyon) County; endemic; 3 (0). The Maguire primrose is a delicate beauty on limestone in mesic sites in Logan Canyon.

Primula parryi **Gray** Parry Primrose. Plants 7–55 cm tall, ill-scented, from thick rhizomatous caudex branches, these clothed with brown marcescent leaf bases; leaves 8–50 cm long, 1–9 cm wide, spatulate to oblanceolate or elliptic, entire to sinuate, glandular and green on both sides; bracts numerous, not saccate at the base, 2–10 mm long, oblong to lance-attenuate; peduncle apex merely glandular; umbels 3- to 20-flowered, the pedicels 0.3–12 cm long; calyx 7–16 mm long, the lobes shorter than or subequal to the tube; corolla reddish purple (or white) with a yellow center, the limb 12–28 mm wide, the lobes shorter than the tube, 6–12 mm wide or more; capsules ovoid, subequal to the calyx; $2n = 44$. Meadows, talus slopes, and stream banks in spruce-fir and alpine tundra communities at 2470 to 3965 m in Beaver, Box Elder, Cache, Daggett, Duchesne, Emery, Garfield, Grand, Iron, Juab, Piute, Salt Lake, San Juan, Sanpete, Sevier, Summit, Tooele, Uintah, Utah, and Wayne counties; Idaho and Montana to Arizona and New Mexico; 64 (vii).

Primula specuicola **Rydb.** Cave Primrose; Easter-flower. Plants 6–28 cm tall, from short, rhizomatous caudex branches, these more or less clothed with marcescent leaf bases; leaves 2–20 cm long, 3–33 mm wide, spatulate to oblanceolate or elliptic, crenate-serrate to serrate- dentate, strongly to slightly white-mealy beneath, green above; bracts 4–8 mm long, lance-attenuate, numerous, swollen at the base, more or less mealy; umbels 5- to 40-flowered, sometimes a second one superposed, the pedicels 8–34 mm long; calyx 5–8.5 mm long, often mealy, the lobes 2–4 mm long (longer to shorter than the tube), diverging; corolla lavender to rose, pink, or white, the tube ca twice as long as the calyx, the limb 10–16 mm wide, the lobes shorter than the tube, rarely double; capsules shorter than the calyx; seeds roughened. Hanging gardens and their margins at 1125 to 1680 m in Garfield, Grand, Kane, San Juan (type from near Bluff), and Wayne counties; Arizona; 38 (xv). This beautiful primrose is an endemic of the canyons of the Colorado. The plants produce flower buds as early as January and complete flowering in early May. Its beauty hardly diminishes following flowering due to the bead-like, glistening, mealy surface and bicolored leaves. It often clings to vertical, wet cliff faces, where its roots are anchored in an algal mat, within the surface of platy exfoliating sandstone, or in sandy detritus. The coincidence of flowering time with Easter, in most years, accounts for the common name applied in southeastern Utah.

Samolus L.

Glabrous, perennial herbs; leaves simple, alternate, entire; flowers small, white, in terminal racemes or panicles; calyx tube adnate to the ovary below, 5-lobed; corolla campanulate, perigynous, 5-lobed; stamens 5, inserted on the corolla tube opposite the lobes, alternating with 5 staminodia; ovary more than half inferior; capsule subglobose, 5-valved; seeds numerous.

Samolus parviflorus **Raf.** Water-pimpernell; Brookweed. Plants 1.5–4.5 dm tall or more; stems simple or branched; leaves mainly 1–15 cm long, 0.5–4 cm wide, obovate to spatulate, oblanceolate, ovate, or oval, sessile or petiolate, rounded to obtuse; racemes (or panicles) sessile or nearly so, the slender axis flexuous; pedicels filiform, mainly 5–20 mm long, bracteate (above to below the middle); calyx lobes triangular to ovate; corolla white, 2–3 mm wide, the lobes oblong, longer than the tube; capsules 2–3 mm wide. Canal banks, seeps and springs, and in riparian communities at 850 to 1345 m in Salt Lake and Washington counties; widespread in the U. S. and southern Canada; Mexico; 8 (ii).

PUNICACEAE Horan.
Pomegranate Family

Small trees or shrubs; branches thorn-tipped; leaves opposite, estipulate; flowers epigynous, perfect; calyx lobes 5–7, persistent; petals 5–7, alternate with the sepals; stamens numerous, the filaments slender; pistil 1, the ovary inferior, 3– to 7 (or more) -loculed, the ovules numerous on all the placentae; style and stigma 1; fruit a thick-skinned, depressed globose berry, the seeds surrounded by juicy pulp; x = 8, 9.

Punica L.

Plants glabrous; leaves leathery; flowers 1–5 on tips of axillary branches; receptacle or hypanthium campanulate or tubular, leathery, topped by the persistent sepals; corolla lobes alternate with the sepals and inserted on the hypanthium, wrinkled; stamens obscuring the inside of the receptacle.

***Punica granatum* L.** Pomegranate. Plants deciduous; stems to 7 m tall; leaves short-petioled, oblong or oval-lanceolate, 2.5–6 cm long, obtuse, glabrous, shiny; flowers orange red, 2.5–4 cm broad; fruit to 10 cm in diameter, brownish to orange red; seeds numerous, the fleshy coat red or pink; 2n = 16, 18. Cultivated for its sweet fleshy seeds, long persisting, and rarely escaping at 720 to 1270 m in Washington County (and previously at Hite, San Juan County, prior to Glen Canyon Dam); introduced from Asia, now widely grown in warm areas of the world; 5 (ii). There are many named varieties, with single or double flowers. Many produce no fruit and are grown for aesthetic value only.

PYROLACEAE Dumort.
Wintergreen Family

Suffrutescent or herbaceous perennials; leaves simple, alternate, opposite, or appearing whorled, evergreen or much reduced and lacking chlorophyll; flowers usually perfect, regular or irregular; calyx with 4 or more or less distinct sepals; corolla with 4 or 5 more or less distinct petals (united in *Pterospora*); stamens twice as many as the petals, the anthers pendulous, opening by apparently terminal pores or by slits, or the anthers erect, awnless or 2-awned; pistil 1, the ovary superior, 4– or 5–loculed; style 1; fruit a capsule; x = 8, 11, 13, 16, 19, 23.

Welsh, S. L. 1980. Pyrolaceae. pp. 53–55. In: Utah flora: Miscellaneous families. Great Basin Nat. 40: 38–58.

1. Plants lacking chlorophyll; leaves reduced and scale-like, reddish, brownish, purple, or yellowish when fresh, often drying dark .... *Pterospora*
— Plants with chlorophyll (rarely without); leaves not reduced to scales, except rarely, commonly evergreen ... 2

2(1). Flowers solitary, the petals rotate or nearly so ... *Moneses*
— Flowers few to several, the petals concave 3

3(2). Stems leafy, though short, the leaves apparently whorled; flowers corymbose; staminal filaments dilated near the base; styles very short or lacking .... *Chimaphila*
— Stems leafy at base only; flowers in elongate racemes; filaments not especially dilated at the base; styles in most species over 2 mm long .... *Pyrola*

Chimaphila Pursh

Low shrubs from creeping rhizomes, the stems erect or ascending; leaves evergreen, leathery, apparently whorled or some alternate; flowers regular, (1) 2 to several, borne in pedunculate, umbellate corymbs; sepals usually 5, distinct nearly to the base, persistent; petals usually 5, distinct, rotate campanulate; stamens usually 10, the filaments dilated and ciliate near the base; anthers awnless, opening by falsely terminal pores on short tubes; ovary superior, 5–lobed and 5–loculed; fruit a loculicidally dehiscent capsule.

1. Leaves lance-ovate to ovate; flowers 1–3, white; plants rare in Washington County .... *C. menziesii*
— Leaves oblanceolate; flowers 3–7, pink; plants uncommon, widely distributed .... *C. umbellata*

***Chimaphila menziesii* (R. Br.) Sprengel** [*Pyrola menziesii* R. Br. ex D. Don]. Plants erect, mainly 1–1.5 dm tall, simple or sparingly branched; leaves alternate or indistinctly whorled, lance-ovate to ovate, 15–30 (35) mm long, serrulate, dark green above, pale beneath, sometimes mottled; peduncles mostly 3.5–5 cm long; bracts obovate, persistent; sepals 4–5 mm long, rounded, erose; petals white, 5–6 mm long; capsules 5–6 mm thick; 2n = 18, 26. Juniper and mountain brush community at ca 1585 m in Washington (Zion National Park) County; British Columbia and Idaho south to California; 1 (0).

***Chimaphila umbellata* (L.) Barton** Pipsissewa; Prince's Pine. [*Pyrola umbellata* L.; *C. occidentalis* Rydb.; *C. umbellata* ssp. *occidentalis* (Rydb.) Hulten]. Plants (1) 1.5–2.5 (3) dm tall, the stems glabrous, only somewhat woody; leaves 1.5–4.5 (6) long, 0.5–1.5 (2) cm wide, elliptic to oblanceolate, cuneate basally, sharply serrate, shining above, pale beneath, glabrous; peduncles 4–7 (10) cm long, glabrous or minutely glandular-puberulent, often suffused with red purple; pedicels glandular-puberulent or merely puberulent; flowers 1–6 or more, umbellate-corymbose; sepals erose-ciliate; petals 5–7 mm long, pink; stamens with expanded bases ciliate; capsules 5–7 mm broad. Ponderosa pine, white fir-Douglas fir, aspen, lodgepole pine, and spruce-fir communities at 2195 to 2900 m in Daggett, Duchesne, Garfield, Juab, Millard, Piute, Salt Lake, Summit, Uintah, and Washington counties; Alaska south to California and Mexico, east to New Mexico and Colorado, and in the eastern U. S.; Eurasia; 14 (0). Our materials are referable to **var. *occidentalis* (Rydb.) Blake**.

Moneses Salisb.

Rhizomatous herbs; leaves with chlorophyll, leathery, persistent, mainly basal, but sometimes opposite or in whorls; flowers solitary, nodding, borne on a long peduncle; sepals usually 5, persistent; petals usually 5, distinct, spreading; stamens usually 10, the filaments tapering to the apex, the anthers awnless, nodding, opening by means of apparently terminal pores; ovary superior, 5–loculed, the stigma borne on an elongate, glabrous style; fruit a loculicidal capsule.

***Moneses uniflora* L.** Single Delight; Wax Flower. [*M. reticulata* Nutt.; *M. uniflora* var. *reticulata* (Nutt.) Blake]. Plants 0.4–1.7 dm tall; leaves (including petioles) 0.8–4 cm long, 0.6–2 cm broad, serrate to crenate-serrate; peduncles 3–15 cm long usually with 1 or 2 bracts along its length; flowers 1.3–2.5 cm broad, white to cream; sepals 1.5–2.5 cm long, ciliate; petals 7–11 mm long, spreading; style 2–4 mm long; capsule 5–8 mm broad; 2n = 24, 26. Moist sites in spruce-fir communities at 2450 to 3355 m in Beaver, Carbon, Duchesne, Emery, Garfield, Grand, Iron, Juab, Salt Lake, San Juan, Uintah, Utah, and Wasatch counties; widely distributed in North America; Eurasia; 19 (ii).

Pterospora Nutt.

Plants herbaceous saprophytes, devoid of chlorophyll, tall, reddish or purplish brown, the stems arising from the bulbous clusters of coralloid roots; leaves alternate, simple, scalelike, colored like the stems; flowers numerous, borne in an elongate raceme, nodding; calyx 5–lobed; corolla urn-shaped, the tube much longer than the lobes; stamens 10, the filaments flattened, tapering to the apex, glabrous, the anthers with 2 recurved awns, dehiscent almost throughout; ovary superior, 5–loculed, the stigma borne on a short thick style; fruit a loculicidal capsule. Note: This genus is often included within the Monotropaceae.

Pterospora andromeda Nutt. Pinedrops. Plants erect, the stems simple, 2–8.5 (10) dm tall, reddish brown, succulent, arising from a cluster of roots to 5 cm in diameter, glandular-hairy, leafy only near the base; racemes 3–35 cm long or more; flowers 5–8 mm long, nodding, axillary; pedicels 5–15 mm long, recurved; sepals oblong, glandular; corolla pale yellow, depressed urn-shaped; capsule 8–12 (14) mm broad, 5–lobed, depressed globose. Pinyon-juniper, ponderosa pine, mountain brush, aspen, and lodgepole pine communities at 1980 to 3355 m in Beaver, Daggett, Duchesne, Garfield, Grand, San Juan, Summit, Uintah, and Washington counties; widely distributed in North America; 21 (iii).

Pyrola L.

Rhizomatous herbs; leaves with chlorophyll, leathery, persistent, all basal or apparently so, or rarely lacking and the plants then partially or completely saprophytic; flowers regular to irregular, borne in terminal racemes; sepals 5, united at the base; petals 5, distinct, usually concave, deciduous; stamens 10, the filaments tapering to the apex, the anthers unawned, pendulous, opening by means of apparently terminal pores; ovary superior, 4–loculed, the stigma borne on a straight or curved style; fruit a loculicidal capsule.

Copeland, H. F. 1947. Observations on structure and classification of the Pyroleae. Madroño 9: 65–102.

1. Style straight or nearly so; pores of the anthers sessile; stigma usually much broader than the style 2
— Style bent or curved; pores of anthers usually borne on short tubes; stigmas only slightly broader than the styles
 3
2(1). Style 2 mm long or less, not (or seldom) exserted from the flower; flowers not secund; petals pinkish to cream
 . . . . P. minor
— Style over 2 mm long, exserted from the flower; flowers secund; petals greenish white . . . . P. secunda
3(1). Flowers pink to purplish; sepals longer than broad . . .
 . . . . P. asarifolia
— Flowers pale, greenish yellow; sepals broader than long
 4

4(3). Leaves marked with white or silvery gray along the veins; plants known from Washington County . . . . P. picta
— Leaves uniformly green throughout; plants of broad distribution, including Washington County . . . . . P. virens

Pyrola asarifolia Michx. Liver-leaf Wintergreen. [*P. rotundifolia* var. *bracteata* (Hook.) Gray; *P. asarifolia* var. *bracteata* (Hook.) Jepson; *P. rotundifolia* var. *purpurea* Bunge; *P. asarifolia* var. *purpurea* (Bunge) Fern.; *P. incarnata* Fisch. in DC.; *P. asarifolia* var. *ovata* Farw.; *P. uliginosa* T. & G. ex Torr.; *P. rotundifolia* var. *uliginosa* (T. & G.) Gray; *P. asarifolia* var. *uliginosa* (T. & G.) Farw.; *P. elata* Nutt.; *P. bracteata* var. *hilli* J. K. Henry]. Plants 1.3–4 dm tall; leaves basal or essentially so, the blades 1.3–7.5 cm long, 1.1–7.3 cm wide, oval, rotund, elliptic, or acute basally, rounded to obtuse or emarginate apically, entire to serrulate; petioles 1–9 cm long; racemes mostly 2–12–flowered; pedicels 3–8 mm long; sepals longer than broad, 1.5–4 mm long; petals pink to purplish, 5–7 mm long; anthers pink, the pores on short tubes; style curved, with a flaring collar below the stigma. Mountain brush, aspen, lodgepole pine, spruce-fir, and bristlecone pine communities at 1675 to 3205 m in Cache, Daggett, Duchesne, Emery, Garfield, Grand, Iron, Juab, Millard, Piute, Rich, Salt Lake, Sanpete, Summit, Uintah, Utah, Washington, and Weber counties; Alaska east to Newfoundland, south to California, New Mexico, South Dakota, and New England; Asia; 50 (iv). Varietal status of Utah material is not clear.

Pyrola minor L. Lesser Wintergreen. [*Amelia minor* (L.) Alef.; *Erxlebenia minor* (L.) Rydb.; *P. minor* var. *conferta* C. & S.; *P. conferta* (C. & S.) Fisch. ex Ledeb.]. Plants 0.8–2.4 dm tall; leaves basal, the blades (0.4) 1.1–3.3 cm long, (0.6) 0.9–2.5 cm broad, oval, elliptic, or ovate, obtuse to rounded apically, crenate to subentire; petioles 0.2–3 cm long; racemes mostly 5– to 13–flowered; pedicels 2–3 mm long sepals 1–1.5 mm long, erose to subentire; petals pale pink to cream, 3.5–4.5 mm long; anthers with pores sessile; style straight, very short, not exserted from the corolla, with a more or less distinctive collar below the stigma; 2n = 46. Aspen, white fir, lodgepole pine, and spruce-fir communities at 2255 to 3420 m in Beaver, Box Elder, Daggett, Duchesne, Garfield, Juab, Piute, Salt Lake, San Juan, Sevier, Uintah, and Washington counties; Alaska and Yukon, east to Greenland and south to California and Colorado; circumboreal; 19 (ii).

Pyrola picta J. E. Sm. Pictureleaf Wintergreen. Plants mainly 1–2 dm tall; leaves basal, the blades 2–7 cm long, 1–3.5 cm wide, lance-ovate to elliptic, white along the veins, obtuse to rounded apically; petioles 0.3–3.5 cm long; racemes mostly 2– to 7–flowered; pedicels 3–8 mm long; sepals 1.8–2 mm long; petals greenish to cream, 7–8 mm long; anthers yellowish, the pores on elongate tubes; style curved, with a flaring collar below the stigma. Ponderosa pine, fir, and aspen communities at ca 2530 to 2900 m in Washington (Pine Valley Mts.) County; British Columbia to Montana, south to California and Wyoming; 2 (0).

Pyrola secunda L. Secund Wintergreen. [*Ramischia secunda* (L.) Garcke; *Actinocyclus secundus* (L.) Klotzsch; *P. secunda* var. *obtusata* Turcz.; *Orthilia secunda* var. *obtusata* (Turcz.) House; *P. secunda* var. *pumila* Paine; *P. secunda* f. *eucycla* Fern.]. Plants 0.6–1.8 (2.1) dm tall; leaves basal or rarely some cauline, or sometimes with a naked stem below the leaves, the blades 1.3–4 (5) cm long, 1–3 cm wide, ovate, oval, elliptic, or orbicular, obtuse or rounded basally, acute to obtuse or rounded apically, crenate-serrate; petioles 0.6–2 cm long; racemes mostly 4- to 15-flowered, the flowers secund; pedicels 2–5 mm long sepals 0.5–1.5 mm long; petals greenish white, 4–6 mm long; anthers with pores sessile; style straight, exserted from the corolla, lacking a collar; 2n = 46. Mountain brush, ponderosa pine, Douglas fir, aspen, lodgepole pine, and spruce-fir communities at ca 2010 to 3420 m in Box Elder, Cache, Carbon, Daggett, Duchesne, Garfield, Juab, Kane, Millard, Piute, Rich, Salt Lake, San Juan, Sanpete, Summit, Uintah, Utah, Wasatch, and Washington counties, broadly distributed in North America; Eurasia; 57 (ix). Segregation of our materials into the various proposed infraspecific categories seems unwarranted.

Pyrola virens Schweigger in Schweigg. & Koerte Greenish Wintergreen. [*P. chlorantha* Swartz; *P. chlorantha* var. *saximontana* Fern.; *P. virens* f. *paucifolia* (Fern.) Fern.; *P. chlorantha* f. *paucifolia* (Fern.) Camp]. Plants 0.9–2.5 dm tall; leaves basal, the blades 0.6–3.5 cm long, 0.5–3 cm broad, elliptic, oval, or obovate, obtuse to rounded basally, rounded to obtuse apically, crenate-serrate to subentire; petioles 0.8–6 cm long; racemes mostly 2- to 9-flowered; pedicels 3–8 mm long; sepals 0.5–1.5 mm long; petals greenish yellow, 5–7 mm long; anthers yellowish, the pores on elongate tubes; style curved, with a flaring collar below the stigma. Ponderosa pine, aspen, lodgepole pine, and spruce-fir communities at 2225 to 3050 m in Daggett, Duchesne, Juab, Piute, Salt Lake, San Juan, Summit, Uintah, and Washington counties; widely distributed in North America; Eurasia; 17 (ii).

RANUNCULACEAE A. L. Juss.
Buttercup Family

Annual or perennial herbs or trailing vines; leaves alternate, opposite, or basal, simple, deeply divided or variously compound; flowers hypogynous, perfect or rarely imperfect, regular or irregular, lacking a hypanthium; sepals 3 to many, often petaloid; petals 3 to many or lacking; stamens several to many; pistils 1 to many, superior; fruit an achene, follicle, or berry; x = 6–10, 13.

1.	Flowers markedly irregular, mostly dark blue or purple	2
—	Flowers regular, seldom dark blue or purple	3
2(1).	Upper sepal spurred at the base; petals usually 4 .... *Delphinium*	
—	Upper sepal not spurred, hooded at apex; petals usually 2 .... *Aconitum*	
3(1).	Flowers with sepals or petals spurred anteriorly	4
—	Flowers not spurred	5
4(3).	Leaves simple, linear to narrowly spatulate; plants diminutive annuals .... *Myosurus*	
—	Leaves ternately once to thrice compound; plants perennial .... *Aquilegia*	
5(3).	Perianth consisting of a single whorl, arbitrarily called sepals, but often petaloid in texture and color	6
—	Perianth of sepals and petals, but the sepals sometimes deciduous and leaving scars below the petals	10
6(5).	Leaves simple, the blades reniform, cordate; perianth white, large, showy; plants of wet meadows at higher elevations .... *Caltha*	
—	Leaves deeply lobed or compound; perianth showy or not; plants of various habitats, and occasionally as above	7
7(6).	Leaves alternate; flowers greenish, inconspicuous	8
—	Leaves opposite or whorled; flowers often showy	9
8(7).	Leaves simple, palmately lobed or parted; flowers perfect; anthers oval or ovate, 1 mm long or less ... . *Trautvetteria*	
—	Leaves compound; flowers perfect or imperfect; anthers mostly more than 1 cm long .... *Thalictrum*	
9(7).	Stems with sessile or short-petioled, whorled leaves; flowers erect .... *Anemone*	
—	Stems with petiolate, opposite leaves; flowers nodding or erect .... *Clematis*	
10(5).	Flowers inconspicuous, numerous, in dense, terminal racemes; pistils 1 per flower; fruit a red or white fleshy berry .... *Actaea*	
—	Flowers solitary or few in cymes; pistils several to numerous; fruit an achene or a follicle	11
11(10).	Plants annual; leaves dissected, the ultimate segments 0.5–1.5 mm wide; petals white to red ... *Adonis*	
—	Plants perennial, or if annual, not otherwise as above	12
12(11).	Petals linear, narrower than the white to cream petaloid sepals; fruit a several-seeded follicle; plants montane .... *Trollius*	
—	Petals mostly as broad as or broader than the usually green sepals, or lacking; fruit usually a 1-seeded achene .... *Ranunculus*	

Aconitum L.

Perennial herbs from tuberous roots; stems simple, erect; leaves alternate, palmately divided; flowers perfect, irregular, large and showy, borne in terminal racemes or panicles; sepals 5, petaloid, the upper one large, helmetlike, the lateral ones oval, broader than the lower 2; petals 2, enclosed within the helmetlike hood, highly modified (with a slender filamentous claw, a pendulous blade, and a saclike or curved spur), sometimes 3 additional scalelike petals present; stamens numerous, the filaments flattened; pistils 3–5, distinct; fruit a follicle.

Aconitum columbianum Nutt. in T. & G. Monkshood. [*A. bakeri* Greene; *A. insigne* Greene; *A. divaricatum* Rydb., type from City Creek Canyon; *A. glaberrimum* Rydb., type from southern Utah?]. Plants mainly 5–15 dm tall, the stems glabrous below, pubescent above, with at least some spreading hairs and often glandular, usually contracted at or below ground level; leaves mostly cauline, 5–25 cm long, the blades 3- to 5-lobed, the sinuses lateral to the terminal lobe extending almost to the base, the terminal lobe 2.5–10 cm long or more; sepals mostly a deep bluish purple, rarely white, yellowish, or greenish; hood 14–25 mm high, only the upper 2 petals usually present, the spur commonly coiled; follicles commonly 3, glabrous or glandular, 10–20 mm long; 2n = 16, 18. Often along streams in mountain brush, sagebrush, aspen, hairgrass-sedge, lodgepole pine, and spruce-fir communities at 1980 to 3355 m in all Utah counties, except Wayne; British Columbia to Montana, south to California, Arizona, and New Mexico; 124 (xvi). Our specimens belong to **var. *columbianum***.

Actaea L.

Perennial, rhizomatous herbs, the stems simple or more usually branched, erect; leaves all cauline, alternate, twice ternately or twice ternate-pinnately compound; flowers perfect, regular, small, borne in terminal racemes; sepals 3–5, caducous; petals 5–10 (rarely lacking); stamens numerous; pistil 1; fruit a red or white berry.

Actaea rubra (Ait.) **Willd.** Baneberry. [*A. spicata* var. *rubra* Ait.; *A. arguta* Nutt. ex T. & G.; *A. rubra* ssp. *arguta* (Nutt.) Hulten]. Plants 3–10 dm tall; stems glabrous or glabrate; leaves all cauline, the segments broad, ovate to lanceolate, oblong, or obovate, 2–10 cm long, sharply toothed and often lobed; sepals 2–3 mm long, whitish (rarely purplish); stamens much surpassing the petals; berries 5–10 mm long, white or red; 2n = 16. Mountain brush, willow-birch, aspen, Douglas fir, limber pine, fir, and spruce-fir communities at 1980 to 3050 m in probably all Utah counties; Alaska and Yukon east to the Atlantic, south to California, Arizona, and New Mexico; 61 (vi). This plant is considered to be poisonous, especially the berries.

Adonis L.

Annual herbs; leaves alternate, bipinnatifid, the ultimate segments linear to subulate; flowers showy, regular, complete, solitary, borne on axillary or terminal pedicels; sepals 5, membranous or chartaceous; petals 5–20, white or red orange, sometimes suffused with dark blue or black basally, lacking a glandular basal scale; stamens numerous; pistils numerous; fruit a crested, sculptured achene.

Adonis aestivalis L. Pheasant-eye. Plants glabrous; stems simple or branched, 2–6 dm tall; leaves 2–6 cm long, dissected into narrow segments; sepals gibbous basally, greenish, straw colored, or reddish; petals 5–12.5 mm long, white or red orange, blue to black basally, narrowly oblong-oblanceolate; stamens blue or brown; achenes hardened, crested, 5–6 mm long (including the beak); 2n = 16, 32. Disturbed sites in fields and in sagebrush, aspen, and spruce-fir communities at 1370 to 2400 m in Cache, Garfield, Salt Lake, and Utah counties; widespread in the U. S.; introduced from Eurasia; 4 (0).

Anemone L.

Perennial herbs from rhizomes or caudices; stems with usually a single whorl of leaves below the inflorescence; basal leaves long-petioled, the blades simple and palmately lobed or ternately compound and often deeply dissected; flowers perfect, regular, showy, borne 1 per peduncle, the peduncles solitary or 2 to several in umbellate cymes; sepals 4–10, petaloid, variously colored; petals lacking; stamens numerous; fruit an achene.

1. Sepals 2–5 cm long, blue to purplish (rarely white); styles plumose, 2–4 cm long in fruit .... *A. patens*
— Sepals 0.5–2 cm long, white, yellow, purplish, or bicolored; styles not plumose, mainly less than 5 mm long in fruit 2
2(1). Plants from tubers; stems constricted below ground level, growing in xeric habitats in southern and western Utah .... *A. tuberosa*
— Plants from a caudex or slender rhizome, not or rarely constricted below ground level, growing in mesic habitats, usually at higher elevations, the distribution various 3
3(2). Cauline whorl of leaves with slender petioles 1–4 cm long, ca 1 mm wide, compound; achenes short-pubescent .... *A. quinquefolia*
— Cauline whorl of leaves sessile or with broadly winged petioles, simple or compound; achenes pilose or woolly hairy 4
4(3). Leaflets of basal leaves merely crenate or cleft to near the middle and with rounded lobes; sepals white within, usually bluish dorsally; flowers solitary ..... *A. parviflora*
— Leaflets of basal leaves dissected into narrowly oblong, elliptic, or lanceolate, acute lobes; sepals purplish, maroon, yellow, or white; flowers 1 to several .. *A. multifida*

Anemone multifida Poir. Cutleaf Anemone. [*A. multifida* var. *globosa* T. & G.; *A. globosa* (T. & G.) Nutt.]. Plants from caudices, 0.4–5 dm tall, the herbage pubescent with long, spreading or ascending hairs; basal leaves 3–22 cm long, 3-foliolate, the leaflets dissected to below the middle into narrowly oblong, elliptic, or lanceolate, acute lobes, the terminal leaflet 1–6 cm long; cauline leaves 1.5–6.5 cm long, in 1 or 2 whorls, silky-hairy to glabrate; peduncles 1–3, 3–15 cm long or more; flowers often showy; sepals purple, maroon, reddish, yellow, or white (often bicolored) 6–12 mm long, silky-hairy without; cluster of achenes ovoid to globose or short-cylindric, mainly 1–2 cm long; achenes 3–4 mm long, silky-hairy, the styles 0.7–1.2 mm long; 2n = 16, 32. Mountain brush, ponderosa pine, sagebrush, aspen, Douglas fir, lodgepole pine, spurce-fir, and alpine tundra communities at 2225 to 3390 mm in Cache, Carbon, Daggett, Duchesne, Emery, Garfield, Iron, Juab, Kane, Piute, Rich, Salt Lake, San Juan, Sevier, Summit, Uintah, Utah, Wasatch, Washington, and Weber counties; Alaska and Yukon to Newfoundland, south to California, New Mexico, and New York; 84 (viii). Our materials apparently belong, at least in large part, to **var. tetonensis** (T. C. Porter) C. L. Hitchc. [*A. tetonensis* T. C. Porter; *A. stylosa* A. Nels., type from Fish Lake, Sevier County]. Possibly recognition of this taxon is inconsequential, since the specimens pass by degree into **var. multifida**. Reports of *A. cylindrica* Gray probably belong here.

Anemone parviflora Michx. Northern Anemone. Plants from elongate rhizomes, mainly 4–15 cm tall (or more), the herbage spreading-villous to glabrous; basal leaves 1.5–10 (17) cm long, 3-foliolate, the leaflets merely lobed or incised to near the middle, the segments or teeth rounded, the terminal leaflet 3–25 (28) mm long; cauline leaves 7–28 mm long, in 1 whorl, silky-hairy to glabrous; peduncles 1 per inflorescence, 2–10 (26) cm long; flowers solitary, large and showy; sepals white to cream, often tinged bluish dorsally; cluster of achenes ovoid, 0.8–1.3 cm long; achenes 2–3 mm long, woolly hairy, the styles 1.2–2 mm long; 2n = 16, 32. Spruce-fir community at 3050 m in Duchesne and Salt Lake counties; Alaska and Yukon, east to the Atlantic, south to Oregon and Colorado; 2 (0).

Anemone patens L. Pasque-flower; Wild Crocus. [*Pulsatilla patens* ssp. *multifida* (Pritz.) Zam.]. Plants from caudices, 1–5.5 dm tall, the herbage more or less grayish villous; basal leaves 8–30 cm long, 3-foliolate, the leaflets dissected into narrowly oblong or lanceolate segments, the terminal leaflet 2.5–6 cm long; cauline leaves 3–6 cm long, in 1 whorl, silky hairy; peduncles 1 per inflorescence; flowers solitary, large and showy; sepals 2–4.5 cm long, blue to purplish (or white), silky hairy dorsally; achenes 4–6 mm long, silky-hairy, the elongate plumed style 2–3.5 cm long; 2n = 16. Aspen, sagebrush, lodgepole pine, and alpine tundra communities at 2560 to 3510 m in Daggett, Duchesne, and Summit counties; Alaska, Yukon, and Mackenzie, south to Washington, Colorado, Texas, and Illinois; 4 (0). Our plants apparently belong to var. *multifida* Pritz.

Anemone quinquefolia L. American Wood-anemone. [*A. oregona* Gray; *A. quinquefolia* var. *oregana* (Gray) Robins.]. Plants from slender scaley rhizomes, mainly 1–2 (3) dm tall, the herbage short-hairy to glabrous; basal leaves 6–20 cm long, 3– to 5-foliolate, the leaflets finely to coarsely serrate or dentate and shallowly cleft, the terminal leaflet 2–4.5 cm long; cauline leaves 3–12 cm long, in 1 whorl, sparingly pilose to subglabrous, ciliate; peduncles 1 per inflorescence; flowers large, showy; sepals 8–22 mm long, white, sometimes tinged bluish or purplish without, glabrous; achenes 3–4 mm long, short-hairy, the cluster subglobose; style ca 1 mm long. Spruce-fir community at 3050 m in Salt Lake County (Cottam et al. s.n. BRY); widespread in the U. S.; 1 (0). Our material belongs to var. *lyallii* (Britt.) Robins. [*A. lyallii* Britt.].

Anemone tuberosa Rydb. Desert Anemone. Plants from fusiform, tubers (tuberous rhizomes), mainly 8–50 cm tall, the herbage sparingly silky-hairy to glabrous; basal leaves 6–16 cm long, 3-foliolate, the leaflets deeply cleft or parted, the ultimate segments obovate to cuneate, the terminal leaflet (including petiolule) 1.3–5.5 cm long; cauline leaves 15–70 mm long, in 1 whorl (rarely 2), sparingly pilose to glabrous; peduncles 1–5 per inflorescence; flowers showy; sepals 9–18 mm long, rose-pink; achenes 3.5–4.5 mm long, woolly-hairy; styles 0.8–1.8 mm long. Blackbrush, creosote bush, hopsage (other warm desert shrub), mixed desert shrub and juniper-cliffrose communities at 800 to 1650 m in Beaver, Juab, Kane, San Juan, Tooele, and Washington counties; Nevada, California, Arizona, and New Mexico; 27 (iv). This is a remarkably pretty plant, with its apparent northern distribution on Stansbury Island in Great Salt Lake, where it occurs with *Astragalus nuttallianus*, far north of its more usual range.

Aquilegia L.

Perennial herbs from caudices and taproots; stems simple or branched above, erect; leaves alternate or mainly basal, the cauline ones much reduced, ternate to triternately compound; flowers regular, showy, solitary or 2 to several in bracteate, long-pedicellate, terminal racemes; sepals 5, petaloid, not spurred; petals 5, the lower part closed and forming a spur with a bulbous, glandular tip; stamens numerous, the inner ones often sterile (staminodia); pistils mostly 5, distinct; fruit a follicle.

Munz, P. A. 1946. *Aquilegia*. The cultivated and wild columbines. Gentes Herb. 7: 1–150.
Payson, E. B. 1918. The North American species of *Aquilegia*. Contr. U. S. Nat. Herb. 20: 133–157.

1. Herbage conspicuously glandular-pubescent; stems mainly 3–7 dm tall; flowers white or suffused with blue or pink; plants of alcoves in southeastern Utah .... *A. micrantha*
— Herbage glabrous, glaucous, or glandular-pubescent (especially in the inflorescence and throughout in some *A. flavescens* and *A. formosa*); stems of various height; flowers variously colored; plants of various distribution, but not, if ever, of hanging gardens (except in Zion Canyon) 2

2(1). Sepals and spurs scarlet, sometimes fading yellowish; flowers nodding 3
— Sepals and spurs white, blue, yellow, or pinkish, but never truly scarlet; flowers erect to spreading 4
3(2). Flowers much longer than wide, appearing subconic; sepals erect-ascending; spurs 3–5 times longer than petal blades .... *A. elegantula*
— Flowers as wide as long or wider; sepals horizontally spreading; spurs mainly 5–8 times longer than petal blades .... *A. formosa*
4(2). Sepals and spurs uniformly yellow (or merely suffused with pink); plants of Zion Canyon or of central to northern Utah 5
— Sepals and spurs bicolored, blue and white, red and yellow, or other, or monochrome, but not yellow; plants variously distributed 6
5(4). Spurs subequal in length to the sepals; flowers mainly 2.5–4 cm long; plants of central and northern Utah .. .. *A. flavescens*
— Spurs much longer than the sepals; flowers mainly 5–7 cm long; plants of Zion Canyon .... *A. chrysantha*
6(4). Sepals and spurs pinkish; stamens exceeding the petal blades by 6–13 mm; herbage glaucous; plants of shale outcrops in Uinta Basin .... *A. barnebyi*
— Sepals and spurs white or blue; stamens shorter than petal blades or exceeding them by 2–6 mm; plants glaucous or not but, if so, not of the Uinta Basin 7
7(6). Ultimate leaf segments overlapping, often thickened and glaucous; plants mainly 0.6–4.5 dm tall; sepals and spurs commonly blue .... *A. scopulorum*
— Ultimate leaf segments not overlapping, not thickened, seldom glaucous except below; plants mainly 2–10 dm tall; sepals and spurs white or sometimes blue .... *A. caerulea*

Aquilegia barnebyi Munz Shale Columbine. Plants mainly 3–8 dm tall; stems glabrous below, glandular pubescent above; leaves mainly basal, 0.5–2.7 dm long, triternate, glaucous, glabrous, not especially bicolored; flowers 1 to several, erect or spreading; sepals horizontally spreading, 12–20 mm long, pink, ovate-lanceolate to elliptic; petals with spurs colored like the sepals, the blades yellowish to cream, 6–10 mm long; spurs 13–27 mm long, the blades ca 1/4 as long, rounded; staminodia ca 8 mm long; stamens exceeding the blades by 6–13 mm; follicles 20–25 mm long, glandular-puberulent. Shale outcrops of the Douglas Creek and Parachute Creek (Mahogany beds) of the Green River and Uinta formations in mixed desert shrub, Salina wildrye, pinyon-juniper, and Douglas fir communities at ca 1675 to 2255 m in Duchesne and Uintah counties; Colorado (a Uinta Basin endemic); 22 (i). This singular columbine is distinguished from *A. flavescens* by the sepals being shorter than the spurs and by the shorter, glaucous tuft of basal leaves.

Aquilegia caerulea James Colorado Columbine. Plants mainly 1.5–10 dm tall; stems glabrous or sparsely puberulent below, glandular-pubescent or hirtellous above; leaves mainly basal, 0.5–4 dm long, biternate or rarely triternate, the blades glabrous or puberulent beneath, often paler beneath; flowers 1 to several, erect; sepals horizontally spreading, 2–4 cm long, pale blue to blue (purplish) or white, oblong-lanceolate to elliptic; petals colored like the sepals, but the blades often paler or white; spurs 3–7.3 cm long, the blades 12–23 mm long, truncate; stamens shorter than or slightly surpassing the blades; follicles 5–10, erect, glandular-puberulent, 2–3 cm long; 2n = 14. Sagebrush, pinyon-juniper, mountain brush, aspen, Douglas fir-white fir, aspen-tall forb, spruce-fir, and alpine tundra communities at 1675 to 3660 m in probably all Utah counties; Idaho and Montana, south to Arizona and New Mexico; 167 (xxvii). The common, strikingly beautiful pale-flowered phase of this species in Utah can be called var. *ochroleuca* Hook. [*A. leptocera* Nutt.; *A. caerulea* var. *albiflora* Gray]. The less common plants of higher elevations belong to var. *caerulea* [*A. pinetorum* Tidestr.; *A. caerulea* var. *pinetorum* (Tidestr.) Payson ex Kearney & Peebles; *A. caerulea* ssp. *pinetorum* (Tidestr.) Payson]. Alpine plants are commonly dwarfed. They approach and pass by degree into *A. scopulorum* within its range. This beautiful columbine is the state flower of Colorado.

Aquilegia chrysantha Gray Golden Columbine. Plants mainly 4–12 dm tall; stems glabrous or glandular-puberulent, forming huge clumps; basal leaves 10–60 cm long or more, mostly triternate, glabrous above and glabrous or finely puberulent and glaucous beneath; cauline leaves well developed; pedicels mostly 3–10 cm long; flowers 1 to several, erect, longer than broad; sepals horizontally spreading, 12–25 mm long, clear golden yellow, lance-ovate to lanceolate; petals with spurs colored like the sepals, the blades golden yellow, 8–16 mm long; spurs 4–7 cm long; stamens exceeding blades by 8–10 mm; follicles 5–7, 2–3 cm long. Hanging gardens, stream and seep margins, and other moist sites at ca 1340 m in Zion Canyon, Washington County; Arizona, Colorado, New Mexico, and Mexico; 3 (iii). This pretty yellow columbine hybridizes with *A. formosa*, resulting in varicolored hybrid swarms in some of the hanging gardens.

Aquilegia elegantula Greene Elegant Columbine. Plants mainly 1–4 dm tall; stems glabrous or glandular to glandular-puberulent; leaves mainly basal, 4–15 (25) cm long, biternate, glabrous or glandular, glaucous at least beneath; flowers 1 to few, nodding, cone-shaped, much longer than wide; sepals ascending, 7–11 (15) mm long, usually scarlet, ovate to lanceolate; petals with spurs colored like the sepals, the blades yellow, 6–11 mm long; spurs 15–35 mm long; stamens exceeding the blades by 6–10 mm; follicles 5, 15–20 mm long. Moist woods or in riparian sites in aspen, Douglas fir, and spruce-fir communities at 1735 to 2745 m in Carbon, Emery, Garfield, Grand, Kane, San Juan, and Washington counties; Colorado; 7 (iii).

Aquilegia flavescens Wats. Yellow Columbine. [*A. chrysantha* authors, not Gray; *A. depauperata* Jones, type from Provo Canyon; *A. flavescens* f. *minor* Tidestr., type from Wasatch Plateau]. Plants mainly 1.5–8 dm tall; stems glabrous or glandular-puberulent; leaves mainly basal, 4–40 cm long, biternate (or almost or quite triternate in some), glabrous or finely puberulent or glandular, glaucous beneath; flowers 1 to several, erect, as broad as long or broader; sepals horizontally spreading, 10–25 mm long, yellow or pinkish, ovate to lanceolate or oblong-elliptic; petals with spurs colored like the sepals, the blades yellow (or pink), 5–13 mm long; spurs 10–22 mm long; stamens exceeding blades by 5–11 mm; follicles 5–7, 20–27 mm long. Two more or less distinctive varieties are present.

1. Leaves, stem, and inflorescence glandular hairy; flowers pink with cream or white petal blades; plants of the east margin of the Wasatch Plateau ....
 .... *A. flavescens* var. *rubicunda*

— Leaves and stem glabrous and glaucous, or the stem puberulent to glandular above; flowers clear yellow or as above; plants variously distributed ....
 .... *A. flavescens* var. *flavescens*

Var. *flavescens* Moist woods or riparian sites in mountain brush, aspen, spruce-fir, and mixed montane communities at 1370 to 3265 m in Box Elder, Cache, Davis, Duchesne, Emery, Juab, Millard, San Juan, Salt Lake, Sevier, Summit, Tooele, Utah, and Weber counties; British Columbia and Alberta, south to Oregon and Colorado; 40 (vi). The type is from the Wasatch Mts. The yellowish, less commonly pink, proportionately short flowers, and moderately exserted stamens, are diagnostic. Specimens intermediate with *A. caerulea* are known.

Var. *rubicunda* (Tidestr.) Welsh [*A. rubicunda* Tidestr., type from "Link Trail," west of Emery in Sevier County]. Ponderosa pine, aspen, and spruce-fir communities at 2255 to 2505 m in Emery and Sevier counties; endemic; 4 (i). This phase of *A. flavescens* has long passed as a synonym under *A. micrantha*, primarily because of the smallish flowers and glandular vesture. It is allopatric with that taxon, but is contiguous to (if not sympatric with) *A. flavescens*.

Aquilegia formosa Fisch. in D.C. Western Columbine. Plants 1.5–10.5 dm tall; stems glabrous or glandular-hirtellous above; leaves mainly basal, 4–38 cm long, biternate, glabrous or villous or glandular, green above, paler beneath; flowers usually 2–4, nodding, as broad as long or broader; sepals horizontally spreading, 14–27 mm long, red, reddish, or rarely yellowish; petals with spurs colored like the sepals, the blades yellow, 2–7 (9) mm long; spurs 15–20 (30) mm long; stamens exceeding the blades by 12–17 mm; follicles usually 5, pubescent, 15–25 mm long; 2n = 14. Commonly in riparian or palustrine habitats in sagebrush, mountain brush, pinyon-juniper, ponderosa pine, aspen, and white fir communities at 1370 to 2685 m in Beaver, Box Elder, Iron, Juab, Kane, Millard, Sanpete, Tooele, Uintah (?), and Washington counties; Alaska and Yukon, south to Baja California, Nevada, and Montana; 41 (xi). The truly scarlet perianth parts (except for the petal blades), short, broad, nodding flowers, short petal blades, and long-exserted stamens are diagnostic for this species. An unusual plant, represented by three collections, is present in eastern Washington County; spurs and blades are longer than usual (the measurements in parentheses in the description), leaf segments are more dissected, and the herbage is glandular-pubesent overall. This later material is designated as var. *fosteri* Welsh (type from Zion Canyon).

Aquilegia micrantha Eastw. Alcove Columbine. [*A. ecalcarata* Eastw., not Maxim. or (Hort.) ex Steudel; *A. pallens* Payson, type from La Sal Creek]. Plants 2–7.5 dm tall (or more); stems glandular pubescent throughout, or rarely glabrous below, or sometimes long-pilose; leaves mainly basal, 10–30 cm long, bi- or triternate, commonly glandular-pubescent on one or both surfaces and sometimes sparingly to densely long-pilose beneath; flowers 1 to several, erect or somewhat nodding, about as broad as long; sepals horizontally spreading, 7–20 mm long, white to cream or pale blue (rarely pink?); petals with spurs colored like the sepals, the blades white or cream, 5–13 mm long; spurs 15–35 mm long; stamens subequal to the blades or exceeding them by 4 or 5 mm; follicles usually 5, glandular-hairy, ca 15–20 mm long. Moist alcoves in hanging garden communities and less commonly in seeps and springs not in alcoves at 1125 to 1895 m in Emery, Garfield, Grand, Kane, San Juan (type from Bluff), and Wayne counties; Colorado and Arizona; 45 (xii). The usually glandular herbage, short sepal to spur length ratio, and small flowers appear to be diagnostic. The stamens are less exserted than in *A. flavescens*.

Aquilegia scopulorum Tidestr. Rock Columbine. [*A. caerulea* var. *calcarea* Jones, type from above Cannonville]. Plants 0.6–4.5 dm tall; stems villous to hispidulous throughout or glabrous below; leaves mainly basal, 2–11 (22) cm long, ternate to biternate, the segments distinctly overlapping and crowded, commonly glabrous, glaucous, and thickened, but sometimes green and thin; flowers 1 to several, erect, somewhat longer than broad; sepals horizontally spreading, 13–27 mm long, blue; petals with spurs colored like the sepals, the blades blue to pale blue or white, 8–15 mm long; spurs 21–42 mm long; stamens subequal to the blades or surpassing them by 1–4 mm; follicles usually 5, glandular-hairy, 12–22 mm long. Pinyon-juniper, ponderosa pine, bristlecone pine, spruce-fir, and alpine tundra communities at 2135 to 3420 m in limestone or on igneous scree slopes, in Garfield, Juab, Kane, Piute, Sanpete (type from Wasatch Peak), and Sevier counties; Nevada; 20 (v). The overlapping leaf segments and usually blue flowers appear to be diagnostic for most specimens, but plants from the Tushar Mts., growing on igneous scree slopes have green leaves and are completely transitional with *A. caerulea*. Possibly the best status for this attractive taxon would be to regard it as *A. caerulea* var. *calcarea*.

Caltha L.

Perennial subrhizomatous herbs with fibrous roots; stems short, erect; leaves basal or alternate, petiolate, the blades simple, cordate to oval-cordate or ovate; flowers perfect, regular, showy, solitary or 2 or 3; sepals 5–12, petaloid; petals lacking; stamens numerous, rarely petaloid and resulting in "double flowers;" pistils 5 to many; fruit of subsessile follicles.

Caltha leptosepala DC. Marsh-marigold. Plants erect, 0.5–5 dm tall; basal leaves 4–30 cm long, the petioles 2–22 cm long, the blades 1.5–8 cm long (from sinus to apex), 1.3–4.5 cm wide, oval- to oblong-cordate, crenate to subentire; stems usually with a single leaf; peduncles 1 or 2, 4–33 cm long, 1-flowered; sepals white, tinged bluish or purplish externally, 2–27 mm long; follicles subsessile, 12–15 mm long; 2n = 48, 96. Wet meadows, streamsides, and around seeps and springs in aspen, lodgepole pine, spurce-fir and alpine tundra communities at 2315 to 3600 m in Beaver, Box Elder, Cache, Daggett, Duchesne, Emery, Garfield, Iron, San Juan, Salt Lake, Sanpete, Sevier, Summit, Uintah, Utah, Wasatch, and Wayne counties; Alaska and Yukon, south to Oregon, Nevada, and Colorado; 67 (xiv). Our material belongs to var. *leptosepala*. The broad-leaved *C. biflora* DC. is reported for Utah (Pacific N. W. Flora 2: 336. 1964), and is approached by some specimens, especially those taken by Lewis (7402; 7603 BRY) from Sanpete County. Specimens transitional to those occur among our herbarium materials and no segregation appears possible. Beyond Utah the species are distinct.

Clematis L.

Perennial herbs or woody vines; stems erect or clambering; leaves opposite, pinnately to ternately compound, some petioles often twining or clasping; flowers small to large, solitary, or few to many in compound cymes, perfect or imperfect (the plants then dioecious or polygamous); sepals 4 or 5, valvate in bud; petals lacking; stamens many, the outer sometimes broadened and petaloid; pistils many, 1-ovuled, forming 1-seeded, long-tailed achenes.

Pringle, J.S. 1971. Taxonomy and distribution of *Clematis*; Sect. Atragene (Ranunculaceae), in North America. Brittonia 23: 361–393.

1. Flowers few to many in cymose clusters; sepals white, 5–8 mm long; plants widespread at lower elevations . . . . *C. ligusticifolia*
— Flowers solitary; sepals yellow to blue or purple; plants of various distribution 2

2(1). Stems erect, not at all vinelike; leaves pinnately compound; sepals commonly densely hairy dorsally, thickened . . . . *C. hirsutissima*
— Stems commonly viney; leaves ternate to biternate; sepals glabrous or sparingly hairy dorsally, thin 3

3(2). Sepals yellow; plants of lower elevations, commonly along roadsides, introduced . . . . *C. orientalis*
— Sepals blue to purple or rarely white; plants of middle to higher elevations, indigenous 4

4(3). Leaves bi- or triternately compound, the ultimate segments usually sharply toothed; plants widespread . . . . *C. columbiana*
— Leaves merely ternate, the leaflets bluntly toothed; plants of northern Utah . . . . *C. occidentalis*

Clematis columbiana (Nutt.) T. & G. [*Atragene columbiana* Nutt.; *C. pseudoatragene* (ssp.) *pseudoalpina* Kuntze; *C. pseudoalpina* (Kuntze) A. Nels. in Coult. & Nels.; *C. pseudoatragene* (ssp.) *wenderothioides* Kuntze, type from Kane County; *C. alpina* (ssp.) *occidentalis* 2. *repens* Kuntze, type from American Fork Canyon; *C. alpina* var. *occidentalis* subvar. *tenuiloba* Gray in Newton & Jenney; *C. tenuiloba* (Gray) C. L. Hitchc.; *C. columbiana* var. *tenuiloba* (Gray) Pringle]. Plants perennial; stems woody, trailing or sometimes short; herbage glabrous to long-hairy; leaves petiolate, bi- or triternate, the ultimate segments usually lobed or toothed; peduncles 3–19 cm long, recurved near the apex; flowers solitary, bell shaped, nodding; sepals 4, violet blue to lilac, fading red purple, lanceolate to ovate, 12–60 mm long, commonly hairy dorsally; staminal filaments petaloid; achenes ca 5 mm long; plumose styles 20–55 mm long; 2n = 16. Sagebrush, mountain brush, ponderosa pine, Douglas fir-white fir, aspen, and spurce-fir communities at 1830 to 3295 m in Cache, Carbon, Duchesne, Emery, Garfield, Grand, Iron, Kane, Piute, Salt Lake, San Juan, Sanpete, Sevier, Summit, Utah, Wasatch, Washington, Wayne, and Weber counties; Montana to North Dakota, south to Arizona and New Mexico; 78 (xv). Our material long passed under the name of *C. pseudoalpina* until problems of typification were resolved by Pringle (1971), who points to the similarity of North American plants to those of Old World *C. alpina* (L.) Miller, accounting in part for the occurrence of that name among the synonyms. Segregation of specimens into two varieties on the basis of much dissected leaves and short aerial stems (var. *tenuiloba*) versus not much dissected leaves and developed aerial stems (var. *columbiana*) seems arbitrarily possible only and does not seem to be geographically correlated.

Clematis hirsutissima Pursh Lions-beard. [*Virona hirsutissima* (Pursh) Heller; *C. wyethii* Nutt.; *C. douglasii* ssp. *jonesii* Kuntze, type from American Fork Canyon, Utah County]. Perennial, erect herbs from a branched caudex; stems 2–7 dm tall, simple or branched above; herbage usually long-hairy; leaves usually 2–6 pair, the lowermost bladeless and sheathing, pinnately compound, the primary divisions again once or twice dissected into narrowly elliptic, oblong, or linear segments; peduncles 2–15 cm long, recurved apically in flower, erect in fruit; flowers solitary (seldom more on axillary upper branches), perfect; sepals 4, thickened, urceolate-campanulate, brownish purple to purple, grayish villous-lanate dorsally, purple within, 2–5 cm long, connate basally, the tips free and recurved; stamens included, the filaments villous; styles plumose, 20–50 mm long. Sagebrush, mountain brush, ponderosa pine, aspen, aspen-tall forb, and spruce-fir communities at 1860 to 3175 m in Cache, Carbon, Daggett, Duchesne, Emery, Grand, Salt Lake, San Juan, Sanpete, Sevier, Summit, Uintah, Utah, and Wasatch counties; British Columbia to Montana, south to Oregon and Colorado; 58 (vii).

Clematis ligusticifolia Nutt. White Virgins-bower. Vigorous, dioecious, woody vines to 10 m long or more; herbage glabrous to strigose or villous; leaves pinnately compound with 3–7 leaflets 2–8 cm long, these lanceolate to ovate, coarsely few-toothed and often lobed or entire; flowers few to many in bracteate cymes; sepals white to cream, 6–11 mm long, elliptic to oblanceolate; staminate flowers lacking pistils; pistillate flowers with sterile stamens; achenes villous; styles 2–4.2 cm long; 2n = 16. Commonly in riparian communities at 1125 to 2380 m in all Utah counties except Davis, Morgan, Salt Lake, Summit, and Wasatch; British Columbia to North Dakota, south to California, Arizona, and New Mexico; 79 (xiii). This plant festoons willows, cottonwoods, and other trees and shrubs in riverbottom communities, but curiously it is missing from the central Wasatch Front.

Clematis occidentalis (Hornem.) DC. Purple Virgins-bower. [*Atragene occidentalis* Hornem.; *C. columbiana* authors, not (Nutt.) T. & G., q.v.]. Plants woody, sprawling or clambering vines, mainly 1–30 dm long; herbage pilose, finally glabrate; leaves ternate, the 3 leaflets lanceolate to ovate or orbicular, 2–11 cm long, entire to crenate, occasionally lobed; peduncles 2.5–15 cm long, recurved near the apex in flower, erect in fruit; flowers solitary, more or less campanulate; sepals 4, blue to violet or lilac, rarely white, lanceolate to elliptic or ovate, 2.5–6 cm long, ciliate and sometimes hairy externally; stamens with petaloid filaments; achenes ca 5 mm long; styles plumose 30–70 mm long. Moist woods and riparian habitats in mountain brush, aspen, and spruce-fir communities at 2010 to 2715 m in Cache, Daggett, Duchesne, Salt Lake, Summit, Uintah, and Utah counties; British Columbia to Montana, south to Oregon and Colorado; 28 (0).

Clematis orientalis L. Oriental Clematis. [*C. aurea* Nels. & Macbr.?]. Vigorous clambering vines to 30 dm long or more; herbage sparingly villous to glabrous; leaves ternate or pinnate-ternate, the segments ovate to lanceolate or elliptic, 1–5 cm long, coarsely toothed to lobed or entire; peduncles 6–12 cm long; flowers solitary, nodding to erect; sepals 4, yellow (or tinged reddish), oblong to lanceolate, 15–25 mm long, pubescent, ultimately spreading; stamens with petaloid, pubescent filaments; achenes 3–5 mm long; styles plumose, 3–10 cm long; 2n = 16, 32. Sagebrush, mountain brush, and ruderal habitats at 1370 to 2290 m in Cache, Juab, Millard, Salt Lake, Tooele, Uintah, and Utah counties; sporadic in the U. S.; introduced from Asia; 18 (0). This plant is now well established in Utah.

Delphinium L.

Annual or perennial herbs from slender or tuberous roots; stems simple (rarely branched), erect; leaves alternate or mostly basal, the blades palmately divided; flowers perfect, irregular, large and showy, borne in terminal racemes or panicles, subtended by a pair of bracts; sepals 5, petaloid, the upper one produced into a prominent spur, the lateral ones often shorter than the lower 2; petals 4, in 2 pair, the upper ones spurred and clawless, the lower ones clawed and with expanded blades; stamens numerous; pistils 3–5; fruit a follicle.

Ewan, J. 1945. A synopsis of the North American species of *Delphinium*. Univ. Colorado Stud. Ser. D. 2: 55–244.

1. Plants annual or short-lived perennial; lowermost leaves often deciduous at anthesis; adventive weedy or cultivated plants .... A. *ajacis*
— Plants perennial; leaves mainly basal and persistent at anthesis or mainly cauline and with broadly cuneate lobes; indigenous 2
2(1). Leaves mainly cauline, the lowermost often withered at anthesis; leaf segments cuneate, often more than 1 cm broad; plants often over 1 m tall, of montane habitats . ... *D. occidentale*
— Leaves mainly basal, much reduced upwards, the lowermost often persistent at anthesis; leaf segments variously shaped, but less than 1 cm wide; plants less than 1 m tall, of xeric to montane habitats 3
3(2). Stems conspicuously narrowed below ground level, easily separable from the fusiform tuberous roots; plants ubiquitous, common .... *D. nuttallianum*
— Stems not especially narrowed below the ground level or easily separable from the caudex; roots slender, elongate, not tuberous or fusiform 4
4(3). Herbage puberulent with curved or crinkly hairs; pedicels spreading-ascending, much longer than the spurs; plants of northeastern Utah .... *D. geyeri*
— Herbage glabrous or essentially so; pedicels erect-ascending, longer to shorter than the spurs; plants of southeastern, southern, and western Utah . *D. andersonii*

Delphinium ajacis L. Rocket Larkspur. Plants annual; stems 1.5–7 dm tall or more, glabrous or puberulent above, simple or branched above; basal leaves often withered at anthesis, the blade divided into 3 sections and these again divided into numerous linear segments; pedicels curved ascending, 1–6 cm long, subtended by a subulate bract and with another pair near the middle; flowers blue to violet, rose, pink, or white, the spurs 10–20 mm long; pistils mainly solitary; follicles usually solitary, pubescent, 12–25 mm long; 2n = 16. Cultivated ornamental, escaped and persistent in Cache, Juab, Millard, Salt Lake, Sevier, and Utah counties; widely grown in the U. S.; introduced from Europe; 11 (0).

Delphinium andersonii Gray Anderson Larkspur. Perennial herbs from fibrous coarse roots and caudices, mainly 10–60 cm tall; stems simple, glabrous or sparingly long hairy in the inflorescence; leaves mainly basal or on lower portion of the stem, usually at least some persistent at anthesis, much reduced upwards; blades mainly 1–6 cm wide, thickened, usually thrice dissected, the ultimate segments to 4 mm wide; inflorescence a raceme or panicle, the main raceme with 1–15 flowers, lower pedicels shorter than or to 4 times longer than the spur or longer, ascending to erect; sepals blue (pale or purplish), 9–15 mm long; follicles 12–25 mm long, erect, glabrous or puberulent. Two morphologically similar geographic races are present in Utah.

1. Plants subscapose, the cauline leaves often reduced to mere rudiments upward; flowers pale blue or sometimes dark blue; distribution in southeastern and southern Utah
    .... *D. andersonii* var. *scaposum*
— Plants not especially subscapose, the cauline leaves reduced upwards, but not especially rudimentary; flowers usually dark blue; distribution in the Great Basin .....
    .... *D. andersonii* var. *andersonii*

Var. *andersonii* [*D. cognatum* Greene; *D. andersonii* ssp. *cognatum* (Greene) Ewan; *D. leonardii* Rydb., type from Garfield, Salt Lake County]. Pinyon-juniper, sagebrush, mixed desert shrub, and mountain brush communities in Beaver, Box Elder, Juab, Millard, Piute, Sanpete, Sevier, Tooele, and Utah counties; Oregon to Montana, south to California and Nevada; 40 (v). Plants of this and the next variety have long been confused. In their extremes they are easily separable, but when dark flowered plants occur within the range of the usually pale flowered variety their identification beyond specific rank is dubious; hence, the presentation herein. In the monograph by Ewan three largely sympatric taxa were recognized, but our material seems better segregated on geographic and morphological basis into two very similar taxa.

Var. *scaposum* (Greene) Welsh [*D. scaposum* Greene; *D. coelestinum* Rydb., type from Washington County; *D. amabile* Tidestr.]. Joshua tree-creosote bush, blackbrush, shadscale, Vanclevea-ephedra, and pinyon-juniper communities in Emery, Garfield, Grand, Kane, San Juan, Washington, and Wayne counties; Colorado, New Mexico, Arizona, and Nevada; 74 (xii).

Delphinium geyeri Greene Geyer Larkspur. Plants 1.5–9 dm tall, from a woody caudex and stout fibrous roots; stems not or seldom fistulous, simple, crisp-puberulent with curved hairs; leaves mainly basal or lower cauline, 4–25 cm long, the blades divided into linear to oblong segments, mainly 2–4 mm wide; inflorescence simple or paniculate, the main axis 7–32 cm long, the ascending to erect pedicels longer or shorter than the spurs; calyx spreading hairy or with curved subappressed hairs, blue, blue purple, or white; lateral sepals 11–19 mm long, elliptic to oblanceolate, obtuse; spurs 11–19 mm long; blade of upper sepal spreading-ascending; follicles hirtellous, 12–15 mm long; 2n = 16, 32. Salt desert shurb, sagebrush, pinyon-juniper, mountain brush, and aspen communities at 1585 to 2745 m in Daggett, Summit, and Uintah counties; Wyoming, Nebraska, and Colorado; 18 (ii).

Delphinium nuttallianum Pritz. Nelson Larkspur. [*D. nelsonii* Greene; *D. menziesii* authors, not DC.; *D. menziesii* var. *utahense* Wats., type from Wasatch Mts.; *D. nelsonii* ssp. *utahense* (Wats.) Ewan; *D. pinetorum* Tidestr.; *D. nelsonii* f. *pinetorum* (Tidestr.) Ewan]. Plants 0.9–9.2 dm tall, from clustered tuberous roots; stems not fistulous (or rarely so), simple, glabrous to crisp puberulent or hirtellous; leaves mainly lower cauline, 2–15 cm long, the blades divided into linear oblong to elliptic lobes mainly 2–5 mm wide; inflorescence simple or paniculate, the main axis 2–35 cm long, the ascending to erect-ascending pedicels longer than to shorter than the spurs; calyx hirtellous, blue purple (rarely pale blue or white); lateral sepals 9–17 mm long, oblong-elliptic to oblanceolate, obtuse to rounded; spurs 12–19 mm long; blade of upper sepal spreading-ascending; follicles hirtellous, puberulent, or glabrous, 9–19 mm long (including the beak); 2n = 16. Blackbrush, mixed desert shrub, pinyon-juniper, mountain brush, sagebrush, ponderosa pine, aspen, grass-forb, Douglas fir, spruce-fir, and meadow communities at 1125 to 3175 m in probably all Utah counties; British Columbia to Alberta, south to California, Arizona, New Mexico, and Nebraska; 203 (xxvii). This is our most common and most ubiquitous larkspur species. Despite its commonness and ubiquity this species has been known under three specific epithets—*menziesii*, *nelsonii*, and *nuttallianum*, and, even now, the nomenclatural problem is not readily resolved. Should our plants be treated as conspecific with *D. menziesii*, with which they are reported to intergrade (Vascular Plants Pacific N. W. 2: 357. 1964), that name (having priority) must be used. In an attempt at a stable nomenclature the name *nuttallianum* is used herein. Recognition of infraspecific categories among our specimens seems unwarranted.

Delphinium occidentale (Wats.) Wats. Western Larkspur. [*D. elatum* var. *occidentale* Wats., type from the Wasatch Mountains; *D. cuculatum* A. Nels.; *D. occidentale* ssp. *cuculatum* (A. Nels) Ewan; *D. occidentale* var. *cuculatum* (A. Nels.) R. J. Davis; *D. abietorum* Tidestr., type from Wasatch Plateau]. Plants perennial, 6–20 dm tall or more; stems often fistulose, simple, often glaucous, glabrous or crisp puberulent below and crisp puberulent to glandular hirtellous above; leaves mostly cauline, the lower ones usually withered at anthesis; blades 3–20 cm wide, cleft to the base into 3 main lobes, the lateral ones again cleft (not as deeply) into 2 main, cuneate, lobed, cleft, or toothed broad segments; inflorescence simple or paniculate, the main axis 10–35 cm long, the pedicels shorter to somewhat longer than the spurs; calyx crisp puberulent to glandular hirtellous or glandular longhairy, blue purple to rose or white; lateral sepals variably shaped, 7–20 mm long, rounded-acute to acute or attenuate; spurs 10–17 mm long; blade of upper sepal erect to ascending or spreading-ascending; follicles glabrous to puberulent, or glandular-hirtellous; 2n = 16. Two varieties are recognized.

1. Axis of racemes and pedicels with lustrous, long spreading glandular hairs; lateral sepals lanceolate to lance-oblong, 13–20 mm long, acute to acuminate; follicles mostly glabrous .... *D. occidentale* var. *barbeyi*

— Axis of racemes and pedicels with crisp puberulent curved or crinkly hairs or sparingly short glandular hairs; lateral sepals ovate, oblong, or lance-elliptic, 7–15 mm long, rounded-acute; follicles puberulent to glandular hirtellous .... *D. occidentale* var. *occidentale*

Var. *barbeyi* (Huth) Welsh [*D. exaltatum* var. *barbeyi* Huth; *D. barbeyi* (Huth) Huth; *D. scopulorum* var. *attenuatum* Jones, type from Bullion Creek; *D. barbeyi* f. *subroseum* (Cockerell) Ewan]. Mountain brush, ponderosa pine, aspen, spruce-fir, alpine meadow, and alpine tundra communities at 2590 to 3450 m in Beaver,

Garfield, Iron, Kane, Piute, San Juan, Sanpete, and Sevier counties; Colorado, New Mexico, and Arizona; 44 (x). Similarity of this attractive poisonous plant to *D. occidentale* has long resulted in confusion as to taxonomic limits of both. They are partially sympatric, overlapping in Sanpete County. Presence of glandular hairs in both taxa compounds the difficulty. The glandular hair in var. *barbeyi* is long, yellowish, lustrous, and usually copious; in var. *occidentale* with glandular pubescence the hairs are short, lustrous or not, yellowish or whitish, and usually sparse. Plants with hairs long and copious or with long spreading sepals or spreading-ascending blades of upper sepals occur beyong the range of var. *barbeyi* as strictly interpreted. Because of these problems *barbeyi* is treated at varietal rank.

Var. *occidentale* Western Larkspur. Mountain brush, sagebrush, Douglas fir-white fir, aspen, grass-tall forb, and spruce-fir communities at 2135 to 3115 m in Box Elder, Cache, Carbon, Daggett, Duchesne, Emery, Juab, Millard, Rich, Salt Lake, Sanpete, Tooele, Uintah, Utah, Wasatch, and Weber counties; Oregon to Montana, south to Nevada and Colorado; 58 (iv). The western larkspur has been regarded as *D. scopulorum* ssp. *occidentale* (Wats.) Abrams, indicating a portion of a broader problem than can be solved on the basis of Utah plants alone.

Myosurus L.

Diminutive tufted, scapose herbs; leaves basal, entire or toothed, linear to narrowly spatulate; flowers minute, greenish or purplish to whitish, solitary; sepals 5 (6 or 7), spurred at the base; petals lacking or 6 or 7; stamens 5 to many; pistils many on a slender tapering axis, this elongate in fruit; achenes apiculate or aristate.

Campbell, G. R. 1952. The genus *Myosurus* in North America. El Aliso 2: 389–403.

1. Achenes cupulate dorsally, the beak arising from the horseshoe-shaped cup .... *M. cupulatus*
— Achenes ridged dorsally, the beak a continuation of the ridge 2
2(1). Achene beak 0.8–2 mm long, usually more or less spreading .... *M. apetalus*
— Achene beak 0.2–0.5 mm long, straight ..... *M. minimus*

Myosurus apetalus C. Gay Mousetail. [*M. aristatus* Benth. ex Hook.; *M. minimus* ssp. *montanus* G. R. Campbell]. Plants slender, 1–6 (10) cm tall; leaves 1–4 (7) cm long, 0.5–1.5 mm wide, linear to narrowly spatulate, entire or toothed; scapes 1 to several; flowers solitary; sepals 5, greenish or purplish, narrowly oblong, 1–3 mm long, spurred basally, the spur 1–3 mm long; petals 5 or lacking, whitish, ca 1–1.5 mm long, caducous; stamens usually 5; achenes 20–50 or more, borne on a receptacle mainly 0.4–2 cm long, ca 2 mm wide; dorsal keel of achene projecting beyond the body as a usually spreading-ascending beak 0.7–2 mm long. Lake and pond margins and along drainages at 2075 to 2950 m in Box Elder, Cache, Daggett, Duchesne, Juab, San Juan, Salt Lake, Sanpete, Sevier, Uintah, Utah, and Weber counties; British Columbia to Montana, south to California and Wyoming; South America; 15 (0).

Myosurus cupulatus Wats. Horseshoe Mousetail. Plants slender, 2–15 cm tall; leaves 0.7–7 cm tall, 0.5–1.5 mm wide, linear to narrowly spatulate, entire; scapes 3 to many; flowers solitary; sepals 5, whitish to greenish or brownish, 1.5–3 mm long, spurred basally, the spur 1–2 mm long; petals 5, ca 2–2.5 mm long, caducous; stamens ca 5; achenes 20–50 or many more; borne on a receptacle 7–40 mm long, ca 2.5–4 mm wide; beak of achene arising from a depression surrounded by a spongy thickened margin, the beak, ca 0.5–0.8 mm long. Mixed desert shrub, mountain brush, and sagebrush communities at 850 to 2135 m in Beaver, Iron, Millard, Tooele, and Washington counties; Arizona and California 9 (i). Note: the closely related *M. nitidus* Eastw. is known from the Kaibab Plateau and from Montezuma County, Colorado. It might occur in Utah adjacent to those regions. It has fewer than 4 scapes per plant and longer (ca 1 mm) slender achene beak from a less pronounced or non-cupulate achene.

Myosurus minimus L. Tiny Moustail. Plants slender, mainly 3–10 cm tall or more; leaves 2–5 mm long, 0.5–1.5 mm wide, narrowly spatulate to linear, entire; scapes 1 to numerous; flowers solitary; sepals 5, whitish to greenish, 2–3 mm long, the spur 1–3 mm long; petals 5, linear, caducous; stamens usually 10; achenes 30–100 or more, borne on a receptacle projecting beyond the body as a beak 0.3–0.5 mm long; $2n = 16$. Pond margins, drainage bottoms, and other moist sites at ca 1735 to 2380 m in Cache, Daggett, Salt Lake, San Juan, and Summit counties; widely distributed in North America; circumboreal; 4 (0). This species is sometimes delimited to include *M. apetalus* (*M. aristatus*) at infraspecific level, but that solves no problem, rather, the problem is then reduced to separation of subspecies or varieties. There do not seem to be any real problems with our Utah materials once ssp. *montanus* G. R. Campbell is placed in synonymy with *M. apetalus*.

Ranunculus L.

Perennial, biennial, or annual aquatic or terrestrial herbs with fibrous to tuberous roots; stems erect, ascending or sometimes prostrate and stoloniferous; leaves basal or cauline and alternate, simple or compound; flowers perfect, regular, solitary or few to several in cymes; sepals 3–5, herbaceous or petaloid, usually deciduous; petals 5–16, rarely more; stamens 5 to numerous, sometimes the outer ones petaloid; pistils 5 to many; fruit an achene.

Benson, L. 1948. A treatise on the North American Ranunculi. Amer. Midl. Nat. 40: 1–264.
Padmore, P. A. 1957. The varieties of *Ranunculus flammula* L. and the status of *R. scoticus* E. S. Marshall and *R. reptans* L. Watsonia 4: 19–27. 1957.
Palmieri, M. D. 1976. A revision of the genus *Ranunculus* (Ranunculaceae) for the state of Utah. Unpublished M. S. thesis; Brigham Young University. 141 pp.

1. Plants aquatic; petals white; submersed leaves divided into linear-filiform segments .... *R. aquatilis*
— Plants terrestrial or aquatic; petals yellow or red; leaves variable, but not with linear-filiform segments .. 2
2(1). Plants tomentose annuals; stamens commonly 5–7; fruit 3-loculed, the lateral locules empty .... *R. testiculatus*
— Plants perennial or, if annual, not as above; stamens commonly 10 or more; fruit 1-loculed 3
3(2). Petals and sepals red purple or lined with red purple; leaves dissected; plants of lower elevation habitats in desert ranges .... *R. andersonii*

	Petals yellow, the sepals often greenish or brownish; leaves various; distribution various, but seldom in lower elevation desert sites 4
4(3).	Leaves all entire or merely crenate, not lobed or dissected or, if 3- to 5-lobed, the lobes rounded 5
—	Leaves, leaflets, or bracts dissected or lobed (the lower sometimes merely crenate) 10
5(4).	Plants stoloniferous, rooting at the nodes, growing in palustrine or riparian habitats 6
—	Plants not stoloniferous, of various habitats, sometimes as above 7
6(5).	Leaf blades entire, elliptic, several times longer than broad; plants of high elevations .... *R. flammula*
—	Leaf blades crenate, cordate-oval, the length subequal to the width; plants seldom of high elevations . ... *R. cymbalaria*
7(5).	Leaf blades with 3–5 broadly rounded teeth or lobes, usually broader than long; plants aquatic or semiaquatic 8
—	Leaf blades entire or essentially so, much longer than broad; plants not aquatic or semiaquatic 9
8(7).	Leaf blades mostly 3–8 mm long; achenes 10–20, receptacle mainly 1.5–3 mm long ..... *R. hyperboreus*
—	Leaf blades mostly 4–15 mm long; achenes 20–60; receptacles 3–5 mm long .... *R. nutans*
9(7).	Petals 5–7 mm broad or more; achenes ca 50–100 in an ovoid cluster 10–12 mm long, 9–10 mm thick; achenes 3–3.2 mm long; plants of Kane and San Juan counties .... *R. oreogenes*
—	Petals 2–5 mm broad; achenes 10–50 in a hemispherical to subglobose cluster 4–7 mm long, 6.5–8.5 mm thick; achenes 1.5–2.5 mm long .... *R. alismifolius*
10(4).	Plants aquatic or semiaquatic; leaves often dimorphic, at least some finely dissected; stems prostrate and rooting at the nodes or erect and fistulous; achenes glabrous, beakless or essentially so 11
—	Plants neither aquatic nor semiaquatic and with finely dissected leaves, but sometimes growing in wet places; stems mostly erect, seldom rooting at the nodes; achenes either hairy or with a beak more than 0.5 mm long 13
11(10).	Plants annual, erect, the stems fistulous; achenes beakless or essentially so .... *R. sceleratus*
—	Plants perennial, the stems usually prostrate and rooting, not fistulous; achenes with a beak 0.3–0.5 mm long 12
12(11).	Leaf blades 2–5 cm long or more; petals 7–12 mm long; achenes 1.8–2 mm long, prominently keeled; beak 1–2 mm long .... *R. flabellaris*
—	Leaf blades 1–2 cm long; petals 3–7 mm long; achenes 1–1.5 mm long; not prominently keeled; beak usually less than 1 mm long .... *R. gmelinii*
13(10).	Plants annual; achenes 5–8 in a globose cluster, compressed, 4–7.5 mm long (including beaks), armed with stout spines; adventive .... *R. arvensis*
—	Plants perennial; achenes commonly more than 10 and less than 4 mm long, unarmed 14
14(13).	Roots tuberous, clavate, 2–5 mm thick; plants low (mainly 2–12 cm tall), flowering in early springtime in sagebrush-mountain brush habitats .... *R. jovis*
—	Roots not especially thickened, tapering or linear, mainly less than 2 mm thick; plants of various height, flowering in springtime or summer, of various habitats 15

15(14).	Basal leaf blades entire or merely crenately toothed or lobed, or doubly so 16
—	Basal leaf blades divided, parted, or cleft, the divisions one-half or more to the base 18
16(15).	Blades of basal leaves mainly 2–4 times as long as wide, entire or shallowly 3-lobed ..... *R. glaberrimus*
—	Blades of basal leaves about as broad as long, crenate to crenate-serrate 17
17(16).	Cauline leaf segments 3–10 mm broad or more; petals mostly 5–7 mm long .... *R. inamoenus*
—	Cauline leaf segments 1–3 mm wide; petals 8–10 mm long or lacking .... *R. cardiophyllus*
18(15).	Plants mainly 5–20 cm tall; leaf blades commonly 1–3 cm long; habitats at high elevations 19
—	Plants commonly 20–60 cm tall; leaf blades mainly 2–8 cm long; habitats various, but sometimes at high elevations 22
19(18).	Basal leaves with 3 main lobes, the lateral lobes again merely toothed or lobed .... *R. eschscholtzii*
—	Basal leaves with 3 main lobes but the lateral lobes again deeply divided or parted into linear or oblong segments 20
20(19).	Stems abruptly horizontally spreading, curved from near the middle; leaf blades 0.5–1 cm long, the ultimate segments oblong; petals 4–6 mm long .. *R. gelidus*
—	Stems erect; leaf blades various, the ultimate segments linear to narrowly oblong; petals various 21
21(20).	Leaf blades 1–4 cm long, the ultimate segments linear to narrowly oblong; petals 6–12 mm long; plants locally common and more widespread ..... *R. adoneus*
—	Leaf blades 0.8–1.2 cm long; petals 4–6 mm long; plants evidently rare in the Uinta Mts. .. *R. pedatifidus*
22(18).	Leaves compound, the primary divisions petiolulate; herbage glabrous .... *R. ranunculinus*
—	Leaves palmatifid or, if compound, the primary divisions sessile; (stalked in *R. repens*); herbage variously hairy 23
23(22).	Basal leaves apparently pinnately 5- to 7-lobed; achene beaks 2–4 mm long .... *R. orthrorhynchus*
—	Basal leaves pinnately ternate, palmatifid, or palmately 3-lobed (the lobes often again lobed); achene beaks mainly 0.3–2 mm long, often hooked or recurved 24
24(23).	Petals not much if at all surpassing the sepals, mostly less than 7 mm long .... *R. macounii*
—	Petals much surpassing the sepals, mostly 7–12 mm long or more 25
25(24).	Basal leaves with stalked primary divisions, compound; stems rooting at lower nodes, often prostrate or decumbent; flowers single or double ..... *R. repens*
—	Basal leaves palmatifid or palmately compound with sessile primary segments; stems not rooting at lower nodes and not prostrate to decumbent 26
26(25).	Basal leaves pentagonal in outline, the blades incised into short acute teeth or lobes; achene body 2–2.5 (3) mm long, the beaks 0.5–0.6 mm long, curved; sepals spreading to reflexed .... *R. acris*
—	Basal leaves not especially pentagonal in outline, the blades incised into broad blunt to acute segments; achene body 2.5–3.5 mm long, the beaks 0.2–0.6 mm long, abruptly curved; sepals reflexed .... *R. acriformis*

Ranunculus acriformis Gray Sharp Buttercup. Terrestrial perennials, mainly 2.5–6 dm tall; stems spreading-hirsutulous, 2–2.5 mm wide, erect (the base sometimes decumbent), not rooting at the nodes; basal leaf blades simple, ternate, twice divided, some segments again lobed or parted, the ultimate segments linear-lanceolate, acute or abruptly obtuse, 2–5 mm wide, reniform in outline and often more or less folded when pressed, 3–5 cm wide, 1–5 cm long; petioles 5–10 cm long; cauline leaves alternate, similar to the lower; bracts linear or with 3 linear lobes; pedicels 5–10 cm long, sparingly long-hairy; sepals 5, yellowish green, 4–5 mm long, dorsally pilose, reflexed, deciduous, petals 5, yellow, 7–11 mm long; achenes 20–40 in a hemispheric cluster 5–7 mm high and 6–8 mm wide, the bodies 3–3.4 mm long, with beaks 0.2–0.6 mm long, abruptly curved; receptacle glabrous. Wet meadows adjacent to cold bogs and springs at 1950 to 2595 m in Carbon, Emery, Sanpete, and Wasatch counties; Idaho and Montana, south to Wyoming and Colorado; 7 (0). Materials from Emery and Wasatch counties apparently are more nearly related to the body of the species to the north and east.

Ranunculus acris L. Tall Buttercup. Terrestrial perennials mostly 2–5 dm tall; stems spreading long-hairy below, the hairs appressed upwards, 1.5–3 mm wide, erect, not rooting at the nodes; basal leaf blades simple, 3-parted, the 3 main segments again lobed or toothed, pentagonal in outline, the ultimate segments mostly 1–7 mm wide; petioles 3–15 cm long; cauline leaves simmilar but more shortly petiolate; bracts with 1–3 linear segments, sessile; pedicels 2–20 cm long or more; sepals 5, greenish, or marginally yellowish, 4–5.5 mm long, pilosulous, spreading deciduous; petals 5, yellow, 8–10 mm long, obovate; achenes glabrous, 2.6–3 mm long, ca 50, in a subglobose cluster, 5.5–6 mm high and 6–7.5 mm wide; beaks ca 0.5–0.6 mm long; receptacle hirtellous; n = 14, 28. Two varieties, one indigenous and the other introduced are present in Utah.

1. Leaf segments narrowly cuneate, the ultimate segments mainly 1–4 mm wide; plants of Garfield County . . . .
 . . . . R. acris var. aestivalis
— Leaf segments broadly cuneate to obtuse basally, the ultimate segments more than 4 mm wide; plants of northern Utah . . . . R. acris var. acris

Var. acris Riparian habitats at 1460 to 2170 m in Box Elder, Cache, and Rich counties; naturalized widely in North America; adventive from the Old World; 3 (0).

Var. aestivalis (L. Benson) Welsh [R. acriformis var. aestivalis L. Benson, type from north of Panguitch].. Sedge-grass meadow at ca 1964 m in Garfield County; endemic; 4 (0). The similarity of this taxon to R. acris, sens. lat., seems to override considerations of close relationship to R. acriformis, with which it shares similar achene characteristics. However, the achene features of both acris and acriformis are hardly convincing as diagnostic characters. The acris complex includes various taxa, often recognized at specific rank in North America, including R. acriformis, and it is hardly surprising that a phase of the species should occur in the western U. S., at the southern margin of the species complex proper. The R. acris complex should be treated monographically, an attempt far beyond the scope of this work.

Ranunculus adoneus Gray Alpine Buttercup. Terrestrial perennials, mostly 0.5–3.2 dm tall; stems glabrous, 0.5–4 mm wide, ascending to erect; basal leaf blades simple, several times divided into usually linear segments 0.7–3 mm wide, the whole 1–5 cm long and about as broad; petals 1–9 mm long; cauline leaves similar to the basal but with progressively shorter petioles; bracts short-petioled to subsessile commonly 5 to many lobed; pedicels 0.5–14 cm long, glabrous; sepals 5, green or brownish to lavender, 3.5–10 mm long, pilose or glabrous, deciduous, spreading; petals 5–12, yellow, 5–15 mm long; achenes glabrous, 2.5–4 mm long, 40–70 in an ovoid cluster 5–10 mm long and 3–9 mm thick; beaks 1.4–1.9 mm long, straight or curved; receptacle glabrous. Moist sites near melting snow in open meadows, talus slopes, and in spruce-fir communities at 2400 to 3660 m in Cache, Daggett, Duchesne, Juab, Salt Lake, Sanpete, Summit, Uintah and Utah counties; Idaho; Wyoming, Nevada, and Colorado; 46 (0). Our material is assignable to var. **alpinus (Wats.) L. Benson** [R. orthrorhynchus var. alpinus Wats., type from Wasatch Mts.], but is probably not significantly different from var. alpinus, from which it has been segregated by its narrow leaf segments and straight beaks. The species has been treated at infraspecific rank as R. eschscholtzii var. alpinus (Wats.) C. L. Hitchc.

Ranunculus alismifolius Geyer Plantain Buttercup. Terrestrial perennials, 0.5–4.6 dm tall, glabrous; stems 1.5–6 mm wide, erect, not rooting at the nodes; basal leaf blades simple, entire or merely dentate, 2–10 cm long, 0.5–3.5 cm wide, lanceolate to ovate-lanceolate; petioles 2–17 cm long; cauline leaves sessile or short-petiolate, alternate or opposite; bracts like the leaves only narrower; pedicels 2–14 cm long glabrous or pubescent; sepals 5, yellowish or greenish, or suffused with lavender, 4.5–7.5 mm long, glabrous, spreading, deciduous; petals 3–10, yellow, 4–14 mm long; achenes glabrous, 1.6–3.2 mm long, 15–45 in a subglobose cluster 4–7 mm long and 6.5–8.5 mm wide; beaks 0.4–1.2 mm long; receptacle glabrous; 2n = 16. Wet meadows, pond margins, and stream banks in sagebrush, sedge-hairgrass, aspen, fir, and spruce-fir communities at 1830 to 3355 m in Beaver, Box Elder, Cache, Davis, Duchesne, Emery, Garfield, Grand, Iron, Salt Lake, San Juan, Sanpete, Sevier, Summit, Utah and Wasatch counties; widespread in the western U. S. and Canada; 47 (v). Specimens in Utah are referable to var. **montanus** Wats. (type from upper Provo River, Uinta Mts.). The variety is distinguished from phases of the species beyond Utah in its more numerous petals (9 or 10, not 5). This species is distinguished from R. oreogenes by its glabrous achenes and receptacle, and from R. glaberrimus by its narrow fruiting head (less than 9 mm) and usually entire cauline leaves.

Ranunculus andersonii Gray Violet Buttercup. [R. andersonii var. tenellus Wats., type from Pilot Rock Point, Great Salt Lake]. Terrestrial perennials, 0.5–3 dm tall, glabrous; stems scapose, 1.5–4 m thick, not rooting at the nodes; blades of basal leaves compound, 1–5 cm long, 1.5–5.5 cm wide, triternately 2–3 times divided, the ultimate segments 0.5–4 mm wide, acute; petioles 1–12 cm long; cauline leaves lacking, or a single bracteate leaf subtending a second pedicel; sepals 5, reddish or purple, 6–15 mm long, glabrous, spreading, persistent; petals 5, red purple, 12–17 mm long, orbicular; achenes 4.5–6.5 mm long, 30–60 in a hemispheric cluster ca 10 mm long and 9–15 mm wide; beaks 0.2–1 mm long; receptacle glabrous. Shadscale, horsebrush, sagebrush, juniper, pinyon-juniper, and ponderosa pine communities at 1095 to 2410 m in Beaver, Box Elder, Juab, Kane, Millard,

Tooele, and Washington counties; Oregon and Idaho, south to California, Nevada, and Arizona; 33 (ii). Our materials have long been regarded at species level as *R. juniperinus* Jones, based mainly on size of the achenes - 4-6 mm versus 9-14 mm for *R. andersonii*. But a cline in size of fruit exists east and west. The *juniperinus* phase was further substantiated by the presence of a second flower subtended by a leafy cauline bract in some specimens. Plants with two flowers occur in Washington County uniformly and less commonly north to Millard County. These are designated as **var. *juniperinus* (Jones) Welsh** [*R. juniperinus* Jones, type from Washington County; *Beckwithia juniperina* (Jones) Heller]. The remainder of the Utah material belongs to **var. *andersonii*.**

Ranunculus aquatilis L. Water Crowfoot. Aquatic perennials, mainly 2-8 dm long, glabrous; stems 1-2.5 mm wide, submersed or floating, rooting at the nodes; leaves all cauline, alternate, submersed ones dissected into filiform segments, the blades 1-3 cm long on petioles 3-18 mm long, the floating ones simple but deeply 3-parted and 1-2.5 cm broad (1-1.5 cm long) on petioles 1-3 cm long; pedicels 1-6 cm long, finally recurved but usually not from the base, glabrous; sepals 5, green or purplish, 3-5 mm long, glabrous, spreading or reflexed, deciduous; petals 5, white, 1.6-2 mm long, 15-30 in a subglobose cluster; 4-6.5 mm high, 4-6 mm wide; beaks 0.2-0.4 mm long or lacking; pericarp transversely ridged; receptacle hairy; $2n = 16, 32, 48$. Four closely allied, and possibly taxonomically insignificant, phases are known.

1. Leaves petiolate, the petiole arising from the summit of the stipular enlargement, commonly collapsing when withdrawn from water 2
— Leaves sessile, the first divisions arising from the summit of the stipular enlargement, often not collapsing when withdrawn from water 3
2(1) Leaves dimorphic, the submersed segments capillary, the floating broadly lobed ... *R. aquatilis* var. *hispidulous*
— Leaves all alike, all submersed and with capillary segments .... *R. aquatilis* var. *capillaceus*
3(2). Pedicels recurved basally in front; petals 2-4 mm wide; stamens 5-10; achenes 30-80, the beak 0.2-0.5 mm long .... *R. aquatilis* var. *subrigidus*
— Pedicels not recurved in front; petals 2.5-6 mm wide; stamens 10-20; achenes 7-25, the beak 0.7-1.1 mm long, slender .... *R. aquatilis* var. *longirostris*

Var. *capillaceus* (Thuill.) DC. [*R. capillaceus* Thuill.; *Batrachium aquatile* var. *capillaceum* (Thuill.) Garrett]. Streams, lakes, ponds, and irrigation canals at 1320 to 3050 m in Box Elder, Cache, Daggett, Duchesne, Garfield, Iron, Salt Lake, Sevier, Uintah, Utah, Wasatch, Washington; and Weber counties; widely distributed in North America; circumboreal (as to the species); 34 (ii). Specimens on mud in drying pond margins have leaves less dissected—with shorter and proportionally (if not actually) broader segments. Possibly these account for the report (Vascular Plants Pacific N. W. 2: 378. 1964) of var. *porteri* Benson for Utah.

Var. *hispidulus* E. Drew Ponds and lakes at 1675 to 2745 m in Beaver, Cache, Salt Lake, and San Juan counties; widespread in western north America; 2 (i).

Var. *longirostris* (Godron) Lawson [*R. longirostris* Godron]. Ponds, reservoirs, lakes, streams, springs, and irrigation canals at 1370 to 1800 m in Cache, Davis, Duchesne, Grand, Salt Lake, Rich, Summit, and Wasatch counties; Saskatchewan to Quebec, south to Nevada, Colorado, New Mexico, Arkansas, Alabama, and Delaware; 8 (0). This is a close ally of both var. *subrigidus* and var. *capillaceus*, with intermediates representing a grading series and with no one feature being useful in segregation of all specimens.

Var. *subrigidus* (E. Drew) Breitung [*R. circinatus* var. *subrigidus* (E. Drew) L. Benson]. Ponds, lakes, and streams at 1065 to 2745 in Cache, Daggett, Garfield, Rich, Salt Lake, Sanpete, Sevier, Summit, Utah, Washington, and Wayne counties; widespread in North America; Europe; 8 (i). This plant is closely allied to both *R. aquatilis* var. *capillaceus* and to *R. aquatilis* var. *longirostris*. The (usually) sessile leaves and short beaked achenes are considered as diagnostic, but these are tenuous characters at best.

Ranunculus arvensis L. Field Buttercup. Terrestrial annuals, mainly 0.7-6 dm tall; stem glabrous to pilosulous, ca 1-2 mm wide, erect, not rooting at the nodes; basal leaf blades simple, cuneate to obovate, apically toothed or 3-parted; petioles 1.5-3.5 cm long; cauline leaves alternate, numerous, deeply divided into 2-5 oblanceolate segments, with petioles progressively shorter upwards; bracts sessile or petiolate, toothed or lobed; pedicels 1-5 cm long, strigose; sepals 5, green to yellow, 3-5.5 mm long, long-hairy dorsally, spreading, finally deciduous; petals 5, yellow, 4-6.5 mm long, obovate; achenes 6.5-8.5 mm long, commonly 8 or fewer in a whorl, this 9-16 mm wide, the beaks stout, 1.8-3 mm long, the surfaces with straight spines ca 1-3 mm long; receptacles pubescent; $2n = 32$. Weedy species of fields and other open habitats at 1370 to 1680 m in Cache, Salt Lake, Utah, and Weber counties; widely naturalized in the U. S.; adventive from Europe; 4 (0).

Ranunculus cardiophyllus Hook. Showy Buttercup. Terrestrial perennials, mainly 1.3-5 dm tall; stems 1-3 mm wide, with long, spreading hairs, erect, not rooting at the nodes; basal leaf blades simple, with crenate to lobed margins, the apical lobe sometimes free 1/2 the blade length, truncate to cordate basally, 2-3.5 cm long, 1.8-3.5 cm wide, oval to orbicular; cauline leaves sessile or essentially so, 3- to 7-parted, the segments 1-5 mm wide; bracts sessile, the lobes 1-3 mm wide; pedicels 2-17 cm long; sepals 5, green or diffused with purple dorsally, 4-5.5 mm long, villous dorsally, spreading, deciduous; petals 5 (or lacking), yellow, 8-15 mm long, obovate; achenes short-hairy, 1.6-3 mm long, 50-100 or more, in a cylindroid cluster, 6-10 mm long and 4.5-7.5 mm wide; beaks 0.6-1 mm long; receptacle hirsute; $2n = 32, 64$. Meadows and tall forb communities in aspen, ponderosa pine, lodgepole pine, and spruce-fir woodlands at 2315 to 3600 m in Beaver, Daggett, Duchesne, Garfield, Iron, Piute, and Uintah counties; British Columbia to Saskatchewan, south to Washington, Arizona, and New Mexico; 9 (ii). This is a poorly understood taxon in Utah. Its gross morphology closely simulates the common *R. inamoenus* (q.v.), but differs in its larger flowers and usually narrower upper bracts segments.

Ranunculus cymbalaria Pursh Marsh Buttercup. [*R. cymbalaria* var. *saximontana* Fern.]. Palustrine or riparian perennials; vegetative stems 0.5-1.5 mm wide, produced into strawberrylike stolons, rooting at the nodes, glabrous to pubescent; flowering stems 0.2-3 dm tall; basal leaves simple, the petioles 1-15 cm long, the blades 0.5-4 mm long, cordate, ovate, or uniform, crenate;

pedicels 2–10 cm long; sepals 5, 2.5–7.5 mm long, glabrous, deciduous; petals 5 or more, yellow, 2.5–8 mm long; achenes ca 1.1–1.5 mm long, 25–200 in a cylindrical cluster, glabrous, straight, the beak to ca 0.3 mm long, straight; 2n = 16. Muddy sites along streams, near pools or lakes, or in marshlands, at 850 to 2745 m in all Utah counties; widely distributed in North America; South America and Eurasia; 143 (xiii). This most common of our buttercup species has been interpreted as belonging to two varieties, but such status does not seem to be warranted.

Ranunculus eschscholtzii Schlecht. Eschscholtz Buttercup. Terrestrial perennials, 0.4–2.6 dm tall, glabrous or sometimes pilose; stems 1.5–3.5 mm wide, erect, not rooting at the nodes; basal leaf blades simple, cleft into 3 main lobes, the midsegment entire or 3–toothed, the lateral segments lobed or toothed, truncate to cordate or obtuse basally, 0.9–5 cm wide, 0.7–2.5 cm long; cauline leaves sessile or subsessile, alternate; bracts sessile, usually 3–lobed; pedicels 0.5–9 cm long; sepals 5, green or purplish, 4–6 mm long, pilose dorsally, reflexed, deciduous; petals 5, yellow, 6–10 mm long, obovate; achenes glabrous or hairy, 1.3–1.8 mm long, 20–50 or more in an elongate cluster 6–12 mm long and 6–7.5 mm wide; beak 0.7–1 mm long; receptacle glabrous or hairy; 2n = 32, 40, 48. Spruce-fir and alpine meadow communities at 2745 to 3205 m in Daggett, Duchesne, Piute, Salt Lake, San Juan, Sanpete, and Summit counties; Alaska and Yukon, south to California, Nevada, and Colorado; 21 (iii). Our material has been segregated into the common **var. eschscholtzii**, with broad lobes and rounded teeth and **var. trisectus** (Eastw.) L. Benson [*R. trisectus* Eastw. in Robins.] with slender more acute lobes. Importance of the distinction is unclear. Specimens from the La Sal Mts. have been designated as **var. eximius** (Greene) L. Benson [*R. eximius* Greene] on the basis of more sharply lobed lower leaves.

Ranunculus flabellaris Raf. Yellow Water-crowfoot. [*R. delphinifolius* Torr. in A. A. Eaton, not H.B.K.]. Aquatic or palustrine perennials 3–8 dm long or more, submersed or on mud, rooting at the nodes; stems 2–8 mm wide; leaves all cauline, alternate, ternately compound, the main segments further divided into linear segments, these 0.3–2 mm wide, the blades 1–6 cm long; petioles 2–12 cm long or more, the aerial leaves simple, 3–parted to -divided and again lobed, with petioles to 20 cm long; pedicels 1–8 cm long; sepals 5, greenish, 6–7.5 mm long, glabrous, spreading, deciduous; petals 5, yellow, 7–10 mm long, obovate to orbicular; achenes not ridged, ca 4 mm long, 40–100 or more in a subglobose cluster 6–10 mm long and 7–11 mm wide; beaks 1.3–2 mm long; receptacles hairy; n = 16. Ponds, marshes, and other wet sites at 1310 to 2685 m in Cache, Duchesne, Salt Lake, and Summit counties; widespread in North America; 1 (0).

Ranunculus flammula L. Creeping Spearwort. [*R. reptans* L.]. Palustrine perennials, with vegetative stems produced into strawberrylike stolons, often rooting at the nodes, glabrous or nearly so; flowering stems 1–10 cm tall; leaves simple, alternate (appearing basal), the petioles 2–13 mm long, the blades 0.5–7.5 cm long, 1–8 mm wide, linear to narrowly elliptic, lanceolate or oblanceolate, acute at both ends, entire; pedicels 2–10 cm long; sepals 5, 2–5 mm long, glabrous or pubescent, greenish, deciduous; petals 5, yellow, 2–8 mm long; achenes 1.2–1.8 mm long, 5–30 in a subglobose cluster, glabrous; beak 0.1–0.7 mm long; receptacle glabrous; 2n = 32. Margins of ponds, lakes, streams and other moist muddy or gravelly sites in aspen, alder, lodgepole pine, spruce-fir, and meadow communities at 2315 to 3355 m in Daggett, Duchesne, Garfield, Iron, Kane, Salt Lake, Sevier, Summit, Uintah, Utah, and Wayne counties; widespread in North America; circumboreal; 26 (ix). Segregation of infraspecific taxa seems to be unnecessary.

Ranunculus gelidus Kar. & Kir. Tundra Buttercup. Terrestrial perennials, from a subterranean root crown and long-sheathing marcescent leaf base; stems abruptly curved above or below the floral bracts, 0.2–1 dm tall, not rooting at the nodes, glabrous or sparsely hairy; basal leaves simple, the petioles 1.5–6 cm long, the blades 0.5–1 (2) cm long, cordate to uniform in outline, 3–parted and again cleft or parted; cauline leaves alternate, sessile or subsessile; pedicels 1.5–6.5 cm long; sepals 5, 2.5–4.5 mm long, spreading or reflexed, pubescent with pale hairs, deciduous; petals 5, 3.5–5 mm long; achenes 2–2.5 mm long, 30–80 in a subglobose to elongate cluster, glabrous; beak 0.4–0.7 mm long; receptacles glabrous; 2n = 16. Smelowskia-sedge community in alpine tundra at 3415 to 3800 m in Duchesne and Summit counties; Alaska and Yukon, south to Colorado; 2 (i). The sheathing marcescent bases, abruptly curved, horizontally spreading stems, and dissected short lower leaf blades are diagnostic for this species, which is disjunct from the arctic.

Ranunculus glaberrimus Hook. Sagebrush Buttercup. Terrestrial perennials, mainly 5–19 cm tall, glabrous; stems 1–2 mm thick, erect, not rooting at the nodes; basal leaves simple, the blades entire or apically 3–lobed, 1–5.2 cm long 0.5–2 cm wide, orbicular to ovate, elliptic, or oblanceolate, rounded to acute or obtuse at both ends; petioles 2–7 cm long; cauline leaves simple or dissected into 2 or 3 linear or lanceolate lobes, bracteate; pedicels 1–8.5 cm long; sepals usually 5, green or suffused with purple, 4.5–7 mm long, pubescent, spreading, deciduous; petals 5, yellow, 6.5–11 mm long, mostly 5–10 mm wide; achenes hairy, 1.3–3.5 mm long, 60–180 in a globose cluster 8–11 mm thick; beaks 0.5–0.7 mm long; receptacle glabrous; 2n = 64, 80. Sagebrush, meadow, ponderosa pine, Douglas fir, aspen, and lodgepole pine communities at 1460 to 3050 m in Beaver, Box Elder, Cache, Carbon, Daggett, Juab, Salt Lake, San Juan, Sevier, Summit, Tooele, and Uintah counties; British Columbia to the Dakotas, south to California and New Mexico; 22 (i). Our material has been recognized as belonging to two varieties; **var. glaberrimus**, with orbicular usually lobed basal leaves and **var. ellipticus** Greene, with narrow unlobed basal leaves. The differences do not seem to represent taxonomic entities.

Ranunculus gmelinii DC. Gmelin Buttercup. [*R. multifidus* var. *repens* Hook., in Wats., type from Weber Valley, Summit County]. Palustrine or aquatic perennials, mainly 2–5 dm long, glabrous or pubescent; stems 1.5–3 mm thick, prostrate or floating, rooting at the nodes; leaf blades simple, 3–parted or divided, the primary lobes again lobed or forked, seldom some dissected into linear segments, cordate basally, orbicular to uniform in outline; pedicels 1–5 cm long; sepals 5, greenish, 2.5–6 mm long, glabrous or pubescent, deciduous; petals 5, yellow, 4–8 mm long; achenes 1–1.5 mm long, 50–70 in an ovoid cluster, glabrous; beak 0.6–0.8 mm long; receptacle hairy. Pond margins and wet places often exposed as water dries, at 1705 to 2930 m in Duchesne, Morgan, Rich, Sanpete, Sevier, Summit, and Weber counties; widely distributed in North America; Asia; 2 (0). *R. gmelinii* is a near congener of *R. flabellaris*, from which it is distinguished mainly by its usually smaller parts, but with which it would probably best be combined at some infraspecific status. Both were published in 1818, and would be difficult to decide priority should they be combined. Recognition of infraspecific taxa within our portion of the species seems unnecessary.

Ranunculus hyperboreus Rottb. Arctic Buttercup. Aquatic (sometimes palustrine) perennials; stems prostrate to erect (sometimes floating), 1–5 dm long, rooting at the nodes, glabrous; leaves mostly cauline, alternate, the petioles 1.5–5 cm long; blades simple, 3–8 (10) mm long, with 3–5 entire, rounded lobes, truncate to rounded or cordate basally; pedicels 0.5–4 cm long; sepals 5, greenish, 2–3 mm long, glabrous, deciduous; petals 5, yellow, 2–4 mm long; achenes 0.6–0.8 mm long, 10–20 in a globose cluster, glabrous; beaks ca 0.1 mm long; receptacle glabrous; 2n = 16, 24, 32, 64. Bogs and wet meadow at 2590 to 2900 m in Daggett, Rich, and Summit counties; widespread in northern North America; circumpolar; 6 (0). This is a near ally of *R. natans*, q.v., and our materials might be more nearly allied to that taxon. More specimens must be examined to determine the exact affinities.

Ranunculus inamoenus Greene Drab Buttercup. [*R. alepeophilus*. A. Nels.; *R. utahensis* Rydb., type from Alta; *R. inamoenus* var. *alepeophilus* (A. Nels.) L. Benson]. Terrestrial perennials, mainly 0.5–4.4 dm tall, hairy or glabrous; stems 1.5–3 mm thick, erect, not rooting at the nodes; basal leaf blades simple, crenate or 3-lobed, truncate to obtuse or acute basally, 1–5.5 cm long, 1–4.5 cm wide; petioles 1.5–15 cm long; cauline leaves sessile or shortly petiole, deeply lobed or divided; bracts sessile usually 3- to 5-lobed, the lobes mainly 3–10 mm wide or more; pedicels 0.5–7 cm long; sepals 5, green or suffused with purple, 4–7 mm long, obovate; petals 5, yellow, mostly 5–7 mm long, elliptic to obovate; achenes hairy or glabrous, 1.6–2.5 mm long, 16–200 in an elongate cluster 7.5–17 mm long and 5.5–8.5 mm thick; beaks 0.2–0.8 mm long, straight or curved; receptacle hairy or glabrous. Sagebrush, ponderosa pine, mountain brush, Douglas fir, aspen, spruce-fir communities, often in meadows, at 1370 to 3540 m in all Utah counties, except Morgan and Rich; British Columbia and Alberta, south to Nevada, Arizona, and New Mexico; 98 (xii). Recognition of infraspecific taxa seems to be inadvisable.

Ranunculus jovis A. Nels. Jupiter Buttercup. [*R. digitatus* Hook., not Gilib.]. Terrestrial perennials, glabrous, 2–11 cm tall; roots tuberous, clustered, clavate, 2–5 mm thick; stems 1–4 mm wide, erect, not rooting at the nodes; basal leaf blades simple but divided into 2–5 lance-oblong to oblanceolate segments 0.8–4 cm long and 2–10 cm wide; petioles 1–8 cm long; cauline leaves sessile, commonly 2 or 3-lobed; pedicels 0.5–6 cm long, abruptly curved-spreading in fruit; sepals 5, green or suffused with purple, 2–7.5 mm long, glabrous, reflexed, deciduous; petals 5, yellow, 4–15 mm long; achenes glabrous or pubescent, 1–2.3 mm long, 50–70 in a globose to elongate cluster 4–6.5 mm long and 4.5–6 mm wide; beaks 0.2–1 mm long; receptacle glabrous. Sagebrush, mountain brush, aspen, and spruce-fir communities at 1525 to 2745 m in Box Elder, Cache, Juab, Morgan, Salt Lake, Summit, Utah, and Wasatch counties; Wyoming, Idaho, and Nevada; 24 (iii). This is a remarkable plant, flowering at edges of snowbanks from low to high elevations.

Ranunculus macounii Britt. Macoun Buttercup. [*R. hispidus* Hook.]. Terrestrial perennials; stems 2–10 dm tall, erect to decumbent, sometimes rooting at the lower nodes, pubescent with long spreading hairs, 3–12 mm in diameter; basal leaves simple or pinnately (subpalmately) 3- to 5- foliolate (rarely double the compound); petioles 5–20 cm long, the blades 3–9 cm long, on petiolules 0.1–2.3 cm long; cauline leaves alternate, similar to the basal; pedicel 0.3–11.5 cm long; sepals 5, often suffused with purple, 3.5–8.5 mm long, reflexed, long-hairy, deciduous; petals 5, yellow, 3–8 mm long; achene 2–3 mm long (excluding beaks), 20–60 in a subglobose cluster 7–14 mm long and 7–15 mm wide; beaks 0.8–1.5 mm long; receptacle glabrous; 2n = 48. Stream margins and moist meadows at 1370 to 2960 m in Box Elder, Cache, Davis, Daggett, Duchesne, Garfield, Grand, Kane, Morgan, Piute, Rich, Salt Lake, San Juan, Sanpete, Sevier, Summit, Uintah, Utah, Washington, and Weber counties; widely distributed in North America; 48 (vii). The short petals proportional to sepal length, and moderately long, straight achene beaks, are diagnostic characteristics allowing segregation of *R. macounii* from other tall buttercup species. Occasional specimens have creeping stems rooting at the nodes as in *R. repens*, the apparent nearest ally. However, that species has very showy flowers and usually curved achene beaks.

Ranunculus natans C. A. Mey. in Ledeb. Floating Buttercup. [*R. hyperboreus* var. *natans* Regel]. Aquatic or palustrine perennials, mainly 0.5–5 dm long; stems prostrate or floating, rooting at the nodes; leaves all cauline, alternate, the petioles 1–6 cm long, the blade 5–15 cm long and usually broader, rounded to cordate basally, 3- to 5-lobed, the lobes broadly rounded; pedicels 2–8 cm long; sepals 5, greenish, 2–4.5 mm long, spreading, deciduous; petals 5, yellow, 3–4.5 mm long; achenes 1–1.2 mm long, glabrous, 25–70 in an ovoid cluster; style beak 0.1–0.3 mm long. Ponds, bogs, and wet meadows at 1920 to 2745 m in Daggett and Rich counties; Alberta to Colorado; Eurasia; 3 (0).

Ranunculus oreogenes Greene Mountain Buttercup. [*R. collomae* L. Benson]. Terrestrial perennial, 5–10.5 cm tall, glabrous; stems 1–2 mm thick, erect but abruptly bent, soon spreading, not rooting at the nodes; blades of basal leaves simple, linear to elliptical, entire, 1.5–5 cm long, 0.2–1.3 cm wide, the petioles 2–3.2 cm long; cauline leaves and bracts simple, not lobed; pedicels 2.5–6 cm long; sepals 5, greenish or suffused with purple, ca 5 mm long, pilose, spreading, deciduous; petals 5, yellow, 7–12 mm long; achenes hairy, 3–3.2 mm long, 50–100 in an ovoid cluster 10–12 mm long and 9–10 mm wide; beaks ca 1 mm long, curved; receptacle pubescent. Pinyon-juniper and ponderosa pine communities at ca 1830 to 2440 m in Kane (Benson 1948) and San Juan counties; Arizona, Colorado, and New Mexico; 3 (0). A specimen from the north side of Navajo Mt. (Atwood 4186 BRY) is peculiar in having slender involute leaves and more or less persistent sepals. This species is identical with *R. glaberrimus* in general aspect, differing in no discernable way except for the entire cauline leaves and bracts. Possibly it would best be treated at varietal level within that taxon, a fact not intended or implied herein.

Ranunculus orthrorhynchus Hook. Straight-beaked Buttercup. [*R. orthorhynchus* var. *platyphyllus* Gray, based on *R. macranthus* Wats., type from Wasatch Mts.]. Terrestrial perennials 2.5–8 dm tall, spreading long-hairy; stems 2.5–8 mm thick, erect, not rooting at the nodes; basal leaves mostly pinnately trifoliolate, the terminal division long-petiolulate and deeply lobed or again compound, the lateral leaflets more shortly petiolulate but sometimes again compound, the blade 4–13 cm long and 4–11 cm wide; petioles 6–31 cm long; cauline leaves similar to the basal, gradually reduced upward to the sessile or short petiolate bracts, these 3–parted with broad, toothed segments; pedicels 2–16 cm long; sepals 5, greenish and suffused with purple, 6.5–12.5 mm long, sparingly long-pilose, reflexed, tardily deciduous; petals 5, yellow, 8.5–14 mm long; achenes 5–7.5 mm long, 25–65 in a globose cluster 8.5–17 mm high and 10–19 mm thick; beaks 2.5–4 mm long; receptacle hirsute; n = 16. Pinyon-juniper, mountain brush, sagebrush-grass, aspen, tall forb, and spruce-fir communities, often in meadows, at 2195 to 3115 m in Cache, Carbon, Daggett, Duchesne, Emery, Grand, Iron, Salt Lake, Sanpete, Sevier, Summit, Utah, and Wasatch counties; British Columbia to Montana, south to California; Nevada, and Wyoming; 30 (ii). Our material is assignable to **var. platyphyllus** Gray. The large flowers, large achenes with long beaks, and definitely pinnnately compound main lower leaves are diagnostic.

Ranunculus pedatifidus J. E. Sm. ex Rees Northern Buttercup. Terrestrial perennials; stems erect, mainly 0.6–2 dm tall (rarely taller), puberulent to glabrous, not rooting at the nodes; basal leaves simple, the petioles 1–4 cm long or more, the blades 0.8–1.2 (3) cm long, usually divided into numerous linear to narrowly oblanceolate lobes; cauline leaves alternate; bracts divided into 3–7 linear lobes, sessile; pedicels 1–10 cm long or more; sepals 5, greenish, 4–6 mm long, spreading, pubescent, deciduous; petals 5, yellow, 4–6 (10) mm long; achenes ca 2 mm long, 25–70 in an elongate cluster, glabrous, the beak 0.6–1 mm long. Alpine tundra at ca 3415 m in Summit County (Ostler 640, 1977 BRY); widespread in Northwestern America; 1 (0). This attractive dwarf plant shares features with both *R. gelidus* (size of petals and leaves) and *R. adoneus* (narrow leaf segments). More collections are necessary to determine the nature of this plant in Utah.

Ranunculus ranunculinus (Nutt.) Rydb. Little Buttercup. [*Cyrtorhyncha ranunculina* Nutt. ex. T. & G.]. Terrestrial perennials 1.5–2.5 dm tall, glabrous; stems 0.8–2.5 mm wide, erect, not rooting at the nodes; blades of basal leaves 4–6 cm wide, 3–5.5 cm long, pinnately ternate, the leaflets again lobed or parted into segments 1–4 mm wide; petioles 7–11 long; cauline leaves and bracts similar to the basal leaves or reduced and becoming sessile; pedicels 1–7 cm long; sepals 5, yellowish, 2–4 mm long, glabrous, spreading, deciduous; petals 5, yellow, 3–6 mm long; achene 2.4–2.8 mm long, 10–30 in a hemispheric cluster ca 5 mm long and 5–7 mm wide; beaks 0.6–1.2 mm long; receptacle glabrous; 2n = 32. Aspen and spruce fir or mountain brush communities, often in crevices in limestone cliffs at ca 2135 to 2595 m in Cache county; Wyoming, Colorado, and New Mexico; 3 (0).

Ranunculus repens L. Creeping Buttercup. Terrestrial perennials, mainly 1–8 dm tall, densely to sparingly spreading hairy; stems 0.8–8 mm thick, prostrate-decumbent to erect often rooting at the nodes; blades of basal leaves trifoliate (rarely 5–foliolate) but pinnately compound, the petiolule of the central leaflet longer than those of lateral ones, 1–11 cm long and about as broad, the leaflets again lobed or toothed, rarely again compound; petioles 3–28 cm long; cauline leaves similar to the basal ones, reduced upwards and transitional to sessile or subsessile bracts; pedicels 1–15 cm long; sepals 5, greenish or suffused with purple, 5–8.5 mm long, hairy, spreading, deciduous; petals 5 to numerous, yellow, 3.5–15 mm long; achenes 2.4–4 mm long, 20–40 in a subglobose cluster 6.5–9.5 mm long and 7–10 mm wide; beaks 0.5–1 mm long, the margins prominent; receptacle hairy. Lawns, gardens, banks of streams, and irrigation canals at 1310 to 1895 m in Cache, Davis, Duchesne, Garfield, Salt Lake, Sevier, Summit, Uintah, and Utah counties; widely established in North America; adventive from Eurasia; 21 (i). Two phases are present; a many-petaled one with buttonlike flowers (**var. pleniflorus** Fern.), and a single flowered one with usually 5 large petals (**var. repens**), They sometimes grow intermixed. Plants with usually 5–petaled flowers are distinguished from the similar *R. macounii* by the proportionately greater petal-sepal ratio and usually curved achene beaks.

Ranunculus sceleratus L. Blister Buttercup. Palustrine or semiaquatic annuals, glabrous or nearly so; stems 2–15 mm thick, fistulous, thin-walled and often succulent, erect, not rooting at the nodes; blades of basal leaves simple, 3–lobed or parted, 0.5–4.5 cm long, 0.5–6 cm wide, the main lobes again lobed, parted, or toothed; cauline leaves similar to the basal ones, alternate, petiolate; petioles 1.5–30 cm long; bracts sessile or shortly petiolate variously narrowly lobed or entire; sepals 5; greenish, 2–5.5 mm long, hairy spreading, deciduous; petals 5, yellow, 0.5–5 mm long; achenes glabrous, 0.8–1.3 mm long, mostly 90–200 or more in a cylindric to subglobose cluster 4–11 mm long and 2.5–8 mm wide; beaks to 0.2 mm long; receptacle sparsely hairy; 2n = 32, 64. Pond, lake, stream, and spring margins and banks, or in other wet places, at 1495 to 2535 m in probably all Utah counties; Alaska and Yukon; 41 (v). Our material is assigneable to **var. multifidus** Nutt. in T. & G. This is the only annual buttercup of wetlands in Utah.

Ranunculus testiculatus Crantz Bur Buttercup. [*Ceratocephalus testiculatus* (Crantz) Roth]. Terrestrial annuals, 1.5–10 cm tall, tomentose; stems scapose, 0.5–1.5 mm thick erect or decumbent, not rooting at the nodes; leaves all basal, simple, 0.3–4 cm long, 0.3–3 cm broad, deeply 3–parted, the lateral segments again lobed into linear segments; petioles 0.5–3 cm long, broadly winged; sepals 5, green, 2.5–8.5 mm long, tomentose, spreading, persistent; petals 2–5, yellow, 3.5–8 mm long; achenes tomentose, 3–loculed, the 2 lateral locules empty, 5–80; in an elongate cluster 7–23 mm long; receptacle tomentose; 2n = 14. Weedy species in most dry land community types at 1155 to 3050 m in probably all Utah counties; introduced from Eurasia, now widely established in Western U. S.; 54 (vi). This species is one of our earliest flowering plants, often opening the first flowers before the scape elongates. It has long been regarded as belonging to the genus *Ceratocephalus* by European workers.

Thalictrum L.

Perennial, glabrous, rhizomatous herbs, the stems simple or branched above, erect; leaves alternate or basal, bi- or triternately compound; flowers perfect or imperfect, regular, inconspicuous, borne in racemes or panicles; sepals 4 or 5, green, whitish, or purplish, caducous; petals lacking; stamens 8 to many; pistils 2 to many; fruit a sessile or stipitate, beaked achene.

Boivin, B. 1944. American Thalictra and their Old World Allies. Rhodora 46: 337–77; 391–445; 453–87.

1. Plants dioecious, the flowers imperfect; peduncles straight below the achene cluster; achenes spreading in all directions; anthers 2.5–3.5 mm long, borne on filiform, usually contorted stamens; plants common .... *T. fendleri*
— Plants with perfect flowers or, if imperfect, the ultimate leaf segments less than 6 mm long (see *T. heliophilum*) . 2

2(1). Plants dioecious, the flowers imperfect; ultimate leaf segments less than 6 mm long; known from western Colorado, and to be sought on the shales of eastern Utah .... *T. heliophilum* Wilkins
— Plants with flowers perfect; ultimate leaf segments various in size, seldom if ever of shales in eastern Utah 3

3(2). Leaves all basal, except for a single small leaf below the stem middle; flowers in racemes; staminal filaments linear; stems mainly less than 2.5 dm tall; plants rare in alpine sites .... *T. alpinum*
— Leaves mainly cauline; flowers in panicles; staminal filaments dilated upwards; stems commonly over 3 dm tall; plants uncommon in montane sites .. *T. sparsiflorum*

Thalictrum alpinum L. Arctic Meadowrue. [*T. duriusculum* Greene, type from near Fish Lake]. Plants 0.3–2.5 (3) dm tall; leaves all basal or with a single leaf near the base, 2–8 (13) cm long, biternate, the segments 3–8 mm long; racemes elongate; pedicels recurved in fruit; flowers perfect; sepals purplish tinged; stamens 8–15, the filaments linear or slightly expanded apically; anthers 1.5–3 mm long; achenes 2–3.5 mm long, subsessile. Wet meadows and cold (often calcareous) bogs in willow-sedge, lodgepole pine, and spruce-fir communities at ca 2440 to 2900 m in Daggett, Duchesne, Emery, Juab, Sevier, Summit, and Wayne counties; Alaska and Yukon, east to the Atlantic, south to California, Nevada, and New Mexico; circumboreal; 9 (i).

Thalictrum fendleri Engelm. ex Gray Fendler Meadowrue. Plants dioecious, perennial; stems mainly 2–8 (10) cm tall, often purplish, commonly branching above; leaves mainly cauline (but larger towards the base), usually biternate, the ultimate segments (5) 8–20 mm long, ovate to obovate or orbicular, thin, dark green above, pale and often glandular beneath; panicles small to large, leafy bracteate; pedicels straight, not abruptly bent below the flowers in fruit; sepals whitish or greenish, 2–4 mm long; stamens 18–25, the anthers apiculate, 2.5–3 mm long, shorter than the usually contorted, filiform (often purplish) filaments; achenes 5–11 mm long, sessile or nearly so, spreading (often appearing as a round cluster when pressed), the body 4–7 mm long, strongly flattened, glabrous or glandular, obliquely elliptic-obovate in outline; styles 2–4 mm long, stigmatic almost the entire length; 2n = 28. Willow, birch, mountain brush, sagebrush-snowberry, boxelder-cottonwood, alder, ponderosa pine, lodgepole pine, aspen-tall forb, and spruce-fir communities at 1370 to 3355 m in probably all Utah counties; Oregon to Wyoming, south to California, Arizona, New Mexico, and Texas; northern Mexico; 136 (xxi). This is our common meadowrue, being practically omnipresent in the aspen zone in Utah. Reports of *T. occi-*

dentale Gray probably belong here. That species has the achenes spreading to reflexed, usually more than twice longer than wide, stigmas often purplish, with the style 3–4 mm long. None of our numerous specimens approaches that species.

Thalictrum sparsiflorum Turcz. ex Fisch. & Mey. Montane Meadowrue. Plants 2.5–10 dm tall, simple or branched above; leaves mainly cauline, usually triternate, the ultimate segments 5–20 mm long, green above, pale and often glandular beneath, obovate to orbicular; inflorescence simple or branched, leafy, with bracts subtending pedicels leafletlike; pedicels abruptly recurved below the flowers in fruit; flowers perfect; sepals whitish or greenish, 2–3.5 mm long; stamens 12–20, the filaments whitish, somewhat dilated upwards, 3–4.5 mm long; anthers 0.5–0.8 mm long; achenes shortly stipitate, 3–4.5 mm long, strongly flattened, glandular, obliquely obovate in outline; styles 1–1.5 mm long, the stigmatic portion ca 1 mm long; 2n = 14, 42. Alder, willow, aspen, lodgepole pine, and spruce-fir communities at 1860 to 3355 m in Beaver, Carbon, Duchesne, Piute, Sevier, Summit, Uintah, and Utah counties; Alaska and Yukon, south to California and Colorado; 13 (iv).

Trautvetteria Fisch. & Mey.

Perennial, rhizomatous herbs; stems erect; leaves basal and cauline, alternate, deeply palmately lobed; flowers in more or less dichotomously branched corymbose panicles, perfect; perianth white or greenish, in 1 series, of 3–7 caducous, concave segments; stamens and pistils numerous; achenes sharply angular, not ribbed, with a hooked, persistent style.

Trautvetteria caroliniensis (Walter) Vail [*Hydrastis caroliniensis* Walter; *T. grandis* Nutt.; *T. media* Greene]. Plants rhizomatous, the stems solitary or few to several, erect, 2.5–10 dm tall, usually glabrous below but puberulent above; leaves mainly basal, long-petiolate, the blades 1–3 dm broad, 5- to 11-lobed, the lobes variously again lobed and serrate-dentate, glabrous above, commonly pubescent and paler beneath; cauline leaves 1 or 2, alternate, short-petiolate; sepals 3–6 mm long, ovate; staminal filaments 5–7 mm long; achenes 3–4 mm long, strongly veined on the angles. Boggy streamsides at ca 2745 to 3000 m in the Abajo Mts., San Juan County; British Columbia south to California and New Mexico; 2 (0).

Trollius L.

Perennial, rhizomatous herbs; stems erect; leaves basal and cauline, alternate, palmately lobed to compound; flowers perfect, regular, solitary or few; sepals 5 to several, petaloid; petals 5–10, linear, glandular (sometimes regarded as staminodia); stamens numerous; pistils few to numerous; fruit of few- to several-seeded follicles.

Trollius laxus Salisb. [*T. americanus* DC.; *T. laxus* var. *albiflorus* Gray; *T. albiflorus* (Gray) Rydb.]. Terrestrial perennials, usually tufted from a caudex and fibrous roots; stems erect, 1 to several, 1–7.5 dm tall; basal leaves with petioles 2–30 cm long, the blade palmately cleft and usually 5–lobed, the lobes again lobed or toothed; cauline leaves usually 2–4, becoming sessile upwards; flowers solitary on pedicels 2–20 cm long; sepals 5–9, 10–20 mm long, white to cream; petals 2–5 mm long; follicles 8–12 mm long, drying blackish. Bogs, wet meadows, and other moist sites in aspen, spruce-fir, and lodgepole pine communities at 2745 to 3300 m in Duchesne, Summit, Uintah, and Wasatch counties; British Columbia east to the eastern U. S., south to Oregon and Colorado; 20 (iii). This is a very attractive large-flowered plant that is often associated with *Caltha leptosepala* in cold wet sites at higher elevations.

RESEDACEAE S. F. Gray
Mignonette Family

Annual or perennial herbs with watery sap; leaves alternate, simple, or pinnately to subpalmately divided; flowers perfect, irregular, borne in terminal racemes; sepals (4) 5–6 (8), distinct; petals (4) 5–6 (8), unequal in size, the upper one the largest, appendaged; stamens 8 or more, borne on the upper side of a rounded disk, the anthers 2–loculed; pistil 1, the ovary superior, with usually 3 (2–6) carpels; style lacking; fruit a capsule, usually open at maturity; x = 6–15.

Reseda L.

Erect or ascending annual or perennial herbs from a taproot; leaves alternate; flowers greenish yellow; sepals subequal; petals unequal; pistils 1, the carpels usually 3, open toward the apex.

Reseda lutea L. Yellow Mignonette. Plants simple or much branched, glabrous; leaves pinnatifid or subpalmately divided; flowers greenish yellow, numerous, borne in elongate racemes; petals usually 6, each commonly with 3 connate or distinct appendages; ovary and capsule usually with 3 apical lobes; 2n = 24, 48. Cultivated ornamental; rarely escaping Utah; introduced from the Old World; 1 (0).

RHAMNACEAE A. L. Juss.
Buckthorn Family

Shrubs or small trees; leaves simple, alternate or opposite, pinnately of palmately veined; stipules small; flowers perigynous, perfect or polygamous, regular or nearly so; sepals 4 or 5, lined with a disk; petals 4 or 5, distinct, or absent; hypanthium present; stamens equaling the petals in number and opposite them; pistil 1, the ovary free or coalescent to the fleshy disk, 2– or 3–loculed, each locule with 1 ovule; style 1, commonly cleft; fruit a capsule or a berry; x = 9–13, 23.

1. Fruit dry, capsular; flowers showy, borne in terminal or axillary clusters .... *Ceanothus*
— Fruit fleshy, drupelike; flowers small, inconspicuous, borne axillary or in axillary clusters 2
2(1). Leaves palmately 3–veined from the base ...... *Ziziphus*
— Leaves pinnately several– to many-veined ..... *Rhamnus*

Ceanothus L.

Shrubs; leaves alternate or opposite, deciduous or persistent, commonly with 3 main veins from the base; flowers small, but showy, white to cream, borne in terminal or lateral, corymbose cymes; sepals 5, somewhat petaloid, united below with the urn shaped receptacle (this filled with a glandular disk surrounding the ovary); petals 5, distinct, clawed; stamens 5, opposite the petals; ovary 3–loculed, 3–lobed; style 3–cleft; fruit a capsule; seeds smooth.

1. Leaves opposite, pinnately veined, with several broad veins; stipules thick, commonly persistent (at least the lower part) .... *C. greggii*
— Leaves alternate, palmately 3–veined from the base; stipules thin, usually deciduous except for a thickened basal scar 2
2(1). Branches thorny; leaves both entire and permanently pubescent beneath .... *C. fendleri*
— Branches unarmed; leaves toothed or entire, but not permanently pubescent beneath 3
3(2). Leaves deciduous, mainly 0.5–2.5 cm long, entire near the base; plants inodorous .... *C. martinii*
— Leaves persistent, leathery, mainly 2.5–6 cm long or more, serrulate throughout; plants with odor of cinnamon .... *C. velutinus*

Ceanothus fendleri Gray Fendler Mountain-lilac. [*C. fendleri* var. *viridis* Jones, type from Elk Ranch, Kane (?) County]. Low, rounded to spreading, thorny (rarely unarmed) shrubs, mainly 2–8 dm tall; leaves alternate, deciduous, short-petiolate, the blades 3–24 mm long, 2–11 (14) mm wide, elliptic to oblong, ovate, acute or acutish, entire or serrulate, pale and persistently short-hairy beneath; inflorescence more or less corymbose; sepals ca 0.3–0.5 mm long; petals hooded, ca 1.5–1.8 mm long, white; fruit 4–5 mm thick, 3–lobed; seeds shiny, brown, 2.5–3 mm long. Ponderosa pine-manzanita, and less commonly in pinyon-juniper, aspen, and spruce communities at 1950 to 2840 m in Beaver, Daggett, Duchesne, Garfield, Iron, Kane, Piute, San Juan, Sevier, Uintah, and Wayne counties; Wyoming to South Dakota, south to Arizona, New Mexico, and Texas; 36 (iii).

Ceanothus greggii Gray Desert Mountain-lilac. Shrubs, mainly 0.2–2 m tall, erect or low and rounded, unarmed, the branchlets tomentose; leaves opposite, more or less persistent, 5–16 mm long, 3–8 mm wide, elliptic to obovate, thick and leathery, entire or toothed, more or less hairy on one or both sides, pinnately veined; inflorescence more or less corymbose; sepals white ca 1–1.5 mm long; petals white or purple, hooded, ca 1.5–3 mm long; fruit 4–5 mm thick; seeds brown, shiny, ca 4 mm long. Mixed desert shrub, pinyon-juniper, and mountain brush communities at 1220 to 2870 m in Grand, Iron, San Juan, Washington, and Wayne counties; Nevada, California, Arizona, New Mexico; Texas, and Mexico; 29 (iii). Specimens from San Juan County are mainly rimrock plants at the margin of pinyon-juniper communities, and form low-rounded bushes mainly 2–5 dm tall, typical of the material in the Colorado Drainage. Plants from Washington County are typically much taller (1–2 m).

Ceanothus martinii Jones Utah Mountain-lilac. [*C. utahensis* Eastw., type from Soldier Summit, Wasatch County]. Low, rounded to spreading, unarmed shrubs, mainly 2–8 dm tall; leaves alternate, deciduous, short-petiolate, the blades 7–30 mm long, 4–22 mm wide, elliptic to oval, ovate or obovate, entire or serrulate except at the base, green on both sides, variously strigose, but especially so on veins beneath; inflorescence more or less corymbose; sepals 0.5–1.2 mm long, petals hooded 1.8–2.4 mm long, white; fruit 4–5 mm thick, 3–lobed; seeds shiny brown, ca 3 mm long. Pinyon-juniper, mountain brush, sagebrush, ponderosa pine, Douglas fir, aspen, and bristlecone pine communities at 1830 to 2930 m in Beaver, Box Elder, Carbon, Davis, Duchesne, Emery,

Garfield, Juab, Kane, Millard, Piute, Salt Lake, Sanpete (type from Manti Canyon), Sevier, Tooele, Utah and Washington counties; Nevada, Arizona, Colorado, and Wyoming; 50 (viii).

Ceanothus velutinus Dougl. Deer-brush. Rounded to spreading unarmed shrubs, mainly 0.5–1.5 m tall; leaves alternate, persistent with petioles 0.6–2.4 cm long, the blades 18–85 mm long, 12–55 mm wide, broadly elliptic to oval or oblong, strongly 3–veined from the base, serrulate throughout, green and shiny above, pale and velvety beneath; inflorescence more or less corymbose; sepals 0.9–1.4 mm long, white; petals hooded, 2–2.5 mm long, white; fruit 3–4 mm thick, 3–lobed; seeds tan, shiny, 2–2.5 mm long; $2n = 24$. Ponderosa pine, sagebrush, mountain brush, aspen, and lodgepole pine communities at 1890 to 2900 m in Box Elder, Cache, Daggett, Davis, Duchesne, Millard, Morgan, Salt Lake, Sanpete, Summit, Tooele, Utah, and Weber counties; British Columbia to South Dakota, south to California, Nevada, and Colorado; 45 (iii). The plants have a pervasive odor of cinnamon that is especially noticeable when one tramples the plant.

Rhamnus L.

Shrubs or small trees; leaves alternate, pinnately veined, deciduous; flowers small, perfect or imperfect, in axillary clusters; calyx with 4 or 5 lobes, circumcissally deciduous following anthesis; petals 4 or 5 or lacking; stamens 4 or 5, with short filaments; pistils 1, the ovary free of the disk, 2– to 4–loculed; fruit a berrylike drupe.

Wolf, C. B. 1938. The North American species of *Rhamnus*. Rancho Santa Ana Bot. Gard. Monogr. Bot. Ser. 1.

1. Leaves mainly with 10–12 lateral veins along each side; plants of the Colorado River Canyon and tributories .. .. *R. betulifolia*
— Leaves with fewer lateral veins on each side (2–7); plants of different distribution 2
2(1). Main lateral veins of leaves 2–4 on each side; flowers 4–merous, the petals usually present; style 4–lobed .... *R. cathartica*
— Main lateral veins of leaves 5–9; flowers 5–merous, the petals lacking; styles 3–lobed 3
3(2). Main lateral veins of leaves 5–7; uncommonly cultivated shrubs, but known to escape and persist .... *R. alnifolia*
— Main lateral veins of leaves 8–9; uncommonly cultivated ornamental, not known to escape or persist .... *R. frangula* L.

Rhamnus alnifolia L'Her. Alder Buckthorn. Shrubs, mainly 1–1.5 m tall; leaves 1–8 cm long, elliptic to ovate, serrulate, glabrous or puberulent on one or both sides; umbels with 1–3 flowers, appearing before the leaves; flowers imperfect, mostly 5–merous; pedicels 2–6 mm long; calyx ca 3 mm long; petals lacking; fruit black, 6–8 mm long, subglobose, 3–seeded. Aspen-fir community at ca 2196 m in Davis and Salt Lake counties; British Columbia to Newfoundland, south to California and Wyoming; 1 (0).

Rhamnus betulifolia Greene Birchleaf Buckthorn. Shrubs, mainly 1–2.5 m tall; leaves alternate, with petioles 0.3–1.8 cm long, the blades 2.5–15 cm long, 1.6–8.5 cm wide, elliptic to obovate, or oblong, obtuse to abruptly short attenuate apically, hirtellous on one or both sides, at least when young, the main lateral veins 7–12 on each side; inflorescence axillary, corymbose; flowers perfect, 5–merous; hypanthium, pedicels, and branches of the inflorescence puberulent; sepals 1.5–2.5 mm long, triangular; petals brownish, ca 1 mm long; style ca 1 mm long; ovary 3–loculed; fruits 7–10 mm long, with 3 seeds. Rock crevices, defiles, monolith bases, and hanging gardens at 1125 to 1895 m in Emery, Garfield, Grand, Kane, San Juan, Washington, and Wayne counties; Arizona, New Mexico, Texas, and Nevada; 52 (xvii).

Rhamnus cathartica L. Common Buckthorn. Shrubs or small trees, mainly 2–5 m tall, often armed with blunt thorns; leaves opposite or subopposite, the petioles mainly 5–25 mm long, the blades mostly 1.5–6 cm long, 1.2–3.5 cm wide, elliptic to obovate, oval or obovate, crenulate, rounded to abruptly short-attenuate, with 2–4 main lateral veins on each side; flowers 1–3, axillary, 4–merous; sepals 1.5–3 mm long; petals brownish, 1–1.3 mm long or shorter; style 4–lobed; fruit black, 5–7 mm thick, often with 4 seeds; $2n = 24$. Cultivated, escaping, and persisting in Salt Lake and Utah counties; introduced from Eurasia; 1(0).

Ziziphus Miller

Shrubs or small trees; leaves alternate, petiolate, markedly 3–veined from the base; flowers in axillary cymes, the hypanthum filled with a massive disk surrounding the pistil but not adherent; petals present, caducous; ovary 2–loculed; fruit drupaceous, with a solitary 2–loculed stone.

Ziziphus jujuba Miller Jujube. Glabrous shrubs or small trees, mainly 3–8 m tall, spiny or unarmed; branchlets often clustered, simulating pinnate leaves often with 2 spines at each node; leaves glossy, commonly 2–6 cm long, ovate to lanceolate, serrate, obtuse to marginate; drupes dark red or brown. Uncommonly cultivated fruit plant, escaping and persisting mainly in Washington County; introduced from Eurasia; 5 (0). The fruit is edible, either dried or preserved, and resembles a small date.

ROSACEAE A. L. Juss.
Rose Family

Annual, biennial, or perennial herbs, shrubs, or trees; leaves alternate or basal (and still alternate) or less commonly opposite, simple or pinnately to palmately compound, mostly deciduous, stipulate or rarely exstipulate; flowers perfect or imperfect, regular, complete or incomplete, perigynous to epigynous, borne singly or in racemose, corymbose, umbellate or cymose clusters; sepals usually 5 (more in some), often bearing bracteoles alternate with the lobes, borne with petals and stamens on margin of a hypanthium; petals usually 5 (lacking or more in some), commonly showy; stamens 5 to numerous; pistils 1 to many, of 1 carpel, or of 5 connate or distinct carpels enclosed in the hypanthium; fruit an achene, follicle, drupe, pome, aggregate, hip, or accessory; $x = 7$–9, 17+. Note: The rose family is both large and complex. The diversity of fruit type reflects the many morphological differences in structure of the gynoecium. Suggestions by some workers that the group should be segregated into more than one family is not without merit. It is held together by the presence of the hypanthium on which the perianth and stamens are displayed. This is a complex structure, with several possible origins, and might fail ultimately as a diagnostic character.

1. Plants annual, biennial, or perennial herbs .... Key 1
— Plants trees, shrubs, or subshrubs .... Key 2

Key 1.

Plants herbaceous.

1. Petals lacking; flowers numerous, borne in dense spikes; leaves pinnately compound .... *Sanguisorba*
— Petals present; flowers not both numerous and borne in spikes 2
2(1). Leaves bi- or triternately dissected into linear segments; petals white; plants of Piute, Beaver, and Sevier counties .... *Chamaerhodos*
— Leaves various, but not bi- or triternately dissected into linear segments; petals white, yellow, or pink 3
3(2). Flowers solitary on scapose peduncles; leaves simple, crenate; fruit of plumose achenes; sepals and petals 8–10 each .... *Dryas*
— Flowers usually more than 1; leaves compound or lobed, rarely simple; fruit not of plumose achenes; sepals and petals usually 5 each 4
4(3). Bractlets lacking between the sepals; flowers with a stalked receptacle; hypanthium funnelform; plants of Washington County .... *Purpusia*
— Bractlets present, alternating with the sepals; flowers with sessile receptacle; hypanthium not funnelform 5
5(4). Leaflets tridentate apically, entire along the sides; stamens 5; plants prostrate or mat-forming, of high elevations .... *Sibbaldia*
— Leaflets variously toothed or lobed, but not regularly tridentate apically; stamens 5, 10, or more; plants of various habit and habitat 6
6(5). Leaves trifoliolate; plants with well-developed stolons; flowers white; receptacle ripening into an accessory fruit .... *Fragaria*
— Leaves mostly with more than 3 leaflets but, if trifoliolate, the lacking stolons; flowers typically yellow (less commonly purple or white); receptacle not ripening 7
7(6). Leaflets very numerous, mostly less than 6 mm long; petals usually clawed .... *Ivesia*
— Leaflets 3–15 (rarely more), commonly much more than 6 mm long; petals sessile 8
8(7). Leaves palmately or pinnately lobed or compound, not lyrate pinnatifid; styles at maturity not elongate and conspicuous .... *Potentilla*
— Leaves pinnately lobed or compound or more usually lyrate-pinnatifid; styles at maturity elongate and conspicuous .... *Geum*

Key 2.

Plants woody.

1. Leaves compound 2
— Leaves simple 7
2(1). Stems and/or leaves armed with prickles or spines ... 3
— Stems and leaves lacking prickles or spines 4
3(2). Pistils several, enclosed within a fleshy hypanthium; fruit a hip; petals very showy .... *Rosa*
— Pistils several to many, on an elongate receptacle; fruit an aggregate; petals not especially showy ... *Rubus*

4(2). Leaves bipinnately compound, the ultimate segments 0.5–1.5 mm long; herbage glandular-stellate, aromatic .... *Chamaebatiaria*
— Leaves once pinnately compound, the leaflets much longer than 1.5 mm; herbage not glandular-stellate .. 5
5(4). Leaflets 3–7; leaves 1.5–3.5 cm long; flowers yellow; low shrub .... *Potentilla*
— Leaflets 7–15 or more; leaves 5–20 cm long or more; flowers white to cream; moderate shrubs to small trees 6
6(5). Ovary superior; stamens 20 or more; leaflets 13–23; cultivated shrubs .... *Sorbaria*
— Ovary inferior; stamens 15–20; leaflets 9–15; indigenous shrubs or cultivated trees .... *Sorbus*
7(1). Leaves opposite; petals lacking; intricately branched, low desert shrubs of southern and southeastern Utah .... *Coleogyne*
— Leaves alternate; petals present (lacking in *Cercocarpus*); plants of various habits and habitats 8
8(7). Shrubs low, mat-forming; flowers solitary or in dense spikes on leafless or merely bracteate scapes 9
— Shrubs or small trees, never mat-forming; flowers various, but neither scapose nor subscapose 10
9(8). Flowers solitary, the sepals and petals mostly 8–10 each; leaves crenate; plants of alpine tundra .... *Dryas*
— Flowers in dense spikes, the sepals and petals commonly 5 each; leaves entire; plants of rock surfaces at low to moderate elevations .... *Petrophytum*
10(8). Pistils superior, the 1 to several ovaries separate or partially connate, not adnate to the hypanthium; fruit a drupe, aggregate, achene, follicle, or capsule 11
— Pistils inferior, the 3– to 5–carpellate ovaries adnate to the hypanthium; fruit a pome 20
11(10). Flowers inconspicuous; petals lacking; leaves entire and evergreen (except in *C. montanus*) .. *Cercocarpus*
— Flowers showy, though small in some; petals present; leaves mainly toothed or lobed, often deciduous 12
12(11). Pistil 1; fruit a drupe; leaves commonly with glands at base of blade or on petiole .... *Prunus*
— Pistils 1 to many; fruit not a drupe; leaves not gland-bearing 13
13(12). Leaves pinnately veined, the lobes, if any, pinnate .. 14
— Leaves palmately veined, the lobes palmately arranged or flabellate 16
14(13). Flowers yellow, solitary, terminating branches of the current year .... *Kerria*
— Flowers white to pink or lavender, borne in corymbs, panicles or racemes 15
15(14). Flowers borne in racemes, 1.5 cm wide or more; petals 6–12 mm long .... *Exochorda*
— Flowers borne in corymbs or panicles, less than 1 cm wide; petals 6–15 mm long .... *Spiraea*
16(13). Flowers large, 2 cm broad or more, in few-flowered cymes; fruit an aggregate .... *Rubus*
— Flowers commonly less than 2 cm broad, solitary or in corymbs or panicles 17
17(16). Flowers numerous, borne in panicles ...... *Holodiscus*
— Flowers borne singly or in few- to many-flowered corymbs 18
18(17). Flowers borne in umbellate corymbs; leaves broad and thin, commonly 1–6 cm wide or more .... *Physocarpus*

	Flowers borne singly or in corymbose racemes; leaves thickish, seldom to 1 cm wide 19
19(18).	Pistils numerous; petals white; leaf lobes tightly revolute; plants of low elevations in southern Utah . *Fallugia*
—	Pistils 1–5 (rarely more); petals white to cream or pale yellowish; leaf lobes not tightly revolute; plants of broad distribution .... *Purshia*
20(10).	Stems armed with thorns or spines 21
—	Stems unarmed 23
21(20).	Leaves evergeen, crenate-serrate; pomes commonly orange; petals white, less than 4 mm long . *Pyracantha*
—	Leaves deciduous, serrate or doubly serrate; pomes variously colored, rarely orange; petals more than 5 mm long 22
22(21).	Shrubs to 2 m tall (generally less); flowers 20–45 mm broad; fruit over 2 cm thick .... *Chaenomeles*
—	Shrubs or small trees to 5 m tall or more; flowers 9–18 mm broad; fruit less than 1.5 cm thick ..... *Crataegus*
23(20).	Leaves entire or essentially so 24
—	Leaves serrate to doubly serrate (see also *Peraphyllum*) 26
24(23).	Leaves ovate to cordate ovate, 1.5–5 cm wide or more; pomes clothed with a villous tomentum .... *Cydonia*
—	Leaves variously shaped, less than 1.5 cm wide; pomes glabrous 25
25(24).	Shrubs to 1.5 m tall or more, indigenous; leaves narrowly elliptic; fruit an acrid pome .... *Peraphyllum*
—	Shrubs of various height, cultivated; leaves ovate to obovate or oblanceolate; pomes mealy, non-acrid .. .. *Cotoneaster*
26(23).	Flowers white, in racemes; plants indigenous, rarely cultivated; leaves prominently toothed towards the apex .... *Amelanchier*
—	Flowers white or otherwise, in corymbs; plants cultivated, sometimes escaping; leaves toothed or lobed throughout 27
27(26).	Leaves deeply or at least prominently lobed ... *Sorbus*
—	Leaves moderately if at all lobed 28
28(27).	Shrubs to 2 m tall (generally less); flower solitary or sessile in corymbose clusters .... *Chaenomeles*
—	Shrubs or trees to 7 m tall or more; flowers pedicellate in corymbose or umbellate clusters 29
29(28).	Flowers in umbels; styles connate at the base; fruit with few if any stone cells, apple-shaped .... *Malus*
—	Flowers in corymbs; styles free; fruit with stone-cells, mostly pear-shaped .... *Pyrus*

Amelanchier Medicus

Shrubs or small trees with unarmed branches; leaves alternate, simple, not lobed; stipules linear, caducous; flowers perfect, regular, borne in racemes; hypanthium short, with a glandular disk on the inner surface; sepals 5, persistent; petals 5, white; stamens usually 10 or more; pistil 1, the ovary inferior, usually 5–loculed (appearing as 10); styles 2–5, the stigmas capitate; fruit a reddish to purplish, often glaucous, pome.

Jones, G. N. 1946. American species of *Amelanchier*. Ill. Biol. Monogr. 20: 1–126.

1. Leaves mainly over 2.5 cm long; petals mostly 9–15 mm long; styles commonly 5 .... *A. alnifolia*
— Leaves mainly less than 2.5 cm long; petals 5–10 mm long; styles 2–4 (rarely 5) .... *A. utahensis*

Amelanchier alnifolia (Nutt.) Nutt. Serviceberry, Shadbush, Saskatoon. [*Aronia alnifolia* Nutt.; *Amelanchier canadensis* var. *pumila* Nutt. in T. & G.; *A. pumila* (Nutt.) Roemer; *A. alnifolia* var. *cusickii* (Fern.) C. L. Hitchc.]. Low shrubs to small trees, mostly 2–5 m tall; leaves petiolate, mainly 20–50 mm long, 15–40 mm broad, oval to oblong, acute to rounded or subcordate basally, rounded to truncate apically, serrate near the apex, glabrous or hairy on one or both sides; flowers in short racemes; sepals 2.8–4.6 mm long; petals 9–15 mm long, 3.3–5.8 mm wide, spatulate-oblanceolate, white to pinkish; styles 5 (or 4); fruit purplish to black purple, glaucous, subglobose, 6–14 mm long, palatable; 2n = 34. Streamsides, meadows, and mountain slopes at 1500 to 2900 m in sagebrush, mountain brush, aspen, and mixed conifer communities at 1220 to 2900 m in all Utah counties; Alaska and Yukon east to Hudson Bay and south to California, Arizona, New Mexico, and Nebraska; 112 (iii). Attempts to segregate the various proposed infraspecific taxa among our Utah materials are fraught with difficulties not easily overcome, even by application of mechanical and arbitrary keys. Pubescence or its absence and the position of that pubescence form the basis of the main proposed segregates. The feature of pubescence seems to be so variable, not only within *A. alnifolia*, but within *A. utahensis* (q.v.), that it might indicate a response to ecological conditions rather than genetic affinities. More work is indicated.

Amelanchier utahensis Koehne Utah Serviceberry. [*A. bakeri* Greene; *A. oreophila* A. Nels.; *A. utahensis* ssp. *oreophila* (A. Nels.) Clokey; *A. florida* var. *oreophila* (A. Nels.) R. J. Davis; *A. utahensis* var. *cinerea* Goodding, type from Washington County]. Low to large shrubs, mostly 0.5–4 m tall, intricately branched, often in dense clumps; leaves petiolate, mainly 10–27 mm long, 6–27 mm wide, oval to ovate, oblong, or elliptic, acute to rounded or subcordate basally, rounded to truncate or less commonly acute apically, serrate near the apex, hairy on one or both sides, rarely glabrous; sepals 1–3 mm long; petals 5.2–10 mm long, 1.8–4.2 mm wide, spatulate-oblanceolate to elliptic, white, cream or pinkish; styles 2–4 (5); fruit purplish or pinkish, 5–12 mm long, palatable or dry and hardly edible. Streamsides, dry slopes, or thickets in sagebrush, grassland, mountain mahogany, mountain brush, pinyon-juniper, aspen, and ponderosa pine communities at 900 to 2800 m in all counties in Utah (type from Leeds, Washington County); Washington to Montana and south to Baja California, Arizona, New Mexico, and Texas; 268 (xxxi). Segregation of all specimens in the *alnifolia-utahensis* complex is difficult if not impossible. Diagnostic features show overlap, and while trends are apparent in the vast amount of material available, the best of characteristics fail singly and often in combination as well. Because of the trends indicated by leaf and petal size, and other features, it seems best to treat *A. utahensis* apart from the tangled morphology of *A. alnifolia*. Additionally, variation within the *utahensis* assemblage is as great as (or greater) than that known to occur in the *alnifolia* materials. Both taxa are hosts to a cedar-apple rust.

Cercocarpus H.B.K.

Shrubs or small trees with unarmed branches and very dense wood; leaves alternate, simple, entire or toothed; stipules small, adnate to petiole; flowers perfect, regular, borne solitary or in small clusters, terminal or axillary; hypanthium trumpetlike, with a deciduous apical portion; sepals 5; petals lacking; stamens 10 or more, borne in 2 or 3 whorls; pistil 1, of 1 carpel; style terminal; fruit an achene, with the elongate plumose style persisting.

Martin, F. L. 1950. A revision of *Cercocarpus*. Brittonia 7: 91–111.

1. Leaves deciduous, toothed, not especially revolute .. .. *C. montanus*
— Leaves evergreen, entire or toothed, decidedly revolute 2
2(1). Leaves (at least some) toothed; a hybrid .... *C. montanus* x *C. ledifolius*
— Leaves entire 3
3(2). Leaves elliptic, commonly 12–30 mm long or more; shrubs or small trees, mostly of middle and higher elevations .... *C. ledifolius*
— Leaves linear to narrowly oblong, usually less than 12 mm long; low intricately branched shrubs of lower middle and lower elevations .... *C. intricatus*

Cercocarpus intricatus Wats. Dwarf Mountain Mahogany. [*C. ledifolius* var. *intricatus* (Wats.) Jones; *C. intricatus* var. *villosus* Schneider, type from Deep Creek, Juab (?) County; *C. arizonicus* Jones]. Shrubs mostly 0.5–2 m tall, intricately branched; leaves 3–18 mm long, 0.8–1.4 mm wide, oblong to linear (rarely elliptic), tightly revolute, glabrous, strigose-pilose, or villous, coriaceous and persistent; flowers 3.2–8.7 mm long; sepals 0.6–1.2 mm long; stamens 10–20; tails of achenes 1–3 cm long. Rimrock, cliffs, and slopes in desert shrub, pinyon-juniper and mountain brush communities at 1370 to 2400 m in Beaver, Box Elder, Cache, Daggett, Emery, Garfield, Grand, Juab, Kane, Millard, San Juan, Sanpete, Sevier, Uintah, Utah (type from American Fork Canyon), Washington, and Wayne counties; Nevada, California, and Arizona; 53 (xiii). Pubsecence of leaves varies from strigose-pilose to crinkly-hairy, or it is lacking. Plants with pilose leaves form the basis of var. *villosus*, but the feature does not seem to be correlated with any other. Leaves are heavily cutinized in plants from Kane and Washington counties.

Cercocarpus ledifolius Nutt. in T. & G. Curl-leaf Mountain Mahogany. [*C. ledifolius* var. *intercedens* Schneider; *C. ledifolius* var. *intercedens* f. *subglaber* Schneider, type from Slate Canyon, Utah County; and f. *hirsutus* Schneider, type from Odgen]. Shrubs or small trees, mainly 2–5 m tall; leaves 10–42 mm long, 2–14 mm wide, elliptic to oblong, the margin only revolute, pubescent to glabrous, coriaceous, persistent; flowers 7–10 mm long; sepals 1.2–2.1 mm long; stamens 20–30; tails of achenes 4.5–8 cm long. Mountain brush, pinyon-juniper, aspen, and spruce-fir communities, often in stands, at 1400 to 3000 m in all Utah counties; Washington to Montana and south to California, Arizona, and Colorado; 89 (xi). Hybrids are known between *C. ledifolius* and *C. montanus*. They are easily discerned by the coriaceous persistent toothed leaves. They occur as scattered individuals in places of contact between the parental types. Similarly, putative hybrids involving *C. ledifolius* and *C. intricatus* are known. Longer very narrow and markedly revolute leaves mark those apparent hybrids.

Cercocarpus montanus Raf. Alder-leaf Mountain Mahogany. [*C. betuloides* Nutt. in T. & G.; *C. betulifolius* Nutt. ex Hook.; *C. parviflorus* var. *glaber* Wats.; *C. parviflorus* var. *betuloides* (Nutt.) Sarg.; *C. montanus* var. *glaber* (Wats.) Martin; *C. parvifolius* var. *minimus* Schneider, type from Utah?; *C. flabellifolius* Rydb., type from Glenwood, Sevier County]. Shrubs, or less commonly, small trees commonly 1.2–4 m tall; leaves short-petiolate, the blade obovate to oblanceolate or orbicular, 6–44 mm long, 5–23 mm wide, crenate-serrate, glabrous above, pubescent beneath (sometimes glabrous), deciduous; flowers 9.5–17.5 mm long; sepals 0.9–1.7 mm long; stamens 25–40, the anthers hairy; tails of achenes 3–10 cm long. Mountain brush, sagebrush, grassland, pinyon-juniper, aspen, and mixed conifer communities at 1400 to 2800 m throughout Utah, except in Box Elder and Weber counties; Oregon to Wyoming and south to Mexico; 91 (xiv). This and other species of *Cercocarpus* are valuable browse plants for wildlife and domestic livestock. They are components of wild seed mixtures in reclamation attempts.

Chaenomeles Lindl.

Shrubs, usually armed with thorns; leaves alternate, simple, serrate; stipules large, deciduous; flowers perfect, regular, solitary or 2–5 or more in sessile clusters; hypanthium short; sepals not persistent; petals 5 (sometimes more), variously colored; stamens 20 or more; pistil 1, the ovary inferior, usually 5-loculed; styles 5, joined at the base; the stigmas capitate; fruit a pome of moderate size.

1. Branchlets with verrucose small scars left by deciduous short hairs; flowers orange scarlet, ca 2.5 cm wide; plants mostly less than 1 m tall .... *C. japonica*
— Branchlets smooth, lacking hair scars; flowers variously colored, over 2.5 cm wide; plants commonly more than 1 m tall .... *C. speciosa*

Chaenomeles japonica (Thunb.) Lindl. ex Spach Japanese Quince. [*Pyrus japonica* Thunb.; *Cydonia japonica* (Thunb.) Pers.; *Cydonia lagenaria* Lois.; *Cydonia japonica* var. *lagenaria* (Lois.) Makino; *Chaenomeles lagenaria* (Lois.) Koidz.]. Shrubs to 1 m tall (rarely more); the short branchlets often modified as thorns, the young branchlets with deciduous short hairs leaving verrucose scars on falling; leaves obovate, 20–50 mm long, 8–35 cm wide, obtusely serrate, obtuse to subacute apically; flowers orange-scarlet, ca 2.5 cm wide; fruit subglobose, ca 3 cm thick. Cultivated ornamental in Carbon, Salt Lake, and Utah counties; introduced from Japan; 4 (i).

Chaenomeles speciosa (Sweet) Nakai Flowering Quince. [*Cydonia speciosa* Sweet; *C. lagenaria* Koidz, not (Lois.) Koidz.]. Shrubs to 2 m tall, the short branchlets often modified as spines, glabrous or with short deciduous hairs leaving no scars on falling; leaves oblong to ovate or lanceolate, 22–65 mm long, 12–35 mm wide, sharply serrate, acute to subacute apically; flowers scarlet to white or red, 2.5–3.5 cm wide; fruit subglobose to pyriform, 2–5 cm thick. Cultivated ornamental in Juab, Salt Lake, and Utah counties; introduced from China; 3 (i).

Chamaebatiaria (T. C. Porter) Maxim.

Aromatic shrubs, unarmed; leaves alternate, bi- or tripinnately compound, the herbage stellate-pubescent; stipules herbaceous, more or less persistent; flowers perfect, regular, showy, borne in terminal panicles; hypanthium turbinate; sepals 5, persistent; petals 5, white; stamens many; pistils 5, more or less connate below, the ovary superior; styles 5; fruit of follicles.

Chamaebatiaria millefolium (Torr.) Maxim. Fern Bush, Desert Sweet. [*Spiraea millefolium* Torr.]. Shrub 8–20 dm tall (rarely more), the stems and herbage glandular and stellate-pubescent when young; leaves 0.9–6.7 cm long, 0.4–1.8 cm wide, oblong to lanceolate in outline, with 8–24 pairs of pinnae, these again pinnate, the tertiary segments again pinnatifid; panicles 3–15 cm long; flowers 0.8–1.5 cm wide; sepals ovate to lanceolate, 3–5 mm long, green; petals white, 2.5–5 mm long and about as broad; follicles 4–6 mm long, few-seeded; n = 9. Sagebrush, mountain brush, aspen, limber pine, and spruce-fir communities at 1495 to 2900 m in Beaver, Box Elder, Garfield, Iron, Juab, Kane, Millard, Piute, Tooele, Washington, and Wayne counties; Oregon, Idaho, Wyoming, and south to California and Arizona; 34 (vii).

Chamaerhodos Bunge

Plants biennial or short-lived perennial herbs; leaves alternate and in a basal rosette, bi- or triternately divided, the segments narrow; stipules foliose, narrowly oblong, simple or divided, persistent; flowers perfect, regular, borne in bracteate, corymbose cymes; hypanthium cup-shaped, long-hairy within; sepals 5; petals 5; stamens 5, borne at the base of the petals; pistils 5 (rarely fewer), distinct, the ovaries superior, each 1-loculed; style 1 per pistil, the stigma capitate; fruit an achene.

Chamaerhodos erecta Bunge in Ledeb. American Chamaerhodos. [*C. erecta* var. *nuttallii* T. & G.; *C. erecta* ssp. *nuttallii* (T. & G.) Hulten; *C. nuttallii* (T. & G.) Rydb.]. Plants erect, mostly 7–28 (30) cm tall, from a taproot and a basal rosette, the stems freely branched above the base; leaves mostly 7–40 mm long, the ultimate segments linear to oblong, sparingly long-hairy; cymes equaling 1/4–1/2 the plant height; flowers short-pedicellate, inconspicuous; sepals 1.2–2.5 mm long, triangular, sparsely hirsute; petals white, equaling or slightly longer than the sepals; achenes 1.2–1.5 mm long, glabrous, grayish; 2n = 14. Igneous gravel and sandy loam in sagebrush, grassland, and alpine tundra at 2745 to 3355 m in Piute, Sevier, and Wayne (?) counties; Alaska and Yukon east to Michigan and south to Colorado and North Dakota; Asia. Our materials are assignable to **var. parviflora (Nutt.) C. L. Hitchc.** [*Sibbaldia erecta* var. *parviflora* Nutt.]; 8 (iv).

Coleogyne Torr.

Shrubs, the stems intricately branched, spinescent; leaves opposite, fasciculate on spur branchlets, entire, coriaceous, persistent; flowers perfect, regular, solitary and terminal on spur branchlets, subtended by paired, trifid bracts; hypanthium cup-shaped, coriaceous, persistent; sepals valvate, 4, persistent; petals 0 or rarely 4 and yellow; stamens 20–40, basally inserted on outside of tubular sheath enclosing ovary; pistil 1, with a lateral, twisted, exserted, persistent style pubescent at base; fruit a glabrous achene.

Coleogyne ramosissima Torr. Blackbrush. Rounded shrubs, 3–12 dm or more tall, with divaricate branches; leaves 3–12 mm long, mainly 0.8–1.5 mm wide, narrowly oblanceolate, obtusish and commonly mucronate apically, strigose with malpighian hairs; sepals 4.5–6.5 (8) mm long, ovate to lanceolate, malpighian hairy and red brown dorsally, glabrous and yellowish ventrally, the margin scarious; petals typically lacking, or rarely 4, yellow, 6–10 mm long; hypanthium membranous, tapering to 5-toothed apex, silky-hairy within, glabrous without; achene ovate, curved, glabrous, 5–8 mm long. Shallow sandy to clay soils in blackbrush and warm desert shrub communities at 760 to 1830 m in Emery, Garfield, Grand, Kane, San Juan, Washington, and Wayne counties; Nevada and Colorado south to California and Arizona; 35 (v). Specimens with petals occur sporadically in the range of the species. The plants are eaten by deer in winter.

Cotoneaster Medicus

Shrubs, unarmed, erect to arcuate or horizontal, deciduous or evergreen; leaves alternate, simple, entire; stipules linear, deciduous; flowers perfect, regular, solitary or in cymes terminating lateral branches; hypanthium short, persistent; sepals 5; petals 5, white or pink; stamens 10–20; pistil 1, the ovary inferior, 2- to 5-loculed, the styles 2–5; fruit a pome. Note: Members of this genus are cultivated widely in Utah as ornamentals. They have potential for use in reclamation and stabilization projects and will probably be maintained as a portion of our introduced flora. The genus has some 50 species distributed in Eurasia, and many more are in cultivation than are treated herein. The species keyed below are merely representative.

1. Leaves mainly 5–12 mm long; flowers usually solitary .. 2
— Leaves mainly 1.5–10 cm long or more; flowers commonly several to many 3
2(1). Petals spreading, white; shrubs spreading; leaves evergreen .... *C. microphylla* Lindl.
— Petals erect, pink; plants depressed-horizontal; leaves semi-evergreen .... *C. horizontalis* Decne.
3(1). Petals erect, obovate, pinkish or white; fruit red or black 4
— Petals spreading, suborbicular, white; fruit red 5
4(3). Fruit black; at least some leaves more than 2.5 cm long .... *C. acutifolia* Turcz.
— Fruit red; leaves less than 2.5 cm long . *C. dielsiana* Pritz.
5(3). Leaves glabrate at maturity .... *C. multiflora* Bunge
— Leaves persistently, white to rusty tomentose beneath . 6
6(5). Leaves white-tomentose beneath, commonly 1–3 cm long .... *C. pannosa* Franch.
— Leaves rusty-tomentose beneath, or finally glabrate, often at least some over 3 cm long .. *C. salicifolia* Franch.

Note: *Cotoneaster* species are not described due to lack of adequate specimens in herbaria. Much work on cultivated plants is needed.

Crataegus L.

Small, deciduous trees or shrubs, commonly armed with thorns; leaves alternate, simple, serrate to doubly serrate, lobed in some; stipules small, adnate to the petiole, glandular-serrate, deciduous; flowers perfect, regular, in corymbose cymes terminating short lateral branchlets; hypanthium short, free above the ovary; sepals 5, tardily deciduous; petals 5, white or pink; stamens (5) 10–25 or more, the filaments filiform; pistil 1, the ovary inferior, 1– to 5–loculed, the styles 1–5; fruit a pome. Only one taxon is widespread in Utah, growing as a portion of the indigenous flora. The following key contains other native and widely cultivated or escaping taxa. Others are present in cultivation but are excluded.

1. Leaves deeply 3– to 7–lobed; styles 1–3; fruits with 1 or 2 seeds; plants ornamental, cultivated and escaping .. .. *C. monogyna*
— Leaves serrate to doubly serrate or somewhat lobed; styles 2–5; plants indigenous 2
2(1). Leaves mostly more than twice long as broad; plants common and widespread; fruit black .... *C. douglasii*
— Leaves mostly less than twice as long as broad; plants rare; fruit red, yellow, or orange 3
3(2). Petioles, at least some, with stalked red glands; teeth of leaves conspicuously tipped with reddish glands; plants known from Cache County .... *C. chrysocarpa*
— Petioles lacking stalked red glands; teeth of leaves not conspicuously red-tipped; plants known from Provo Canyon .... *C. succulenta*

Crataegus chrysocarpa Ashe Yellow Hawthorn. [*C. rotundifolia* Moench, not Lam.; *C. doddsii* Ramaley]. Shrub or small tree with rounded crown, 2–4 m tall or more, with thorns 1–5 cm long or more; leaf blades 1.8–7 cm long, 1–7.2 cm wide, orbicular to obovate, acute apically, acute to broadly obtuse basally, the margins sharply doubly serrate (the serrations red-tipped) and commonly lobed as well; petioles often glandular (at least some); inflorescence more or less villous at anthesis; sepals lance-attenuate to triangular, serrate; petals white, 6–8.5 mm long and about as broad; stamens 5–10; styles 2–4; fruit red to yellow orange; "seeds" not pitted or deeply concave ventrally. Streamsides at ca 1617 m in Blacksmiths Fork, Cache County; Alberta and Manitoba south to Colorado and Nebraska; 1 (0).

Crataegus douglasii Lindl. River Hawthorn. Shrubs or small trees, with rounded crowns, mainly growing as thickets, 2.5–5 (7) m tall, with thorns 1.2–3.5 cm long; leaf blades 1.3–9.5 cm long, 0.8–5.8 cm wide, lanceolate to elliptic, oblanceolate, or obovate, acute to obtuse apically, cuneate basally, serrate to doubly serrate, seldom lobed; petioles often with a pair of raised sub-basal glands; inflorescence glabrous; sepals triangular-attenuate, entire or serrulate, 1.5–3.5 mm long; petals white, 3–7.8 mm long and about as broad or broader; stamens 10–20; styles normally 5; fruit blackish, 6–12 mm thick. The species occurs from southeastern Alaska east to Michigan. Two rather well defined varieties are present.

1. Petals 3–3.8 mm long, 2–4.2 mm wide; leaves slender, at least some 2–4 times longer than broad; fruit 6–8 mm thick when dried; plants of the Uinta Basin .... *C. douglasii* var. *duchesnensis*
— Petals 4.5–8 mm long, 5–7.8 mm wide; leaves commonly less than twice longer than broad; fruit 8–12 mm thick when dried .... *C. douglasii* var. *rivularis*

Var. duchesnensis Welsh Stream courses at 1800 to 2450 m in Duchesne (type from NW of Duchesne) and Uintah counties; endemic; 8 (ii).

Var. rivularis (Nutt.) Sarg. River Hawthorn. [*C. rivularis* Nutt. in T. & G.]. Terraces, flood plains, alluvial fans and canal banks and other moist sites at 1370 to 2135 m in Beaver, Box Elder, Cache, Daggett, Davis, Juab, Millard, Piute, Salt Lake, San Juan, Sanpete, Sevier, Summit, Tooele, Utah, Wasatch, Washington, and Weber counties; Wyoming and Idaho, south to Arizona and New Mexico; 34 (iv).

Crataegus monogyna Jacq. One-seeded Hawthorn; English Hawthorn. [*C. oxyacantha* var. *monogyna* (Jacq.) Loudon; *C. oxyacantha* L. nom. illeg.]. Shrub or small tree to 6 m tall, with thorns to 2 cm long; leaf blades 1.5–5.5 cm long and about as wide, broadly ovate to orbicular in outline, deeply 3– to 7–lobed, obtuse to truncate basally; petioles lacking glands; inflorescences glabrous or glabrate; sepals broadly triangular, 1.5–2.5 mm long; petals pink or white, 3–4 mm long, 4–5 mm wide; stamens commonly 20; styles 1 (2); fruit reddish or orange, 5.5–8 mm thick; $2n = 34$. Cultivated ornamental in Cache, Davis, Juab, Salt Lake, Utah, Weber and perhaps in other counties, escaping in some; introduced from the Old World; 13 (0).

Crataegus succulenta Schrader ex Link. Red Hawthorn. Shrubs or small trees, mainly 2–4 m tall with thorns to 4.5 cm long; leaf blades 1.8–7 cm long, 1–4.5 cm broad, elliptic to obovate, serrate to doubly serrate and often lobed; attenuate to cuneate basally, abruptly acuminate apically; petioles lacking glands; inflorescence sparingly villous to glabrate; petals white 5–7 mm long, 5.8–7.7 mm wide; stamens 10–20; styles 2–4; fruit red, 7–12 mm thick when dry; $2n = 34+$, 51. Indigenous in riparian habitats, mainly in Provo Canyon, Utah County; Colorado eastward to Pennsylvania and southeastern Canada; 3 (i). Note: **Crataegus mollis** (T. & G.) Scheele and/or *C. crus-gallii* L., and perhaps other species of hawthorn occur in cultivation in Utah. The extent is not known, at present.

Cydonia Miller

Small trees, unarmed; leaves alternate, simple, entire; stipules foliose, glandular-margined, deciduous; flowers perfect, regular, solitary, terminal on leafy shoots; hypanthium short; sepals 5, persistent; petals 5, white or pale pink; stamens 15–20; pistil 1, the ovary inferior, commonly 5-loculed; styles 5; fruit a tomentose pome.

Cydonia oblonga Miller Quince. [*Pyrus cydonia* L.; *C. vulgaris* Pers.]. Trees to 6 m tall; leaves petiolate, the blades 1.1–9.5 cm long, 0.8–7 cm wide, ovate to ovate-oblong, villous-tomentose beneath, tomentose above when young, becoming glabrate; flowers solitary; sepals foliose, 6–9 mm long, tomentose (especially within), glandular-margined; petals 13–25 mm long, 8–17 mm wide, obovate to obcordate, white to pale pink; fruit 6–10 cm in diameter, yellow, densely tomentose, fragrant, broadly pyriform; $2n = 34$. Cultivated ornamental and botanical curiosity in Utah, Washington, and possibly other counties; introduced from the Middle East; 2 (0).

Dryas L.

Shrubs or subshrubs, with stoloniferous branches; leaves alternate, simple, crenate to entire, sometimes incised at the base, evergreen; stipules narrowly lanceolate, adnate to the petiole, persistent; flowers perfect (rarely imperfect), regular, solitary; hypanthium saucer-shaped, with an internal glandular disk; sepals 8–10, persistent; petals 8–10; stamens numerous; pistils numerous, distinct, the ovaries superior, each 1-loculed; style 1 per pistil, much elongated in fruit; fruit an achene with a long-plumose style.

Hulten, E. 1959. Studies in the genus *Dryas*. Svensk Bot. Tidskr. 5: 507–542.
Porsild, A. E. 1947. The genus *Dryas* in North America. Canad. Field-Naturalist 61: 175–192.

Dryas octopetala **L.** Mat-forming shrubs; leaves petiolate, the petioles glabrous to sparingly villous and often glandular; leaf blades mostly 1–4 cm long, 3–10 mm broad, lanceolate to lance-oblong, obtuse apically, obtuse to subcordate basally, crenate, green and glabrous to pubescent above, tomentose below, commonly with stipitate glands on the midrib below, often revolute; scapes 1–11 cm long, tomentose and stipitate- glandular; petals white (fading yellowish) or rarely yellowish, 9–15 mm long; staminal filaments glabrous; style plumose, in fruit to 4 cm long; $2n = 18$. Moraines, slopes, and ridge crests in alpine tundra and meadows at 3500 to 3965 m in Daggett, Duchesne, Summit, and Uintah counties; widespread in northern North America; circumboreal; 18 (ii). Our material is assignable to var. **hookeriana** (**Juz.**) **Breitung** (*D. hookeriana* Juz.; *D. octopetala* ssp. *hookeriana* (Juz.) Hulten].

Exochorda Lindl.

Shrubs, deciduous, unarmed; leaves alternate, simple, entire or serrate; stipules none; flowers more or less imperfect (polygamo-dioecious), borne in terminal racemes; hypanthium flaring, with a broad disk internally; sepals 5; petals 5, white; stamens 15–25; pistils 5, connate except for the 5 free styles, the ovary superior, 5-loculed; fruit a bony capsule.

Exochorda racemosa (**Lindl.**) **Rehder** Pearl Bush. (*Amelanchier racemosa* Lindl.; *E. grandiflora* Hook.). Slender shrubs with spreading crowns, to 2.5 m tall or more, the herbage glabrous; leaf blades 1.2–6.5 cm long, 0.5–3.5 cm wide, elliptic to oblong or obovate, cuneate basally, mucronate apically, entire or some serrate in the upper half; racemes 3- to 10-flowered; flowers very showy; sepals 1–2 mm long, broadly rounded, erose apically, chartaceous; petals 10–17 mm long. Introduced ornamental in Utah County and in other low elevation urban areas; native to Asia; 2 (0).

Fallugia Endl.

Shrubs, deciduous, unarmed; leaves alternate, pinnately dissected; stipules adnate to the petiole, triangular-subulate, persistent; flowers mainly perfect, regular, terminal and solitary or in few-flowered cymes; hypanthium hemispheric, persistent, hairy within; sepals 5, alternating with slender bractlets; petals 5, white; stamens numerous; pistils numerous, the ovaries superior, of 1 carpel each; style terminal; fruit an achene, tipped by the plumose style.

Fallugia paradoxa (**D. Don**) **Endl.** Apache Plume. (*Sieversia paradoxa* D. Don). Shrubs to 1.5 m tall, the herbage stellate hairy, the bark scaly; leaves mainly 4–16 mm long, cuneate-flabellate, 3- to 5-lobed, green and lepidote above, rusty-lepidote beneath; pedicels 2–18 mm long; sepals 4–7 (11) mm long, broadly ovate, abruptly acuminate-cuspidate apically; petals 11–14 mm long, 8–15 mm wide, white; pistils numerous; style plumose, 2–4 cm long in fruit. Wash bottoms in mixed desert shrub and pinyon-juniper communities at 940 to 2290 m in Emery, Garfield, Iron, Kane, San Juan, Washington, and Wayne counties; Nevada and California east to Texas and south to Mexico; 32 (vii).

Fragaria L.

Herbaceous, rosulate perennials, commonly stoloniferous; leaves compound, with 3 serrate leaflets; stipules adnate to base of elongate petiole; flowers more or less imperfect (polygamo-dioecious), solitary or in scapose cymes; hypanthium widely spreading; sepals 5, alternating with bractlets; petals 5, white or pinkish; stamens 20, sometimes abortive; pistils numerous, on a pulpy receptacle, superior; fruit of achenes, on a fleshy accessory receptacle.

Rydberg, P. A. 1908. *Fragaria*. N. Amer. Fl. 2: 356–365.

1. Petioles spreading-hairy; terminal tooth of leaflets relatively well developed, commonly surpassing the adjacent lateral teeth; inflorescence usually as long as or longer than the leaves .... *F. vesca*
— Petioles with hairs ascending to appressed-ascending; terminal tooth of leaflets small, commonly surpassed by the adjacent lateral teeth; inflorescence usually shorter than the leaves .... *F. virginiana*

Fragaria vesca **L.** Starvling Strawberry. Stoloniferous herbs, with stems, petioles, and peduncles pubescent with slender spreading to somewhat ascending hairs; petioles 0.8–17.5 cm long (rarely longer); leaflets 3, the terminal one 1.3–6.5 cm long, 1–4.2 cm wide, thin, elliptic to oblong or obovate, coarsely serrate, silky pilose, subsessile or indistinctly petiolulate; scapes ultimately equaling or surpassing the leaves; cymes 3- to 15-flowered; sepals 3.5–7.2 mm long, acuminate to caudate; bracteoles 1.6–5.8 mm long, often bilobed; petals 5–10.5 mm long, white or pinkish; fruit to 1 cm thick, succulent, and palatable; $2n = 14$. Stream banks, terraces, and slopes, broad-leaved deciduous and coniferous woods and brushlands at 1800 to 3200 m in Box Elder, Cache, Carbon, Davis, Duchesne, Salt Lake, San Juan, Sanpete, Summit, Tooele, Utah, and Weber counties; British Columbia and Alberta south to California and New Mexico; 34 (ii). Our material is referable to var. **bracteata** (**Heller**) **R. J. Davis** [*F. bracteata* Heller; *F. vesca* ssp. *bracteata* (Heller) Staudt; *F. helleri* Holz.].

Fragaria virginiana **Duchesne** Mountain Strawberry. Stoloniferous herbs with stems, petioles, and peduncles with appressed to ascending hairs; petioles 2–15 cm long; leaflets 3, the terminal one 1.1–4.4 cm long, 0.5–2.2 cm wide, thickish, obovate to elliptic, coarsely serrate, silky pilose to glabrate, commonly petiolulate; scapes shorter than to surpassing the leaves; cymes 2- to 12-flowered; sepals 3.1–6.5 mm long; bracteoles 1.8–4.5 mm long, not or seldom bilobed; petals 3.5–10 mm long, white or rarely pinkish; fruit to 1 cm thick or more, succulent and palatable; $2n = 56$. Meadows, deciduous and coniferous woods at 2280 to 3300 m in Beaver, Duchesne, Emery, Garfield, Iron, Kane, Piute, Sanpete, Summit, Tooele, Utah, and Wayne counties; Alaska to Northwest Territories, south to Colorado and California; 22 (vi). Our material has been treated as belonging to two rather weak and intergrading varieties. The phase with large petals (i.e., more than 6 mm long) is supposedly more densely pubescent with spreading hairs. That phase is known as var. *platypetala* (Rydb.) Hall [*F. platypetala* Rydb.]. It apparently intergrades completely with the small-flowered supposedly scantily pubescent phase with appressed hairs, known as var. *glauca* Wats. [*F. vesca* var. *americana* Rydb., not T. C. Porter; *F. glauca* (Wats.) Rydb.; *F. virginiana* ssp. *glauca* (Wats.) Staudt]. In a broad sense, as herein interpreted, all of our material is best referred to a single taxon. The oldest available epithet appears to be var. *glauca* Wats. (type from Parleys Park, Summit County).

Geum L.

Perennial rhizomatous herbs; leaves alternate or opposite or mainly basal, pinnatifid to lyrate-pinnatifid; stipules foliose (at least on cauline leaves); flowers perfect, regular, solitary or in open cymes; hypanthium campanulate to saucer-shaped; sepals 5, persistent, alternating with 5 bractlets; petals 5, usually yellow (sometimes pinkish or purplish); stamens numerous; pistils numerous, the ovaries superior, each 1-carpellate; style straight to bent or strongly geniculate and jointed, in some elongate in fruit and in some then with a deciduous terminal segment, in others plumose and persistent; fruit an achene.

1. Stems decidedly leafy; plants often more than 3.5 dm tall; sepals reflexed at anthesis; style strongly geniculate and jointed, the persistent base hooked apically 2
— Stems subscapose; plants commonly less than 3.5 dm tall; sepals ascending to erect at anthesis; style neither geniculate nor jointed 4
2(1). Persistent style base glandular-pubescent; terminal segment of basal leaves much larger than the lateral lobes, mostly rounded or subcordate at base; our common meadow and woodland *Geum* .... *G. macrophyllum*
— Persistent style base glabrous or hirsute, not glandular; terminal segment of basal leaves only somewhat larger than the lateral lobes, cuneate at base; plants uncommon 3
3(2). Petals equal to or shorter than the sepals; receptacle pubescent with coarse hairs; stem leaves with lobes or leaflets often about as broad as long, tapering to a rounded or acute apex; achenes ca 70 ...... *G. urbanum*
— Petals longer than the sepals; receptacle minutely hairy; stems leaves with lobes or leaflets distinctly longer than broad and mostly tapering to an acute apex; achenes 200 or more .... *G. aleppicum*
4(1). Cauline leaves opposite; petals white, pink, or only yellow-tinged, erect or convergent; style much elongate and plumose in fruit .... *G. triflorum*
— Cauline leaves alternate; petals yellow, spreading; style about as long as the achene, glabrous .... *G. rossii*

Geum aleppicum Jacq. Erect Avens. [*G. canadense* Murray, not Jacq.; *G. strictum* Ait.; *G. aleppicum* var. *strictum* (Ait.) Fern.; *G. aleppicum* ssp. *strictum* (Ait.) Clausen]. Plants shortly rhizomatous, 4.5–8 (10) dm tall, the stems and petioles spreading-hirsute; basal leaves 8–23 cm long, lyrate-pinnatifid, main lobes 5–9, all cuneate-obovate, strongly cleft and toothed, the terminal lobe larger but similarly shaped; cauline leaves several; flowers 2 to several; sepals soon reflexed, 4–8 mm long; petals yellow, spreading, about equaling the sepals; stamens 60 or more; achenes 3–4.5 mm long, tipped by persistent style; style strongly geniculate above the middle, the lower segment hirsute to glabrous, not glandular near the base, persistent and hooked apically; $2n = 42$. Wet to dryish meadows at 1400 to 2300 m in Carbon, Grand, Salt Lake, San Juan, Summit, Utah, and Wasatch counties; widespread in North America and Eurasia; 5 (0).

Geum macrophyllum Willd. Large-leaved Avens. [*G. urbanum* ssp. *oregonense* Scheutz; *G. oregonense* (Scheutz) Rydb.; *G. macrophyllum* var. *rydbergii* Farw.]. Plants shortly rhizomatous, 2.3–11.5 dm tall, the stems erect, with spreading hairs; basal leaves 4–28 cm long or more, long-petiolate, the leaflets 9–25 or more, the apical lobe 2.5–13 cm long, 2–15 cm wide, acute to truncate or subcordate basally, dentate and conspicuously lobed, glabrate to glabrous above, hairy along the veins beneath; cauline leaves numerous; sepals 2.5–4.8 mm long, reflexed at anthesis; bracteoles 0.5–3.5 mm long, linear to lanceolate or lacking; petals yellow, 3–5.8 mm long; style elongate with an S-shaped curve above the middle, glabrous or hairy above the bend, glandular (often sparingly so, or the glandular hairs deciduous) below the bend, hooked at apex, the apical section deciduous; $2n = 42$. Aspen, spruce-fir, birch-willow, and grass-sedge communities at 1280 to 3050 m throughout Utah; widespread in North America; Asia; 94 (xiii). Our materials, as described above, belong to a poorly differentiated variety, the var. *perincisum* (Rydb.) Raup [*G. perincisum* Rydb.; *G. macrophyllum* ssp. *perincisum* (Rydb.) Hulten].

Geum rossii (R. Br.) Ser. Ross Avens. [*Sieversia rossii* R. Br.; *Acomastylis rossii* (R. Br.) Greene; *S. scapoidea* A. Nels. in Coult. & Nels, type from W of Marysvale, Piute County]. Herbs, shortly rhizomatous, the rhizomes and stem bases with a persistent thatch of marcescent petioles and stems; stems 0.3–1.8 dm tall, erect, often with 1–3 greatly reduced leaves below the inflorescence; basal leaves 2.2–12 cm long, short-petiolate, pinnatifid, with 15–31 entire to several-toothed or -lobed lateral divisions, the apical lobes similar to the lateral ones except smaller, glabrous to pubescent along veins below, ciliate; sepals 2.9–5.1 mm long, ovate to ovate-lanceolate; bracteoles 2–3.9 mm long, lanceolate; petals yellow, 6.4–8.6 mm long; styles straight, elongate, glabrous except at base, erect; $2n = 70$. Alpine meadows, rock stripes, and talus slopes at 3050 to 3400 m in Beaver, Box Elder, Daggett, Duchesne, Garfield, Grand, Iron, Juab, Piute, Salt Lake, San Juan, Summit, Uintah, Utah, and Wayne counties; Alaska and Yukon south to Oregon, Nevada, Arizona, and New Mexico; Asia; 71 (viii). The specimens from Utah are referable to var. **turbinatum** (**Rydb.**) **C. L. Hitchc.** [*Potentilla nivalis* Torr., not Lapeyr.; *G. turbinatum* Rydb.; *Sieversia turbinata* (Rydb.) Greene; *Acomastylis turbinata* (Rydb.) Greene; *G. sericeum* Greene; *S. sericea* (Greene) Greene], which is only weakly separable from the larger typical phase that occurs in the arctic.

Geum triflorum Pursh Purple Avens; Old Man's Beard. [*Sieversia triflora* (Pursh) R. Br.; *Erythrocoma triflora* (Pursh) Greene; *G. ciliatum* var. *triflorum* (Pursh) Jepson; *E. brevifolia* Greene, type from Panguitch Lake, Garfield County]. Plants shortly rhizomatous, clothed basally with persistent leaf bases; stems 1–4 dm tall, with 1–2 pairs of opposite leaves; basal leaves 2–18.5 cm long, pinnatifid, with mostly 15–31, cleft to lobed divisions, puberulent to pilose; flowers 1–7 (9); sepals showy, reddish to pink or purplish, 7–10 mm long; bracteoles linear to narrowly elliptic; petals yellowish to pink or suffused with crimson, 7–11.2 mm long; styles straight to moderately geniculate, plumose, 2–4 cm long at maturity. Oak-sagebrush, aspen, aspen-fir, and mixed conifer communities, often in meadows, at 1980 to 3500 m in Beaver, Box Elder, Cache, Daggett, Duchesne, Garfield, Iron, Juab, Kane, Millard, Piute, Rich, Salt Lake, Sanpete, Sevier, Summit, Tooele, Wasatch, and Wayne counties; British Columbia east to Newfoundland, south to Nevada, Arizona, New Mexico, Nebraska, and Illinois; 76 (vi). Our specimens are assignable to the weakly differentiated var. *ciliatum* (Pursh) Fassett [*G. ciliatum* Pursh; *Sieversia ciliata* (Pursh) D. Don; *Erythrocoma ciliata* (Pursh) Greene].

Geum urbanum L. Plants shortly rhizomatous, (3) 6–10 dm tall, the stems and petioles spreading-hirsute; basal leaves mainly 2–10 cm long, pinnatifid, the main lobes 3–5 (10), cuneate-obovate, rounded to acute at the apex, the terminal one rhombic-ovate and slightly larger than the lateral ones; cauline leaves several to many; sepals spreading to reflexed, 4–7 mm long; petals yellow, spreading, somewhat shorter than the sepals; achenes 3–6 mm long, tipped by the persistent style; style strongly geniculate above the middle. Disturbed sites in Salt Lake County (Arnow 5546, UT); adventive from Eurasia; 1 (0).

Holodiscus Maxim.

Shrubs unarmed; leaves simple, alternate, toothed, deciduous; stipules lacking; flowers perfect, regular, each closely subtended by 1–3 bracteoles, numerous, borne in panicles; hypanthium saucer-shaped, lined with a disk; sepals 5, persistent; petals 5, white to cream or less commonly pinkish; stamens ca 20; pistils 5, the ovaries superior, each 1-carpellate; style terminal, persistent; fruit a short-stipitate, villous achene.

Ley, A. 1943. A taxonomic revision of the Genus *Holodiscus* (Rosaceae). Bull. Torrey Bot. Club 70: 275–288.

Holodiscus dumosus (Nutt.) Heller Mountain Spray. [*Spiraea dumosa* Nutt. in T. & G.; *S. discolor* var. *dumosa* (Nutt.) Wats.; *Schizonotus argenteus* var. *dumosus* (Nutt.) Kuntze; *H. discolor* var. *dumosa* (Nutt.) Dippel; *Schizonotus dumosus* (Nutt.) Koehne; *S. discolor* var. *dumosus* (Nutt.) Rehder; *Sericotheca dumosa* (Nutt.) Rydb.; *H. microphyllus* Rydb., type from Alta; *S. microphylla* (Rydb.) Rydb.; *H. discolor* var. *microphyllus* (Rydb.) Jepson; *S. concolor* Rydb.]. Shrubs, densely to intricately branched, mainly 0.5–1.5 m tall; main foliage leaves on spur branches, the blades 0.5–3.2 cm long, 0.2–2.3 cm wide, obovate to oblanceolate or elliptic, cuneate basally, prominently toothed or lobed, villous to glabrate or glabrous on one or both surfaces, pale beneath; inflorescence 3–15 cm long; sepals 1.3–1.7 mm long, villous, sometimes pinkish; petals 1.9–2.2 mm long, white, cream, or pinkish; achenes somewhat flattened, villous-hirsute. Ubiquitous in numerous plant communities, where especially common on rock outcrops, slickrock plateau margins, and at bases of cliffs, or in talus slopes at 1280 to 3550 m in all Utah counties; Oregon east to Wyoming and south to California, Nevada, Arizona, and New Mexico; 60 (xii). Attempts at recognition of infraspecific taxa are fraught with difficulties which are likely ecological rather than genetic reflections. A conservative approach is indicated.

Ivesia T. & G.

Perennial herbs, from a caudex; leaves pinnately compound, alternate or primarily basal; stipules of cauline leaves foliose; flowers perfect, regular, borne in compact to open cymes; sepals 5, alternating with 5 bracteoles; hypanthium saucer- to cup-shaped, lined with a disk; petals 5, yellow or white; stamens 5 or 20; pistils 1–15, the receptacle hairy, the ovaries superior; style subterminal; fruit of achenes.

Keck, D. D. 1938. Revision of *Horkelia* and *Ivesia*. Lloydia 1: 75–142.

1.	Leaflets usually fewer than 20; plants of western Utah .. 2
—	Leaflets commonly 22–80 or more; plants various 3
2(1).	Leaflets mainly 3–10 mm long or more; leaves often over 10 cm long; plants known from the Deep Creek Mts. .... *I. setosa*
—	Leaflets mainly 1–3 mm long; leaves commonly less than 8 cm long; plants known from western Beaver County .... *I. shockleyi*
3(1).	Petals white; stems more or less radiate-decumbent 4
—	Petals yellow; stems erect or ascending 5
4(3).	Stamens 5; hypanthium cup-shaped; plants of Utah, Salt Lake, and Summit counties .... *I. utahensis*
—	Stamens 20; hypanthium saucer-shaped; plants of Beaver, Garfield, Sevier, and Tooele counties ... *I. kingii*
5(3).	Hypanthium cup-shaped; flowers in dense, elongate cymes; plants widespread in central and northern Utah, our most common *Ivesia* .... *I. gordonii*
—	Hypanthium saucer-shaped; flowers in open cymes; plants of Beaver, Garfield, Kane, and Washington counties .... *I. sabulosa*

Ivesia gordonii (Hook.) T. & G. Gordon Ivesia. [*Horkelia gordonii* Hook.; *Potentilla gordonii* (Hook.) Greene]. Plants erect, 7–30 cm tall, from a thick woody caudex clothed with persistent leaf bases; herbage puberulent, glandular-puberulent, or glabrous; basal leaves 1–25 cm long; leaflets 20–50 or more, 2–17 mm long, divided to base; cymes congested, many-flowered, 1, few, or several; hypanthium campanulate; sepals 2.6–5 mm long, erect at anthesis, triangular-subulate; bracteoles 1.2–2.8 mm long; stamens 5; pistils 1–6; style not glandular; achenes 1.7–2.1 mm long. Ponderosa pine, spruce-fir, and mixed conifer woods and upwards in alpine sites, often in rocky meadows, at 2050 to 3660 m in Beaver, Cache, Davis, Duchesne, Garfield, Juab, Millard, Morgan, Piute, Salt Lake, San Juan, Sanpete, Sevier, Summit, Tooele, Uintah, Utah, Wasatch, and Weber counties; Washington to Montana, south to California and Colorado; 69 (iv). Specimens from the Wasatch Mts. are more robust than those of other places, but do not seem worthy of taxonomic recognition.

Ivesia kingii Wats. King Ivesia. [*Potentilla kingii* (Wats.) Greene; *Horkelia kingii* (Wats.) Rydb.; *P. eremica* Cov.; *H. eremica* (Cov.) Rydb.; *P. kingii* var. *incerta* Jones; *I. halophila* Heller]. Plants decumbent, the stems 5–22 cm long, radiating from a thickened caudex clothed with blackish persistent leaf bases; herbage glabrous or pubescent, not glandular; basal leaves 0.5–12 cm long or more; leaflets 24–60 or more, 1–6 mm long, entire or ternately divided; cymes dichotomously divided, open; hypanthium saucer-shaped; sepals 2.5–3.1 mm long, spreading at anthesis, lance-attenuate; bracteoles 1.3–1.9 mm long, ovate-lanceolate; petals white, 3–4.2

mm long, 2.5–3 mm wide, clawed; stamens 20; pistils 2–9; style not glandular; achenes 1.8–2.2 mm long. Saline meadows and pans in rabbitbrush, saltgrass, shadscale, greasewood, and sedge communities at 1460 to 2380 m in Beaver, Garfield, Iron, Sevier, and Tooele counties; Nevada and California; 10 (vii). The plants tend to blend with the pale substrate of the saline pans, perhaps accounting for the paucity of our records from Utah.

Ivesia sabulosa (Jones) Keck Sevier Ivesia. [*Potentilla sabulosa* Jones, types from head of Sevier River, "25 miles south of Panguitch"; *Comarella sabulosa* (Jones) Rydb.; *Horkelia sabulosa* (Jones) L. O. Williams; *H. mutabilis* Brandegee; *I. mutabilis* (Brandegee) Rydb.]. Plants erect, the stems 10–42 (50) cm tall, from a woody caudex clothed with persistent leaf bases; herbage glabrous to villous and glandular; basal leaves 3.5–23 (30) cm long, the petioles often suffused red purple; leaflets 30–80, paired, 1–13 mm long, usually divided to the base; cymes branched, open; hypanthium saucer-shaped; sepals 3.3–5.5 mm long, triangular-acuminate, spreading at anthesis; bracteoles 1.3–2.7 mm long, lanceolate; petals yellow, 2.1–3 mm long, 0.3–0.5 mm wide; stamens 5; pistils 1–5; style glabrous or nearly so; achenes 1.7–2.2 mm long. Sagebrush, pinyon-juniper, pygmy sagebrush, ponderosa pine, and spruce communities, commonly on limestone at 1735 to 2745 m in Beaver, Garfield, Kane, Sevier, and Washington counties; Arizona and Nevada; 20 (vii). The type locality of *I. sabulosa* is imprecisely known. Jones states (see Leafl. West. Bot. 10: 216. 1965) that ". . .And 4 miles below Ranch I got 6031 to 33f." The type of *I. sabulosa* is his number 6032. Jones further states that "Ranch is the name of the post office at the head of the Sevier and serves a little farming area there." Presumably "Ranch" and the collecting site are both in Garfield County.

Ivesia setosa (Wats.) Rydb. [*Horkelia baileyi* var. *setosa* (Wats.) Rydb.; *I. baileyi* ssp. *setosa* (Wats.) Keck]. Plants 4–25 cm tall, from a woody caudex clothed with persistent leaf bases; herbage glandular-pubescent; basal leaves 3.5–12 cm long; leaflets 10–20, 3–15 mm long, parted to divided; hypanthium discoid; sepals 2.5–3.2 mm long, ovate-lanceolate; bracteoles lance-oblong to ovate; petals white or cream, subequal to the sepals; stamens 5; pistils 3–7; style glandular; achenes 1.6–2.3 mm long. Mountain slopes at 1700 to 3100 m in Tooele County (Deep Creek Mts.); eastern Nevada; 1 (0).

Ivesia shockleyi Wats. [*Horkelia shockleyi* (Wats.) Rydb.; *Potentilla shockleyi* (Wats.) Jepson]. Plants with stems decumbent to erect, from a woody caudex, this clothed with marcescent leaf bases; herbage pale, densely glandular-puberulent; basal leaves 2–7 cm long, the leaflets in 7–10 crowded pairs, 1–3 mm long, divided into 2–5 oblanceolate to orbicular segments; cymes open, few-flowered; hypanthium yellow, disciform, finally pentagonal; sepals broadly lanceolate, 1.5–3.5 mm long; petals pale yellow, shorter than the sepals; stamens 5, the filaments linear-subulate; pistils typically 3; style scarcely glandular; achenes brown, 2–2.5 mm long. Quartzitic outcrop in pinyon-jnuniper and ponderosa pine communities at 1950 to 2410 m in western Beaver County; Nevada and California; 4 (0).

Ivesia utahensis Wats. Utah Ivesia. [*Potentilla utahensis* (Wats.) Greene; *Horkelia utahensis* (Wats.) Rydb.]. Plants decumbent or ascending, the stems radiating from a thickened caudex clothed with brownish leaf bases; herbage glandular-viscid or glandular-pubescent; basal leaves 1.2–9 cm long; leaflets 30–40, paired, 1–4 mm long, divided to base; cymes capitate or in congested corymbs; hypanthium campanulate; sepals 1.5–2.7 mm long, narrowly triangular; bracteoles 1–1.3 mm long, oblong; petals white, 1.8–2.7 mm long, 1–1.7 mm wide; stamens 5; pistils commonly 2; style not glandular; achenes 1.7–1.9 mm long. Alpine tundra and krummholz communities, often in talus, at 3200 to 3600 m in Salt Lake (type from Bald Mt.), Summit, Utah, and Wasatch counties; endemic; 5 (0).

Kerria DC.

Cultivated shrubs, unarmed, deciduous; leaves alternate, simple, doubly serrate; stipules lance-linear, deciduous; flowers perfect, regular, solitary, terminating short lateral branchlets of the current year; hypanthium short, lined by a disk; sepals 5, small, entire; petals 5, yellow; stamens numerous; pistils 5–8, the ovaries superior; style slender; fruit an achene (seldom produced).

Kerria japonica (L.) DC. Japanese Kerria. [*Rubus japonicus* L.]. Shrubs with slender green arching branches; leaves 1–7 cm long, 0.7–3.5 cm wide, ovate to lanceolate, long-acuminate, bright green; pedicels 0.5–2.5 cm long; sepals ca 3.5 mm long, broad-ovate, membranous; petals yellow, 12–20 mm long; achenes 4–5 mm long. Widely cultivated ornamental, persisting but not spreading, observed in Box Elder, Cache, Davis, Salt Lake, Utah, and Weber counties; introduced from Japan and China; 2 (i). Double flowered phases are common.

Malus Miller

Trees with mainly unarmed branches; leaves alternate, simple, not or rarely lobed; stipules linear, caducous; flowers perfect, regular, borne in umbels; hypanthium short; sepals 5, persistent or deciduous; petals 5, white or pink; stamens usually 15–50; pistil 1, the ovary inferior, usually 5–loculed; styles 2–5, connate at the base; fruit variously shaped, applelike, the flesh usually lacking stone-cells. Note: The apples are all introduced into our flora. Mostly they are cultivated and persist following cultivation, but some escape and are established widely in Utah, especially the pomological varieties. The following key is tentative; our cultivars often represent hybrid derivatives involving two to several of the species types. Other taxa probably are present in the state.

1. Leaves on elongated shoots lobed or notched; a crabapple .... *M. ioensis*
— Leaves on elongated shoots neither lobed nor notched .. 2
2(1). Mature leaves glabrous, crenate-serrate, but not sharply so; fruit mainly 1.5–2.5 cm in diameter; a crabapple .... *M. sylvestris*
— Mature leaves more or less tomentose on one or both surfaces or, if glabrous, the pedicels very long and the fruit smaller 3
3(2). Pedicels 3–4.5 cm long, very slender; fruit mainly 0.8–1.2 cm in diameter; a crabapple ...... *M. hupehensis*
— Pedicels mainly shorter than 3 cm long, slender to moderately thickened; fruit size various 4
4(3). Leaf margins sharply serrate; flowers ordinarily pink; flowering ornamental crabapple .... *M. floribunda*
— Leaf margins merely crenate-serrate to serrate; flowers various 5

5(4). Flowers commonly pink; fruit 0.8–1.2 cm in diameter; a crabapple .... M. baccata

— Flowers commonly white within (pink sometimes on dorsal surface; apple of commerce .... M. pumila

Malus baccata (L.) Borkh. Siberian Crab. [*Pyrus baccata* L.]. Small trees to 5 m, the branchlets glabrous; leaves ovate to oblong, 2–8 cm long, serrate, glabrous to puberulent on one or both sides; petioles 1.5–5 cm long; petals white or pink, 12–18 cm long; fruit 0.8–1.2 cm in diameter, the calyx deciduous. Cultivated ornamental tree in lower elevation portions of Utah; introduced from Asia; 4 (0).

Malus floribunda Sieb. ex Van Houtte Showy Crab. [*Pyrus floribunda* Sieb.]. Small trees to 8 m, the branchlets pubescent; leaves ovate to oblong, 2–7 cm long, sharply serrate, usually tomentose on one or both sides; petioles mostly 1.5–5 cm long; petals rose pink, 15–20 mm long or more; fruit mostly 0.5–1 cm in diameter. Cultivated ornamental trees in lower elevation portions of Utah; introduced from Japan; 5 (0).

Malus hupehensis (Pamp.) Rehder Tea Crab. [*Pyrus hupehensis* Pamp.; *M. theifera* Rehder]. Small trees to 5 m tall, the branchlets glabrous or essentially so; petioles 0.8–4 cm long; leaf blades oblong-elliptic to ovate, 2.5–7.8 cm long, minutely tomentulose on both sides at maturity; peduncles 3–4.5 cm long; sepals 4–6 mm long; petals 15–20 mm long, commonly white; fruit 0.8–1.2 cm in diameter. Cultivated ornamental trees of lower elevations in Utah; introduced from Asia; 7 (0).

Malus ioensis (Wood) Britt. Iowa Crab. [*Pyrus ioensis* (Wood) Bailey]. Small trees to 9 m tall, the branchlets tomentose; leaves ovate to oblong, 2.5–10 cm long, tomentose on both sides, at least when young; petals usually white (sometimes pinkish), 12–25 mm long; fruit 2–3 cm in diameter. The Iowa crab is grown occasionally in lower elevation portions of Utah; introduced from the north-central states; 4 (0).

Malus pumila Miller Common Apple. [*M. domestica* Borkh.]. Small to moderate trees to 10 m tall, the branchlets tomentose when young, becoming glabrous; leaves ovate to oblong or elliptic, 1.5–10 cm long, tomentose on one or both sides (even in age); petals usually white within, often pink dorsally, 12–25 mm long; fruit mainly 2.5–12 cm in diameter, red, reddish purple, or yellow. This is the apple of commerce, and it is widely cultivated in Utah; it persists and occurs as established trees throughout the state; introduced from Eurasia; 18 (ii).

Malus sylvestris Miller Crabapple. Small to moderate trees to 10 m tall, the branchlets glabrous or puberulent when young, often somewhat thorny; leaf blades 2–6 cm long, ovate to elliptic, crenate-serrate, acuminate or cuspidate; petals 8–20 mm long, white or pink; fruit mainly 1.5–2.5 cm in diameter, sour. Introduced cultivated trees, persisting and escaping in Utah; native to Eurasia; 6 (i).

Peraphyllum Nutt. in T. & G.

Shrubs, unarmed, deciduous; leaves alternate, simple, entire or nearly so; stipules adnate to petioles, triangular, minute, deciduous; flowers perfect, regular, solitary or few on lateral branchlets of the season; hypanthium campanulate, disk-lined; sepals 5, spreading to reflexed, persistent; petals 5, white or pink; stamens 15–20; pistil 1, the ovary inferior, 2- to 3-carpellate but falsely 4- to 6-loculed by intrusion of parietal septa; styles 2 or 3, the stigma capitate; fruit a fleshy applelike pome.

Peraphyllum ramosissimum Nutt. in T. & G. Squaw-apple. Shrubs 4–15 (20) dm tall, intricately branched; leaves alternate, mainly on short lateral spurs, 1.1–3.9 cm long, 0.4–0.9 cm wide, oblanceolate, abruptly acute, appressed puberulent (especially beneath), entire or minutely serrulate; pedicels 4–13 mm long, with 1–3 caducous bractlets; sepals 2.9–4 mm long, triangular-acuminate, serrulate to entire; petals white to pink, 6.5–9 mm long, 5–8.5 mm wide; pomes 8–18 mm thick, yellow orange, the flavor bad when ripe. Oak-sagebrush, pinyon-juniper, mountain brush, and ponderosa pine communities at 1500 to 2500 m in Beaver, Garfield, Grand, Iron, Juab, Kane, Millard, Piute, San Juan, Sanpete, Sevier, Uintah, and Washington counties; Oregon and Idaho south to California and Colorado; 27 (iv). The fruit is attractive when ripe but the flavor is not agreeable. Perhaps, when cooked with sugar it might be better?

Petrophytum (Nutt.) Rydb.

Shrubs, prostrate and mat-forming, conforming to the rock substrate; leaves alternate, commonly appearing rosulate, simple, entire, stipules lacking; flowers perfect, regular, borne in compact spikelike panicles on scapose, bracteate peduncles; pedicels with 1 or more bractlets; hypanthium cup-shaped, lined with a disk; sepals 5, erect at anthesis; petals 5, white; stamens numerous; pistils commonly 5, the ovaries superior, each 1–loculed; styles slender, exserted from the flower; fruit of usually 5 follicles.

Petrophytum caespitosum (Nutt.) Rydb. Rock Spiraea. [*Spiraea caespitosa* Nutt. in T. & G.; *Eriogyna caespitosa* (Nutt.) Wats.; *Luetkea caespitosa* (Nutt.) Kuntze]. Mat-forming shrubs to 10 dm broad or more; leaves 3–17 mm long, 1.5–4.5 mm wide, spatulate to oblanceolate or obovate, pilose on one or both surfaces, rarely almost or quite glabrous; peduncles 0.5–12 cm long, with bractlike leaves much reduced upward; panicles spikelike, 0.5–2.5 cm long, often branched at base or with axillary panicles along axis of peduncle; pedicels 0.5–3 mm long, bracteolate; sepals 1.2–2.1 mm long, narrowly triangular; petals 1.3–2.5 mm long, 0.4–0.8 mm wide, white; fruit 1.5–2.1 mm long. On limestone or granitic outcrops or gravels from sagebrush upward to spruce-fir communities at 1375 to 3050 m, and on sandstone (Entrada, Navajo, Kayenta, Cedar Mesa, etc.) often in hanging gardens at 1125 to 1895 m in Beaver, Box Elder, Cache, Daggett, Garfield, Grand, Iron, Juab, Kane, Millard, Piute, Salt Lake, Summit, Tooele, Uintah, Utah, Washington, and Wayne counties; Oregon east to South Dakota and south to California, Arizona, New Mexico and Texas; 73 (xvii). This beautiful dwarf shrub flowers in late summer and autumn. Major variations involve the tendency to glabrous leaves of some Great Basin specimens, and a tendency to short peduncles in some materials from the hanging gardens of southeastern Utah. More materials are required to adequately access the variation.

Physocarpus Maxim.

Shrubs, unarmed, deciduous, with exfoliating bark; leaves alternate, simple, palmately lobed and veined, usually with at least some stellate hairs; stipules membranous, deciduous; flowers perfect, regular, borne in terminal corymbs; hypanthium cup-shaped, lined with a disk;

sepals 5; petals 5, white or pink; stamens 20–40, inserted with petals at edge of disk; pistils 1–5, the ovaries superior and partially connate; styles slender, the stigmas capitate; fruit of one or more follicles, each several seeded.

Howell, J. T. 1931. A Great Basin species of *Physocarpus*. Proc. Calif. Acad. Sci. IV. 20: 129–134.

1. Pistil and style solitary; leaves less than 2 cm long; staminal filaments of two alternating and markedly unequal lengths .... *P. alternans*
— Pistils and styles 2 or 3, or the carpels connate below; leaves various but commonly over 2 cm long; staminal filaments subequal or somewhat unequal 2
2(1). Leaves mainly 0.7–2.5 cm long; mature carpels swollen, not flattened; plants evidently rare in Utah .... *P. monogynus*
— Leaves mainly 2–8 cm long; mature carpels flattened; plants common in south central to northern Utah .... *P. malvaceus*

Physocarpus alternans (Jones) J. T. Howell Dwarf Ninebark. [*Niellia monogyna* var. *alternans* Jones; *Opulaster alternans* (Jones) Heller]. Shrubs commonly 4–12 dm tall, and about as broad; twigs stellate pubescent and sometimes glandular; bark shreddy on older twigs; leaf blades 0.3–2 cm long, 0.3–2.2 cm wide, oval-ovate to ovate, cordate to subcordate basally, more or less 3-lobed, doubly crenate, pubescent to sparingly pubescent on both sides; inflorescence subumbellate, (1) 2– to 12 (17) -flowered; pedicels 2–10 mm long; hypanthium stellate-hairy; sepals 1.3–3.2 mm long, oval to suboblong; petals 1.8–3.2 mm long, 1.5–3 mm wide, white or suffused with red pink; follicle solitary, densely stellate, 4–5 mm long. Rock outcrops, ledges, and cliff faces in desert shrub, pinyon-juniper, oak, and ponderosa pine communities at 1980 to 2750 m in Box Elder, Cache, Carbon, Daggett, Duchesne, Emery, Garfield, Grand, Juab, Millard, Salt Lake, San Juan, Tooele, Utah, Washington, and Wayne counties; Idaho, Nevada, and Colorado; 24 (iv). Jones (Zoe 4: 38–44. 1893) discussed the problems within the genus *Physocarpus* then called *Niellia*. The basic problems are still as Jones outlined them eight decades ago. The genus is still in need of a definitive revision.

Physocarpus malvaceus (Greene) Kuntze Mallow-leaved Ninebark. [*Niellia malvacea* Greene; *Opulaster malvaceus* (Greene) Kuntze; *N. monogyna* var. *malvacea* (Greene) Jones; *Spiraea opulifolia* var. *pauciflora* T. & G.; *Opulaster pubescens* Rydb.; *O. cordatus* Rydb.] Shrubs, mainly 8–20 dm tall, rarely more, and often as broad; twigs glabrous to minutely stellate; bark shreddy on older branchlets; leaf blades (0.8) 2.2–8 cm long, (1.2) 2–8.2 cm wide, ovate to broadly ovate, cordate basally, 3-lobed, doubly crenate, glabrous above, stellate-pubescent to glabrous beneath; inflorescence corymbose, 5– to 30 (+) -flowered; pedicels 0.7–2.3 cm long; hypanthium stellate-hairy; sepals 2.2–4.6 mm long, ovate to lance-oblong; petals 3.3–6.7 mm long, 1.5–4.8 mm wide, white; follicles paired, connate to the middle or above, substipitate, densely stellate, 4.9–6 mm long. Moist slopes and streamsides in mountain brush, aspen, and mixed conifer woodlands at 1600 to 3300 m in Cache, Emery, Garfield, Juab, Millard, Salt Lake, San Juan, Sanpete, Summit, Tooele, Utah, Wasatch, and Weber counties; British Columbia east to Alberta and south to Oregon and Wyoming; 69 (vii). Two other large-leaved ninebark species are grown in Utah. They are **P. opulifolius** (L.) Raf. [*Spiraea opulifolia* L.] and **P. capitata** (Pursh) Kuntze [*Spiraea capitata* Pursh]. They both possess 3–5 glabrous pistils connate only at the base. They differ in that the leaves of *P. opulifolius* are commonly glabrous beneath while those of *P. capitatus* are ordinarily densely stellate beneath. The extent of these species in cultivation is not known.

Physocarpus monogynus (Torr.) Coult. Mountain Ninebark. [*Spiraea monogyna* Torr.; *Opulaster monogynus* (Torr.) Kuntze]. Shrubs, mainly 4–20 dm tall; twigs glabrous to densely stellate; bark shreddy on older branchlets; leaf blades 0.5–2.8 cm long, 0.6–3.2 cm wide, ovate to orbicular, cordate basally, commonly 3-lobed, doubly crenate, glabrate or glabrous on both sides or sometimes stellate-hairy especially below; inflorescence corymbose, 9– to 25–flowered (sometimes more); pedicels 0.3–0.8 cm long; hypanthium stellate-hairy; sepals 2.3–3 mm long, ovate; petals 2.2–3.9 mm long, 2.6–3.5 mm wide, white; follicles paired, connate to the middle or above, substipitate, densely stellate, 3–4.5 mm long. Canyon bottoms and moist slopes in mountain brush, aspen, and Douglas fir communities at 1650 to 2150 m in Carbon and Utah counties; South Dakota and Wyoming south to Texas and Arizona (?); 4 (i). Our material seems to fit well within the range of variation of materials from Colorado, New Mexico, and South Dakota. The plants are smaller in all features from the similar *P. malvaceus* (q.v.).

Potentilla L.

Annual, biennial, or perennial herbs (a shrub in *P. fruticosa*); leaves alternate or basal, palmately or pinnately compound; stipules lanceolate to ovate, sometimes sheathing; flowers perfect, regular, borne solitary or in cymes; hypanthium saucer- to cup-shaped; sepals 5, alternating with bractlets; petals 5, yellow to ochroleucous (fading white in some), broadly obovate and often emarginate; stamens 10–25; pistils numerous, on a hemispheric to conical receptacle, superior; fruit of achenes, the styles terminally, medially, or basally attached, jointed, finally deciduous, glabrous or papillose. Note: This is a rather difficult genus, with many interpretations, due partially to hybridization, which obscures differences between taxa, and because of different weight given to portions of the genus as providing basis of segregation into several genera. Segregate genera *Argentina*, *Drymocallis*, and *Pentaphylloides* are cited in synonymy. Synonymy is restricted to basionyms and those involving Utah plant types.

Hitchcock, C. L. 1961. Rosaceae. In: Hitchcock, C. L., A. Cronquist, M. Ownbey, and J. W. Thompson. Vascular plants of the Pacific Northwest. Univ. Washington Publ. Biol. 17(3): 89–194.
Keck, D. D. 1940. *Potentilla glandulosa* and its allies: Taxonomy. pp. 128–137. In: Clausen, J., D. D. Keck, and W. M. Hiesey. Environmental studies on the nature of species. I. Effect of varied environments on western north American plants. Carnegie Inst. Washington Publ. 520: 31–194. *Potentilla gracilis* and its allies.
Rydberg, P. A. 1898. Monograph of the North American Potentillae. Mem. Dept. Bot. Columbia Univ. No. 2.
———. 1908. Potentilleae. N. Amer. Fl. 22(3/4): 268–377.

1. Flowers reddish purple to dark red or maroon; petals less than half as long as the sepals .... *P. palustris*
— Flowers white, cream, or yellow, not at all red or purple 2

2(1).	Plants woody shrubs; styles laterally attached to the ovary; ovaries and achenes hairy .... *P. fruticosa*			
—	Plants herbaceous; styles basal, terminal, or lateral; ovaries and achenes glabrous 3			
3(2).	Plants strongly stoloniferous; styles laterally attached to the ovary; leaves evidently pinnate, strongly bicolored, silvery white beneath .... *P. anserina*			
—	Plants without stolons, or rarely somewhat stoloniferous; styles basal or terminal; leaves usually not evidently pinnate and strongly bicolored 4			
4(3).	Plants annual, biennial, or perennial from a taproot; basal leaves poorly developed; leaves green on both sides, never tomentose 5			
—	Plants perennial, commonly with well-developed caudices; basal leaves well developed, tomentose or greene 8			
5(4).	Leaflets 5–9; stamens commonly more than 20; plants of low elevations .... *P. paradoxa*			
—	Leaflets commonly 3 (rarely 5 on some leaves); stamens 10–20; plants of various elevations 6			
6(5).	Stems pubescent with multicellular, glandular hairs; sepals glandular .... *P. biennis*			
—	Stems pubescent with hirsute or crinkly simple hairs; sepals not glandular 7			
7(6).	Stems softly pubescent with crinkly, villous hairs; lowermost leaves sometimes 5–foliolate ..... *P. rivalis*			
—	Stems rough hairy with spreading, hirsute hairs; leaves all 3–foliolate .... *P. norvegica*			
8(4).	Leaves with 3 leaflets; plants of high elevations, La Sal Mts. .... *P. nivea*			
—	Leaves with more than 3 leaflets (if rarely with 3 leaflets not from the La Sal Mts.) 9			
9(8).	Leaves palmately compound 10			
—	Leaves pinnately compound 13			
10(9).	Plants mainly 0.5–2 dm tall; leaflets villous or villous-tomentose on one or both sides; mainly alpine and subalpine 11			
—	Plants mainly 2–5 dm tall; leaflets variously pubescent to subglabrous; typically montane but sometimes alpine 12			
11(10).	Leaflets deeply divided into oblong lobes, pilose above, tomentose beneath; plants of high elevations in the Uinta Mts. .... *P. rubricaulis*			
—	Leaflets subentire to toothed or lobed, the lobes (if any) tapering to the apex, pilose above, tomentose and pilose beneath; plants widespread, but not known from high elevations in the Uinta Mts. .... *P. concinna*			
12(10).	Leaflets toothed from somewhat below the middle or above, the base long-cuneate, green or gray green on both sides; anthers mostly 0.4–0.7 mm long .... *P. diversifolia*			
—	Leaflets commonly toothed to well below the middle, the base cuneate or not, commonly bicolored, often tomentose or paler beneath; anthers mainly 0.8–1.3 mm long .... *P. gracilis*			
13(9).	Styles lateral near the base of the achene; leaflets broadly obovate to ovate, green on both sides 14			
—	Styles subterminal or borne above the midlength of the achene; leaflets various, but never both broadly obovate to ovate and green on both sides 15			
14(13).	Inflorescence narrow and erect, often much elongate, the branches suberect; sepals equaled or exceeded by the pale yellow petals; plants uncommon .... *P. arguta*			
—	Inflorescence open, the branches often diverging; sepals surpassing the yellow to pale yellow petals; plants common .... *P. glandulosa*			
15(13).	Styles shorter than mature achene, glandular-thickened at base; leaflets toothed 1/2 the distance to the middle or more; stipules often cleft 16			
—	Styles exceeding the mature achene, not glandular-thickened at the base; leaflets various; stipules entire or only shallowly lobed 17			
16(15).	Leaflets 5–11 (more than 5 on some leaves); plants widely distributed .... *P. pensylvanica*			
—	Leaflets 3–5; plants of the Uinta Mts. ... *P. rubricaulis*			
17(15).	Plants mainly 0.5–1.5 dm tall; leaflets deeply dissected, or 5 or fewer and silky pilose-tomentose 18			
—	Plants mainly 1.8–5 dm tall; leaflets various but, if deeply dissected, the plants larger and otherwise different 20			
18(17).	Leaflets commonly 3–5, subpalmate, densely silky-villous .... *P. concinnus*			
—	Leaflets commonly 7–21, pinnate, densely silky-pilose to pilose or strigose and green on both surfaces . 19			
19(18).	Leaflets silky-sericeus on both surfaces, 7–21; stems ascending to erect .... *P. ovina*			
—	Leaflets green on both surfaces, 7–11; stems decumbent to ascending .... *P. plattensis*			
20(17).	Leaflets mostly 5–7, often subpalmate and deeply cleft, green on both sides .... *P. diversifolia*			
—	Leaflets mostly 5–13, pinnate (subpalmate in some *P. gracilis*), tomentose on one or both sides 21			
21(20).	Leaflets toothed to the base, mainly 5–9, often subpalmate; anthers over 0.8 mm long .... *P. gracilis*			
—	Leaflets toothed from below the middle or only at the apex, mainly 7–13, pinnate; anthers less than 0.8 mm long 22			
22(21).	Leaflets shallowly toothed near the apex, usually entire below the middle, pilose-sericeus beneath, seldom also tomentose .... *P. crinita*			
—	Leaflets toothed from below the middle, definitely tomentose beneath the pilose-sericeus hairs .... *P. hippiana*			

Potentilla anserina L. Common Silverweed. [*Argentina anserina* (L.) Rydb.]. Perennial herbs, with long strawberrylike stolons; leaves 2–10 cm long or more, pinnately compound with 5–25 main leaflets interspersed by smaller ones, the terminal leaflet 0.5–5.5 cm long, 0.3–2.6 cm wide, oval to oblong or oblanceolate to obovate, coarsely serrate, green and glabrous to pilose above, pale and villous over a tomentum beneath; scapes 1.5–15 cm long or more, villous to densely so, leafless; sepals ovate, 3–10 mm long, pubescent to glabrous, erect, enlarging in fruit; petals yellow, 7.5–16 mm long; achenes 1.5–2 mm long; 2n = 28, 42. Meadows, lake shores, terraces, and floodplains, especially where wet part of the season, at 1300 to 2600 m in all Utah counties, except Davis, Grand, Iron, Morgan, Washington, and Weber; widespread in North America; circumboral; 55 (iv). The worldwide review by Rousi (Ann. Bot. Fenn. 2: 47–112. 1965) indicates that formal varieties are not warranted.

Potentilla arguta Pursh Acute Cinquefoil. Perennial, glandular-pubescent herbs, 2.5–6 (8) dm tall, from a caudex; basal leaves 6–30 mm long, pinnately compound with 5–11 leaflets, the terminal one 15–60 mm long (or more), 12–40 mm wide, oval to elliptic or obovate, doubly dentate to somewhat lobed, green and glandular-pubescent on both surfaces; flowers several to many, showy; sepals 4–8 mm long, lance-ovate, longer in fruit; bracteoles 2–6 mm long, oblong to narrowly lanceolate; petals yellow to cream or white, mostly 5–8 mm long; receptacle sparsely hairy; achenes 1–1.5 mm long; styles basal, ca 1 mm long, deltoid. Mountain brush, aspen, and spruce-fir communities, often in meadows, at 1950 to 3360 m in Box Elder, Cache, Carbon, Daggett, Davis, Iron, Juab, Morgan, Salt Lake, San Juan, Sanpete, Summit, Utah, Washington, and Weber counties; the species is widespread in North america; 24 (iii). Our material is referable to **var. convallaria (Rydb.) F. T. Wolf** [*P. convallaria* Rydb.], and is transitional to *P. glandulosa* (q.v.).

***Potentilla biennis* Greene** Greene Cinquefoil. [*P. lateriflora* Rydb., type from Utah]. Annual or biennial herbs, mostly 1–6 (7) dm tall, from taproots; leaves mainly cauline, palmately 3 (4) -foliolate, the terminal leaflet 1–5 cm long, 1–3 cm wide, obovate to oblanceolate, crenate-serrate, pubescent with spreading to appressed hairs and multicellular glandular ones; flowers several to numerous, inconspicuous; sepals mostly 2–4 mm long, ovate to lance-ovate; bracteoles 2–3 mm long, ovate-lanceolate to oblong; petals yellow, 1.5–3 mm long; achenes numerous, ca 1 mm long; styles terminal, ca 1 mm long, basally thickened. Meadows, streamsides, springs, and seeps at 1525 to 2324 m in Beaver, Box Elder, Cache, Daggett, Davis, Duchesne, Garfield, Iron, Juab, Millard, Salt Lake, Sevier, Wasatch, Washington, and Weber counties; widespread in western North America; 7 (v).

***Potentilla concinna* Richards.** Pretty Cinquefoil. Perennial herbs, the stems decumbent- spreading to ascending, 0.1–1 dm tall; leaves mainly basal, palmately to pinnately 5– to 7 (9) -foliolate, the terminal leaflet 0.3–3.8 cm long, 1–10 mm wide, obovate to oblanceolate, toothed only at apex or along the length, often folded, not markedly bicolored, pilose and tomentose beneath, pilose to glabrous (less commonly tomentose) above; cauline leaves 1 or 2; cymes (1) 2– to 7–flowered, the flowers showy; sepals 2.2–4.8 mm long, triangular-ovate; bracteoles 1.3–3.5 mm long, oblong to lanceolate; petals yellow, 2.7–7.3 mm long; achenes numerous, 1.6–2 mm long; style subapically attached, smooth or glandular, not basally thickened, sometimes clavate. Three largely sympatric, variable, and often weak varieties are present in Utah. It seems likely that some, if not most, of the variation can be accounted for by intermediacy with other taxa beyond the circumscription of *P. concinna*. The trends are separable as follows:

1. Leaflets pinnately disposed, often 7, the lowermost often scattered or reduced; stems usually trailing, longer than the basal leaves; pubescence translucent-brownish; plants of western Utah ..... *P. concinna* var. *proxima*

— Leaflets palmately disposed, commonly 5–9; stems shorter than the basal leaves; pubescence white 2

2(1). Leaflets toothed from the middle or below (at least some), commonly flat and green above; plants widespread .... *P. concinna* var. *modesta*

— Leaflets toothed at apex only, rarely minutely crenate also, commonly folded, the obscured upper surface greenish or not .... *P. concinna* var. *bicrenata*

Var. *bicrenata* (Rydb.) Welsh & Johnston [*P. bicrenata* Rydb.]. Pinyon-juniper, grassland, sagebrush, ponderosa pine, and spruce-fir communities at 2050 to 2875 m in Beaver, Duchesne, Emery, Garfield, San Juan, Sevier, and Wayne counties; Colorado, Wyoming, and New Mexico; 18 (iv).

Var. *modesta* (Rydb.) Welsh & Johnston [*P. modesta* Rydb.]. Sagebrush, meadow, aspen, spruce-fir, and Douglas fir communities at 2280 to 3480 m in Carbon, Emery, Garfield, Piute (type from Mt. Barrett, near Marysvale), Sanpete, Sevier, Wasatch, and Wayne counties; endemic; 22 (v). The *modesta* phase of the pretty cinquefoil occurs generally above the range of var. *bicrenata*, with which it forms intermediates. Specimens from Emery and Sanpete counties approach *P. gracilis*.

Var. *proxima* (Rydb.) Welsh & Johnston [*P. proxima* Rydb., type from Mt. Belknap Peak, Piute County; *P. beanii* Clokey; "*P. quinquefolia*" sensu Rydb.]. Pinyon-juniper, sagebrush, ponderosa pine, ans spruce-fir communities at 2200 to 3000 m in Beaver, Carbon, Garfield, Iron, Piute, and Wayne counties; Nevada; 40 (v).

***Potentilla crinita* Gray** Hair-tuft Cinquefoil. [*Ivesii lemmonii* Wats.; *P. lemmonii* (Wats.) Greene; *P. crinita* var. *lemmonii* (Wats.) Kearney & Peebles]. Perennial herbs, the stems ascending to erect, 1.3–4.3 dm tall; leaves mainly basal, pinnately 7– to 13–foliolate, the terminal leaflet 0.6–3 cm long, 0.2–0.9 cm wide, elliptic-oblong to narrowly oblanceolate, toothed at the apex only (somewhat below on some leaves), not markedly bicolored, pilose and sometimes slightly tomentose beneath, strigose-pilose to glabrate above; cauline leaves 2 or 3; cymes several- to many-flowered, the flowers showy; sepals 3–5.4 mm long, oblong-lanceolate, acuminate; petals yellow, 5–6.5 mm long; achenes several, 1.3–1.8 mm long; styles subapical, filiform, 1.8–2.1 mm long. Sagebrush, mountain brush, pinyon-juniper, ponderosa pine, and aspen communities at 1890 to 2590 m in Garfield and Wayne counties; Arizona, Nevada, New Mexico, and Colorado; 14 (iii). The varieties are connected by intermediates. Two varieties have been distinguished on the basis of pubescence types and purported leaflet tooth number differences. They are sympatric and wholly confluent. Their segregation appears to be moot. This taxon is distinguished from *P. hippiana* by the narrow, few-toothed leaflets, strigose-glandular pubescence, and openly branched inflorescence with smaller flowers. Intermediates are known, and possibly *P. crinita*, which is wholly within the range of *P. hippiana*, would best be represented at infraspecific rank within that species, a fact not proposed herein.

***Potentilla diversifolia* Lehm.** Wedge-leaf Cinquefoil. Perennial herbs, the stems ascending to erect, 0.6–3.2 dm tall; leaves mainly basal, palmately to less commonly pinnately (3) 5– to 7–foliolate, the terminal leaflet 1.2–4 cm long, 0.3–1.8 cm wide, obovate to oblanceolate, toothed mainly above the middle, not bicolored, green on both sides, strigose to pilose on both sides, seldom tomentose below; cauline leaves 1–3; cymes several-flowered, the flowers showy; sepals 5–6.5 mm long, triangular to triangular-atttenuate; bracteoles 2.1–4.5 mm long, oblong to lanceolate; petals yellow, 5.6–9.5 mm long; achenes numerous, 1.5–1.9 mm long; style subapical, 1.9–2.6 mm long, smooth or glandular, not basally thickened. Two varieties are present:

1. Leaves palmately or subpalmately disposed, merely toothed or lobed; plants common and widespread .... *P. diversifolia* var. *diversifolia*
— Leaves pinnately disposed (on at least some leaves), the leaflets dissected almost to the midvein; plants of the Deep Creek Mts. .... *P. diversifolia* var. *multisecta*

Var. *diversifolia* Dry to wet meadows, lake margins, stream banks, forest margins, alpine tundra, and rocky ridges at 2745 to 3500 m in all Utah counties, except Carbon, Davis, Kane, Millard, Morgan, Rich, Washington, and Weber; Alaska and Yukon, south to California, Arizona, New Mexico, and South Dakota; 89 (xii). This is a highly variable taxon, transitional to both *P. gracilis* and *P. ovina*. Reports of var. *perdissecta* (Rydb.) C. L. Hitchc. from the Uinta Mts. appear to be based on plants with pinnately disposed leaflets intermediate between var. *diversifolia* and other taxa. The following variety is the most pronounced of regional ecotypes, but is difficult to distinguish in all cases.

Var. *multisecta* Wats. [*P. dissecta* var. *multisecta* (Wats.) Wats.; *P. multisecta* (Wats.) Rydb.]. Ponderosa pine, spruce-fir, and limber pine communities at 2135 to 3050 m in Juab County; Nevada, Idaho, and Montana; 7 (ii).

Potentilla fruticosa L. Shrubby Cinquefoil; Yellow Rose; Tundra Rose. [*Dasiphora riparia* Raf.; *D. fruticosa* (L.) Rydb.; *P. floribunda* Pursh; *Pentaphylloides floribunda* (Pursh) A. Love]. Shrubs to 1 m tall or more; bark shreddy; leaves 1–5 cm long, pinnately 3- to 7-foliolate, the terminal leaflet 0.5–2.5 cm long, 0.2–1 cm broad, oblong to elliptic, entire, green and sparsely hairy to glabrate above, grayish and silvery hairy below, somewhat revolute; flowers 1 to several, conspicuous; sepals 3.5–9 mm long, ovate-lanceolate; bracteoles 4–13 mm long, lanceolate to elliptic; petals yellow, 6–14 mm long, rounded; receptacle hairy; achenes 1.5–2 mm long, white-hairy; 2n = 14, 28. Meadows, sagebrush, aspen, lodgepole, ponderosa pine, and spruce-fir communities often on floodplains or stream banks at 1700 to 3500 m in all Utah counties, except Beaver, Davis, Millard, Morgan, and Tooele; Alaska east to Newfoundland, south to California, New Mexico, Iowa and New Jersey; Eurasia; 111 (xiv). This handsome plant is known in cultivation in Utah, Summit, and Salt Lake counties at elevations below 1900 m, and should be grown widely in the state.

Potentilla glandulosa Lindl. Glandular Cinquefoil. [*Drymocallis glandulosa* (Lindl.) Rydb.] Perennial glandular-pubescent herbs, 0.8–6 (7) dm tall, from a caudex; basal leaves 3–22 cm long, pinnately compound, with 5–9 leaflets, the terminal one 0.7–6.8 cm long, 0.6–4.2 cm wide, obovate to elliptic, doubly dentate or lobed, green and variously pubescent or glandular on both surfaces, flowers several to many, showy or inconspicuous; sepals 5.5–9.3 mm long, lance-ovate, often acuminate, longer in fruit; bracteoles 4.5–7.2 mm long, oblong to narrowly lanceolate; petals mainly yellow, mostly 4–7.5 mm long; receptacle sparsely hairy; achenes numerous, 1–1.2 mm long; styles from below the middle, 0.8–1 mm long. The glandular cinquefoil is widespread and common in much of Utah, where it consists of a series of intergrading populations varying by degree from each other and only arbitrarily separable from *P. arguta* (q.v.), with which it probably should be combined. In that case, the name would be *P. arguta*, since that epithet has priority. The complex was summarized by Keck (Carnegie Institution Washington 520: 26–124. 1940). He recognized four morphological phases from Utah (i.e., the subspecies *arizonica*, *micropetala*, *glabrata*, and *pseudorupestris*). Examination of a fairly large series of specimens from Utah demonstrates that all, except that designated as *micropetala*, are connected by a series of intermediates, and might best be regarded as belonging to a single polymorpic and highly plastic var. *intermedia*. The following arbitrary key will serve to differentiate most specimens of those two varieties.

1. Petals much shorter than the sepals, 4–5 mm long, 2–4 mm broad .... *P. glandulosa* var. *micropetala*
— Petals shorter than to slightly exceeding the sepals, mainly 5–7.5 mm long and about as broad .... *P. glandulosa* var. *intermedia*

Var. *intermedia* (Rydb.) C. L. Hitchc. [*Drymocallis pseudorupestris* var. *intermedia* Rydb.]. Mountain brush, ponderosa pine, lodgepole pine, aspen, and spruce-fir communities, often in meadows at 1890 to 3200 m in Beaver, Box Elder, Cache, Carbon, Daggett, Duchesne, Garfield, Juab, Millard, Piute, Rich, Salt Lake, Sanpete, Sevier, Summit, Tooele, Uintah, Utah, Wasatch, Washington, and Weber counties; British Columbia and Alberta south to Oregon, Arizona, and Wyoming; 86 (viii).

Var. *micropetala* (Rydb.) Welsh & Johnston [*Drymocallis micropetala* Rydb., type from City Creek Canyon; *P. glandulosa* ssp. *micropetala* (Rydb.) Keck]. Sagebrush, mountain brush, upwards to alpine meadows at 1430 to 3050 m in Cache, Salt Lake, Sanpete, Sevier, Summit, and Weber counties; Idaho and Wyoming; 6 (i). This small-petaled phase resembles *P. norvegica* in flower size and in general conformation, but differs inter alia in pistil features and leaflet number.

Potentilla gracilis Dougl. ex Hook. Slender Cinquefoil. Perennial, variably pubescent herbs from a caudex; stems ascending to erect, 0.4–6 (8) dm tall; basal leaves 3–30 cm long, or more, palmately or pinnately compound, with 5–9 leaflets, the terminal one 1.3–10.7 cm long, 0.4–3.7 cm wide, obovate to oblanceolate, crenate, serrate, or toothed to dissected, commonly bicolored, but not uncommonly green on both sides; flowers several to numerous, showy; sepals 3.5–9.5 mm long, lanceolate to lance-ovate, acute to attenuate-acuminate; bracteoles 2.4–8.5 mm long, lance-oblong; petals yellow, 5.6–8 (10) mm long; achenes numerous, 1.3–1.6 mm long; styles subapical, 2–2.3 mm long, filiform or thickened to ca 1/2 the length, tapered to stigma. Several intergrading phases are recognizable at varietal level in this most common and widespread of our cinquefoil species. Variants tend to represent recombinant types of several recurrent features, i.e., glands on calyx teeth, tomentum on lower leaflet surface, and depth of incision of leaflet margin. There are four more or less intergrading varieties separable as follows:

1. Leaflets dissected almost or quite to the midrib, or some teeth, at least, more than 5 mm long; plants uncommon, often of marshes at lower to moderate elevations .... *P. gracilis* var. *elmeri*
— Leaflets seldom dissected as much as 1/2 the distance to the midrib; teeth seldom to 5 mm long; plants more common to abundant, often at moderate to higher elevations 2

2(1). Leaves prominently discolored, green above, white-tomemtose beneath, often subpinnate to pinnate .... *P. gracilis* var. *pulcherrima*

— Leaves less prominently discolored, pilose to glabrate, less commonly somewhat tomentose beneath 3

3(2). Calyx lobes and often the leaflets glandular and stiffly pilose; plants often drying brownish .... *P. gracilis* var. *brunnescens*

— Calyx lobes and leaflets without glandular hairs; plants usually green .... *P. gracilis* var. *glabrata*

Var. *brunnescens* (Rydb.) C. L. Hitchc. [*P. brunnescens* Rydb.]. Meadows, mountain brush, aspen, spruce-fir, and alpine tundra at 2300 to 3180 m in Box Elder, Cache, Duchesne, Emery, Salt Lake, Sanpete, Sevier, Summit, Uintah, and Utah counties; Washington and Montana south to Nevada and Wyoming; 15 (iii). This variety is freely transitional to var. *pulcherrima* and to *P. diversifolia* where they occur together.

Var. *elmeri* (Rydb.) Jepson [*P. elmeri* Rydb.; *P. pectinisecta* Rydb., type from Salt Lake City; *P. pecten* Rydb.]. Stream banks, meadows, and seeps at 1370 to 2680 m in Box Elder, Duchesne, Juab, Rich, Salt Lake, Sanpete, Sevier, Summit, Uintah, Utah, Wasatch, Washington and Weber counties; Oregon to Montana and south to California and New Mexico; 27 (0). The variety is transitional to var. *glabrata*.

Var. *glabrata* (Lehm.) C. L. Hitchc. [*P. nuttallii* var. *glabrata* Lehm.]. Mountain brush, sagebrush, aspen, and spruce-fir communities at 1675 to 2745 m in Daggett, Emery, Piute, Sanpete, Sevier, Summit, Wasatch, and Weber counties; British Columbia and Alberta south to California, New Mexico, and Nebraska. Included in our limited materials are those specimens with spreading hairs on petioles and stems, passing as var. *permollis* (Rydb.) C. L. Hitchc. [*P. permollis* Rydb.]. They seem to be transitional completely with var. *glabrata*; 13 (i).

Var. *pulcherrima* (Lehm.) Fern. [*P. pulcherrima* Lehm.; *P. hippiana* var. *pulcherrima* (Lehm.) Wats.]. Sagebrush, mountain brush, aspen parkland, and spruce-fir communities often in meadows at 1890 to 3450 m in all Utah counties, except Kane, Millard, Morgan, Tooele, Wayne, and Weber; British Columbia east to Saskatchewan, south to Nevada, Arizona, New Mexico, and South Dakota; 125 (vii). This phase of *P. gracilis*, especially, forms intermediates with *P. hippiana* and *P. concinna*. It is likewise freely transitional to the other varieties of *P. gracilis*.

Potentilla hippiana Lehm. Hipp Cinquefoil; Woolly Cinquefoil. [*P. wardii* Greene, type from Thousand Lake Mt., Wayne County]. Perennial, variably pubescent herbs from a caudex, the stems ascending to erect, 1.1–4.8 (5.5) dm tall; basal leaves 2.5–19 cm long or more, pinnately compound with 7–11 leaflets, the terminal one 0.9–4.7 cm long, 0.4–1.9 cm wide, oblanceolate to oblong or elliptic, serrate or toothed less than half way to midrib, the teeth from below the middle, grayish tomentose to pilose on one or both surfaces; flowers several to numerous, showy; sepals 4.3–6.5 mm long, lance-ovate, acute to acuminate; bracteoles 2.4–5.5 mm long, lance-oblong; petals yellow, 6–7.7 (9.5) mm long; achenes several to numerous, 1.5–1.9 mm long; styles subapical, 1.8–2.3 mm long. Meadow, aspen, spruce-fir, and alpine tundra communities at 2250 to 3450 m in Beaver, Daggett, Duchesne, Emery, Garfield, Grand, Iron, Kane, Piute, Salt Lake, San Juan, Sevier, Summit, Uintah (?), Wasatch, and Wayne counties; British Columbia east to Michigan and south to New Mexico, Arizona, and Nebraska; 66 (iv). Apparent intermediates are known between the Hipp cinquefoil and *P. gracilis*, and especially var. *pulcherrima*, but also with var. *brunnescens*.

Potentilla nivea L. Perennial non-glandular herb, the stem decumbent to ascending, 0.7–2 dm long, from a caudex; stems and petioles densely tomentose; basal leaves 2–9 cm long, trifoliolate, the terminal leaflet 0.8–2.7 cm long, 0.7–1.7 cm wide, obovate to elliptic or oblanceolate, coarsely toothed to near the base, green and silky-pubescent above, densely snow-white tomentose below, strongly bicolored; flowers 1–15 in an open inflorescence, showy; sepals 2.5–5 mm long, lanceolate; bracteoles 1.8–4.5 mm long, oblong to lanceolate; petals yellow, 4–7 mm long; achenes several, 1–1.5 mm long; style subterminal, ca 1 mm long, uniformly thickened; 2n = 14, 28, 49, 56, 63, 70. Alpine tundra at 3050 to 3965 m in Grand, Juab, San Juan, and Summit counties; Alaska east to Greenland, south in the Rocky Mountains to New Mexico and southeastern Utah; Eurasia; 8 (iii).

Potentilla norvegica L. Rough Cinquefoil. [*P. monspeliensis* var. *norvegica* (L.) Farw.; *P. monspeliensis* L.; *P. norvegica* ssp. *monspeliensis* (L.) Asch. & Graebn.]. Annual or biennial (short-lived perennial?) herbs, the stems erect, 0.7–5 dm tall or more, from a taproot, the stems and petioles sparsely stiff-hairy; leaves mostly cauline, palmately (or subpinnately) compound with 3 (rarely 5) leaflets, the terminal one 1–8 cm long, 0.6–2.7 cm wide, obovate to oblanceolate, coarsely toothed to near the base or entire and cuneate in the lower part, green and sparsely stiff-hairy to glabrous above, paler and stiff-hairy beneath, especially along the veins; flowers several to many, inconspicuous; sepals 4–6 mm long, ovate-lanceolate, enlarging in fruit; bracteoles 3–6 mm long, oblong to elliptic or lanceolate; petals yellow or whitish, 2.5–3.6 mm long; achenes numerous, 0.8–1.1 mm long; style subterminal, 1.6–0.9 mm long; 2n = 56, 63?, 70. Floodplains, wet meadows, lake shores, and other moist sites at 1370 to 2930 m in Beaver, Cache, Duchesne, Emery, Garfield, Grand, Iron, Kane, Morgan, Salt Lake, Sevier, Summit, Uintah, Utah, Wasatch, and Wayne counties; widely distributed in the nothern hemisphere; 16 (ii).

Potentilla ovina Macoun Perennial herbs, from a caudex, the stems decumbent to ascending, 1–3.5 (5) dm tall; leaves mainly basal, pinnately compound with (5) 9–18 leaflets, the terminal mostly 0.6–2.1 mm long, 0.4–1.1 cm wide, deeply pinnately dissected or apically few-toothed, conspicuous; calyx 6–8 (10) mm long, lobes ovate-lanceolate to lance-attenuate; bracteoles 1.8–4 mm long, oblong to lance-oblong; petals yellow, 5–6.4 mm long; achenes many, 1.1–1.7 mm long; style subterminal, 1.8–2.7 mm long, filiform. There are two fairly distinct varieties in Utah:

1. Leaflets densely to uniformly sericeous, grayish green or gray, often with a lower layer of sparse tomentum, 5–10 mm long, often with more than 6 teeth; leaf rachis 2–6 cm long .... *P. ovina* var. *ovina*

— Leaflets glabrous to sparsely sericeous-strigose, usually green, never tomentose, 10–20 mm long, often with 3–5 teeth; leaf rachis 3–12 cm long .... *P. ovina* var. *decurrens*

Var. *decurrens* (Wat.) Welsh & Johnston [*P. dissecta* var. *decurrens* Wats., type from Uinta Mts., Summit

County]. Meadows and rocky ridges, ponderosa pine, lodgepole, spruce-fir, aspen, and (less commonly) alpine tundra communities, often on siliceous substrates, at 2200 to 3500 m in Beaver, Box Elder, Cache, Daggett, Davis, Duchesne, Garfield, Grand, Piute, Salt Lake, San Juan, Sanpete, Summit, Tooele, Uintah, and Wayne counties; Colorado to Nevada and Wyoming, north to Alberta; 57 (iv). This is a characteristic form of montane meadows in the Uinta Mts. It forms intermediates with the following and with *P. diversifolia*, sometimes simulating *P. diversifolia* var. *multisecta* (q.v.).

Var. *ovina* [*P. diversifolia* var. *pinnatisecta* Wats., type from Uinta Mts., Summit County]. Meadows and rocky ridges, openings in spruce-fir and alpine tundra communities, often on limestone, at 2800 to 3900 m in Box Elder, Cache, Daggett, Davis, Duchesne, Grand, Juab, Piute, Salt Lake, San Juan, Sanpete, Summit, Uintah, Utah, and Wasatch counties; Alberta to Oregon and Wyoming; 21 (i). This taxon has been confused with *P. plattensis*, but its tomentulose leaflets and erect-ascending pedicels in fruit appear to be diagnostic. Hybrids between *P. ovina* (usually var. *decurrens*) and *P. gracilis* var. *pulcherrima* are frequently encountered in Utah (Box Elder, Cache, Daggett, Piute, and Summit counties); leaves are pinnate or subpalmate with 7-15 leaflets that are usually bicolored and moderately tomentose beneath, and stems are ascending, contrasting with the usually decumbent *ovina* and the erect-ascending *pulcherrima*.

Potentilla palustris (**L.**) Scop. Marsh Cinquefoil. [*Comarum palustre* L.]. Perennial, rhizomatous, palustrine herbs, the stems prostrate to ascending, purplish, glabrous to hairy, mostly 1-10 dm long; leaves mostly cauline, the lower ones the largest, mostly 0.5-2 dm long, pinnately (3) 5- to 7-foliolate, the terminal leaflet 1.8-8.5 cm long, 0.7-3.2 cm broad, oblanceolate to elliptic, coarsely serrate (sometimes doubly so), green and hairy to glabrous above, paler and hairy beneath; flowers few to several, showy; sepals 7-15 mm long, ovate-lanceolate, enlarging in fruit, purplish; bracteoles 3.5-11 mm long, lanceolate to linear; petals purplish or reddish purple, 2.5-5 mm long; receptacle glabrous; achenes numerous, 1-1.5 mm long. Wet meadows and bogs at 2865 to 2930 m in the Uinta Mts., Uintah County; Alaska and Yukon east to Labrador, south to California and Wyoming; circumboreal; 4 (0). This remarkable potentilla species was discovered in Utah by Mr. Joel Tuhy.

Potentilla paradoxa Nutt. in T. & G. Contrary Cinquefoil. Annual, biennial or short-lived perennial herbs, the stems decumbent to ascending or erect, 0.8-9 dm tall, sparsely to moderately villous; leaves mainly cauline, all pinnately compound with 5-11 leaflets, the terminal one 1.2-4.5 cm long, 0.4-2.5 cm wide, often dissected into 3 main confluent lobes, obovate to oblanceolate, toothed to below the middle, cuneate to the base, green on both sides, strigose to strigulose; flowers few to many, not especially showy; sepals 3-5.2 mm long, ovate, shorter than the bracteoles, enlarging in fruit; bracteoles 4-6.5 mm long, lanceolate; petals yellow or whitish, 3.2-4.3 mm long; achenes numerous, 0.8-1 mm long, laterally enlarged along ventral suture; style subterminal, 0.5-0.7 mm long. Beaches, marshes, and lake shores at 1350 to 1650 m in Salt Lake and Utah counties; widespread in North America; Asia; 6 (0).

Potentilla pensylvanica L. Perennial, glandular to nonglandular herbs from a caudex, the stems ascending to erect, 0.5-4.5 dm tall or more; leaves mostly basal, 2-18 (25) cm long, erect, pinnately compound with 5-17 leaflets, the terminal one 0.9-4.5 (6) cm long, 0.4-1.2 cm wide, elliptic to oblong or oblanceolate, coarsely toothed to narrowly lobed, the sinuses extending more than half way to the midrib, conspicuously revolute-margined, green and somewhat hairy above, white- or more commonly greenish- or yellowish tomentose below; flowers few to many in a glomerate inflorescence, showy; sepals 3.5-7.5 mm long, ovate-lanceolate; bracteoles 3-6 mm long, narrowly lanceolate; petals yellow, 5-8 mm long; receptacle glabrous; achenes numerous, 0.9-1.2 mm long; style subterminal, 0.9-1.0 mm long, conic, with a conspicuously thickened and often pappillose base. This is a highly variable species throughout its large range in North America; the taxonomic problems in this species are mirrored in the whole section (sect. Multifidae), to which *P. rubricaulis* and *P. multifida* belong. There is seemingly only one entity in Utah which merits separation from typical *P. pensylvanica*:

1. Leaflets 5-7, subdigitate (10-30% of the rachis occupied with leaflets); pubescence silvery or yellowish white, often of long entangled hairs densely matted; plants alpine
.... *P. pensylvanica* var. *paucijuga*
— Leaflets 7-17, often pinnate (30-60% of rachis occupied); pubescence olive-greenish, dull, not silvery, usually composed of short curly hairs sparsely matted; plants widespread, sometimes alpine ....
.... *P. pensylvanica* var. *pensylvanica*

Var. *paucijuga* (Rydb.) Welsh & Johnston [*P. paucijuga* Rydb., type from La Sal Mts.]. Grassy tundra and alpine communities at 3200 to 3800 m in Duchesne, Grand, Juab, Piute, San Juan, and Summit counties; endemic to Utah; 3 (0). In the La Sal Mts. this form may hybridize with *P. nivea*; the taxa are usually distinguished by the glomerate inflorescence with glandular pubescence, revolute-margined leaflets, and pinnate leaves of var. *paucijuga*.

Var. *pensylvanica* Sagebrush, sagebrush-grass, and meadow communities at 2200 to 3725 m in Beaver, Box Elder, Carbon, Daggett, Duchesne, Emery, Garfield, Grand, Iron, Piute, San Juan, Sevier, Wayne, and Wasatch counties; Hudson Bay to Alaska, south to northern Mexico, Texas, and Arizona. 55 (iv).

Potentilla plattensis Nutt. Perennial herbs with stems 0.5-2.5 dm long, decumbent or creeping through meadow herbs and grasses; leaves basal and cauline, 3-11 cm long, pinnately compound with 11-23 leaflets, these verticillate or subverticillate; terminal and lower leaflets all pinnately toothed with 5-8 teeth cutting 70-90% to midrib, glabrous to sparsely strigose, green to grayish green on both surfaces; stems with 3-15 flowers on recurved pedicels in fruit; calyx (5) 6-8 mm long, strigose or puberulent; achenes numerous; styles subterminal, 1.5-3.0 mm long, filiform. Wet meadows, bogs, and valley bottoms at 1820 to 2700 m in Kane and Wayne counties; Alberta and Manitoba south through Wyoming and Colorado to central Arizona and New Mexico; 2 (0). This species has been confused with *P. ovina* in the past, but the habitats are usually sharply distinct, and the combination of trailing stems, pinnately-dissected leaflets, and recurved pedicels are diagnostic for *P. plat*-

tensis. Sometimes there are two forms at the same site; a lax trailing form in moist portions of a meadow, and a more compact form with prostrate stems on better-drained soil in the same meadow.

Potentilla rivalis Nutt. [*P. millegrana* Engelm. ex Lehm; *P. leucocarpa* Rydb. in Britt. & Br.]. Annual or biennial herbs from a taproot, the stems 2–6 dm tall, spreading to suberect, pubescent with fine villous or villous-tomentose hairs; leaves mainly cauline, palmately to subpinnately compound with 3–5 leaflets, the terminal 1–4 cm long, 0.4–1.3 cm wide, obovate to oblong to oblanceolate, coarsely toothed (at least above the middle), green and minutely villous on both sides; flowers numerous, inconspicuous; sepals 2.3–4.2 mm long, ovate; bracteoles 3.3–4 mm long, lanceolate; petals yellow, 1.5–1.9 mm long, not enlarged laterally; style subterminal, 0.6–0.8 mm long, fusiform. Reservoir shores and other mesic sites at 1400 to 2320 m in Carbon, Salt Lake, Sevier, and Uintah counties; widespread in North America; 4 (i).

Potentilla rubricaulis Lehm. [*P. hookeriana* Lehm.; *P. saximontana* Rydb.; *P. pedersenii* Rydb.; *P. furcata* Porsild]. Perennial herbs, stems ascending from a caudex, leaves mostly basal, eglandular, palmate to subpinnate with 5–7 leaflets, the terminal one 1.0–2.5 cm long and 0.5–1.0 cm wide, few-toothed with narrow-crenate or serrate teeth, usually deeply divided 50–90%, often revolute-margined, densely grayish tomentose below, sparsely tomentose to puberulent above, strongly bicolored, often pilose as well below with tufts of hairs at the tips of ultimate segments, the petioles pilose-sericeous with subsidiary tomentum; cymes 1– to 10–flowered, the flowers showy but small; sepals 4–6 mm long; bracteoles 3–6 mm long; style subapical, 0.9–1.1 mm long, conic, usually conspicuously thickened and papillose below. Alpine tundra meadows and rocky ridges at 2410 to 3780 m in Duchesne, Grand, San Juan, Summit, and Uintah counties; Greenland to Alaska, south to British Columbia, sparsely and rarely through Montana into Wyoming, and Colorado; 9 (0). This name is very often misapplied, but here it is taken for a strictly alpine species with cone-shaped, thickened style ca 1 mm long, more or lesss digitate leaves with 5–7 leaflets, and petioles with long straight hair in addition to tomentum. It has affinities with sect. Multijugae (see *P. pensylvanica*), but may be allied to *P. nivea* and its relatives as well, with whom it may hybridize.

Prunus L.

Shrubs or small trees with unarmed branches; leaves alternate, simple, entire to serrate or serrulate, rarely lobed, commonly with glands on petioles or blade bases; stipules lance-attenuate to linear, caducous; flowers perfect, regular, solitary, in umbellate clusters, or in racemes; sepals 5, borne atop a cup-shaped to turbinate hypanthium; petals 5, white, pink, rose, or red; stamens numerous; pistil 1, free from the usually deciduous hypanthium; style 1, elongate, with capitate stigma; fruit a drupe. Note: The genus is represented by only three indigenous species in Utah, *P. emarginata*, *P. fasciculata*, and *P. virginiana*. However, the numerous introduced species persist following cultivation, and many escape. They are treated herein for those reasons.

1.	Leaves entire or serrate only near the apex, borne in fascicles, less than 2 cm long and 5 mm wide; shrubs with divaricate branches, of low elevations, indigenous in Beaver, Millard, and Washington counties . . . . *P. fasciculata*	
—	Leaves various, usually larger; shrubs to trees of moderate size, variously distributed, but not of desert shrublands	2
2(1).	Flowers borne in pedunculate, elongate or corymbose clusters	3
—	Flowers borne singly or in sessile umbellate clusters from winter buds	8
3(2).	Flowers borne in corymbose racemes, the pedicels subtended by persistent bracts	4
—	Flowers borne in elongate racemes, the pedicels without persistent bracts	6
4(3).	Flowers to 20 mm wide or more, the petals often pink or double or both; bracts 3–8 mm long, cuneate, the truncate apex fringed-serrulate; flowering cherry . . . . *P. serrulata*	
—	Flowers ca 12 mm wide, the petals white, single; bracts ca 1 mm long, ovate, serrulate	5
5(4).	Corymbs subumbellate; fruit red; leaves much longer than wide; plants indigenous, uncommon . . . . *P. emarginata*	
—	Corymbs elongate; fruit black; leaves about as wide as long; plants introduced, escaping (rootstock) . . . . *P. mahaleb*	
6(3).	Leaves leathery thickened, more or less evergreen, entire; plants cultivated shrubs . . . . *P. laurocerasus*	
—	Leaves not especially thickened, serrate to serrulate; shrubs or small trees, cultivated or not	7
7(6).	Hypanthium pubescent within; racemes pendulous or spreading in anthesis; cultivated ornamental . . . . *P. padus*	
—	Hypanthium glabrous within; racemes erect or ascending in anthesis; indigenous or less commonly cultivated . . . . *P. virginiana*	
8(2).	Leaves cordate-ovate to broadly ovate; cultivated apricot, rarely escaping . . . . *P. armeniaca*	
—	Leaves lanceolate to oblong, or less commonly oblanceolate to broadly elliptic; cultivated ornamental and fruit trees and shrubs	9
9(8).	Plants shrublike, mainly 2 m tall or lower	10
—	Plants shrub or treelike, mainly 2–5 m tall or more .	12
10(9).	Leaves glabrous, less than 2 cm broad; flowers commonly 2–4 per bud . . . . *P. besseyi*	
—	Leaves hairy, at least along veins beneath; flowers solitary (rarely 2) per bud	11
11(10).	Leaves long-villous beneath, sparingly villosulose above, broadly elliptic to oblong; flowers single; cultivated for ornament and for fruit . . . . *P. tomentosa*	
—	Leaves sparingly villous on both sides; flowers mostly double; cultivated ornamental . . . . *P. triloba*	
12(9).	Flowers solitary, or sometimes 2 or 3 per bud	13
—	Flowers 3 or more per bud	16
13(12).	Axillary buds on fruiting branchlets borne in 3's, the lateral ones producing flowers	14
—	Axillary buds on fruiting branchlets borne singly	15
14(13).	Leaves pubescent beneath; pedicels pubescent; cultivated plum . . . . *P. domestica*	
—	Leaves glabrous beneath, except on midrib; pedicels glabrous; cultivated flowering plum . . . . . *P. cerasifera*	

15(13). Leaves sharply serrate; fruit not leathery and splitting, exposing the stone at maturity; cultivated peach, commonly escaping .... *P. persica*

— Leaves crenate-serrate; fruit leathery, splitting and exposing stone at maturity; sparingly cultivated almond, not escaping .... *P. dulcis*

16(12). Plants forming shrubby thickets; flowers 14 mm across or less; fruit an astringent plum ... *P. americana*

— Plants treelike or small trees, not forming thickets; flowers 10–30 mm across or more; fruit a cherry 17

17(16). Inner flowerbud scales reflexed or spreading; petals 8–14 mm long; sweet cherry .... *P. avium*

— Inner flower bud scales erect; petals 6–9 mm long; tart cherry .... *P. cerasus*

***Prunus americana* Marshall** American Plum; Pottawattami Plum. Shrubs, often forming thickets, rarely treelike, to 5 m tall; branchlets sometimes thornlike, glabrous; leaves 2.5–7 cm long, 0.5–3 cm wide, elliptic to ovate or lanceolate, sharply serrate, long-attenuate apically, acute to obtuse basally, glabrous or pubescent along veins beneath; petioles usually glandless; flowers 1–4, in sessile or subsessile umbels, from lateral buds, appearing before the leaves, in ours mainly 14 mm broad or less; petals white, 5–7 mm long, 2.5–3 mm wide; sepals spreading, puberulent; hypanthium puberulent without and within; fruit a yellow to red plum. Cultivated fruit plant, and probably indigenous in portions of Utah, but also introduced by pioneers to all areas of the state. The plants spread underground and form thickets which persist in Cache, Duchesne, Piute, Uintah, Salt Lake, Utah, Washington, and Wayne counties; introduced from central U. S.; 12 (ii).

***Prunus armeniaca* L.** Apricot. Small trees to 8 m tall or taller; branchlets green to brown, armed or unarmed; leaf blades mainly 1.5–7 cm long, 1–6 cm wide on mature branches, cordate-ovate to ovate, obtusely serrate, abruptly attenuate, obtuse to cordate basally, glabrous, or hairy along veins beneath; petioles usually with glands; flowers solitary, appearing before leaves; petals white (rarely pinkish), 8–12 mm long, obcordate to orbicular; sepals glandular, the hypanthium glabrous except basally; pedicels villosulose; fruit pubescent, fleshy and edible. Apricot trees are grown in the lower elevation portions of the state, where they escape and persist as waifs along canals, roadsides, and fence rows. Specimens examined are from Beaver, Carbon, Millard, Utah, and Wayne counties; introduced from China; 10 (iii).

***Prunus avium* L.** Sweet Cherry. Trees to 8 m tall or taller; branchlets soon brown, unarmed; leaf blades mainly 3–10 cm long, 2.5–8 cm wide, broadly oblanceolate to obovate or elliptic, obtusely serrate to doubly serrate, abruptly short-attenuate, obtuse to rounded basally, glabrous or long-hairy especially along veins beneath; petioles commonly with glands; flowers 2–4 per bud, appearing with early leaves; petals white (rarely pink), 8–14 mm long, obcordate to orbicular; sepals, hypanthium, and pedicels glabrous; fruit glabrous, red to almost black, edible. The Bing, Lambert, Royal Ann, and other cultivars of the sweet cherry are grown commercially and in garden orchards at low elevations in Utah, where frost is not prohibitive. Trees escape from cultivation along canals and in riverbottoms. Many are grown as grafted stock; some on rootstocks of other *Prunus* species, especially on *P. mahaleb* (q.v.). Specimens examined are from Carbon, Utah, and Washington counties; introduced from Eurasia; 9 (i).

***Prunus besseyi* Bailey** Western Sand Cherry. Shrubs to 1.3 m tall; branchlets soon brownish, not thornlike; leaf blades mainly 1.5–5 cm long, 0.5–1.8 cm wide, elliptic to oblanceolate, obtusely serrate, acute to cuspidate apically, cuneate basally; glabrous; petioles commonly lacking glands; flowers commonly 2–4 per bud, appearing with early leaves; petals white, 4–6 mm long, 2.5–3.5 mm wide, elliptic; sepals, petals, and hypanthium glabrous; fruit glabrous, black, commonly astringent. The western sand cherry is used as a dwarfing rootstock for peaches, cherries, and other *Prunus* species. The plants persist and escape in orchard areas of the state; erosion control and wildlife plantings are probable; specimens were examined from Utah and Wasatch counties; introduced from the Great Plains; 3 (i).

***Prunus cerasifera* Ehrh.** Cherry Plum; Flowering Plum. Small trees to 6 m tall, rarely taller; branchlets soon brown, not thornlike; leaf blades mainly 1.8–6.5 cm long, 1.2–4 cm wide, ovate to elliptic, serrate to doubly serrate, acute apically, acute to rounded basally, villous along the veins beneath; petioles seldom with glands; flowers solitary, or less commonly 2 or 3 per bud; petals pink to violet or white, 6–9 mm long, 4.5–7 mm wide, suborbicular; sepals, hypanthium, and pedicels glabrous, except along marginal insertion of filaments; fruit red to purple, edible; 2n = 16. Cultivated ornamentals of streets, lawns, and gardens, persisting and in lower elevation portions of the state; specimens examined from Utah County; introduced from Asia; 19 (i). The hybrid between *P. cerasifera* and *P. mume* Sieb. & Zucc., the Japanese apricot is also cultivated in Utah, under the name **P. x blireana Andre.** The hybrid has broad purple leaves.

***Prunus cerasus* L.** Sour Cherry; Pie Cherry. Small trees to 5 m tall, rarely more; branchlets soon brown, not thornlike; the leaf blades mainly 3–10 cm long, 1.2–5 cm wide, oblanceolate to obovate or elliptic, doubly serrate, abruptly acuminate, acute to rounded basally, glabrous except hairy along veins beneath; petioles bearing glands; flowers commonly 3 per bud (less commonly fewer); petals white, 7–9 mm long and about as broad; sepals crenate-serrate, glabrous; hypanthium and pedicels glabrous; fruit red, soft, sour. This is the tart cherry of commerce, widely cultivated in Utah and a favorite food of robins; specimens examined from Utah County; introduced from Eurasia; 2 (0). The species flowers later than does the sweet cherry, and fruit is harvested in mid- to late July, after the sweet cherries have been harvested. The trees long persist.

***Prunus domestica* L.** Common or European Plum; Italian Prune. Small trees to 5 m tall, with grayish or ashy bark; branchlets pubescent when young, not commonly thornlike; leaf blades mainly 2–10 cm long and 1–6 cm wide, ovate to obovate, coarsely serrate, rough and thinly pubescent above, pubescent beneath; petioles commonly with glands; flowers solitary or sometimes 2 or 3 per bud; petals white, ca 10 mm long; sepals and hypanthium pubescent or glabrous; pedicels commonly pubescent; fruit commonly blue purple, glaucous; 2n = 48. Included here are the yellow fleshed plums known as Damson and green-gage, other plums, and the Italian prune. The species is widely cultivated in Utah, where it persists and less commonly escapes. The red-fleshed plums, called Japanese or Satsuma, belong to *P. salicina* Lindl. and are cultivated in the area too, but less commonly; specimens were examined from Beaver and Utah counties; introduced from Eurasia; 5 (i).

Prunus dulcis (Miller) D. A. Webb Almond. Small trees to 5 m tall or taller; branchlets pale, not thornlike; leaf blades 2–7 cm long, 0.7–2 cm long on mature branches, oblong-lanceolate, crenate-serrate, abruptly short-apiculate, glabrous; petioles usually bearing glands; flowers solitary, appearing before or with early leaves; petals pink; sepals villous at margin; fruit pubescent, splitting at maturity and exposing the stone; $2n = 16$. Sparingly cultivated ornamental, botanical curiosity, and "nut" tree, known from Utah and Washington counties; persisting, but not escaping (?); introduced from the Old World; 9 (i).

Prunus emarginata (Dougl.) Walp. Bitter Cherry. [*Cerasus emarginata* Dougl.]. Shrubs or small trees to 4 m tall or more, typically with glabrous, red, shining branchlets; leaves mainly 1.5–5 cm long, the blades oblong-obovate to elliptic, acutish to obtuse, finely serrulate, sparingly pubescent to glabrous, short-petioled; corymbs mainly 2–4 cm long, 3- to 10-flowered; hypanthium campanulate, glabrous, ca 3 mm long; sepals 1.4–2 mm long; petals 5–7 mm long, obovate, white; fruit red, bitter, 6–8 mm long, the stone ellipsoid. Mixed chaparral and pinyon-juniper communities at ca 1525 m in Washington County (Warrick 1474 BRY); British Columbia to Idaho, south to California; 1 (0).

Prunus fasciculata (Torr.) Gray Desert Peach. [*Emplectocladus fasciculatus* Torr.]. Low intricately branched shrub to 1.5 (2) m tall; branchlets pubescent, ashy or grayish, more or less thornlike; leaves mainly 0.5–2.5 cm long, 0.1–0.6 cm wide, cuneate-spatulate, entire or few toothed near the apex, apiculate or cuspidate, sessile or nearly so; not bearing glands at leaf base, more or less puberulent on both sides; flowers mainly 1 per bud; petals cream to white, 3.5–5 mm long, spatulate to elliptic; sepals, hypanthium, and pedicels glabrous; fruit hairy, thin-fleshed, inedible. The desert peach occurs in mixed desert shrub, chapparal, and lower pinyon-juniper communities at 625 to 1770 m in Beaver, Millard, Iron (?), and Washington counties; Arizona, Nevada and California; 42 (viii). This species is now being tested as a rootstock for grafting of other *Prunus* species.

Prunus laurocerasus L. Cherry-Laurel. Evergreen shrub to 2 m, rarely more; branchlets pallid, glabrous, not at all thornlike; leaf blades mainly 3–10 (13) cm long, and 1–3 cm wide, entire or remotely crenate-serrate, elliptic to oblong, attenuate to abruptly so, cuneate to acute basally, leathery-thickened, glabrous; petioles lacking glands; racemes not leafy, commonly 3–10 cm long, many-flowered, the pedicels 0.5–1.5 mm long, subtended by large caducous bracts; flowers ca 1 cm wide or less; petals white, 3–4 mm long, obovoid; sepals fringed, glabrous; hypanthium and pedicel glabrous; fruit black, inedible. Cultivated ornamental; specimens from Davis, Utah and Weber counties; introduced from Eurasia; 13 (0).

Prunus mahaleb L. Mahaleb; St. Lucie Cherry. Small trees to 6 m tall or taller; branchlets pale brown, copiously pubescent, not thornlike; leaf blades mainly 2–6 cm long, 1.5–4.5 cm wide, oval to elliptic or ovate, finely crenulate, abruptly acute, acute to rounded basally, commonly glabrous on both sides; petioles sometimes with glands; racemes with 3–12 flowers, corymbose, the axis short, leafy bracted at base; petals white, 4–6 (8) mm long, oblong-oblanceolate; sepals, hypanthium, and pedicels glabrous; fruit black. The Mahaleb Cherry is used as rootstock for other cherry cultivars. It persists and escapes, becoming established along canals and in riverbottom forests in Salt Lake and Utah counties; introduced from Asia; 8 (ii).

Prunus padus L. European Bird Cherry; May Day Cherry. Small trees to 8 m tall, rarely taller; branchlets soon brown, glabrous or puberulent, not thornlike; leaf blades 1.5–20 cm long, 0.7–3.5 cm wide, elliptic to oblanceolate or obovate, serrate, abruptly acuminate, acute to truncate basally, glabrous except sometimes on veins beneath; petioles usually bearing glands; racemes with leafy peduncles, commonly 7–12 cm long, 15- to 25-flowered, the pedicels 5–17 mm long, subtended by caducous bracts; flowers 12–20 mm broad; petals white, 5–7.5 mm long, almost as broad; sepals fringed, glabrous; hypanthium and pedicels glabrous; fruit black, bitter, astringent; $2n = 32$. Cultivated ornamental of yards and other plantings; Utah County; introduced from Eurasia; 1 (0).

Prunus persica (L.) Batsch Peach. Small trees to 3 (4) m tall, seldom more; branchlets green to pallid, becoming ashy in age, glabrous, not thornlike; leaf blades mainly 3–15 cm long (or more), 0.5–5.5 cm wide, oblong-lanceolate to lanceolate, serrate to crenate-serrate, attenuate, obtuse to acute basally, glabrous; petioles usually bearing glands; flowers solitary, appearing before the leaves; petals pink, white, or red; sepals villous at margin; fruit pubescent, fleshy at maturity, edible. Cultivated fruit and ornamental trees at lower elevations in the state, widely escaping along roads; specimens seen from Carbon, Salt Lake, Utah, and Washington counties; introduced from China; 23 (ii). Numerous horticultural forms are grown in the state.

Prunus serrulata Lindl. Flowering Cherry. Trees to 10 m tall or more; branchlets pallid, becoming brown, glabrous, not thornlike; leaf blades mainly 3–15 cm long, 1.2–8 cm wide, ovate to lance-ovate, abruptly long-acuminate, aristate-serrate, acute to obtuse basally; petioles usually bearing glands; racemes corymbose, 3- to 5-flowered, naked at the base, the pedicels 1.5–30 mm long, each subtended by a cuneate bract fringed at the truncate apex; petals 12–20 mm long, oval, white, rose, pink, often double; sepals glabrous, acuminate or toothed; hypanthium and pedicels glabrous; fruit small, black. This flowering cherry is gaining in popularity, especially in grafted or pendulous forms; specimens examined from Utah County; introduced from Japan; 9 (0).

Prunus tomentosa Thunb. Bush Cherry. Shrubs to 2 m tall (commonly lower); branchlets brown, copiously pubescent, not thornlike; leaf blades mainly 2–6 cm long, 1.5–3 cm wide, obovate to elliptic, abruptly acuminate, doubly serrate and obscurely lobed (in some), puberulent above, long-villous beneath; petioles pubescent, not bearing glands; flowers 1 (rarely 2) per bud, appearing before the leaves, sessile or subsessile; petals white or pink, 7–9 mm long, oval; sepals serrulate, pilose; hypanthium glabrous below; fruit red, sparingly pubescent, sour, edible. Cultivated ornamental, escaping and persisting in Utah County; introduced from Asia; 4 (0).

Prunus triloba Lindl. Flowering Almond. Shrubs to 2 m (rarely taller); branchlets brown, glabrous; leaf blades commonly 2–5 cm long, 1.5–4 cm broad, ovate to obovate, sharply doubly serrate, pubescent on both sides; flowers 1 or 2 per bud, short-pedicellate; petals pink or white, 8-15 mm long, oval, commonly double; sepals serrulate, glabrous; hypanthium glabrous; pedicels puberulent; fruit red, seldom produced. Sparingly cultivated shrubs, in Salt Lake and Utah counties; introduced from China; 1 (0).

Prunus virginiana L. Chokecherry. Shrubs or small trees to 8 m tall with ashy bark; branchlets brown, glabrous, not thornlike; leaf blades 2-10 cm long, 1.5-7 cm wide, elliptic to oblong-ovate, serrate, abruptly acuminate, acute to rounded basally, glabrous or sometimes pubescent beneath; petioles usually bearing glands; racemes 4-20 cm long, the peduncles usually leafy, 2-8 cm long; flowers 10-20 mm wide, numerous, the pedicels 4-17 mm long, each subtended by a caducous bract; petals white, 4-6 mm long, suborbicular; sepals fringed, glabrous; hypanthium and pedicels glabrous; fruit black when ripe, astringent, edible. Sagebrush, pinyon-juniper, mountain brush, and aspen communities at 1370 to 3050 m in probably all Utah counties; widely distributed in North America; 192 (xxv). Chokecherry fruit has been gathered since early days for making of jelly and syrup, and prior to that by indigenous peoples as a component of pemican. Our material has been assigned to var. *melanocarpa* (A. Nels.) Sarg. [*Cerasus demissa* var. *melanocarpa* A. Nels.; *P. melanocarpa* (A. Nels.) Rydb.; *P. demissa* var. *melanocarpa* (A. Nels.) A. Nels.]. Reports of *P. demissa*, or of *P. virginiana* var. *demissa* (Nutt.) Torr., for Utah belong to var. *melanocarpa*. That variety is widely distributed in western North America.

Purpusia Brandegee

Plants perennial, caespitose, glandular herbs, arising from a caudex; leaves mainly basal, pinnately compound; flowers perfect, regular, borne in few-flowered cymes; hypanthium campanulate to turbinate, usually lacking bractlets; sepals 5; petals 5, white to yellowish; stamens 5, opposite the sepals; pistils mostly 6-12, on a stalked receptacle; fruit of achenes.

Purpusia saxosa Brandegee [*P. arizonica* Eastw.; *P. osterhoutii* A. Nels.; *Potentilla osterhoutii* (A. Nels.) J. T. Howell]. Perennial glandular herbs from a caudex, the stems 0.5-2 dm tall; basal leaves pinnate with 5-11 leaflets, the leaflets 0.5-1.5 cm long, deeply toothed or cleft; sepals 2.5-3 mm long, lance-ovate, acuminate; petals yellow or white, 3-4 mm long, oblanceolate, acuminate; achenes on a receptacle 1.5-2 mm long. This species is included in Utah on the basis of a collection reported by Meyer (1976) taken at Kolob Reservoir, Washington County; Arizona, Nevada, and California; 0 (0).

Purshia DC. ex Poir.

Shrubs or small trees with unarmed branches; leaves alternate, simple, pinnatifid or apically 3-toothed, usually glandular; stipules triangular-attenuate, persistent; flowers perfect, regular, solitary, on short lateral spur branchlets; sepals 5, borne atop a turbinate-funnelform persistent hypanthium; petals 5, white to cream or yellow; stamens ca 25; pistils 1 to 12 (typically 1 or 5), borne on a stipe at base of hypanthium; fruit an achene, this plumose-tailed or not. The genus, as treated here, includes the species traditionally placed in *Cowania* D. Don.

1. Pistils 4-12 (typically 5); styles persistent as apical plumes on the achene .... *P. mexicana*
— Pistil 1 (2 or 3); styles not plumose on the achene apex .. 2
2(1). Leaves conspicuously punctate, with depressed glands, glabrous above; plants of Washington County ....
 .... *P. glandulosa*
— Leaves lacking punctate, depressed gands, puberulent above; plants widespread .... *P. tridentata*

Purshia glandulosa Curran Shrubs, much branched, 1-2 (3) m tall; branchlets prominently glandular; leaves 3-10 mm long, 1- to 4-mm wide, cuneate, glabrous above, slightly tomentose beneath, the margins revolute; hypanthium glabrous to tomentose, 2.5-5 mm long, funnelform; petals 4-8 mm long, spatulate, creamy white to yellowish; achenes oblique, ca 2 cm long, including style, puberulent. Blackbrush, chaparral, and pinyon-juniper communities at 1065 to 1375 m in Washington County; Nevada, Arizona, and California; 10 (ii). This taxon is presumed to have arisen as a product of hybridization betweeen *P. mexicana* and *P. tridentata*. It resembles the former in vegetative characteristics and the later in pistil number (usually 1, but sometimes 2 or 3). and non-plumose styles.

Purshia mexicana (D. Don) Welsh Cliff-rose. [*Cowania mexicana* D. Don]. Much branched shrubs or small trees, mainly 0.6-3.5 m tall with shreddy bark and glandular branchlets; leaves 3-15 mm long, cuneate-flabellate, mainly 5-lobed, glandular-punctate and green above, white-tomentose beneath; pedicels 2-8 mm long; sepals 4-6 mm long, ovate; petals 5-9 mm long, white to cream or yellowish; pistils commonly 5; styles plumose 2-6 cm long or more in fruit. Blackbrush, live oak, pinyon-juniper, ponderosa pine, desert peach, mixed grass-desert shrub, and mountain brush communities at 975 to 2745 m in Beaver, Box Elder, Cache, Carbon, Emery, Garfield, Grand, Iron, Juab, Kane, Millard, Salt Lake, San Juan, Sanpete, Sevier, Tooele, Utah, Washington, and Wayne counties; Nevada, Colorado, Arizona, California, New Mexico, and Mexico; 179 (xxx). Our materials belong to var. **stansburiana** (Torr.) Welsh [*Cowania stansburiana* Torr., type from Great Salt Lake; *C. mexicana* var. *stansburiana* (Torr.) Jepson; *C. mexicana* ssp. *stansburiana* (Torr.) Murray]. This species forms introgressants with *Purshia tridentata* (q.v.). The problem has been investigated by Stutz and Thomas (1964. Evolution 18: 183-195).

Purshia tridentata (Pursh) DC. Bitterbrush. [*Tigarea tridentata* Pursh]. Shrubs, much branched, to 2 m tall (rarely taller); branchlets brown, tomentulose; leaves mainly 4-20 mm long, 2-12 mm wide, cuneate, apically 3 (5)-toothed, tomentose but green above, grayish tomentose beneath, the margins more or less revolute; hypanthium tomentose and stipitate glandular; sepals mainly 4-6 mm long, ovate-oblong, entire; petals 5-9 mm long, oblong to obovate or spatulate, yellow; achenes obliquely ovoid, 1-2 cm long, the beak ca 1/3 the length, puberulent. Bitterbrush is a plant of sagebrush, mountain brush, pinyon-juniper, and ponderosa pine communities at 1220 to 2775 m in all counties in Utah; British Columbia east to Montana and south to California, Arizona, and New Mexico; 159 (ix). This species forms hybrids with *P. mexicana*; such plants are distinguished by the more lobed leaves, and longer and more numerous achenes.

Pyracantha Roemer

Evergreen shrubs, armed with thorns; leaves alternate, simple, crenate-serrate, petiolate; stipules minute, caducous; flowers perfect, regular, borne in simple or branched corymbs; sepals 5; petals 5, white; stamens 20, the filaments subulate; pistil 1, the ovary inferior, 5-loculed; styles usually 5; fruit a pome.

Pyracantha coccinea Roemer Fire-thorn. Shrubs to 3 m tall or more, with thorns 0.5–1.5 cm long or more; leaf blades 0.8–4.5 cm long, 0.5–1.8 cm wide, elliptic to oblanceolate, acute apically, cuneate to acute basally, crenate-serrate; petioles puberulent; inflorescence pilosulous at anthesis; sepals broadly triangular; petals white, 3–4 mm long, 2–3 mm wide; pomes red orange, persistent. Cultivated ornamental, persisting, and escaping rarely in Salt Lake and Utah counties; introduced from Eurasia; 6 (0).

Pyrus L.

Trees with unarmed branches; leaves alternate, simple, not or seldom lobed; stipules deciduous; flowers perfect, regular, borne in corymbs; hypanthium short; sepals 5, persistent; petals 5, white; stamens many; pistil 1, the ovary inferior, usually 5-loculed; styles 3–5, separate to the base; fruit usually pear-shaped, the flesh with stone-cells.

Pyrus communis L. Common pear. Small trees to 6 m tall, the branchlets commonly glabrous; leaves 2–8 cm long, ovate to oblong or elliptic, glabrous or glabrate, leathery, crenate-serrulate to subentire; flowering with the leaves; petals white, mainly 12–18 mm long; fruit pear shaped to almost spherical; 2n = 34. Cultivated fruit plant at low to moderate elevations in Garfield, Salt Lake, Utah, Washington, and Wayne counties; introduced from Eurasia; 15 (iii). This the pear of commerce is widely grown in Utah, with Bartlet being the most common cultivar. The Callery or Bradford pear, *P. calleryana* Decne., is now widely grown in Utah as a shade and ornamental tree. It differs in having ovate to orbicular leaves, and very small fruit, resembling some crabapples.

Rosa L.

Shrubs, deciduous; stems armed with prickles or spines, rarely unarmed; leaves alternate, pinnately 3- to 9-foliolate; stipules conspicuous, adnate to petioles; flowers perfect, solitary or in corymbs; hypanthium urn-shaped to globose or ellipsoid, red, red orange, yellow, or purplish, fleshy at maturity; sepals 5; petals 5 (or numerous in double forms); stamens numerous, inserted on margin of ringlike disk; pistils few to numerous; styles exserted through or to orifice of disk; fruit of achenes, enclosed in the fleshy hypanthium (hip). The cultivated roses are largely of hybrid derivation, and are not well represented in herbaria. The key includes both indigenous and cultivated taxa because of the propensity of some species to escape and of others to persist for long periods following cultivation. The key is tentative at best, because of the propensity of all roses to hybridize.

1. Stipules deeply fringed or pectinate, appearing as lateral projections of petiole base; flowers white (or pink in some hybrids keying here); cultivated and escaping .. .. R. multiflora
— Stipules entire, or rarely fringed, but not cut to the petiole; flowers variously colored 2
2(1). Flowers mostly 5–9 cm broad; styles long-exserted from hypanthium; cultivated and persisting .... R. odorata
— Flowers mostly less than 5 cm broad; styles not exserted, forming a dense headlike stopper in orifice of hypanthium 3
3(2). Leaflets stipitate-glandular beneath; sepals strongly stipitate-glandular, erect or spreading in fruit; cultivated and escaping .... R. eglanteria
— Leaflets not at all or only sparingly stipitate-glandular beneath; sepals various in fruit; cultivated and escaping or indigenous 4
4(3). Sepals reflexed and finally deciduous after flowering; stipitate glands sparse on lower midveins and sepals; cultivated and escaping .... R. canina
— Sepals erect and persistent following flowering; stipitate glands rarely present; plants indigenous 5
5(4). Sepals 1.5–4 cm long; petals 2.2–4 cm long; hips 1–2 cm long at maturity .... R. nutkana
— Sepals 1–2.2 cm long; petals 1–2.5 cm long; hips 0.6–1.5 cm long at maturity .... R. woodsii

Rosa canina L. Dog Rose. Shrub 1–3.5 (4) m tall; stems sometimes clambering, armed with scattered strongly curved to straight spines and no prickles, glabrous; stipules entire; leaves 3–8 cm long or more, with 3–7 leaflets, the terminal leaflet 0.6–2.2 cm long, 0.3–1.2 cm broad, glabrous to sparingly villous and stipitate-glandular along the veins beneath, serrate to doubly serrate; flowers single, solitary or 2–5 on short glabrous or glandular pedicels; sepals 1.2–1.5 cm long, sparingly stipitate-glandular, reflexed in fruit, soon shattering; petals white or pink, 1–1.8 cm long; hips ovoid, ca 1 cm thick, red; 2n = 35. Cultivated, persisting, and rarely escaping in lower elevation portions of Utah, specimens examined from Salt Lake and Utah counties; introduced from Europe; 3 (0).

Rosa eglanteria L. Sweetbriar. [*R. rubiginosa* L.]. Shrubs 0.5–2 m tall; stems erect or ascending, armed with distinctive flattened infrastipular spines and often with straight internodal prickles; leaves 3.5–10 cm long, with 5–7 (9) leaflets, the terminal one 1.2–3.5 cm long, 0.8–3 cm wide, conspicuously stipitate glandular on one or both surfaces; doubly serrate; flowers solitary or 2–4; sepals 1–2 cm long; petals 0.8–2 cm long, pink to white; hips 1–1.5 cm long, ellipsoid to subglobose. Cultivated, persisting, and escaping in Cache, Salt Lake, Tooele, and Uintah counties; introduced from Europe; 4 (0).

Rosa multiflora Thunb. Multiflora Rose. Shrub to 2 or 3 m tall or more; stems sometimes clambering, armed with prickles or rarely unarmed, glabrous; stipules about half as long as the petiole, pectinate, cut almost to petiole; leaves 4–9 cm long or more, 5- to 11-foliolate, the terminal leaflet 1.5–3 cm long, glabrous or puberulent and rarely with some glands beneath, serrate; flowers numerous (to 50 or more), white (pink in some hybrids), single (double); sepals reflexed in fruit, pubescent; petals 5, or numerous, 0.6–0.8 cm long; hips ovoid, ca 6 mm thick, brownish red; 2n = 14. Herbarium specimens are very few for this species in Utah, and those involve hybrids with other species and are only tentatively placed herein. Specimens examined are from Uintah and Utah counties; introduced from Japan; 3 (0).

Rosa nutkana Presl Nutka Rose. [*R. spaldingii* Crepin]. Shrubs 0.3–2 m tall, rarely more; stems erect or ascending, armed with distinctive infrastipular spines (or rarely unarmed), the internodal prickles lacking or few and different from the infrastipular spines; leaves 6–13 cm long, with 5–7 (9) leaflets, the terminal one 1.5–7 cm long, 0.8–3.5 cm broad, pubescent to glabrous, rarely stipitate-glandular beneath, serrate to doubly serrate; flowers solitary (rarely 2 or 3); sepals 1.5–4 cm long, 3–6 mm wide; petals 5, pink, 2.2–4 cm long; hips ellipsoid to subglobose, 1–2 cm long and thick, red orange to purplish. Oak-maple, aspen, mountain brush, sagebrush, Douglas fir, cottonwood, and spruce-fir communities at 1525 to 3355 m in most if not all Utah counties; Alaska south to California, Nevada, and Colorado; 45 (viii). The features of larger fruit, longer and broader sepals, and usually solitary flowers are diagnostic when taken in combination for most specimens. However, intermediate specimens between Nootka and Woods rose are known. Our material has been assigned to var. *hispida* Fern.; 48 (vii).

Rosa odorata Sweet Tea Rose; Hybrid Rose. Shrubs, 0.3–6 m long or more; stems erect or ascending to clambering, armed with infrastipular and/or internodal spines or prickles, or unarmed; leaves 4–20 cm long, with usually (3) 5 leaflets, the terminal 1–7 cm long, 0.8–4 cm wide, once or twice serrate, glabrous or pilose and somewhat stipitate glandular; flowers solitary or usually few to numerous; petals 5, or more commonly numerous, of various colors and sizes; hips various in size, shape, and color. Included within this catch-all name are the hybrid tea roses of commerce, grown widely in Utah for ornament. The cultivars are mainly complex hybrids, with ancestors involving *R. moschata* J. Herrmann, *R. foetida* J. Herrmann, *R. gallica* L., *R. chinensis* Jacq., and *R. multiflora* Thunb. (see Flora Europaea, 1968, p. 26 for a more complete discussion). The hybrid roses persist and are present around abandoned farmsteads in Utah; introduced from Eurasia; 3 (i).

Rosa woodsii Lindl. Woods Rose. [*R. fendleri* Crepin; *R. neomexicana* Cockerell; *R. manca* Greene; *R. arizonica* Rydb.; *R. californica* var. *ultramontana* Wats.; *R. macounii* Greene; *R. chrysocarpa* Rydb., type from Abajo Mts.; *R. puberulenta* Rydb., type from Montezuma Canyon]. Shrubs 0.1–2.5 (3) m tall; stems armed with infrastipular spines and/or internodal prickles or spines, or unarmed; leaves 1.5–13 cm long, 0.4–2.5 cm broad, pubescent to glabrous and rarely stipitate glandular, serrate to doubly serrate; flowers solitary or 2 to several; sepals 1–2.2 cm long, 2–3.5 cm wide; petals 5, pink (rarely white), 1–2.5 cm long; hips ellipsoid to subglobose, 0.6–1.5 cm long and thick, red orange to yellow. Streamsides, irrigation canals, marsh lands, lake shores and hillsides in palustrine, lacustrine, and riparian habitats in mountain brush, juniper, aspen, and spruce-fir communities at 850 to 3355 m in all Utah counties; Alaska and MacKenzie east to Hudson Bay and south to California, Texas, Missouri, and Wisconsin; 183 (xviii). The species is represented in Utah by a variable assemblage, which has been included within the concept of **var. *ultramontana*** (Wats.) Jepson. There are plants from Garfield and San Juan counties, especially, which have very coarse internodal as well as infrastipular spines. These would key to *R. neomexicana* Cockerell, but gradient specimens tie these striking exceptions to the mass of variation within the complex of forms. The unusual *R. stellata* Wooton, might occur in southern Utah. The young stems of that species are stellate-pubescent or copiously glandular hispidulous or both, and the older stems are armed with numerous long, nearly straight prickles.

Rubus L.

Shrubs; stems armed with prickles or bristles, or unarmed; leaves alternate, pinnately compound or palmately veined and lobed; stipules various, usually persistent; flowers perfect or imperfect, regular, solitary or few to numerous in cymes; hypanthium short, saucerlike, lined with a glandular disk; sepals usually 5, lacking bracteoles; petals the same number as the sepals; stamens 15 to numerous, linear-subulate; pistils several to many, the ovaries superior, each 1–loculed; styles 1 per pistil, the stigma capitate; fruit of separate drupelets, or the drupelets coherent and free of the receptacle, hence an "aggregate" fruit.

Bailey, L. H. 1941. Species Batorum. The genus *Rubus* in North America. Gentes. Herb. 5: 1–932.

1. Leaves simple, palmately veined and lobed; stems unarmed 2
— Leaves compound pinnately 3– to 5– foliolate; stems armed 3

2(1). Leaves mainly less than 6 cm wide, the lobes rounded in general outline, green above, white tomentose beneath; flowers mainly solitary, plants rare, known only from low elevations in San Juan County . *R. neomexicanus*

— Leaves mainly 6–30 cm wide, the lobes acute to attenuate in general outline; flowers borne in clusters of 2–6 or more; plants locally common, montane, widespread .... *R. parviflorus*

3(1). Main prickles straight, slender, retrorsely disposed along stem; fruit red when ripe .... *R. idaeus*
— Main prickles flattened, curved or straight, retrorse or retrorsely curved 4

4(3). Receptacle fleshy, the druplets adhering, not slipping free when ripe; stems usually strongly armed, trailing or clambering; cultivated, persisting, and escaping .... *R. discolor*

— Receptacle not fleshy, the druplets slipping free when ripe; stems arching, but not trailing or clambering; indigenous and with cultivated phases ..... *R. leucodermis*

Rubus discolor Weihe & Nees Himalayan Blackberry. [*R. procerus* Muell.-Arg.]. Shrubs, often clambering or sprawling, the stems to several meters long, armed with strong, straight, flattened spines; stipules linear, entire; leaves 7–20 cm long, pinnately to palmately compound, with 3–5 leaflets, the terminal leaflet 3–12 cm long, 2–8 cm wide, green and glabrous above, tomentose beneath; flowers usually perfect, conspicuous, mainly 3–20 in clusters; sepals 6–10 mm long, lanceolate; petals white (rarely reddish), 10–15 mm long; staminal filaments linear; pits to 3 mm long, the drupelets adherent to receptacle, numerous, the flavor agreeable. Roadsides, field margins, and abandoned farmsteads at 850 to 1525 m in Juab, Utah, and Washington counties (likely elsewhere); introduced from the Old World; 5 (i). This is a vigorous, strongly armed shrub that is locally established. The fruit is of excellent quality, but difficult to pick, due to the spiny nature of the plants.

Rubus idaeus L. Raspberry. [*R. strigosus* Michx.; *R. idaeus* var. *strigosus* (Michx.) Maxim.; *R. sachalinensis* H. Levl.; *R. idaeus* ssp. *sachalinensis* (H. Levl.) Focke; *R. idaeus* var. *canadensis* Richards.; *R. melanolasius* Dieck]. Shrubs, 2–15 (20) dm tall, the stems, petioles, and veins on lower leaf surfaces with glandular pricklelike processes or prickles, or both; stipules linear; leaves 2–20 cm long, pinnately compound with 3–5 leaflets, the terminal leaflet 1.2–10 cm long, 0.6–7.5 cm broad, green and glabrous to hairy above, white- or gray-hairy to glabrate and greenish beneath; flowers perfect, not conspicuous, solitary or 1 to few in clusters; sepals 4–12 mm long, lanceolate; petals white, 4–7 mm long; staminal filaments slender, often somewhat clavate; pits 2–2.5 mm long, the drupelets coherent, red, several to many, the flavor agreeable; $2n = 14$. Riparian sites and talus slopes in aspen and mixed conifer communities at 2135 to 3420 m in probably all Utah counties; Alaska east to the Atlantic and south to California, Mexico, Iowa, and North Carolina; Eurasia; 60 (xii). Our indigenous material belongs to ssp. *melanolasius* (Dieck) Focke. Cultivated phases belong mainly to ssp. *idaeus*.

Rubus leucodermis Dougl. ex T. & G. Black Raspberry. Shrubs, mainly 1–3 m long, the stems, petioles, and some veins on lower leaf surfaces armed with retrorsely curved, flattened, catclawlike prickles; stipules linear; leaves 6–14 cm long, pinnately compound, with 3–5 leaflets, the terminal leaflet 3–7.5 cm long, 2–6 cm wide, green and almost or quite glabrous above, white-tomentose beneath; flowers usually perfect, not conspicuous, mainly 2–10 in clusters; sepals 6–12 mm long, lance-acuminate; petals white, shorter than the sepals; staminal filaments slender, linear-subulate; pits to 2.5 mm long, the druplets coherent, several to many, the flavor agreeable. Dry open slopes in mountain brush and in riparian communities at 1220 to 2200 m in Box Elder, Emery, Iron, Millard, Salt Lake, Utah, and Washington counties; British Columbia east to Montana and south to California and Nevada; 6 (0). The plants are rare in collections, and the distribution is probably wider than indicated.

Rubus neomexicanus Gray Shrubs, 0.5–1.5 m tall, the stems, petioles, and leaves unarmed, merely villous-puberulent and sometimes minutely glandular; stipules lance-ovate, entire or serrate; leaves palmately lobed and veined, simple, the blades 1.2–4.2 cm long (from sinus to apex), 1.5–5.5 cm wide, green above, pale green beneath, puberulent on one or both sides; flowers usually perfect, solitary, showy; sepals 10–14 mm long, lance-ovate, entire or serrate; petals white, 2–17 mm long, the drupelets not especially coherent, red, several to many, thinly-fleshed, hardly palatable. Hanging garden with *Ostrya knowltonii*, at 1130 to 1160 m in Ribbon, Knowles, and Cataract canyons, San Juan County; Arizona and New Mexico; 8 (vi). This is a truly attractive species, with startlingly large white roselike flowers.

Rubus parviflorus Nutt. Thimbleberry. Shrubs, 0.5–2 m tall, rarely taller, the stems, petioles, and leaves unarmed, stipitate-glandular; stipules lanceolate, entire or serrate; leaves palmately lobed and veined, simple, the blades 4.5–15 cm long (from sinus to apex), 5.5–20 cm wide, green above, pale beneath, puberulent on one or both sides, or glabrate above; flowers usually perfect, in clusters of 2–7, showy; sepals 8–19 mm long, ovate, the apex caudate-attenuate, entire; petals white, 13–18 (20) mm long, or more; staminal filaments linear-subulate; pits to 3 mm long, the drupelets coherent as an aggregate, red, numerous, thinly fleshy, almost dry at maturity, palatable. Riparian habitats in aspen, spruce, fir, lodgepole, Douglas fir, and mountain brush communities at 1435 to 2745 m in Box Elder, Cache, Davis, Duchesne, Emery, Salt Lake, San Juan, Sanpete, Summit, Tooele, Uintah, Utah, Wasatch and Weber counties; Alaska east to Great Lakes, and south to California, Arizona, New Mexico and the Dakotas; 40 (ii). Our plants belong to var. *parviflorus*.

Sanguisorba L.

Perennial herbs, from a branching caudex; leaves basal and alternate, pinnately compound; stipules adnate to the petioles, persistent; flowers mostly imperfect, regular, numerous in short to elongate, dense spikes; hypanthium subglobose, restricted near the apex; sepals 4, petaloid; petals lacking; stamens numerous; pistils 1–3, the ovary superior, 1–loculed; styles 1 per pistil, the stigma capitate, fringed; fruit an achene, enclosed by the usually 4–angled to 4–winged hypanthium.

Sanguisorba minor Scop. Burnet. Plants mainly 2–5 dm tall; caudex clothed with persistent stipules and petioles; basal leaves 4–18 cm long, with mostly 9–17 oval to obovate-oblong leaflets, 0.6–1.8 cm long, coarsely serrate; spikes subglobose to cylindroid, 8–40 mm long; bractlets ovate; flowers mainly imperfect, the lower staminate and the upper pistillate; calyx greenish or pinkish; hypanthium cone-shaped in fruit, woody; stamens numerous, the filaments filiform, long-exserted; $2n = 28, 56$. Introduced revegetation and erosion control plant at 1525 to 2135 m elevation in Cache, Davis, Duchesne, Emery, Garfield, Grand, Iron, Millard, Salt Lake, San Juan, Tooele, Utah, and Washington counties; introduced from Europe; 15 (ii).

Sibbaldia L.

Perennial herbs from a caudex; leaves basal or cauline and alternate, long-petioled, palmately 3–foliolate; stipules adnate to petioles, persistent, lanceolate; flowers perfect, regular, borne in leafy-bracted cymes; hypanthium short, saucer-shaped, lined with a glandular disk; sepals 5, alternating with 5 sepaloid bracteoles; petals 5; stamens usually 5; pistils 5–20, distinct, the ovaries superior; styles 1 per pistil, the stigmas capitate; fruit an achene.

Sibbaldia procumbens L. Sibbaldia. Plants low, mat-forming, the flowering stems 0.4–1.4 dm tall; leaves 2–12 cm long, the 3 leaflets oblanceolate to obovate, 3 (rarely 5)-toothed apically, the terminal leaflet 11–32 mm long, 7–18 mm broad, stiffly hairy on both surfaces; flowers inconspicuous; sepals 2.5–5 mm long; petals pale yellow, 1.5–3 mm long; achenes stipitate, ca 1 mm long; $2n = 14$. Alpine tundra, krummholz, spruce-fir, meadow, and lodgepole pine communities, often in talus or gravel at 2745 to 3660 m in Beaver, Box Elder, Daggett, Duchesne, Garfield, Grand, Iron, Juab, Piute, Salt Lake, San Juan, Sevier, Summit, Uintah, Utah, and Wayne counties; Alaska east to Newfoundland and south to California, Colorado, Quebec, and New Hampshire; circumboreal; 47 (vi).

Sorbaria (Ser.) A. Br.

Shrubs with unarmed branches; leaves alternate, pinnately compound; stipules persistent; flowers perfect, regular, borne in terminal panicles; hypanthium short, lined with a glandular disc; sepals 5, persistent; petals 5, white; stamens 20–50; pistils 5, somewhat connate basally; styles 1 per pistil, the stigmas capitate; fruit of follicles.

Sorbaria sorbifolia (L.) A. Br. Sorbaria. [*Spiraea sorbifolia* L.]. Shrubs to 2 m tall, sometimes taller; leaves

8–20 cm long or more; leaflets 11–23, lanceolate to oblong-lanceolate, serrate or doubly so, long-acuminate, glabrous or puberulent, the hairs stellate; inflorescence 10–25 cm long; flowers white, ca 8 mm wide; hypanthium glabrous; fruit glabrous; 2n = 36. Cultivated ornamental in Davis, Salt Lake and Utah counties, and probably elsewhere; introduced from Asia; 7 (0).

Sorbus L.

Shrubs or small trees with unarmed branches; leaves alternate, pinnately lobed or compound; stipules persistent or deciduous; flowers perfect, regular, numerous in corymbose cymes; hypanthium short, lined with a glandular disk; sepals 5, persistent; petals 5, cream to white; stamens 15–20; pistil 1, the ovary inferior, 2– to 5–loculed; styles 2–5, the stigmas capitate; fruit a pome.

Jones, G. N. 1939. A synopsis of the North American species of *Sorbus*. J. Arnold. Arb. 20: 1–43.

1. Leaves simple, lobed or pinnatifid; petioles and branchlets of inflorescence densely white villous-tomentose . . . . *S. hybrida*
— Leaves compound; petioles and branchlets of inflorescence sparingly tomentose 2

2(1). Winter buds densely white-villous, the surface obsured by the hairs; maximum leaflet number commonly 15; plants cultivated . . . . *S. aucuparia*
— Winter buds sparingly hairy, the shiny surface not at all obscured by the hairs; maximum leaflet number 13; plants indigenous . . . . *S. scopulina*

Sorbus aucuparia L. European Mountain-ash. Trees, mostly 3–6 m tall, with grayish or yellowish green smooth bark; winter buds densely white-villous; leaves pinnately compound; leaflets 11–15, 3–5 cm long, 1–1.8 cm broad, the margins coarsely serrate except at the base; petioles and branches of inflorescence sparingly white-hairy at least in flower; stipules persistent; flowers 8–10 mm broad; sepals triangular; petals white to cream, orbicular, 3–4.5 mm long; fruit 9–11 mm long, scarlet, drying purplish; 2n = 34. Cultivated ornamental, persisting, and escaping (?) in Salt Lake, Summit, and Utah counties; introduced from Europe; 14 (0).

Sorbus hybrida L. Trees, mostly 3–6 m tall, with grayish or yellowish green smooth bark; winter buds white villous-tomentose; leaves simple or pinnatifid, usually with at least 1 pair of lobes free at base of blade, the lobes coarsely serrate or doubly so; petioles and branches of inflorescence densely white villous-tomentose; stipules deciduous; flowers 10–14 mm broad; sepals triangular; petals white to cream, broadly elliptical 5–6 mm long; fruit 10–12 mm long, globose, red. Cultivated ornamental, persisting, in Cache, Salt Lake, and Utah counties; introduced from Europe; 8 (0).

Sorbus scopulina Greene Rock Mountain-ash. Shrubs, 1–4 m tall, with grayish red or yellowish bark; winter buds glutinous and glossy, white-hairy to glabrous; leaves pinnately compound; leaflets 7–13, 2–9 cm long, 0.7–3 cm broad, sharply serrate almost to the base; branches of inflorescence sparingly to rather densely pubescent with white hairs; stipules persistent or tardily deciduous; flowers 8–12 mm broad; sepals triangular; petals white to cream, oval, 4–6 mm long; fruit 5–10 mm long, scarlet to orange, drying purplish. Aspen, spruce-fir, white fir, Douglas fir, and ponderosa pine communities at 2075 to 2900 m in Cache, Carbon, Davis, Duchesne, Morgan, Salt Lake, San Juan, Sanpete, Summit, Utah, Wasatch, Washington, and Weber counties; Alaska south to California, New Mexico, and the Dakotas; 28 (vi).

Spiraea L.

Deciduous shrubs with unarmed branchlets; leaves alternate, simple; stipules obsolete; flowers perfect, regular, borne in terminal corymbs; hypanthium cup-shaped; sepals 5, persistent; petals 5; stamens 25 or more; pistils 3–7 (usually 5), distinct, the ovaries superior, each 1–loculed; styles 1 per pistil, the stigmas capitate; fruit a few-seeded follicle.

Spiraea x vanhouttei (**Briot**) **Zabel** [*S. aquilegifolia vanhoutei* Briot]. Shrubs to 2 m tall; stems finally arching; leaves 0.8–3.5 cm long, 0.4–1.7 cm wide, cuneate-obovate, serrate to doubly serrate at the apex and often 3–5 lobed; inflorescences pedunculate, terminal on short lateral branches; petals white, 3.5–4.5 mm long, oval; follicle to 5 mm long (including styles). Commonly cultivated ornamental, persisting in Carbon, Salt Lake, and Utah counties; introduced from Eurasia; 8 (0). This plant of hybrid derivation from **S. cantoniensis** Lour. x **S. trilobata** L.

RUBIACEAE A. L. Juss.
Madder Family

Herbs or subshrubs; leaves opposite or whorled, simple, entire; stipules sheath- or leaflike; flowers sympetalous, perfect or dioecious, regular, usually in paniculate or cymose clusters, less commonly solitary; calyx with 4–8 lobes, persistent; corolla funnelform, salverform, or rotate; stamens 3–5, inserted on the corolla tube; pistil 1, the ovary inferior or partly so, crowned by a disk, 1– to several-loculed; style filiform, divided above; fruit a capsule, berry, drupe, or schizocarp; x = 6–17.

1. Leaves opposite; corolla tube elongate 2
— Leaves whorled; corolla tube short 3

2(1). Ovules and seeds several in each carpel; fruit a 2–loculed capsule, lacking hooked hairs . . . . . . *Houstonia*
— Ovules and seeds solitary in each carpel; fruit didymous, hispid with hooked hairs . . . . *Kelloggia*

3(2). Corolla with 3–4 lobes; fruit dry, glabrous, or hairy with straight or hooked bristles . . . . *Galium*
— Corolla 5–lobed; fruit fleshy, glabrous . . . . *Rubia*

Galium L.

Annual or perennial herbs, or occasionally subshrubs; stems 4–angled, slender, erect or clambering; leaves whorled or apparently so (the stipules large and leaflike in some); flowers small, perfect or imperfect, in cymes or panicles; sepals minute or lacking; corolla rotate, usually with 4 lobes; stamens 4 (3); ovary 2–lobed, 2–loculed, 2–seeded; styles 2; fruit didymous, of 2 indehiscent dry carpels.

Dempster, L. 1959. A reevaluation of *Galium multiflorum* and related taxa. Brittonia 11: 105–122.
Dempster, L and F. Ehrendorfer. 1965. Evolution of the *Galium multiflorum* complex in western North America. II. Critical taxonomic revision. Brittonia 17: 289–334.
Ehrendorfer, F. 1956. Survey of the *Galium multiflorum* complex in western North America. Contr. Dudley Herb. 5: 1–36.

1. Leaves 5–8 per whorl 2
— Leaves 2–4 per whorl 5
2(1). Ovary and fruit glabrous; leaves obtuse or rounded apically .... *G. trifidum*
— Ovary and fruit bristly hairy; leaves acute or bristle-tipped 3
3(2). Flowers several on each branch, borne in a terminal paniculate cluster; fruit ca 1 mm thick, with bristles less than 0.3 mm long .... *G. mexicanum*
— Flowers mostly 2 or 3 in axillary cymules; fruit more than 1 mm thick 4
4(3). Plants perennial; leaves ovate-oblong to broadly obovate, 7–13 mm broad .... *G. triflorum*
— Plants annual; leaves linear to narrowly oblong or oblanceolate, mainly 3–5 (7) mm broad ...... *G. aparine*
5(1). Plants herbaceous, not at all woody at the base 6
— Plants woody at the base, at least in the previous years growth 8
6(5). Plants perennial; fruit glabrous .... *G. trifidum*
— Plants annual; fruit with hooked hairs 7
7(6). Fruiting pedicels 3–30 mm long; plants glabrous; leaves unequal in the whorl .... *G. bifolium*
— Fruiting pedicels to 1 mm long; plants hispidulous; leaves subequal in the whorl .... *G. proliferum*
8(5). Mature fruits glabrous or merely puberulent; flowers numerous in terminal paniculate clusters ..... *G. boreale*
— Mature fruits densely bristly, with hairs almost equaling body of fruit 9
9(8). Plants shrubby; leaves rigid and acerose; known from Washington County .... *G. stellatum*
— Plants herbaceous to suffrutescent; leaves not especially rigid or acerose; distrbution various ..... *G. multiflorum*

Galium aparine L. Cleavers. [*G. vaillantii* DC. in Lam.]. Clambering annuals, 1–20 dm tall or more; stems with angles retrorsely hispid; leaves 2–5 cm long, 2–7 mm wide, 6–8 per whorl, linear to linear-oblong, or oblanceolate, cuspidate, 1-veined; flowers perfect, in 1- to 3-flowered cymes, in upper leaf axils; pedicels well developed; corolla white to greenish white; fruit 2–5 mm wide, with short uncinate bristles or subglabrous. Sagebrush, mountain brush, pinyon-juniper, aspen, and spruce-fir communities at 915 to 3050 m in Beaver, Box Elder, Cache, Davis, Juab, Kane, Millard, Morgan, Salt Lake, San Juan, Sanpete, Tooele, Utah, Wasatch, Washington, and Weber counties; widely distributed in North America; Eurasia; 48 (viii). Our materials belong to **var. echinospermum** (Wallr.) Farw. [*G. agreste* var. *echinospermon* Wallr.].

Galium bifolium Wats. Twinleaf Bedstraw. Erect or ascending annual herbs, 0.5–2 dm tall, glabrous; leaves 1–2.5 cm long, 1–5 mm broad, with 2–4 per whorl, when 4, dimorphic, with 2 smaller and 2 larger ones, thin, linear-elliptic, obtuse to acutish, 1-veined; flowers perfect, solitary, on peduncles 5–30 mm long in fruit; corolla white; fruit 2.5–3.5 mm thick, with uncinate hairs. Mountain brush, sagebrush, ponderosa pine, meadow, lodgepole pine, aspen, and spurce-fir communities in 1400 to 3235 m in all Utah counties except Grand, Piute, San Juan, Summit, and Wayne; Montana to British Columbia, south to California and Colorado; 42 (ix). The type is from the Wasatch Mts.

Galium boreale L. Northern Bedstraw. [*G. utahense* Eastw., type from Soldier Summit]. Perennial herbs, from rhizomes, the stems mostly 2–8 (10) dm tall, erect or ascending, rough to smooth along the angles, not retrorsely bristly; leaves 1–6.5 cm long, 2–12 mm wide, narrowly lanceolate to oblong or linear, borne in whorls of 4, glabrous or roughened marginally and on the veins beneath, 3-veined, rounded apically; flowers several to numerous, borne in terminal cymose panicles; corolla white to cream, 3–7 mm broad, 4-lobed; fruit 1.5–2 mm long, pubescent with straight or curled hairs, rarely glabrous; $2n = 44, 64?, 66$. Sagebrush, mountain brush, lodgepole pine, aspen, meadow, and spruce-fir communities at 1650 to 3100 m in Cache, Carbon, Daggett, Duchesne, Emery, Grand, Salt Lake, San Juan, Sanpete, Sevier, Summit, Uintah, Utah, Wasatch, and Weber counties; widely distributed in North America; circumboreal; 53 (v).

Galium mexicanum H.B.K. Rough Bedstraw. [*G. filipes* Rydb., type from City Creek Canyon]. Clambering or prostrate perennial herbs, 2–8 dm tall; stem angles retrorsely scabrous; leaves 1.5–4 cm long, 2–12 mm wide, 4–5 (6) per whorl, elliptic to narrowly linear, cuspidate, 1-veined, retrorsely scabrous or glabrous; flowers perfect, in loose, leafy-bracteate cymose panicles; corolla white; fruit ca 1 mm thick, with very short (0.3 mm) hooked bristles; $2n = 22, 44$. Mountain brush, aspen, and fir communities at 2000 to 2700 m in Cache, Morgan, Salt Lake, and Tooele counties; Washington to Montana, south to Mexico and Central America; 3 (i). Our materials belong to **var. asperulum** (Gray) Dempster [*G. asperrimum* var. *asperulum* Gray]. The more compact **var. asperrimum** (Gray) Higgins & Welsh [*G. asperrimum* Gray] occurs near the Utah boundary in Colorado, and should be sought in eastern or southeastern Utah.

Galium multiflorum Kellogg Erect or ascending, clump-forming perennial herbs or subshrubs; stems herbaceous or woody below, glabrous to hispid; leaves 5–20 mm long, 1–8 mm wide, usually 4 per whorl, linear to ovate, acute to cuspidate, the midrib evident and with lateral veins evident near the base; flowers imperfect, dioecious, in narrow panicles; corolla greenish yellow, to 4 mm wide; fruit, including hairs, 4–5.5 mm wide, the hairs straight or curved, flattened, 1.5–2.5 mm long. This complex of forms has been interpreted as belonging to a series of species and infraspecific taxa recognizeable only arbitrarily by use of morphological features and differences in chromosome levels. For the purposes of practical taxonomy they are treated more conservatively herein as belonging to two variable and intergrading taxa.

1. Leaves mainly 5–10 times longer than broad, mostly 1–2 mm wide and 10–20 mm long, glabrous or nearly so; plants of the Colorado drainage system (and in the Virgin River area) .... *G. multiflorum* var. *coloradoense*
— Leaves mainly 1.5–6 times longer than broad, mostly 1–5 mm wide and 3–15 mm long, glabrous or hispid; plants widely distributed .... *G. multiflorum* var. *multiflorum*

Var. coloradoense (Wight) Cronq. [*G. coloradoense* Wight]. Warm desert shrub, shadscale, sagebrush, mountain brush, and pinyon-juniper communities at 1500 to 2300 m in Grand, San Juan, Uintah, and Washington counties; Colorado, Arizona, and New Mexico; 27 (vii).

Var. multiflorum [*G. multiflorum* var. *watsonii* Gray, type from Wasatch Mts.; *G. munzii* Hilend & J. T. Howell; *G. hypotrichium* ssp. *utahense* Ehrend., type from Utah County; *G. hypotrichium* ssp. *scabriusculum* Ehrend. (including *G. scabriusculum* (Ehrend.) Dempster & Ehrend., type from Calf Springs Wash, Emery County, not Dalla Torre & Sarnth., and *G. emeryense* Dempster & Ehrend., type from San Rafael Swell; *G. scabriusculum* ssp. *protoscabriusculum* Dempster & Ehrend., type from Castle Gate); *G. desereticum* Dempster & Ehrend., type from Fish Springs]. Blackbrush, sagebrush, mountain brush, riparian poplar-willow, and pinyon-juniper communities at 1125 to 2700 m in Beaver, Box Elder, Cache, Carbon, Daggett, Emery, Garfield, Grand, Iron, Juab, Kane, Millard, Salt Lake, Sanpete, Sevier, Tooele, Uintah, Utah, Washington, and Weber counties; Washington to Idaho, south to California, New Mexico, and Arizona; 65 (xxii).

Galium proliferum Gray Erect annual herbs, 0.5–3 dm tall, simple or branched, glabrous or hispidulous; leaves 5–8 mm long, 0.5–2 mm wide, 4 per whorl, unequal, lanceolate to ovate; flowers perfect, solitary, subsessile in axil of leafy bracts on axillary peduncles, apparently lateral; corolla white, minute; fruit 1.5–3 mm thick, uncinate-hairy. Blackbrush, creosote bush, and other warm desert shrub communities at 885 to 1130 m in Kane and Washington counties; Nevada and California, east to Texas and Mexico; 7 (iii).

Galium stellatum Kellogg Stellate Bedstraw. Dioecious shrubs, mainly 2–7 dm tall; stems much branched, scabrous to hispidulose, with exfoliating white epidermis; leaves 4–10 mm long, 1–3 mm wide, 4 per whorl, lance-acerose, hispidulous, revolute, acuminate-cuspidate, 1-veined; staminate flowers in dense panicles, the pistillate solitary at branchlet ends, short-pedicellate; corolla greenish yellow, bristly hairy externally; fruit 2–3 mm thick, with dense white or tawny hairs 1–2 mm long. Larrea, blackbrush, buddleja, and other warm desert shrub communities, often on limestone or dolomite outcrops, at 850 to 1100 m in Washington County (where rare); Nevada, Arizona, and California; 8 (iii).

Galium trifidum L. Small Bedstraw. [*G. brandegei* Gray; *G. subbiflorum* (Wieg.) Rydb.; *G. trifidum* var. *subbiflorum* Wieg.; *G. tinctorium* var. *subbiflorum* (Wieg.) Fern.]. Clambering to ascending perennial herbs with fibrous roots from short stolons, 0.5–3 dm long or more, retrorsely scabrous on stem angles to subglabrous; leaves 3–15 mm long, 0.8–3 mm wide, 4–6 per whorl, elliptic-oblong to linear, obtuse, 1-veined, glabrous to obscurely scabrous on margins and vein; flowers perfect, 1–3 on axillary peduncles 5–18 mm long; corolla whitish or greenish, mostly 3-merous; fruit glabrous, 1–1.5 mm thick; 2n = 24. Bogs, stream banks, and lake and pond margins in mountain brush, ponderosa pine, aspen, lodgepole pine, spruce-fir, and hairgrass-sedge communities at 1370 to 3355 m in Beaver, Box Elder, Cache, Daggett, Duchesne, Emery, Garfield, Grand, Kane, Piute, Salt Lake, Sanpete, Sevier, Summit, Uintah, Utah, Wasatch, and Washington counties; widespread in North America; circumboreal; 39 (xii).

Galium triflorum Michx. Sweet-scented Bedstraw. Clambering to ascending perennial herbs, mainly 2–8 dm tall; stems retrorsely scabrous on the angles; leaves 1.5–6 cm long, 4–15 mm wide, usually 6 (4 or 5) per whorl, elliptic to oblong-ovate, thin, cuspidate, glabrous or thinly strigulose below and on the veins; flowers perfect, on trifurcate 3-flowered peduncles, the pedicels divergent; corolla greenish white or cream; fruit ca 2 mm thick, densely bristly with hooked hairs 0.5–1 mm long. Moist sites, commonly in riparian areas, in mountain brush, ponderosa pine, Douglas fir-white fir, and aspen communities at 1220 to 2500 m in Box Elder, Cache, Daggett, Duchesne, Garfield, Grand, Juab, Piute, Salt Lake, San Juan, Sanpete, Tooele, Uintah, Utah, Washington, and Weber counties; widespread in North America; circumboreal; 28 (ix).

Houstonia L.

Perennial herbs; leaves opposite, stipulate, mostly linear; flowers perfect, showy, 4-merous, heterostylous or homostylous; corolla salverform; style slender; stigmas 2; ovary inferior or partly so, 2-loculed; fruit a capsule, loculicidally dehiscent; seeds several.

Houstonia rubra Cav. Bluets. [*Hedyotis rubra* (Cav.) Gray; *Houstonia saxicola* Eastw., type from San Juan County]. Caespitose herbs from a slender taproot; stems short, from a multicipital caudex, ca 0.5–1 dm long; leaves 1–3 cm long, 1–2.5 mm wide, linear, revolute; flowers subsessile in the axils, in fruit with stout recurved pedicels 1–3 mm long; calyx lobes 2–4 mm long; corolla deep rose to purple, rarely white, 1.5–3 cm long; style elongate; ovary inferior, becoming half inferior in fruit; seeds several in each locule, black. Desert shrub, blackbrush, and juniper communities at ca 1280 to 1400 m in San Juan County; Arizona and New Mexico, south to Mexico; 4 (i).

Kelloggia Torr.

Rhizomatous perennial herbs; stems simple or sparingly branched, glabrous; leaves opposite, lanceolate, with interpetiolar stipules; flowers in loose terminal forked cymes, 4- to 5-merous; corolla pinkish white, small, funnelform; stamens exserted; ovary inferior, 2-loculed, style filiform, exserted; stigmas 2, linear-clavate; fruit dry, uncinate-hispid.

Kelloggia galioides Torr. Stems 1–6 dm tall; leaves 2–5 cm long, 2–15 mm wide, lanceolate to linear, glabrous; pedicels filiform, 1–12 cm long, divergent; corolla 3–7 mm long, mostly 4-lobed; fruit broadly clavate, 3–4 mm long. Pinyon-juniper, sagebrush, mountain brush, aspen, ponderosa pine, and spruce-fir communities at 1500 to 3000 m in Beaver, Garfield, Grand, Iron, Kane, Piute, Utah, and Washington counties; Washington to Idaho, south to California, Nevada, and Arizona; 12 (iv).

Rubia L.

Coarse, clambering perennial herbs; leaves whorled, coarsely scabrous on margins and midvein; flowers small in axillary or terminal cymes; calyx minute or obsolete; corolla 5-lobed; ovary glabrous; style bifid; stigmas capitate; fruit fleshy.

Rubia tinctoria L. Madder. Stems mainly 5–15 dm long, with hooked prickles on the angles; leaves 3–11 cm long, 0.5–3 cm wide, 4–6 per whorl, lance-elliptic to oblong-elliptic, or ovate; corolla pale yellow, broadly funnelform-campanulate; fruit dark green, turning blackish at maturity, 6–8 mm thick. Weedy plants of roadsides, ditchbanks, and other disturbed areas below 1600 m in Cache, Garfield, Iron, Juab, Kane, Salt Lake, Sanpete,

Tooele, Utah, and Washington counties; widespread in North America; adventive from Europe; 16 (iv).

RUTACEAE A. L. Juss.
Citrus Family

Shrubs or small trees; leaves alternate, simple or compound, glandular-punctate with aromatic oil glands; stipules lacking; flowers commonly perfect, in axillary or terminal clusters; sepals and petals of 3–5 separate or fused parts; stamens 4–8, inserted on a glandular disk; pistil 1, the ovary and fruit superior; carpels mainly 2 or 3; stigma simple or lobed; fruit a capsule or samara; x = 7–11+.

1. Leaves compound, thin; large shrubs or small trees; plants indigenous in southern Utah and cultivated widely .... *Ptelea*
— Leaves simple, thickish; low ill-scented indigenous shrubs in Washington County .... *Thamnosma*

Ptelea L.

Unarmed, aromatic, polygamo-dioecious shrubs or small trees, the bark greenish to whitish; leaves alternate, trifoliolate; leaflets variable in size and shape; flowers in terminal panicles, greenish white; sepals 4 or 5 (6), imbricate in bud, broadly elliptic to ovate or linear-oblong; stamens 4–6, alternate with the petals; disk lobed, serving as a gynophore; ovaries compressed; fruit a samara, with a thin wing around the margin.

Bailey, V. L. 1962. Revision of *Ptelea*. Brittonia 14: 1–45.

***Ptelea trifoliata* L.** Hoptree. Large shrubs or small trees; leaves with 3 ovate to elliptic or linear-lanceolate leaflets 1.3–10 cm long and 3–50 mm wide, petiolate, the leaflets subsessile, entire or finely crenulate, glabrous or variously pubescent; flowers 10 mm wide or less, greenish white, in corymbs 2–5 cm broad; fruit 10–25 mm wide, oblong-orbicular to orbicular, the thin broad wing very veiny. There are two groups of this species present in our area, an indigenous taxon that occurs rarely in southern Utah and a cultivated introduced taxon of dubious origin growing widely in the state.

1. Twigs pubescent, greenish; leaflets ovate to broadly elliptic; plants cultivated .... *P. trifoliata* ssp. *trifoliata*
— Twigs glabrous, pale; leaflets linear-lanceolate to oblong-lanceolate .... *P. trifoliata* ssp. *pallida*

Ssp. *pallida* (Greene) V. L. Bailey [*P. pallida* Greene; *P. neglecta* Greene, type from Kane County]. Along canyons in Garfield (?), Kane, and Washington counties; Arizona; 1 (0). Our material is reported as belonging to **var. *lutescens* (Greene) V. L. Bailey** (*P. lutescens* Greene], a variable taxon at best that intergrades with other taxa in Arizona. Reports of this species from Garfield County (Baxter & Kraus sn 1926 MICH and Siler 1552h MO are likely from Kane County, as is Wetherill sn 1897 CAS). The plant should be sought near Kanab, and possibly it persists still in Glen Canyon along the shores of Lake Powell.

Ssp. *trifoliata* Cultivated ornamentals in Cache, Salt Lake, Utah, and Weber counties; widespread in the U. S.; 5 (0).

***Thamnosma* Torr. & Frem.**

Ill scented low shrubs, with punctate glands; leaves alternate, narrow, simple, early deciduous; flowers cymosely arranged, but apparently racemose, perfect; calyx 4–lobed, persistent; petals 4, erect; stamens 8, inserted on the disk; ovary mostly 2–loculed and 2–lobed, stipitate, the style filiform; stigma capitate; fruit a leathery capsule, 2–lobed, dehiscent at the apex; seeds reniform, 4–6.

***Thamnosma montana* Torr. & Frem.** Stems branching, broomlike, yellowish green, 3–6 dm tall; leaves 5–15 mm long, linear-oblong to oblanceolate; sepals ovate to orbicular, 3–4 mm long, greenish or purplish; petals ovate to oblong, 8–12 mm long, purple (rarely white); stamens 8, 4 long and 4 short, with filiform filaments; style usually exceeding petals; capsule with 2 subglobose glandular lobes, each ca 5 mm thick; seeds 1–3 per locule, whitish, ca 4 mm long. Creosote, Joshua tree, and other warm desert shrub communities at 760 to 1300 m in Kane (near Glen Canyon) and Washington counties; New Mexico, Arizona, and California; 22 (v).

SALICACEAE Mirbel
Willow Family

Dioecious dwarf shrubs to large trees; leaves alternate, simple, entire, serrate, crenate, rarely lobed, usually stipulate, but the stipules often readily deciduous; flowers borne in aments (catkins), without a perianth, each subtended by a small, scalelike bract (commonly referred to as a scale); staminate flowers of (1) 2 to many stamens; pistillate flowers of a single pistil with 2–4 carpels and as many stigmas; placentation parietal or basal; fruit a sessile or stipitate capsule with 2–4 valves; seeds numerous, small, covered with long white hairs, dispersed easily by wind; x = 11, 12, 19.

1. Trees with pendulous aments; leaf-buds covered by several, usually resinous scales .... *Populus*
— Trees, shrubs, or dwarf shrubs with mostly ascending to erect aments; leaf buds covered by a single nonresinous scale .... *Salix*

Populus L.

Small to large trees; leaf buds covered by several overlapping scales, resinous in most taxa; aments pendulous, mostly appearing before the leaves, and often soon deciduous, the scalelike bracts caducous, deeply lobed to laciniate, often dilated; each flower subtended by a cuplike disk; stamens 6–60 or more, the filaments free; inserted on the disk; capsules pedicellate, with 2–4 valves, glabrous (except in *P. balsamifera*).

Eckenwalder, J. E. 1977. North American cottonwoods (*Populus*, Salicaceae) of sections Abaso and Aiegiros. J. Arnold Arboretum 58: 194–208.
Hitchcock, C. L., and A. Cronquist. 1964. Vascular plants of the Pacific Northwest. Part 2: Salicaceae to Saxifragaceae. Univ. Washington Publ. Biol. Vol. 17. 597 pp.
Rehder, A. 1951. Manual of cultivated trees and shrubs hardy in North America. Macmillan Co., New York. 996 pp.

1. Leaves deeply 3–5 lobed and aceriform (at least some), often densely tomentose beneath; plants introduced, cultivated, and escaping .... *P. alba*

— Leaves not deeply lobed, not aceriform, merely toothed, glabrous or nearly so; plants indigenous or cultivated 2

2(1). Bark white and smooth, covered with a whitish powdery bloom, furrowed and gray only in great age; plants not confined to water courses .... *P. tremuloides*

— Bark soon turning gray or brown and roughly furrowed on older trunks; plants mostly cultivated or growing along water courses or edges of lakes 3

3(2). Leaves 0.6–1.3 times longer than wide, deltoid to rhombic or ovate; petioles compressed laterally 4

— Leaves (1) 1.2–7 (10) times longer than wide, ovate to lanceolate; petioles terete or dorsiventrally compressed 6

4(3). Bud scales and twigs of the season pubescent; leaf blades commonly with 4–10 (15) fine to coarse teeth on each side; plants native, sometimes cultivated, common along the drainages of the Colorado River system, and sporadic elsewhere .... *P. fremontii*

— Bud scales and twigs typically glabrous; leaf blades with 15–25 (30) fine teeth on each side; plants introduced, cultivated, sometimes persisting 5

5(4). Leaf blades rhombic-ovate, cuneate at the base, seldom over 7 cm long; capsules 2–valved; trees columnar (in those planted in our area) .... *P. nigra*

— Leaf blades more or less deltoid or broadly ovate, broadly cuneate at the base, some regularly over 7 cm long; capsules 2– to 4–valved; trees not columnar ... . *P. x canadensis*

6(3). Leaf blades distinctly darker above than beneath, very strongly resinous especially when young, the petiole terete or nearly so; ovary and young fruit hairy or glabrous; stamens 30–60 .... *P. balsamifera*

— Leaf blades about equally yellow green on both sides, ovary and young fruit glabrous; stamens mostly 12–30 .. 7

7(6). Leaf blades (1.8) 2.5–6 (9.5) times longer than wide; petioles 1/5–1/3 (2/5) as long as the blades, dorsiventrally compressed; carpels 2 .... *P. angustifolia*

— Leaf blades 1–2.4 times as long as wide; petioles 1/5–3/4 as long as the blades, subterete or somewhat flattened; carpels 2 or 3; plants hybrids, intergrading with *P. angustifolia*, *P. fremontii*, and others ..... *P. acuminata*

Populus acuminata Rydb. (hybrid) Lanceleaf Cottonwood. This is a series of hybrids between *P. angustifolia*, *P. fremontii*, and other taxa with broad leaves, having features intermediate between the parents and intergrading into *P. angustifolia* on one hand and into the broadleaved parent on the other; petioles commonly (1.5) 2.5–5.5 (6.5) cm long, 1/5–3/4 times as long as the blade; leaf blades 1–2.4 times longer than wide. Along streams and rivers and around ponds and lakes, often in mouths of canyons where the parental types come together, also probably cultivated, at 1370 to 1920 m in Box Elder, Cache, Duchesne, Emery, Garfield, Iron, Kane, Salt Lake, San Juan, Sevier, Uintah, Utah, Wasatch, Washington, and Wayne counties; Colorado and Arizona; 29 (0). The name *P. acuminata* in the strict sense is applied to crosses of *P. angustifolia* and *P. deltoides* Marshall var. *occidentalis* Rydb. It is used here in a broad sense to include crosses with other broadleaved taxa.

Populus alba L. White Poplar. Trees spreading by root sprouts, to ca 30 m tall, the trunk to 1 m or more in diameter, the branches usually spreading, the crown more less rounded; bark gray green to whitish and smooth on upper parts of the trunk and branches, rough and furrowed and turning blackish on lower parts of old trunks; twigs tomentose or glabrous; buds tomentose; petioles terete, 1–5 cm long, 0.2–0.6 times as long as the blade; leaf blades longer than wide, deltoid-ovate in outline, undulate-toothed to deeply palmalely 3– to 5–lobed and aceriform, the lobes serrate or crenate, the 2 primary lateral lobes sometimes hastately lobed, dark green above, silvery white-tomentose beneath or glabrous; aments appearing before or with the leaves, the rachis pilose-tomentose, the bracts entire to toothed, not laciniate, ciliate fringed with long pilose hairs, very quickly deciduous; staminate aments 8 cm long or more, the flowers with 6–10 stamens; pistillate aments 4–9 cm long; capsules 2–5 mm long, glabrous, 2 or 3 valved, the pedicels ca 1 (2) mm long; stigmas 2, each 2–lobed, the lobes linear, not dilated; $2n = 38$. Introduced from Eurasia, cultivated, escaping, and more or less naturalized, in populated areas, along fencelines, ditchbanks, and abandoned homesteads and fields, at ca 1370 to 1980 m, to be expected in all counties of the state; widely distributed in the U. S.; introduced from Europe; 21 (ii). Trees with leaves densely white-tomentose beneath are referable to var. *alba*. Those with leaves and twigs glabrous or glabrate and fastigiate crowns are referable to var. *bolleana* Lauche, the Lombardy poplar. These may be hybrids between *P. alba* and some other species.

Populus angustifolia James Narrowleaf Cottonwood. Trees ca 7–15 (20) m tall, the trunk 3–6 (8) dm in diameter, the branches erect-ascending, the crown more or less pyramidal; bark pale green to whitish when young, furrowed and grayish on old trunks, twigs glabrous or pubescent; buds ovoid-conic, pointed, strongly resinous, reddish brown, glabrous or pubescent; petioles semiterete or horizontally flattened and channeled above, especially near the blade, 3–25 mm long, ca 0.3–0.4 times as long as the blade; leaf blades 4–14 cm long, 0.7–2.5 (4) cm wide, (1.8) 2.5–6 (9.5) times longer than wide, lanceolate or occasionally narrow elliptical or ovate, glabrous or nearly so, usually acute at the apex, rounded at the base, the margins finely to coarsely serrate; aments often developing with the leaves, the rachis glabrous or nearly so, the bracts broadly obovate, deeply and irregularly lacrate; staminate aments 2–6 cm long, the flowers with 12–20 stamens; pistillate aments 6–10 cm long; capsules 3–6 (7) mm long, 2–valved, glabrous, the pedicels about 2–10 mm long; stigmas 2, dilated, irregularly lobed. Along water courses, often in canyons, from ca 1525 to 2440 m in all counties of the state; Washington and Alberta to Montana, South Dakota, and Nebraska, south to Nevada, Arizona, New Mexico, and Mexico; 79 (i). This species crosses rather freely with broadleaved species of the genus (see *P. acuminata*).

Populus balsamifera L. Balsam Poplar; Black Cottonwood. Tree 15–30 (50) m tall, the trunk mostly 0.6–1 (1.5) m in diameter, bark furrowed and grayish on older trunks; buds large, the scales very resinous, glabrous or inconspicuously puberulent; petioles more or less terete, 2–6.2 cm long, 1/4–3/4 as long as the blade; leaf blades 4.3–11 cm long, 3.2–8 cm wide, 1.3–2.6 times longer than wide, ovate-acuminate, cuneate to cordate at the base, the margins crenulate, sometimes short ciliate, strongly resinous, glabrous at maturity on both sides, the upper side dark green, the lower side distinctly paler and often rufus tinged in dried specimens; bracts of aments lacerate-fringed, otherwise glabrous or sometimes with minute hairs, these to 0.5 mm long; staminate aments 2–3 (5) cm long, readily deciduous; stamens commonly 30–60; pistillate aments 8–20 cm long; capsules 5–8 mm long, glabrous or pubescent, subsessile; stigmas broadly dilated; 2n = 38. Along streams, mostly in canyons, and cultivated at 1370 to 2350 m in Cache, Juab, Salt Lake, Sevier, Utah, Wasatch and Wayne counties; widespread in North America from Newfoundland south to New York and west to Alaska (ssp. *balsamifera*), and from Alaska south to Baja California (ssp. *trichocarpa*); 9 (0). The native trees of our area are assignable to ssp. *trichocarpa* (T. & G.) Brayshaw [*P. trichocarpa* T. & G.] with mostly pubescent and 3 (rarely 2–4) carpellate capsules. Some of the cultivated trees might be ssp. *balsamifera* with mostly glabrous and 2 (rarely 3–4) carpellate capsules.

Populus x canadensis Moench Carolina Poplar; Gray Poplar. Cultivated and persisting, rarely escaping, to 35 m tall, the trunk 0.5–1.5 (2) m in diameter; bark deeply furrowed and grayish on old trunks; buds large, the scales glabrous but resinous; petioles laterally flattened 3.5–8.5 cm long, 1/3 to as long as the blade; leaf blades mostly 3.5–11.5 cm long, 3.5–11 cm wide, or much larger on stump sprouts, 0.9–1.3 (rarely to 1.5) times as long as wide, deltoid-ovate, acuminate at the apex, mostly broadly cuneate or truncate at the base, the margin crenate-serrate, glabrous and equally green on both sides; staminate aments 4–7 cm long; stamens 15–25; pistillate aments unknown. Cultivated for shade tree in Utah, probably in all counties; widely grown in North America; 15 (iii). This plant probably originated in France as a cross between *P. deltoides* Marshall and *P. nigra* (Rehder, 1951). *P. deltoides*, one of the putative parents of *P.* x *canadensis*, might also be expected in the state as an introduced tree from the eastern U. S., but no specimens were seen that are clearly assignable to that taxon. The original Carolina poplar is *P. deltoides*, but for many years the nursery stock distributed under that name has been *P.* x *canadensis* (Hitchcock and Cronquist, 1964).

Populus fremontii Wats. Fremont Cottonwood. Trees 10–25 m tall with broad rounded crowns, the crown often as broad or broader than the tree is high, the trunk 0.5–2 m in diameter; bark smooth and whitish on young trees and on twigs and young branches, deeply furrowed and grayish or brownish on old trunks; petioles (0.8) 3–9.5 cm long, 1/2 to as long as the blade, flattened, rarely with 2 glands at the summit; leaf blades (2) 4–10 cm long, (15) 4.5–12.5 cm wide, or much larger on sterile sprouts, 0.6–1.2 times as long as wide, deltoid, ovate, rarely nearly rhombic, with truncate, cuneate, or cordate base, acuminate apically, coarsely to finely crenate or serrate with ca 8–11 (15) glandular teeth, glabrous, greenish or yellow green on both sides, turning yellow in autumn; staminate aments 4–10 cm long, the flowers with a broad oblique disk and 50–80 stamens with dark red anthers; pistillate aments 5–15 cm long, the flowers with a cup shaped disk, this to 5 mm wide in fruit; capsules 7–10 (12) mm long, to 8 mm wide, ovoid to subglobose, 3- to 4-valved, glabrous, the stipes 2–6 (10) mm long; stigmas strongly dilated and irregularly lobed; 2n = 38. Along flood plains of rivers and along washes, irrigation ditches, and occasionally cultivated, from 760 to 1860 m, in Cache, Duchesne, Garfield, Grand, Iron, Kane, Salt Lake, San Juan, Sevier, Tooele, Uintah, Utah, Washington, Wayne, and Weber counties; Colorado, Arizona, New Mexico, and Nevada; 98 (ii). This tree is part of a transcontinental complex, of which *P. arizonica* Sarg., *P. deltoides*, *P. sargentii* Dode, and *P. wislizenii* (Wats.) Sarg. are portions. *P. arizonica* and *P. wislizenii* have generally been considered closely allied to *P. fremontii* and they have by some authors been included as varieties of or as synonymous with *P. fremontii*. Specimens having capsules with stipes up to 6 or even 20 mm long are found in Emery County and other points along the Colorado River system. These trees have been referred to as ***P. fremontii* var. *wislizenii* Wats.** Based on the long stipes, they have recently been assigned to *P. deltoides* var. *wislizenii* (Wats.) Eckenwalder (Eckenwalder, 1977). However, the plants are similar to *P. fremontii* in the lack of glands at the junction of petiole and blade and in the few, broad, and coarse teeth on leaf margins, and it seems best to regard them as part of *P. fremontii*.

Populus nigra L. Black Poplar. Tree to 30 m tall; bark deeply furrowed and grayish on old trunks; bud scales glabrous, resinous; petioles flattened laterally, slender 1–4.5 cm long, 0.4–0.8 times as long as the blade; leaf blades 2.2–6.5 cm long, 1.8–8 cm wide, occasionally larger 0.8–1.2 (rarely 1.4) times as long as wide, very often as wide or wider than long, rhombic ovate, or orbicular, usually strongly accuminate at the apex, cuneate at the base, glabrous, equally green on both sides or a little darker above, the margin crenate-serrate, not ciliate; bracts of aments laciniate; staminate aments 4–6 cm long; stamens 20–30; pistillate aments not seen; 2n = 38. Introduced, cultivated for shade and wind breaks, in Beaver, Salt Lake, and Utah counties, but to be expected throughout the state; previously widely grown in western North America; 6 (0). Most of the trees in our area, known as Lombardy poplar or Mormon tree, are from a staminate clone with strongly ascending branches that produced a narrow, often nearly cylindrical crown. Trees of this clone have been assigned to **var. *italica* Duroi**, Lombardy poplar. The foxtail shaped trees were a mark of Mormon cities and towns, especially in the first century following colonization.

Populus tremuloides Michx. Aspen; Quaking Aspen; Quakey. [*P. aurea* Tidestrom]. Colonial tree 10–15 (20) m tall, seldom taller; trunk seldom over 40 cm in diameter; bark white and smooth, covered with a powdery white bloom, turning black and rough where scarred and at the base of very old trunks; branches usually spreading, the crown usually rounded; bud scales shiny but hardly resinous; petioles laterally flattened, 2–5.5 cm long, (1/2) 3/4 to nearly as long as the blade; leaf blades 2–6.5 cm long, 1.8–6.5 cm wide, or much larger on stump sprouts, 3/4–1 1/3 times longer than wide, ovate to reniform-cordate, the margin subentire to serrate or undulate, ciliate, glabrous on the surfaces at maturity; bracts of the aments more or less persistent, especially the staminate ones, 3- to 7-lobed or -cleft, silky-pilose ciliate, the hairs up to 2 mm long; staminate aments 2–4 cm long, readily deciduous; stamens 6–14; pistillate aments 4–12 cm long, to 13 mm wide; capsules 4–6 mm long, the stipes 1–2 mm long, subtended by a cuplike disk ca 2 mm across; carpels 2; stigmas 2, each deeply cleft into 2 or more slender lobes; n = 19, 28, 29. Along water courses and forming clones and aggregates of clones in canyons and on mountain sides at 1400 to 3200 m in all Utah counties; widespread in North America from Labrador to Alaska and south to Tennessee and northern Mexico; 96 (i). Aspen is cultivated as a reclamation, shade, or ornamental tree. In recent years, nursery stock has become readily available from commercial nurseries. The plant is consistently used by beaver for food and dam material.

Salix L.

Depressed, mat-forming dwarf shrubs to large trees; buds covered with 1 nonresinous scale; aments erect to spreading, rarely drooping, developing before (precocious), with (coetaneous) or after (serotinous) the leaves, the bracts mostly entire, occasionally with a slightly toothed apex, flowers with 1, occasionally 2 minute glands near the base; stamens (1) 2–8 (12), the filaments free or united toward the base, inserted on the base of the bract; capsules sessile or stipitate, glabrous or pubescent. Note: Difficulty in identification of the willows is compounded by unisexual plants, aments that are sometimes precocious and mostly caducous, and variation among the usually smaller leaves of the flowering branches, which often lack or have inconspicuous stipules, and the usually much larger leaves and stipules of vegetative branchlets or vigorous young shoots. Thus, herbarium specimens of each species often bear 3 or 4 phases (pistillate, staminate, flowering twigs with or without the deciduous aments, and vegetative twigs). Vigorous young shoots sometimes add a fifth dimension. At times whole plants in the field bear only one or two of the various phases. To facilitate identification of plants of the different phases, pistillate, staminate, and vegetative features have been included in many of the leads in the key. Thus, some of the leads are rather long and features not applicable to a particular specimen will need to be skipped. An alternative approach to lengthy leads is separate keys for the different sexual and vegetative phases. Many such keys have been written, but these sometimes also contain a mixing of vegetative and sexual features. To have an adequate basis for a staminate key, many more staminate specimens are required.

Archer, W. A., and E. E. Little. 1965. Salicaceae of Nevada. Contr. Flora Nevada 50: 10–59.
Argus, G. W. 1965. The taxonomy of the *Salix glauca* complex in North America. Contr. Gray Herb. 196: 1–142.
———. 1973. The genus *Salix* in Alaska and the Yukon. Nat. Mus. Canad. Publ. Bot. 2: 1–279.
———. 1980. The typification and identity of *S. eriocephala* Michx. (Salicaceae). Brittonia 32(2): 170–177.
Cronquist, A. 1964. Salicaceae. In: Vascular plants of the Pacific Northwest. Part 2: Salicaceae to Saxifragaceae. Univ. Washington Publ. Biol. Vol. 17. 597 pp.
Dorn, R. D. 1975. A systematic study of *Salix* section *Cordatae* in North America. Can. J. Bot. 53: 1491–1522.
———. 1976. A synopsis of American *Salix*. Can. J. Bot. 54: 2769–2789.
———. 1977. Willows of the Rocky Mountain states. Rhodora 79: 390–429.
Rehder, A. 1951. Manual of cultivated trees and shrubs hardy in North America. Macmillan Co., New York. 996 pp.

1. Shrubs or dwarf shrubs not over 1 (1.5) m tall, subalpine to alpine 2
— Shrubs or trees, mostly over 1.5 m tall, of valleys to montane 3
2(1). Shrubs depressed, dwarf, 1–10 (20) cm tall mostly alpine, often forming mats, the stems creeping on or below the ground surface .... Key 1
— Shrubs (1) 2–10 dm tall or taller, subalpine or alpine, not forming mats on the ground, the stems ascending to erect .... Key 2
3(1). Leaves 10–30 times longer than wide; plants colonial, spreading underground and forming patches, our most common and widespread low-land willow ..... *S. exigua*
— Leaves less than 8 times as long as wide 4
4(3). Bracts persistent, dark brown to blackish or, if pale green or pale brown in age, silky pilose and with the hairs exceeding the bract by 1–2 mm and the capsules pubescent (rarely glabrous); stamens 2, the filaments glabrous or pilose in a few species; shrubs or occasionally trees, mostly native .... Key 3
— Bracts of at least the pistillate aments caducous, pale green or yellowish tan in age, short pubescent, the hairs not exceeding the bract by more than 1 mm, if at all; capsules glabrous; stamens more than 2 or, if only 2, the filaments pilose; plants mostly trees or treelike except in *S. lasiandra* 5
5(4). Plants native; stamens 3–9 per flower; stipes of capsules mostly 1–2 mm long, obviously longer than the gland .... Key 4
— Plants introduced; stamens 2, except in *S. pentandra*; capsules sessile or the stipes mostly less than 1 mm long and hardly longer than the gland .... Key 5

Key 1.

Depressed mat-forming dwarf shrubs, 1–10 (20) cm tall, at or above timberline.

1. Leaves elliptic to orbicular, 1.4–2.6 times longer than wide, glaucous and strongly reticulate-veined beneath, the tips mostly rounded or obtuse .... *S. reticulata*
— Leaves elliptic or narrow elliptic, (1.2) 2.3–4.7 times longer than wide, glaucous or not, not strongly reticulate-veined beneath, the tips mostly pointed 2
2(1). Leaves 2–5 (7) mm wide, 2–4.7 times longer than wide, sessile or the petiole to 3 mm long; plants seldom over 3 cm tall; aments 0.5–2.2 cm long .... *S. cascadensis*
— Leaves 5–20 mm wide, mostly 2–3 times longer than wide, with petiole 3–13 mm long; plants mostly 5–10 (20) cm tall; aments (1) 2–4 cm long .... *S. arctica*

Key 2.

Low shrubs (1) 2–10 (30) dm tall, subalpine or alpine.

1. Capsules glabrous, the style and stigma together less than 1 mm long; leaves permanently pubescent on both sides, the lower surface not glaucous but often more densely pubescent and thus lighter than the upper surface .... *S. wolfii*
— Capsules pubescent at least until mature or style and stigma together over 1 mm long; leaves often glaucous beneath, glabrous or pubescent 2
2(1). Mature leaves glabrous, dark green and shiny above, strongly glaucous and glabrous or with a few hairs beneath; twigs of the season glabrous or very scattered pubescent, dark chestnut to lustrous purplish black .... *S. planifolia*
— Mature leaves pubescent on both sides, but sometimes glabrate or glabrous in age; twigs of the current season densely pubescent; aments coetaneous or subserotinous, born on stalks to 2 (4) cm long, these usually bearing and subtended by bractlike leaves 3
3(2). Bracts of aments pale green when young, tan in age; capsules 3–5 mm long, pubescent even in age, crowded and nearly sessile so as to mostly conceal the rachis at the center of the aments, the stipes seldom over 0.5 mm long; pistillate aments 0.8–2 (2.5) cm long, 8–10 mm wide .... *S. brachycarpa*

— Bracts of aments brown to blackish, sometimes light brown to whitish tan but not green even when young; capsules (4) 5–7 (8) mm long, sometimes glabrate in age, dense but often not so crowded as to conceal the rachis at the center of the ament, the stipes 0.5–2 mm long; pistillate aments (1.8) 2.5–5 cm long, 11–15 mm wide .... *S. glauca*

Key 3.

Native (or some introduced) shrubs or small trees; aments mostly with dark bracts; stamens 2; capsules glabrous or pubescent.

1. Capsules glabrous; leaves not both glaucous and pubescent on the lower surface when fully expanded . 2

— Capsules mostly pubescent except in *S. lasiolepis*; leaves both glaucous and pubescent on the lower surface when fully expanded 6

2(1). Leaves glaucous beneath, not or scarcely pubescent when fully expanded 3

— Leaves not glaucous beneath, although sometimes lighter colored due to hairs; pubescent at least in part on both sides when fully expanded, but sometimes glabrate in age 5

3(2). Leaves mostly entire, often slightly revolute; twigs dark chestnut to lustrous purplish black, essentially glabrous; plants often less than 1.5 m tall .. *S. planifolia*

— Leaves serrate, serrulate, or entire, not at all revolute; twigs variously colored, glabrous or those of the current season more often pubescent; plants often over 1.5 m tall 4

4(3). Styles 0.7–1.5 (1.8) mm long; leaves of fertile and vegetative twigs often less than 3 times longer than wide, evidently crenulate-serrate or subentire; bark of older twigs not ashy gray or whitish; plants apparently uncommon, in the eastern and central part of the state, mostly montane .... *S. monticola*

— Styles 0.2–0.7 mm long; leaves of vegetative twigs 2–5 times longer than wide, serrulate or entire; bark of older twigs usually ashy gray or white; plants widespread, mostly of valleys and lower montane ... *S. lutea*

5(2). Aments (1.5) 2–5 cm long, with dense crisped-villous, tangled hairs; leaves subglabrate in age, with inconspicuous hairs, entire or sometimes serrulate; plants over 2 m tall .... *S. boothii*

— Aments 0.8–1.5 (3) cm long, with hairs straight or nearly so; leaves permanently pubescent throughout on both sides even in age, the hairs readily conspicuous with a 10 x lens, entire; plants 0.6–1.5 (2) m tall *S. wolfii*

6(1). Twigs strongly blue glaucous, the bloom sometimes evanescent; capsules densely pubescent 7

— Twigs not glaucous or those of the current season often pubescent, or leaves not sericeous; capsules pubescent or glabrous 8

7(6). Pistillate aments 2–5 cm long; capsules sessile or the stipes to 1 mm long, the style and stigmas together 0.8–1.3 mm long; staminate aments ca 2 cm long, the filaments glabrous .... *S. drummondiana*

— Pistillate aments 1–2 cm long; capsules stipitate, the stipes 2–3 mm long, the style and stigmas together ca 0.5 mm long; staminate aments 8–15 mm long, the filaments pilose on the lower 1/2 .... *S. geyeriana*

8(6). Shrubs 0.6–3 m tall, midmontane to above timberline, the stems less than 4 cm thick; leaves mostly less than 2 cm wide, elliptic to narrowly lanceolate .. Key II

— Shrubs or small trees, commonly 3–4 m tall or taller, of valleys or montane; stems of mature plants often 4–10 cm thick or thicker; leaves sometimes over 2 cm wide, oblong, obovate, oblanceolate, or elliptic 9

9(8). Capsules glabrous; filaments about 3–5 mm long; bracts of aments blackish or purplish black, about as wide as long and rounded at the apex, densely pilose-tomentose, the hairs exceeding the bracts by ca 1 mm; plants of Great Basin and Virgin River drainages .... *S. lasiolepis*

— Capsules pubescent; filaments longer or bracts not as dark; bracts of aments of lighter color or, if blackish, with hairs exceeding the bracts by ca 2 mm, pointed or somewhat rounded; plants of various distribution . 10

10(9). Twigs of the second and current year and the dark red bud scales velvety villous; lower surface of leaves densely velvety villous throughout the season; twigs with longitudinal ridges beneath the bark; aments precocious; plants introduced, cultivated . *S. cinerea* L.

— Twigs of the second year glabrous, those of the current season villous or with appressed hairs; lower surface of leaves villous at first but usually rather scattered-villous to glabrate in age; aments various; plants native 11

11(10). Capsules long-beaked, loosely arranged so as to expose much of the rachis; filaments of stamens 3–6 mm long; leaves mostly elliptic, occasionally lanceolate or obovate .... *S. bebbiana*

— Capsules not long beaked, densely arranged and mostly concealing the rachis; filaments ca 10 mm long; leaves obovate or oblanceolate .... *S. scouleriana*

Key 4.

Native, tall shrubs or trees; bracts pale green or yellow, at least the pistillate ones, deciduous; stamens 3–8 (12); capsules glabrous with a stipe 1–2 mm long.

1. Bracts 3–4 mm long; bud scales fused, without free overlapping margins; styles 0.5–1 mm long; plants mostly multi-stemmed, large shrubs from large root crowns, rarely trees, widespread in the state . *S. lasiandra*

— Bracts of aments 1–2 mm long; bud scales with free overlapping margins; styles 0.1–0.2 mm long; plants mostly trees with solitary or few trunks, of various distribution 2

2(1). Leaf blades not glaucous beneath, (2.5) 4–7 times longer than wide; twigs whitish or grayish yellow; plants of the southern 1/2 of the state .... *S. gooddingii*

— Leaf blades glaucous beneath, variously shaped; twigs and distribution various 3

3(2). Twigs reddish or reddish brown, often pubescent at least near the nodes, horizontal or spreading; some of the leaf blades usually 4–5 times longer than wide, shiny dark green above; plants of San Juan and Washington counties .... *S. laevigata*

— Twigs ashy gray or yellowish when fresh, glabrous, tending to droop; leaf blades usually not over 3 times longer than wide, not shiny dark green above; plants widespread in the state, mostly north of the counties listed above .... *S. amygdaloides*

Key 5.

Small to rather large trees, introduced, cultivated, sometimes escaping and persisting; bracts pale green or yellowish, at least the pistillate ones deciduous; capsules glabrous, sessile or nearly so. (Note: The cultivated species of this key, except S. fragilis, are not described due to lack of adequate specimens in herbaria.)

1. Stamens 3–12; leaves with glands on upper part of petiole and lower margins of blade, the blade seldom over 3 times longer than wide, usually glabrous, except along the midrib above, lighter beneath but not glaucous (bay willow) .... S. pentandra L.
— Stamens 2, leaf blades 3–5 times longer than wide, usually glaucous beneath, glabrous or variously pubescent 2

2(1). Pistillate aments 1–2.5 (3) cm long, the capsules 1–2.5 mm long; staminate aments to 4 cm long; petioles glandless; trees weeping, with very slender, greatly elongate, pendulous branches or, if not weeping, the twigs more or less contorted 3
— Pistillate aments mostly over 3 cm long, the capsules 3–6 mm long; petioles sometimes with small glands near the base of the blade; trees not weeping, with upright branches; twigs spreading, not contorted 5

3(2). Trees not weeping; twigs not pendulous; aments 1–1.5 cm long; trees with twigs and branches twisted and contorted are referable to f. *tortuosa* Rehder (corkscrew willow), those with broad umbrella-shaped or semiglobose crowns are f. *umbraculifera* Rehder (umbrella or globe willow) .... S. matsudana Koidz.
— Trees weeping; twigs pendulous, very straight; aments sometimes longer than above 4

4(3). Leaves mostly 3–15 mm wide, mostly deciduous in October; twigs often bright yellow; capsules sessile (weeping willow) .... S. babylonica L.
— Leaves 15–22 mm wide, often persisting into December; twigs greenish or yellow-green; capsules with stipe exceeding the gland; plants hybrids of S. babylonica x S. fragilis (Niobe or Wisconsin weeping willow) .... S. x blanda Anderss.

5(2). Leaves glabrous when unfolded, the margin of mature leaves usually serrate with 4–8 teeth per cm; twigs glabrous or nearly so; stipe of capsules 0.5–0.8(1) mm long; plants common, cultivated and escaping .. S. fragilis
— Leaves sericeous, or glabrous when unfolded, the margin of mature leaves finely serrulate with 9–10 teeth per cm; twigs sometimes pubescent; capsules sessile or subsessile; plants not known outside of cultivation (white willow) .... S. alba L.

Salix amygdaloides Anderss. Peach-leaf Willow. Mostly small trees, rarely shrublike, 4–10 (12) m tall, often with 2–4 ascending trunks; twigs whitish, yellowish, or ashy-gray, rarely reddish, glabrous except when very young; stipules minute and soon deciduous; petioles (3) 5–15 (25) mm long; leaf blades, (1.8) 2.3–6 (7.5) cm long, (7) 12–19 (23) mm wide, or up to 10.5 cm and 3.2 cm wide on vigorous young shoots, elliptical to lanceolate, entire or serrulate, glabrous except when very young, glaucous beneath, green above; aments coetaneous, rarely subprecocious, on leafy or bracteate twigs of the season, 1.5–4 cm long; bracts 1–2 mm long, at least the pistillate ones soon deciduous, pale green, orbicular, the dorsal side woolly-pilose below and along the margins, but mostly glabrous toward the apex, the ventral surface woolly-villous, the hairs seldom exceeding the bract by more than 0.5 mm; staminate aments 2–10 cm long, 7–11 mm wide; stamens 4–7, the filaments pilose on the lower half; pistillate aments (1.5) 2.5–8 cm long, 13–20 mm wide; capsules 4–7 mm long, glabrous, the stipe 1.2–3 mm long, the style ca 0.2 mm long, not surpassing the stigmas, $2n = 38$. Riparian and palustrine habitats, and in neglected fields and pastures, at 1070 to 1710 m in Beaver, Box Elder, Cache, Daggett, Davis, Duchesne, Emery, Grand, Juab, Salt Lake, San Juan, Tooele, Uintah, Utah, and Washington (?) counties; southern Canada and widespread in the U. S., except the southern part; 63 (vi).

Salix arctica Pallas Arctic Willow. [S. *anglorum* var. *antiplasta* Schneid.]. Depressed shrubs with stems creeping on or under the ground, seldom more than 10 (20) cm tall, tending to form mats; stipules minute or lacking; petioles 2–12 mm long; leaf blades (5) 11–47 mm long, (4) 6–16 mm wide, elliptical, narrow elliptical, obovate, or oblanceolate, entire, slightly paler beneath than above but not strongly glaucous, pilose-sericeous when young, sparingly pubescent or glabrous in age; aments coetaneous, leafy or naked pedunculate, the peduncles 7–35 mm long; bracts persistent, dark brown, pinkish purple basally, pilose-sericeous on both sides, sometimes less so dorsally than ventrally, the hairs to 1 mm longer than the bract; staminate aments 15–25 mm long, 7–9 mm wide; stamens 2, the filaments glabrous, to ca 7 mm long; pistillate aments 1.5–7 cm long, 10–12 mm wide, with 25–75 fruits; capsules 4–7 mm long, pubescent, the stipe ca 1 mm long, the style and stigmas together 1–2 mm long; $2n = 76, 114$. Snowbanks, meadows, shores of lakes, and rocky slopes near or a little above timberline at 2775 to 3600 m in Beaver, Cache, Duchesne, Piute, Salt Lake, Summit, and Utah counties; circumboreal and south to California and New Mexico; 16 (ii). Our plants are var. *petraea* Anderss. They intergrade with S. *cascadensis* in the Uinta Mts.

Salix bebbiana Sarg. Bebb Willow. Shrubs, occasionally treelike, (2) 4–6 (8) m tall, with 1 to several stems; branchlets glabrous, puberulent or densely pubescent; stipules usually inconspicuous and caducous; petioles (2) 3–8 (10) mm long, reddish or pale; leaf blades 1–4 cm long, 1.2–2 cm wide or to 7 cm long and 3 cm wide on vigorous shoots, 2.2–2.8 times longer than wide, typically elliptical, occasionally obovate or oblanceolate, entire to slightly undulate-crenate, dark green above, glaucous beneath, pubescent when young on both sides; fully expanded leaves glabrous above, usually with a few hairs beneath near the midrib; aments coetaneous on a bracteate peduncle 3–15 mm long; bracts persistent, pale green to pale brown in age, sometimes reddish at the apex, particularly in staminate aments, silky pubescent, the hairs exceeding the bract by ca 1 mm; staminate aments 1.5–2 cm long, to 13 mm wide; stamens 2, the filaments 3–6 mm long, glabrous or sparingly pilose at the base; pistillate aments 1.5–4 (5) cm long, to 2 cm wide; capsules 6–8 (10) mm long, rostrate, with a rounded basal portion 1–2 mm wide and a long slender beak, pubescent, rather loosely arranged and not concealing the rachis, the stipe 2–3.5 mm long, the style 0.1–0.2 mm long; stigmas 0.3–0.5 mm long, bilobed to the base; $2n = 76$. Riparian communities and occasionally along irrigation ditches at 1370 to 2710 m in Box Elder, Cache, Daggett, Davis,

Garfield, Grand, Juab, Kane, Morgan, Rich, Salt Lake, San Juan, Sevier, Summit, Uintah, Utah, Wasatch, Washington, and Wayne counties; widespread in the U. S. and Canada; 87 (xx). Our plants with leaves sparsely appressed pubescent and soon glabrous beneath and rather weakly raised reticulate-veiny are often referred to as var. *perrostrata* (Rydb.) Schneider, but the separation is probably taxonomically insignificant.

Salix boothii Dorn Booth Willow. [*S. pseudocordata* (Anderss.) Rydb., misapplied]. Shrubs (1.5) 2–4 m tall; young twigs finely hairy, stipules small, inconspicuous and caducous or larger and leaflike on vigorous shoots; petioles mostly 2–5 mm long or to 2 cm long on vigorous shoots; leaf blades (0.8) 2.5–6 cm long, (4) 8–22 mm wide, or to 11.2 cm long and 4 cm wide, elliptical, lanceolate, or almost linear, rarely oval, entire or serrulate, not glaucous beneath, sparingly to moderately pubescent on both sides or glabrate in age, subequally pubescent apically and basally, finally coriaceous; aments subprecocious or coetaneous, sessile or on a naked or bracteate peduncle to 8 mm long; bracts persistent, dark brown to purplish black at the apex, often with a lighter base; aments sericeus-pilose at first but soon crisped-villous, the hairs somewhat entangled, usually exceeding the bracts by 1–2 mm, sometimes deciduous; staminate aments 1–2.5 cm long; stamens 2, the filaments ca 5 mm long, glabrous; pistillate aments (1) 2–4 (6) cm long; capsules 3–6 mm long, glabrous, the stipe 1.5–2 mm long; styles 0.3–1 (1.5) mm long. Riparian and wet meadow communities at 2075 to 3050 m, particularly common on the plateaus of central Utah, but from all counties of the state, except Millard, Morgan, Rich, San Juan, Tooele, Washington, Wayne, and Weber; Colorado west to California and north to Alberta and British Columbia; 139 (xii). Our plants are closely related to **S. myrtillifolia** Anderss. of Alaska and Canada. They vary from those of Alaska and Canada by either taller stature, pubescent leaves (or both), and longer stipules, which are more sharply acute at the apex. The plants might be treated best as a variety of *S. myrtillifolia*, but no new combination is proposed here. They have been referred to *S. pseudocordata*, but that name is synonymous with *S. myrtillifolia* (Dorn, 1975). Plants occasionally trend toward *S. wolfii* ir pubescence of leaves and sometimes are difficult to distinguish from that species vegetatively. This species is similar to *S. lutea*, with which it is is sometimes confused, but the leaves are coriaceous in age, more persistently pubescent, not glaucous beneath, and the plants occur generally at higher elevations.

Salix brachycarpa Nutt. Barren-ground Willow. Shrubs 2.5 6–15 dm tall, rarely taller; internodes with epidermis breaking in translucent flakes, branchlets dark or reddish beneath the dense pubescence; stipules inconspicuous, deciduous; petioles 1–4 mm long, usually not longer than the bud, often reddish; leaf blades (0.6) 1.5–4 cm long, (3) 5–18 mm wide, or to 7 cm long and 3 cm wide on sterile branches, 2–4 (5) times longer than wide, elliptical, broadly lanceolate, or almost linear, entire, thinly to moderately sericeous to nearly glabrous on both sides, strongly glaucous beneath, the midrib sometimes reddish; aments coetaneous or serotinous, nearly sessile or more often on leafy peduncles; bracts pale green, tan, or light brown in age, rarely pink or pale reddish at apex, scattered to densely pilose on both sides, the hairs exceeding the bract by ca 1 mm or less; staminate aments (6) 8–10 (12) mm long, 5–6 mm wide; stamens 2, the filaments 2.5–5 mm long, densely pilose at base and scattered pilose for 1/3–3/4 the length, the pubescent portion sometimes exceeding the scale, anthers 0.3–0.5 (0.6) mm long, orbicular, yellowish; pistillate aments 8–25 mm long, 3–10 mm wide; capsules 3–5 mm long, contiguous and mostly concealing the rachis, sessile or on stipes to 0.5 (1) mm long, pubescent, the hairs persistent even on over-wintering capsules, the style 0.5–1 mm long, the stigmas ca 0.5 mm long, bilobed to the base; 2n = 38. Riparian sites, wet meadows, dry rocky and talus slopes, and rocky open (mostly basic) ground in mountains at 2070 to 3230 m in Cache, Duchesne, Emery, Garfield, Grand, Iron, Juab, Kane, Salt Lake, Sanpete, Sevier, Summit, Utah, and Wasatch counties; Alaska, south to Oregon and Colorado; 86 (xi). Our plants are assignable to **var. brachycarpa** with bracts greenish at anthesis and subspherical or short cylindrical, densely flowered pistillate aments, leaves coarsely pubescent on both sides, and with comparatively tall stature. This willow is closely related to and, often confused with, *S. glauca*. In addition to the features given in the key, *S. brachycarpa* differs from *S. glauca* in having twigs with more numerous aments, distal leaves of fertile twigs often considerably larger than the 3 or 4 proximal ones, and reddish as well as yellowish petioles with the reddish color sometimes extending onto the midrib. Although most of our plants seem quite distinct, apparently there is widespread introgression with *S. glauca* in the Rocky Mt. Region and particularly southward in Colorado (Argus 1965); see discussion under *S. glauca*.

Salix cascadensis Cockerell Cascade Willow. Depressed, mat-forming subshrubs, 1–3 cm tall, from a tap root and much-branched rhizomatous caudex; petiole lacking or to 3 mm long; leaf blades 6–18 mm long, 1.5–4 mm wide, 2–4.7 times longer than wide, linear or narrow elliptical, entire, pilose-sericeous when young, soon glabrous and green on both sides or slightly paler below, some persistent for 1 or more years; aments coetaneous, terminal on short leafy lateral branches, these ca 8–22 mm long; bracts persistent, black or purplish black, reddish purplish basally, 1–2 mm long, 1 mm wide, pilose on both sides, but less so to nearly glabrous at the base ventrally, the hairs ca 1 mm long; staminate aments 3–12 mm long, 5–8 mm wide; stamens 2, separate to the base, the filaments ca 3–4 mm long, glabrous, the anthers reddish or purplish; pistillate aments 5–22 mm long, 5–11 mm wide; capsules 3–4 mm long, pubescent, sessile or the stipe less than 1 mm long, the style and stigmas together ca 1.5 mm long. Alpine tundra in the Uinta Mts. at 3350 to 3935 m in Daggett, Duchesne, Summit, and Uintah counties; British Columbia to Montana, south to Colorado; 15 (iv).

Salix drummondiana Barratt in Hook. Drummond Willow. [*S. subcoerulea* Piper]. Shrubs (1) 2–3 (4) m tall; twigs glabrous or puberulent when very young, heavily glaucous, the bloom persisting into the second year, yellow brown to blackish purple beneath the bloom; stipules narrow, small and deciduous, or larger and more persistent on vigorous shoots; petioles 4–12 mm long; leaf blades 2.2–8 cm long, (5) 13–20 mm wide, or to 14 cm long and 3 cm wide on juvenile shoots, lanceolate or narrowly elliptical, rarely oblanceolate, entire, sometimes with slightly revolute margins, dark green and glabrous or thinly pubescent above, densely silvery white pubescent beneath with short appressed or spreading and slighty tangled hairs, pale glaucous beneath the pubescence; aments precocious or subcoetaneous; bracts persistent, purplish black or purplish brown, pilose on both sides, the longest hairs exceeding the bract by 1.5–2 mm; staminate aments 19–22 mm long, 3–10 mm wide, sessile or on a peduncle to 3 mm long; stamens 2, the filaments, 4–9 mm long, glabrous; pistillate aments 2–4.5 cm long, 3–12 mm wide; capsules 3–6 mm long, pubescent, sessile or the stipe to 1 mm long, the style 0.5–0.7 mm long, the stigmas 0.3–0.6 mm long, 2n = 38, 76. Riparian habitats, wet meadows, and other wet sites at 2135 to 3290 m in Beaver, Box Elder, Cache, Daggett, Davis, Duchesne, Emery, Grand, Piute, Salt Lake, Sanpete, Sevier, Summit, Uintah, Utah, and Wasatch counties; British Columbia and Alberta south to California and New Mexico; 84 (xxv).

Salix exigua Nutt. Coyote, Dusky, or Narrowleaf Willow. Colonial shrub (1) 2–3 m tall or rarely treelike and to 8 m tall; stems ashy gray, branches often reddish, twigs of the season greenish, pubescent; leaves (1) 2–11 cm long, (0.1) 0.2–1 cm wide, sessile or with petiole 1–3 mm long, or to 17.5 cm long and 1.6 cm wide with petiole up to 12 mm long on juvenile shoots, linear, entire or serrulate-dentate with glandular teeth, glabrate to densely white sericeous; aments coetaneous or serotinous, on slender leafy peduncles or twigs of the season, these 0.5–14 cm long; bracts ca 2 mm long and 1 mm wide, pale green or yellowish, deciduous, pubescent on both sides but often glabrate or glabrous dorsally especially toward the apex, occasionally only ciliate ventrally; staminate aments 1.5–4.5 cm long, 0.5–1 cm wide; stamens 2, the filaments pilose on the lower 1/2; pistillate aments 1.5–6 cm long, 8–16 mm wide; capsules 4–7 mm long, mostly glabrous, sometimes pubescent, sessile or the stipe up to 0.8 mm long, the style obsolete. Riparian habitats and other moist sites, often on saline soils, at 825 to 2590 m in all Utah counties; widespread in North America; 170 (ii). Most of our specimens have glabrous capsules and belong to **ssp. exigua var. stenophylla** (Rydb.) **Schneider**. Some specimens from the northern part of the state have somewhat pubescent capsules and might be referable to var. exigua. The closely related **S. melanopsis** Nutt. has been reported for Utah, but I have seen no materials.

Salix fragilis L. Crack Willow. Large trees to 20 m tall, the trunks to 1.3 m in diameter, solitary or few, erect or ascending, with thick furrowed gray or blackish gray bark; branches ascending, often large; branchlets spreading, not pendulous, very brittle; leaf blades lanceolate to narrow elliptic, (2.5) 3–17 cm long, (7) 10–32 mm wide, acute or accuminate, serrate, glaucous or glaucescent beneath, glabrous or sericeous when young and glabrous when mature; aments coetaneous on branchlets of season, these 1–2.5 cm long, with (1) 2–3 (4) reduced leaves, the leaves like those of the nonfloriferous twigs but sometimes oblanceotate; bracts pale green, pale yellow green, or greenish white, tan or very pale brown upon drying, the pistillate deciduous by capsule maturity, sericeous, with the hairs exceeding the bract by about 1–1.5 mm; staminate aments 3.5–7 (9) cm long, 9–12 mm wide; stamens 2; filaments 3–6 mm long, yellow, pilose toward the base, the pilose portion subequal to or shorter than the subtending bract; pistillate aments (2.5) 4–7 cm long, 10–13 mm wide; capsules 4–6 mm long, crowded but the rachis usually apparent, glabrous, the stipes ca 1 mm long; styles 0.5–1 mm long, the stigmas 0.2–0.3 mm long; 2n = 38, 76. Cultivated shade and windbreak trees, persisting, escaping and naturalized along irrigation and natural waterways and lake margins, solitary to forming groves, at 1370 to 2075 m in Beaver, Box Elder, Cache, Duchesne, Juab, Rich, Salt Lake, Sanpete, Sevier, Summit, Tooele, Uintah, Utah, and Wasatch counties; widely grown in the U. S.; introduced from Eurasia; 30 (xlv).

Salix geyeriana Anderss. Geyer Willow. Shrubs 1.5–4.5 m tall; twigs glabrous or sparingly puberulent, strongly glaucous, the bloom sometimes evanescent; stipules minute and deciduous; petioles 3–10 mm long; leaves (1) 2–4.5 cm long (4) 8–12 mm wide, elliptical or narrow elliptical to narrow lanceolate, entire or nearly so, glaucous beneath, sericeous when unfolding, sparsely to moderately sericeous at maturity especially below, the hairs white or a few pale reddish; aments subprecocious to coetaneous; peduncles of aments leafy or bracteate, the staminate 2–5 mm long, the pistillate 3–10 mm long; bracts persistent, sericeous-pilose on both sides, or glabrate or glabrous ventrally, especially in age, the hairs exceeding the bract by 0.5–1 mm, the staminate light brown when young, turning reddish to purplish black at the tips, the pistillate greenish brown to brown; staminate aments 7–15 mm long, 5–9 mm wide; stamens 2, the filaments ca 4 mm long, pilose to about midlength, the pilose portion subequal to or exceeding the bract; pistillate aments 1–2 cm long, 6–15 mm wide; capsules 4–7 mm long, pubescent, the stipe (1) 2–3 mm long, the style 0.2–0.3 mm long; stigmas ca 0.2 mm long; 2n = 38. Along streams and rivers and in other wet places at 2195 to 2895 m, common in the Uinta Mts. to Strawberry Valley and occasional elsewhere, in Beaver, Cache, Daggett, Duchesne, Emery, Garfield, Kane, Rich, Salt Lake, Sevier, Summit, Uintah, Utah, Wasatch, Washington, and Wayne counties; British Columbia south to California and east to Montana and Colorado; 67 (xix). Our plants are assignable to var. **geyeriana**.

Salix glauca L. Glaucous Willow; Grayleaf Willow. [S. pseudolapponum Seemann in Engler]. Mostly low shrubs (0.1) 0.3–1 (3) m tall; branchlets sometimes glaucous, the epidermis of internodes exfoliating in translucent flakes, reddish beneath whitish pubescence, occasionally glabrate, often with a tuft of pilose hairs at the node; stipules mostly small and caducous; petioles (1) 2–6 (18) mm long, mostly yellowish or greenish, the color often extending onto the midrib; leaf blades 2–55 mm long, 7–22 mm wide, or to 9 cm long and to 5 cm wide on juvenile stems, elliptical, pubescent when young to glabrate or glabrous in age, entire or rarely serrate; aments coetaneous, subsessile on twigs or on leafy peduncles; bracts persistent, pale brown to blackish, pilose; staminate aments 1.5–4 cm long; stamens 2, the filaments free or rarely united, glabrous or sparsely pilose at the base, the anthers 0.5–0.8 mm long; pistillate aments 1.5–5 cm long, 11–15 mm wide; capsules (4) 5–7 (9) mm long, densely pubescent to glabrate or glabrous in age, crowded but the rachis usually apparent, the stipes 0.5–2 mm long; style 0.6–1 mm long, the stigmas ca 0.5 mm long; 2n = 76, 114, 144. Along streams, other wet sites, on talus slopes, and in snowflush areas and dry alpine tundra at 2775 to 3660 m in Cache, Daggett, Duchesne, Salt Lake, Sanpete, Summit, Uintah and Wasatch counties; circumboreal, south to New Mexico; 57 (xii). Plants of the Uinta Mts. are comparable to those that have passed under the name of *S. pseudolapponum*. On windswept alpine summits, they approach the stature of *S. arctica*, but the stems are still ascending to erect. The Uinta Mt. materials tend more toward glabrescence in the capsules and have darker scales than is typical of those in the Bear River Range. The twigs are quite persistently pubescent, and the leaves are seldom over 5 cm long or over 2 cm wide. Plants from the Bear River and Wasatch ranges have densely and persistently pubescent capsules, pale brown to dark brown to occasionally pinkish tan or rarely whitish tan scales, with twigs often soon glabrate and some of the leaves are frequently over 5 cm long and over 2 cm wide. Specimens from Horseshoe Flat area of the Wasatch Plateau have glabrous or pubescent capsules, mostly dark scales, and glabrate and unusual distinctly serrate leaves. The variability in *S. glauca* almost encompasses *S. brachycarpa*. However, I prefer to follow Argus (1965) in keeping them separate.

Salix gooddingii Ball Black Willow. [*S. nigra* Marshall sens. lat.]. Trees or occasionally shrubs (2) 6–10 (24) m tall; branchlets yellowish, glabrous or initially finely pubescent; stipules to 8 mm long, more or less glandular, usually caducous; petioles 3–7 mm long; leaf blades 2–7.5 cm long, 6–16 mm wide or to 10.2 cm long and 18 mm wide with petiole to 15 mm long on juvenile shoots, narrowly to broadly lanceolate, short to long acuminate, entire or more often glandular-serrulate, greenish on both sides, pubescent initially but finally glabrate; aments coetaneous; peduncles 1–6 cm long, with 3–6 leaves; floral bracts pale green or pale yellow, soon fading tan, and the pistillate ones deciduous, pubescent on both sides or glabrous toward the apex, entire or with 1–3 minute, rounded teeth; staminate aments 2.5–6.5 cm long, 5–10 mm wide; stamens 3–6, the filaments pilose to about midlength; pistillate aments 1.5–6 cm long, 10–17 mm wide; capsules 4–7 mm long, glabrous, the rachis apparent between them, the stipe 1–2 mm long; style 0.1–0.3 mm long. Along rivers and other drainages at 825 to 1585 m in Garfield, Grand, Kane, San Juan, and Washington counties; Kansas to California, south to Mexico; 25 (ii). *S. gooddingii* is a part of the *S. nigra* complex. The relationship of these taxa is in need of further study.

Salix laevigata Bebb. Red Willow. Shrubs or trees 2–15 m tall; branchlets reddish brown or dull brown, ashy red or ashy gray; stipules inconspicuous or to 6 mm long on juvenile shoots, usually deciduous; petioles stout, 4–14 mm long; leaf blades (1) 1.8–4 (6) cm long, 5–20 cm wide, or to 19 cm long and 4 cm wide on vigorous young shoots, narrowly to broadly lanceolate, glandular-serrulate, somewhat revolute, usually thick and firm, dark green and glabrous above, glaucous and glabrous or pubescent toward the base and along the midrib; aments subprecocious to coetaneous, on leafy peduncles; bracts 1–2 mm long, at least the pistillate deciduous, pale yellow, crinkly pilose on both sides or often glabrous dorsally, entire or erose to dentate at apex; staminate aments 3–6 cm long, ca 1 cm wide; stamens 3–7, pilose on lower half; pistillate aments 4–8 (11) cm long, to 1.5 mm wide; capsules 4–5 (6) mm long, glabrous, the stipes 1.5–2.5 mm long, styles 0.1–0.2 mm long, equaling the bilobed stigmas. Along drainages at 700 to 1370 m in Kane (?), San Juan, and Washington counties; Arizona, California, Nevada, and Baja California; 16 (0). Perhaps the red willow is not distinct from *S. bonplandiana* H.B.K., with which it was synonymized by Dorn (1977).

Salix lasiandra Benth. Whiplash Willow; Caudate Willow. Shrubs or small trees (2) 3–6 (12) m tall; twigs glabrous or finely hairy when young; stipules often well developed, broadly rounded, gland-toothed, 2–10 mm long, eventually deciduous; petioles 3–15 (25) mm long, often bearing 2 or more glands on the upper side at or near the base of the blade; leaf blades (2.2) 5.5–11.5 cm long, (5) 12–21 mm wide, or to 26 cm long and 5.5 cm wide on juvenile shoots, lanceolate, elliptical or narrow elliptical, gradually long acuminate, closely serrulate, glabrous except in youth; aments coetaneous, on leafy peduncles 1–3.5 cm long, these with 3–5 leaves to 6.5 cm long and 1.2 cm wide, deciduous after the fruit matures; bracts deciduous by capsule dehiscence, 3–4 mm long, glabrous or nearly so on the upper 1/2, pubescent toward the base, entire or minutely toothed apically with a few rounded teeth, the staminate yellow, the pistillate pale greenish; staminate aments 1.8–4.5 cm long, 3–12 mm wide; stamens 3–8, usually 5, the filaments pilose; pistillate aments 2–7 cm long, 11–18 mm wide; capsules 4–8 mm long, glabrous, the stipe 1–2 mm long, the style 0.5–1 mm long, the stigmas to 0.5 mm long; $2n = 76$. Riparian and palustrine habitats at 1525 to 2625 m in Beaver, Box Elder, Cache, Carbon, Daggett, Davis, Duchesne, Emery, Garfield, Juab, Millard, Morgan, Piute, Rich, Salt Lake, San Juan, Sanpete, Sevier, Summit, Uintah, Utah, Wasatch, and Washington counties; Alaska and Yukon to California and New Mexico; 48 (xxi). Our plants belong to var. **caudata** (Nutt.) Sudw. [*S. pentandra* var. *caudata* Nutt.].

Salix lasiolepis Benth. Arroyo Willow. Shrubs or small trees mostly 4–6 m tall in our range; twigs yellowish olive to reddish, usually soft puberulent when young; stipules minute, soon deciduous or lacking, occasionally well developed on juvenile shoots; petioles 3–15 mm long; leaf blades 1.5–4.2 cm long, 6–13 mm wide, or to 11 cm long and 2.5 cm wide on vigorous shoots, usually oblanceolate or oblong, occasionally elliptical, entire, rarely minutely toothed, somewhat revolute, dark green and glabrous above, at maturity glaucous beneath, more or less coriaceous, rather densely soft pubescent on both sides when unfolding, more or less hairy beneath at maturity; aments precocious to subcoetaneous on bracteate or naked peduncles 3–6 mm long; bracts persistent, purple black, obovate, broadly rounded apically, densely villous, obscured by the hairs; staminate aments 2.2–4.5 cm long; stamens 2, the filaments glabrous; pistillate aments (1.8) 2.2–4.5 cm long, 10–12 mm wide; capsules 3–4 (5) mm long, glabrous, the stipe 1–2 cm long; style ca 0.5 mm long, the stigmas 0.2–0.3 mm long. Along streams, ditches, and washes at 1460 to 2330 m in Beaver, Iron, Juab, Kane, Millard, Sevier, Tooele, Utah, and Washington counties; British Columbia south to Baja California and east to Idaho, Texas, and northern Mexico; 40 (xx).

Salix lutea Nutt. Yellow Willow. [*S. lutea* var. *platyphylla* Ball, type from the La Sal Mts.; *S. lutea* var. *watsonii* (Bebb) Jepson]. Shrubs or rarely small trees, (2) 3–5 (9) m tall; branchlets slender, yellowish to reddish initially, often pale on one side and red purple on the other, glabrous; older twigs and smaller branches often grayish white; stipules small and inconspicuous or to 1 cm long or more and foliose on vigorous shoots, usually deciduous; petioles 1–11 (20) mm long; leaf blades (1) 2–5.5 cm long, (4) 9–21 mm wide or to 10.7 cm long and 3 cm wide on vigorous young shoots, elliptical or lanceolate, rarely linear, entire or occasionally serrulate, glaucous beneath and usually glabrous at maturity, the lower surface glabrous or less pubescent than above; aments precocious or subprecocious; peduncles 1–7 cm long, naked or with 1–3 bracts; rachis and usually peduncle covered with a tangle of crisped-villous white hairs; bracts persistent, pubescent with crisped-villous, entangled hairs, sometimes only moderately pilose-woolly toward the base or near the apex ventrally, the dorsal side usually glabrous apically and often throughout as the crinkly hairs are readily deciduous; staminate aments 2–5 cm long, ca 1 cm wide; stamens 2, the filaments glabrous, the anthers yellowish or turning purple; pistillate aments 2–7 cm long, to 2 cm wide; capsules 3–6 mm long, glabrous, compactly to subcompactly disposed; the stipe (1) 1.3–3 (4) mm long; style 0.2–0.7 mm long, the stigmas often scarcely bilobed; 2n = 38. Along streams and ditches in valleys and canyons and occasionally on mountains at 1340 to 2350 m in all Utah counties, except Carbon, Davis, Iron, Morgan, and Rich; New Mexico to California, north to Alberta; 160 (xxx). Our plants are closely related to and possibly a part of the *S. eriocephala* Michx. complex. They have been referred to as *S. rigida* Muhl., but Argus (1980) has placed *S. rigida* in synonymy under *S. eriocephala*. He did not place *S. lutea* in synonymy, but suggested that more study is needed. Until such a study is made, it seems best to retain the traditional name for our plants. *S. lutea* var. *ligulifolia* Ball (*S. ligulifolia* (Ball) Ball has been reported for southern Utah, but I have seen no material that fits the description exactly. Hence, the taxon is not treated formally herein.

Salix monticola Bebb ex Coult. Shrubs 1.5–4 m tall; twigs yellowish when fresh, drying blackish, initially puberulent; stipules small and inconspicuous or foliose on vigorous shoots; petioles 5–10 (15) mm long; leaf blades 2–5 cm long, 0.7–1.5 mm wide or up to 11 cm long and 4 cm wide on juvenile shoots, mostly elliptical or elliptic-obovate, crenate-serrate or subentire, slighty pubescent when very young, more so above than beneath, usually glabrous when fully expanded, glaucous beneath; aments precocious or coetaneous, subsessile or on peduncles to 1 cm long, often subtended by bracts; floral bracts persistent, dark brown to blackish, pilose, or soon crisped-villous, the hairs exceeding the bract by ca 2 mm, more or less tangled; staminate aments 2–3.5 cm long, 1–1.5 cm wide; filaments 2, glabrous; pistillate aments 2–6 cm long, 1–1.5 cm wide; capsules 4–7 mm long, glabrous, subsessile, the stipe less than 1 mm long; style 0.7–1.8 mm long, longer than the stigmas; 2n = 38, 114. Along streams and other wet places at 2195 to 3200 m in Beaver, Garfield, Piute, San Juan, Sanpete, Sevier, Uintah and Wasatch counties; Wyoming, Colorado, Arizona, and New Mexico; 12 (0). This willow is closely allied to both *S. boothii* and *S. lutea* and rather easily confused with them. Separation from *S. boothii* is often compounded by the lack of glaucescence on young leaves. More specimens are needed to gain a better understanding of this plant in the state. All materials examined to date could pass for broadleaved specimens of *S. boothii*.

Salix planifolia Pursh Plainleaf Willow. [*S. phylicifolia* ssp. *planifolia* (Pursh) Hiit.]. Shrubs 0.5–1.5 (4) m tall; internodes often with epidermis exfoliating in translucent flakes or strips, branchlets typically glabrous and lustrous black or purplish black, rarely glaucous in part; stipules small and usually deciduous; petioles 2–10 mm long; leaf blades 1.2–3.8 (8) cm long, 4–13 (30) mm wide, or to 5 (13) cm long and 2 (5) cm wide on juvenile twigs, elliptical or narrow elliptical, soon glabrous and dark green above, glaucous and glabrous to sparingly pubescent below, entire or rarely with minute teeth; aments precocious (at least the staminate) to coetaneous, nearly sessile or rarely on a naked peduncle to 0.5–1 cm long; bracts persistent, blackish, scattered to densely villose to pilose, the hairs usually exceeding the bract by ca 2 mm; staminate aments 10–25 mm long; stamens 2, the filaments glabrous, ca 6 mm long; pistillate aments 2–4 cm long, 1–1.5 cm wide; capsules 3–7 mm long, pubescent, at least near the base, or glabrate in age, the stipe mostly less than 1 mm long; style and stigmas together more than 1.5 mm long. Streamside meadows, margins of lakes and ponds and other wet places at 2255 to 3660 m in Daggett, Duchesne, Garfield, Iron, Salt Lake, Sanpete, Sevier, Summit, Uintah, and Wasatch counties; circumboreal, south to California and New England; 40 (x). Our plants mostly fall well within the concept of var. *monica* (Bebb) Jepson [*S. monica* Bebb]. However, some taller plants having larger leaves at moderate elevations in the Uinta Mts. are apparently var. *planifolia*. The differences do not seem to be taxonomically significant.

Salix reticulata L. [*S. nivalis* Hook.; *S. nivalis* var. *saximontana* (Rydb.) Schneider]. Caespitose dwarf shrubs; stems creeping at or just below the ground surface, the slender aerial twigs rarely more than 2–3 cm long, usually prostrate; stipules minute and deciduous or none; petioles 1–8 (15) mm long; leaf blades 0.5–3 cm long, 0.3–2 cm wide, ovate, obovate, orbicular or occasionally broadly elliptical, entire, glabrous, green above, glaucous beneath, strongly reticulate veined; aments subcoetaneous, but mostly serotinous, on the ends of current shoots; bracts persistent, pale green or yellowish, sometimes with reddish tops, spatulate or obovate, glabrous or sparsely pubescent ventrally, especially toward the margin with short hairs that extend less than 1 mm beyond the bract; staminate aments 0.5–2 cm long, slender, the flowers loose and not concealing the puberulent rachis, on a slender glabrous peduncle ca 10–12 mm long; stamens 2; filaments 1.5–2 mm long, glabrous or pilose toward the base; anthers soon reddish or purple; pistillate aments 5–15 mm long, 5–8 mm wide, on a slender peduncles 1–2 cm long; capsules 1.5–3 mm long, pubescent, or glabrous in age, sessile or the stipe to 0.5 mm long; style obsolete or to 0.2 mm long, the stigmas 0.1–0.2 mm long; 2n = 38. Open rocky slopes and ridges and alpine tundra at 2985 to 3965 m in Daggett, Duchesne, Grand, Salt Lake, San Juan, Summit, and Utah counties; circumboreal, south to California, New Mexico, Utah, and Colorado; 33 (0). Most of our plants are referable to var. *saximontana* (Rydb.) Kelso, which may not be distinct from var. *reticulata*. A few specimens approach var. *nivalis* (Hook.) Anderss. The diagnostic features seem to be poorly correlated in our plants.

Salix scouleriana Barratt in Hook. Scouler Willow. Shrubs or small trees 3–7 m tall; stipules small and inconspicuous or large and leaflike on juvenile shoots, finally deciduous; petioles 2–11 mm long; leaf blades 2–6 cm long, (0.8) 1–3 cm wide or to 11.5 cm long and 4 cm wide on vigorous shoots, obovate to oblanceolate, rounded to acute or sometimes accuminate at the apex, entire or finely serrate, or often coarsely crenate or serrate on larger leaves of innovations, densely crisp-hairy or sericeous especially beneath initially, the mature ones dark green and glabrous above, except sometimes puberulent along the midrib, the lower side strongly glaucous, sparsely puberulent with translucent whitish or rusty minute hairs, or occasionally densely felty-villous; bracts of aments blackish or purplish-black nearly throughout, reddish or pale at the very base, sericeous-pilose on both sides, the hairs at the apex usually exceeding the bract by 1.5–2 mm; staminate aments 15–35 mm long, nearly as wide as long, strictly precocious, nearly sessile or on thickened bracteate peduncles to 7 mm long, the bracts 3–4 mm long, about 2 mm wide, pale green to whitish, sericeous; stamens 2, the filaments to 11 mm long at maturity, glabrous; pistillate aments 2–6 cm long, 13–17 mm wide, precocious or subcoetaneous, nearly sessile or on thickened bracteate peduncles to 17 mm long, the bracts to 7 mm long and 2 mm wide, not at all leaflike; capsules (5) 6–9 mm long, pubescent, subsessile or with a stipe 1–3 mm long; style 0.3–0.4 mm long, rarely shorter, the stigmas 0.5–1 mm long; 2n = 76, 114? Around springs, along streams, and on well-drained slopes in aspen and conifer woods at 1400 to 3355 m in all Utah counties except Beaver, Emery, Kane, Morgan, Piute, and Wayne; Alaska and Yukon to California, Arizona and New Mexico; 69 (xi). Scouler willow is most closely allied to S. *humilus* Marshall and to S. *discolor* Muhl. of eastern U. S. and Canada. The staminate phase of S. *discolor* might be cultivated in the state as pussy willow, but all specimens of pussy willow examined prior to this publication belong to the Eurasian S. *cinerea*. Occasionally specimens of Scouler willow have leaves densely pubescent beneath. Arnow et al. (1980) attributed this to hybridization with S. *drummondiana*.

Salix wolfii Bebb in Rothr. Wolf Willow. Shrubs 0.6–1.5 (2) m tall; twigs yellow to orange when young, chestnut brown in age, those of the season thinly villous-puberulent; stipules 1–5 mm long, often glandular-serrulate, eventually deciduous; petioles 2–10 mm long; leaf blades 1.2–4.2 cm long, 5–13 mm wide or to 5.3 cm long and 16 mm wide towards the ends of juvenile twigs, narrow elliptical, linear-lanceolate or occasionally oblanceolate, entire, sparsely to densely sericeous-to-mentose on both sides, even in age, or finally glabrate; aments coetaneous or subserotinous, subsessile or on bracteate peduncles to 1 cm long; floral bracts persistent, blackish or pale at the very base, pilose-sericeous on both sides, the hairs exceeding the bract by ca 1 mm; staminate catkins 10–15 mm long, 8–10 mm wide; stamens 2, the filaments about 3–4 mm long, glabrous; pistillate aments 8–20 (30) mm long, 6–10 mm wide; capsules 3–5 mm long, glabrous or rarely pubescent, the stipe less than 1 mm long; style ca 0.5 mm long, the stigmas about 0.2 mm long; 2n = 38. Along streams and lakes and ponds margins at 2470 to 3290 m in Cache, Daggett, Duchesne, Emery, Summit, Uintah, and Wasatch counties; Oregon to Montana, south to Nevada, and Colorado; 44 (ix). Our plants belong **var. wolfii**. One specimen (B. Maguire, D. Hobson, & R. Maguire 14104 UT) from White Pine Lake, Cache County has pubescent capsules and leaves that are larger than others from the state. This specimen is like S. *wolfii* var. *idahoensis* Ball, which is known from well north and west of Utah. Other specimens from the vicinity of Whitepine Lake and other places in the Bear River Range have glabrous capsules, and I prefer not to list var. *idahoensis* for the state based on this one specimen. The plants from the Bear River Range with pistillate aments 15–30 mm long do, however, seem intermediate toward var. *idahoensis* when compared to those of the Uinta Mts. with pistillate aments 8–15 mm long. The specimen with pubescent capsules and somewhat larger leaves is probably the basis of reports of S. *commutata* Bebb for Utah.

SANTALACEAE R. Br.
Sandalwood Family

Perennial parasitic herbs; leaves alternate, short-petiolate to nearly or quite sessile, entire, simple, green; flowers perfect, at least some, borne in terminal or axillary corymbose clusters, white to greenish or pinkish; sepals usually 5; petals lacking; stamens as many as the sepals and opposite them, inserted on a fleshy disk; pistil 1, the ovary inferior, 1–loculed; style 1; fruit drupaceous; x = 5, 6, 7, 12, 13+.

Comandra Nutt.

Glabrous perennials; stems striate; leaves alternate, sessile; flowers small, perfect; sepals usually 5; ovary 1–loculed; stigma capitate; fruit a drupe.

Comandra umbellata (L.) Nutt. Bastard Toadflax. [*Thesium umbellatum* L.; C. *linearis* Rydb., type from Green River]. Stems erect, 0.8–3.4 dm tall, simple or branched; leaves 1–3.3 cm long, 1–10 mm wide, linear to narrowly elliptic or lanceolate, pale green and glaucous; sepals oblong to lanceolate, 2.5–5 mm long, persistent; fruit 6–9 mm thick, globose to ovoid, somewhat fleshy or dry when old, purplish to brown; n = 26. This species is present in numerous plant communities from 850 to 2780 m in most if not all Utah counties; widely distributed in North America; 110 (v). Our material belongs to **var. pallida** (**A. DC.**) Jones [C. *pallida* A. DC. in DC.].

SAPINDACEAE A. L. Juss.
Soapberry Family

Trees or large shrubs; leaves alternate, pinnately or bipinnately compound; flowers perfect or imperfect, irregular, borne in panicles; sepals 4 or 5, distinct or connate at the base; petals 4 or 5, inserted below an annular fleshy disk, with 2 upturned appendages at the base of the blade; disk borne between the petals and the stamens; stamens 8 or fewer, the filaments distinct; pistil 1, the ovary superior, 3–loculed, with 2 ovules per locule, the placentation axile; style 1; fruit a loculicidal capsule; x = 10–16.

Koelreuteria Laxmann

Showy ornamental umbrella shaped trees; leaves large, odd-pinnate to bipinnate, the leaflets serrate; flowers yellow, borne in summer; capsules bladdery, often persisting through winter.

Koelreuteria paniculata Laxmann Goldenrain Tree. Trees or shrubs, mainly 2.5–7 m tall, with the canopy spreading; bark fissured, grayish; leaves 10–40 cm long, with (3) 5–15 leaflets, these irregularly ovate to oblong or lance-ovate, coarsely and irregularly lobed to doubly crenate-serrate, acute to acuminate apically, rounded to obtuse basally, green and glabrous above, paler and usually hairy along the veins beneath; flowers numerous, mainly

8–12 mm long, bright yellow, borne in July and August; capsules mainly 4–6 cm long, the valves soon separating at the margins, cordate to ovate in outline, bladdery inflated, often persisting through the autumn and winter. Commonly grown ornamental tree, especially of streets and byways in the major population centers, producing seeds abundantly, escaping and tending to be weedy, mainly below 1525 m in Carbon, Grand, and Utah counties; introduced from Asia; 17 (i).

SAURURACEAE Meyer
Lizard-tail Family

Erect more or less aromatic perennial herbs from elongate stolons; Leaves mostly basal, alternate, simple, usually petioled; stipules adnate to the petiole; flowers perfect, in congested spikes or lax racemes subtended by conspicuous petaloid bracts; perianth lacking; stamens 6–8, free or adnate to ovary base, or epigynous; anthers 2-loculed; pistils 3 or 4, distinct or united basally; seeds several; x = 11, 12(?).

Anemopsis H. & A.

Colonial herbs; stems nodose, scapelike; leaves mostly basal, minutely punctate; spike conic, with a persistent white involucre of several bracts; ovary sunk in axis of spike; stigmas 3–4; fruit a capsule.

Anemopsis californica (**Nutt.**) **H. & A.** Yerba Mansa. [*Anemia californica* Nutt.]. Plants 1–5 dm tall, pubescent, with a broadly clasping ovate leaf above the middle; flowering stems arising from thick simple caudices; basal leaves 3.5–50 cm long or more, the blades elliptic-oblong, the petioles as long or longer than the blades, entire, cordate basally; spikes 1.5–4 cm long; involucral bracts at base of inflorescence 1–3 cm long, white or suffused with red, each flower subtended by a white obovate clawed bract; ovules 6–10 on each placenta; n = 22. Wet, marshy, usually saline areas adjacent to seeps, springs, and streams in Utah and Washington counties; California to Colorado and Texas, south to Mexico; 16 (iii).

SAXIFRAGACEAE A. L. Juss.
Saxifrage Family

Perennial herbs or shrubs; leaves basal, alternate, or opposite, with or without stipules; flowers perfect, regular, solitary to many and racemose or cymose; sepals, petals, and stamens borne on a floral cup or hypanthium; hypanthium saucer- or cup-shaped or tubular, sometimes small or essentially lacking; sepals 4 or 5, often appearing as lobes of the hypanthium, petallike in *Ribes*; petals 4 or 5, distinct, alternate with the sepals; stamens (4), 5, 8, 10, or more; pistil 1 (the carpels sometimes almost distinct), the ovary superior, partly inferior, or inferior; style 1 or more; fruit a capsule, follicle, or berry; x = 6–15, 17, 30. This is an extremely variable family. As treated herein it includes: Grossulariaceae (*Ribes*), Hydrangiaceae (*Philadelphus*), and Parnassiaceae (*Parnassia*).

1. Plants shrubs, woody well above ground level 2
— Plants herbaceous, sometimes with woody caudices at or below ground level 7

2(1). Leaves alternate, lobed and toothed; petals shorter than the sepals, not over 4 mm long, often similar in texture and color to the sepals; ovary inferior; fruit a berry .... *Ribes*
— Leaves opposite, entire or toothed; petals longer than the sepals, over 4 mm long (except in *Fendlerella*), of contrasting texture and color from the sepals; ovary not completely inferior; fruit a capsule 3

3(2). Stamens more than 10; petals 4 (or sometimes more in cultivated plants) .... *Philadelphus*
— Stamens 8 or 10; petals 4 or 5 4

4(3). Leaves toothed, with petioles 2–10 mm long, or rarely sessile; petals 5, 4–13 mm long; stamens 10 ... 5
— Leaves entire, sessile, or the petiole 1–2 mm long; petals 4 and 13–20 mm long, or 5 and only 2–4 mm long; stamens 8 or 10 6

5(4). Flowers rather numerous, paniculate; ovaries and sometimes the petals stellate-pubescent; filaments often petaloid and bifid apically; plants introduced, cultivated .... *Deutzia scabra* Thunb.
— Flowers few, in small cymes; ovaries sericeous-canescent; petals glabrous; filaments hardly petaloid, sometimes dilated basally, not bifid apically; leaves sericeous-canescent; plants indigenous .... *Jamesia*

6(4). Petals 4, 13–20 mm long; stamens 8; styles 4; leaf blades 9–30 mm long, the lower surface without pustulate hairs; shrubs ca 1–2 m tall, of Grand and San Juan counties .... *Fendlera*
— Petals 5, 2–4 mm long; stamens 10; styles 3; leaf blades 4–12 mm long, the lower surface sometimes with pustulate hairs; shrubs to ca 1 m tall, of other distribution .... *Fendlerella*

7(1). Leaves all basal, distinctly and often abruptly petioled, not lobed more than 1/2 the distance to the midrib; flowers on naked scapes or these with a solitary bract; stamens 5 or 10 8
— Leaves not all basal, or sometimes so in depauperate plants, but then the blades not distinctly petioled or else lobed more than 1/2 the distance to the midrib; flowers often not scapose; stamens 10 12

8(7). Flowers solitary and terminal on long scapes with a solitary bract; petals 6–14 mm long; leaves not toothed, sometimes cordate, otherwise not lobed; fertile stamens 5 .... *Parnassia*
— Flowers not solitary; scapes without bracts; petals 2–4 mm long; leaves either toothed or lobed or both and sometimes cordate as well; fertile stamens various 9

9(8). Leaves peltate, the blades 5–40 cm wide, cupped in the center; petioles and scapes to 10 dm long or more .... *Peltiphyllum*
— Leaves not peltate, the blades to 10 cm wide, not cupped in the center; petioles not over 2.5 dm long; scapes less than 7.5 dm long 10

10(9). Flowers in simple, very narrow, elongate, spikelike, ebracteate racemes; petals parted or divided into filiform segments; leaves toothed and lobed .... *Mitella*
— Flowers not in spikelike racemes or, if so, bracteate; petals entire 11

11(10). Leaves crenate-toothed and lobed; stamens 5; plants often of dry rocky places .... *Heuchera*
— Leaves subentire, crenate, or very coarsely dentate but not lobed; stamens 10; plants mostly of dry meadows or wet places .... *Saxifraga*

12(7). Leaves parted or divided to the midrib, the basal ones abruptly constricted into slender petioles 0.5–10 cm long; petals deeply lobed or cleft .... *Lithophragma*

— Leaves entire, toothed, or lobed, but not divided more than 1/2 the distance to the midrib or, if so, sessile or nearly so; petals entire 13

13(12). Leaf blades of basal and stem leaves crenate-toothed (the teeth more than 10), also shallowly lobed, reniform or orbicular; petals pink to deep red; floral cup 4–7 mm long, often reddish or purplish; plants of the Bear River Range .... *Boykinia*
— Leaf blades entire, toothed, or lobed, (the teeth or lobes 8 or fewer); petals yellow or white, sometimes with purple markings; floral cup 1–3 mm long, greenish or purplish; plants variously distributed .. *Saxifraga*

Boykinia Nutt.

Caulescent, glandular herbs from woody branched caudices and thick scaly rootstocks; leaves petioled, basal and alternate, with membranous stipules; flowers perfect, regular, borne in compact, few-flowered, bracteate cymes; hypanthium calyxlike; sepals 5; petals 5; stamens 10; ovary ca 1/2 inferior; styles 2, free or connate below; fruit a capsule; seeds several.

Boykinia jamesii (Torr.) Engler James Saxifrage. [*Saxifraga jamesii* Torr.; *Telesonix jamesii* (Torr.) Raf.]. Perennial caulescent herbs, 6.5–20 cm tall; caudices clothed with broad marcescent leaf bases; stems hirsute and stipitate-glandular; petioles 1.5–5.5 cm long, glandular; leaf blades 12–60 mm wide and about as long, reniform or orbicular, truncate or cordate at the base, crenate and more or less shallowly lobed; bracts similar to the upper leaves but smaller and less toothed, the upper ones usually entire; hypanthium 4–7 mm long, campanulate, glandular to pilose-glandular, often reddish or purplish; sepals 3–4.5 mm long, glandular; petals subequal to the sepals or shorter, pinkish or reddish. Crevices, often in limestone, at 2680 to 2990 m in the Bear River Range, Cache County; Idaho and Montana to Colorado, and southern Nevada (Spring Mts.); 6 (0). Our plants are referable to var. ***heucheriformis*** (Rydb.) Engler [*Therefon heucheriforme* Rydb.].

Fendlera Engelm. & Gray

Shrubs with opposite, nearly sessile; leaves without stipules, deciduous; flowers perfect, showy; hypanthium calyxlike; sepals 4; petals 4; stamens 8, the filaments flattened, lobed at the apex; ovary inferior at the base, 4–loculed; styles 4; stigmas minute; fruit a capsule, over 1/2 superior, septicidal; seeds few in each locule.

Fendlera rupicola Gray Fendlerbush. [*F. tomentosa* Thornb.]. Much branched shrubs 1–2 m tall; bark of twigs longitudinally ridged and grooved, reddish or straw colored, turning gray; leaves opposite or appearing fasciculate-opposite, 9–30 mm long, 2–7 mm wide, lance-linear, linear, elliptic, or less often ovate, entire, sometimes slightly revolute, sparingly strigose on both sides, the midrib prominent, grooved above, ridged beneath; flowers solitary or 2–3 together at the ends of short branches; hypanthium 2–3 mm long; sepals 3–5 mm long, to 8 mm in fruit, persistent, strigose beneath, tomentose-villous above; petals 13–20 mm long, constricted to a narrow claw, the blade to 11 mm wide, white; staminal filaments ca 6–8 mm long, to 2 mm wide at the base, 2–lobed apically, the lobes 2–3 mm long, the anther shorter or longer than the lobes; styles 4, appearing as 2 at anthesis, glabrous or with multicellular hairs; capsules 8–15 mm long. Blackbrush, mixed desert shrub, and pinyon-juniper communities at 1370 to 1710 m in Grand and San Juan counties; Colorado and Arizona to Texas; 11 (0).

Fendlerella Heller

Shrubs with opposite, sessile, or nearly sessile leaves, lacking stipules; flowers perfect, in small compound cymes; hypanthium calyxlike; sepals 5; petals 5; stamens 10, the filaments dilated below the narrow apex; ovary ca 1/2 inferior, 3–loculed; styles 3; fruit a septicidal capsule; seeds 1 in each locule.

Fendlerella utahensis (Wats.) Heller Utah Fendlerella. [*Whipplea utahensis* Wats., type from near Kanab, Kane County]. Sprawling or ascending, much branched shrubs to 1 m tall; bark of twigs strigose, whitish, exfoliating in milky or translucent strips or flakes; leaves 4–12 mm long, 1–6 mm wide, linear-oblanceolate, linear, elliptic, or less commonly ovate, entire, slightly revolute, strigose, the hairs sometimes pustulate, especially on the lower surface; flowers in small compound cymes; hypanthium inconspicuous at first, finally to 2 mm long in fruit, turbinate-campanulate; sepals 1–1.5 mm long; petals 2–4 mm long, white; staminal filaments dilated and petaloid just below the narrow apex, white; styles ca 1.5–2 mm long; capsules 3–4 mm long. Sagebrush, pinyon, juniper, and mountain brush communities, mostly on sandstone and sandy soil, at 1480 to 2745 m in Garfield, Kane, Millard, Uintah, Utah, and Washington counties; northwestern Colorado to Arizona and west to California; 17 (0).

Heuchera L.

Perennial scapose herbs from scaly, somewhat woody branched caudices or rootstocks; leaves basal, stipulate; flowers paniculate or racemose or nearly spicate, bracteate, perfect, regular; hypanthium calyxlike; sepals 5; petals 5, small, entire, usually clawed; stamens 5; ovary partly inferior, 1–loculed; styles 2; fruit a capsule, opening between the 2 more or less divergent stylar beaks; seeds many.

Rosendahl, C. O., F. K. Butters, and O. Lakela. 1936. A monograph on the genus *Heuchera*. Minnesota Stud. Pl. Sci. 2: 1–180.

1. Stamens shorter than the sepals; petals ca 1–2 mm long; sepals not reddish or pinkish; pedicels 1–2 mm long or nearly obsolete .... *H. parvifolia*
— Stamens exserted 1–4 mm beyond the sepals; petals 3–4 mm long; sepals often pinkish or reddish; pedicels (1) 2–7 mm long .... *H. rubescens*

Heuchera parvifolia Nutt. in T. & G. Littleaf Alumroot. [*H. utahensis* Rydb., type from Salt Lake County.]. Scapose glandular herbs, 12–71 cm tall; the caudex branches clothed with marcescent leaf bases; scapes stipitate-glandular; petioles 1–13 cm long, with stipules glandular-puberulent or glabrate; leaf blades (1) 1.5–6.5 cm wide, wider than long, orbicular or reniform, cordate, palmately lobed with 3–7 primary lobes, these crenate and usually again shallowly lobed, commonly with scattered stipitate glands below, glabrate above and sometimes throughout; flowers in open or congested panicles or sometimes racemose or spicate, bracteate, the bracts toothed or fimbriate; inflorescence 2–35 mm long; pedicels obsolete or 1–2 mm long; hypanthium 2.5–3.5 mm long or to 5 mm in fruit; sepals 0.5–1 mm long, sometimes yellowish; petals ca 2 mm long, white; stamens shorter than the sepals; capsules 4–7 mm long. Pinyon, juniper, sagebrush, mountain brush, ponderosa pine, aspen, grass-forb, Douglas fir, and white fir communities, often in rocky places, at 1675 to 3200 m in all Utah counties, except Kane, Morgan, and Sevier; Alberta to New Mexico and west to Idaho and Nevada; 96 (v). Our plants are referable to var. *utahensis* (Rydb.) Garrett.

Heuchera rubescens Torr. in Stansbury Red Alumroot. Scapose perennial herbs, 5–30 (53) cm tall, the stem bases clothed with marcescent leaf bases; herbage glabrous to glandular-puberulent or hirsute; petioles 1–6 (8) cm long, glabrate, the stipular bases often fimbriate; leaf blades 0.7–4.3 cm wide, slightly longer than wide to wider than long, orbicular to broadly ovate, cordate or truncate, palmately lobed, the primary lobes sometimes again shallowly lobed, dentate or crenate, hirsute-ciliate; flowers in racemose or spicate panicles, the inflorescence 3–18 (31) cm long; pedicels (1) 2–7 mm long; hypanthium 2–4 mm long, campanulate, pinkish, lavender, or whitish; sepals 1–2.5 mm long, or to 3.5 mm long in fruit, mostly pinkish or lavender; petals 3–4 mm long, white; stamens exserted ca 1–4 mm beyond the sepals; capsules ca 4–6 mm long. There are two more or less morphologically and geographically correlated varieties.

1. Petioles hirsute, the spreading hairs to 2 mm long .....
   .... *H. rubescens* var. *versicolor*
— Petioles merely puberulent ....
   .... *H. rubescens* var. *rubescens*

Var. *rubescens* [*H. versicolor* Greene f. *pumila* Rosend, Butters, & Lakela, type from Utah?]. Crevices of rock outcrops, and other rocky places, in pinyon, juniper, sagebrush, mountain brush, ponderosa pine, limber pine, bristlecone pine, aspen, fir, and spruce communities at (1525) 1825 to 3355 m in Beaver, Box Elder, Cache, Garfield, Grand, Millard, Piute, Salt Lake, Sanpete, Tooele (type from Stansbury Island), Utah, Wayne, and Weber counties; Oregon and Idaho to California, east to Arizona, and Colorado; 59 (viii).

Var. *versicolor* (Greene) M. G. Stewart [*H. versicolor* Greene]. Pinyon, juniper, mountain brush, and sprucefir communities, often on rock outcrops, at 1370 to 2590 m in Garfield, Iron, Kane, San Juan, and Washington counties; Arizona, western New Mexico, and southeastern Nevada; 16 (0). In addition to the hirsute petioles, var. *versicolor* is reported to have filaments not noticeably flattened toward the base and attached at or slightly below the point of attachment of the petals. These features appear to be weakly, if at all, correlated in Utah materials. Our plants might represent intermediates between var. *rubescens* and the more southern var. *versicolor*, or perhaps the distinction is so tenuous as to be of little taxonomic value.

Jamesia T. & G.

Shrubs; leaves opposite, mostly petioled, without stipules; flowers perfect, regular, cymose; hypanthium calyxlike; sepals 5; petals 5, white; stamens 10; filaments narrow; ovary ca 1/2 inferior, partly 3– to 5–loculed at first, finally 1-loculed; styles 3–5, distinct; fruit a capsule, ca 1/3 inferior, with 3–5 valves, these with slender beaks; seeds numerous.

Jamesia americana T. & G. Cliff Jamesia. [*Edwinia americana* (T. & G.) Heller; *E. macrocalyx* Small, type from American Fork Canyon]. Shrubs 3–15 dm tall; foliage and young twigs pubescent with multicellular hairs; bark of twigs reddish or whitish, exfoliating in long whitish or translucent strips; petioles 2–10 mm long, canescent or sericeus-tomentose, sometimes lacking; leaf blades 7–40 mm long, (3) 6–32 mm wide, ovate or elliptic, serrate or dentate, rarely nearly entire, green and sparingly strigose on the upper surface, sericeus-canescent below; flowers few in small cymes, hypanthium 1–3 mm long in fruit, sericeus-canescent; sepals 3–4 mm long in flower, 5–6 mm long in fruit; petals 5–11 mm long, white, clawed, somewhat pubescent subapically; filaments gradually dilated basally; capsules 4–5 mm long. Mountain brush and spruce-fir communities, mostly on cliffs and other rocky places at 1220 to 3200 m in Juab, Millard, Salt Lake, Utah, and Washington counties; Wyoming to New Mexico, Nevada, and California; 25 (0).

Lithophragma Nutt.

Small, usually stipitate-glandular herbs from fibrous roots and very slender rhizomes, bearing ricelike bulblets; leaves basal and alternate, variously palmately cleft or divided; flowers in racemes, perfect, regular; hypanthium calyxlike; sepals 5; petals 5, variously palmately lobed or cleft, with a narrow claw and expanded blade; stamens 10; ovary 1–loculed; styles 3; fruit a 3–valved capsule; seeds numerous.

Taylor, R. L. 1965. The genus *Lithophragma* (Saxifragaceae). Univ. California Publ. Bot. 37: 1–122.

1. Plants with few to several purple bulblets in the inflorescence and usually in axils of the upper leaves; lower pedicels 1.5–3 times longer than the hypanthium ....
   .... *L. glabra*
— Plants without bulblets in inflorescence or in leaf axils; lower pedicels ca 0.5–1.5 times longer than hypanthium 2

2(1). Petals mainly 3–lobed, 5–9 mm long, excluding the claw, white; ovary inferior or nearly so; plants of north-central mountains .... *L. parviflora*
— Petals 5– to 7–lobed, 3–6 mm long excluding the claw, pinkish or white; ovary ca 1/2 inferior or less; plants of western and southern Utah .... *L. tenella*

Lithophragma glabra Nutt. in T. & G. Fringecup. [*L. bulbifera* Rydb.; *L. glabra* var. *bulbifera* Jepson]. Plants 5–35 cm tall, glandular-pubescent, the gland-tips mostly dark purple; petioles of lower leaves 1.2–4.8 cm long; leaf blades 5–24 mm long, 8–48 mm wide, orbicular or reniform in outline, parted or divided and trifoliate or palmate, the main divisions again lobed to parted, reduced above and often with purple bulblets in the axils; inflorescence simple or branched, purple, 1–5 (13) cm long, with (1) 2–4 normal flowers and others reduced to bulblets; pedicels 2–10 mm long; hypanthium 2–5 mm long, campanulate; sepals ca 1 mm long; petals pinkish, white, or pale lavender, the claw 1–2 mm long, the blades 4–7 mm long, 3– to 5–lobed or parted; ovary ca 1/4 inferior; seeds muricate. Aspen, oak-maple, sagebrush, pinyon-juniper, mountain brush, riparian, ponderosa pine, spruce, fir, lodgepole pine, and rarely in greasewood communities at 1310 to 3050 m in Beaver, Box Elder, Cache, Carbon, Daggett, Davis, Duchesne, Grand, Juab, Iron, Millard, Morgan, Rich, Salt Lake, San Juan, Sanpete, Sevier, Summit, Tooele, Uintah, Utah, Wasatch, Washington, and Weber counties; British Columbia to California, east to Alberta, the Dakotas, and Colorado; 82 (vi). Our plants are referable to **var. ramulosa (Suksd.) Goodrich** [*L. tenella* var. *ramulosa* Suksd.]

***Lithophragma parviflora* (Hook.) Nutt. in T. & G.** Smallflower Woodlandstar. [*Tellima parviflora* Hook.]. Perennial herbs from fibrous roots and slender rootstocks bearing small bulblets, 9–47 cm tall, pubescent with mostly glandular hairs, rather densely so in the inflorescence; petioles of basal leaves 2–11 cm long; leaf blades 9–35 mm long, 11–50 mm wide, orbicular or reniform in outline, 3-parted or 3-foliolate, with the divisions again lobed, cleft, or parted; cauline leaves similar to the basal ones but sometimes reduced upwards; racemes congested, 1–2 (3) cm long in flower, to 7.5 cm long in fruit, 4– to 7-flowered, the pedicels 1–5 mm long in flower, to 8 mm long in fruit; hypanthium (3.5) 4–5 mm long in flower, to 6 mm long in fruit, wedge-shaped or funnelform basally and gradually tapered to the pedicel; sepals 0.7–1.4 mm long; petals white, clawed, the claws 2–3.5 mm long; ovary inferior or nearly so; seeds smooth. Oak-maple, sagebrush, aspen, riparian, meadow, lodgepole pine, spruce, and fir, communities at 1450 to 2590 (3050) m in Box Elder, Cache, Carbon, Davis, Emery, Juab, Millard, Morgan, Salt Lake, Sanpete, Summit, Tooele, Uintah, Utah, Wasatch, and Weber counties; British Columbia to northern California, east to Alberta, South Dakota, and Colorado; 88 (i).

***Lithophragma tenella* Nutt. in T. & G.** Slender Woodlandstar. [*L. australis* Rydb.]. Plants 10–43 cm tall, pubescent, the hairs mostly glandular, the gland-tips whitish or pale purplish; petioles 0.5–4.5 cm long; leaf blades 4–25 mm long, (5) 9–30 mm wide, orbicular or ovate, parted or 3-foliolate, the main divisions again lobed or parted; racemes ca 1–3 cm long in flower, to 15 cm long in fruit, with 4–13 flowers; pedicels 2–7 mm long; hypanthium campanulate, 2–3 mm long; sepals ca 1 mm long; petals pinkish or whitish, the claws 1–2 mm long, the blades 3–6 mm long, with 5–7 lobes; ovary ca 1/2 inferior; seeds smooth. Pinyon-juniper, sagebrush, mountain brush, ponderosa pine, aspen, riparian, meadow, and spruce communities at (1310) 2075 to 3050 m in Beaver, Box Elder, Cache, Garfield, Grand, Iron, Juab, Kane, Piute, Salt Lake, San Juan, Sevier, Tooele, and Washington counties; western Washington south to Arizona, east to Montana and New Mexico; 50 (ii).

Mitella L.

Scapose glandular-puberulent herbs from rhizomes; leaves basal, palmately lobed, cordate; flowers in racemes, perfect, regular; hypanthium calyxlike; sepals 5; petals 5, pinnately or palmately lobed, small and soon withered; stamens 5; ovary ca 1/2 inferior, 1-loculed; style 1; fruit a capsule, dehiscent by ventral suture and appearing almost circumscissile; seeds numerous.

1. Petals with (2) 3 lobes; racemes secund; pedicels lacking or to 3 mm long; sepals whitish or purplish; leaf margin often ciliate .... *M. stauropetala*
— Petals pinnately divided; racemes not especially secund; pedicels mainly 2–8 mm long; sepals greenish; leaf margin seldom if ever ciliate .... *M. pentandra*

***Mitella pentandra* Hook.** Fivestar Miterwort. Scapose perennial herbs, 10–35 cm tall; petioles 1.8–10 cm long, glabrous or sparingly pilose; leaf blades 1.5–5.7 cm wide, ovate to orbicular, cordate basally, with 5–9 shallow lobes, the lobes dentate, the teeth mucronate, seldom if at all ciliate, glabrous, or more often hirsute to pilose; scapes glandular above; racemes 3–16 cm long, 4– to 20-flowered; pedicels 2–8 mm long, shorter above; hypanthium saucer-shaped to campanulate, 1–2.5 mm long; sepals less than 1 mm long; petals 2–3 mm long, greenish, pinnately dissected into 4–10 filiform segments; capsules ca 2 mm long. Meadows and other moist sites, often where shaded, in aspen, conifer, and willow communities at 1740 to 3650 m in Box Elder, Daggett, Davis, Duchesne, Emery, Salt Lake, Sanpete, Summit, Uintah, Utah, and Wasatch counties; Alaska south to northern California, east to Alberta and Colorado; 33 (iii).

***Mitella stauropetala* Piper** Smallflower Miterwort. Perennial scapose herbs 15–50 cm tall or more; petioles 1.2–13 cm long, glandular and sometimes hirsute; leaf blades 1–7.8 cm wide and about as long, orbicular or broadly ovate, cordate, 5– to 9-lobed, the lobes crenate, the margin ciliate, glabrous or minutely glandular and sparingly hirsute; scapes glandular to the base; racemes 4.5–18 (24) cm long, mostly strongly secund, 7– to 24-flowered, mainly with 1 flower per node; pedicels obsolete or to 2 or 3 mm long; hypanthium ca 1–2 mm long, cup-shaped; sepals ca 1.5 mm long, oblong, whitish or purplish; petals 2–4 mm long, 3 (2) -lobed or rarely entire, the lobes filiform; capsules ca 2 mm long. Mountain brush, aspen, ponderosa pine, fir, and spruce communities at 1615 to 3050 m in Box Elder, Cache, Carbon, Duchesne, Emery, Juab, Millard, Salt Lake, San Juan, Sanpete, Summit, Tooele, Utah, Wasatch, and Weber counties; eastern Washington and Oregon to Montana, south to Colorado; 65 (ii). Our plants are referable to **var. stenopetala (Piper) Rosend.** [*M. stenopetala* Piper, type from the Wasatch Mts.].

Parnassia L.

Scapose perennial glabrous herbs from short rootstocks; leaves basal, entire; scapes with a solitary, entire bract or bractlike leaf; flowers solitary and terminal, perfect, regular; hypanthium calyxlike, sometimes nearly obsolete; sepals 5; petals 5, white; stamens 5, alternating with clusters of gland-tipped staminodia; ovary superior or slightly inferior, 1-loculed; styles lacking; stigmas 3 or 4, sessile; fruit a capsule, loculicidal at the apex; seeds numerous.

1. Petals fringed below the middle; petioles (1.5) 3–16 cm long; leaf blades cordate to truncate basally, 12–45 mm wide, broader than long; bract mostly borne above the middle of the scape .... *P. fimbriata*
— Petals entire; petioles 0.7–4 cm long; leaf blades cuneate or obtuse basally (rarely truncate or cordate), 5–20 mm wide, mostly longer than broad; bract borne below the middle of the scape .... *P. palustris*

***Parnassia fimbriata* Koenig** Fringed Grass-of-Parnassus. Perennial, scapose glabrous herbs, 14–43 cm tall; petioles (1.5) 3–16 cm long; leaf blades 12–45 mm wide, about as long, reniform, orbicular, or broadly ovate, cordate or truncate at the base; bracts of scapes ovate to orbicular, 7–17 mm long, mostly clasping, borne at or above the middle of the scape; flowers solitary and terminal; hypanthium nearly obsolete; sepals 4–7 mm long, 2–4 mm wide, sometimes slightly fimbriate; petals 8–15 mm long, including the narrow clawlike base, strongly fimbriate below the middle; staminodia thickened and scalelike, flared above the middle and usually with a central subterminal larger lobe and 7-9 marginal ones; capsules to 1 cm long. Moist habitats in aspen, spruce, fir, and mountain brush communities at 2010 to 3355 m in Cache, Duchesne, Salt Lake, San Juan, Summit, and Utah counties; Alaska south to California, east to Alberta and New Mexico; 45 (v). Our plants are referable to var. *fimbriata*.

Parnassia palustris L. Perennial scapose glabrous herbs, 8-44 cm tall; petioles 0.7-4 cm long; leaf blades 7-27 mm long, 5-20 mm wide, ovate to nearly orbicular, mostly cuneate or obtuse at the base, rarely truncate; bracts of scapes linear to ovate, (3) 5-25 mm long, sessile and sometimes clasping, borne mostly below the middle of the scape; flowers solitary and terminal; hypanthium ca 2 mm long; sepals 3-10 mm long; petals 6-14 mm long, including the narrow claw, entire, white; staminodia with a thickened scalelike base, flared upwards and divided into (5) 7-11 (or many) slender filamentous segments, these terminating in capitate knobs; capsule ovoid 8-10 (12) mm long. Wet meadows and other moist sites, sometimes in woods, at (1375) 1830 to 3415 m in Beaver, Box Elder, Cache, Duchesne, Emery, Garfield, Grand, Iron, Juab, Kane, Piute, Sanpete, Summit, Uintah, Utah, Washington, and Wayne counties; arctic America east to Quebec and south to Colorado, Nevada (Spring Mts.), and California; Eurasia; 64 (0). Our plants are assignable to var. *montanensis* (Fern. & Rydb.) C. L. Hitchc. [*P. montanensis* Fern. & Rydb.]. Many of our specimens have been identified previously as *P. parviflora* DC., which is known from well to the north of our area, and which may be only a small-flowered phase of *P. palustris* not worthy of taxonomic recognition.

Peltiphyllum Engler

Scapose perennial herbs from thick rhizomes; leaves developing after the scapes; petioles long; blades peltate, very large; flowers in bractless, paniculate-corymbose cymes, showy, regular; hypanthium very short, adnate to the ovary; sepals 5; petals 5; stamens 10; carpels 2, free above the hypanthium, tapering to the discoid-capitate stigma; fruit a follicle, fully dehiscent; seeds cellular-rugulose.

Peltiphyllum peltatum (Torr.) Engler in Engler & Prantl Shieldleaf. [*Saxifraga peltata* Torr. ex Benth.]. Robust perennial herbs; rhizomes fleshy but tough, to 5 cm thick; petioles to 1 m tall or more, hirsute; leaf blades 5-40 cm broad, peltate, depressed at the center above the point of attachment to the petiole, nearly orbicular, 10- to 15-lobed, the divisions again lobed and serrate-dentate; scapes 3-6 dm tall or more, hirsute-glandular, naked, or with a small bract; sepals 2.5-3.5 mm long, reflexed; petals 4.5-7 mm long, white to bright pink; filaments flattened and broad at the base; follicles 6-10 mm long, purplish, fused basally. Known in Utah from a small colony growing along a cold mountain stream on the east side of Mt. Timpanogos, Utah County; coastal southwestern Oregon and northern California; 4 (0). The widely disjunct colony may be from an introduction.

Philadelphus L.

Shrubs; leaves opposite, subsessile or on short petioles; flowers perfect, regular, in few-flowered cymes at the ends of leafy branches; hypanthium calyxlike; sepals 4 (5); petals 4 (5), white or nearly white; stamens many (ca 20-60); ovary at least 2/3 inferior, with 3-5 locules; styles 3-5, distinct or united; fruit a loculicidal capsule, leathery or woody; seeds numerous. Note: In addition to the taxa included in the key, *P. verrucosus* Schrader and *P.* x *virginalis* Rehder are occasionally cultivated in Utah. These differ from those treated below in having sepals pubescent on the back, not just on the margins. *P. verrucosus* has single flowers and *P.* x *virginalis* has double flowers. Possibly *P. lewisii* Pursh of the Northwest is also cultivated in the state.

1. Leaves entire, 4-26 mm long; plants native . . . .
 . . . . *P. microphyllus*
— Leaves toothed, sometimes larger than above; plants introduced, to 3 m tall, our common cultivated mockorange . . . . *P. coronarius* L.

Philadelphus microphyllus Gray Littleleaf Mockorange. [*P. nitidus* A. Nels.]. Shrubs 8-20 dm tall, with opposite leaves; branchlets appressed pubescent; petioles ca 1 mm long; leaf blades 4-26 mm long, 2-13 mm wide, mostly elliptic or ovate to lanceolate or linear, entire, slightly revolute, sparingly to moderately strigose or glabrate; pedicels to 3 mm long, strigose-sericeous; hypanthium ca 2-3 mm long, pubescent like the pedicels; sepals 2-5 mm long; petals 8-15 mm long, white; stamens subequal to the sepals; capsules 6-8 mm long. Pinyon-juniper, mountain brush, ponderosa pine, aspen, and fir communities, mostly on sandstone at 1220 to 2650 m in Beaver, Carbon, Daggett, Emery, Garfield, Grand, Iron, Juab, Millard, Piute, San Juan, Sevier, Tooele, Uintah, Washington, and Wayne counties; Wyoming and Utah to Texas; 80 (iii). Plants from the southern part of the state have leaves that are slightly larger on the average than those from more northern areas, but the difference does not seem to warrant taxonomic recognition.

Ribes L.

Shrubs with or without bristles and spines; leaves alternate, palmately lobed, crenate or dentate; stipules none or adnate to the petiole; flowers perfect, regular, in racemes, or rarely solitary; pedicels subtended by bracts and usually with 2 bractlets about midlength; hypanthium mostly corollalike; sepals 5, mostly petaloid; petals 5 (4), often smaller than the sepals; stamens 5 (4-6); ovary completely inferior, 1-loculed; styles 2, united or distinct; fruit a berry, crowned by the withered flowers; seeds several to many. Note: The taxa with jointed and disarticulating pedicels, several flowers per raceme, and stipitate-glandular or glabrous berries are referred to as currants. The currants in our area are without spines or bristles except for *R. lacustre* and *R. montigenum*. The taxa with nonjointed and persistent pedicels and with only (1) 2 or 3 flowers per raceme and without stipitate-glandular hairs on the berries are referred to as gooseberries. The gooseberries of Utah all have nodal spines.

Berger, A. 1924. A taxonomic review of currents and gooseberries. New York State Agr. Exp. Sta. Tech. Bull. 109: 1-118.

1. Branchlets armed with nodal spines, sometimes also with internodal bristles 2
— Branchlets unarmed 8
2(1). Racemes with 3-15 flowers; ovaries and berries with setose-stipitate glands; hypanthium saucer-shaped . . . 3

—	Racemes with 1–3 flowers; ovaries and berries glabrous or pubescent, but not with setose-stipitate glands; hypanthium tubular or cup-shaped 4	11(9).	Ovaries and berries stipitate-glandular or with crystalline, yellowish, sessile glands; lower leaf surface with sessile glands; berries blackish; flowers (5) 8–30 per raceme; styles glabrous 12
3(2).	Racemes with 3–8 flowers; berries red; leaf blades cleft 3/4 of the way to nearly all the way to the base; plants common and widespread .... *R. montigenum*	—	Ovaries and berries glabrous or, if with sessile nonglandular spots, the flower 1–3 per raceme and styles pubescent; berries red or reddish purple 15
—	Racemes with (5) 7–15 flowers; berries black or purple black; leaf blades lobed 2/3 of the way to the base or less; plants of the Uinta and Wasatch mts. .... *R. lacustre*	12(11).	Lower surface of leaves and floral parts with scattered, crystalline yellow glands; racemes many-flowered (to 30); plants strongly aromatic 13
4(2).	Leaves 7–20 mm wide; styles connate to near the apex; free hypanthium and sepals pubescent externally; plants of desert ranges of western and southern Utah 5	—	Leaves and/or floral parts with stipitate glands, or variously glandular, but not with yellow crystalline glands; racemes with (5) 8–16 flowers; plants not strongly aromatic 14
—	Leaves over 20 mm wide (at least some), and/or styles lobed or cleft 1/4–1/2 the length; free hypanthium and sepals mostly glabrous externally; plants of various distribution 6	13(12).	Sepals longer than the free hypanthium; racemes erect or ascending; plants indigenous . *R. hudsonianum*
5(4).	Free hypanthium 4–5.5 mm long; sepals 4–6 mm long; ovaries and berries mostly glabrous .... *R. leptanthum*	—	Sepals shorter than the free hypanthium; racemes spreading to pendulous; plants cultivated. Black or Bedbug Currant .... *R. nigrum* L.
—	Free hypanthium 1–2 mm long; sepals 2–4 mm long; ovaries and berries glabrous to hairy ..... *R. velutinum*	14(12).	Bracts of racemes 3–4 mm long, ca 1/2 as long as the pedicels, oblong; flowers whitish or greenish; stipitate hairs of racemes whitish or pale greenish, rarely purplish, often sparse except on ovary and berry; racemes crowded, the upper internodes seldom over 3 mm long; plants of the La Sal Mts. and portions of central Utah .... *R. wolfii*
6(4).	Hypanthium pubescent within; berries greenish or yellowish (ripening purplish); plants cultivated, persisting, and rarely escaping. English Gooseberry .. .. *R. grossularia* L.		
—	Hypanthium glabrous within, but the styles pubescent; berries reddish or blackish; plants indigenous 7	—	Bracts of racemes 1–2 mm long, less than 1/2 as long as the pedicels, linear to acute; flowers pinkish or purplish; stipitate hairs of racemes mostly with purple tips, usually well developed on rachis and pedicels as well as on ovaries and berries; racemes loose, 1 or more of the upper internodes regularly over 3 mm long, plants of the Deep Creek Mts. ... . *R. laxiflorum*
7(6).	Free hypanthium 2–3.5 mm long; berries reddish purple; nodal spines 1 (rarely 3) or lacking; branchlets usually glabrous; plants common and widespread .. .. *R. inerme*		
—	Free hypanthium 4–5 mm long; berries purple black; nodal spines usually 3; branchlets finely puberulent and sometimes with internodal bristles; plants rather rare, except along the south slope of the Uinta Mts. .... *R. setosum*	15(11).	Flowers 1–3 per raceme, white; sepals longer than wide; free hypanthium ca 2–3.5 mm long; styles pubescent; plants indigenous .... *R. inerme*
8(1).	Flowers bright yellow, often reddish in part in age, glabrous; ovaries and berries glabrous; leaves glabrous, not cordate, with 3 (rarely 5) primary lobes, the primary lobes seldom with more than 3 teeth or lobes .... *R. aureum*	—	Flowers 8–20 per raceme, greenish; sepals wider than long; hypanthium to 2 mm long; styles glabrous; plants cultivated and escaping. Red Currant .... *R. sativum* Syme
—	Flowers not yellow, often pubescent or glandular; ovaries and berries with sessile or stipitate glands (except in *R. sativum* and *R. inerme*); leaves often pubescent, mostly cordate, with (3) 5–7 primary lobes, these with usually more than 3 teeth or lobes .. 9		
9(8).	Leaf blades, flowers, and fruit stipitate-glandular or, if sparingly so, flowers pinkish and berries red; free hypanthium 4–11 mm long, campanulate to cylindric; anthers glandular apically 10		
—	Leaf blades not stipitate-glandular, glabrous or with hairs mostly on veins, or with sessile glands; flowers rarely pinkish; berries red only in *R. inerme*; free hypanthium mostly less than 4 mm long, variously shaped; anthers eglandular 11		
10(9).	Flowers (1) 2 or 3 per raceme, pinkish, the hypanthium less than 3 mm wide; berries red; leaf blades 7–30 (44) mm wide, not pilose-hirsute ..... *R. cereum*		
—	Flowers 4–12 per raceme, greenish white to cream, the hypanthium 3–6 mm wide; berries blackish; leaf blades 30–100 mm wide, at least some, pilose-hirsute and stipitate-glandular .... *R. viscossissimum*		

Ribes aureum Pursh Golden Currant. Shrubs, 1–3 m tall, unarmed; branchlets glabrous; petioles 0.5–2.5 (3) cm long; leaf blades 0.6–4.7 cm long, 1–6.7 cm wide, orbicular, reniform, obovate, cuneate to truncate basally, strongly 3–lobed, the lobes entire or crenate to lobed, glabrous; racemes with (3) 6–9 flowers; bracts 3–12 mm long, entire; pedicels to 3 mm long; free hypanthium cylindric, yellow, or often reddish in age, corollalike; sepals 4–6 mm long yellow, spreading; petals ca 2 mm long, yellow, cream, or reddish; stamens subequal to the petals, the anthers longer than the filaments; styles united to near the apex; berries 8–12 mm long, black, red, orange, or translucent-golden, glabrous. Riparian and palustrine habitats in greasewood-shadscale, sagebrush, pinyon-juniper, mountain brush, ponderosa pine, and Douglas-fir communities at 1340 to 2590 m in all Utah counties, except Grand and San Juan; Washington to Saskatchewan, south to California and New Mexico; 150 (iii). The golden currant is similar to **R. odoratum** Wendl. of the eastern U. S. and Canada, which may also be cultivated in the state. The berries are the most palatable of any species of *Ribes* in Utah, and for that reason the plants are widely grown in cultivation in Utah.

Ribes cereum Dougl. Wax or Squaw Currant. Shrubs, (0.2) 0.5–1.5 (2) m tall, unarmed; branchlets pilose-villous and stipitate-glandular; petioles 0.4–2.2 (2.9) cm long; leaf blades 0.5–2.5 (3.4) cm long, 0.7–3 (4.4) cm wide, orbicular, reniform, rarely ovate, cordate or truncate basally, with 3–7 shallow lobes, the lobes crenate or dentate, puberulent and stipitate glandular, or glabrous except on margins and along veins beneath; racemes with 2 or 3 flowers, the axis very short; bracts 2–5 mm long ciliate, fringed or lacerate, glandular; free hypanthium 4–11 mm long, pinkish, pilose, sometimes also stipitate-glandular; sepals ca 2 mm long, spreading to deflexed, whitish or pinkish; petals ca 1 mm long, whitish; staminal filaments subequal to the anthers; styles united to near the apex; ovaries stipitate-glandular; berries 6–8 mm long, reddish, sparingly stipitate-glandular, rarely glabrate. Mountain brush, sagebrush, pinyon-juniper, riparian, ponderosa pine, aspen, limber pine, spruce-fir, krummholz, alpine, and less commonly desert shrub communities at 1520 to 3260 m in all but Davis County; British Columbia to Montana, south to California and New Mexico; 230 (vi). Most of our plants are referable to var. *inebrians* (Lindl.) C.L. Hitchc. [*R. inebrians* Lindl.]. Occasional specimens have fan-shaped bracts that are truncate to broadly rounded and prominently toothed or with as many as 6 lobes. These specimens might be referable to var. *cereum*, but the distinction is tenuous at best. Intergradation of the morphological features is common and no geographical correlation is apparent in Utah.

Ribes hudsonianum Richards. Northern Black Currant. Shrubs, 0.5–1.5 (2) m tall, rather strongly scented, unarmed; herbage bearing crystalline, yellowish glands, usually puberulent; petioles (1) 2.5–8.5 cm long; leaf blades (2.1) 2.7–8.2 cm long, (2) 5–12.2 cm wide, orbicular or ovate in outline, strongly 3–lobed and doubly dentate, the lateral lobes again lobed, cordate basally; racemes ca 20– to 30–flowered, to 12 cm long in fruit; bracts ca 0.5–1.5 mm long, awl-shaped, soon deciduous; free hypanthium to ca 1 mm long, whitish; sepals 2.5–4 mm long, white; petals ca 1/2 as long as sepals; styles united for more than 1/2 their length; berries 7–12 mm long, blackish, glabrous except for sessile glands. Streamsides, usually in aspen, lodgepole pine, or spruce-fir communities at 1830 to 2590 m in Box Elder, Duchesne, Salt Lake, Summit, Utah, and Wasatch counties; Alaska to Hudson Bay, south to California, Wyoming, and Minnesota; 11 (i). Our plants are referable to var. **petiolare** (Dougl.) Jancz. [*R. petiolare* Dougl.].

Ribes inerme Rydb. Whitestem Gooseberry; Wine Gooseberry. Shrubs 7.5–20 dm tall; branchlets often whitish, glabrous, armed at the nodes with 1 (3) spines, or the spines lacking; internodal bristles mostly lacking or few and sparse; petioles (0.3) 0.5–4.5 cm long, sometimes with 1 to few pilose gland-tipped hairs; blades (0.8) 1.5–9 cm long, orbicular or nearly so, cordate to truncate with 3–5 main lobes, these again lobed and crenate-dentate toothed, the major sinuses cut 1/3–2/3 to the base, glabrous, glabrate with hairs mostly along the veins or occasionally with moderately dense strigose, not glandular, paler beneath than above; racemes 1– to 4–flowered, the axis to ca 12 mm long; bracts 1–2 mm long, greenish, glabrous or glandular ciliolate and puberulent; pedicels 2–5 mm long; free hypanthium 2–3.5 mm long, cylindric to narrowly campanulate, greenish or greenish cream, sometimes purplish tinged, densely pilose to villous-wooly inside; sepals ca 3 mm long, colored as the hypanthium; petals about 1–1.5 mm long, obovate to narrowly fan-shaped, white; stamens (1.5) 2–2.5 times longer than the petals, 2–4 (5) mm long; styles subequal to or slightly longer than the stamens, cleft 1/2–2/3 to the base, rather densely pilose on the lower 1/2 or more; berries 7–10 mm long, reddish or reddish purple, succulent, more or less edible. Pinyon-juniper, mountain brush, aspen, willow, Douglas fir, spruce-fir, tall forb, and meadow communities at 1830 to 3100 m, in Cache, Carbon, Daggett, Duchesne, Garfield, Grand, Juab, Piute, San Juan, Salt Lake, Sanpete, Sevier, Summit, Tooele, Uintah, Utah, Wasatch, and Weber counties. British Columbia to California, Montana to New Mexico; 56 (vii).

Ribes lacustre (Pers.) Poir. in Lam. Swamp Black Gooseberry. [*R. oxyacanthoides* var. *lacustre* Pers.]. Shrubs, 7.5–15 dm tall; branchlets armed with internodal prickles and nodal spines, puberulent, eglandular; petioles 0.3–5 cm long, glabrous or with scattered stipitate-glandular hairs; leaf blades (0.6) 1.5–5.6 cm long, (1) 2–8 cm wide, orbicular in outline, cordate at the base, usually 5–lobed, the lobes again lobed and doubly crenate-dentate, glabrous or sparingly hairy along the veins; racemes rather loosely 5– to 15–flowered, the axis to 4.5 cm long in fruit, stipitate-glandular with reddish or purplish glands, puberulent; bracts 2–3 mm long, ciliate-glandular; pedicels 3–8 mm long; free hypanthium less than 1 mm long, saucer-shaped, yellowish green, pinkish, or reddish; sepals 2.5–3 mm long, yellow green, pinkish, or reddish; petals shorter than the sepals, pinkish; stamens subequal to the petals; styles parted to the base; berries 6–8 mm long, dark purple, coarsely stipitate-glandular. Moist sites, often in conifer and aspen woods, at 2100 to 3350 m in the Uinta and Wasatch mts., in Duchesne, Salt Lake, Summit, and Wasatch counties; Alaska to Newfoundland, south to California, Colorado, South Dakota, and Michigan; 20 (iv).

Ribes laxiflorum Pursh Western or Trailing Black Currant. Shrubs to ca 0.7 m tall, the stems sprawling or ascending, unarmed; branchlets and some older branches puberulent; petioles (0.5) 2–4.5 cm long, puberulent, sometimes short stipitate-glandular near the blade; leaf blades (1) 2–5 cm long, 1.5–6.5 cm wide, orbicular or nearly so, cordate, with 3–5 primary lobes, these again lobed and crenate-dentate toothed, the major sinuses cut ca 1/3–1/2 to the base, glabrate or with some puberulent and stipate glandular hairs, especially toward the base on veins beneath, slightly paler beneath with translucent crystalline sessile glands; racemes 5–10 flowered, stipitate glandular and puberulent, the axis 2–4 cm long; bracts 1–2 mm long, linear or narrowly triangular, greenish; pedicels 4–10 mm long, jointed just below the ovary, some usually persisting at least until fruit is nearly mature; free hypanthium less than 1 mm long; sepals 2–3 mm long, pinkish or purplish; petals ca 1 mm long, 1–1.3 mm wide, broadly fan-shaped with concave margins; stamens subequal to the petals; styles cleft 1/3–2/3 their length; berries to ca 1 cm long, blackish, stipitate-glandular, the glands and stalks mostly purplish. Known in Utah from a single small population in the Deep Creek Mts., Juab County, where abundant in wet shady places among downed timber and boulders, in a spruce-aspen community; Alaska to Washington and along the coast to southwest California, east to Alberta and northen Idaho, widely disjunct in Utah; 2 (i). The closely related *R. coloradense* Cov. of Colorado and New Mexico has been reported for

Utah, but I have seen no specimens. It is separable, reportedly, from *R. laxiflorum* by glandular hairs on the back of sepals (a feature that occasionally shows up in plants as far away as Alaska), by petals nearly twice as broad as long, and by berries without bloom. The separation appears rather tenuous at the species level. The combination *R. laxiflorum* var. *coloradense* (Cov.) Jancz. is available. The amazingly disjunct population in Utah seems more closely allied to the coastal and northern plants than to those of Colorado.

Ribes leptanthum Gray Trumpet Gooseberry. Shrubs, 0.5–2 m tall; branchlets armed at the nodes with 1–3 spines, usually lacking internodal bristles, puberulent; petioles 0.2–1.2 cm long; leaf blades 0.5–1.6 cm long, 0.7–2 cm wide, orbicular, cordate basally, mostly 5-lobed, the main lobes again shallowly lobed or toothed, glabrous or less commonly puberulent and rarely glandular; racemes 1- to 3-flowered, the axis very short; bracts glabrous except glandular-ciliate or -toothed; pedicels ca 1 mm long; free hypanthium 4–5.5 mm long, whitish, pilose or short-villous; sepals 4–6 mm long, whitish; petals 2.5–3 mm long, whitish; stamens subequal to the petals; anthers shorter than the filaments; styles glabrous, apically notched; berries ca 6–10 mm long, blackish, glabrous. Pinyon-juniper, mountain brush, ponderosa pine, aspen, spruce-fir, and meadow communities at 1830 to 2590 m in Beaver, Emery, Garfield, Grand, Kane, Piute, and San Juan counties; Colorado, New Mexico, and Arizona; 22 (i).

Ribes montigenum McClatchie Gooseberry Currant. [*R. lacustre* var. *lentum* Jones, type from Henry Mts.]. Shrubs, 2–7 dm tall; branchlets armed with 1–5 nodal spines, the internodal bristles lacking or present, puberulent or glabrous; petioles 0.4–4 cm long; leaf blades 0.4–3.7 cm long, 0.6–5 cm wide, orbicular in outline, cordate at the base, usually 5-lobed, the primary lobes again lobed or toothed, the vesture of glandular or nonglandular hairs and sometimes with sessile crystalline glands; racemes 3- to 8-flowered, the axis to 3.5 cm long in fruit; bracts 2–3 mm long, glandular-pubescent; pedicels 1–3 mm long; free hypanthium ca 1 mm long, saucer-shaped, lined with a yellowish or pinkish disk, glandular-hairy; sepals ca 3 mm long, pinkish lavender to whitish; petals ca 1 mm long, pinkish or pink purple; stamens subequal to the petals; filaments longer than the anthers; style divided to near the base, glabrous; berries mostly 5–10 mm long, red, stipitate-glandular. Spruce-fir, Douglas fir, lodgepole pine, bristlecone pine, aspen, krummholz, sagebrush-snowberry, and sedge-grass communities, and often in talus and scree slopes at 2135 to 3660 m in all Utah counties except Davis, Millard, Morgan, and Rich; British Columbia to southern California, east to Montana and New Mexico; 151 (v).

Ribes setosum Lindl. Missouri Gooseberry. Shrubs 7.5–20 dm tall; branchlets sometimes armed with internodal bristles, the nodes with 1–3 spines; petioles 1–4.8 cm long, often puberulent to pubescent, sometimes with a few long glandular seta; blades (1.5) 2–5.7 cm long, orbicular or nearly so, truncate to cordate, with ca 5 main lobes that are again lobed and crenate-dentate, the major sinuses cut a little less than 1/2 to 2/3 to the base of the blade, scattered to moderately pubescent with short curved hairs or rarely glabrate, the hairs rarely confined to the nerves, often stipitate glandular and/or with sessile translucent glandular dots especially beneath, slightly paler beneath; racemes 1–4 flowered, the axis to ca 5 mm long; bracts 1–3 mm long, greenish, minutely glandular ciliate; pedicels 2–5 mm long; free hypanthium 4–5 mm long, cylindric, greenish white or pinkish tinged, pilose to villous-woolly within; sepals 3–4 mm long, the color of the hypanthium; petals ca 2 mm long, obovate, white; stamens subequal to the petals, shorter than the erect to ascending sepals; styles subequal to the sepals, cleft 1/2–2/3 to the base, densely pilose below; berries 10–12 mm long, blackish or deep purple black, succulent, more or less edible. With aspen, alder, birch, and willows in riparian or palustrine habitats at 2130 to 2750 m in Box Elder, Duchesne, Juab, and Uintah counties; Alberta east to the Dakotas and Michigan; 15 (iii). A few Utah specimens seem intermediate to *R. inerme*.

Ribes velutinum Greene Desert Gooseberry. Shrubs, mostly 1–2 m tall, with 1 (2 or 3) nodal spine (s), lacking internodal bristles; branchlets usually pubescent; petioles 0.3–1.5 cm long; leaf blades 0.5–1.4 cm long, 0.7–2 mm wide, orbicular, 3– to 5-lobed, the main lobes usually again lobed, pubescent or glabrous; racemes 1– to 3-flowered, the axis very short; bracts 2–3 mm long, puberulent; pedicels 1–2 mm long; free hypanthium ca 1–2 mm long, pubescent externally, whitish or pinkish; sepals 2–4 mm long, whitish; stamens subequal to the petals; filaments subequal to the anthers; style simple, subequal to the stamens; styles united to the apex; berries 5–8 mm long, reddish (?), glabrous or hirsute, not very fleshy. Pinyon-juniper, mountain brush, and sagebrush-desert shrub communities at 1340 to 2321 m in Beaver, Kane, Millard, Tooele, and Washington counties; Washington to California, east to Idaho and Arizona; 23 (ii). Our plants are referable to var. ***velutinum***.

Ribes viscosissimum Pursh Sticky Currant. Aromatic shrubs, mostly 1–2 m tall, unarmed; branchlets pilose-hirsute and stipitate-glandular; petioles 0.6–7 cm long; leaf blades 0.9–6.6 cm long, 1.3–10 cm wide, orbicular, rarely ovate, cordate basally, 3– to 7-lobed, the main lobes crenate or dentate and sometimes again lobed, glandular-hairy and often pilose or hirsute; racemes 4– to 12-flowered, the axis 5–30 mm long, glandular; bracts 3–10 mm long, entire to toothed, glandular; pedicels 3–17 mm long; free hypanthium 5–9 mm long, whitish or pale green, stipitate-glandular and pilose-hirsute; sepals 3–5.5 mm long, white or yellow green, or occasionally pinkish; petals 2–3 mm long, whitish; stamens subequal to the petals; filaments longer than the anthers; styles simple, glabrous or nearly so; berries 10–13 mm long, black, rather dry, stipitate-glandular. Commonly growing in shade of aspen, fir, Douglas fir, lodgepole pine, and spruce woods, and less commonly in mountain brush, meadows, and openings at 1965 to 2925 m in Box Elder, Cache, Carbon, Daggett, Duchesne, Kane, Millard, Morgan, Salt Lake, Sanpete, Summit, Tooele, Uintah, Utah, Wasatch, and Washington counties; British Columbia, south to California, east to Montana and Arizona. This plant is unique among our species in the dense stipitate-glandular and nonglandular hairs of the herbage and in the rather broad and long hypanthia. Our plants are referable to var. ***viscosissimum***; 42 (ii).

Ribes wolfii Rothr. Rothrock Currant. [*R. mogollonicum* Greene]. Shrubs, 0.5–3 m tall, unarmed; branchlets glabrous or puberulent; petioles 0.7–4.5 cm long, glabrous or puberulent; leaf blades 1.2–5.7 cm long, 1.2–8 cm wide, orbicular, cordate basally, 3 (5) -lobed, the main lobes again lobed and variously 1– or 2–crenate or -dentate, glabrous except for sessile, clear crystalline glands; racemes 8– to 16–flowered, glandular, the axis ca 1–4 cm long; bracts 3–6 mm long, mostly entire; pedicels 1–5 (7) mm long; free hypanthium 0.7–1.5 mm long, green, bowl-shaped, glabrous or puberulent; sepals 2–3 mm long, whitish; petals ca 1.5 mm long, white; styles free or united below the middle; berries 6–10 mm long, blackish, not very fleshy, stipitate-glandular. Mountain brush, aspen, Douglas fir, and spruce- fir communities, usually in shade, at 1645 to 3350 m in Carbon, Emery, Grand, Juab, Millard, Salt Lake, San Juan, Sanpete, Sevier, Tooele, and Utah counties; Colorado, Arizona, and New Mexico, and in Washington and Idaho; 80 (iv). Vegetative specimens of *R. wolfii* are sometimes confused with *R. hudsonianum*. Both have sessile crystalline glands on the lower surface of the leaves. In *R. wolfii*, these glands are more clear than yellow, smaller, and they are seldom noticeable on the often puberulent petioles and young twigs. In *R. hudsonianum* the glands are yellowish, larger, and more conspicuous, and they often extend down the less puberulent petioles and twigs.

Saxifraga L.

Perennial herbs; leaves alternate or basal; flowers perfect; hypanthium obsolete to well developed; sepals 5, erect to deflexed; petals 5, clawless or clawed, deciduous or persistent; stamens 10, the filaments subulate, linear, or flattened; carpels 2–5, connate only at the base or to near the tip; stigmas capitate; fruit capsular and dehiscent across the top or follicular; seeds numerous.

1. Leaves all basal, the blades subentire to coarsely toothed but not lobed, commonly over 15 mm long, usually distinctly petioled; flowers mostly more than 10; plants common and widespread 2
— Leaves cauline, at least in part or, if all basal, less than 15 mm long, not distinctly petioled or, if so, the blades lobed but not toothed; flowers 1–5 or rarely more; plants mainly of high mountains 3

2(1). Inflorescence open; leaf blades orbicular to reniform; petioles 1.2–23 cm long, usually longer than the blades; plants 16–67 cm tall .... *S. odontoloma*
— Inflorescence congested, occasionally interrupted; leaf blades rhombic, obovate or ovate, cuneate at the base; petioles 0.3–2.5 cm long, usually shorter than the blades; plants 3–20 (30) cm tall .... *S. rhomboidea*

3(1). Petioles evident, 0.5–4.5 cm long, usually longer than the blades; blades orbicular or reniform, palmately lobed or trilobate, cordate or truncate basally, with some usually over 5 mm wide 4
— Petioles not especially evident; blades linear to oblanceolate, entire or 3– to 7–lobed, not cordate or truncate basally, 0.5–5 mm wide 5

4(3). Bulblets present in upper leaf-axils and inflorescence, not at petiole bases of lower leaves; hypanthium ca 1 mm long, not turbinate .... *S. cernua*
— Bulblets not present in upper leaf-axils and inflorescence, often present at petiole bases of lower leaves; hypanthium 2–3 mm long, turbinate .... *S. debilis*

5(3). Leaves toothed (sometimes obscurely) or lobed; petals white; sepals erect or somewhat spreading, ca 1/2 as long as the hypanthium; plants glandular 6

— Leaves entire; petals yellow (white in *S. bronchialis*), but fading whitish; sepals spreading to reflexed, mostly more than 1/2 as long as the hypanthium 7

6(5). Plants depressed-caespitose; leaves lobed, those of the stem generally less deeply lobed than the basal ones; petals gradually narrowed to the base, clawless or very shortly clawed; filaments longer than the sepals .... *S. caespitosa*
— Plants not depressed caespitose; leaves merely toothed, those of the stems often more prominently toothed than the basal ones; petals abruptly narrowed to a short claw; filaments shorter than the sepals .... *S. adscendens*

7(5). Plants stoloniferous; sepals ascending; hypanthium ca 1/2 to as long as sepals; herbage stipitate-glandular; leaves ciliate; petals yellow .... *S. flagellaris*
— Plants not stoloniferous; sepals spreading to reflexed or the petals white; hypanthium obsolete or very short; herbage not stipitate-glandular (or the stem sparingly so); leaves glabrous or, if ciliate, the petals white 8

8(7). Petals white; racemes often more than 3–flowered; leaves ciliate; sepals erect to spreading .... *S. bronchialis*
— Petals yellow when fresh; flowers mostly solitary; leaves not ciliate; sepals finally reflexed 9

9(8). Petals 4–5 mm long, short-clawed; sepals 2–3 mm long, glabrous; leaves 4–8 mm long, 0.5–1.5 mm wide; stems with minute stipitate glands, not pilose; plants 2–6 cm tall .... *S. chrysantha*
— Petals 7–10 (15) mm long, not clawed; sepals 4–5 mm long, often pilose; leaves often over 8 mm long and some over 1.5 mm wide; stems often rusty pilose; plants 6–20 cm tall .... *S. hirculus*

Saxifraga adscendens L. Wedge-leaf Saxifrage. Plants short-lived perennials, 3–10 cm tall, from a small simple caudex, strongly glandular-pubescent; leaves 5–15 mm long, sessile or gradually narrowed to a petiolelike base, entire, or 3 (5) -toothed or shallowly lobed at the apex, obovate; sepals about 1–2 mm long, usually reddish purple; petals white, 3–6 mm long, narrowed abruptly to a claw; filaments slender; ovary inferior or nearly so; capsules 3.5–5 mm long. Alpine tundra and rocky slopes in Grand, San Juan, and Summit counties; northern Rocky Mts. and Cascades to Colorado, and Europe; 2 (0). Our material is referable to **var. *oregonensis* (Raf.) Breitung** [*Ponista oregonensis* Raf].

Saxifraga bronchialis L. Spotted Saxifrage. Plants (1) 3–15 cm tall, arising from a caudex with a taproot; caudex branches prostrate to ascending, clothed with densely imbricate persistent leaves; flowering stems glandular-pubescent, with few to several alternate leaves; leaves 3–15 mm long, 1–3 mm broad, sessile, entire, leathery, elliptic to oblong or spatulate, setose-ciliate and spinulose-tipped; flowers showy, yellowish or whitish; sepals greenish, ovate, 2–3 mm long, erect to spreading, glabrous, not ciliate; petals whitish or yellowish, yellow- to red-spotted, 3–6 mm long, indistinctly 3–veined; stamens shorter to longer than the petals, the filaments subulate; ovary only slightly inferior; capsules 4–6 mm long. Open rocky slopes at ca 3050 to 3931 m in the La Sal Mts., Grand and San Juan counties; Alaska and Yukon, south to Oregon, Idaho, and New Mexico; 9 (0). Our plants are apparently referable to **var. *austromontana* (Wiegand) G. N. Jones** [*S. austromontana* Wiegand].

Saxifraga caespitosa L. Tufted Saxifrage. [*S. caespitosa* ssp. *exaratoides* var. *purpusii* Engl. & Irm., type from the La Sal Mts.]. Plants 1–17 cm tall, from a caudex and a taproot; caudex branches prostrate to ascending, clothed with densely imbricate leaves; flowering stems glandular-pubescent, often densely so, with 1 to several alternate leaves; basal leaves 3–15 mm long, 2–7 mm wide, with 3–5 (7) apical, triangular to lanceolate or narrowly oblong lobes, cuneate, glandular-ciliate, not spinulose-tipped; cauline leaves often entire; flowers solitary or 2–4, moderately showy; sepals often purplish, nearly linear to ovate, 1–2 mm long, erect, glandular-pubescent and ciliate; petals white, cream, or yellowish, 2–3 mm long, 3-veined; stamens shorter than the petals, longer than the sepals, the filaments subulate; ovary almost completely inferior; capsule 3–4 (7) mm long. Spruce and alpine tundra communities, fell fields, and rocky slopes at 2985 to 3990 m in Beaver, Duchesne, Grand, Piute, Salt Lake, San Juan, Summit, and Utah counties; circumboreal, south to Nevada, Arizona, and New Mexico; 29 (0). Specimens from above 3660 m on Kings Peak have many entire basal leaves and have sometimes been confused with *S. adscendens*. Fresh petals of some plants (Franklin 955 BRY) are yellowish. Our plants seem to be referable to **var. minima Blank.**

Saxifraga cernua L. Nodding Saxifrage. Plants 6–15 cm tall, arising from fibrous roots; stems sparsely to densely glandular-villous, with several alternate leaves; basal leaves with petioles 0.5–3 cm long, the blades (3) 5- to 7-lobed, reniform, 4–15 mm long, 6–15 mm wide; cauline leaves becoming smaller, fewer lobed, and shorter petioled upwards, at least the upper ones bearing purplish bulblets in the axils; flowers showy, solitary (rarely 2) at the apex of the inflorescence, the others replaced by bulblets; sepals green to dark reddish purple, ovate to lanceolate, 2.5–3 mm long, erect, sparsely to densely glandular, somewhat ciliate; petals white, 5–8 (12) mm long, 3- to 5-veined; stamens longer than the sepals, much shorter than the petals; ovary slightly if at all inferior; capsules rarely developing. Rocky places at 3445 to 3960 m in the Uinta and La Sal mts., Grand, San Juan and Summit counties; circumboreal, south to New Mexico; 10 (0).

Saxifraga chrysantha Gray Golden Saxifrage. Plants 2–6 cm tall, arising from slender rhizomes, sometimes mat-forming; stems glandular; leaves imbricate at the stem base and scattered upwards, 4–8 mm long, 0.5–1.5 mm wide, narrowly spatulate to narrowly oblong, entire, glabrous or occasionally sparsely glandular; flowers solitary; sepals 2–3 mm long, strongly reflexed, glandular externally, green or purplish; petals 4–5 mm long, yellow; stamens longer than the sepals but shorter than the petals; ovaries slightly inferior; capsules 6–8 mm long, ovoid. Alpine tundra, fell fields, and rock stripes at 3415 to 3960 m in the Uinta Mts., Duchesne and Summit counties; Wyoming south to New Mexico; 6 (0).

Saxifraga debilis Engelm. Pygmy Saxifrage. Plants 2–15 cm tall, arising from fibrous roots; stems usually glandular, rarely pilose; basal leaves sometimes with whitish or purple-tipped bulbils in the axils; petioles 0.5–6 cm long; leaf blades 3–15 (20) mm wide, mostly wider then long, reniform or orbicular, with 3–7 lobes; flowers solitary and terminal or 2 or 3 on naked or bracteate, slender pedicels, 0.5–15 mm long; sepals 1.5–2 mm long; petals 2–5 mm long, white; stamens exceeding the sepals, but shorter than the petals; ovaries only slightly inferior; capsules about 5–8 mm long. Alpine tundra, cirque basins, and Engelmann spruce communities, often near melting snowbanks at 2743 to 3960 m in Beaver, Daggett, Duchesne, Garfield, Grand, Iron, Piute, Salt Lake, San Juan, Summit, Tooele, Uintah, and Utah counties; British Columbia south to California, east to Montana and Arizona; 30 (i). Some plants from Kings Peak are rather densely pilose and have slightly shorter and relatively broader sepals than others from the state. These plants seem more like those from the northern Rocky Mts. and Cascades than others from the state.

Saxifraga flagellaris Willd. Flagellate Saxifrage. Plants 3–10 (12) cm tall, arising from slender rhizomes and with naked flagellate stolons; stems erect, stipitate-glandular villous, with several cauline leaves; basal leaves 5–17 mm long, 1–5 mm broad, entire, cuneate-oblanceolate to spatulate, entire, setose-ciliate; cauline leaves similar to the basal ones except becoming glandular-ciliate and smaller above; flowers 1 to few, showy; sepals greenish to reddish purple, oblong to lanceolate or ovate, 2.5–4.5 mm long, stipitate-glandular, the glands purplish; petals bright yellow, but faded in dried specimens, 6–12 mm long, 7– to 9–veined; stamens longer than the sepals, the filaments subulate; ovary only slightly to 1/4 inferior; capsules 4–5 mm long. Alpine tundra at 3350 to 3962 m in the Uinta and La Sal mts., Duchesne, Grand, San Juan, and Summit counties; circumboreal, south to Arizona and New Mexico; 7 (0).

Saxifraga hirculus L. Yellow Marsh Saxifrage. Plants 6–20 cm tall; flowering stems erect, brownish or yellowish tomentose, at least above, arising from a basal rosette with fibrous roots, with usually several cauline leaves; basal leaves 1.2–3.5 cm long, 1–3 (4) mm wide, entire, linear-oblanceolate to spatulate, glabrous or glabrate; cauline leaves similar to the basal only smaller above; flowers 1 to few, showy; sepals greenish to reddish, oblong to lanceolate, 2.5–5.5 mm long, spreading, glandular-villous to glabrate, ciliate; petals bright yellow, 7–12 mm long, 5– to 7–veined; stamens longer than the sepals, the filaments subulate; ovary only slightly inferior; capsules 7–12 mm long. Wet meadows in the Uinta Mts., Daggett County; circumboreal, south to Colorado; 3 (ii).

Saxifraga odontoloma Piper Brook Saxifrage. [*S. arguta* D. Don, misapplied]. Plants scapose, 16–78 cm tall, from rootstocks with fibrous roots; leaves all basal, glabrous or nearly so; petioles 1.2–23 cm long; blades 1–9.5 cm wide and about as long, orbicular to reniform, cordate or truncate at the base, coarsely dentate or crenate-dentate; scapes glabrous below, glandular above; inflorescence a spreading cymose panicle, several-flowered, stipitate-glandular; sepals 2–3 mm long, strongly reflexed in anthesis, purplish; petals 2.5–4 mm long, white, spreading; stamens equaling or exceeding the petals; capsules 5–9 mm long. Along streams, springlets, and about ponds and lakes, and in other moist sites in montane plant communities at 1830 to 3350 m in all Utah counties except Daggett, Davis, Kane, Millard, Morgan, Rich, Washington, Wayne, and Weber; Alaska to California, east to Alberta and New Mexico; 104 (ii).

Saxifraga rhomboidea Greene Diamondleaf Saxifrage. Plants scapose, 3–20 (30) cm tall, from short rootstocks and fibrous roots; leaves all basal; petioles 0.3–2.5 cm long, usually pilose-ciliate or fringed, rather broad and flat, gradually differentiated from the blade; blades 9–35 mm long, 3–20 mm wide, rhombic, obovate, or ovate, dentate or crenate to subentire, not lobed, glabrous or ciliate; inflorescence cymose-paniculate, mostly very congested, globose or headlike, glandular; flowers several to many; sepals 1–2 mm long, more or less triangular, not reflexed; petals 2–4 mm long, white; stamens longer than the sepals, subequal to the petals; capsules 4–6 mm long excluding the stylar beaks. Meadows in alpine tundra and sometimes in ponderosa pine, Douglas fir, and spruce communities at 2075 to 3965 m in Beaver, Box Elder, Cache, Daggett, Duchesne, Garfield, Grand, Iron, Juab, Millard, Piute, Salt Lake, San Juan, Sevier, Summit, Tooele, Uintah, Utah, Washington, and Wayne counties; British Columbia to Colorado; 111 (iii). *S. integrifolia* Hook. has been discovered in Cache County (Thorne 4831 BRY). It differs in having entire or merely denticulate leaves.

SCROPHULARIACEAE A. L. Juss.
Figwort Family

Annual, biennial, or perennial herbs or shrubs or vines, some parasitic or hemiparasitic; leaves alternate, opposite, all basal or sometimes whorled, simple to pinnatifid; inflorescence of spikes, panicles, thyrsoid panicles or in some the flowers solitary on leaf axils or on long slender pedicels; bracts foliaceous to reduced; calyx lobes 4 or 5 (rarely 2), distinct or united; corolla sympetalous, regular or bilabiate, some saccate at the base or spurred, 4- or 5-lobed, commonly 2-lipped, tubular, rotate to campanulate; stamens epipetalous, alternate with the corolla lobes, 4 and didynamous, or 4 fertile and 1 sterile, or 5 fertile (*Verbascum*), the anthers 2-celled, equal or unequal; ovary superior, 2-locular; style 1, sometimes forked into 2 stigmatic lobes; fruit a capsule; seeds usually many and small or rarely few, winged or wingless; $x = 6+$.

1.	Leaves opposite or sometimes whorled in *Collinsia*	2
—	Leaves alternate (at least above) or mostly all basal	9
2(1).	Fertile stamens 2	3
—	Fertile stamens 4	4
3(2).	Corolla tubular, bilabiate; stigmas distinct; calyx 5-lobed; capsule not compressed	*Gratiola*
—	Corolla subrotate, regular to slightly irregular; stigmas completely united; calyx 4-lobed; capsule compressed	*Veronica*
4(2).	Plants annual	5
—	Plants perennial	7
5(4).	Flowers with a fifth sterile stamen; lower corolla lip keeled-saccate and forming a pouch that encloses the stamens and style	*Collinsia*
—	Flowers lacking a sterile stamen	6
6(5).	Calyx strongly 5-angled (ribs prominent), the lobes equal or, if unequal, less than half the tube length; anthers all well developed; capsule glabrous	*Mimulus*
—	Calyx 5-sulcate but not angled (ribs obscure), the lobes unequal, the anterior lobes at least half the calyx length; lower anther pairs reduced or wanting; capsule glandular puberulent	*Mimetanthe*
7(4).	Style forked; stigmatic lobes 2, distinct	*Mimulus*
—	Style entire with 1 stigma	8
8(7).	Corolla small, 8.5–14 mm long, urceolate, not showy; staminode flabellate, attached high on the corolla tube	*Scrophularia*
—	Corolla mostly longer, tubular to strongly ventricose, ampliate, mostly showy; staminode filiform, attached well down in the corolla tube	*Penstemon*
9(1).	Stamens 5 and usually all bearing anthers; basal leaves forming a rosette, the cauline alternate, simple, sessile	*Verbascum*
—	Stamens 2 or 4; leaves various	10
10(9).	Basal leaves well developed; cauline leaves bractlike, sessile, alternate	11
—	Leaves mostly cauline	13
11(10).	Scapose, stoloniferous perennial; flowers solitary on slender pedicels arising from the basal leaves; plants submerged or on muddy shores	*Limosella*
—	Subscapose perennials with fibrous roots; flowers borne in a dense spike or spikelike raceme; moist alpine meadows and open ridges	12
12(11).	Blade of the basal leaves crenate to crenate-serrate; corolla wanting, vestigial or, if present, strongly bilabiate	*Besseya*
—	Blade of the basal leaves laciniate to pinnatifid; corolla present, campanulate to subrotate	*Synthyris*
13(10).	Upper lip of the corolla forming a hood that encloses the anthers	14
—	Upper lip not forming a hood, usually distinctly lobed, if present	17
14(13).	Anther cells alike, equal; galea rounded dorsally; leaves mostly pinnatifid or bipinnatifid or crenate to serrate	*Pedicularis*
—	Anther cells dissimilar, obliquely positioned; galea mostly straight; leaves entire to pinnately-lobed	15
15(14).	Plants perennial or, if annual (*Castilleja exilis*), the leaves and bracts entire and the galea longer than the lower lip	*Castilleja*
—	Plants all annual; leaves, bracts, and galeas various	16
16(15).	Calyx forming 2-lobes, cleft to or nearly to the base in front, bractlike	*Cordylanthus*
—	Calyx forming 4 lobes, cleft 2/5 the calyx length, tubular-campanulate	*Orthocarpus*
17(13).	Corolla spurred at the base	*Linaria*
—	Corolla not spurred at the base, gibbous or pouched at the base	18
18(17).	Corolla 4–5 cm long, typically spotted on the lower lip and within; plants cultivated and persisting	*Digitalis*
—	Corolla typically less than 4 cm long, seldom if ever with the lower lip maculate; plants mainly indigenous (cultivated in *Antirrhinum majus*)	19
19(18).	Plants perennial; leaves coarsely toothed or hastately lobed, palmately veined and lobed	*Maurandya*
—	Plants annual; leaves entire, not palmately veined	20
20(19).	Plants densely hairy from ground level to cotyledons and above; flowers either yellow and peduncles twining or pale lavender and axillary	*Antirrhinum*
—	Plants not densely hairy at the base; flowers yellow, axillary, lacking twining peduncles	*Mohavea*

Antirrhinum L.

Plants annual or perennial herbs; stems erect or twining; leaves opposite, or the upper alternate, entire, lanceolate to linear, sessile; flowers few to several in a bracteate raceme; calyx 5-lobed; corolla yellow or blue to purple or white, strongly bilabiate; stamens 4, didynamous, filament sometimes dilated apically; stigmas capitate, united; capsule woody, asymmetrical or symmetrical, irregularly dehiscent or regularly so; style deflexed, persistent; seeds many with longitudinal corky, tuberculate ridges. Note: The segregate genus *Neogaerrhinum* consists of only two species, which are likely more closely related to species within *Antirrhinum* than they are to each other. Thus, *A. filipes* is here treated traditionally. The garden snapdragon, *A. majus* L., is commonly grown in Utah. The large multicolored flowers serve to distinguish it from the slender and diminutive flowered indigenous species.

1. Pedicels prehensile; capsules lacking pores ..... *A. filipes*
— Pedicels not prehensile; capsules with pores ..... *A. kingii*

Antirrhinum filipes Gray Twining Snapdragon. [*Neogaerrhinum filipes* (Gray) Rothm.]. Stems densely hairy from root apex to first foliage leaves, mostly 3–8 dm long, simple or branched, clambering or with twining pedicels; leaves 15–30 mm long, 4–8 mm wide below, 1–2 mm wide on the upper stem, ovate to lanceolate or linear respectively; flowers yellow; calyx 3–5 mm long, the segments subequal; corolla 10–15 mm long, bilabiate, the tube yellow, maculate with brown on the lower lip, glandular-hairy, the palate hairy; capsule 3.8–5.5 mm long. Creosote bush, Joshua tree, and other warm desert shrub communities at ca 700 to 900 m in Washington County; California, Nevada, and Arizona; 5 (i). The species is rare in collections, probably because it is so inconspicuous. The slender stems blend into the vegetation on which the plants climb. The flowers are individually beautiful.

Antirrhinum kingii Wats. King Snapdragon. Slender, erect annual, 1–3 dm high; stems simple or branched, often flowering to the base; herbage densely woolly below, glabrous upward from the woolly area; leaves 0.6–2.5 cm long, narrowly lanceolate to linear, reduced upward; flowers solitary in the leaf axils and branches, 2–3.5 mm long; sepals irregular; corolla white with purple veins, 5–7 mm long, bilabiate, the palate closing the throat; capsule 2.5–4 mm long, subglobose; seeds 0.7–0.8 mm long, ovoid, tuberculate. Gravelly slopes and flats, wash banks in creosote bush, Joshua tree, other warm desert shrub, sagebrush, and shadscale communities at 1525 to 1830 m in Beaver, Juab, Millard, Sevier, Tooele (type from Stansbury Island) and Washington counties; Arizona, Nevada, California, and Oregon; 8 (i). This plant has evidently suffered great reduction of range since it was initially taken in Utah in 1850. The plants are small and difficult to see among other vegetation, and that might account, in part, for the paucity of modern records.

Besseya Rydb.

Plants subscapose, herbaceous perennials; basal leaves well developed, petiolate, the blades rounded to cordate or cuneate to truncate, crenate-serrate; cauline leaves bracteate, sessile, much smaller than the basal, alternate; inflorescence a terminal, dense spike; sepals 2–4, distinct or united; corolla vestigial or present and bilabiate, colored; stamens 2; stigma capitate; capsule loculicidal; seeds numerous, flat.

1. Corolla present, violet purple; filaments inconspicuously colored, southeastern Utah .... *B. alpina*
— Corolla lacking; filaments conspicuously colored, Uinta and Raft River mts. .... *B. wyomingensis*

Besseya alpina (Gray) Rydb. Alpine Kittentails. [*Synthyris alpina* Gray]. Subscapose perennial herbs, 0.5–2 dm tall; stems simple, erect; herbage woolly to glabrate; basal leaves petiolate, the petiole 1–1.5 cm long, blades 3–5 cm long, 1.5–4 cm wide, elliptic to ovate, crenate; cauline leaves alternate, bracteate, sessile; inflorescence dense, villous, the bracts spatulate to obovate; sepals 4, 4–6 mm long, white-villous; corolla violet to purple, 5–8 mm long; filaments inconspicuously colored; capsule 3–5 mm long, orbicular, the style 3–4 mm long; seeds flat, orbicular, 1–1.5 mm long. Moist rocky alpine meadows 3290 to 3660 m in the La Sal Mts., Grand and San Juan counties; Colorado, New Mexico, and Wyoming; 2 (i).

Besseya wyomingensis (A. Nels.) Rydb. Wyoming Kittentails. [*Synthris wyomingensis* Heller]. Subscapose perennial herb, 1–2.5 dm tall; stems simple, erect; herbage woolly to glabrate; basal leaves long petiolate, the blade 2.5–11.5 cm long, 1.5–5 cm wide, broadly lanceolate to ovate, crenate to serrate; cauline leaves much reduced, bracteate, alternate, sessile to short-petiolate; inflorescence dense, subspicate, villous, calyx 3–5 mm long, 2 or sometimes 3–lobed; corolla lacking; filaments violet or purple, 5–10 mm long; capsule 3.5–6 mm long, the style 4–6.5 mm long; seeds 1–1.5 mm long, flat, orbicular. Moist open areas on high ridges and in meadow areas, 2684 to 3355 m in Box Elder (Raft River Mts.) and Duchesne (Uinta Mts.) counties; north in the Rocky Mts. to British Columbia, east to South Dakota; 7(ii).

Castilleja Mutis ex L. f.

Perennial or rarely annual herbaceous plants; stems usually several, erect or decumbent; leaves alternate, entire or pinnately divided, sessile, all cauline; inflorescence of terminal spikes, these prominently bracteate; calyx 4–cleft, tubular, variously colored; corolla narrowly tubular, bilabiate, upper lip (galea) beaklike, its lobes united to the apex and enclosing the anthers, the lower lip 3–toothed; stamens 4, didynamous; stigma capitate, entire or 2–lobed; capsule loculicidal; seeds numerous.

1. Plants annual, growing in alkaline meadows, seeps, springs, and marshes, usually in the valleys; stems mostly erect, solitary; leaves and bracts all entire, linear-lanceolate, the upper 1/3–1/2 of the bracts red .... *C. exilis*
— Plants perennial growing in various habitats; stems usually 2 or more in a cluster, often decumbent to ascending or erect; leaves and bracts various 2

2(1). Plants low with decumbent to ascending stems 3–9 cm long; corolla short, 9–14 mm long, the lobes small, yellow to pink tipped; plants of the Deep Creek Mts. above 3050 m .... *C. nana*
— Plants larger or, if small, with erect stems and larger corollas; corolla 1.3–4.3 cm long, the lobes mostly showy and variously colored 3

3(2). Calyx divided more deeply below than above 4
— Calyx equally or subequally divided above and below 5

4(3). Inflorescence mostly yellow, rarely reddish; corolla 15–22 mm long .... *C. flava*
— Inflorescence mostly red to orange red, rarely yellow; corolla 2.5–5.2 cm long .... *C. linariifolia*

5(3). Inflorescence predominantly yellow (rarely tinged with purple in *C. occidentalis*) or yellow-green 6
— Inflorescence rarely yellow, typically pink, orange yellow, red to scarlet or crimson 10

6(5). Upper leaves and bracts deeply divided 7

—	Leaves and bracts mostly entire or only shallowly divided near the apex	8
7(6).	Inflorescence hispid-villous, yellow or white; lateral leaf lobes 1 mm wide or more	*C. angustifolia*
—	Inflorescence viscid-villous, sometimes tinged with purple	*C. pulchella*
8(6).	Stems decumbent to ascending, 2.2 dm tall or less; inflorescence often tinged with purple; plants from the La Sal Mts. above timberline at 3264 to 3599 m	*C. occidentalis*
—	Stems erect, mostly taller than 2 dm; inflorescence yellow, not tinged with purple; mostly below timberline in meadows and open areas in conifer or conifer-aspen communities	9
9(8).	Plants with 1 to few stems; 1.2–2.6 dm tall; leaves linear, appressed-ascending; plants of Garfield County	*C. aquarensis*
—	Plants with few to several stems, 1.5–6.8 dm tall; leaves linear to lance-ovate, appressed to spreading; plants of the La Sal Mts., Wasatch Plateau, and northward	*C. rhexifolia*
10(5).	Lower portion of the stems devoid of chlorophyll and with bracteate leaves; stems 0.5–2.2 dm tall; upper portion of the corolla (galea) often completely exserted from the calyx	*C. scabrida*
—	Stems green and the leaves mostly well-developed; galea equaling the calyx or slightly longer or, if well exserted, the plants well over 2.2 dm tall	11
11(10).	Leaves (at least the upper) deeply cleft into linear, spreading lobes; herbage hispid to glandular pubescent; calyx a little more deeply cut in back than in front	12
—	Leaves entire or the upper leaves sometimes shallowly lobed at the tip; herbage glabrous to puberulent or villous above	14
12(11).	Plants glandular-pubescent; leaves wavy-margined	*C. applegatei*
—	Plants finely puberulent to hispid-villous, rarely glandular; leaves not wavy-margined	13
13(12).	Inflorescence orange to yellow orange or whitish, rarely yellow or red; western Utah; Millard County northward to Box Elder County	*C. angustifolia*
—	Inflorescence orange red to bright red, rarely yellow; throughout the state, except infrequently in western Utah	*C. chromosa*
14(11).	Galea 5–8 mm long; stems 0.9–2.1 dm long; southwestern central Utah	*C. parvula*
—	Galea 8–18 mm long; stems mostly over 2.1 dm long	15
15(14).	Bracts lanceolate to ovate, shallowly lobed, magenta or purple, rarely red; galea 8–12 mm long	*C. rhexifolia*
—	Bracts narrowly lanceolate to lanceolate, mostly divided to below the middle, red, or reddish-orange; galea 1.2–1.8 cm long	*C. miniata*

Castilleja angustifolia (Nutt.) G. Don [*Euchroma angustifolia* Nutt.]. Herbaceous perennial, 0.6–3.4 dm tall; stems erect, mostly in clusters; herbage hispid-villous; leaves 1–5 cm long, the lower usually entire, linear to lanceolate, the upper leaves with 1 or 2 pair of lateral lobes; inflorescence villous-hispid, yellow, yellow orange, white, or pink; bracts lanceolate, mostly with 1 or 2 pairs of lateral lobes; calyx 1.4–2.3 cm long, unequally cleft; corolla 1.4–2.7 cm long, the galea 7–13 mm long, the tube 7–17 mm long; capsule 8–14 mm long. Sagebrush and pinyon-juniper communities at 1400 to 3050 m in Beaver, Box Elder Iron, Juab, Millard, and Tooele counties; adjacent Nevada, Idaho, Wyoming, Oregon, and Montana; 38 (i). Our plants belong to var. *flavescens* (Pennell) N. Holmgren [*C. flavescens* Pennell ex Edwin], which is morphologically similar to *C. chromosa* A. Nels.

Castilleja applegatei Fern. Wavy-leaf Paintbrush. Herbaceous perennial; 0.7–3.3 dm tall; stems clustered, erect or ascending from a woody caudex, glandular-villous, sometimes hirsute; leaves glandular-puberulent, usually wavy-margined, the lower mostly entire, linear-lanceolate, upper usually with a pair of lateral lobes 1–4.5 cm long; inflorescence pale red to bright red, ours rarely yellowish; bracts deeply 3– to 5–parted, glandular-villous; calyx 1.5–2.3 cm long, deeply and subequally cleft; corolla 1.3–3.2 cm long, galea equalling to or slightly shorter than the tube, pubescent; capsule 6–14 mm long, finely pubescent; $2n = 24$. Dry mountain ridges to subalpine slopes and meadows 2440 to 3416 m in Box Elder, Cache, Duchesne, Salt Lake, Sanpete, Summit, Tooele, Uintah, Utah, and Wasatch counties; Oregon, California, Idaho, Wyoming and Nevada; 60 (iv). Our plants belong to var. *viscida* (Rydb.) Ownbey [*C. viscida* Rydb., type from Big Cottonwood Canyon, Salt Lake County].

Castilleja aquariensis N. Holmgren Aquarius Plateau Paintbrush. Herbaceous perennial, 1.2–2.6 dm tall; stems erect, unbranched, 1 to several, pubescence short, irregular in length, often blue purple; leaves 1–4.5 cm long, 1–5 mm wide, linear-lanceolate, finely and evenly pubescent, mostly cauline and longer than the internodes; inflorescence villous-glandular, pale yellow; bracts broadly lanceolate to ovate, lower ones entire, upper with 1–2 pair of short lateral lobes near the apex; calyx 1.5–1.8 cm long, subequally cleft, deeper in front than the back; corolla 1.3–1.6 cm long, the galea 5–8 mm long; capsule 7–10 mm long; seeds ca 100. Sagebrush-mixed forb meadow openings in spruce-fir stands at 2985 to 3355 m on the Aquarius Plateau (type locality), Garfield County; endemic; 19 (v).

Castilleja chromosa A. Nels. Common Paintbrush. [*C. collina* A. Nels.; *C. angustifolia* (f.) *collina* Garrett]. Herbaceous perennial, 1–5.5 dm tall; stems clustered, ascending to erect, finely puberulent and hispid; leaves puberulent and hispid, lower ones linear-lanceolate, mostly entire, upper with 1–3 pair of lateral lobes; inflorescence villous to hispid; bright red to orange red, sometimes yellowish; calyx 1.5–2.7 cm long, the primary lobes subequally cleft and divided into 1–4 mm long acute to rounded segments; corolla 2–3.6 cm long, galea 9–18 mm long and equaling the tube in length, the lower lip much reduced; capsule 9–18 mm long; $2n = 12, 24$. Creosote bush, blackbrush, sagebrush, and pinyon-juniper communities at 915 to 2610 m throughout Utah; Arizona, New Mexico, and Colorado, north to Wyoming, west to Oregon, California, and Nevada; 288 (xxiv). This species is closely related to *C. angustifolia*, particularly var. *flavescens*, which occurs in western Utah.

Castilleja exilis A. Nels. Annual Paintbrush. [*C. minor* misapplied, not (Gray) Gray]. Annual, 1.9–9 dm tall; herbage glandular-pilose; stems erect, usually single and unbranched, sometimes branched above; leaves linear-lanceolate, attenuate, entire, 3–9.5 cm long; inflorescence a narrow raceme, greatly elongated in fruit, the fruiting calyces remote, bracts foliose, the upper ones scarlet, showy; calyx 1.5–2.2 cm long, subequally cleft, the primary lobes divided again into linear, acute segments; corolla yellowish, 1.5–2.5 cm long, usually exceeding the calyx, the galea 5–10 mm long, blunt, puberulent, the tube 1.1–1.9 cm long; capsule 1–1.7 cm long, glabrous; n = 12. Wet, often saline places in flood plains, meadows, seeps, springs, bogs, and lake shores at 850 to 2245 m in all Utah counties except Cache, Davis, Morgan, and Wasatch counties; Washington and Oregon, east to Montana and Wyoming, south to Arizona and New Mexico; 73 (i).

Castilleja flava Wats. Yellow Paintbrush. Herbaceous perennial, 8–42 cm tall; stems usually several in a cluster, erect to ascending, often branched above, retrorsely cinereous pubescent (glabrous forms in Sevier County); leaves 1.6–6.7 cm long, linear, the lower entire, the upper with a pair of lateral lobes, retrorsely cinereous pubescent; inflorescence compact, villous, mostly yellow but sometimes reddish; bracts lanceolate, with 1–2 pair of lateral lobes; calyx 12–19 mm long, unequally cleft, the lobes acute; corolla 15–22 mm long, the galea shorter than the tube; capsule 9–14 mm long; n = 24. Sagebrush and mountain brush communities at 1980 to 3205 m in Box Elder, Carbon, Daggett, Duchesne, Emery, Garfield, Grand, Rich, Sevier, Summit, Uintah, Utah, and Wasatch counties; Colorado, Wyoming, Nevada, Idaho, and Oregon; 51 (iv). The Utah specimens are referable to var. *flava*. This taxon closely resembles *C. cusickii* from southern Idaho, which is not known from Utah but should be expected in Box Elder, Cache, or Rich counties.

Castilleja linariifolia Benth. Narrowleaf Paintbrush. [*C. arcuata* Rydb., type from Fish Lake, Sevier County]. Herbaceous perennial, 1.9–8 dm tall; stems simple or branched, glabrous to puberulent; leaves 1–10 cm long, linear, filiform to narrowly lanceolate, mostly entire but sometimes a few with a pair of lateral lobes, usually glabrous; inflorescence conspicuous bright red, scarlet, rarely yellow, the lower flowers remote and usually pedicellate, villous; bracts mostly 3-lobed, shorter than the calyx (usually); calyx 1.7–4 cm long, the primary lobes unequally cleft, deeply cleft in front for 9–24 mm, only 2–10 mm in back; corolla 2.5–5.2 cm long, the galea and lower lip projecting through the front of the deeply cleft calyx, usually arcuate; capsule 9–17 mm long, dark brown, glabrous; n = 12, 24. Sagebrush-grass, pinyon-juniper, mountain brush, and aspen-conifer communities at 1140 to 3145 m in all Utah counties; Arizona, New Mexico, California, Oregon, Idaho, and Montana; 241 (xv).

Castilleja miniata Dougl. ex Hook. Scarlet Paintbrush. [*C. variabilis* Rydb., type from Big Cottonwood Canyon, Salt Lake County]. Herbaceous perennial, 3–7 dm tall; stems few, erect or ascending, often branched above, woody basally; leaves 3–7 cm long, narrowly lanceolate to lanceolate, mostly entire, the upper leaves sometimes lobed, glabrous to puberulent; inflorescence conspicuous, scarlet or bright red, sometimes crimson, rarely yellow, elongating in fruit, villous to viscid-villous; bracts lanceolate, with 1–2 pairs of deeply divided lobes, the lobes acute, rarely entire, puberulent and villous; calyx 1.6–2.8 cm long, deeply and subequally cleft; corolla 2–4.5 cm long, the galea 1.2–1.8 cm long, the tube 1.4–2.2 cm long; capsule 1–1.5 cm long, glabrous; n = 12, 24, 36, 48, 60. Meadows, along streams and lakes, and aspen-conifer communities at 1765 to 3450 m in all Utah counties except Millard and Morgan; Alaska, south to New Mexico, Arizona, and California; 130 (vii).

Castilleja nana Eastw. Dwarf Paintbrush. [*C. lapidicola* Heller]. Herbaceous perennial, 3–9 cm tall; stems decumbent to ascending, glandular and villous, several; leaves 0.4–1.5 cm long, linear to narrowly lanceolate, entire to lobed (uppermost), glandular pubescent and short villous; inflorescence short, dense, glandular pubescent and villous; bracts spreading, the tips dull yellow to pink or white, broader than the leaves, 3- to 5-lobed; calyx 7–12 mm long, deeply and subequally cleft into 4 linear lobes, tips yellow to pink or white; corolla 8–14 mm long, the lobes yellow to pink-tipped; capsule 6–7 mm long, glabrous. Alpine meadows and windswept ridges in granitic gravelly sand at 3200 to 3600 m in the Deep Creek Mts., Juab County; Nevada and California; 13 (i).

Castilleja occidentalis Torr. Western Paintbrush. Herbaceous perennial, 0.2–2.2 dm tall; stems erect or ascending, several to many in a cluster, puberulent, the hairs partly retrorse; leaves 1–4 cm long, lanceolate, entire or the upper leaves rarely 3–lobed, puberulent; inflorescence short and dense in flower, pale yellow or tinged with purple; bracts broad, 1.5–2 cm long, 0.5–1 cm wide, ovate, entire or sometimes with a pair of lateral lobes near apex, greenish yellow or streaked with red; calyx 1.5–2 cm long, subequally cleft; corolla mostly exceeding the calyx, greenish, the galea shorter than the corolla tube, finely puberulent; capsule 7–10 mm long; n = 12, 24. Alpine tundra, often in talus or rock stripes, at 3260 to 3600 m in the La Sal Mts., Grand and San Juan counties; Colorado and New Mexico, north to Montana and Canada; 8 (iii). This paintbrush is related to *C. rhexifolia* var. *sulphurea*. The latter occurs at or below timberline in the La Sal Mts. and elsewhere in the state.

Castilleja parvula Rydb. Tushar Paintbrush. Herbaceous perennial, 0.9–2.1 dm tall; stems erect to ascending, 1 to several per cluster, old stems persisting, puberulent below to villous above; leaves entire, glabrous to retrorsely-puberulent, the lower leaves small and scale-like, linear, the main stem leaves narrowly lanceolate to broadly lanceolate; inflorescence dense, viscid-villous, crimson to magenta; bracts broadly lanceolate, entire or lobed near the tip; calyx 1.2–2.1 cm long, the lobes unequally cleft; corolla 1.5–2.5 cm long, the galea slightly shorter than the tube; capsule 1 cm long, glabrous. Two varieties are separable as follows:

1. Stems mostly solitary but sometimes 2–3 per plant, usually purplish; growing on the Wasatch Limestone Formation in Garfield, Iron, and Kane counties, rare . . . . *C. parvula* var. *revealii*

— Stems mostly several, the old stems persisting, rarely purplish; usually growing on igneous gravel in Beaver, Garfield, and Piute counties, rare . *C. parvula* var. *parvula*

Var. *parvula* [*C. parvula* Rydb., type from the Tushar Mts., Piute County]. Alpine ridgetops and talus slopes above timberline on tertiary igneous sandy gravel at 3050 to 3584 m in Beaver, Garfield, and Piute counties; endemic; 10 (v).

Var. *revealii* (N. Holmgren) Atwood [*C. revealii* N. Holmgren, type from Bryce Canyon]. n = 12. Bristlecone and ponderosa pine communities on the Wasatch Limestone Formation at 2285 to 3050 m in Garfield Iron, and Kane counties; endemic; 5 (iv).

Castilleja pulchella Rydb. Pretty Paintbrush. Herbaceous perennial, 0.5–1.4 dm tall; stems clustered, decumbent to erect, simple, viscid-villous; leaves 0.5–4 cm long, the lower usually linear-lanceolate, entire, the middle and upper leaves often 3–lobed, glandular-pubescent; inflorescence relatively short and broad, viscid-villous and puberulent, mostly yellowish, rarely purplish (in ours); bracts broader than the leaves, lanceolate to ovate, usually with a pair of lateral lobes; calyx yellow green, 1.5–2 cm long, deeply and subequally cleft above and below; corolla 1.7–2.3 cm long, the galea shorter than the tube; capsule 7–12 mm long; n = 12. Rocky alpine tundra and fell fields at 3355 to 3800 m in the Uinta Mts., Daggett, Duchesne, Summit, and Uintah counties; Wyoming, Montana, and Idaho; 16 (ii).

Castilleja rhexifolia Rydb. Rhexia-leaved Paintbrush. [*C. leonardii* Rydb., type from American Fork Canyon]. Herbaceous perennial, 1.2–8.2 cm tall; stems mostly erect, or sometimes ascending from a woody base, occasionally branched above, few to several per plant, glabrous or retrorsely puberulent and villous; leaves 2–7 cm long, entire, linear-lanceolate, or sometimes broader, glabrous to puberulent; inflorescence conspicuous, short and dense, villous, sometimes viscid, purple to crimson, rarely yellow; bracts lanceolate to ovate, entire or shallowly lobed near the tip; calyx 1.5–2.5 cm long, subequally cleft; corolla 1.5–3.5 cm long, the galea shorter than the tube; capsule 10–12 mm long; 2n = 12, 24, 48. Two varieties, differing largely in flower color, are present in Utah.

1. Inflorescence yellow, rarely suffused with other colors
.... *C. rhexifolia* var. *sulphurea*
— Inflorescence violet, red purple, purple, dark red; plants transitional to the above variety ....
.... *C. rhexifolia* var. *rhexifolia*

Var. ***rhexifolia*** Moist alpine or subalpine meadows and slopes or open aspen-conifer communities at 2135 to 3660 m in Box Elder, Cache, Carbon, Daggett, Davis, Duchesne, Emery, Grand, Juab, Morgan, Salt Lake, San Juan, Sanpete, Sevier, Summit, Tooele, Uintah, Utah, Wasatch, and Weber counties; Canada, south to New Mexico and Oregon; 106 (ii).

Var. ***sulphurea*** (Rydb.) Atwood [*C. sulphurea* Rydb.; *C. septentrionalis* authors, not Lindl.]. Meadows and openings in aspen and conifer communities or with sagebrush at 2070 to 3175 m in Cache, Carbon, Daggett, Duchesne, Emery, Grand, Juab, San Juan, Sanpete, Summit, Tooele, Uintah, Utah, and Wasatch counties; New Mexico, Colorado, Idaho, Montana, and Canada; 79 (xi). This variety is allied with and not always easily segregated from *C. occidentalis* Torr., with which it is sympatric in the La Sal Mts.

Castilleja scabrida Eastw. Eastwood Paintbrush. Herbaceous perennial, 0.5–2.2 (2.7) dm tall; stems from a woody caudex, erect to decumbent, several per plant, cinereous-hirsute, the hairs spreading to reflexed, the lower portion devoid of chlorophyll and with bracteate leaves; leaves 1.5–4 cm long, linear to lanceolate or the upper deeply 3– to 5–lobed; inflorescence conspicuous, hirsute to villous, mostly bright red or orange red, rarely yellow; bracts deeply lobed, the lower leaflike; calyx 1.8–2.8 cm long, unequally cleft, the lobes lanceolate, acute; corolla 2.3–4.3 cm long, the galea 0.9–2.0 cm long and mostly exserted from the calyx, the stigma and upper part of the style exserted from the tip of the galea; capsule 1–1.5 cm long; n = 12. Two varieties are recognized as follows:

1. Plants from western Utah in Beaver and Millard counties; limestone gravels and rock crevices; stems often purplish, moderately pubescent .... *C. scabrida* var. *barnebyana*
— Plants from southern, east central, and eastern Utah; sandy soils, all sandstones, rarely on clay soils; stems greenish, densely pubescent .... *C. scabrida* var. *scabrida*

Var. ***barnebyana*** (Eastw.) N. Holmgren [*C. barnebyana* Eastw.]. Gravelly, rocky ridges, slopes, and rock crevices of rhyolite and dolomite limestone, in mountain brush, pinyon-juniper, and black sagebrush communities at 1890 to 2565 m in Beaver, Juab, and Millard counties; Nevada; 10 (0).

Var. ***scabrida*** [*C. scabrida* Eastw.; *C. zionis* Eastw., type from Zion National Park]. This is the common phase of the species which probably grows on most of the sandstone formations or sandy soils south and east of the Wasatch Mts., in pinyon-juniper, sagebrush, mountain brush, and ponderosa pine communities at 1340 to 2535 m in Carbon, Duchesne, Emery, Garfield, Grand, Kane, San Juan, Sevier, Uintah, Washington, and Wayne counties; Colorado and New Mexico; 99 (vi).

Collinsia Nutt.

Herbaceous annual; stems erect, leaves opposite or partly whorled, entire to crenate or toothed; inflorescence racemose, flowers solitary to several in the axils of the upper leaves; calyx with 5 subequal lobes; corolla bilabiate, white to pale blue, upper lip 2–lobed, lower lip 3–lobed, the tube keeled-saccate; stamens 4 in 2 pairs, unequally inserted in the corolla tube; staminode gland-like; stigma capitate or slightly 2–lobed; capsule 4–valved, the walls thin.

Collinsia parviflora Dougl. Blue-eyed Mary. Plants (1) 5–9 (23) cm tall; stems simple or branched, ascending to erect, glabrous to puberulent below and glandular above; leaves 1–3.5 cm long, entire, the upper sometimes whorled, sessile to clasping, the lower small, petiolate, orbicular; flowers long-pedicellate, the upper 2–7 per cluster, the lower often solitary in the axils; calyx 3–7 mm long, the slender lobes longer than the tube; corolla 4–7 mm long, blue with a white upper lip, the tube gibbous dorsally near the base; capsule 3–5 mm long, ellipsoid; seeds brown, 2 per locule; n = 7, 14. Dry foothills to moderately moist areas from sagebrush, pinyon-juniper, oak-maple, and conifer communities at 1250 to 2745 m in all Utah counties except Kane, Morgan, Piute, Rich, and Wayne; New Mexico, Arizona, and California, north to Alaska and Canada; 98 (vi).

Cordylanthus Nutt. ex Benth.

Caulescent annual herbs; stems single, erect, mostly branched; leaves deeply 3– to 5–parted or entire, alternate, all cauline; inflorescence spicate, or flowers sometimes solitary, the bracts foliose; calyx cleft to the base (or nearly so) forming dorsal and ventral segments; corolla bilabiate, tubular, the upper lip narrow, forming a head which encloses the anthers, lower entire to 3–lobed; fertile stamens 2 or 4, in pairs, anthers unequal, the outer attached at its middle, the inner at the apex, sometimes reduced or obsolete; style swollen and hooked near the tip; capsule loculicidal, glabrous. A genus of 20–25 species, native to western North America.

Chuang, T. I. and L. R. Heckard. 1986. Systematics and evolution of *Cordylanthus* (Scrophulariaceae-Pedicularieae) (including the taxonomy of Subgenus Cordylanthus). Systematic Botany Monogr. 10: 1–105.

1. Leaves all entire, narrowly lanceolate to more broadly so; plants of alkaline areas .... *C. maritimus*
— Leaves (some or all) deeply divided and mostly filiform or linear; plants not in alkaline areas 2
2(1). Stamens 2; filaments glabrous; plants of northeastern Nevada and southwestern Idaho; expected for northern and western Box Elder County, Utah in sagebrush and pinyon-juniper areas .... *C. capitatus* Nutt. ex Benth.
— Stamens 4; filaments puberulent to villous 3
3(2). Plants cinereous-puberulent; veins of the bracts and calyx prominent, raised, the pubescence often forming parallel lines; outer bracts subtending each flower; plants of northern Utah .... *C. ramosus*
— Plants glabrous to finely glandular or glandular villous; veins of the bracts evident but not raised and prominent, the pubescence evenly distributed 4
4(3). Plants glabrous to finely glandular, the hairs short, 0.1 mm long; southeastern Utah from Emery and Grand counties south .... *C. wrightii*
— Plants densely glandular or glandular villous, the hairs longer than 0.1 mm; plants of southwestern, northwest central, or northeastern Utah 5
5(4). Calyx 1.2–1.6 cm long; corolla 1.3–1.9 cm long; inner bracts entire; plants of southwestern Utah and Kane and Garfield counties .... *C. parviflorus*
— Calyx 1.5–2.5 cm long; corolla 1.5–2.7 cm long; inner bracts divided; southern Millard County, southwestern Utah and northeast central and northeastern Utah ... . *C. kingii*

Cordylanthus kingii Wats. King Birdsbeak. [*Adenostegia kingii* (Wats.) Greene]. Plants 4–25 (40) cm tall; stems diffusely branched, glandular-pubescent; leaves 1.5–4 cm long, entire or more commonly pinnately 3– to 5–lobed, the divisions filiform; inflorescence a capitate spike; flowers sessile; bracts leaflike, the outer (when present) subtending the spike and mostly 3– to 5–lobed, the inner bracts subtending each flower 3– to 7–lobed or the upper bracts sometimes entire, violet to purple; calyx 1.5–2.7 cm long, bidentate, narrowly lanceolate; corolla 1.5–2.7 cm long, the lower lip purple, upper one violet, retrorsely villous; stamens 4, unequally 2–loculed, the filaments puberulent. Two weakly separable and possible neglegible taxa are present in Utah:

1. Central inflorescence compact, typically with more than 8 flowers and 4–5 cm long; plants of Carbon, Duchesne, Emery, and Uintah counties ..... *C. kingii* var. *densiflorus*
— Central inflorescence not especially compact, typically with 8 or fewer flowers and mainly less than 4 cm long; plants of Garfield, Iron, Kane, and Washington counties .... *C. kingii* var. *kingii*

Var. densiflorus (Chuang & Heckard) Atwood stat. nov. [based on: *Cordylanthus kingii* ssp. *densiflorus* Chuang & Heckard Syst. Bot. Monogr. 10: 81. 1986]. Salt desert shrub, sagebrush, bitterbrush, pinyon-juniper, and Douglas fir communities at 1525 to 2320 m in Carbon, Duchesne, Emery, and Uintah counties; endemic; 32 (ii).

Var. kingii Black sagebrush, rabbitbrush, winterfat, big sagebrush, pinyon-juniper, and ponderosa pine communities at 1250 to 2380 m in Iron, Kane, Millard, and Washington counties; Nevada; 9 (ii).

Cordylanthus maritimus Nutt. ex Benth. Alkali Birdsbeak. [*Adenostegia maritima* Greene; *C. parryi* Wats. in Parry, type from St. George; *C. maritimus* var. *parryi* (Wats.) Jepson]. Plants (6) 1–25 (34) cm tall; stems branched, corymbose, glabrate to viscid-villous; leaves 1–22 (3) cm long, lanceolate, entire, viscid-villous to glabrous with age; inflorescence a spike; bracts foliose, each subtending a flower, all entire; calyx 1.3–1.9 cm long, lanceolate, bidentate, broader than the leaves; corolla 18–20 mm long, the tips unequal, lower lip pale with purple lines, upper lip (galea) yellow; stamens 4, unequally 2–loculed, the filaments glabrous to puberulent; n = 15. Alkaline flats and meadows at 835 to 1465 m in Box Elder, Juab, Millard, Salt Lake, Tooele, Utah, Washington, and Weber counties; Oregon, California, and Nevada; 14 (0). Our plants belong to **var. canescens (Gray) Jepson** [*C. canescens* Gray; *C. maritimus* ssp. *canescens* (Gray) Chuang & Heckard].

Cordylanthus parviflorus (Ferris) Wiggins Smallflower Birdbeak. [*Adenostegia parviflorus* Ferris]. Plants 11–25 (45) cm tall; stems branched, glandular and often purple; leaves 0.5–1.8 (3.2) cm long, the lower mostly 3–lobed, upper linear, entire, glandular; inflorescence a globular spike, this 1– to 2–flowered and terminal on each branch; outer bracts subtending the flower cluster, 3–lobed, the inner bracts subtending each flower and entire; calyx 12–16 mm long, entire to slightly bidentate, narrowly lanceolate; corolla 1.3–1.9 cm long, pink to violet, upper lip (galea) longer than the lower lip, lower lip with reflexed lobes and a yellow tip; stamens 4, unequally 2–loculed, the filaments villous. Warm desert shrub, blackbrush, sagebrush, and pinyon-juniper communities at 975 to 2135 m in Beaver, Garfield, Iron, Juab, Kane, and Washington counties; Nevada and California; 15 (iii).

Cordylanthus ramosus Nutt. ex Benth. Branched Birdbeak. [*Adenostegia ramosa* Greene]. Plants 8–27 (35) cm tall; stems much branched, cinereous-puberulent; leaves 1–3.5 (4.5) cm long, entire or more commonly dissected into filiform segments, cinereous-puberulent; inflorescence in headlike spikes, 1– to 5–flowered; bracts foliose, 5– to 7–lobed, subtending each flower; calyx 1.0–1.7 cm long, entire or slightly bidentate; corolla 10–21 mm long, dull yellow or the lower lip purplish; stamens 4, unequally 2–loculed, the filaments villous. Sagebrush, juniper, and pinyon-juniper communities at 965 to 2505 m in Box Elder, Cache, Daggett, Duchesne, Morgan, Rich, Salt Lake, Summit, Uintah, and Wasatch counties; Colorado to Montana, Idaho, and California; 23 (iii).

Cordylanthus wrightii Gray Wright Birdbeak. [*Adenostegia wrightii* (Gray) Greene]. Plants 13–35 (51) cm tall; stems much branched, finely glandular, rarely retrorsely puberulent; leaves 1–2.7 (3.5) cm long, entire or 3– to 5–lobed into filiform divisions; inflorescence in capitate spikes, flowers rarely solitary; outer bracts subtending each spike 3– to 5–lobed, inner bracts subtending each flower, mostly entire or 3– to 5–lobed; calyx 16–21 mm long, linear to lanceolate, shallowly cleft to entire; corolla 15–27 mm long, the lips subequal, yellow to purple, sparsely villous; stamens 4, unequally

2-loculed, the filaments villous; n = 13. Sagebrush, Vanclevea, juniper, and pinyon-juniper communities, often in sandy soils, at 1125 to 2135 m in Emery, Garfield, Grand, Kane, San Juan, and Wayne counties; Colorado, New Mexico, and Arizona; Mexico; 24 (viii).

Digitalis L.

Biennial (perennial) herbs; stems erect, typically simple; leaves alternate, simple, petiolate (at least below); flowers in terminal, bracteate, usually secund racemes; calyx 5-lobed, much shorter than the corolla tube; corolla ampliate, constricted at the base, the throat open; limb more or less 2-lipped, the lips erect to somewhat spreading, the upper shorter than the lower; stamens 4; capsule ovoid; seeds numerous.

Digitalis purpurea L. Foxglove. Herbs mainly 6–12 dm tall, typically lanate; basal leaves long-petiolate, the blades 8–20 cm long and 4–9 cm wide or more, lanceolate to ovate or elliptic, obtuse to attenuate apically, tapering basally to the broadly winged petiole, the margin serrate to crenate-serrate, white tomentose below, puberulent and green above; racemes simple or less commonly branched, mainly 1–3 dm long; pedicels shorter than the calyx; calyx lobes ovate-lanceolate; corolla 4–6 cm long, purple, pink, or white, usually spotted within and on the lower lip; capsule subequal to the calyx. Rather commonly grown ornamental in lower elevation portions of Utah; introduced from Europe; 1 (0).

Gratiola L.

Small annual herbs; stems simple or branched; leaves opposite, sessile, entire or toothed, lanceolate to ovate; flowers, long pedicellate, solitary or paired in the leaf axils; calyx of 5 more or less distinct segments, subtended by 2 bractlets; corolla tubular, bilabiate, the upper lip entire or shallowly lobed, the lower 3-lobed; fertile stamens 2, included, the others wanting or merely a vestigial pair of filaments near the base of the corolla tube, the anther sacs separated by a broad connective; stigmas 2, flattened; capsule dehiscing by four valves.

Gratiola neglecta Torr. Common Hedge-hyssop. Annual, with fibrous roots, 5–20 cm tall, glandular, at least above; stems erect or diffuse, simple or branched; leaves 1–5 cm long, linear to oblong, entire or finely toothed above, sessile; pedicels 8–22 mm long; calyx 3.2–5.5 mm long (up to 7 mm long in fruit), the lobes subequal, united basally; corolla 7–12 mm long, the tube yellow, the limb white to pale colored; stamens included. Stream and pond margins, often rooting in mud, in Cache County; Canada, south to California, east to Alabama and Georgia; 1 (0).

Limosella L.

Herbaceous, scapose, aquatic or semi-aquatic, glabrous, perennial; leaves mostly long-petiolate, the blades narrow, entire, elliptic to spatulate; flowers small, solitary on the peduncles; calyx campanulate, with 5 equal lobes; corolla rotate-campanulate, white or pinkish, essentially regular, 5-lobed, the lobes shorter than the tube; stamens 4, nearly equal, the anthers confluent; stigmas united, capitate.

Limosella aquatica L. Diminutive, caespitose, usually stoloniferous, scapose, glabrous perennials; leaves 0.4–2.3 cm long, elliptic, entire, long-petiolate; petioles to 9 cm long; scapes slender, bearing a single flower; peduncle 1–5.3 cm long, shorter than the leaves; calyx 1–3 mm long, campanulate, the lobes triangular, equal; corolla 2.5–4 mm long, white to pink, regular or nearly so; stamens 4 with flattened filaments; stigma capitate; 2n = 40. Muddy flats and shallow water of seeps, ponds, lakes, and streams at 1370 to 3050 m in Beaver, Cache, Daggett, Davis, Duchesne, Garfield, Grand, Kane, Millard, Morgan, Rich, Salt Lake, Sevier, Uintah, Utah, Wasatch, and Washington counties; Alaska to Newfoundland, south in much of the U. S.; circumboreal; South America; 23 (i).

Linaria Miller

Annual or perennial, herbs; stems erect, leafy, simple or branched; leaves alternate (the lower sometimes opposite), entire, sessile; inflorescence of terminal racemes; calyx of 5 more or less distinct sepals; corolla yellow, blue or white, strongly bilabiate, the tube spurred ventrally; stamens 4, didynamous, included; stigmas capitate, united.

1. Plants annual; corolla blue or whitish ..... *L. canadensis*
— Plants perennial; corolla yellow 2
2(1). Stem leaves cordate clasping, broad, over 6 mm wide; calyx 6–8 mm long (9–13 mm in fruit) ..... *L. dalmatica*
— Stems leaves not cordate clasping, narrow, less than 6 mm wide; calyx 3.5–5 mm long .... *L. vulgaris*

Linaria canadensis (L.) Dum.-Cours. Blue Toadflax. [*Linaria texana* Scheele]. Slender annual from a taproot; stems 1–5 dm tall, glabrous, erect or with prostrate stems present; leaves of erect (main) stems 1–3 cm long, alternate, linear, those on short basal stems opposite or whorled, wider; flowers in slender glabrous to glandular pubescent racemes; calyx 2.8–3.4 mm long, the lobes broadly lanceolate; corolla 7–13 mm long, exclusive of the 5–11 m long spur, bilabiate, light blue with paler palate; capsule 2.7–3.5 mm long, subglobose. Creosote bush and mesquite communities at 610 to 1650 m in Washington County (?); British Columbia to Quebec, south to Mexico; 0 (0). This species is not definitely known to occur in Utah, but is to be expected.

Linaria dalmatica (L.) Miller Dalmation Toadflax. [*Antirrhinum dalmaticum* L.; *L. genistifolia* ssp. *dalmatica* (L.) Maire & Petit]. Herbaceous perennial, from horizontal rootstocks; stems stout, 3.2–11.7 dm tall, branched above, glabrous and glaucous; leaves 1.4–3.2 (5) cm long, ovate or ovate-lanceolate, sessile, cordate-clasping below, crowded, alternate, entire; calyx 6–8 mm long, 9–13 mm in fruit; sepals subequal, lanceolate, acute; corolla 1.2–2.3 cm long, yellow, spurred, the spur 1–1.7 cm long, the palate bearded, the hairs white to orange; capsule 6–7 mm long, subglobose; 2n = 12. Oak, aspen, sagebrush, mountain brush, and riparian communities at 1340 to 3050 m in Beaver, Cache, Emery, Garfield, Grand, Rich, Salt Lake, Summit, Uintah Utah, and Wasatch counties; sporadic in the U. S.; introduced from Europe; 33 (0). The species was evidently introduced as an ornamental in Provo Canyon in the 1930's.

Linaria vulgaris Hill Butter-and-eggs. [*Antirrhinum linaria* L.; *L. linaria* Karsten]. Herbaceous perennial from creeping roots; stems 2.5–6.8 dm tall, glabrous; leaves 2–4.5(5.5) cm long, narrow, linear to narrowly lanceolate, mostly alternate; racemes dense in flower, elongating in fruit; calyx 3.5–5 mm long; corolla 13–17 mm long, bright yellow, with a bearded orange palate, the spur 10–12 mm long, mostly straight; capsule 5–9 mm long, subglobose; 2n = 12. Waste places, pastures, and roadsides at 1950 to 2810 m in Duchesne, Garfield, Millard, Piute, Salt Lake, Sanpete, Summit, Wasatch, Wayne, and Weber counties; scattered over temperate North America; native of Eurasia; 9 (i).

Maurandya Ortega

Perennial herbs; stems twining or climbing; leaves alternate, petiolate, coarsely toothed to hastately lobed; calyx 5–parted; corolla gibbous or saccate basally, bilabiate, usually with internal plaits, with a palate; stamens 4, didynamous; filaments with 2 rows of tackshaped glands; capsule scarcely oblique, irregularly dehiscent; seeds oblong in outline, with irregular corky ridges and tubercles.

Maurandya antirrhiniflora Humb. & Bonpl. ex Willd. [*Antirrhinum maurandioides* Gray]. Stems mainly 2–8 dm long, typically much branched from the base, twining and clambering, glabrous; leaves triangular, hastate to 5-lobed, mainly 1–2.5 dm long, on long flexuous petioles; flowers solitary, axillary, on slender pedicels to 2 cm long; calyx 10–12 mm long; corolla 25–30 mm long, the tube pale, the lobes pink purple to purple, the palate yellowish or whitish, with dark lines, hairy; stamens 17–19 mm long; capsules 7–8 mm long, subglobose; seeds ca 1 mm long. Warm desert shrub community at ca 700 to 900 m in Washington County; California to Texas; Mexico; 2 (0). Evidently the species has not been collected in Utah since taken by C. C. Parry near St. George in the 1870's.

Mimetanthe Greene

Herbaceous annual with long villous hairs; leaves narrow, opposite, sessile; inflorescence racemose, flowers in the axils of the opposite foliose bracts, in pairs on long ebracteate pedicels; calyx not angled, somewhat 5-sulcate, deeply cleft; corolla small, yellow, bilabiate, upper lobes external in bud; stamens 4, didynamous, the lower pair tending to be reduced or wanting; stigmas distinct; capsule loculicidal, membranous, glandular-puberulent.

Mimetanthe pilosa (Benth.) Greene Hairy Mimetanthe. [*Herpestis pilosa* Benth.; *Mimulus pilosus* (Benth.) Wats.]. Small herbaceous, unpleasantly scented annuals 6–32 cm tall; stems erect, simple or much-branched, glandular and spreading villous; leaves mostly entire, sessile, 1.2–3.5 (4.9) cm long, narrowly elliptic to oblanceolate; calyx 5–8 mm long, unequal, upper lobe the longest; corolla 5–10 mm long, yellow, usually with maroon dots on the palate, long pedicellate, the pedicel 9–20 mm long; capsule finely glandular, 5–7 mm long. Sandy and gravelly wet soils at 760 to 1740 m in Iron, Kane, and Washington counties; Arizona, east to Baja California, north to Washington; 9 (ii).

Mimulus L.

Annual or perennial herbs; stems leafy, glabrous to more often glandular-pubescent or pilose; leaves opposite or basal, entire to dentate, sessile or petioled; inflorescence of solitary axillary flowers or in leafy terminal racemes; calyx strongly 5-angled, tubular or campanulate, the lobes shorter than the tube; corolla mostly yellow, but also blue or red to purple, bilabiate to nearly regular, 5–lobed, the tube funnelform to cylindrical; stamens 4, in pairs, all fertile; stigmas mostly distinct, lamelliform; capsule loculicidal, membranous to coriaceous, glabrous.

1. Flowers long-pedicellate, the pedicels longer than the calyx, caducous 2
— Flowers with short pedicels up to as long as the calyx, marcescent 12
2(1). Corolla large, 3.3–6 cm long, scarlet, red to pink or violet 3
— Corolla mostly less than 3 cm long, yellow 5
3(2). Stems prostrate with leafy stolons; corolla crimson or red; pedicels mostly 1–2.5 cm long (4 cm); plants of southeastern Utah .... *M. eastwoodiae*
— Stems erect or decumbent, rhizomatous; corolla scarlet, pink or magenta to violet; pedicels 3–10 cm long; plants of southwestern or northern Utah 4
4(3). Stigma and anthers exserted; corolla scarlet; plants of Washington County .... *M. cardinalis*
— Stigma and anthers included; corolla pink to rose red; plants of northern Utah .... *M. lewisii*
5(2). Calyx lobes distinctly unequal, the upper lobe much longer; fruiting calyx strongly inflated 6
— Calyx lobes more or less equal, not strongly inflated in fruit, cylindrical to campanulate 8
6(5). Corolla throat open; lateral and lower calyx lobes blunt and relatively short or nearly obsolete; corolla 10–20 mm long .... *M. glabratus*
— Corolla throat closed by the well-developed palate; lateral and lower calyx teeth more or less acute; corolla 9–30 mm long 7
7(6). Stems mostly 1- to 5-flowered; corolla mostly over 2 cm long; plants usually at middle or lower elevations .... *M. guttatus*
— Stems often over 5-flowered; corolla mostly less than 2 cm long; plants mostly above 2440 m elevation ... . *M. tilingii*
8(5). Corolla 14–25 mm long; plants perennial 9
— Corolla mostly less than 14 mm long; plants annual .. 10
9(8). Leaves palmately veined, crowded at or near the base; pedicels 4–7 times longer than the flowers; plants of southwestern Utah .... *M. primuloides*
— Leaves pinnately veined; plants leafy stemmed; pedicels 2–3 times longer than the flowers; plants of northcentral and eastern Utah .... *M. moschatus*
10(8). Leaves petiolate, ovate to lanceolate ... *M. floribundus*
— Leaves sessile (at least above), linear to narrowly lanceolate 11
11(10). Plants compressed, much branched, the internodes shorter than the leaves; fruiting pedicels widely spreading, sigmoid, 2–7 mm long .... *M. suksdorfii*
— Plants open, simple or few-branched, the internodes mostly longer than the leaves; pedicels arched, 4–16 mm long .... *M. rubellus*
12(1). Corolla small, 4–8 mm long; calyx 3.5–5 mm long; plants from Sevier and Tooele counties northward .... *M. breweri*
— Corollas larger, 10–23 mm long; calyx 5–12 mm long; plants of Washington County 13
13(12). Leaves cuspidate, ovate to nearly suborbicular, densely crowded, thick .... *M. spissus*

— Leaves not cuspidate and densely crowded or thick . 14
14(13). Plants glandular-viscid and villous; corolla lobes purple or violet, the tube lighter .... M. *bigelovii*
— Plants puberulent; corolla mostly yellow but sometimes violet .... M. *parryi*

Mimulus bigelovii (Gray) Gray Bigelow Monkey-flower. [*Eunanus bigelovii* Gray in Whipple]. Annual, 0.3–1 dm tall; stems simple or branched below, glabrous to finely puberulent; leaves sessile to short-petiolate, 5–10 mm long, elliptic, entire; calyx 6–9 mm long in flower, the ribs reddish, the lobes unequal; corolla 18–23 mm long, glabrous, magenta to violet, funnelform, bilabiate, the lobes subequal, the palate with yellow spots; anthers and style included. Pinyon-juniper, sagebrush, creosote bush, and Joshua tree communities at 1065 to 1525 m in Washington County (reported from the Beaver Dam Mts.); Arizona, Nevada, California, and Baja California; 0 (0).

Mimulus breweri (Greene) Cov. Brewer Monkey-flower. [*Eunanus breweri* Greene; *M. rubellus* Gray var. *breweri* (Greene) Jepson]. Annual, 2.5–11 cm long; stems simple or branched, glandular-pubescent; leaves 6–12 mm long, linear to narrowly-oblanceolate, sessile, entire; calyx 3.5–5 mm long; finely glandular, the lobes nearly equal, 0.5–1.5 mm long; corolla small, 4–8 mm long, persistent, violet to magenta, the lobes short, ca 1 mm long; anthers included, glabrous; capsule 5–6 mm long, oblanceolate. Mountain brush, ponderosa pine, aspen, Douglas fir, and alpine fir communities at 2560 to 2745 m in Box Elder, Sevier, Summit, and Tooele counties; British Columbia south to California and Nevada; 4 (0).

Mimulus cardinalis Dougl. ex Benth. Cardinal Monkey-flower. [*M. verbenaceus* Kearney & Peebles]. Perennial, 2.2–6 dm tall; stems erect or decumbent, glabrous below to viscid-villous above; leaves 4–9.5 cm long, 1.5–6 cm wide, obovate, tapering basally to a clasping base, palmately veined, irregularly toothed; calyx 2–3 cm long, glabrous to pubescent, the lobes 4–7 mm long, subequal, acuminate tipped; corolla 3.6–5 cm long, scarlet, strongly bilabiate, the tube 2.5–3.5 cm long, the palate with pale yellow hairs; pedicels 3–10 cm long; anthers and stigma exserted, the anthers hairy; capsule 12–16 cm long, oblong, acuminate; n = 8. Hanging gardens and margins of seeps and streams at 1250 to 1530 m in Zion National Park, Washington County; Oregon to Baja California, east to Nevada, Arizona, New Mexico, and adjacent Mexico; 6 (0). This beautiful plant flowers in springtime.

Mimulus eastwoodiae Rydb. Eastwood Monkey-flower. Perennial with leafy stolons; stems 7–3 (43) cm tall, glandular-puberulent or glandular-villous; leaves all sessile, dentate, palmately veined, the lower reduced in size, 0.5–2 cm long, fan shaped, upper leaves larger, 2–5 (7) cm long, elliptic to lanceolate or oblanceolate to obovate; calyx 15–21 cm long, the lobes subequal 4–7 mm long, lanceolate, acuminate; corolla 25–35 (45) cm long, crimson or red, deciduous, strongly bilabiate, the tube 20–30 mm long; pedicels 1–2.5 (4) cm long; anthers and style exserted, the anthers villous; capsule 6–10 mm long, elliptic; n = 8. Moist seeps and hanging gardens in sandstone cliffs at 1125 to 1375 m in the Canyonlands of the Colorado River, in Garfield, Grand, Kane, and San Juan (type from near Bluff) counties; Arizona and Colorado; 25 (v). Flowers are displayed in late summer and autumn.

Mimulus floribundus Dougl. ex Lindl. [*M. membranaceus* A. Nels.; *M. floribundus* var. *membranaceus* (A. Nels.) A. L. Grant]. Annual, 3–15 (24) cm tall; stems erect to subprostrate, simple or branched, glandular to viscid-villous; leaves petiolate, the petiole to 12 mm long, the blade 4–19 mm long, deltoid to ovate or lanceolate, thin, entire to dentate; calyx 3–6 mm long (up to 9 mm in fruit), glandular, the lobes nearly equal, 1–2 mm long, the tube cylindric; corolla 5–10 (14) mm long, yellow, tubular, somewhat bilabiate, soon deciduous, mostly long-pedicellate; pedicels 5–20 mm long; stamens and style included; capsule 4–7 mm long, elliptic to ovoid. Pinyon-juniper, hanging garden, aspen, spruce-fir, and alpine meadow communities, typically where wet, at 1370 to 2595 m in Box Elder, Cache, Daggett, Emery, Grand, Juab, Kane, Millard, Salt Lake, Sevier, Tooele, Utah, Washington, and Weber counties; British Columbia to Montana and South Dakota, south to California, Nevada, Arizona, and Mexico; 10 (iii).

Mimulus glabratus H.B.K. Glabrous Monkey-flower. [*M. glabratus* var. *fremontii* A. L. Grant; *M. glabratus* ssp. *utahensis* Pennell, type from Millard County]. Perennial, rhizomatus, creeping and rooting at the nodes, glabrous or nearly so, 1–6 dm tall; leaves short-petiolate below to sessile above, 1–3 cm long, palmately 3–7 veined, broadly ovate to orbicular, dentate or sometimes entire; inflorescence of solitary flowers from the upper leaf axils; pedicels (0.8) 2–3.5 cm long; calyx campanulate, 6–15 mm long, glabrous, the lobes unequal, the teeth blunt and relatively short or nearly lacking; corolla yellow, 10–20 mm long, soon deciduous, distinctly bilabiate, often maculate, the throat more or less open; n = 14, 15, 30, 31, 46. Pinyon-juniper, mountain brush, aspen, and other communities, typically in springs, seeps, and streams, at 820 to 2440 m in Davis, Duchesne, Juab, Kane, Piute, Rich, Salt Lake, Sevier, Tooele, Utah, Wasatch, Washington, and Weber counties; Manitoba and Ontariao and widely distributed in the U. S.; Mexico and Central and South America; 28 (ii). Two rather weak subspecies have been recognized for Utah: ssp. *fremontii* (**Benth.**) **Pennell** [*M. jamesii* var. *fremontii* Benth. in A. DC.] has small flowers, pedicels 1–3 cm long, a calyx 5–10 mm long and suborbicular leaves (it enters our area in southwestern Utah); and ssp. *utahensis* Pennell with longer pedicels (2–6 cm long), a calyx 7–16 mm long, and oval leaves. Most of our plants belong to ssp. *utahensis*.

Mimulus guttatus DC. Common Monkey-flower. [*M. guttatus* DC. var. *depauperatus* (Gray) A. L. Grant]. Fibrous rooted annuals or perennials with stout stolons, 0.5–9 dm tall, glabrous or pubescent; stems stout, erect or decumbent, simple or branched, often very succulent; leaves irregularly dentate, broadly ovate to obovate or reniform-cordate, palmately or subpalmately 5–9 veined; petiolate below to sessile above; inflorescence of terminal, long-pedicellate racemes or sometimes solitary; calyx accrescent, campanulate, 6–16 mm long in flower, longer and much inflated in fruit, lateral calyx teeth more or less acute and tending to fold inward in fruit; corolla yellow, 9–30 mm long, soon deciduous, strongly bilabiate, the throat flaring, closed by the well developed palate, maculate; 2n = 28. Marshy areas, seeps, springs, and riverbanks in many plant communities at 1000 to 3000 m in all Utah counties, except Davis, Morgan, and San Juan; Alaska and Yukon, south to New Mexico; 158 (ix). This is our most common species of *Mimulus*.

Mimulus lewisii **Pursh** Lewis Monkey-flower. Herbaceous perennial from a rootstock; stems 1.7–9.2 dm tall, erect, stout, glandular to glandular-villous; leaves (2) 3.5–7.5 (9) cm long, elliptic, ovate to ovate-lanceolate, subentire to irregularly dentate, palmately veined, sessile; flowers solitary or few to several, the pedicels 3–7 (8.5) cm long; calyx 1.3–2.5 cm long, glandular, the lobes subequal, 3–5 (7) mm long, acuminate; corolla 3.3–4.6 cm long, pink to rose red, soon deciduous, somewhat bilabiate, the lobes broad and rounded, 9–14 mm long, the tube much longer than the calyx, often lined with red dots, pinkish to yellowish, 2.6–3.4 cm long; anthers and style included, the anthers villous; capsule 13–16 mm long, narrowly oblong; n = 8. Mountain brush, aspen, Douglas fir, spruce-fir, and lodgepole pine communities, typically along stream banks, at 2135 to 3390 m in Box Elder, Daggett, Davis, Duchesne, Salt Lake, San Juan (?), Summit, Tooele, and Weber counties; Alaska, British Columbia, and Alberta, south to California, Nevada, and Colorado; 20 (iv).

Mimulus moschatus **Dougl. ex Lindl.** Musk Monkey-flower. [*M. guttatus* var. *moschatus* (Dougl.) Prov.]. Herbaceous musk-scented perennial; stems 2–30 cm tall, prostrate to decumbent, mostly branched or simple, glandular-villous, rooting at lower nodes; leaves petiolate, the petiole 2–10 mm long, the blade 1.2–5 cm long, ovate, entire to sparingly dentate, pinnately veined; pedicel 5–14 mm long; calyx 6–13 mm long, campanulate, the lobes subequal, lanceolate, 1.5–4 mm long; corolla 14–20 mm long, yellow, soon deciduous, tubular-funnelform, the tube with red lines extending to the throat, 10–16 mm long, the palate bearded; anthers and style included, the anthers bearded; capsule 3–7 mm long, ovate; n = 16. Sagebrush, aspen, spruce-fir, lodgepole pine, and meadow communities, often where wet and shady at 2135 to 2745 m in Cache, Daggett, Duchesne, Morgan, Salt Lake, Utah, and Wasatch counties; Colorado, Nevada, and California, north to Canada; sporadic in the eastern U. S.; 16 (i).

Mimulus parryi **Gray** Parry Monkey-flower. [*Eunanus parryi* (Gray) Greene]. Annual; stems 2–16 cm high, simple or branched from the base, glandular-puberulent; leaves entire, (2) 10–27 mm long, the lower usually short-petiolate or tapering into a petiole, ovate or spatulate, the upper sessile, linear to oblanceolate; flowers few to many, pedicellate, the pedicel 1–5 mm long; calyx 6–12 mm long, campanulate, the ribs dark green or red, paler between the ribs, the lobes unequal, the upper longer and wider; corolla yellow, rarely magenta, 10–23 mm long, persistent, bilabiate, funnelform, the palate bearded; anthers and style included, glabrous; capsule 5–8 mm long, narrow. Creosote bush, Joshua tree, and blackbrush communities at 610 to 1070 m in Washington (type from near St. George) County; Arizona and Nevada; 19 (iii). The magenta phase of this species can and has been confused with *M. bigelovii*, which is viscid-villous and glandular-pubescent rather than finely glandular.

Mimulus primuloides **Benth.** Primrose Monkey-flower. Herbaceous, usually mat-forming perennial with flagelliform rhizomes; stems 0.5–6 cm long; leaves sessile or the base narrow and tapering into a petiole, crowded toward the base, 7–25 mm long, entire to dentate, oblanceolate to ovate, glabrous to villous, palmately veined; flowers 1–2 (3) per stem, the pedicels slender, 2–10 cm long; calyx narrow, tubular, 4–8 mm long, the ribs dark, the lobes nearly equal, ciliate; corolla yellow, funnelform, maroon dotted, 14–20 mm long, nearly regular, the palate bearded; anthers and style included, glabrous; capsule 5–7 mm long, ovate; n = 17. Moist areas at 2620 to 2810 m in Beaver and Duchesne counties, and reported from the Pine Valley Mts., Washington County; Arizona and California, north to Washington and Idaho; 2 (0).

Mimulus rubellus **Gray** [*M. gratioloides* Rydb.]. Herbaceous, glandular-puberulent, slender annual, 1–22 cm tall; stems simple or branched from base; leaves 3–15 mm long, sessile or the lowest subpetiolate linear to narrowly lanceolate; flowers pedicellate, the pedicel 7–20 mm long; calyx 4–7 mm long, tubular, the ribs dark, the lobes nearly equal, ciliate; corolla yellow, sometimes reddish, 7–9 mm long, the palate puberulent, the lobes red dotted, deciduous; anthers and style included, glabrous; capsule 4–6 mm long, ovoid. Cool desert shrub, sagebrush, and pinyon-juniper communities at 1555 to 2745 m in Beaver, Cache, Duchesne, Emery, Iron, Juab, Kane, Millard, Piute, Salt Lake, San Juan, Sevier, Tooele, Uintah, Utah, and Washington counties; California and Nevada to Colorado, Texas, and New Mexico; 30 (vi).

Mimulus spissus **A. L. Grant** Viscid-villous, mephitic annual, 2–20 cm tall; stems simple or branched throughout; leaves sessile or the lower leaves tapering into a petiole, densely crowded especially in larger plants, 8–17 mm long, and nearly as wide, obovate to suborbicular, cuspidate, tapering to the base, entire, thick; flowers few to many in larger plants and usually flowering the full length of the stems; pedicels 1–2 mm long; calyx 5–9 mm long, campanulate, the lobes unequal, curved out, subulate to lanceolate, the upper lobe longer, larger; corolla 12–18 mm long, magenta to pink, persistent, funnelform, bilabiate, the palate and corolla tube densely short pubescent; anthers included, glabrous; style and stigma included, the style glandular, the stigma ciliolate; capsule 9–13 mm long, narrowly lanceolate. Warm desert shrub communities at 915 to 1830 m in Beaver Dam Wash, Washington County; Nevada and California; 7 (0).

Mimulus suksdorfii **Gray** Suksdorf Monkey-flower. Compact annual, 1–7 cm tall; stems finely glandular-puberulent, usually much branched; leaves 5–13 mm long, mostly longer than the internodes, sessile or the lower short petiolate, linear to oblanceolate, entire; flowers few to several along the entire stem; pedicels 2–7 mm long; calyx 3–6 mm long, cylindrical, glandular-puberulent, reddish, the ribs a little darker, the lobes ca 1 mm long, the margins not ciliate; corolla 4–6 mm long, yellow, deciduous, narrowly-funnelform, the lobes subequal, emarginate, the palate and tube puberulent; anthers and style visible in the throat, glabrous; capsule 4–5 mm long, ovoid. Mixed desert shrub, sagebrush, pinyon-juniper, mountain brush, ponderosa pine, and windswept ridge communities at 1310 to 2820 m in Beaver, Box Elder, Juab, Kane, Millard, Morgan, San Juan, Uintah, Wasatch, and Weber counties; Washington to Wyoming, south to California and Arizona; 18 (ii).

Mimulus tilingii **Regel** Subalpine Monkey-flower. Low perennial, from creeping rhizomes or stolons, 5–26 cm tall, glabrous or sometimes puberulent; stems mostly simple or rarely branched; leaves petiolate below to sessile above, the blade 0.8–3 cm long, rhombic to ovate, dentate, palmately 3- to 5-veined, pedicels mostly less than 1 (3.5) cm long; calyx campanulate, 7–14 mm long in flower, to 21 mm long in fruit and strongly inflated, pale green to reddish with dark spots, the lobes unequal; corolla yellow, red dotted, 15–30 mm long, soon deciduous, broadly funnelform, the palate densely yellow hairy; n = 14, 28. Streambanks, springs, and other wet places in ponderosa pine, aspen, Douglas fir, and spruce-fir communities at 2285 to 3235 m in Beaver, Box Elder, Davis, Duchesne, Juab, Kane, Piute, Salt Lake, Sanpete, Utah, and Wasatch counties; Canada, south to California, Nevada, Arizona, and New Mexico; 17 (iv).

Mohavea Gray

Annual herbs; stems erect, often branched from near the base; leaves alternate, entire, short-petioled; flowers axillary, subtended by foliose bracts; calyx 5-parted, campanulate; corolla bilabiate, the tube short, merely gibbous at the base, the lips fan-shaped, the lower with a hairy palate; fertile stamens 2, connivent, the other 3 abortive; capsule ovoid, thin; seeds discoid, each surrounded by a wing.

Mohavea breviflora Cov. Plants simple or few-branched, viscid-pubescent, mainly 0.5–2 dm tall,; leaves 1–4 cm long, ovate-lanceolate; pedicels 2–5 mm long; calyx 10–12 mm long; corolla 15–18 mm long, bright lemon yellow; stamens glabrous; capsules 8–10 mm long; seeds ca 2 mm long. Warm desert shrub at ca 850 m in Washington County; Arizona, Nevada, and California; 3 (i). This is another Mohavean plant that encroaches on Utah in the Beaverdam Wash vicinity. It was taken on 17 April 1986 by N. D. Atwood, G. I. Baird, and K. H. Thorne.

Orthocarpus Nutt.

Annual herbs; stems erect, branching above; leaves alternate, entire to dissected, sessile or nearly so; inflorescence of leafy-bracted spikes; calyx 4-cleft, tubular-campanulate; corolla bilabiate, yellow, white or purple, the tube elongate, narrow, the lower lip entire, somewhat 3-saccate apically, upper lip beaklike, the lobes united, enclosing the anthers; stamens 4, attached near top of corolla tube; anthers dissimilar, the outer attached at the middle, the inner attached at its apex; stigma entire, penicillate.

1. Corolla 15–20 mm long, violet and white; plants of southeastern Utah .... O. purpureo-albus
— Corolla 9–16 mm long, yellow (rarely violet); plants of various distribution 2
2(1). Galea longer than the lower lip, the tip terminating in a short hook; stems often much branched above .. O. tolmiei
— Galea subequal to the lower lip, the tip not hooked; stems mostly simple .... O. luteus

Orthocarpus luteus Nutt. Yellow Owl-clover. Stems 1–4 dm tall, erect, mostly simple, sometimes branched above, spreading-hairy; leaves 1–4 cm long, linear to linear-lanceolate, mostly entire, sometimes trifid, with pubescence shorter than the stem hairs; spikes many-flowered, narrow, partly glandular; bracts 10–15 mm long, 3- to 5-cleft, leaflike but shorter and more cleft above; calyx 5–8 mm long, subequally 4-lobed, the lobes 1–2 mm long; corolla 9–12 mm long, golden yellow, puberulent, galea short and broad, 2.5–4 mm long; anthers pubescent; capsule 4–7 mm long, elliptical; n = 14. Sagebrush, meadows, ponderosa pine, and aspen communities at 2225 to 2930 m in Cache, Carbon, Daggett, Duchesne, Garfield, Grand, Kane, Morgan, Piute, Rich, Salt Lake, San Juan, Sanpete, Sevier, Summit, Uintah, Wasatch, and Washington counties; British Columbia, south to California, Colorado, and New Mexico, east to Michigan and Minnesota; 32 (i).

Orthocarpus purpureo-albus Gray Stems 1–4 dm tall, erect, mostly simple or branched above, purplish, glandular pubescent; leaves 1.5–3.5 cm long, filiform to linear-lanceolate, entire or 3-lobed, the lobes filiform; spike few- to many-flowered; bracts 10–20 mm long, 3-lobed; calyx 6–10 mm long, glandular, the nerves dark green; corolla 13–20 mm long, white or purplish, lower lip wider and shorter than galea; anthers puberulent; capsule 6–7 mm long, elliptical. Sagebrush and ponderosa pine communities at 2000 to 2700 m in Garfield and San Juan counties; Colorado, Arizona, and New Mexico; 3 (0).

Orthocarpus tolmei H. & A. Tolmie Owl-clover. Stems erect, mostly branched above, 1–3 dm tall, puberulent; leaves 2–4 cm long, narrowly lanceolate, entire; spikes short, compact; flowers short-pedicellate; bracts mostly 3-lobed, sometimes entire, green to yellow, finely glandular; calyx 5–7 mm long, unequally divided, the upper longer and more deeply cut; corolla 10–15 cm long, yellow or violet to purple, glabrous to puberulent; anthers usually ciliate; capsule 3–5 mm long, obovate. Sagebrush-grass, mountain brush, spruce-fir, and alpine communities at 2195 to 3265 m in Cache, Carbon, Davis, Duchesne, Millard, Morgan, Rich, Salt Lake, Sanpete, Sevier, Summit, Utah, and Weber counties; Idaho and Wyoming; 19 (ii).

Pedicularis L.

Herbaceous perennials, mostly caulescent or subacaulescent; stems simple or branched at base, erect; leaves alternate, opposite or basal, toothed to bipinnatifid; inflorescence of bracteate spikes or spicate racemes; calyx 2- to 5-lobed, accrescent, the lobes unequally cleft; corolla strongly bilabiate, yellow, purple, red, or white, the galea hooded, the lower lip 3-lobed; stamens 4, didynamous, glabrous; stigmas united, capitate; capsule loculicidal, flattened, asymmetrical, glabrous.

1. Leaves simple, serrate to crenate; calyx 2-lobed 2
— Leaves pinnatifid or bipinnatifid; calyx 5-lobed 3
2(1). Corolla white or pale yellow, 10–16 mm long, the galea beaked and curved downward; leaves serrate or sometimes crenate, not cartilaginous or revolute .. P. racemosa
— Corolla rose to purple, 20–25 mm long, the galea beakless; leaves double crenate, the crenulations white-cartilaginous, revolute; western Colorado and Nevada, to be expected in Utah in wet meadows or along streams .... P. crenulata Benth.
3(1). Galea prolonged into a slender curved beak over 8 mm long (curved out and up), the galea and beak resembling the head and trunk of an elephant .... P. groenlandica
— Galea beakless or with a short straight beak or incurved and short (5 mm or less long) 4
4(3). Stems less than 1 dm tall; inflorescence a dense compact spicate-raceme, 4–6 cm long; leaves surpassing the stems and usually the inflorescence; plants of low elevations .... P. centranthera
— Stems over 1 dm tall; inflorescence an elongate spicate-raceme over 6 cm long and exceeding the leaves; plants of middle and higher elevations 5

5(4). Corolla 20–35 mm long; bracts simple or serrate to toothed apically 6
— Corolla 10–22 mm long; bracts pinnately to palmately lobed 7
6(5). Corolla 21–26 mm long; bracts simple or serrate apically; calyx 7–10 mm long .... *P. bracteosa*
— Corolla 25–35 mm long; bracts entire or toothed apically, calyx 11–16 mm long .... *P. procera*
7(5). Corolla 16–22 mm long; calyx lobes apparently dark purplish to the naked eye; leaves distinctly lighter ventrally, the leaf segments up to 7 mm long; plants mainly from the Uinta and Wasatch mts. and Wasatch Plateau .... *P. parryi*
— Corolla less than 16 mm long; calyx lobes not apparently dark purple; leaves not distinctly lighter ventrally, at least some of the leaf segments over 7 mm long; plants of Raft River Mts. .... *P. contorta*

Pedicularis bracteosa Benth. in Hook. Stems 3–9 dm tall, leafy, erect, glabrous; leaves 5–15 cm long, pinnatifid, linear to oblong-lanceolate, mostly cauline, the cauline sessile to short-petiolate, the basal long-petiolate; inflorescence densely flowered, spicate; bracts 1–2 cm long, lanceolate, simple, the margins villous and entire to serrate; calyx 8–10 mm long, 5-lobed, unequal, villous and sometimes glandular; corolla 20–26 mm long, yellow to yellowish, the galea 8–10 mm long, erect, curved apically, not beaked; anthers 1.8–2.5 mm long; capsule 10–12 mm long, asymmetrical; n = 16. Lodgepole pine and spruce-fir communities at 2745 to 3390 m in Daggett, Duchesn, Grand, San Juan, Summit, Uintah, Wasatch, and Uintah counties; Colorado, Wyoming, Idaho, and Montana; 38 (vi). Our materials belong to **var. *paysoniana*** (Pennell) Cronq. [*P. paysoniana* Pennell].

Pedicularis centranthera Gray Pinyon-juniper Lousewort. Stems 4–7 cm tall from a tuberous root, shorter than the leaves, glabrous; leaves 6–15 cm long, petiolate, pinnatifid, the segments double crenate to dentate; inflorescence a short spicate raceme; lower bracts foliose, reduced upward; pedicels 1–4 mm long; calyx 15–20 mm long, the lobes 5, 6–9 mm long, ciliate; corolla 30–35 mm long, purple or yellowish, glabrous; capsule 10–13 mm long, ovoid. Pinyon-juniper, juniper, mountain brush, and ponderosa pine communities at 1645 to 2745 m in Beaver, Box Elder, Carbon, Duchesne, Emery, Garfield, Grand, Iron, Kane, Millard, San Juan, Sevier, Uintah, Washington, and Wayne counties; New Mexico, Arizona, Colorado, Wyoming, Nevada, and Oregon; 45 (x).

Pedicularis contorta Benth. Stems mainly 1.5–4.5 dm tall, from a caudex, glabrous; leaves mostly basal, 6–16 cm long, pinnatifid, the segments narrow, serrate, the cauline ones reduced upward; inflorescence a spicate raceme, mainly 5–20 cm long, bracts narrow, pinnately or subpalmately lobed; calyx 7–9 mm long, the lobes 1.5–3 mm long, ciliate; corolla 10–13 mm long, white or pale yellow, often purple maculate, the tube curved; galea 4–5.5 mm long, prolonged into a beak 6–8 mm long, the lower lip 5–6 mm long, the lateral lobes rounded, often enfolding the galeate beak; anthers 2–3 mm long; capsule ca 10 mm long. Sagebrush, limber pine, and other alpine communities at ca 2867 m in the Raft River Mts., Box Elder County; British Columbia to Alberta, south to California and Wyoming; 3 (ii).

Pedicularis groenlandica Retz. Elephant-head. Stems 1.5–7 dm tall, mostly clustered, glabrous; leaves mostly basal, 5–25 cm long, pinnatifid, the segments narrow, dentate to crenate, petiolate, the cauline leaves smaller, reduced upward, petioled to sessile; inflorescence spicate, quite dense, glabrous; bracts mostly shorter than the flowers, the lower foliose, sometimes cleft, the upper smaller, cleft to entire; calyx 5–7 mm long, 5-lobed, the lobes ca 1 mm long, entire, subequal, broadly triangular, the veins prominent, ciliate; corolla 10–15 mm long, violet to purple, strongly hooded, prolonged into an upturned beak, the lobes deflexed, spreading, the whole flower resembling the head and trunk of an elephant; capsule 7–9 mm long, asymmetrical. Lodgepole pine, spruce-fir, alpine tundra, and sedge-grass communities, typically where wet, at 2255 to 3815 m in Beaver, Cache, Daggett, Duchesne, Garfield, Iron, Kane, Piute, Rich, Salt Lake, Sanpete, Sevier, Summit, Uintah, Utah, and Wasatch counties; boreal North America south to New Mexico and California; 78 (vii).

Pedicularis parryi Gray Parry Lousewort. Stems numerous from a thickened fibrous root, 0.5–3 dm tall, glabrous below the inflorescence; leaves mostly basal, pinnatifid, the segments toothed or incised, peteiolate, the cauline few, reduced; inflorescence spicate-racemose, 5–25 cm long; bracts often trifid to subpinnately cleft, glabrous to villous; calyx 7–10 mm long, 5-lobed, unequally cleft, the lower deepest cut; corolla pink to pale yellow, 16–22 mm long, the galea curved downward at tip, 6–10 mm long, the beak straight; capsule 10–14 mm long, asymetrical. There are two geographically correlated varieties in Utah.

1. Corolla rose or pink; plants of Cache County .... *P. parryi* var. *purpurea*
— Corolla pale yellow (rarely white), northeastern to southern Utah .... *P. parryi* var. *parryi*

Var. *parryi* Sagebrush, spruce-fir, lodgepole pine, sedge-forb, and alpine tundra communities at 2315 to 3815 m in Beaver, Daggett, Duchesne, Emery, Garfield, Iron, Piute, Sanpete, Sevier, Summit, and Wasatch counties; Colorado, Arizona, and New Mexico; 48 (v).

Var. *purpurea* Parry Open slopes and meadows from moderate to high elevations in Cache County; Montana and Idaho; 1 (0).

Pedicularis procera Gray Gray Lousewort. [*P. grayi* A. Nels.]. Stout perennial, 5–11 dm tall, pubescent above and in the inflorescence; leaves basal and cauline, glabrous, pinnatifid with irregularly serrate segments, the basal long-petiolate, the stem leaves short-petiolate; inflorescence spicate, 15–35 cm long, villous to shortly villous; bracts linear, entire to toothed, the lower ones longer than the flowers; calyx narrowly linear, 1–1.5 cm long, 5-lobed; corolla 2.5–3.4 cm long, pale yellow, sometimes streaked with red, galea 9–16 mm long, beakless but with 2 lateral teeth below apex, the lower lip 7–12 mm long; capsule 10–16 mm long, ovoid; n = 16. Aspen, meadow, and spruce-fir communities at 2680 to 3295 m in Grand, San Juan, and Sevier counties; Wyoming, Colorado, New Mexico, and Arizona; 4 (ii).

Pedicularis racemosa Dougl. ex Benth. Leafy Lousewort. Stems mostly clustered, from a woody caudex, 1.5–5 dm tall, glabrous or nearly so; leaves mostly cauline, 4–7 cm long, the lowermost much reduced, lanceolate, simple, serrate to crenate, short-petiolate; inflorescence a few-flowered, short, spicate raceme or some flowers in the upper leaf axils; bracts foliose below, reduced upwards; calyx 2–lobed, these obliquely ovate, the tip acuminate; corolla white, pale yellow, or purple, 9–15 mm long, the galea 5–8 mm long, strongly arched and tapering into a slender decurved beak, the beak 5–7 mm long, the lower lip prominent, deflexed-spreading; capsule 10–16 mm long, asymmetrical; n = 8. Lodgepole pine and spruce-fir communties at 2445 to 3295 m in Cache, Daggett, Duchesne, Grand, Piute, Rich, Salt Lake, San Juan, Sanpete, Summit, Uintah, and Wasatch counties; Alberta to British Columbia, south to New Mexico, and California; 43 (v). Our plants belong to var. *alba* (**Pennell**) Cronq. [*P. racemosa* ssp. *alba* Pennell].

Penstemon Mitch.

Contributed by
Elizabeth C. Neese

Glabrous, pubescent, or glandular-pubescent perennial herbs or subshrubs; leaves basal and opposite (or sometimes all cauline), simple, entire or toothed, the lower petioled, becoming sessile upward, reduced and more or less bractlike in the inflorescence; inflorescence of verticillate axillary cymes, these sometimes (rarely) reduced to a single flower; calyx 5–cleft, the lobes often scarious-margined; corolla tubular to funnelform or ventricose-ampliate, bilabiate, rarely obscurely so; fertile stamens 4, didynamous, glabrous to pubescent; a fifth sterile filament (staminode) present, glabrous or variously bearded; stigma globose; capsule 2–valved, septicidal; seeds few to numerous. Note: Most of our taxa in *Penstemon* are distinguished by small but well-marked differences. Measurements given in the following keys and descriptions are of well pressed, fully open flowers, of calyces at anthesis (not in fruit), and of anther sacs following natural anthesis, not immature ones that have opened during drying. Presence or absence of anther pubescence as noted is for the surface only (suture margins are usually more or less spiculate or minutely stiff-hairy). Color is given for the usual color range but occasional pink or white individuals occur in most species. *Penstemon* is a large and complex genus. Critical garden tests over many years by Dr. Glen Moore and others have demonstrated an absence of barriers to hybridization among even distantly related taxa. Nonetheless, although often subtly differentiated morphologically, most species are sharply separated geographically and edaphically, and hybrids in nature appear to be uncommon.

Holmgren, N. H. 1984. *Penstemon*. pp. 370–455. *In*: Cronquist, A. et al. Intermountain Flora. Vol. 4. New York Botanical Garden, New York.

1. Anther sacs opening only across the top, the distal free ends indehiscent, remaining saccate, the anthers permanently horseshoe-shaped .... Key 1
— Anthers sacs dehiscing throughout their length or, if partially, then dehiscing from the distal ends and remaining indehiscent at the apex, if horseshoe-shaped the distal portion never remaining closed and saccate ... 2
2(1). Anther sacs opening partially or, if the full length, at least the connective indehiscent, the valves hardly parting .... Key 2
— Anther sacs opening throughout and across the connective, the valves often widely spreading 3
3(2). Plants caespitose, forming mats, the stems creeping or, if flowering stems erect, these arising from older, rooting, prostrate stems; leaves always entire generally none over 4 cm long or 5 mm wide .... Key 3
— Plants neither caespitose nor mat-forming; leaves entire or toothed, the larger regularly either over 4 cm long or 5 mm wide 4
4(3). Inflorescence axis and corolla both regularly pubescent or glandular externally .... Key 4
— Inflorescence axis or corolla or both glabrous externally (rarely a few long hairs may be present on the corolla lobes in *P. rydbergii*) .... Key 5

Key 1.

Anther sacs opening partially, across the apex.

1. Corolla red .... *P. rostriflorus*
— Corolla some shade of blue, lavender, or pink 2
2(1). Leaves coarsely and sharply serrate; summits of filaments and anther bases villous with long white hairs .
.... *P. venustus*
— Leaves entire; filaments and anthers glabrous 3
3(2). Calyx 2.1–3.2 mm long, the lobes broadly ovate or obovate, obtuse to acute .... *P. sepalulus*
— Calyx 3.4–7 mm long, the lobes lance-ovate to lanceolate, acuminate to attenuate 4
4(3). Flowers mostly 14–20 mm long, the corolla blue with lavender throat, or violet to lavender (Washington County); anther sacs 1–1.4 mm long .... *P. leonardii*
— Flowers mostly 20–30 mm long, lavender; anther sacs 1.3–2.5 mm long 5
5(4). Middle stem leaves mostly greater than 8 mm wide, elliptic, sharply acute, noticeably firm and plane; plants mostly over 3 dm tall; plants of Davis, Duchesne, Salt Lake, and Weber counties .... *P. platyphyllus*
— Middle stem leaves mostly less than 7 mm wide, (ob)lanceolate to linear, many of them obtuse; plants mostly less than 3 dm tall; plants of Juab, Millard, and Tooele counties .... *P. leonardii*

Key 2.

Anther sacs opening partially, from the distal end but not across the connective.

1. Corolla red 2
— Corolla not red, some shade of blue, lavender, or pink
.... 3
2(1). Corolla strongly 2–lipped, the lower lip reflexed ..
.... *P. barbatus*
— Corolla not or obscurely 2–lipped, nearly tubular, the lobes more or less equally projecting .... *P. eatonii*
3(1). Anthers glabrous on the surface (margin of suture may be scabrous or ciliate-toothed) 4
— Anthers pubescent on the surface, sometimes sparsely so 13
4(3). Stems often fistulose; herbage and inflorescence papillate-glutinous, the surfaces obscured by adhering sand grains; leaves (most) prominently undulate or crisped; staminode pale, dotted with dark purplish blue papillae on upper surface; plants of blow sand in Garfield, Kane, and Washington counties ....
.... *P. ammophilus*
— Stems never fistulose; herbage neither glutinous nor coated with tightly packed sand grains; leaves rarely crisped; staminode glabrous or variously pubescent but not as above 5
5(4). Herbage below the inflorescence glabrous 6
— Herbage below the inflorescence pubescent 9

6(5). Calyx and corolla glandular-puberulent externally; plants of southern Utah .... *P. leiophyllus*
— Calyx and corolla glabrous externally, if (rarely) obscurely glandular the plants from northwestern Utah 7

7(6). Corolla 18–23 mm long; anther sacs to 1.8 mm long; plants of Navajo Mt., San Juan County ..... *P. navajoa*
— Corolla 23–35 mm long; anther sacs longer than 1.8 mm; plants of southwestern or northwestern Utah ... 8

8(7). Staminode glabrous; plants of the Deep Creek and Raft River ranges in northwestern Utah ... *P. speciosus*
— Staminode yellow-bearded; plants of southwestern Utah .... *P. laevis*

9(5). Cauline leaves linear or nearly so; anthers to 1.5 mm long; plants of Garfield and Piute counties 10
— Cauline leaves seldom linear; anthers 1.5–3.3 mm long (rarely shorter); plants not from Garfield and Piute counties 11

10(9). Calyx and corolla glandular-puberulent ..... *P. parvus*
— Calyx and corolla glabrous .... *P. pseudoputus*

11(9). Corolla 16–21 mm long; anthers 1.5–1.8 mm long . ... *P. tidestromii*
— Corolla 23–33 mm long; anthers 1.8–2.4 mm long .. 12

12(11). Leaves, stems and inflorescence branches and bracts densely cinereous puberulent; corolla glandular-puberulent; plants of Sanpete and Sevier counties ... . *P. wardii*
— Herbage glabrous to sparingly and minutely puberulent, not cinereous; corolla glabrous; plants of the Deep Creek and Raft River ranges in western Utah (*P. x jonesii* of Kane, and Washington counties may key here — see discussion under *P. laevis*) .... *P. speciosus*

13(3). Anthers pubescent with long tangled hairs as long or longer than the length of the sac, more or less woolly 14
— Anthers pubescent with straight or flexuous hairs shorter than the length of the sac, hardly woolly 15

14(13). Corolla pale blue or pale lavender blue; inflorescence relatively broad, the lower peduncles often exceeding 2 cm and somewhat spreading, the inflorescence thus not markedly secund; anther-sacs usually densely woolly with long tangled hairs that obscure the surface .... *P. comarrhenus*
— Corolla blue to dark blue; inflorescence narrow, the peduncles mostly shorter than 1.5 cm, erect to appressed, the inflorescence strongly secund; anther-sacs often more sparingly woolly, the surface scarcely obscured .... *P. strictus*

15(13). Anthers pubescent with slender, flexuous hairs about as long or longer than the width of the sac 16
— Anthers pubescent with relatively stiff, straight hairs shorter than the width of the sac 19

16(15). Calyx 2.5–6 mm long, the lobes ovate, obtuse to acute or sometimes shortly acuminate 17
— Calyx 5–11 mm long, the lobes lanceolate, acuminate to attenuate 18

17(16). Staminode bearded .... *P. cyanocaulis*
— Staminode glabrous .... *P. navajoa*

18(16). Inflorescence glabrous; calyx lobes narrowly scarious-margined; plants from southeastern Utah .... *P. strictiformis*
— Inflorescence glandular-puberulent; calyx lobes with prominent, broad, usually erose, scarious margins; plants from central and northeastern Utah . *P. scariosus*

19(15). Leaves, stem, and inflorescence densely cinereus-puberulent throughout; plants of the Uinta Basin . . .. *P. fremontii*
— Herbage glabrous or nearly so, green or, if (rarely) cinereus-puberulent, plants not of the Uinta Basin .. 20

20(19). Leaf margins (at least some) undulate to crisped; inflorescence glabrous; plants of southeastern Utah .. .. *P. cyanocaulis*
— Leaf margins not noticeably crisped; inflorescence glabrous or pubescent; plants not of southeastern Utah 21

21(20). Pedicels, calyx, and corolla glabrous (calyx rarely glandular); anther sacs diverging at about 90–120 degrees, dehiscing in the distal 2/3–3/4, the line of dehiscence somewhat lateral; inflorescence cylindrical to secund .... *P. cyananthus*
— Pedicels, calyx, and corolla finely glandular (sometimes obscurely so—then best seen on backs of corolla lobes of unopened flowers); anther sacs diverging at nearly 180 degrees, dehiscing fully (but not across the connective), the line of dehiscence median; inflorescence secund 22

22(21). Plants mostly less than 15 (rarely to 18) cm tall; verticellasters 1–4; plants from near and above timberline, of the central and eastern crest of the Uinta Mts. .... *P. uintahensis*
— Plants mostly much taller; verticellasters 4–11; plants seldom if ever above treeline, of wider distribution .... *P. subglaber*

Key 3.

Anthers sacs opening throughout; plants caespitose.

1. Plants forming compact mats not over 3 cm tall, the leaves densely clustered in evidently acaulescent tufts; plants of Daggett County .... *P. acaulis*
— Plants mat-forming or not; plants caulescent, the leaves distributed along an evident stem; distribution various . 2

2(1). Leaves linear, at least 5 times longer than wide; plants not from Grand or San Juan County 3
— Leaves broader, narrowly lanceolate to obovate or spatulate, the broader ones less than 5 times longer than wide (or, if sometimes narrower, plants from Grand or San Juan County) 5

3(2). Flowering stems prostrate .... *P. caespitosus*
— Flowering stems ascending to erect 4

4(3). Staminode densely bearded with a tuft of long yellow hairs at the apex, only sparsely bearded proximally; anthers bicolorous (purplish black, with a white band along the suture); plants of southwestern Utah .... *P. linarioides*
— Staminode rather equitably and densely bearded most of its length with golden orange hairs; anthers rather uniformly purplish black, without a pronounced white band along the suture; plants of Sevier and Utah counties .... *P. abietinus*

5(2). Pubescence of leaves of flattened, appressed, scalelike hairs 6
— Pubescence of leaves of terete, more or less spreading hairs 7

6(5). Stems ascending to erect, the plants rather shrubby; leaves moderately pubescent, the surface not or scarcely obscured, often glabrate below; plants from above 2590 m in Beaver (Tushar Mtns.), Iron, and Piute counties .... *P. caespitosus* var. *suffruticosus*

— Stems prostrate or very short and forming tufts, scarcely shrubby; leaves densely cinereous-pubescent on both surfaces, the surface largely obscured; plants from below 2165 m in Beaver, Iron, and Kane counties .... *P. thompsoniae*

7(5). Leaves (the larger) regularly over 1.5 cm long, glabrous to sparsely scaberulous-puberulent; Grand and San Juan counties .... *P. crandallii*

— Leaves 0.4–1 (2) cm long, densely cinereous-puberulent with retrorsely curved hairs; distribution various, but not from Grand or San Juan counties .. *P. caespitosus*

Key 4.

Anther sacs opening thoughout; plants not mat-forming; both inflorescence and corolla pubescent or glandular.

1. Leaves prominently and regularly toothed (rarely entire in *P. palmeri* but the upper leaves connate-perfoliate) 2

— Leaves (most) entire or undulate, rarely some with a few teeth; upper leaves never connate-perfoliate ... 10

2(1). Corolla white, sometimes slightly brown- or lilac-tinged; plants of western Box Elder County . *P. deustus*

— Corolla pink, lavender, blue, or purple; plants with distribution various but not of Box Elder County 3

3(2). Small subshrubs of limestone crevices or higher elevation talus; leaves all cauline 4

— Plants herbaceous, of various habitats but not restricted as above; basal leaves present 5

4(3). Corolla 24–40 mm long; anthers densely woolly with long white hairs; plants known in Utah from the Wasatch plateaus and mts. .... *P. montanus*

— Corolla 13–17 mm long; anthers essentially glabrous; plants known in Utah from southwestern Washington County .... *P. petiolatus*

5(3). Corolla 20–40 mm long 6

— Corolla 8–20 mm long 7

6(5). Upper cauline leaves connate-perfoliate; corolla (25) 27–40 mm long, strongly inflated, the throat when pressed over 11 mm wide, the lower lip reflexed; plants of low to middle elevations .... *P. palmeri*

— Upper cauline leaves not perfoliate; corolla 20–27 (30) mm long, the throat when pressed less than 11 mm wide, the lower lip projecting; plants montane to alpine .... *P. whippleanus*

7(5). Calyx 3–6 mm long; anther sacs 0.4–0.8 mm long, about as broad as long .... *P. humilis*

— Calyx or anther sacs regularly longer 8

8(7). Leaves densely cinereous-puberulent throughout . ... *P. nanus*

— Leaves, at least the lower, glabrous or essentially so .. 9

9(8). Anthers becoming peltate-explanate, the sacs less than 0.9 mm long, about as broad as long; plants of Beaver and Millard counties .... *P. concinnus*

— Anther sacs not explanate, not divaricate but remaining parallel, over 0.9 mm long, longer than broad; plants of Iron County .... *P. pinorum*

10(1). Herbage and inflorescence glutinous, the surface obscured with adhering sand grains, otherwise glabrous; leaves (most) prominently undulate or crisped; stems often fistulose; plants of sand dunes in Garfield, Kane, and Washington counties .... *P. ammophilus*

— Herbage and inflorescence not glutinous, not with obscuring sand grains; plants otherwise differing ... 11

11(10). Leaves glabrous or essentially so (petioles of lower leaves may be sparingly puberulent) 12

— Leaves pubescent 22

12(11). Stamens very short, scarcely if at all exserted from the tube; corolla funnelform, not bilabiate, the throat flaring gradually, the throat and limb scarcely distinguishable 13

— Stamens well-exserted from the tube, the longer pair nearly reaching or exceeding the orifice; corolla bilabiate, the throat and limb usually abruptly differentiated 14

13(12). Stems glabrous below the inflorescence; staminode densely bearded with golden orange hairs, lighter yellow at the tip; plants of Duchesne and Uintah counties .... *P. goodrichii*

— Stems puberulent; staminode more sparsely bearded, the hairs all about the same color; plants of Carbon and Emery counties .... *P. marcusii*

14(12). Corollas mostly over 20 mm long 15

— Corollas mostly less than 20 mm long 16

15(14). Plants 2–8 dm tall, of montane to alpine communities; corolla not over 30 mm long ...... *P. whippleanus*

— Plants not over 2 dm tall, of shale barrens in the pinyon-juniper zone; corolla 30–38 mm long .... *P. grahamii*

16(14). Calyx mostly less than 5 mm long 17

— Calyx mostly over 5 mm long 18

17(16). Inflorescence secund; anthers dehiscing nearly or quite to the connective but the valves hardly parting, the sacs over 1 mm long, longer than broad; plants of Aquarius Plateau, Garfield and Wayne counties ... . *P. parvus*

— Inflorescence not secund; anthers dehiscing the full length and across the connective, the valves spreading, the sacs less than 1 mm long, about as broad; plants widespread .... *P. humilis*

18(16). Corolla less than 14 mm long; plants not of Grand or San Juan counties 19

— Corolla over 14 mm long or, if shorter, the plants of Grand and San Juan counties 20

19(18). Stem puberulent; anthers becoming peltate-explanate, the sacs broadly ovate; plants of Beaver, Juab, and Millard counties .... *P. concinnus*

— Stem mainly glabrous; anthers not becoming explanate, the sacs oblong; plants of Garfield and Kane counties .... *P. atwoodii*

20(18). Anther sacs oblong, scarcely explanate; throat generally purple-veined on the palate only; corolla lobes about equally spreading; staminode included .... *P. moffattii*

— Anther sacs broadly ovate, becoming peltate-explanate; throat longitudinally purple-veined nearly throughout 21

21(20). Staminode exserted; corolla ventricose-ampliate, upper lobes projecting-spreading, the lower ones spreading-reflexed; plants of wide distribution in southeastern Utah .... *P. ophianthus*

| | Staminode included or merely showing at the orifice; corolla tubular funnelform to somewhat ampliate, the lobes about equally spreading; plants uncommon, of San Juan County .... *P. breviculus* |

22(11). Staminode prominently exserted; corolla ventricose-ampliate, the orifice constricted and prominently bearded with long white hairs; stems mostly less than 2 dm tall; plants of Daggett and Summit counties .. .. *P. cleburnei*

— Staminode, corolla, and stems not as above in all respects 23

23(22). Calyx 3–6 mm long; anther sacs 0.4–0.8 mm long, about as broad as long .... *P. humilis*

— Calyx or anther sacs regularly longer 24

24(23). Plants dwarf, the stems usually shorter than 1.5 dm; stems and leaves densely cinereus-puberulent 25

— Stems mostly taller; herbage more sparingly puberulent, scarcely cinereus 28

25(24). Anthers permanently horseshoe-shaped to sagittate, the sacs remaining parallel; inflorescence densely glandular-puberulent with spreading hairs; plants of western Utah .... *P. nanus*

— Anther sacs becoming divaricate; inflorescence densely cinerous-puberulent or, if glandular-pubescent with spreading hairs, not of western Utah 26

26(25). Inflorescence glandular-pubescent with spreading hairs .... *P. moffattii*

— Inflorescence densely cinereus-puberulent with retrorse hairs, scarcely if at all glandular 27

27(26). Corolla lobes orbicular, overlapping at the base, widest near the middle, 4–7 mm wide when pressed; staminode densely and rather evenly orange-bearded in the distal half, gradually widening to a moderately dilated apex less than 1 mm wide, the tip straight or slightly curved; plants of Duchesne County .... *P. duchesnensis*

— Corolla lobes mostly oblong, widest at the base, not overlapping, 3–5 mm wide when pressed; staminode abruptly dilated apically, the tip over 1 mm wide, strongly curved to curled, the expanded tip bearded with a tuft of golden yellow hairs, more sparsely bearded proximally; plants of west-central Utah ... . *P. dolius*

28(24). Leaves principally cauline, the basal ones absent or poorly developed; corolla lobes blue, the throat blue to blue violet dorsally, whitish and 2-ridged ventrally, the lower lip projecting, exceeding the upper; plants of the northern tier of counties ..... *P. radicosus*

— Basal leaves well developed; corolla throat not prominently ridged, not white ventrally, the lobes about equally spreading; plants of central and southeastern Utah 29

29(28). Anther sacs oblong, scarcely explanate; throat generally purple-veined on the palate only, the lobes about equally spreading; staminode mostly included .... *P. moffattii*

— Anther sacs broadly ovate, becoming peltate-explanate; throat longitudinally purple-veined nearly throughout 30

30(29). Staminode included or merely showing at the orifice; corolla tubular-funnelform to moderately ampliate, the lobes about equally spreading .... *P. breviculus*

— Staminode usually conspicuously exserted; corolla ventricose-ampliate, the upper lobes projecting-spreading, the lower spreading-reflexed .... *P. ophianthus*

Key 5.
Anther sacs opening throughout;
Inflorescence or corolla or both glabrous externally.

1. Leaves green, not particularly glaucous, thin 2

— Leaves strongly glaucous, thick and leathery, somewhat fleshy when fresh 4

2(1). Corolla 7–12 mm long, the throat nearly tubular, mostly 2–3 mm wide when pressed; anther sacs 0.4–0.7 mm long, suborbicular to ovate, about as broad as long, becoming explanate; inflorescence interrupted, strongly fasciated, the fascicle(s) densely flowered, globular, mostly 2–3 cm in diameter .... *P. procerus*

— Corolla 10–20 mm long, the throat somewhat ampliate, mostly 3–5 mm wide when pressed; anther sacs 0.6–1.1 mm long, ovate to oblong, longer than wide, not fully explanate; inflorescence often less prominently fasciated, the fascicle(s) less densely flowered, mostly 3–4 cm in diameter 3

3(2). Calyx 3–9 mm long, the lobes long-acuminate to caudate, the broadly scarious margins usually prominently erose to lacerate; basal tufts of leafy sterile shoots often present .... *P. rydbergii*

— Calyx mostly 2–3.5 mm long, the lobes triangular to ovate, obtuse, acute, or abruptly short-acuminate, the margins scarious but seldom prominently erose; leaves predominantly cauline .... *P. watsonii*

4(1). Leaves toothed; corolla 25–40 mm long, the throat markedly inflated, when pressed 12–20 mm in diameter .... *P. palmeri*

— Leaves entire; corolla smaller, less than 25 mm long, the throat to 10 mm wide 5

5(4). Corolla densely glandular internally near the orifice, not bearded on the palate; staminode glabrous or merely papillate, only slightly dilated apically; leaves lanceolate, usually narrowly so 6

— Corolla not evidently glandular within, the palate often with at least a few long hairs; staminode bearded (rarely papillate-pubescent); leaves various 7

6(5). Corolla red to carmine pink, retaining the red coloration on drying; guidelines usually absent; calyx mostly 2.5–4.5 mm long; plants of Washington County and east of the central Utah plateaus .... *P. utahensis*

— Corolla rose pink, lavender pink or purplish red, drying pink violet to bluish; guidelines often present (may be faint); calyx mostly 4–6 mm long; plants of Washington County and west of the Utah Plateaus . ... *P. confusus*

7(5). Plants from Daggett, Duchesne, or Uintah counties .. 8

— Plants not from Daggett, Duchesne, or Uintah counties 12

8(7). Leaves linear to narrowly (ob)lanceolate throughout, to 12 mm wide; lower bracts of the inflorescence with broad clasping bases, abruptly narrowed to a caudate or acuminate tip; plants of sand dunes or loose sand; plants of Daggett and the northern half of Uintah counties .... *P. angustifolius* var. *vernalensis*

— Leaves broader, broadly lanceolate to spatulate, the basal ones when present with broadly expanded blades; plants seldom of especially sandy places 9

9(8). Basal leaves absent or poorly developed; flowers true soft-pink, retaining the pink color on drying; staminode bearding of papillate hairs to 0.1 mm long; plants endemic to the Roosevelt/Myton area on either side of the Duchesne/Uintah county line .. *P. flowersii*

— Basal leaves well developed; flowers lavender pink to blue or blue purple, usually drying violet to bluish; staminode with hairs 0.2–2 mm long 10

10(9). Staminode densely bearded with long, tangled hairs that obscure the apex, these 1–2 mm long ....
.... *P. pachyphyllus*

— Staminode bearded with hairs 0.2–1 mm long 11

11(10). Corolla glabrous externally; the throat summit and lobes prominently veined within on all sides with wine red guidelines; plants widespread in the Uinta Basin; staminode hairs mostly 0.5–1 mm long .....
.... *P. pachyphyllus*

— Corolla finely glandular-puberulent externally, the throat and lobes without prominant guidelines within; staminode hairs 0.2–0.5 mm long; plants restricted in the Uinta Basin to the Tavaputs Plateau .
.... *P. carnosus*

12(7). Plants dwarf; stems 3–10 (15) cm tall from a rhizomatous caudex; plants of multi-hued limestone barrens in western Garfield County .... *P. bracteatus*

— Plants larger; distribution various 13

13(12). Staminode densely bearded with long, tangled hairs 1–2 mm long; corolla usually blue violet to blue purple .... *P. pachyphyllus*

— Staminode bearded with more or less straight hairs less than 1 mm long; corolla usually lavender pink to pale blue violet or blue 14

14(13). Flowers pink to lavender pink; plants of Millard, Juab, and Tooele counties 15

— Flowers lavender to violet blue, or sometimes pinkish lavender; distribution otherwise 16

15(14). Basal leaves poorly developed; lower leaves linear-lanceolate to lanceolate, rarely over 1 cm wide, the blade gradually narrowed to the petiole; plants of sand dunes in eastern Juab and Millard counties ...
.... *P. angustifolius* var. *dulcis*

— Basal leaves well developed; lower leaves ovate to broadly oblanceolate, the larger well over 1 cm wide, the blade abruptly narrowed to the petiole; plants usually growing in clay or silty soils; known from near the Nevada border in Juab, Millard, and Tooele counties .... *P. immanifestus*

16(14). Leaves principally cauline, the lower ones sessile or gradually narrowed to the petiole, lanceolate, narrower than the upper, in the inflorescence becoming broad-based, clasping, and abruptly narrowed to an acuminate, and often conduplicate and recurved tip
.... *P. angustifolius*

— Basal leaves well developed, spatulate, ovate or broadly oblanceolate, the blade abruptly narrowed to the petiole; upper leaves and bracts broad-based and clasping, but usually more gradually narrowed to the obtuse, acute, or mucronate tip 17

17(16). Corolla often finely glandular-puberulent externally; stems mostly 2–3.5 dm tall, the inflorescences compact or sometimes interrupted, cylindrical; peduncles usually less than 1 cm long, the cymes densely-flowered, congested; plants not of San Juan County
.... *P. carnosus*

— Corolla glabrous; stems mostly 3–7 dm tall, the inflorescence more or less elongate, interrupted, secund; peduncles (at least the lower) usually 1–2 cm long, evident, the cymes few-flowered; plants of San Juan County .... *P. lentus*

Penstemon abietinus Pennell Firleaf Penstemon. Low suffrutescent mat-forming perennial; old stems prostrate to decumbent, rooting at the nodes; flowering stems 0.5–1.6 (2.3) dm tall, puberulent, ascending to erect, leafy; leaves 0.5–2 (2.5) cm long, 0.4–2 mm wide, linear, acute, entire, often involute, glabrous or sparsely puberulent near the base, scarcely reduced in the inflorescence; inflorescence secund, the cymes usually only 1 per node, 1– to 2 (3) -flowered, the peduncles and pedicels puberulent and with a few gland-tipped hairs, the pedicels bracteate; calyx 4–7 mm long, glandular-pubescent, the lobes abruptly caudate above their broad, scarious-margined bases; corolla glandular-pubescent externally, 12–18 mm long, the tube purple, the throat lavender blue, bulged above, flattened and longitudinally 2–ridged and lighter in color below, the lobes darker blue, projecting-spreading, the palate bearded with pale yellow hairs; staminode included, densely bearded for most of its length with golden orange hairs; stamens ca reaching the orifice or the longer pair exserted, the anthers purplish black, the sacs 0.8–1.8 mm long. Pinyon-junipr, oak, and sagebrush communities at 1750 to 2300 m in Emery, Sevier (type from Salina Canyon), and Utah counties; 16 (ii). This taxon is known from Salina Canyon and the Fishlake Plateau, and from a disjunction in Spanish Fork Canyon (Markham 8958 BRY). Perhaps it is better treated at varietal level within the widespread *P. linarioides*, but no new combination is here proposed.

Penstemon acaulis L. O. Williams Stemless Penstemon. [*P. yampaensis* Pennell]. Diminutive mat-forming, long-lived perennial from a multicipital, fibrous-rooted caudex; plants evidently acaulescent, the old stems rooting, more or less stoloniferous, buried in duff and soil, and clothed with the remnants of previous leaves and flowers; current years growth tufted, the leaves 0.3–2.8 cm long, 0.3–4 mm wide, oblanceolate to linear, entire, obtuse to acute, papillate-scabrous, viscid when fresh; flowers usually solitary, sessile, borne among the leaves; calyx 5–8 mm long, densely glandular-puberulent, the lobes lance-linear, narrowly acuminate; corolla 12–15 mm long, glandular-pubescent, ampliate, moderately bilabiate, the lobes about equally spreading, blue, the palate paler and sparsely bearded, the throat rounded below; staminode reaching the orifice, densely orange-bearded; anthers blue black, glabrous, included to barely reaching the orifice, the sacs 0.7–0.9 mm long; capsules persistent on the old stems, indehiscent, retaining the seeds for several years, the seeds apparently released by eventual decomposition of the capsule. *Penstemon acaulis* is an unusual penstemon of unparalleled dwarf habit. Plants at the eastern end of the distribution have slightly larger leaves and flowers and have been recognized as *P. yampaensis* Penl., but specimens from Brown's Park, Utah, near the Colorado border are geographically and morphologically transitional. The variation seems best recognized at the varietal level as follows:

1. Leaves linear, less than 1.6 mm wide and 2 cm long; plants of Brown's Park and west to Manila and adjacent Sweetwater County, Wyoming ..... *P. acaulis* var. *acaulis*

— Leaves, at least some, larger, more broadly linear to narrowly oblanceolate; plants of Brown's Park, Utah and east to near Greystone, Moffat County, Colorado ....
.... *P. acaulis* var. *yampaensis*

Var. acaulis Semibarren substrates in pinyon-juniper and sagebrush-grass communities at 1790 to 2000 m in Daggett County; Sweetwater County, Wyoming and Moffat County, Colorado; 20 (vi).

Var. yampaensis (Penl.) Neese [*P. yampaensis* Penl.]. Semibarren, pale substrates in pinyon-juniper and sagebrush-grass communities at ca 1780 m in Daggett County; Moffatt County, Colorado; 2 (ii).

Penstemon ambiguus Torr. Bush Penstemon; Gilia Penstemon. [*Leiostemon purpureum* Raf.; *L. ambiguus* (Torr.) Greene]. Woody-based, freely branched, bushy perennial 2.5–6 (8) dm tall; stems glabrous, branching candelabralike, the branches slender, divaricate near their bases, otherwise ascending to erect, the lower relatively long, the upper shorter, the flowers thus clustered toward the summit of the plant; leaves filiform, 0.4–3.8 (5) cm long, 0.2–1.7 mm wide, glabrous, entire, acute, involute, appearing terete; inflorescence of 2–11 verticellasters, the peduncles mostly 4–12 mm long, filiform, the cymes 1- to 3-flowered; calyx 2–3 (4) mm long, the lobes ovate, obtuse, acute or mucronate, broadly scarious-margined; corolla phloxlike, not 2-lipped, salverform, the tube and throat slightly arched, together 10–14 mm long, glabrous externally, pubescent internally toward the summit and at the orifice, this only 1–3 mm in diameter, the limb 10–16 cm wide, obliquely flat-spreading, glabrous, white or pale pinkish ventrally, pink to lavender pink dorsally; staminode included, the tip not expanded, glabrous or with a few hairs; anthers 5–8 mm long, included, glabrous, blackish, the sacs opening their full length and across the connective, finally explanate; capsule ca 6–7 mm long; 2n = 16. Common or locally abundant in sandy soil in creosote bush, sand sagebrush, mixed desert shrub, live oak, manzanita-ponderosa pine, and juniper-sagebrush communities at 760 to 1950 m in Emery, Garfield, Kane, San Juan, and Washington counties; Nevada and Arizona, east to Colorado, Kansas, Oklahoma, Texas, and Mexico; 34 (i). Utah specimens belong to **var. laevissimus (Keck) N. Holmgren** [*P. ambiguus* ssp. *laevissimus* Keck in Kearney & Peebles].

Penstemon ammophilus N. Holmgren & L. Shultz Sandloving Penstemon. Herbaceous perennial from a branched caudex; stems clumped, 5–32 cm tall or more, decumbent to erect, sometimes elongating and developing fibrous roots in response to burying, usually fistulose and swollen above and in the inflorescence (except in depauperate stems); herbage without trichomes, papillate-glutinous when young, the surface obscured by tightly packed adhering sand grains; leaves 1.5–7 cm long, the basal ones petiolate, the petiole equaling or exceeding the blade, this suborbicular to oblanceolate, rounded to obtuse, the cauline leaves oblanceolate to narrowly lanceolate, becoming sessile above, undulate to prominently crisped; inflorescence of (2) 4–12 bracteate verticellasters, these congested or the lower ones remote, the cymes 2- to 8-flowered, the flowers subsessile; calyx 4–8 mm long, the lobes narrowly lanceolate, attenuate; corolla 14–17 mm long, lavender blue, moderately ampliate, scarcely bilabiate, the lobes projecting to spreading, without trichomes, glutinous and with adhering sand grains dorsally near the base of the lobes; anther sacs ca 1 mm long, included, dehiscing from the tip to the connective; staminode reaching the orifice or slightly exserted, the tip not expanded, pale, the upper surface dotted with dark blue purple papillae. In blow sand derived from the Navajo Sandstone, where the long-lived clumps act as sand stabilizers, in ponderosa pine and mixed shrub communities at 1800 to 2200 m in Garfield, Kane, and Washington (type from Canaan Mt.) counties; endemic; 4 (i).

Penstemon angustifolius Pursh Narrowleaf Penstemon. Perennial herb 2–6 dm tall; herbage glabrous, glaucous; stems ascending to erect, more or less woody at the base, 1 or few from a thick caudex; leaves 2–9.5 cm long, 1.5–25 (35) mm wide, leathery, entire, the basal tuft usually poorly developed or lacking, the basal and lower cauline ones (ob)lanceolate (usually narrowly so), narrowed imperceptibly to a petiolar base, the middle and upper ones usually wider and sessile, in the inflorescence becoming caudate-acuminate with broad and clasping bases; inflorescence glabrous, interrupted, of (3) 7–17 verticellasters, the fascicles usually densely flowered, the peduncles and pedicels short, mostly obscured by bracts and flowers; calyx 4–7.5 mm long, glabrous, the lobes narrowly lanceolate, acuminate, scarious-margined; corolla 15–21 (23) mm long, moderately ampliate, glabrous externally and internally, pink, lavender, blue lavender, or blue, the lobes about equally spreading; anthers included, the sacs 0.9–1.4 mm long, glabrous, dehiscing the full length and across the connective, not explanate; staminode included, dilated and yellow-bearded distally with hairs 0.5–1 mm long. There are three geographically discontinuous varieties in Utah, all growing in sandy sites.

1. Corolla pink to rose, not becoming violet or bluish on drying; lower and middle cauline leaves mostly linear-lanceolate, seldom over 8 mm wide; plants of sand dunes in Juab and Millard counties .... *P. angustifolius* var. *dulcis*
— Corolla lavender to blue less commonly pink, usually drying somewhat bluish; leaves linear or not; plants of eastern and south-central Utah 2

2(1). Corolla blue; basal and lower cauline leaves usually linear to linear-lanceolate, seldom over 8 mm wide; plants of Daggett and Uintah counties .... *P. angustifolius* var. *vernalensis*
— Corolla lavender to blue violet or sometimes pink basal and lower cauline leaves usually broader, the larger usually over 8 mm wide; plants of south-central and southeastern Utah .... *P. angustifolius* var. *venosus*

Var. dulcis Neese Four-wing saltbush, sagebrush-eriogonum, and juniper communities at 1400 to 1650 m in Juab and Millard (type from the west base of the Canyon Mts.) counties; endemic; 9 (0).

Var. venosus (Keck) N. Holmgren [*P. angustifolius* ssp. *venosus* Keck in Kearney & Peebles; *P. venosus* (Keck) Reveal]. Blackbrush, sand sagebrush, rabbitbrush-Indian ricegrass-ephedra, and sagebrush-juniper communities at 1200 to 2200 m in Garfield, Kane, San Juan, and Wayne counties; Arizona and New Mexico; 26 (ii). This entity is, perhaps, not distinct from var. *caudatus* (Heller) Rydb., which occurs to the south and east of Utah.

Var. vernalensis N. Holmgren In ephedra-rabbitbrush and sagebrush-juniper communities at 1500 to 1800 m in Daggett and Uintah (type from 3 mi N of Maeser) counties; Colorado; 26 (xi). This is a well-marked taxon geographically isolated from other members of the *P. angustifolius* complex. It would not be inconsistent within *Penstemon* to treat it at specific rank. *Penstemon arenicola* A. Nels., a somewhat similar Wyoming species, has been reported from Daggett and Uintah counties and may occur here, but I have seen no specimens referrable to that taxon. The report may have been based on misinterpretation of depauperate specimens of *P. angustifolius* var. *vernalensis*.

Penstemon atwoodii Welsh Atwood Penstemon. Perennial herb 1.3–3.8 (5.2) dm tall; stems 1 to few, from a sparingly branched caudex and taproot, glabrous below, becoming glandular-pubescent near the inflorescence, ascending to erect; leaves 1.5–8 cm long, 2–12 mm wide, entire or irregularly few-toothed, the basal ones petioled, spatulate to narrowly oblanceolate, the cauline ones narrowly (ob)lanceolate, becoming sessile and clasping upward, glabrous; inflorescence densely glandular-pubescent throughout with spreading hairs, the thyrse narrow, interrupted, of 4–8 densely flowered verticellasters; calyx 5–9 mm long, densely glandular-pubescent, the lobes narrowly lanceolate, herbaceous, cyaneous; corolla 11–14 mm long, densely glandular-puberulent externally, only moderately ampliate, bilabiate, the lobes reflexed, the limb blue, the throat lavender with darker guidelines on the palate; anthers about reaching the orifice, glabrous, the sacs divaricate, dehiscing the full length and across the connective but not widely explanate; staminode exserted, golden yellow bearded. Sandy to clay soils, often on the "Blues" (gray clay badlands of the Kaiparowits Formation), pinyon-juniper woodland at 1650 to 2100 m in Garfield and Kane (type from the south end of Horse Mt.) counties; endemic; 12 (0).

Penstemon barbatus (Cav.) Roth. Beardlip Penstemon; Scarlet Penstemon. [*Chelone barbata* Cav.]. Perennial herb 3–11 dm tall; stems few from a stout, short-branched caudex, glabrous, ascending to erect, the internodes often remote; leaves 2–10 cm long, 1–20 mm wide, entire, glabrous (usually), glabrate, or the lower puberulent, the basal ones spatulate to broadly oblanceolate, petioled, the upper sublinear to filiform, sessile; inflorescence glabrous, secund, of 3–7 (12) verticellasters, these rather remote and the inflorescence thus wandlike, the pedicels slender, ascending, the cymes 1- to 2 (4) -flowered; calyx 3–5 (7) mm long, glabrous, the lobes ovate, obtuse to acute, sometimes shortly apiculate, mostly entire, more or less scarious-margined; corolla 25–35 mm long, scarlet, glabrous externally, sometimes long-pubescent on the palate, geniculate at the base, the flower thus oriented horizontally, markedly bilabiate, the upper lip projecting and ca 1 cm long, the lower lip reflexed; anthers long-exserted, the sacs glabrous (or villous in var. *trichander*), 1.5–2 mm long, widely spreading, dehiscing partially, the proximal portion near the connective remaining closed; staminode included, glabrous. Recognition as var. *torreyi* (Benth.) Gray has been accorded some of our material based on the presence of yellow palate pubescence; I can detect only the following infraspecific taxa:

1. Anthers villous; palate of the corolla glabrous ....
   .... *P. barbatus* var. *tricander*

— Anthers glabrous; palate glabrous to sparsely bearded ..
   .... *P. barbatus* var. *barbatus*

Var. *tricander* Gray Sandstone cliffs, near Church Rock, at ca 1800 m (Cottam & Hutchings 2331), also possibly in the Abajo Mts., San Juan County; Arizona, New Mexico; 1 (0). It has been postulated that the villous anthers are an expression of introgression from *P. strictus* Benth.

Var. *barbatus* Ponderosa woodland, often in sandy soil at 1700–2600 m in Garfield, Kane, and Wayne counties; Arizona, New Mexico, Texas, Mexico; 7 (0).

Penstemon bracteatus Keck Platy Penstemon. Dwarf perennial herb 0.3–1.2 (1.5) dm tall; herbage and inflorescence glabrous; stems 1 or few from a rhizomatous caudex, ascending to erect; leaves 0.3–4 cm long, 1.5–12 (15) mm wide, entire, thick, glaucous, spatulate to (ob)ovate or broadly (ob)lanceolate, the lower narrowed to a petiole, the upper sessile and clasping; inflorescence secund, of 1–4 (6) verticellasters, the cymes 1- to few-flowered; calyx 3–5 (6) mm long, the lobes ovate to broadly lanceolate, glabrous, entire, scarious-margined; corolla (12) 14–17 (19) mm long, moderately ampliate, blue to blue violet, the lobes spreading, glabrous externally, the palate sparsely white-bearded; stamens included, the sacs 0.8–1.2 mm long, glabrous, dehiscing the full length and across the connective, not explanate; staminode included or reaching the orifice, densely golden bearded in the distal third. Pink and White Limestone members of the Wasatch Formation, in ponderosa pine, limber pine, or bristlecone pine-manzanita communities at 2105 to 2535 m in Garfield (type from Red Canyon) County; endemic; 13 (i). The species is endemic to the Bryce Canyon-Red Canyon vicinity.

Penstemon breviculus (Keck) Nisbet & R. Jackson [*P. jamesii* subsp. *breviculus* Keck]. Perennial herb 0.8–3 (3.5) dm tall; stems clustered from a branched caudex and tap root, ascending to erect, puberulent (usually) to glabrous; leaves 2–8 cm long, 4–18 mm wide, spatulate to (ob)lanceolate, the lower and basal narrowed to a petiolar base, the upper sessile, glabrous or sparingly puberulent on the midrib near the base, entire, crisped, or the cauline ones sometimes irregularly toothed; herbage somewhat viscid when fresh; inflorescence glandular-pubescent, of 3–7 verticellasters, the cymes several-flowered; calyx 5–7 (8) mm long, glandular-pubescent, the lobes lanceolate, scarious-margined near the base; corolla (12) 14–(18) 20 mm long, glandular-pubescent, funnelform to moderately ampliate, bluish purple with darker longitudinal veins, the palate bearded with long, pale yellow hairs, the lobes about equally spreading or the lower sometimes spreading-reflexed; stamens included, becoming peltate-explanate, the sacs (0.6) 0.8–1 mm long, about as broad, glabrous; staminode included to shortly exserted, bearded with long golden orange hairs. Sagebrush and pinyon-juniper communities at 1600 to 2000 m in Grand and San Juan counties; Colorado and New Mexico; 5 (1). This taxon is very similar to *P. ophianthus*, differing in the mostly smaller, less ventricose-ampliate flowers. Some of our few collections seem fully transitional, and the two taxa are more or less sympatric. Additional collections may dictate treatment at infraspecific level within *P. ophianthus*.

Penstemon caespitosus Nutt. ex Gray Mat Penstemon. Plants perennial, more or less caespitose, 0.2–1.4 dm tall, herbaceous to suffruticose; herbage pubescent with terete to flattened retrorse hairs; flowering stems 2–11 (20) cm long, prostrate and mat-forming with tips ascending, or (in var. *suffruticosus*) ascending to erect, proliferating from previous years stems, these eventually rooting, semi-woody, and prostrate; leaves 0.2–15 mm long, 1–7 mm wide, linear to obovate or spatulate, narrowed to a short petiole, entire; inflorescence leafy, the flowers borne among the leaves or overtopping the leafy mat, obscurely secund; cymes often only 1 per node, 1–2 (3) flowered; calyx 3–6 mm long, the lobes herbaceous or with very narrow scarious margins, linear-lanceolate, pubescent; corolla 11–20 mm long, glandular-pubescent externally, ampliate, bulged above, flattened and 2-ridged below, the lobes projecting-spreading, the lower lobes slightly longer than the upper, blue to blue violet, the palate lighter, sparsely light yellow bearded; stamens shortly exserted, the anther sacs 0.6–1.2 (1.4) mm long, glabrous, blue, dehiscing the full length and across the connective, not fully explanate; staminode reaching the orifice, golden yellow to orange bearded. There are four essentially allopatric varieties:

1. Leaves linear to narrowly oblanceolate, mostly 1–2 mm wide, more than 5 times longer than wide, green, scabrous-puberulent but only slightly cinereus; plants of northeastern Utah near the Uinta Mts. ....
   .... *P. caespitosus* var. *P. caespitosus*

— Leaves narrowly oblanceolate to obovate or spatulate, mostly 2–4 mm wide, only 3–4 times longer than wide; leaves cinereus-puberulent, at least below; plants of more southerly distribution 2

2(1). Pubescence of leaves mostly flattened, appressed, scalelike; flowering stems ascending to erect; plants somewhat suffruticose, hardly mat-forming, of the Tushar, Sevier, and Markagunt plateaus ....
   .... *P. caespitosus* var. *suffruticosus*

— Pubescence of leaves mostly terete, retrorsely spreading; plants mat-forming, the flowering stems mostly prostrate; distribution otherwise 3

3(2). Anther sacs mostly more than 9 mm long; plants of the Aquarius and Paunsaugunt plateaus ....
   .... *P. caespitosus* var. *desertipicti*

— Anther sacs mostly less than 9 mm long; plants of the Tavaputs and Wasatch plateaus ....
   .... *P. caespitosus* var. *perbrevis*

Var. *caespitosus* In sagebrush, pinyon-juniper, mountainbrush, and forb-grass openings in aspen-conifer communites at 2100 to 2900 (3250) m in Daggett, Duchesne, Rich, Uintah, and Wasatch counties; Wyoming and Colorado; 20 (viii).

Var. *desertipicti* (A. Nels.) N. Holmgren [*P. desertipicti* A. Nels; *P. caespitosus* subsp. *desertipicti* (A. Nels.) Keck]. In sandy clay or gravelly soil, often on multi-hued semi-barrens of the Wasatch Limestone Formation, in pinyon-juniper, sagebrush, and ponderosa communities at 2000 to 2800 m in Garfield, Kane, and Wayne counties; Arizona; 17 (iii).

Var. *perbrevis* (Pennell) N. Holmgren [*P. caespitosus* subsp. *perbrevis* Pennell]. In sparsely vegetated sagebrush-grass openings of pinyon-juniper, mountain brush, and aspen-conifer communities in the vicinity of the Tavaputs Plateau at 2500 to 2650 (2900) m in Carbon (type from Castle Gate), Duchesne, Grand, Sanpete, and Uintah counties; Colorado; 42 (ix).

Var. *suffruticosus* [*P. caespitosus* subsp. *suffruticosus* Keck; *P. tusharensis* N. Holmgren]. In conifer-manzanita, black sagebrush, and aspen parkland communities, in rocky or gravelly openings at 2500 to 3360 m on the Tushar, Sevier, and Markagunt plateaus in Beaver (type from the Tushar Mts.), Garfield, Iron, Kane (?), and Piute counties; 22 (0). These more or less clump-forming plants from high elevation in the vicinity of the Tushar Mts. have been variously treated infraspecifically within the *P. caespitosus* complex; the phase was elevated to specific rank (*P. tusharensis*) by Noel Holmgren (1979), who believed it more closely allied with *P. thompsoniae*. Recent collections to the south and east of the Tushar Mountains have shown var. *suffruticosus* to be transitional morphologically, geographically, and elevationally with *P. caespitosus* var. *desertipicti*, and less similar to the nearby *P. thompsoniae* var. *thompsoniae* (q.v.). The strong tendency in var. *suffruticosus* toward the scalelike leaf pubescence which characterized *P. thompsoniae* is indicative of the close relationship between the two species.

Penstemon carnosus Pennell Fleshy Penstemon. Perennial herb 1.5–3.5 dm tall; herbage glabrous, glaucous; stems ascending to erect, 1 or few from a thick, woody sparingly branched caudex; leaves 2–9 (11) cm long, 5–32 (40) mm wide, thick and fleshy, entire, spatulate to (ob)ovate or (ob)lanceolate, the basal tuft well developed, the lowermost petioled, the upper sessile, clasping, obtuse (usually); inflorescence glabrous, the thyrse cylindrical, congested or sometimes moderately elongate, of 3–10 verticellasters, the cymes 1– or few-flowered, the peduncles rarely exceeding 1 cm; calyx 4–8 mm long, glabrous, the lobes ovate to broadly lanceolate, acute to acuminate, broadly scarious-margined; corolla (16) 18–21 (24) mm long, moderately ampliate, finely and obscurely glandular-puberulent externally (specimens from Garfield and Wayne counties sometimes glabrous), lavender pink to blue violet (usually drying blue violet), the throat rounded above and below, the lobes about equally spreading, the palate glabrous or sparsely white-bearded, guidelines absent or obscure; stamens included or shortly exserted, the sacs 1.1–1.5 mm long, glabrous, dehiscing fully and across the connective, not explanate; staminode included to reaching the orifice, dilated (the tip 1–1.5 mm wide) and curved distally, sparsely to moderately bearded with short, yellow to yellow brown hairs, these 0.2–0.5 mm long. Salt desert shrub, sagebrush, pinyon-juniper (usually), and mountain brush communities, often in clay or shaly soil at 1500 to 2260 (2500) m in Carbon, Emery (type from the San Rafael Swell), Garfield, Uintah, and Wayne counties; endemic; 41 (vi). Specimens from Wayne and Garfield counties are problematical, sharing variously characters of *P. carnosus*, *lentus*, and *pachyphyllus*.

Penstemon cleburnei Jones Cleburne Penstemon. Perennial herb 7–22 cm tall; stems from a tap-rooted caudex, 1 to few, ascending to erect; herbage densely puberulent, becoming glandular in the inflorescence; leaves 1.5–6 cm long, 4–13 mm wide, entire, broadly spatulate to narrowly (ob)lanceolate, the basal ones narrowed to a petiolar base, the upper sessile; thyrse leafy, congested, of 3–7 verticellasters; calyx 6–9 mm long, glandular-pubescent, the lobes lanceolate, herbaceous; corolla 18–24 mm long, glandular-puberulent, ventricose-ampliate, the throat rounded above and below, constricted at the orifice, bilabiate, the lips spreading, lavender blue, the lower lip lighter and veined with wine purple guidelines, the palate densely bearded with conspicuous white hairs; stamens included, the anthers glabrous, peltate-explanate, the sacs 0.8–1.1 mm long, broadly ovate; staminode bearded for most of its length with long yellow hairs (more densely bearded proximally and at the tip than in the central portion), exserted, the tip curved downward; in sandy, gravelly, or (often) clay soils, sagebrush-grass and cool-desert shrub communities at 1800 to 2500 m in Daggett and Summit counties; Wyoming; 7 (iv).

Penstemon comarrhenus Gray Dusty Penstemon. Robust perennial herb 3–8 (12) dm tall; herbage glabrous (rarely sparingly puberulent); stems 1 to few from a short-branched, fibrous-rooted caudex, ascending to erect, puberulent below; leaves 2.5–13 cm long, 1–20 (30) mm wide, entire, rounded to obtuse, the basal ones broadly to narrowly (ob)lanceolate, petioled, the cauline ones linear to narrowly oblanceolate, becoming sessile upwards, often folded; inflorescence glabrous, moderately or scarcely secund, of 5–16 verticellasters, the cymes 1- to 4-flowered, the lower peduncles spreading, 13–40 mm long, or the lower axils sometimes producing branched, many-flowered, secondary inflorescences; calyx 3–7 (10) mm long, glabrous, the lobes ovate, blunt or apiculate, scarious margined; corolla 24–37 mm long, glabrous, strongly ventricose-ampliate and bilabiate, the tube nearly white, the throat pale lavender blue, often paler ventrally, the lobes pale blue, the upper ones projecting-spreading, the lower spreading-reflexed; anthers exserted, the sacs 2–2.7 mm long, spreading at about a 90 degree angle, remaining indehiscent in the proximal 1/4–1/6, white-villous with tangled hairs longer than the sacs, usually densely so; staminode included or reaching the orifice, moderately dilated apically, glabrous (usually) or with a few long hairs. Openings in Douglas fir-aspen, mountain brush, sagebrush-grassland, and pinyon-juniper communities at (1000) 1700 to 2750 m in Beaver, Carbon, Emery, Garfield (type from the Aquarius Plateau), Grand, Iron, Kane, Millard, Piute, San Juan, Sanpete, Sevier, Washington, and Wayne counties; Colorado, Arizona, Nevada; 123 (xiii). See note under *P. strictus*.

Penstemon concinnus Keck Elegant Penstemon. Perennial herb 0.6–2.3 dm tall from a compact, much-branched caudex and taproot; stems finely retrorsely puberulent, usually suffused with purple; leaves 1.8–5 (7) cm long, 1.5–6 (9) mm wide, entire to toothed, glabrous (sometimes sparingly puberulent near the base or on the midrib), somewhat glaucous, the basal ones well developed, petioled, linear to (ob)lanceolate and becoming sessile above; inflorescence glandular-pubescent with spreading hairs, the thyrse narrow, leafy, the lowermost bracts usually wider than the cauline leaves, of (2) 4–7 (9) verticellasters, the cymes densely flowered, the peduncles and pedicels short; calyx 5–8 mm long, densely glandular-pubescent with spreading hairs, the lobes narrowly lanceolate, herbaceous, usually suffused with purple; corolla 8–10 (12) mm long, moderately ampliate, the lower lip usually reflexed, glandular-pubescent externally, blue violet, the palate veined with dark purple guidelines and copiously bearded with long white hairs; anthers included or shortly exserted, glabrous, the sacs 0.5–0.9 mm long, peltate-explanate, about as broad as long, blue black; staminode exserted, curved at the tip, bearded, the hairs pale yellow to white. Calcareous or igneous gravels, usually on pale, limestone-derived soil, in pinyon-juniper and sagebrush communities at (1600?) 1900 to 2300 m in western Beaver, closely adjacent Millard (type from Tunnel Springs Mts.), and Iron counties; Nevada; a Great Basin endemic; 29 (ii).

Penstemon confusus Jones Mistaken Penstemon. Perennial herb 0.8–5 (8.5) dm tall; herbage glabrous and glaucous; stems erect, single or few from a thick crown; leaves thick and leathery, entire, rounded to obtuse apically, the basal ones 2–7 cm long, 4–11 mm wide, (ob)lanceolate (sometimes narrowly so) or rarely spatulate, the cauline ones 0.8–6.5 cm long 1.5–15 mm wide, narrowly (ob)lanceolate to oblong, sessile, few, more or less reduced above; inflorescence glabrous, often lax and wandlike, of 3–10 (16) verticellasters, the cymes 1- to 3 (5)-flowered, not congested; calyx 3.5–6 mm long, glabrous, the lobes ovate to lanceolate, obtuse to acute or shortly-acuminate, broadly scarious-margined; corolla 15–20 (22) mm long, tubular-funnelform to moderately ampliate, the lobes about equally spreading, glandular to glabrous externally, densely glandular within near the orifice, the limb lavender pink, violet, or magenta, the throat and tube lighter, usually marked with reddish guidelines, drying lavender to blue violet or rarely reddish pink; stamens included or reaching the orifice, the anthers glabrous, pale or (often) cyaneus, dehiscing fully and across the connective, the sacs ca 1 mm long, more or less explanate but scarcely peltately so; staminode included, slightly dilated apically, glabrous or papillate. Sandy or gravelly, often sparsely vegetated places or in clay, in mixed desert shrub, sagebrush, juniper-black sage, sagebrush-oak, and juniper-mountain brush communities at (1000) 1370 to 2200 m in Beaver, Iron, Juab (type from Detroit), Millard, Sanpete, Sevier, and Washington counties; Nevada; 86 (vi). This species is closely related to the more eastern *P. utahensis* and transitional to it in Washington County and the Tushar Mts.

Penstemon crandallii A. Nels. Crandall Penstemon. Herbaceous or obscurely suffrutescent perennial; old stems prostrate, rooting; flowering stems 3–12 dm long, decumbent to ascending, puberulent; leaves 0.5–2.5 (3.8) cm long, 1–5 mm wide, spatulate, obovate, or oblanceolate, becoming narrowly oblanceolate above, entire, sparingly scaberulous- puberulent to subglabrous, usually floriferous for most of their lengths, the flowers 1–several per node, usually only 1 axil per node bearing flowers; calyx 4–6 mm long, the lobes lanceolate, acute to acuminate, usually scarious-margined in the lower half; scaberulous-puberulent, obscurely glandular; corolla 14–21 mm long, glandular-pubescent externally, blue to blue lavender, flattened and 2-ridged below, 2-lipped, the lobes spreading, the palate sparsely bearded; stamens reaching the orifice or the longer exserted, the sacs 1–1.2 mm long, dehiscing fully but the sacs not fully explanate; staminode included, densely golden orange-bearded most of its length. $2n = 16$. Our material may be referred to two poorly differentiated varieties:

1. Leaves narrowly oblanceolate; stems ascending; plants of the Abajo Mts. .... *P. crandallii* var. *crandallii*
— Leaves spatulate, obovate, or oblanceolate; stems more or less prostrate; plants of the La Sal Mts. ....
 .... *P. crandallii* var. *atratus*

Var. *atratus* (Keck) N. Holmgren [*P. crandallii* ssp. *atratus* Keck, type from the La Sal Mts.]. Pinyon-juniper, sagebrush, and mountain brush communities at 2135 to 2595 m in the La Sal Mts., Grand and San Juan counties; endemic; 3 (0).

Var. *crandallii* Pinyon-juniper, mountain brush, and ponderosa pine communities at 2135 to 2440 m in the Abajo Mts., San Juan County; Colorado; 4 (0).

Penstemon cyananthus Hook. Wasatch Penstemon. Perennial herb 0.6–9 dm tall; herbage glabrous (usually) to cinereous-puberulent; stems decumbent to erect; inflorescence of 2–9 verticellasters, secund to cylindrical, the cymes 1– to 7–flowered; calyx 3–8.5 mm long, glabrous or (in var. *compactus*) obscurely glandular, the lobes ovate to lanceolate, acute to caudate, more or less erosely scarious-margined; corolla 15–32 mm long, glabrous, blue, moderately ventricose-ampliate, the lobes spreading; palate glabrous; stamens (the longer pair) exserted; anther sacs 1.3–2.2 mm long, pubescent with straight to flexuous hairs mostly shorter than the width of the sac, diverging at about a 90–120 degree angle, dehiscing in the distal 2/3–4/5, the suture lateral; staminode white or sometimes blue-tipped, bearded with yellow (usually) hairs, sometimes sparsely so, or rarely glabrous; 2n = 16. *P. cyananthus* is a variable and difficult group, unified by uniform and consistent anther characteristics, whose members have been treated at various taxonomic ranks. Names applied to taxa in our flora include *compactus*, *holmgrenii*, *longiflorus*, and *subglaber* (of Gray, applied at varietal level, not *P. subglaber* Rydb., a different species). A reticulate array of morphological and geographic intermediates of these as well as other unnamed races occurs. Most variation within the complex may be recognized at varietal level as follows:

1. Plants relatively dwarf, mostly 1–2.5 dm tall, the stems decumbent to ascending: leaves commonly folded; calyx obscurely glandular; inflorescence reduced, usually of 2–4 verticellasters; plants of rocky limestone outcrops, high elevation in the northern Wasatch Range, Cache County .... *P. cyananthus* var. *compactus*
— Plants taller, usually over 3 dm tall, the stems ascending to erect; leaves seldom folded; calyx glabrous; inflorescence usually of 5 or more verticellasters; plants of wider distribution 2
2(1). Leaves linear-lanceolate to lanceolate, the cauline ones mostly 4 times or more longer than wide; lower leaves usually cinereous-puberulent; plants of Box Elder County .... *P. cyananthus* var. *subglaber*
— Leaves lanceolate to ovate, the cauline ones less than 4 times longer than wide; lower leaves sparingly puberulent to glabrous; plants of other distribution (rarely occuring in eastern Box Elder County) 3
3(2). Inforesence secund, the cymes mostly 1–3 flowered; corollas usually over 25 mm long; herbage usually puberulent; plants of mountain ranges bordering the east edge of the Bonneville Basin ....
 .... *P. cyananthus* var. *longiflorus*
— Inflorescence more or less cylindrical, usually densely flowered, the cymes mostly 5– to 7–flowered; flowers often smaller; herbage glabrous or nearly so; plants of the Wasatch Mts. .... *P. cyananthus* var. *cyananthus*

Var. compactus (Crosswhite) Neese [*P. compactus* Crosswhite, type from the Bear River Range; *P. cyananthus* subsp. *compactus* (Crosswhite) Keck]. High elevations near Tony Lake and Mt. Naomi in Cache County; Idaho, a northern Wasatch endemic; 9 (0).
Var. cyananthus Common in the Wasatch Mts. and foothills; a very showy species; in Box Elder, Cache, Davis, Duchesne, Juab, Morgan, Salt Lake, Sanpete, Summit, Utah, Wasatch, and Weber counties; Idaho and Wyoming; 92 (iii).
Var. longiflorus (Pennell) Neese [*P. cyananthus* subsp. *longiflorus* Pennell, type from near Beaver; *P. longiflorus* (Pennell) S. Clark]. Tushar and Pavant ranges, at 1800 to 2500 m, passing into var. *subglaber* northward in the small ranges of the east Bonneville Basin, and into var. *cyananthus* in the southwestern part of the Wasatch Range; in Beaver, Juab, Millard, Piute, Sevier, and Tooele counties; endemic; 18 (i).
Var. subglaber (Gray) N. Holmgren [*P. fremontii* var. *subglaber* Gray; *P. cyananthus* subsp. *subglaber* (Gray) Pennell; *P. holmgrenii* S. Clark]. Mountain brush and sagebrush communities at 1800 to 3300 m in Box Elder County; Idaho and Wyoming; 23 (ii). This variety passes into var. *cyananthus* in eastern Box Elder County, and into var. *longiflorus* southward.

Penstemon cyanocaulis Payson Bluestem Penstemon. Herbaceous perennial 2–4.5 dm tall; stems glabrous, ascending to erect, 1 to several, from a short-branched woody caudex; leaves 1.5–12 cm long, 2–20 (30) mm wide, the margins commonly crisped, the basal and lower cauline ones obovate to (ob)lanceolate, narrowed to a petiolar base, the upper oblong or (ob)lanceolate, sessile, often somewhat clasping, mostly obtuse, not folded; inflorescence of (5) 7–9 (12) verticellasters, the cymes 1–4 (5) flowered, glabrous; calyx 3.5–5 (6) mm long, glabrous, the lobes ovate to broadly lanceolate, acute, scarcely if at all acuminate or caudate, the margins narrowly scariousmargined, entire or somewhat erose; corolla 16–24 mm long, glabrous, ventricose-ampliate, bilabiate, the upper and lower lobes spreading about equally or the upper somewhat reflexed, the lobes blue, the tube and throat lavender blue; anthers (the longer pair) about reaching the orifice, the sacs pubescent, the hairs mostly straight and shorter than the width of the sac, dehiscing from the distal end, the line of dehiscence not reaching the connective; staminode sparsely yellow-bearded, included to slightly exserted. Pinyon-juniper communities, mostly in sandy soil at 1500 to 2300 m in Carbon, Emery, Grand, and San Juan counties; Colorado; 26 (iii). Bluestem penstemon is very similar to *P. strictiformis*, a taxon of similar habit that occurs slightly to the south of *P. cyanocaulis*. Although some specimens are problematical, *P. cyanocaulis* usually may be distinguished by the relatively broader, mostly obtuse rather than acute, more crisped leaves, somewhat shorter more hispid pubescence of the anthers, and ovate, broadly acute (not lance-acuminate) calyx lobes.

Penstemon deustus Dougl. ex Lindl. Hotrock Penstemon; Scabland Penstemon. Subshrub 1.5–3 dm tall; herbage glabrous to glandular- puberulent; stems slender, decumbent, ascending, or erect, numerous, from a freely branched woody base; sterile leafy shoots borne among the taller, flowering ones; leaves 1–2.5 (5) cm long, 2.5–8 (18) mm wide, serrate to dentate, the lower ones spatulate to oblanceolate, narrowed to a short petiolar base, the upper elliptic to lanceolate, sessile; inflorescence of 3–9 (14) verticellasters, the cymes 1– to few– flowered; calyx and pedicels sparingly glandular-puberulent, the calyx 2–5 (6) mm long, the lobes lanceolate, acute, narrowly scarious-margined near the base; corolla 10–14 mm long (ours), sparingly glandular- puberulent externally, white, the throat sometimes brown or lilac-tinged externally, the upper lobes arched-projecting, faintly blotched with brownish purple internally, the lower lobes longer than the upper, spreading, veined with reddish purple; stamens exserted, the sacs ca 0.5 mm long and equally broad, glabrous, dehiscing the full length and across the connective, explanate; staminode included or reaching the orifice, glabrous or sparsely bearded apically, slender, not dilated apically; 2n = 16. Rocky places in the pinyon-juniper zone at 2000 to 2700 m in western Box Elder County; Washington, Idaho, Oregon, California, Wyoming; 2 (0). Utah specimens are var. *pedicellatus* Jones.

Penstemon dolius Jones ex Pennell Jones Penstemon. Perennial herb 4–17 dm tall; herbage densely cinereous-puberulent with retrorse hairs; stems tufted, several to many from a compactly branched caudex, spreading or the marginal ones more or less prostrate, often reddish purple; leaves 1–4 cm long (to 6 cm at the base of the inflorescence), 2–11 mm wide, basal and cauline, entire, spatulate to oblanceolate, the lower ones narrowed to a broad petiolar base; inflorescence cinereus-puberulent, essentially non-glandular, leafy-bracteate, congested, of (1) 3–8 verticellasters, the cymes 1– to 3 (5) -flowered, often only one of the axil pairs floriferous, the flowers subsessile; calyx cinereus-puberulent with non-glandular hairs, 5–9 mm long, the lobes lanceolate to narrowly lanceolate-attenuate, herbaceous; corolla 13–21 mm long, glandular-pubescent with spreading hairs, tubular-funnelform to moderately ampliate; corolla lobes about equally spreading or the upper arched and projecting, 3–5 mm wide when well pressed, usually widest at the base and not overlapping, pale blue to blue or blue violet, the tube and throat lighter, the palate sparsely bearded and sometimes marked with red violet guidelines; anthers included, the sacs 1–1.2 mm long, glabrous, dehiscing the full length and across the connective, becoming divaricate, not explanate, blue black; staminode included or reaching the orifice, bearded with a tuft of golden yellow hairs distally, more sparsely bearded proximally, abruptly dilated to a broad tip, this more than 1 mm wide and strongly curved to coiled. Calcareous gravels or on rocky or sandy semibarrens, often in clay or silt of old playas or alluvial deposits, in shadscale, mixed desert shrub, sagebrush-grass, or juniper communities at 1370 to 2000 m in Beaver, Juab, Millard, Sanpete, Sevier, and Tooele (type from the Deep Creek Mts.) counties, Nevada; 58 (iv).

Penstemon duchesnensis (N. Holmgren) Neese Duchesne Penstemon. [*P. dolius* var. *duchesnensis* N. Holmgren, type from 17 km east of Duchesne]. Low perennial herb 2.5–12 cm tall; herbage densely cinereous-puberulent with retrorse hairs; stems ascending to erect, few to several from a branched caudex; leaves 0.8–4 cm long, 1.5–8 mm wide, basal and cauline, entire, narrowed to a petiolar base, the blade broadly elliptic to (ob)lanceolate; inflorescence leafy-bracteate, cinereous-puberulent, of 1–4 (6) more or less congested verticellasters, not secund, the cymes 1 to several-flowered, often only one of the axil pairs floriferous, the flowers subsessile; calyx cinereus-puberulent with non-glandular hairs, 6–11 mm long, the lobes lanceolate to narrowly lanceolate-attenuate, acute; corolla 16–22 (25) mm long, tubular-funnelform to ampliate, glandular pubescent externally with spreading hairs, blue to blue purple, the palate sparsely bearded, guidelines absent or obscure; lobes about equally spreading, orbicular, overlapping at the base, widest at the middle, 4–7 mm wide when pressed; stamens included, the anthers glabrous, the sacs 1.1–1.3 mm long, dehiscing the full length and across the connective, becoming divaricate, not explanate; staminode included or reaching the orifice, densely and rather evenly bearded with orange hairs in the distal half, gradually widened to a dilated apex, the tip less than 1 mm wide and straight or only slightly curved. On gravelly or silty sand or clay semibarrens in open pinyon-juniper woodlands at 1640 to 1830 m near Duchesne, Duchesne County; endemic; 15 (vii). *Penstemon duchesnensis*, similar to *P. dolius* and previously included within that taxon at varietal level, is a much showier plant and is distinguished from the latter by a series of characters as indicated in the keys and descriptions; the two are geographically separated by about 130 km.

Penstemon eatonii Gray Eaton Penstemon; Scarlet-bugler Penstemon. Perennial robust herb 2.5–10 dm tall; stems few to several from a stout caudex, ascending to erect; herbage glabrous throughout, or puberulent in var. *undosus*; leaves entire, crisped (usually) or plane, the basal and lower cauline 3–19 cm long, 8–40 (55) mm wide, the blade (ob)lanceolate or broadly obovate, narrowed to a petiolar base, the upper 1.5–11 cm long, (4) 8–40 mm wide, ovate to broadly lanceolate (rarely narrowly lanceolate), sessile, clasping, often cordate; inflorescence secund, of 4–12 (18) verticellasters, wandlike, little-congested, the cymes 1– to 3 (6) -flowered, the peduncles and pedicels ascending to erect, glabrous or puberulent; calyx (2.6) 3.2–5.5 (6.5) mm long, glabrous or glandular puberulent, the lobes ovate, obtuse, acute or (rarely) shortly acuminate, the margins moderately scarious-erose; corolla (15) 20–30 (33) mm long, red, glabrous, somewhat geniculate at the base (the corolla thus horizontal or declining), nearly tubular, the tube 2.8–3.5 mm in diameter, widening imperceptibly distally, the throat at the orifice 5–7.5 mm wide, the lobes projecting or only slightly spreading, the upper slightly longer; staminode white, glabrous, included; anthers finely spiculate-puberulent, included to exserted, the sacs 1.5–2.5 mm long, opening in the distal 2/3's. Mixed desert shrub, sagebrush, mountain brush, and aspen communities at 900 to 3400 m in all but the northern quarter of Utah; California, Nevada, Arizona, New Mexico, and Colorado; 231 (xiv). A loose geographic correlation between presence or absence of pubescence may be recognized at the varietal level as follows:

1. Herbage glabrous; plants often of rocky places, mostly in central and western Utah .... *P. eatonii* var. *eatonii*
— Herbage finely puberulent; plants often of sandy places, mostly in southern and southeastern Utah ....
.... *P. eatonii* var. *undosus*

Var. *eatonii* Plants of numerous habitats in many vegetative types in Beaver, Carbon, Duchesne, Emery, Garfield, Iron, Juab, Kane, Millard, Piute, Salt Lake, San Juan, Sevier, Tooele, Utah (type from Provo Canyon),, and Wasatch counties; Nevada, California, Colorado.

Var. *undosus* Jones Plants of mixed warm and cool desert shrub, pinyon-juniper and other plant communities in Garfield, Grand, Iron, Kane, San Juan, Washington (type from St. George), and Wayne counties; Nevada, California, Arizona, New Mexico. Occasional plants bear yellow flowers.

Penstemon flowersii Neese & Welsh Flowers Penstemon. Perennial herb 8–25 (32) dm tall; herbage glabrous and glaucous; stems few to several, arising from a branched, woody caudex, ascending to erect; leaves (1) 2–7 cm long, 4–25 mm wide, all cauline, entire, thick, blue-glaucous, spatulate to (ob)ovate or broadly (ob)lanceolate, the lower narrowed to a petiolar base, the upper sessile, becoming clasping and broadly ovate in the inflorescence; inflorescence glabrous, the thyrse compact, cylindric, of 4–9 verticellasters, the cymes several-flowered; calyx 5–7 mm long, the lobes ovate to broadly lanceolate, acute to attenuate, broadly scarious-margined; corolla 15–20 (22) mm long, the throat moderately ampliate, the lobes more or less spreading, glabrous within and without, rose pink (drying pink not blue), the palate inconspicuously marked with darker lines; stamens ca reaching the orifice or the longer pair exserted, the sacs 0.9–1.3 mm long, opening the full length and across the connective, glabrous, not fully explanate; staminode included, the tip moderately dilated, pubescent, with very short yellow hairs to 0.1 mm long. Shadscale and salt desert shrub communities at 1520 to 1650 m in Duchesne and Uintah (type from 5.6 km W of Randlett) counties; 10 (vii). The species is endemic to clay badlands in the vicinity of Roosevelt.

Penstemon fremontii T. & G. Fremont Penstemon. [*P. glaber* var. *fremontii* (T. & G.) Jones, type on the Uinta plains, Fremont]. Perennial herb (0.8) 1.5–4.2 dm tall; stems erect, one or few, from a crown and taproot; herbage evenly retrorsely cinereus-puberulent; leaves (1.5) 3–12 cm long, 5–27 mm wide, entire, obtuse, the basal and lower cauline ones elliptic or obovate to (ob)lanceolate, narrowed to a petiolar base, the upper ones ovate (rarely) or more often oblong to lanceolate, usually narrowly so, sessile; inflorescence relatively congested, of 4–11 verticellasters, the cymes (1) 2– to 4 (7)-flowered, not markedly secund, puberulent to sparingly glandular or glabrous; calyx 4–6.5 mm long, glabrous or sparingly glandular, the lobes ovate to lanceolate, acute to (usually) acuminate, scarious-margined, moderately erose; corolla (12) 18–23 (25) mm long, medium to dark blue, glabrous, moderately ampliate or nearly tubular, the lobes moderately spreading, about equal or the lower somewhat larger and projecting; anthers included, sparsely hispid, the hairs shorter than the width of the sac, straight to somewhat flexuous, the sacs 1.3–1.8 mm long, dehiscing from the distal end 2/3–4/5 their length, moderately divergent near the connective but with tips flared; staminode sparsely bearded with golden hairs, dilated apically, included. Common in the Uinta Basin on arid benches and slopes, in shadscale, mixed desert shrub-grassland, and pinyon-juniper communities at 1500 to 2500 m in Duchesne and Uintah counties; Wyoming and Colorado; 82 (xxxiv). The type was taken "on Uinta plains" by John Charles Fremont on his second expedition in the spring of 1844. One specimen (Collins & Harper 146 BRY) from near Nephi keys here on the basis of pubescent rather than glabrous anthers. It has, otherwise, the features and distribution of *P. tidestromii* Pennell.

Penstemon goodrichii N. **Holmgren** Goodrich Penstemon. Perennial herb 1.2–4 dm tall; stems several to many from a freely branched caudex and taproot, ascending to erect, glabrous, becoming glandular-pubescent in the inflorescence; leaves basal and cauline, 2–6 (8) cm long, 2–3 mm wide, narrowly (ob)lanceolate to linear, entire or sometimes with a few teeth, glabrous; inflorescence lax, of 3–6 (9) verticellasters, the cymes 1– to 3–flowered; flowers short-pediceled; calyx 4–7 mm long, glandular-pubescent, the lobes lanceolate, obtuse, herbaceous; corolla 11–16 mm long, nearly regular, funnelform, flaring gradually from the summit of the tube to the apex, glandular-puberulent, blue to blue lavender, the throat pale, striate internally with violet guidelines; stamens very short, included in the tube or shortly exserted into the throat; anthers glabrous, peltate-explanate, the sacs ca 0.8 mm long; staminode densely golden bearded, included, slightly exceeding the stamens; style very short, included in the tube. Blue gray to reddish, clay-impregnated badlands of the Duchesne river Formation in shadscale and juniper-mountain mahogany communities at 1705 to 1895 m in Duchesne and Uintah (type from near Lapoint) counties; 16 (v). This species is known only from the Lapoint-Tridell-Whiterocks area.

Penstemon grahamii **Keck** Graham Penstemon. Perennial herb 0.5–2 dm tall from a sparingly branched, taprooted caudex; stems ascending to erect, puberulent; leaves 1.5–5 cm long, 3–18 mm wide, thick, entire or rarely with a few inconspicuous teeth, glabrous (sometimes with a few hairs near the base), becoming glandular-puberulent in the inflorescence, the basal ones ovate, spatulate, or broadly oblanceolate, narrowed to a petiolar base, the upper oblong or (ob)lanceolate, sessile, clasping; inflorescence glandular-pubescent, of 1–5 crowded verticellasters, the cymes 1– to 3–flowered; calyx 5–9 mm long, glandular-pubescent, the lobes lanceolate, harbaceous; corolla 25–37 mm long, glandular-pubescent, abruptly and conspicuously ampliate, conspicuously bilabiate, the upper lobes projecting, the lower reflexed, the throat and tube about equal, the throat rounded above and below, pinkish lavender, the palate lighter and marked with wine red guidelines; stamens exserted, the anthers glabrous, peltate-explanate, the sacs 1–1.4 mm long; staminode conspicuously exserted and the tip curved downward, evenly and densely bearded on all sides with short golden orange hairs. Uncommon in sparsely vegetated desert shrub and pinyon-juniper communities, usually growing with shadscale and *Elymus salina* on shale ledges and talus of the Green River Formation at 1400 to 2060 m in Carbon and Uintah (type from near Sand Wash) counties; Colorado; 29 (xii). This is one of our most distinctive penstemons, often growing with *Astragalus lutosus*, *Aquilegia barnebyi*, *Cryptantha barnebyi*, and other rare species on peculiar, oil-rich, substrates.

Penstemon humilis **Nutt. ex Gray** Low Penstemon. Perennial herb 0.7–3.5 (5) dm tall; basal sterile shoots well developed, leafy, semi-matforming; flowering stem decumbent to erect, puberulent, numerous from a freely branched caudex, the caudex branches eventually rooting; leaves entire or rarely with a few teeth, glabrous to cinereous-puberulent; basal and lower leaves 1–9 cm long, 2–20 (32) mm wide, petioled, the blades ovate to elliptic or lanceolate, abruptly to gradually narrowed to the petiole; middle and upper cauline leaves 0.8–5.5 cm long, 2–12 mm wide, ovate to (usually) narrowly (ob)lanceolate, oblong, or pandurate, sessile, the bases often clasping; inflorescence of (2) 3–9 verticellasters, the cymes (1) 3– to 9–flowered, glandular-pubescent; calyx 3–5 (6) mm long, glandular- pubescent, the lobes lanceolate, acute to caudate, the tips often squarrose; corolla 8–19 (21) mm long, glandular-pubescent, nearly tubular or the throat ampliate, the tube violet, the throat and limb blue, the lobes usually flat-spreading; palate pale, veined with purple guidelines, yellow- to white-bearded, usually inconspicuously so; stamens included or the longer pair short-exserted, the anthers glabrous, purplish black dorsally, usually pale ventrally, the sacs 0.4–0.8 mm long, broadly ovate, divaricate, not fully explanate; staminode included, densely bearded distally with golden yellow hairs; capsule 4–8 mm long; 2n = 16. Widespread in mountainous areas in many plant communities, from desert shrub to above timberline at 1500 to 3400 m in most of the northwestern two-thirds of the state; Washington to Wyoming, south to California, Nevada, and Colorado. (xxix). *Penstemon humilis* is a highly variable taxon as regards height, flower size, and degree of pubescence, and correlation in Utah of evident morphological features with geographic distribution is imperfect at best. The following may be used to identify, in part, the variation encompassed in our material:

1. Leaves (some) toothed; largest basal leaves usually 15–25 mm wide; plants from the vicinity of Zion National Park, Washington County .... *P. humilis* var. *obtusifolius*
— Leaves (usually all) entire; basal leaves usually narrower 2

2(1). Corolla mostly 12 mm or less long; leaves glabrous or essentially so, at least distally; plants principally of the Wasatch Range and northwestern Utah .... *P. humilis* var. *brevifolius*
— Corolla mostly at least 12 mm long; leaves usually puberulent, sometimes densely so; plants widespread in mountains but usually outside the distributions given above .... *P. humilis* var. *humilis*

Var. *brevifolius* Gray Sagebrush, mountain brush, aspen, and spruce-fir to alpine communities at ca 1765 to 3390 m in Box Elder, Cache, Juab, Morgan, Rich, Salt Lake (type from Cottonwood Canyon), Sanpete, Sevier, Tooele, Utah, Wasatch, and Weber counties; endemic; 90 (vii).

Var. *humilis* Sagebrush, pinyon-juniper, mountain brush, aspen-fir, spruce-fir, and alpine communities at 1525 to 3355 m Beaver, Daggett, Duchesne, Iron, Millard, Morgan, Summit, Tooele, Utah, Wasatch, and Weber counties; Washington, Oregon, California, Nevada, Idaho, Wyoming, and Colorado; 125 (xxiv).

Var. *obtusifolius* (Pennell) Reveal [*P. obtusifolius* Pennell, type from Springdale; *P. humilis* subsp. *obtusifolius* (Pennell) Keck]. Pinyon-juniper, mountain brush and ponderosa pine communities at 1580 to 2450 m in Washington county; endemic; 10 (0).

Penstemon immanifestus N. Holmgren Perennial herb 1.2–3.5 tall; herbage glabrous, strongly glaucous; stems ascending to erect, 1 or few from a woody caudex; basal leaves present or (usually) poorly developed; leaves 2–7 cm long, 5–25 (30) mm wide, thick and fleshy, entire, the basal ones when present petiolate, spatulate to broadly oblanceolate, the cauline ones ovate to broadly (ob)lanceolate, usually rounded to obtuse, becoming sessile, broad-based, and clasping upward; inflorescence glabrous, thyrsoid, cylindrical, usually congested, of 4–9 (11) verticellasters, the fascicles densely flowered, the peduncles and pedicels short, obscured by the bracts and flowers; calyx 5–9 mm long, glabrous, the lobes ovate to lanceolate, acute, broadly scarious-margined; corolla 16–21 mm long, glabrous externally and internally or with a few long hairs on the palate, ampliate, the lobes about equally spreading or the upper somewhat shorter and spreading-projecting, lavender pink to pale pink; anthers included, the sacs 1–1.5 mm long, glabrous, dehiscing the full length and across the connective, not explanate; staminode yellow-bearded with hairs 0.5–1 mm long. In greasewood and shadscale communities, usually on playas or clay soil at 1500 to 1700 m near the Nevada state line in Juab, Millard, and Tooele counties; Nevada; 9 (0). Specimens from Nevada are often bluish lavender while ours are uniformly pink to pink lavender, and sometimes are difficult to distinguish from nearby *P. angustifolius* var. *amoenus* (q.v.). The two are, however, allopatric and have strikingly different habitat preferences.

Penstemon x jonesii Pennell This is the name applied to putative hybrids between *P. laevis* (q.v.) and *P. eatonii* var. *undosus*. The plants have been taken in Zion National Park, Washington (type from Sprindale) and Kane counties; 2 (i).

Penstemon laevis Pennell Smooth Penstemon. Perennial herb (1.5) 3–10 dm tall; stems glabrous, erect, 1 or few from a stout crown, often robust; herbage glabrous throughout, glaucous; leaves 2–20 cm long, 3–35 mm wide, entire or undulate-margined, rather thick, the basal and lower cauline ones narrowed to a petiolar base, spatulate or obovate to oblanceolate, the upper ones lanceolate, sessile, clasping, obtuse to acute; inflorescence glabrous, of (2) 7–12 (18) verticellasters, lax and more or less secund or (when cymes congested and many-flowered) scarcely so, the cymes (1) 3–8 flowered; calyx 4–7 (8)mm long, glabrous, the lobes ovate, acute mucronate, or shortly-acuminate, erose, scarious margined; corolla (20) 23–35 mm long, glabrous externally and internally, bilabiate, the lobes blue, the throat lavender blue and ventricose-ampliate (to 12 mm wide when well-pressed), the palate lighter with wine-red guidelines; staminode yellow-bearded apically, included; anthers glabrous, at least the longer pair exserted, the sacs 1.8–2.5 mm long, dehiscing partially, the line of dehiscence nearly reaching the connective; capsules 8–14 mm long. Sandy places, in pinyon-juniper, ponderosa-manzanita, and mountain brush communities at 1500 to 2150 m in Kane and Washington (type from Springdale) counties; Arizona; 25 (v). A specimen (Burkey 165 BRY) taken near Bryce Canyon, has distinctly puberulent lower stems and petioles. It is intermediate with the similar but disjunct *P. speciosus*. Anomalous specimens from Zion National Park (Foster & Foster 3947, 0.5 mi east E of tunnel, and Atwood & Neese 7791, Checkerboard Mesa, both BRY) represent apparent hybrids between *P. laevis* and *P. eatonii* var. *undosus*. Such specimens have been given the name *P. jonesii* Pennell. They are morphologically similar to *P. laevis* but have puberulent stems and reddish purple, scarcely ampliate corollas with projecting rather than spreading lobes.

Penstemon leiophyllus Pennell Markagunt Penstemon. Perennial herb 2–9 dm tall; stems 1–several, glabrous, erect to ascending from a branched caudex or crown; leaves 1–13 cm long, 1.2–30 mm wide, entire, glabrous, glaucous; basal and lower cauline leaves petioled, spatulate to (ob)lanceolate, the upper ones sessile, clasping, narrowly to broadly lanceolate, sometimes cordate; inflorescence secund, of 3–13 verticellasters, the cymes 1– to 4 (6) -flowered, the peduncles and pedicels glandular-puberulent, erect to ascending; calyx 4–8 mm long, glandular-puberulent, the lobes lanceolate, acute to acuminate, narrowly scarious-margined, mostly entire; corolla 20–33 mm long, glandular puberulent externally and at the base of the upper lip internally, blue to blue purple, the throat ventricose-ampliate, bilabiate, the upper lip projecting to spreading, the lower spreading, the lobes blue to purplish blue, the tube and throat bluish violet, the palate glabrous or with a few white hairs, two-ridged, the ridges usually white; staminode glabrous, included, anthers glabrous, white or cyaneous, included or the longer pair reaching the orifice, the sacs (1.2) 1.5–2 mm long, opening partially, the line of dehiscence not reaching the connective. In pinyon-juniper, mountain brush, ponderosa pine-manzanita, bristlecone-limber pine, sagebrush, aspen-conifer, and high elevation meadow and ridgetop communities at 2000 to 3500 m in Garfield (type from Markagunt Plateau), Iron, Kane, and Washington counties; endemic; 66 (iii).

Penstemon lentus Pennell Abajo Penstemon. Robust perennial herb (2) 3–5 (7) dm tall; herbage glaucous, glabrous; stems ascending to erect, 1–few, from a coarse, sparingly branched caudex; leaves 2–11 cm long, 5–30 (35) mm wide, fleshy, entire, the basal ones petioled, spatulate, the cauline ones sessile and clasping or the lower narrowed to a short petiolar base, (ob)ovate to broadly (ob)lanceolate; inflorescence rather narrow and wandlike, occasionally somewhat congested, moderately secund, of 5–10 verticellasters, glabrous, the lower peduncles often elongated and exceeding 1 cm in length the cymes 1– to (usually) several-flowered; calyx 4–8 mm long glabrous, the lobes ovate to lanceolate, acute to acuminate, scarious-margined; corolla 17–23 mm long, moderately ampliate glabrous externally, lavender to blue violet or white (var. *albiflorus*), or occasionally pink, the throat rounded above and below, the lobes about equally spreading, the palate sparingly white-bearded to glabrous, guidelines absent or obscure; stamens included or reaching the orifice, glabrous, the sacs 1.1–1.5 mm long, dehiscing fully and across the connective, not explanate; staminode included or reaching the orifice, moderately dilated and bearded distally with yellow hairs. There are two varieties:

1. Corolla lavender pink to blue or blue violet, plants from eastern San Juan County, eastward from the Abajo Mts.
   .... *P. lentus* var. *lentus*
— Corolla white (sometimes suffused with very pale pink or blue); plants from the vicinity of Natural Bridges National Monument and the west side of the Abajo Mts. ....
   .... *P. lentus* var. *albiflorus*

Var. *albiflorus* (Keck) Reveal [*P. l.* ssp. *albiforus* Keck, type from 8 mi W of Blanding]. Pinyon-juniper community at ca 1830 to 1895 m in San Juan County; endemic; 11 (i).

Var. *lentus* Sagebrush, pinyon-juniper, and mountain brush-ponderosa pine communities, usually in sandy soil, at 1500 to 2600 m in San Juan County; Colorado and Arizona; a Colorado Plateau endemic; 16 (iii).

Penstemon leonardii Rydb. Leonard Penstemon. Herbaceous woody-based perennial (7) 10–35 (45) cm tall from a much-branched caudex; stems numerous, clustered, sparingly puberulent at least below, erect to decumbent; leaves 1.5–6.5 cm long, 1–10 mm wide, entire, glabrous, elliptic to more often lanceolate to spatulate, especially below, acute to obtuse or rounded, not especially firm or plane; inflorescence of 3–10 verticellasters, the cymes 1– to several-flowered, the pedicels to 2.5 mm long, granular-puberulent; calyx (3) 4–6.5 mm long, glabrous, the lobes lanceolate, attenuate to caudate; corolla 14–26 mm long, deep blue with violet to lavender tube, or lavender to violet throughout, glabrous, bilabiate, slightly to moderately ventricose-ampliate, the lobes subequal, spreading or the lower reflexed; anthers 1–1.6 (1.9) mm long, glabrous or sparingly hirsutulose on the inner margin, the suture margins finely spinulose. Three geographically correlated varieties are present:

1. Corolla mostly more than 20 mm long; anther sacs mostly over 1.4 mm long; plants of west-central Utah .
   .... *P. leonardii* var. *patricus*
— Corolla mostly less than 20 mm long; anther sacs mostly less than 1.4 mm long; plants of other distribution 2
2(1). Corolla lobes blue or blue violet; plants of central and north-central Utah .... *P. leonardii* var. *leonardii*
— Corolla lavender when fresh; plants of the vicinity of the Pine Valley Mts., Washington County ....
   .... *P. leonardii* var. *higginsii*

Var. *higginsii* Neese Pinyon-juniper, mountain brush, ponderosa pine-manzanita, and aspen conifer communities at 2000 to 2750 m in the Pine Valley (type locality) and Bull Valley mts. and Kolob Plateau, Washington County; endemic; 18 (i). Earlier reports of *P. thurberi* Torr. from Washington County belong here.

Var. *leonardii* Rocky openings in mountain brush, sagebrush-grassland, conifer-aspen, and lodgepole pine communities at 1830 to 3050 m in Cache, Davis, Juab, Millard, Morgan, Rich, Salt Lake, Sanpete, Summit, Utah, Wasatch, and Weber counties; endemic; 63 (iii). The type is from the Wasatch Mts.

Var. *patricus* (N. Holmgren) Neese [*P. patricus* N. Holmgren, type from the Deep Creek Mts., Juab County]. Limestone or (more usually) granitic rocks in mountain brush, sagebrush-grassland, and aspen conifer communities in the Deep Creek and House Ranges, Juab, Millard, and Tooele counties; Nevada; 8 (0). This larger flowered phase from western Utah passes into the typical phase through intermediate populations in the Canyon Range in eastern Millard County.

Penstemon linarioides Gray Suffrutescent perennial 9–38 cm tall; stems puberulent with terete, tapered, retrorsely spreading hairs, ascending to erect from a much branched subwoody base, the old stems sometimes prostrate, rooting and forming mats; leaves 8–35 mm long, linear, entire, puberulent with terete tapered hairs or subglabrous; inflorescence secund, compact to elongate, the nodes usually bearing a single cyme, the cymes 1– or 2 (4) -flowered, the pedicels bracteate, glandular-pubescent; calyx 5–7 mm long, the lobes ovate to broadly lanceolate, shortly caudate, scarious-margined; corolla 13–18 mm long, ampliate, bulged above, flattened and 2–ridged below, lavender blue, the lobes darker, the throat and palate lighter and veined with reddish purple, the palate sparsely bearded; staminode short-exserted, densely bearded apically with a tuft of long yellow hairs, more sparsely bearded with shorter hairs proximally; stamens reaching the orifice or the longer well-exserted, the anthers glabrous, purplish black with a prominent white band along the suture, the sacs 1–1.4 mm long, opening the full length and across the connective, not explanate. Sagebrush, pinyon-juniper, chaparral, mountain brush, and ponderosa pine communities at 1500 to 3000 m in Beaver, Garfield, Iron, Kane (?, type of var. *sileri* from Osmer), and Washington counties; a disjunct station in Boxelder County,(Albee 5773, 5882 UT) may represent an introduction; Nevada, Arizona, New Mexico, and Colorado; 53 (v). Utah plants belong to var. *sileri* Gray [*P. linarioides* ssp. *sileri* (Gray) Pennell; *P. l.* var. *viridis* Keck], and perhaps they are more closely allied to *P. abietinus* than to the typical variety, which occurs to the southeast of our area. Also, plants at the eastern margin of the Utah distribution tend to be more compact and less erect than the more westerly ones and may deserve taxonomic recognition.

Penstemon marcusii (Keck) N. Holmgren Price Penstemon. [*P. pseudohumilis* Jones, not Rydb.; *P. moffattii* ssp. *marcusii* Keck, type from near Price]. Perennial herb 12–26 cm tall; stems ascending to erect, from a branched caudex and taproot, puberulent; leaves basal and cauline, (2) 2.5–8 cm long, 3–10 (15) mm wide, linear-lanceolate to lanceolate, petioled below, becoming sessile near the inflorescence, essentially glabrous (sometimes sparingly pubescent near the inflorescence), entire; inflorescence glandular-pubescent, narrow, the thyrse interrupted, of 3–5 (7) verticellasters, the cymes 2- to 5-flowered, short-petioled; calyx 4–7 mm long, glandular-pubescent with spreading hairs, the lobes narrowly lanceolate, more or less obtuse, herbaceous; corolla 10–16 mm long, glandular-puberulent externally, nearly regular, funnelform, flaring gradually from tube to apex, the lobes subequal, spreading, blue to blue lavender or bluish purple, the throat lighter and veined with purple; stamens very short, included or only slightly exserted from the tube, the anthers glabrous, peltate-explanate, the sacs ca 0.8 mm long; staminode moderately bearded with golden hairs, included in the throat, shortly exceeding the anthers. Shadscale, mat-atriplex, sagebrush, and salt desert shrub-juniper communities, usually in gravelly places on Mancos Shale-derived clay at 1700 to 2000 m in Carbon and Emery counties; endemic; 8 (0). The Jones penstemon occurs in a small area at the east side of the Wasatch Plateau between Price and Emery. It is very similar to *P. goodrichii* of the Uinta Basin, but is separated by small but consistent differences and their edaphic habitats are distinct.

Penstemon moffattii Eastw. Moffatt Penstemon. Perennial herb 0.8–3 dm tall; stems ascending to erect, clustered from a branched caudex and taproot, puberulent; leaves 1.5–6.5 cm long, 2–15 mm wide, entire, glabrous or puberulent, the basal ones petioled and spatulate to broadly obovate or oblanceolate, the upper ones sessile, lanceolate; inflorescence glandular-pubescent (viscid when fresh), rather congested, the stems floriferous for most of their length, the cymes several- flowered, the peduncles and pedicels short; calyx 6–8 (9) mm long, glandular-pubescent, the lobes lanceolate, acute, herbaceous; corolla (14) 15–20 (21) mm long, glandular-pubescent externally, blue violet, moderately ampliate but neither ventricose nor constricted at the orifice, the throat rounded above and below, not noticeably veined, the palate sparsely white-bearded and veined with violet purple, the lobes about equally spreading; stamens included, the anther sacs dehiscing completely and across the connective, the sacs 1–1.4 mm long, white to pale blue, becoming divaricate but not fully explanate, about twice as long as wide; staminode included, sparsely yellow-bearded. Blackbrush, shadscale, mat-atriplex, other salt desert shrub, sagebrush, and pinyon-juniper communities, usually in clay soils, at 1300 to 1900 m in Duchesne, Emery, Garfield, Grand, San Juan, Utah, and Wayne counties; endemic; 13 (ii).

Penstemon montanus Greene Cordroot Penstemon. Perennial, forming loose mats; stems woody below, sprawling or decumbent and often partially buried in rocks and talus, those of the current year erect, 6–35 cm tall, arising severally from the suffrutescent base; herbage puberulent with spreading hairs; leaves 6–51 mm long, 3–26 mm wide, spatulate, elliptic, ovate, or broadly lanceolate, all cauline, the basal ones narrowed to a short petiolar base, the middle ones the largest, the upper sessile and clasping, sharply and prominently toothed, becoming bractlike and more or less entire in the inflorescence; inflorescence of 1–6 verticellasters, the cymes 1-flowered, the peduncles, pedicels, bracts, and sepals glandular puberulent; calyx 10–18 mm long, the lobes narrowly lanceolate to nearly linear, long-attenuate; corolla 24–40 mm long, glabrous externally, pink violet (drying bluish purple) only moderately ventricose-ampliate, longitudinally ridged below, ca 10 mm wide at the orifice, the lobes projecting, the palate bearded with crinkled white hairs; staminode glabrous or rarely with a few white hairs, included; anthers densely white-woolly with long hairs, flat opening, included or reaching the orifice. Mountain brush, aspen, and spruce-fir communities, on talus and cliffs, often in crevices of limestone, at 2400 to 3400 m in Duchesne, Juab, Salt Lake, Sanpete, and Utah counties; Idaho, Montana, and Wyoming; 18 (i).

Penstemon nanus Keck Dwarf Penstemon. Perennial herb 4–15 cm tall; herbage evenly and densely cinereus-puberulent; stems spreading to ascending, few to numerous from a thick, short-branched, taprooted caudex; leaves 1–5.5 cm long, 2–8 mm wide, entire or sometimes irregularly toothed, the basal ones numerous, petioled, spatulate to narrowly oblanceolate, the cauline ones few, (ob)lanceolate (sometimes narrowly so), becoming sessile above; inflorescence glandular-pubescent, usually leafy-bracteate below, the bracts clasping, of 2–5 (6) mostly crowded verticellasters, the cymes several-flowered, the peduncles and pedicels short; calyx 4–8 mm long, glandular-pubescent with spreading hairs, the lobes narrowly ovate to lanceolate, herbaceous, suffused with purple; corolla (10) 12–17 (20) mm long, tubular-funnelform, the throat rounded above and below, glandular-puberulent externally, the palate sparsely bearded, the lobes subequal, spreading, rose purplish (drying blue), albino or pink forms rarely present; anthers included or reaching the orifice, horseshoe-shaped to sagittate, oblong, the sacs not divaricate, 1.4–2 mm long, white to pale blue, dehiscing the full length and across the connective, glabrous; staminode about reaching the orifice, densely bearded with short golden orange hairs. Mixed desert shrub, sagebrush, and pinyon-juniper communities on calcareous or dolomitic gravels at 1580 to 2140 m in Beaver, Iron (?), and Millard (type from ca 10 mi E of Garrison) counties; endemic; 19 (i).

Penstemon navajoa N. Holmgren Navajo Mt. Penstemon. Short-lived perennial herb 2–4.5 dm tall; stems few, glabrous, slender, ascending to erect, wandlike, the leaf pairs remote, the inflorescence lax; leaves glabrous, entire, 2.5–9 cm long, 1–15 mm wide, rounded to acute, the basal ones oblanceolate, narrowed to petiolar bases, the cauline ones narrowly oblanceolate to linear, sessile; inflorescence glabrous, of 2–7 verticellasters, the cymes 1- or 2 (3) -flowered; calyx 3.5–4.5 mm long, glabrous, the lobes ovate, obtuse to acute, scarious-margined; corolla 18–21 (23) mm long, glabrous externally, ventricose-ampliate, the lobes bluish, the upper erect to spreading, the lower spreading or reflexed, the throat and tube pale blue to white, the palate sparsely white-bearded; staminode glabrous, slightly expanded apically; anthers exserted, the sacs (1.2) 1.4–1.8 mm long, glabrous or sparsely long-villous, dehiscing from the distal end to the connective, the valves little-spreading, the suture line straight, spiculose-margined. Ponderosa pine and conifer-aspen woods, in open rocky places, at 2500 to 3160 m on Navajo Mt.(type locality), San Juan County; endemic; 7 (0).

Penstemon ophianthus Pennell [*P. jamesii* ssp. *ophianthus* (Pennell) Keck]. Perennial herb (1) 1.5–4 dm tall; stems clustered from a branching caudex and taproot, ascending to erect, glabrous to sparingly puberulent, more or less viscid when fresh, entire to undulate or the upper sometimes sinuate-dentate, spatulate to narrowly oblanceolate, the basal and lower cauline ones narrowed to a petiolar base, the upper ones sessile; inflorescence glandular-pubescent, of (2) 4–7 verticellasters, the cymes several-flowered, more or less congested; calyx 5–10 mm long, glandular-pubescent, the lobes lanceolate, herbaceous or narrowly scarious-margined near the base; corolla 15–21 (24) mm long, glandular-pubescent, ventricose-ampliate, constricted at the orifice, bluish to pale blue lavender, longitudinally veined with purple stripes nearly throughout, the palate bearded with long white to pale yellow hairs, the lower lobes spreading-reflexed, the upper projecting-spreading; stamens included or the longer pair sometimes exserted, becoming peltate-explanate, the sacs 0.7–1 (1.2) mm long, about as broad, glabrous; staminode exserted, bearded with long, yellow hairs. Mixed desert shrub, pinyon-juniper, sagebrush, and ponderosa pine communities, often in sandy soil, at 1500 to 2400 m in Emery, Garfield, Grand, Kane, Sevier, and Wayne (type from near Loa) counties; Colorado, Arizona, and New Mexico; 28 (iii). This species passes into *P. breviculus* (Keck) Nisbet & R. Jackson (q.v.).

Penstemon pachyphyllus Gray ex Rydb. Thickleaf Penstemon. [*P. nitidus* var. *major* Benth. in A. DC., type from "Hillsides of Du Chesne Fork," Duchesne (?) County;]. Perennial herb (1) 1.5–5.7 dm tall; herbage glabrous, strongly glaucous; stems ascending to erect, 1 or few from a woody caudex and taproot; leaves thick and fleshy, entire, the basal ones petiolate, 1.8–18 cm long, 5–45 (53) mm wide, spatulate to broadly lanceolate, the cauline ones 1.5–8 cm long, 7–32 mm wide, sub-orbicular, (ob)ovate or (ob)lanceolate, rounded to acute, occasionally mucronate, sessile and clasping or the lower narrowed to a short petiolar base; inflorescence glabrous, the thyrse congested to interrupted, of (2) 4–10 verticellasters, the cymes (1) 3– to 6–flowered; calyx 4–8 mm long, glabrous, the lobes ovate to lanceolate, acute to acuminate, scarious-margined; corolla 15–23 mm long, ampliate, or sometimes tubular-salverform glabrous externally and internally or with a few long white hairs on the palate, the limb when well pressed usually 12–20 mm wide, the lobes about equally spreading or the lower longer, blue, blue violet, or purple to dark blue purple, the throat with or without prominent guidelines; stamens included or reaching the orifice, the anthers glabrous, the sacs (0.7) 1–1.5 mm long, dehiscing the full length and across the connective, not explanate; staminode reaching the orifice or shortly exserted, dilated distally, the tip 0.5–2.2 mm wide, sparsely to densely bearded distally with pale yellow to golden brown hairs 0.5–2 mm long. Three intergrading phases are recognized at varietal level:

1. Staminode sparsely bearded with hairs 0.2–1 mm long, these scarcely tangled; plants predominantly of central and northern Uintah and Daggett County ....
   .... *P. pachyphyllus* var. *mucronatus*
— Staminode densely bearded with long tangled hairs that obscure the apex, these 1–2 mm long; distribution various 2
2(1). Staminode broadly expanded at the apex, 1–2 mm wide; corolla usually with definite guidelines; plants principally of the Tavaputs Plateau and Duchesne County ..
   .... *P. pachyphyllus* var. *pachyphyllus*
— Staminode not broadly expanded at the apex, less than 1 mm wide; corolla lacking prominent guidelines; plants of Sanpete and Emery counties south and southwestward .... *P. pachyphyllus* var. *congestus*

Var. **congestus** (**Jones**) **Holmgren** [*P. acuminatus* var. *congestus* Jones, type from near Rockville; *P. p.* ssp. *congestus* (Jones) Keck in Kearney & Peebles; *P. congestus* (Jones) Pennell]. Beaver, Emery, Garfield, Iron, Kane, Millard, Piute, Sanpete, Sevier, Washington, and Wayne counties; Nevada and Arizona; 113 (viii).

Var. **mucronatus** (**N. Holmgren**) **Neese** [*P. mucronatus* N. Holmgren, type from 4 mi S of Manila]. In various substrates with a clay component in salt desert shrub, sagebrush, and pinyon-juniper communities at 1650 to 2150 m in Daggett, Duchesne, and Uintah counties; Wyoming and Colorado; 80 (xxv). Var. *mucronatus* is a well-marked taxon in the vicinity of the type collection, but is fully transitional to var. *pachyphyllus* in Duchesne and southern Uintah counties. In addition to characters given in the key, var. *mucronatus* commonly possesses a relatively broad limb and sky-blue (when dried) rather than blue violet or dark blue purple corolla. The corolla throat and the lobes near their base are usually prominently lined on all sides within with wine-red guidelines. This is one of our most beautiful penstemons.

Var. **pachyphyllus** Salt desert shrub, sagebrush-grass, pinyon-juniper, mountain brush, and conifer communities at 1340 to 3200 m in Carbon, Duchesne, Uintah, Utah, and Wasatch, counties; endemic; 56 (xiii). Var. *pachyphyllus* shares with var. *congestus* the possession of a densely wooly staminode apex; it shares with var. *mucronatus* the characters of a broadly dilated staminode, cauline leaves more often rounded-mucronate than lanceolate, and a bright blue, broad corolla limb with prominently veined lobes. Members of the *pachyphyllus* complex in the Uinta Basin have been variously interpreted, with the name *P. osterhoutii* frequently misapplied.

Penstemon palmeri Gray Palmer Penstemon. Plants robust, perennial, herbaceous to suffrutescent; stems ascending to erect, clustered from a branched woody caudex, glabrous, glaucous, often becoming glandular-puberulent in the inflorescence; leaves 2–13 cm long, 5–48 (55) mm wide, glabrous, thick and usually strongly glaucous, the lower ones petioled, becoming sessile upward, the bases progressively broader, finally broadly connate-perfoliate near the inflorescence, usually sharply serrate, rarely subentire; inflorescence tall, wandlike, more or less secund, of 4–17 or more verticellasters, the cymes 1– to 3 (5) -flowered, the pedicels ascending to appressed, the axis, peduncles, and pedicels glandular-pubescent or glabrous; calyx mostly 4–6 mm long, densely glandular-pubescent to glabrous, broadly ovate, acute, scarious-margined; corolla 24–37 mm long, finely glandular-puberulent, abruptly broadly inflated from the short tube, ventricose, the throat 11–20 mm wide when pressed, constricted at the orifice, prominently bilabiate, the upper lip arching over the reflexed lower one, the upper lobes spreading, the lower reflexed, creamy white to pink externally, pink to dark rosy pink internally, the palate prominently veined with wine red guidelines on a white background; stamens about reaching orifice, the sacs divaricate, 1.7–2.4 mm long, about twice as long as wide, dehiscing the full length and across the connective, cupped within the upper lip, white, glabrous; staminode long-exserted, filiform proximally, moderately dilated and bearded with a dense tuft of long yellow hairs distally, curved apically. Two partially sympatric varieties occur:

1. Inflorescence axis, peduncles, and pedicels glandular pubescent .... *P. palmeri* var. *palmeri*
— Inflorescence axis, peduncles, and pedicels glabrous . ... *P. palmeri* var. *eglandulosus*

Var. **palmeri** Warm desert shrub, cool desert shrub, pinyon-juniper, mountain brush, and ponderosa pine communities at 800 to 2750 m in Beaver, Cache, Davis, Garfield, Iron, Juab, Kane, Millard, San Juan, Sanpete, Washington, and Wayne counties; Nevada, California, and Arizona (introduced in Idaho); 62 (iv). Probably restricted originally in our area to the southern half or less of Utah, Palmer penstemon has been included widely in seeding mixtures and has become established in many places beyonds its natural distribution.

Var. **eglandulosus** (Keck) N. Holmgren [*P. p.* ssp. *eglandulosus* Keck, type from 2.5 mi N of Kanab]. Mixed desert shrub, mountain brush, ponderosa pine, and pinyon-juniper communities at 1500 to 2600 m in Garfield, Iron, Kane, Washington, and Wayne counties; Arizona; 23 (ii).

Penstemon parvus Pennell Little Penstemon. Perennial herb 7–20 (25) cm tall; stems slender, decumbent to erect, arising from slender, fibrous-rooted caudex branches; herbage puberulent; leaves 0.7–6 cm long, 1–6 mm wide, entire, often involute or folded, the basal ones spatulate to narrowly oblanceolate, narrowed to a petiolar base, rounded to obtuse, the cauline ones mostly linear, sessile, obtuse to acute; inflorescence secund, glandular puberulent at least above, of 3–8 verticellasters, the cymes 1– to 2–flowered; calyx 3.5–5 mm long, glandular, usually cyaneous, the lobes ovate to broadly lanceolate, obtuse to shortly acuminate (rarely truncate), the margins herbaceous and entire to narrowly scarious-erose; corolla 14–20 mm long, glandular-puberulent externally, the palate glabrous, blue to dark blue, moderately ampliate and bilabiate, the lobes spreading; staminode white, glabrous, somewhat dilated apically, reaching the orifice or slightly exserted; anthers glabrous, the longer pair exserted, the sacs 1–1.5 mm long, dehiscing the full length from distal end to connective, the valves little-spreading. Sagebrush-grassland communities at 2500 to 3100 m in Garfield (type from the Aquarius Plateau), Piute, and Wayne (?) counties; endemic; 8 (ii).

Penstemon petiolatus Brandegee Crevice Penstemon. Suffruticose perennial; stems freely branching to form a low rounded shrub (0.5) 0.9–2 (4) dm tall, at first green to blue-glaucous, puberulent, eventually becoming trunk-like, gnarled, and blackish near the base; leaves thick, 7–40 mm long, abruptly petiolate, the blade sharply dentate, 4–27 mm wide, 5–26 mm long, mostly conduplicate, green or (usually) blue-glaucous, suborbicular to broadly ovate, obtuse, the petiole and sometimes the blade puberulent to finely papillate; inflorescence relative short and broad, with 2–4 verticellasters; calyx 4–7 mm long, glandular-pubescent, suffused with purple, the lobes herbaceous or very narrowly scarious-margined, narrowly lanceolate to elliptic, acute, the cymes 2– to 4–flowered, the peduncles and pedicels glandular; corolla carmine pink, 13–17 mm long, glandular pubescent, narrowly ampliate, the palate sparsely bearded; staminode short-bearded with yellow hairs, slightly exserted; stamens included or reaching the orifice, the anthers glabrous, peltate-explanate, the sacs 0.6–0.9 mm long. Warm desert shrub and pinyon-juniper communities, on cliffs of limestones, at 900 to 1300 m in the Beaverdam Mts., Washington County; Nevada; 12 (ii).

Penstemon pinorum L. Shultz & J. Shultz Pinyon Penstemon. Perennial herb from a branching caudex; stems puberulent to glabrate below, usually glabrous above or glandular-puberulent near the inflorescence; leaves ca 2–8 cm long, 2–10 mm wide, sharply serrate (rarely entire), narrowly (ob)lanceolate, rounded to acute, the basal ones narrowed to a petiolar base, this puberulent to glabrate, the blades glabrate or glabrous, the cauline leaves, clasping; inflorescence conspicuously glandular-pubescent, moderately congested to lax (shade forms), not secund; verticellasters 5–7, the cymes 2– to 7–flowered; calyx ca 6 mm long, glandular-pubescent, the lobes lanceolate, acute, cyaneous; corolla 10–15 mm long, only moderately ampliate, blue to bluish lavender, the exterior sparsely glandular-pubescent, the palate white- to yellow-bearded, with purple guidelines; staminode exserted, bearded with flattened golden yellow hairs most of the length; anthers glabrous, the sacs ca 1 mm long, dehiscing the full length and across the connective, not explanate, remaining nearly parallel, the longer pair exserted. Pinyon-juniper community at 1700 to 2000 m in Iron (type from 8 km SW of Newcastle) County; endemic; 3 (i).

Penstemon platyphyllus Rydb. Broadleaf Penstemon. [*P. heterophyllus* var. *latifolius* Wats., type from Cottonwood Canyon, Salt Lake County]. Suffrutescent perennial 3–6 dm tall from a branching caudex; flowering stems mostly unbranched, erect, numerous, herbaceous, from decumbent or prostrate woody bases of old stems, puberulent or glabrate; leaves 2–5.6 (6) cm long, 3–15 (20) mm wide, all cauline, firm, plane, glabrous or minutely scabrous on the margins, glaucous, the middle and upper ones elliptic, tapering to both ends, the lower and those of vegetative shoots often oblanceolate or spatulate; inflorescence of 5–10 verticellasters, the cymes 1– to 3–flowered, the pedicels 0.5–6 mm long, glabrous or obscurely granular-puberulent at the summit; calyx 4–6 mm long, glabrous, the lobes lanceolate, acuminate to attenuate, recurved; corolla (18) 20–30 mm long, lavender or reddish violet, glabrous, bilabiate, the tube 3–3.5 mm in diameter, gradually widening to the ventricose-ampliate throat, the lobes subequal, the upper projecting to spreading, the lower spreading to reflexed; stamens (the longer pair) exserted, the anthers 1.5–2.5 mm long, bluish purple, glabrous on the sides, sparsely pubescent on the inner margins, the suture spiculose-margined, 1/3–1/2 the length of the sac; staminode slightly exserted, glabrous, white- or blue-tipped. Mountain brush communities at 1600 to 2700 m in the Wasatch Mts., Davis, Salt Lake, Utah, and Weber counties (also in Indian Canyon, Duchesne County, Harrison 406 BRY); endemic; 15 (iii).

Penstemon procerus Dougl. ex Graham Small-flowered Penstemon. [*P. rydbergii* var. *rydbergii* in part, of authors, not A. Nels.]. Perennial herb 0.8–4 (5) dm tall; stems numerous from a freely branched, fibrous-rooted caudex, puberulent to glabrate, the flowering ones few, decumbent to erect, slender, overtopping the sometimes matlike basal clusters of sterile shoots; leaves 0.8–8 cm long, 2–17 (30) mm wide, entire, glabrous to sparingly puberulent, elliptic to (ob)lanceolate, petioled below, the middle and upper ones sessile, ascending to appressed; inflorescence interrupted, fascicled, glabrous to puberulent, of 1–4 (or rarely as many as 8) verticellasters, if solitary the inflorescence appearing capitate, the cymes densely flowered, mostly 2–3 cm wide, usually remote, somewhat globular, the peduncles usually obscured; calyx 1.7–5.5 mm long, glabrous to obscurely puberulent, the lobes ovate to lanceolate, truncate-apiculate to caudate, scarious-margined at the base, often broadly so and conspicuously erose to lacerate; corolla 6–12 mm long, spreading to deflexed, dark blue to blue violet, glabrous externally, sparsely to densely bearded with yellowish hairs on the palate; throat tubular to moderately widened to the limb, 2–3.5 (4) mm in diameter when well pressed, often slightly arched downward; corolla lobes small, 1–2.5 mm wide, widely spreading; stamens about reaching the orifice, the sacs 0.4–0.7 mm long, glabrous, suborbicular to ovate, about as broad as long, dehiscing the full length and across the connective, becoming explanate, blue black; staminode included, slender, bearded at the very tip with golden brown hairs; $2n = 16$. Dry to usually moist places in loamy soil, in sagebrush, sedge, willow, and grass-forb meadows in mountain brush, aspen, spruce-fir and alpine communities at 2200 to 3600 m in Beaver, Cache, Carbon, Daggett, Duchesne, Emery, Garfield, Grand, Kane, Piute, Salt Lake, Sanpete, Sevier, Summit, Uintah, Wasatch, and Wayne counties; Alaska and Yukon south to California, Nevada, and Colorado; 111 (xxii). The phase in the Uinta and Wasatch mts. with calyx lobes 3–5 mm long, corolla 8–12 mm long, and plants mostly 2–4 cm tall is assignable to **var. procerus**; that from the southern Wasatch Mts. and the Wasatch Plateau with calyx lobes mostly 2–3 mm long, corolla 6–9 mm long, and plants mostly 1–3 dm tall is assignable to **var. aberrans (Jones) A. Nels.** [*P. confertus* var. *aberrans* Jones, type from Soldier Summit; *P. p.* ssp. *aberrans* (Jones) Keck].

Penstemon pseudoputus (Crosswhite) N. Holmgren [*P. virgatus* ssp. *pseudoputus* Crosswhite]. Perennial herb to 4 dm tall; stems slender, erect, one to few from a branched fibrous-rooted caudex; herbage finely puberulent to glabrate; leaves entire, linear or the basal ones narrowly oblanceolate, 1–5 (8) cm long, 1–3 mm wide, involute or folded; inflorescence secund, narrow, strict, of about 5 verticellasters, the cymes 1 (2) -flowered; calyx 2–3 mm long, glabrous, the lobes ovate, obtuse, scarious-margined, entire to erose; corolla 15–22 mm long, glabrous, moderately ventricose-ampliate, the upper lobes projecting-spreading, the lower reflexed, blue purple to blue lavender, the palate lighter with purple guidelines, glabrous; anther sacs 1.2–1.5 mm long, glabrous, the longer pair much exserted, the shorter reaching the orifice, dehiscing from the distal end to the connective, the suture scabrous-margined; staminode white, glabrous, slightly expanded apically, about reaching the orifice. Sagebrush community at 2500 m in Garfield County; Arizona; 1 (0). Our solitary record (Foster & Foster 4489) is from near Panguitch Lake from "loamy soil, rolling open sage community."

Penstemon radicosus A. Nels. Matroot Penstemon. Perennial herb 2–4 dm tall; stems ascending to erect, slender, few to numerous from a woody caudex, puberulent with retrorsely curved hairs; leaves all cauline, 1.5–5.5 cm long, 1.5–11 mm wide, narrowly (ob)lanceolate or elliptic, sessile or the lower with short petiolar bases, ascending, pubescent to glabrate, entire; inflorescence glandular-pubescent, of 2–5 verticellasters, the cymes 1- to few-flowered, not congested, the lower peduncles sometimes elongate; calyx 5–8 mm long, glandular-pubescent, the lobes ovate to lanceolate, the margins below the attenuate tip scarious, erose to denticulate; corolla (15) 18–21 (23) mm long, glandular-pubescent, only slightly ampliate, the upper lip short and the lobes reflexed, the lower lip projecting and the lobes spreading, blue to blue violet dorsally, whitish and longitudinally 2-ridged ventrally, the palate white-bearded; stamens included or the longer pair reaching the orifice, the sacs 0.8–1.3 mm long, blue black, glabrous, opening the full length and across the connective, not explanate; staminode about reaching the orifice, yellow-bearded; $2n = 16$. Juniper, sagebrush-grass, mixed mountain brush, and aspen-fir communities at 1370 to 2400 (3000) m in Box Elder, Daggett, and Rich counties; Nevada, Idaho, Montana, Wyoming, and Colorado; 11 (ii).

Penstemon rostriflorus Kellogg Bridges Penstemon; Beaked Penstemon. [*P. bridgesii* Gray; *P. bridgesii* var. *amplexicaulis* Monnet]. Perennial, mostly 3–9 dm tall; stems few to many, erect or ascending, glabrous or obscurely puberulent below, arising from an open, freely branched caudex or subshrubby base; leaves entire, all cauline, 1.5–9 cm long, 1–2 (14) mm wide, (ob)lanceolate or the lower sometimes spatulate, the upper usually linear, rounded to obtuse or acute, sessile or the lower narrowed to a petiolar base, glabrous (usually) or the lower obscurely puberulent to glabrate; inflorescence of 3–13 verticellasters, the cymes 1- to 3-flowered, the pedicels ascending, the flowers spreading to somewhat declined, the lower axils sometimes producing elongate, several-flowered, racemiform branches, the inflorescence then lax and open; pedicels and calyx glandular-pubescent, the calyx 4–6 mm long, the lobes ovate or lanceolate, acute, entire, the margins herbaceous; corolla 23–31 (33) mm long, sparingly glandular puberulent externally, scarcely ampliate, red to orange red or rarely yellow, the tube widening imperceptibly to the limb, strongly 2-lipped, the upper lobes scarcely divided, projecting beaklike, ca 1 cm long, the lower 3 narrowly oblong, reflexed; anthers glabrous, narrowly horseshoe-shaped, prominently exserted, sometimes equaling or even exceeding the projecting upper lip, the sacs ca 2 mm long, subparallel or the distal ends slightly divergent, dehiscing only across the curved proximal end, the suture line spiculose-margined; staminode glabrous, slender, not expanded apically, exserted equaling or exceeding the upper lip; $2n = 42$. Creosote bush, blackbrush, mixed desert shrub, pinyon-juniper, sagebrush, mountain brush, ponderosa pine, aspen-conifer, and alpine communities at 1000 to 3300 m in Beaver, Garfield, Iron, Kane, Millard, Piute, San Juan, Sevier, Washington, and Wayne counties; California to Colorado, Arizona, and New Mexico; 67 (viii).

Penstemon rydbergii A. Nels. Rydberg Penstemon. [*P. aggregatus* Pennell; *P. rydbergii* ssp. *aggregatus* (Pennell) Keck; *P. rydbergii* var. *aggregatus* (Pennell) N. Holmgren]. Perennial herb; basal leafy sterile shoots usually present; flowering stems 0.2–5 dm tall, puberulent or rarely glabrate or glabrous, ascending to erect from a woody crown, the old caudex branches eventually fibrous-rooted; leaves 1.5–15 cm long, 3–33 (35) mm wide, glabrous or rarely puberulent near the base, elliptic to (ob)lanceolate, the middle cauline ones tapered to the scarcely clasping base (usually not so broadly clasping as in *P. watsonii*); inflorescence interrupted, fascicled, of 1–4 (6) verticellasters (if verticellasters 1 or 2 the inflorescence subcapitate), the fascicles mostly 3–4 cm wide, many-flowered but less densely so than in *P. procerus*, the peduncles usually evident; inflorescence and calyx glabrous or (usually) puberulent to pubescent with retrorse-spreading hairs; calyx 3–9 mm long, the lobes ovate-lanceolate to lanceolate, acuminate to caudate, often ciliate, broadly scarious-margined, erose to lacerate; corolla (10) 12–17 (22) mm long, glabrous externally or with a few long hairs on the backs of the lobes, moderately ampliate, the throat 3–4.5 mm in diameter at the summit, the lobes spreading, mostly 2–3.5 mm wide, bearded with white to yellowish hairs on the palate and near the sinuses of the lobes; stamens reaching the orifice or the longer pair exserted, the sacs 0.6–1.1 mm long, ovate, usually longer than wide, opening the full length and across the connective, not fully explanate, blue black, glabrous; staminode bearded distally with golden yellow hairs or rarely subglabrous; 2n = 32. Mountain brush, sagebrush, and aspen-fir-spruce parkland communities at (2100) 2500–3700 m in Box Elder, Cache, Carbon, Duchesne, Garfield, Iron, Juab, Morgan, Rich, Salt Lake, Sanpete, Sevier, Summit, Uintah, Utah, and Wasatch counties; Washington, Oregon, Idaho, Montana, Wyoming, California, Nevada, Arizona, and New Mexico; 86 (viii). Separation of most materials of *P. rydbergii* and *P. procerus* is unequivocal, but in the Uinta Mts. some populations are intermediate. Plants referred previously to *P. rydbergii* var. *rydbergii* are often transitional to or indistinguishable from *P. procerus* var. *procerus* and may represent introgressants. Occasional individuals of *P. rydbergii* are transitional to *P. watsonii*. Intermediacy of *P. rydbergii* suggests an origin from hybrid derivatives of *P. watsonii* and *P. procerus*. Resolution of relationships requires additional study.

Penstemon scariosus Pennell Plateau Penstemon. Perennial herb 1.5–5 dm tall; stems several, clustered from a thick crown, decumbent to (usually) ascending or erect, glabrous to puberulent; leaves 3–17 cm long, 2–23 mm wide, linear to (ob)lanceolate, usually involute or channeled above, entire, glabrous, the lower tapered to a winged petiolar base, the upper sessile; inflorescence glandular-pubescent or rarely glabrous, congested, secund, of 3–9 verticellasters, the cymes (1) 2– to 5-flowered; calyx 3–12 mm long, glandular-pubescent or rarely glabrous, the lobes ovate to lanceolate, acute to caudate, scarious-margined (usually broadly and conspicuously so), entire to (usually) erose; corolla (15) 22–30 (33) mm long, ventricose-ampliate, rounded above, slightly flattened and 2-ridged below, prominently 2-lipped, the upper lobes spreading, the lower spreading-reflexed and larger than the upper, suborbicular, the bases overlapping, pale lavender blue or blue purple, glabrous to glandular-pubescent externally, the palate glabrous and without guidelines; stamens exserted, the anthers sparsely to moderately white-bearded with slender flexuous hairs about as long as the width of the sac, dark blue, the sacs 1.3–2.6 mm long, dehiscing nearly the full length but not across the connective, divaricate at about a 90 degree angle; staminode shortly exserted, white, moderately dilated apically, sparsely yellow-bearded. As noted by Holmgren (in Cronquist et al. 1984, q.v.), *P. scariosus* exhibits a wide range of variability that sorts into weakly differentiated varieties. In addition to those recognized below there exist other races, each differing in subtle ways, growing on disjunct peculiar calcareous, sandy, or shaly barrens. The mosiac of variation seems to dictate against recognition of further infraspecific taxa. Most of our material can be referred to the following varieties:

1. Corolla mostly over 25 mm long and calyx over 8 mm long; plants of the Wasatch Plateau and Sevier River Valley .... *P. scariosus* var. *scariosus*
— Corolla and calyx mostly less than 25 mm and 8 mm long respectively; plants of northeastern Utah and the southern Wasatch Mts. 2

2(1). Corolla conspicuously glandular externally, usually blue purple; stems and calyx often cyaneous; stems decumbent to ascending; plants of sandstone crevices, summit of Blue Mountain Plateau, Uintah County ... . *P. scariosus* var. *cyanomontanus*
— Corolla usually glabrous, blue to pale blue lavender; stems ascending; plants of other distribution 3

3(2). Leaves linear to linear-lanceolate, rarely over 7 mm wide; corolla usually pale blue lavender; plants of the Green River Formation, southern Uintah County ... . *P. scariosus* var. *albifluvis*
— Leaves lanceolate, mostly over 7 mm wide; corolla blue; plants of the Uinta Mts., Tavaputs Plateau, and southern Wasatch Mts. .... *P. scariosus* var. *garrettii*

Var. albifluvis (England) N. Holmgren [*P. albifluvis* England, type from along the White River, Uintah County]. Mixed desert shrub and pinyon-juniper communities, on sparsely vegetated shale slopes of the Green River Formation, southeast of Bonanza, Uintah County; Colorado; 42 (xvi). This phase passes into var. *garrettii* in the Hill Creek/Willow Creek area.

Var. cyanomontanus Neese Sandstone crevices in sagebrush-grass communities on the Blue Mountain Plateau summit near the Colorado/Utah border, Uintah (type from Blue Mt.) County; Colorado; a Uinta Basin endemic; 13 (vi). The species was previously reported from the region as *P. cyanocaulis* Payson (Welsh 470 BRY).

Var. garrettii (Pennell) N. Holmgren [*P. garrettii* Pennell, type from Midway, Wasatch County]. In many plant communities on the southern and eastern Uinta Mts., and in the southern Wasatch Mts., Carbon, Daggett, Duchesne, Grand, Uintah, and Wasatch counties; endemic; 40 (xi). This variety is transitional to var. *scariosus* on the Tavaputs Plateau.

Var. **scariosus** In many plant communities in Carbon, Emery, Grand, Juab, Piute, Sanpete (type from Musinea Peak), Sevier, and Wayne counties; endemic; 38 (vi).

Penstemon sepalulus A. Nels. Littlecup Penstemon. [*P. azureus* var. *ambiguus* Gray (not *P. ambiguus* Torr.), type from Provo Canyon]. Woody-based perennial herb 4.5–8 (9) dm tall; herbage glabrous and glaucous; stems much-branched, erect or somewhat decumbent, from a woody caudex; leaves all cauline, 1.5–9 cm long, 2–10 mm wide, entire, sessile, narrowly elliptic to linear, moderately reduced upward, often only subopposite and then bases neither confluent nor connected by a line; inflorescence of 3–14 bracteate verticellasters, the cymes 1- to 2 (3) -flowered, the lower nodes sometimes bearing thyrsoid branches, the pedicels 0.4–1.9 cm long, strongly ascending, glabrous; calyx 1.9–3.2 mm long at anthesis, the lobes mostly about as broad as long, rounded to acute or mucronate, glabrous; corolla 20–29 mm long, moderately ampliate, bilabiate, the lobes spreading, violet, the tube paler and 2–2.5 mm in diameter, ca 1/4 the length of the corolla; staminode white, expanded at the tip, about equaling the throat, glabrous; stamens included or the longer pair reaching the orifice; anthers 1.5–2 mm long, about as broad, horseshoe-shaped, bluish purple, opening only across the confluent apex of the sacs, the suture about 1/2–1/3 the length of the sac, its margin spiculose. Sagebrush, mountain brush, aspen, and Douglas fir communities at 1600 to 2800 m in Juab, Utah, and Wasatch counties; endemic; 35 (ii). A single collection from Washington County (Woodbury s.n. 1925 BRY), may represent an introduction or a mislabeled specimen.

Penstemon speciosus Dougl. ex Lindl. Showy Penstemon. Perennial herb mostly 15–25 cm tall; stems several, ascending, from a branched caudex, usually puberulent below; leaves 1.5–7 cm long, 1–12 mm wide, entire, glabrous or puberulent, rounded to acute, the basal and lower cauline ones oblanceolate and petiolate, the upper narrowly (ob)lanceolate and sessile, often folded; inflorescence secund, of 2–7 rather crowded verticellasters, the flowers ascending to spreading; calyx glabrous (rarely glandular-puberulent), 5–7 mm long, the lobes ovate to broadly lanceolate, acute to shortly acuminate, the margins scarious, erose; corolla 23–35 mm long, light blue, the tube and throat lighter, lavender blue, ventricose-ampliate, bilabiate, the upper lobes projecting-spreading, the lower spreading to reflexed; staminode included or reaching the orifice, glabrous or obscurely short-bearded, little expanded apically; anthers glabrous, the sacs ca 2 mm long, pale, partially dehiscent, the suture line not reaching the connective, sigmoid, often blue-margined. Pinyon-juniper and sagebrush-grass communities at 1800 to 2900 m in Box Elder, Juab, and Tooele counties; Washington, Oregon, California, Idaho, and Nevada; 7 (0). This is a highly variable taxon beyond the limited Utah portion of its range.

Penstemon strictiformis Rydb. [*P. strictus* ssp. *strictiformis* (Rydb.) Keck in Kearney & Peebles]. Herbaceous perennial (1) 2–5 (7) dm tall; stems erect, glabrous, 1 to few from a branched caudex; leaves 2–9 cm long, 3–15 (17) mm wide, entire, glabrous, seldom crisped, mostly narrowly (ob)lanceolate, the basal and lower cauline ones narrowed to a petiolar base, the upper sessile, mostly acute, often folded; inflorescence glabrous, of (3) 5–9 (12?) verticellasters, the cymes (1) 3- to 5-flowered; calyx mostly 5–8 mm long, glabrous, the lobes lanceolate, acuminate or caudate, the tip often recurved; corolla (12) 16–22 (30?) mm long, ventricose-ampliate, bilabiate, the lobes spreading, bluish purple to (usually) pale bluish lavender; staminode included, yellow-bearded or glabrous (?); anthers sparsely pubescent with slender flexuous hairs that about equal the length of the sac, the sacs 1.4–1.6 (2) mm long, dehiscing from the distal end, the line of dehiscence not reaching the connective. Pinyon-juniper communities at 1500 to 2100 m in San Juan County; Colorado, New Mexico, and Arizona; 20 (ii). The name *strictiformis* is tentatively applied to the Utah materials, which have substantially smaller size than given by Holmgren (in Cronquist et al. 1984). Some specimens approach *P. cyanocaulis* (q.v.).

Penstemon strictus Benth. Rocky Mountain Penstemon. [*P. strictus* ssp. *angustus* Pennell]. Herbaceous perennial 3–8 dm tall; stems 1 to few from a fibrous-rooted branched caudex, ascending to erect; herbage glabrous (rarely puberulent); leaves 2–11 (15) cm long, 2–17 (20) mm wide, linear to (ob)lanceolate (usually narrowly so), the basal and lower ones petioled, the upper sessile, often folded, entire, rounded to obtuse; inflorescence glabrous, wandlike, strongly secund, of 3–10 verticellasters, the cymes 1- to 4-flowered, the peduncles mostly 3–15 mm long and erect-appressed; calyx 2.5–5 mm long, glabrous, the lobes ovate, blunt or sometimes apiculate, scarious-margined, usually cyaneous; corolla 23–32 mm long, glabrous, ventricose-ampliate, the upper lobes projecting-spreading, the lower ones spreading-reflexed, blue to dark blue or blue purple; anthers well-exserted (the longer pair), the sacs ca 2 mm long, spreading at about a 90 degree angle, remaining indehiscent in the proximal 1/4–1/6, white-villous with hairs mostly longer than the sacs; staminode included or reaching the orifice, sparsely villous to glabrous. Pinyon-pine, mountain brush, snowberry, sagebrush, and aspen-conifer communities at 2100 to 3200 m in Daggett, Duchesne, Emery, Garfield, Grand, San Juan, Sevier, Uintah, Utah, Wasatch, and Wayne counties; Wyoming, Colorado, New Mexico, and Arizona; 95 (xxviii). Materials from the Tavaputs Plateau southward are sometimes transitional to *P. comarrhenus* (q.v.).

Penstemon subglaber Rydb. Smooth Penstemon. [*P. glaber* var. *utahensis* Wats., type from the Uinta Mts.]. Robust perennial herb 2–10 dm tall; stems ascending to erect, few to numerous from a stout caudex; herbage glabrous; leaves 1.5–15 cm long, 3–35 mm wide, entire, elliptic, (ob)lanceolate (usually narrowly so), the basal and lower cauline ones narrowed to a petiolar base, the upper sessile, the middle and upper cauline ones mostly 4–7 times longer than wide; inflorescence markedly secund, of 4–11 verticellasters, the cymes 2- to 4 (6) -flowered, the peduncles and pedicels and sometimes the axis glandular-puberulent; calyx 4–8 mm long, glandular-puberulent, often (in dried specimens) dotted with a white exudate, the lobes ovate to broadly lanceolate, obtuse (rarely) to acute or (usually) acuminate, scarious-margined, sometimes broadly so; corolla 21–31 mm long, mid to dark blue (occasional white or pink forms present), ventricose-ampliate, glandular-puberulent externally (sometimes obscurely so); staminode sparsely bearded with short golden hairs, included or reaching the orifice; anthers short-hispid near the attachment of the filaments, usually glabrous laterally toward the suture line, the longer pair exserted, the sacs 1.5–2.1 (2.5) mm long, dehiscing mostly or quite to the connective, mostly opposing-divaricate, the suture line nearly median. Sagebrush, mountain brush, snowberry-tall forb, and sagebrush-grass communities, and in openings in aspen and conifer woods, at 1830 to 3355 m in Carbon, Daggett, Duchesne, Emery, Juab, Piute, Sanpete, Sevier, Uintah, Utah, and Wasatch counties; endemic; 115 (xiv).

Penstemon thompsoniae (Gray) Rydb. Thompson Penstemon. [*P. pumilus* var. *thompsoniae* Gray, type from near Kanab; *P. caespitosus* var. *thompsoniae* (Gray) A. Nels.; *P. c.* ssp. *thompsoniae* (Gray) Keck in A. Nels; *P. c.* var.? *incanus* Gray, type from SW Utah?; *P. pumilus* var. *incanus* (Gray) Gray; *P. incanus* (Gray) Tidestr.]. Low, often mat-forming perennial herb, often obscurely suffrutescent; old stems freely branched, leafless, more or less buried in surface soil and litter, rooting; current stems prostrate to erect, densely leafy, the leafy (above ground) portion 0.4–7.5 (13) cm long; herbage white-cinereous throughout with flattened, appressed, scalelike hairs; leaves 0.4–2 cm long, 1–5 mm wide, narrowed to a petiolar base, spatulate to oblanceolate, thick, crowded; stem and inflorescence scarcely differentiated, the flowers or flower clusters appearing axillary, usually only 1 per axil per node floriferous, the flowers 1 to several per node, when stems elongate all but the lowermost nodes flower-bearing, when stems compact flower-bearing nodes few and the flowers borne within a tuft of leaves that obscures the stem; calyx 4–6 mm long, white-cinereous with appressed scales, the lobes lanceolate, herbaceous; corolla 12–20 mm long, glandular-pubescent externally, 2–lipped, the lips spreading, the tube and throat violet to blue violet, the lobes blue, the throat 2–ridged ventrally, the palate sparsely bearded; anthers about reaching the orifice or the longer pair exserted, the sacs 0.8–1.3 mm long, dehiscing the full length and across the connective but not explanate; staminode golden-bearded, about reaching the orifice. Sagebrush and pinyon-juniper communities, often on calcareous semibarrens, at 1670 to 2030 (3000?) m in Beaver, Iron, Kane, and Washington counties; 13 (0). Holmgren (in Cronquist et al. 1984), reports *P. thompsoniae* only from Kane County in Utah. Additional collections from Kane, as well as from Beaver, Iron, and Washington counties reveal two morphologically and geographically distinct races:

1. Stems tufted, very short, more or less erect, the inflorescence not elongating, the flowers thus borne within the basal tuft; larger leaves 1.2 or more cm long, 3.5 mm or more wide; plants of south-central Kane County ......
   .... *P. thompsoniae* var. *thompsoniae*
— Stems elongating in the inflorescence, more or less prostrate and mat-forming, the flowers borne beyond the basal tuft; leaves, even in the largest, usually smaller; plants from western Beaver and Iron counties ....
   .... *P. thompsoniae* var. *desperatus*

Var. *desperatus* Neese Sagebrush and pinyon-juniper communities at 1800 to 2075 m in Beaver (type from Indian Peak Range) and Iron counties; Nevada; 6 (0). One specimen tentatively assigned here (Atwood 5151 BRY) has stems unusually long, and is probably more closely allied to materials from the Mohave Strip region of Arizona. Possibly it represents still another taxon worthy of naming. Other infraspecific taxa may occur within the species beyond Utah.

Var. *thompsoniae* Juniper-sagebrush and pinyon-juniper communities at 1675 to 1895 m in Kane County; Arizona; 5 (0).

Penstemon tidestromii Pennell Tidestrom Penstemon. [*P. leptanthus* Pennell, type from Twelve-mile Creek, Sanpete County]. Perennial herb 2–6 dm tall; stems several from a branched caudex, finely retrorsely puberulent, ascending to erect; leaves 1.5–7 (10) cm long, 2–18 (25) mm wide, entire, puberulent, obtuse to acute, the basal ones oblanceolate and narrowed to a petiolar base, the cauline ones lanceolate to oblanceolate, the upper sessile and more or less clasping; inflorescence strict, the stems, pedicels, and bracts sparingly and finely puberulent to glabrate, more or less secund, of 4–11 (15) verticellasters, the cymes (1) 2– to several-flowered; calyx 3–5 (6) mm long, glabrous, the lobes ovate to lanceolate, obtuse or more often acute to acuminate, the margins narrowly scarious-margined, entire to irregularly denticulate; corolla 13–21 (22) mm long, moderately ampliate, the lobes blue to deep blue, glabrous, nearly equal, spreading, the tube and throat bluish lavender to violet; stamens included or reaching the orifice, the sacs 1.2–1.8 mm long, glabrous, dehiscing partially, the suture line not reaching the connective; staminode yellow-bearded apically and for ca 3/4's its length, reaching the orifice. Desert shrub, sagebrush, and pinyon-juniper communities at 1630 to 2500 m in Juab, Sanpete (type from the Sanpitch Mts.), and Utah counties; endemic; 12 (ii). The range of flower and anther size reported above, as well as of remaining morphological features, encompasses that designated for *P. leptanthus* Pennell, known definitely only from the type, which was taken from within or closely adjacent to the narrow geographic range of *P. tidestromii*. I can detect only one taxon in our material. One specimen (Collins & Harper 146 BRY) from near Nephi is anomalous in possessing hispid anthers.

Penstemon uintahensis Pennell Uinta Penstemon. Perennial herb (4) 6–20 cm tall; stems 1 to few, slender, glabrous, arising from a compact branched caudex; leaves 1–7 (10) cm long, 1–8 (12) mm wide, entire, glabrous, often folded, narrowly spatulate or (ob)lanceolate to linear, rounded or (usually) obtuse, the lower narrowed to a petiolar base, the upper sessile; inflorescence racemose, secund, glandular-puberulent, of 1–4 (7) verticellasters, the cymes 1– or 2 (3) –flowered; calyx 4–7 mm long, glandular-puberulent, the lobes ovate, acute, or shortly acuminate, the margins scarious, erose; corolla relatively large, showy, 18–25 mm long, glandular-puberulent (sometimes sparingly so), blue to blue violet, ampliate, the lobes spreading, the tube ca 4 mm in diameter; staminode expanded at the tip, golden-bearded, included or barely reaching the orifice; anthers included or reaching the orifice, sparsely hispid on the sides, the hairs much shorter than the width of the sac; sacs ca 1 mm long, dehiscing from the distal end nearly to the connective, becoming divaricate; capsule 8–11 mm long. Spruce-fir and alpine tundra communities at 3200 to 3815 m in Daggett, Duchesne, Summit, and Uintah (type from the Uinta Mts.) counties; endemic; 36 (xii).

Penstemon utahensis Eastw. Utah Penstemon. [*P. eastwoodiae* Heller, type from near Monticello]. Perennial herb (0.8) 1.5–7 dm tall; herbage glabrous, glaucous; stems erect, single or clustered, from a woody crown; leaves thick and leathery, entire, rounded to obtuse apically, the basal ones 2–10 cm long, 4–22 mm wide, spatulate, (ob)ovate to (ob)lanceolate, narrowed to a petiolar base, the cauline ones 1–8 cm long, 2–20 mm wide, lanceolate, sessile or the lower petioled, few, becoming reduced and bractlike above; inflorescence glabrous, lax, more or less wandlike, of 3–11 (15) verticellasters, the cymes 1– to 3 (5) -flowered, not congested, the peduncles and pedicels usually well developed, sometimes branched from the lower axils; calyx 2–4.5 mm long, glabrous, the lobes broadly ovate, obtuse to acute or shortly acuminate, usually tinged with red, broadly scarious-margined; corolla 16–25 mm long, salverform or tubular-funnelform, the lobes flat spreading or the limb slightly oblique, glandular externally, densely glandular near the orifice, crimson, carmine, or sometimes carmine pink, retaining the reddish coloration on drying, without guidelines or these obscure; stamens included, the anthers glabrous, pale, the sacs 0.6–1.1 mm long, broadly ovate, becoming more or less peltate-explanate; staminode included, pale, slightly dilated apically, glabrous to papillate. Mixed desert shrub, blackbrush-juniper, pinyon-juniper, and mountain brush communities at 1200 to 2500 m in Carbon, Emery, Garfield, Grand, Kane, San Juan (type from N of Monticello), Washington, and Wayne counties; Nevada and California; 102 (vii).

Penstemon venustus Dougl. Suffrutescent perennial 4–10 dm tall; stems ascending to erect, sparingly puberulent to glabrate or glabrous, clustered from a branched woody base; leaves (2) 4–8 (12) cm long, 5–20 (35) mm wide, all cauline, glabrous, lanceolate, sharply serrate; inflorescence subsecund, glabrous or the axis sparingly puberulent, of 4–10 verticellasters, the cymes (1) 3– to 6-flowered, the peduncles well developed, ascending, mostly 1–3 cm long; calyx 3–6 mm long, glabrous, the lobes lanceolate, acuminate, scarious-margined; corolla 23–30 (38) mm long, glabrous except for the ciliate lobes, blue to blue violet, strongly ventricose-ampliate, the throat (when well pressed) 10–12 mm wide, bilabiate, the upper lobes projecting-spreading, the lower ones spreading to reflexed; filaments hirsute in the distal portion; anthers 1.3–2 mm long, glabrous or hirsute near the filament, nearly circular in outline, the sacs dehiscing across the confluent apices, the distal ends remaining saclike; staminode flattened, only moderately dilated, bearded with long white hairs; capsule 7–9 mm long. Our only record (Goodrich 19243 BRY) is from a rocky road embankment near Mountain Home, Duchesne County (apparently an introduction); Washington, Oregon, and Idaho; 1 (0).

Penstemon wardii Gray Ward Penstemon. [*P. glaber* var. *wardii* (Gray) Jones]. Herbaceous perennial 1.5–4.3 dm tall; stems erect, clustered, from fibrous-rooted caudex branches; herbage densely cinereus-puberulent (also usually coated with dust from clay substrates); leaves entire, the basal ones 1.5–9 cm long, 3–21 mm wide, spatulate to oblanceolate, rounded to obtuse, narrowed to a petiolar base, the cauline ones 2.2–6.5 cm long, 3–15 mm wide, narrowly lanceolate to oblanceolate, of 2–4 pairs, the upper sessile; inflorescence leafy-bracteate below, of (3) 5–10 verticellasters, the cymes 1– to several-flowered, the pedicels glandular-puberulent; calyx 6–10 mm long, glandular-puberulent, the lobes lanceolate, acute to acuminate, narrowly to broadly scarious-margined, the margins sometimes erose; corolla 24–32 (35) mm long, glandular-puberulent, moderately ventricose-ampliate, the lobes widely flaring to reflexed, the lower ones the larger, light blue to violet blue, the throat and palate lighter, with purple red guidelines; staminode glabrous, included; anthers white, the longer pair exserted, the sacs 2–2.4 mm long, the suture sigmoidally curved, often blue-margined. Semibarren, white to gray, fine-textured (often calcareous or gypsiferous) substrates (principally the Arapien Shale Formation) in desert shrub and pinyon-juniper communities at 1675 to 2075 m in Sanpete and Sevier (type from near Glenwood) counties; endemic; 22 (iv).

Penstemon watsonii Gray Watson Penstemon. [*P. phlogifolius* Greene, type from Castle Gate]. Perennial herb 2.5–6.5 dm tall; stems ascending to erect, clustered from a freely-branched caudex, puberulent or sometimes glabrate above; leaves 1.5–7.5 cm long, 2.5–23 mm wide, all cauline, elliptic to lanceolate, the lower ones petiolate, sessile and clasping above, the middle ones the largest and abruptly narrowed to a clasping base, glabrous (usually) to finely puberulent; inflorescence loosely thyrsoid to moderately fasciculate, of (2) 3–6 (8) verticellasters, the cymes 3– to 7 (10) –flowered, the flowers not particularly crowded, the peduncles and pedicels usually evident, the axis and pedicels puberulent to glabrate; calyx 1.8–3.2 (3.7) mm long, glabrous or essentially so, the lobes triangular to ovate, acute, apiculate, or abruptly short-acuminate, the bases broad, overlapping, scarious-margined, moderately erose; corolla 12–17 mm long, usually spreading horizontally, glabrous externally, the palate white-bearded, the tube violet, the throat and limb rich blue to blue violet, often paler beneath, gradually flaring from the narrow tube, the throat when well pressed 4–5 mm wide at the summit, the lobes about equally spreading; stamens reaching the orifice or the longer pair exserted, the sacs ca 1 mm long, longer than wide, glabrous, dehiscing fully and across the connective, divaricate, not ampliate; staminode slender, reaching the orifice or shortly exserted, yellow-bearded in the distal 3–5 mm; $2n = 16$. Aspen, mountain brush, juniper-sagebrush, sagebrush-grass, pinyon-mountain mahogany, and open conifer communities at (1650) 2000 to 3300 m in mountains and higher plateaus of Beaver, Carbon, Duchesne, Emery, Garfield, Grand, Iron, Juab, Millard, Piute, Sanpete, Sevier, Summit, Tooele, Uintah, Utah, Wasatch, and Wayne counties; Nevada, Idaho, Wyoming, and Colorado; 181 (xxix). Occasional specimens are encountered that appear to be hybrids with *P. rydbergii*.

Penstemon whippleanus Gray Whipple Penstemon. Perennial herb (1) 3–6 (10) dm tall; herbage glabrous below, becoming glandular- pubescent with spreading white hairs in the inflorescence; stems slender, ascending to erect, clustered from a branched, fibrous-rooted caudex; basal tuft of leaves usually well developed; leaves 1.3–13 cm long, 3–35 (40) mm wide, thin, entire to denticulate, rounded to obtuse, lanceolate, the basal ones abruptly narrowed to a slender petiole, the upper sessile; inflorescence somewhat lax, of (1) 2–4 (6) verticellasters, the cymes 1– to 5 –flowered; flowers spreading-reflexed to pendulous; calyx 6–11 mm long, densely glandular-pubescent with spreading white hairs, the lobes narrowly lanceolate, attenuate, blunt, narrowly scarious-margined near the base; corolla 20–30 mm long, glandular-pubescent, the throat abruptly tubular-ampliate from a short tube, the upper lip spreading, the lower one projecting, exceeding the upper, conspicuously bearded with long white hairs, dark blue or wine purple to dull lavender or dingy blue white; stamens about reaching the orifice or the longer pair exserted, the sacs 1–1.5 mm long, glabrous, blue, dehiscing throughout and across the connective, becoming opposite and explanate; staminode exserted, slender, white, bearded apically with a tuft of pale hairs; 2n = 16. Alpine tundra, talus slopes, alpine meadows, krummholz, sagebrush-snowberry-ribes grassland, lodgepole, and aspen-conifer communities, in almost all our higher mountain ranges (apparently absent from the Raft River and Pine Valley mts.) at (1800) 2300 to 3500 m in Beaver, Cache, Carbon, Daggett, Davis, Duchesne, Garfield, Grand, Iron, Juab, Millard, Piute, Salt Lake, Sanpete, Sevier, Summit, Uintah, and Wasatch counties; Montana, Idaho, Wyoming, Colorado, Arizona, New Mexico; 137(xiv). This is one of our most distinctive species of beardtongue.

Scrophularia L.

Coarse, caulescent perennials; stems 4–angled; leaves opposite, petiolate; flowers in terminal paniculate cymes; calyx 5–parted; corolla greenish to brownish or purplish, strongly bilabiate, the upper lip 2–lobed, erect, the lower shorter and with erect lateral lobes, the middle lobe deflexed; stamens 5, 4 fertile, the fifth scalelike on the corolla tube; stigmas usually united, capitate; capsule septicidal; seeds numerous, oblong-ovoid, turgid.

Scrophularia lanceolata Pursh Figwort. [*S. utahensis* Gand., type from Cache (?) County]. Plants 5–20 dm tall; stems glabrate to puberulent below and glandular in the inflorescence; leaves 5–15 cm long, 3–6 cm wide, lanceolate to lance-ovate, the base typically cordate or rounded to truncate, the apex acuminate, the margins serrate to doubly so; inflorescence narrow and elongate; calyx 2–4 mm long, the lobes ovate with scarious, slightly erose margins; corolla 6–10 mm long, the tube urceolate, the throat somewhat constricted; staminode flabellate, yellow green (sometimes purplish); capsule 5–10 mm long, acuminate, ovoid; n = 48. Sagebrush, mountain brush, aspen, and spruce-fir communities at 1705 to 3175 m in Cache, Carbon, Davis, Duchesne, Grand, Juab, Millard, Morgan, Rich, Salt Lake, San Juan, Sanpete, Summit, Tooele, Uintah, Utah, Wasatch, Washington, and Weber counties; British Columbia to Montana, south to California and New Mexico; northeastern U. S.; 54 (0).

Synthris Benth.

Subscapose herbaceous perennials; leaves mostly basal, long- petiolate, the blades ovate, serrate to pinnatifid; cauline leaves much reduced, bracteate, sessile, alternate; inflorescence a dense terminal raceme; calyx 4–lobed, the segments distinct; corolla mostly blue, less commonly pink or white, campanulate, subequally 4–lobed; stamens 2; stigmas capitate; capsule loculicidal, notched apically or entire; seeds 2 to many per locule.

Synthris pinnatifida Wats. Stems 8–22 (28) cm tall, simple, erect, villous to glabrate above to sparsely woolly below; basal leaves long petiolate, the blade 4–6 cm long, ovate in outline, laciniate or pinnatifid to pinnately compound; petioles 2–10 cm long; cauline leaves bracteate, alternate, sessile; raceme dense, the pedicels 1–3 (6) mm long; calyx 2.5–5.5 mm long, the lobes elliptic to oblong, ciliate; corolla 5.5–7.5 mm long, dark blue; capsule 4–6 mm long, subglobose, notched apically, puberulent; seeds 1–2 mm long. Two varieties are present.

1. Leaves pinnatifid to pinnately compound; plants of northern Utah County and northward ....
   .... *S. pinnatifida* var. *pinnatifida*
— Leaves simple, laciniately cleft; plants from Mt. Nebo and the Deep Creek Mts., southward ....
   .... *S. pinnatifida* var. *laciniata*

Var. *laciniata* Gray [*S. laciniata* (Gray) Rydb.; *S. laciniata* ssp. *ibaphensis* Pennell, type from the Deep Creek Mts.]. Spruce-fir and alpine tundra communities at 2775 to 3630 m in Beaver, Garfield, Juab, Iron, Sanpete, Sevier, Utah, and Wayne counties; endemic; 43 (iv).

Var. *pinnatifida* [*Wulfenia pinnatifida* (Gray) Greene]. Spruce-fir, lodgepole pine, alpine tundra, and meadow communities at 2560 to 3390 m in Cache, Daggett, Duchesne, Salt Lake, Summit, Tooele, and Utah (type from American Fork Canyon) counties; Idaho, Wyoming, Montana, and Washington; 29 (0).

Verbascum L.

Biennial or possibly perennial herbs; stems tall, erect; leaves basal and cauline, the basal in a rosette, the cauline alternate, simple, sessile, clasping to somewhat decurrent; inflorescence a raceme or spicate panicle; bracts reduced; calyx deeply lobed, the lobes equal; corolla yellow (rarely white), rotate, almost regular, 5–lobed, the upper pair shorter than the lower 3, the tube short; stamens 5, all fertile, exserted, the filaments commonly villous or the lower ones glabrous or essentially so; capsules globular to ovoid-oblong, septicidal, mostly 2–cleft apically; seeds numerous, longitudinally rugose.

1. Leaves densely woolly; inflorescence spicate, the flowers crowded; filaments with yellow hairs or glabrous .
   .... *V. thapsus*
— Leaves glabrous; inflorescence a laxly flowered raceme; filaments with violet hairs 2
2(1). Leaves glabrous; pedicels 10–25 mm long ... *V. blattaria*
— Leaves puberulent; pedicels 2–5 mm long ... *V. virgatum*

Verbascum blattaria L. Moth Mullein. Biennial; stems 4–12 dm tall, erect, glabrous below, glandular above; cauline leaves 1.5–15 cm long, glabrous, green, lanceolate, dentate, incised or lobed; inflorescence a lax interrupted raceme, elongating to 5 dm; pedicels 5–20 mm long; calyx 5–8 mm long, the lobes lanceolate, glandular; corolla 25–30 mm wide, yellow (rarely white); lower 2 staminal filaments with violet hairs, the upper ones with white or violet hairs; capsule 6–8 mm long, globose; seeds 0.6–1 mm long; 2n = 18, 30. Waste places in salt grass and greasewood communities, or in other open sites, at 1220 to 2075 m in Box Elder, Utah, and Weber counties; naturalized from Europe; widespread in North America; 8 (0).

Verbascum thapsus L. Woolly Mullein. Robust biennial; stems stout, erect, 5–15 dm tall or more, densely woolly, simple or branched; basal leaves in a rosette, oblong-ovate to obovate or lanceolate, 6–50 cm long, petiolate; cauline leaves elliptic-lanceolate, reduced upward, decurrent, entire to crenate; panicle spicate, densely flowered, to 3 dm long or more; pedicels ca 2 mm long; calyx 5–10 mm long, the lobes lanceolate, subequal to the capsules; corolla 15–30 mm wide, mostly yellow (rarely white); lower 2 stamens glabrous or sparingly hairy, the upper ones with yellow hairs; capsule 6–10 mm long, stellate, broadly ovoid; seeds 0.7–0.8 mm long; 2n

= 36. Open sites, especially along roadsides, at 1220 to 2745 m in probably all Utah counties; adventive from Eurasia, not widely established in North America; 39 (i).

Verbascum virgatum Stokes Wand Mullein. Biennial; stems 7–12 dm tall, erect; leaves 2–24 cm long, lanceolate, crenate to dentate, glandular-puberulent and hispid; racemes lax, interrupted, elongating to 4.5 dm; pedicels 2–5 mm long; calyx 5–8 mm long, the lobes lanceolate, glandular; corolla 30–40 mm wide, yellow; lower pair of stamens with violet hairs, the upper ones with white to violet hairs; capsule 6–9 mm long, globose, glandular; seeds 0.6–1 mm long; 2n = 64, 66. Roadsides and other open sites in Box Elder, Davis, and Utah counties; naturalized from Europe; widely distributed in North America; 4 (0).

Veronica L.

Annual or perennial herbs; stems erect, ascending, decumbent, or prostrate; leaves all opposite, cauline, entire to serrate; inflorescence of terminal or axillary racemes or some flowers solitary in the upper axils; bracts foliose to reduced, mostly alternate; calyx 4–lobed; corolla blue, violet to pink or white, subrotate, the tube very short, irregularly 4–lobed, the upper lobe the largest, the lower lobe the smallest; stamens 2; stigmas united, capitate (style measurement includes the stigma); capsule mostly wider than long, compressed, or ovoid, notched or lobed at the tip, often heart-shaped; seeds few to many, plano-convex, flattened, smooth or roughened, small. Note: Two cultivated species are represented in our collections, V. prostrata L. and V. longifolia L., but are not included in the following treatment.

1. Racemes terminal or flowers solitary and axillary; plants often in wet places but not true aquatics 2
— Racemes axillary and opposite, never terminating the main stem; plants aquatic or semi-aquatic 8
2(1). Flowers and fruit long pedicellate, 1.5–2.5 cm long; fruit broadly notched, the lobes divergent, 5.5–8 mm wide .... V. persica
— Flowers and fruit mostly short pedicellate, rarely over 1.5 cm long; fruit not broadly notched and with divergent lobes; capsule mostly less than 5.5 mm wide 3
3(2). Leaves petiolate, at least some palmately 3– to 5–lobed; calyx lobes conspicuously long-ciliate; seeds 1–2 per locule .... V. hederaefolia
— Leaves various but not palmately lobed; calyx lobes glabrous or pubescent but not long-ciliate; seeds few to many per locule 4
4(3). Racemes compact, headlike; capsule longer than broad, 5.4–8 mm long, ellipsoid, shallowly notched; flowers dark blue, large, 5–10 mm wide; plants perennial ... . V. wormskjoldii
— Racemes not headlike; flowers light blue, violet or white; capsules usually wider than long or equal in width and length, often deeply notched; plants perennial or annual 5
5(4). Plants rhizomatous perennials; style 2–3.5 mm long . ... V. serpyllifolia
— Plants tap-rooted annuals; style less than 2 mm long 6
6(5). Flowers white or nearly so; style very short, 0.1–0.3 mm long; leaves narrowly oblong to oblanceolate .... V. peregrina
— Flowers light blue or violet; style over 0.3 mm long; leaves broadly elliptic, ovate or broadly deltoid 7

7(6). Capsules deeply notched to the middle or below, 2–3.5 mm deep; style much shorter than the lobes; flowers 3–4.5 mm wide; capsule 3–5.8 mm wide ...... V. biloba
— Capsules notched 0.5–0.8 mm deep, less than half way to the middle; style 0.4–1 mm long, subequal to the lobes; flowers 2–2.5 mm wide; capsule 3–4 mm wide .... V. arvensis
8(1). Leaves all short-petiolate .... V. americana
— Leaves of the middle and upper flowering stems, at least, sessile and clasping 9
9(8). Leaves 3–5 times as long as wide, mostly entire or inconspicuously toothed; calyx 2.5–3.5 mm long, as long as the capsule; flowers light blue, pink or white, 3–5 mm wide .... V. catenata
- Leaves not more than 3 times as long as wide, mostly distinctly toothed; calyx 3–6 mm long, sometimes longer than the capsule; flowers blue, 5–10 mm wide .... V. anagallis-aquatica

Veronica americana Schwein. ex Benth. American Brooklime. [V. beccabunga var. americana (Schwein.) Raf.]. Rhizomatous perennial, 0.5–4.8 dm tall, glabrous throughout; stems erect or decumbent, mostly branched, rooting at the lower nodes; leaves 1.5–6 cm long, at least twice as long as wide, petiolate, lanceolate to ovate, widest near base; flowers in axillary racemes, 3– to 26–flowered; pedicellate, the pedicel 3–9 (12) mm long; calyx 2–5 (6) mm long, acute; corolla 5–10 mm wide, blue; capsule wider than long, 2–4 mm long, 3–5 mm wide, slightly notched apically; style 1.7–3 mm long; seeds 0.5–0.7 mm long, brownish, plano-convex; 2n = 36. Mostly in moist meadows, seeps, peat bogs, streamsides, hanging gardens, and aspen communities at 1370 to 3205 m in all Utah counties except Davis, Grand, Morgan, and San Juan; Alaska south to California, Arizona, New Mexico, Texas, and Mexico; east to North Carolina; 98 (iv).

Veronica anagallis-aquatica L. Water Speedwell. Aquatic perennial, 0.4–8.3 dm tall, glabrous throughout; stems erect or ascending, simple to much branched below; leaves elliptic, sessile, or short petiolate below, clasping, crenate-serrate or entire, 1–7 cm long; racemes axillary, pedunculate, few- to many-flowered; calyx 3–6 mm long, the lobes lanceolate; corolla 3–5 mm across, blue; fruiting pedicels mostly strongly ascending, 2–7.5 mm long; capsule 2.8–3.8 mm long, slightly longer than wide, scarcely notched, turgid; style 0.9–2.5 mm long; seeds 0.5–1 mm long, plano-convex, brown; 2n = 18, 34–36, 54. Slow moving streams or ditches, sand bars, and wet meadows at 1220 to 2505 m in all Utah counties except Box Elder, Cache, Carbon, Iron, Juab, Morgan, and Tooele counties; widely distributed in North and South America; adventive from Europe; 71 (iii).

Veronica arvensis L. Corn Speedwell. Annual from a taproot, villous-hirsute to puberulent, 0.4–3 dm tall; stems erect or subprostrate, simple or branched; leaves ovate, crenate-serrate, occasionally subcordate below; racemes terminal, 1– to 30–flowered; bracts alternate, larger below than above; calyx 2.5–4.5 mm long, the lobes lanceolate, unequal; corolla small, 2–2.5 mm wide, blue violet; pedicels 0.5–2 mm long; capsule 3 mm long, obcordate, glandular-pubescent marginally, strongly flattened; style 0.4–1 mm long; seeds 0.7–1 mm long, plano-convex; 2n = 16. Open fields and gardens at ca 1370 to 1525 m in Utah County; adventive from Eurasia, but now widely naturalized in North America; 1 (0).

***Veronica biloba* L.** Bilobed Speedwell. Taprooted annual, (2) 5–25 (32) cm tall, villous-hirsute to glandular-puberulent; stems simple or branched below, erect; leaves short-petiolate, the blade lance-ovate or elliptic, mostly 8–20 (40) mm long, coarsely toothed; racemes terminal, elongate, 1– to 25–flowered, glandular, the bracts alternate, reduced; calyx 4–8 mm long, the lobes prominent, mostly veiny, broadly lanceolate; corolla 3–4.5 mm wide, blue, inconspicuous, pedicels elongating to 1.8 cm long in fruit; capsule 2–5.8 mm long, glandular, notched to the middle or below, 2–3.5 mm deep; style 0.9–1.3 mm long; seeds 1.4–2 mm long, pale yellow, tear-shaped and with transverse ridges; 2n = 14. Mostly in disturbed ground at 1310 to 2900 m in Box Elder, Davis, Duchesne, Garfield, Juab, Millard, Morgan, Salt Lake, Sanpete, Sevier, Summit, Tooele, Uintah, Utah, and Weber counties; Colorado, Nevada, north to Idaho and Montana, native of Asia; 37 (ii).

***Veronica catenata* Pennell** Rhizomatous aquatic perennial, 1–6 dm tall, glabrous; stems erect or ascending, branched throughout; leaves lanceolate, entire or subentire, 5–15 mm long, sessile and clasping; racemes axillary, pedunculate, 15– to 25–flowered, the bracts much reduced; calyx 2.5–3.5 mm long, the lobes broadly lanceolate to ovate; corolla 3–5 mm wide, white to pink or light blue; capsule 2.5–3 mm long, yellow brown, plano-convex; 2n = 36. Wet places, slow moving water, ditchbanks, ponds, and lake shores at the lower elevations. *Veronica catenata* is reported for Utah (Intermountain Flora, Vol. 4) along the Wasatch front but I have seen no specimens. This species is closely allied with *V. anagallis-aquatica*.

***Veronica hederifolia* L.** Ivy-leaved Speedwell. Taprooted annual, 2–12 dm tall, sparsely to moderately spreading-hirsute; stems weakly ascending or prostrate, branched below; leaves petiolate, the blade suborbicular to ovate, 4–10 mm long, palmately veined, toothed to somewhat lobed; raceme terminal or a single flower in the upper leaf axils; calyx 3–5 mm long, accrescent, the lobes broadly triangular, conspicuously ciliate; corolla 3–6 mm wide, pale blue, the pedicel to 10 mm long in fruit; capsule 2.5–3.5 mm long, subglobose, shallowly notched (0.2–0.3 mm deep); style 0.6–0.9 mm long; seeds 2.3–3.1 mm long, blackish, cup-shaped, transversely ridged; 2n = 28, 36, 54. Waste places and canyon bottoms in Davis and Salt Lake counties; native to Europe and widely naturalized in the northeastern U.S., but rarely collected in Utah; 2 (0).

***Veronica peregrina* L.** Purslane Speedwell. Taprooted annual, 4–26 cm tall, glandular-pubescent; stems erect or curved below, simple or branched especially below; leaves linear-oblong to oblanceolate, sessile or the lower with a petiolar base, 5–18 (35) mm long, entire or irregularly toothed; racemes terminal, elongate, lax; bracts foliose below and reduced upward, subtending a single flower; calyx 2.3–4 mm long, the lobes narrowly elliptic to lanceolate, unequal; corolla 2–3 mm wide, inconspicuous, whitish; fruiting pedicels 1–2 (2.6) mm long; capsule 2.5–4 mm long, 3.5–4.5 mm wide, obcordate, notch evident (0.2–0.5 mm deep); style 0.1–0.3 mm long, mostly shorter than the notch; seeds 0.5–0.8 mm long, plano-convex, pale yellow; 2n = 52. On moist lake shores, edges of ponds, streambanks, and meadows at 1370 to 3235 m in Beaver, Cache, Daggett, Duchesne, Emery, Garfield, Iron, Juab, Millard, Salt Lake, Sevier, Tooele, Uintah, Utah, and Washington counties; wide-spread in temperate parts of North and South America; 35 (iv). Our plants belong to var. **xalapensis (H.B.K.) St. John & Warren** [*V. xalapensis* H.B.K; *V. peregrina* ssp. *xalapensis* (H.B.K.) Pennell].

***Veronica persica* Poir.** Persian Speedwell. Taprooted annual, 4–28 cm tall, pilose; stems simple or branched, especially below, loosely ascending, often rooting at the lower nodes; leaves short-petiolate, the blade broadly ovate or suborbicular, crenate-serrate, sometimes deeply so, 10–27 mm long; raceme terminal, lax, elongate, 1– to 26–flowered, the bracts similar to the leaves, reduced upward and subtending a single flower, the pedicel 15–25 mm long; calyx 4.5–7 mm long, the lobes prominent, usually veiny, broadly lanceolate to elliptic, ciliate; corolla 7.7–11 mm wide, blue; capsule 2.3–4 mm long, 5.5–8 mm wide, broadly notched (0.7–1.1 mm deep), the lobes divergent; style 1.6–3.0 mm long; seeds 1.5–2.2 mm long, straw colored, plano-convex, elliptic to ovate, transversely rugose; 2n = 28. Weedy species of lawns and gardens, waste places and sagebrush foothills at 850 to 1405 m in Davis, Salt Lake, Utah, and Washington counties; native to Eurasia, naturalized in North America; 15 (0). The flowers open in the morning, but the corollas are deciduous by early afternoon.

***Veronica serpyllifolia* L.** Thyme-leaf Speedwell. Rhizomatous perennial, 4–30 cm tall, glabrous to finely puberulent to glandular above; stems ascending or decumbent below; leaves short-petiolate below, the blade elliptic to broadly ovate, 4–20 mm long, entire to obscurely crenate; raceme terminal, elongate, 1– to 20–flowered, bracts lanceolate, leaflike below; calyx 2.5–3.2 mm long, the lobes ovate to oblong; corolla 3.5–5 mm wide, blue or white; pedicels 2.5–7.5 mm long; capsule 2.6–3.4 mm long, 3.9–4.1 mm wide, glandular-pubescent, notched (0.3–0.8 mm deep); style 2–3.5 mm long; seeds 0.7–0.8 mm long, straw colored, plano-convex; 2n = 14, 28. Riparian habitats, meadow, and streambanks at 1920 to 3355 m in all Utah counties, except Carbon, Davis, Juab, Millard, Morgan, Rich, and Salt Lake counties; Alaska and Newfoundland, south to California, Nevada, Arizona, New Mexico, and Mexico, east to Vermont; Eurasia; South America; 55 (iv).

***Veronica wormskjoldii* R. & S.** Wormskjold Speedwell. [*V. alpina* var. *wormskjoldii* (R. & S.) Hook.]. Rhizomatous perennial, 8–29 (39) cm tall, sparsely to densely villous-hirsute to glandular; stems mostly simple, erect to ascending, leaves elliptic to broadly lanceolate, 10–30 (37) mm long, sessile, entire to slightly crenate; raceme terminal, compact, somewhat elongate in fruit; calyx 4–5.5 mm long, the lobes lanceolate; corolla 5–10 mm wide, deep blue; pedicels 2–6 mm long; capsule broadly notched, 5.4–8 mm long, 4.2–5.5 mm wide, longer than wide, glandular; style 0.8–1.2 mm long; seeds 0.8–0.9 mm long, straw colored, strongly flattened; 2n = 18. Moist streambanks, meadows, near ponds and lakes and alpine slopes at 2835 to 3050 m in Beaver, Daggett, Duchesne, Garfield, Juab, Piute, Salt Lake, San Juan, Sevier, Summit, Uintah, Utah, Wasatch, and Washington counties; Alaska to Greenland, south to California, New Mexico and New Hampshire; Asia; 55 (vi).

SIMAROUBACEAE DC.

Quassia Family

Malodorous trees or shrubs, usually with bitter bark containing oil sacs; leaves pinnate, alternate; stipules usually none or minute; flowers regular, typically imperfect and often dioecious, but sometimes perfect or polygamous, axillary, panicled or racemose; sepals 3–8, distinct or partly united basally; petals 3–5; stamens 2–10 in 1 or 2 series; pistil 1, the ovary superior, of 2–8 united carpels; style 1, typically lobed apically; fruit of samaras; x = 8–13+.

Ailanthus Desf.

Trees or large shrubs with gray bark; leaves large, odd-pinnately compound; flowers small, in large terminal panicles, 5-merous; stamens 10 or 2 or 3; ovary 2– to 5-cleft; fruit a samara.

Ailanthus altissima (Miller) Swingle Tree-of-heaven. [*Toxicodendron altissimum* Miller]. Trees or large shrubs to 20 m tall or more, with smooth gray bark, polygamo-dioecious; stems with large pith; leaves to 6 dm long, with 9–25 leaflets; leaflets lanceolate to elliptic, acuminate, to 15 cm long and 4.5 cm wide, subentire or with 1 or more rounded teeth, especially near the base, these typically each with a circular gland; flowers small, greenish, in large terminal panicles, the staminate malodorous; sepals 5, imbricated; petals 5, spreading; stamens 10 in staminate flowers, 2 or 3 in perfect ones; disk lobed; ovary 2– to 5-cleft, with flat, 1-loculed divisions; ovules solitary in each locule; fruit of 1–5 linear or oblong samaras 3–5 cm long; 2n = 64. Cultivated, escaping, and established weedy trees at 780 to 1800 m in Carbon, Juab, Salt Lake, Sevier, Utah, and Washington counties; introduced from China; 15 (i).

SOLANACEAE A. L. Juss.

Potato Family

Herbs, shrubs, or trees; leaves alternate or fascicled, occasionally opposite, entire to odd-pinnate; flowers in umbels, cymes, panicles, or solitary, perfect, regular or nearly so, 4– to 6-merous; calyx usually 5-lobed or -cleft, rotate, campanulate, or tubular, usually persistent; corolla usually 5-lobed, tubular, campanulate, or rotate, the lobes valvate or imbricate and usually plicate in bud; stamens 5, distinct or slightly united by the anthers; filaments distinct, inserted on the corolla tube alternate with the lobes; anthers opening by slits or pores; ovary superior, usually 2-loculed; style 1, the stigma entire or 2-lobed; fruit a berry or capsule; x = 7–12.

1.	Corolla rotate to broadly campanulate	2
—	Corolla salverform to funnelform or urn-shaped	6
2(1).	Calyx accrescent in fruit, enclosing the berry entirely or nearly so	3
—	Calyx usually not noticeably accrescent in fruit, spiny if investing the fruit	4
3(2).	Calyx becoming much enlarged and membranous-inflated to entirely and permanently enclosing the berry, reticulate-veiny; corolla usually lacking tomentose pads on the lower part of the lobes	*Physalis*
—	Calyx herbaceous and closely investing the berry or most of it; corolla with tomentose pads alternating with the filaments	*Chamaesarcha*
4(2).	Anthers not connivent around the style	*Capsicum*
—	Anthers connivent around the style	5
5(4).	Anthers opening by a terminal pore or slit; plants indigenous and cultivated	*Solanum*
—	Anthers opening by a longitudinal slit from base to apex; plants cultivated	*Lycopersicon*
6(1).	Stamens in 2 sets of 2, the fifth one by itself and often smaller, abortive, or lacking; plants cultivated	7
—	Stamens not differentiated in sets, all alike and fertile	8
7(6).	Stamens all functional, all producing pollen	*Petunia*
—	Stamens not all functional, only 4 producing pollen	*Salpiglossis*
8(6).	Plants shrubs or vines, the branches armed with thorns; fruit a berry	*Lycium*
—	Plants herbaceous; fruit a capsule	9
9(8).	Capsule circumscissile near the apex; corolla ca 2 cm long, slightly irregular	*Hyoscyamus*
—	Capsule opening by longitudinal valves; corolla 2–15 cm long or more	10
10(9).	Capsule completely included in the calyx, not spiny, less than 15 mm long; flowers in terminal racemes or panicles, mostly less than 5 cm long	*Nicotiana*
—	Capsules not included in the calyx, spiny, to 40 mm long or more; flowers solitary in upper stem axils, mainly 8–15 cm long or more	*Datura*

Capsicum L.

Annual herbs; stems erect, much branched, glabrous; leaves ovate, elliptic, or narrowly lanceolate, simple, entire; flowers white or greenish white, pedicellate, solitary or 2 or 3, erect or reflexed; calyx short, truncate or with very short lobes; corolla rotate or nearly so, 5-lobed; stamens 5, mostly bluish; anthers opening longitudinally; ovary 2- or 3-loculed (more in cultivated plants); style simple, the stigma capitate; fruit a many-seeded berry, widely varying in size, shape, and color.

Capsicum frutescens L. Pepper. Annual herbs in our area, or perennial and shrubby under glass; leaves mainly lanceolate to ovate, 1.5–14 cm long, petiolate, usually acuminate; corolla 1–2 cm wide; fruit very diverse in cultivated varieties; 2n = 24. Cultivated ornamental and food plants in gardens throughout Utah, sometimes escaping but not persisting; introduced from the Old World (?); 1 (0).

Chamaesaracha Gray

Perennial herbs; stems leafy, decumbent or prostrate, branched; leaves alternate, entire, repand, or pinnatifid, sessile to petiolate, the petioles margined; flowers solitary on slender peduncles or sometimes 2 or 3 per node, the peduncles recurved in fruit; calyx campanulate, 5-lobed, not becoming bladderlike but closely investing the fruit; corolla rotate, yellow, greenish white or purplish, 5-angled; filaments adnate to corolla base; anthers opening longitudinally; fruit a berry; seeds flattened and more or less wrinkled.

Chamaesaracha coronopus (Dunal) Gray [*Solanum coronopus* Dunal]. Plants diffusely branched from a perennial caudex; stems prostrate or decumbent, subglabrous to densely and finely stellate-pubescent; leaves sessile or shortly petiolate, linear to lanceolate 1.5–6 cm long, tapering at the base, sinuately lobed to occasionally subentire or pinnatifid; peduncles elongate; calyx 2.5–4 mm long, the triangular lobes obtuse to acute, densely pubescent; corolla 6–10 (15) mm wide, with prominent contiguous white-tomentose appendages closing the throat; fruit 4–8 mm thick; n = 12, 24, 36. Creosote bush, blackbrush, Joshua tree, and other warm desert shrub communities at 760 to 1525 m in Kane, San Juan, and Washington counties; Kansas to Texas, New Mexico, Arizona, and Mexico; 8 (ii).

Datura L.

Rank, ill-scented, herbs; stems stout, mostly erect, branched; leaves large, ovate to elliptic, petioled; flowers large, showy, white to lavender, solitary in axils of branches; calyx prismatic or cylindrical, funnelform, 5–lobed; corolla funnelform to tubular with a spreading, 5– to 10–toothed, plaited limb; stamens 5, included; ovary 2–loculed or falsely 4–loculed; fruit a large, globose or ovoid, normally spiny capsule; seeds flat. Note: All species are poisonous due to presence of alkaloids.

1. Capsules erect, ovoid, opening by 4 valves; corollas 5–10 cm long . . . . D. stramonium
— Capsules pendulous, globose, opening irregularly; corollas 12–20 cm long 2

2(1). Plants glabrous, cultivated . . . . D. fastuosa
— Plants puberulent and somewhat glaucous, indigenous and sometimes cultivated . . . . D. wrightii

Datura fastuosa L. [D. metel L.; D. cornucopia Herter]. Annual, glabrous herbs; stems 1–2 m tall; leaves ovate-lanceolate, acute to acuminate, unequal at the base, sinuate-toothed or repand, glabrous, alternate or the upper in pairs with one larger than the other, 12–20 cm long; flowers erect, the calyx tubular-angulate, purplish, 4–5 cm long, 5–toothed, the teeth acuminate; corolla white, violet, to yellowish externally, 13–15 cm long, with 5–6 long cuspidate lobes, or the flowers double; stamens 5–6; capsules nodding, globose, short-spiny, irregularly dehiscent; n = 12. This is the most common garden datura; the color is variable. The plants are sporadically grown in Utah; 0 (0).

Datura stramonium L. Jimpson Weed. Annual herbs; stems 3–10 dm tall, glabrous or nearly so; leaves 1–2 dm long, ovate to oblong, repand to coarsely sinuately toothed, nearly lobed, glabrous or nearly so; calyx 3–6 cm long, the lobes 3–7 mm long, triangular-lanceolate; corolla 5–11 cm long, white to violet; fruit erect, ovoid, regularly dehiscent into 4 valves; 2n = 24. Waste places and cultivated land, often in barnyards and corrals at 915 to 1065 m in Garfield, Salt Lake, and Washington counties; widespread in North America; 2 (0).

Datura wrightii Regel Indian-apple; Angels-trumpet. [D. meteloides authors, not Dunal]. Annual or perennial herbs; stems 3–10 dm tall, erect, canescent-puberulent with dense, fine, gray hairs; leaves 5–25 cm long, the blades ovate, repand-dentate to entire, often uneven at the base, canescent-puberulent; calyx 7–13 cm long, the lobes lanceolate, ca 2 cm long; corolla 15–23 cm long, whitish to violet; fruit nodding, subglobose, breaking irregularly at maturity; 2n = 24. Creosote bush, blackbrush, Joshua tree, sagebrush, and pinyon-juniper communities at 700 to 1870 m in Garfield, Grand, Kane, San Juan, Washington, and Wayne counties; Colorado to Texas, west to California; Mexico; 38 (xvi).

Hyoscyamus L.

Annual or biennial herbs; stems leafy, erect; leaves alternate, lobed or pinnatifid; flowers solitary in upper leaf axils; inflorescence secund; calyx urn-shaped to campanulate, 5–cleft, accrescent and becoming reticulate in fruit, enclosing the capsule; corolla greenish yellow to whitish with dark rose or purple veins, funnelform, with slightly oblique, 5–lobed limb; stamens exserted; anthers opening longitudinally; ovary 2–loculed; fruit a capsule, circumscissile above the middle.

Hyoscyamus niger L. Henbane. Stems 3–15 dm tall, stout, viscid, short-villous; leaves 6–20 cm long, the blades oblong, ovate, or lanceolate, irregularly lobed, cleft, or pinnatifid, sessile or the upper ones clasping, viscid or short-villous; calyx 2–2.5 cm long in fruit; corolla ca 1.5–2 cm long; capsule 10–14 mm long; 2n = 34. Along roadsides and waste places (often in sheep bedgrounds) in sagebrush, mountain brush, aspen, and spruce-fir communities at 1675 to 2440 m in Beaver, Box Elder, Cache, Carbon, Daggett, Duchesne, Grand, Rich, Sanpete, Summit, and Weber counties; Nova Scotia to Montana, south to New York and Colorado; adventive from Europe; 20 (iv). The plant is notoriously poisonous to livestock and humans, due to the presence of atropine and hyocyamine.

Lycium L.

Shrubs or vines, the branchlets often modified as thorns; leaves mostly fascicled, entire or minutely dentate, glabrous and glaucous to glandular or pubescent; flowers mostly axillary, solitary or in small clusters of 2–4; calyx campanulate to tubular, irregularly toothed or cleft into 4–6 lobes; corolla whitish to purplish or greenish purple, regular, tubular-funnelform or salverform, the limb 4– to 7–lobed; stamens 4–5, the anthers affixed near their middle; berries fleshy or dry, globose to ovoid, subtended by the persistent calyx; seeds 2 to many.

1. Calyx lobes 2/3 as long as the tube, or at least 2 mm long 2
— Calyx lobes less than 2/3 as long as the tube, or less than 2 mm long 3

2(1). Plants glaucous; fruit not transversely grooved near the middle . . . . L. pallidum
— Plants not glaucous, densely pubescent; fruit with 2 transverse grooves above the middle . . . . L. cooperi

3(1). Plants clambering shrubs or vines, adventive near habitations and cemetaries . . . . L. barbarum
— Plants erect or ascending shrubs, indigenous in xeric habitats 4

4(3). Corolla lobes lanate-ciliate; leaves over 5 mm wide, at least some . . . . L. torreyi
— Corolla lobes glabrous to ciliolate; leaves less than 5 mm wide . . . . L. andersonii

Lycium andersonii Gray Anderson Wolfberry. Much branched, rounded shrubs; stems 1–2 (3) m tall, subglabrous to sparsely pubescent; thorns slender, needle-like; leaves spatulate, thickened, 3–15 mm long, 1–3 mm wide; flowers 1 or 2 per leaf axil; pedicels 3–9 mm long; calyx 1.5–3 mm long, cup-shaped, glabrous; corolla whitish lavender, tubular-funnelform, the tube 10–16 mm long, the lobes 1.5–2.5 mm long, entire, ciliolate; stamens exserted 2–3 mm; fruit ellipsoid to ovoid, red, fleshy, 4–8 mm long. Creosote bush, blackbrush, Joshua tree, mixed desert shrub, sagebrush, and pinyon-juniper communities at 730 to 2130 m in Beaver, Garfield, Juab, Kane, Millard, San Juan, Tooele, and Washington counties; New Mexico, Arizona, Nevada, California, and Mexico; 30 (ix). The palatable fruit is abundantly produced in some years.

Lycium barbarum L. Matrimony Vine; Teavine. [*L. halamifolium* Miller]. Sprawling, clambering, or climbing shrubs or vines, mainly 1–4.5 m long; stems erect or spreading, with arching, scrambling branches 1–6 dm long, glabrous, pale green, short-petiolate; calyx ca 4 mm long, cupulate, the lobes triangular, obtusish, from half as long as the tube to subequal to the tube; corolla rotate-campanulate, dull lilac purple, 3–7 mm long; stamens ca subequal to the corolla lobes, exserted; fruit ovoid, salmon to red, 10–15 mm thick; $2n = 24, 48$. Disturbed sites, homesites, cemetaries, and other open places at 930 to 2130 m in Box Elder, Cache, Duchesne, Grand, Millard, Salt Lake, Sevier, Summit, Uintah, Utah, Washington, and Wayne counties; widely established in the U. S., adventive from Eurasia; 16 (i). This plant was grown by Utah pioneers to protect graves in cemeteries; all succeeding generations have worked to undo the handiwork of those now residing permanently where their decendents labor.

Lycium cooperi Gray Cooper Wolfberry. Densely leafy, compact, thorny shrubs; stems 1–1.5 (2) m tall, glandular puberulent; leaves oblanceolate to spatulate, 1–3 cm long, 5–10 mm wide; flowers drooping, the pedicels 8–15 mm long; calyx saucer-shaped, 8–15 mm long, the lobes half as long as the tube; corolla greenish white, funnelform, 8–12 mm long and almost as broad; stamens equaling the corolla tube; fruit ovoid, dry, greenish, constricted above, 6–10 mm long; seeds several. Creosote bush, blackbrush, and other warm desert shrub communities on dry slopes and mesas at 760 to 1065 m in Washington County; Arizona, Nevada, California, and Mexico; 6 (ii).

Lycium pallidum Miers Pale Wolfberry; Tomatilla. Intricately branched shrub 1–2 m tall; stems erect but with spreading branches, thorny, glabrous and glaucous; leaves 1–4 cm long, glaucous, green, glabrous, oblong-spatulate to oblanceolate or elliptic; calyx 5–8 mm long, the lobes as long as or somewhat longer than the tube; corolla 15–20 mm long, narrowly funnelform, greenish or tinged with purple; stamens usually somewhat exserted; fruit ovoid, ca 1 cm long, red to reddish blue. Blackbrush, sagebrush, mesquite, greasewood, mountain brush, and pinyon-juniper communities at 1000 to 1870 m in Garfield, Iron, Kane, San Juan, and Washington counties; Colorado, Texas, New Mexico, Arizona, and Mexico; 47 (vii). This plant is often associated with ruins of the Anasazi occupation in southeastern Utah.

Lycium torreyi Gray Torrey Wolfberry. Spreading, much-branched shrubs; stems 1–3 m tall, subglabrous, heavily spined; leaves broadly spatulate, 1–5 cm long, 3–10 mm wide; flowers mostly in small fascicles, the pedicels 5–20 mm long; calyx 2.5–4 mm long, cupulate to short-cylindric, the lobes 0.5–2 mm long, ciliolate; corolla lavender purple, tubular clavate, 10–15 mm long, the lanceolate to ovate lobes lanate-ciliate, 3–4 mm long; stamens subequal to the corolla lobes; fruit a juicy, red, ovoid, many-seeded berry 7–10 mm long; $n = 12$. Tamarix, Baccharis, greasewood, and shadscale communities at 760 to 1160 m in Garfield and Washington counties; Texas, New Mexico, Arizona, Nevada, California, and Mexico; 9 (iv).

Lycopersicon Miller

Allied with *Solanum*, but always unarmed; leaves always pinnate or pinnatifid; flowers yellow; anthers projected into a narrow or sharp sterile tip and dehiscing throughout; fruit red or yellow, with 2 or more locules.

Lycopersicon esculentum Miller Tomato. Annual (perennial under glass) herbs; stems mainly 3–15 dm long, spreading, glandular-hairy and strong smelling; leaves odd-pinnate to pinnatifid, mainly 8–40 cm long; flowers 3–7 (or more) per cluster, nodding, ca 2 cm wide, on jointed pedicels; calyx 5–parted to the base; corolla yellow, the 5 lanceolate lobes recurved-reflexed; fruit a pulpy berry, mostly 2–10 cm wide; seeds numerous; $2n = 24$. Cultivated food plant throughout Utah; introduced from South America; occasionally escaping from cultivation; 3 (0).

Nicotiana L.

Annual or perennial, viscid-puberulent herbs; stems erect, leafy; leaves sessile or petioled, large, alternate, entire or repand; flowers in terminal racemes or panicles; calyx tubular-campanulate, 5–cleft, persistent; corolla funnelform, salverform, or tubular, the limb spreading and shallowly 5–lobed; stamens included, the anthers opening longitudinally; ovary 2–loculed, the stigma capitate; fruit a capsule; seeds numerous.

1. Corolla pink to red; plants cultivated .... *N. tabacum*
— Corolla whitish or greenish yellow; plants indigenous .. 2
2(1). Leaves petiolate, not cordate or auriculate-clasping at the base; corolla externally glabrous or only sparingly pubescent .... *N. attenuata*
— Leaves cordate and sessile or auriculate-clasping basally; corolla very pubescent externally .... *N. trigonophylla*

Nicotiana attenuata Torr. ex Wats. Coyote Tobacco. Annual herbs; stems erect, simple or branched, glandular-pubescent to glabrate, 3–16 dm tall; leaves 5–15 cm long, ovate to lance-ovate, mostly petioled; inflorescence racemose or paniculate; calyx campanulate, 6–8 mm long, the teeth deltoid; corolla white, 2.5–3 cm long, ca 1 cm broad; capsule 8–12 mm long; seeds brown, 0.6–0.7 mm long; $2n = 24$. Plants ubiquitous, usually in disturbed sites, at 830 to 2850 m in all Utah counties, except Box Elder, Cache, Duchesne, Rich, Wasatch, Wayne and Weber; Idaho and Washington to Colorado, Texas, Arizona, and California; 90 (v).

Nicotiana tabacum L. Tobacco. Annual herbs; stems erect, branching above, mainly 1–2.5 m tall, viscid-pubescent; leaves oblong-lanceolate, sessile, apically acuminate, mainly 8–40 cm long; flowers pedicellate, bracteate, in panicled racemes; calyx oblong, the lobes lanceolate, unequal, acute; corolla funnelform, rose colored with a red limb, woolly without; capsules ca 2 cm long; $2n = 23, 46, 48, 77, 96$. Cultivated ornamental and tobacco plants, rarely if at all escaping; widely grown in the U. S. and elsewhere; 2 (i). Another species, *N. sylvestris* Spegaz., is grown in Utah occasionally. It has flowers of various colors, white, yellow, bronze, or red, on plants mainly less than 1 m tall.

Nicotiana trigonophylla Dunal in DC. Desert Tobacco. Annual or perennial herbs; stems viscid-pubescent, 2–8 dm tall, erect, simple or branched, often clump-forming; lowermost leaves petiolate, the others sessile, auricled, and oblong-ovate to lanceolate, mainly 2–8 cm long; inflorescence loosely paniculate-racemose; pedicels 5–10 mm long; calyx campanulate, 6–12 mm long, with lance-subulate lobes as long as the tube; corolla greenish white, 18–22 mm long, constricted at the throat, the limb 8–10 mm wide; capsule 8–10 mm long; seeds dark brown, ca 0.6 mm long. Creosote bush, blackbrush and mixed desert shrub communities, often at cliff bases or in crevices, at 760 to 1660 m in Garfield, Kane, San Juan, and Washington Counties; Nevada, California, east to Texas, and south to Mexico; 32 (xii).

Petunia Juss.

Herbaceous annual or perennial herbs; stems branching, viscid-pubescent, weak and sprawling; leaves simple, entire, alternate or the upper ones opposite; flowers variously colored, solitary, axillary or terminal; calyx 5–cleft, the lobes oblong to linear, obtuse; corolla funnelform or salverform, the tube fitting loosely in the calyx, the limb broad and usually 5–lobed; stamens 5, 4 in pairs, the fifth smaller or essentially lacking; capsule 2–loculed; seeds numerous, small.

1. Corolla tube ca 6 mm long; leaves 5–12 mm long; plants indigenous in Beaver Dam Wash, Washington County . . . . P. parviflora
— Corolla tube 2–5 cm long; leaves 20–60 mm long or more; plants cultivated, rarely escaping . . . . P. hybrida

Petunia hybrida Vilm. Common Petunia. Highly variable cultingens, annual or perennial; stems erect or finally sprawling; leaves softly viscid-pubescent, ovate to ovate-oblong; corollas 5–9 cm long, with the limb as wide or wider, funnelform, varicolored, and single or double; n = 7. Cultivated ornamental, rarely escaping but not persisting, in all Utah counties; introduced from South America; 6 (0). The cultivated petunias have apparently been derived from hybridization involving both **P. axillaris** B.S.P. and **P. violacea** Lindl. of South America. The former is reported to have dull white flowers with a slender tube, and the latter has small rose or violet flowers.

Petunia parviflora Juss. Streamside Petunia. Plants annual, prostrate, rooting along the stem, the circular-spreading stems 3–40 cm long or more; herbage glandular-puberulent; leaves 5–12 mm long, 1–2 mm wide, linear-oblong to spatulate, fleshy; flowers solitary, lateral on very short pedicels; calyx 5–parted to below the middle; sepals 6–11 mm long in fruit, 1–1.5 mm wide, linear-oblanceolate to spatulate, obtuse; corolla funnelform, purple to violet, the yellowish to whitish tube 5–6 mm long; capsules 3–4 mm long, ovoid-ellipsoid, 1–loculed, bivalved; seeds numerous. Braided stream gravels along perennial water at ca 750 to 885 m in Beaver Dam Wash, Washington County; California to Virginia and Florida, south to tropical America; 8 (iv). This minute-flowered petunia was first discovered in Utah, growing in some abundance along Beaver Dam Wash, in the summer of 1985.

Physalis L.

Annual or perennial herbs; stems leafy; leaves entire to sinuate-dentate; flowers solitary in leaf axils, or in clusters of 2–5; pedicels slender; calyx campanulate to tubular-campanulate, 5–toothed, enlarging and bladdery-inflated in fruit; corolla rotate or open-campanulate, obscurely 5–lobed, yellowish or whitish to purplish; stamens 5, inserted near the base of the corolla tube; style slender; stigma faintly 2–lobed; fruit a berry; seeds few to many.

1. Corolla rotate or rotate-campanulate, purple or whitish . . . . P. lobata
— Corolla campanulate, yellow or yellowish, usually campanulate, yellow or yellowish, usually with a brownish or purplish center 2

2(1). Pubescence at least partly of forked or stellate hairs . . . . P. hederaefolia
— Pubescence of simple hairs or glabrate 3

3(2). Plants annual; herbage and calyx strongly pubescent with glandular or eglandular hairs; leaf blades broadly ovate, cordate or rounded at the base . . . . . . P. pubescens
— Plants perennial, often rhizomatous; leaf blades various . 4

4(3). Leaf blades narrowly lanceolate to oblong-lanceolate; plants sparsely pubescent to hairy above, never glandular 5
— Leaf blades wide, ovate to oblong or ovate, the base rounded, truncate, or cordate; plants often glandular-pubescent 7

5(4). Calyx tube minutely strigose, chiefly in 10 narrow, longitudinal stripes, with minute, appressed, simple, or obscurely septate hairs as much as 0.5 mm long; leaves glabrous or minutely hairy only on the main veins; plants common . . . . P. longifolia
— Calyx tube hirtellous over its entire surface, with spreading, septate hairs 0.5–1.5 mm long; leaves sparsely hairy over the surface on both sides; plants rare 6

6(5). Hairs of the upper part of the stem mostly or all of them decurved; plants reported from Utah, but specimens not seen . . . . P. virginiana Miller
— Hairs of the upper part of the stem ascending or ascending-spreading; plants reported from Utah, but specimens not seen . . . . P. hispida (Waterfall) Cronq.

7(4). Plants densely long-villous with spreading, simple, gland-tipped hairs of 2 lengths, some multicellular and 1–2 mm long, the others much shorter and finer; leaves broadly cordate-ovate, the larger ones at least 3.5 cm broad; anthers mostly 3.5–4.5 mm long . . P. heterophylla
— Plants mostly either non-villous or eglandular; leaves rarely cordate but, if so, less than 3.5 cm wide; anthers less than 3 mm long 8

8(7). Pedicels at anthesis not much, if any, longer than the flowers; filaments mostly strongly flattened; stems typically sparingly branched and not obviously flexuous . . . . P. hederaefolia
— Pedicels at anthesis often much longer (1.5–6 times) than the flowers; filaments nearly terete; stems much-branched and mostly flexuous, rather brittle . . . . P. crassifolia

Physalis crassifolia Benth. Thickleaf Ground-cherry. Diffusely and intricately branched perennial herbs; stems 2–5 dm tall, viscid-puberulent; leaf blades variable, ovate, deltoid, or cordate, 1–3 cm long, entire or shallowly sinuate, on petioles of equal length; pedicels slender, 1–2 cm long; calyx campanulate, 3–5 mm long at anthesis, in fruit becoming 15–25 mm long, ovoid, obscurely angled; corolla pale tawny-yellow, 10–15 mm wide; berry greenish. Creosote bush, blackbrush, and other warm desert shrub communities at 760 to 1335 m in San Juan and Washington counties; Nevada, Arizona, and California; 5 (iii).

Physalis hederifolia Gray Ivy-leaved Ground-cherry. Perennial from a somewhat woody base, to 7 or 8 dm tall, strongly pubescent with simple to forked or stellate, glandular to eglandular hairs; leaf blades ovate to lanceolate, entire to sinuate-toothed, with acute to rounded, cordate or reniform base, usually somewhat grayish green; pedicels rather stout, mostly 5–10 mm long at anthesis and shorter than the flowers, but up to 15 (20) mm long in fruit; calyx 7–10 mm long at anthesis, up to 30 mm in fruit, the lobes triangular to lanceolate; corolla yellow, often brownish spotted, 12–15 mm broad; filaments much flattened and broadly clavate in outline, the anthers 1.5–3.5 (4.3) mm long. Three rather distinctive varieties are present in Utah.

1. Pubescence partially or wholly of long multicellular hairs, these sometimes branched; anthers (2.5) 3–4.3 mm long .... *P. hederifolia* var. *hederifolia*
— Pubescence short, without long, multicellular hairs; anthers 1.5–3 mm long 2

2(1). Hairs mostly forked to dendritic, not glandular .... *P. hederifolia* var. *fendleri*
— Hairs all simple, gland-tipped, very short .... *P. hederifolia* var. *palmeri*

Var. *fendleri* (Gray) Cronq. [*P. fendleri* Gray; *P. fendleri* var. *cordifolia* Gray, type from St. George]. Sagebrush, oak-maple, pinyon-juniper, and mountain brush communities at 1660 to 2500 m in Beaver, Garfield, Grand, Iron, Kane, San Juan, Sevier, Washington, and Wayne counties; California to New Mexico, Oklahoma, Texas, and Mexico; 19 (iv).

Var. *hederifolia* Sagebrush, pinyon-juniper, oakbrush, and mountain brush communities at 1600 to 2000 m in Millard and Washington counties; Texas to California and Mexico; 0 (0). I have not seen any specimens of this taxon from Utah, but it is reported in the Intermountain Flora.

Var. *palmeri* (Gray) C. L. Hitchc. [*P. palmeri* Gray]. Creosote bush, cottonwood, sagebrush, pinyon-juniper, and mountain brush communities at 900 to 2500 m in Beaver, Kane, Piute, and Washington counties; California and Arizona; 10 (iv).

Physalis heterophylla Nees Perennial herbs from a deep rootstock; stems mostly erect, 1.5–9 dm tall, simple or branching, pubescent with short viscid-glandular hairs and some long jointed ones 1–2 mm long; leaves ovate, mostly 5–10 cm long, 3.5–6 cm wide, with petioles 3–6 cm long; pedicels slender, 1.5–4 cm long; calyx 7–12 mm long at anthesis, in fruit becoming 25–30 mm long, greatly inflated; corolla 10–18 mm wide, yellow, maculate, but not strongly so; berry greenish yellow. Our one record is from the Salt Lake Valley at ca 1535 m in Salt Lake County; eastern U. S., southern Canada, and northern Rocky Mountains; 1 (0).

Physalis lobata Torr. Perennial herbs; stems procumbent and spreading, with pubescence of scurfy, crystalline vesicles; leaves ovate-lanceolate to linear-lanceolate, 4–10 cm long, 5–30 mm wide, narrowing to a cuneately winged petiole, the margin entire to pinnatifid; pedicels 1–3 (5) mm long; calyx at anthesis 3–4 mm long, in fruit becoming 15–20 mm long, pentagonal-ovoid, inflated, the pedicels becoming 10–30 mm long; corolla blue to violet or white, rotate, 15–20 mm wide; berry greenish; n = 12. Pinyon-juniper community at ca 1830 m in Grand and San Juan counties (Atwood et al. 8738; Goodrich and Atwood 20393 BRY); Kansas and Colorado to New Mexico, Arizona, Texas, and Mexico; 2 (0).

Physalis longifolia Nutt. Common Ground-cherry. Perennial herbs from thick rootstocks; stems erect, stout, 2–8 dm tall, branched above, glabrous or with a few flat long hairs, especially above; leaves lanceolate, oblanceolate, or linear, 4–5 times longer than broad, entire to repand; flowers usually solitary; calyx at anthesis campanulate, 4–8 mm long, in fruit becoming 20–40 mm long, ovoid, the fruiting pedicels 1–2 cm long; corolla yellow, maculate, 10–20 mm broad; berry greenish yellow; n = 12. Riparian, sagebrush, pinyon-juniper, and mountain brush communities at 930 to 2670 m in Box Elder, Cache, Carbon, Daggett, Duchesne, Emery, Garfield, Grand, Iron, Juab, Morgan, Salt Lake, San Juan, Tooele, Uintah, Utah, Wasatch, Washington, Wayne, and Weber counties; eastern U. S., Canada, and Montana south to Arizona, Texas, and Mexico; 48 (v).

Physalis pubescens L. Annual herbs; stems stout, seldom branching from the base, 0.8–9 dm tall, villous, sometimes viscid to nearly glabrous; leaves ovate to nearly orbicular, 3–9 cm long, 2–4 cm wide, margins irregularly toothed to nearly entire, acuminate apically, cordate basally, the petioles 2–7 cm long; calyx at anthesis 4–10 mm long, the pedicels 3–6 mm long; fruiting calyx strongly 5-angled, acuminate apically, 18–30 mm long, on pedicels 5–13 mm long, soft hairy; corolla yellowish, dark, maculate, matted-hairy below the maculations, 10–15 mm broad; berry 10–18 mm broad, greenish yellow; n = 12. Disturbed sites at ca 855 m in Washington County (Galway 6247); widespread in the U. S., and south to Panama; 1 (0).

Salpiglossis Ruiz & Pavon

Annual herbs; stems erect, glandular-hairy; leaves alternate, simple, entire to sinuate-dentate or pinnatifid; flowers large, few, long-pedicellate; calyx tubular, 5-toothed; corolla funnelform, with a broad throat, the limb with 5 emarginate lobes; stamens 4, didynamous, included, the fifth reduced to a staminodium or lacking; fruit an oblong or ovoid capsule, the valves 2-cleft; seeds small.

Salpiglossis sinuata Ruiz & Pavon Poormans-orchid. Stems branching, 3–7 dm tall; leaves elliptic to lanceolate or linear, sinuately toothed to pinnatifid or nearly entire, subsessile; flowers 5–6 cm long and about as broad, pale yellow to scarlet or blue, with much variation in venation and coloration. Widely grown ornamentals in lower elevation portions of Utah; introduced from South America; 0 (0).

Solanum L.

Herbs or shrubs, glabrous to pubescent or tomentose, often glandular, sometimes clambering or twining, armed or unarmed; leaves simple and entire to lobed or parted; flowers mostly in umbels or cymes, white or yellow to blue or purple; calyx 5-cleft or -toothed, rotate to campanulate; corolla 5-angled or -lobed, plaited in bud; stamens 5, inserted on the corolla tube; filaments short; anthers connivent around the style, dehiscent by a terminal pore or short slit; ovary 2-loculed; stigma small, capitate or bilobed; fruit a globose berry with several to many flattened seeds.

1. Flowers lavender; tubers large; plants seldom producing fruit, cultivated for edible tubers .... *S. tuberosum*

— Flowers variously colored; tubers lacking or small; plants typically producing fruit, indigenous, adventive, or rarely cultivated, but not grown for edible tubers 2

2(1). Plants armed with prickly, stiff spines 3
— Plants unarmed 5

3(2). Plants annual; leaves pinnatifid; flowers yellow; calyx enlarged, armed with long, straight spines .. *S. rostratum*
— Plants perennial; leaves entire, sinuate, or lobed, but not pinnatifid; flowers white or bluish; calyx not as above 4

4(3). Leaves entire to repand-dentate, permanently gray canescent, at least above .... *S. elaeagnifolium*
— Leaves lobed or deeply dentate, not permanently gray canescent on the upper surface .... *S. carolinense*

5(2). Leaves entire or shallowly toothed 6
— Leaves, at least in part, pinnately compound, pinnatifid, or hastate 7

6(5). Stems and leaves glabrate, puberulent or strigose; berries black when ripe .... *S. nigrum*
— Stems and leaves viscid-villous; berries yellow when ripe .... *S. sarrachoides*

7(5). Plants sprawling, perennial vines; leaves 3-lobed or hastate; corolla 12–16 mm wide, bright violet or blue purple; fruit red when ripe .... *S. dulcamara*
— Plants neither sprawling nor vinelike; leaves pinnately compound or pinnatifid; corolla white 8

8(7). Plants perennial from globose tubers; leaves pinnately compound or essentially so; corolla 12–18 mm wide .. .. *S. jamesii*
— Plants annual, lacking tubers; leaves entire, sinuate-dentate, or pinnatifid; corolla 6–12 mm wide .... *S. triflorum*

Solanum carolinense L. Coarse perennial herbs from creeping underground rhizomes; stems erect, branched, mainly 3–10 dm tall, spiny and loosely pubescent throughout with 4- to 8-rayed, stellate hairs; leaves ovate to ovate-elliptic, to 14 cm long, the blades rounded basally, usually with several large teeth or shallow lobes on each side, more or less spiny along the main veins, the petiole to 3 cm long; flowers several, racemose; calyx 5–7 mm long, the lobes lance-acuminate; corolla pale violet to white, 2–3 cm wide; fruit globose, yellow at maturity, 1–2 cm wide; $2n = 48$. Collected once at the University of Utah, growing in a flower bed, at ca 1650 m in Salt Lake County; eastern U. S., and adventive elsewhere; 1 (0).

Solanum dulcamara L. European Bittersweet. Plants perennial; stems woody, clambering, to 3 m long, sparsely puberulent to nearly glabrous; leaves ovate, acuminate, with or without auricled or hastate bases, 5–12 cm long; petioles 1–4 cm long; cymes several-flowered; pedicels ca 1 cm long; calyx 3–4 mm long; corolla bright blue to purple, 12–16 mm wide, deeply 5-cleft; anthers ca 5 mm long, connivent; berry red, ovoid or ellipsoid, 8–12 mm long; seeds rounded, ca 2 mm wide, minutely patterned; $2n = 24$. Fence rows, canal banks, and other moist sites at 1330 to 2170 m in Cache, Davis, Juab, Salt Lake, Sanpete, Summit, Uintah, Utah, Wasatch, and Weber counties; widely distributed in North America; introduced from Eurasia; 37 (ii). The berries are reported to be poisonous when eaten in quantity.

Solanum elaeagnifolium Cav. Silverleaf Nightshade; White Horsenettle. Plants perennial; stems mainly 3–10 dm tall, silvery canescent, with dense, stellate hairs and sparse to abundant small prickles; leaves oblong to linear or oblong-lanceolate, to 15 cm long, obtuse apically, the margin entire to sinuate-repand; cymes terminal, short-peduncled, few-flowered; pedicels rather long, recurved or reflexed in fruit; calyx 5–angled, with slender lobes subequal to the tube; corolla violet or sometimes white, slightly 5-lobed, 2–2.5 cm wide, the lobes triangular-ovate; ovary white-tomentose; berry globose, to 15 mm in diameter, yellowish or eventually blackish; $2n = 24$. Waste places and other disturbed sites at 900 to 1300 m in Grand, Kane, San Juan, Washington, and Wayne counties; Missouri and Kansas, south to Louisiana, Texas, and Arizona; adventive elsewhere; 10 (ii).

Solanum jamesii Torr. James Potato. Perennial herbs with rootstocks bearing nearly globose tubers ca 1 cm thick; stems erect or spreading, bushy, mainly 1–5 dm tall, glabrous to sparingly pilose; leaves odd-pinnate, to 15 cm long, the leaflets 7–11, linear-oblong to lanceolate; flowers cymose-paniculate, the pedicels 1–2 cm long, articulate near the middle; calyx 4–8 mm long, irregularly lobed to near the middle; corolla white, stellate, 12–28 mm broad, the lobes ovate-lanceolate to triangular-lanceolate, acute; fruit globose, ca 1 cm broad; $n = 12$. Pinyon-juniper and mountain brush communities at ca 1800 m in Garfield and San Juan counties; Nebraska, Colorado, Arizona, New Mexico, Texas, and Mexico; 2 (0).

Solanum nigrum L. Black Nightshade. [*S. americanum* Miller]. Plants annual; stems slender, usually divergently much-branched, mainly 1.5–10 dm tall, glabrous or nearly so; leaves petioled, the blades ovate to oval or ovate-lanceolate, entire to sinuate-dentate, pale green, to 10 cm long; umbels with 2–4 flowers, on slender peduncles to 3 cm long; pedicels reflexed; calyx of unequal acutish to obtuse spreading lobes 1–1.5 mm long; corolla white or purplish tinged, the lobes 4.5–7 mm long; berry black, glossy, 5–9 mm wide. Roadsides, gardens, and other cultivated or open lands at 900 to 2000 m in Cache, Kane, Salt Lake, Sevier, Uintah, Utah, Washington, and Weber counties; adventive from Europe and now widespread in the U. S.; 10 (ii). This species is often mistaken for the more common *S. sarrachoides* (q.v.).

Solanum rostratum Dunal Buffalobur. Plants annual; stems to 7 dm tall, somewhat hoary or yellowish with copious, wholly stellate pubescence, also abundantly armed with straight prickles; leaves 1- to 2-pinnatifid; racemes with ascending pedicels; calyx nearly hidden by the numerous, spinelike prickles; corolla yellow, 20–25 mm broad, the short lobes broadly ovate; stamens and style much declining, the lowermost anther much longer and exceeding the others, and with an incurved beak; berry wholly enclosed by the investing calyx; seeds coarsely undulate-rugose; $2n = 24$. Occasional weed of cultivated and other disturbed lands at 930 to 1870 m in Cache, Davis, Garfield, Grand, Millard, Rich, Salt Lake, San Juan, Tooele, Uintah, Utah, and Washington counties; indigenous in the central United States, adventive elsewhere; 13 (ii).

Solanum sarachoides Sendt. ex Martius [*S. villosum* Miller, *S. nigrum* var. *villosum* (Miller) Miller]. Plants annual; stems much-branched, ascending to decumbent, 1–5 dm long, shortly viscid-villous; leaves ovate, 2.5–6 cm long, gradually to abruptly narrowed basally, apically acute to obtuse, entire to sinuately toothed, the petioles 1–1.5 cm long; peduncles 5–10 (20) mm long; calyx 2–2.5

mm long at anthesis, accrescent in fruit; corolla white, 3–5 mm wide, the lanceolate lobes villous outside; berry 6–7 mm wide, globose, yellow when ripe; 2n = 24. Weedy plants of fields, roadsides, gardens, and other open sites at 850 to 2250 m in Cache, Duchesne, Piute, Salt Lake, Sevier, Uintah, Utah, Washington, and Weber counties; adventive from South America; widely established in North America; 15 (ii).

Solanum triflorum Nutt. Cutleaf Nightshade. Annual herbs; stems much branched from near the base, decumbent, 1–4 dm long; leaves oblong, deeply pinnatifid, with rounded sinuses, 2.5–4 cm long, with petioles mainly 5–15 mm long; cymes 1- to 3-flowered, the pedicels soon reflexed; calyx 2.5–3 mm long, the lobes lance-ovate; corolla white, 7–9 mm broad; berry globose, green, translucent, 7–15 mm thick; seeds many, pale, 2.5–3 mm wide; n = 12. Creosote bush, blackbrush, sagebrush, pinyon-juniper, mountain brush, and other plant communities at 920 to 2830 m in probably all Utah counties; British Columbia to California, east to Minnesota, Kansas, and Texas; 54 (v).

Solanum tuberosum L. Potato. Perennial herbs from large tubers; stems branched, glabrate to pubescent, mainly 4–12 dm long; leaves odd-pinnate, 8–30 cm long or more; flowers in branching cymes on long peduncles, usually violet; calxy lobes linear-lanceolate; corolla rotate; berry greenish or yellowish to 2 cm thick, seldom produced; 2n = 24, 36, 48. Cultivated food plant throughout Utah; introduced from South America; 1 (0). This is one of the most important carbohydrate sources in the temperate regions of the earth.

TAMARICACEAE Link.
Tamarisk Family

Shrubs or small to moderate trees; leaves alternate, scalelike, estipulate, entire; flowers mostly perfect, regular, borne in spikelike racemes arranged in panicles; sepals 4 or 5; petals 4 or 5, overlapping, arising from the base of a nectiferous disk; stamens usually as many as or twice as many as the petals, the anthers 2–loculed; pistil 1, the ovary superior, unilocular, usually 3 or 5 carpelled, the placentation basal; stigmas 2–5, separate; ovules 2 per locule; fruit a capsule, the seeds comose.

Baum, B. R. 1967. Introduced and naturalized tamarisks in the United States and Canada (Tamaricaceae). Baileya 15: 19–25.

Tamarix L.

Deciduous or evergreen shrubs or trees, the branchlets deciduous; leaves clasping or sheathing; flowers small, shortly pedicelled; petals white to pink or lavender, inserted below the disk; capsules dehiscent by 3–5 valves.

1. Leaves sheathing; trees evergreen, of moderate size, restricted to Washington County .... *T. aphylla*
— Leaves not sheathing, at most merely clasping; deciduous trees of small size or merely shrubs of broad distribution 2
2(1). Flowers 4-merous, or the stamens sometimes more than 4; stamens emerging gradually from the disk-lobes; plants uncommon both in cultivation and as escapes .. .. *T. parviflora*
— Flowers 5-merous, or the stamens sometimes more than 5; stamens inserted under disk near the margin between the emarginate lobes; plants abundant, cultivated and otherwise .... *T. ramosissima*

Tamarix aphylla (L.) Karsten Athel Tamarisk. [*Thuja aphylla* L.]. Trees to 10 m tall and 6 dm in diameter or more, the bark reddish brown to gray; branchlets jointed; leaves sheathing, minute, evergreen; bracts longer than the pedicels; flowers 5–merous; sepals entire, the inner ones slightly larger; petals elliptic-oblong to ovate, 2–2.2 mm long, early deciduous or with 1–2 persisting; staminal filaments inserted between the disk lobes. Cultivated sparingly in Washington County, where it seldom flowers; native to Africa and the Middle East; introduced in California, Nevada, Arizona, and Texas; 2 (i).

Tamarix parviflora DC. Small-flowered Tamarisk. Shrubs or small trees to 5 m tall; bark brown to deep purple; branchlets not jointed; leaves merely sessile, not sheathing, deciduous with branchlets; bracts longer than the pedicels, more or less translucent; flowers 4–merous; sepals erose-denticulate, the outer 2 keeled and acute, the inner flat or slightly keeled and obtuse; petals oblong to ovate, 1.9–2.3 mm long, persistent; staminal filaments arising gradually from disk-lobes. Cultivated and naturalized along streams and seeps at 850 to 1710 m in Emery, Kane, Salt Lake, Utah, and Washington counties, and to be expected elsewhere; introduced from southern Europe and now widespread in Canada and the U. S.; 7 (i).

Tamarix ramosissima Ledeb. Tamarisk; Salt Cedar. [*T. gallica* authors, not L.; *T. pentadra* authors, not Pallas]. Shrubs or small trees to 6 m tall, or rarely taller; bark reddish brown; branchlets not jointed; leaves merely sessile, not sheathing, deciduous with the branchlets; bracts longer than the pedicels, scarious but scarcely translucent; flowers 5–merous; sepals erose-denticulate, the outer 2 narrower than the inner, all more or less acute; petals obovate, 1–1.8 mm long, persistent; filaments inserted under the disk near the margin between the emarginate lobes; 2n = 24. Cultivated and naturalized along seeps, streams, and reservoirs, almost throughout Utah, except in Iron, Morgan, Piute, Rich, and Summit counties; introduced from Eurasia, now widespread in the southern U. S.; 99 (xix).

TILIACEAE A. L. Juss.
Linden or Basswood Family

Trees; leaves alternate, simple, serrate to obscurely lobed, usually oblique, stipulate; flowers regular, perfect, borne in cymes; sepals 5, distinct or more or less connate; petals 5, alternate with the sepals; stamens numerous, the filaments free or conate in clusters; pistil 1, the ovary 5–loculed, the style 1; fruit drupaceous; x = 7–41.

Tilia L.

Cultivated trees; leaves long petioled, the blades obliquely cordate, serrate or doubly so, sometimes obscurely lobed; flowers in long pedunculed cymes, the peduncle adnate at its base to a ligulate bract; sepals 5; petals 5; stamens numerous, distinct or in 5 clusters, sometimes bearing petaloid staminodia opposite the petals; ovary 5-loculed, the stigma 5-lobed; fruit subglobose, 1- to 3-seeded.

1. Branchlets and petioles densely white-hairy; leaf blades white stellate hairy beneath .... *T. tomentosa*
— Branchlets and petioles glabrous or nearly so; leaf blades variously pubescent or glabrous 2
2(1). Leaf blades hairy (sometimes thinly so) over the lower surface and usually along the veins beneath 3
— Leaf blades glabrous beneath, except in vein axils 5
3(2). Leaf blades with densely white or brown stellate hairs .... *T. heterophylla*
— Leaf blades variously hairy but the surface not obscured by hairs 4
4(3). Hairs of lower leaf surface stellate, at least somewhat; flowers with staminodes .... *T. neglecta*
— Hairs of lower leaf surface all simple; flowers without staminodes .... *T. platyphyllos*
5(2). Leaf blades definitely glaucous beneath, usually less than 8 cm long; flowers lacking staminodes .... *T. cordata*
— Leaf blades green or merely pale beneath, the largest usually more than 8 cm long; flowers with or without staminodes 6
6(5). Flowers with staminodes; leaves serrate to doubly serrate with long-acuminate teeth, the largest blades on flowering stems to 10 cm long .... *T. americana*
— Flowers without staminodes; leaves serrate with short acute teeth, the largest blades on flowering stems usually less than 10 cm long .... *T. europaea*

Tilia americana L. American Linden. Moderate to large trees of streets and other ornamental plantings; common in Salt Lake, Utah, and Weber counties; indigenous to the eastern states and Canada; 9 (0).

Tilia cordata L. Small-leaved European Linden. Small to large trees of ornamental plantings; common in Box Elder, Cache, Juab, Salt Lake, Utah, and Weber counties; widely cultivated in North America; introduced from Europe; 12 (0).

Tilia x europaea L. Common or European Linden. Moderate to large trees of ornamental plantings, uncommon in Utah; indigenous to Europe. This tree is reputed to be a hybrid derivative of *T. cordata* x *T. platyphyllos*; 2 (0).

Tilia heterophylla Vent. White Basswood. Large ornamental trees, uncommon in Utah; indigenous to the eastern U. S.; 2 (0).

Tilia neglecta Spach. Moderate to large ornamental trees, uncommon in Utah; indigenous to the eastern U. S. and Canada. This taxon resembles, and apparently intergrades with, *T. americana*, with which it is very closely allied; 2 (0).

Tilia platyphyllos Scop. Large-leaf Linden. Moderate to large ornamental trees, moderately common in Salt Lake, Utah, and Weber counties; indigenous to Europe; 8 (0).

Tilia tomentosa Moench Silver Linden. Moderate to large ornamental trees, moderately common in Cache, Juab, Salt Lake, and Weber counties; indigenous to eastern Europe and Asia Minor; 5 (0).

TROPAEOLACEAE DC.
Tropaeolum Family

Annual or perennial herbs; leaves alternate, digitately angled or peltate, sometimes lobed or dissected; flowers perfect, irregular, usually solitary; sepals 5; petals 5, or sometimes fewer, often cut or fringed, the upper ones unlike the others and usually smaller and inserted in the opening of a spur; stamens 8, unequal; pistil 1, the ovary superior, 3-lobed and 3-loculed; style 1, apical; stigmas 3; fruit separating at maturity into three 1-seeded indehiscent segments; x = 12–14.

Tropaeolum L.

Herbs with acrid watery sap, spreading or climbing by coiling petioles; flowers showy, mostly yellow, orange, or red; sepals 5, the upper one produced into a spur; petals 5, the upper 2 differing from the lower 3; stamens 8, in 2 whorls.

Tropaeolum majus L. Garden Nasturtium. Annual herbs, the stems usually sprawling, glabrous and more or less succulent; leaf blades orbicular to subreniform, mostly 3–8 cm wide (or more), entire or nearly so; flowers shades of orange, yellow, or red and sometimes striped or spotted, mainly 5–7 cm across. Rather commonly cultivated ornamental in Utah; widely grown in North America; introduced from South America; 2 (0). The striking *T. peregrinum* L., the canary-bird flower, is sometimes grown, mainly in hanging baskets. The plants are climbing and the flowers are canary yellow.

ULMACEAE Mirbel
Elm Family

Trees or shrubs, with pith finely chambered at nodes; leaves alternate, simple, unequal at the base, deciduous, somewhat palmately veined; flowers perfect or imperfect, axillary, solitary or in small clusters; perianth 4- to 6-parted; stamens 4–5; ovary superior, 1-loculed; styles none; stigmas 2; fruit a drupe with thin flesh and hard-pitted seed or a samara; x = 10, 11, 14.

1. Fruit a samara; flowers on last year's branches, perfect, appearing before the leaves (except in *U. parvifolia*) . ... *Ulmus*
— Fruit a drupe; flowers on new growth with the leaves or after 2
2(1). Calyx campanulate, 4- to 5-lobed; style not central; leaves with 7 or more pairs of parallel veins, the lowest pair not prominent; winter buds somewhat spreading .... *Zelkova*
— Calyx of distinct sepals, 5- to 6-parted; style central; leaves 3-veined at the base, pairs of veins typically fewer than 6; winter buds appressed .... *Celtis*

Celtis L.

Deciduous trees; bark smooth or reticulate; leaves 3-veined at the base, serrate or entire; flowers greenish, appearing with the leaves; calyx 5- or 6-parted; staminate flowers in cymose clusters; pistillate flowers solitary or in few-flowered clusters in upper axils; ovary ovoid; fruit a drupe.

1. Leaves serrate; plants cultivated and persisting .... *C. occidentalis*
— Leaves entire or rarely with a few teeth; plants indigenous, low trees .... *C. reticulata*

Celtis occidentalis L. Hackberry. Small to large trees; bark finally deeply furrowed and reticulate or warty; branchlets typically hairy; blades of fertile shoots mostly

3–12 cm long, 1.2–8 cm wide, obliquely ovate, coriaceous, scabrous, serrate, those of leading shoots often larger; drupe orange red to brownish, on pedicels to 15 mm long, spherical, 8–10 mm thick; stone 7–9 mm long, 5–8 mm thick. Uncommonly cultivated ornamental and shade tree of Salt Lake, Utah, and Weber counties; introduced from the eastern U. S.; 13 (0). The sugarberry, *C. laevigata* Willd., is sparingly grown in Utah also. It differs from *C. occidentalis* in the entire leaves, and from *C. reticulata* in the sharply acute to acuminate leaves.

Celtis reticulata Torr. Netleaf Hackberry. [*C. villosula* Rydb., type from Utah]. Small trees (or sometimes shrubby) to ca 5 m tall, and with a spreading canopy almost as broad; bark finally reticulate and with corky ridges; branchlets typically hairy; blades 2–8 cm long, 1.5–4 cm wide, obliquely ovate to lanceolate, obtuse to acute, rounded to cordate at the base, entire or occasionally sparingly serrate, coriaceous, scabrous, reticulately veined and usually pale beneath, typically infected with insect galls; fruit spherical, reddish to orange or dark red, the flesh thin and sweet, 8–9 mm long, on pedicels 6–12 mm long. Warm desert shrub, sagebrush-grass, mixed desert shrub, and mountain brush communities at 915 to 1525 m in Box Elder, Cache, Davis, Emery, Garfield, Grand, Kane, Millard, Salt Lake, San Juan, Tooele, Uintah, Utah, Washington, Wayne, and Weber counties; Washington and Idaho, south to California, Texas, and Mexico; 60 (xi). The thinly fleshed fruits are eaten by coyotes and fox, who aid in their dispersal.

Ulmus L.

Small to large trees; buds with rounded brown scales imbricated in 2 series; leaves simply or doubly serrate, often oblique; stipules linear to obovate; flowers in clusters or cymes, appearing prior to the leaves or much after; calyx campanulate, 4- to 9-lobed; stamens 4–9, the filaments long and slender; fruit a 1-loculed and 1-seeded, compressed nutlet surrounded by a broad or narrow membranous wing (a samara). Note: All species of *Ulmus* in Utah are introduced, but at least one is established in our flora. The following key is to the most commonly grown species, but there are several others represented in the cultivated flora.

1. Leaves once serrate (obscurely twice serrate in *U. pumila*); plants widely cultivated and established or uncommon 2
— Leaves twice serrate 3

2(1). Flowers borne in late summer and autumn; bark with orange, corky lenticels; trees mainly less than 8 m tall, uncommonly cultivated ornamentals ...... *U. parvifolia*
— Flowers borne in early springtime, prior to the leaves; bark with inconspicuous lenticels; trees often more than 8 m tall, widely grown, escaping, and established .... *U. pumila*

3(1). Trees umbrella-shaped, with ascending branches; samaras ciliate; flowers pendulous; commonly grown shade tree of great beauty .... *U. americana*
— Trees more or less excurrent; samaras glabrous marginally; flowers erect; uncommonly grown tree .... *U. procera*

Ulmus americana L. American Elm. Trees to 30 m tall and with trunks 1–1.5 m thick; bark, gray, scaly and deeply fissured with broad scaly ridges; branchlets pubescent initially or almost glabrous; buds ovoid, obtuse to acute, glabrous or only slightly pubescent with whitish hairs; leaves with petioles 4–8 mm long, borne in 2 ranks on the twigs, ovate-oblong, 7–15 cm long, 3–7.5 cm wide, unequal at the base, acuminate apically, doubly serrate, glabrous or scabrous above, pubescent or subglabrous beneath; flowers borne on elongated, unequal pedicels 1–2 cm long; calyx 5- to 8-lobed; stamens 7 or 8, exserted; stigmas white; samaras elliptic, flat, ca 1 cm long, deeply notched apically, ciliate; $2n = 56$. Cultivated shade and specimen trees at 925 to 1895 m in Cache, Davis, Salt Lake, Summit, Uintah, Washington, and Weber counties; introduced from the eastern U. S.; 36 (0). Possibly the greatest of the silent tragedies in the forests of the U. S. involves the plight of the American elm, which has been essentially eradicated by Dutch elm disease, spread by beetles, in much of its native range. Perhaps the last vestiges of the species will be street trees in some remote communities in the West.

Ulmus parvifolia Jacq. Chinese Elm. Small trees, mainly less than 7 m tall, with trunks 1–5 dm thick; bark gray, with prominent, orange, corky lenticels; branchlets pubescent; leaf blades elliptic to ovate, 2–7 cm long, 1–2.5 cm wide, slightly unequal basally, acute apically, once serrate or essentially so, glabrous above, glabrous to slightly pubescent in vein axils beneath; flowers produced in late summer and early autumn; samaras ca 6 mm long, notched apically, elliptic to ovate; $2n = 28$. Cultivated shade and specimen trees in Salt Lake, Utah, and Wasatch counties; introduced from Asia; 8 (0).

Ulmus procera Salisb. English Elm. Trees to 25 m tall or more, with trunks 5–10 dm thick or more; bark deeply fissured; branchlets pubescent; leaf blades 3–8 cm long, 2–5 cm wide, oval to ovate, short-acuminate, the base unequal, coarsely doubly serrate, green and scabrous above, pubescent beneath; flowers short-stalked, in clusters; samaras ca 25 mm long. Cultivated shade trees in Carbon, Tooele, Utah, and Washington counties; introduced from Eurasia; 18 (0).

Ulmus pumila L. Siberian Elm; Chinese Elm. Trees to 25 m tall or more, with trunks 5–15 dm thick; bark finally fissured; branchlets slender, glabrous or glabrate, or hairy in youth; buds essentially glabrous; leaves in 2 rows on branchlets, narrowly elliptic to lanceolate, 2.5–7.5 cm long, acute, only slightly oblique basally, once serrate, thick, smooth and dark green above, becoming glabrous beneath; flowers small, greenish, in clusters, borne in springtime; stamens 8; samaras ca 10–13 mm wide, glabrous, obovate to rotund; $2n = 24, 28$. Rapidly growing and widely cultivated shade tree at 850 to 1895 m in most if not all Utah counties; introduced from Asia; 42 (iii). This elm has escaped from cultivation and is now a part of the established flora of Utah, especially along stream courses and around lakes at lower elevations. It is a vigorous, brittle tree, often infected with organisms that result in slime flux on the trunk. The trees shed branches with great ease, during even mild cyclonic disturbances, providing work when none is needed. Seeds are produced in great profusion, and each germinates somewhere, with seedlings established in every conceivable place in lower elevation portions of the state.

Zelkova Spach

Trees; leaves short-petiolate, the blades serrate; flowers polygamous, borne on branchlets of the season; staminate flowers clustered, axillary; calyx campanulate, 4- or 5-lobed; stamens 4 or 5; pistillate or perfect flowers solitary or few in upper leaf axils; fruit a 1-seeded drupe, with an eccentric style.

***Zelkova serrata* (Thunb.) Makino** [*Corchorus serrata* Thunb.]. Trees to 25 m or more, with rounded crown; branches slender; leaves 2-10 cm long, ovate to oblong or lanceolate, acuminate, rounded or subcordate basally, sharply and coarsely serrate, somewhat scabrous above, pubescent beneath; fruit subglobose. Cultivated shade and specimen tree at ca 1310 to 1375 m in Salt Lake and Utah counties; introduced from Japan; 5 (0).

UMBELLIFERAE A. L. Juss.
Parsley Family

Annual biennial or perennial acaulescent or caulescent herbs from taproots, rhizomes, fibrous or tuberous roots, or caudices; leaves simple to decompound, petioles typically sheathing basally or the upper leaves reduced to dilated sheaths; inflorescence of compound umbels, the primary umbels with or without a subtending involucre of bracts, the secondary umbels (umbellets) with or without a subtending involucel of bractlets; flowers mostly regular, perfect or some of them staminate or sterile; sepals 5 or lacking; petals 5, small, usually inflexed at the tip, white, yellow, or purple; stamens 5, small, alternate with the petals; pistil 1, the ovary inferior, bicarpellate, 2-loculed, with 1 ovule per locule, the two styles with or without a conical base (stylopodium); fruit a schizocarp of 2 mericarps united by their faces (the commmissure) nearly terete, dorsally or laterally compressed; mericarps separating at maturity and apically attached to and pendulous on a fine wirelike entire or bifid to divided carpophore or remaining adherent and then the carpophore usually lacking or poorly developed and usually adnate to the commissural faces, each mericarp usually 5-nerved, 3 of the nerves dorsal and 2 on the lateral margins, the nerves filiform to winged, or obscure or lacking, the internerve areas commonly with 1 or more oil-tubes, the commissural faces often with 2 or more oil-tubes; x = 4-12 [Apiaceae Lindl.].

Mathias, M. E. and L. Constance. 1944-1945. Umbelliferae. N. Amer. Fl. 28B: 43-297.

1. Leaves peltate, simple, orbicular; flowers in a verticellate spikelike inflorescence; plants rhizomatous, of Washington County .... *Hydrocotyle*
—   Leaves not as above; flowers in compound umbels or globose heads (in a few taxa of *Cymopterus*); plants rarely rhizomatous 2
2(1). Plants caulescent; pseudoscape lacking; peduncles few to several, mostly shorter than the leafy stem; styles rarely over 1 mm long; stylopodium present and petals white in most native taxa 3
—   Plants acaulescent, the leaves sometimes whorled atop a pseudoscape or, if subcaulescent, the usually solitary peduncle longer than the short leafy stem, and lateral umbels if any typically borne on the lower 1/3 of the plant; styles often over 1 mm long; stylopodium lacking or present; petals yellow, white, or purple 4
3(2). Leaves simple, pinnate or ternate; leaflets mostly sessile .... Key 1
—   Leaves various; leaflets usually petiolulate, at least the primary ones .... Key 2
4(2). Leaves ternate or biternate with 3-9 leaflets or rarely a few simple, usually only 2-3 per plant; petals white .. .. *Orogenia*
—   Leaves and leaflets not as above or, if so, the plants mostly taller and/or petals yellow 5
5(4). Stylopodium low conic; plants of the Raft River and Uinta mts., mostly above 2440 m; involucels lacking or of 1-2 linear bractlets; fruit 3-6 mm long; petals white .... *Ligusticum*
—   Stylopodium lacking except in *Podistera*; plants not as above in all features; involucels mostly present; fruit mostly longer or petals yellow 6
6(5). Key to plants with mature fruits .... Key 3
—   Key to plants in flower or with young fruits .... Key 4

Key 1.

Plants caulescent; peduncles and umbels mostly shorter than the stem; stylopodium usually present; leaves simple, pinnate, or ternate; leaflets sessile.

1. Leaflets entire, linear or linear-elliptic 2
—   Leaflets toothed and/or lobed, not linear 4
2(1). Leaves soon withering, with some leaflets often over 1.5 cm long; stylopodium present; plants from a tuberous root or fascicle of tuberous roots, the stem readily detached from the tuberous base, from northern Utah .... *Perideridia*
—   Leaves more persistent, the leaflets not over 1.5 cm long or, if so, the plants of San Juan and Wayne counties; stylopodium lacking; plants from a taproot and a branched crown or caudex 3
3(2). Petals and stamens yellow when fresh; leaflets mostly 2-5 cm long; fruit 6-8 mm long; plants of San Juan and Wayne counties .... *Cymopterus beckii*
—   Petals and stamens white; leaflets 0.3-2 cm long; fruit 2-4 mm long; plants of Cache County ...... *Musineon*
4(1). Basal leaves mostly simple, shallowly toothed, cordate at the base; stem leaves usually ternate, not over 3 cm long; petals bright yellow .... *Zizia*
—   Leaves pinnate or, if ternate or upper ones simple, over 3 cm long; petals white or yellow 5
5(4). Leaves ternate, the upper ones sometimes simple, the 3 leaflets 8-36 cm long, about as wide; plants 1-2 m tall or taller, villous-woolly at least on some of the nodes; petals 4-8.5 mm long (at least some) . *Heracleum*
—   Leaves pinnate, the leaflets less than 8 cm long and much narrower; plants shorter or not villous-woolly; petals smaller 6
6(5). Umbels sessile or nearly so; leaflets ovate to suborbicular, 3-lobed to near the middle; fruit ca 1.5 mm long; plants cultivated and rarely escaping except in Washington County (celery) ..... *Apium graveolens* L.
—   Umbels not sessile except sometimes the terminal one; leaflets variously shaped, but not ovate to suborbicular and lobed to near the middle; fruit over 1.5 mm long except in *Berula*; distribution various 7
7(6). Involucre and involucels well developed, sometimes spreading or deflexed, the bracts 1-6, the bractlets (2) 4-12; fruit 1.5-3 mm long, the ribs not winged; plants of very wet places, often growing in water, from fibrous roots 8
—   Involucre lacking or infrequently of 1 or 2 bracts; involucels often lacking; fruit over 3 mm long or else the ribs winged; plants of various habitats, from tap or tuberous roots 9

8(7). Stems often sprawling, sometimes stoloniferous; leaves with (3) 5–15 opposite pairs of leaflets, these 0.3–4 (6.5) cm long; rays 4–16; ribs of the fruit obscure .... *Berula*

— Stems erect, not stoloniferous; leaves with 4–6 opposite pairs of leaflets, these 2–8 (15) cm long; rays 11–24; ribs of the fruit prominently corky .... *Sium*

9(7). Umbels often more than 7 per stem; fruit strongly flattened dorsally, 5–8 mm long, 3–6 mm wide, the lateral ribs slightly winged, the dorsal ones filiform; petals greenish yellow or reddish; plants introduced, cultivated, or established .... *Pastinaca*

— Umbels fewer than 7 per stem; fruit not strongly flattened dorsally or, if so, 3–5 mm long; petals white or greenish; plants indigenous 10

10(9). Fruit over 1 cm long; leaves rarely all pinnate; peduncles mostly not subtended by dilated, bladeless sheaths or these greatly reduced .... *Osmorhiza*

— Fruit 3–5 mm long; leaves mostly once-pinnate; peduncles often with subtending dilated sheaths 11

11(10). Fruit strongly flattened, the dorsal ribs filiform, the lateral ribs conspicuously winged; plants with tuberous roots, of the Abajo and La Sal mts. .... *Oxypolis*

— Fruit rounded in cross-section, the dorsal and lateral ribs with small wings; plants from taproots, widespread .... *Angelica*

Key 2.

Plants caulescent; peduncles and umbels mostly shorter than the stems; stylopodium present; leaves more than once-compound; primary leaflets not sessile.

1. Ultimate leaf segments over 2 cm long (at least some), toothed or lobed, but not entire or pinnatifid 2

— Ultimate leaf segments less than 2 cm long or, if longer, entire or pinnatifid 5

2(1). Plants from creeping rhizomes, cultivated and rarely escaping; lower leaves long-petioled, often biternate with 9 leaflets but sometimes irregularly compound (ground elder) .... *Aegopodium podagraria* L.

— Plants not from creeping rhizomes, seldom cultivated; leaves various 3

3(2). Involucels of ca 6 bractlets, 1–4 mm long; umbels 6–20 or more per stem, the rays 15–26, 1.5–4 cm long; fruit 2–4 mm long, the ribs corky .... *Cicuta*

— Involucels mostly lacking; umbels often fewer than 6 per stem and/or the rays either fewer or longer than above or both; fruit 4–25 mm long, the ribs various .. 4

4(3). Fruit (10) 12–25 mm long, bristly pubescent (except in *O. occidentalis*) the dorsal ribs not prominent; leaflets often hirtellous; dilated sheaths seldom subtending the peduncles .... *Osmorhiza*

— Fruit 4–5 mm long, not bristly pubescent, the dorsal ribs with small wings; leaflets glabrous; peduncles often subtended by dilated bladeless or nearly bladeless sheaths .... *Angelica*

5(1). Fruits and ovaries with bristly hairs; involucre often of pinnatifid or compound bracts; plants annual or biennial 6

— Fruits and ovaries without bristly hairs; involucre mostly of entire bracts; plants mostly biennial or perennial 8

6(5). Involucre lacking or of 1 entire bract; plants with appressed hispid hairs, from Washington County .. .. *Torilis*

— Involucre of few to several pinnatifid to compound bracts; plants glabrous or with spreading hairs 7

7(6). Bracts of the involucre leaflike, pinnately compound; rays 1–7 (9), 1.5–10 cm long, some much longer than the involucres, some often nearly as long as the peduncles; inflorescence open; bristly hairs of the fruit hooked; plants annual, of Washington County ... *Yabea*

— Bracts of the involucre pinnatifid; rays mostly 10–60 or more, seldom over 3 cm long or, if longer, plants biennial, often not much exceeding the involucres, rarely longer than the peduncles; inflorescence congested; bristly hairs of the fruit glochidiate at apex; plants biennial, widespread .... *Daucus*

8(5). Involucel and involucre lacking 9

— Involucel and often involucre present 13

9(8). Petals yellow; plants introduced, cultivated and adventive, ultimate segments of leaves filiform, 1–40 mm long, ca 0.5 mm wide 10

— Petals white or yellow (in *Lomatium*) and the plants native; ultimate segments various, often over 0.5 mm wide 11

10(9). Plants annual, not glaucous, widely cultivated; leaves not especially crowded toward the stem base, the petiolules of the lowest pair of primary leaflets mostly less than 2 cm long, the ultimate segments 4–20 mm long (dill) .... *Anethum graveolens* L.

— Plants perennial, glaucous, occasionally adventive; leaves sometimes crowded toward the stem base, the petiolules of the lowest pair of primary leaflets often over 2 cm long, the ultimate segments 4–40 mm long .... *Foeniculum*

11(9). Plants biennial from taproots, introduced; umbels often 6–12, or more per stem; fruit 3–4 mm long, the ribs filiform, not at all winged .... *Carum*

— Plants perennial from taproots or caudices, native; umbels rarely more than 8 per stem; fruit 3–14 mm long, the lateral and sometimes the dorsal ribs winged 12

12(11). Petals white; fruit 3–8 mm long, rounded, the dorsal and lateral ribs narrowly winged; stylopodium low conic; ultimate leaflets over 50 per leaf, 1–10 (15) mm long .... *Ligusticum*

— Petals yellow when fresh; fruit 8–14 mm long, dorsally flattened, the dorsal ribs filiform, the lateral ribs winged; stylopodium lacking; ultimate leaflets usually 3–45 per leaf, 0.3–9 cm long .... *Lomatium*

13(8). Petals yellow or greenish yellow; plants cultivated, rarely escaping (parsley) .... *Petroselinum crispum* (Mill.) A. W. Hill

— Petals white; plants not cultivated 14

14(13). Stems often purple-spotted, usually much branched, mostly with 10–30 or more umbels; plants 5–30 dm tall, naturalized, weedy in moist or wet places in valleys and foothills, occasionally montane; involucre of 2–6 bracts, 2–6 (15) mm long .... *Conium*

— Stems not purple-spotted with few branches, with (1) 3–7 (12) umbels; plants to 10 dm tall, native, often montane, involucre lacking or seldom as above 15

15(14). Ultimate leaflets 2–6 cm long (at least some) and entire; leaves often withering shortly after anthesis; plants from a tuberous root or fascicle of tuberous roots, these easily detached from the stem, and seldom collected .... *Perideridia*

— Ultimate leaflets not over 2 cm long or, if so, not entire; leaves more persistent than above; plants from a taproot or a cluster of tuberous roots 16

16(15). Involucels usually with more than 3 bractlets; fruit slightly compressed dorsally; root-crown mostly simple, without marcescent petiole bases; plants rather rare in eastern Utah .... *Conioselinum*

— Involucels lacking or rarely with more than 3 bractlets; fruit terete or slightly compressed laterally; root-crown simple or branched, usually with fibrous marcescent petiole bases; plants common, widespread .... *Ligusticum*

Key 3.

Plants typically acaulescent; styles often over 1 mm long; stylopodium mostly lacking.

1. Fruit strongly flattened dorsally; dorsal ribs filiform, not winged, the lateral ribs more or less winged; body 8–15 (20) mm long or, if shorter, the plants usually pubescent (note: *L. cous*, *L. minimum*, and *L. scabrum* have small fruits and glabrous herbage); involucre lacking . *Lomatium*

— Fruit not strongly flattened or, if so, the dorsal ribs winged, the body usually less than 8 mm long, the wings sometimes to 12 (15) mm long especially in plants with an involucre; plants glabrous to hirtellous 2

2(1). Stylopodium conic; leaves once compound with palmatifid leaflets; bractlets of involucels with 2 or 3 or more teeth; plants of high elevations in the La Sal Mts., where apparently rare .... *Podistera*

— Stylopodium lacking; leaves either more than once compound or leaflets not palmatifid; bractlets entire or plants not of high elevations in the La Sal Mts. (except *Oreoxis bakeri*) 3

3(2). Ribs of fruit not winged or at most with low corky wings; carpophore well developed; leaves pinnate, a few leaflets sometimes pinnatifid and nearly bipinnate; plants of Cache, Garfield, and San Juan counties 4

— Ribs of the fruit with papery wings or, if with low corky wings, the carpophore lacking; leaves usually more than once compound; plants of broad distribution 5

4(3). Ribs of fruit not winged; terminal umbel often subtended by a smaller one; petals and stamens white; plants often subcaulescent, of the Bear River Range, Cache County .... *Musineon*

— Ribs of fruit with low, corky wings; umbel solitary; petals and stamens yellow; plants strictly acaulescent, of Garfield and San Juan counties .... *Aletes*

5(3). Fruit slightly compressed laterally, with low corky wings, 2–5 mm long and with rather conspicuous, persistent calyx teeth, the carpophore lacking; plants from branched caudices, strictly acaulescent, not strongly aromatic, mostly hirtellous or scabrous throughout or the bractlets toothed, 1–10 (15) cm tall, montane, mostly above 2440 m (except *O. trotteri*) .... *Oreoxis*

— Fruit compressed dorsally with prominent, papery or corky wings, the calyx teeth obsolete or not persisting or the carpophore well developed; plants from fibrous often tuberous taproots or, if from branched caudices, with one or more cauline leaves and from lower elevations, or strongly aromatic, or glabrous, or regularly over 10 cm tall; bractlets often not toothed ... *Cymopterus*

Key 4.

Plants with features of Key 3, but in flower or with immature fruits.

1. Leaves pinnate, pinnatifid, or palmatifid, rarely trifid with linear entire segments 2

— Leaves either more than once-compound or else ternate or ternately divided and with toothed to lobed leaflets 7

2(1). Petals and stamens white; terminal umbel sometimes subtended by a smaller axillary umbel; plants of Cache County .... *Musineon*

— Petals and stamens yellow when fresh; umbel solitary; plants of the southern 1/2 of the state 3

3(2). Bractlets of the involucel usually with 3 or more teeth or lobes, linear-elliptic or oval to obovate; stylopodium present or lacking; plants of the La Sal Mts. 4

— Bractlets of the involucel mostly entire, linear or narrowly elliptic; stylopodium lacking; plants not of the La Sal Mts. 5

4(3). Leaflets more or less palmatifid, the major segments again trilobate to palmatifid; stylopodium conspicuous; base of plant with few if any persistent leaf-bases .... *Podistera*

— Leaflets pinnatifid or trifid; stylopodium lacking; base of plant clothed with persistent leaf-bases .... *Oreoxis bakeri*

5(3). Leaflets entire, 0.5–2 mm wide, linear-filiform to very narrowly elliptic; plants of Emery, Garfield, Iron, Sevier, and Wayne counties .... *Lomatium*

— At least some of the leaflets lobed or, if all entire, some over 2 mm wide and elliptic; distribution various 6

6(5). Leaflets.3–12 mm long; rays 4–8, 2–10 mm long; involucels 4–5 mm long; plants of Garfield and central San Juan counties .... *Aletes*

— Some of the leaflets regularly over 1.2 cm long; rays (4) 6–13, 5–20 mm long; involucels 2–15 mm long; plants of Grand and San Juan counties .... *Lomatium latilobum*

7(1). Plants pubescent, not more so just below the umbel than elsewhere (see also *Oreoxis alpina*) .... *Lomatium*

— Plants glabrous or scabrous, sometimes hirtellous just below the umbel and then with glabrous leaves 8

8(7). At least some of the ultimate leaflets over 2 cm long and entire or at most toothed 9

— Ultimate leaflets less than 2 cm long or, if longer, lobed 10

9(8). Peduncles hirtellous just below the umbel, glabrous below; leaves glabrous, lowest pair of primary leaflets sessile or on petiolules less than 2 cm long, the ultimate leaflets to 2.5 cm long, 1–3 (4) mm wide; plants not aromatic, from the southern 1/2 of Utah .... *Cymopterus lemmonii*

— Plants glabrous or, if scabrous, not more so just below the umbel than elsewhere; lowest pair of primary leaflets either with petiolules longer than above or some of the ultimate leaflets mostly longer or wider than above; plants of the northern 1/2 of the state or else strongly aromatic .... *Lomatium*

10(8). Plants from a taproot, this sometimes enlarged and tuberlike, the crown simple or few-branched, with few if any marcescent leaf bases; pseudoscape (at least a subterranean one) often conspicuous; leaf blades sometimes with confluent portions wider than the ultimate teeth or lobes 11

— Plants from a simple or more typically branched often woody caudex, this often clothed with marcescent leaf bases; pseudoscape lacking; leaf blades finely and completely dissected so that the ultimate segments the widest undivided portions of the blade 12

11(10). Root abruptly tuberous; plants strongly aromatic and from Salt Lake north to Cache County or, not aromatic, and of northwestern Box Elder County (*L. ambiguum* and *L. cous*) .... *Lomatium*
— Root gradually if at all tuberous; plants not aromatic, distribution various (note: rare glabrous specimens of *Lomatium junipernum* will key here) ..... *Cymopterus*

12(10). Leaves with ca 2–4 opposite pairs of primary leaflets; plants of mountains, mildly if at all aromatic 13
— Leaves (at least some) with 5–11 opposite pair of primary leaflets; plants of mountains and deserts, aromatic or not 14

13(12). Primary leaflets 4–14 mm long, sessile; leaf blades 1–3.5 cm long; plants to about 12 cm tall, scabrous-hirtellous throughout or else bractlets of the involucel toothed .... *Oreoxis*
— Primary leaflets (at least the lowest pair) usually 15–35 mm long, sessile or on petiolules to 15 mm long; plants 8–50 cm tall, glabrous except hirtellous on the peduncle below the umbel and sometimes scabrous or hirtellous in the umbel; bractlets of the involucel entire .... *Cymopterus lemmonii*

14(12). Lowest pair of primary leaflets seldom over 1/4 as long as the leaf blade, sessile or on petiolules to 18 mm long; leaves pinnately compound, the blades more or less oblong in outline 15
— Lowest pair of primary leaflets (1/4) 1/3–3/4 as long as the leaf blade, on petiolules over 18 mm long; leaves more or less ternate-pinnately compound, the blades often ovate in outline 16

15(14). Plants strongly aromatic, mostly known from above 2350 m, glabrous, widespread; ultimate segments of leaves 1–12 mm long; some bractlets commonly exceeding the flowers .... *Cymopterus hendersonii*
— Plants not strongly aromatic or from lower elevations, scabrous if from above 2350 m, from the southern 1/2 of Utah; bractlets rarely exceeding the flowers (*L. parryi* and *L. scabrum*) .... *Lomatium*

16(14). Calyx teeth lacking or to 0.3 mm long; ultimate segments of leaves 0.2–0.3 mm wide ..... *Lomatium grayi*
— Calyx teeth 0.5–0.9 mm long; ultimate segments of leaves 0.5–1 (1.5) mm wide ... *Cymopterus terbinthus*

Aletes Coult. & Rose

Perennial, acaulescent, glabrous to pubescent herbs; leaves pinnate or bipinnate, petiolate, the leaflets distinct or confluent, often lobed and spinulose-dentate or entire; umbels compound; involucre lacking; rays few to several, spreading to reflexed; involucel of free or united bractlets; calyx teeth conspicuous, deltoid-ovate; stylopodium lacking; carpophore divided to the base, sometimes readily deciduous or possibly lacking; fruit oblong to ovoid-oblong, slightly compressed laterally or subterete, the ribs subequal, prominently corky-winged or obscure.

Aletes macdougalii Coult. & Rose Plants 7–20 cm tall, acaulescent, glabrous or scabrous, from a branched caudex, this more or less clothed with persistent leaf bases; leaves pinnate or some of the leaflets pinnatifid and nearly bipinnate, with 2–6 opposite pair of lateral leaflets, petioles 1.5–7 cm long, blades 1–5 cm long; leaflets 3–12 mm long, sessile, narrowly elliptic and entire or obovate and with 1–3 (5) teeth or lobes; peduncles 5–15 cm long; umbel solitary; rays 4–8, 2–10 mm long; bractlets of the involucel ca 4–6, 4–5 mm long, linear or linear-elliptic, more or less united at the base; pedicels 1–2 mm long; calyx teeth ca 1–1.5 mm long, narrowly to broadly deltoid; petals yellow when fresh; styles 1.5–2.5 mm long; fruit 4–6 mm long, the ribs with small, more or less corky wings, the lateral ones ca 1 mm wide, the dorsal ones smaller. Rock crevices, rocky slopes, and sandy ground in pinyon-juniper and limber pine-bristlecone pine communities at 1280 to 2740 m in Garfield and San Juan counties; Arizona and Colorado; 8 (0). Our plants are referable to ssp. ***breviradiatus*** Theobald & Tseng. The genus could reasonably be included within an expanded *Cymopterus*.

Angelica L.

Perennial, caulescent, single stemmed herbs from a stout taproot; leaves pinnately to ternately 1–3 times compound, with broad leaflets; lower blades on elongate petioles, the middle ones often arising directly from a dilated sheath, the upper ones often much reduced or lacking and the leaves reduced to a dilated sheath; umbels compound; involucre and involucel lacking or of narrow scarious or foliaceous bracts or bractlets; calyx teeth minute or obsolete; petals white, seldom pink or yellow; stylopodium broadly conic; carpophore divided to the base; fruit elliptic-oblong to orbicular strongly compressed dorsally, the lateral and dorsal ribs with small but obvious wings, or the ribs all corky-thickened and scarcely winged.

1. Leaflets lanceolate to linear, mostly over 3 times as long as wide; leaf blades oblong in outline; umbels 2–7, with 7–20 rays 2
— Leaflets ovate or broader or, if lanceolate and over 3 times as long as wide, the umbels mostly more than 7 and rays 20–40 3

2(1). Leaves ternate-pinnate, the lowest pair of primary leaflets on petiolules 3.5–12 cm long; leaflets coarsely toothed to lobed, the margins with 1–3 teeth or lobes per cm; plants of the Deep Creek Mts. .... *A. kingii*
— Leaves pinnate or scarcely ternate-pinnate, the lowest pair of primary leaflets sessile or on petiolules to 1.5 cm long; leaflets finely to rather coarsely toothed, the margins with ca 3–7 teeth per cm; plants not from the Deep Creek Mts. .... *A. pinnata*

3(2). Plants less than 1 m tall, of rocky places above 3050 m; umbels 1–3; leaflets 1–5 cm long, serrate-dentate, rarely lobed; involucels of 1–3 or more linear bractlets 3–10 mm long; ovaries and fruit glabrous or scabrous . ... *A. roseana*
— Plants 1–2 m tall, of wet places below 3050 m; umbels several; leaflets 3–16 cm long, some usually lobed as well as toothed; involucels lacking; ovaries and young fruit hispid to hirsute .... *A. wheeleri*

Angelica kingii (Wats.) Coult. & Rose Great Basin Angelica. [*Selinum kingii* Wats.]. Plants (3) 4–12 dm tall, glabrous, except scabrous to short-hispid in the inflorescence, from a taproot, lacking persistent leaf bases; leaves ternate-pinnate with 4–5 (6) opposite pair of lateral primary leaflets, the lower pairs pinnate; lower petioles to 25 cm long, dilated at the base, the upper ones reduced; lower blades to 40 cm long, oblong in outline, the upper ones reduced; lowest pair of primary leaflets 2/3–3/4 as long as the leaf blade, ascending and more or less parallel to the primary rachis, on petiolules 3.5–12 cm long; leaflets 2–14 cm long, 4–15 (40) mm wide, lanceolate to nearly linear, coarsely toothed or lobed, the margins with 1–3 teeth or lobes per cm or rarely entire; peduncles mostly 4–17 cm long; umbels 3–7; involucre lacking; rays 11–20, 1.5–9.5 cm long, scabrous; involucels lacking; pedicels 1–6 mm long, scabrous or short-hispid; petals white, sometimes marked with purple in age; stamens white; styles to ca 1.5 mm long; fruit 4–5 mm long, densely hispid, the ribs slightly winged, the lateral wings a little wider than the dorsal ones, n = 22. Aspen fir and streamside communities at 2130 to 2380 m in the Deep Creek Mts., Juab County; Nevada, California, and Idaho; 5 (0).

Angelica pinnata Wats. Small-leaved Angelica [*A. leporina* Wats., type from Rabbit Valley, Wayne County]. Plants 4.5–10 (15) dm tall, glabrous or nearly so, except scabrous to hirtellous in the inflorescence, without persistent leaf bases, from a taproot and sometimes branched crown; leaves pinnate or partly bipinnate with 3 (4) opposite pair of leaflets, the lowest pair sometimes bipinnate or partly bipinnate, the upper ones pinnate, lower petioles 5–26 cm long, gradually expanded into a dilated partly sheathing base, reduced and the blades sometimes sessile on the dilated sheath; blades (5) 9–21 cm long, more or less oblong in outline; leaflets 1.5–13 cm long, 4–37 mm wide, sessile, lanceolate, elliptic, or ovate, serrate, the margins with ca 3–7 teeth per cm; peduncles 3.5–14 cm long; umbels (1) 2–5; involucre lacking; rays 7–14, 2–8.5 cm long, scabrous to hirtellous; involucels lacking or very rarely of 1 or more green to scarious, linear or nearly linear bractlets 3–13 mm long; pedicels 3–7 mm long, glabrous or scabrous; petals white; styles to ca 1 mm long; ovary glabrous to hirtellous; fruit 4–5 mm long, glabrous or sparsely hirtellous, the lateral wings ca 1 mm wide, the dorsal wings about 0.5 mm wide. Tall forb, oak, maple, aspen, Douglas fir, spruce-fir, willow, and wet meadow communities, very often along streams or around seeps and springs at 1520 to 3290 m in all Utah counties except Beaver, Box Elder, Carbon, Emery, Millard, Morgan, Rich, and Summit (type from the Uinta Mts.); eastern Idaho to western Montana, south to Utah and Colorado; 65 (xv).

Angelica roseana Henderson Rock Angelica. Plants 30–75 cm tall, strongly aromatic, glabrous, or scabrous in the inflorescence, from stout taproots; stems stout, hollow, 1–2 cm in diameter; leaves ternate-pinnate with 3–4 opposite pair of lateral primary leaflets, the lower ones bipinnate or ternate and petiolulate, the upper pinnate and sessile; petioles to 8 cm long or, lacking on the upper leaves, and the blades sessile on a dilated sheath; blades 5–17 cm long, ovate in outline, the upper ones reduced or lacking and leaves reduced to dilated sheaths; lowest pair of primary leaflets ca 3/4 as long as the leaf blade, on petiolules 2.5–5.2 cm long, the blades (1) 2–5 cm long, ovate to orbicular, sharply serrate-dentate, rarely lobed; peduncles 4–17 cm long, the terminal one about as thick as the stem, the lateral ones partly enveloped in bladeless, dilated sheaths; umbels 1–3; involucre lacking or occasionally of 1–2 linear bracts to 1.5 cm long; rays 15–30, 3.5–12 cm long, scabrous; bractlets of the involucel 1–3 (rarely more), 3–10 mm long, 0.2–0.5 mm wide, separate, linear; pedicels 4–9 mm long, glabrous or scabrous; petals white; stamens whitish; styles ca 2 mm long; ovary glabrous or at most scabrous; fruit ca 5 mm long, the ribs with wings about 1 mm wide. Talus slopes, boulder fields, and rock strips, above timberline or upper spruce zone at 3050 to 3570 m in Daggett, Duchesne, Summit, Uintah, and Utah counties, Montana to Idaho, south to Colorado and Utah; 14 (iii).

Angelica wheeleri Wats. Utah Angelica. [*A. dilatata* A. Nels. in Coult. & Rose, type from City Creek Canyon, Salt Lake County]. Robust plants 1–2 m tall or taller, glabrous except in the inflorescence, mildly if at all aromatic, from stout rootcrowns with large fibrous roots; stems hollow, to 3 cm in diameter; lower leaves ternate-pinnately compound, with 3–5 opposite pair of lateral primary leaflets, the lower ones bipinnate or tripinnate and petiolulate, the upper often pinnate and sessile; petioles to 45 cm long, often dilated; blades to 40 cm long, ovate in outline; lowest pair of primary leaflets to 21 cm long, ca 1/2 as long as the leaf blade, on petiolules to 5 cm long; blades of leaflets 3–16 cm long, 2–8 cm wide, lanceolate to ovate, serrate and some usually lobed; peduncles 2–29 cm long, often subtended by bladeless or nearly bladeless dilated sheaths 2–20 cm long; umbels several; involucres lacking or sometimes of 1–2 linear bracts to 2 cm long; rays 20–45, 5–10 cm long, scabrous; involucels none; pedicels 5–12 mm long, glabrate to scabrous-hirsute; petals white; stamens whitish; styles ca 1 mm long; ovary and young fruit sparingly to densely hispid to hirsute; fruit 4–5 mm long, densely hispid, the lateral and dorsal ribs commonly winged. Boggy or very wet areas often in riparian communities or in seeps and springs at 1950 to 3050 m in Cache, Juab, Piute, Salt Lake, Sevier, and Utah counties; endemic to Utah; 7 (ii). The type is from northern or central Utah. *A. arguta* Nutt. in T. & G. has been reported for Utah, but I have not seen a specimen and suspect that reports are based on *A. wheeleri*. It is apparently different from *A. wheeleri* only in the glabrous ovaries and fruit.

Berula Hoffm.

Perennial, caulescent, glabrous herbs from fibrous roots, often stoloniferous; leaves pinnately compound or the submerged ones sometimes with filiform-dissected blades; umbels compound; involucre and involucel usually well developed; calyx teeth minute or obsolete; stylopodium conic; carpophore divided to the base, inconspicuous, adnate to the mericarps; fruit elliptic to orbicular, somewhat compressed laterally, glabrous, the ribs inconspicuous.

Berula erecta (Hudson) Cov. Cutleaf Water-parsnip. [*Sium erectum* Huds.]. Stems 1.5–10 dm long or longer, from numerous fibrous roots; leaves pinnate with (3) 5–15 opposite pair of lateral leaflets, or the submerged leaves (if present) often with filiform-dissected blades; petioles to 32 cm long or upper blades sessile on a dilated sheath; blades 2–31 cm long; leaflets 0.3–4 (6.5) cm long, sessile, nearly linear to lanceolate or ovate in outline, toothed to incised or occasionally a few entire; peduncles 1.5–8 cm long; umbels 3–20 or more; bracts of the involucre 1–6, 2–15 (25) mm long, linear or elliptic, entire, toothed, or rarely pinnatifid; rays 4–16, 0.5–2.5 (4) cm long; bractlets of the involucels ca 4–7, 1–7 mm long, linear or elliptic, entire; pedicels 2–7 mm long; petals white; stamens white; styles less than 1 mm long; fruit ca 2 mm long, the ribs obscure. In mud and water of streams, seeps, springs, marshes, swamps, margins of ponds and lakes, and in wet hanging gardens at 850 to 2130 m in all counties of the state except Beaver, Cache, Carbon, Daggett, Emery, Grand, Iron, Morgan, San Juan, Summit, and Wayne; widespread in Europe, Mediterranean regions, and North America. The American plants are referable to **var. incisa** (Torr.) Cronq. [*Sium ? incisum* Torr.]; 64 (vi).

Carum L.

Perennial, caulescent, glabrous herbs from taproots; leaves pinnately compound; inflorescence of compound umbels; involucre and involucel lacking or of a few inconspicuous bracts or bractlets; calyx teeth obsolete; stylopodium low conic; carpophore divided to the base; fruit oblong to broadly elliptic-oblong, somewhat compressed laterally, evidently ribbed.

Carum carvi L. Caraway. Plants 3–6 (10) dm tall; leaves 2–3 times pinnate and then often pinnatifid, with 6–11 opposite or offset pairs of lateral primary leaflets; petioles to 15 cm long, the upper ones reduced and the blades sometimes sessile on a dilated sheath; blades 5–16 cm long, oblong in outline; primary leaflets from less than 1/4–1/2 as long as the leaf blade, sessile, the ultimate segments 2–8 (15) mm long, 0.5–2 mm wide, linear and entire or obovate and toothed to lobed; peduncles 4–12 cm long, usually subtended by a dilated sheath; umbels 6–12 or more; involucre lacking or inconspicuous; rays 6–12 (14), 1.5–8 cm long; involucels lacking or of minute scarious teeth; pedicels (5) 8–20 mm long; petals white; filaments white, the anthers pale green or whitish; styles 0.5–0.9 mm long; fruit 3–4 mm long, the ribs filiform; 2n = 20, 22. Cultivated, the fruits used in flavoring, escaping, and established in mountain brush, meadow, and aspen communities, at 1375 to 2640 m in Box Elder, Cache, Daggett, Davis, Duchesne, Salt Lake, Sanpete, Sevier, and Summit counties; native to Eurasia, now widespread across the U. S.; 9 (iii).

Cicuta L.

Perennial, caulescent, glabrous, violently poisonous herbs, from clusters of fibrous roots, some of these commonly tuberous-thickened; base of stem thickened, with hollow chambers separated by transverse septae; internodes of stems hollow; leaves 1–3 times pinnate or ternate-pinnate, with well-developed leaflets; umbels several, compound; involucre wanting or of a few inconspicuous narrow bracts; involucel of several narrow bractlets or rarely lacking; petals white or greenish; calyx teeth evident; stylopodium depressed or low-conic; carpophore divided to the base, deciduous; fruit ovate or orbicular, compressed laterally, the ribs usually prominent and corky.

Cicuta maculata L. Water Hemlock. [*C. douglasii* (DC.) Coult. & Rose, misapplied]. Plants, 6–21 dm tall or taller, with clusters of fibrous roots surmounted by a thickened crown; stems 5–15 mm or more in diameter; leaves pinnate or ternate-pinnate with 4–7 opposite pair of lateral primary leaflets, the lower ones again pinnate, the upper once pinnate and sessile, the lower petioles 5–40 cm long, the the upper ones reduced and the blades often sessile on dilated sheaths, the lowest pair of petiolules 1–3 cm long, leaflets 2–11 cm long, 3–25 mm wide, narrowly lanceolate to lanceolate or linear, finely to coarsely serrate; peduncles (2) 4–15 cm long; umbels 6–30 or more; involucre lacking or of 1 or few linear bracts to 1 cm long; rays 15–26, 1.5–4 cm long; bractlets of the involucels ca 6, 1–4 mm long, linear or narrowly deltoid, pale yellow green or purplish, scarious-margined; pedicels 3–10 mm long; calyx teeth ca 0.5 mm long, often pale green with whitish margins; petals white; stamens white; styles 0.5–1 mm long; fruit 2–4 mm long, oval to globose, the ribs prominent, more or less corky, green, often wider than the darker (often purple) intervals; 2n = 11, 22. Along streams, rivers, ditches, canals, margins of pond and lakes, in wet meadows and marshes at 1370 to 2320 m in Beaver, Cache, Daggett, Kane, Millard, Piute, Salt Lake, Sanpete, Summit, Tooele, Uintah, Utah, Wasatch, Wayne, and Weber counties; widespread in North America; 46 (v.) Some of our plants have leaflets less than 5 times as long as wide (a feature of **var. maculata**, which is found mostly east of Utah), but in these specimens, as well as others from the state, the styles are not more than 1 mm long. All Utah specimens I have seen belong to **var. angustifolia** Hook., the common phase in western North America. This plant is poisonous, with a yellow orange resinol, cicutoxin, concentrated in the chambered root crown and less concentrated elsewhere in the plant.

Conioselinum Hoffm.

Perennial more or less caulescent herbs from a taproot or cluster of fleshy-fibrous roots, sometimes with a caudex; leaves pinnately or ternate-pinnately decompound; umbels compound; involucre lacking or of a few narrow or leafy bracts; involucels of well developed, narrow, often scarious bractlets; calyx teeth obsolete; petals white; stylopodium conic; carpophore divided to the base or nearly so; fruit elliptic or elliptic-oblong, slightly dorsally compressed, glabrous, the lateral ribs evidently thin-winged, the dorsal ribs less so and corky.

Conioselinum scopulorum (Gray) Coult. & Rose [*Ligusticum scopulorum* Gray]. Plants perennial 3–10 dm tall, glabrous except in the inflorescence, from a fusiform taproot with simple or very sparingly branched crown, without persistent leaf bases or these few and weakly persisting; leaves pinnate or ternate-pinnate with (3) 4–5 opposite pair of lateral primary leaflets, the lower ones 2–3 times pinnate and petiolulate, the upper pinnate, pinnatifid, and sessile or nearly so; petioles 3–23 cm long; blades 3.5–19 cm long, ovate in outline, the lowest pair of primary leaflets 1/2–2/3 as long as the leaf blade, on petiolules (0.5) 1–3.5 cm long, the ultimate segments 2–15 mm long, 1–5 mm wide; peduncles 3–21 cm long, often subtended by a dilated sheath, this usually with a reduced sessile blade; umbels 1–3; involucre lacking or of 1 or few linear bracts to 1 cm long; rays 9–15, 1.5–5 cm long; involucels of 3–6 linear or linear-filiform bractlets 2–8 mm long; pedicels 4–12 mm long; petals white; stamens white; styles to ca 1.3 mm long; fruit 4–6 mm long, with lateral ribs narrowly corky winged, the dorsal ones not winged. Apparently rare, along streams at 2560 to 3200 m in Daggett, Grand, Garfield, Piute, San Juan, Summit, and Wayne counties; Wyoming to Arizona and New Mexico; 18 (iii). Plants of *C. scopulorum* are often confused with *Ligusticum porteri*. The two taxa differ in the following subtle ways, with features of *L. porteri* in parentheses: fruit dorsally flattened (nearly terete); bractlets of the involucel often 3 or more (0–2, rarely more); terminal umbel solitary or subtended by alternate lateral umbels (often subtended by opposite or whorled umbels); and plants from a taproot, with a mostly simple crown and with few if any persisting fibrous leaf bases (the crown simple or branched and often with numerous, persistent, fibrous leaf bases). In addition, the rays average shorter and the ultimate segments of the leaves are less conspicuously veined than in those of *L. porteri*.

Conium L.

Biennial caulescent glabrous herbs from stout taproots with purple-spotted, freely branching hollow stems; leaves pinnately or ternate-pinnately dissected; umbels compound, several to numerous; involucre and involucels of small, lanceolate to ovate bracts or bractlets; calyx teeth obsolete; petals white; stylopodium depressed conic; carpophore entire; fruit broadly ovoid, somewhat laterally compressed, with prominent, raised, often wavy slightly winged ribs.

Conium maculatum L. Poison Hemlock. Plants 5–30 dm tall, glabrous; leaves pinnate or ternate-pinnately decompound with 6–9 opposite pair of lateral primary leaflets, the lower ones usually twice or more pinnate and then pinnatifid, petiolulate, the upper once pinnate, pinnatifid, and sessile; petioles of larger leaves 4–18 cm long; larger leaf blades to 30 cm long, reduced upwards and sessile on dilated sheaths, ovate in outline; lowest pair of primary leaflets less than 1/2–2/3 as long as the leaf blade, on petiolules 1–5.5 mm long or shorter upward; ultimate leaflets pinnatifid, the lobes entire or toothed, the widest confluent portions 2–5 (10) mm wide; peduncles 2–7.5 cm long; umbels many; involucral bracts 2–6, 2–6 (15) mm long, entire and ovate or deltoid, caudate to cuspidate, green with scarous margins, or rarely pinnatifid; rays 9–16, 1–4 cm long; bractlets of the involucels 4–6, 1–3 mm long, shaped like the involucral bracts; pedicels 2–6 mm long; petals white; stamens white; styles ca 0.5 mm long; fruit 2–2.5 mm long, the ribs prominently ridged, narrower than the intervals; $2n = 22$. Along ditches, streams, rivers, roadsides, and fence lines, in wet and boggy meadows and moist waste places at 1400 to 2135 (2990) m, in Box Elder, Cache, Davis, Duchesne, Juab, Rich, Salt Lake, Sanpete, Summit, Tooele, Uintah, Utah, Washington, and Weber counties; introduced from Eurasia, now widespread in North America; 34 (iv). This plant is deadly poisonous, due to alkaloids that can cause paralysis of respiratory muscles.

Cymopterus Raf.

Perennial, acaulescent or subcaulescent, glabrous or scabrous herbs from slender to greatly enlarged and tuberlike taproots to branching woody caudices; leaves all basal (these sometimes elevated on an aerial pseudoscape) or basal and 1 to few cauline mostly on the lower 1/2 of the stems, ternate to pinnate or ternate-pinnately compound, rarely simple and ternately cleft; umbels solitary to several, open or reduced to globose heads; involucres lacking or developed; involucels of separate or united bractlets; pedicels obsolete or developed; calyx teeth obsolete to conspicuous; petals white, yellow, or purple; stylopodium lacking; carpophore lacking, inconspicuous and adhering to the inner faces (commissure) of the mericarps or present, persistent on the pedicel, and divided to the base; fruit ovoid to oblong, somewhat flattened dorsally, the lateral and usually 1 or more of the dorsal ribs prominently winged. The strongly aromatic members of the group with woody branched caudices and greenish acute conspicuous calyx teeth have been included in the genus *Pteryxia*. Most of those in *Cymopterus* are ternate or have only 2–6 opposite pairs of lateral primary leaflets. However, the caudex and sometimes the number of primary leaflets are repeated in *C. bipinnatus* Wats., *C. aboriginum* Jones, and in other taxa long included in *Cymopterus*. If high volatile oil content is unique to taxa of the *Pteryxia* group, that group might stand at generic rank. Chemical studies might prove useful in resolving this problem.

1. Leaves 1– or 2–pinnate or a few merely ternate, with entire (rarely bifid) linear or linear-elliptic leaflets 0.5–4 (5.5) cm long and 1–2 (3) mm wide; plants caulescent, of San Juan and Wayne counties, rare .. .. *C. beckii*
— Leaves not as above in all features; plants acaulescent or subcaulescent with 1–3 leaves mostly on the lower 1/3 of the stem 2
2(1). Peduncles rather densely hirtellous just below the umbel, mostly glabrous elsewhere .... *C. lemmonii*
— Peduncles not hirtellous just below the umbel, sometimes scabrous but then not more so just below the umbel than elsewhere 3
3(2). Plants strongly aromatic, from a branched more or less woody caudex, mostly clothed at the base with marcescent leaf bases and sometimes stem bases, often of rocky places; calyx teeth rather prominent, ca 0.5–1 mm long, acute, greenish (*Pteryxia* group) 4
— Plants not strongly aromatic, from fibrous taproots with simple or sparingly branched crowns, without or with few persisting leaf bases, not specific for rocky places; calyx teeth to ca 0.5 mm long, rarely acute ... 5
4(3). Lowest pair of primary leaflets (1/4) 1/2–3/4 or more the length of the leaf blade, mostly 3–9 cm long, several times longer than the upper pairs, on petiolules 2–4 cm long; plants mostly of lower elevations .... *C. terebinthinus*
— Lowest pair of primary leaflets 1/4 or less the length of the leaf blade, to 2.7 cm long, often not more than twice as long as some of the upper pairs, sessile or on petiolules to 1 cm long; plants mostly of high elevations .... *C. hendersonii*
5(3). Involucels scarious, purplish or whitish with purple nerves, the bractlets mostly over 3 mm wide, sometimes united to midlength 6
— Involucels greenish or the bractlets very narrow and divided to the base or nearly so 8
6(5). Rays 1–3.5 cm long, usually at least some exceeding the well-developed to obsolete involucre, not obscured by the dense mature fruits; plants of the Colorado drainage .... *C. bulbosus*
— Rays 0.3–1 cm long, rarely longer, not exserted beyond the usually well-developed involucre; plants of the Great Basin and Colorado drainage 7
7(6). Lobes of involucels and usually of the involucres with more than 3 parallel purplish nerves extending to or near the tip; involucre sometimes reduced to a ring; plants of Kane and Washington counties .... *C. multinervatus*
— Lobes of involucels and involucres with a midnerve extending to the tip and sometimes 1 or 2 lateral shorter ones extending to near the middle; involucre present; plants widespread .... *C. purpurascens*
8(5). Involucels green and foliose, seldom scarious-margined, the bractlets 1.5–4 mm wide; plants obscurely viscid and with adhering grains of sand 9
— Involucels rarely wholly green, not foliose, often scarious-margined and/or the bractlets linear or narrowly elliptic and not over 1.5 mm wide; plants not viscid and with adhering sand grains 10

9(8). Leaves once ternate, the 3 leaflets ternately lobed or cleft, the blades with confluent portions 5–35 mm wide; outer rays 1–3.3 cm long; bractlets of involucel entire or rarely tridentate; pseudoscape lacking; plants of the southern Utah .... *C. newberryi*

— Leaves 2–3 times pinnate with 2 (3) opposite pairs of lateral primary leaflets, some rarely ternate, the blades with confluent portions 1–7 (12) mm wide, rays to 1.3 cm long; bractlets of the involucel often with 2–3 teeth; pseudoscape often present; plants widespread .... *C. acaulis*

10(8). Rays obsolete or short and concealed in the very dense fruits of a globose headlike inflorescence; styles less than 1 mm long, or, if rays evident (to 17 mm long) and styles to 2 mm long, the leaves ternate, without a rachis, and with lobes sharply dentate-serrate; plants endemic in the Great Basin 11

— Rays short or rather long, not concealed in the dense inflorescence; styles mostly 1–3 mm long; leaves 2–3 times pinnate or, if ternate, usually with a rachis and the lobes not dentate-serrate; plants of broad distribution 13

11(10). Leaves pinnate, with 2–3 pairs of lateral primary leaflets, rarely some ternate, the blades narrowly ovate to oblong in outline; rays and pedicels obsolete; inflorescence a dense globose head; wings of fruit more or less spongy thickened; anthers white .... *C. globosus*

— Leaves ternate or occasionally simple and ternately cleft, the blades reniform, orbicular, to ovate in outline; rays and sometimes pedicels more or less evident when young, inflorescence various; wings of fruit papery; anthers yellowish or purplish 12

12(11). Leaves 2, opposite, rarely 3, the ultimate lobes crenate; pseudoscape solitary, subterranean; peduncles solitary; rays 3–10 mm long, hidden at maturity in the very dense globose headlike umbel; styles ca 0.4 mm long; plants of Juab, Sanpete, Sevier and Tooele counties .... *C. coulteri*

— Leaves often more than 2, the ultimate lobes dentate; pseudoscapes or stems rarely solitary; peduncles 1–4 (6); rays 8–17 mm long, not hidden as above; the umbel not globose and headlike; styles 1.5–2 mm long; plants of Beaver and Millard counties .... *C. basalticus*

13(10). Pseudoscape rather quickly developing, (3.5) 5–24 cm long; leaf blades with 4–6 opposite or offset pairs of lateral primary leaflets; umbels sometimes nodding on recurved peduncles; petals white or yellow seldom turning light purple; plants mostly montane in central and western Utah 14

— Pseudoscape lacking or mostly subterranean, the aerial portion not over 3 cm long; leaf blades with 2–4 opposite pair of lateral primary leaflets, or ternate; umbels not nodding; petals white or yellow often turning dark purple 15

14(13). Leaves 3-pinnate, finely and completely dissected, with the ultimate widest undivided portions of the blades, these to 2 mm long and to 1 mm wide; upper primary leaflets not tending to be confluent with the rachis; petals white; anthers purple; plants of western Utah .... *C. ibapensis*

— Leaves 2 (3) -pinnate, the ultimate lobes or teeth 1–5 mm long, 0.5–3 mm wide, these often not as wide as the confluent portions of the blade which are up to 12 mm wide; upper primary leaflets tending to be confluent with the rachis, and pinnatifid or only lobed; petals and anthers yellow or white; plants of northern and central Utah .... *C. longipes*

15(13). Leaves once pinnate with 2 opposite pairs of lateral primary leaflets, or a few ternate or rarely biternate, glaucous, confluent portions of the blades (3) 6–25 (40) mm wide; petals and stamens bright yellow when fresh, fading to cream or white; plants of the Uinta Basin .... *C. duchesnensis*

— Leaves ternate or 2–3 times pinnately compound with up to 4 opposite pair of lateral primary leaflets, glaucous or not, the confluent portions mostly 1–4 mm wide or, if wider, the leaves ternate; petals yellow, purple, or white when fresh, if yellow, turning dark purple in age; plants of broad distribution 16

16(15). Petals creamy pink to pale purple when fresh, finally pale to moderately purple; rays 2–18 mm long; pedicels to 3 mm long; blades 1–3 cm long, pinnately dissected, the confluent portions rarely over 3 mm wide; plants scabrous, of Garfield, Iron, and Kane counties, at moderate to high elevations ... *C. minimus*

— Petals yellow or purplish when fresh, ultimately dark purple; rays, pedicels, and leaves mostly longer than above or the leaves mostly ternate with confluent portions often 5–21 mm wide; plants widespread .. .. *C. purpureus*

Cymopterus acaulis (Pursh) Raf. Plains Spring-Parsley. [*Selinum acaule* Pursh]. Plants 5–18 (27) cm tall, from a simple or rarely branched, deep seated, nearly linear or slightly to much enlarged fibrous taproot; herbage often more or less viscid and dotted with sand grains; pseudoscapes 1–2 (3) per plant, 0.5–5.5 cm long, often partly or wholly subterranean; leaves basal, or more often whorled and subtending the peduncles atop the pseudoscape, occasionally 1 or 2 on a stem, 2–3 times pinnate, with 1–3 opposite pairs of lateral primary leaflets; petioles 2–8 (11) cm long, blades (1) 2–5.5 (7) cm long, the confluent portions 1–7 (12) mm wide, oblong, ovate, to nearly linear in outline; primary leaflets 5–35 mm long, gradually reduced upwards, pinnate to bipinnatifid with few to several rounded to narrow lobes, the ultimate teeth or lobes to 10 (16) mm long and 2 mm wide; peduncles 1–14, (1.5) 3–14 (19) cm long; involucres lacking; rays 6–9, 1–13 mm long; bractlets of the involucel 3–8 (11) mm long, ca 1.5–4 mm wide, more or less united at the base, entire or with 2–3 teeth or lobes, green or purple in age, of the texture of the leaves; pedicels to 2 mm long; calyx teeth ca 0.2 mm long, greenish; petals white, yellow, or purple; stamens the color of the petals; styles ca 2.5 mm long; carpophore lacking; fruit 5–10 mm long, the wings slightly longer than the body, to 2 mm wide, slightly corky, some of the dorsal ones sometimes obsolete. There are 4 more or less intergrading varieties in the state.

1. Petals and stamens yellow when fresh, sooner or later fading to white or cream when dried 2

— Petals and stamens white or purple when fresh 3

2(1). Peduncles mostly shorter than the leaves, to ca 4 cm long; wings of the fruit mostly strongly wavey and often erose, to 7 mm long; leaf blades to ca 4 cm long; plants seldom over 7 cm tall, of the Great Basin .... *C. acaulis* var. *parvus*

— Peduncles equaling or exceeding the leaves, to 14 (19) cm long; wings of the fruit straight or slightly wavey, mostly entire or obscurely erose, to 10 mm long; leaf blades to 7 cm long; plants often over 7 cm tall, of the Colorado Basin .... *C. acaulis* var. *fendleri*

3(1). Petals and stamens purple; peduncles mostly exceeding the leaves; plants of Kane County . . . . *C. acaulis* var. *higginsii*
— Petals and stamens white; peduncles mostly shorter than or equaling the leaves; plants of the Uinta Basin . . . . *C. acaulis* var. *acaulis*

Var. *acaulis* Desert shrub, sagebrush, and juniper communities at 1432 to 1980 m in Duchesne and Uintah counties; Saskatchewan and Minnesota west to Oregon and south to Texas and northern Utah; 20 (v). Plants with white flowers grow among those with yellow flowers in the Uinta Basin where it is difficult if not impossible to recognize two taxa. Even when fresh the white flowers do not seem as bright as those of Wyoming, and the Uinta Basin materials seem transitional to var. *fendleri*.

Var. *fendleri* (Gray) Goodrich comb. nov. [based on: *C. fendleri* Gray Mem. Amer. Acad. II. 4: 56. 1849; *C. decipiens* Jones, type from Cisco, Grand County]. Desert shrub, blackbrush, sagebrush, and pinyon-juniper communities often on sandy soil at 1885 to 2890 m in Duchesne, Carbon, Emery, Garfield, Grand, Kane, San Juan, Uintah, and Wayne counties; Utah and Arizona; 74 (v). This taxon has long been recognized at specific level. The rather recent discovery of intermediate plants in the Uinta Basin and of other yellow flowered varieties (var. *parvus* from the Great Basin and var. *greeleyorum* Grimes & Packard of Oregon) greatly weaken the case for recognition at specific level.

Var. *higginsii* (Welsh) Goodrich [*C. higginsii* Welsh, type from north of Glen Canyon]. Desert shrub communities, often on sandy alluvium of Tropic Shale at about 1525 m in Kane County; endemic; 4 (0). The color of the petals persists as a bright purple long following collection.

Var. *parvus* Goodrich Desert shrub, sagebrush, and juniper communities, often on aeolian sand, at 1400 to 1585 m in Millard and Tooele counties; endemic; 15 (xii). This variety is similar to var. *acaulis* in the short scape, but differs in the yellow flowers. It is like var. *fendleri* in the yellow flowers, but differs in the short scapes. The plant is similar to extralimital **var. *greeleyorum* Grimes & Packard** in the short scape and yellow flowers. It is apparently most closely related to the latter taxon, from which it differs in the strongly undulate erose wings of the smaller fruit.

***Cymopterus basalticus* Jones** Dolomite Spring-parsley. Plants 4–15 cm tall, from a taproot with simple or branched crown, glabrous, glaucous, with 1–few mostly etiolated, subterranean, short scapose stems; pseudoscape mostly lacking or, if present, short and enveloped in bladeless sheaths, the crown sometimes with a few marcescent leaf bases; leaves basal, (2) 3–9 per plant, ternately divided without a rachis, or occasionally simple and ternately cleft; petioles 1.5–5 cm long; blades 1–3.5 cm long, orbicular to reniform, the confluent portions 4–32 mm wide; leaflets 5–30 mm long, sessile, orbicular, ternately lobed, the major lobes again lobed, mostly ternately so, the ultimate lobes coarsely dentate; peduncles 1–4 (6) per plant, 3.5–8 (14) cm long; involucre lacking; rays 6–14, 8–17 mm long, usually evident in fruit; bractlets of the involucel 6–8, 2–5 mm long, more or less united at the base, white, pink, or purplish with white scarious margins; pedicels obsolete or nearly so; petals white or purplish; stamens yellowish or purplish; styles 1.5–2 mm long; carpophore lacking; body of fruit 3–6 mm long, the wings 4–7 mm long, 1–2 mm wide, whitish, shrub communities, gravelly hills and alluvial fans mostly on dolomite, at 1705 to 1985 m in western Beaver (type from Wah Wah) and Millard counties; White Pine County; Nevada, a Great Basin endemic; 21(v). The orbicular to reniform leaf blades without a rachis are unique in the genus.

***Cymopterus beckii* Welsh & Goodrich** Pinnate Spring-parsley. Plants 0–4 dm tall, glabrous, weakly if at all aromatic; caulescent, the leaves extending up the stem, from a taproot with a simple or sparingly branched crown, often clothed at the base with marcescent leaf bases; leaves 1– or 2–pinnate, with 2–3 opposite pairs of lateral leaflets, or the upper ones ternate; petioles 2–13 cm long; blades 2–10 cm long; leaflets 3–7, 0.5–4 cm long, or the terminal one to 5.5 cm long, 1–2 (3) mm wide, sessile, linear or linear-elliptic, entire or rarely a few bifid; peduncles 4–8 (19) cm long; umbels 1–3 per stem; involucres lacking; rays 6–11, 0.6–1.4 cm long; bractlets of the involucels ca 5, 1–5 mm long, to 1 mm wide, greenish or with narrow scarious margins, mostly separate; pedicels 1–3 mm long; petals and stamens bright yellow when fresh, fading whitish when dried; styles 1.2–2.2 mm long; carpophore weak, adhering to the mericarps; fruit 6–8 mm long, oblong, the lateral wings to ca 1 mm wide, the dorsal ones narrower, some often obsolete. Sandy or stoney places, pinyon-juniper-mountain brush communities at 1700 to 2150 m in San Juan and Wayne (type from Fruita) counties; endemic; 8 (iv). The Beck spring-parsley is apparently closely allied to *C. lemmonii*, but differs in entire leaflets, glabrous peduncles and rays, and the slightly longer fruit.

***Cymopterus bulbosus* A. Nels.** Onion Spring-parsley. Plants 8–27 cm tall, glabrous, glaucous, from a stout, thickened, often bulbous, fibrous taproot with a simple or sparingly branched crown; pseudoscapes obsolete or 1–3, to 6.5 cm long and often partly or wholly subterranean, enveloped in dilated bladeless sheaths; leaves few to several, basal or whorled atop the pseudoscape with the peduncles, rarely 1 or 2 cauline, (1) 2–3 times pinnate, with (2) 3–6 opposite or offset pairs of lateral primary leaflets, the upper pairs often once-pinnate and more or less confluent; blades 2–10 cm long, ovate to oblong or nearly linear, confluent portions 1–5 mm wide, lowest pair of primary leaflets to 4 cm long, sessile or on petiolules to 2 (5) mm long, the other primary leaflets progressively reduced upwards, the ultimate lobes and teeth 1–8 (12) mm long, ca 1–4 mm wide, more or less rounded; peduncles (1) 3–8 (11) per pseudoscape, 4–18 cm long; involucre obsolete or reduced to a ring or cup, or the bracts present and to 13 mm long, translucent, white, more or less united, with a green midrib and occasionally 1 or 2 lateral nerves that extend to ca 1/3 the length of the bract; rays 5–15, 1–3.5 cm long, usually exceeding the involucre; involucels 3–10 mm long, the bractlets more or less united at the base, similar in texture and color to the involucre, with a green center and midrib or the midrib sometimes purple, rarely with 1 or 2 lateral nerves extending to near midlength; pedicels 3–9 mm long; calyx teeth 0.5–1 mm long, scarious, white, like the involucel, with a green midrib to about midlength; petals white, sometimes purplish in age; stamens white, or purple especially in age; styles ca 2–4 mm long; carpophore divided to the base, more or less persistent on the pedicel after the mericarps have fallen; fruit 6–11 mm long, to 2 (3) mm wide, the wings (7) 9–13 mm long, 1.7–3 (4) mm wide. Desert shrub and juniper communities at 1220 to 2005 m in the Colorado drainage in Carbon, Duchesne, Emery, Garfield, Grand, San Juan, Uintah, and Wayne counties; Wyoming to New Mexico and Arizona; 95 (xi). This species is similar to *C. purpurascens* (q.v.).

Cymopterus coulteri (Jones) Mathias Two-leaf Spring-parsley. [*C. corrugatus* var. *scopulicola* Jones, type from Sevier Bridge; *C. corrugatus* var. *coulteri* Jones; *Rhysopterus jonesii* Coult. & Rose, type from Juab]. Plants 4–11 cm tall, glabrous from a slightly to much enlarged fibrous taproot with simple crown, this giving rise to a solitary, mostly subterranean pseudoscape 2–6 cm long; leaves 2 (rarely 3 or 4 and the third and fourth ones usually smaller), opposite, borne at or near ground level, ternate or rarely simple and ternately cleft, petioles 1–3 cm long, blades 2–4 cm long, ovate to nearly orbicular in outline, confluent portions (5) 8–38 mm wide; leaflets 7–35 mm long, the lateral ones sessile mostly ternately lobed, the terminal on a winged more or less confluent rachis to 1 cm long, ternately cleft, the main lobes again lobed or crenate-toothed; peduncle solitary, 2–7 cm long; umbel globose, headlike, 1.5–5 cm across in pressed fruiting specimens; involucre lacking; rays 7–14 or perhaps more, 3–10 mm long, somewhat evident at anthesis, but hidden by the dense mature fruits; bractlets of the involucel 2–4 mm long, linear to narrowly ovate, green or purplish in age, often 3-nerved, with whitish or purplish scarious margins; pedicels shorter than the bractlets; calyx teeth minute but white and of the texture of the petals, deciduous; petals white; filaments white, anthers purple; styles (including the stigmas) ca 0.4 mm long; carpophore lacking; body of fruit 5–7 mm long, the wings 7–10 mm long, to ca 2 mm wide, papery; n = 11. Desert shrub, black sagebrush, and juniper communities, often on Arapien Shale and other clayey and gravelly barrens or semibarrens at 1540 to 1700 m in Juab, Sanpete, Sevier, and Tooele counties; endemic; 28 (vii). The strong tendency for plants to have only two leaves is unique in the genus.

Cymopterus duchesnensis Jones Uinta Basin Spring-parsley. Plants 7–23 cm tall, from a slender or more often enlarged bulbous taproot with simple or branched crown, glabrous and glaucous, not or weakly aromatic; stems short, often branched; pseudoscape lacking; leaves basal or 1–3 or more cauline, pinnate with 2 pairs of lateral leaflets, or occasionally a few ternate, or rarely biternate; petioles 2.2–11 cm long; blades (2) 3–10.5 mm long, ovate to oblong in outline, the confluent portions (3) 6–25 (40) mm wide; leaflets 1–5, on petiolules 2–32 mm long, ternately cleft or divided, the major lobes to 3 cm long, (3) 5–15 (20) mm wide, often again toothed or lobed and mostly ternately so, the ultimate teeth or lobes 1–8 mm wide; peduncles 1–3 per stem, 7–17 cm long; involucres lacking; rays 6–17, 1.5–4.4 cm long; bractlets of the involucel lacking or more often 1–7, 1–5 mm long, more or less united at the base, linear; pedicels (2) 4–9 mm long; petals and stamens bright yellow when fresh, fading to cream or greenish in herbarium specimens, not turning purple; styles 2–2.2 mm long; carpophore divided to the base, more or less persistent on the pedicel; body of fruit 5–9 mm long, the wings to 11 mm long, 2–2.5 mm wide, undulate to corrugated, more or less papery and not corky. Desert shrub, sagebrush, and juniper communities, sandy clay and clay semibarrens of Duchesne River,

Mancos Shale, Morrison, Uinta, and Wasatch formations at 1430 to 1860 m in Uintah and Duchesne (type from Myton) counties; Moffat and Rio Blanco counties, Colorado; a Uinta Basin endemic; 47 (xiii).

Cymopterus globosus (Wats.) Wats. Golfball Spring-parsley. [*C. montanus* var. *globosus* Wats.]. Plants 4–10 cm tall, from a slender or thickened fibrous taproot; pseudoscape (1) 2–6 cm long, all or nearly all subterranean, often loosely enveloped in dilated bladeless sheaths; leaves 1 or 2 atop the pseudoscape, and some usually arising directly from the fibrous root and then with etiolated subterranean petioles, pinnate or bipinnate and then trifid or pinnatifid, with 2–3 opposite pairs of sessile lateral primary leaflets, or rarely ternate; petioles 1.5–6 cm long; blades 2–5 cm long, narrowly ovate to oblong, confluent portions 5–10 mm wide, lowest pair of primary leaflets 10–18 mm long, the ultimate lobes to 4 mm long, to 2 mm wide, mostly toothed; peduncles 1 or 2, 3–6 cm long; involucre lacking; umbel a globose head, the rays and pedicels obsolete; involucels concealed in the dense flowers and fruits; petals white; stamens white; styles 0.5–0.8 mm long; carpophore lacking; body of the fruit ca 6 mm long, the wings ca 9 mm long, to 2.8 mm wide, wider towards the outside of the head, spongy thickened. Desert shrub communities at 1400 to 1525 m in Box Elder, Juab, Millard and Tooele counties; eastern California, Nevada, and western Utah; a Great Basin endemic; 6 (0).

Cymopterus hendersonii (Coult. & Rose) Cronq. Mountain Rock-parsley [*Pseudoteryxia longiloba* Rydb., type from the Abajo Mts.; *Pterxia hendersonii* (Coult. & Rose) Math. & Const.; *Pseudocymopterus hendersonii* Coult. & Rose]. Plants (3) 5–34 cm tall, glabrous, strongly aromatic, from a branched woody caudex, clothed at the base with old petiole and peduncle bases, these sometimes persisting for a few or several years without shredding; leaves basal, bipinnate or occasionally partly tripinnate with 5–10 opposite or offset pairs of lateral primary leaflets, petioles (1) 2–14 cm long, blades (1) 1.5–10 cm long, oblong in outline, finely dissected so that the ultimate segments are the widest undivided parts of the blade, the lowest pair of primary leaflets less than 1/4 as long as the leaf blade, 5–27 mm long, sessile or on petiolules to 1 cm long, upper primary leaflets gradually reduced, the ultimate segments 1–12 mm long, 0.3–1.4 mm wide, acute, with a usually whitish tiny mucro; peduncles 7–30 cm long; umbels compact; involucres lacking; rays 6–16, 0.5–2.4 cm long, the inner ones shorter than the outer ones and often abortive; bractlets of the involucel 2–6, 2–10 mm long, linear, acute; pedicels 1–5 mm long; calyx teeth ca 1 mm long, persistent in fruit, greenish, often reddish tinged, acute; petals and stamens bright yellow when fresh, fading whitish in herbarium specimens; styles to 2 (2.5) mm long; carpophore divided to the base; fruit 4–8 mm long, the wings to ca 1 mm wide, some of the dorsal ones sometimes obsolete. Talus, cliffs, ledges, rocky spruce-fir, limber pine, and alpine communities at (2285) 2740 to 3660 m in Beaver, Box Elder, Cache, Daggett, Duchesne, Grand, Juab, Piute, Salt Lake, San Juan, Sevier, Summit, Tooele, Uintah, and Utah counties; southwestern Montana and central Idaho, south to New Mexico; 66 (x).

Cymopterus ibapensis Jones Ibapah Spring-parsley. [*C. watsonii* (Coult. & Rose) Jones]. Plants 7–25 cm tall, glabrous or granular-scabrous, not or weakly aromatic, from a linear taproot, this hardly if at all swollen, with a simple or occasionally branched crown; pseudoscapes 1 or 2 (5) per root, the aerial portion 3.5–10 cm long, commonly enveloped at the base by scarious dilated bladeless sheaths; leaves whorled atop the pseudoscape, rarely some arising directly from the root, tripinnate, with 5–6 opposite or offset pairs of lateral primary leaflets; petioles (1) 1.5–3.5 cm long; blades (2.5) 4–11 cm long, ovate in outline, completely dissected so that the ultimate segments are the widest undivided portions of the blade; lowest pair of primary leaflets 1/2–3/4 as long as the leaf blade, sessile or on petiolules to 2 cm long, with 4–6 (8) opposite or offset pairs of secondary leaflets, the ultimate segments to 2 mm long, to ca 1 mm wide; peduncles (2) 4–8 per pseudoscape, 2–15 cm long; umbels and peduncles occasionally nodding or recurved; involucre lacking; rays 10–18, 5–20 mm long; bractlets of the involucels to 4 mm long, to 0.5 mm wide, distinct or nearly so, green with a purple midrib and narrow scarious margins; pedicels 4–6 mm long; calyx teeth to 1 mm long, greenish; petals white; filaments white, the anthers purple; styles 1–2 mm long; carpophore divided to the base; body of fruit 5–8 mm long, the wings 6–9 mm long, to 2 mm wide, some of the dorsal ones sometimes reduced. Greasewood-sagebrush, sagebrush-grass, and pinyon-juniper communities at 1520 to 2755 m in Beaver, Box Elder, Iron, Millard, Piute, either or perhaps both Juab and Tooele (type from Deep Creek Valley), and Washington counties; southeastern Oregon, and Nevada; 30 (viii).

Cymopterus lemmonii (Coult. & Rose) Dorn Lemmon Spring-parsley. [*Pseudocymopterus lemmonii* (Gray) Coult. & Rose; *P. montanus* (Gray) Coult. & Rose; *P. tidestromii* Coult. & Rose, type from Mt. Terrill, Sevier County; *P. versicolor* Rydb., type from Aquarius Plateau; *Thaspium montanum* Gray; *Ligusticum montanum* (Gray) Gray]. Plants 8–50 cm tall, glabrous except on the peduncle and in the inflorescence, not or weakly aromatic, from a taproot with simple or branched crown, more or less clothed at the base with shredded persisting leaf bases; pseudoscape lacking; leaves basal and sometimes 1 or 2 cauline ones on the lower 1/3 of the stem or 1 above the middle, mostly bipinnate and then often bifid or pinnatifid in the lower part, with 2–4 opposite or offset pairs of lateral primary leaflets, rarely pinnate in part with entire leaflets; petioles 1–13 cm long, dilated basally; blades (1) 2–8 cm long, confluent portions to 4 mm wide; lowest pair of primary leaflets 1/4–2/3 as long as the leaf blade, sessile or on petiolules to 15 mm long, the ultimate segments 2–20 mm long, linear or narrowly elliptic; peduncles 1–9, (4) 9–28 cm long, rather densely hirtellous below the umbel; involucre lacking or rarely of 1 or 2 small bracts; rays (5) 9–18, 0.8–2.5 cm long, glabrous, scabrous or hirtellous; bractlets of involucels 5–11, to 5.5 mm long, linear or narrowly elliptic, separate or united at the base, green or sometimes with a scarious or purplish margin; pedicels obsolete or to 2 mm long; calyx teeth less than 0.5 mm long, deciduous; petals and stamens bright yellow when fresh, pale or purplish in age; styles ca 2 mm long; carpophore apparently lacking to well developed and divided to the base; fruit mostly 3–6 mm long, the wings ca 1.5 mm wide, some of the dorsal ones sometimes obsolete; n = 11. Grass-forb, aspen, Douglas fir, and spruce-fir communities, and windswept ridges and raw escarpments especially in limestone, at 2375 to 3600 m in Beaver, Emery, Garfield, Grand, Iron, Piute, San Juan, Sanpete, Sevier, Washington, and Wayne counties; southeastern Wyoming to Arizona and Mexico; 138 (vii). Occasional specimens have been confused with *Lomatium juniperinum*.

Cymopterus longipes Wats. Long-stalk Spring-parsley [*Peucedanum lapidosum* Jones, type from Echo, Summit County; *C. lapidosus* (Jones) Jones; *Lomatium lapidosum* (Jones) Garrett; *C. lapidosus* var. *deserti* Jones]. Plants 7–30 (50) cm tall, glabrous, not aromatic, from a thickened fibrous taproot with a simple or sparingly branched crown; pseudoscapes 1–3, 4–24 cm long, mostly aerial, or partially subterranean, more or less enveloped by dilated bladeless sheaths; leaves whorled atop the pseudoscape, rarely any rising from the fibrous root, 1–3 times pinnately compound, with mostly 4–6 opposite or offset pairs of lateral primary leaflets, the upper pairs often more or less confluent and merely pinnatifid, petioles 1–5 cm long, blades 3–8.5 cm long, oblong to ovate in outline, the confluent portions 2–12 mm wide; lowest pair of primary leaflets 1–5 cm long, sessile, or on petiolules to 5 mm long, the ultimate lobes or teeth 1–5 mm long, 0.5–3 mm wide; peduncles 3–18 per pseudoscape, 4–24 cm long; umbels and peduncles sometimes nodding or recurved; involucre lacking; rays 4–11, 0.5–3.3 cm long; bractlets of the involucel to 7 mm long and less than 1 mm wide, mostly separate, green with narrow scarious margins; pedicels 1–12 mm long; calyx teeth 0.2–0.5 mm long; petals yellow or white when fresh, when yellow fading to white in herbarium specimens; stamens colored like the petals; styles ca 2 mm long; carpophore divided to the base; body of the fruit 4–6 mm long, the wings 5–8 mm long, 1–2 mm wide. Sagebrush-grass, pinyon-juniper, and mountain brush communities at 1340 to 3155 m in Box Elder, Cache, Carbon, Daggett, Davis, Duchesne, Juab, Morgan, Rich, Salt Lake (type from near Salt Lake City), Sanpete, Summit, Tooele, Uintah, Utah, Wasatch, and Weber counties; Idaho and Colorado; 160 (xxi). White flowered specimens are common in the Bear River Range and occasional to the central part of the Wasatch Range. The yellow petals of more southern specimens turn whitish in the herbarium, and the ranges of the two color variants are difficult to determine from herbarium specimens. Other than the color difference, there seems to be no way to tell the phases apart. A third phase, in which the fruits are like a *Lomatium* (dorsal ribs not or scarcely winged), apparently has white flowers. This phase, known from Summit County, is referable to *C. lapidosus* (Jones) Jones and may be worthy of the specific status given it by Jones. More work is indicated.

Cymopterus minimus (Mathias) Mathias Least Spring-parsley. [*Aulospermum minimum* Mathias, type from Cedar Breaks, Iron County]. Plants 3–10 (10) cm tall, acaulescent, scabrous, from a slender to much enlarged often deep-seated taproot, with few to several frequently soboliferous branches; stems mostly subterranean and etiolated; pseudoscapes lacking or short and below ground; leaves 2–3 times pinnately dissected, with 3–4 opposite pairs of lateral primary leaflets, petioles 0.5–2 cm long or sometimes much longer with etiolated subterranean portions, blades 1–3 cm long, the confluent portions 1–6 (10) mm wide, the primary leaflets sessile or the lowest pair on petiolules to 0.5 cm long, the ultimate segments to 3 mm long, to ca 2 mm wide; peduncles 1.5–14 cm long; rays of the umbel mostly 5–10, 2–18 mm long; bractlets of the involucel ca 3–4, 2–4 mm wide;

pedicels lacking or to 3 mm long; calyx teeth minute or lacking; petals cream pink or pale purple (reputedly white) with whitish margins or with moderately purple markings in herbarium specimens; stamens whitish; styles ca 2 mm long; carpophore divided to the base; fruit 4–8 mm long, the wings to 1 mm wide. Ponderosa pine, bristlecone pine, spruce-fir, and perhaps pinyon-juniper communities at (2190) 2440 to 3170 m in Garfield, Iron, and Kane counties; endemic; 23 (i). Occasional specimens are intermediate to *C. purpureus* and more work is needed to establish the range of the small plants with cream pink or pale purple petals. These plants have been mistaken for *C. purpureus* var. *rosei* and they are similar in the short rays, short pedicels, small fruit, small leaves, and in scabrosity, but the leaves are not ternate, and they are more finely dissected than those of *C. purpureus* var. *rosei*. In some features of the leaves and in distribution this taxon is more closely allied with *C. purpureus* var. *purpureus*. At the extreme (Cedar Breaks) these plants are very different, but through a series of recent collections from the Markagunt, Paunsaugunt, and Table Cliff plateaus, and Escalante Mts., a rather close relationship to *C. purpureus* var. *purpureus* is evident. Perhaps this is only a part of the *C. purpureus* complex, and could be treated as a variety, but no such combination is proposed herein. The color of the petals seems to be diagnostic.

Cymopterus multinervatus (Coult. & Rose) Tidestr. Purple-nerved Spring-parsley. [*Phellopterus multinervatus* Coult. & Rose]. Plants 10–15 cm tall, glabrous and glaucous, from a linear or slightly to much enlarged fibrous taproot; pseudoscapes lacking or solitary, to 7.5 cm long, partly or mostly subterranean, often enveloped by dilated, scarious bladeless sheaths; leaves basal or whorled, with the peduncles atop the pseudoscape, 2–3 times pinnately compound, with 3–5 opposite pairs of lateral primary leaflets; petioles 1–7.5 cm long; blades 1–7 cm long, ovate in outline, lowest pair of primary leaflets to 4.5 cm long, sessile or on petiolules to 5 mm long, the ultimate lobes to 4 (7) mm long, to 2 mm wide, sometimes with small rounded teeth; peduncles 1–8, 4–10 cm long; involucre to 1 (1.5) cm long, the bracts more or less united, sometimes forming a cup or reduced to a ring, greenish basally and centrally, with broad, white scarious margins, and with many purplish almost parallel nerves; rays (3) 5–11, 0.3–1 cm long, obscured by the dense fruits, included in or exserted beyond the involucre; involucels 5–10 mm long, like the involucre in color, texture, and venation, but never reduced to a ring; pedicels to ca 4 mm long, included in the involucel; calyx teeth 0.5–1 mm long or somewhat enlarged and simulating the involucre; petals white or purple; stamens white or purple; styles 2–3 (4) mm long; carpophore lacking; body of the fruit 7–10 mm long, the wings 12–13 mm long, 5–7 mm wide. Desert shrub and sagebrush communities at 1220 to 1525 m in Kane and Washington counties; California to Texas and Mexico; 7 (0).

Cymopterus newberryi (Wats.) Jones Sweetroot Spring-parsley. [*Peucedanum newberryi* Wats.; *Coloptera jonesii* Coult. & Rose, type from Milford]. Plants 7–18 cm tall, sparingly viscid and often dotted with adhering sand grains, from a slender to tuberous taproot with simple or rarely branched crown; pseudoscape lacking; leaves arising directly from the root crown, ternate, rarely simple and ternately cleft; petioles 3.5–10.5 cm long, (1.6) 2.5–5 times as long as the blades, often partly subterranean, blades 1–4 (5.5) cm long, confluent portions 5–35 mm wide; leaflets 1–3.5 cm long, mostly ternately cleft or divided, the major lobes again lobed or toothed, the ultimate lobes or teeth to 6 mm long and 5 mm wide; peduncles 1–10, 5–17 cm long, often partly subterranean; involucre lacking; rays 5–16, the central ones often greatly reduced or obsolete, the outer ones 1–3.3 cm long; bractlets of the involucels 3–12 mm long, to 3 mm wide, entire, green, or sometimes purplish in age, with texture of the leaves; pedicels ca 1 mm long; calyx teeth ca 0.5 mm long, deciduous; petals and stamens yellow when fresh, fading to cream or greenish in age; styles 2–3 mm long; carpophore lacking; body of fruit 5–8 mm long, the wings 6–10 mm long and 1.5–1 mm wide, more or less corky, some of the dorsal ones obsolete. Desert shrub, blackbrush, sand sagebrush, desert grassland, and juniper communities, mostly on very sandy soil, at 850 to 1830 m in Beaver, Garfield, Grand, Kane, Millard, San Juan, Washington, and Wayne counties; Arizona; 47 (x).

Cymopterus purpurascens (Gray) Jones Widewing Spring-parsley [*C. montanus* var. *purpurascens* Gray]. Plants 5–15 cm tall, glabrous and glaucous, from a mostly tuberous root with simple or sparingly branched crown, this usually with few to several persistent shredded leaf bases; pseudoscapes lacking or to 3 per plant and to 6 cm long, mostly subterranean, usually enveloped by scarious dilated bladeless sheaths; leaves basal or more or less whorled atop the pseudoscape, 2–3 times pinnately compound, with 3–6 opposite pairs of lateral primary leaflets, the pairs gradually reduced upwards; petioles 0.6–5 cm long, sometimes longer including etiolated subterranean portions; blades 1.2–7 cm long, oblong to ovate in outline, confluent portions to 3 (5) mm wide, lowest pair of primary leaflets (0.4) 1–2 (4) cm long, sessile or on petiolules to 3 mm long, the ultimate lobes or teeth rounded, mostly with narrow-scarious margins; peduncles 1–3 per pseudoscape, 3–9 cm long; involucre 8–14 mm long, more or less united at the base and sometimes to about midlength, whitish, scarious, the lobes with a greenish or purplish midnerve extending to the tip, and usually 1 or 2 parallel lateral much shorter nerves; rays ca 4–7, rarely longer than 1 cm, mostly shorter than the involucre, hidden in the dense broadly winged fruits; involucels like the involucre but shorter (ca 5–7 mm) and usually with the lateral nerves over 1/2 as long as the midnerve, the nerves occasionally branched; pedicels to 5 mm long, mostly concealed in the involucels and in the dense fruits; calyx teeth less than 0.5 mm long, rounded; petals white or purplish with a green or purplish midvein; filaments white, the anthers purple; styles ca 2 mm long; carpophore lacking or filiform and very thin, not persisting on the pedicel; body of fruit 6–11 mm long, the wings 9–16 mm long and 3–6.5 mm wide. Desert shrub, sagebrush, pinyon-juniper, bullgrass, and ponderosa pine communities, on aeolian sand to heavy clay at 1065 to 2745 m in all Utah counties except Daggett, Davis, Grand, Morgan, Rich, Summit, Wasatch, and Weber; southeastern Idaho to southeastern California and northwestern New Mexico; 106 (xxi). *Cymopterus purpurascens* is often confused with *C. bulbosus*, but that species has mostly 1–nerved lobes of the involucre and involucel, the involucre is sometimes reduced to a ring or cup, the plants flower later (often a month or so), and it is confined to lowlands mostly in heavy soil. Rare specimens (Neese

7169 and Thorne et al. 1707) show the broad wings of fruit typical of *C. purpurascens*, have rays well over 1 cm long that are exserted beyond the involucre, and at least some of the fruits have a well developed carpophore. Perhaps these specimens indicate hybridization of these two taxa.

Cymopterus purpureus Wats. Variable Spring-parsley. Plants 5-26 cm tall, from a slender to tuberous root with a simple or branched crown; stems solitary to several, arising at or just below ground level; pseudoscape lacking, or less than 2 cm long and then usually mostly subterranean; leaves basal or nearly so, ternate or (1) 2-3 times pinnately compound, with up to 4 opposite pairs of lateral primary leaflets; petioles 1-7 cm long; blades 1.5-13 cm long, mostly ovate in outline, the lowest pair of primary leaflets mostly over 1/2 and to 3/4 as long as the leaf blade, sessile or on petiolules to 32 mm long, the ultimate lobes or teeth acute or rounded; peduncles 1-5, 3-21 cm long; involucre lacking; rays 5-22, 0.2-9.5 cm long; bractlets of the involucel 4-8, 2-4 mm long, separate or united at the base, acute to acuminate, entire; pedicels 1-10 mm long; calyx teeth less than 0.5 mm long, deciduous; petals yellow when fresh, drying dark purple in age; stamens yellow when fresh, remaining yellowish or cream or at least pale in age; styles 2-3 mm long; carpophore divided to the base; body of fruit ca 4-8 mm long, the wings 5-10 (12) mm long, 1.5-4 mm wide, often marked with purple. With 3 more or less intergrading varieties in the State.

1. Fruiting rays 5-8 (15), 0.2-2 (3) cm long; fruiting pedicels 1-5 (7) mm long; wings of fruit 5-8 mm long, to 2 mm wide; leaf blades mostly ternate; plants of lower to midmontane areas, mostly of central Utah . . . . *C. purpureus* var. *rosei*

— Fruiting rays (8) 12-22, (2) 2.5-7 (9.5) cm long; fruiting pedicels 5-10 mm long; wings of fruit 8-10 (12) mm long, (2) 2.5-4 mm wide; leaf blades pinnately compound, rarely ternate; plants of deserts and lower montane, widespread 2

2(1). Some wings of fruit apparently thickened and spongy; ultimate teeth of leaves acute; plants conspicuously glaucous, of Beaver, Iron, and Washington counties . . . . *C. purpureus* var. *jonesii*

— Wings of fruit mostly thin and papery; ultimate teeth of leaves acute to rounded; plants not conspicuously glaucous, not of the above counties except Washington . . . . *C. purpureus* var. *purpureus*

Var. jonesii (Coult. & Rose) Goodrich [*C. jonesii* Coult. & Rose, type from Frisco, Beaver County; *Aulospermum jonesii* (Coult. & Rose) Coult. & Rose]. Sagebrush, pinyon-juniper, and mountain brush communities at 1520 to 1905 m in Beaver, Iron, and Washington counties; Nevada; a Great Basin endemic; 10 (i).

Var. purpureus Desert shrub, sagebrush, pinyon-juniper, mountain brush, ponderosa pine, and rarely aspen-fir communities in sandy to heavy clay soils at 1100 to 2375 (2880) m in Carbon, Duchesne, Emery, Garfield, Grand, Kane, San Juan, Uintah, Washington, and Wayne counties; eastern and southern Utah, western Colorado, northern Arizona, and northwestern New Mexico; 134 (v). Specimens with rather broad leaflets from the Uinta Basin (Neese et al. 7273, and White and Neese 123) indicate a close relationship to and possible hybridization with *C. duchesnensis*.

Var. rosei (Jones) Goodrich [*Aulospermum rosei* Jones in Coult. & Rose., type from Richfield; *C. rosei* (Jones) Jones]. Pinyon-juniper, sagebrush, mountain brush, bull grass, limber pine, white fir, and rarely desert shrub communities, in marly limestone, shaley slopes, and clay or sandy-clay soils, at 1615 to 2290 (2650) m in Duchesne, Juab, Millard, Sanpete, Sevier, and Wasatch counties; endemic to central Utah; 28 (x).

Cymopterus terebinthinus (Hook.) T. & G. Rock Parsley. [*Selinum terebinthinum* Hook.; *Pteryxia terebinthina* (Hook.) Coult. & Rose]. Plants (12) 15-35 (40) cm tall, glabrous, strongly aromatic, from a heavy subligneus root and a branched caudex, the caudex clothed with marcescent non-shredded leaf bases; pseudoscape lacking; leaves basal and often 1-3 on the lower part of the stem, mostly 2-4 times pinnately or ternate-pinnnately compound, with (4) 6-10 opposite or offset pairs of lateral primary leaflets, petioles 2-13 cm long, blades 1.5-14 cm long, finely and completely dissected so that the ultimate segments are the widest undivided part of the blade, the lowest pair of primary leaflets mostly 3-9 cm long, (1/5) 1/2-3/4 as long as the leaf blade, on petiolules 2-4 cm long, the ultimate segments 1-5 (7) mm long, 0.5-1 (1.5) mm wide; peduncles 10-34 cm long; involucre lacking; rays 7-13, 0.7-5 (8) cm long; bractlets of the involucel (0) 1-5, 2-5 mm long, separate or united at the base, linear or linear-subulate; pedicels 2-5 (10) mm long; calyx teeth 0.5-0.9 mm long, acute, rather persistent; petals and stamens bright yellow when fresh, fading whitish and rarely yellow for more than 2 years in herbarium specimens; fruiting styles (2-5) 3-4 mm long, mostly curved or coiled; carpophore divided to the base, persisting on the pedicel; body of fruit 5-8 mm long, the wings 6-9 mm long, 0.5-1.5 (2.5) mm wide, some of the dorsal ones often reduced or obsolete. There are two varieties in Utah.

1. Lateral pairs of primary leaflets all longer than their internodes; lowest pair of primary leaflets with the lower secondary leaflets sometimes petiolulate; plants of the Uinta Basin and northern Utah . . . . *C. terebinthinus* var. *calcareus*

— Upper 4-6 pairs of lateral primary leaflets equal or shorter than their internodes; lowest pair of primary leaflets with sessile secondary leaflets; plants of eastern Utah south of the Uinta Basin . . . . *C. terebinthinus* var. *petraeus*

Var. calcareus (Jones) Cronq. [*Cymopterus calcareus* Jones]. Desert shrub, sagebrush, pinyon-juniper, and mountain brush communities, often in talus, colluvium, and crevices of rock outcrops at 1445 to 2320 (2560) m in Box Elder, Cache, Daggett, Juab, Rich, and Uintah counties; Montana to Colorado, and west to southern Idaho and northeastern Nevada; 36 (vii). Some of the Uintah County materials are transitional to plants of var. *petraeus*. Plants of var. *calcareus* are very similar to those of **var. albiflorus (T. & G.) Jones,** from Montana and Wyoming, which were originally described as having white flowers. Dried flowers soon turn white or whitish in all of the *C. terebinthinus* complex, and the two varieties are both reported from the same regions of Montana and Wyoming. Unfortunately, the misnomer var. *albiflorus* has priority by many years, and our plants might belong to that taxon.

Var. petraeus (Jones) Goodrich Skeletonleaf Rockparsley. [*C. petraeus* Jones]. Desert shrub, blackbrush, and pinyon-juniper communities, often in talus, colluvium, crevices of rock outcrops, and in sandy to clayey soil at 1400 to 2075 m in Emery, Grand, and San Juan counties; Nevada, Idaho and Arizona; 16 (i).

Daucus L.

Annual or biennial caulescent herbs from taproots; leaves pinnately dissected; umbels compound; involucre of pinnatifid bracts or lacking; involucel of toothed or entire bracts or lacking; calyx teeth evident to obsolete; petals white or those of the central flower of the umbel or umbellet often purple or rarely all the flowers pink or yellow; stylopodium conic; carpophore entire or bifid at the apex; fruit oblong to ovoid, slightly compressed and evidently ribbed dorsally, with 2 ribs on the commissure, beset with stout spreading glochidiate or barbed ribs.

1. Plants biennial, cultivated or established, widespread; bracts of the involucre mostly pinnatifid into mostly entire rather rigid elongate segments .... *D. carota*
— Plants annual, native, not cultivated, known from the Virgin Narrows in Arizona, to be expected in Washington County; bracts of the involucre pinnatifid into often lobed or toothed nonrigid segments .... *D. pusillus* Michx.

Daucus carota L. Carrot. Plants 6–10 dm tall, from a taproot; herbage glabrous or hirsute; leaves in rosettes and cauline, mostly 1–2 times pinnate and then pinnatifid, with ca 4–9 opposite or offset pair of lateral primary leaflets, basal and lower cauline petioles to 15 cm long, basal and lower blades 5–15 cm long or more, the upper ones reduced and sessile on dilated sheaths, the lowest pair of primary leaflets 1/3–1/2 as long as the leaf blade, on petiolules 4–15 mm long, ultimate segments 1–10 mm long, 0.5–2 mm wide, elliptic, narrowly deltoid, or linear, often acute; peduncles mostly 8–30 cm long; umbels 4–10 or more; involucre of pinnatifid bracts 1–5 cm long, the segments linear and narrow; rays ca 15–60 or more, (0.5) 1–6 cm long; involucels similar to the involucre but smaller, or the bractlets entire, 2–16 mm long; fruit 3–4 mm long, bristly hirsute in rows, the hairs or bristles ca 2 mm long, minutely glochidate apically, the intervals often with shorter simple hairs; $2n = 18$. Cultivated in all counties of the state, or widely established; introduced from Eurasia; 16 (iv). The wild plants (ssp. *carota*) differ from the cultivated plants [ssp. *sativus* (**Hoffm.**) Arcangeli] primarily in the size and flavor of the root.

Foeniculum Adanson

Biennial or perennial, caulescent herbs with strong odor of anise, glabrous, glaucous, from a taproot; leaves pinnately dissected with filiform ultimate segments; umbels compound; involucre and involucel lacking; calyx teeth obsolete; petals yellow; stylopodium conic; carpophore divided to the base; fruit oblong, subterete, or slightly compressed laterally, with prominent ribs.

Foeniculum vulgare Miller Sweet Fennel. Short-lived perennial herbs 0.5–2 m tall, from a taproot; stems solitary, branched above; leaves to 3 times ternate-pinnately compound with 6–9 opposite pair of lateral primary leaflets; petioles to 15 cm long, rather abruptly expanded into a dilated sheathing base or lacking and blades arising directly from the sheath; larger blades to 30 or 40 cm long, ovate in outline, finely and completely dissected, the elongated filiform ultimate segments 4–40 mm long and less than 1 mm wide, the lowest pair of primary leaflets on petiolules often over 2 cm long; peduncles 1.5–6.5 cm long; umbels several; rays 10–40, 2–8 cm long; petals yellow; styles 0.3–0.4 mm long; fruit 3.5–4 mm long; $n = 8, 11, 13, 15, 22$. Roadsides and waste places at 850 to 1465 m in Utah and Washington counties; introduced from Europe, now widespread in much of the U. S.; 3 (0).

Heracleum L.

Biennial or perennial herbs from tap or fascicled fibrous roots; leaves ternately or pinnately compound, with broad toothed or cleft leaflets; umbels compound; involucre lacking or of a few deciduous bracts; involucel lacking or of slender bractlets; flowers of the marginal umbellets generally irregular, the outer petals enlarged and often deeply bilobed; calyx teeth obsolete or minute; stylopodium conic; carpophore divided to the base; fruit orbicular to obovate or elliptic, strongly flattened dorsally, usually pubescent, the dorsal ribs narrow, the lateral ribs broadly winged.

Brummit, R. K. 1971. Relationship of *Heracleum lanatum* Michx. of North America to *H. sphondylium* of Europe. Rhodora 73: 578–584.

Heracleum lanatum Michx. Cow parsnip. [*H. sphondylium* ssp. *lanatum* (Michx.) Love & Love]. Stout single-stemmed perennial herbs 8–25 dm tall, from a taproot or cluster of fibrous roots, glabrate or thinly to densely villous or villous-hirsute below to villous-woolly above especially on the nodes; leaves ternate or the upper ones simple, petioles to 25 cm long or longer, or lacking on upper leaves with the petiolules and rachis arising directly from a dilated sheath, blades to 40 cm long or longer, ovate to orbicular; leaflets 8–36 cm long or longer, ovate to orbicular, usually with 3 major lobes that are again lobed and coarsely toothed; peduncles 5–24 cm long; involucre lacking or of few mostly linear entire bracts to 2 cm long; rays 12–25, 3.5–12 cm long; involucels of 3–5 linear, subulate or caudate bractlets to 15 mm long; pedicels 6–26 mm long; petals white (2) 4–8.5 mm long at least some deeply bilobed; filaments white, the anthers whitish to dark green or yellow with pollen; styles ca 1 mm long, the stigmas incurved; fruit 8–12 mm long, obovate to obcordate, strongly flattened, the lateral ribs with wings 1–1.5 mm wide, the dorsal ribs filiform. Aspen, tall forb, fir, oak-maple, willow, streamside, and wet meadow communities at 1430 to 2930 m in Box Elder, Cache, Carbon, Davis, Duchesne, Juab, Salt Lake, Sanpete, Sevier, Summit, Tooele, Uintah, Utah, Wasatch, and Weber counties; Eurasia and across much of North America; 61 (iii).

Hydrocotyle L.

Perennial herbs; stems creeping or floating, rooting at the nodes; leaves petiolate, often peltate; inflorescences sessile, or borne on axillary peduncles; involucres small or lacking; petals white, greenish, or yellow; calyx minute or lacking; stylopodium conic to depressed; fruit orbicular to ellipsoid, more or less flattened laterally, the dorsal surfaces rounded or acute, the ribs obsolete or narrow and acute; carpophore lacking.

Hydrocotyle verticillata Thunb. Plants glabrous, with slender creeping stems; leaves peltate, suborbicular, 0.5–6 cm wide, shallowly lobed and often crenate; petioles slender, 3–20 cm long or longer; peduncles slender, axillary; flowers apparently verticillate in few to several well separated whorls; petals pale, small; fruits subsessile, subtruncate at the base, 1.5–2 mm long, 2–3 mm wide. Moist ground or in water at 850 to 1005 m in Washington County; South America north to Massachussetts and California; 4 (0).

Ligusticum L.

Perennial caulescent or acaulescent herbs from taproots; leaves ternately or ternate-pinnately compound or dissected, the lower ones with well-developed petioles, the upper ones with blades arising directly from dilated sheaths; umbels compound; involucre and involucel lacking or of a few narrow bracts or bractlets; calyx teeth evident or obscure; petals white; stamens white; stylopodium low-conic; carpophore divided to the base; fruit oblong to ovate or suborbicular, subterete or slightly compressed laterally, the ribs evident, often winged.

1. Ultimate leaf segments more or less linear or narrowly elliptic, mostly 0.5–3 mm wide, entire 2
— Ultimate leaf segments (at least some) elliptic or broader, some usually over 3 mm wide, sometimes toothed or lobed 3
2(1). Umbels mostly solitary, occasionally 2, rarely 3, never opposite; rays 0.5–3.6 cm long; petioles 1.2–13.5 cm long; leaf blades 3–19 cm long; plants 10–45 (64) cm tall, of the Uinta Mts. .... *L. tenuifolium*
— Umbels 2–5 or more the lateral ones occasionally opposite or whorled; rays 2.5–6.5 (8) cm long; petioles 8–32 cm long; leaf blades (9) 12–30 cm long; plants (4) 6–10 dm tall, of central Utah and western Uinta Mts. .... *L. filicinum*
3(1). Umbels 2 or 3, the lateral 1 or 2 alternate, subtended by much reduced leaves; plants of the Raft River Mts. .... *L. grayi*
— Terminal umbel subtended by often opposite or whorled umbels and 1–3 or more alternate umbels from the axils of reduced or well-developed leaves; plants widespread .... *L. porteri*

Ligusticum filicinum Wats. Fernleaf Ligusticum. Plants (4.5) 6–13 dm tall, aromatic, glabrous, from a heavy taproot with a simple or branched crown, the crown clothed with fibrous persisting petiole bases; leaves basal and 1–3 cauline, ternate-pinnately 3 times dissected, with 5–6 (7) opposite pair of lateral primary leaflets; basal petioles 8–32 cm long; blades (9) 12–30 cm long, ovate in outline, the lowest pair of primary leaflets 1/2–3/4 as long as the leaf blade, on petiolules 2.5–10 cm long; ultimate leaf segments 1–18 mm long, 0.7–2.5 (3) mm wide, linear, narrowly elliptic or narrowly deltoid, entire, bifid, or trifid; peduncles (5) 10–17 (23) cm long; terminal umbel subtended by 1–3 smaller umbels, the lateral ones arising from axils of leaves and alternate or the upper ones not from leaf-axils and opposite or in whorls of 3; involucre lacking; rays 7–27, 2.5–6.5 (8) cm long; involucels of 1–3 linear separate usually deciduous bractlets to 5 mm long; pedicels ca 1–4 mm long; petals white; stamens whitish; styles ca 0.5 mm long; fruit 5–8 mm long. Tall forb, aspen, sagebrush-grass, forb-grass, Douglas fir, and spruce-fir communities at 1920 to 3110 m in Cache, Duchesne, Juab, Morgan, Sanpete, Summit (type from the Uinta Mts.), Tooele, Utah and Wasatch counties; Idaho, Montana, Utah and Wyoming; 46 (xx). This taxon is rather easily confused with *L. porteri* (q.v.).

Ligusticum grayi Coult. & Rose Grays Ligusticum. Plants 3–6 (9.5) dm tall, glabrous, aromatic, from a stout taproot with simple or branched crown, the crown clothed with fibrous persistent petiole bases; leaves basal and usually 1–3 much reduced cauline ones, ternate-pinnately twice compound and then pinnatifid with (2) 3–5 opposite pairs of lateral primary leaflets; petioles (2.5) 4–34 cm long; blades 4–26 cm long, ovate in outline, the lowest pair of primary leaflets 1/2–3/4 as long as the blade, on petiolules 1–6.5 cm long, the larger secondary leaflets pinnatifid with the larger lobes again bi- or trilobate; peduncles 2–55 (90) cm long; terminal umbel subtended by 1–2 alternate umbels arising from the axils of much reduced leaves; involucre lacking or rarely of 1 linear mostly deciduous bract to ca 1 cm long; rays 8–18, 1.2–4 cm long; involucels lacking or of 1–5 linear bractlets to 4.5 mm long; pedicels 4–10 mm long; petals white; stamens whitish; styles 0.8–1.1 mm long; fruit 4–6 mm long. Forb-grass and fir communities and snowflush areas at 2650 to 2900 m in the Raft River Mts., Box Elder County; Washington to California and east to Idaho; 3 (iii).

Ligusticum porteri Coult. & Rose Southern Ligusticum. [*L. brevilobum* Rydb., type from the Aquarius Plateau]. Similar to *L. filicinum*, but leaves with broader ultimate segments, these (1.5) 3–8 mm wide, and with the terminal umbel often subtended by a whorl of 3–8 lateral umbels, and occasionally with up to 12 or more umbels, but sometimes with the lateral umbels only 2 and opposite, but not alternate; n = 11. Sagebrush, oak, aspen, Douglas fir, spruce, fir, and occasionally in open forb-grass communities at 2255 to 3171 m in Beaver, Carbon, Duchesne, Garfield, Grand, Iron, Juab, Kane, Millard, Piute, San Juan, Sanpete, Sevier, Uintah, and Utah counties; southern Wyoming to northern Mexico, west to Idaho and Arizona; 58 (x). Plants of this taxon are sometimes mistaken for *Conioselinum scopulorum* (q.v.). The separation of *L. porteri* from *L. filicinum* is made difficult by a rather extensive overlap in distribution and lack of definitive morphology in plants from Utah County south to Sevier County. Otherwise the ranges of the two taxa are essentially discrete in Utah, but occasional specimens from scattered locations throughout the state would be difficult to place without location data.

Ligusticum tenuifolium Wats. Small Ligusticum; Slender-leaf Ligusticum. [*L. filicinum* var. *tenuifolium* (Wats.) Math. & Const.]. Plants 11–64 cm tall, glabrous mildly aromatic from a taproot, the crown more or less covered by short shredded marcescent bases; leaves basal and sometimes 1 or 2 cauline, ternate and then 2–3 times pinnate with 5–7 pairs of lateral primary leaflets; petioles 1.2–13.5 cm long; blades 3–19 cm long, completely dissected, ovate in outline; lowest pair of primary leaflets 1/2–2/3 as long as the blade, on petiolules (0.5) 1–4 cm long, the upper primary leaflets progressivelly reduced, the ultimate segments 2–9 mm long, 0.5–1.5 (2.5) mm wide; scapes or peduncles 10–45 (61) cm long; involucre lacking; umbel solitary or the terminal one sometimes subtended by 1 or 2 lateral ones that usually arise from the axil of a reduced leaf; rays 6–15, 0.5–3.6 cm long; involucels lacking or of 1–3 filiform-linear bractlets to 3 mm long; pedicels 2–4 mm long; calyx obsolete; petals ca 1 mm long, white, sometimes tinged with light purple in age; stylopodium evident, conic; styles 0.5–0.8 mm; fruit about 3–5 mm long. Moist and wet meadows, along streams in lodgepole pine and Engelmann spruce woods at 2440 to 3420 m, common across the Uinta Mts., in Daggett, Duchesne, Summit, Uintah, and Wasatch counties; Oregon to Montana and south to Colorado; 37 (xviii).

Through a series of features (none of which are exclusive), plants of *L. tenuifolium* are readily distinguished from those of *L. filicinum*. The two taxa are sympatric in the western Uinta Mts., where some intermediate specimens occur, but the range of overlap is small.

Lomatium Raf.

Plants perennial, acaulescent or caulescent, occasionally with a short pseudoscape, glabrous or pubescent, from a slender taproot with sometimes 1 or more tuberlike segments, or from a thickened woody branching caudex, sometimes clothed at the base with marcescent leaf bases; stems simple or rarely branched, the peduncles and umbels mostly solitary; leaves pinnate or pinnately to ternate-pinnately compound, the sheaths often dilated; petioles developed, and distinct or confluent with and poorly differentiated from the sheath, or lacking and the petiolules arising directly from the sheath, the ultimate segments extremely variable; involucre lacking or inconspicuous; rays few to many, spreading to ascending, the central ones often shorter and sterile; involucel mostly of separate or partly united bractlets, rarely lacking; pedicels slender or stout, the central ones often shorter and sterile; petals small, yellow, white, greenish yellow or purplish; calyx teeth obsolete or small, or conspicuous in some species; styles slender, often curved or coiled; stylopodium lacking; carpophore divided to the base; fruit linear to orbicular or obovate, flattened dorsally, glabrous or pubescent, dorsal ribs filiform or obsolete or occasionally with rudimentary wings at the base. Note: The genus is closely related to *Cymopterus*, and the filiform wingless dorsal ribs of the fruit seem to be the only consistent difference from *Cymopterus*. The dependability of this separation is somewhat weakened by the tendency for lack of dorsal wings in some *Cymopterus* taxa.

1. Leaves once-pinnate and/or the ultimate segments over 15 mm long and fewer than 50 per leaf; plants glabrous and/or petals yellow when fresh 2
— Leaves more than once-compound, the ultimate segments not over 15 mm long and mostly more than 50 per leaf or, if a few ultimate segments over 15 mm long or fewer than 50 per leaf, the plants pubescent and petals white 9
2(1). Leaves once-pinnate or partly bipinnate, the leaflets sessile and more or less confluent with the rachis; plants from stout, more or less woody caudices, clothed at the base with marcescent leaf bases, from southern Utah 3
— Leaves more than once-compound, the primary leaflets mostly with well-developed petiolules, not confluent with the rachis; (plants from taproots or small caudices, not much if at all clothed at the base with marcescent leaf bases, from northern Utah 6
3(2). Leaflets lanceolate to elliptic, 2–12 mm wide, some always over 5 mm wide; plants of Grand and northern San Juan counties . . . . *L. latilobum*
— Leaflets linear, not over 4 mm wide; distribution not as above 4
4(3). Leaves with 1–7 elongate terete leaflets that simulate the rachis in diameter and shape, these 1–18 cm long, at least some commonly over 5 cm long in each leaf; calyx teeth greenish, acute, ca 1 mm long; plants of Emery, Garfield, Sevier, and Wayne counties . . . . *L. junceum*
— Leaves either with more than 7 leaflets and/or leaflets less than 5 cm long, more or less flattened and, wider than the rachis; calyx teeth not over 0.6 mm long, scarious or greenish; plants of various distribution . . . 5
5(4). Plants 2–12 (17) cm tall, of Garfield and Iron counties; fruit 4–7 mm long, leaflets 3–13 per leaf, 0.2–1.5 (2) cm long . . . . *L. minimum*
— Plants 15–30 cm tall or taller, not known from Garfield and Iron counties; fruit 5–15 mm long; leaflets sometimes more than 13 per leaf, sometimes over 2 cm long . . . . *L. kingii*
6(2). Ultimate leaflets less than 3 times as long as wide, at least some dentate; rays (4) 8–19 cm long; peduncle often swollen just beneath the umbel; plants of the Deep Creek Mts. and western Box Elder County . . . . *L. nudicaule*
— Ultimate leaflets 3 or more times longer than wide, entire; rays 0.5–10 cm long; peduncle not swollen just beneath the umbel; distribution not as above 7
7(6). Plants caulescent, glabrous; peduncles to 13 cm long; involucel lacking . . . . *L. ambiguum*
— Plants acaulescent or, if caulescent, the peduncles mostly over 13 cm long and plants puberulent; involucels present, to 1 cm long 8
8(7). Plants strongly aromatic; caudex often clothed with marcescent leaf bases; leaves strictly basal; ultimate leaflets 0.3–5 (6.5) cm long, 0.5–2 (4) mm wide; lateral wings of the fruit to 1 mm wide . . . . *L. kingii*
— Plants not strongly aromatic; marcescent leaf bases lacking or weakly persisting; cauline leaves sometimes 1–3; ultimate leaflets 1–13 cm long, 1–6 (15) mm wide; lateral wings of the fruit 1–2 mm wide . . . . *L. triternatum*
9(1). Larger mature leaves with blades (10) 15–30 cm long, ternate-pinnately compound, the larger ultimate segments 2–3 mm wide; plants 3–13 dm tall; peduncles fistulose, (3) 4–6 (10) mm thick at the base . . . . *L. dissectum*
— Larger mature leaves with blades 2–11 cm long or, if longer, either not at all ternate or with ultimate segments not over 1 mm wide; plants rarely over 50 cm tall; peduncles fistulose or not, often less than 4 mm thick 10
10(9). Plants pubescent 11
— Plants glabrous or at most scabrous 15
11(10). Ovaries and fruit glabrous or somewhat scabrous; plants mainly of central and eastern Utah 12
— Ovaries and young fruit densely pubescent, the older fruit sometimes glabrous but often retaining some hairs; plants of various distribution 13
12(11). Bractlets of the involucel ca 10, the longer ones 4–10 mm long, pubescent; herbage more or less villous . . . . *L. macrocarpum*
— Bractlets of the involucel 1–5, 1–4.5 mm long, glabrous; herbage glabrate to puberulent . . . . *L. juniperinum*
13(11). Longer ultimate leaf segments 5–27 mm long, some often over 1 mm wide; petals white . . . . . . *L. nevadense*
— Longer ultimate leaf segments 1–5 mm long, to 1 mm wide; petals yellow, rarely white 14
14(13). Petals and anthers yellow; leaves often conspicuously ternate-pinnately compound; plants common . . . . *L. foeniculaceum*
— Petals white; anthers purple or whitish; leaves pinnately or scarcely ternate-pinnately compound; plants apparently rare, known only from western Millard County . . . . *L. ravenii*

15(10). Lowest pair of primary leaflets less than 1/3 as long as the leaf blade 16
— Lowest pair of primary leaflets 1/3–3/4 as long as the leaf blade 18
16(15). Mature pedicels 1–10 mm long; fruit 4–9 mm long; leaves and peduncles scabrous, the blades 2–7 cm long; plants of the Great Basin and Washington County .... *L. scabrum*
— Mature pedicels (at least the longer) 10–20 mm long; fruit (6) 8–20 mm long; leaves and peduncles glabrous or plants of the Colorado Basin 17
17(16). Leaf blades 3–7 cm long, the ultimate segments 2–4 mm long; fruit 8–10 mm long; rays 4–6, 1–3 cm long; plants densely scabrous, 10–15 cm tall, of western Colorado, to be expected in Utah in extreme eastern Grand and San Juan counties .... *L. eastwoodiae* (Coult. & Rose) Macbr.
— Leaf blades 7–24 cm long, the ultimate segments 1–15 mm long; fruit to 20 mm long; plants glabrous, glaucous, 8–40 cm tall, of the Colorado Basin and Washington County .... *L. parryi*
18(15). Bractlets of the involucel broadly elliptic to obovate, to 3 mm wide; pedicels 1–2 mm long; ultimate segments of leaves 2–13 mm long, 0.5–4 mm wide, dimorphic; plants of northwestern Box Elder County .... *L. cous*
— Bractlets of the involucel linear to subulate, not over 1 mm wide; pedicels 2–18 mm long; ultimate segments of leaves 1–7 mm long, 0.2–1.5 mm wide, not much if at all dimorphic 19
19(18). Petals white or cream; ultimate segments of leaves 0.5–1.5 mm wide; plants not aromatic; pedicels 3–16 mm long; leaves with 3–6 opposite pairs of lateral primary leaflets; rare glabrous forms ... *L. juniperinum*
— Petals yellow when fresh; ultimate segments of leaves 0.2–0.6 mm wide; plants strongly aromatic, glabrous; pedicels various; leaves various 20
20(19). Fruit 2–4 mm wide, the wings 0.4–0.6 mm wide; pedicels 2–5 mm long; rays very unequal .... *L. bicolor*
— Fruit 5–8 mm wide, the wings ca 1.5–2 mm wide; pedicels 5–18 mm long; rays subequal .... *L. grayi*

Lomatium ambiguum (Nutt.) **Coult. & Rose** Wyeth Biscuitroot. [*Eulophus ambiguus* Nutt.]. Plants caulescent, 1–4 dm tall, glabrous, lacking persistent leaf bases; root very slender, or sometimes with 1 or more globose or elongate tuberous segments; leaves ternately or ternate-pinnately compound; petioles to 2 cm long or lacking and blades arising from a dilated sheath 1.5–4 cm long; blades 4–15 cm long, ovate in outline, the lowest pair of primary leaflets mostly over 1/2 as long as the blade, with petiolules 1–4 cm long, the ultimate segments 15–45, 0.3–9 cm long, 1–4 mm wide, often very unequal in the same leaf; peduncle 2.5–13 cm long; involucre lacking; rays 0.5–6.5 cm long, very unequal in the same umbel; involucel lacking; pedicels 2–12 mm long; petals and stamens yellow, fading in age; styles ca 1 mm long; fruit 8–10 mm long, 2–3 mm wide, the lateral wings ca 0.5 mm wide, the dorsal ribs filiform; n = 11. Sagebrush and mountain brush communities at 1525 to 1980 m in Cache, Salt Lake, Utah, and Weber counties; Washington and British Columbia to Montana, south to Wyoming; 14 (0). The larger leaf segments are conspicuously different from the very narrow and shorter ones of *L. bicolor*.

Lomatium bicolor (**Wats.**) **Coult. & Rose** Wasatch Biscuitroot. [*Peucedanum bicolor* Wats., type from Parleys Park, Summit County]. Plants 10–50 cm tall, acaulescent or caulescent, aromatic, glabrous, lacking persistent leafbases or these few and weakly persisting; from a slender taproot, this often with one or more tuberous branches; leaves ternate-pinnately decompound; petioles mostly lacking and the blades arising directly from a dilated sheath; blades 4–12 cm long, ovate in outline, finely and completely dissected, the lowest pair of primary leaflets mostly over 1/2 as long as the leaf blade, with petiolules 2.5–6 cm long, each with 5–6 opposite or offset pairs of secondary leaflets, the ultimate segments mostly more than 300, 1–4 (6) mm long, 0.2–0.6 mm wide; peduncle 10–28 cm long; rays 3–12 (20), 1–8 (11) cm long, very unequal in the same umbel; involucel lacking or of 1–8 linear distinct bractlets; pedicels 2–5 mm long; petals and stamens yellow; styles ca 1 mm long; fruit 8–11 mm long, 2–4 mm wide, congested; lateral wings 0.4–0.6 mm wide, dorsal ribs filiform; n = 11. Sagebrush, mountain brush, aspen, and meadow communities at 1525 to 2438 m in Cache, Morgan, Rich, Salt Lake, and Weber counties; Idaho; 25 (0). This taxon has been included in *L. leptocarpum* (T. & G.) Coult. & Rose. Perhaps they should be combined but, if so, *L. bicolor* is the proper name.

Lomatium cous (**Wats.**) **Coult. & Rose** Cous Biscuitroot. [*Peucedanum cous* Wats.]. Plants 5–15 (25) cm tall, not or weakly aromatic, glabrous, from a globose or fusiform tuberous root, this sometimes deep seated and giving rise to a subterranean pseudoscape; leaves basal and sometimes 1 or 2 cauline on the lower 1/3–1/2 of the stem, 2–3 times pinnately or ternate-pinnately compound; blades 4–8 (11) cm long, mostly borne on dilated sheaths with the petioles obsolete or short, or some leaves arising from the roots and then with etiolated petioles to 11 cm long, the lowest pair of primary leaflets 1/2–3/4 as long as the leaf blade, sessile or with petiolules to 42 mm long, ultimate segments or lobes 2–13 mm long, 0.5–4 mm wide, as many as 200 or more, linear to elliptic; peduncles 1–7, 3–18 cm long; involucre lacking or of a solitary bract to 7 (10) mm long; rays 6–15, 0.4–5 cm long, strongly dimorphic in the same umbel; bractlets of the involucel 6–10, 3–5 mm long, to 3 mm wide, broadly elliptic, ovate or obovate, greenish, sometimes with yellowish or scarious margins; pedicels 1–2 mm long; calyx teeth obsolete; petals and stamens yellow when fresh, fading to white in age; styles ca 1.5 mm long; fruit 6–9 mm long, the lateral wings to ca 1 mm wide, the dorsal ribs filiform or obscurely winged. Sagebrush-grass communities at 2440 to 2560 m, in the Grouse Creek and Raft River mts., Box Elder County; Oregon to Montana, south to Nevada; 2 (ii).

Lomatium dissectum (**Nutt.**) **Math. & Const.** Giant Lomatium. [*Leptotaenia dissecta* Nutt.]. Plants 3–13 dm tall, mostly short caulescent, puberulent or rarely glabrous, from a woody thickened taproot or caudex, without marcescent leaf bases or these soon shredding and deciduous; leaves pinnately or ternate-pinnately decompound, with 5–9 opposite or offset pairs of primary leaflets, or the upper cauline leaves much reduced; petioles 3–20 cm long, often lacking on cauline leaves and then the blades sessile on a dilated sheath; blades 10–30 cm long or smaller on cauline leaves, ovate in outline, the lowest pair of primary leaflets usually over 1/2 as long as the leaf blade, with petiolules 2.5–12 cm long, ultimate segments numerous, 1–12 mm long, 0.5–3 mm wide;

peduncles 15–50 (90) cm long; involucre lacking or rarely of 1–3 deciduous bracts; rays 9–27, 2–7 (12) cm long; bractlets of the involucel 3–6 mm long, or occasionally much longer and foliaceous; pedicels 3–10 (15) mm long; petals and stamens yellow, yellow-green, or purplish; styles ca 1.5 mm long; fruit 9–15 (20) mm long, 6–10 mm wide, the lateral wings ca 1–2 mm wide, the dorsal ribs filiform. Sagebrush, pinyon-juniper, oak-maple, aspen-fir, riparian, and rarely greasewood-desert shrub communities at 1280 to 2650 (3170) m in Beaver, Box Elder, Cache, Duchesne, Iron, Juab, Millard, Morgan, Rich, Salt Lake, Sanpete, Summit, Tooele, Uintah, Utah, Washington, and Weber counties; British Columbia and Alberta south to California, Arizona and Colorado; 91 (xiii). Utah materials are referable to **var. eatonii** (Coult. & Rose) Cronq. [*Leptotaenia eatonii* Coult. & Rose; type from Utah]. Sometimes the leaves are mistaken for those of *Ligusticum porteri*, but the mostly solitary umbel is strikingly different from the usually opposite or whorled lateral umbels in addition to the terminal one in the *Ligusticum*.

Lomatium foeniculaceum (Nutt.) **Coult. & Rose** Desert-parsley. [*Ferula foeniculacea* Nutt.]. Plants 5–25 (38) cm tall, acaulescent, densely pubescent throughout, from a more or less branched caudex and deep taproot, often clothed at the base with persistent leaf-bases; leaves ternate-pinnately dissected, with 6–8 opposite pairs of lateral primary leaflets; petioles to 2.5 cm long or lacking and the blade arising from a dilated sheath; blades 2–13 cm long, completely and finely dissected, ovate in outline, the lowest pair of primary leaflets over 1/2 as long as the blade, sessile or with petiolules to 5 cm long, the ultimate segments numerous, often over 500, 1–3 (5) mm long, 0.5–1 (2.5) mm wide; peduncles 4–30 cm long; rays 5–20, 0.2–7 cm long; bractlets of the involucel 2–5 (6) mm long, separate or united at the base, linear; pedicels 2–12 mm long; petals and anthers yellow (rarely white) when fresh and mostly remaining yellow or occasionally turning purplish; styles 1.5–2 mm long; fruit 5–10 mm long, 3–7 mm wide, lateral wings 1–2 mm wide, dorsal ribs filiform. Sagebrush (mostly black sagebrush), pinyon-juniper, and mountain brush communities at 1250 to 2635 m in Beaver, Box Elder, Daggett, Emery, Juab, Kane, Millard, Sanpete, and Tooele counties; Manitoba to Missouri and Texas, west to Oregon and California; 71 (xvi). Utah materials are generally referable to **var. macdougalii** (Coult. & Rose) Cronq. [*L. macdougalii* Coult. & Rose; *L. jonesii* Coult. & Rose, type from Ireland Ranch, Sevier County]. However, some plants from western Utah have ciliolate petals and are referable to **var. fimbriatum** (Theob.) **J. Boivin** [*L. foeniculaceum* ssp. *fimbriatum* Theobald], but the diagnostic feature is not always reliable.

Lomatium grayi (Coult. & Rose) **Coult. & Rose** Milfoil Lomatium. [*Cogswellia grayi* Coult. & Rose; *Peucedanum millefolium* Wats, type from Antelope Island; *L. millefolium* (Wats.) Macbr.]. Plants (8) 15–40 (80) cm tall, acaulescent or subcaulescent, strongly aromatic, glabrous, from a simple or branched caudex and thick taproot often clothed at the base with marcescent, mostly shredded, fibrous leaf bases; leaves ternate-pinnately dissected, with 7–10 opposite pairs of lateral primary leaflets; petioles to 14 cm long or lacking and the blades arising from a dilated sheath 1–16 cm long; blades 7–16 (2) cm long, finely and completely dissected, ovate in outline, the lowest pair of primary leaflets from 1/2 to as long as the blade, with petiolules 1–7.5 cm long, the ultimate segments very numerous, extremely minute, 1–3 (6) mm long, 0.2–0.3 mm wide; peduncle 10–45 (70) cm long; rays 10–26, 1.5–6 (8) cm long; bractlets of the involuces 3–5 mm long, linear, separate or united at the base; pedicels 5–13 (18) mm long; petals and stamens yellow when fresh, soon fading whitish when dried; styles 1.5–2.5 mm long; fruit 6–12 mm long, 5–8 mm wide, the lateral wings ca 2 mm wide, the dorsal ribs filiform; n = 11. With 2 intergrading but more or less geographically correlated varieties.

1. Fruit 6–9 (10) mm long, the lateral wings to ca 1.5 mm wide; leaves rather openly dissected, with a few hundred ultimate segments; plants 8–20 (35) cm tall, of the western tier of counties from Box Elder south to Beaver County .... *L. grayii* var. *depauperatum*

— Fruit 8–12 mm long, the lateral wings to ca 2 mm wide; leaves with congested and numerous ultimate segments, these several hundred to a thousand or more; plants 15–40 (80) cm tall, of more easterly distribution and only in the eastern 1/4 of the western tier of counties where more or less transitional with the preceeding variety ... . *L. grayii* var. *grayi*

Var. depauperatum (Jones) Mathias [*Cogswellia millefolia* var. *depauperata* Jones, type from Dugway, Tooele County]. Desert shrub, pinyon-juniper, and mountain brush communities at 1525 to 2835 m in Beaver, Box Elder, Juab, Millard, and Tooele counties; western Utah and adjacent Nevada; 58 (xiii).

Var. grayi Sagebrush, pinyon-juniper, mountain brush, ponderosa pine, and Douglas fir communities at 1340 to 2745 m in Box Elder, Cache, Daggett, Davis, Duchesne, Grand, Juab, Morgan, Rich, Salt Lake, San Juan, Sanpete, Summit, Tooele, Uintah, Utah, and Weber counties; Washington to Nevada and east to Idaho and Colorado; 121 (xxiii).

Lomatium junceum Barneby & N. Holmgren Rush-lomatium. Plants (6) 10–37 cm tall, acaulescent, glabrous, from a simple to much branched woody caudex, clothed at the base with persistent petioles; leaves rushlike, trifid or pinnatifid or rarely reduced to a petiole and a linear bladeless rachis, with 1–7 linear segments; petioles 3–15 mm long with a short sheath at the base; blades 3–17 cm long, the segments 1–18 cm long, 1–2 mm wide, terete and similar to the rachis and petioles in diameter; peduncles 5–25 cm long; rays of umbels 6–13, 1.5–3 cm long; bractlets of the involucel 1.5–3 mm long, distinct or connate at the base, linear; pedicels 4–11 mm long; calyx teeth to ca 1 mm long, acutish, somewhat persistent; petals and stamens bright yellow or cream, quickly fading to white when frozen or dried; styles 2–3 mm long; fruit 8–12 mm long, 5–7 mm wide, the lateral wings 1–2 mm wide, the dorsal ribs filiform. Desert shrub, sagebrush, pinyon-juniper, ponderosa pine, and Douglas fir communities at 1615 to 2485 m in Emery (type from San Rafael Swell), Garfield, Sevier, and Wayne counties; endemic; 21 (iii).

Lomatium juniperinum (Jones) **Coult. & Rose** Juniper Lomatium. [*Peucedanum juniperinum* Jones, type from Coalville]. Plants 8–32 cm tall, acaulescent or with 1–3 leaves on the lower part of the stems, more or less hirtellous to glabrate and glabrous, often with a short pseudoscape, from a taproot with simple or sparingly branched crown, lacking marcescent leaf bases or these weakly persistent; leaves ternate-pinnately dissected, with (3) 4–6 opposite or offset pairs of lateral primary leaflets; petioles to 8 cm long or lacking and the blades arising directly from dilated sheaths 1–4 cm long; blades 2.5–8 (11) cm long, ovate in outline, the lowest pair of lateral primary leaflets 1/3 as long to as long as the leaf blade, with petiolules to 3 cm long, the ultimate segments 50–400, 1–7 mm long, 0.7–1.5 mm wide; peduncles 6–29 cm long; rays of the umbel 3–12, 1–8 cm long; bractlets of the involucel 1–5, 1–4.5 mm long, linear, distinct or connate at the base; pedicels 3–16 mm long; petals white, cream, or yellow; anthers white, ochroleucus, purple, or yellow; styles 1–2 mm long; fruit 5–8 (11) mm long, 3–6 mm wide, glabrous, or scabrous to sparsely hirtellous, especially when young; lateral wings 0.5–1.5 mm wide, the dorsal ribs filiform. Sagebrush, pinyon-juniper, forb-grass, aspen, Douglas fir, and alpine communities at 1830 to 3230 m in Carbon, Daggett, Duchesne, Grand, Juab, Sanpete, Summit, Uintah, Utah, and Wasatch counties; Wyoming, Idaho, and Colorado; 52 (xx). This species is variable as to color of petals and anthers. Plants with yellow petals and anthers are known only from the west and north side of the Uinta Mts. and West Tavaputs Plateau. In one population from the east end of the Uinta Mts., the flower color is white, cream, and yellow. Those with white petals and white to purplish anthers are found in the Wasatch Mts., south slope of the Uinta Mts., Tavaputs Plateau and north end of the Wasatch Plateau. The plants are sometimes confused with *L. nevadense* and *Cymopterus lemmonii*.

Lomatium kingii (Wats.) Cronq. Stinking Lomatium. Plants 1.5–5 dm tall, acaulescent, glabrous, strongly aromatic, from a branched caudex, this clothed with persistent leaf-bases; leaves pinnatifid to bipinnate or ternate-pinnately compound; petioles gradually expanded into a dilated sheath, with the sheath 2–21 cm long; blades 2–15 cm long, usually oblong in outline, ultimate leaflets or segments about 7–30, 0.3–6.5 cm long, 0.5–2 (4) mm wide; peduncles 12–47 cm long; rays of the umbel (3) 5–12, 1–5 cm long; bractlets of the involucel 3–10 mm long; pedicels 2–10 mm long; calyx teeth ca 0.5 mm long, rather scarious; petals and stamens yellow, soon turning pale to white in age; styles 2–3 mm long; fruit 5–15 mm long, 3–5 mm wide, the lateral wings 0.5–1 mm wide, the dorsal ribs filiform or somewhat prominent with rudimentary wings. There are two geographically correlated varieties that are somewhat transitional morphologically.

1. Fruit 5–8 mm long, the lateral wings ca 0.5 mm wide; umbels with only 3–6 rays; leaves once pinnatifid with sessile segments or some of the lower pairs of segments bipinnatifid; plants of Washington and western Millard counties . . . . *L. kingii* var. *alpinum*
— Fruit 10–15 mm long, the lateral wings ca 1 mm wide; umbels with up to 12 rays; leaves pinnatifid to ternate-pinnately compound, with the lowest pair of primary leaflets on petiolules (1) 2.5–9 cm long; plants of Sevier and eastern Millard counties and northward . . . .
. . . . *L. kingii* var. *kingii*

Var. alpinum (Wats.) Cronq. [*Peucedanum graveolens* var. *alpinum* Wats.; *P. graveolens* Wats., type from the Wasatch Mts.]. Pinyon-juniper and mountain brush communities at 2225 to 2440 m in Millard and Washington counties; western Nevada and southwestern Utah; 4 (0).

Var. kingii [*Peucedanum kingii* Wats., type from the Wasatch and Uinta mts.; *L. nuttallii*, misapplied, not (Gray) Macbr.]. Sagebrush, bullgrass, mountain brush, Douglas fir, limber pine, and spruce-fir communities, often in rocky places, mostly on limestone and other basic substrates, sometimes in raw snowflush areas, at 1980 to 3200 m in Cache, Davis, Duchesne, Millard, Rich, Salt Lake, Sanpete, Sevier, Summit, Tooele, Utah, Wasatch, and Weber counties; western Nevada, and western Wyoming; 85 (xviii).

Lomatium latilobum (Rydb.) Mathias Canyonlands Lomatium. [*Cynomarathrum latilobum* Rydb., type from Wilson Mesa, Grand County]. Plants (6) 10–30 cm tall, acaulescent, glabrous, from a branched woody caudex, clothed with persistent leaf-bases; leaves pinnate with 3–4 (5) pair of lateral leaflets; petioles 2–16 cm long; blades 1–10 cm long, oblong in outline; leaflets 1–4 cm long, 2–12 mm wide, sessile, entire or a few bifid or trifid; peduncles 4–27 cm long; rays of the umbel 4–13, 0.5–2 cm long; bractlets of the involucel 2–15 mm long, 0.5–2 mm wide, linear or elliptic, separate; pedicels 1–4 mm long; calyx teeth 1–1.5 mm long, acute; petals yellow when fresh, drying white; styles 2–3 mm long; fruit 8–12 mm long, 3–7 mm wide, the lateral wings ca 1 mm wide, the dorsal ribs filiform. Pinyon-juniper and desert shrub communities, mainly in Entrada Sandstone, at ca 1525 m in Grand and San Juan counties; Mesa County, Colorado; a Navajo Basin endemic; 17 (i).

Lomatium macrocarpum (H. & A.) Coult. & Rose Bigseed Lomatium. [*Ferula macrocarpa* H. & A.]. Plants 12–30 cm tall, acaulescent or subcaulescent with leaves mostly on the lower 1/4 of the stem, more or less tomentose-villous or glabrate, from a thickened taproot with a simple or sparingly branched crown with few or no persistent leaf bases; leaves pinnately or ternate-pinnately dissected, with ca 4 opposite pair of lateral primary leaflets; petioles often long tapering into a dilated sheath, the petiole and sheath 3–6 cm long; blades 3–6 cm long, ovate in outline, the lowest pair of primary leaflets 1/2–3/4 as long as the leaf blade, sessile or with petiolules to 1 cm long, ultimate segments 30–300 or more, 1.5–6 mm long, 0.5–2 mm wide, elliptic or linear; peduncles 8–26 cm long; rays of the umbel 6–18, 1–4 (6.5) cm long; bractlets of the involucel ca 10, 2–10 mm long, separate or united at the base, pubescent; pedicels 2–5 mm long; calyx teeth to ca 0.5 mm long; petals white or purplish in age; anthers white; styles ca 2–3 mm long; fruit 9–12 (15) mm long, 4–5 mm wide, glabrous, the lateral wings 1–1.5 mm wide, the dorsal ribs filiform; 2n = 22. Desert shrub, sagebrush, and pinyon-juniper communities at 1480 to 2550 m in Daggett, Juab, Millard, Sanpete, Tooele, and Uintah counties; southern British Columbia to California and east to Manitoba and Colorado; 29 (xi).

Lomatium minimum (Mathias) Mathias Least Lomatium. [*Cogswellia minima* Mathias, type from Bryce Canyon]. Plants 2–12 (17) cm tall, acaulescent, glabrous or scabrous, from a branched caudex, the caudex branches clothed with persistent leaf bases; leaves once-pinnatifid or rarely trifid, with (3) 5–9 (13) segments; petioles to 2 cm long; blades 1–2.5 cm long, the segments 2–15 (20) mm long, 0.5–2 mm wide; peduncles to 10 (16) cm long; rays of the umbel 3–6, 0.3–2.3 (3.2) cm long; bractlets of the involucel 2–4 mm long, linear-subulate, separate; pedicels 1–3 mm long; calyx teeth to 0.6 mm long, acute, greenish or purplish in age with scarious margins; petals and stamens yellow, drying to cream; styles 1.5–2 mm long; fruit 4–7 mm long, 3–4 mm wide, the lateral wings 0.5–1 mm wide, the dorsal ribs mostly filiform. Forb-grass, ponderosa pine, and bristlecone pine communities, often on exposed ridges and raw escarpments, often on limestone at 2165 to 3170 m in Garfield, Iron, and Kane counties; endemic; 20 (0). This appears much like a diminutive form of *L. kingii*.

Lomatium nevadense (Wats.) Coult. & Rose Nevada Lomatium. [*Peucedanum nevadense* Wats.]. Plants 10–36 cm tall, acaulescent or with 1 or 2 leaves on the lower part of stems, more or less pubescent throughout, from a slender root frequently with a fusiform tuberous segment, with or without persistent leaf bases; leaves 2–3 times pinnately compound, with ca 4 opposite pair of lateral primary leaflets; petioles to 7.5 cm long or often lacking and the blade sessile on a dilated sheath; blades 2.5–9 cm long, ovate in outline, the lowest pair of primary leaflets ca 1/2 to nearly as long as the blade, sessile or on petiolules to 5 mm long, the ultimate segments 20–80, 1–27 mm long, 0.5–3 mm wide; peduncles 7–33 cm long; rays of the umbel 7–12, sometimes with as few as 3 of them fertile, 1.5–4 cm long; bractlets of the involucel 2–3 mm long, lanceolate, linear-elliptic, or narrowly obovate; pedicels 4–11 mm long; petals and stamens white; styles ca 1–1.5 mm long; fruit 5–10 mm long, 3–7 mm wide, densely puberulent or glabrate to glabrous; lateral wings 0.8–2 mm wide, the dorsal ribs filiform. Desert shrub, sagebrush, pinyon-juniper, mountain brush, and ponderosa pine communities at 1524 to 2285 m in Beaver, Garfield, Iron, Kane, Millard, and Washington counties; Oregon and California east to Colorado and Arizona; 44 (viii). Our materials are perhaps referable to **var. parishii** (Coult. & Rose) Jepson [*Peucedanum parishii* Coult. & Rose]. This variety has been keyed as having glabrous fruits, and the Utah materials with pubescent fruits have been referred to as **var. nevadense**. However, the Utah plants have dimorphic ultimate leaf segments 1–27 mm long. This is a feature of var. *parishii*. The ultimate leaf segments of var. *nevadense* are only 2–3 mm long. A specimen from Navajo Mt. (Albee 4463 UT), San Juan County, has uniformly small leaf segments and glabrous fruits.

Lomatium nudicaule (Pursh) Coult. & Rose Nakedstem Lomatium. [*Smyrnium nudicaule* Pursh]. Plants 2–4.5 dm tall, acaulescent, glabrous, from a taproot, without persistent leaf bases or these few and weakly persisting; leaves ternate or biternate, with 3–11 distinct leaflets; petioles to 6 cm long or obsolete and the blades arising from a dilated sheath; blades 4–10 cm long, ovate in outline, the leaflets 2–5 cm long mostly 1–5.5 cm wide, ovate or orbicular to reniform, coarsely toothed toward the apex; peduncles 15–27 cm tall, sometimes swollen at the apex; rays of the umbel 7–27, 8–10 cm long; involucel lacking; pedicels 3–10 mm long; petals white; styles ca 1–2 mm long; fruit 8–12 mm long, 2–5 mm wide, the lateral wings ca 0.5 mm wide, the dorsal ribs filiform; n = 11. Sagebrush, pinyon-juniper, and mountain brush communities at 1585 to 2530 m in Box Elder, Juab, and Tooele counties; British Columbia to California and east to Alberta and Idaho; 6 (iii).

Lomatium parryi (Wats.) Macbr. Parry Lomatium. [*Peucedanum parryi* Wats., type from southern Utah; *Cogswellia cottami* Jones, type from Beaverdam Mts.]. Plants 8–40 cm tall, acaulescent, glabrous, from a branched caudex, clothed at the base with persistent leaf bases; leaves bi- or partly tripinnatifid, with mostly 7–9 opposite pair of primary leaflets or the upper leaflets simple; petioles 3–16 cm long, terete, often persisting for some years without shredding; blades 7–24 cm long, the lowest pair of lateral primary leaflets less than 1/4 as long as the leaf blade, sessile or with petiolules to 1.2 cm long, the ultimate segments mostly 50–150, 1–15 mm long, 1–2 mm wide, acute; peduncles 5–32 cm long; rays of the umbel 8–13, 1–5 cm long; bractlets of the involucels 3–10 mm long, entire, tridentate or rarely pinnatifid, spreading to reflexed in age; pedicels 1–2 cm long; petals yellow, turning white in age; styles 2–4 mm long; fruit 6–20 mm long, 5–10 mm wide, the lateral wings 1–3 mm wide, the dorsal ribs filiform. Desert shrub, blackbrush, pinyon-juniper, and mountain brush communities at 975 to 2320 m in Emery, Garfield, Grand, Iron, Kane, San Juan, and Washington counties; Utah to eastern California; 65 (v).

Lomatium ravenii Math. & Const. Raven Lomatium. Plants 4–23 cm tall, acaulescent, densely hirtellous throughout, from a taproot, with a simple or branched crown, usually clothed at the base by shredded leaf-bases; leaves ternate-bipinnate or 2–3 times pinnately dissected, with 5–7 (8) opposite pair of primary leaflets; petioles to 6 cm long or lacking and blades arising from dilated sheaths to 2 cm long; blades 1.5–8 cm long, finely and completely dissected, the lowest pair of primary leaflets usually over 1/2 as long as the leaf blade, sessile or with petiolules to 1.5 cm long, the ultimate segments 300–600 or more, 1–5 mm long, 0.5–1 mm wide; peduncles 2.5–21 cm long; rays of the umbel nearly obsolete or to 3.7 cm long; bractlets of the involucel 1–3 mm long, linear, pubescent; pedicels 1–8 mm long; petals white; anthers purple; styles 1–2 mm long; fruit 6–10 mm long, 3–6 mm wide, pubescent, the lateral ribs with wings 0.5–1 mm long, the dorsal ribs filiform. Pinyon-juniper-mahogany communities, at ca 3440 m in western Millard County; Oregon and Idaho to California and Nevada; 1 (i). Except for the white petals, purple anthers and sometimes slightly less pubescent foliage, plants of this taxon could pass for plants of *L. foeniculaceum*. Fruiting specimens may be difficult to distinguish.

Lomatium scabrum (Coult. & Rose) Mathias Rough Lomatium; Cliff Lomatium. [*Cynomarathrum scabrum* Coult. & Rose, type from Frisco, Beaver County]. Plants 6–25 (34) cm tall, acaulescent, mostly scabrous, from a branched caudex, clothed at the base by persistent leaf bases; leaves bi- to tripinnately dissected, with (5) 7–11 opposite pairs of lateral primary leaflets; petioles 1–7 (10) cm long; blades (1.5) 2–11 cm long, lowest pair of primary leaflets less than 1/3 as long as the leaf blade, sessile or nearly so, the ultimate segments 50–400 or more, 1–4 mm long, 0.4–2 mm wide; peduncles 5–25 (32) cm long; rays of the umbel 4–11, 0.5–2 (3) cm long; bractlets of the involucel 1–4 mm long, linear; pedicels 1–5 (10) mm long; petals and stamens mostly yellow or occasionally white when fresh, fading white in age; styles ca 2–3 mm long; fruit 4–7 mm long, 3–4 mm wide, the lateral wings to 1 mm wide, the dorsal ribs filiform or sometimes with a rudimentary wing at the base. There are two intergrading varieties.

1. Leaves mostly bipinnately dissected, with ca 50–110 (140) ultimate segments; fruit 4–8 mm long; plants mostly found above 1615 m .... *L. scabrum* var. *scabrum*

— Leaves tripinnately dissected, with ca 150–400 or more ultimate segments; fruit 6–9 mm long; plants mostly found below 1615 m .... *L. scabrum* var. *tripinnatum*

Var. *scabrum* Desert shrub, pinyon-juniper, mountain brush, and white fir communities, mostly on limestone and dolomite outcrops at 1615 to 2684 m in Beaver, Iron, Juab, and Millard counties; Nevada; a Great Basin endemic; 55 (xii). Some specimens, especially from Iron County, are wholly transitional to the following variety.

Var. *tripinnatum* Goodrich Blackbrush and pinyon-juniper communities, often on sandstone or in sandy places at 792 to 2170 m in Washington County; Arizona; a Virgin-Mohave endemic; 29 (ii).

***Lomatium triternatum* (Pursh) Coult. & Rose** Ternate Lomatium. [*Seseli triternatum* Pursh]. Plants 2–7 dm tall, acaulescent or subcaulescent, mostly hirtellous throughout except the fruit, from a taproot with simple or sparingly branched crown, not clothed at the base with persistent leaf bases or only weakly so; leaves ternate-pinnately compound with (3) 9–21 leaflets or segments; petioles up to 23 cm long including the dilated sheathing base, or reduced to the sheath; blades 4–20 cm long, ovate in outline, the lowest primary leaflets often over 1/2 as long as the leaf blade, the ultimate leaflets or segments 1–13 cm long, 1–15 mm wide; peduncles 15–55 cm long; rays of the umbel 4–20, 2–10 cm long; bractlets of the involucel 6–10, 1–10 mm long, 0.1–0.5 mm wide; pedicels 2–7 mm long; petals and stamens bright yellow when fresh but fading to white in age; styles 1–1.5 mm long; fruit 8–15 mm long, 4–11 mm wide, the lateral wings 1–2.5 (4) mm long, the dorsal ribs filiform; n = 11. There are two subspecies.

1. Ultimate leaflets or segments linear, over 10 times as long as wide, to 13 cm long, 1–6 (10) mm wide; fruit broadly elliptic, the mature wings as broad or nearly as broad as the body .... *L. triternatum* ssp. *platycarpum*
— Ultimate leaflets elliptic, 3–9 times as long as wide, 2–6 cm long, (3) 6–15 mm wide; fruit rather narrowly elliptic to nearly linear, the mature wings seldom more than half as wide as the body ...... *L. triternatum* ssp. *triternatum*

Ssp. *platycarpum* (Torr.) Cronq. [*Peucedanum triternatum* var. *platycarpum* Torr. in Stansbury, type from Great Salt Lake; *P. simplex* Nutt. ex Wats.; *L. simplex* (Nutt.) Macbr.]. Sagebrush-grass, pinyon-juniper, mountain brush, ponderosa pine, lodgepole pine, and dry meadow communities at 1310 to 2895 m in Box Elder, Cache, Daggett, Duchesne, Morgan, Rich, San Juan, Summit, Uintah, Utah, Wasatch, and Weber counties; British Columbia and Montana to Idaho and Colorado; 116 (viii).

Ssp. *triternatum* Mountain brush and aspen communities, sometimes on heavy clay soils with *Wyethia* at 1580 to 2590 m in Weber and Summit counties; southern Alberta and British Columbia to Utah; 8 (0). Utah specimens are referable to **var. *anomalum*** (Jones) Cronq. [*L. anomalum* Jones].

Musineon Raf.

Perennial plants with leaves mostly at or near the base, from a thickened taproot with a simple or branched crown or caudex; leaves 1 or more times pinnately or ternate-pinnately compound; umbel compound; involucre usually lacking; involucel of several separate or basally united bractlets; calyx teeth present, ovate; petals and stamens white or yellow; stylopodium lacking; carpophore entire to deeply cleft; fruit ovoid to linear oblong, somewhat laterally compressed, evidently ribbed.

***Musineon lineare* (Rydb.) Mathias** Rydberg Musineon. [*Aletes tenuifolia* Coult. & Rose, type from Cache County; *Daucophyllum lineare* Rydb., type from near Logan]. Plants 5.5–25 cm tall, caulescent or subcaulescent, glabrous, from a mostly branched caudex, more or less clothed at the base with persistent leaf bases; leaves mostly on the lower 1/3 of the plants, ternate or more often pinnate, with 2–4 opposite pair of lateral leaflets; petioles 0.5–6 (14) cm long; blades 1–5.3 cm long; leaflets, sessile, entire or bifid, trifid or rarely pinnatifid, the ultimate leaflets or lobes 3–20 mm long; peduncles 5–22 cm long, slender; umbel solitary; rays 5–10, 1–5 mm long; involucels of ca 3 linear or narrowly elliptic bractlets 4–10 mm long; pedicels ca 1 mm long; calyx teeth ca 0.5 mm long, greenish or purplish with scarious margins; petals and stamens white; styles ca 1 mm long; fruit 2–4 mm long, minutely scabrous, the ribs evident but not winged. Limestone cliffs in the Bear River Range, Cache County; endemic; 4 (0).

Oreoxis Raf.

Caespitose, acaulescent herbs from branched woody caudices, these usually clothed with marcescent leaf bases; leaves pinnate or bipinnate; umbels compound; involucre mostly lacking; bractlets of the involucel more or less united at the base, usually exceeding the flowers; calyx teeth conspicuous; petals and stamens yellow at least when fresh; stylopodium lacking; carpophore lacking; fruit oblong to ovoid-oblong, slightly compressed laterally, the ribs corky-winged. Plants of the genus could reasonably be included in *Cymopterus* and with the recent discovery of the low elevation *O. trotteri* such inclusion might be necessary.

1. Bractlets obovate, toothed at the apex, usually purplish, plants of the La Sal Mts. .... *O. bakeri*
— Bractlets linear or narrowly elliptic, entire, acute to acuminate; plants of wide or other distribution 2

2(1). Plants pulvinate-caespitose, forming clumps to 30 cm wide, from low elevations in Grand County; caudex clothed with a thatch of terete leaf bases; ultimate segments elliptic to cuneate-ovate .... *O. trotteri*
— Plants caespitose but hardly pulvinate, from high elevations, widespread; caudex clothed with short, more or less flattened leaf bases; ultimate segments of leaves linear to linear-elliptic .... *O. alpina*

***Oreoxis alpina* (Gray) Coult. & Rose** Alpine Oreoxis. [*Cymopterus alpinus* Gray]. Plants 2.5–11.5 cm tall, scabrous-hirtellous throughout, from a branched caudex, the caudex clothed with persistent leaf bases; leaves all basal, mostly bipinnate, with ca 4 opposite pair of sessile or nearly sessile lateral primary leaflets, the upper ones and those of smaller leaves sometimes once pinnate and then trifid or pinnatifid; petioles 0.5–2.5 cm long; blades 1–3.5 cm long oblong in outline, the lowest pair of primary leaflets 4–14 mm long, the ultimate segments 1–6 mm long, 0.4–1.5 mm wide, linear to narrowly elliptic; peduncles 2–10.5 cm long; umbel solitary; involucre lacking; rays 4–7, 1–6 mm long; involucels of 5–9 bractlets 1–4 mm long, united at the base; pedicels obsolete or to ca 0.3 mm long; calyx teeth 0.6–1 mm long, green; petals and stamens yellow when fresh, fading to white or cream or purple tinged in age; styles 1.7–2 (3) mm long; fruit 4–5 mm long, the ribs with low corky wings to ca 0.7 mm wide; 2n = 60. Forb-grass, limber pine, spruce, and alpine communities, and raw escarpments and barren ridges at 2440 to 3475 m in Duchesne, Garfield, Grand, San Juan, Sanpete, Summit, and Wayne counties; Wyoming to New Mexico and Arizona; 27 (vi).

Oreoxis bakeri Coult. & Rose Plants 1–12 cm tall, slightly puberulent at base of umbels and rays; leaves basal, bipinnate for the most part or pinnate with pinnatifid or trifid leaflets, with 3–4 opposite pair of lateral primary leaflets, the petioles 0.8–2.5 cm long; blades 0.8–5 cm long, the lowest pair of primary leaflets to ca 1 cm long, sessile or nearly so, the ultimate segments to 7 mm long, to 1 mm wide; peduncles 1–11 cm long; umbels solitary, involucre lacking; rays 3–8, 3–5 mm long; bractlets of the involucel united at base, 3–5 mm long, nearly linear-elliptic to obovate, usually 3-toothed at the apex; petals and stamens yellow at least when fresh; styles to ca 1 mm long; fruit 2–4 mm long, the ribs with low corky wings to 0.75 mm wide. Alpine forb-grass communities at ca 3660 m in the La Sal Mts., Grand and San Juan counties; Colorado and New Mexico; 4 (0).

Oreoxis trotteri Welsh & Goodrich Plants pulvinate-caespitose, forming clumps to 30 cm wide, 4–8 cm tall, scabrous and more or less glandular, from a branching caudex, this clothed with a thatch of persistent, terete leaf bases and peduncles; leaves all basal, bipinnate, with ca 4 opposite pair of sessile, lateral, primary leaflets, the upper ones and those of the smaller leaves sometimes once-pinnate and then trifid or pinnatifid; petioles 1–3.5 cm long; blades 1.5–2.3 cm long, oblong in outline, the lowest pair of primary leaflets 3.5–5 mm long, the ultimate segments 1–3.5 mm long, 1–3 mm wide, elliptic to cuneate-ovate; peduncles 4–7.5 cm long; umbel solitary; involucre lacking; rays 5–7, 3–5 mm long; involucels of 4–7 linear-subulate bractlets 2–3.5 mm long, distinct or essentially so; pedicels obsolete or to ca 1 mm long; calyx teeth ca 1 mm long, green or purplish; petals and stamens yellow; styles 1–1.2 mm long; fruit 2.8–4.8 (5) mm long, the ribs with low, corky wings to 0.7 mm wide. Mixed juniper and warm desert shrub community at ca 1464 m in Grand (type from NW of Moab) County; endemic; 3 (i).

Orogenia Wats.

Perennial acaulescent glabrous low herbs from a fusiform or globose root; leaves ternate or biternate with linear entire leaflets; umbel compound; involucre lacking or of a few linear minute scarious bractlets; calyx teeth obsolete; petals and stamens white or purplish; stylopodium lacking; carpophore lacking; fruit oblong to oval, nearly round in cross section, the dorsal ribs evident or obsolete, the lateral ones corky winged but inflexed into the commissure, a corky riblike projection also running the length of the commissural faces of each mericarp.

Orogenia linearifolia Wats. Indian Potato. Plants 5–10 (13) cm tall, glabrous, not aromatic, from a globose or fusiform root, with a fragile etiolated subterranean scapose stem easily detached from the tuberous root; leaves borne at ground level or a few arising from the tuberous root with etiolated petioles, ternate or biternate blades 3–8 (12.5) cm long, the 3–9 leaflets 1.5–5 (11.5) cm long, 1–11 mm wide, linear, entire, the lowest pair of petiolules to 2 cm long; peduncles 3–8 cm long, usually a little longer than the subterranean stem; involucre lacking; rays 3–12, but rarely more than 5 of them fertile, 0.3–3 cm long; involucel proper apparently lacking but some of the pedicels usually bearing a linear bractlet to 4 mm long; pedicels nearly obsolete or to 2 mm long; petals white; filaments white, anthers pale or dark purple; styles ca 1 mm long; fruit 4–6 mm long; dorsal ribs filiform. Sagebrush-grass, mountain brush, aspen, ponderosa pine, white fir, and rarely desert shrub communities, mostly flowering at the edge of melting snow at 1370 to 2805 m in Beaver, Box Elder, Juab, Millard, Morgan, Salt Lake, San Juan, Sanpete, Sevier, Summit (type from Parleys Park), Tooele, Uintah, Utah, Wasatch, Washington, and Weber counties; Washington to Montana, south to Colorado; 50 (viii).

Osmorhiza Raf.

Perennial caulescent usually pubescent herbs from taproots with simple or branched crowns; leaves ternately or pinnately 1–3 times compound with well marked leaflets; umbels compound; involucre lacking or of 1 or a few narrow foliaceous bracts; involucel lacking or of several foliaceous reflexed bractlets; calyx teeth obsolete; petals and stamens white, greenish white, yellow, pink, or purple; stylopodium, conic to depressed; carpophore bifid less than 1/2 its length; fruit linear or clavate, somewhat compressed laterally, bristly hispid to glabrous, the ribs narrow.

1. Ovaries and fruit glabrous, generally obtuse at both ends; petals and stamens yellow or greenish yellow; leaves (1) 2 times pinnately or ternate-pinnately compound; plants strongly aromatic, usually with more than 2 stems .... *O. occidentalis*
— Ovaries and fruit bristly hispid, with long, pointed bristly hispid tails; petals white or greenish white; leaves biternate; plants not strongly aromatic, often with solitary stems 2

2(1). Mature fruit including tails mostly 16–25 mm long, the apex concavely pointed into a beak 1–2 mm long; fruiting pedicels mostly ascending-spreading .... *O. chilensis*
— Mature fruit including tails mostly 13–18 mm long, the apex convex and obtuse; fruiting pedicels horizontally spreading to ascending .... *O. depauperata*

Osmorhiza chilensis H. & A. [*O. nuda* Torr.]. Stems often solitary, 18–75 cm tall, from a taproot, without marcescent leaf bases; herbage not strongly aromatic; leaves basal and 2–3 cauline, biternate, usually with 9 distinct leaflets; petioles 3–16 cm long or cauline leaves sessile; blades 5–15 cm long, the lateral primary leaflets nearly as long as the central one, with petiolules (1) 2–5.5 cm long; blades of leaflets 1–4 (5) cm long, elliptic to ovate, lobed to cleft, and toothed, ciliate and often pubescent on nerves below and sometimes scattered pubescent between the nerves; peduncles 5–34 cm long; umbels 1–5; involucre lacking; rays 3–7, 2.5–9 (13) cm long, ascending or spreading-ascending, glabrous to hirtellous; involucels lacking; pedicels 5–22 (30) mm long, ascending; petals and stamens greenish white; styles less than 0.5 mm long; fruit including the tails 16–25 mm long, linear-clavate, bristly hispid, the beak concavely pointed, 1–2 mm long, the concave beak usually evident in young fruits; n = 11. Mountain brush, aspen, Douglas fir, white fir, narrowleaf cottonwood, and riparian communities at 1520 to 2680 m in Box Elder, Cache, Daggett, Davis, Duchesne, Juab, Millard, Salt Lake, Sanpete, Tooele, Uintah, Utah, Wasatch, Washington, and Weber counties; Alaska to California, east to Alberta and Arizona, also in Great Lakes region; Argentina and Chile; 54 (viii).

Osmorhiza depauperata Phil. Blunt-fruit Sweet-cicely. [*O. obtusa* (Coult. & Rose) Fern.]. Stems mostly solitary, 14–63 (77) cm tall, often with a slight ring of hairs at the nodes, from a taproot, without persisting leaf-bases; herbage not strongly aromatic; leaves basal and 1–3 cauline, biternate, usually with 9 distinct leaflets, or the upper cauline ones once-ternate; petioles (1) 3–17 cm long, often with dilated, ciliate bases; blades (2) 4–11 cm long, the lateral primary leaflets almost equal to the central one or a little shorter, with petiolules (0.5) 1–4 cm long, blades of leaflets 1–4 (5.5) cm long, elliptic to ovate, lobed to cleft and toothed, ciliate and often pubescent on nerves below and sometimes scattered pubescent between them; peduncles 3.5–15 (22.5) cm long; umbels 3–6; involucre lacking, or rarely of a solitary bract to 12 mm long; rays 3–5, 1.5–8.5 cm long, spreading to divaricate; involucels lacking or infrequently of 1 or 2 separate ciliolate bractlets to 3 mm long; pedicels 5–20 mm long, spreading to divaricate; petals greenish white; styles ca 0.2 mm long; fruit including the tails (11) 13–18 mm long, linear-clavate, the beak convex-obtuse; n = 11. Mountain brush, aspen, ponderosa pine, Douglas fir, lodgepole pine, spruce-fir, riparian, and rarely pinyon-juniper and sagebrush communities at 1980 to 3200 m in all Utah counties, except Cache, Davis, Emery, Morgan, Piute, Rich, and Wayne; Alaska to California, east to South Dakota and New Mexico; also in the Great Lakes Region; Chile and Argentina; 110 (xv).

Osmorhiza occidentalis (Nutt.) Torr. Western Sweet-cicely. [*Glycosma occidentalis* Nutt. ex T. & G.; *G. maxima* Rydb., type from Mt. Nebo]. Plants 6–13 dm tall, from a taproot, with few or no persistent leaf bases, strongly aromatic; leaves (1) 2 times pinnate or the upper cauline ones ternate pinnately compound, with 3–4 pairs of opposite lateral primary leaflets; petioles of lower leaves 4–30 cm long or longer, the upper ones reduced; lower blades to 25 cm long or longer, the upper ones much reduced, the lowest pair of primary leaflets usually again pinnate, usually over 1/2 as long as the leaf blade, with petiolules 1–3.5 cm long, the ultimate leaflets 1–9 cm long 0.5–4 cm wide, lanceolate to lance-elliptic or ovate coarsely toothed and some often lobed; peduncles 6–20 cm long; umbels 3–5; involucre lacking or occasionally of 1–2 linear or filiform bracts to 16 mm long; rays 7–13, 2–6.5 cm long; involucels lacking; pedicels 2–7 mm long; calyx obsolete; petals greenish white or greenish yellow, 1–2 mm long; stylopodium low; styles 0.7–1 mm long; carpophore divided to the base; fruit 16–20 mm long, 2–3 mm wide, linear, glabrous. Tall forb, aspen, oak-maple, spruce-fir, riparian, and infrequently in sagebrush communities at 1765 to 3170 m in Box Elder, Cache, Carbon, Duchesne, Iron, Juab, Millard, Morgan, Salt Lake, San Juan, Sanpete, Sevier, Summit, Tooele, Utah, Wasatch, Washington, and Weber counties; British Columbia and Alberta south to California and Colorado; 67 (xvi).

Oxypolis Raf.

Perennial, caulescent, glabrous herbs from fascicled tuberous roots; leaves pinnate; umbels compound; involucre and involucel lacking; rays ascending; calyx teeth conspicuous; petals white to purple; stylopodium conic; carpophore divided to the base; fruit oblong to oval, strongly flattened dorsally, with dorsal ribs filiform and lateral ribs broadly winged.

Oxypolis fendleri (Gray) Heller [*Archemora fendleri* Gray]. Plants 6–8 dm tall, without persistent leaf bases; leaves pinnate with 2–5 pairs of opposite lateral leaflets, the upper ones sometimes reduced to bladeless or nearly bladeless sheaths, the petioles (3) 5–15 cm long or the upper blades sessile on a dilated sheath; blades 7–17 cm long, oblong in outline, the leaflets sessile, 2–5 cm long, ovate to orbicular, shallowly to deeply crenate-dentate or serrate or rarely incised, or those of the upper leaves lanceolate to linear and sometimes entire; peduncles (1) 4–20 cm long; umbels usually 4 or more per stem; involucre lacking; rays 5–14, 1–5 (7) cm long, ascending; involucels lacking; pedicels 3–10 mm long; petals and stamens white; styles mostly less than 1 mm long; fruit 3–5 mm long. Streambanks on the Abajo and La Sal mts. in San Juan County; Wyoming south to New Mexico; 4 (0).

Pastinaca L.

Biennial or perennial caulescent herbs from large taproots; leaves pinnately compound, with broad toothed to pinnatifid leaflets; umbels compound; involucre and involucel usually lacking; calyx teeth obsolete; petals yellow or red; stylopodium depressed-conic; carpophore divided to the base; fruit elliptic to obovate, strongly flattened dorsally, the dorsal ribs filiform, the lateral ones narrowly winged.

Pastinaca sativa L. Parsnip. Biennial caulescent aromatic herbs 8–15 dm tall, from a taproot; leaves pinnate or partly bipinnate in some of the lower leaflets, with 3–6 opposite or offset pair of lateral leaflets; petioles 3–15 mm long or lacking and the blade sessile on a dilated sheath; blades 12–35 cm long or longer, oblong in outline; leaflets sessile and sometimes confluent or the lower ones sometimes on petiolules to 1.7 cm long, the blades 2.5–12 cm long, lanceolate to ovate, coarsely serrate, and often lobed; umbels 6–15 or more, the terminal one sessile or pedunculate but shorter than the 2 immediately lateral ones, the lateral umbels alternate or opposite or on opposite branches supporting 2 or more umbels; involucre lacking or of 1 to few linear entire or occasionally toothed or lobed bracts to 2 (4) cm long; rays 9–25, 0.8–8.5 cm long; involucels lacking or infrequently of 1–2 linear bractlets to 2 mm long; pedicels 4–20 mm long; petals greenish yellow or reddish; styles 0.4–1 mm long; fruit 5–8 mm long, 3–6 mm wide, broadly elliptic to orbicular or obovate, strongly flattened dorsally the dorsal ribs filiform and the lateral ones slightly winged. Ditchbanks, roadsides, fencelines, gardens, fields, margins of ponds and lakes, and moist flood plains at 1370 to 2365 m, cultivated in all counties of the state, escaping and persisting, introduced from Europe, now widely established in North America; 23 (x). The cultivated plants, ssp. ***sativa***, differ from the wild plants ssp. ***sylvestris*** (Miller) Rouy & Camus, in having larger roots. Some of the wild plants might be recent escapes from cultivation.

Perideridia Reichenb.

Perennial, caulescent herbs from a fusiform or tuberous root, these often deep-seated and easily detached from the rather fragile etiolated subterranean portion of the stem, and often lacking in herbarium specimens; leaves ternate, pinnate, or ternate-pinnately compound, the upper ones sometimes reduced to a simple, linear rachis; petioles sheathing; umbels compound; involucre lacking or of mostly few more or less scarious bracts;

involucel lacking or of 1 to few bractlets; calyx teeth inconspicuous, of the texture and color of the petals; petals white; stamen white; stylopodium conic or low conic; carpophore divided to the base; fruit linear-oblong to orbicular, scarcely compressed or lightly so at right angles to the commissure, with filiform ribs.

1. Bractlets of the involucel scarious, as wide or to 5 times as wide as the pedicels, 3–5 mm long; longest rays rarely over 2 cm long; lower leaves with ultimate divisions 10–50 or more .... *P. bolanderi*
— Bractlets of the involucel not scarious or with narrow scarious margins, only about as wide as the pedicels, to 3 mm long; longest rays 2–3 (4) mm long; leaves commonly with 3–5 leaflets .... *P. gairdneri*

Perideridia bolanderi (Gray) **Nels. & Macbr.** Yampah. [*Podosciadium bolanderi* Gray]. Plants 23–40 cm tall, glabrous, without marcescent leaf bases; leaves often crowded on the lower part of the stem, ternate-pinnately 2 or more times compound, with petiolulate primary leaflets, the upper ones reduced and sometimes simple and linear, often withered before or shortly after anthesis; petioles to 4 cm long or lacking and the petiolules arising directly from a dilated sheath; blades 4–12 cm long, the ultimate leaflets strongly dimorphic, 0.2–8 cm long, mostly 10–50 or more per leaf on the lower leaves; peduncles (2) 5–14 cm long; umbels 2–6 per stem; involucre lacking or usually of 1–4 scarious bracts to 5 mm long; rays 4–12, 1–2 cm long; bractlets of the involucels 4–8, 3–5 mm long, to 2.5 mm wide, linear to ovate and often caudate, with pale yellow green midrib, this often flanked on either side by purple and then by conspicuous scarious margins; pedicels 3–6 mm long; petals white; styles 1–2 mm long, spreading to recurved; fruit 3–4 (5) mm long, some of the ribs usually conspicuously ridged; n = 19. Sagebrush, juniper, mountain brush, and stream side communities, sometimes in snow flush areas at 1524 to 2320 m in western Box Elder County and Deep Creek Mts., Juab County; Oregon and Idaho south to California; 12 (iii). This plant was an important source of food for Indians and early explorers.

Perideridia gairdneri (H. & A.) **Mathias** False Yarrow. [*Atenia gairdneri* H. & A.; *Carum garrettii* A. Nels. in Coult. & Rose, type from the Wasatch Mts.]. Plants 1.5–7.5 dm tall, glabrous, without marcescent leaf bases; leaves 2–5 per stem, ternate or pinnate, with 3–5, or rarely more, sessile leaflets, the upper ones reduced and often simple and linear; leaflets to 13 cm long, mostly confluent with the rachis, linear and hardly wider than the petiole, occasionally expanded and to 11 mm wide; peduncles (1) 2–5 (7) cm long; umbels 2–5 per stem; involucre lacking or occasionally of 1 or 2 linear bracts to 6 mm long; rays 7–16, 0.7–4 cm long; bractlets of the involucels lacking or more often 1–6, 1–3 mm long, linear or linear-subulate, hardly if at all wider and conspicuously shorter than the pedicels, not marked with purple or, if so, the whole bractlet mostly purple; pedicels 3–5 mm long; petals white or turning purplish; styles to 1 mm long, recurved; fruit 2–3 mm long, orbicular, the ribs obscure; n = 19. Sagebrush, forb-grass-sagebrush, meadow, oak, maple, aspen, and willow communities at 1680 to 2685 m in Box Elder, Cache, Daggett, Juab, Salt Lake, Sanpete, Summit, Utah, and Wasatch counties; British Columbia to California, east to Saskatchewan and New Mexico; 25(v). Our plants belong to ssp. *borealis* Chuang & Const.

Podistera Wats.

Perennial acaulescent glabrous plants from taproots or branched caudices; leaves pinnate with deeply lobed leaflets; umbel solitary, compound, compact; involucre wanting; involucel of toothed bractlets; calyx teeth conspicuous, ovate; petals greenish yellow; stylopodium conic; carpophore stout, undivided; fruit oval, slightly flattened laterally, the ribs filiform to prominent.

Podistera eastwoodiae (Coult. & Rose) **Mathias & Const.** [*Ligusticum eastwoodae* Coult. & Rose]. Plants 7–20 (30) cm tall, acaulescent, without or with few marcescent leaf bases; leaves pinnate, with 4–6 pairs of sessile lateral leaflets; petioles 1.5–7 cm long; blades 2.5–7.5 cm long, oblong in outline; leaflets 1–2 cm long, ovate to obovate in outline, ternately or palmately lobed or cleft, the larger lobes again toothed or lobed; peduncles (7) 10–20 (30) cm tall; involucre lacking; rays 5–8, 2–8 mm long; bractlets of the involucel 4–6 mm long, often exceeding the flowers and fruit, ovate or obovate, with 2–3 teeth or lobes, with the texture and color of the leaves; pedicels 1–2 mm long; petals greenish yellow, turning purple; styles ca 1 mm long; fruit 3–4 mm long, the ribs evident but not winged. Apparently rare at upper elevations of the La Sal Mts., Grand and San Juan counties; Colorado and New Mexico; 1 (0).

Sium L.

Perennial, caulescent herbs from fascicles of fibrous roots; leaves mostly pinnately compound or decompound, with well marked, toothed to pinnatifid leaflets; umbels compound; involucre of entire or incised, often reflexed bracts; involucel of narrow bractlets; calyx teeth minute or obsolete; petals white; stylopodium depressed or rarely conic; carpophore divided to the base (but threadlike and adnate to the faces of the mericarps); fruit elliptic to orbicular, slightly compressed laterally and somewhat constricted at the commissure, the subequal ribs prominent and corky but hardly winged.

Sium suave Walter Hemlock Water-parsnip. Plants 5–10 dm tall; leaves pinnate or occasionally partly bipinnate, with 4–6 opposite pairs of sessile lateral leaflets, the lower petioles to 25 cm long, often septate, the upper ones smaller and sometimes reduced to a dilated sheath; lower blades 14–32 cm long, the upper ones reduced; leaflets 2–8 (15) cm long, (1) 3–8 (20) mm wide, linear to lanceolate, sharply and uniformly serrate to pinnatifid with linear segments; peduncles 4–10 cm long; umbels 3–11 or more per stem; involucre of 1–6 separate, often reflexed bracts 2–9 mm long; rays 11–24, 1.5–3 cm long; involucels of (2) 5–12 separate bractlets 2–5 mm long; pedicels 2–8 mm long; petals and stamens white; styles ca 1 mm long; fruit 2–3 mm long, the ribs prominent; 2n = 6, 12, 22. Mud flats, marshlands, wet meadows, along streams and shorelines, and in ponds and lakes at 1365 to 2990 m in Garfield, Piute, Rich, Salt Lake, Sanpete, Sevier, Utah, and Wayne counties; British Columbia to Newfoundland, south to California and Virginia; 15 (i). This species is often mistaken for *Cicuta*, and is frequently found with that genus in herbaria, but differs conspicuously in the merely pinnate leaves.

Torilis Adanson

Annual caulescent hispid or pubescent herbs from slender taproots; leaves 1–2 times pinnate or pinnately decompound; petioles sheathing; umbels compound, capitate or open, sessile or pedunculate; involucre lacking or of a few small bracts; involucel of several linear or filiform bractlets; calyx teeth evident to obsolete; petals white; stylopodium thick, conic; carpophore bifid or cleft ca 1/3–1/2 its length; fruit ovoid or oblong, flattened laterally, tuberculate or prickly, the primary ribs filiform, setulose, the lateral ribs displaced onto the commissural surface, the intervals covered with glochidiate prickles or tubercles.

Torilis arvensis **(Hudson) Link** Hedge Parsley. [*Caucalis arvensis* Hudson]. Plants 3–10 dm tall, divaricately branched, appressed-hispid throughout, retrorsely so on the stems and antrorsely so on the leaves and rays; leaves 2–3 times pinnate, or the upper ones once-pinnate, the ultimate leaflets 5–60 mm long, 2–20 mm wide, ovate to linear lanceolate, acute or acuminate, regularly incised or divided; peduncles 2–12 cm long; involucre lacking or of a single small bract; rays 2–10, 0.5–2.5 cm long; involucel of several subulate bractlets longer than the pedicels; pedicels 1–4 mm long; petals white; styles short; fruit ovoid-oblong, 3–5 mm long, the mericarps densely covered with straight glochidiate prickles with minute retrorse barbs, these spreading almost at right angles and about as long as the fruit is wide; 2n = 12. The species was taken from an orchard (Barnum 1316 BRY) in Washington County; adventive; introduced from southern and central Europe; 4 (0). This plant was observed growing abundantly in an orchard in La Verkin in 1983.

Yabea Kozo-Polj.

Annual caulescent herbs from taproots; leaves pinnate or dissected; umbels compound; involucre of a few entire or dissected usually somewhat scarious bractlets; calyx teeth evident; petals white; stylopodium thick and conic; carpophore entire or bifid at the apex; fruit oblong or ovoid, somewhat compressed laterally, with spreading uncinate prickles along alternating ribs and bristly-hairy on the other ribs.

Yabea microcarpa **(H. & A.) Kozo-Polj.** California Hedge-parsley. [*Caucalis microcarpa* H. & A.]. Plants annual, caulescent, 8–40 cm tall, pubescent with spreading hispid hairs, from a slender taproot; leaves 2–3 (4) times pinnate or ternate-pinnate, with ca 3–4 opposite pairs of lateral primary leaflets; blades 1–5 cm long, oblong or ovate in outline, on petioles 1–4.5 cm long or the upper ones sessile, the lowest pair of primary leaflets almost 1/2 as long as the leaf blade, sessile or petiolulate, the ultimate segments 1–8 mm long, 0.5–2 mm wide; peduncles 3–10 cm long; umbels 1–4; involucre resembling the upper leaves or a little smaller; rays (1) 2–7 (9), 1.5–10 cm long, often about as long as the peduncles; involucels similar to the involucre, but usually reduced to much reduced and the bractlets merely pinnatifid or entire; pedicels 5–15 cm long; petals white; stamens white; styles very short; carpophore bifid for ca 1/5 its length; fruit 3–7 mm long. The one specimen seen (Atwood 4871 BRY) is from the Pine Valley Mts., Washington County; British Columbia south to Baja California, east to Idaho and Arizona; 1 (0).

Zizia Koch

Perennial glabrous or subglabrous herbs with basal and cauline leaves, from a short caudex and a cluster of fleshy-fibrous roots; leaves simple or ternate, with toothed blades or leaflets; umbels compound; involucre lacking or obsolete; involucel of a few inconspicuous bractlets; calyx teeth well developed; petals bright yellow; stylopodium lacking; carpophore bifid about half its length; fruit oblong or broadly elliptic, somewhat laterally compressed, the ribs prominent but not winged.

Zizia aptera **(Gray) Fern.** [*Thaspium trifoliatum* var. *apterum* Gray]. Perennial, caulescent, glabrous herbs 15–50 cm tall, from a taproot or fascicle of roots, without marcescent leaf-bases; basal leaves simple, rarely ternate; petioles 3–18 cm long; blades 1.5–5 cm long, ovate to nearly orbicular, cordate, crenate-serrate; cauline leaves ternate, not over 3 cm long, the leaflets sessile or on petiolules to 4 mm long; peduncles 6–12 cm long; umbels 1 or 2 per stem; involucre lacking or obsolete; rays 10–17, 0.5–2 cm long; involucels of 4–6 bractlets, to ca 2 mm long, separate or united at the base; pedicels 1–3 mm long; petals yellow; stamens yellow; styles ca 1 mm long; fruit ca 2 mm long, the ribs prominent. Willow-streamside and meadow communities at 2130 to 2440 m in Sanpete, Sevier, Summit, Utah, and Wasatch counties; widespread in the U. S. and Canada; 12 (iii).

URTICACEAE A. L. Juss.
Nettle Family

Annual or perennial herbs; leaves opposite or alternate, simple, with or without stinging hairs; plants monoecious or dioecious; flowers hypogynous, imperfect, inconspicuous, arranged in spicate cymes or small cymose clusters; staminate flowers with 3–6 sepals and 3–6 stamens; pistillate flowers with 4 or 5 sepals or the perianth lacking; pistil 1, the ovary superior, 1-loculed; style 1; stigma 1; fruit an achene; x = 6–14.

1. Leaves alternate, entire; plants lacking stinging hairs .. .. *Parietaria*
— Leaves opposite, toothed; plants with stinging hairs . *Urtica*

Parietaria L.

Annual, monoecious or perfect herbs, lacking stinging hairs but often otherwise pubescent; leaves simple, alternate, entire, estipulate; flowers borne in bracteate axillary clusters, the perianth 4-lobed; stamens 4; ovary superior; stigma tufted; fruit an ovoid achene.

Parietaria pensylvanica **Muhl. ex Willd.** Hammerwort. Plants sprawling to erect, simple or branched, 4–50 cm tall; leaves petiolate, the blades 4–50 mm long or more, 3–25 mm wide, lanceolate to ovate or elliptic, strigulose, hirsute (the hairs often with pustular bases), or glabrous, ciliate; flowers in bracteate clusters; perianth 1.5–2.2 mm long, more or less cylindric-campanulate, green to brownish; achenes smooth, shining, tan, 0.9–1.3 mm long; 2n = 16. Warm desert shrub, hanging garden, and ruderal habitats at 850 to 1740 m in Box Elder, Cache, Daggett, Kane, Millard, Salt Lake, Uintah, and Washington counties; British Columbia to the Atlantic, south to California, Mexico, and Alabama; 18 (ii). Use of relative length of bracts versus calyx, absolute length of calyx, and degree of harshness of pubescence are not apparently reliable for segregation of our specimens into the proposed species identified from Utah. Our plants have passed under the names *P. floridana* Nutt., *P. occidentalis* Rydb., and *P. obtusa* Rydb., and some of those might be valid elsewhere, but our materials belong to a single variable taxon, fitting into a continuous graded series.

Urtica L.

Perennial or annual, monoecious to dioecious herbs armed with stinging hairs (and otherwise pubescent); leaves simple, opposite, toothed, stipulate; flowers in axillary spicate cymes, the perianth 4–lobed; staminate flowers with 4 stamens and a rudimentary pistil; ovary superior, the stigmas tufted; fruit a lenticular achene.

Fernald, M. L. 1926. Urtica gracilis and some related North American species. Rhodora 28: 191–99.
Woodland, D. W. 1982. Biosystematics of the perennial North American taxa of Urtica. II. Taxonomy. Syst. Bot. 7: 282–290.

1. Plants rhizomatous perennials; stipules more than 5 mm long; plants widespread .... *U. dioica*
— Plants taprooted annuals; stipules less than 5 mm long; plants known from nearby in Nevada .... *U. urens* L.

Urtica dioica L. Stinging Nettle. Perennial rhizomatous herbs; stems 6–20 dm tall, bearing stinging hairs and otherwise hairy to subglabrous; leaves 4–18 cm long, the petioles 1–6 cm long, the blades 3–15 cm long and 1–8 cm wide, linear-lanceolate or lanceolate to ovate, coarsely serrate, acute apically, the base cordate to truncate or acute; stipules 5–15 cm long; flowers inconspicuous, the perianth 1–2 mm long, greenish; 2n = 26. Two geographically correlated varieties are present in Utah.

1. Lower leaf surface glabrous or essentially so, except for occasional stinging hairs; stems not obscured by pubescence; plants of eastern Utah
.... *U. dioica* var. *procera*
— Lower leaf surface more or less densely hirtellous; stems usually obscured by pubescence; plants mainly of the western 2/3 of Utah .... *U. dioica* var. *occidentalis*

Var. occidentalis Wats. [*U. holosericea* Nutt.; *U. dioica* var. *holosericea* (Nutt.) C. L. Hitchc.; *U. dioica* ssp. *holosericea* (Nutt.) Thorne; *U. brewerii* Wats.]. Riparian and palustrine habitats but also in xeric sites in mountain brush, sagebrush, aspen and Douglas fir communities at 1370 to 2595 m in Cache, Carbon, Daggett, Davis, Duchesne, Garfield, Juab, Millard, Piute, Salt Lake, Summit, Tooele, Uintah, Utah, Wasatch, Washington, Wayne, and Weber counties; Washington to Montana, south to Mexico and Colorado; 43 (iii). A specimen from Utah County (Allred 450 BRY) is intermediate with the following in pubescence features.

Var. procera (Muhl.) Wedd. [*U. procera* Muhl. ex Willd.; *U. gracilis* Ait.; *U. dioica* var. *gracilis* (Ait.) Taylor & MacBryde; *U. dioica* ssp. *gracilis* (Ait.) Selander; *U. lyallii* Wats.; *U. dioica* var. *lyallii* (Wats.) C. L. Hitchc.]. Mountain brush, sagebrush, aspen, and spruce communities (often in moist sites) at 2135 to 3050 m in Grand, San Juan, and Uintah counties; widely distributed in North America; 7 (i).

VALERIANACEAE Batsch
Valerian Family

Annual or perennial herbs; leaves opposite, estipulate; flowers epigynous, small, perfect or unisexual, borne in cymes or heads; calyx a ring, reduced, or lacking, variously toothed, often inrolled in flower and forming a leathery pappus in fruit; corolla funnelform to rotate or nearly salverform, 4– or 5–lobed, the base saccate or spurred; stamens 1–4, inserted near the base of the corolla tube; pistil 1, the ovary superior, mostly 3–loculed, with 2 of the locules sterile; ovule 1, pendulous; fruit dry, achenelike, indehiscent, sometimes plumose or winged; x = 7–12.

1. Stamen 1; corolla with a long spur, usually magenta or red, rarely white, more than 1 cm long; plants perennial, cultivated and persisting .... *Centranthus*
— Stamens 3; corolla gibbous or spurred at the base, white, pink, or bluish to purple, less than 1 cm long 2
2(1). Calyx limb of plumose bristles in fruit; plants perennial, mostly from rhizomes or taproots .... *Valeriana*
— Calyx limb lacking; plants annual, from taproots 3
3(2). Flowers in cymose clusters, forming a more or less flat-topped inflorescence; stems dichotomously branched; plants annual, introduced .... *Valerianella*
— Flowers densely clustered in capitate or interrupted spikelike inflorescences; plants annual, indigenous ...
.... *Plectritis*

Centranthus DC.

Glaucous annual or perennial herbs; stems erect; leaves entire, dentate, or pinnatifid; flowers small, red to white; calyx dissected into 5–15 linear lobes, inrolled at anthesis, in fruit spreading and enlarging; corolla tube slender, spurred basally, the limb 5–lobed; stamen 1; fruit 1–loculed, narrow, crowned with the pappuslike calyx, 1–seeded.

Centranthus ruber (L.) DC. [*Valeriana rubra* L.]. Plants glabrous, glaucous, 3–7 dm tall; leaves sessile, 3–10 cm long, lance-ovate to ovate, entire; flowers fragrant, red to pink or white, densely clustered; corolla tube 10–13 mm long, the spur elongate, slender, the limb spreading; stamen exserted; fruit 3–4 mm long; 2n = 32. Cultivated ornamental, rarely escaping at 930 to 1533 m in Davis, Salt Lake, Utah, and Washington counties; widely grown in the U. S.; introduced from Europe; 5 (0).

Plectritis DC.

Annual herbs, glabrous or nearly so; stems erect, sparingly branched, with tufts of hair at the nodes; leaves entire or few-toothed, short-petiolate or sessile; flowers in capitate or interrupted spicate clusters; bracts linear-subulate; calyx obsolete; corolla small, bilabiate, 5–lobed, funnelform, spurred at base; stamens 3; stigma usually bilobed; ovary inferior, 1–loculed; fruit an achene.

Plectritis macrocera T. & G. Plants 1–6 dm tall; stems often glandular-pubescent above; leaves 1–6 cm long, 4–25 (40) mm wide, obovate to oblong or linear, the lower ones short-petiolate, becoming sessile upward; corolla 2–5 mm long, white to pale pink, irregular to nearly regular, with 5 short lobes, the spur thick, to 1.5 mm long; fruit 2–3 mm long, yellowish, with broad wings, glabrous or puberulent; 2n = 32. Sagebrush, pinyon-juniper, mountain brush, and ponderosa pine communities at 1533 to 2330 m in Box Elder, Davis, Millard, Salt Lake, Tooele, Utah, and Weber counties; British Columbia to Montana, south to California; 17 (ii).

Valeriana L.

Perennial, strongly scented herbs from rhizomes or caudices; stems leafy or subscapose; leaves opposite, entire to pinnatifid; flowers cymose or in branched clusters; calyx inconspicuous, involute, finally spreading, the limb sessile, hyaline, becoming setose, the setae plumose and pappus-like; corolla rotate to funnelform, the tube somewhat gibbous at the base, throat more or less hairy, the 5 lobes subequal; stamens 3; ovary inferior, basically 3-loculed, maturing 1 pendulous ovule; style 1; stigma 3-lobed; fruit an achene.

1. Leaves mostly ligulate-spatulate, gradually tapering to clasping bases, the cauline ones often pinnatifid and more or less decurrent; plants from vertical, usually forked taproots .... *V. edulis*
— Leaves mostly pinnatifid, generally petiolate; plants rhizomatous 2
2(1). Corolla 7-13 mm long, tubular-funnelform, the tube commonly longer than the limb; basal leaves broadly elliptic, ovate, or suborbicular, with bases rounded, subcordate, or short-cuneate .... *V. arizonica*
— Corolla 7 mm long or less, funnelform-campanulate; leaves various 3
3(2). Corolla tube short, the limb widely flaring (rotate or nearly so), 2-3.5 mm long; leaves mostly oblong in outline, the lateral segments of stem leaves commonly broadly lanceolate to elliptic, obtuse to acute ....
.... *V. occidentalis*
— Corolla funnelform, 4-6 mm long; leaves ovate to spatulate in outline, the lateral segments of cauline leaves usually narrowly lanceolate and acuminate .. *V. acutiloba*

Valeriana acutiloba Rydb. [*V. capitata* ssp. *acutiloba* (Rydb.) F. G. Meyer; *V. pubicarpa* Rydb., type from Mt. Nebo; *V. acutiloba* var. *pubicarpa* (Rydb.) Cronq.; *V. puberulenta* Rydb., type from near Marysvale; *V. utahensis* Gand., type from Utah]. Plants from a stout rhizome or caudex with fibrous roots; stems erect, 1-6 dm tall, sparingly spreading hairy; leaves mainly 3-8 cm long and 1.5-3.7 cm wide, petiolate, the basal ones mostly simple, obovate to oblong or ovate, entire; cauline leaves entire to pinnatifid; calyx segments 10-17, plumose; corolla white to pinkish, 4-6 mm long, funnelform, the lobes 1-2 mm long; stamens exserted; fruit 3.5-5 mm long, puberulent to glabrous. Mountain brush, aspen, spruce-fir, and alpine tundra communities at 2240 to 3920 m in Beaver, Box Elder, Cache, Carbon, Duchesne, Juab, Kane, Morgan, Piute, Salt Lake, Summit, Tooele, Utah, Wasatch, and Weber counties; Oregon to Montana, south to California, Arizona, and New Mexico; 49 (vi). Treatment of our material as a portion of an expanded *V. capitata* is not without merit.

Valeriana arizonica Gray Plants from slender rhizomes; stems mostly 20-30 cm tall, glabrous; leaves petiolate, mostly basal, subentire or sometimes pinnatifid, 4-17 cm long, 15-40 mm wide, ovate to suborbicular; calyx segments 10-12; corolla white or pinkish, 7-13 mm long, tubular-funnelform, the lobes 2-2.5 mm long; stamens and style exserted; fruit 2-5 mm long, glabrous. Moist canyon, west rim of Zion National Park, Washington County (Cottam 6975 BRY); Arizona to Texas and Mexico; 1 (0).

Valeriana edulis Nutt. ex T. & G. Plants from taproot and caudex, this clothed with black, marcescent leaf bases; stems erect, 1-10 dm tall, glabrous or nearly so; leaves thick, the basal ones 7-40 cm long, 7-50 mm wide, linear to obovate, entire or essentially so; cauline leaves 1-3 pairs, sessile or nearly so, usually pinnatifid, with 3-7 lanceolate to linear lobes; inflorescence elongate, open; calyx segments plumose, 9-13; corolla 1.5-3 mm long, rotate, yellowish to whitish; fruit 2.5-5 mm long, glabrous or pubescent; 2n = 64. Sagebrush, aspen, lodgepole pine, spruce-fir, and alpine tundra communities at 1500 to 3700 m in all Utah counties, except Davis, Juab, Millard, Morgan, Rich, Tooele, Washington, and Weber; British Columbia to Ohio, south to Mexico; 64 (xiii). This is perhaps the most ill-scented of the valerian species in Utah.

Valeriana occidentalis Heller Plants rhizomatous; stems erect, 3-8 dm tall, glabrous or puberulent; basal leaves mainly 3-25 cm long, petiolate, elliptic to lanceolate or spatulate, entire or with 1 or 2 pairs of lateral lobes; cauline leaves sessile or short-petiolate, ovate to lanceolate, pinnatifid, with 3-9 lobes, these lanceolate to elliptic; calyx segments 9-15, plumose; corolla 2-3.5 mm long, rotate or campanulate, whitish, the lobes subequal to the tube; fruit 4-5 mm long, pubescent to glabrous. Sagebrush, mountain brush, aspen, spruce-fir, sedgegrass, and sedge- or grass-forb, and alpine tundra communities at 1830 to 3720 m in all Utah counties, except Kane, Morgan, Rich, and Washington; Oregon to North Dakota, south to California, Arizona, and Colorado; 83 (xi).

Valerianella Miller

Annual or biennial herbs; stems erect, branched above; leaves somewhat succulent, opposite, mostly entire; flowers small, in cymose clusters, whitish or pale bluish, bracteate; calyx limb obsolete or toothed; corolla funnelform or salverform, often gibbous or minutely spurred, the limb equally 5-lobed; stamens 3; ovary 3-loculed; stigma 3-lobed; fruit 3-loculed, but 2 of them sterile.

Valerianella locusta (**L.**) **Betcke** [*Valeriana locusta* L.]. Stems dichotomously branched above, mainly 1-3.5 dm tall, pubescent; lower leaves petiolate, spatulate, entire, the upper ones oblong-ovate, sessile, entire or few toothed, all ciliate, 1-7 cm long, 3-15 mm wide; flowers in clusters 5-13 mm wide; corolla 1-2 mm long, funnelform, white with bluish lobes; fruit 2-4 mm long, glabrous to puberulent, the central locule with a thick corky mass dorsally, the lateral locules sterile, narrow, with a shallow groove between them; 2n = 14, 16. Sagebrush and mountain brush communities at ca 1533 m in Salt Lake County; introduced from Europe, now widely established in the U. S.; 1 (0).

VERBENACEAE St. Hil.

Vervain Family

Herbs or shrubs; leaves mostly opposite, simple, estipulate; inflorescence cymose, racemose, or of spikes or panicles; flowers sessile or pedicellate, perfect or sometimes imperfect, more or less irregular; calyx persistent, 2- to 4 (5)-toothed or -lobed; corolla regular or irregular, funnelform or salverform, usually with a well-developed tube, the limb commonly 4- or 5-lobed; stamens 4 and didymous or rarely 2 or 5, inserted on the corolla tube; pistil 1, the ovary superior, 2- to 4-loculed, the ovules 1 per locule; style simple, with 1 or 2 stigmas; fruit dry, of 2–4 nutlets; x = 5–12.

1. Calyx and corolla 5–lobed or 5–toothed; nutlets 4; plants common .... Verbena
— Calyx 2– to 4–lobed; corolla 4–lobed; nutlets 2; plants uncommon 2

2(1). Plants shrubs; flowers scattered, in slender elongate spikes or spikelike racemes; bracts narrow, not closely subtending the flowers .... Aloysia
— Plants herbaceous; flowers in dense, short-spicate or subcapitate spikes; bractlets broad, imbricate, closely subtending the flowers .... Phyla

Aloysia Ortega

Sweet-scented aromatic shrubs; leaves opposite, simple, the blades toothed; flowers small, in open leafy panicles composed of slender elongate spikes or spicate racemes; bractlets narrow, inconspicuous to conspicuous; calyx tubular-campanulate, usually spreading hairy or villous, not inflated at maturity, the tube angled, not flattened, 4–lobed; corolla salverform, 2–lipped; somewhat unequal; stamens 4, didymous, included; ovary 1– or 2–loculed; fruit a small, dry schizocarp.

Aloysia wrightii (Torr.) Heller ex Abrams [*Lippia wrightii* Gray]. Shrubs to 15 dm tall; stems slender, brittle, the wood bright yellow, the bark readily peeling in long thin filamentous strips, the young twigs densely grayish puberulent; leaves small, ovate to rotund, 2–15 (30) mm long, 2–17 mm wide, rounded, crenulate or serrulate, strigulose-scabrous above, very densely tomentulous beneath with resinous-glandular, grayish or yellowish hairs, the venation deeply impressed above; inflorescence exceeding the leaves; spikes 1–4 cm long, densely many-flowered, the bractlets 1.5–2 mm long, lanceolate, long-acuminate, densely puberulent; corolla white, ca 2.5 mm long, puberulent externally, the limb ca 2 mm wide. Rare plants of rocky slopes, ledges, and limestone hills with creosote bush, blackbrush, yucca, and various cacti at ca 1200 m in Washington County; Nevada and Arizona, to New Mexico, Texas, and Mexico; 4 (ii). The plants flower in late summer and early autumn.

Phyla Lour.

Perennial, procumbent or prostrate herbs with trailing or ascending stems, sometimes woody near the base, glabrate to canescent-strigulose; leaves opposite, dentate apically; inflorescence a dense spike, axillary, cylindric, elongate in fruit; flowers small, sessile, borne singly in axils of cuneate-obovate bractlets, not at all 4–ranked.

1. Plants densely matted, more or less densely strigose-canescent throughout; leaves very small and cuneiform; known from Washington County .... P. nodiflora
— Plants forming open mats, finely strigulose; leaves 1–1.5 cm long or more, narrowly cuneate-spatulate, not of Washington County .... P. cuneifolia

Phyla cuneifolia (Torr.) Greene Wedge-leaf Frogfruit. [*Zapania cuneifolia* Torr.; *Lippia cuneifolia* (Torr.) Steudel in Marcey]. Stems prostrate to procumbent, branching from a woody base, often rooting at the nodes, somewhat flexuous, with short, erect branches at the nodes; leaves sessile, thick-textured, linear-oblanceolate or cuneiform, 1–5 cm long, 2–8 mm wide, acute with 2–8 teeth above the middle, gradually cuneate basally, appressed-strigulose on both surfaces; inflorescence shorter than or slightly longer than the leaves; peduncles 0.8–5 cm long; heads globose at first, finally elongating and cylindrical, to 2 cm long and 8–12 mm wide; bractlets conspicuous, obovate, ca 5 mm long, abruptly long-acuminate; corolla whitish to reddish or purplish, the tube 4–5 mm long. Gravel bars and sandy terraces at 1190 to 1550 m in Carbon, Grand, Millard, Uintah, Utah, and Wayne counties; South Dakota and Wyoming to Arizona, Texas, and Mexico; 10 (iv). The only known report of this species from Millard County is from near Flowell, where it served as the host plant for *Cuscuta warneri* Yuncker. The parasite has not been relocated since the initial (and only) collection in 1957.

Phyla nodiflora (L.) Greene [*Verbena nodiflora* L.]. Plants matted, more or less carpetlike, somewhat woody basally, cinereous-strigulose; leaves pale green, narrowly oblanceolate to obovate, entire to toothed from below the middle, mainly 4–15 mm long, acutish apically; peduncles mostly 1.5–3 cm long; spikes ovoid, mostly 5–8 (10) mm long, 6 mm broad; bracts ovate; calyx ca 1 mm long; corolla rose to white, 4–5 mm long; 2n = 36. Cultivated ground cover, escaping along ditch banks and roadsides below 1200 m in Washington County; California to Texas and South America; 3 (iii).

Verbena L.

Annual or perennial herbs; stems and branches procumbent, ascending, or erect, glabrous or pubescent; leaves mostly opposite; flowers small or medium, bracteate, in terminal corymbose or paniculate spikes; calyx usually tubular, 5–ribbed, 5–toothed; corolla salverform or funnelform, its tube straight or curved, the limb spreading, 5–lobed, regular or slightly 2–lipped; stamens 4, didynamous, included; ovary entire or somewhat 4–lobed apically, 4–loculed, each locule with 1 ovule; fruit dry, enclosed in the calyx, separating into 4 nutlets at maturity.

1. Spikes mostly broad and dense; flowers showy; corolla tube 8 mm long or more; calyx 7–13 mm long 2
— Spikes mostly slender and elongate following anthesis; flowers not especially showy; corolla tube 6 mm long or less; calyx 2.5–5 mm long 3

2(1). Corolla tube only slightly longer than the calyx; herbage conspicuously villous; corolla limb 8–9 mm wide .... V. gooddingii
— Corolla tube 1.3–5 times longer than the calyx; herbage densely hispid-hirsute; corolla limb 8–10 mm wide .. .. V. bipinnatifida

3(1). Plants branching at the base, the branches decumbent or ascending; bracts conspicuous, at least twice as long as the calyx .... V. bracteata
— Plants rarely branching at the base; stems erect or nearly so; bracts inconspicuous, less than 1.5 times longer than the calyx 4

4(3). Leaf blades thin, not reticulate, often hastately lobed at base; fruiting spikes narrow, less than 7 mm wide .... V. hastata
— Leaf blades thick, rugose-reticulate, not hastate basally; fruiting spikes thick, over 7 mm wide .... V. macdougalii

Verbena bipinnatifida Nutt. Dakota Vervain. [*Glandularia bipinnatifida* (Nutt.) Nutt.]. Stems diffusely branched from the base, ascending, rooting at the nodes, hispid-hirsute; leaves bi- or tripinnately parted, with the divisions again bipinnatifid, 2–6 cm long, the lobes linear or oblong, appressed hirsute on both surfaces; spikes compact at anthesis, elongating in fruit; bractlets usually longer than the calyx, linear-subulate, hirsute-hispid, ciliate; calyx 8.5–10 mm long, the lobes subulate, hispid ciliate; corolla pink to lavender, the limb 8–10 mm wide, the lobes emarginate; n = 5. Dry roadsides, pastures and vacant lots below 1600 m in Utah and Washington counties; South Dakota to Missouri and eastward, south to Arizona and Mexico; 2 (i). This species is known from both cultivated and established material. It apparently is assignable to var. **latiloba** Perry.

Verbena bracteata Lag. & Rodr. Prostrate Vervain. [V. bracteosa Michx.]. Annual to perennial herbs; stems diffusely branched, decumbent or ascending, 1–5 dm long, more or less hirsute; leaves oblong to cuneate-obovate in outline, 1–4 cm long, pinnately 3-parted or -lobed, the middle lobe the largest and incisely toothed or cleft, more or less hirsute on both surfaces; spikes sessile; bracts conspicuous, much longer than the calyx; calyx 3–4 mm long; corolla pale blue to purple, the tube 3–5 mm long, the limb 2.5–3 mm wide. Roadsides, fields, and other disturbed sites at 930 to 2660 m in nearly all Utah counties; widely distributed in North America; Mexico; 94 (viii).

Verbena gooddingii Briq. Gooding Vervain. [*Glandularia gooddingii* (Briq.) Solbrig]. Perennial herbs; stems several, ascending, 2–4.5 dm tall, more or less villous-hirsute and glandular; leaves rounded in outline, 1–2 cm long, palmately 3-parted, then pinnately cleft, with petioles 5–15 mm long, villous-hirsute on both surfaces; spikes capitate, somewhat elongating in fruit; bracts lance-linear, ca 8 mm long; calyx 7–10 mm long; corolla purplish to rose, the tube 8–12 mm long, the limb 8–10 mm wide, with retuse lobes; n = 15. Blackbrush, pinyon-juniper, and ponderosa pine communities at 1100 to 2030 m in Washington County; Texas to California and Mexico; 11 (iii).

Verbena hastata L. Blue Vervain. Erect perennial herbs; stems 4–15 dm tall, branched above only, strigose-hispidulous; leaves lanceolate to ovate-lanceolate, 5–15 cm long, the lower sometimes hastate, acute, serrate or incised-dentate, shortly petiolate, rough pubescent on both surfaces; spikes straight, usually numerous in an upright panicle, not over 7 mm wide; bracts shorter than the calyx; calyx 2.5–3 mm long; corolla blue to purplish or pink, the tube ca 3 mm long, the limb 2.5–4 mm wide. Wet meadows, marshes, and bogs in palustrine and riparian habitats at 1300 to 2000 m in Box Elder, Cache, Davis, Salt Lake, Tooele, Utah, and Weber Counties; widely distributed in the U. S. and southern Canada; 18 (ii).

Verbena macdougalii Heller New Mexico Vervain. Perennial herbs; stems 3–10 dm tall, erect, stout, simple or sparingly branched above, ashy green, hirsute; leaves oblong-elliptic to ovate-lanceolate, 6–10 cm long, shortly petiolate, coarsely and irregularly serrate-dentate, hirtellous, rugose above, prominently veined below; spikes solitary or few, thick, over 7 mm wide; bracts ca 1–2 mm longer than the calyx; calyx 4–5 mm long; corolla blue to purple, the tube 4–5 mm long, scarcely exceeding the calyx, the limb 6 mm wide. Ponderosa pine forest at ca 2160 to 2600 m in Garfield County; Wyoming and Colorado to New Mexico and Arizona; 1 (0).

VIOLACEAE Batsch
Violet Family

Annual or perennial herbs; leaves basal or cauline and alternate, simple or pedatifid; flowers perfect, irregular, or sometimes cleistogamous, solitary; sepals 5, distinct or nearly so; petals 5, the lowermost spurred; stamens 5; pistils 1, the ovary superior, 1-loculed, 3- to 5-carpelled; style 1; stigma usually lobed; fruit a loculicidal capsule; x = 6–13, 17, 21, 23.

Baker, M. S. 1935. Studies in western violets -I. Madrono 3: 51–56.
———. 1940. Studies in western violets -III. Madrono 5: 218–231.
———. 1949. Studies in western violets -IV. Leafl. W. Bot. 5: 141–156.

Viola L.

Plants with caudices, rhizomes, or stolons, acaulescent or caulescent; leaves with stipules and petioles, the blades pedate or not; flowers on axillary peduncles, the early ones open-flowered, the later ones often cleistogamous; sepals persistent; petals purple, pink, violet, white, or varicolored; stamens with short filaments, the connective extended beyond the anther apex as a broad membranous appendage; capsule explosively dehiscent.

1. Leaf blade pedately compound, the main segments again deeply divided into slender segments; plants rare along the Wasatch Front .... V. beckwithii
— Leaf blades entire or merely toothed, not as above; plants common or uncommon 2
2(1). Stipules foliaceous, toothed or pinnatifid basally 3
— Stipules not foliaceous, entire or merely toothed 4
3(2). Plants annual; flowers small, not especially showy .. .. V. arvensis
— Plants annual or perennial; flowers large and showy . ... V. tricolor
4(2). Petals yellow, sometimes suffused with purple dorsally 5
— Petals blue, purple, or white 6
5(4). Leaf blades commonly toothed or lobed (rarely entire), about as broad as long, usually purplish veined .... V. purpurea
— Leaf blades entire or remotely toothed, much longer than broad, not purplish veined .... V. nuttallii
6(4). Plants caulescent, at least 1 internode apparent on flowering stems, the peduncles visibly axillary 7
— Plants acaulescent or at most stoloniferous; peduncles arising from a basal tuft of leaves 8
7(6). Petals white or suffused with pale lavender; spur short; stems usually with 2 or more apparent internodes .. .. V. canadensis
— Petals violet; spur elongate; stems mainly with 1 or 2 apparent internodes .... V. adunca
8(6). Plants stoloniferous 9
— Plants acaulescent 10
9(8). Flowers white to pale violet; plants uncommon in moist sites, usually in higher elevations, indigenous . ... V. palustris
— Flowers deep violet purple, fragrant; plants cultivated and persisting .... V. odorata
10(8). Leaf blades acute apically, as long as broad or longer; plants cultivated and persisting .... V. papilionacea

— Leaf blades rounded apically, usually broader than long; plants indigenous .... *V. nephrophylla*

Viola adunca **J. E. Sm. ex Rees** Blue Violet. [*V. mamillata* Greene, type from Uintah County; *V. oxysepala* Greene, type from Sanpete County; *V. tidestromii* Greene, type from Sanpete County]. Plants with short to elongate rhizomes and more or less developed caudex, usually stemless early in the season, stemmed as the season advances, often with 1–3 apparent internodes; stipules 3–12 (15) mm long, entire or toothed; petioles 1–12 cm long, glabrous or hairy; leaf blades cordate to ovate or reniform, crenate, 0.4–4.5 cm wide, 0.6–6 cm long, hairy or glabrous, the apex bluntly acute to rounded; peduncles 0.6–5 cm long, with 2 bracts borne usually at or above the middle; flowers 7–15 (21) mm long, the spur to half as long as the lowest petal (or longer); petals blue to violet or white, the lateral pair bearded; style bearded; 2n = 20, 30, 40. Mountain brush, aspen, ponderosa pine, lodgepole pine, spurce-fir, meadow, and alpine tundra communities at 1525 to 3450 m, known in all Utah counties, except Davis, Iron, Rich, and Washington; Alaska and Yukon, east to Newfoundland, south to California, New Mexico, and the Great Lakes; 151 (xxv). A dwarf, high elevation phase of this species from the Wasatch and Uinta mts. has been identified as **var. *bellidifolia* (Greene) Harrington** [*V. bellidifolia* Greene; *V. bellidifolia* ssp. *valida* Baker, type from Brighton], but it is transitional in all features with the common phase of the species that adjoins it.

Viola arvensis **Murray** Field Pansy. Plants annual, the stems simple or branched, mainly 1–3 dm tall, with reflexed hairs on the angles; petioles 0.5–2 cm long (or lacking); leaf blades variable, the lower ones ovate to orbicular, the upper ones oblong to narrowly elliptic, crenate, blunt or acutish apically; peduncles 2–8 cm long; flowers white to yellow or blue, or with the tips purplish, 10–15 mm long, 0.7–10 mm wide; sepals as long as the petals or longer, lanceolate; style head hairy; 2n = 34. Weedy species, widely grown in gardens, escaping and persisting in Cache County (Maguire sn 1932 BRY); sporadic in the U. S.; adventive from Europe; 1 (0).

Viola beckwithii **T. & G.** Beckwith Violet. [*V. beckwithii* var. *cachensis* C. P. Sm., type from Cache County?; *V. bonnevillensis* Cottam, type from Salt Lake City]. Plants arising from subterranean, vertical, root crowns, the leaves and stems pale and thin below ground level, above ground portion mainly 5–12 cm tall; herbage puberulent (especially the leaves); stipules membranous, 0.8–3 cm long above ground, adnate to the petiole; petioles 2–15 cm long; leaf blades ternately 2–3 times compound, the ultimate segments linear to narrowly oblong; peduncles shorter than to longer than the leaves; flowers 12–18 mm long; sepals obtuse to rounded, rarely acute; upper paired petals red purple, the lower ones violet, the lateral pair bearded; style head hairy. Foothills, in bluebunch wheatgrass and mountain brush communities, at ca 1370 to 1680 m in Box Elder, Cache, Salt Lake, and Utah counties; Oregon to Idaho, Nevada, and California; 6 (0). The name *V. bonnevillensis* Cottam, is based on plants more or less intermediate with *V. purpurea*. Possibly, the type of the species came from Utah (between Great Salt Lake and the Sierra Nevada). This handsome violet has been displaced largely from its range in Utah by housing subdivisions and commercial developments. It has not been collected in Utah County, to my knowledge, since 1933.

Viola canadensis **L.** Canada Violet. [*V. rugulosa* Greene; *V. canadensis* var. *rugulosa* (Greene) C. L. Hitchc.]. Plants with short to elongate rhizomes and sometimes with stolons (seldom collected) as well; aerial stems 5–40 cm tall, usually with 1–4 elongate internodes; stipules 8–20 mm long, entire; petioles 2–25 cm long, glabrous or puberulent; leaf blades cordate, acute to acuminate, puberulent to glabrous, 3–10 cm long, 2–10 cm broad; peduncles 1–4 cm long, arising from upper axils, the bracts from below to above the middle; flowers 10–15 mm long, the spur short; petals white, yellow at the base, the lower ones with purple lines, the lateral pair bearded, purplish dorsally; style bearded. Mountain brush, ponderosa pine, sagebrush, aspen, and spurce-fir communities at 1830 to 3085 m in Beaver, Carbon, Emery, Garfield, Grand, Iron, Kane, Piute, Rich, Salt Lake, San Juan, Sanpete, Sevier, Utah, and Washington counties; British Columbia east to the Atlantic, south to Oregon, Arizona, and New Mexico; 51 (xi). Segregation of varieties from among our specimens seems unwarranted.

Viola nephrophylla **Greene** Bog Violet. [*V. clauseniana* Baker, type from Zion Canyon]. Plants 0.4–2 dm tall, with thickened, elongate rhizomes, the leaves arising from vertical rhizome apices; stipules 5–12 mm long, entire; petioles 3–25 cm long, glabrous; leaf blades ovate to cordate or reniform, rounded to obtusely cuspidate apically, crenate-dentate, 2–7 cm wide, 2–7 cm long from sinus to apex, glabrous to minutely puberulent along the veins of the upper surface; peduncles 2–25 cm long, mostly surpassing the leaves, with bracts from below to above the middle; flowers 10–22 mm long, the spur 2–5 mm long; petals bluish violet, the lower 3 whitish basally and bearded, the upper pair beardless or bearded; style glabrous. Blackbrush communities upward to alpine forests at 1155 to 3175 m in all Utah counties, except Rich, Sanpete, and Summit; British Columbia to Newfoundland, south to California, Arizona, New Mexico, and Wisconsin; 55 (v).

Viola nuttallii **Pursh** Nuttall Violet [*V. praemorsa* Dougl.; *V. lingaefolia* Nutt. in T. & G.]. Plants acaulescent or short caulescent, typically 5–25 cm tall, with thickened mostly vertical rhizomes; herbage glabrous to densely hairy; stipules 5–20 mm long, entire or toothed; petioles 0.5–15 cm long; leaf blades ovate to elliptic, lanceolate, or oblong, rounded to obtuse apically, irregularly toothed to subentire, mostly 1–7 cm long, 0.5–4 cm wide, cuneate to truncate or subcordate basally; peduncles shorter than or longer than the leaves; flowers 8–17 mm long; petals yellow, the upper ones often brownish or purplish dorsally, the lower 3 lined purple, the lateral pair bearded; style head bearded. Sagebrush, mountain brush, aspen, and mixed conifer communities at 1370 to 3205 m in Beaver, Box Elder, Cache, Carbon, Daggett, Davis, Duchesne, Emery, Juab, Millard, Morgan, Salt Lake, Sanpete, Sevier, Summit, Uintah, Utah, and Weber counties; British Columbia to Saskatchewan, south to California, Arizona, Colorado, and Kansas; 119 (x). The Nuttall violet is known to form intermediates with *V. purpurea* (q.v.), but most specimens can be segregated easily. Our materials have been treated as belonging to two varieties, based mainly on leaf shape; **var. *vallicola* (A. Nels.) St. John** [*V. vallicola* A. Nels.], with usually truncate to subcordate leaf bases, and **var. *major* Hook.**, with cuneate leaf bases. Only arbitrary separation seems possible.

Viola odorata L. Sweet Violet; English Violet. Plants mostly 4–12 cm tall, from slender, creeping rhizomes and elongate strawberrylike stolons; herbage sparingly puberulent throughout, except retrorsely hairy on peduncles; stipules mostly 5–12 mm long, entire or serrate; petioles 2–9 cm long, arising from the summit of an erect, thickened rhizome; leaf blades 1.3–6 cm long from sinus to apex, 1.5–8 cm wide, cordate to subreniform, crenate, obtuse to rounded apically; peduncles mostly 4–12 cm long, the bracts near the middle; flowers 12–22 mm long, fragrant; petals typically deep violet, varying to paler shades or white, the lateral petals bearded; spur 2.5–4 mm long; style head glabrous; 2n = 20. Lawns, gardens, irrigation canal banks, and other moist sites, usually near habitations, at ca 1340 to 1710 m in Salt Lake and Utah counties; naturalized in many places in the U. S.; introduced from Europe; 5 (0). This is a commonly grown vigorous ornamental that spreads and persists.

Viola palustris L. Marsh Violet. Plants 3–19 cm tall, from slender elongate rhizomes and creeping stolons; herbage glabrous; stipules 5–11 mm long, entire; petioles 0.7–13 cm long; leaf blades 0.5–3.5 cm long from sinus to apex, 0.5–4 cm wide, peduncles 2–15 cm long, often surpassing the leaves, the bracts from below to above the middle; flowers 9–16 mm long, the spur 2–3 mm long; petals lavender to almost white, the lower petals with purple lines, the lateral pair sparsely bearded to glabrous; style glabrous; 2n = 48. Stream banks, wet meadows, and talus slopes, in willow, aspen, spruce, and alpine tundra communities, at 2375 to 3355 m in Box Elder, Duchesne, Garfield, Piute, Salt Lake, Sevier, Summit, and Wayne counties; Alaska east to Labrador, south to California and Colorado; circumboreal; 15 (i).

Viola papilionacea Pursh Meadow Violet. Plants mainly 6–30 cm tall, from thick horizontal and verical rhizomes; herbage sparingly puberulent to spreading long-hairy or glabrous; stipules mainly 10–30 mm long, entire or with long gland-tipped teeth; petioles 3–24 cm long; leaf blades 1.2–10 cm long from sinus to apex, 1.5–13 cm wide, cordate to orbicular, crenate-serrate, rounded-acutish apically; peduncles mostly 5–12 cm long, the bracts often well below the middle; flowers mostly 18–25 mm long, the spur 1–3 mm long; petals white to pink, violet, or lavender, with blue purple lines, the lateral petals bearded; style head glabrous. Cultivated ornamentals, spreading and long persisting in Utah County; widely distributed in the eastern U. S.; 3 (0).

Viola purpurea Kellogg Pine Violet. [*V. utahensis* Baker & Clausen, type from Providence Canyon]. Plants mainly 4–15 cm tall, from a thick subrhizomatous caudex; herbage scaberulous to subglabrous; stipules mostly 4–10 mm long, entire or toothed; petioles 1–12 cm long; leaf blades 0.6–4 cm long and about as broad or broader, ovate, oval, or orbicular, mostly obtuse to acute (rarely subcordate) basally, crenate to entire or shortly lobed; peduncles 1–9 cm long, the bracts often below the middle; flowers mostly 5–12 mm long; petals yellow, purple veined and commonly purple dorsally, the lateral ones bearded or glabrous; spurs 1–2 mm long; style head bearded. Sagebrush, mountain brush, pinyon-juniper, aspen, and spruce-fir communities at 1525 to 3175 m in Beaver, Box Elder, Cache, Davis, Duchesne, Juab, Kane, Millard, Morgan, Piute, Salt Lake, Sanpete, Sevier, Summit, Tooele, Utah, Wasatch, Washington, and Weber counties; Washington to Montana, south to California, Arizona, and Colorado; 77 (v). The bulk of our material belongs to **var. venosa (Wats.) Brainerd** [*V. nuttallii* var. *venosa* Wats.], but material with entire leaves from Washington County is referable to **var. charlestonensis (Baker & Clausen) Welsh & Reveal** [*V. charlestonensis* Baker & Clausen]. The pine violet forms an intergrading series with *V. nuttallii*, a portion of that variation forms the basis for recognition of *V. utahensis*, but most specimens can be assigned readily to one or the other of the species.

Viola tricolor L. Pansy. Plants caulescent, mainly 10–45 cm long, annual (or short-lived perennial) from taproots; herbage glabrous or pubescent; petioles short or lacking; stipules foliose and pinnately lobed; leaf blades 1–8 cm long, 1–2.5 cm wide, orbicular to lanceolate or elliptic, crenate; peduncles 2–12 cm long; flowers 1–6 cm wide; petals variously colored, marked with yellow, purple, bronze, or white, the upper ones usually darker, the lateral pair bearded; 2n = 26, 28, 42, 46. Cultivated ornamentals, escaping, and occasionally persisting in Salt Lake, Uintah, and Utah counties; widely grown elsewhere; native to the Old World; 7 (0). The large flowered phase has been derived through hybridization and is known as *V. x. wittrockiana* Gams.

VISCACEAE Miers
Mistletoe Family

Plants parasitic on branches of trees and shrubs; stems brittle (especially when dry), with swollen jointed nodes, often freely branched; leaves opposite, simple, entire (often scale-like), leathery and persistent; flowers imperfect, inconspicuous, apetalous, in axillary spikes or cymes or clustered at the nodes (sometimes solitary); perianth calyxlike with 2–5 sepals, persistent; staminate flowers with 2–5 stamens inserted on the lower portion of the sepals, the anthers mostly 1- to 2-loculed; pistillate flowers with a more reduced perianth; ovary inferior, 1-loculed; style 1; fruit a berry, fleshy, 1 (2–3)-seeded; x = 10–15.

Hawksworth, F. G. and D. Wiens. 1972. Biology and classification of dwarf mistletoes (*Arceuthobium*). U.S.D.A., Forest Service, Agriculture Handbook No. 401. 234 pp.

1. Perianth of the pistillate flower 2-lobed; fruits compressed, on recurved pedicels; anthers 1-loculed; parasitic on conifers except junipers .... *Arceuthobium*

— Perianth of the pistillate flower 3-lobed; fruits globose to subglobose, sessile; anthers 2-loculed; parasitic on junipers and deciduous trees and shrubs ..... *Phoradendron*

Arceuthobium Bieb.

Parasitic on conifers; stems yellowish to greenish, often strongly divaricate or erect, glabrous, jointed, terete to angled with short, swollen internodes; leaves reduced, scalelike, opposite; flowers axillary, 2 to several per node or terminal, dioecious; staminate flowers mostly 3- or 4-merous, the segments somewhat fleshy and each lobe with an adnate anther; pistillate flowers with 2 perianth lobes adnate to the ovary; stigma 1, entire to slightly lobed; style lacking. Note: The taxonomy in this treatment follows that of Hawksworth and Wiens (1972).

1. Branches, flowers, and fruit in whorls; internodes ca 10 times longer than wide; parasitic mostly on *Pinus contorta* (occassional on *P. ponderosa*) ...... *A. americanum*
— Branches flabellate but not whorled; flowers in decussate pairs, never whorled; parasitic on various conifers .. 2

2(1). Shoots olive green to brownish; parasitic on *Pinus edulis* and *P. monophylla* .... *A. divaricatum*
— Shoots variously colored; parasitic on other hosts 3

3(2). Shoots orange to reddish brown, not bright; plants mostly parasitic on ponderosa pine and Engelmann spruce (occassional on bristle cone pine and limber pine) .... *A. vaginatum*
— Shoots yellow to yellow green or olive green, not orange to reddish brown; principal host *Abies*, *Pseudotsuga*, *Pinus longaeva*, and *P. flexilis* 4

4(3). Shoots olive green; pistillate plants less than 4 cm tall; plants parasitic mostly on *Pseudotsuga* (occassional on *Abies concolor*) .... *A. douglasii*
— Shoots yellow green to yellow; pistillate plants over 4 cm tall; plants parasitic mostly on *Abies concolor*, *Pinus flexilis*, and *P. longaeva* 5

5(4). Plants 10 cm high or more (up to 2.2 dm), mostly yellowish or yellowish green; parasitic on *Abies concolor* .... *A. abietinum*
— Plants 3–5 dm tall, yellow green; parasitic on pines, *P. flexilis* and *P. longaeva* ours .... *A. cyanocarpum*

Arceuthobium abietinum Engelm. ex Munz Fir Dwarf-mistletoe. [*A. douglasii* var. *abietinum* Engelm. in Wats., *A. occidentale* var. *abietinum* Engelm. in Wats., *Razoumofskya douglasii* var. *abietinum* (Engelm.) Howell, *R. abietina* (Engelm.) Abrams, *A. campylopodium* f. *abietinum* (Engelm.) Gill]. Plants 8–22 cm high with yellowish or yellowish green shoots; main stems 1.5–2 (6) mm wide basally; staminate flowers ca 2.5 mm in diameter; perianth 3 (4) –merous, acute apically, yellowish to yellow green, lobes 1.2 mm long, 1 mm wide; mature fruit ca 4 mm long, 2 mm wide. Growing on *Abies concolor* in Kane County; Washington to California, Nevada, and Arizona; 1 (0). Our plants belong to f. sp. **concoloris** Hawksworkth & Wiens. This mistletoe is known to parasitize *Abies lasiocarpa* in northern Arizona where that species grows with *A. concolor* but occurrence on submontane fir is not presently known in Utah.

Arceuthobium americanum Nutt. ex Engelm. in Gray Lodgepole Pine Dwarf-mistletoe. [*Razoumofskya americana* (Nutt. ex Engelm.) Kuntze]. Stems yellowish to olive green 2–9 (30) cm long, the branches whorled; staminate flowers on short lateral stems, 2 to several per node; flowers 2 mm long; perianth mostly 3–merous (sometimes 4), the lobes 1.1 mm long; pistillate flowers 2 to several, 2–merous, whorled at the nodes, short-pedicellate; mature fruit 3.5–4 mm long, maturing the second summer. Lodgepole pine and other pine and spruce species in Cache, Daggett, Duchesne, San Juan, Summit, Uintah, and Wasatch counties; Canada, southern California and central Colorado; 6 (0).

Arceuthobium cyanocarpum Coult. & Nels. Limber Pine Dwarf-mistletoe. [*Razoumofskya cyanocarpum* Nels., *A. campylopodium* Engelm. f. *cyanocarpum* (Nels.) Gill]. Plants with yellow green shoots, 3–5 cm long, in dense, flabellate branched clusters; staminate flowers 3 mm in diameter; perianth 3–merous with acute apex; mature fruit 3–5 mm long. Parasitic mostly on *Pinus flexilis* and *P. longaeva* in Beaver, Box Elder, Cache, Daggett, Duchesne, Garfield, Iron, Kane, San Juan, Tooele, and Wasatch counties; Montana, Colorado, southwest to California; 12 (0). This species is on ponderosa and lodgepole pines also.

Arceuthobium divaricatum Engelm. Pinyon Dwarf-mistletoe. [*Razoumofskya divaricatum* Cov., *A. campylopodium* f. *divaricatum* (Engelm.) Gill]. Shoots olive green to brown, 5–13 cm long, the branches flabellately arranged; staminate flowers 2.5 mm long; perianth 3–merous, the lobes 1.1 mm long; mature fruit 3.5 mm long. Apparently parasitic only on pinyon pines in Beaver, Carbon, Duchesne, Emery, Garfield, Grand, Iron, Juab, Kane, Millard, Piute, San Juan, Sanpete, Uintah, and Washington counties; California, Baja California; southwestern Colorado, Arizona, New Mexico, and west Texas; 14 (0).

Arceuthobium douglasii Engelm. Douglas Fir Dwarf-mistletoe. [*Razoumofskya douglasii* (Engelm.) Kuntze]. Stems short, 1–7 cm long, olive green, flabellately arranged; staminate flowers 2 mm long, borne in axillary flowers; perianth mostly 3 (4) –merous, the lobes rounded apically and with a reddish to purple inner surface, ca 1 mm long; pistillate flowers also in axillary pairs, 1.5 mm long; mature fruits olive green, 3–4 mm long, maturing the second season. On Douglas fir in Beaver, Cache, Daggett, Duchesne, Garfield, Grand, Iron, Kane, Millard, Piute, Salt Lake, San Juan, Sanpete, Sevier, Summit, Tooele, and Utah counties; British Columbia south to California, Idaho, Montana, Wyoming, Nevada, Colorado, Arizona, New Mexico, and Mexico; 19 (0). The species is occassional on *Abies concolor*, *A. lasiocarpa*, *Picea pungens*, and *P. engelmanii*, when these species are associated with Douglas-fir.

Arceuthobium vaginatum (Willd.) Presl in Bercht. Southwestern Dwarf-mistletoe. [*A. vaginatum* f. *cryptopodium* (Engelm.) Gill]. Shoots orange to reddish brown, flabellately arranged, the main shoots 2 (4–8) 10 mm wide at base; staminate flowers light green to yellowish, 2.5–3 mm long, axillary; perianth lobes ca 1.3 mm long; pistillate flowers axillary, the lateral ones in pairs; fruit 4.5–5.5 mm long. On ponderosa pine and Engelmann spruce in Garfield, Kane, San Juan, Sanpete, Sevier, and Wayne counties; Colorado, south to New Mexico, Arizona, west Texas and Mexico; 7(i). The species is occassionally found on lodgepole, bristlecone, and limber pines. Our plants belong to ssp. **cryptopodium** (Engelm.) Hawksworth & Wiens. [*A. cryptopodium* Engelm.].

Phoradendron Nutt.

Parasitic, dioecious; sometimes woody plants with brittle, branched, usually jointed stems; leaves scalelike, opposite, entire; flowers in short jointed spikes, sessile and sunk in the jointed rachis; perianth 1– to 4–merous (usually 3–merous), the lobes triangular, rounded; staminate flowers usually in clusters, with an adnate 2–loculed anther at the base of each perianth lobe; pistillate flowers with the perianth adnate to the ovary; ovary inferior, 1–loculed; fruit a sessile drupe, ovoid to globose. The plants are parasitic on some conifers and various woody deciduous plants.

1. Scalelike leaves strongly connate; pistillate inflorescence with 1 segment and 2 flowers; stems crowded, with stout branches, the internodes 5–16 mm long; mostly parasitic on juniper species .... *P. juniperinum*
— Scalelike leaves slightly connate; pistillate inflorescence with 2 or more segments, some segments with more than 2 flowers; stems not crowded, the branches slender and flexuous, with internodes usually 1.3–2.8 cm long; mostly parasitic on *Acacia greggii* .... *P. californicum*

Phorodendron californicum Nutt. Acacia Mistletoe. Stems slender, terete, the branches long and slender, clustered, 1–5 cm long, with the internodes 13–28 mm long, commonly reddish; leaves scalelike, ca 1 mm long, entire; flowers in axillary spikes; staminate spikes 2- to 6-jointed; pistillate spikes 2- to 7-jointed; fruit white to red, globose, 3–4 mm in diameter. Parasitic on *Acacia greggii*, and much less commonly on *Larrea tridentata*, in Beaver Dam Wash, Washington County; Nevada, Arizona, California, and Mexico; 12 (i). One collection from Beaver Dam Wash is reported as growing on Fremont cottonwood.

Phorodendron juniperinum Gray Juniper Mistletoe. Stems usually stout, swollen at the nodes, the internodes short 5–16 mm long, glabrous; leaves scalelike, spreading, 1–2 mm long, united at base, triangular; green to yellowigh green (sometimes turning dark when pressed); staminate spike with 1–2 segments, each with 5–9 flowers; sepals (2) 3 (4), triangular, free; stamen attached to base of each sepal, anther 2–loculed; pistillate spike with 1 segment and 2 flowers; ovary inferior, 1–loculed, the style 1, the sepals fused to the ovary; fruit a drupe, 1–seeded, on a recurved pedicels. Mostly parasitic on Utah juniper in Beaver, Davis, Emery, Garfield, Grand, Iron, Kane, Piute, Salt Lake, San Juan, Sanpete, Tooele, Utah, Washington, and Wayne counties; Oregon, south to California, Nevada, Colorado, New Mexico, Arizona, west Texas, and Mexico; 34(ii). Our plants belong to **ssp. juniperinum**. Curiously the species is apparently lacking in the Uinta Basin, a principal locality for the host plant.

VITACEAE A. L. Juss.
Grape Family

Shrubs or woody vines, usually supported by tendrils opposite the leaves or on peduncles, sympodial; leaves alternate usually palmately 3- to 5-lobed or compound; stipules deciduous; inflorescence terminal, appearing opposite the leaves; flowers often imperfect and perfect on the same plant, small, regular, greenish; sepals 4 or 5, minute; petals 4 or 5; stamens as many as the petals and opposite them; style short or none; pistil 1, the ovary 2–loculed, 1- to 4-ovuled; fruit a berry (grape); x = 11–20.

1. Leaves merely palmately veined and lobed; plants indigenous in Washington County, widely cultivated there and elsewhere .... *Vitis*
— Leaves palmately compound; plants indigenous in southeastern Utah, and cultivated elsewhere ..... *Parthenocissus*

Parthenocissus Planchon

Woody vines, supported by tendrils that clasp or afix themselves by adhesive disks; leaves palmately compound; leaflets 3–7, coarsely serrate; inflorescence cymose; flowers perfect or imperfect; calyx minutely 5–toothed; petals distinct; disk lacking; berries thinly fleshy.

1. Tendrils with 3–8 branches, these terminating in adhesve disks; inflorescence with the axis usually well developed; plants cultivated, rarely escaping .... *P. quinquefolia*
— Tendrils with 3–5 branches, these lacking adhesive disks; inflorescence dichotomously branched; plants indigenous in southeastern Utah, seldom cultivated .... *P. vitacea*

Parthenocissus quinquefolia (**L.**) Planchon Virginia Creeper. [*Hedera quinquefolia* L.]. Vines to 5 m tall or more, supported by tendrils with 3–8 branches, these terminated by adhesive disks; leaflets with distinct petiolules, 3–15 cm long, 1.6–8 cm wide, oblanceolate to obovate, lanceolate or ovate, coarsely serrate, pale green and dull above, glaucous beneath; cymes usually with a definite central axis with flowers clusters arising laterally from that axis; berries 5–7 mm thick. Cultivated ornamental in much of Utah, escaping and established in Daggett, Salt Lake, Uintah, and Utah counties; introduced from eastern U. S.; 4 (0). The brightly colored leaves in autumn are beautiful.

Parthenocissus vitacea (**Knerr**) **A. S. Hitchc.** Thicket Creeper. [*Ampelopsis quinquefolia* var. *vitacea* Knerr; *P. inserta* (Knerr) Fritsch, misapplied]. Vines clambering or trailing, rarely weakly climbing, supported by tendrils with 3–5 branches with slender nonadhesive tips; leaflets shortly petiolulate, mainly 3–13 cm long, 1.5–6 cm wide, obovate to oblanceolate, coarsely serrate, lustrous above, green and not especially glaucous beneath; inflorescence mainly dichotomously branched, without a main central axis; berries 8–10 mm thick; $2n = 40$. Hanging gardens, canyon bottoms, and near seeps and springs in Garfield, Kane, and San Juan counties; Montana to Quebec, south to Wyoming and Texas; 11 (v).

Vitis L.

Clambering shrubs; leaves palmately veined and usually lobed; flowers small; calyx minute, the limb unlobed; inflorescence a compound thyrsoid cyme opposite the leaf; petals 5, usually coherent apically; disk of 5 glands alternate with the stamens; ovary 2–loculed; style short; berry usually 2- to 4-seeded.

1. Leaves persistently and densely tomentose on the lower surface; plants cultivated and long persisting . *V. labrusca*
— Leaves thinly if at all hairy or the hairs confined to lower vein axils; plants cultivated or indigenous 2
2(1). Bark shreddy, hanging loose between the nodes; skin of fruit slipping free from the pulp; plants indigenous in Washington County .... *V. arizonica*
— Bark not especially shreddy; skin of fruit not slipping free from the pulp; plants widely grown ...... *V. vinifera*

Vitis arizonica Engelm. Canyon Grape. Sprawling or clambering woody vines, mainly 2–6 m long; herbage thinly tomentose to subglabrous; leaf blades 3–12 cm long, 4–14 cm wide, cordate to cordate-ovate or uniform; inflorescence 5–10 cm long, with slender peduncle, paniculiform; grapes 6–10 mm thick; juicy, palatable, but tending towards bitter. Canyon bottoms and talus slopes, often where moist, at ca 850 to 1350 m in Washington County; Nevada, Arizona, New Mexico, Texas, and Mexico; 13 (i).

***Vitis labrusca* L.** Fox Grape; Concord Grape. Sprawling or clambering woody vines, mainly 2–15 m long; herbage tomentose, densely and persistently so on lower leaf surfaces; leaf blades mainly 3–15 cm long and 4–20 cm wide, cordate to uniform; inflorescence 5–12 cm long, paniculiform; grapes 6–15 mm thick, juicy, palatable, though often laden with tannins; n = 19. Widely grown fruit plant, long persisting following cultivation, and more or less established in Washington County (at least); introduced from the eastern U. S.; 2 (i). This is the common slip-skin grape of commerce. Fruit color varies from greenish to purplish or purplish black. Many horticultural varieties are grown.

***Vitis vinifera* L.** Wine Grape. Sprawling or clambering woody vines, mainly 3–6 m long; herbage subglabrous or more or less hairy in lower vein axils; leaf blades mainly 4–13 cm long and 5–18 cm wide, orbicular, cordate, or oval in outline; inflorescences paniculiform, mainly 4–13 cm long and 5–18 cm wide, orbicular, cordate or oval in outline; inflorescences paniculiform, mainly 6–30 cm long (or more); grapes 6–30 mm long (or more), juicy, palatable, seeded or seedless; 2n = 38, 76. Widely grown dessert and wine grape in Utah, persisting but not known from naturalized occurrences; introduced from Eurasia; 1 (i). The history of this grape is closely tied to Eurasian history, grapes having been grown since antiquity. Popular seedless and seeded hardy or semihardy varieties are grown in warmer portions of the state.

ZYGOPHYLLACEAE R. Br. in Flinders

Caltrop Family

Annual or perennial herbs or shrubs; leaves usually opposite, pinnately compound or digitately 2 or 3 (7)-foliolate, the leaflets entire; flowers perfect, regular or nearly so; sepals 5, distinct or united at the base; petals usually 5; stamens usually 10; pistil 1, the ovary superior, usually 4- or 5-loculed; style 1; fruit splitting into several nutlets; x = 6, 8–13+.

1. Flowers purple; stipules spiny; leaflets palmately 3-foliolate .... *Fagonia*
— Flowers yellow; stipules not spiny; leaves bifoliolate or pinnately plurifoliolate 2
2(1). Plants woody shrubs; leaves with 2 leathery resinous leaflets; fruit globose, villous .... *Larrea*
— Plants prostrate herbs; leaves pinnately compound with few to many leaflets; fruit not as above 3
3(2). Fruit armed with spines, splitting into 5, or fewer, 2-spined nutlets .... *Tribulus*
— Fruit not armed with spines, splitting into 10, merely tuberculate nutlets .... *Kallstroemia*

Fagonia L.

Perennial herbs, woody below; leaves opposite, usually 3- foliolate, the leaflets more or less spinose-tipped; stipules spinose; pedicels axillary, solitary, reflexed in fruit; sepals 5, caducous; petals 5, caducous; stamens 10; ovary 5-loculed, with 2 ovules per locule; fruit ovoid, deeply 5-grooved, separating into 5 ultimately dehiscent nutlets.

***Fagonia laevis* Standley** [*F. californica* authors, not Benth.?; *F. clutensis* var. *laevis* (Standley) Johnston]. Plants clump-forming, much branched, 1–4 dm tall; stems angled, smooth or scabrous above; stipules subulate, 1.5–4 mm long, spreading; leaves petiolate, the leaflets 3–13 mm long, 1–3 mm wide, lanceolate to oblanceolate, mucronate; pedicels 3–15 mm long; sepals 2–4 mm long; petals 5–9 mm long, purplish; fruit 3–5 mm long, reticulate, pubescent, tipped by a stylar beak 1.5–2 mm long. Creosote bush community at ca 915 m in Washington County (US); Arizona; 1 (0).

Kallstroemia Scop.

Annual herbs; stems prostrate; leaves even-pinnate; stipules membranous, deciduous; flowers solitary, axillary, 5-merous; stamens 10, the filaments slender; pistil 1, the carpels typically 5; fruit breaking into twice as many indehiscent, 1-seeded tuberculate, not spiny nutlets and leaving a more or less persistent axis.

***Kallstroemia californica* (Wats.) Vail** [*Tribulus californica* Wats.]. Stems prostrate to decumbent, branched, mostly 1.5–8 dm long, whitish hairy; stipules ca 2 mm long; leaflets in 2–7 pairs, oblong to elliptic, mainly 3–8 mm long and 1–3 mm wide; pedicels 0.5–3 cm long in fruit; sepals 3–4 mm long, deciduous; petals 3–5 mm long, yellow; fruit strigulose, ovoid-globose, ca 3 mm long, with sharp tubercles on the back of the carpels; beak glabrous, slender, 2–3 mm long. Mixed shrub and juniper communitiy at ca 1280 m in eastern Kane County; Arizona, California, and Mexico; 3 (ii).

Larrea Cav.

Resinous, aromatic, evergreen shrubs; leaves opposite, with 2, spreading, sessile, asymmetrical olive green leaflets; flowers solitary, yellow, showy; sepals 5, unequal, deciduous; petals 5, yellow; stamens 10, borne on a 10-lobed nectary disk; pistil 5-lobed and 5-carpelled; style slender; stigmas 5; fruit globose, hairy, separating into 5 indehiscent, 1-seeded carpels.

***Larrea tridentata* (DC.) Cov.** Creosote bush. [*Zygophyllum tridentatum* DC.; *L. glutinosa* Engelm.; *L. divaricata* authors, not Cav.]. Shrubs, mainly 1–3 m tall; stems flexuous, arranged in broadly rounded clumps; nodes with dark bands; stipules brown, persistent; leaflets 3–10 mm long, obliquely lance-ovate; pedicels 4–10 mm long; sepals 5–8 mm long, broadly elliptic, villous-strigose; petals yellow, 5–12 mm long; fruit 4–5.5 mm long, villous with long, white to tawny hairs; style persistent 5–9 mm long; 2n = 26, 52? Creosote bush and other warm desert shrub communities at 670 to 1130 m in Washington County; California, Nevada, Arizona, Texas, and Mexico; 34 (i). There are reports of the species from Iron County, but I have not seen specimens from there. The name *L. divaricata* Cav. has been applied to our material, but that name properly lies with South American plants, which are considered currently as distinct from the North American taxon. This species is a principal component of the Mohavean warm desert vegetative zone. The leaves have been used medicinally under the name "chaparral tea."

Tribulus L.

Annual, prostrate herbs; leaves even-pinnate with several to many leaflets; stipules membranous; flowers axillary, solitary usually 5-merous; stamens usually 10, with slender filaments; petals 5, bright yellow; sepals 5, caducous; ovary 5-lobed, 5-loculed; fruit 5-angled, horizontally compressed, pubescent, separating into 5, indurate, 3- to 5-seeded, indehiscent, 7-spined, nutlets.

Tribulus terrestris L. Puncture vine; Goat-head; Caltrop. Plants with prostrate spreading stems mainly 1–10 cm long; leaflets 3–8 pairs, 5–14 mm, oblong to elliptic, acute apically, pilose; petals 3–5 mm long, yellow; obovate; fruit segments crested, sculptured into elongate spinose protruberances, these spinulose-aristate, and bearing 2 puberulent spines mainly 3–6 mm long; 2n = 12, 24, 36, 48. Gardens, roadsides, sidewalks, and other open sites at 850 to 2135 m in Cache, Davis, Duchesne, Emery, Garfield, Grand, Kane, Millard, Salt Lake, San Juan, Uintah, Washington, and Weber counties; widespread in the U. S.; adventive from the Old World; 28 (ii). This tribulation of the earth is a vicious weed, leaving in its wake a refuse heap of punctured tires and painfully injured feet. This tribulation of the earth is adequately named scientifically.

CLASS LILIOPSIDA
The Monocots

Annual, winter-annual, or perennial herbs, or shrubs or trees; leaves alternate, spiral, or whorled, typically simple and parallel-veined; flower parts 3–merous; embryo with 1 cotyledon.

Key to the Families.

1. Plants woody or with leaves and/or stems persisting above ground throughout the year 2
— Plants not woody or the leaves and/or stems dying each year or persisting below ground only 3
2(1). Leaves sword-shaped or curved, apically spinose, laterally typically spinose or filiferous ... **Agavaceae** p. 647
— Leaves with palmately lobed, plaited blades, sometimes filiferous, and the petioles toothed or pronged, but not as above .... **Palmae** p. 817
3(1). Plants small, free-floating aquatics, without true stems .... **Lemnaceae** p. 799
— Plants with stems and leaves, terrestrial or aquatic but not free-floating 4
4(3). Perianth lacking or reduced and inconspicuous, the parts often bristlelike or scalelike but not petallike in color and texture 5
— Perianth well developed, at least the inner segments petaloid in color and texture 15
5(4). Flowers sessile in axils of chaffy or husklike scales; leaves with sheathing bases 6
— Flowers not in axils of chaffy bracts, sessile or pedicellate; leaves with or without sheathing bases 7
6(5). Leaf sheaths split lengthwise on side opposite the blade; leaves 2–ranked; stems mostly hollow and terete; anthers versatile; flowers subtended by 2 bracts .... **Gramineae** p. 684
— Leaf sheaths not split; leaves 3–ranked; stems mostly solid and triangular in cross section; anthers basifixed; flowers subtended by 1 bract .... **Cyperaceae** p. 653
7(5). Plants floating or submerged, usually not emergent .. 8
— Plants terrestrial or in shallow water, usually leaves and flowers both emergent 12
8(7). Flowers in spikes or heads 9
— Flowers axillary and solitary or few in clusters 10
9(8). Flowers imperfect, in globose heads, the lower pistillate, the upper staminate .... **Sparganiaceae** p. 822
— Flowers perfect, in peduncled or axillary spikes; sepals 4 .... **Potamogetonaceae** p. 818
10(8). Leaves alternate .... **Ruppiaceae** p. 822
— Leaves opposite 11
11(10). Leaves 3–10 cm long or more; carpels 2 or more ... . **Zannichelliaceae** p. 824
— Leaves mostly less than 3 cm long; carpel 1 .... **Najadaceae** p. 811
12(7). Inflorescence a dense, elongate spike 13
— Inflorescence in subglobose heads, racemes, or otherwise, but not in dense spikes 14
13(12). Inflorescence a double spike, staminate above and pistillate below; plants usually over 1 m tall .... **Typhaceae** p. 823
— Inflorescence a single spike (or spicate racmeme), the sexes intermingled or the flowers perfect; plants usually less than 1 m tall .... **Juncaginaceae** p. 798
14(12). Flowers imperfect, the lower heads pistillate; perianth inconspicuous, of chaffy scales .... **Sparganiaceae** p. 822
— Flowers perfect, usually not in heads; perianth usually 6–parted, in 2 whorls .... **Juncaceae** p. 791
15(3). Pistils several to many, 1–loculed and 1–ovuled, maturing into a cluster or whorl of achenes .... **Alismaceae** p. 650
— Pistil 1 per flower, 3– to 12–loculed, maturing into a capsule or a berry 16
16(15). Ovary superior 17
— Ovary inferior 18
17(16). Sepals green, rarely reddish; petals colored; flowers in umbels .... **Commelinaceae** p. 652
— Sepals and petals colored alike or, if unlike, the flowers not in umbels .... **Liliaceae** p. 800
18(16). Plants aquatic, with mostly submersed, whorled leaves .... **Hydrocharitaceae** p. 788
— Plants terrestrial, the leaves neither submersed nor whorled ' 19
19(18). Inflorescence a scapose umbel, sometimes reduced to a single flower, subtended by more or less membranous bracts; stamens 6 ...... **Amaryllidaceae** p. 651
— Inflorescence various but not an umbel, if subtended by a bract this not usually membranous (except in Iridaceae) 20
20(19). Fertile stamens 3; flowers regular, erect .... **Iridaceae** p. 789
— Fertile stamens 1 or 2; flowers irregular, typically nodding or bent .... **Orchidaceae** p. 812

AGAVACEAE Endl.
Agave Family

Plants succulent or subsucculent, xerophytic, acaulescent to arborescent, suckering, erect, simple or branched perennials; leaves basal and cauline, alternate (appearing spirally arranged), thick or thin and leathery, spine-tipped, the margins with spines or fibers, linear to ovate or oblanceolate; flowers mostly perfect or polygamous, small to large and often showy, white to cream, greenish, or yellowish orange, 3–merous, the perianth segments distinct or connate basally, similar in color and texture, fleshy; stamens 6; pistil 1, the ovary superior or inferior, 3–loculed, with usually numerous ovules; placentae axile or parietal; style 1, thick, often short; stigma 3–lobed or capitate; fruit a dry or fleshy capsule or berry; x = 16–30+.

1. Ovary inferior; perianth yellow, the segments connate basally into a short tube; stamens exserted 10–12 mm .... *Agave*
— Ovary superior; perianth white to cream or greenish, the segments distinct or nearly so; stamens included ... 2
2(1). Flowers small, less than 1 cm long, dryish; leaves linear, long and flexible, without an apical spine .... *Nolina*
— Flowers larger, mostly 3 cm long or more, fleshy; leaves with an apical spine .... *Yucca*

Agave L.

Rosettiform perennials, forming offsets at the base; roots hard, fibrous; stems thick, very short, simple or branched; leaves thick, spine-tipped, the margins armed with teeth or unarmed; flowering stems tall, bracteate, scapose; inflorescence spicate, racemose, or paniculate; flowers showy; perianth tubular to shallowly funnelform, the segments erect to curved, similar or dimorphic; stamens 6, exserted; filaments inserted on the tube or at the segment bases; ovary inferior, 3–loculed, succulent; pistil elongate; stigma 3–lobed, papillate-glandular; fruit a loculicidal capsule; n = 30.

Gentry, H. S. 1982. Agaves of contintental North America. Univ. Arizona Press. Tucson, Arizona.

***Agave utahensis* Engelm.** Utah Century-plant. [*A. newberryi* Engelm.; *A. scaphoidea* Greenm. & Roush, type from near St. George; *A. utahensis* var. *discreta* Jones]. Plants caespitose, the rosettes compact, pale grayish to yellowish green, 18–30 cm tall; leaves (8) 15–30 cm long, 1.5–3 cm wide, linear-lanceolate, stiff, straight or falcate, concave above, teeth brown-ringed around base, thickish, light gray, 2–4 mm long, the apical spine 20–40 mm long; inflorescence spicate, racemose, or paniculate, 1.2–4 m tall; flowers in clusters of 2–8, sessile or on short lateral branches, 25–30 mm long, urceolate, the segments 9–12 mm long, erect; stamens inserted near base of tube, 18–20 mm long, exserted from the perianth; capsule ovoid to oblong, 1–2 cm long. Dry slopes on limestone and sandstone in creosote bush, blackbrush, other warm desert shrub, and in the pinyon-juniper communities at 1000 to 1550 m in Washington County; Arizona and Nevada; 15 (v). The type was collected by Edward Palmer in 1870 near St. George. The species is very common, with thousands of individuals on the west slope of the Beaver Dam Mts. and also on the south slope of the Pine Valley Mts. In Paria Canyon, Coconino County, Arizona the species is known from within a short distance of the Utah boundary. That phase of the species is known as **var. *kaibabensis* (McKelvey) Breitung** [*A. kaibabensis* McKelvey]. Further exploration might demonstrate its existence in Kane County. Following the production of an inflorescence the rosette dies, and the offsets yield subsequent inflorescences, each in turn, dying.

Nolina Michx.

Perennials with a conspicuous caudex or stem; leaves numerous, linear, entire or with filamentous hairs or teeth on the margins, the apex acute but not spinelike; inflorescence a broad, racemose panicle, the main branches and pedicels subtended by scarious bracts; flowers very small, perfect and imperfect, whitish to yellowish white; perianth segments 6, all alike, distinct, oblanceolate to obovate; stamens 6; ovary superior, 3–loculed, with 2 ovules per locule; style short; stigma 3–lobed; capsule splitting or not, broadly 3–winged; n = 19.

***Nolina microcarpa* Wats.** Beargrass. Flowering stems stout, to ca 1.8 m tall, arising from a short caudex; leaves rather thick and somewhat keeled, 0.5–1 m long, 6–12 mm wide, serrulate to scabrous-denticulate, the apex fibrous-lacerate; panicle rather narrow, branched at the base, the branches to 3 dm long; bracts small; fruit thin and inflated, notched at both ends, ca 6 mm wide, on fruiting pedicels 4–6 mm long. Rocky slopes and ridges at ca 1200 m in Washington County (Cottam 5368 BRY); Arizona, New Mexico, and northern Mexico; 1 (0).

Yucca L.

Plants rosettiform from a caudex or arborescent, rhizomatous; leaves numerous, spirally arranged at stem or branch tips, elongate, thin and flaccid or thick and rigid, spine-tipped, with tough or fibrous margins; inflorescence a terminal raceme or panicle; flowers usually large and showy, numerous, perfect; perianth segments 6, thick, white to cream or greenish, oval to oblong or lanceolate; stamens 6; stigmas 3; fruit dry or fleshy, dehiscent or indehiscent; seeds flat, blackish; n = 30. Note: *Yucca filamentosa* L. (Adams-needle) is rather widely cultivated in Utah. It has somewhat flaccid leaves to 8 dm long and 2.5 cm wide. It flowers profusely in most years.

McKelvey, S. D. 1938. Yuccas of the Southwestern United States. Part I. Arnold Arboretum. Jamaica Plain, Mass. 150 pp.
———. 1947. Yuccas of the southwestern United States. Part II. Arnold Arboretum. Jamaica Plain, Mass. 150 pp.
Webber, J. M. 1953. Yuccas of the Southwest. USDA Agric. Monographs 17: 1–97.
Reveal, J. L. 1977. Yucca. In: Cronquist, A. et al. Intermountain Flora 1: 526–538.

1. Fruit indehiscent, fleshy or spongy and dry; plants treelike and with short non-filiferous leaves, or caespitose and with large, broad, swordlike leaves 2 cm or more wide 2
— Fruit dehiscent, dry; plants acaulescent or short-caulescent, with narrow, filiferous margined leaves often less than 2 cm wide 4
2(1). Plants treelike, usually branched; leaf blades seldom more than 2.5 dm long, minutely denticulate; fruit spongy and dry at maturity .... *Y. brevifolia*
— Plants treelike or not; leaf blades mostly more than 6 dm long, entire or filiferous; fruit fleshy at maturity 3
3(2). Plants treelike, with a well-defined stem; pistils 1.8–4 cm long; rare on the Beaver Dam Slope .... *Y. schidigera*
— Plants acaulescent, or with short, decumbent stems; pistils 4.5–9 cm long; common on the Beaver Dam Slope and east through southern Utah .... *Y. baccata*
4(1). Leaf blades concavo-convex, 0.6–5 dm long, lanceolate; inflorescence racemose, 3.5–7 dm long, on a short peduncle 1–4 dm long; style green, 9–13 mm long; fruit constricted .... *Y. harrimaniae*
— Leaf blades plano-convex, linear to acicular; style white to pale green; fruits constricted or not; peduncles and inflorescence much longer 5
5(4). Inflorescence paniculate; scape elongate, the first flower well above the foliage leaves; plants usually caulescent, with a short simple or branched stem to 1.3 m tall, in Washington County .... *Y. utahensis*

— Inflorescence racemose, or if paniculate, the distribution otherwise; peduncle short or elongate; plants acaulescent or only rarely short-caulescent 6

6(5). Peduncle short, less than 1 dm long; inflorescence subequal to the leaves or shorter, these together usually less than 1 m long; fruits pendulous, not constricted; plants of Garfield, Grand, Kane, and San Juan counties .... *Y. baileyi*

— Peduncle much longer; inflorescence elevated to well above the leaves; fruits erect or spreading 7

7(6). Capsules 3–4.2 cm long, deeply constricted; scape and inflorescence together reaching to 1.8 m tall; plants widespread .... *Y. angustissima*

— Capsules 4.5–7.5 cm long, moderately constricted; peduncle and inflorescence together mainly 2–3.5 m tall; plants rather restricted 8

8(7). Flowers 3–4.5 (5.2) cm long; pistil 1.5–3.2 cm long; peduncle 1.2–3.8 m tall, much surpassing the leaves; inflorescence commonly with 1–15 lateral branches; plants of Glen Canyon, Kane, Garfield, and San Juan counties .... *Y. toftiae*

— Flowers 5.5–6.5 cm long; pistil 3–3.5 cm long; peduncles 0.5–1 m long, usually shorter than the leaves; inflorescence simple or with 1 to few lateral branches; plants of sandy sites in Kane and Washington counties .... *Y. kanabensis*

Yucca angustissima Engelm. ex Trel. Narrow-leaved Yucca. [*Y. angustissima* var. *avia* Reveal, type from Loa Pass, Piute County]. Plants acaulescent or rarely short-stemmed, solitary or clumped with numerous rosettes; leaves plano-convex, linear, flexible, 1.5–4.5 dm long, 4–8 mm wide, the margins white or greenish white, in age eventually filiferous with fine curled fibers; inflorescence to 1.8 m tall, racemose, occasionally with a few branches, the peduncle much surpassing the leaves; flowers campanulate to globose, pendant, greenish white, the segments elliptic, 3–5 cm long, 2.5–3.5 cm wide, tinged dorsally with rose or rose purple; ovary 1–2.8 cm long; style white to pale green, 10–13 mm long; capsule 3–4.5 cm long, commonly with a deep constriction. Typically in sandy sites in warm desert shrub, sagebrush, pinyon-juniper, and ponderosa pine communities at 1100 to 2675 m in Garfield, Grand, Kane, Piute, San Juan, Sevier, Washington, and Wayne counties; Colorado to New Mexico, Arizona, and Nevada; 38 (xxvi). The var. *avia* Reveal of Piute county is at the northern limits of the species, and seemingly represents a high altitude form at the end of a cline. There are intermediates between narrow-leaved yucca and *Y. baileyi*, *Y. kanabensis*, and *Y. toftiae*. This plant was used prehistorically as a source of fiber for cordage and textiles. Both Basketmaker and Pueblo cultures depended on this and other species of yucca for many aspects of their civilization.

Yucca baccata Torr. Datil Yucca. Plants simple or clumped at ground level, rarely caulescent with 2–6 short, procumbent stems; leaves broadened near the middle, straight or incurved, occasionally twisted, rather deeply concavo-convex, rigid, 3–10 dm long, (1.5) 2.5–5 (6) cm wide, the margins usually with coarse fibers; inflorescence 3.5–6.5 dm long, paniculate, hidden by the leaves or only somewhat surpassing them; flowers campanulate, pendent, white to cream colored, commonly tinged with purple, the segments lanceolate to oblanceolate, 4–8 (11.5) cm long, 1.4–3 cm wide; ovary 3.8–7 cm long, narrowly ovoid and tapering to the style, this 4–10 mm long, tapered; capsule 8–17 cm long or more, 3–6 cm wide, fleshy and sweet at maturity. Two rather distinctive varieties occur in Utah.

1. Leaf blade straight, dark green, glabrous; scape and pedicels green, typically surpassing the leaves; plants broadly distributed .... *Y. baccata* var. *baccata*
— Leaf blade falcate, blue green, glaucous; scape and pedicels reddish purple, typically shorter than the leaves; plants of Kane (?) and Washington counties .... *Y. baccata* var. *vespertina*

Var. *baccata* Dry, sandy or rocky slopes and mesas in pinyon-juniper, sagebrush, mountain brush, and ponderosa pine communities at 1200 to 2400 m in Beaver, Iron, Kane, San Juan, and Washington counties; California, Nevada, Arizona, Colorado, New Mexico, Texas, and Mexico; 8 (iii).

Var. *vespertina* McKelvey Creosote bush, Joshua tree, blackbrush, and other warm desert shrub and pinyon-juniper communities at 750 to 1700 m in Kane (?) and Washington counties; California, Nevada, and Arizona; 9 (iv).

Yucca baileyi Woot. & Standl. Bailey Yucca. [*Yucca standleyi* McKelvey; *Y. navajoa* Webber]. Plants acaulescent or rarely with a short, ascending stem, solitary or clumped; leaves linear, plano-convex, flexible, often falcate, 2–6 dm long, 3–8 mm wide, pale green or yellow green, the margin white, soon filiferous; inflorescence racemose, 3.5–12 dm long, unbranched, usually not much surpassing the foliage, densely flowered; flowers campanulate to globose, greenish white, tinged with purple, the segments thin, ovate to obovate, 4–6 cm long, 2–3 cm wide; ovary 2.3–2.9 cm long, oblong to obovoid; style white to pale green, 7–9 mm long; capsule oblong, seldom constricted, 4–7 cm long, 2–5 cm wide, pendulous at maturity. Grassland, Mormon tea, Vanclevea, mixed warm desert shrub, pinyon-juniper, and ponderosa pine communities at 1200 to 2200 m in Garfield, Grand, Kane, and San Juan counties; Colorado, Arizona, and New Mexico; 9 (i).

Yucca brevifolia Engelm. Joshua Tree. [*Y. draconis* var. *arborescens* Torr.; *Y. arborescens* (Torr.) Trel.]. Plants arborescent, with 1 to several large stems, mainly 1–4 m tall, the branches erect or ascending (or pendulous in age); leaves numerous, linear-acicular, spinulose apically, plano-convex to triquetrous, rigid, 1.5–3.5 dm long, 0.7–1.5 cm wide, the margin thin, horny, minutely denticulate; inflorescence a compact panicle, 2–4 dm long, dense, ovoid; flowers ellipsoid to globose, scarcely expanding, greenish-white to cream, the segments thick and succulent, oblong to lanceolate, 2.5–4.5 (5) cm long, 1–2 cm wide; ovary 2–3.5 cm long, tapering from base to apex, pale green; style 0–4 mm long; capsule ellipsoid, 6–8.5 cm long, 3–4.5 cm wide, dry and spongy at maturity. Bajadas and slopes in creosote bush, blackbrush, other warm desert shrub, and pinyon-juniper communities at 800 to 2200 m in Washington County; California, Nevada, and Arizona; Mexico; 7 (v). A single Joshua tree waif stands near the Red Mound in the Monroe Hot Springs area of Sevier County. Its status as indigenous or introduced is recondite. The species is successfully grown as far north as Salt Lake City.

Yucca harrimaniae Trel. Harriman Yucca. [*Y. harrimaniae* var. *gilbertiana* Trel., type from Juab County; *Y. gilbertiana* (Trel.) Rydb.]. Plants acaulescent, forming dense small clumps of 1–22 rosettes; leaves falcate or straight, lanceolate to spatulate-lanceolate, concavo-convex, deeply striate, rather thick and rigid, pale green, pungent apically, 1–5 dm long, 0.7–4 cm wide, the margin white or brown, in age filiferous, the fibers somewhat coarse and curly; inflorescence 3.5–7 dm tall, racemose or rarely with a few short branchlets, extending from within the foliage to well above; flowers broadly campanulate, pendent, yellowish or greenish yellow to cream, tinged with purple, the segments broad, 4–5 (6) cm long, 1.6–3.5 cm broad; ovary 1.5–2 cm long, pale green; style 9–11 mm long, bright green; capsule cylindric, with a short attenuate beak, 3.7–5 (6) cm long, usually deeply constricted toward the center and flaring open when dried. Warm desert shrub, grasslands, sagebrush, pinyon-juniper, and mountain brush communities at 1200 to 2700 m in Beaver, Carbon (type from near Helper), Duchesne, Emery, Garfield, Grand, Iron, Juab, Millard, Piute, San Juan, Sevier, Uintah, and Wayne counties; Nevada, Colorado, Arizona, and New Mexico; 42 (xii). This species, with both Great Basin and Colorado Drainage distribution, is the most widely distributed of our species. It forms apparent intermediates with *Y. angustissima* in eastern Utah. Plants from western Utah have been regarded as *Y. harrimaniae* var. *gilbertiana*. Except for the more robust size of most individuals, the Great Basin materials seem not to differ substantially. In the lower elevation portions of the Uinta Basin there is a strongly rhizomatous phase of the species with flaccid leaves that tend to sprawl on the surface of the ground. Leaf margins tend to be less filiferous than in the species proper. These plants have been collected in flower, but are not known to produce fruit. They are designated as **var. sterilis Neese & Welsh** (type from Uintah County).

Yucca kanabensis McKelvey Kanab Yucca. [*Y. angustissima* var. *kanabensis* (McKelvey) Reveal]. Plants acaulescent or short-caulescent, forming small to large clumps; leaves linear, 4.5–8 dm long, 7–15 (20) mm wide, concavo-convex, stiff but flexible, the margin white, soon with straight or curled fibers; inflorescence mostly 1.5–3 m tall, racemose or with 3–4 short branchlets, the short peduncle and lower portion of inflorscence usually included among the leaves; flowers large, campanulate to globose, white to greenish white or cream, often tinged purplish dorsally, the segments elliptic to nearly orbicular (4) 5.5–6.5 cm long, 2.3–4 cm broad; ovary oblong, 2.3–3 cm long; style short, 3–7 mm long; capsule cylindric, oblong, 6.5–7.5 cm long, sometimes constricted, roughened. Sandy soils in sagebrush, mountain brush, pinyon-juniper, and ponderosa pine communities at 1300 to 2300 m in Kane (type from 8 mi nw of Kanab) and Washington counties; Arizona; 16 (vi). The Kanab yucca occurs mainly from Zion Canyon east to Johnson Canyon, with the greatest concentration northwest of Kanab, in the vicinity of the Coral Pink Sand Dunes. It is truely one of our most striking species. It forms intermediates westward with *Y. utahensis* and eastward with *Y. angustissima*.

Yucca schidigera Roezl ex Ortega Plants shrublike or arborescent, single or clumped; stems erect, mainly 0.5–2.5 m tall, simple or branched; leaves linear to linear-lanceolate, 3–10 dm long, 2.5–5 cm broad, concavo-convex, thick, very rigid, yellow green, the margin thick, usually with coarse curved fibers; inflorescence paniculate, 5–13 dm long, mostly contained within the foliage; flowers globose, white or cream, tinged with purple, the segments lanceolate to broadly lanceolate, 2.4–4.5 cm long, 0.6–1 cm wide; ovary 1.8–2.5 cm long; style 1–2 mm long; fruit cylindrical, 7–10.5 cm long, fleshy, sweet and succulent. Bajadas in mixed warm desert shrub and Joshua trees at 900 to 1500 m in Washington County; California, Nevada, and Arizona; 5 (ii). The plant occurs as an introduction at Castle Cliffs, and is native on the bajadas along the road to Beaver Dam Wash west of Castle Cliffs.

Yucca toftiae Welsh Toft Yucca. [*Y. angustissima* var. *toftiae* (Welsh) Reveal]. Plants acaulescent or with stems to 1 m long, solitary or clumped; leaves linear 2–8 dm long, 4–17 mm wide, plano-convex to concavo-convex, green to yellow green, the margins white, soon filiferous; inflorescence mostly 1.2–3.4 m tall, racemose or paniculate (with 2–18 lateral branches), the peduncle much surpassing the foliage; flowers campanulate to globose, white to cream, the segments elliptic to lance-ovate, acute, 2.5–5.2 cm long; ovary 1.5–3.2 cm long; style (3) 7–11 mm long; capsule erect, cylindrical, woody, and somewhat constricted near the middle, 4.6–5.5 (6) cm long. Sandstone outcrops and hanging gardens in warm desert shrub communities along Glen Canyon (type locality) and the San Juan River arms of Lake Powell and the southern portion of the Kaiparowits Plateau in Garfield, Kane, and San Juan counties; endemic; 5 (iii). McKelvey indicates that this plant was described to her in a letter received from Miss E. U. Clover, who noted the plant in her exploration of the canyons of the Colorado.

Yucca utahensis McKelvey Utah Yucca. [*Y. elata* var. *utahensis* (McKelvey) Reveal]. Plants usually caulescent, the stems to 1.3 m tall, forming clusters of up to 15 rosettes; leaves linear, (2) 4–7.5 dm long, 7–20 mm wide, broadest near the middle, concavo-convex, flexible, yellow green, the margins white, with a few fibers; inflorescence paniculate, 1.2–2 m long, the peduncle produced well above the foliage; flowers campanulate, cream, the segments broadly elliptic or ovate, 3.5–5 cm long, 2–2.5 cm wide; ovary 2.5–3 cm long; style white, very slender, 5–10 mm long; capsule oblong-cylindrical, 4–7 (8) cm long, not or only slightly constricted. Sandy soils in creosote bush, blackbrush, sand sagebrush, other warm desert shrub, and pinyon-juniper communities at 930 to 1850 m in Washington (type from northwest of St. George) County; Nevada and Arizona; 11 (iv). This entity is disjunct from *Y. elata*. It is in contact, instead, with *Y. kanabensis*, with which it forms occasional intermediates, and with whom its alliance is as great or greater than with the allopatric *Y. elata*.

ALISMACEAE Vent.

Water-plantain Family

Herbaceous aquatic or marsh perennials from a thickened rootstock and fibrous roots; leaves basal, long-petioled, erect or floating, the veins longitudinal and prominent; flowers perfect or imperfect, subtended by a whorl of bracts, in racemes or panicles; sepals 3, green; petals 3, white or pink, often large and showy; stamens 6 or more, distinct, hypogynous, the anthers 2–loculed; pistils several to many; style terminal, persistent; $x = 6, 7, 8, 11$.

1. Leaves sagittate; flowers dioecious, monoecious, or sometimes polygamous .... *Sagittaria*
— Leaf blades elliptic, ovate, or rarely cordate, but not sagittate; flowers perfect 2

2(1). Stamens 6 (rarely 9); flowers rather small and inconspicuous; petals 2–6 mm long; achenes arranged in a single whorl on the receptacle .... *Alisma*
— Stamens 12; flowers rather showy; petals 5–10 mm long; achenes in a burlike head .... *Echinodorus*

Alisma L.

Scapose perennial herbs with fibrous roots; leaves basal, erect or floating; flowers in compound, whorled panicles, perfect and somewhat inconspicuous; sepals 3, broad, persistent; petals 3, white or pink, deciduous; stamens 6 (rarely 9); carpels 10–25 in a single whorl, each distinct; fruit of achenes.

1. Leaves broad, usually 3–8 cm wide, broadly elliptic to ovate or oblong; achenes with a central groove along the center of the outer edge and top, 2–2.8 mm long, somewhat longer than broad; fruiting pedicels ascending to erect, at least some over 2 cm long ... *A. plantago-aquatica*
— Leaves narrow, mostly 0.3–2 cm wide, narrowly elliptic to linear; achenes with a central ridge and a groove on the center of the outer edge and top, ca 2 mm long, suborbicular; pedicels spreading to recurved, mostly less than 2 cm long .... *A. gramineum*

Alisma gramineum Lej. [*A. geyeri* Torr. ex Nicollet]. Perennial herbs 0.6–5.2 dm tall; leaves all basal, erect or floating, often extending above the inflorescence, long-petiolate, the blade 2.5–7 (9) cm long, 0.5–2 cm wide, narrowly elliptic to linear, the petiole sheathing, equaling the blade in length or more often much longer; scapes bracteate, paniculately compound, erect, mostly shorter than the leaves; bracts lanceolate, papery; pedicels spreading to recurved, mostly less than 2 cm long; sepals green, 1.5–3 mm long, persistent; petals white or pinkish, 2–4 mm long; stamens 6–9; pistils 10–25 in a single whorl; style curved, 0.4–0.5 mm long; achenes ca 2 mm long; $2n = 14, 16$. Ponds, reservoirs, and marshes, growing in mud or submersed, at 1400 to 2260 m in Duchesne, Millard, Rich, and Uintah counties; widely distributed in North America; Eurasia; 14 (ii).

Alisma plantago-aquatica L. [*A. triviale* Pursh; *A. brevipes* Greene]. Perennial herbs 2–8.9 dm tall; leaves all basal, erect, often shorter than the inflorescence, long-petiolate, the blade 4–17 cm long, 2–8 cm wide, broadly elliptic to ovate or oblong, the petiole 2.5–32 cm long, sheathing; scapes erect, 1 to few and mostly paniculately compound, bracteate, longer than the leaves; bracts lanceolate, papery, 2–3 mm long; pedicels ascending to erect, 0.5–4 cm long; sepals greenish, 3–4 mm long, persistent; stamens 6; pistils 10–22, in a single whorl; style usually straight, 0.5–0.7 mm long; achenes 2–2.8 mm long; $2n = 12, 14, 24, 28$. Mud flats, open water and wet meadows at 1340 to 2320 m in Cache, Daggett, Rich, Utah, Washington, and Weber counties; widely distributed in the U. S.; Canada; Eurasia, Africa, Australia, and South America; 19 (i).

Echinodorus Rich.

Submersed or amphibious annuals from fleshy rhizomes; leaves erect or ascending, emergent or floating, all basal; flowers several, bracteate, in panicles or racemes, perfect, 3-merous; sepals persistent, green; petals deciduous; stamens 12; pistils in a dense cluster, numerous; achenes flattened.

Echinodorus rostratus (Nutt.) Engelm. ex Gray [*Alisma rostrata* Nutt.; *E. rostratus* var. *lanceolatus* Engelm. ex Wats. & Coult.]. Annual herbs, mainly 1.1–3.7 dm tall; leaves basal, erect to ascending or floating, the blades elliptic to broadly lanceolate, 3–9 cm long, 0.5–4.5 cm wide, sheathing, petioled, the petioles subequal to the blade or somewhat longer, shorter than the inflorescence; scapes erect, few, the flowers clustered; bracts linear-lanceolate with hyaline margins; pedicels spreading to ascending, to 1.4 cm long; sepals broad, persistent, dark green, 2–4 mm long; petals rhombic, 5–9 mm long, white to greenish white; stamens 1–2 mm long; pistils many, compact; style longer than the ovary; achenes 2–3 mm long, ribbed, the beak stout, 1–2 mm long, persistent, the fruiting head burlike. Mud flats, lake shores, and marshes at ca 1400 to 1450 m in Millard and Uintah counties; California to Texas, east to Illinois, and south to Mexico; 5 (i).

Sagittaria L.

Perennial aquatic herbs; leaves erect to spreading or floating, long-petioled, sagittate; inflorescence a long, verticillate panicle; flowers mostly monoecious; sepals persistent, greenish; petals deciduous, white, longer than the sepals; stamens many; pistils many on an enlarged receptacle; fruit an achene.

Sagittaria cuneata Sheldon [*S. arifolia* Nutt., *S. arifolia* var. *cuneata* Lunell]. Submersed to emergent, glabrous, perennial herbs, mainly 1.1–6 dm tall; leaves basal, the blades typically sagittate with the two basal lobes narrower and shorter than the upper one, this 2–11 cm long, 1–7 cm wide; petioles sheathing, to 4.5 dm long and much broader near the base; scapes simple or branched, erect, with 2–7 whorls of flowers; bracts lanceolate, membranous, 6–20 mm long; pedicels mostly ascending, 5–33 mm long; staminate flowers usually above and the pistillate below; sepals 4–8 mm long, persistent, ovate; petals 6–14 mm long, white, with a narrow claw and an expanded blade; stamens 15–22; pistils many; fruiting heads 1–1.5 cm thick, globose; achenes obovate, flattened, 2–3 mm long, with a short, erect, straight or curved beak and dorsal wing. Streams, marshes, lakes, and mudflats at 1370 to 2745 m in Beaver, Cache, Daggett, Garfield, Grand, Piute, Rich, Sanpete, Sevier, Uintah, Utah, and Weber counties; rather widely distributed in the U. S.; Canada; 30 (i). Another species, *S. latifolia* Willd., might occur in Utah, but no specimens were seen by me. It is distinguished from *S. cuneata* in having achenes with beaks extending at right angles to the upper body, the leaf lobes more or less equal in length, and bracts mostly 5–10 mm long.

AMARYLLIDACEAE J. St. Hil.
Amaryllis Family

Perennial acaulescent herbs from corms or bulbs; leaves basal and alternate, entire; flowers perfect, the perianth of 6 parts; stamens 6; stigma 3–lobed; ovary inferior, 3–loculed; fruit a loculicidal capsule or a berry; seeds mainly flat or subglobose, in 2 rows per locule; $x = 5, 7, 9, 10–15$. Note: The members of this family keyed below are all known from introduced, cultivated individuals. However, some persist for long periods of time following cultivation and others occur as waifs along roadsides and at refuse dumps.

1. Flowers white, nodding, borne in late winter and early spring, as the snow melts .... *Galanthus*
— Flowers yellow, cream, pink, scarlet, red, candy-striped, or white, borne in spring or under glass in autumn, winter, and spring 2
2(1). Flowers with a definite corona, this forming a tube or ring between the perianth and stamens; perianth typically yellow or cream, less commonly white; plants typically grown outside .... *Narcissus*
— Flowers with corona inconspicuous, if formed at all; perianth commonly red, pink, or white; plants typically grown inside .... *Hippeastrum*

Galanthus L.

Bulbous, scapose herbs, the bulbs with membranous outer scales; leaves typically 2, linear to narrowly oblong; flowers solitary; perianth segments distinct, the 3 inner ones shorter, overlapping and appearing tubular; fruit a berry.

***Galanthus nivalis* L.** Snowdrop. Plants from small bulbs; leaves 2, appearing with the flowers, linear, mainly 6–10 cm long and 6 mm wide; flowers solitary, nodding, white, subtended by a papery spathe, this split on one side, 20–25 mm wide, the outer segments white, spreading, the inner erect, white, with green markings around the sinus apically; fruit green; 2n = 18, 24, 36. Commonly grown ornamental, flowering optimistically as early as mid-February, while snow persists on the beds, in most if not all Utah counties; introduced from Europe; 2 (0).

Hippeastrum Herbert

Bulbous, scapose herbs, the bulbs with papery outer scales; leaves linear to oblong, usually several; umbels 1- or more-flowered, usually declined, subtended by distinct spathaceous bracts; perianth tube short; stamens declined, unequal, of several different lengths; style declined, the stigma 3-lobed; fruit a loculicidal capsule.

***Hippeastrum striatum* (Lam.) H. Moore** Amaryllis. [*Amaryllis striata* Lam.]. Leaves strap-shaped, 10–30 cm long or more, 1–4 cm wide; scapes 1.5–4 dm tall, fistulose; spathes green or whitish, or suffused with anthocyanins; umbels 1- to 4-flowered; flowers large and showy, crimson, pink, white, scarlet, or salmon, mainly 5–15 cm long and as broad or broader; fruit seldom formed. Cultivated ornamental of startling beauty, grown mainly under glass, and offered for sale mainly in winter and early spring; sometimes grown outside in summer; introduced from South America; 1 (0). Material included in the above description is probably of hybrid origin, involving ***H. aulicum* (Ker Gawler) Herbert, *H. elegans* (Sprengel) H. Moore, *H. reginae* (L.) Herbert,** and others.

Narcissus L.

Scapose bulbous herbs; leaves linear, usually several; flowers 1 to several, borne in umbels subtended by a single bracteate spathe; perianth yellow or white, with a long and tubular or short ringlike corona separate from the filaments; fruit a capsule with subglobose black seeds.

1. Leaves nearly terete, narrowly channeled on the face; corona much shorter than the perianth segments; flowers yellow, 2–6 per umbel .... *N. jonquilla*
— Leaves flat or nearly so; corona shorter to longer than the perianth segments; flowers yellow, cream, or white, solitary to many per umbel 2
2(1). Corona equaling or surpassing the perianth segments, these spreading; our most common species ....
    .... *N. pseudo-narcissus*
— Corona less than half as long as the perianth segments; plants common to uncommon 3
3(2). Corona with the margin red to red orange, shallowly cup-shaped; flowers mostly solitary .... *N. poeticus*
— Corona not margined with red, deeply cup-shaped to shortly tubular; flowers often 4 or more per umbel ...
    .... *N. tazetta*

***Narcissus jonquilla* L.** Jonquil. Leaves mainly 10–30 cm long, narrow; flowers mostly 2–6, yellow, fragrant; perianth tube 2–2.5 cm long; corona margin undulate, less than half as long as the spreading perianth segments; 2n = 14, 21. Commonly grown ornamental, long persisting in lower elevation portions of Utah; introduced from Europe and Africa; 1 (0).

***Narcissus poeticus* L.** Poets Narcissus. Leaves mainly 10–30 cm long, 4–7 mm wide; flowers mostly solitary, white, fragrant; perianth tube mainly 20–25 mm long; corona short, with undulate red margin, much shorter than the segments; 2n = 14, 21, 28. Commonly grown ornamental of charm and beauty, long persisting in lower elevation portions of Utah; introduced from Europe; 1 (0).

***Narcissus pseudo-narcissus* L.** Daffodil. Leaves mainly 1.5–30 cm long and 10–20 mm wide; flowers solitary, spreading or deflexed, yellow, with greenish perianth tube mainly 12–20 mm long; corona undulate-crisped, bright yellow, subequal to or surpassing the perianth segments; 2n = 14, 15, 21, 26–29, 30, 35+. Commonly grown ornamental, often available as cut flowers, in much of lower elevatión portions of Utah; introduced from Europe; 5 (ii).

***Narcissus tazetta* L.** Polyanthus Narcissus. Leaves mainly 10–30 cm long and 15–20 mm wide; flowers typically 4–8, white, fragrant; perianth tube mainly 20–25 mm long; corona pale yellow or white, much shorter than the segments; 2n = 20, 30, 32. Occasionally grown ornamental, but commonly forced, and offered for sale as early as late autumn; introduced from Europe; 2 (0).

COMMELINACEAE R. Br.
Spiderwort Family

Herbaceous perennials; stems erect, mostly glabrous, stout; leaves typically cauline, alternate, narrow, distinctly parallel-veined; flowers perfect, showy, 3-merous, differentiated into sepals and petals; sepals distinct, imbricate, green; petals free, alternate with the sepals and colored; stamens 6, in 2 series, the filaments flattened basally; anthers basifixed; ovary superior, sessile, 3-loculed, with 1 or 2 ovules per locule; style 1; fruit a loculicidal capsule, enclosed by fleshy sepals; seeds 1–6; x = 4–19.

Tradescantia L.

Plants with thickened, fleshy roots; leaves alternate, linear-lanceolate, sessile, sheathing at the base; flowers blue, violet, or magenta, subtended by leafy bracts, pedicellate; cymes terminal; sepals 3, distinct, persistent, glandular; petals purple, homomorphic, separate; stamens 6, the filaments pilose; ovary with the stigma capitate; seeds small.

Tradescantia occidentalis (Britt.) Smyth Spiderwort. [*T. virginiana* var. *occidentalis* Britt. in Britt. & Br.]. Erect perennial, 2–4 (5) dm tall; stems strict or ascending, glabrous, glaucous; leaves linear-lanceolate, membranaceous, long-acuminate, stiff, divaricate, 0.7–3.7 dm long, 0.2–1 cm wide, the uppermost the smallest, swollen at juncture with the node and forming a sheath; flowers in umbellate clusters, terminating the main stem and axillary, the clusters subtended by 2 or 3 leafy, divaricate bracts up to 2 dm long; flowers long-pedicellate, the pedicels erect or spreading to deflexed, glandular, 1–2 cm long, broadly campanulate; sepals green or tinged with purple, elliptic, glandular-puberulent or glabrous; petals broadly ovate, 7–15 mm long, blue to magenta; filaments densely pilose; capsules oblong, pubescent, 4–7 mm long; seeds 2–4 mm long, compressed; 2n = 24. Mostly in sandy substrates of the lower ponderosa pine community and mixed desert shrub and shrub-grass communities at 1250 to 2135 m in Garfield, Kane, San Juan, and Washington counties; Arizona east to Texas, north to North Dakota; 33 (iii). The ornamental, *T. virginiana* L., is occasionally cultivated in Utah. It is distinguished by its glabrous to villous, but not glandular sepals.

CYPERACEAE A. L. Juss.

Sedge Family

Grasslike, perennial or annual herbs; culms (stems) triangular, rounded, or flattened in cross section, not jointed; leaves simple, linear, entire, 3-ranked, sometimes reduced to bladeless sheaths, these mostly closed; inflorescence spicate, racemose, or umbellate, of 1 to many spikes or spikelets; flowers much reduced, imperfect or perfect, sessile or nearly so, subtended by a very small bract (lemma), referred to as a scale; perianth lacking or reduced to bristles; stamens usually 3, rarely 1 or 2, exserted at anthesis, the anther basifixed; ovary superior; stigmas 2 or 3, the achenes accordingly lenticular or trigonous; x = 5–60+.

Cronquist, A., A. H. Holmgren, N. H. Holmgren, J. L. Reveal, and P. K. Holmgren. 1977. Intermountain flora, Vol. 6, Columbia University Press, New York, 584 pp.

1. Achenes enclosed by or folded within a closed or open sac or small bract as well as subtended by a scale; perianth bristles lacking; flowers imperfect 2
— Achenes exposed, neither enveloped by nor enfolded in a sac or bract, merely subtended by a scale and usually by perianth bristles (except in *Cyperus*); flowers mostly perfect (except in *Cladium*) 3

2(1). Achene completely enclosed in a sac (perigynium), this closed except at the apex, through which the stigmas are exserted, concealing the attachment of the style to the achene .... *Carex*
— Upper part of achene usually exposed, the subtending bract with open margins, exposing the attachment of the style to the achene .... *Kobresia*

3(1). Perianth bristles more than 10, much exceeding the scales, appearing as cottonlike tufts; plants of the Uinta Mts. .... *Eriophorum*
— Perianth bristles fewer than 10, equal to or shorter than the scales, not forming cottonlike tufts; plants widespread 4

4(3). Inflorescence a solitary spike, not subtended by leafy bracts; leaves reduced to bladeless sheaths (*Scirpus caespitosus* is also keyed here) .... *Eleocharis*
— Inflorescence usually with few to many spikelets, these sometimes umbellate, subtended by 1 or more leafy bracts; leaves various but some commonly with conspicuous blades 5

5(4). Lower scales empty; flowers in each spike all staminate except the terminal perfect one; plants 1–2 m tall, with leafy stems and foliose bracts, of canyonlands in Kane and San Juan counties .... *Cladium*
— Scales typically all subtending flowers, these mostly perfect; plants various, but seldom if ever as above 6

6(5). Scales arranged in 2 vertical ranks; perianth lacking .. .. *Cyperus*
— Scales spirally arranged; perianth various 7

7(6). Perianth lacking; styles thickened toward the base; plants rather rare .... *Fimbristylis*
— Perianth commonly of bristles; styles not thickened toward the base; plants often common .... *Scirpus*

Carex L.

Plants perennial, monoecious or occasionally dioecious; stems tufted or arising singly or few together from creeping rhizomes; spike solitary or few to several, sessile, congested or remote, pedunculate, short to elongate, staminate, pistillate, androgynous (staminate flowers borne above the pistillate ones), or gynaecandrous (pistillate flowers borne above the staminate ones), each with few to many unisexual, sessile or short-stipitate flowers; perianth lacking; staminate flowers with 2 or 3 stamens, the pistillate of a pistil, this enveloped in a closed scale or saclike perigynium with the stigmas exserted from the apex.

Hermann F. J. 1970. Manual of the Carices of the Rocky Mountains and Colorado Basin. U. S. D. A. Agr. Handbook No. 374. 397 pp.
Lewis M. E. 1958. *Carex*—its distribution and importance in Utah, Brigham Young Univ. Sci. Bull., Biol. Ser. 1, No. 2, Provo, Utah. 43 pp.
Mackenzie K. K. 1931–35. *Carex* L. Fl. N. Amer. 18: 9–478.

1. Spike solitary, the perigynia attached directly to the rachis .... Key 1
— Spikes more than 1, the perigynia attached to a rachilla and the rachilla attached to the rachis 2

2(1). Spikes all sessile, aggregated into a capitate or spicate inflorescence, the terminal one not staminate and the lower ones commonly not all pistillate except in imperfect specimens 3
— Spikes peduncled (at least some) or, if sessile, well separated from the upper ones or the terminal one staminate; lower 1–5 spikes typically all pistillate; plants monoecious 6

3(2). Stigmas 3; achenes trigonous; inflorescence with 3–5 spikes, often subtended by a leaflike bract; terminal spike gynaecandrous, the lateral ones all pistillate; pistillate scales black or blackish purple (section Atratae) .... Key 5
— Stigmas 2; achenes lenticular; inflorescence commonly with more than 5 spikes, or the pistillate scales paler than above, not subtended by a leaflike bract (except in *C. athrostachya*); spikes typically all androgynous or all gynaecandrous 4

4(3). Spikes typically androgynous or the plants with well-developed creeping rhizomes .... Key 2
— Spikes gynaecandrous; plants caespitose; rhizomes lacking or very short 5

5(4). Perigynia round-margined, not winged, not conspicuously flattened, mostly less than 3.5 mm long (longer in *C. deweyana*); scales pale green to brown; inflorescence commonly less than 2 cm long and/or less than 1 cm wide; plants mostly of wet places .... Key 3
— Perigynia wing-margined, often conspicuously flattened, (2.5) 3.5–7.5 (8) mm long; scales commonly brownish to dark brown, often with a green midrib; inflorescence often longer or wider than above; plants of dry or wet places (Section Ovales) .... Key 4

6(2). Terminal spike gynaecandrous, the lateral ones all pistillate or gynaecandrous with very few staminate flowers; pistillate scales dark brown purple to black; plants tufted, lacking rhizomes or these very short (except in *C. buxbaumii*) .... Key 5
— Terminal spike and sometimes 1 or more lateral ones staminate or androgynous (the terminal spike occasionally gynaecandrous in a few taxa but the scales then pale green or, if brown, the culms from slender rhizomes) ... 7

7(6). Stigmas 2; achenes lenticular; scales often black or black purple or with blackish lines flanking a greenish or pale midstripe, often contrasting with the greenish or stramineous perigynia; plants commonly of wet places .... Key 6
— Stigmas 3; achenes trigonous; scales greenish or brownish or, if blackish, usually of the same color as the perigynia (except in *C. raynoldsii* of well-drained soil) .. 8

8(7). Spikes all androgynous; staminate flowers few, inconspicuous and exceeded by the upper perigynia; lower pistillate scales leaflike, greatly exceeding the perigynia, commonly 1–7 cm long; achenes ca 3 mm long ... . *C. backii*
— Spikes rarely all androgynous; pistillate scales not leaflike, much shorter than above; achenes mostly smaller .... Key 7

Key 1.

Spike solitary.

1. Spike unisexual; scales very dark brown puple to black 2
— Spike androgynous; scales green to dark brown 4

2(1). Perigynia glabrous or nearly so; unusual specimens .... *C. parryana*
— Perigynia pubescent 3

3(2). Plants densely caespitose, lacking rhizomes or these very short, 21–40 cm tall, of canyonlands in Kane and San Juan counties .... *C. curatorum*
— Plants rhizomatous, loosely to not at all caespitose, 4–34 cm tall, montane in the northern 1/2 of Utah . ... *C. scripoidea*

4(1). Plants densely caespitose, often fasciculate; rhizomes lacking or not evident; leaf blades rarely over 1 mm wide 5
— Plants caespitose or not, but with conspicuous rhizomes, or if rhizomes lacking, the leaves over 1 mm wide 9

5(4). Perigynia blunt at the retuse apex, not conspicuously beaked; leaves and culms lax to flaccid; plants rare, of very wet sites at ca 2200 m in the Uinta Mts . *C. leptalea*

— Perigynia beaked; leaf blades and stems rather strict; plants mostly of dry places and/or of higher elevations 6

6(5). Rachilla lacking; perigynia shortly stipitate, somewhat to widely spreading; pistillate scales deciduous in age, about as wide as the perigynia; leaves flat or slightly channeled; plants rare .... *C. pyrenaica*
— Rachilla subequal to or longer than the achene; perigynia sessile, appressed or nearly so; pistillate scales persistent, usually wider than the perigynia; leaves involute; plants variously abundant 7

7(6). Perigynia minutely hirtellous at least in the upper 1/2 (use 15–20x magnification); spikes 1.3–2.5 cm long, the staminate portion exceeding the uppermost perigynia by ca 8–13 mm; plants below 2300 m .... *C. filifolia*
— Perigynia glabrous or serrulate-ciliolate or with a few minute hairs near the beak; spikes 0.5–1.8 cm long, the staminate portion exceeding the uppermost perigynia by 2–7 mm; plants above 2800 m 8

8(7). Perigynia with conspicuous flattened margins, these commonly serrulate-ciliolate; spikes equaling or included in the leaves, the staminate part inconspicuous, exceeding the uppermost perigynia by ca 2–3 mm; scales conspicuously greenish or pale along the midrib, subequal to or a little wider than the perigynia; stigmas 2 or 3 (see also *C. capitata*) .... *C. nardina*
— Perigynia with rounded or somewhat involute margins, glabrous, or rarely serrulate-ciliolate, or with a few minute hairs near the beak; spikes equaling or exceeding the leaves, the staminate part usually conspicuous and exceeding the uppermost perigynia by ca 3–7 mm; scales mostly not conspicuously green or pale along the midrib, wider than the perigynia; stigmas 3 .... *C. elynoides*

9(4). Stigmas 2; achenes lenticular; leaves to ca 1 mm wide 10
— Stigmas 3; achenes trigonous; leaves various 11

10(9). Perigynia plump, widely spreading at maturity, conspicuously nerved dorsally, the wall thick and spongy; rachilla obsolete; culms arising singly from very slender, long rhizomes .... *C. dioica*
— Perigynia flattened, ascending to loosely spreading, very finely nerved dorsally, the wall thin; rachilla over 1/2 as long as the achene; culms arising from stout, very short rhizomes .... *C. capitata*

11(9). Spikes with only 1 or 2 perigynia 5–6 mm long, the staminate portion 1–2 (3) cm long .... *C. geyeri*
— Perigynia either more numerous or smaller than above; staminate portion of spike usually less than 1 cm long 12

12(11). Rachilla exserted beyond the orifice of the perigynium; perigynia soon strongly reflexed, these and the scales early deciduous, greenish or pale brown; plants rare and restricted .... *C. microglochin*
— Rachilla lacking or included in the perigynia; perigynia ascending and persistent or, if strongly spreading and deciduous, these and the scales black or blackish purple at least in part; plants various 13

13(12). Leaves (at least some) exceeding the spikes, often strongly curved to completely curled at the tip; rachilla lacking; plants alpine on the Uinta Mts. ... . *C. rupestris*
— Leaves mostly exceeded by the spikes, straight or at least not curled; rachilla subequal to the achene (except in *C. nigricans*) 14

14(13). Rachilla lacking; perigynia narrowed to a substipitate base, ascending at first, these and the scales widely spreading and deciduous in age .... *C. nigricans*

— Rachilla subequal to the achene; perigynia sessile or inconspicuously stipitate, these and the scales ascending and persistent 15

15(14). Perigynia firm, mostly 1–6 per spike; achenes filling the perigynia; plants occasional or locally common . ... *C. obtusata*

— Perigynia thin, mostly more than 6 per spike; achenes various; plants rare (Section Inflatae) 16

16(15). Perigynia 4–7.5 mm long, mostly over 1.8 mm wide; spike 6–12 mm wide, 1.2–2.5 times as long as wide; achenes much smaller than the perigynia; plants of alpine well-drained sites .... *C. brewerii*

— Perigynia 2.5–3 (4) mm long, to 1.5 mm wide; spike commonly 4–6 mm wide, 2.5–3 times as long as wide; achenes subequal to the perigynia; plants of subalpine wet places .... *C. subnigricans*

Key 2.

Spikes few to many, sessile, androgynous or unisexual; stigmas 2; achene lenticular.

1. Culms arising singly or few together from creeping rhizomes 2

— Culms caespitose; rhizomes lacking or short 10

2(1). Spikes mostly widely spaced from each other, with 1–3 staminate and pistillate flowers each, ca 5 mm long; scales and perigynia green .... *C. disperma*

— Spikes congested into a headlike or continuous spikelike inflorescence, mostly with more flowers and longer than above; scales and perigynia brownish or blackish 3

3(2). Spikes closely aggregated into a compact, nearly globose head, not evident without dissecting the head; plants rare, mostly 5–25 cm tall, subalpine 4

— Spikes not so closely aggregated, the lower ones and sometimes the upper readily evident without dissecting the head 6

4(3). Perigynia widest below the middle, 2.9–3.8 mm long, 0.9–1.4 mm wide, hardly (if at all) winged, the marginal nerves not much more conspicuous than the facial ones (see lead 11 below) .... *C. jonesii*

— Perigynia widest near the middle or above, 3.3–4.7 mm long, 1–2.6 mm wide, conspicuously flattened, winged or at least the marginal nerves more conspicuous than the facial ones 5

5(4). Perigynia 1.9–2.6 mm wide, 1.5–2 times as wide as long, almost twice as wide as the achenes, not stipitate; leaves 1–2 mm wide, commonly equal to or surpassing the culms; plants from above timberline on the La Sal Mts. .... *C. perglobosa*

— Perigynia 0.1–1.6 mm wide, (2) 2.5–4.5 times as long as wide, mostly less than 1.5 times as wide as the achene, stipitate; leaves 2–4 mm wide, shorter than the culms; plants known from subalpine meadows on the Uinta Mts. .... *C. foetida*

6(3). Perigynia winged, deeply bidentate, serrulate to below the middle; lateral spikes androgynous or staminate, the upper one often wholly pistillate; plants 19–36 cm tall; rhizomes pale brown .... *C. siccata*

— Perigynia not winged or scarcely so, bidentate or not, if serrulate, only on the upper half; spikes androgynous or imperfect 7

7(6). Perigynia 1.7–2.6 mm long, yellow green or yellow brown, thick-walled and firm especially toward the base; beak of perigynium 0.2–0.5 mm long, slightly winged and serrulate at the confluence of the beak and the body only; plants of boggy meadows .... *C. simulata*

— Perigynia or their beaks longer than above, the walls rather thin, mostly brownish, wings and serrulations lacking or inconspicuous except in *C. praegracilis* and then usually extending onto the body; plants of dry or wet places 8

8(7). Rhizomes averaging 2–4 mm thick, blackish or dark brown; lower leaves reduced to bladeless sheaths, blackish or dark brown; plants 10–70 cm tall, often of wet places; perigynia narrowly winged, serrulate distally and sometimes medially .... *C. praegracilis*

— Rhizomes less than 2 mm thick, brownish or tan; lower leaves with blades or, if reduced to bladeless sheaths, these pale brown or greenish; plants 8–28 cm tall, mostly of well-drained soil; perigynia wingless, not serrulate or inconspicuously so 9

9(8). Plants dioecious; beak of perigynium 1–1.5 mm long; inflorescence 1.5–5.5 cm long, with the most conspicuous coloration due to the hyaline margins of the scales .... *C. douglasii*

— Plants monoecious, the spikes androgynous, the staminate portion usually conspicuous; inflorescence 0.8–2 cm long, with the most conspicuous coloration coming from the brown or dark brown body of the scales .... *C. stenophylla*

10(1). Inflorescence simple, ovoid or ovoid-oblong or, if linear, the plants of well-drained soil; stature various 11

— Inflorescence compound (some of the lower spikes borne on branched rachillas), usually oblong to linear; plants mostly of wet places, (2) 3–12 dm tall 15

11(10). Perigynia abruptly contracted to a beak or conspicuously serrulate; plants of well-drained soil (Section Bracteosae) 12

— Perigynia tapered to a beak, not or inconspicuously serrulate; plants mostly of wet places (Section Vulpinae in part) 14

12(11). Perigynia entire or serrulate on the beak only, somewhat rounded on the margins, with the marginal nerves displaced onto the ventral surface, pale green or straw colored, ca 3–7 per spike, the beak not (or very slightly) bidentate .... *C. vallicola*

— Perigynia serrulate on the beak and usually on the upper 1/2 of the slightly winged margins of the body, with the marginal nerves on the wing-edges or slightly displaced onto the ventral surface, green to dark brown, ca 4–12 per spike, the beak bidentate .. 13

13(12). Spikes closely congested into an ovoid or ovoid-oblong head, rarely any of the lower ones noticeably separated, the internodes obscure .... *C. hoodii*

— Spikes loosely congested into an oblong to linear head or spikelike inflorescence, the lower ones generally noticeably separated, the lowest internode conspicuous, generally 2–7 mm long .... *C. occidentalis*

14(11). Leaves crowded basally; sheaths relatively short, mostly less than 3 cm from one collar to the next, the ventral surface not conspicuously cross-rugose, usually truncate or concave at the mouth and not cartilaginous-thickened; culms 15–30 (40) cm tall, ca 2 mm thick at the base .... *C. jonesii*

— Leaves more widely spaced; sheaths relatively long, ca 3–5 cm long from one collar to the next, the ventral surface exposed and usually conspicuously cross-rugose, convex or prolonged and often cartilaginous at the mouth; culms 20–80 cm tall, ca 3.5 mm thick at the base .... *C. neurophora*

15(10). Sheaths conspicuously cross-rugose ventrally; perigynia very gradually tapering to a slender beak, without a keeled dorsal suture .... *C. stipata*

— Sheaths not cross-rugose ventrally; perigynia more abruptly tapered to the beak, this often with a keeled dorsal suture 16

16(15). Leaf sheaths not copper colored at the mouth, not red-dotted; inflorescence 2–3.5 (5) cm long, not interrupted or but slightly so; perigynia 2–3 mm long, not concealed by the scales; leaves 1–2.5 (3) mm wide .... *C. diandra*

— Leaf sheaths copper colored at the mouth, often red-dotted; inflorescence 3–8 cm long interrupted; perigynia 2.7–4 mm long, nearly concealed by the scales; leaves 3–5 mm wide; plants reported for Utah .... *C. cusickii* Mack.

Key 3.

Spikes more than 1, sessile, gynaecandrous; stigmas 2; achenes lenticular; perigynia with rounded (wingless) margins.

1. Perigynia widely spreading at maturity, the lower ones sometimes reflexed (Section Stellulatae) 2

— Perigynia appressed 3

2(1). Beak 1/4–1/3 (1/2) the length of the body, up to ca 1 mm long, inconspicuously bidentate with broad short teeth .... *C. interior*

— Beak 1/2 or more the length of the body, conspicuously bidentate, the teeth narrower .... *C. muricata*

3(1). Perigynia mostly 3.8–4.6 mm long, the beak sharp-margined and conspicuously serrulate; inflorescence 4–6.5 cm long............................. *C. deweyana*

— Perigynia less than 3 mm long or, if longer, the inflorescence less than 3 cm long and the beak with mostly rounded margins, serrulate or not 4

4(3). Perigynia and scales black or dark brown at maturity; spikes in an ovoid head, 0.8–1.4 cm long, closely congested and not or but slightly evident without dissecting the head .... *C. illota*

— Perigynia greenish or light brown and/or the inflorescence more open with the spikes evident (Section Heleonastes in part) 5

5(4). Pistillate scales dark brown, equal to the body of the perigynium but often surpassed by the beak; perigynia 2.4–3.4 mm long; spikes 2–4, all approximate; plants rare, alpine on the Uinta Mts .... *C. bipartita*

— Pistillate scales greenish to dark brown, shorter than the body of the perigynium; perigynia 1.7–2.5 mm long; spikes 4–8, the lower ones sometimes separated; plants various 6

6(5). Lower spikes 4–8 mm long, equal or shorter than the internode of the rachis (to 8 times shorter), with 5–10 (15) perigynia .... *C. brunnescens*

— Lower spikes sometimes larger, usually equal to or longer than the internodes of the rachis, with (10) 15–32 perigynia 7

7(6). Pistillate scales brown to dark brown; perigynia slightly granular-roughened, not or sparingly serrulate distally, the dorsal suture encroaching onto the body; plants 10–31 cm tall .... *C. praeceptorum*

— Pistillate scales greenish or pale, occasional with very light brown markings; perigynia conspicuously granular-roughened and serrulate distially (at 20x or greater magnification), the dorsal suture evident only on the beak, if at all; plants 18–50 cm tall .... *C. canescens*

Key 4.

Spikes more than 1, sessile, gynaecandrous; stigmas 2; achenes lenticular; plants caespitose; perigynia winged, usually flattened (Section Ovales).

1. Bracts, at least the lowermost and sometimes 1 or more upper ones, as long as or longer than the inflorescence .... *C. athrostachya*

— Bracts lacking or shorter than the inflorescence or rarely equaling it 2

2(1). Pistillate scales as long and as wide as the perigynia; inflorescence (except in *C. araphoensis*), pale green to stramineous or light reddish brown, rarely bicolored and/or tending to be spicate with the internodes of the rachis evident; spikes 1–7 3

— Pistillate scales shorter and narrower than the perigynia; inflorescence variously colored but often dark brown to blackish or bicolored, with the perigynia lighter than the scales, tending to be capitate, with the internodes of the rachis concealed; spike number various 8

3(2). Perigynia 2.9–4.3 mm long, 1–1.5 mm wide; plants 11–28 cm tall, often of drying pools or ponds and lakes shores, known from the western Uinta Mts. .... *C. leporinella*

— Perigynia 4–8 mm long, 1.2–2.8 mm wide; plants of various stature and habitat, but often of drier places, the distribution various 4

4(3). Perigynium with a flattened, winged, serrulate, and often ill-defined beak; plants mostly of southern Utah 5

— Perigynium with a narrowly margined more or less distally terete beak, this serrulate in the proximal portion only; plants widespread 6

5(4). Inflorescence a compact head, the first 2 internodes collectively 1.5–6 mm long; plants mostly 15–30 (40) cm tall, commonly found above timberline .... *C. arapahoensis*

— Inflorescence spicate, the first 2 internodes collectively (8) 10–18 mm long; plants commonly 30–70 cm tall, mostly found below timberline ...... *C. xerantica*

6(4). Perigynia (5.7) 6–7.5 (8) mm long; plants lower montane to subalpine .... *C. petasata*

— Perigynia 4–6 mm long; plants of various habitats 7

7(6). Spikes all or mostly 2 or more times longer than the internodes of the rachis; inflorescence rather strict, 1.4–3 cm long; plants subalpine and more commonly alpine .... *C. phaeocephala*

— Spikes, at least the lower 2 or 3, only 1–2 times longer than the internodes of the rachis; inflorescence rather flexuous, sometimes over 3 cm long; plants lower montane to subalpine .... *C. praticola*

8(2). Perigynium with a flattened, winged, serrulate, often ill-defined beak 9

— Perigynium with a narrow, more or less terete beak, this not serrulate in the distal portion 13

9(8). Perigynia 2.3–3.7 mm long, 1–1.2 mm wide; plants known from wet places in Uintah County .... *C. bebbii*

— Perigynia 4.5–8 mm long, 1.6–3.5 mm wide, plants usually of well-drained soil, of various distribution .. 10

10(9). Perigynia plano-convex; scales reddish brown with broad hyaline margins; plants of the Bear River and Uinta Mts. .... *C. multicostata*

— Perigynia strongly flattened except where distended by the achene; scales often brownish with green markings; plants of various distribution 11

11(10). Inflorescence spicate, the first 2 internodes collectively (8) 10–18 mm long; plants of southern and eastern Utah .... *C. xerantica*

— Inflorescence capitate, the first 2 internodes collectively only 4–7 (9) mm long; distribution various 12

12(11). Perigynia 2.5–3.5 mm wide, commonly over 5.6 mm long, up to 45 or more per spike; spikes 7–12 mm wide; anthers 2.5–3.7 mm long; plants (15) 30–90 cm tall, of rather broad distribution .... *C. egglestonii*

— Perigynia 1.6–2.5 mm wide, not over 5.6 mm long, to ca 20 per spike; spikes 5–7 mm wide; anthers 1.5–2 mm long; plants 24–45 cm tall, known from the central Wasatch and western Uinta mts. .... *C. straminiformis*

13(8). Perigynia 2.5–3.2 (3.5) mm long; inflorescence 0.7–1.8 cm long (or longer in *C. subfusca*) 14

— Perigynia 3.2–7.1 mm long; inflorescence sometimes longer 16

14(13). Perigynia sharp-edged but hardly wing-margined, not serrulate, 2.5–3 mm long; inflorescence 0.8–1.4 cm long, very dark brown to blackish; plants of wet places in the Uinta and central Wasatch mts. ... *C. illota*

— Perigynia wing-margined, serrulate, 2.5–3.5 mm long; inflorescence and distribution various 15

15(14). Inflorescence 0.7–1.5 (1.8) cm long, compact head, the rachis obscured by the close spikes, which are scarcely evident except by dissecting the head; plants known from the northern half of Utah ... *C. limnophila*

— Inflorescence 1–3.5 cm long, capitate or spicate, the rachis internodes sometimes conspicuous, the spikes more or less evident without dissecting the inflorescence; plants known from southern Utah .... *C. subfusca*

16(13). Perigynia plano-convex, the perigynal cavity nearly filled by the plump achene; scales greenish, light brown or reddish brown; plants rather rare and somewhat restricted 17

— Perigynia strongly flattened except where distended by the relatively small achene, or else the scales dark brown to blackish; plants generally more common and widespread 19

17(16). Perigynia 2.7–3.2 (3.6) mm long; plants of Sevier, Washington, and Wayne counties .... *C. subfusca*

— Perigynia 3.5–6.2 mm long; plants known from the Uinta and Wasatch mts. 18

18(17). Pistillate scales equaling or slightly shorter than the appressed perigynia, at least some with broad, shining, white hyaline margins 0.1–0.3 mm wide, the scales scarcely shorter than the appressed perigynia .... *C. multicostata*

— Pistillate scales conspicuously shorter than the spreading perigynia, generally lacking white hyaline margins or these less than 0.1 mm wide .... *C. pachystachya*

19(16). Perigynia 5.2–7.1 mm long (at least some), 4–6.4 times longer than wide; spikes averaging over 10 mm long, the lower 1 or 2 regularly 11–13 (15) mm long, often elliptic, smooth, with the perigynia appressed; inflorescence often cuneate at the base ...... *C. ebenea*

— Perigynia either shorter than above or else 1.6–4 times as long as wide; spikes averaging less than 10 mm long, the lower 1 or 2 typically less than 11 mm long, usually ovate, with the perigynia more or less spreading; inflorescence usually truncate at the base 20

20(19). Perigynia 1.3–2.3 mm wide, the body green or brownish in age, the beak often darker brown; scales brownish; inflorescence often conspicuously bicolored from the green or paler perigynia contrasting with the brownish scales .... *C. microptera*

— Perigynia (1.7) 2.2–2.8 mm wide, usually averaging over 2.2 mm wide, the body brown to dark brown, sometimes the winged margins greenish; scales brown to dark brown or blackish; inflorescence usually not conspicuously bicolored .... *C. haydeniana*

Key 5.

Spikes 3–5, the terminal one gynaecandrous, the lateral pistillate; pistillate scales very dark brownish purple to blackish; stigmas 3; achene trigonous (Section Atratae except *C. misandra* and *C. paupercula*).

1. Spikes sessile, congested into a head, the internodes of the rachis obsolete or at least very short and hidden by the closely aggregated heads 2

— Spikes sessile or peduncled, at least the lowest spike separated from the others by a conspicuous rachis internode 3

2(1). Perigynia 1.1–1.8 mm wide, usually somewhat inflated, without flattened margins or these generally not conspicuous, nearly filled by the mature achene, papillate at least apically (use 20x or greater magnification) .... *C. nelsonii*

— Perigynia 2–3 mm wide, strongly flattened with conspicuous broad margins, wider than the mature achene, more or less glossy on the faces, smooth or sometimes ciliolate apically .... *C. nova*

3(1). Lowest bract with a closed sheath 7–45 mm long; perigynia ciliolate-serrulate; spikes all on slender peduncles (Section Ferrugineae) .... *C. misandra*

— Lowest bract not sheathing or, if so, the sheath closed for less than 5 mm; perigynia glabrous or papillate; spikes various 4

4(3). Culms arising singly or few together from long creeping rhizomes, these covered with a yellowish feltlike tomentum; lateral spikes spreading or drooping, on slender peduncles; scales equaling or longer than the perigynia .... *C. paupercula*

— Culms tufted and rhizomes lacking or short or, if long, the spikes strongly ascending to erect or, if these spreading or drooping, the scales shorter than the perigynia 5

5(4). Pistillate scales aristate, the awn 1–2 mm long; culms arising singly or few together from long rhizomes; leaves all on the culms, the lower ones reduced to bladeless sheaths; plants rare, of wet places .... *C. buxbaumii*

— Pistillate scales acute to acuminate but not aristate; culms tufted; some leaves basal; plants various 6

6(5). Spikes not over 1.3 cm long; perigynia 2–2.8 mm long, green at maturity and strongly contrasting with the dark scales; intermediate spikes sessile or nearly so .... *C. norvegica*

— Spikes (at least the largest) commonly over 1.3 cm long and/or some of the perigynia over 2.8 mm long, these sometimes as dark as the scales at least distally or, if mostly green, the intermediate spikes commonly pedunculate 7

7(6). Upper spikes not particularly more crowded than the lower ones, all sessile or nearly so; inflorescence not over 1 cm wide; perigynia 1.9–3.6 mm long .... *C. parryana*

— Upper spikes generally more crowded than the lower ones or at least the lowest one commonly pedunculate; inflorescence sometimes over 1 cm wide; perigynia 2.7–5 mm long 8

8(7). Perigynia whitish green, rarely with red markings, strongly contrasting with the shorter, dark scales; lateral spikes ascending, spreading, or drooping at maturity, each usually on a slender peduncle, at least some in each inflorescence commonly gynaecandrous .... *C. bella*

— Perigynia olive green or commonly marked with dark red, purple, or black, often about as dark as the scales distally; lateral spikes ascending to erect, the upper ones commonly sessile or short-pedunculate, all pistillate or infrequently gynaecandrous in *C. atrata* 9

9(8). Perigynia 2.7–3.5 mm long, 1.6–2.1 mm wide, consistently papillate, sometimes completely black or black purple at maturity; lowest rachis internode 0.2–1 cm long; lowest spike sessile or on a peduncle to 0.3 cm long, the upper ones sessile or subsessile; plants 10–28 cm tall .... *C. albonigra*

— Perigynia mostly more than 3.5 mm long or over 2.1 mm wide, rarely papillate, sometimes greenish at maturity; lowest internode of the rachis ca 0.3–7 cm long; lowest spike sessile or on a peduncle to 4.5 cm long; plants (2) 2.5–10 dm tall .... *C. atrata*

Key 6.

Terminal spike staminate;
stigmas 2; achenes lenticular; lateral spikes
pedunclate and/or separated from the upper one.

1. Scales pale green; lowest peduncle enveloped at the base by a closed sheath 2–6 mm long, often originating near the base of the plant or on the lower 1/2 of the culm .... *C. aurea*

— Scales black or blackish purple, sometimes with a green or paler midstripe; lowest peduncle not in a sheath or, if so, the sheath less than 2 mm long, mostly on the upper 1/2 of the culm 2

2(1). Perigynia about as dark as the scales at maturity, slightly inflated; stigmas 2 or 3; style continuous with, persistent on, and of the same firm texture as the achene, strongly bent at maturity .... *C. saxatilis*

— Perigynia greenish or at least paler than and strongly contrasting with the dark scales, not inflated; stigmas 2; style jointed to, deciduous from, and of softer texture than the achene, straight (Section Acute) 3

3(2). Leaves 1–2.6 mm wide; plants caespitose, the larger roots covered with yellowish or yellow brown feltlike hairs; pistillate spikes 3–4 mm wide; perigynia rather conspicuously stipitate, soon deciduous after maturity .... *C. lenticularis*

— Leaves (at least some) over 2.6 mm wide or the perigynia nerved only on the margins; plants rhizomatous; fibrous roots lacking feltlike hairs or, if present, whitish; perigynia not or scarcely stipitate 4

4(3). Perigynia nerved on both sides and marginally, the beak (0.2) 0.4–0.6 mm long and bidentate and/or ciliolate; midrib of scales conspicuous throughout and at least some excurrent as a mucro 0.5–1 (2) mm long; plants widespread .... *C. nebrascensis*

— Perigynia nerveless except marginally, the beak less than 0.3 mm long, entire or obliquely cleft, not ciliolate; midrib of scales often inconspicuous toward the apex, not excurrent; plants low to high montane 5

5(4). Lowest bract shorter than the inflorescence; mature perigynia usually about as dark as the scales at least apically; pistillate spikes 5–10 mm wide; plants of the Deep Creek, Raft River, and Uinta mts. . *C. scopulorum*

— Lowest bract usually exceeding the inflorescence; mature perigynia not so dark as the scales; pistillate spikes 3–5 mm wide; plants of the central and eastern part of Utah .... *C. aquatilis*

Key 7.

Terminal spike staminate; stigmas 3; achenes trigonous;
the lower spikes peduncled and/or usually
separated from the terminal one.

1. Perigynia pubescent 2
— Perigynia glabrous 7

2(1). Pistillate spikes with 1–4 perigynia, the lower ones arising near the culm base; plants 5–35 cm tall, densely caespitose, usually growing on dry or well-drained soil; staminate spikes 5–12 mm long .. *C. rossii*

— Pistillate spikes with more than 4 perigynia, arising from the upper half of the culm; plants often taller, caespitose or not, mostly of moist places; staminate spikes (at least some) over 12 mm long 3

3(2). Perigynia densely pubescent throughout; staminate spikes (1) 2–3 (4) 4

— Perigynia sparsely pubescent; staminate spike solitary 6

4(3). Leaf sheaths pubescent at least ventrally toward the summit; style continuous with and persistent on the achene; plants to be expected in northwestern Box Elder County .... *C. sheldonii* Mack.

— Leaf sheaths glabrous; style jointed with the achene, deciduous; plants of various distribution 5

5(4). Leaves involute or folded at least distally, often appearing terete, 1–1.5 (2) mm wide; plants of ponds and marshes on subalpine moraines in the eastern Uinta Mts. .... *C. lasiocarpa*

— Leaves flat or nearly so, the larger ones commonly 2–6 mm wide; plants widespread in the state .... *C. lanuginosa*

6(3). Pistillate spikes on slender peduncles, these ca 1/2 enveloped in the closed sheath of the subtending bract .... *C. luzulina*

— Pistillate spikes sessile or nearly so, the peduncles, if present, not enveloped in a closed sheath .... *C. parryana*

7(1). Style continuous with, persistent on, and of the same firm texture as the achene, not withering; staminate spikes more than 1 and/or usually 2.1–8.5 cm long ... 8

— Style jointed to, deciduous from, and of softer texture than the achene, withering; staminate spikes mostly 1, 0.3–2.1 cm long 13

8(7). Leaf sheaths pubescent; style straight; plants known from Sevier and Utah counties .... *C. atherodes*

— Leaf sheaths glabrous; style curved to strongly bent in age; plants of various distribution 9

9(8). Pistillate scales narrowed to a serrulate-ciliolate awn .... *C. hystricina*

— Pistillate scales not awned (Section Vesicariae) 10

10(9). Pistillate scales soon purple black or black, except at the white hyaline acute tip; perigynia with the 2 marginal nerves evident; styles 2 and 3; achenes lenticular or trigonous .... *C. saxatilis*

— Scales green, stramineous, reddish brown or, if (rarely) dark brown, usually acuminate or caudate-acuminate; perigynia conspicuously nerved dorsally; styles 3; achenes trigonous 11

11(10). Inflorescence 5–13 cm long; spikes crowded, the pistillate ones longer than the rachis internodes; lowest bract 2–3 times longer than the inflorescence; perigynia 7–10 mm long; plants uncommon, known from Wasatch and Weber counties .... *C. retrorsa*

— Inflorescence commonly over 13 cm long; spikes not crowded, the pistillate one shorter to longer than the rachis internodes; lowest bract 1–2 times as long as the inflorescence; perigynia 4–7 mm long; plants of various distribution 12

12(11). Perigynia spreading at maturity, the ellipsoid to subglobose body more or less abruptly contracted to a conspicuous beak; plants with robust, long-creeping rhizomes, widespread, common .... *C. rostrata*

— Perigynia appressed or slightly ascending, the lanceolate to lance-ovate body gradually tapering to the often poorly defined beak; plants with rather short rhizomes, known from the Uinta Mts., uncommon . ... *C. vesicaria*

13(7). Peduncle of the lowest and often the upper pistillate spikes enveloped in a closed sheath 0.4–2 (4) cm long or longer 14

— Peduncles not enveloped in a sheath or this closed for less than 0.4 cm 18

14(13). Plants with slender creeping rhizomes; lowest spike often borne on the lower half of the culm 15

— Plants caespitose; rhizomes lacking or very short; lowest spike usually borne on the upper half of the culm 16

15(14). Bracts not exceeding the inflorescence; stigmas 3; achenes trigonous; larger pistillate spikes commonly with over 20 perigynia; plants known from a marl bog in Kane County .... *C. crawei*

— Bracts (at least one) commonly exceeding the inflorescence; stigmas 2 (3); achenes lenticular; pistillate spikes with 4–18 perigynia; plants widespread .... *C. aurea*

16(14). Pistillate spikes 1–3 cm long; leaf blades 3–9 mm wide; perigynia 3–5 mm long .... *C. luzulina*

— Pistillate spikes 0.5–1.1 cm long; leaf blades 1–3 mm wide; perigynia 2.2–3.5 mm long 17

17(16). Pistillate spikes sessile or on erect or ascending peduncles, all crowded or the lowest one sometimes separate; inflorescence 2–5 (10) cm long; staminate spike 0.7–2.1 cm long .... *C. oederi*

— Pistillate spikes borne on capillary, spreading or drooping peduncles, remote; inflorescence 5–30 cm long; staminate spike 0.3–0.9 cm long .... *C. capillaris*

18(13). Pistillate spikes 4–6 mm long, sessile; perigynia widely spreading, greenish; scales greenish or pale; plants caespitose .... *C. interior*

— Pistillate spikes longer and/or at least one on a conspicuous peduncle; perigynia and/or scales brownish, blackish, or black purple 19

19(18). Pistillate spikes borne on slender, spreading to drooping peduncles; perigynia pale green; scales greenish brown to purple brown; culms arising singly or few together from long, creeping rhizomes; roots covered with a feltlike tomentum 20

— Pistillate spikes sessile or on erect peduncles or, if the lower peduncles spreading, the perigynia and/or the scales often blackish or black purple in part; culms tufted or arising singly or few together; roots not with a feltlike tomentum 21

20(19). Terminal spike 9–12 mm long, occasionally monoecious; lateral spikes often staminate at the base; leaves mostly with well-developed blades ..... *C. paupercula*

— Terminal spike (10) 15–21 mm long, staminate; lateral spikes sometimes staminate at the apex; lower leaves usually bladeless .... *C. limosa*

21(19). Perigynia 1.9–3 (3.6) mm long; pistillate spikes 4–5 mm wide, commonly all sessile or nearly so .... *C. parryana*

— Perigynia (3) 3.3–5.2 mm long; pistillate spikes 5–8 mm wide, at least the lowest one commonly pedunculate 22

22(21). Perigynia green or olive green, strongly contrasting with the dark scales; plants uncommon in the Uinta Mts., more common elsewhere on well-drained soil, tufted, the rhizomes lacking or short ..... *C. raynoldsii*

— Perigynia as dark, or nearly as dark, as the blackish or black purple scales; plants most common in the Uinta Mts., mostly of wet places, commonly rhizomatous . 23

23(22). Perigynia slightly inflated, not flattened; leaf blades 1.5–3 mm wide; styles continuous with and persistent on the achene, contorted in age; stigmas 2 or 3; staminate spikes 1–2 (3); plants mostly of wet places .... *C. saxatilis*

— Perigynia flattened, not inflated; leaf blades 2–6 mm wide; styles jointed with, and deciduous from, the achene, straight; stigmas ; staminate spikes solitary; plants commonly of well-drained soil ..... *C. paysonis*

Carex albonigra Mack. in Rydb. White and Black Sedge. Plants (10) 15–28 cm tall, the culms tufted; rhizomes lacking or short; leaves basal and on the lower culms, the blades 1–5 mm wide, typically flat, the lower sheaths bladeless, often dark purple; lowest bract shorter than the inflorescence, this (1) 1.5–3 cm long, surpassing the leaves, the lowest internode 2–10 mm long; spikes 3 (2–4), 8–15 mm long, 3–6 mm wide, elliptic to narrowly ovate, the terminal gynacandrous, often a little larger than the lateral pistillate ones, the lowest one sessile or on a peduncle to 3 mm long, the others sessile or subsessile; pistillate scales subequal to or a shorter than the perigynia, black or blackish purple, with white hyaline margins, the midrib sometimes greenish or pale; perigynia 2.7–3.5 mm long, 1.6–2.1 mm wide, elliptic to nearly orbicular, papillate (at least apically), black or blackish purple throughout; stigmas 3; achenes trigonous, sessile or nearly so. High mountains, in fell fields, dry and wet meadows, and alpine tundra at 3430 to 3810 m in Duchesne, Grand, Juab, San Juan, Summit, Uintah, and Wasatch counties; Alaska and Mackenzie south to California and Arizona; 13 (ii). *Carex albonigra* appears to be allied to *C. atrata*, but it is consistently smaller, always of high elevations, and does not show the variation found in

C. atrata. The perigynia are always papillate but only occasionally and randomly papillate in *C. atrata*. Although apparently a distinct taxon, *C. albonigra* seems to pass into both *C. atrata* and *C. nova* (q.v.).

Carex aquatilis Wahl. Water Sedge. [*C. interimus* Maguire, type from Tony Grove Lake, Cache County]. Plants (0.6) 1.5–9.5 dm tall, from thick long rhizomes; culms tufted or arising singly or few together; leaves on the lower 1/2 of the culms, the blades 1.5–5.5 mm wide, shorter than to exceeding the culms; bracts leaflike, the lowest shorter than, subequal to, or surpassing the inflorescence, this 3–19 cm long, with 1–2 (3) staminate spikes above 2–3 (5) pistillate spikes, the transitional ones often androgynous; terminal staminate spike 1–3.5 cm long, the first lateral one 0.8–1.8 cm long, the second lateral or, if present, mostly larger than the first; pistillate spikes (0.7) 1.5–4.5 (7) cm long, 3–5 mm wide; pistillate scales blackish or black purple, shorter to longer and narrower than the perigynia, lanceolate to ovate, acute to accuminate or rather rounded at the tip, entire, the midrib green or pale or colored like the scale, not raised or excurrent; perigynia numerous, 2–3.3 mm long including the beak, elliptic to obovate, more or less strongly flattened, marginally nerved only, light green or stramineous in age, often speckled or suffused with reddish brown, minutely granular-muricate, the beak 0.1–0.3 mm, entire or oblique but not bidentate or ciliolate; stigmas 2; achenes lenticular, shorter and usually narrower than the perigynia; 2n = 76. Wet and boggy meadows, along streams, and margins of ponds and lakes at 2195 to 3475 m in Beaver, Cache, Daggett, Duchesne, Emery, Garfield, Grand, Iron, Kane, Salt Lake, San Juan, Sanpete, Sevier, Summit, Uintah, Utah, Wasatch, Washington, and Wayne counties; circumboreal, south in North America to California and New Jersey; 82 (xv). Our members of section Acutae are not easily distinguished. The primary distinction (ribs of the perigynia) between *C. aquatilis* and *C. nebrascensis* (q.v.) hardly seems to warrant segregation at the specific level. However, this feature is correlated with other morphological and with ecological trends. The distinction between *C. aquatilis* and *C. scopulorum* (q.v.) seems to be tenuous, but the taxa have quite different ranges. The separation from *C. lenticularis* (q.v.) is real but difficult. **Carex bigelowii** Torr. has been reported for Utah. Plants of that taxon have narrow spikes similar to those of *C. aquatilis* but have the lowest bract shorter than the inflorescence as in those of *C. scopulorum*. Occasional specimens with slender spikes and lowest bract shorter than the inflorescence occur sporadically in Utah within the range of *C. aquatilis*. The significance of such plants in Utah is not clear and they are referred here to *C. aquatilis*.

Carex arapahoensis Clokey Arapaho Sedge. Plants 12–30 cm tall, densely caespitose; rhizomes lacking or very short; leaves basal and on the lower culms, the blades 1.5–4 mm wide, flat or nearly so; lowest bract lacking; inflorescence 1–2 cm long, headlike, equaling or surpassing the leaves, the lowest internode of the rachis ca 2–6 mm long; spikes usually 3–6, gynaecandrous, 8–12 mm long, sessile, closely aggregated but more or less evident; pistillate scales equaling or longer than and as wide or wider than the perigynia, brown, with a pale midrib and hyaline margins; perigynia 4.5–5.5 mm long, 2–2.5 mm wide, more or less elliptic to ovate or elliptic-obovate, flattened, with thin winged margins, the wing extending onto the beak, serrulate at least in the distal 1/4–1/2, finely nerved on both sides, dark reddish brown; stigmas 2; achenes lenticular. Alpine communities of the La Sal Mts. at 3660 to 3960 m in Grand and San Juan counties; Wyoming and Colorado; 2 (0).

Carex atherodes Sprengel Awned Sedge. Plants 4–15 dm tall; culms arising singly or few together from robust, creeping rhizomes; leaves distributed along the culms, the sheaths villous-hirsute, or sometimes pubescent only toward the ventral summit, the blades 4–12 mm wide; inflorescence to 45 cm long with well-spaced spikes mostly shorter than the internodes or the upper ones longer; upper 2–4 spikes staminate, 2.5–5 cm long, sessile or nearly so, except the terminal on a peduncle to 3 cm long; lower spikes pistillate or the intermediate ones sometimes androgynous, cylindrical, 5.5–9 cm long, ca 10 mm wide, on peduncles 1–11 cm long, the lower ones enveloped in a closed sheath up to ca 9 cm long; pistillate scales aristate-awned, the awn 1–5 mm long and subequal to the perigynia, the body pale, rather scarious, narrower than the perigynia; perigynia 7–10 mm long, lanceolate or lance-ovate, inflated below, conspicuously ribbed, ascending, with a bidentate beak, the teeth 1.5–3 mm long and often more or less divergent; stigmas 3; achenes trigonous; style continuous with, persistent on, and of the same firm texture as the achene, more or less contorted in age; 2n = 74. Margins of ponds and lakes, and along streams, at 1370 to 2695 m in Sevier and Utah counties; circumboreal, south in America to New York, Colorado, and Oregon; 1 (0).

Carex athrostachya Olney Slender-beaked Sedge. [*C. macloviana* var. *pachystachya* f. *involucrata* Kukenthal in Engl.] Plants 10–52 cm tall, densely caespitose; rhizomes lacking or very short; leaves basal and on the lower culms, the blades 1–3 mm wide, flat or channeled; lowest bract (at least) subequal to or surpassing the inflorescence, but not uncommonly shorter, or broken and appearing shorter; inflorescence 1–2.5 cm long, 8–15 mm wide, headlike, usually exserted beyond the leaves, the lowest internode 1–3.5 mm long; spikes ca 3–15, gynaecandrous, 5–10 mm long, ca 5 mm wide, sessile, tightly aggregated, much longer than and concealing the internodes of the rachis, not, or hardly, evident without dissecting the inflorescence or the lower ones conspicuous; pistillate scales subequal to and almost as wide as the perigynia, brown with a green or pale midstripe and with or without narrow hyaline margins; perigynia 3.5–4.5 mm long, 1–1.5 mm wide, lanceolate or narrowly ovate, flattened, with winged margins, slightly nerved dorsally and sometimes ventrally, greenish, serrulate at least in the distal 1/2, tapered into a more or less terete, often brownish beak; stigmas 2; achenes lenticular. Drying mud of ephemeral pools, margins of ponds and lakes, and lodgepole pine, aspen-spruce, forb-grass, and wet meadow communities at 1825 to 3475 m in Beaver, Cache, Daggett, Duchesne, Emery, Garfield, Kane, Salt Lake, Sanpete, Sevier, Summit, Uintah, Utah, Wasatch, Washington and Weber counties; Alaska to Saskatchewan, south to California and Colorado; 35 (x).

Carex atrata L. Blackened Sedge. [*C. heteroneura* W. Boott]. Plants 2.5–10 dm tall; culms tufted; rhizomes lacking or short; leaves basal and on the lower culms, the blades 1.5–10 mm wide, flat; lowest bract often foliose, mostly subequal to or shorter than the inflorescence, this 2–10 cm long; first internode of the rachis (0.5) 1–7 cm long or almost obsolete, shorter than to surpassing the spike; second internode ca 1–8 mm long; spikes 3–5, 0.8–2.5 cm long, 4–8 mm wide, narrowly oblong to linear, the terminal one gynaecandrous, the lateral ones pistillate or infrequently gynaecandrous, the lowest one typically on a peduncle to 4.5 cm long, the upper pedunculate or sessile; pistillate scales shorter than to much longer than the perigynia, blackish, dark brownish purple, or blackish purple; perigynia 3–5 mm long, 1.7–4 mm wide, oblong, ovate, or nearly orbicular, olive green, yellow brown, or greenish marginally but often suffused with dark purple or purple black; stigmas 3; achenes trigonous, commonly short-stipitate; $2n = 54$. There are three varieties in Utah.

1. Perigynia not flattened, lanceolate or narrowly elliptic, olive green, becoming yellow brown at maturity, seldom suffused with dark color, except for the beak, not much wider than the mature achene; plants rare in Duchesne County .... *C. atrata* var. *atrosquama*
— Perigynia flattened, broadly elliptic to ovate or obovate, mostly conspicuously suffused with dark purple or black, wider than the mature achene; plants widespread 2

2(1). Pistillate scales equal to or longer than the perigynia; plants mosty of the southern half of the state, but also of the Wasatch Mts. .... *C. atrata* var. *chalciolepis*
— Pistillate scales equal to or shorter than the perigynia; plants rather widely distributed .... *C. atrata* var. *erecta*

Var. atrosquama (Mack.) Kelso [*C. atrosquama* Mack.]. Engelmann spruce woods and openings and alpine communities at 2800 to 3370 m in the Uinta Mts. in Duchesne County; British Columbia and Montana south to Colorado; 6 (vi). Unlike the following two varieties this seems to be a well marked taxon.

Var. chalciolepis (H. T. Holm) Kukenthal [*C. chalciolepis* H. T. Holm]. Oak-ponderosa pine-fir, aspen, spruce-fir, and alpine communities at 2470 to 3660 m in Beaver, Duchesne (Harrison 7705A), Grand, Iron, Salt Lake, San Juan, Utah, and Wayne counties; Wyoming, Colorado, Utah, and Arizona; 37 (x).

Var. erecta W. Boott [*C. epapillosa* Mack. in Rydb, type from Piute County; *C. heteroneura* var. *epapillosa* F. Hermann]. Lodgepole pine, spruce-fir, moist meadow, and alpine communities, sometimes in talus or boulder fields, at 2285 to 3870 m in Cache, Daggett, Duchesne, Emery, Juab, Rich, Salt Lake, Sevier, Summit, Tooele, Uintah, Wasatch, and Washington counties; Washington to California, east to Montana and Utah; 55 (vi). Our plants show a continuum in the relative length of the scales, and to segregate all Utah specimens of var. *erecta* and var. *chalciolepis* becomes an exercise in futility. However the trends in morphology, as well as geography, seem sufficient to maintain a distinction at varietal level. The shape of the perigynia and length of the peduncles are features that have been used to distinguish these two taxa, but these are poorly if at all correlated with the length of the scales. Random plants from throughout the range of the species in Utah have glabrous and granular-roughened perigynia (at least on the margins toward the beak), but this feature appears to have no taxonomic merit.

Carex aurea Nutt. Golden sedge. [*C. garberi* Fern.; *C. hassei* Bailey]. Plants 4–40 (65) cm tall; culms tufted or arising singly or few together from slender pale rhizomes; leaves borne on the lower culms, the blades 1–4 mm wide, flat, surpassing or shorter than the bracteate inflorescence; bracts leaflike, with sheathing bases, some usually equaling or exceeding the inflorescence, the lowest with a closed sheath 2–6 mm long or longer; terminal spike staminate or often gynaecandrous, rarely androgynous, 0.4–1.5 (2) cm long, 2–3 mm wide, with peduncles 0.4–2 cm long; lateral spikelets 1–3 (4), pistillate, 0.5–2.5 cm long, 3–5 mm wide, usually well spaced and pedunculate; peduncles filiform or capillary, commonly 0.3–7 cm long, or the lowest one arising at or near the culm base and to 11 cm long, this often enclosed in a long sheath; pistillate scales greenish or brown with a green or pale center, shorter than the perigynia; perigynia 4–18, 1.7–3 mm long, ellipsoid to obovoid-globose, pale green and whitish papillose when young, sometimes turning golden and somewhat fleshy in age; stigmas 2, rarely 3 in some perigynia; achenes lenticular; $2n = 52$. Around seeps and springs, along streams, wet meadows, and hanging gardens at 1140 to 3350 m in all Utah counties except Box Elder, Carbon, Davis, Morgan, Wayne, and Weber; Alaska to Newfoundland southward to New Mexico and Pennsylvania; 75 (vii). Tall specimens from the canyonlands section of Utah are similar in appearance to *C. aquatilis*. Such specimens might be referred to *C. hassei*. However, the numerous specimens seen show a continuum in size from very small to quite large, and the color (golden versus green) of the perigynia seems poorly if at all correlated with any other feature. Our plants are probably no more than a subspecies of **C. bicolor** All. of Eurasia and Alaska, but no combination is proposed here.

Carex backii F. Boott in Hook. [*C. saximontana* Mack.]. Plants ca 15–30 cm tall, densely caespitose, without rhizomes; culms slender, becoming capillary, or toward the base shorter than the leaves; leaves mostly basal, the blades 2–6 mm wide, flat; inflorescence less than 1 cm long, terminal and basal or nearly so, the terminal one with 2 or 3 spikes subtended (and greatly exceeded) by a leaflike bract, the basal ones of a solitary spike and bractless but with the scales bracteate; spikes androgynous but with the staminate part short, inconspicuous, and exceeded by the upper perigynia, sessile or nearly so; upper pistillate scales shorter than the perigynia, the lower ones bracteate and surpassing the perigynia, 1–7 cm long; perigynia 4–5.4 mm long, green or greenish, inconspicuously nerved, rhombic-ovoid and obscurely beaked or subglobose and conspicuously beaked; stigmas 3; achenes rounded-trigonous, ca 3 mm long; $2n = 64, 66$. Oak-maple communities at 1720 to 1980 m in Cache, Daggett, Juab, Salt Lake, and Utah counties; British Columbia to Quebec and south to Oregon and New York; 2 (0).

Carex bebbii Olney ex Fern. Plants 3–7 dm tall, densely tufted; rhizomes lacking; leaves borne on the lower culms, the lowermost reduced to bladeless sheaths or with short blades, the upper ones with blades 2–4 mm wide; lowest bract lacking; inflorescence 1–3 cm long, 5–12 mm wide, leaflike, subequal to or surpassing the leaves, the lowest internode of the rachis 3–4 mm long; spikes usually 4–12, gynaecandrous, 5–9 mm long, sessile, closely aggregated but more or less evident without dissecting the inflorescence, equal to or to 3 times as long as the rachis internodes; pistillate scales shorter and narrower than the perigynia, stramineous to light brown, with a greenish or pale midrib, the margins hyaline or not;

perigynia 2.3–3.7 mm long, 1–1.2 mm wide, pale green to straminous or tan, flattened, with thin winged margins, the wings extending to near the ill-defined, flattened beak apex, densely crowded in the spike and spreading-ascending, serrulate; stigmas 2; achenes lenticular; 2n = 68. Riparian communities, margins of beaver ponds, and along ditches at 1800 to 2225 m in Uintah County (Whiterocks drainage); British Columbia to Newfoundland and south to Oregon and New Jersey, and Colorado; 4 (iv).

Carex bella Bailey Beautiful Sedge. Plants 20–68 cm tall, tufted, rhizomes lacking or short; leaves basal and on the lower culms, the blades 1–5 mm wide, flat; lower bract usually leaflike, typically shorter than or subequal to the inflorescence, or infrequently to 2.5 cm longer than the inforescence, with a sheathing base ca 3–4 mm long, the scarious margins of the sheath free or nearly so; inflorescence somewhat flexuous and tending to nod, 4–7 cm long or longer; spikes (2) 3–4, commonly gynaecandrous, pedunculate, more or less cylindric, 1–3.2 cm long, 3–6 mm wide, the uppermost usually with a conspicuous but short staminate portion, subsessile to pedunculed; the lateral ones usually with only a few stamens and infrequently all pistillate; peduncles slender flexuous, the lower ones shorter to longer than the spikes, the lowest one rarely arising near the base of the culm; pistillate scales shorter than the perigynia, blackish or black purple, the pale midrib obscure or slightly keeled, especially apically and sometimes excurrent as a mucro; perigynia 2.8–4 mm long, broadly oval to oblong-oval, flat but swollen by the ripening achene, green and strongly constrasting with the dark scales; stigmas 3; achenes trigonous. Ponderosa pine-oak, aspen, and more commonly lodgepole pine, spruce, and fir communities and openings, and alpine slopes at 2650 to 3340 m in Beaver, Carbon, Daggett, Duchesne, Emery, Garfield, Grand, Iron, Piute, San Juan, Sevier, Summit, Uintah, and Wayne counties; Colorado and South Dakota, south to Arizona, New Mexico, and Mexico; 44 (x).

Carex bipartita All. Plants 5–20 cm tall, tufted; rhizomes lacking or short; leaves basal and on the lower culms, the blades 0.5–2 mm wide; lowest bract lacking or shorter than the first spike; inflorescence 1–1.5 cm long, 0.5–1 cm wide, surpassing the leaves; spikes 2–4, gynaecandrous, 0.5–1 cm long, 3–5 mm wide, sessile, approximate, longer than the internodes of the rachis, but evident in the inflorescence; pistillate scales subequal to and as wide as the perigynia, largely concealing them, except for the beak, brown to blackish purple brown, with hyaline margins and usually a pale midstripe; perigynia 2.4–3.4 mm long, elliptic to obovate, more or less concavely tapered to a short beak to 0.7 mm long, the dorsal suture extending from the beak to the distal portion of the body, more or less flattened but with rounded margins, finely several-nerved on each side, yellow green or brown green with the nerves often brown red, finely granular-roughened, ascending, ca 10–30 per spike; stigmas 2; achenes lenticular. Wet alpine meadows at 3370 m in the Uinta Mts., Summit County; circumboreal, south in the U./S. to Colorado; 4 (i). With brownish scales, reddish nerved perigynia, approximate spikes, and alpine habitat, the plants of *C. bipartita* are more like those *C. praeceptorm* than they are to those of *C. canescens* or *C. brunescens*.

Carex breweri F. Boott [*C. engelmanii* Bailey]. Plants 5–30 cm tall; culms arising singly or few together from creeping rhizomes; leaves basal and on the lower culms, the blades ca 1 mm wide, involute or deeply channeled, more or less wiry like the culms; bract lacking; spike solitary, erect, androgynous, 1–2 cm long, 6–12 mm wide, 1.2–2.5 times as long as wide, more or less shiny, ellipsoid or ovoid, the staminate portion rather inconspicuous; scales brownish or dark brown, the pistillate ones subequal to the perigynia; perigynia numerous, 4–7.5 mm long, 1.8–4.5 mm wide, ovate or broadly elliptic, very thin walled, abruptly or gradually short-beaked; rachilla ca 1/2 to as long as the achene; stigmas 3; achenes trigonous, only 1/4–1/2 as long and as wide as the perigynia. Wet or dry places, sometimes among limestone rocks or talus, at 3050 to 3350 m in Cache, Duchesne, and Salt Lake counties; British Columbia to California, east to Wyoming and Colorado; 4 (i). Our plants have a single nerve on the pistillate scales and are referred to var. *paddoensis* (Suksd.) Cronq. [*C. paddoensis* Suksd.].

Carex brunnescens (Pers.) Poir. in Lam. Brownish Sedge. Plants (1) 2–6 dm tall, tufted; rhizomes lacking or short; leaves basal and on the lower culms, the blades 1–2.5 mm wide; lowest bract inconspicuous or awned and surpassing the lowest spike but not the inforescence, this 1–3.5 cm long, ca 4–6 mm wide, usually surpassing the leaves, spikelike, typically interrupted; spikes 4–9, gynaecandrous, 4–8 mm long, ca 3 mm wide, sessile, approximate and longer than the internodes or more often the lower ones widely remote and only 1/8–1/3 as long as the internodes; lowest internode sometimes to 2.5 (4) cm long; pistillate scales usually much shorter than the perigynia, mostly scarious, except for the green or rarely brownish midstripe; perigynia 1.7–2.5 mm long, elliptic or elliptic-ovate, more or less compressed but with rounded (not winged) margins, finely nerved on both sides or nearly nerveless ventrally, green or brownish distally in age, ca 5–10 (15) per spike, usually slightly spreading, so that the beaks interrupt the outline of the spike; dorsal suture extending from the beak onto the body, this often with a narrow hyaline overlapping flap; stigmas 2; achenes lenticular; n = 27, 28. Wet meadows and other wet places at 2225 to 2956 m in the Uinta Mts., Summit and Uintah counties; circumboreal, south in America to Oregon and Tennessee; 2 (i). This taxon is obviously related to *C. canescens* and *C. praeceptorum*, but the culms are more lax and other features of the spikes and perigynia are quite different.

Carex buxbaumii Wahl. Plants 3–5.5 dm tall, the culms arising singly or 2 or 3 together and well spaced on long creeping rhizomes; leaves all cauline, borne on the lower culms, the lower ones reduced to bladeless sheaths, the blades 1–3 mm wide; lowest bract leaflike, shorter than to exceeding the inflorescence by 2 cm; inflorescence 3.5–6.5 cm long, mostly surpassing the leaves; spikes 3–4, 0.7–2.2 cm long, 4–8 mm wide, ovoid to nearly cylindric, erect or strongly ascending, the terminal one gynaecandrous, longer than its peduncle, the lower ones pistillate, sessile or longer than their peduncles, each slightly shorter or exceeding the internode it subtends; pistillate scales conspicuously aristate, the awn-point ca 1–2 mm long, the body usually shorter than the perigynia, but with the awn exceeding it, brownish purple or blackish purple, with a greenish midrib; perigynia 2.5–4 mm long, elliptic or elliptic-obovate, faintly nerved; stigmas 3; achenes trigonous. Wet and boggy meadows and seeps at 2255 to 2900 m in the Uinta Mts.,

Daggett, Duchesne, Summit, and Uintah counties; circumboreal, south in America to North Carolina and California; 5 (iv).

Carex canescens L. Pale Sedge. [*C. canescens* var. *dubia* Bailey, type from Bear River Canyon, Summit County]. Plants 18–50 cm tall, densely caespitose; rhizomes lacking or very short; leaves basal and on the lower culms, the blades 1.5–3.1 mm wide; lowest bract lacking or mostly shorter than the first spike and never exceeding the second one; inflorescence 1.4–5.2 cm long, 0.5–1 cm wide, spikelike, equaling or surpassing the leaves; spikes 4–8, androgynous, 0.5–1.4 cm long, 2.5–5 mm wide, sessile, more or less cylindric, greenish or silver, approximate or the lower ones spaced, equal to or longer than the internodes or the lowermost rarely shorter; pistillate scales usually conspicuously shorter than the perigynia, pale green or silvery, mostly hyaline except for the green midstripe, sometimes pale brown in age; perigynia 1.8–2.5 mm long, 1–1.2 mm wide, elliptic to elliptic-ovate, more or less compressed but with rounded (wingless) margins, greenish, yellow green, or grayish green, sometimes brownish toward the beak, finely nerved on both sides, the nerves green or rarely pale brown, (10) 15–32 per spike, quite noticeably granular-roughened distally, very short-beaked, the beak granular-roughened and often serrulate (use high magnification), the dorsal suture evident only on the beak, if at all; stigmas 2; achenes lenticular; n = 28, 36. Wet and boggy meadows, and on the margins of ponds and lakes, streams, and seeps and springs at 1920 to 3475 m in Beaver, Daggett, Duchesne, Emery, Garfield, Piute, Salt Lake, San Juan, Sevier, Summit, and Uintah counties; circumboreal, south in America to California and Virginia; 29 (xi).

Carex capillaris L. Hair Sedge. Plants 5–70 cm tall, densely caespitose from short rhizomes; culms slender, rather weak; leaves basal and on the lower culms, or 1 above the middle, the blades 1–3 mm wide; lowest bract foliose, shorter than the inflorescence, strongly sheathing, the sheath closed for 9–20 mm long; inflorescence 5–30 cm long, surpassing the leaves; terminal spike staminate, rarely gynaecandrous, 3–9 mm long, borne on a capillary peduncle 1–5 cm long; lateral spikes pistillate, 6–11 mm long, with 5–25 perigynia, borne on capillary, spreading to drooping peduncles to 3.5 cm long, each peduncle enveloped basally by the closed sheath of the subtending bract, the lowest spike and peduncle together much shorter than the accompanying internode of the rachis; pistillate scales 1/2–3/4 as long as the perigynia, with broad hyaline margins and greenish centers or greenish midstripe flanked by light reddish brown stripes; perigynia 2.4–3.3 mm long, more or less elliptic or lance-ovate, more or less rounded, the marginal nerves rather prominent, green at first, turning a glossy brown; stigmas 3; achenes trigonous, not filling the distal part of the perigynia; 2n = 50?, 52, 54. Wet meadows and along streams at 2440 to 2835 m in Cache, Beaver, Daggett, Duchesne, Iron, Juab, Uintah, and Utah counties; circumboreal, south in America to New Mexico and New York; 16 (iv).

Carex capitata L. Capitate Sedge. Plants 5–20 cm tall; culms closely spaced on short rhizomes; leaves borne on the lower 1/4 of the culms, the lower ones often reduced to bladeless sheaths, the upper ones with blades to ca 1 mm wide, involute, filiform; bract lacking; spike solitary, erect, androgynous, 4–10 mm long, ca 3–7 mm wide, the staminate part short or elongate; pistillate scales shorter and narrower than the perigynia, dark brown with broad hyaline margins; perigynia 6–25, 2–3.5 mm long, ovate to suborbicular, conspicuously beaked; rachilla about reaching the beak of the perigynium; stigmas 3; achenes trigonous, not filling the perigynia; 2n = 50. Moist or dry alpine tundra of the Uinta Mts., Duchesne County; British Columbia and Alberta south to California and Colorado, and reputedly to South America; 2 (0).

Carex crawei Dewey Plants 1–3 dm tall; culms arising singly or few together from creeping rhizomes; leaves borne toward the base of the culms, the blades 1.5–3 mm wide; inforescence 4–18 cm long with the spikes well-spaced and shorter than the internodes, the lowest spike sometimes borne near the base of the culm; terminal spike staminate, 1.2–2 cm long; lateral spikes pistillate, 1–2 cm long, on peduncles 0.2–4.2 cm long, these short above, longer below, each enclosed in the sheath of the subtending bract, the closed sheath 2–25 mm long, the bracts shorter than the inflorescence; pistillate scales shorter than the perigynia, with a greenish or pale, more or less 3-nerved, narrow, triangular midstripe and scarious but brownish or brown purple margins; perigynia (5) 10–15, 2.3–3.8 mm long, elliptic or ovate, greenish to straw colored or light brown, inconspicuously beaked, faintly many-nerved; stigmas 3; achenes trigonous, loosely filling the lower part of the perigynia; 2n = 38. The one specimen seen (Maguire 18828) is from a marl bog at ca 2680 m in Kane County; British Columbia to Quebec and south to Missouri; 1 (0).

Carex curatorum Stacey Canyonlands Sedge. [*C. scirpoidea* var. *curatorum* (Stacey) Cronq.]. Plants 21–40 cm tall, dioecious, densely caespitose, without rhizomes or these very short; leaves basal and on the lower culms; blades 1–4 mm wide; bract mostly lacking or rarely to 1.5 cm long; spike solitary, erect or nearly so, 1.8–4.5 cm long, ca 5 mm thick; scales reddish brown with a pale midstripe and hyaline margins, often ciliolate, somewhat erose, sparsely pubescent dorsally, the pistillate ones shorter and narrower than the perigynia, these 3.2–4 mm long, broadly elliptic, obovate to nearly orbicular, blackish purple distally, greenish below, conspicuously pubescent with translucent hairs; stigmas 3; achenes trigonous. Hanging gardens and canyons along the Colorado and San Juan rivers at 1155 to 1340 m in Kane and San Juan counties; northern Arizona; 7 (0). Some plants of *C. scirpoidea* from Emery (Scad Valley), Garfield, and Iron counties are very close to this taxon, and possibly should be placed here.

Carex deweyana Schwein. [*C. bolanderi* Olney]. Plants 40–75 cm tall, tufted; rhizomes lacking or short; leaves borne on the lower culms, the blades 1–3 mm wide, flat; lowest bract lacking or present and shorter than or subequal to the inforescence, this 4–6.5 cm long, ca 5–10 mm wide, spikelike, equaling the leaves or surpassing them; spikes 4–6 (10), gynaecandrous, 0.7–1.6 cm long, ca 4–6 mm wide, pale green, sessile, the lowest usually remote, subequal to or to 2.5 times shorter than the internode, the upper ones remote or approximate but individually evident without dissecting the inforescence, the staminate flowers few, their scales rather closely appressed to a pedunclelike portion of the rachis; pistillate scales as long as the body of the perigynia, pale green or slightly stramineous, often largely hyaline, with a darker green midrib, this sometimes excurrent; perigynia 3.8–4.6 mm long, 1–1.2 mm wide, lanceolate to lance-elliptic, pale green, the margins not winged or scarcely so distally, the marginal nerves prominent, raised, sometimes slightly extended onto the ventral surface, the dorsal ribs conspicuous, the ventral lacking or obscure, the body tapering to a serrulate bidentate beak with teeth to 0.4 mm long; stigmas 2; achenes lenticular, 1.5–2 mm long; 2n = 54. Oak, aspen, and white fir communities, mostly along streams, at 2195 to 2950 m in Cache, Juab, Millard, Piute, and Salt Lake counties (Deep Creek, Canyon, and Wasatch mts.); British Columbia to Newfoundland and south to California and Pennsylvania; Asia; 5 (ii). Utah materials are more or less referable to var. *bolanderi* (Olney) W. Boott [*C. bolanderi* ssp. *leptopoda* (Mack.) Calder & Taylor].

Carex diandra Schrank Lesser Panicled Sedge. Plants 3–10 dm tall; caespitose with short rhizomes; leaves borne on the lower culm, the lowermost reduced to bladeless sheaths, the upper with elongate blades 1–2.5 (3) mm wide, these subequal to or surpassing the inflorescence, flat or nearly so; inflorescence 1–6 cm long, 7–10 mm wide, compound but headlike or spikelike; spikes crowded, scarcely evident, ca 3–5 mm long, few-flowered, androgynous; pistillate scales rather hyaline, brownish or stramineous, the midrib firmer and sometimes projected as an aristate awn; perigynia 2.4–3 mm long, lance-ovate or lance-truncate, short-stipitate, abruptly to gradually tapered to a serrulate beak, the dorsal suture sometimes extended into a membranous flap or keel, this usually conspicuous toward the apex, the body thick-walled, dark brown, shiny; stigmas 2; achenes lenticular. Wet meadows and sphagnum bogs, rare, at ca 2285 m in Duchesne and Garfield (Intermountain Flora report) counties; circumboreal, south in America to California and Pennsylvania; 1 (i). Possibly reports of the closely related *C. cusickii* Mack. in Piper & Beattie for Utah are based on specimens of *C. diandra*.

Carex dioica L. Yellow Bog Sedge. [*C. gynocrates* Wormsk.]. Plants 2–20 cm tall, culms arising singly from long, very slender rhizomes, these to ca 0.5 mm thick; leaves borne on the lower culms, the blades 0.4–0.9 mm wide, narrowly involute; bract lacking; spike solitary, androgynous or nearly all staminate to nearly all pistillate, 1–1.5 cm long, the staminate part only 1–2 mm wide; pistillate scales usually shorter but broader than the perigynia, light brown, persistent; perigynia 3–3.5 mm long, plump, prominently ribbed dorsally, strongly spreading to descending, commonly chestnut brown at maturity, thick-walled, spongy at the base (especially ventrally), abruptly contracted to a short beak; rachilla obsolete; stigmas 2; achenes lenticular, filling the perigynial cavity; 2n = 48, 50, 70. Wet and boggy meadows at 2740 to 3130 m in Daggett, Duchesne, Emery, Garfield, Iron, Summit, and Uintah counties; circumboreal, south in America to Oregon, Colorado, and Pennsylvania; 10 (vii). Our materials are referable to var. *gynocrates* (Wormsk.) Ostenf.

Carex disperma Dewey Softleaved Sedge. Culms 10–45 cm long, densely tufted, from short slender rhizomes, rather flaccid or flexuous; leaves basal and on the lower culms, the blades 0.5–2 mm wide; inflorescence 0.8–4 cm long, 3–5 mm wide, usually interrupted, spikelike, sometimes subtended by a narrow bract, this shorter than the inflorescence; spikes sessile, 3–5, to ca 5 mm long, about as wide as long, androgynous, with 1 or 2 (3) staminate flowers and 1–3 perigynia, the staminate flowers inconspicuous and the spikes sometimes appearing pistillate; pistillate scales shorter than to equaling the perigynia, pale green, scarious; perigynia 2–3 mm long, elliptic, abruptly contracted to a very short beak; stigmas 2; achenes lenticular, filling the perigynium; style semipersistent on the achene; 2n = 70. Wet and boggy meadows and along streams, mostly in shade of woods, at 2285 to 3050 m in Beaver, Cache, Daggett, Duchesne, Garfield, Grand, Juab, Piute, Salt Lake, San Juan, Sanpete, Summit, Uintah, Utah, Wasatch, Washington, and Wayne counties; circumboreal, south in America to New Mexico and New Jersey; 34 (xv).

Carex douglasii F. Boott in Hook. Plants 8–28 cm tall, mostly dioecious; culms arising singly or few together from long slender brownish rhizomes less than 2 mm in diameter; leaves basal and on the lower culms, the blades 0.5–2 mm wide; lowest bract lacking or sometimes present and foliose; inflorescence 1.5–5.5 cm long, (5) 10–20 mm wide, headlike, exceeded by the leaves or surpassing them; spikes ca 10–25, generally all staminate or all pistillate, rarely androgynous, densely crowded, over twice as long as the internodes of the rachis, more or less evident without dissecting the head or the upper ones sometimes not, the lower 0.8–2 cm long, ovoid or more commonly the larger ones elongate and nearly linear or linear-elliptic; pistillate scales longer than the perigynia and concealing them, these and usually the staminate ones with a sometimes excurrent, light green midstripe, this flanked on both sides by a narrow stramineous or pale brown stripe and then the broad hyaline margins, very conspicuous in the spikes; perigynia 2.2–4.7 mm long, 1.1–1.8 mm wide, ovate to elliptic, brownish, the marginal nerves prominent and distended toward the ventral surface, otherwise obscurely nerved; beak of the perigynium 1–1.5 mm long, obliquely cleft, often hyaline, the dorsal suture often conspicuous and with a hyaline flap, this sometimes encroaching onto the body of the perigynium; stigmas 2; achenes lenticular; 2n = 60. Pinyon-juniper, sagebrush-grass, mountain brush, lodgepole pine, and spruce-fir communities, often along roads and near seeps and springs where trampling of livestock has compacted the soil, at 1525 to 3110 m in all Utah counties except Carbon, Davis, Grand, Kane, Piute, and San Juan; British Columbia to Manitoba and south to California and Iowa; 45 (vii). Some of the Utah specimens have very small perigynia (only 2.2 mm long) but they do not appear to differ in any other way from plants with larger perigynia.

Carex ebenea Rydb. Ebony Sedge. Plants 19–59 cm tall, densely caespitose; rhizomes lacking or short; leaves basal and on the lower culms, the blades 1–4.5 mm wide, flat or nearly so; lowest bract usually lacking, or if present, mostly shorter than the inflorescence, this a head 1.3–2.8 cm long and about as wide, or more commonly slightly narrower (in unpressed specimens), ovate or broadly elliptic, truncate or more commonly cuneate at the base, the internodes of the rachis 0.5–2 mm long; spikes (3) 6–12, gynaecandrous, commonly 9–15 mm long, averaging over 10 mm long, sessile, densely congested, usually slightly more discernable than in *C. microptera*, ca 5–10 times longer than the internodes of the rachis; pistillate scales slightly shorter and usually narrower (at least distally) than the perigynia, dark brown to blackish brown, with or without hyaline margins; perigynia (4.8) 5.2–7.1 mm long, 0.9–2 mm wide, the better developed ones commonly 4–6.4 times longer than wide, brown to dark brown in the center and commonly with narrow greenish wings, faintly nerved on both sides, serrulate except near the apex of the prominent slender beak; stigmas 2; achenes lenticular, 1.5–1.8 mm long. Lodgepole pine, Engelmann spruce, meadow, and alpine communities at 2895 to 3570 m in Daggett, Duchesne, Garfield, Grand, Iron, San Juan, Sevier, Summit, Tooele, and Wayne counties; Wyoming to Arizona and New Mexico; 33 (xi). Plants of C. ebenea are more similar to C. haydeniana than to C. microptera (q.v.) especially in the darker colored heads, fewer spikes on the average, and somewhat larger perigynia. They are also more common at higher elevations than is C. microptera. However, the perigynia in C. microptera are sometimes quite narrow and, except for mostly being shorter, could be mistaken for those C. ebenea.

Carex egglestonii Mack. Plants (1.5) 3–9 dm tall, caespitose; rhizomes lacking or very short; leaves borne on the lower culms, the blades 1.5–5 mm wide, flat; lowest bract lacking, shorter than, or rarely equaling the inflorescence, this 1.5–2.7 cm long, 1–3 cm wide, headlike, surpassing the leaves, the lowest internode to ca 5 mm long; spikes 3–6, gynaecandrous, 10–17 mm long, 7–12 mm wide, ovate to nearly globose, sessile, closely aggregated, but more or less evident without dissecting the head, to ca 3 or more times longer than the rachis internodes, with up to 45 or more perigynia; pistillate scales conspicuously shorter and narrower than the perigynia, brownish, with a green or pale brown midstripe and hyaline margins, acute; perigynia (5) 6–8 mm long, 2.5–3.5 mm wide, ovate, elliptic, or lanceolate, greenish or brownish, strongly flattened, with thin broadly winged margins, finely serrulate distally and frequently below, gradually or rather abruptly tapered to a bidentate, more or less flattened, winged beak; anthers 2.5–3.7 mm long; stigmas 2; achenes lenticular. Sagebrush-grass, aspen, forb-grass, Douglas fir, spruce-fir, lodgepole pine, and alpine communities at 2440 to 3475 m in Beaver, Carbon, Duchesne, Emery, Garfield, Grand, Iron, Kane, Millard, Piute, San Juan, Sanpete, Sevier, Summit, Uintah, Utah, and Wasatch counties; Wyoming and Colorado; 48 (xx).

Carex elynoides H. T. Holm Kobresia Sedge. Plants 5–16 cm tall, densely caespitose, lacking rhizomes, the culms and leaves fasciculate; leaves basal and on the lower 1/4 of the rounded culms, mostly equal to, or surpassed by, the inflorescence, the blades 0.3–0.5 mm wide, strongly folded or involute and nearly terete; bract lacking; spike solitary, erect, androgynous, 1–1.8 cm long, the staminate part shorter to longer than the pistillate part, exceeding the perigynia by 3–7 mm; pistillate scales brown to dark brown, the centers usually not noticeably green or pale, the margins hyaline, wider to much wider and mostly longer than the perigynia, these usually 4–12, 2.5–4.5 mm long, oblong or ellipsoid, filled by the mature achene, glabrous or occasionally very sparingly serrulate-ciliolate near the beak; rachilla subequal to the achene; stigmas 3; achenes trigonous. Krummholz and alpine communities at 2865 to 3905 m in Beaver, Box Elder, Duchesne, Garfield, Grand, Iron, Juab, Piute, Salt Lake, San Juan, Sevier, Summit, Utah, Wasatch and Wayne counties; Montana to Colorado and Nevada; 34 (ix). C. elynoides is separated from C. nardina by a series of features, but these are easily confused. It is more closely allied to C. filifolia, and except for the glabrous or nearly glabrous perigynia and different habitat, it would be almost impossible to distinguish them. It is also easily mistaken for **Kobresia bellardii** (q.v.), which grows in the same kind of habitat.

Carex filifolia Nutt. Threadleaf Sedge. Plants 8–38 cm tall, densely caespitose, without rhizomes, the culms and leaves fasciculate; leaves basal or on the lower 1/4 of the rounded culms, equaling or exceeded by the inflorescence, the blades 0.3–0.5 mm wide, strongly folded to involute and nearly terete; bract lacking; spike solitary, erect, androgynous, 1.3–2.5 cm long, the staminate part usually equaling or longer than the pistillate part, commonly exceeding the perigynia by 8–13 mm; scales light brown or brown with broad white hyaline margins, the pistillate ones wider and longer than the perigynia, these usually 5–15, 3–4.5 mm long, obovoid to ellipsoid, filled by the mature achene, minutely hirtellous at least on the upper 1/2; rachilla subequal to the achene; stigmas 3; achenes trigonous; $2n = 50$. Desert shrub, sagebrush, and pinyon-juniper communities at 1675 to 2225 m in Daggett, Emery, and Wayne counties; Yukon to Manitoba southward to California and Texas; 7 (i).

Carex foetida All. [C. vernacula Bailey]. Plants 5–25 cm tall; culms arising singly or 2 or 3 together from rhizomes; leaves basal and on the lower culms, the blades 2–4 mm wide, flat or nearly so; inflorescence a globose or ovoid head 8–16 mm long, surpassing the leaves; lowest bract lacking or shorter than the inflorescence; spikes usually 10 or more, androgynous, ca 4–7 mm long, sessile, very densely crowded and mostly not evident in the head, concealing the very short internodes of the rachis; pistillate scales subequal to the perigynia, dark brown with paler midrib and hyaline margins; perigynia 3.3–4.7 mm long, 1–1.6 mm wide, lanceolate to narrowly elliptic, usually brownish, often with paler greenish margins, flattened, with the marginal nerves prominent, faintly nerved toward the base on both sides, conspicuously beaked; stigmas 2; achenes lenticular. Wet meadow at 3020 to 3200 m in Wasatch County; Washington to Wyoming, south to California and Colorado; 1 (0). Utah materials are referable to **var. vernacula (Bailey) Kukenthal.** The one specimen (Hayward 9948 BRY) seen is rather young and tentatively identified. Other reports of this species in the state are likely based on specimens of C. jonesii (q.v.).

Carex geyeri F. Boott Elk Sedge. Plants 15–30 (50) cm tall, more or less caespitose and with short to elongate rhizomes; leaves basal and cauline, the lower ones often reduced to bladeless sheaths, the blades 1–3 mm wide, flat or nearly so, evergreen, the tips often turning yellow or brown but the lower parts remaining green through at least one winter; spike solitary, erect, androgynous, the staminate part 1–2 (3) cm long, linear, the pistillate usually with 1 or 2 plump perigynia, these proximal to or somewhat remote from the staminate part; bract lacking or in robust specimens to 1.5 cm long; scales light brown, with broad hyaline margins, the pistillate ones usually equaling or to 4 mm longer than the perigynia; perigynia solitary or, if 2, rather remote, with the internode subequal to or longer than the perigynia, 5–6 mm long, 2–ribbed, greenish or light brown, ellipsoid or obovoid; rachilla obsolete or less than 1/2 as long as the achene; stigmas 3; achenes trigonous, filling the perigynium. Mountain brush, ponderosa pine, aspen, Douglas fir, lodgepole pine, and spruce-fir communities at 1830 to 3300 m in Cache, Carbon, Daggett, Duchesne, Grand, Juab, Millard, Morgan, Salt Lake, San Juan, Sanpete, Summit, Tooele, Uintah, Utah, and Wasatch counties; Alberta and British Columbia south to California and Colorado; 47 (xii).

Carex haydeniana Olney in Wats. Cloud Sedge. [*C. nubicola* Mack.]. Plants 9–33 cm tall, densely caespitose; rhizomes lacking or very short; leaves basal and on the lower culm, the blades 1–4 mm wide; lowest bract lacking or, if present, not over 1/2 as long as the inflorescence; internodes of the rachis ca 0.5–1 mm long; inflorescence a head 1–1.5 cm long, ovoid to nearly orbicular in outline, usually truncate at the base, surpassing the leaves; spikes 3–8, gynaecandrous, 7–11 mm long, ovate or nearly so, sessile, densely congested, hardly if at all discernable without dissecting the head, 7–10 times longer than the internodes of the rachis; pistillate scales shorter and narrower than the perigynia, dark brown to blackish, with a lighter brown or green midrib, the margins narrowly hyaline or not; perigynia 4–6 (6.3) mm long, (1.7) 2.2–2.8 mm wide, 1.6–2.9 times longer than wide, ovate or lance-ovate to elliptic, brown, dark brown, or blackish, nearly as dark as the scales or somewhat lighter and sometimes the margins green, faintly nerved dorsally, nerveless or faintly nerved ventrally, strongly flattened except where distended by the achene, the margins thin and winged, serrulate except on the subterete beak tip; stigmas 2; achenes lenticular, commonly 1.5–2 mm long. Spruce-fir, limber pine, Engelmann spruce, krummholz, and alpine communities at (2650) 2890 to 3965 m in Beaver, Box Elder, Cache, Duchesne, Garfield, Salt Lake, Sanpete, Summit, Tooele, Uintah, Utah, and Wayne counties; British Columbia and Alberta, south to California and Colorado; 45 (x). This species is similar to and probably transitional with *C. microptera* (q.v.).

Carex hoodii F. Boott in Hook. Plants (20) 30–62 cm tall, densely caespitose; rhizomes lacking or very short; leaves basal and on the lower culm; lowest bract lacking or shorter than the inflorescence, this 1–2 cm long, 0.7–1.5 cm wide, headlike, oblong to nearly orbicular; spikes all tightly congested or the lowest slightly separated, 4–12, androgynous, not over 5 mm long, with ca 4–12 perigynia, sessile; pistillate scales equal to or slightly shorter than the perigynia, brown with a green midstripe and scarious margins; perigynia 3.5–4.6 mm long, 1.5–2 mm wide, narrowly elliptic to ovate, the body brown to dark brown, with greenish serrulate (at 20x magnification) slightly winged margins and beak, (the wings more pronounced than in *C. occidentalis*), the beak bidentate (more so than in *C. occidentalis*), the teeth ca 0.3–0.5 mm long, the marginal nerves on the winged edges or marginally on the ventral face, the dorsal and ventral faces nerveless or faintly so; stigmas 2; achenes lenticular. Sagebrush-grass, mountain brush, aspen, tall forb, forb-grass, spruce, and fir communities at 2105 to 3050 m in Beaver, Box Elder, Cache, Carbon, Daggett, Davis, Duchesne, Grand, Juab, Millard, Morgan, Piute, Rich, Salt Lake, Sanpete, Sevier, Summit, Tooele, Uintah, Utah, Wasatch, and Weber counties; British Columbia to Saskatchewan and south to California and Colorado; 55 (x). This species is not easily distinguished from *C. occidentalis* (q.v.).

Carex hystricina Muhl. ex Willd. Bottlebrush Sedge. Plants 3–6 (10) dm tall; culms clustered on rhizomes; leaves basal and on the lower culm, the lowermost leaves sometimes reduced to bladeless sheaths, the upper ones with flat blades 2–9 mm wide; inflorescence 6–10 cm long; terminal spike staminate or rarely gynaecandrous, 2–3 cm long; lateral spikes 1–3, pistillate, (1) 1.5–3 cm long, 12–15 mm wide, ascending to strongly spreading, on slender peduncles to 5 cm long or the upper one sessile or subsessile, the lowest subtended by a leafy bract usually exceeding the inflorescence, this usually sheathless or the sheath closed for up to 14 mm; pistillate spikes also with leafy bracts; pistillate scales equaling or much shorter than the perigynia, pale green or straw colored in age, aristate-awned, the awn 2–6 mm long and serrulate-ciliolate; perigynia 5–7 mm long, pale green, prominently nerved, lanceolate or lance-ovate, slightly inflated, with a conspicuous narrow beak ca 2 mm long, strongly spreading and giving a bristly appearance to the spikes; stigmas 3; achenes trigonous, much smaller than the perigynia; style continuous with, persistent on, and of the same firm texture as the achene, straight or bent in age; $2n = 58$. Riparian and wet meadow communities of the Colorado and Green rivers drainages at 1370 to 1920 m in Carbon, Garfield, Grand, Kane, San Juan, Uintah, and Washington counties; Washington to New Brunswick and south to California and Virginia; 10 (i).

Carex illota Bailey Sheep Sedge. Plants 10–42 cm tall, densely tufted; rhizomes lacking or short; leaves basal and on the lower culm, the blades 0.5–2.8 mm wide, flat; bracts not much larger than the scales; inflorescence 0.8–1.4 cm long, nearly or fully as wide as long, headlike, much surpassing the leaves; spikes 3–6, gynaecandrous, ca 4–6 mm long, sessile, closely aggregated and scarcely evident from one another in the dense head, with 5–15 ascending or spreading perigynia; pistillate scales ca 1/2–3/4 as long as the perigynia, blackish or black brown, often with a greenish or yellow brown midstripe and with or nearly without hyaline margins; perigynia 2.5–3 mm long, 1–1.4 mm wide, lanceolate to narrowly ovate, yellowish or greenish brown to blackish, dorsally and sometimes ventrally nerved, the nerves sometimes reddish, the margins entire, rounded or rather sharp, sometimes with very small wings in the distal portion, gradually to rather abruptly tapered to an ill-defined or conspicuous beak, the beak usually blackish, obliquely cleft, the dorsal suture conspicuous through the length of the beak and onto the distal part of the body, usually with an overlapping involute flap; stigmas 2; achenes lenticular. Wet and boggy meadows, flooded pond and lake margins, and along streams of the Uinta and central Wasatch mts. at 2285 to 3505 m in Daggett, Duchesne, Salt Lake, Summit, Uintah, and Wasatch counties; British Columbia to Montana, south to California and Colorado; 30 (xi). *C. illota* has about as many, if not more, affinities with section Stellulatae as it does with section Ovales.

Carex interior Bailey Inland Sedge. Plants 17–45 cm tall, densely caespitose; rhizomes lacking; leaves borne on the lower 1/3 of the culms, the blades 1–2 mm wide; lowest bract lacking; inflorescence 0.8–2 cm long, 4–6 mm wide, included in or exserted beyond the leaves; spikes (2) 3 or 4, sessile, gynaecandrous, the terminal one often clavate with a conspicuous slender staminate portion, rarely all staminate, 7–10 mm long, the lateral ones ca 4–6 mm long and about as wide, with only a few staminate flowers and (1) 5–10 perigynia, rather closely spaced but readily discerned, subequal to or a little longer than the internodes of the rachis; pistillate scales shorter than the perigynia, pale green, stramineous or marked with brown, with broad hyaline margins; perigynia widely spreading, beaked but not conspicuously so as in *C. muricata* and the spikes not so bristly, 2.2–3.2 mm long including the shallowly bidentate serrulate beak, this 1/4–1/2 as long as the body, not winged or thin-margined, the marginal nerves prominent and slightly ridged, these displaced toward the flattened ventral surface, the dorsal surface convex; stigmas 2; achenes lenticular; $2n = 54$. Wet often calcareous boggy meadows at 1890 to 3170 m in Beaver, Cache, Duchesne, Emery, Garfield, Iron, Kane, Salt Lake, Sanpete, Summit, and Uintah counties; British Columbia to Labrador and south to Mexico and Pennsylvania; 12 (ii).

Carex jonesii Bailey Plants 15–60 cm tall, caespitose or the culms commonly scattered along conspicuous rhizomes; leaves clustered on the lower 1/4 of the culm, the lower ones often bladeless, the upper ones with flat blades 1.5–3 mm wide, the sheaths neither cross-rugulose nor thickened at the mouth; inflorescence a head, 0.8–2 cm long, 7–11 mm wide, much surpassing the leaves; spikes 4–8, less than 1 cm long, androgynous, the staminate flowers inconspicuous; pistillate scales shorter than the perigynia but as wide, brownish to blackish with hyaline margins; perigynia 3–4 mm long, 1.2–1.5 mm wide, widest below the middle, ovate-lanceolate to lanceolate, gradually tapering to a smooth or slightly serrulate beak, stramineous to dark brown, widely spreading, nerved on both sides, the marginal nerves not (or hardly) more evident than the facial nerves; stigmas 2; achenes lenticular. Meadows and along streams at 2500 to 3050 m in Beaver, Juab, and Salt Lake counties; Washington to California and east to Colorado; 8 (iii). In manuals this taxon is often keyed as being caespitose, either without rhizomes or with very short ones. Our specimens commonly have conspicuously prolonged rhizomes and their presence has probably resulted in some confusion with materials of *C. foetida*.

Carex lanuginosa Michx. Woolly Sedge. [*C. lasiocarpa* var. *lanuginosa* Kukenthal]. Plants 24–80 cm tall or taller; culms arising singly or more often few together or in small tufts from thick creeping rhizomes; leaves mostly borne on the lower 1/2 of the culms, the lower ones often reduced to bladeless sheaths, the upper ones with flat blades 2–6 mm wide; inflorescence 8–21 cm long, leafy bracteate, the first and sometimes the second bracts equaling to much surpassing the inflorescence, mostly not sheathing but the lowest one sometimes with a closed sheath to 17 mm long; terminal spike staminate, 1.7–4.5 cm long, cylindrical, sessile or subsessile or on peduncles to 7 cm long, often subtended by 1–2 smaller staminate spikes; lateral pistillate spikes 1–3 (4), 1–4.8 cm long, 5–7 mm wide, the lower ones well spaced and mostly shorter than their internodes, sessile or nearly so, or on peduncles up to 8 cm long; pistillate scales usually shorter and narrower than the perigynia, acute or awn-tipped, brownish or purplish, typically with a green or pale midstripe; perigynia densely velutinous or velutinous-sericeous, 3.3–5 mm long, ovoid to subglobose, with an abrupt conspicuous bidentate beak, many-ribbed, thick-walled; stigmas 3; achene trigonous, loosely filling the perigynium; $2n = 78$. Along streams, rivers, ponds and lakes, and other wet places at 1250 to 2805 m in all Utah counties except Millard, Morgan, and Tooele; British Columbia to New Brunswick, south to California and Arkansas; 71 (vii).

Carex lasiocarpa Ehrh. Slender Sedge. Plants 3–8 cm tall or taller; culms reddish or purplish at the base, loosely caespitose, from long creeping robust rhizomes; leaves mostly on the lower 1/2 of the culms, the lower ones reduced to bladeless sheaths, the upper ones with blades 2 mm wide or less, these flattish at the base, strongly involute apically and long-attenuate; inflorescence 12–21 cm long; terminal spike staminate, 3–5 cm long, closely or remotely subtended by a smaller staminate spike; pistillate spikes 2 or 3, below the staminate ones, rarely with a few staminate flowers apically, 1–2.5 (3) cm long, more or less cylindric, sessile or on peduncles to ca 4 mm long, well spaced, and shorter than the internodes of the rachis, each subtended by a leafy bract, these from shorter to longer than the inflorescence; pistillate scales aristate-awned, the body narrower than the perigynia, the awn serrulate-ciliolate, exceeding the lower perigynia, but shorter than the upper perigynia; perigynia pubescent as in *C. lanuginosa*, 2.5–5 mm long, broadly obovoid to ovoid, abruptly contracted to a bidentate beak; stigmas 3; achenes trigonous, loosely filling the perigynia; $2n = 56$. Swampy meadows and ponds (in water to ca 30 cm deep) on subalpine moraines at 2950 to 3000 m in the Uinta Mts., Daggett and Uintah counties; Alaska to New Brunswick, south to California and Tennessee; 5 (iv).

Carex lenticularis Michx. [*C. kelloggii* W. Boott]. Plants 15–61 cm tall, caespitose from fibrous roots, the larger roots covered with feltlike yellowish or yellow brown hairs; rhizomes lacking or short; leaves mostly on the lower 1/4 of the culms, the blades 1–2.6 mm wide; lowest bract foliose, shorter than or exceeding the inflorescence, this 4–13 cm long, with a solitary (or 2) staminate spike (s) 1–3 cm long above 2–4 cylindrical pistillate ones, these 0.6–4 cm long and 3–4 mm wide, the uppermost pistillate one sometimes reduced; pistillate scales shorter and narrower than the perigynia, blackish or black purple with a green midrib, this shorter than or equaling the tip of the scale and neither excurrent nor ridged; perigynia 2–3 mm long, lanceolate, elliptic to ovate, more or less faintly nerved dorsally and ventrally, deciduous at maturity, usually with a well-defined stipe 0.1–0.4 mm long, green except for the often blackish beak, this 0.1–0.4 mm long, entire; stigmas 2; achenes lenticular; $n = 34, 44$. Stream and river banks, wet meadows, and pond, lake, seep, and spring margins, where sometimes partly submersed, at 1860 to 3265 m in Duchesne, Salt Lake, Summit, Tooele, Uintah, Utah, and Wasatch counties; Alaska to Labrador south to California and Michigan; 42 (vii). This taxon is distinquished from others in the section Atratae by the very narrow leaves, narrow spikes, caespitose habit, numerous fibrous roots often covered with yellowish or yellow brown feltlike hairs, and by the quite stipitate and rather readily deciduous perigynia. The scales and beaks of the perigynia are like those in *C. aquatilis*, while the perigynial nerves indicate an alliance with *C. nebrascensis*.

Carex leporinella Mack. Sierra-hare Sedge. Plants 11–28 cm tall, densely caespitose; rhizomes lacking or very short; leaves basal and on the lower culm, the blades 1–2 mm wide, flat or channeled; lowest bract lacking or shorter than the inflorescence, this 12–25 mm long, 5–12 mm wide, headlike, included in the leaves or surpassing them, the lowest internode usually 1–3 mm long; spikes

3–6, gynaecandrous, 5–11 mm long, 3–5 mm wide, sessile, densely aggregated or the lower ones somewhat separated, with 8–22 perigynia; pistillate scales subequal to and as wide as the perigynia, brown, with a green or pale midrib and hyaline margins; perigynia 2.9–4.3 mm long, 1–1.5 mm wide, elliptic-lanceolate to elliptic, greenish to brownish, sharp-edged but scarcely winged or the wings very narrow, serrulate, evidently nerved dorsally, not nerved or faintly so ventrally, gradually tapering to an ill-defined beak; stigmas 2; achenes lenticular. Wet and drying meadows, often in drying mud of ephemeral pools or around ponds and lakes at 2835 to 3200 m in the central and western Uinta Mts., Duchesne, Summit, and Wasatch counties; Washington to Idaho, south to California; 8 (v).

Carex leptalea Wahl. Flaccid Sedge. Plants 16–35 cm tall, densely caespitose, with slender rhizomes, the rhizomes obscured by the matted habit; culms slender, rather flaccid, usually exceeding the numerous fine leaves; leaves basal or nearly so, rather lax, the blades 0.4–1 mm wide, flat or somewhat folded; bract lacking; spike solitary, androgynous, 4–6 mm long, the staminate part included in or slightly exceeding the pistillate part; scales greenish or pale brown, more or less hyaline, the pistillate ones mostly shorter than the perigynia and soon deciduous; perigynia 1–3, 2.5–3.5 mm long, pale green, retuse or notched at the beakless tip, elliptic or nearly so, finely striate, substipitate and more or less spongy at the base, empty in the upper 1/2 or nearly so; rachilla lacking; stigmas 3; achenes trigonous; 2n = 52. Shady seepy bogs at 2195 to 2230 m in Uinta and Whiterocks Canyons, Uinta Mts., Duchesne and Uintah counties; Alaska to Labrador, south to northern California and Florida; 4 (iv).

Carex limnophila F. Hermann Pond Sedge. Plants 20–45 cm tall, caespitose; rhizomes lacking or very short; leaves borne on the lower 1/4 of the culms, the lower ones sometimes reduced to bladeless sheaths, the upper ones with blades 1–3 mm wide, these flat or nearly so; lowest bract lacking or shorter than the inflorescence, this headlike, 7–15 (18) mm long, 5–10 mm wide, usually surpassing the leaves; spikes 4–8, gynaecandrous, 4–7 mm long, densely aggregated and hardly or not evident without dissecting the head, mostly concealing the very short internodes of the rachis; pistillate scales subequal to the perigynia, brown with hyaline margins and sometimes a pale midrib; perigynia 2.5–3.2 (3.5) mm long, 1.1–1.7 mm wide, lanceolate to ovate, plano-convex, brown or blackish brown and faintly nerved dorsally, brownish and nerveless ventrally but often with 1 or 2 transverse folds near the middle or on the lower 1/2 of the body, tapered or abruptly contracted to a short, obliquely cleft beak, doubly serrulate to below the middle; stigmas 2; achenes lenticular. Wet meadows at 2165 to 2750 m in Daggett, Duchesne, Emery, Juab, and Uintah counties; Washington to Alberta and south to Nevada and Colorado; 4 (i). Perhaps *C. limnophila* represents a diminutive phase of *C. pachystachya*. One specimen (Goodrich 19593 BRY) has perigynia 3.5 mm long (the minimum for *C. pachystachya* and the maximum for *C. limnophila*), and a few of the perigynia have transverse folds on the ventral side as in *C. limnophila*, but the inflorescence is 15–20 mm long, well within the range of *C. pachystachya* but beyond that listed for *C. limnophila*.

Carex limosa L. Mud Sedge. Plants 13–26 cm tall, often glaucous; culms arising singly or 2 or 3 together from slender rhizomes; roots covered with a yellowish or yellow brown feltlike tomentum; leaves mostly on the lower 1/2 of the culms, the lower ones usually reduced to bladeless sheaths, the upper ones with blades 0.5–1.5 mm wide, more or less flat, the midrib strongly keeled below, often sharply serrulate; lowest bract from shorter to longer than the inflorescence, this mostly 3.5–6.5 cm long; terminal spike staminate, (1) 1.5–2.1 cm long; lateral spikes 1 or 2, pistillate or occasionally with 1 or 2 staminate flowers apically, 8–18 mm long, spreading or drooping, on flexuous peduncles 5–22 mm long, or rarely the lowest spike on an elongate peduncle originating near the culm base; pistillate scales shorter than to slightly longer than and nearly as wide as the perigynia, greenish brown to dark reddish brown, with a green center, the margins not or scarcely hyaline, obtuse or acute to cuspidate; perigynia 2.9–3.5 mm long, 1.2–1.9 mm wide, substipitate, with a very short, more or less conspicuous beak, pale green and densely white-granular (the grains visible at high magnification); stigmas 3; achenes trigonous; 2n = 56, 64. Floating bogs and boggy meadows at 2955 to 3200 m in the Uinta Mts., Duchesne, Summit, and Uintah counties; circumboreal, south in America to California; 6 (iv).

Carex luzulina Olney Woodrush Sedge. [*C. ablata* Bailey; *C. fissuricola* Mack.; *C. luzulina* var. *ablata* F. Hermann]. Plants 15–80 cm tall, tufted; rhizomes lacking or short; leaves basal and 1–3 cauline, these with closed sheaths 3–5 cm long or longer, the blades 3–9 mm wide, flat; lowest bract with a closed sheath 1–4.5 cm long, shorter than the inflorescence and mostly not exceeding the next lowest spike; inflorescence 6.5–17 cm long, surpassing the leaves; terminal spike staminate or with a few perigynia at the base, 1–2 cm long, sessile or on a short peduncle; lateral spikes 4–5, 1–3 cm long, the upper 1 or 2 crowded with the staminate spike, sessile or nearly so, pistillate or with a few stamens at the apex, the lower ones pistillate and well separated, borne on slender elongate peduncles, the lowest peduncle to 8 cm long, ca 1/2 of it enclosed in the sheath of the bract, the next lowest peduncle ca 1–2 cm long and 1/2 or more of it enclosed in the sheath of the second bract; pistillate scales dark brown with broad hyaline, sometimes minutely frimbiate or ciliate margins, subequal to the perigynia but not covering the beak; perigynia 3–5 mm long, 1–1.5 mm wide, lanceolate or lance-ovate, compressed, ciliolate-serrulate; stigmas 3; achenes trigonous. Engelmann spruce and meadow communities, often along streams at 2590 to 3350 m in Beaver, Piute, and Salt Lake counties; British Columbia to Montana and south to California; 2 (0).

Carex microglochin Wahl. Plants (5) 15–25 cm tall; culms arising few together from slender rhizomes, the few leaves mostly toward the base; leaf blades ca 0.5–0.8 mm wide, channeled or nearly terete; bract lacking; spike solitary, androgynous, ca 0.5–1 cm long, 3–5 mm wide, erect; scales brownish or tinged with purple, the pistillate ones almost half as long as the perigynia, quickly deciduous; perigynia subulate or linear lanceolate, with the rachilla conspicuously exserted from the orifice, 3–5 mm long including the rachilla, ca 1 mm wide at the base, strongly spreading, soon deflexed and deciduous; stigmas 3; achenes elongate, nearly cylindric or obtusely trigonous; 2n = 58. Calcareous boggy meadows at 2800 to 2830 m in Duchesne and Emery counties; Greenland to Alaska, southward to Quebec, Washington and Colorado; 4 (iii).

Carex microptera Mack. Small-wing Sedge. [*C. macloviana* var. *microptera* (Mack.) J. Boivin; *C. festivella* Mack.]. Plants 2–8 dm tall, densely caespitose; rhizomes lacking or very short; leaves basal and on the lower culm, the blades 1–5 mm wide, flat or nearly so; lowest bract mostly lacking, rarely to ca 1/2 as long as the inflorescence, this a head, 1.2–2.5 cm long and as wide, orbicular or nearly so, usually truncate at the base, the internodes of the rachis ca 0.5–2 mm long or the lower one to 3 (4) mm long; spikes (2) 5–15 (21), gynaecandrous, commonly 5–11 mm long, sessile, densely congested, hardly if at all discernable without dissecting the head, much longer than the internodes of the rachis; pistillate scales shorter and narrower than the perigynia, brownish; perigynia 3.2–5 mm long, 1.3–2.2 mm wide, 2–3 (4) times longer than wide, greenish or brownish to dark brown in age, with a green or pale midstripe, lightly or obscurely nerved on both sides, strongly flattened, except where distended by the mature achene, the margins thin, winged, serrulate except on the more or less terete beak-tip, not hyaline or very narrowly so; stigmas 2; achenes lenticular, commonly 1.1–1.5 mm long. Moist places in many montane plant communities at 1525 to 3415 m in all Utah counties except Davis and Rich; British Columbia to Saskatchewan, south to California and New Mexico; 215 (xvii). This is one of the most common and widespread sedges in the state. It is often confused with several others; among which are *C. hoodii* that is distinguished by its androgynous spikes. *C microptera* forms a series of intergrading forms with *C. ebenea*, *C. haydeniana*, *C. pachystachya*, and *C. subfusca*. All of the above listed species except *C. ebenea* have been reduced previously to varietal status under *C. macloviana* Urv. The comments by Cronquist et al. (1977) are pertinent.

Carex misandra R. Br. Shortleaf Sedge. Plants 10–30 cm tall, densely caespitose; rhizomes lacking or very short; leaves basal and on the lower culm, all bearing flat blades 1–4 mm wide; lowest bract with a closed sheath 7–20 mm long, this subequal to or longer than the blade, the second bract also sheathing; inflorescence ca 2–5 cm long, usually surpassing the leaves, with 3 or 4 spikes 1–1.5 cm long, the terminal one gynaecandrous or occasionally pistillate, the lateral ones pistillate, often spreading or drooping, on long slender peduncles, these enveloped basally in the closed sheath of the bracts, the exserted portion mostly longer than the spikes; pistillate scales shorter but commonly wider than the perigynia, brown to dark purple, with hyaline margins and sometimes a paler midrib; perigynia 3.3–5 mm long, ca 1 mm wide, lanceolate, more or less flattened, especially distally, sometimes stipitate at the base, very gradually tapering to a beak, ciliolate-serrulate in the distal 1/2, nerveless or nearly so except at the margins, brown or purple distally, pale where covered by the scale; stigmas 3; achenes trigonous; 2n = 40. Alpine tundra communities at 3600 to 3960 m in the Uinta Mts., Duchesne and Summit counties; circumboreal, and south in America to Alberta and Quebec, disjunct in Colorado; 3 (0).

Carex multicostata Mack. Many-ribbed Sedge. Plants 15–70 cm tall, caespitose; rhizomes lacking or very short; leaves borne on the lower 1/4 of the culms, the lower ones bladeless or nearly so, the upper ones with blades 2–4 mm wide; lowest bract lacking or shorter than the inflorescence, this 1.5–4 cm long, 14–20 mm thick, headlike, surpassing the leaves, the internodes mostly concealed by the congested spikes, ca 1–2 mm long or the lowest one slightly longer; spikes 5–10, gynaecandrous, 6–10 mm long, sessile, closely aggregated but generally more readily discernable than in *C. microptera*; pistillate scales subequal to the length and width of the perigynia or somewhat shorter, light reddish brown or brown, with a pale 3-nerved midstripe and white hyaline margins 0.1–0.3 mm wide; perigynia 4.5–6.2 mm long, 1.6–2.4 mm wide, narrowly to broadly ovate, plano-convex, obscurely to conspicuously many-nerved on both sides or nerveless or few-nerved ventrally, winged-margined and serrulate, tapering to the prominent slender or more or less flattened beak, this margined and serrulate to near the apex or terete and entire for up to 0.6 (1) mm; stigmas 2; achenes lenticular. Moist and wet meadows and along streams at 2625 to 2835 m in Cache, Daggett, Summit, and Uintah counties; Washington to California and east to Idaho; 5 (iv).

Carex muricata L. [*C. anqustior* Mack.]. Plants 1–3 dm tall, densely caespitose; rhizomes lacking or very short; leaves borne on the lower 1/3 of the culms, the blades 1–2.2 mm wide, flat; lowest bract lacking or usually shorter than the first spike; inflorescence 1–2.3 cm long, ca 7 mm wide, exceeded by the leaves or surpassing them; terminal spike gynaecandrous, usually with more staminate flowers than pistillate ones or sometimes all staminate, 7–10 mm long, often clavate, with a slender staminate portion and wide pistillate portion; lateral spikes (1) 2 or 3, gynaecandrous, with only a few staminate flowers and 5–10 perigynia, 5–7 mm long and nearly as wide, subequal to or a little longer than the internodes, rather closely spaced but readily evident in the spike; pistillate scales shorter than the perigynia, pale green or pale stramineous with a green midstripe and broad hyaline margins; perigynia widely spreading and conspicuously beaked and giving a bristly appearance to the spikes, 2.8–3.5 (4) mm long, including the prominent serrulate beak 1.1–1.6 mm long, this 1/2 or more as long as the body, neither winged nor thin-margined but the marginal nerves prominent and sometimes ridged, these displaced toward the flattened ventral surface, the dorsal surface convex; stigmas 2; achenes lenticular. Wet meadows and most common in cold boggy places at 2250 to 3180 m in the Uinta Mts., Daggett, Duchesne, Summit, Uintah, and Wasatch counties; circumboreal, south in America to California and North Carolina; 11 (vi). Utah specimens are from areas with quartzitic, noncalcareous substrates. In contrast, the closely related *C. interior* seems to be more common on calcareous or at least basic substrates.

Carex nardina Fries Spikenard Sedge. [*C. hepburnii* F. Boott]. Plants 3–14 cm tall, densely caespitose, without rhizomes, the culms and leaves fasciculate; leaves basal and on the lower rounded culm, mostly surpassing the inflorescence, the blades 0.3–0.7 mm wide, folded to involute and nearly terete; bract lacking; spike solitary, erect, androgynous, 0.5–1 cm long, the staminate part shorter than the pistillate part, exserted beyond the perigynia by 2–3 mm; scales brown with green or pale center and hyaline margins, the pistillate ones as wide or a little wider than the perigynia, these ca 5–15, 3.5–4.5 mm long, elliptic, lanceolate or oblong-obovate, with conspicuous more or less flattened margins, the margins usually serrulate-ciliolate, not wholly filled by the achene; rachilla subequal to the achene; stigmas 2 or 3, with the achenes accordingly lenticular or trigonous; 2n = 68.

Alpine communities at 3050 to 3505 m in Beaver, Duchesne, Piute, Salt Lake, and Summit counties; circumboreal, south to Washington and Colorado; 9 (ii). The tendency for 3 stigmas in our plants might reflect introgression with *C. elynoides*.

Carex nebrascensis Dewey Nebraska Sedge. Plants (1) 2–11.5 dm tall or taller, from robust, long-creeping, scaly rhizomes; culms somewhat tufted or more often arising singly or 2–3 together; leaves borne on the lower 1/4 of the culms, the blades 3–11 mm wide; bracts foliose (at least some), often exceeding the inflorescence, this 4–25 cm long, mostly exceeding the leaves, with 1–3 staminate spikes above 2–4 pistillate spikes, the transitional ones occasionally androgynous; terminal staminate spike 1.5–5 cm long, the lateral, if present, to 2.5 cm long; pistillate spikes (1.5) 2–6 (10) cm long, 5–8 mm wide, cylindrical, the lower one pedunculate, the upper sessile or short-pedunculate; pistillate scales mostly longer and narrower than the perigynia, lanceolate or narrowly awl-shaped, glabrous or occasionally ciliolate apically, blackish or black purple, with a narrow green or at least pale line in the center, this extending with the midrib to the apex; midrib prominent, sometimes slightly ridged dorsally, sometimes excurrent as a mucro or aristate point, this glabrous or occasionally antrorsely ciliolate; perigynia 2.6–4.1 mm long including the beak, elliptic to obovate, with some conspicuous dorsal and ventral nerves plus the marginal nerves, not as flattened as in *C. aquatilis*, greenish at first, yellowish brown in age, sessile or nearly so, persistent, the beak (0.2) 0.4–0.6 mm long, more or less bidentate and ciliolate; stigmas 2; achenes lenticular. Wet meadows, swamps, streams and ditches, seeps, springs, ponds, and lakes, where tolerant of some alkali, at 1365 to 3065 m in all Utah counties; Washington to South Dakota, south to California and New Mexico; 136 (viii). Nebraska sedge is the most widespread member of section Acutae in Utah. Herbarium specimens are frequently mixed with those of *C. aquatilis* (q.v.).

Carex nelsonii Mack. Plants 10–35 cm tall, caespitose; rhizomes lacking or short; leaves basal and on the lower culm, the blades not over 12 cm long, 1–2.5 mm wide, flat; bract lacking or shorter than the inflorescence, this headlike, 8–15 mm long, 5–12 mm wide, mostly much surpassing the leaves; spikes 2 or 3, sessile, congested and sometimes appearing as 1, 5–10 mm long, 4–8 mm wide, ovate to oblong, the terminal one gynaecandrous, usually larger than the lower pistillate ones; pistillate scales shorter than or equal to the perigynia, blackish or black purple, occasionally with a pale midrib, the margins sometimes narrowly hyaline; perigynia 3–3.8 mm long, 1.1–1.8 mm wide, elliptic, narrowly ovate or obovate, subsessile or with a stipe to 1 mm long, rather gradually tapered to a beak 0.5 mm long, cellular-striate on the faces and minutely papillate, at least apically; stigmas 3; achenes sharply trigonous, the angles more or less nerved, the nerves slightly ridged. Wet and dry meadows, seeps, streams, and in rather dry alpine tundra and on slopes at 2835 to 3870 m in the Uinta Mts., Duchesne, Uintah, and Wasatch counties; Wyoming and Colorado; 16 (v). *C. nelsonii* might better be considered as a variety of *C. nova* (q.v.).

Carex neurophora Mack. Alpine Nerved Sedge. [*C. vernacula* var. *hobsonii* Maguire, type from Bear River Mts.]. Plants 3–7 dm tall, caespitose, with short or long rhizomes; leaves borne on the lower 1/2–3/4 of the culms, not so clustered toward the base as in *C. jonesii*, the lower ones bladeless, the upper with flat blades 3–3.5 mm wide, the sheaths conspicuously cross-rugulose ventrally and thickened at the mouth; inflorescence a head 15–25 mm long, 8–12 mm thick, surpassing the leaves; spikes 5–10, less than 1 cm long, androgynous, the staminate flowers inconspicuous; pistillate scales ca 1/2 as long and as wide as the perigynia, brownish with a green midrib; perigynia 3.2–4 mm long, ca 1.5 mm wide, light to dark brown, finely nerved on both sides, gradually tapering to a minutely serrulate beak, widely spreading; stigmas 2; achenes lenticular. Wet meadows and along streams at 2500 to 2960 m in Beaver, Cache, Emery, Salt Lake and Wasatch counties; Washington to Montana, south to Nevada and Colorado; 2 (0). A specimen of this species (Lewis 495 BRY) from near Puffer Lake, Beaver County was growing sympatrically with *C. jonesii* (Lewis 489 BRY). The two specimens are clearly marked with features of the respective taxa. The plants of both taxa have been placed erroneously with *C. foetida*.

Carex nigricans C. A. Mey. Black Alpine Sedge. Plants 7–21 cm tall; culms arising few together from short to elongate rhizomes, with few to several leaves mostly borne toward the base; leaf blades 1–3 mm wide, flat or slightly channeled; bract lacking; spike solitary, erect, 0.8–1.8 cm long, 5–8 mm wide, androgynous, the staminate portion ca 1/4–1/2 as long as the pistillate, the scales blackish purple with scarious margins, the staminate ones erect, persistent, the pistillate ones spreading with the perigynia and soon deciduous; perigynia 3–4.5 mm long, lanceolate, often greenish except blackish purple on the beak, soon strongly spreading to declined, often deciduous after maturity, stipitate, the stipe ca 0.5 mm long,; rachilla obsolete; stigmas 3; achenes trigonous; $2n = 72$? Dry to moist or boggy meadows, along streams, and in moist woods at 2430 to 3380 m in Daggett, Duchesne, Salt Lake, Summit, Uintah and Wasatch counties; Alaska and the Siberian Coast, south to California and Colorado; 15 (viii).

Carex norvegica Retz. Scandinavian Sedge. [*C. media* R. Br.]. Plants 12–68 cm tall, densely caespitose, rhizomes lacking or very short; leaves basal and on the lower culm, the blades 1–4 mm wide, flat; lowest bract shorter than or longer than the inflorescence, this (1) 1.5–3.5 (6) cm long, mostly surpassing the leaves; spikes mostly 3 (2–4), 0.5–1.3 cm long, ca 4 mm wide, ovoid to nearly cylindrical, erect or strongly ascending, the terminal one gynaecandrous, equal to or longer than its peduncle, the lower spikes pistillate, mostly longer than their slender peduncles, or the lowest 1 rarely on an elongate peduncle arising on the lower 1/3 of the culm, the intermediate 1 or 2 often nearly or quite sessile, crowded toward the terminal one, and longer than their internodes, with the lowest shorter to longer than its internode; pistillate scales shorter than the perigynia, blackish, or blackish purple, the midrib inconspicuous and rarely pale, sometimes papillate or ciliate, the margins often white hyaline; perigynia 2–2.8 mm long, including the short beak, elliptic to narrowly obovate, green, except for the dark beak, strongly contrasting with the dark scales, becoming greenish brown in age, cellular-striate; stigmas 3; achenes trigonous, the angles slightly rounded, not strongly nerved; $2n = 54, 56$. Aspen, lodgepole pine, spruce, fir, and wet meadow communities, often along streams or around seeps and springs, at 2255 to 3370 m in Beaver,

Daggett, Duchesne, Garfield, Iron, Salt Lake, San Juan, Sanpete, Summit, Sevier, Uintah, and Wasatch counties; circumboreal, south in America to Quebec and New Mexico; 27 (x). Specimens of this species are sometimes misidentified as *C. nelsonii* (q.v.).

Carex nova Bailey [*C. pelocarpa* F. Hermann, type from Lamotte Peak, Summit County]. Plants (6) 16–72 cm tall, densely caespitose; rhizome lacking or very short; culms erect to somewhat flexuous and nodding; leaves basal and on the lower culm, the blades 1–5.5 mm wide; bract lacking or more often present and subtending the inflorescence or to 3 cm below it and not surpassing the inflorescence, this a head, 0.9–2.5 cm long, 9–20 mm wide, with 3 or 4 spikes, usually surpassing the leaves; spikes mostly sessile or on peduncles to 3 mm long, 5–15 mm long, 3–8 mm wide, ovoid to oblong, the terminal one gynaecandrous, usually slightly larger than the lateral pistillate ones; pistillate scales shorter than or equaling the perigynia, much narrower, black or purple black, rather glossy; perigynia 3.5–4.9 mm long, 2–3 mm wide, broadly elliptic, obovate, to ovate-orbicular, flattened, with broad margins, conspicuously wider than the achene, greenish at first, but soon blackish or blackish purple, more or less glossy, glabrous or the upper margins ciliolate; stigmas 3; achenes trigonous. Lodgepole pine, Engelmann spruce, meadow, and alpine communities, often along streams but usually in well-drained soil, and in talus, at 2790 to 3690 m in Box Elder, Cache, Daggett, Duchesne, Grand, Piute, Salt Lake, San Juan, Summit, Tooele, Uintah, Utah, and Wasatch counties; Oregon to Montana, south to Nevada and New Mexico; 48 (xxxiii). Specimens of this taxon are occasionally confused with those of *C. nelsonii* (q.v.). Morphological intermediates occur between *C. nova* and *C. albonigra* (q.v.).

Carex obtusata Lilj. Blunt Sedge. Plants 7–20 cm tall; culms usually arising singly from and well spaced (commonly 1–5 cm apart) along very dark brown to purplish black creeping rhizomes; leaves basal or nearly so, the blades ca 1–1.5 mm wide, flat or loosely folded; spike solitary, erect, androgynous, 5–17 mm long, linear to nearly ovoid, the staminate portion longer or shorter than the pistillate portion; scales brownish with hyaline margins, the pistillate ones mostly shorter and narrower than the perigynia, these usually 1–6, 3–4 mm long, narrowly ovoid or ellipsoid, thick-walled, brown or blackish brown, glossy, obscurely to conspicuously ribbed, conspicuously beaked; rachilla subequal to or longer than the achene, often with a flattened, hyaline, apical appendage; styles 3; achenes trigonous, filling the perigynia; 2n = 52. Sagebrush, mountain brush, aspen, dry meadow, and spruce communities at 1980 to 3050 m in Daggett, Duchesne, Emery, Garfield, Sanpete, Sevier, Uintah, Utah, and Wayne counties; Yukon to Manitoba, south to New Mexico and South Dakota; Eurasia; 17 (vii).

Carex occidentalis Bailey Western Sedge. Plants 2–8 dm tall, densely caespitose; rhizomes lacking or very short; leaves basal and on the lower culm, the blades 1–3 mm wide; lowest bract lacking or less than 1/2 as long as the inflorescence, this (1) 1.5–3 cm long, ca 5–10 mm wide, spikelike to headlike, oblong to linear-oblong or linear, at least the lower spikes usually separated, the first internode of the rachis ca 2–7 mm long; spikes ca 4–10, androgynous, not over 5 mm long, sessile, with ca 4–12 perigynia; pistillate scales equal to and nearly concealing the perigynia, brownish with a green midstripe and hyaline margins; perigynia 3–3.8 mm long, 1.6–2 mm wide, elliptic to ovate or ovate-orbicular, the body greenish or brownish with greenish, slightly winged, serrulate (use 20x) margins at least in the upper 1/2 and beak, this bidentate, the teeth ca 0.3 mm long, the marginal nerves on the winged edges or scarcely displaced to the ventral surface, the faces faintly nerved or essentially nerveless; stigmas 2; achenes lenticular. Mountain brush, ponderosa pine, aspen, lodgepole pine, and spruce-fir communities at 1830 to 3230 m in Beaver, Box Elder, Cache, Daggett, Duchesne, Garfield, Grand, Iron, Juab, Rich, San Juan, Salt Lake, Sevier, Uintah, Utah, Wasatch, and Washington counties; Wyoming to New Mexico, west to Nevada and Arizona; 39 (xv). The separation of *C. occidentalis* from *C. hoodii* at the species level is rather tenuous. Although the extremes in shape and congestion of the inflorescence are striking, a continuum is seen in the numerous specimens available for examination in the state. Perhaps the perigynia of *C. hoodii* turn dark sooner and are ultimately darker on the average. The perigynia are more strongly margined and more conspicuously serrulate, but these features do not seem convincing at the species level. Plants of the two taxa occupy different ranges in Utah, but there is considerable overlap from Sevier County northward, with *C. hoodii* having a more northerly and westerly distribution.

Carex oederi Retz. Green Sedge. Plants 12–33 cm tall, densely caespitose; rhizomes lacking; leaves basal and on the lower culm, the blades 1–3 mm wide, flat or nearly so; inflorescence 2–5 (10) cm long, exceeded by the leafy bracts; terminal spike 7–21 mm long, staminate or with a few perigynia at the base or intermixed with the staminate flowers; lateral spikes 2–4, pistillate, 0.5–1 cm long, ca 5 mm wide, the upper ones aggregated, closely subtending the terminal spike, and sessile or nearly so, the lowest one approximate and sessile to remote (the internode 5–12 cm long) and on a peduncle to 2.3 cm long, this enclosed in the sheath (to 2 cm long) of the subtending bract; pistillate scales much shorter than the perigynia, pale green, mostly hyaline; perigynia 2.2–3.5 mm long, including the conspicuous beak, pale green or yellow green, strongly ribbed, obovoid; stigmas 3; achenes trigonous, filling the lower part of the perigynia only; 2n = 70, 72. Marl bogs and wet meadows, mostly with basic substrate, and riparian communities at 1760 to 2745 m in Duchesne, Garfield, Kane, Summit, and Uintah counties; circumboreal, south in America to California and New Jersey; 7 (iii). Our plants are referable to **var. viridula** (Michx.) Kukenthal [*C. viridula* Michx.].

Carex pachystachya Cham. ex Steudel Chamisso Sedge. [*C. macloviana* var. *pachystachya* (Cham. ex Steudel) Kukenthal]. Plants 3–10 dm tall, densely caespitose; rhizomes lacking or very short; leaves borne on the lower 1/2 of the culms, the lower ones reduced to bladeless sheaths, the upper ones with essentially flat blades 2–6 mm wide; lowest bract lacking; inflorescence 15–23 cm long, 13–16 mm wide, usually oblong, but sometimes ovoid, surpassing the leaves or sometimes subequal to or surpassed by them, the internodes of the rachis concealed by the dense spikes or apparent and 2–4 mm long; spikes 4–12, gynaecandrous, 6–9 mm long, closely aggregated, hardly evident without dissecting the head but usually more conspicuous than in *C. microptera*; pistillate scales shorter and narrower than the perigynia, brown, with or without white hyaline margins to 0.1 mm wide; perigynia

3.5–4.3 (5) mm long, 1.1–2 mm wide, ovate, plano-convex to concavo-convex, lightly nerved dorsally, nerveless or lightly nerved and infrequently with a transverse fold ventrally, greenish and soon stramineous or brownish to copper colored, usually widely spreading at maturity and often deciduous, the winged margins rather narrow, serrulate, rather abruptly narrowed into a short, darkened, more or less terete, smooth beak; stigmas 2; achenes lenticular, 1.5–2 mm long, nearly filling the perigynal cavity. Streams, seeps and springs, and in boggy meadows at 1430 to 2470 m in Daggett, Duchesne, Salt Lake, Summit, Uintah, Wasatch, and Weber counties; Alaska to California and east to Alberta and Colorado; 17 (xi). Several specimens from the state are clearly assignable to this taxon, and entire populations (e.g. in Strawberry Valley) from very wet places are typical of the species. However, the plants seem to merge with *C. limnophylla* and *C. microptera* (q.v.). They differ in general from *C. microptera* in being rather rare and restricted to wetter places. The perigynia of *C. pachystachya* are plano-convex with the achene more nearly filling the cavity, and they are not so strongly flattened as in *C. microptera*. Also, they are more widely spreading to deflexed and much sooner deciduous, and the achenes are commonly larger (1.6–2 mm long versus 1.1–1.5 mm).

Carex parryana Dewey [*C. aboriginum* Jones misapplied; *C. parryana* var. *brevisquama* F. Hermann; *C. hallii* Olney]. Plants 18–50 cm tall; culms densely tufted or arising a few together from short to rather elongate rhizomes; leaves basal and on the lower culm, the blades 1–4 mm wide; lowest bract shorter than or subequal to the inflorescence, this 1.8–6 cm long, 5–10 mm wide, with (1) 2–4 (6) spikes, the lowest internode of the rachis ca 0.5–2 times as long as the lowest spike, the upper ones usually equal to or shorter than the spikes; terminal spike gynaecandrous with a short staminate portion or occasionally nearly or entirely staminate or all pistillate, 1.2–3 cm long, 3–5 mm wide, subsessile or on a peduncle to 1.5 cm long; lateral spikes pistillate, 0.7–2.4 cm long, ca 4–5 mm wide, ovoid to cylindric, sessile or essentially so, or the lowest on a peduncle to 4 cm long; pistillate scales shorter than or subequal to the perigynia and narrower apically, blackish or dark brown purple, with greenish or pale midrib, sometimes ciliolate apically; perigynia 1.9–3 (3.6) mm long, more or less obovate, glabrous to papillate, or occasionally ciliolate to strigose-hirtellous, the dorsal surface greenish marginally and basally and usually blackish purple in the center near the beak; stigmas commonly 3 but sometimes 2; achenes mostly trigonous; 2n = 54. Silver sagebrush-meadow and wet meadow communities, and also around seeps and springs in desert shrub and pinyon-juniper communities, at 1220 to 3110 m in Cache, Emery, Juab, Sanpete, San Juan, Utah, and Wasatch counties; Mackenzie to Manitoba, south to Nevada and Colorado; 18 (vii). Plants with the terminal spike all staminate or all pistillate are found in the same populations with those that have gynaecandrous spikes. I see no basis for recognition of infraspecific taxa based on these features.

Carex paupercula Michx. Poor Sedge. Plants 15–30 cm tall; culms arising singly or few together from short to long rhizomes; roots covered with a yellowish or yellow brown feltlike tomentum; leaves basal and cauline, mostly bearing blades, the blades usually 1–2 mm wide, flat, sometimes serrulate-ciliolate, especially apically, the midrib more or less keeled below, this sometimes serrulate; lowest bract shorter than to longer than the inflorescence, this 2–5 cm long; terminal spike staminate or sometimes with 1 to several perigynia at or near the apex, 0.9–1.2 cm long; lateral spikes 1–3, pistillate or often with 1–3 staminate flowers at the base, 0.7–1.6 cm long, spreading- ascending to drooping, on slender flexible peduncles 1–2.5 cm long; pistillate scales longer and slightly narrower than the perigynia, acuminate, brown or dark purple brown; perigynia 2.2–2.9 (3.5) mm long, 1.2–1.7 mm wide, sessile or substipitate, with the beak nearly obsolete, densely white-granular on a pale green background; stigmas 3; achenes trigonous; n = 29. Wet and boggy meadows and occasionally along well-drained banks of streams at 2970 to 3125 m in the Uinta Mts., Duchesne, Summit, Uintah, and Wasatch counties; circumboreal, south to Colorado; 8 (vi). *C. paupercula* is very similar to *C. limosa* (q.v.).

Carex paysonis Clokey [*C. podocarpa* var. *paysonis* (Clokey) J. Boivin; *C. tolmiei* F. Boott misapplied]. Plants 22–40 cm tall, caespitose, with short or somewhat prolonged rhizomes; leaves basal and on the lower culms, or occasionally 1 on the upper culm, the blades 2–6 mm wide, flat; lowest bract from shorter than to as long as the inflorescence, sometimes foliose; inflorescence 3.5–6 cm long; terminal spike staminate, 1.2–1.8 cm long, 4–8 mm wide, on a peduncle 5–10 mm long; lateral spikes 2–5, pistillate, or with 1 or 2 staminate flowers at the apex, (0.8) 1.2–2 cm long, 5–8 mm thick, rather closely aggregated or the lower separated, erect to spreading-ascending or widely spreading, on slender peduncles to 15 mm long or the upper one sessile; pistillate scales subequal to or shorter than the perigynia, black or purple black, the midrib sometimes green or pale, the margins not or scarcely hyaline; perigynia 3.3–4.4 mm long, broadly elliptic to subrotund, abruptly contracted to the beak, rather fragile and thin, strongly flattened, conspicuously wider than and not filled by the achene, soon blackish or dark brown purple or at least strongly suffused with these colors; stigmas 3; achenes trigonous; 2n = 32?, 60, 62, 64, 66. Alpine tundra and rocky slopes at 3290 to 3660 m in the Uinta Mts., Duchesne and Summit counties; Alberta to Nevada and Oregon; 11 (v).

Carex perglobsa Mack. Mount Baldy Sedge. Plants 6–20 cm tall, loosely caespitose from slender creeping rhizomes; leaves clustered at the base of the culms, the blades 0.7–1.5 mm wide; inflorescence a globose head ca 1 cm in diameter, usually surpassing the leaves; spikes 6–15, androgynous, not evident in the crowded head, less than 1 cm long, the staminate flowers inconspicuous; bracts lacking; pistillate scales subequal to the perigynia, brownish with hyaline margins; perigynia 4–4.75 mm long, 1.7–2.3 mm wide, ovate-elliptic, inflated, gradually tapered to a bidentate beak; stigmas 2; achenes lenticular. Alpine communities at 3688 to 3960 m in the La Sal Mts., Grand and San Juan counties; Colorado; 1 (0). Some specimens of this species from the La Sal Mts. share features of *C. foetida* var. *vernacula* (Cronquist et al. 1977).

Carex petasata Dewey Liddon Sedge. Plants 2–7 dm tall, caespitose; rhizomes lacking or very short; leaves basal and on the lower culm, the blades 1–4 mm wide, flat or channeled; lowest bract lacking; inflorescence 2.2–5 cm long, 8–12 mm wide, spikelike, surpassing the leaves, the lowest internode of the rachis 4–7 mm long, the second lowest subequal; spikes 3–6, gynaecandrous, 10–22 mm long, 5–7 mm wide, more or less clavate, sessile, loosely aggregated and evident without dissecting the inflorescence, ca 2–3 times longer than the internodes of the rachis; pistillate scales largely concealing, subequal to, and as wide as the perigynia, stramineous or light reddish brown, with a green or pale midstripe and hyaline margins; perigynia (5.7) 6–7.5 (8) mm long, 1.5–2.8 mm wide, lanceolate, narrowly ovate or nearly elliptic, greenish or light brown, plano-convex, with thin winged margins, serrulate at least in the distal 1/2, gradually tapered to an obliquely cleft beak, conspicuously nerved on both sides; stigmas 2; achenes lenticular. Sagebrush-grass, mountain brush, aspen, and grass-forb communities at 2070 to 3050 m in all Utah counties except Carbon, Davis, Grand, Iron, Kane, Morgan, Summit, Washington, Wayne, and Weber; British Columbia to Saskatchewan and south to California and Arizona; 40 (vii). This plant passes into *C. praticola* (q.v.), and probably into *C. phaeocephala*. Further, it is not easily distinguished from *C. xerantica* (q.v.). Additional study of this complex is indicated.

Carex phaeocephala Piper Dunhead Sedge. [*C. eastwoodiana* Stacey]. Plants (10) 15–30 (55) cm tall, densely caespitose; rhizomes lacking; leaves basal and on the lower culm, the blades ca 1–3 mm wide, flat, channeled, or involute; lowest bract lacking or occasionally present and subequal to the inflorescence or longer; inflorescence 1.4–3 cm long, rarely shorter, 5–15 mm wide, equaling or surpassing the leaves; spikes (1) 2–5 (7), gynaecandrous, 8–15 mm long, 5–7 mm thick, sessile, approximate but mostly evident without dissecting the head, mostly twice or more as long as the internodes; lowest internode of the rachis commonly 4–7 mm long; pistillate scales as long and almost as wide as the perigynia, largely concealing them (sometimes pressed to one side in herbarium specimens), brown to dark brown, with green or pale midstripe and usually conspicuous hyaline margins; perigynia 4–5.5 mm long, 1.5–2.2 mm wide, lanceolate to ovate-elliptic, brownish or stramineous, with greenish stramineous thin winged margins, finely but conspicuously nerved dorsally and often ventrally, gradually or rather abruptly tapered to a bidentate beak; stigmas 2; achenes lenticular. Spruce-fir, lodgepole pine-Engelmann spruce, krummholz, and alpine communities, often in rocky places, at 2830 to 3500 m in Beaver, Duchesne, Grand, Iron, Juab, Piute, San Juan, Sanpete, Sevier, Summit, Tooele, Uintah, Utah, and Wasatch counties; British Columbia to Alberta and south to California and Colorado; 40 (xi). This species apparently grades into both *C. petasata* and *C. xerantica* (q.v.).

Carex praeceptorum Mack. Plants 10–31 cm tall, densely to loosely tufted, with very short or somewhat prolonged rhizomes; leaves basal and on the lower culm, the blades 1.5–3 mm wide; lowest bract lacking or shorter than the lowest spike; inflorescence 1–2.7 cm long, ca 5–8 mm wide, usually equaling or surpassing the leaves; spikes 4–6, gynaecandrous, 5–12 mm long, 3–4 mm wide, sessile, approximate or the lower ones somewhat separate, but usually twice as long as the internodes or occasionally equaling them, all usually evident; pistillate scales shorter than the perigynia, brown to dark brown, with a green midstripe and very narrow to rather broad hyaline margins; perigynia 1.7–2.2 (2.5) mm long, 1–1.2 mm wide, elliptic to elliptic-ovate, more or less flattened, but with rounded (wingless) margins, yellow green or brownish green, (10) 15–25 per spike, slightly granular-roughened, infrequently serrulate on the beak, finely nerved on both sides, the nerves frequently reddish or brown red, sometimes only faintly so, the dorsal suture conspicuous on the beak and extending barely onto the body; stigmas 2; achenes lenticular. Wet and boggy meadows, margins of lakes, ponds, seeps, and springs at 3050 to 3115 m in Beaver, Duchesne, Piute, Salt Lake, Summit, Uintah, and Wasatch counties; Washington to Wyoming, south to Nevada and Colorado; 15 (v). Plants of *C. praeceptorum* are very similar to *C. canescens*. A combination has been made [*C. canescens* var. *praeceptorum* (Mack.) Bailey], and perhaps recognition at the varietal level best represents this taxon. Of the two taxa, *C. praeceptorum* is more restricted and less common. However, they are essentially sympatric in Utah. Only on the La Sal Mts. and Aquarius Plateau is *C. canescens* known in the absence of *C. praeceptorum*.

Carex praegracilis W. Boott Blackcreeper Sedge. Plants (1) 1.5–7 dm tall, dioecious (and forming clones) or monoecious; culms arising singly or few together or occasionally tightly clustered, from robust blackish or dark brown rhizomes 2–4 mm thick, these covered with scales that often become filamentous in age; lower leaves reduced to bladeless sheaths, these often dark brown or blackish, the upper leaves borne on the lower 1/4 of the culms, with blades 1–3.5 mm wide; lowest bract lacking or ca 1/2 as long as the inforescence, this 1.5–4.3 cm long, ca 5–15 mm wide, simple or compound, capitate or spike-like, linear to oblong, usually surpassing the leaves; spikes ca 6–25, androgynous, staminate or pistillate, to ca 1 cm long, closely aggregated, or the lower one somewhat separate, all longer than the rachis internodes, sessile; pistillate scales mostly as long and as wide as the perigynia and largely concealing them, these and the staminate ones brown with a green or paler midrib and hyaline margins; perigynia 2.8–3.5 mm long, 1.1–1.7 mm wide, ovate or lance-ovate to elliptic, usually short-stipitate, the margins rather thin when young, firm in age and slightly turned toward the ventral surface; beak of the perigynium conspicuous, 0.6–1.3 mm long, obliquely cleft, serrulate, the minute teeth often extending on to the body of the perigynium; stigmas 2; achenes lenticular; 2n = 60. Wet meadows, along streams and ditches, margins of lakes, ponds, seeps and springs, often where saline, or sometimes in rather dry places, in sagebrush-grass and dry meadow communities at 850 to 2965 m in all Utah counties except Iron; British columbia to Michigan, south to California and Mexico; 96 (xii).

Carex praticola Rydb. Meadow Sedge. Plants 30–80 cm tall, densely caespitose; rhizomes lacking or very short; leaves basal and on the lower culm, the blades 2–4 mm wide, flat; lowest bract lacking or less commonly present and shorter than the inflorescence, this (2.5) 3–4 cm long, 1–1.5 cm wide, spicate, more or less flexuous, sometimes more or less secund, overtopping the leaves, the lowest internode ca 5–12 mm long, the second one as long or nearly so; spikes (2) 4–7, gynaecandrous, (8) 10–15 mm long, ca 5–7 mm wide, usually clavate, sessile, loosely aggregated and evident without dissecting the inflorescence, ca 1–2 times as long as the internodes, or the lower one slightly shorter, and the upper ones to ca 3 times as long; pistillate scales subequal to and as wide as the perigynia, largely concealing them, dull reddish brown with a green midstripe and broad hyaline margins; perigynia (4) 4.5–5.7 (6) mm long, 1.2–2.1 mm wide, plano-convex, with thin winged margins, serrulate, gradually tapered to a shallowly bidentate beak, finely nerved dorsally, faintly nerved ventrally, the beak more or less terete and not serrulate, at least in the distal portion; stigmas 2; achenes lenticular; 2n = 64, 70? Aspen, Douglas fir, lodgepole pine, and Engelmann spruce woods and openings at 2740 to 3140 m in Daggett, Duchesne, Sevier, and Summit counties; Alaska to Labrador, south to California and Quebec; 10 (x). Plants of this taxon pass imperceptably into those of *C. petasata*. Reduction to infraspecific status does not seem unreasonable, but no combination is intended here.

Carex pyrenaica Wahl. Pyrenaean Sedge. Plants 5–15 cm tall, densely caespitose, without rhizomes, the culms and leaves often fasciculate; leaves 2–4 per culm, the lower bladeless, the upper with blades 0.5–1.5 mm wide, flat or loosely to strongly channeled; bract lacking; spike solitary, erect, androgynous, 0.7–2 cm long, 4–7 mm thick, the staminate portion ca 2–4 mm long; pistillate scales tan to dark brown, the margins hyaline, subequal to the perigynia, deciduous in age; perigynia usually (5) 10 or more, ca 2.5–4 mm long, lanceolate, somewhat stipitate and jointed to the rachis, ascending or sometimes reflexed in age; rachilla lacking; stigmas 3; achenes trigonous. Moist places in alpine communities at 3655 to 3690 m in the Uinta Mts. (where rare), Duchesne County; British Columbia to Mackenzie, south to Colorado and California; Eurasia; 2 (0).

Carex raynoldsii Dewey Plants 1.9–8 dm tall, densely caespitose; rhizomes lacking or short; leaves basal and on the lower culm, the blades 2–7 mm wide, flat; lowest bract shorter than to somewhat longer than the inflorescence or sometimes foliose; inflorescence 2–11 cm long, usually surpassing the leaves, the first internode ca 1–7 cm long, the others shorter; terminal spike staminate, 1–1.5 cm long, sessile or nearly so; lateral spikes pistillate, 1–4 (5) 1–2.5 cm long, 6–8 mm wide, the lowest on a peduncle 0.2–2 (5) cm long, the upper one(s) sessile or short-pedunculate; pistillate scales shorter than the perigynia, brownish black or purplish black, sometimes with very narrow hyaline margins; perigynia (3) 3.3–4.4 mm long, elliptic or elliptic-obovoid, somewhat inflated, green or olive green and strongly contrasting with the dark scales, the beak usually purple black; stigmas 3; achenes trigonous. Sagebrush-grass, aspen, tall forb, spruce, and fir communities at 2195 to 3170 m in Beaver, Box Elder, Cache, Carbon, Daggett, Duchesne, Juab, Morgan, Salt Lake, Sanpete, Sevier, Summit, Tooele, Uintah, Utah, and Wasatch counties; British Columbia to Alberta and south to California and Colorado; 32 (x).

Carex retrorsa Schwein. Knotsheath Sedge. Plants 3–10 dm tall; culms densely clustered on short rhizomes; leaves mostly borne on the culms, the blades 4–10 mm wide; inforescence 5–13 cm long, with 1 staminate, gynaecandrous, or androgynous spike above 4–6 pistillate spikes, these 1.5–4.5 cm long, ca 1.5–2 cm thick, cylindric, closely spaced and much longer than the internodes of the rachis, or the lower one widely spaced but still produced beyond its internode, each subtended by leaves that much surpass the inflorescence; pistillate scales shorter and narrower than the perigynia, greenish or stramineous, narrowly acute, acuminate, or short-awned; perigynia widely spreading, giving a bristly appearance to the spikes, 1–10 mm long, shiny, greenish, conspicuously nerved, the body inflated, ellipsoid to subglobose, narrowed to a conspicuously bidentate beak 2–3 (4) mm long; stigmas 3; achenes trigonous, much smaller than the perigynia, the style continuous with, persistent on, and of the same firm texture as the achene, contorted in age; 2n = 70. Along rivers and margins of ponds and reservoirs at 1400 to 1830 m in Wasatch and Weber counties; British Columbia to Quebec and south to Oregon, Colorado, and New Jersey; 3 (0).

Carex rossii F. Boott Ross Sedge. [*C. brevipes* W. Boott]. Plants 5–32 cm tall, densely caespitose, with or without short rhizomes, often deep reddish purple basally; leaves basal and on the lower culm, mostly overtopping the inflorescence, the blades 0.5–3 mm wide; inflorescence essentially the length of the culms, with the lowest spike borne on a slender peduncle arising near the culm base and much removed from the rest of the inflorescence; terminal spike staminate, 5–12 mm long; lateral spikes mostly 2–5, very short with only 1–4 perigynia, bracteate or leafy bracteate, the lower bracts (excluding those of the basal spike) usually exceeding the inflorescence; pistillate scales shorter than the perigynia, with a greenish, keeled, often scabrescent-ciliolate midrib, this flanked by stripes of brown or brownish purple and then by broad hyaline margins; perigynia sparsely to densely puberulent, 2.7–4.5 mm long including the oblique beak, this 0.7–1.8 mm long, the body plump, with 2 prominent marginal nerves; stigmas 3; achenes trigonous, filling the perigynia; 2n = 36. Sagebrush, pinyon-juniper, mountain brush, ponderosa pine, aspen, lodgepole pine, spruce-fir, dry meadow, and alpine communities at 1340 to 3445 m in all Utah counties except Davis, Emery, Morgan, and Weber; Yukon to California and Arizona east to Michigan; 79 (xii). **Carex geophila** Mack. [*C. pitophila* Mack.] of Arizona, Colorado and New Mexico has been reported for Utah, and would key here. That taxon is reported to have firm leaves, the bract of the lowest nonbasal spike not exceeding the staminate spike, and the perigynium beak 0.4–1.2 mm long.

Carex rostrata Stokes ex With. Beaded Sedge. Plants 6–14 dm tall; culms arising singly or 2–3 together from robust, scaley, long-creeping rhizomes, often spongy thickened and to 1 cm thick near the base; leaves mostly borne on the lower 1/2 of the culm, the lower sheaths septate-nodulose, the blades 3–12 mm wide, flat; ligule tapering across the ventral surface of the blade at a very rounded angle, not noticeably projected into a point; inflorescence 15–45 cm long, with 2–4 staminate spikes borne above 2–3 (5) pistillate ones, the intermediate ones sometimes androgynous; terminal staminate spike (2) 3–8.5 cm long, the lateral staminate ones somewhat to much shorter, approximate to widely separated; pistillate spikes 2.5–10.5 cm long, 6–13 mm wide, cylindric, sessile or on peduncles to 3 (7) cm long, rarely compound at the base in very large spikes, well spaced and much shorter to longer than the rachis internodes, each subtended by a foliose bract, the bracts often equaling or exceeding the inflorescence; pistillate scales shorter than or subequal to and narrower than the perigynia, greenish, turning red brown to purple brown, with a paler median, narrowly acute to acuminate or short-awned; perigynia strongly spreading, glossy, light to very dark brown, 4–7 mm long, the body inflated, broadly ellipsoid to subglobose, rather abruptly contracted to a prominent, conspicuously bidentate beak 1–2 mm long, the teeth 0.2–1 mm long; stigmas 3; achenes trigonous, much smaller than the perigynia, the style continuous with, persistent on, and of the same firm texture as the achene, twisted or abruptly bent; 2n = 70?, 72, 76. Ponds and lakes (in water to ca 45 cm deep), swampy meadows, sometimes along streams but mostly in or near quiet water at 1830 to 3200 m in all Utah counties, except Carbon, Davis, Juab, Millard, Tooele, and Washington; circumboreal, south in America to California and Delaware; 90 (iv). One specimen (N. Holmgren et al. 2328 BRY) has sparsely pubescent perigynia and appears to have been influenced by *C. lanuginosa*.

Carex rupestris All. Curly Sedge. [*C. drummondiana* Dewey]. Plants 3–10 (15) cm tall, caespitose, with very short rhizomes; leaves basal or nearly so, the blades 1–2 mm wide, flat or strongly folded, the midrib sometimes conspicuously keeled on the lower side, curved or strongly curled toward the apex in age; spike solitary, erect, 1–2 cm long, 2–4 mm wide, androgynous, the pistillate portion with ca 3–10 (15) perigynia, subequal to or shorter than the staminate portion; scales brownish or purplish, with hyaline margins, the pistillate ones mostly concealing the perigynia, mostly obtuse to rounded; perigynia 3–4 mm long, elliptic, gradually short-stipitate at the base, very short truncate-beaked at the apex, conspicuously nerved; rachilla lacking; stigmas 3; achenes trigonous, purplish black at maturity. Krummholz and alpine tundra communities, often in rocky places, at 3200 to 3660 m in Duchesne, Summit, Uintah, and Wasatch counties; circumboreal, south in North America to Quebec; 10 (vii).

Carex saxatilis L. Russet Sedge. [*C. physocarpa* Presl]. Plants 21–40 (56) cm tall; culms arising singly or rather closely spaced on creeping rhizomes; leaves mostly borne on the lower 1/2 of the culms and sometimes higher, the blades 1.5–3 mm wide, flat; inflorescence 4–9 cm long, with 1–2 (3) staminate spikes 1–3 cm long borne above 1–2 pistillate spikes 1–2.6 cm long and ca 5–8 mm wide, the lateral staminate ones well spaced, subequal to or shorter than the rachis internodes, the pistillate ones sessile or on slender, spreading or drooping peduncles to 3 cm long, each subtended by a leafy bract, the lowest bract usually overtopping the inflorescence; pistillate scales slightly to conspicuously shorter than the perigynia, acute, very dark brown, purple black or black, sometimes with a pale midstripe, the margins whitish hyaline at least toward the often erose tip; perigynia 3.2–5.2 mm long, 1.5–3.3 mm wide, lanceolate to nearly orbicular, slightly inflated, soon black or blackish purple at least where exposed beyond the scale, with marginal or sometimes dorsal nerves conspicuous, nerveless ventrally, tapering to a short beak, the beak nearly entire or with inconspicuous teeth, these not over 0.5 mm long and easily broken off; styles 2 or occasionally 3, sometimes both in the same spike and achenes accordingly lenticular or trigonous, the style continuous with, persistent on, and of the same firm texture as the achene, often strongly bent or recurved back against the achene in age; 2n = 78?, 80. Wet and boggy meadows and margins of ponds and lakes at 2740 to 3810 m in Duchesne, Garfield, Kane, Summit, Uintah, and Wasatch counties; circumboreal and south in America to Colorado; 33 (x). Utah materials might be referable to *var. major* Olney. Robust specimens are sometimes confused with *C. vesicaria* q.v.

Carex scirpoidea Michx. Plants 4–34 cm tall, unisexual; culms arising singly or few together from rather short but conspicuous rhizomes; leaves basal and on the lower culm, the blades 1.5–3.5 cm long, 3–7 mm wide, linear; scales mostly rounded at the apex, blackish purple with hyaline, ciliolate margins, sometimes sparsely pubescent dorsally, the pistillate ones mostly longer and wider than the perigynia; anthers 3–5 mm long; perigynia obovoid, 2–4 mm long including the beak, pubescent; rachilla obsolete. Dry subalpine meadow, lodgepole pine-Engelmann spruce, krummholz, and alpine tundra communities at 2895 to 4100 m in Cache, Daggett, Duchesne, Garfield, Grand, Iron, San Juan, Summit, and Uintah counties; Labrador to Alaska south to New Hampshire and California; 41 (v). Our materials are referable to *var. pseudoscirpoidea* (Rydb.) Cronq. [*C. pseudoscirpoidea* Rydb.]. Some of the specimens from Iron, Garfield, and Emery counties are similar to *C. curatorum*.

Carex scopulorum H. T. Holm Rock Sedge. [*C. campylocarpa* H. T. Holm; *C. campylocarpa* ssp. *affinis* Maquire & Holmgren, type from Indian Farm Creek, Deep Creek Mts.]. Plants 22–67 cm long; culms arising singly or few together or somewhat tufted, from robust scaley rhizomes; leaves basal and mostly on the lower culm, the blades (2) 3–7 mm wide; lowest bract foliose but shorter than the inflorescence, this 2.5–9 cm long, with 1 or 2 staminate spikes 1–2.2 cm long above 2–4 cylindrical or ovoid pistillate spikes 1–3 cm long and 5–10 mm wide, the intermediate spikes occasionally androgynous; pistillate scales shorter to longer than the perigynia and mostly narrower at least toward the apex, black or purple black; perigynia 1.8–3.3 mm long, elliptic to broadly obovate, nerveless except for the 2 marginal nerves, blackish to dark reddish brown at least distally, sessile or nearly so, the beak nearly obsolete; stigmas 2, and achenes lenticular, or very rarely stigmas 3 and achenes trigonous. Wet meadows and along streams, Deep Creek, Raft River, and Uinta mts. at 2370 to 2790 m in Box Elder, Daggett, and Juab counties; British Columbia to California and east to Colorado; 10 (viii). The perigynia are the darkest of any among Utah members of section Acutae, and the pistillate spikes are the widest for their length. Occasional specimens from the Wasatch Plateau appear to be intermediate between *C. scopulorum* and *C. aquatilis* (q.v.), with the bract shorter than the inflorescence but with rather narrow pistillate spikes. However, these are referred herein to *C. aquatilis*.

Carex siccata Dewey Silvertop Sedge. [*C. foenea* sensu Svenson, not Willd.]. Plants 19–36 cm tall; culms arising singly or few together and loosely to rather closely clustered on stout rhizomes; leaves clustered on the lower 1/4 of the culms, the lower ones often reduced to bladeless sheaths, the upper ones with blades 1–2 mm wide, flat, the longer ones subequal to or shorter than the inflorescence, the lowest bract subequal to the lowest spike; inflorescence 1–3 (6) cm long, 5–10 mm wide, linear or clavate or occasionally orbicular; spikes sessile, congested or interrupted, ca 5–12 mm long, the lower ones gynaecandrous or pistillate, some (especially the middle ones) staminate, the upper one often pistillate and closely subtended by a staminate spike and thus falsely appearing to be gynaecandrous; scales light brown with broad hyaline margins, the pistillate ones subequal to the body of the perigynia but surpassed by the beak; perigynia 4.5–6.2 mm long including the prominent beak, flattened, serrulate to near the middle and sometimes below, slightly wing-margined below, the body more or less elliptic; stigmas 2; achenes lenticular, filling the perigynia, sometimes rupturing at maturity; $2n = 70$. Sagebrush, oak, ponderosa pine, aspen, lodgepole pine, Engelmann spruce, and forb-grass communities at 915 to 3170 m in Daggett, Garfield, Grand, San Juan, Uintah, and Washington counties; Maine to the Mackenzie, south to Arizona; 14 (vi).

Carex simulata Mack. Plants 22–45 cm tall, dioecious (and commonly forming clones) or monoecious, the culms mostly arising singly from, and closely to remotely spaced on long, whitish or light brown rhizomes; leaves cauline, mostly borne on the lower 1/4 of the culms, the lower ones reduced to bladeless sheaths, the upper ones with blades 1.5–4 (6) mm wide; bracts subequal to the spikes or the lower 1–3 from 1/2 as long to equaling the inflorescence, long acuminate-caudate; inflorescence 1.8–3.5 cm long, 7–25 mm wide, simple or occasionally compound, headlike, linear to oblong or occasionally ovate, subequal to or surpassing the leaves; spikes commonly 10–20, unisexual or androgynous, crowded and usually concealing the rachis, more or less evident without dissecting the inflorescence; pistillate scales longer than the perigynia, brownish, with narrow to rather broad pale hyaline margins, the midrib sometimes green; perigynia 1.7–2.6 mm long, elliptic-ovate, or nearly orbicular, shining brown, sessile or occasionally very short-stipitate, the margins rounded, not thin, the dorsal surface with a few raised nerves extending to ca 1/2 or more to apex or occasionally nerveless, the ventral nerves like the dorsal ones but mostly shorter and more often lacking, the wall thick and coriaceous, the beak 0.2–0.6 mm long, inconspicuously winged and serrulate near the confluence with the body; stigmas 2; achenes lenticular. Boggy meadows, quaking bogs, and streamside communities, commonly in places with calcareous or at least basic substrate, at 2286 to 2865 m in Box Elder, Daggett, Duchesne, Emery, Garfield, Kane, Sanpete, Sevier, Summit, Uintah, and Wasatch counties; Washington and Alberta, south to California and New Mexico; 35 (xx).

Carex stenophylla Wahl. Narrowleafed Sedge. [*C. eleocharis* Bailey]. Plants 8–28 cm tall; culms arising singly or few together from slender brownish or pale colored rhizomes less than 2 mm in diameter, and with numerous fibrous roots; leaves basal and on the lower culm, the blades 0.5–1.5 mm wide; lowest bract lacking; inflorescence 0.8–2 cm long, 5–10 mm wide, headlike; spikes ca 3–10, sessile, androgynous or rarely all unisexual, 4–9 mm long, longer than the internodes of the rachis and closely aggregated, disernable or not without dissecting the inflorescence; pistillate scales subequal to or slightly longer than the perigynia, brownish, with a dark green or brownish often keeled midrib, and with narrow or broad scarious margins; perigynia 2.6–3.5 mm long, rather narrowly fusiform-elliptic to broadly elliptic-ovate, brownish, the marginal nerves prominent, otherwise obscurely nerved; beak of the perigynium, obliquely cleft, 0.5–1.2 mm long, the dorsal suture inconspicuous; stigmas 2; achenes lenticular; $2n = 60$. Sagebrush-grass, pinyon-juniper, and windswept montane communities at 1920 to 3270 m in Daggett, Duchesne, Garfield, San Juan, Sevier, Uintah, Utah, Wasatch, and Wayne counties; mostly of the Great Plains, but from Iowa to California and south to Arizona; 13 (iv).

Carex stipata Muhl. in Willd. Prickly Sedge. Plants 28–75 cm tall or taller, caespitose; rhizomes lacking or very short; leaves basal and on the lower 1/2 of the culms, the lower culm leaves reduced to bladeless sheaths, the upper ones with blades 4–8 (11) mm wide, flat, the ventral side of the sheath membranous, usually cross-corrugated; inflorescence (2.5) 3–5.5 cm long, 1–2 cm wide, compound but spikelike, with many, short, crowded spikes, appearing bristly from the widely spreading, narrowly beaked perigynia; lowest bract not much if any longer than the lowest branch of the inflorescence, more or less awnlike, usually broken off in older specimens; individual spikes scarcely evident, androgynous, to ca 1 cm long and nearly as wide; pistillate scales mostly shorter than the perigynia, pale or brownish, rather scarious, with the more or less greenish or pale midrib sometimes exserted as an awn-tip; perigynia (3.6) 4–5.2 mm long, widely spreading, lance-triangular or lance-ovate, broadest at the sometimes abruptly truncate, somewhat spongy thickened base, gradually tapered to a long, minutely serrulate beak, conspicuously nerved at least dorsally, subsessile to conspicuously stipitate; stigmas 2; achenes lenticular; $2n = 52$. Wet places at 1525 to 1830 m in Davis, Salt Lake, and Weber counties; Alaska to Newfoundland and south to California and Florida; 3 (0). Utah materials are referable to var. *stipata*.

Carex straminiformis Bailey Mt. Shasta Sedge. Plants 24–45 cm tall, densely caespitose; rhizomes lacking or very short; leaves basal and on the lower culm, the blades 1–3 mm wide, flat or channeled; lowest bract lacking or shorter to slightly longer than the inflorescence, this 1.5–2.5 cm long, 1–2 cm wide, headlike, usually much overtopping the leaves, the lowest internode to ca 5 mm long; spikes 3–9, gynaecandrous, 7–15 mm long, 5–7 mm wide, ovate to nearly globose, sessile, closely aggregated but more or less evident without dissecting the head, ca 3 or more times longer than the internodes of the rachis, with up to 20 perigynia; pistillate scales shorter than or subequal to the perigynia and a little narrower, brownish with a green midstripe and hyaline margins, acute; perigynia 4–5.6 mm long, 1.6–2.5 mm wide, ovate-elliptic, greenish or light brown, strongly flattened with conspicuous thin-winged margins, finely serrulate in the distal 1/2 and sometimes below, gradually or rather abruptly tapered to a bidentate, more or less flattened beak; anthers 1.5–2 mm long; stigmas 2; achenes lenticular. Sagebrush-grass, forb-grass, and Engelmann spruce communities, open slopes and ridges, and rocky cirques at 2740 to 3140 m in Duchesne, Salt Lake, Tooele, Utah, and Wasatch counties; Washington to Montana, and south to California; 12 (vi).

Carex subfusca W. Boott in Wats. Rusty Sedge. [*C. macloviana* var. *subfusca* (W. Boott) Kukenthal]. Plants 20–65 cm tall, densely caespitose; rhizomes lacking or very short; leaves on the lower 1/4 of the culms, the blades 2–4 mm wide, flat; lowest bract lacking or shorter than the inflorescence, this 1–3.5 cm long, 7–12 mm wide, headlike or spikelike, the internodes of the rachis sometimes conspicuous; spikes 4–12, gynaecandrous, ca 5–9 cm long, sessile, the lower approximate and ca 2–3 times as long as the rachis internodes, the upper usually more crowded; pistillate scales shorter and narrower than the perigynia, reddish brown with lighter midstripe and margins; perigynia 2.7–3.2 (3.6) mm long, 1–1.5 mm wide, ovate, plano-convex, many-nerved on both sides or nerveless or nearly so ventrally, the margin narrowly winged, serrulate, tapering or contracted into a serrulate, more or less flattened beak; stigmas 2; achenes lenticular. Dry meadows at ca 2900 m in Sevier, Washington, and Wayne counties; British Columbia to California and east to Arizona; 3 (0). *C. subfusca* is apparently closely related to *C. microptera* (q.v.) through *C. pachystachya*.

Carex subnigricans Stacey Dark Alpine Sedge. [*C. rachillis* Maquire, type from Gilbert Peak, Summit County]. Plants 5–20 cm tall, sod-forming, with culms arising few together along rhizomes with short internodes; leaves basal or on the lower culm, the blades to ca 1 mm wide, involute or deeply channeled, more or less wiry like the culms; bract lacking; spike solitary, erect, androgynous, 1–2 cm long, 3–6 mm thick, 2.5–3 times as long as thick, the staminate portion very short or quite conspicuous; pistillate scales tan to dark brown, hyaline at the margins, subequal to the perigynia, these (5) 10 or more, 2.5–4 mm long, to 1.5 mm wide, elliptic to ovate; stigmas 3; achenes trigonous, almost filling the perigynia. Moist and boggy meadows at 2965 to 3355 m in Duchesne, Piute, Summit, Uintah, and Wayne counties; Oregon to Wyoming and south to Nevada; 4 (iii).

Carex vallicola Dewey Valley Sedge. Plants 15–56 cm tall, densely caespitose; rhizomes lacking or very short; leaves basal and on the lower culms, the blades 0.5–2.5 mm wide; lowest bract lacking or shorter than the inflorescence, this 0.8–3 cm long, 5–9 mm wide, slightly to much overtopping the leaves, spicate or capitate, shortly interrupted or more often continuous, with the individual spreading perigynia more evident than the individual spikes; spikes ca 2–10, to 5 mm long, with ca 2–7 perigynia, sessile, androgynous, the staminate portion inconspicuous; pistillate scales shorter than the perigynia, with greenish or brownish centers and hyaline margins; perigynia 3.3–3.75 mm long, 1.7–2.3 mm wide, oblong-elliptic, with more or less rounded wingless margins, pale green or straw colored, the marginal nerves somewhat displaced onto the ventral surface, otherwise nerveless or nearly so, the body entire, rather abruptly contracted to a very finely serrulate beak 0.6–1 mm long, this slightly or not at all bidentate; stigmas 2; achenes lenticular, filling the perigynia and sometimes rupturing it at maturity. Sagebrush-grass, mountain brush, ponderosa pine, Douglas fir, aspen, spruce-fir, and dry meadow communities at 1675 to 3010 m in all Utah counties, except Carbon, Davis, Kane, Morgan, San Juan, Wayne, and Weber; California to South Dakota, south to Mexico; 52 (xii).

Carex vesicaria L. Blister Sedge. [*C. exsiccata* Bailey]. Plants 30–75 cm tall, caespitose; rhizomes short or somewhat prolonged; leaves mostly on the lower 1/2 of the culms, the blades 2–4 (8) mm wide, flat, the lower sheaths septate-nodulose; ligules tapering across the ventral surface of the blade at a rather sharp angle and prolonged into a point; inflorescence 9–21 cm long, with 2–3 staminate spikes borne above 2–3 pistillate ones; staminate spikes 2–7 cm long, the terminal one the largest, the lateral spaced or closely subtending the terminal one; pistillate spikes 2–3 (7) cm long, ca 1 cm thick, cylindric, sessile or on peduncles to 2.3 cm long, spaced, subequal to much shorter than the rachis internodes, each subtended by a leafy bract, these often much exceeding the inflorescence; pistillate scales shorter and narrower (at least distally) than the perigynia, stramineous to reddish brown, rarely dark brown, commonly acuminate; perigynia ascending, not spreading, 5–6 mm long, inflated, lanceolate or lance-ovate, gradually tapering to a bidentate beak, glossy; stigmas 3; achenes trigonous, much smaller than the perigynia, the style continuous with, persistent on, and of the same firm texture as the achene, strongly bent or contorted with age; 2n = 82. Ponds, lakes, swamps, and boggy meadows, in standing water to ca 30 cm deep, at 3000 to 3050 m in Duchesne, Iron, Sevier, Summit, and Uintah counties; circumboreal, south in America to California and Delaware; 7 (i). Utah materials are apparently referable to **var. vesicaria**. The few specimens I have examined have 3 stigmas and perigynia 5–6 mm long, with a bidentate beak with teeth not over 0.5 mm long. The perigynia, as well as the scales, are sometimes dark brown when mature and often well before maturity. These features do not exclude *C. saxatilis*, and the dark color of the perigynia and scales strongly indicate that taxon. The plants are taller and have larger spikes that are more ascending than in most Utah specimens of *C. saxatilis*. They are apparently caespitose, and with perigynia gradually tapered to the beak, and with conspicuous dorsal ribs. However, they look as much (or more) like typical specimens of *C. saxatilis* from Utah than they look like Idaho specimens of *C. vesicaria*. Cronquist et al. (1977) noted some anomalous features that seem to align cordilleran plants of *C. saxatilis* with *C. vesicaria* (i.e., the larger perigynia with better developed nerves). More work is indicated.

Carex xerantica Bailey Dry land Sedge. Plants 20–45 cm tall, caespitose, from short stout rhizomes; leaves borne on the lower 1/4 of the culms, the lower ones bladeless or nearly so, the upper ones with blades 2–4 mm wide; lowest bract lacking or shorter than the inflorescence, this 2–4 cm long, 7–15 mm wide, more or less spikelike, usually much surpassing the leaves, the internodes of the rachis obscured or sometimes discernable, the lower 2 collectivelly 8–18 mm long; spikes 3–6, gynaecandrous, 8–17 mm long, sessile, approximate but usually readily evident without dissecting the inflorescence, ça twice as long as the internodes of the rachis; pistillate scales shorter to longer than the perigynia, reddish brown with a greenish midstripe and hyaline margins; perigynia 4.2–7 (7.4) mm long, 1.9–2.8 mm wide, elliptic or ovate, slightly to strongly plano-convex, lightly to evidently many-nerved on both sides, or the ventral side nerveless or with few nerves, wing-margined to the tip of the ill-defined flattened beak, serrulate, greenish, stramineous, or pale brown; stigmas 2; achenes lenticular; 2n = 68. Forb-grass communities and open woods at 2620 to 3505 m in Beaver, Emery, Garfield, Grand, San Juan, Sevier, and Washington counties; Alberta to Minnesota and south to Arizona and New Mexico; 9 (0). Although plants of this taxon have been separated from those of *C. petasata* (q.v.) at the subsection level (Foeneae versus Specificae), they may not be so distantly related. They have broadly overlapping geographic ranges, and morphological differences are few and overlapping. The flattened versus terete beak of the perigynium is a subtle difference in this case and perhaps does not warrant segregation at the subsection level. Indeed, reduction to infraspecific status might not extend reality, but no combination is proposed here. I am not certain that all specimens placed here or in *C. petasata* are correctly determined, but in general the perigynia are smaller on the average than in *C. petasata*, and, in Utah, the plants are more commonly found in open exposed places and not so much with sagebrush and aspen. In these respects, *C. xerantica* simulates *C. phaeocephala*, and specimens of these two taxa are sometimes mixed in herbaria. These three taxa together with *C. praticola*, form a complex that is in need of further study.

Cladium R. Br.

Plants perennial from robust scaly rhizomes, with hollow, leafy stems; inforescence a compound umbel with numerous spikelets, these few-flowered; scales spirally arranged, the lower ones empty or staminate; perfect flower solitary; perianth bristles lacking; stamens 2; style 2 or 3 cleft.

Cladium californicum (Wats.) O'Neill in Tidestr. & Kittell Saw-grass. [*C. mariscus* var. *californicum* Wats.]. Culms 1–2 m tall, subterete to subtriangular, ca 1–1.5 mm thick; leaves 0.8–1.5 cm wide, flat, serrulate, the teeth cartilaginous; umbels 3–10 or more, with several rays and numerous spikes 3–4 mm long; scales few, reddish brown, the lowest one empty, the middle ones empty or subtending staminate flowers, the terminal one subtending a perfect flower; achenes ovoid, ca 2 mm long, without a tubercle. Hanging gardens along Lake Powell at 1125 to 1150 m in Kane and San Juan counties; California to Arizona, south to Mexico; 7 (0). Most of the sites discovered for this plant have been drowned by the rising water of Lake Powell. Leaves at the base of an accumulation of marcescent remains of this plant in Driftwood Canyon were dated at 440 years B.P.

Cyperus L.

Annual or perennial plants with mostly solid, triangular stems; spikelets several to many, arranged in one or usually several capitate to spicate clusters, the terminal cluster commonly sessile or nearly so, the others from rays originating in the axils of sheathless leafy involucral bracts; scales of the spikelets 2–ranked in vertical rows, the lowest one empty and somewhat modified; flowers perfect; perianth lacking; stamens (1–2) 3; styles 2 or 3 or with 2 or 3 branches; achenes lenticular or trigonous.

1. Plants perennial, from creeping rhizomes; stamens 3 ... 2
— Plants annual or short-lived perennial; rhizomes lacking; stamens 1, 2, or 3 3
2(1). Spikelets with 5–8 flowers; scales 3–4 mm long; rhizomes short, thick, not ending in a terminal tuber; plants of sandy places in Kane County ... *C. schweinitzii*
— Spikelets with ca 12–28 flowers; scales 2–3 (4) mm long; rhizomes slender, ending in a terminal tuber; distribution various .... *C. esculentus*
3(1). Scales 3–6 mm long, not especially crowded, usually fewer than 10 per spikelet, linear or narrowly lanceolate; stamens 3; plants perhaps to be expected in Utah .... *C. strigosus* L.
— Scales 1–2 (2.5) mm long, more or less crowded, usually more than 10 per spikelet, lanceolate to ovate or obovate; stamens various 4
4(3). Stigmas 2; scales blunt, not mucronate or awn-tipped, more or less strongly anthocyanic with a pale midrib; plants perhaps to be expected in Utah .. *C. rivularis* Kunth
— Stigmas 3; scales tapered or blunt but then usually mucronate or awn-tipped, variously colored but not as above 5
5(4). Stamens 3; scales broadly rounded or blunt apically, sometimes abruptly mucronate, with broad yellow gold or yellow brown scarious margins, the nerves not conspicuously raised .... *C. erythrorhizos*
— Stamens 1 or 2; scales gradually tapered to an acuminate or awned tip, the margins variously colored, the nerves not conspicuously raised 6
6(5). Scales with (5) 7–9 nerves, the slender awn-tip squarrose .... *C. aristatus*
— Scales with 3 nerves, the acute to acuminate tips not squarrose .... *C. acuminatus*

Cyperus acuminatus Torr. & Hook. Plants annual, 5–25 cm tall; leaves few, borne near the culm base, 1–2 (3) mm wide; involucral bracts unequal, the longest ones to 10 cm long or longer and much surpassing the inflorescence; spikelets in very dense rounded clusters; spikelets 3–7 mm long, strongly flattened; scales ca 2 mm long, 3–nerved, mostly hyaline except for the raised nerves, strongly overlapping, deciduous in age, the tip tending to be shortly recurved-acuminate; rachilla wingless, persistent; stamen solitary; styles trifid for ca 3/4 their length; achenes trigonous. Mud flats and drying shores at ca 1375 m in Millard County; North Dakota to Georgia and west to the Pacific; 4 (0).

Cyperus aristatus Rottb. Bearded Flat-sedge. Plants annual, 2–15 cm tall, tufted, with fibrous roots; leaves borne near the base of the culms, mostly 0.5–2.5 mm wide; involucral bracts commonly 2–6, foliose, some or all exceeding the inflorescence; spikelets in headlike clusters, the terminal cluster sessile, the others, if present, on slender rays to 2 cm long; spikelets 4–10 mm long, flattened; scales ca 1–2 mm long, with (5) 7–9 more or less raised nerves, deciduous with age, the narrowed apex awnlike, 0.3–1 mm long, spreading to somewhat recurved; rachilla wingless, persistent; stamen solitary; style 3–branched; achenes trigonous, 0.6–1 mm long. Seeps, springs, disturbed ground in meadows, margins of ponds and lakes, and flood plains at 1220 to 2015 m in Duchesne, Emery, Garfield, Grand, Salt Lake, San Juan, Uintah, Utah, Washington, and Weber counties; more or less cosmopolitan except at high latitudes; 13 (iii).

Cyperus erythrorhizos Muhl. Redroot Flat-sedge. Plants annual, 10–73 cm tall; culms sharply triangular, arising singly or few to several together; leaves crowded toward the base of the culms, the blades elongate, 2–10 mm wide; involucral bracts commonly 4–10, unequal, the larger ones to 40 cm long, sometimes wider than the leaves; inflorescence compact, or the open and umbelliform clusters of spikelets elongate, sometimes cylindrical, 1–3 cm long, the terminal sessile, the others borne on rays to 9 cm long; spikelets 3–12 mm long, 1–1.5 mm wide, spirally arranged; rachilla narrowly hyaline-winged, the wings readily deciduous in short segments at maturity; scales ca 1–1.5 mm long, broadly rounded or blunt apically, with 3–5 faint lateral nerves, closely overlapping, eventually deciduous, the midrib scarcely excurrent as a blunt mucro; stamens 3; styles 3–branched for ca 1/4–1/2 their length; achenes trigonous, pale, shiny, 0.7–1 mm long. Margins of ponds and lakes and on sand bars along rivers at 1504 to 1820 m in Garfield, Grand, San Juan, Utah, and Washington counties; U. S. and Canada; 18 (0).

Cyperus esculentus L. Chufa Flat-sedge. Plants perennial, 3–7 dm tall; culms sharply triangular, usually arising singly, from numerous fiberous roots and slender rhizomes which terminate in small tubers; leaves few to many, clustered toward the base of the culms, the blades elongate, 3–8 mm wide; involucral bracts ca 3–6, unequal, from shorter than to much exceeding the inflorescence, this umbelliform; clusters of spikelets short and congested to open and elongate, the terminal sessile, the others on rays to 10 cm long; spikelets slender, 0.5–2 cm long, 1–2 mm wide, with ca 12–28 flowers; rachilla narrowly hyaline-winged; scales 2–3 (4) mm long, severalnerved, overlapping ca 1/2 their length, deciduous at maturity; stamens 3; styles 3-branched ca 1/2–3/4 their length; achene unequally trigonous, 1.3–2 mm long; n = 48?, 54, 104. Weedy in areas of cultivation at ca 1230 m in Davis and Grand counties; widespread in tropical and warm temperate regions; 2 (0).

Cyperus schweinitzii Torr. Plants perennial, 10–34 cm tall, from short, thick rhizomes; culms triangular, somewhat bulbous-based, commonly antrorsely scabrous especially above; leaves few to several, mostly on the lower 1/4 of the culm, the blades elongate, ca 1–4 mm wide; involucral bracts commonly 3 or 4, unequal, to 10 cm long; terminal cluster of spikelets sessile, the others on rays to 7 cm long, the clusters ovoid to oblong; spikelets 5–10 mm long, compressed, ca 2 mm wide, with ca 5–8 flowers; scales 3–4 mm long, striate nerved, each one overlapping the next directly above ca 1/2 its length, deciduous at maturity, the midrib commonly excurrent as a tiny mucro; rachilla narrowly hyaline-margined; stamens 3; style 3-branched to near the base; achenes equally trigonous, 2–2.5 mm long; 2n = 80. Dunes and sandy soil in rabbitbrush-yucca and black sagebrush-juniper communities at 1735 to 1980 m in Kane County (near Kanab); Ontario to Texas and Mexico, and northern Idaho; 6 (0).

Eleocharis R. Br.

Annual or perennial plants; culms angular, flattened, or terete; leaves reduced to bladeless sheaths, or the sheaths bristle-tipped, basal or nearly so; spikelet solitary and terminal; involucral bracts lacking; scales spirally arranged, the lower empty or with flowers (in the *E. acicularis* group); flowers perfect; perianth lacking or of bristles, these up to ca 10; stamens 3 (or fewer); styles 2 or 3 or with 2 or 3 branches, thickened toward the base, the thickened part persistent on the achene as a tubercle, this confluent with or sharply differentiated from the body; achenes lenticular to more or less trigonous.

1. Stigmas 2 2
— Stigmas 3 3
2(1). Plants perennial, our most common and widespread spikerush; spike linear-lanceolate; achenes with a distinct tubercle, this constricted at the base and as high as wide or higher .... *E. palustris*
— Plants annual, rare; spike ovate-lanceolate; achenes with a flattened tubercule, this with a broad base and wider than high .... *E. ovata*
3(1). Tubercle constricted at the base, more or less forming a cap on the achene or wider than high 4
— Tubercle not constricted at the base, confluent with and not forming a cap on the achene, as high as or higher than wide 7
4(3). Achenes whitish or pale gray, longitudinally manyribbed and with numerous cross corrugations forming ladderlike configurations; spikelets 2.5–9 mm long, with 3–15 flowers; culms very slender, ca 0.2 mm thick, not over 12 cm tall 5
— Achenes yellowish, golden brown, or blackish in age, 3-ribbed, cellular roughened but without cross corrugations; spikelets sometimes over 9 mm long and/or with more than 15 flowers; culms slender but mostly over 0.2 mm thick, 5–30 cm tall 6
5(4). Plants perennial, with very slender rhizomes, widespread in Utah; anthers 0.7–1.3 mm long; spikes 2.5–9 mm long .... *E. acicularis*
— Plants annual, without (rarely with) rhizomes, known from Washington County where rarely collected; anthers 0.2–0.4 mm long; spikelets 1.5–3 mm long. . *E. bella*
6(4). Tubercle wider than high, not noticably constricted at the base, but forming a broad low apiculate crown on the achene; spikelet ca 2–3 times longer than wide, 3–8 mm long; leaf sheaths with an entire collar; plants rarely collected, mostly known from the northeastern and east-central part of the state .... *E. bolanderi*
— Tubercle as high as or higher than wide, noticeably constricted at the base and forming a cap on the achene; spikelet ca 3–7 times longer than wide, 5–16 mm long; leaf sheaths, at least some, usually with a mucronate collar, the mucro to 0.8 mm long; plants widespread in the Great Basin and southern part of the state . *E. parishii*
7(3). Achenes 0.9–1.3 mm long; scales 1.5–2 (2.5) mm long plants 2–6 (10) cm tall, of low elevations; spikelet 2.5–4 (6) mm long .... *E. parvula*
— Achenes 1.9–2.8 mm long; scales 2.5–5.5 mm long; plants various; of various elevations; spikelet diverse ... 8
8(7). Plants (1) 2–8 dm tall or taller, from ascending or vertical stout rhizomes; larger culms 1–2 mm thick, some of them arching and rooting from an apical bulbil (proliferous); spikelet 5–13 mm long, with (5) 10–20 (25) flowers .... *E. rostellata*
— Plants 5–15 (20) cm tall, either with slender rhizomes or these lacking; culms to ca 1 mm thick, neither arching nor proliferous; spikelet 4–8 mm long, usually with 3–9 flowers 9
9(8). Leaf sheaths entire; scales entire, soon dark purple to purple black; perianth bristles minutely retrorsely barbellate; culms arising singly or few together from slender, creeping rhizomes; plants widespread . *E. pauciflora*
— Leaf sheath, at least the uppermost, of some culms with a bristlelike leafblade 4–6 mm long; lowest scale with a blunt awn 1–3 mm long, the scales greenish, pale in age; perianth bristles smooth; culms densely tufted; rhizomes lacking; plants of the Uinta Mts. .... *Scripus caespitosus* (q.v.)

Eleocharis acicularis (L.) R. & S. Plants perennial, 2–15 (20) cm tall, culms 0.3–0.5 mm thick, tufted, from very slender rhizomes, usually with a solitary sheath ca 3–15 mm long; spikelet 2–8 mm long, with ca 4–25 flowers; scales (1.3) 1.5–2.2 mm long, blackish purple, typically with a green midstripe and hyaline margins; perianth bristles commonly 3 or 4, equaling or surpassing the achene or sometimes reduced or lacking; anthers 0.7–1.3 mm long; stigmas 3; achenes white to pale gray, rounded-trigonous, 0.7–1.1 mm long including the lowconic or more or less triangular-conic, basally constricted tubercle, with 8–18 longitudinal ribs, and with numerous cross corrugations connecting the ribs; 2n = 20. In mud and water, along streams, margins of ponds and lakes, and bogs and wet meadows, at (1310) 2190 to 3200 m in Beaver, Cache, Daggett, Duchesne, Garfield, Kane, Rich, Salt Lake, Sevier, Summit, Uintah, Utah, and Wasatch counties; circumboreal, south in America to Florida and Mexico; 23 (vii).

***Eleocharis bella* (Piper) Svenson** Pretty Spikerush. [*E. acicularis* var. *bella* Piper]. Plants annual but forming dense tufts and appearing perennial, rarely with slender rhizomes, 1–8 cm tall; culms ca 0.2 mm thick, usually with a solitary short sheath; spikelet 1.5–3 mm long, with 3–15 flowers; scales 1–1.3 mm long, with a green or pale midstripe and hyaline margins and sometimes with dark purplish strips flanking the green midstripe; perianth bristles lacking; anthers 0.2–0.4 mm long; stigmas 3; achenes white or nearly so, rounded trigonous, with several longitudinal ribs and numerous cross corrugations connecting the ribs, these forming ladderlike configurations, 0.6–0.8 mm long including the low-conic basally constricted tubercle. Margins of ponds and lakes and other wet places in a pinyon-juniper community at ca 1830 m in Washington County (south of Enterprise, Atwood & Higgins 5901 BRY); Washington to Idaho, south to California and New Mexico; 1 (0).

***Eleocharis bolanderi* Gray** Plants perennial, 8–30 cm tall; culms ca 0.5 mm thick, closely clustered on short rhizomes, often purplish at the base, usually with 1 or 2 sheaths, the upper sheath extending 1.5–3 cm on the culm; spikelet 3–8 mm long, ca 2–3 times longer than wide, with 10–25 flowers; scales usually 1.8–2.7 mm long, blackish or black purple, with a pale hyaline tip, lacking a pale midstrip or this rather faint; perianth bristles 3 or 4, reddish brown, minutely retrorsely barbellate, shorter than the achene; stigmas 3; achenes unequally trigonous, ca 1.5 mm long, greenish yellow at first, golden yellow at maturity, perhaps blackish in age, minutely cellular roughened, the tubercle not constricted at the base, depressed conic and apiculate. Moist and wet meadows at 2475 to 2590 m in Cache, Daggett, Duchesne, Grand, San Juan, Summit, and Washington counties; Oregon to California east to Idaho; 11 (v).

***Eleocharis flavescens* (Poir.) Urban** Yellow Spikerush. [*Scirpus flavescens* Poir.]. Plants perennial, tufted, with slender rhizomes, 3–15 cm tall; leaf sheaths membranous, whitish, prolonged, many of them free from the culms; spikelet 2–6 mm long, ovate, with fewer than 20 flowers; scales purplish brown with a paler midvein; bristles 6–7, white, equaling or slightly longer than the achene; achenes glossy, purplish brown, the surface minutely punctate, ca 1 mm long; tubercle conic, acute. Moist places in Cache County; South America, West Indies, and north to the southern U. S., introduced in California and isolated stations in Wyoming and Montana; 1 (0).

***Eleocharis ovata* (Roth) R. & S.** [*Scirpus ovatus* Roth]. Plants annual; culms few to several from fibrous roots, ca 0.5–1 mm thick, several-ribbed, with 1 or 2 sheaths, the upper sheath extending ca 4.5 cm up the culm; spikelet (2) 5–13 mm long, ovoid, usually with more than 40 flowers; scales usually 1.7–2.5 mm long, purplish or brownish, with a greenish midstripe and hyaline margins; bristles 6 or 7, brownish, retrorsely barbellate, subequal to or exceeding the achene, or reduced or lacking; anthers 0.3–0.8 mm long; stigmas 2 or occasionally 3; achenes lenticular, 1–1.5 mm long, smooth, straw colored to dark brown, the tubercle depressed conic, not conspicuously constricted at the base. Shore of Utah Lake, Utah County (Brotherson 2756 BRY); widespread in the Northern Hemisphere; 1 (0).

***Eleocharis palustris* (L.) R. & S.** Creeping Spikerush. [*Scirpus palustris* L.; *E. calva* Torr.; *E. macrostachya* Britt. ex Small]. Plants perennial, 1–7 (10) dm tall; culms arising singly or few together or more or less tufted, from slender creeping rhizomes, 0.5–3 (4) mm thick, usually with 2 sheaths, at least the lower one usually reddish or dark purple, the upper one extending up to ca 2–15 cm on the culm; spikelet 5–25 mm long, linear to lance-ovate in outline, with several to many flowers; scales 2–4.5 mm long, brownish purple, with a greenish midstripe and hyaline margins, the lower 1 or 2 empty; perianth bristles 4 (5 or 6), retrorsely barbellate, a little longer than the achene or sometimes reduced or lacking; anthers 1.3–2.5 mm long (dry); stigmas 2; achenes lenticular, green yellow, yellow, or medium brown, finely cellular-roughened, ca 1.5–2.5 mm long including the tubercle; tubercle constricted at the base, rather elongate-conic, 0.4–0.7 mm long; 2n = 14, 16, 18, 38. Margins of ponds and lakes, along streams, in swampy meadows, and other wet places at 1135 to 3200 m in all Utah counties; widespread in the Northern Hemisphere; 140 (x).

***Eleocharis parishii* Britt.** Plants perennial, 7–40 cm tall; culms ca 0.3–1 mm thick, tufted, arising singly, or few together, from slender rhizomes; sheaths usually 2, often reddish or purplish at least in part, the upper one extending (1) 1.5–4.5 cm up the culm, usually at least a few of them ending in a mucro to 0.8 mm long; spike 5–16 mm long, ca 3–7 times longer than wide, with ca 20–40 flowers; scales usually 2.4–3 mm long, purplish or brownish purple, usually with a green or pale midstripe and hyaline margins, the lowest one tending to be shorter, broader, and empty; perianth bristles usually 6 or 7, white or brown, retrorsely barbellate, longer or shorter than the achene; stigmas 3; achene plano-convex or unequally trigonous, greenish yellow when young, yellow or brownish to blackish when mature, 1–1.2 mm long, the tubercle more or less elongate-conic, much like that of *E. palustris* but perhaps not quite so noticeably constricted at the base. Margins of ponds, lakes, seeps and springs, wet and swampy meadows, along streams, and other wet places at 775 to 2200 (2700) m in Beaver, Cache, Davis, Juab, Kane, Millard, Salt Lake, San Juan, Sanpete, Tooele, Utah, Washington, and Weber counties; Oregon south to California and New Mexico; 22 (i).

***Eleocharis parvula* (R. & S.) Link ex Bluff & Fingerh.** [*Scirpus parvulus* R. & S.]. Plants perennial, from very slender inconspicuous rhizomes, 2–7 (10) cm tall; culms ca 0.5–1 mm thick, densely tufted, sometimes forming mats, usually with only 1 sheath, this to ca 1 cm long; spikelet 2.5–5 mm long, with 2–9 (20) flowers; scales 1.5–2 (2.5) mm long, the lowest one empty; stigmas 3; achenes 0.9–1.3 mm long, usually unequally trigonous; tubercle confluent with the achene, not constricted, very short. Usually in mud of drying margins of ponds and lakes or on flood plains at 1250 to 1770 m in Millard, Salt Lake, Uintah, Utah, and Weber counties; British Columbia to Newfoundland, south to South America; Europe; 10 (iii).

***Eleocharis pauciflora* (Light.) Link** [*E. quinqueflora* (Hartman) Swartz]. Plants perennial, 7–20 (30) cm tall; culms 0.2–1 mm thick, arising singly or few together or occasionally tufted, from slender rhizomes, usually with 1 or 2 sheaths, these entire, pale or reddish, the upper one extending 1–2 cm up the culm; spikelet 4–8 mm long, with ca 3–9 flowers; scales 2.5–5.5 mm long, commonly purplish to purple black, with hyaline margins, often without a green or pale midstripe, the lowest 2 or 3 from

1/2–3/4 as long as the spikelet or longer; perianth bristles equaling or exceeding the achene or sometimes reduced, finely retrorsely barbellate; stigmas 3; achenes 1.9–2.6 mm long, equally or unequally trigonous or plano-convex, broadest above the middle; tubercle confluent with the achene, not constricted, higher than wide; n = 68. Wet and boggy meadows and along streams at 2080 to 3385 m in Box Elder, Cache, Daggett, Duchesne, Emery, Garfield, Juab, Kane, Salt Lake, Sanpete, Sevier, Summit, Uintah, Utah, Wasatch, and Wayne counties; circumboreal, south in America to California and New Jersey; 40 (xv). This species is sympatric with *Scripus caespitosus* (q.v.) in the Uinta Mts.

Eleocharis rostellata (Torr.) Torr. [*Scirpus rostellatus* Torr.]. Plants perennial, (1.5) 4–8 (10) dm tall; culms usually 1–2 mm thick, few together or densely clustered, from a thick, ascending or nearly vertical rhizome, some of the culms often arching and rooting from an apical bulbil (proliferous); sheaths usually 2 per culm, brownish or occasionally reddish or purple, the upper one extending ca 2–8 cm up the culm; spikelet (5) 8–11 mm long, with (5) 10–20 (25) flowers; scales purplish to black purple, with broad pale hyaline margins, with or without a green or pale midstripe, ca 3–4 mm long; lowest scale a little larger than the others, rarely over 1/2 as long as the spikelet, empty; perianth bristles somewhat shorter to a little longer than the achene, retrorsely barbellate; stigmas 3; achenes 1.9–2.8 mm long, rounded trigonous to plano-convex, smooth or slightly cellular-roughened, greenish, the tubercle confluent with the body of the achene, not constricted, medium to high conic. Wet meadows, marshes, ditches and streams, and seeps and springs (often where alkaline) at 1280 to 2255 m in Beaver, Cache, Duchesne, Emery, Garfield, Iron, Juab, Kane, Millard, Salt Lake, San Juan, Sanpete, Summit, Tooele, Utah, and Wasatch counties; British Columbia to Nova Scotia, south to South America; 37 (v). The arching vegetative stems rooted at both ends, and catching one's feet when walking, are unique among the Utah flora.

Eriophorum L.

Perennial grasslike plants with mostly solid culms; leaf sheaths closed, the blades grasslike or the upper ones sometimes lacking; bracts subtending the inflorescence scalelike to leaflike; spikelets 1 to several in a terminal inflorescence; scales of spikelets spirally arranged, not awned; perianth of numerous persistent bristles, these elongate at least in fruit and forming a cottonlike tuft; stamens 3; styles 3; achenes unequally trigonous, with a usually slender apiculate tip.

1. Spikelets (2) 3–6 (8), mostly on slender peduncles; lowest bract more or less leaflike, 2.5–5.5 cm long; leaf blades well developed, mainly 2–6 mm wide; anthers commonly 2.5–4 mm long .... *E. polystachion*
— Spikelet solitary, sessile; bracts scalelike, less than 1 cm long; leaf blades ca 1 mm wide, often reduced or lacking; anthers commonly 0.5–1 mm long .... *E. scheuchzeri*

Eriophorum polystachion L. Tall Cottongrass. [*E. angustifolium* Honck.]. Plants perennial, 20–60 cm tall; culms mostly arising singly, from long creeping rhizomes; leaf blades usually well developed, mainly 2–6 mm wide, flat or narrowed and channeled to triangular distally; lowest involucral bract more or less leaflike, 2.5–5.5 cm long, often dark purple, at least toward the base, the upper bracts much smaller; spikelets (2) 3–6 (8), mostly on slender peduncles to 2.5 cm long, or rarely sessile or nearly so, in an umbellate inflorescence; scales greenish, brownish, or purplish, with hyaline margins; perianth bristles numerous, white, forming cottonlike tufts 2–3.5 cm long; achenes blackish, 2–3 mm long, oblanceolate to obovate; 2n = 58, 60. Wet meadows and bogs in spruce-fir and lodgepole pine forest openings at 2740 to 3385 m in Daggett, Duchesne, and Uintah counties; circumboreal, south in America to Oregon, New Mexico, and New York; 7 (iv).

Eriophorum scheuchzeri Hoppe White Cottongrass. Plants perennial, 13–40 cm tall; culms mostly arising singly from long creeping rhizomes; leaf blades to 8 cm long and 1 mm wide, channeled or triangular, often much reduced or lacking especially the upper one; involucral bracts scalelike, 3–8 mm long, blackish or black purple; fertile scales blackish, with broad white hyaline margins, acute; perianth bristles numerous, white, forming cotton-like tufts 2–3.5 cm long; achenes brown or blackish ca 2 mm long. Wet meadows, margins of lakes, and other wet places in lodgepole pine, spruce-fir, and alpine tundra at 3350 to 3810 m in Duchesne, Summit, and Uintah counties; circumboreal, south in America to Colorado, and Newfoundland; 10 (i).

Fimbristylis Vahl

Grasslike plants with triangular culms; leaves basal or nearly so; spikelets several to many in a terminal, simple or compound umbellate cymes subtended by sheathless bracts; scales spirally arranged; flowers perfect; perianth lacking; stamens 1–3; style lobed, enlarged toward the base.

Fimbristylis spadicea (L.) Vahl [*Scirpus spadiceus* L.]. Plants perennial, 20–90 cm tall; leaves basal and on the lower 1/5 of the culms, the sheaths more or less closed, glabrous or pubescent, the ventral side membranous, often with minute brownish or reddish spots, the blades 1–4 mm wide; inflorescence simple or more often compound, with the terminal spikelet sessile and subtended by an involucral bract, the lateral spikelets on rays or peduncles mainly 1–7 cm long; spikelets 8–23 mm long, 3–5 mm thick, elliptic-cylindrical, many-flowered; scales 3–5 mm long, usually ciliate to pubescent over the back, reddish brown or grayish, the midrib lighter and prolonged into a mucro or short awn; stamens 3; style 2–lobed ca 1/4 the length; achenes ca 1.5 mm long, obovate, minutely apiculate, finely many-ribbed and cross-rugulose; n = 10. Hanging gardens, wet meadows, margins of ponds and lakes, and along streams, tolerant of alkali, at 850 to 1375 m in Kane, Millard, San Juan, Utah, and Washington counties; tropical America north to California and New York; 6 (0).

Kobresia Willd.

Plants perennial, densely caespitose, grasslike; stems obtusely triangular, solid; leaves with closed sheaths and narrow blades; inflorescence of 1 to several spikes with few to several spikelets; scales spirally arranged; flowers imperfect, without a perianth, arranged in small spikelets; achenes subtended by 2 small bracts, the lower bract corresponding to the lower bracteate scale in *Carex* and the inner bract corresponding to the perigynium but with unsealed margins and exposing the achene; stamens 3; stigmas 3.

1. Spike solitary, 1–2 cm long; plants 5–20 cm tall, of exposed, alpine ridges at or above 3200 m ...... *K. bellardii*
— Spikes (1) 3–12, 0.5–1.5 cm long; plants 25–35 cm tall, of wet or boggy, subalpine meadows below 3200 m .... *K. simpliciusula*

Kobresia bellardii (All.) Degl. [*Carex bellardii* All.; *C. myosurioides* Vill.; *K. mysosurioides* (Vill.) Fiori & Paoli]. Plants 5–20 cm tall, caespitose; leaves basal and on the lower stem, the blades tightly involute or channeled, wiry, ca 0.4–0.8 mm wide; spike solitary, 1–2 cm long, 2–3 mm wide, the terminal flower staminate, the lateral ones mostly androgynous, occasionally pistillate; scales 3–4 mm long, brownish, largely scarious; achenes ca 2.5 mm long, brownish with a darker, persistent, apiculate style base; $2n = 36?, 58$. Alpine tundra communities, rock stripes, and talus at 3200 to 3630 m in Daggett, Duchesne, and Summit counties; circumboreal, south in North America to California and Newfoundland; 15 (i). This plant and *Carex elynoides* (q.v.) are habitually and ecologically similar, and they are rather easily mistaken.

Kobresia simpliciuscula (Wahl) Mack. [*Carex simpliciuscula* Wahl]. Plants perennial, 25–35 cm tall, densely caespitose, without rhizomes; leaves basal and on the lower 1/5 of the stems, the sheaths open and with scarious margins, the blades flat or strongly folded, 0.5–1.2 mm wide; spikes (1) 3–12, 5–15 mm long; sessile or occasionally short-pedunculate; spikelets usually with a pistillate and a staminate flower; scales 3–3.5 mm long, brownish or greenish brown, largely scarious; achenes 2–3 mm long; $2n = 72–76$. Boggy, mostly calcareous meadows at 2590 to 2805 m in Daggett, Duchesne and Emery counties; circumboreal and south in western North America to Idaho and Colorado; 4 (ii).

Scirpus L.

Annual or perennial plants; leaves reduced to bladeless sheaths or with well-developed blades; involucre of 1 to many scalelike or leaflike bracts; spikelets solitary to numerous in a capitate, umbellate, or paniculate inflorescence; scales of spikelets spirally arranged, with or without an excurrent awn; flowers perfect; perianth of 1–6 bristles, these sometimes obsolete; stamens 3 (rarely fewer); stigmas 2 or 3; achenes lenticular or trigonous, with or without a stylar apiculus, but without a tubercle.

1. Spikelet solitary; involucre of 2 or 3 modified empty scales less than 1 cm long; plants 4–20 cm tall, perennial, densely tufted, of the Uinta Mts. .... *S. caespitosus*
— Spikelets 2 to many; involucre of 1 to several more or less leaflike bracts; plants various but often taller, of lower elevations 2
2(1). Plants annual, tufted, from fibrous roots, 3–25 cm tall; achenes dark gray brown to blackish, conspicuously cross-ridged .... *S. supinus*
— Plants perennial, from rhizomes, often over 25 cm tall; achenes mostly lighter, not cross-ridged 3
3(2). Involucre with 2 or more foliose, spreading bracts 4
— Involucre with only 1 foliose bract, this erect or nearly so 7
4(3). Spikelets mostly 3–6 mm long and over 100 in an open umbellate inflorescence 5
— Spikelets mostly 10–35 mm long and 3–40 (rarely more) in a compact or somewhat open umbellate inflorescence 6
5(4). Stigmas 2; midrib of scales abruptly contracted into a short mucro; leaf sheaths mostly anthocyanic; plants widespread .... *S. microcarpus*
— Stigmas 3; midrib of scales tapered into a short awn; plants of the Uinta Basin and Utah County .. *S. pallidus*
6(4). Stigmas 2; spikelets 3–25, all in a sessile cluster or also with 1–4 additional clusters borne on peduncles to 6 cm long; plants common, widespread ..... *S. maritimus*
— Stigmas 3; spikelets commonly 10–40, at least a few usually borne singly, the inflorescence typically umbellate with few to several raylike peduncles to 7 cm long; plants rare, known from Box Elder, Cache, and Daggett counties .... *S. fluviatilis*
7(3). Inflorescence open, with conspicuous branches; spikelets numerous, more than 20; culms terete, commonly 1 cm or more in diameter, 8–30 dm tall; leaves reduced to bladeless sheaths or the blades short and erect or nearly so 8
— Inflorescence headlike, with the sessile spikelets in a sessile cluster; spikelets 3–15; culms triangular or, if subterete, usually with well-developed ascending-spreading leaf blades, commonly less than 1 cm in diameter, 1–16 dm tall 9
8(7). Spikelets appearing dull orange or reddish brown, the scales with striolae about the same color as the rest of the scale; scales mostly 2–3 (3.5) mm long; plants rather uncommon .... *S. validus*
— Spikelets appearing dull gray brown, the scales with prominent red brown striolae that contrast with the gray white color of the rest of the scale; larger scales mostly 3.5–4 mm long; plants common ...... *S. acutus*
9(7). Achenes entire at the apex, not apiculate; scales mostly entire at the apex, the midrib hardly (if at all) exserted; culms subterete .... *S. nevadensis*
— Achenes apiculate; scales notched at the apex, the midrib extending between the notch as an awn or awn-point; culms obtusely to sharply triangular 10
10(9). Culms sharply triangular, ridged or winged on the angles, conspicuously concave on the sides, (4) 5–16 dm tall; leaf blades to 10 cm long, rarely longer; scalelike bracts lacking; plants mainly of the Great Basin. .... *S. americanus*
— Culms subterete to sharply triangular but not concave on the sides, commonly less than 5 dm tall or, if taller, the longest leaf blade mostly over 10 cm long; scalelike bracts 1–2; plants widespread .... *S. pungens*

***Scirpus acutus* Muhl. ex Bigelow** Hardstem Bulrush. Plants perennial, commonly 10–30 dm tall, from robust rhizomes; culms terete, ca 5–20 (30) mm thick; leaves borne on the lower 1/4 of the culms, reduced to bladeless sheaths, or the upper one with a blade up to 12 cm long; inflorescence compact, subumbellate, subtended by greenish involucral bract that simulates a continuation of the culm, this commonly shorter than or equaling the inflorescence, rarely longer, also with additional inconspicuous bracts; spikelets usually numerous, more or less grayish or gray brown, commonly 8–15 mm long, solitary or 2 to several; lower scales (3) 3.5–4 mm long, mostly gray hyaline except for the greenish midrib, the gray hyaline background sharply contrasting with reddish brown striolae, the margins ciliolate or lacerate; perianth bristles retrorsely barbellate, subequal to or slightly exceeding the achene; styles 2-branched for ca 3/4 their length; achenes 2.2–2.5 mm long, plano-convex, more or less completely hidden by the scales; $2n = 36$. Margins of ponds and lakes, marshes and swamps, seeps, springs, washes, and flood plains at 1097 to 2200 m in all Utah counties except Beaver, Iron, Morgan, Summit, and Wasatch; temperate North America; 75 (viii). This species apparently intergrades with S. *validus* q.v.

Scirpus americanus Pers. Olney Threesquare. [S. *olneyi* Gray]. Plants perennial, (4) 5–16 dm tall, from robust rhizomes; culms very sharply triangular, ridged or winged on the 3 angles, conspicuously concave on the sides, easily flattened in pressing, mostly thicker than in S. *pungens*, commonly 5–10 mm thick toward the base; leaves commonly 1–4, borne on the lower 1/3 of the culms, the lower ones often reduced to bladeless sheaths, the upper ones with blades to 10 cm long, the blades usually strongly folded, to 6 mm wide when pressed, very gradually tapered; involucral bract solitary, 1–3.5 (5) cm long, commonly 1–3 times longer than the inflorescence, appearing as a continuation of the culm; involucral bracts lacking; spikelets 2–15, sessile, usually 6–15 mm long, scales yellowish brown, reddish brown, or purplish brown, rather hyaline, the midrib firm distally and commonly exserted as a mucro, the mucro subequal to the notch of the scale; perianth bristles 4 (5–6), usually retrorsely barbellate; styles 2; achenes 2–3 mm long including the apiculus, 1.4–1.7 mm wide, plano-convex; n = 39? Swamps, marshes, seeps, springs, margins of ponds and lakes, wet alkaline meadows, and saltgrass-greasewood communities at 730 to 1510 m in Box Elder, Cache, Davis, Juab, Millard, Salt Lake, Sanpete, Tooele, Utah, Washington, Wayne, and Weber counties; Nova Scotia to Washington, south to South America; 33 (iii). This species is much less common than S. *pungens*, for which it is occasionally mistaken.

Scirpus caespitosus L. Deerhair Bulrush. Plants perennial, 4–20 cm tall, densely tufted, with very short rhizomes; culms subterete, ca 1 mm thick, clothed at the base with persistent leaf sheaths; basal leaves reduced to bladeless sheaths, the upper borne on the lower 1/5 of the culm, this commonly with a rudimentary blade 4–7 mm long; inflorescence of a solitary, terminal spikelet 4–6 mm long, this light to medium brown; involucre of 2 or 3 empty scales, these usually deciduous in age, the lowest scale with a blunt awn 1–3 mm long, the awn subequal to the spikelet; perianth bristles 6, white, smooth, a little longer to twice as long as the achene; achenes trigonous, ca 1.5 mm long; n = 52? Boggy and peaty meadows at 2895 to 3415 m in Duchesne, Summit, Uintah, and Wasatch counties; circumboreal, south in North America to North Carolina and Oregon; 15 (viii). This bulrush is sympatric with *Eleocharis pauciflora* in the Uinta Mts., but perhaps it is less tolerant of permanently saturated soil.

Scirpus fluviatilis (Torr.) Gray River Bulrush. [S. *maritimus* var. ? *fluviatilis* Torr.]. Plants perennial, commonly 6–15 dm tall, from robust rhizomes; culms sharply triangular, often with concave sides; leaves commonly 3–5, well distributed on the lower 1/2–3/4 of the culm, mostly with well-developed flat blades to 50 cm long and 6–15 mm wide; leaflike involucral bracts usually 3–5, to 35 cm long and 7 mm wide, flat; inflorescencee of few to several raylike peduncles to 7 cm long, each with a cluster of (1) 2–5 spikelets; some spikelets with empty scalelike bracts at the base, these often entire apically; spikelets commonly 10–40, 10–18 mm long, 6–10 mm thick; scales tan or light brown, scarious, hirtellous on the back, the midrib excurrent as an awn, the awn commonly 2–5 times longer than the apical notch of the scales, often recurved; perianth bristles 6, retrorsely barbellate, often equaling the achene; styles 3-branched for ca 1/3 their length; achenes trigonous, 3.5–5 mm long; minutely cellular-reticulate; 2n = 94. Riparian and marsh communities at 1382 to 1650 m in Box Elder, Cache, and Daggett counties; widespread but irregularly distributed in Canada and the U. S.; 5 (0).

Scirpus maritimus L. Alkali Bulrush. [S. *paludosus* A. Nels.]. Plants perennial, 2–15 dm tall, from robust rhizomes, these commonly bearing tubers; culms sharply triangular, often with concave sides, 3–13 mm thick toward the base; leaves commonly 4–8, borne on the lower 1/2 (3/4) of the culm, mostly all with well-developed folded or flat blades (5) 10–40 cm long or longer and to 1 cm wide, the lower 1 or 2 sometimes reduced to bladeless sheaths; leaflike involucral bracts 2 or 3 (1–4), the longest one usually much exceeding the others and to 34 cm long; scalelike involucral bracts usually few to several, the larger ones commonly with the green midrib exserted as a caudate awn; spikelets 3–25 or rarely more, 10–35 mm long, 5–10 mm thick, all sessile in a compact cluster or 1–4 additional clusters borne on peduncles to 6 cm long; scales tan or light brown, rarely dark brown, scarious, minutely hirtellous on the back, the firm midrib exserted as a mucronate awn, this commonly 1–3 times longer than the apical notch of the scale; perianth bristles few, minutely retrorse-barbellate, ca 1/4–1/2 (3/4) as long as the achenes; styles 2-branched for ca 1/3 the length; achenes lenticular, 2.5–4 mm long, minutely cellular reticulate; n = 55; 2n = 40, 90? Marshes, lakes and ponds, ditches, streams and rivers, mud flats, seeps, and alkaline meadows at 840 to 2075 m in all Utah counties, except Beaver, Daggett, Iron, Morgan, Piute, Wasatch, and Wayne; widespread in the Northern Hemisphere; 80 (xiii).

Scirpus microcarpus Presl Panicled Bulrush. Plants perennial, (3) 6–15 dm tall, from robust rhizomes; culms obtusely triangular, 6–15 (20) mm thick toward the base; leaves well distributed on the stems, the sheaths often reddish purple in part, the blades well developed, flat, 15–60 cm long, 6–20 mm wide; leaflike involucral bracts 3–5, to 27 cm long; inflorescence a compound, umbellate, terminal cyme; spikelets very numerous, 3–6 (8) mm long, borne in small clusters of mostly 5–20 at the ends of rather slender branches or some of the clusters sessile; scales with a green midrib and broad scarious margins, flecked with dark purple and appearing greenish black, the midrib hardly, if at all, exserted as a mucro; perianth bristles 4–6, rather sparsely retrorsely barbellate, subequal to the achene; styles 2-branched for ca 1/2–3/4 their length; achenes lenticular, ca 1 mm long; n = 33; 2n = 67. Banks of streams and canals, and in meadows, at 1370 to 2890 m in Box Elder, Cache, Daggett, Davis, Duchesne, Garfield, Juab, Piute, Salt Lake, San Juan, Sevier, Uintah, Utah, Wasatch, Washington, and Weber counties; Canada south to California and West Virginia; 39 (viii).

Scirpus nevadensis Wats. Nevada Bulrush. Plants perennial, 1–5 dm tall, with creeping rhizomes; culms subterete; leaves few to several, clustered near the base of the culms, the lower ones sometimes reduced to bladeless sheaths, the upper ones with channeled or flat blades to 20 cm long and 1–3 mm wide; leaflike involucral bract solitary, 1.5–7 cm long; scalelike involucral bracts 1 or more, sometimes with the firm midrib exserted as an awn shorter than the body; spikelets (1) 2–10, sessile in a compact cluster, 1–2 cm long; scales shining brown, with white, hyaline, sometimes ciliolate margins, the apex entire, the midrib firmer than the body and greenish, not exserted as a murco; perianth bristles 1–3 (4), retrorsely barbellate, mostly less than 1/2 as long as the achene; styles 2-branched for ca 1/2 their length or less; achenes plano-convex, ca 2 mm long, cellular-reticulate, not at all apiculate. Moist alkaline meadows at ca 1315 m in Juab County; Washington to Saskatchewan, south to California and Wyoming; Argentina; 1 (i).

Scirpus pallidus (**Britt.**) **Fern.** [*S. atrovirens* var. *pallidus* Britt.]. Plants perennial, 4–15 dm tall, from robust rhizomes; culms triangular, 6–15 (20) mm thick toward the base; leaves cauline, the sheaths not marked with reddish purple, the blades well developed, flat, commonly 20–60 cm long, 6–20 mm wide; leaflike involucral bracts commonly 3–5, mostly 3–15 cm long or longer; inflorescence a compound, umbellate, terminal cyme; spikelets very numerous, 3–4 mm long, borne in sessile or pedunculate clusters averaging larger than in *S. microcarpus*, ca (20) 40 or more per cluster; scales with a green midrib and scarious margins flecked with dark purple and appearing greenish black, the midrib exserted as a short mucro; perianth bristles mostly 6, minutely retrorse-barbellate above the middle, subequal to or shorter than the achene; achene trigonous, with the ventral side the widest, ca 1 mm long. Along ditches, canals, and streams at 1370 to 1710 m in Uintah and Utah counties; Washington to Minnesota, south to Texas and Missouri; 11 (vii).

Scirpus pungens **Vahl** Common Threesquare. [*S. americanus* Pers. misapplied]. Plants perennial, 1.3–11.6 dm tall, from robust rhizomes; culms subterete to sharply triangular, the sides not concave, 2–5 (7) mm thick; leaves commonly 2–4, borne on the lower 1/3 of the culms, the lower ones often reduced to bladeless sheaths, the upper ones with blades commonly (5) 8–25 (38) cm long, the blades flat to involute, 0.5–4 mm wide, linear, well developed; leaflike involucral bract solitary, rarely 2, 2–11 cm long, commonly 3–7 times longer than the inflorescence, more or less appearing as a continuation of the culm, accompanied by 1 or 2 smaller, empty, scalelike bract(s), these often blackish purple and usually with an awn from ca 1/3 to as long as the body; spikelets 1–6, sessile or essentially so in a compact cluster, commonly 7–20 mm long; scales yellowish brown to reddish brown, some (especially the lower ones) often blackish purple, rather scarious, the midrib prominent and exserted as a mucro, the mucro equal to or a little longer than the apical notch of the scale; perianth bristles 4–6, retrorsely barbellate; styles 2 or 3; achenes lenticular or trigonous, 2.2–3.3 mm long including the conspicuous apiculus, 1.6–2.3 mm wide; n = 39?; 2n = 74. Tolerant of alkali, along flood plains, ditches, and streams, seeps, springs, margins of ponds and lakes, and in marsh, swamp, and lowland meadow communities at 850 to 2290 m in all Utah counties, except Morgan and Summit; Canada to South America, Europe, Australia, and New Zealand; 95 (xiii). Unfortunately, the name *S. americanus* has generally been misapplied to this species. While plants of the two taxa are apparently closely related, they are strikingly different in the field. Our plants are referable to var. *longispicatus* (**Britt.**) **Cronq.** [*S. americanus* var. *longispicatus* Britt.].

Scirpus supinus **L.** Sharpscale Bulrush. Plants annual, tufted, 2–25 cm tall, from fibrous roots; culms decumbent to erect; leaves few, mostly on the lower 1/2 of the culms, reduced to bladeless or bristle-tipped sheaths, or the upper ones with short blades or the blades rarely elongate and to 5 cm long; principal involucral bract solitary, erect to somewhat incurved, ca 2–10 cm long, more or less simulating a continuation of the culm, smaller bracts borne at the base of the spikelets, these inconspicuous or to 3 cm long; spikelets commonly 2–7, sessile or nearly so or some conspicuously short-pedunculate; scales hyaline except for the greenish, short-excurrent midrib; perianth bristles present or lacking; style branches 3; achenes unequally trigonous, the ventral side the widest, ca 1.4 mm long, dark gray brown to blackish, conspicuously rugulose; n = 14. Shores of drying lakes and ponds, mud flats, and flood plains at 1380 to 1417 m in Millard, Salt Lake, and Uintah Counties; almost cosmopolitan, but rare and irregularly distributed; 4 (0).

Scirpus validus **Vahl** Softstem Bulrush; Tule. Plants perennial, commonly 8–12 dm tall, from robust rhizomes; culms terete, ca 5–10 (15) mm thick; leaves borne on the lower 1/4 of the culms, reduced to bladeless sheaths or the upper one with a mostly erect blade to 9 cm long and 4 mm wide; inflorescence commonly umbellate, subtended by a greenish involucral bract that simulates a continuation of the culm, this shorter or longer than the inflorescence, also with additional inconspicuous bracts; spikelets usually numerous, more or less orange brown or reddish brown, commonly 6–10 (15) mm long, borne singly or 2 or 3 together; scales (2) 2.5–3 (3.5) mm long, with red brown striolae that hardly, if at all, contrast with the red brown back ground color of the scales except on the sometimes pale hyaline margins, the margins entire or shortly fringed-ciliolate to somewhat lacerate; perianth bristles retrorsely barbellate, subequal to the achenes, styles 2-branched for ca 3/4 their length; achenes ca 1.8–2.3 mm long, plano-convex, the margins usually not covered by the scales; 2n = 42. Streams, ditches, canals, margins of ponds and lakes, flood plains of rivers, and marshlands, sometimes emergent in shallow water at 1220 to 1880 m in Box Elder, Carbon, Daggett, Duchesne, Emery, Grand, Kane, Uintah, Utah, and Wayne counties; temperate North America and into tropical America; 23 (iv). A number of our specimens apparently show intergradation with *S. acutus*, but the plants are quite distinct in the field.

GRAMINEAE A. L. Juss.

Grass Family

Contributed and edited by Lois A. Arnow

Perennial or annual herbaceous plants (more or less woody only in giant reeds and bamboos); stems (culms) terete or flattened, simple or branched, with swollen or depressed nodes that often differ from the internodes in color, the internodes hollow or less commonly pithy to solid; leaves alternate, 2-ranked, the lower portion (sheath) closed and tubular or more commonly open with membranous, usually overlapping margins, the lateral margins at the summit occasionally prolonged as small, acute to rounded appendages (auricles), the upper portion (blade) narrowly elongate, parallel-veined, flat or folded to involute, often scabrous due to the presence of sharp-pointed siliceous spicules, the junction of the sheath and blade on the abaxial surface (collar) usually marked by thickened tissues and a different coloration, on its adaxial surface nearly always giving rise to an appendage (ligule) consisting of a band of membranous tissue or a ring of hairs. Flowers (florets) small and inconspicuous, commonly wind-pollinated, bisexual or, in a few species, unisexual (sometimes proliferating as bulblets or some florets sterile or rudimentary), borne in few to numerous spikelets, these sessile or pedicelled at each node of a primary axis (rachis) or secondary axis (branch), the ultimate inflorescence a spike, raceme, or panicle based on the arrangement of the spikelets (not flowers) with the most mature spikelets toward the apex; each spikelet normally consisting of a pair of subopposite bracts (glumes) subtending 1–50 individual florets, these alternating on opposite sides of an axis (rachilla), the first glume (lower) sometimes reduced or lacking, rarely both glumes absent; each floret typically consisting of a pair of subopposite, usually dissimilar bracts, the margins of the outer bract (lemma) partially or completely enfolding the inner one (palea), these subtending 2 (3) minute scales (lodicules), the stamens, and the pistil; lemma either flattened to convex on the back (dorsiventrally compressed) or more or less folded (keeled) along the midnerve (laterally compressed), in some species the base of the lemma (or floret) prolonged and hardened to form a blunt to sharply acute, often bearded structure (callus); palea nerveless or (1) 2 (3) –nerved and thought by some to represent fused members of an ancestral trio of petals, the third member entirely suppressed; stamens usually (1 or 2) 3; pistil 1, the ovary 2 (3) –carpellate, superior, 1-loculed, 1-ovuled, glabrous or occasionally hairy, the styles 2 (3), each with a more or less featherlike stigma. Fruit a caryopsis having the pericarp fused with the seed coat, or sometimes are achene having the pericarp free of the seed coat, the seeds with a hard or soft, fairly large starchy endosperm and a small to large embryo lying toward one side. [Poaceae Barnh.]. Note: See additonal grass references pp. 10–12.

Allred, K. 1982. Describing the grass inflorescence. J. Range Management 35: 672–675.
Sutherland, D. 1986. Poaceae, in Great Plains Flora Association, Flora of the Great Plains. Univ. Press of Kansas. Lawrence, Kansas.

Note: The following key to the genera occasionally employs apparent rather than actual characters. Data pertaining to lemmas are taken from those of the lowermost floret in the spikelet, and that pertaining to ligules from those of the uppermost culm leaf. Nerves of the glumes and lemmas may be more accurately assessed from the ventral than the dorsal surface.

1. Inflorescence a small headlike cluster largely obscured by subtending leaf fascicles; plant a low, stoloniferous annual; leaves typically narrowly white-margined, flat, and 1–3 mm wide .... *Munroa*
— Inflorescence not headlike or, if so, not largely obscured by the leaves; plants not otherwise as above 2

2(1). Spikelets enclosed within a spiny bur or cuplike structure, not usually visible 3
— Spikelets not so enclosed, usually exserted at flowering . 4

3(2). Spikelets enclosed within a spiny bur; plant a widely distributed weedy annual .... *Cenchrus*
— Spikelets enclosed within a lobed, cuplike structure; plant a rare, strongly stoloniferous perennial .... *Buchloe*

4(2). Spikelets dorsiventrally compressed (glumes or lemmas flattened to convex on the back), 1–flowered or in some species 2–flowered (the lower floret then reduced and the lemmas dissimilar), disarticulating below (and sometimes also above) the glumes, in a few species the rachis disarticulating below the sessile spikelet of a pair or trio .... Key 1, p. 685
— Spikelets terete to laterally compressed (the lemma rounded on the back to strongly keeled along the midnerve) or, if dorsiventrally compressed, the spikelets either more than 2–flowered or disarticulating above the glumes 5

5(4). Spikelets sessile, the inflorescence either a solitary terminal spike or made up of 1 to many, usually simple, 1–sided, spikelike branches, these variously arranged along the main axis [in *Hordeum* the lateral spikelets of a trio short-pedicelled; in *Bouteloua curtipendula* the branches very short (spikelike) and pendulous] .... Key 2, p. 686
— Spikelets pedicelled, the inflorescence a spikelike to open panicle (reduced to a single spikelet in *Danthonia unispicata*), the pedicels 0.2–50 mm long (plants with spikelike panicles in which some of the spikelets are sessile will key here) 6

6(5). Spikelets with a single well-developed floret, in several species with 1 or 2 (3) staminate or neuter to rudimentary florets above or below the perfect one, the lemmas then dissimilar .... Key 3, p. 687
— Spikelets with 2 to many florets, rarely the florets replaced by bulblets; lemmas often progressively smaller upward, otherwise similar 7

7(6). Glumes (at least the longer one) consistently equal to or longer than the body of the lowermost lemma, in some species surpassing all of the florets .... Key 4, p. 689
— Glumes (both) averaging shorter than the body of the lowermost lemma, sometimes only slightly so .... Key 5, p. 690

Key 1.

Spikelets dorsiventrally compressed (glumes or lemmas flattened to convex on the back), 1–flowered or occasionally 2–flowered (the lower floret then neuter or staminate and the lemmas dissimilar), disarticulating below (and occasionally also above) the glumes, in a few species the rachis disarticulating below the sessile spikelet of a pair or trio.

1. Spikelets in 3s at each node of a spicate inflorescence, the lateral spikelets often reduced; rachis disarticulating at each node (except in a cultivated annual); glumes awnlike throughout or nearly so ..... *Hordeum*
— Spikelets diversely arranged on the rachis (occasionally in 3s at tips of branches); either the rachis continuous or the glumes not awnlike 2

2(1). Glumes thinner than the lemma of the fertile floret, the first glume sometimes reduced or lacking, the second then resembling the lemma of the lower sterile floret 3
— Glumes firmer than the lemma, in some species only in the sessile spikelet of a pair or trio 9

3(2). Spikelets subtended by 1 to many scaberulous bristles .... *Setaria*
— Spikelets not subtended by bristles 4

4(3). Ligule lacking; spikelets in irregular clusters along one side of the panicle branches; glumes and sterile lemma with short stiff hairs on the nerves . *Echinochloa*
— Ligule present; plants not otherwise as above 5

5(4).	Inflorescence an open to compact panicle, the primary branches regularly rebranched .... *Panicum*		
—	Inflorescence of chiefly simple, 1–sided, spikelike branches 6		
6(5).	Panicle of erect-appressed, spikelike branches mostly less than 5 cm long and racemosely arranged along the main axis 7		
—	Panicle of mostly widely spreading, often elongate, spikelike branches mostly digitately arranged 8		
7(6).	Plants annual; second glume minutely awn-tipped; ligule a ring of hairs .... *Eriochloa*		
—	Plants perennial; second glume and lower lemma abruptly acute to rounded at the apex; ligule membranous .... *Panicum obtusum*		
8(6).	Plants annual; spikelets to ca 1 mm wide; panicle branches 2–16 .... *Digitaria*		
—	Plants perennial, stoloniferous; spikelets 1.2–2 mm wide; panicle branches 2 (4) .... *Paspalum*		
9(2).	Inflorescence typically 2.5–6 dm long; spikelets enveloped in long hairs arising chiefly from the base of the spikelet; culms mostly 2–4 m tall ...... *Saccharum*		
—	Inflorescence less than 2 dm long (occasionally longer in two rare species); spikelets glabrous or hairy, in a few species partially obscured by long hairs arising from the rachis and pedicels; culms diverse in height 10		
10(9).	Inflorescence a compact to open panicle with primary branches rebranched 11		
—	Inflorescence of 1 to numerous, slender, subdigitately arranged racemes or racemelike branches terminating culms and (when present) culm branches .. 12		
11(10).	Spikelets all alike, each subtended by 1 or 2 short-hairy, nonspikelet-bearing pedicels ..... *Sorghastrum*		
—	Spikelets dimorphic, the sessile broader than the pedicelled ones .... *Sorghum*		
12(10).	Culm and its branches terminating in a solitary raceme; pedicelled spikelets typically no wider than the pedicels, the latter ciliate with white hairs to ca 3 (4) mm long .... *Schizachyrium*		
—	Culm and (when present) its branches terminating in (1) 2 to numerous, loosely digitate racemelike branches; pedicelled spikelets diverse, in one species represented only by the pedicels; pedicels copiously pubescent with white or yellow hairs mostly 2–9 mm long 13		
13(12).	Pedicels and rachis segments (at least those toward the apex of the branches) vertically grooved, the central portion membranous; first glume of the sessile spikelet usually pubescent below midlength .... *Bothriochloa*		
—	Pedicels and rachis segments flat to rounded, not grooved down the center; first glume of the sessile spikelet glabrous or scaberulous below midlength . ... *Andropogon*		

Key 2.

Spikelets sessile, 1– to many-flowered; lemmas usually laterally compressed; inflorescence either a solitary terminal spike or made up of 1 to many, usually simple, 1–sided, spikelike branches.

1.	Inflorescence a solitary terminal spike; spikelets arising on alternate sides of or appearing to encircle the rachis 2
—	Inflorescence a 1–sided terminal spike or of (1) 2 to many, usually simple, spikelike branches, the latter in a few species spikeletlike and pendulous; spikelets arising from one side of a more or less flattened rachis 11
2(1).	Glumes longer and broader than the solitary floret, strongly compressed and stiffly ciliate on the keel; spike dense, ovoid to cylindrical .... *Phleum*
—	Glumes not both larger than a solitary floret and ciliate on the keel; spike diverse 3
3(2).	First glume lacking (except in the terminal spikelet); spikelets strongly compressed, placed edgewise to and alternating on either side of the rachis ..... *Lolium*
—	First glume nearly always present, sometimes much reduced; spikelets not otherwise as above 4
4(3).	Spikelets in clusters of 3 at each node of the rachis, the clusters hairy at the base (the hairs mostly 2–5 mm long) and falling from the rachis as a unit ... *Hilaria*
—	Spikelets 1–7 at each node of the rachis, neither hairy at the base nor falling from the rachis as a unit 5
5(4).	Spikelets 3 at each node of a readily disarticulating rachis, (continuous in the annual cultivar, *H. vulgare*), each 1–flowered, the central spikelet of the trio sessile, the lateral ones pedicelled (all sessile in *H. vulgare*) .... *Hordeum*
—	Spikelets 1–7 per node of the continuous or disarticulating rachis, all sessile or nearly so and more than 1–flowered 6
6(5).	Spikelets crowded on a continuous rachis, the internodes averaging less than 3 (0.2–3) mm long near midlength of the spike 7
—	Spikelets diversely arranged on a continuous or disarticulating rachis, the internodes averaging at least 3 mm long near midlength of the spike 8
7(6).	Plants annual; uppermost sheath typically inflated; internodes of the rachis 0.2–1 mm long .. *Eremopyrum*
—	Plants perennial; uppermost sheath not inflated; internodes of the rachis mostly 1–3 mm long . *Agropyron*
8(6).	Plants perennial, tufted or rhizomatous, occupying native or disturbed habitats .... *Elymus*
—	Plants annual or biennial, solitary or in small tufts, occupying disturbed habitats 9
9(8).	Spike cylindrical, to ca 5 mm wide; spikelets about as wide as the rachis and appressed to its concave aspect (sometimes slightly spreading at flowering) ... *Aegilops*
—	Spike flattened or more or less 4–angled in cross section, nearly always more than 5 mm wide; spikelets broader than the rachis, erect to spreading 10
10(9).	Glumes subulate to narrowly lanceolate, 1 (3) –nerved; lemmas coarsely ciliate on keel and exposed margins .... *Secale*
—	Glumes broad, 3– to several-nerved; lemmas not coarsely ciliate .... *Triticum*
11(1).	Inflorescence with spikelike branches in 1 or more whorls, in some species 1 to several branches loosely spaced below the terminal whorl 12
—	Inflorescence either a solitary, 1–sided spike or a panicle with 1 to many, erect to pendulous spikelike branches racemosely arranged along the rachis 15
12(11).	Glumes or lemmas with awns at least 1 mm long 13
—	Glumes and lemmas awnless 14

13(12). Panicle branches 3–20 in 1–5 whorls, each branch typically terminating in a floret; spikelets 2-flowered, the reduced lemma of the upper floret truncate and awned .... *Chloris*
— Panicle branches 2–6 in a single whorl, each branch terminating in a naked, sharp point 1–7 mm long; spikelets 3 (5) –flowered, the florets all well developed, the lemmas awn-tipped to short-awned; known in Utah from a single 1953 collection (Flowers 2317 UT), not persistent, not included in the text .... *Dactyloctenium aegyptium* (L.) Beauv.
14(12). Spikelets 1 (2) –flowered; both glumes 1-nerved; plants perennial, rhizomatous or stoloniferous .... *Cynodon*
— Spikelets 3– to 9-flowered; second glume 3– to 7-nerved; plants annual, usually branched from the base .... *Eleusine*
15(11). Spikelets obovate; glumes typically inflated to some degree, enclosing the floret except for the sharply pointed tip of the lemma .... *Beckmannia*
— Spikelets linear or lanceolate to elliptical; glumes not inflated, the florets often visible 16
16(15). Glumes (at least the conspicuous one) more than 1-nerved 17
— Glumes (both) 1-nerved 19
17(16). Plants perennial, strongly rhizomatous; spikelike branches 2–50 .... *Spartina*
— Plants annual; spike or spikelike branch solitary 18
18(17). Lemmas obtuse to truncate apically; spikelets disarticulating below the glumes .... *Sclerochloa*
— Lemmas acute to mucronate apically; spikelets disarticulating above the glumes .... *Eleusine*
19(16). Spikelike branches 0.5–2 cm long, readily disarticulating near the base and falling as a unit 20
— Spikelike branch or branches 0.5–15 cm long, persistent 21
20(19). Spikelike branches 1–5 per culm, erect to slightly spreading; florets staminate; plants stoloniferous, rare in Utah .... *Buchloe*
— Spikelike branches 2–80 per culm, mostly ascending to widely spreading or pendulous; lowermost florets perfect; plants rarely stoloniferous, general in distribution .... *Bouteloua*
21(19). Spikelets disarticulating below the glumes; glumes nearly always short-ciliate on the prominent keel; plants coarse, strongly rhizomatous, occupying wet places .... *Spartina*
— Spikelets disarticulating above the glumes; glumes diverse, not short-ciliate on the keel; plants mostly slender-stemmed, often annual, occupying diverse habitats 22
22(21). Spikelets 1-flowered, about as broad as and mostly appressed to the concave surface of the branches; branches linear, 2–20 cm long and stiffly spreading from the main axis .... *Schedonnardus*
— Spikelets more than 1-flowered (upper florets often reduced or rudimentary), either broader than or not strictly appressed to the branches; branches nonlinear, 0.5–16 cm long 23
23(22). Spikelets with 2–12 well-developed florets . *Leptochloa*
— Spikelets with 1 perfect floret below 1 or 2 (4) rudimentary ones 24
24(23). Upper florets consisting entirely or in part of 1–3 awns; lemma of the perfect floret awned or unawned; spikelike branches 0.5–5 cm long, usually about equally spaced along the main axis, occasionally solitary .... *Bouteloua*

— Upper floret more or less club-shaped with a short solitary awn; lemma of the perfect floret with an awn 4–12 mm long; spikelike branches 2–16 cm long, usually at least some of them in closely spaced groups .... *Chloris*

Key 3.

Spikelets pedicelled, each with a single perfect or pistillate floret, several species having 1 or 2 (3) staminate or neuter to rudimentary florets above or below the fertile one, the lemmas then dissimilar.

1. Glumes lacking; spikelets strongly flattened, mostly short-pedicelled and appressed to one side of the panicle branches; lemmas minutely ciliate on the keel; plants tall, occupying wet areas .... *Leersia*
— Glumes (at least one) present or, if obsolete, the plants not otherwise as above 2
2(1). Reduced florets regularly present in each spikelet (in one species absent in the lowermost spikelet of each panicle branch) 3
— Reduced florets not regularly present in each spikelet (or not recognizable as such) 13
3(2). Spikelets with 1–3 rudimentary florets above the perfect one 4
— Spikelets with all reduced florets below (or appearing lateral to) the perfect one, in some species the spikelets appearing to have 3 glumes 9
4(3). Inflorescence of 1 to many, simple, spikelike, 1-sided branches 5
— Inflorescence a spikelike to loose panicle, the latter with branches mostly rebranched and not 1-sided ... 6
5(4). Branches of the inflorescence 2–16 cm long, arranged in whorls, occasionally some of the branches solitary or racemosely arranged below the terminal whorl .. .. *Chloris*
— Branches of the inflorescence 0.5–5 cm long, racemosely arranged, occasionally solitary ... *Bouteloua*
6(4). Lowermost lemma termintaing in 9 plumose awns . ... *Enneapogon*
— Lowermost lemma awnless or no more than 3-awned 7
7(6). Lemmas leathery and glossy, the lowermost typically awnless, the upper with a straight to hooked awn to 2 mm long .... *Holcus*
— Lemmas thin or dull, the lowermost awned, the upper diverse 8
8(7). Lowermost lemma with a stout, geniculate awn 10–15 mm long arising from the back ... *Helictotrichon*
— Lowermost lemma with the midnerve free as a plumose awn 3–5 mm long .... *Blepharidachne*
9(3). Lower lemma with an awn 1–2 cm long .... *Arrhenatherum*
— Lemmas awnless 10
10(9). Sterile lemma resembling the second glume in size and texture .... *Panicum*
— Sterile lemma not resembling the second glume 11
11(10). Panicle open, the branches naked at the base; spikelets bronze in color; lower (lateral) florets broader than the fertile one .... *Hierochloe*
— Panicle compact or, if open, the branches spikelet-bearing to the base or nearly so; spikelets whitish, green, or tawny; rudimentary florets much narrower than the fertile one 12
12(11). Glumes with hairs mostly 8–15 mm long .... *Imperata*
— Glumes glabrous to short-hairy .... *Phalaris*

13(2). Glumes (at least one) averaging shorter than the lemma; glumes and lemmas all 1-nerved, neither awned nor awn-tipped 14
— Plants not as above in every detail 17
14(13). Ligule membranous .... *Muhlenbergia*
— Ligule a ring of hairs 15
15(14). Lemmas bearded on the callus with hairs at least half as long as the lemma .... *Calamovilfa*
— Lemmas glabrous or at most scaberulous 16
16(15). Panicle spikelike, 0.3–8 cm long; plants annual and often mat-forming .... *Crypsis*
— Panicle open or, if spikelike, more than 8 cm long; plants perennial or rarely annual but not mat-forming .... *Sporobolus*
17(13). Spikelets disarticulating below (and sometimes also above) the glumes, in a few species disarticulation occurring in the rachis, the pedicels, or at the base of short panicle branches 18
— Spikelets disarticulating above the glumes, not regularly breaking off below the glumes or in the rachis, pedicels, or panicle branches 32
18(17). Rachis of the inflorescence readily disarticulating; spikelets short-pedicelled in groups of 3 at each node; glumes awnlike throughout their length or nearly so .... *Hordeum*
— Rachis of the inflorescence remaining intact; plants not otherwise as above 19
19(18). Spikelets enveloped by hairs 4–12 mm long arising chiefly from the base of the spikelet; panicles 10–60 cm long 20
— Spikelets not enveloped by long hairs, or the hairs not arising chiefly from the base of the spikelet; panicles diverse 21
20(19). Lemmas awned .... *Saccharum*
— Lemmas awnless .... *Imperata*
21(19). Ligule a ring of hairs 22
— Ligule membranous 23
22(21). Plants annual, often mat-forming with decumbent to prostrate culms; sheaths glabrous or minutely ciliate below the summit; panicles spikelike, 0.3–6.5 (8) cm long .... *Crypsis*
— Plants perennial or, if annual, not otherwise as above .... *Sporobolus*
23(21). First glume typically 2-nerved, bifid or awned at the apex; branches of the panicle disarticulating at the base 24
— First glume not both 2-nerved and either bifid or awned at the apex; branches of the panicle not regularly disarticulating 25
24(23). Plants annual; first glume unawned, typically bifid apically; spikelets 1–3 per branch, the pedicels distinctly unequal in length .. *Muhlenbergia depauperata*
— Plants perennial; first glume with 2 awns typically 1–5 mm long; spikelets borne on paired, subequal pedicels .... *Lycurus*
25(23). Glumes averaging shorter than the body of the lemma, conspicuously soft-hairy on the keel .... *Alopecurus*
— Glumes (at least the second) equal to or longer than the body of the lemma, glabrous or hairy on the keel 26

26(25). First glume lacking, the second one firm and completely enclosing the thin lemma; plant a low, strongly rhizomatous and often stoloniferous lawn-grass or weed, rare in Utah .... *Zoysia*
— First glume present; plant either tall or without rhizomes 27
27(26). Glumes with awns averaging more than 1 (1–10) mm long; inflorescence a dense to somewhat interrupted, mostly spikelike panicle 28
— Glumes awnless or with an awn tip to ca 1 mm long; inflorescence diverse 29
28(27). Awn of the glumes noncapillary, to 4 mm long; glumes strongly compressed, the prominent keel ciliate with spreading hairs mostly at least 0.5 mm long .... *Phleum*
— Awn of the glumes hairlike or nearly so, 1–10 mm long; glumes not strongly compressed or, if so, the keel only minutely hairy .... *Polypogon*
29(27). Panicle nodding, 1–3 dm long; glumes narrowly lanceolate, scabrous to short-ciliate on the sharp keel .... *Cinna*
— Panicle erect or plants not otherwise as above 30
30(29). Inflorescence of multiple, 1-sided, spikelike branches; glumes glabrous .... *Beckmannia*
— Inflorescence a terminal, spikelike to more or less interrupted panicle; glumes scabrid-roughened to variously hairy 31
31(30). Spikelets strongly compressed, the glumes hairy on the keel or throughout .... *Alopecurus*
— Spikelets subterete, the glumes roughened to scabrid-puberulent over the back (at 10x magnification) .... *Polypogon semiverticillatus*
32(17). Lemma with a terminal or dorsal awn averaging at least 1 mm long (in some species the awn jointed to the lemma and readily deciduous, the lemmas then terete or nearly so and firmer than the glumes) 33
— Lemmas awnless or with an awn tip averaging less than 1 mm long 45
33(32). Awn of the lemma 3-branched .... *Aristida*
— Awn of the lemma simple 34
34(33). Awn arising from the back of the lemma 35
— Awn terminal on the lemma 40
35(34). Callus bearded, the longer hairs at least 1/3 as long as the lemma .... *Calamagrostis*
— Callus glabrous or bearded with hairs less than 1/3 as long as the lemma 36
36(35). Lemmas 3 mm long or longer 37
— Lemmas less than 3 mm long 38
37(36). Awn of the lemma arising below midlength .... *Calamagrostis*
— Awn of the lemma arising above midlength (spikelets normally 2-flowered, plant will rarely key here) ... . *Trisetum*
38(36). Rachilla of the spikelet prolonged behind the palea and conspicuously hairy ..... *Calamagrostis rubescens*
— Rachilla sometimes minutely prolonged but not long-hairy 39
39(38). Lemma thinner than the glumes, glabrous, the awn rarely as much as 5 mm long .... *Agrostis*
— Lemmas slightly firmer than the glumes, regularly scaberulous above midlength, the awn (4) 6–16 mm long .... *Apera*
40(34). Awns deciduous 41

	Awns persistent 42
41(40).	Body of the lemma with hairs 1–4 mm long, the callus 0.5–1 mm long and obtuse to sharply acute ...... *Stipa*
—	Body of the lemma glabrous or with hairs less than 1 mm long, the callus less than 0.5 mm long and blunt .... *Oryzopsis*
42(40).	Awn jointed to the lemma, the junction between awn and lemma clearly visible unless obscured by hairs arising near the apex of the lemma .... *Stipa*
—	Awn confluent with the lemma, no joint visible; lemma glabrous to scabrous or, if hairy, the hairs not prolonged beyond the apex 43
43(42).	Pedicels distinctly short-hairy (at 5x magnification), mostly about equal to or slightly longer than the body of the lemma; second glume 4.5–7 mm long .... *Festuca dasyclada*
—	Pedicels glabrous to scabrid-puberulent (at 10x magnification), averaging shorter or much longer than the body of the lemma; glumes rarely to 4.5 mm long ... 44
44(43).	Second glume slightly exceeding the floret; lemma with a hairlike, flexuous awn 4–16 mm long .... *Apera*
—	Second glume shorter than the floret or, if longer, the lemma long-hairy, awn-tipped, or with a stout rigid awn .... *Muhlenbergia*
45(32).	Lemmas laterally compressed, firm and glossy, subtended laterally by minute, closely appressed, bristlelike structures that often appear to be pubescence on the lemma .... *Phalaris*
—	Lemmas not as above in every particular 46
46(45).	Lemmas yellow brown, more or less fleshy, conspicuously 3–nerved, apically truncate and erose .. *Catabrosa*
—	Lemmas not as above 47
47(46).	Ligule of hairs 48
—	Ligule a membrane 50
48(47)	Lemmas bearing a basal tuft of hairs to 2 mm long .. .. *Redfieldia*
—	Lemmas glabrous to minutely hairy 49
49(48)	Lemmas awn-tipped to short-awned .... *Muhlenbergia*
—	Lemmas obtuse to acute, neither awn-tipped nor awned .... *Sporobolus*
50(47)	Lemmas 3–nerved, each nerve hairy over at least the lower 1/3, otherwise glabrous; glumes smooth throughout .... *Blepharoneuron*
—	Lemma diversely nerved, glabrous or the hairs not restricted to the nerves; glumes often scabrous, at least on the midnerve 51
51(50).	Callus bearded with hairs 1/4 the length of the lemma or longer .... *Calamagrostis*
—	Callus glabrous or minutely hairy 52
52(51).	Lemmas less than 3 mm long (including awn tip when present) 53
—	Lemmas 3 mm long or more (including awn tip when present) 55
53(52).	Glumes (at least one) averaging shorter than the lemma, rarely both longer and then the lemma conspicuously hairy; spikelets greenish white to dark gray .... *Muhlenbergia*
—	Glumes both longer than the lemma (sometimes barely so); lemma glabrous or minutely hairy on the callus; spikelets green or purple-tinged to dark purple 54
54(53).	Glumes stiffly ciliate on the prominent keel, abruptly tapered at the apex into a stout awn 0.6–4 mm long .... *Phleum*
—	Glumes glabrous or scabrous, awnless or at most minutely awn-tipped .... *Agrostis*
55(52).	Lemmas green and often purple-tinged; rachilla prolonged behind the palea and pubescent with hairs at least 0.5 mm long .... *Trisetum wolfii*
—	Lemmas whitish or more often gray green to black (in one species red purple to brownish); rachilla either not prolonged or not long-hairy .... *Muhlenbergia*

Key 4.

Spikelets pedicelled, 2– to many-flowered, disarticulating above or below the glumes; florets perfect or occasionally unisexual; lemmas variously compressed, all similar as to shape and texture (except in *Hierochloe*); at least the longer glume equal to or exceeding the body of the lowermost lemma.

1.	Lemmas awnless or with an awn tip not more than 1 mm long 2
—	Lemmas (at least 1 per spikelet) awned, the awns terminal or dorsal, averaging more than 1 mm long . 15
2(1).	Ligule a ring of hairs, sometimes with a short membranous base; lemmas pubescent 3
—	Ligule membranous; lemmas glabrous or pubescent . 5
3(2).	Lemmas more than 3–nerved; plants annual . *Schismus*
—	Lemmas 3–nerved; plants perennial 4
4(3).	Panicle to 4.5 cm long, about as broad as long; leaf blades rarely as much as 6 cm long; lemmas awn-tipped .... *Erioneuron*
—	Panicle 4.5–20 cm long, longer than broad; leaf blades mostly 3–25 cm long; lemmas usually shallowly notched, not awn-tipped .... *Tridens*
5(2)	Glumes 1.6–3 cm long; plants annual, cultivated or escaped .... *Avena fatua* var. *sativa*
—	Glumes no more than 1.4 cm long; plants either perennial or native 6
6(5)	Lemmas typically more than 5–nerved, the nerves prominent; culms either bulbous-based or the spikelets 6–24 mm long and falling entire. ...... *Melica*
—	Lemmas no more than 5–nerved or the nerves faint (lemmas obscured by the long hairs in *Arundo*); culms not bulbous-based; spikelets either disarticulating above the glumes or less than 6 mm long 7
7(5).	Spikelets ovate; glumes soon turning bronze-membranous throughout; florets subsessile, the narrow central floret often obscured by the broad lateral ones .... *Hierochloe*
—	Spikelets lanceolate to elliptical or the glumes not both bronze in color and membranous, the florets not dimorphic as above 8
8(7).	Glumes markedly dissimilar, the first linear, the second oblanceolate to obovate, 3–4 times as wide as the first; spikelets disarticulating below the glumes .... *Sphenopholis*
—	Glumes similar in shape or not dissimilar as above; spikelets disarticulating above the glumes 9
9(8).	Lemmas pubescent with hairs 6–9 mm long ... *Arundo*
—	Lemmas glabrous or pubescent with hairs less than 6 mm long 10
10(9).	Rachilla pubescent with hairs to ca 1 mm long; glumes equal or nearly so, averaging more than 5 (4.5–8.5) mm long .... *Trisetum wolfii*
—	Rachilla glabrous to minutely hairy; glumes often unequal, at least one less than 5 mm long 11

11(10). Spikelets to 2.8 mm long, whitish to pale or dark gray (spikelets typically 1–flowered, plant will rarely key here) .... *Muhlenbergia asperifolia*
— Spikelets either at least 3 mm long or not whitish to dark gray 12

12(11). Panicle spikelike or nearly so, the rachis and pedicels of the panicle minutely soft-hairy throughout (at 10x magnification); paleas shiny-hyaline throughout, prominent at flowering .... *Koeleria*
— Panicle open or rarely spikelike, the rachis and pedicels glabrous to scabrid-puberulent; paleas not hyaline throughout, at least the nerves green 13

13(12). Lemmas minutely awn-tipped .... *Festuca*
— Lemmas abruptly to sharply acute but not awn-tipped 14

14(13). Panicle of simple, spikelike, 1–sided branches; second glume 1–nerved; plants reported only from Washington County .... *Leptochloa filiformis*
— Panicle of regularly rebranched branches; second glume 3– to 5–nerved; plants general in distribution .... *Poa*

15(1). Awn arising from the tip of the lemma or produced in the sinus of a mostly bilobed apex 16
— Awn arising from the back of the lemma, below the entire to toothed or lobed apex 22

16(15). Lemmas terminating in 9 plumose awns ... *Enneapogon*
— Lemmas with fewer awns 17

17(16). Lemmas pubescent with hairs 6–9 mm long; panicle 20–70 cm long .... *Arundo*
— Lemmas glabrous or pubescent with hairs less than 6 mm long; panicles often less than 20 cm long 18

18(17). Lemmas prominently 3–nerved; ligule a ring of hairs 19
— Lemmas either more than 3–nerved or the nerves obscure; ligule diverse 20

19(18). Spikelets 4–flowered; uppermost lemma reduced to a 3–awned rudiment, the lower lemmas lobed to midlength or below, the midnerve free as a plumose awn .... *Blepharidachne*
— Spikelets mostly (4) 5– to 18–flowered; uppermost lemma smaller than those below but not rudimentary, the lower lemmas entire or lobed for less than half their length, the awn not plumose .... *Erioneuron*

20(18). Awn (at least the central one) flattened near the base; spikelets 9–24 mm long; ligule a ring of hairs .... *Danthonia*
— Awn terete or nearly so; spikelets often less than 9 mm long; ligule membranous 21

21(20). Awns avaeraging more than 5 (4–10) mm long; panicle branches glabrous or scabrous to puberulent ... . *Bromus*
— Awn 1–5 mm long; panicle branches typically short-hairy .... *Festuca dasyclada*

22(15). Glumes 1.6–3 cm long; plants coarse annuals of the valleys .... *Avena*
— Glumes less than 1.6 cm long; plants diverse 23

23(22). Florets dissimilar, at least as to the awns 24
— Florets similar 26

24(23). Lower floret typically awnless; upper floret with a hooked or occasionally straight awn to ca 2 mm long .... *Holcus*
— Lower floret with a geniculate awn; upper floret awnless or with a straight, twisted, or geniculate awn usually more than 2 mm long 25

25(24). Plants to 2 dm tall, rare, restricted to high elevations .... *Helictotrichon*
— Plants (6) 7–18 dm tall, occupying disturbed sites from low to moderately high elevations .... *Arrhenatherum*

26(23). Lemmas at maturity smooth and shiny, typically blunt and minutely toothed to erose apically, the awn arising near midlength or below .... *Deschampsia*
— Lemmas dull, often scaberulous, acute and often minutely bristle-toothed apically, the awn diverse .. 27

27(26). Callus bearded with hairs more than 1 mm long or the awn arising below midlength of the lemma (florets typically 1–flowered, plant will rarely key here) ... . *Calamagrostis*
— Callus glabrous or bearded with hairs no more than 1 mm long, the awn arising above midlength of the lemma .... *Trisetum*

Key 5.

Spikelets pedicelled, 2– to many-flowered, disarticulating above or rarely below the glumes; florets perfect or occasionally unisexual, rarely replaced by bulblets; both glumes averaging shorter than the body of the lowermost lemma.

1. Plants mostly 2–4 m tall, occurring in low wet places; panicles 1.5–5 dm long; spikelet rachilla with spreading hairs exceeding the narrowly lanceolate lemmas .... *Phragmites*
— Plants either less than 2 m tall or not otherwise as above 2

2(1). Spikelets in dense, 1–sided clusters at the tips of otherwise naked branches; glumes awn-tipped, the keel coarsely ciliate to occasionally merely scabrous; sheaths closed at least half their length, often compressed and keeled .... *Dactylis*
— Plants not as above in every particular 3

3(2). Lemmas (1 or more per spikelet) with an awn more than 0.5 mm long 4
— Lemmas awnless or with an awn tip no more than 0.5 mm long 11

4(3). Spikelets 2–flowered; lemma of the lower floret with a dorsifixed, geniculate awn mostly 1–2 cm long; lemma of upper floret awnless or with a terminal awn shorter and more slender than that of the lower floret .... *Arrhenatherum*
— Spikelets more than 2–flowered or the awns of the lemmas not as above 5

5(4). Lemmas prominently 3–nerved, the lateral nerves and usually the midnerve hairy (often near the base), otherwise glabrous, the awn 0.5–3.5 mm long ... 6
— Lemmas either more than 3–nerved or not with pubescence restricted to the nerves, the awn variable in length 7

6(5). Panicle 7–40 cm long; plants of mostly wet places .. .. *Leptochloa*
— Panicle to 4.5 cm long; plants of dry places . *Erioneuron*

7(5). Sheaths of culm leaves closed at least half their length, sometimes mechanically split and then the margins not overlapping; lemmas 6–17 mm long 8
— Sheaths of the culm leaves open throughout or nearly so, the margins typically overlapping; lemmas in some species less than 6 mm long 9

8(7). Callus of the lowermost lemma bearded with hairs 1–3 mm long .... *Schizachne*
— Callus glabrous or minutely hairy .... *Bromus*

9(7). Awn of the lemma arising 0.5–2.5 mm below the entire to bristle-toothed apex; rachilla of the spikelet pubescent with hairs mostly 0.5–1 mm long .. *Trisetum*

— Awn of the lemma terminal or arising less than 0.5 mm below the acute apex; rachilla of the spikelet glabrous or the hairs minute 10

10(9). Inflorescence a usually glistening, spikelike panicle; spikelets 2 (3–5) -flowered; glumes and lemmas strongly keeled, broadly membranous-margined, the awns to ca 1 mm long .... *Koeleria*

— Inflorescence not spikelike; spikelets 2– to 20–flowered; glumes and lemmas mostly rounded on the back, not broadly membranous-margined, the awns often more than 1 mm long .... *Festuca*

11(3). Pedicels hairlike, mostly 1–5 cm long, much longer than the 4–8 mm long spikelets; rhizomatous perennial of sand dunes .... *Redfieldia*

— Pedicels not both long and hairlike or the spikelets less than 4 mm long; plants of diverse habitats 12

12(11). Lemmas prominently 3–nerved, the lateral nerves sometimes very near the margins 13

— Lemmas either more than 3–nerved or the nerves not regularly conspicuous 18

13(12). Lemmas averaging more than 6 (6–14) mm long, glabrous or, if puberulent, the hairs not confined to the nerves 14

— Lemmas averaging less than 6 (2–6) mm long or rarely to 6 mm, the 3 nerves then conspicuously hairy 15

14(13). Spikelets to 11 mm long; panicles usually nodding .... *Festuca sororia*

— Spikelets 12–40 mm long; panicles usually erect .... *Bromus*

15(13). Lemmas firm to somewhat fleshy, typically yellow brown, erose to irregularly toothed at the truncate apex; plants rhizomatous perennials of wet places .. .. *Catabrosa*

— Lemmas not as above; plants either annual or without rhizomes, occupying diverse habitats 16

16(15). Ligules membranous; sheaths glabrous or rarely long-hairy throughout .... *Leptochloa*

— Ligules of short hairs, sheaths often ciliate with longer hairs lateral to the ligule 17

17(16). Glumes 1–3 mm long; lemmas with nerves glabrous or scaberulous .... *Eragrostis*

— Glumes (the longer one) 4–6 mm long; lemmas with nerves hairy .... *Tridens*

18(12). Glumes dissimilar, the first linear, the second oblanceolate to obovate and 3–4 times as broad as the first; lemmas obscurely nerved, glabrous or scaberulous .... *Sphenopholis*

— Glumes similar in shape or not dissimilar as above; lemmas diverse 19

19(18). Lemmas no more than 3 (1–3) mm long 20

— Lemmas averaging more than 3 (3–17) mm long 24

20(19). Lemmas 1.2–2 mm long, 3–nerved, whitish to dark gray green, occasionally purple-tinged; plant a rhizomatous perennial occurring below 2135 m (spikelets normally 1–flowered, plant will rarely key here) .... *Muhlenbergia asperifolia*

— Lemmas to 3 mm long, often more than 3–nerved; plants otherwise diverse 21

21(20). First glume more than 2 mm long or, if shorter, the lemmas compressed and keeled, glabrous to conspicuously hairy, 3– to 5–nerved, sometimes obscurely so .... *Poa*

— First glume averaging less than 2 (0.4–2.2) mm long; lemmas rounded on the back, glabrous or inconspicuously hairy near the base, 5– to 7–nerved, obscurely or prominently so 22

22(21). Lemmas obscurely nerved .... *Puccinellia*

— Lemmas prominently (5) 7– to 9–nerved 23

23(22). Sheaths closed to the top or nearly so; second glume 1–nerved; lemmas mostly obtuse to rounded apically .... *Glyceria*

— Sheaths open throughout, the membranous borders overlapping; second glume 3–nerved; lemmas mostly truncate apically .... *Puccinellia pauciflora*

24(19). Inflorescence a spikelike panicle (the short branches often spreading at flowering), the rachis and pedicels minutely soft-hairy (at 10x magnification); palea shiny-hyaline throughout, usually prominent at flowering .... *Koeleria*

— Inflorescence diverse but, if spikelike, the rachis glabrous or scabrous, not minutely soft-hairy; plants not otherwise as above in every particular 25

25(24). Rachilla pubescent with hairs to ca 1 mm long; lower lemma 4–6 mm long, keeled; anthers 0.6–1.8 mm long .... *Trisetum wolfii*

— Rachilla glabrous to minutely hairy; plants otherwise diverse 26

26(25). Sheaths of the culm leaves closed about half their length or more, occasionally mechanically split, the margins then not overlapping 27

— Sheaths of culm leaves regularly open more than half the length to throughout, the margins typically overlapping 31

27(26). Lemmas mostly 7– to 15–nerved 28

— Lemmas with fewer than 7 nerves or the nerves obscure 30

28(27). Lemmas to ca 4 mm long; plants of wet places .... *Glyceria borealis*

— Lemmas 6–18 mm long; plants of mostly mesic to dry sites 29

29(28). Culms bulbous at the base or the spikelets borne on pubescent pedicels and disarticulating below the glumes .... *Melica*

— Culms fibrous-rooted; spikelets not borne on hairy pedicels or, if so, disarticulating above the glumes . ... *Bromus*

30(27). Spikelets at maturity averaging more than 12 (10–50) mm long; lemmas at least 6 mm long; ovary or fruit bearing an apical tuft of short hairs .... *Bromus*

— Spikelets at maturity averaging no more than 12 (2–12) mm long; lemmas often less than 6 mm long; ovary or fruit lacking an apical tuft of hairs .... *Poa*

31(26). Panicle 1–sided, spikelike (occasionally branched near the base), to 4 (5) cm long and 12 mm wide; spikelets disarticulating below the glumes and falling entire; plant a low, glabrous, weedy annual .... *Sclerochloa*

— Panicle either not spikelike or not secund, often more than 5 cm long and 12 mm wide; spikelets disarticulating above the glumes; plants otherwise diverse ... 32

32(31). Lemmas 7- to 13-nerved, the nerves visible at 10x magnification on all but the most mature florets; florets unisexual; leaf blades at culm base typically scalelike; plants strongly rhizomatous, growing in alkaline-saline soils (occasionally weedy in neutral soils of the valleys) .... *Distichlis*

— Lemmas not more than 5-nerved, or the plants not otherwise as above in every particular 33

33(32). Lemmas distinctly roughened to scabrous throughout (at 10x magnification), firmer than the largely or entirely membranous glumes, (4.5) 5–10 mm long; florets typically unisexual, often subterete; panicle branches and pedicels smooth or nearly so ... *Leucopoa*

— Lemmas not both scabrous throughout and conspicuously firmer than the glumes, variable in length; florets diverse; panicle branches smooth or scabrous to hairy 34

34(33). Lemmas acute to awn-tipped, sometimes minutely so, rounded on the back or, if somewhat keeled, the lemmas often averaging more than 6 (6–10) mm long; sheaths in some species with well-developed auricles .... *Festuca*

— Lemmas rounded to acute but not awn-tipped at the apex, compressed and keeled, or if rounded on the back, rarely averaging more than 6 (2–7) mm long; sheaths lacking auricles (in *P. bulbosa* the florets replaced by bulblets) .... *Poa*

Aegilops L.

Annuals; sheaths open, with or without auricles; blades usually flat; ligule membranous, short; inflorescence a spike, the rachis disarticulating either at the nodes or near the base and then the spike falling entire; spikelets 2- to 8-flowered, solitary and sessile at each node, terete, and fitting closely into a shallow concavity in the rachis, the lowermost 1 or 2 spikelets vestigial; glumes oblong to obovate, leathery to hardened, subequal, more than half as long as the opposing lemma, flat to rounded on the back, many nerved, awned or awnless; lemma oblong to lance-oblong, usually rounded on the back, 5- to 13-nerved, awned or awnless; callus very short, obtuse; palea subequal to the lemma; stamens 3; caryopsis free of or adherent to the lemma and palea; embryo to about half as long as the grain; $x = 7$.

Aegilops cylindrica Host Jointed Goatgrass. [*Triticum cylindricum* (Host) Cesati]. Annual; culms mostly 2–6 dm tall, erect or geniculate at the lower nodes; lower sheaths sparsely hairy, the upper glabrous with auricles small or lacking; blades flat to somewhat involute, 2–5 mm wide, often sparsely hairy; ligule to ca 0.5 mm long; spike cylindrical, 4–12 cm long, 3–5 mm wide, disarticulating at the base and ultimately between the florets; spikelets 2- to 5-flowered, sessile and solitary at each node of the rachis, subequal to the internodes, terete and strongly appressed to the more or less concave joint of the rachis or occasionally slightly spreading at flowering; glumes hardened, 6–10 (14) mm long, mostly 6- to 9-nerved, the apex laterally toothed or awned; awns of the terminal spikelet mostly 3–8 cm long, those of the lower spikelets 0.1–2 cm long; lemma membranous below, hardened above, rounded on the back, 5- to 7-nerved, those of upper spikelets with awns mostly 3–7 cm long, those of lower spikelets with awns to ca 2 cm long; $2n = 28$. Disturbed sites, especially along roadsides and in fallow fields at 1070 to 2130 m in Box Elder, Cache, Davis, Duchesne, Garfield, Iron, Juab, Kane, Millard, Salt Lake, San Juan, Sanpete, Tooele, Uintah, Utah, Wasatch, Washington, and Weber counties; native to Eurasia, introduced to the U. S. about 1930 and now widely established; 51 (vi). According to Tsvelev (1984), jointed goatgrass is a good source of forage.

Agropyron Gaertner

Perennials; culms solitary or tufted; sheaths with or without auricles, those of the cauline leaves open for more than 2/3 their length, those of the basal leaves closed throughout or nearly so; blades flat to involute; ligule membranous; inflorescence a spike, the rachis continuous, the internodes much shorter than the spikelets; spikelets 3- to 10-flowered, sessile and solitary at each node of the rachis, strongly overlapping to widely spreading, compressed, disarticulating above the glumes and between the florets; glumes lanceolate to lance-ovate, somewhat asymmetrical, unequal, 1- to 4-nerved, often obscurely so, typically keeled, acute or short-awned; lemma lanceolate to lance-oblong, 5-nerved, laterally compressed and usually keeled, with or without a short awn; callus to 0.2 mm long, rounded, glabrous or minutely bearded; palea equal or subequal to the lemma, keeled on each of the 2 nerves, 2-toothed at the apex; stamens 3; caryopsis ellipsodial, typically somewhat adherent to the palea and often to the lemma, the embryo 1/6–1/2 as long as the grain; $x = 7$. The tribe Triticeae, of which *Agropyron* is a member, consists of a series of closely allied, interbreeding taxa about whose generic alignments considerable disagreement exists. Barkworth et al. (1983) and Dewey (1984), on the basis of genome homology as revealed by the level of chromosome pairing in artificially produced hybrids, would redistribute among seven genera (*Elytrigia*, *Leymus*, *Pascopyrum*, *Pseudoroegneria*, *Psathyrostachys*, *Thinopyrum*, and a redefined *Elymus*) those species traditionally included in *Agropyron*, *Elymus*, and *Sitanion*. In applying the genomic method of classification to the Triticeae of North America, however, Barkworth and Dewey (1985) find it necessary to describe plants belonging to the genus *Leymus* as having one set of characters when growing inland and a diametrically opposite set when growing along the coast. Thus, a classification based on genomic data violates the concept that a genus is composed of taxa universally recognizable by a unique cluster of characters. I have followed the treatment of Gould (1947) and the persuasive logic of Estes and Tyrl (1982) in adding *Sitanion* and all of *Agropyron* except the crested wheatgrasses to the genus *Elymus*. A guide to some of the more common hybrids produced by the often freely interbreeding members of the tribe Triticeae is provided following the key to the genus *Elymus*.

Barkworth, M. E. and D. R. Dewey. 1985. Genomically based genera in the perennial Triticeae of North America: Identification and membership. Amer. J. Bot. 72: 767–776.

_____. D. R. Dewey, and R. E. Atkins. 1983. New generic concepts in the Triticeae (Gramineae) of the Intermountain Region: Keys and comments. Great Basin Naturalist 43: 561–572.

Dewey, D. R. 1982. Genomic and phylogenetic relationships among North American Triticeae. *In*: J. R. Estes et al. (eds.), Grasses and grasslands, pp. 51–88. Univ. Oklahoma Press, Norman.

_____. 1983a. New nomenclatural combinations in the North American perennial Triticeae (Gramineae). Brittonia 35: 30–33.

_____. 1983b. Historical and current taxonomic perspectives of *Agropyron*, *Elymus* and related genera. Crop Sci. 23: 637–642.

_____. 1984. The genomic system of classification as a guide to intergeneric hybridization within the perennial Triticeae. *In*: J. P. Gustafson (ed.), Gene manipulation in plant improvement, pp. 209–280. Plenum Press, New York.
Estes, J. R. and R. J. Tyrl. 1982. The generic concept and generic circumscription in the Triticeae: an end paper. *In*: J. R. Estes, R. J. Tyrl, and J. N. Brunken (eds.), Grasses and Grasslands, pp. 145–164. Univ. Oklahoma Press, Norman.
Gould, F. W. 1947. Nomenclatural changes in *Elymus* with a key to the California species. Madroño 9: 120–128.
Stebbins, G. L., J. I. Valencia, and R. M. Valencia. 1946. Artificial and natural hybrids in the Gramineae, tribe Hordeae. I. *Elymus*, *Sitanion*, and *Agropyron*. Amer. J. Bot. 33: 338–351.

Agropyron cristatum (L.) Gaertner Fairway or Crested Wheatgrass. [*Bromus cristatus* L.; *A. desertorum* (Link) Schultes; *A. fragile* (Roth) Candargy; *A. pectiniforme* R. & S.; *A. sibiricum* (Willd.) Beauv.]. Perennial; culms tufted, (1) 2–10 dm tall, erect or geniculate at the lower nodes, sheaths glabrous or the lower ones sparsely hairy, often with well-developed auricles; blades flat or loosely involute upon drying, 2–5 (7) mm wide, entirely glabrous or scaberulous to short-hairy; ligule to ca 1 mm long; spike narrowly oblong to lanceolate or oval, (1.5) 2–10 (15) cm long, 5–23 mm wide, the rachis typically puberulent, the internodes mostly 1–3 mm long; spikelets 3- to 10-flowered, 5–15 mm long (excluding awns), strongly ascending and closely overlapping to widely divergent; glumes thin to somewhat thickened, subequal or unequal, ovate to lance-elliptical, (2) 3–6 (7) mm long, asymmetrical, the midnerve often prolonged to a short awn, glabrous or the keel scabrous to sparsely ciliate; lemma 4–8.5 mm long, glabrous or occasionally short-hairy, acute to more often awn-tipped or awned, the awn to 3 (4) mm long; palea scabrous on the keels; anthers 2.5–4 mm long; 2n = 14, 28, 42. Usually in disturbed or revegetated sites along roadways, on open hillsides, and in salt desert shrub, sagebrush, pinyon-juniper, mountain brush, and ponderosa pine communities at 910 to 2740 m in all Utah counties except Summit; native to Eurasia, introduced to North America from Siberia in 1898, now widely distributed in western Canada, much of the western U. S., and the Great Plains; 135 (v). Various treatments have been proposed whereby at least three species may be supposedly recognized in contemporary material, but of the three only two are widely planted. The following key may be helpful in separating these two cultivars, especially in the field:

1. Spikelets ultimately divergent from the rachis at a wide angle; culms often less than 4 dm tall, typically variable in height and erect or more often geniculate at the lower nodes; leaves both basal and cauline ....
 .... *A. cristatum* (fairway wheatgrass)
— Spikelets at maturity strongly ascending; culms often more than 4 dm tall, typically about equal in height and strictly erect; leaves chiefly basal; plants adapted to slightly drier sites than *A. cristatum* ....
 .... *A. desertorum* (standard wheatgrass)

Agropyron cristatum (sensu amplo), is known to hybridize in nature with *Elymus repens* and *E. trachycaulus* (see guide to hybrids p. 722). All strains of crested wheatgrass, being long-lived and resistant to both cold and drought, are useful in arid areas of the Intermountain Region for hay, pasture, range seeding, and erosion control. Standard wheatgrass can be grown where the mean annual precipitation is as low as 23 cm, whereas fairway requires a minimum of 30 cm. Both strains are easily established and in favorable years make good fall regrowth (Thornburg 1982). The species is moderately alkali-tolerant and does well on all but hard clay and coarse, sandy soils. Crested wheatgrass provides good spring forage but is not as desirable on summer range as other introduced wheatgrasses (Vallentine 1961).

Agrostis L.

Annuals or perennials; sheaths glabrous, open to the base or nearly so; ligule membranous; inflorescence a spikelike to open panicle; spikelets laterally compressed to terete, 1-flowered or rarely a few spikelets within a panicle 2 (3) -flowered, the rachilla disarticulating above the glumes and occasionally prolonged behind the palea as a short stub or bristle; glumes equal or unequal, lanceolate to ovate, obtuse to awn-tipped, the first exceeding the second, 1–nerved, the second 1 (3) –nerved, typically equal to or longer than the lemma; lemma thinner than or occasionally similar to the glumes in texture, whitish to dark purple, somewhat compressed laterally, glabrous or the upper portion scabrous, obscurely 3– to 5–nerved, acute to truncate and often minutely erose apically, occasionally with a straight to bent awn arising from the back; callus glabrous or laterally bearded with minute tufts of hair; palea hyaline, shiny, well developed to rudimentary or lacking; stamens 3; caryopsis free within the lemma and palea, broadly ellipsoidal, the embryo less than half as long as the grain; x = 7. Difficulties in identifying taxa in *Agrostis* may be due in part to the frequency of interspecific hybridization, ample evidence for which exists in our populations.

Bjorkman, S. O. 1960. Studies in *Agrostis* and related genera. Symb. Bot. Upsal. 17: 1–112.
Carlbom, C. G. 1967. A biosystematic study of some North American species of *Agrostis* L. and *Podagrostis* (Griseb.) Scribn. & Merr. Oregon State Univ., Eugene. Dissertation. 223 pp.
Lamson-Scribner, F. and E. D. Merrill. 1910. The grasses of Alaska. Contr. U. S. Natl. Herb. 13: 47–92.
Philipson, W. R. 1937. A revision of the British species of the genus *Agrostis*. J. Linn. Soc. Bot. 51: 73–151.
Simpson, D. R. 1967. A study of species complexes in *Agrostis* and *Bromus*. Univ. Washington, Seattle. Dissertation. 88 pp.
Swallen, J. R. 1948. *Agrostis variabilis* Rydb., a valid species. Leafl. W. Bot. 5: 123–125.
Widen, K. 1971. The genus *Agrostis* L. in eastern Fennoscandia. Taxonomy and distribution. Fl. Fennica 5: 1–209.

1. Palea lacking or less than half as long as the lemma 2
— Palea at least half as long as the lemma 5
2(1). Panicles narrow and green or occasionally purple-tinged, the branches scabrous and mostly spikelet-bearing to the base or nearly so; glumes long-tapered and distinctly longer than the lemma; plants (1) 2–15 dm tall; leaf blades flat, 1.5–11 mm wide .... *A. exarata*
— Panicles linear to open and diffuse, often purple-tinged to dark purple, the branches smooth or scabrous and often naked near the base; glumes diverse; plants to 7 dm tall; leaf blades flat to involute, 0.5–2 mm wide 3
3(2). Plants with any combination of characters cited in the leads in couplet 4 (the name a catchall for various indeterminate elements) .... *A. idahoensis*
— Plants matching in all details one of the leads in the following couplet 4
4(3). Panicles (8) 10–30 cm long, the branches mostly at least 5 cm long, naked for more than half their length, mostly strongly scabrous and rebranching above midlength, at maturity widely spreading to deflexed; glumes typically unequal, long-tapered to sharply acute or aristate tips, averaging at least 0.5 mm longer than the lemmas ...
 .... *A. scabra*

— Panicles 2–10 cm long, the branches to ca 3 cm long, often naked for less than half their length, typically smooth or sparingly scabrous, rebranching at or below midlength, at maturity erect to spreading; glumes mostly equal or very nearly so, often abruptly acute, averaging less than 0.5 mm longer than the lemmas .. .. *A. variabilis*

5(1). Anthers at maturity 1 mm long or longer; plants rhizomatous and often stoloniferous 6

— Anthers 0.5–0.8 mm long; plants with culms solitary or tufted, neither rhizomatous nor stoloniferous 7

6(5). Ligule of upper culm leaves 2.5–7 mm long; pedicels usually densely scabrous; primary panicle branches (at least the lowermost whorl) typically subtended by much shorter ones .... *A. stolonifera*

— Ligule of upper culm leaves 0.5–1.5 (2) mm long; pedicels smooth or sparingly scabrous; primary panicle branches not regularly subtended by much shorter ones .... *A. capillaris*

7(5). Spikelets ultimately disarticulating below the glumes and falling entire, the panicle branches bare in age; glumes usually averaging less than 2 (1–2.2) mm long; plants of valleys and foothills ....
.... *Polypogon semiverticillatus* (q.v.).

— Spikelets disarticulating above the glumes; glumes persistent on the plant, averaging at least 2 (1.5–2.8) mm in length; plants of midmontane or above 8

8(7). Glumes mostly elliptical and barely exceeding the lemmas; lemmas ultimately whitish; panicles 1.5–15 cm long, the branches mostly 0.5–5 cm long, typically more or less angled and scaberulous; leaf blades 1–4 mm wide; culms 0.5–6 dm tall .... *A. thurberiana*

— Glumes mostly lanceolate to ovate and averaging ca 0.2 mm longer than the lemmas; lemmas often purple or purple-tinged; panicle 1–6 cm long, the branches mostly 0.5–2 cm long, typically terete and smooth or nearly so; leaf blades 0.3–2 mm wide; culms 0.3–2.6 dm tall .... *A. humilis*

Agrostis capillaris L. Colonial Bentgrass. [*A. tenuis* Sibth.]. Rhizomatous, sometimes stoloniferous perennial strongly resembling *A. stolonifera* but differing as follows: culms (1) 2–10 dm tall, the blades to 6 mm wide; ligule of upper culm leaves 0.5–1.5 (2) mm long; panicle to 2 dm long, erect to lax, the primary branches in the lowermost whorl not regularly subtended by distinctly shorter ones, the branches and pedicels capillary, glabrous or sparingly scabrous; glumes abruptly acute or gradually tapered to a sharply acute tip; lemma not infrequently awned, the awn arising below midlength, typically bent and often exceeding the glumes; palea 1/2–2/3 as long as the lemma. Native of Eurasia, introduced to the U. S. for use in lawns and for erosion control, grown primarily in the coastal regions of the Northeast and Northwest where Kentucky bluegrass is not well adapted, occasionally included in lawn seed sold locally, readily hybridizing with *A. stolonifera* (Widen 1971). Several specimens with characters suggesting intergradation between *A. capillaris* and *A. stolonifera* have been collected from disturbed areas below 1980 m in Salt Lake and Utah counties. I follow Widen (1971) and Tutin et al. (1980) in applying the binomial *A. capillaris* to this species; 1 (i).

Agrostis exarata Trin. Spike Bentgrass. Tufted perennial, often flowering the first year, occasionally developing slender rhizomes; culms (1) 2–10 (15) dm tall, usually erect; blades flat, (1.5) 2–11 mm wide, mostly 5–30 cm long; ligule (2) 3–9 (14) mm long; panicle linear to narrowly elongate, somewhat loose to dense, (1.5) 3–20 (30) cm long and 0.6–3 cm wide, sometimes interrupted, the branches erect to slightly spreading at flowering, scabrous, mostly spikelet-bearing to the base or nearly so; spikelets green, occasionally purple-tinged; glumes narrowly lanceolate, equal or subequal, 1.8–4 mm long, typically long-tapered to a sharply acute or minutely awn-tipped apex, scabrous on the keel, usually at least some of the glumes within a panicle scaberulous or roughened over the back; lemma 1.2–2.5 (4) mm long, mostly 1/2–4/5 as long as the glumes, occasionally awned from about midlength, the awn to ca 5 mm long; palea ranging from rudimentary to 1/3 the length of the lemma; anthers 0.2–0.6 mm long; $2n = 28, 42, 56$. Streamside, margins of seeps in rocky areas, and in mesic to moist places in pinyon-juniper, aspen, spruce-fir, ponderosa pine, lodgepole pine, and meadow communities at 1520 to 3200 m in Beaver, Box Elder, Cache, Daggett, Duchesne, Emery, Garfield, Grand, Iron, Juab, Millard, Piute, Salt Lake, San Juan, Sanpete, Sevier, Summit, Tooele, Uintah, Utah, Wasatch, and Washington counties; Alaska to Saskatchewan, south throughout the western U. S. into Mexico, east to South Dakota, Nebraska, Oklahoma, and Texas; Siberia; 121 (viii). *Agrostis exarata* is here treated as a polymorphic species without recognizable infraspecific taxa following Simpson (1967). Occasional specimens (e.g., Stanton 594 UT; Genz 8750 BRY) appear to combine the characters of *A. exarata* and *A. scabra*, others (e.g., Arnow 5189 and 6134 UT) those of *A. exarata* and *A. stolonifera*. Spike bentgrass is an important source of montane forage, remaining green and palatable throughout the summer.

Agrostis humilis Vasey Alpine Bentgrass. [*Podagrostis humilis* (Vasey) Bjorkman]. Tufted perennial; culms 0.3–2.6 dm tall, erect; leaves chiefly basal; blades flat to involute, 0.3–2 mm wide, glabrous or scabrous to puberulent; ligule 0.5–3 (4) mm long; panicle linear to narrowly elongate or occasionally somewhat pyramidal, 1–6 cm long and to ca 1.5 cm wide when pressed, the branches mostly 0.5–2 cm long, erect or at flowering spreading, the pedicels often spreading, both branches and pedicels smooth or nearly so throughout; spikelets purple or purple-tinged, rarely green; glumes lanceolate to ovate or less often elliptical, equal or subequal, 1.5–2.5 (2.8) mm long, averaging ca 0.2 mm longer than the lemmas, often gradually tapered to a sharply acute to aristate apex, the margins generally ragged at 30x magnification; lemma typically green or purple-tinged to dark purple, 1.5–2.2 mm long, obscurely to distinctly nerved, the nerves often purple-stained at the obtuse, erose apex, the midnerve occasionally prolonged as a minute awn to ca 1 mm long; palea 2/3–3/4 the length of the lemma, the rachilla either not prolonged behind the palea or bristle-like and to 0.5 (0.7) mm long, rarely bearing a rudimentary floret; anthers 0.5–0.8 mm long; $2n = 14$. Wet meadows and other moist places, in openings in spruce-fir and lodgepole pine communities, and on open, rocky slopes chiefly at 2740 to 3350 m in Daggett, Duchesne, Juab, Salt Lake, Summit, Uintah, and Wasatch counties; Alaska and British Columbia, south throughout most of the western U. S.; 36 (iii). The differences between *A. humilis* and *A. thurberiana*, because they are chiefly quantitative, appear to be related to habitat or to stage of growth. It was found, however, that fruiting plants with their roots essentially entwined could be readily segregated according to the characters provided in the key. Thus, the differences are neither a function of maturity nor of microhabitat, and the reality of two species is upheld. A few herbarium specimens display evidence of intermediacy, however, suggesting that hybridization does occasionally occur.

Agrostis idahoensis Nash Idaho Bentgrass. The binomial *A. idahoensis* has long been employed in umbrella-like fashion to cover a number of otherwise indeterminate elements. The name, which may be readily applicable to a single collection confined to a herbarium sheet, does not appear to apply to populations. In Utah, *A. idahoensis* is reported only from areas where *A. scabra*, *A. exarata*, or *A. variabilis* are also growing, and may represent intergradation among these species. It is relatively infrequent and, to my knowledge, does not form pure stands. The taxon is also reported from British Columbia and Alberta, south throughout the western U. S., as are *A. exarata* and *A. scabra*.

Agrostis scabra Willd. Ticklegrass. [*A. hyemalis* of authors, not (Walter) B.S.P.; *A. hyemalis* var. *tenuis* (Tuckerman) Gleason]. Short-lived perennial; culms solitary or tufted, 0.5–6 (7) dm tall, erect; blades 0.5–2 mm wide, flat to involute, essentially smooth or more often scabrous, occasionally pubescent above with short stiff hairs; ligule (1.2) 2–7 mm long, averaging at least 2 mm long; panicle elliptical to pyramidal, (8) 10–30 cm long and 0.5–20 cm or more wide, the branches to 12 cm long, erect to ultimately widely spreading or deflexed, typically strongly scabrous and rebranching at or above midlength, naked below; spikelets mostly purple at maturity; glumes subequal to unequal, the first longer than the second, 1.5–3 (4) mm long, mostly sharply acute to aristate, scabrous on the midnerve and often laterally roughened near the tip, exceeding the lemma by 1/3–1/2 their length; lemma (1) 1.5–2 mm long, occasionally awned from about midlength, the awn 0.5–2 mm long; palea lacking or rudimentary; anthers 0.3–0.7 mm long; 2n = 28, 42, 43. Openings in mountain brush, aspen, spruce-fir, ponderosa pine, lodgepole pine, and meadow communities, rarely in oak-sagebrush and pinyon-juniper communities, at (1825) 2440 to 3350 m in Beaver, Cache, Carbon, Daggett, Duchesne, Emery, Garfield, Grand, Iron, Juab, Kane, Morgan, Rich, Salt Lake, San Juan, Sanpete, Sevier, Summit, Uintah, Utah, Wasatch, Washington, and Wayne counties; Alaska to Greenland, throughout most of the U. S. and in Mexico; Asia; naturalized in Europe and adventive elsewhere; 188 (xii). *Agrostis scabra* is here maintained as a species distinct from *A. hyemalis* of the eastern U. S. in accordance with recent treatments. *Agrostis scabra* appears to hybridize with *A. stolonifera* and *A. exarata* (q.v.).

Agrostis stolonifera L. Redtop; Carpet Bentgrass. [*A. alba* of authors, not L.; *A. gigantea* Roth; *A. palustris* Hudson]. Rhizomatous and often stoloniferous perennial, sometimes mat-forming; culms 0.2–1.5 m tall, erect or decumbent; blades mostly (1.5) 2–10 (12) mm wide, flat or folded, smooth to strongly scabrous; ligule of upper culm leaves (2) 2.5–7 mm long; panicles narrowly to broadly lanceolate or elliptical, 4–40 cm long and 1–15 cm wide, the branches erect to ultimately widely spreading, subglabrous to strongly scabrous, at least those in the lowermost whorl typically subtended by distinctly shorter branches; spikelets green or reddish purple; glumes narrowly to broadly lanceolate or elliptical, equal or subequal, 1.5–3.5 mm long, averaging ca 0.3 mm longer than the lemma, abruptly acute and sometimes submucronate to short-tapered at the apex, usually scabrous on the midnerve and otherwise smooth to somewhat roughened; lemma 1.2–2.8 mm long, awnless or rarely short-awned from above midlength; palea 1/2–3/4 as long as the lemma; anthers 1–2 (2.3) mm long; 2n = 28–46, 56. Along waterways (often partially or totally submersed in shallow running water), in marshes, wet meadows, and near seeps in greasewood, pinyon-juniper, aspen, and spruce-fir communities at 980 to 3050 m in all Utah counties except Morgan; native to Eurasia and North Africa, introduced to North America as a lawn and pasture grass prior to 1750 (Vallentine 1961) and now circumboreal; 272 (xiii). Some workers treat rhizomatous plants with panicles remaining open after flowering as *A. gigantea*, and stoloniferous ones with panicles contracted after flowering as *A. stolonifera*. In Utah, plants with panicles contracted and others with panicles open after flowering occasionally occur within the same tuft (e.g., Cottam 13470 and Arnow 4699 UT). Although Widen (1971) states that the presence or absence of stolons or rhizomes in this complex is habitat-related, he recognizes two species, separating them, in what appears to be an arbitrary break in a continuum of variation, by the occurrence in *A. gigantea* of rhizomes with more than three scale leaves. I have followed Scoggan (1978) and Sutherland (1986) in assigning *A. gigantea* to synonymy. *Agrostis stolonifera* is known to form hybrids with *A. capillaris* (Widen 1971) and, based on specimens examined, is evidently hybridizing with both *A. scabra* (Harrison 12360 UT; Gierisch 238 and 533 UTC; Goodrich s.n. BRY) and *A. exarata* (Harrison 12680 BRY). According to Tutin et al. (1980) and Tsvelev (1984), *A. stolonifera* also hybridizes with *Polypogon monspeliensis* (q.v.) and possibly with *P. semiverticillatus* (Arnow 5960 UT). Redtop is extensively used in turf culture, especially on golf courses. On moist sites it produces good forage throughout the growing season but is less productive and less palatable than most introduced grasses (Vallentine 1961).

Agrostis thurberiana A. S. Hitchc. Thurber Bentgrass. [*Podagrostis thurberiana* (A. S. Hitchc.) Hulten]. Tufted perennial, often with short rhizomes; culms 0.5–6 dm tall; leaves chiefly basal or, in robust plants, the cauline leaves conspicuous; blades flat, 0.5–4 mm wide and to ca 10 cm long, glabrous or the margins scaberulous; ligule 1–4 mm long; panicle linear to narrowly oblong, 1.5–10 (15) cm long and mostly less than 2 cm wide, sometimes nodding, the branches capillary, mostly 0.5–5 cm long or more, erect to moderately spreading, naked below, the pedicels remaining appressed, the branches and pedicels typically scaberulous; spikelets pale green to purple; glumes mostly narrowly elliptical and equal or subequal, 1.5–2.5 (2.8) mm long, barely exceeding the lemma or occasionally a few of them distinctly longer, mostly abruptly tapered to an acute apex, often smooth on the midnerve, the margins typically smooth at 30x magnification; lemma greenish or ultimately white, sometimes faintly purple-tinged near the apex, 1.5–2.2 mm long, often more or less distinctly nerved, usually appearing acute unless spread out, awnless; palea 2/3–3/4 the length of the lemma, the rachilla often prolonged behind the palea as a bristlelike structure to 0.5 mm long, rarely bearing a rudimentary floret; anthers 0.4–0.7 mm long;

$2n = 14$. Wet meadows and other wet sites above 2620 m in Duchesne, Salt Lake, Uintah, and Wasatch counties; Alaska south throughout most of the western U. S.; 13 (ii).

Agrostis variabilis **Rydb.** Mountain Bentgrass. [*A. rossiae* of authors, not Vasey]. Tufted perennial; culms 0.6–3 (4) dm tall, erect; leaves chiefly basal; blades 0.3–2.5 (3) mm wide, flat to involute, smooth or scaberulous to puberulent; ligule 0.5–2.5 (3) mm long; panicle linear to narrowly elongate, 2–10 cm long and rarely more than 2 cm wide, erect, the branches to ca 3 cm long, erect to somewhat spreading, typically smooth or sparingly scabrous, rebranching at or below midlength; spikelets 1-flowered, green or more often purple-tinged to purple; glumes typically equal or subequal, 1.7–2.5 (3) mm long, slightly longer than the lemma, mostly abruptly acute apically, sometimes sharply so but rarely aristate, glabrous or the midnerve scabrous, sometimes scaberulous lateral to the midnerve near the tip; lemma whitish or some of them lightly purple-tinged, 1.5–2.2 mm long, 3- to 5-nerved, occasionally with a short awn arising from the back at about midlength or above; palea lacking or rudimentary; anthers 0.3–0.7 mm long; $2n = 28$. Mountain meadows, talus, streamside, and other mostly moist sites in spruce-fir and lodgepole pine communities at 2930 to 3570 m in Beaver, Duchesne, Garfield, Grand, San Juan, Summit, Uintah, and Washington counties; British Columbia and Alberta, south throughout most of the western U. S.; 53 (i). For many years *A. variabilis* was confused with *A. rossiae* Vasey, an annual endemic to Yellowstone National Park (Swallen 1948). Occasionally plants within populations of *A. variabilis* resemble *A. borealis* in being awned, but the panicle branches are not widely spreading as in the latter. *Agrostis borealis* is reported for Utah by Hitchcock (1951), but I have seen no Utah materials with both open panicles and consistently awned lemmas.

Alopecurus L.

Annuals or perennials; sheaths open; blades usually flat; ligule membranous; inflorescence a spikelike cylindrical panicle; spikelets 1-flowered, short-pedicelled, flattened, strongly overlapping, lanceolate to elliptical, disarticulating below the glumes; glumes laterally compressed, equal or nearly so, slightly shorter to barely longer than the floret, 3-nerved, the margins often fused near the base, the keel conspicuously ciliate; lemma strongly laterally compressed, the margins free or fused to some degree, awned from the back, the awn included or exserted; palea small or lacking; stamens 3; caryopsis free within the lemma and palea, the embryo 1/4–1/2 as long as the grain; $x = 7$.

1. Glumes sharply acute at the apex, averaging more than 3 (2.5–6) mm long; awn typically exserted 2 mm or more beyond the glumes; anthers 1.5–3.5 mm long 2
— Glumes obtuse at the apex, averaging no more than 3 (1.5–3) mm long; awn included within the glumes or variously exserted; anthers 0.3–1.8 mm long 3

2(1). Glumes pubescent throughout with more or less crinkly hairs, those along the keel mostly 1–2 mm long; panicles 1–4 (5.5) cm long .... *A. alpinus*
— Glumes pubescent chiefly on the keel with hairs less than 2 mm long; panicles 2–10 cm long ..... *A. pratensis*

3(1). Awn included within the glumes or the exserted portion averaging less than 1.5 mm long; anthers 0.5–0.8 mm long .... *A. aequalis*

— Awn exserted, the exserted portion typically averaging more than 2 mm long; anthers various 4

4(3). Anthers 0.3–0.5 mm long; plants annual . *A. carolinianus*
— Anthers 1–2 mm long; plants perennial ... *A. geniculatus*

Alopecurus aequalis **Sobol.** Shortawn Foxtail. Perennial, sometimes flowering the first year; culms 0.8–6 (7) dm tall, erect or geniculate and occasionally rooting at the lower nodes; blades flat, 0.6–6 mm wide, glabrous or scaberulous; ligule (2) 4–8 mm long; panicle spikelike, cylindrical, 1.5–8 (10) cm long and 3–6 (7) mm wide; spikelets strongly flattened, elliptical, 1.5–2.7 (3) mm long and to ca 1 mm wide, pale green, occasionally purple-tinged; glumes fused basally, obtuse apically, ciliate on the keel and often appressed-hairy on the lateral nerves, with or without shorter hairs over the back; lemma lance-ovate, 1.6–2.5 (3) mm long, obtuse, glabrous, the awn delicate, arising above or below midlength of the lemma, mostly 0.5–2.5 mm long, straight or slightly bent, included within the glumes or the exserted portion averaging less than 1.5 mm long; anthers 0.5–0.8 mm long; $2n = 14$. Along waterways, in mostly wet, sometimes saline meadows and in other moist sites (occasionally partly submersed) in aspen, spruce-fir, lodgepole pine, and ponderosa pine communities at 1280 to 3200 m in all Utah counties except Carbon, Morgan, and Tooele; circumboreal, south in the U. S. throughout the West, most of the Great Plains, and east through New England; Argentina; Eurasia; 139 (iv).

Alopecurus alpinus **J. E. Sm.** Alpine Foxtail. Rhizomatous perennial; culms 1–8 dm tall, erect or somewhat geniculate below, solitary or in small tufts; blades usually flat, 1–7 mm wide, glabrous or scabrous; ligule 1–4 mm long; panicle spikelike, ovoid to cylindrical, 1–4 (5.5) cm long and to ca 16 mm wide; spikelets flattened, elliptical, (2.5) 3–5 (6) mm long and to ca 2 mm wide; glumes fused basally, acute apically, the keel and the lateral nerves typically outlined in dark purple, otherwise pale or suffused to some degree with purple, ciliate on the keel and densely pubescent throughout with more or less crinkly hairs 1–2 mm long or longer; lemma broadly elliptical, scaberulous to puberulent near the often purple-tinged apex, glabrous and pale below, the awn arising below or rarely above midlength of the lemma, straight or slightly bent, 3.5–7 mm long, the ultimately exserted portion averaging more than 2 mm long; anthers 1.8–2.8 mm long; $2n = 98$, ca 100, 105, 110, 112, 113. Riparian, mesic to wet meadows, aspen, lodgepole pine, and ponderosa pine communities at (2320) 2740 to 3230 m in Daggett, Summit, and Uintah counties (Uinta Mts.); Alaska to Labrador, south in the Rocky Mts. to Colorado; Eurasia; 27 (0).

Alopecurus carolinianus **Walter** Carolina Foxtail. Annual; culms erect or geniculate at the lower nodes, 1–5 dm tall, solitary or tufted; blades flat, 1–4 (5) mm wide; ligule (1) 3–7 mm long; panicle spikelike, cylindrical, 2–6 (8) cm long and to ca 6 mm wide; spikelets flattened, elliptical, 1.8–2.5 mm long and to ca 1 mm wide, mostly pale green, occasionally with minute purple markings at the tips; glumes equal or nearly so, fused basally, obtuse apically, short-ciliate on the keel and usually hairy to some degree laterally; lemma lance-elliptical, obtuse, glabrous, the awn arising below midlength, typically geniculate, 3–5 mm long, the ultimately exserted portion averaging more than 2 (1–3) mm long; anthers 0.3–0.5

mm long; 2n = 14. Wet, often saline meadows and along waterways below 1520 m in Cache and Davis counties; Canada south throughout most of the U. S.; 3 (0).

Alopecurus geniculatus L. Marsh or Water Foxtail. Perennial, sometimes flowering the first year; culms erect or geniculate below, occasionally rooting at the lower nodes, 1–6 dm tall; blades flat, 2–6 (8) mm wide, glabrous or scaberulous; ligule 2–6 (8) mm long; panicle spikelike, cylindrical, 2–8 cm long and to ca 6 mm wide; spikelets strongly flattened, blunt-elliptical, 1.8–3 (3.5) mm long, 1–1.5 mm wide, green and usually purple-tinged; glumes fused basally, obtuse apically, short-ciliate on the keel and often appressed-hairy on the lateral nerves, with or without shorter hairs over the back; lemma lance-elliptical, obtuse, glabrous, the awn arising at or below midlength of the lemma, straight or at maturity geniculate and often twisted, 1.5–4 (6) mm long, the ultimately exserted portion averaging at least 1.5 mm long; anthers 1–2 mm long; 2n = 14, 28. Wet meadows or seeps (sometimes partially submersed) in aspen and spruce-fir communities at 1320 to 3200 m in Cache, Garfield, Iron, Salt Lake, Sevier, Summit, and Wayne counties; native to Eurasia, sporadically adventive from Alaska to Greenland or possibly circumboreal, south in the U. S. through most of the West and the Northeast; 20 (i). Utah plants are unusual in having awns shorter than those of typical *A. geniculatus*. The species is often sympatric with the short-awned *A. aequalis*, and evidence of intergradation, which could explain the shorter awns, exists in specimens combining awns longer than is typical for *A. aequalis* with the very short anthers of that species. Hybrids between *A. geniculatus* and *A. pratensis* have been reported (J. Linn. Soc., Bot. 83: 285. 1981).

Alopecurus pratensis L. Meadow Foxtail. Perennial, occasionally with short rhizomes; culms tufted, erect or occasionally geniculate at the lower nodes, 2–12 dm tall; blades flat, 2–10 mm wide; ligule 1.5–6 mm long; panicle spikelike, cylindrical, (2) 3–10 cm long and to ca 14 mm wide; spikelets flattened, elliptical, (3.5) 4–6 mm long and 1.2–2.5 mm wide; glumes fused basally, sharply acute apically, the keel and the lateral nerves typically outlined in green or purple, the remainder pale or suffused to some degree with purple, ciliate on the keel with hairs to ca 1.5 mm long, otherwise glabrous or the lateral nerves appressed-hairy; lemma elliptical, acute, glabrous or occasionally pubescent on the upper portion of the keel, the awn arising near the base of the lemma, geniculate, 5–8 (10) mm long, the exserted portion 2–6 mm long; anthers 1.5–3.5 mm long; 2n = 14, 28, 42. Mesic to wet sites along roads, in pastures, meadows, and aspen-conifer communities at 1310 to 3230 m in Cache, Carbon, Daggett, Emery, Garfield, Salt Lake, Sanpete, Sevier, Summit, Tooele, Utah, and Wasatch counties; native to Eurasia, introduced to North America in the mid-1800s, now essentially circumboreal, south in the U. S. through much of the West and the Northeast; 44 (vii). Meadow foxtail has been used in Utah since about 1960 for range rehabilitation and erosion control. The closely related *A. arundinaceus* Poir. (creeping meadow foxtail) has reportedly been used for range seeding in Utah, but I have seen no specimens. It differs from *A. pratensis* chiefly in being more strongly rhizomatous and in having short, rarely exserted awns, and truncate rather than acute glumes and lemmas.

Andropogon L.

Perennial; culms erect, simple or branched above the base; sheaths open; blades flat, folded, or loosely involute; ligule membranous; inflorescence a panicle with (1) 2 to several, more or less digitately arranged spikelike branches; spikelets dorsiventrally compressed, 2-flowered, borne in pairs, one sessile, the other short-pedicelled (or lacking and represented only by the pedicel), the rachis disarticulating below the sessile spikelet, the spikelet pair falling as a unit together with the associated pedicel and rachis segment, the latter structures usually rounded on the abaxial side, not medially grooved; sessile spikelet with the lower floret neuter (often rudimentary) and the upper floret perfect or pistillate; pedicelled spikelet well developed to rudimentary or lacking; glumes lanceolate, 2– to several-nerved, at least the first more or less leathery to hardened, exceeding both florets, firmly clasping the often thinner, shorter, more or less keeled second glume; lemma hyaline, that of the upper floret (sessile spikelet) entire or 2–lobed apically, unawned or with a straight or geniculate, twisted awn arising from the apex or from between apical lobes; palea hyaline or lacking; stamens 1–3; caryopsis free of the lemma and palea, remaining enclosed within the glumes, the embryo about half the size of the grain; x = 9, 10. *Andropogon* and the closely related genera *Bothriochloa* nad *Schizachyrium* are treated as a single genus (*Andropogon*) by Sutherland (1986).

Gould, F. W. 1957. New North American andropogons of subgenus *Amphilophus* and a key to those species occurring in the United States. Madrono 14: 18–29.

———. 1967. The grass genus *Andropogon* in the United States. Brittonia 19: 70–76.

1. Sessile spikelets 6–12 mm long, the pedicelled spikelets averaging more than 3 (3–8) mm long .... *A. gerardii*

— Sessile spikelets 3–4.5 mm long, the pedicelled spikelets rudimentary or represented only by a pedicel ....
 .... *A. glomeratus*

Andropogon gerardii Vit. Big Bluestem. [*A. furcatus* Muhl. in Willd.]. Perennial, occasionally with rhizomes; herbage often glaucous; culms usually tufted, 5–20 dm tall, sparingly branched above; sheaths glabrous or occasionally hairy along the margins near the ligule; blades (3) 5–10 mm wide, flat or loosely involute toward the long-tapered filiform tips; ligule of culm leaves 0.5–2.5 mm long, occasionally surpassed by long hairs from the base of the blade, a ligule lacking in blades subtending branches of the inflorescence; panicle of (1) 2–7, usually erect, loosely whorled, spikelike branches, these 4–11 cm long, the rachis internodes and pedicels sparsely to densely pubescent with ascending-spreading, white to yellow hairs mostly 2–6 mm long; sessile spikelet narrowly lanceolate, 6–12 mm long, the glumes dimorphic, the first ultimately hardened, dorsiventrally compressed, narrowly lanceolate, longitudinally concave to deeply grooved, glabrous or more often scabrous to some degree, the second thinner, shorter, and somewhat keeled, the upper floret usually perfect, the lemma awnless or with a geniculate, basally twisted awn to 2 cm long; pedicelled spikelet often broader than the pedicel, mostly 3–8 mm long (excluding the awn tip when present), the upper floret neuter or staminate, the lemma short-awned or awnless; anthers 3–5 mm long; 2n = 20, 40, 60, 80–86,

According to Gould (1975), two intergrading and often sympatric varieties are present in Utah:

1. Hairs of the rachis internodes copious, typically 3–4 mm long and whitish or yellow; rhizomes well developed .. .. A. *gerardii* var. *chrysocomus*
— Hairs of rachis internodes sparse to copious, mostly 1–2 mm long and whitish; rhizomes short or lacking .... A. *gerardii* var. *gerardii*

Var. chrysocomus (Nash) Fern. [A. *chrysocomus* Nash; A. *hallii* Hackel]. Dry sandy sites in desert shrub and pinyon-juniper communities at ca 1400 to 1770 m in Duchesne (a recent roadside introduction), Emery, Garfield, Kane, San Juan, and Wayne counties; Montana and the Dakotas, south to Arizona, New Mexico, Texas, and Mexico; 27 (0).

Var. gerardii Dry to moist sites in creosote bush, live oak, mountain brush, pinyon-juniper, and ponderosa pine communities at 850 to 1890 m in Emery, Garfield, San Juan, Washington, and Wayne counties; Saskatchewan to Quebec, and south through all except the westernmost states of the U. S.; 13 (0).

***Andropogon glomeratus* (Walter) B.S.P.** Bushy Bluestem. [*Cinna glomerata* Walter]. Perennial; herbage often glaucous; culms densely tufted, mostly 7–15 dm tall, erect, typically much branched above; sheaths glabrous or long-hairy on the margins and sometimes near the collar, those of the upper branches ultimately reddish brown; blades 2.5–6 (8) mm wide, flat or folded, often becoming loosely involute toward the long-tapered, filiform tips; ligule a firm membrane to 1 mm long, anterior to and occasionally surpassed by long hairs arising at the base of the blade; inflorescence of mostly paired, slender branches 1.5–3 cm long terminating each of the numerous culm branchlets, and partially enclosed by the subtending sheaths, the whole leafy, broomlike, and mostly 1–3 dm long, rachis internodes and pedicels conspicuously pubescent with ascending-spreading, silvery hairs mostly 5–8 mm or more long; sessile spikelet narrowly lanceolate, 3–4.5 mm long, the glumes firm, the first dorsiventrally compressed and concave on the back, glabrous throughout or scabrous on the upper margins, the lemma of the upper floret with a slender, straight awn 1–2 cm long; pedicelled spikelet rudimentary or represented only by the pedicel; anthers 0.5–1.5 mm long; 2n = 20. Hanging gardens and along streams at 1125 to 1160 m along Lake Powell, in Garfield, Kane, and San Juan counties; throughout the southern U. S., in northern Mexico, and the Atlantic coastal states south to Massachusetts; 8 (0).

***Apera* Adanson**

Annuals; sheaths open; blades flat or involute; ligule membranous; inflorescence a panicle; spikelets flattened to subterete, 1–flowered, disarticulating above the glumes; glumes unequal, lanceolate, the first 1–nerved, the second 3–nerved and slightly exceeding the floret; lemma laterally compressed to subterete, firmer than the glumes, obscurely nerved, bearded on the callus, awned from just below the apex; palea equal to or slightly shorter than the lemma; stamens 3; caryopsis free within the lemma and palea, ellipsoidal, the embryo less than half as long as the grain; x = 7.

***Apera interrupta* (L.) Beauv.** Italian Windgrass. [*Agrostis interrupta* L.]. Annual; culms solitary or tufted, 1–5 (7) dm tall, usually erect; sheaths smooth; blades lax, 1–4 mm wide, flat or involute when dry, glabrous or scaberulous; ligule 1.5–5 mm long; panicle narrowly elongate, (2) 3–20 cm long and 0.3–1.5 cm wide, compact to somewhat interrupted, especially below, the branches erect or slightly spreading, often spikelet-bearing to the base; spikelets 1–flowered, 2–3 mm long, green or purple-tinged; glumes lanceolate, laterally compressed to rounded on the back, membranous-margined, acute to minutely awn-tipped apically, scaberulous, especially toward the apex, the first 1.5–2.2 mm long, 1–nerved, the second 2–2.8 mm long, 3–nerved and slightly exceeding the floret; lemma at maturity firmer than the glumes, lanceolate, 1.6–2.5 mm long, minutely bearded on the callus, laterally compressed to rounded on the back, the margins completely enclosing the palea, scaberulous above midlength, the awn terminal or subterminal, hairlike or nearly so, (4) 6–16 mm long, straight or flexuous, anthers 0.3–0.5 mm long; 2n = 14, 28. On open, mostly sandy sites, in greasewood communities or as a lawn weed below 1525 m in Cache, Salt Lake, and Tooele counties; native to Eurasia, now sporadic throughout much of the U. S. except in the South; South America; 11 (v). *Apera interrupta* is occasionally host to the smut fungus *Tilletia decipiens* (Pers.) Koern.

***Aristida* L.**

Annuals or perennials, tufted or rhizomatous; sheaths open; blades narrow, usually involute; ligule a ring of hairs or a short, ciliate membrane; inflorescence an open to contracted panicle, raceme, or spike; spikelets terete to compressed, 1–flowered, disarticulating above the glumes; glumes equal or unequal, lanceolate, acute to acuminate or awn-tipped, 1– to 3–nerved, the second glume usually slightly exceeding the body of the lemma; lemma at maturity firmer than the glumes, lanceolate, rounded on the back with overlapping margins, 3–nerved, long-tapered toward the apex and terminating in 3 awns, these often fused at the base to form a slender awn column, the lateral awns sometimes partially or rarely totally reduced; callus hardened, more or less sharply pointed, usually bearded; palea entirely enclosed within and small relative to the lemma; stamens (1) 3; caryopsis free of but firmly enclosed within the lemma, ellipsoidal to narrowly cylindrical, the embryo 1/3–2/3 the length of the grain; x = 11. Species of *Aristida* are the only known host for the smut *Sorosporium consanguineum* Ellis & Everhart. This treatment of *Aristida* was reviewed by Kelly W. Allred.

Allred, K. W. 1984. Morphologic variation and classification of the North American *Aristida purpurea* complex (Gramineae). Brittonia 36: 382–395.

1. Plants annual; sheaths glabrous at the summit; awns not twisted at the base .... A. *adscensionis*
— Plants perennial; sheaths often hairy at the summit; awns ultimately twisted near the base 2

2(1). Glumes distinctly unequal, the second averaging at least 3–4 mm longer than the first; awns 1.5–12 cm long, equal or nearly so .... A. *purpurea*
— Glumes mostly equal or subequal; central awns 1–3.5 cm long, usually exceeding the lateral ones .. A. *arizonica*

Aristida adscensionis L. Sixweeks Threeawn. [*A. oligantha* authors, not Michx.]. Annual or occasionally perennial; culms 0.8–8 dm tall, branched from the base and from the lower nodes, erect or geniculate at nodes near the base; sheaths glabrous; blades to ca 10 cm long and 2.5 mm wide, involute, flexuous to stiffly ascending, usually puberulent; ligule a membranous-based ring of hairs to ca 1 mm long; panicle narrow, 2–22 cm long, compact to loose, the branches short, erect or slightly divergent; spikelets often purplish; glumes thin, unequal, narrowly lanceolate, 1-nerved, acute to rarely awn-tipped, the first (4) 5–9 mm long, the second 7–12 mm long; lemma narrowly lanceolate, 6–13 mm long, tapered toward the apex, an awn column short or lacking, the central awn 6–16 (25) mm long, the lateral ones slightly to occasionally much shorter, about equally divergent at maturity, often flattened but not twisted at the base; callus sharp, bearded, 0.5–1.2 mm long; anthers to ca 1 mm long; 2n = 22. Known in Utah by a single collection (Harrison 12717 BRY) from a desert shrub community at ca 1280 m in Kane County; southwestern U. S., Texas, Kansas, Oklahoma, Missouri, and Mexico; South America; Africa and Eurasia; 1 (0).

Aristida arizonica Vasey Arizona Threeawn. Perennial; culms in small tufts, mostly 2–10 dm tall, stiffly erect; sheaths glabrous throughout or hairy at the summit; blades mostly 6–25 cm long and 0.5–3 (4) mm wide, flat or involute, the basal blades often loosely to strongly recurved, glabrous or scaberulous to scabrid-puberulent; ligule a membranous-based ring of hairs to ca 0.5 mm long; panicle narrow, 5–25 cm long, erect to somewhat nodding, loosely few-flowered, the branches erect, the lowermost mostly 2–6 cm long; spikelets often reddish to purple; glumes equal or subequal, narrowly lanceolate, 8–18 mm long, acute to awn-tipped; lemma narrowly lanceolate, 10–16 mm long, tapered to an ultimately twisted awn column 3–6 mm long, the central awn 1–3.5 cm long, usually longer than the lateral ones; callus sharp, bearded, to ca 0.7 mm long; 2n = 22. Dry sandy soil in desert shrub communities below 1520 m in Kane and San Juan counties; southwestern U. S., Texas, and Mexico; 3 (0).

Aristida purpurea Nutt. Purple Threeawn. [*A. fendleriana* Steudel; *A. glauca* (Nees) Walp.; *A. longiseta* Steudel; *A. wrightii* Nash]. Perennial; culms densely tufted, 0.4–8 (10) dm tall, erect; leaves often chiefly basal; sheaths often hairy at the summit, otherwise glabrous or nearly so; blades 2–25 cm long and 0.5–2 mm wide, involute or occasionally flat, stiffly ascending to strongly recurved, glabrous or scaberulous to scabrous; ligule a membranous-based ring of hairs to ca 0.5 mm long; panicle narrow, (2) 5–20 (30) cm long, erect or nodding, loosely flowered, the branches appressed to ascending or the lowermost somewhat spreading; spikelets often reddish to purple; glumes unequal, narrowly lanceolate, acute to short-awned, the first (4) 7–14 mm long, the second (8) 11–25 mm long; lemma narrowly lanceolate, (7) 10–15 (17) mm long, tapered to a straight or somewhat twisted awn column mostly 1–6 mm long, the awns equal or nearly so, (1.5) 2–8 (12) cm long, terete or flattened near the base, often purple; callus sharp, bearded, to ca 1 mm long; anthers 1–2.5 mm long; 2n = 22, 44, 66, 88. Creosote bush, salt desert shrub, oak-sagebrush, and pinyon-juniper communities at 820 to 2320 m in all Utah counties except Morgan, Rich, Summit, and Wasatch; southwestern Canada, south in the U. S. throughout the western and most of the central states, and in Mexico; 385 (vi). Cronquist et al. (1977) first recognized the overall pattern of continuous variation in the *A. purpurea* complex, and treated it as a single species. A demonstration of polymorphism within closely spaced populations of the species corroborated their findings (Arnow et al. 1980). Based on a statistical analysis of the morphological variation within the complex, Allred (1984) recognized a single species with seven varieties. Much of our material, however, does not fit into the infraspecific categories described in his treatment; moreover, specimens that do, often occur within a single population. It would appear that in Utah, at least, the factors that render unsatisfactory recognition of more than one species also militate against recognition of infraspecific taxa. Where locally abundant, purple threeawn provides a moderate amount of forage during winter and early spring; but once the sharply pointed, coarsely awned lemmas begin to mature (and until they are shed), the grass is a source of irritation and trauma to the mouth parts of grazing animals (Vallentine 1961). In fact, one of the common names applied to this plant is "no eatum."

Arrhenatherum Beauv.

Perennials; sheaths open; blades flat; ligule membranous; inflorescence a panicle; spikelets subterete to compressed, usually 2-flowered, the florets nearly opposite, dissimilar, the rachilla prolonged beyond the upper floret as a bristlelike structure and disarticulating above the glumes; glumes unequal, laterally compressed and more or less keeled, acute; lemma firmer than the glumes, laterally compressed to rounded on the back, acute, 5- to 9-nerved, the callus or pedicel base glabrous or bearded, the lemma of the lower floret with a stout awn arising from below midlength, that of the upper floret awnless or with a shorter, more slender, subterminal awn; palea hyaline, shorter than the lemma; stamens 3; caryopsis free within the lemma and palea, narrowly oblong, the embryo not more than 1/3 the length of the grain; x = 7.

Arrhenatherum elatius (L.) Presl Tall Oatgrass. [*Avena elatior* L.]. Short-lived perennial; culms occasionally bulblike at the base, loosely tufted, erect, (6) 7–18 dm tall; sheaths glabrous or sparsely hairy; blades flat, (1) 2–10 mm wide, scaberulous on the margins and occasionally sparsely hairy; ligule 0.5–3 mm long, truncate to obtuse; panicle narrowly oblong, (5) 10–30 cm long, loose or more or less compact; spikelets narrowly lanceolate, (6) 7–10 mm long, glistening, 2-flowered, the florets dissimilar, the lower staminate or neuter with an awned lemma, the upper pistillate or perfect with an awnless or bristle-tipped lemma, the two florets falling as a unit; glumes membranous, the first 3.5–7 mm long, 1–nerved, the second 7–10 mm long, 3–nerved, shorter than the lower floret to longer than the upper one; lemma 7–10 mm long, mostly 7- to 9-nerved, scabrous or sometimes sparsely short-hairy, the callus bearded or occasionally glabrous, some of the nerves occasionally prolonged as minute teeth, the lemma of the lower floret with a stout, twisted, geniculate awn 1–2 cm long arising just below midlength, that of the upper floret awnless or with a subterminal awn shorter and more slender than that of the lower lemma; anthers 4–5 mm long; 2n = 14, 28, 42. Roadsides, streambanks, fields, and seeded areas in oak-sagebrush, aspen, spruce-fir, ponderosa pine, and meadow communities at 1310 to 3110 m in Beaver, Box Elder, Cache, Carbon, Davis, Duchesne, Juab, Millard, Morgan, Piute, Salt Lake, San Juan, Sanpete, Sevier, Summit, Tooele, Utah, Wasatch, and Weber counties; native to Eurasia, now circumboreal, south throughout the western and northeastern U. S.; South America; 43 (vi). Tall oatgrass was introduced in the U. S. as a pasture grass in 1807 (Vallentine 1961). It is widely used for revegetation in mountain brush, aspen, and subalpine meadows.

Arundo L.

Tall, rhizomatous perennials; culms stout, hollow; blades flat; ligule a ring of hairs or a ciliate membrane; inflorescence a large, much branched panicle; spikelets 2- to 7-flowered, terete or compressed, the rachilla glabrous, disarticulating above the glumes and between the florets; glumes membranous, subequal, about as long as the florets, 3 (5) -nerved; lemma membranous, rounded on the back, 3- to 5-nerved, pubescent with long silky hairs; palea hyaline; stamens (2) 3; caryopsis free of and loosely enclosed by the lemma and palea, oblong, terete; x = 12.

***Arundo donax* L.** Giant Reed. Perennial from thick knotty rhizomes; culms subligneous in age, densely tufted, mostly 2–6 m tall and 1–4 cm in diameter; sheaths glabrous; leaves chiefly cauline; blades flat, (1) 2–6 cm wide, long-tapered to a sharply acute tip, scabrous on the margins; ligule a minutely ciliate membrane to ca 1 mm long; panicle plumelike, (2) 3–7 dm long and to ca 1 dm wide, much branched, densely flowered, tawny or greenish and often purple-tinged; spikelets subterete to compressed, narrowly lanceolate, 8–15 mm long, mostly 2- to 4-flowered, the florets all perfect or the uppermost neuter, the rachilla glabrous, the florets progressively smaller upward, all terminating in approximately the same plane; glumes membranous, subequal, lanceolate, 10–14 mm long, more or less keeled to rounded on the back, tapered to a sharply acute tip, the whole about equal to the body of the uppermost lemma; lemma membranous, lanceolate, mostly rounded on the back, to ca 13 mm long, obscurely 3- to 9-nerved, acute at the apex or the nerves prolonged, the lateral as minute bristlelike teeth, the midnerve as an awn 0.5–3 mm long, the back pubescent over at least the lower half with soft hairs 6–9 mm long; palea ca 2/3 as long as the lemma; fertile caryopses seldom developing. Disturbed sites and margins of gardens at 850 to 1250 m in Washington County; native to Eurasia, widely cultivated as an ornamental, and established from California to New Mexico, Mexico, and occasional eastward; 2 (0).

Avena L.

Annuals; sheaths open; blades flat, broad; ligule membranous; inflorescence a large, open, nodding panicle or raceme; spikelets large, pendulous, 2- to 6-flowered, disarticulating above the glumes and between at least the lowermost florets; glumes lanceolate, equal or nearly so, some laterally compressed, 3- to 11-nerved, longer than the lower floret and often exceeding the uppermost; lemma firm, lanceolate to lance-ovate, rounded on the back, (5) 7-nerved, notched at the apex, awnless (in cultivars) or the awn arising from the back; callus 1–7 mm long, usually acute, glabrous or bearded; palea shorter than the lemma; stamens 3; caryopsis adherent to the lemma and palea or, in cultivated forms, free-threshing; x = 7.

Baum, B. R. 1968. On some relationships between *Avena sativa* and *A. fatua* (Gramineae) as studied from Canadian material. Canad. J. Bot. 46: 1013–1022.

***Avena fatua* L.** Oats. Annual; culms solitary or in small tufts, (2) 3–19 dm tall; sheaths glabrous or sparsely long-hairy; blades flat, (1) 3–15 mm wide, usually scabrous, at least on the margins, sometimes sparsely hairy; ligule 2–6 (7) mm long; inflorescence a panicle or in depauperate plants a raceme 0.4–4 dm long; spikelets mostly 8–30 per inflorescence, 1.5–5 cm long, borne on slender drooping pedicels to ca 5 cm long, 2- to 3 (4) -flowered, the florets above the first two usually greatly reduced; glumes broadly lanceolate, subequal, 1.6–3 cm long, rounded on the back, usually long-tapered to an acute membranous tip, 7- to 11-nerved, surpassing the lower and often the uppermost florets; lemma ultimately hardened, rounded on the back, 1.4–2 cm long, notched at the membranous apex, glabrous throughout or hairy on the lower half, awnless or the awn arising near midlength, varying from slender, short, and more or less straight to stout, 2–5 cm long, and geniculate; anthers 4–5 mm long. Two intergradient varieties are recognized:

1. Spikelets usually 3- or 4-flowered; lemmas pubescent with stiff, usually reddish brown hairs, bearing a stout geniculate awn mostly 2–5 cm long .... *A. fatua* var. *fatua*
— Spikelets usually 2-flowered; lemmas glabrous, awnless or the awn poorly developed .... *A. fatua* var. *sativa*

Var. *fatua* Wild Oats. Roadsides, ditchbanks, wet meadows, fallow fields, and among annual crops at 850 to 2620 m in Cache, Carbon, Duchesne, Salt Lake, Sanpete, Sevier, Utah, Wasatch, Washington, and Wayne counties, and to be expected wherever cereal crops are grown; native to Eurasia, now circumboreal, south through much of the U. S.; South America; 23 (i).

Var. *sativa* (L.) Hausskn. Oats. [*A. sativa* L.]. Widely grown in Utah as food for humans and livestock, escaping and common as a waif in agricultural regions; native to Eurasia; widely cultivated in temperate regions of the world; 14 (0). The varieties of *A. fatua* are often treated at the specific level. The cultivated variety is derived from *A. fatua*, however, and, because the taxa intergrade freely, the characters used to distinguish them are by no means consistently correlated. The treatment of Gould and Shaw (1983) is followed here. The stout awn of var. *fatua* may cause mechanical injury to mouth parts of grazing animals. Moreover, oat hay is one of the most common sources of plant-related nitrate poisoning, evidently resulting from moisture accumulation on outdoor haystacks. Oats are also among the crops that can produce mineral imbalance in cattle, giving rise to the syndrome known as grass tetany (Kingsbury 1964).

Beckmannia Host

Annuals or perennials; culms moderately stout; sheaths open; blades broad and flat; ligule membranous; inflorescence a panicle consisting of racemosely arranged, erect to ascending, simple or sparingly rebranched, spikelike branches, the sessile to subsessile, obovate spikelets closely overlapping in two rows along one side of the rachis, 1- or 2-flowered, disarticulating below the glumes and falling entire; glumes laterally compressed but typically somewhat inflated, equal, slightly shorter than the spikelet, 3-nerved, rounded to an obtuse to abruptly apiculate apex; lemma lance-fusiform, 5-nerved, rounded on the back, tapered to a slender, acute, included or exserted tip; palea slightly shorter than the lemma; stamens 3; caryopsis free within the lemma and palea, to ca 2 mm long, 3-angled, the embryo 1/4-1/3 the length of the grain; x = 7.

Beckmannia syzigachne (Steudel) Fern. American Sloughgrass. [*Panicum syzigachne* Steudel; *B. eruciformis* of American authors, not (L.) Host]. Annual; herbage light green; culms 3-12 dm tall, solitary or loosely tufted; blades 3-12 mm wide, flat, scabrous; ligule 4-11 mm long; panicle narrow, 0.7-3.5 dm long, the 1-sided, spikelike branches erect or strongly ascending, (0.7) 1-4 (5) cm long, simple or a few of them rebranched, spikelet-bearing to the base; spikelets pale green, obovate, 2-3.2 mm long (excluding the awn tip) and about as wide as long, sessile or nearly so and closely overlapping in 2 rows arising from one side of the rachis, 1-flowered, occasionally with a reduced second floret; glumes equal, boat-shaped, typically inflated to some degree, rounded to an apiculate apex; lemma narrow, subterete, the slender, sharply pointed tip to ca 0.8 mm long, exserted beyond the glumes for ca 1 mm; anthers 0.4-1 mm long; 2n = 14. Wet, rarely saline meadows and marshes, and along waterways, at 1220 to 2740 m in Box Elder, Cache, Carbon, Daggett, Duchesne, Emery, Garfield, Morgan, Piute, Rich, Salt Lake, Sanpete, Sevier, Summit, Uintah, Utah, Wasatch, Wayne, and Weber counties; Alaska to Greenland, throughout the western U. S., east to Kansas, Iowa, and the Great Lakes region; Eurasia; 65 (ii).

Blepharidachne Hackel

Monoclinous or monoecious annuals or perennials; culms low, tufted; leaves crowded near the base; sheaths open; blades firm, flat to involute; ligule a ring of short hairs or lacking; inflorescence a short compact panicle, often overtopped by the subtending leaves; spikelets subsessile or short-pedicelled, laterally compressed, 4-flowered, disarticulating above and often below the glumes but not between the florets, the first and second florets generally neuter or staminate, the third pistillate or perfect, the fourth reduced to a 3-awned rudiment; glumes thin, subequal, 1-nerved, generally keeled; lemmas dimorphic, the lower three slightly firmer than the glumes, rounded on the back, 3-nerved, deeply lobed to about midlength or below, the midnerve largely free as an awn and the lateral nerves often prolonged as minute awns, the uppermost lemma a 3-awned, bristlelike rudiment; palea of the neuter or staminate florets greatly reduced, that of the perfect or pistillate floret about equal to the body of the lemma; stamens (1) 2 or 3; caryopsis free of the lemma and palea, oblanceolate, laterally compressed; x = 7(?).

Hunziker, A. and A. M. Anton. 1979. A synoptical revision of *Blepharidachne* (Poaceae). Brittonia 31: 446-453.

Blepharidachne kingii (Wats.) Hackel King Desertgrass. [*Eremochloe kingii* Wats.]. Monoclinous perennial; culms densely tufted, erect to spreading, 0.2-1 (1.4) dm tall, not rooting at the nodes; leaf sheaths short and broad, membranous-margined, often hairy near the base and minutely to conspicuously so at the summit; blades firm, arcuate, folded or involute and in that state less than 1 mm wide, pungent-tipped, more or less scaberulous; ligule a ring of hairs to ca 0.5 mm long, longer hairs often present along the sheath margins; panicle headlike or nearly so, 1-2.5 cm long, subtended by 1 or 2 leaves, ultimately at least partially emergent from the sheaths, often exceeded by the upper culm blades; spikelets 6-8.5 mm long, pale or occasionally purple-tinged, 4-flowered, the internodes of the rachilla very short, the florets forming a basally hairy, fan-shaped cluster; glumes thin, broadly lance-elliptical, sharply acute to awn-tipped, 6-8.5 mm long overall, typically keeled, usually scaberulous to minutely hairy near the base, especially on the midnerve, as long as or longer than the florets, often slightly exceeded by the awns of the lemma; lemmas dimorphic: the first two 3.4-5.8 mm long, lobed to about midlength, the midnerve largely free as a plumose awn 3-5 mm long, the lateral nerves prolonged beyond the lobes or not, the margin conspicuously ciliate; third lemma similar to the lower 2 but often glossy and not ciliate laterally below midlength; fourth lemma a 3-awned, bristlelike rudiment; palea of the neuter florets about half as long as the lemmas, very narrow, the keels ciliate, the palea of the fertile floret subequal to the lemma; stamens 2, the anthers 1.5-2 mm long; 2n = 14 (Reeder, Amer. J. Bot. 64: 104. 1977). Greasewood, desert shrub, and sagebrush communities at 1070 to 1830 m in Beaver, Box Elder, Juab, Millard, and Tooele counties; California and Nevada; 43 (i). *Blepharidachne kingii* outwardly resembles *Erioneuron pulchellum* (q.v.), but the structure of the spikelet is distinctive.

Blepharoneuron Nash

Perennials; sheaths open; ligule membranous; inflorescence a panicle; spikelets 1-flowered, disarticulating above the glumes, typically dark gray; glumes thin, 1- to 3-nerved; lemmas 3-nerved; palea about equal to the lemma; stamens 3; caryopsis free within the lemma and palea, the embryo about half the length of the grain; x = 8. *Blepharoneuron* is a monotypic North American genus.

Blepharoneuron tricholepis (Torr.) Nash Hairy or Pine Dropseed. [*Vilfa tricholepis* Torr.]. Tufted, sometimes ring-forming perennial; culms erect, 1.5-7 (8.5) dm tall; leaves chiefly basal; sheaths glabrous; blades mostly 0.5-2 mm wide, flat to more often involute, scabrous; ligule mostly 0.5-2 mm long; panicle narrowly oblong to pyramidal, 4-20 (22) cm long, loosely contracted to open, the primary branches smooth or scabrous, the hairlike pedicels smooth, mostly 2-9 mm long; spikelets 1-flowered, subterete, 2-3.8 mm long, gray green to blackish and sometimes obscurely purple-tinged; glumes thin, oblong-elliptical, equal or unequal, typically rounded on the back, glabrous, abruptly acute to rounded at the apex, obscurely 1- to 3-nerved, the first 1.5-2.5 mm long, the second 1.7-2.8 (3.2) mm long, averaging shorter than the lemma; lemma firmer than or similar to the glumes in texture, elliptical, 2-3.2 (3.8) mm long, acute to obtuse at the apex, 3-nerved, each nerve densely hairy over at least the lower 1/3, the hairs appressed to spreading, the midnerve often minutely prolonged; palea densely hairy between the 2 nerves, appearing 1-nerved; anthers 1.4-2 mm long; 2n = 16. Openings in ponderosa pine, lodgepole pine, aspen-spruce-fir communities, occasionally in sagebrush, mountain brush, and meadow communities, at 1980 to 3200 m in Beaver, Daggett, Duchesne, Garfield, Iron, Juab, Piute, San Juan, Sevier, Summit, Uintah, Wasatch, Washington, and Wayne counties; Colorado, Arizona, New Mexico, Texas, and Mexico; 69 (i). Pine dropseed is a good source of forage in those areas of the Uinta Mts. and the high plateaus where it is the dominant grass (Cronquist et al. 1977).

Bothriochola Kuntze

Perennials; culms simple or sparingly to much branched above, glabrous or hairy at the nodes; sheaths open, occasionally somewhat keeled; blades flat to loosely involute; ligule membranous; inflorescence a panicle of 2 to numerous, digitate to closely spaced, spikelike, disarticulating branches, these simple or in some species rebranched, at least the lower ones often partially included in the upper sheath, the spikelets arranged on the rachis in pairs, one in each pair sessile, the other short-pedicelled, the rachis disarticulating below the sessile spikelet, the latter falling together with the associated spikelet, pedicel, and rachis segment, the pedicels and at least the terminal rachis segments flat and vertically grooved, the central portion membranous, the margins more or less thickened; spikelets lanceolate, 2–flowered, the sessile spikelet with the lower floret neuter (often rudimentary) and the upper floret perfect or pistillate, the pedicelled spikelet rudimentary or well developed, with the upper floret staminate or neuter; glumes firm, narrowly lanceolate, exceeding both florets, prominently to obscurely several-nerved, the first equal to or slightly longer than the second; lemma hyaline, that of the well-developed floret of the sessile spikelet entire or 2–lobed apically and usually bearing a geniculate, twisted awn arising from the apex or from between the apical lobes; palea hyaline or lacking; stamens 3; caryopsis free of but often remaining enclosed within the lemma and palea, the embryo 1/2–2/3 the length of the grain; x = 10. This treatment was reviewed by Kelly W. Allred.

Allred, K. W. and F. W. Gould. 1983. Systematics of the *Bothriochloa saccharoides* complex (Poaceae: Andropogoneae). Syst. Bot. 8: 168–184.

1. Pedicelled spikelets about equal in size to the sessile ones; internodes of the panicle branch with hairs averaging less than 3 mm long .... *B. ischaemum*
— Pedicelled spikelets smaller and narrower than the sessile ones; internodes of the panicle branch with hairs averaging more than 3 mm long 2

2(1). Sessile spikelets averaging less than 4 (2.5–4.5) mm long; awns typically averaging less than 15 (8–18) mm long .... *B. laguroides*
— Sessile spikelets (4) 5–7.3 mm long; awns 15–30 mm long 3

3(2). Main axis of the panicle less than 5 cm long, nearly always shorter than the branches; branches 2–9 per panicle, 4–7 cm long; upper culm nodes with hairs mostly 3–7 mm long .... *B. springfieldii*
— Main axis of the panicle often more than 5 cm long, mostly equal to or longer than the surrounding branches; branches often more than 9 per panicle, 2–9 cm long; upper nodes of culm with hairs mostly 1–2 (3) mm long .... *B. barbinodis*

Bothriochloa barbinodis (Lag.) Herter Cane Bluestem. [*Andropogon barbinodis* Lag.]. Perennial; herbage often glaucous; culms densely tufted, mostly 6–15 dm tall, erect or sometimes geniculate at the lower nodes, usually becoming decumbent and much branched below in age, sparingly or not at all branched above, the nodes pubescent with erect to spreading hairs mostly 1–3 mm long; sheaths glabrous; blades flat, mostly 2–7 (10) mm wide, often becoming loosely involute toward the long-tapered tips, glabrous or scabrous, the base of the blade usually long-hairy and the collar often puberulent, the uppermost culm blades typically greatly reduced; ligule 1–2.5 mm long, often surpassed by the few to numerous, long stiff hairs arising from the base of the blade; panicle whitish to tan, 4.5–13 cm long, contracted to somewhat open, often partially included in the upper sheath, the few to numerous, suberect to spreading branches mostly 2–9 cm long, the main axis mostly equal to or longer than the branches, the rachis segments and the pedicels pubescent along the thickened margins with dull white hairs mostly 5–9 mm long, at least those rachis segments and pedicels near the tips of the branches with a membranous-based medial groove; sessile spikelet lanceolate, (4) 5–7.3 mm long, typically partially obscured by the long hairs of the rachis, the glumes firm, the first flat to centrally concave, usually sparsely hairy below midlength, occasionally with a centrally located glandular pit or depression, the lemma of the well-developed floret bearing a geniculate, basally twisted awn 15–30 mm long; pedicelled spikelet rudimentary, not much wider than the pedicel, 3–5 mm long, often early deciduous; anthers 0.5–1.3 mm long; 2n = 70, 80, 90, 120, 180. Chiefly in dry, sandy, gravelly, or rocky sites in desert shrub, pinyon-juniper, and ponderosa pine communities at 920 to 1830 m in Garfield, Grand, Kane, San Juan, Washington, and Wayne counties; southwestern U. S., Oklahoma, Texas, and into Mexico; South America; 23 (0).

Bothriochloa ischaemum (L.) H. Keng Yellow Bluestem. [*Andropogon ischaemum* L.]. Perennial; herbage yellow green; culms tufted, mostly (3) 4–10 dm tall, erect or sometimes decumbent to prostrate at the base, simple or sparingly branched above the base, the nodes glabrous or short-hairy; leaves chiefly basal; sheaths glabrous; blades flat, often becoming loosely involute toward the long-tapered tips, mostly 1–4 mm wide, usually sparsely long-hairy with papilla-based hairs at least along the margins near the ligule, the uppermost culm blades greatly reduced; ligule to ca 1 mm long, minutely ciliate; panicle 4–10 cm long, reddish purple, well exserted above the uppermost leaf, the 2–10 spikelike branches digitate or nearly so, ascending to spreading, mostly 3–9 cm long and infrequently rebranched, the pedicels and rachis segments pubescent along the thickened margins with hairs to ca 3 mm long, at least those rachis segments and pedicels near the tips of the racemes with a medial groove; sessile spikelet lanceolate to elliptical, 3–5 mm long, the glumes firm, the first flat or shallowly concave, never with a glandular pit or depression, glabrous or scabrous above and often short-hairy below midlength, the lemma of the well-developed floret bearing a geniculate, basally twisted awn mostly 10–15 mm long; pedicelled spikelet staminate, about equal in size to the sessile one, awnless, the first glume glabrous or hairy below midlength; anthers ca 2 mm long; 2n = 40, 50, 60. Known in Utah by a single population (Welsh & Neese 23066 BRY) from a roadside area at ca 1400 m in Utah County; native to Eurasia, introduced in the southwestern U. S. and in Mexico as a pasture grass, and occasionally escaping cultivation; 2 (0).

Bothriochloa laguroides (DC.) Herter Silver Bluestem. [*Andropogon laguroides* DC.; *A. saccharoides* authors, not Swartz; *A. saccharoides* var. *torreyanus* (Steudel) Hackel; *B. saccharoides* var. *torreyana* (Steudel) Gould]. This species is closely related to but differs from *B. barbinodis* as follows: plants less robust with more slender culms, leaves more basally clustered, leaf blades glabrous; panicle branches mostly 1–4 cm long and erect to strongly ascending; sessile spikelet 2.5–4 (4.5) mm long, lacking a glandular pit, the awns 8–16 mm long; pedicelled spikelet about as wide as the pedicel and 2.5–3 (4) mm long; 2n = 60. Dry, usually sandy sites, often along roadsides below 1520 m in Kane (Welsh 11881 BRY) and Grand (Allred 1765 BRY) counties; Great Plains, with isolated stations in Utah, Missouri, Georgia, and Alabama; Mexico; South America; 2 (0). Allred and Gould (Syst. Bot. 8: 168–184. 1983) recognize several species within the *B. sacchariodes* complex, separating *B. laguroides* from the remainder of the complex on the bases of chromosome number (*B. sacchariodes*, 2n = 120), leaf pubescence, width of the medial groove of rachis and pedicels, and the presence or absence of axillary pulvini. Our plants are ssp. *torreyana* (Steudel) Allred & Gould, differing from the southern element of the species in having spikelets visible through the subtending hairs, sheaths glabrous at the summit, and culm nodes nearly always bearded (Allred & Gould 1983).

Bothriochloa springfieldii (Gould) Parodi Springfield Bluestem. [*Andropogon springfieldii* Gould]. This species is similar to and commonly identified as *B. barbinodis* but differs from that species as follows: culms 3–8 dm tall, the nodes densely hairy with spreading hairs mostly 3–7 mm long; sheaths often densely hairy on the margins below the summit; leaf blades 2–3 (4.5) mm wide, sometimes sparsely hairy on the ventral surface; panicles densely white-hairy, with 2–9 spikelike branches 4–7 cm long, the main axis of the panicle less than 5 cm long, often surpassed by the branches, these infrequently once rebranched near the base, the rachis segments and pedicels densely pubescent with bright white hairs 5–10 mm long; sessile spikelets 5.5–8.5 mm long. Sandstone cliffs and ledges, along waterways, and in desert shrub communities at 1130 to 1770 m in Garfield, Grand, Kane, San Juan, and Wayne counties; Arizona, New Mexico, Texas, Louisiana, and northern Mexico; South America; 16 (0).

Bouteloua Lag.

Annuals or perennials; culms often sparingly branched above the base; leaves chiefly basal; sheaths open; blades flat to involute; ligule a ring of hairs or a short ciliate membrane; inflorescence a panicle of 1 to numerous, short, spikelike, secund branches, these racemosely ranged along a main axis, either disarticulating at the base and falling entire or persistent on the main axis, each branch with 1 to numerous, sessile or subsessile spikelets strongly to loosely overlapping in 2 closely spaced rows along one aspect of a subterete to flattened axis, the branch axis sometimes prolonged beyond the terminal spikelet as a naked, needlelike tip; spikelets with 1 perfect floret and usually with 1 or 2 (3) reduced florets (rudiments) above the perfect one, the rachilla disarticulating above the glumes or remaining intact in those species with deciduous branches; glumes subequal or unequal, lanceolate to elliptical, 1-nerved, the second (outer) shorter to slightly longer than the lowermost lemma; lemma of the perfect floret lanceolate, usually minutely bearded on the callus, rounded on the back to laterally compressed and keeled, 3-nerved, the lateral nerves marginal or nearly so, frequently the midnerve and often the lateral ones prolonged as awns; palea of the perfect floret usually well developed, the 2 nerves sometimes prolonged as minute awns; lemma of the reduced florets present or absent, that of the first rudiment usually 3–awned or reduced to awns, that of the upper rudiment(s) often scalelike and awnless; stamens 3; caryopsis free within the lemma and palea, the embryo about half the length of the grain; x = 10.

Gould, F. W. 1979. The genus *Bouteloua*. Ann. Missouri Bot. Gard. 66: 348–416.

1. Branches of the inflorescence readily disarticulating at the base and falling entire, each branch with 1–8 (13) spikelets 2
— Branches of the inflorescence persistent on the culm, each branch with 8 to numerous spikelets 3
2(1). Plants annual; panicle branches 4–15 (20) per culm, mostly ascending-spreading .... *B. aristidoides*
— Plants perennial; panicle branches typically (13) 20–80 per culm, spreading to pendulous .... *B. curtipendula*
3(1). Culms white-woolly, especially on the lower internodes .... *B. eriopoda*
— Culms glabrous or scabrid-puberulent, not at all woolly . 4
4(3). Rachis prolonged beyond the terminal spikelet as a hard, needlelike tip 5–20 mm long .... *B. hirsuta*
— Rachis terminating in a rudimentary or well-developed floret 5
5(4). Plants annual 6
— Plants perennial 7
6(5). Inflorescence typically a single spikelike branch; second glume 3.5–5.5 mm long .... *B. simplex*
— Inflorescence of (1) 2–9 spikelike branches; second glume 1.3–3 mm long .... *B. barbata*
7(5). Rachis terminating in one or more rudimentary spikelets; rudimenary floret long-beared from the base .... *B. gracilis*
— Rachis terminating in a well-developed spikelet; rudimentary floret not long-bearded from the base .... *B. trifida*

Bouteloua aristidoides (H.B.K.) Griseb. Needle Grama. [*Dinebra aristidoides* H.B.K.]. Annual; culms tufted, slender, 0.6–3 (5) dm tall, erect or geniculate to decumbent below; sheaths glabrous or more often ciliate with long hairs near the summit; blades 0.5–2 mm wide, flat to involute, glabrous or sparsely long-hairy; ligule a minute ciliate membrane or entirely of hairs; panicle of 4–15 (20) slender, loosely spaced, appressed to widely spreading or occasionally pendulous, spikelike branches mostly 1–2 cm long, the branches disarticulating at the base and falling as a unit, the axis densely pubescent at the base, prolonged beyond the point of origin of the terminal spikelet as a flattened, naked tip 5–11 mm long; spikelets 1–6 per branch, mostly 5–7 mm long including the awn tips, terete or nearly so, typically closely appressed to the branch rachis, only the lowermost spikelet on each branch usually without a rudimentary floret above the perfect one; glumes relatively firm, very unequal, the outer slightly shorter to longer than the fertile lemma; lemma of the lower floret firm, rounded on the back, the callus sharp, bearded, the body to ca 4 mm long, acute or awn-tipped to long-awned, the anthers ca 2 mm long; all but the lowermost spikelet of each branch with a solitary rudiment consisting of a short stalklike base terminating in 3 awns 2–6 mm long; 2n = 40. Dry open slopes, in rock crevices, along washes and roadsides in desert shrub and Joshua tree communities at 1070 to 1830 m in Garfield, Kane, and Washington counties; California to New Mexico, Texas, and Mexico; South America; 8 (0).

Bouteloua barbata Lag. Sixweeks Grama. [*B. rothrockii* authors, not Vasey]. Annual; culms tufted, slender, 0.2–3 (6) dm tall, erect or geniculate to sprawling or prostrate, occasionally rooting at the lower nodes, glabrous; leaves well distributed along the culm; sheaths entirely glabrous or with tufts of hair lateral to the ligule; blades 0.5–2 (3) mm wide, flat to loosely involute, scabrous and occasionally sparsely hairy with papilla-based hairs; ligule a ring of hairs to ca 1 mm long; panicle of (1) 2–9 erect to widely spreading, persistent, spikelike branches, these 0.5–3 cm long, the spikelets ultimately widely spreading, the axis straight or arcuate, scabrous, often puberulent at the base, terminating in a well-developed spikelet; spikelets 6–50 per branch, disarticulating above the glumes, mostly subterete, 2–5 mm long including the short awns, each with 1 or 2 rudimentary florets above the perfect one; glumes unequal, the first membranous, to ca 2 mm long, the second firm, rounded on the back, mostly broadly elliptical, 1.3–3 mm long, glabrous, typically notched and awn-tipped apically, pale or purple-tinged, surpassed by the fertile lemma or its awns; lemma of the lower floret rounded on the back, to ca 2.5 mm long, conspicuously hairy on the margins and often on the lower half of the midnerve, 3-lobed and 3-awned, the central lobe bifid, the awns 0.5–3 mm long, the anthers ca 0.5 mm long; first rudiment with the lemma 0.5–1 mm long, often cleft to the base along the nerves, the 3 nerves then free throughout and 0.5–3 mm long; second rudiment (when present) minute, fan- to club-shaped, awnless; 2n = 20, 40. Dry to moist sites in desert shrub and creosote bush communities at 850 to 1830 m in Emery, Garfield, Grand, Kane, Millard, San Juan, Washington, and Wayne counties; California to Colorado, Texas, and Mexico; Argentina; 64 (0). Our plants are **var. barbata**, differing from the more southern variety in being consistently annual, without stolons, and with mostly geniculate culms rarely more than 3 dm tall (Gould 1979).

Bouteloua curtipendula (Michx.) Torr. Sideoats Grama. [*Chloris curtipendula* Michx.]. Perennial, often with slender to stout, scaly rhizomes; culms solitary to densely tufted, 0.2–1 m tall, simple or occasionally sparingly branched above the base, decumbent to stiffly erect, glabrous; leaves blue green; sheaths glabrous or sparsely long-hairy, the basal sheaths of previous years persistent; blades 2–7 mm wide, flat to loosely involute, scaberulous, sometimes sparsely ciliate with coarse, papilla-based hairs near the ligule; ligule a ciliate membrane or a ring of hairs mostly less than 1 mm long; panicle often secund with 13–80 mostly pendulous, readily disarticulating, short-peduncled, spikelike branches, these typically elliptical to V-shaped, 0.6–2 cm long, the axis flattened, glabrous or scabrous to puberulent, prolonged beyond the point of origin of the terminal spikelet, the prolonged portion rarely surpassing the terminal spikelet; spikelets (1) 2–8 (13) per branch, subterete, closely appressed to the short axis, with or without a single rudimentary floret above the perfect one; glumes unequal, the first small and membranous, the second firm, lance-olate, 3.8–7.5 mm long, sharply acute to awn-tipped, glabrous or scabrous, often purple-tinged, varying from shorter to longer than the body of the fertile lemma; lemma of the lower floret rounded on the back to somewhat keeled, 3.6–6 mm long, the lateral nerves often free toward the acute tip to form short awns, these subequal to the tip, the anthers red, orange, or yellow, ca 2 mm long; rudiment (when present) with 1–3 awns, the longest to ca 7 mm; 2n = 20, 28, 35, 40–66, 69–103. Desert shrub, sagebrush, pinyon-juniper, and ponderosa pine communities, occasionally on open slopes, at 980 to 2440 m in Beaver, Duchesne, Garfield, Emery, Grand, Kane, San Juan, Washington, and Wayne counties; south-central Canada, most of the U. S., and in Mexico; South America; 77 (0). Gould (1975) cites as doubtful one record for *B. uniflora* from southern Utah. That taxon differs from *B. curtipendula* chiefly in having only one spikelet per branch of the inflorescence, a condition occasionally encountered within populations of *B. curtipendula*. Sideoats grama is a major source of forage over large areas of the U. S., but in Utah it is not sufficiently abundant to be of significance. The species is included in contemporary seed mixtures for use in reclamation projects, and at least the Duchesne County record is thought to represent such an origin.

Bouteloua eriopoda (Torr.) Torr. Black Grama. [*Chondrosium eriopodum* Torr.]. Tufted perennial from a hard, knotty, woolly base, sometimes becoming stoloniferous; culms 2–6 (7) dm tall, sprawling to ascending, densely white-woolly, especially on the lower internodes; sheaths glabrous or ciliate at the summit; blades 0.5–2.2 mm wide, flat to involute, smooth or scaberulous, often ciliate with papilla-based hairs near the ligule; ligule a ring of hairs less than 1 mm long; panicle of 2–8 ascending to occasionally widely spreading, persistent, spikelike branches, these (1) 2–4 (5) cm long, the branch axis densely white-woolly at the base and usually hairy to some degree at the base of each spikelet, prolonged beyond the point of origin of the terminal spikelet as a needlelike tip; spikelets 8–20 per branch, disarticulating above the glumes, subterete, strongly ascending, 6–10 mm long excluding the awns, each with a single rudimentary floret above the perfect one; glumes very unequal, the first membranous, 2–4.5 mm long, the second firm, lance-elliptical, 6–10 mm long, long-tapered to a sharply acute tip, pale or purple-tinged, exceeding the body of the fertile lemma; lemma of the lower floret bearded on the callus, rounded to somewhat keeled on the back, 4.5–7 mm long, glabrous or sparsely hairy, tapered to a terminal awn 1.5–3 mm long, the lateral nerves often prolonged as awn tips, the anthers 1.5–3.5 mm long; rudiment bearded at the short, stalklike base, its 3 awns mostly 4–8 mm long; 2n = 20, 21, 28. Sagebrush, desert shrub, and pinyon-juniper communities at 850 to 2440 m in Garfield, Grand, Kane, San Juan, Washington, and Wayne counties; California and Arizona, and from Wyoming south to northern Mexico, Kansas, Oklahoma and Texas; 39 (i). Black grama is a palatable and nutritious source of forage in summer and winter. It is also drought resistant and some strains have exceptional seedling vigor and good seed production (Thornburg 1982).

Bouteloua gracilis (**H.B.K.**) **Lag. ex Steudel** Blue Grama. [*Chondrosium gracile* H.B.K.]. Perennial, often with short, stout rhizomes, sometimes mat-forming; culms tufted, 1–6 (7) dm tall, erect or decumbent to geniculate-sprawling, glabrous or the nodes minutely hairy; leaves chiefly basal; sheaths glabrous or sparsely long-hairy, sometimes with tufts of hair lateral to the ligule; blades 0.5–2.2 mm wide, flat to involute, often recurved, scaberulous to short-hairy; ligule a ring of hairs, sometimes membranous-based, the whole less than 1 mm long; panicle of 1–3 (4) persistent, ascending to spreading, spikelike branches, these 1–4.5 (5) cm long, pectinate, the axis generally arcuate, usually pubescent at the base and glabrous or scabrous above, terminating in one or more rudimentary spikelets; spikelets densely crowded (30–90 per branch), at maturity widely spreading, disarticulating above the glumes, subterete to strongly keeled, generally 4–6 mm long excluding the awns, each with 1 or 2 rudimentary florets above the perfect one; glumes unequal, the first membranous and small, the second firm, lanceolate, 3.5–6 (6.5) mm long, slightly shorter than the fertile lemma, long-tapered to a sharply acute tip, pale or purple-tinged, glabrous or the midnerve scabrous to pubescent with simple or papilla-based hairs; lemma of the lower floret rounded or keeled on the back, 3–6 (6.5) mm long, pubescent at least below and bearded on the callus, the midnerve prolonged between 2 slender lobes as an awn 1–3 mm long, the lateral nerves often prolonged as small bristles, the anthers mostly 2.5–3.5 mm long; first rudiment with a 3-lobed lemma surpassed by the long hairs at the summit of the rachilla and terminating in 3 awns 2.5–5 mm long, the second rudiment (when present) obovate, 1–2 mm long and awnless; 2n = 20, 21, 28, 35, 40, 42, 60–62, 77, 84. Salt desert shrub, desert shrub, grass-sagebrush, pinyon-juniper, and ponderosa pine communities at 980 to 2960 m in Beaver, Carbon, Daggett, Duchesne, Emery, Garfield, Grand, Iron, Kane, Millard, Piute, San Juan, Sanpete, Sevier, Uintah, Utah, Wasatch, Washington, and Wayne counties; southwestern Canada, throughout the southwestern and central U. S. to Mexico, rare in the eastern U. S.; 210 (iii). Blue grama is highly palatable as forage, and is resistant to cold, drought, and grazing. It forms a dense sod, and is useful for erosion control in arid and semiarid regions. Although its productivity is relatively low in Utah, it does form dense stands in some areas of southern and eastern Utah. The grass is unusual in that plants remain dormant throughout the spring and summer, growing and producing flowering culms at a rapid rate following late summer rains.

Bouteloua hirsuta Lag. Hairy Grama. This species is similar to *B. gracilis*, differing as follows: culms densely tufted, not rhizomatous; sheaths and blades glabrous or more often long-hairy with papilla-based hairs; spikelike branches with the rachis prolonged beyond the terminal spikelet as a needlelike tip 5–20 mm long; second glume 3–5 mm long, distinctly shorter than the fertile lemma, the midnerve often ciliate with papilla-based hairs 1–2 mm long and prolonged as an awn tip; first rudiment with a glabrous rachilla; 2n = 12, 20–22, 24, 28, 36, 37, 40, 42, 46, 50. Known in Utah by a single 1946 collection (Booth s.n. UTC) from an open dry, gravelly site in Zion National Park, Washington County; Wyoming south throughout the southwestern and central states into Mexico, east to Wisconsin, Illinois, and Louisiana, disjunct in Florida; 1 (0).

Bouteloua simplex Lag. Mat Grama. [*B. procumbens* (Durand) Griffiths]. Tufted annual; culms 0.5–2 (3.5) dm tall, erect or geniculate below to prostrate; sheaths glabrous or long-hairy near the ligule; blades 0.5–1.5 mm wide, flat to involute, scaberulous or sparsely hairy, often ciliate above the ligule; ligule a ciliate membrane ca 0.2 mm long; inflorescence a persistent, solitary, spikelike branch 0.8–2.5 cm long, ultimately pectinate and arcuate, the axis essentially glabrous or minutely hairy at the base, terminating in a rudimentary spikelet; spikelets crowded (10–80 per branch), at maturity widely spreading, disarticulating above the glumes, mostly flattened, 4–6 mm long including the short awns, each with 1 or 2 rudimentary florets above the perfect one; glumes unequal, keeled, the first 2–3 mm long, the second lanceolate, 3.5–5.5 mm long, sharply acute at the tip, usually pale, glabrous or the midnerve scaberulous, ultimately exceeded by the awn of the fertile lemma; lemma of the lower floret generally laterally compressed and keeled, 2.5–3.5 mm long, usually bearded on the callus and pubescent on the nerves, the latter prolonged as awns 1.5–2.5 mm long, the anthers to ca 0.5 mm long; first rudiment a minute membranous body with 3 awns to 4 mm long, the second rudiment (when present) an obovate scale; 2n = 20, 40. In rocky or sandy-gravelly soils, roadsides and along waterways, in greasewood and sagebrush communities at 1520 to 2440 m in Beaver, Garfield, Iron, Piute, Sevier, and Washington counties; Wyoming and Colorado to Arizona, New Mexico, Texas, and Mexico; Argentina; 9 (0).

Bouteloua trifida Thurber in Wats. Red Grama. Tufted perennial from a somewhat hard, often short-rhizomatous base; culms 0.4–3 (4) dm tall, wiry, erect or occasionally geniculate below, glabrous or scabrid-puberulent; leaves chiefly basal; sheaths glabrous or scabrid-puberulent; blades 0.5–1.5 mm wide, usually involute, scaberulous to puberulent or sometimes with a few papilla-based hairs; ligule a ring of hairs less than 1 mm long; panicle of 1–7 persistent, erect-ascending, spikelike branches, these 0.7–2.5 (4) cm long, the axis mostly straight or only slightly arcuate, subglabrous to puberulent, terminating in a perfect spikelet; spikelets loosely spaced, 8–32 per branch, at maturity spreading-ascending, disarticulating above the glumes, subterete, mostly 3–5 mm long excluding the awns, each with a single rudimentary floret above the perfect one; glumes subequal to unequal, lanceolate, acute to slightly bifid and mucronate apically, often purple-tinged, glabrous, the first 1.5–4 mm long, about as wide as the second, the latter mostly (2) 3–5 mm long, typically exceeding the body of the perfect lemma; lemma of the lower floret 2–4 mm long, glabrous to short-bearded on the callus, rounded on the back, glabrous to sparsely pubescent, the 3 nerves prolonged to form subequal or unequal awns 1–7 mm long, the anthers ca 0.5 mm long; rudiment usually consisting of a short stalklike base terminating in 3 awns 3.5–6 mm long; 2n = 20, 28. Our few specimens from cliffs and rocky hillsides at 920 to 1220 m in San Juan and Washington counties; California and Nevada to Texas and Mexico; 2 (0).

Bromus L.

Annuals or perennials; culms tufted, rarely rhizomatous; herbage glabrous or pubescent (mostly retrorsely so); sheaths closed at least half their length, in a few species bearing auricles at the summit; blades flat or occasionally loosely involute; ligule membranous, often brownish, erose to lacerate, pubescent or glabrous; inflorescence a compact to open, erect to nodding, panicle or raceme, the branches 1–8 per node; spikelets compressed to terete, at least 12 mm long, 1– to many-flowered, the uppermost floret sometimes reduced and sterile, the rachilla disarticulating above the glumes and between the florets; glumes lanceolate, rounded on the back to strongly laterally compressed and keeled; usually shorter than the lowermost lemma, awnless or rarely awn-tipped, the first 1– to 7–nerved, the second 3– to 9–nerved; lemma lanceolate to lance-ovate or obovate with a small glabrous or laterally puberulent callus, the body rounded on the back to strongly laterally compressed and keeled, 3– to 13–nerved, narrowly to broadly membranous-margined, awnless or more often with a straight to divergent, mostly subterminal awn, often arising between the 2 lobes or teeth of a bifid apex; palea shorter or barely longer than the lemma; stamens (2) 3, the anthers minute in cleistogamous florets, larger in chasmogamous ones; caryopsis usually adherent to the lemma and palea (one or both sometimes falling away at maturity), linear-oblong to elliptical, bearing at the summit a persistent tuft of short hairs, the embryo 1/8–1/3 the length of the grain; x = 7. Tsvelev (1984) and Weber (1976) elevate to the rank of genus three of the subgeneric sections of *Bromus* as recognized by Hitchcock (1951): *Ceratochloa* comprised of those species with strongly compressed-keeled glumes and lemmas, *Bromopsis* of perennial species, and *Bromus* of annual ones. I have followed the more conservative approach of other contemporary North American workers and of Tutin et al. (1980). Most perennial bromes provide good forage, but the fruiting awns of some annuals are capable of causing mechanical injury to eyes and mouths of grazing animals.

Elliott, F. C. 1949. *Bromus inermis* and *B. pumpellianus* in North America. Evolution 3: 142–149.
Wagnon, H. K. 1952. A revision of the genus *Bromus*, section *Bromopsis*, of North America. Brittonia 7: 415–480.

Note: In the following key, spikelet length excludes awns; lemma length includes apical teeth (when present) but excludes awns; and awns are measured from the point at which they arise on the lemma.

1.	Lemmas with awns averaging at least 10 mm long; plants annual	2
—	Lemmas awnless or with awns averaging less than 10 mm long (rarely to 12 mm in the perennial *B. vulgaris*); plants annual or perennial	7
2(1).	Awns of the lemmas mostly 30–60 mm long; first glume 13–25 mm long	*B. diandrus*
—	Awns of the lemmas averaging less than 30 mm long; first glume averaging less than 13 mm long	3
3(2).	First glume 3– to 5–nerved; lemmas averaging more than 1.5 mm wide midnerve to margin	*B. japonicus*
—	First glume 1–nerved; lemmas averaging not more than 1.5 mm wide midnerve to margin	4
4(3).	Awns geniculate and twisted near the base; lemma with apical teeth often bristlelike; second glume typically equal to or exceeding the lowermost lemma; plant reported only from Washington County	*B. trinii*
—	Awns straight or nearly so, not twisted below; lemma with apical teeth membranous; second glume often shorter than the lowermost lemma; plants more general in distribution	5
5(4).	Panicles dense, erect, mostly wedge-shaped in outline, 0.3–1 dm long including the awns, the branches mostly very short and erect-ascending; apical teeth of the lemma mostly 2.5–5 mm long	*B. rubens*
—	Panicles loose, often either nodding or more than 1 dm long, the branches often elongate, flexuous, and widely spreading; apical teeth of the lemma to 3 mm long	6
6(5).	Lemmas 9–13 (14) mm long, the awns 7–20 mm long; primary panicle branches typically bearing more than 3 spikelets (except in very young plants)	*B. tectorum*
—	Lemmas averaging more than 13 (13–20) mm long, the awns 15–30 mm long; primary panicle branches bearing 1 or 2 (3) spikelets	*B. sterilis*
7(1).	Spikelets strongly compressed, the glumes and lemmas sharply keeled (except at fruiting); lemmas long-tapered to an acute tip; first glume (3) 5– to 9–nerved	8
—	Spikelets terete to moderately flattened, the glumes and lemmas not regularly sharply keeled; lemmas often obtuse to rounded apically; first glume 1– to 5–nerved	9
8(7).	Lemmas awnless or with awns averaging no more than 3 (0.5–3) mm long; first glume 5– to 9–nerved; plants rare, weedy	*B. catharticus*
—	Lemmas with awns averaging more than 3 (2–8) mm long; first glume 3 (5) –nerved; plants widespread, occupying chiefly native habitats	*B. carinatus*
9(7).	Plants annual (occasionally biennial); second glume 5– to 7–nerved	10
—	Plants perennial; second glume 3–nerved (occasionally 5–nerved in the rhizomatous *B. inermis*)	13
10(9).	Palea equal to or exceeding the lemma, the palea tip visible at the apex of at least some of the lemmas; lemmas at maturity usually shiny-indurate with the margins incurved around the grain	*B. secalinus*
—	Palea shorter than the lemma; lemmas at maturity neither shiny-indurate nor with margins inrolled	11
11(10).	Lemmas awnless or with an awn tip less than 2 mm long, at maturity 3–4.5 mm wide midnerve to margin	*B. briziformis*
—	Lemmas regularly awned, rarely as much as 3 mm wide midnerve to margin	12
12(11).	Pedicels mostly shorter than the spikelets, the panicles compact, erect; lemmas hairy or less often glabrous, at maturity the area between the nerves often concave	*B. hordeaceus*
—	Pedicels mostly longer than the spikelets; panicles loose, erect or nodding; lemmas glabrous or merely scabrous, the area between the nerves flat	*B. japonicus*
13(9).	Lemmas glabrous or obscurely puberulent; awns lacking or 1–3 (4) mm long; panicle branches mostly erect or ascending; plants rhizomatous	*B. inermis*
—	Lemmas distinctly hairy, the awns 1–10 (12) mm long; panicle branches drooping; plants without rhizomes	14
14(13).	Awns averaging no more than 3 (1–4) mm long; glumes often pubescent, the first mostly 3–nerved; leaf blades 1–5 (6) mm wide	*B. anomalus*
—	Awns averaging more than 3 (2–12) mm long; glumes glabrous or pubescent, the first 1–nerved (sometimes 3–nerved, usually only at the base); leaf blades often more than 6 (3–15) mm wide	15

15(14). Awns averaging less than 6 (2–6) mm long; second glume glabrous or scabrous, not distinctly hairy; culm nodes typically glabrous; ligule to 2 mm long .... *B. ciliatus*

— Awns averaging at least 6 (5–12) mm long; second glume often hairy near the margins; culm nodes typically pubescent; ligule 2–5 mm long .... *B. vulgaris*

Bromus anomalus Rupr. ex Fourn. Nodding Brome. [*Bromopsis anomala* (Rupr.) Holub; *B. porteri* (Coult.) Nash; *Bromus frondosus* (Shear) Woot. & Standl.; *B. porteri* Nash]. Perennial; culms solitary or in small tufts, 3–10 (12) dm tall, typically puberulent, at least near the nodes; sheaths glabrous or hairy, without auricles; blades 1–5 (6) mm wide, flat to loosely involute, scabrous and occasionally long-hairy; ligule 0.2–1 (2) mm long; panicle 0.5–1.8 dm long, open, nodding; spikelets typically 1 or 2 per primary branch, 13–35 mm long, subterete to moderately compressed, (5) 7- to 11–flowered; glumes often pubescent, the first elliptical or less often lanceolate, 5–8.5 mm long, typically widest at about midlength, obtuse to acute apically, (1) 3–nerved (usually some 1–nerved within any one panicle), the second similar to the first in shape, 6–11 mm long, 3–nerved; lemma narrowly elliptical, slightly compressed to rounded on the back, (8) 9–13 mm long and 1–2 mm wide midnerve to margin, 5- to 7–nerved, conspicuously hairy throughout, sometimes more densely so along the margins, acutish to rounded and occasionally shallowly notched at the apex, the awn (1) 1.5–3 (4) mm long; anthers 2–3 (3.7) mm long; $2n = 14, 28$. Sagebrush, mountain brush, pinyon-juniper, ponderosa pine, aspen, and spruce-fir communities, on open slopes and in meadows at higher elevations, at 1680 to 3350 m in Beaver, Box Elder, Cache, Carbon, Daggett, Duchesne, Emery, Garfield, Grand, Iron, Kane, Millard, Piute, Rich, Salt Lake, San Juan, Sanpete, Sevier, Summit, Uintah, Utah, Wasatch, Washington, and Wayne counties; British Columbia to Saskatchewan, south throughout most of the western U. S. and in Mexico, east to the Dakotas and Texas; 157 (iv). Segregation of *B. frondosus* from *B. anomalus* has been based on glabrous sheaths and panicles more than 1 dm long, characters not necessarily correlated in our material. According to Gould (1951), *B. frondosus* is probably only a robust form of *B. anomalus*. See comments under *B. ciliatus* concerning the close relationship of that species to *B. anomalus*. Nodding brome provides palatable forage and its presence in abundance is evidence of good range condition.

Bromus briziformis Fisch. & Mey. in Fisch., Mey., & Trautv. Rattlesnake Chess. Annual; culms 1–6 (8.5) dm tall, solitary or in small tufts; sheaths densely soft-hairy, without auricles; blades ascending to lax, 1–6 (8) mm wide, flat or occasionally somewhat involute, scabrous to more often pubescent; ligule 0.5–3 mm long; panicle 0.4–1.7 (2) dm long, nodding; spikelets 1–3 per primary branch, moderately compressed, elliptical to oblong, 12–32 mm long and 7–15 mm wide, 7- to 15–flowered; glumes broad, rounded to U-shaped on the backs, obscurely multinerved, the first 4–6 mm long, the second 5–9 mm long, shorter than the lowermost lemma; lemma broadly diamond-shaped, angled on the margins, at maturity more or less inflated, 6.5–10 mm long and (2.8) 3–4.5 mm wide midnerve to margin, obscurely multinerved, glabrous, abruptly acute to obtuse and entire to emarginate apically, awnless or with an awn tip to ca 1 mm long; palea shorter than the lemma; anthers to ca 0.8 mm long; $2n = 14$. Salt desert shrub, sagebrush, mountain brush, and juniper communities, and on open slopes, at 1280 to 1955 m in Box Elder, Cache, Daggett, Davis, Millard, Salt Lake, Sevier, Summit, Tooele, Uintah, and Utah counties; native to Eurasia, introduced in Alaska, British Columbia, Ontario, most of the western U. S., Mexico, the Great Plains, and the Northeast; 65 (v). The spelling *"brizaeformis"* is corrected to *briziformis* in accordance with the International Code of Botanical Nomenclature (Rhodora 69: 451–455. 1967). A population of *B. briziformis* adjacent to one of *B. japonicus*, included a few plants with the broad spikelets of the former and lemmas awned as in the latter (Arnow 2895 UT).

Bromus carinatus H. & A. Mountain or California Brome. [*B. breviaristatus* Buckley; *B. marginatus* Nees ex Steudel; *B. polyanthus* Scribn. ex Shear; *Ceratochloa carinata* (H. & A.) Tutin]. Perennial, often flowering the first season; culms solitary or tufted, usually erect, (2) 6–18 dm tall, glabrous or minutely hairy; sheaths glabrous or short- to long-hairy, evidently lacking auricles in our material; blades lax, 2–12 (15) mm wide, flat or rarely folded to involute, glabrous or more often scabrous to variously pubescent; ligule 1–4 mm long; panicle narrowly elongate to broadly pyramidal, compact to open, (0.5) 1–3 dm long, the branches mostly 3–15 cm long, erect or somewhat divergent at flowering, at fruiting often drooping under the weight of the spikelets; spikelets usually 1–4 per primary branch, strongly compressed, 15–50 (60) mm long, 4- to 16–flowered; glumes lanceolate, strongly keeled, sharply acute at the apex, glabrous or scabrous to short-hairy, the first 6.5–11 mm long, 3 (5) –nerved, the second 9–13 mm long, 5- to 7 (8) –nerved; lemma lanceolate, sharply keeled (becoming less so as the caryopsis matures), (10) 13–17 mm long and 2–3 mm wide midnerve to margin, long-tapered to a sharply acute, entire or minutely notched apex, more or less prominently 7- to 9–nerved, glabrous or scabrous to densely pubescent throughout, the awn essentially terminal, (2) 3–7 (8) mm long; anthers (1.5) 2–8 (9) mm long or 0.4–1 mm long in cleistogamous florets; $2n = 28, 42, 56, 70$. Sagebrush, mountain brush, aspen, spruce-fir, and subalpine meadow communities, at (920) 1830 to 3200 m in all Utah counties except Daggett and Wayne; Alaska to Alberta and Ontario, the western U. S, Texas, and Mexico; naturalized in Europe; 277 (ix). The facultatively cleistogamous *B. carinatus* is part of a highly successful polymorphic complex consisting of a number of closely related, possibly conspecific taxa widespread in western North America and in parts of Central and South America. Mountain brome is a palatable, short-lived bunchgrass valuable for grazing and erosion control in areas with a mean annual precipitation of 46 cm or more (Thornburg 1982). It is better adapted for broadcast seeding than most grasses but is somewhat less nutritous than many other mountain grasses at comparable stages of growth (Vallentine 1961).

Bromus catharticus Vahl Rescue Grass. [*Festuca unioloides* Willd.; *B. unioloides* (Willd.) H.B.K.; *B. willdenowii* Kunth; *Ceratochloa unioloides* (Willd.) Beauv.]. Annual to short-lived perennial; culms solitary to densely tufted, erect to somewhat decumbent, (1) 3–12 dm tall; sheaths glabrous or pubescent, without auricles; blades flat, 2–10 mm wide, glabrous or scabrous to variously pubescent; ligule 2–5 mm long; panicle narrowly elongate to broadly pyramidal, compact to open, 0.5–3 dm long, erect or ultimately nodding; spikelets mostly 1–6 per primary branch, strongly compressed, 10–45 mm long, 4– to 14–flowered; glumes lanceolate, strongly keeled, sharply acute at the apex, the first 6–12 mm long, 5– to 7 (9) –nerved, the second 7–13 mm long, 7– to 9–nerved; lemma lanceolate, sharply keeled, 9–15 (20) mm long and 1.5–2.5 mm wide midnerve to margin, long-tapered to a sharply acute, rarely emarginate apex, usually prominently 7– to 13–nerved, glabrous or scabrous to pubescent, awnless or awn-tipped, rarely with an awn to 3 mm long; palea shorter than the lemma; anthers 2–4 mm long, much shorter in cleistogamous florets; 2n = 42. Along waterways, in waste places, fallow fields, and as a weed of cultivated sites, at 850 to 1890 m in Carbon, Garfield, Iron, Kane, Piute, Utah, Washington, and Wayne counties; native to South America, cultivated in the southern U. S. for forage, escaping and adventive northward; naturalized in Europe; 23 (0). This plant was referred to *B . unioloides* when the original name, *B . catharticus*, was rejected as a *nomen confusum*. Recently, Pinto-Escobar restored *B . catharticus* as the valid name (Caldasia 11: 9–16. 1976). Raven (Brittonia 12: 219–221. 1960) differentiated between *B . unioloides* and *B . willdenowii*, but subsequent workers including Pinto-Escobar (Bot. Jahrb. Syst. 102: 445–457. 1981) have found no basis for recognizing more than one species. Rescue grass is a plant of subtropical regions, which "rescues" livestock with its lush winter growth.

Bromus ciliatus L. Fringed Brome. [*Bromopsis ciliata* (L.) Holub; *Bromus richardsonii* Link]. Perennial; culms in mostly small loose tufts, 5–12 dm tall, usually glabrous; sheaths without auricles, glabrous or hairy, the lowermost sometimes densely pubescent, the upper more sparsely so, the margins often ciliate near the ligule; blades lax, 3–15 mm wide, scabrous or sparsely long-hairy; ligule 0.3–2 mm long; panicle narrowly elongate, 0.7–2.5 dm long, nodding; spikelets typically 1–4 per primary branch, terete to moderately compressed, 15–30 mm long, 5– to 9-flowered; glumes glabrous or scabrous to occasionally scabrid-puberulent on the nerves, the first 5–10 mm long, tapered from below midlength to an acute apex, 1– or basally 3–nerved, the second lanceolate to broadly elliptical, 6–12 mm long, 3–nerved; lemma narrowly elliptical, rounded on the back, 8–15 mm long and 1–2 mm wide from midnerve to margin, 5– to 7–nerved, conspicuously hairy along the margins or sometimes throughout, entire or shallowly bifid apically, the awn (2) 3–6 mm long, mostly straight; anthers 1–3 (4) mm long; 2n = 14, 28, 56. Aspen and spruce-fir communities and occasionally under ponderosa pine and lodgepole pine, in meadows, and on open slopes, at 1520 to 3510 m in Beaver, Box Elder, Cache, Carbon, Daggett, Duchesne, Emery, Garfield, Grand, Iron, Juab, Kane, Piute, Salt Lake, San Juan, Sanpete, Sevier, Summit, Tooele, Uintah, Utah, Wasatch, and Washington counties; Alaska, across southern Canada, south throughout the western U. S., and in much of the remainder of the U. S.; Mexico; 181 (x). Although *B . ciliatus* and *B . anomalus* are closely related and occupy the same type of habitats, they appear in Utah, at least, to be maintaining their integrity as distinct species. Only about 3 percent of all specimens examined (e.g., Arnow 6009 UT) display evidence of intermediacy. Baum (Canad. J. Bot. 45: 1845–1852.

1967) retypified Kalm's specimens, assigning the name *B . ciliatus* to *B . latiglumis* (Shear) A. S. Hitchc. of the northeastern U. S. and referring what we have called *B . ciliatus* to *B . canadensis* Michx. Considerable confusion surrounds the definition and distribution of these and several other closely related taxa, however, and I have followed other contemporary workers (e.g., Cronquist et al. 1977) in retaining the more familiar names. Fringed brome is highly palatable to all classes of livestock and to deer and elk. Although it does not form dense stands, the species is sufficiently common to constitute a moderately good source of forage on some forest ranges.

Bromus diandrus Roth Ripgut Brome. [*B . rigidus* Roth]. Annual; culms 1–9 (10) dm tall, solitary to loosely tufted, usually erect; sheaths without auricles, more or less densely pubescent with moderately long, spreading hairs; blades (1) 2–8 mm wide, flat or involute, scabrous and typically hairy; ligule (1) 2–6 mm long; panicle compact to open, to 2.5 dm long, the branches erect or spreading to drooping; spikelets usually 1 or 2 per primary branch, compressed, 30–50 mm long, (4) 5– to 8–flowered; glumes narrow, glabrous or scabrous on the keel, tapered to sharply acute tips, the first 13–25 mm long, mostly 1–nerved, the second 20–32 mm long, mostly 3–nerved; lemma narrowly lanceolate, 20–36 mm long and 1–1.5 mm wide midnerve to margin, rounded on the back, 5– to 7–nerved, typically scabrous, bidentate apically, the teeth 4–7 mm long, the awn mostly 30–60 mm long; palea shorter than the lemma; anthers 0.8–1.5 mm long; 2n = 28, 42, 56, 70. Creosote bush, pinyon-juniper, and mountain brush communities, mostly in the wake of disturbance, at 975 to 1985 m in Davis, Millard, Salt Lake, Washington, and Weber counties; native to Eurasia, adventive to established from British Columbia south through much of the western U. S. to Mexico, rare in the East; South America; 37 (iv). The binomial *B . rigidus* was applied by Roth in 1790 to a plant differing only slightly from the one he named *B . diandrus* in 1787. Gould and Shaw (1983) refer North American material to *B . diandrus*, and Tsvelev (1984) likewise reduces *B . rigidus* to synonymy.

Bromus hordeaceus L. Soft Chess. [*B . mollis* L.; *B . racemosus* L.]. Annual or biennial; culms solitary or in small tufts, 1–9 (10) dm tall, erect or ascending, occasionally geniculate at the lower nodes; sheaths densely soft-hairy, without auricles; blades 0.5–5 (7) mm wide, flat or occasionally involute, hairy or less often glabrous; ligule 0.2–2 mm long; panicle to 1 (1.3) dm long, erect, compact, the pedicels averaging shorter than the spikelets; spikelets moderately compressed, (8) 10–22 mm long, 5– to 12–flowered; glumes pubescent or glabrous, the first lanceolate to elliptical, 5–8 mm long, 3– to 5–nerved, the second elliptical, 6–9 mm long, 5– to 7–nerved; lemma ovate to obovate or diamond-shaped and then angled on the margins, rounded on the back to slightly compressed, 6–9 (10) mm long and 1.5–2.6 (3) mm wide midnerve to margin, prominently 7– to 9–nerved, the area between the nerves often concave, glabrous or pubescent, the awn arising 0.3–1 (1.5) mm below the entire to bifid tip, the awns of the upper lemmas 4–9 (10) mm long, straight to occasionally divergent and basally twisted; palea shorter than the lemma; anthers 0.5–2 mm long; 2n = 14, 28. Salt desert shrub and sagebrush communities, occasionally on open or wooded slopes, at 1280 to 2530 m in Box Elder, Cache, Davis, Garfield, Juab, Millard, Salt Lake, Summit, Utah, and Weber counties; native to Eurasia, now essentially circumboreal, south through much of the U. S. and in Mexico; South America; 46 (xii). Linnaeus applied the name *B. hordeaceus* to a depauperate grass later shown to be conspecific with *B. mollis*. In consequence, most contemporary European authors (e.g., Tutin et al. 1980) have adopted the earlier epithet as the correct name for the species. *Bromus hordeaceus* occasionally intergrades with *B. japonicus* (q.v.).

Bromus inermis Leysser Smooth or Hungarian Brome. [*B. pumpellianus* Scribn.; *Bromopsis inermis* (Leysser) Holub]. Rhizomatous perennial; culms (1) 2–15 dm tall, solitary or in loose tufts, glabrous or hairy; sheaths glabrous or scabrous to long-hairy, with auricles (when present) to ca 1 mm long; blades 2–15 (19) mm wide, flat, scabrous to variously hairy; ligule 0.5–2 (2.5) mm long; panicle mostly oblong and erect, 0.5–2 (3) dm long, narrow and compact to broad and open, the branches appressed to ascending, usually widely spreading at flowering, sometimes the lower ones drooping in age; spikelets 1–5 per primary branch, terete to somewhat compressed, (12) 15–40 mm long, 5– to 13–flowered; glumes lanceolate to elliptical, rounded on the backs to moderately keeled, sharply acute to rounded apically, the first (4) 5–8 mm long, 1– or occasionally faintly 3–nerved, the second (5) 6–10 mm long, 3– to 5–nerved; lemma elliptical, rounded on the back, (7) 9–13 (14) mm long and to 1.6 mm wide midnerve to margin, acute or obtuse to rounded at the entire to emarginate apex, more or less prominently 3–nerved, often with additional fainter nerves, glabrous or obscurely puberulent only near the base to nearly throughout, awnless or with awns 1–3 (4) mm long; anthers 2.5–6 mm long; $2n = 28, 42, 49, 54$–$58, 70$. Along roads and waterways, in fallow fields and other waste places, and in openings in mountain brush, pinyon-juniper, aspen, spruce-fir, ponderosa pine, lodgepole pine, and meadow communities, at 1280 to 3235 m in all Utah counties except Grand; native to Eurasia, introduced to North America in 1884 (Elliott 1949), now circumboreal, south throughout the U. S. except in the Southeast; 186 (x). Elliott (1949) suggested that due to extensive introgression between the introduced *B. inermis* and the native *B. pumpellianus*, the latter might well be reduced to a subspecies of *B. inermis*, distinguished from the typical element of the species by its pubescent versus glabrous culm nodes and lemmas. Elliott cited a single locality for *B. pumpellianus* in Utah, whereas Wagnon (1952) described ssp. *pumpellianus* as occurring entirely to the east of the State. In any case, I have seen no material that could be identified as belonging to the native strain. Smooth brome, of major economic importance as a sodforming grass, is used for pasture, hay, silage, and erosion control throughout much of the U. S. In the West it is extensively used in seeding rangeland and roadsides below 2800 m in areas with mean annual precipitation of 38 cm or more (Thornburg 1982). The closely related *B. biebersteinii* R. & S., native to Asia, is included in some contemporary seed mixes that have been used in range seedings in our area. This species differs from *B. inermis* in having racemelike panicles, consistently auricled sheaths, and short-awned lemmas.

Bromus japonicus Thunb. ex Murray Japanese or Meadow Chess. [*B. commutatus* Schrader]. Annual; culms solitary or in small tufts, 2–8 (12) dm tall, erect or ascending, often geniculate at the lowermost nodes; sheaths soft-hairy, without auricles; blades ascending to lax, 1–7 (8) mm wide, flat or involute, usually soft-hairy and sometimes scabrous; ligule 0.5–1.5 mm long; panicle to ca 2 dm long, erect to ultimately nodding, the lowermost branches 1–15 cm long; spikelets usually 1 or 2 per primary branch, but the lower branches occasionally bearing as many as 8, moderately compressed at maturity, 12–30 (40) mm long, 5– to 18–flowered; glumes glabrous or scabrous, acute to obtuse at the apex, the first lanceolate, (3.5) 4–6 (7) mm long, 3– to 5–nerved, the second mostly elliptical, 5–8.5 mm long, 5– to 7 (9) –nerved; lemma more or less diamond-shaped to broadly elliptical, rounded on the back to laterally compressed but not sharply keeled, (6.5) 7–12 mm long and 1.5–3 mm wide midnerve to margin, more or less prominently 7– to 9–nerved, glabrous throughout or scaberulous above, the awn arising 0.5–3 mm below the entire to bifid apex of the lemma, those of the lowermost lemmas often poorly developed, those of the upper ones mostly 8–15 mm long, straight or at maturity divergent and often twisted at the base (hygroscopic); palea distinctly shorter than the lemma; anthers 0.5–1.7 mm long; $2n = 14, 28, 56$. In mostly disturbed sites in greasewood, sagebrush, mountain brush, and juniper communities, in meadows and on open slopes, at 760 to 2440 m in Box Elder, Cache, Carbon, Davis, Duchesne, Emery, Garfield, Grand, Iron, Juab, Kane, Millard, Morgan, Rich, Salt Lake, San Juan, Sanpete, Sevier, Summit, Tooele, Utah, Wasatch, Washington, and Weber counties; first collected in Utah in 1909 (Muhlenbergia 6: 63. 1910); native to Eurasia, now circumboreal, south throughout the U. S. and in Mexico; South America; 167 (xxix). As described in most contemporary floras, *B. japonicus* and *B. commutatus* represent extremes in what can be seen in nature as a continuous range of variation. The condition of panicle branches (relatively long, flexuous, and widely spreading in *B. japonicus* versus shorter and stiffly ascending in *B. commutatus*) is frequently cited as diagnostic. Both conditions are usually found within a single population, the length of panicle branches evidently related either to microhabitat or to maturity. That the illustration accompanying Thunberg's original description of *B. japonicus* depicts the panicle as having relatively short, strongly ascending branches also tends to eliminate this character as a criterion for recognizing two species. Numerous additional features, have been employed as diagnostic characters, but none is consistently correlated with the nature of the panicle or with any other character. The variety of characters employed by different authors as diagnostic in their attempt to comply with the concept that two species exist, is itself indicative of the absence of any real distinction. The taxa have the same chromosome number ($2n = 14$), occupy identical habitats, and have essentially the same geographic range. Gould (1951) and Voss (1972) note the lack of any clear distinction between them. *Bromus japonicus* hybridizes with both *B. briziformis* (Arnow 2895 & 4835 UT) and *B. hordeaceus* (Harrison 8370 BRY, Arnow 3964, 5919 UT).

Bromus rubens L. Red Brome; Foxtail Brome. Annual; culms 1–6 (8) dm tall, erect, solitary or in small tufts; sheaths without auricles, pubescent with short soft hairs or glabrous in part; blades 1–5 mm wide, flat to loosely involute, puberulent to pubescent; ligule 1–4 mm long; panicle 0.3–1 dm long (including awns), erect, often turning reddish purple, typically obovoid to wedge-shaped in outline, dense, the short branches only occasionally visible, mostly erect to slightly spreading; spikelets subterete to somewhat compressed, 18–30 mm long, 3- to 8-flowered; glumes narrowly lanceolate, tapered to finely acute tips, scabrous to hairy or occasionally glabrous, the first 5–10 mm long, 1-nerved, the second 8–13 mm long, 3-nerved; lemma lanceolate, 12–17 mm long and 1–1.3 mm wide midnerve to margin, more or less keeled to rounded on the back, 3- to 5-nerved, scaberulous (at least above) to distinctly hairy, bidentate at the sharply acute apex, the teeth membranous, mostly 2.5–5 mm long, the awn 10–24 mm long, straight to slightly divergent; anthers 0.4–0.9 mm long; 2n = 28. Blackbrush, creosote bush, and other mixed desert shrub communities at 700 to 1710 m in Beaver, Box Elder, Garfield, Juab, Kane, Millard, San Juan, Tooele, and Washington counties; first reported in Utah in 1935; native to Eurasia, now introduced from British Columbia south throughout most of the western U. S. and in Mexico; 89 (0).

Bromus secalinus L. Rye Chess or Cheat. Annual or biennial; culms solitary or in small tufts, 2–12 dm tall, erect; sheaths without auricles, glabrous or sparsely pubescent with long soft hairs; blades lax to ascending, 1–7 (10) mm wide, flat or involute, short- to long-hairy and usually scabrous; ligule 0.5–4 mm long; panicle to 2 dm long, erect to ultimately nodding, compact or open, the branches erect-appressed to spreading, the pedicels shorter to longer than the spikelets; spikelets mostly 1 or 2 (3) per primary branch, subterete to moderately compressed, 12–40 mm long, 4- to 12-flowered, slow to disarticulate; glumes mostly elliptical, abruptly acute to rounded at the apex, glabrous, rarely short-awned, the first 4–6 mm long, 3(5)–nerved, the second 5–9 mm long, 5- to 7-nerved; lemma broadly elliptical to diamond-shaped, rounded on the back, 6–9 mm long and 1.8–2.3 mm wide midnerve to margin, 5- to 7-nerved, becoming hardened except for the membranous, ultimately inrolled margin, at maturity usually smooth and shiny, rarely hairy, the awn arising 1–2 mm below the obtuse, entire to barely notched apex, the awns of the upper lemmas 0.5–8 (9) mm long, straight or flexuous, erect or divergent, occasionally lacking; palea equal to or more often slightly longer than the lemma; anthers 0.8–2 mm long; 2n = 14, 28. Disturbed sites below 1525 m in Cache and Salt Lake counties; native to Eurasia, introduced from Alaska to Alberta, sporadically elsewhere in Canada, and through most of the U. S.; South America; 6 (0). Owing to improved methods of freeing cereal grains from impurities, this farmland weed has become uncommon or rare; since 1909 it has been collected in Utah only once (in 1979).

Bromus sterilis L. Poverty Brome. Annual; culms 2–10 dm tall, solitary or in small tufts, erect or occasionally geniculate at the lower nodes; sheaths without auricles, more or less uniformly puberulent, at least below, occasionally the hairs long and spreading; blades 2–5 (8) mm wide, flat to loosely involute, obscurely to conspicuously hairy; ligule 1.5–4 mm long; panicle 1–2 (2.4) dm long, open, the branches ascending to drooping, with spikelets usually 1 or 2 (3) per primary branch, these subterete to somewhat compressed, (15) 20–30 (36) mm long, 4- to 9-flowered; glumes narrowly lanceolate, glabrous, tapered to awnlike tips, the first 7–12 (14) mm long, 1-nerved, the second 9–17 (20) mm long, 3-nerved; lemma narrowly lanceolate, 13–20 mm long and 0.8–1.3 mm wide midnerve to margin, 5- to 7-nerved, glabrous or more often scaberulous to scabrid-puberulent, bidentate at the sharply acute tips, the teeth membranous, very slender, mostly 0.5–2 (3) mm long, the awn 15–30 mm long; palea shorter than the lemma; anthers 0.8–1.2 (1.6) mm long; 2n = 14, 28. Usually in disturbed sites below 1830 m in Davis, Salt Lake, Utah, and Washington counties; native to Eurasia, sparingly introduced in Canada and in the western U. S., adventive in Texas, widespread in the East; 26 (xi).

Bromus tectorum L. Cheatgrass; Downy Chess. Annual or winter annual; culms 0.5–8 (11) dm tall, solitary to tufted, usually erect; sheaths without auricles, more or less densely soft-hairy, often with some longer spreading hairs; blades 0.5–6 (8) mm wide, flat to loosely involute, usually densely soft-hairy and sometimes scaberulous, often with longer spreading hairs, especially along the margins near the ligules; ligule 1–3 (4) mm long; panicle to 1.5 (2) dm long, loose to open and ultimately nodding, the branches often flexuous; spikelets 1–8 or more per primary branch, subterete to somewhat compressed, 10–24 mm long, 3- to 7-flowered, green or purple-tinged to purple; glumes narrowly lanceolate, tapered to sharply acute tips, pubescent or occasionally glabrous, the first (4) 5–8 (9) mm long, 1-nerved, the second 8–11 (13) mm long, 3-nerved; lemma narrowly lanceolate, 9–13 mm long and 1–1.5 mm wide midnerve to margin, more or less obscurely 5- to 7-nerved, variously pubescent or occasionally glabrous, bidentate at the apex, the teeth membranous, 1–3 mm long, the awn (7) 10–18 (20) mm long; anthers 0.5–0.9 mm long; 2n = 14. Open slopes, salt desert shrub, sagebrush, pinyon-juniper, and less commonly aspen and conifer communities, at 850 to 2590 m in all Utah counties; native to Eurasia, now circumboreal, south throughout the U. S.; South America; 212 (iv). Cheatgrass, first collected in Utah by Marcus E. Jones in 1894, is now an integral part of our flora, turning whole hillsides red brown as it matures. Often regarded as a noxious weed, the grass has been shown to have considerable economic importance as forage and to provide fair protection to soil and watersheds (Ecology 30: 58–74. 1949). These advantages are offset, however, by the fire-hazard potential and the threat of injury to animals ingesting the harsh awns and sharp-tipped lemmas of fruiting plants. Moreover, in drought situations the presence of *B. tectorum* causes rapid depletion of soil moisture that may retard or prevent the establishment of perennial grasses.

Bromus trinii Desv. in Gay Chilean Chess. Annual; culms 2–7 (10) dm tall, erect, solitary or in small tufts; sheaths sparsely to densely pubescent with long spreading hairs, occasionally with auricles to 0.8 mm long; blades 2–10 mm wide, flat, scabrous to variously hairy; ligule 1–4 mm long; panicle narrowly elongate, 0.8–2 dm long, loosely erect, the branches mostly strongly ascending to somewhat spreading; spikelets 1–4 or more per primary branch, compressed, 14–25 mm long, 5- to 9-flowered; glumes narrowly lanceolate, tapered to sharply acute, often awnlike tips, glabrous or scabrous, the first 7.5–12 mm long, 1-nerved, the second 8–15 mm long, typically equal to or exceeding the lowermost floret, 3-nerved; lemma narrowly lanceolate, 9–16 mm long and 1–1.3 mm wide midnerve to margin, 5-nerved, densely short-hairy, bidentate apically, the teeth often bristlelike, 2–5 mm long, the awn 10–24 mm long, twisted near the base and geniculate; palea often exceeding the lemma; anthers 1.5–2.5 mm long; 2n = 42. Rock crevices at 940 to 1220 m in Washington County; California, Nevada, Arizona, Colorado and Mexico; Chile and Argentina; 9 (0). This species is sometimes regarded as introduced in North America (Gould & Moran 1981).

Bromus vulgaris (Hook.) Shear Columbia Brome. [*B. purgans* var. *vulgaris* Hook.; *Bromopsis vulgaris* (Hook.) Holub]. Perennial; culms loosely tufted, (6) 8–12 dm tall, nodes typically pubescent; sheaths without auricles, glabrous or more or less sparsely pubescent with long straight hairs, the margins often ciliate near the ligule; blades lax, 4–12 mm wide, flat, glabrous or scaberulous to occasionally hairy; ligule 2–5 mm long, often brownish; panicle narrowly elongate, 1–2 dm long, nodding; spikelets mostly 1–3 per primary branch, subterete to somewhat compressed, 12–30 mm long, 3- to 7 (10) –flowered; glumes narrowly lanceolate, sharply acute at the tips, often hairy near the margins, the first 5–9 mm long, 1–nerved, the second 8–11 mm long, 3–nerved; lemma narrowly elliptical to lanceolate, 8–14 mm long and 1–1.5 mm wide midnerve to margin, rounded on the back, 5- to 7–nerved, short-hairy, especially on the keel and along the margins or occasionally only near the base, entire to minutely emarginate at the sharply acute to subacute apex (sometimes split and appearing apically toothed), the awn terminal or subterminal, 5–10 (12) mm long, straight or sinuous; anthers 2–4 mm long; 2n = 14. Aspen-spruce-fir communities at 1825 to 2300 m in Salt Lake County; British Columbia and Alberta south to California, Idaho, and Wyoming; 9 (iii).

Buchloe Engelm.

Low. dioecious or occasionally monoecious, stoloniferous perennial; sheaths open; ligule a ring of hairs; staminate inflorescence a panicle of racemosely arranged, spikelike branches; pistillate inflorescence a panicle of several headlike clusters on a common peduncle, the whole scarcely emergent from the uppermost sheaths; spikelets dimorphic, sessile or nearly so. *Buchloe* is a monotypic North American genus.

Buchloe dactyloides (Nutt.) Engelm. Buffalo Grass. [*Sesleria dactyloides* Nutt.]. Plant mat-forming; culms 0.5–3 dm tall, densely tufted from subsurface branches or in small tufts from the nodes of elongate stolons; sheaths glabrous or more often pubescent with long spreading hairs; blades flat to involute, 1–2.5 mm wide, 0.2–1 dm long, usually sparsely pubescent with hairs mostly 1–2 (3) mm long; ligule a ring of hairs to ca 1 mm long; staminate panicle elevated above the basal leaf blades on slender culms to ca 2.5 dm tall, composed of 1–5 racemosely arranged, 1–sided, spikelike branches 0.5–1.5 cm long, each disarticulating at its base and falling as a unit; pistillate panicle exceeded by the leaves, made up of 2–7 small headlike clusters borne on a short common peduncle, the panicle partially included in the somewhat inflated upper sheaths, each branch disarticulating at the base and falling as a unit, the hardened upper portion of the rachis united with the modified second glumes of the spikelets to form a cuplike structure; staminate spikelets 2–flowered, 4–6 mm long, the glumes lanceolate, unequal, shorter than the lemmas, somewhat laterally compressed to rounded on the back, 1–nerved, the lemma membranous, 3–nerved, acute, the anthers 2–3.5 mm long; pistillate spikelets 1–flowered, sessile or subsessile, (2) 3–5 (7) in each headlike cluster, the first glume usually reduced, the second broad, hardened, rounded on the back and apically 3–lobed, the lemma thin but firm, 3–nerved, each of the nerves prolonged as awn tips; 2n = 20, 40, 60. Known by a single collection (Welsh & Moore 18709 BRY) from the margin of a maintained grassy campground at ca 2044 m in Daggett County; Montana through Wyoming to Mexico and through most of the central states; 1 (0). Buffalo grass is a dominant species of the shortgrass prairie of the western Great Plains, and our single population may represent an introduction. The species is an excellent source of forage, curing on the ground and providing nutritious feed during the winter. It is drought resistant, adapted to grazing, and provides erosion control on heavy soils.

Calamagrostis Adanson

Perennials; sheath open; blades flat to involute; ligule membranous; inflorescence an open to spikelike panicle; spikelets compressed to subterete, narrowly lanceolate to elliptical, 1 (2) –flowered, the rachilla disarticulating above the glumes and prolonged behind the palea as a short, often hairy, bristlelike structure; glumes equal or nearly so, as long as or longer than the floret, the first 1–nerved, the second 1– or 3–nerved; lemma often membranous, laterally flattened to subterete, 3– to 5–nerved, the lateral nerves often prolonged as 4 minute bristlelike teeth, a slender awn arising anywhere from near the base to just below the apex, occasionally the awn lacking; callus well developed, with lateral tufts of hair exceeding or not less than 1/6 the length of the lemma; palea subequal to or shorter than the lemma; stamens 3; caryopsis free within the lemma and palea, the embryo less than half as long as the grain; x = 7. Complexity within this genus derives from the frequency of hybridization among closely related species and possibly from the existence of apomixis in some species complexes. According to Nygren (1954, 1958), apomixis occurs in three of our species: *C. canadensis*, *C. purpurascens*, and *C. stricta* (as *C. inexpansa*). *Calamagrostis epigejos* (L.) Roth, native to Eurasia and introduced in the Great Plains, has been grown in experimental plots in Utah but is not otherwise known for the State. The plant is readily distinguished from native species by its long, linear-subulate glumes and by the abundant callus hairs, these at least one and a half times as long as the lemma.

Nygren, A. 1954. Investigations on North American *Calamagrostis* I. Hereditas 40: 377–397.
———. 1958. Investigations on North American *Calamagrostis* II. Lantbrukshogskolans Ann. 24: 363–368.
Stebbins, G. L. Jr. 1930. A revision of some North American species of *Calamagrostis*. Rhodora 32: 35–57.

1. Awns 2–9 mm long, at maturity geniculate and generally twisted below the bend, typically arising below midlength of the lemma; callus hairs mostly less than half as long as the lemma 2

— Awns to ca 3 mm long, straight or bent but not usually twisted below, arising above or below midlength of the lemma, fragile, often obscure, sometimes lacking; callus hairs from 1/3 to as long as the lemma 3

2(1). Glumes averaging less than 5 (3–5.5) mm long; awn 2–4.5 mm long; panicle typically pale greenish to tawny .... *C. rubescens*

— Glumes averaging more than 5 (5–9) mm long; awn 4–9 mm long; panicle usually purple or purple-tinged .... *C. purpurascens*

3(1). Pedicels mostly smooth; anthers 2–3 mm long ......
.... C. scopulorum
— Pedicels regularly scabrous; anthers 1–3 mm long 4
4(3). Callus hairs as long as the lemma or nearly so 5
— Callus hairs not more than 2/3 as long as the lemma 6
5(4). Glumes gradually long-tapered, often to nearly awnlike tips; panicles ultimately open and often more than 3 cm wide; blades typically dark to bright green and lax, 2–10 mm wide and flat .... C. canadensis
— Glumes mostly abruptly acute, sometimes with minute apiculate tips; panicles contracted or open at flowering but rarely more than 3 cm wide when pressed; blades often light green and mostly stiffly ascending, 1–5 (6) mm wide, often folded to involute .... C. stricta
6(4). Glumes averaging at least 4 (4–6.5) mm long; anthers 2–3 mm long; herbage typically pale blue green .....
.... C. scopulorum
— Glumes averaging less than 4 (2–4.2) mm long; anthers 1–2 mm long; herbage often yellow green ..... C. stricta

Calamagrostis canadensis (Michx.) Beauv. Bluejoint Reedgrass. [*Arundo canadensis* Michx.; *C. langsdorfii* (Link) Trin.; *C. purpurea* (Trin.) Trin.]. Rhizomatous perennial; culms 0.4–1.8 m tall; herbage typically bright to dark green; sheaths glabrous or scaberulous; blades 2–8 (10) mm wide, usually flat and lax, scabrous; ligule 3–10 (12) mm long; panicle open or occasionally dense and narrow, 6–25 cm long, scabrous throughout; spikelets 1 (2) –flowered, the prolongation of the rachilla minute, long-bearded; glumes lanceolate, 3–5 (6) mm long, long-tapered to sharply acute, often awnlike tips to ca 1 mm long, scabrous throughout or only on the keel, often purple-tinged to purple; lemma usually thinner than the glumes, 2.5–4 mm long, the awn fragile, straight, arising slightly below to slightly above mid-length, 1.2–3 mm long; callus hairs abundant and about equal to the lemma; anthers 1–2 mm long, typically without pollen development; 2n = 28, 42–66. Meadows and other moist sites in aspen, spruce-fir, and lodgepole pine communities, and occasionally in alpine meadows, at 1950 to 3355 m in Beaver, Cache, Daggett, Duchesne, Emery, Garfield, Grand, Juab, Morgan, Rich, Salt Lake, San Juan, Sanpete, Sevier, Summit, Uintah, Utah, Wasatch, and Washington counties; circumboreal, south in the U. S. through all but the Southeast; Eurasia; 89 (v). Although *C. canadensis* has long been regarded as a North American species distinct from the Eurasian *C. purpurea*, N. N. Tsvelev of the U.S.S.R. found representative specimens of our material (e.g., Dick s.n. UT) indistinguishable from *C. purpurea*. The basionym *Arundo canadensis* Michx. (1803), being older than *A. purpurea* (1820), *C. canadensis* remains the correct name for this circumboreal species. *Calamagrostis canadensis* appears to hybridize freely with *C. stricta* (e.g., Taye 2836 UT) and with *C. scopulorum* (e.g., Harper s.n. BRY). The grass forms dense stands in moist places in the mountains but, because of the coarseness of the leaves and stems, is only moderately palatable to grazing animals. Only on the high mountains where little else grows does it acquire some significance as a source of forage.

Calamagrostis purpurascens R. Br. in Richards. Purple Reedgrass. Perennial, often with short rhizomes; culms strongly tufted, 1.5–10 dm tall; herbage green or glaucous; sheaths glabrous or scaberulous; blades (1) 2–5 mm wide, flat to involute, smooth or scabrous to minutely hairy; ligule 1–6.5 mm long; panicle narrow, often spike-like, 4–17 cm long and mostly 0.8–1.5 cm wide, dense to occasionally somewhat interrupted, often purple, the branches mostly short and erect (spreading at flowering), scabrid-puberulent to puberulent or rarely scabrous; spikelets 1 (2) –flowered, the prolongation of the rachilla long-hairy; glumes lanceolate, (5) 6–8 (9) mm long, long-tapered to a sharply acute tip, glabrous or scaberulous to puberulent, usually purple or purple-tinged; lemma scarcely thinner than the glumes, 4–7 mm long, glabrous or scaberulous, the awn relatively stout, especially below, ultimately twisted and geniculate, typically arising from the lower 1/3, mostly 4–9 mm long, often exceeding the spikelet by 2 mm or more, rarely lacking; callus hairs less than half as long as the lemma, occasionally lacking; anthers 1.5–3 mm long; 2n = 28, 40–58, 84. Open rocky slopes, aspen-spruce-fir, meadow, and lodgepole pine communities at (2290) 2740 to 3965 m in Daggett, Duchesne, Grand, Iron, Juab, Piute, Salt Lake, San Juan, Sevier, Summit, and Uintah counties; subarctic America south through most of the western U. S., east to South Dakota and Minnesota; Siberia; 78 (v). A specimen of *C. scopulorum* (Harrison 11301 UT), collected near a population of *C. purpurascens* (Harrison 11282 BRY), is atypical in having the yellow green foliage and scabrid-puberulent panicle branches of *C. purpurascens*, suggesting that hybridization between the two species is occurring.

Calamagrostis rubescens Buckley Pinegrass. Rhizomatous perennial; culms (4) 7–11 dm tall; herbage yellow green to dark green; sheaths glabrous or pubescent, often hairy on the collar; blades flat or folded, 1.5–5.5 mm wide, scabrous to hairy; ligule 1–4 (5) mm long; panicle narrow, mostly 6–20 cm long and to ca 3 cm wide at flowering, loose to compact with branches mostly less than 2 cm long, erect or somewhat spreading at flowering, typically pale greenish to tawny, the branches and pedicels scabrous to more often scabrid-puberulent; spikelets 1-flowered, the prolonged rachilla conspicuously hairy; glumes lanceolate, (3) 4–5.5 mm long, mostly gradually tapered to an acute apex, smooth throughout or scabrous on the keel, occasionally scaberulous over the back, usually whitish to pale green or tawny; lemma similar to the glumes in texture, 2.5–4 mm long, glabrous or scaberulous, the awn relatively stout, arising below midlength, at maturity geniculate and twisted below the bend, 2–4.5 mm long; callus hairs to ca 1 mm long, usually sparse; anthers 1.5–2.8 mm long; 2n = 28, 42, 56. Ponderosa pine, aspen, and more commonly lodgepole pine communities, commonly intermixed with *Carex geyeri*, at 2440 to 2745 m in Daggett, Rich, and Uintah counties (Uinta Mts.); British Columbia to Manitoba, south to California, Idaho, and Colorado; 4 (0). Pinegrass is shade tolerant but rarely flowers except in cutover areas. It is important as a sod-former and, although unpalatable at maturity when its herbage becomes tough and harsh, is highly palatable during the spring (Vallentine 1961).

Calamagrostis scopulorum Jones Jones Reedgrass. [*C. scopulorum* var. *lucidula* Kearney, type from the Wasatch Range]. Perennial, often with short rhizomes; culms usually loosely tufted, 3–9 dm tall; herbage pale blue green or rarely yellow green to dark green; sheaths glabrous or scaberulous; blades 1–7 (11) mm wide, flat or folded, mostly firm and erect-ascending, generally scaberulous; ligule 2–5 mm long; panicle narrow, mostly

5-20 cm long and to ca 3 cm wide at flowering, typically compact with branches erect to slightly spreading at flowering, the branches and pedicels smooth or sparingly scabrous; spikelets 1 (2) -flowered, the prolongation of the rachilla minute, long-hairy; glumes lanceolate to elliptical, 4-6 (6.5) mm long, abruptly acute to long-tapered at the apex, glabrous or more often the keel scaberulous, usually whitish to pale green or brownish purple, occasionally dark purple; lemma scarcely thinner than the glumes, (3) 3.5-5 mm long, glabrous or scaberulous, the awn delicate (occasionally lacking) erect or rarely spreading, not twisted near the base, arising near or above midlength, to ca 2 mm long; callus with the longer hairs from 1/3-2/3 as long as the lemma, rarely longer; anthers 2-3 mm long; 2n = 28. Hanging gardens and other rocky sites, streamside and in meadow communities, occasionally in conifer woodlands and krummholtz, at 1070 to 3510 m in Daggett, Duchesne, Emery, Garfield, Grand, Iron, Juab, Kane, Salt Lake, San Juan, Sanpete, Summit, Uintah, Utah, Wasatch, and Washington (type from Springdale) counties; Montana south to Arizona and New Mexico; 107 (0). *Calamagrostis scopulorum* apparently hybridizes with *C. purpurascens* (q.v.), *C. canadensis* (e.g., Welsh, Neese, & Atwood 19075 BRY), and *C. stricta* (e.g., Neese 14881 BRY).

***Calamagrostis stricta* (Timm) Koeler** Slimstem or Northern Reedgrass. [*Arundo stricta* Timm; *C. inexpansa* Gray; *C. neglecta* authors, not (Ehrh.) Gaertn., Mey. & Scherb.]. Rhizomatous perennial; herbage typically yellow green, occasionally glaucous to dark green; culms 3-13 dm tall; sheaths glabrous or scaberulous; blades 1-5 (6) mm wide, flat to involute, mostly firm and ascending, smooth or scabrous to puberulent; ligule 0.7-7 (10) mm long; panicles dense and narrow or rarely open, 6-25 cm long and mostly less than 3 cm wide, the branches erect to strongly ascending or more or less spreading at flowering, regularly scabrous throughout; spikelets 1 (2) -flowered, the prolongation of the rachilla minute, long-bearded; glumes lanceolate to elliptical, 2-4.2 mm long, abruptly acute at the apex or occasionally some of them long-tapered, glabrous or scaberulous, green but soon turning golden brown, occasionally purple-tinged; lemma similar to the glumes, 1.8-3.2 mm long, scaberulous or rarely glabrous, the awn straight or rarely bent but not twisted below, arising from above or below midlength, mostly 1-2.5 mm long; callus hairs rather sparse, from 2/3 to as long as the lemma; anthers 1.3-2 mm long; 2n = 28, 56, 58, 70, 84-105. Moist sites, often streamside, in ponderosa pine, spruce-fir, lodgepole pine, and meadow communities at 1280 to 3230 m in Beaver, Cache, Daggett, Duchesne, Emery, Garfield, Grand, Iron, Kane, Rich, Salt Lake, San Juan, Sanpete, Sevier, Summit, Uintah, Utah, Wasatch, Washington, and Wayne counties; circumboreal, south throughout the western and northeastern U. S.; South America; Eurasia; 118 (ii). I have followed Tutin et al. (1980), Love (Taxon 19: 299-300. 1970), and Voss (Michigan Bot. 11: 28-29. 1972) in referring this taxon to *C. stricta*. See *C. canadensis* and *C. scopulorum* for comments concerning possible hybrid formation.

Calamovilfa Hackel

Rhizomatous perennials; culms usually tall and stout; sheaths open; blades firm, flat to involute; ligule a ring of short hairs; inflorescence an open or contracted panicle; spikelets compressed to subterete, 1-flowered, disarticulating above the glumes; glumes unequal, keeled to rounded on the back, 1-nerved; lemma becoming leathery, more or less rounded on the back, 1-nerved, awnless; callus bearded; palea about equal to the lemma; caryopsis free of the pericarp; x = 10.

Reeder, J. R. and M. E. Ellington. 1960. *Calamovilfa*, a misplaced genus of Gramineae. Brittonia 12: 71-77.

***Calamovilfa gigantea* (Nutt.) Scribn. & Merr.** Big Sandreed. [*Calamagrostis gigantea* Nutt.]. Tall, coarse, rhizomatous perennial; sheaths glabrous except occasionally near the collar; culms mostly solitary at each rhizome node, 1-2 m tall; ligule a ring of hairs 1-2 mm long; blades firm, flat to involute, 5-12 mm wide, tapering into a long, ultimately filiform tip; panicle mostly 3-6 dm long, pale green to tawny, open, the branches to ca 2.5 dm long, mostly solitary at each node and somewhat distantly spaced, stiffly erect to widely spreading, naked at the base; spikelets short-pedicelled, narrowly lanceolate, mostly 7-9 mm long; glumes usually firm, unequal, 1-nerved, obtuse to acute apically, the second shorter than or slightly exceeding the lemma; lemma narrowly lanceolate, 1-nerved, acute, hairy on the back above the base; callus with a dense tuft of silky hairs 3-5 mm long; anthers 3-5.5 mm long; 2n = 60. Sand dunes and sandy soil in sagebrush, pinyon- juniper, and ponderosa pine communities below 1830 m in Kane, San Juan, and Washington counties; Arizona and Colorado, east in the southern Great Plains; 13 (0). Big sandreed is a good sand binder but is too coarse to be of value as forage.

Catabrosa Beauv.

Perennial; sheaths closed 1/2-3/4 the length; blades lax, flat; ligule membranous; inflorescence an open panicle; spikelets subterete, (1) 2 (3) -flowered, disarticulating above the glumes and between the florets; glumes unequal, linear to obovate, shorter than the lowermost lemma; lemma rounded on the back, prominently 3-nerved, the nerves parallel or nearly so; palea similar to the lemma in texture and size; caryopsis free of the lemma and palea, the hilum ovate, 1/4 as long as the grain; stamens 3; x = 5. *Catabrosa* is an amphi-Atlantic monotypic genus.

***Catabrosa aquatica* (L.) Beauv.** Brookgrass. [*Aira aquatica* L.]. Perennial, often stoloniferous or rhizomatous; culms 0.5-7 (8) dm tall, erect or decumbent, freely rooting at the lower nodes; sheaths glabrous; blades lax, flat, 2-13 mm wide, typically prow-shaped at the tips, essentially smooth; ligule 1.5-8 mm long; panicle open, ovate to oblong, 0.7-3 dm long, erect, the branches smooth, ascending to widely spreading or descending; spikelets terete to somewhat dorsiventrally compressed, 2 (3) -flowered, blunt-elliptical, 1.6-3.5 (4) mm long; glumes thin, often pale, linear to fan-shaped, obscurely nerved, much shorter than the lowermost floret, the first glume often much reduced, the second 1.2-2.5 mm long, acute to rounded or truncate and erose to irregularly toothed apically; lemma somewhat fleshy, typically yellow brown to greenish, 1.5-3.5 mm long, rounded on the back, glabrous, with 3 prominent, nearly parallel nerves, the apex membranous, truncate and erose; anthers 0.7-1.8 mm long; 2n = 20, 42. Wetland or aquatic grass associated with streams, springs, wet meadows, and marshes, in oak-maple, aspen-spruce-fir, and ponderosa pine communities at 1310 to 3110 m in all Utah counties except Daggett, Millard, San Juan, and Weber; Alberta to Newfoundland, south through most of the western U. S., the Dakotas, Nebraska, and Wisconsin; Eurasia; Africa; introduced in South America; 144 (vi). Brookgrass is a fairly good pasture plant but can produce severe acid poisoning in cattle if consumed in large quantities (Tsvelev 1984).

Cenchrus L.

Annuals or perennials; culms erect or more often decumbent to prostrate; sheaths open, often compressed and more or less keeled; blades flat to involute; ligule a membranous-based fringe of hairs or lacking; inflorescence a spikelike panicle consisting of few to numerous, subsessile, spiny burs (formed by the partial fusion of modified sterile branchlets) arranged along a short unbranched rachis, each bur partially enclosing 1–3 (8) spikelets and falling intact at maturity; spikelets sessile within the bur, 2–flowered, the lower floret staminate or neutral, the upper one perfect; glumes membranous, unequal, ovate to lanceolate, the first 1– to 3–nerved, sometimes reduced or lacking, the second 1– to 7–nerved, nearly as long as the spikelet; lemma of the sterile lower floret similar in texture and size to the second glume, 3– to 7–nerved, the palea smaller than the lemma; lemma of the upper, fertile floret usually firm, smooth, and obscurely nerved, the margins inrolled around the palea; caryopsis free of but firmly enclosed within the lemma and palea, the embryo at least half the length of the grain; x = 9.

DeLisle, D. G. 1963. Taxonomy and distribution of the genus *Cenchrus*. Iowa State J. Sci. 37: 259–351.

Cenchrus longispinus (Hackel) Fern. Field Sandbur.

[*C. echinatus* f. *longispinus* Hackel; *C. pauciflorus* authors, not Benth.]. Annual, often forming large clumps with culms ascending to decumbent or prostrate, 0.4–9 dm long; sheaths compressed, open and loose, glabrous or occasionally long-hairy; blades mostly flat, (1.5) 3–9 mm wide and to ca 20 cm long, scabrous to sparsely long-hairy; ligule a ciliate membrane, 0.7–1.7 mm long; inflorescence spikelike, (1.5) 3–10 cm long; burs subglobose, subglabrous to densely pubescent, the body 6–12 mm long and 3–7 mm wide, cleft on two sides, the spines 45–75, irregularly spaced, 1–6 (7) mm long and 0.1–1.4 mm wide at the terete to flattened base, spreading or reflexed, several to many of the numerous spines near the base of the bur bristlelike; spikelets 2–3 (4) per bur, 6–8 mm long; first glume 1.5–3.8 mm long, 1–nerved; second glume 4.4–6 mm long, 3– to 5–nerved; sterile lower lemma similar to but sometimes slightly longer than the second glume, the palea usually smaller than the lemma; fertile upper lemma firm and smooth, 5.8–7.6 mm long, obscurely nerved; anthers 0.6–2 mm long; 2n = 34, 36. Fields, gardens, and other disturbed sites chiefly below 1830 m in Box Elder, Davis, Garfield, Grand, Kane, Salt Lake, San Juan, Uintah, Utah, Washington, Wayne, and Weber counties; Ontario, south through most of the U. S.; Mexico to Venezuela; introduced in Europe; 28 (iii).

Chloris Swartz

Annuals or perennials; sheaths open, rounded to flattened and keeled; blades flat or folded; ligule membranous or sometimes lacking; inflorescence a panicle with 2–20 spikelike branches digitately or racemosely arranged, the spikelets sessile or short-pedicelled and overlapping to remote along one side of the slender, more or less 3–angled rachis; spikelets flattened to subterete, 2– to 4–flowered, disarticulating above the glumes, pale to brownish, the lower floret perfect, the upper 1–3 florets staminate or rudimentary, rarely perfect; glumes membranous or nearly so, subequal or unequal, lanceolate, acute to long-tapered, 1–nerved, the second shorter to longer than the florets; lemma of the perfect lower floret laterally compressed and keeled to subterete, 1– to 5–nerved, glabrous or variously pubescent to scabrous on the nerves and margins, often minutely bearded on the callus, awnless or more often awned from the tip or from between two small apical teeth; lemma of the upper sterile floret(s) frequently reduced to a club-shaped, awned or awnless rudiment; paleas shorter than the lemmas; caryopsis free within the lemma and palea, the pericarp only partially fused with the outer seed coat, the embryo to ca 3/4 the length of the grain; x = 10.

Anderson, D. E. 1975. *Chloris* Swartz. pp. 316–333. *In*: F. W. Gould, The grasses of Texas. Texas A. & M. University Press, College Station.

1. Lower lemma densely long-hairy on the margin just below the apex, the hairs mostly 2–3 mm long; plants annual
   .... *C. virgata*

— Lower lemma pubescent with very short hairs; plants perennial............................. *C. verticillata*

Chloris verticillata Nutt. Tumble Windmillgrass.

Tufted perennial; culms 4 dm tall, erect or decumbent at the base, often rooting at subsurface nodes; sheaths glabrous or long-hairy at the summit; blades flat or folded, 1–4 mm wide, abruptly acute to obtuse at the tips, usually scabrous; ligule membranous, mostly ca 1 mm long, occasionally subtended at the base of the blade by hairs to 3 mm or more long; panicle of 10–20 ultimately divergent, slender, spikelike branches, these mostly 5–16 cm long, in 2–5 more or less closely spaced whorls, some of the branches occasionally racemosely arranged, the spikelets in 1 row, sessile to short-pedicelled, appressed to the rachis and usually slightly overlapping or more or less remote, especially near the base of the branches; spikelets 2–4 mm long excluding awns, 2–flowered, the lower floret perfect, the upper typically reduced; glumes unequal, narrowly lanceolate, 1–nerved, awn-tipped, the second glume to 4 mm long including the awn tip, slightly longer than the upper floret; lemma of the perfect lower floret firm, strongly laterally compressed, lanceolate to elliptical, 2–3.5 mm long, pale green or whitish to tawny, occasionally purple-tinged, 3–nerved, glabrous or minutely hairy on the keel, short-hairy on the marginal nerves, acute or nearly so at the entire to minutely toothed apex, with an awn 4–9 mm long arising just below the apex; lemma of the upper floret oblong to broadly club-shaped, 1–2.3 mm long, often somewhat inflated, glabrous, truncate apically, with an awn 3–7 mm long; 2n = 40. Weed of cultivated areas below 1830 m in Emery and Salt Lake counties; most of the southwestern U. S., east to Indiana and Louisiana, introduced in Maryland; 9 (0).

Chloris virgata Swartz Feather Fingergrass. Annual;

culms usually tufted, mostly 2–10 dm tall, sparingly branched above the base, often geniculate at the lower nodes; sheaths glabrous, occasionally long-hairy near the ligule; blades flat, 2–15 mm wide, tapered to sharply acute tips, glabrous or scabrous to long-hairy; ligule membranous, 0.2–1 mm long, occasionally lacking; panicle of 3–20 erect to strongly ascending, spikelike branches 2–10 cm long in a terminal whorl on the culm, the whole breaking away at maturity, the spikelets in 2 rows, mostly subsessile, closely overlapping throughout the length of the branch; spikelets 2.5–4.2 mm long excluding awns, 2 (3) –flowered, the lower floret perfect, the upper 1 (2) typically reduced; glumes unequal, narrowly lanceolate, 1–nerved, the second glume short-awned, to 4.3 mm long including the awn, usually equal to or slightly longer than the upper floret; lemma of the perfect lower floret firm, strongly laterally compressed, skewed- elliptical, 2.5–4.2 mm long, pale green or whitish to tawny, often purple-tinged, 3–nerved, the keel glabrous or densely short-hairy, the marginal nerves densely hairy along the upper 1/3 with hairs mostly 2–3 mm long and obscurely hairy below, acute or nearly so at the entire to minutely toothed apex with a subterminal awn 5–12 mm long; lemma of the upper 1 (2) florets narrowly club-shaped, 1.4–3 mm long, glabrous, truncate apically, with an awn 3–9.5 mm long; $2n = 14, 20, 26, 30, 40$. Along waterways and roadsides, and as a weed of irrigated sites chiefly below 1830 m in Grand, Juab, Kane, and Washington counties; southwestern and south-central U. S., into Mexico, and along most of the east coast from Maine to Florida; South America; Asia and Africa; 21 (0).

Cinna L.

Perennials; sheaths open; blades flat; ligule membranous; inflorescence a panicle; spikelets laterally compressed, 1–flowered, disarticulating below and sometimes above the glumes; glumes about equal to the floret, narrowly lanceolate, 1– to 3–nerved; lemma similar to the glumes, 3–nerved, awnless or a short straight awn arising just below the tip; palea nearly equal to the lemma; stamens 1–3; caryopsis free within the lemma and palea, the hilum oval, 1/7–1/6 the length of the grain; $x = 7$.

Cinna latifolia (Trev. ex Goeppert) Griseb. Drooping Woodreed. [*Agrostis latifolia* Trev. ex Goeppert]. Rhizomatous perennial; culms loosely tufted, 0.5–2 m tall; sheaths glabrous or scabrous; blades lax, flat, (5) 7–17 (20) mm wide, scabrous; ligule brownish, (2) 3–8 mm long; panicle open, nodding, light green, 1–3 dm long; spikelets 2–4 (4.5) mm long and to ca 1 mm wide, disarticulating below and sometimes also above the glumes, the rachilla prolonged behind the palea as a bristlelike structure ca 0.5 mm long; glumes equal or nearly so, narrowly lanceolate, 2–4 (4.5) mm long, equal to or exceeding the floret, 1–nerved, typically strongly keeled along the midnerve, tapered to a sharply acute to awnlike apex, strongly scabrous to short-ciliate on the keel and often scaberulous over the back; lemma obscurely 3–nerved, glabrous or more often scaberulous, awnless or with a subterminal awn to 1.2 mm long; anthers 0.6–1 mm long; $2n = 28$. Chiefly streamside in mountain brush, aspen, and spruce-fir communities at 1710 to 2595 m in Juab, Millard, Salt Lake, Summit, Tooele, Uintah, Utah, and Wasatch counties; circumboreal, south through most of the western U. S., the northern Great Plains, and the Northeast; Eurasia; 50 (viii).

Crypsis Ait.

Annuals; culms prostrate to erect, usually much branched from the base and often from leaf axils; sheaths open; blades mostly short, flat to involute; ligule a ring of short hairs; inflorescence a spikelike panicle; spikelets 1–flowered, strongly compressed, disarticulating below and sometimes above the glumes; glumes subequal or unequal, 1–nerved, keeled, typically shorter than or occasionally equal to the lemma, acute; lemma keeled, 1–nerved, acute or mucronate to short-awned; palea equal to or slightly shorter than the lemma, (1) 2–nerved; stamens 2 or 3; caryopsis ellipsoidal to oblong, the pericarp swelling upon becoming wet and usually ejecting the seed, the embryo almost as long as the seed; $x = 9$.

Hammel, B. E. and J. R. Reeder. 1979. The genus *Crypsis* (Gramineae) in the United States. Syst. Bot. 4: 267–280.

1. Panicles mostly less than 5 (2.5–8) mm wide, at flowering free of the upper, unexpanded sheath; spikelets often black-tinged; lemma averaging less than 3 (1.5–3) mm long .... *C. alopecuroides*
— Panicles mostly more than 5 (5–15) mm wide, typically remaining partially enclosed within the expanded upper sheath; spikelets pale to purple-tinged; lemmas often averaging 3 (2.4–4) mm or longer .... *C. schoenoides*

Crypsis alopecuroides (Piller & Mitterp.) Schrader Pricklegrass. [*Phleum alopecuroides* Piller & Mitterp.; *Heleochloa alopecuroides* (Piller & Mitterp.) Host]. Annual; herbage frequently purple to nearly black; culms prostrate to ascending, 0.3–2 (7) dm long, branched from the base, rarely above, often geniculate at the lower nodes; sheaths glabrous or some of them minutely ciliate below the summit; blades flat to loosely involute, mostly 1–12 cm long and 1–2.5 mm wide, glabrous or scaberulous, often hairy near the ligule; ligule a ring of hairs 0.2–1 mm long; panicle spikelike, cylindrical, 0.5–6.5 (8) cm long and 2.5–8 mm wide, at flowering borne above the unexpanded upper sheath; spikelets 1–flowered, strongly overlapping, often black-tinged; glumes subequal or unequal, 1–nerved, glabrous or scaberulous to minutely ciliate on the keel, the first narrowly lanceolate, 1–1.7 (2) mm long, the second slightly broader than the first, 1.2–2 (2.4) mm long, both usually shorter than the lemma; lemma lanceolate, 1.5–2.4 (3) mm long, 1–nerved, keeled, glabrous; anthers 0.5–1 mm long; $2n = 18$. Margins of lakes, ponds, and waterways, and in drying reservoir bottoms, below 1830 m in Cache and Millard counties; native to Eurasia; introduced in the U. S. from Washington to California, also reported from Idaho, Nevada, Wyoming, Montana, and Pennsylvania; 3 (0).

Crypsis schoenoides (L.) Lam. Common Pricklegrass. [*Phleum schoenoides* L.; *Heleochloa schoenoides* (L.) Host ex Roemer]. Annual; herbage frequently purple-tinged; culms prostrate to erect, 0.3–4 (7) dm long, branched from the base, occasionally above, often matforming; sheaths glabrous or some of them minutely ciliate below the summit, the uppermost inflated; blades flat or often involute, especially toward the long-tapered tips, mostly 2–10 cm long and 1–4 mm wide, glabrous or short-hairy; ligule a ring of hairs 0.5–1 mm long; panicle spikelike, ovoid to cylindrical, 0.3–5 (7.5) cm long and 5–15 mm wide, typically remaining partially enveloped by the spathelike upper sheath; spikelets 1–flowered, strongly overlapping, pale to purple-tinged; glumes subequal, narrowly lanceolate, 1.8–2.7 mm long, usually shorter than the lemma, 1–nerved, glabrous or scaberulous to minutely ciliate on the keel; lemma lanceolate, 2.4–3 (4) mm long, 1–nerved, glabrous or the keel scaberulous; anthers 0.7–1 mm long; $2n = 32, 36$. Drying margins of lakes, ponds, and waterways, and in reservoir drawdown areas below 1525 m in Box Elder and Washington counties; native to Eurasia, adventive in California, Oregon, and Nevada, and sporadic in the East; 8 (0).

Cynodon Rich.

Perennials with stolons or rhizomes; culms simple to much branched; sheaths open; blades short, flat to involute; ligule membranous; inflorescence a panicle of 2–8 slender spikelike branches in one or more whorls at the summit of the culm, the spikelets sessile or nearly so, typically in 2 rows along one side of a 3–angled rachis; spikelets 1 (2) –flowered, laterally compressed, the rachilla disarticulating above the glumes and prolonged behind the palea as a slender bristlelike structure, sometimes bearing a second, usually rudimentary floret; glumes subequal or unequal, 1–nerved, keeled, shorter than the lemma; lemma firmer than the glumes, laterally compressed, obscurely 3–nerved, the lateral nerves submarginal, the keel ciliate; palea subequal to the lemma, much narrower; stamens 3; caryopsis free within the lemma and palea, the embryo ca 1/2 the length of the grain; $x = 9$.

de Wet, J. M. J. and J. R. Harlan. 1970. Biosystematics of *Cynodon* L. C. Rich. (Gramineae). Taxon 19: 565–569.

***Cynodon dactylon* (L.) Pers.** Bermuda Grass. [*Panicum dactylon* L.; *Capriola dactylon* (L.) Kuntze]. Perennial with stolons or rhizomes; culms; decumbent to prostrate, 2–8 dm long, leafy, rooting at the lower nodes and producing erect flowering shoots 1–5 dm tall; sheaths smooth below, often hairy near the summit; blades flat or folded, 1.5–5 (15) mm wide, glabrous or scaberulous to sparsely hairy; ligule a ciliate membrane less than 0.5 mm long, subtended at the base of the blade by hairs 1–3 mm long or longer; panicle of (2) 3–7 (10) spikelike branches, these 2–8 (10) cm long, in a single whorl at the apex of the culm or rarely in 2–4 whorls, the spikelets sessile, appressed to the rachis, closely overlapping and alternating in 2 rows (often appearing to be a single row) on one side of a 3–angled rachis; spikelets 1.8–2.8 mm long, typically 1–flowered; glumes 1–nerved, keeled, the first narrowly crescent-shaped or lanceolate, 0.7–1.8 mm long, the second narrowly lanceolate, 1.2–2.2 mm long and scarcely if at all wider than the first, typically shorter than the lemma; lemma firmer than the glumes, boat-shaped, 1.8–2.5 mm long, smooth, keeled, usually pale green to whitish or purple-tinged, glabrous or scabrous to pubescent on the keel and submarginal lateral nerves; anthers 0.6–1 mm long; caryopsis almond-shaped, to ca 1.4 mm long; $2n = 18, 27, 30, 36, 40, 54$. A weed of lawns, and waste places, and along waterways below 1525 m in Cache, Salt Lake, Utah, Washington, and Weber counties; native to Africa, widely introduced in the U. S. and Mexico; South America; Eurasia; 28 (i). Bermuda grass, introduced to the U. S. as early as 1751, is frequently cultivated in the South as a pasture and lawn grass. The species is cited by Holmgren and Maguire (Utah State Circ. 123: 7. 1949) as having become a serious pest in Utah sometime during the 1940s. Its presence as a lawn weed has increased appreciably in some areas of the state in recent years.

Dactylis L.

Perennial; sheaths closed throughout or open to half their length; ligule membranous; inflorescence a panicle with spikelets in dense clusters on short to relatively long, erect to reflexed branches; spikelets laterally compressed, 2– to 5–flowered; glumes subequal to unequal, keeled, 1– to 3–nerved; lemmas firm, keeled, 5–nerved, usually obscurely so; palea shorter than the lemma; stamens 3; caryopsis free within or slightly adherent to the lemma and palea; $x = 7$. *Dactylis* is a widespread monotypic genus.

Stebbins, G. L., Jr. and D. Zohary. 1959. Cytogenetic and evolutionary studies in the genus *Dactylis*. Univ. Calif. Publ. Bot. 31: 1–40.

***Dactylis glomerata* L.** Orchard Grass. Perennial; culms tufted, erect, to 12 dm tall, often with short stout rhizomes; sheaths smooth or scaberulous, often compressed and keeled; blades flat or folded, 1.5–11 mm wide, nearly smooth to strongly scabrous; ligule 2.5–11 mm long; panicle 4–20 cm long, mostly interrupted, the usually few primary branches 1 or 2 (3) at each node of the rachis, ca 0.5–12 cm long or longer, erect to widely spreading or reflexed, naked below and bearing near the tips 1 to several, dense, secund clusters of subsessile to short-pedicelled spikelets; spikelets 5–9 mm long, 2–' to 5 (8) –flowered, occasionally purple-tinged, ultimately disarticulating above the glumes and between the florets; glumes subopposite, lanceolate, often asymmetrical, keeled to rounded on the back, acute to awn-tipped apically, glabrous or scabrous to coarsely ciliate on the keel, otherwise glabrous or occasionally pubescent, the first often longer and firmer than the second, 5–6.5 mm long, the second usually membranous, 3–6 mm long, shorter than the lowermost lemma; lemma mostly firm, lanceolate, (3) 5–6 mm long, keeled, acute or more often tapered to an awnlike tip to ca 2 mm long, coarsely ciliate or rarely only scabrous on the upper portion of the keel, otherwise glabrous or scaberulous; anthers 2–4 mm long; $2n = 14, 21, 27-31, 42$. Sagebrush, mountain brush, ponderosa pine, aspen, spruce-fir, and occasionally desert shrub communities, often in the wake of disturbance, at 850 to 3235 m in all Utah counties except Box Elder, Carbon, and Morgan; native to Eurasia and Africa, now circumboreal, naturalized in temperate regions of the western Hemisphere; 121 (iv). Stebbins and Zohary (1959) agree with most European authors that *Dactylis* is best treated as a monotypic genus. Orchard grass was introduced to the north-central U. S. as a hay and pasture grass sometime prior to 1760 (Vallentine 1961). In the West it grows in irrigated areas or in sites with a minimum of 45 cm annual rainfall, although drought-resistant strains have been developed in recent years (Hanson 1972). The grass has been successfully used in range seedings in both mountain brush and aspen-fir communities. Orchard grass produces an abundance of palatable and nutritious forage during spring and summer and is noted for its resistance to close grazing (Valentine 1961).

Danthonia DC.

Tufted perennials; culms often disarticulating at the nodes in age; sheaths open; blades narrow; ligule a ring of hairs, sometimes with a short membranous base; inflorescence a panicle or raceme, in one species reduced to a single spikelet; spikelets subterete to laterally compressed, (2) 4- to 10-flowered, the florets progressively smaller upward, the rachilla disarticulating above the glumes and between the florets; cleistogamous, 1- or 2-flowered spikelets occasionally present in the axils of the lower leaves; glumes equal or subequal, 1- to 5-nerved, often obscurely so, equal to or longer than the body of the uppermost lemma; lemma broad, somewhat laterally compressed but not keeled, several-nerved, usually obscurely so, bilobed apically, the midnerve typically prolonged between the sometimes short-awned lobes as an elongate, stout, basally flattened, twisted to geniculate awn; callus typically well developed and often bearded; palea nearly as long as the lemma; anthers well developed or in cleistogamous florets minute; caryopsis free within the lemma and palea, ellipsoidal to obovoid, the embryo to ca 3/5 the length of the grain; $x = 9$.

Clay, K. 1983. Variation in the degree of cleistogamy within and among species of the grass *Danthonia*. Amer. J. Bot. 70: 835–843.

1. Panicle typically reduced to a single spikelet; sheaths (at least the upper ones) usually conspicuously hairy; lemma with apical lobes acute or, if awnlike, mostly less than 2 mm long; plants rarely more than 3 dm tall .... *D. unispicata*
— Panicle of multiple spikelets or rarely reduced to one, the sheaths then glabrous or inconspicuously hairy, the lemma with apical lobes mostly awnlike and often more than 2 mm long, and the plants mostly more than 3 dm tall 2

2(1). Panicle branches erect or nearly so, glabrous or scabrid-puberulent; lowermost lemmas (including lateral awns) 6–10 mm long .... *D. intermedia*
— Panicle branches erect to widely spreading, often flexuous, typically hairy; lowermost lemma (including lateral awns) 9.5–15 mm long .... *D. californica*

Danthonia californica Bolander California Oatgrass. Perennial; culms densely tufted, 3–7 (9) dm tall, erect; sheaths glabrous or with more or less conspicuous tufts of hair lateral to the ligule; blades (0.5) 2–4 (5) mm wide, flat to loosely involute, usually scabrous, occasionally sparsely long-hairy; ligule ca 0.5 mm long; panicle of (1) 2–5 spikelets, the pedicels mostly 0.5–3 cm long, erect to widely spreading, often flexuous, typically pubescent; spikelets 5- to 8-flowered; glumes lanceolate, (10) 14–18 (20) mm long (including the often awnlike tip), exceeded only by the awns of the upper florets; lemma 9.5–13 (15) mm long, hairy on the margin near midlength, otherwise glabrous, the apical lobes usually awnlike, often more than 2 mm long, the terminal awn 5–12 mm long, flattened and twisted at the base; callus densely short-hairy laterally; $2n = 36$. Meadows, open grassy or rocky slopes, and pinyon-juniper, aspen-spruce-fir, and ponderosa pine communities at 1580 to 3050 m in Cache, Daggett, Duchesne, Morgan, Summit, Uintah, Utah, and Wasatch counties; southwestern Canada to Saskatchewan, and throughout the western U. S.; Chile; 29 (ii).

Danthonia intermedia Vasey Timber Oatgrass. [*D. cusickii* (Williams) A. S. Hitchc.]. Perennial; culms densely tufted, 1–5 (7) dm tall, erect; sheaths glabrous or with a tuft of hair on either side of the ligule, occasionally sparsely long-hairy throughout; blades 1–4 mm wide, flat or involute, glabrous or sparsely long-hairy on one or both surfaces; ligule of hairs to ca 1 mm long; panicle narrow, often secund, 3–6 (9) cm long, with 4–9 (13) spikelets, the branches short and strictly erect or the lowermost somewhat spreading, the pedicels and branches glabrous or scabrid-puberulent; spikelets 3- to 6-flowered; glumes broadly lanceolate, (9) 11–17 (19) mm long, acute, exceeded only by the awns of the upper florets; lemma (6) 7–10 mm long, hairy on the margin near midlength, the apical lobes acute or with awnlike tips 1.5–2.5 mm long, the terminal awn 6–10 mm long, flattened and ultimately twisted at the base; callus long-hairy laterally; $2n = 18$, 36. Dry to moist meadows in spruce-fir and lodgepole pine communities, and on exposed rocky ridges at higher elevations, at 2440 to 3660 m in Beaver, Daggett, Duchesne, Garfield, Grand, Iron, Salt Lake, San Juan, Sanpete, Sevier, Summit, Tooele, Utah, Uintah, and Washington counties; Alaska and across southern Canada, south throughout the western U. S., in South Dakota and Michigan; 70 (ii).

Danthonia unispicata (Thurber) Munro ex Macoun Oneside Oatgrass. [*D. californica* var. *unispicata* Thurber]. Perennial; culms densely tufted to matted, 1–3 (3.5) dm tall, slender and erect to curved-ascending; sheaths (at least some of them) conspicuously pubescent with long, spreading, often papilla-based hairs, the hairs lateral to the ligule usually 2–5 mm long; blades mostly 1–2.5 mm wide, flat to loosely involute, spreading-hairy or rarely glabrous; ligule of hairs to ca 1 mm long; inflorescence typically reduced to a single spikelet, often with the elongate, minutely hairy pedicel jointed to the culm, rarely 1 or 2 mostly smaller spikelets below the terminal one; spikelets 3- to 6-flowered; glumes lanceolate, (9) 12–23 mm long, the second slightly shorter than the first, long-tapered to an acute or awn-tipped apex, exceeded only by the awns of the upper florets; lemma (7) 8–14 mm long, marginally hairy along the lower 2/3, the apical lobes acute or with awnlike tips rarely more than 2 mm long, the terminal awn 3–9 mm long, flattened and twisted, sometimes geniculate at the base; callus minutely bearded; $2n = 36$. Dry to wet sites, in mountain brush, aspen-fir, ponderosa pine, lodgepole pine, and meadow communities, and at higher elevations on open rocky slopes, at 2130 to 3050 m in Cache, Daggett, Davis, Juab, Millard, Salt Lake, Summit, and Uintah counties; British Columbia and Alberta, south through most of the western U. S.; 30 (v).

Deschampsia Beauv.

Annuals or perennials; sheaths open; blades flat to involute, mostly narrow; ligules membranous; inflorescence an open to contracted panicle, often glistening, the branches commonly subcapillary; spikelets lanceolate to oblong, laterally flattened to subterete, 2 (3) -flowered, the rachilla usually hairy, disarticulating above the glumes and between the florets, sometimes prolonged beyond the uppermost floret and then occasionally bearing a rudimentary floret; glumes lanceolate, subequal or unequal, keeled to rounded on the back, broadly membranous-margined, acute to long-tapered at the apex, 1- to 3-nerved, equal to the lower floret, the second usually equal to or exceeding the uppermost floret; lemma thin to firm, smooth and glossy, narrowly lanceolate to ovate, rounded on the back, 5-nerved, often obscurely so, blunt and erose or bilobed to variously toothed apically, a slender, straight or geniculate awn arising from the back;

callus bearded; palea equal to or shorter than the lemma; stamens 3; caryopsis free of but enclosed within the lemma and palea, spindle-shaped, the embryo not more than 1/3 the length of the grain; x = 7.

1. First glume averaging shorter than the upper floret; anthers 1.2–3 mm long; panicles ultimately open, often nodding; blades 1–5 mm wide .... *D. cespitosa*
— First glume equal to or exceeding the upper floret; anthers 0.2–0.7 mm long; panicles sublinear to open, erect; blades 0.5–1.5 mm wide 2
2(1). Awn straight or nearly so; panicle sublinear, the branches nearly always strictly erect; plants perennial .... *D. elongata*
— Awn geniculate; panicle narrow or ultimately open, the branches spreading-ascending; plants annual .... *D. danthonioides*

Deschampsia cespitosa (L.) Beauv. Tufted Hairgrass. [*Aira cespitosa* L.]. Densely to loosely tufted perennials, 0.4–12 dm tall; leaves chiefly basal; sheaths glabrous or scabrous; blades somewhat stiff, flat to involute, 1–3 (5) mm wide, typically scabrous; ligule 3–11 (15) mm long; panicle narrowly elongate to pyramidal, 0.3–4 (5) dm long, often nodding, becoming silvery or tawny and to some degree purple, the branches erect to ultimately widely spreading, usually naked below; spikelets 2 (3) –flowered, 3–6 (9) mm long, the rachilla hairy; glumes ovate to narrowly lanceolate, keeled to rounded on the back, acute to blunt and often erose apically, at maturity usually tawny or silvery, the midportion purple or purple-tinged, the broad membranous margins glistening, the first glume (2) 3–5 (6) mm long, 1–nerved, averaging shorter than the upper floret, the second (2.5) 3–5.5 (7) mm long, obscurely 3–nerved, occasionally exceeding the upper floret; lemma thin, smooth and shiny, 1.7–4.5 mm long, pale to more often purple in a band below the tawny margin, truncate to blunt and erose to 4–toothed apically, the awn arising at or below midlength, 1–5 (6) mm long, straight or twisted and occasionally geniculate; callus with short hairs; anthers 1.2–2.5 (3) mm long; 2n = 24–28, 52. Mesic to wet meadows and occasionally along streams under spruce-fir, lodgepole pine, ponderosa pine, and in forest openings or rarely on open rocky slopes, at (1370) 1525 to 3810 m in all Utah counties except Davis, Tooele, and Washington; circumboreal, south throughout the western U. S. into Mexico; sporadic eastward; Eurasia; Africa; introduced in Argentina; 294 (xiii). According to Voss (Rhodora 68: 440. 1966), alteration of the spelling of the epithet *cespitosa*, as originally published by Linnaeus (Sp. Pl. 64. 1753), is unjustified. Tufted hairgrass is a good and often abundant source of forage throughout its growing season. Although it is typically associated with mesic meadows, Thornburg (1982) reports that the U. S. Forest Service has had success with ecotypes adapted to dry, windblown, disturbed sites at high elevations.

Deschampsia danthonioides (Trin.) Munro ex Benth. Annual Hairgrass. [*Aira danthonioides* Trin.]. Annual; culms solitary or tufted, 0.5–7 (8) dm tall; leaves few, chiefly basal; sheaths glabrous; blades to 1.5 mm wide, usually involute; ligule (1) 3–6 mm long; panicle sublinear to open, occasionally reduced to a raceme, erect, 0.3–2.5 dm long, pale green to purple, becoming tawny, the branches erect to ultimately spreading, usually naked below; spikelets 2 (3) –flowered, 4–8 mm long, the rachilla densely hairy; glumes generally equal or nearly so, lanceolate, (4) 5–8 mm long, exceeding the upper floret, gradually tapered to a sharply acute apex, more or less obscurely 3–nerved, occasionally purple-tinged; lemma at maturity relatively firm, smooth and shiny, 1.8–3 mm long, often purple-tinged, truncate and more or less unequally 4–toothed apically, the awn arising at or below midlength, 3–7 mm long, geniculate; callus bearded with hairs less than half as long as the lemma; anthers 0.3–0.5 mm long; 2n = 26. Salt desert shrub and saline meadow communities, pastures and marshes below 1525 m in Cache, Davis, Salt Lake, and Weber counties; Alaska to Alberta, south throughout most of the western U. S. and in Mexico; 18 (v).

Deschampsia elongata (Hook.) Munro ex Benth. Slender Hairgrass. [*Aira elongata* Hook.]. Perennial; culms slender, erect, densely tufted, 1–10 (12) dm tall; leaves chiefly basal and short relative to the culm; sheaths usually glabrous; blades soft but ascending, flat to involute-filiform, 0.5–1.5 mm wide, glabrous or scaberulous; ligule 2.5–8 (9) mm long; panicle typically sublinear, 0.5–3 dm long, mostly erect, green, becoming tawny, occasionally purple-tinged or purple, the branches erect-appressed or rarely slightly spreading, the naked portion of the branches usually appressed to the rachis and inconspicuous; spikelets 2 (3) –flowered, mostly 4–7 mm long, somewhat flattened to subterete, the rachilla sparsely hairy; glumes equal or nearly so, lanceolate, mostly 4–7 mm long, exceeding the upper floret, gradually tapered to a sharply acute apex, obscurely to conspicuously 3–nerved; lemma relatively firm, smooth and shiny, 1.5–3 mm long, occasionally purple to some degree, truncate and unequally 4–toothed at the apex, the awn arising just below the midlength, 1.5–6 mm long, nearly straight; callus bearded with hairs often at least half as long as the lemma; anthers 0.2–0.7 mm long; 2n = 26. Dry to moist sites, streamside and in maple, aspen, spruce-fir, lodgepole pine, and meadow communities, occasionally on open slopes and rocky ridges, at 1830 to 3140 m in Box Elder, Cache, Davis, Duchesne, Juab, Morgan, Rich, Salt Lake, Summit, Utah, and Wasatch counties; Alaska to Alberta, south throughout most of the western U. S. and into Mexico; Argentina; 76 (xii).

Digitaria Heister

Annuals or perennials; culms erect to prostrate; sheaths open; blades flat; ligule membranous; inflorescence a panicle of 2–20 slender, spikelike, secund branches, these simple or rarely branched near the base, whorled or occasionally distributed along a short axis, the spikelets arising from one side of a somewhat 3–angled winged rachis, usually 2 or 3 per node, borne on appressed pedicels of different lengths, the lower pedicel very short, the second ca 0.8–4 mm long, the third (when present) slightly longer, often fused with the rachis in the upper part; spikelets dorsiventrally compressed, disarticulating below the glumes, 2–flowered, the lower floret reduced to an empty lemma, the upper floret perfect; glumes unequal, the first minute or lacking, the second ranging from much shorter to fully as long as the lemma, conspicuously 3– to 7–nerved; sterile lemma similar to the second glume in texture and nervation, but nearly to fully as long as the fertile lemma; fertile lemma leathery and more or less glossy, dorsiventrally compressed, acute to acuminate at the apex, the membranous margins curved around the palea; palea similar to the fertile lemma in texture; stamens 3; caryopsis free of but firmly enclosed within the lemma and palea, the embryo 2/5–1/2 the length of the grain; x = 9.

1. First glume triangular, often minute but opaque and readily visible under low magnification; second glume narrower and distinctly shorter than the spikelet; herbage pubescent with spreading, often papilla-based hairs .... *D. sanguinalis*

— First glume transparent and obscure or lacking; second glume about as long as the spikelet; herbage glabrous or inconspicuously hairy, occasionally with longer hairs near the ligule .... *D. ischaemum*

Digitaria ischaemum (Schreber) Schreber ex Muhl. Smooth Crabgrass. [*Panicum ischaemum* Schreber in Schweigger]. Annual; culms 1–4 dm long, erect or decumbent to prostrate; sheaths usually long-hairy near the ligule, otherwise glabrous or the lower ones inconspicuously hairy; blades flat, 2–4.5 (6) mm wide, glabrous or scabrous; ligule 1–2 mm long; panicle of 2–6 spikelike branches 3–10 (11) cm long in a terminal whorl, occasionally with 1 or 2 branches below the terminal ones, the rachis flat or nearly so and rather broadly winged; spikelets borne in groups of 3 or some of them in pairs, elliptical, 1.7–2.5 mm long and to ca 1 mm wide, often purple or purple-tinged; first glume lacking or present as a minute transparent scale, the second thin and often pale, lanceolate to ovate, equal to or not less than 4/5 as long as the spikelet, rounded on the back, prominently 3-nerved, minutely crisp-hairy, the hairs typically enlarged apically (30x magnification); sterile lemma relatively firm and often purple, essentially flat, equal to or slightly shorter than the fertile lemma, prominently 5- to 7-nerved, minutely crisp-hairy; fertile lemma leathery, at maturity dark brown to purple black, glossy and finely striate; anthers 0.5–0.7 mm long; 2n = 36, 45. Weed of lawns below 1525 m in Cache, Salt Lake, and Utah counties; native to Eurasia, widely distributed in southern Canada and through most of the U. S.; 9 (i).

Digitaria sanguinalis (L.) Scop. Hairy Crabgrass. [*Panicum sanguinale* L.; *Syntherisma sanguinalis* (L.) Dulac]. Annual; culms 2–10 dm tall, erect or ascending or prostrate, often rooting at the lowermost nodes and mat-forming; sheaths conspicuously pubescent with spreading, often papilla-based hairs mostly 1–3 mm long; blades flat, 2.5–12 mm wide, scabrous or sparsely long-hairy, the hairs usually papilla-based; ligule mostly 0.5–2.5 mm long; panicle of 2–16 branches 3–20 cm long in 1–3 whorls, occasionally some of the branches solitary, the rachis 3-angled, each angle narrowly winged; spikelets borne in pairs, narrowly lanceolate to elliptical, 2.2–3.5 mm long and to ca 1 mm wide, often purple or purple-tinged; first glume triangular, 0.1–0.5 mm long, firm or at least opaque (easily seen at low magnification), the second glume narrowly wedge-shaped, typically pale, 0.8–2.5 mm long, 1/3–3/5 the length of the spikelet, 3-nerved, ciliate to densely puberulent throughout; sterile lemma often purple, essentially flat, equal to the fertile lemma or nearly so, prominently 5- to 7-nerved, the lateral nerves closely spaced near the incurved ciliate margin and scabrous to pubescent with appressed or spreading hairs; fertile lemma leathery, at maturity pale to dark gray, glossy and finely striate; anthers 0.5–0.8 mm long; 2n = 18, 28, 36–48, 76. Weed of lawns, gardens, and waste places below 1525 m in Box Elder, Cache, Duchesne, Kane, Salt Lake, Utah, Washington, and Weber counties; native to Eurasia, introduced in southern Canada, south throughout the U. S.; Mexico; South America; 29 (ii). *Digitaria adscendens* (H.B.K.) Henrard, reported from Washington County by Ebinger (Brittonia 14: 248–253. 1962), is synonymous with *D. ciliaris* (Retz.) Koeler (Gould 1975), a Eurasian species very closely related to *D. sanguinalis*. As described by Gould, the taxon differs from *D. sanguinalis* chiefly in having essentially glabrous leaves and the lateral nerves of the sterile lemma smooth rather than scaberulous. I have seen no Utah plants in which these characters are correlated.

Distichlis Raf.

Dioecious (rarely monoecious) perennials; leaves chiefly cauline; sheaths open; blades firm; ligule membranous; inflorescence a compact panicle or raceme; spikelets laterally compressed, few- to many-flowered, disarticulating above the glumes and between the florets; glumes equal or unequal, keeled, the first 3- to 5-nerved, the second 5- to 7 (9) –nerved, the nerves sometimes obscure; lemma broad, 7- to 13-nerved, keeled, those of the pistillate spikelets leathery; palea subequal to the lemma in length and width; stamens 3; caryopsis free within but enclosed by the lemma and palea, the embryo less than 1/2 as long as the grain; x = 10.

Beetle, A. A. 1943. The North American variations of *Distichlis spicata*. Bull. Torrey Bot. Club 70: 638–650.
Reeder, J. R. 1943. The status of *Distichlis dentata*. Bull. Torrey Bot. Club 70: 53–56.

Distichlis spicata (L.) Greene Desert Saltgrass. [*Uniola spicata* L.; *D. dentata* Rydb.; *D. stricta* (Torr.) Rydb.; *D. spicata* var. *stricta* (Torr.) Scribn.; *D. maritima* var. *laxa* H. T. Holm, type from Lake Park, Uintah Co.]. Dioecious, strongly rhizomatous or sometimes stoloniferous perennial; culms leafy, 1–4 (5) dm tall, stiffly erect from a more or less decumbent base; sheaths short, glabrous or with hairs at the summit 0.5–3 mm long; blades of the middle and upper culm leaves conspicuously 2-ranked, stiffly ascending-spreading, 1–4 mm wide, loosely involute on drying, long-tapered to a sharply acute tip, glabrous or the upper surface scaberulous to long-hairy; blades near the culm base typically scalelike; ligule a ciliate membrane 0.2–0.6 mm long; panicle (raceme) mostly narrowly oblong, compact, 2–9 cm long, those of mature staminate plants usually exserted above the leaves, those of pistillate plants usually equal to or shorter than the blades; spikelets mostly strongly laterally compressed, 5- to 16 (20) –flowered, 6–22 (25) mm long, glabrous, light green and occasionally purple-tinged; glumes ovate to lanceolate, equal or unequal, keeled, 3- to 9-nerved, sometimes obscurely so, glabrous throughout or obscurely scabrous on the keel, the first glume (1.5) 2–6 (7.8) mm long, the second 2.5–8.5 mm long; lemma ovate to lanceolate, keeled to more or less rounded on the back, 7- to 13-nerved, sometimes obscurely so, glabrous, membranous-margined, the abruptly to sharply acute tips often incurved; lemmas of pistillate florets 3–10 mm long, those of staminate florets mostly 3–7 mm long; palea about equal to the lemma; anthers 2–4 mm long; 2n = 38, 40, 42, 72. Moist sites in salt desert shrub communities and occasionally around seeps and springs in sagebrush, oak, and pinyon-juniper communities, sometimes invading cultivated areas as a weed, at 1010 to 2290 m in all Utah counties except Morgan, Summit, and Wasatch; British Columbia to Manitoba, south throughout the western and central U. S. into Mexico, and along the Atlantic Coast and the Gulf of Mexico; South America; 177 (vii). This species is commonly considered to consist of two varieties with plants distributed along the Atlantic and Pacific coastal regions and Gulf of Mexico being referred to var. *spicata*, and inland plants to var. *stricta*. Authors do not agree, however, as to how the two varieties differ. According to Beetle (1943), who recognized a number of infraspecific taxa, var. *stricta* in various parts of its range "approaches in certain characters all other varieties and even typical material from the eastern seaboard." Finding no correlation between morphological variation and geographical distribution, I choose to treat the complex as a single taxonomic unit. Desert saltgrass is not exceptionally palatable but, in areas of saline soil, is often grazed for lack of other forage. Because of its rhizomatous nature, saltgrass provides excellent protection against erosion by wind and water. Seed is not usually available commercially, however, because of low seed production by the plant.

Echinochloa Beauv.

Annuals or perennials; sheaths open; blades flat; ligule a ring of hairs or lacking; inflorescence a panicle of simple or rebranched, densely flowered, secund branches with spikelets subsessile in clusters or in regular rows; spikelets dorsiventrally compressed, disarticulating below the glumes, 2–flowered, the lower floret staminate or neuter, the upper one perfect; glumes membranous, unequal, the first clasping the base of the spikelet, the second acute to awned at the apex; lemma of the lower floret similar to the second glume in size, shape, and texture, both structures closely sheathing the upper floret; lemma of the perfect floret plano-convex, firm, smooth and glossy, the firm margins incurved around the equally firm palea; stamens 3; caryopsis free of but firmly enclosed within the lemma and palea, broadly ellipsoidal to spherical, the embryo 2/3–3/4 the length of the grain; x = 9.

Gould, F. W., M. A. Ali, and D. E. Fairbrothers. 1972. A revision of *Echinochloa* in the United States. Amer. Midl. Naturalist 87: 36–59.

***Echinochloa crus-galli* (L.) Beauv.** Barnyard Grass. [*Panicum crus-galli* L.; *E. muricata* var. *microstachya* Wieg.; *E. crus-galli* var. *zelayensis* (H.B.K.) A. S. Hitchc.]. Coarse annual; culms solitary or in small tufts, erect to ascending, 1–10 (18) dm tall, sheaths glabrous or occasionally with a few long hairs near the collar; blades elongate, flat, 3–30 mm wide, glabrous or scabrous to sparsely long-hairy; ligule lacking; panicle narrowly elongate, 0.5–2.5 dm long, erect to nodding, the branches densely flowered, erect-ascending to widely spreading, occasionally the longer ones rebranched; spikelets plano-convex, 2.8–4 mm long; glumes unequal, the first 1.2–1.6 mm long, sometimes awn-tipped, the second as long as the spikelet, usually coarsely stiff-hairy on the crowded lateral nerves and variously scabrous to puberulent over the back, rarely subglabrous, acute to short-awned; lemma of the lower floret similar to the second glume but occasionally with an awn 0.5–5 cm long; lemma of the upper floret leathery or hardened, broadly elliptical, rounded on the back, smooth and glossy, unawned or awn-tipped; anthers 0.4–1.2 mm long. Weed of gardens, fields, and other open sites, especially along waterways, at 820 to 2135 m in Beaver, Box Elder, Cache, Carbon, Davis, Emery, Garfield, Grand, Iron, Juab, Kane, Millard, Salt Lake, San Juan, Sanpete, Tooele, Uintah, Utah, Washington, Wasatch, Wayne, and Weber counties; southern Canada, south through most of the U. S. and in Mexico; South America; Eurasia; 100 (v). Polymorphism in *E. crus-galli* has led to the segregation of varieties, but the existence of morphological intermediates and the lack of geographical or ecological disjunction among the proposed taxa suggest that they do not merit taxonomic recognition (Cronquist et al. 1977). Several species of *Echinochloa* are cultivated as food and fodder plants in Asia and Africa.

Eleusine Gaertner

Annuals or perennials; culms more or less compressed; sheaths open; blades flat to loosely involute; ligule membranous; inflorescence of 1 to several, spikelike, digitately arranged branches, often with 1 or 2 branches below the terminal whorl, the spikelets sessile in 2 closely overlapping rows along one side of a flattened rachis, the rachis not prolonged beyond the terminal spikelet, the latter scarcely if at all reduced; spikelets strongly laterally compressed, mostly 2– to 10–flowered, disarticulating above the glumes and between the florets; glumes keeled, unequal, the first mostly 1–nerved, the second 3– to 7–nerved, shorter than or equal to the lowermost lemma; lemma keeled, 3 (5) –nerved; palea shorter than the lemma; stamens 3; caryopsis free of the lemma and palea, at maturity loosely enclosed by or free of the pale membranous pericarp, 3–angled, horizontally striate, the embryo about half the length of the grain; x = 9.

***Eleusine indica* (L.) Gaertner** Goosegrass. [*Cynosurus indicus* L.]. Annual; culms typically much branched from the base, ascending to prostrate, 0.5–3 (8) dm long; sheaths often keeled, glabrous or pubescent near the summit with hairs 2–6 mm long; blades flat to loosely involute, 1–8 mm wide, usually glabrous; ligule 0.3–1.2 mm long, ciliate, the membranous portion longer than the cilia; panicle of (1) 2–6 (12) spikelike branches digitately or racemosely arranged, 3–10 (15) cm long and mostly 3–6 mm wide, the sessile spikelets closely appressed to one side of a flattened rachis, the latter ca 1 mm wide; spikelets strongly flattened, elliptical, 3– to 9–flowered, ovate to lanceolate-elliptical, 5–8 mm long; glumes firm, lanceolate, keeled, acute, the first 2–3 mm long, 1 (2) –nerved, the second 2.7–4 mm long, 3– to 7–nerved, shorter than or equal to the lowermost lemma; lemma membranous except near the midrib, keeled, lanceolate, 2.5–4 mm long, (3) 5– to 7–nerved, abruptly acute to mucronate at the apex, glabrous; anthers 0.3–0.7 mm long; 2n = 18, 36. Weed, chiefly of lawns, below 1525 m in Cache, Salt Lake, and Washington counties; native to the tropics of Eurasia, adventive in Canada and established through much of the U. S. and Mexico; South America; 7 (ii). Utah material has sometimes been interpreted as *E. tristachya* (Lam.) Lam., but that taxon differs from *E. indica* in having spikelike branches no more than 2.5 cm long and lemmas at least 4.2 mm long (Tsvelev 1984).

Elymus L.

Perennials, strongly rhizomatous to densely tufted; sheaths open, with or without auricles; blades flat to involute; ligule membranous; inflorescence a terminal, solitary (rarely compound) spike, the rachis continuous or in a few species readily disarticulating; spikelets 1–8 or more per node, sessile, usually more or less overlapping, at least near midlength of the spike, subterete to laterally flattened, 2– to 11–flowered, the rachilla disarticulating above the glumes and between the florets or, in a few species, below the glumes; occasionally some of the spikelets, especially those near the base of the spike, reduced to 1 or 2 glumes; glumes subequal to unequal, awnlike throughout to subulate or elliptical, occasionally asymmetrical, 0– to 5 (6) –nerved, acute to truncate, awned or awnless, shorter to longer than the lowermost lemma; lemma lanceolate, either the rounded back or the sides visible between the glumes (often variable within a species) (3) 5– to 7–nerved, the nerves often visible only near the apex, awned or awnless; palea well developed; stamens 3; caryopsis adherent to the palea and often to the lemma, ellipsoidal, the embryo 1/6–1/2 as long as the grain; x = 7. See discussion of the tribe Triticeae under *Agropyron*. Species of *Elymus* (especially *E. cinereus*, *E. canadensis*, and *E. virginicus*) are subject to infestation with the parasitic fungus *Claviceps purpurea* (Fr.) Tul., commonly known as ergot. Consumption by animals of grasses so infected can cause serious injury or even death if consumed in sufficient quantity (Kingsbury 1964).

Atkins, R. J., M. E. Barkworth, and D. R. Dewey. 1984. A taxonomic study of *Leymus ambiguus* and *L. salinus* (Poaceae: Triticeae). Syst. Bot. 9: 279–294.
Barkworth, M. E. and R. J. Atkins. 1984. *Leymus* Hochst. (Gramineae: Triticeae) in North America: Taxonomy and distribution. Amer. J. Bot. 71: 609–625.
Barkworth, M. E., D. R. Dewey, and R. J. Atkins. 1983. New generic concepts in the Triticeae of the Intermountain Region: key and comments. Great Basin Naturalist 43: 561–572.
Dewey, D. R. 1984. The genomic system of classification as a guide to intergeneric hybridization within the perennial Triticeae. pp. 209–279. *In*: J. P. Gustafson (ed), Gene manipulation in plant improvement. Plenum Publishing Corp., New York.
Gould, F. W. 1947. Nomenclatorial changes in *Elymus* with a key to the Californian species. Madrono 9: 120–128.
———. 1949. Nomenclatorial changes in Arizona grasses. Madrono 10: 94.
Stebbins, G. L. Jr., J. I. Valencia, and R. M. Valencia. 1946a. Artificial and natural hybrids in the Gramineae, tribe Hordeae. I., *Elymus*, *Sitanion* and *Agropyron*. Amer. J. Bot. 33: 338–351.
———. 1946b. Artificial and natural hybrids in the Gramineae, tribe Hordeae. II. *Agropyron*, *Elymus*, and *Hordeum*. Amer. J. Bot. 33: 579–586.
Wilson, F. D. 1963. Revision of *Sitanion* (Triticeae, Gramineae). Brittonia 15: 303–323.

Note: A plant that cannot be identified using the key below may be a hybrid; see guide to hybrids, p. 722.

1.	Rachis readily disarticulating	2
—	Rachis continuous	4
2(1).	Lemmas acute or with an awn to 3 mm long; glumes subulate, acute to awn- tipped; spikelets (2) 3 per node; anthers 3–5 mm long	*E. junceus*
—	Lemma awns mostly 10–85 mm long; glumes awnlike throughout or awned at the apex; spikelets 1 or 2 per node; anthers 1–2 mm long	3
3(2).	Glumes awnlike throughout or only slightly expanded near the base, mostly 1– or 2–nerved, 20–85 mm long; spikelets 2 per node; plants general in distribution	*E. elymoides*
—	Glumes linear to narrowly lanceolate, 1– to 5–nerved, with an awn (4) 8–30 mm long; spikelets 1 (2) per node; plants mostly above 2740 m	*E. scribneri*
4(1).	Glumes subulate to narrowly lanceolate, 0– to 1 (3) –nerved, usually obscurely so (the nerves not raised), acute or with an awn to ca 3 mm long	5
—	Glumes variously shaped, conspicuously 2– to 7–nerved (the nerves raised), truncate to acute, awnless or with an awn to 20 mm long	9
5(4).	Spikelets (2) 3–7 at each node of the rachis; glumes (when flattened) to ca 0.5 mm wide near the base	6
—	Spikelets 1 or 2 per node, rarely 3 at a few nodes; glumes (when flattened) often more than 0.5 (0.3–1.5) mm wide near the base	7
6(5).	Ligules of lower culm leaves averaging more than 1.5 (1–7) mm long; blades 3–20 mm wide; spikelets mostly 3– to 7–flowered; glumes 5–20 mm long, often more than 3/4 the length of the lower lemma; old sheath bases papery, not disintegrating into fibers	*E. cinereus*
—	Ligules of lower culm leaves averaging less than 1.5 (0.2–1.5) mm long; blades 1–5 mm wide; spikelets 2 (3) –flowered; glumes not more than 9 mm long, mostly less than 3/4 the length of the lower lemma; old sheath bases fibrillose	*E. junceus*
7(5).	Plants strongly tufted, occasionally rhizomatous, mostly of mountain sides and plateaus, rarely in low-lying alkaline sites; spikelets 1 per node or occasionally 2 at some or all nodes; leaf blades to 4 mm wide; sheaths nearly always glabrous	*E. salinus*
—	Plants strongly rhizomatous, mostly of low-lying, often alkaline sites; spikelets (1) 2 per node; blades 1–10 mm wide; sheaths glabrous or pubescent	8
8(7).	Lemmas gradually tapered to an awn 2–7 mm long; sheaths glabrous; plants known from Daggett County	*E. simplex*
—	Lemmas acute to abruptly awn-tipped or with an awn to 3 mm long, typically shiny and darkened toward the base; sheaths usually hairy (sometimes only minutely so) along the margins; plants widespread	*E. triticoides*
9(4).	Spikelets 1 per node or occasionally 2 at some nodes	10
—	Spikelets usually 2 at most nodes, rarely 3 or 4 at some nodes	17
10(9).	Glumes truncate or obtuse to abruptly acute or coarsely mucronate, (1.6) 2–3 mm wide, thickened and indurate, glabrous or hairy; sheaths typically ciliate on at least one margin	11
—	Glumes acute to awned, rarely obtuse, diverse in width but not thickened and indurate; sheaths rarely ciliate	12
11(10).	Plants with rhizomes	*E. hispidus*
—	Plants densely tufted	*E. elongatus*
12(10).	Glumes with an awn mostly 8–30 mm long; culms rarely as much as 4 dm tall, often decumbent to curved-ascending; plants mostly above 2740 m	*E. scribneri*
—	Glumes awnless or with an awn to 5 mm long; culms 3–15 dm tall and erect; plants general in distribution	13
13(12).	Anthers 0.6–2.2 mm long (usually retained in many of the spikelets even in age); spikelets at midlength of the spike about twice as long as the internodes of the rachis; glumes conspicuously (3) 5– to 9–nerved, the second at least 2/3 as long as the lowermost lemma	*E. trachycaulus*

— Anthers (2.5) 3–7 mm long (usually falling after flowering); spikelets at midlength of the spike variable; glumes diverse 14

14(13). Plants densely tufted, rarely with short rhizomes; spikelets at midlength of the spike not exceeding the internodes of the rachis by more than 1/3 their length; lemmas glabrous to scabrous, acute or more often awned, the awn 3–25 mm long and at maturity strongly divergent .... *E. spicatus*

— Plants strongly rhizomatous; spikelets at midlength of the spike variable; lemmas glabrous to puberulent, awnless or with a mostly straight awn to 15 mm long 15

15(14). Leaf blades flat, (2) 5–15 mm wide, mostly dark green; anthers (3) 4–7 mm long .... *E. repens*

— Leaf blades either involute or mostly less than 5 mm wide, typically glaucous; anthers (2.5) 3–5 mm long . 16

16(15). Glumes broadest at or above midlength, abruptly acute or awn-tipped, the body of the first glume mostly 2–6 mm shorter than the adjoining lemma . ... *E. lanceolatus*

— Glumes broadest below midlength, gradually tapered from near the base and passing imperceptibly into an awn-tip, the whole typically as long as to longer than the body of the lowermost lemma .... *E. smithii*

17(9). Spikelets disarticulating below the glumes; glumes subterete at the base, the basal portion nerveless and hardened, usually curved outward; plants rare, reported only from north-central Utah ..... *E. virginicus*

— Spikelets disarticulating above the glumes; glumes not both subterete and bowed outward at the base; plants general in distribution 18

18(17). Plants strongly rhizomatous, typically glaucous; glumes gradually long-tapered from below midlength to a sharply acute or awn-tipped apex; leaf blades 1–3 (5) mm wide .... *E. smithii*

— Plants tufted, yellow green to glaucous; glumes broadest at or above midlength, acute to long-awned; leaf blades (3) 4–20 mm wide 19

19(18). Glumes with awns averaging more than 5 (4–20) mm long; lemmas short-hairy or rarely merely scabrous, the awns mostly 10–40 mm long, ultimately spreading; spikes usually flexuous to nodding .. *E. canadensis*

— Glumes awnless or the awns averaging no more than 5 (1–5) mm long; lemmas glabrous to minutely scabrous, the awns 1–30 mm long, usually erect; spikes stiffly erect .... *E. glaucus*

Note: Extensive hybridization and introgression among species in the tribe Triticeae help to explain the occurrence of anomalous plants that cannot be identified using the foregoing key. Some frequently recurring hybrids were originally mistaken for species and given names accordingly, e.g., *Agropyron pseudorepens* Scribn. & Sm., *A. saundersii* (Vasey) A. S. Hitchc., *A. saxicola* (Scribn. & Sm.) Piper, *Elymus macounii* Vasey, and *Sitanion hansenii* (Scribn.) J. G. Sm. Current names for these plants are given in the following guide to some of the more common hybrids of the tribe Triticeae:

1. Internodes of the rachis averaging no more than 3 mm long; glumes unawned 2

— Internodes of the rachis averaging more than 3 mm long or, if shorter, the glumes awned 3

2(1). Anthers more than 2.5 mm long; plants often rhizomatous ...... *Agropyron cristatum* x *Elymus repens*

— Anthers to 2 mm long; plants tufted .... *Agropyron cristatum* x *Elymus trachycaulus*

3(1). Glumes awnless or merely awn-tipped; rachis continuous 4

— Glumes with awns at least 3 mm long; rachis continuous or disarticulating 12

4(3). Spikelets 2 or more at some or all nodes; plants with or without rhizomes 5

— Spikelets solitary at each node; plants often shortly rhizomatous 7

5(4). Glumes elliptical to lanceolate, (2) 3– to 9–nerved; lemmas awned; plants tufted; anthers not usually more than 2 mm long .... *Elymus glaucus* x *E. trachycaulus*

— Glumes narrowly lanceolate to subulate, 1 (2) –nerved; lemmas unawned or awn-tipped; plants tufted or rhizomatous; anthers more than 2 mm long . 6

6(5). Plants sometimes rhizomatous; lemmas typically brownish .... *Elymus triticoides* x *E. salinus*

— Plants tufted; lemmas green .... *Elymus salinus* x *E. cinereus*

7(4). Glumes mostly subulate, 1–nerved; anthers more than 2 mm long .... *Elymus salinus* x *E. lanceolatus*

— Glumes broader, 2– to 9–nerved; anthers often no more than 2 mm long 8

8(7). Spikelets at midlength of the rachis (at least some) less than twice as long as the internodes; anthers more than 2 mm long; lemmas awn-tipped or with a divergent awn to 15 mm long .... *Elymus spicatus* x *E. lanceolatus*

— Spikelets at midlength of the rachis usually at least twice the length of the internodes; anthers at maturity often less than 2 mm long; lemmas unawned or awned 9

9(8). Plants of subalpine to alpine .... *Elymus trachycaulus* x *E. scribneri*

— Plants mostly at lower elevations 10

10(9). Leaves often more than 5 mm wide; internodes of the rachis short throughout the spike .... *Elymus trachycaulus* x *E. repens*

— Leaves no more than 5 mm wide; internodes of the rachis variable in length 11

11(10). Plants rhizomatous, often glaucous .... *Elymus trachycaulus* x *E. lanceolatus* [*E.* x *pseudorepens* (Scribn. & Sm.) Barkworth & D. R. Dewey]

— Plants lacking rhizomes, mostly yellow green .... *Elymus trachycaulus* x *Agropyron cristatum*

12(3). Internodes of the rachis (2) 3–4 mm long throughout the length of the spike or nearly so; glumes variable in width from awnlike throughout to basally expanded, scaberulous throughout their length .... *Elymus trachycaulus* x *Hordeum jubatum* [x *Elyhordeum macounii* (Vasey) Barkworth & D. R. Dewey]

— Internodes of the rachis generally more than 4 mm long at most of the nodes; glumes variable as to width, not finely scaberulous throughout 13

13(12). Spikelets mostly 2 or more per node; plants with or without rhizomes 14

— Spikelets mostly solitary at each node; plants lacking rhizomes 17

14(13). Glumes 2– to 4–nerved at the base; lemmas green, with awns mostly at least 3 cm long .... *Elymus elymoides* x *E. glaucus* [*E.* x *hansenii* Scribn.]

— Glumes mostly 1–nerved or nerveless; awn of lemma less than 3 cm long 15

15(14). Lemmas brownish and smooth; plants often rhizomatous .... *Elymus triticoides* x *E. elymoides*

— Lemmas green and glabrous to minutely hairy; plants not rhizomatous 16

16(15). Lemmas minutely hairy; leaf blades (some of them) more than 4 mm wide; spikelets generally more than 2 at some of the nodes .. *Elymus cinereus* x *E. elymoides*

— Lemmas glabrous; leaf blades narrow and often involute; spikelets 2 per node .... *Elymus salinus* x *E. elymoides*

17(13). Glumes awnlike throughout their length or nearly so, mostly nerveless or 1–nerved; rachis usually disarticulating .... *Elymus salinus* x *E. elymoides*

— Glumes mostly at least 1 mm wide near the base, (2) 3– to 7–nerved; rachis diverse 18

18(17). Rachis continuous; glumes consistently 3– to 7–nerved 19

— Rachis often disarticulating; glumes occasionally less than 3–nerved 20

19(18). Glumes with awns to 4 mm long; lemmas mostly glabrous to scabrous, the awns straight; spikes erect .... *Elymus glaucus* x *E. trachycaulus*

— Glumes with awns more than 4 mm long; lemmas often short-hairy, the awns recurved; spikes sometimes flexuous to nodding .... *Elymus canadensis* x *E. trachycaulus*

20(18). Anthers (or some of them) more than 2 mm long; internodes at midlength of the rachis often more than half the length of the spikelets; awns of the lemmas often recurved .... *Elymus spicatus* x *E. elymoides* [*E.* x *saxicolus* Scribn. & Sm.]

— Anthers to 2 mm long; internodes at midlength of the rachis not more than half the length of the spikelets; awns of the lemma divergent but not recurved 21

21(20). Culms often decumbent; spikelets 1 per node; rachis internodes averaging 4 mm or less; plants of high elevations .... *Elymus scribneri* x *E. trachycaulus*

— Culms erect; spikelets sometimes 2 at a few nodes; rachis internodes diverse; plants of valleys to alpine .... *Elymus trachycaulus* x *E. elymoides* [*E.* x *saundersii* Vasey]

Elymus canadensis L. Canada Wildrye. Perennial; culms tufted, 8–20 (28) dm tall, erect or decumbent at the base; leaves chiefly cauline; sheaths glabrous or sometimes long-hairy, usually with well-developed auricles; blades flat or folded, (3) 4–15 (20) mm wide, glabrous or scaberulous to scabrous, rarely pubescent; ligule 0.2–1.5 (2) mm long; spike 7–20 (25) cm long, usually nodding, the rachis continuous, the internodes at midlength mostly (4) 5–8 mm long; spikelets 2–4 per node, (2) 3– to 5–flowered, mostly 10–18 mm long; glumes narrowly elliptical, mostly 0.6–1.3 mm wide at the widest point, conspicuously (2) 3– to 5–nerved, narrowed below and tapered to an awn at the apex, the body 6–15 mm long, the awn 4–20 mm long, mostly spreading, the whole usually exceeding the spikelet; lemma with the rounded back visible between the glumes, mostly 8–15 mm long, glabrous or more often scabrous to short-hairy, 5– to 7–nerved, tapered to an ultimately spreading awn 10–40 (50) mm long; anthers 2.5–4 mm long; $2n = 28, 42$. Along waterways and in wet, sometimes saline meadows at 1220 to 2440 m in Box Elder, Cache, Carbon, Daggett, Emery, Garfield, Grand, Iron, Juab, Kane, Millard, Piute, Salt Lake, San Juan, Sanpete, Sevier, Uintah, Utah, Wasatch, Washington, and Wayne counties; British Columbia to Nova Scotia, south through the U. S except in the extreme Southeast; adventive in Eurasia; 104 (iv). *Elymus canadensis* hybridizes with *E. trachycaulus* (see guide to hybrids p. 722).

Elymus cinereus Scribn. & Merr. Great Basin Wildrye. [*E. condensatus* authors, not Presl; *Leymus cinereus* (Scribn. & Merr.) Love]. Robust perennial, often with short rhizomes; culms mostly 7–25 dm tall, densely tufted, often forming clumps to 1 m or more in diameter; herbage bright green to strongly glaucous; sheaths glabrous or pubescent, often with well-developed auricles, old sheath bases papery, breaking off, not disintegrating into fibers; blades flat or nearly so, (1) 3–15 (20) mm wide, glabrous or scabrous to pubescent; ligule of lower culm leaves 1.5–7 mm long; spike stiffly erect, 7–25 (29) cm long and to 20 mm wide, rarely branching or a few spikelets short-pedicelled, the rachis continuous, often hairy, especially on the angles, the internodes at midlength of spike mostly (3) 4–7 mm long; spikelets (2) 3–6 (7) per node, disarticulating above or below the glumes, 9–20 mm long (excluding awns), mostly 3– to 6 (7) –flowered; glumes subulate to narrowly lanceolate, to ca 0.5 mm wide near the base and tapered to an awnlike tip, the whole 5–14 (20) mm long, shorter to longer than the lowermost lemma, 1 (3) –nerved, often obscurely so, glabrous or scaberulous; lemma rounded on the back, the back or occasionally the sides visible between the glumes, (6.5) 7–13 (15) mm long, glabrous or more often scaberulous to short-hairy, 5– to 7–nerved, acute to awn-tipped or with an awn to 5 (7) mm long; anthers 3–7 mm long; $2n = 28, 42, 56$. Moist to ultimately dry sites along waterways and roadsides, on open grassy slopes, in wet meadows, and in openings in sagebrush, juniper, pinyon-juniper, mountain brush, aspen, spruce-fir, ponderosa pine, and lodgepole pine communities at 790 to 2900 m in all Utah counties except Emery, Kane, Morgan, and Wayne; throughout the western U. S., rare in the Great Plains; 137 (iv). This species forms natural hybrids with *E. elymoides* and *E. triticoides* (see guide to hybrids p. 722). Great Basin wildrye is a moderately palatable, long-lived species valuable for range seeding and erosion control, especially in alkaline-saline soils.

Elymus elongatus (Host) Runem. Tall Wheatgrass. [*Triticum elongatum* Host; *Agropyron elongatum* (Host) Beauv.; *Elytrigia elongata* (Host) Nevski; *E. pontica* (Podp.) Holub; *Thinopyrum elongatum* (Host) D. Dewey; *T. ponticum* (Podp.) Barkworth & D. Dewey]. Perennial; herbage green or glaucous; culms densely tufted, erect, 5–20 dm tall; sheaths typically ciliate on at least one margin, otherwise glabrous, the auricles often well developed; blades flat to involute, 2–8 mm wide, glabrous or more often at least the upper surface scabrous or pubescent with short stiff hairs, often minutely hairy near the ligule; ligule to ca 0.7 mm long; spike erect to lax, 8–40 cm long, the rachis continuous, the internodes mostly 7–20 mm long at midlength of the spike; spikelets solitary at each node, at maturity compressed, 13–35 (40) mm long, 5– to 18–flowered, glabrous or hairy; glumes thickened and hardened, oblong, conspicuously 5– to 7 (9) –nerved, 6–11 mm long and (1.5) 2–3 mm wide, rounded on the back, truncate or occasionally slightly mucronate apically; lemma thickened and hardened, oblong to lanceolate, (8) 9–13 mm long, the sides visible between the glumes, 5–nerved, truncate to acutish at the apex, typically glabrous; anthers 4–7 mm long; 2n = 14, 28, 42, 56, 70. Along roadways and in salt desert shrub, sagebrush, aspen, and lower montane grassland communities, rarely under spruce-fir, at 1220 to 2740 m in Box Elder, Cache, Carbon, Daggett, Davis, Duchesne, Garfield, Juab, Kane, Millard, Salt Lake, Sanpete, Sevier, Summit, Tooele, Utah, Wasatch, Washington, Wayne, and Weber counties; native to Eurasia, introduced in the U. S. through much of the West, and in the central states as far south as Kansas and Iowa; South America; 54 (xi). Tall wheatgrass was first introduced in Utah in the 1930s. It is especially useful for seeding wet or subirrigated alkaline-saline soils, and for open foothill sites where it requires more than 30 cm mean annual precipitation. The grass is an excellent source of forage and cures well when left standing, thus providing a source of winter forage. Some animals find it unpalatable, however, because of its coarseness (Vallentine 1961).

Elymus elymoides (Raf.) Swezey Squirreltail. [*Sitanion elymoides* Raf.; *Aegilops hystrix* Nutt.; *S. hystrix* (Nutt.) J. G. Sm.; *S. breviaristatum* J. G. Sm.; *S. cinereum* J. G. Sm.; *S. insulare* J. G. Sm., type from Carrington Island, Great Salt Lake; *S. longifolium* J. G. Sm.]. Perennial; culms solitary to densely tufted, 0.5–8 dm tall, erect or occasionally spreading; herbage glabrous to long-hairy throughout; sheaths with auricles mostly inconspicuous or lacking; blades flat to involute, 1–4 (6) mm wide; ligule mostly less than 0.5 mm long; spike erect to flexuous or nodding, 2–15 (17) cm long (excluding awns) and at maturity to ca 12 cm or more wide (including awns), the rachis readily disarticulating, the internodes mostly 2–12 mm long at midlength of the spike; spikelets (1) 2 (4) per node, (1) 2– to 6–flowered, mostly loosely overlapping; lowermost floret occasionally reduced to an awn; glumes typically 1– or 2 (3) –nerved, 0.2–1 mm wide near the base, tapering to an ultimately widely spreading, scabrous awn, the whole 2–10 (12) cm long, occasionally some of the glumes cleft into 2 (3) awns of unequal lengths, these occasionally with 1 to several, bristlelike lateral appendages, the awns often becoming purple; lemma lanceolate to elliptical, rounded on the back, the back or sides visible between the glumes, 6–12 mm long, glabrous or scabrous to pubescent, 3– to 5–nerved, the midnerve prolonged into an ultimately recurved, scabrous awn mostly 2–8.5 cm long; anthers 1–2 mm long; 2n = 28. Dry to moist sites, occurring in every vegetative type from salt desert shrub to alpine grasslands at 1070 to 3500 m, in all Utah counties except Morgan; British Columbia to Saskatchewan, south throughout the western and central U. S. and in Mexico; 424 (x). Bentham and Hooker (*Genera Plantarum* 1883) included *Sitanion* as a section of *Elymus*. Although most workers have followed Hitchcock (1935) in recognizing *Sitanion* as a separate genus, cytological evidence assembled by Dewey (Bot. Gaz. 128: 11–16. 1967) and by Stebbins (1946a), in addition to the ease with which it hybridizes with other species of *Elymus* and findings from genome analysis demonstrate that it is more reasonably grouped with other species of *Elymus*. An occasional plant in our area, with some glumes 3–cleft, has been referred to *E. multisetus* (J. G. Sm.) Jones [*Sitanion jubatum* J. G. Sm.]. Wilson (1963) claims that in general this species is restricted to California, parts of Nevada, and the three northwestern states. The sporadic occurrence in Utah of such plants may be the result of introgressive hybridization between *E. glaucus* and *E. elymoides*, a phenomenon acknowledged by Wilson (1963). On the other hand, Cronquist et al. (1977) suggest that *E. multisetus* (as *S. jubatum*) is very similar to and probably conspecific with *E. elymoides* (as *S. hystrix*). *Elymus elymoides* readily crosses in nature with other species of *Elymus* and infrequently with those in *Hordeum*. Squirreltail is a native invader of disturbed areas in deserts, valleys, foothills, and mountain grasslands where overgrazing has eliminated more desirable range grasses. Where locally abundant, it produces fair forage during spring and early summer. At maturity the bristlelike awns may cause injury to grazing animals, but new growth after fall rains is highly palatable and remains so after it has dried (Vallentine 1961).

Elymus glaucus Buckley Blue Wildrye. Perennial; culms erect, 6–18 dm tall, forming small, mostly loose tufts; herbage glaucous or dark green; leaves chiefly cauline; sheaths glabrous or puberulent to long-hairy, mostly with well-developed auricles; blades lax, flat to loosely involute, (3) 4–19 mm wide, glabrous or scaberulous to sparsely long-hairy; ligule 0.3–1.5 mm long; spike stiffly erect, 6–16 (20) cm long (excluding the awns), the rachis continuous, the internodes at midlength of spike mostly 3–8 mm long; spikelets 2 or rarely 3 per node or solitary at upper and lower nodes, 2– to 4–flowered, 10–16 mm long, disarticulating above the glumes; glumes elliptical, conspicuously 3– to 7–nerved, (5) 6–19 mm long and mostly 1–1.5 mm at the widest point, slightly narrowed below, but the pair usually concealing the base of the lowermost floret, gradually or abruptly tapered to an acute or awn-tipped apex, occasionally with an awn 1–5 mm long; lemma rounded on the back, the back or the sides visible between the glumes, 8–15 mm long, glabrous or scaberulous to scabrid-puberulent, 5– to 7–nerved, awn-tipped or more often tapered to a straight or occasionally recurved awn mostly 8–30 mm long; anthers 1.5–3 mm long; 2n = 28. Mountain brush, aspen, ponderosa pine, spruce-fir, and lodgepole pine communities in open areas or in shade and often streamside at (1310) 1520 to 3200 m in Beaver, Box Elder, Cache, Carbon, Davis, Duchesne, Emery, Grand, Iron, Juab, Millard, Piute, Rich, Salt Lake, San Juan, Sanpete, Sevier, Summit, Tooele, Uintah, Utah, Wasatch, Washington, and Weber counties; Alaska to Ontario, south throughout the western U. S. and into Mexico, rare in the northern Great Plains and eastward; 128 (viii). *Elymus glaucus* forms hybrids with *E. elymoides* and *E. trachycaulus* (see guide to hybrids p. 722). Blue wildrye provides good forage during summer months but is intolerant of heavy grazing (Valentine 1961).

Elymus hispidus (Opiz) Meld. Intermediate Wheatgrass. [*Agropyron hispidum* Opiz; *A. intermedium* (Host) Beauv.; *A. trichophorum* (Link) Richter; *Elytrigia intermedia* (Host) Nevski; *Thinopyrum gentryi* (Meld.) D. Dewey; *T. intermedium* (Host) Barkworth & D. Dewey]. Rhizomatous perennial; herbage glaucous to dark green; culms 4–15 dm tall; sheaths typically ciliate on at least one margin, occasionally short-hairy throughout, usually with well-developed auricles; blades flat or involute, 2–10 (12) mm wide, glabrous or more often scabrous or pubescent with short stiff hairs; ligule to ca 0.5 mm long; spike stiffly erect, 6–20 (30) cm long, the rachis continuous, the internodes 6–20 mm long at midlength of the spike; spikelets solitary at each node, at maturity compressed, (8) 9–20 mm long, (1) 3– to 8–flowered, glabrous or hairy; glumes thickened and hardened, oblong to lanceolate, 4–9 mm long and (1.5) 2–3 mm wide at midlength, rounded to slightly keeled on the back, conspicuously 3– to 7–nerved, truncate to obtuse or occasionally acutish to abruptly mucronate or coarsely awn-tipped; lemma firm to hardened, oblong to lanceolate, 6–12 mm long, the sides visible between the glumes, 3– to 5–nerved, truncate to acutish or mucronate at the apex, rarely with an awn 1–10 mm long; anthers 3–5 (8) mm long; 2n = 42. Dry to mesic sites, chiefly along roadsides and in other waste places, in sagebrush, mountain brush, pinyon-juniper, aspen, ponderosa pine, and spruce-fir communities at 1280 to 3050 m in all Utah counties except Box Elder, Iron, and Piute; native to Eurasia, introduced in much of the western U. S., sporadically in the Great Plains and in the Northeast; 101 (xiii). Intermediate wheatgrass, introduced from the U.S.S.R. in the 1930s, is used for pasture, hay, range seedings, and erosion control. A sod-former, adapted to well-drained soils where the mean annual rainfall is 30–36 cm or more, it is only slightly inferior to *Agropyron cristatum* in persistence, drought tolerance, and winter hardiness (Hanson 1972). The grass is highly palatable to all grazing animals throughout the spring and summer (Vallentine 1961).

Elymus junceus **Fisch.** Russian Wildrye. [*Psathyrostachys juncea* (Fisch.) Nevski]. Perennial; culms densely tufted, (2) 4–11 dm tall, erect or decumbent at the base; sheaths usually glabrous, often with well-developed auricles, old sheath bases more or less persistent, often ultimately shredding into fibers; blades flat to involute, 1–5 mm wide, glabrous or scaberulous; ligule 0.2–0.8 (1) mm long; spike 3–16 cm long and 5–17 mm wide, erect, the rachis at maturity disarticulating, often tardily so, the internodes mostly 2–5 mm long at midlength of the spike; spikelets (2) 3 per node, disarticulating above or below the glumes, 7–10 (12) mm long (excluding awns) and strongly overlapping, 2 (3) –flowered, the upper 1 or 2 florets often only slightly exceeding the lowermost one, sometimes reduced or obsolete; glumes subulate, (4) 5–9 mm long and to ca 0.5 mm wide near the base, obscurely 1–nerved, scabrous or puberulent, acute to awn-tipped, distinctly shorter than the lowermost lemma; lemma lance-elliptical, the rounded back visible between the glumes, 6–10 mm long, scabrous to densely short-hairy, tapered to a sharply acute tip or with an awn to ca 3 mm long; anthers 2.5–5 mm long; 2n = 14. Salt desert shrub, sagebrush, pinyon-juniper, mountain brush, aspen, and ponderosa pine communities at 1280 to 2870 m in Box Elder, Daggett, Duchesne, Emery, Garfield, Iron, Juab, Kane, Millard, Piute, Salt Lake, Sevier, Tooele, Uintah, Washington, and Weber counties; native to the U. S. S. R., introduced in Canada, and from Montana to Arizona, in the Dakotas and Nebraska; 47 (vi). Russian wildrye was first introduced into the northern Great Plains in 1927 as a pasture grass and for use in erosion control. The species is especially useful in the Intermountain West because of its drought resistance and salt tolerance. Moreover, it remains green into August, thus providing a longer period of grazing than crested wheatgrass, which matures earlier. Unfortunately, it has relatively low seedling vigor and poor seed production, although a strain with improved seed production is now available (Thornburg 1982).

Elymus lanceolatus **(Scribn. & Sm.) Gould** Thickspike Wheatgrass. [*Agropyron lanceolatum* Scribn. & Sm.; *A. albicans* Scribn. & Sm.; *Elymus lanceolatus* ssp. *albicans* (Scribn. & Sm.) Barkworth & D. Dewey; *A. dasystachyum* (Hook.) Scribn.; *A. elmeri* Scribn.; *A. griffithsii* Scribn. & Sm. ex Piper; *A. riparium* Scribn. & Sm.; *Elymus subvillosus* (Hook.) Gould; *Elytrigia dasystachya* (Hook.) Love & Love]. Perennial, strongly rhizomatous; herbage typically glaucous; culms mostly erect, 3–13 dm tall, solitary or tufted; sheaths glabrous or rarely ciliate on the margins to puberulent or long-hairy, often with well-developed auricles; blades involute or occasionally flat and then 1–3.5 (5) mm wide, usually stiffly ascending, glabrous or scaberulous to long-hairy; ligule to ca 1 mm long; spike stiffly erect, 6–22 cm long, the rachis continuous, the internodes 5–16 mm long at midlength of the spike; spikelets solitary at each node of the rachis or occasionally in pairs at a few nodes, 3– to 12–flowered, 7–20 (24) mm long, mostly closely overlapping except sometimes near the base of the spike; glumes subequal or unequal, lance-elliptical to oblanceolate, mostly conspicuously (2) 3–5 (7) –nerved, glabrous to puberulent, 4–10 (11) mm long and 1–2 mm wide at the widest point, tapered from midlength or above to an acute or awn-tipped apex, the body mostly 2–6 mm shorter than the body of the lowermost lemma; lemma rounded on the back, the sides visible between the glumes, 7–14 mm long, glabrous or scaberulous to short-pubescent, (3) 5– to 7–nerved, subacute to awn-tipped at the apex; anthers (2.5) 3–5 mm long; 2n = 28. Roadsides and other disturbed sites, sagebrush, juniper, mountain brush, aspen, and meadow communities, at 1220 to 3350 m in all Utah counties except Beaver; Alaska and across Canada, south throughout the western U. S. and in Mexico, in the Great Plains south to Kansas, and in the East as far south as West Virginia; 240 (ix). The name *Agropyron albicans* was originally applied to a plant differing from *E. lanceolatus* in having lemmas with divergent awns to ca 15 mm long. Dewey (Amer. J. Bot. 57: 12–18. 1970) demonstrated its origin from a cross between *E. lanceolatus* (as *A. dasystachyum*) and *E. spicatus* (as *A. spicatum*). The hybrid has been classified as a subspecies of *E. lanceolatus* (Barkworth et al. 1983); but, because as hybrids awned plants can be expected to occur sporadically, I see no advantage in obscuring their origin by assigning them subspecific status (see guide to hybrids p. 722). Thickspike wheatgrass is low in productivity and has only fair forage value (Vallentine 1961). Because of its sod-holding ability, however, it is being widely used for revegetating mine spoils and other disturbed areas.

Elymus repens **(L.) Gould** Quackgrass; Couchgrass. [*Triticum repens* L.; *Agropyron repens* (L.) Beauv.; *Elytrigia repens* (L.) Nevski]. Rhizomatous perennial; herbage dark green or occasionally glaucous; culms erect or occasionally decumbent at the base, 5–16 dm tall, solitary or loosely tufted; sheaths glabrous or the lowermost sparsely to densely pubescent with soft spreading hairs, often with well-developed auricles; blades mostly lax to stiffly ascending, flat or loosely involute on drying,

(2) 5–15 mm wide, at least some of the blades of any one plant 5 mm or more wide, scabrous or hairy; ligule to ca 0.5 mm long; spike erect, 5–20 cm long, the rachis continuous, the internodes mostly 3–7 (8) mm long at midlength of the spike, the spikelets usually no less than twice as long as the internodes throughout the spike; spikelets solitary at each node of the rachis or occasionally in pairs at a few nodes, 8–17 (20) mm long, 3– to 6 (8) –flowered; glumes lanceolate to lance-elliptical, usually asymmetrical, more or less keeled, 6–11 mm long, ca 2/3 as long to slightly longer than the lowermost lemma, mostly 1.3–3 mm wide at midlength, rounded to somewhat keeled on the back, conspicuously 3– to 9–nerved, glabrous or scabrous on the midnerve, gradually long-tapered to an acute or awn-tipped apex, sometimes abruptly so, occasionally with an awn to 5 mm long; lemma rounded on the back to somewhat keeled, the sides visible between the glumes, 6–12 mm long, glabrous or scaberulous near the tip, 5–nerved, tapered to an acute to awn-tipped apex or with an awn to 10 (15) mm long; anthers (3) 4–7 mm long; $2n = 42$. Chiefly in disturbed, mesic to moist sites, often as a weed of cultivated lands, along waterways and in meadows, occasionally in mountain brush and conifer communities, at 1220 to 3050 m in Box Elder, Cache, Carbon, Daggett, Duchesne, Garfield, Juab, Kane, Millard, Rich, Salt Lake, Sanpete, Uintah, Utah, Wayne, and Weber counties; native to Eurasia, now circumboreal, south through the U. S. except in the southern states; South America; 88 (x). *Elymus repens* occasionally forms hybrids with *Agropyron cristatum* (see guide to hybrids, p. 722). Quackgrass provides fairly good spring forage, is a rapid invader, and quickly stabilizes moist, eroding soils. It is considered a noxious weed in cultivated areas, however, and is not recommended for seeding (Vallentine 1961).

Elymus salinus Jones Salina Wildrye; Bullgrass. [*E. ambiguus* var. *salina* (Jones) C. L. Hitchc.; *E. ambiguus* var. *salmonis* C. L. Hitchc.; *Leymus salinus* (Jones) Love; *L. salinus* ssp. *salmonis* (C. L. Hitchc.) Atk.]. Perennial, often short-rhizomatous, occasionally with rhizomes well developed; herbage green or more often glaucous; culms densely tufted, mostly erect, 4–14 dm tall; sheaths glabrous or scabrous to rarely pubescent, often with well-developed auricles, old sheath bases persistent, more or less papery, usually straw-colored but sometimes darkened, typically becoming loose, flattened, and forming a conspicuous feature; blades firm, flat or more often strongly involute, 2–3 (4) mm wide, glabrous or scabrous to puberulent, sometimes ciliate near the base; ligule 0.2–1.5 mm long; spike erect, 4–15 cm long and 2.5–8 mm wide, the rachis continuous, the internodes (3) 5–13 mm long at midlength of the spike; spikelets usually solitary at most of the nodes, occasionally paired at some or rarely most of the nodes, 9–17 (21) mm long, 3– to 6–flowered, light green or tinged with lavender; glumes subulate, subequal to distinctly unequal, nerveless or obscurely 1–nerved, keeled or rounded on the back, glabrous or scabrous, 1–12 mm long and 0.4–1 mm wide at the base, occasionally a spikelet reduced to a solitary glume or the first glume much reduced or lacking, the second glume typically shorter than the lowermost lemma; lemma lance- elliptical, rounded on the back, the sides or the back visible between the glumes, 7–12.5 mm long, 5– to 7–nerved, sometimes obscurely so, occasionally lavender-tinged, typically dull but occasionally shiny, glabrous or rarely scabrid-puberulent, awnless or tapered to an awn to 2.6 mm long; anthers 3–7.5 mm long; $2n = 28, 42, 56$. Chiefly on dry, sandy, gravelly, or rocky sites in salt desert shrub, desert shrub, sagebrush-grass, and pinyon-juniper communities, less frequently in mountain brush, ponderosa pine, and aspen-fir communities, at 1520 to 3050 m in Box Elder, Carbon, Daggett, Duchesne, Emery, Garfield, Grand, Kane, Millard, San Juan, Sanpete, Sevier (type from Salina Pass), Tooele, Uintah, Utah, Wasatch, and Wayne counties; Idaho, Nevada, California, Colorado, Wyoming, and Arizona; 304 (iii). Based on other plant names published at the same time, Atkins et al. (1984) conclude that Jones intended the epithet for this species to be *salinus* rather than *salina*. Barkworth and Atkins (1984) recognize two subspecies in *E. salinus* (as *Leymus salinus*), based on correlation between herbage pubescence or lack thereof and the number of spikelets per node. Examination of hundreds of Utah specimens of *E. salinus* revealed no consistent correlation between these characters. The one specimen that did combine pubescent sheaths with paired spikelets at most nodes of the rachis (Taye 1326 UT) appeared to be a hybrid between *E. salinus* and *E. triticoides*. *Elymus triticoides* is usually readily distinguished from *E. salinus* by its habitat preference, conspicuous rhizomes, darkened, papery, usually pubescent sheaths, broader leaves, and mostly paired spikelets. Each of these characters can occasionally be seen in *E. salinus*, however, perhaps due to intergradation between the two taxa. *E. salinus* also appears to form hybrids with *E. elymoides*, *E. cinereus*, and *E. lanceolatus* (see guide to hybrids, p. 722). *Elymus ambiguus* (as *Leymus ambiguus*) has been reported from Utah, but according to Atkins et al. (1984), is a distinct taxon restricted to the Colorado Front Range. Salina wildrye provides a moderate amount of fair quality forage during the growing season but is unpalatable when mature and dried (Vallentine 1961).

Elymus scribneri (**Vasey**) Jones Scribner or Spreading Wheatgrass. [*Agropyron scribneri* Vasey]. Perennial; culms tufted, 0.4–4 (6) dm long, prostrate to curved-ascending, often geniculate at the lower nodes or flexuous throughout; leaves often chiefly basal; sheaths glabrous to short-hairy with auricles present or more often lacking, the sheath bases of previous years persistent, pale and papery, often a conspicuous feature; blades flat to involute, 1–3 (4) mm wide, glabrous to puberulent, the ligule to ca 0.5 mm long; spike often nodding or flexuous, 3–7 cm long (excluding awns), the rachis ultimately disarticulating, the internodes 2–4 (5) mm long at midlength of the spike; spikelets solitary at each node of the rachis or rarely in pairs at a few nodes, (2) 3– to 6–flowered, 9–14 mm long; glumes linear to narrowly lanceolate, the body 4–8 mm long and 0.3–1.5 mm wide, 1– to 3 (5) –nerved, the midnerve prolonged as a usually divergent, scabrous awn (4) 8–30 mm long, a short lateral awn sometimes present; lemma rounded on the back, the sides visible between the glumes, 7–10 mm long, glabrous to scabrid-puberulent, 5–nerved, the midnerve prolonged as a divergent to recurved, scabrous awn 10–30 (40) mm long; anthers 1–1.7 mm long; $2n = 28$. Open, often rocky slopes and exposed alpine ridges, rare in openings in mountain brush and ponderosa pine communities at (2740) 3050 to 3810 m in Beaver, Cache, Daggett, Duchesne, Garfield, Grand, Iron, Juab, Piute, Salt Lake, San Juan, Sanpete, Sevier, Summit, Uintah, and Utah counties; British

Columbia and Alberta, south in the mountains of the western U. S.; 106 (iii). *Elymus scribneri* hybridizes with *E. trachycaulus* (see guide to hybrids, p. 722).

Elymus simplex Scribn. & Will. Alkali Wildrye. [*E. triticoides* var. *simplex* (Scribn. & Will.) A. S. Hitchc.; *Leymus simplex* (Scribn. & Will.) D. Dewey]. Perennial, strongly rhizomatous; herbage glaucous; culms mostly erect, 1.5–9 dm tall; sheaths glabrous; auricles lacking or sometimes well developed, the old sheath bases papery; blades firm, usually involute, 1–4 mm wide, glabrous or scaberulous to puberulent; ligule 0.3–0.5 mm long; spike erect, 6–16 (20) cm long and to ca 8 mm wide, the rachis continuous, the internodes 5–10 mm long at midlength of the spike; spikelets solitary or occasionally in pairs at some or all nodes of the rachis, 11–20 (25) mm long, 3– to 6 (10) –flowered, light green or rarely tinged with lavender; glumes subulate to lanceolate, mostly subequal, rounded to somewhat keeled on the back, nerveless or obscurely 1 (3) –nerved, smooth or occasionally scabrous, 5–16 mm long and mostly 0.3–1.5 mm wide near the base, awn-tipped or with awns to ca 3 mm long; lemma more or less firm, lanceolate, 7–11 mm long, rounded on the back, the sides usually visible between the glumes, obscurely nerved, glabrous or occasionally puberulent, gradually tapered to an awn 2–7 mm long; anthers 3–5 mm long; 2n = 28. Known in Utah only from along the Green River in Daggett County; Wyoming and Colorado; 1 (0). *Elymus simplex* is said to differ morphologically from *E. triticoides* chiefly in having spikelets mostly solitary rather than mostly paired at each node of the rachis and in having awns of the lemma to 7 mm long. Barkworth and Atkins (1984) treat *E. simplex* (as *Leymus simplex*) at the level of species but note that additional data may show that it would be more appropriately treated at an infraspecific rank.

Elymus smithii (Rydb.) Gould Western Wheatgrass. [*Agropyron smithii* Rydb.; *Elytrigia smithii* (Rydb.) Love; *Pascopyrum smithii* (Rydb.) Love]. Perennial, strongly rhizomatous; herbage glaucous, rarely yellow green; culms erect, mostly 2–11 dm tall; sheaths glabrous to rarely densely puberulent, frequently with well-developed auricles, old sheath bases often pale and more or less conspicuous; blades firm, involute to flat, 1–5 (6) mm wide, scaberulous to variously hairy; ligule to ca 1 mm long; spike stiffly erect, (3) 6–20 cm long, the rachis continuous, the internodes 4–16 mm long at midlength of the spike; spikelets solitary at each node of the rachis or not infrequently in pairs at some nodes, (3) 5– to 12–flowered, 13–26 (30) mm long; glumes firm, subequal to unequal, linear-lanceolate to lanceolate, typically prominently ribbed with 3–7 nerves, glabrous or scaberulous to puberulent, 8–15 mm long, 1.3–2 mm wide near the base and tapered from below midlength to a sharply acute, often awnlike apex, the second glume typically at least as long as the body of the lowermost lemma; lemma lanceolate, 8–14 mm long, rounded or somewhat keeled on the back, the sides visible between the glumes, 5–nerved, glabrous or scabrous to occasionally pubescent, sharply acute to more often awn-tipped or with an awn to ca 5 mm long; anthers (2.5) 3–5 mm long; 2n = 28, 42, 56. Chiefly on dry, open, often disturbed sites in salt desert shrub, sagebrush, mountain brush, and pinyon-juniper communities, rarely on open slopes at higher elevations, at 1220 to 2130 (2740) m in Beaver, Box Elder, Cache, Carbon, Daggett, Davis, Emery, Garfield, Grand, Iron, Millard, Morgan, Salt Lake, San Juan, Sanpete, Sevier, Summit, Tooele, Uintah, Utah, Wasatch, Washington, and Weber counties; Alaska and British Columbia to Quebec, south throughout the western and central U. S., sparingly introduced in the eastern U. S. and in Europe; 105 (x). *Elymus smithii*, the only octoploid species in the Triticeae, rarely if ever hybridizes with other members of the tribe (Dewey 1984). Western wheatgrass is the dominant grass in some areas of our region. Although its palatability is variable, it ranks as a valuable range grass, becoming coarse and stiff during the summer, but curing well and providing a good source of winter forage. The grass is widely planted for reclamation of surface mines and in erosion control, but is not regularly used for range seeding because it is normally less palatable than intermediate, tall, and even crested wheatgrass (Vallentine 1961). It is best adapted to a mean annual precipitation of 36–50 cm and to silt or clay soils, although it performs satisfactorily on sandy soils and will tolerate even strongly saline ones (Thornburg 1982).

Elymus spicatus (Pursh) Gould Bluebunch Wheatgrass. [*Festuca spicata* Pursh; *Agropyron spicatum* (Pursh) Scribn. & Sm.; *A. inerme* (Scribn. & Sm.) Rydb.; *A. spicatum* var. *inerme* (Scribn. & Sm.) Heller; *Elytrigia spicata* (Pursh) D. Dewey; *Pseudoroegneria spicata* (Pursh) Love]. Perennial, rarely with short rhizomes; herbage green or glaucous; culms densely tufted, 3–10 dm tall, erect or nearly so; sheaths glabrous or puberulent to soft-hairy, usually with well-developed auricles; blades flat to more often involute, 1–3.5 (4.5) mm wide, glabrous or more often puberulent to variously pubescent; ligule to ca 1 mm long; spikes typically erect, 6–20 cm long, the rachis continuous, the internodes 7–17 mm long at midlength of the spike; spikelets rarely more than 1 per node, 5– to 8–flowered, mostly 10–20 mm long, those at midlength of the spike rarely exceeding the internodes by more than 1/3 their length; glumes mostly subequal, oblong-elliptical to somewhat oblanceolate, conspicuously 3– to 7–nerved, 4.5–12 mm long, 1–2.2 mm wide at midlength, glabrous or scaberulous, obtuse to acute or rarely awn-tipped, from ca 1/2 to as long as the lowermost lemma; lemma lance-elliptical, usually rounded on the back, the sides visible between the glumes, the body 7–13 mm long, 5–nerved, glabrous or rarely scaberulous, acute or more often terminating in an ultimately widely divergent awn (3) 6–25 mm long; anthers 4–6 mm long; 2n = 14, 28. Sagebrush, pinyon-juniper, mountain brush, aspen, ponderosa pine, and openings in spruce-fir communities at 1370 to 2900 m in all Utah counties except San Juan and Wayne; Alaska to Saskatchewan, south to California, Arizona, New Mexico, and Texas; 294 (vi). The awnless form of *E. spicatus*, at one time referred to *Agropyron inerme*, occurs at random essentially throughout the range of the species, and is not deemed deserving of formal taxonomic recognition. *Elymus spicatus* hybridizes in nature with *E. lanceolatus* and with *E. elymoides* (see guide to hybrids, p. 722). Bluebunch wheatgrass is the dominant species in some grassland areas of the Pacific Northwest and in the Intermountain Region, forming up to 60 percent of foliage cover in some localities (D. D. Collins, dissertation, Montana State University, 1965), and is considered one of the most important forage grasses on western rangelands. A cultivar of the awnless form of the species tolerates drought and is adapted to the same soils and climate as crested wheatgrass (Thornburg 1982). Cultivars thus far developed do not, however, match crested wheatgrass in seedling vigor.

Elymus trachycaulus (Link) Gould ex Shinners Slender Wheatgrass. [*Triticum trachycaulum* Link; *Agropyron trachycaulum* (Link) Malte; *A. caninum* authors, not (L.) Beauv.; *A. latiglume* (Scribn. & Sm.) Rydb.; *A. pauciflorum* (Schwein.) sensu A. S. Hitchc., not Schur; *A. subsecundum* (Link) A. S. Hitchc.; *A. tenerum* Vasey; *A. violaceum* authors, not (Hornem.) Lange; *A. violaceum* var. *andinum* Scribn. & Sm.; *Elymus subsecundus* (Link) Löve & Löve]. Perennial, occasionally with short rhizomes; culms tufted, 3–10 (15) dm tall, erect or decumbent at the base; sheaths glabrous or retrorsely puberulent, the auricles short or lacking; blades stiffly ascending to somewhat lax, usually flat, 2–7 (8) mm wide, glabrous or scabrous to short-hairy; ligule to ca 0.5 mm long; spike slender, 4–20 (25) cm long, stiffly erect, the rachis continuous, the internodes at midlength mostly 4–9 mm long; spikelets solitary at each node of the rachis, 3– to 7–flowered, 9–20 mm long, those at midlength of the rachis usually about twice as long as the internodes; glumes oblong-elliptical, conspicuously (3) 5– to 7 (9) –nerved, glabrous or scaberulous, (5) 6–15 mm long and 1–3 mm wide at midlength, gradually or abruptly tapered to an acute or awn-tipped apex or with an awn to ca 5 mm long, the second glume at least 2/3 as long as the body of the adjoining lemma, sometimes slightly exceeding it; lemma rounded on the back, the sides visible between the glumes, 7.5–13 mm long, glabrous or scabrous to minutely hairy, 3– to 5–nerved, acute to awn-tipped or with an awn 2–20 (30) mm long; anthers 0.6–2.2 mm long; 2n = 28. Chiefly in grass-sagebrush, mountain brush, aspen, ponderosa pine, spruce-fir, and lodgepole pine communities, in alpine meadows and on exposed rocky slopes, at (1280) 1370 to 3660 m, in all Utah counties; Alaska, through much of Canada, south in the U. S. throughout the western, central, and northeastern states and in northern Mexico; Siberia; introduced in temperate Eurasia; 595 (xix). The small- anthered, largely self-fertilized complex of grasses to which *E. trachycaulus* belongs is widespread throughout the northern hemisphere. Because of a strong resemblance between the two species, North American plants were referred to *Agropyron caninum* (L.) Beauv. by C. L. Hitchcock et al. (1969), but examination of Eurasian material at the Komarov Botanical Institute revealed that the plant now known as *Elymus caninus* (L.) L. is the forest phase of the Eurasian portion of the complex for which there is no North American counterpart. *Elymus trachycaulus* hybridizes with *Hordeum jubatum*, *Agropyron cristatum*, *Elymus lanceolatus*, *E. scribneri*, *E. elymoides*, *E. glaucus*, *E. canadensis*, and possibly *E. repens* (see guide to hybrids, p. 722). Slender wheatgrass is useful as a short-lived species in reclamation plantings and range seedings. Adapted strains may be grown in areas with a minimum mean annual precipitation of 36 cm (Thornburg 1982).

Elymus triticoides Buckley Beardless or Creeping Wildrye. [*Leymus triticoides* (Buckley) Pilger]. Perennial, strongly rhizomatous; herbage glaucous or green; culms erect, 1.5–12.5 dm tall; sheaths typically hairy to some degree, often with well-developed auricles, the old sheath bases usually papery and darkened, rarely if ever disintegrating into fibers; blades moderately stiff, flat to involute, (1) 2–7 (10) mm wide, typically scaberulous to hairy; ligules mostly less than 1 mm long; spikes erect, 3–20 cm long and 5–15 mm wide, the rachis continuous, the internodes 2–10 mm long at midlength of the spike; spikelets usually in pairs at each node of the rachis, occasionally solitary at some or rarely all nodes, (6) 8–22 mm long, 3– to 6 (8) –flowered, mostly glaucous or brownish, sometimes faintly purple-tinged; glumes subulate to narrowly lanceolate, subequal to unequal, flat or rounded to somewhat keeled on the back, nerveless or obscurely 1 (3) –nerved, 4–16 mm long and 0.3–1.5 mm wide near the base, acute to awn-tipped apically, shorter to longer than the lowermost lemma, occasionally the first glume lacking or both glumes of the upper spikelets reduced or obsolete; lemma typically leathery, elliptical to lanceolate, rounded on the back, the sides or backs visible between the glumes, the body 5–12 mm long, 5– to 7–nerved, usually obscurely so, generally darkened below the apex, shiny, glabrous or pubescent, acute to abruptly awn-tipped or with a stout awn to ca 3 mm long; anthers (2.5) 3–6 mm long; 2n = 28. Saline meadows, salt desert shrub, and juniper communities at 1280 to 1830 m in Box Elder, Daggett, Davis, Juab, Millard, Rich, Salt Lake, San Juan, Sanpete, Sevier, Tooele, Utah, and Weber counties; throughout the western U. S., in Texas and Mexico; 41 (iii). *Elymus triticoides* forms natural hybrids with *E. cinereus* and *E. salinus* (see guide to hybrids, p. 722). Plants tentatively identified as *E. flavescens* Scribn. & Sm. have been collected from Tooele County (Taye 746 UTC); but they are unlike typical specimens of that species in having spikelets solitary rather than paired at each node of the rachis, the lemmas rounded on the back rather than mostly laterally compressed, and awn-tipped rather than acute. The Utah material appears to be a pubescent phase of *E. triticoides*. Beardless wildrye is a salt-tolerant, sod-forming grass, especially valuable for soil stabilization because of its ability to grow in a wide range of soil types.

Elymus virginicus L. Virginia Wildrye. Perennial; culms tufted, 6–12 dm tall; sheaths glabrous, usually with well-developed auricles; blades lax, flat, (2) 4–10 (15) mm wide; ligule to ca 0.5 mm long; spike stiffly erect, 4–17 cm long, the rachis continuous; spikelets mostly 2 per node, strongly overlapping, 4– to 5–flowered, disarticulating below the glumes; glumes subulate to narrowly lanceolate, 7–22 mm long, the usually yellowish base nerveless, subterete, hardened, and typically curved outward (exposing the base of the lowermost floret), broadened and conspicuously 2– to 5–nerved above the base, mostly 1–2 mm wide at about midlength, awn-tipped or with an awn to ca 20 mm long; lemma lance-elliptical, 6–9 mm long, the rounded back usually visible between the glumes, glabrous to minutely hairy, 5–nerved, tapered to an awn to 30 mm long or in ours (**var. submuticus Hook.**) merely awn-tipped; anthers 1.5–3 mm long; 2n = 28. A weed of cultivated areas, below 1830 m in Cache, Utah, Wasatch, and Weber counties; British Columbia to Newfoundland, south through most of the U. S.; introduced in Europe; 3 (0). Where their ranges overlap, *E. virginicus* and *E. canadensis* are known to hybridize.

Enneapogon Desv. ex Beauv.

Annuals or perennials; culms tufted; sheaths open almost to the base; blades flat to involute; ligule a ring of hairs; inflorescence a compact to spikelike panicle; spikelets subterete, 3– to 6–flowered, some of the upper florets reduced, the rachilla disarticulating above the glumes and tardily between the florets; glumes membranous, (3) 7– to 9–nerved, longer than the body of the lemma; lemma broad, conspicuously 9–nerved, the nerves prolonged as (3) 7–9 equal or subequal awns; stamens 3; caryopsis free within the lemma and palea, the embryo 3/4–5/6 the length of the grain; $x = 10$.

Enneapogon desvauxii Beauv. Spike Pappusgrass. [*Pappophorum wrightii* Wats.]. Perennial, sometimes flowering the first year; culms tufted, 1–5 dm tall, often geniculate, ultimately disarticulating at the nodes; sheaths open, usually pubescent; blades filiform, 0.5–2 mm wide, mostly 2–12 cm long, folded to involute, hairy to some degree; ligule a ring of hairs to ca 1 mm long; panicle spikelike, 2–9 cm long, gray green to gray, bristly; spikelets mostly 5–7 mm long including the awns, typically 3–flowered, the lowermost floret perfect, the middle floret perfect, staminate, or neuter, the uppermost typically reduced to multiple awns; cleistogamous, 1–flowered spikelets often developing in the lower leaf sheaths; glumes lanceolate, usually sparsely puberulent, subequal, 3– to 9–nerved, mostly 3–5 mm long, longer than the body of the uppermost lemma; lemma broad, pubescent on the rounded back, the body 1.3–2.5 mm long, with 9 strong nerves prolonged as subequal plumose awns 3–4.5 mm long; anthers 0.2–0.6 mm long; $2n = 20$. Dry, sandy or gravelly to rocky soil on open desert flats and slopes below 1800 m in Beaver, Emery, Garfield, Grand, Kane, and Millard counties; California, Arizona, New Mexico, Texas, and Mexico; South America; Asia; 16 (0).

Eragrostis N. M. Wolf

Annuals or perennials, tufted or in a few species rhizomatous; sheaths open, often long-hairy at the summit, otherwise glabrous or hairy; blades flat or folded to involute; ligule a ring of hairs or rarely membranous; inflorescence an open to spikelike panicle; spikelets subterete to laterally compressed, (2) 3– to 40–flowered, pale green or gray to nearly black, either the lemmas falling and the paleas persistent on an intact rachilla, or the rachilla disarticulating above the glumes and between the florets; glumes usually membranous, lanceolate or ovate, subequal or unequal, keeled or rounded on the back, 1 (3) –nerved, obtuse to sharply acute at the apex, shorter than the lowermost lemma; lemma usually membranous, lanceolate to broadly ovate, keeled or rounded on the back, 3–nerved, the nerves nearly parallel and typically conspicuous, the apex obtuse to sharply acute, awnless; palea variable in length; stamens 2 or 3; caryopsis falling free of the lemma and palea, rounded or shallowly to deeply grooved on the ventral side, the embryo 2/5–2/3 as long as the grain; $x = 10$.

Koch, S. 1974. The *Eragrostis pectinacea-pilosa* complex in North and Central America (Gramineae: Eragrostoideae). Illinois Biol. Monogr. 48: 1–74.

――――. 1978. Notes on the genus *Eragrostis* (Gramineae) in the southeastern United States. Rhodora 80: 390–403.

1. Plants perennial; spikelet rachilla disarticulating, all but the lowermost florets falling with the rachilla internodes 2

— Plants annual; spikelet rachilla remaining intact, the lemmas falling, the paleas usually persistent 3

2(1). Basal leaf blades often more than 30 cm long and flexuous or arching toward the ground; pedicels remaining appressed to the branchlets .... *E. curvula*

— Basal leaf blades mostly 2–10 cm long, stiffly ascending to spreading; pedicels ultimately spreading .... *E. lehmanniana*

3(1). Craterlike glands or glandular bands on pedicels or leaf blade margins 4

— Craterlike glands not present on panicle branches or leaf margins 5

4(3). Mature spikelets averaging more than 2 (2–4) mm wide at midlength; lemmas frequently bearing craterlike glands on the midnerve, averaging at least 0.8 mm wide midnerve to margin .... *E. cilianensis*

— Mature spikelets averaging not more than 2 (1.5–2.2) mm wide at midlength; lemmas rarely glandular on the midnerve, averaging less than 0.8 mm wide midnerve to margin .... *E. minor*

5(3). Culms decumbent to prostrate and often rooting at the nodes, rarely erect; panicles mostly less than 4 (1–6) cm long; palea to about half as long as the lemma .... *E. hypnoides*

— Culms erect to occasionally prostrate but not rooting at the nodes; panicles mostly more than 4 cm long; palea more than half as long as the lemma 6

6(5). Spikelets 0.6–1.2 mm wide; caryopses flattened and shallowly to deeply grooved on the ventral side; panicles 15–40 cm long, the branchlets and pedicels ultimately widely spreading to reflexed .... *E. mexicana*

— Spikelets at maturity averaging more than 1.2 (1–2.5) mm wide; caryopses terete, not grooved; panicles 5–35 cm long, the branchlets and pedicels diverse 7

7(6). Primary panicle branches mostly spikelet-bearing to within 1–3 mm of the base, the branchlets and pedicels ultimately spreading; lemmas often obtuse at the apex; culms and rachis of the panicles usually bearing a complete or partial ring of yellow glandular tissue below the nodes .... *E. barrelieri*

— Primary panicle branches mostly naked for more than 3 mm above the base; branchlets and pedicels at maturity typically appressed to the branches; lemmas acute; culms and panicles without glandular tissue below the nodes (which may be yellow) .... *E. pectinacea*

Eragrostis barrelieri Daveau Mediterranean Lovegrass. Annual; culms tufted, 1–6 dm long, erect to decumbent-ascending or prostrate, branching from the lower nodes, usually with a complete or partial ring of yellow glandular tissue below the panicle and below the nodes; sheaths long-hairy at the summit, otherwise glabrous; blades flat to involute, 1–5 mm wide and mostly 2–10 cm long, glabrous or scaberulous, occasionally sparsely hairy; ligule a ring of hairs; panicle open, ovoid to narrowly oblong, 5–16 cm long and to ca 7.5 cm wide, the primary branches ultimately stiffly spreading, branchlets or pedicels smooth or scaberulous, typically arising within 1–3 (4) mm of the base, ultimately widely spreading; spikelets compressed, lance-linear, 5–11 (15) mm long, 1.2–1.5 mm wide, 6– to 15 (20) –flowered, pale to dark gray green, the rachilla remaining intact at fruiting; glumes subequal, 0.8–1.8 mm long, 1–nerved; lemma lanceolate to elliptical, 1.8–2.5 mm long, 3–nerved, acute to more often obtuse at the apex; palea at least 2/3 as long as the lemma; anthers 0.2–0.5 mm long; caryopsis subglobose to broadly ellipsoidal, to ca 1 mm long, not ventrally grooved; 2n = 60. Gardens and disturbed, often sandy sites below 1520 m in Salt Lake, San Juan, and Washington counties; native to the Mediterranean region of Europe, introduced in North America through most of the southwestern U. S., east to Texas and Kansas, naturalized in Mexico (Koch 1978); South America; 5 (i).

Eragrostis cilianensis (All.) Mosher Stinkgrass. [*Poa cilianensis* All.; *E. megastachya* (Koeler) Link]. Annual; herbage with few to many craterlike glands, malodorous when fresh; culms tufted, 1–6(8) dm tall, ascending or spreading to prostrate, often branching from the lower nodes, usually with a ring of glands below the nodes; sheaths long-hairy at the summit, often bearing scattered, craterlike glands; blades flat or folded, 2–7 (8) mm wide, glabrous or scaberulous to occasionally long-hairy, with craterlike glands especially numerous along the margins; ligule a ring of hairs; panicle compact to somewhat open, ovoid to oblong, 1–16 (20) cm long and 1–5 (7) cm wide, the branches ascending-spreading, the pedicels at maturity spreading, 0.3–7 mm long, averaging less than 2 mm long, often gland-dotted, otherwise smooth or scaberulous; spikelets compressed, ovate to lanceolate or oblong, 7– to 40– flowered, 4–15 (20) mm long and at maturity 2–3 (4) mm wide at midlength, pale to dark gray green, occasionally tinged with dull purple, the rachilla remaining intact at fruiting; glumes subequal, 1–2.5 mm long, 1– (3)–nerved, the midnerve often beset with craterlike glands; lemma slightly firmer than the glumes, ovate, 1.5–2.8 mm long, 3–nerved, often bearing craterlike glands along the midnerve, abruptly acute at the apex; palea at least 2/3 as long as the lemma; anthers 0.2–0.5 mm long; caryopsis globose to ovoid or ellipsoidal, 0.6–0.8 mm long, not ventrally grooved; 2n = 20, 40. Weed of gardens and dry to mesic, disturbed sites at 850 to 2320 m in Box Elder, Cache, Carbon, Davis, Grand, Kane, Salt Lake, Tooele, Uintah, Utah, Washington, and Weber counties; native to Eurasia, introduced in Canada and widespread throughout the U. S., in northern Mexico; South America; 49 (iv).

Eragrostis curvula (Schrader) Nees Weeping Lovegrass. [*Poa curvula* Schrader; *E. chloromelas* Steud.]. Perennial; culms tufted, (3) 6–12 (18) dm tall, erect, occasionally branching at the lower nodes; sheaths glabrous or hairy, sometimes with long hairs at the summit, the lowermost often hard and yellowish, densely appressed-hairy near the base; blades involute and in that state mostly less than 1.4 mm wide, long-tapered to filiform tips, the basal leaves often more than 30 cm long and flexuous or arching toward the ground; ligule a ring of short hairs; panicle variable, contracted to open, oblong, 6–40 cm long and to mostly 5–12 cm wide, erect or nodding, the branches strongly ascending to spreading, the pedicels remaining appressed to the branches; spikelets subterete to moderately compressed, 4– to 15–flowered, linear, 4–10 mm long and 1–1.6 mm wide, pale to dark gray green, the rachilla disarticulating between the florets above, the rachilla internodes falling with the florets, the lower internodes often persistent; glumes somewhat unequal, the first 1–2 mm long, 1–nerved, the second 1.5–3 mm long, 1– to 3–nerved; lemma lanceolate, 1.8–3 mm long, 3–nerved, glabrous or scaberulous, acute to rounded at the apex, often variable within any one panicle; palea about as long as the lemma; anthers 1–1.3 mm long; caryopsis elliptical, 1.1–1.7 mm long and to 0.7 mm wide, not ventrally grooved; 2n = 20, 40, 50, 60. Desert shrub and juniper communities, usually on disturbed sites, below 1520 m in Grand and Washington counties; native to South Africa, introduced into much of the southern U. S.; South America; 5 (0). Weeping lovegrass is used throughout much of the southern U. S. for erosion control and as a pasture grass.

Eragrostis hypnoides (Lam.) B.S.P. Teal or Creeping Lovegrass. [*Poa hypnoides* Lam.]. Annual, often matforming; culms tufted, 1–3 dm long, branching at the base and above, decumbent to prostrate or rarely erect, rooting at the nodes; sheaths long-hairy at the summit and occasionally sparsely so below; blades flat or folded, mostly 0.7–2 mm wide, to about 4 cm long, usually hairy on one or both surfaces; ligule a ring of hairs; panicle contracted to open, ellipsoidal to ovoid, 1–4 (6) cm long and to ca 3 cm wide, the branches ascending to spreading, the branches and pedicels usually with at least a few minute hairs; spikelets compressed, 7–to 22 (40)–flowered, linear-lanceolate, 2.5–15 (20) mm long, mostly pale green, sometimes purple-tinged, glabrous or minutely hairy, the rachilla remaining intact at fruiting; glumes 1–nerved, more or less unequal, the first 0.5–1 mm long, the second 1–1.8 mm long; lemma narrowly lanceolate, 1.5–2 mm long, 3–nerved, often fluted, acute; palea to about half as long as the lemma; anthers to 0.3 mm long; caryopsis compressed, ovate to elliptical, to about 0.6 mm long; 2n = 20. Along waterways, in marshes and beds of drying ponds below 1520 m in Box Elder, Cache, Utah, and Wasatch counties; Canada, south through most of the U. S. and in Mexico; South America; 22 (0).

Eragrostis lehmanniana Nees Lehmann Lovegrass. Perennial; culms tufted, 4.5–7.5 dm tall, erect or decumbent to prostrate, the latter often rooting at the nodes, typically geniculate, sometimes branched above the base; sheaths glabrous or sometimes with a few long hairs at the summit, the lowermost papery, whitish, and often short-hairy at the very base; blades involute, to ca 1 mm wide and mostly 2–10 cm long, mostly stiffly ascending to spreading; ligule a ring of short hairs; panicle open, narrowly oblong to lanceolate, 7–18 cm long and 3–8 cm wide, erect, the branches stiffly ascending to moderately spreading, 4–8 cm long, the pedicels ultimately spreading, much shorter than the spikelets; spikelets slightly compressed, narrowly oblong, mostly 5– to 8 (12)–flowered, 5–14 mm long and 0.8–1 mm wide, gray green to dark gray, the upper rachilla internodes falling with the florets, the lower internodes persistent; glumes unequal, the first 1–1.6 mm long, 1–nerved, the second 1.4–1.8 mm long, 1–nerved; lemma oblong, 1.8–2.2 mm long, mostly conspicuously 3–nerved, glabrous, abruptly acute to more often obtuse at the apex; palea as long as or slightly longer than the lemma; anthers 0.8–1 mm long; caryopsis ellipsoidal, 0.6–1 mm long and to ca 0.5 mm wide. Known by several collections (e.g., Higgins 16693 BRY) from along roadsides in creosote bush communities in Washington County; native to South Africa, introduced into Arizona, New Mexico, Texas, and Oklahoma; 3 (0). Lehmann lovegrass is a readily established, drought-resistant grass introduced to the U. S. in 1932 for range seeding in warm semi-deserts of the Southwest (Hanson 1972). University of Arizona research has shown that this grass produces more forage with less water than any other known forage species (Thornburg 1982).

Eragrostis mexicana (Hornem.) Link Mexican Lovegrass. [*Poa mexicana* Hornem.; *E. orcuttiana* Vasey; *E.*

virescens Presl]. Annual; culms tufted, 4–10 dm tall, erect or geniculate and sometimes branching at the lower nodes, occasionally with a band of glandular tissue below the nodes; sheaths typically hairy at the summit, with or without glandular depressions; blades usually flat, gradually tapered from a broad base to a more or less filiform tip, (1) 2–17 mm wide and 5–25 cm long, usually scaberulous along the margins; ligule a ring of hairs; panicle open, ovoid to narrowly oblong, 15–40 cm long and to ca 14 cm wide, the primary branches naked near the base, the branchlets and pedicels smooth or scaberulous, ascending to ultimately widely spreading or occasionally reflexed, the pedicels shorter or longer than the spikelets; spikelets subterete to somewhat compressed, ovate to linear, 4– to 15–flowered, 4–9.5 mm long and 0.6–1.2 mm wide, pale green to dull gray green, occasionally purple, the rachilla remaining intact at fruiting; glumes subequal, 0.7–1.8 mm long, 1–nerved; lemma lanceolate, 1.5–2.2 mm long, 3–nerved, glabrous, acute; palea as long as the lemma or nearly so; anthers 0.3–0.5 mm long; caryopsis subpyriform to oblong, 0.6–1 mm long, flattened and shallowly to deeply grooved on the ventral side. Weed of gardens and waste places below 1520 m in Davis, Duchesne, Salt Lake, Uintah, and Utah counties; California and Nevada; disjunct in South America; 22 (vi). According to Koch and Sanchez (Phytologia 58: 377–381. 1985), our material is ssp. *virescens* (Presl) **S. D. Koch & I. Sanchez V.**, differing from the typical element of the species in having spikelets less than 1.5 mm wide.

Eragrostis minor Host Minor Lovegrass. [*E. poaeoides* Beauv. ex R. & S.]. Annual; herbage with few to many craterlike glands; culms tufted, 0.3–4 (5) dm tall, erect or decumbent-based to prostrate, branching from the lower nodes, often with a ring of glands below the nodes; sheaths long-hairy at the summit, with or without craterlike glands; blades flat or folded, 1–5 mm wide and mostly 1–15 cm long, smooth or scaberulous and sometimes sparsely hairy, usually with craterlike glands along the margins; ligule a ring of hairs; panicle open, ovoid to narrowly oblong, mostly 2–15 cm long and to 10 (15) cm wide, the branches, branchlets, and pedicels ascending to widely spreading, the pedicels mostly 1–10 mm long, usually averaging more than 2 mm long, typically bearing 1 or more glands or bands of yellow glandular tissue, otherwise smooth or scaberulous; spikelets compressed, 5– to 12 (20) –flowered, 4–11 mm long and 1.5–2.2 mm wide, pale to dark gray green, often purple-tinged, the rachilla remaining intact at fruiting; glumes subequal, 1.2–1.8 mm long, 1–nerved, occasionally 1 or 2 glands on the midnerve; lemma usually slightly firmer than the glumes, broadly lanceolate to ovate, 1.3–2.2 mm long, glabrous or laterally scaberulous, 3–nerved, rarely with 1 or 2 obscure glands on the midnerve, subacute to rounded at the apex; palea at least 2/3 as long as the lemma; anthers 0.2–0.4 mm long; caryopsis globose to broadly ellipsoidal, 0.5–0.8 mm long, not ventrally grooved; 2n = 20, 22, 30, 40, 60. Weed of gardens and waste places, mostly below 1520 m in Duchesne, Salt Lake, and Washington counties; native to Eurasia, adventive in Canada and in most of the southwestern U. S., established in much of the East; South America; 9 (iv). Proof that the generic name *Eragrostis* was validly published prior to the publication of the epithet *minor* necessitates rejection of the later name, *E. poaeoides* (Koch 1978). *Eragrostis lutescens* Scribn., reported from all of the states surrounding Utah (Hitchcock 1951), differs from *E. minor* in being less robust (to 2 dm tall), with panicles narrow (the branches remaining appressed at maturity), and lemmas acute at the apex.

Eragrostis pectinacea (Michx.) Nees Tufted Lovegrass. [*Poa pectinacea* Michx.; *E. caroliniana* authors, not (Sprengel) Scribn.; *E. diffusa* Buckley]. Annual; culms tufted, 1–6 (10) dm tall, erect or decumbent-ascending, often branched at the base and occasionally above, lacking yellow glandular tissue below the nodes; sheaths hairy at the summit, otherwise glabrous; blades flat or folded, 0.5–4.5 mm wide and to 15 (30) cm long, glabrous or scaberulous; ligule a ring of hairs; panicle open, pyramidal or ovoid to narrowly oblong, 5–30 (35) cm long and to about 15 cm wide, the primary branches at maturity spreading or rarely reflexed, mostly naked near the base, the branchlets and pedicels smooth or scaberulous, remaining appressed to the branches; spikelets moderately compressed, 5– to 15 (22)–flowered, ovate to lanceolate, 4–11 mm long and (1) 1.2–2.5 mm wide, light green to dark gray green, sometimes purple-tinged, the rachilla remaining intact at fruiting; glumes 1–nerved, the first 1–1.5 mm long, the second 1.2–1.8 mm long; lemma lance-ovate, 1.5–2.2 mm long, 3–nerved, glabrous throughout or scabrous on the keel and often scaberulous laterally, acute; palea persistent, about equal to the lemma; anthers 0.2–0.4 mm long; caryopsis subglobose to oblong or subpyriform, mostly 0.6–1.1 mm long, not ventrally grooved; 2n = 40, 60. Chiefly a weed of cultivated places, also occurring in moist, usually sandy sites along waterways and in waste places below 1520 m in Cache, Davis, Kane, Salt Lake, San Juan, Uintah, Utah, Washington, and Wayne counties; British Columbia to Nova Scotia, south throughout the U. S. and in Mexico; South America; sparingly introduced in western Europe; 28 (vii).

Eremopyrum (Ledeb.) Jaub. & Spach

Annuals; sheaths open or those of lowermost leaves initially closed to the summit or nearly so, usually bearing auricles; ligules membranous; inflorescence a short, dense, oblong to ovate spike, the rachis disarticulating throughout or only at its base, the internodes mostly to about 1 mm long; spikelets compressed, sessile, solitary at each node and alternating on opposite sides of the rachis, 3– to 6–flowered, the rachilla disarticulating above the glumes and between the florets or remaining intact; glumes firm, often fused at the base, 1– to 5–nerved, prominently keeled, usually asymmetrical, acute or awned; lemma leathery, lanceolate, keeled or rounded on the back, 5–nerved, acute or awned; stamens 3; caryopsis often adherent to the lemma and palea, the embryo not more than half the length of the grain; x = 7. In their native desert and semidesert habitats in Central Asia, these annual members of the tribe Triticeae have forage value only in early spring when large concentrations develop.

Eremopyrum triticeum (Gaertn.) **Nevski** Annual Wheatgrass. [*Agropyron triticeum* Gaertn.]. Annual; culms solitary or in small tufts, 4–40 cm tall, erect or geniculate at some of the nodes, densely puberulent below the spike; sheaths of culm leaves open about 2/3 their length, glabrous or puberulent, the uppermost typically inflated, the auricles small or lacking; blades soft, flat or loosely involute, 1–4 (6) mm wide, smooth or scabrous to sparsely puberulent; ligule 0.2–1 mm long; spike ovate-elliptical, 0.8–2 cm long and to about 1.5 cm wide, the internodes of the rachis 0.2–1 mm long, the rachis ultimately disarticulating at its base, the spike falling intact; spikelets widely divergent, 6–10 mm long including awn tips; glumes becoming thickened and hardened, laterally compressed, lanceolate, 4–7 mm long, more or less saccate near the base of the broad keel, the margins membranous (often inrolled and appearing thickened), the apex acute to awn-tipped; lemma similar to or the upper portion thinner than the glumes in texture, lanceolate, 5–7.5 mm long, rounded on the back, glabrous or basally puberulent, awn-tipped or with an awn to 7 mm long; anthers to about 1 mm long; $2n = 14$. Disturbed sites in salt desert shrub, sagebrush, and juniper communities, mostly below 1520 m in Box Elder, Carbon, Emery, Grand, Juab, Salt Lake, San Juan, Tooele, and Uintah counties; native to central Asia, sparingly introduced from Alberta to Manitoba, south in the U. S. through most of the West; 28 (v).

Eriochloa H. B. K.

Annuals or sometimes perennials; culms often branching at the lower nodes; sheaths open; ligule an often obscurely membranous-based ring of hairs; inflorescence a contracted to open panicle, the spikelets sessile or short-pedicelled, solitary or in pairs in (1) 2 rows on erect to slightly spreading, one-sided, spikelike branches; spikelets dorsiventrally compressed, 2-flowered, the lower floret neuter or staminate, the upper perfect, the rachilla disarticulating near the base, the segment between the glumes expanded to form a cuplike structure, the first glume nearly always vestigial and adherent to the cuplike rachilla base, the second glume thin, about equal to the spikelet; lemma of the lower floret resembling the second glume, lemma of the upper floret leathery, obtuse to acute or short-awned; caryopsis free of but firmly enclosed within the lemma and palea, the embryo to ca 3/4 the length of the grain; $x = 9$.

Eriochloa contracta A. S. Hitchc. Prairie Cupgrass. [*Helopus mollis* C. Muell.]. Annual, often pubescent throughout with short soft hairs; culms branched at and often above the base, (2) 3–7 (8) dm tall, erect or geniculate and spreading; blades mostly 2–7 mm wide, flat to involute, ligule a dense ring of hairs 1–2 mm long; panicle 6–15 cm long, typically with 6–15 erect, slender, spikelike branches 1–2 cm long; spikelets 3.5–5 mm long including the awn tip; first glume vestigial (indistinguishable), second glume and lemma of the lower floret similar, as long as the spikelet, sparsely appressed-hairy, with awn tips to ca 1 mm long; lemma of the upper floret leathery, finely striate, obtuse at the apex and terminating in a scabrous, included awn 0.5–0.8 mm long; $2n = 36$. Known in Utah by a single collection (Higgins 16084 BRY) from a disturbed site at 950 m in the Mohave Desert extension into Washington County; Arizona, Colorado, Nebraska to Texas, Missouri, Louisiana, and Mexico; 2 (0).

Erioneuron Nash

Low, tufted annuals or perennials; leaves chiefly basal; sheaths open; blades narrow, flat to involute; ligule a ring of short hairs, sometimes membranous at the base; inflorescence a short, compact raceme or panicle; spikelets 4- to 18- flowered, disarticulating above the glumes and between the florets, the florets progressively reduced upward, the uppermost sometimes rudimentary; glumes 1-nerved; lemma keeled to rounded on the back, entire or bilobed at the apex, 3-nerved, each nerve densely hairy at least below, the lateral nerves occasionally prolonged as short mucros, the midnerve as an awn-tip or a short awn; palea subequal to the lemma; stamens 3; caryopsis falling free of the lemma and palea, ovoid, translucent to opaque, smooth and glossy, the embryo about half as long as the grain; $x = 8$.

Tateoka, T. 1961. A biosystematic study of *Tridens* (Gramineae). Amer. J. Bot. 48: 565–573.

1. Inflorescence often partially enclosed by the subtending sheath and typically overtopped by the leaf blades; glumes as long as the spikelet or nearly so. . . . *E . pulchellum*
— Inflorescence borne above the blades; glumes much shorter than the spikelet. *E. pilosum*

Erioneuron pilosum (Buckley) Nash Hairy Tridens. [*Uralepis pilosa* Buckley; *Tridens pilosus* A. S. Hitchc.] Perennial with a few slender fibrous roots; culms densely tufted, erect or nearly so, 0.2–3 dm tall; sheaths usually hairy at the summit, otherwise glabrous; blades firm, flat or folded, 1–2 mm wide and mostly 1–6 cm long, typically sparsely short-hairy, the whitish margins thickened; ligule a ring of hairs, with or without a membranous base, the whole to about 0.5 mm long; panicle or raceme typically borne above the basal leaf cluster, ovate-oblong, 1–4.5 cm long; spikelets mostly 10–16 mm long with 6–12 (18) closely overlapping florets, pale or more often strongly purple-tinged; glumes ovate to lanceolate, sharply acute to awn-tipped, 3–6 (8) mm long overall, typically keeled on the solitary nerve, glabrous, the second glume about equal to the lowermost lemma; lemma ovate to lanceolate, 4–7.5 mm long, typically keeled, gradually tapered to an acute, entire to minutely notched apex, 3-nerved, the lateral nerves rimming the margin, the midnerve prolonged as an awn 0.2–2 mm long, the lemma body densely pubescent with hairs mostly 1–2 mm long on the nerves near the base and sometimes between, also pubescent along the margin to above midlength or nearly throughout; anthers (of specimens examined) 0.2–1 mm long; $2n = 16, 32$. Creosote bush, desert shrub, sagebrush, and pinyon-juniper communities at 910 to 2130 m in Beaver, Emery, Garfield, Grand, Kane, Millard, San Juan, Tooele, Uintah, Washington, and Wayne counties; southwestern U. S. from California east to Kansas, Oklahoma, and Texas, south into Mexico; 94 (0).

Erioneuron pulchellum (H.B.K.) Tateoka Fluffgrass. [*Triodia pulchella* H.B.K.; *Tridens pulchellus* (H.B.K.) A.S. Hitchc.]. Perennial or in some areas chiefly annual or biennial, with or without slender stolons; culms densely tufted, 0.2–1.5 dm tall; leaves fascicled at each culm node; sheaths short and broad, typically hairy at the base and at the summit; blades firm, strongly involute and in that state to about 1 mm wide, rarely more than 6 cm long, pungent-tipped, more or less scaberulous; ligule a ring of hairs with or without a membranous base, the whole about 0.5 mm long; panicle a headlike cluster mostly 1.5–4 cm long, subtended by multiple leaves, often partially enclosed in the sheaths and typically surpassed by the blades; spikelets mostly 6–13 mm long, pale or occasionally purple-tinged, (4) 5- to 8 (14)-flowered, the florets progressively reduced upward; glumes broadly lance-elliptical, sharply acute to awn-tipped, 6–8 mm long overall, keeled to rounded on the back, usually glabrous throughout, the second equal to or slightly shorter than the uppermost lemma; lemma lanceolate, mostly 2.5–5 mm long, keeled to rounded on the back, lobed nearly to midlength, 3-nerved, densely long-hairy along the lateral nerves, sometimes almost to the apex, the midnerve prolonged as a glabrous awn nearly equal to or just surpassing the apical lobes, mostly 1–2.5 mm long, often recurved in age; anthers 0.3–0.5 mm long; $2n = 16$, 32. Creosote bush, desert shrub, sagebrush, and pinyon-juniper communities at 820 to 2130 m in Beaver, Emery, Garfield, Grand, Juab, Kane, Millard, San Juan, Washington, and Wayne counties; southwestern U.S. east to Texas and south into Mexico; 72 (0). According to Beatley (Madrono 20: 330–331. 1970), *E. pulchellum* has limited success in becoming established as a perennial where precipitation is variable from season to season. In such areas, most plants are annual or biennial and germinate following spring rains. *Erioneuron pulchellum* occasionally serves as host to what appears to be the same species of woolly aphid that commonly infests *Munroa squarrosa*. See *Blepharidachne kingii*.

Festuca L.

Annuals or perennials; sheaths open or the lowermost entirely to partially closed, with or without auricles; blades flat to involute; ligule membranous, often formed by the continuation of the membranous margin of the sheath summit across the ventral surface of the narrow blade, the lateral portion then usually higher than the center, ligules of the basal leaves sometimes longer than those of the culm leaves; inflorescence a contracted to open, erect to nodding, often more or less secund panicle or raceme, the branches and pedicels subterete to 3-angled, erect to widely spreading or reflexed, glabrous or scabrous to short-hairy; spikelets (1) 2- to 13 (20) -flowered, the upper 1–3 florets sometimes greatly reduced, the rachilla disarticulating above the glumes and between the florets; glumes mostly unequal, subulate to broadly ovate, the first 1 (3)-nerved and acute, the second usually subulate to lanceolate, 3 (5)-nerved, acute to short-awned, usually shorter but occasionally equal to or longer than the lowermost lemma; lemmas laterally or dorsiventrally compressed, lanceolate to ovate, mostly ultimately rounded on the back, 3- to 7-nerved, often obscurely so, gradually long-tapered from near midlength or below to an acute, rarely minutely bifid apex, awnless or with a straight to somewhat sinuous terminal awn; palea about equal to the lemma; stamens 3 in perennial species, 1–3 in annual ones; caryopsis free of or fused with the palea, bearing a very short apical appendage, the embryo 1/6–1/3 as long as the grain; $x = 7$. According to Lonard and Gould (1974), the separation of *Vulpia* from *Festuca* is based on a correlation between a primarily annual life form with the production of mostly cleistogamous flowers in *Vulpia*, versus a perennial life form with chasmogamous flowers in *Festuca*. European taxonomists, however, include in *Vulpia* some perennial species with chasmogamous flowers. Except for these overlapping characters, the two groups appear to be morphologically and cytologically inseparable. The morphological similarity between the annual species and the perennial *Festuca ovina*, for example, is greater than that between *F. ovina* and other perennial species of *Festuca*, such as *F. sororia* or *F. subulata*. This close relationship between the annual and perennial species is totally obscured when each is placed in a separate genus. Under such circumstances, the rational for assigning the annual species to *Vulpia* eludes me; and I have followed Hitchcock (1951), Dorn (1977), and Sutherland (1968) in recognizing only *Festuca*.

Lonard, R. I. and F. W. Gould. 1974. The North American species of *Vulpia* (Gramineae). Madroño 22: 217–230.
Malik, C. P. 1967. Hybridization of *Festuca* species. Canad. J. Bot. 45: 1025–1029.
Terrell, E. E. 1968. Notes on *Festuca arundinaceae* and *F. pratensis* in the United States. Rhodora 70: 564–568.

1.	Leaf blades mostly flat, averaging more than 3 (2–12) mm wide	2
—	Leaf blades flat to involute, averaging less than 3 mm (0.5–3) wide	5
2(1).	Lemma awns averaging more than 4 (5–17) mm long	*F. subulata*
—	Lemmas awnless or with awns to 3 (4) mm long	3
3(2).	Panicles open; sheath auricles lacking; spikelets with 1–4 well developed florets; plants of moist wooded areas, mostly at or above midmontane	*F. sororia*
—	Panicles narrow (branches often somewhat spreading at flowering); sheath auricles usually present; spikelets (4) 5- to 13-flowered; plants mostly occupying open sites below midmontane	4
4(3).	Lemmas averaging less than 7 (4–7.5) mm long; leaf blades rarely more than 7 (2–9) mm wide; auricles typically glabrous; panicle branches 2 at the lowermost node, together rarely bearing more than 9 spikelets	*F. pratensis*
—	Lemmas usually averaging at least 7 (6–10) mm long; leaf blades 2–12 mm wide; auricles typically ciliate; panicle branches 2 or 3 at the lowermost node, together usually bearing 10–30 spikelets	*F. arundinacea*
5(1).	First glume 0.3–1.5 (2) mm long; awns of the lemmas 5–22 mm long	*F. myuros*
—	First glume more than 2 mm long or rarely shorter and then awns of the lemmas averaging less than 5 (0.5–6) mm long	6
6(5).	Ligule 2.5–9 mm long; plants 4–12 dm tall, occupying montane habitats	*F. thurberi*
—	Ligule less than 2.5 mm long; plants mostly less than 4 dm tall, rarely montane	7
7(6).	Plants annual; stamens 1–3	8
—	Plants perennial; stamens consistently 3	10
8(7).	Spikelets at maturity 5- to 17-flowered, the internodes of the rachilla averaging about 0.5 (0.3–1) mm long; lemma glabrous to pubescent with minute, stiffly spreading hairs, the awn 0.5–5 (6) mm long	*F. octoflora*
—	Spikelets 2- to 4 (8)-flowered, the internodes of the rachilla averaging 1 mm long; lemma glabrous or scabrous, the awn 3–20 mm long	9
9(8).	Pedicels and usually the branches of the panicle erect and remaining so at maturity, pulvini generally lacking in the axils; pedicels averaging more than 2 (1–6 or more) mm long; glumes glabrous	*F. bromoides*
—	Pedicels and usually the branches (at least the lowermost) widely spreading to reflexed at maturity, pulvini usually present in the axils; pedicels averaging less than 2 (0.5–6) mm long; glumes sometimes hairy	*F. microstachys*

10(7). Spikelets 1- or 2 (3)-flowered; panicles ultimately open, the branches widely spreading to reflexed, distinctly pubescent (at 10x magnification) .. *F. dasyclada*

— Spikelets (2) 3- to 10-flowered; panicles mostly narrow, the branches spreading only at flowering, glabrous to scabrid-puberulent 11

11(10). Plants densely tufted, lacking rhizomes, occupying native habitats or rarely escaping from cultivation; leaves chiefly basal, numerous, the blades filiform, averaging about 1 (0.3–2) mm wide .... *F. ovina*

— Plants solitary or loosely tufted, usually decumbent-based and rhizomatous, in cultivation or escaped therefrom; leaves basal and cauline, few, the blades often flat, 0.5–3 mm wide .... *F. rubra*

Festuca arundinacea Schreber Tall Fescue. [*F. elatior* ssp. *arundinacea* Hackel]. Differs from *F. pratensis* chiefly in being more robust, with culms 0.5–2 m tall, sheath auricles ciliate or rarely glabrous, leaf blades (2) 3–12 mm wide, panicle to 35 cm long, spikelets mostly 12–15 mm long, first glume (3) 4–6 mm long, second glume 4–7 mm long, and lemmas (6) 7–10 mm long; 2n = 28, 42, 56, 63, 70. Along roadways, in salt desert shrub, sagebrush, and meadow communities, and as a weed of cultivated areas, at 1280 to 1830 m in Davis, Duchesne, Garfield, Morgan, Salt Lake, Tooele, Utah, and Wayne counties; native to Europe, now cultivated from Alaska through southern Canada and much of the U. S., escaping and becoming widely established; South America; 16 (viii). The characters by which Terrell (Rhodora 70: 564–568. 1968) separates *F. arundinacea* from *F. pratensis* (as listed in part in the key) occur at random in more than half the specimens examined. Although the genomic constitution of the cultivar growing in our area may have been altered through selective breeding, both morphological and seed protein studies indicate that in natural populations *F. pratensis* and *F. arundinacea* do retain independent specific status (Pl. Syst. & Evol. 149: 135–140. 1985). The name *F. arundinacea* is retained here following European, Asian, and nearly all American workers. Tall fescue is valuable for pasture, hay, and erosion control on irrigated sites or in areas receiving a mean annual precipitation of about 46 cm or more. The grass is salt-tolerant and does well on heavy alkaline soils of low-lying valleys where it has been widely used in recent years. It also produces good forage yields in mountain meadows and in openings in aspen-conifer woodlands, especially when planted in pure stands (Vallentine 1961). Unfortunately, the plant is a noxious invader of lawns. Moreover, ingestion of large amounts of tall fescue may lead to lameness and ultimately gangrene of one or both hind feet in susceptible cattle (Kingsbury 1964).

Festuca bromoides L. Brome Sixweeks Fescue. [*F. dertonensis* (All.) Asch. & Graebn.; *Vulpia bromoides* (L.) S.F. Gray]. Annual; culms solitary or in small tufts, 0.5–6 dm tall, usually erect; sheaths glabrous or short-hairy, without auricles; blades usually involute, 0.5–2.5 mm wide, glabrous or hairy; ligule 0.2–0.6 mm long; panicle (rarely a raceme) narrow, 1–15 cm long, the short branches erect to slightly spreading, pulvini usually lacking in branch axils, pedicels 1–6 mm or longer, remaining appressed; spikelets 2– to 4 (7) -flowered, 6.5–11.5 mm long (excluding awns), the internodes of the rachilla usually at least 1 mm long; glumes subequal to unequal, glabrous, acute to minutely awned at the tips, the first 2.5–5 (6) mm long, the second 4.5–7 mm long, slightly shorter to longer than the body of the lowermost lemma; lemma ultimately rounded on the back, (4) 5–8 (9) mm long, glabrous or scabrous, long-tapered to an awn 3–12 mm long; stamens 1–3, the anthers 0.2–3 mm long; 2n = 14, 42. Dry, disturbed sites at 1310 to 1830 m in Cache, Millard, Salt Lake, and Washington counties; native to Eurasia, introduced in British Columbia and along the west coast of the U. S. to northern Mexico, infrequent eastward; South America; 5 (0). *Festuca bromoides* closely resembles and is possibly not specifically distinct from *F. microstachys*.

Festuca dasyclada Hackel ex Beal Utah Fescue. [*Argillochloa dasyclada* (Hackel ex Beal) W. A. Weber]. Tufted perennial; culms erect to curved-ascending, 1.5–4 dm tall; sheaths as in *F. ovina*; blades 0.5–2 mm wide, folded to involute, glabrous or more often minutely hairy; ligule to ca 1 mm long; panicle ultimately open, to ca 12 cm long, branches mostly solitary or in pairs, the branches and pedicels ultimately widely spreading to abruptly reflexed, distinctly pubescent (at 10x magnification), the pedicels about equal to or slightly longer than the body of the lemma; spikelets 1– or 2 (3) –flowered, 5–7 (8) mm long; glumes unequal, long-tapered to sharply acute tips, the first subulate to narrowly lanceolate, 2.6–4.5 mm long, the second lance-elliptical, (4.5) 5–7 mm long, often scaberulous, shorter than the lowermost to as long as the second floret; lemmas mostly rounded on the back, (4) 5–6 mm long, scabrous to scabrid-puberulent, 5-nerved, the midnerve prolonged as an awn 1–5 mm long; palea about equal to the lemma, the 2 nerves prolonged as minute bristles; stamens 3, the anthers 1.5–2 mm long; 2n = 28. Chiefly in sagebrush, mountain brush, and juniper communities at 2130 to 3048 m in Emery (type from Joes Valley), Garfield, Sanpete, and Wasatch counties; western Colorado; 19 (0). Weber (Phytologia 55: 1–2. 1984) places *F. dasyclada* in a newly erected monotypic genus (*Argillochloa*), stating that the taxon differs strikingly from *Festuca* by its rigidly divaricate secondary branches with pulvini in the axils. Widely spreading secondary branches with pulvini also occur in *F. thurberi*, *F. subulata*, and *F. sororia*. The pulvini-based branches of *F. microstachys* are stiffly spreading to reflexed at maturity. Other features cited by Weber as unusual are the long second glume (present in *F. bromoides* and one variety of *F. ovina*), the 2-flowered spikelets (occurring in some forms of *F. microstachys* and *F. ovina*), and the nonsecund nature of the inflorescence (found in *F. subulata*, *F. thurberi*, and other perennial species of *Festuca*). I find that *Festuca dasyclada* is a distinctive species but is well within the generic circumscription of *Festuca*. *Festuca dasyclada* is one of four species of *Festuca* endemic to the western U. S.

Festuca microstachys Nutt. Small Fescue. [*F. arida* Elmer; *F. pacifica* Piper; *F. reflexa* Buckley; *Vulpia microstachys* (Nutt.) Benth.]. Annual; culms solitary or in small tufts, 0.5–4 (6) dm tall, usually erect; sheaths glabrous or pubescent, without auricles; blades mostly involute, rarely as much as 1.5 mm wide; ligule 0.2–1 mm long; panicle (rarely a raceme) initially narrow, open at maturity, 3–13 cm long, pedicels 0.5–6 mm long, mostly less than 2 mm long, the branches and pedicels at first erect, pulvini forming in axils, the pedicels and at least the lower branches ultimately spreading to reflexed; spikelets 1– to 6-flowered, the internodes of the rachilla usually at least 1 mm long; glumes mostly unequal, acute to minutely awnlike at the tips, glabrous or pubescent, the first 1.7–5.5 mm long, the second 3.5–7.5 mm long, equal to or more often shorter than the body of the lowermost lemma; lemma ultimately rounded on the back, (3.5) 4–8 mm long, glabrous or scabrous to pubescent, long-tapered to an awn (3) 6–20 mm long; stamens 1–3, the anthers 0.2–3 mm long; 2n = 28. Along waterways and roadsides and on dry open slopes below 1830 m in Washington County; British Columbia, south through most of the western U. S. and into Mexico; 8(i). Lonard and Gould (1974) recognize four varieties as comprising *F. microstachys* (as *Vulpia microstachys*) based on variations in the indument of the spikelets. Utah plants, with spikelets glabrous or scabrous, are referred by them to **var. pauciflora** Scribn. ex Vasey, the most common element within the species, occurring throughout its range. See *F. bromoides*.

Festuca myuros L. Myur Fescue. [*F. megalura* Nutt.; *Vulpia myuros* (L.) C.C. Gmel.]. Annual; culms solitary or in small tufts, 0.8–7 dm tall, usually erect; sheaths glabrous or rarely the lowermost puberulent, without auricles; blades involute or folded, 0.5–3 mm wide; ligule 0.2–0.6 mm long; panicle narrow, often rather dense, 3–25 cm long, the branches mostly erect to strongly ascending; spikelets 3– to 5 (8) –flowered; glumes unequal, the first 0.3–1.5 (2) mm long, the second 2.5–6.5 mm long, at least twice as long as the first, shorter than the body of the lowermost lemma; lemma typically rounded on the back, 4.5–7.5 mm long, usually scabrous, at least toward the apex, sometimes ciliate on the narrowly membranous, ultimately inrolled margins, long-tapered to an awn 5–22 mm long; stamens 1–3, the anthers 0.2–1.5 mm long; 2n = 14, 42, 63. Dry to moist sites, saline meadows and salt desert shrub communities, on open slopes and along roadsides, chiefly below 1830 m in Davis, Salt Lake, Tooele, Utah, Washington, and Weber counties; native to Eurasia, now circumboreal, south throughout most of the western U. S., east in most of the coastal states from Texas to Maine, and in Mexico; South America; 7(i). *Festuca megalura* was long thought to be native to North America and distinct from the closely related European introduction, *F. myuros*, in having ciliate lemmas. Plants with ciliate lemmas also occur in Europe, however, where they are referred to a variety of the species. In our area, ciliate and non-ciliate lemmas occur on plants within the same population and even within a single panicle, rendering formal recognition of the variation of doubtful significance. A cultivar of myur fescue (foxtail fescue), because of its extensive fibrous root system and excellent seedling vigor, is proving valuable in controlling erosion in areas with a mean annual precipitation as low as 25 cm (Thornburg 1982).

Festuca octoflora Walter Sixweeks Fescue. [*F. octoflora* var. *hirtella* (Piper) Henrard; *Vulpia octoflora* (Walter) Rydb.]. Annual; culms solitary or in small tufts, 0.2–3 (6) dm tall, erect or occasionally geniculate at the lower nodes; sheaths glabrous or retrorsely puberulent, without auricles; blades involute, to ca 1.3 mm wide; ligule 0.2–1 mm long; panicle 1–10 (20) cm long, the short branches mostly erect-appressed; spikelets 5– to 15 (17) –flowered, the internodes of the rachilla 0.3–1 mm long, averaging ca 0.5 mm long; glumes unequal, the first subulate to narrowly lanceolate, (1.7) 2–5.5 mm long, acute, the second lance-elliptical, 3.5–7 mm long, acute or with an awn tip mostly less than 1 mm long, the whole shorter than the body of the lowermost lemma; lemma ultimately rounded on the back, 2.7–7 mm long, glabrous or scabrous to pubescent with stiffly spreading hairs, narrowly or not at all membranous on the ultimately inrolled margins, long-tapered to an awn 0.5–5 (6) mm long, the awn rarely longer than the body of the lemma; stamens 1–3, the anthers 0.2–0.7 (1.3) mm long; 2n = 14. Salt desert shrub, creosote bush, desert shrub, pinyon-juniper, and sagebrush communities, occasionally on grassy slopes and in association with ponderosa pine, at 760 to 2290 m in Beaver, Box Elder, Cache, Daggett, Davis, Emery, Garfield, Grand, Iron, Juab, Kane, Millard, Piute, Salt Lake, San Juan, Sanpete, Sevier, Tooele, Uintah, Utah, Washington, Wayne, and Weber counties; British Columbia to Quebec, south throughout the U. S., and in Mexico; 115 (v). Plants with glabrous or slightly scabrous lemmas have been referred to var. *octoflora*, those with scabrid-puberulent to distinctly hairy lemmas to var. *hirtella* (Piper) Henrard. The extremes are not geographically segregated, however, and do in fact occur within a single population and even on the same plant.

Festuca ovina L. Sheep Fescue. [*F. arizonica* Vasey; *F. brachyphylla* Schultes; *F. brevifolia* R. Br.; *F. brevifolia* var. *utahensis* St. Yves (type from Alta, Salt Lake County); *F. calligera* (Piper) Rydb.; *F. idahoensis* Elmer; *F. saximontana* Rydb.]. Tufted perennial; herbage bright green to glaucous; culms 0.2–11 dm tall, erect to rarely decumbent at the base, sometimes rooting at subsurface nodes and appearing short-rhizomatous; leaves chiefly basal, usually numerous and filiform, strongly ascending, rarely more than half the length of the culm (except in **var. brevifolia**); sheaths glabrous to minutely hairy, the outer sheath bases (sometimes missing in herbarium specimens) closed to the summit, mostly brown, thin, and soon shredding, the inner sheath bases open, often to the base or nearly so, pale to dark gray, opaque and persistent, typically broader than the blade, the margins narrowly membranous and usually overlapping, lacking auricles (sheath margins occasionally slightly prolonged at the summit and resembling auricles); blades folded or involute and in that state mostly 0.2–1 mm wide, rarely flat and then to 1.5 (2) mm wide, smooth or scabrous to minutely hairy; ligule ranging from a mere ridge to 1.5 (2) mm long, higher on the sides than in the center; panicle 1–12 (15) cm long, often more or less secund, the short branches scabrous to scabrid-puberulent, erect to strongly ascending or at flowering widely spreading to reflexed; spikelets (2) 3– to 7 (9) –flowered, 3–12 (14) mm long, green or purple-tinged; glumes narrowly to broadly lanceolate, glabrous throughout or scabrous to scabrid-puberulent, especially distally, the first 1.5–4.5 mm long, the second 2.5–5.5 (6) mm long, shorter than (or in var. *brevifolia* sometimes equal to) the body of the lowermost lemma; lemma laterally to dorsiventrally compressed, rounded on the back to slightly keeled toward the apex, 3–7 mm long, 3– to 5-nerved, often obscurely so, glabrous or distally scabrous to minutely hairy, the narrow membranous margins ultimately curved around the palea, the apex long-tapered to an awn 0.5–4 (5) mm long; stamens 3, the anthers 0.5–4 mm long; 2n = 14, 21, 28, 35, 36, 42, 44, 49, 56, 70. Often the dominant grass, occupying mostly mesic sites, chiefly in sagebrush, aspen-spruce-fir, ponderosa pine, lodgepole pine, and montane grass communities, and on exposed rocky slopes and ridges, occasionally with pinyon-juniper and mountain brush, at 1950 to 3960 m, rarely an escape from cultivation at lower elevations, in all Utah counties except Morgan; circumboreal, south in the U. S. throughout the West, in the Great Plains, the Northeast, and in Mexico; South America and Eurasia; 431 (xiii). *Festuca ovina* is a widespread polymorphic grass with many habitat-related forms, some of which have been treated as species on the basis of quantitative characters that form a continuum of variation. The robust forms referred to *F. idahoensis* and *F. arizonica* by Hitchcock (1935) were first described as varieties of *F. ovina* by Beal (Grasses of North America, 1896). Gould and Shaw (1983) and Cronquist et al. (1977) concede that the two forms can with reason be treated at the varietal level. The following key permits assignment to varietal status of some of the more commonly occurring native forms, although the many intermediate plants will not satisfactorily fit into any category.

1. Culms mostly more than 4 dm tall; panicles typically more than 7 cm long; spikelets 8–14 mm long 2
—   Culms 0.2–4 (5) dm tall; panicles rarely more than 7 cm long; spikelets 4–8 (9) mm long 3
2(1). Awn of the lemma usually less than 1 mm long; plants reported from Navajo Mt. in San Juan County, occurring in dry meadows or in openings in aspen-fir (*F. arizonica*) .... *F. ovina* var. *arizonica* Hackel ex Beal
—   Awn of lemma usually at least 1 mm long; widespread, plants of foothills to subalpine, often in sagebrush-grass communities (*F. idahoensis*) ....
    .... *F. ovina* var. *ingrata* Hackel ex Beal
3(1). Culms over 2.5 dm tall, at least twice the height of the basal leaves; anthers more than 1 mm long; plants of foothills to alpine .... *F. ovina* var. *rydbergii* St. Yves
—   Culms 0.2–2.5 dm tall, usually less than twice the height of the basal leaves; anthers to 1 mm long; plants of subalpine and alpine (*F. brachyphylla*) ....
    .... *F. ovina* var. *brevifolia* (R. Br.) Wats.

Native strains of *F. ovina* are an important source of forage and will thrive in areas with a mean annual precipitation from 20–36 cm (Thornburg 1982). Commercial strains used for turf and for erosion control have been developed from plants originating in Europe, e.g., var. *ovina* and var. *duriuscula* (L.) Koch. They are adapted to areas receiving mean annual precipitation of approximately 36 cm or more. Blue fescue, a low, usually strongly glaucous form of *F. ovina* grown for ornament, is generally referred to var. **glauca** Koch.

Festuca pratensis Hudson Meadow Fescue. [*F. elatior* L., nom. ambig.]. Tufted perennials; culms rarely more than 1 m tall, erect to basally curved-ascending, sometimes rooting at the lower nodes and appearing to be short-rhizomatous; sheaths mostly open, the outermost often dark and soon shredding, auricles usually present, glabrous or rarely minutely ciliate; blades (2) 3–9 mm wide, flat to occasionally involute, smooth or more often scabrous; ligule 0.2–1 (2) mm long; panicles mostly 6–20 cm long, erect or somewhat nodding, the branches erect or at flowering spreading, often spikelet-bearing throughout their length or nearly so; spikelets (4) 5– to 13–flowered, 10–15 (17) mm long; glumes subequal or unequal, lanceolate, acute, the first 2.5–4 mm long, 1– to 3–nerved, the second 3.5–5 mm long, 3– to 5–nerved, shorter than the body of the lowermost lemma; lemma laterally or dorsiventrally compressed, lanceolate to elliptical, (4) 5–7 (7.5) mm long, more or less obscurely 5– to 7–nerved, smooth or somewhat scabrous, membranous-margined, acute or with an awn to 3 (4) mm long; stamens 3, the anthers 1.5–3.5 mm long; $2n = 14, 28, 42, 70$. Moist to dry sites, chiefly along roadsides and waterways, in meadows, fallow fields, and other disturbed sites, occasionally in aspen-spruce-fir, ponderosa pine, and lodgepole pine communities and on open slopes, at 1310 to 2900 m in all but Morgan and Piute counties; native to Eurasia, introduced in Canada and throughout most of the U. S.; South America; 105 (v). *Festuca pratensis* readily hybridizes with *Lolium perenne*. The sterile progeny of such a cross, described as having short-pedicelled spikelets and rudimentary first glumes, are referred to x *Festulolium loliaceum* (Hudson) P. Fourn. (Tutin et al. 1980). Meadow fescue (Hanson 1972) is used as a pasture grass and for erosion control in the humid parts of the northern U. S., but it is neither as high in yield nor so persistent as *F. arundinacea* (q.v.).

Festuca rubra L. Red Fescue. *Festuca rubra* differs from *F. ovina* chiefly in nearly always having rhizomes. Our few specimens also differ from the strongly tufted *F. ovina* in having more or less conspicuously hairy sheaths, leaf blades mostly flat and 0.5–1.5 (3) mm wide, and spikelets from 7.5–17 mm long. These distinctions may represent a habitat-related extreme of variation; or they may be the product of selective breeding, since in Utah this combination of characters apparently exists only in plants under cultivation or escaped therefrom; $2n = 14, 21, 28, 42, 46, 49, 53, 56, 64, 70$. Cultivated as a lawn grass, rarely escaping, reported from Cache, Salt Lake, Summit, Utah, and Washington counties; native races circumboreal, south through the U. S., except in the Great Plains, the Southeast, and parts of the Southwest; South America; 13 (iii).

Festuca sororia Piper Ravine Fescue. Perennial; culms loosely tufted, 4–10 dm tall, erect or basally curved-ascending, sheaths glabrous, otherwise as in *F. ovina*; blades thin, flat, 2.5–7 mm wide, glabrous or scaberulous; ligule to 1 (1.5) mm long; panicle loose to open, 1–2 dm long, usually nodding, the branches solitary or in pairs, in some plants to 8 cm or more long, ultimately widely spreading to reflexed, scabrous, spikelet-bearing over the distal 1/2–2/3; spikelets 2– to 4 (5) –flowered, 8–11 mm long; glumes unequal, long-tapered to sharply acute tips, the first subulate to narrowly lanceolate, 2–4 mm long, the second lanceolate, 3.5–5.5 mm long, shorter than the lowermost lemma; lemma rounded on the back to more or less laterally compressed, 6–8 (8.5) mm long, with 3–5 evident nerves, scaberulous or nearly smooth, long-tapered to an acute or awn-tipped apex; stamens 3, the anthers 1.5–2.5 mm long. Spruce-fir communities in more or less dense shade at 2440 to 3050 m in Emery, Iron, San Juan, and Utah counties; Arizona, New Mexico, and Colorado; 7 (0).

Festuca subulata Trin. Bearded Fescue. [*F. jonesii* Vasey, type from Salt Lake County]. Perennial; culms mostly small, loose tufts, 5–10 dm tall, erect or curved-ascending near the base, often rooting at the lowermost nodes, occasionally stoloniferous; leaves chiefly cauline; sheaths glabrous or scaberulous, otherwise as in *F. ovina*; blades thin, flat, 3–10 mm wide, smooth or scaberulous; ligule to ca 1 mm long; panicle loose to open, 10–40 cm long, frequently nodding, the branches 1–5 (6) per node, often to 8 cm or more long, ultimately widely spreading to occasionally abruptly reflexed, scabrous or glabrous, mostly spikelet-bearing over the distal 1/2–2/3; spikelets usually with 1–4 perfect florets below 1–3 partially developed to vestigial ones, (6) 7–9 (10) mm long; glumes unequal, sharply acute to awn-tipped, the first subulate, 1.5–4 mm long, the second subulate to narrowly lanceolate, 2–6 mm long, shorter than the body of the lowermost lemma; lemma rounded to keeled on the back, 4–7 mm long, mostly 3-nerved, glabrous or scaberulous, long-tapered to a slender awn (4) 5–17 mm long; stamens 3, the anthers 1.1–3 mm long; 2n = 28. Mesic to moist sites, usually in dense shade within maple-aspen, aspen-mountain brush, and aspen-spruce-fir communities at 1680 to 2320 m in Cache, Davis, Salt Lake, and Utah counties; Alaska to Alberta, south to California, east to Idaho, Montana, and Wyoming; 28 (viii).

Festuca thurberi Vasey Thurber Fescue. [*Poa festucoides* Jones, type from Mt. Ellen, Henry Mts.]. Tufted perennial; culms erect, relatively stout, 4–9 (12) dm tall; sheaths glabrous or obscurely scabrous, without auricles, the persistent old bases loose, opaque, and often conspicuous; blades firm, involute or rarely flat, to ca 2 mm wide, glabrous or scabrous on the lower surface, minutely hairy above; ligule 2.5–8 (9) mm long; panicle loosely contracted to open, 8–17 cm long, frequently nodding, the branches mostly solitary at each node, to 8 cm or more long, ultimately widely spreading, scabrid-puberulent to minutely hairy, mostly spikelet-bearing over the distal 1/2–2/3; spikelets 3– to 6 (7) –flowered, (8) 10–14 mm long, terete to somewhat compressed; glumes usually subequal, narrowly to broadly lance-elliptical, 3–5.5 mm long, broadly membranous-margined, sharply acute to obtuse at the apex, shorter than the lowermost lemma; lemma lance-elliptical, 6–10 mm long, rounded to somewhat keeled on the back, 5-nerved, often conspicuously so, glabrous or scaberulous, acute to minutely awn-tipped; stamens 3, the anthers 1.5–4.5 mm long; 2n = 28, 42. Mesic to moist sites, chiefly in dense shade within aspen-spruce-fir communities, occasionally in sagebrush-grass, mountain brush, ponderosa pine, and meadow communities, and on open slopes, at 2130 to 3350 m in Emery, Garfield, Grand, San Juan, Sevier, Uintah, Utah, and Wayne counties; Wyoming, Colorado, New Mexico, and Arizona; 63 (i). Thurber fescue is a palatable grass that may ultimately prove valuable for range seeding.

Glyceria R. Br.

Perennials; culms erect to prostrate; sheaths closed throughout or over at least 2/3 the length, the lowermost often cross-veined; blades flat to loosely involute; ligule membranous; inflorescence a narrow to open, erect to nodding panicle or raceme; spikelets linear to ovoid, flattened or subterete, 3– to 15 (20) –flowered, disarticulating above the glumes and between the florets; glumes unequal, ovate to lanceolate, shorter than the lowermost lemma, 1-nerved; lemmas broadly ovate or obovate to lance-elliptical, rounded to somewhat flattened on the back, with 7–9 usually prominent parallel nerves, obtuse to acutish apically; palea shorter to slightly longer than the lemma; stamens 2 or 3; caryopsis free of the lemma and palea, narrowly to broadly ovoid to obovate, the embryo 1/5–2/5 as long as the grain; x = 10.

1. Lemmas 3.3–4 mm long; spikelets linear (except at flowering), 9–15 (18) mm long .... *G. borealis*
— Lemmas 1.4–3 mm long; spikelets ovate to oblong, to 8 mm long 2

2(1). Second glume 3-nerved; sheaths open throughout; lemmas mostly truncate at the broadly membranous apex *Puccinellia pauciflora*
— Second glume 1-nerved; sheaths closed nearly to the summit; lemmas obtuse to rounded at the firm or narrowly membranous apex 3

3(2). Lemmas membranous-margined and broadest at the apex, 1.4–2.3 mm long .... *G. striata*
— Lemmas firm throughout, broadest at midlength or below, 2–2.8 mm long .... *G. grandis*

Glyceria borealis (Nash) Batch. Northern Mannagrass. [*Panicularia borealis* Nash]. Rhizomatous perennial; culms 6–15 dm tall, erect or decumbent at the base; sheaths often open 1–4 cm below the summit, glabrous; blades flat or folded, 2–7 mm wide, glabrous throughout or scaberulous to papillose on the upper surface; ligule 4–12 mm long; panicle usually narrow and erect, 2–4 (5) dm long, the primary branches relatively few and widely spaced, to ca 13 cm long, appressed to the main axis or occasionally more or less spreading, the spikelets borne on slender, mostly appressed pedicels; spikelets linear and subterete (lemmas spreading at flowering), 9–15 (18) mm long, 8– to 12–flowered; glumes lance-elliptical to oblanceolate, obscurely nerved, broadly membranous-margined, acute to rounded at the apex, the first 1.1–2.2 mm long, the second 2.2–3.5 mm long, shorter than the lowermost lemma; lemma lance-elliptical, 3.3–4 mm long, rounded or more or less flattened on the back, prominently (5) 7–nerved, green or tawny, obscurely scaberulous on the nerves, otherwise essentially glabrous, broadly membranous and often erose at the acute to rounded apex; palea slightly shorter than the lemma; stamens 3, the anthers 0.5–0.8 mm long; 2n = 20. Wet places, occasionally partially submersed, in meadows, streamside, along lake margins, and in areas of ground-water seepage in mountain brush, aspen-spruce-fir, and lodgepole pine communities at 1890 to 3140 m in Duchesne, Garfield, Iron, San Juan, Sevier, Summit, Uintah, and Utah counties; Alaska to Newfoundland, south in the U. S. throughout most of the West, and from the Dakotas to Maine, south to Illinois and Pennsylvania; 36 (0).

Glyceria grandis Wats. ex Gray American Mannagrass. [*G. maxima* ssp. *grandis* (Wats.) Hulten; *Panicularia grandis* (Wats.) Nash]. Rhizomatous perennial; culms 5–20 dm tall; sheaths closed throughout or nearly so, glabrous or scaberulous; blades flat to loosely involute, 4.5–12(15) mm wide, glabrous or scaberulous; ligule (2) 4–9 mm long; panicle ultimately open, 2–4 dm long, the branches often drooping; spikelets mostly oblong, 4–6.5 mm long, 4– to 9-flowered; glumes membranous, pale, elliptical to ovate, acute at the apex, the first 1.2–2 mm long, the second 1.5–2.5 mm long, shorter than the lowermost lemma; lemma elliptical to ovate, 2–2.8 mm long, rounded on the back, prominently (5) 7-nerved, typically purple or purple-tinged, obscurely scaberulous, gradually tapered from about midlength to the obtuse, firm, usually entire apex; palea mostly equal to or slightly longer than the lemmas; stamens (2) 3, the anthers 0.5–0.9 mm long; 2n = 20. Wet sites, occasionally partially submersed, along waterways, in meadows, and in areas of ground-water seepage at 1310 to 2440 m in

Cache, Duchesne, Morgan, Piute, Rich, Salt Lake, Summit, Uintah, Utah, Wasatch, and Weber counties; Alaska to Nova Scotia, south in the U. S. through most of the West, and from the northern Great Plains to Maine, south to Tennessee; sparingly introduced in northern Europe; 47 (ii). *Glyceria grandis* is a link in a circumboreal chain of very closely related taxa and was for a time reduced by Hulten (*Flora of Alaska* 1968) to infraspecific status under the Eurasian *G. maxima* (Hartman) Holmberg. Our material differs from *G. maxima*, however, in having smaller anthers, smaller, less acute glumes, and smaller lemmas that are firm throughout rather than membranous at the tip.

Glyceria striata (Lam.) A. S. Hitchc. Fowl Mannagrass. [*Poa striata* Lam.; *G. elata* (Nash) Jones; *G. striata* var. *stricta* (Scribn.) Fern.; *Panicularia elata* Nash; *P. nervata* (Willd.) Kuntze]. Rhizomatous perennial; culms (1.5) 3–10 (15) dm tall, erect or decumbent at the base; sheaths closed throughout or nearly so, glabrous or scabrous; blades flat to loosely involute, (1) 2–10 (12) mm wide, glabrous or scaberulous; ligule 1–4 (6) mm long; panicle open, 0.5–3 dm long, the branches usually drooping; spikelets ovate to oblong, 2–4 (5) mm long, 3– to 7–flowered; glumes membranous except along the midnerve, pale or purple-tinged, mostly lance-elliptical to obovate, obtuse to rounded and usually erose at the apex, the first 0.4–1.3 mm long, the second 0.7–1.5 mm long, shorter than the lowermost lemma; lemma obovate, 1.4–2.3 mm long, rounded on the back, prominently 7– to 9–nerved, green or purple-tinged, scaberulous, broadest near the rounded, narrowly membranous, often erose apex; palea mostly equal to or longer than the lemma; stamens 2, the anthers 0.3–1 mm long; $2n = 20, 28$. Wet meadows and in other moist sites in aspen-spruce-fir, ponderosa pine, and lodgepole pine communities, chiefly between 1220 and 3200 m in all but Carbon County; Alaska to Newfoundland, south throughout the U. S. and in Mexico; introduced in Europe; 267 (xv). *Glyceria striata* ranges across the entire North American continent, exhibiting considerable variation, particularly in the height of the culm and width of blades. In the western states, robust forms have been referred to *G. elata* and those less robust to *G. striata*. Plants referred to *G. elata* in the West, however, are identical to what is called *G. striata* var. *striata* in the East. Moreover, characters traditionally used to separate two taxa are quantitative, overlapping, and by no means consistently correlated. Under these circumstances, formal recognition of more than one entity appears unwarranted.

Helictotrichon Besser

Tufted perennials; sheaths open to at least midlength; blades flat to involute; ligule membranous; inflorescence a more or less narrow panicle; spikelets large, subterete, 2– to 7–flowered, the lower 1–4 perfect, the upper 1–4 neuter, the rachilla long-hairy, disarticulating above the glumes and usually between the florets; glumes lanceolate, broadly membranous-margined, 1– to 5–nerved; lemma lanceolate, firmer than the glumes, membranous-margined and often cleft at the apex, keeled to rounded on the back, 5– to 7–nerved, at least that of the lower lemma awned from about midlength, the awn stout, geniculate, the lower portion terete and twisted; palea shorter than the lemma; stamens 3; caryopsis free of but enclosed within the lemma and palea, the embryo not more than 1/3 as long as the grain; $x = 7$.

Helictotrichon mortonianum (Scribn.) Henrard Alpine Oat. [*Avena mortoniana* Scribn.]. Perennial; culms densely tufted, 0.5–2 dm tall; sheaths open; ligule membranous, to about 1 mm long; blades usually involute and in that state mostly less than 1 mm wide; panicle narrow, compact to loose, 2–7 cm long, the branches short, erect or nearly so; spikelets (1) 2 (3) –flowered, 7–12 mm long, typically solitary at the tip of each branchlet, the lower floret perfect, the upper 1 (2) typically reduced and neuter, the rachilla long-hairy; glumes equal or nearly so, equal to or longer than the uppermost floret, pale or purplish, broadly membranous-margined, long-tapered to a sharply acute tip, the first 1–nerved, the second 3–nerved; lemma of the perfect lower floret 6–9 mm long, bearded on the callus with hairs mostly 1–2 mm long, firm and obscurely nerved except near the sharply acute, often cleft, membranous apex, the body scaberulous, at least toward the apex, awned from midlength or above, the awn 10–15 mm long, stout, geniculate, the lower portion twisted; lemma of the upper floret variable in size, usually reduced, awnless or with a straight or geniculate awn more slender than that of the lower floret; anthers 1.5–2.5 mm long. Grass-sedge communities, along lake margins, and on open slopes above timberline, at 3050 to 3720 m in Daggett, Duchesne, Summit, and Uintah counties (Uinta Mts.); southern Rocky Mts. of Colorado to New Mexico; 9(0).

Hierochloe R. Br.

Perennials; culms erect; sheaths open; blades soft, flat to involute; ligule membranous; inflorescence a compact to ultimately open panicle; spikelets compressed, 3–flowered, the 2 lower florets staminate with 3 stamens, the upper floret perfect with 2 stamens, all 3 florets closely occupying nearly the same plane, the rachilla disarticulating above the glumes and the florets falling as a unit; glumes membranous, ovate, equal or the second slightly longer than the first, 1– or 3–nerved, about as long as the uppermost floret; lemmas laterally compressed, usually obscurely 5–nerved, those of the staminate florets entire to short-awned from a bifid tip, the lemma of the perfect floret ultimately hardened, sometimes awned; caryopsis free of but enclosed by the lemma and palea, ellipsoid, the embryo 1/4–1/2 as long as the grain, often undeveloped; $x = 7$.

Weimarck, G. 1971. Variation and taxonomy of *Hierochloe* (Gramineae) in the Northern Hemisphere. Bot. Not. 124: 129–175.

Hierochloe odorata (L.) Beauv. Sweetgrass. [*Holcus odoratus* L.; *Torresia odorata* (L.) A. S. Hitchc.]. Rhizomatous perennial with an aroma of coumarin; culms solitary or loosely tufted, 1.5–6 (7.5) dm tall; basal sheaths brownish or reddish; blades flat, (2) 3–6 (8)mm wide, scabrous and sometimes long-hairy, those of the culm leaves often much reduced; ligule (0.3) 1.5–6 (8) mm long; panicle mostly pyramidal, compact to open, 3–10 (12) cm long, the branches smooth, ascending to ultimately drooping; spikelets compressed to somewhat inflated, ovate, mostly 3–6.5 mm long and 2–4 mm wide, greenish tan or at maturity bronze, 3–flowered, the florets nearly sessile, the perfect central one flanked by and falling with the larger staminate ones; glumes soon becoming membranous throughout, ovate, 2.5–7.5 mm long, obscurely nerved, slightly shorter to slightly longer than the uppermost floret; lemmas entire or rarely bifid and awn-tipped, otherwise dimorphic: those of the lateral, staminate florets ovate, 3–5 mm long, more or less keeled, scabrous to appressed short-hairy on the back with longer hairs along the margins; that of the perfect floret lanceolate, 2.2–4 mm long, usually rounded on the back, minutely hairy above midlength, glabrous below; anthers 1.2–2.5 mm long; 2n = 28, 42, 56. Wet meadows and along streams, lake shores, and margins of spruce-fir and lodgepole pine communities at 2130 to 3500 m in Daggett, Duchesne, Emery, Garfield, Summit, Uintah, Wasatch, and Wayne counties; circumboreal, south throughout most of the western U. S., in the Dakotas, the Great Lake States, and New England; Eurasia; 50 (0). With few exceptions, populations of *H. odorata* are largely infertile, reproducing largely by rhizomes or by apomixis or both (Reeder & Norstag, Bull. Torrey Bot. Club 88: 77. 1961). Weimarck (1971) treats *H. odorata* as a complex comprised of several cytological races.

Hilaria H.B.K.

Perennials; culms stiffly erect or decumbent at the simple to much-branched base; sheaths open; blades flat to involute; ligule membranous or rarely entirely of hairs; inflorescence a more or less slender spike, the superficially uniform spikelets in appressed, often overlapping clusters of 3 at each node of a zigzag rachis, the clusters falling as a unit; the central spikelet 1- or 2-flowered, the lateral spikelets 2- or 3-flowered; glumes membranous to firm, typically laterally compressed, often asymmetrical with the keel near one margin, 2- to 9-nerved, variable within the cluster as to size, shape, and development of awns, shorter to slightly longer than the lemmas; lemma membranous, laterally compressed, 3-nerved, awnless or awned; palea similar to the lemma in texture, varying from slightly shorter to longer than the lemma; stamens (2) 3; caryopsis not seen; x = 9.

1. Sheaths woolly-hairy to some degree; ligule often chiefly or entirely of hairs............................ *H. rigida*
— Sheaths essentially glabrous or long-hairy only at the summit, not at all woolly-hairy; ligule a ciliate membrane 1–4 mm long, the hairs mostly shorter than the membranous portion............................ *H. jamesii*

Hilaria jamesii (Torr.) Benth. Galleta or Curlygrass. [*Pleuraphis jamesii* Torr.]. Strongly rhizomatous or stoloniferous perennial; herbage generally glaucous; culms 1–5 (6.5) dm tall, stiffly erect or sometimes decumbent at the simple to much-branched base, usually puberulent below the inflorescence and long-hairy at the nodes; leaves firm, generally recurved when dry; sheaths glabrous or scaberulous, often with a few long hairs at the summit; blades firm, 1–3 (4) mm wide and mostly 1–16 cm long, flat to involute, long-tapered to a sharply acute tip, glabrous or scaberulous to minutely hairy; ligule membranous, 1–4 mm long, ciliate, the hairs usually shorter than the membranous portion; spike 2–9 cm long, the pale to purple-tinged spikelets in clusters of 3, the clusters 6–11 mm long (including awns) and subtended by a dense tuft of hairs mostly 2–5 mm long; central spikelet about equal to or slightly exceeding the lateral ones, 1- or 2-flowered, the lower floret typically perfect, the upper (when present) usually staminate, the minutely ciliate glumes more or less fan-shaped and irregularly cleft, the (2) 3–9 nerves mostly free at varying levels as awns 1–6 (7) mm long, occasionally 1 or more awns arising from the back, the body of the glumes shorter than the lemma; lemma lanceolate, 4–9 mm long, apically obtuse to acutish and notched, 3-nerved, the midnerve prolonged as an awn 1–2.5 mm long, the anthers to ca 5 mm long; lateral spikelets 2- or 3-flowered, the florets typically staminate or the upper ones neuter, the glumes narrowly lanceolate to elliptical, shorter to longer than the florets, minutely ciliate with hairs rarely more than 0.5 mm long, tapered to an acute to obtuse tip, the first usually asymmetrical (one side undeveloped), with a straight to ultimately recurved awn 1–6 (7) mm long arising at or above midlength, occasionally lobed or with a second awn, rarely awnless, the second glume often asymmetrical and usually awnless or only minutely awn-tipped, the lemma lanceolate, obtuse to acutish at the tip, more or less obscurely nerved laterally, awnless, the anthers to ca 4 mm long; 2n = 36, 38. Desert flats to dry foothills, in salt desert shrub, creosote bush, desert shrub, sagebrush, and pinyon-juniper communities chiefly below 2130 m in Beaver, Box Elder, Carbon, Daggett, Duchesne, Emery, Garfield, Grand, Iron, Juab, Kane, Millard, San Juan, Sevier, Tooele, Uintah, Washington, and Wayne counties; southwestern U. S., Wyoming, and Texas; 286(v). Galleta is moderately palatable as forage during the growing season but is harsh and unpalatable after growth ceases. It has the advantage of being drought resistant and tolerant of heavy grazing (Vallentine 1961).

Hilaria rigida (Thurber) Benth. ex Scribn. Big Galleta. [*Pleuraphis rigida* Thurber]. Strongly resembling *H. jamesii*, differing as follows: culms 3–8(10) dm tall, often stout (to 5 mm or more in diameter), woolly-hairy; sheaths (some of them) typically conspicuously woolly-hairy; blades firm to rigid, often woolly-hairy to some degree; ligule ciliate with the membranous portion short or consisting only of a dense ring of woolly hairs; central spikelet typically 1-flowered; lateral spikelets mostly 2-flowered with glumes narrowly oblanceolate to elliptical and obtuse to rounded or truncate at the often erose to shallowly lobed apex, usually ciliate with hairs often 1 mm or more long; 2n = 18, 36, 108. Creosote bush and shadscale communities below 1220 m in the Mohave desert extension into Washington County; California, Nevada, Arizona, and northern Mexico; 26 (0). Reeder (Amer. J. Bot. 64: 102–110. 1977) presents cytological and morphological evidence of hybridization and subsequent introgression between *H. rigida* and *H. jamesii*.

Holcus L.

Perennials or annuals; sheaths open or closed to above midlength; blades flat; ligule membranous; inflorescence a contracted panicle; spikelets compressed, 2 (3) –flowered, the lower floret perfect, the upper usually smaller and staminate, the rachilla disarticulating above the glumes and often at fruiting between the florets; glumes membranous, subequal, strongly keeled, the first 1-nerved, the second 3-nerved, exceeding the florets; lemmas more or less leathery, shiny, laterally compressed, obscurely nerved, obtuse to abruptly acute apically, lemma of the lower floret awnless, that of the upper floret awned from just below the apex (ours), rarely both awned; palea equal to or slightly shorter than the lemma; stamens 3; caryopsis free of or only slightly adherent to the lemma and palea, the embryo 1/3–2/5 as long as the grain; x = 7.

Holcus lanatus L. Yorkshirefog Velvetgrass. [*Notholcus lanatus* (L.) Nash]. Perennial, often short-lived; culms solitary or tufted, mostly (2) 3–10 dm tall, erect or ascending; sheaths open to midlength or below, soft-hairy; blades lax, flat, 3–10 (12) mm wide, soft-hairy thoughout; ligule 1–3 (4.2) mm long; panicle compact to loose, 4–15 (20) cm long, whitish to pale green or variously pink- to purple-tinged, the branches erect to ascending-spreading, spikelet-bearing nearly to the base, the rachis and pedicels minutely soft-hairy throughout; spikelets oblong-elliptical, strongly compressed, (3.5) 4–5 (6) mm long, 2 (3) -flowered, the lower floret typically perfect, the upper 1(2) staminate or neuter, usually smaller than the lower one; glumes thin, whitish or purple-tinged, subequal, 3.5–6 mm long, exceeding both florets, scaberulous to puberulent, often awn-tipped, the first glume 1-nerved, the second 3-nerved; lemma more or less leathery, shiny, lanceolate, the lowermost to ca 2.5 mm long and typically awnless, the upper usually smaller with a relatively stout, hooked or occasionally straight, subterminal awn to ca 2 mm long, the awn included within the glumes or exserted; anthers 1.3–2.5 mm long; $2n = 14$. Weed of cultivated places, occasionally in meadows and oak-maple communities, mostly below 1520 m in Cache, Davis, Salt Lake, Utah, and Washington counties; native to Eurasia and Africa, now circumboreal, south in the U. S. throughout most of the West, essentially absent from the Great Plains, well established in much of the East; South America; 10 (i).

Hordeum L.

Annuals or perennials; culms erect, solitary or tufted; sheaths with or without auricles, the membranous sheaths of the lowermost leaves initially closed to the summit or nearly so, those of the culm leaves open; blades flat to involute; ligule membranous; inflorescence a dense terminal spike or spikelike raceme with 3 spikelets at each node of the rachis, the central spikelet perfect and sessile or, in a few species, short-pedicelled, the lateral spikelets short-pedicelled (sessile in the cultivar *H. vulgare*) with rudimentary, staminate or occcasionally perfect florets, the rachis at maturity readily disarticulating at the nodes (continuous in *H. vulgare*), the rachis segment falling with the spikelet trio; spikelets 1–flowered, the rachilla of the well-developed florets prolonged behind the palea as a bristle and sometimes bearing a rudimentary second floret; glumes firm, borne in front of the floret and fused at the base, obscurely 1 (3) -nerved, awnlike throughout or in a few species some of the glumes dilated or winged for a short distance above the base; lemma of the central spikelet firm, lanceolate, laterally to somewhat dorsiventrally compressed, obscurely 5-nerved, awned from the apex; palea slightly shorter than the lemma; stamens 3; caryopsis free of or adherent to the palea and sometimes to the lemma, the hilum linear and as long as the grain; $x = 7$. *Hordeum* is a member of the tribe Triticeae (see discussion under *Agropyron*). Species of *Hordeum* are known to hybridize with species in other genera within the Triticeae (Bowden 1967). These hybrids can often be detected, in plants that otherwise have the characters of *Hordeum*, by the presence of more than one floret per spikelet (see guide to hybrids, p. 722).

Baum, B. R. and L. G. Bailey. 1984. Taxonomic studies in wall barley (*Hordeum murinum* sensu lato) and sea barley (*Hordeum marinum* sensu lato). 2. Multivariate morphometrics. Canad. J. Bot. 62: 2754–2764.
Bothmer, R., N. Jacobson, R. B. Jorgensen, and E. Nicora. 1982. Revision of the *Hordeum pusillum* group. Nordic J. Bot. 2: 307–321.
Bowden, W. M. 1967. Taxonomy of the intergeneric hybrids of the tribe Triticeae from North America. Canad. J. Bot. 45: 711–724.
Mitchell, W. W. 1967. On the *Hordeum jubatum*—*H. brachyantherum* question. Madrono 19: 108–110.
———. and A.C. Wilton. 1964. The *Hordeum jubatum*—*caespitosum*—*brachyantherum* complex in Alaska. Madrono 17: 269–280.
Rajhathy, T. and J. W. Morrison. 1959. Cytogenetic studies in the genus *Hordeum*. Canad. J. Genet. Cytol. 1: 124–132.

1. Rachis remaining intact; spikelets all sessile; lemma with an awn 60–150 mm long and often 1 mm wide near the base (or apically 3–lobed but awnless); plants cultivated, occasionally escaping .... *H. vulgare*
— Rachis ultimately disarticulating; lateral spikelets short-pedicelled; lemmas with awns variable in length but less than 1 mm wide at the base; plants native or weedy 2
2(1). Glumes of at least the central spikelet ciliate along the basally expanded portion, the hairs to ca 1 mm long; lateral spikelets about equal to or longer than the central one .... *H. murinum*
— Glumes not ciliate below; lateral spikelets typically smaller than the central one 3
3(2). Glumes dimorphic, at least the inner glumes of the lateral spikelets slightly to markedly dilated below, the outermost glumes of the spikelet trio awnlike throughout 4
— Glumes uniformly tapered from the base upward, none abruptly expanded near the base 5
4(3). Spikes to about 1 cm wide including the awns; awn of the central spikelet typically averaging no more than 8 (2–9) mm long; uppermost sheath typically inflated, spathelike .... *H. pusillum*
— Spikes at maturity more than 1 cm wide including the awns; awn of the central spikelet typically averaging more than 8 (8–20) mm long; uppermost sheath not dilated and spathelike .... *H. marinum*
5(3). Glumes 30–100 mm long, ultimately widely spreading; spike at maturity mostly 4–15 cm wide including the awns .... *H. jubatum*
— Glumes 6–26 mm long, mostly ascending; spike at maturity to about 3 cm wide including the awns 6
6(5). Plants perennial, occupying chiefly native habitats; culms 1–10 dm tall, mostly erect; awn of the central lemma 2–8 (11) mm long .... *H. brachyantherum*
— Plants annual, weedy; culms to 5 dm tall, frequently geniculate at the lower nodes; awn of the central lemma mostly 8–20 mm long .... *H. marinum*

Hordeum brachyantherum Nevski Meadow Barley. [*Critesion brachyantherum* (Nevski) Barkworth & D. Dewey; *H. nodosum* of American authors, not L.]. Perennials; culms tufted, erect or occasionally geniculate at some of the lowermost nodes, 1–10 dm tall; sheaths glabrous or variously hairy, lacking auricles; blades flat to loosely involute, 1–8 mm wide, scabrous to long-hairy; ligule 0.2–0.7 mm long; spike 2–10 cm long and at maturity to ca 1.5 cm wide including the awns, erect, green or purplish, the rachis disarticulating, the internodes at mid-length of the spike 1–2 mm long; spikelets 3 per node, the central spikelet sessile, the much reduced (rarely staminate) lateral ones borne on curved pedicels 0.7–1 mm long; glumes awnlike throughout, mostly 7–15 mm long, scaberulous to the base, straight or nearly so, erect to slightly divergent; lemma of the central spikelets 5–10 mm long, glabrous throughout or scaberulous near the apex, tapering to an awn 2–8 (11) mm long; anthers 0.8–1.8 mm long; 2n = 28. Mesic to wet sites in the spruce-fir zone mostly between 1830 and 3200 m, infrequently to as low as 3000 m, from all but Daggett, San Juan, Wayne, and Weber counties; Alaska south throughout the western U. S. and in Mexico; sparingly introduced in the eastern U. S. and Canada; Siberia; 142 (x). The largely North American *H. brachyantherum* was thought to be a phase of the European *H. nodosum* [a nomen confusum now replaced by *H. secalinum* Schreber (Tutin et al. 1980)] until Nevski recognized the North American plant as distinct from *H. secalinum* in having no auricles, shorter awns, and smaller anthers. *Hordeum brachyantherum* occasionally forms natural hybrids with *H. jubatum* (q.v.) and with *Elymus trachycaulus*. See discussion following *H. jubatum* on the relationship between *H. brachyantherum* and *H. jubatum*. Meadow barley is palatable but rarely present in sufficient quantity to constitute a good source of forage.

Hordeum jubatum L. Foxtail Barley. [*Critesion jubatum* (L.) Nevski]. Perennials, short-lived, sometimes flowering the first year; herbage glabrous to densely softpubescent throughout; culms tufted, 0.7–5 (7) dm tall, erect, or sometimes decumbent at the base, occasionally geniculate at some of the nodes; sheaths with auricles lacking or less than 0.5 mm long; blades flat to involute, 1–5 mm wide; ligule 0.2–1 mm long; spikes ultimately nodding, 4–15 cm long and at maturity often nearly as wide (including the awns), pale green or purple, the rachis readily disarticulating; spikelets 3 per node, the central spikelet sessile, the much reduced lateral ones borne on curved pedicels to about 1 mm long; glumes awnlike throughout, mostly 3–8 (10) cm long, ultimately widely spreading; lemma of the central spikelet 5.5–8 mm long, tapered to an ultimately widely spreading awn mostly 2–8 cm long; anthers 1–1.5 mm long; 2n = 14, 28, 42. Along roadsides, in saline meadows, and sporadically in every type of plant community from salt desert shrub at 1280 m to aspen-conifer communities as high as 3350 m, also occupying dry weedy areas and open rocky slopes, known from all but Morgan and Piute counties; throughout most of temperate North America; Siberia; adventive in many other parts of the world; 186 (v). On the bases of interfertility and chromosome homology, Rajhathy and Morrison (1959) conclude that *H. jubatum* and *H. brachyantherum* are conspecific; but Mitchell and Wilton (1964) cite differences in morphology, habitat preference, and overall geographic distribution favoring the continued recognition of two species. In our area, these taxa form distinct, largely isolated populations, with only an occasional plant appearing to be a hybrid (e.g., Brotherson 2774 and Neese 14921, BRY). Plants with awns of the lemmas intermediate in length between those of *H. jubatum* and *H. brachyantherum* (1.5–3 cm long) have been referred to *H. caespitosum* Scribn. or to *H. jubatum* var. *caespitosum* (Scribn.) A. S. Hitchc. According to Mitchell and Wilton (1964), such plants are the result of hybridization between the two species and should be referred to *H.* x *caespitosum* Scribn. *Hordeum jubatum* also hybridizes occasionally with *Elymus trachycaulus* (see guide to hybrids, p. 722). Foxtail barley is a native plant that moves into disturbed areas in the manner of an introduced weed. Although it is rated fair as forage for cattle prior to seed-head development, mature fruiting heads can cause injury to eyes and mouth parts of all grazing animals. The species is widely cultivated in Europe and the U.S.S.R. as an ornamental.

Hordeum marinum Hudson Mediterranean Barley. [*Critesion marinum* (Hudson) Love; *H. geniculatum* All., nom. ambig.; *H. gussoneanum* Parl.; *H. hystrix* Roth]. Annual; culms solitary or tufted, mostly 1–5 dm tall, erect or frequently geniculate at some of the lower nodes; herbage gray green; sheaths glabrous or short-hairy, the auricles obscure or lacking; blades flat to loosely involute, 1–3.5 mm wide, glabrous or scaberulous to short-hairy; ligule 0.2–0.5 mm long; spikes erect, mostly 2–5 cm long and to ca 3 cm wide including the stiffly ascending to moderately spreading awns, the rachis readily disarticulating, the internodes 0.8–1.5 mm long; spikelets 3 per node, the central spikelet sessile, the moderately to greatly reduced lateral ones borne on straight to slightly curved pedicels 0.7–1.5 mm long; glumes mostly (6) 9–22 (26) mm long, typically smooth or nearly so at the base, scaberulous above, all awnlike or the inner glumes of the lateral spikelets dilated below, the expanded portion 0.2–0.6 (1.4) mm wide; lemma of the central spikelet 5–6.5 mm long, glabrous or rarely scaberulous, with an awn 8–20 mm long; anthers 0.5–1 mm long; 2n = 14, 28. Salt desert shrub, along roadsides, and as a weed of other dry waste places chiefly below 1370 m but occasionally to 1650 m, in Box Elder, Cache, Davis, Salt Lake, Utah, and Weber counties; native of Eurasia, adventive in British Columbia and Ontario, south in the U. S. through much of the West and in Mexico; South America; 32 (vii). As described by Tsvelev (1984), the inner glume of the lateral spikelets of *H. marinum* varies from awnlike throughout to markedly dilated below. Some authors (e.g., Tutin et al. 1980) treat those plants with the inner glume not at all or only slightly dilated as *H. hystrix* Roth, and those with the inner glume "lanceolate or winged below" as *H. marinum*. European specimens seen at the National Herbarium exhibit a continuum of variation, however, the lateral spikelets with inner glumes ranging from 0.2–1.4 mm wide, occasionally the entire range of variation occurring within a single population. For this reason, and because no other character is consistently correlated with the width of the inner glume, I have followed the treatment of Tsvelev (1984), according to which our plants, with the inner glumes of the lateral spikelets only moderately dilated, are ssp. *gussoneanum* (Parl.) Thell.

Hordeum murinum L. Rabbit Barley. [*Critesion murinum* (L.) Love; *H. glaucum* Steudel; *H. leporinum* Link; *H. stebbinsii* Covas]. Annual; culms solitary or loosely to densely tufted, 0.5–8 (10) dm tall, frequently geniculate at some of the nodes; sheaths usually glabrous, often with slender auricles; blades flat or folded, 2–10 mm wide, glabrous or short-hairy; ligule to ca 1 mm long; spikes erect, 2–12 cm long and to ca 3 cm wide at midlength including the awns, the rachis readily disarticulating, the internodes 1.5–3 mm long; spikelets 3 per node, about equal in length or the lateral ones longer than the central one, the central spikelet sessile or borne on a pedicel to ca 2 mm long, the sterile or staminate lateral ones borne on erect pedicels 1–2 mm long; glumes of the central spikelet mostly 12–30 mm long, expanded for a short distance above the base to ca 0.5 mm wide at the widest point, the expanded portion ciliate with hairs to ca 1 mm long; glumes of the lateral spikelets dimorphic, the outer awnlike or nearly so throughout the length, subglabrous to scaberulous, the inner slightly to distinctly dilated below, the expanded portion often ciliate; lemma of the central spikelet 6–12 mm long, glabrous or scaberulous, tapered to an awn 1–5 (6) cm long; lemmas of the lateral spikelets 7–18 mm long, each tapered to an awn about equal to that of the central one; anthers 0.2–1.4 mm long; $2n = 14, 28, 42$. Chiefly a weed of roadsides, cultivated areas, and dry waste places, rare on saline soils and in sagebrush and mountain brush-conifer communities, mostly at 810 to 1830 m, occasionally as high as 3000 m, in Cache, Davis, Grand, Kane, Millard, Salt Lake, San Juan, Sevier, Tooele, Utah, Washington, and Weber counties; native to Europe, adventive in British Columbia, now established throughout the western U. S., in Texas, and northern Mexico; South America and Asia; 91 (ix). *Hordeum murinum* sensu lato has been treated as consisting of two or more species based chiefly upon a single character or quantitative differences, the range of variation in each character either overlapping or forming a continuum. Consistent correlation between even two of six features used by various authors as distinguishing characteristics was lacking in 19 of 35 specimens examined. I have, therefore, followed Tsvelev (1984) and Tutin et al. (1980) in treating the complex as a single species.

Hordeum pusillum Nutt. Little Barley. [*Critesion pusillum* (Nutt.) Love; *H. pusillum* var. *pubens* A. S. Hitchc., type from LaVerkin, Washington County]. Annual; culms solitary or loosely tufted, 1–4 (6) dm tall, erect to ascending or geniculate at some of the nodes; sheaths glabrous to long-hairy, without auricles, the uppermost inflated; blades flat, 1–5 mm wide, glabrous or scabrous to long-hairy; ligule 0.2–0.8 mm long; spikes erect, 2–8 (9) cm long and to ca 1 cm wide including the awns, the rachis readily disarticulating, the internodes to ca 2 mm long; spikelets 3 per node, the central spikelet sessile, the usually reduced lateral ones borne on curved pedicels 0.3–2.3 mm long; glumes mostly 7–17 mm long, usually scaberulous, the inner glumes of the lateral spikelets and often those of the central spikelet dilated below, the expanded portion 0.5–1.8 mm wide, the outermost glumes of the spikelet trio awnlike throughout; lemma of the central spikelet 4–7 (8) mm long, glabrous to pubescent, with an awn 2–9 mm long; anthers 0.5–1.3 mm long; $2n = 14$. Roadsides, in salt desert shrub, desert shrub, and pinyon-juniper communities, at 1280 to 1980 m in Garfield, Grand, Salt Lake, San Juan, Uintah, and Washington counties; British Columbia, Alberta, and Ontario, south throughout most of the U. S., and in Mexico; South America; 20(i).

Hordeum vulgare L. Common Barley. Annual; culms moderately stout, erect, 5–14 dm tall; sheaths glabrous, typically with auricles to 6 mm long; blades flat, 5–16 mm wide, glabrous or scabrous; ligule 0.5–3 mm long; spikes erect to nodding, 2–10 cm long excluding awns the rachis continuous; spikelets 3 per node, all sessile and fertile; glumes subulate, to about 8 mm long, nerveless or 1– to 3–nerved, glabrous or scaberulous to appressed-hairy, tapering to a slender, smooth to scabrous awn mostly 4–20 mm long; lemma (6.5) 8–12 mm long, glabrous and often glaucous, tapering to a flattened, smooth to scabrous awn to 15 cm long and to ca 1 mm wide near the base (in some cultivars the lemma 3–lobed but awnless at the apex); anthers 2–2.5 mm long; $2n = 14, 28$. Barley is native to Eurasia and widely cultivated as a cereal throughout the temperate and subtropical regions of the world. It occasionally escapes from cultivation but is not known to persist; 5 (ii).

Imperata Cirillo

Rhizomatous perennials; culms tufted, mostly tall; sheaths open; blades flat; ligule membranous; inflorescence a terminal, contracted, whitish panicle, the spikelets borne on closely spaced, short pedicels arising in groups of 2 or 3 from erect-ascending branches; spikelets terete or subterete, 2–flowered, disarticulating below the glumes, the lower floret neuter, the upper perfect, the upper floret much shorter than the long white hairs arising from the base of the spikelet and from the glumes; glumes membranous, subequal, about equal to the upper floret, 3– to 9–nerved; sterile lemma and the fertile lemma and palea hyaline; stamens 1 or 2; caryopsis free of but usually remaining enclosed within the lemma and palea, the embryo about 1/2–3/5 as long as the grain; $x = 10$.

Imperata brevifolia Vasey Satintail. [*I. hookeri* Rupr. ex Hackel]. Perennial with short, scaly rhizomes; culms tufted, erect, 0.7–1.5 m tall; sheaths glabrous; blades flat, (3) 4–18 mm wide and to ca 5 dm long, tapered at both ends, glabrous throughout or more often long-hairy near the base of the blade; ligule 1–2 mm long, usually truncate; panicle contracted, dense, 1–3 dm long and to ca 3 cm wide, silvery white to grayish, speckled with the brown of anthers and stigmas, the slender branches more or less elongate, erect-ascending and spikelet-bearing to the base or nearly so; spikelets 2–flowered, subterete, awnless, largely obscured by silky white hairs mostly 8–15 mm long arising from the spikelet base and from the glumes; glumes membranous, narrowly lanceolate, 5– to 7–nerved, long-hairy, the first 2.3–3.5 mm long, the second to 4 mm long, about equal to the upper floret; first floret consisting of a neuter lemma to ca 2.5 mm long, the upper floret perfect with a lanceolate lemma to ca 3.5 mm long, the palea broadly ovate; anthers 1.4–2.5 mm long; caryopsis obovoid, ca 1 mm long, dark brown. Streamside and in other moist places below 1220 m in San Juan County; California east to New Mexico, Texas, and Mexico; 4 (0).

Koeleria Pers.

Annuals or tufted perennials; sheaths open; blades narrow, flat to involute; ligule membranous; inflorescence a more or less glistening, spikelike panicle; spikelets laterally compressed, lanceolate to elliptical, 2– to 5–flowered, the rachilla disarticulating above the glumes and between the florets, prolonged beyond the uppermost floret as a short, usually scabrous bristle or bearing a rudimentary floret; glumes equal or unequal, keeled, the first 1–nerved, the second 3– to 5–nerved, slightly shorter than the lowermost floret to barely exceeding the uppermost one; lemma lanceolate to ovate-lanceolate, keeled, 3– to 5–nerved, acute or occasionally short-awned; palea hyaline throughout, equal to or shorter than the lemma; stamens 3; caryopsis free within the lemma and palea, the embryo not more than 1/3 the length of the grain; $x = 7$.

Greuter, W. 1968. Notulae nomenclaturales et bibliographicae 1–4. Candollea 23: 81–108.

Shinners, L. H. 1956. Illegitimacy of Persoon's species of *Koeleria* (Gramineae). Rhodora 58: 93–96.

***Koeleria macrantha* (Ledeb.) Schultes** Junegrass. [*K. cristata* (L.) Pers.; *K. gracilis* Pers.; *K. nitida* Nutt.; *K. pyramidata* of American authors, not (Lam.) Beauv.]. Loosely tufted perennial, sometimes with short rhizomes; culms 1–7 (8) dm tall; leaves chiefly basal; sheaths glabrous or scabrous to more often retrorsely puberulent; blades flat to involute, 0.5–2 (3) mm wide, more or less prow-shaped at the long-tapered tips, smooth or minutely to strongly scabrous, occasionally short-hairy; ligule 0.2–2 mm long, truncate; panicle spikelike, (2)3–16 cm long and 0.5–1.5 (2) cm wide when pressed, occasionally somewhat interrupted or the branches spreading at flowering, the lower branches rarely more than 2 cm long, densely flowered with the spikelets mostly pale and glistening (especially at flowering when the hyaline paleas are exposed), the rachis, pedicels, and branches minutely soft-hairy (at 10x magnification); spikelets compressed, 3–6 (6.3) mm long, 2 (3–4) –flowered, the rachilla minutely hairy (at 10x magnification), sometimes prolonged beyond the uppermost floret when 2–flowered; glumes narrowly lanceolate to oblanceolate-elliptical, subequal to and unequal, strongly keeled, broadly membranous and glistening on the margins, abruptly to sharply acute and occasionally minutely awn-tipped at the apex, subglabrous to scaberulous, the first (2) 2.5–5 (6) mm long, narrower than the second, 1–nerved, the second 3–6 (6.3) mm long, obscurely to distinctly 3– to 5–nerved, slightly shorter than the lower floret to barely exceeding the upper one; lemmas similar to the glumes in texture and color, lanceolate, often narrowly so, (3) 3.5–6 (6.3) mm long, strongly laterally compressed, obscurely 5–nerved, minutely hairy on the callus, otherwise glabrous to scaberulous, acute to awn-tipped or with a subterminal awn to ca 1 mm long; palea shiny-hyaline throughout, usually slightly shorter than the lemma, spreading at flowering; anthers 1.2–2.5 (3) mm long; 2n = 14–16, 28, 42, 70. Chiefly in sagebrush, mountain brush, and pinyon-juniper communities, and on open slopes, or occasionally in moist meadow communities, at 1370 to 3480 m in Beaver, Box Elder, Cache, Carbon, Daggett, Davis, Duchesne, Emery, Garfield, Grand, Kane, Morgan, Piute, Rich, Salt Lake, San Juan, Sanpete, Sevier, Summit, Uintah, Utah, Wasatch, Washington, and Wayne counties; circumboreal, south throughout the western half of the U. S., in much of the East, and into Mexico; Eurasia; 209 (xvi). The tortuous nomenclatural history of this plant is briefly described by Arnow et al. (1980). Although Shinners (1956) arrived at the name *K. macrantha* as the correct one for our plants, Gould (1975), without explanation, refers North American plants to *K. pyramidata*. European taxonomists assign a more robust phase of the complex (plants having spikelets (5) 6–8 mm long) to *K. pyramidata* and smaller plants (having spikelets 2–5 (6) mm long) to *K. macrantha*. Although North American populations of *K. macrantha* occasionally contain some plants with a few spikelets more than 6 mm long, none match in all details the specimens of *K. pyramidata* seen at the Komarov Botanical Institute. Until the complex as it occurs throughout its range is better understood, however, the name *K. macrantha* should be considered tentative. Two collections of *K. macrantha* (Harrison 12518 UT and Harrison 12527 BRY) provide evidence of hybridization between this species and *Trisetum spicatum*. One sheet contains two plants, one of which is "good" *K. macrantha*, whereas the other has with the awned lemmas, long-hairy rachilla, and broad leaves of *T. spicatum* in combination with the exposed paleas and longer anthers of *K. macrantha*. The second sheet is similar to the first except that the anthers are short as in *T. spicatum*. *Koeleria macrantha* can be distinguished from two outwardly similar species with which it is occasionally sympatric (*Poa fendleriana* and *Trisetum wolfii*), by the minutely hairy panicle axis and generally smaller spikelets.

Leersia Sol. ex Swartz

Perennials, mostly rhizomatous; herbage glabrous to strongly scabrous; sheaths open; blades mostly long-tapered to a sharply acute tip; ligule membranous; inflorescence an open panicle with the spikelets subsessile or short-pedicelled, in strongly overlapping, and mostly single rows along branches and branchlets; spikelets mostly flat, 1–flowered, disarticulating below the spikelet and, in most species, falling entire; glumes lacking or reduced; lemma membranous or firm, laterally compressed and sharply keeled, 5–nerved, the marginal nerves sometimes indistinct; palea narrow, usually 3–nerved, the lateral nerves marginal; stamens 1–6; caryopsis free within the lemma and palea, the embryo typically ca 2/5 the length of the grain; x = 12.

***Leersia oryzoides* (L.) Swartz** Rice Cutgrass. [*Phalaris oryzoides* L.; *Homalocenchrus oryzoides* (L.) Mieg]. Rhizomatous perennial; culms 0.5–1.5 (2) m tall, solitary or loosely tufted, tapered to a usually weak base and supported by surrounding vegetation, simple or branched at nodes above the base; sheaths usually strongly scabrous, the collar often puberulent; blades (3) 6–15 mm wide and to ca 3 dm long, flat or folded, long-tapered to a sharply acute tip, usually scabrous, strongly so on the margins; ligules firm, to ca 1 mm long, truncate; panicles 1–2 (3) dm long, ultimately open and nodding or more or less erect in age, the spikelets overlapping in a single row and appressed to one side of the slender, often flexuous branches; panicles of culm branches remaining at least partially included within the sheath, the included florets cleistogamous; spikelets flat, asymmetrically elliptical, 1.5–2 mm wide; glumes lacking; lemma semicircular, 3.7–5.5 mm long, ciliate on the sharp keel with minute stiff hairs, more or less scabrous on the 5 nerves and on the area between; palea equal to or slightly longer and much narrower than the lemma, stiffly ciliate on the sharp keel; stamens 3, the anthers to ca 2 mm long; 2n = 48, 60. Wet, heavily vegetated sites along waterways or in marshes below 1400 m in Davis, Utah, and Weber counties; Canada, throughout the U. S. and in Mexico; Eurasia; 19 (ii).

Leptochloa Beauv.

Tufted annuals and perennials; sheaths open; blades flat to involute; ligule membranous; inflorescence a panicle with few to numerous spikelike or racemose branches, these 1–4 or more per node, the spikelets sessile or short-pedicelled and remote to closely overlapping; spikelets 2– to 12–flowered, subterete to distinctly compressed, disarticulating above the glumes and between the florets, the uppermost florets in some species staminate or neuter; glumes membranous, laterally compressed and more or less keeled to rounded on the back, the first 1–nerved, the second 1– to 3–nerved, shorter to longer than the lowermost lemma; lemma similar to the glumes in texture and color, laterally compressed and more or less keeled to dorsiventrally compressed, 3–nerved, acute to truncate and entire or erose to bifid at the apex, awnless or mucronate to awned; palea well developed; stamens 3; caryopsis free within the lemma and palea, the embryo to about half the length of the grain; x = 10. The characters upon which McNeill (1979) based the separation of *Leptochloa* into *Diplachne* and *Leptochloa* are not consistently correlated throughout the complex. Gould and Shaw (1983), without explanation, follow Hitchcock (1951) in treating *Leptochloa* as a single genus.

McNeill, J. 1979. *Diplachne* and *Leptochloa* (Poaceae) in North America. Brittonia 31: 399–404.

1. Spikelets 1.5–3.5 mm long; sheaths usually papillose-hairy .... *L. filiformis*
— Spikelets 4–13 mm long; sheaths glabrous or scabrous . . 2

2(1). Lemmas averaging more than 3 (3–6) mm long, mostly sharply acute, the midnerve often prolonged as an awn 0.5–3.5 mm long; second glumes sharply acute ......
.... *L. fascicularis*
— Lemmas averaging no more than 3 (1.8–3) mm long, abruptly acute to truncate and usually mucronate; second glume abruptly acute to rounded ...... *L. uninervia*

Leptochloa fascicularis (Lam.) Gray Bearded Sprangletop. [*Festuca fascicularis* Lam.; *Diplachne fascicularis* Beauv.]. Tufted annual; culms 1–7 (10) dm tall, erect to ascending or rarely prostrate, often branched at nodes above the base; sheaths glabrous; blades flat to loosely involute, 1–5 (7) mm wide, long-tapered to sharply acute or filiform tips, smooth or scabrous to occasionally sparsely hairy, those of the culm often equal to or exceeding the inflorescence; ligule 2–7 mm long; panicle 7–30 (40) cm long, typically partially enclosed by the uppermost sheath, the few to numerous, stiffly erect to widely spreading, spikelike branches mostly 4–12 cm long; spikelets mostly on pedicels less than 1 mm long, appressed, overlapping to remotely spaced, 5– to 12–flowered, linear, subterete, 4–13 mm long, initially whitish or pale green, becoming grayish at maturity, rarely purple-tinged; glumes unequal, 1–nerved, mostly sharply acute and sometimes minutely awn-tipped, the first subulate to narrowly lanceolate, 1.3–4 mm long, the second lanceolate to elliptical, 2.5–4 (5) mm long, shorter than the lowermost lemma; lemma similar to the glumes in texture and color, rounded on the back, mostly lanceolate to elliptical, 3–5 (6) mm long, 3–nerved, appressed-hairy, at least near the base of the lateral nerves, the entire or minutely bifid apex acute, the midnerve usually prolonged as a mucro or as an awn to ca 3.5 mm long, the lateral nerves often prolonged as minute teeth; anthers 0.2–0.5 mm long; 2n = 20. Wet or drying margins of waterways, ponds, and lakes, occasionally partially submersed in shallow water, at 1310 to 2740 m in Cache, Grand, Duchesne, Salt Lake, San Juan, Uintah, Utah, and Washington counties; throughout most of the U. S.; Mexico; South America; 26 (iii).

Leptochloa filiformis (Lam.) Beauv. Red Sprangletop. [*Festuca filiformis* Lam.]. Annual; culms solitary or tufted, 0.8–8 (10) dm tall, erect to decumbent-based and spreading, usually sparingly branched at nodes above the base; sheaths typically pubescent with papilla-based, soft hairs to 6 mm long; blades flat or folded, mostly 2–13 mm wide, long-tapered to sharply acute or filiform tips, smooth or scaberulous, sometimes papillose-hairy near the base; ligule 0.5–2 mm long, ciliate and sometimes appearing to be of hairs; panicle 10–40 cm long, usually 1/3–1/2 the height of the plant, with few to numerous, ascending to spreading, spikelike branches 2–15 cm long; spikelets sessile or occasionally short-pedicelled, appressed to the rachis, slightly overlapping to more or less remote, 2– to 4–flowered, lance-elliptical, 1.5–3.5 mm long, green or more often reddish purple; glumes equal or subequal, 1–nerved, acute to long-tapered at the apex, the first subulate, 1–2.5 mm long, the second slightly broader, 1.5–2.7 mm long, shorter to longer than the first glume, slightly shorter to distinctly longer than the lowermost lemma, sometimes both glumes exceeding all the florets; lemma similar to the glumes in texture and usually in color, mostly compressed laterally and slightly keeled along the midnerve, 0.7–1.7 mm long, 3–nerved, abruptly acute to truncate and often erose to minutely bifid at the apex, usually appressed-hairy along each nerve, at least below; anthers to ca 0.3 mm long; 2n = 20. Known in Utah by a single collection (Higgins 823 BRY) from a wet site near a spring below 1520 m in a creosote bush community in Washington County; southwestern U. S. east through Texas to Mississippi; rare in the eastern U. S.; Mexico; South America; 1 (0).

Leptochloa uninervia (Presl) Hitchc. & Chase Mexican Sprangletop. [*Megastachya uninervia* Presl]. Tufted annual; culms mostly 4–10 dm tall, usually strictly erect, often branched at nodes above the base; sheaths glabrous or scabrous; blades flat to loosely involute, 1–4 mm wide, elongate, long-tapered to filiform tips, smooth or more often scabrous; ligule 2–6 mm long; panicle narrowly ovoid to oblong, 10–30 cm long, tardily emergent from the uppermost sheath, the usually numerous, stiffly erect to moderately spreading, spikelike branches mostly 2–7 cm long; spikelets on pedicels mostly less than 1 mm long, typically appressed to the rachis and overlapping, 6– to 10–flowered, subterete to oblong, 4–10 mm long, pale gray green to dark gray; glumes unequal, 1–nerved, the first subulate to linear, 0.8–1.7 mm long, acute to obtuse, the second oblong to oblanceolate, 1.5–2.6 mm long, abruptly acute to rounded and often minutely mucronate, shorter than the lowermost lemma; lemma similar to the glumes in texture, usually darker in color, rounded to somewhat flattened on the back, 1.8–3 mm long, 3–nerved, abruptly acute to truncate and usually mucronate, the lateral nerves rimming the margins and often inconspicuous, sparsely hairy near the base of each nerve, the lateral nerves sometimes protruding as minute teeth; anthers 0.2–0.6 mm long; 2n = 20. Known in Utah by a single collection (Menzies 8001 BRY) from a moist site in a grass community at 1370 m in Utah County; California east to Colorado and from Arizona to Mississippi, sparingly introduced along the eastern seaboard; Mexico; South America; 1 (0).

Leucopoa Griseb.

Imperfectly dioecious perennials; sheaths open; ligule membranous; inflorescence a loose to contracted panicle; spikelets subterete to flattened, 3– to 7–flowered, the rachilla disarticulating above the glumes and between the florets; glumes soon membranous throughout, pale, obscurely nerved; lemmas pale green, soon membranous, laterally compressed and somewhat keeled to rounded on the back, 5-nerved; stamens 3; caryopsis at maturity typically adherent to lemma and palea, bidentate apically, the embryo less than half as long as the grain; x = 7. This North American species was included in *Festuca* by Hitchcock (1935) and removed by Chase to *Hesperochloa* (Hitchcock 1951). Weber (1966) found the genus *Hesperochloa* inseparable from the largely Asian genus *Leucopoa*, but meantime Tsvelev (1984) has returned *Leucopoa* to subgeneric rank under *Festuca*.

Weber, W. A. 1966. Additions to the flora of Colorado IV. Univ. Colorado Stud. Ser. Biol. 23: 2.

Leucopoa kingii (Wats.) W. A. Weber Spike Fescue. [*Poa kingii* Wats.; *Festuca confinis* Vasey; *Festuca kingii* (Wats.) Cassidy; *Hesperochloa kingii* (Wats.) Rydb.]. Incompletely dioecious perennial, often with short rhizomes; herbage green or glaucous; culms usually densely tufted, (2) 3–10 dm tall; sheaths smooth, the old sheath bases straw colored and persistent; blades firm, erect-ascending or occasionally recurved, flat or loosely involute upon drying, mostly 2–10 mm wide, long-tapered to an acute tip, smooth or occasionally scaberulous; ligules 0.5–4 mm long; panicle narrow, (5) 10–20 cm long, the branches erect or slightly spreading, smooth or occasionally sparingly scabrous on the angles; spikelets soon turning tawny, subterete to somewhat compressed, 6–10 (12) mm long, 3– to 5 (6) –flowered, the florets mostly functionally unisexual; glumes initially green along the midnerve, becoming membranous throughout, broadly lanceolate to ovate, rounded on the back to more or less keeled, glabrous, the first glume (2) 3–5.5 mm long, 1-nerved, the second 4–7 mm long, shorter than the lowermost lemma, 3-nerved, the lateral nerves obscure; lemma firmer than the glumes, lanceolate to ovate, (4.5) 5–9 (10) mm long, rounded on the back to more or less keeled, obscurely to distinctly 5-nerved, typically roughened to scabrous throughout (at 10x magnification), acute to minutely awn-tipped at the apex; palea about as long as the lemma; pistillate florets with anthers to ca 1 mm long, the staminate ones with anthers 2.5–6 mm long; 2n = 56. Chiefly grassy or lightly wooded slopes in oak-sagebrush and aspen-spruce-fir zones and on exposed rocky slopes at high elevations, occasionally in pinyon-juniper, ponderosa pine, and meadow communities, at 1370 to 3660 m in Box Elder, Cache, Carbon, Daggett, Davis, Duchesne, Emery, Juab, Morgan, Piute, Rich, Salt Lake, Sanpete, Tooele, Utah, Wasatch, and Weber counties; in most of the western U. S., in the Great Plains as far south as Kansas; 161 (vi). In our area, staminate plants appear to outnumber pistillate ones, but the range of variation in size of spikelets and lemmas in the plants I have seen is the same for pistillate as in staminate florets. *Poa fendleriana* differs from *L. kingii*, which it superficially resembles, in having sheaths closed at least near the base, mostly involute leaves rarely more than 3 mm wide, and lemmas no firmer than the glumes and often pubescent on the keel and marginal nerves, rarely scabrous throughout. Spike fescue is palatable as forage in the spring but becomes harsh and unpalatable as it matures. Because it occurs chiefly as scattered clumps, its production per acre is usually low (Vallentine 1961).

Lolium L.

Annuals or perennials; sheaths open, with or without auricles; blades flat to more or less involute; ligule membranous; inflorescence a solitary terminal spike (rarely branched), the spikelets solitary and sessile at each node of the rachis, narrowly oblong to elliptical, compressed, 2– to 22–flowered, the rachilla disarticulating above the glumes and between the florets; first glume lacking except on the terminal spikelet, the second glume 3– to 9–nerved, awnless, shorter to longer than the lowermost florets; lemma membranous to hardened, mostly lanceolate, laterally compressed, rounded on the back, 5– to 9–nerved, awnless or awned; palea usually subequal to but as broad as the lemma; stamens 3; caryopsis somewhat adherent to the lemma and palea, with an apical appendage, the embryo 1/6–1/3 as long as the grain; x = 7. When taxonomic considerations are based on characteristic of the flower and fruit rather than those of the inflorescence, *Lolium perenne*, traditionally included in the tribe Triticeae, is seen to be more closely related to *Festuca pratensis* (q. v.) than to other species of *Lolium*, and *F. pratensis* is likewise more closely allied to *L. perenne* than to other representative species of *Festuca* (Stebbins, Evolution 10: 235–236. 1956). The ease with which the two species hybridize supports this contention. Accordingly, Gould and Shaw (1983) include the genus *Lolium* in the tribe Poeae.

Lolium perenne L. Ryegrass. [*L. multiflorum* Lam.; *L. perenne* var. *italicum* Parnell]. Annual or perennial; culms tufted, mostly erect, 3–7 (10) dm tall; sheaths with auricles often well developed; blades flat or somewhat involute, 2–8 (10) mm wide; ligule to 1.5 mm long; spike 7–25 cm long; spikelets (3) 4– to 15 (20) –flowered, 7–20 (23) mm long, solitary and sessile at each node, placed edgewise to and alternating on opposite sides of the rachis; first glume (adaxial) lacking except on the terminal spikelet, the second glume membranous or hardened, lanceolate, (4) 6–12 mm long, mostly 5– to 9–nerved, 1/2–3/4 as long as the spikelet; lemma membranous or hardened, (4) 5–7.5 (8) mm long, rounded on the back with margins inrolled at maturity, mostly 5–nerved, the lateral nerves typically conspicuous, the remainder more or less obscure, glabrous or scaberulous, acute or with awns to 8 mm long; anthers (2) 3–4 mm long; 2n = 14. An escape from cultivation, in dry to moist sites, along waterways, in waste places, fallow fields, and on open grassy slopes at 760 to 2440 m in Cache, Davis, Duchesne, Garfield, Grand, Salt Lake, Sanpete, Tooele, Utah, Wasatch, Washington, and Weber counties; native to Eurasia and Africa, now circumboreal, south throughout most of the U. S. and in Mexico; South America; 58 (v). *Lolium multiflorum* is often treated as a species distinct from *L. perenne* in being annual and in having larger spikelets, more numerous florets, and awned rather than awnless lemmas. As it occurs in this hemisphere, the grass is a mixture of annual and perennial forms from which improved varieties have been developed for commercial use. In the absence of consistent correlation between even two characters, recognition of more than one species becomes impractical. Although the awned form may be referred to var. *aristatum* Willd. [var. *italicum*], individual panicles may bear both awned and awnless lemmas, rendering unsatisfactory the recognition of variation even at the infraspecific level (Gould 1975). Ryegrass is thought to be the first meadow grass to have been cultivated in Europe, records of its use in England dating back to 1681. It was being grown in North America as early as 1782 (Gould and Shaw 1983) and is now widely used for pasture, hay, lawns, and erosion control in most of North America and in temperate South America. See *Festuca pratensis* for discussion of hybrids between that species and *L. perenne*.

Lycurus H.B.K.

Perennials; culms tufted, erect; sheaths open; blades narrow, flat or folded; ligule membranous; inflorescence a cylindrical, spikelike panicle with spikelets borne on mostly paired pedicels, these united at the base and disarticulating as a unit; spikelets narrowly lanceolate, subterete, 1–flowered, not or only tardily disarticulating, the floret borne on the shorter pedicel staminate or neuter, that borne on the longer pedicel perfect; glumes subequal, the first 2 (3) –nerved, the second 1–nerved, both with the nerves prolonged as slender awns, the body of the glumes shorter than the lemma; lemma 3–nerved, the midnerve prolonged as an awn; palea about equal to or shorter than the lemma; stamens 3; caryopsis free of the lemma and palea; $x = 10$. According to Reeder (1985), who treats the North American genus *Lycurus* as consisting of three species, plants north of the Trans-Pecos region of Texas [*L. setosus* (Nutt.) C. Reeder] are distinct from the remainder of the complex in having at least the upper leaves terminating in a slender seta to 10 mm long or more, ligules (3) 5–10 mm long, and culms erect. In contrast, the plants south of the Trans-Pecos are described as having leaves without setae or with only short mucros, ligules 0.5–2 (3) mm long, and culms decumbent or ascending to widely spreading. Those plants I have seen (all from north of the Trans-Pecos) do indeed have the mostly longer ligules and seta-tipped leaves described by Reeder, although some of them have decumbent-based, spreading culms. I suggest that variation of this nature, in which some overlapping occurs, might more reasonably be given recognition at the infraspecific rather than at the specific level.

Reeder, C. G. 1985. The genus *Lycurus* (Gramineae) in North America. Phytologia 57: 283–291.

Lycurus phleoides H.B.K. Wolftail. Perennial; culms tufted, 1.5–6 dm tall, simple or sparingly branched at the nodes, erect or decumbent near the base, often geniculate at nodes below midlength, the culms then ascending to widely spreading; sheaths generally keeled, glabrous or puberulent; blades flat or folded, 1–2 (3) mm wide and mostly 3–15 cm long, usually scabrous on the narrow white margins, terminating in a sharply acute to bristle-like tip; ligule (1.8) 2–6 (10) mm long, acute or 3–lobed; panicle spikelike, cylindrical, (2) 3–10 (13) cm long and 3–8 mm wide, densely flowered, the short branches erect or nearly so, rebranched immediately above the base to form two subequal pedicels averaging less than 2 mm long, each branch disarticulating at the base of the paired pedicels, the spikelets and attached pedicels usually falling as a unit; spikelets subterete, 1–flowered, borne on short, mostly paired, erect pedicels, the florets perfect, staminate or neuter, that on the shorter pedicel more frequently reduced than that on the longer one; glumes membranous, narrow, subequal, 1–1.8 mm long,

the first 2–nerved and 2–awned, the second 1–nerved with a solitary awn, the awns mostly 1–5 mm long, the body of the glumes shorter than the lemma; lemma similar to or slightly firmer than the glumes in texture, narrowly lanceolate, rounded on the back, 2.7–4.2 mm long, light green to grayish, often purple-tinged toward the apex, short-hairy along the margins over the back, 3–nerved, the midnerve prolonged as an awn mostly 2–4 mm long; palea about as long as the lemma; anthers 1–2 mm long; $2n = 40$. Desert shrub and juniper communities, on open slopes, and in rock crevices at 1520 to 2130 m in Garfield, Kane, San Juan, and Washington counties; Arizona east to Colorado, Oklahoma, Texas, and in Mexico; 12 (0). *Muhlenbergia wrightii* resembles *L. phleoides* but can readily be distinguished from *Lycurus* by its 1–nerved first glume. See *M. depauperata* for a discussion of its possible relationship to *Lycurus*.

Melica L.

Perennials; culms solitary to densely tufted, some species arising from bulblike corms; sheaths closed throughout or nearly so; blades flat to involute; ligule membranous, sometimes ventrally fused; inflorescence a contracted to open panicle or raceme; spikelets lanceolate to elliptical and subterete to moderately flattened, mostly 3– to 10–flowered, disarticulating above or below the glumes; florets becoming progressively smaller upward, the lower 1–6 perfect, the upper 1–4 reduced to a mostly club-shaped rudiment consisting of one or more empty lemmas; glumes unequal or subequal, ovate to oblong-elliptical, laterally compressed and somewhat keeled to rounded on the backs, membranous at least near the apex, the first (1) 3– to 7–nerved, the second (3) 5– to 9–nerved, shorter or longer than the lowermost floret; fertile lemmas firmer than the glumes, laterally compressed, slightly keeled to rounded on the back, (5) 7– to 15–nerved (the weaker nerves often alternating with more pronounced ones), broadly membranous-margined, awnless (ours) or awned from an entire or bifid apex; callus glabrous; palea shorter than the lemma, often markedly so; stamens 3; caryopsis free within the lemma and palea, the embryo 1/5–2/5 as long as the grain; $x = 9$. Utah species of *Melica* are highly palatable, but populations of even the relatively common *M. bulbosa* are too well dispersed to be of much significance as forage.

Boyle, W. S. 1945. A cytotaxonomic study of the North American species of *Melica*. Madroño 8: 1–26.

1. Spikelets disarticulating above the glumes; pedicels mostly erect-ascending, glabrous to scabrid-puberulent on the upper portion; culms arising from bulblike corms 2

— Spikelets disarticulating below the glumes; pedicels often spreading or recurved, hairy at the summit; culms not distinctly bulblike at the base 3

2(1). Corms "threaded" along a rhizome, when detached the rhizome fragments forming "tails" on each corm; first glume averaging less than 6 (3.5–7) mm long, typically becoming membranous throughout, ovate to broadly lanceolate and acute apically .... *M. spectabilis*

— Corms tightly clustered, not "tailed"; first glume averaging at least 6 (5–11) mm long, mostly opaque, at least in the center, lance-elliptical and obtuse to abruptly acute at the apex .... *M. bulbosa*

3(1). First glume averaging no more than 6 (4–6) mm long; plants rhizmatous, loosely tufted .... M. porteri

— First glume averaging more than 6 (6–16) mm long; plants densely tufted, not rhizomatous .... M. stricta

Melica bulbosa Geyer ex Port. & Coult. Oniongrass. [*Bromelica bulbosa* (Port. & Coult.) W. A. Weber; *M. bella* Piper]. Perennial; culms solitary or in mostly small tufts, 3–6 (10) dm tall, each usually arising from a swollen, bulblike corm, these typically tightly clustered; sheaths glabrous or scabrous to short-hairy; blades flat to involute, 2–5 mm wide, often short-hairy; ligule 2–5 (7) mm long; panicle sublinear, (7) 10–16 (20) cm long, the short branches mostly erect-appressed; spikelets mostly erect, (6) 9–18 (24) mm long, disarticulating above the glumes and ultimately between the florets, the fertile florets 3–6, the rudiment 1.5–5 mm long; glumes subequal to unequal, lance-elliptical, obtuse to abruptly acute at the pale to tawny or purplish, broadly membranous apex, mostly scaberulous, at least on the nerves, the first (5) 6–9 (11) mm long, 1– to 4–nerved, the second 6–13 mm long, 5– to 7–nerved, typically shorter than the lowermost lemma; lemma (lowermost) lance-elliptical, 6–12 mm long, (5) 7– to 15–nerved, firmer and more conspicuously scabrous than the glumes, usually with a purple band below the pale to brownish membranous tip; anthers 1.3–4 mm long; 2n = 18. Open, grassy or rocky slopes, in sagebrush, oak, pinyon-juniper, mountain brush, aspen, and ponderosa pine communities, and at upper elevations in montane meadows and on exposed rocky ridges, at 1520 to 3200 m in Beaver, Box Elder, Cache, Carbon, Daggett, Davis, Duchesne, Juab, Millard, Piute, Rich, Salt Lake, Sanpete, Sevier, Summit, Tooele, Uintah, Utah, Wasatch, Washington, and Weber counties; British Columbia and Alberta south through most of the western U. S.; 136 (v).

Melica porteri Scribn. Porter Melic. Rhizomatous perennial; culms loosely tufted, 0.5–1 m tall; sheaths typically glabrous; blades flat to somewhat involute, mostly 2–5 mm wide, smooth or scabrous to hairy; ligule 3–7 mm long; panicle green or tawny, 13–25 cm long, narrow, the branches few, erect or nearly so, often racemose and usually secund, the spikelets loosely spaced, the pedicels erect or abruptly recurved at the puberulent summit; spikelets ascending to drooping, mostly elliptical, 8–16 mm long, disarticulating below the glumes and falling entire, the fertile florets 3–5, the rudiment 2–5 mm long; glumes subequal, subovate to oblanceolate, acute to obtuse at the apex, broadly membranous-margined, glabrous or scaberulous, the first 4–6 mm long, 3– to 5–nerved, the second 5–8 mm long, 5– to 7–nerved, shorter than the uppermost floret; lemma (lowermost) 6–10 mm long, slightly firmer and usually more scabrous than the glumes, pale or green, mostly 7– to 9–nerved, occasionally with a faint purple band below the membranous apex; anthers to ca 2 mm long; 2n = 18. Known in Utah by a single collection (Thompson s.n. BRY) from an aspen-mountain brush community at 2440 m in San Juan County; Arizona east to Texas and from Colorado south into Mexico; 1 (0).

Melica spectabilis Scribn. Purple Oniongrass. [*Bromelica spectabilis* (Scribn.) W. A. Weber]. Perennial; culms solitary or in mostly small tufts, 2.5–12 dm tall, each arising from a bulblike corm, the corms spaced along a rhizome, usually at intervals of 1–3 cm, a portion of the rhizome typically forming a "tail" on any corm becoming detached; sheaths mostly glabrous or scabrous; blades flat to involute, 1–5 mm wide, smooth or scabrous; ligule 1–3.2 mm long; panicle usually sublinear, 5–26 cm long, the short branches or pedicels often flexuous, erect-appressed to slightly spreading, scaberulous; spikelets mostly erect, 7–20 mm long, disarticulating above the glumes and ultimately between the florets, the fertile florets 3–7, the rudiment 1.5–3.5 mm long; glumes subequal to unequal, ovate to broadly lanceolate, abruptly acute to acuminate apically, typically becoming membranous and brownish to purple throughout, rarely pale, glabrous or scaberulous on the nerves and sometimes over the back, the first glume 3.5–6 (7) mm long, 1– to 3–nerved, the second 5–7.5 mm long, 7– to 9–nerved, equal to or shorter than the lowermost lemma; lemma (lowermost) ovate to broadly lance-elliptical, 6–9 mm long, firmer and more conspicuously scabrous than the glumes, 7– to 9–nerved, usually purple-banded below the brownish membranous tip or occasionally the tip purplish; anthers 1.8–2.6 mm long; 2n = 18 (Amer. J. Bot. 64: 103. 1977). Dry to moderately moist sites, chiefly under aspen and in meadows in the spruce-fir zone, less frequent in sagebrush or juniper communities and on open grassy slopes, at 1980 to 3200 m in Cache, Carbon, Daggett, Duchesne, Grand, Rich, Salt Lake, San Juan, Sanpete, Summit, Uintah, Utah, and Wasatch counties; British Columbia and Alberta, south through most of the western U. S.; 45 (0).

Melica stricta Bolander Rock Melic. Perennial; culms densely tufted, 2–6 (8.5) dm tall, more or less thickened but not bulblike at the base; sheaths typically scaberulous, occasionally pubescent; blades flat to somewhat involute, 1.5–3 (5) mm wide, smooth or scabrous to more often short-hairy; ligule 3–5.5 mm long; panicle or raceme mostly pale green or whitish, 3–30 cm long, narrow, the branches few, erect or nearly so, often racemose and more or less secund, the spikelets loosely spaced, the pedicels erect throughout or more often abruptly recurved at the puberulent summit; spikelets spreading to drooping, becoming broadly V-shaped, (8) 10–20 mm long, disarticulating below the glumes and falling entire, the fertile florets 2–5, the rudiment 2–7 mm long; glumes mostly subequal, oblanceolate-elliptical, often broadly so, acute at the apex, pale-membranous and glistening marginally, initially greenish and occasionally purple-tinged medially, glabrous or scaberulous, 5– to 7–nerved, the first 6–16 mm long, the second 7–15 (18) mm long, typically narrowed below to a pedicellike base, the whole often as long as or longer than the uppermost floret; lemma (lowermost) 8–16 mm long, 9–nerved, slightly firmer and usually more scabrous than the glumes, typically pale green to whitish throughout; anthers 1–3 mm long; 2n = 18. Dry to moist sites, chiefly rocky slopes and open woods at 1520 to 2740 m in Beaver, Box Elder, Juab, and Tooele counties; Oregon, California, and Nevada; 11 (0).

Muhlenbergia Schreber

Annuals or perennials; culms simple or sparingly to freely branched at nodes above the base; sheaths open; blades flat or folded to involute; ligule membranous, rarely obsolete or consisting chiefly or entirely of hairs; inflorescence an open or contracted to spikelike panicle; spikelets whitish or pale to dark gray or less often reddish purple, terete to somewhat laterally compressed, 1–flowered (occasionally 2– or 3–flowered in *M. asperifolia*, rarely so in other species), disarticulating above the glumes (except in *M. depauperata*); glumes usually membranous, mostly subequal, shorter to longer than the lemma, rounded to acute or short-awned apically, the first 1 (2) –nerved (rarely obsolete), the second 1– to 3–nerved; lemma similar to or firmer than the glumes, mostly rounded on the back, 3 (5) –nerved, often obscurely so, glabrous or variously pubescent, sometimes bearded on the callus, obtuse to acute or awned from the tip, rarely from between minute apical lobes, the awn often flexuous; palea similar to the lemma in texture, shape, and often in length; stamens 3; caryopsis free of but usually remaining enclosed within the lemma and palea, the embryo less than half as long as the grain; x = 10. Note: This treatment of *Muhlenbergia* has been reviewed by C. M. Morden.

Pohl, R. W. 1969. *Muhlenbergia*, subgenus Muhlenbergia (Gramineae) in North America. Amer. Midl. Naturalist 82: 512–542.

Rehder, C. G. 1949. *Muhlenbergia minutissima* (Steudel) Swallen and its allies. J. Wash. Acad. Sci. 39: 363–367.

1. Panicles at maturity open; pedicels averaging longer than the spikelets (excluding awns) 2
— Panicles narrow, compact or occasionally loose to open, and then the pedicels averaging shorter than the body of the lemma 5

2(1). Lemmas terminating in an awn 3–20 mm long; plant perennial, tufted from a hard, knotty base .. *M. porteri*
— Lemmas awnless or with an awn to 2 mm long; plants annuals or strongly rhizomatous perennials 3

3(2). Lemmas 4–6 mm long including the short awn; basal sheaths (sometimes only the inner ones) woolly-hairy; leaf blades typically pungent-tipped ...... *M. pungens*
— Lemmas 2 mm long or less; basal sheaths glabrous or scaberulous; leaf blades soft or firm but not pungent-tipped 4

4(3). Annual with fibrous roots; glumes mostly blunt, often minutely ciliate at the apex or short-hairy over the back .... *M. minutissima*
— Perennial from scaly rhizomes; glumes mostly acute and glabrous or scabrous .... *M. asperifolia*

5(1). First glume conspicuously 2–nerved and usually bifid at the apex, the second glume 1–nerved; panicle branches disarticulating at the base and falling with the spikelets; plants annual, much branched from the lower nodes .... *M. depauperata*
— First glume 1 (3) –nerved, rarely obsolete; plants not otherwise as above 6

6(5). Second glume 3 (5) –nerved, typically 3–toothed at the apex; ligule 4–20 mm long .... *M. montana*
— Second glume 1 (3) –nerved but not 3–toothed; ligule rarely more than 3.5 mm long 7

7(6). Panicles both terminal and axillary; glumes with an awn-tip to 1.5 mm long; anthers to 0.5 mm long ... . *M. mexicana*
— Panicles all terminal on culms or culm branches; plants otherwise variable 8

8(7). Lemmas awned, the awns averaging more than 1 (1–25) mm long 9
— Lemmas awnless or with awn-tips averaging no more than 1 mm long 13

9(8). Glumes averaging not more than 1 mm long (often rudimentary or obsolete in *M. schreberi*) 10

— Glumes averaging more than 1 mm long 11

10(9). Awns mostly 10–20 mm long; panicles loose to open, often purple .... *M. microsperma*
— Awns 1–5 mm long; panicles contracted, typically green .... *M. schreberi*

11(9). Lemma densely bearded from the base, the hairs about equal to the body; glumes equal to or exceeding the body of the lemma; leaf blades often flat, 1.5–5 mm wide, mostly 5–12 cm long .... *M. andina*
— Lemma glabrous to variously pubescent, the hairs shorter than the body; glumes shorter than the body of the lemma or, if longer, the leaf blades to 1.5 (2) mm wide and mostly 1–3 cm long (the two taxa below sometimes intergrading) 12

12(11). Sheaths glabrous; glumes averaging more than 2/3 as long as the lemma; ligule averaging more than 1 (1–5) mm long; leaf blades mostly 1–15 cm long and strongly ascending; lemmas glabrous or short-hairy, the awns averaging at least 6 (5–25) mm long .... *M. pauciflora*
— Sheaths typically pubescent; glumes averaging more than 2/3 as long as the lemma; ligule 0.2–0.8 mm long; leaf blades typically 1–3 cm long and stiffly spreading; lemmas pubescent with hairs to ca 1.5 mm long, the awns usually less than 6 (0.8–10) mm long .... *M. thurberi*

13(8). Glumes awned or awn-tipped 14
— Glumes awnless, obtuse to acute at the apex 16

14(13). Glumes with an awn almost as long as the body, the whole 3–8 mm long .... *M. racemosa*
— Glumes with an awn distinctly shorter than the body, the whole to 3.8 mm long 15

15(14). Lemmas pubescent with hairs to ca 1.5 mm long ... . *M. thurberi*
— Lemmas glabrous to minutely hairy ...... *M. wrightii*

16(13). Lemmas pubescent with hairs mostly ca 1.5 mm long .... *M. thurberi*
— Lemmas glabrous to minutely hairy 17

17(16). Plants annual (rarely perennial) from hairlike fibrous roots; anthers 0.2–1 mm long; culms not distinctly nodulose .... *M. filiformis*
— Plants perennial from slender, knotty rhizomes; anthers at maturity more than 1 mm long, sometimes only slightly so; culms typically nodulose at least above (at 20x magnification) 18

18(17). Lemmas averaging more than 2.7 (2.5–4) mm long (including the awn-tip); glumes 1.5–3 mm long, typically more than half as long as the lemma; plants known from a single, low-elevation population in Kane County .... *M. repens*
— Lemmas averaging no more than 2.7 (1.8–3) mm long; glumes 0.6–1.8 mm long, mostly not more than half as long as the lemma; plants of montane areas, widely distributed .... *M. richardsonis*

Muhlenbergia andina (Nutt.) A. S. Hitchc. Foxtail Muhly. [*Calamagrostis andina* Nutt.]. Perennial from somewhat coarse, scaly rhizomes; culms 2.5–6 (10) dm tall, erect, simple or sparingly branched; leaves chiefly cauline, green or glaucous; sheaths glabrous or minutely hairy; blades flat to loosely involute, mostly 5–15 cm long and 1.5–4 (5) mm wide, scaberulous; ligule 0.5–1.3 mm long, truncate; panicle spikelike to narrowly oblong, compact or sometimes loose, often interrupted or lobed, 4–16 cm long and to ca 2 cm wide, the primary branches mostly 0.5–5 cm long, erect-appressed to slightly spreading, the pedicels mostly shorter than the body of the lemma; spikelets 1–flowered, 2.5–4.5 mm long (excluding the awn), whitish to gray green, occasionally purple-tinged; glumes narrowly lanceolate, subequal, the larger 2.5–4.5 mm long, equal to or exceeding the body of the lemma, prominently 1–nerved, gradually tapered to a sharply acute to awnlike tip; lemma similar to the glumes in texture and shape, 1.8–3.2 mm long, 3–nerved, tapered to a hairlike awn 2–10 mm long, the callus densely bearded with silky hairs about as long as the body of the lemma; palea equal to or more often slightly exceeding the lemma; anthers 0.4–0.7 mm long; 2n = 20. Streamside, moist sandy soil, and wet rock faces at 1370 to 2900 m in Cache, Carbon, Daggett, Garfield, Grand, Iron, Kane, Rich, San Juan, Sanpete, Sevier, Summit, Uintah, Utah, Washington, and Wayne counties; throughout the western U. S., rare in Texas; 46 (i).

Muhlenbergia asperifolia (Nees & Mey.) Parodi Scratchgrass. [*Sporobolus asperifolius* Nees & Mey.]. Yellow green to glaucous perennial from long, slender, scaly rhizomes, occasionally stoloniferous; culms 1–5 (10) dm tall, slender, decumbent-ascending, often branched, especially near the base; leaves chiefly cauline; sheaths glabrous; blades ascending to spreading, flat or folded, mostly 2–8 (14) cm long and 1–2.5 (3) mm wide, essentially glabrous or more often scaberulous to scabrid-puberulent; ligule 0.2–1 mm long, truncate; panicle open, mostly 5–25 cm long and nearly as wide, branches and hairlike pedicels ultimately widely spreading, the pedicels much longer than the spikelets, the entire panicle usually breaking off at maturity; spikelets 1 (2 or 3) –flowered, 1.2–2 mm long, pale to dark gray, often purple-tinged; glumes subequal, the larger 0.7–1.7 mm long, considerably shorter to slightly longer than the lemma, 1–nerved, sometimes obscurely so, mostly sharply acute at the apex, glabrous or more often scabrous on the nerve and sometimes over the back; lemma lanceolate to elliptical or ovate, 1.2–2 mm long, 3–nerved, glabrous, obtuse to minutely awn-tipped at the apex; palea as long as the lemma or nearly so; anthers to 1.2 mm long; 2n = 20, 22, 28. Wet or drying, often alkaline or saline sites, along waterways and in meadows, in salt desert shrub, sagebrush, and pinyon-juniper communities, often on disturbed sites, at 1000 to 2130 m in all Utah counties except Daggett, Davis, Iron, Morgan, Sanpete, and Wasatch; British Columbia to Manitoba, south throughout the western half of the U. S. and into Mexico, sporadic in the central states east to Ohio; South America; 144 (vi). The caryopsis of *M. asperifolia* is frequently infected by a smut fungus (*Tilletia asperifolia* Ellis & Everhart), producing a globose black body.

Muhlenbergia depauperata Scribn. Sixweeks Muhly. Annual; herbage glabrous or scabrous to puberulent; culms 2–15 cm tall, much branched from the lower nodes, erect to ascending; sheaths glabrous or puberulent; blades flat or loosely involute on drying, 0.5–1.5 mm wide and to ca 3 cm long; ligule 1.5–4.5 mm long, generally split near the margins, the lateral portions auriclelike; panicle spikelike, often partially included in the uppermost sheath, 1–6 (7) cm long and to ca 5 mm wide, each short branch appressed to the rachis, bearing 1–3 spikelets, and disarticulating at the base; spikelets 1–flowered, 2.5–3.5 mm long (excluding awn when present), disarticulating (if at all) above the glumes; glumes narrowly lanceolate, subequal, typically scaberulous, the first 2–nerved and usually bifid at the apex, 2.5–3.5 mm long, the second 1–nerved and acute or tapered to an awn tip 0.5–2 mm long, to 3.8 mm long overall, often equal to or exceeding the body of the lemma; lemma lanceolate, 2.5–3.5 mm long, conspicuously 3–nerved, usually minutely pubescent with appressed hairs between the nerves, the apex acute or the midnerve prolonged as an awn to 10 mm long; palea about as long as the lemma, usually minutely hairy between the nerves; anthers 0.5–0.8 mm long; 2n = 20. Sagebrush and pinyon-juniper communities at 1520 to 2130 m in Beaver and San Juan counties, first collected in Utah in 1983 (Welsh 22448 BRY); Arizona, New Mexico, Texas; Mexico; 3 (0). Because *M. depauperata* has the same distinctive morphology and the same basic chromosome number as *Lycurus phleoides*, inclusion of this species in the genus *Lycurus* where it was placed (as *L. schaffneri*) by C. Mez (Repert. Sp. Nov. Regni Veg. 17: 212. 1921) would seem a more rational treatment than this, the currently accepted one.

Muhlenbergia filiformis (Thurber) Rydb. Pullup Muhly. [*Vilfa depauperata* var. *filiformis* Thurber, type from Uinta Mts., Summit County]. Annual or occasionally a short-lived, fibrous-rooted perennial; culms slender, 0.2–3 (4) dm tall, simple to freely branched from the base, sparingly so above, erect to ascending, not distinctly nodulose, often rooting at the lower nodes; sheaths glabrous; blades flat or involute near the tips, mostly 0.3–2 mm wide and rarely more than 3 cm long, scaberulous on the margins and often puberulent; ligule 1–3.5 mm long, obtuse to acute; panicle linear to narrowly oblong, 0.5–5 (9) cm long, usually interrupted, the short branches erect to slightly divergent, the pedicels much shorter to slightly longer than the spikelets; spikelets 1–flowered, 1.5–2.5 mm long, greenish or more often pale to dark gray, sometimes mottled; glumes ovate to fan-shaped, subequal, 0.5–1.8 (2) mm long, much shorter than the lemma, both 1–nerved, often scaberulous near the acute to rounded, entire or erose apex; lemma lanceolate, 1.5–2.5 mm long, 3–nerved, often conspicuously so, sometimes appearing fluted, usually sparsely short-hairy laterally or along the midrib and scabrous toward the acute to mucronate or awn-tipped apex; palea as long as the body of the lemma or nearly so; anthers 0.2–1 mm long; 2n = 18. Wet meadows or other moist to mesic places in sagebrush-grass, aspen, spruce-fir, and lodgepole pine communities at 1280 to 3200 m in Cache, Duchesne, Emery, Garfield, Iron, Juab, Kane, Salt Lake, Sanpete, Sevier, Summit, Tooele, Uintah, Utah, and Wasatch counties; British Columbia south throughout the western U. S. and into Mexico, occassional in the Great Plains; 121 (vi). *Muhlenbergia filiformis* is occasionally mistaken for *M. filiculmis* Vasey, which differs from *M. filiformis* in being perennial and in having the second glume 3–nerved and 3–toothed as in *M. montana* (q.v.). I have seen no Utah plants that can be identified as *M. filiculmis*.

Muhlenbergia mexicana (L.) Trin. Mexican Muhly. [*Agrostis mexicana* L.; *M. ambigua* Torr.; *M. foliosa* ssp. *ambigua* (Torr.) Scribn.]. Perennial from stout, scaly rhizomes; culms 3–10 dm tall, erect or ascending, simple below and branching from the middle and upper nodes; sheaths glabrous or nearly so; blades lax, flat, 1.5–6 mm wide and mostly 4–15 cm long, scaberulous; ligule 0.4–1.5 mm long, truncate to acute; panicles both terminal and axillary, narrow, often interrupted, 3–21 cm long and 2–12 mm wide, often partly included in the uppermost sheath, the branches typically erect, densely flowered to the base, the pedicels much shorter than the spikelets; spikelets 1–flowered, 2–4.4 mm long, pale green, sometimes purple-tinged; glumes narrowly lanceolate to elliptical, subequal, 2–4.4 mm long, 1–nerved, the nerve prolonged as an awn 0.5–1.5 mm long, the body of the glume equal to or slightly shorter than the body of the lemma; lemma narrowly lanceolate, the body 2–4.4 mm long, 3–nerved, the callus bearded with hairs 0.2–1.5 mm long, the apex sharply acute to awn-tipped or with an awn to about 6 mm long; palea about as long as the body of the lemma; anthers 0.3–0.5 mm long; $2n = 40$. Along waterways and in other moist, often disturbed sites below 1520 m in Utah and Grand counties; Alaska and British Columbia to Nova Scotia, sporadic in the U. S. except in the Southeast; 4 (0). *Muhlenbergia mexicana* is sometimes confused with *M. frondosa* (Poir.) Fern. of the eastern states. In *M. frondosa*, however, the internodes of the culm are shiny and glabrous rather than dull and scabrous to pubescent as in our plants (Gould 1975).

Muhlenbergia microsperma **(DC.) Kunth** Littleseed Muhly. [*Trichochloa microsperma* DC.]. Annual or short-lived perennial; culms 0.5–6 (7) dm tall, much branched at the lower nodes, sparingly so above, erect or decumbent, often geniculate at the nodes; sheaths loose, glabrous or scaberulous; blades often deciduous, flat or loosely involute, 1–6 (10) cm long and 1–1.5 (2) mm wide, scaberulous to short-hairy; ligule 0.5–2 mm long, truncate, erose, and soon splitting vertically; panicle narrow, loosely flowered, 2–20 cm long, the branches mostly less than 3 cm long, erect to more or less spreading, the pedicels rather stout, scabrous to scabrid-puberulent, averaging shorter than the spikelets; spikelets 1–flowered, 2–4 mm long, often purple; glumes unequal or subequal, ovate to oblong, averaging no more than 1 mm long, broadly acute to rounded at the apex, 1(2) –nerved, glabrous or the midnerve scaberulous; lemma narrowly lanceolate, 2–3.5 (4) mm long, 3–nerved, sometimes obscurely so, short-hairy on the lower 1/4–1/3, scabrous above, the midnerve prolonged from between minute teeth as a hairlike awn mostly 10–20 mm long; palea slightly shorter than the lemma; anthers 0.3–1 mm long. Dry rocky sites, often in the shelter of desert shrubs or cacti, in creosote bush communities at about 1070 m in Washington County; California, Nevada, Arizona, and Mexico; South America; 7 (0). Cleistogamous spikelets are usually present in the axils of the lowermost culm branches of *M. microsperma*. They develop within a short, hardened, reduced sheath, lack glumes, have a short-awned lemma, and a caryopsis more rounded than those of the chasmogamous florets.

Muhlenbergia minutissima **(Steudel) Swallen** Annual Muhly. [*Agrostis minutissima* Steudel]. Annual; culms slender, mostly 0.5–3.5 dm tall, simple or branched from the base, erect or ascending; leaves often chiefly basal; sheaths essentially glabrous; blades flat to weakly involute near the tips, mostly 1–6 cm long and 0.3–1.5 (2) mm wide, scaberulous to puberulent; ligule 1–2.5 (3) mm long, truncate, erose, and soon splitting vertically; panicle ultimately open, 2–18 cm long, usually 1/2–3/4 the height of the plant, and to about 4 cm wide, the branches soon spreading, the pedicels hairlike, ascending-spreading, averaging longer than the spikelets; spikelets 1–flowered, 0.8–1.8 mm long, greenish to straw colored, often purple-tinged; glumes subequal, elliptical to lanceolate or ovate, 0.6–1 mm long, shorter than the lemma, 1–nerved, acutish to rounded and often minutely ciliate at the apex, usually sparsely short-hairy over the back; lemma lanceolate to elliptical, 0.8–1.8 mm long, obscurely nerved, glabrous or more often sparsely short-hairy below, obtuse to acute or awn-tipped; palea about equal to the lemma; anthers 0.3–0.6 mm long; $2n = 60, 80$. Dry to mesic sites in openings in sagebrush, pinyon-juniper, and aspen-spruce communities at 1830 to 2740 m in Beaver, Iron, and Sevier counties, also reported [as *Sporobolus confusus* (Fourn.) Vasey] from Tooele County by Flowers (Bot. Gaz. 95: 405. 1934); throughout the western U. S., and in Texas and Mexico, rare in the Great Plains; 9 (0).

Muhlenbergia montana **(Nutt.) A. S. Hitchc.** Mountain Muhly. [*Calycodon montanum* Nutt.]. Tufted perennial; culms slender to moderately robust, 0.6–8 (9) dm tall, erect or sometimes decumbent at the much branched base; leaves chiefly basal, often glaucous; sheaths glabrous or scaberulous, old sheath bases persistent, often broad and loose, becoming conspicuously flattened; blades typically stiffly erect or ascending, flat or more often involute, mostly 6–25 cm long and 1–2 mm wide, more or less scaberulous; ligule 4–14 (20) mm long, acute, entire; panicle sublinear to loose and narrowly oblong, 2–25 cm long, the branches mostly 1–9 cm long, erect to moderately spreading, the pedicels averaging shorter than the body of the lemma; spikelets 1–flowered, 2.5–4.5 mm long excluding awns, greenish to tan or pale to dark gray, sometimes whitish or purple-tinged; glumes equal or unequal, usually scabrous on the keel and sometimes across the back, the first lanceolate to ovate, 1–3 mm long, 1 (3) –nerved, acute to mucronate or awn-tipped, the second ovate to elliptical, 1.5–3.2 mm long, 3 (5) –nerved, 3–toothed or some of them erose and then more or less truncate, both glumes shorter than the lemma; lemma firmer than the glumes, lanceolate, 2.5–4 (4.5) mm long, 3–nerved, usually obscurely so, often mottled with dark gray, the callus short-bearded, the body variously hairy on the lower portion or laterally nearly to the apex, typically scabrous above, gradually tapered to a slender flexuous awn 2–12 (20) mm long, or occasionally sharply acute to awn-tipped; palea about equal to the body of the lemma; anthers 1.5–2.3 mm long; $2n = 40$. Dry to moist sites on grassy slopes and in openings in sagebrush, aspen, spruce-fir, and ponderosa pine communities at 1830 to 3295 m in Beaver, Garfield, Grand, Iron, San Juan, Washington, and Wayne counties; sporadic in the western U. S.; 81 (0). Vasey (Contr. U.S. Natl. Herb. 1: 267–280. 1893) separated *M. montana* Vasey from *M. filiculmis*, with which Utah plants have occasionally been confused, on purely quantitative bases: culms filiform, ligules to 2 mm long, lemma awns about 1 mm long. As currently interpreted, however, the range of variation in the two taxa is continuous or overlapping. The type specimen of *M. filiculmis* (US) reveals no basis for distinguishing it from *M. montana* other than the quantitative characters mentioned above. In view of the random nature of the variation and the existence of numerous intermediates, it is possible that size is related to microhabitat, and that the name *M. filiculmis* has been applied to depauperate specimens of *M. montana*. Mountain muhly is a good source of forage early in the year but becomes unpalatable at maturity unless fully grazed throughout the growing season (Valentine 1961). The species appears to have potential for use in land reclamation.

Muhlenbergia pauciflora Buckley New Mexican Muhly. Tufted perennial, rarely short-rhizomatous; culms 1–7 dm long from a hard, knotty base, erect or ascending to spreading, sometimes geniculate and rooting at the lower nodes, simple or sparingly to freely branched chiefly below midlength; leaves chiefly cauline; sheaths glabrous; blades typically involute-filiform or rarely flat and to ca 1.5 mm wide, mostly 1–15 cm long and strongly ascending, scaberulous or scabrous to occasionally short-hairy; ligule 1–5 mm long, often split, the lateral, auriclelike projections exceeding the central portion; panicle narrow but frequently loose, mostly 2–15 cm long, the primary branches 0.5–6 cm long, erect-appressed to slightly spreading, the pedicels erect, scabrid-puberulent, averaging shorter than the body of the lemma; spikelets 1 (2) –flowered, (3) 3.5–5.5 mm long, often reddish purple, occasionally greenish white; glumes lanceolate, subequal, (1.5) 2–3.5 mm long, the longer glume 1/2–2/3 as long as the lemma, 1–nerved, acute to awn-tipped, rarely subacute to obtuse, entire to minutely erose at the apex; lemma narrowly lanceolate, (3) 3.5–5.5 mm long, glabrous or scabrous to sparsely hairy, the hairs usually minute, often arising just above the callus or along the lower lateral margins (lemma pubescence variable within a panicle), gradually tapered to a hairlike awn (5) 7–25 mm long; palea about equal to or slightly exceeding the body of the lemma; anthers 1.5–2.7 mm long. Desert shrub, sagebrush, juniper, pinyon-juniper, and ponderosa pine communities at 1520 to 2130 m in Beaver, Garfield, Kane, Millard, San Juan, Washington, and Wayne counties; California, Colorado, Arizona, New Mexico, western Texas, and Mexico; 21(0). *Muhlenbergia arsenei* is reported for Utah by Hitchcock (1951). As originally described (Proc. Biol. Soc. Wash. 41: 161. 1928), *M. arsenei* differs from *M. pauciflora* in having a ligule only 1 mm long and glumes "somewhat hispidulous toward the awnless and rounded tips." The paratype cited by Hitchcock (Rydberg and Garrett 9498 UT) is, however, a specimen of *M. pauciflora* with ligules longer than 1 mm and glumes sharply acute. Both *M. pauciflora* (Welsh & Moore 2508 BRY) and *M. thurberi* (Rydb. & Garrett 9511 UT) have been collected from the paratype locality for *M. arsenei*. Under these circumstances, it appears possible that *M. arsenei*, with the short leaves and rounded glumes of *M. pauciflora* and the awns of the lemmas intermediate between those of *M. thurberi* and *M. pauciflora*, may have been based on a hybrid between the two taxa. *Muhlenbergia polycaulis* Scribn., reported for Utah by Tidestrom (1925), differs from *M. pauciflora* in being consistently rhizomatous and in having shorter ligules (0.5–1 mm long), glumes with an awn to 1.5 mm long, and smaller lemmas (2.3–3.5 mm long) densely hairy over the lower 1/2–2/3 (Gould 1975). I have seen no Utah material matching Gould's description of *M. polycaulis*. See *M. thurberi* for discussion of the possibility of hybridization between that species and *M. pauciflora*.

Muhlenbergia porteri Scribn. ex Beal Bush Muhly. [*M. texana* Thurber]. Tufted perennial from a hard knotty base; culms mostly 1.5–10 dm tall, sparingly branched above the base, ascending-spreading, typically geniculate at some or all the numerous, often swollen nodes; leaves chiefly cauline; sheaths glabrous or scabrous, those below loose and becoming more or less flattened; blades soft, flat, mostly 1.5–6 cm long and 0.5–2 mm wide, scaberulous; ligule 1–2 mm long, often vertically split laterally; panicle ultimately open, mostly 4–10 cm long and at maturity nearly as broad, the branches and hairlike pedicels ultimately widely spreading, the latter averaging longer than the body of the lemma; spikelets 1-flowered, 2–5 mm long (excluding awns), often reddish purple; glumes narrowly lanceolate, equal or unequal, 1 (3) –nerved, acute or the midnerve of at least the second glume often prolonged as an awn-tip or as an awn to about 3 mm long, the body of the glumes 1.5–3 mm long, shorter than the lemma; lemma narrowly lanceolate, 2–4 (5) mm long, 3–nerved, sparsely pubescent between the nerves and laterally, terminating in a hairlike awn 3–20 mm long; palea about equal to the body of the lemma; anthers 1.6–2.5 mm long; 2n = 20, 40. Dry, mostly rocky or sandy sites on open hillsides, in creosote bush and desert shrub communities, at 760 to 1310 m in San Juan, Washington, and Wayne counties; southwestern U. S. from California east to Colorado, Texas, and Mexico, rare in Oklahoma; 21 (0). Bush muhly is highly palatable to all classes of livestock but is not sufficiently abundant in Utah to provide a significant source of forage.

Muhlenbergia pungens Thurber in Gray Sandhill Muhly. Glaucous perennial from coarse scaly rhizomes, often forming large rounded clumps and dying out in the center; culms 1–6 (7) dm tall, erect or more often decumbent-ascending, usually much branched just above the base; sheaths mostly woolly-hairy at the base, essentially glabrous above; blades rigid, involute, typically pungent-tipped, 2–6 (8) cm long and to ca 1 mm wide, glabrous to puberulent; ligule 0.2–1 mm long, often with membranous lateral lobes, occasionally appearing to consist of a ring of hairs; panicle ultimately open, (7) 8–15 (18) cm long and 2–8 cm wide, the branches and hairlike pedicels ultimately widely spreading, the pedicels to ca 1.5 cm long, averaging much longer than the spikelets; spikelets 1 (2) –flowered, 4–6 mm long (including awns), mostly red purple to brownish; glumes lanceolate to ovate, subequal, 1 (3) –nerved, entire or erose to 3–toothed, often awn-tipped, the whole 1.2–4 mm long, considerably shorter than the lemma, scaberulous above or nearly throughout; lemma narrowly lanceolate, gradually tapered to a short awn 1–1.5 (2) mm long, the whole 4–6 mm long, glabrous throughout or sparsely short-hairy near the base; palea equal to or slightly longer than the body of the lemma, the 2 nerves prolonged as awn-tips; anthers 1.3–2.5 mm long; 2n = 42, 60. Desert shrub, sagebrush-juniper, and pinyon-juniper communities at 1070 to 1980 m in Emery, Garfield, Grand, Kane, San Juan, and Wayne counties; Wyoming, Colorado, Arizona, New Mexico and Texas, east to South Dakota and Nebraska; 98 (iii).

Muhlenbergia racemosa (Michx.) B.S.P. Green Muhly. [*Agrostis racemosa* Michx.]. Perennial from long scaly rhizomes; culms 2.5–10 (13) dm tall, typically erect, simple or more often branching at or above the base, often from the nodes at midlength, the internodes glabrous to puberulent; leaves chiefly cauline; sheaths glabrous or scaberulous; blades soft, flat or loosely involute, to ca 15 cm long and mostly 2–6 mm wide, scabrous; ligule 0.2–1 (1.5) mm long, truncate; panicle spikelike to narrowly oblong, 4–18 cm long and 0.3–0.7 (1.5) cm wide, often lobed or interrupted, the branches 0.5–2 cm or more long, erect to slightly spreading, the lowermost branch often somewhat remote, the pedicels shorter than the spikelets; spikelets 1–flowered, 2.2–4.5 mm long including the awn-tip, usually greenish, occasionally purple-tinged; glumes narrowly lanceolate, equal or subequal, the solitary scabrous nerve prolonged as an awn often as long as the body or nearly so, the whole 3–7.5 (8) mm long, exceeding the lemma; lemma narrowly lanceolate, acute or awn-tipped, the whole 2.2–4 (4.5) mm long, 3–nerved, often mottled with black, pubescent on the lower half and sometimes along the margin almost to the apex, the hairs to ca 1 mm long; palea about equal to the lemma; anthers 0.4–1 (1.5) mm long; $2n = 20, 40$. Dry to moist sites, on open slopes, in hanging gardens, in mountain brush, aspen, ponderosa pine, and meadow communities, and a weed of gardens, at 1220 to 3050 m in Duchesne, Garfield, Grand, Piute, Salt Lake, San Juan, Uintah, Utah, Washington, Wayne, and Weber counties; Alaska and British Columbia to Newfoundland, south in the U. S. except in the Southeast; 22(i). Pohl (1969) recognizes two species in this widespread complex, separating them as follows:

1. Internodes of the culm dull, puberulent; ligules 0.2–0.6 mm long; anthers 0.8–1.5 mm long; lemma pilose at the base and along the margins; species of wet sites .... *M. glomerata* (Willd.) Trin.
— Internodes of the culm smooth and polished except near the summit, ligules 0.6–1 mm long; anthers 0.5–0.8 mm long; lemma pilose at the base only; species of dry open sites .... *M. racemosa*

According to Pohl (1969), both species are present in Utah. Although the total range of variation for both forms can be seen in our plants, the characters as outlined above are not consistently correlated, either in Utah material or in numerous North American specimens examined at the National Herbarium. Under these circumstances, the validity of recognizing *M. glomerata* as a distinct species appears to need additional scrutiny.

Muhlenbergia repens (Presl) A. S. Hitchc. Creeping Muhly. [*Sporobolus repens* Presl]. Perennial with hard shiny rhizomes; culms often loosely tufted, wiry, freely branched above the base, 0.6–3.5 dm tall, the sterile culms decumbent, the flowering ones erect, usually smooth below and nodulose to some degree above; leaves chiefly cauline; sheaths usually glabrous; blades flat or soon involute, mostly 1–2 mm wide, those of the sterile shoots to as much as 9 cm long, those of the flowering culms 1–3 (5) cm long and often recurved; ligule 0.5–1 mm long, acute to obtuse; panicles terminal and axillary, narrow, 1–6 cm long, the spikelets loosely spaced along the rachis on erect or slightly spreading pedicels averaging shorter than the spikelets; spikelets 1 (2) –flowered, 2.5–4 mm long, pale green to gray; glumes lanceolate to ovate, subequal, 1.5–3 mm long, averaging at least half as long as the body of the lemma, 1–nerved, typically acute, glabrous or scaberulous on the midnerve; lemma typically a darker gray than the glumes, lanceolate, 2.5–4 mm long including the awn tip; 3–nerved, the lateral nerves sometimes obscure, glabrous or scaberulous, acute or tapered to an awn tip to ca 0.5 mm long; palea about as long as the lemma; anthers 1.2–1.5 mm long; $2n = 72$. This low-elevation plant is represented in Utah by a single collection (Morden 649 UT) from a small population in a sagebrush community at 1615 m in Kane County; Nevada and Arizona to Texas; Mexico; 1 (0).

Muhlenbergia richardsonis (Trin.) Rydb. Mat Muhly. [*Vilfa richardsonis* Trin.; *M. squarrosa* (Trin.) Rydb.]. Mat-forming perennial from slender knotty rhizomes; culms slender, 0.5–6 (7) dm tall, erect or decumbent-ascending, often geniculate at the lower nodes, typically much branched at or near the base, usually nodulose (at 10–20x magnification); leaves chiefly cauline; sheaths glabrous; blades flat or more often involute, 0.5–1.8 (2) mm wide, 1–5 (9) cm long, scaberulous to puberulent; ligule 1–3 mm long, acute; panicles terminal and often axillary, narrow, (1) 2–7 (10) cm long, mostly exserted from the upper sheath, often somewhat interrupted, the short branches erect or slightly divergent, the pedicels averaging shorter than the spikelets; spikelets 1–flowered, 1.8–3 mm long including the awn tips, greenish or more often pale to dark gray; glumes ovate, subequal, 0.6–1.5 (1.8) mm long, much shorter than the lemma, obtuse to acute, occasionally minutely erose at the apex, glabrous or scaberulous on the midnerve, the first 1–nerved, the second 1 (3) –nerved; lemma lanceolate, 1.8–2.8 (3) mm long, 3–nerved, usually obscurely so, glabrous or scaberulous, acute or mucronate to awn-tipped; palea about as long as the lemma; anthers 1–1.8 mm long; $2n = 40$. Dry to moist sites in sagebrush, mountain brush, pinyon-juniper, aspen-spruce-fir, ponderosa pine, and meadow communities, occasionally on open slopes, at 1680 to 3200 m in Beaver, Box Elder, Cache, Carbon, Daggett, Duchesne, Emery, Garfield, Grand, Iron, Piute, Rich, Salt Lake, San Juan, Sanpete, Sevier, Summit, Uintah, Utah, Wasatch, Washington, and Wayne counties; Alaska and Yukon to New Brunswick, south throughout the western U. S. and into Mexico, east in the Great Plains as far south as Nebraska, rare in the eastern states; 110 (vii).

Muhlenbergia schreberi J. F. Gmel. Nimblewill Muhly. Perennial, often appearing annual, sometimes stoloniferous; culms lax, 1–6 dm long, erect or ascending to prostrate and rooting at the nodes, simple or branched above the base; leaves evenly distributed along the culm; sheaths glabrous or with a few long hairs near the summit; blades soft, flat, mostly 1–3 (4) mm wide and 3–8 cm long, glabrous or sparsely hairy near the ligule; ligule to 0.5 mm long, truncate, sometimes evidently lacking or consisting of short hairs; panicle linear to somewhat loose, mostly 4–15 cm long and to about 6 mm wide, often nodding, the short branches erect to slightly spreading, the pedicels erect, averaging shorter than the body of the lemmas; spikelets 1–flowered, 2–2.5 mm long excluding awns, usually greenish; glumes obsolete or the second to about 0.4 (1) mm long; lemma narrowly lanceolate, 2–2.5 mm long, often mottled with black, 3–nerved, typically puberulent near the base, terminating in a hairlike awn 1–5 mm long; palea about equal to the body of the lemma; anthers 0.1–0.4 mm long; $2n = 40$. Known in Utah by a single collection (Deming s.n. UTC) at about 1200 m from a dry hillside in Washington County; Colorado, Arizona, Texas, and Mexico, and from Ontario south throughout the eastern U. S.; South America; adventive in many subtropical countries; 1 (0).

Muhlenbergia thurberi Rydb. Thurber Muhly. [*M. curtifolia* Scribn., type from Kane County]. Glaucous perennial from slender rhizomes; culms wiry, mostly 1–6 dm tall, solitary or tufted, erect or decumbent-based, often loosely branched above the base; leaves chiefly cauline; sheaths typically pubescent; blades soft or more often pungent-tipped, flat or involute toward the tips, 1–1.5 (2) mm wide and mostly 1–3 cm long, usually widely spreading, often puberulent; ligule 0.2–0.8 mm long, lacking lateral projections; panicle sublinear to narrow, 2–8 (9) cm long, often more or less interrupted, the short branches erect or slightly divergent, the pedicels scabrid-puberulent, averaging shorter than the spikelets; spikelets 1 (2) –flowered, 2.5–4.5 mm long excluding the awn, greenish white, occasionally tinged with rose or purple; glumes lanceolate, subequal, (2) 2.5–4 mm long, often nearly as long as the lemma, usually purple-tinged, acute or tapered to awnlike tips, essentially glabrous or scaberulous on the upper half of the midnerve, the first 1–nerved, the second 1– to 3–nerved; lemma narrowly lanceolate, the body 2.5–4.5 mm long, 3–nerved, pubescent in part or throughout with hairs to ca 1.5 mm long, sharply acute or more often awned, the awn 0.8–7 (10) mm long; palea about equal to or slightly longer than the body of the lemma; anthers 1–2 mm long. Dry to moist, often sandy or rocky sites, in hanging gardens, desert shrub, pinyon-juniper, and ponderosa pine communities at 1130 to 2500 m in Garfield, Grand, Kane, San Juan, Washington, and Wayne counties; Nevada, Arizona, and New Mexico; 29(0). *Muhlenbergia curtifolia* is reduced to synonymy by Cronquist et al. (1977), who found the asserted distinctions between *M. curtifolia* and *M. thurberi* difficult to detect. Although *M. thurberi* and *M. pauciflora* are usually easily separable according to the characters cited in the key, occasional specimens (e.g., Welsh 21384 BRY and Albee 3414 UT) combine the characters of both, suggesting that hybridization is occurring where their ranges overlap.

Muhlenbergia wrightii Vasey ex Coult. Spike Muhly. Tufted perennial from a usually knotty base, occasionally with elongate, decumbent tillers and appearing rhizomatous; culms wiry, 1–5 (6) dm tall, erect or occasionally somewhat decumbent at the base, simple or more often sparingly branched above the base, sometimes obscurely nodulose; sheaths glabrous or scabrous to puberulent; blades soft, typically folded or involute, 0.5–2 mm wide and to about 15 cm long, scaberulous to puberulent; ligule 0.8–2 (3) mm long, truncate or obtuse to acute; panicle spikelike, 1–15 cm long, 0.2–1 cm wide, often interrupted, the short branches erect, the lowermost often more or less remote, the pedicels much shorter than the spikelets; spikelets 1 (2) –flowered, 3–4.5 (5) mm long including the awn tips, greenish white or more often pale to dark gray; glumes ovate to lanceolate, subequal, mostly 0.6–2 mm long, the solitary scabrous nerve generally prolonged as an awn tip 0.2–1 mm long, the whole shorter than the lemma; lemma narrowly lanceolate, gradually tapered to a short awn tip averaging less than 1 mm long, mostly 3–5 mm long overall, 3–nerved, sometimes obscurely so, glabrous or scabrous to puberulent; palea about as long as the body of the lemma; anthers 1.2–2.2 mm long. Desert shrub, sagebrush, mountain brush, pinyon-juniper, and ponderosa pine communities, occasionally roadside or in other disturbed places, at 1370 to 2740 m in Emery, Garfield, Millard, and Sevier counties; Colorado, Arizona, and New Mexico; Mexico; 15 (0). Spike muhly is palatable to all classes of livestock and is an excellent soil binder recommended for revegetation of mined lands.

Munroa Torr.

Low, often stoloniferous, gynomonoecious annuals; culms branching at and above the decumbent base; leaves appearing fascicled at culm nodes; sheaths open, hairy at the summit; blades narrow, mostly flat or folded; ligule a ring of hairs; inflorescence a small headlike cluster of sessile to short-pedicelled spikelets closely subtended by and partially hidden within the fascicled leaf blades; spikelets (2) 3– to 6 (10) –flowered, the lower florets pistillate or perfect, the terminal one rudimentary, the rachilla continuous or disarticulating above the glumes and between the florets; glumes narrowly lanceolate, 1–nerved, usually shorter than the lowermost lemma, sometimes one or both reduced or lacking; lemma obscurely to prominently 3–nerved, mucronate or short-awned, those of the lower spikelets leathery, those of the upper spikelets typically membranous; palea shorter and narrower than the lemma; stamens 2 or 3, well developed or rudimentary; caryopsis free of the lemma and palea, the embryo to about half the length of the grain; $x = 8$.

Anton, A. M. and A. T. Hunziker. 1978. El genero *Munroa* (Poaceae): Sinopsis morfologica y taxonomica. Bol. Acad. Nac. Ci. 52: 229–252.

Munroa squarrosa (Nutt.) Torr. False Buffalograss. [*Crypsis squarrosa* Nutt.]. Low, mat-forming annual; culms 0.1–1.5 (3) dm long, branched at the base and at the nodes above the base, prostrate to ascending or erect, sometimes rooting at the nodes, a pair of short-sheathed leaves arising at each node, these closely subtending several leafy culm branches with greatly shortened internodes, the lowermost sheaths in the clusters greatly reduced, sometimes fused, the blades awnlike; sheaths of unreduced leaves mostly less than 1 cm long, hairy at the base and sometimes along the margins, often with tufts of hairs 1–3 mm long at the summit; blades of unreduced leaves firm or occasionally soft, generally white-margined and long-tapered to an often needlelike tip, flat or folded, mostly 1–2 (3) mm wide and 1–4 cm long, essentially glabrous to strongly scabrous or scabrid-puberulent; ligule a ring of hairs 0.2–1 mm long; panicle headlike, composed of 2–4 spikelets borne in the axils of leaf clusters, much shorter than and often obscured by the leaves; spikelets sessile or short-pedicelled, subterete to somewhat compressed, 6–8 mm long, 3– to 5–flowered, all but the lowermost spikelet usually disarticulating above the glumes and between the florets, the lower 1 or 2 florets often pistillate, the uppermost rudimentary; glumes membranous, narrow, variable in size but shorter than the lowermost lemma, 1–nerved, the first sometimes lacking (reduced leaf blades often appearing to be glumes); lemmas rounded on the back, (3) 3.5–5 mm long, that of the lowermost spikelet in each panicle ultimately hardened, more or less obscurely nerved, scaberulous or otherwise roughened and with or without lateral pubescence, those of the upper spikelets often more or less membranous and conspicuously 3–nerved, truncate, minutely notched and awn-tipped, glabrous to hairy over the back, the lateral nerves pubescent below midlength with hairs to ca 1 mm long, palea shorter than the lemma; stamens 3, the anthers either rudimentary or 1–2 mm long; $2n = 16$. Salt desert shrub, desert shrub, sagebrush, and pinyon-juniper communities, usually in disturbed sites, at 1280 to 2930 m in Beaver, Carbon, Duchesne, Emery, Garfield, Iron, Kane, Millard, San

Juan, Sevier, Uintah, Washington, and Wayne counties; most of the southwestern U. S., Montana to Colorado and throughout the Great Plains, south to Texas and northern Mexico; 86 (i). Plants of this species are often infected by a woolly aphid, with the frequently abundant remains of egg cases deposited in leaf axils and over leaf surfaces as white-woolly tufts.

Oryzopsis Michx.

Tufted perennial; sheaths open; blades flat to involute; ligule membranous; inflorescence a narrow to open panicle, occasionally reduced to a raceme; spikelets terete to dorsally or laterally compressed, 1-flowered, disarticulating above the glumes; glumes green to pale and translucent, elliptical to ovate, shorter to more often longer than the body of the lemma, usually 3- to 9-nerved, rounded to short-tapered apically; lemma at maturity firm to hardened, elliptical to obovate, terete or somewhat compressed, glabrous or hairy, the callus short and blunt, the apex awned, the awn readily deciduous, straight, curved, or geniculate, sometimes basally twisted; palea similar to the lemma in texture, color, and length; stamens 3; caryopsis free of or adherent to the palea, firmly enclosed within the hardened lemma and palea, the embryo 1/6–1/3 as long as the grain; $x = 11$. *Oryzopsis miliacea* (L.) Benth. & Hook., native to the Mediterranean Region and distinguished by its broad, flat leaf blades and glabrous lemmas, has been introduced in scattered localities in the U. S., but I have seen no specimens from Utah.

1. Blades of basal leaves (1.5) 2–10 mm wide; blades of cauline leaves typically much reduced; lemmas 5–7 mm long, the callus densely short-bearded .... *O. asperifolia*
— Blades of basal and cauline leaves well developed but rarely more than 2 mm wide; lemmas to 5 mm long, not more conspicuously hairy on the callus than above 2

2(1). Lemma averaging more than 3 (3–5) mm long, the awn once geniculate; panicle branches strictly erect .... *O. exigua*
— Lemma 1.8–3 mm long, the awn straight or flexuous; panicle branches erect or ultimately spreading to deflexed. .... *O. micrantha*

Oryzopsis asperifolia Michx. Roughleaf Ricegrass. Perennial; culms tufted, erect to prostrate, 2–7 dm long; sheaths glabrous; blades of the basal leaves firm and erect-ascending, often exceeding the culms, (1.5) 2–10 mm wide, flat or rarely loosely involute, tapered at both ends, scabrous on the margins, dark green and smooth to scaberulous on the lower surface, glaucous and typically scabrid-puberulent above; blades of the culm leaves much reduced, mostly less than 3 cm long or lacking; ligule 0.1–0.7 mm long, truncate; panicle or raceme narrow, 3.5–8 cm long, the short branches appressed or at flowering somewhat spreading; glumes broadly elliptical to obovate, subequal, 5–8.5 mm long, 5- to 9-nerved, membranous-margined, abruptly acute at the tip, shorter than or more often at least the second equal to or slightly exceeding the lemma; lemma at maturity leathery to hardened, lanceolate, terete to somewhat flattened, 5–7 mm long and to ca 2.5 mm wide, pale green to white or yellowish, sometimes purple-tinged, glossy or dull, sparsely puberulent over the body, the short, blunt callus densely short-bearded, the awn terminal or nearly so, 5–14 mm long, usually flexuous, readily deciduous; palea similar to the lemma; anthers 2.7–3.5 mm long, with an apical tuft of hairs; $2n = 46, 48$. Moist to mesic sites under aspen, ponderosa pine, and lodgepole pine at 2250 to 2770 m in Daggett, Duchesne, Juab, and Uintah counties (Uinta and Deep Creek mts.); across Canada, south into Idaho and Montana to Colorado and New Mexico, in the Dakotas and Nebraska, and through most of the Northeast; 12 (0).

Oryzopsis exigua Thurber in Torr. Little Ricegrass. Perennial; culms tufted, slender, stiffly erect, 1–4.3 dm tall; sheaths glabrous or scaberulous; leaves chiefly basal; blades involute-filiform or occasionally flat and then to ca 2 mm wide, typically stiffly ascending and scabrous; ligule 1.5–4 mm long, acute; panicle narrow, 3–9 (11) cm long, the few-flowered, short branches appressed to the rachis; glumes membranous throughout or green near the base, occasionally purple-tinged, elliptical to obovate, subequal, 3–6 mm long, obscurely 3- to 5(7)-nerved, smooth or scaberulous, abruptly acute to rounded at the erose or entire tip, shorter than to barely exceeding the lemma; lemma at maturity pale to brown, leathery to hardened, terete to somewhat compressed, ellipsoidal, (3) 3.5–5.5 mm long and to ca 1 mm wide, sparsely appressed-puberulent and minutely bearded above the short, blunt callus, the awn terminal or nearly so, 3–7 mm long, stout, once geniculate, usually loosely twisted below, readily deciduous; palea similar to the lemma; anthers 1.5–2 mm long; $2n = 22$. Dry, open, often rocky sites and in aspen-fir and lodgepole pine communities at 1820 to 2900 m in Cache, Daggett, Duchesne, Juab, San Juan, and Summit counties; British Columbia and Alberta, south in much of the western U. S. and in South Dakota; 29 (0).

Oryzopsis micrantha (Trin. & Rupr.) Thurber Littleseed Ricegrass. [*Urachne micrantha* Trin. & Rupr.]. Perennial; culms tufted, slender, erect-ascending, 3–7 (9) dm tall; sheaths glabrous to puberulent; blades flat to loosely involute, 0.5–2 (3) mm wide, more or less soft, glabrous or more often scabrous; ligule 0.2–2 mm long, usually truncate; panicle 5–20 cm long, narrow to open, the branches erect to widely spreading or deflexed, the spikelets on short, appressed pedicels; glumes thin-membranous throughout or occasionally greenish near the base, lance-elliptical, 2.7–4 (5) mm long, equal or nearly so, mostly 5- to 7-nerved, often obscurely so, abruptly acute at the tip, both glumes exceeding the lemma; lemma at maturity pale to dark brown, leathery, terete to somewhat compressed, ellipsoidal, 1.8–3 mm long and to ca 1 mm wide, glossy, glabrous or sparsely puberulent, the callus very short and blunt, glabrous or minutely hairy, the awn essentially terminal, 4–9.5 mm long, hairlike, straight or flexuous, readily deciduous; palea similar to the lemma; anthers to ca 1 mm long; $2n = 22, 24$. Dry to moist, often sandy or rocky sites in sagebrush, pinyon-juniper, mountain brush, and ponderosa pine communities, occasionally in openings in or under aspen and spruce-fir, at 1380 to 2930 m in Beaver, Carbon, Daggett, Duchesne, Emery, Garfield, Grand, Iron, Juab, Millard, San Juan, Sevier, Uintah, Washington, and Wayne counties; British Columbia to Manitoba, south in most of the states west of the Mississippi; 85 (i).

Panicum L.

Annuals or perennials; sheaths open; blades mostly flat; ligule a ring of hairs, often membranous at the base, occasionally entirely membranous or lacking; inflorescence an open to contracted panicle or raceme; spikelets plano-convex to subterete, disarticulating below and in some species above the glumes, 2-flowered, the lower floret staminate or neuter, the upper floret perfect; glumes often markedly unequal, usually prominently several-nerved, the first commonly clasping the base of the spikelet, the second often somewhat hooded or beaked at the apex; lemma of the lower floret generally similar to the second glume in size, shape, and texture, the lemma and second glume typically sheathing the upper floret; lemma of the upper floret dorsiventrally compressed, firm or leathery, acute to rounded at the apex, at maturity often smooth and shiny or finely striate, the firm margins incurved around the edges of the equally firm palea; stamens 3; caryopsis free of but firmly enclosed within the lemma and palea, the embryo at least half the length of the grain; x = 9, 10. *Panicum bulbosum* H.B.K., reported from southern Utah by Cronquist et al. (1977), apparently reaches its northernmost limits in Arizona and New Mexico (Gould 1975). The manuscript for this genus was reviewed by Gerrit Davidse.

Gould, F. W. and C. A. Clark. 1978. *Dichanthelium* (Poaceae) in the United States and Canada. Ann. Missouri Bot. Gard. 65: 1088–1132.
Hitchcock, A. S. and A. Chase. 1910. The North American species of *Panicum*. Contr. U. S. Natl. Herb. 15: 1–396.
Lelong, M. 1984. New combinations for *Panicum* subgenus *Panicum* and subgenus *Dichanthelium* (Poaceae) of the southeastern United States. Brittonia 36: 262–273.

1. Glumes pubescent (at 10 x magnification) 2
— Glumes glabrous or scabrous, not hairy 3
2(1). Spikelets averaging no more than 2.5 (0.8–2.7) mm long
   .... *P. acuminatum*
— Spikelets averaging more than 2.5 (2.5–4) mm long ..
   .... *P. oligosanthes*
3(1). Second glume and sterile lemma acute to acuminate at the apex, distinctly longer than the hardened upper floret 4
— Second glume and sterile lemma abruptly acute to rounded at the apex, only slightly exceeding to shorter than the hardened upper floret 8
4(3). Plants perennial, occupying mostly native habitats 5
— Plants annual, typically occupying disturbed habitats ... 6
5(4). Plants with stout scaly rhizomes; culms 6–30 dm tall; first glume 2–4 mm long, nearly always averaging more than 2.5 mm long .... *P. virgatum*
— Plants tufted, not rhizomatous; culms 2–7.5 dm tall; first glume 1.7–2.5 mm long, usually averaging less than 2.5 mm long .... *P. hallii*
6(4). Sheaths glabrous; first glume not more than 1/3 as long as the spikelets .... *P. dichotomiflorum*
— Sheaths pubescent, the hairs usually papilla-based; first glume mostly more than 1/3 as long as the spikelet 7
7(6). Spikelets 4–6 mm long .... *P. miliaceum*
— Spikelets 2–3.5 mm long .... *P. capillare*
8(3). First glume about equal to the spikelet; ligule membranous, without cilia .... *P. obtusum*
— First glume 1/3–1/2 as long as the spikelet; ligule a ring of hairs or a membrane with cilia 9
9(8). Plants to 6 dm tall; panicles to 13 cm long; pedicels mostly longer than the spikelets .... *P. oligosanthes*
— Plants 5–30 dm tall; panicles 12–25 cm long; pedicels consistently much shorter than the spikelets ....
   .... *P. antidotale*

Panicum acuminatum Swartz Bundle Panic. [*Dichanthelium acuminatum* (Swartz) Gould & Clark; *D. lanuginosum* (Ell.) Gould; *Panicum huachuacae* Ashe; *P. lanuginosum* Ell.; *P. occidentale* Scribn.; *P. tennesseense* Ashe; *P. thermale* Bolander]. Perennial; culms 1–7 (8) dm tall, tufted, erect to ascending, unbranched during initial growth, culms appearing late in the season freely branched above the base, the ultimate branchlets forming short, dense, leafy fascicles; herbage glabrous or sparsely to densely long-hairy throughout, usually at least some of the hairs papilla-based; blades flat, initially chiefly cauline, 3.5–12 mm wide and 4–10 cm long, the blades of the autumnal fascicled branchlets mostly 2–4 cm long, a basal rosette of blades shorter and broader than those of the culm usually appearing in late summer or early fall; ligule of hairs 1–5 mm long; panicle open to somewhat contracted, terminal and axillary, the latter usually developing after the terminal ones have flowered; terminal panicles 2–12 cm long, the very slender, flexuous branches ascending to spreading, the pedicels usually averaging longer than the spikelets; axillary panicles similar but rarely more than 3 cm long, usually partially included in the sheaths; spikelets plano-convex, ovate to obovate-elliptical, 0.8–2.7 mm long; florets of the terminal panicles chasmogamous, those of the axillary panicles cleistogamous; glumes pubescent with minute spreading hairs, the first 1/5–1/3 as long as the spikelet, broadly acute to somewhat rounded apically, the second glume and the pubescent sterile lemma about equal, as long as or slightly shorter than the glossy fertile lemma; 2n = 18. Chiefly in moist sandy sites, in hanging gardens and along waterways at 1070 to 2230 m in Emery, Garfield, Grand, Kane, Salt Lake, San Juan, Uintah, Utah, and Washington counties; sporadic in Canada, throughout the U. S., and in Mexico; South America; adventive in the U.S.S.R.; 109 (0). According to Gould and Clark (1978), the species commonly known as *P. lanuginosum* is considered to be synonymous with *P. acuminatum*. As support for elevation of the subgenus *Dichanthelium* to the rank of genus, Gould and Clark (1978) provide a table comparing 24 features within *Panicum* (sensu stricto) with similar features in the dichanthelia. The table demonstrates that with respect to only one of these features, namely the Kranz syndrome, can the two groups be consistently separated. Most species of dicanthelia develop a rosette of short broad basal leaves and a terminal inflorescence during the cool season, and small clusters of axillary cleistogamous flowers on much branched, reduced lateral shoots during the warm season. These seasonal variations, the only readily observable morphological features by which the two groups may be distinguished, are not consistently present throughout the geographic range of the dicanthelia. Perhaps largely for this reason, no consensus exists among agrostologists as to the merit of recognizing two genera (Pohl, Madrono 30: 197. 1981). I have, therefore, followed Lelong (1984) in retaining the dicanthelia within *Panicum*.

Panicum antidotale Retz. Blue Panicgrass. Stoloniferous perennial from an elongate, stout rootstock; herbage often glaucous; culms 5–20 (30) dm tall, erect, becoming much branched at and above the base in age; sheaths glabrous or the collar puberulent; blades elongate, flat, mostly 3–12 mm wide, glabrous or puberulent; ligule a membranous-based ring of hairs to ca 1 mm long; panicle contracted to open, mostly 12–25 cm long, the branches many flowered, ascending to horizontally spreading, the pedicels consistently shorter than the spikelets; spikelets dorsiventrally to laterally compressed, lanceolate to ovate, 2–3 mm long; glumes glabrous, the first more or less clasping, to about half as long as the spikelet, abruptly acute, the second glume acute to obtuse, about equal to the membranous sterile lemma, both about equal to the glossy fertile one; 2n = 18. First reported for Utah in 1985 (Higgins 16086 BRY) from a creosote bush-blackbrush community at 950 m in Washington County; native to India, becoming established in Arizona, Texas, and Mexico; 1 (0). Blue panicgrass is an excellent sand-binder, introduced into the Southwest for erosion control. It is not tolerant of cold and its value as forage is questionable.

Panicum capillare L. Witchgrass. [*P. barbipulvinatum* Nash in Rydb.; *P. capillare* var. *occidentale* Rydb.]. Annual; culms 1–13 dm tall, typically branched from the base and often sparingly so above, erect to more often decumbent-ascending; sheaths conspicuously pubescent with long, spreading, usually papilla-based hairs; blades flat or folded, 4–15 (20) mm wide, usually long-hairy to some degree, the hairs commonly papilla-based; ligule a membranous-based ring of hairs 0.5–2 (3) mm long; panicle often at least half as long as the total length of the plant and about as broad as long, emerging tardily from the upper sheaths, much branched, the branches slender and ultimately widely spreading, the spikelets solitary at the tips of numerous, short to elongate, hairlike pedicels, the entire panicle ultimately breaking off at maturity; spikelets plano-convex to subterete, narrowly lanceolate to elliptical, 2–3.5 mm long, green or purple; glumes smooth or scabrous, the first broadly ovate, 1/3–2/3 as long as the spikelet, acute to abruptly short-tapered, the second short- to long-tapered at the apex, usually longer than the similarly shaped sterile lemma, both distinctly longer than the glossy fertile lemma; 2n = 18. Moist to drying, often sandy sites in salt desert shrub and desert shrub communities, in rock crevices, along waterways, roadsides, and in other waste places, less frequently in meadows and in mountain brush communities, at 850 to 1830 m in Cache, Carbon, Daggett, Davis, Emery, Garfield, Grand, Iron, Kane, Salt Lake, San Juan, Sanpete, Sevier, Uintah, Utah, Washington, and Wayne counties; across southern Canada, south throughout most of the U. S. and in Mexico; South America; naturalized in much of Eurasia; 89 (ix). *Panicum flexile* **(Gattinger) Scribn.**, which resembles *P. capillare*, is represented in the Utah flora by a solitary collection (Cottam 206 BRY) made in 1925 at about 1370 m in Utah County. This species differs from *P. capillare* in having a narrower, oblong, relatively few-flowered panicle, typically with a single, long-pedicelled spikelet terminating each branch and branchlet. *Panicum flexile* has not been reported since the initial collection and is otherwise known only from the eastern half of the U. S.

Panicum dichotomiflorum Michx. Spreading or Fall Panicum. Annual; culms 1–20 dm long, branched from the base and often sparingly so above, erect to decumbent-ascending; sheaths glabrous; blades flat, 3–20 mm wide, glabrous or scaberulous to sparsely long-hairy on the upper surface; ligule a membranous-based ring of hairs to ca 2 mm long; panicle terminal and often axillary, variable in size and shape, the terminal one to 40 cm long, emerging tardily from the upper sheaths, the primary branches usually stiffly ascending, naked at the base, the branchlets relatively short and mostly appressed to the branch axis; spikelets short-pedicelled, plano-convex to subterete, narrowly lanceolate, 2–4.3 mm long; glumes glabrous, the first broadly ovate, 0.5–1.5 mm long, 1/5–1/3 as long as the spikelet, abruptly acute to rounded at the apex, the second acute, about equal to the membranous sterile lemma, the two distinctly longer than the glossy fertile lemma; 2n = 54. Apparently first reported from Utah in 1919 (Garrett s.n. UT), occurring chiefly in cracks in curbing and along waterways below 1520 m in Salt Lake County; native to the eastern U. S. and southeastern Canada, sporadically introduced in the West and the Great Plains; Eurasia; 11 (vii).

Panicum hallii Vasey Hall Panicum. Tufted perennial, often glaucous; culms strictly erect, 2–6 (7.5) dm tall; sheaths glabrous or pubescent with spreading, often papilla-based hairs; blades flat, typically curling in age, 1.5–6 (7) mm wide, glabrous or long-hairy, especially near the ligule, the hairs occasionally papilla-based; ligule an inconspicuously membranous-based ring of hairs to 1.5 mm long; panicle variable in size and shape, 6–20 (25) cm long, ultimately well exserted, the primary branches relatively few and stiffly ascending, naked at the base, the branchlets appressed to the branch axis; spikelets short-pedicelled, plano-convex, lanceolate to subovate, 2.2–4.2 mm long; glumes glabrous, the first broadly ovate, 1.7–2.5 mm long, about half as long as the spikelet, the second acute to acuminate at the apex, about equal to the membranous sterile lemma; both distinctly longer than the glossy fertile lemma; 2n = 18. Known in Utah by a single collection (Hutchings & Stahmann 20 UT) from a juniper-grass community at 1680 m in Beaver County; Colorado, Arizona, New Mexico, Oklahoma, and Texas, south to Mexico; 2 (0). *Panicum lepidulum* Hitchc. & Chase, reported from Utah by Hitchcock (1951), was originally separated from *P. hallii* as having "more evenly flowered," narrower panicles with larger spikelets, and a "greater amount of pubescence" (Hitchcock and Chase 1910). Narrow panicles are not, however, necessarily associated with larger spikelets (Gould 1951). Other small quantitative distinctions in vegetative features noted by Hitchcock (1951) are bridged by Gould's (1975) description of *P. hallii* as it occurs in Texas. Cronquist et al. (1977) noted the weak nature of the distinction between the taxa and did not include *P. lepidulum* as part of the Intermountain Flora.

Panicum miliaceum L. Broomcorn Panicum; Hog Millet. Annual; culms 2–10 (12) dm tall, erect or decumbent-ascending, simple or branching from the base or above; sheaths loose, pubescent with long, usually papilla-based hairs; blades flat, 6–15 (20) mm wide, variously hairy to subglabrous, the margins often undulate; ligule an obscurely membranous-based ring of hairs 1–3 mm long; panicle more or less compact, mostly narrowly oblong, 5–30 cm long, emerging tardily from the upper sheaths, the branches erect-appressed and usually drooping near the tips, the spikelets short-pedicelled and crowded along the branches; spikelets plano-convex, broadly ovate, 4–6 mm long; glumes glabrous, the first 2.5–4.7 mm long, 1/2–3/4 as long as the spikelet, acute or acuminate apically, the second glume and sterile membranous lemma about equal, acute or acuminate apically, typically distinctly longer than the glossy fertile lemma; 2n = 18, 36, 40, 42, 49, 54, 72. Escaping from cultivation and occurring as a weed of gardens and waste places mostly below

1830 m in Cache, Salt Lake, Sanpete, Utah, and Washington counties; native to China and central Asia, cultivated in temperate regions of the world since prehistoric times either as a cereal or for fodder, escaped from cultivation in North America and reported from parts of Canada and throughout the western and northeastern U. S.; South America; Eurasia; 12 (iii). Broomcorn panicum is currently used in the U. S. as feed grain, in birdseed mixtures, and as a food crop for game birds.

Panicum obtusum **H.B.K.** Vine Mesquite. Perennial from a base with stolons to ca 2 m long, these with crowded, swollen, densely hairy nodes; herbage glaucous; culms mostly 2–6 (8) dm tall, erect or decumbent-based, simple or branched above the base; sheaths glabrous or coarsely hairy; blades firm, elongate, 2–7 mm wide, flat to involute, often scabrous, usually with a few long hairs near the ligule; ligule membranous, not ciliate, 0.5–2 mm long; panicle narrow, to 15 cm long, the branches 2–6, spikelike and secund, racemosely arranged along the main axis, mostly 1–5 cm long, erect to slightly spreading; spikelets sessile to short-pedicelled, dorsiventrally to somewhat laterally compressed or subterete, narrowly oblong to obovate, 2.8–4 mm long and 1.5–2.3 mm wide; glumes subequal, 2.2–4.1 mm long, 5- to 7-nerved, acute to rounded apically, glabrous, the second glume and sterile membranous lemma about equal, the two about as long as or barely exceeding the glossy fertile lemma; 2n = 20, 36, 40. First reported from Utah in 1941 (Harrison 10391 BRY), generally in sandy sites along waterways, on open hillsides, and along sandstone ledges from 1010 to 1830 m in Garfield, Grand, Kane, San Juan, Washington, and Wayne counties; Arizona east to Missouri, Arkansas, and Texas, and to southern Mexico; 20 (0).

Panicum oligosanthes **Schultes** Scribner Panicum. [*Dicanthelium oligosanthes* (Schultes) Gould; *P. scribnerianum* Nash]. Perennial; culms 1.5–8.5 dm tall, loosely to densely tufted, unbranched during initial culm growth, the autumnal culms freely branched above the base, the ultimate branchlets forming short, dense, leafy fascicles; sheaths glabrous or more often conspicuously pubescent with papilla-based hairs; blades flat, glabrous to short-hairy or (outside Utah) tomentose, initially chiefly cauline, 3–15 mm wide and 4–14 cm long, the blades of the autumnal fascicled branchlets to ca 7 cm long, a small basal rosette of blades usually appearing in late summer or early fall; ligule of hairs 0.5–4.2 mm long; panicles open, terminal and axillary, the latter developing after the terminal ones have flowered; terminal panicles typically pyramidal, 3.5–9 cm long, the flexuous branches ascending to spreading, the pedicels shorter to much longer than the spikelets, the axillary panicles smaller than the terminal ones and usually partially included in the sheaths; spikelets plano-convex, ovate to obovate-elliptical, 2.8–4 mm long; glumes glabrous or pubescent, the first to about half as long as the spikelet, acute to obtuse or occasionally toothed apically, the second glume usually slightly shorter than the sterile lemma and the glossy fertile one; 2n = 18. Moist sandy soil or rock crevices in hanging garden and ponderosa pine communities at 850 to 2130 m in Garfield and Washington counties; southern Canada, south through most of the U. S. and in Mexico; 3 (0). According to Gould and Clark (1978), the only phase of the species west of Colorado is var. **scribnerianum** (**Nash**) **Fern.** The variety differs from the typical element of the species in having a ligule less than 1.6 mm long and the abaxial leaf surface glabrous to long-hairy, not tomentose.

Panicum virgatum **L.** Switchgrass. Perennial with stout, scaly rhizomes; herbage green or more often glaucous; culms 6–20 (30) dm tall, erect, simple or occasionally branched above the base; sheaths glabrous throughout or ciliate along the upper margin; blades elongate, flat, 3–15 mm wide, glabrous or scabrous to occassionally long-hairy; ligule a membranous-based ring of hairs (1) 1.5–5 mm long; panicle open, mostly 10–50 cm long, the mostly few-flowered branches often elongate, ascending to more or less widely spreading, the spikelets borne on slender pedicels and scattered along or clustered near the tips of primary or slender secondary branches; spikelets dorsiventrally to somewhat laterally compressed, more or less ovate, mostly 3–5 mm long; glumes glabrous, the first clasping, broadly ovate, 2–4 mm long, 2/3–3/4 as long as the spikelet, sharply acute to acuminate at the tip, the second glume 3.5–5 mm long, with an erect to spreading, acuminate tip that often slightly exceeds the similarly shaped, membranous, sterile lemma, the glume and sterile lemma distinctly longer than the glossy fertile lemma; 2n = 18, 21, 25, 30, 32, 36, 54–65, 70, 72, 90, 108. Chiefly in moist, sandy, or rocky places along waterways and in hanging gardens, rarely in meadows, pinyon-juniper communities, or on dry hillsides, below 1830 m in Emery, Garfield, Grand, Kane, San Juan, Utah, and Wayne counties; southern Canada from Saskatchewan to Nova Scotia, south through much of the U. S., and in Mexico; introduced in Asia; 34 (i). Switchgrass is an important grass of the central and southern Great Plains, valuable as forage and pasture and for erosion control. It has been introduced in central Asia for fodder and as an ornamental.

Paspalum L.

Annuals or perennials; culms usually tufted or branched from the base; sheaths open; blades flat or folded, often broad; ligule membranous or a ring of short hairs; inflorescence a panicle of 1–many, spikelike, secund branches, these scattered along the axis or paired at the apex of the culm, the spikelets subsessile or short-pedicelled, solitary or paired, in 1 or 2 rows from one side of a flattened to 3-angled, often winged axis; spikelets orbicular to oblong, flattened to plano-convex, disarticulating below the glumes, 2-flowered, the lower floret staminate or neuter, the upper one perfect; first glume very short or lacking (the spikelet then appearing to have 2 glumes and 1 perfect floret); second glume and the sterile lemma of the lower floret membranous, usually equal or nearly so, closely sheathing the fertile floret; lemma of the perfect floret dorsiventrally compressed, leathery, glossy, minutely striate; palea leathery like the lemma, partially enclosed by the incurved, firm lemma margins; stamens 3; caryopsis free of but firmly enclosed within the lemma and palea, the embryo at least half the length of the grain; x = 10.

Paspalum distichum **L.** Knotgrass. [*P. distichum* var. *indutum* Shinners]. Stoloniferous perennial; culms 2–10 dm tall, erect to decumbent-ascending, the nodes glabrous or pubescent; sheaths essentially glabrous or more often ciliate on the margins and otherwise pubescent to some degree with long, spreading, mostly papilla-based hairs; blades flat or folded, 1.5–6 (7) mm wide, scabrous on the margins and often sparsely pubescent with papilla-based hairs; ligule membranous, 1–2.5 mm long; panicle of paired or closely spaced, spikelike branches 1.5–7 cm long, occasionally with 1 or 2 branches below the terminal pair, the axis of the branch flat and narrow with 2 rows of solitary or paired, subsessile to short-pedicelled spikelets arising from one side, one row on either side of the midrib; spikelets elliptical to narrowly ovate, plano-convex, 2.5–4 mm long, 1.2–2 mm wide, pale or rarely purple-tinged; first glume lacking or present in variable form on some of the spikelets; second glume thin, broadly elliptical, about equal to the spikelet or nearly so, rounded on the back, 3– or obscurely 5-nerved, acute, minutely crisp-hairy; sterile lemma slightly firmer than the second glume and essentially flat, barely exceeding the fertile lemma, 3 (5) –nerved, acute, glabrous or puberulent; fertile lemma leathery, dorsiventrally compressed, pale, glossy, finely striate; anthers 1.3–1.8 mm long; $2n = 20, 30, 40, 48, 60$. Saline meadows, along waterways, and in desert shrub communities below 1520 m in Utah and Washington counties; most of the western U. S., throughout the South, and into Mexico; South America; southern Eurasia; 22 (0). The status of the name *P. distichum* has recently been the subject of much debate, the consensus appearing to lie with the view expressed by Fosberg (Taxon 26: 201–202. 1977) that the Linnaean type of *P. distichum* is the plant generally known as *P. vaginatum* Swartz. A proposal to reject the name *P. distichum* L. was, however, declined with the recommendation that the traditional use of the name be retained (Taxon 32: 281. 1983). Allred (Great Basin Naturalist 42: 103–104. 1982) finds no basis for bestowing taxonomic recognition upon the pubescent forms of *P. distichum* referred by Shinners to var. *indutum*.

Phalaris L.

Annuals or perennials; sheaths open, often with short auricles; blades flat; ligules membranous; inflorescence an ovoid to narrowly elongate, usually compact, sometimes spikelike panicle; spikelets usually 3–flowered, the terminal floret perfect, the lower 1 or 2 sterile, generally greatly reduced, occasionally lacking, the rachilla disarticulating above the glumes, the florets falling together as a unit; glumes laterally compressed, equal or subequal, longer than the terminal floret, sometimes winged on the keel; lemmas of the lower florets lacking or rudimentary and usually pubescent, the lemma of the perfect floret leathery, often glossy, laterally compressed; palea membranous, shorter and narrower than the lemma; stamens 3; caryopsis free of but at maturity usually firmly enclosed by the lemma and palea, the embryo 1/4–1/2 as long as the grain; $x = (6) 7$.

Anderson, D. E. 1961. Taxonomy and distribution of the genus *Phalaris*. Iowa State J. Sci. 36: 1–96.

1. Plants annual; spikelets obovate; glumes broadly winged .... *P. canariensis*
— Plants perennial and rhizomatous; spikelets lanceolate to elliptical; glumes wingless or rarely slightly winged on the keel .... *P. arundinacea*

Phalaris arundinacea L. Reed Canarygrass; Ribbongrass. Rhizomatous perennial; culms erect, 5–20 dm tall; sheaths without auricles; blades flat, 6–25 mm wide, scabrous throughout to essentially smooth except near the long-tapered tips; ligule 2–10 mm long; panicle narrowly elongate, sometimes more or less lobed near the base or throughout, more or less open at flowering, (5) 7–30 (40) cm long, the branches strictly erect or at flowering divergent to widely spreading; spikelets lanceolate to elliptical, pale green or sometimes tinged with pinkish purple; glumes subequal, narrowly lanceolate, (3.5) 4–7.5 mm long, strongly keeled, rarely slightly winged on the keel, long-tapered to sharply acute tips, more or less prominently 3–nerved, glabrous or scaberulous, exceeding the solitary perfect floret; sterile lower lemmas 2, reduced to linear, hairy scales 1–2.3 mm long, these closely subtending the perfect floret (often appearing to be pubescence on the fertile lemma); lemma of the perfect floret leathery and glossy, lanceolate, 2.7–4.5 mm long, pale green to yellowish, at maturity straw colored to dark gray brown, glabrous or sparsely appressed-pubescent, ultimately enclosing the similar but smaller palea; anthers 2.5–3.5 mm long; $2n = 14, 27–42, 56$. Along waterways and in wet meadows chiefly at 1290 to 2745 m in Box Elder, Cache, Carbon, Daggett, Davis, Duchesne, Emery, Morgan, Rich, Salt Lake, Sanpete, Sevier, Summit, Uintah, Utah, Wasatch, Wayne, and Weber counties; circumboreal, south through the U. S. (except in the South) and in northern Mexico; South America; Eurasia; 78 (viii). The ribbon grass of gardens is *f. variegata* (**Parnell**) Druce [var. *picta* L.], differing from the typical element of the species in having white-striped leaves. The cultivar rarely escapes but was found growing along a stream in Park City, Utah. Reed canarygrass provides good forage prior to maturity and is useful in controlling erosion in wet areas. Large populations of the grass can, however, greatly reduce the carrying capacity of irrigation and drainage ditches (Thornburg 1982).

Phalaris canariensis L. Canary Grass. Annual; culms 2–6 (10) dm tall; sheath of uppermost culm leaf inflated, sometimes bearing minute auricles; blades flat, 3–10 mm wide; ligule 3–8 mm long; panicle ovate to oblong, 1.5–4 cm long, to 2 cm wide, compact; spikelets short-pedicelled, obovate, closely overlapping; glumes subequal, 5–10 mm long, abruptly acute, pale with dark green lateral stripes, broadly winged on the keel, glabrous or sparsely appressed-pubescent; sterile lower lemmas 2, linear to narrowly lanceolate, green or chaffy, 2.5–5 mm long, sparsely hairy, one appressed to each side of the perfect floret; lemma of the perfect floret firm and glossy, lanceolate, 4–6.8 mm long, appressed-hairy; $2n = 12$. Moist to mesic, disturbed sites below 1850 m, reported from Salt Lake and San Juan counties; native to northwest Africa and the Canary Islands, a casual weed throughout most of the U. S. and in parts of Mexico; South America; Eurasia; 4 (ii). Canary grass is widely cultivated as a component of birdseed.

Phleum L.

Annuals or perennials; culms solitary or tufted; sheaths open, with or without minute rounded auricles; blades usually flat; ligule membranous; inflorescence a dense, spikelike panicle, the branches very short, often fused with the rachis all along their length (the spikelets then appearing sessile); spikelets 1–flowered, disarticulating above the glumes and at maturity below; glumes laterally compressed, keeled, equal or nearly so, longer than the lemma, gradually tapered to an acute tip or (ours) obliquely truncate and mucronate to short-awned, mostly 3-nerved, usually ciliate on the prominent keel; lemma laterally compressed, ovate, mostly 3- to 7-nerved; palea equal to the lemma or nearly so; stamens 3; caryopsis free within the lemma and palea, the embryo to about half the length of the grain; $x = 7$.

1. Sheath of uppermost leaf typically inflated; panicles to 4(5) cm long; awns usually averaging more than 1.5 (1.5-4) mm long; culms 1-5 (6) dm tall, not bulblike at the base .... *P. alpinum*
— Sheath of uppermost leaf not regularly inflated; panicles often more than 4 (2-30) cm long; awns usually averaging no more than 1.5 (0.6-2) mm long; culms 2-15 dm tall, typically bulblike at the base .... *P. pratense*

Phleum alpinum L. Alpine Timothy. Perennial; culms solitary or in small loose tufts, 1-5 (6) dm tall, erect to decumbent, not bulblike at the base; sheaths glabrous, with or without small rounded auricles, sheath of the uppermost culm leaf inflated; blades flat, 2-7 (8) mm wide, subglabrous to scabrous; ligule 0.5-4.5 mm long; panicle ovoid to cylindrical, mostly 1.5-5 cm long and to about 12 mm wide when pressed, often purple; spikelets 1-flowered, oblong, 3.5-8 mm long including awns; glumes equal or nearly so, broadly membranous-margined, the apex obliquely truncate with the midnerve prolonged as an awn 1.5-4 mm long, the body of the glume generally 2-4 mm long, the keel ciliate with mostly spreading hairs to ca 1 mm long; lemma membranous, to ca 2/3 as long as the glumes, truncate-erose at the apex; anthers 1-2 mm long; $2n = 14, 28$. Chiefly streamside, in mesic to wet meadows, and in openings in aspen-spruce-fir and lodgepole pine communities, occasionally in grass-sagebrush and sagebrush-aspen ecotones, and on exposed rocky ridges at high elevations, mostly from 2135 to 3660 m in Beaver, Box Elder, Cache, Carbon, Daggett, Duchesne, Emery, Garfield, Grand, Iron, Juab, Piute, Rich, Salt Lake, San Juan, Sanpete, Sevier, Summit, Tooele, Uintah, and Utah counties; Alaska to Alberta, south throughout the western U. S. and into Mexico, east to northern Michigan and the mountains of New England; South America; Eurasia; 192 (v).

Phleum pratense L. Timothy. Perennial; culms loosely to densely tufted, 2-15 dm tall, erect, typically with 1 or 2 (3) bulblike nodes at the base; sheaths glabrous, with or without small rounded auricles, none regularly inflated; blades flat, 3-8 (10) mm wide, at least the margins scabrous; ligule 1-6 mm long; panicle cylindrical, (2) 3-20 (30) cm long and 5-8 (12) mm wide when pressed, occasionally purple-tinged; spikelets 1-flowered, oblong, 2-5.2 mm long including awns; glumes equal or nearly so, broadly membranous-margined, the apex obliquely truncate with the midnerve prolonged as an awn 0.6-2 mm long, the body of the glume generally 2-3.2 mm long, the keel ciliate with mostly spreading hairs to about 1 mm long; lemma membranous, 2/3-3/4 as long as the glumes, truncate-erose at the apex; anthers 1.2-2.3 mm long; $2n = 14, 21, 28-84$. Roadsides and along waterways, in dry to wet meadows, in aspen-spruce-fir, and occasionally in oak-sagebrush, pinyon-juniper, mountain brush, and ponderosa pine communities, at 1310 to 3200 m in all Utah counties except Wayne; native to Eurasia, now circumboreal, south throughout the U. S.; South America; 134 (iv). Timothy, introduced to America from Europe prior to 1747 (Vallentine 1961) because of its palatability as a hay and pasture grass, is now naturalized in most of the temperate regions of the world. It is both cold and shade tolerant, but is vulnerable to drought, high temperatures, and close grazing.

Phragmites Adanson

Tall, coarse perennials with rhizomes or stolons; sheaths open; blades broad, usually flat; ligule membranous or a ring of hairs, the latter sometimes membranous-based; inflorescence a large, dense to somewhat lax panicle; spikelets subterete to laterally compressed, (1) 3- to 10-flowered, the rachilla densely long-hairy, disarticulating above the glumes and between the florets; florets successively smaller upward, the lowermost 1 or 2 staminate or neuter, the uppermost 1 or 2 rudimentary, the remainder pistillate or perfect, occasionally all spikelets abortive; glumes unequal, the first 1- to 3-nerved, the second 3 (5)-nerved, shorter than the lowermost floret; lemma 3-nerved; palea to about half as long as the lemma; stamens 1-3, the number variable within the spikelet; caryopsis free of and only loosely enclosed by the lemma and palea, the embryo 1/2-2/3 as long as the grain, the grain rarely reaching maturity; $x = 12$.

Clayton, W. D. 1968. The correct name for the common reed. Taxon 17: 168-169.

Phragmites australis (Cav.) **Trin. ex Steudel** Common Reed. [*Arundo australis* Cav.; *P. communis* Trin.]. Rhizomatous and often stoloniferous perennial; culms mostly 2-4 m tall, to 2 cm or more in diameter, usually unbranched; blades flat, to 5 dm long and mostly 1-5 cm wide, strongly scabrous on the margins, otherwise glabrous, long-tapered at the apex, readily breaking off at the base in age due to an abscission layer in the vicinity of the ligule; ligule membranous, minutely ciliate, 0.5-3 mm long; panicle plumelike, oblong to obovoid, mostly 1.5-5 dm long and to about 2 dm wide, densely flowered, initially gray purple, becoming whitish as spikelets mature; spikelets narrowly lanceolate, 10-16 mm long, (1) 2- to 10-flowered, (occasionally entire panicles with all florets abortive), the rachilla with spreading silky hairs 6-11 mm long; glumes thin, elliptical to lanceolate, unequal, the first 3-7 mm long, the second 5.5-10 mm long, shorter than the lowermost floret; lemma thin, linear-lanceolate, the lowermost (8) 9-15 mm long, those above progressively smaller, the uppermost seldom exceeding the lowermost; $2n = 36, 44, 46, 48, 49-52, 54, 72, 84, 96$. Along waterways and in saline or freshwater marshes at 760 to 1980 m, known from all Utah counties except Beaver, Iron, Morgan, Piute, Summit, and Wasatch; circumboreal, south throughout the U. S. but less frequent in the southeast and in Mexico; South America; Eurasia; 83 (ii). *Arundo donax*, which rarely escapes from cultivation in Utah and is similar in general aspect to *P. australis*, can readily be distinguished from the latter by its hairy lemmas and glabrous rachilla. The common reed, the most widely distributed of all angiosperms (Good, Geography of Flowering Plants. 1974), is currently being used for erosion control of shore lines (Thornburg 1982).

Poa L.

Annuals or perennials, tufted or rhizomatous, occasionally stoloniferous, some species dioecious or gynomonoecious, one typically viviparous; culms usually erect;

sheaths persistent or shredding, open throughout or closed nearly to the summit; blades flat or folded to involute, mostly prow-shaped at the tips, in some species obscurely so; ligule membranous, glabrous or the abaxial surface scabrous to pubescent; inflorescence an open to compact, erect to nodding panicle; spikelets laterally compressed to less often terete or nearly so, 2– to 12 (15) –flowered, disarticulating above the glumes and between the florets; glumes subequal to unequal, lanceolate to ovate or elliptical, keeled to rounded on the back, the first glume 1– to 3–nerved, the second 3– to 5–nerved, equal to or in a few species slightly exceeding the lowermost lemma; lemma ovate to narrowly lanceolate, keeled to rounded on the back, with 3–5 (7) obscure to conspicuous nerves converging toward the rounded to acute but awnless apex; callus in some species with a tuft of mostly crumpled, cobweblike hairs, these usually longer than any on the body of the lemma; palea glabrous or scabrous to variously pubescent; palea shorter than the lemma; stamens 3; caryopsis free of or somewhat adherent to the lemma and palea, ellipsoidal, often somewhat compressed ventrally, the embryo 1/6–1/3 as long as the grain; x = 7. The frequency of apomixis in species of *Poa* has been viewed as contributing to problems in classification within what is considered a "difficult" genus. Of the species growing in our area, apomixis is known to occur in: *Poa arctica* (Engelbert 1940), *P. alpina*, *P. bulbosa*, *P. compressa*, *P. glauca*, *P. nervosa*, *P. palustris*, *P. pratensis* (Nygren 1954), *P. secunda* (Nygren 1951) and possibly also in *P. fendleriana* (Hitchcock et al. 1969). Although Kellogg (1985) found no correlation between morphological complexity and apomixis in the large and variable *P. secunda* complex, a number of apomictic species are pseudogamous, producing unfertilized seed only after pollination (Nygren 1954, Engelbert 1940, Kellogg 1985). As suggested by Nygren, this need for pollination may lead to hybridization among related species in natural populations. Knobloch (1963), in fact, cited 82 known interspecific hybrids for the genus *Poa*, by far the largest number encountered in all but one of the 64 genera reported upon by him. Evidence of hybridization in our material exists in the field as well as in numerous herbarium specimens, especially among plants growing at high elevations. Intergradation, if it does exist on a large scale, would clearly help to explain the complexity within the genus. Hybrids usually retain sufficient characters of one parent to be classified as such, but vary in one or two characters by which the taxon is generally recognizable. Awareness of this potential for variation should be useful to anyone attempting to identify species of *Poa*. Soreng (1985) reports *Poa stenantha* Trin. as rare in northern Utah, but I have seen no evidence that populations with the characters generally attributed to this little understood taxon actually occur south of Idaho. This treatment of *Poa* was reviewed by Elizabeth Kellogg.

Akerberg, E. 1942. Cytogenetic studies in *Poa pratensis* and its hybrid with *Poa alpina*. Hereditas 28: 1–126.
Clausen, J. 1961. Introgression facilitated by apomixis in polyploid poas. Euphytica 10: 87–94.
Engelbert, V. 1940. Reproduction in some *Poa* species. Canad. J. Res. Sect. C-D. 18: 518–521.
Halperin, M. 1933. The taxonomy and morphology of bulbous bluegrass, *Poa bulbosa vivipara*. J. Amer. Soc. Agron. 25: 408–413.
Kellogg, E. 1985. A biosystematic study of the *Poa secunda* complex. J. Arnold Arbor. 66: 201–242.
Marsh, V. L. 1952. A taxonomic revision of the genus *Poa* of United States and southern Canada. Amer. Midl. Naturalist 47: 202–250.

Nygren, A. 1951. Embryology of *Poa*. Carnegie Inst. Wash. Year Book 50: 113–115.
_____. 1954. Apomixis in the angiosperms II. Bot. Rev. 20: 577–621.
Soreng, R. J. 1985. *Poa* L. in New Mexico, with a key to middle and southern Rocky Mountain species (Poaceae). Great Basin Naturalist 45: 395–422.

1. Culms bulbous at the base; florets proliferated into bulblets with leafy shoots, occasionally some, rarely all the florets floriferous .... *P. bulbosa*
— Culms not bulbous-based; florets not producing bulblets 2

2(1). Lemmas averaging no more than 3 (1.8–3.1) mm long, typically firm throughout and green (rarely purple-tinged) below a distinct terminal band of bronze or purple (rarely the tip pale), obscurely nerved, the lateral margins soon curving around the palea; panicle branches erect to spreading, not deflexed 3
— Lemmas averaging more than 3 mm long or not otherwise as above 4

3(2). Panicles 10–35 cm long, at maturity often more than 3 cm wide, the branches naked below; ligule (1.8) 2–6 mm long; plants 1.5–13 dm tall, rarely appearing short-rhizomatous .... *P. palustris*
— Panicles often less than 10 (2–14) cm long, rarely more than 3 cm wide, many or all of the branches spikelet-bearing to the base or nearly so; ligule 0.5–2 mm long; plants 1–6 (7) dm tall, strongly rhizomatous .... *P. compressa*

4(2). Lemmas with a tuft of crumpled hairs arising from the callus, i.e., webbed at the base 5
— Lemmas not webbed at the base 13

5(4). Plants annual; lemmas either glabrous except for the web or densely pubescent to near the apex; plants from below 2450 m 6
— Plants perennial, not otherwise as above 7

6(5). Lemmas densely pubescent to near the apex on keel and marginal nerve and usually to some degree between; panicle branches to 6 cm long, at maturity erect to slightly divergent .... *P. bigelovii*
— Lemmas glabrous or scaberulous; panicle branches few and remote, linear, to 11 cm long, ultimately stiffly spreading to reflexed .... *P. bolanderi*

7(6). Panicle spikelike, 1–5 cm long and 1 cm wide; second glume averaging more than 3.5 (3.5–5) mm long; glumes and lemmas strongly keeled; plants mostly less than 1.5 dm tall and occurring above 3050 m .. .. *P. pattersonii*
— Panicle not spikelike, somtimes small and narrow but then not otherwise as above 8

8(7). Body of the lemma glabrous or sparsely pubescent only on the prominent keel, typically conspicuously 5–nerved; first glume often crescent-shaped; ligules 3–10 mm long; plants mostly more than 3 (3–10) dm tall and occurring below 2440 m .... *P. trivialis*
— Body of the lemma hairy on keel and marginal nerves or, if glabrous, the plants not otherwise as above 9

9(8). Glumes mostly incurved at the sharply acute tips and generally scabrous (at 10x magnification) on the upper keel, the second glume usually angled on the margin; lemmas to 3.5 (4) mm long, typically conspicuously 5–nerved, copiously webbed, the hairs (when extended) longer than the lemma; plants strongly rhizomatous, common from low to high elevations .... *P. pratensis*

— Glumes not both incurved at the tips and conspicuously scabrous on the keel; lemmas often more than 4 mm long or only sparsely webbed; plants either tufted or rarely occurring below 2440 m 10

10(9). Anthers averaging less than 1 (0.2–1) mm long; panicle axis usually nodding, at maturity the branches widely spreading to deflexed 11

— Anthers averaging at least 1 (1–2.5) mm long; panicle axis erect, the branches erect to spreading, in *P. arctica* occasionally deflexed 12

11(10). First glume typically long-tapered to a nearly awnlike tip; lemmas 3–5 mm long, glabrous between keel and marginal nerves; palea generally ciliate on the keels with minute stiff hairs .... *P. leptocoma*

— First glume typically abruptly acute or tapered but not at all awnlike at the tip; lemmas averaging no more than 3.5 (2–3.5) mm long, at least some of them sparsely hairy between keel and marginal nerves; palea ciliate on the keels with crinkled hairs, sometimes obscurely so .... *P. reflexa*

12(10). Lemmas often rounded on the back, averaging more than 3.5 (3.4–6) mm long, typically densely pubescent (often to above midlength) on midnerve and marginal nerves and conspicuously so between; ligules 1–4 mm long; plants with short to well-developed rhizomes .... *P. arctica*

— Lemmas mostly keeled, averaging no more than 3.5 (2–4) mm long, densely to sparsely pubescent on keel and marginal nerves, glabrous or sparsely hairy between; ligules 0.5–2 (2.5) mm long; plants densely tufted, occasionally rooting at subsurface nodes ...
.... *P. glauca*

13(4). Florets pistillate, the anthers rudimentary 14

— Florets perfect or staminate, the anthers sometimes small or sterile but not colorless 16

14(13). Panicle internodes elongate, the lowermost averaging at least 3 (1.5–13) cm long, at flowering most of the branches horizontally spreading to deflexed; sheaths glabrous or scaberulous; plants usually bearing both perfect and pistillate florets .... *P. curta*

— Panicle internodes variable, the lowermost averaging less than 3 (1–4) cm long, the branches erect to spreading or occasionally the lower ones drooping; sheaths glabrous or hairy; plants usually bearing only pistillate florets 15

15(14). Panicles at flowering typically open; sheaths (at least some of them) retrorsely puberulent, sometimes obscurely so, the sheath bases from previous years often turning purple or dark brown and ultimately shredding in age, not usually a conspicuous feature; plants generally rhizomatous .... *P. nervosa*

— Panicles typically narrowly oblong and more or less compact; sheaths glabrous or scabrous, occasionally scabrid-puberulent, the sheath bases from previous years typically straw colored or gray and papery (not disintegrating into fibers), often a conspicuous feature; plants densely tufted, rarely with a few short rhizomes per tuft .... *P. fendleriana*

16(13). Spikelets broadly rounded at the base and averaging at least 2 mm wide at midlength (florets nondivergent) 17

— Spikelets narrowly rounded to wedge-shaped at the base, usually averaging less than 2 mm wide (florets nondivergent) 18

17(16). Panicles often pyramidal, 2–6 (9) cm long; second glume to 3 (3.5) mm long, usually incurved at the tip and scabrous on the upper portion of the keel; basal leaf blades to 8 (10) cm long and at least some of them 2 mm or more wide; plants above 2440 m .... *P. alpina*

— Panicles mostly narrowly oblong, 2–15 cm long; second glume 3–5.3 mm long, not usually incurved at the tip, glabrous or obscurely scaberulous on the keel; basal leaf blades often more than 10 cm long and less than 2 mm wide; plants of low to high elevations
.... *P. fendleriana*

18(16). Second glumes typically angled on the margin; lemma generally pubescent on all 5 nerves (rarely entirely glabrous); panicles usually short-pyramidal, the branches smooth or nearly so; plants annual or occasionally perennial and mat-forming ..... *P. annua*

— Second glume not angled on the margin or the plants not otherwise as above in every particular 19

19(18). Anthers averaging less than 1 (0.3–1) mm long; panicles to 5 cm long and sublinear to oblong or spikelike; low plants occurring at or above 3050 m 20

— Anthers averaging at least 1 (1–4) mm long; panicles often more than 5 cm long; plants otherwise diverse . 21

20(19). Lemmas glabrous (rarely pubescent), 2–3.5 mm long; glumes mostly rounded on the back and appearing somewhat inflated .... *P. lettermanii*

— Lemmas pubescent to some degree, (3) 3.5–5 mm long; glumes strongly compressed and sharply keeled
.... *P. pattersonii*

21(19). Sheaths (at least some) retrorsely puberulent; panicles open; plants usually rhizomatous ...... *P. nervosa*

— Sheaths glabrous or scabrous, rarely scabrid-puberulent and then panicles compact and plants densely tufted 22

22(21). Lemmas hairy (often to above midlength) on keel and marginal nerves and typically conspicuously so between, averaging more than 3.5 (3.4–6) mm long; panicles loose, the branches naked below and smooth or nearly so; plants rhizomatous, restricted to high elevations .... *P. arctica*

— Lemma not hairy as above or, if so, either the lemmas averaging less than 3.5 mm long or the plants densely tufted, otherwise diverse 23

23(22). Lowermost internode of the panicle averaging at least 3 (1.5–13) cm long, most of the branches horizontally spreading or deflexed by flowering .... *P. curta*

— Lowermost internode of the panicle averaging less than 3 cm long, those of individual plants rarely to 4 cm long and then most of the branches erect to spreading, rarely the lowermost drooping 24

24(23). Lemmas averaging no more than 3.5 (2–4) mm long; ligules 0.5–2.5 mm long; anthers 1–2 mm long ....
.... *P. glauca*

— Lemmas averaging more than 3.5 (3–7) mm long; ligules 0.5–12 mm long; anthers 1.2–4 mm long 25

25(24). Spikelets subterete or occasionally compressed, and then the upper florets very narrow and often terete or nearly so; panicles nearly always narrowly elongate and loose; lemmas glabrous or more often pubescent with minute hairs more or less evenly distributed over the lower portion .... *P. secunda*

— Spikelets compressed with all florets more or less strongly keeled; panicles open to compact; lemmas glabrous or pubescent with hairs restricted to (or most pronounced on) the keel and marginal nerves .. 26

26(25). Panicles open, the branches spreading, mostly naked below; glumes and lemmas typically narrowly membranous-margined; sheath bases from previous years typically brown and disintegrating into fibers, not usually a conspicuous feature; plants generally rhizomatous .... *P. nervosa*

— Panicles mostly oblong and densely flowered, the short branches erect to ascending, mostly spikelet-bearing to the base; glumes and lemmas typically broadly membranous-margined; sheath bases from previous years papery (not fibrillose), straw colored or gray, usually a conspicuous feature; plants densely tufted, rarely with a few short rhizomes per tuft ...
.... *P. fendleriana*

Poa alpina L. Alpine Bluegrass. Perennial; culms tufted, erect or more or less flexuous, 0.4–4 (6) dm tall, each arising from a short, thickened caudex, the latter coalesced beneath the surface of the ground; leaves chiefly basal; sheaths open to the base or nearly so, glabrous or occasionally the innermost membranous sheaths minutely hairy; blades (1) 2–4.5 (5) mm wide and to 8 (10) cm long, flat or folded, gradually to abruptly tapered to a prow-shaped tip; ligule 1–4 (5) mm long, truncate to obtuse, glabrous abaxially; panicle oblong to more often pyramidal, compact to open, 2–6 (9) cm long, the branches mostly 1 or 2 per node, 0.5–3 (4) cm long, naked below, smooth or some of them scaberulous, ascending to widely spreading or sometimes reflexed; spikelets strongly compressed, broadly lanceolate to ovate, (3) 4–6 (7) mm long, rounded to subcordate at the base, mostly 2–3 mm wide (florets nondivergent), generally purple-tinged to some degree, 3– to 6 (9) –flowered, the florets perfect or occasionally viviparous; glumes typically ovate, 1.5–3 (3.5) mm long, sharply keeled, acute at the mostly incurved tips, the keel scabrous, at least above, the second glume rounded or angled on the margins, shorter than to nearly as long as the lowermost lemma; lemma 3–5 mm long, usually sharply keeled, abruptly acute to rounded at the broadly membranous, mostly incurved tip, generally obscurely nerved, sparsely to densely pubescent on the keel and marginal nerves and usually sparsely hairy between, not webbed at the base; palea minutely stiff-ciliate on the upper portion of the keels, typically with longer hairs at or below midlength; anthers (1) 1.2–2 mm long; 2n = 14, 21–74. Meadows, along streams and lake margins, in spruce-fir and lodgepole pine communities, and on exposed rocky slopes above timberline, at 2440 to 3960 m in Daggett, Duchesne, Salt Lake, Sanpete, Summit, Uintah, and Utah counties; circumboreal, south in North America to Oregon, Idaho, Wyoming, Colorado, and Michigan; Eurasia; 67(vii). *Poa alpina* is known to form hybrids with *P. pratensis* (Akerberg 1942, Clausen 1961).

Poa annua L. Annual Bluegrass; Winter Bluegrass. Annual or short-lived perennial, rooting at subsurface nodes and often forming dense mats; herbage bright green; culms erect or decumbent, terete or compressed, 0.2–3 (4) dm tall; sheaths often compressed, open 1/2–3/4 the length, glabrous; blades 0.8–3 (4) mm wide, flat or folded, acute to distinctly prow-shaped at the tips, occasionally scaberulous on the margins, otherwise glabrous, often transversely wrinkled, those of the uppermost culm leaves 0.5–8 cm long; ligule 0.7–4 (5) mm long, truncate to rounded, glabrous abaxially; panicle typically pyramidal, (1) 2–6 (8) cm long, the branches 1 or 2 at each node, to ca 3 (5) cm long, naked below, smooth or nearly so, ultimately widely spreading to deflexed, the lowermost internode of the rachis rarely as much as 2 cm long; spikelets compressed, ovate to lance-elliptical, 2.5–7 (10) mm long, rarely purple-tinged, (2) 3– to 6 (10) –flowered, the florets perfect; glumes moderately to sharply keeled, the keels glabrous, the first glume narrowly lanceolate, 1.3–2.5 (3) mm long, the second mostly diamond-shaped and angled on the margin, 2–3 (4) mm long, shorter than the lowermost lemma; lemma 2–3 (4) mm long, pale membranous at the acute or more often obtuse or rounded tip, 5-nerved, sometimes obscurely so, sparsely to densely appressed-hairy on all the nerves and glabrous between or rarely completely glabrous, not webbed at the base; palea typically ciliate on the keels with minute soft hairs; anthers 0.3–1.2 mm long; 2n = 24–26, 28, 52. An invader of lawns and moist, often waste places, occasionally in salt desert shrub, sagebrush, oak-maple, aspen-fir, lodgepole pine, and meadow communities, at 1220 to 2960 m in Beaver, Box Elder, Cache, Daggett, Davis, Duchesne, Emery, Juab, Millard, Piute, Salt Lake, Tooele, Uintah, Utah, Wasatch, and Washington counties; native to Eurasia and western North America, now cosmopolitan (Clausen 1961); 67 (v). In our area this plant is an undesirable invader of lawns, turning yellow except under cool, moist conditions.

Poa arctica R. Br. Arctic Bluegrass. [*Poa grayana* Vasey]. Perennial with short to well-developed rhizomes; culms solitary or loosely tufted, erect or decumbent-based, 0.7–4 (6) dm tall; sheaths open to midlength or below, glabrous or occasionally scaberulous; blades 0.5–3 (4) mm wide, folded or flat, generally distinctly prow-shaped at the abruptly acute tips, the blades of the uppermost culm leaves 1–5 (7) cm long; ligule 1–3 (4) mm long, abruptly acute to rounded or truncate, glabrous or abaxially scaberulous; panicle oblong to loosely pyramidal, (2) 4–8 cm long, the branches 1–4 per node, mostly 1–5 cm long, naked below, smooth or obscurely scaberulous, ascending to widely spreading or occasionally deflexed; spikelets subterete to moderately compressed, elliptical, 4–7 (8) mm long, pale green streaked with purple or purple throughout, 2– to 5 (6) –flowered, the florets perfect; glumes lanceolate, equal or nearly so, (2.5) 3–5 (6) mm long, weakly to moderately keeled, abruptly acute or more often long-tapered to sharply acute tips, glabrous or nearly so on the keels, shorter than to as long as the lowermost lemma; lemma (3.4) 4–6 mm long, keeled to more or less rounded on the back, acute to obtuse at the broadly membranous apex, distinctly to obscurely 5–nerved, typically densely pubescent (often to above midlength) on the keel and marginal nerves and usually densely so between, rarely pubescent on all the nerves and glabrous between, webbed or not webbed at the base; palea minutely to profusely crisp-hairy on the keels and between; anthers 1.2–2.5 mm long; 2n = 36–106. Open slopes, streamside, in aspen-spruce-fir, lodgepole pine, and meadow communities at 2560 to 3960 m in Beaver, Duchesne, Garfield, Grand, Iron, Juab, Kane, San Juan, Sanpete, Sevier, Summit, and Wasatch counties; circumboreal, south in the mountains of Oregon, Nevada, and New Mexico; Eurasia; 52 (0). Finding that variation within *P. arctica* does not appear to be correlated with either geography or habitat, Welsh (1974) treats the species as a single polymorphic entity. Atypically robust plants from Iron County (Harrison 9853 UT) with panicles more than 6 cm long and lemmas less pubescent than usual, appear to be hybrids with *P. secunda*. Such plants are also known from Colorado (San Juan Mts.) and New Mexico (Wheeler Peak) where both *P. arctica* and *P. secunda* are known to occur. Otherwise, our plants are well within the range of *P. arctica* as it occurs in Eurasia.

Poa bigelovii Vasey & Scribn. in Vasey Bigelow Bluegrass. Annual; herbage yellow green, occasionally purple-tinged, glabrous; culms solitary or in small tufts, erect or occasionally geniculate at the lowermost node, 1–7 dm tall; leaves chiefly cauline; sheaths open half their length or more, glabrous or nearly so; blades lax, 1–6 mm wide, mostly flat or folded, long-tapered to slender, mostly obscurely prow-shaped tips, the blades of the uppermost culm leaves to ca 8 cm long; ligule 1–6 mm long, acute, glabrous abaxially; panicle sublinear to narrowly oblong, 4–20 cm long, the branches mostly 2 or 3 at the lower nodes, to ca 6 cm long, smooth, strictly erect or slightly divergent at flowering; spikelets strongly compressed, elliptical to lance-ovate, 4–6 (7) mm long, 3– to 6 (7) -flowered, the florets perfect; glumes equal or nearly so, 2–4 mm long, sharply keeled, the tips acute and often incurved, smooth or scaberulous on the upper portion of the keel, the second glume rounded to somewhat angled on the margins, shorter than or equal to the lowermost lemma; lemma 2.7–4.5 (5) mm long, sharply keeled, incurved at the typically acute, membranous-margined apex, usually obscurely nerved, densely pubescent to near the apex on keel and marginal nerves and usually to some degree between, webbed at the base; palea pubescent on the keels and between with minute soft hairs; anthers 0.4–1 mm long; 2n = 28, 29. Desert shrub, juniper, and creosote bush communities, in moist washes and on open, sandy or rocky slopes below 1520 m in Kane, San Juan, and Washington counties; southern California east to Colorado, New Mexico, Oklahoma, and Texas, south into Mexico; 21 (0).

Poa bolanderi Vasey Bolander Bluegrass. Annual; culms solitary or in small tufts, erect, 2–5 (8) dm tall; leaves chiefly cauline; sheaths terete or compressed, open 1/3–1/2 the length, glabrous; blades lax, 2–4 (5) mm wide, usually flat, abruptly or gradually tapered to a prow-shaped tip, essentially glabrous, the blades of the uppermost culm leaves to ca 10 cm long; ligule 1.5–4 (5) mm long, rounded to long-tapered, glabrous abaxially; panicle ultimately open, 4–20 cm long, the branches few, remote, linear, initially erect, ultimately stiffly spreading to reflexed, smooth or scaberulous, the lowermost 2–11 cm long; spikelets compressed, elliptical, 3.5–6 (7) mm long, 2– or 3–flowered, the florets perfect; glumes keeled, the first 1.5–2.8 mm long, the second 2–3.5 mm long, mostly shorter than the lowermost lemma; lemma 1.5–4.5 mm long, keeled, not regularly incurved at the acute to obtuse, membranous tip, obscurely to more or less distinctly nerved, webbed at the base, otherwise glabrous; palea glabrous throughout or scaberulous on the keels; anthers 0.5–1.5 mm long; 2n = 28. Dry to moist, open or wooded habitats from ca 1525 to 2620 m in Cache and Garfield (Eggleston 8219 US, originally determined as *P. occidentalis* Vasey) counties; Washington to southern California, east to Idaho and Nevada; 1 (0).

Poa bulbosa L. Bulbous Bluegrass. Short-lived perennial, in our climate nearly always viviparous to some degree; herbage green or glaucous; culms solitary to densely tufted, 1–5 (7) dm tall, erect or geniculate at one or two of the lower nodes, arising from more or less pear-shaped bulbs to ca 1 cm long; sheaths open from just below midlength nearly to the base, glabrous or occasionally minutely hairy; blades 0.5–3 (4.5) mm wide, flat to loosely involute, obscurely to distinctly prow-shaped at the tips, the blades of the uppermost culm leaves 2–10 cm long; ligule (1) 1.5–5 (8) mm long, obtuse to long-tapered, abaxially glabrous; panicle narrowly oblong to ovoid, (1.5) 2–13 cm long, erect or nodding, compact or more or less open at flowering, the branches 1–5 per node, usually scaberulous, mostly 0.5–3 cm long; spikelets typically borne in more or less dense clusters at tips of short branches, the 3–6 (8) florets mostly proliferating into a single, ultimately purple-based bulblet, the lowermost 1 or 2 lemmas typically empty, the succeeding one basally thickened and elongating, closely subtending the remaining lemmas, these modified to form the bulblet, the apical portions of the upper lemmas assuming the typical leaf characters of sheath, ligule, and blade, the latter often reaching 3 cm or more in length while still attached to the plant; floriferous spikelets strongly compressed, elliptical or lanceolate to ovate, (2.5) 3–7 mm long, 3– to 7–flowered; glumes equal or nearly so, (1.5) 2–4 mm long, sharply keeled, incurved at the acute tips, the keels scabrous on at least the upper portion, the second glume distinctly shorter than the lowermost lemma; lemma (2) 3–5 (6) mm long, strongly keeled, typically incurved at the acute membranous tip, obscurely 5–nerved, pubescent on keel and marginal nerves or only on the keel, occasionally the body completely glabrous (often variable within a single panicle), webbed or not at the base; palea keels glabrous or scabrous to ciliate with minute, stiff hairs; anthers 1–2.5 mm long; 2n = 14, 21, 28, 39–58. Moderately dry to moist sites, as a weed of cultivated and waste areas, in sagebrush, oak-maple, pinyon-juniper, mixed conifer, meadow, and streamside communities at 1200 to 3000 m in Box Elder, Cache, Daggett, Davis, Duchesne, Juab, Millard, Morgan, Salt Lake, San Juan, Tooele, Uintah, Utah, Washington, and Weber counties; native to Eurasia and North Africa, now circumboreal, south in the U. S. to California, east to Montana, Wyoming, and Colorado, infrequent to absent elsewhere in the U. S.; 115 (xi). *Poa bulbosa* is the only member of the Gramineae arising from true bulbs (Halperin 1933). Bulbous bluegrass produces palatable forage soon after spring snow melt, but total production is low and the growing period short (Vallentine 1961). The species has been used to some extent for pasture and for erosion control in parts of the West, having been adapted for use as an understory grass in range seedings of wheatgrass and other dryland grasses at elevations of less than 1200 m (Hanson 1972).

Poa compressa L. Canada Bluegrass. Perennial, rhizomatous and sometimes stoloniferous; herbage often glaucous; culms solitary to loosely tufted, erect or decumbent-based, often geniculate at one or two of the lower nodes, 1–6 (7) dm tall, usually strongly compressed, at least in the upper portion; leaves chiefly cauline; sheaths typically compressed and keeled, open nearly or quite to the base, glabrous or scabrid-puberulent; blades 1–4 (5) mm wide, flat or folded, at least those of the innovations distinctly prow-shaped at the tips, the blades of the uppermost culm leaves rarely more than 10 cm long; ligule 0.5–2 mm long, truncate to rounded, glabrous or abaxially scaberulous; panicle at maturity erect, sublinear to narrowly oblong, 2–10 (14) cm long, compact to open but rarely if ever more than 3 cm wide, the branches 1–4 (5) per node, smooth or scaberulous, erect-ascending or at flowering moderately spreading, rarely as much as 4 (7) cm long, mostly spikelet-bearing to the base or nearly so; spikelets moderately compressed, elliptical or lanceolate to ovate, 3–6 (8) mm long, (2) 3– to 8 (10) –flowered, the florets perfect; glumes subequal, elliptical to lanceolate, 1.8–3.1 mm long, moderately to strongly keeled, abruptly acute at the tips, the second shorter than the lowermost lemma; lemma firmer than the glumes, 1.8–3.1 mm long, keeled but not strongly compressed (boat-shaped), narrowly membranous-margined, the lateral margins soon curving around the palea, rounded to abruptly acute at the entire, bronze- or purple-tipped (rarely pale) apex, obscurely nerved, pubescent below midlength on keel and marginal nerves and glabrous between to glabrous over the entire body, copiously to sparsely or not at all webbed at the base; palea subglabrous or roughened to scabrid-puberulent on the keels; anthers 0.8–2 mm long; 2n = 14, 35, 39, 42, 45–50, 56, 84. Mesic to wet sites in pastures, along waterways, and in open places in sagebrush, juniper, oak-maple, aspen, ponderosa pine, and lodgepole pine communities, often in the wake of disturbance, at 1300 to 2600 m in Cache, Carbon, Duchesne, Juab, Iron, Rich, Salt Lake, San Juan, Summit, Tooele, Uintah, Utah, Wasatch, Washington, and Weber counties; native to Eurasia, now cosmopolitan (Clausen 1961); 46 (viii). In the more humid areas of the U. S., Canada bluegrass is employed as a pasture grass and for erosion control. It adapts well to poor soils but does not withstand heavy grazing (Hanson 1972).

Poa curta Rydb. Wasatch Bluegrass. Perennial, usually rhizomatous, typically gynomonoecious; herbage dark green to glaucous; culms solitary or in small loose tufts, erect or decumbent-based, 1.5–8 dm tall, terete or frequently compressed to some degree; sheaths terete or compressed, occasionally keeled, open from about half to nearly the entire length, glabrous or scabrous; blades 1.5–5 (7.5) mm wide, flat, glabrous or scaberulous, obscurely to distictly prow-shaped at the tips, the blades of the uppermost culm leaves 2.5–13 cm long; ligule 0.5–4 mm long, truncate to acute, the abaxial side glabrous or scaberulous to minutely hairy; panicle pyramidal to oblong, (5) 6–22 cm long, mostly erect, ultimately open, the nodes of the rachis remote, each with (1) 2 or 3 (5) branches, the lowermost internode (1.5) 2.5–13 cm long, in any one population averaging at least 3 cm long, the branches smooth or scabrous, usually all but a few of the uppermost branches horizontally spreading or more often deflexed by flowering, the longer ones typically spikelet-bearing over the distal half, the lowermost branches (1) 3–10 (12) cm long; spikelets moderately compressed, elliptical to lanceolate, (4) 5–9 (11) mm long, green or occasionally faintly purple-tinged, (1) 2– to 4 (6) –flowered, the florets perfect or pistillate, in most populations both present within each panicle, an occasional population with plants mostly pistillate, rarely some plants mostly perfect-flowered; glumes moderately keeled, narrowly lanceolate and sharply acute at the tips to subovate or obovate with tips abruptly acute to rounded, mostly narrowly membranous-margined, the keels typically glabrous or merely roughened, the first glume 1.5–4.6 mm long, the second 2–5 (5.3) mm long, shorter than the lowermost lemma; lemma (2.8) 3–6 (6.5) mm long, narrowly to broadly membranous-margined, not regularly incurved at the sharply acute to rounded, entire to erose tip, obscurely to prominently 5–nerved, glabrous or scabrous to puberulent with short crinkly hairs, often sparsely crisp-hairy on the keel and over the center back near the base with the marginal nerves glabrous, less frequently sparsely crisp-hairy over the entire lower third of the lemma (variable within a population), not webbed at the base; palea keels scaberulous (at 30x magnification); anthers of perfect florets (1.3) 2–3 (3.6) mm long, those of pistillate florets colorless and mostly less than 1 mm long. Under oak-maple, aspen, and spruce-fir, rare in meadows and on open grassy slopes at higher elevations, at 1680 to 3200 m in Cache, Carbon, Davis, Duchesne, Emery, Juab, Morgan, Salt Lake, Sanpete, Sevier, Summit, Utah, Wasatch, and Weber counties; British Columbia, Idaho, Oregon, and Wyoming; 95 (xv). The holotype of this taxon (NY) is a specimen of *P. nervosa*; the name *P. curta* is retained here on a tentative basis. See *P. nervosa* for discussion of the relationship between *P. curta* and *P. nervosa*.

Poa fendleriana (Steudel) Vasey Muttongrass. [*Eragrostis fendleriana* Steudel; *P. cusickii* Vasey; *P. epilis* Scribn.; *P. longiligula* Scribn. & Williams, type from Silver Reef, Washington County; *P. eatonii* Wats., type from Cottonwood Canyon, Salt Lake County; *P. scabriuscula* Williams, type from south of Glenwood, Sevier County]. Perennial, occasionally with a few short rhizomes, usually dioecious; herbage pale to dark green or glaucous; culms tufted, erect, 1–6 (7.5) dm tall; sheaths terete or compressed, open 1/3–3/4 the length, glabrous or scaberulous to retrorsely scabrid-puberulent, the sheath bases of previous years typically straw colored or gray, papery (not disintegrating into fibers), forming a conspicuous feature; blades 1–3 (5) mm wide, flat to involute and then often appearing filiform, distinctly to obscurely prow-shaped at the tips, scabrous to puberulent on one or both surfaces or less often glabrous throughout, the blades of the uppermost culm leaves to ca 7 cm long, often very short or lacking; ligule 0.5–12 mm long, truncate to long-tapered, glabrous or abaxially scaberulous; panicle mostly oblong, 2–10 (15) cm long, densely flowered or occasionally somewhat loose at flowering, the short branches 1–5 per node, scaberulous or smooth, erect to ascending or at flowering more or less divergent, spikelet-bearing to the base or over more than half the length; spikelets compressed, elliptical or lanceolate to ovate, 4–12 (14) mm long, 2–4 (5) mm wide (florets nondivergent) or narrower in very young plants, often tinged with purple or turning pinkish, appearing shiny because of the broad membranous margins of glumes and lemmas, (2) 3– to 8 (10) –flowered, the florets evidently all pistillate in ca 92 percent of plants seen, the remainder with at least some staminate or perfect florets; glumes ovate to narrowly lanceolate, mostly sharply keeled, abruptly to sharply acute at the tips, glabrous to merely roughened on the keel, typically broadly membranous-margined, the first (2) 3–5 mm long, the second 3–5.3 mm long, much shorter than to as long as the lowermost lemma; lemma (3) 4–6 (7) mm long, keeled, typically broadly membranous-margined, rounded to sharply acute at the tip, usually obscurely nerved, glabrous or scabrous to more often sparsely to densely pubescent on keel and marginal nerves and glabrous or rarely sparsely hairy between, not webbed at the base; palea keels scabrous or densely scabrid-puberulent to ciliate with relatively long, soft hairs; anthers of staminate or perfect florets (1.5) 2–4 mm long, those of pistillate florets colorless and less than 1 mm long; 2n = 29, 56. Commonly in openings in sagebrush, oak, pinyon-juniper, mountain brush, ponderosa pine, and spruce-fir communities, occasionally in desert shrub communities, in talus and meadows, fringing aspen, streamside, and on exposed alpine ridges, at 910 to 3660 m in all Utah counties; British Columbia to Manitoba, south throughout the western U. S., east to the Dakotas, Nebraska, and western Texas; 787 (xiii). *Poa fendleriana* appears to intergrade with *P. nervosa* by means of occasional staminate or perfect florets (e.g., Arnow 6075, Harrison 9472 and 13004, Cottam 9451 UT) and with *P. secunda* (e.g., Harrison 12479 and Maguire 16381 UT). Because of its ability to hybridize with other species and to occupy diverse habitats throughout a broad altitudinal range, *P. fendleriana* forms an exceptionally polymorphic complex that would appear to comprise more than one species. On the basis of variation in lemma pubescence pattern and leaf width, *P. cusickii* and *P. epilis* have traditionally been segregated from *P. fendleriana* as distinct species. The transition from densely hairy to glabrous lemmas is continous, however, and the plants with glabrous lemmas grow side by side with those having hairy ones. Plants with glabrous lemmas appear to occur in areas where *P. nervosa* also grows, and because such plants often have other characters of *P. nervosa* (such as its dark green color, broader leaves, unreduced upper blades, less persistent, puberulent sheath bases, etc.) it seems reasonable to assume that they result from intergradation between the two species. In any case, I have been unable to detect even one character that is consistently correlated with pubescence pattern of the lemma and thus find no basis for recognizing more than one species. *Poa arida* Vasey is a grass native to the Great Plains, differing from *P. fendleriana*, at least in part, in being strongly rhizomatous and in having glaucescent, thickened, stiffly erect basal leaves and lemmas pubescent between the nerves. *Poa glaucifolia* Scribn. & Williams and *P. juncifolia* Scribn. (a form with glabrous or scabrous lemmas) are both closely related to, if not synonymous with, *P. arida*. I have seen no evidence that any member of the *P. arida* complex occurs in Utah. Muttongrass provides palatable and nutritious spring forage for cattle and horses, and good to fair feed for sheep, elk, and deer (Vallentine 1961). The species is drought-resistant, adapted to a wide variety of soil types, and holds up well under moderately heavy grazing. Its use in range seeding is restricted, however, because the plant produces relatively small amounts of viable seed.

Poa glauca Vahl Greenland Bluegrass. [*P. glauca* ssp. *rupicola* (Nash) W. A. Weber; *P. glaucantha* Gaudin; *P. interior* Rydb.; *P. nemoralis* ssp. *interior* (Rydb.) Butters & Abbe]. Perennial; herbage bright green to glaucous or anthocyanic; culms densely tufted, occasionally rooting at subsurface nodes, 0.6–5 (6) dm tall, slender, more or less wiry, erect or nearly so; sheaths open nearly to the base, glabrous or occasionally scaberulous to scabrid-puberulent; blades stiffly erect to somewhat lax, 0.5–2.5 (3) mm wide, flat to involute, tapered to slender, sometimes obscurely prow-shaped tips, the blades of the uppermost culm leaves 1–12 cm long; ligule (0.5) 1–2 (2.5) mm long, truncate to obtuse, usually abaxially scaberulous; panicle sublinear to narrowly oblong, (0.8) 1–13 (17) cm long, compact to loose, the branches (1) 2 or 3 (5) per node, usually naked below, closely scabrous throughout to smooth or nearly so, erect or ascending, occasionally widely spreading, the lower branches 0.3–9 cm long; spikelets compressed to occasionally subterete, elliptical to ovate, (2.5) 3–6 (7) mm long, often purple-tinged or purple throughout, (1) 2– to 5 (6) –flowered, the florets perfect; glumes lanceolate to ovate, at least moderately keeled, narrowly membranous at the abruptly to sharply acute tips, these often distinctly spreading, the first glume 1.5–3 (3.5) mm long, the second 2–3.5 (4) mm long, shorter to occasionally slightly longer than the lowermost lemma; lemma 2–3.5 (4) mm long, moderately to strongly keeled or occasionally rounded on the back, narrowly to broadly membranous-margined at the acute to obtuse tip, often banded below the pale membranous tips with bronze or purple or both, usually obscurely nerved, sparsely to densely hairy on keel and marginal nerves, glabrous or variously hairy between, sparsely or not at all webbed at the base; palea glabrous or scaberulous to densely and minutely scabrid-puberulent on the keels; anthers 1–2 mm long; 2n = 28, 42, 43, 44, 56–58, 62, 70–72, 75, 78. Variation within *P. glauca* as it occurs throughout its range is very great, and plants indistinguishable from the various forms that grow in our area, including the type of *P. interior* (NY), occur in Europe, in the U.S.S.R., and throughout Canada, where the species has long been known as *P. glauca*. I find no basis for treating *P. interior* as being distinct in North America. Some workers follow Hitchcock (1951) in recognizing more than one species within the *P. glauca* complex as it occurs in North America. Only about half the Utah plants examined can be satisfactorily segregated by means of the characters given below, and correlation among these and other features is even less consistent outside Utah. I have followed Boivin (Naturaliste Canad. 94: 527. 1967) in treating the complex as comprising a single species with two intergradient, partially sympatric varieties:

1. Panicles 0.8–7 cm long, the branches often smooth, at least in part; glumes about equal, mostly abruptly acute at the tips; lemmas (at least some) pubescent between keel and marginal nerves, not webbed at the base; plants to 2.5 (3) dm tall .... *P. glauca* var. *rupicola*
— Panicles (3) 4–15 cm long, the branches distinctly and uniformly scabrous throughout; first glume narrower than the second and subaristate at the apex; lemmas glabrous between keel and marginal nerves, sparsely webbed at the base; plants 2–6 dm tall .... *P. glauca* var. *glauca*

Var. glauca Chiefly on open, often rocky slopes and in openings in aspen-spruce-fir, ponderosa pine, and lodgepole pine at 2290 to 3735 m in Beaver, Carbon, Daggett, Duchesne, Emery, Garfield, Grand, Iron, Juab, Kane, Millard, Piute, Salt Lake, San Juan, Sanpete, Sevier, Summit, Uintah, Utah, Wasatch, Washington and Wayne counties; circumboreal, south in the Sierra Nevadas of California, in the Rocky Mts. to Arizona and New Mexico, in the Great Plains to South Dakota, and to Minnesota and Vermont; South America; Eurasia; 317 (ix).

Var. rupicola (Nash) J. Boivin [*P. rupicola* Nash in Rydb.]. Among scattered conifers, in alpine meadows, and on exposed rocky ridges at 2745 to 3735 m in Daggett, Duchesne, Emery, Garfield, Grand, Juab, Piute, San Juan, Sanpete, Summit, and Uintah counties; through most of the western U. S.; 58 (0). *Poa glauca* var. *glauca* apparently forms hybrids with *P. secunda* (e.g., Welsh & Christensen 2658 and Harrison 12526 UT), *P. lettermanii* (e.g., Rydberg and Garrett 9054 NY), and *P. pattersonii* (e.g., Seiler s.n. UT). *Poa glauca* differs from the outwardly similar *P. palustris* in occupying mostly open sites at elevations generally above 2290 m and in having lemmas with pale, often erose, membranous tips, ligules mostly less than 2 mm long, and panicles 0.8–15 cm long, in contrast to *P. palustris*, which is nearly always streamside and often below 2290 m, and has entire, mostly bronze- or purple-tipped lemmas, ligules 2–6 mm long, and panicles 10–35 cm long.

Poa leptocoma Trin. Bog Bluegrass. Perennial; culms solitary or in small tufts, erect or weak-stemmed and more or less decumbent below, 1–7.5 (10) dm tall, often rooting at subsurface nodes; sheaths open 1/2–3/4 the length, terete or compressed, glabrous or occasionally scaberulous to retrorsely scabrid-puberulent; blades 1–4.5 mm wide, flat or occasionally folded, smooth or scaberulous, those of the basal leaves mostly distinctly prow-shaped at the tips, those of the culm leaves generally obscurely so, the uppermost blade (2) 6–14 cm long; ligule 1–4.2 mm long, truncate to obtuse, glabrous abaxially; panicle sublinear to ultimately oblong-pyramidal, 3–17 cm long, often nodding, the branches in more or less remote whorls of (1) 2 (4), essentially smooth or sparingly scaberulous, rarely scabrous throughout, erect-ascending to widely spreading or drooping to deflexed, mostly naked below midlength, the lowermost to ca 9 cm long; spikelets compressed, elliptical to lanceolate, (3) 4–7 (8) mm long, usually purple-tinged to some degree, 2– to 4 (6) –flowered, the florets perfect; glumes strongly to moderately keeled, the keel scaberulous (at 20x magnification), the first narrowly lanceolate, 2–4 mm long, usually tapered to a minutely awnlike tip, the second often considerably broader than the first, 2.5–5 mm long, abruptly acute to long-tapered at the apex, much shorter than to as long as the lowermost lemma; lemma 3–4.5 (5) mm long, typically tapered to a sharply acute, membranous, bronze or purple tip, 5-nerved, sometimes distinctly so, pubescent on the keel and marginal nerves and glabrous between, or occasionally the body entirely glabrous, webbed at the base, sometimes very sparsely so; palea keels subglabrous or more often ciliate with minute, mostly stiff, often upwardly curved hairs; anthers 0.3–0.9 (1.1) mm long; 2n = 42. Under aspen-spruce-fir and lodgepole pine, in meadows, along streams, and on moist open slopes or wet cliffs at 2135 to 3570 m in Beaver, Cache, Carbon, Daggett, Duchesne, Emery, Garfield, Grand, Iron, Juab, Piute, Rich, Salt Lake, San Juan, Sanpete, Sevier, Summit, Tooele, Uintah, Utah, Wasatch, and Wayne counties; Alaska to Alberta, south through Washington to California and Nevada, and in nearly all the Rocky Mt. states; Siberia; 84 (iv). The chief morphological difference between the outwardly similar *P. leptocoma* and *P. reflexa* lies in the shape and size of the glumes and lemmas, the longer, more consistently flat leaves, and the minute stiff cilia of the palea keels in *P. leptocoma* versus the relatively short, mostly folded leaves, and soft, wavy cilia of the palea keels in *P. reflexa*. The taxa also appear to differ ecologically in that *P. leptocoma* is more frequently found in wet places than is *P. reflexa*. The two species are occasionally sympatric, but intermediacy was observed in only 0.05 percent of the specimens examined. Vasey (Contr. U. S. Natl. Herb. 1: 267–280. 1893), in his original description of *Poa occidentalis*, cites Utah collections made by L. F. Ward. The one such collection I have seen (Ward 284 US) proved to be a robust form of *P. leptocoma*. *Poa occidentalis*, as treated by Soreng (1985), is not reported for Utah.

Poa lettermanii Vasey Letterman Bluegrass. Perennial with numerous hairlike fibrous roots; herbage glabrous or nearly so; culms densely tufted, erect or lax, 1–12 (15) cm tall; sheaths open to the base or nearly so, glabrous or some of the inner sheaths scaberulous, outer sheaths at the base of the plant often disintegrating into fibers; blades 0.3–1.8 mm wide and typically less than 3 cm long, folded or rarely flat, mostly gradually tapered to slender, prow-shaped tips; ligule 1–5 mm long, truncate to long-tapered, typically glabrous abaxially; panicle sublinear to narrowly oblong or occasionally dense and spikelike, 0.6–3.5 cm long and to about 0.5 cm wide, the branches erect-appressed or slightly spreading at flowering, smooth or obscurely scaberulous, 1–3 per node, the lowermost branch to about 1.5 cm long; spikelets short-pedicelled, moderately compressed to subterete, elliptical, 2–5 mm long, usually purple to some degree, 2– to 4-flowered, the florets perfect, the uppermost about half as long as the lower and subterete; glumes typically elliptical, mostly rounded on the back and appearing somewhat inflated, abruptly or occasionally sharply acute at the narrowly to broadly membranous tips, glabrous, the first glume 2–3 mm long, the second 2–3.5 (4) mm long, about equal to or longer than the lowermost lemma, often enclosing all the florets or nearly so; lemma 2–3.5 mm long, moderately keeled to rounded on the back, abruptly acute to obtuse at the mostly broadly membranous tip, obscurely nerved, glabrous or rarely pubescent, not webbed at the base; palea subglabrous or scaberulous to scabrous on the keels, at least above midlength; anthers 0.3–0.8 mm long. Meadows and on exposed, often rocky slopes, usually in association with *Poa glauca*, *P. secunda*, or *P. pattersonii*, at or above 3050 m in Duchesne, Grand, San Juan, and Summit counties; British Columbia and Alberta, south in the mountains of Washington, Idaho, Montana, California, Nevada, and Wyoming; 8 (0). *Poa lettermanii* appears to hybridize with *P. secunda* (Humphrey 47 NY), *P. glauca* (Rydberg and Garrett 9054 NY and Seiler s.n. UT), and *P. pattersonii* (Lackschewitz 9697 UT, Henderson & Brunsfeld 4845 NY), the resulting medley of character combinations leading to understandable confusion. *Poa lettermanii* most closely resembles the circumboreal *P. abbreviata* R. Br. Both are very small plants occupying only arctic or alpine habitats and are morphologically similar in having short compact panicles, spikelets only slightly compressed, glumes almost as long as the 4-flowered spikelets, and anthers less than 1 mm long. *Poa lettermanii* is, however, distinct in having flat to folded rather than strongly involute leaf blades, longer ligules, consistently smaller spikelets, and (typically) glabrous rather than pubescent lemmas.

Poa nervosa (Hook.) Vasey Wheeler Bluegrass. [*Festuca nervosa* Hook.; *P. wheeleri* Vasey in Rothr.]. Perennial, usually rhizomatous, incompletely dioecious; herbage green or occasionally glaucous; culms loosely to densely tufted, erect or decumbent-based, (2) 3–8 (9) dm tall; sheaths terete or compressed, open 1/5–1/2 the length, sheaths retrorsely puberulent (sometimes only the upper ones) or rarely glabrous, the lowermost often purple-tinged, usually turning dark brown and shredding in age; blades 1–5 mm wide and (1) 2–10 cm long, flat or folded, mostly long-tapered to sharply acute or obscurely prow-shaped tips, glabrous or the upper surface minutely hairy; ligule 1–3 (4) mm long, truncate to acute, opaque or occasionally translucent, densely to sparsely puberulent or rarely glabrous on the abaxial side; panicle narrowly to broadly oblong or pyramidal, 4–13 cm long, relatively compact to open, the nodes of the rachis more or less remote, each with 1–5 branches, the lowermost internode 1–3 (4) cm long, the branches scabrous or smooth, ascending to widely spreading or occasionally those near the base drooping to slightly deflexed, usually naked below, the lower branches 1–6 (7) cm long, ranging from much shorter to occasionally slightly longer than those above; spikelets compressed, elliptical to lanceolate, (4) 5–10 (11) mm long, (2) 4– to 6 (8) –flowered, the florets evidently all pistillate in approximately 90 percent of plants seen, the remainder with some anther development; glumes lanceolate, moderately to sharply keeled, the keels smooth or scabrous, the first glume (2) 2.5–4.5 mm long, the second 3–5.2 mm long, distinctly shorter than the lowermost lemma; lemma (3.7) 4–6 (6.5) mm long, keeled, acute to obtuse, and usually narrowly membranous-margined at the apex, sometimes purple-banded below the margin, obscurely to distinctly 5–nerved, glabrous or scabrous to occasionally puberulent on keel and marginal nerves and rarely between, not webbed at the base; palea scaberulous on the keels; anthers of perfect florets 2–3 mm long, those of pistillate florets rudimentary, colorless and not usually more than 1 mm long; 2n = 28, 29, 56, 61–91. Under aspen, spruce-fir, and lodgepole pine, in openings in wooded areas, in association with krummholz and on exposed rocky slopes at higher elevations, infrequent in sagebrush, mountain brush, and meadow communities, at (1900) 2500 to 3660 m in Beaver, Box Elder, Cache, Carbon, Daggett, Davis, Duchesne, Emery, Iron, Juab, Kane, Piute, Rich, Salt Lake, Sanpete, Sevier, Summit, Tooele, Uintah, Utah, Wasatch, Washington, and Weber counties; British Columbia and Alberta, south throughout most of the western U. S.; 191 (xx). According to Soreng (1985), our material is **var. wheeleri (Vasey) C. L. Hitchc.**, differing from the typical element of the species (found only west of the Cascade Mts.), in lacking long hairs on the margins of the sheath collars and in reproducing only by agamospermy. Plants with perfect or staminate florets do, however, occur sporadically in our area, enabling *P. nervosa* to form hybrids with *P. secunda* (Goodrich 2752 a and b UT), *P. fendleriana* (Harrison 9472 and 12479 UT), *P. pratensis* (e.g., Harrison 12840 and 12605, Arnow 5989, Goodrich 2624, UT), and with *P. curta* (Arnow 6110 UT). Hitchcock et al. (1969) suggest that *P. curta* might realistically be treated as a perfect-flowered race of *P. nervosa*. Over much of their range, the taxa are distinct in reproductive strategy, morphological characteristics, and flowering time. The two species generally form discrete populations but do intergrade to some degree when sympatric. Plants from eastern Oregon and Washington, however, appear to be less sharply defined than those in other areas of the West.

Poa palustris L. Fowl Bluegrass. [*P. crocata* Michx.]. Perennial; culms loosely tufted, erect or decumbent-based, rooting at the subsurface nodes and rarely giving rise to short, slender, rhizomelike structures, (1.5) 3–13 dm tall, terete to distinctly compressed, often branching at the lower nodes; leaves chiefly cauline; sheaths terete or occasionally compressed, open to the base or nearly so, glabrous or scabrous to puberulent; blades strongly ascending or lax, (0.5) 1–4 (5) mm wide, flat or folded, glabrous or scabrous, long-tapered to slender, prow-shaped tips, blades of the uppermost culm leaves to about 20 cm long; ligule (1.8) 2–5 (6) mm long, truncate to acute, glabrous or abaxially scaberulous; panicle initially narrow, ultimately open and oblong to narrowly pyramidal, 10–35 cm long, nodding, the branches in rather distant whorls of 2–6, ascending-spreading, naked below, generally scaberulous, the lowermost 2–13 cm long; spikelets moderately compressed, elliptical, 2.5–4 (5) mm long, (1) 2– to 5 (6) –flowered, the florets perfect; glumes subequal, narrowly lanceolate, 2.2–3.2 mm long, keeled, gradually long-tapered to sharply acute, erect to spreading tips, the second generally about as long as the lowermost lemma; lemma becoming firmer than the glumes, 2–3.1 mm long, keeled but not strongly compressed (boat-shaped), narrowly membranous-margined, the lateral margins soon curving around the palea, acute, entire, and nearly always green below the bronze or purple (rarely pale) tip, obscurely nerved, pubescent below midlength on keel and marginal nerves, usually glabrous between, webbed at the base, often scantily so; palea keels subglabrous or roughened to obscurely scaberulous; anthers 0.8–1.6 mm long; 2n = 28, 30, 32, 42. Chiefly along waterways, less often in meadows or damp woods, at 1370 to 2895 m in Beaver, Box Elder, Cache, Carbon, Daggett, Davis, Duchesne, Morgan, Piute, Rich, Salt Lake, Sanpete, Sevier, Summit, Uintah, Utah, Wasatch, and Weber counties; circumboreal (Gleason and Cronquist 1963), south throughout the western, north-central, and northeastern U. S.; introduced in South America; 77 (xii). Although *P. palustris* most closely resembles *P. compressa* in spikelet morphology, it is more frequently confused with *P. glauca* (q.v.).

Poa pattersonii Vasey Patterson Bluegrass. [*P. abbreviata* ssp. *pattersonii* (Vasey) Love, Love, & Kapoor]. Perennial; culms tufted, 0.5–1.5 (2) dm tall, robust plants typically tapered to a narrowly elongate, conspicuously sheath-clad, often decumbent base; leaves chiefly basal; sheaths terete or compressed, open to the base or nearly so, glabrous or some of the innermost basal sheaths minutely hairy, old sheath bases persistent, loose, papery, pale, ultimately disintegrating into fibers; blades 0.5–2.5 (3) mm wide and mostly less than 8 cm long, flat or folded, mostly gradually tapered to slender, obscurely prow-shaped tips; ligule (0.5) 1–4 (8) mm long, truncate to acute, glabrous or occasionally scaberulous abaxially; panicle spikelike, 1–4.5 (5) cm long and to ca 1 cm wide, the short branches typically erect-appressed, smooth or less often scaberulous, 1–3 (4) at the lowermost node, the longest mostly less than 2 cm (including spikelets); spikelets subsessile to short-pedicelled, typically strongly compressed, mostly elliptical, 3–7 (7.5) mm long, purple or purple-tinged, 2– to 4–flowered, the florets perfect, the uppermost more than half as long as the lower and compressed; glumes mostly lanceolate and sharply keeled, broadly membranous-margined, long-tapered to sharply acute tips, the first 3–4.8 mm long, the second (3.5) 4–5 mm long, shorter than to slightly exceeding the lowermost lemma; lemma (3) 3.5–5 mm long, keeled, typically broadly membranous-margined, sharply acute to occasionally obtuse at the apex, usually mottled with purple or more often purple throughout, obscurely nerved, pubescent on the keel and marginal nerves, glabrous or variously pubescent between the nerves, very rarely the body of the lemma entirely glabrous, webbed or not webbed at the base; palea roughened to obscurely scaberulous on the upper portion of the keel; anthers 0.7–1 (1.3) mm long. Exposed rocky slopes at or above 3350 m in Beaver, Duchesne, Grand, Piute, and San Juan counties; Alaska, Alberta, and through most of the western U. S.; 9(0). *Poa pattersonii* apparently hybridizes with *P. secunda* (Humphrey 47 NY), *P. glauca* (Seiler s.n. UT), and *P. lettermanii* (q.v.). Adding to the confusion generated by hybridization, the line drawing of *Poa pattersonii* in Hitchcock et al. (1969) (also used by Cronquist et al. 1977) was evidently made from a specimen of *P. glauca* var. *rupicola*. It does not portray the dense, spikelike panicle with the sharply acute glumes and lemmas of *P. pattersonii*. *Poa pattersonii* is part of a circumboreal complex involving *P. laxa* Haenke of Europe and a number of other North American taxa. *Poa pattersonii* differs from *P. laxa* only in having slightly larger spikelets and somewhat more compact panicles and could very reasonably be treated as a subspecies of that taxon. In fact, our material so strongly resembles a collection from New Hampshire (NY) and several from Spitzbergen, Norway (NY), that it may well have a much wider distribution than it is currently thought to have. Although *P. pattersonii* has been assigned to infraspecific status under the circumboreal *P. abbreviata*, the latter is much more closely allied morphologically to *P. lettermanii* (q.v.) than to *P. pattersonii*.

Poa pratensis L. Kentucky Bluegrass. [*P. aggassizensis* J. Boivin & D. Love]. Perennial, strongly rhizomatous; culms erect, (2) 4–10 (14) dm tall, smooth; sheaths terete or compressed, sometimes keeled, open 1/3–3/4 the length, glabrous or occasionally scaberulous to puberulent; blades 1–6 (7) mm wide, flat or folded, distinctly to obscurely prow-shaped at the tips, smooth or scabrous on the margins and sometimes on the midnerve below, occasionally short-hairy on the upper surface, especially above the ligule and on the margins near the collar, blades of the uppermost culm leaves 1–12 (14) cm long; ligule (0.7) 1–2.5 (3) mm long, truncate, glabrous or abaxially scaberulous; panicle narrowly oblong and compact to pyramidal and diffuse, 2–16(20) cm long, erect to more or less nodding, the branches (1) 2–5 (9) per node, mostly naked below, smooth or scaberulous, ascending to widely spreading or occasionally drooping, the lowermost ranging from very short to about 10 (12) cm long; spikelets strongly compressed, mostly elliptical to ovate, 3–6.5 (7) mm long, usually purple-tinged to some degree, (2) 3– to 6 (7) –flowered, the florets perfect; glumes strongly keeled, the long-tapered, sharply acute tips typically incurved, the keels mostly scabrous on at least the upper half (at 10x magnification or less), the first glume narrowly lanceolate, 1.8–3.5 mm long, the second broader than the first, 2–3.8 mm long, often angled on the margin, shorter than to as long as the lowermost lemma; lemma 2.5–3.5 (4) mm long, strongly keeled, narrowly to broadly membranous-margined and generally banded with purple or bronze just below the usually sharply acute, mostly incurved tip, typically distinctly 5–nerved, pubescent on keel and marginal nerves, glabrous or occasionally sparsely hairy between, copiously webbed at the base; palea keels glabrous or obscurely scaberulous; anthers (0.8) 1–2 mm long; $2n = 21$–147. Mesic to moist sites, in gardens, pastures, marshes, along roadways and waterways, in greasewood, grass-sagebrush, pinyon pine, oak-maple, aspen, ponderosa pine, lodgepole pine, aspen-spruce-fir, and meadow communities at 1280 to 3290 m in all Utah counties except Morgan; circumboreal (Kellogg 1985), south throughout the U. S. except in the Southeast, and throughout most of the remainder of the world except in desert, tropical, and subtropical regions; 266 (x). *Poa pratensis* is capable of reproducing sexually as well as apomictically by apospory (development of the gametophyte from somatic cells) although pseudogamous apomictic forms are predominant in nature (Akerberg 1942). The species is known to form natural hybrids with *P. arctica*, *P. alpina* (Akerberg 1942), and *P. secunda* (as *P. ampla*, Clausen 1961). In our area *P. pratensis* also appears to hybridize rather frequently with *P. nervosa* (q.v.) and less often with *P. reflexa* (Harrison 12406 UT) and *P. palustris* (Taye 2455 BRY). Kentucky bluegrass is said to be the most important perennial sod-forming grass cultivated in North America. It is extensively used for pasture, recreational turf, and erosion control through most of the northern U. S., southward to southern California, and in the mountains to northern Georgia (Hanson 1972). As forage it is highly palatable and nutritious for all classes of livestock and for deer and elk. Although it can withstand continued heavy grazing better than almost any other grass, it is not drought tolerant and is, therefore, satisfactory for range seeding only in mountainous regions (Vallentine 1961).

Poa reflexa Vasey & Scribn. Nodding Bluegrass. Perennial; culms solitary or loosely tufted, mostly erect, 0.5–5 (6) dm tall; sheaths open from 1/3–2/3 the length, terete or compressed, glabrous; blades 1–4 (5) mm wide, folded or sometimes flat, those of the basal leaves mostly distinctly prow-shaped at the tips, those of the culm leaves sometimes obscurely so, blades of the uppermost culm leaves 1–9 cm long; ligule 1–3.5 mm long, truncate to obtuse, abaxially glabrous; panicle sublinear to ultimately broadly oblong-pyramidal, 2–15 cm long, often nodding, the nodes of the rachis more or less remote, each with 1–2 (3) branches, these essentially smooth or rarely scaberulous to some degree, initially erect-ascending but soon spreading to drooping or deflexed, generally spikelet-bearing above midlength, the lower mostly 2–8 cm long; spikelets compressed, elliptical or ovate to lanceolate, 3–5 (6) mm long, often purple-tinged or entirely purple, 2– to 4 (5) –flowered, the florets perfect; glumes lanceolate to ovate, sharply to moderately keeled, the keel scaberulous (at 20x magnification), the first glume lanceolate to elliptical, 1.5–3.2 mm long, abruptly acute to obtuse or occasionally some of them tapered to a slender tip, the second 2–3.5 mm long, mostly abruptly acute to obtuse, much shorter than to as long as the lowermost lemma; lemma (2) 2.5–3.5 mm long, sharply acute to rounded at the narrowly to broadly membranous, often purple- or bronze-tipped apex, obscurely to distinctly 5–nerved, typically short- to long-hairy on the keel and marginal nerves and often minutely so on the intermediate nerves, webbed at the base; palea keels ciliate with minute soft hairs, somtimes obscurely so; anthers 0.2–0.8 (1) mm long; $2n = 28$. Chiefly in spruce-fir communities and on open, often rocky slopes, infrequently under aspen and lodgepole pine, in meadows, and streamside, mostly from (2550) 2750 to 3660 m in Beaver, Daggett, Duchesne, Emery, Grand, Juab, Piute, Salt Lake, San Juan, Sanpete, Sevier, Summit, Tooele, Uintah, Utah, and Wasatch counties; in the Rocky Mts. from British Columbia and Alberta south to New Mexico, also in eastern Oregon, Nevada, and Arizona; 105 (xiv). *Poa reflexa* is closely related to but morphologically and ecologically distinct from *P. leptocoma* (q.v.). *Poa reflexa* can be readily distinguished from *P. curta* by its webbed lemmas. The species occasionally serves as a host for the smut *Tilletia fusca* Ellis & Everhart.

Poa secunda Presl Sandberg Bluegrass. [*P. ampla* Merr.; *P. brachyglossa* Piper; *P. canbyi* (Scribn.) Howell; *P. confusa* Rydb.; *P. gracillima* Vasey; *P. juncifolia* authors, not Scribn.; *P. nevadensis* Vasey; *P. sandbergii* Vasey; and *P. scabrella* (Thurber) Benth.; *Atropis laevis* var. *rigida* Beal, type from Lake Point, Tooele County]. Tufted perennial, occasionally with a few short rhizomes; herbage yellow to dark green or strongly glaucous (sometimes within the same population), plants of dry habitats often anthocyanic to varying degrees; culms loosely to more often densely tufted, mostly erect, (0.5) 1–12 dm tall; sheaths open at least 3/4 the length, sometimes closed only near the base, glabrous or occasionally scaberulous, old sheath bases not consistently a conspicuous feature; blades 0.4–3.5 (4) mm wide, mostly folded or involute upon drying, tapered to a sometimes obscurely prow-shaped tip, usually scabrous to some degree, blades of the uppermost culm leaves 0.2–22 cm long; ligule 0.5–7 mm long, truncate to acute, usually scaberulous to some degree on the abaxial side; panicle sublinear to narrowly elongate or in an occasional plant pyramidal, 2–27 cm long, loose or rarely compact, the branches (1) 2–5 (8) per node, scabrous or smooth, erect-ascending or spreading, the latter chiefly during flowering, the lowermost branches from very short to ca 6 cm long, some of them spikelet-bearing to the base or nearly so; spikelets subterete or occasionally compressed (but then the upper florets very narrow and often terete or nearly so), usually narrowly elliptical, (3.5) 4–12 mm long, typically averaging at least 4 times as long as wide (florets nondivergent), often bronze- or purple-tinged, (2) 3- to 6-flowered, the florets perfect; glumes narrowly to broadly lanceolate or rarely ovate, mostly either rounded on the back or not keeled to the base, not regularly incurved at the abruptly acute or obtuse to rounded tips, the first glume 2–5 mm long, the second (2.5) 3–5 (6) mm long, distinctly shorter to slightly longer than the lowermost lemma; lemma 3–6 mm long, rounded on the back or occasionally keeled, typically narrowly membranous-margined at the acute to rounded, often bronze- or purple-banded apex, 5-nerved, usually obscurely so, commonly evenly puberulent over the lower third but varying from densely hairy over the lower 2/3 to scabrous or entirely glabrous, not webbed at the base; palea keels scabrous to scabrid-puberulent; anthers 1.2–3 (3.8) mm long; 2n = 44, 56, 61–72, 74, 78, 81–106. Often the dominant grass, occurring on dry to mesic, neutral to strongly saline soils, in openings in greasewood-iodinebush, sagebrush, oak, pinyon-juniper, mountain brush, aspen, aspen-fir, lodgepole pine, spruce-fir, and meadow communities at 1280 to 3660 m in every Utah county; Alaska, across much of Canada to Quebec, south in the western U. S. and into Mexico, in the Great Plains states as far south as Nebraska, and in Michigan; South America; 771 (xliv). *Poa secunda* apparently hybridizes with *P. glauca*, *P. fendleriana*, *P. lettermanii*, *P. nervosa*, *P. pattersonii*, and *P. pratensis* (q.v.). Examination of numerous specimens of South American *P. secunda* revealed that plants of the southern hemisphere are morphologically and ecologically within the range of variation displayed by the group of western North American plants referred by Hitchcock et al. (1969) to the *P. sandbergii* complex (Arnow, Syst. Bot. 6: 412–421. 1981). *Poa secunda*, being the older epithet, replaces *P. sandbergii* as the correct name for the complex. Occurring as they do in an extraordinarily wide range of habitats, members of the complex exhibit an unusual number of outwardly distinct forms, many of which have been segregated as species. Kellogg (1985), in a comprehensive biosystematic review of the complex, found, as did Marsh (1952), that the only defensible taxonomic treatment for the group involves recognition of two species: *P. curtifolia*, a narrow endemic restricted to the Wenatchee Mts. in central Washington and *P. secunda*, the wide ranging species described above. Of the many forms taken by *P. secunda*, that with the open panicle referred to *P. gracillima* is one of the most distinctive. In a few plants collected from a hybrid swarm between *P. secunda* and *P. nervosa* (Goodrich 27526 UT), all characters of *P. nervosa* are suppressed except for the looseness of the panicles. Because *P. gracillima* occurs only sporadically (does not form populations) and because it occupies habitats where *P. nervosa* may also be present, one wonders whether hybridization with introgression may be the explanation for this distinctive entity wherever it occurs. Sandberg bluegrass is variable as to palatability but provides good, though relatively coarse, early spring forage in areas where it receives ample moisture. It has been successfully seeded in areas with as little as 30 cm mean annual precipitation and is useful in controlling erosion because volunteer plants fill the spaces between larger bunchgrasses (Thornburg 1982).

Poa trivialis L. Rough-stalked Bluegrass. Perennial, sometimes stoloniferous; culms tufted, often rooting at subsurface nodes and appearing short-rhizomatous, erect or decumbent-based, 3–10 dm tall, terete or compressed; sheaths terete or compressed, open at least half the length, usually glabrous; blades lax, 1–6 (8) mm wide, flat, long-tapered to slender, obscurely prow-shaped tips, typically scabrous at least on the margins, blades of the uppermost culm leaves mostly 4–15 cm long; ligule (3) 4–10 mm long, truncate to acute, abaxially glabrous; panicle narrowly oblong to pyramidal, (4) 10–23 cm long, relatively compact to open, the branches 1–6 (8) per node, mostly naked at the base, typically strongly scabrous, ascending to widely spreading, the lowermost to about 8 cm long; spikelets strongly compressed, elliptical to ovate, 2.5–5 mm long, 2- to 4-flowered, the florets perfect; glumes sharply keeled, the sharply acute tips mostly incurved, the keels scabrous, the first glume narrowly lanceolate to crescent-shaped, 1.7–3.2 mm long, the second broader than the first, 2–3.5 mm long, often angled on the margin, shorter than to nearly as long as the lowermost lemma; lemma 2–3.5 (4) mm long, generally strongly keeled, regularly incurved at the usually narrowly membranous, sharply acute tips, conspicuously 5-nerved, the body glabrous or occasionally sparsely hairy on the keel, conspicuously to sparsely webbed at the base; palea keels glabrous throughout or merely roughened; anthers 1–2 mm long; 2n = 14, 15, 28, 42. Along waterways, in wet meadows, and in moist woods at 1280 to 2440 m in Cache, Garfield, Grand, Morgan, Salt Lake, Utah, and Weber counties; native to Eurasia, now circumboreal, sporadic in the northwestern U. S., south to New Mexico and east to South Dakota, common throughout the Northeast; South America; 12 (ii).

Polypogon Desf.

Annuals or perennials; sheaths open; culms tufted; blades flat; ligules membranous; inflorescence a dense to somewhat interrupted, spikelike panicle; spikelets laterally compressed to subterete, 1–flowered, disarticulating below a short, calluslike segment of the pedicel, the spikelet and pedicel segment usually falling together, the rachilla ultimately disarticulating above the glumes; glumes equal or the first slightly the longer, lanceolate to oblanceolate, exceeding the floret, 1–nerved, keeled to rounded on the back, acute to obtuse, entire or bifid at the awned or awnless apex; lemma hyaline and shiny, ovate, subterete, obscurely 5–nerved, the lateral nerves often prolonged as minute teeth, awnless or the awn subterminal, delicate, straight, and readily deciduous; callus short, glabrous; palea slightly shorter than the lemma; stamens 1–3; caryopsis free of or somewhat adherent to the lemma and palea, the embryo less than half the length of the grain; $x = 7$.

Bjorkman, S. O. 1960. Studies in *Agrostis* and related genera. Symb. Bot. Upsal. 17: 1–112.

1. Glumes awnless .... *P. semiverticillatus*
— Glumes awned 2
2(1). Glumes obtuse and shallowly lobed at the apex, the awn 2–10 mm long .... *P. monspeliensis*
— Glumes acute and entire to minutely bifid at the apex, the awn 1–3(5) mm long .... *P. interruptus*

Polypogon interruptus H.B.K. Ditch Polypogon. [*P. lutosus* of American authors, not (Poir.) A. S. Hitchc.]. Tufted annual or perennial; culms 2–8 (10) dm tall, erect to decumbent, often geniculate at the lower nodes; blades lax to ascending, flat, 2–12 mm wide; ligule 3–10 mm long; panicle 2–10 (15) cm long, compact to open, the short to relatively long branches mostly erect to strongly ascending or at flowering widely spreading, spikelet-bearing to the base or nearly so; spikelets green or purple-tinged, disarticulating only above the glumes or tardily so below a calluslike segment of the pedicel; glumes nearly equal, narrowly lanceolate to elliptical, 1.5–2.8 (3) mm long, scabrous only on the keel or more often scaberulous to puberulent throughout, acute and entire to minutely bifid at the apex, with a terminal awn 1–3 (5) mm long; lemma shiny, transparent or opaque, ca 1/2–3/4 as long as the glumes, awnless or with a delicate subterminal awn 0.2–2 mm long; anthers usually sterile, to ca 0.8 mm long; $2n = 28, 42$. Disturbed moist sites in pastures and along waterways below 1520 m in Garfield, Salt Lake, San Juan, and Utah counties; British Columbia south throughout the southwestern U.S. and into Mexico, rare in the Great Plains; South America; Eurasia; 10(iii). In Eurasia this largely sterile, relatively rare plant is referred to x *Agrostis littoralis* (Smith) C. E. Hubbard in recognition of its status as a naturally occurring hybrid between *Agrostis stolonifera* and *Polypogon monspeliensis* (Tutin et al. 1980, Tsvelev 1984). In our area, two large hybrid swarms have been observed wherein the parent plants were growing in close association (Arnow 5801 and 5960 UT). The hybrid is unlike *Polypogon* in that the spikelets do not (or do not as readily) disarticulate below the glumes, and is unlike *Agrostis* in having awned glumes. Of 30 or more panicles examined, only 2 had developing caryopses in a few florets.

Polypogon monspeliensis (L.) Desf. Rabbitfoot Grass. [*Alopecurus monspeliensis* L.]. Tufted annual; culms 0.5–6 (7) dm tall, erect to ascending, often geniculate at the nodes; blades lax, flat, 0.5–13 mm wide; ligule (2) 4–10 (12) mm long; panicle more or less cylindrical, appearing softly long-hairy, mostly (0.5) 1–15 cm long, and 0.5–3 (4) cm wide when pressed, compact or somewhat lobed to interrupted, the short branches more or less spreading at flowering, spikelet-bearing to the base or nearly so; spikelets pale green or at maturity tawny, 1–flowered, readily disarticulating below a calluslike segment of the pedicel; glumes nearly equal, narrowly oblanceolate, 1.4–2.5 mm long, minutely hairy and more or less scabrous on the midnerve, shallowly lobed at the apex, a delicate, scabrous awn 2–9 (10) mm long arising between the minute, rounded lobes; lemma shiny, about half as long as the glumes, awnless or with a fragile awn to about 1.5 mm long; palea nearly as long as the lemma; anthers 0.2–0.5 mm long; $2n = 14, 28$. Mostly along waterways and in marshy, sometimes saline sites at 1220 to 2130 m in Beaver, Box Elder, Cache, Carbon, Daggett, Davis, Emery, Garfield, Grand, Juab, Kane, Millard, Salt Lake, San Juan, Sanpete, Sevier, Tooele, Uintah, Utah, Washington, Wayne, and Weber counties; native to Eurasia and Africa, now circumboreal, south throughout the western U. S., from South Dakota to Louisiana, in all the Atlantic coastal states, and in Mexico; South America; 190 (vii).

Polypogon semiverticillatus (Forsskal) Hylander Water Polypogon. [*Phalaris semiverticillata* Forsskal; *Agrostis semiverticillata* (Forsskal) C. Chr.; *A. verticillata* Vill.]. Tufted or stoloniferous perennial; culms 1.5–8 dm tall, erect to decumbent-based, often geniculate and freely rooting at the lower nodes; blades lax or ascending, flat, 1.5–8 mm wide; ligule 1–3 (7) mm long; panicle 1.5–16 cm long, compact to interrupted and lobed, the branches erect to widely spreading at flowering, spikelet-bearing to the base or nearly so; spikelets green or occasionally purple-tinged, 1–flowered, readily disarticulating below a calluslike segment of the pedicel; glumes nearly equal, elliptical, 1–2.2 mm long, scabrous on the keel and roughened to scabrid-puberulent over the back, rounded to abruptly acute at the apex, awnless; lemma shiny, about 1/2–2/3 as long as the glumes, awnless; anthers 0.2–0.5 mm long; $2n = 14, 28$. Moist places, especially along waterways from 610 to 2130 m in Garfield, Iron, Kane, Millard, Salt Lake, San Juan, Sevier, Tooele, Utah, Washington, and Wayne counties; native to Eurasia and North Africa, introduced in the U. S. from Washington to California, in Arizona, New Mexico, Oklahoma, and Texas, south to Mexico; South America; 70 (iii). Because the glumes are awnless, many North American workers include *P. semiverticillatus* in the genus *Agrostis*. The species is, however, closer to *Polypogon* than to *Agrostis* in having the spikelets disarticulating below the glumes, the lemma considerably shorter than the glumes, and the callus consistently glabrous. The transfer of this species to *Polypogon* was effected by Hylander (Uppsala Univ. Arsskr. 7: 74, 1945); the validity of his work confirmed by Bjorkman (1960); and his treatment followed by Tutin et al. (1980) and Tsvelev (1984). Tutin et al. refer the species to an earlier name, *P. viridis* (Gouan) Breistr., whereas Tsvelev retains *P. semiverticillatus*.

Puccinellia Parl.

Annuals or perennials, culms erect or geniculate to decumbent-spreading at the base, sheaths open throughout or nearly so, blades flat to involute, ligule membranous; inflorescence a narrow to open panicle; spikelets terete or subterete, (2) 3– to 9–flowered, disarticulating above the glumes and between the florets; glumes unequal to subequal, lanceolate to ovate, rounded on the back, the first 1 (3) –nerved, the second 3– to 5–nerved, shorter than the lowermost lemma; lemma obovate or ovate to oblong, rounded or rarely keeled on the back, prominently to more often obscurely (3) 5– to 7–nerved, the nerves parallel or nearly so, the membranous-margined apex obtuse to acute and often erose; palea as long as the lemma or nearly so; stamens 3; caryopsis free of but enclosed within the lemma and palea, broadly ovoid to fusiform, embryo 1/6–1/3 as long as the grain; $x = 7$. In spite of their abundance in some areas, most species of *Puccinellia* have little forage value because they cannot withstand grazing pressure.

Church, G. L. 1949. A cytotaxonomic study of *Glyceria* and *Puccinellia*. Amer. J. Bot. 36: 155–165.
———. 1952. The genus Torreyochloa. Rhodora 54: 197–200.
Clausen, R. T. 1952. Suggestion for the assignment of *Torreyochloa* to *Puccinellia*. Rhodora 54: 42–45.
Davis, J. I. 1983. Phenotypic plasticity and the selection of taxonomic characters in *Puccinellia* (Poaceae). Syst. Bot. 8: 341–353.

1. Annuals mostly less than 1 (0.1–1.2) dm tall; panicles sublinear, 1–8 cm long .... *P. simplex*
— Perennials mostly more than 1 (1–14) dm tall; panicles narrow to ultimately open, 2–30 cm long 2

2(1). Lemmas prominently nerved; plants strongly rhizomatous .... *P. pauciflora*
— Lemmas obscurely nerved; plants solitary to tufted, not rhizomatous 3

3(2). Panicle branches typically spikelet-bearing to the base or nearly so, mostly narrowly divergent at maturity; lemmas typically glabrous throughout, firmer than the glumes and somewhat shiny .... *P. fasciculata*
— Panicle branches naked below, ultimately widely spreading or deflexed; lemmas typically puberulent near the base (sometimes obscurely so), similar to the glumes in texture 4

4(3). Plants with yellow green herbage and mostly erect culms; panicle branches to 15 cm long, widely spreading, rarely deflexed at maturity .... *P. nuttalliana*
— Plants with blue green herbage and mostly geniculate to decumbent-based culms; panicle branches to about 6 (8) cm long, at least some of them deflexed at maturity .. .. *P. distans*

Puccinellia distans (L.) Parl. Weeping Alkaligrass. [*Poa distans* L.]. Perennial; herbage typically blue green; culms solitary or more often tufted, sometimes mat-forming, usually decumbent-based, often geniculate at the lower nodes, 1–7 dm tall; sheaths glabrous; blades flat, often becoming involute, 0.5–4.5 mm wide, glabrous or scaberulous; ligule 0.9–2.2 mm long; panicle open, narrowly lanceolate to broadly pyramidal, 6–20 cm long, the branches loosely spaced, 2–6 per node, to ca 6 (8) cm long, ascending to ultimately widely spreading or often deflexed, usually spikelet-bearing over the distal half, glabrous or scabrous; spikelets ovate to narrowly oblong, 3–8 mm long, 3– to 9–flowered, often purple-tinged; glumes ovate to narrowly lanceolate, acute to obtuse and entire or erose at the membranous-margined apex, the first 0.5–1.5 mm long, the second 1–2.5 mm long; lemma similar to the glumes in texture, ovate to obovate or oblong, 1.4–3 mm long, rounded on the back, faintly nerved, the lowermost mostly truncate to obtuse at the minutely erose-ciliate, membranous-margined apex, minutely hairy near the base, sometimes obscurely so; anthers 0.3–1 mm long; $2n = 14, 28, 42$. Moist sites, chiefly in salt desert shrub communities, occasionally along roads and waterways and in pastures and meadows, or in mixed conifer communities, at 1070 to 2620 m in Box Elder, Cache, Daggett, Davis, Duchesne, Emery, Garfield, Grand, Juab, Kane, Millard, Rich, Salt Lake, San Juan, Sanpete, Sevier, Tooele, Uintah, Utah, Wasatch, Washington, Wayne, and Weber counties; native to Eurasia, adventive and becoming naturalized in North America from Alaska across Canada to Nova Scotia and southward throughout most of the western U. S., in the Great Plains as far south as Nebraska, and occasional in the Northeast; 110 (xvi). See *P. nuttalliana*.

Puccinellia fasciculata (Torr.) Bicknell Torrey Alkaligrass. [*Poa fasciculata* Torr.]. Tufted perennial; herbage usually glaucous; culms erect to somewhat decumbent at the base, 1–6 (8) dm tall; sheaths glabrous; blades flat to involute, 1–5 mm wide, scabrous; ligule 0.5–2.5 mm long; panicle narrow and contracted, 2–18 cm long, the branches strictly erect or at maturity narrowly divergent to occasionally widely spreading, to ca 7 cm long, at least the shorter branches spikelet-bearing over the distal 2/3 or more, usually scabrous or scaberulous; spikelets ovate to narrowly oblong, (2) 2.5–6 mm long, 2– to 5 (8) –flowered, rarely faintly purple-tinged; glumes ovate, acute to obtuse at the subentire, broadly membranous-margined apex, the first 0.5–1.5 mm long, 1–nerved, the second 1–2 mm long, 3–nerved; lemma typically firmer than the glumes and somewhat shiny, ovate to oblong, 1.5–2 (2.5) mm long, rounded on the back or slightly keeled toward the tip, faintly nerved, mostly acute to obtuse at the minutely erose membranous-margined apex, typically glabrous throughout; anthers 0.5–1 mm long; $2n = 28$. Moist sites in salt desert shrub communities below 1370 m in Davis, Salt Lake, Tooele, and Utah counties; native to Europe, introduced in Nevada and Arizona and along the east coast from Nova Scotia to Virginia; 22 (ix).

Puccinellia nuttalliana (Schultes) A. S. Hitchc. Nuttall Alkaligrass. [*Poa nuttalliana* Schultes; *Puccinellia airoides* (Nutt.) Wats. & Coult.]. Perennial; herbage yellow green; culms slender, solitary or in small tufts, nearly always strictly erect, (1) 3–10 dm tall; sheaths glabrous; blades flat, often becoming involute, 0.5–4 mm wide, glabrous or scabrous; ligule 1–3.1 mm long; panicle at maturity open, pyramidal to oblong, 6–30 cm long, the branches loosely spaced, 2–6 per node, the lowermost to ca 15 cm long, ascending to ultimately widely spreading or infrequently deflexed, usually spikelet-bearing on the distal half, glabrous or scabrous; spikelets ovate to narrowly oblong, 4–8 mm long, 3– to 7–flowered, often purple-tinged; glumes ovate to narrowly lanceolate, acute to obtuse, entire or erose at the membranous-margined apex, the first 0.5–1.8 mm long, the second 1–2.5 mm long; lemma similar to the glumes in texture, ovate to obovate or oblong, 1.8–3.5 mm long, rounded on the back, faintly nerved, the lowermost mostly acute to obtuse at the minutely erose-ciliate, membranous-margined apex, minutely hairy near the base, sometimes obscurely so; anthers 0.6–1.4 (2) mm long; 2n = 42, 56. Moist sites, chiefly in salt desert shrub communities, in meadows and along waterways at 1280 to 2620 m in Box Elder, Cache, Daggett, Davis, Duchesne, Emery, Garfield, Grand, Rich, Salt Lake, Sanpete, Sevier, Tooele, Uintah, Utah, Wayne, and Weber counties; Alaska to Manitoba, south in the U. S. throughout the West, south in the Great Plains to Kansas, and in Minnesota and Wisconsin, rare in the East; 92 (ix). According to McNeill and Dore (Naturaliste Canad. 103: 536. 1976), the earliest legitimate epithet for this species is *Poa nuttalliana* Schultes, published in 1824 as a replacement for *Poa airoides* Nutt. (1818), the later being a homonym of the validly published *Poa airoides* Koeler (1802). Although workers usually describe the native North American *Puccinellia nuttalliana* as having glumes and lemmas larger than those of the introduced *P. distans*, plants that in the field otherwise resemble *P. distans* exhibit nearly the same range of variation in these characters as does *P. nuttalliana*. All other characters by which the two species have traditionally been separated are likewise overlapping, although the range of variation for each is distinct. The reality of two species, however, is readily observed in the field. Populations of *P. nuttalliana*, with its yellow green herbage, its erect, slender culms in loosely spaced small tufts, and its mostly ascending-spreading panicle branches, present a distinct contrast to populations of *P. distans* with its blue green herbage and ultimately purple gray panicles, its decumbent-ascending, tufted, somewhat coarser culms (often mat-forming in wet places), and its frequently deflexed panicle branches. Identification becomes a problem when these distinctions are lost in the process of converting the plants to herbarium specimens. Moreover, the two species are not infrequently sympatric and, being outcrossers (Smith, J. Agric. Res. 68: 93. 1944), may occasionally hybridize.

Puccinellia pauciflora (Presl) Munz Weak Mannagrass. [*Glyceria pauciflora* Presl; *Torreyochloa pauciflora* (Presl) Church]. Rhizomatous perennial; culms erect or decumbent-based, 2.5–14 dm tall; sheaths smooth, the lowermost often cross-veined; blades lax, flat, 3–15 mm wide, scaberulous; ligule 3–9 mm long; panicle 10–30 cm long, open, often nodding, the more or less remote branches ascending to drooping, to 8 cm long or longer, mostly spikelet-bearing over the distal half, glabrous or scaberulous; spikelets laterally compressed, ovate to narrowly oblong, 2.5–8 mm long, 3– to 7–flowered; glumes mostly subequal, lanceolate to ovate or obovate, acute to obtuse at the entire to subentire, broadly membranous-margined apex, the first 0.8–1.3 mm long, 1 (3) –nerved, the second 1.2–2 mm long, 3–nerved; lemma broadly oblong, 1.8–3 mm long, rounded on the back, prominently 5– to 7–nerved, mostly truncate at the minutely erose apex, usually purple-banded below the membranous margin, typically scaberulous, at least on the nerves; anthers 0.5–0.8 mm long; 2n = 14. Chiefly streamside in conifer forests and in wet meadows at 2130 to 3230 m in Daggett, Duchesne, San Juan, Sevier, Summit, and Uintah counties; Alaska to southwestern Canada, south throughout the western U. S.; 16 (0). Church (1949) removed several species (including *P. pauciflora*) from *Glyceria*, placing them in a new genus, *Torreyochloa*, based primarily on differences in basic chromosome number, condition of leaf sheaths, and nervation of the second glume. Clausen (1952) pointed out that *Torreyochloa* is aligned with *Puccinellia* in the above characters and recommended that it be included in that genus as a section, distinguished from other sections in *Puccinellia* primarily by the prominent nerves of the flowering glumes. Munz (1968), Hitchcock et al. (1969), Voss (1972), and Cronquist et al. (1977) have all followed Clausen's recommendation.

Puccinellia simplex Scribn. Annual; culms tufted or solitary, erect to stiffly ascending or prostrate, 0.1–1.2 dm tall; sheaths broad and loose, glabrous; blades mostly folded to involute, 0.5–1.5 mm wide, scaberulous; ligule 1–2.5 mm long; panicle sublinear, 1–8 cm long, stiffly erect, often remaining partially enclosed within the sheath at maturity, the short branches appressed to the rachis; spikelets narrowly oblong, 5–8 mm long, 3– to 6–flowered, green or faintly purple-tinged; glumes mostly unequal, lanceolate, acute or obtuse at the nearly entire, membranous-margined apex, the first 1–1.5 (2) mm long, 1– to 3–nerved, the second 1.5–3 (3.5) mm long, 3–nerved; lemma at maturity firmer than the glumes, lanceolate, 2.5–3.5 mm long, rounded on the back below, more or less keeled toward the tip, obscurely nerved, acute to obtuse at the entire to subentire, membranous-margined apex, minutely hairy below midlength or throughout; anthers to ca 0.5 mm long; 2n = 56. Known in Utah from a single population in a heavily grazed greasewood community at 1310 m in Weber County (Arnow 3901, 5907 UT); California; 8 (ii). This grass was first seen in 1974 and was recollected from a population covering many acres in 1982, but not seen on subsequent visits.

Redfieldia Vasey

Perennial; sheaths open; ligule of hairs; inflorescence a panicle; spikelets 1– to 6–flowered, disarticulating above the glumes and between the florets; glumes subequal, 1–nerved; lemma 3–nerved, the callus long-hairy; stamens 3; caryopsis free of but retained within the lemma and palea, oblong, the hilum punctate, the embryo nearly half as long as the grain; x = 10? A monotypic genus of the west-central U. S.

Reeder, J. R. 1976. Systematic position of *Redfieldia* (Gramineae). Madrono 23: 434–438.

Redfieldia flexuosa (Thurber) Vasey Blowout Grass. [*Graphephorum flexuosum* Thurber]. Perennial with long, slender rhizomes; culms 5–12 dm tall, often branched above the base; sheaths glabrous, shredding in age; blades 2–5 mm wide, loosely involute, glabrous or scaberulous; ligule a ring of hairs to ca 1.5 mm long; panicle diffuse, oblong, open, 2–5 dm long, the branches and branchlets slender to hairlike, widely spreading, the pedicels hairlike, flexuous, mostly 1–5 cm long; spikelets lanceolate to ovate, (1) 2– to 6–flowered, (4) 5–8 mm long, gray green to brownish; glumes subequal, narrowly lanceolate, 2.5–5 mm long, shorter than the lowermost lemma; lemma somewhat firmer than the glumes, laterally compressed, lanceolate, 4–6.5 mm long, distinctly 3–nerved, the lateral nerves sometimes slightly prolonged beyond the lemma, the midnerve often prolonged as an awn-tip to ca 0.5 mm long, the callus pubescent with hairs to ca 2 mm long; palea less than 2/3 as long as the lemma; anthers 2–3 mm long; 2n = 25 (Reeder, Amer. J. Bot. 64: 103. 1977). Sand dunes in desert shrub and pinyon-juniper communities at 1490 to 1830 m in Kane and Uintah counties; Arizona, New Mexico, Colorado, and from North Dakota to Texas; 17 (0). Blowout grass is of value only as a sand binder.

Saccharum L.

Perennials; culms erect, to 3 (6) m tall; sheaths open to the base or nearly so; leaf blades flat or folded; ligule membranous or of hairs arising from a short membranous base; inflorescence a large, usually dense panicle, the primary branches persistent, the branchlets fragile, bearing 2 (3) spikelets, one sessile and usually falling with the rachis internode and pedicel(s) of the other spikelet(s), the remaining spikelet(s) short-pedicelled and disarticulating below the glumes; spikelets all alike, dorsally compressed to subterete, enveloped in long hairs arising from the base, 2-flowered, the lower floret reduced and neuter, the upper perfect; glumes membranous or firm, equal or nearly so, (1) 3– to 5–nerved, surpassing the florets; lemma of both florets membranous, that of the upper one awned or awnless; palea of the lower floret reduced or lacking, that of the upper floret shorter than the lemma; stamens 3; caryopsis free of the lemma and palea, ellipsoidal, the embryo about half as long as the grain; x = 10.

Saccharum ravennae (L.) Murray Ravenna Grass. [*Andropogon ravennae* L.; *Erianthus ravennae* (L.) Beauv.]. Coarse, densely tufted perennial; culms mostly 2–4 m tall; sheaths short-hairy or smooth; blades firm, to ca 12 mm wide, strongly scabrous on the margins and usually densely long-hairy at the base; ligule membranous, 0.3–1.2 mm long, anterior to the hairs of the blade and appearing to be composed of hairs; panicle plumelike, 2.5–6 dm long, the primary branches to about 20 cm long and more or less spreading, the numerous ultimate branchlets 1–2 cm long, disarticulating at the base at maturity; spikelets lanceolate, 3.5–7 mm long excluding awns, enveloped in hairs to about 7 mm long arising chiefly from the base; glumes firmer than the lemma, equal to or exceeding the florets, whitish to tawny or purple-tinged, long-tapered to sharply acute, often awnlike tips, usually hairy on the back, as well as from the base; lemma of the perfect floret membranous, 3–5 mm long, with a terminal slender awn 3–8 mm long; anthers ca 2 mm long; 2n = 20. An escape from cultivation and persisting; hanging gardens, along waterways, and on open slopes at 610 to 910 m in Washington County; native to Eurasia, cultivated as an ornamental and occasionally escaping in parts of the western U. S.; 8 (0). Tutin et al. (1980) have submerged *Erianthus* in *Saccharum*, the two having been separated solely according to whether or not the fertile lemma is awned or awnless.

Schedonnardus Steudel

Perennial; leaves chiefly basal; sheaths open; ligule membranous; inflorescence a panicle, the primary branches simple, spikelike, usually few and remotely spaced, the spikelets sessile, well spaced and appressed along 2 sides of a sometimes obscurely 3–angled rachis, at maturity the entire panicle breaking off at the base and becoming a tumbleweed; spikelets slender, 1–flowered, disarticulating above the glumes; glumes unequal, 1–nerved; lemma narrow, 3–nerved; palea subequal to the lemma; stamens 3; caryopsis free of lemma and palea, subcylindrical, the embryo about half as long as the grain; x = 10. A monotypic genus of the western hemisphere.

Schedonnardus paniculatus (Nutt.) Trel. Tumblegrass. [*Lepturus paniculatus* Nutt.]. Perennial, sometimes flowering the first year; herbage subglabrous to scabrous or scabrid-puberulent; culms slender, wiry, solitary or in small tufts, 0.8–5 (7) dm long, erect or decumbent-spreading; sheaths compressed and keeled, broadly membranous-margined; blades 2–12 cm long, 0.6–2 (3) mm wide, flat or folded, narrowly whitish-margined, becoming spirally twisted on drying; ligule 1–3.5 (4) mm long; panicle 1/2–3/4 the height of the plant, composed of 2 to several, linear, stiffly spreading, spikelike branches 2–10 (20) cm long, these widely spaced along an erect to loosely S-curved main axis; spikelets sessile, about as broad as and mostly strictly appressed to the concave surface of the branches and along the terminal portion of the main axis, subterete to laterally compressed, 3–6 mm long, the glumes more or less firm, narrowly lanceolate, keeled, sharply acute to short-awned, the second glume shorter than or equal to the solitary floret; lemma firm, narrowly lanceolate, 3–5 mm long, typically appressed-hairy near the base, acute to minutely awn-tipped; anthers to ca 1.2 mm long; 2n = 20, 30. Disturbed sites, chiefly below 1830 m in Cache, Carbon, Emery, and Uintah counties; central North America from Saskatchewan south into Mexico; South America; 14 (0).

Schismus Beauv.

Annuals but occasionally perennating; culms tufted; sheaths open; ligule a usually membranous-based ring of hairs; inflorescence a panicle; spikelets subterete to moderately flattened, 4– to 10–flowered, the uppermost more or less reduced, the rachilla disarticulating above the glumes and between the florets; glumes equal or subequal, 3– to 7–nerved, about equal to the terminal floret; lemma lance-ovate to obovate, laterally compressed but at maturity rounded on the back, 9–nerved, broadly membranous-margined, the midnerve often extended as a minute bristle; palea slightly shorter than the lemma; stamens 3; caryopsis free of the lemma and palea, glossy and translucent, ellipsoidal, the embryo 3/5–2/3 the length of the grain; x = 6.

Conert, H. J. and A. M. Turpe. 1974. Revision der Gattung *Schismus* (Poaceae: Arundinoideae: Danthonieae). Abh. Senckenberg. Naturf. Ges. 532: 1–81.

1. Lemmas apically lobed (the lobes 0.5–1 mm long), hairy over the back with hairs mostly 1 mm long; glumes typically averaging more than 5 (5–7) mm long .... *S. arabicus*

— Lemmas subentire to emarginate at the apex (the notch less than 0.5 mm long), subglabrous to hairy chiefly on the margins or minutely hairy on the back; glumes typically averaging 5 (3.5–6) mm long or less .... *S. barbatus*

Schismus arabicus Nees Arabian Mediterranean Grass. This species is similar to *S. barbatus*, differing as follows: leaf blades 0.5–2 mm wide, often flat or merely folded, frequently hairy; panicles ovate to elliptical and compact, occasionally loose, 1–5 cm long; glumes 5–7 mm long; lemmas 2.2–3.5 mm long, lobed apically to the base of the membranous margin, the lobes mostly 0.5–1 mm long, pubescent over the back with hairs to ca 1 mm long; palea mostly sharply acute apically; 2n = 12. Desert shrub and creosote bush communities and open slopes below 1370 m in Kane and Washington counties, first reported in Utah in 1941 (Harrison 10229 BRY); native to the Mediterranean Region and Asia, introduced in the southwestern U. S.; South America; 7 (0). Although *S. arabicus* and *S. barbatus* have essentially the same distribution in Utah and are sometimes sympatric, both species are inbreeding and do not appear to be intergrading.

Schismus barbatus (L.) Thell. Mediterranean Grass. [*Festuca barbata* L.]. Annual; culms tufted, 0.5–3.5 dm tall, erect to widely spreading; sheaths glabrous or with scattered long hairs, typically with long hairs lateral to the ligule at the summit; blades mostly 0.5–1 mm wide and strongly involute, glabrous or occasionally long-hairy; ligule an obscurely membranous-based ring of hairs to ca 2 (3) mm long; panicle narrowly oblong, mostly 1–4 cm long, loose to compact; spikelets subterete to laterally compressed, 3.5–7 mm long, often purple-tinged, 5- to 7-flowered; glumes about equal, lanceolate, 3.5–5.5 (6) mm long, equal to or slightly shorter than the terminal floret, 5- to 7-nerved; lemma obovate, 1.5–2.5 mm long, rounded on the back, conspicuously 9-nerved, minutely emarginate at the broad apex, the notch occupying only the upper portion of the broad, membranous margin, typically hairy on the lateral margins and often with a few hairs near the middle of the back; palea rounded to acute at the apex; anthers 0.2–0.5 mm long; 2n = 12. Desert shrub, pinyon-juniper, and creosote bush communities below 1370 m in Washington County; native to the Mediterranean region and Asia, introduced into the southwestern U. S.; South America; 14 (0).

Schizachne Hackel

Perennials; sheaths closed; blades flat to loosely involute; ligule membranous; inflorescence a usually narrow, often nodding, loose panicle or raceme; spikelets slender, 3- to 7-flowered, the rachilla disarticulating above the glumes and between the florets; glumes membranous, unequal, rounded on the back, the first (1) 3- to 5-nerved, the second 5-nerved; lemma firmer than the glumes, lanceolate, rounded on the back, (5) 7- to 13-nerved, glabrous or scaberulous, bearded on the callus, awned from below a bifid apex; palea shorter than the lemma, densely puberulent along the keeled nerves; stamens 3; caryopsis free of the lemma and palea, smooth and shiny, ellipsoidal, the embryo 1/5–2/5 as long as the grain; x = 10.

Koyama T. and S. Kawano. 1964. Critical taxa of grasses with North American and eastern Asiatic distribution. Canad. J. Bot. 42: 859–884.

Schizachne purpurascens (Torr.) Swallen False Medic. [*Trisetum purpurascens* Torr.]. Tufted perennial; culms erect or decumbent at the base, 3–11 dm tall; sheaths glabrous or rarely scaberulous; blades flat to loosely involute, 1–5 mm wide, glabrous or occasionally sparsely long-hairy; ligule 0.3–2 (3) mm long, initially united in front, often splitting and the outer margins forming auriclelike remnants on the sheaths; panicle (raceme) (5) 6–17 cm long, with 5–20 spikelets, the slender branches ascending to drooping, each bearing 1–3 spikelets; spikelets terete or nearly so, narrowly elliptical, 1–2 cm long, 3- to 7-flowered; glumes broadly lanceolate, acute to obtuse at the apex, pale or purple-tinged below the broad membranous margin, the first 4–6.5 mm long, (1) 3 (5) –nerved, the second 6–9 mm long, 3- to 5-nerved, shorter than the lowermost lemma; lemma lanceolate, 7.5–12 mm long, rounded on the back, distinctly 7- to 9 (13) -nerved, glabrous or scaberulous, the callus of the lowermost lemma densely bearded with hairs mostly 1–2 (3) mm long, the apex bifid, the acute teeth 1.2–2 mm long, the awn arising between the teeth, 8–15 mm long, ultimately recurved; palea distinctly shorter than the lemma, densely puberulent on the keeled nerves; anthers 1.4–2 mm long; 2n = 20. Mesic to moist sites, in aspen, lodgepole, and ponderosa pine communities, streamside and in wet meadows, at 2030 to 2500 m in Daggett, Duchesne, and Uintah counties (Uinta Mts.); Alaska to Newfoundland, south in the U. S. from Montana to New Mexico, and from North Dakota and Nebraska to Maine and Maryland; Siberia; 13 (0).

Schizachyrium Nees

Annuals or perennials; culms sparingly to much branched above the base; sheaths open, terete or compressed and keeled; blades flat to loosely involute; ligule membranous; inflorescence of solitary spikelike racemes terminating the culm and each of its branches, often partially enclosed by the uppermost sheath; spikelets plano-convex, lanceolate, 2–flowered, arranged on the rachis in pairs, one sessile, the other short-pedicelled, the rachis disarticulating below the sessile spikelet, the spikelet pair falling as a unit together with the associated pedicel and rachis internode, the latter structures nearly flat to rounded on the abaxial side, not medially grooved; sessile spikelet with the lower floret neuter, often rudimentary, the upper floret perfect or pistillate; pedicelled spikelet much reduced or occasionally lacking; glumes firm to hardened, narrowly lanceolate, longer than the florets, the first (abaxial) obscurely several-nerved, slightly shorter to longer than the second; lemma membranous, that of the upper floret of the sessile spikelet entire or 2–lobed at the apex, awnless or with a geniculate twisted awn arising from the apex or from between the apical lobes; palea hyaline or lacking; stamens 1–3; caryopsis free of the lemma and palea, the floret remaining enclosed within the glumes, the embryo large; x = 10. See *Andropogon*.

Schizachyrium scoparium (Michx.) Nash in Small Little Bluestem. [*Andropogon scoparius* Michx.]. Perennial, occasionally producing short rhizomes; herbage often glaucous or purplish; culms tufted, usually erect, 7–15 dm tall, the upper half sparingly to much branched; sheaths often flattened and more or less keeled, glabrous or less often hairy; blades 1.5–5 (6) mm wide, flat or loosely involute toward the long-tapered, filiform tips, scabrous and sometimes sparsely long-hairy; ligule 1–3 mm long; racemes 2.5–5 (6) cm long, solitary at the tips of the culm and its branches, the rachis segments and the ultimately recurved pedicels ciliate with widely spreading white hairs 2–3 (4) mm long, the abaxial surface of the pedicels visible between the rows of hairs; sessile spikelet narrowly lanceolate, 5–9 mm long, short-bearded at the base, the glumes ultimately firm to hardened, dorsiventrally compressed, the first nearly flat or dorsally shallowly grooved, glabrous or scaberulous, the lemma of the upper floret with a geniculate, basally twisted awn 6–16 mm long; pedicelled spikelet typically no wider than the pedicel, 2–5 (6) mm long (excluding the awn-tip when present); anthers 3–5 mm long; 2n = 40. Along waterways, in rock crevices, and in desert shrub, pinyon-juniper, ponderosa pine, and hanging garden communities at 1070 to 2290 m in Emery, Garfield, Grand, Kane, San Juan, Washington, and Wayne counties; British Columbia to Nova Scotia, south through most of the western U. S. and in the Great Plains; 35 (0). Occasional plants with but one raceme terminating the culm and its branches (e.g., Welsh 19345 BRY) otherwise resemble *Andropogon gerardii* in having copious yellow hairs to 4 or 5 mm long within the inflorescence. See *Andropogon*.

Sclerochloa Beauv.

Annuals; stems erect to prostrate; sheaths open half the length to nearly throughout; ligule membranous; inflorescence a 1–sided, mostly spikelike panicle; spikelets 3– to 8–flowered, disarticulating below the glumes and falling entire; glumes firm, unequal; lemma leathery, membranous-margined, keeled; palea hyaline, shorter than the lemma, truncate; stamens 3; caryopsis free of the lemma and palea, narrowly triangular-ovoid, the hilum basal, punctiform; x = 7. A monotypic genus native to southern Europe.

Swallen, J. R. 1931. *Crassipes*, a new grass genus from Utah. Amer. J. Bot. 18: 684–685.

Sclerochloa dura (L.) Beauv. Hardgrass. [*Cynosurus durus* L.; *Crassipes annuus* Swallen, type from "between Salt Lake City and Ogden"]. Low glabrous annual; culms 0.3–1.5 (2) dm long, densely tufted, typically prostrate to ascending; sheaths more or less compressed, open nearly to the base; blades flat or occasionally folded, 1–3 mm wide, to ca 6 cm long, somewhat prow-shaped at the tips; ligule 0.5–2 mm long; panicle secund, spikelike or occasionally branched near the base, 1–4 (5) cm long and 6–12 mm wide, with strongly overlapping, sessile to short-pedicelled spikelets, often exceeded by the upper leaf blades; spikelets narrowly oblong, compressed, 6–9 (11) mm long, 3– to 5–flowered, the lower 2 or 3 fertile and the upper 1–3 staminate or sterile; glumes firm, broadly lanceolate to ovate, strongly keeled, typically obtuse apically, the first 1.5–3.5 mm long, (1) 3–nerved, the second 2.5–5 mm long, 5– to 7–nerved, shorter than the lowermost lemma; lemma leathery, linear-oblong, strongly keeled, (3) 4–6 mm long, prominently 5– to 7–nerved (the nerves parallel or nearly so), obtuse to truncate at the apex; anthers 0.8–1.5 mm long; 2n = 14. Apparently first collected in Utah in 1928 (Fallas s.n. US), occurring in chiefly dry, disturbed sites below 1520 m in Cache, Davis, Juab, Salt Lake, Utah, and Weber counties; native to Eurasia, adventive to well established in the northwestern U. S., Colorado, and Texas; 28 (iv).

Secale L.

Annuals or perennials; sheaths open, typically with relatively well-developed auricles; blades flat or involute; ligule membranous, short; inflorescence a somewhat flattened, compact, terminal spike, the rachis continuous or disarticulating; spikelets solitary and sessile at each node of the rachis, 2 (3) –flowered; glumes subulate to narrowly lanceolate, usually 1–nerved, acute to awn-tipped, shorter than the spikelet; lemma lanceolate, laterally compressed, sharply keeled, 5–nerved, long-awned; palea well developed; stamens 3; caryopsis free of the lemma and palea, subcylindrical, the embryo not more than half as long as the grain; x = 7. See discussion of the tribe Triticeae under *Agropyron*.

Secale cereale L. Cultivated Rye. Annual or occasionally biennial; culms solitary to tufted; culms mostly 6–18 dm tall, erect; sheaths glabrous to hairy, the auricles (when present) to 1 mm long; blades flat, 3–10 mm wide, glabrous or scabrous to stiffly hairy; ligule to ca 2 mm long; spike 4–15 (17) cm long (excluding awns), more or less 4–angled in cross section and to ca 1 cm wide, often somewhat nodding, the rachis continuous or ultimately disarticulating, the internodes 2–5 mm long at midlength of the spike, marginally hairy; spikelets solitary at each node, disarticulating above the glumes or the spikelet falling entire with the rachis segment, the 2 lower florets well-developed and arising at nearly the same level from a broad, contracted rachilla, the rachilla sometimes prolonged and bearing a third rudimentary floret; glumes subulate to narrowly lanceolate, equal or nearly so, 5–12 (17) mm long, strongly keeled, often asymmetrically so, 1 (3) –nerved, acute to awn-tipped, shorter than the florets; lemma firm, broad, (10) 12–16 mm long, more or less asymmetrically compressed, the sides visible between the glumes, 5–nerved, ciliate along the keel and the inner upper margin with coarse bristlelike hairs to ca 1 mm long, tapered to a straight, scabrous awn 0.5–7 cm long; anthers 7–9 mm long; 2n = 14, 16, 27–29. Cultivated, commonly escaping, often along roadsides, in fallow fields and other waste places, occasionally on open grassy slopes and in desert shrub, sagebrush, and mountain brush communities at 1280 to 2900 m in Beaver, Box Elder, Cache, Davis, Garfield, Grand, Juab, Kane, Millard, Salt Lake, Sanpete, Tooele, Utah, and Washington counties; native to Eurasia, widely cultivated throughout Canada, the U. S., and to a lesser extent in Mexico; South America; 54 (ii). *Secale cereale*, as it occurs in the Northern Hemisphere, is the rye of commerce, derived from and possibly incorporating features of several taxa native to Asia. See *Triticum* for a discussion of the artificially produced hybrid commonly known as triticale. *Secale montanum* Gussone, a tufted perennial native to Asia, has been used for range seeding programs in Davis County. It differs from *S. cereale* in being perennial and in having a readily disarticulating panicle rachis.

Setaria Beauv.

Annuals or perennials; culms sometimes branching at the lower nodes; sheaths open, often compressed and keeled; blades usually flat; ligule a ciliate membrane or a ring of hairs; inflorescence a cylindrical, spikelike panicle with numerous very short branches, each branch with 1–numerous scabrous bristles (modified sterile branchlets) subtending the closely clustered, subsessile spikelets; spikelets plano-convex to subterete, lanceolate to elliptical, 2–flowered, typically disarticulating below the glumes, the lower floret staminate or neuter, with or without a palea, the upper floret perfect; glumes membranous, subequal to unequal, the first clasping the base of the spikelet, generally less than half its length, 1– to 3–nerved, the second shorter than or equal to the upper floret, 5– to 7–nerved; lemma of the lower floret similar in shape, texture, and often in size to the second glume; lemma of the upper floret firm or hardened, dorsiventrally compressed, faintly 5–nerved, rounded at the apex, smooth or more often faintly to strongly transversely wrinkled, the firm margins incurved around the edges of the equally firm, similarly textured palea; stamens 3; caryopsis free of but remaining enclosed within the firm lemma and palea, ellipsoidal to subglobose, the embryo more than half the length of the grain; x = 9.

Rominger, J. M. 1962. Taxonomy of *Setaria* (Gramineae) in North America. Illinois Biol. Monogr. 29: 1–132.

1. Upper lemma at maturity strongly cross-wrinkled, to about twice as long as the second glume; spikelets averaging more than 2.5 (2.5–3.8) mm long, each subtended by 4–20 greenish to yellow or orange bristles .. S. *glauca*
— Upper lemma nearly smooth to obscurely wrinkled, about equal to or only slightly exceeding the second glume; spikelets averaging less than 2.5 (1.8–2.7) mm long, each subtended by 1–4 (5) greenish or purple bristles 2

2(1). Bristles antrorsely scabrous, mostly 5–11 mm long ... . S. *viridis*
— Bristles retrorsely scabrous, typically averaging less than 5 mm long .... S. *verticillata*

Setaria glauca (L.) Beauv. Yellow Bristlegrass. [*Panicum glaucum* L.; *Chaetochloa lutescens* (Weigel) Stuntz; *Setaria lutescens* (Weigel) Hubb.]. Annual; culms 2–8 (13) dm tall, occasionally sparingly branched above the base, erect or geniculate at the lower nodes; sheaths glabrous; blades flat or twisted in a loose spiral, 3–10 mm wide, long-hairy above the ligule and often scaberulous; ligule an obscurely membranous-based ring of hairs 0.2–1.2 mm long; panicle cylindrical, (1) 3–15 cm long, densely flowered, the axis puberulent, each spikelet subtended by 4–20 ultimately yellow to orange, antrorsely scabrous bristles 3–6 (9) mm long, these at maturity imparting a yellowish color to the panicle; spikelets plano-convex, broadly elliptical to ovate, mostly 2.5–3.8 mm long; glumes green to whitish, subequal, the second to about half the length of the spikelet; lemmas of the reduced lower floret similar to the second glume in texture and shape, about equal to the lemma of the upper floret; lemma of the upper floret at maturity hardened and strongly transversely wrinkled; $2n = 18, 36, 72$. Mostly in dry, waste places below 1520 m in Cache, Emery, Salt Lake, Tooele, Uintah, Utah, and Washington counties; native to Eurasia, now circumboreal, south throughout the contiguous U. S., in Mexico; South America; 18 (iv). Controversy has long revolved around the application of the name *Setaria glauca*, but the most recent analyses (Taxon 25: 297–304. 1976 and Naturaliste Canad. 103: 564. 1976) establish that *Panicum lutescens* Weigel, the basionym for S. *lutescens*, is illegitimate, and that Linnaeus himself clearly indicated how he wished the name *P. glaucum*, the basionym for S. *glauca*, applied. These conclusions appear to resolve the controversy, establishing S. *glauca* as the correct name for the species. Yellow bristlegrass in relatively small amounts in hay can produce extensive ulceration in oral tissues of livestock (Kingsbury 1964).

Setaria verticillata (L.) Beauv. Bur Bristlegrass. [*Panicum verticillatum* L.; *Chaetochloa verticillata* (L.) Scribn.]. Annual; culms 3–10 (15) dm tall, occasionally branched above the base, erect or geniculate at the lower nodes; sheaths glabrous or the margins ciliate near the summit; blades flat, 5–15 mm wide, strongly scabrous on the margins, often sparsely long-hairy; ligule a ciliate membrane 0.7–2 mm long; panicles cylindrical, (2) 5–10 (15) cm long, densely to loosely flowered, often somewhat interrupted, the axis more or less retrorsely scabrid-puberulent, each spikelet subtended by a solitary, greenish, retrorsely scabrous bristle 2–5 (7) mm long; spikelets plano-convex, oblong-elliptical, 2–2.5 mm long; glumes greenish, unequal, the second subequal to the spikelet; lemma of the reduced lower floret similar to the second glume, about equal to the lemma of the perfect floret; lemma of the upper floret at maturity hardened, glossy, minutely papillose to obscurely cross-wrinkled; $2n = 18, 36, 54$. Dry to moist waste places below 1520 m in Cache, Salt Lake, Tooele, Utah, Washington, and Weber counties; native to Eurasia and Africa, widely introduced from British Columbia to Ontario and in the U. S.; 23 (i).

Setaria viridis (L.) Beauv. Green Bristlegrass. [*Panicum viride* L.; *Chaetochloa viridis* (L.) Scribn.]. Annual; culms 1–10 dm tall, occasionally branched above the base, erect or geniculate at the lower nodes; sheaths often ciliate on the margins above, occasionally hairy across the collar, otherwise glabrous or hairy; blades flat to folded, 3–10 (18) mm wide, strongly scabrous; ligule a ciliate membrane 1–2.5 mm long; panicle cylindrical, rarely interrupted, 2–9 (16) cm long, the axis antrorsely scabrid-puberulent to more often spreading-hairy, each spikelet subtended by 1–4 (5) greenish or purple, antrorsely scabrous bristles (4) 5–12 (14) mm long; spikelets plano-convex, elliptical, 1.8–2.7 mm long; glumes greenish, unequal, the second subequal to the spikelet; lemma of the reduced lower floret similar to the second glume, about equal to the lemma of the upper perfect floret; lemma of the upper floret at maturity hardened, glossy, minutely papillose to obscurely cross-wrinkled and often brown-spotted at maturity; $2n = 18, 35, 36$. Weed of dry to moist sites at 1280 to 2130 m in Box Elder, Cache, Carbon, Duchesne, Garfield, Grand, Kane, Salt Lake, Summit, Tooele, Uintah, Utah, Washington, and Weber counties; native to Eurasia, now circumboreal, south throughout the U. S. and in Mexico; South America; 64 (iii).

Sorghastrum Nash

Perennials; sheaths open; blades flat; ligule membranous, often thickened; inflorescence a narrow, contracted to loose panicle with spikelets borne mostly in pairs on slender, fragile branchlets; one spikelet sessile, the other pedicelled, the latter reduced, often represented only by the pedicel (spikelets at tips of branches in 3s, one sessile, two pedicelled); spikelets lanceolate, (1) 2–flowered, disarticulating below the glumes and falling with the attached pedicel(s) and rachis internode, the lower floret rudimentary or lacking, the upper one perfect; glumes firm or hardened, dorsiventrally compressed, subequal, exceeding the florets, glabrous or hairy; lemma and palea of the lower floret hyaline or lacking, that of the perfect, upper floret hyaline, the midnerve prolonged as a well-developed awn; stamens 3; caryopsis free of but retained within the lemma and palea, the embryo large; $x = 10$.

Sorghastrum nutans (L.) Nash in Small Indiangrass. [*Andropogon nutans* L.; S. *avenaceum* (Michx.) Nash]. Perennial with short stout rhizomes; culms tufted, erect, 0.8–2.5 m tall, the nodes pubescent with stiffly erect hairs; sheaths glabrous or sparsely long-hairy, the margins at the summit usually projected upward as stiff, awl-shaped structures lateral to the ligule; blades long, flat, the lower ones mostly 5–10 mm wide, narrowed at the base and long-tapered to a filiform tip, scabrous, occasionally ciliate above the ligule; ligule (1.5) 2–5 mm long; panicle mostly narrowly oblong, loose, soon turning yellowish to reddish brown, 1–3 (4.5) dm long, the primary branches and slender branchlets mostly glabrous, with a solitary sessile spikelet 6–8 mm long at each node, each spikelet subtended by a plumose, nonspikelet-bearing pedicel, the terminal spikelet by 2 such pedicels (rarely a vestigial spikelet present); glumes firm to hardened, often darkening with age, hairy at the base and otherwise glabrous or pubescent with hairs 1–2 mm long, the first glume flattened, the second more or less laterally compressed, the awn of the perfect floret (7) 8–20 mm long, twisted, geniculate, and persistent; anthers 3–4 mm long; $2n = 20, 40, 80$. Hanging gardens, along washes, and in disturbed sites at 1200 to 2130 m in Grand, Kane, and San Juan counties; native to the Great Plains, introduced from Wyoming to Arizona, south to Mexico and throughout the eastern U. S.; 8 (ii). Baum (Canad. J. Bot. 45: 1845–1852. 1967) retypifies S. *nutans*, but Gould (1975) rejects his conclusions, which would substitute the name S. *avenaceum* (Michx.) Nash for S. *nutans*.

Sorghum Moench

Annuals or perennials; culms often tall and stout; sheaths open; blades long and flat, usually broad; ligule membranous; inflorescence a usually large, sometimes nodding panicle with spikelets mostly borne in pairs at each node of the branchlets, 1 spikelet sessile and well developed and 1 (2 at branch tips) pedicelled and usually more or less reduced, at maturity usually disarticulating below the sessile spikelet and falling with or without the adjacent internode and pedicel(s); spikelets lanceolate to ovate, 2-flowered, the lower floret reduced, the upper perfect or that of the pedicelled spikelet(s) staminate or neuter, the rachilla not disarticulating; sessile spikelets with glumes subequal, leathery or less often papery, pale turning dark brown, the first flat to somewhat rounded, the second more or less laterally compressed, slightly exceeding the florets, the lower lemma hyaline, the palea lacking, the upper lemma hyaline and awnless or with a geniculate, twisted awn, the palea hyaline; pedicelled spikelets with glumes papery, the lemmas and paleas hyaline or lacking and the spikelets reduced to 1 or 2 glumes; stamens 3; caryopsis free of the lemma and palea, the floret usually retained within the firm glumes, broadly ellipsoidal to subglobose, the embryo 1/2–2/3 as long as the grain; $x = 10$. Since prehistoric times members of the genus *Sorghum* have been cultivated in Africa and Asia as a source of food and forage. Hybridization and selection have produced many different strains that are, in our area, best treated as belonging to an annual and a perennial species. In their native habitats the two species can be differentiated by several characters, but the plants as they occur in the western hemisphere are so thoroughly introgressed that only the annual and perennial habit serve to distinguish them with any degree of certainty. Various strains of *Sorghum* have a potential for producing dangerous levels of cyanide compounds leading to their prohibition by law as noxious weeds in several southern states. Currently, however, strains with low cyanide potential that pose no danger regardless of environmental or management factors are available (Kingsbury 1964, p. 492).

de Wet, J. M. J. 1978. Systematics and evolution of *Sorghum* sect. Sorghum (Gramineae). Amer. J. Bot. 65: 477–484.
Warwick, S. I., B. K. Thompson, and L. D. Black. 1984. Population variation in *Sorghum halepense*, Johnson grass, at the northern limits of its range. Canad J. Bot. 62: 1780–1790.

1. Plants rhizomatous perennials . . . . S. *halepense*
— Plants annual, sometimes tillering . . . . S. *bicolor*

Sorghum bicolor (L.) Moench Milo or Grain Sorghum. [*Holcus bicolor* L.; S. *vulgare* Pers.]. *Sorghum bicolor* is morphologically indistinguishable from S. *halepense* except for being annual and without rhizomes, and in having a broader range of variation as follows: culms 1–3 m tall; leaf blades 1.2–5 cm wide; spikelet-bearing branchlets not usually shattering; sessile spikelets mostly ovoid to subglobose, the glumes leathery or papery; $2n = 20$. Occasionally cultivated in Utah, escaping and sometimes persisting along ditchbanks and roadways below 1500 m in Grand, Millard, Uintah, Washington, and Wayne counties; native to Eurasia, cultivated in many of the warmer regions of the world, over the last several decades becoming an important crop in the U. S. and in Mexico; 6 (0). *Sorghum bicolor* is a conserved species name (Taxon 24: 176. 1975). encompassing a number of cultivated strains, a widely distributed wild African complex, and the many weedy strains derived from introgression among domesticated grain sorghums and their wild relatives.

Sorghum halepense (L.) Pers. Johnson Grass; Millet. [*Holcus halepensis* L.; S. *miliaceum* (Roxb.) Snowden]. Strongly rhizomatous perennial; culms 0.5–1.5 m tall; sheaths essentially glabrous, sometimes hairy at the summit and across the collar; blades mostly flat, to 9 dm long and 0.4–4 cm wide, usually scabrous on the margins and sometimes densely hairy at the base of the blade; ligule 1.5–3 (5) mm long, ciliate with hairs shorter than the membranous portion, typically densely hairy on the abaxial side; panicle elliptical to lanceolate in outline, 1–6 dm long, compact to open; spikelets in pairs or at branch tips in 3s; sessile spikelet lanceolate to ovate, (3) 4–6.5 mm long, the glumes leathery, glossy, subglabrous to densely hairy, the first glume flat or nearly so, multi-nerved near the minutely toothed apex, the second somewhat laterally compressed and not toothed, the lower floret reduced to a membranous lemma, the upper one perfect with a membranous, ciliate lemma to ca 5 mm long and awnless or terminating in a twisted, geniculate, readily deciduous awn to ca 16 mm long; pedicelled spikelets narrowly lanceolate, 4–7 mm long, the pedicels often densely ciliate, the glumes papery, multi-nerved, the florets staminate or neuter, the lemmas membranous, ciliate; anthers 2.2–3 mm long; $2n = 20, 40, 41, 42, 43$. Weed of ditchbanks and mostly mesic, waste places from 850 to 1500 m in Box Elder, Kane, Piute, San Juan, Salt Lake, Utah, Washington, and Wayne counties; native to Eurasia, cultivated and escaping or introduced as a weed in the southern half of the U. S., in the Northeast, and in warm temperate regions throughout the remainder of the world. In 1937 Maguire reported that *Sorghum halepense* was becoming a common weed in Washington County (Leafl. W. Bot. 2: 23).

Spartina Schreber

Perennials, tufted or rhizomatous, often tall with stout culms; sheaths open; blades firm, flat or involute; ligule a ring of hairs; inflorescence a panicle with few to numerous, appressed to spreading, spikelike branches arising at intervals along a primary axis with the spikelets sessile and strongly overlapping in 2 closely spaced rows along a 3-angled rachis; spikelets narrowly elliptical to lanceolate, strongly compressed laterally, 1 (2) –flowered, disarticulating below the glumes; glumes leathery to hardened, unequal, strongly keeled, the first 1–nerved, the second usually slightly exceeding the floret, 1– to 3 (6) –nerved, awnless or the midnerve prolonged as a mostly short awn; lemmas ultimately firm with a broad membranous margin, strongly keeled, 1– to 6–nerved, usually narrowly obtuse to rounded apically; palea thinner than the lemma, slightly shorter to longer than the lemma; stamens 3; caryopsis retained within the lemma and palea, the embryo nearly as long as the grain; x = 10.

Mobberley, D. G. 1956. Taxonomy and distribution of the genus *Spartina*. Iowa State Coll. J. Sci. 30: 471–574.

1. Second glume with an awn (1) 2–8 mm long, overall length 10–25 mm; plants mostly more than 1 m tall; culm blades 6–15 mm wide .... S. *pectinata*

— Second glume acute or occasionally awn-tipped, overall length 6–10 (11) mm; plants to ca 1 m tall; culm blades 1.5–6 (8) mm wide .... S. *gracilis*

Spartina gracilis Trin. Alkali Cordgrass. Strongly rhizomatous perennial; culms 0.3–1 m tall, erect; sheaths glabrous; blades firm, 1.5–6 (8) mm wide, to ca 3 dm long, flat or involute upon drying, gradually tapered to more or less filiform tips, mostly scabrous; ligule a ring of hairs 0.5–1.5 mm long; panicle mostly 0.5–2.5 dm long, the 2–10 (12) spikelike branches 1.5–6 (8) cm long, spikelet-bearing to the base or nearly so, mostly appressed to the main axis; spikelets 10–30 per branch, ovate to lanceolate, essentially flat and strongly overlapping, 1–flowered; glumes usually ciliate on the keel, otherwise glabrous or scaberulous to sparsely hairy, the first mostly crescent-shaped, 3–6 (7) mm long, 1–nerved, apically acute, the second narrowly elliptical, 6–10 (11) mm long, shorter to slightly longer than the floret, (2) 3–nerved, the lateral nerves conspicuous on only one side (close to the midnerve and best seen on the ventral surface), acutish or the midnerve prolonged as an awn tip to ca 1 mm long; lemma elliptical to lanceolate, 6–9 mm long, 1–nerved, ciliate on the keel, at least toward the apex; palea thinner in texture and slightly shorter to distinctly longer than the lemma; anthers 2.5–6 mm long; 2n = 40, 42. Moist to wet, often saline soils, along waterways, in meadows, and in hanging gardens, at 1220 to 1980 m in Beaver, Box Elder, Cache, Daggett, Emery, Garfield, Grand, Juab, Kane, Rich, Salt Lake, Sanpete, Tooele, Uintah, Utah, Wasatch, Washington, and Wayne counties; British Columbia to Ontario and north to the Mackenzie Territory, south throughout the western U. S. and in the Great Plains to Kansas; 63 (ii).

Spartina pectinata Link Prairie Cordgrass or Slough Grass. Strongly rhizomatous perennial; culms mostly 1–2.5 m tall, erect; sheaths glabrous; blades firm, 6–11 (15) mm wide at the base, to ca 6 dm long, flat or involute upon drying, gradually tapered to often filiform tips, typically scabrous on the margins; ligule a ring of hairs 1–3 mm long; panicle 1–4 (5) dm long, the 4–30 spikelike branches mostly 4–10 cm long, spikelet-bearing to the base or nearly so, appressed to the rachis or slightly spreading; spikelets 10–80 per branch, narrowly lanceolate, essentially flat and strongly overlapping, 1–flowered; glumes strongly scabrous to ciliate with short stiff hairs on the keel and sometimes on the lateral nerves, otherwise glabrous or scabrous, the first glume mostly narrowly crescent-shaped, 5–11 mm long, 1–nerved, acute to awn-tipped, the second glume narrowly lanceolate, 10–25 mm long including the awn, 3–nerved, the lateral nerves closely paralleling the midnerve, the awn (1) 2–7 (8) mm long; lemma narrowly lanceolate, (6.5) 7–9.5 mm long, 1–nerved, coarsely ciliate on the keel at least above; palea subequal to ca 1 mm longer than the lemma; anthers 3–6 mm long; 2n = 40, 42, 70, 80, 84. Wet, often saline sites along waterways and in meadows below 1520 m in Box Elder, Cache, Grand, and Uintah counties; British Columbia to Nova Scotia and through much of the U. S.; 10 (i).

Sphenopholis Scribn.

Annuals or short-lived perennials; culms solitary or tufted; sheaths open; blades lax, flat; ligule membranous; inflorescence a compact to loose, erect or nodding panicle; spikelets flattened, (1) 2– or 3–flowered, disarticulating below the glumes and falling as a unit, the rachilla prolonged beyond the uppermost floret as a slender bristle, sometimes bearing a rudimentary floret; glumes laterally compressed, more or less keeled, dissimilar in size and shape, the first narrow, 1 (3) –nerved, the second broad, obovate to oblanceolate, 3 (5) –nerved, broadly membranous-margined, usually shorter than the lowermost lemma; lemma firmer than the glumes, laterally compressed, rounded on the back to somewhat keeled, obscurely 5–nerved, awnless, or occasionally an awn arising just below the apex; palea hyaline throughout, smaller than the lemma; stamens 3; caryopsis free within the lemma and palea, oblong, terete or slightly compressed; x = 7.

Erdman, K. S. 1965. Taxonomy of the genus *Sphenopholis* (Gramineae). Iowa State J. Sci. 39: 289–336.

Sphenopholis obtusata (**Michx.**) **Scribn.** Prairie Wedgegrass. [*Aira obtusata* Michx.; *S. intermedia* (Rydb.) Rydb.; *S. longiflora* (Vasey) A. S. Hitchc.; *S. pallens* sensu Scribn.]. Annual or short-lived perennial; culms 2–10 (15) dm tall; sheaths glabrous or scabrous to pubescent; blades flat, 1.5–8 (12) mm wide, typically scabrous; ligule 0.5–3.5 mm long; panicle spikelike to open, (3) 5–20 (25) cm long, 0.5–2 cm wide; spikelets flattened, 1.5–5 mm long, (1) 2– or 3–flowered; glumes dissimilar, the first linear, 1–3 (4) mm long, 1–nerved, acute or nearly so, narrowly membranous-margined, scaberulous on the keel, the second oblanceolate to obovate, 3–4 times as wide and usually slightly longer than the first, 3– to 5–nerved, rounded above to faintly mucronate, broadly membranous-margined, the lateral nerves often scaberulous, from slightly shorter to as long as the lowermost lemma; lemma lanceolate, 1.5–3 (4.4) mm long, obscurely nerved, glabrous or minutely pustulose to scaberulous, especially toward the awnless apex; anthers 0.3–0.7 mm long; 2n = 14. Periodically moist sites, chiefly along waterways, in meadows, near seepage in rock formations, and in aspen-spruce-fir communities, rarely on open hillsides, at 1220 to 2740 m in Beaver, Cache, Duchesne, Garfield, Grand, Juab, Kane, Millard, Rich, San Juan, Tooele, Uintah, Utah, Washington, and Wayne counties; Alaska to Newfoundland, south throughout the U. S. and in Mexico; 44 (0).

Sporobolus R. Br.

Annuals or perennials; sheaths open; blades flat, folded, or involute; ligule a ring of hairs or a short, ciliate membrane; inflorescence a panicle, terminal and occasionally axillary, spikelike to open and diffuse, often partially enclosed in the subtending sheath; spikelets 1-flowered, subterete to strongly laterally compressed, disarticulating above and occasionally below the glumes, pale to dark gray green; glumes thin to moderately firm, subequal to unequal, 1-nerved, keeled to rounded on the back, the second shorter to slightly longer than the lemma; lemma usually similar to the glumes in texture, lanceolate to boat-shaped, 1-nerved, acute to subacute apically, glabrous or scaberulous; palea shorter to slightly longer than the lemma, sometimes apically split at maturity and appearing to be a second lemma; stamens 3; caryopsis with seed free of the pericarp at maturity and ejected from the floret when the pericarp swells upon becoming wet, ovate to obovate, somewhat asymmetrical and compressed, the surface sculptured, the embryo 1/2–3/5 as long as the seed; x = 9.

Riggins, R. 1977. A biosystematic study of the *Sporobolus asper* complex (Gramineae). Iowa State J. Res. 51: 287–321.

1. Panicles at maturity pyramidal, 1–8 cm long; lemmas to 2 mm long; plants usually annual and less than 2 (0.5–3) dm tall .... S. *pulvinatus*
— Panicles at maturity spikelike to open, either the panicles more than 8 cm long or the lemmas averaging more than 2 mm long; plants nearly always perennial and more than 2 dm tall 2

2(1). Spikelets 3.5–7 mm long; lemmas abruptly acute to obtuse; panicles narrow 1.5–3 mm long . S. *asper*
— Spikelets averaging less than 3.5 (1.5–4) mm long; either the lemmas sharply acute or the panicles open; anthers 0.2–1.5 (1.7) mm long 3

3(2). Panicles spikelike, rarely the lower portion somewhat loose, the branches spikelet-bearing to the base; glumes and lemmas typically keeled 4
— Panicles ultimately open, the branches not consistently spikelet-bearing to the base; glumes and lemmas keeled or rounded on the backs 5

4(3). Culms to 10 (12) dm tall and to 5 mm thick at the base; glumes typically roughened to scaberulous on the keel; caryopses rarely more than 1.2 mm long ... S. *contractus*
— Culms 6–20 dm tall and 2–12 mm thick at the base; glumes typically smooth or nearly so on the keel; caryopses 1.3–1.7 mm long .... S. *giganteus*

5(3). Glumes and lemmas mostly rounded on the back and typically smooth on the midnerve (at 20x magnification); mature anthers averaging more than 1 (1–1.7) mm long; primary panicle branches 3–15 cm long; old sheath bases persistent, thickened, straw colored and slick, typically a conspicuous feature .... S. *airoides*
— Glumes and lemmas keeled and typically scaberulous on the midnerve (at 20x magnification); anthers 0.2–1 mm long; primary panicle branches mostly less than 6 cm long; old sheath bases not a consistently conspicuous feature 6

6(5). Pedicels and usually the branchlets appressed to the rachis, the primary branches thus more or less spikelike; plants general in distribution .... S. *cryptandrus*
— Pedicels and branchlets soon widely spreading, the primary branches loosely flowered, not at all spikelike; plants restricted to the southern half of Utah (Millard and Emery counties southward) .... S. *flexuosus*

Sporobolus airoides (Torr.) Torr. Alkali Saccaton. Perennial; culms typically densely tufted, often forming large clumps and dying out in the center, (3) 4–20 dm tall, erect or decumbent-based; sheaths glabrous throughout or long-hairy on the margins at the summit, old sheath bases persistent, typically broad, loose, straw colored and slick, forming a conspicuous feature; blades flat or involute, 2–4 (6) mm wide and to ca 4 dm long, typically scaberulous, often long-hairy at the base; ligule a short ring of hairs, often anterior to long hairs at the base of the blade; panicle open, pyramidal, 10–45 cm long, the branches loosely flowered, naked below, mostly 3–15 cm long, the branches, branchlets, and pedicels typically smooth and ascending to widely spreading; spikelets subterete, 1.3–3 mm long, disarticulating above and often below the glumes, pale green to brownish or dull purple; glumes thin, unequal, rounded on the back, glabrous, the first 0.4–2 mm long, acute, the second 1.2–2.8 mm long, acute or obtuse, shorter or in some spikelets slightly longer than the lemma; lemma thin, 1.7–3 mm long, acute or rarely obtuse, glabrous; palea about equal to the lemma; anthers (1) 1.2–1.5 (1.7) mm long; caryopsis ca 1 mm long; 2n = 80, 108, 126. According to Gould and Moran (1981), the species is comprised of two varieties:

1. Culms to ca 1 m tall; panicle branches in age widely spreading, the branchlets naked below, the pedicels ultimately spreading; plants with the range of the species .. .. S. *airoides* var. *airoides*
— Culms commonly 1–2 m tall; panicle branches in age ascending, the branchlets spikelet-bearing to the base, the pedicels mostly appressed in age; plants restricted to Washington County .... S. *airoides* var. *wrightii*

Var. airoides [*Agrostis airoides* Torr.]. Dry to moist sites, in salt grass, salt desert shrub, desert shrub, sagebrush, and pinyon-juniper communities, along waterways, roadsides, and occasionally on open slopes, at 800 to 2350 m in all Utah counties except Davis, Iron, Morgan, Rich, Summit, and Wasatch; throughout the western U. S., in the Great Plains, Texas, Missouri, and Mexico; 172 (v).

Var. wrightii (Munro ex Scribn.) Gould [*Sporobolus wrightii* Munro ex Scribn.]. Rocky slopes and open ground, often in saline soils, below 1220 m in Washington County; southwestern U. S., east to Oklahoma, south into Mexico; 2 (0). Alkali sacaton provides an abundance of moderately palatable forage in spring, particularly in areas dominated by greasewood (Vallentine 1961). It is drought-tolerant, performing well in areas with a mean annual precipitation of 30–46 cm (Thornburg 1982).

Sporobolus asper (Michx.) Kunth Tall Dropseed. [*Agrostis aspera* Michx.]. Perennial, occasionally short-rhizomatous; culms solitary or tufted, erect, 2–13 dm tall; sheaths glabrous throughout or ciliate at the summit; blades flat to involute and often flexuous, mostly 1–5 mm wide and to ca 5 dm long, scabrous to pubescent or rarely long-hairy; ligule a ring of very short hairs, usually with lateral tufts of hairs to 4 mm long or longer; panicle narrow and compact, 5–30 cm long and to ca 1.6 cm wide when pressed, generally partially included within the sheath at maturity, the branches and pedicels appressed-erect or some of them slightly spreading, the plants often bearing shorter, mostly included axillary panicles with cleistogamous spikelets; spikelets strongly compressed, 3.5–7 (8) mm long, disarticulating above the glumes, pale green or whitish to dark gray green or dull purple; glumes thin to moderately firm, unequal or subequal, mostly keeled, acute to obtuse at the apex, glabrous or more often scaberulous on at least the upper part of the keel, the first 1.7–4.7 mm long, the second 2.5–6.2 mm long, mostly distinctly shorter than the lemma; lemma similar to or firmer than the glumes in texture, boat-shaped, (3.5) 4–6 (7) mm long, glabrous or scabrous on the keel, abruptly acute or obtuse; palea about equal to the lemma; anthers 1.5–3 mm long; caryopses mainly 1.6–2 mm long; $2n = 54, 88, 108$. Dry, often sandy sites in juniper communities and in fallow fields below 1530 m in Grand, Uintah, and Utah counties; sporadic in the western U. S., throughout the remainder of the states except in the Southeast; 4 (0).

Sporobolus contractus **A. S. Hitchc.** Spike Dropseed. This species strongly resembles *S. giganteus*, differing only as follows: culms to 10 (12) dm tall and to ca 5 mm thick at the base; blades flat or involute throughout upon drying, 1.5–8 mm wide and to ca 3.5 dm long; ligule a dense ring of hairs mostly 0.4–0.7 mm long; panicle 15–35 (50) cm long and 0.3–0.8 (1) cm wide; spikelets 1.7–3.2 mm long; glumes roughened to scaberulous (at 20x magnification) on the keel and usually laterally, especially toward the apex; caryopses to 1.2 (1.3) mm long. Dry or periodically moist, mostly sandy, occasionally saline soils, chiefly in desert shrub-grass and pinyon-juniper communities below 1980 m in Beaver, Carbon, Daggett, Emery, Garfield, Grand, Kane, Millard, San Juan, Tooele, Uintah, Washington, and Wayne counties; throughout the southwestern U. S., east to Oklahoma, south to Texas and in Mexico; 43 (0). In brief, *S. contractus* is less robust than *S. giganteus*, has a wider distribution, and less frequently occupies roadside areas. Although the two species can usually be distinguished as outlined in the key, they are known to grow side by side, and an occasional plant can be only arbitrarily assigned one name or the other. Plants with panicles bearing large numbers of sterile spikelets suggest that hybridization between the two species is occurring. *Sporobolus cryptandrus* differs from *S. contractus* chiefly in having spreading branches. Because the lower branches of *S. contractus* may sometimes be somewhat spreading and because those of *S. cryptandrus* are initially appressed, an occasional herbarium specimen may be difficult to identify. On a population basis, however, these taxa are readily differentiated. Spike dropseed is useful for dune stabilization, especially in areas of the Southwest where the mean annual precipitation is only 18–30 cm.

Sporobolus cryptandrus **(Torr.) Gray** Sand Dropseed. [*Vilfa cryptandra* Torr.]. Perennial, sometimes appearing annual; culms solitary or in mostly small tufts, often from a hard knotty base, (2) 3–10 (12) dm tall, erect or curved-ascending to prostrate; sheaths ciliate at the summit and often along the upper margins, the hairs 1–4 mm long, sometimes extending across the collar, persistent old sheath bases often present but not a consistently conspicuous feature; blades flat to involute toward the long-tapered tips, 1–5 (8) mm wide and to ca 3 dm long, more or less scaberulous; ligule a dense ring of hairs to 0.5 mm long; panicle ultimately open, pyramidal to oblong, 1–3 (4) dm long, usually remaining partly included within the subtending sheath, the primary branches 1 or 2 per node, ascending to widely spreading or occasionally deflexed, mostly 1–6 cm long, naked near the base, the short pedicels and usually the branchlets erect-appressed, smooth or scaberulous, the pulvini at base of panicle branches glabrous or short-hairy; spikelets subterete to compressed, 1.5–2.5 mm long, disarticulating above and sometimes below the glumes, pale to dark gray green or dull purple; glumes thin, unequal, keeled, acute, typically scaberulous on the keel and often laterally, the first 0.6–1 mm long, the second 1.4–2.5 mm long, slightly shorter to longer than the lemma; lemma thin, 1.5–2.5 mm long, acute, scaberulous on the upper midnerve and sometimes laterally (at 20x magnification); palea about equal to the lemma; anthers 0.2–1 mm long; caryopses ca 1 mm long; $2n = 18, 36, 38, 72$. A weed of roadsides and other waste places, in desert shrub, sagebrush, and pinyon-juniper communities, occasionally in salt desert shrub, aspen-fir, and ponderosa pine communities, streamside, and on grassy slopes, at 850 to 2870 m in all but Morgan, Rich, Summit, and Wasatch counties; British Columbia to Quebec, south through the U. S. (except in the Southeast), and in Mexico; 218 (vi). Sand dropseed is a relatively good source of spring forage. Although the palatability of the mature plant is low, growth in sandy desert sites is often renewed in the fall, the green basal leaves providing forage into the winter. The plant is useful for range seeding on dry, sandy lowlands receiving less than 22.5 cm annual precipitation (Vallentine 1961). Being a prolific seed producer, the plant increases on depleted ranges and wastelands.

Sporobolus flexuosus **(Thurber) Rydb.** Mesa Dropseed. [*Vilfa cryptandra* var. *flexuosa* Thurber]. Perennial, sometimes appearing annual; culms solitary or more often in small tufts, typically erect, 3–10 (12) dm tall; sheaths densely long-hairy at the summit and often ciliate along the margins; blades flat or folded, becoming involute, at least toward the long-tapered tips, 1–4 (6) mm wide and to ca 2 dm long, smooth or the margins scaberulous; ligule a dense ring of short hairs; panicle open, oblong, 10–30 cm long, often remaining partially included within the subtending sheath, the primary branches ascending to widely spreading or deflexed, mostly 1–6 cm long, typically rebranched near the base, the branchlets and usually the pedicels soon becoming widely spreading to deflexed, the pulvini at the base of the lower panicle branches often long-hairy; spikelets indistinguishable from those of *S. cryptandrus*; caryopses ca 1 mm long; $2n = 36, 38$. Desert shrub, sagebrush, and pinyon-juniper commuities from 850 to 1710 m in Emery, Garfield, Grand, Kane, Millard, San Juan, Washington, and Wayne counties; most of the southwestern U. S., east to Texas, and in Mexico; 93 (ii). Mesa dropseed is valuable for stabilization of loose sandy soils. It survives and reseeds readily in areas with as little as 15–18 cm mean annual precipitation (Thornburg 1982). The tangled masses formed from adjacent panicles are characteristic of this plant.

Sporobolus giganteus **Nash** Giant Dropseed. Perennial; culms solitary or tufted, erect, 6–20 dm tall and 2–12 mm thick near the base; sheaths typically ciliate along the margins, densely so at the summit, the hairs variable in length, occasionally extending onto the collar, old sheath bases sometimes present but not a consistently conspicuous feature; blades flat or folded, becoming involute toward the tips, 2.5–10 mm wide and to ca 5 dm long, smooth or the margins scaberulous; ligule a dense ring of hairs 0.5–1.5 mm long; panicle spikelike or occasionally somewhat interrupted below, 25–70 cm long and 0.5–2 cm wide, typically remaining partly or sometimes entirely included within the subtending sheath, the branches spikelet-bearing to the base, appressed to the main axis or occasionally those of the lower portion of the panicle spreading, those above remaining erect or nearly so; spikelets compressed, 2.5–3.5 (4) mm long, disarticulating above and ultimately below the glumes, pale to gray green; glumes thin, unequal, narrow, acute, essentially smooth on the prominent keel, glabrous to slightly scaberulous laterally (at 20x magnification), the first 0.7–2 mm long, the second 2–3.5 (4) mm long, slightly shorter to barely longer than the lemma; lemma thin, 2.5–3.5 (4) mm long, similar to the second glume; palea slightly shorter than the lemma; anthers 0.5–1 mm long; caryopses (1.3) 1.5–1.7 mm long; 2n = 36. Sandy soil in desert shrub and juniper communities, often on sand dunes or in roadside areas below 1830 m in Carbon, Emery, Garfield, Grand, Kane, San Juan, Utah, Washington, and Wayne counties; through most of the southwestern U. S. and in Texas, Kansas, Oklahoma, and northern Mexico; 16 (ii). See *S. contractus*. Giant dropseed is useful for dune stabilization in areas where the mean annual precipitation is 18–30 cm (Thornburg 1982).

Sporobolus pulvinatus Swallen Cushion Dropseed. Annual or short-lived perennial; culms solitary or tufted, 0.5–1.5 (3) dm tall, erect to spreading or geniculate at the lower nodes; sheaths glabrous throughout or sparsely long-hairy at the summit; blades flat, 1–5 mm wide and to ca 7 cm long, typically scaberulous and often sparsely long-hairy, the uppermost culm blade often much reduced; ligule a ring of hairs to ca 0.5 mm long; panicle open, pyramidal, 1–8 cm long, at maturity free of the subtending sheath, the lowermost primary branches whorled, naked at the base, the short pedicels erect-appressed; spikelets compressed, 1.5–2 mm long, pale to dark gray green; glumes unequal, more or less keeled, the first 0.3–0.7 mm long, the second 1.3–2 mm long, about equal to the lemma or nearly so; lemma 1.3–2 mm long, abruptly acute, glabrous or the upper midnerve scaberulous; palea about equal to the lemma; anthers to ca 0.3 mm long; caryopses to 1 mm long; 2n = 24, 36, 54. Known in Utah by a single collection (Harrison 12183 BRY, UTC) from sparsely vegetated desert pavement at about 1460 m in San Juan County; Colorado, Arizona, New Mexico, Texas, and northern Mexico; 2 (0).

Stipa L.

Tufted perennials, rarely annuals; culms generally erect; leaves often chiefly basal; sheaths open, the basal portion from previous years frequently persistent and conspicuous; blades involute or less often flat to folded; ligule membranous or rarely a ring of hairs; inflorescence a spikelike to open panicle; spikelets terete or slightly compressed, 1-flowered, disarticulating above the glumes; glumes moderately firm to largely or entirely membranous, lanceolate to lance-elliptical or occasionally ovate, equal or unequal, the first often longer than the second, both exceeding the body of the lemma, 1- to 5 (9) -nerved, acute to more often long-tapered to a sharply acute tip; lemma at maturity stiffly membranous to leathery or hardened, subglobose to narrowly elliptical, (3) 5-nerved, mostly obscurely so, generally hairy to some degree on the body, the apex awned, the junction of awn and lemma distinct, the awn persistent or in a few species readily deciduous, usually once or twice abruptly bent, the lower 1 or 2 segments typically more or less twisted; callus well developed, obtuse to sharply acute, bearded; palea hyaline to stiffly membranous, 2-nerved, obscured by the overlapping margins of the lemma except at flowering in a few species; stamens 3, the anthers in some species bearing at the apex a minute tuft of hairs; caryopsis closely invested by the lemma and palea, the embryo 1/6–1/3 as long as the grain; x = 11. The treatment of *Stipa* was reviewed by M. E. Barkworth. Species of *Stipa* hybridize more or less freely among themselves (Gould and Shaw 1983) and with the species traditionally known as *Oryzopsis hymenoides* (Johnson 1945, 1972), here included in the genus *Stipa*. Hybrids involving *S. hymenoides* have been particularly well documented, possibly because they are easiest to recognize (see *S. x bloomeri*).

Barkworth, M. E. 1979. Proposal to reject *Stipa columbiana* (Poaceae) and nomenclatural changes affecting three western North American species of *Stipa* (Poaceae). Taxon 28: 621–625.

———, J. McNeil, and J. Maze. 1979. A taxonomic study of *Stipa nelsonii* (Poaceae) with a key distinguishing it from related taxa in western North America. Canad. J. Bot. 57: 2539–2553.

Johnson, B. L. 1945. Natural hybrids between *Oryzopsis hymenoides* and several species of *Stipa*. Amer. J. Bot. 32: 599–608.

———. 1972. Polyploidy as a factor in the evolution and distribution of grasses. pp. 23–34. In: *The biology and utilization of grasses*. Edited by V. B. Youngner and C. M. McKell. Academic Press, New York.

Note: In the following key and descriptions, panicle length excludes the awns; lemma measurements include the callus but neither the apical hairs (when present) nor the awn.

1.	Awn deciduous, the lemma pubescent throughout with hairs 1–4 mm long, sometimes ultimately glabrate	2
—	Awn persistent, the lemma variously pubescent	3
2(1).	Panicle at maturity open, the elongate branches mostly dichotomously branched and divaricate; lemmas less than 3 times as long as wide, the awn 3–6 mm long	*S. hymenoides*
—	Panicles contracted to somewhat open, the short to long branches erect or ascending and variously rebranched; lemmas at least 3 times as long as wide, the awn 7–20 mm long	*S. x bloomeri*
3(1).	Awns of the lemmas averaging more than 50 (53–240) mm long	4
—	Awns of the lemmas averaging no more than 50 mm long	6
4(3).	Lemma 4–6 mm long	*S. arida*
—	Lemma 8–19 mm long	5
5(4).	Terminal awn segment pubescent with hairs 1.5–3 mm long	*S. neomexicana*
—	Terminal awn segment scabrous or puberulent with hairs to about 1 mm long	*S. comata*
6(3).	Awn pubescent near the base with hairs averaging more than 0.5 (0.5–9) mm long	7
—	Awn glabrous or scabrous throughout or, if pubescent, the hairs averaging less than 0.5 mm long	8
7(6).	Lower awn segment with hairs 3–9 mm long	*S. speciosa*
—	Lower awn segment with hairs 0.5–2 mm long	*S thurberiana*
8(6).	Awn (35) 40 mm or longer, flexuous or weakly once-geniculate, the upper portion hairlike or nearly so	*S. arida*
—	Awn averaging less than 35 (3–35) mm long, typically once or twice geniculate, the upper portion variable	9
9(8).	Lemma densely hairy with apical or lateral hairs at least 2 mm long	10

—	Lemma glabrous above or sparsely pubescent with hairs nearly always less than 2 mm long 12
10(9).	Sheaths at base of plant terete and mostly less than 1 mm wide, often grayish and usually inconspicuous; blades typically less than 1 mm wide; callus sharply acute .... *S. pinetorum*
—	Sheaths at base of plant loose, often flattened, more than 1 mm wide, usually straw colored, more or less thickened, papery, and typically forming a conspicuous feature; blades 1–7 mm wide; callus blunt (at 10x magnification) 11
11(10).	Awn 8–14 mm long, once geniculate; palea long-hairy throughout, the hairs surpassing the tip .... *S. coronata* var. *depauperata*
—	Awn 12–25 mm long, once or twice geniculate; palea glabrous or sparsely hairy below, the hairs not surpassing the tip .... *S. coronata* var. *parishii*
12(9).	Joint between awn and lemma oblique, not obscured by hairs, the apical margin of the lemma thickened, forming a minute unilateral "hump" on the mature fruit; palea nearly as long as the lemma .... *S. lemmonii*
—	Joint between awn and lemma horizontal, in some species obscured by hairs, the apical margins of the lemma either membranous or not prolonged; palea not more than 2/3 as long as the lemma 13
13(12).	Palea less than half as long as the lemma, glabrous; lemma at maturity narrowed toward the apex into a whitish "neck" to about 1 mm long, the joint between lemma and awn not obscured by the scant apical hairs; plants introduced, occupying disturbed habitats .... *S. viridula*
—	Palea either at least half as long as the lemma or pubescent to some degree; lemma at maturity not consistently white at the apex, the joint between lemma and awn often obscured by the apical hairs; plants native, generally forming large populations in native habitats (the next two species occasionally intergrading) 14
14(13).	Palea at least 2/3 as long as the lemma; awns typically averaging no more than 22 (10–23) mm long; callus to about 0.7 mm long, the glabrous tip blunt; plants usually less than 6 dm tall; leaf blades mostly involute, 0.5–2 mm wide. .... *S. lettermanii*
—	Palea about half as long as the lemma; awns averaging more than 20 mm long; callus 0.7–1.5 (3) mm long, the glabrous tip acute and usually curved; plants to 15 dm tall; leaf blades often flat, frequently at least some of them more than 2 mm wide .... *S. nelsonii*

Stipa arida Jones Mormon Needlegrass. [*S. mormonum* Mez, type from near Milford, Beaver County]. Tufted perennial; culms relatively slender, 3–8 (10) dm tall; sheaths glabrous or scabrous, occasionally minutely hairy just below the collar or ciliate near the summit, old sheath bases persistent; blades involute-filiform, those of the culm rarely flat and then 0.8–2.5 mm wide, scabrous and often puberulent; ligule 0.2–1.5 mm long; panicle narrowly oblong, 0.4–3 dm long, the branches to ca 6 cm long and erect to slightly divergent, the base of the inflorescence typically remaining enclosed in the upper sheath; glumes ultimately membranous throughout or nearly so, pale and shiny, narrowly lanceolate, unequal or occasionally equal or nearly so, long-tapered to sharply acute tips, 3 (5)-nerved, mostly obscurely so, glabrous or sparsely scaberulous, the first (8) 9–15 mm long, the second 7–10 (11) mm long; lemma pale or at maturity dark brown and papery, 4–6 mm long, lacking free apical margins, more or less sparsely short-pubescent throughout or glabrous above, the hairs less than 0.7 mm long, the awn (35) 40–75 mm long, slender, sometimes weakly once geniculate, more or less twisted on the lower 1–2 cm, straight to flexuous and hairlike or nearly so above, scaberulous below or glabrous throughout; callus to ca 1 mm long, curved, with a sharply acute, abaxially glabrous tip to ca 0.5 mm long; palea about half as long as the lemma, hairy; anthers 2–3 mm long. Chiefly on rocky sites in blacksage, shadscale, sagebrush, and pinyon-juniper communities from 1050 to 2300 m in Beaver, Emery, Garfield, Grand, Juab, Kane, Millard, Piute (type from near Marysvale), San Juan, Tooele, Washington, and Wayne counties; California east to Colorado and south through Arizona and New Mexico to Texas; 70 (ii).

Stipa x bloomeri Bolander Bloomer Ricegrass. [*Oryzopsis bloomeri* (Bolander) Ricker; *x Stiporyzopsis bloomeri* (Bolander) B. L. Johnson]. Tufted perennial; culms erect, (1) 3–9.5 dm tall; sheaths glabrous or scaberulous to variously pubescent, the old sheath bases typically persistent; blades flat to involute, to ca 2 mm wide, often long-tapered to filiform tips, smooth or scaberulous; ligule mostly 1–4 mm long; panicle contracted to somewhat open, 0.6–3.5 dm long, the branches mostly 2–7 cm long or longer, erect or more or less divergent; glumes greenish below or membranous throughout, occasionally purple-tinged, subequal or unequal, (4.5) 5–11 mm long, broad below, long-tapered to a finely acute or awnlike tip, obscurely or conspicuously nerved, glabrous or scaberulous; lemma at maturity leathery, mostly 4–5.8 mm long and at least 3 times as long as wide, sparsely to densely pubescent with spreading to appressed hairs mostly 1–3 mm long, the awn more or less readily deciduous, 7–20 mm long or longer, usually once geniculate, the lower segment twisted and scabrous to pubescent; callus to ca 1 mm long with a blunt to acute, glabrous or hairy tip; anthers 1.2–5.5 mm long, often bearing a minute apical tuft of hairs, the pollen typically undeveloped. Dry, mostly open sites, at 2130 to 2750 m in Garfield, Iron, Utah, and Washington counties (and likely to occur throughout the range of *S. hymenoides*); throughout the western U. S. and in South Dakota; 37 (i). Although Johnson (1945) originally applied the name x *Stiporyzopsis bloomeri* only to the hybrid between *Stipa hymenoides* (as *Oryzopsis hymenoides*) and *S. occidentalis*, additional hybrids between *S. hymenoides* and 10 other *Stipa* species are sufficiently similar to the hybrid with *S. occidentalis* to collectively bear the name, here revised to *S. x bloomeri*. Of those species occurring in our area, the following are reported to have formed hybrids with *S. hymenoides*: *S. viridula*, *S. thurberiana*, *S. neomexicana*, *S. speciosa*, *S. pinetorum*, and *S. nelsonii* (Johnson 1972).

Stipa comata Trin. & Rupr. Needle-and-thread Grass. Tufted perennial, often glaucous; culms 3–13 dm tall; sheaths glabrous or scabrous, the old bases persistent; blades flat or involute, 1–2 (3) mm wide, smooth or scabrous to puberulent; ligule 1.5–9 mm long; panicle contracted to somewhat open, 0.7–4 dm long, often nodding, the branches few flowered, appressed to somewhat divergent; glumes ultimately membranous throughout or nearly so, equal or subequal, 15–40 mm long, long-tapered to filiform tips 5–20 mm or more long, the first 5-nerved, the second 5- to 7-nerved, usually conspicuously so; lemma brown and leathery at maturity, 8–17 mm long, sparsely short-hairy throughout or glabrate near the apex, the awn (53) 60–240 mm long, once or twice geniculate, sometimes weakly so, the lower half twisted, scabrous to pubescent with hairs to ca 1 mm long, the upper segment mostly flexuous and scaberulous, occasionally the awn short-hairy throughout, the hairs of the lower portion longer than those above (unlike *S. neomexicana*); callus (2) 3–6 mm long with a sharply acute, glabrous tip 0.5–0.8 mm long; palea subequal to the lemma, firm, glabrous or sparsely hairy; anthers to ca 8 mm long; 2n = 38, 44–46. Chiefly in salt desert shrub, desert shrub, sagebrush, pinyon-juniper, and mountain brush communities, occasionally in openings in aspen, lodgepole pine, and ponderosa pine, from 1060 to 3050 m in all Utah counties except Iron, Morgan, Summit, and Weber; Alaska to Ontario, south throughout the western U. S. and the Great Plains to Texas, sporadic in the eastern U. S.; 263 (vii). The Utah material I have seen is var. *comata* with the terminal segment of the awn flexuous and usually well over 50 mm long, and with the lower branches of the panicle generally included in the sheath. See *S. neomexicana* for a discussion of its relationship to *S. comata*. Needle-and-thread grass produces highly palatable early spring forage and is a good source of fodder in fall and winter, although on the summer range mature fruit frequently causes mechanical injury to grazing animals (Vallentine 1961). The grass grows in areas with mean annual precipitation of less than 25 cm and does well on sandy or gravelly sites (Thornburg 1982).

Stipa coronata Thurber Crested Needlegrass. The *S. coronata* complex is comprised of tufted perennials characterized by light green to glaucous herbage; papery, persistent sheath bases; firm, erect to recurved leaf blades long-tapered to involute tips; a short (to 2 mm long) ligule; relatively firm, conspicuously nerved glumes, a lemma pubescent with hairs 2–5 mm long, and a palea 1/2–2/3 as long as the lemma. As described by Hitchcock (1951), the species consists of two varieties: var. *coronata* with culms 10–20 dm tall, panicles 3–4 dm long, and lemmas with a twice-geniculate awn 40–50 mm long, the plants ranging from southern California to Arizona (?), south to northern Mexico; and var. *depauperata* (Jones) A.S. Hitchc. with culms 3–5 dm tall, panicles 1–1.5 dm long, and lemmas with a once-geniculate awn to about 2.5 cm long, the plants ranging from southern California to Nevada, Utah, and Arizona. In addition to the typical element of the species, which does not enter Utah, the complex as it occurs in the state appears to be composed of two previously undifferentiated enties:

1. Culms 2–7 (8) dm tall, erect; sheaths ciliate with a small tuft of hairs lateral to the ligule; blades 2–6 dm long and 1.4–6 mm wide; panicles 1–2.6 dm long; lemma 6–8.5 mm long, the apical hairs 2–3 mm long, the awn (12) 17–25 mm long and once or weakly twice geniculate; callus with a minute, often obscure, knoblike glabrous tip; palea glabrous or sparsely hairy below, the hairs not surpassing the tip; anthers 3.5–5 mm long .... *S. coronata* var. *parishii*

— Culms 1–4 (5) dm tall, erect to decumbent-based; sheaths glabrous or minutely ciliate, lacking longer tufts of hairs lateral to the ligule; blades to ca 1.5 dm long and 1–2.5 mm wide; panicles 0.4–1.5 dm long; lemma 5–7(8) mm long, the apical hairs 3–5 mm long, the awn 8–14 mm long and once geniculate; callus with an oblong, abaxially glabrous tip to ca 0.3 mm long; palea hairy throughout, the hairs extending beyond the tip; anthers 2–3.5 mm long .... *S. coronata* var. *depauperata*

Var. *depauperata* (Jones) A. S. Hitchc. [*S. parishii* var. *depauperata* Jones, type from Juab County]. Salt desert shrub, desert shrub, sagebrush, and pinyon-juniper communities at 1500 to 2700 m in Beaver, Juab, Millard, Tooele, and Washington counties; Nevada; 41 (0).

Var. *parishii* (Vasey) A. S. Hitchc. [*S. parishii* Vasey]. Desert shrub, sagebrush, and pinyon-juniper communities, from 1585 to 1830 m in Garfield, Kane, and San Juan counties; California, Nevada, and Arizona; 25 (0). The complex, as it occurs throughout its range, is currently being studied (M. E. Barkworth, pers. comm. 1986); and one or both of the taxa described above may prove to merit species status. In any case, the names applied here are tentative. *Stipa scribneri* Vasey, which differs from the *S. coronata* complex in combining a sharply acute callus with a palea less than half as long as the lemma, is reported from Utah by Johnson (1972); but I have seen no material with this combination of characters and suggest that the plants reported as *S. scribneri* were actually *S. coronata* var. *parishii*. Two collections from the western Great Basin that outwardly resemble *S. coronata* var. *parishii* may prove to be a previously undescribed species.

Stipa hymenoides R. & S. Indian Ricegrass. [*Oryzopsis hymenoides* (R. & S.) Ricker]. Densely tufted perennial; culms erect, 2–7 (8.5) dm tall; sheaths glabrous to puberulent, the margins often ciliate, with or without tufts of longer hairs lateral to the ligule, the old bases of previous years thickened and persistent, the uppermost sheath often partially enclosing the panicle; blades elongate, firm, involute and in that state ca 1 mm wide; ligule 2.5–8 mm long; panicle open, 5–20 cm long, the branches slender, often flexuous, usually dichotomously branched, the branchlets mostly 0.5–4 cm long, nearly hairlike and divaricate; glumes initially greenish, soon becoming white-membranous, broadly ovate and saccate, equal or nearly so, 4–8 mm long, 3– to 5–nerved, puberulent to glabrous or nearly so, abruptly tapered to a slender tip mostly 3–6 mm long; lemma at maturity hardened, glossy brown to black, subglobose to fusiform, 2.5–5 mm long and to ca 3 mm wide, less than 3 times as long as wide, densely pubescent with hairs 2–4 mm long or sometimes glabrate, the awn readily deciduous, 3–6 mm long, stout, straight or nearly so; callus 0.5–1 mm long, obtuse to sharply acute, long-hairy throughout; palea similar to the lemma in texture, color, and length; anthers 0.8–1.5 mm long, with an apical tuft of hairs; 2n = 28, 48, 65, 130. Chiefly dry, open, often sandy sites, in greasewood, creosote bush, shadscale, sagebrush, mountain brush, pinyon-juniper, and ponderosa pine communities, occasionally at margins of aspen-spruce-fir communities, at 750 to 2750 m in all Utah counties except Morgan; British Columbia to Manitoba, south throughout the western U. S., east through the Great Plains and in Texas; in arid regions of the Pacific slope in Chile; 233 (ii). Originally included in the genus *Stipa*, *S. hymenoides* was transferred to *Oryzopsis* largely because of its short plump caryopsis and its readily deciduous awn. In 1945, Johnson (Bot. Gaz. 107: 1–32) transferred *O. webberi* (Thurb.) Benth. ex Vasey to *Stipa*, thus eliminating the deciduous awn as a character restricted to *Oryzopsis*. An evaluation of floral development in a number of species of *Oryzopsis* led Kam and Maze (Bot. Gaz. 135: 227–247. 1974) to recommend that *O. hymenoides* also be placed in *Stipa*. Subsequently, in a study of the embryological characters and taxonomy of the Stipeae, Barkworth (Taxon 31: 233–243. 1982) found *S. hymenoides* (as *Oryzopsis hymenoides*) to be the most stipoid in its embryological characters of any of the seven species of *Oryzopsis* examined. These data, added to the morphological resemblance of *O. hymenoides* to *Stipa* (e.g., the long callus, lanceolate glume tips, and long-hairy lemma), and its ability to hybridize with ten or more species of *Stipa* (Johnson 1972), make a most convincing case for the return of *O. hymenoides* to the genus *Stipa*. Indian ricegrass is dominant over large desert and semidesert areas where it forms an important source of forage for all classes of livestock. The grass cures exceptionally well and the seeds, which remain on the plant, are high in protein. Strains with an unusually low percentage of nongerminating seeds are recommended for range rehabilitation throughout the Intermountain West (Thornburg 1982).

Stipa lemmonii (Vasey) Scribn. Lemmon Needlegrass. [*Stipa pringlei* var. *lemmonii* Vasey]. Tufted perennial; culms 3–8 dm tall; sheaths glabrous or pubescent, old sheath bases persistent; blades involute-filiform or flat and then 1–2 mm wide, glabrous or scaberulous to puberulent; ligule 1–3 mm long; panicle narrow, 0.5–1.2 dm long, the branches mostly erect; glumes ultimately membranous throughout or nearly so, occasionally purple-tinged, equal or the first slightly shorter than the second, 7–11 mm long, somewhat abruptly tapered to sharply acute or hairlike tips, more or less obscurely (3) 5-nerved; lemma pale or at maturity brown and leathery to hardened, 5.5–7 mm long, sparsely to moderately pubescent with appressed to spreading hairs to ca 1 mm long, the few hairs at the apex typically much shorter, the joint between the awn and the lemma body oblique, the free apical margin of the lemma thickened, forming a firm, unilateral "hump" at the tip of the mature fruit, the awn 10–30 (35) mm long, mostly twice geniculate, the lower 2/3 twisted and puberulent; callus to about 1 mm long with a very short, blunt, glabrous tip 0.1–0.25 mm long; palea firm, nearly as long as the lemma, abruptly acute at the apex, glabrous or inconspicuously hairy; $2n = 34, 36$. Grass and sagebrush communities and open woodlands from 1830 to 2280 m in Cache and possibly Salt Lake (Wasatch Range, Jones s.n., GH) counties; southwestern Canada south through much of the western U. S.; 2 (0). Barkworth and Linman (Madrono 31: 48–56. 1984) found that morphological and distribution data for *S. lemmonii* do not support recognition of infraspecific taxa.

Stipa lettermanii Vasey Letterman Needlegrass. [*S. minor* (Vasey) Scribn.]. Tufted perennial; culms 2–7 (10) dm tall; sheaths glabrous or scaberulous, old sheath bases persistent; blades involute and linear to filiform, rarely flat and then to ca 2 mm wide, smooth or scaberulous to puberulent; ligule 0.2–2 mm long; panicle sublinear to narrowly oblong, mostly 0.4–2 dm long, the branches erect or nearly so; glumes ultimately membranous throughout or nearly so, green or purple-tinged, equal or the second slightly longer than the first, (5) 6–10 mm long, tapered to hairlike tips (these sometimes broken off), 3-nerved, often obscurely so, glabrous or scaberulous; lemma at maturity stiffly membranous, light brown, 4.5–6 mm long, sparsely pubescent throughout, the apical hairs mostly 1–1.7 mm long, equal to or longer than those on the body, the joint between the awn and lemma body horizontal, usually obscured by the apical hairs, the free margins of the lemma often prolonged at the apex as minute membranous lobes mostly less than 1 mm long, the awn 10–23 mm long, mostly twice geniculate, the two lower segments twisted, glabrous to scabrid-puberulent, the terminal segment glabrous; callus to about 0.7 mm long with a blunt, glabrous tip ca 0.1 mm long; palea exposed at flowering, thin, about 2/3 as long as the lemma, tapered to an acute tip, pubescent; anthers 1.7–2.5 mm long; $2n = 32, 66, 68$. Openings in sagebrush, mountain brush, pinyon-juniper, ponderosa pine, lodgepole pine, aspen, spruce-fir, and occasionally in mountain meadow communities, from 1500 to 3570 m in all Utah counties; throughout most of the western U. S.; 162 (xvii). Discrepancies in the treatment accorded *S. lettermanii* by authors of western manuals (e.g., Gould 1951, Hitchcock et al. 1969, and Harrington 1954) attest to the variation in populations. Of Utah specimens examined, 12 percent failed in at least one varying character to conform with those used in the key herein to separate *S. lettermanii* from the closely related, often sympatric *S. nelsonii*. Specimens from adjoining western states, however, appear to combine characters of both species in a much more confusing pattern of intermediacy. Although the taxa appear unequivocally distinct in Utah despite occasional intergradation, Weber (1976) finds Colorado specimens doubtfully so. Letterman needlegrass provides fair to good forage for cattle, depending upon its abundance (Vallentine 1961).

Stipa nelsonii Scribn. Nelson Needlegrass. [*S. columbiana* authors, not Macoun; *S. columbiana* var. *nelsonii* (Scribn.) A. S. Hitchc.; *S. occidentalis* var. *minor* sensu C. L. Hitchc.; *S. williamsii* Scribn.]. Tufted perennial; culms 1–20 dm tall; sheaths glabrous or sparsely to densely pubescent, occasionally ciliate near the summit, old sheath bases more or less persistent; blades flat to involute, 0.5–5 (7) mm wide, smooth or scabrous on the lower surface, glabrous to scaberulous; ligule 0.2–2 mm long; panicle narrow, 0.7–3.6 dm long, 0.5–2 cm wide, the branches erect or strongly ascending; glumes ultimately membranous throughout or nearly so, occasionally purple-tinged, equal or subequal, (6.4) 7–13 mm long, long-tapered to hairlike tips (these sometimes broken off), prominently to obscurely 3 (5)–nerved, glabrous or scaberulous; lemma at maturity stiffly membranous, brown, 4.9–7.2 mm long, appressed-pubescent throughout, the apical hairs mostly 0.5–1.5 mm long, equal to or longer than those on the body, the joint between awn and lemma horizontal, usually obscured by the apical hairs, the free apical margins of the lemma often prolonged as membranous lobes to about 0.5 mm long, the awn mostly 20–30 mm long, usually twice geniculate, the lower segments twisted, some portion of the middle section scaberulous to pubescent with hairs to ca 0.4 mm long, rarely the entire awn essentially glabrous; callus 0.7–3 mm long with an acute (ours), usually curved, abaxially glabrous tip mostly 0.2–0.9 mm long; palea exposed at flowering, thin, about half as long as the lemma, blunt to acutish at the apex, hairy at least near the tip; anthers 2–4 mm long; $2n = 36, 44$. Chiefly in openings in sagebrush, mountain brush, pinyon-juniper, and aspen-spruce-fir, occasionally in ponderosa pine, lodgepole pine, and meadow communities or on exposed alpine ridges, at 1500 to 3350 m in all Utah counties except Kane and Piute; Alaska and western Canada, south through most of the western U. S., east to Texas; 258 (xviii). According to

Barkworth (pers. comm. 1986), the only phase of the species to enter our area is ssp. *nelsonii*, characterized by an acute rather than blunt callus. Hitchcock et al. (1969) treated the *S. occidentalis* complex, including *S. nelsonii* (as *S. columbiana*), as a single species. Barkworth et al. (1979) segregate *S. occidentalis* as a species distinct from the remainder of the complex in having the lower awn segment pubescent with hairs 1–2 mm long. Although Hitchcock (1951) reported *S. occidentalis* (in the sense of Barkworth et al.) from Utah, I have seen no plants in our area with awn pubescence longer than 0.5 mm. *Stipa nelsonii* is known to form hybrids with *S. hymenoides* (see *S. x bloomeri*) and also intergrades to some extent with *S. lettermanii* (q.v.). Nelson needlegrass is variable in palatability, but is usually rated high for cattle and horses on summer range (Vallentine 1961).

Stipa neomexicana (Thurber) Scribn. New Mexico Feathergrass. [*S. pennata* var. *neomexicana* Thurber]. Tufted perennial, typically glaucous; culms 4–10 dm tall; sheaths glabrous or occasionally puberulent, rarely ciliate with longer hairs; blades involute-filiform, glabrous or puberulent; ligule of upper culm blades 0.5–3 mm long; panicle mostly 1–3 dm long, narrow; glumes ultimately membranous throughout or nearly so, subequal, (20) 30–60 mm long, long-tapered to sharply acute or hairlike tips, 5- to 7-nerved, glabrous; lemma at maturity brown and leathery, mostly 12–18 (19) mm long, generally short-hairy, the awn 120–220 mm long, once or occasionally twice geniculate, the lower segment(s) twisted and glabrous to minutely hairy, the longer upper segment densely pubescent throughout or nearly so with hairs mostly 1.5–3 mm long; callus to 6 mm long with a curved, sharply acute, abaxially glabrous tip to 1 mm long; palea similar to the lemma in length and texture, glabrous to sparsely hairy; $2n = 44$. Desert shrub and pinyon-juniper communities at 915 to 2000 m in Emery, Garfield, Kane, San Juan, Washington, and Wayne counties; Wyoming south to Arizona, New Mexico, and Texas, adventive in Oklahoma; 23 (0). Gould (1975) suggests that *S. neomexicana* may not be specifically distinct from *S. comata*. Indeed, the only distinction appears to lie in the awn indument: scabrous to puberulent in *S. comata* and densely pubescent with relatively long hairs in *S. neomexicana*. Moreover, the occurrence of intermediate specimens with awns short-hairy throughout is not uncommon. *Stipa neomexicana* is known to hybridize with *S. hymenoides* (see *S. x bloomeri*) and appears to intergrade with *S. comata* where their ranges overlap.

Stipa pinetorum Jones Pinewoods Needlegrass. Tufted perennial, often forming large circular clumps, usually glaucous; culms slender, 1–6 (8) dm tall; sheaths typically glabrous throughout, the old sheath bases terete, short, slender (mostly less than 1 mm wide), usually grayish, often inconspicuous; blades involute-filiform, typically less than 1 mm wide, smooth or scaberulous to puberulent; ligule 0.1–0.6 mm long; panicle sublinear to narrow, 0.5–2 dm long, the branches erect to slightly divergent; glumes ultimately membranous throughout or nearly so, equal or subequal, 7–12 mm long, long-tapered to sharply acute or hairlike tips, 1- to 3-nerved, glabrous or scaberulous; lemma at maturity papery and light brown, with membranous apical margins to ca 1 mm long, the body 3.5–5 (6) mm long, densely pubescent with hairs mostly 2.5–4 mm long, the length usually more or less uniform throughout, the awn 10–25 mm long, typically twice geniculate, the two lower segments twisted, essentially glabrous or scaberulous to scabrid-puberulent; callus to ca 0.8 mm long, with a sharply acute abaxially glabrous tip to ca 0.3 mm long; palea at least 2/3 as long as the lemma, acute at the apex, pubescent like the lemma, the hairs at the tip exceeding the body of the lemma; anthers 1.5–3 mm long; $2n = 32$. Dry, often rocky slopes, shale barrens, and sagebrush, mountain brush, pinyon-juniper, and ponderosa pine communities at 2000 to 2900 m in Beaver, Daggett, Emery, Garfield (type locality), Juab, Millard, Sevier, Utah, Wasatch, Washington, and Wayne counties; Oregon east to Wyoming, south to California, Nevada, and Colorado; 22 (0). *Stipa pinetorum* is known to form hybrids with *S. hymenoides* (see *S. x bloomeri*).

Stipa speciosa Trin. & Rupr. Desert Needlegrass. Tufted perennial, the herbage typically glaucous; culms 3–6 (7.5) dm tall; sheaths firm, glabrous throughout or minutely hairy along the margins, especially near the summit, the old sheath bases persistent, forming a conspicuous feature; blades involute-filiform, mostly stiffly erect and pungent at the tips, smooth or scabrous; ligule of lower blades short, consisting chiefly of a ring of hairs, those of the uppermost blades typically membranous, 3–10 mm long, glabrous or variously hairy, often densely ciliate; panicle 0.7–2 dm long, narrow, the branches erect or strongly ascending; glumes ultimately membranous throughout or nearly so, equal or subequal, 14–25 mm long, long-tapered to sharply acute tips, glabrous or scaberulous, the first 1- to 3-nerved, the second 3- to 7-nerved; lemma at maturity brownish and firm, 7–10 mm long, more or less densely short-pubescent throughout or glabrate toward the apex, the awn 30–50 mm long, typically once geniculate, the lower segment twisted and densely pubescent with hairs 3–9 mm long, the terminal segment scaberulous and straight or nearly so; callus 1–2.5 mm long, with a curved, sharply acute, abaxially glabrous tip 0.5–1 mm long; palea thin, about half as long as the lemma, obtuse to acutish, glabrous; anthers 1.5–4 mm long; $2n = 60, 64, 66$. Joshua tree, desert shrub, sagebrush, and pinyon-juniper communities, and open, often rocky slopes, from 800 to 1900 m in Beaver, Emery, Garfield, Grand, Juab, Kane, Millard, San Juan, Tooele, Uintah, and Washington counties; southwestern U. S. and northern Mexico; South America; 102 (i). *Stipa speciosa* is known to form occasional hybrids with *S. hymenoides* (see *S. x bloomeri*).

Stipa thurberiana Piper Thurber Needlegrass. Tufted perennial, typically glaucous; culms slender, 1.5–6 (8) dm tall; sheaths glabrous or occasionally scabrous to puberulent below, not hairy at the summit; blades flexuous, involute-filiform, usually scabrous or soft-pubescent; ligule 2–7 mm long, glabrous or scaberulous on the abaxial surface; panicle mostly 0.5–1.5 dm long, narrow to somewhat open, the branches few flowered; glumes greenish or purple-tinged, ultimately membranous throughout or nearly so, equal or subequal, 8–16 mm long, long-tapered to sharply acute, often hairlike tips, 3 (5) -nerved, glabrous or nearly so; lemma pale, leathery at maturity, 6–8.7 mm long, sparsely short-hairy, often glabrate toward the apex, the awn 30–50 mm long, twice geniculate, the 2 lower segments twisted and pubescent with hairs 0.5–1.5 (2) mm long near the base and progressively shorter upward; callus to ca 1 mm long, the sharply acute, abaxially glabrous tip to ca 0.3 mm long; palea subequal to the lemma, otherwise similar to it in texture and pubescence; anthers 2–4 mm long; 2n = 34. Sagebrush, oak, and pinyon-juniper communities at 1650 to 1980 m in Box Elder and Juab counties; through much of the western U. S.; 5 (0). *Stipa thurberiana* is known to hybridize with *S. hymenoides* (see *S. x bloomeri*).

Stipa viridula **Trin.** Green Needlegrass. Tufted perennial; culms 5–11 dm tall; sheaths glabrous throughout or hairy along the margins and often at the summit, the old sheath bases persistent; blades flat to involute, 2–6 mm wide, smooth or scabrous; ligule 0.5–2 (3) mm long; panicle narrowly oblong, 1–2.5 dm long and to ca 2 cm wide, the branches erect or only slightly divergent; glumes ultimately membranous throughout or nearly so, equal or subequal, 8–13 mm long, long-tapered to sharply acute or hairlike tips, prominently 3 (5) –nerved, glabrous or the midnerve scaberulous; lemma at maturity brown and leathery, usually 5–6.5 mm long, the body narrowed toward the apex into a glabrate, whitish "neck" 0.5–1 mm long, sparsely appressed-pubescent below the "neck" and with a tuft of apical hairs generally less than 1 mm long, the apical margins not prolonged, the awn 20–35 mm long, mostly twice geniculate, the 2 lower segments twisted, subglabrous to scaberulous; callus to ca 1 mm long with a very short, blunt, glabrous tip; palea stiffly membranous, less than half as long as the lemma, acute to rounded at the apex, glabrous; anthers 2–3 mm long or vestigial and scarcely 1 mm long; 2n = 82, 88. Along roads and in grass and sagebrush communities, at 1370 to 2150 m in Iron, Rich, Salt Lake, Sanpete, and Uintah counties; native to the Great Plains north of Kansas, introduced in Canada from British Columbia to Manitoba, south through the Rocky Mt. states; 8 (ii). *Stipa viridula* is known to hybridize with *S. hymenoides* (see *S. x bloomeri*). Green needlegrass is native to the northern and central Great Plains and is a fairly recent and infrequent introduction in Utah. The plant has been adapted for seeding on a wide range of soil textures, but does especially well on clayey and clay soils. It can survive in areas receiving as little as 25 cm mean annual precipitation (Thornburg 1982).

Tridens R. & S.

Perennials, rarely with short rhizomes; culms tufted; sheaths open; blades flat to involute; ligule a ring of hairs or a ciliate membrane; inflorescence an open to contracted panicle or raceme; spikelets mostly subcylindrical, 3– to 14–flowered, disarticulating above the glumes and between the florets; glumes subequal to unequal, the first 1–nerved, the second 1– to 3 (7) –nerved; lemma rounded on the back, mostly notched at the apex, 3–nerved, usually more or less pubescent below midlength; palea subequal to the lemma, often minutely hairy; stamens 3; caryopsis falling free of the lemma and palea, the embryo ca 2/5 as long as the grain; x = 10.

Tateoka, T. 1961. A biosystematic study of *Tridens* (Gramineae). Amer. J. Bot. 48: 565–573.

Tridens muticus **(Torr.) Nash** Slim Tridens. [*Tricuspis mutica* Torr.; *Triodia mutica* (Torr.) Scribn.]. Perennial; herbage often glaucous, glabrous or scaberulous to hairy; culms densely tufted, mostly 1.5–6.5 dm tall, erect; leaf blades flat or involute, 1–3 mm wide and mostly 3–25 cm long; ligule a ring of short hairs, sometimes membranous-based, the whole less than 1 mm long; panicle or raceme sublinear to narrowly elongate, 4.5–20 cm long and mostly less than 1 cm wide, branches and pedicels erect or slightly divergent at flowering, the pedicels averaging much shorter than the spikelets; spikelets terete or nearly so, 5– to 11–flowered, (6) 7–12 (14) mm long, often purple-tinged; glumes lanceolate to elliptical, acute to obtuse at the apex, glabrous, the first 2–5 (6) mm long, 1–nerved, the second 4–6 mm long, 1 (3) –nerved, shorter than to nearly as long as the lowermost lemma; lemmas broadly elliptical to obovate, (3.5) 4.5–6 mm long, obtuse to rounded and usually shallowly notched and often mucronate at the apex, conspicuously 3–nerved, densely soft-hairy, the hairs to ca 1.5 mm long, appressed or spreading, and confined chiefly or entirely to the nerves from below to slightly above midlength; palea shorter and narrower than the lemma; anthers 0.3–2 mm long; 2n = 32–40. Creosote bush, desert shrub, sagebrush, and pinyon-juniper communities at 610 to 1750 m in Beaver, Garfield, Millard, San Juan, and Washington counties; much of the southwestern U. S., the southern Great Plains and Texas, south to central Mexico; 29 (0). Slim tridens is palatable to all classes of livestock, but the plants are too scattered to be of significance as a source of forage.

Trisetum Pers.

Annuals or tufted perennials; sheaths open; blades mostly flat; ligule membranous; inflorescence a spikelike to open but usually narrow, more or less glistening panicle; spikelets narrowly oblong, flattened, (1) 2– to 3 (5) –flowered, the rachilla typically prolonged beyond the uppermost floret, conspicuously hairy, disarticulating above the glumes and between the florets or sometimes below the glumes; glumes lanceolate, equal or unequal, keeled, membranous-margined, acute at the apex, the first 1 (3) –nerved, the second 3 (5) –nerved and ranging from slightly shorter than the lowermost lemma to slightly longer than the uppermost one; lemma similar to or firmer than the glumes in texture, lanceolate to narrowly ovate, laterally compressed and keeled to ultimately rounded on the back, obscurely 5–nerved, sharply acute to narrowly rounded and entire or more often cleft at the membranous apex into 2 often bristlelike teeth, awnless or with an awn arising from above midlength on the back; callus minute, glabrous or short-bearded; palea membranous throughout, shorter to longer than the lemma; stamens 3; caryopsis free within the lemma and palea, fusiform, the embryo to ca 1/3 as long as the grain; x = 7.

1. Lemmas awnless or with awns to 2 mm long .... *T. wolfii*
— Lemmas with awns 3–14 mm long 2
2(1). Lemmas averaging at least 6 (6–8) mm long (including apical teeth), the awns averaging at least 7 (7–14) mm long; leaf blades 1–12 mm wide .... *T. canescens*
— Lemmas averaging less than 6 (3–6) mm long (including apical teeth), the awns averaging less than 7 (3–7) mm long; leaf blades 1–5 (6) mm wide 3
3(2). Panicles spikelike, compact or occasionally interrupted, the branches mostly less than 1 cm long (rarely the lowermost to 4 cm) and appressed to somewhat spreading .... *T. spicatum*
— Panicles loose, the branches 2–6 cm long and erect to spreading .... *T. montanum*

Trisetum canescens **Buckley** Tall Trisetum. [*T. cernuum* var. *canescens* (Buckley) Beal; *T. cernuum* ssp. *canescens* (Buckley) Calder & Taylor]. Perennial; culms solitary or loosely tufted, (3) 5–12 (14) dm tall, erect or decumbent-based; sheaths long-hairy to occasionally merely scabrous; blades mostly flat, (1) 2–12 mm wide, long-tapered and involute at the finely acute tips, essentially smooth to long-hairy, often scabrous on the margins; ligule 1.5–5 mm long; panicle narrow but loose, (6) 8–25 cm long, erect to drooping, the rachis, branches, and pedicels glabrous or scaberulous; spikelets 1– or 2 (3) –flowered (the third floret often greatly reduced), (5) 7–10 mm long, the rachilla pubescent with hairs ca 1 mm long; glumes unequal, the first narrowly lanceolate, (3) 4–5 mm long, membranous except along the midnerve, long-tapered to a mostly aristate tip, the second broadly lanceolate, 4.5–7.5 mm long, membranous except along the 3 (5) nerves, acute to minutely aristate at the apex, consistently shorter than the lowermost floret; lemma at maturity slightly firmer than the glumes, scaberulous, 6–8 mm long including the bristlelike apical teeth, these mostly ca 1 mm long, the awn arising below the bifid apex, that of the first lemma (7) 8–14 mm long, typically twisted below and geniculate; callus bearded with hairs to ca 1 mm long; anthers 1.3–2.2 mm long; 2n = 28, 42. Under maple, aspen, and spruce-fir, and on open slopes, from 1980 to 2440 m in Cache and Salt Lake counties; British Columbia and Alberta (as *T. cernuum* Trin.), south in much of the western U. S.; 9 (vi). *Trisetum cernuum* of Canada and the northwestern U. S. has a distribution similar to that of *T. canescens*, but differs morphologically chiefly in having shorter, apically erose second glumes to 4.5 mm long and shorter anthers to ca 1 mm long. The two elements occupy the same type of habitat and, where their ranges overlap, intermediate specimens are encountered. A number of contemporary authors, consequently, refer *T. canescens* to varietal status under *T. cernuum*.

Trisetum montanum **Vasey** Rocky Mountain Trisetum. [*T. canescens* var. *montanum* (Vasey) A. S. Hitchc.]. *Trisetum montanum* is a puzzling element that has been allied with both *T. spicatum* (Cronquist et al. 1977) and *T. canescens* (Proc. Biol. Soc. Wash. 41: 160. 1928). The taxon as originally described (Bull. Torrey Bot. Club 13: 118. 1886), does appear to be intermediate between these two species, with spikelet morphology within the range of variation found in *T. spicatum* and the panicle loose with longer, spreading branches as in *T. canescens*. The distribution of specimens referable to *T. montanum* (Southwest) and of *T. canescens* (Northwest), however, militates against any theory that *T. montanum* is a product of hybridization. According to the original description, *T. montanum* differs from *T. spicatum* only in having panicles loose rather than spikelike. Occasionally, within populations of otherwise typical *T. spicatum*, however, plants with spreading branches as much as 4 cm long do occur. Although panicle branches for the species average much shorter in length, such plants, occurring as they do within populations of *T. spicatum* and differing in no other way, must be treated as being within the range of variation of that species. A few collections from the Abajo Mts. of Utah (e. g., Rydberg and Garrett 9861 UT), however, have spreading panicle branches that are mostly 2–6 cm long. These specimens appear to be anomalous *T. spicatum* plucked from a normal population represented by collections of Rydberg and Garrett 9860 (UT). Unless additional information concerning these plants ultimately proves otherwise, their presence within populations of typical *T. spicatum*, the infrequency with which they occur, their similarity to *T. spicatum*, and the presence of intermediates suggest that they do not constitute a distinct species. *Trisetum montanum* has been reported from much of the southwestern U. S., but plants so identified usually prove to be loose-panicled *T. spicatum*.

Trisetum spicatum **(L.) Richter** Spike Trisetum. [*Aira spicata* L.]. Perennial; culms solitary or tufted, (0.4) 1–8 dm tall; sheaths glabrous or short- to long-hairy, often densely so; blades flat or folded, 1–5 (6) mm wide, long-tapered and involute to prow-shaped at the tips, scabrous or puberulent to densely long-hairy; ligules 0.5–3 mm long; panicle spikelike or occasionally interrupted and somewhat loose, (2) 3–20 cm long, green and silvery to tawny or purple-tinged, the branches mostly less than 1 cm long or occasionally the lowermost to as much as 4 cm long, strictly appressed or occasionally somewhat spreading at flowering, the rachis, branches, and pedicels glabrous to densely pubescent; spikelets (1) 2 (3) –flowered, 4–7.5 mm long, the rachilla pubescent with hairs typically less than 1 mm long; glumes lanceolate, nearly equal to distinctly unequal, glabrous or hairy, the first 3–5 mm long, much narrower to nearly as broad as the second, membranous except along the midnerve, 1 (3) –nerved, acute to aristate at the apex, the second 4.5–6.2 mm long, broadly membranous-margined, 3 (5) –nerved, acute, slightly shorter than the lowermost floret to slightly longer than the upper one; lemma similar in texture to the glumes or at maturity slightly firmer, 3–6 mm long, glabrous or scaberulous, acute or bifid at the apex into bristlelike teeth to ca 1 mm long, the awn arising from the upper 1/3 of the back, that of the lower lemma 3–6 (7) mm long, straight or flexuous to geniculate, often widely spreading in age; callus glabrous or sparsely short-bearded; anthers 0.7–1.2 (1.5) mm long; 2n = 14, 28, 42. Dry to wet sites, in aspen-spruce-fir, lodgepole pine, and meadow communities, and on exposed rocky slopes at higher elevations, occasionally in sagebrush and ponderosa pine communities, at 1830 to 4000 m in all Utah counties except Box Elder, Kane, and Weber; circumboreal, south throughout the western U. S., essentially absent from the Great Plains, sporadic in the Northeast and in Mexico; South America; Eurasia; 391 (xii). The widely distributed *T. spicatum* occupies a variety of climates and a wide range of habitats. On the premise that such plants "cannot be identical in all places," Hulten (Svensk Bot. Tidskr. 53: 203–228. 1959) divided the species into 14 subspecies, 3 of which were reported as occurring in Utah. The races are, however, weakly differentiated, their ranges overlap, and their formal recognition serves no useful purpose. See *Koeleria macrantha* for discussion of a possible hybrid with *T. spicatum*. Spike trisetum provides good forage throughout its growing season and late into the fall. It seldom grows in dense stands, however, and is of only moderate importance as a source of forage (Vallentine 1961).

Trisetum wolfii **Vasey** Wolf Trisetum. Perennial, sometimes short-rhizomatous; culms solitary or tufted, mostly 2–10 dm tall; sheaths glabrous or rarely puberulent to sparsely long-hairy; blades flat, (1) 2–6 mm wide, long-tapered and involute to prow-shaped at the tips, essentially smooth or scabrous to rarely sparsely long-hairy; ligule 1–3 mm long; panicle narrow, spikelike to interrupted or somewhat loose, erect, 4–16 cm long, the lowermost branches from very short to 6 cm or more long, the rachis and branches glabrous or scabrous; spikelets (1) 2– or 3 (4) –flowered, 4.2–8 mm long, the rachilla pubescent with hairs to ca 1 mm long; glumes equal or nearly so in length and usually in width, lanceolate to elliptical, (4.5) 5–8 (8.5) mm long, slightly shorter to longer than the lemmas, broadly membranous-margined, the margins often becoming tawny and to some degree purple, the apex abruptly to sharply acute, the first 1 (3) –nerved, the second 1– to 3–nerved; lemma at maturity usually slightly firmer than the glumes and scaberulous, 4–6 mm long, acute to blunt at the entire to minutely bifid apex, awnless or with an awn to 2 mm long arising below the tip; callus glabrous to minutely hairy; anthers 0.6–1.3 mm long. Aspen-spruce-fir and meadow communities at 2600 to 4000 m in Beaver, Daggett, Duchesne, Emery, Salt Lake, San Juan, Summit, Uintah, Utah, and Wasatch counties; southwestern Canada and throughout the western U. S.; 50 (iii).

Triticum L.

Annuals or biennials; culms erect; sheaths open, typically with well-developed auricles; blades mostly flat; ligule membranous; inflorescence a somewhat flattened, compact, terminal spike, the rachis continuous or disarticulating; spikelets sessile and solitary at each node, 2– to 5– (9) –flowered (the uppermost 1 or 2 usually neuter), disarticulating above the glumes and between the florets; glumes firm or rarely membranous, laterally compressed and more or less keeled, often asymmetrically so, equal or subequal, 3– to several–nerved, the apex truncate and mucronate to notched or with 1–several awns; lemma firm, broad, moderately compressed laterally and usually more or less keeled, with several nonconvergent nerves, abruptly acute to awned; stamens 3; caryopsis free-threshing, the embryo to about half as long as the grain; x = 7. See discussion of the tribe Triticeae under *Agropyron*.

Triticum aestivum L. Wheat. Annual; culms erect, often branched at the base, 3–15 dm tall; sheaths glabrous, the auricles typically well developed; blades flat, 2–16 (20) mm wide; ligule to ca 1 mm long; spike 3–18 cm long (excluding awns), stiffly erect to lax, the rachis continuous, the internodes glabrous, 4–8 mm long at about midlength of the spike; spikelets solitary at each node, subterete to moderately compressed; glumes firm, equal or subequal, lanceolate to ovate, 6–11 mm long, somewhat compressed laterally and keeled at least near the apex, usually asymmetrically so, truncate and mucronate to notched or awned, shorter than the lowermost florets; lemma firm, broad, 7–12 mm long, more or less asymmetrically compressed laterally, keeled to rounded on the back, acute or awn-tipped to awned, the awn 0.2–16 cm long; anthers to ca 3 mm long; 2n = 42. Widely grown in Utah and often escaping; native to Eurasia, widely cultivated in temperate regions of the world. The numerous cultivars are variously treated as varieties or as distinct species. Surprisingly, Hanson (1972) reports that awnlessness in wheat is not compatible with maximum wheat production. Triticale (x *Triticosecale* Wittm.), an artificially produced hybrid between wheat and rye, has created considerable excitement in the world of agriculture, especially during the last twenty years. Although much progress has been made toward production of a "manmade" cereal of commercial value, several serious problems remain to be solved before utilization of the hybrid can become widespread (Advances in Genetics 21: 256–345. 1982).

Zoysia Willd.

Low perennials, rhizomatous and stoloniferous; culms erect or ascending; sheaths open throughout or nearly so; blades short, stiffly spreading-ascending, often more or less pungently acute at the tips; ligule a short-ciliate membrane; inflorescence a slender, spikelike raceme or panicle to ca 4 cm long, the short-pedicelled spikelets appressed to the rachis; spikelets 1–flowered, laterally compressed, mostly 2–3 mm long, disarticulating below the glumes; first glume lacking, the second firm, soon becoming hardened, asymmetrically compressed, smooth and glossy, obscurely 1– to 3–nerved, acute or with an awn to ca 1.2 mm long; lemma membranous, broadly lanceolate, 1–nerved, with 2 rounded apical lobes; stamens 3, the anthers to 1.8 mm long; caryopsis free of the lemma and palea, the embryo 1/2 as long as the grain; x = 10. Three species of this Asian genus (including *Z. japonica* Steudel, *Z. matrella* (L.) Merr., and *Z. tenuifolia* Willd. ex Trin.) are known to have been introduced as lawn grasses into the warmer parts of the U. S., one as early as 1895. Cultivars derived from hybrids among the species have proved unsatisfactory when introduced in our area because they become straw colored or brown at first frost and do not green up until late in May. Moreover, once introduced, the cultivar is difficult to eradicate because of its rhizomatous nature.

HYDROCHARITACEAE A. L. Juss.
Frogbit Family

Plants herbaceous, submerged perennials; leaves opposite or whorled; flowers regular, dioecious or polygamous, mostly imperfect, sessile or on scapelike peduncles in a spathe; sepals mostly 3, valvate; petals 3 or lacking, imbricate; stamens mostly 3–9, free; anthers 2–loculed; stigmas 3–12; pistil 1, the ovary inferior, 1–loculed; fruit indehiscent; x = 7–12.

Elodea Michx.

A single genus with characteristics of the family.

1. Upper leaves in whorls of 3, 6–13 mm long, the margins entire .... *E. canadensis*
— Leaves opposite, 16–32 mm long, the margins serrate .... *E. longivaginata*

Elodea canadensis Michx. [*Anacharis canadensis* Planchon; *Philotria canadensis* Britt.; *Udora canadensis* Nutt.]. Herbaceous aquatic perennial, 2–5 dm high; stems slender, sparsely branched; leaves whorled at least above (opposite below), linear-oblong, crowded, the blade 4–17 cm long, 1–2 mm wide, dark green, finely serrulate, apex rounded; flowers in the upper leaf axils; staminate perianth 4–5 mm long, remaining attached to the long pedicel; stamens 9; pistillate sepals 2–2.4 mm long; petals 2.5–3 mm long; fruit a capsule, 5–6 mm long; seeds 3–4, 4.5 mm long. In ponds, canals, reservoirs, and streams at 1370 to 2745 m in Beaver, Cache, Sevier,

Uintah, and Utah counties; Canada, south to California, Nevada, Arizona, New Mexico, Arkansas, Alaska, and North Carolina; 5 (0).

Elodea longivaginata St. John Herbaceous aquatic perennials; stems slender, little branched; leaves opposite (rarely in 3s), 16–32 mm long, 1–3 mm wide, acute or blunt apically, finely serrulate; staminate sepals 3.5–4 mm long, petals ca 5 mm long; stamens 9; pistillate sepals 2.5–3 mm long; staminodia 3, style slender, stigmas 3, not lobed; capsule 8–10 mm long; seeds usually 6, 5–6 mm long. In ponds, sloughs, and lakes at ca 2290 m in Wasatch County; Alberta and North Dakota, south to New Mexico; 1 (0).

IRIDACEAE A. L. Juss.
Iris Family

Perennial, from short rhizomes or corms; leaves narrow, distichous, with a sheathing base and equitant blade; flowers solitary or few to several, in terminal clusters, each subtended by a bract and typically all subtended by spathaceous bracts; perianth of 6 segments in 2 series of 3 each; stamens 3, the filaments often connate below into a tube; pistil 1, 3–carpellate, the ovary inferior; style 3–lobed, the branches sometimes again divided and petaloid; fruit a loculicidal capsule; x = 3–19+.

1. Plants lacking an aerial scape, blooming in autumn or early spring; flowers sessile on the corm that is several cm below ground level .... *Crocus*
— Plants with an aerial scape; flowers on elongate stems, not arising below ground 2

2(1). Flowers irregular or the perianth oblique and curved, secund; stamens usually disposed at one side ... *Gladiolus*
— Flowers regular, not curved or secund; stamens regularly disposed 3

3(2). Style branches large and petaloid; flowers usually over 5 cm wide; sepals and petals unlike .... *Iris*
— Style branches not petaloid; flowers less than 3 cm wide; sepals and petals alike .... *Sisyrinchium*

Crocus L.

Cormous herbs; leaves linear, grasslike; peduncles and ovary subterranean, the ovary subtended by colorless spathaceous bracts; flowers varicolored; perianth segments 6, subequal; stamens 3, not connate, inserted on the perianth tube, included; styles 3–branched; stigmas lacerate (or branched).

1. Corm scales reticulately fibrous; flowers white to lilac or purple; staminal filaments white .... *C. vernus*
— Corm scales with parallel fibers or smooth; flowers typically yellow (sometimes otherwise); staminal filaments various 2

2(1). Corm scales with parallel fibers .... *C. flavus*
— Corm scales smooth .... *C. chrysanthus*

Crocus chrysanthus (Herbert) Herbert Corm scales brownish to tan, smooth; bladeless sheaths 4 or 5 below ground; leaves of flowering stems 4–6, shorter to longer than the flowers; flowers typically yellow, the segments 2–4 cm long, often marked and veined with brown or purple; anthers with black sagittate bases; style branches lacerate. Commonly grown and long-persisting ornamental in much of Utah; introduced from Eurasia; 2 (0). This and other species of *Crocus* are among the earliest flowering ornamentals, forming bright patches of color in the drabness of the late winter and early spring seasons.

Crocus flavus Weston [*C. aureus* J. G. Sm.]. Corm scales dark brown to brownish purple, with parallel fibers (at least ultimately); leaves of flowering stems 4–6, shorter to longer than the flowers; flowers typically yellow, the segments 2.5–4 cm long, sometimes marked or veined with darker colors; anthers with pale sagittate bases; style branches lacerate. Commonly cultivated ornamental, long-persisting in much of Utah; introduced from Eurasia; 2 (0).

Crocus vernus (L.) Hill Corm scales reticulately fibrous; leaves of flowering stems 4–6, shorter than or surpassing the flowers; flowers white, lilac, or purple, often marked with dark purple outside, the segments 2.5–7 cm long; anthers with yellow sagittate bases; style branches lacerate. Commonly grown ornamental in much of Utah, long persisting; introduced from Europe; 2 (0).

Gladiolus L.

Cormous herbs; stems leafy, swordlike; flowers sessile, secund, several to many, each subtended by a spathaceous bract, typically curved; perianth segments oblong to elliptic, unequal; stamens inserted below the throat; style slender; fruit a loculicidal capsule.

Gladiolus carneus Delaroche Plants to 1 m (or more) tall; stems slender, from corms 2–8 cm wide; leaves 1–4 dm long and 1–2.5 cm wide; flowers several to many, showy, variously colored, the segments 4–6 cm long. Commonly grown ornamental, occuring rarely as a waif, but frost sensitive and not persistent except by reintroduction, in much of Utah; widely grown in the U. S.; introduced from the Old World; 1 (i). Our material represents hybrid derivatives involving several species and the name applied here is only one of them. No inclusive binomial is available.

Iris L.

Rhizomatous herbs; leaves mainly basal, equitant, linear or swordlike; flowers showy, solitary or few to several, borne in 2–bracted spathes; perianth segments 6, dimorphic, the outer 3 (sepals or falls) usually spreading, sometimes bearded, the inner 3 (petals or standards) usually erect; stamens 3, the filaments not united, inserted at the base of the sepals; style branches 3, bifid or crested beyond the stigmas, petaloid, covering the stamens; fruit a capsule. Note: Many species and hybrids are grown as ornamentals in Utah, including *I. pumila* L., *I. reticulata* Bieb., *I. stolonifera* Maxim., *I. korolkowii* Regel, *I. variegata* L., their hybrids and other hybrid combinations. Those cited below are either indigenous or are known to occur spontaneously in our flora, the strictly cultivated ones are not treated.

1. Leaves 5 mm wide; rhizomes ca 5–10 mm thick; spathaceous bracts subopposite, 2–5 mm wide .... *I. pariensis*
— Leaves linear to swordlike, mainly 6–15 mm wide or more; rhizomes mostly more than 10 mm thick; spathaceous bracts alternate or opposite, mostly 5–20 mm wide or more 2

2(1). Spathaceous bracts opposite, mainly less than 12 mm wide; plants indigenous and widespread . *I. missouriensis*
— Spathaceous bracts alternate, typically more than 12 mm wide; plants introduced and persisting 3

3(2). Flowers pale yellow to almost orange; plants often more than 20 dm tall, growing in wet places ... *I. pseudacorus*

— Flowers mainly purple, blue purple, or white; plants typically less than 10 dm tall, growing in gardens or dry places .... *I. germanica*

Iris germanica L. (hybrid). Flag; Fleur-de-lis. Rhizomes typically 20–30 mm thick, not especially clothed with leaf bases; leaves few to several, glaucous, 1.2–3.5 cm wide, 2–5 dm long, pale to stramineous basally; flower stems 4–8 dm tall, often branched; flowers 2 or 3 per spathe; spathaceous bracts alternate, scarious or partially herbaceous, lance-ovate to oblong, 15–25 mm wide, 2.5–7 cm long, erect or ascending; ovary 1–2 cm long; perianth tube 2–3 cm long; sepals 5–8 cm long, usually purple (violet), less commonly white; petals 5–8 cm long, often colored like the sepals; style branches 2.5–4.5 (5) cm long; style crests 10–15 mm long; anthers 14–18 mm long; 2n = 24, 36, 44, 48. Reclamation and ornamental plants, persisting and escaping in much of Utah; widely grown and established in the U. S.; introduced from Europe; 2 (i).

Iris missouriensis Nutt. Missouri Iris. Rhizomes typically 10–20 mm thick, clothed with brownish scaly bases; leaves glaucous, few to many, 0.6–7.5 dm long, 3–10 (12) mm wide, stramineous or brown basally; flower stems 2–8 dm tall, branched or simple, with 1–3 flowers; spathaceous bracts opposite, scarious or greenish, lanceolate to ovate, 2.5–7 cm long, 6–12 (15) mm wide; ovary 15–30 mm long; perianth tube 0.8–1.2 cm long; sepals 4–6 cm long, the blades obovate to ovate; pale lilac to whitish with purple veins; petals 4–5 cm long, oblanceolate, often more brightly colored than the sepals; style branches 1.5–2.5 cm long, the crests 5–8 mm long; anthers 10–15 mm long; capsules 3–6 cm long; 2n = 38. Moist meadows at 1430 to 2900 m in Box Elder, Carbon, Daggett, Duchesne, Garfield, Grand, Iron, San Juan, Tooele, Uintah, and Utah counties; British Columbia to North Dakota, south to New Mexico and California; 50 (iv).

Iris pariensis Welsh Paria Iris. Rhizomes less than 10 cm thick, clothed with shredded fibrous leaf bases; leaves several to many on both fertile stems and innovations, 2–5 mm wide, (4) 7–24 cm long, stramineous to brownish or purplish basally; flower stem 4 cm tall, with several sheathing leaves free almost throughout; flower 1; spathaceous bracts linear-attenuate, 2–3 mm wide, 5–6 cm long, parallel, subopposite, herbaceous; ovary ca 12 mm long; perianth apparently white, the tube 15 mm long; sepals ca 6 cm long and 1 cm wide; petals narrowly oblanceolate, ca 6 cm long and 8 mm wide; style branches ca 2.8 cm long, the crests ca 7 mm long; anthers ca 13 mm long; capsule unknown. Grass-shrub community at 1400 m in Kane County; endemic; 1 (0).

Iris pseudacorus L. Yellow Flag. Rhizomes 20–30 mm thick or more; leaves few to several, linear, 2.5–9 dm long, mostly 10–20 mm wide, flowering stems 8–12 dm tall, leafy, usually branched, few- to several-flowered; flowers pale to deep yellow or almost orange, purple lined; perianth tube ca 12 mm long; sepals 4–6 cm long, obovate, the petals shorter and narrower; style branches 20–25 mm long, the crests ca 10 mm long; capsules 5–8 mm long; 2n = 24, 32, 34. Irrigation canal banks and pond margins at 1280 to 1315 m in Salt Lake County; eastern U. S.; introduced from Europe; 1 (0).

Sisyrinchium L.

Caespitose perennial herbs, seldom shortly rhizomatous; leaves equitant, sheathing at the base, the blade flattened and grasslike; stems simple or branched, flattened and wing-margined; spathes terminal or terminal on umbellately disposed rays from a primary (or secondary) foliose bract; flowers 1 or few, umbellate within the ultimate spathes, the pedicels usually further subtended by bractlets within the spathe; perianth rotate, the 6 segments alike, shortly connate basally, blue, rose purple, or white; staminal filaments connate basally and sometimes almost throughout; style branches slender; ovary 3-loculed; capsules loculicidally dehiscent.

1. Perianth typically rose purple, the segments 1.3–2.5 cm long or more; filaments united at the base; plants rare (in Tooele County) .... *S. douglasii*

— Perianth typically blue or purple (or white), less than 1.5 cm long; filaments united almost to the tips 2

2(1). Stems terminated by the paired spathes of mostly unequal bractlets; plants seldom of saline substrates .... *S. idahoense*

— Stems terminated by 2 or more umbellately disposed rays arising from within the sheathing base of a foliose bract, the rays each with a pair of subequal bractlets at the apex; plants often of saline substrates ... *S. demissum*

Sisyrinchium demissum Greene Blue-eyed Grass. [*S. radicatum* Bicknell, type from St. George]. Caespitose, the stems erect, 0.7–4 dm tall or more, typically with a sheathing foliose bract subtending 2 or more umbellately disposed rays terminated by paired subequal to unequal spathes or rarely some of the rays again branched; leaves 1–32 cm long, 0.7–5 mm wide, entire or denticulate; rays 2–14 cm long; spathes paired, the outer mainly 11–22 mm long, connate at the base for 2.5–8 mm, the inner mainly 12–24 mm long; flowers 1–7 on glabrous or glandular-puberulent pedicels, these typically enclosed within the bracts; perianth pale to dark blue (or white) with a yellow center, the segments apiculate, 6–12 mm long; staminal filaments united for 3.5–5.5 mm; capsules globose to obovoid, 2.4–6.2 mm long; n = 16. Seeps, springs, wet meadows, and stream banks, often where saline, at 850 to 2380 m in Beaver, Carbon, Duchesne, Garfield, Juab, Kane, Millard, Piute, San Juan, Tooele, Uintah, Utah, and Washington counties; Nevada, Arizona, New Mexico, and Texas; Mexico; 40 (vii). All features regarded as diagnostic for this entity overlap with those of *S. idahoense* and it is not always possible to place a given specimen within either of the species with certainty.

Sisyrinchium douglasii A. Dietr. Purple-eyed Grass. [*S. grandiflorum* Dougl. ex Lindl., not Cav.]. Caespitose, the stems erect, 10–38 cm tall; leaves clasping basally, shorter than the stem, 1.8–4 mm wide, entire; spathes terminal on the stem, the bracts unequal, the inner 22–40 mm long or more, the outer 35–60 mm long or more, surpassing the flowers, these 1–4; pedicels subequal to the inner spathaceous bract; perianth rose purple, pink, or white, the segments 13–25 mm long, the segments apiculate; filaments connate in the basal half; anthers 3–9 mm long; style slender; capsules spherical to broadly obovoid, 5–10 mm long; n = 32. Gravelly north slope at ca 1585 m in Tooele County (Stockton Bar, Cottam 7483 BRY); British Columbia to Idaho, California, and Nevada; 1 (0).

Sisyrinchium idahoense Bicknell [*S. occidentale* Bicknell; *S. segetum* Bicknell; *S. oreophilum* Bicknell]. Caespitose, the stems usually unbranched, mainly 1–5 dm tall; leaves 1–30 cm long, 1–3 (4.5) mm wide, entire or denticulate; spathe typically terminal, rarely borne on rays from a foliose sheathing bract, the spathaceous bracts usually unequal, the outer one 15–65 mm long and connate basally for 3–8 mm, the inner one 13–33 mm long; flowers 1–4; pedicels glabrous or glandular, shorter to longer than the outer bract; perianth blue to purple, commonly with a yellow center, the segments 6–15 mm long, apiculate; filament column 3–6.5 mm long; capsules 3–6 mm long, globose to ovoid; n = 32, 48. Seeps, springs, wet meadows and stream banks at 1310 to 2745 m in all Utah counties, except Carbon, Davis, Millard, Morgan, and Washington; British Columbia to Montana, south to Nevada and Colorado; 53 (v).

JUNCACEAE A. L. Juss.
Rush Family

Perennial or annual grasslike herbs; stems terete or flattened, not jointed, caespitose or arising singly or few together from rhizomes; leaves sheathing, alternate or all basal, mostly 2-ranked, the blades linear, sometimes much reduced or lacking; inflorescence headlike to open paniculate, subtended by an involucral bract; branches, heads, and pedicels often subtended by bractlets; flowers perfect, sometimes subtended by bracteoles borne at pedicel apices, directly below the perianth; perianth segments 6, the petals and sepals essentially alike, membranous, rather scalelike, greenish or brownish; stamens (3) 6; pistil 1; ovary superior, with 1 or 3 locules; fruit a capsule; x = 3–36.

1. Seeds numerous in each capsule; leaves glabrous, the sheaths open; bracteoles subtending the flowers entire or lacking . . . . *Juncus*
— Seeds 3 per capsule; leaf blades pubescent at least on the margins near the base (except sometimes in *L. parviflora*), the sheaths closed; bracteoles subtending the flowers entire to lacerate . . . . *Luzula*

Juncus L.

Perennial or annual grasslike herbs; stems terete or flattened; leaf blades flat, strongly folded, or terete, when terete sometimes hollow with cross membranes at intervals (septae) or reduced to a bristle or lacking; flowers as described in the family; seeds numerous, minute, usually apiculate or tailed.

1. Plants annual 2
— Plants perennial 3

2(1). Plants 5–20 mm tall, the scapose stem with 1 flower; stamens 3; leaves not over 5 mm long . . . . . . *J. bryoides*
— Plants 2–30 cm tall, not scapose, the inflorescence with 1–20 flowers; stamens 6; leaves 0.5–10 cm long . . . . *J. bufonius*

3(1). Flowers (1) 2–5 in a solitary terminal head; leaves basal or nearly so, hollow, septate; plants densely tufted, without rhizomes, 3–19 cm tall, infrequent in the Uinta Mts. . . . . *J. triglumis*
— Flowers either more numerous or not in a solitary terminal head or plants otherwise different from above 4

4(3). Stems with 0–2 (rarely more) leaf blades; blades borne on the lower 1/5 of the plant, neither hollow nor septate; sheaths sometimes ending in a rudimentary bristle instead of a blade; flowers borne singly, each subtended by 2 hyaline bracteoles; rhizomes lacking or short and plants caespitose (except in *J. arcticus*) 5
— Stems with 2 or more well-developed leaf blades, at least the uppermost blade borne above the lower 1/3 of the plant or else hollow and septate; flowers borne in 1 or more heads and not individually subtended by bracteoles; plants mostly rhizomatous 13

5(4). Leaves all reduced to bladeless sheaths, the upper ones sometimes with a bristle-tip, this not over 5 mm long; inflorescence with 5–75 or more flowers; seed not tailed 6
— Leaves of most stems, at least the uppermost, with a well developed leaf blade well over 10 mm long or, if leaves all reduced to bladeless sheaths (*J. drummondii*), the inflorescence with only 1–3 flowers and seeds tailed 7

6(5). Involucral bract subequal to or longer than the stem and the inflorescence appearing at or below the middle of the plant; stems seldom over 1 mm thick, somewhat tufted; plants of the Uinta Mts., above 2950 m, rather rare . . . . *J. filiformis*
— Involucral bract mostly shorter than the stem and the inflorescence appearing above midlength of the plant; stems often over 1 mm thick, mostly arising singly or few together from robust dark rhizomes; plants widespread . . . . *J. arcticus*

7(5). Seeds tailed at each end, the tails from 1/2 to longer than the body; inflorescence with 1–6 flowers; stem leaves with blades lacking or reduced to a bristle, or the uppermost 1 (2) with a well-developed blade; plants mostly found above 2620 m 8
— Seeds apiculate but not tailed; inflorescence with (1) 6–50 or more flowers; stem leaves (at least some lower ones) commonly with well-developed blades; plants of various elevations 10

8(7). Stems with bladeless leaves, the uppermost and often the lower sheaths tipped with a bristle, this not over 1 cm long; tails of seeds equal to or longer than the body; perianth 5–8 mm long; capsules blunt and more or less retuse, equal or a little shorter than the perianth . . . . *J. drummondii*
— Most of the stems with a well-developed leaf blade on at least the uppermost sheath, the lower sheaths often tipped with a bristle; tails of seeds equal to or shorter than the body; perianth and capsules various 9

9(8). Capsules ovoid, retuse at the apex; perianth 4–5 (5.5) mm long; anthers less than 1 mm long, the filaments longer than the anthers . . . . *J. hallii*
— Capsules oblong, pointed; perianth 5–8 mm long; anthers 1.5–2 mm long, the filaments only ca 0.3 mm long . . . . *J. parryi*

10(7). Perianth segments, at least the outer ones, with incurved or hooded tips, rather obtuse, 1.5–2.5 mm long; uppermost leaf often borne above midlength of the stem 11
— Perianth segments with acute to acuminate erect tips, 3–5 mm long; uppermost leaf borne on the lower 1/3 of the stem 12

11(10). Anthers ca 3 times longer than the filaments; capsule ellipsoid-ovoid, equal to or slightly exceeding the perianth; plants sometimes over 40 cm tall, of Salt Lake County . . . . *J. gerardii*

— Anthers scarcely longer than the filaments; capsule globose-ovoid, distinctly exserted; plants 2–4 dm tall, known from floodplains of the Green and Colorado rivers .... *J. compressus*

12(10). Capsules retuse at the apex, completely 3-loculed; panicles mostly less than 2 cm long; perianth segments with hyaline margins extending to the apex of the acute tip; plants montane, mostly above 2380 m .... *J. confusus*

— Capsules blunt but not retuse, incompletely 3-loculed; panicles various but often over 2 cm long; perianth segments, at least the outer, with the hyaline margins not extending to the acuminate or acuminate-attenuate tip; plants found mostly below 2380 m .... *J. tenuis*

13(4). Flowers borne singly, each subtended by 2 bracteoles; pedicels sometimes also subtended by bractlets; perianth segments with incurved or hooded tips, 1.5–2.5 mm long 14

— Flowers borne in 1 to many heads, not subtended immediately by bracteoles, the pedicels usually subtended by bractlets; perianth segments, at least the outer, with erect or spreading tips, as short as or longer than above 15

14(13). Anthers ca 3 times longer than the filaments; capsule ellipsoid-ovoid, equal to or slightly exceeding the perianth; plants sometimes over 40 cm tall, of Salt Lake County .... *J. gerardii*

— Anthers scarcely longer than the filaments; capsule globose-ovoid, distinctly exserted; plants 20–40 cm tall, known from flood plains of the Green and Colorado rivers .... *J. compressus*

15(13). Leaf blades flat or strongly folded and appearing flat at least toward the base, neither terete nor hollow, the sheaths with hyaline margins; capsules not exserted beyond the perianth 16

— Leaf blades terete and hollow, if only toward the tip, the sheaths without hyaline margins and capsules conspicuously exserted beyond the perianth 18

16(15). Leaves strongly folded, the narrow edge oriented toward the flattened stem, the scarious margins of the sheaths extending well beyond the juncture with the stem, gradually tapering to inconspicuous auricles or the auricles lacking, the margins of the blade more or less united beyond the scarious margins ... *J. ensifolius*

— Leaf blades flat, the flat surface oriented toward the terete stem, the scarious margins of the sheaths not extending beyond the juncture with the stem 17

17(16). Seeds tailed, the tails as long as or longer than the body; perianth segments granular-papillate on the back; heads sometimes with more than 10 flowers; plants known from Duchesne, Salt Lake, and Wasatch counties .... *J. regelii*

— Seeds apiculate but not tailed; perianth segments smooth on the back; heads with 3–10 flowers; plants widespread .... *J. longistylis*

18(15). Leaf blades folded to enrolled toward the base, becoming terete and hollow distally, the sheaths without hyaline margins; auricles lacking; capsules conspicuously exceeding the perianth; seeds long-tailed, stamens 6; plants rare, from above timberline .... *J. castaneus*

— Leaf blades terete and hollow from the collar and outward, septate, the scarious margins of sheaths projected into auricles; capsules not much if any longer than the perianth and seeds not tailed or else the stamens 3; plants widespread 19

19(18). Seeds tailed; stamens 3; capsules conspicuously exceeding the perianth; plants known from Box Elder County .... *J. tweedii*

— Seeds not tailed; stamens 6; capsules various; plants widespread 20

20(19). Capsules tapered almost from the base into a mostly nondehiscent, conspicuous stylar beak, often divergent in all directions in the mature, globose or hemispheric head; heads rarely solitary, greenish or light brown; rhizomes sometimes swollen and tuberous at the nodes 21

— Capsules rather abruptly narrowed above into a dehiscent short or inconspicuous stylar bank, ascending to slightly spreading in the heads or, if spreading in all directions, the heads solitary; heads light or deep brown to blackish purple; rhizomes not as above 22

21(20). Auricles 1.5–5 mm long; perianth 4–5 mm long with rigid long-acuminate or subulate tips; mature heads 10–15 mm wide; capsules shorter or scarcely longer than the perianth; stems to 6 mm thick ...... *J. torreyi*

— Auricles 0.2–1 mm long; perianth segments 2.5–3.5 mm long, acuminate, the tips not so rigid as above; mature heads 5–12 mm wide; capsules to ca 1 mm longer than the perianth; stems 1–2 mm thick .... *J. nodosus*

22(20). Perianth segments 3–5 mm long or, if smaller, the inflorescence with 1 or rarely 2 heads, equal to or conspicuously exceeding the capsules 23

— Perianth segments 1.5–2.8 mm long, shorter than the capsules; inflorescence with (1) 4–25 heads 24

23(22). Heads 1 or rarely 2, globose or nearly so, with 5–40 or more flowers; anthers 0.5–1 mm long, shorter than the filaments; perianth purplish black .. *J. mertensianus*

— Heads (1)2–13, not or hardly globose, with 3–13 flowers; anthers 1–2 mm long, longer than the filaments; perianth brown to purplish black . *J. nevadensis*

24(22). Outer perianth segments obtuse, mostly longer than the inner ones; branches of the inflorescence stiffly erect; capsule rather abruptly rounded at the tip; plants of low elevations and montane .... *J. alpinus*

— Outer perianth segments acute, subequal to or shorter than the inner ones; capsules rather gradually tapering to the tip; branches of the inflorescence spreading to nearly divaricate; plants rather rare, apprently not montane .... *J. articulatus*

Juncus alpinus Vill. Northern Rush. Perennial plants 5–40 cm tall; stems loosely tufted on creeping rhizomes, with 1–2 leaves mostly borne on the lower 1/3 or 1/2; leaves mostly with well-developed blades, the scarious margins of the sheaths prolonged into auricles, these 0.5–1 mm long, the blades terete, hollow, septate, 1–1.5 mm in diameter; involucral bract 1–7 (10) cm long, sometimes leaflike; inflorescence 0.8–8.5 cm long, rather openly branched, the branches mostly erect or strongly ascending, with (1) 3–25 small heads or headlike glomerules, the heads 3–4 (5) mm long, with 2–6 flowers; bractlets subtending the branches and heads scarious; pedicels obsolete or less than 1 mm long; bracteoles lacking; perianth 1.5–2.5 mm long, brownish to dark purple, the outer segments usually slightly longer than the usually more obtuse inner ones; stamens 6, the filaments ca 0.5 mm long, the anthers 0.2–0.5 mm long; styles with stigmas ca 1 mm long; capsules 1-loculed, 2.5–3 mm long, rather abruptly narrowed to a stylar beak to 0.3 mm long; $2n = 40$. Seeps, bogs, margins of lakes and ponds, and along streams, often on limestone or other basic substrates at 1524 to 2800 m in Box Elder, Daggett, Duchesne, Garfield, Kane, Uintah, and Wayne counties; Alaska to Newfoundland and south to Washington, Colorado, and Quebec; 21 (vi). Some specimens are rather difficult to distinguish from those of *J. articulatus*.

Juncus arcticus Willd. Wiregrass [*J. balticus* Willd.; *J. balticus* var. *montanus* Engelm.; *J. balticus* var. *vallicola* Rydb.]. Perennial plants 2–10.5 dm tall; stems terete or slightly compressed, 1–5 mm in diameter, arising singly, 2–3 together, or rather tightly clustered, from robust dark brown or blackish rhizomes to 5 mm in diameter; leaves reduced to bladeless sheaths, the sheaths confined to the lower 1/5 of the plant, entire or occasionally tipped with a tiny bristle; involucral bract (2.5) 4–23 cm long, appearing as a continuation of the stem; inflorescence 0.5–15 cm long, congested and headlike with very few (ca 5) flowers to open-paniculate with up to 75 or more flowers; bractlets subtending the branches and pedicels scarious; pedicels nearly obsolete or to 8 mm long; bracteoles subtending the flowers ovate or nearly so, scarious; perianth segments 3.5–5 mm long, acute to acuminate, pale to dark brown, the outer ones often a little longer, more sharply pointed and with darker margins than the usually more rounded, often broadly scarious-margined inner ones; stamens 6, the filaments ca 0.5 mm, the anthers 1.4–2.1 mm long; styles and stigmas to ca 3 mm long; capsules 3–4 mm; seeds ca 0.6–0.8 mm long; $2n = 80, 84$. Margins of ponds and lakes, along streams and rivers, in saline to nonsaline meadows, seeps, springs, marshes, and swamps at 850 to 3050 m in all Utah counties, except Morgan; circumboreal, widespread in western North America, Eurasia; 150 (i). *Juncus arcticus* forms a highly variable complex. It seems best to await further monographic work for the disposition of infraspecific taxa.

Juncus articulatus L. Jointed Rush. Perennial plants 17–36 cm tall; stems loosely to rather densely tufted, from short stout or prolonged rootstocks, with 2–4 leaves, the upper leaf often on the upper 1/2–3/4 of the stem; leaves mostly bearing well-developed blades, the sheaths with scarious margins prolonged into auricles ca 1 mm long, the blades 0.5–2 mm wide, terete, septate; involucral bract 1–3.7 cm long; inflorescence 1.5–7 cm long, openly branched, the branches ascending to divaricate, with 4–25 small heads, the heads with (2) 5–10 flowers; bractlets subtending the branches and heads scarious; pedicels obsolete or less than 1 mm long; bracteoles lacking; perianth 2–2.8 mm long, mostly all acute, the segments equal or the inner ones slightly longer than the outer ones, greenish or purplish with conspicuous scarious margins, often minutely granular scabrous on the back; stamens 6, the filaments ca 0.5 mm long, the anthers 0.5–0.7 mm long; styles and stigmas ca 1 mm long; capsules 1-chambered, 2.8–3.8 mm long, gradually tapered to a stylar beak to 0.3 mm long; seeds 0.4–0.5 mm long, apiculate. Along streams, sand bars of rivers, around ponds, and in wet lowland meadows at 1220 to 1710 m in Grand, San Juan, Tooele, and Utah counties; British Columbia to Newfoundland and south to Arizona and West Virginia; Eurasia; 5 (i).

Juncus bryoides F. Hermann Minute Rush. [*J. kelloggii* Engelm. misapplied]. Annual plants ca 0.5–2 cm tall, the scapose stems capillary, 0.1–0.2 mm thick; leaves basal or nearly so, mostly less than 1/2 as long as the stems, flat or channeled, with scarious margins, not over 1 mm wide; inflorescence a solitary terminal flower, this subtended by mostly 2 scarious bracteoles; perianth segments 1.5–2 mm long, acute, about equal; stamens 3, the filaments ca 0.4 mm long, the anthers ca 0.2 mm long; capsules shorter than the perianth; seeds 0.3–0.4 mm long, obscurely apiculate. Ponderosa pine, aspen, and mountain brush communities apparently mostly on moist or spring-fed sandy soil or sandstone or quartzite at 2400 to 2550 m in Daggett, Salt Lake, Sevier, and Uintah counties; California to western Colorado; 5 (0). This very small plant is easily overlooked, it is probably more widespread than the few collections indicate.

Juncus bufonius L. Toad Rush. [*J. bufonius* var. *occidentalis* F. Hermann; *J. sphaerocarpus* misapplied]. Tufted annual plants, 2–30 cm tall, from few to numerous fibrous roots; stems few to many; leaves basal and cauline or all basal in very small plants, some of the basal ones bladeless, margins of sheaths hyaline but not projected into auricles, the blades 0.2–1 mm wide, involute or channeled; involucral bract to 3.5 cm long, filiform and leaflike, or scarious, much reduced and hardly different from the bracteoles in very small plants; inflorescence 0.6–12 cm long, often 1/4–4/5 the height of the plant, with 1–20 flowers, the flowers borne singly and sometimes rather remote; bractlets of the nodes scarious, often aristate; pedicels obsolete or less than 1 mm long; bracteoles subtending the flowers scarious, usually ovate; perianth segments 3–6 (8) mm, lanceolate, acute to acuminate, the inner ones shorter than and not as pointed as the outer; stamens 6, the filaments 0.7–1 mm long, the anthers 0.3–0.8 mm long; styles less than 1 mm long; capsules incompletely 3-loculed, 3–4 mm long, ca 1/2 as long to nearly equaling the perianth, subglobose to cylindric-ovoid; seeds 0.3–0.5 mm long, scarcely apiculate; $2n = 34, 54?, 70, 80, 100–110, 160$. Moist or wet soil of rocky drainages, ephemeral pools, along streams and rivers, margins of ponds and lakes, hanging gardens, and wet meadows, in many plant communities at 1135 to 2850 m in all Utah counties, except Iron; cosmopolitan except in the tropics and extreme arctic; 93 (viii).

Juncus castaneus J. E. Sm. Chestnut Rush. Perennial plants, stems arising singly or 2 together from slender, rather long rhizomes, 7–22 cm tall; leaves extending up to the upper 1/3–4/5 of the stem, the sheaths without hyaline margins, lacking auricles, the blades strongly folded below, becoming channeled toward the middle and terete, hollow, and septate toward the tip; involucral bract 1.5–3.5 cm long, scarious and broadly expanded below, prolonged into an involute tip; inflorescence 1–7 cm long, of 1–4 compact heads, the heads ca 5–10 flowered, bractlets subtending the pedicels scarious like the involucral bract but not with a prolonged involute tip; pedicels 2–4 mm long; bracteoles subtending the flowers lacking; perianth segments 4–7 mm long at anthesis, linear-lanceolate, acute or the inner ones somewhat obtuse, chestnut colored or purplish brown; stamens 6, ca 1 mm shorter than the perianth at late anthesis, the filaments ca 3 mm long, the anthers ca 1 mm long; styles ca 2–3 mm long; capsules 1-loculed, 6–10 mm long, elongate, much exceeding the perianth, very gradually tapering into a stylar beak ca 1 mm long; seeds 2.5–4 mm long, strongly tailed, the tails longer than the body; $2n = 60$. Wet alpine communities in the Wasatch Plateau and Uinta Mts. in Sanpete and Summit counties; circumboreal and south in mountains of western North America to

New Mexico; 2 (i). The leaf blades are folded toward the base and are thus similar to those of *J. ensifolius*, but the sheaths are without the scarious margins that are so prominent in those of *J. ensifolius*.

Juncus compressus Jacq. Caespitose perennial plants from thickened short rootstocks and fibrous roots, 20–40 (60) cm tall; leaves on the lower 1/2 of or extending well up on the stems, the scarious margins of sheaths terminating in scarious aricles, the blades 0.5–1.2 mm wide, more or less channeled; involucral bract 2–15 cm long, leaflike; inflorescence 3–15 cm long, congested to open paniculate, with ca 12–80 flowers, these borne singly; bractlets subtending the branches of the inflorescence similar to but smaller than the involucral bract, or the upper ones scarious and much reduced; pedicels obsolete or to 3 mm long, bracteoles of the flowers scarious, rounded to broadly acute; perianth segments 1.5–2.2 mm long, slightly coriaceous and incurved at the tip, especially the outer set; stamens with filaments ca 1 mm long, and anthers ca 0.5–0.7 mm long; styles ca 1 mm long; capsules 2–2.5 mm long, subglobose, obtuse, exceeding the perianth; seeds ca 0.4 mm long, apiculate at both ends. Sand bars, mud flats, swamps, and riparian communities along the flood plain of the Green and Colorado rivers at 1220 to 1675 m in Daggett, Emery, Grand, and Uintah counties; Montana to Utah and east to Nova Scotia, and Eurasia, most likely introduced in the western U. S. and probably in the East; 8 (i).

Juncus confusus Cov. Densely caespitose, perennial plants from fibrous roots, 8–52 cm tall; leaves basal or borne on the lower 1/4 of the stems, the upper 2–4 with well-developed blades, these 0.5–1.3 mm wide, mostly narrower than the stems, channeled or involute, the hyaline margins of the sheaths projected into rounded auricles; involucral bract 1–13 cm long, less than 1 mm wide, very slender, with a scarious decurrent auricle; inflorescence 0.8–2 cm long, with (1) 3–40 flowers, congested, but the flowers borne singly and not in heads; bractlets subtending the pedicels scarious or the lowest one similar to but much smaller than the involucral bract; pedicels nearly obsolete or to ca 2 mm long; bracteoles 2, or these subtended by additional smaller ones borne on the pedicel; perianth segments 3–5 mm long, elliptic-lanceolate or lanceolate, acute to broadly acute, with conspicuous hyaline margins extending to the tip, subequal or the inner set slightly shorter; stamens to ca 1.7 mm long, the anthers 0.5–0.6 mm long; styles ca 1 mm long; capsules 3-loculed, 3–4 mm long, subequal to the perianth, retuse at the apex; body of seeds ca 0.4 mm long, more or less blunt at the apex with a bent apiculate tip from one side. Ponderosa pine, aspen, lodgepole pine, Engelmann spruce, dry and wet meadow, and willow communities at 2070 to 3110 m in Box Elder, Cache, Daggett, Duchesne, Garfield, Iron, Juab, Morgan, Salt Lake, Sanpete, Sevier, Summit, Uintah, Wasatch, and Washington counties; British Columbia to Saskatchewan and south to California and New Mexico; 50 (xv).

Juncus drummondii E. Mey. in Ledeb. Densely caespitose perennial plants from fibrous roots, 8–40 cm tall; stems terete; leaves mostly on the lower 1/4 of the stem, reduced to sheaths or the upper most sheath mostly with a bristle-tip (much reduced blade), this 1–10 mm long or rarely longer, the upper sheath slightly bilobed and scarious at the apex; involucral bract 1–4 cm long, shorter than or exceeding the inflorescence by 2 cm, terete, with scarious decurrent auricles; inflorescence 0.8–2 (3) cm long, with (1) 2 or 3 flowers, the flowers borne separately on pedicels 1–15 (20) mm long, the pedicels often subtended by and partly enveloped in scarious bractlets, the lowest bractlet sometimes similar to but shorter than the involucral bract, each flower subtended by 2 scarious bracteoles 2–4 mm long, these rounded to broadly acute; perianth segments 5–8 mm long, equal or the outer set slightly longer, narrowly acute; stamens ca 2 mm long, the anthers ca 2–3 times longer than the filaments; styles ca 2 mm long; capsules 3-loculed, 5–7 mm long, ca as long or a little shorter than the sepals, retuse at the tip; body of seeds ca 1.5–2 mm long, appendaged on both ends, the appendages as long or slightly longer than the body. Lodgepole pine, spruce-fir, wet and dry meadow, and alpine communities, in wet to dry rocky places at 2940 to 3475 m in Box Elder, Daggett, Duchesne, Garfield, Grand, Juab, Piute, Salt Lake, Sevier, Summit, Uintah, Utah, and Wasatch counties; Alaska to New Mexico; 76 (iv).

Juncus ensifolius Wikstrom Swordleaf Rush. Perennial plants 21–72 cm tall; stems flattened, arising singly or loosely to rather densely tufted, from creeping rhizomes; leaves clustered on the lower 1/2 of the stems, but the upper one usually borne near or above midlength of the stem, strongly folded and flattened, the narrow ventral edge facing the stem, the broad scarious margins of the sheaths extending onto the blade, hardly if at all projected into auricles, the blades more or less closed above the scarious margin, partly to completely septate in the closed position; involucral bract much reduced to leaflike and up to 5 cm long, but shorter than the inflorescence, this 1–17 cm long, with (2) 3–90 or perhaps more heads, the heads light green to nearly black, with 4–25 flowers; bractlets subtending the heads and pedicels scarious; pedicels to ca 1 mm long; bracteoles lacking; perianth segments 2.3–4 mm long, the inner ones slightly shorter and slightly less pointed than the acute outer ones; stamens 3 or 6, filaments ca equal the anthers, these 0.4–1.3 (1.5) mm long, styles 0.5–1.5 mm long; capsules 3-loculed, rounded toward the tip, equal or a little longer than the perianth, the body of the seeds 0.4–0.6 mm long, with or without tailed appendages to 0.2 mm long at either end. There are three varieties as follows:

1. Stamens 3; plants rare in Utah . *J. ensifolius* var. *ensifolius*
— Stamens 6; plants common 2

2(1). Heads (5) 10–60 (90) or perhaps more, mostly 3–8 mm thick, with ca 4–12 flowers; seeds with apiculate tails, or the tails rarely lacking (in Utah specimens); plants common in the Canyonlands section of the state, and wholly intergrading into the following variety throughout much of the state . . . . *J. ensifolius* var. *brunnescens*

— Heads 2–10 or rarely more, mostly 8–15 mm thick, with ca 10–15 (25) flowers; seeds with or without apiculate tails; plants state wide and over a wide elevational range, but more common in mountains and in the Great Basin than the preceding variety . . . .
. . . . *J. ensifolius* var. *montanus*

Var. brunnescens (Rydb.) Cronq. [*J. brunnescens* Rydb.; *J. saximontanus* f. *brunnescens* (Rydb) F. Hermann; *J. tracyi* Rydb., type from Ogden; *J. tracyi* f. *utahensis* (Martin) F. Hermann; *J. utahensis* Martin, type from Ashley National Forest, Summit (?) County]. Along rivers, streams, ditchbanks, around seeps, springs, ponds, lakes, and in hanging gardens, marshes, meadows, and bogs at 1065 to 2450 (2740) m in Beaver, Duchesne, Emery, Garfield, Grand, Iron, Kane, Rich, San Juan, Sanpete, Sevier, Summit, Uintah, Utah, Washington, and Wayne counties, and intergrading into var. *montanus* in nearly all counties of the state; nearly throughout the range of var. *montanus*, but more common southward especially in Arizona and the only phase in Texas; 145 (v). See discussion under var. *montanus*.

Var. *ensifolius* Wet places in mountains in Daggett, Salt Lake, Tooele, and Uintah counties; Alaska to northern Mexico and east to Alberta and Arizona, 4 (i). The few specimens from isolated stations in Utah with only 3 stamens per flower could be nothing more than odd specimens of var. *montanus*. However, to the north of our area this is a common phase.

Var. *montanus* (Engelm.) C. L. Hitchc. [*J. xiphioides* var. *montanus* Engelm.; *J. saximontanus* A. Nels.]. In meadows, along streams and rivers, about seeps and springs and other wet places at (853) 1830 to 3100 m in all counties of the state; Alaska to southern California and east to Saskatchewan and New Mexico; 128 (v). A study of the numerous specimens from Utah revealed the following trends: 1) Most plants of the lower elevations in southeastern Utah are rather easily assigned to var. *brunnescens* (they mostly have apiculate-tailed seeds); 2) Plants of the Great Basin are often referrable to var. *montanus* (they have apiculate but rarely tailed seeds); and 3) Throughout the plateaus and mountains of central Utah and the Uinta Mts., there are plants of both varieties and numerous intermediate plants (seeds are commonly with or without tails in both varieties as well as in intermediate plants). Color phases of the inflorescence (pale green to purplish black) are also found in both varieties and in intermediate plants. Perhaps outside Utah, the situation is not so complex, but this seems to be where the two varieties overlap. More than a quarter of the specimens examined appear to be intermediate.

Juncus filiformis L. Plants perennial, 5–40 cm tall; stems arising singly or in tufts from creeping rhizomes, terete, rarely over 1 mm in diameter; leaves reduced to bladeless sheaths, the uppermost one often tipped with a tiny bristle, confined to the lower 1/5 of the plant; involucral bract 10–27 cm long, appearing as a continuation of the stem, as long as or to over 4 times longer than the stem; inflorescence appearing lateral and on the lower 1/2–1/10 of the plant, 0.5–1 (2) cm long, compact, with ca 5–20 flowers, the flowers borne singly; bractlets subtending the branches and pedicels scarious, the lower ones sometimes aristate; pedicels nearly obsolete or to 4 mm long; bracteoles scarious, ovate or oblong; perianth segments 3–4.5 mm long, greenish, lanceolate, acute to acuminate, subequal or the outer ones slightly longer; stamens 6, the filaments ca 0.6 mm, the anthers 0.4–0.6 mm long; styles and stigmas less than 1 mm long; capsules 3-loculed, ca 2–3 mm long, greenish, ovoid to obovoid, abruptly tapered to a very short stylar beak; seeds 0.4–0.6 mm long, minutely winged-apiculate at both ends; $2n = 70, 80, 84$. Wet subalpine meadows and along streams in the Uinta Mts. at 2990 to 3200 m in Summit, Uintah, and Wasatch counties; Alaska to Labrador and south to Utah and West Virginia; 10 (iv).

Juncus gerardii Lois. Black Grass; Mud Rush. Perennial plants 15–80 cm tall; stems somewhat tufted on slender dark rhizomes; leaves rather scattered on the stems, the upper one usually borne on the upper 1/2 of the stem, the lower ones bladeless or with reduced blades, the upper blades flat, 1.5–3 mm wide; inflorescence with several to many flowers, the flowers borne singly and subtended by scarious bracteoles, nearly sessile to long-pediceled; perianth segments 2–3.5 mm long, dark brown with a greenish midstripe, blunt and usually hooded at the tip; stamens 6, the anthers ca 1.5 mm long, ca 2–3 times longer than the filaments; capsules ovoid to obovoid, rounded, subequal to but a little shorter than the perianth; seeds 0.5–0.6 mm long, slightly apiculate at the tapered end, nearly truncate-apiculate at the other end. Becks Hot Spring in Salt Lake County (Flowers sn 24 Sept. 1924 UT); Atlantic and Pacific coasts, sporadic inland; Eurasia; 1 (0). The Salt Lake County population is probably eradicated, since the area is now part of Interstate 15.

Juncus hallii Engelm. Halls Rush. Perennial caespitose plants 20–40 cm tall, from fibrous roots, rhizomes lacking; stems terete, to ca 1.5 mm thick; leaves basal and on the lower 1/5 of the plant, usually only the uppermost cauline leaf bearing a well-developed blade, the lower stem leaves bladeless or tipped with a short bristle, the innovations sometimes with well-developed blades, the blades terete, the upper side more or less channeled toward the base, not or very inconspicuously channeled toward the tip, less than 1 mm wide; involucral bract 0.7–2.5 (7.5) cm long, scarious and caudate to awned, or elongate and leaflike with scarious margins projected into auricles; inflorescence to 1.7 cm long, with (2) 3–6 flowers, the flowers rather congested, but borne singly; bractlets subtending the pedicels scarious, attenuate to caudate; pedicels 1–8 mm long; bracteoles subtending the flowers hyaline, ovate to nearly orbicular; perianth segments 4–5 (5.5) mm long, subequal or the inner ones a little shorter, lanceolate, acute, usually with greenish centers flanked by purple and with hyaline margins; stamens 6, the filaments 1–1.5 mm long, the anthers 0.5–0.7 mm long; styles and stigmas not over 1 mm long; capsules 3-loculed, equaling or ca 1 mm longer than the perianth, triquetrous, retuse at the apex, dark brown to purplish black; body of the seeds 0.6–0.7 mm long, tailed at each end, the tails ca 1/2 as long as the body. Dry, wet, and boggy meadows, margins of ponds and lakes, and along streams at 2956 to 3350 m in Beaver, Daggett, Duchesne, Garfield, Summit, Uintah, and Wasatch counties; Montana to Colorado; 13 (xi).

Juncus longistylis Torr. in Emory Longstyle Rush. Perennial plants 20–63 cm tall; stems arising singly or few together from creeping rhizomes, terete; leaves somewhat crowded on the lower 1/2 of the stem, but the uppermost one often on the upper 1/2–3/4 of the stem, the scarious margins of the sheaths prolonged into auricles to ca 1 mm long, the blades flat, not hollow, not septate, 1–3 mm wide; involucral bract 1–4 cm long, ca equaling or shorter than the inflorescence, mostly scarious, rarely leaflike, narrowly attenuate to caudate; inflorescence 1–7.5 cm long, usually with (1) 3–13 heads, the heads with 3–10 flowers; bractlets subtending the heads and pedicels scarious, acute to caudate; pedicels to ca 2 mm long, concealed in the scarious bractlets; perianth segments (4) 5–6 mm long, acute to acuminate, often purplish with greenish center and broad whitish or silvery hyaline margins; stamens 6, the filaments 0.5–1 mm long, the anthers (1) 1.3–2 mm long; styles 1–2 mm long;

stigmas ca 2 mm long; capsules 3–loculed, shorter than or rarely equaling the perianth, rather abrupt to retuse at the stylar beak, brownish or purplish black; seeds ca 0.5 mm long, apiculate at each end; $2n = 40$. Wet meadows, along streams and rivers, about seeps and springs and other wet places, occasionally in saline places at 1380 to 3350 m in Box Elder, Cache, Daggett, Duchesne, Emery, Garfield, Grand, Kane, Millard, Morgan, Salt Lake, San Juan, Sanpete, Sevier, Summit, Tooele, Uintah, Utah, Wasatch, Washington and Wayne counties; southern Canada, Washington to South Dakota and south to California and New Mexico; 100 (viii).

Juncus mertensianus Bong. Mertens Rush. Perennial plants 13–42 cm tall; stems arising singly or loosely to tightly clustered on long creeping or short rhizomes, the rhizomes sometimes very short and stout and plants caespitose with numerous fibrous roots; leaves basal and on the upper 1/4–3/4 or higher on the stems, scarious margins of the sheaths projected into ligulelike auricles 0.5–2 mm long; blades terete, channeled above, hollow, septate, 0.5–2 mm wide when pressed; involucral bract 0.8–3.2 cm long, rarely leaflike, often caudate; inflorescence 0.5–1.5 (3.5) cm long, with 1 (2) head(s), the heads with 5–40 or more flowers, to ca 1.5 cm thick; bractlets subtending the heads and pedicels scarious, acute to caudate; pedicels to ca 1 mm long; bracteoles lacking; perianth segments 2.5–4 mm long, acute to acuminate, blackish purple; stamens 6, filaments 1–1.3 mm long, anthers 0.5–1 mm long, shorter than the filaments; styles to 1 mm long; stigmas to 1 mm long; capsules 1–chambered, triquetrous, slightly to conspicuously shorter than the perianth, abruptly tapered to or slightly retuse at the stylar beak, often as blackish purple as the perianth; seeds 0.5–0.7 mm long, apiculate but hardly tailed; $2n = 20, 40$. Wet meadows, along streams, about seeps and springs, margins of lakes and ponds, and Engelmann spruce-lodgepole pine, and tundra communities at 2435 to 3415 m in Box Elder, Cache, Duchesne, Garfield, Iron, Juab, Piute, Salt Lake, San Juan, Sanpete, Sevier, Summit, Uintah, Utah, and Wasatch counties; Alaska and Yukon south to southern California and South Dakota; 83 (vi).

Juncus nevadensis Wats. Nevada Rush. [*J. badius* Suksd.]. Perennial plants 12–35 (53) cm tall; stems more or less terete, arising singly or a few together from creeping rhizomes; leaves basal and on the upper 1/4–3/4 or higher on the stems, with well-developed blades, the scarious margins of the sheaths projected into ligulelike auricles 1.5–3 mm long, blades terete, hollow, septate, somewhat channeled above, 0.5–2 mm wide; involucral bract 1–3 (8.5) cm long, bractlike to leaflike, seldom much exceeding the inflorescence; inflorescence 1–8 cm long, with (1) 2–13 heads, the heads with (3) 6–13 or more flowers; bractlets subtending the heads membranous, attenuate to caudate; pedicels to ca 1 mm long; bracteoles lacking; perianth segments (3) 3.5–5 mm long, brown to purplish black; anthers 1–2 mm long, longer than the filaments; styles to 3 mm long, the stigmas 1–2 mm long; capsules 1–loculed, equal to or conspicuously shorter than the perianth, triquetrous, rounded or rarely slightly retuse at the apex; seeds 0.5–0.6 mm long, apiculate but not tailed. Dry meadow, wet meadow, silver sagebrush-meadow, and lodgepole pine communities at 2286 to 3050 m in Box Elder, Cache, Daggett, Duchesne, Emery, Garfield, Grand, Kane, Rich, Sanpete, Summit, Uintah, Wasatch, and Washington counties; southern British Columbia and Alberta south to California and New Mexico; 43 (xvii). The inflorescence is sometimes similar to those of *J. ensifolius* and *J. longistylis*, and plants of these taxa are sometimes confused. The leaves are different in each of these. Occasional plants with only 1 or 2 heads are rather easily mistaken for those of *J. mertensianus*.

Juncus nodosus L. Jointed Rush. Perennial plants 17–58 cm tall; stems terete, 1–2 mm thick, arising singly to densely clustered on creeping rhizomes, the rhizomes sometimes with small tuberlike segments; leaves usually extending well up the stems, those of the stem with well-developed blades, those of the innovations often with blades, the scarious margins of the sheaths prolonged into very short auricles 0.3–1 mm long, the blades terete or channeled above, hollow, septate, 0.5–1.5 mm wide when pressed; involucral bract 2.5–12 cm long, more or less leaflike; inflorescence 1.5–7 cm long, congested or rather open, with 3–12 globose or nearly globose heads, the heads sessile or pedunculate, with (5) 10–25 flowers, 5–12 mm wide, the flowers widely spreading to divergent; bractlets subtending the heads scarious, acute to cuspidate; pedicels to ca 1 mm long; perianth segments 2.5–3.5 mm long, subequal, acuminate, the acuminate tips shorter than and not so rigid as in those of *J. torreyi*; stamens 6, the filaments ca 0.8 mm long, the anthers 0.6–0.8 mm long; styles to about 3 mm long, the stigmas ca 1 mm long; capsules incompletely 3–loculed, to ca 1 mm longer than the perianth, slender, gradually prolonged into a tardily dehiscent stylar beak, sharply triangular in cross section; seeds 0.4–0.5 mm long, apiculate. Wetlands along streams and rivers and in wet and boggy meadows at (1250) 1640 to 2320 m in Cache, Duchesne, Garfield, Piute, Rich, Summit, Uintah, Washington, and Wayne counties; southern Canada and northern U. S., south to California and Texas; 36 (xiii).

Juncus parryi Engelm. Parry Rush. Perennial caespitose plants (10) 15–30 cm tall, from fibrous roots, lacking rhizomes; stems terete, about 1 mm thick; leaves basal and borne on the lower 1/5 of the stems, usually only the uppermost one with a well-developed blade, the lower sheaths usually tipped with a bristle or much reduced blade, and the uppermost one sometimes reduced on a few of the stems, the scarious margins of the sheaths barely prolonged into auricles less than 0.5 mm long, the blades less than 1 mm thick, terete, channeled to strongly involute below, obscurely channeled above, not septate; involucral bract 1.5–6 (9) cm long, leaflike, terete, more or less simulating a continuation of the stem, often auriculate; inflorescence 0.7–2.2 cm long, with 1–4 flowers, the flowers borne singly; bractlets subtending the pedicels scarious, acute to caudate, or one of them often similar to the involucral bract but smaller; pedicels 1–20 mm long, the abaxial ones often much longer than the adaxial; bracteoles subtending the flowers ovate to lance-ovate, rounded to acute or acuminate-attenuate; perianth segments 5–8 mm long, the inner ones to 1 mm shorter than the outer and somewhat less pointed and more scarious; stamens 6, the filaments ca 0.3 mm long, anthers 1.5–2 mm long; styles ca 1.5 mm long, stigmas to 3.5 mm long; capsules a little shorter to a little longer than the perianth; body of seeds 0.6–0.7 mm long, with tails a little shorter than or to 0.1 mm longer than the body. Engelmann spruce, lodgepole pine, meadow, and alpine communities, on wet to dry rocky ground, sometimes in rocky snowflush areas at 2620 to 3420 m in Cache, Duchesne, Iron, Juab, Salt Lake, Summit, Uintah, and Wasatch counties; British Columbia to Alberta and south to California; 50 (v).

Juncus regelii Buch. Regels Rush. [*J. jonesii* Rydb., type from Alta]. Perennial plants 10–60 cm tall; stems arising singly or few together from creeping rhizomes; leaves basal and extending well up on the stems, the scarious margins of the sheaths prolonged into inconspicuous or short auricles, the blades flat, 2–4 mm wide, neither hollow nor septate; involucral bract 1–4 cm long; inflorescence with 1–5 globose or hemispherical heads, the heads 8–20 mm across; bractlets subtending the heads scarious; bracteoles lacking; pedicels ca 1–2 mm long; perianth segments 4–6 mm long, papillose-roughened on the back, with a greenish midstripe flanked by dark brown and with scarious margins, the inner ones slightly shorter and slightly less pointed than the outer ones; stamens 6, the anthers 1–1.5 mm long, subequal to the filaments; capsules 3–loculed, subequal to the perianth, oblong-ovoid, truncate to retuse; body of seeds about 0.5 mm long, tailed at each end, the tails about as long or longer than the body. Meadows and along streams at 2750 to 3060 m in Duchesne, Salt Lake, and Wasatch counties; southern Washington to northern California and east to Montana; 8 (i). Much like *J. longistylis* but distinct in the tailed seeds and more or less marked by papillose-roughened perianth.

Juncus tenuis Willd. Poverty Rush. [*J. dudleyi* Wieg.; *J. interior* Wieg.; *J. tenuis* var. *dudleyi* (Wieg.) F. Hermann; *J. tenuis* var. *congestus* Engelm.] Perennial, caespitose plants 22–65 cm tall, with fibrous roots; rhizomes lacking; stems terete, to 1.8 mm wide; leaves basal and cauline, borne on the lower 1/5 of the plant, those of the stem mostly with well-developed blades, some of the basal ones with blades reduced to bristles, scarious margins of the sheaths projected into auricles to ca 0.75 mm long, the blades flat but soon moderately to strongly involute, not hollow, not septate, to ca 2 mm wide; involucral bract 2–18 cm long, leaflike; inflorescence (0.7) 1.5–8.5 cm long, congested to rather open, with (4) 10–50 or more flowers, the flowers borne singly; bractlets subtending branches and pedicels scarious, caudate-acuminate or awned, or the lower ones leaflike and similar to the involucral bract; pedicels obsolete or to 5 mm long; bracteoles subtending the flowers, scarious, ovate to lanceolate, acute to caudate; perianth segments 4–5 mm long, subequal or the outer ones a little longer than the inner, the outer ones narrowly acuminate or acuminate-attenuate with the hyaline margins mostly not extending on to the acuminate tip, the inner ones mostly acute to slightly acuminate with the hyaline margins often extending to the tip; stamens 6, the filaments 0.6–1 mm long, the anthers 0.5–0.8 mm long; styles and stigmas ca 1.5 mm long; capsules imperfectly 3–loculed, 1–2 mm shorter than the perianth, obtuse to truncate; body of seeds 0.3–0.4 mm long, with apiculate ends to ca 0.1 mm long; $2n = 40, 80, 84$. Along streams, washes, ditchbanks, rivers, margins of ponds and reservoirs, about seeps and springs, and in meadows and hanging gardens at 1135 to 2380 m in Cache, Daggett, Duchesne, Garfield, Grand, Millard, Rich, San Juan, Uintah, Utah, Wasatch, Washington, and Wayne counties; widespread in North America and introduced in temperate regions elsewhere in the world; 41 (iv). Three intergradient phases are present in our plants: var. *dudleyi* with cartilaginous, often yellow to brown auricles ca 0.5 mm long, and vars. *congestus* and *tenuis* with membranous, usually greenish or whitish auricles often over 0.5 mm long, the former with congested panicles mostly less than 3 cm long, the latter with open panicles mostly over 3 cm long. The morphological differences are minor at best and the apparent taxa are more or less sympatric. However, plants with features of var. *tenuis* do seem to be more common in the southern half of Utah. Perhaps the recognition of these varieties serves more to waste time than to achieve any purpose.

Juncus torreyi Cov. Torrey Rush. Perennial plants (1) 2–8 (10) dm tall; stems terete, to 6 mm in diameter near the base, arising singly or a few together from robust creeping rhizomes, the rhizomes often with swollen tuberlike segments; leaves well distributed up the stem, the scarious margins of the sheaths prolonged into auricles, (1.5) 2–5 mm long, the blades terete sometimes channeled on the upper side, hollow, septate, 1–3 mm thick; involucral bract 1.5–17 cm long, more or less leaflike; inflorescence 1.5–7 cm long, congested, with (1) 3–13 more or less globose and sometimes burlike heads, the heads 6–15 mm across, with 10–50 or more flowers, the flowers widely spreading, with some usually reflexed; pedicels very short and hidden in the compact heads; bracteoles lacking; perianth segments 4–5 mm long or the inner ones slightly shorter, long-acuminate and rigid at the tip; stamens 6, the filaments 0.7–1 mm long, the anthers 0.5–0.8 mm long; styles ca 0.25 mm long, the stigmas ca 1 mm long; capsules incompletely 3–loculed, slender, triquetrous, equal to or slightly longer than the perianth, the slender stylar beak tardily dehiscent; seeds 0.4–0.5 mm long, apiculate but not tailed. Along streams, rivers, washes, and ditchbanks, at margins of ponds and lakes, about seeps and springs, and in saline or alkaline, moist to wet meadows, marshes and swamps at 850 to 2010 m in all Utah counties, except Iron, Millard, Piute, Summit, and Wasatch; southern Canada to northern Mexico; 112 (iv).

Juncus triglumis L. Three-flowered Rush. Perennial plants 4–19 cm tall, densely caespitose; leaves basal or on the lower 1/4 of the stems, those of the stem with well-developed blades, scarious margins of sheaths projected into auricles to ca 0.8 mm long, the blades ca 0.5 mm wide, hollow, sepatate; involucral bract 5–10 mm long, often purplish; inflorescence a solitary head, this 5–8 mm long, with (1) 2–5 flowers, the bractlets subtending the pedicels similar to but somewhat smaller than the involucral bract; pedicels to ca 1 mm long; perianth segments 3–4 mm long, acute, cream, yellowish or greenish yellow, and often marked with purple; stamens 6, as long as the perianth or ca 1 mm shorter, the filaments to 2 mm long, the anthers 0.5–0.8 mm long; stigmas and styles ca 1 mm long; capsules shorter or ca 1 mm longer than the perianth, abruptly tapered to blunt or subtruncate at the tip, blackish purple; trigonous-cylindric body of seeds ca 0.7–1 mm long, tailed at both ends, each tail ca 1/2 to as long as the body, more or less flattened, scarious; $n = 22$; $2n = 132$. Wet meadows and bogs at 2800 to 3810 m in the Uinta Mts., Daggett, Duchesne, Summit, and Uintah counties; circumboreal, south in the Rocky Mts. to New Mexico; 10 (v). Utah plants are referrable to var. **albescens** (**Fern.**) **Lange** [*J. albescens* Fern.].

Juncus tweedyi Rydb. [*J. canadensis* var. *kuntzei* Buch., type from near Corinne, Box Elder County].

Perennial plants 20–40 cm tall, the stems clustered, terete; rhizomes apparently lacking; leaves basal and cauline, the scarious margins of the sheath projected into auricles (0.5) 1–2 mm long, the blades 1–2.5 mm thick, terete or nearly so, hollow, septate; involucral bract shorter than or somewhat longer than the inflorescence, this with 2–8 heads, these with 3–8 flowers, 3–8 mm wide, brown, bractlets subtending the heads and pedicels scarious; bracteoles lacking; perianth segments 3–4 mm long, gradually acute, the inner ones equal to or a little longer than the outer; stamens 3, the anthers 0.5–0.7 mm long, shorter than the filaments; capsule slightly longer than the perianth, triquetrous, more or less acute, imperfectly 3–loculed; seeds cylindrical with tails ca 1 mm long at each end. Near Corinne in Box Elder County (Kuntze 3133, NY). Wet places about hot springs, Yellowstone National Park, Wyoming; 1 (0).

Luzula DC.

Perennial grasslike herbs generally with long, spreading hairs along the margins of leaf blades at least when young; leaves sheathing, the sheaths closed, the blades flat; inflorescence headlike or spicate to open-paniculate; flowers subtended by bracteoles; perianth segments 6; stamens 6; capsules 1–loculed, with 3 seeds, dehiscent along the midribs of the carpels; seeds with or without caruncles, sometimes comose with extremely fine hairs.

1. Inflorescence an open panicle, the panicle sometimes drooping; leaves glabrous or nearly so at maturity, the blades 3–11 mm wide; plants 27–77 cm tall . *L. parviflora*

— Inflorescence of few to several congested or remote spikes or spikelike racemes, sometimes headlike; margins of leaves pubescent with long hairs especially near the collar, 1–6 mm wide; plants 5–42 cm tall 2

2(1). Flowers borne in a terminal compound spikelike or headlike inflorescence; leaves 1–3 mm wide; seeds without or with an inconspicuous caruncle; plants widespread .... *L. spicata*

— Inflorescence with 1 or more lateral spikes, some of the lateral ones often borne on peduncles to 3 (5.5) cm long; leaves 2–6 mm wide; seeds with a conspicuous caruncle; plants known from the Uinta Mts. ...... *L. campestris*

Luzula campestris (L.) DC in Lam. & DC. Hairy Woodrush. [*Juncus campestris* L.; *L. multiflora* (Retz.) Lej.; *L. intermedia* (Thuill.) A. Nels.] Plants 13–42 cm tall; stems more or less tufted; leaves basal and cauline, the blades flat, 2–6 mm wide, the margins with scattered to moderately dense villous hairs ca 2–7 mm long or longer; involucral bract 3–9 mm long, leaflike, some of the bractlets subtending branches and peduncles herbaceous, others scarious; inflorescence 1.5–5 cm long, usuallly conspicuously branched, with 3–12 spikes, these 5–12 mm long, sessile or on peduncles to 3 (5.5) cm long, with 5–15 or more flowers; bracteoles subtending the flowers hyaline, entire or ciliate to fringed; perianth segments 2–3.5 mm long, greenish to brownish, acute or acuminate to scarcely caudate; anthers ca 0.5–1 mm long, the filaments subequal to the anthers; capsules equal to or shorter than the perianth; seeds 1.4–1.7 mm long, with a whitish caruncle ca 0.3–0.6 mm long; $2n = 12, 24, 36$. Lodgepole pine, Engelmann spruce, and meadow communities at 2440 to 3110 m in the Uinta Mts., Daggett, Duchesne, Summit, Uintah, and Wasatch counties; widespread in temperate regions of the world; 10 (vii).

Utah plants apparently belong to var. *multiflora* (Ehrh.) Celak. [*Juncus multiflorus* Ehrh.].

Luzula parviflora (Ehrh.) Desv. Millet Woodrush. [*Juncus parviflorus* Ehrh.; *L. wahlenbergii* Rupr., misapplied]. Plants 27–77 cm tall, stems solitary or few together from short rhizomes and fibrous roots; leaves basal and cauline, glabrous or with a few scattered long-villous hairs especially near the collar, the blades flat, 2–11 mm wide; involucral bract scarious and as short as 1 cm or leaflike and to 4 (7) cm long, shorter than the inflorescence, sometimes sheathing at the base for up to 1 cm; inflorescence 3.5–16 cm long, open-paniculate, with slender flexuous branches; bractlets subtending the branches mostly scarious, sometimes fimbriate toward the apex; flowers borne singly on slender pedicels to 10 mm long or 2 or 3 together on very short pedicels; bracteoles subtending the flowers hyaline, ovate-acute, entire to incised; perianth segments 1.5–2.5 mm long, brownish and partly hyaline, acute to acuminate, subequal; anthers 0.3–0.4, the filaments ca 0.5 mm long; capsules slightly longer than the perianth, blackish purple, shiny; seeds 1.2 mm long, with obsolete or inconspicuous caruncles at each end; $2n = 22, 24, 36$. Ponderosa pine, aspen, lodgepole pine, spruce-fir, willow-streamside, and wet meadow communities at 2300 to 3475 m in Beaver, Box Elder, Carbon, Daggett, Duchesne, Garfield, Grand, Iron, Juab, Piute, Salt Lake, San Juan, Sanpete, Sevier, Summit, Uintah and Wasatch counties; circumboreal, extending south in western North America to California and New Mexico; 84 (v).

Luzula spicata (L.) DC. Spike Woodrush. [*Juncus spicatus* L.]. Plants 5–40 cm tall, stems more or less caespitose from fibrous roots, rhizomes apparently lacking; leaves basal and cauline, the blades flat and 1–3 mm wide or involute and to only 0.5 mm wide, the margins with scattered to moderately dense, long-villous hairs; involucral bract 0.7–4 cm long, bractlike or occasionally leaflike, shorter than or equal to the inflorescence or occasionally longer; inflorescence 1–3 cm long, often nodding, of ca 4–10 or perhaps more sessile or subsessile heads or short spikes that are congested into a continuous or basally interrupted compound spike, the individual heads or spikes with few to several flowers; bractlets subtending the spikes or heads scarious and bractlike, or rarely leaflike; bracteoles subtending flowers scarious, fimbriate-ciliate, acuminate-caudate; perianth segments 2–3 mm long, dark brown or partly hyaline, acuminate or acuminate caudate, the inner ones a little shorter than the outer; anthers 0.3–0.5 mm long, ca 1/2 as long to as long as the filaments; capsules a little shorter than the perianth; seeds 1–1.3 mm long, the caruncle obsolete or inconspicuous, not over 0.2 mm long; $2n = 24, 36$. Lodgepole pine, spruce, fir, dry meadow, wet meadow, streamside-meadow, alpine tundra, and rarely aspen and oak-ponderosa pine communities, also in talus and fell fields at 2470 to 3810 m in Daggett, Duchesne, Grand, Juab, Piute, Salt Lake, San Juan, Summit, Tooele, and Uintah counties; circumboreal, south in western North America to California and Arizona; 65 (vi).

JUNCAGINACEAE Rich.
Arrowgrass Family

Herbs of marshes and wet meadows, perennial, rhizomatous; leaves 2–ranked, mainly basal, linear, sheathing at the base; inflorescence a terminal spike; flowers perfect or imperfect, regular; perianth segments usually 6, in 2 series of 3; stamens commonly 6; pistil 1, the ovary 3- to 6-carpellate; fruit capsular; x = 6, 8, 9.

Triglochin L.

Wandlike perennials; flowers few to numerous; anthers oblong to elliptic, sessile or nearly so; carpels 3 or 6, at length separating and sometimes only the inner 3 fertile and each 1-seeded.

1. Stigmas 3; fruit linear to clavate, 5-8 mm long, the carpels tapering to the base .... *T. palustris*
— Stigmas 3; fruit cylindrid to ovoid-angled, 4-6 mm long, the carpels not tapering at the base 2
2(1). Ligules 1-5 mm long; leaf blades 1-4 mm wide; stems tufted, mainly 3-10 dm tall .... *T. maritima*
— Ligules 0.5-1 mm long; leaf blades 0.5-1.5 mm wide; stems solitary along the rhizome, mainly 1.5-4 dm tall .... *T. concinna*

Triglochin concinna Davy Low Arrowgrass. Plants mainly 1.5-4 dm tall, the stems borne along a slender rhizome; leaves linear, 8-20 cm long, typically much overtopped by the spike, the blade 0.5-1.5 mm wide, subterete, the ligule bilobed, 0.5-1 mm long; flowers few to many, spaced along the spicate raceme; pedicels 2-4 mm long; perianth 6-segmented, 1-1.5 mm long, green; stigmas 6; fruit short-cylindroid, 4-5 mm long, the carpels all fertile, rounded at the base, finally deciduous from the axis. Saline seeps and marshes at 1280 to 2105 m in Beaver, Box Elder, Cache, Daggett, Garfield, Iron, Juab, Millard, Rich, San Juan, Sanpete, Tooele, and Washington counties; British Columbia to the Dakotas, south to Baja and Arizona; 11 (i). Our material, which is not always separable from dwarf specimens of *T. maritima*, belongs to var. *debilis* (Jones) J. T. Howell [*T. maritima* var. *debilis* Jones, type from Johnson, Kane County; *T. debilis* (Jones) Love & Love].

Triglochin maritima L. Maritime Arrowgrass. Plants 0.7-10.5 dm tall (rarely more) from a thick woody rhizome; leaves linear, 2-50 cm long, flattened or channeled, mostly 2-4 mm broad, obtuse apically, the sheath prominently hyaline-margined; racemes several- to many-flowered, the flowers not subtended by bracts; pedicels 1-4 mm long; perianth segments greenish or yellowish, 1-2.2 mm long; fruit ovoid-oblong, mostly 4-6 mm long, deciduous, the axis terete; 2n = 24, 36, 48. Wet meadows and other moist sites in salt desert shrub, mixed desert shrub, sagebrush, mountain brush, aspen, ponderosa pine, and spruce-fir communities at 1310 to 2685 m in all Utah counties except Beaver, Millard, and Morgan; widespread in North America; circumboreal; 79 (ix). This plant occurs in meadows that are cut for wild hay. Drying plants yield hydrocyanic acid, which causes poisoning of livestock.

Triglochin palustris L. Marsh Arrowgrass. Plants 0.8-6.5 dm tall from short, ascending rhizomes; leaves linear-filiform, 2-28 cm long, flattish, 0.3-2 mm broad, acutish apically, the sheath narrowly hyaline-margined; racemes with few to many flowers, the flowers not subtended by bracts; pedicels 1-6 mm long; perianth segments yellowish or greenish, often suffused with purple, 1.5-2 mm long; fruit linear-clavate, commonly 8-10 mm long, the carpels separating from the base and remaining attached apically; 2n = 24, 28, 36. Wet meadows and other moist sites in ponderosa pine, aspen, spruce-fir, and lodgepole pine communities at 1830 to 2960 m in Daggett, Duchesne, Garfield, Grand, Kane, Piute, Salt Lake, Sanpete, Sevier, Summit, and Uintah counties; Alaska to the Atlantic, south to California, New Mexico, Iowa, and New York; circumboreal; Southern Hemisphere; 11 (ii).

LEMNACEAE S. F. Gray

Duckweed Family

Plants small, free-floating or submersed aquatic perennials with flattened, thalluslike stems (fronds), these rootless or with 1 or more simple roots borne on the lower surface; leaves lacking; reproducing vegetatively; flowers rarely produced, borne in a sac- or flask-shaped spathe; staminate flowers consisting of a single stamen (rarely 2); pistillate flowers of a single carpel with 1 to several ovules; perianth lacking; fruit a utricle; x = 8, 10, 11, 21. This is the simplest and smallest of the flowering plants, reproducing mainly by budding.

1. Roots lacking; fronds veinless, ellipsoid .... *Wolffia*
— Roots 1 or more per plant; fronds with 1 to several veins 2
2(1). Root 1 per frond; plant body with 1-5 veins ...... *Lemna*
— Roots several per frond; plant body with 5-11 veins .... *Spirodela*

Lemna L.

Fronds flattened, free-floating or somewhat submersed, 1- to 5-veined, orbicular to obovate, each frond with 1 root; flowers 3, 1 pistillate and 2 staminate, borne in a pouch; anthers 2-loculed; ovary 1-loculed and 1- to several-ovuled. Note: Specimens from Utah have been annotated by F. Landolt with an array of names whose taxonomic delimitations are not understood by this worker. A future revision of the genus seems likely.

1. Fronds long-stalked, 6-12 mm long, elliptic to somewhat lanceolate .... *L. trisulca*
— Fronds not stalked or the stipe very short, oblong to ovate 2
2(1). Dorsal surface of fronds flat and smooth, veinless or with 1 vein; fronds usually elliptic, 1.5 mm wide or less 3
— Dorsal surface of fronds with evident protuberances, faintly 3-veined; fronds obovate to orbicular, mostly over 1.5 mm wide 4
3(2). Fronds 2.5-5 mm long, elliptic to obovate, in colonies of 2 or more .... *L. valdiviana*
— Fronds 2.5 mm long or less, ovoid to elliptic, solitary or in pairs .... *L. minuta*
4(2). Fronds mainly 3.5-6 mm long and 2.5-5 mm wide, obovate to orbicular, the dorsal surface mottled yellowish green, the 3 nerves distinct; fruit winged, usually 2-seeded (at least) .... *L. gibba*
— Fronds various in size but, if as above, the dorsal surface uniformly green and the three nerves indistinct; fruit not winged, usually 1-seeded 5
5(4). Fronds 1.5-3.5 mm long, 1-2.5 mm wide, oblong to obovate .... *L. obscura*
— Fronds 3-6 mm long, 1.5-4 mm wide, obovate to suborbicular .... *L. minor*

Lemna gibba **L.** Fronds obovate to orbicular, 1–4 per colony, 3.5–6 mm long, 2.5–5 mm wide, usually 3-veined, mottled or green above, flat to somewhat convex, strongly inflated to gibbous below at maturity; fruits winged and with 1 or more seeds; 2n = 40, 50, 60?, 80. Ponds, lakes, and sluggish streams at ca 1370 m in Utah County; widespread in temperate and tropical regions of the world; 1 (0).

Lemna minor **L.** Lesser Duckweed. Fronds solitary or in colonies of a few floating fronds, ovate to suborbicular, 3–6 mm long, 1.5–4 mm wide, symmetrical or nearly so, thickish, obscurely 3-nerved, upper surface sometimes keeled and with a row of papillae along the midrib, surfaces green or purplish; fruits not winged, 1–seeded; 2n = 20, 30, 40, 42, 50, 80. Common species of ponds, lakes, and sluggish streams at 1060 to 2440 m in Box Elder, Davis, Cache, Garfield, Kane, Morgan, Rich, Salt Lake, Sevier, Utah, and Wasatch counties; widespread in North America; almost cosmopolitan; 11 (0).

Lemna minuta **H.B.K.** [*L. minima* Kunth; *L. minuscula* Herter; *L. valdiviana* var. *minima* (Kunth) Hegelm.] Fronds solitary or in pairs, floating, ovoid to elliptic, 2.5 mm long or less, 1–veined or veinless, flat or slightly convex above and below, usually with a row of papillae along the nerves on the upper surface; fruits not winged and with 1 seed. Ponds, lakes, and sluggish streams at 1370 to 2625 m in Cache, Davis, Duchesne, Kane, Millard, Salt Lake, Utah, Wasatch, Washington, and Weber counties; California and Nevada to Texas; South America; 12 (i).

Lemna obscura **(Austin) Daubs** [*L. minor* var. *obscura* Austin in Gray]. Fronds obovate to oblong, 1.5–3.5 mm long, 1–2.5 mm wide, solitary or in groups of 2 or 3, floating, upper surface darker green, lower surface slightly reddish and inflated, obscurely 3-veined; fruits wingless, 1–seeded. Sluggish streams and ponds at 1430 to 2685 m in Beaver, Daggett, Garfield, Sevier, Uintah, and Utah counties; widely distributed in the U.S.; 11 (0).

Lemna trisulca **L.** Ivy-leaf Duckweed. Fronds usually floating just below the surface, with several generations attached to each other, 2.5–8.5 mm long, 0.9–3.2 mm wide, elliptic to lanceolate, obscurely 3–veined, denticulate at the apex; fruit asymmetrical; seeds ribbed; 2n = 20, 40, 44, 60, 80. Springs, streams, ponds, and lakes at 1525 to 2592 m in Cache, Garfield, Millard, Morgan, Salt Lake, Sevier, Summit, and Utah counties; widely distributed in North America; circumboreal; 10 (0).

Lemna valdiviana **Phil.** Fronds 2.5–5 mm long, elliptic to obovate, borne in colonies of 2 or more, floating or submersed, strongly asymmetrical at the base, the tip symmetrical, more or less flat on both surfaces and pale green, obscurely 1–veined; fruit wingless, 1–seeded. Ponds, lakes, and sluggish water at 820 to 2745 m in Cache, Utah, Washington, and Weber counties; widely distributed in North America; South America; 5 (0).

Spirodela Schleiden

Fronds flattened, colonial by elongate stipes, obovate to orbicular, with several roots per frond, greenish or reddish; flowers monoecious in pouches, with 2 or 3 staminate and 1 pistillate per lateral pouch; stamens 1 per flower; pistils 2–ovuled.

Spirodela polyrhiza **(L.) Schleiden** [*Lemna polyrhiza* L.]. Floating colonial or solitary plants, the fronds rounded-obovate, flat, dark green above and usually purplish beneath (sometimes also above), 4–7 mm long, 2–6 mm wide, with 5–11 nerves and 5–12 rootlets; fruits somewhat winged. Wet meadows, sluggish streams, and ponds at 1280 to 1405 m in Cache, Davis, Salt Lake, and Utah counties; widely distributed in North America; Eurasia, Africa, and Australia; 7 (0).

Wolffia Horkel ex Schleiden

Small, floating, rootless plants, globular, solitary, and greenish brown; flowers unisexual, 2 per spathe from the upper surface, 1 staminate and 1 pistillate; anthers 1–loculed; ovary 1–loculed and with 1 ovule; fruit a utricle, 1–seeded.

Wolffia punctata **Griseb.** Fronds 0.7–1.2 mm long, 0.4–0.7 mm wide, solitary, rootless, globular, bearing stomata on the upper surface, punctate both ventrally and dorsally. Open water in Cache, Morgan, and Weber counties; widespread in North America; 3 (0).

LILIACEAE A. L. Juss.
Lily Family

Herbaceous perennials from bulbs, corms, or rhizomes; stems annual, dying to below ground level each year; flowers regular or nearly so, perfect; sepals 3; petals 3, distinct or connate; stamens 6; pistil 1, the ovary superior, 3–loculed; style 1 or 3; stigma 3–lobed or 3; fruit a capsule or a berry; x = 3–19+.

1.	Perianth segments not alike, the sepals of different color or texture from the petals; plants indigenous . . . . *Calochortus*	
—	Perianth segments alike, the sepals and petals both petaloid in color and texture	2
2(1).	Inflorescence umbellate, or subumbellate and nearly sessile	3
—	Inflorescence a raceme, panicle, or corymb, or the flowers solitary and yet not umbellate	7
3(2).	Perianth 3–8 cm long, the lobes white; umbel subsessile, arising from below ground . . . . *Leucocrinum*	
—	Perianth less than 2 cm long, variously colored; umbels scapose, borne above ground	4
4(3).	Perianth segments distinct or nearly so; plants with odor of onion or garlic . . . . *Allium*	
—	Perianth segments united below the middle; plants not smelling of onion or garlic	5
5(4).	Staminal filaments distinct; styles 2–4 mm long; capsules stipitate . . . . *Triteleia*	
—	Staminal filaments united; styles 4–6 mm long; capsules sessile	6
6(5).	Stamens in 2 series, the anthers dissimilar; leaves 2–5; capsules 4–6 mm long; plants mostly 3–6 dm tall or more; flowers blue . . . . *Dichelostemma*	
—	Staminal filaments in 1 series, the anthers alike; leaves often more than 5; capsules 10–15 mm long; plants 1–3 dm tall; flowers whitish to greenish, with purple veins . . . . *Androstephium*	
7(2).	Stems much-branched, from thick, tuberous roots; leaves minute, scalelike, subtending filiform, leaflike branchlets (phyllodia) . . . . *Asparagus*	
—	Stems or scapes simple or sparingly branched; underground parts various, but usually not consisting of thick, tuberous roots; leaves neither scalelike nor subtending phyllodia	8

8(7). Plants 10–20 dm tall; leaves 15–30 cm long and at least some of them over 8 cm wide; flowers borne in panicles, these usually more than 20 cm long and 10 cm wide .... *Veratrum*

— Plants mainly less than 10 dm tall; leaves less than 15 cm long or less than 8 cm wide or both; flowers solitary, in racemes, or panicles but, if the latter, usually less than 20 cm long 9

9(8). Flowers borne separately from the foliage leaves, either at different seasons or on separate stems; plants cultivated 10

— Flowers terminating leafy branches of the season, or in leaf axils 11

10(9). Flowers 5–8 cm wide, usually pink to lavender, erect, appearing in autumn, the leaves borne in springtime .... *Colchicum*

— Flowers 4–6 mm wide, white, nodding, appearing in springtime, the leaves borne on separate branches . ... *Convallaria*

11(9). Flowers solitary of few, the perianth segments 2–4 cm long or more, sometimes the calyx not distinctly petaloid; plants cultivated and escaping .... *Tulipa*

— Flowers few to many or, if solitary, smaller than above 12

12(11). Leaves definitely cauline; peduncles or pedicels less than 10 cm long 13

— Leaves definitely basal or, if on lower stem, the peduncles or scapes often more than 1 dm long 18

13(12). Flowers solitary in ovate to lanceolate leaf axils .... *Streptopus*

— Flowers terminal, either solitary, paired, or in racemes or panicles; leaves various 14

14(13). Flowers 1 or few, nodding; perianth white to cream, 8–12 mm long; stems dichotomously branched .... *Disporum*

— Flowers solitary or numerous, borne in racemes or panicles, not nodding or, if so, the perianth larger .. 15

15(14). Flowers numerous, borne in racemes or panicles; plants arising from rhizomes .... *Smilacina*

— Flowers solitary or few; plants bulbous 16

16(15). Flowers white, the segments with purple veins; plants of alpine tundra .... *Lloydia*

— Flowers variously colored, spotted in some; plants seldom if ever of alpine tundra 17

17(16). Flowers often reflexed and with tips of petals reflexed, mainly 3–7 cm long or more; plants cultivated and persisting .... *Lilium*

— Flowers reflexed or not, the petals erect to spreading, mainly 1–2 cm long; plants indigenous ...... *Fritillaria*

18(12). Perianth segments united, at least below 19

— Perianth segments distinct or nearly so 21

19(18). Flowers orange red, the segments 7–12 cm long; plants cultivated .... *Hemerocallis*

— Flowers variously colored, the segments less than 3 cm long 20

20(19). Perianth typically blue purple, 5–9 mm long, urn-shaped, not opened at the throat; lobes very short . ... *Muscari*

— Perianth variously colored, 1.5–2.5 cm long, bell-shaped, open at the throat; lobes long and spreading .... *Hyacinthus*

21(18). Plants shortly rhizomatous, with thickened roots; flowers white, numerous in contracted, racemose panicles; growing in sandy deserts in southeastern Utah .... *Eremocrinum*

— Plants bulbous; flowers variously colored, solitary, few, or numerous, borne in open racemes or panicles; diverse in habitat and distribution 22

22(21). Flowers bright yellow, nodding, the segments reflexed, solitary or few .... *Erythronium*

— Flowers white, cream, or blue, nodding or erect, the segments erect or spreading, solitary to numerous .. 23

23(22). Perianth white to cream 24

— Perianth blue, or blue and white (rarely white in some) 25

24(23). Flowers in corymbose racemes; plants cultivated and more or less weedy .... *Ornithogalum*

— Flowers in racemes or panicles, not corymbose; plants indigenous .... *Zigadenus*

25(23). Flowers several to many in racemes; plants indigenous .... *Camassia*

— Flowers solitary or few in contracted racemes; plants cultivated 26

26(25). Flowers nodding, blue throughout .... *Scilla*

— Flowers erect, the lobes blue apically, white at the base .... *Chionodoxa*

Allium L.

Perennial herbs with odor of onion or garlic, scapose from tunicate bulbs or less commonly from rhizomes; leaves 1 to several, linear to narrowly oblanceolate or elliptic, flat to terete, hollow or solid, often withered or deciduous at anthesis; flowers perfect, borne in spathaceous umbels or else subcapitate, sometimes replaced by bulblets; sepals 3, petaloid; petals 3, petaloid, distinct or nearly so; stamens 6, more or less connate at the base and adherent to the perianth segments; ovary sessile, 3-lobed, sometimes 3- or 6-crested, 3-loculed, with 2 to several ovules per locule; style 1, the stigma capitate to 3-lobed; capsule loculicidal, usually 6-seeded.

Cronquist, A. 1977. *Allium*. pp. 508–522. In: Cronquist et al. Intermountain Flora. Vol. 6. Columbia University Press, New York. 584 pp.

1. Outer bulb coat traversed by coarse, interwoven fibers 2
— Outer bulb coat not coarsely fibrous 5

2(1). Spathaceous bracts 3- to 5-nerved; ovary usually strongly crested; leaves typically 2 per scape; plants of southeastern Utah .... *A. macropetalum*
— Spathaceous bracts typically 1-nerved; ovary low crested to crested; plants of broad or other distribution 3

3(2). Leaves flat, curved, mostly 3–6 mm wide, 2 or 3 per scape; perianth pink; plants of Box Elder County .. .. *A. passeyi*
— Leaves semi-cylindric or grooved, not especially curved, 2 or more per scape; plants broadly distributed 4

4(3). Leaves 3 or more per scape; perianth pink (or white); plants of moist montane sites mainly east of the mountain and plateau axis, except in western Garfield county .... *A. geyeri*
— Leaves typically 2 per scape; perianth white; plants of lower elevation sites in eastern Utah .... *A. textile*

5(1). Leaves hollow, terete or channeled; plants cultivated 6

— Leaves seldom hollow, typically flattened or keeled; plants indigenous 7

6(5). Scapes inflated .... *A. cepa*

— Scapes not inflated .... *A. schoenoprasum*

7(5). Inflorescence nodding at anthesis; ovary strongly crested; stamens exserted; plants of eastern Utah .. .. *A. cernuum*

— Inflorescence erect at anthesis; ovary crested or not; stamens various 8

8(7). Bulbs borne on short, stout rhizomes, elongate; outer bulb scales striate with elongate cells arranged in regular, vertical rows .... *A. brevistylum*

— Bulbs not, or seldom, borne on rhizomes, ovoid to subglobose; outer bulb scales variously ornamented, but not as above 9

9(8). Leaves 1 per scape, the leaf terete or essentially so .. 10

— Leaves 2 or more per scape, the leaves flat or channeled 11

10(9). Outer bulb scales not reticulate; perianth typically dark rose purple (at least when dry); plants of Beaver, Kane, and Millard counties .... *A. atrorubens*

— Outer bulb scales reticulate; perianth typically rose pink, often drying pale pink; plants widely distributed .... *A. nevadense*

11(9). Outer bulb scales with walls of cellular reticulations thin, minutely sinuous, striate above; ovary strongly 6-crested, the crests triangular and minutely serrulate; plants mainly of the western half of Utah, less commonly in Kane and San Juan counties .... *A. biceptrum*

— Outer bulb scales with walls of reticulum not both thin and sinuous, or not reticulate, not especially striate above; ovary variously crested 12

12(11). Leaves 2–4, linear, subterete or grooved, flexuous, shorter than the scape; bulb coats with prominent, thick-margined, cellular reticulations; plants broadly distributed .... *A. acuminatum*

— Leaves 2 (or 1), flat, curved, usually surpasing the scape; bulb coats not especially reticulate 13

13(12). Perianth 7–10 mm long; ovary crested; plants of central western Utah .... *A. parvum*

— Perianth 5–8 mm long; ovary not crested; plants common from Millard and Sanpete counties northward, less common elsewhere .... *A. brandegei*

Allium acuminatum Hook. Bulbs 1–1.5 cm long and as wide or wider, buried (2) 2.5–10 (12) cm, the scales reticulate with evidently thickened ridges, not fibrous-shredded; leaves 2–4, linear, concave-convex, shorter than the scape, typically with sheathing, purplish bases; scapes (6) 10–30 cm tall (above ground); spathaceous bracts 2, 10–25 mm long; umbel (3) 15– to 38–flowered, the pedicels straight or curved; perianth 7–12 mm long, rose to pink purple or pink (white), the segments dimorphic, with recurved tips; stamens included; ovary minutely crested and with 3 minute, bilobed processes; 2n = 14. Sagebrush, mountain brush, pinyon-juniper, and ponderosa pine communities at 1370 to 2685 m in all Utah counties except Carbon, Emery, Iron, Kane, Morgan, Piute, and Wayne; British Columbia to Idaho and Wyoming, south to California, Nevada, and Arizona; 104 (ix).

Allium atrorubens Wats. Bulbs 0.8–1.5 cm long, 0.6–1.2 cm thick, buried 3–11 cm, the scales not reticulate (merely cellular), not fibrous-shredded; leaves solitary, typically surpassing the scape, linear, terete to semiterete, the sheathing bases usually hyaline; scapes 2–10 cm long (above ground); spathaceous bracts 2 or 3, 5–13 mm long; umbels 8– to 25–flowered, the pedicels usually straight; perianth 7–12 mm long, deep rose purple to less commonly, pinkish or white, the segments subequal, with tips spreading; stamens included; ovary crested, the crests thin, entire or toothed. Shadscale, sagebrush, and pinyon-juniper communities at 1430 to 1925 m in Beaver, Kane, and Millard counties; Nevada and California; 8 (i).

Allium biceptrum Wats. [*A. palmeri* Wats.; *A. biceptrum* var. *palmeri* (Wats.) Cronq.; *A. biceptrum* var. *utahense* Jones, type from City Creek Canyon, Salt Lake County]. Bulbs 1–1.5 cm long and about as thick, buried 1–9 cm below ground, the scales cellular-reticulate, the reticulum subrectangular, with minutely sinuous vertical walls or irregular and with all sides sinuous; leaves 2–4, shorter than the scape, flat or broadly channeled, 2–13 mm wide, usually persistent at anthesis, the sheathing bases hyaline to purplish; scapes 10–40 cm tall (above ground); spathaceous bracts 2, distinct, 3– to 5–veined; umbels 8– to 36–flowered, the pedicels straight to curved; perianth pink purple to pink or white, 7–12 mm long, the segments subequal, the tips usually erect; stamens included; ovary strongly crested, with 6 triangular, flattened, denticulate-papillose crests; 2n = 18. Sagebrush, pinyon-juniper, mountain brush, aspen, grass-forb, and spruce-fir communities at 1370 to 3145 m in Beaver, Box Elder, Cache, Davis, Iron, Juab, Kane, Millard, Morgan, Piute, Salt Lake, San Juan, Sevier, Summit, Tooele, Utah, Washington, and Weber counties; Oregon and Idaho, south to California, Arizona, and New Mexico; 42 (iv). Our materials have been treated as belonging to two varieties, separable on a series of overlapping and difficultly discernable features. Their recognition seems to be moot. Reports of *A. campanulatum* Wats. probably belong with this taxon.

Allium brandegei Wats. [*A. tribracteatum* var. *diehlii* Jones, type from Parleys Canyon; *A. diehlii* (Jones) Jones]. Bulbs 0.7–1.7 cm long and about as thick, typically purple and shiny beneath the older scales, buried 2–15 cm below ground, the scales minutely reticulate with rectangular cells; leaves 2 per scape, linear, channeled to flattened, 1–6 mm wide, usually persistent at anthesis, the sheathing bases usually hyaline; scapes 1–12 cm tall (above ground); spathaceous bracts 2 or 3, several-veined; umbels 5– to 35–flowered, the pedicels curved or straight; perianth white or less commonly pink, 5–11 mm long, the segments subequal, the tips usually erect; stamens included; ovary not crested; 2n = 14. Sagebrush, mountain brush, grass-forb, lodgepole pine, aspen, spruce-fir, and alpine tundra communities at 1830 to 3355 m in Box Elder, Cache, Davis, Duchesne, Juab, Millard, Salt Lake, Sanpete, Sevier, Summit, Uintah, Wasatch, Washington, and Weber counties; Oregon to Montana, south to Nevada and Colorado; 50 (v). A collection from Washington County, included here within *A. brandegei*, consists of unusually robust plants with bright pink flowers, but these seem to be properly placed.

Allium brevistylum Wats. Bulbs 0.8–2 cm thick, 1–4 cm long, tapering from base upward, the roots appearing lateral from a short, rhizomatous base, buried 2–10 cm, the scales minutely striate with elongate cells in vertical rows, not fibrous-reticulate; leaves 2–5, usually shorter than the scape, linear, flat, 2–8 mm wide; scapes 15–68 cm long (above ground); spathaceous bracts 2, connate basally, 3- to 5-veined; umbels 5- to 17-flowered, the pedicels typically curved; perianth deep rose pink (at least when dried), 10–13 mm long, subequal, the tips recurved; stamens included; ovary not crested; 2n = 14. Aspen, lodgepole pine, spruce-fir and sedge-forb or grass-forb communities, often in moist sites, at 2285 to 3235 m in Daggett, Duchesne, Garfield, Sanpete, Sevier, Summit (type from Uinta Mts.), Uintah, and Wasatch counties; Idaho and Montana to Colorado; 39 (ii). This species has been reported for the La Sal Mts. also.

Allium cepa L. Onion. Biennial or perennial herbs (grown often as annuals) with well-developed bulbs of many shapes, colors, and sizes; leaves hollow, glaucous; umbels borne on scapes 3–10 dm tall; flowers pink to white, mostly 4–7 mm long, on pedicels to ca 25 mm long; capsules about as long as the perianth; 2n = 16, 32, 54. Cultivated onion of gardens and commerce, widely grown in Utah; introduced from Eurasia; 3 (0).

Allium cernuum Roth Bulbs 1–2.5 cm thick, 3–9 cm long, tapering from base to ground level, the roots central or sometimes from an eccentric, subrhizomatous base, the scales typically purple or purplish, membranous, minutely striate with elongate cells in vertical rows; leaves several to many per scape, shorter than the scape, linear, flat or concave-convex, 1–6 mm wide; scapes 10–60 cm tall (above ground), abruptly recurved near the tip, the umbel nodding; spathaceous bracts connate, membranous, fragile and soon fragmented; umbels 6- to 40-flowered, the pedicels straight or curved; perianth campanulate, 4–6 mm long, pink or white; stamens exserted; ovary strongly 6-crested; 2n = 14. Mountain brush, sagebrush, ponderosa pine, Douglas fir, aspen, and lodgepole pine communities at 1980 to 3205 in Carbon, Daggett, Duchesne, Emery, Garfield, Grand, San Juan, Sanpete, Sevier, Uintah, and Wayne counties; British Columbia to New York, south to Mexico, Texas, and Georgia; 46 (iv).

Allium geyeri Wats. Bulbs 0.7–2 cm thick and about as long or longer, ovoid, buried 1–8 cm below ground, the scales fibrous-reticulate; leaves 2 or 3 per scape, shorter than the scape, 1–5 mm wide, concave-convex; scapes 15–50 cm tall (above ground); spathaceous bracts 2 or 3, mostly 1-veined; umbels 5- to 37-flowered, the pedicels straight or curved; perianth pink or white, 6–10 mm long, the segments subequal, with tips more or less recurved, often replaced by bulblets; stamens included; capsule inconspicuously low-crested. Pinyon-juniper, ponderosa pine, sagebrush, aspen, grass-forb, and spruce-fir communities, typically in moist sites, at 2105 to 3540 m in Cache, Daggett, Duchesne, Emery, Garfield, Rich, San Juan, Summit, Uintah, and Wasatch counties; British Columbia and Alberta, south to Oregon, Nevada, Arizona, New Mexico, and Texas; 24 (i). A peculiar specimen from San Juan County (Welsh et al. 22371 BRY) keys to this species, but is anomalous in having many layers of fibrous reticulum and other peculiar features.

Allium macropetalum Rydb. [*A. reticulatum* var. *deserticola* Jones, type from Cisco, Grand County; *A. deserticola* (Jones) Woot. & Standl.]. Bulbs 1.2–2.5 cm thick and about as long, buried 3–15 cm below ground, the scales fibrous-reticulate, long-persistent and sometimes many layers thick; leaves typically 2, semiterete, usually surpassing the scape, 1–3 mm wide; scapes 3–18 cm tall (above ground); spathaceous bracts 2 or 3, mostly 3- to 5-veined; umbels 4- to 29-flowered, the pedicels often curved; perianth pink to whitish with purple midvein, the segments subequal, straight or more or less recurved; stamens included; capsule conspicuously 6-crested. Shadscale, other salt desert shrub, and pinyon-juniper communities at 1125 to 1925 m in Emery, Garfield, Grand, Kane, San Juan, and Wayne counties; Colorado, Arizona, New Mexico, and Texas; 57 (xii).

Allium nevadense Wats. [*A. cristatum* Wats., type from near St. George; *A. nevadense* ssp. *criatatum* (Wats.) Ownbey]. Bulbs 0.8–1.5 cm thick and about as long, buried 3–11 cm below ground, the scales with contorted reticulations; leaf solitary (very rarely a second poorly developed), terete or nearly so, surpassing the scape, often coiled apically, mainly 2–4 mm thick; scape 2–10 cm tall (above ground); spathaceous bracts 2 or 3, mostly 3- to 7-veined; umbels 3- to 25-flowered, the pedicels straight or curved; perianth 7–13 mm long, pink, white, or rose, the segments subequal, the tips spreading or recurved; stamens included; ovary distinctly crested, the 6 crests entire or toothed; 2n = 14. Blackbrush, shadscale, other salt desert shrub, sagebrush, wildrye, mountain brush, and pinyon-juniper communities at 850 to 2440 m in Beaver, Box Elder, Carbon, Duchesne, Emery, Garfield, Grand, Juab, Kane, Millard, San Juan, Sanpete, Sevier, Tooele, Uintah, Washington, and Wayne counties; Oregon and Idaho to Colorado, California, Nevada, and Arizona; 71 (vii).

Allium parvum Kellogg Bulbs 1.2–2.5 cm thick and about as long, buried 5–11 cm below ground, the scales merely cellular; leaves 1 or 2 per scape, flat, 2–5 mm wide, surpassing the scape, usually strongly curved; scapes 1.5–4 cm tall (above ground); spathaceous bracts 2, suffused with deep purple, many-veined; umbels 3- to 15-flowered, the short pedicels often recurved; perianth 7–12 cm long, white to pink or purple, with dark purple midveins, the outer often broader than the inner; stamens included; capsules with 3, low, rounded crests; 2n = 14. Pinyon-juniper and mountain brush communities at 1675 to 2930 m in Juab and Millard counties; Oregon to California and Nevada; 5 (0). The species has been reported for Tooele County also.

Allium passeyi N. & A. Holmgren Bulbs 1–2.2 cm thick and as long or somewhat longer, buried 5–7 cm below ground, the scales fibrous-reticulate; leaves 2 or 3 per scape, flat, curved, shorter than to somewhat surpassing the scape, 3–7 mm wide; scapes 10–17 cm tall (above ground); spathaceous bracts 2–4, usually 1-veined; umbels 8- to 27-flowered, the pedicels straight or curved; perianth segments mostly 6–9 mm long, pink to white, dimorphic, the inner ones narrower; stamens included; capsules with 3 low crests. Sagebrush community at 1460 to 1650 m in Box Elder (type from Howell Valley) County; endemic; 7 (0).

Allium schoenoprasum L. Chive. Tufted biennial or short-lived perennial herbs from poorly developed, clustered bulbs; leaves hollow, very narrow; umbels capitate on uninflated scapes 1.5–7.5 dm tall; pedicels shorter than the flowers; flowers bright rose to purplish, 7–12 mm long; capsule included in the perianth; 2n = 16, 32. Cultivated ornamental and food plant in Utah; indigenous in Eurasia and widely grown across North America; 2 (i).

Allium textile Nels. & Macbr. Bulbs 1–2.5 cm thick and about as long, buried 3–13 cm below ground, the scales fibrous-reticulate; leaves 2–4 per scape, concave-convex, 1–5 mm thick, shorter than to surpassing the scape; scapes 3–25 cm tall (above ground); spathaceous bracts 3, typically 1-veined; umbels 5– to 54–flowered, the pedicels straight or curved; perianth segments 5–9 mm long, white or pale pink, dimorphic; stamens included; capsules with tiny crests toward the style; 2n = 14. Shadscale, mat-saltbush, other salt desert shrub, sagebrush, and pinyon-juniper communities at 1430 to 2290 m in Carbon, Daggett, Duchesne, Emery, Garfield, Grand, and Uintah counties; Alberta and Saskatchewan to Nevada and New Mexico; 50 (viii).

Androstephium Torr.

Perennial herbs from fibrous-reticulate cormous bulbs; leaves linear, channeled; flowers inconspicuous, though relatively large, pedicellate, borne in umbels, these subtended by spathaceous bracts; sepals 3, petaloid; petals 3, petaloid, united below and adnate to the sepals; stamens 6, the filaments united, with bifid lobes between the basifixed anthers; ovary 3–loculed, several-ovuled, sessile; seeds flat.

Androstephium breviflorum **Wats.** [*Brodiaea breviflora* (Wats.) Macbr.; *B. paysonii* A. Nels.]. Bulbs 1–3.5 cm thick and about as long, buried 3–9 cm below ground, the scales fibrous-reticulate; leaves 1–3 per scape, linear, surpassing the scapes or subequal to them, usually straight or only somewhat curved; spathaceous bracts connate basally, usually 5-veined; umbels 3– to 8–flowered, the pedicels typically straight; perianth 12–21 mm long, a dirty greenish or purplish white, with the midvein often blue purple, the segments connate ca 1/3 of the length; stamens included; capsules 10–15 mm long and about as broad, 3–lobed; seeds 7–9 mm long, black. Creosote bush, Joshua tree, blackbrush, salt desert shrub, sagebrush, and pinyon-juniper communities at 820 to 1985 m in Beaver, Carbon, Daggett, Duchesne, Emery, Garfield, Grand, Iron, Juab, Kane (type from near Kanab), Millard, San Juan, Sevier, Uintah, Utah, Washington, and Wayne counties; Nevada, Arizona, and Colorado; 87 (xvii).

Asparagus L.

Perennial herbs from elongate tuberous roots; stems erect, diffusely branched; leaves alternate, scalelike, subtending flattened, photosynthetic phyllodia; flowers perfect or functionally dioecious, borne in pairs along the stem; pedicels jointed; sepals 3; petals 3, distinct or nearly so; stamens 6; ovary 3–loculed, with 2 or few ovules per locule; berries globose.

Asparagus officinalis **L.** Asparagus. Plants mainly 10–20 dm tall; stems glaucous; leaves scalelike; phyllodia 3–20 mm long, linear-filiform; flowers pendulous, the pedicels filiform, jointed, 4–20 mm long; perianth campanulate, greenish white, 4–7 mm long; berries red, 5–8 mm thick. Cultivated food plant, escaping and established along canal banks, in orchards, and among indigenous, riparian vegetation at 760 to 1895 m in lower elevation sites in much of Utah; introduced from Eurasia; 24 (iv).

Calochortus Pursh

Perennial herbs from tunicate bulbs; stems leafy, erect or sprawling, straight or flexuous, often with axillary bulbs below ground level; leaves alternate, linear to narrowly attenuate, reduced upwards; bracts opposite to subopposite or alternate; flowers showy, solitary or 2–6 in an umbellate or racemose cluster, erect; perianth dimorphic, the sepals more slender and differing in color and texture, the petals more brightly colored and glandular near the base, the area around the bearded gland variously yellow and/or purple striped or mottled and often variously bearded; stamens 6, the anthers basifixed; ovary 3–loculed; capsules 3–angled, septicidal; seeds numerous, flattened.

Ownbey, M. A. 1940. A monograph of the genus *Calochortus*. Ann. Missouri Bot. Gard. 27: 371–560.
_____. 1969. *Calochortus*. Univ. Washington Publ. Biol. 17(1): 765–779.
Reveal, J. L. 1977. *Calochortus*. pp. 496–504. In: Cronquist, A. et al. Intermountain Flora. Vol. 6. Columbia Univ. Press, New York.

1. Flowers yellow or yellow orange 2
— Flowers white to cream or lavender to purplish 3
2(1). Base of petals with a field of purplish black surrounding the gland, the dark field extending to petal margins; petals golden yellow to yellow orange; plants evidently rare .... *C. kennedyi*
— Base of petals not wholly purplish black, mainly with a purple crescent or inverted "V" above the gland; petals bright yellow; plants locally common in southern Utah .... *C. aureus*
3(1). Stems soon sprawling and flexuous; flowers 1–6 in a flexuous raceme, purplish; glands surficial, not surrounded by a membrane; plants of southern Utah .... *C. flexuosus*
— Stems erect; flowers mostly 1–4 (6) in an umbellate cluster, variously colored; glands depressed and surrounded by a membrane 4
4(3). Petals usually 2–3 times longer than broad, with a longitudinal greenish median; plants uncommon in western and northern Utah .... *C. bruneaunis*
— Petals usually 1–2 times longer than broad, the median sometimes faintly purplish but not usually green; plants diverse in abundance and distribution 5
5(4). Anthers acute; petal base with a broad purple band above the gland extending across the width; plants of eastern Utah .... *C. gunnisonii*
— Anthers obtuse; petal base with a purple inverted "V" or crescent above the gland (typically separated by a yellow stripe), the purple marking not extending to the blade margins 6
6(5). Gland on petal circular, shield-shaped, or longitudinally elongate; petal hairs simple, not branched or enlarged apically; plants common and widespread .... *C. nuttallii*
— Gland on petal transversely oval in outline; petal hairs bilobed or conspicuously enlarged apically; plants of Washington County .... *C. ambiguus*

Calochortus ambiguus **(Jones) Ownbey** [*C. watsonii* var. *ambiguus* Jones]. Plants erect; stems 1–5 dm tall, usually simple; leaves several, alternate; bracts narrowly lance-attenuate, opposite or subopposite, 3–8 cm long; flowers 1–4; sepals lanceolate, subequal to or shorter than the petals; petals 3–4.5 cm long, rounded to obtuse or abruptly acuminate, the base with a transversely oval gland, this depressed, fringed, and densely hairy, surrounded by yellow and this topped by a purple stripe, otherwise white; anthers pinkish to maroon, obtuse; capsules narrowly oblong, 3–angled, 4–7 cm long. Mountain brush at ca 1525 m in Washington County; Arizona and New Mexico; 1 (0).

Calochortus aureus Wats. [*C. nuttallii* var. *aureus* (Wats.) Ownbey]. Plants erect; bulbs often with very shaggy, persistent scales; stems 9–35 cm tall (above ground), simple or rarely branched; leaves typically 3, alternate; bracts lance-attenuate to lanceolate, opposite (or whorled), 1.6–8 cm long; flowers 1–5; sepals elliptic to oblong, shorter than the petals, somewhat suffused with yellow green and longitudinally veined dorsally, yellowish and often with a purple blotch above the base ventrally; petals golden yellow, 2.8–4.5 cm long, truncate-rounded to abruptly short-apiculate apically, the base with a subcircular gland, this depressed and densely hairy, surrounded by yellow and overtopped by a narrow to broad purple crescent, sometimes with a purple blotch below as well, and often with scattered long hairs outward from the gland; anthers cream to yellow, obtuse; capsules 3–5 cm long, 3–angled, ellipsoid. Blackbrush, shadscale, other salt desert shrub, and juniper communities at 1250 to 1770 m in Garfield, Kane (type from near Kanab), San Juan, Washington, and Wayne counties; Arizona (a Colorado Plateau endemic); 24 (iii).

Calochortus bruneaunis Nels. & Macbr. [*C. nuttallii* var. *bruneaunis* (Nels. & Macbr.) Ownbey]. Plants erect; stems 12–40 cm tall (above ground), simple; leaves 2 or 3, alternate; bracts opposite, linear from a clasping base, 1.5–6 cm long; flowers 1 or 2; sepals narrowly lanceolate, shorter than or subequal to the petals, greenish dorsally, whitish to pale lavender ventrally and often with a purple crescent above the base; petals 2–3 (4) cm long, obtuse to short-acuminate apically, more than twice longer than broad, whitish to pale lavender and with a median longitudinal greenish stripe, the base with a subcircular to elonagte gland, this depressed and densely hairy, surrounded by yellow and often purple-blotched above and below, the surface away from the gland somewhat long-hairy as well; anthers purple, obtuse; capsules ellipsoid, 3–6 cm long, 3–angled. Sagebrush, grass, pinyon-juniper, white fir, and aspen communities at 2135 to 2900 m in Box Elder, Juab, and Rich counties; Idaho and Montana, south to California and Nevada; 5 (ii).

Calochortus flexuosus Wats. Plants decumbent, sprawling, or less commonly erect; stems 8–50 (60) cm long or more, flexuous or sinuous, often branched; leaves 1 or few, often coiled on the ground surface; bracts subopposite to alternate, lance-attenuate to linear-subulate; flowers 1–8; sepals ovate-elliptic to elliptic, much shorter than the petals, greenish externally, purplish and with yellowish and purplish markings above the base ventrally; petals 2.2–4.5 cm long, truncate to broadly rounded apically, somewhat longer than broad, purplish to lavender, the base with a transversely elliptic to crescent-shaped gland, this not depressed or with a surrounding membrane, densely hairy, surrounded by yellow or yellow orange and with bordering purple blotches, often with scattered hairs away from the gland; anthers purplish to cream, obtuse; capsule 2.5–3.5 cm long, 3–angled. Creosote bush, Joshua tree, other warm desert shrub, blackbrush, shadscale, other salt desert shrub, and pinyon-juniper communities at 885 to 2015 m in Beaver, Garfield, Kane (type from near Kanab), Millard, San Juan, and Washington counties; Colorado, New Mexico, Arizona, Nevada, and California; 43 (vii).

Calochortus gunnisonii Wats. Plants erect; stems 20–60 cm tall (above ground), simple; leaves 2–4, alternate; bracts subopposite, lance-linear to lance-attenuate; flowers 1–4; sepals lanceolate, shorter than the petals, greenish and many-veined dorsally, pale lavender to purplish ventrally, typically with a pair of purple blotches above the base; petals 2.7–4.1 cm long, truncate to broadly rounded or obtusish apically, only somewhat longer than broad, purplish to lavender, the base with a transverse, elongate, densely hairy gland, this typically bordered by purple on one or both sides, the purple banding extending to the petal margins, also typically hairy away from the gland with bulbous or bifid tipped long hairs; anthers purplish to cream, acute to apiculate; capsules ellipsoid, 3–angled, 3–5 cm long. Sagebrush, pinyon-juniper, mountain brush, and aspen communities at 2375 to 2900 m in Carbon, Duchesne, Grand, San Juan, and Uintah counties; Montana and South Dakota, south to Arizona and New Mexico; 10 (ii).

Calochortus kennedyi T. C. Porter Plants erect; stems mainly 1–4 dm tall, simple; leaves 2–4, alternate; bracts subopposite, 1.5–6 cm long, lance-attenuate to lance-linear; flowers 1 or 2; sepals lanceolate, shorter than the petals, greenish to purplish dorsally, pale and subbasally blotched with purple ventrally; petals 3–3.7 cm long, truncate to broadly rounded apically, somewhat longer than broad, yellow orange to bright orange, the base with a circular, depressed, densely hairy gland in a field of purple extending to the petal edges, and often with additional long hairs away from the gland; anthers pinkish to purple, abruptly apiculate; capsules not seen. Two specimens belonging to this species were collected in "foothills" at Kanab, Kane County (Snow sn 1925 BRY); Arizona, Nevada, and California; 2 (0). A specimen from Wire Mesa in Washington County, tentatively identified as *C. aureus*, approaches this species.

Calochortus nuttallii T. & G. in Beckwith Sego Lily. [*C. luteus* Nutt.; *C. watsonii* Jones; *C. rhodothecus* Clokey]. Plants erect; stems 8–50 cm tall (above ground), simple; leaves typically 3, alternate; bracts subopposite, 1.5–9 cm long, lance-attenuate to lance-linear; flowers 1–5; sepals lanceolate to ovate, variously greenish to purplish externally, pale within and often blotched with yellow and or purple above the base; petals 2.5–6.3 cm long, truncate-rounded to obtuse or abruptly acuminate apically, 1–2 times longer than broad, white, cream, or lavender (fading purple), the base with a circular, shield-shaped, or longitudinally elongate, depressed, densely hairy gland in a field of yellow often bordered by a crescent or inverted V-shaped stripe or blotch of purple, the outer surface sometimes with a median purplish stripe; anthers cream to pink or purplish, obtuse; capsules 3–5 cm long, lance-ellipsoid, 3–angled. Shadscale, greasewood, other salt desert shrub, sagebrush, pinyon-juniper, mountain brush, ponderosa pine, and aspen communities at 1005 to 3050 m in all Utah counties; Idaho to North Dakota, south to Nevada, Arizona, New Mexico, and Nebraska; 175 (xvii). This beautiful plant is the state flower of Utah.

Camassia Lindl.

Perennial scapose herbs from tunicate bulbs; leaves basal, 3 to several; flowers several to many, borne in bracteate racemes; perianth segments 6, alike in color and texture, distinct; stamens 6, inserted on the receptacle; anthers versatile; ovary superior, 3–loculed; style 1, the stigma 3–lobed; capsules loculicidal.

Gould, F. W. 1942. A systematic treatment of the genus *Camassia* Lindl. Amer. Midl. Naturalist 28: 712–742.

Camassia quamash (Pursh) Greene Camas. [*Phalangium quamash* Pursh; *C. quamash* ssp. *utahense* Gould, type from Blacksmith Canyon, Cache County]. Plants erect, mainly 3–7 dm tall; leaves flat, shorter than the inflorescence, 1–4 dm long, 0.6–2 cm wide, 3–6 from the stem base; racemes 0.8–2.5 dm long; pedicels jointed, 0.7–2 cm long; perianth blue (or white), the segments 1.5–3 cm long; stamens with slender filaments, the anthers yellow or blue; capsules 1–2 cm long. Sagebrush, mountain brush, aspen, and grass-forb communities, often where wet in springtime, at 1890 to 2410 m in Box Elder, Cache, Davis, Summit, Wasatch, and Weber counties; British Columbia and Alberta, south to California and Wyoming; 18 (i). Recognition of infraspecific taxa among the Utah materials seems moot. This plant was important as a food staple for the Indians of the Pacific Northwest.

Chionodoxa Boiss.

Perennial bulbous scapose herbs; leaves basal, usually 2 per scape; flowers erect, solitary or 2 or 3 in a raceme; perianth lobes rotate, the 6 segments connate at the base, forming a campanulate tube; stamens 6, inserted at the base of the tube, the anthers dorsifixed; capsule loculicidal, few-seeded.

Chionodoxa luciliae Boiss. Glory-of-the-snow. Plants erect, from bulbs mainly 2–9 cm below ground; scapes 4–12 cm tall above ground; leaves usually 2 per scape, linear, 2–7 mm wide, subequal to or shorter than the scape; flowers 1–3, in minutely bracteate racemes; perianth blue with a white center, 12–22 mm long, the segments connate in the lower 1/4; stamens exserted beyond the tube, the filaments broad. Cultivated ornamental, persisting and rarely escaping in lower elevation portions of Utah; introduced from Asia Minor; 4 (i).

Colchicum L.

Perennial tunicate cormous herbs; flowers in umbellate, subsessile clusters, arising below ground; perianth 6-lobed, united basally into a long, pedicellike tube; stamens 6, inserted near the base of the segments; anthers versatile; ovary subterranean, the styles 3; capsule septicidal; seeds numerous.

Colchicum autumnale L. Corms 2.5–6 cm long and 2–4 cm thick or more, ovoid to subglobose, the prolonged tunicate scales dark brown, leathery; leaves developed in springtime, 3–5, mostly 1–3.5 dm long and 2–7 cm wide, glabrous; flowers 1–10 (or more), the perianth segments oblong to elliptic or obovate, 2–6 cm long and 8–15 mm wide, pink, appearing in autumn; staminal filaments 10–15 mm long; anthers 5–8 mm long, yellow; styles curved apically, with decurrent stigmas; capsules maturing at or near ground level. Uncommonly grown ornamental herbs, long persisting in lower elevation portions of Utah; introduced from Europe; 2 (i).

Convallaria L.

Perennial rhizomatous herbs; leaves with convolute, sheathing bases, simulating a stem with alternate leaves, each such aggregation consisting of ca 5 bladeless, membranous scales and 1–3 green foliage leaves; scapes arising from the clustered, membranous blades; leaves, usually appearing prior to the foliage leaves; flowers white, borne in racemes, deflexed; perianth campanulate, 6-lobed, connate to above the middle; stamens 6; anthers attached near the base; style 1; fruit a berry.

Convallaria majalis L. Lily-of-the-valley. Plants 10–30 cm tall, the scapes with purplish, membranous, sheathing scales; foliage leaves with sheaths mainly 5–20 cm long, the blades 5–20 cm long and 2–10 cm wide, elliptic to lanceolate, acute to acuminate; flowers fragrant, nodding; perianth 5–10 mm long, white; fruit red; $2n = 38$. Commonly grown ornamental plants, long persisting at lower elevations in Utah; introduced from Europe; 1 (0). The plants are noted for their medicinally active components.

Dichelostemma Kunth

Plants scapose perennial herbs from fibrous-reticulate bulbous corms; leaves 2, basal, linear; flowers lilac to blue, borne in umbellate to subcapitate clusters, subtended by membranous bracts; perianth segments 6, similar in color and texture, connate at the base; stamens 6, dimorphic, adnate to the perianth; anthers basifixed; ovary 3-loculed, few-ovuled; capsule loculicidal.

Dichelostemma pulchellum (Salisb.) Heller [*Hookera pulchella* Salisb.; *Brodiaea pulchella* (Salisb.) Greene; *B. capitata* Benth.]. Plants erect from bulbs 4–12 cm below ground; leaves 2, linear, shorter than to surpassing the scape, straight or curved; scapes mainly 3–9 dm tall above ground; spathaceous bracts distinct, usually severalveined; umbels 3– to 10-flowered, the pedicels often shorter than the flowers; perianth blue to lilac, 14–18 mm long, the segments connate ca 1/3 the length; stamens included; capsules 4–6 mm long, 3-lobed; seeds 2.5–4 mm long. Creosote bush, Joshua tree (other warm desert shrub), pinyon-juniper, mountain brush, and ponderosa pine communities at 850 to 2600 m in Iron, Kane, and Washington counties; California, Nevada, Arizona, and New Mexico; 27 (iii).

Disporum Salisb. ex D. Don

Perennial herbs from rhizomes; stems simple or dichotomously branched, typically flexuous; leaves alternate, ovate to lanceolate or elliptic, sessile and somewhat cordate-clasping; flowers white to greenish, solitary or in terminal, umbellate pairs, nodding on slender recurved pedicels; perianth segments 6, alike, connate near the base; stamens 6, inserted on base of the perianth; anthers subversatile; ovary 3-loculed, with 2–6 ovules per locule; style 1, with 3 stigmatic lobes; berries yellowish to red; seeds subglobose.

Jones, Q. 1951. A cytotaxonomic study of the genus *Disporum* in North America. Contr. Gray Herb. 173: 1–39.

Disporum trachycarpum (Wats.) Benth. & Hook. Fairybells. [*Prosartes trachycarpa* Wats., type from Parleys Park, Summit County]. Plants erect or with spreading branches, mainly 2–6 dm tall; leaves 3–10 cm long, 1.2–9 cm wide, ovate to lanceolate or elliptic, ciliate; flowers 1 or 2, white to greenish, 10–15 mm long, campanulate; stamens subequal to the perianth or somewhat longer; ovary papillate; berries yellowish to red, 7–13 mm wide, strongly papillate; $2n = 22$. Mountain brush, Douglas fir-white fir, ponderosa pine, limber pine, and spruce-fir communities at 1250 to 2990 m in Beaver, Box Elder, Cache, Carbon, Duchesne, Iron, Juab, Millard,

Salt Lake, San Juan, Sanpete, Sevier, Summit, Tooele, Uintah, Utah, Wasatch, Washington, and Weber counties; British Columbia to Manitoba, south to Arizona and New Mexico; 34 (iii).

Eremocrinum Jones

Perennial scapose herbs from short stout rhizomes and fleshy yellowish roots; leaves 3–8 or more, the bases membranous-sheathing, persistent and ultimately fibrous; flowers white to greenish, borne in racemes or in contracted panicles; perianth segments 6, alike, connate and pedicellike at the base; stamens 6, inserted at the ovary base, the anthers basifixed; ovary 3–loculed, with few ovules; style 1, the stigma discoid; capsules loculicidal, 3–lobed, reticulately veined.

***Eremocrinum albomarginatum* (Jones) Jones** [*Hesperanthes albomarginata* Jones, type from near Green River]. Plants mainly 1–3.5 dm tall above ground, erect, arising from a fibrous-sheathed base terminating a short thickened rhizome and thickened yellowish roots; leaves linear, often surpassing the scapes, but commonly coiled or reclining on the ground; flowers 0.8–1.5 cm long, subtended by membranous, ovate-attenuate, white bracts; stamens included; capsules 4–6 mm long; seeds black, 2–3 mm long. Blackbrush, Vanclevea, sand sagebrush, purple sage, ephedra, ricegrass-galleta, shadscale, and other sand and salt desert shrub communities at 1125 to 1895 m in Emery, Garfield, Grand, Kane, San Juan, and Wayne counties; Arizona (a Colorado Plateau endemic); 32 (vii).

Erythronium L.

Perennial, scapose or subscapose herbs from elongate bulbous corms; leaves 2, basal, elliptic to oblong or lanceolate to oblanceolate; flowers yellow, solitary or 2–5 in loose racemes; perianth segments 6, similar in color and texture, distinct; stamens 6, in 2 unequal sets, borne on the receptacle; anthers basifixed; ovary 3–loculed, several-ovuled; style 1, 3–lobed apically; capsules loculicidal.

Applegate, E. I. 1935. The genus *Erythronium*: A taxonomic and distributional study of the western North American species. Madrono 3: 58–113.

***Erythronium grandiflorum* Pursh** Dogtooth-violet. [*E. utahense* Rydb., type from Great Salt Lake]. Bulbous corms mainly 1–2 cm thick, tapering to the stem exsertion, 2.5–8 cm long, the scales cellular-membranous, buried 8–19 cm below ground; leaves 2, the blades 8–20 cm long, 1–5 cm wide, lanceolate to elliptic, oblong, or oblanceolate; flowers solitary or usually 2 or 3 in a bractless raceme; pedicels 2–12 cm long (or more) and recurved subapically in flower, even longer and erect in fruit; perianth segments 2–3.5 cm long, 3–7 mm wide, narrowly lanceolate, spreading to reflexed at anthesis; stamens exserted; anthers 6–12 mm long, reddish or purplish; capsules ellipsoid to obovoid, 2–4 (6) cm long. Mountain brush, sagebrush, aspen, lodgepole pine, and spruce-fir communities at 1675 to 3115 m in Box Elder, Cache, Davis, Duchesne, Emery, Juab, Millard, Morgan, Salt Lake, Sanpete, Sevier, Summit, Tooele, Utah, Wasatch, and Weber counties; British Columbia and Alberta, south to California and Colorado; 74 (v).

Fritillaria L.

Plants perennial glabrous subscapose or caulescent herbs from bulbs and with usually numerous reduced bulblets; leaves alternate or the lower ones subopposite or some almost or quite whorled; flowers yellow to brownish purple and often mottled, usually nodding, campanulate to subrotate, solitary or 2–7 in terminal, bracteate racemes; perianth segments 6, subequal, distinct; stamens 6, inserted on the receptacle; anthers subversatile; ovary 3–loculed, with numerous ovules; capsule 6–angled or –winged, loculicidal.

Beetle, D. E. 1944. A monograph of the North American species of *Fritillaria*. Madrono 7: 133–159.

1. Flowers yellow, narrowly campanulate, sometimes fading purple; style 1 .... *F. pudica*
— Flowers brownish purple, mottled, or sometimes yellow with purplish mottlings; styles 3 .... *F. atropurpurea*

***Fritillaria atropurpurea* Nutt.** Leopard-lily. Plants erect, mainly 1.2–4.5 dm tall above ground, arising from bulbs 2–15 cm below ground; stems leafless for 2–25 cm above ground; leaves numerous, linear, 3–12 cm long, 1.5–7 mm wide; flowers 1–7, borne on recurved pedicels, rotate-campanulate; perianth segments brownish purple to yellowish, mottled, 10–25 mm long, ovate to lanceolate or elliptic; anthers 2–4 mm long; styles connate only at the base; capsules barrel-shaped, 10–22 mm long and about as broad, with 6 entire, acutely angled ridges. Shadscale, pinyon-juniper, mountain brush, sagebrush, white fir, Douglas fir, aspen, and spruce-fir communities at 1645 to 3205 m in most Utah counties (except Carbon, Emery, Grand, and Sanpete); Oregon to North Dakota, south to California, Nevada, Arizona, and New Mexico; 79 (v).

***Fritillaria pudica* (Pursh) Sprengel** Yellow-bell. [*Lilium pudicum* Pursh; *F. lucella* Gand., *F. dichroa* Gand., *F. utahensis* Gand, types of the three latter names are from Cache County]. Plants erect, mainly 6–15 cm tall above ground, arising from bulbs 2–12 cm below ground; stems with leaves almost or quite at ground level, 2 to many, linear to narrowly oblong or elliptic, 3–15 cm long, 2–14 mm wide; flowers solitary or less commonly 2 or 3, borne on recurved bractless pedicels (or peduncles) mainly 3–10 cm long; perianth segments bright yellow or streaked with purple, sometimes fading purple, narrowly campanulate, 12–22 mm long, 4–10 mm wide; style 1, with a discoid stigma; capsules erect, 15–30 mm long and 12–20 mm wide or more, seldom collected. Sagebrush and mountain brush communities at 1525 to 2565 m in Box Elder, Cache, Davis, Juab, Morgan, Salt Lake, Summit, Tooele, Utah, and Weber counties; British Columbia and Alberta, south to California, Nevada, and Colorado; 23 (0).

Hemerocallis L.

Plants perennial scapose herbs from short coarse rhizomes; leaves basal, linear, keeled; flowers yellow, orange, or reddish, large and showy, ascending-spreading, few to many, borne in bracteate racemes or panicles; perianth segments 6, more or less dimorphic, united into a long pedicellike base; stamens 6, long-exserted, inserted on the segments; anthers subversatile; ovary 3–loculed; style long, slender; capsules loculicidal, few-seeded.

Hemerocallis fulva (L.) L. Day-lily. Plants mainly 8–12 dm tall, forming clumps; leaves basal, linear, typically 2–5 dm long and 15–30 mm wide, narrowly elliptic to linear; flowers large and showy, short-pedunculate, but with a pedicellike, connate base enclosing the ovary, the perianth expanded and more or less funnelform above the base, yellow, orange, reddish, and often variously striped, the segments 5–9 cm long beyond the throat; staminal filaments often curved to one side and appearing irregular; capsules seldom formed; 2n = 22, 33, 36. Commonly cultivated ornamentals, long persisting and escaping at lower elevations in of Utah; introduced from Eurasia; 6 (i). Our plants are variable in flower color and size, and might involve horticultural selections of hybrids between this and other species in the genus. The lack of fruit suggests that our material might consist of at least partially sterile triploids.

Hyacinthus L.

Perennial scapose herbs from tunicate bulbs; leaves few to numerous, basal, linear to oblong, flowers in minutely bracteate racemes; perianth funnelform to campanulate, the segments united to near the middle, the lobes spreading to reflexed; stamens 6, inserted on the perianth base; anthers versatile; ovary 3–loculed; style 1, the stigma capitate; capsules loculicidal.

Hyacinthus orientalis L. Common Hyacinth. Bulbs 3–6 cm thick or more, the outer scales cellular-membranous; scapes 6–30 cm tall above ground; leaves 15–30 cm long or more, mostly 1–3 cm wide, subequal to or shorter than the scapes at anthesis; racemes few- to many-flowered; pedicels 1–8 mm long, subtended by membranous bracts; flowers 2–3 cm long, blue, purple, red, yellow, or white, fragrant; capsules seldom produced; 2n = 16. Commonly cultivated ornamental herbs, long persisting and occasionally escaping in Utah; introduced from Asia Minor; 3 (i). This traditional ornamental marks spring with its brightly colored flowers and permeating odor.

Leucocrinum Nutt. ex. Gray

Perennial acaulescent herbs from short stout rhizomes and fleshy grayish to tan roots; leaves 4–8 (or more), the bases membranous-sheathing, persistent but not especially fibrous; flowers white, arising in subumbellate clusters from below ground; perianth segments 6, united into a long pedicellike tube basally, the lobes spreading; stamens 6, the filaments inserted on the perianth tube, subversatile; ovary 3–loculed; styles slender, 3–lobed; capsule loculicidal.

Leucocrinum montanum Nutt. ex Gray Star-lily. Plants caespitose, mainly less than 1 dm tall; leaves linear, mainly 3–20 cm long, 1.5–6 mm wide, the sheathing bladeless bases 3–8 cm long; flowers white, few to several, mostly 5–12 cm long, arising from subterranean pedicels; anthers 4–6 mm long, often coiled or curved; capsules formed below ground, 5–8 mm long; seeds 3–4 mm long; 2n = 11, 13, 14. Sagebrush, pinyon-juniper, and ponderosa pine communities at 1800 to 2290 m in Garfield, Iron, Kane, and Washington (?) counties; Oregon to Montana, south to California, Nevada, and New Mexico; 11 (0).

Lilium L.

Perennial caulescent herbs from scaly bulbs or short rhizomes; leaves alternate or whorled, sessile; flowers showy, white to orange or red and typically spotted, nodding to erect, funnelform to campanulate, solitary or few to several in leafy-bracted racemes, or subumbellate; perianth segments 6, similar, distinct, each with a nectary gland near the base; stamens 6, borne on the receptacle, the anthers versatile; ovary 3–loculed; style slender, 3–lobed; capsule loculicidal.

1. Flowers white, funnelform, the segments reflexed only at the tips .... *L. longiflorum*
— Flowers orange to yellow orange, and usually spotted, campanulate, often reflexed below the tips .. *L. bulbiferum*

Lilium bulbiferum L. Bulbs white, subglobose, 5–8 mm thick; stems 5–12 dm tall, floccose at the nodes and along the stem upward; leaves mainly 6–10 cm long, 6–18 mm wide, lanceolate to narrowly elliptic, often bulbiferous in the upper axils; pedicels 2–5 cm long, spreading; flowers yellow to orange, spotted with brown purple, 1–4, opening widely, the segments 5–10 cm long; staminal filaments straight to curved, but not divaricate, the anthers brownish; styles much longer than the ovary; capsules 2.5–4 cm long; 2n = 24. Cultivated ornamental, persisting in lower elevation portions of the state; introduced from Europe; 3 (ii). Our material is more or less intermediate to **L. maculatum** Thunb.

Lilium longiflorum Thunb. White-trumpet Lily; Easter Lily. Bulbs white, subglobose, 2–5 cm thick; stems 5–8 (10) dm tall, glabrous, green or reddish below; leaves mainly 10–18 cm long, narrowly lanceolate to elliptic, acuminate, not bulbiferous in the axils; flowers white, 1 to few (several), slender, 10–18 cm long, flaring apically, fragrant; staminal filaments straight; style long, slender. Common greenhouse and garden ornamental, which persists at lower elevations in of Utah; introduced from Japan; 1 (i).

Lloydia Salisb.

Perennial bulbous herbs, arising from rhizomes; stems erect, leafy, the leaves reduced upward, arising from a basal, usually sheathing cluster of marcescent leaf bases; flowers 1 to few, in a terminal raceme; perianth segments 6, distinct, each with a transverse gland at the base; stamens 6; pistil 1, with a single style, the stigma 3–lobed; fruit a 3–loculed, loculicidal capsule.

Lloydia serotina (L.) Wats. Alp Lily. [*Bulbocodium serotinum* L.]. Plants erect, 5–15 (20) cm tall; bulbs small, clothed with a brownish coat; stems slender; basal leaves 2–8 cm long, mostly 1–2 mm wide; cauline leaves alternate, reduced upward, 1–4 cm long; flowers creamy white, 8–13 mm long, the segments purple-veined and tinged with rose on back; capsules ovoid, ca 8 mm long; 2n = 24, 36, 48. Alpine tundra and krummholz at 2288 to 3815 m in Daggett, Duchesne, Salt Lake, Summit, and Uintah counties; Alaska and Yukon, south to Washington, Nevada, and Colorado; Eurasia; 10 (ii).

Muscari Miller

Perennial herbs from a truncate bulb; leaves few to several, linear; flowers small, numerous, borne in terminal racemes (rarely paniculate), nodding, typically blue, the apical ones commonly sterile, the lower ones fertile; bracts of pedicels small; perianth connate, urn-shaped, the 6 segments recurved only apically; stamens 6, inserted on the perianth; anthers versatile; pistil 3–loculed, sessile, the style filiform, short; capsules 3–angled, loculicidal.

Muscari botyroides (L.) Miller Grape Hyacinth. [*Hyacynthus botyroides* L.]. Plants scapose, erect, 10–35 cm tall; bulbs ovoid, clothed with membranous scales; leaves all basal, 10–40 cm long; flowers blue purple, ovoid, 5–6 mm long, borne in more or less compact racemes; capsules ovoid; $2n = 18, 36$. Commonly cultivated ornamental in Utah; escaping and persisting; introduced from Europe; 5 (ii).

Ornithogalum L.

Bulbous perennial herbs; leaves all basal; flowers white to greenish, borne in bracteate, scapose, corymbose racemes; perianth segments 6, distinct or connate at the base only; stamens 6, the filaments broadly expanded, membranous; anthers subversatile; fruit a loculicidal capsule.

Ornigthogalum umbellatum L. Bulb surrounded by tufts of leaves arising from offsets; leaves 10–40 cm long or more, 2–8 mm wide; scape glabrous, subequal to the leaves or shorter; bracts membranous or the tips herbaceous, subequal to or shorter than the pedicels; flowers few to numerous; perianth segments 15–22 mm long, lanceolate to oblong-lanceolate or -oblanceolate, white with a green stripe; ovary ovoid to subcylindric, longer than the style; $2n = 18, 27, 28, 35, 36, 44, 45, 52, 54, 77, 90, 108$. Introduced ornamental, escaping and established in Salt Lake and Utah counties; introduced from Europe; 2 (0).

Scilla L.

Bulbous, scapose, perennial herbs; leaves all basal; flowers blue or purple, solitary or racemose, nodding; bracts reduced; perianth segments distinct; stamens 6, the filaments distinct, subulate; anthers subversatile; fruit a subglobose, 3–lobed capsule.

Scilla sibirica Haw. in Andrews Squill. Bulbs ovoid, whitish beneath dark outer scales; leaves 2–4, mainly 6–20 cm long, 10–20 mm wide; scapes 10–25 cm long, subequal to the leaves; pedicels usually shorter than the perianth; flowers 1–5, nodding; bracts 1–1.5 mm long; perianth segments 12–15 mm long, deep blue; $2n = 12, 18, 30$. Cultivated spring-flowering ornamental, persisting at lower elevations in Utah; introduced from U.S.S.R.; 2 (ii).

Smilacina Desf.

Rhizomatous perennial herbs; stems erect or ascending, simple; leaves alternate, many-nerved, sessile or somewhat clasping; flowers in terminal racemes or panicles; perianth 6–segmented, white or greenish white, distinct or nearly so; stamens 6; ovary 3–loculed; style short; stigma typically 3–lobed; fruit a globose berry.

1. Flowers in panicles; perianth segments ca 2 mm long .... *S. racemosa*
— Flowes in racemes; perianth segments 4–7 mm long .... *S. stellata*

Smilacina racemosa (L.) Desf. False Solomon-seal. [*Convallaria racemosa* L.; *S. amplexifolius* Nutt.]. Stems 3–10 dm tall from fleshy rhizomes; leaves 5–9 or more per stem, oblong-lanceolate, sessile or short-petioled, 7–20 cm long, pubescent below with short stiff hairs, the margins minutely ciliate; panicle densely many flowered, 4–10 cm long; perianth segments ca 2 mm long; fruit mottled, becoming red at maturity, 4–6 mm in diameter; $2n = 36, 72$. Mountain brush, pinyon-juniper, Douglas fir, limber pine, aspen, and spruce-fir communities at 1370 to 3235 m in Box Elder, Cache, Carbon, Davis, Duchesne, Grand, Juab, Salt Lake, San Juan, Sanpete, Sevier, Summit, Uintah, Utah, Wasatch, Washington, and Weber counties; widely distributed in North America; 50 (iii).

Smilacina stellata (L.) Desf. [*Convallaria stellata* L.]. Stems 3–5 dm tall or more, from a slender rhizome; leaves 5–9 or more per stem, sessile, pubescent to glabrous beneath, lanceolate, 5–15 cm long, 2–4 cm wide; racemes 3–7 cm long, several-flowered; perianth segments 4–7 mm long; berries mottled, becoming red at maturity, 7–10 mm in diameter; $2n = 36$. Hanging garden, pinyon-juniper, sagebrush, mountain brush, ponderosa pine, aspen, lodgepole pine, bristlecone pine, and spruce-fir communities at 1125 to 3115 m in all Utah counties; widspread in North America; 156 (xv).

Streptopus Michx.

Rhizomatous perennial herbs with leafy branching or simple stems; leaves alternate, thin, many-nerved, sessile or auriculate-clasping; flowers solitary or in pairs, borne laterally on the stem from the upper leaf axils; perianth campanulate, yellowish white (rarely pinkish), the tips at length spreading, 6–segmented; stamens 6; pistil 1, the ovary 3–loculed; fruit a red berry.

Streptopus amplexifolius (L.) DC. Cucumber-root; Clasping twisted-stalk. [*Uvularia amplexifolia* L.]. Plants with stems usually branched, 3–10 dm tall; leaves several, auriculate-clasping, ovate-lanceolate, 5–15 cm long, 2.5–5 cm wide; peduncles 1–5 cm long, 1– or 2–flowered; perianth 8–12 mm long, the segments recurved or spreading apically; stigma entire; fruit ovoid-ellipsoid, yellowish white to red, 10–18 mm long; $2n = 32$. Aspen, lodgepole pine, and spruce-fir communities at 2135 to 3235 m in Beaver, Cache, Daggett, Duchesne, Garfield, Grand, Piute, Salt Lake, San Juan, Sevier, Summit, Uintah, and Wasatch counties; widespread in North America; Eurasia; 19 (vi). Our material belongs to var. *chalazatus* Fassett. The berries are reported to be edible, and have been used in making preserves.

Triteleia Dougl. ex Lindl.

Plants scapose perennials from fibrous-reticulate bulbs; leaves 1 or 2, basal, linear; flowers blue or white, borne in umbellate clusters, subtended by a papery involucre of separate bracts; perianth segments 6, alike, connate at the base; stamens 6, dimorphic, the filaments distinct, inserted on the perianth; anthers versatile; ovary 3–loculed, few-ovuled; capsule loculicidal.

Triteleia grandiflora Lindl. [*Brodiaea douglasii* Wats.]. Plants erect from bulbs 8–20 cm below ground; leaves 1 or 2, linear, shorter than or surpassing the scapes, straight; scapes mainly 2–7 dm tall (above ground); spathaceous bracts distinct, usually several-veined; umbels 5– to 16-flowered, the pedicels shorter to longer than the flowers; perianth blue, 14–27 mm long, the segments connate ca 1/2–2/3 the length; stamens included; capsules 5–10 mm long, subterete, stipitate, the stipe 3–5 mm long; seeds ca 2 mm long. Sagebrush, mountain brush, aspen, and Douglas fir communities at 1400 to 2850 m in Cache, Morgan, Salt Lake, Summit, Utah, Wasatch, and Weber counties; British Columbia to Montana, south to Oregon, Idaho, and Wyoming; 16 (i).

Tulipa L.

Perennial caulescent or subscapose herbs from tunicate bulbs; leaves 2 to several, basal or one or more cauline, ovate to elliptic or linear; flowers typically erect; perianth segments 6, alike or unlike in color and texture, distinct; stamens 6, borne on the receptacle; anthers basifixed; ovary 3–loculed, the stigma sessile, 3–lobed; capsules loculicidal. This is a large genus of Old World plants. Several ornamental species are known in cultivation in Utah. They persist following cultivation and occur as waifs along roadsides and in refuse dumps. At least the common ones are treated here to provide answers to those who discover waifs, especially.

1. Staminal filaments hairy at the base; leaves lance-linear .... *T. sylvestris*
— Staminal filaments glabrous throughout; leaves ovate to lanceolate or broadly elliptic 2
2(1). Perianth segments mostly 8–10 cm long, typically scarlet .... *T. fosteriana*
— Perianth segments mostly 3–6 cm long, variously colored .... *T. gesneriana*

Tulipa fosteriana W. Irv. Empress Tulip. Outer bulb scales dark brown, silky hairy within; leaves 3 or 4, ovate to elliptic or lanceolate, 10–22 cm long, 2–10 cm wide; stems 15–25 cm long; flowers solitary, campanulate, becoming explanate, brilliant scarlet with black blotches basally, these margined with yellow; segments 8–10 cm long, elliptic to obovate, emarginate and ciliate apically; stamens ca 1/4 as long as the segments or shorter. Commonly grown ornamental through much of Utah; introduced from Asia; 1 (0).

Tulipa gesneriana L. Common Tulip. Outer bulb scales brown, appressed hairy near the apex; leaves 3–5, lanceolate to ovate or elliptic, 7–18 cm long, 2–10 cm wide; stems 10–40 cm long or more; flowers solitary or 2 or 3, campanulate, at length opening widely, variously brightly colored or white; segments mainly 3–6 cm long, obovate to elliptic or oblanceolate, obtuse to rounded or emarginate and ciliate apically; stamens ca 1/3 as long as the segments, often purple; 2n = 24, 25, 26, 36, 48. The common tulip is grown widely in Utah; introduced from Asia; 2 (0). The plants grown in Utah consist of a series of hybrid derivatives, mainly involving *T. gesneriana* L. and *T. praestans* Hort.

Tulipa sylvestris L. Outer bulb scales brown, hairy at the apex within; leaves 2 or 3, lance-linear to narrowly elliptic, 8–24 cm long, 0.8–2 cm wide; stems 8–25 cm long; flowers solitary (or 2), campanulate, yellow or orange within, greenish to purplish without, 2.5–5 cm long, the inner segments fringed basally; stamens ca 1/2 as long as the segments, hairy basally. Uncommonly cultivated "species" tulip, rather widely grown in Utah; introduced from Europe; 1 (0). *T. clusiana* DC. is also known to occur in Utah.

Veratrum L.

Tall stout leafy poisonous perennials from stout rhizomes; leaves broad, strongly veined and plaited, the bases sheathing; flowers greenish yellow to whitish, often bicolored, in large, terminal panicles; perianth segments 6, distinct, more or less glandular at the base; stamens 6; pistil 1, the styles 3, persistent; fruit a 3–loculed, septicidal capsule.

Veratrum californicum Durand False Hellebore; Skunk-cabbage. Plants mostly 15–20 dm tall, typically glabrous below, becoming tomentose above; leaves numerous, 1–3 dm long, 6–20 cm wide, ovate to elliptic, or lanceolate, sheathing basally, reduced upward; panicle 2–6 dm long, many-flowered, erect, the branches spreading to ascending; flowers subsessile to shortly pedicellate; perianth greenish yellow to white, the segments abruptly narrowed and clawed basally, 8–15 (17) mm long, entire to erose or denticulate; stamens shorter than the perianth; ovary glabrous or hairy; capsules 2–3 cm long; seeds broadly winged, 10–15 mm long; n = 16. Meadows and stream banks in aspen, mixed conifer, and spruce-fir communities at 1830 to 3115 m in Box Elder, Cache, Davis, Duchesne, Millard, Rich, Salt Lake, Sanpete, Sevier, Summit, Tooele, Uintah, Utah, Wasatch, Washington, and Weber counties; Washington to Montana, south to California, New Mexico, and Mexico; 31 (v). Consumption of this plant by pregnant sheep results in abortion or production of monstrosities.

Zigadenus Michx.

Perennial bulbous poisonous plants; leaves linear, mainly basal, those of the stem reduced upward; flowers in terminal racemes or panicles, perfect or polygamous; perianth withering and persisting, bearing 1 or 2 glands just above the narrowed base, 6–segmented; pistil 1; styles 3, distinct; fruit a 3–locular septicidal capsule.

Preese, S. J. 1956. A. cytotaxonomic study of the genus *Zigadenus* (Liliaceae). Ph.D. Dissertation. Washington State Univ. 167 pp.

1. Perianth segments 4–7 mm long; ovary superior; glands of perianth ovate or semicircular 2
— Perianth segments (8) 9–11 mm long; ovary partly inferior; glands of perianth obcordate 3
2(1). Perianth segments usually less than 4 mm long, acute or acuminate; sepals clawless; flowers borne in panicles (rarely in racemes); plants common and widespread .. .. *Z. paniculatus*
— Perianth segments 5–7 mm long, obtuse to acute; sepals clawed or clawless; flowers in long-pedicellate racemes or less commonly in panicles; plants uncommon, though widespread .... *Z. venenosus*
3(2). Flowers cream to greenish, typically more than 15 mm wide, borne in racemes or less commonly in panicles; plants of higher elevations .... *Z. elegans*
— Flowers white, typically less than 15 mm wide, typically borne in panicles; plants of hanging gardens in southeastern Utah .... *Z. vaginatus*

Zigadenus elegans Pursh Elegant Death Camas. [*Anticlea elegans* (Pursh) Rydb.]. Plants 1.2–8 dm tall; basal leaves 8–30 cm long, 3–15 mm wide, slightly keeled; flowers greenish to cream, borne in racemes or less commonly in panicles; perianth segments 8–11 mm long, not clawed, each bearing a large, obcordate gland at the base; capsule 1.5–2 cm long; 2n = 32. Meadows and streambanks in aspen, lodgepole pine, and spruce-fir communities and in alpine tundra at 2150 to 3510 m, and less commonly downward along streams to 1980 m, in Beaver, Cache, Daggett, Duchesne, Garfield, Grand, Iron, Juab, Kane, Piute, Salt Lake, San Juan, Sanpete, Sevier, Summit, Tooele, Uintah, Utah, Wasatch, and Washington counties; Alaska to the Great Lakes, south to Oregon, Nevada, Arizona, New Mexico, Texas, and northern Mexico; 85 (xi).

Zigadenus paniculatus (Nutt.) Wats. Foothill Death Camas. [*Helonias paniculata* Nutt.]. Plants 2–7 dm tall; leaves mostly below the stem middle, sheathing, 1–3.5 dm long, 3–13 mm broad, much reduced upward; flowers all perfect, or the upper ones perfect and the lower ones staminate or functionally staminate, borne in panicles, these mostly 1–3 dm long; pedicels 5–25 mm long; perianth white or cream, broadly campanulate, the segments unequal, the sepals clawless or nearly so, broadly ovate, 3–4 (4.5) mm long, acute to acuminate, the inner ones longer, with a claw 0.5–1 mm long, the gland broadly ovate; stamens usually 1–2 mm longer than the perianth segments; styles ca 3 mm long; capsules 1.5–2 cm long. Blackbrush, other warm desert shrub, pinyon-juniper, sagebrush, mountain brush, ponderosa pine, Douglas fir, and grassland communities at 850 to 2685 m in all Utah counties; Washington to Montana, south to California, Nevada, Arizona, and New Mexico; 88 (x).

Zigadenus vaginatus (Rydb.) Macbr. Alcove Death Camas. [*Anticlea vaginata* Rydb., type from Natural Bridges National Monument]. Plants 3–10 dm tall or more; leaves mostly below the stem middle, sheathing or the upper ones sheathless, 20–75 cm long, 6–18 mm wide, reduced upward; flowers perfect, borne in panicles, these 15–43 cm long; pedicels 6–23 cm long; perianth white, rotate, the subequal segments clawless, obovate to spatulate or elliptic, 6–7 mm long, rounded to obtuse, each bearing a large obcordate gland at the base; stamens shorter than the perianth segments; styles ca 2–3 mm long; capsules 10–15 mm long. Hanging garden communities in seeps and alcoves in Grand, Kane, and San Juan counties; endemic (?); 9 (vi). The plants flower in late August to the end of October. This species simulates the larger flowered *Z. elegans*, which occurs mainly at high elevations and flowers in spring and early summer, but the alcove death camas evidently has affinities southward with *Z. volcanicus* Benth. of Guatemala.

Zigadenus venenosus Wats. Meadow Death Camas. [*Toxicoscordion venenosum* (Wats.) Rydb.]. Plants 1.5–7 dm tall; leaves mostly at the stem base, 1–3.5 dm long, 2–6 (10) mm wide, keeled, reduced upward; flowers in racemes or less commonly in panicles, these rarely over 1.5 dm long; pedicels ascending, 5–30 mm long; perianth white or cream, campanulate, the segments dimorphic, the sepals 4–5 mm long and with or without a claw, the inner segments somewhat longer and with a claw to 1 mm or more long, the gland usually broader than long; stamens subequal to or slightly surpassing the perianth; styles 2–3 mm long; capsules 8–15 mm long. Wet and drying meadows and streamsides in sagebrush, pinyon-juniper, mountain brush, and white fir communities at ca 2600 to 2990 m in Beaver, Daggett, Duchesne, Millard, Sanpete, Sevier, Uintah, Utah, and Washington counties; British Columbia to Saskatchewan, south to California, Colorado, Nebraska, and Mexico; 13 (0). A specimen from southern Uintah County approaches **var. *gramineus*** (Rydb.) Walsh ex Peck [*Z. gramineus* Rydb.], in the sheathing upper stem leaves and greatly reduced claws or clawless sepals. More material is necessary to adequately interpret varietal status.

NAJADACEAE A. L. Juss.
Water-nymph Family

Submersed annuals with fibrous roots; stems slender, elongate or short (in *N. caespitosa*), branched; leaves small, subopposite or verticillate, sessile, the base dilated and sheathing, linear, entire or toothed; flowers unisexual, small and concealed by the leaves, solitary, axillary; staminate flowers with 2–lipped perianth and 1 stamen, the perianth entire to 4–lobed; pistillate flowers usually naked; pistil 1–loculed and 1–ovulate; stigmas 2–4; fruit sessile, indehiscent; x = 6, 7.

Najas L.

Plants with characters of the family.

1. Leaves and stems spinulose-toothed, the teeth 0.5–1.5 mm long and evident without a hand lens .... *N. marina*

— Leaves entire or minutely toothed, the teeth not evident without a hand lens; stems entire 2

2(1). Plants acaulescent or nearly so, 0.2 dm high; seed coat not coarsely pitted, shiny; leaves gradually tapering to a slender point .... *N. caespitosa*

— Plants caulescent, 1.5–6 dm tall; seed coat coarsely pitted and dull; leaves abruptly tapering to a rounded or obtuse point .... *N. guadalupensis*

Najas caespitosa (Maguire) Reveal Fish Lake Naiad. [*N. flexilis* ssp. *caespitosa* Maguire, type from Pelican Point, Fish Lake, Sevier County]. Plants acaulescent or nearly so, compact, 2–4 cm high; leaves narrowly linear, entire (rarely few toothed), 3–12 mm long, 0.5–1.5 mm wide, expanded into a sheath basally; staminate flowers 2–2.5 mm long; anther 1, 1–loculed; pistillate flowers 2–2.5 mm long, stigmas mostly 3 (2); fruit 2–2.5 mm long; seed 1, shiny, finely reticulate with 45–70 rows of small areolae. In shallow water 3 dm deep at Pelican Point, Fish Lake, at 2623 m in Sevier County; endemic, rare, and known only from the type collection, 3 August 1940 (Maguire 19888, UTC, BRY); 1 (0).

Najas guadalupensis (Sprengel) Morong [*Caulina guadalupensis* Sprengel; *N. flexilis* var. *guadalupensis* A. Br.; *N. microdon* var. *guadalupensis* A. Br.]. Plants caulescent, branching throughout; stems slender, smooth or occasionally minutely toothed; 3–6 dm long; leaves narrowly linear, 7–20 mm long, 0.3–0.8 mm wide, toothed, the teeth not visible without a hand lens, rarely completely entire, expanding into a sheath below; staminate flowers 2–3 mm long, anthers 4–loculed; pistillate flowers 2–3 mm long, stigmas 2–3; fruits 2.5–3 mm long, pitted with 10–20 rows of areolae, dull; 2n = 24. In fresh, slow moving streams, or in ponds or lakes, at 1330 m in Cache and Salt Lake counties; over most of North America and South America; 1 (0).

Najas marina L. Caulescent plants 1–4 dm high; stems branched, stout and spinulose-toothed; leaves stiff, coarsely toothed, narrowly linear, 1–4 cm long, 1–3 mm wide, the teeth 0.5–1.5 mm long and obvious to the naked eye, the basal sheath broadly expanded, 5–6 mm long; staminate flowers 3–4 mm long; anther solitary, 4–loculed; pistillate flowers 3–4 mm long, stigmas 3; fruits 4–5 mm long; seed solitary, coarsely pitted, dull; 2n = 12. In fresh slow moving streams, and in lakes and ponds, in Cache, Juab, and Tooele counties; introduced from the Old World and scattered in the U. S. from California eastward; 1 (0).

ORCHIDACEAE A. L. Juss

Orchid Family

Perennial herbs from coralloid rhizomes, corms, or tubers; stems erect, short or much elongated, slender to somewhat thickened, naked to bracteate or very leafy; leaves simple, linear to orbicular; inflorescence terminal or axillary; flowers small and inconspicuous to large and showy, irregular; sepals 3; petals 3, free or united and adnate to the inferior ovary, one petal (the lip) differing from the other two in form, size, and coloration, often forming a spur or nectary from the base; stamens and pistils adnate to form a column; anthers mostly 2–loculed, producing masses of pollen (pollinia), the pollen powdery, waxy, or mealy; fruit a dry capsule or fleshy pod; seeds numerous; $x = 6-29+$.

Correll, D. S. 1950. Native Orchids of North America. Stanford Univ. Press. Stanford, California.
Leur, C. A. 1975. The Native Orchids of the United States and Canada excluding Florida. New York Botanical Garden. Bronx, N.Y.

1. Lip an inflated sac-like pouch; fertile stamens 2, borne laterally on the column .... *Cypripedium*
— Lip not an inflated sac-like pouch; fertile stamen 1 2
2(1). Leaves absent at flowering time 3
— Leaves present at flowering time 4
3(2). Roots coralloid; plant without chlorophyll .. *Corallorhiza*
— Roots not coralloid; plants with chlorophyll ... *Spiranthes*
4(2). Lip prolonged at base into a prominent saccate or slender spur .... *Habenaria*
— Lip not forming a spur 5
5(4). Flowers 1 or occassionally 2, the lip saccate; leaf single; plant from a small corm .... *Calypso*
— Flowers several; leaves 2 or more 6
6(5). Leaves 2, opposite, borne near middle of the stem; lip wedge-shaped .... *Listera*
— Leaves alternate, lip saccate or concave 7
7(6). Roots arising from a creeping rhizome; leaves mottled, evergreen, mostly in a basal rosette .... *Goodyera*
— Roots fleshy, fascicled; leaves not mottled and evergreen, mostly cauline 8
8(7). Lip saccate at base; flowers racemose, brownish purple; leaves lanceolate to ovate .... *Epipactis*
— Lip concave near the base; flowers in a spirally twisted spike, white or whitish; leaves linear to linear-lanceolate .... *Spiranthes*

Calypso Salisb.

Plant terrestrial, small, with a corm and fleshy roots or a coralloid rhizome; leaf solitary; scape 1–flowered; sepals and petals alike, linear-oblong to lanceolate, spreading; lip larger than the rest of the flower, deeply saccate, dilated at the anterior margin of the orifice and forming an apron that is adorned with hairs at the base; column broadly winged, having the operculate anther just below the apex; pollinia 2, bipartite, waxy; capsule erect, ellipsoid-cylindric.

Calypso bulbosa (L.) Oakes Fairy Slipper. [*Cypripedium bulbosum* L.]. Stems erect, glabrous, 6–20 cm tall; leaf solitary, arising from the summit of the corm, 2–6 cm long, cordate-ovate to elliptic; flower solitary, rarely 2, showy, pendent, purplish to pinkish white; sepals and petals 1.2–2.3 cm long, ca 2.7 mm wide, lip pendent, calceolate, ovate-oblong in outline, 1.5–2.3 cm long, 7–13 mm wide; sac expanded in front and forming a whitish apron, vividly marked on the inner surface with reddish spots and lines; apron bearded at base with 3 longitudinal rows of white, yellow or brown spotted hairs; column petaloid, suborbicular, inverted over the orifice of the lip, 7–12 mm long; capsule erect, 2–3 cm long; n = 28. Sagebrush, lodgepole pine, spruce-fir, and aspen communities at 2700 to 3200 m in Cache, Daggett, Duchesne, Grand, San Juan, Summit, Uintah, and Wasatch counties; Alaska to Labrador, south to California, Arizona, Colorado and New York; Eurasia; 18 (0).

Corallorhiza Chatel.

Non-green scapose saprophytic herbs, with short to elongated rhizomes that are much branched, toothed and coralloid; stems brownish, yellowish or purplish, clothed with several membranaceous sheathing bracts; inflorescence terminating the naked scape, a lax or dense raceme of yellowish, brownish, or purplish flowers; sepals subequal, the lateral ones united at the base and forming a short mentum more or less adnate to the ovary; lip simple to 3-lobed, slightly adhering to the base of the column; column compressed; anther terminal; pollinia 4, waxy, free; capsule ovoid to ellipsoid, pendent.

1. Sepals always 1–nerved; lip 3–5 mm long; petals 5.5 mm or less long; plant small, mostly yellowish .. *C. trifida*
— Sepals usually 3–nerved; lip and petals 5.5 mm or more long; plant larger, 1.5 dm or more tall, usually reddish or purplish 2
2(1). Perianth segments prominently striped with reddish brown or purple; lip linguiform .... *C. striata*
— Perianth segments not prominently striped; lip not linguiform, with flat margins 3
3(2). Lip with an evident though small lateral lobe on each side, spotted with purplish red .... *C. maculata*
— Lip without lateral lobes, not spotted ...... *C. wisteriana*

Corallorhiza maculata Raf. Spotted Coralroot. Stems erect, 2–5.5 dm tall, purplish to yellowish, stout to slender; flowers 10–30, minutely bracteate, erect but reflexed in age; sepals linear to oblanceolate, 6–8 mm long, 3.5–5 mm wide, reddish purple; petals slightly shorter than the sepals, often with a few dark reddish spots; lip white, spotted with red, 5–8 mm long, 3.5–5 mm wide, deeply but unequally 3–lobed, the lateral lobes small, 1.3 mm long, the middle lobe nearly orbicular, more or less crenate; column 4–5 mm long, strongly curved, yellow with magenta spots on the ventral surface; capsule ovoid, 1.5–2.5 cm long; 2n = 42. Mountain brush, aspen, spruce-fir, and lodgepole pine communities at 1830 to 3800 m in Box Elder, Cache, Daggett, Duchesne, Garfield, Grand, Juab, Kane, Millard, Piute, Salt Lake, San Juan, Sanpete, Sevier, Summit, Tooele, Uintah, Utah, Washington, and Wayne counties; Alaska to Guatemala; 48 (iv).

Corallorhiza striata Lindl. Striped Coralroot. Stems erect, 1.5–5 dm tall, reddish purple to brownish purple, stout; flowers 7–25 in a minutely bracted raceme, arcuate, pinkish yellow or whitish tinged and conspicuously striped with reddish purple; sepals oblong-elliptic to linear-lanceolate, 3- to 5-nerved, with the nerves reddish purple, 6.5–16 mm long, 2.3–5 mm wide; petals 6–15 mm long, 2.5–5.5 mm wide, 3- to 5–nerved, with the nerves reddish purple, the lip white, striately veined with reddish purple, 6–12 mm long, to 8.5 mm wide, somewhat reflexed near the base, entire, broadly elliptic, fleshy, with thickened involute margins and 2 median callosities near the base; column 4–7 mm long, arcuate, thickened at the base; capsule ellipsoid, becoming reflexed, 1.2–2 cm long; 2n = 42. Mountain brush, aspen, and spruce-fir communities at 2200 to 2900 m in Box Elder, Cache, Davis, Duchesne, Garfield, Juab, Salt Lake, Tooele, Uintah, Utah, Wasatch, and Washington counties; British Columbia to Quebec, south to Mexico through the western states; 14 (i).

Corallorhiza trifida Chatel. Early Coralroot. Stems erect, 0.4–3 dm tall, pale yellow, greenish yellow or deep yellow, very slender; flowers 3–20, minutely bracteate, small, erect-spreading, yellowish white to greenish or dull purple; sepals linear-oblong to linear-oblanceolate, obtuse, 1–nerved, 4.5–6.5 mm long, 1–1.8 mm wide; petals 4.5–5.5 mm long, 1.4–2 mm wide, slightly smaller than the sepals, the lip white or nearly so, occasionally purple-spotted, 3.5–4.5 mm long, 2.3 mm wide, oblong-quadrate, usually with a small triangular upcurved tooth on each side near the base; column broad arcuate, 3.5–7 mm long; capsule 8–12 mm long; 2n = 38, 42. In moist meadows in aspen, spruce, or lodgepole pine forests at 2000 to 3050 m in Cache, Duchesne, Garfield, Piute, Salt Lake, Sanpete, Summit, Uintah, Wasatch, and Wayne counties; Alaska to Labrador and south to Oregon, Colorado, Indiana, and New Jersey; 12 (0).

Corallorhiza wisteriana Conrad Wister Coralroot. Plant much like *C. maculata* but smaller in size, to 4 dm tall; sepals and petals greenish yellow, 1- to 3–nerved, the lip white, spotted with red, 5.5–7 mm long, 4–5 mm wide, short-clawed, truncate, often retuse at apex, erose-denticulate or undulate, the lateral lobes obsolete. Moist alpine meadows and lodgepole pine communities at 2800 to 3200 m in Duchesne, Garfield, Rich, Summit, Uintah, Wasatch, and Washington counties; throughout the eastern U. S., west to Montana, Wyoming, Arizona, and Mexico; 6 (0).

Cypripedium L.

Terrestrial caulescent herbs with fibrous roots from a short rhizome; leaves sheathing, 2 or more, plicate and prominently ribbed; flowers 1 to several, showy, bracteate; sepals spreading, free or united; petals spreading, free, smaller and narrower than the sepals, lip sessile, inflated, saccate; column declined, with 2 lateral fertile stamens and a thick dorsal staminode; stigma terminal, slightly 3–lobed; ovary 1–loculed; capsule obovoid to ellipsoid.

1. Leaves 2, opposite, near the summit of the stem; flowers corymbose, 2–4; lip purplish .... *C. fasciculatum*
— Leaves 3 or more, scattered; flower usually solitary; lip deep yellow to white, marked with purple about the orifice .... *C. calceolus*

Cypripedium calceolus L. Lady's Slipper. Stems erect, herbaceous, to 5 dm tall, glandular-pubescent throughout; leaves (2) 3–5, elliptic to ovate or ovate-lanceolate, acute, many-nerved, 5–20 cm long, 4–10 cm wide; flowers 1 or 2 at the stem apex, each subtended by a foliaceous bract to 10 cm long and 4 cm wide; sepals and petals greenish yellow to purplish; dorsal sepal twisted, lanceolate, 3–4 cm long, ca 1 cm wide, the lower pair completely connate or with a slight apical notch, petals longer and narrower than the sepals, spirally twisted, 4–9 cm long; lip strongly pouched, firm, yellow with purple dots around the orifice, 2–3 cm long; staminode yellow, spotted with purple, triangular to oblong-linear, auriculate at the base, to 10 mm long; ovary ribbed; capsule ellipsoid. River and canal banks in the valleys or to spruce-fir forests at 1470 to 3000 m in Cache, Grand, Salt Lake, Utah, and Weber counties; British Columbia to Washington, New York, and Louisiana; Eurasia; 9 (0).

Cypripedium fasciculatum Kellogg ex Wats. Stems slender, to 3 dm tall, viscid-villous, with scarious sheaths near the base; leaves 2, opposite or nearly so at the stem apex, oblong-elliptic to orbicular-ovate, nearly glabrous to puberulent, 4–11 cm long, 2.5–7.5 cm wide; peduncle short, elongating in age; bracts ca 3.5 cm long, 6–13 mm wide; flowers small, 2–4; sepals lanceolate, acuminate, 1.5–2.5 cm long, 3–6 mm wide, lower pair connate; petals broadly ovate, similar to the sepals; lip small, globose, greenish yellow, 8–14 mm long, with the purplish margin deeply infolded; staminode smaller than the stigma, 2.5–3 mm long; capsule obovoid-ellipsoid, 1.5–2 cm long; 2n = 20. Usually in duff in spruce-fir or lodgepole pine forests at 2650 to 3000 m in Daggett, Salt Lake, Summit (?), and Uintah counties; Montana, Idaho, Wyoming, and Colorado, west to Washington, Oregon, and California; 9 (0).

Epipactis Swartz

Stems erect, leafy, and arising from a short, creeping rhizome; leaves variable, orbicular to linear-lanceolate; inflorescence a conspicuously bracted, many-flowered raceme; floral bracts foliose, greatly exceeding the flowers; flowers small to medium sized, greenish to purplish; sepals free, subequal; petals similar to sepals but smaller; lip sessile on base of column, fleshy, saccate at base, enlarged above into a flat lamina, constricted or distinctly 3–lobed above; column short; anther sessile, behind the stigma on a slender jointed base; pollinia 4, mealy-granulose; capsule obovoid to ellipsoid, pendent or spreading.

Epipactis gigantea Dougl. ex Hook. Helleborine. Stems erect, stout, glabrous or nearly so, often tinged with purple at the base, leafy, 3–14 dm tall; leaves clasping the stem or with a short tubular petiole, broadest on the lower part of stem, ovate to ovate-elliptic or narrowly lanceolate, 5.5–20 cm long, 2–7 cm wide; raceme 2–12 or more flowered, elongate, lax; bracts leaflike, to 15 cm long and 2 cm wide; flowers showy; sepals greenish to rose, with purple or dull red nerves, deeply concave, 1.3–1.7 cm long, 6–8 mm wide; petals similar to sepals but thinner and more brownish purple; lip strongly veined and marked with red or purple, sessile on base of column, complex, deeply and unequally 3–lobed, 8–9 mm long, 7–8 mm wide; column short, erect, stout, 8–10 mm long; capsule ellipsoid, pendent, 2–2.5 cm long; 2n = 40, 60. Marshy and shady areas along rivers, streams, meadows, seeps, and hanging gardens from warm desert shrub to spruce communities at 830 to 2905 m in Cache, Carbon, Duchesne, Emery, Garfield, Grand, Juab, Kane, Millard, San Juan, Sanpete, Tooele, Uintah, Utah, Washington, Wayne, and Weber counties; British Columbia to Baja California and most of the western U. S. to South Dakota and Mexico; 85 (ii).

Goodyera R. Br.

Plants scapose herbs with creeping rootstocks bearing several thick fibrous roots; leaves basal, ovate to lanceolate, dark green or bluish green, reticulate-veined or variegated with white, the sheaths somewhat inflated; inflorescence a lax to densely flowered secund-spicate raceme; flowers small, white to pink or tinged with yellow or green, the oblique petals connivent with the dorsal sepal and forming a hood over the column and lip; lip sessile, deeply concave and saccate at the base, straight or recurved at the apex, entire or rarely lobed; column short, the anther borne on the back; pollinia 2; capsule erect, ovoid to ellipsoid.

Goodyera oblongifolia Raf. Rattlesnake-plantain. Stems scapose, densely glandular-pubescent above, 2–3.5 dm tall; leaves in a basal rosette, spreading, with broad petioles, oblong-elliptic to lanceolate, dark green or bluish green, concolorous or reticulate-veined and whitish along the midvein, 4–11 cm long, 1.5–3.5 cm wide; raceme densely flowered, strongly secund or in a loose spiral, tapering to the apex, 6–14 cm long; floral bracts ovate to ovate-lanceolate, 8–12 mm long; flowers white, tinged or streaked with green; dorsal sepal triangular-lanceolate, 6.5–10.5 mm long, the lateral ones obliquely ovate-lanceolate, 5–8 mm long; petals 6.5–10 mm long; lip deeply concave-cymbiform, with a long beak and involute margins, 5–8 mm long, saccate portion 1.5–2.5 mm deep and up to 4.5 mm wide, the beak sulcate, 2–3 mm long; column 4–5.5 mm long, with a slender beak; capsule obovoid-elliptic, ca 1 cm long; 2n = 22, 30. Mountain brush, maple, aspen, spruce, and fir communities at 1870 to 3100 m in Cache, Daggett, Iron, Rich, Salt Lake, San Juan, Sanpete, Tooele, Utah, and Washington counties; Alaska to Nova Scotia, south to Maine, Michigan, South Dakota, New Mexico, Arizona, and California; 16 (iii).

Habenaria Willd.

Erect herbs with fleshy or tuberous roots; stems glabrous, simple, leafy or bracteate; leaves 1 or more, basal or cauline, sessile, sheathing the stem; flowers relatively small, racemose; sepals free, alike or dissimilar, the dorsal sepal erect or incurved to form a hood over the column, the lateral sepals spreading or reflexed; petals free, erect, usually connivent with the dorsal sepal, simple or bipartite; lip pendent to upcurved, entire or lobed, extended at the base into a spur, this saccate-scrotiform to elongate-filiform, shorter to much longer than the pedicellate ovary; column short; stigmas with or without papillae; anthers 2-loculed, the sacs separated by a broad connective; pollen granular; capsule narrowly cylindrical to ellipsoid.

1. Leaves 1 or rarely 2, broadly obovate to oblanceolate; lip linear to narrowly triangular-lanceolate; spur tapering from a rather broad base, subequal to the lip .... *H. obtusata*
— Leaves several to many, basal and cauline 2
2(1). Leaves 2–4, always clustered at or near base of stem, usually withering before anthesis; flowers subsessile; sepals 1-nerved; spur subequal to slightly longer than the lip .... *H. unalascensis*
— Leaves many, scattered on the stem, persistent; flowers pedicellate; sepals usually more than 1-nerved 3
3(2). Lip unequally 3-lobed at apex, the lateral lobes the longest; bracts leaflike, more than 2 cm long ... *H. viridis*
— Lip without lobes at apex; bracts mostly shorter than 2 cm 4
4(3). Lip rhombic-lanceolate to broadly lanceolate, dilated at the base; flowers whitish .... *H. dilatata*
— Lip linear to elliptic-lanceolate, never dilated at the base; flowers greenish 5
5(4). Spur saccate to clavate, didymous near the tip, about half as long as the lip .... *H. saccata*
— Spur not saccate, subequal to the lip or greatly exceeding it 6
6(5). Spur 1.5–2 times as long as the lip .... *H. zothecina*
— Spur shorter than or exceeding the lip only slightly 7
7(6). Flowers usually scattered in elongate racemes; lip usually linear and with a fleshy ridge in the center .... *H. sparsiflora*
— Flowers usually in dense cylindrical racemes; lip usually lanceolate and without calli below the middle .... *H. hyperborea*

Habenaria dilitata (Pursh) Hook. White Bog-orchid. [*Orchis dilitata* Pursh]. Plants erect, glabrous, 1.5–12 dm tall; stems slender or stout, leafy; leaves lanceolate to oblanceolate, to 30 cm long and 5.5 cm wide; inflorescence a many-flowered cylindrical laxly or densely flowered raceme to 45 cm long and 3.5 cm wide; floral bracts lanceolate, exceeding the flowers; flowers white, or yellowish white or tinged with green; dorsal sepal ovate to elliptic, connivent with the petals to form a hood over the column, 3-nerved, 3–7 mm long, 2.5 mm wide, the lateral sepals elliptic-lanceolate to narrowly lanceolate, 3-nerved, 4–9 mm long; petals ovate-lanceolate to linear-lanceolate, falcate, obliquely dilated at the base, 1- or 2-nerved, 4–8.2 mm long, the lip variable, rhombic-lanceolate to broadly lanceolate or with a suborbicular base and linear anterior part, strongly dilated at the base, 5–10 mm long, 2–5 mm wide at the base, the spur cylindrical, shorter to twice as long as the lip; n = 21.

1. Spur less than 2/3 the length of the lip, thickly cylindrical .... *H. dilitata* var. *albiflora*
— Spur as long as the lip to twice as long, slenderly cylindrical 2
2(1). Spur ca 1 1/2–2 times as long as the lip and strongly curved; plants of higher elevations .... *H. dilitata* var. *leucostachys*
— Spur subequal to the lip and not strongly curved; plants of middle elevations .... *H. dilitata* var. *dilitata*

Var. *albiflora* (Cham.) Correll [*H. borealis* var. *albiflora* Cham.; *Platanthera dilitata* var. *albiflora* (Cham.) Ledeb.; *Limnorchis dilitata* ssp. *albiflora* (Cham.) A. Love & Simmon]. Moist rocky slopes, meadows, woods, and along streams at 1950 to 3200 m in Box Elder, Cache, Garfield, Salt Lake, Summit, and Wasatch counties; Alaska to Colorado; 12 (0).

Var. *dilitata* [*H. borealis* var. *dilitata* (Pursh) Cham.; *Platanthera dilitata* (Pursh) Beck; *Limnorchis dilitata* (Pursh) Rydb.]. Moist areas in canyons, meadows or along seeps and streams to 2500 m in Cache, Beaver, Duchesne, Juab, and Salt Lake counties; Greenland to Alaska, south to Pennsylvania, Colorado, New Mexico, and Oregon; 18 (vi).

Var. **leucostachys** (Lindl.) Ames [*Platanthera leucostachys* Lindl.; *Limnorchis leucostachys* (Lindl.) Rydb.]. Alpine meadows and springs in lodgepole, aspen, and spruce-fir communities at 2600 to 3500 m elevation in Box Elder, Cache, Daggett, Davis, Duchesne, Garfield, Iron, Juab, Kane, Piute, Salt Lake, Sevier, Summit, Tooele, Uintah, Utah, Washington, and Weber counties; Alaska to Nevada; 75 (vii).

Habenaria hyperborea (L.) R. Br. Northern Bog-orchid. [*Orchis hyperborea* L.; *Platanthera hyperborea* (L.) Lindl.; *Limnorchis hyperborea* (L.) Rydb.; *Habenaria borealis* var. *viridiflora* Cham.]. Plants erect, slender to stout, 1.5–10 dm tall; stems leafy throughout or only near the base, glabrous; leaves variable, linear, oblong-elliptic to oblanceolate, reduced to bracts above, 4.5–30 cm long, 0.8–4.5 cm wide; inflorescence a densely to laxly flowered, spicate raceme 3–25 cm long and 1–2.5 cm wide; floral bracts lanceolate to linear-lanceolate, the lower to 3 cm long and greatly exceeding the flowers; flowers small, often fragrant, green or yellowish green; dorsal sepal suborbicular to elliptic ovate, erect and connivent with the petals to form a hood over the column, 3-nerved, 3–7 mm long, 1.3–4 mm wide, lateral sepals ovate to elliptic-lanceolate, 3–9 mm long; petals usually fleshy, ovate-lanceolate to lanceolate, 1- or 2-nerved; lip fleshy, lanceolate to linear, 3- to 5-nerved, 3–9 mm long, 1.5–2.5 mm wide; spur cylindrical, slender to somewhat clavate, 2.5–7.5 mm long, usually shorter than the lip or subequal to it; capsule erect, ellipsoid. Sagebrush, pinyon-juniper, oak brush, aspen, lodgepole, spruce, and fir communities at 1390 to 3545 m in all Utah counties, except Box Elder, Davis, Kane, Morgan, Rich, Tooele, Wayne, and Weber; Alaska to Newfoundland, Greenland, Iceland, south to California, Nevada, Arizona, New Mexico, and Pennsylvania; 75 (iv).

Habenaria obtusata (Banks) Richards. Small Bog-orchid. [*Orchis obtusata* Banks ex Pursh; *Platanthera obtusata* (Banks) Lindl.]. Small glabrous plants 8–35 cm tall; stems naked or with a linear bract about the middle, 4-angled; leaf solitary or rarely 2 at the base of stem, obovate to linear-lanceolate, 4–15 cm long, 1–4.5 cm wide; inflorescence a short, few-flowered raceme 2.5–17 cm long, 1.5–2 cm wide; floral bracts lanceolate, exceeding the flowers, to 2 cm long; flowers greenish-white; dorsal sepal orbicular-ovate, concave, 3-nerved, connivent with the petals to form a hood over the column, the lateral sepals strongly reflexed, elliptic-lanceolate; petals triangular-ovate, falcate, 1-nerved, 4–5.5 mm long, 1.5–2.2 mm wide at the dilated base; lip fleshy, linear to narrowly triangular-lanceolate, acute, strongly pendent, with a small sulcate callus in the center at the base, usually with the lateral margins somewhat revolute, 6–10 mm long, 1–2 mm wide; spur tapering to a somewhat broadened base, curved, 3–8 mm long; column 2 mm long; capsule erect, ellipsoid, 7–10 mm long. Aspen, spruce, and fir communities at 2850 to 3200 m in Duchesne and Summit counties; Labrador to Alaska and south to British Columbia, Oregon, Colorado, Wisconsin, and New York; 4(ii).

Habenaria saccata Greene Slender Bog-orchid. [*Platanthera saccata* (Greene) Hulten; *Limnorchis saccata* (Greene) A. Love & Simmon; *Habenaria stricta* Rydb; *H. neomexicana* Tidestr.]. Plants erect, slender or stout, light green, 1.5–10 dm tall; stems leafy; leaves narrowly elliptic to linear-lanceolate, 4–14 cm long, 1–4 cm wide; inflorescence a much elongated, laxly many-flowered, slender, spicate raceme to 42 cm long and 8–20 mm wide; floral bracts linear-lanceolate, the lower ones to 6 cm long and greatly exceeding the flowers; flowers small, green, scattered on the elongated rachis; sepals thin, 3-nerved; dorsal sepal suborbicular to ovate or ovate-elliptic, connivent with the petals and forming a hood over the column, 3–5 mm long, 3–3.5 mm wide at the base, the lateral sepals 4–6 mm long, 2–3 mm wide; petals triangular-lanceolate to elliptic-lanceolate, falcate, usually purplish, 1- or 2-nerved, 3–5 mm long, 1.5–2.2 mm wide; lip linear, usually purplish, 4–7.5 mm long, 1–2 mm wide; spur broadly cylindric to scrotiform, occasionally didymous at the tip, 1/3–2/3 the length of the lip; capsule erect, ca 1 cm long; 2n = 42. Moist or wet meadows, bogs, and thickets or along streams in cottonwood, oakbrush, maple, willow, and aspen communities at 1800 to 2550 m in Duchesne, Juab, Millard, Salt Lake, Summit, and Wasatch counties; Alaska southward to California, New Mexico, Colorado, and Montana; 10 (0).

Habenaria sparsiflora Wats. Watson Bog-orchid. [*Platanthera sparsiflora* (Wats.) Schlecht.; *Limnorchis sparsiflora* (Wats.) Rydb.]. Plants erect, glabrous, light green, 1.5–7.5 dm tall; stems more or less leafy; leaves variable, scattered, oblong-elliptic to linear-lanceolate, 6.5–30 cm long, 1–5 cm wide; inflorescence a laxly few- to many-flowered spicate raceme 1–4.5 dm long and 1–3 cm wide; floral bracts narrowly lanceolate, subequal to the flowers or the lower ones to 4 cm long; flowers light green, usually scattered, the lowermost often remote; dorsal sepal suborbicular to ovate or ovate-elliptic, connivent with the petals, 6–7.5 mm long, 4.5–6 mm wide, the lateral sepals strongly reflexed, 6–10 mm long; petals rather fleshy, triangular-lanceolate to lanceolate, 6–8 mm long; lip fleshy, large for the flower, pendent, linear to linear-lanceolate, 6–14 mm long, 1.5–3 mm wide; spur cylindrical, filiform, equal to or slightly exceeding the lip; capsule ellipsoid, to 1.5 cm long.

1. Column large, with a broad connective; lip ca 8 mm long; spur ca 10 mm long, cylindric .... *H. sparsiflora* var. *sparsiflora*
— Column small, usually without a broad connective; lip ca 6 mm long; spur ca 8 mm long, slightly clavate, mostly higher elevations .... *H. sparsiflora* var. *laxiflora*

Var. **laxiflora** (Rydb.) Correll [*Limnorchis laxiflora* Rydb.]. Moist or wet meadows, springs or streamsides in aspen, Douglas fir, spruce, lodgepole pine, and alpine tundra communities up to 3600 m in Daggett, Duchesne, Summit, and Uintah counties; British Columbia to California, eastward to New Mexico, Colorado, Wyoming, and Montana; 8 (0).

Var. **sparsiflora** Moist or wet meadows, bogs, seeps, or along streams in poplar, mahogany, juniper, aspen, ponderosa pine, lodgepole pine, and spruce-fir communities at 1850 to 3600 m in Cache, Emery, Garfield, Grand, Iron, Juab, Kane, Millard, Piute, Salt Lake, San Juan, Utah, Washington, and Wayne counties; Baja California to Nevada, Colorado, Arizona, and New Mexico; 40 (iii).

Habenaria unalascensis (Sprengel) Wats. Alaska Bog-orchid. [*Spiranthes unalascensis* Sprengel; *Platanthera unalascensis* (Sprengel) Kuntz]. Plant strict, scapose, glabrous, 2.5–7 dm tall; stem straw colored or purplish brown, leafy only near the base, bracteate above; leaves

2–4 in a basal cluster, oblanceolate to narrowly linear-lanceolate or rarely obovate, pale green, and withering before or during anthesis, 7.5–15 cm long, 1–3 cm wide; inflorescence a narrowly cylindrical, elongate, densely or laxly flowered spicate raceme, 1–3.5 cm long and 7–15 mm wide; floral bracts ovate to linear-lanceolate, prominently 1-nerved, 3–8 mm long, much shorter than the flowers; flowers white, greenish, or yellowish green, small, numerous, and fragrant; sepals thin, elliptic to broadly ovate-elliptic, 1-nerved, 2–4 mm long, 0.8–1.5 mm wide; petals fleshy, sometimes tinged with purple, ovate to elliptic-lanceolate, 1-nerved, 1–4 mm long, 0.8–1.5 mm wide; lip elliptic with a small lobe on each side at the base, fleshy-thickened through the center below the middle, 3-nerved, 2.5–4.5 mm long, 1.2–3 mm wide at the base; spur cylindrical, slender, curved, 3–4.5 mm long, subequal to the lip; capsule suberect, obliquely ellipsoid, 6–10 mm long. Somewhat dry to moist slopes in sagebrush, mountain brush, juniper, ponderosa pine, aspen, and lodgepole pine communities at 1870 to 3250 m in Cache, Carbon, Daggett, Duchesne, Grand, Juab, Millard, Rich, Salt Lake, Summit, Tooele, Uintah, Utah, and Weber counties; Quebec to Alaska and south to California, Colorado, and North Dakota; 33 (0).

Habenaria viridis (**L.**) **R. Br.** Long-bracted Bog-orchid. [*Satyrium viride* L.; *Orchis virescens* Muhl. ex Willd.]. Plants stout, glabrous, 0.6–6 dm tall; stems leafy; leaves variable, obovate to oblong or lanceolate, 4–15 cm long, 1–6.5 cm wide; inflorescence a dense to laxly flowered raceme to 20 cm long; floral bracts linear-lanceolate, 1.5–5.5 (7) cm long, 2–4 times as long as the flowers; flowers green, with stout pedicels; dorsal sepal ovate-orbicular to oblong-elliptic, 3–6 mm long, 2–3.5 mm wide; lateral sepals ovate-oblong; petals linear lanceolate to linear-oblong, 3–5 mm long; lip narrowly oblong-spatulate or narrowly cuneate, 2- or 3-toothed at the apex, the middle tooth short and obscure, 5–10 mm long, 2–4 mm wide, occasionally tinged with reddish brown, 2–3 times as long as the short saccate spur; capsule ellipsoid, 7–10 mm long. Moist or wet meadows or dryer hillsides in coniferous forests at about 2900 m in Iron (?) and Summit counties; Alaska to Newfoundland south to Washington, Colorado, West Virginia, and North Carolina; 1 (0). Our plants belong to **var. bracteata** (**Muhl.**) **Reichenb. ex Gray** [*Orchis bracteata* Muhl. ex Willd.].

Habenaria zothecina **Higgins & Welsh** Alcove Bog-orchid. Plant erect, glabrous, 1.5–6 dm tall; stem leafy below, reduced upward, oblong-elliptic to linear-oblong, 5–25 cm long, 0.8–6 cm wide; inflorescence laxly 5- to 20-flowered, 0.4–2 dm long, 1–3 cm wide; bracts lanceolate to linear-lanceolate, 9–20 mm long, 1–7 mm wide; flowers yellowish green to greenish; dorsal sepal orbicular or nearly so, connivent with the petals and forming a hood over the column, 4–6 mm long, 3.5–4.5 mm wide, the lateral sepals strongly reflexed, 5–8 mm long; petals triangular-lanceolate to lanceolate, 5–6.5 mm long; lip yellowish, linear to linear-lanceolate, 7–10 mm long, 2–3 mm wide; spur cylindrical, 1 1/2–2 times as long as the lip, curved outward; capsule ellipsoid, erect, 1–1.8 cm long. Moist stream banks, seeps, and hanging gardens in mixed desert shrub, pinyon-juniper, and oakbrush communities at 1330 to 2650 m in Emery, Garfield, Grand (type from near Moab), San Juan, and Uintah counties; endemic; 12 (iv).

Listera **R. Br.**

Plants small terrestrial herbs; stems slender, glandular-pubescent above; leaves 2, opposite or subopposite, sessile, inserted near middle of stems; inflorescence racemose; flowers small, greenish to greenish purple; sepals and petals free, subequal, alike; lip longer than the sepals and petals, rounded to deeply lobed at the apex, variously toothed, auricled, lobed or entire on each side at the base; column wingless; stigma with a rounded beak; anther borne on the back of column near the apex; pollinia 2, powdery; capsule small, slender, pedicellate.

1. Lip linear-oblong, cleft halfway to the base into linear-filiform to narrowly lanceolate lobes .... *L. cordata*
— Lip oblong to narrowly-cuneate, not deeply cleft 2
2(1). Lip shallowly bilobed, not much narrowed to the auriculate base; leaves lanceolate to elliptic-ovate .. *L. borealis*
— Lip retuse, abruptly narrowed to a short claw at the base; leaves broad, suborbicular ...... *L. convallarioides*

Listera borealis **Morong** Northern Twayblade. Plant small, glabrous below, minutely glandular-puberulent above, 5–18 cm tall; leaves 2, subopposite, narrowly elliptic to ovate-elliptic, obtuse, 1.3–6 cm long, 0.7–3 cm wide; raceme few-flowered, lax, 2–9 cm long; flowers pale green or yellowish green, on pedicels 3.5–7 mm long; dorsal sepal elliptic-lanceolate to linear-elliptic, 4–6 mm long, lateral sepals narrowly elliptic to oblong-lanceolate, 4.5–7 mm long; petals linear to narrowly oblong, 4–5.5 mm long; lip broadly oblong, somewhat narrowed at the middle, ciliate on the margins, 7–12 mm long, 4.2–6.5 mm wide above the middle, cleft at the dilated apex into 2 oblong or semiorbicular lobes to 3 mm long, prominently dilated at the base and with conspicuous auricles; column stout, arcuate, 3–4 mm long; $2n = 56$. Moist or wet mossy spruce-fir forests, along streams or alpine meadows to 3300 m in Cache, Duchesne, Salt Lake, San Juan, Sevier, and Summit counties; Alaska to Hudson Bay and south to Washington and Colorado; 7 (ii).

Listera convallarioides (**Swartz**) **Torr.** Broad-leaved Twayblade. [*Epipactis convallarioides* Swartz]. Plant slender or stout, glabrous below, densely and minutely whitish glandular-pubescent above, 6–37 cm tall; leaves 2, opposite, broadly ovate to elliptic, obtuse to apiculate, glabrous, 2–7 cm long, 1.5–5.8 cm wide; raceme laxly many-flowered, 2–12 cm long; flowers yellowish green, on slender pedicels 4–7 mm long; dorsal sepal ovate-lanceolate, 4.5–5 mm long, the lateral sepals lanceolate, strongly falcate-recurved, 4.5–5.5 mm long; petals linear-falcate, 4–5 mm long; lip with a short slender claw, narrowly cuneate, shallowly notched at the apex, with the lobes rounded and minutely toothed in the sinus, and with a short triangular tooth on each side near the base, 8–13 mm long, 5–7 mm wide near the apex; column slender, slightly recurved, 2.5–3 mm long; $2n = 36$. Moist or damp leaf mold, along streams, or in aspen, pinyon pine, mountain mahogany, ponderosa pine, and spruce-fir communities at 2450 to 3000 m in Box Elder, Davis, Duchesne, Juab, Piute, Salt Lake, Sevier, Utah, and Washington counties; Alaska to Newfoundland and south to the Carolinas, Tennessee, Arizona, and California; 46 (iv).

Listera cordata (**L.**) **R. Br.** Heart-leaved Twayblade. [*Ophrys cordata* L.]. Plant slender, glabrous, 6.5–25 cm tall; leaves 2, opposite, broadly to narrowly ovate-cordate or deltoid, mucronate, 0.9–4 cm long, 0.7–3.8 cm wide; raceme slender, laxly to densely flowered, 2–10 cm long; flowers small, purplish or greenish yellow, on pedicels 1–4 mm long; dorsal sepal ovate-oblong to elliptic-oblong, obtuse, 2–3 mm long, the lateral sepals similar; petals elliptic to oblong-linear, obtuse to truncate, 1.5–2.5 mm long; lip linear-oblong, cleft 1/2–2/3 its length, into 2 linear-lanceolate lobes, with a subulate tooth on each side near the base; column short and thick, ca 0.5 mm long; $2n = 34+, 36, 38, 40, 42$. Mossy, damp, spruce-fir, aspen, or sphagnum bog communities at 3000 to 3500 m in Duchesne, Summit, and Wasatch counties; Alaska to Greenland, south to California, Nevada, New Mexico, and the Carolinas; 6 (ii).

Spiranthes Rich.

Small to coarse terrestrial herbs with tuberous or rarely fibrous roots; leaves mostly basal, narrow, reduced above to sheathing bracts; flowers in a congested, terminal, spirally-twisted spike, mostly white or creamish, sordid; sepals in ours connivent with the 2 petals and forming a hood enclosing the column and most of the lip or the lateral sepals free and not forming part of the hood; lip sessile or with a short claw, concave and grooved near the base and partly enclosing the column, simple or lobed, spreading or recurved distally; column short in ours, producing a single anther on the back, tipped by a 2–toothed, viscid rostellum; pollinia 2; capsule erect, ellipsoid to ovoid.

1. Flowers widely spreading or nodding, only the lateral petals more or less connivent with the upper sepal; lateral sepals somewhat loose and spreading; rare in our area .... S. cernua
— Flowers ascending or outcurved; sepals and lateral petals more or less connivent to form a hood opposite the lip 2
2(1). Lip definitely narrowed below the tip (pandurate), its basal callosities obsolete .... S. romanzoffiana
— Lip not narrowed below the tip (ovate), its basal callosities developed .... S. porrifolia

Spiranthes cernua (L.) Rich. [*Ophrys cernua* L.]. Plant erect, glabrous below, downy-pubescent above, 1–5.5 dm tall; roots quite slender, fleshy; leaves mostly basal, linear to linear-lanceolate, 4.5–25 cm long, 0.6–2 cm wide; inflorescence moderately dense, up to 15 cm long; flowers white, usually fragrant, widespreading or somewhat recurved so as to appear nodding, the perianth somewhat downy-pubescent on the outer surface; sepals and petals not connivent and forming a hood over the column, only the upper sepal and lateral petals somewhat connivent; the lateral sepals free; lip ovate-oblong, 6–10 mm long, 3–6 mm wide, arcuate-recurved, with crisped or erose margins, the basal callosities prominent; $2n = 46, 60, 61$. Moist or damp places at low elevations in Salt Lake County; eastern and central U. S. west to New Mexico; 2 (0).

Spiranthes porrifolia Lindl. [*Spiranthes romanzoffiana* var. *porrifolia* (Lindl.) Ames & Correll; *S. diluvialis* Shev.]. Plant erect, glabrous below, somewhat glandular-pubescent above, up to 5 dm tall; roots fascicled, long, fleshy; leaves 3 or 4, elliptic-lanceolate, 5–20 cm long, 0.5–2.5 cm wide, green, sometimes soon withering; flowers small, yellowish white; bracts lanceolate, 8–12 mm long, to 3 mm wide; sepals and petals connivent and forming a hood over the column; sepals 7–9 mm long, 1.5–2.5 mm wide near the base; petals linear-lanceolate to subfalcate, 7–8 mm long, 1–2 mm wide; lip ovate, concave to conduplicate, 7–8 mm long, 3–4 mm wide, the apex not dilate but papillose, the base with prominent callus, the tip spreading, not sharply recurved. Along streams, bogs, and open seepage areas in cottonwood, tamarix, willow, and pinyon-juniper communities at 1340 to 2075 m in Daggett, Duchesne, Garfield, Utah, and Wayne counties; Washington, Oregon, and California; 8 (0).

Spiranthes romanzoffiana Cham. Hooded Ladies-tresses. Plant erect, stout to slender, glabrous below, pubescent above, 0.8–5 dm tall; roots fleshy, fasciculate; leaves 3–6, basal and cauline, linear- lanceolate, 5–25 cm long, 0.5–1.5 cm wide; flowers creamy white; bracts lanceolate, acuminate, 15–25 mm long, somewhat sheathing; sepals and petals connivent and forming a hood over the column; sepals lanceolate, 8–13 mm long, 3–4 mm wide, the lateral somewhat falcate; petals linear, obtuse, 8–12 mm long, 1–2.5 mm wide; lip pandurate, 8–11 mm long, ca 5 mm wide, constricted above the middle, the apex erose, dilated and without papillae, the basal callosities minute, the tip sharply deflexed; $2n = 30$. Bogs to open woods in lodgepole pine, aspen, spruce-fir, and alpine tundra communities at 2700 to 3600 m in Daggett, Duchesne, Juab, Salt Lake, Summit, Tooele, Uintah, Utah, Wasatch, Washington, and Weber counties; Alaska to Newfoundland and southward to California, Arizona, New Mexico, Michigan, and Pennsylvania; 29 (i).

PALMAE A. L. Juss.
Palm Family

Shrubs or trees with terminal clusters of spirally arranged leaves; leaves petiolate, the blades fan-shaped, plaited and palmately divided into segments; spadix elongate, branched, with a spathe at the base of the inflorescence and at the base of each inflorescence branch; flowers perfect or imperfect, regular; sepals 3; petals 3; stamens 6–12; pistil 1, the ovary superior, 3–carpelled; fruit drupaceous; $x = 13–18$. [Arecaceae Schultz-Schultzenstein].

1. Petioles with long fibers at the base, toothed along the margin above the base .... *Trachycarpus*
— Petioles not with long fibers at the base, coarsely toothed along the margin, the teeth often pronglike .. .. *Washingtonia*

Trachycarpus Wendl.

Dioecious or rarely monoecous or polygamous, small to moderate palms; trunks often covered with persistent fibrous leaf sheaths; leaves palmate, divided into 1–ribbed segments, the petiole bases bearing long blackish fibers, the margins armed with small blunt teeth; inflorescence among the leaves, with several thin bracts on the peduncle paniculate; branches not subtended by sheathing bracts; flowers in clusters; sepals 3, basally imbricate; petals 3, imbricate, the male flowers with 6 stamens and 3 small pistillodes; female flowers with 6 staminodes and a pistil; perfect flowers with 6 stamens and a pistil; fruit subglobose.

Trachycarpus wagnerianus Becc. Fan-palm. Shrubs or small trees, mainly less than 3 m tall; leaves to 1 m long or more, the blades mainly less than 5 dm long, rigid and outstanding, not drooping; seldom flowering. Plants cultivated in Washington County, though frost sensitive and with leaf kill in most years; widely grown in the southern U. S.; origin unknown, though probably Asia; 1 (i).

Washingtonia Wendl.

Small to moderately large palms; trunks covered with persistent leaf bases; leaves palmate, divided into 1-ribbed segments, the petioles armed with spinose teeth or prongs along the margin; inflorescence among the leaves and exceeding them, with several bracts on the peduncle, with bracts subtending each of the several branches; flowers with a tubular calyx; petals tardily deciduous; stamens 6; pistil of 3 carpels, separate except for the united style; fruit 1-seeded, ellipsoid to globose.

Washingtonia filifera (L. Linden) Wendl. Desert Fanpalm; Petticoat Palm. [*Pritchardia filifera* L. Linden]. Trees mainly 3-10 m tall; trunks clothed with hanging persistent leaves (or these trimmed, displaying the leaf bases); leaves to 2 m long or more, the blades green, filiferous; occasionally flowering. Cultivated ornamental in Washington County; Arizona and California; 1 (i). This handsome palm suffers from frost damage in some years, but has been successfully grown for many decades. It should be expected to escape into the seeps and springs around St. George.

POTAMOGETONACEAE Dumort.

Pondweed Family

Aquatic perennial rhizomatous fresh water herbs; stems terete or somewhat flattened, simple or freely branched; leaves submersed or some of them floating, alternate or opposite, simple, entire or minutely toothed, the submersed blades mostly narrower than the floating ones; stipules conspicuous, sheathing, free or fused with the leaf base, deciduous or not; flowers perfect or imperfect, ebracteate, sessile or nearly so, in whorls or on axillary peduncles; perianth of 4 short-clawed distinct segments; stamens 4, attached to the perianth segments; pistils 4, distinct; style short, usually persisting as a short beak; fruit a somewhat flattened achene, with dorsal and lateral keels or obscurely keeled in some species; x = 7?, 13-15.

Potamogeton L.

Aquatic herbs; leaves often dimorphic, the floating often broader and leathery in texture; stipules conspicuous, sometimes sheathing; flowers sessile in pedunculate spikes, usually in whorls.

Correll, D. S. and H. B. Correll. 1972. Aquatic and wetland plants of southwestern United States. Environmental Protection Agency, Washington, D. C.
Cronquist, A., et al. 1977. Intermountain Flora. Vol. 6. Columbia University Press.
Mason, H. L. 1957. A flora of the marshes of California. Univ. California Press, Berkeley.
Hitchcock, C. L., et al. 1969. Vascular plants of the Pacific Northwest. Univ. Washington Publ. Biol. 17(1): 1-914.

1. Stipules adnate to the leaf base, forming a sheath enfolding the stem; leaf appearing as though arising from top of sheath 2
— Stipules axillary and free from the leaf base, either free from one another or united to form a cylinder around the stem, or the margins enfolding the stem not united 5
2(1). Leaves with auricles at the base, the margins entire, minutely serrulate toward the tip, 3-8 mm wide, 20- to 60-nerved; plants usually sterile ...... *P. robbinsii*
— Leaves without auricles, the margins entire, to 5 mm wide, 1- to 3-nerved 3
3(2). Leaves with a long tapering tip, filiform, 0.2-1 mm broad; style present, short; fruit 2.6-4.2 mm long, beaked .... *P. pectinatus*
— Leaves with obtuse or rounded apex, to 5 mm broad; style lacking; fruit beak wartlike or lacking 4
4(3). Sheaths inflated at the base, usually subtending 2 or more branches, the free stipule tips not producing an obvious ligule; leaves filiform to narrowly linear, 1-2 mm wide; spikes with equally spaced whorls of flowers; plants rare in Utah .... *P. vaginatus*
— Sheaths not inflated, usually subtending 1 or 2 branches, the free stipular tips projecting upward into a distinct ligule; leaves filiform to linear, 0.2-5 mm wide; spikes with mostly unequally spaced whorls of flowers; plants common .... *P. filiformis*
5(1). Leaves all submersed and essentially alike, the petiole short or lacking 6
— Leaves dimorphic, submersed and floating, the floating with broad blades and long petioles 14
6(5). Leaves crispate and serrate, 3-12 mm wide; achenes with a curved beak 2-3 mm long .... *P. crispus*
— Leaves entire; achene beak various 7
7(6). Leaves lanceolate, oblong, or ovate 8
— Leaves linear-filiform or setaceous 10
8(7). Leaves sessile or only slightly clasping, usually reddish; beak of fruit 1-1.3 mm long .... *P. alpinus*
— Leaves clasping 1/3-3/4 around the stem, rarely if ever reddish; beak of fruit not over 1 mm long 9
9(8). Stipules conspicuous, 3-10 cm long, persistent; leaves mostly over 10 cm long, the apex usually boat-shaped; fruit over 4 mm long and 3 mm wide at maturity .... *P. praelongus*
— Stipules inconspicuous, not over 2 cm long and soon reduced to shreds; leaves less than 10 cm long, the apex not boat-shaped; fruit less than 4 mm long and 3 mm wide .... *P. richardsonii*
10(7). Stems strongly flattened and winged; leaves 9- to 35-nerved, 3-5 mm wide; fruits 3-5.5 mm long .. .. *P. zosteriformis*
— Stems terete or nearly so, not winged; leaves less than 10-nerved, not over 3 mm wide; fruits less than 3 mm long 11
11(10). Leaves without basal glands; peduncle stout, 0.3-1.5 (2) cm long; spikes 2-5 mm long, capitate; fruit compressed, with a thin toothed keel .... *P. foliosus*
— Leaves often with 2 basal glands; peduncle slender, 1-9 cm long, the spike often over 5 mm long; fruit plump, the keel obscure or lacking 12
12(11). Stipules obscure, 0.3-1.4 cm long, not fibrous; stems very slender or capillary and much branched; leaves firm, 3-nerved, 0.3-2 mm wide, with 2 small globose basal glands .... *P. pusillus*
— Stipules obvious, 0.7-2 cm long, strongly fibrous; stems slender and simple or sparsely branched; plants rare in northern Utah 13

13(12). Leaves thin and translucent, (3) 5– to 7–nerved, 1.5–3.5 mm wide, with 2 yellowish glands at the base, the apex rounded; stipules 0.7–1.1 (2) cm long .... *P. friesii*

— Leaves firm, often revolute, 3–nerved, 0.5–2 (2.5) mm wide, glandless, the apex gradually tapering to a slender tip; stipules 0.8–2 cm long ..... *P. strictifolius*

14(5). Submerged leaves thread- or ribbonlike, the sides parallel, less than 8 mm wide 15

— Submerged leaves broader, the sides curving in near the ends, not parallel, over 8 mm wide (except in some *P. gramineus*) 18

15(14). Stipules of submerged leaves fused with the bases of the leaves; inflorescences dimorphic, in spikes in the axils of the floating leaves and in small heads in the axils of submerged leaves; plants not definitely known for Utah .... *P. diversifolius* Raf.

— Stipules not joined with the leaf base; inflorescences alike 16

16(15). Floating leaves over 2.5 cm wide, usually cordate or rounded at the base; submerged leaves less than 2.5 mm wide .... *P. natans*

— Floating leaves less than 2.5 cm wide, usually cuneate at the base; submerged leaves over 2.5 mm wide ... 17

17(16). Submerged leaves with a conspicuous reticulated medial band; fruit 2.5–3.5 mm long, definitely keeled; floating leaves usually opposite .. *P. epihydrus*

— Submerged leaves without a reticulated medial band; fruit 1.7–2.5 mm long, slightly if at all keeled; floating leaves usually alternate .... *P. gramineus*

18(14). Floating leaves delicate and translucent, short-petiolate, not much differentiated from submerged leaves, except wider, reddish tinged .... *P. alpinus*

— Floating leaves coriaceous, opaque, the base rounded or cuneate but distinct from the petiole, the blade with 13 or more nerves 19

19(18). Submerged leaves all petiolate, the longest petioles over 4 cm long; fruit 3.5–4 mm long when mature, usually reddish .... *P. nodosus*

— Submerged leaves over 1.5 cm wide, sessile or on petioles to 4 cm long, seldom over 5 times longer than wide, the nerves usually more than 9 .... *P. illinoensis*

Potamogeton alpinus Balbis Northern Pondweed. Plants partially submerged; stems reddish brown, 1–2 mm thick, terete, simple or rarely branched above, 3–10 dm long; leaves somewhat dimorphic, the submerged ones oblong-linear to linear-lanceolate, 0.5–2 cm wide, 4–18 cm long, rather thin, usually with 7 nerves, delicate, translucent, sessile or slightly clasping, obtuse or rarely acutish, entire; floating leaves elliptical, oblanceolate to obovate or even oblong-linear, 1–2 cm wide, often poorly developed, thin, translucent, tapering, with no sharp distinction to a petiole, 1–3 cm long; stipules thin, membranous, free from the leaf, 1.5–4 cm long, soon withering and deciduous; peduncles 3–10 cm long, usually as thick as the stems; spikes in fruit 1.5–3.5 cm long, dense and compact, with 5–9 whorls; flowers large, the perianth greenish red to red, 1.5–2.5 mm long; style short; fruit 3–4 mm long, obovate, 1– to 3–nerved, the beak curved, 1–1.3 mm long; $2n = 26$. Streams, ponds, and lakes at 2400 to 3850 m in Daggett, Duchesne, Salt Lake, Sevier, Summit, Uintah, and Utah counties; Alaska to Nova Scotia, south to California, Colorado, and the northeastern states; Eurasia; 18 (0). *P. alpinus* is similar to *P. praelongus* but without the whitish flexuous stems.

Potamogeton crispus L. Crisped Pondweed. Plants wholly submerged; stems greenish, compressed, with channeled sides, freely branching above, 4–8 (15) dm long; leaves all submerged, oblong, half-clasping, 2–8 cm long, 5–12 mm wide, 3– to 5 (7) –nerved, rounded apically, the margin crisped and serrulate; stipules axillary, scarious, splitting and becoming fibrillose early, 0.3–0.8 mm long, whitish, often united around the stem to form a sheath; peduncles stout, 3–5 (7) cm long; spikes few-flowered, 1–2 cm long; flowers yellowish, the segments 1.8–3 mm long; fruit ovoid, 3–keeled, 5–6 mm long, with a stout curved beak 2–3 mm long; $2n = 26, 36, 42, 50, 52, 76, 78$. Open water in ponds, streams, and ditches at 1500 to 2330 m in Cache, Daggett, Rich, Salt Lake, Summit, Utah, Wasatch, and Weber counties; widely introduced throughout the world, in the western U. S. as far south as California and Nevada; 10 (0).

Potamogeton epihydrus Raf. Nuttall Pondweed. Stems somewhat flattened, simple or few-branched, 5–20 dm long; leaves dimorphic, the submerged ones linear, sessile, flaccid, (5) 10–20 cm long, 3–10 mm wide, the midvein bordered by several rows of air chambers, the leaves soon decomposing; floating leaves with flattened petioles shorter than the blades, these firm, elliptic to oblong-elliptic, 3–8 cm long, (0.5) 1–2 cm wide; stipules free, not connate, membranous, 1–3 cm long, mostly blunt; peduncles about as thick as the stem, 3–8 cm long; spikes densely flowered, 2–4 cm long in fruit; achenes obliquely broadly ovate, 3–4 mm long, the ventral keel sharp, the lateral ones evident but not as sharp; beak short; $2n = 26$. Shallow to deep slow moving streams, lakes, or ponds, at 2800 to 3000 m in Salt Lake County; Alaska south to California, Idaho, and Colorado, east to Newfoundland and the east-central U. S.; 1 (0).

Potamogeton filiformis Pers. Filiform Pondweed. Plants wholly submerged, arising from creeping, slender, branched rootstock with terminally thickened tubers; stem usually short and much-branched, especially below, 2–6 (9) dm long, subterete, whitish to greenish; leaves all submerged, filiform to linear, rather blunt, 3– to 5–nerved, 5–12 cm long, 0.8–4 (5) mm wide, sheaths 0.5–2 cm long, closed, not swollen, greenish to whitish with a conspicuous hyaline ligule 1–7 mm long; peduncles filiform, 3–10 (15) cm long, with 2–8 whorls of crowded to remote flowers, these small, 1–2.5 mm long, brownish; style short to almost lacking; fruit obovoid, 2–4 mm long, 1.5–3 mm wide, the dorsal keel low and rounded, the 2 lateral keels low and obsolete or lacking; fruiting style a short wartlike beak less than 0.5 mm long. Three varieties are present.

1. Leaves 2–4 (5) mm wide; fruits 3–4 mm long ....
   .... *P. filiformis* var. *latifolius*

— Leaves 0.2–2 mm wide; fruits 2–3 mm long 2

2(1). Leaves 0.5–2 mm wide; plants tall, the stems mostly 3–5 (9) dm long; plants common ....
   .... *P. filiformis* var. *occidentalis*

— Leaves 0.2–0.5 mm wide; plants short, the stems mostly 1–3 dm long; plants rare .. *P. filiformis* var. *alpinus*

Var. alpinus (Blytt) Asch. & Graebn. [*P. marinus* var. *alpinus* Blytt]. Ponds, lakes, and streams at 2850 m and above in Summit County; Eurasia and North America as far south as Idaho and Colorado; 2 (0).

Var. *latifolius* (Robb.) Reveal [*P. pectinatus* var. *latifolius* Robb.; *P. latifolius* (Robb.) Morong]. Streams, ponds, and lakes at 1500 to 2500 m in Box Elder County; California to Nevada and north to Alaska; 6 (0). At its morphological extreme, var. *latifolius* is easily distinguished from var. *occidentalis*. However, in northern Utah the leaf width of the two blend wholly together and a clear distinction is not always possible.

Var. *occidentalis* (Robb.) Morong [*D. marinus* var. *occidentalis* Robb.]. Shallow ponds, lakes, and slow moving streams at 1500 to 3100 m in Box Elder, Duchesne, Emery, Garfield, Kane, Piute, Rich, Salt Lake, Sanpete, Sevier, Summit, Uintah, Utah, Wasatch, and Washington counties; Alaska to Greenland, south to California, Arizona, Colorado, New Mexico, Michigan, and Pennsylvania; 59 (iii).

Potamogeton foliosus Raf. Leafy Pondweed. Plants wholly submerged; stems flattened, leafy, slender, freely branched, 2–10 dm long; leaves all submerged, linear, 3–10 cm long, 1–2.5 mm wide, 3– to 5–nerved, without basal glands, the apex acute to cuspidate, tapering and sessile at the base, attached directly to the stem; stipules at first connate, later free, soon rupturing and deciduous, fibrillose, the fibers persistent in some, 15–25 mm long; peduncles 0.5–2 cm long; spikes subglobose to subcapitate, congested, 0.1–0.5 cm long, few-flowered; flowers minute, greenish, 0.5–1 mm long; style short; fruit broadly ovoid, 2–2.5 mm long, the dorsal keel prominent, undulate-dentate, the 2 lateral keels obscure; beak 0.2–0.4 mm long; $2n = 26$. Two varieties occur in Utah.

1. Stipules delicate, tending to completely disintegrate, the fibers not persistent; plants common ....
   .... *P. foliosus* var. *foliosus*
— Stipules firm, strongly fibrous, the fibers free at the tip of the stipule, becoming longer as the membranous portion of the stipule disintegrates, persistent; plants uncommon
   .... *P. foliosus* var. *fibrillosus*

Var. *fibrillosus* (Fern.) Haynes & Reveal [*P. fibrillosus* Fern.]. Shallow pools, lakes, and slow moving streams at ca 2700 m in Grand and Summit counties; California to Washington, Idaho, and Wyoming; 3 (0).

Var. *foliosus* Shallow pools, lakes, and streams at 1400 to 3050 m in Box Elder, Cache, Daggett, Emery, Duchesne, Grand, Kane, Salt Lake, San Juan, Summit, Uintah, Utah, Wasatch, and Washington counties; British Columbia to Quebec, south to California and Florida; Central America; 30 (ii).

Potamogeton friesii Rupr. Fries Pondweed. Plants wholly submerged; stems slender, simple below, freely branched above, green, flattened, 3–10 dm long; leaves all submerged, narrowly linear, 3–8 cm long, 1.5–3.5 mm wide, mostly 5–nerved, apically rounded but with a mucronate tip, tapering and sessile basally, with 2 yellowish glands just below the point of attachment; stipules at first connate but soon opening, not forming a sheath, 0.7–2 cm long, white, strongly fibrillose and tardily shredded, free from the blade; peduncles somewhat flattened and slightly winged, 1.5–5 cm long; spikes congested, 0.7–2 cm long, with 3–4 whorls of 2–3 flowers each; flowers small, greenish, 1.5–2.5 mm long, the style short; fruit ovoid to obovoid, 1.5–2.5 mm long, the dorsal keel low and rounded, the 2 lateral keels obscure; beak curved, persistent, 0.3–0.8 mm long; $2n = 26$. Shallow lakes and ponds at 2600 to 3100 m in Sevier and Summit counties; Eurasia and North America; Alaska to Newfoundland, south to Washington, South Dakota, Indiana, and Virginia; 3 (0).

Potamogeton gramineus L. Grass Pondweed. [*P. gramineus* var. *maximus* Morong; *P. gramineus* var. *myriophyllus* Robb.]. Plants only partially submerged; stems simple to much branched, greenish, terete, 2–8 (15) dm long; leaves dimorphic, the submerged leaves linear to lanceolate or oblanceolate, 1.5–9 cm long, 1–14 (16) mm wide, rather thin, 3– 9 (11) –nerved, apically acute, tapering to a sessile base; floating leaves usually present, coriaceous, ovate to elliptical, the blades 2–5 (7) cm long, 1–2.5 cm wide, 13– to 20–nerved, the apex obtuse or somewhat mucronate, the base cuneate or rounded to a petiole 2–10 cm long; stipules free, persistent, 0.5–3 cm long; peduncles 2–10 cm long, with numerous flowers, these large, greenish, 2–3 mm long, the style short; fruits obovoid, 2–2.5 mm long, the dorsal keel sharp and prominent, the 2 lateral keels rounded and obscure, the beak 0.3–0.5 mm long, slightly curved. Slow moving streams, shallow pools, lakes, and swampy areas at 1430 to 3500 m in Beaver, Box Elder, Cache, Daggett, Emery, Garfield, Iron, Salt Lake, San Juan, Sanpete, Sevier, Summit, Uintah, and Utah counties; Alaska to Greenland, south to California, Arizona, Colorado, and the northeastern states; Eurasia; 38 (ii). Our material can be separated into two weak varieties; specimens with submerged leaves 3–8 mm wide are var. *gramineus* and those with leaves 6–16 mm wide are placed in var. *maximus* Morong ex Bennett. However, the morphology is a continuum and the recognition of infraspecific taxa seems to be moot. *P. gramineus* is known to hybridize with *P. illinoensis*.

Potamogeton illinoensis Morong Illinois Pondweed. Stems simple to much branched, terete, to 20 dm long; leaves somewhat dimorphic but transitional; submerged leaves thin, lanceolate to oblong-elliptic or oblanceolate, 6–20 cm long, 2–5 cm wide, subsessile or tapered gradually to the petiolar base, conspicuously 9– to 19–nerved; floating leaves (if any) rather firm, gradually transitional to the submerged ones, the blades mostly broadly elliptic or oblong-elliptic, 4–12 cm long, 2–6 cm wide, the petiole much shorter than the blade, 13– to 29–nerved; stipules 2.5–7 cm long; peduncles mostly thicker than the stem, 4–12 cm long; spikes dense, mostly 3–6 cm long; achenes broadly and obliquely obovoid, ca 3.5 mm long, the dorsal keel conspicuous and sharp, the lateral keels also conspicuous but not so sharp; beak ca 0.5 mm long; $2n = 104$. Ponds, lakes, and other quiet, deep water at ca 2330 m in Cache County; British Columbia to the northeastern U. S., south to California, Nevada, and Colorado; 1 (0). This species is closely allied with *P. nodosus* and *P. gramineus*, and intermediates are known.

Potamogeton natans L. Floating Pondweed. Plants partially submerged; stems simple or nearly so, terete, 1–2 mm thick, 5–18 dm long; leaves dimorphic, the submerged ones coriaceous, linear, 10–20 cm long, 1–2 mm wide, obscurely nerved, sessile; floating blades 4–9 cm long, 2.5–6 cm wide, on long petioles, the blade coriaceous, ovate to oblong, ovate-elliptic, basally cordate or rounded, rarely tapering, apically rounded or obscurely mucronate, with 23–37 nerves; stipules persistent, clasping the stem, whitish, fibrous, 4–11 cm long; peduncles 3–8 cm long; spikes compact, especially in age, 3–5 cm long; flowers large, greenish, 1.5–3 mm long, the style short; fruit 3.5–5 mm long, obovoid, lacking keels, the beak 0.7–1 mm long; 2n = 52. Shallow pools or lakes or slow moving streams at 1500 to 3070 m in Rich, Uintah, and Utah counties; Alaska to Newfoundland, south to California, Arizona, New Mexico, and the northeastern U. S.; Eurasia; 7 (0).

Potamogeton nodosus Poir. [*P. fluitans* Roth]. Plants partially submerged; stems terete, simple or branched near the top, 1–1.5 mm thick, 4–20 dm long; leaves dimorphic, the submerged ones thin, lance-linear to lance-elliptic, the blades 9–20 cm long, 1–3.5 cm wide, 7- to 15-nerved, gradually tapering into a long petiole 2–13 cm long, apically acute, the floating blade elliptic to oblong-elliptic, 5–12 cm long, 1.5–4 cm wide, coriaceous, 10- to 20-nerved, apically obtuse, basally cuneate to somewhat rounded, on a long stout petiole arising well below the surface of the water; stipules brownish, often decaying early, linear, 3–6 (9) cm long; peduncles thicker than the stems, 3–15 cm long; spikes dense, with 10–15 whorls, 3–6 cm long in fruit; flowers large, greenish to greenish brown, 1.5–2.5 mm long; style short; fruit obliquely obovoid, 3.5–4 mm long, prominently keeled, brown or reddish when mature; beak erect, short, 0.3–0.5 mm long; 2n = 26. Ponds and streams or in deeper lakes at 1420 to 3330 m in Box Elder, Cache, Kane, Salt Lake, Sanpete, Summit, Uintah, Utah, and Weber counties; nearly cosmopolitan; 12 (0). This is a distinctive species because of its usually very long petiolate submerged leaves.

Potamogeton pectinatus L. Fennel-leaf Pondweed. Plants wholly submerged, arising from creeping slender branched rootstocks with terminally thickened tubers; stems filiform, ca 1 mm thick, much branched, terete, 3–15 dm long; leaves all submerged, filiform, mostly 1-nerved, attenuate apically, 3–15 cm long, 0.2–0.8 (1) mm wide; sheaths 2–5 cm long, only slightly thicker than the stem, often whitish, with a short hyaline ligule; peduncles filiform, 5–25 cm long, with 2–6 unequally spaced whorls of flowers, these small, obliquely obovoid, 1–1.5 mm long, brownish green; style very short; fruit obliquely ovoid, 2.5–4.2 mm long, 2–3.5 mm wide, the dorsal keel low and rounded, the 2 lateral keels inconspicuous; fruiting style slender, incurved, forming a beak; 2n = 42, 78. Streams, lakes, ponds, and ditches at 1420 to 2950 m in Box Elder, Cache, Carbon, Davis, Emery, Garfield, Grand, Kane, Millard, Rich, Salt Lake, Sanpete, Sevier, Summit, Uintah, Utah, and Weber counties; Alaska to Newfoundland, south to California, Nevada, Arizona, Colorado, and Florida; Eurasia; Africa; 89 (i). The sharply pointed filiform leaves are distinctive for this species.

Potamogeton praelongus Wulfen Whitestem Pondweed. Plants mostly submerged; stems simple or occasionally branched, whitish or olive green, 10–30 dm long, 1.5–4 mm thick; leaves all submerged, oblong-lanceolate, 10–30 (35) cm long, 1–3 cm wide, 13- to 25-nerved, rounded and hooded apically, basally cordate or rounded and clasping the stem; stipules free, white, oblong-linear to ovate-lanceolate, 5–10 cm long, usually persistent; peduncles clavate, stout, 15–60 cm long; spikes congested, with 6–12 whorls of flowers, 3–5 cm long in fruit; flowers large, greenish, 2–4 mm long, the style short; fruit obovoid, rounded on the back, cuneate at the base, 4–5 mm long, the dorsal keel acute and well developed; beak short, thick, erect, 0.5–0.8 mm long; 2n = 52. Deep cold water in lakes and slow moving streams at 1500 to 3120 m in Box Elder, Cache, Duchesne, Garfield, Sevier, and Utah counties; Alaska to Newfoundland, south to California, Colorado, Nebraska, Indiana, and New Jersey; Eurasia; 19 (0). The flexuous white stems are distinctive.

Potamogeton pusillus L. Dwarf Pondweed. [*P. berchtoldi* Fieber in Bercht.; *P. pusillus* var. *vulgaris* subvar. *interruptus* Robb., type from Silver Creek, Parleys Park]. Plants wholly submerged; stems capillary, terete or nearly so, usually much-branched, 2–10 dm long; leaves linear, 2–6 cm long, 0.5–2 mm wide, firm, 3-nerved, apically acute to rounded, often mucronate, sessile at the base with 2 small globose glands just below the point of attachment; stipules connate for at least half their length, free from the leaves, hyaline, 6–15 mm long; peduncles filiform, 5–8 cm long; spikes 0.6–1.2 (1.5) cm long, few-whorled, interrupted; flowers small, greenish, 1–2 mm long, the style very short; fruit obliquely ovoid, 2–3 mm long, rounded on the back, obscurely keeled; beak persistent, 0.3–0.5 mm long, short and curved. Shallow pools and lakes and slow moving streams at 1400 to 3330 m in Beaver, Box Elder, Cache, Daggett, Davis, Rich, Salt Lake, Sevier, Summit, Utah, Wasatch, and Weber counties; widespread in North America; Eurasia; Mexico; 15 (0).

Potamogeton richardsonii (Bennett) Rydb. Richardson Pondweed. [*P. perfoliatus* var. *richardsonii* Bennett]. Plants mostly submerged; stems stoutish, terete, sparingly if at all branched, 3–8 dm long; leaves submerged, mostly ovate-lanceolate, 3–10 cm long, 1–2 cm wide, 17- to 29-nerved, cordate and clasping at the base, apically acute to rounded, the margins somewhat crispate; stipules free, whitish, lanceolate to ovate, 1–2 cm long, soon disintegrating into fibers; peduncles stout, enlarged upward, 1.5–20 cm long; spikes congested, with 6–12 whorls, 1.5–4 cm long in fruit; flowers large, greenish, 2.5–4 mm long, the style short; fruit obovate, 2.5–3.5 mm long, the dorsal keel low and rounded, the lateral keels lacking, the persistent beak 0.5–0.8 mm long. Shallow ponds, lakes, and slow moving streams at 1430 to 3000 m in Garfield, Salt Lake, Sevier, Summit, Uintah, and Washington counties; Alaska to Labrador and south to California, Colorado, Nebraska, Indiana, Pennsylvania, and New York; 12 (0).

Potamogeton robbinsii Oakes Robbins Pondweed. Plants wholly submerged; stems stout, much-branched, 1–5 (20) dm long; leaves all submerged, linear, auricled at the base, 2-ranked, stiffly divergent, 4–10 cm long, 4–8 mm wide, the margins thickened and finely serrulate; sheaths 1–1.5 (2) cm long, white, adnate to the base of the leaf, persistent, the upper portion forming a long hyaline ligule, this ultimately fibrous; peduncles reddish, 3–10 cm long; spikes stiff, interrupted, 0.7–2 cm long, loosely flowered; flowers small, reddish to greenish yellow, 1–1.8 mm long, the style short; fruit obliquely obovoid, 3.5–4.5 mm long, the dorsal keel sharp and prominent, the 2 lateral keels low and rounded, the beak subapical, curved, and 0.7–0.9 mm long; 2n = 52. Quiet deep water in lakes and streams at 2900 m in Rich (?) and Sevier counties; British Columbia to Labrador, south to California, Wyoming, Indiana, and Pennsylvania; 1 (i).

Potamogeton strictifolius Bennett Plants wholly submerged; stems slender, simple or few-branched, greenish, 1–10 dm long; leaves all submerged, the blade linear, 2–7 cm long, 0.5–2 (2.5) mm wide, firm or rigid,

3-nerved, apically obtuse, often mucronate, tapering and sessile basally, attached directly to the stem, without basal glands; stipules connate into a tubular sheath 0.8–2 cm long, this soon rupturing and deciduous, fibrillose, free from the blade; peduncles filiform, 1–9 cm long; spikes 0.8–1.5 cm long, interrupted, with 3 or 4 remote whorls of flowers, these small, greenish, 1.3–1.8 mm long, the style short; fruit obliquely ovoid, 2–3 mm long, the dorsal keel prominent but low and rounded, the lateral keels obscure; beak curved to almost erect, 0.3–0.7 mm long, positioned on the ventral margin. Lakes and slow moving streams, known in Utah from a single collection (taken in 1869) along the Bear River (Summit County?); eastern Canada and the northeastern U. S., west to Nebraska; 0 (0). Our material has been assigned to var. *rutiloides* Fern.

Potamogeton vaginatus Turcz. Sheathed Pondweed. Plants wholly submerged; stems freely branched, terete, greenish, 2–7 dm long; leaves all submerged, the blade filiform to linear, 2–12 (15) cm long, 1–2 mm wide, mostly 1–nerved, apically rounded or obtuse; stipules adnate to the base of the leaf and forming a closed, clasping sheath around the stem, the sheath and main stem usually brownish and swollen, especially at the base, enclosing several branches, 2–6 cm long, with a short to rather conspicuous ligule; peduncles slender, 3–15 cm long; spikes moniliform (1) 3–8 cm long, with 4–9 rather evenly spaced whorls of flowers to ca 1 cm apart; flowers small, brownish, 1.5–2 mm long, the style lacking; fruit obovoid, 3–3.5 mm long, the dorsal keel low and inconspicuous, the 2 lateral keels lacking; fruiting stigma broad, sessile, persistent on a wartlike projection well to the side of the fruit; 2n = 78. Ponds, pools, streams, and lakes at 1450 to 2730 m in Garfield and Utah counties; Alaska to Newfoundland, south to Oregon, Idaho, Wyoming, Wisconsin, and New Jersey; Eurasia; 2 (0).

Potamogeton zosteriformis Fern. Eelgrass Pondweed. [*P. compressus* authors, not L.]. Plants wholly submerged; stems flattened and somewhat winged, (1) 2–3 mm broad, constricted at the nodes, freely branched, to 6 dm long; leaves all submerged, linear, to 20 cm long, (2) 3–5 mm wide, many-nerved, narrowed slightly at the base, but not at all clasping; stipules free, 1–3 (4) cm long, often whitish, firm, but eventually shredding; peduncles stout, flattened, to 10 cm long; spikes 1–2.5 cm long, the whorls of flowers continuous; achenes obliquely oblong, 3.5–4.5 mm long, with a narrow sharp somewhat undulate to dentate dorsal keel, the lateral keels obscure; beak persistent, slightly curved, ca 0.7 mm long. Usually in lakes at ca 2970 m in Sevier County; British Columbia to California and east to Montana and the northeastern U. S. and Canada; 3 (0).

RUPPIACEAE Hutch.

Ditchgrass Family

Slender submersed branching aquatic perennial herbs; leaves alternate or rarely opposite, with sheathing adnate stipules; flowers perfect, enclosed in sheathing leaf bases, small, typically 2 per terminal spike; perianth lacking; stamens commonly 2, the anthers with a broad connective; pistils 4, the stigma broad and flat; fruit a drupelet; n = 8, 10.

Ruppia L.

Stems and leaves filiform; peduncles elongating and spirally coiled at maturity; flowers with 2 sessile anthers and 4 pistils sessile in flower and elevated on elongated stipes in fruit.

Fernald, M. L. and K. M. Weigand. 1914. The genus *Ruppia* in eastern North America. Rhodora 16: 119–27.

Ruppia maritima L. Ditchgrass. Stems terete or nearly so, to 8 dm long; leaves 2–20 cm long or more, linear to filiform, sessile atop the sheathing adnate stipules, these membranous, commonly 2 to several times broader than the stem, entirely adnate to the leaf base or the tips free for 1–2 mm, enclosing the flowers; flower spikes axillary, the peduncles elongating as the fruits develop and finally 3–20 cm long or more and straight or coiled; drupelets 1.5–3 mm long, obliquely ovoid; 2n = 20. Brackish water of ditches, ponds, and marshes at 1280 to 1525 m in Box Elder, Juab, Millard, Salt Lake, Tooele, Uintah, Utah, and Weber counties; almost cosmopolitan; 15 (0).

SPARGANIACEAE Rudolphi

Bur-reed Family

Perennial aquatic herbs from creeping rhizomes; stems erect or floating, simple or branched; leaves alternate, linear, entire, sheathing at the base; flowers imperfect in dense globular heads on the upper portion of the stem, the staminate above the pistillate, the pistillate subtended by leafy bracts; stamens of the staminate flowers 3 or more, subtended by 3–5 chaffy bracts, the filaments longer than the anthers and bracts; pistil 1, the ovary mostly 1–loculed, with 3–6 short chaffy bracts; fruit nutlike, hardened, 1–seeded (or 2 when 2–loculed); x = 15.

Sparganium L.

A single genus with characaters of the family.

1. Style branched (often breaking off leaving only the beak); stigmas 2; mature achenes truncate apically; inflorescence often compound .... *S. eurycarpum*
— Style not branched; stigma 1; mature achenes gradually tapering into the beak; inflorescence not branched 2

2(1). Staminate heads mostly 1 or 2; beak of achene 1–1.5 mm long (including stigma); anthers at least half as wide as long; leaves mostly 3–6 mm wide .... *S. minimum*
— Staminate heads mostly 2 or more; anthers 2–3 times longer than wide 3

3(2). Leaves mostly less than 5 mm wide, the margins not scarious; achene beak and stigma together 2 mm long or less .... *S. angustifolium*
— Leaves mostly over 5 mm wide, the margins often scarious at the base; achene beak and stigma together 3–4 mm long .... *S. emersum*

Sparganium angustifolium Michx. Narrowleaved Bur-reed. [*S. natans* var. *angustifolium* Pursh; *S. simplex* var. *angustifolium* Torr.]. Herbaceous perennial, 2–4 dm tall; stems erect or floating; leaves linear, 10–80 cm long, 1–8 mm wide, flat but rounded dorsally, dilated below; inflorescence simple, with 2–5 staminate heads and 2–4 pistillate heads; perianth scalelike, clawed, 1–2 mm long; anthers and stigma ca 1 mm long; the stigma solitary; fruit of achenes in globose heads, the beak 2 mm long or less; 2n = 30. Ponds, lake shores, bogs, and potholes, usually in shallow water, at 2135 to 3160 m in Cache, Daggett, Duchesne, Emery, Garfield, Grand, Iron, Summit, Uintah, and Utah counties; over much of North America; Eurasia; 22 (0).

Sparganium emersum Rehmann Emersed Burreed. [*S. simplex* var. *emersum* Asch. & Graebn.; *S. simplex* var. *multipedunculata* Morong; *S. multipedunculatum* Rydb.; *S. emersum* var. *pedunculatum* Reveal]. Perennial herbs with erect stems 2–4 (6) dm tall; leaves linear, mostly 20–50 cm long, 5–12 mm wide, much longer than the inflorescence, the margin scarious below where expanded; inflorescence simple (usually), the pistillate heads (1) 2–4 (5), staminate heads (1) 2–4; perianth scales somewhat broadened at apex and erose, 2–4 mm long; stigma 1; fruit short-stipitate, 1–3 cm broad, 1-seeded, the beak stout, 3–5 mm long; $2n = 30$. Shallow water of ponds, marshy meadows, and lakes at 2135 to 2930 m in Daggett, Davis, Duchesne, Garfield, Grand, Morgan, Sanpete, Sevier, Summit, Uintah, and Wayne counties; over much of North America; Eurasia; 9 (0).

Sparganium eurycarpum Engelm. Erect herbaceous perennials with stout, branching stems, 6–12 dm tall; leaves linear, 3–8 dm long, 7–14 mm wide; inflorescence branched, 20–30 cm long, staminate heads 5–10 (12), pistillate heads 1–2, at maturity 15–25 mm thick; perianth scales 3–7 mm long, expanded apically; anthers 1–1.5 mm long; style branched, the 2 branches 2 mm long; fruits truncate, abruptly narrowed to the beak, this 3–4 mm long (not including the style branches); $2n = 30$. Swamps, bogs, and marshes at 1373 to 1525 m in Box Elder, Cache, Davis, Garfield, Salt Lake, and Utah counties; British Columbia, south to Baja, east to Newfoundland and Florida; 7 (0).

Sparganium minimum Fries Small Burreed. Aquatic herbaceous perennials with weak, emersed, slender stems, 1–3 dm high (longer when floating); leaves flat, floating or erect, mostly 3–6 mm wide, equaling or a little exceeding the inflorescence, this simple, 2–6 cm long, flower heads 2–4, upper one staminate or with both male and female flowers, the lower ones mixed or pistillate, the pistillate 6–12 mm thick; anthers at least half as long as broad, 0.3–0.6 mm long; perianth scales small, ca 1 mm long; styles simple, the stigma up to 1 mm long; fruits ellipsoid, the body 2–3 mm long,, abruptly contracted into a short beak 0.6–1 mm long (including the stigma); $2n = 30$. Shallow ponds and lakes, often in muck, at ca 3000 m in Duchesne, Summit, and Uintah counties; Labrador to Alaska, south to California, Montana, Tennessee, and New Jersey; Eurasia; 4 (0).

TYPHACEAE A. L. Juss.

Cattail Family

Tall herbaceous semiaquatic or aquatic perennials from creeping, thick rhizomes; stems erect, simple, terete, pithy; leaves alternate, long, linear, flat, sheathing; flowers densely crowded in terminal, cylindrical, spikelike inflorescences, unisexual; true perianth lacking; staminate flowers above, sessile, subtended by bristles, the stamens 2–5; pistillate flowers composed of fertile and sterile ones, the fertile with a stipitate ovary and an elongate style and stigma, the sterile on slender stipes, hairy below and clubshaped apically; ovary superior, 1-loculed; fruit dry, tardily dehiscent; $x = 15$.

Typha L.

The family consists of a single genus, with characteristics of the family.

1. Pistillate portion of spike (the lower part) dark brown, velvety, typically contiguous with the staminate (or upper part); pistillate flowers lacking subtending bracts; stigmas dilated apically, dark brown .... *T. latifolia*
— Pistillate portion of the spike pale to cinnamon brown, typically separated by an interval from the staminate spike; pistillate flowers with subtending bracts; stigmas not dilated apically, pale brown 2

2(1). Leaf sheaths auriculate, closed at the throat; spikes green at anthesis; plants known from Cache County .. .. *T. x glauca*
— Leaf sheathes not auriculate, open at the throat; spikes brown at anthesis; plants widely distributed .... *T. domingensis*

Typha domingensis Pers. Common Cattail. [*T. angustifolia* var. *domingensis* (Pers.) Griseb.]. Plants mainly 1.5–4 m tall; leaves 6–9 per stem, equaling the spikes, 3–14 mm wide, plano-convex, pale yellow green; pistillate spikes 7.5–33.5 cm long, 0.5–2.8 cm thick, pale cinnamon brown; bracts brownish; staminate and pistillate spikes typically separated by a sterile portion 1.3–2.5 cm long; stigma brownish, linear; staminate spikes 11–22 cm long, 0.8–1 cm thick, with golden yellow pollen and brownish simple bracts; filaments branched; $2n = 30$. Seeps, springs, canyon bottoms, and wet meadows at 1065 to 1830 m in Box Elder, Cache, Davis, Emery, Garfield, Grand, Juab, Kane, Millard, San Juan, Tooele, Uintah, Utah, Washington, and Wayne counties, and probably in all other counties; Oregon to Oklahoma, south to California, Nevada, Arizona, New Mexico, Texas, Mexico, and South America; 22 (ii).

Typha glauca Gordon (hybrid). Plants typically 1.5–3 m tall; leaves 8–12 per stem, usually exceeding the spikes, dark bluish green, the sheath closed above and auriculate; pistillate spikes 10–25 cm long, 1.8–2.5 cm wide, red brown, the flowers usually lacking bracts, short-pedicellate, the stigmas red brown, linear-lanceolate, staminate spikes pale brown. Marshes and in shallow water, reported from Cache County, near Logan (Cronquist et al. 1977); widely distributed in North America; 0 (0). This hybrid is the result of crosses between *T. latifolia* and *T. domingensis*, and is probably more widely distributed in Utah.

Typha latifolia L. Broad-leaved Cattail. [*Massula latifolia* (L.) Dulac]. Plants mainly 1–3 m tall; leaves 12–16 per stem, equaling the spikes, 7–16 mm wide, nearly flat, green, the sheaths open, cylindrical, not auriculate; pistillate spikes 6–18 cm long, 1.8–3.5 cm wide, dark brown, the flowers lacking bracts, but pedicellate; stigmas medium to dark brown, lance-ovate; interval between the spikes typically lacking; staminate spikes pale brown, lacking bracts; stamens on branched filaments; $2n = 30$. Marshy areas and slow moving water at 1280 to 2105 m in Box Elder, Cache, Daggett, Davis, Emery, Juab, Kane, Millard, San Juan, Salt Lake, Sevier, Tooele, Uintah, Utah, Wasatch, and Wayne counties; throughout North America; Mexico; Eurasia; Africa; 31 (i).

ZANNICHELLIACEAE Dumort.
Horned Pondweed Family

Perennial submersed aquatic monoecious herbs with creeping rhizomes; stems slender, usually branched; leaves opposite or crowded at the nodes, with adnate sheathing and usually ligulate stipules; flowers minute, imperfect, axillary, solitary or cymose; perianth of 3 scales or lacking; stamens 1; pistils 1–9, the ovary 1–loculed, 1–carpelled, the stigmas 1; fruit an achene.

Zannichellia L.

Stems slender; leaves opposite; flowers axillary; staminate flowers with 1 stamen, the filament slender; pistillate flowers sessile or short pedunculate subtended by a short hyaline bract, mostly with 3–5 pistils.

Zannichellia palustris L. Horned Pondweed. Submersed herbs with slender stems to 4 dm long or more; leaves opposite, 2–10 cm long, linear-filiform, 1–veined; staminate flowers with slender filaments; pistillate flowers sessile or short-pedunculate; fruit short-stipitate to subsessile, the body compressed, lunately curved, the keel often denticulate; styles less than half as long as the body. Ponds, streams, and irrigation canals at 1310 to 2595 m in Davis, Millard, Morgan, Salt Lake, San Juan, Sevier, Tooele, Uintah, and Utah counties; widely distributed in North America; Eurasia; Africa; 24 (iii).

AUTHOR ABBREVIATIONS

Abrams Leroy Abrams (1874–1956)
Achey Daisy Bird Achey (b. 1906)
Adams Michael Friedrich Adams (1780–1829/32)
Adanson Michel Adanson (1727–1806)
Aellen Paul Aellen (1896–1973)
Agardh Carl Adolph Agardh (1785–1859)
Agardh f. Jacob Georg Agardh (1813–1901)
Ahles Harry E. Ahles (b. 1924)
Airy-Shaw Herbert Kenneth Airy-Shaw (b. 1902)
Ait. William Townsend Aiton (1766–1849)
Alef. Friedrich George Christoph Alefeld (1820–1872)
All. Carl Allioni (1725–1804)
Allen Caroline Kathryn Allen (b. 1904)
Al-Shebaz Ihsan A. Al-Shebaz
L. C. Anderson Loren C. Anderson (b. 1936)
Anderss. Nils Johan Andersson (1821–1880)
Andrews Henry C. Andrews (1794–1830)
Andrz. Antoni Lukianowicz Andrezejowski (1784–1868)
Angstrom Johan Angstrom (1813–1879)
Arcangeli Giovanni Arcangeli (1840–1921)
Arnold (possibly a pseudonym ca 1785)
Arv.-Touv. Jean Maurice Casimir Arvet-Touvet (1841–1913)
Asch. Paul Friedrich August Ascherson (1834–1913)
Asch. & Graebn. Paul Friedrich August Ascherson (1834–1913) and Karl Otto Robert Peter Paul Graebner (1871–1933)
Asch. & Mag. Paul Friedrich August Ascherson (1834–1913) and Paul Wilhelm Magnus (1844–1914)
Asch. & Schweinf. Paul Friedrich August Ascherson (1834–1913) and Georg August Schweinfurth (1836–1925)
Ashe William Willard Ashe (1872–1932)
Atk. George Francis Atkinson
Atwood Nephi Duane Atwood (b. 1938)
Austin Coe Finch Austin (1831–1880)
Babc. Ernest Brown Babcock (1877–1954)
Babc. & Stebbins Ernest Brown Babcock (1877–1954) and George Ledyard Stebbins Jr. (b. 1906)
Babington Charles Cardale Babington (1808–1895)
Bailey Liberty Hyde Bailey (1858–1954)
D. K. Bailey David Kenneth Bailey (b. 1931)
V. L. Bailey Virginia Edith Bailey (b. 1908)
Baker John Gilbert Baker (1834–1920)
M. Baker Milo Samuel Baker (1868–1961)
Baker & Clausen Milo Samuel Baker (1868–1961) and Jens Christian Clausen (1891–1969)
Balbis Giovanni Battista Balbis (1765–1831)
Balf. John Hutton Balfour (1808–1884)
Ball Carleton Roy Ball (1873–1958)
Banks Joseph Banks (1743–1820)
Banks & Sprengel Joseph Banks (1743–1820) and Kurt Polycarp Joachim Sprengel (1766–1833)
Barbey William Barbey (1842–1914)
Barkley Fred Alexander Barkley (b. 1908)
T. M. Barkley Theodeore Mitchell Barkley (b. 1934)
Barkworth Mary Elizabeth Barkworth (b. 1941)
Barneby Rupert Charles Barneby (b. 1911)
Barneby & Holmgren Rupert Charles Barneby (b. 1911) and Noel Herman Holmgren (b. 1937)
Barratt Joseph Barratt (1796–1882)
Barton William Paul Gillon Barton (1786–1856)
Bartling Friedrich Gottlieb Bartling (1798–1875)
Batch. Frederick William Batchelder (1838–1911)
Batsch August Johann Georg Karl Batsch (1761–1802)
Baumg. Johann Christian Gottlob Baumgarten (1765–1843)
Baxter Edgar Martin Baxter (b. 1903)
Beal William James Beal (1833–1924)
Beaman John H. Beaman (b. 1929)
Beauv. Baron Ambroise Marie Francois Joseph Palisot de Beauvois (1752–1820)
Bebb M. S. Bebb (1833–1895)
Becc. Odoardo Beccari (1843–1920)
Beck Lewis Caleb Beck (1798–1853)
Beckwith Edward Griffin Beckwith (1818–1881)
Beetle Alan Ackerman Beetle (b. 1913)
Behr Hans Hermann Behr (1818–1904)
Beissner Ludwig Beissner (1843–1927)
Bennett Arthur Bennett (1843–1929)
L. Benson Lyman David Benson (b. 1909)
Benson & Walkington Lyman David Benson (b. 1909) and D. L. Walkington (b. 1930)
Benth. George Bentham (1800–1884)
Benth. & Hook. George Bentham (1800–1884) and Joseph Dalton Hooker (1817–1911)
Bercht. Friedrich von Berchtold (1781–1876)
Berger Ernst Friedrich Berger (1814–1853)
Bergeret Jean Pierre Bergeret (1751–1813)
Bernh. Johann Jacob Bernhardi (1774–1850)
Besser Wilbert Swibert Joseph Gottlieb Besser (1784–1842)
Bessey Charles Edwin Bessey (1845–1915)
Betcke Ernst Friedrich (1815–1865)

Beyrich Heinrich Carl Beyrich (1796–1834)
Bicknell Eugene Pintard Bicknell (1859–1925)
Bieb. Friedrich August Marschall von Bieberstein (1768–1826)
Bigelow Jacob Bigelow (1787–1879)
J. Bigelow John Milton Bigelow (1804–1878)
Bjorkman Sven Oscar Bjorkman (1920–1956)
Blake Sidney Faye Blake (1892–1959)
Blank. Joseph William Blankinship (1862–1938)
Bluff & Fingerh. Mathias Joseph Bluff (1805–1837) and Karl Antoine Fingerhuth (1802–1876)
Blume Carl Ludwig von Blume (1796–1862)
Blytt Mathias Numsen Blytt (1789–1862)
Boeck. Johann Otto Boeckeler (1803–1899)
Boiss. Pierre Edmond Boissier (1810–1885)
Boissev. & Davids. Charles Hercules Boissevain (1893–1946) and Carol Davidson
Boissevain Charles Hercules Boissevain (1893–1946)
J. Boivin Joseph Robert Bernard Boivin (b. 1916)
Bolander Henry Nicholas Bolander (1831–1897)
Bong. August Heinrich von Bongard (1786–1839)
Bonpl. Aime Jacques Alexandre (ne Goujaud) Bonpland (1773–1858)
F. Boott Francis Boott (1792–1863)
W. Boott William Boott (1805–1887)
Borbas Vincze von Borbas (1844–1905)
Boreau Alexandre Boreau (1793–1875)
Borkh. Moritz Balthazar Borkhausen (1760–1806)
Borner Carl Julius Bernard Borner [also Boerner] (b. 1880)
Bornm. Joseph Friedrich Nicolaus Bornmuller (1862–1948)
Bory Jean Baptiste Georges (Genevieve) Marcellin Bory de Saint-Vincent (1778–1846)
Botsch. Victor P. Botschantzev (b. 1910)
Bowden Wray Merrill Bowden (b. 1914)
Brack. William Dunlop Brackenridge (1810–1893)
Brainerd Ezra Brainerd (1844–1924)
Brand August Brand (1863–1930)
K. Brandegee Mary Katharine (nee Curran) Brandegee (1844–1920)
Brandegee Townsend Stith Brandegee (1843–1925)
Branner & Coville John Casper Branner (1850–1922) and Frederick Vernon Coville (1867–1937)
A. Br. Alexander Carl Heinrich Braun (1805–1877)
Brayshaw T. C. Brayshaw (publ. 1965)
Breistr. Maurice A. F. Breistroffer (b. 1906/10)
Breitung August J. Breitung (b. 1913)
Brenckle & Cottam Jacob Frederick Brenckle (b. 1875) and Walter Page Cottam (b. 1894)
Brewer William Henry Brewer (1828–1910)
Brewer & Wats. William Henry Brewer (1828–1910) and Sereno Watson (1826–1892)
Briot Pierre Louis Briot (1804–1888)
Briq. John Isaac Briquet (1870–1931)
Britt. Nathaniel Lord Britton (1859–1934)
Britt. & Br. Nathaniel Lord Britton (1859–1934) and Addison Brown (1830–1913)
Britt. & Rose Nathaniel Lord Britton (1859–1934) and Joseph Nelson Rose (1862–1928)
Britt. & Rusby Nathaniel Lord Britton (1859–1934) and Henry Hurd Rusby (1855–1940)
Britt. & Shafer Nathaniel Lord Britton (1859–1934) and John Adolph Shafer (1863–1918)

B.S.P. Nathaniel Lord Britton (1859–1934), Emerson Ellick Sterns (1846–1926) and Justus Ferdinand Poggenburg (1840–1893)
Broome C. Rose Broome (publ. 1976)
R. Br. Robert Brown (1773–1858)
Brummitt Richard Kenneth Brummitt (b. 1937)
Buch.-Ham. Francis Buchanan-Hamilton (1762–1829)
Buch. Franz Georg Phillipp Buchenau (1831–1906)
Buchholz John Theodore Buchholz (1888–1951)
Buckley Samuel Botsford Buckley (1809–1884)
Bunge Alexander Andrejewitsch von Bunge (1803–1890)
Burgsd. Friedrich August Ludwig Burgsdorf (1747–1802)
Burman Johannes Burman (1706–1779)
Burman f. Nicolaas Laurens Burman (1734–1793)
Butters Frederic King Butters (1878–1945)
Butters & Abbe Frederic King Butters (1878–1945) and Ernst Cleveland Abbe (b. 1905)
Butters & St. John Frederic King Butters (1878–1945) and Harold St. John (b. 1892)
Calder & Taylor J. A. Calder (b. 1915) and Raymond Leech Taylor (b. 1901)
Camp Wendell Holmes Camp (1904–1963)
Campbell Douglas Houghton Campbell (1859–1953)
G. R. Campbell Gloria R. Campbell (publ. 1952)
Camus Edmond Gustave Camus (1852–1915)
Canby William Marriot Canby (1831–1904)
Candargy Paleologos C. Candargy (b. 1870)
Carr. Elie Abel Carriere (1818–1896)
Carruth James Harrison Carruth (1807–1896)
Cassidy James Cassidy (1844–1889)
Cassini Alexandre Henri Gabriel Cassini (Comto de) (1781–1850)
Cav. Antonio Jose Cavanilles (1745–1804)
Cesati Vincenzo Cesati (1806–1883)
Chaix Dominique Chaix (1731–1800)
C. & S. Ludolf Karl Adalbert von Chamisso (1781–1838) and Diederich Franz Leonhard von Schlectendal (1794–1866)
Cham. Ludolf Karl Adalbert von Chamisso (1781–1838)
Chatel. Jean Jacques Chatelain (1736–1822)
Chatterley Louis Matthew Chatterley (b. 1951)
Chaudhri Mohammad Nazeer Chaudri (b. 1932)
Chia Liang-Chi Chia (fl. 1955)
Chiov. Emilio Chiovenda (1871–1941)
Choisy Jacques Denys Choisy (1799–1859)
C. Chr. Carl Frederick Albert Christensen (1872–1942)
Chrtek Jindrich Chrtek (b. 1930)
Chuang & Heckard Tsan-Lang Chuang (b. 1933) and Lawrence R. Heckard (b. 1923)
Church George Lyle Church (b. 1905)
Cirillo Domenico Maria Leone Cirillo (also Cyrillo) (1730/39–1799)
Clapham Arthur Roy Clapham (b. 1904)
S. Clark Stephen L. Clark (b. 1940)
C. B. Clarke Charles Baron Clarke (1832–1906)
Clausen Jens Christian Clausen (1891–1969)
R. T. Clausen Robert Theodore Clausen (b. 1911)
Clayton John Clayton (1686–1773)
Clokey Ira Waddell Clokey (1878–1950)
Clover Elzada Urseba Clover (b. 1897)
Clover & Jotter Elzada Urseba Clover (b. 1897) and Mary Lois Jotter (b. 1914)
Clute Willard Nelson Clute (1869–1950)
Cockerell Theodore Dru Alison Cockerell (1866–1948)
Congdon Joseph Whipple Congdon (1834–1910)

Conrad Solomon White Conrad (1779–1831)
Constance Lincoln Constance (b. 1909)
Copel. Edwin Bingham Copeland (1873–1964)
Correll Donovan Stewart Correll (b. 1908)
Cory Victor Louis Cory (b. 1880)
Cosson Ernest Saint-Charles Cosson (1819–1889)
Cottam Walter Page Cottam (b. 1894)
Coult. John Merle Coulter (1851–1928)
Coult. & Nels. John Merle Coulter (1859–1928) and Aven Nelson (1859–1952)
Coult. & Rose John Merle Coulter (1851–1928) and Joseph Nelson Rose (1862–1928)
Coulter Thomas Coulter (1793–1843)
Covas Guillermo Covas (b. 1915)
Cov. Frederick Vernon Coville (1867–1937)
B. Cox Billy J. Cox
Craig Thomas Theodore Craig (b. 1907)
Crantz Heinrich Johann Nepomuk von Crantz (1722–1797)
Crepin Francois Crepin (1830–1903)
Critchf. William B. Critchfield
Cronq. Arthur John Cronquist (b. 1919)
Cronq. & Keck Cronquist, Arthur John (b. 1919) and David Daniels Keck (b. 1903)
Crosswhite Frank Samuel Crosswhite (b. 1940)
Curran Mary Katherine Curran (1844–1920)
Cutler Hugh Carson Cutler (b. 1912)
Czernj. Vassili Matveievich Czernjaew (1796–1871)
Dalle Torre Karl Wiilhelm von Dalle Torre (1850–1928)
Dandy James Edgar Dandy (1903–1976)
Daniels Francis Potter Daniels (1869–1947)
Danser Benedictus Hubertus Danser (1891–1943)
Darby John Darby (1804–1877)
Darlington Josephine Darlington (b. 1905)
Daston J. S. Daston (publ. 1946)
Daubs Edwin Horace Daubs (publ. 1965)
Daveau Jules Alexandre Daveau (1852–1929)
A. Davidson Anstruther Davidson (1860–1932)
R. J. Davis Ray Joseph Davis (b. 1895)
Davy Joseph Burtt Davy (1870–1940)
Day Alva Day (b. 1920)
Dayton William Adams Dayton (1885–1958)
DC. Augustin Pyramus de Candolle (1778–1841)
A. DC. Alphonse Louis Pierre Pyramus de Candolle (1806–1893)
A. & C. DC. Alphonse Louis Pierre Pyramus de Candolle (1806–1893) and Anne Casimir Pyramus de Candolle (1836–1918)
Decne. Joseph Decaisne (1807–1882)
Degl. Jean Vincent Yves Degland (1773–1841)
Delaroche Daniel Delaroche (1743–1812)
Dempster Lauramay Tinsley Dempster (b. 1905)
Dempst. & Ehrend. Lauramay Tinsley Dempster (b. 1905) and Friedrich Ehrendorfer (b. 1927)
Desf. Rene Louiche Desfontaines (1750–1833)
Desmarais Yves Desmarais (publ. 1952)
Desr. Louis Auguste Joseph Desrousseaux (1753–1838)
Desv. Auguste Nicaise Desvaux (1784–1856)
Detl. LeRoy Ellsworth Detling (1909–1967)
Dewey Chester Dewey (1784–1867)
D. Dewey Douglas R. Dewey
Dieck Georg Dieck (1847–1925)
Diels Friedrich Ludwig Emil Diels (1874–1945)
Dietr. Friedrich Gottlieb Dietrich (1768–1850)
A. Dietr. Albert Gottfried Dietrich (1795–1856)

D. Dietr. David Nathaniel Friedrich Dietrich (1800–1888)
Dippel Leopold Dippel (1827–1914)
Dode Louis-Albert Dode (1875–1943)
Dole William Herbert Dole (b. 1869)
Domin Karel Domin (1882–1953)
D. Don David Don (1799–1841)
G. Don Georg Don (1798–1856)
Donn James Donn (1758–1813)
Dorn Robert D. Dorn (publ. 1977)
Dougl. David Douglas (1798–1834)
Drejer Solomon Thomas Nicolai Drejer (1813–1842)
Drew William Brooks Drew (b. 1908)
E. Drew Elmer Reginald Drew (1865–1930)
Druce George Claridge Druce (1850–1932)
Drury W. H. Drury (fl. 1952)
Duchesne Antoine Nicolas Duchesne (1747–1827)
Dugle Janet R. Dugle (publ. 1966)
Duhamel Henri Louis Duhamel du Monceau (1700–1781)
Dulac Joseph Dulac (1827–1897)
Dum-Cours. George Louis Marie Dumont de Courset (1746–1824)
Dumort. Barthelemy Charles Joseph Dumortier (1797–1878)
Dunal Michel Felix Dunal (1780–1856)
Dunn David Baxter Dunn (b. 1917)
Dunn & Harmon David Baxter Dunn (b. 1917) and William E. Harmon
Durand Elias Magloire Durand (1794–1873)
Dur. & Hilg. Elias Magloire Durand (1794–1873) and Theodore Charles Hilgard (1828–1875)
Durazz. Antonio Durazzini
Durazzo Ippolito Durazzo (1750–1818)
Durieu Michel Charles Durieu de Maisonneuve (1796–1878)
Duroi Johann Philipp Duroi (1741–1785)
Dusen Per Karl Hjalmar Dusen (1855–1926)
DuTour DuTour de Salvert (fl. 1803–1815)
Dziek. & Dunn Chester T. Dziekanowski and David Baxter Dunn (b. 1917)
Earle W. Hubert Earle (b. 1906)
Eastw. Alice Eastwood (1859–1953)
A. A. Eaton Alvah Augustus Eaton (1865–1908)
D. C. Eaton Daniel Cady Eaton (1834–1895)
Eckl. & Zeyh. Christian Frederick Ecklon (1795–1868) and Carl Ludwig Philipp Zeyher (1799–1858)
Edwards Sydenham Teast Edwards (1769–1819)
Edwin Gabriel Edwin (b. 1926)
Ehrend. Friedrich Ehrendorfer (b. 1927)
Ehrh. Friedrich Ehrhart (1742–1795)
Eichler Hansjoerg Eichler (b. 1916)
Elkan Louis (Ludwig) Eklan (1815–1851)
Ell. Stephen Elliott (1771–1830)
Ellis John Ellis (1710–1776)
Ellison William L. Ellison (b. 1923)
Elmer Adolph Daniel Edward Elmer (1870–1942)
Emory William Hernsley Emory (1811–1887)
Endl. Stephan Friedrich Ladislaus Endlicher (1804–1849)
Engelm. Georg Engelmann (1809–1884)
Engelm. & Bigel. Georg Engelmann (1809–1884) and John Milton Bigelow (1804–1878)
England Larry John England (publ. 1982)
Engler Heinrich Gustav Adolph Engler (1844–1930)

Engl. & Irm. Heinrich Gustav Adolph Engler (1844–1930) and Edgar Irmscher (1887–1968)
Engler & Prantl Heinrich Gustav Adolph Engler (1844–1930) and Karl Anton Eugen Prantl (1849–1893)
Erskine David S. Erskine (b. 1900)
Eschsch. Johann Friedrich Gustav von Eschscholtz (1793–1831)
Ewan Joseph Andorfer Ewan (b. 1909)
Farw. Oliver Atkins Farwell (1867–1944)
Fassett Norman Carter Fassett (1900–1954)
Fedde Friedrich Karl Georg Fedde (1873–1942)
Feinbrun Naomi Feinbrun (b. 1907)
Fenzl Eduard Fenzl (1808–1879)
A. Ferg. Alexander McGowan Ferguson (b. 1874)
Fern. Merritt Lyndon Fernald (1873–1950)
Ferris Roxana Judkins Ferris (b. 1895)
Fieber Franz Xaver Fieber (1807–1872)
Fiori Adriano Fiori (1865–1950)
Fisch. Friedrich Ernst Ludwig von Fischer (1782–1854)
Fisch. & Mey. Friedrich Ernst Ludwig von Fischer (1782–1854) and Carl Anton Andrievic Meyer (1795–1855)
Fisch. & Trautv. Friedrich Ernst Ludwig von Fischer (1782–1854) and Ernst Rudolph Trautvetter (1809–1889)
Fisch., Mey., & Trautv. Friedrich Ernst Ludwig von Fischer (1782–1854), Carl Anton Andrievic Meyer and Ernst Rudolph Trautvetter (1809–1889)
Fleak Sam Fleak (fl 1971)
Flinders Matthew Flinders (1774–1814)
Florin Carl Rudolf Florin (1894–1965)
Flous M. Fernande Flous (b. 1908)
Flowers Seville Flowers (1900–1968)
Focke Wilhelm Olbers Focke (1834–1922)
Forb. & Hemsl. A. E. E. Forberg (b. 1851) and William Botting Hemsley (1843–1924)
Forbes James Forbes (1773–1861)
Forsskal Pehr (Peter) Forsskal (1732–1763)
Forster A. Forster (1810–1884)
Fosberg Francis Raymond Fosberg (b. 1908)
Foug. Auguste Denis Fougeroux de Bondaroy (1732–1789)
Fourn. Eugene Pierre Fournier (1834–1884)
P. Fourn. Paul Victor Fournier (1877–1964)
Fowler Robert Lawrence Fowler (b. 1910)
Franchet Adrien Rene Franchet (1834–1900)
Franco Joao Manuel Antonio do Amaral Franco (b. 1921)
Franklin John Franklin (1786–1847)
Fraser John Fraser (1750–1811)
Frem. John Charles Fremont (1813–1890)
Fresen. Johann Baptist Georg Wolfgang Fresenius (1808–1866)
Fries Elias Magnus Fries (1794–1878)
Fritsch Karl F. Fritsch (1864–1934)
Gaertner Joseph Gaertner (1732–1791)
Gaertn., Mey. & Scherb. Joseph Gaertner (1732–1791), Bernhard Meyer (1767–1836) and Johannes Scherbius (1769–1813)
Gagnebin Abraham Gagnebin (1707–1800)
Gaill. Charles Gaillardot (1814–1883)
Galloway Leo A. Galloway (publ. 1975)
Gams Helmut Gams (b. 1893)
Gand. Michel Gandoger (1850–1926)
Garcke Christian August Friedrich Garcke (1819–1904)
Garrett Albert Osbun Garrett (1870–1948)

Gates Reginald Ruggles Gates (1882–1962)
Gattinger Augustin Gattinger (1825–1903)
Gaudin Jean Francois Gottlieb Phillippe Gaudin (1766–1833)
C. Gay Claude Gay (1800–1873)
Gentry Howard Scott Gentry (b. 1903)
J. Gentry Johnnie L. Gentry
Geyer Carl Andreas Geyer (1809–1853)
Gilg Ernst Friedrich Gilg (1867–1933)
Gilib. Jean Emmanuel Gilibert (1741–1814)
Gill Jiri Gill (b. 1936)
J. M. Gillett John Montagu Gillett (b. 1918)
Giseke Paul Dietrich Giseke (1745–1796)
Glad Judith B. Glad (publ. 1971)
Gleason Henry Allan Gleason (1882–1975)
C. C. Gmel. Carl Christian Gmelin (1762–1837)
J. F. Gmel. Johann Friedrich Gmelin (1748–1804)
S. G. Gmel. Samuel Gottlieb Gmelin (1745–1774)
Godron Dominique Alexandre Godron (1807–1880)
Goepp. Johann Heinrich Rober Goeppert (1800–1884)
Goldie John Goldie (1793–1886)
Goodding Leslie Newton Goodding (b. 1880)
Goodman George Jones Goodman (b. 1904)
Goodm. & Hitchc. George Jones Goodman (b. 1904) and Charles Leo Hitchcock (b. 1902)
Goodrich Sherel Goodrich (b. 1943)
Gopp. Heinrich Robert Goppert (1800–1884)
Gord. & Glend. George Gordon (1806–1879) and Robert Glendinning (fl. 1844–1858)
Gouan Antoine Gouan (1733–1821)
Gould Frank W. Gould (b. 1913)
Gould & Clark Frank W. Gould (b. 1913) and C. A. Clark (publ. 1978)
Gould & Kapadia Frank W. Gould (b. 1913) and Zarir Kapadia (b. 1935)
Grab. Heinrich Emanuel Grabowski (1792–1842)
Graebner Karl Otto Robert Peter Paul Graebner (1871–1933)
Graf Siegmund Graf (1801–1838)
Graham Robert Graham (1786–1845)
Grant Alva Day Grant (b. 1920)
A. & V. Grant Alva Day Grant (b. 1920) and Verne Edwin Grant (b. 1917)
A. L. Grant Adele Lewis Grant (1881–1969)
V. Grant Verne Edwin Grant (b. 1917)
Gray Asa Gray (1810–1888)
Gray, Wats. & Robins. Asa Gray (1810–1888), Sereno Watson (1826–1892) and Benjamin Lincoln Robinson (1864–1935)
S. F. Gray Samuel Frederick Gray (1766–1828)
Greene Edward Lee Greene (1843–1915)
Greenman Jesse More Greenman (1867–1951)
Greenm. & Roush Jesse More Greenman (1867–1951) and Eva Myrtelle Roush (b. 1886)
Gren. Jean Charles Marie Grenier (1808–1875)
Griffith John William Griffith (1819–1901)
Griffiths David Griffiths (1867–1935)
Griscom Ludlow Griscom (1890–1959)
Griseb. August Heinrich Rudolf Grisebach (1814–1879)
Gueldenst. Anton Johann (von) Gueldenstaedt (1745–1781)
Guerke Robert Louis August Maximilian Guerke (1854–1911)
Guimple Friedrich Guimple (1774–1839)
Hackel Eduard Hackel (1850–1926)
Hadac Emil Hadac (b. 1914)

Haenke Thaddaeus Peregrinus Xaverius Haenke (1761–1817)
Halacsy Eugene von Halacsy (1842–1913)
Hall Harvey Monroe Hall (1874–1932)
H. & C. Harvey Monroe Hall (1874–1932) and Frederic Edward Clements (1874–1945)
Hand.-Mazz. Heinrich von Handel-Mazzetti (1882–1940)
Hanks & Small Lenda Tracy Hanks (1879–1944) and John Kunkel Small (1869–1938)
C. A. Hanson Craig Alfred Hanson (b. 1935)
Harmon William E. Harmon (fl 1968)
Harms Vernon L. Harms
Harrington Harold David Harrington (b. 1903)
Hartman Carl Johan Hartman (1790–1849)
Harv. & Gray William Henry Harvey (1811–1866) and Asa Gray (1810–1888)
Hauman Lucien Hauman (1880–1965)
Hausskn. Heinrich Carl Haussknecht (1838–1903)
Hawksworth & Wiens F. G. Hawksworth (fl. 1964) and Delbert Weins (b. 1935)
Haw. Adrian Hardy Haworth (1768–1833)
Hayden Ferdinand Vandeveer Hayden (1829–1887)
Hayne Friedrich Gottlob Hayne (1763–1832)
Haynes Robert R. Haynes (publ. 1973)
Hegelm. Christoph Friedrich Hegelmaier (1833–1906)
Heil Kenneth D. Heil (b. 1941)
Heimerl Anton Heimerl (1857–1942)
Heiser Charles Bixler Heiser (b. 1920)
Heister Lorenz Heister (1683–1758)
Heller Amos Arthur Heller (1867–1944)
Hemsley William Botting Hemsley (1843–1924)
Henckel Leo Victor Felix Henckel von Donnersmarck (1785–1861)
Henderson Louis Forniquet Henderson (1853–1942)
Henrard Jan Theodoor Henrard (b. 1881)
J. K. Henry Joseph Kaye Henry (1866–1930)
L. Henry Louis H. Henry (1853–1903)
Herbert William Herbert (1778–1847)
Herder Ferdinand Gottfried Theobald Herder (1828–1896)
F. Hermann Frederick Joseph Hermann (b. 1906)
Herrmann Johann Herrmann (1738–1800)
Herter William Gustav Franz Herter (1884–1958)
Hess & Dunn Loyd W. Hess and David Baxter Dunn (b. 1917)
Hieron. Georg Hans Emmo Wolfgang Hieronymus (1846–1921)
Higgins Larry C. Higgins (b. 1936)
Hiit. Henrik Ilmari Augustus Hiitonen (b. 1898)
Hildebr. Friedrich Hermann Gustav Hildebrand (1835–1915)
Hilend Martha Luella Hilend (b. 1902)
Hilend & Howell Martha Luella Hilend (b. 1902) and John Thomas Howell (b. 1903)
Hill John Hill (1716–1775)
A. W. Hill Arthur William Hill (1875–1941)
A. S. Hitchc. Albert Spear Hitchcock (1865–1935)
Hitchc. & Chase Albert Spear Hitchcock (1865–1935) and Mary Agnes Chase (1869–1963)
C. L. Hitchc. Charles Leo Hitchcock (b. 1902)
Hitchc. & Maguire Charles Leo Hitchcock (b. 1902) and Bassett Maguire (b. 1904)
E. Hitchc. Edward Hitchcock (1793–1864)
Hoch Peter C. Hoch (fl 1977)
Hochst. Christian Ferdinand Hochstetter (1787–1860)

Hoffm. George Franz Hoffman (1760–1826)
K. Hoffm. Kathe Hoffmann
Hoffsgg. Johann Centurius von Hoffmannsegg (1766–1849)
H. T. Holm Herman Theodor Holm (1854–1932)
A. & N. Holmgren Arthur Hermann Holmgren (b. 1912) and Noel Herman Holmgren (b. 1937)
Holmgren, Shultz & Lowrey Arthur Hermann Holmgren (b. 1912), Leila M. Shultz (b. 1946) and Timothy K. Lowrey
N. Holmgren Noel Herman Holmgren (b. 1937)
N. & P. Holmgren Noel Herman Holmgren (b. 1937) and Patricia (nee Kern) Holmgren (b. 1940)
P. Holmgren Patricia (nee Kern) Holmgren (b. 1940)
Holub Josef Holub (b. 1930)
Holz. John Michael Holzinger (1853–1929)
Honck. Gerhard August Honckeny (1724–1805)
Hook. f. Joseph Dalton Hooker (1817–1911)
Hook. & Baker Joseph Dalton Hooker (1817–1911) and John Gilbert Baker (1834–1920)
Hook. William Jackson Hooker (1785–1865)
H. & A. William Jackson Hooker (1785–1865) and George Arnott Walker Arnott (1799–1868)
Hook. & Grev. William Jackson Hooker (1785–1865) and Robert Kaye Greville (1794–1866)
Hopkins Milton Hopkins (b. 1906)
Hoppe David Heinrich Hoppe (1760–1846)
Horan. Paul Fedorowitsch Horaninow (1796–1865)
Horkel Johann Horkel (1769–1846)
Hornem. Jens Wilken Hornemann (1770–1841)
Horton James H. Horton (b. 1931)
Hort. Hortulanorum, means of the gardeners.
Host Nicolaus Thomas Host (1761–1834)
House Homer Doliver House (1878–1949)
J. T. Howell John Thomas Howell (b. 1903)
Howell Thomas Jefferson Howell (1842–1912)
Hu & Cheng Shiu-ying Hu (b. 1910) and Ching-yung Joyce Cheng (b. 1919)
C. E. Hubbard Charles Edward Hubbard (b. 1900)
Hubb. Frederic Tracy Hubbard (1875–1962)
Hudson William Hudson (1730–1793)
Hulten Oskar Eric Gunnar Hulten (1894–1981)
H.B.K. Friedrich Wilhelm Heinrich Alexander von Humboldt (1769–1859), Aime Jacques Alexandre Bonpland (1773–1858) and Karl Sigismund Kunth (1788–1850)
Humb. Friedrich Wilhelm Heinrich Alexander von Humboldt von (1769–1859)
Hutch. John Hutchinson (1884–1972)
Huth Ernest Huth (1845–1897)
Hylander Nils Hylander (1904–1970)
Iltis Hugh Hellmut Iltis (b. 1925)
Irmscher Edgar Irmscher (1887–1968)
W. Irv. Walter Irving (1867–1934)
Isely Duane Isely (b. 1918)
Ives Joseph Christmas Ives (1828–1868)
A. B. Jackson Albert Bruce Jackson (1876–1947)
R. Jackson Raymond C. Jackson (b. 1928)
Jacq. Nicolaus Jacquin (1727–1817)
Jaeger Hermann Jaeger (1815–1890)
James Edwin James (1797–1861)
Jancz. Eduard Janczewski von Glinka (1846–1918)
Jarm. A. V. Jarmolenko (1905–1944)
Jaub. Hyppolyte Francois Jaubert (1798–1874)
Jepson Willis Linn Jepson (1867–1946)

Jepson & Bailey Willis Linn Jepson (1867–1946) and Liberty Hyde Bailey (1858–1954)
B. L. Johnson B. Lennart Johnson (b. 1909)
Johnston Ivan Murray Johnston (1898–1960)
J. R. Johnston John Robert Johnston (b. 1880)
M. C. Johnston Marshall Conring Johnston (b. 1930)
Jones Marcus Eugene Jones (1852–1934)
G. Jones George Neville Jones (1904–1970)
Q. Jones Quentin Jones (b. 1920)
Jonsel Bengt Edvard Jonsel (b. 1936)
Jordan Alexis Jordan (1814–1897)
Juss. Christophe de Jussieu (1685–1758)
A. L. Juss. Antoine Laurent de Jussieu 1748–1836
Juz. Sergei Vasilievic Juzepczak (1893–1959)
Kapoor Brij Mohan Kapoor (b. 1936)
Kar. & Kir. Grigorij Siliyc Karelin (1801–1872) and Ivan Petrovic Kirilow (1821 or 1822–1842)
Karsten Carl Wilhelm Gustav Hermann Karsten (1817–1908)
Kass Ronald J. Kass
Kaulf. Georg Friedrich Kaulfuss (1786–1830)
Kearney Thomas Henry Kearney (1874–1956)
Kearney & Peebles Thomas Henry Kearney (1874–1956) and Robert Hibbs Peebles (1900–1955)
Keck David Daniels Keck (b. 1903)
Kellogg Albert Kellogg (1813–1887)
Kelso Leon Hugh Kelso (b. 1907)
H. Keng Hsuan Keng (b. 1923)
Kenn. Patrick Beveridge Kennedy (1874–1930)
Kenney Patricia Kenney (fl 1977)
Ker John Bellenden Ker (previously John Gawler) (1764–1842)
Kerner Anton Joseph Kerner von Marilaun (1831–1898)
King George King (1840–1909)
Kiob. (publ. ca 1770)
Kit. Paul Kitaibel (1757–1817)
Kittell Marie Teresa Kittell (b. 1892)
W. Klein William Klein (publ. 1962)
Klotzsch Johann Friedrich Klotzsch (1805–1860)
Knerr Ellsworth Brownell Knerr (1861–1942)
Knuth Reinhard Gustav Paul Knuth (1874–1957)
K. Koch Karl Heinrich Emil Ludwig Koch (1809–1879)
Koch Wilhelm Daniel Joseph Koch (1771–1849)
Koehne Bernard Adalbert Emil Koehne (1848–1918)
Koeler George Ludwig Koeler (1765–1807)
Koenig Carl Dietrich Eberhard Koenig (1774–1851)
Koen. & Sims Carl Dietrich Eberhard Koenig (1774–1851) and John Sims (1740–1831)
Koern. Friedrich August Koernicke (1828–1908)
Koidz. Genichi Koidzumi (1883–1953)
Komarov Vladimir Leontjevic Komarov (1869–1945)
Kozo-Polj. Boris Mikhailovic Kozo-Poljansky (1890–1957)
Krap. Antonio Krapovickas (publ. 1970)
Krause Ernst Hans Ludwig Krause (1859–1942)
Kuhn Friedrich Adalbert Maximilian Kuhn (1842–1894)
Kukenthal Georg Kukenthal (1864–1955)
Kunth Karl Sigismund Kunth (1788–1850)
Kuntze Carl Ernst Otto Kuntze (1843–1907)
Kunze Gustav Kunze (1793–1851)
Kurz Wilhelm Sulpiz Kurz (1834–1878)
L'Her. Charles Louis de Brutelle L'Heritier (1746–1800)
Lag. Mariano Lagasca y Segura (1776–1839)

Lag. & Rodr. Mariano Lagasca y Segura (1776–1839) and Jose Demetris Rodriguez (1780–1846)
Lakela Olga Korhoven Lakela (b. 1890)
Lam. Jean Baptiste Antoine Pierre de Monnet de Lamarck (1744–1829)
Lam. & DC. Jean Baptiste Antoine Pierre de Monnet de Lamarck (1744–1829) and Augustin Pyramus de Candolle (1778–1841)
Lam. & Poir. Jean Baptiste Antoine Pierre de Monnet de Lamarck (1744–1829) and Jean Louise Marie Poiret (1755–1834)
Lambert Aylmer Bourke Lambert (1761–1842)
Lange Johan Martin Christian Lange (1818–1898)
Lapeyr. Philippe Picot de Lapeyrouse (1744–1818)
Lauche Friedrich Wilhelm Georg Lauche (1827–1882)
Lawson Charles Lawson (1794–1873)
Laxmann Erik G. Laxmann (1737–1796)
Ledeb. Carl Friedrich von Ledebour (1785–1851)
Lehm. Johann Georg Christian Lehmann (1792–1860)
Leiberg John Bernhard Leiberg (1853–1913)
Lej. Alexandre Louis Simon Lejeune (1779–1858)
Lellinger David Bruce Lellinger (b. 1937)
Lem. Charles Antoine Lemaire (1801–1871)
Lemmon John Gill Lemmon (1832–1908)
Lepage Ernest Lepage (b. 1905)
Less. Christian Friedrich Lessing (1809–1862)
H. Levl. Augustin Abel Hector Leveille (1863–1918)
Lewis & Szweykowski Frank Harlan Lewis (b. 1919) and Jerzy Szweykowski (b. 1925)
Leysser Friedrich Wilhelm von Leysser (1731–1815)
Light. Lightfoot, John (1735–1788)
Lilj. Samuel Liljeblad (1761–1815)
L. Linden Lucien Linden (1851–1940)
Lindl. John Lindley (1799–1865)
Lindl. & Gord. John Lindley (1799–1815) and George Gordon (1806–1879)
Lindl. & Paxt. John Lindley (1799–1865) and Joseph Paxton (1803–1865)
Lindsay George Edmund Lindsay (b. 1916)
Link Johann Heinrich Friedrich Link (1767–1851)
L. Carl Linnaeus (1707–1778)
L. f. Carl von Linne (fil.) (1741–1783)
Little Elbert Luther Little Jr. (b. 1907)
Litv. Dmitrij Ivanovitsch Litvinov (1854–1929)
Lodd. Conrad Loddiges (1738–1826)
Lois. Jean Louis August Loiseleur-Deslongchamps (1774–1849)
Long Stephen Harriman Long (1784–1864)
Loudon John Claudius Loudon (1783–1843)
Lour. Joao de Loureiro (1717–1791)
Love Askell Love (b. 1916)
Love & Love Askell Love (b. 1916) and Doris Benta Maria (nee Wahlen) Love (b. 1918)
Lund Peter Wilhelm Lund (1801–1880)
Lunell Joel Lunell (1850–1920)
Ma Yu-Chuan Ma (b. 1916)
McArthur E. Durant McArthur (b. 1941)
Macbr. James Francis Macbride (1892–1976)
Macbr. & Payson James Francis Macbride (1892–1976) and Edwin Blake Payson (1893–1927)
Macbryde Bruce Macbryde (b. 1941)
McClatchie Alfred James McClatchie (b. 1906)
McClelland John McClelland (1805–1883)
McClintock & Epling Elizabeth May McClintock (b. 1912) and Carl Clawson Epling (1894–1968)

McDermott Laura Frances McDermott (fl 1908)
Macfady. James Macfadyen (1798–1850)
MacGregor Donald MacGregor (1877–1933)
McGregor Ronald Lighton McGregor (b. 1919)
Mack. Kenneth Kent Mackenzie (1877–1934)
McKelvey Susan Delano McKelvey (b. 1883)
Mack. & Bush Kenneth Kent Mackenzie (1877–1934) and Benjamin Franklin Bush (1858–1937)
Macmillan Conway Macmillan (1867–1929)
McMinn Howard Earnest McMinn (1891–1963)
Macoun John Macoun (1831–1920)
McVaugh Rogers McVaugh (b. 1909)
Maguire Bassett Maguire (b. 1904)
Maguire & Cronq. Bassett Maguire (b. 1904) and Arthur John Cronquist (b. 1919)
Maguire & Holmgren Bassett Maguire (b. 1904) and Arthur Herman Holmgren (b. 1912)
Maguire & Woodson Bassett Maguire (b. 1904) and Robert Everard Woodson (1904–1963)
Maire Rene Charles Joseph Ernest Maire (1878–1949)
Makino Tomitaro Makino (1862–1957)
Malte Oscar Malte (1880–1933)
Manetti Giuseppe Manetti (1831–1858)
Mansfeld Rudolf Mansfeld (1901–1960)
Marcy Randolph Barnes Marcy (1812–1887)
Marshall Humphry Marshall (1722–1801)
W. T. Marshall William Taylor Marshall (1886–1957)
J. Martin James Stillman Martin (b. 1914)
Martin Robert F. Martin (b. 1910)
Martius Carl Friedrich Phillipp von Martius (1794–1868)
Mason Herbert Louis Mason (b. 1896)
Mathias Mildred Esther Mathias (b. 1906)
Math. & Const. Mildred Esther Mathias (b. 1906) and Lincoln Constance (b. 1909)
Mattf. Johannes Mattfeld (1895–1951)
Maxim. Carl Johann Maximowicz (1827–1891)
Maxon William Ralph Maxon (1877–1948)
Mayr Heinrich Mayr (1856–1911)
Medicus Friedrich Casimir Medicus (Medikus) (1736–1808)
L. Medicus Ludwig Wallrad Medicus (1771–1850)
Meissner Carl Daniel Friedrich Meissner (1800–1874)
Meld. Alexander Melderis (b. 1909)
Merriam Clinton Hart Merriam (1855–1942)
Merr. Elmer Drew Merrill (1876–1956)
Mett. Georg Heinrich Mettenius (1823–1866)
Mey. & Scherb. Bernhard Meyer (1767–1836) and Johannes Scherbius (1769–1813)
C. A. Mey. Carl Anton Andreevic von Meyer (1795–1855)
E. Mey. Ernst Heinrich Friedrich Meyer (1791–1858)
F. G. Meyer Frederic Gustav Meyer (b. 1917)
Mez Carl Christian Mez (1866–1944)
Michx. Andre Michaux (1746–1802)
Michx. f. Francois Andre Michaux (fil.) (1770–1855)
Miers John Miers (1789–1879)
Miki Sigeru Miki (1901–1974)
Milde Carl August Julius Milde (1824–1871)
Miller Phillip Miller (1691–1771)
G. N. Miller C. N. Miller
Milliken Jessie Milliken (b. 1887)
Millsp. Charles Frederick Millspaugh (1854–1923)

Miq. Friedrich Anton Wilhelm Miquel (1811–1871)
Mirbel Charles Francois Brisseau de Mirbel (1776–1854)
Mitchell John Mitchell (1676–1768)
Moench Conrad Moench (1744–1805)
Mold. Harold Norman Moldenke (b. 1909)
Molina Giovanni Ignazio Molina (1740–1829)
Moore Thomas Moore (1821–1887)
G. Moore Glen Moore (b. 1917)
H. Moore Harold Emery Moore (b. 1917)
Moq. Christian Horace Benedict Alfred Moquin-Tandon (1804–1863)
Moretti Giuseppe Moretti (1782–1853)
Morong Thomas Morong (1827–1894)
Morot Louis Rene Marie Francois Morot (1854–1915)
Morris Edward Lyman Morris (1870–1913)
Morton Conrad Vernon Morton (1905–1972)
Mosher Edna Mosher (publ. 1915)
Mosquin Theodore Mosquin (b. 1932)
C. Muell. Johann Karl (Carl) August Friedrich Wilhelm Mueller (1817–1899)
Muell.-Arg. Jean Mueller-Argoviensis (1828–1896)
Muenchh. Otto von Muenchhausen (1716–1774)
Muhl. Gotthilf Henry Ernest Muhlenberg (1753–1815)
Mulligan Brian O. Mulligan (b. 1907)
Munro William Munro (1818–1889)
Munz Philip Alexander Munz (1892–1974)
Munz & Klein Philip Alexander Munz (1892–1974) and William McKinley Klein (b. 1933)
Murray John Andreas Murray (1740–1791)
E. Murray Edward Murray
Mutis Jose Celestino Mutis (1732–1808)
Nakai Takenoshin Nakai (1882–1952)
Nash George Valentine Nash (1864–1921)
Necker Noel Martin Joseph de Necker (1729–1793)
Nees Christian Gottfried Daniel Nees von Esenbeck (1776–1858)
Nees & Mey. Christian Gottfried Daniel Nees von Esenbeck (1776–1858) and Franz Julius Ferdinand Meyen (1804–1840)
Neese Elizabeth Janet (nee Chase) Neese (b. 1934)
Neese & Welsh Elizabeth Janet Neese (b. 1934) and Stanley Larson Welsh (b. 1928)
A. Nels. Aven Nelson (1859–1952)
Nels. & Kennedy Aven Nelson (1859–1952) and Patrick Beveridge Kennedy (1874–1930)
Nels. & Macbr. Aven Nelson (1859–1952) and James Francis Macbride (1892–1976)
E. Nels. Elias Emanuel Nelson (1876–1949)
Nesom Guy L. Nesom (publ. 1976)
Neuwied Maximilian Alexander Philipp Wied-Neuwied (1782–1867)
Nevski Sergei Arsenjevic Nevski (1908–1938)
Newberry John Strong Newberry (1822–1892)
Newton A. Newton (fl. 1971)
Nicollet Jean Nicholas Nicollet (1786–1843)
Nieuwl. Julius Arthur Nieuwland (1878–1936)
Nisbet Gladys T. Nisbet (publ. 1960)
Northstrom & Welsh Terry Edward Northstrom (b. 1945) and Stanley Larson Welsh (b. 1928)
J. B. S. Norton John Bitting Smith Norton (1872–1966)
Nutt. Thomas Nuttall (1786–1859)
Nyman Carl Fredrik Nyman (1820–1893)
O'Neill Hugh Thomas O'Neill (1894–1969)
Oakes William Oakes (1799–1848)

Oeder Georg Christian von Oeder (1728–1791)
Olney Stephen Thayer Olney (1812–1878)
Opiz Philipp Maximilian Opiz (1787–1858)
Ortega Casimiro Gomez Ortega (1740–1818)
Ortgies Karl Eduard Ortgies (1829–1916)
Ostenf. Carl Emil Hansen Ostenfeld (1873–1931)
Osterh. George Everett Osterhout (1858–1937)
Ottley Alice Maria Ottley (b. 1882)
Otto Christoph Friedrich Otto (1783–1856)
Ownbey Francis Marion Ownbey (b. 1910)
G. B. Ownbey Gerald Bruce Ownbey (b. 1916)
Paine John Alsop Paine (1840–1912)
Pallas Peter Simon von Pallas (1741–1811)
Pamp. Renato Pampanini (1875–1949)
Paoli Guido Paoli (1881–1947)
Parish Samuel Bonsall Parish (1838–1928)
Parke John Grubb Parke (publ. 1855)
Parker Kitty Lucille Parker (b. 1910)
Parl. Filippo Parlatore (1816–1877)
Parnell Richard Parnell (1810–1882)
Parodi Lorenzo Raimundo Parodi (1895–1966)
Parry Charles Christopher Parry (1823–1890)
Pasq. Giuseppe Antonio Pasquale (1820–1893)
Patze Carl August Patze (1808–1892)
Pax Ferdinand Albin Pax (1858–1942)
Pax & K. Hoffm. Ferdinand Albin Pax (1858–1942) and Karl August Otto Hoffmann (1853–1909)
Paxton Joseph Paxton (1801–1865)
Payne Willard William Payne (b. 1934)
Payson Edwin Blake Payson (1893–1927)
Pease & Moore Arthur Stanley Pease (1881–1964) and Albert Hanford Moore (b. 1883)
Peck Morton Eaton Peck (1871–1959)
Peebles Robert Hibbs Peebles (1900–1956)
Penl. Charles William Theodore Penland (b. 1899)
Pennell Francis Whittier Pennell (1886–1952)
Perry Matthew Calbraith Perry (1794–1858)
Pers. Christiaan Hendrick Persoon (1761–1836)
Petit Felix Petit
Petrak Franz Petrak (1886–1973)
Phil. Rudolf Amandus Philippi (1808–1904)
Phillips Lyle L. Phillips (b. 1923)
Pilger Robert Knud Friedrich Pilger (1876–1953)
Piller & Mitterp. Mathias Piller (1733–1788) and Ludwig Mitterpacher von Mitterburg (1734–1814)
Pilz G. E. Pilz (publ. 1978)
Piper Charles Vancouver Piper (1867–1926)
Piper & Beattie Charles Vancouver Piper (1867–1926) and Rolla Kent Beattie (1875–1960).
Planchon Jules Emile Planchon (1823–1888)
Plenck Joseph Jacob von Plenck (1738–1807)
Podp. Josef Podpera (1878–1954)
Poir. Jean Louis Marie Poiret (1755–1834)
Polunin Nicholas (Vladimir) Polunin (b. 1909)
Porsild Alf Erling Porsild (b. 1901)
C. L. Porter Charles Lyman Porter (b. 1889)
T. C. Porter Thomas Conrad Porter (1822–1901)
Port. & Coult. Thomas Conrad Porter (1822–1901) and John Merle Coulter (1859–1928)
D. Post Douglas Manners Post (b. 1920)
Prain David Prain (1857–1944)
Prantl Karl Anton Eugen Prantl (1849–1893)
Presl Carel Boriwag Presl (1794–1852)
J. & C. Presl Jan Swatopluk Presl (1791–1849) and Carel Boriwag Presl (1794–1852)
Pringle James S. Pringle (b. 1937)
Pritz. Georg August Pritzel (1815–1874)
Purpus Joseph Anton Purpus (1860–1932)
Pursh Frederick Traugott Pursh (1774–1820)
Rabenh. Gottlob Ludwig Rabenhorst (1806–1881)
Raf. Constantine Samuel Rafinesque-Schmaltz (1783–1840)
Rafn Carl Gotlieb Rafn (1769–1808)
Ramaley Francis Ramaley (1870–1942)
Rattan Volney Rattan (1840–1915)
Raup Hugh Miller Raup (b. 1901)
Rauschert Stephen Rauschert (b. 1931)
Raven Peter Hamilton Raven (b. 1936)
Rech. Karl Rechinger (1867–1952)
Rech. f. Karl Heinz Rechinger (b. 1906)
Red. Pierre Joseph Redoute (1761–1840)
C. Reeder Charlotte Olive (nee Goodding) Reeder (b. 1916)
Rees Abraham Rees (1743–1825)
Regel Eduard August von Regel (1815–1892)
Rehder Alfred Rehder (1863–1949)
Rehmann Anton Rehmann (1840–1917)
Reichenb. Heinrich Gottlieb Ludwig Reichenbach (1793–1879)
Rein. Joseph Reiner (1765–1797)
Retz. Anders Jahan Retzius (1742–1821)
Reuter Georges Francois Reuter (1805–1872)
Reveal James Lauritz Reveal (b. 1941)
Reveal & Brotherson James Lauritz Reveal (b. 1941) and Jack D. Brotherson (b. 1938)
Reveal, Broome, & Beatley James Lauritz Reveal (b. 1941), C. Rose Broome and Janice C. Beatley
A. Rich. Achille Richard (1794–1852)
Rich. Louis Claude Marie Richard (1754–1821)
A. Richards. Alfred Richardson (publ. 1976)
Richards. John Richardson (1787–1865)
Richter Karl Richter (1855–1891)
Ricker Percy Leroy Ricker (1878–1973)
Riddell John Leonard Riddell (1807–1865)
Robb. James Watson Robbins (1801–1879)
Robins. Benjamin Lincoln Robinson (1864–1935)
Robins. & Fern. Benjamin Lincoln Robinson (1864–1935) and Merritt Lyndon Fernald (1873–1950)
Robins. & Greenm. Benjamin Lincoln Robinson (1864–1935) and Jesse More Greenman (1867–1951)
Roehl. Johann Christoph Rohling (Roehling) (1757–1813)
Roemer Johann Jacob Roemer (1763–1819)
R. & S. Johann Jacob Roemer (1763–1819) and Josef August Schultes (1773–1831)
Roezl Benito Roezl (1824–1885)
Rogler George A. Rogler (publ. 1960s)
Rohrb. Paul Rohrbach (1847–1871)
Rollins Reed Clark Rollins (b. 1911)
Rollins & Shaw Reed Clark Rollins (b. 1911) and Elizabeth Anne Shaw (b. 1938)
Rose Joseph Nelson Rose (1862–1928)
Rose & Painter Joseph Nelson Rose (1862–1928) and Joseph Hannum Painter (1879–1908)
Rosend. Carl Otto Rosendal (1875–1956)

Rosend., Butters, & Lakela Carl Otto Rosendal (1875–1956), Frederic King Butters (1878–1945) and Olga Korhoven Lakela (b. 1890)
Rostkov Friedrich Wilhelm Gotlieb Rostkov (1770–1848)
Roth Albrecht Wilhelm Roth (1757–1834)
Rothm. Werner Hugo Paul Rothmaler (1908–1962)
Rothr. Joseph Trimble Rothrock (1839–1922)
Rottb. Christen Friis Rottboell (1727–1797)
Rousseau Jean-Jacques Rousseau (1712–1778)
Rouy Georges C. Chr. Rouy (1851–1924)
Rowley Gordon D. Rowley (b. 1921)
Roxb. William Roxburgh (1751–1815)
Royle John Forbes Royle (1779–1858)
Rudd Velva E. Rudd (b. 1910)
Rudolphi Karl Asmund Rudolphi (1771–1832)
Ruiz & Pavon Hipolito Ruiz (1754–1815) and Jose Antonio Pavon (1750–1844)
Rumpler Theodor Rumpler (1817–1891)
Runem. Hans Runemark (b. 1927)
Rupr. Franz Josef Ruprecht (1814–1870)
Rydb. Per Axel Rydberg (1860–1931)
St. John Harold St. John (b. 1892)
J. St. Hil. Jean Henri Jaume Saint-Hilaire (1840–1912)
St.-Yves Alfred Saint-Yves (1855–1933)
Salisb. Richard Anthony Salisbury (1761–1829)
Sarg. Charles Sprague Sargent (1841–1927)
Sarnth. Ludwig von Sarnthein (1861–1914)
Sauer Jonathan Sauer (b. 1918)
Savi C. Gaetano Savi (1769–1844)
Schaffner Wilhelm Schaffner (1830–1882)
J. H. Schaffn. John Henry Schaffner (1866–1939)
Schauer Johann Conrad Schauer (1813–1848)
Scheele George Heinrich Adolf Scheele (1808–1864)
Scheutz Nils Johan Wilhelm Scheutz (1836–1889)
Schinz & R. Keller Hans Schinz (1858–1941) and Robert Keller (1854–1939)
Schinz & Thell. Hans Schinz (1858–1941) and Albert Thellung (1881–1928)
Schischkin Boris Konstantinovich Schischkin (1886–1963)
Schkuhr Christian Schkuhr (1741–1811)
Schlecht. Diederich Franz Leonhard von Schlechtendal (1794–1866)
Schleicher Johann Christoph Schleicher (1768–1834)
Schleiden Matthias Jacob Schleiden (1804–1881)
Schmidt Franz Schmidt (1751–1834)
F. W. Schmidt Franz Wilibald Schmidt (1764–1796)
Schmidel Casimir Christoph Schmidel (1718–1792)
Schneider Camillo Karl Schneider (1876–1951)
Schnitzl. Adalbert Carl Friedrich Schnitzlein (1814–1868)
Schoener Carol Susan Schoener (b. 1946)
Schott Heinrich Wilhelm Schott (1794–1865)
Schrader Heinrich Adolph Schrader (1767–1836)
Schrank Franz von Paula von Schrank (1747–1835)
Schreber Johann Christian Daniel von Schreber (1739–1810)
B. Schreiber Beryl Olive Schreiber (b. 1911)
Schrenk Alexander Gustav von Schrenk (1816–1876)
Schultes Josef August Schultes (1773–1831)
Schult. & Schult. Josef August Schultes (1773–1831) and Julius Hermann Schultes (1804–1840)
Schultz-Bip. Carl Heinrich Schultz-Bipontinus (1805–1867)
Schultz-Schultzens. Karl Heinrich Schultz-Schultzenstein (1798–1881)
Schulz Otto Eugen Schulz (1874–1936)
K. Schum. Karl Moritz Schumann (1851–1904)
Schur Philipp Johann Ferdinand Schur (1799–1878)
Schweigger August Friedrich Schweigger (1783–1821)
Schweigg. & Koerte August Friedrich Schweigger (1783–1821) and Franz Koerte (1782–1845)
Schwein. Ludwig David von Schweinitz (1780–1834)
Scop. Giovanni Antonio Scopoli (1723–1788)
Scribn. Frank Lamson Scribner (1851–1938)
Scribn. & Merr. Frank Lamson Scribner (1851–1938) and Elmer Drew Merrill (1876–1956)
Scribn. & Sm. Frank Lamson Scribner (1851–1938) and Jared Gage Smith (1866–1925)
Scribn. & Will. Frank Lamson Scribner (1851–1938) and Thomas Albert Williams (1865–1900)
Seemann Berthold Carl Seeman (1825–1871)
Selander Nils Sten Edward Selander (1891–1957)
Sellow Friedrich Sellow (1789–1831)
Sendt. Otto Sendtner (1813–1859)
Sennen & Pau Frere E. C. Sennen (1861–1937) and Carlos Pau y Espanol (1857–1937)
Ser. Nicolas Charles Seringe (1776–1858)
Seub. Moritz August Seubert (1818–1878)
W. Sharp Ward McClintic Sharp (b. 1904)
E. Shaw Elizabeth A. Shaw (b. 1938)
Shear Cornelius Lott Shear (1865–1956)
Sheldon Edmund Perry Sheldon (b. 1869)
Sherff Earl Edward Sherff (1886–1966)
Shev. Charles J. Sheviak
Shinn. Lloyd Herbert Shinners (1918–1971)
L. Shultz Leila M. Shultz (b. 1946)
J. Shultz John Shultz (b. 1943)
Sibth. John Sibthorp (1758–1796)
Sieb. & Zucc. Philipp Franz Siebold (1796–1866) and Joseph Gerhard Zuccarini (1797–1848)
Silliman Benjamin Silliman (1779–1864)
Silva Tarouca Ernst (Graf) Silva Tarouca (1860–1936)
Silveus William Arento Silvens (b. 1875)
Sitgr. Lorenzo Sitgreaves (d. 1888)
Slosson Margaret Slosson (b. 1872)
Small John Kunkel Small (1869–1938)
Small & Cronq. John Kunkel Small (1869–1938) and Arthur John Cronquist (b. 1919)
Smiley Frank Jason Smiley (b. 1880)
C. P. Sm. Charles Piper Smith (1877–1955)
F. G. Sm. Frank G. Smith
J. E. Sm. James Edward Smith (1759–1828)
J. G. Sm. Jared Gage Smith (1866–1925)
Smyth Bernard Bryan Smyth (1843–1913)
Snowden Joseph Davenport Snowden (1886–1973)
Sobol. Gregorius Fedorovitch Sobolevski (1741–1807)
Sol. Daniel Carl Solander (1733–1782)
Solbrig Otto Thomas Solbrig (b. 1930)
Soul.-Bod. Etienne Soulange-Bodin (1774–1846)
Spach Edouard Spach (1801–1879)
Spegaz. Carlo Luigi Spegazzini (1858–1926)
Sprengel Kurt Polycarp Joachim Sprengel (1766–1833)
Stacey John William Stacey (1871–1943)
Standley Paul Carpenter Standley (1884–1963)
Stansb. Howard Stansbury (1806–1863)
Stapf Otto Stapf (1857–1933)
Staudt Gunther Staudt (publ. 1961)
Stebbins George Ledyard Stebbins (b. 1906)
Sternb. Caspar Maria (Graf) von Sternberg (1761–1838)

Steudel Ernst Gottlieb von Steudel (1783–1856)
Steven Christian von Steven (1781–1863)
Stewart M. G. Stewart (publ. 1934)
Steyerm. Julien Alfred Steyermark (b. 1909)
Stockhouse Stockhouse, W. (fl 1984)
Stockwell William Palmer Stockwell (1898–1950)
Stokes Susan Gabriella Stokes (1868–1954)
Strother John Lance Strother (b. 1941)
Stuckey Ronald Lewis Stuckey (b. 1938)
Sturm Jakob W. Sturm (1771–1848)
Stutz Howard Stutz (b. 1918)
Sudw. George Bishop Sudworth (1864–1927)
Suksd. Wilhelm Nikolaus Suksdorf (1850–1932)
Svenson Henry Knute Svenson (b. 1897)
Swallen Jason Richard Swallen (b. 1903)
Swartz Olof Peter Swartz (1760–1818)
Sweet Robert Sweet (1783–1835)
Swezey Godwin Deloss Swezey (1851–1934)
Sydow Hans Sydow (1879–1946)
Syme John Thomas Irvine Boswell Syme (1822–1888)
Tateoka Tuguo Tateoka (b. 1931)
Taub. Paul Hermann Wilhelm Taubert (1862–1897)
Tausch Ignaz Friedrich Tausch (1793–1848)
Taylor Thomas Taylor (1775–1848)
Ten. Michele Tenore (1780–1861)
Thell. Albert Thellung (1881–1928)
Theobald William Louis Theobald (b. 1936)
Theobald & Tseng William Louis Theobald (b. 1936) and Chiao C. Tseng (publ. 1964)
Thieb. Arsenne Thiebaud de Berneaud (1777–1850)
H. J. Thompson Henry Joseph Thompson (b. 1921)
Thompson & Roberts Henry Joseph Thompson (b. 1921) and Roberts
Thompson Zadock Thompson (1796–1856)
Thornber John James Thornber (1872–1962)
Thorne Robert Folger Thorne (b. 1920)
K. Thorne Kaye (nee Hugie) Thorne (b. 1939)
Thuill. Jean Louis Thuillier (1757–1822)
Thunb. Carl Peter Thunberg (1743–1828)
Thurber George Thurber (1821–1890)
Tidestr. Ivar Frederick Tidestrom (1864–1956)
Tidestr. & Kittell Ivar Frederick Tidestrom (1864–1956) and Sister Marie Teresa Kittell (b. 1892)
Timm Joachim Christian Timm (1734–1805)
Toft & Welsh Catherine Ann Toft (b. 1950) and Stanley Larson Welsh (b. 1928)
Tomb Andrew Spencer Tomb (b. 1943)
Torr. John Torrey (1796–1873)
Torr. & Frem. John Torrey (1796–1873) and John Charles Fremont (1813–1890)
T. & G. John Torrey (1796–1873) and Asa Gray (1810–1888)
Torr. & Hook. John Torrey (1796–1873) and William James Hooker (1785–1865)
Trautv. Ernst Rudolf von Trautvetter (1809–1889)
Trel. William Trelease (1857–1945)
Trev. Ludolf Christian Treviranus (1779–1864)
Trew Christoph Jakob Trew (1695–1769)
Trin. Carl Bernhard von Trinius (1778–1844)
Trin. & Rupr. Carl Bernhard von Trinius (1778–1844) and Franz Josef Ruprecht (1814–1870)
Tuckerman Edward Tuckerman (1817–1886)
Tuckey James Kingston Tuckey (1776–1816)
Tul. Edmond Louis Rene Tulasne (1815–1885)
Turcz. Porphir Kiril Nicolas Stepanovich Turczaninow (1796–1864)
Turner Billie Lee Turner (b. 1925)
Turra Antonio Turra (1730–1796)
Tutin Thomas Gaskel Tutin (b. 1908)
Underw. Lucien Marcus Underwood (1853–1907)
Urban Ignatz Urban (1848–1931)
Urv. Jules Sebastian Cesar Dumont d' Urville (1790–1842)
Vahl Martin Hendriksen Vahl (1749–1804)
Vail Anna Murray Vail (b. 1863)
Van Houtte Louis Van Houtte (1810–1876)
L. Van Houtte Louis Benoit Van Houtte (1898–1952)
Vasey George Vasey (1822–1893)
Vasey & Scribn. George Vasey (1822–1893) and Frank Lamson Scribner (1851–1938)
Veitch John Gould Vietch (1839–1870)
Velen. Josef Velenovsky (1858–1949)
Vell. Jose Mariano da Conceicao Velloso (1742–1811)
Vent. Etienne Pierre Ventenat (1757–1808)
Vict. Alexandre Marie Victorin (Conrad A. Kirouac) (1885–1944)
Vill. Dominique Villars (1745–1814)
Vilm. Pierre Louis Francois Leveque de Vilmorin (1816–1860)
Vit. Joanna Vitasey
Vitman Fulgenzio Vitman (1728–1806)
Voss Andreas Voss (1857–1924)
Wagenkn. Rudolfo Wagenknecht (fl. 1955)
Wagner Warren Herbert Wagner (b. 1920)
W. L. Wagner Warren L. Wagner
Wahl Herbert Alexander Wahl (b. 1900)
Wahl. Georg Wahlenberg (1780–1851)
Waldst. & Kit. Franz de Paula Adam von Waldstein-Wortemburg (1759–1823) and Paul Kitaibel (1757–1817)
Wallich Nathaniel Wallich (1786–1854)
Wallr. Carl Friedrich Wilhelm Wallroth (1792–1857)
Walp. Gerhard Wilhelm Walpers (1816–1863)
Walsh Robert Walsh (1772–1852)
Walter Thomas Walter (1740–1789)
Wangenh. Friedrich Adam Julius von Wangenheim (1747–1800)
Ward Lester Frank Ward (1841–1913)
G. Ward G. H. Ward (publ. 1953)
Warder John Aston Warder (1812–1883)
Warm. Johannes Eugen Bulow Warming (1841–1924)
Warren Fred Adelbert Warren (b. 1902)
Waterfall Umaldy Theodore Waterfall (1910–1971)
E. E. Watson Elba Emanuel Watson (1871–1936)
Wats. Sereno Watson (1826–1892)
Wats. & Coult. Sereno Watson (1826–1892) and John Merle Coulter (1851–1928)
T. J. Watson T. J. Watson (publ. 1977)
Webb Philip Barker Webb (1793–1854)
D. A. Webb David Allardice Webb (b. 1912)
Webber John Milton Webber (b. 1897)
Weber Georg Heinrich Weber (1752–1828)
Weber & Mohr Georg Heinrich Weber (1752–1828) and Daniel Mathias Heinrich Mohr (1779–1808)
W. A. Weber William Alfred Weber (b. 1918)
Wedd. Hugh Algernon Weddell (1819–1877)
Weigel Christian Ehrenfried von Weigel (1748–1831)
Weihe & Nees Carl Ernst August Weihe (1779–1834) and Christian Gottfried Daniel Nees von Esenbeck (1776–1858)

Weinm. Johann Anton Weinmann (1782–1858)
Welsh Stanley Larson Welsh (b. 1928)
Welsh & Atwood Stanley Larson Welsh (b. 1928) and Nephi Duane Atwood (b. 1938)
Welsh & Barneby Stanley Larson Welsh (b. 1928) and Rupert Charles Barneby (b. 1911)
Welsh & Goodrich Stanley Larson Welsh (b. 1928) and Sherel Goodrich (b. 1943)
Welsh & Johnston Stanley Larson Welsh (b. 1928) and Barry C. Johnston
Welsh & Moore Stanley Larson Welsh (b. 1928) and Glen Moore (b. 1917)
Welsh & Reveal Stanley Larson Welsh (b. 1928) and James Lauritz Reveal (b. 1941)
Welsh, Atwood, & Reveal Stanley Larson Welsh (b. 1928), Nephi Duane Atwood (b. 1938) and James Lauritz Reveal (b. 1941)
Wendl. Hermann A. Wendland (1823–1903)
Wendt Albert Wendt (1887–1958)
Weston Richard Weston (1733–1806)
Wheeler George Montague Wheeler (b. 1842)
L. Wheeler Louis Cutter Wheeler (b. 1910)
Wherry Edgar Theodore Wherry (b. 1885)
Whipple Amiel Wicks Whipple (1818–1863)
White Theodore Greeley White (1872–1901)
Wiegand Karl McKay Wiegand (1873–1942)
Wiegand & Backeberg Karl McKay Wiegand (1873–1942) and Curt Backeberg (1984–1966)
Wiesl. Albert Everett Wieslander (b. 1890)
Wiggers Friedrich Heinrich Wiggers (1746–1811)
Wiggins Ira Loren Wiggins (b. 1899)
Wight William Franklin Wight (1874–1954)
Wikstr. Johan Emanuel Wikstrom (1789–1856)
Wilkes Charles Wilkes (1798–1877)
Willd. Carl Ludwig von Willdenow (1765–1812)
E. W. Williams Elizabeth W. Williams (publ. 1957)
F. Williams Frederick Newton Williams (1862–1923)
L. O. Williams Louis Otho Williams (b. 1908)
Williams Thomas Albert Williams (1865–1900)
F. D. Wilson Frank D. Wilson (publ. 1963)
Wimmer Christian Heinrich Wimmer (1803–1868)
Winkler Hubert J. P. Winkler (1875–1941)
Winward Alma H. Winward (b. 1937)
Wisliz. Friedrich Adolph Wislizenus (1810–1889)
With. William Withering (1741–1799)
Wittm. Marx Carl Ludewig Wittmack (1839–1929)
F. T. Wolf Franz Theodor Wolf (1841–1924)
S. J. Wolf & Packer S. J. Wolf and John G. Packer (b. 1929)
Wood Alphonso Wood (1810–1881)
Woodruff Dorde (nee Wright) Woodruff
Woodson Robert Everard Woodson (1904–1963)
Wooton Elmer Ottis Wooton (1965–1945)
Woot. & Standl. Elmer Ottis Wooton (1865–1945) and Paul Carpenter Standley (1884–1963)
Wormsk. Morten Wormskiold (1883–1845)
Wulfen Franz Xavier Wulfen (1728–1805)
Yates Harris Oliver Yates (b. 1934)
York Harlan Harvey York (b. 1875)
A. Young Alven Young (1903–1965)
Yuncker Truman George Yuncker (1895–1964)
Zabel Hermann Zabel (1932–1912)
Zam. Aleksander Zamels (1897–1943)
Zenker Jonathan Carl Zenker (1799–1837)
Zucc. Joseph Gerhard Zuccarini (1797–1848)

GLOSSARY

A- (prefix). Meaning without, lacking, as in acaulescent.
Abaxial. On the side away from the axis.
Aberrant. Atypical, different from the usual.
Abortive. Not developing; rudimentary; barren.
Abrupt. Terminating suddenly.
Abscission. Falling away due to breakdown of thin-walled cells at the base of a structure, as of a leaf or petiole.
Acaulescent. Without a stem, the leaves all basal.
Accessory fruit. Fruit with a large and succulent receptacle, as in the strawberry.
Accrescent. Enlarging with age, as with the calyx in many flowers.
Accumbent. Lying against, as in the cotyledons against the radicle.
Acerose. Needle-shaped, like pine leaves.
Achene. A small dry, hard, 1-loculed, 1-seeded indehiscent fruit.
Acicular. Needle-shaped.
Acorn. The leathery fruit of an oak, containing a single large seed and enclosed basally in a cup formed from bract.
Acrid. Sharp or bitter tasting, as with the sap of many plants.
Actinomorphic. Radially symmetric, as in a regular flower.
Aculeate. Armed with prickles, as with the stem of a rose; sharply pointed.
Acumen. Apex.
Acuminate. Gradually tapering to a point, the sides somewhat concave.
Acute. Sharp; tapering to the apex with straight sides.
Ad-. Latin prefix, meaning to or toward.
Adaxial. Located on the side nearest the axis.
Adherent. Sticking together.
Adnate. Union of 2 unlike parts, as with stamens attached to the petals, or sepals to petals.
Adventitious. Organs developing in an unusual position, as with roots developing from stems or leaves.
Adventive. Applied to an introduced plant that has spread or is spreading into a new locality or region.
Agamospermy. Lacking sexual union; an embryo developing from the innermost layer of the nucellus.
Aggregate fruit. Fruit having several separate pistils of a single flower attached to a common receptacle, as in the raspberry.
Alkaline. Of or pertaining to, or having the properties of alkali; basic, having a pH greater than 7.0.
Allopatric. Occupying different geographical areas.
Alluvial. Pertaining to or composed of alluvium as sand, gravel, or similiar detrital material deposited by running water or wind.
Alpine. Applying to plants growing at or above timber line.
Alternate. Borne between, not in front of, as with stamens when between the petals.
Alveolate. Honeycombed, with pits in the surface as in some seeds and receptacles.

Ament. Catkin; a usually deciduous dense spike or raceme with bracteate apetalous unisexual flowers, as in Salicaceae.
Amplexicaul. Clasping the stem, as a leaf base or stipule.
Ampliate. Enlarged.
Androecium. Collectively, the entire set of stamens in a flower.
Androgynous. With staminate flowers borne above the pistillate ones, as in many species of *Carex*.
Angiosperm. A group of flowering plants with the ovules enclosed in an ovary.
Annual. Plants growing from seed and producing flowers and seeds and dying the same year.
Annular. In the form of a ring.
Annulus. In ferns, the organ along the side of a sporangium, which functions in bursting the sporangium at maturity.
Anterior. Front; on the front side; the side away from the axis.
Anther. The pollen-bearing part of the stamen.
Antheridium. Sperm producing structure in the gametophyte generation, as in ferns.
Antheriferous. Anther-bearing.
Anthesis. Flowering; the period when the flower is completely expanded and typically functional.
Anthocyanic. Containing anthocyanins—blue, purple, or red pigments.
Antrorsely. Directed forward or upward, opposite of retrorse.
Apetalous. Without petals.
Aphyllous. Leafless.
Apical. Located at the tip.
Apiculate. Ending in an abrupt slender tip.
Apiculus. A small sharp point.
Apomictic. Producing seeds without any form of sexual union or fertilization.
Apospory. Production of gametophytes directly from somatic cells.
Appendage. A secondary attached part.
Appressed. Lying close and flat against another part.
Approximate. Close together but not united.
Aquatic. Living in water.
Arachnoid. With cobwebby or entangled, very fine hairs.
Arborescent. Treelike.
Arcuate. Arching or curved like a bow.
Areola (pl. areolae). A small space on or beneath the surface, as the area between the small veins; in cacti, the structure which bears flowers, spines, or glochids, or all three.
Argillaceous. Clayey; growing in clay; clay colored.
Aril. An appendage growing at or above the hilum of a seed.
Aristate. With an awn or stiff bristle.
Armature (armed). Spines, barbs, prickles, or thorns (or bearing these).
Articulate. Jointed, and separating at maturity with a clean scar.
Ascending. Growing obliquely upward, often curving.
Asexual. Sexless; reproducing vegetatively.
Atropurpurea. Dark purple; blackish purple.
Attenuate. Gradually narrowing to a tip or base.
Auricle. An ear-shaped appendage.
Auriculate. With one or more auricles.
Awl-shaped. Narrowly triangular, short, sharp-pointed like an awl.
Awn. A bristle like appendage, as on the tips of glumes and lemmas of many grasses.
Axil. The upper (ventral) angle between a leaf and a stem.
Axile. Belonging to or situated in the axis, as with a placenta situated in the axis of a pistil.
Axillary. Located in or arising in an axil.
Axis. The longitudinal central line around which organs are borne, as with the stem supporting the inflorescence.
Banner. Upper petal of a papilionaceous flower, as in the sweet pea.
Barbate. Bearded with long stiff hairs; diminutive of barbed.
Barbed. Bearing sharp, rigid, reflexed points like the barb of a fishhook.
Barbellate. With short, usually stiff hairs.
Bark. Outer layer of a woody stem, usually including all tissues external to the cambium.
Basal. Related to or located at the base.
Basifixed. Attached by the base, as an anther attached by its base to the filament.
Basionym. The name on which a binomial (plant name) or other taxonomic category is based.
Beak. A prolonged, usually narrowed tip of a thicker organ, as in some fruits and petals.
Beaked. Bearing a beak.
Bearded. Bearing long hairs, usually in tufts.
Berry. A pulpy indehiscent fruit, usually with several to may seeds (no stone), as the tomato.
Bi-. Latin prefix signifying two, twice, or doubly, as in bidentate (two-toothed).
Bidentate. Having two teeth.
Biennial. Plants living two years, usually flowering and fruiting the second year (i.e., germinating one spring and producing a rosette that overwinters prior to bolting, flowering, and fruiting in the spring of the second year, thereafter dying).
Bifid. Two-cleft to about the middle.
Bifurcate. Forked, as in Y-shaped hairs.
Bilabiate. With two lips, as in many irregular sympetalous flowers.
Bilocular. With two locules, as an ovary.
Binate. In pairs.
Binomial. The names of the genus and species taken collectively, e.g. *Astragalus utahensis*.
Bipinnate. Doubly or 2–pinnate, as in many compound leaves.
Bipinnatifid. Twice pinnately cleft.
Bisected. Completely divided into two parts.
Biseriate. In two rows or series, one above the other, as in the seeds of some mustards.
Bladdery. Thin and inflated.
Blade. The expanded part of a leaf or petal.
Brackish. Somewhat saline.
Bract. A reduced leaf subtending a flower, usually associated with the inflorescence.
Bracteate. With bracts.
Bracteolate. Provided with bracteoles, as at the base of a flower near the apex of a pedicel.
Bracteole. A small bract, especially on a floral axis; also called bractlet.
Bractlet. See bracteole.
Bridge. The band of tissue connecting adjacent corolla scales in *Cuscuta*.
Bristly. Bearing stiff hairs.

Bud. An undeveloped leaf or flower shoot, often enclosed by specialized leaves called bud-scales.
Bulb. An underground leaf bud with thickened scales, as in the onion.
Bulblet. A small bulb, usually axillary and above ground.
Bulbose. Bulblike.
Bur. A structure or part armed with spines or appendages, typically dispersed by animals.
Caducous. Falling off very early or prematurely.
Caespitose (also cespitose). Growing in tufts.
Callosity. A hardened thickening.
Callus. The thickened extension at the base of the lemma in some grasses.
Calyptra. A lid or hood.
Calyx. Outer whorl of flowering parts; collective term for all the sepals of a flower.
Calyx lobe. In a gamosepalous calyx, one of the free parts.
Calyx tube. The tube of a gamosepalous calyx.
Cambium. A tissue composed of actively or potentially actively dividing cells, typically situated between the xylem and phloem.
Campanulate. Bell-shaped.
Canescent. Covered with gray white or hoary, fine hairs.
Capillary. Hairlike, very fine and slender.
Capitate. Head-shaped, or in a head, as the stigma of many flowers or the inflorescence of many Compositae.
Capsule. A dry fruit of more than 1 carpel that opens to release the seeds.
Carinate. Keeled, with a sharp longitudinal ridge.
Carpel. A simple pistil, or 1 of the modified leaves forming a compound pistil; a megasporophyll.
Carpophore. That part of the receptacle prolonged between the carpels.
Cartilaginous. Like cartilage, tough and firm.
Caruncle. A protuberance or outgrowth near the hilum of a seed.
Caruncular. Of or pertaining to a caruncle.
Caryopsis. The grain or fruit of grasses.
Catkin. A deciduous dense spike or raceme with bracteate, apetalous unisexual flowers as in Salicaceae; ament.
Caudate. Bearing a tail or a slender taillike appendage.
Caudex. The woody base of an otherwise herbaceous perennial.
Caulescent. With a definite leafy stem.
Cauline. Belonging to or on the stem.
Cell. A cavity of an anther containing the pollen, or of an ovary with ovules; a locule.
Ceraceous. Waxy in appearance or color.
Cernuous. Nodding, pendulous, drooping.
Cespitose. See caespitose.
Chaff. Thin dry scales, especially the bracts on the receptacle of heads in composites.
Channeled. Deeply groved longitudinally.
Chaparral. An evergreen vegetation type with a dense growth of leathery leaved shrubs.
Chartaceous. With the texture of paper, usually not green.
Chasmogamous. Said of open, cross-pollinated flowers.
Ciliate. Fringed with marginal hairs.
Cilium (pl. cilia). A slender, hairlike process.
Cinereous. Ash colored; light gray.
Circinate. Coiled from the top downward with the apex as the center, as in the developing leaves of ferns.
Circumboreal. Occuring around the earth in the northern part of the northern hemisphere south of the circumpolar region.
Circumpolar. Occurring all the way around the pole, usually meant to be the north pole as with many arctic plants.
Circumscissile. Dehiscing by a transverse line around the fruit or anther, the top falling as a lid.
Cladophyll. A flattened leaflike stem, functioning as a leaf.
Clasping. Leaf partly or wholly surrounding the stem.
Clavate. Club-shaped, gradually thickened toward the tip.
Claw. A narrowed base of the petal in some flowers.
Cleft. Split nearly to the middle.
Cleistogamous. Flowers self-fertilizing without opening, as in the violet.
Clone. Vegetatively produced individuals from a common parent.
Coalescent. Grown together to form a single unit.
Coetaneous. With the flowers developing the same time as the leaves.
Coherent. Two or more similar organs or parts touching and adhering but usually not joined by fusion of tissues.
Collar. The area on the outside of a grass leaf at the junction of the blade and the sheath.
Colonial. In colonies; usually a reference to plants occurring in clumps in which the individuals are connected by rhizomes or other underground organs.
Column. Body formed by the union of stamens and pistils, as in orchids, mallows and milkweeds.
Coma. A tuft of hairs, as in the seeds of milkweeds.
Commissure. The face by which two carpels cohere, as in the Umbelliferae.
Comose. Bearing a tuft of hairs.
Complete. Having all the parts belonging to it, as a flower with sepals, petals, stamens, and pistil(s).
Compliate. Folded together.
Compound. Having two or more similar parts in one organ.
Compound leaf. A leaf with two or more leaflets.
Compressed. Flattened.
Concave. Hollowed out.
Conduplicate. Folded together lengthwise, as the leaves of many grasses.
Conspecific. Of the same species.
Cone. A strobilus, the fertile aggregation of microsporophylls or megasporophylls and associated axis and bracts, as in *Pinus*.
Coniferous. Bearing cones or strobili.
Confluent. Blending of one part into another.
Congested. Crowded together.
Conglomerate. Densely clustered.
Conic. Cone-shaped, with the point of attachment at the broad base.
Connate. The union of like structures.
Connective. Portion of the filament connecting the two cells of an anther.
Connivent. Converging but not united.
Consimilar. Like one another.
Constricted. Drawn in, narrowed and then expanded.
Contorted. Twisted or bent.
Contracted. Narrowed in a particular place.
Convex. Rounded on the surface.
Convolute. Rolled up longitudinally.

Copious. Abundant, plentiful.
Coralloid. Corallike.
Cordate. Heart-shaped.
Coriaceous. Leathery.
Corm. A short, bulblike, underground stem, with only papery scale leaves.
Corolla. Inner whorl of floral parts; collective name for petals.
Corona. A crown; the whorl of structures between petals and stamens; cuplike in *Narcissus* or of separate parts as in *Asclepias*.
Corrugated. Strongly and usually regularly wrinkled or folded.
Cortex. Rind or bark; parenchymatous tissue between epidermis and pericycle.
Corymb. A flat-topped or convex racemose flower cluster, the lower or outer pedicels longer, their flowers opening first.
Corymbose. Arranged in corymbs.
Costa (pl. costae). A rib, as in a thickened midvein (costate—ribbed).
Cotyledon. The primary leaf or leaves of the embryo.
Creeping. Growing along the ground or just under the surface and producing roots at the nodes.
Crenate. Having the margins with rounded teeth.
Crenulate. Crenate, but the teeth themselves small.
Crested. With a ridge or elevated rib on the summit or back.
Creosote. A colorless, oily, aromatic liquid, as in reference to *Larrea tridentata*, an odoriferous shrub.
Crisped. Curled; wavey.
Cristate. Crested or with a terminal tuft.
Crown. The persistent base of a herbaceous perennial; the top of a tree; a corona.
Cruciform. Cross-shaped.
Crustaceous. Of brittle texture when dry, as in the leaf margin of some plants.
Cucullate. Hood-shaped.
Cucullus. A partial to complete seed covering external to the seed coat in *Nemophila*.
Culm. The type of hollow or pithy slender stem found in grasses.
Cultivar. A race or variety originating under cultivation.
Cuneate. Wedge-shaped, triangular, with the narrow part at point of attachment.
Cupulate. Furnished with or subtended by a small cup.
Cuspidate. Tipped with a cusp or a sharp, short, rigid point.
Cyathium. The inflorescence of *Euphorbia*, consisting of a 3-loculed ovary and a cuplike involucre bearing male flowers with solitary stamens.
Cylindric. Elongate and circular in cross section.
Cymbiform. Boat-shaped.
Cyme. A flat-topped or convex paniculate flower cluster, with the central flowers opening first.
Cymose. With the flowers in a cyme.
Cymule. A small cyme.
Deciduous. Falling off; not evergreen.
Declined. Curved downward.
Decompound. More than once divided or compounded.
Decumbent. Resting on the ground, but with tip of the stem ascending.
Decurrent. Extending down the stem below the insertion, as with leaves or stipules.
Decussate. Opposite pairs, usually applied to leaves, alternating at right angles with those above and below.

Deflexed. Turned abruptly downward.
Dehiscent. Opening spontaneously when ripe to discharge the contents, as an anther or fruit.
Deltoid. Equilaterally triangular; shaped like the Greek letter delta.
Dendritic. Branched like a tree, as some hairs in the Cruciferae
Dentate. Having the margins cut with sharp teeth that are not directed forward.
Denticulate. Minutely dentate or toothed.
Depauperate. Small or poorly developed, usually due to environmental conditions.
Depressed. Low and flattened from above.
Determinate. Said of an inflorescence in which the terminal (central) flower opens first.
Diadelphous. Stamens united by their filaments into two sets.
Diaphanous. Transparent.
Dichotomous. Repeatedly forked in pairs.
Dicotyledon. Plant group with two seed leaves; the Magnoliopsida.
Didymous. Twin; found in pairs, as the fruit halves in the genus *Physaria*.
Didynamous. Having 4 stamens disposed in pairs, 2 long and 2 short, as in many Labiatae.
Diffuse. Widely and diffusely spreading.
Digitate. Fingered; shaped like an open hand; palmate.
Dilated. Flattened and broadened, as an expanded filament.
Dimorphic. Having two forms.
Dioecious. Having staminate and pistillate flowers on different plants.
Diploid. Having two complete sets of chromosomes in each cell.
Disarticulating. Separating joint from joint at maturity, as leaves at the petiole base.
Discoid. Disklike; having disk flowers.
Disk. A fleshy development of the receptacle about the base of an ovary; in Compositae the tubular flowers of the head as distinct from the ray; also spelled disc.
Dissected. Deeply divided into numerous fine segments.
Distal. Opposite the point of attachment; apical; away from the axis.
Distichous. In 2 vertical rows or ranks.
Distinct. Separate; not united with parts in the same whorl.
Diurnal. Opening during the day.
Divaricate. Widely divergent.
Divergent. Spreading.
Divided. Separated to the base.
Dolabriform. Pick-shaped as in two-armed hairs attached at or near the middle and tapering to the ends.
Dorsal. Pertaining to the back; the surface turned away from the axis; abaxial.
Dorsifixed. Attached to the back.
Dorsiventral. Having an upper and lower surface.
Double. Said of flowers when floral parts are more numerous than usual.
Drooping. Erect at the base but bending downward above.
Drupe. Fleshy fruit in which the inner layer of the ovary wall becomes hard, as in the peach.
Drupelet. One fruit of an aggregation of such fruits (aggregate fruit), as in a raspberry.
E-, Ex-. Latin prefix, meaning without, out of or from.

Ebracteate. Without bracts.
Echinate. With stout bluntish prickles.
Ecotone. The area of intergradation between vegetative types.
Ecotype. Those individuals adadpted to one kind of environmental condition.
Edaphic. Pertaining to, or influenced by, soil conditions.
Ellipsoid. An elliptic solid.
Elliptic. In the form of a flattened circle, more than twice as long as broad, widest in the center and the two ends equal.
Emarginate. With a small notch at the apex.
Emersed. Raised above the water; emergent.
Enation. An outgrowth on the surface of an organ.
Endemic. Restricted to a geographic region, topographic unit, or edaphic situation.
Endocarp. The inner layer of the pericarp.
Endosperm. The food storage tissue surrounding the embryo of a seed and formed in the embryo sac, resulting from fertilization of the polar nuclei by a sperm (typically 3n or 5n).
Enfold. To fold inward or toward one another.
Ensiform. Sword-shaped, as the leaves of an *Iris*.
Entire. Undivided; the margin continuous, not incised or toothed.
Ephemeral. Lasting for short time.
Epidermis. The cellular covering layer of an organ, as a leaf, stem, or root.
Epigynous. Attached at the top of the ovary or apparently so.
Epithet. A name given to a plant, e. g., *utahensis* in *Astragalus utahensis*.
Equitant. Overlapping in 2 ranks; as the leaves of an *Iris*.
Erect. Upright in relation to the ground, or sometimes perpendicular to the surface of attachment.
Erose. Irregularly toothed as if gnawed.
Estipulate. Without stipules (also exstipulate).
Etiolated. White through failure to develop chlorophyll and usually elongate and spindly.
Evanescent. Soon disappearing; lasting only a short time.
Even pinnate. A compound leaf lacking a terminal leaflet.
Evergreen. Remaining green over winter; not deciduous altogether.
Excurrent. Extending beyond the margin, as in a leaf base extending down the stem.
Exfoliate. To come off in scales or flakes, as the bark of some trees.
Exocarp. The outer layer of the pericarp.
Explanate. Spread out flat.
Exserted. Protruding, as stamens projecting from the corolla; not included.
Extrorse. Facing inward; said of anther dehiscence.
Exudate. Anything excreted, often forming a deposit on the surface of plant part.
Falcate. Sickle-shaped.
Falls. Outer whorl of perianth parts in an *Iris*; the sepals.
Farinaceous. Having a mealy texture or surface.
Farinose. Covered with meallike particles as in some taxa of *Chenopodium*.
Fasciated. Much flattened or convoluted due to a teralogical widening of its stem, as in cacti.
Fascicle. A close cluster or bundle of flowers, leaves, stems, or roots.
Fasciculate. Connected or drawn into a bundle.
Fastigiate. Clustered, parallel, erect branches.

Faveolate (alveolate). Honeycombed, as with the surface of some *Phacelia* seeds and the receptacle of many Compositae.
Fenestrate. Perforated with openings or translucent areas, as in schizocarps of Malvaceae.
Fertile. Said of pollen-bearing stamens and seed-bearing fruits.
Festucoid. Like a *Festuca*; a plant belonging to the grass group Festucoidea.
Fetid. Having a disagreeable odor.
Fibrillose. With or bearing fibers.
Filament. A thread, especially the stalk of an anther.
Filiform. Threadlike.
Fimbriate. Fringed.
Fistulose. Hollow, often rather enlarged.
Flabellate. Fanlike.
Flaccid. Weak, limp or flabby.
Fleshy. Thick and juicy; succulent.
Flexuose. Zigzag.
Floccose. Bearing tufts of soft woolly hair.
Floral tube. A somewhat elongate tube of the perianth or other floral parts.
Floret. Lemma and palea of the small included flower of a grass; also the small flower of the Compositae.
Floriferous. Bearing or producing many flowers.
Flower. A fertile branch, typically consisting of four whorls, i.e. sepals, petals, stamens, and pistils, but one or more of the whorls often lacking.
Foliaceous. Leaflike.
Foliolate. Having leaflets.
Foliose. Closely clothed with leaves; leaflike.
Follicle. A dry, dehiscent fruit, consisting of a single carpel which opens only on the ventral suture, as in the milkweed.
Forma. The ultimate category in plant classification, typically based on minor morphological differences; abbreviation = f.
Fornices. A set of small appendages in the throat of the corolla in the Boraginaceae.
Fornix. One of a set of small appendages in the corolla throat in the Boraginaceae.
Foveate. Pitted.
Foveolate. Pitted, the diminutive of foveate.
Free. Not joined to other organs.
Free-central. Said of an axile placenta standing free in the center of an ovary with one locule.
Frond. Leaf of a fern.
Fruit. The ripened pistil with all of its accessory parts.
Fruticose. Shrubby or shrublike, also frutescent.
Frutescent. Shrubby or bushy in the same sense of being woody.
Funiculus. The stalk by which the ovule is attached to the placenta.
Funnelform. Gradually widening upwards, like a funnel.
Fusiform. Spindle-shaped, thickest near the middle and tapering toward each end.
Galea. The upper lip in certain 2–lipped corollas, as in *Castilleja*.
Galeate. Hollow and vaulted, as in many lipped corollas.
Gametophyte. The gamete-bearing generation in the alternation of generations (see sporophyte); greatly reduced in seed plants.
Gamopetalous. Corolla with petals united; also sympetalous.
Gamosepalous. Calyx with united sepals; also synsepalous.

Geniculate. Abruptly bent.
Genome. The chromosome compliment.
Genus (pl. genera). A grouping usually consisting of two or more closely related species (less commonly a single species) marked by morphological and often other distinctions from other genera in a family.
Gibbous. Swollen on one side; ventricose.
Glabrate. Becoming glabrous in age.
Glabrous. Without hairs.
Gland. A depression, protuberance, or appendage which secretes a usually sticky fluid.
Glandular. Bearing glands.
Glaucescent. Slightly glaucous; becoming glaucous.
Glaucous. Covered or whitened with a bloom, as a cabbage leaf.
Globose. Spherical or nearly so.
Glochid. A barbed hair or bristle.
Glochidiate. Pubescent with barbed prickles or heads.
Glomerate. Densely compacted in clusters.
Glomerulate. Arranged in small, compact clusters.
Glomerule. A compact capitate cyme.
Glume. A chaffy bract; especially one of the pair of bracts at the base of a grass spikelet.
Glutinous. Sticky; with a sticky exudation.
Graduate. Marked with small regular distances.
Granular. Covered with very small grains or granules; minutely mealy.
Gymnosperm. A cone-bearing vascular plant, e.g., a conifer.
Gynaecandrous. Having staminate and pistillate flowers in the same spikelet, the latter above the former.
Gynandrium. A column bearing stamens and pistils.
Gynandrous. Stamens adnate to the pistil.
Gynobase. An elongation of the receptacle supporting the pistil(s) or nutlets, as in the Boraginaceae.
Gynoecium. Collectively, the pistils of a flower.
Gynophore. The stalk of the pistil of recptacular origin, as in some *Astragalus* species.
Habit. General appearance of a plant.
Habitat. The part of the environment where a plant grows.
Halophyte. A plant adapted to salty soil.
Hanging garden. A vegetated, moist alcove or bedding plane in sandstone, typical of the canyons of the Colorado River.
Hastate. Arrowhead-shaped but with the basal lobes turned outward.
Head. A dense globular cluster of sessile or subsessile flowers arising from a common receptacle.
Helicoid cyme. A one-sided cymose inflorescence that appears to uncoil from the end, as in Boraginaceae.
Hemi-. Greek prefix meaning half.
Hemispheric. Half spherical.
Herb. A plant without a persistent woody stem, at least above ground.
Herbaceous. Pertaining to an herb; opposed to woody; having the texture or odor of a foliage leaf; dying to the ground each year.
Herbarium. A collection of dried plants; the place where such a collection is housed.
Hetero-. Greek prefix meaning unlike or different.
Heterogamous. Having flowers of different sex, as in many Compositae.
Heteromorphic. Of more than one kind or form.
Heterosporous. Producing spores of two kinds, microspores and megaspores.
Hexaploid. Having six complete sets of chromosomes per cell.
Hilum. A scar on a seed, as on a bean, marking the point of attachment.
Hirsute. Rough, with coarse, stiff hairs.
Hirsutulous. Minutely hirsute; hirtellous.
Hirtellous. Minutely hirsute.
Hispid. Rough, with stiff or bristly hairs.
Hispidulous. Minutely hispid.
Holosericeous. Covered with fine and silky hairs.
Holotype. The one specimen on which the name of a species or other taxon is based.
Homomorphous. All alike, as the nutlets in some Boraginaceae; having perfect flowers of only one type.
Homonym. A taxonomic name rejected because it duplicates a previously and validly published name at the same rank.
Homosporous. With spores all alike in size and shape.
Homostylous. With anthers and style the same length.
Hood. A covering; the arching upper petal in *Aconitum*.
Host. A plant that nourishes a parasite.
Hyaline. Translucent, when viewed in transmitted light.
Hybrid. A cross between two species, subspecies, varieties, or genera.
Hydrophyte. An aquatic or water plant.
Hypanthium. A cup-shaped structure on which the calyx, corolla, and often the stamens are inserted; in perigyny, the "calyx tube".
Hypocotyl. The axis of a plant embryo or seedling below the cotyledon.
Hypogeous. Growing or living below the surface of the ground.
Hypogynous. Borne on the receptacle below or free of the pistil, as with sepals, petals, and stamens.
Igneous. Specifically, formed by volcanic action or intense heat, as with igneous rock.
Imbricate. Overlapping as shingles on a roof.
Immersed. Growing under water.
Imparipinnate. Unequally pinnate; odd-pinnate.
Incised. Cut more or less deeply and sharply, intermediate between toothed and lobed.
Included. Not protruding beyond the surrounding organ or envelope.
Incumbent. Lying upon, as an anther against the inner face of its filament.
Incurved. Curved inward.
Indehiscent. Not splitting open at maturity.
Indeterminate. Not terminated absolutely, as in an inflorescence that continues to flower at the distal end of the axis.
Indigenous. Native to the area.
Indurate. Hardened.
Indusium. In ferns, the epidermal outgrowth that covers or invests the sorus.
Inferior. Lower or beneath, as with the ovary when it is below the other parts of a flower.
Inflated. Blown up; bladdery.
Inflorescence. The flower cluster of a plant.
Infraspecific. A taxonomic category below the rank of species, i.e., subspecies, variety, or form.
Innate. Borne on the apex of the support; in an anther the antithesis of adnate.
Inrolled. Involute; curled or curved inward.
Inserted. Attached to or growing upon.

Integument. The outer covering of an ovule.
Intercostal. Between the ribs or nerves.
Internode. The portion of stem between two nodes.
Interrupted. Not continuous.
Introduced. Brought in (usually purposefully) from another area.
Introgression. Infusion of genetic material from different taxa, resulting in intermediate hybrid progeny capable of transferring genetic material to both parental types.
Introrse. Faced or turned toward the axis.
Involucel. A secondary involucre, as the bracts subtending the secondary umbels in the Apiaceae.
Involucre. A whorl of bracts subtending a flower cluster, as in the heads of Compositae.
Involute. With the edges rolled inward toward the upper side; not revolute.
Irregular. Showing a lack of uniformity; bilaterally symmetric, as a zygomorphic flower.
Isotype. Duplicate of the holotype.
Joint. The node of a grass stem or cactus as in *Opuntia*.
Jointed. With nodes or places of real or apparent articulation.
Keel. A prominent dorsal ridge, analogous to the keel of a boat; the two lower united petals of a papillonaceous corolla.
Key. A graphic divice wherein series of alternative choices are presented, the solution of which allows one to determine the name of the taxonomic rank of any unknown plant.
Krummholz. Literally wind-forest; the low wind-deformed forest at tree or timberline.
Labiate. Lipped; a member of the Labiatae.
Lacerate. Appearing irregularly torn or cleft.
Laciniate. Cut into narrow lobes or segments.
Lamina. The blade or expanded part of a leaf, petal, etc.
Lanate. Woolly; densely clothed with long entangled hairs.
Lanceolate. Lance-shaped; much longer than broad, tapering from below the middle to the apex and to the base.
Lateral. On or at the side.
Latex. Milky sap.
Lax. Loose, with component parts separated.
Leaflet. A segment of a compound leaf.
Legume. A superior 1–loculed fruit derived from a simple pistil, usually dehiscent into two valves having the seeds attached along the ventral suture; a leguminous plant.
Lemma. In grasses, the lower of the two bracts immediately enclosing the floret.
Lenticels. Corky spots on young bark, arising in relation to the epidermal stomates.
Lenticular. Lens-shaped.
Lepidote. Covered with fine scurfy scales.
Ligneous. Woody; of or resembling wood.
Ligulate. Strap-shaped; furnished with a ligule.
Ligule. The strap-shaped part of a ray corolla in Compositae; the thin collarlike appendage on the inside of the blade at the junction with the sheath in grasses.
Liliopsida. Moncotyledoneae or monocots.
Limb. The expanded flat part of an organ, especially the expanded part of a gamopetalous corolla.
Linear. Resembling a line; long and narrow, of uniform width, as the leaf blades of grasses.

Lip. One of two segments of an irregular corolla or calyx, as in many Labiatiae; a labium.
Lobe. A division or segment of an organ, as of a leaf.
Lobulate. Made up of lobules.
Lobule. A small lobe.
Locally abundant. Abundant where found, but not occurring uniformly.
Locule. The "cell" or cavity of an organ, applied to pistils and stamens.
Loculicidal. Dehiscent longitudinally through the middle of the back of a pericarp, directly into the locule.
Lodicule. Paired rudiments at the base of the ovary in grass flowers, which spread the palea and lemma by swelling at anthesis.
Loment. A legume which is constricted between the seeds.
Longitudinal. Along the long axis.
Lyrate. Lyre-shaped; pinnatifid, with the terminal lobe large and rounded, the lower lobes small.
Machaerantheroid. With squarrose or recurved type on involucral bracts, as in *Machaeranthera*.
Macro-. Greek prefix meaning large or long.
Macrophyll. A large leaf, as the complex, usually few- to many-veined leaves of higher vascular plants; see microphyll.
Macrosporangium. The organ in which macrospores are produced; see megasporangium.
Macrospore. The larger of two kinds of spores in Selaginellaceae; see megaspore.
Macrosporophyll. The modified leaf that bears the macrosporangium; see megasporophyll.
Maculate. Blotched or spotted.
Magnoliopsida. Dicotyledoneae or dicots.
Malpighian hair. Straight appressed hairs attached above the base and tapering to the free tips; pick-shaped.
Mammillate. Having nipples, or nipplelike protruberances.
Marcescent. Withering without falling off, as the persistent leaves at the bases of some plants.
Marginate. Distinctly margined.
Medial. Of the middle.
Megasporangium. The spore case bearing the megaspore.
Megaspore. A synonym of macrospore; the female spore.
Megasporophyll. A fertile, modified leaf bearing the megasporangium; a carpel.
Membranous. Of the nature of a membrane; thin, soft, and pliable.
-merous. A suffix denoting parts or numbers, as 3–merous.
Mericarp. A portion of a schizocarp, as in Umbelliferae.
Mesocarp. The middle layer or coat of a fruit.
Mesophyte. A plant growing where moisture conditions are moderate.
Micro-. Greek prefix meaning small.
Microphyll. A small leaf, as the simple, usually 1–veined leaves of *Lycophyta* and *Equisetophyta*.
Microsporangium. The organ producing microspores.
Microspore. The smaller of the two kinds of spores, as in Selaginellaceae; the male spore.
Microsporophyll. A fertile leaf bearing microsporangia; a stamen.
Midrib. The central rib of a leaf or other ogran.
Monadelphous. Stamens united by their filaments into a tube surrounding the gynoecium, as in malvaceous flowers.

Monocarpic. Bearing fruit but once and then dying; an annual or biennial.
Monoclinous. Having stamens and pistils in the same flower.
Monocotyledon. Plant group with one seed leaf; Liliopsida.
Monecious. Having staminate and pistillate flowers on the same plant, but not perfect ones.
Monotypic. Having one representative, as a family with one genus.
Montane. Growing in the mountains.
Moraines. A mass of rocks, gravels, or sand deposited by a glacier at the terminal end or along the sides.
Mottled. Marked with colored spots.
Mucilaginous. Moist and viscid; slimy; composed of mucilage.
Mucro. A small and short abrupt tip of an organ.
Mucronate. Possessing a short and straight point (mucro), as some leaves.
Multi-. Latin prefix, meaning many.
Multifid. Cleft into many narrow lobes or segments.
Muricate. Rough with short and firm sharp excrescences.
Naked. Nude.
Naturalized. Said of plants introduced from elsewhere but now established.
Nectariferous. Having nectar.
Nectary. An organ which secretes nectar.
Nerve. A simple vein or slender rib of a leaf or bract.
Neuter. Lacking functional stamens or gynoecium.
Nocturnal. Opening at night, as in flowers of *Oenothera*.
Node. The joint of a stem; the point of insertion of a leaf or leaves.
Nodding. Bent to one side.
Nodulose. Having minute nodules; finely knobby.
Nut. A hard, indehiscent, usually 1-seeded fruit, produced from a compound ovary.
Nutlet. A small nut.
Ob-. A Latin prefix usually signifying inversion.
Obconic. Conical, but attached at the narrower end.
Obcordate. Inversely cordate.
Oblanceolate. Inversely lanceolate.
Oblique. With unequal sides, slanting.
Oblong. Much longer than broad, with nearly parallel sides.
Obovate. Shaped like the longitudinal section of an egg, but with the broadest part toward the tip.
Obovoid. Inversely ovoid.
Obsolete. Rudimentary or not evident, as an organ that is almost entirely suppressed; vestigial.
Obtuse. Blunt to almost rounded at the end.
Ocrea. A sheath around the stem derived from the stipules; chiefly in the Polygonaceae.
Odd-pinnate. Having a terminal leaflet instead of a tendril or pair of leaflets.
Odoriferous. Having an odor.
Operculum. A lid, as with the deciduous cap of a circumscissle fruit.
Opposite. Set against, as leaves when two at a node; one part in front of another, as a stamen in front of a petal.
Orbicular. Approximately circular in outline.
Orchroleucous. Yellowish white.
Orifice. The mouthlike opening of a tubular corolla at the tube opening.
Outcross. To exchange pollen with other plants; opposite of selfing.

Oval. Broadly elliptic.
Ovary. The part of the pistil that contains the ovules.
Ovate. With the outline of an egg in longitudinal section, the bradest end downward.
Ovoid. Egg-shaped.
Ovule. The megasporangium and associated integuments of a seed plant; the part within an ovary that becomes a seed.
Ovuliferous. Bearing ovules.
Palate. An appendage in the throat of an irregular flower partly or completely closing the throat.
Palea. One of the chafflike scales on the receptacle of many Compositae; the inner bract off a grass floret, often partly surrounded by the lemma.
Paleaceous. Covered with or resembling chaffy scales.
Palmate. Lobed or veined where the branches arise from a common point, like the fingers of a hand.
Palmatifid. Palmately lobed or cleft.
Pandurate. Fiddle-shaped.
Panicle (adj. paniculate). A compound racemose inflorescence; (said of such inflorescences).
Papilionaceous. Applied to the butterflylike corolla of the pea, with banner, wings, and keel.
Papillae. Soft superficial glands or protuberances.
Papillate. Having papillae.
Pappillose. Bearing minute rounded or swollen projections.
Pappose. Pappus-bearing.
Pappus. The modified calyx limb on Compositae, consisting of a crown of bristles or scales at the summit of the achene.
Parallel veined. With veins parallel with the leaf axis, or arising from the midrib and parallel to each other to the margin.
Parasite. An organism that uses energy or water at the expense of another organism (the host).
Parietal. Attached to the wall of the ovary, instead of to the axis.
Parted. Deeply cleft nearly to the base.
Pectinate. With narrow closely set divisions like the teeth of a comb.
Pedate. Palmate, with the lateral lobes 2-cleft.
Pedicel. The stalk of a single flower in a flower cluster, or of a spikelet in grasses.
Pedicellate. Having a pedicel, as in many flowers.
Peduncle. The stalk of a flower or of a flower cluster.
Pedunculate. Having a peduncle, as a flower cluster.
Pellucid. Clear, transparent or nearly so.
Peltate. Shield-shaped; a flat body having a stalk attached to the lower surface instead of at the base or margin.
Pendulous. Hanging downward; pendent.
Pepo. An indehiscent, fleshy, 1-loculed or falsely 3-loculed, many-seeded berry, usually with a hard rind, as in the cucumber.
Perennate. To renew itself, as with lateral shoots from a caudex.
Perennial. Of three or more years duration.
Perfect. A flower having both stamens and pistils.
Perfoliate. With the leaf entirely surrounding the stem.
Perforate. With holes (see *Thysanocarpus curvipes* Cruciferae).
Perianth. The floral envelopes; collectively, the calyx and corolla, especially when they are alike.
Pericarp. The ripened wialls of the ovary, referring to the fruit.

Perigynium (pl. perigynia). The scalelike organ surrounding the pistil in *Carex*.
Perigynous. Borne around the ovary in contrast to beneath it, as when the stamens, corolla, and sepals are inserted on the floral tube.
Pernicious. Highly injurious or destructive in character; deadly.
Persistent. Remaining attached, as a calyx on the fruit.
Petal. One of the second whorl of floral parts or parts of a corolla, usually colored.
Petaloid. Resembling a petal.
Petiole. A leaf stalk.
Petiolule. The petiole of a leaflet of a compound leaf.
Phloem. The usually food conductive tissue; bark.
Phyllode. A leaflike petiole with no blade, as in some *Astragalus* species.
Pilose. Bearing soft and straight hairs.
Pilosulose. Bearing very small, soft, straight hairs.
Pinna (pl. pinnae). A leaflet or primary division of a pinnate leaf.
Pinnate. A compound leaf, having leaflets arranged on each side of a common petiole; featherlike.
Pinnatifid. Pinnately cleft into narrow lobes not reaching the midrib.
Pinnatisect. Pinnately cut to the midrib.
Pinnule. The secondary pinna, as in thrice compound leaves of ferns.
Pistil. The ovule-bearing organ of a flower, consisting of stigma and ovary, usually with a style between; gynoecium; a megasporophyll.
Pistillate. Provided with pistils and without stamens; female (carpellate).
Pith. The parenchymatous tissue in the stem center in Magnoliopsida.
Pitted. Having small pits or depressions.
Placenta (pl. placentae). The ovule-bearing part of an ovary.
Placentation. The arrangement or orientation of the placenta.
Plano-convex. Flat on one side and convex on the other.
Plicate. Plaited; folded as a fan.
Plumbeous. Lead-colored.
Plumose. Feathery; having fine hairs on each side as a plume.
Pluricipital. Many headed, as in a branched caudex.
Pod. Any dry, dehiscent fruit, especially a legume.
Pollen. The male microspores of seed plants located in the anther.
Pollinum (pl. pollinia). The pollen masses of the orchids and milkweeds.
Poly-. Greek prefix, meaning many.
Polychrome. Many colored.
Polygamous. Bearing unisexual and bisexual flowers on the same plant.
Polymorphic. Of many forms.
Polypetalous. With a corolla of separate petals.
Polyploid. With three or more sets of chromosomes in each cell.
Pome. A fleshy fruit made up mostly of a modified floral tube, an apple, pear, or haw.
Poricidal. Opening by means of pores, as in the capsules of poppies.
Porrect. Like a parrot beak.

Posterior. On the side toward the axis; the upper side of the flower; at or toward the back.
Precocious. Flowering before the leaves appear.
Prickle. Sharp outgrowth of the bark or epidermis.
Primary. First.
Primary leaflet. The apical leaflet; the only one produced on some early leaves.
Procumbent. Trailing on the ground, but not rooting.
Proliferous. Bearing offshoots or bulblets.
Prostrate. Lying flat upon the ground.
Pruinose. Having a waxy powdery surface; covered with whitish dust or bloom.
Prothallus. The gametophyte in the alternation of generations (see sporophyte), especially in ferns.
Psammophyte. A plant growing in sand.
Pseudoscape. A false scape, as in a tulip where not all leaves are basal.
Puberulent. Minutely pubescent.
Pubescent. Covered with short, soft hairs; downy.
Pulvinate. Cushion-shaped.
Pulvinus (pl. pulvini). An enlargement at a leaf or leaflet base.
Punctate. Dotted with punctures or with translucent pitted glands or with colored dots.
Pungent. Ending in a rigid, sharp point or prickle; acrid (flavor).
Pustulate. Bearing irregular blisterlike swellings or pustules, mostly at the bases of hairs.
Pyramidal. Shaped like a pyramid.
Pyriform. Pear-shaped.
Pyxis. A capsule whose top comes off as a lid.
Raceme. A simple, elongated, indeterminate inflorescence with each flower subequally pedicelled.
Racemose. Having racemes; racemelike.
Rachilla. A small rachis, specifically the axis of a grass spikelet.
Rachis. The axis of a spike or raceme or of a compound leaf.
Radiate. Spreading from a common center; bearing rays.
Radical. Pertaining to roots.
Radicle. That portion of the embryo below the cotyledons.
Raphe. A ridge along the side of a seed adjacent to the hilum.
Ray. A primary branch of an umbel; in Compositae, the ligule of a ray flower.
Receptacle. That portion of the floral axis upon which the flower parts are borne; in Compositae, that which bears the flowers in the head.
Recurved. Curved backward or downward.
Reduced. With parts diminished; diminished.
Reflexed. Bent downward.
Regular. Said of a flower having radial symmetry, with the parts in each series alike.
Relict. A species with a disjunct distribution left over from an earlier geological period.
Remote. Distantly spaced.
Reniform. Kidney-shaped.
Repand. With an undulating margin, less strongly wavy than sinuate.
Repent. Prostrate and rooting.
Replicate. Folded backward.
Replum. Partition between the two locules of siliques or silicles; see Cruciferae.
Resinous. Bearing resin.

Reticulate. With a network; net-veined.
Retrorse. Bent backward or downward.
Retuse. Notched shallowly at a rounded apex.
Revolute. Rolled backward from both margins toward the underside.
Rhizomatous. Having rhizomes.
Rhizome. An underground stem or rootstock, with scales, leaves, and buds at the nodes.
Rhombic. Diamond-shaped.
Rhomboidal. Approaching a rhombic outline; quadrangular, with the lateral angles obtuse; diamond-shaped.
Riparian. Growing along stream banks.
Root. The usually below ground axis of a plant, lacking nodes, scalelike leaves, and regularly placed buds.
Rosette. A crowded cluster of radiating leaves arising from a shortened stem at or near ground level.
Rostrate. With a short, stout beak.
Rosulate. With a collection of clustered leaves; a rosette.
Rotate. Wheel-shaped; said of a sympetalous corolla with obsolete tube and with a flat and circular limb.
Rotund. Rounded in outline.
Rudimentary. Imperfectly developed; vestigial.
Rufus. Reddish-brown in color.
Ruga (pl. rugae). A wrinkle or fold.
Rugose. Wrinkled.
Rugulose. Somewhat wrinkled.
Runcinate. Sharply pinnatifid or incised, the lobes pointing downward.
Runner. A slender stolon or trailing stem rooting at the nodes or tip.
Sac. The cavity of an anther.
Saccate. Bag-shaped.
Sagittate. Arrowhead-shaped, with the basal lobes turned downward.
Salverform. A corolla with slender tube, abruptly expanding into a flat limb, as in *Phlox*.
Samara. An indehiscent, winged fruit.
Saprophyte. A plant living on dead organic matter, without chlorophyll and not parasitic.
Scaberulose. Intermediate between scabrous and minutely pilose.
Scabrid. Minutely scabrous.
Scabrous. Rough to the touch, owing to the structure of the epidermis or to the presence of short stiff hairs.
Scale. Any thin, scarious bract, usually a vestigial leaf.
Scandent. Climbing.
Scape. A leafless peduncle rising from the ground in acaulescent plants.
Scapose. With the flowers borne on a scape.
Scarious. Thin, dry, and membraneous, not green.
Schizocarp. An indehicent fruit that splits at maturity into two or more one-seeded segments (mericarps).
Scorpioid. A one-sided inflorescence with a circinately coiled axis like a scorpion's tail.
Scurfy. Clothed with small branlike scales.
Secondary leaflet. The leaflet below the terminal one.
Secund. Arranged on one side only; unilateral.
Seed. The ripened ovule; the integuments, embryo, and endosperm (when present).
Seep. A moist location where underground water comes to the surface.
Segment. A division or part of a leaf or other organ.
Segregate. To separate.
Seleniferous. Bearing selenium.

Selenophyte. A plant that takes up selenium from the soil, and typically grows only on selenium bearing soils.
Semi-. (prefix). Literally half but used as meaning partly or nearly.
Sepal. A segment of a calyx, or a segment of a perianth where only one whorl is present.
Sepaloid. Sepallike.
Septicidal. Dehiscence of a capsule through the septa and between the locules.
Septate. Bearing septae or partitions.
Septum. A partition between the cavities, as in an ovary.
Seriate. Disposed in rows.
Sericeous. Silky with long, slender, soft, more or less appressed hairs.
Serotinous. Late coming; late to leaf, flower, or appear.
Serrate. Saw-toothed, the sharp teeth pointed forward.
Serrulate. Minutely serrate.
Sessile. Attached directly by the base, not stalked, as a leaf without a petiole.
Seta. (pl. setae) A bristle.
Setose. Beset with bristles.
Sheath. The tubular basal part of the leaf that encloses the stem, as in the grasses and sedges.
Sheathing. Enclosed with a sheath, as the sheath of a grass leaf that surrounds the stem.
Shrub. A woody plant smaller than a tree and with several to many stems.
Sigmoid. Doubly curved, as in the letter S; S-shaped.
Silicle. A short silique, typically less than twice as long as wide.
Silique. A many-seeded capsule of the Cruciferae, with two valves splitting from the bottom and leaving the placentae with the false partition (replum) between them, typically more than twice as long as wide.
Silky. Like covered with silk; sericeous.
Simple. Unbranched, as a stem or hair; not compound, as a leaf; single, as a pistil of one carpel.
Sinuate. With a strongly wavy margin.
Sinuous. Sinuate, with a deeply wavy margin.
Sinus. The cleft or recess between two lobes of an expanded organ such as a leaf.
Smooth. Not rough to the touch.
Soboliferous. Bearing sobols; elongate caudex branches.
Solitary. Alone, borne singly.
Sonoran habitat. Warm to hot desert habitat, as at St. George, Washington County.
Sordid. Of a dull or dirty hue.
Sorus (pl. sori). The fertile portion of a fern frond; the place where sporangia are borne.
Spathaceous. Spathe bearing; like a spathe.
Spathe. A bract or pair of bracts enclosing a flower cluster.
Spatulate. Like a spatula, a blade rounded above and gradually tapering to the base.
Species. An agglomeration of related individuals with a similiar mode of reproduction at a given point of time.
Spicate. Having the form of or arranged in a spike.
Spike. An elongated rachis of sessile flowers or spikelets.
Spikelet. A secondary spike; the ultimate flower cluster in grasses, consisting of two glumes and one or more florets, also in sedges.
Spine. A sharp-pointed, stiff, woody body, arising from below the epidermis; commonly the counterpart of a stipule.
Spinulose. Having spinules.

Sporangium (pl. sporangia). A spore case or sac.
Spore. A reproductive cell, typically haploid, resulting from meiotic cell division in specialized reproductive structures (sporangia).
Sporocarp. The receptacle containing sporangia or spores.
Sporophyll. A spore-bearing leaf.
Sporophyte. The spore-bearing generation in alternation of generations (see gametophyte), the usual plant seen in seed plants, ferns, and other vascular plants.
Spreading. Diverging almost to the horizontal; nearly prostrate.
Spur. A slender, saclike, nectariferous process from a petal or sepal.
Squama. A scale, as applied to some types of pappus in the Compositae.
Squarrose. Rough or scurfy with spreading and outstanding processes; recurved at the tips, as the involucral of bracts (see machaerantheroid).
Stalk. The stem of an organ.
Stamen. The male organ of the flower that bears pollen; the microsporophyll.
Staminate. Having stamens but not pistils; said of a flower or plant that is male, hence not seed-bearing.
Staminode. A sterile stamen, or what corresponds to a stamen.
Standard. Banner or uppermost petal in leguminous flowers.
Stellate. Star-shaped.
Stem. The usually above ground axis of a plant bearing nodes, leaves, end buds.
Sterile. Infertile or barren, as a stamen lacking an anther.
Stigma. The receptive part of the pistil on which the pollen germinates.
Stipe. The stalk beneath an ovary that is inserted upon the receptacle (see gynophore).
Stipel. The stipule of a leaflet.
Stipitate. With a stipe or stalk.
Stipulate. Bearing stipules.
Stipule. One of the pair of usually foliaceous appendages found at the base of the petiole in many plants.
Stolon. A modified stem bending over and rooting at the tip or creeping and rooting at the nodes; a horizontal stem that gives rise to a new plant at its tip.
Stoloniferous. Having stolons.
Stomate. A breathing pore or aperture in the epidermis.
Stone. The hardened, bony endocarp of a drupe enclosing the seed.
Stramineous. Strawlike as color or terture.
Striate. Marked with fine longitudinal lines or furrows.
Strict. Very straight and upright, not at all lax or spreading.
Strigillose. Minutely strigose.
Strigose. Clothed with sharp and stiff appressed straight hairs.
Strigulose. Intermediate between strigose and pilose.
Strobilus. Conelike aggregation of sporophylls; a cone.
Style. The contracted portion of the pistil between the ovary and the stigma.
Stylopodium. An enlargement or disklike expansion at the base of the style as in the Apiaceae.
Sub-. A prefix usually signifying somewhat, slightly, rather, or almost.
Submersed. Submerged.
Subspecies. A unit of classificaation below the rank of species consisting of two or more varieties.
Subtend. To be below and close to, as the leaf subtends the shoot borne in its axis.
Subulate. Awl-shaped.
Succulent. Juicy; fleshy (and sometimes soft).
Suffrutescent. Obscurely shrubby; very slightly woody, but not necessarily low.
Suffruticose. Woody; diminutively shrubby.
Sulcate. Longitudinally grooved, furrowed, or channeled.
Superior. Growing above, as an ovary that is free from the other floral organs.
Suture. The line of dehiscence of fruits or anthers; the line of a natural union or division between coherent parts.
Syconium. Fruit of a fig; inflorescence with inverted hollow recptacle bearing flowers internally, the whole ripening.
Symmetric. Said of a flower having the same number of parts in each circle.
Sympatric. Growing together, or having the same distribution.
Sympetalous. With petals united in a one-piece corolla; gamopetalous.
Syn-. Greek prefix, meaning united.
Synonym. An outmoded systematic name, as with a species published superflously.
Talus. The accumulation of detritus, usually boulders and smaller particles.
Taproot. The primary root from which smaller secondary roots originate.
Tawny. Dull brownish yellow.
Taxa. Two or more taxonomic entities as with two species.
Taxon. Any taxonomic entity, division, class, order, genus, species, etc.
Taxonomy. Classification; principals of classification.
Tendril. a slender, coiling or twining organ by which a climbing plant grasps its support.
Tepals. A segment of a perianth, either sepal or petal, as in the Polygonceae.
Terete. Cylindric; round in cross section.
Ternate. In 3's, as a leaf consisting of three leaflets.
Terrestrial. Growing on ground as opposed to in water; not aquatic.
Tesselate. Checkered.
Testa. The outer seed coat.
Tetradynamous. Having four long and two short stamens.
Tetraploid. With four complete sets of chromosomes in a nucleus; a plant with such a condition.
Thorn. A short, hard, woody, modified stem with a sharp point.
Throat. The orifice of a gamopetalous corolla; the expanded portion between the limb and tube proper.
Thyrse. A compact, ovate panicle; strictly, a panicle with the main axis indeterminate, but other axes cymose.
Tomentose. With tomentum; covered with a rather short, densely matted, soft white wool.
Tomentulose. Slightly tomentose.
Tomentum. A cobwebby covering of hairs.
Tooth. Any small marginal lobe.
Torose. With swellings at intervals.
Tortuous. Bent or twisted in different directions.

Torulose. Constricted between the seeds.
Tracheophyta. The vascular plants.
Trailing. Creeping, said of plants with weak very long stems, as in *Convolvolus*.
Translucent. Nearly transparent.
Transverse. At a right angle (across) to the longitudinal axis.
Tree. A woody perennial usually with a single main trunk or stem.
Tri-. Prefix, meaning three as in trifoliate (with three leaflets).
Tribe. Subdivision of a family.
Trichome. A hair.
Trichotomous. Three-forked.
Tridentate. Three-toothed.
Trifid. Three-cleft to about the middle.
Trifoliate. With three leaflets.
Trigonous. Three-angled.
Tripartite. Three-parted.
Triquetrous. Three-edged.
Triternate. Thrice or three times ternate.
Truncate. As if cut squarely at the end.
Tube. A hollow structure, as in the fused (lower) part of a united corolla or calyx.
Tuber. A thickened, solid, short underground stem with nodes bearing buds.
Tubercle. A small tuberlike prominence or nodule; the persistent base of the style in some Cyperaceae.
Tuberculate. Bearing tubercles.
Tuberous. Producing or resembling a tuber.
Tubular. Shaped like a tube or cylinder.
Tufted. In tufts; caespitose.
Turbinate. Top-shaped.
Turgid. Swollen; inflated.
Turions. Winter buds or over wintering young shoots, as in *Epilobium*.
Type (type specimen). The specimen on which the name of a plant is based.
Umbel. A flat or convex flower cluster in which the pedicels arise from a common point, like the rays of an umbrella.
Umbellate. In umbels, or appearing as if an umbel.
Uncinate. Hooked at the tip.
Undulate. Wavy; repand; with less pronounced waves than sinuate.
Unguiculate. Clawed.
Unilateral. One-sided, or turned to one side of an axis; secund.
Uni-. Latin prefix, meaning one.
Unilocular. With a single cavity or compartment, as in many fruits.
Unisexual. Male or female but not both, as in a staminate (male) flower.
Urceolate. Pitcherlike; hollow and contracted at the mouth like an urn or pitcher.
Urn-shaped. Like an urn; pitcher-shaped.
Utricle. A small, bladdery, 1-seeded usually indehiscent fruit.
Vaginate. Sheathed.
Valvate. Opening by valves, as in most dehiscent fruits and some anthers.
Valve. One of the segments into which a dehiscent capsule or legume separates.
Variety. A taxonomic subunit of a species or subspecies.
Vascular. Conductive; with conductive tissue (i.e., xylem and phloem).
Vascular plant. A plant with xylem and phloem.
Vein. A vascular bundle of a leaf or other flat organ.
Velum. The membranous indusium (covering) in the genus *Isoetes*.
Velutinous. Velvety.
Venation. The arrangement of the veins of a leaf; nervation.
Ventral. Relating to the inward surface of an organ, in relation to the axis, as in the usually upper surface of a leaf.
Ventricose. Swollen or inflated on one side, as in some corollas.
Vernal. Spring; as opposed to autumnal.
Vernation. The arrangement of foliage leaves within the bud.
Verrucose. Warty.
Versatile. An anther attached near the middle.
Verticil. A whorl, or circular arrangement of similiar parts about the same point on an axis.
Verticillate. Whorled; arranged in whorls.
Vespertine. Opening in the evening.
Vestigial. Reduced to a vestige, or trace of a part or organ at one time more perfectly formed.
Villosulose. Diminutive of villous.
Villous. Bearing long, soft, and unmatted hairs; shaggy.
Virgate. Wandlike, erect and straight.
Viscid. Sticky; glutinous.
Viviparous. Forming bulblets in the inflorescence.
Weed. A plant with aggressive tendencies, the product or beneficiary of mans agriculture.
Whorl. A ring of similiar organs radiating from a node; a verticil.
Wing. A thin, usually dry extension bordering an organ; a lateral petal of a papilionaceous flower.
Winter annual. A plant whose seed germinates in late summer or autumn and completes flowering and fruiting before dying in late spring or summer of the year following.
Woolly. Having long, soft, entangled hairs; lanate.
Xeric. Of dry places.
Xerophyte. A plant adapted to live in dry areas, such as desert plants.
Xylem. The usually dead water conducting cells (vessels or trachieds); wood.
Zygomorphic. Bilaterally symmetric; that which can be bisected only in one plane into similar halves.

INDEX

Abies
 balsamea, 29
 lasiocarpa, 29
 concolor, 29
 engelmannii, 31
 grandis, 29
  concolor, 29
 lasiocarpa, 29
 menziesii, 33
 subalpina, 29
 taxifolia, 33
Abronia
 argillosa, 426
 bakeri, 426
 carnea, 430
 elliptica, 426
 fallax, 426
 fragrans, 426
  elliptica, 426
  fragrans, 426
  pterocarpa, 426
 glabra, 426
 lanceolata, 427
 mellifera, 427
 micrantha, 430
  pedunculata, 430
 nana, 427
  harrisii, 427
  nana, 427
  pedunculata, 430
 pumila, 426
 salsa, 426
 turbinata, 426
  marginata, 426
 villosa, 427
Absinthe, 146
Abutilon
 parvulum, 419
 theophrasti, 419
Acacia
 catclaw, 339
 greggii, 339
 arizonica, 339
Acamptopappus
 sphaerocephalus, 137
Acanthochiton
 wrightii, 44
Acanthogonum
 rigidum, 471
Acanthopanax
 sieboldianus, 50
Acer
 campestre, 41
 circinatum, 41
 diffusum, 42
 ginnala, 42
 glabrum, 42
  diffusum, 42
  glabrum, 42
  neomexicanum, 42
  tripartitum, 42
 grandidentatum, 42
  interior, 42
  kingii, 42
 negundo, 42
  interior, 42
  neomexicanum, 42
 palmatum, 42
 platanoides, 42
 pseudoplatanus, 42
 rubrum, 42

 saccharinum, 42
 saccharum, 43
  grandidentatum, 42
  trilobum, 42
  tripartitum, 42
Aceraceae, 41
Acerates
 asperula, 52
 decumbens, 52
 erecta, 52
 rusbyi, 54
Achillea
 filipendulina, 137
 lanulosa, 137
 millefolium, 137
  alpicola, 137
  lanulosa, 137
  millefolium, 137
Achyranthes
 lanuginosa, 45
Achyronichia
 cooperi, 102
Aciphyllaea
 acerosa, 178
Acomastylis
 rossii, 526
 turbinata, 526
Aconitum
 bakeri, 503
 columbianum, 503
  columbianum, 503
  divaricatum, 503
  glaberrimum, 503
  insigne, 503
Acourtia
 wrightii, 217
Acrolasia
 montana, 415
Acroptilon
 repens, 162
Acrostichum
 septentrionale, 19
Actaea
 arguta, 504
  arguta, 504
  rubra, 504
  spicata, 504
  rubra, 504
Actinella
 biennis, 207
 cooperi, 207
 depressa, 207
 grandiflora, 208
 torreyana, 208
Actinocyclus
 secundus, 502
Actinolepis
 lanosa, 192
Adder'stongue, 16
Adelia
 neomexicana, 431
 parvifolia, 431
Adenophyllum
 cooperi, 178
Adenostegia
 kingii, 572
  kingii, 572
  maritima, 572
  parviflorus, 572
  ramosa, 572
  wrightii, 572

Adiantum
 capillus-veneris, 18
 modestum, 18
 rimicola, 18
 modestum, 18
 pedatum, 18
  aleuticum, 18
  rimicola, 18
Adonis
 aestivalis, 504
Adoxa
 moschatellina, 43
Adoxaceae, 43
Aegilops
 cylindrica, 692
 hystrix, 724
Aegochloa
 intertexta, 466
Aesculus
 arguta, 315
 carnea, 315
  arguta, 315
  glabra, 315
  hippocastanum, 315
  octandra, 315
Agastache
 pallidiflora, 329
 urticifolia, 329
Agavaceae, 647
Agave
 family, 647
 kaibabensis, 648
 newberryi, 648
 scaphoidea, 648
 utahensis, 648
  discreta, 648
  kaibensis, 648
Agoseris
 agrestis, 138
 annual, 138
 arizonica, 137
 aurantiaca, 137
  aurantiaca, 137
  purpurea, 138
 caudata, 138
 confinis, 138
 glauca, 138
  agrestis, 138
  dasycephala, 138
  glauca, 138
  laciniata, 138
  parviflora, 138
  pumila, 138
 gracilens, 137
 grandiflora, 138
 heterophylla, 138
 isomeris, 138
 longirostris, 137
 orange, 137
 pale, 138
 parviflora, 138
 pumila, 138
 purpurea, 138
 retrorsa, 138
 retrorse, 138
 scorzoneraefolia, 138
 taracifolia, 138
 taraxacoides, 138
 villosa, 138
Agropyron
 caninum, 728

 cristatum, 693
 dasystachyum, 725
 desertorum, 693
 elmeri, 725
 elongatum, 723
 fragile, 693
 griffithsii, 725
 hispidum, 724
 intermedium, 724
 inerme, 727
 lanceolatum, 725
 latiglume, 728
 pauciflorum, 728
 pectiniforme, 693
 repens, 725
 riparium, 725
 scribneri, 726
 sibiricum, 693
 smithii, 727
 spicatum, 727
  inerme, 727
 subsecundum, 728
 tenerum, 728
 trachycaulum, 728
 trichophorum, 724
 triticeum, 731
 violaceum, 728
  andinum, 728
Agrostis
 airoides, 779
 alba, 695
 aspera, 779
 capillaris, 694
 exarata, 694
 hyemalis, 695
  tenuis, 695
 humilis, 694
 idahoensis, 695
 interrupta, 698
 latifolia, 715
 mexicana, 749
 minutissima, 750
 palustris, 695
 racemosa, 751
 rossiae, 696
 scabra, 695
 semiverticellata, 770
 stolonifera, 695
 tenuis, 694
 thurberiana, 695
 variabilis, 696
 verticillata, 770
Ailanthus
 altissima, 604
Aira
 aquatica, 713
 cespitosa 718
 danthonoides, 718
 elongata, 718
 obtusata, 778
 spicata, 787
Aizoaceae, 43
Ajuga
 reptans, 329
Albizia
 julibrissin, 339
Alcea
 rosea, 419
Aletes
 macdougallii, 616
 breviradiatus, 616

tenuifolia, 633
Alfalfa, 395
  yellow, 394
Algarobia
  glandulosa, 401
Alisma
  brevipes, 651
  geyeri, 651
  gramineum, 651
  plantago-aquatica, 651
  rostrata, 651
  triviale, 651
Alismaceae, 650
Alkaligrass
  Nuttall, 771
  Torrey, 771
  Weeping, 771
Alkali-heath
  family, 306
Alkanet, 61
Allenrolfea
  occidentalis, 117
Alliaria
  alliaria, 247
  officinalis 247
Allionia
  decipiens, 428
  incarnata, 427
  linearis, 428
  multiflora, 429
  nyctaginea, 429
  pumila, 429
Allioniella
  oxybaphoides, 429
Allium
  acuminatum, 802
  atrorubens, 802
  biceptrum, 802
    palmeri, 802
    utahense, 802
  brandegei, 802
  brevistylum, 802
  campanulatum, 802
  cepa, 803
  cernuum, 803
  cristatum, 803
  deserticola, 803
  diehlii, 802
  geyeri, 803
  macropetalum, 803
  nevadense, 803
    cristatum, 803
  palmeri, 802
  parvum, 803
  passeyi, 803
  reticulatum, 803
    deserticola, 803
  schoenoprasum, 803
  textile, 803
  tribracteatum, 802
    diehlii, 802
Allocarya
  cognata, 83
  cusickii, 83
  hispidula, 83
  nitens, 83
  orthocarpa, 83
  penicillata, 83
  scouleri, 83
  versicolor, 83
Almond, 538
  flowering, 538
Alnus

incana, 57
  incana, 57
  rugosa, 57
    occidentalis, 57
    tenuifolia, 57
    occidentalis, 57
    tenuifolia, 57
Alopecurus
  aequalis, 696
  alpinus, 696
  carolinianus, 696
  geniculatus, 697
  monspeliensis, 770
  pratensis, 697
Aloysia
  wrightii, 640
Alsine
  media, 114
  rubella, 105
Alsinella
  saginoides, 109
Alsinopsis
  obtusiloba, 105
  propinqua, 105
  pusilla, 105
Althaea
  rosea, 419
Alumroot
  littleleaf, 558
  red, 559
Alyssum, 247
  alyssioides, 247
  desert, 247
  desertorum 247
  incanum, 254
  ludovicianum, 276
  maritimum, 277
  minus, 247
  saxatile, 247
  sweet, 247, 277
  Szowits, 248
  szowitsianum, 248
Amaranth
  family, 44
  California, 44
  grain, 44
  pale, 44
  Palmer, 45
Amaranthaceae, 44
Amaranthus
  acanthochiton, 44
  albus, 44
  blitoides, 44
  blitum, 45
  californicus, 45
  caudatus, 45
  graecizans, 44
  hybridus, 45
  hypochondriacus, 45
  leucocarpus, 45
  lividus, 45
  fimbriatus, 45
  palmeri, 45
  retroflexus, 45
Amaryllidaceae, 651
Amaryllis, 652
  family, 651
  striata, 652
Amauria
  dissecta, 157
Amberboa
  moschata, 162
Ambrosia

acanthicarpa, 139
artemisiifolia, 139
coronopifolia, 139
dumosa, 139
elatior, 139
eriocentra, 139
psilostachya, 139
tomentosa, 139
trifida, 139
Amelanchier
  alnifolia, 521
    cusickii, 521
  bakeri, 521
  canadensis, 521
    pumila, 521
  florida, 521
    oreophila, 521
    pumila, 521
  racemosa, 525
  utahensis, 521
    cinerea, 521
    oreophila, 521
Amelia
  minor, 502
Amellus
  spinulosus, 201
  villosus, 204
Amesia
  gigantea, 813
Ammannia
  coccinea, 417
  robusta, 417
  robusta, 417
Ammogeton
  scorzoneraefolius, 138
Amorpha
  canescens, 339
  fruticosa, 339
Ampelopsis
  quinquefolia, 545
  vitacea, 545
Amphiachyris
  fremontii, 140
  spinosus, 140
Amphipappus
  fremontii, 140
  spinosus, 140
Amsinckia
  arvensis, 60
  campestris, 60
  collina, 61
  conica, 61
  eatonii, 61
  helleri, 61
  hispidissima, 84
  intactilis, 60
  intermedia, 60
  menziesii, 61
  parviflora, 61
  pustulata, 61
  retrorsa, 61
  rugosa, 61
  tessellata, 61
  utahensis, 61
  velens, 60
Amsonia
  brevifolia, 48
  eastwoodiana, 48
  jonesii, 48
  latifolia, 48
  tomentosa, 48
  stenophylla, 48

Anabasis
  glomerata, 127
Anacardiaceae, 46
Anacharis
  canadensis, 788
Anagallis
  arvensis, 497
Anaphalis
  margaritacea, 140
Anastatica
  syriaca 269
Anchusa
  azurea, 61
  officinalis, 61
Andropogon
  barbinodis, 702
  chrysocomus, 698
  furcatus, 697
  gerardii, 697
    chrysocomus, 698
    gerardii, 698
    glomeratus, 698
    hallii, 698
  ischaemum, 702
  laguroides, 702
  nutans, 776
  ravennae, 773
  saccharoides, 702
  scoparius, 774
  springfieldii, 703
Androsace
  capillaris, 497
    carinata, 497
  chamaejasme, 497
    carinata, 497
    lehmanniana, 497
  filiformis, 497
    lehmanniana, 497
  occidentalis, 498
  septentrionalis, 498
  simplex, 498
Androstephium
  breviflorum, 804
Anemia
  californica, 557
Anemone
  cutleaf, 504
  cylindrica, 504
  desert, 505
  globosa, 504
  lyallii, 505
  multifida, 504
    globosa 504
    multifida, 505
    tetonensis, 504
  northern, 504
  oregona, 505
  parviflora, 504
  patens, 504
  quinquefolia, 505
    lyallii, 505
    oregana, 505
  stylosa, 504
  tetonensis, 504
  tuberosa, 505
Anemopsis
  californica, 557
Anethum
  graveolens, 614
Angelica
  arguta, 617
  dilatata, 617
  Great Basin, 616

kingii, 616
leporina, 617
pinnata, 617
rock, 617
roseana, 617
small-leaved, 617
wheeleri, 617
Utah, 617
Angels-trumpet, 605
Anisolotus
　denticulatus, 388
　numularius, 388
　longebracteatus, 388
　rigidus, 389
　wrightii, 389
Anogra
　albicaulis, 445
　deltoides, 447
　pallida, 447
Anotites
　jonesii, 111
Antennaria
　alpina, 140
　　media, 140
　anaphaloides, 140
　aprica, 141
　arida, 141
　austromontana, 140
　carpathica, 141
　　pulcherrima, 141
　concinna, 141
　corymbosa, 141
　dimorpha, 141
　　macrocephala, 141
　dioica, 142
　luzuloides, 141
　marginata, 141
　media, 140
　microphylla, 141
　nardina, 141
　neglecta, 141
　oblanceolata, 141
　obtusata, 141
　parvifolia, 141
　pulcherrima, 141
　rosea, 141
　rosulata, 141
　umbrinella, 142
Anthemis
　cotula, 142
　tinctoria, 142
Anthyllis
　vulneraria, 339
Anticlea
　elegans, 810
　vaginata, 811
Antirrhinum
　dalmaticum, 573
　filipes, 568
　kingii, 568
　linaria, 573
　majus, 568
　maurandioides, 574
Apera
　interrupta, 698
Aphyllon
　cooperi, 449
Apinus
　flexilis, 32
Apium
　graveolens, 613
Apocynaceae, 47
Apocynum

androsaemifolium, 48
　androsaemifolium, 48
　pumilum, 48
　cannabinum, 48
　　glaberrinum, 48
　　floribundum, 49
　lividum, 49
　medium, 49
　pubescens, 48
　sibiricum, 49
　suksdorfii, 48
Apple
　common, 529
Applebush, 127
Apricot, 537
Aquifoliaceae, 50
Aquilegia
　barnebyi, 505
　caerulea, 505
　　abiflora, 505
　　caerulea, 506
　　calcarea, 506
　　ochroleuca, 506
　　pinetorum, 506
　chrysantha, 506
　depauperata, 506
　elegantula, 506
　ecalcarata, 506
　flavescens, 506
　　flavescens, 506
　　minor, 506
　　rubicunda, 506
　formosa, 506
　　formosa, 506
　fosteri, 506
　leptocera, 506
　micrantha, 506
　pallens, 506
　pinetorum, 506
　rubicunda, 506
　scopulorum, 507
　　calcarea, 505
Arabidopsis
　stenocarpa, 269
　thaliana, 248
　virgata, 269
Arabis
　albertina, 249
　angulata, 252
　angustifolia, 249
　aprica, 249
　arcuata, 251
　　perennans, 252
　　secunda, 251
　　subvillosa, 253
　arida, 252
　armerifolia, 251
　beckwithii, 252
　brachycarpa, 250
　bracteolata, 251
　brebneriana, 269
　bridgeri, 252
　caduca, 251
　canescens, 251
　　latifolia, 251
　confinis, 250
　　brachycarpa, 250
　connexa, 249
　consanguinea, 251
　demissa, 249
　　russeola, 249
　densa, 251
　densicaulis, 252

diehlii, 249
divaricarpa, 250
drummondii, 249
　alpina, 251
　brachycarpa, 250
　connexa, 249
　lyallii, 251
　oreophila, 251
　oxyphylla, 249
　egglestonii, 251
　eremophila, 252
　exilis, 251
　falactoria, 252
　fendleri, 250
　　spatifolia, 250
　fernaldiana, 253
　formosa, 253
　glabra, 250
　　furcatipilis, 250
　　glabra, 250
　gracilenta, 252
　hirsuta, 250
　　glabrata, 250
　　laevis, 252
　　ovata, 250
　　pycnocarpa, 250
　holboellii, 250
　　fendleri, 250
　　pendulocarpa, 251
　　pinetorum, 251
　　retrofracta, 251
　　secunda, 251
　inyoensis, 253
　kockii, 251
　lasiocarpa, 251
　latifolia, 251
　lemmonii, 251
　lignifera, 251
　lignipes, 251
　　impar, 252
　longirostris, 284
　lyallii, 251
　macdougalii, 251
　macella, 252
　macounii, 252
　macrocarpa, 250
　microphylla, 251
　　macounii, 252
　　microphylla, 252
　multiceps, 251
　nevadensis, 249
　nuttallii, 252
　oreophila, 251
　oxyphylla, 249
　pallens, 253
　pendulina, 249
　pendulocarpa, 251
　peramoena, 253
　perelegans, 253
　perennans, 252
　perfoliata, 250
　philonipha, 249
　pinetorum, 251
　puberula, 252
　pulchra, 252
　　duchesnensis, 253
　　munciensis, 252
　　pallens, 253
　pycnocarpa, 250
　　glabra, 250
　recondita, 252
　reptans, 267
　retrofracta, 251

rhodantha, 251
rugocarpa, 249
rupestris, 250
sabulosa, 252
　frigida, 252
　colorata, 252
schistacea, 253
secunda, 251
selbyi, 253
setulosa, 249
shockleyi, 253
sparsiflora, 253
　peramoena, 253
　secunda, 251
　sparsiflora, 253
　subvillosa, 253
spathulata, 252
spatifolia, 250
stokesiae, 250
stricta, 249
subpinnatifida, 252
　beckwithii, 252
　impar, 252
　tenuicola, 252
　tenius, 251
　thalliana, 248
　vivariensis, 253
Arachis
　hypogaea, 340
Aragallus
　besseyi, 396
　bigelovii, 397
　knowltonii, 397
　lambertii, 397
　　sericeus, 398
　majusculus, 398
　metcalfei, 397
　minor, 397
　multiceps, 397
　　minor, 397
　oreophilus, 397
　parryi, 398
　patens, 397
　ventosus 396
Aralia
　racemosa, 50
　bicrenata, 50
　spinosa, 50
Araliaceae, 50
Arborvitae
　American, 27
　Oriental, 28
Arbutus
　uva-ursi, 297
Arceuthobium
　abietinum, 644
　americanum, 644
　campylopodium, 644
　　abietinum, 644
　　cyanocarpum, 644
　　divaricatum, 644
　　cryptopodium, 644
　douglasii, 644
　　abietinum, 644
　　occidentale, 644
　　abietinum, 644
　　vaginatum, 644
　　cryptopodium, 644
Archemora
　fendleri, 635
Arctium
　lappa, 142
　minus, 142

Arctomecon
  californica, 450
  humilis, 450
  merriamii, 450
Arctostaphylos
  obtusifolia, 297
  officinalis, 297
  parryana, 297
  pinetorum, 297
  patula, 297
  incarnata, 297
  pinetorum, 297
  platyphylla, 297
  pringlei, 297
  procumbens, 297
  pungens, 297
    platyphylla, 297
  uva-ursi, 297
    adenotricha, 297
    coactilis, 297
Arenaria
  aculeata, 103
    uintahensis, 103
  calycantha, 103
  cephaloidea, 102
  congesta, 102
    congesta, 102
    cyphaloidea, 102
    lithophylla, 102
    subcongesta, 103
  eastwoodiae, 103
  fendleri, 103
    aculeata, 103
    brevicaulis, 103
    brevifolia, 103
    diffusa, 103
    eastwoodiae, 103
    fendleri, 103
    glabrescens, 103
    porteri, 103
    subcongesta, 103
    tweedyi, 103
  filiorum, 103
  hookeri, 104
    desertorum, 104
  jamesiana, 113
  kingii, 103
    glabrescens, 103
    plateauensis, 103
    uintahensis, 103
  lanuginosa, 104
  lateriflora, 104
  macradenia, 104
    macradenia, 104
    ferrisiae, 104
  macrophylla, 104
  nuttallii, 104
    gracilipes, 104
  obtusa, 104
  obtusiloba, 105
  propinqua, 104
  pusilla, 105
  rossii, 105
  rubella, 105
  rubra, 112
    marina, 112
  sajanensis, 104
  saxosa, 104
  serpyllifolia, 105
  tweedyi, 103
  uintahensis, 103
  verna, 105
    rubella, 105

Argemone
  corymbosa, 451
    arenicola, 451
  munita, 451
  rotundata, 451
    rotundata, 451
Argentina
  anserina, 531
Argillochloa
  dasyclada, 734
Aristida
  adscensionis, 699
  arizonica, 699
  fendleriana, 699
  glauca, 699
  longiseta, 699
  oligantha, 699
  purpurea, 699
  wrightii, 699
Armoracia
  armoracia, 254
  rusticana, 254
Arnica
  arachniodea, 144
  broadleaf, 144
  caudata, 144
  Chamisso, 143
  chamissonis, 143
    foliosa, 143
    incana, 143
    longinodosa, 144
  cordifolia, 143
  diversifolia, 143
  foliosa, 143
    incana, 143
  fulgens, 143
  gracilis, 144
  hairy, 144
  heartleaf, 143
  jonesii, 144
  latifolia, 144
    gracilis, 144
    longifolia, 144
    longifolia, 144
  meadow, 145
  mollis, 144
  orange, 143
  ovata, 144
  parryi, 144
  pedunculata, 143
  rayless, 143
  Rydberg, 144
  rydbergii, 144
  sororia, 145
  varying, 143
Aronia
  alnifolia, 521
Arrhenatherum
  elatius, 699
Arrowgrass
  family 798
  low, 799
  maritime, 799
  marsh, 799
Arrowleaf, 205
Arrowweed, 219
Artemisia
  abrotanum, 146
  absinthium, 146
  albula, 148
  annua, 146
  arbuscula, 146
    nova, 149

aromatica, 147
biennis, 146
bigelovii, 146
campestris, 147
  borealis, 147
  pacifica, 147
  scouleriana, 147
cana, 147
  viscidula, 147
candicans, 148
carruthii, 147
caudata, 147
discolor, 148
dracunculus, 147
  glauca, 147
drancunculoides, 147
filifolia, 147
forwoodii, 147
frigida, 148
glauca, 147
gnaphaloides, 148
incompta, 148
longiloba, 148
ludoviciana, 148
  albula, 148
  candicans, 148
  gnaphaloides, 148
  incompta, 148
  latiloba, 148
  ludoviciana, 148
  mexicana, 148
mexicana, 148
michauxiana, 148
microcephala, 148
norvegica, 149
  piceetorum, 149
nova, 149
pacifica, 147
parryi, 149
purshianus, 148
pygmaea, 149
rothrockii, 150
scopulorum, 149
spiciformis, 150
  longiloba, 148
spinescens, 150
tridentata, 150
  arbuscula, 146
  nova, 149
  pauciflora, 150
  spiciformis, 150
  tridentata, 150
  vaseyana, 150
  wyomingensis, 150
tripartita, 150
vulgaris, 147
  wrightii, 147
wrightii, 147
Arundo
  australis, 759
  canadensis, 712
  donax, 700
  stricta, 713
Asclepiadaceae, 51
Asclepias
  asperula, 52
    asperula, 52
    capricornu, 52
    cryptoceras, 52
    cryptoceras, 52
  curassavica, 52
  cutleri, 52
  engelmanniana, 54

  rusbyi, 54
  erosa, 52
  fascicularis, 53
  galioides, 54
  hallii, 53
  incarnata, 53
  involucrata, 53
    tomentosa, 54
  labriformis, 53
  latifolia, 53
  leucophylla, 52
  macrosperma, 54
  obtusifolia, 53
    latifolia, 53
  rusbyi, 54
  ruthiae, 54
  speciosa, 54
  subverticillata, 54
  tuberosa, 54
    terminalis, 55
  verticillata, 54
    subverticillata, 54
  welshii, 55
Ash
  black, 432
  blue, 432
  European, 432
  flowering, 432
  green, 432
  red, 432
  singleleaf, 432
  velvet, 432
  white, 432
Aspalathus
  caragana, 381
Asparagus, 804
  officinalis, 804
Aspen, 548
  quaking, 548
Asperugo
  procumbens, 61
Aspidium
  aculeatum, 23
  scopulinum, 23
Aspidotus
  densa, 18
Asplenium
  adiantum-nigrum, 18
  andrewsii, 18
  resiliens, 19
  septentrionale, 19
  trichomanes, 19
  viride, 19
Astephanus
  utahensis, 55
Aster
  abatus, 290
  adscendens, 152
    parryi, 153
  alpigenus, 154
  alpinus, 155
  alkali, 155
  angustus, 152
  arenarioides, 183
  arenosus, 211
  bellus, 211
  bigelovii, 212
  Bigelow, 212
  blueleaf, 153
  brachyactis, 152
  campestris, 152
  canescens, 213

aristatus, 213
chilensis, 152
  adscendens, 152
cichoriaceus, 213
commutatus, 152
crag, 155
Eaton, 152
eatonii, 152
elegans, 155
  engelmannii, 155
Engelmann, 152
engelmannii, 152
elongate, 155
ericoides, 155
exscapa, 235
falcatus, 153
foliaceus, 153
  apricus, 153
  canbyi, 153
  eatonii, 152
  frondeus, 153
  parryi, 153
frondosus, 153
glabriuscula, 240
glaucodes, 153
  glaucodes, 154
  pulcher, 154
glaucus, 156
  wasatchensis, 156
gumweed, 213
halophilus, 152
hesperius, 154
hirtifolius, 211
hoary, 213
integrifolius, 154
junciformis, 152
King, 154
kingii, 154
  barnebyana, 154
laevis, 154
laetivirens, 154
leafy, 153
leafybract, 153
leucanthemifolius, 213
leucelene, 211
leucopsis, 152
linearis, 213
Markagunt, 156
meadow, 152
multiflorus, 153
  commutatus, 153
  pansus, 155
New England, 154
novae-angliae, 154
Nuttall, 155
occidentalis, 155
oregonus, 152
Pacific, 152
pansus, 155
parvulus, 214
pauciflorus, 155
perelegans, 155
scopulorum, 155
sibiricus, 155
Siberian, 155
Siskiyou, 154
smooth, 154
spinosus, 156
subspicatus, 153
tanacetifolius, 214
tansyleaf, 214
thermalis, 155
thickstem, 154

venustus, 240
wasatchensis, 156
western, 155
Astragalus
  aboriginorum, 355
  adanus, 376
  adsurgens, 353
    robustior, 353
  agrestis, 353
    polyspermus, 353
  alpinus, 353
  amphioxys, 354
    amphioxys, 354
    cymbellus, 360
    melanocalyx, 354
    vespertinus, 354
  ampullarius, 354
  angustus, 357
    ceramicus, 357
    pictus, 357
  anserinus, 354
  annuus, 364
  araneosus, 367
  arctus, 374
  aretioides, 354
  argillosus, 363
  argophyllus, 354
    argophyllus, 355
    cnicensis, 355
    martinii, 355
    panguicensis, 355
    pephragmenoides, 355
  arietinus, 358
  arrectus, 362
    eremiticus, 362
  artemisiarum, 357
  artipes, 372
  asclepiadoides, 355
  australis, 355
  barnebyi, 355
  beckwethii, 355
    beckwithii, 356
    purpureus, 356
  bigelovii, 370
    thompsonae, 370
  bisulcatus, 356
    bisulcatus, 356
    haydenianus, 356
    major, 356
  bisontum, 397
    minor, 397
  bodinii, 356
    yukonis, 356
  brachycarpus, 376
  brachylobus, 378
  brandegei, 356
  brevicaulis, 357
  bryantii, 367
  caespitosus, 376
  callithrix, 357
  calycosus, 357
    calycosus, 357
    mancus, 357
    scaposus, 357
  campestris, 396
    diversifolius, 361
  canadensis, 357
    brevidens, 357
    canadensis, 357
  candicans, 357
  canovirens, 359
  carltonii, 370
  carolinianus, 357

  castaneiformis, 359
    consobrinus, 359
  ceramicus, 357
    ceramicus, 358
    jonesii, 357
  chamaeleuce, 358
    panguicensis, 355
  chamaemeniscus, 358
  chloodes, 358
  cibarius, 358
  cicadae, 358
    laccoliticus, 358
  cicer, 359
  coltonii, 359
    aphyllus, 359
    coltonii, 359
    foliosus, 359
    moabensis, 359
  confertiflorus, 363
    flaviflorus, 363
  consobrinus, 359
  convallarius, 359
    convallarius, 359
    diversifolius, 361
    finitimus, 359
  cottamii, 359
  coulteri, 367
  cronquistii, 360
  curtilobus, 378
  cymboides, 360
  dasyglottis, 353
  debilis, 356
  decumbens, 370
    oblongifolius, 370
  deflexus, 396
  desereticus, 360
  desperatus, 360
    conspectus, 355
    desperatus, 360
    petrophilus, 361
  detritalis, 361
  diehlii, 364
  diversifolius, 361
    junceus, 359
    roborum, 359
  dodgianus, 380
  drummondii, 361
  duchesnensis, 361
  eastwoodae, 361
  emoryanus, 362
  ensiformis, 362
    gracilior, 362
  episcopus, 362
    lancearius, 366
  equisolensis, 362
  eremiticus, 362
    ampullarioides, 362
  eucosmus, 376
  eurekensis, 363
  fallax, 364
  falcatus, 363
  familicus, 364
  filipes, 363
  flavus, 363
    argillosus, 363
    candicans, 363
    flavus, 363
  flexuosus, 364
    diehlii, 364
    flexuosus, 364
  fremontii, 367
  fucatus, 364

  garrettii, 370
  geyeri, 364
  gilviflorus, 364
  glareosus, 375
  goniatus, 353
  gracilentus, 364
    fallax, 364
  hallii, 364
    fallax, 364
  hamiltonii, 364
  harrisonii, 365
  haydenianus, 356
    major, 356
  henrimontanensis, 365
  holmgreniorum, 365
  humistratus, 365
    humivagans, 365
  hylophilus, 370
    oblongifolius, 370
  hypoglottis, 353
    polyspermus, 353
  ibapensis, 361
  impensus, 366
  inflexus, 375
    glareosus, 375
  iodanthus, 365
    diaphanoides, 365
  iselyi, 366
  jejunus, 366
  jessiae, 366
  junceus, 359, 361
    attenuatus, 359
    diversifolius, 361
    orthocarpus, 361
  junciformis, 359
  kaibensis, 362
  kentrophyta, 366
    coloradoensis, 366
    elatus, 366
    impensus, 366
    implexus, 366
    jessiae, 366
    rotundus, 366
  lambertii, 397
    bigelovii, 397
  lancearius, 366
  lentiginosus, 367
    araneosus, 367
    chartaceus, 367
    diphysus, 367
    fremontii, 367
    palans, 367
    platyphyllidius, 368
    pohlii, 368
    salinus, 368
    scorpionis, 368
    stramineus, 368
    ursinus, 367
    vitreus, 368
    wahweapensis, 368
  limnocharis, 368
    montii, 370
    limnocharis, 368
    tabulaeus, 368
  loanus, 368
  lonchocarpus, 368
    hamiltonii, 364
  lutosus, 369
  macer, 368
  malacoides, 369
  mancus, 357
  marcus-jonesii, 354
  marianus, 373

megacarpus, 369
  caulescens, 372
  prodigus, 369
  parryi, 369
minthorniae, 362
  gracilior, 362
miser, 369
  oblongifolius, 370
  tenuifolius, 370
missouriensis, 370
  amphibolus, 370
  moencoppensis, 370
mollissimus, 370
  thompsonae, 370
montii, 370
monumentalis, 370
  cottamii, 359
mortonii, 357
  brevidens, 357
montanus, 366
  coloradoensis, 366
  impensus, 366
  tegetarius, 366
munzii, 397
multicaulis, 374
multiflorus, 378
musiniensis, 371
  newberryoides, 371
nelsonianus, 371
newberryi, 371
  castoreus, 371
  newberryi, 371
  wardianus, 368
nidularius, 371
nitidus, 353
  robustior, 353
nuttallianus, 371
  imperfectus, 372
  micranthiformis, 372
oophorus, 372
  artipes, 372
  caulescens, 372
  lonchocalyx, 372
oreophilus, 397
palans, 367
  araneosus, 367
panguicensis, 355
pardalinus, 372
parryanus, 398
pattersonii, 372
  praelongus, 374
paucijugus, 370
peabodianus, 375
pectinatus, 371
  platyphyllus, 371
perianus, 372
pictus, 357
  angustus, 357
  foliolosus, 357
  magnus, 357
pinonis, 373
piscator, 373
piutensis, 373
platytropis, 373
praelongus, 373
  ellisiae, 374
  lonchopus, 374
  praelongus, 374
preussii, 374
  arctus, 374
  culteri, 374
  eastwoodae, 361
  latus, 374

laxiflorus, 374
laxispicatus, 374
preussii, 374
sulcatus, 374
procerus, 374
pubentissimus, 374
  peabodianus, 375
  pubentissimus, 375
purshii, 375
  glareosus, 375
  interior, 375
  purshii, 375
pygmaeus, 358
  laccoliticus, 358
racemosus, 375
  treleasei, 375
rafaelensis, 375
robbinsii, 376
sabinarum, 355
sabulonum, 376
sabulosus, 376
salinus, 368
saurinus, 376
scobatinatulus, 356
scoposus, 357
scopulorum, 376
sericoleucus, 354
  aretioides, 354
serotinus, 359
  campestris, 359
serpens, 377
sesquiflorus, 377
  brevipes, 376
shortianus, 354
  brachylobus, 378
  minor, 378
sileranus, 378
  caraicus, 378
simplicifolius, 376
  spathulatus, 376
simplex, 376
spatulatus, 377
  simplex, 376
  uniflorus, 376
spectabilis, 361
stocksii, 365
straturensis, 377
striatiflorus, 377
striatus, 353
subcinereus, 378
  basalticus, 378
  subcinereus, 378
syrticolus, 370
tegetarius, 365
  elatus, 366
  implexus, 366
  rotundus, 366
tenellus, 378
  strigulosus, 378
tephrodes, 378
  brachylobus, 378
tetrapterus, 378
thompsonae, 370
toanus, 379
triphyllus, 364
uintensis, 355
uncialis, 379
ursinus, 362,367
utahensis, 379
vespertinus, 354
virgineus, 376
viridis, 366
  impensus, 366

wardii, 379
webberi, 358
  cibarius, 358
welshii, 380
wetherillii, 380
wingatanus, 380
  dodgeanus, 380
woodruffii, 380
yukonis, 356
zionis, 380
Atelphragma
  ibapense, 361
  straturense, 376
Atenia
  gairdneri, 636
Athyrium
  alpestre, 19
    americanum, 19
  cyclosorum, 19
  distentifolium, 19
    americanum, 19
  felix-femina, 19
    cyclosorum, 19
Athysanus
  pusillus, 246
Atragene
  columbiana, 507
  occidentalis, 508
Atrichoseris
  platyphylla, 156
Atriplex
  argentea, 118
    caput-medusae, 122
  bonnevillensis, 119
  canescens, 118
    garrettii, 120
  gigantea, 118
  occidentalis, 118
  caput-medusae, 122
  carnosa, 121
  collina, 118
  confertifolia, 118
  cornuta, 122
  corrugata, 118
  cuneata, 119
    introgressa, 119
  elegans, 119
  falcata, 119
  gardneri, 119
    bonnevillensis, 119
    cuneata, 119
    falcata, 119
    gardneri, 119
    tridentata, 119
    welshii, 120
  garrettii, 120
  graciliflora, 120
  hastata, 121
  heterosperma, 120
  hortensis, 120
  hymenelytra, 120
  lentiformis, 120
  nuttallii, 119
    falcata, 119
    utahensis, 119
  obovata, 120
  patula, 120
    patula, 121
    triangularis, 121
  pleiantha, 121
  powellii, 121
  rosea, 121
  rydbergii, 118

  saccaria, 121
  saccaria, 122
  caput-medusae, 122
  semibaccata, 122
  subdecumbens, 122
  subspicata, 121
  tenuissima, 122
  torreyi, 122
  triangularis, 121
  tridentata, 119
  truncata, 122
  welshii, 120
  wolfii, 122
Atropis
  laevis, 769
  rigida, 769
Audibertia
  dorrii, 334
Audibertiella
  argentea, 334
Aulospermum
  jonesii, 625
  minimum, 623
  rosei, 625
Avena
  elatior, 699
  fatua, 700
    fatua, 700
    sativa, 700
    mortoniana, 738
  sativa, 700
Avens
  erect, 526
  large-leaved, 526
  purple, 526
  Ross, 526
Azolla
  mexicana, 24
Babysbreath, 107
Baccharis
  emoryi, 156
  glutinosa, 156
  salicina, 157
  sarothroides, 156
  sergilloides, 157
  viminea, 157
  wrightii, 157
Bahia
  dissecta, 157
  desertorum, 219
  integrifolia, 219
  nudicaulis, 219
  oblongifolia, 219
  ourolepis, 219
  wallacei, 219
Baileya
  desert, 158
  multiradiata, 158
  pauciradiata, 157
  pleniradiata, 158
Balsamita
  major, 165
Balsamorhiza
  hirsuta, 158
    neglecta, 158
  hispidula, 158
  hookeri, 158
    hispidula, 158
    neglecta, 158
  macrophylla, 158
  sagittata, 158
Balsamroot
  arrowleaf, 158

cutleaf, 158
Hooker, 158
Banalia
  occidentalis, 128
Baneberry, 504
Barbarea
  americana, 254
  orthoceras, 254
    dolichocarpa, 254
  vulgaris, 254
Barberry
  common, 56
  family, 55
  Fendler, 56
  Thunberg, 56
Barley
  common, 742
  foxtail, 741
  little, 742
  meadow, 740
  Mediterranean, 741
  rabbit, 741
Bartonia
  multiflora, 416
  pumila, 416
Bassia
  hyssopifolia, 122
Basswood
  family, 610
  white, 611
Batidophaca
  desperata, 360
  humivagans, 356
  petrophila, 361
  sabinarum, 355
  sesquiflora, 377
Batis
  vermiculata, 129
Batrachium
  aquatile, 513
    capillaceum, 513
Bean
  kidney, 400
Bearberry, 297
Bearclaw-poppy
  low, 450
Beard
  old-man's, 526
Beargrass, 648
Beauty-bush, 98
Bebbia
  juncea, 159
Beckmannia
  eruciformis, 701
  syzygachne, 701
Beckwithia
  juniperina, 513
Bedstraw
  northern, 544
  rough, 544
  small, 545
  stellate, 545
  sweet-scented, 545
  twinleaf, 544
Beebalm
  plains, 332
Beech
  American, 304
  European, 304
  family, 304
Beet
  red table, 122
  sugar, 122

Beeplant
  Rocky Mountain, 97
  yellow, 97
Beggarticks
  devil's, 159
Bellflower
  family, 94
  Parry, 95
Bellis
  perennis, 159
Bentgrass
  alpine, 694
  carpet, 695
  colonial, 694
  Idaho, 695
  mountain, 696
  spike, 694
  Thurber, 695
Berberidaceae, 55
Berberis
  aquifolium, 56
  fendleri, 56
  fremontii, 56
  julianae, 56
  repens, 56
  sargentiana, 56
  thunbergii, 56
  verruculosa, 56
  vulgaris, 56
Bergia
  texana, 296
Berteroa
  incana, 254
Berula
  erecta, 617
  incisa, 617
Besseya
  alpina, 568
  wyomingensis, 568
Beta
  vulgaris, 122
Betony
  marsh, 336
  woolly, 336
Betula
  alnus, 57
    incana, 57
  fontinalis, 58
  glandulosa, 57
  occidentalis, 58
  papyrifera, 58
  pendula, 58
Betulaceae, 57
Bidens
  bigelovii, 159
  cernua, 159
  comosa, 159
  connata, 159
    comosa, 159
  frondosa, 159
  megapotamica, 233
  tripartita, 159
Bigelovia
  albida, 165
  douglasii, 169
    stenophylla, 169
  glareosa, 167
  graveolens, 167
    glabrata, 167
  howardii, 168
    attenuata, 168
  juncea, 167
  leiosperma, 167

    abbreviata, 167
    menziesii, 201
    scopularum, 201
  paniculata, 168
  pulchella, 168
  turbinata, 168
  vaseyi, 169
Bignonia
  linearis, 59
  radicans, 59
Bignoniaceae, 58
Billberry
  dwarf, 298
Bindweed, 241
  Black, 489
Birch
  European white, 58
  family, 57
  glandular, 57
  swamp, 57
  water, 58
Bird-of-paradise, 381
Birdsbeak
  alkali, 572
  branched, 572
  King, 572
  smallflower, 572
  Wright, 572
Biscuitroot
  Cous, 629
  Wasatch, 629
  Wyeth, 629
Bistort
  alpine, 490
  American, 489
Bitterbrush, 539
Bittercress
  Brewer, 256
  heartleaf, 256
  Muhlenberg, 257
  Nuttall, 257
  smallflower, 257
Bitterroot, 494
Bittersweet
  European, 609
Blackberry
  Himalayan, 541
Blackbrush, 523
Bladder-fern
  bulbet, 21
Bladderpod
  alpine, 275
  Arizona, 275
  breaks, 277
  Colorado, 276
  Fendler, 275
  Garrett, 275
  King, 276
  kodachrome, 277
  Rich, 276
  Rollins, 277
  silver, 276
  Skyline, 275
  slender, 277
  Utah, 277
  Ward, 277
  Wasatch, 276
  Watson, 275
  western, 276
Bladder-senna, 382
Bladderwort
  common, 411
  family, 411

  flatleaf, 411
  lesser, 411
Blanketflower, 193
  Arizona, 194
  Hopi, 194
Blepharidachne
  kingii, 701
Blepharipappus
  glandulosus, 211
Blepharoneuron
  tricholepis, 701
Blitum
  capitatum, 124
  chenopodioides, 128
  nuttallianum, 128
Bluebell
  Flagstaff, 80
  lanceleaf, 81
  mountain, 80
  spindle, 80
  tall, 80
  Wasatch, 80
  western, 81
Bluebells of Scotland, 95
Blue-eyed-Mary, 571
Blue-eyes
  low, 76
Bluegrass
  alpine, 762
  annual, 762
  arctic, 762
  Bigelow, 763
  bog, 766
  Bolander, 763
  bulbous, 763
  Canada, 763
  fowl, 767
  Greenland, 765
  Kentucky, 768
  Letterman, 766
  nodding, 768
  Patterson, 767
  rough-stalked, 769
  Sandberg, 769
  Wasatch, 764
  Wheeler, 766
  winter, 762
Bluestem
  big, 697
  bushy, 698
  cane, 702
  little, 774
  silver, 702
  Springfield, 703
  yellow, 702
Bluets, 545
Boechera
  demissa, 249
  drummondii, 249
  fendleri, 250
  holboellii, 250
  pendulina, 249
  perennans, 252
  retrofracta, 251
  selbyi, 253
Boerhaavia
  spicata, 427
    torreyana, 427
  torreyana, 427
Bog-orchid
  Alaska, 815
  Alcove, 816
  long-bracted, 816

northern, 815
slender, 815
small, 815
Watson, 815
white, 814
Boisduvalia
  densiflora, 134
  glabella, 434
Bolelia
  laeta, 95
Bolophyta
  ligulata, 217
Bonset
  false, 209
Borage
  common, 62
  family, 59
Borago
  officinalis, 62
Boraginaceae, 59
Bothriochloa
  barbinodis, 702
  ischaemum, 702
  laguroides, 702
  saccharoides, 703
    torreyana, 703
  springfieldii, 703
Botrychium
  boreale, 16
  crassinervium, 16
    obtusilobum, 16
  lanceolatum, 16
  lunaria, 16
    minganense, 16
    onondagense, 16
  minganense, 16
  onondagense, 16
  simplex, 17
  virginianum, 17
    simplex, 17
Bottle-gourd, 219
Bounding-bet, 109
Bouteloua
  aristoides, 703
  barbata, 704
    barbata, 704
  curtipendula, 704
  eriopoda, 704
  gracilis, 704
  hirsuta, 705
  procumbens, 705
  rothrockii, 704
  simplex, 705
  trifida, 705
Boxelder, 42
Boykinia
  jamesii, 558
    heucheriformis, 558
Brachyactis
  angusta, 152
  frondosa, 153
Brachylobus
  hispidus, 281
Brachyris
  microcephala, 197
Brassica
  alba, 255
  arvensis, 255
  campestris, 255
  caulorapa, 2545
  hirta, 255
  juncea, 255
  kaber, 255

napobrassica, 254
napus, 254
nigra, 255
oleracea, 255
  botrytis, 255
  capitata, 255
  gemifera, 255
  italica, 255
  orientalis, 259
rapa, 255
Brassicaceae, 247
Breadroot
  Kane, 399
  large-flowered, 399
  Paradox, 399
  Paria, 400
  skunk, 399
Brickellbush
  California, 160
  longleaf, 160
  Mohave, 161
  rough, 160
  spiny, 160
Brickellia
  atractyloides, 160
  californica, 160
  grandiflora, 160
  longifolia, 160
  linifolia, 161
  microphylla 160
    scabra, 160
    watsonii, 161
  oblongifolia, 161
    linifolia, 161
  scabra, 161
  watsonii, 161
Bristlecone
  western, 32
Bristlegrass
  bur, 776
  green, 776
  yellow, 776
Brittle-fern, 21
Broadbean, 410
Brodiaea
  breviflora, 804
  capitata, 806
  douglasii, 809
  paysonii, 804
  pulchella, 806
Bromopsis
  anomala, 707
  ciliata, 708
  inermis, 709
  porteri, 707
  vulgaris, 711
Bromelica
  bulbosa, 747
  spectabilis, 747
Brome
  California, 707
  Columbia, 711
  foxtail, 709
  fringed, 708
  Hungarian, 709
  mountain, 707
  nodding, 707
  poverty, 710
  red, 709
  ripgut, 708
  smooth, 709
Bromus
  anomalus, 707

breviaristatus, 707
briziformis, 707
carinatus, 707
catharticus, 707
ciliatus, 708
commutatus, 709
cristatus, 698
diandrus, 708
frondosus, 707
hordeaceus, 708
inermis, 709
japonicus, 709
marginatus, 707
mollis, 707
polyanthus, 707
porteri, 707
pumpellianus, 709
purgans, 711
  vulgaris, 711
racemosus, 708
richardsonii, 708
rigidus, 708
rubens, 709
secalinus, 710
sterilis, 710
tectorum, 710
trinii, 710
unioloides, 707
vulgaris, 711
willdenowii, 707
Brookgrass, 713
Brooklime
  American, 602
Brookweed, 500
Broom
  Scots, 382
  Spanish, 404
Broom-flax, 412
Broomrape
  family, 448
Broomshrub
  Nevada, 211
Bryonia
  alba, 289
Buchloe
  dactyloides, 711
Buckbean
  family, 424
Buckeye
  Ohio, 315
  sweet, 315
Buckthorn
  alder, 519
  birchleaf, 519
  common, 519
  family, 518
Buckwheat
  arid, 487
  Bailey, 474
  Bateman, 475
  bent, 479
  birdnest, 482
  coin, 482
  Comb Wash, 477
  corymb, 477
  cream, 486
  Cronquist, 478
  cushion, 483
  Darrow, 479
  Davidson, 479
  desert, 476
  Dinosaur, 482
  Duchesne, 477

Dugway, 483
dwarf, 476
Eastwood, 481
Ellen, 485
Elsinore, 475
Ephedra, 476
family, 470
flat top, 478
Frisco, 485
Gate Canyon, 478
golden, 478
Gordon, 479
Grand, 477
Gray, 487
Heermann, 479
hermit, 475
Howell, 480
Ibex, 482
James, 480
leafy, 484
little-butterfly, 481
Logan, 476
longleaf, 481
low, 484
mat, 477
Mojave, 479
Mowry, 482
nodding, 477
obscure, 483
Palmer, 483
Panguitch, 483
Parry, 475
Porter, 486
pretty, 475
Red Creek, 484
redroot, 484
Reveal, 478
rimrock, 478
sand, 481
San Juan, 478
Sevier, 485
Shockley, 485
shortstem, 475
skeletonweed, 479
slender, 482
slenderleaf, 481
slender-stipe, 486
smooth, 484
son, 485
spreading, 479
spotted, 482
Stokes, 485
sulfur, 486
Thomas, 485
unique, 480
varying, 476
Watson, 480
Westwater, 484
Wetherill, 487
whorled, 480
Widtsoe, 474
winged, 474
wirestem, 483
Woodside, 486
Wright, 487
Zion, 484
Buddleja
  davidii, 85
  utahensis, 85
Buddlejaceae, 84
Budsage, 150
Buffaloberry
  roundleaf, 295

silver, 295
Buffalobur, 609
Buffalograss
  false, 753
Bugloss, 61
Bugseed, 126
Bulbocodium
  serotinum, 808
Bulbostylis
  annua, 220
  californica, 160
  microphyllus, 160
Bugleweed
  American, 331
  rough, 331
Bull-bay, 418
Bullgrass, 726
Bulrush
  alkali, 683
  deerhair, 683
  hardstem, 682
  Nevada, 683
  panicled, 683
  river, 683
  sharpscale, 684
  softstem, 684
Bupthalmium
  sagittatum, 158
Burdock, 142
  great, 142
Bur-marigold, 159
Bursa
  bursa-pastoris, 256
  pastoris, 256
Bur-sage, 139
Burnet, 542
Bur-reed
  emersed, 823
  family, 822
  narrowleaved, 822
  small, 823
Burrobush, 205
Bush
  burning, 115
  creosote, 646
  fern, 523
  iodine, 117
  paperbag, 323
  pearl, 525
  strawberry, 115
Bush-clover
  Thunberg, 387
Butter-and-eggs, 573
Buttercup
  alpine, 512
  Arctic, 515
  blister, 515
  bur, 515
  creeping, 516
  drab, 515
  Eschscholtz, 514
  family, 503
  field, 513
  floating, 515
  Gmelin, 514
  Jupiter, 515
  little, 515
  Macoun, 515
  marsh, 513
  mountain, 515
  northern, 515
  plantain, 512
  sagebrush, 514

sharp, 511
showy, 513
straight-beaked, 515
tall, 512
tundra, 514
violet, 512
Butterfly-weed, 54
Butternut, 290, 327
Butterweed, 226
Cactaceae, 85
Cactus
  barrel, 87
  family, 85
  fragilis, 90
  Gypsum, 92
  viviparus, 86
Caesalpinia
  gilliesii, 381
  repens, 381
Calais
  lindleyi, 215
Calamagrostis
  andina, 748
  canadensis, 712
  gigantea, 713
  inexpansa, 713
  langsdorfii, 712
  neglecta, 713
  purpurascens, 712
  purpurea, 712
  rubescens, 712
  scopulorum, 712
  lucidula, 712
  stricta, 713
Calamovilfa
  gigantea, 713
Calandrina
  grayi, 494
  nevadensis, 494
  pygmaea, 494
Callichroa
  platyglossa, 211
Calligonum
  canescens, 118
Callisteris
  arizonica, 459
Callitriche
  anceps, 94
  hermaphroditica, 94
  heterophylla, 94
  palustris, 94
  verna, 94
Callitrichaceae, 94
Calocedrus
  decurrens, 25
Calochortus
  ambiguus, 804
  aureus, 805
  bruneaunis, 805
  flexuosus, 805
  gunnisonii, 805
  kennedyi, 805
  luteus, 805
  nuttallii, 805
    aureus, 805
    bruneaunis, 805
  rhodothecus, 805
  watsonii, 805
    ambiguus, 804
Caltha
  biflora, 507
  leptosepala, 507
    leptosepala, 507

Calycodon
  montanum, 750
Caltrop, 647
  family, 646
Calycoseris
  parryi, 161
  wrightii, 161
Calylophus
  lavandulifolia, 434
Calypso
  bulbosa, 812
Calyptridium
  monandrum, 493
  parryi, 493
  umbellatum, 494
    caudicifera, 494
Calystegia
  longipes, 240
  sepium, 240
Camas, 806
  alcove death, 810
  elegant death, 811
  foothill, 811
  meadow, 811
Camassia
  quamash, 806
    utahensis, 806
Camelina
  microcarpa, 255
Camissonia
  andina, 435
  atwoodii, 436
  boothii, 436
    alyssoides, 436
    condensata, 436
    villosa, 436
  breviflora, 436
  brevipes, 436
    pallidula, 436
  chamaenerioides, 436
  clavaeformis, 437
    clavaeformis, 437
    integrior, 437
    purpurascens, 437
  eastwoodiae, 437
  gouldii, 437
  megalantha, 436
  minor, 437
  multijuga, 437
  parryi, 438
  parvula, 438
  pterosperma, 438
  pusilla, 438
  refracta, 438
  scapoidea, 438
    brachycarpa, 438
    scapioidea, 439
    utahensis, 439
    subacaulis, 439
    walkeri, 439
    tortilis, 439
Campanula
  parryi, 95
  perfoliata, 96
  rapunculoides, 95
  rotundifolia, 95
  uniflora, 95
Campanulaceae, 94
Campe
  orthoceras, 254
Camphor-weed, 219
Campion
  Maguire, 111

Menzies, 111
  moss, 110
  Oregon, 11
Campsis
  radicans, 59
Canaigre, 492
Cannabaceae, 96
Cannabis
  sativa, 97
    indica, 97
    sativa, 97
Canarygrass
  reed, 758
Cancerroot
  cluster, 449
  one-flower, 449
Cantua
  aggregata, 458
  longiflora, 461
  pungens, 464
Caper
  family, 97
Capnodes
  brachycarpum, 307
Capnorea
  pumila, 317
  watsoniana, 317
Caprifoliaceae, 98
Capparaceae, 97
Capsella
  bursa-pastoris, 256
  rubella, 256
  thracica, 256
Capsicum
  frutescens, 604
Caragana
  arborescens, 381
  caragana, 381
  inermis, 381
  sibirica, 381
Caraway, 618
Cardamine
  breweri, 256
  cordifolia, 256
    cordifolia, 257
    pubescens, 256
  flexuosa, 257
  pensylvanica, 257
  hirsuta, 257
    pensylvanica, 257
  infausta, 256
  oligosperma, 257
  palustris, 281
    jonesii, 281
  parviflora, 257
  pensylvanica, 257
  uintahensis, 256
  vallicola, 256
Cardaria
  chalepensis, 257
  draba, 257
    chalepensis, 257
  pubescens, 257
  elongata 258
  repens, 257
Cardinal-flower, 96
Carduus
  eatonii, 172
  lacerus, 174
  nutans, 161
  olivescens, 174
  pulchellus, 172
  scopulorum, 174

tracyi, 174
undulatus, 174
vulgaris, 175
Carex
 ablata, 668
 aboriginum, 672
 albonigra, 659
 anqustor, 669
 aquatilis, 660
 arapahoensis, 660
 atherodes, 660
 athrostachya, 660
 atrata, 660
  atrosquama, 661
  chalciolepis, 661
  erecta, 661
 atrosquama, 661
 aurea, 661
 backii, 661
 bellardii, 681
 bebbii, 661
 bella, 662
 bicolor, 661
 bigelowii, 660
 bipartita, 662
 bolanderi, 663
  leptopoda, 664
 brevipes, 674
 breweri, 662
  paddoensis, 662
 brunnescens, 662
 buxbaumii, 662
 campylocarpa, 675
  affinis, 675
 canescens, 663
  dubia, 663
  praceptorum, 673
 capillaris, 663
 capitata, 663
 chalciolepis, 661
 crawei, 663
 curatorum, 663
 deweyana, 663
  bolanderi, 664
  leptopoda, 664
 diandra, 664
 dioica, 664
  gynocrates, 664
 disperma, 664
 douglasii, 664
 drummondiana, 675
 eastwoodiana, 673
 ebenea, 664
 egglestoni, 665
 eleocharis, 676
 elynoides, 665
 engelmannii, 662
 epapillosa, 661
 exsiccata, 677
 festivella, 669
 filifolia, 665
 fissuricola, 668
 foenea, 675
 foetida, 665
  vernacula, 665
 garberi, 661
 geophila, 674
 geyeri, 665
 gynocrates, 664
 hallii, 672
 hassei, 661
 haydeniana, 666
 hepburnii, 669
 heteroneura, 661
  epapillosa, 661
 hoodii, 666
 hystricina, 666
 illota, 666
 interimus, 660
 interior, 666
 jonesii, 667
 kelloggii, 667
 lanuginosa, 667
 lasiocarpa, 667
  lanuginosa, 667
 lenticularis, 667
 leporinella, 667
 leptalea, 667
 limnophila, 668
 limosa, 668
 luzulina, 668
  ablata, 668
 macloviana, 669
  microptera, 669
  pachystachya, 671
  involucrata, 660
  subfusca, 676
 media, 670
 microglochin, 668
 microptera, 669
 misandra, 669
 multicostata, 669
 muricata, 669
 myosuriodes, 682
 nardina, 669
 nebrascensis, 670
 nelsonii, 670
 neurophora, 670
 nigricans, 670
 norvegica, 670
 nova, 671
 nubicola, 666
 obtusata, 671
 occidentalis, 671
 oederi, 671
  viridula, 671
 pachystachya, 671
 paddoensis, 662
 parryana, 672
  brevisquama, 672
 paupercula, 672
 paysonis, 672
 pelocarpa, 671
 perglobosa, 672
 petasata, 672
 phaeocephala, 673
 physocarpa, 675
 pityophila, 674
 podocarpa, 672
  paysonis, 672
 praeceptorum, 673
 praegracilis, 673
 praticola, 673
 pseudoscripoidea, 675
 pyrenaica, 674
 rachillis, 677
 raynoldsii, 674
 retrorsa, 674
 rossii, 674
 rostrata, 674
 rupestris, 675
 saxatilis, 675
  major, 675
 saximontana, 661
 scirpoidea, 675
  curatorum, 673
 pseudoscerpoidea, 675
 scopulorum, 675
 siccata, 675
 simpliciuscula, 682
 simulata, 676
 stenophylla, 676
 stipata, 676
  stipata, 676
 straminiformis, 676
 subfusca, 676
 subnigricans, 677
 tolmiei, 672
 vallicola, 677
 vernacula, 665
  hobsonii, 670
 vesicaria, 677
  vesicaria, 677
 viridula, 671
 xerantica, 677
Carnation, 107
Carpet-bugle, 329
Carpetweed, 43
 family, 43
Carphephorus
 juncea, 159
Carpinus
 caroliniana, 58
Carrot, 626
Cartiera
 cordata, 285
Carum
 garrettii, 636
 carvi, 618
Carya
 illinoensis, 327
 ovata, 327
Caryophyllaceae, 101
Cashew
 family, 46
Castanea
 dentata, 304
Castilleja
 angustifolia, 569
  collina, 569
  flavescens, 569
 applegatei, 569
  viscida, 569
 aquariensis, 569
 arcuata, 570
 barnebyana, 571
 chromosa, 569
 cusickii, 570
 exilis, 569
 flava, 570
 flavescens, 569
 lapidicola, 570
 leonardii, 571
 linariifolia, 570
 miniata, 570
 minor, 569
 nana, 570
 occidentalis, 570
 parvula, 570
  parvula, 570
  revealii, 570
 pulchella, 571
 revealii, 570
 rhexifolia, 571
  rhexifolia, 571
  sulphurea, 571
 scabrida, 571
  barnebyana, 571
  scabrida, 571
 septentrionalis, 571
 sulphurea, 571
 variabilis, 570
 viscida, 569
 zionis, 571
Castor-bean, 304
Catabrosa
 aquatica, 713
Catalpa
 bignonioides, 59
 Chinese, 59
 common, 59
 family, 58
 ovata, 59
 showy, 58
 speciosa, 59
Catchfly
 Drummond, 108
 night-flowering, 111
Catchweed, 61
Catnip, 333
Cat's-ears, 209
Cattail
 broad-leaved, 823
 common, 823
 family, 823
Caucalis
 arvensis, 637
 microcarpa, 637
Caulanthus
 cooperi, 258
 crassicaulis, 258
  crassicaulis, 258
  glaber, 258
  major, 258
 divaricatus, 286
  glaber, 258
  hastatus, 259
 lasiophyllus, 258
  utahensis, 258
 major, 258
 pilosus, 259
 procerus, 258
 senilis, 258
Caulina
 guadalupensis, 811
Ceanothus
 fendleri, 518
  viridis, 518
 greggii, 581
 martinii, 518
 utahensis, 518
 velutinus, 519
Cedar
 Atlas, 30
 incense, 25
 Port Orford, 25
 red, 27
 white, 27
 salt, 610
 Utah, 26
Cedrus
 atlantica, 30
 lebani, 30
  atlantica, 30
Celastraceae, 114
Celandine, 451
Celosia
 argentea, 45
 cristata, 45
 cristata, 45
Celtis
 laevigata, 612

occidentalis, 611
reticulata, 612
villosula, 612
Cenchrus
  echinatus, 714
    longispinus, 714
  longispinus, 714
  pauciflorus, 714
Centaurea
  Brownscale, 162
  calcitrapa, 162
  cyanus, 162
  jacea, 162
  maculosa, 162
  melitensis, 162
  montana, 162
  moschata, 162
  picris, 162
  repens, 162
  scabiosa, 163
  solstitialis, 163
  spotted, 162
  virgata, 163
Centaurium
  calycosum, 308
    arizonica, 308
  exaltatum, 308
  namophilum, 308
  nevadense, 308
  nuttallii, 308
Centaury
  Buckley, 308
  Great Basin, 308
Centranthus
  ruber, 638
Century-plant
  Utah, 648
Centrostegia
  thurberi, 471
Cerastium
  arvense, 106
  beeringianum, 106
  brachypodum, 106
  caespitosum, 106
  fontanum, 106
  nutans, 106
    brachypodum, 106
  tomentosum, 106
  variable, 106
  vulgatum, 106
Cerasus
  demissa, 539
    melanocarpa, 539
Ceratocephalus
  testiculatus, 516
Ceratochloa
  carinata, 707
  unioloides, 707
Ceratophyllaceae, 115
Ceratophyllum
  demersum, 116
Ceratoides
  lanata, 123
    lanata, 123
    ruinina, 123
    subspinosa, 123
    subspinosa, 123
Cercis
  canadensis, 381
  occidentalis, 382
    orbiculata, 382
    orbiculata, 382
Cercocarpus

arizonicus, 522
betulifolius, 522
betuloides, 522
flabellifolius, 522
intricatus, 522
  villosus, 522
  intricatus, 522
  ledifolius, 522
    intricatus, 522
    intercedens, 522
      hirsutus, 522
      subglaber, 522
  montanus, 522
    glaber, 522
    parviflorus, 522
      betuloides, 522
      glaber, 522
      minimus, 522
Cereus
  coccineus, 87
    melanacanthus, 87
    engelmannii, 86
      chrysocentrus, 86
      variegatus, 87
    mojavensis, 87
Chaenactis
  achilleifolia, 163
  alpina, 163
    alpina, 163
    rubella, 163
  leucopsis, 163
  brachiata, 163
    stansburiana, 163
  carphoclina, 163
  douglasii, 163
    achilleifolia, 163
    alpina, 163
    montana, 163
  fremontii, 164
  leucopsis, 163
  macrantha, 164
  rubella, 163
  stevioides, 164
Chaenomeles
  japonica, 522
  lagenaria, 522
  speciosa, 522
Chaetochloa
  lutescens, 776
  verticillata, 776
  viridis, 776
Chaff-bush, 139
Chamaebatiaria
  millefolium, 523
Chamaechaenactis
  scaposa, 164
    parva, 164
Chamaecyparis
  lawsoniana, 25
  nootkatensis, 25
Chamaenerion
  angustifolium, 441
    abbreviatum, 441
    platyphyllum, 441
Chamaerhodos
  American, 523
  erecta, 523
    nuttallii, 523
  parviflora, 523
  nuttallii, 523
Chamaesaracha
  coronopus, 604
Chamaesyce

setiloba, 303
Chamomile, 164
  yellow, 142
Chamomilla
  resuctita, 164
  suaveolens, 164
Chard
  Swiss, 122
Charlock, 255
Cheat, 710
Cheatgrass, 710
Cheeses, 420
Cheilanthes
  covillei, 20
  eatonii, 20
  feei, 20
  gracillima, 20
  parryi, 22
  siliquosa, 18
Cheiranthus
  asper, 268
  asperrimus, 268
  argillosus, 268
  bakeri, 268
  capitatus, 268
  cheiranthoides, 268
  elatus, 268
  inconspicuus, 268
  nivalis, 268
  amoenus, 268
Cheirinia
  amoena, 268
  brachycarpa, 268
  cheiranthoides, 268
  elata, 268
  inconspicua, 268
  repanda, 269
  syrticola, 268
  wheeleri, 268
Chelodonium
  majus, 451
Chelone
  barbata, 585
Chenopodina
  linearis, 130
  nigra, 130
Chenopodiaceae, 116
Chenopodium
  album, 124
    album, 124
    berlandieri, 124
    leptophyllum, 126
  ambrosioides, 124
  atrovirens, 124
  berlandieri, 124
    zschackei, 124
  botrys, 124
  calceoliforme, 130
  capitatum, 124
    capitatum, 124
    parvicapitatum, 124
  chenopodioides, 124
  dessicatum, 125
  fremontii, 125
    atrovirens, 124
    fremontii, 125
    incanum, 125
  gigantospermum, 125
  glaucum, 125
    salinum, 125
  hians, 124
  hybridum, 125
    gigantospermum, 125

  incanum, 125
  incognitum, 124
  leptophyllum, 126
    dessicatum, 125
    oblongifolium, 125
  murale, 126
  overi, 124
  petiolare, 125
    leptophylloides, 125
    pratericola, 125
    oblongifolium, 125
  rubrum, 126
  salinum, 125
  scoparium, 128
  spinosum, 127
  subspicatum, 121
  watsonii, 125
  zschackei, 124
Cherry
  bitter, 538
  bush, 538
  European Bird, 538
  flowering, 538
  May Day, 538
  pie, 537
  sour, 537
  St. Lucie, 537
  sweet, 537
  western sand, 537
Cherry-laurel, 538
Chess
  Chilean, 710
  downy, 710
  Japanese, 709
  meadow, 709
  rattlesnake, 707
  rye, 710
  soft, 708
Chia, 334
Chickweed
  Bering, 106
  common, 114
  field, 106
  mouse-ear, 106
  nodding, 106
Chicory, 169
  California, 221
  desert, 221
Chilopsis
  linearis, 59
Chimaphila
  menziesii, 501
  occidentalis, 501
  umbellata, 501
    occidentalis, 501
Chinaberry, 424
Chinch-weed, 217
Chionodoxa
  luciliae, 806
Chivalry-grass, 488
Chive, 803
Chloris
  curtipendula, 704
  verticillata, 714
  virgata, 714
Chlorocrambe
  hastatus, 259
Chokecherry, 539
Chondrosium
  eriopodum, 704
  gracile, 704
Cholla
  buckhorn, 88

pale, 89
sand, 91
Whipple, 91
Chorispermum
tenellum, 259
Chorispora
tenella, 259
Chorizanthe
brevicornu, 471
rigida, 471
thurberi, 471
Chrysanthemum
balsamita, 165
leucanthemum, 165
parthenium, 165
Chrysoma
pumila, 218
Chrysopsis
acaulis, 198
alpina, 155
caespitosa, 198, 204
foliosa, 204
hispida, 204
jonesii, 204
villosa, 204
scabra, 204
viscida, 204
cinerascens, 204
Chrysothamnus
affinis, 168
albidus, 165
axillaris, 169
consimilis, 167
depressus, 166
greenei, 166
lanceolatus, 169
linifolius, 166
marianus, 169
nauseosus, 166
abbreviata, 167
albicaulis, 167
arenarius, 167
bigelovii, 167
consimilis, 167
glabratus, 167
glareosus, 167
gnaphaloides, 167
graveolens, 167
hololeucus, 167
iridis, 167
junceus, 167
leiospermus, 167
nitidus, 167
psilocarpus, 167
salicifolius, 168
turbinatus, 168
uintahensis, 168
paniculatus, 168
parryi, 168
affinis, 168
attenuatus, 168
howardii, 168
nevadensis, 168
parryi, 168
pulchellus, 168
baileyi, 168
salicifolius, 168
speciosus, 167
albicaulis, 167
gnaphaloides, 167
vaseyi, 169
viscidiflorus, 169
axillaris, 169

lanceolatus, 169
puberulus, 169
pumilus, 169
stenophyllus, 169
viscidiflorus, 169
zionis, 167
Chylisma
brevipes, 436
claviformis, 437
scapoidea, 438
seorsa, 438
walkeri, 439
Cicendia
exaltata, 308
Cichorium
intybus, 169
Cicuta
douglasii, 618
maculata, 618
angustifolia, 618
maculata, 618
Cigartree, 59
Cinquefoil
acute, 531
contrary, 535
glandular, 533
green, 532
hair-tuft, 532
Hipp, 534
marsh, 535
pretty, 532
rough, 534
shrubby, 533
slender, 533
wedge-leaf, 532
woolly, 534
Cinna
glomerata, 698
latifolia, 715
Circaea
alpina, 439
pacifica, 439
pacifica, 439
Cirsium
acaule, 174
americanum, 174
acaulescens, 174
arizonicum, 171
arizonicum, 171
nidulum, 171
arvense, 171
mite, 171
horridum, 171
barnebyi, 171
bipinnatum, 172
calcareum, 171
bipinnatum, 172
calcareum, 172
pulchellum, 172
centaureae, 172
clavatum, 172
coloradoense, 174
davisii, 174
drummondii, 174
eatonii, 172
eatonii, 172
harrisonii, 173
murdockii, 173
eriocephalum, 172
leiocephalum, 172
foliosum, 174
lactucinum, 173
neomexicanum, 173

neomexicanum, 173
utahense, 173
nidulum, 171
ochrocentrum, 172
ownbeyi, 173
parryi, 174
pulchellum, 172
glabrescens, 172
rothrockii, 173
rydbergii, 173
scariosum, 174
scariosum, 174
thorneae, 174
scopulorum, 174
subniveum, 174
tioganum, 174
tweedyi, 173
undulatum, 174
albescens, 175
tracyi, 174
undulatum, 175
albescens
utahense, 173
virginensis, 175
vulgare, 175
wheeleri, 175
Citrullus
lanatus, 289
vulgaris, 289
Citrus
family, 546
Cladium
californicum, 678
mariscus, 678
californicum, 678
Cladothrix
oblongifolia, 46
Cladrastis
kentukea, 382
lutea, 382
tinctoria, 382
Clammy-weed, 98
Claretcup, 87
Clarkia
broad-leaved, 439
rhomboidea, 439
Claytonia
arctica, 494
megarhiza, 494
chamissoi, 495
cordifolia, 495
grayana, 494
lanceolata, 494
linearis, 495
megarhiza, 494
multicaulis, 494
multiscapa, 494
perfoliata, 495
rosea, 494
triphylla, 495
Clay-verbena, 426
Cleavers, 544
Clematis
alpina, 507
occidentalis, 507
repens, 507
teniuloba, 507
aurea, 508
columbiana, 507
tenuiloba, 507
douglasii, 508
jonesii, 508
hirsutissima, 508

ligusticifolia, 508
occidentalis, 508
oriental, 508
orientalis, 508
pseudoalpina, 507
pseudoatragene, 507
pseudoalpina, 507
wenderothioides, 507
tenuiloba, 507
wyethii, 508
Cleome
dodecandra, 98
integrifolia, 97
angusta, 97
lutea, 97
pinnata, 284
serrulata, 97
angusta, 97
serrulata, 97
Cleomella
cornuta, 98
montrosae, 98
nana, 98
palmeriana, 98
goodrichii, 98
plocasperma, 98
Cliff-brake
Brewer, 22
slender, 20
spiny, 22
Suksdorf, 22
ternate, 22
Cliff-rose, 539
Clinopodium
vulgaris, 535
Clubmosses, 13
Clusiaceae, 314
Cloak-fern
border, 21
Jones, 21
Parry, 21
Clomenocoma
cooperi 178
Clover
Alsike, 407
Andean, 406
Beckwith, 407
bur, 395
dwarf, 408
Frisco, 406
holy, 395
hop, 394
King, 408
lean, 408
Nuttall, 407
Parry, 408
red, 409
Rydberg, 408
strawberry, 407
tundra, 408
Uinta, 407
variegated, 409
white, 409
woolly. 407
Wormskjold, 409
Clypeola
alyssioides, 247
maritima, 277
minor, 247
Cnemidophacos
argillosus, 363
confertiflorus, 363
flavus, 363

moencoppensis, 370
rafaelensis, 375
toanus, 379
Cnicus
  arizonicus, 171
  benedictus, 175
  calcareus, 171
  clavatus, 172
  drummondii, 172
    acaulescens, 174
    bipinnatum, 172
  nidulus, 171
  rothrockii, 173
    diffusus, 173
  wheeleri, 175
Cochlearia
  armoracia, 254
  draba, 257
Cockle
  white, 108
Cocklebur, 239
Cockscomb, 45
Coffee-tree
  Kentucky, 385
Cogswellia
  cottamii, 632
  grayi, 630
  millefolia, 630
    depauperata, 630
  minima, 631
Colchicum
  autumnale, 806
Coldenia
  canescens, 84
    subnuda, 84
  hispidissima, 84
  nuttallii, 84
Coleogyne
  ramosissima, 523
Coleosanthus
  californicus, 160
  garrttii, 160
  longifolia, 160
Collinsia
  parviflora, 571
Collomia
  debilis, 456
  gilioides, 460
  glandiflora, 456
  gracilis, 466
    humilior, 466
  linearis, 436
  tenella, 456
  tinctoria, 456
Coloptera
  jonesii, 624
Columbine
  alcove, 506
  Colorado, 505
  elegant, 506
  golden, 506
  rock, 507
  shale, 505
  western, 506
  yellow, 506
Colutea
  arborescens, 382
Comandra
  linearis, 556
  pallida, 556
  umbellata, 556
    pallida, 556
Combseed

bent, 82
flattened, 82
saucer, 82
unequal, 82
Comarella
  sabulosa, 528
Comarum
  palustre, 535
Commelinaceae, 652
Common fern
  family, 17
Compositae, 131
Conanthus
  demissus, 318
  hispidus, 318
  parviflorus, 318
Coneflower
  cutleaf, 222
  Prairie, 221
  western, 222
Conioselinum
  scopulorum, 618
Conium
  maculatum, 619
Conringia
  orientalis, 259
Convallaria
  majalis, 806
  racemosa, 809
  stellata, 809
Convolvulaceae, 240
Convolvulus
  arvensis, 241
  batatas, 241
  equitans, 241
  longipes, 240
  purpureus, 241
  sepium, 240
Conyza
  bonariensis, 175
  canadensis, 176
    glabrata, 176
  coulteri, 176
Copperweed, 216
Coralberry, 101
Coralroot
  early, 813
  spotted, 812
  striped, 812
  Wister, 813
Corallorhiza
  maculata, 812
  striata, 812
  trifida, 813
  wisteriana, 813
Corchorus
  serrata, 613
Cordgrass
  alkali, 778
  prairie, 778
Cordylanthus
  canescens, 572
  kingii, 571
    densiflorus, 571
    kingii, 571
  maritimus, 572
    canescens, 572
    parryi, 572
  parryi, 572
  parviflorus, 572
  ramosus, 572
  wrightii, 572
Corispermum

emarginatum, 126
hyssopifolium, 126
imbricatum, 126
marginale, 126
nitidum, 126
simplicissimum, 126
villosum, 126
Cornaceae, 241
Cornflower, 162
Cornus
  mas, 242
  sericea, 242
    occidentalis, 242
    sericea, 242
    stolonifera, 242
  stolonifera, 242
Coronilla
  varia, 382
Corydalis
  aurea, 307
  caseana, 307
  engelmannii, 307
    exaltata, 307
  golden, 307
Corylus
  avellana, 58
Coryphantha
  chlorantha, 86
  marstonii, 85
  missouriensis, 85
    marstonii, 85
  vivipara, 86
    arizonica, 86
    desertii, 86
    vivipara, 86
Costmary, 165
Cotinus
  coggygria, 46
Cotoneaster
  acutifolia, 523
  dielsiana, 523
  horizontalis, 523
  microphylla, 523
  multiflora, 523
  pannosa, 523
  salicifolia, 523
Cottongrass
  tall, 481
  white, 481
Cottonwood
  black, 547
  Fremont, 548
  lanceleaf, 547
  narrowleaf, 547
Couchgrass, 725
Coulterina
  newberryi, 280
Cowania
  mexicana, 539
    stansburiana, 539
  stansburiana, 539
Crab
  Iowa, 529
  showy, 529
  Siberian, 529
  Tea, 529
Crabapple, 529
Crabgrass
  hairy, 719
  smooth, 719
Cranesbill
  Bicknell, 312
  Carolina, 313

slender, 313
Crassipes
  annuus, 775
Crassulaceae, 242
Crataegus
  chrysocarpa, 524
  crus-galli, 524
  doddsii, 524
  douglasii, 524
    duchesnensis, 524
    rivularis, 524
  mollis, 524
  monogyna, 524
  oxyacantha, 524
    monogyna, 524
  rivularis, 524
  rotundifolia, 524
  succulenta, 524
Creeper
  thicket, 645
  Virginia, 645
Crepidium
  glaucum, 178
Crepis
  acuminata, 176
  atrabarba, 176
  barbigera, 177
  capillaris, 176
  intermedia, 177
  modocensis, 177
  nana, 177
  occidentalis, 177
    costatus, 177
    gracilis, 176
    occidentalis, 177
    pumila, 177
    pumila, 177
  runcinata, 177
    alpicola, 178
    glauca, 178
    hispidulosa, 178
    runcinata, 178
Cress
  mouse-ear, 248
  water, 248
Cressa
  erecta, 241
  truxillensis, 241
Creutzfeldt-flower, 67
Crinitaria
  viscidiflorus, 169
Cristaria
  coccinea, 421
Critesion
  brachyantherum, 740
  jubatum, 741
  marinum, 741
  murinum, 741
  pusillum, 742
Crocus
  aureus, 789
  chrysanthus, 789
  flavus, 789
  vernus, 789
  wild, 504
Crookneck
  winter, 290
Croton
  californicus, 299
    longipes, 299
  longipes, 299
  Mohave, 299
  setigerus, 299

texensis, 299
Crowfoot
  water, 513
Crownbeard, 237
Crown-vetch, 382
Cruciferae, 243
Crustweed, 195
Crypsis
  alopecuroides, 715
  schoenoides, 715
  squarrosa, 753
Cryptanth
  ally, 64
  Baja, 72
  Baker, 65
  Barneby, 65
  Basin, 70
  bearded, 65
  beguiling, 68
  Carbon, 71
  desert, 73
  dwarf, 69
  dye, 71
  erect, 74
  Fish Lake, 74
  golden, 67
  Graham, 69
  Greene, 68
  head, 66
  James, 66
  Johnston, 70
  Jones, 66
  Kelsey, 70
  long-flower, 71
  low, 64
  mound, 67
  narrowleaf, 65
  Nevada, 71
  opening, 67
  Osterhout, 72
  Paradox, 72
  Pipe Springs, 73
  plains, 67
  recurved, 73
  Rollins, 73
  short-flower, 65
  silky, 73
  slender, 74
  tall, 68
  Torrey, 74
  tubercle, 73
  tufted, 66
  unequal, 70
  Utah, 74
  Virgin, 75
  Watson, 75
  Wilkes, 64
  wing-nut, 72
  yellow, 68
  yellow-eye, 69
  yellow-hair, 69
  yellowish, 71
Cryptantha
  abata, 64
  affinis, 64
  ambigua, 64
    robustior, 64
  angustifolia, 65
  bakeri, 65
  barbigera, 65
  barnebyi, 65
  breviflora, 65
  caespitosa, 66

  capitata, 66
  cinerea, 66
    abortiva, 66
    arenicola, 66
    cinerea, 66
    jamesii, 66
    multicaulis, 66
  circumscissa, 67
  compacta, 67
  confertiflora, 67
  confusa, 64
  crassisepala, 67
    elachantha, 67
  creutzfeldtii, 67
  decipiens, 68
  depressa, 64, 67
  dumetorum, 68
  eastwoodiae, 64
  elata, 68
  fendleri, 68
  flava, 68
  flavoculata, 69
  fulvocanescens, 69
    echinoides, 69
  geminata, 64
  gracilis, 69
    hillmannii, 69
  grahamii, 69
  hillmannii, 69
  humilis, 69
    commixta, 69
    nana, 70
    ovina, 69
    shantzii, 69
  inaequata, 70
  interrupta, 70
  jamesii, 66
    abortiva, 66
    cinerea, 66
    disticha, 66
    multicaulis, 66
    pustulosa, 67
    setosa, 66
  johnstonii, 70
  jonesiana, 70
  kelseyana, 70
  leptophylla, 71
  longiflora, 71
  mensana, 71
  micrantha, 71
  multicaulis, 64
  nana, 70
    ovina, 69
  nevadensis, 71
  ochroleuca, 71
  osterhoutii, 72
  paradoxa, 72
  pattersonii, 70
  polycarpa, 64
  pterocarya, 72
    cycloptera, 72
    pterocarya, 72
  pustulosa, 67
  racemosa, 72
  ramulosissima, 68
  recurvata, 73
  rollinsii, 73
  rugulosa, 73
  scoparia, 73
  semiglabra, 73
  sericea, 73
  setosissima, 74
  stricta, 74

  tenuis, 74
  torreyana, 74
  utahensis, 74
  virginensis, 75
  watsonii, 75
  wetherillii, 75
  wyomingensis, 68
Cryptogramma
  acrostichoides, 20
  crispa, 20
    achrosticoides, 20
  densa, 18
  stelleri, 20
Cucumber
  squirting, 291
  stinking, 290
  wild, 291
Cucumber-root, 809
Cucumber-tree, 418
Cucumis
  melo, 290
  sativus, 290
Cucurbita
  californica, 290
  foetidissima, 290
  maxima, 290
  moschata, 290
  palmata, 290
  pepo, 290
    melopepo, 290
  siceraria, 291
Cucurbitaceae, 289
Cudweed
  cottonbatting, 195
  lowland, 196
  marsh, 196
  viscid, 196
  Wright, 196
Cupgrass
  prairie, 732
Cupressaceae, 25
Cupressus
  arizonica, 25
  disticha, 34
  lawsoniana, 25
Curlygrass, 739
Currant
  golden, 562
  gooseberry, 564
  northern black, 563
  Rothrock, 564
  squaw, 562
  sticky, 564
  trailing black, 563
  wax, 562
  western, 563
Cuscuta
  alata, 292
  anthemi, 292
  applanata, 292
  approximata, 292
  arvensis, 293
    calycina, 293
    pentagona, 293
    verrucosa, 293
  californica, 293
    squamigera, 294
  campestris, 293
  cephalanthii, 292
  cupulata, 292
  curta, 293
  cuspidata, 292
  decora, 293

    indecora, 293
    pulcherrima, 293
    subnuda, 293
  denticulata, 292
  glabrior, 293
    pedicellata, 293
  gracilis, 292
  gronovii, 293
    curta, 293
    hispidula, 293
  indecora, 293
    bifida, 293
    hispidula, 293
    neuropetala, 293
    subnuda, 293
  megalocarpa, 293
  neuropetala, 293
    littoralis, 293
    minor, 293
  nevadensis, 294
  occidentalis, 293
  pentagona, 293
    calycina, 293
    verrucosa, 293
  planiflora, 292
    approximata, 292
    pulcherrima, 293
  salina, 294
    apoda, 294
    squamigera, 294
  squamigera, 294
  tenuiflora, 292
  umbosa, 293
  urceolata, 292
  veatchii, 294
    apoda, 294
    verrucosa, 293
    glabrior, 293
    hispidula, 293
  warneri, 294
Cuscutaceae, 291
Cushaw, 290
Cutgrass
  rice, 743
Cutleaf, 157
Cycladenia
  humilis, 49
  jonesii, 49
  venusta, 49
  jonesii, 49
Cycloloma
  atriplicifolium, 126
Cydonia
  japonica, 522
    lagenaria, 522
  lagenaria, 522
  oblonga, 524
  speciosa, 522
  vulgaris, 524
Cymopterus
  acaulis, 620
    acaulis, 621
    fendleri, 621
    greeleyorum, 621
    higginsii, 621
    parvus, 621
  alpinus, 633
  basalticus, 621
  beckii, 621
  bulbosus, 621
  calcareus, 625
  corrugatus, 622
  coulteri, 622

scopulicola, 622
coulteri, 622
decipiens, 621
duchesnensis, 622
fendleri, 621
globosus, 622
hendersonii, 622
higginsii, 621
ibapensis, 622
jonesii, 625
lapidosus, 623
deserti, 623
lemmonii, 623
longipes, 623
minimus, 623
montanus, 622
globosus, 622
purpurascens, 624
multinervatus, 624
newberryi, 624
petraeus, 625
purpurascens, 624
purpureus, 625
jonesii, 625
purpureus, 625
rosei, 625
rosei, 625
terebinthinus, 625
albiflorus, 625
calcareus, 625
petraeus, 625
watsonii, 622
Cynanchum
utahense, 55
Cynodon
dactylon, 716
Cynoglossum
officinale, 75
Cynomarathrum
latilobum, 631
scabrum, 631
Cynosurus
durus, 775
indicus, 720
Cyperaceae, 653
Cyperus
acuminatus, 678
aristatus, 678
erythrorhizos, 678
esculentus, 678
schweinitzii, 679
Cypress
Arizona, 25
bald, 34
family, 25
swamp, 34
Cypripedium
bulbosum, 812
calceolus, 813
fasciculatum, 813
Cyrtorhyncha
ranunculina, 516
Cystium
araneosum, 367
cicer, 359
fremontii, 367
platyphyllidium, 368
platytrope, 373
salinum, 368
scorpionis, 368
stramineum, 368
Cystopteris
bulbifera, 21

fragilis, 21
Cytissus
laburnum, 386
scoparius, 382
Dactylis
glomerata, 716
Dactyloctenium
aegypticum, 687
Daffodil, 652
Daisy
Abajo, 183
alcove, 192
Awapa, 184
Basin, 189
bear, 191
Canaan, 184
Carrington, 184
Coulter, 185
Cronquist, 185
darkhead, 188
dwarf, 188
Eaton, 185
European, 159
Engelmann, 186
fern-leaf, 185
glaber, 187
Garrett, 186
Goodrich, 187
Greene, 190
hairy, 183
hoary, 184
Jones, 187
Kachina, 187
La Sal, 188
longleaf, 187
Maguire, 188
Marysvale, 188
mound, 184
mountain, 185
needleleaf, 188
Oregon, 190
Payson, 192
pretty, 184,186
professor, 189
religious, 189
silver, 184
smooth, 187
splendid, 191
spreading, 185
strange, 189
tall, 186
thin, 191
threadleaf, 186
trailing, 186
tufted, 184
Untermann, 191
Utah, 191
vernal, 191
Wah Wah, 192
Wasatch, 183
Yellow, 187
Zion, 190
Dalea
arborescens, 402
amoena, 402
pubescens, 402
candida, 383
oligophylla, 383
epica, 383
flavescens, 383
epica, 383
flavescens, 383
fremontii, 402

minutifolia, 402
pubescens, 402
johnsonii, 402
lanata, 383
lanata, 383
terminalis, 383
nummularia, 403
oligophylla, 383
polydenia, 403
searlsiae, 384
thompsonae, 403
terminalis, 383
whitingii, 403
woolly, 383
Dandelion
alpine, 232
common, 232
rough, 232
Danthonia
californica, 717
unispicata, 717
cusicki, 717
intermedia, 717
unispicata, 717
Dasiphora
fruticosa, 533
riparia, 533
Datura
cornucopia, 605
fastuosa, 605
metel, 605
meteloides, 605
stramonium, 605
wrightii, 605
Daucopappus
linearis, 187
Daucophyllum
lineare, 633
Daucus
carota, 626
carota, 626
sativus, 626
pusillus, 626
Day-lily, 808
Dead-nettle, 330
Deer-brush, 519
Delphinium
abietorum, 509
ajacis, 508
amabile, 509
andersonii, 508
andersonii, 509
cognatum, 509
scaposum, 509
barbeyi, 509
subroseum, 509
coelestinum, 509
cognatum, 509
cuculatum, 509
elatum, 509
occidentale, 509
exaltatum, 509
barbeyi, 509
geyeri, 509
leonardii, 509
menziesii, 509
utahense, 509
nelsonii, 509
utahense, 509
pinetorum, 509
nuttallianum, 509
occidentale, 509
barbeyi, 509

cuculatum, 509
occidentale, 510
pinetorum, 510
scaposum, 509
scopulorum, 509
attenuatum, 509
occidentale, 509
Densecress, 271
Deschampsia
caespitosa, 718
danthonioides, 718
elongata, 718
Descurainia
andrenarum, 261
brachycarpa, 261
nelsonii, 261
californica, 260
canescens, 261
halictorum, 261
andrenarum, 261
osmiarum, 261
incisa, 261
leptophylla, 261
intermedia, 260
longipedicellata, 260
glandulosa, 260
pinnata, 260
filipes, 260
glabra, 260
halictorum, 261
intermedia, 260
nelsonii, 261
osmiarum, 261
paysonii, 261
richardsonii, 260
brevipes, 261
incisa, 261
macrosperma, 261
procera, 261
sonnei, 261
viscosa, 261
rydbergii, 261
eglandulosa, 260
serrata, 261
sophia, 261
Desertgrass
King, 701
Desert-parsley, 630
Desmanthus
illinoensis, 384
Desmodium
thunbergii, 387
Devilweed
Mexican, 156
Dianthus
armeria, 107
barbatus, 107
caryophyllus, 107
chinensis, 107
deltoides, 107
Dicentra
spectabilis, 307
uniflora, 307
Dichanthelium
acuminatum, 755
lanuginosum, 755
oligosanthes, 759
Dichelostemma
pulchellum, 806
Dicoria
brandegei, 178
canescens, 178
clarkae, 178

paniculata, 178
wetherillii, 178
Dicots, 35
Didiplis
   diandra, 417
   linearis, 417
Dieteria
   gracilis, 200
Digitalis
   purpurea, 573
Digitaria
   adscendens, 719
   ciliaris, 719
   ischaemum, 719
   sanguinalis, 719
Diholcus
   bisulcatus, 356
   haydenianus, 356
   scobatinatulus, 356
Dimorphocarpa
   wislizenii, 262
Dinebra
   aristidoides, 703
Diospyros
   virginiana, 294
Diotis
   lanata, 123
Diplachne
   fascicularis, 744
Diplopappus
   hispidus, 204
Diplotaxis
   muralis, 261
Dipsacaceae, 294
Dipsacus
   fullonum, 294
      sylvestris, 294
   sylvestris, 294
Disella
   hederacea, 421
Dishwater-grass, 488
Disporum
   trachycarpum, 806
Distichlis
   dentata, 719
   maritima, 719
      laxa, 719
   spicata, 719
      stricta, 719
   stricta, 719
Ditchgrass
   family, 822
Dithyrea
   wislizenii, 262
Dock
   beach, 492
   bitter, 492
   curled, 491
   golden, 492
   western, 492
Dodder
   family, 292
   field, 293
   largefruit, 293
   plain, 293
   salt, 294
   slender, 292
   slenderflower, 292
   smalltooth, 292
   smooth, 293
   toothed, 292
   Warner, 294
   western, 293

wing, 292
Dodecatheon
   alpinum, 498
      majus, 498
   dentatum, 498
   jeffreyi, 498
      alpinum, 498
      redolens, 499
   meadia, 498
      alpinum, 498
      pauciflorum, 498
   pauciflorum, 498
      monanthum, 498
   pulchellum, 498
      pulchellum, 498
      monanthum. 498
      zionense, 499
   radicatum, 498
   redolens, 499
   zionemse, 499
Dogbane
   family, 47
   spreading, 48
Dogtooth-violet, 807
Dogweed, 179
Dogwood
   American, 242
   family, 241
   red-osier, 242
Donia
   lanceolata, 200
   squarrosa, 197
   uniflora, 200
Dondia
   ramosissima, 130
Doveweed, 299
Downingia
   laeta, 95
Draba
   Alaska, 267
   ammophila, 264
   andina, 266
   arctic, 265
   apiculata, 265
      daviesiae, 265
   asprella, 263
      zionensis, 263
   aurea, 263
      aureiformis, 263
      decumbens, 263
      luteola, 263
   aureiformis, 263
      leiocarpa, 263
   brachystylis, 264
   breaks, 267
   caeruleomontana, 265
      piperi, 265
   cana, 265
   caroliniana, 267
      stellifera, 267
   chrysantha, 264
      crassa, 264
   coloradoensis, 267
   crassa, 264
   crassifolia, 264
      parryi, 264
   creeping, 267
   cuneifolia, 264
      cuneifolia, 264
      helleri, 264
      leiocarpa, 264
      platycarpa, 264
      decumbens, 263

deflexa, 267
densifolia, 264
   apiculata, 265
   daviesiae, 265
   densifolia, 265
   nelsonii, 265
dictyota, 266
douglasii, 265
   douglasii, 265
   doublecomb, 266
Douglas, 265
dwarf, 266
fladnizensis, 265
glacialis, 265
   pectinata, 265
globosa, 265
   sphaerula, 265
golden, 263
hairy, 264
helleri, 264
incerta, 265
   laevicapsula, 265
   juniperina, 266
kassii, 265
laevicapsula, 265
lanceolata, 265
lonchocarpa, 265
   dasycarpa, 266
   exigua, 266
   lonchocarpa, 266
   semitonsa, 266
   vestita, 266
luteola, 263
Maguire, 266
maguirei, 266
   burkei, 266
   mccallae, 263
   micrantha, 267
montana, 266
mountain, 266
mulfordae, 265
nelsonii, 265
nemorosa, 266
   stenoloba, 267
nitida, 267
   nana, 267
   praelonga, 267
nivalis, 266
   elongata, 266
   exigua, 266
oligosperma, 266
   andina, 266
   juniperina, 266
   leiocarpa, 266
   microcarpa, 266
   oligosperma, 266
   pectinipila, 266
   saximontana, 266
oriebata, 267
parryi, 264
pattersonii, 265
   dasycarpa, 265
   hirticaulis, 265
pectinata, 265
platycarpa, 264
ramulosa, 267
rectifructa, 266
reptans, 266
   micrantha, 267
   reptans, 267
   stellifera, 267
rockcress, 254
saximontana, 266

sobolifera, 267
   uncinalis, 267
spectabilis, 267
   glabrescens, 267
   spectabilis, 267
sphaerula, 265
splendid, 267
spring, 267
stenoloba, 267
subalpina, 267
surculifera, 263
thick, 264
tundra, 267
uber, 263
uncinalis, 267
valida, 265
ventosa, 267
   ventosa, 267
verna, 268
viperensis, 264
Wasatch, 264
woods, 266
Zion, 265
zionensis, 263
Dracocephalum
   nuttallii, 333
   parviflorum, 329
   thymiflorum, 329
   virginianum, 333
Dropseed
   cushion, 781
   giant, 780
   hairy, 701
   mesa, 780
   pine, 701
   sand, 780
   spike, 780
   tall, 779
Dryas
   hookeriana, 525
   octopetala, 525
   hookeriana, 525
Drymocallis
   glabrata, 533
   glandulosa, 533
   micropetala, 533
   pseudorupestris, 533
      intermedia
Dryopteris
   filix-mas, 21
Duckweed
   family, 800
   ivy-leaf, 800
   lesser, 800
Dudleya
   arizonica, 242
   pulverulenta, 242
   arizonica, 242
Dunebroom
   narrow-leaf, 398
Dugaldia
   hoopesii, 398
Dusty-maiden
   alpine, 163
   Douglas, 163
   Fremont, 164
   showy, 164
   Stevia, 164
Dwarf-mistletoe
   fir, 644
   limber pine, 644
   lodgepole pine, 644
   pinyon, 644

southern, 644
Dyssodia
  acerosa, 178
  cooperi, 178
  papposa, 179
  pentachaeta, 179
    belinidium, 179
  thurberi, 179
Easter-flower, 500
Eastwood-plant, 164
Ebenaceae, 294
Ebony
  family 294
Ecballium
  elaterium, 291
Echinocactus
  acanthodes, 87
  glaucus, 93
  johnsonii, 87
  lecontei, 87
  polycephalus, 86
    polycephalus, 86
    xeranthemoides, 86
  pubispinus, 93
  sileri, 92
  simpsonii, 92
    minor, 92
  whipplei, 93
    spinosior, 93
Echinocereus
  engelmannii, 86
    chrysocentrus, 86
    purpureus, 87
    variegatus, 87
  phoeniceus, 87
    inermis, 87
  triglochidiatus, 87
    inermis, 87
    melanacanthus, 87
    mojavensis, 87
Echinochloa
  crus-galli, 720
    zelayensis, 720
    muricata, 720
    microstachya, 720
Echinocystis
  lobata, 291
Echinodorus
  rostratus, 651
    lanceolatus, 651
Echinomastus
  johnsonii, 87
Echinopsilon
  hyssopifolius, 122
Echinospermum
  collinum, 78
  floribundum, 76
  redowskii, 78
    cupulatum, 78
    occidentalis, 78
Echium
  vulgare, 76
Eddya
  hispidissima, 84
Edwinia
  americana, 559
  macrocalyx, 559
Ehretia
  hispida, 84
Elaeagnaceae, 294
Elaeagnus
  angustifolia, 295
  argentea, 295

commutata, 295
Elatinaceae, 295
Elatine
  brachysperma, 296
  californica, 296
  gracilis, 296
  rubella, 296
  triandra, 296
Elderberry
  blue, 100
  red, 100
Elecampane, 209
Eleocharis
  acicularis, 679
    bella, 680
  bella, 680
  bolanderi, 680
  calva, 680
  flavescens, 680
  macrostachya, 680
  ovata, 680
  palustris, 680
  parishii, 680
  parvula, 680
  pauciflora, 680
  quinqueflora, 680
  rostellata, 681
Elodea
  canadensis, 788
  longivaginata, 789
Elephant-head, 578
Eleusine
  indica, 720
  tristachya, 720
Elkweed, 311
Ellisia
  micrantha, 316
Elm
  American, 612
  Chinese, 612
  English, 612
  family, 611
  Siberian, 612
Elymus
  ambiguus, 726
    salina, 726
    salmonis, 726
  canadensis, 723
  cinereus, 723
  condensatus, 723
  elongatus, 723
  elymoides, 724
  glaucus, 724
  hanseni, 722
  hispidus, 724
  junceus, 725
  lanceolatus, 725
    albicans, 725
  macounii, 722
  pontica, 723
  pseudorepens, 722
  repens, 725
  salinus, 726
  saundersii, 723
  saxicolus, 723
  scribneri, 726
  simplex, 727
  smithii, 727
  spicatus, 727
  subsecundus, 728
  subvillosus, 725
  trachycaulus, 728
  triticoides, 728

simplex, 727
virginicus, 728
submuticus, 728
Elytrigia
  dasystachya, 725
  elongata, 723
  intermedia, 724
  pontica, 723
  repens, 725
  spicata, 727
  smithii, 727
Emmenanthe
  foliosa, 325
  lutea, 325
  penduliflora, 316
  salina, 325
  scopulina, 325
Emplectocladus
  fasciculatus, 538
Encelia
  bush, 179
  farinosa, 179
  frutescens, 179
    frutescens, 179
    resinosa, 179
    virginensis, 179
  microcephala, 202
  nudicaulis, 180
  nutans, 180
  virginensis, 179
Enceliopsis
  argophylla, 179
  nudicaulis, 180
  nutans, 180
Enneapogon
  desvauxii, 729
Ephedra
  coryi, 29
    viscida, 29
  cutleri, 29
  family, 28
  fasciculata, 28
  green, 28
  Mohave, 28
  Nevada, 28
  nevadensis, 28
  Torrey, 28
  torreyana, 28
  viridis, 28
    viridis, 29
    viscida, 29
Ephedraceae, 28
Epilobium
  adenocaulon, 441
    americanum, 441
    occidentale, 441
  alpinum, 440
    alpinum, 440
    clavatum, 441
    lactiflorum, 442
    nutans, 442
  americanum, 441
  anagallidifolium, 440
  angustifolium, 441
    abbreviatum, 441
    canescens, 441
    circumvagum, 441
    platyphyllum, 441
  brachycarpum, 441
  brevistylum, 441
  canum, 448
    garrettii, 448
  ciliatum, 441

  glandulosum, 441
  watsonii, 441
  clavatum, 441
  glaberrimum, 441
  glandulosum, 441
    adenocaulon, 441
  halleanum, 442
  hornemannii, 442
    hornemannii, 442
    lactiflorum, 442
  lactiflorum, 442
  latifolium, 442
  leptophyllum, 442
  nevadense, 442
  nutans, 442
  oliganthum, 442
    gracile, 442
  palmeri, 441
  paniculatum, 441
  saximontanum, 442
  tracyi, 441
  watsonii, 441
Epipactis
  convallarioides, 816
  gigantea, 813
Epithymum
  arvense, 293
  cephalanthii, 292
  indecorum, 293
Equisetaceae, 14
Equisetum
  arvense, 15
  funstonii, 15
  hyemale, 15
  kansanum, 15
  laevigatum, 15
  variegatum, 15
Eragrostis
  barrelieri, 729
  caroliniana, 731
  chloromelas, 730
  cilianensis, 730
  curvula, 730
  diffusa, 731
  fendleriana, 764
  hypnoides, 730
  lehmanniana, 730
  megastachya, 730
  mexicana, 730
  minor, 731
  orcuttiana, 730
  pectinacea, 731
  poaeoides, 731
  virescens, 731
Erectcress, 273
Eremalche
  exile, 421
Eremocarpus
  setigerus, 299
Eremocarya
  muricata, 71
Eremochloe
  kingii, 701
Eremocrinum
  albomarginatum, 807
Eremopyrum
  triticeum, 731
Erianthus
  ravennae, 773
Eriastrum
  diffusum, 457
  eremicum, 457
  sparsiflorum, 457

Ericaceae, 296
Ericameria
    nana, 201
Erigeron
    abajoensis, 183
    acris, 183
        asteroides, 183
        debilis, 183
    annuus, 183
    aphanactis, 183
        aphanactis, 183
        congestus, 183
    arenarioides, 183
    argentatus, 184
    awapensis, 184
    bellidiastrum, 184
    bonariensis, 175
    caespitosus, 184
        laccoliticus, 184
        nauseosus, 188
    callianthemus, 189
    camphoratum, 219
    canaani, 184
    canadensis, 176
        glabrata, 176
    canus, 184
    carringtonae, 184
    chrysopsidis, 184
    cinereus, 185
        aridus, 185
    compactus, 184
        compactus, 185
        consimilis, 185
    compositus, 185
        glabratus, 185
    concinnus, 183
        aphanactis, 183
    congestus, 183
    consimilis, 185
    controversus, 186
    corymbosus, 185
    coulteri, 185
    cronquistii, 185
    divaricatus, 185
    divergens, 185
        cinereus, 185
        divergens, 185
    eatonii, 185
        molestus, 185
    elatior, 186
    engelmannii, 186
    filifolius, 186
    flagellaris, 186
        trilobatus, 189
    florifer, 235
    formosissimus, 186
    frucetorum, 186
    garrettii, 186
    glabellus, 187
        mollis, 190
    goodrichii, 187
    grandiflorus, 186,190
        elatior, 186
    inamoenus, 188
    jonesii, 187
    kachinensis, 187
    leiomerus, 187
    leiophyllus, 190
    linearis, 187
    lonchophyllus, 187
    macranthus, 190
    maguirei, 188
        harrisonii, 188

mancus, 188
melanocephalus, 188
minusculus, 187
nanus, 188
nauseosus, 188
nematophyllus, 188
peregrinus, 189
    callianthemus, 189
        callianthemus, 189
        eucallianthemus, 189
        scaposus, 189
    pinnatisectus, 188
        insolens, 188
    proselyticus, 189
    pulcherrimus, 189
        pulcherrimus, 189
        wyomingia, 189
pulvinatus, 184
pumilis, 189
    concinnoides, 189
        condensatus, 189
        subglaber, 189
    intermedius, 189
        gracilior, 189
        euintermedius, 189
        intermedius, 189
    regalis, 189
    religiosus, 189
    salsuginosus, 189
        scaposus, 189
    simplex, 190
    sionis, 190
    sparsifolius, 189
    speciosus, 190
        macranthus, 190
        mollis, 190
        speciosus, 190
        uintahensis, 191
    stenophyllus, 183
        tetrapleuris, 191
    strigosus, 180
    subtrinervis, 190
    superbus, 191
    tener, 191
    uintahensis, 191
    uniflorus, 188
        melanocephalus, 188
    untermannii, 191
    ursinus, 191
    utahensis, 191
        sparsifolius, 191
        utahensis, 191
        tetrapleuris, 191
    vagus, 192
    wahwahensis, 192
    zothecinus, 192
Eriocarpum
    grindelioides, 213
Eriochloa
    contracta, 732
Eriodictyon
    angustifolium, 316
Eriogonum
    alatum, 474
        triste, 474
    ammophilum, 482
    angulosum, 482
        maculatum, 482
    aretioides, 474
    aridum, 486
    aureum, 478
        glutinosum, 478
        azaleastrum, 487

baileyi, 474
    davidsonii, 479
    tomentosum, 483
batemanii, 475
    batemanii, 475
    eremicum, 475
    ostlundii, 475
bicolor, 475
biumbellatum, 487
brachypodium, 475
brevicaule, 475
    aureum, 476
    brevicaule, 476
    cottamii, 476
    desertorum, 476
    ephedroides, 476
    laxifolium, 476
    loganum, 476
    nanum, 476
    promiscuum, 476
    pumilum, 476
    viridulum, 477
    wasatchense, 476
caespitosum, 477
campanulatum, 476
cernuum, 477
    tenue, 472
    umbraticum, 477
    viminale, 477
chrysocephalum, 476
    alpestre, 483
    angustum, 476
    desertorum, 476
    loganum, 476
clavellatum, 477
confertiflorum, 476
    stansburyi, 476
contortum, 477
corymbosum, 477
    albogilvum, 477
    aureum, 478
    corymbosum, 478
    cronquistii, 478
    davidsei, 478
    divaricatum, 478
    erectum, 478
    glutinosum, 478
    halophilum, 478
    humivagans, 478
    hylophium, 478
    matthewsiae, 485
    orbiculatum, 478
    velutinum, 478
    revealianum, 478
    smithii, 478
crispum, 478
cronquistii, 478
cupreum, 487
darrovii, 479
davidsonii, 479
deflexum, 479
    brachypodum, 475
    deflexum, 479
    exaltatum, 480
    hookeri, 480
    gilvum, 480
    insigne, 480
    nevadense, 479
    parryi, 475
    ultrum, 483
    desertorum, 476
    divergens, 478
    divaricatum, 479

duchesnense, 477
dudleyanum, 483
effusum, 477
    contortum, 477
    corymbosum, 478
    durum, 478
    foliosum, 482
    leptocladon, 481
    leptophyllum, 481
    orbiculatum, 478
    pallidum, 481
    shandsii, 481
ephedroides, 476
eremicum, 475
exaltatum, 480
fasciculatum, 479
    polifolium, 479
ferrissii, 487
filicaule, 485
filiforme, 487
flexum, 479
    ferronis, 479
    friscanum, 482
    fusiforme, 480
    fruticosum, 478
    glutinosum, 478
    glaberrimum, 487
    aureum, 487
gordonii, 479
grayii, 476
heermannii, 479
    subracemosum, 480
    sulcatum, 480
heracleoides, 480
    subalpinum, 486
    utahense, 480
hookeri, 480
howellii, 480
    subracemosum, 480
howellianum, 480
humivagans, 478
hylophilum, 478
inflatum, 480
    fusiforme, 480
    inflatum, 480
insigne, 480
intermontanum, 482
jamesii, 480
    flavescens, 481
    rupicola, 481
juncinellum, 479
kearneyi, 483
kingii, 476
    laxifolium, 476
lancifolium, 478
leptocladon, 481
    leptocladon, 481
    papiliunculi, 481
    ramosissimum, 481
    leptophyllum, 481
    loganum, 476
lonchophyllum, 481
    lonchophyllum, 482
    saurinum, 482
longilobum, 485
luteum, 487
maculatum, 482
marginale, 487
medium, 476
microthecum, 482
    bicolor, 475
    foliosum, 482
    laxiflorum, 482

leptocladon, 481
leptophyllum, 481
minutiflorum, 485
molestum, 479
  davidsonii, 479
nanum, 476
natum, 485
neglectum, 487
nelsonii, 482
nidularium, 482
nivale, 483
nudicaule, 476, 485
  angustum, 476
  garrettii, 476
  ochroflorum, 485
  parleyense, 476
  pumilum, 476
nummulare, 482
  ammophilum, 482
  nummulare, 483
nutans, 483
ochrocephalum, 476
  angustum, 476
ostlundii, 475
ovalifolium, 483
  multiscapum, 483
  nivale, 483
  ovalifolium, 483
  utahense, 483
palmeri, 483
palmeranum, 483
panguicense, 483
  alpestre, 483
parryi, 475
pauciflorum, 483
  panguicense, 483
perfoliatum, 487
pharnaceoides, 483
  cervinum, 483
plumatella, 483
  palmeri, 483
polifolium, 479
polycladon, 484
porteri, 486
puberulum, 484
pulvinatum, 485
pusillum, 484
racemosum, 484
  racemosum, 484
  zionis, 484
  coccineum, 484
ramosissimum, 481
reniforme, 484
  pusillum, 484
revealianum, 478
rubiflorum, 483
rydbergii, 487
salsuginosum, 484
saurinum, 482
scabrellum, 484
sessile, 487
shockleyi, 485
  longilobum, 485
  shockleyi, 485
simpsonii, 482
smithii, 478
soredium, 485
spathulatum, 485
  natum, 485
  spathulatum, 485
  panguicense, 483
spathuliforme, 475
subalpinum, 486

subreniforme, 485
sulcatum, 480
tenellum, 476
  cottamii, 476
  grandiflorum, 482
thomasii, 485
thompsonae, 485
  albiflorum, 486
  thompsonae, 486
trichopes, 486
trichopodum, 486
  minus, 486
triste, 474
tumulosum, 486
umbellatum, 486
  aridum, 486
  aureum, 487
  deserticum, 486
  dichrocephalum, 486
  ferrissii, 487
  glabratum, 487
  intectum, 487
  majus, 486
  porteri, 486
  subalpinum, 486
  subaridum, 487
  umbellatum, 487
  umbelliferum, 487
umbelliferum, 487
villiflorum, 487
  tumulosum, 486
vimineum, 479, 484
  polycladon, 484
  juncinellum, 479
  viridulum, 476
  wasatchense, 476
wetherillii, 487
wrightii, 487
zionis, 484
  coccineum, 484
Eriogyna
  caespitosa, 529
Erioneuron
  pilosum, 732
  pulchellum, 732
Eriophorum
  angustifolium, 681
  polystachion, 681
  scheuchzeri, 681
Eriophyllum
  lanatum, 192
  integrifolium, 192
  lanosum, 192
  wallacei, 192
Eritrichium
  angustifolium, 65
  aretioides, 76
  elongatum, 76
  barbigerum, 65
  californicum, 83
    subglochidiatum, 83
  canescens, 83
  arizonicum, 83
  crassisepalum, 67
  elongatum, 76
  paysonii, 76
  hispidum, 68
  leiocarpum, 68
  holopterum, 74
  submolle, 74
  micranthum, 71
  muriculatum, 64
  ambiguum, 64

  nanum, 76
  elongatum, 76
  pterocaryum, 72
  pectinatum, 72
  racemosum, 72
  setosissimum, 74
Erodium
  cicutarium, 312
  texanum, 312
Ervum
  multiflorum, 378
Erxlebenia
  minor, 502
Erysimum
  alliaria, 247
  aridum, 268
  asperrimum, 268
  asperum, 268
  amoenum, 268
  inconspicuum, 268
  purshii, 268
  bakeri, 268
  barbarea, 254
  capitatum, 268
  argillosum, 260
  cheiranthoides, 268
  elatum, 268
  glaberrimum, 282
  inconspicuum, 268
  nuttallii, 252
  oblanceolatum, 268
  officinale, 283
  parviflorum, 268
  pinnatum, 260
  puberulum, 252
  repandum, 269
  syrticolum, 268
  wheeleri, 268
Erythraea
  calycosa, 308
  arizonica, 308
  nuttallii, 308
Erythremia
  grandiflora, 212
Erythrocoma
  brevifolia, 526
  ciliata, 527
  triflora, 526
Erythronium
  grandiflorum, 807
  utahense, 807
Erythrostemon
  gilliesii, 381
Escheveria
  arizonica, 242
  pulverulenta, 242
Eschscholzia
  californica, 451
  glyptosperma, 452
  ludens, 452
  mexicana, 452
  minutiflora, 452
Espeletia
  amplexicaulis, 238
Eucephalus
  elegans, 155
Euchroma
  angustifolia, 569
Euclidium
  syriacum, 269
Eucnide
  urens, 413
Eucrypta

desert, 316
micrantha, 316
Euklisia
  cordata, 285
Eulophus
  ambiguus, 629
Eunanus
  bigelovii, 575
  breweri, 575
  parryi, 576
Euonymus
  alata, 115
  americana, 115
  atropurpurea, 115
  bungeana, 115
  europaea, 115
  japonica, 115
Eupatorium
  ageratifolium, 193
  herbaceum, 193
  bruneri, 193
  grandiflorum, 160
  herbaceum, 193
  maculatum, 193
  bruneri, 193
  occidentale, 193
  western, 193
Euphorb
  Fendler, 301
Euphorbia
  albomarginata, 300
  arenicola, 302
  brachycera, 300
  cyathophora, 301
  cyparissias, 301
  dentata, 301
  esula, 301
  fendleri, 301
  glyptosperma, 301
  heterophylla, 301
  hyssopifolia, 302
  maculata, 302
  marginata, 302
  micromera, 302
  montanum, 300
  robusta, 300
  myrsinites, 302
  nephradenia, 302
  ocellata, 302
  arenicola, 302
  parryi, 302
  peplus, 303
  podagrica, 302
  prostrata, 303
  pulcherrima, 300
  revoluta, 303
  robusta, 300
    interioris, 301
  serpyllifolia, 303
  setiloba, 303
  spathulata, 303
  supina, 302
Euphorbiaceae, 299
Euploca
  convulvulacea, 77
Eurotia
  lanata, 123
  subspinosa, 123
Euthamia
  occidentalis, 228
Eutoca
  franklinii, 323
  glandulosa, 323

heterophylla, 324
lutea, 325
sericea, 356
Evening primrose
  annual, 446
  biennial, 445
  Bridges, 447
  bronze, 447
  early, 448
  family, 434
  Hooker, 447
  Munz, 446
  Nuttall, 448
  pale, 447
  rootstock, 446
  Watson, 446
  whitestem, 445
  yellow, 447
Evening-snow, 465
Everlasting
  pearly, 140
Evolvulus
  nuttallianus, 241
  pilosus, 241
Exinia
  pulchella, 498
Exochorda
  grandiflora, 525
  racemosa, 525
Fabaceae, 336
Fagaceae, 304
Fagonia
  californica, 646
  clutensis, 646
  laevis, 646
  laevis, 646
Fagus
  grandiflora, 304
  sylvatica, 304
Fairybells, 806
Fallugia
  paradoxa, 525
Falseflax, 255
False-mermaid, 411
Fan-palm, 818
  desert, 818
Feathergrass
  New Mexico, 785
Felwort, 311
Fendlerbush, 558
Fendlera
  rupicola, 558
  tomentosa, 558
Fendlerella
  utahensis, 558
  Utah, 558
Fennel
  sweet, 626
Fern
  bracken, 23
  goldback, 23
  maidenhair, 18
  male, 23
  northern maidenhair, 18
Ferns, 15
Ferocactus
  acanthodes, 87
  lecontei, 87
  johnsonii, 87
Ferula
  foeniculacea, 630
  macrocarpa, 631
Fescue

bearded, 736
brome, 734
meadow, 736
Myur, 735
ravine, 736
red, 736
sheep, 735
sixweeks, 735
small, 734
spike, 734
tall, 734
thurber, 737
Utah, 734
Festuca
  arida, 734
  arizonica, 735
  arundinacea, 734
  barbata, 774
  brachyphylla, 735
  brevifolia, 735
    utahensis, 735
  bromoides, 734
  calligera, 735
  confinis, 745
  dasyclada, 734
  dertonsis, 734
  elatior, 736
    arundinacea, 734
  fascicularis, 744
  filiformis, 744
  idahoensis, 735
  jonesii, 736
  kingii, 744
  megalura, 735
  microstachys, 734
    pauciflora, 735
  myuros, 735
  nervosa, 677
  octoflora, 735
    octiflora, 735
    hirtella, 735
  ovina, 735
    brevifolia, 736
    duriuscula, 736
    glauca, 736
    ovina, 736
    rydbergii, 736
  pacifica, 734
  pratensis, 736
  reflexa, 734
  rubra, 736
  saximontana, 735
  sororia, 736
  subulata, 736
  thurberi, 736
  unioliodes, 707
Ficus
  carica, 425
Filago
  californica, 193
Fiddleneck
  medium, 60
  Menzies, 61
  rough, 61
Fieldcress
  Austrian, 281
Fig
  common, 425
Figwort
  family, 567
Filbert
  European, 58
Fimbristylis

spadicea, 681
Fingergrass
  feather, 714
Fir
  Douglas, 33
  subalpine, 29
  white, 29
Fire-thorn, 540
Fireweed, 441
  Dwarf, 442
Fishhook
  basin, 93
  Whipple, 93
  Wright, 94
Flag, 790
  yellow, 790
Flat-sedge
  bearded, 678
  chufa, 678
  redroot, 678
Flaveria
  campestris, 193
Fleur-de-lis, 790
Flax
  blue, 412
  broom, 412
  family, 412
  small yellow, 412
  Utah yellow, 413
Fleabane
  annual, 183
  bitter, 183
Fleece-flower, 489
Floerkea
  proserpinacoides, 411
Flower-of-an-hour, 419
Fluffgrass, 732
Foeniculum
  vulgare, 626
Footcactus
  Despain, 92
  Simpson, 92
  Winkler, 92
Forestiera
  neomexicana, 431
  pubescens, 431
Forget-me-not, 87
Forsellesia
  meionandra, 115
  nevadensis, 115
Forsythia
  suspensa, 431
Four-o'clock
  Bigelow, 428
  family, 426
  large, 429
  spreading, 429
  Standley, 429
  trailing, 427
  Watson, 428
  wild, 429
Foxglove, 573
Foxtail
  alpine, 696
  Carolina, 696
  marsh, 697
  meadow, 697
  shortawn, 696
  water, 697
Fragaria
  bracteata, 525
  glauca, 526
  helleri 526

platypetala, 526
vesca, 525
  americana, 526
  bracteata, 525
  virginiana, 525
    glauca, 526
    platypetala, 566
Frankenia
  jamesii, 306
  pulverulenta, 306
Frankeniaceae, 306
Franseria
  acanthicarpa, 139
  discolor, 139
  dumosa, 139
  eriocentra, 139
  tamentosa, 139
Frasera
  albomarginata, 311
  gypsicola, 311
  paniculata, 311
  speciosa, 311
  utahensis, 311
Fraxinus
  americana, 432
  anomala, 432
  coriacea, 432
  cuspidata, 432
  dipetala, 432
  excelsior, 432
  lanceolata, 432
  nigra, 432
  ornus, 432
  pennsylvanica, 432
    lanceolata, 432
    velutina, 432
  quadrangulata, 432
  velutina, 432
    coriacea, 432
Fremontia
  vermicularis, 129
Fringecup, 559
Fritillaria, 807
  atropurpurea, 807
  dichroa, 807
  leucella, 807
  pudica, 807
  utahensis, 807
Frogbit
  family, 788
Frog-fruit
  wedgeleaf, 640
Fumaria
  officinalis, 307
Fumariaceae, 306
Fumitory
  family, 306
Funastrum
  heterophyllum, 55
Gaillardia
  acaulis, 207
  aristata, 193
  arizonica, 194
  crassifolia, 194
  flava, 194
  gracilis, 194
  mearnsii, 194
  parryi, 194
  pinnatifida, 194
  spathulata, 194
  straminea, 194
Galanthus
  nivalis, 652
Galega

officinalis, 384
Galium
  agreste, 544
    echinospermum, 544
  aparine, 544
    echinospermum, 544
  asperrimum, 544
    asperulum, 544
  bifolium, 544
  boreale, 544
  brandegei, 545
  coloradoense, 544
  desereticum, 545
  emeryense, 545
  filipes, 544
  hypotrichium, 545
    scabriusculum, 545
    utahense, 545
  mexicanum, 544
    asperrimum, 544
    asperulum, 544
  multiflorum, 544
    coloradoense, 544
    multiflorum, 545
    watsonii, 545
  munzii, 545
  proliferum, 545
  scabriusculum, 545
    protoscabriusculum, 545
  stellatum, 545
  subbiflorum, 545
  tinctorium, 545
    subbiflorum, 545
  trifidum, 545
    subbiflorum, 545
  triflorum, 545
  utahense, 544
  vaillantii, 544
Galleta, 739
  big, 739
Galpinsia
  lavandulifolia, 434
Garrya
  flavescens, 307
Garryaceae, 307
Gastrolychnis
  drummondii, 108
Gaultheria
  humifusa, 297
  myrsinites, 297
Gaura
  coccinea, 443
  parviflora, 443
  scarlet, 443
  willow, 443
Gayophytum
  decipiens, 443
  diffusum, 443
    parviflorum, 444
  humile, 444
  intermedium, 444
  lasiospermum, 444
  nuttallii, 444
  racemosum, 444
  ramosissimum, 444
Genista
  juncea, 404
Gentian
  arctic, 308
  barbellate, 310
  Engelmann, 309
  explorer, 309
  family, 307

Jones, 310
Lapland, 310
meadow, 310
moss, 309
northern, 309
Parry, 309
Rocky Mountain, 308
Gentiana
  acuta, 309
  affinis, 308
  algida, 308
  amarella, 309
    acuta, 309
    heterosepala, 309
  barbellata, 309
  bracteosa, 309
  calycosa, 309
    asepala, 309
  detonsa, 310
  elegans, 310
  heterosepala, 309
  parryi, 309
  plebeia, 309
  prostrata, 309
  romanzovii, 308
  strictiflora, 308
  tenella, 310
  thermalis, 310
  tortuosa, 309
Gentianaceae, 307
Gentianella
  amarella, 309
    acuta, 309
    heterosepala, 309
  barbellata, 310
  detonsa, 310
    elegans, 310
  heterosepala, 309
  tenella, 310
  tortuosa, 310
Gentianopsis
  barbellata, 310
  detonsa, 310
  elegans, 310
Geraea
  canescens, 194
Geraniaceae, 311
Geranium
  atropurpureum, 313
  bicknellii, 312
  caespitosum, 312
    parryi, 313
  carolinianum, 313
  cicutarium, 312
  common, 311
  family, 311
  fremontii, 313
    parryi, 313
  marginale, 312
  pusillum, 313
  Richardson, 313
  richardsonii, 313
  small-leaf, 313
  sticky, 313
  viscosissimum, 313
    nervosum, 313
  nervosum, 313
Germander
  American, 336
Geum
  allepicum, 526
    strictum, 526
  canadense, 526

ciliatum, 527
  triflorum, 526
macrophyllum, 526
  perincisum, 526
  rydbergii, 526
  oregonense, 526
  perincisum, 526
rossii, 526
  turbinatum, 526
strictum, 526
  sericeum, 526
triflorum, 526
  ciliatum, 527
turbinatum, 526
urbanum, 526
  oregonense, 526
Gilia
aggregata, 458
  aggregata, 459
  attenuata, 458
    utahensis, 458
  arizonica, 459
  maculata, 459
  microsiphon, 458
arenaria, 460
  leptantha, 460
  rubella, 460
arizonica, 459
aurea, 465
bigelovii, 465
breweri, 466
brecciarum, 460
burleyana, 459
caespitosa, 459
calcarea, 460
capillaris, 459
clokeyi, 460
congesta, 459
  burleyana, 459
  congesta, 459
  crebrifolia, 459
  frutescens, 459
  nuda, 462
  paniculata, 459
  palmifrons, 459
crebrifolia, 459
dactylophyllum, 465
debilis, 456
demissa, 465
depressa, 459
eremica, 457
filifolia, 457
  diffusa, 457
filiformis, 460
flavocincta, 460
  australis, 460
floribunda, 464
  arida, 464
frutescens, 459
gilioides, 460
gracilis, 465
  spirillifera, 465
  nana, 465
Gunnison, 460
gunnisonii, 460
harknessii, 465
Hayden, 460
haydenii, 460
hutchinsifolia, 460
inconspicua, 460
  inconspicua, 461
  sinuata, 461
  variegata, 460

latifolia, 461
leptomeria, 461
  leptomeria, 461
  micromeria, 461
  rubella, 460
  tridentata, 461
longiflora, 461
mcvickerae, 461
micromeria, 461
nuttallii, 464
ochroleuca, 460
  transmontana, 460
opthalmoides, 460
  australis, 460
parvula, 463
pinnatifida, 461
polycladon, 461
pumila, 462
Rabbit Valley, 459
roseata, 462
scarlet, 458
scopulorum, 462
  deformis, 462
sinuata, 460
sparsiflora, 457
spicata, 462
  spicata, 462
  tridactyla, 463
spiny, 461
stellata, 462
stenothyrsa, 462
straminea, 460
subnuda, 463
superba, 463
tenerrima, 463
tenuituba, 459
tridactyla, 463
triodon, 461
watsonii, 464
yellow, 460
Ginkgo, 29
  biloba, 29
  family, 29
Ginkgoaceae, 29
Ginseng
  family, 50
Gladiolus
  carneus, 789
Glandularia
  bipinnatifida, 640
  gooddingii, 641
Glandweed
  Cooper, 178
  pappose, 178
  scale, 178
Glaucocarpum
  suffrutescens, 283
Glaux
  maritima, 499
Glecoma
  hederacea, 329
Gleditsia
  triacanthos, 384
Globemallow
  common, 421
  Jane, 423
  Jones, 421
  Moore, 423
  Munroe, 423
  Nelson, 424
  Psoralea, 424
Glory-of-the-snow, 806
Glossopetalon

meionandrum, 115
nevadense, 115
Glyceria
  borealis, 737
  elata, 738
  grandis, 737
  maxima, 737
    grandis, 737
  pauciflora, 772
  striata, 738
    striata, 738
    stricta, 738
Glycine
  floribunda, 411
Glycosma
  maxima, 635
  occidentalis, 635
Glycyrrhiza
  glabra, 385
  lepidota, 384
Glyptopleura
  marginata, 195
  setulosa, 195
Gnaphalium
  alpinum, 140
  chilense, 195
  dimorphum, 141
  exilifolium, 195
  grayi, 195
  luteo-album, 195
  margaritaceum, 140
  microcephalum, 195
  palustre, 196
  uliginosum, 196
  viscosum, 196
  wrightii, 196
Goatgrass
  jointed, 692
Goat-head, 647
Goatsrue, 384
Goldenaster
  hairy, 204
  Jones, 204
  sand, 204
Golden-bell, 431
Goldenbush
  Antelope, 199
  Cedar Breaks, 207
  cobwebby, 200
  larchleaf, 200
  low, 201
  Mohave, 200
  Pine Valley, 199
  spindly, 201
  Watson, 201
Golden-chain, 386
Goldeneye
  hairy, 237
  showy, 238
  Tropic, 238
Goldenhead, 137
Goldenrod
  alcove, 228
  coast, 229
  dwarf, 228
  low, 228
  Missouri, 228
  Nevada, 229
  Parry, 228
  rock, 218
  western, 228
Goldenweed
  Clement, 199

meadow, 200
slender, 201
slight, 200
spiny, 201
stemless, 198
thrifty, 199
Goodyera
  oblongifolia, 814
Gooseberry
  desert, 564
  Missouri, 564
  swamp black, 563
  trmpet, 564
  whitestem, 563
  wine, 563
Goosefoot
  desert, 125
  family, 116
  Fremont, 125
  mapleleaf, 125
  mountain, 124
  narrowleaf, 126
  nettleleaf, 126
  oakleaf, 125
  red, 126
Goosegrass, 720
Gourd
  California, 290
  family, 289
  stinking, 290
Grama
  black, 704
  blue, 704
  hairy, 705
  mat, 705
  needle, 703
  red, 705
  sideoats, 704
  sixweeks, 704
Gramineae, 684
Grammica
  campestris, 293
  cuspidata, 292
  denticulata, 292
  indecora, 293
    neuropelata, 293
  occidentalis, 293
  pentagona, 293
  salina, 294
Graphephorum
  flexuosum, 772
Grape
  canyon, 645
  Concord, 646
  family, 645
  fox, 646
  Oregon, 56
  wine, 646
Grapefern
  lanceleaf, 16
  little, 17
  northern, 16
Grass
  Arabian Mediterranean, 773
  barnyard, 720
  Bermuda, 716
  black, 796
  blowout, 772
  blue-eyed, 790
  buffalo, 711
  canary, 758
  family, 684
  Johnson, 777

Mediterranean, 774
needle-and-thread, 782
orchard, 716
purple-eyed, 790
rabbitfoot, 770
Ravenna, 773
rescue, 707
slough, 778
Grass-of-Parnassus, 560
  fringed, 560
Gratiola
  neglecta, 573
Gravel-ghost, 156
Graylocks, 208
Grayia
  brandegei, 130
  polygaloides, 127
  spinosa, 127
Greasebush
  Nevada, 115
  Utah, 115
Greasewood, 129
Greenthread, 233
Grindelia
  aphanactis, 196
  brownii, 196
  fastigiata, 196
  laciniata, 196
  nana, 196
    brownii, 196
  serrulata, 197
    depressa, 197
  squarrosa, 197
    serrulata, 197
    squarrosa, 197
  stylosa, 237
Gromwell, 78
Ground-cherry
  common, 608
  ivy-leaved, 608
  thickleaf, 607
Groundsel
  alpine, 223
  arrowleaf, 227
  balsam, 226
  Bigelow, 223
  black, 223
  broom, 226
  desert, 224
  different, 224
  Douglas, 224
  Fendler, 224
  foetid, 225
  Fremont, 225
  Gray, 223
  Hart, 225
  manyface, 228
  montane, 227
  roundhead, 226
  saffron, 224
  saw, 226
  thick, 224
  Uinta, 225
  water, 225
Grouseberry, 299
Gruvelia
  setosa, 82
Guilandina
  dioica, 385
Guillenia
  cooperi, 258
  lasiophylla, 258
Gumweed

curly, 197
erect, 196
low, 196
Gutierrezia
  microcephala, 197
  petradoria, 197
  pomariensis, 197
  sarothrae, 107
    microcephala, 197
    pomariensis, 197
Guttiferae, 314
Gymnocladus
  dioica, 385
Gymnogramma
  trianglare, 23
Gymnolomia
  hispida, 237
    ciliata, 237
    linearis, 238
    longifolia, 237
    multiflora, 237
    annua, 238
    nevadensis, 238
Gymnosperms, 24
Gymnosteris
  parvula, 463
Gypsophila
  paniculata, 107
  scorzonerifolia, 107
Habenaria
  borealis, 814
    albiflora, 814
    dilitata, 814
    viridiflora, 815
  dilatata, 814
    albiflora, 814
    dilatata, 814
    leucostachys, 815
  hyperborea, 815
  neomexicana, 815
  obtusata, 815
  saccata, 815
  sparsiflora, 815
    laxiflora, 815
    sparsiflora, 815
    stricta, 815
  unalaschcensis, 815
  viridis, 816
    bracteata, 816
    zothecina, 816
Hackberry, 611
  netleaf, 612
Hackelia
  floribunda, 76
  ibapensis, 77
  jessicae, 77
  micrantha, 77
  patens, 77
    harrisonii, 77
    patens, 77
Hairgrass
  annual, 718
  slender, 718
  tufted, 718
Hakmatack, 30
Halimolobos
  virgatus, 269
Halogeton, 127
  glomeratus, 127
Haloragaceae, 314
Halostachys
  occidentalis, 117
Hamamilidaceae, 314

Hammerwort, 637
Hamosa
  atratiformis, 377
  calycosa, 357
  emoryana, 362
  ensiformis, 362
  imperfecta, 372
  manca, 357
  scaposa, 357
Haplopappus
  acaulis, 198
    acaulis, 199
    glabratus, 199
  apargioides, 199
  armerioides, 199
    armerioides, 199
    gramineus, 199
  cervinus, 199
  clementis, 199
  crispus, 199
  croceus, 200
  drummondii, 200
  falcatus, 199
  gracilis, 200
  integrifolius, 199
  interior, 200
  lanceolatus, 200
    subviscosus, 200
    tenuicaulis, 200
  laricifolius, 200
  linearifolius, 200
    interior, 200
  macronema, 200
  nanus, 201
  nuttallii, 213
    depressus, 214
  parryi, 228
    minor, 228
  racemosus, 201
    prionophyllus, 201
    sessiliflorus, 201
  rydbergii, 201
  scopulorum, 202
    hirtellus, 201
  sphaerocephalus, 137
  spinulosus, 201
  tenuicaulis, 200
  tortifolius, 240
  uniflorus, 199, 200
  watsonii, 201
    rydbergii, 201
  zionis, 201
Hardgrass, 775
Harebell
  arctic, 95
Hashish, 97
Hawksbeard
  Gray, 177
  meadow, 177
  Modoc, 177
  mountain, 176
  slender, 176
  thread, 176
  western, 177
Hawkweed
  Fendler, 205
  Houndstongue, 205
  slender, 205
  white, 204
Hawthorn
  English, 524
  one-seeded, 524
  red, 524

river, 524
yellow, 524
Hazelnut, 58
Heal-all, 333
Heath
  family, 296
Hedeoma
  dentata, 330
  nana, 330
  drummondii, 330
  incana, 333
  nana, 330
Hedge-hyssop
  common, 573
Hedge-parsley
  California, 637
Hedera
  helix, 51
  quinquefolia, 645
Hedyotis
  rubra, 545
Hedysarum
  boreale, 385
    boreale, 385
    gremiale, 385
    obovatum, 385
    utahense, 385
    cinerascens, 385
  canescens, 385
  carnulosum, 385
  gremiale, 385
  lancifolium, 385
  mackenziei, 385
  pabulare, 385
  marginastrum, 385
  occidentale, 385
    canone, 386
    occidentale, 386
  onobrychis, 385
  pabulare, 385
  rivulare, 385
  uintahense, 385
  utahense, 385
Helebore
  false, 810
Heleniastrum
  hoopesii, 202
Helenium
  autumnale, 202
  hoopesii, 202
Heleochloa
  alopecuroides, 715
  schoenoides, 715
Helianthella
  microcephala, 202
  multicaulis, 202
  quinquenervis, 202
  uniflora, 202
Helianthus
  annuus, 203
    lenticularis, 203
    macrocarpus, 203
  anomalus, 203
  bracteatus, 203
  deserticolus, 203
  giganteus, 203
    utahensis, 203
  lenticularis, 203
  macrocarpus, 203
  niveus, 203
  nuttallii, 203
  petiolaris, 203
  fallax, 203

quinquenervis, 202
tuberosus, 203
uniflorus, 202
utahensis, 203
Helicotrichon
  mortoniana, 738
Heliomeris
  hispida, 237
  longifolia, 237
    annua, 238
    multiflora, 237
      hispida, 237
    multiflora, 230
    nevadensis, 238
  soliceps, 238
Heliotrope
  salt, 77
  showy, 77
Heliotropium
  convulvulaceum, 77
  curassavicum, 77
    obovatum, 78
    oculatum, 78
  oculatum, 78
Helleborine, 813
Helonias
  paniculata, 811
Helopus
  mollis, 811
Hemerocallis
  fulva, 808
Hendecandra
  texensis, 299
Hemlock
  poison, 619
  water, 618
Hemp, 96
  family, 96
Henbane, 605
Henbit
  red, 330
Heracleum
  lanatum, 626
  spondylium, 626
    lanatum, 626
Hercules-club, 50
Hermidium
  alipes, 428
  pallidum, 428
Herniaria
  glabra, 107
Herpestis
  pilosa, 574
Hesperanthes
  albomarginata, 807
Hesperis
  matronalis, 269
  virgata, 269
Hesperochiron
  californicus, 317
  pumilus, 317
Hesperochloa
  kingii, 745
Hesperonia
  glutinosa, 428
Heterotheca
  grandiflora, 204
  jonesii, 204
  psammophila, 204
  subaxillaris, 204
  villosa, 204
    foliosa, 204
    hispida, 204

villosa, 204
Heuchera
  parvifolia, 558
  utahensis, 558
  rubescens, 558
    rubescens, 559
    versicolor, 559
  versicolor, 559
  pumila 559
  utahensis, 558
Heyderia
  decurrens, 25
Hibiscus
  syriacus, 420
  trionum, 420
Hickory
  shagbark, 327
Hieracium
  albiflorum, 204
  cynoglossoides, 205
  fendleri, 205
  gracile, 205
  griseum, 205
  runcinatum, 177
  scouleri, 205
  utahense, 205
Hierochloe
  odorata, 738
Hilaria
  jamesii, 739
  rigida, 739
Hippocastanaceae, 315
Hippeastrum
  aulicum, 652
  elegans, 652
  reginae, 652
  striatum, 652
Hippophae
  argentae, 295
  canadensis, 295
Hippuridaceae, 315
Hippuris
  vulgaris, 315
Hispidcress, 272
Hoffmanseggia
  repens, 381
Hofmeistera
  pluriseta, 205
Holcus
  halepensis, 777
  lanatus, 740
  odoratus, 738
Holly
  desert, 120
  family, 50
Holly-fern
  Krukeberg, 23
  rock, 23
Hollyhock, 419
  wild, 420
Holodiscus
  discolor, 527
    dumosa, 527
    microphyllus, 527
  dumosus, 527
  microphyllus, 527
Holosteum
  umbellatum, 107
  umbellatum, 107
Homalobus
  brachycarpus, 377
  caespitosus, 377
  campestris, 359

canescens, 377
canovirens, 359
coltonii, 359
detritalis, 361
episcopus, 362
flexuosus, 364
humilis, 370
junceus, 359
junciformis, 359
lancearius, 366
macrocarpus, 368
multiflorus, 378
oblongifolius, 370
orthocarpus, 361
paucijugus, 370
strigulosus, 378
tegetarius, 366
tenellus, 378
tenuifolius, 370
uniflorus, 377
wingatanus, 380
woodruffii, 380
Homalocenchrus
  oryzoides, 743
Homopappus
  racemosus, 201
Honesty, 278
Honeysuckle
  family, 98
  fragrant, 99
  Japanese, 99
  tatarian, 99
  Utah, 99
Hookera
  pulchella, 806
Hop
  American, 97
  European, 97
Hophornbeam
  western, 58
Hopiweed, 44
Hopsage, 127
Hoptree, 546
Hordeum
  brachyantherum, 740
  caespitosum, 741
  geniculatum, 741
  glaucum, 741
  gussonianum, 741
  hystrix, 741
  jubatum, 741
    caespitosum, 741
  leporinum, 741
  marinum, 741
    gussoneanum, 741
  murinum, 741
  nodosum, 740
  pusillum, 742
    pubens, 742
  stebbinsii, 741
  vulgare, 742
Horehound
  common, 331
Horkelia
  baileyi, 528
    setosa, 528
  eremica, 527
  gordonii, 527
  kingii, 527
  mutabilis, 528
  sabulosa, 528
  shockleyi, 528
  utahensis, 528

Hornbeam, 58
Horned pondweed
  family, 824
Hornwort
  common, 116
  family, 115
Horsebrush
  Gray, 233
  Nuttall, 233
  littleleaf, 233
  longspine, 232
  thorny, 233
Horsechestnut
  common, 315
  family, 315
  red, 315
Horse-nettle, 329
Horse-radish, 254
Horsetail
  family, 14
  meadow, 15
Horsetails, 14
Horseweed, 176
Hosackia
  denticulata, 388
  plebeia, 388
  rigida, 389
    nummularia, 388
  wrightii, 389
Houndstongue, 75
Houstonia
  rubra, 545
  saxicola, 545
Huckleberry
  dwarf, 298
  mountain, 298
  western, 298
Huegelia
  eremica, 457
Hulsea
  heterochroma, 205
Hummingbird-flower, 448
Humulus
  americanus, 97
  lupulus, 97
Hutchinsia
  calycina, 284
    americana, 284
  procumbens, 270
Hyacinth
  common, 808
  grape, 809
Hyacinthus
  botyroides, 809
  orientalis, 808
Hyalineherb, 205
Hydrastis
  caroliniensis, 517
Hydrocharitaceae, 788
Hydrocotyle
  verticillata, 626
Hydrophyllaceae, 316
Hydrophyllum
  albifrons, 317
    fendleri, 317
  alpinum, 317
  capitatum, 317
    alpinum, 317
    capitatum, 317
    laxum, 317
    pumilum, 317
    densum, 317
  densifolium, 317

  fendleri, 317
  macrophyllum, 317
    occidentale, 317
  occidentale, 317
    fendleri, 317
    watsonii, 317
    pumilum, 317
    watsonii, 317
Hymenatherum
  acerosum, 178
  belinidium, 179
  pentachaetum, 179
  thurberi, 179
Hymenoclea
  salsola, 205
Hymenolobus
  divaricatus, 270
  erectus, 270
Hymenopappus, 205
  cinereus, 206
  douglasii, 163
  eriopodus, 206
  filifolius, 205
    alpestris, 206
    cinereus, 206
    eriopodus, 206
    lugens, 206
    luteus, 206
    megacephalus, 206
    nanus, 206
    nudipes, 206
    panciflorus, 206
    tomentosus, 206
  lugens, 206
  luteus, 206
  nanus, 206
  niveus, 206
  nudipes, 206
    alpestris, 206
    pauciflorus, 206
    tomentosus, 206
Hymenophysa
  pubescens, 257
Hymenoxys
  acaulis, 207
    acaulis, 207
    arizonica, 207
    caespitosa, 207
    cooperi, 207
    ivesiana, 207
  Cooper, 207
  cooperi, 207
  depressa, 207
  grandiflora, 208
  helenoides, 208
  ivesiana, 207
  lapidicola, 208
  Lemmon, 208
  lemmonii, 208
    greenei, 208
  low, 208
  richardsonii, 208
    floribunda, 208
    richardsonii, 208
    utahensis, 208
  rock, 208
  subintegra, 208
  torreyana, 208
Hyoscyamus
  niger, 605
Hypericum
  anagalloides, 314
  formosum, 314

  scouleri, 314
Hypochaeris
  radicata, 209
Ilex
  aquifolium, 50
  myrsinites, 115
  opaca, 50
Iliamna
  rivularis, 420
Imperata
  brevifolia, 742
  hookeri, 742
Incienso, 179
Indian-apple, 605
Indiangrass, 776
Indigo
  bastard, 339
  false, 339
Indigo-bush
  beauty, 402
  Fremont, 402
  glandular, 403
  Thompson, 403
Inula
  ericoides, 211
  helenium, 209
Ipomoea
  batatas, 241
  purpurea, 241
Ipomopsis
  aggregata, 458
    attenuata, 458
  congesta, 459
  depressa, 459
  frutescens, 459
  gunnisonii, 460
  inconspicua, 450
  longiflora, 461
  polycladon, 461
  pumila, 462
  roseata, 462
  spicata, 462
  tenuituba, 459
Iridaceae, 789
Iris
  family, 789
  germanica, 790
  Missouri, 790
  missouriensis, 790
  Paria, 790
  pariensis, 790
  pseudacorus, 790
Isatis
  tinctoria, 270
Isoetaceae, 13
Isoetes
  bolanderi, 13
  echinospora, 13
  howellii, 13
  lacustris, 13
Iva
  acerosa, 216
  axillaris, 209
  xanthifolia, 209
Ivesia
  baileyi, 528
    setosa, 528
  Gordon, 527
  gordonii, 527
  halophila, 527
  King, 527
  kingii, 527
  lemmonii, 522
  mutabilis, 528

sabulosa, 528
setosa, 528
Sevier, 528
shockleyi, 528
Utah, 528
utahensis, 528
Ivy
  English, 51
  poison, 47
Jacobsladder
  blue, 469
  leafy, 469
  pretty, 469
  viscid, 470
Jamesia
  americana, 559
  cliff, 559
Jerusalem-oak, 124
Jonesiella
  arcta, 374
  asclepiadoides, 355
  ellisiae, 374
  pattersonii, 372
  praelonga, 372
  sabulosa, 376
Jonquil, 652
Judas-tree, 381
Juglandaceae, 386
Juglans
  cinerea, 327
  illinoensis, 327
  major, 327
  nigra, 327
  regia, 327
  rupestris, 327
    major, 327
Jujube, 519
Juncaceae, 791
Juncaginaceae, 798
Juncus
  ablescens, 796
  alpinus, 792
  arcticus, 793
  articulatus, 793
  badius, 796
  balticus, 793
    montanus, 793
    vallicola, 793
  brunnescens, 794
  bryoides, 793
  bufonius, 793
    occidentalis, 793
  campestris, 798
  canadensis, 796
    kuntzii, 796
  castaneus, 793
  compressus, 794
  confusus, 794
  drummondii, 794
  dudleyi, 797
  ensifolius, 794
    brunnescens, 794
    ensifolius, 795
    montanus, 795
  filiformis, 795
  gerardii, 795
  hallii, 795
  interior, 796
  jonesii, 797
  kelloggii, 793
  longistylis, 795
  mertensianus, 796
  multiflorus, 798
  nevadensis, 796
  nodosus, 796
  parryi, 796
  parviflorus, 798
  regelii, 797
  saximontanus, 795
    brunnescens, 794
  sphaerocarpus, 793
  spicatus, 798
  tenuis, 797
    congestus, 796
    dudleyi, 796
  torreyi, 797
  tracyi, 794
    utahensis, 794
  triglumis, 797
    albescens, 796
  tweedyi, 797
  utahensis, 794
  xiphioides, 795
    montanus, 795
Junegrass, 743
Juniper
  Chinese, 26
  common, 26
  creeping, 26
  needle, 27
  one-seed, 27
  Pfitzer, 26
  Rocky Mountain, 27
  Utah, 26
  Virginia, 27
Juniperus
  californica, 27
    utahensis, 27
  chinensis, 26
  communis, 26
    depressa, 26
    montana, 26
    sibirica, 26
  horizontalis, 26
  mexicana, 26
    monosperma, 26
  monosperma, 26
  occidentalis, 26
    monosperma, 26
    utahensis, 27
  osteosperma, 26
  rigida, 27
  sabina, 27
  scopulorum, 27
  sibirica, 26
  tetragona, 26
    osteosperma, 26
  utahensis, 27
  virginiana, 27
    montana, 27
    scopulorum, 27
Jussiaea
  subacaulis, 438
Kallstroemia
  californica, 646
Kalmia
  glauca, 298
    microphylla, 298
  microphylla, 298
  polifolia, 298
    microphylla, 298
Kelloggia
  galioides, 545
Kentrophyta
  coloradoensis, 366
  canyon, 366
  Jessie, 366
  mountain, 366
  tall, 366
Kerria
  Japanese, 528
  japonica, 528
Kinnikinnick, 242, 297
Kittentails
  alpine, 568
  Wyoming, 568
Knapweed
  Russian, 162
Knotgrass, 757
Knotweed, 488
  broadleaf, 490
  bushy, 490
  Douglas, 489
  Kellogg, 489
  Sawatch, 489
Kobresia
  bellardii, 682
  mysosurioides, 682
  simpliciuscula, 682
Kochia
  americana, 127
    vestita, 127
  hyssopifolia, 122
  iranica, 128
  prostrata, 127
  scoparia, 128
    trichophylla, 128
    trichophylla, 128
    vestita, 127
Koeleria
  gracilis, 743
  macrantha, 743
  nitida, 743
  pyramidata, 743
Koelreuteria
  paniculata, 556
Kolkwitzia
  amabilis, 98
Krameria
  glandulosa, 328
  grayi, 327
  parvifolia, 328
    glandulosa, 328
    imparata, 328
Krameriaceae, 327
Kraunia
  floribunda, 411
Krynitzkia
  affinis, 64
  ambigua, 64
  angustifolia, 65
  barbigera, 65
  cycloptera, 72
  depressa, 64
  decipiens, 68
  dumetorum, 68
  echinoides, 69
  fendleri, 68
  glomerata, 75
    acuta, 75
    virginensis, 75
  leucophaea, 67
    alata, 67
  mensana, 71
  mixta, 65
  multicaulis, 66
    setosa, 66
  sericea, 73
  torreyana, 74
  utahensis, 74
  watsonii, 75
Kuhnia
  chlorolepis, 209
Kuhnistera
  candida, 383
    occidentalis, 383
  flavescens, 383
    occidentalis, 383
  oligophylla, 383
  searlsiae, 384
Labiatae, 328
Laburnum
  anagyroides, 386
  vulgare, 386
Lacefern, 20
Lace-vine
  silver, 488
Lactuca
  biennis, 210
  canadensis, 210
  ludoviciana, 210
  sativa, 210
  scariola, 210
  serriola, 210
  serriola, 210
  tatarica, 210
    pulchella, 210
Ladies-tresses
  hooded, 817
Lady-fern, 19
  alpine, 19
Ladysthumb, 490
Lagenaria
  siceraria, 291
Lagerstroemia
  indica, 417
Lamiaceae, 328
Lambsears, 336
Lambsquarter, 124
Lamium
  amplexicaule, 330
  purpureum, 330
Langloisia
  schottii, 463
  setosissima, 463
    campyloclados, 463
Laphamia
  palmeri, 218
  tenella, 218
  stansburyi, 218
Lappula
  collina, 78
  echinata, 78
  jessicae, 77
  micrantha, 77
  occidentalis, 78
    cupulata, 78
  squarrosa, 78
  texana, 78
Lapsana
  communis, 210
Larch
  European, 30
  Montana, 30
  mountain, 30
  western, 30
Larix
  decidua, 30
  europaea, 30
  occidentalis, 30
Larkspur
  Anderson, 508

Geyer, 509
Nelson, 509
rocket, 508
western, 509
Larrea
  divaricata, 646
  glutinosa, 646
  tridentata, 646
Lathyrus
  arizonicus, 387
  brachycalyx, 386
    brachycalyx, 386
    eucosmus, 386
    zionis, 386
  bradfieldianus, 387
  coriaceus, 387
  dissitifolius, 410
  eucosmus, 386
  laetivirens, 387
  lanszwertii, 387
    arizonicus, 387
    laetivirens, 387
    lanszwertii, 387
  latifolius, 387
  leucanthus, 387
    laetivirens, 387
  linearis, 410
  odoratus, 387
  pauciflorus, 387
    utahensis, 387
  sylvestris, 387
  utahensis, 387
  zionis, 386
Laurel
  bog, 298
Lavauxia
  flava, 447
  howardii, 447
Layia
  glandulosa, 211
  platyglossa, 211
    breviseta, 211
    campestris, 211
Lead-plant, 339
Ledum
  glandulosum, 298
Leersia
  oryzoides, 743
Leguminosae, 336
Legume
  family, 336
Leiostemon
  ambiguus, 584
  purpureum, 584
Lemna
  gibba, 800
  minima, 800
  minor, 800
    obscura, 800
  minuta, 800
  minuscula, 800
  obscura, 800
  polyrhiza, 800
  trisulca, 800
  valdiviana, 800
    minima, 800
Lemnaceae, 799
Lentibulariaceae, 411
Leontodon
  ceratophorus, 232
  lyratus, 232
Leonurus
  cardiaca, 330

Leopard-lily, 807
Lepargyraea
  argentea, 295
  canadensis, 295
  rotundifolia, 295
Lepidium
  albiflorum, 273
  alyssioides, 273
    jonesii, 273
    stenocarpum, 273
  barnebyanum, 271
  bourgeauanum, 271
  brachybotryum, 273
  campestre, 271
  chalepense, 257
  corymbosum, 273
  crandallii, 273
  crenatum, 273
  densiflorum, 271
    bourgeauanum, 271
    densiflorum, 271
    macrocarpum, 271
    pubicarpum, 271
    pubicaule, 271
    ramosum, 271
  dictyotum, 271
    macrocarpum, 271
  draba, 257
    chalepense, 257
  fremontii, 271
  georginum, 272
  heterophyllum, 273
  hirsutum, 274
  integrifolium, 272
    heterophyllum, 273
  intermedium, 274
    pubescens, 274
  repens, 257
  jonesii, 273
  lasiocarpum, 272
    georginum, 272
    lasiocarpum, 272
    palmeri, 272
  lasiophyllum, 272
  latifolium, 272
  montanum, 272
    alpinum, 273
    alyssioides, 273
    canescens, 273
    demissum, 271
    eastwoodiae, 273
    heterophyllum, 273
    integrifolium, 272
    jonesii, 273
    montanum, 273
    neeseae, 273
    spathulatum, 273
    stellae, 273
    stenocarpum, 273
  neglectum, 271
  ostleri, 273
  oxycarpum, 273
    strictum, 273
  palmeri, 272
  perfoliatum, 273
  philonitrum, 273
  procumbens, 270
  pubescens, 273
  pubicarpum, 271
  ramosum, 271
  repens, 257
  reticulatum, 273
  scopulorum, 273

  nanum, 273
  spathulatum, 273
  spathulatum, 273
  strictum, 273
  tortum, 273
  utahense, 272
  utaviense, 273
  vaseyanum, 273
  virginicum, 273
    pubescens, 274
  zionis, 272
Lepidospartum
  latisquamum, 211
Leptochloa
  fascicularis, 244
  filiformis, 244
  uninervia, 244
Leptodactylon
  brevifolium, 464
  caespitosum, 464
  nuttallii, 464
  pungens, 464
  watsonii, 464
Leptoseris
  sonchoides, 214
Leptotaenia
  dissecta, 629
  eatonii, 630
Lepturus
  paniculatus, 773
Lespedeza
  thunbergii, 387
Lesquerella
  alpina, 275
    alpina, 275
    condensata, 275
    intermedia, 275
    laevis, 275
    parvula, 275
    spatulata, 275
  argentea, 276
  arizonica, 275
    nudicaulis, 275
  barnebyi, 276
  condensata, 275
  curvipes, 275
  fendleri, 275
  folicacea, 275
  garrettii, 275
  goodrichii, 276
  gordonii, 277
    sessilis, 277
  hemiphysaria, 275
    hemiphysaria, 275
    lucens, 275
  hitchcockii, 277
    rubicundula, 277
    tumulosa, 277
  intermedia, 275
  kingii, 276
    cordiformis, 276
    latifolia, 276
    parvifolia, 276
  latifolia, 276
  ludoviciana, 276
  multiceps, 276
  occidentalis, 276
    cinerascens, 276
    parvifolia, 276
  palmeri, 276
  parvula, 275
  praecox, 275

  prostrata, 276
  rectipes, 276
  rubicundula, 277
  spatulata, 275
  stenophylla, 275
  subumbellata, 277
  tenella, 277
  tumulosa, 277
  utahensis, 277
  wardii, 277
Lettuce, 210
  blue, 210
  prickly, 210
Leucanthemum
  parthenium, 165
  vulgare, 165
Leucelene
  arenosa, 211
  ericoides, 211
Leucocrinum
  montanum, 808
Leucopoa
  kingii, 745
Lewisia
  brachycalyx, 494
  brachycarpa, 494
  pygmaea, 494
    nevadensis, 494
  rediviva, 495
  triphylla, 495
Leymus
  cinereus, 723
  salinus, 726
    salmonis, 726
  simplex, 727
  triticoides, 728
Libocedrus
  decurrens, 25
Licorice, 387
Ligularia
  amplectens, 223
  bigelovii, 223
    hallii, 223
  holmii, 223
  pudica, 226
Ligusticum
  brevilobum, 627
  eastwoodae, 636
  fernleaf, 627
  filicinum, 627
    tenuifolium, 627
  Grays, 627
  grayi, 627
  montanum, 623
  porteri, 627
  scopulorum, 619
  slender-leaf, 627
  small, 627
  southern, 627
  tenuifolium, 627
Ligustrum
  vulgare, 423
Lilac
  Amur, 433
  Chinese, 433
  common, 434
  Peking, 433
  Persian, 433
Liliopsida, 647
Liliaceae, 800
Lilium
  bulbiferum, 808
  longiflorum, 808
  maculatum, 808

pudicum, 807
Lily
 Alp, 808
 Easter, 808
 family, 800
 white-trumpet, 808
Lily-of-the-valley, 806
Limnathaceae, 411
Limnia
 utahensis, 495
Limnorchis
 dilitata, 814
  albiflora, 814
 hyperborea, 815
 laxiflora, 815
 leucostachys, 815
 saccata, 815
 sparsiflora, 815
Limosella
 aquatica, 573
Linaceae, 412
Linanthastrum
 nuttallii, 464
Linanthus
 aureus, 465
 bigelovii, 465
 demissus, 465
 dactylophyllum, 465
 dichotomus, 465
 harknessii, 465
 nuttallii, 464
 septentrionalis, 465
Linaria
 canadensis, 573
 dalmatica, 573
 genistifolia, 573
  dalmatica, 573
 linaria, 573
 texana, 573
 vulgaris, 573
Linden
 American, 611
 common, 611
 European, 611
 family, 610
 large-leaf, 611
 silver, 611
 small-leaved European, 611
Linnaea
 americana, 99
 borealis, 99
  americana, 99
 longiflora, 99
  longiflora, 99
Linosyris
 bigelovii, 167
 drummondii, 200
 howardii, 168
  nevadensis, 168
 parryi, 168
 serrulata, 169
 viscidiflorus, 169
  puberula, 169
Linum
 aristatum, 412
  subteres, 413
 australe, 412
 kingii, 412
  pinetorum, 412
  sedoides, 412
 lewisii, 412
 perenne, 412
  lewisii, 412

puberulum, 413
 rigidum, 413
  puberulum, 413
 subteres, 413
 usitatissimum, 413
Lipfern
 Coville, 20
 Eaton, 20
 Fee, 20
Lippia
 cuneifolia, 640
 nodiflora, 640
 wrightii, 640
Liquidambar
 striaciflua, 315
Liriodendron
 tulipifera, 418
Listera
 borealis, 816
 convallarioides, 816
 cordata, 816
Lithophragma
 australis, 560
 bulbifera, 559
 glabra, 559
  bulbifera, 559
 ramulosa, 560
 parviflora, 561
 tenella, 561
  ramulosa, 561
Lithospermum
 arvense, 78
 circumscissum, 67
 incisum, 79
 multiflorum, 79
 ruderale, 79
Live-oak
 turbinella, 306
Lizard-tail
 family, 557
Lloydia
 serotina, 808
Loasaceae, 413
Lobelia
 cardinalis, 96
 graminea, 96
 carnulosa, 96
 erinus, 96
  graminea, 96
 scarlet, 96
 splendens, 96
Lobularia
 maritima, 277
Locoweed
 Bessey, 396
 Lambert, 397
 silky, 397
 viscid, 397
 white, 397
 woolly, 370
 yellow, 397
Locust
 black, 404
 honey, 384
 New Mexico, 403
Loeflingia
 squarrosa, 108
Logania
 family, 84
Lolium
 multiflorum, 745
 perenne, 745
 aristatum, 745

italicum, 745
Lomatium
 ambiguum, 629
 anomalum, 633
 bicolor, 629
 big-seed, 631
 canyonlands, 631
 cliff, 632
 cous, 629
 dissectum, 629
  eatonii, 630
 foeniculaceum, 630
  fimbriatum, 630
  macdougalii, 630
 giant, 629
 grayi, 630
  grayi, 630
  depauperatum, 630
 jonesii, 630
 junceum, 630
 juniper, 630
 juniperinum, 630
 kingii, 631
  alpinum, 631
  kingii, 631
 lapidosum, 632
 latilobum, 631
 least, 631
 leptocarpum, 631
 macdougalii, 630
 macrocarpum, 631
 milfoil, 630
 millefolium, 630
 minimum, 631
 nakedstem, 632
 Nevada, 632
 nevadense, 632
  nevadense, 632
 parishii, 632
 nudicaule, 632
 nuttallii, 631
 Parry, 632
 parryi, 632
 Raven, 632
 ravenii, 632
 rough, 632
 rush, 630
 scabrum, 632
  scabrum, 633
  tripinnatum, 633
 simplex, 633
 stinking, 631
 ternate, 633
 triternatum, 633
  platycarpum, 633
  triternatum, 633
  anomalum, 633
Lonchophaca
 duchesnensis, 361
 kaibensis, 362
 macra, 368
 macrocarpa, 368
Lonicera
 fragrantissima, 99
 involucrata, 99
 ledebourii, 99
 japonica, 99
 tatarica, 99
 utahensis, 99
Looking-glass
 Venus, 96
Loosestrife
 family, 417

fringed, 499
 tufted, 499
Lophanthus
 urticifolius, 329
Lotodes
 ellipticum, 402
Lotus
 corniculatus, 388
 tenuifolius, 389
 denticulatus, 388
 humistratus, 388
 longebracteatus, 388
 neomexicanus, 388
 numularius, 388
 nummulus, 388
 oroboides, 388
  nanus, 388
 plebeius, 388
 rigidus, 389
 subpinnatus, 388
 tenuifolius, 389
 tenuis, 389
 utahensis, 389
 wrightii, 389
Lousewort
 Gray, 578
 leafy, 578
 Parry, 578
 pinyon-juniper, 578
Lovegrass
 creeping, 730
 Lehmann, 730
 Mediterranean, 729
 Mexican, 730
 minor, 731
 Orcutt, 730
 teal, 730
 tufted, 731
 weeping, 730
Lucern, 395
Luetkea
 caespitosa, 529
Lunaria
 annua, 278
Lupine
 broad-leaved, 392
 dwarf, 393
 elegant, 392
 Jones, 394
 King, 392
 Marysvale, 394
 Mohave, 394
 Robinson, 393
 rolled, 392
 sand, 393
 shortstem, 392
 showy, 393
 silky, 393
 silvery, 390
 sink, 394
 spur, 390
 spurred, 391
 stemless, 392
 white-leaved, 393
 yellow-eye, 392
Lupinus
 aduncus, 391
 aegra-ovium, 394
 alpestris, 391
 ammophilus, 393
 arbustus, 390
 calcaratus, 390
 arcticus, 393

prunophilus, 393
argenteus, 390
　argenteus, 391
　argophyllus, 391
　boreus, 391
　fulvomaculatus, 391
　moabensis, 391
　palmeri, 391
　parviflorus, 391
　rubricaulis, 391
　tenellus, 391
argillaceus, 392
argophyllus, 391
aridus, 392
　utahensis, 392
argillaceus, 392
barbiger, 394
brevicaulis, 391
caespitosus, 392
　utahensis, 392
calcaratus, 390
capitatus, 392
caudatus, 391
　argophyllus, 391
　cutleri, 392
　rubricaulis, 391
　utahensis, 392
columbianus, 392
concinnus, 392
　orcuttii, 392
cutleri, 392
decumbens, 391
　argentatus, 391
　argophyllus, 391
　eatonanus, 393
　egglestonianus, 394
　flavoculatus, 392
flexuosus, 394
fulvomaculatus, 391
garrettianus, 391
hillii, 391
helleri, 391
holosericeus, 392
　utahensis, 392
huffmanii, 394
humicola, 393
jonesii, 394
kingii, 392
　argillaceus, 392
　kingii, 392
larsonanus, 394
latifolius, 392
　columbianus, 392
　leucanthus, 392
laxiflorus, 391
　argophyllus, 391
　calcaratus, 390
laxus, 391
lepidus, 392
　aridus, 392
　caespitosus, 392
　utahensis, 392
leucanthus, 392
leucophyllus, 393
　lupinus, 392
lucidulus, 391
maculatus, 391
marianus, 394
micensis, 392
macounii, 391
maculatus, 391
odoratus, 392
　flavoculatus, 392

orcuttii, 392
palmerii, 391
parviflorus, 391
　fulvomaculatus, 391
polyphyllus, 393
　ammophilus, 393
　humicola, 393
　prunophyllus, 393
prunophilus, 393
pulcher, 391
puroviridis, 394
pusillus, 393
　intermontanus, 393
　pusillus, 393
　rubens, 393
quercus-jugi, 394
rickeri, 394
rubens, 393
　flavoculatus, 392
rubricaulis, 391
salinensis, 394
sericeus, 393
　barbiger, 394
　flexuosus, 394
　huffmanii, 394
　jonesii, 394
　marianus, 394
　sericeus, 394
sileri, 392
sparsiflorus, 394
spathulatus, 391
　boreus, 391
stenophyllus, 391
tenellus, 391
tooelensis, 393
volutans, 392
utahensis, 392
watsonii, 392
wyethii, 393
　prunophilus, 393
Luzula
　campestris, 798
　multiflora, 798
　intermedia, 798
　multiflora, 798
　parviflora, 798
　spicata, 798
　wahlenbergii, 798
Lychnis
　alba, 108
　apetala, 108
　　kingii, 108
　　montana, 108
　drummondii, 108
　　striata, 108
　kingii, 108
　montana, 108
　nodding, 108
　striata, 108
Lycium
　andersonii, 605
　barbarum, 606
　cooperi, 606
　halimifolium, 606
　pallidum, 606
　torreyi, 606
Lycopersicon
　esculenta, 606
Lycopodiophyta, 13
Lycopus
　americanus, 331
　asper, 331
　lucidus, 331

sinuatus, 331
Lycurus
　phleoides, 746
Lygodesmia
　arizonica, 212
　dianthopsis, 212
　doloresensis, 212
　entrada, 211
　exigua, 220
　grandiflora, 212
　　arizonica, 212
　　dianthopsis, 212
　　grandiflora, 212
　　stricta, 212
　juncea, 212
　　dianthopsis, 212
　spinosa, 231
Lysimachia
　ciliata, 499
　thyrsiflora, 499
Lythraceae, 417
Lythrum
　californicum, 418
　salicaria, 418
Machaeranthera
　bigelovii, 212
　canescens, 213
　　aristata, 213
　　canescens, 213
　　commixta, 212
　　latifolia, 213
　　leucanthemifolia, 213
　　vacans, 213
　commixta, 212
　glabriuscula, 240
　　confertifolia, 239
　grindelioides, 213
　　depressa, 214
　　grindelioides, 214
　kingii, 154
　　barnebyana, 154
　latifolia, 213
　leptophylla, 213
　leucanthemifolia, 213
　linearis, 213
　mucronata, 212
　paniculata, 213
　parviflora, 214
　pulverulenta, 213
　　vacans, 213
　rubicaulis, 213
　rubrotinctus, 213
　tanacetifolia, 214
　tephrodes, 213
　tortifolia, 240
　　imberbis, 240
　　tortifolia, 240
Maclura
　pomifera, 425
Macronema
　discoideum, 200
　obovatum, 201
Macrorhynchus
　glaucus, 138
　　laciniatus, 138
　heterophyllus, 138
　retrorsus, 138
Madder
　family, 543
Madia
　glomerata, 214
　gracilis, 214
Madronella

oblongifolia, 332
parvifolia, 332
sessilifolia, 332
Magnolia
　acuminata, 418
　denudata, 418
　family, 418
　grandiflora, 418
　liliflora, 418
　showy, 418
　soulangeana, 418
Magnoliaceae, 418
Magnoliophyta, 35
Magnoliopsida, 35
Mahaleb, 538
Mahoberberis, 56
Mahogany
　alder-leaf mountain, 522
　family, 524
　curl-leaf mountain, 522
　dwarf mountain, 522
Mahonia
　aquifolium, 56
　creeping, 56
　Fremont, 56
　fremontii, 56
　repens, 56
　shining, 56
Mairania
　uva-ursi, 297
Malacothrix
　californica, 214
　　glabrata, 214
　Cleveland, 214
　clevelandii, 214
　coulteri, 215
　glabrata, 215
　snakeshead, 215
　sonchoides, 215
　　torreyi, 215
　torreyi, 215
Malcolmia
　africana, 278
Mallow
　alkali, 421
　curled, 420
　family, 419
　high, 420
Malus
　baccata, 529
　domestica, 529
　floribunda, 529
　hupehensis, 529
　ioensis, 529
　pumila, 529
　sylvestris, 529
　theifera, 529
Malva
　coccinea, 421
　hederacea, 421
　californica, 421
　leprosa, 421
　munroana, 423
　neglecta, 420
　parviflora, 420
　rivularis, 420
　rotundifolia, 420
　sylvestris, 420
　verticillata, 420
　crispa, 420
Malvaceae, 419
Malvastrum
　coccineum, 421

dissectum, 421
  elatum, 421
  cockerellii, 423
   dissectum, 421
   elatum, 421
   exile, 421
   grossulariifolium, 423
   letophyllum, 423
   micranthum, 423
   munroanum, 423
Malvella
  leprosa, 421
Malveopsis
  exile, 420
Mammillaria
  arizonica, 86
  chlorantha, 86
  deserti, 86
  missouriensis, 85
  tetrancistra, 87
  vivipara, 86
Mannagrass
  American, 737
  fowl, 738
  northern, 737
  weak, 772
Mansa
  yerba, 557
Manzanita
  greenleaf, 297
  Mexican, 297
  pink-bracted, 297
Maple
  Amur, 42
  bigtooth, 42
  family, 41
  hedge, 41
  Japanese, 42
  Rocky Mountain, 42
  Norway, 42
  red, 42
  silver, 42
  sugar, 43
  Sycamore, 43
  vine, 42
Marestail
  common, 315
  family, 315
Marigold
  marsh, 507
Marilaunidium
  hispidum, 318
  tenue, 318
Marrubium
  vulgare, 331
Marsh-elder, 209
Marsilea
  mucronata, 16
  oligospora, 16
  vestita, 16
Marsileaceae, 16
Martynia
  louisianica, 453
  parviflora, 453
Massula
  latifolia, 453
Mat-atriplex, 118
Mat-saltbush, 118
Matricaria
  chamomilla, 164
  maritima, 215
  matricarioides, 164
  parthenium, 165

resutita, 164
Maurandya
  antirrhiniflora, 574
Mayweed, 142
Meadowfoam
  family, 411
Meadowrue
  arctic, 517
  Fendler, 517
  montane, 517
Medic
  false, 774
Medicago
  falcata, 394
  hispida, 395
  lupulina, 394
  polymorpha, 395
  sativa, 395
Medick
  black, 394
Megastachya
  uninervia, 744
Melandrium
  album, 108
  apetalum, 108
  drummondii, 108
Melia
  azedarach, 424
Meliaceae, 424
Melic
  Porter, 747
  rock, 747
Melica
  bella, 747
  bulbosa, 747
  porteri, 747
  spectabilis, 747
  stricta, 747
Melilotus
  albus, 395
  indicus, 395
  officinalis, 395
Mengea
  californica, 45
Menodora
  scabra, 433
  spinescens, 433
Mentha
  aquatica, 331
  arvensis, 331
   glabrata, 331
  borealis, 331
  canadensis, 331
   glabrata, 331
  citrata, 331
  piperita, 331
   citrata, 331
  spicata, 332
Mentzelia
  acuminata, 415
  albicaulis, 414
   integrifolia, 415
   gracilis, 414
  argillosa, 414
  californica, 414
  cronquistii, 414
  dispersa, 415
   obtusa, 415
  integra, 416
  laevicaulis, 415
  marginata, 415
  montana, 415
  multicaulis, 415
  librina, 415

  multicaulis, 415
  multiflora, 415
  integra, 416
   multiflora, 416
  nuda, 416
   rusbyi, 416
  obscura, 416
  ornata, 415
  pterosperma, 416
  pumila, 416
   lagarosa, 416
   multicaulis, 415
   multiflora, 416
   pumila, 416
   rusbyi, 416
  rusbyi, 416
  shultziorum, 416
  speciosa, 416
  thompsonii, 417
  tricuspis, 417
  urens, 416
Menyanthaceae, 424
Menyanthes
  trifoliata, 424
Merimea
  texana, 296
Mermaid
  false, 411
Mertensia
  arizonica, 80
   leonardii, 80
   subnuda, 80
   amoena, 81
  brevistyla, 80
  cana, 81
  canescens, 81
  ciliata, 80
  coriacea, 81
   dilitata, 81
  franciscana, 80
  fusiformis, 80
  lanceolata, 81
   lanceolata, 81
   nivalis, 81
   viridis, 81
  leonardii, 80
  oblongifolia 81
   amoena, 81
   nevadensis, 81
  paniculata, 81
   nivalis, 81
   praecox, 81
   sampsonii, 80
   toyabensis, 80
   subnuda, 80
   viridis, 81
   cana, 81
   dilatata, 81
Mesquite
  honey, 401
  vine, 757
Metasequoia
  glyptostroboides, 34
Mexican-tea, 124
Micropuntia
  barkleyana, 91
  brachyropalia, 91
  spectatissima, 91
Microseris
  lindleyi, 215
  linearifolia, 215
  nutans, 216
Microsteris

  gracilis, 465
  humilior, 466
Mignonette
  family, 518
Milkvetch
  alkali, 375
  alpine, 353
  ashen, 378
  Barneby, 355
  Beckwith, 355
  Bicknell, 359
  birds-nest, 371
  Bishop, 362
  Bodin, 357
  Brandegee, 357
  broad-keeled, 373
  broad-leaved, 368
  browse, 358
  Callaway, 357
  Canada, 357
  canoe, 360
  chickpea, 359
  cicada, 358
  Cisco, 376
  clay, 363
  cobweb, 367
  Colton, 359
  crescent, 354
  Cronquist, 360
  Cottam, 359
  Dana, 365
  debris, 361
  Deseret, 360
  Dinosaur, 376
  Draba, 376
  Dragon, 369
  Drummond, 361
  Duchesne, 361
  Eastwood, 361
  egg, 372
  Emory, 362
  escarpment, 376
  Eureka, 363
  Ferron, 371
  field, 353
  Fisher, 373
  Fort Wingate, 380
  four-wing, 378
  freckled, 367
  Fremont, 367
  Garrett's weedy, 370
  Geyer, 364
  glass, 369
  Goose Creek, 354
  gravel, 376
  great bladdery, 369
  great rushy, 368
  Green River, 374
  groundcover, 365
  ground-crescent, 355
  gumbo, 354
  Hall, 364
  Hamilton, 364
  Harrison, 365
  Heliotrope, 370
  hermit, 362
  Holmgren, 365
  Hopi, 376
  Horseshoe, 362
  Humboldt River, 365
  Isely, 366
  Kaiparowits, 369
  lancer, 366

lesser rushy, 359
Loa, 368
meadow, 354
mesic, 361
milkweed, 355
Missouri, 370
Moenkopi, 370
monument, 370
Navajo Lake, 368
Nelson, 371
Newberry, 371
Pagumpa, 362
painted, 357
pallid egg, 372
panther, 372
Patterson, 372
pink egg, 372
pinyon, 373
plateau, 376
Polh, 368
prairie, 364
Preuss, 374
pulse, 378
Pursh, 375
rimrock, 360
Robbins, 376
Rocky Mountain, 376
Russian sickle, 363
Rydberg, 372
Rydberg weedy, 370
St. George, 363
salt, 363
sandstone, 376
San Rafael, 375
scorpion, 368
Sevier, 363
silver, 378
Silver Reef, 376
small-flowered, 371
standing, 353
starvling, 366
stinking, 373
straggling, 367
straw, 368
subarctic, 355
Toana, 379
Torrey, 357
two-grooved, 356
Utah, 379
Ward, 379
weedy, 369
Welsh, 380
Wetherill, 380
Woodruff, 380
yellow, 363
Zion, 380
Milkweed
 broadleaf, 53
 climbing, 55
 Cutler, 52
 desert, 52
 dwarf, 53
 Eastwood, 54
 family, 51
 Hall, 53
 Jones, 53
 Mexican, 53
 orange, 54
 pallid, 52
 Rusby, 54
 Ruth, 54
 showy, 54
 spider, 52
 swamp, 53
 tropics, 52
 Welsh, 55
 whorled, 54
Milkwort
 cushion, 470
 family, 470
 sea, 499
 thorny, 470
Miller
 dusty, 106
Millet, 777
 hog, 756
Miltitzia
 foliosa, 325
 lutea, 325
  scopulina, 325
 salina, 325
 scopulina, 325
Mimetanthe
 hairy, 574
 pilosa, 574
Mimosa
 Illinois, 384
 illinoensis, 384
Mimulus
 bigelovii, 575
 breweri, 575
 cardinalis, 575
 eastwoodiae, 575
 floribundus, 575
  membranaceus, 575
 glabratus, 575
 fremontii, 575
 utahensis, 575
 gratioloides, 575
 guttatus, 575
  depauperatus, 575
  moschatus, 576
 jamesii, 575
  fremontii, 575
 lewisii, 576
 membranaceus, 575
 moschatus, 576
 parryi, 576
 pilosus, 574
 primuloides, 576
 rubellus, 576
  breweri, 575
 spissus, 576
 suksdorfii, 576
 tilingii, 576
 verbenaceus, 575
Miners-lettuce, 495
Mint
 Bergamot, 331
 family, 328
 field, 331
 lemon, 331
Minuartia
 obtusiloba, 105
 pusilla, 105
 rubella, 105
Mirabilis
 alipes, 428
 bigelovii, 428
  retrorsa, 428
 glabra, 428
 glutinosa, 428
 jalapa, 428
 linearis, 428
  decipiens, 428
  linearis, 429
 multiflora, 429
  glandulosa, 429
  multiflora, 429
  pubescens, 429
 nyctaginea, 429
 oxybaphoides, 429
 pumila, 429
 retrorsa, 428
Mistletoe
 acacia, 645
 family, 643
 juniper, 645
Mitella
 pentandra, 560
 stauropetala, 560
 stenopetala, 560
 stenopetala, 560
Miterwort
 fivestar, 560
 smallflower, 560
Mockorange
 littleleaf, 561
Moehringia
 lateriflora, 104
 macrophylla, 104
Mohavea
 breviflora, 577
Moldavica
 parviflora, 329
 thymifolia, 329
Mollugo
 cerviana, 43
Molly
 gray, 127
Molucella
 laevis, 332
Momordica
 elaterium, 291
 lanata, 289
Monarda
 fistulosa, 332
  menthifolia, 332
  menthifolia, 332
  pectinata, 332
Monardella
 odoratissima, 332
Monesis
 reticulata, 501
 uniflora, 501
  reticulata, 501
Monkey-flower
 Bigelow, 575
 Brewer, 575
 cardinal, 575
 common, 575
 Eastwood, 575
 glabrous, 575
 Lewis, 576
 musk, 576
 Parry, 576
 primrose, 576
 subalpine, 576
 Suksdorf, 576
Monkshood, 503
Monocots, 647
Monolepis
 chenopodioides, 128
 nuttalliana, 128
 pusilla, 128
Monoptilon
 bellidiforme, 216
 belliodes, 216
Montia
 chamissoi, 495
 cordifolia, 495
 linearis, 495
 perfoliata, 495
  perfoliata, 496
  utahensis, 495
Moonpod, 429
Moonwort, 278
Moparia
 repens, 381
Moraceae, 425
Morning-glory, 241
Morning glory
 family, 240
Morning-lily, 445
Mortonia
 scabrella, 115
  utahensis, 115
  utahensis, 115
Morus
 alba, 425
 nigra, 425
 rubra, 425
Moschatel
 family, 43
Mosquitofern, 24
Mother-of-millions, 496
Motherwort, 330
Mountain-ash
 European, 543
 rock, 543
Mountain-lilac
 desert, 518
 Fendler, 518
 Utah, 518
Mountain-lover, 115
Mousetail
 horseshoe, 510
 tiny, 510
Muhlenbergia
 ambigua, 749
 andina, 748
 arsenei, 751
 asperifolia, 749
 curtifolia, 752
 depauperata, 749
 filiculmis, 749
 filiformis, 749
 foliosa, 749
  ambigua, 749
 frondosa, 750
 glomerata, 752
 mexicana, 749
 microsperma, 750
 minutissima, 750
 montana, 750
 pauciflora, 751
 porteri, 751
 pungens, 751
 racemosa, 751
 repens, 752
 richardsonis, 752
 schreberi, 752
 squarrosa, 752
 texana, 751
 thurberi, 752
 wrightii, 753
Muhly
 annual, 750
 bush, 751
 creeping, 752
 foxtail, 748
 green, 751
 littleseed, 750

mat, 752
Mexican, 749
mountain, 750
New Mexican, 751
nimblewill, 752
pullup, 749
sandhill, 751
sixweeks, 749
spike, 753
Thurber, 752
Mulberry
  black, 425
  family, 425
  red, 425
  white, 425
Mule-fat, 157
Mulesears, 238
  rough, 238
Mullein
  moth, 610
  wand, 602
  woolly, 601
Munroa
  squarrosa, 753
Muscari
  botryoides, 809
Musineon
  lineare, 633
  Rydberg, 633
Musk-mustard, 259
Muskmelon, 290
Mustard
  African, 278
  black, 255
  garlic, 247
  family, 243
  hares-ear, 259
  hedge, 283
  Indian, 255
  Jim Hill, 283
  tower, 250
  tumbling, 283
  white, 255
Muttongrass, 764
Myagrum
  argenteum, 276
Myosotis
  micrantha, 81
  scorpioides, 81
  scouleri, 83
  squarrosa, 78
  tenella, 83
Myosurus
  apetalus, 510
  aristatus, 510
  cupulatus, 510
  minimus, 510
    montanus, 510
  nitidus, 510
Myriophyllum
  exalbescens, 314
  spicatum, 314
    exalbescens, 314
  verticillatum, 314
Myrtle
  grave, 50
Myzorhiza
  corymbosa, 449
Naiad
  Fishlake, 811
Najadaceae, 811
Najas
  caespitosa, 811
  flexilis, 811

caespitosus, 811
guadalupensis, 811
guadalupensis, 811
marina, 811
microdon, 811
  guadalupensis, 811
Nama
  compact, 318
  demissum, 318
    deserti, 318
  densum, 318
  hispidum, 318
  retrorsum, 318
Narcissus
  jonquilla, 652
  poeticus, 652
  Poets, 652
  polyanthus, 652
  pseudo-narcissus, 652
  tazetta, 652
Nasturtium
  armoracea, 254
  austriacum, 281
  garden, 611
  hispidum, 281
  linifolium, 282
  obtusum, 281
    alpinum, 281
    spaerocarpum, 282
  officinale, 278
  palustre, 281
    glabrum, 281
    hispidum, 281
  pumilum, 282
  sinuatum, 281
  sphaerocarpum, 282
Navarretia
  breweri, 466
  intertexta, 466
  propinqua, 466
  schottii, 463
  setosissima, 463
Needle
  Spanish, 216
Needle-and-thread, 282
Needlegrass
  crested, 783
  desert, 785
  green, 786
  Lemmon, 784
  Letterman, 784
  Mormon, 782
  Nelson, 784
  pinewoods, 785
  Thurber, 785
Neltuma
  glandulosa, 401
Nemacladus
  glanduliferus, 96
  orientalis, 96
  rubescens, 96
Nemophila
  austinae, 319
  breviflora, 318
  explicata, 318
  inconspicua, 318
  parviflora, 318
    austinae, 319
    typica, 318
      inconspicua, 318
  petrophila, 318
Neogaerrhinum
  filipes, 568

Neolloydia
  johnsonii, 87
Nepeta
  cataria, 333
  glecoma, 329
  hederacea, 329
Nerium
  oleander, 49
Nest-straw
  desert, 231
Nettle
  family, 637
  stinging, 638
Nicotiana
  attenuata, 606
  sylvestris, 606
  tabacum, 606
  trigonophylla, 606
Niellia
  malvacea, 530
  monogyna, 530
    alternans, 530
    malvacea, 530
Nightshade
  black, 609
  cutleaf, 610
  enchanters, 439
  silverleaf, 609
Ninebark
  dwarf, 530
  narrow-leaved, 530
  mountain, 530
Nipplewort, 210
Niterwort, 128
Nitrophila
  occidentalis, 128
Noddinghead, 180
Nolina
  microcarpa, 648
Norta
  altissima, 283
  irio, 283
Notholaena
  fendleri, 21
  jonesii, 21
  limitanea, 21
  parryi, 22
Notholcus
  lanatus, 740
Nuphar
  luteum, 430
    polysepalum, 430
  polysepalum, 430
Nuttallia
  lobata, 416
  marginata, 415
  multicaulis, 415
  munroana, 423
Nyctaginaceae, 426
Nyctelea
  pinetorum, 316
Nymphaea
  odorata, 430
Nymphaeaceae, 430
Oak
  bur, 306
  English, 306
  Gambel, 305
  red, 306
  swamp-white, 305
  shinnery, 305
  white, 305
Oat
  alpine, 738

Oats, 700
  wild, 700
Oatgrass
  California, 717
  oneside, 717
  tall, 699
  timber, 717
Obedient-plant, 333
Obione
  confertifolia, 118
  elegans, 119
  gardneri, 119
  hymenelytra, 120
  lentiformis, 120
  truncata, 122
Odora, 219
Oenothera
  albicaulis, 445
    californica, 446
    decumbens, 447
    runcinata, 447
  alyssoides, 437
    minutiflora, 437
    villosa, 436
  acutissima, 447
  alyssioides, 437
    minutiflora, 437
    villosa, 436
  ambigua, 447
  andina, 435
  avita, 446
  biennis, 445
  boothii, 436
  breviflora, 436
  brevipes, 436
    pallidula, 436
    parviflora, 439
  bufonis, 448
  caespitosa, 445
    caespitosa, 445
    crinita, 445
    jonesii, 445
    macroglottis, 446
    marginata, 446
    navajoensis, 446
  californica, 446
    avita, 446
    cavernae, 446
    chamaenerioides, 436
    claviformis, 437
      intergrior, 437
    coronopifolia, 446
    decorticans, 436
      condensata, 436
  deltoides, 446
    ambigua, 447
    decumbens, 447
    deltoides, 447
    densiflora, 434
    eastwoodiae, 437
  elata, 447
  flava, 447
    acutissima, 447
    flava, 447
  gauriflora, 436
    hitchcockii, 436
  glabella, 434
  gouldii, 437
  heteranther, 439
    taraxifolia, 439
  hookeri, 447
    angustifolia, 447
  howardii, 447

johnsonii, 448
lavandulifolia, 434
longissima, 447
marginata, 446
  purpurea, 445
minor, 437
montana, 445
multijuga, 437
orientalis, 439
pallida, 447
  californica, 446
  latifolia, 447
  leptophylla, 447
  trichocalyx, 447
parryi, 438
parviflora, 439
parvula, 438
primiveris, 448
  bufonis, 448
pterosperma, 438
refracta, 438
scapoidea, 438
  claviformis, 437
  eastwoodiae, 437
  purpurascens, 437
  tortilis, 439
  utahensis, 438
speciosa, 448
subacaulis, 438
  taraxifolia, 439
tenuissima, 438
trichocalyx, 447
triloba, 447
  ecristata, 447
villosa, 445
walkeri, 439
  tortilis, 439
Ohio buckeye
Old-man's-beard
Oleaceae, 431
Oleander, 49
Oleaster
  family, 294
Olive
  desert, 431
  family, 431
  Russian
Onagraceae, 434
Onion
Oniongrass, 434
  purple, 747
Onobrychis
  onobrychis, 395
  sativa, 395
  viciifolia, 395
Onopardum
  acanthium, 216
Ophioglossaceae, 16
Ophrys
  cernua, 817
  cordata, 816
Opulaster
  alternans, 530
  cordatus, 530
  malvaceus, 530
  monogynus, 530
  pubescens, 530
Opuntia
  acanthocarpa, 88
    acanthocarpa, 88
    coloradensis, 88
  aurea, 89
  barbata, 91

  gracillima, 91
  basilaris, 88
    aurea, 89
    basilaris, 89
    heilii, 89
    woodburyi, 90
    brachyarthra, 90
  chlorotica, 89
  compressa, 90
  discata, 90
  echinocarpa, 89
  engelmannii, 90
  littoralis, 90
  erinacea, 89
    aurea, 89
    erinacea, 89
    hystricina, 89
    martiniana, 90
    ursina, 89
    utahensis, 89
    xanthostema, 89
  eriocentra, 90
    martiniana, 90
  fragilis, 90
    brachyarthra, 90
    denudata, 90
  humifusa, 90
  hystricina, 89
  juniperina, 91
  littoralis, 90
    martiniana, 90
  macrorhiza, 90
  missouriensis, 91
    rufispina, 91
    trichophora, 91
    multigeniculata, 91
  nicholii, 90
  phaeacantha, 90
    discata, 90
    major, 90
    phaeacantha, 91
  polyacantha, 91
    juniperina, 91
    polyacantha, 91
    rufispina, 91
    trichophora, 91
    watsonii, 91
  pulchella, 91
  rhodantha, 89
  rubrifolia, 89
  sphaerocarpa, 89
    utahensis, 89
  ursina, 89
  utahensis, 89
  whipplei, 91
    multigeniculata, 91
    whipplei, 91
  xanthostemma, 89
Orach
  Blue Valley, 120
  Four-corners, 121
  garden, 120
  medusa-head, 122
  Powell, 121
  silver, 118
  slender, 122
  stalked, 121
  tumbling, 121
  two-seed, 120
  wedge, 122
  wheelscale, 119
Orange
Osage

Orchid
  family, 812
Orchidaceae, 812
Orchis
  bracteata, 816
  dilitata, 814
  hyperborea, 815
  obtusata, 815
  virescens, 816
Oreobroma
  brachycalyx, 494
  pygmaea, 494
Oreocarya
  abortiva, 66
  bakeri, 65
  breviflora, 65
  caespitosa, 66
  capitata, 66
  cinerea, 66
  commixta, 69
  confertiflora, 67
  disticha, 66
  dolosa, 69
  elata, 68
  flava, 68
  flavoculata, 69
  humilis, 69
  interrupta, 70
  jonesiana, 70
  longiflora, 71
  nana, 70
  osterhoutii, 72
  paradoxa, 72
  pustulosa, 66
  rugulosa, 73
  shantzii, 69
  stricta, 74
  tenuis, 74
  torva, 68
  wetherillii, 75
  williamsii, 74
Oreoxis
  alpina, 633
  alpine, 633
  bakeri, 634
  trotteri, 634
Ornithogalum
  umbellatum, 809
Orobanche
  cooperi, 449
  corymbosa, 449
  fasciculata, 449
  lutea, 449
  ludoviciana, 449
    araneosa, 449
    cooperi, 449
  multiflora, 449
    araneosa, 449
  uniflora, 449
Orobanchaceae, 448
Orogenia
  linearifolia, 634
Orophaca
  cushion, 354
  plains, 364
  triphylla, 364
Orthilia
  secunda, 502
  obtusata, 502
Orthocarpus
  luteus, 577
  purpureoalbus, 577
  tolmei, 577

Oryzopsis
  asperifolia, 754
  bloomeri, 782
  exigua, 754
  hymenoides, 783
  micrantha, 754
  miliacea, 754
Osmorhiza
  chilensis, 634
  depauperata, 635
  nuda, 634
  obtusa, 635
  occidentalis, 635
Osmunda
  crispa, 20
  lanceolata, 16
  lunaria, 16
Ostrya
  knowltonii, 58
Owl-clover
  Tolmie, 577
  yellow, 577
Oxalidaceae, 449
Oxalis, 449
  corniculata, 449
  dillenii, 450
  stricta, 450
Oxeye-daisy, 165
Oxybaphus
  angustifolius, 429
    viscidus, 429
  glaber, 429
  linearis, 428
  multiflorus, 429
  nyctagineus, 429
  pumilus, 429
Oxypolis
  fendleri, 635
Oxyria
  digyna, 487
Oxytenia
  acerosa, 216
Oxytheca
  perfoliata, 487
Oxytrope
  Jones, 397
  mountain, 397
  Parry, 398
  Rocky Mountain, 397
  stemmed, 396
Oxytropis
  besseyi, 396
    obnapiformis, 396
    ventosa, 396
  bilocularis, 396
  campestris, 396, 397
    cusickii, 396
    viscida, 397
  cusickii, 396
  deflexa, 396
    deflexa, 397
    sericea, 397
  jonesii, 396
  lambertii, 397
    bigelovii, 397
    sericea, 397
  minor, 397
  multiceps, 397
    minor, 397
  nana, 396
    obnapiformis, 396
    ventosa, 396
  obnapiformis, 396

oreophila, 397
 juniperina, 397
 oreophila, 397
parryi, 398
patens, 396
sericea, 398
viscida, 398
 viscida, 398
Oyster-plant, 237
Pachylophus
 canescens, 445
 crinitus, 445
 macroglottis, 446
Pachypodium
 integrifolium, 288
 sagittatum, 286
Pachystima
 myrsinites, 115
Paeonia
 brownii, 450
 lactiflora, 450
Paeoniaceae, 450
Pagoda-tree
 Japanese, 404
Paintbrush, 569
 annual, 569
 Aquarius Plateau, 569
 common, 569
 dwarf, 570
 Eastwood, 571
 narrowleaf, 570
 pretty, 571
 rhexia-leaved, 571
 scarlet, 570
 Tushar, 570
 wavy-leaf, 569
 western, 570
 yellow, 570
Palafoxia
 arida, 216
 linearis, 216
Palm
 family, 817
 petticoat, 818
Palmae, 817
Panic
 bundle, 755
Panicgrass
 blue, 755
Panicularia
 borealis, 737
 elata, 738
 nervata, 738
Panicum
 acuminatum, 755
 antidotale, 755
 barbipulvinatum, 756
 broomcorn, 756
 bulbosum, 755
 capillare, 755
  occidentale, 756
 crus-galli, 720
 dactylon, 716
 dichotomiflorum, 756
 fall, 756
 flexile, 756
 glaucum, 776
 Hall, 756
 hallii, 756
 huachucae, 755
 ischmaeum, 719
 lanuginosum, 755
 lepidulum, 756

miliaceum, 756
obtusum, 757
occidentale, 755
oligosanthes, 757
 scribnerlanum, 757
sanguinale, 719
Scribner, 757
spreading, 756
syzigachne, 701
tennesseensis, 755
thermale, 755
verticillatum, 776
virgatum, 757
viride, 776
Pansy, 643
 field, 642
Papaver
 glaucum, 452
 kluanense, 452
 orientale, 452
 pygmaeum, 452
 radicatum, 452
  pygmaeum, 452
 rhoeas, 452
 somniferum, 452
Papaveraceae, 450
Paperflower
 greenstem, 221
 whitestem, 221
 woolly, 221
Pappophorum
 wrightii, 729
Pappusgrass
 spike, 729
Parietaria
 floridana, 637
 occidentalis, 637
 obtusa, 637
 pennsylvanica, 637
Parkinsonia
 aculeata, 398
Parnassia
 fimbriata, 560
  fimbriata, 560
 palustris, 561
 montanensis, 561
 parviflora, 561
Paronychia
 pulvinata, 109
 longiaristata, 109
 sessiliflora, 109
Parosela
 amoena, 402
 arborescens, 402
 fremontii, 402
 johnsonii, 402
 minutiflora, 402
 pubescens, 402
 lanata, 383
 polydenia, 403
 thompsonae, 403
Parrya
 platycarpa, 278
 rydbergii, 278
 uinta, 278
Parryella
 filifolia, 398
Parsley
 family, 613
 hedge, 637
 rock, 625
Parsley-fern, 20
Parsnip

cow, 626
Parthenium
 alpinum, 217
 ligulatum, 217
 incanum, 217
 ligulatum, 217
Parthenocissus
 inserta, 645
 quinquefolia, 645
 vitacea, 645
Pascopyrum, 692
 smithii, 727
Paspalum
 distichum, 757
 indutum, 757
 vaginatum, 758
Pasque-flower, 504
Passiflora
 caerulea, 453
Passifloraceae, 453
Passion-flower
 family, 453
Pastinaca
 sativa, 635
 sativa, 635
 sylvestris, 635
Paxistima, 115
Pea
 garden, 400
 golden, 405
 yellow, 405
Peach, 538
 desert, 538
Peanut, 340
Pear
 common, 540
Peartree, 381
Pecan, 327
Pectis
 papposa, 217
Pectocarya
 heterocarpa, 82
 penicillata, 82
  heterocarpa, 82
 platycarpa, 82
 recurvata, 82
 setosa, 82
 aperta, 82
 holoptera, 82
Pedaliaceae, 453
Pedicularis
 bracteata, 578
 paysoniana, 578
 centranthera, 578
 contorta, 578
 grayi, 578
 groenlandica, 578
 parri, 578
 parryi, 578
 purpurea, 578
 paysoniana, 578
 procera, 578
 racemosa, 578
 alba, 579
Pediocactus, 92
 despainii, 92
 hermannii, 92
 sileri, 92
 simpsonii, 92
 minor, 92
 winkleri, 92
Pediomelum, 398
 aromaticum, 399

tuhyi, 399
castoreum, 399
epipsilum, 399
megalanthum, 399
mephiticum, 399
pariense, 400
retrorsum 399
Pelargonium
 hortorum, 314
Pellaea
 breweri, 22
 glabella, 22
 jonesii, 21
 longimucronata, 22
 limitanea, 21
 suksdorfiana, 22
 ternifolia 22
 truncata, 22
Peltiphyllum
 peltatum, 561
Pennellia
 micrantha 287
Pennyroyal
 mock, 330
Penstemon
 Abajo, 592
 abietinus, 583
 acaulis, 583
  acaulis, 584
  yampaensis, 584
 acuminatus, 594
  congestus, 594
  albifluvis, 597
 ambiguus, 584
  laevissimus, 584
 ammophilus, 584
 angustifolius, 584
  dulcis, 584
  venosus, 584
  vernalensis, 584
 arenicola, 584
 Atwood, 585
 atwoodii, 585
 azureus, 598
  ambiguus 598
 barbatus, 585
  barbatus, 585
  trichander, 585
  torreyi, 585
 beardlip, 585
 bluestem, 588
 bracteatus, 585
 breviculus, 585
 Bridges, 596
 bridgesii, 596
  amplexicaulis, 596
 broadleaf, 595
 bush, 584
 caespitosus, 586
  caespitosus, 585
  desertipicti, 586
  incanus, 599
  perbrevis, 586
  suffruticosus, 586
  thompsoniae, 599
 carnosus, 586
 Cleburn, 586
 cleburni, 586
 comarrhenus, 587
 compactus, 588
 concinnus, 587
 confusus, 587
 congestus, 594

cordroot, 593
Crandall, 587
crandallii, 587
    atratus, 587
crevice, 595
cyananthus, 587
    compactus, 588
    cyananthus, 588
    longiflorus, 588
    subglaber, 588
cyanocaulis, 588
desertipicti, 586
deustus, 588
    pedicellatus, 589
dolius, 589
    duchesnensis, 589
Duchesne, 589
duchesnensis, 589
dusty, 587
dwarf, 593
eastwoodiae, 599
Eaton, 589
eatonii, 589
    eatonii, 589
    undosus, 589
elegant, 587
firleaf, 583
fleshy, 586
Flowers, 589
flowersii, 589
Fremont, 590
fremontii, 590
    subglaber, 588
garrettii, 597
Gilia, 584
glaber, 590
    fremontii, 590
    utahensis, 598
    wardii, 600
Goodrich, 590
goodrichii, 590
Graham, 590
grahamii, 590
heterophyllus, 595
    latifolius, 595
holmgrenii, 588
hotrock, 588
humilis, 590
    brevifolius, 591
    humilis, 591
    obtusifolius, 591
immanifestus, 591
jamesii, 594
    breviculus, 585
    ophianthus, 594
Jones, 589
jonesii, 591
laevis, 591
leiophyllus, 591
lentus, 592
    albiflorus, 592
    lentus, 592
Leonard, 592
leonardii, 592
    higginsii, 592
    leonardii, 592
    patricus, 592
leptanthus, 599
linarioides, 592
    sileri, 593
    viridis, 593
little, 595
longiflorus, 588

low, 590
marcusii, 593
Markagunt, 591
mat, 585
matroot, 596
mistaken, 587
Moffatt, 593
moffatii, 593
    marcusii, 593
    montanus, 593
    mucronatus, 594
    nanus, 593
    narrowleaf, 584
    Navajo Mountain, 593
    navajoa, 593
    nitidus, 594
        major, 594
    obtusifolius, 591
    ophianthus, 594
    pachyphyllus, 594
        congestus, 594
        mucronatus, 594
        pachyphyllus, 594
    Palmer, 594
    palmeri, 594
        eglandulosus, 595
        palmeri, 595
    parvus, 595
    patricus, 592
    petiolatus, 595
    phlogifolius, 600
    Pinyon, 595
    pinorum, 595
    Plateau, 597
    platy, 585
    platyphyllus, 595
    Price, 593
    procerus, 596
        aberrans, 596
        procerus, 596
        pseudoputus, 596
    pumilus, 599
        incanus, 599
        thompsoniae, 599
    radicosus, 596
    Rocky Mountain, 598
    rostriflorus, 596
    Rydberg, 597
    rydbergii, 597
        aggregatus, 597
        rydbergii, 597
    sandloving, 584
    scariosus, 597
        albifluvis, 597
        cyanomontanus, 597
        garrettii, 597
        scariosus, 598
    scarlet-bugler, 589
    sepalulus, 598
    showy, 598
    small-flowered, 596
    smooth, 591
    speciosus, 598
    stemless, 583
    strictiformis, 598
    strictus, 598
        angustus, 598
        strictiformis, 598
        subglaber, 598
    thickleaf, 594
    Thompson
    thompsoniae, 599
        desperatus, 599

thompsoniae, 599
Tidestrom, 599
tidestromii, 599
tusharensis, 586
Uinta, 599
uintahensis, 599
Utah, 599
utahensis, 599
venosus, 584
venustus, 600
virgatus, 596
    pseudoputus, 596
Ward, 600
wardii, 600
Wasatch, 587
Watson, 600
watsonii, 600
Whipple, 600
whippleanus, 600
yampaensis, 584
Pentaphylloides
    floribunda, 533
Peony, 450
    family, 450
Peplis
    diandra, 417
Pepper, 604
Peppergrass, 273
Peppermint, 331
Pepperplant
    Ostler, 273
Pepperwort
    family, 16
Peraphyllum
    ramosissimum, 529
Perezia
    wrightii, 217
Perideridia
    bolanderi, 636
    gairdneri, 636
    borealis, 636
Perityle
    emoryi, 218
    palmeri, 218
    specuicola, 218
    stansburyi, 218
    tenella, 218
Persimmon
    common, 294
Petalonyx
    nitidus, 417
    parryi, 417
    parryi, 417
    thruberi, 417
    nitidus, 417
Petalostemon
    candidus, 383
        oligophyllus, 383
    gracilis, 383
        oligophyllus, 383
    flavescens, 383
    occidentale, 383
    oligophylla, 383
    searlsiae, 384
    sonorae, 383
    truncatus, 383
Peteria
    nevadensis, 400
    Thompson, 400
    thompsonae, 400
Petradoria
    graminea, 218
    pumila, 218

pumila, 218
    graminea, 218
Petrophytum
    caespitosum, 529
Petroselinium
    crispum, 614
Petunia
    axillaris, 607
    common, 607
    hybrida, 607
    parviflora, 607
    streamside, 607
    violacea, 607
Peucedanum
    bicolor, 629
    cous, 629
    graveolens, 631
        alpinum, 631
        juniperinum, 632
    kingii, 631
    lapidosum, 623
    millefolium, 630
    nevadense, 632
    newberryi, 624
    parishii, 632
    parryi, 632
    simplex, 633
    triternatum, 633
        platycarpum, 633
Peucephyllum
    schottii, 218
Phaca
    ampullaria, 354
    annua, 364
    artemisiarum, 355
    artipes, 372
    beckwithii, 355
    bisulcata, 356
    bodinii, 356
    caespitosa, 364
    convallaria, 359
    eastwoodae, 361
    flexuosa, 364
    glareosa, 375
    jejuna, 366
    laxiflora, 374
    macrocarpa, 368
    megacarpa, 369
    mollissima, 379
        utahensis, 379
    pardalina, 372
    picta, 357
    platytropis, 373
    preussii, 374
    pubentissima, 374
    purshii, 375
    pygmaea, 358
    robbinsii, 376
    sabulonum, 376
    salsula, 405
    serpens, 377
    silerana, 378
    subcinerea, 378
    wardii, 379
    wetherillii, 380
    yukonsis, 356
Phacelia
    affinis, 320
    alba, 321
    alpina, 323
    ambigua, 321
    anelsonii, 321
    argillacea, 321

austromantana, 321
Bluff, 324
brittle, 322
cephalotes, 321
clay, 321
coerulea, 321
constancei, 322
corrugata, 322
crenulata, 322
ambigua, 321
angustifolia, 322
corrugata, 322
crenulata, 322
cryptantha, 322
curvipes, 322
demissa, 322
demissa, 323
heterothricha, 323
minor, 323
foetida, 325
foliosa, 325
franklinii, 323
fremontii, 323
glandulifera, 324
glandulosa, 323
eglandulosa, 321
elatior, 321
hastata, 323
hastata, 323
leucophylla, 323
heterophylla, 323
alpina, 323
howelliana, 323
incana, 323
indecora, 324
integrifolia, 324
palmeri, 325
ivesiana, 324
glandulifera, 324
ivesiana, 324
laxiflora, 324
lemmonii, 324
leucophylla, 323
limestone, 326
linearis, 324
lutea, 325
scopulina, 325
mammillarensis, 325
micrantha, 316
neomexicana, 321
alba, 321
coulteri, 321
foliosissima, 321
Nipple, 325
nudicaulis, 323
orbicularis, 322
Palmer, 325
palmeri, 325
foetida, 325
parishii, 320
peirsoniana, 325
perityloides, 324
laxiflora, 324
pinetorum, 316
pulchella, 325
luteola, 325
gooddingii, 325
pulchella, 325
sabulonum, 325
rafaelensis, 325
rotundifolia, 326
salina, 325
saxicola, 320

scopulina, 325
sericea, 326
ciliosa, 326
splendens, 319
tetramera, 326
utahensis, 326
vallis-mortae, 326
Phacomene
artemisiarum, 356
beckwithii, 356
Phacopsis
pattersonii, 372
praelongus, 374
Phalangium
quamash, 806
Phalaris
arundinacea, 758
picta
variegata, 758
canariensis, 758
oryzoides, 743
semiverticillata, 770
Pharnaceum
cervianum, 43
Phaseolus
coccineus, 400
coronarius, 561
limensis, 400
vulgaris, 400
Pheasant-eye, 504
Phelipaea
lutea, 449
Phellopterus
multinervatus, 624
Phellosperma
tetrancistra, 87
Philadelphus
coronarius, 561
lewisii, 561
microphyllus, 561
nitidus, 561
verrucosus, 561
virginalis, 561
Philotria
canadensis, 788
Phleum
alopecuroides, 715
alpinum, 759
pratense, 759
schoenoides, 715
Phlox
austromontana, 467
austromontana, 467
jonesii, 467
lutescens, 467
prostrata, 467
caesia, 467
caespitosa, 468
pulvinata, 468
canescens, 468
carpet, 467
Cedar Canyon, 467
cluteana, 467
desert, 467
family, 455
garden, 468
gladiformis, 467
gracilis, 465
grahamii, 468
grayleaf, 467
griseola, 467
hoodii, 467
canescens, 468

jonesii, 467
longifolia, 468
brevifolia, 468
gladiformis, 467
stansburyi, 468
longleaf, 468
moss, 468
multiflora, 468
muscoides, 468
Navajo Mountain, 467
nivalis, 468
paniculata, 468
pulvinata, 468
speciosa, 468
stansburyi, 468
stansburyi, 467
subulata, 468
tumulosa, 469
Phoradendron
californicum, 645
juniperinum, 645
juniperinum, 645
Phragmites
australis, 759
communis 759
Phyla
cuneifolia, 640
nodiflora, 640
Phymosia
rivularis, 420
Physalis
crassifolia, 607
fendleri, 608
cordifolia, 608
hederifolia, 608
fendleri, 608
hederifolia, 608
palmeri, 608
heterophylla, 608
lobata, 608
longifolia, 608
palmeri, 608
pubescens, 608
Physaria
acutifolia, 279
acutifolia, 279
purpurea, 279
stylosa, 279
australis, 279
chambersii, 279
chambersii, 280
membranacea, 280
sobolifera, 280
didymocarpa, 280
newberryi, 280
floribunda, 280
grahamii, 280
lepidota, 280
newberryi, 280
racemosa, 280
repanda, 279
stylosa, 279
Physocarpus
alternans, 530
capitata, 530
malvaceus, 530
monogynus, 530
opulifolius, 530
Physolepidium
repens, 257
Physostegia
parviflora, 333
virginiana, 333

Picea
abies, 30
concolor, 29
engelmannii, 31
glauca, 30
engelmannii, 31
pungens, 31
Picradenia
helenioides, 208
lemmonii, 208
richardsonii, 208
Pie-plant, 491
Pigmy-cedar, 218
Pigweed, 124
prostrate, 44
redroot, 45
winged, 126
Pilosella
stenocarpa, 269
virgata, 269
Pimpernel
scarlet, 497
Pinaceae, 29
Pine
Austrian, 33
black, 33
eastern white, 33
family, 29
Himalayan white, 33
Japanese red, 31
Jeffrey, 32
limber, 32
lodgepole, 31
mountain, 33
Mugo, 33
ponderosa, 33
red, 33
Scots, 33
western white, 32
western yellow, 33
white, 33
yellow, 33
Pinedrops, 502
Pinegrass, 712
Pink
family, 101
Pinophyta, 24
Pinus, 31
atlantica, 30
cembroides, 32
edulis, 32
monophylla, 32
contorta, 31
latifolia, 31
murrayana, 31
densiflora, 32
umbraculifera, 32
divaricata, 31
latifolia, 31
edulis, 32
monophylla, 32
flexilis, 32
griffithii, 33
jeffreyi, 32
lasiocarpa, 29
longaeva, 32
monophylla, 32
edulis, 32
montana, 33
monticola, 32
mugo, 33
murrayana, 31
nigra, 33

ponderosa, 33
strobus, 33
　　monticola, 32
sylvestris, 33
wallichiana, 33
Pinyon
　singleleaf, 32
　two-needle, 32
Pipsissewa, 501
Pisophaca
　diehlii, 364
　familica, 364
　flexuosa, 364
　pinonis, 373
Pistacia
　atlantica, 46
　vera, 46
Pistacio, 46
Pisum
　sativum, 400
Pityrogramma
　triangularis, 23
Plagiobothrys
　arizonicus, 83
　asper, 83
　cusickii, 83
　cognatus, 83
　hispidulus, 83
　humifusus, 83
　jonesii, 83
　kingii, 83
　　haknessii, 83
　leptocladus, 83
　nelsonii, 83
　nitens, 83
　parvulus, 83
　scopulorum, 83
　scouleri, 83
　　penicillatus, 83
　tenellus, 83
　　parvulus, 83
　　　humifusus, 83
Planetree
　oriental, 455
Plant
　obedient, 333
Plantaginaceae, 453
Plantago
　elongata, 454
　eriopoda, 454
　fastigiata, 454
　insularis, 454
　lanceolata, 454
　major, 454
　　pachyphylla, 454
　myosuroides, 454
　patagonica, 454
　　breviscapa, 455
　　gnaphaloides, 455
　　spinulosa, 455
　purshii, 454
　tweedyi, 455
Plantain
　broadleaf, 454
　family, 453
Plantanaceae, 455
Platanus
　acerifolia, 455
　occidentalis, 455
　orientalis, 455
Plants
　flowering, 35
Platanthera

dilitata, 814
　albiflora, 814
　hyperborea, 815
　leucostachys, 815
　obtusata, 815
　saccata, 815
　sparsiflora, 815
　unalascensis, 815
Platycladus
　orientalis, 28
Platyschkuhria
　desertorum, 219
　integrifolia, 219
　　desertorum, 219
　　integrifolia, 219
　　oblongifolia, 219
　　ourolepis, 219
Platystemon
　californicum, 453
　remotus, 453
　rigidulus, 453
　terminii, 453
Plectritis
　macrocera, 638
Pleurahis
　jamesii, 739
　rigida, 739
Pleurophragma
　gracilipes, 288
　integrifolium, 288
　lilacinum, 288
　platypodum, 288
　rhomboideum, 287
Pluchea
　camphorata, 219
　sericea, 219
Plum
　American, 537
　cherry, 537
　common, 537
　European, 537
　flowering, 537
　Pottawattami, 537
Plume
　Apache, 525
　desert, 284
　prince's, 284
Poa
　abbreviata, 767
　　pattersonii, 767
　aggassizensis, 768
　alpina, 762
　ampla, 769
　annua, 762
　arctica, 762
　bigelovii, 763
　bolanderi, 763
　brachyglossa, 769
　bulbosa, 763
　cilianensis, 730
　compressa, 763
　confusa, 769
　crocata, 767
　curta, 764
　curvula, 730
　cusickii, 764
　distans, 771
　eatonii, 764
　epilis, 764
　fasciculata, 771
　fendleriana, 764
　festucoides 737
　glauca, 765

glauca, 765
rupicola, 765
glaucantha, 765
glaucifolia, 765
gracillima, 769
grayana, 762
hypnoides, 730
interior, 765
juncifolia, 769
kingii, 745
leptocoma, 766
lettermanii, 766
longiligula, 764
mexicana, 730
nemoralis, 765
　interior, 765
nervosa, 766
　wheeleri, 767
nevadensis, 769
nuttalliana, 771
occidentalis, 766
palustris, 767
pattersonii, 767
pectinacea, 731
pratensis, 768
reflex, 768
rupicola, 765
sandbergii, 769
scabrella, 769
scabriuscula, 764
secunda, 769
stenantha
striata, 738
trivialis, 769
wheeleri, 766
Podagrostis
　humilis, 694
　thurberiana, 695
Podistera
　eastwoodae, 636
Podosciadium
　bolanderi, 636
Poinciana
　gilliesii, 381
Poinsettia
　annual, 301
　dentata, 301
Polanisia
　dodecandra, 98
　trachysperma, 98
　trachysperma, 98
Polemoniaceae, 455
Polemonium
　albiflorum, 469
　brandegei, 469
　caeruleum, 469
　　amygdalinum, 469
　　pterospermum, 469
　　occidentale, 469
　delicatum, 470
　foliosissimum, 469
　　albiflorum, 469
　　alpinum, 469
　　foliosissimum, 469
　　micranthum, 469
　　occidentale, 469
　　　amygdalinum, 469
　pulcherrimum, 470
　　delicatum, 470
　viscosum, 470
Poliomintha
　incana, 333
Polygala

acanthoclada, 470
acanthoclada, 470
intricata, 470
intermontana, 470
subspinosa, 470
verticillata, 470
Polygalaceae, 470
Polygonaceae, 470
Polygonum, 487
amphibium, 488
argyrocoleon, 490
aubertii, 488
aviculare, 488
bistorta, 489
　oblongifolium, 489
　linearifolium, 489
bistortoides, 489
coccineum, 487
convolvulus, 489
cuspidatum, 489
douglasii, 489
　douglasii, 489
　johnstonii, 489
　engelmannii
　hydropiper, 489
　kelloggii, 489
lapathifolium, 490
nodosum, 490
minimum, 490
pensylvanicum, 490
persicaria, 490
ramosissimum, 490
sawathchense, 489
scabrum, 490
utahense, 489
viviparum, 490
zuccarinii, 489
Polypappus
　sericeus, 219
Polypodiaceae, 17
Polypodiophyta, 15
Polypodium
　bulbiferum, 21
　felix-femina, 19
　felix-mas, 21
　fragile, 21
　hesperium, 23
　lonchitis, 23
Polypody
　western, 23
Polypogon
　ditch, 770
　interruptus, 770
　lutosus, 770
　monspeliensis, 770
　semiverticillatus, 770
　water, 770
Polystichum
　krukebergii, 23
　lonchitis, 23
　scopulinum, 23
Ponista
　oregonensis, 565
Pomegranate, 501
　family, 501
Pondlily
　yellow, 430
Pondweed
　crisped, 819
　dwarf, 821
　eelgrass, 822
　family, 818
　fennel-leaf, 821

filiform, 819
floating, 820
Fries, 820
grass, 820
horned, 824
Illinois, 820
leafy, 820
northern, 819
Nuttall, 819
Richardson, 821
Robbins, 821
sheathed, 822
whitestem, 821
Poormans-orchid, 608
Popcornflower
  hairy, 83
  Jones, 83
  Scouler, 83
  slender, 83
Poplar
  balsam, 547
  black, 548
  Carolina, 548
  gray, 548
  Lombardy, 547
  white, 547
  yellow, 418
Poppy
  California, 451
  corn, 452
  family, 450
  opium, 452
  oriental, 452
  tulip, 452
Populus
  acuminata, 547
  alba, 547
    bolleana, 547
  angustifolia, 547
  arizonica, 548
  aurea, 548
  balsamifera, 547
    trichocarpa, 548
  deltoides, 548
    occidentalis, 547
    wislizenii, 548
  canadensis, 548
  fremontii, 548
    wislizenii, 548
  nigra, 548
    italica, 548
  sargentii, 548
  tremuloides, 548
  trichocarpa, 548
  wislizenii, 548
Porophyllum
  gracile, 219
Porterella
  carnosula, 96
Portulaca
  grandiflora, 496
  mundula, 496
  oleracea, 496
  parvula, 496
  retusa, 496
Portulaceae, 493
Potamogeton
  alpinus, 819
  berchtoldi, 821
  compressus, 822
  crispus, 819
  epihydrus, 819
  filiformis, 819

alpinus, 819
latifolius, 820
occidentalis, 820
fluitans, 821
foliosus, 820
  fibrillosus, 820
  foliosus, 820
friesii, 820
gramineus, 820
  maximus, 820
  myriophyllus, 820
illinoensis, 820
latifolius, 820
marinus, 819
  alpinus, 819
  occidentalis, 820
natans, 820
nodosus, 821
pectinatus, 821
  latifolius, 820
perfoliatus, 821
  richardsonii, 821
praelongus, 821
pusillus, 821
  vulgaris, 821
    interruptus, 821
richardsonii, 821
robbinsii, 821
strictifolius, 821
  rutiloides, 822
vaginatus, 822
zosterformis, 822
Pot, 97
Potamogetonaceae, 818
Potato, 610
  family, 604
  Indian, 634
  James, 609
  sweet, 241
Potentilla
  anserina, 531
  arguta, 531
  bicrenata, 532
  biennis, 532
  brunnescens, 534
  candida, 534
  concinna, 532
    modesta, 532
    bicrenata, 532
    proxima, 532
  crinita, 532
    lemmonii, 532
    convallaria, 532
    dissecta, 532
    decurrens, 534
    multisecta, 532
  diversifolia, 532
    diversifolia, 533
    multisecta, 533
    pinnatisecta, 534
  elmeri, 534
  eremica, 527
    floribunda, 533
  fruticosa, 533
  furcata, 536
  glandulosa, 533
    arizonica, 533
    glabrata, 533
    intermedia, 533
    micropetala, 533
    pseudorupestris, 533
    gordonii, 527

gracilis, 533
  brunescens, 534
  elmeri, 534
  glabrata, 534
  permollis, 534
  pulcherrina, 534
hippiana, 534
  pulcherrima, 534
hookeriana, 536
kingii, 527
  incerta, 527
lateriflora, 532
lemmonii, 532
leucocarpa, 536
millegrana, 536
modesta, 532
monspeliensis, 534
  norvegica, 534
multisecta, 532
nivalis, 526
nivea, 534
norvegica, 534
  monspellensis, 534
nuttallii, 534
  glabrata, 534
osterhoutii, 539
ovina, 534
  decurrens, 534
  ovina, 535
palustris, 535
paradoxa, 535
paucijuga, 535
pecten, 534
pectinisecta, 534
pedersensii, 536
pensylvanica, 535
  paucijuga, 535
  pensylvanica, 535
permollis, 534
plattensis, 535
proxima, 532
pulcherrima, 534
quinquefolia, 532
rivalis, 536
rubricaulis, 536
sabulosa, 528
saximontana, 536
shockleyi, 528
utahensis, 528
wardii, 534
Poverty-weed, 128
Prairie-clover
  Hole-in-the-Rock, 383
  Kanab, 383
  Searls, 384
  western, 383
Prenanthella
  exigua, 220
Prenanthes
  exigua, 220
  juncea, 212
  pauciflora, 230
  tenuifolia, 231
Pricklegrass, 715
  common, 715
Pricklypear
  berry, 90
  brittle, 90
  central, 91
  pancake, 89
  plains, 90
Prickly-poppy
  armed, 451

San Rafael, 451
Primrose
  cave, 500
  family, 497
  House Range, 499
  Maguire, 500
  Parry, 500
  silvery, 500
Primula
  domensis, 499
  incana, 500
  maguirei, 500
  nevadensis, 499
  parryi, 500
  specuicola, 500
Primulaceae, 497
Princes feather, 45
Prince's-pine, 501
Pritchardia
  filifera, 818
Privet
  common, 433
Proboscidea
  lousianica, 453
  parviflora, 453
Professor-weed, 384
Prosartes
  trachycarpa, 806
Prosopis
  glandulosa, 401
  torreyana, 401
  juliflora, 401
    glandulosa, 401
    torreyana, 401
  odorata, 401
  pubescens, 401
Prune
  Italian, 537
Prunella
  pensylvanica, 333
  lanceolata, 333
  vulgaris, 333
    lanceolata, 333
Prunus
  americana, 537
  armeniaca, 537
  avium, 537
  besseyi, 537
  blireana, 537
  cerasifera, 537
  cerasus, 537
  demissa, 539
    melanocarpa, 539
  domestica, 537
  dulcis, 538
  emarginata, 538
  fasciculata, 538
  laurocerasus, 538
  mahaleb, 538
  melanocarpa, 539
  mume, 537
  padus, 538
  persica, 538
  salicina, 537
  serrulata, 538
  tomentosa, 538
  triloba, 538
  virginiana, 539
    demissa, 539
    melanocarpa, 539
Psathyrostachys
  juncea, 692
Psathyrotes

annua, 220
pilifera, 220
schottii, 218
Pseudocymopterus
 hendersonii, 622
 lemmonii, 623
 montanus, 623
 tidestromii, 623
 versicolor, 623
Pseudopteryxia
 longiloba, 622
Pseudoroegneria
 spicata, 727
Pseudotsuga
 douglasii, 33
 glauca, 33
 globosa, 33
 menziesii, 33
 glauca, 33
 taxifolia, 33
Psilocarpus
 brevissimus, 220
Psilostrophe
 cooperi, 221
 sparsiflora, 221
 tagetina, 221
Psoralea
 aromatica, 399
 bigelovii, 402
 castorea, 400
 elliptica, 402
 epipsila, 399
 floribunda, 402
 juncea, 401
 lanceolata, 402
  stenophylla, 402
  stenostachys, 402
 laxiflora, 402
 megalantha, 399
 mephitica, 399
  retrosa, 399
 micrantha, 402
 obtusiloba, 402
 pariensis, 400
 rafaelensis, 399
  magna, 399
 retrorsa, 399
 stenophylla, 402
 stenostachys, 402
 tenuiflora, 402
  bigelovii, 402
Psoralidium
 bigelovii, 402
 floribundum, 402
 junceum, 401
 lanceolatum, 401
  lanceolatum, 402
  stenophyllum, 402
  stenostachys, 402
 micranthum, 402
 stenophyllum, 402
  stenostachys, 402
 tenuiflorum, 402
Psorodendron
 amoenum, 402
 arborescens, 402
 fremontii, 402
 johnsonii, 402
 pubescens, 402
Psorothamnus
 arborescens, 402
  pubescens, 402
 fremontii, 402

polyadenius, 403
 jonesii, 403
 polyadenius, 403
 thompsonae, 403
  thompsonae, 403
  whitingii, 403
Ptelea
 lutescens, 546
 neglecta, 546
 pallida, 546
 trifoliata, 546
  pallida, 546
   lutescens, 546
   trifoliata, 546
Pterchiton
 occidentale, 118
Pteridium
 aquilinum, 23
  pubescens, 23
Pteris
 aquilina, 23
 ternifolia, 22
Pterophacos
 tetrapterus, 378
Pterospora
 andromeda, 502
Pterostegia
 drymarioides, 490
Pteryxia
 hendersonii, 622
 terebinthina, 625
Ptilocalais
 macrolepis, 216
Puccinellia
 airoides, 771
 distans, 771
 fasciculata, 771
 nuttalliana, 771
 pauciflora, 772
 simplex, 772
Pulmonaria
 ciliata, 80
 lanceolata, 81
 oblongifolia, 81
Pulsatilla
 patens, 504
  multifida, 504
Pumpkin
 field, 290
Punica
 granatum, 501
Punicaceae, 501
Purpusia
 arizonica, 539
 osterhoutii, 539
 saxosa, 539
Purse
 shepherds, 256
Purshia
 glandulosa, 539
 mexicana, 539
  stansburiana, 539
 tridentata, 539
Purslane
 dwarf, 496
 family, 493
 horse, 43
Pusley, 496
Pussytoes
 alpine, 140
 breaks, 141
 common, 141
 field, 141

low, 141
 mountain, 142
 pearly, 140
 plains, 141
 rosy, 141
 rush, 141
 showy, 141
Pygmyweed, 243
Pyracantha
 coccinea, 540
Pyrethrum
 parthenium, 165
Pyrola
 asarifolia, 502
  bracteata, 502
  ovata, 502
  purpurea, 502
  uliginosa, 502
 bracteata, 502
  hillii, 502
 chlorantha, 503
  saximontana, 503
  paucifolia, 503
 conferta, 502
 elata, 502
 incarnata, 502
 menziesii, 501
 minor, 502
  conferta, 502
 picta, 502
 rotundifolia, 502
  bracteata, 502
  purpurea, 502
  uliginosa, 502
 secunda, 502
 eucycla, 502
 obtusata, 502
 pumila, 502
 uliginosa, 502
 umbellata, 501
 uniflora, 501
 virens, 503
  paucifolia, 503
Pyrolaceae, 501
Pyrrocoma
 chieranthifolia, 199
 clementis, 199
 lapathifolia, 199
 prionophylla, 201
 sessiliflora, 201
 subcaesia, 199
 subviscosa, 200
Pyrus
 baccata, 529
 calleryana, 540
 communis, 540
 cydonia, 524
 floribunda, 529
 hupehensis, 529
 ioensis, 529
 japonica, 522
Quackgrass, 725
Quakey, 548
Quamoclidion
 multiflorum, 429
 oxybaphoides, 429
Quassia
 family, 604
Quercus
 alba, 305
 bicolor, 305
 borealis, 306
 eastwoodiae, 305

gambelii, 305
 havardii, 305
 macrocarpa, 306
 pauciloba, 306
 robur, 306
 rubra, 306
 stellata, 305
  utahensis, 305
 turbinella, 306
 undulata, 305
  utahensis, 305
Quillwort
 Bolander, 13
 family, 13
 Howell, 13
 lake, 13
 spiny, 13
Quince
 flowering, 522
 Japanese, 522
Rabbitbrush
 alkali, 165
 dwarf, 166
 Greene, 166
 Mohave, 168
 Parry, 168
 rubber, 166
 southwest, 168
 spreading, 166
 Vasey, 169
 viscid, 169
 white, 165
Radicula
 alpina, 281
 austriaca, 281
 armoracia, 254
 curvipes, 281
 glabra, 287
 hispida, 281
 integra, 281
 obtusa, 281
  sphaerocarpa, 282
 palustris, 281
  glabra, 281
  hispida, 281
 sinuata, 281
 spaerocarpa, 282
 tenerrima, 282
 underwoodii, 281
Radish, 280
Rafinesquia
 californica, 221
 neomexicana, 221
Ragweed
 bur, 139
 common, 139
 giant, 139
 low, 139
 western, 139
Ramischia
 secunda, 502
Ranunculaceae, 503
Ranunculus, 510
 acriformis, 511
  aestivalis, 512
 acris, 512
  acris, 512
  aestivalis, 512
 adoneus, 512
 alpinus, 512
 alepeophilus, 515
 alismifolius, 512
  montanus, 512

andersonii, 512
 andersonii, 512
 juniperinus, 513
 tenellus, 512
aquatilis, 513
 capillaceus, 513
 hispidulus, 513
 longirostris, 513
 porteri, 513
 subrigidus, 513
arvensis, 513
 capillaceus, 513
cardiophyllus, 513
circinatus, 513
 subrigidus, 513
collomae, 151
cymbalaria, 513
 saximontana, 513
delphinifolius, 514
digitatus, 515
eschscholtzii, 514
 alpinus, 512
 eschscholtzii, 514
 eximius, 514
 trisectus, 514
eximius, 514
flabellaris, 514
flammula, 514
gelidus, 514
glaberrimus, 514
 ellipticus, 514
 glaberrimus, 514
gmelinii, 514
hispidus, 515
hyperboreus, 515
 nutans, 515
inamoenus, 515
 alpeophilus, 515
jovis, 515
juniperinus, 513
longirostris, 513
macounii, 515
macranthus, 515
multifidus, 514
 repens, 514
nutans, 515
oreogenes, 515
orthrorhynchus, 515
 alpinus, 512
 platyphyllus, 515
pedatifidus, 516
ranunculinus, 516
repens, 516
 repens, 516
 glabratus, 516
 pleniflorus, 516
sceleratus, 516
 multifidus, 516
testiculatus, 516
trisectus, 514
utahensis, 515
Rape, 255
Raphanus
 raphanistrum, 280
 sativa, 280
 tenellus, 259
Raspberry, 541
 black, 542
Ratany
 family, 327
 range, 328
 white, 327
Ratibida

columnaris, 221
columnifera, 221
Rattlesnake-plantain, 814
Rattlesnake-weed, 300
Razoumofskya
 abietina, 644
 americana, 644
 cyanocarpum, 644
 divaricatum, 644
 douglasii, 644
 abietinum, 644
Redbud
 American, 381
 Western, 382
Redfieldia
 flexuosa, 772
Redtop, 695
Redwood
 dawn, 34
Reed
 common, 759
 giant, 700
Reedgrass
 bluejoint, 712
 Jones, 712
 northern, 713
 purple, 712
 slimstem, 713
Reseda
 lutea, 518
Resedaceae, 518
Resinbush, 237
Retama, 398
Reverchonia
 arenaria, 303
Rhamnaceae, 518
Rhamnus
 alnifolia, 519
 betulifolia, 519
 cathartica, 519
 frangula, 519
Rheum
 rhaponticum, 491
 rhubarbarum, 491
Rhodiola
 integrifolia, 243
 rosea, 243
Rhubarb, 491
Rhus
 aromatica, 46
 simplicifolia, 47
 trilobata, 47
 canadensis, 47
 simplicifolia, 47
 cismontana, 47
 copallina, 47
 glabra, 47
 nitens, 47
 radicans, 47
 rydbergii, 47
 rydbergii, 47
 toxicodendron, 47
 rydbergii, 47
 trilobata, 47
 anisophylla, 47
 simplicifolia, 47
 typhina, 47
 utahensis, 47
Rhysopterus
 jonesii, 622
Ribbongrass, 758
Ribes
 aureum, 562

 cereum, 562
 inebrians, 563
 cereum, 563
 coloradense, 563
 grossularia, 562
 hudsonianum, 563
 petiolare, 563
 inebrians, 563
 inerme, 563
 lacustre, 563
 lentum, 564
 laxiflorum, 563
 coloradense, 564
 leptanthum, 564
 mogollonicum
 montigenum, 564
 odoratum, 562
 oxyacanthoides, 563
 lacustre, 563
 nigrum, 562
 petiolare, 563
 sativum, 562
 setosum, 564
 velutinum, 564
 velutinum, 564
 viscosissmum, 564
 viscosissmum, 564
 wolfii, 564
Ricegrass
 Bloomer, 782
 Indian, 783
 little, 754
 littleseed, 754
 roughleaf, 754
Ricinus
 communis, 304
Riddellia
 cooperi, 221
 tagetina, 221
 sparsiflora, 221
Ridgecress, 271
Robinia
 breviloba, 403
 caragana, 381
 hispida, 403
 luxurians, 403
 neomexicana, 403
 luxurians, 403
 subvelutina, 403
 pseudoacacia, 404
Rochelia
 patens, 77
Rock-brake, 20
Rockcress
 common, 252
 hairy, 250
 Holboell, 250
 Kass, 265
 Lemmon, 251
 Lyall, 251
 Nuttall, 252
 park, 253
 pretty, 252
 puberulent, 252
 schist, 253
 Selby, 253
 Shockley, 253
 sickle, 253
 small-leaf, 251
 Wasatch, 251
Rock-daisy
 alcove, 218
 Emory, 217

 Jones, 218
 Stansbury, 218
Rocket
 sweet, 269
Rocknettle, 413
Rock-parsley, 625
 mountain, 622
 skeletonleaf, 625
Roemeria
 refracta, 453
Rorippa
 alpina, 281
 armoracia, 254
 austriaca, 281
 curvipes, 281
 alpina, 281
 curvipes, 281
 integra, 281
 curvisiliqua, 281
 integra, 281
 islandica, 281
 glabra, 281
 hispida, 281
 nasturtium-aquaticum, 278
 obtusa, 281
 alpina, 281
 integra, 281
 spaerocarpa, 282
 palustris, 281
 glabra, 281
 hispida, 281
 sinuata, 281
 sphaerocarpa, 281
 sylvestris, 282
 tenerrima, 282
 underwoodii, 287
Rosa
 arizonica, 541
 californica, 541
 ultramontana, 541
 canina, 540
 chinensis, 541
 chrysocarpa, 541
 eglanteria, 540
 fendleri, 541
 foetida, 541
 gallica, 541
 macounii, 541
 manca, 541
 moschata, 541
 multiflora, 540
 neomexicana, 541
 nutkana, 540
 hispida, 541
 odorata, 541
 puberulenta, 541
 rubiginosa, 540
 spaldingii, 540
 stellata, 541
 woodsii, 541
 ultramontana, 541
Rosaceae, 519
Rose-acacia, 403
Rose
 dog, 540
 family, 519
 hybrid, 541
 multiflora, 540
 Nutka, 540
 tea, 541
 tundra, 533
 Woods, 541
 yellow, 533

Rose-heath, 211
Rose-of-Sharon, 420
Roseroot, 243
Rosettes, mealy, 220
Rubia
  tinctoria, 545
Rubiaceae, 543
Rubberweed
  Colorado, 208
Rubus
  discolor, 541
  idaeus, 541
    canadensis, 541
    idaeus, 542
    melanolasius, 542
    sachalinensis, 541
    strigosus, 541
  japonicus, 528
  leucodermis, 542
  melanolasius, 542
  neomexicanus, 542
  parviflorus, 542
    parviflorus, 542
  procerus, 541
  sachalinensis, 541
  strigosus, 541
Rudbeckia
  columnifera, 221
  laciniata, 222
  montana, 222
  occidentalis, 222
    montana, 222
Rumex
  acetosella, 491
  crispus, 491
  dentatus, 492
  digynus, 487
  fueginus, 492
  hymenosepalus, 492
  maritimus, 492
    athrix, 492
    fueginus, 492
  mexicanus, 493
  obtusifolius, 492
  occidentalis, 492
  patientia, 492
  paucifolius, 492
  salicifolius, 492
    mexicanus, 493
    montigenitus, 493
    triangulivalvis, 493
  stenophyllus, 493
  subalpinus, 492
  triangulivalvis, 493
  utahensis, 493
  venosus, 493
Ruppia
  maritima, 822
Ruppiaceae, 822
Rush
  chestnut, 793
  family, 791
  Halls, 795
  jointed, 793
  longstyle, 795
  Mertens, 796
  minute, 793
  mud, 795
  Nevada, 796
  northern, 792
  Parry, 796
  poverty, 796
  Regels, 796

swordleaf, 794
three-flowered, 796
toad, 793
Torrey, 796
Rush-lomatium, 630
Rush-pea
  creeping, 381
Rushpink
  Entrada, 211
  showy, 212
Russian-thistle, 129
  barbwire, 129
Rutaceae, 546
Rydbergiella
  arcta, 374
  pattersonii, 372
  praelonga, 374
  preussii, 374
Rye
  cultivated, 775
Ryegrass, 745
Sacaton
  alkali, 779
Saccharum
  ravennae, 773
Sage
  bladder, 333
  Dorr, 334
  garden, 334
  purple, 333
Sagebrush
  big, 150
  Bigelow, 146
  black, 149
  common, 150
  fringed, 148
  garden, 146
  longleaf, 148
  low, 146
  old-man, 147
  Osterhout, 150
  pygmy, 149
  sand, 147
  silver, 147
  threetip, 150
  Vasey, 150
  Wyoming, 150
Sagewort
  dwarf, 149
  prairie, 148
Sagina
  linnaei, 109
  procumbens, 109
  saginoides, 109
Sagittaria
  arifolia, 651
    cuneata, 651
  cuneata, 651
  latifolia, 651
Sainfoin, 395
St. Johnswort
  family, 314
Salazaria
  mexicana, 333
Salicaceae, 546
Salicornia
  europaea, 128
    rubra, 128
    trona, 128
  pacifica, 129
    utahensis, 129
    rubra, 128
    trona, 128

utahensis, 129
virginica, 129
Salix
  alba, 551
  amygdaloides, 551
  anglorum, 551
    antiplasta, 551
  arctica, 551
    petraea, 551
  babylonica, 551
  bebbiana, 551
    perrostrata, 551
  blanda, 551
  boothii, 552
  bonplandana, 554
  brachycarpa, 552
  cascadensis, 552
  commutata, 556
  discolor, 556
  drummondiana, 552
  eriocephala, 555
  exigua, 553
    exigua, 553
    stenophylla, 553
  fragilis, 553
  geyeriana, 553
    geyeriana, 553
  glauca, 553
  gooddingii, 554
  humilis, 556
  laevigata, 554
  lasiandra, 554
    candata, 554
  lasiolepis, 554
  ligulifolia, 555
  lutea, 554
    ligulifolia, 555
    platyphylla, 554
    watsonii, 554
  matsudana, 551
  tortuosa, 551
  umbraculifera, 551
  melanopsis, 553
  monica, 555
  monticola, 555
  myrtillifolia, 552
  nigra, 554
  nivalis, 555
    saximontana, 555
  pentandra, 551
    caudata, 554
  phylicifolia, 555
    planifolia, 555
  planifolia, 555
    monica, 555
    planifolia, 555
  pseudocordata, 552
  pseudolapponum, 553
  reticulata, 555
    nivalis, 555
    reticulata, 555
    saximontana, 555
  rigida, 555
  scouleriana, 555
  subcoerulea, 552
  wolfii, 556
    wolfii, 556
    idahoensis, 556
Salpiglossis
  sinuata, 608
Salsify, 237
Salsola
  atriplicifolia, 126

collina, 129
hyssopifolia, 122
iberica, 129
iranica, 128
kali, 129
paulsenii, 129
pestifer, 129
prostrata, 127
Saltbush
  Australia, 122
  basin, 119
  big, 120
  Bonneville, 119
  Castle Valle, 119
  fat-hen, 120
  four-wing, 118
  Gardner, 119
  Garrett, 120
  Jones, 119
  New Mexico, 120
  Torrey, 122
  Welsh, 120
Saltgrass
  desert, 719
Salvia
  aethopsis, 334
  azurea, 334
  carnosa, 334
    pilosa, 334
  columbariae 334
  dorrii, 334
    argentea, 334
    dorrii, 334
    lanceolata, 334
  officinalis, 334
  pilosa, 334
  reflexa, 334
  sclarea, 335
  splendens, 335
Salviniaceae, 24
Sambucus
  caerulea 100
  canadensis, 100
  glauca, 100
  melanocarpa, 100
  nigra, 100
  racemosa, 100
    melanocarpa, 100
    microbotrys, 100
Samolus
  parviflorus, 500
Samphire
  annual, 128
Sandalwood
  family, 556
Sandberry, 297
Sandbur, field, 714
Sandspurge, 303
Sandplant
  Brandegee, 178
  Gray, 178
Sandpuffs, 430
Sandreed
  big, 713
Sand-verbena
  fragrant, 426
  honey, 427
  low, 427
  sticky, 427
  Wooton, 430
Sandwort
  beach, 103
  bluntleaf, 104

dwarf, 105
Fendler, 103
head, 102
Hooker, 105
Nuttall, 104
reddish, 105
Ross, 105
Rydberg, 105
shrubby, 104
thymeleaf, 105
Sanguisorba
  minor, 542
Santalaceae, 556
Santolina
  suaveolens, 164
Sapindaceae, 556
Saponaria
  officinalis, 109
  segetalis, 114
  vaccaria, 114
Sarcobatus
  vermiculatus, 129
    vermiculatus, 129
Sarcostemma
  cynanchoides, 55
Saskatoon, 521
Satin-flower, 278
Satintail, 742
Savin, 27
Satureja
  vulgaris, 335
Satyrium
  viride, 816
Saururaceae, 557
Saw-grass, 678
Saxifraga
  adscendens, 565
    oregonensis, 565
  arguta, 565
  austromontana, 565
  bronchialis, 565
    austromontana, 565
  caespitosa, 565
    exoratoides, 565
    purpusii, 565
  minima, 566
  cernua, 566
  chrysantha, 566
  debilis, 566
  flagellaris, 566
  hirculus, 566
  integrifolia, 566
  jamesii, 558
  odontoloma, 556
  peltata, 561
  rhomboidea, 566
Saxifragaceae, 557
Saxifrage
  brook, 566
  diamondleaf, 566
  family, 557
  flagellate, 556
  golden, 556
  James, 558
  nodding, 566
  pygmy, 566
  spotted, 565
  tufted, 565
  wedge-leaf, 565
  yellow marsh, 566
Schedonnardus
  paniculatus, 773
Schismus

arabicus, 773
barbatus, 774
Schizachne
  purpurascens, 774
Schizachyrium
  scoparium, 774
Schizonotus
  argenteus, 527
    dumosus, 527
  discolor, 527
    dumosus, 527
Schkuhria
  integrifolia, 219
  oblongifolia, 219
Schmaltzia
  affinis, 47
Schoberia
  occidentalis, 130
Schoencrambe
  argillacea, 282
  barnebyi, 282
  decumbens, 283
  linifolia, 282
    pinnata, 283
    pinnata, 283
    pygmaea, 283
  suffrutescens, 283
Scilla
  sibirica, 809
Scirpus
  acutus, 682
  americanus, 683
    longispicatus, 684
    atrovirens, 684
    pallidus, 684
  caespitosus, 683
  flavescens, 680
  fluviatilis, 683
  maritimus, 683
    fluviatilis, 683
  microcarpus, 683
  nevadensis, 683
  olneyi, 683
  ovatus, 680
  pallidus, 684
  paludosus, 683
  palustris, 680
  parvulus, 680
  pungens, 684
    longispicatus, 684
  rostellatus, 681
  spadiceus, 681
  supinus, 684
  validus, 684
Sclerocactus
  contortus, 93
  glaucus, 93
  havasupaiensis, 93
    roseus, 93
  intermedius, 93
  parviflorus, 93
    intermedius, 93
  polyancistrus
  pubispinus, 93
    pubispinus, 93
    spinosior, 93
  spinosior, 93
  terrae-canyonae, 93
  whipplei, 93
    glaucus, 93
    intermedius, 93
    roseus, 93

wrightiae, 94
Sclerocarpus
  gracilis, 214
Sclerochloa
  dura, 775
Scorzonella
  nutans, 216
Scouringrush
  common, 15
  smooth, 15
  variegated, 15
Scratchgrass, 749
Screwbean, 401
Scrophularia
  lanceolata, 601
  utahensis, 601
Scrophulariaceae, 567
Scurfpea
  dune, 401
  plains, 402
  prairie, 402
  rush, 401
Scutellaria
  antirrhinoides, 335
  galericulata, 335
  nana, 335
    sapphirina, 335
Seapurslane, 43
Secale
  cereale, 775
  montanum, 775
Sedge
  alpine nerved, 670
  Arapaho, 660
  awned, 660
  beaded, 674
  beautiful, 662
  black alpine, 670
  blackcreeper, 673
  blackened, 660
  blister, 677
  blunt, 671
  bottlebrush, 665
  brownish, 662
  Canyonlands, 662
  capitate, 662
  Chamisso, 671
  cloud, 666
  curly, 675
  dark alpine, 677
  dry land, 677
  dunhead, 673
  ebony 664
  elk, 665
  family, 653
  flaccid, 668
  green, 671
  golden, 660
  hair, 662
  inland, 666
  knotsheath, 674
  Kobresia, 665
  lesser panicled, 664
  Liddon, 672
  many-ribbed, 669
  meadow, 673
  Mount Baldy, 672
  Mt. Shasta, 676
  mud, 668
  narrowleaf, 676
  Nebraska, 670
  pale, 662
  pond, 668

poor, 672
prickly, 676
Pyrenaean, 674
rock, 675
Ross, 674
russet, 675
rusty, 676
Scandinavian, 670
sheep, 666
shortleaf, 669
sierra-hare, 667
silvertop, 675
slender, 667
slender-beaked, 660
small-wing, 669
softleaved, 664
spikenard 669
threadleaf, 665
valley, 677
water, 660
western, 671
white and black, 659
woolly, 667
woodrush, 668
yellow bog, 664
Sedum
  album, 242
  debile, 242
  integrifolium, 243
  lanceolatum, 243
  meehanii, 243
  rhodanthum, 243
  rosea, 243
    integrifolium, 243
  stenopetalum, 243
Seepweed
  broom, 130
  Torrey, 130
  western, 130
Seepwillow
  Emory, 156
  Rio Grande, 157
  Sticky, 156
Selaginella
  densa, 14
  fendleri, 14
  mutica, 14
  rupestris, 14
  densa, 14
  scopulorum, 14
  underwoodii, 14
  utahensis, 14
  watsonii, 14
Selaginellaceae, 13
Selinocarpus
  diffusus, 429
    diffusus, 429
    nevadensis, 429
  nevadensis, 429
Selinum
  acaule, 620
  kingii, 616
  terebinthinum, 625
Senecio
  admirabilis, 226
  ambrosioides, 224
  amplectens, 223
    holmii, 223
  aquariensis. 227
  atratus, 223
  aureus, 227
    werneriifolius, 227
  bigelovii, 223

hallii, 223
blitoides, 225
canus, 223
cernuus, 226
convallium, 223
crassulus, 224
crocatus, 224
dimorphophyllus, 224
    dimorphophyllus, 224
    intermedius, 224
douglasii, 224
    longilobus, 224
    monoensis, 224
eremophilus, 224
    kingii, 224
fendleri, 224
filifolius, 224
    jamesii, 224
foetidus, 225
fremontii, 225
    blitoides, 225
    fremontii, 225
hartianus, 225
holmii, 223
hydrophyllus, 225
incurvus, 226
integerrimus, 225
jonesii, 227
kingii, 224
lapidum, 225
leonardii, 227
longilobus, 224
lugens, 226
    hookeri, 226
malmstenii, 227
monoensis, 224
multicapitatus, 226
multilobatus, 225
mutabilis, 226
neomexicanus, 225
    mutabilis, 226
pammelii, 227
pauperculus, 226
platylobus, 227
pudicus, 226
purshianus, 223
rubricaulis, 227
    aphanactis, 227
serra, 226
    admirabilis, 226
    serra, 226
spartioides, 226
    multicapitatus, 226
    spartioides, 226
sphaerocephalus, 226
streptanthifolius, 227
triangularis, 227
uintahensis, 227
vulgaris, 227
wardii, 227
werneriifolius, 227
Sequoia
    giant, 34
Sequoiadendron
    giganteum, 34
Sericotheca
    concolor, 527
    dumosa, 527
    microphylla, 527
Serratula
    arvensis, 171
Seriviceberry
    Utah, 521

Sesame
    family, 453
Seseli
    triternatum, 633
Sesleria
    dactyloides, 711
Sesuvium
    sessile, 43
    verrucosum, 43
Setaria
    glauca, 776
    lutescens, 776
    verticillata, 776
    viridis, 776
Shadbush, 521
Shadscale, 118
Shepherdia
    argentea, 295
    canadensis, 295
    rotundifolia, 295
Shieldleaf, 561
Shooting-star
    alpine, 498
    pretty, 498
Sibbaldia
    erecta, 523
        parviflora, 523
    procumbens, 542
Sicyos
    lobata, 291
Sida
    dissecta, 421
    grossularifolia, 422
    hederacea, 421
    nervata, 421
    oregana, 421
Sidalcea
    candida, 421
        candida, 421
        glabrata, 421
    crenulata, 421
    neomexicana, 421
        crenulata, 421
        neomexicana, 421
    oregana, 421
        oregana, 421
Sieversia
    ciliata, 527
    paradoxa, 525
    rossii, 526
    sericea, 526
    triflora, 526
    turbinata, 526
    scapoidea, 526
Silene
    acaulis, 110
        subacaulescens, 110
    alba, 108
    andersonii, 112
    antirrhina, 110
    apetala, 108
        vacarifolia, 110
    armeria, 110
    douglasii, 110
    drummondii, 108
    menziesii, 111
    noctiflora, 111
    oregana, 111
    parryi, 111
    petersonii, 111
        minor, 111
    scouleri, 111
        hallii, 112

    pringlei, 112
    verecunda, 112
        andersonii, 112
    wahlbergella, 108
Silk-tassel
    family, 307
Silk-tree, 339
Siltbush, 130
Silverberry, 295
Silverweed
    common, 531
Simaroubaceae, 604
Simsia
    frutescens, 179
Sinapsis
    alba, 255
    arvensis, 255
    juncea, 255
    kaber, 255
    nigra, 255
Single-delight, 501
Sisymbrium
    alliaria, 247
    altissimum, 283
    amphibium, 281
        palustre, 281
    aureum, 286
    canescens, 261
        brevipes, 261
        major, 261
    decumbens, 283
    elegans, 286
    glandifera, 260
    gracilis, 260
    helictorum, 261
    hispidum, 281
    incisum, 260
    filipes
        sonnei
    intermedium, 260
    irio, 283
    islandicum, 281
    lasiophyllum, 258
    leptophyllum, 261
    linifolium, 282
        decumbens, 283
        pinnatum, 283
    longipedicellata, 260
        glandulosa, 260
    murale, 261
    nasturitum-aquaticum, 278
    officinale, 283
    palustre, 281
    pinnatum, 260
    procerum, 261
    pygmaeum, 282
    richardsonii, 261
    sophia, 261
    sylvestre, 282
    virgatum, 269
    viscosum, 261
Sisyrinchium
    demissum, 790
    douglasii, 790
    grandiflorum, 790
    idahoensis, 791
    occidentale, 791
    oreophilum, 791
    radicatum, 790
    segetum, 791
Sitanion
    breviaristatum, 724
    cinereum, 724

    elymoides, 724
    hystrix, 724
    insulare, 724
    jubatum, 724
    longifolium, 724
Sium
    erectum, 617
    incisum, 617
    suave, 636
Skullcap, 335
Skunkbush, 47
Skunk-cabbage, 810
Slenderweed, 270
Slipper
    fairy, 812
    lady's, 813
Sloughgrass
    American, 701
Smartweed
    water, 488
Smelowskia
    americana, 284
    californica, 260
    calycina, 284
        americana, 284
        lineariloba, 284
        virescens, 284
Smilacina
    amplexifolius, 809
    racemosa, 809
    stellata, 809
Smyrnium
    nudicaule, 632
Snakeweed
    broom, 197
    goldenrod, 197
    orchard, 197
    thread, 197
Snapdragon
    King, 568
    twining, 568
Sneezeweed
    common, 202
    orange, 202
Snowberry
    long-flower, 100
    mountain, 101
    white, 100
Snowdrop, 652
Snow-in-cement, 106
Snow-on-the-mountain, 302
Soapberry
    family, 556
Soapwort, 109
Solanaceae, 604
Solanum
    americanum, 609
    carolinense, 609
    coronopus, 604
    dulcamara, 609
    elaeagnifolium, 609
    jamesii, 609
    nigrum, 609
        villosum, 609
        virginicum, 609
    rostratum, 609
    sarachoides, 609
    triflorum, 610
    tuberosum, 610
    villosum, 609
Solidago
    altissima, 228
    canadensis, 228

ciliosa, 228
decumbens, 229
elongata
  garrettii, 228
  guiradonis, 229
    spectabilis, 229
  humilis, 229
  nana, 229
lepida, 228
missouriensis, 228
mollis, 229
multiradiata, 228
  neomexicana, 229
  scopulorum, 228
nana, 228
nemoralis, 229
occidentalis, 228
parryi, 228
radulina, 228
sarothrae, 197
sparsiflora, 228
spathulata, 229
  nana, 229
  neomexicana, 229
spectabilis, 229
Solomon-seal
  false, 809
Sonchus
  arvensis, 229
    uliginosus, 230
  asper, 229
    glanduliferus, 299
  biennis, 210
  ludovicianus, 210
  oleraceus, 229
    asper, 224
  pulchellus, 210
  uliginosus, 229
Sonnea
  jonesii, 83
Sophia
  andrenarum, 261
    osmiarum, 261
  brevipes, 261
  filipes, 260
  glabra, 260
  halictorum, 261
  incisa, 261
  intermedia, 260
  leptostylis
    leptophylla, 261
  nelsonii, 261
  parviflora, 261
  pinnata, 260
  procera, 261
  purpurascens, 261
  richardsoniana, 261
  serrata, 261
  sonnei, 261
  sophia, 261
  viscosa, 261
Sophora
  japonica, 404
  kentukea, 382
  nuttalliana, 404
  sericea, 404
  silky, 404
  silvery, 404
  stenophylla, 404
Sorbaria
  sorbifolia, 542
Sorbus
  aucuparia, 543
  hybrida, 543

scopulina, 543
Sorghastrum
  arenaceum, 776
  nutans, 776
Sorghum
  bicolor, 777
  grain, 777
  halepense, 777
  miliaceum, 777
  milo, 777
  vulgare, 777
Sorrel
  alpine, 492
  mountain, 487
  sheep, 491
Sow-thistle
  common, 229
  field, 229
  meadow, 229
  spiny, 229
Sparganiaceae, 822
Sparganium
  angustifolium, 822
  emersum, 823
    pedunculatum, 823
  eurycarpum, 823
  minimum, 823
  multipedunculatum, 823
  natans, 822
    angustifolium, 822
  simplex, 822
    angustifolium, 822
    emersum, 823
    multipedunculata, 823
Spartina
  gracilis, 778
  pectinata, 778
Spartium
  junceum, 404
  scoparium, 382
Spatterdock, 430
Spearmint, 332
Spearwort
  creeping, 514
Spectacle-pod, 262
Speedwell
  bilobed, 603
  corn, 602
  ivy-leaved, 603
  Persian, 603
  thyme-leaf, 603
  water, 602
  Wormskjold, 603
Spergula
  saginoides, 109
Specularia
  perfoliata, 96
Spergularia
  marina, 112
  media, 112
  rubra, 112
  salina, 112
Spergulastrum
  lanuguosum, 104
Sphaeralcea
  ambigua, 422
    ambigua, 422
  arizonica, 424
  caespitosa, 422
  coccinea, 422
    dissecta, 422
    elata, 423
  digitata, 423

dissecta, 422
elata, 422
exile, 421
grossulariifolia, 423
  grossulariifolia, 423
  pedata, 423
  moorei, 423
  pedata, 423
janeae, 423
leptophylla, 423
  janeae, 423
marginata, 424
munroana, 423
  subrhomboidea, 423
parvifolia, 424
pedata, 423
psoraloides, 424
rivularis, 420
rusbyi, 424
subrhomboidea, 423
Sphaeromeria
  capitata, 230
  diversifolia, 230
  ruthiae, 230
Sphaerophysa
  salsula, 405
Sphaerostigma
  chamaeneroides, 436
  minor, 437
  utahense, 436
Sphenopholis
  intermedia, 778
  longiflora, 778
  obtusata, 778
  pallens, 778
Spiderling, 427
Spiderwort, 653
  family, 652
Spiesia
  lambertii, 397
    sericea, 398
  multiceps, 397
  oreophila, 397
  parryi, 398
  viscida, 398
Spikemoss
  family, 13
  Rydberg, 14
  Underwood, 14
  Utah, 14
  Watson, 14
Spikenard
  American, 50
Spikerush
  creeping, 680
  pretty, 680
  yellow, 680
Spinach, 129
Spinacia
  oleracea, 129
Spindlestem, 258
Spindletree
  Bunge, 115
  European, 115
  Japanese, 115
Spineflower
  rigid, 471
  short, 471
  Thurber, 471
Spiraea
  aquilegifolia, 543
    vanhoutii, 543
  caespitosa, 529

elatior, 529
cantoniensis, 543
capitata, 530
discolor, 527
  dumosa, 527
dumosa, 527
millefolium, 523
monogyna, 530
opulifolia, 530
  pauciflora, 530
prunifolia
rock, 529
sorbifolia, 542
trilobata, 543
vanhoutei, 543
Spiranthes
  cernua, 817
  diluvialis, 817
  porrifolia, 817
  romanzoffiana, 817
    porrifolia, 817
  unalascensis, 815
Spirodela
  polyrhiza, 800
Spleenwort
  black, 18
  ebony, 19
  green, 19
  maidenhair, 19
Sporobolus
  airoides, 779
    airoides, 779
    wrightii, 779
  asper, 779
  asperifolius, 749
  contractus, 780
  cryptandrus, 780
  flexuosus, 780
  giganteus, 780
  pulvinatus, 781
  repens, 752
  wrightii, 779
Spraguea
  umbellata, 494
Sprangletop
  bearded, 744
  Mexican, 744
  red, 744
Spray
  mountain, 527
Spring-beauty, 494
Spring-parsley
  dolomite, 621
  Dorn, 623
  golfball, 622
  Ibapah, 622
  least, 623
  Lemmon, 623
  long-stalk, 623
  mountain, 622
  onion, 621
  pinnate, 621
  plains, 620
  purple-nerved, 624
  sweetroot, 624
  two-leaf, 622
  Uinta Basin, 622
  variable, 625
  widewing, 624
Spruce
  blue, 31
  Engelmann, 31

Norway, 30
Spurge
　blue, 302
　cypress, 301
　eyed, 302
　family, 299
　fringed
　leafy, 301
　Parry, 302
　petty, 303
　prairie, 303
　prostrate, 303
　revolute, 303
　ridge-seeded, 301
　shorthorn, 300
　spotted, 302
　thyme-leaved, 303
　toothed, 301
　Utah, 302
Squash
　fall, 290
　winter, 290
Squaw-apple, 529
Squill, 809
Squirreltail, 724
Stachys
　albens, 336
　asperrima, 336
　byzantina, 336
　lanata, 336
　olympica, 336
　palustris, 336
　　pilosa, 336
　pilosa, 336
　rothrockii, 336
Stafftree
　family, 114
Stainplant, 83
Stanleya
　albescens, 284
　arcuata, 284
　canescens, 284
　fruticosa, 284
　glauca, 284
　　latifolia, 284
　heterophylla, 284
　integrifolia, 284
　pinnata, 284
　　gibberosa, 284
　　integrifolia, 284
　　inyoensis, 284
　　pinnata, 284
　　pinnatifida, 284
　　　integrifolia, 284
　viridiflora, 284
Stanleyella
　wrightii, 288
　　tenella, 288
Star-lily, 808
Star-thistle
　yellow, 161
Starwort
　long-leaved, 113
　long-stalked, 113
Steershead, 307
Stegnocarpus
　canescens, 84
　leiocarya, 84
Steironema
　ciliatum, 499
Stellaria
　calycantha, 113
　jamesiana, 113

kingii, 103
longifolia, 113
longipes, 113
　altocaulis, 113
　longipes, 113
　monantha, 113
　stricta, 113
　vestita, 113
media, 114
monantha, 113
nitens, 114
obtusa, 114
stricta, 113
umbellata, 114
vestita, 113
Stenactis
　speciosa, 190
Stenogonum
　flexum, 479
　salsuginosum, 484
Stenophragma
　virgatum, 269
Stenotus
　armerioides, 199
　falcatus, 199
　latifolius, 199
Stephanomeria
　exigua, 230
　parryi, 230
　pauciflora, 231
　runcinata, 231
　spinosa, 231
　tenuifolia, 231
　　tenuifolia, 231
　　uintahensis, 231
Stickleaf
　Arapien, 414
　Cronquist, 414
　family, 413
Stickseed
　Deep Creek, 77
　European, 78
　pale, 77
　small-flower, 77
　showy, 77
　western, 78
Stinkgrass, 730
Stipa
　arida, 782
　bloomeri, 782
　columbiana, 784
　　nelsonii, 784
　comata, 782
　coronata, 783
　　depauperata, 783
　　parishii, 783
　hymenoides, 783
　lemmonii, 784
　lettermanii, 784
　minor, 784
　mormonum, 782
　nelsoni, 784
　neomexicana, 785
　occidentalis, 784
　　minor, 784
　parishii, 783
　　depauperata, 783
　pennata, 785
　　neomexicana, 785
　pinetorum, 785
　pringlei, 784
　　lemonii, 784
　speciosa, 785

thurberiana, 785
viridula, 786
williamsii, 784
Stiporyzopsis
　bloomeri, 782
Stonecrop
　common, 243
　family, 242
　opposite, 242
　pink, 243
Stoneseed
　contra, 79
　pretty, 79
　showy, 79
Storksbill, 312
Strawberry
　mountain, 525
　starvling, 526
Strawberry-spinach, 124
Streptanthella
　longirostris, 285
Streptanthus
　angustifolius, 249
　cordatus, 285
　crassicaulis, 258
　crassifolius, 285
　longirostris, 284
　pilosus, 259
　sagittatus, 286
　virgatus, 251
　wyomingensis, 286
Streptopus
　amplexifolius, 809
　chalazatus, 809
Stripe
　green, 44
Strombocarpa
　odorata, 401
　pubescens, 401
Stylocline
　micropoides, 231
Stylopappus
　grandiflorus, 138
Suaeda
　calceoliformis, 130
　depressa, 130
　erecta, 130
　intermedia, 130
　occidentalis, 130
　torreyana, 130
　　ramosissima, 130
　　torreyana, 130
Suida
　stolonifera, 242
Sumac
　shining, 47
　smooth, 47
　staghorn, 47
　velvet, 47
Summer-cypress, 128
Sunflower
　common, 203
　desert, 194
　family, 131
　fivenerve, 202
　Nuttall, 203
　onehead, 202
　prairie, 203
　sand, 203
　smallhead, 202
Swainsonia
　salsula, 405
Swallow-wort, 55

Sweetbriar, 540
Sweetbush, 159
Sweet-cicely
　blunt-fruit, 635
　western, 635
Sweet-clover
　India, 395
　white, 395
　yellow, 395
Sweetgrass, 738
Sweetgum, 315
Sweetpea
　Lanszwert, 387
　perennial, 387
　Rydberg, 386
　Scots, 387
　seemly, 386
　Utah, 387
　Zion, 386
Sweetvetch
　northern, 385
　western, 385
Swertia
　albomarginata, 311
　fritillaria, 311
　gypsicola, 311
　perennis, 311
　radiata, 311
　Utah, 311
　utahensis, 311
　white-margined, 311
　White River, 311
Switchgrass, 757
Sycamore
　American, 455
　family, 455
Syria-weed
Symphoricarpos
　albus, 100
　longiflorus, 100
　occidentalis, 100
　orbiculatus, 101
　oreophilus, 101
　parshii, 101
　utahensis, 101
　parishii, 101
　racemosus, 100
　　laevigatus, 100
　rotundifolius, 101
　　vaccinoides, 101
　tetonensis, 101
　utahensis, 101
　vaccinioides, 101
Syntherisma
　sanguinalis, 719
Synthris
　alpina, 568
　laciniata, 601
　　ibahpensis, 601
　pinnatifida, 601
　　laciniata, 601
　　pinnatifida, 601
　wyomingensis, 568
Syntrichopappus
　fremontii, 231
Syringa
　amurensis, 433
　chinensis, 433
　pekinense, 433
　persica, 433
　suspensa, 431
　villosa, 434
　vulgaris, 434

Tackplant
  pale, 161
  purple, 161
Tagetes
  papposa, 129
Talinum
  brevifolium, 496
  parviflorum, 496
  pygmaeum, 494
  thompsonii, 497
  validulum, 497
Tamarack, 30
  western, 30
Tamaricaceae, 610
Tamarisk
  Athel, 610
  family, 610
  small-flowered, 610
Tamarix
  aphylla, 610
  gallica, 610
  parviflora, 610
  pentandra, 610
  ramosissima, 610
Tanacetum
  capitatum, 230
  diversifolium, 230
  parthenum, 165
  vulgare, 232
Tansy, 232
Tansy-mustard
  California, 260
  pinnate, 260
  Richardson, 261
Tarweed, 214
Taraxacum
  ceratophorum, 232
  lyratum, 232
  officinale, 232
Taraxia
  breviflora, 436
  subacaulis, 438
Tasselflower, 160
Taxaceae, 34
Taxodiaceae, 34
Taxodium
  distichum, 34
  family, 34
Taxus
  baccata, 34
  cuspidata, 34
Tea
  Brigham, 28
  Mormon, 28
  Trapper's, 298
Teasel, 294
  family, 294
Teavine, 606
Telesonix
  jamesii, 558
Tellima
  parviflora, 560
Terragon, 147
Tessaranthium
  radiatum, 311
Tessaria
  sericea, 219
Tetradymia
  axillaris, 232
    longispina, 233
  canescens, 233
  glabrata, 233
  linearis, 233

nuttallii, 233
spinosa, 233
  longispina, 233
Tetraneuris
  acaulis, 207
  caespitosa, 207
  arizonica, 207
  epunctata, 207
  ivesiana, 207
Teucrium
  canadense, 336
    occidentale, 336
  occidentale, 336
Thalictrum
  alpinum, 517
  duriusculum, 517
  fendleri, 517
  occidentale, 517
  sparsiflorum, 517
Thamnosma
  montana, 546
Thaspium
  montanum, 623
  trifoliatum, 637
    apterum, 637
Thelesperma
  megapotamicum, 233
  subnudum, 233
    alpinum, 233
    subnudum, 234
Thelycrania
  sericea, 242
Thelypodiopsis
  ambigua, 285
  argillacea, 282
  aurea, 286
  barnebyi, 282
  divaricata, 286
  elegans, 286
  nuttallii, 286
  sagittata, 286
    ovalifolia, 286
    sagittata, 286
  torulosa, 286
  vermicularis, 286
  wyomingensis, 286
Thelypodium
  offine, 287
  ambiguum, 285
  amplifolium, 286
  aureum, 286
  cooperi, 258
  crenatum, 273
  elegans, 286
  flexuosum, 287
  gracilipes, 288
  integrifolium, 287
    affine, 287
    complanatum, 287
    gracilipes, 288
    integrifolium, 288
  laciniatum, 288
    milleflorum, 288
  lasiophyllum, 258
    utahense 258
  laxiflorum, 288
  lilacinum, 288
    subumbellatum, 288
  longirostris, 284
  macropetalum, 286
  micranthum, 287
  millefolium, 288
  nuttallii, 286

ovalifolium, 286
  palmeri, 286
  rhomiboideum, 287
    gracilipes, 288
    rollinsii, 288
  sagittatum, 286
    ovalifolium, 286
    vermicularis, 286
  suffrutescens, 283
  torulosum, 286
  utahensis, 258
  wrightii, 288
    tenellum, 288
Therefon
  heucheriforme, 558
Thermopsis
  montana, 405
    ovata, 405
  ovata, 405
  pinetorum, 405
  rhombifolia, 405
    montana, 405
    ovata, 405
Thesium
  umbellatum, 556
Thimbleberry, 542
Thinopyrum
  elongatum, 723
  gentryi, 724
  intermedium, 724
  ponticum, 723
Thistle
  Arizona, 171
  Barneby, 171
  blessed, 175
  bull, 175
  carmine, 174
  Cainville, 171
  Canada, 171
  creeping, 171
  Eaton, 172
  Fish Lake, 172
  fringed, 172
  gray, 174
  meadow, 174
  musk, 161
  New Mexico, 173
  nodding, 161
  Ownbey, 173
  Rothrock, 173
  Rydberg, 173
  Scotch, 216
  Virgin, 175
  Wheeler, 175
Thlaspi
  alpestre, 288
    glaucum, 288
    purpurascens, 288
  arvense, 288
  bursa-pastoris, 256
  campestre, 271
  cochleariforme, 288
  coloradoense, 288
  fendleri, 289
    coloradoense, 289
    hesperium, 289
    tenuipes, 289
  glaucum, 288
    hesperium, 288
    pedunculatum, 288
  hesperium, 289
  montanum, 288
  fendleri, 289

montanum, 289
  nuttallii, 288
  prolixum, 289
  purpurascens, 288
Thoroughwort
  white, 193
Threeawn
  Arizona, 699
  purple, 699
  sixweeks, 699
Threesquare
  common, 684
  Olney, 683
Thuja
  aphylla, 610
  decurrens, 25
  occidentalis, 27
  plicata, 28
  orientalis, 28
Thyme, 336
Thymophylla
  acerosa, 178
  pentachaeta, 179
    belinidium, 179
Thymus
  serpyllum, 336
  vulgaris, 336
Thysanocarpus
  curvipes, 289
  eradiatus, 289
  trichocarpus, 289
Ticklegrass, 695
Tickseed, 126
Tidestromia
  lanuginosa, 45
  oblongifolia, 46
Tidytips, 211
Tigarea
  tridentata, 539
Tilia
  americana, 611
  cordata, 611
  europaea, 611
  heterophylla, 611
  platyphyllos, 611
  tomentosa, 611
Tiliaceae, 610
Tillaea
  aquatica, 243
Tiletia
  decipiens, 698
Timothy, 759
  alpine, 759
Tiquilia
  brevifolia, 84
  canescens, 84
  hispidissima, 84
  hairy, 84
  latior, 84
  nuttallii, 84
  parviflora, 84
Tissa
  salina, 112
Tithonia
  argophylla, 180
Tithymalus
  cyparissias, 301
  esula, 301
  robusta, 301
Tium
  atratiforme, 377
  desperatum, 360
  drummondii, 361

palans, 367
scopulorum, 376
Toad-lily, 495
Toadflax
  blue, 573
  Dalmatian, 573
Tobacco, 606
  coyote, 606
  desert, 606
Tobacco-weed, 156
Tomatilla, 606
Tomato, 606
Toutera
  multicaulis, 415
Torilis
  arvensis, 637
Torresia
  odorata, 738
Torreyochloa
  pauciflora, 772
Townsendia
  alpigena, 236
    minima, 236
  annua, 234
  aprica, 234
  arizonica, 235
  condensata, 234
  dejecta, 236
  exscapa, 235
  florifer, 235
    communis, 235
  hookeri, 235
  incana, 235
    ambigua, 235
    prolixa, 236
  jonesii, 235
    jonesii, 235
    lutea, 235
  leptotes, 235
  mensana, 236
    jonesii, 235
    minima, 236
  montana, 236
    caelilinensis, 236
    minima, 236
    montana, 236
  scapigera, 236
    ambigua, 235
  sericea, 235
    leptotes, 235
  strigosa, 236
  watsonii, 235
Toxicoscordion
  venenosum, 810
Toxicodendron
  altissimum, 604
  longipes, 47
  radicans, 47
    rydbergii, 47
  rydbergii, 47
Toxylon
  pomiferum, 425
Tracheophyta, 12
Trachycarpus
  wagnerianus, 817
Tradescantia
  occidentalis, 652
  virginiana, 652
    occidentalis, 652
Tragacantha
  ampullaria, 354
  beckwithii, 356
  bisulcata, 356

caespitosa, 377
calycosa, 357
campestris, 359
drummondii, 361
episcopa, 362
flaviflora, 363
glareosa, 375
haydeniana, 356
iodantha, 365
jejuna, 366
juncea, 359
lonchocarpa, 368
magacarpa, 369
multiflora, 388
pattersonii, 372
picta, 357
platytropis, 393
preussii, 374
pubentissima, 374
purshii, 375
pygmaea, 358
scopulorum, 376
sesquiflora, 377
tegetaria, 366
thompsonae, 370
triphylla, 364
utahensis, 379
Tragia
  ramosa, 304
  stylaris, 304
Tragopogon
  dubuis, 236
  porrifolius, 237
  pratensis, 237
Trautvetteria
  caroliniensis, 517
  grandis, 517
  media, 517
Treacle, 268
Tree
  big, 34
  Goldenrain, 556
  Joshua, 649
  maidenhair, 29
  smoke, 46
  tulip, 418
Tree-of-heaven, 604
Trefoil
  birds-foot, 388
  bush, 389
  long-bracted, 388
  low, 388
  Mohave, 388
  slender, 389
  Utah, 389
  Wright, 389
Trianthema
  portulacastrum, 43
Tribulus
  californica, 646
  terrestris, 647
Tricardia
  watsonii, 326
Trichochloa
  microsperma, 750
Trichophyllum
  integrifolium, 192
Tricuspis
  mutica, 786
Tridens
  hairy, 732
  muticus, 786
  pilosus, 732

pulchellus, 732
Trifolium
  andersonii, 406
    friscanum, 406
  andinum, 406
  beckwithii, 407
  brachypus, 408
  confusum, 408
  corniculata, 388
  cyathiferum, 406
  dasyphyllum, 407
    uintense, 407
  eriocephalum, 407
    villiferum, 407
  fimbriatum, 409
  fragiferum, 407
  gymnocarpon, 407
    plummerae, 407
    subcaulescens, 407
  heterodon, 409
  hybridum, 407
  inequale, 409
  involucratum, 409
    fimbriatum, 409
  kingii, 408
    macilentum, 408
  longipes, 408
    brachypus, 408
    pygmaeum, 408
    reflexum, 408
    rusbyi, 408
  macilentum, 408
  (Melilotus) corniculata, 409
  melilotus-indica, 395
  melilotus-officinalis, 395
  montanense, 409
  nanum, 408
  nemorale, 407
  oreganum, 408
    brachypus, 408
    rusbyi, 408
    rydbergii, 408
  parryi, 408
    montanense, 409
  plummerae, 407
  pratense, 409
  repens, 409
  rusbyi, 409
  rydbergii, 408
  spinulosum, 409
  subcaulescens, 407
  uintense, 407
  variegatum, 409
  villiferum, 407
  willdenovii, 409
    fimbriatum, 409
  wormskjoldii, 409
Triglochin
  concinna, 799
  debilis, 799
  debilis, 799
  maritima, 799
    debilis, 799
  palustris, 799
Trigonella
  corniculata, 409
Triodanis
  perfoliata, 96
Triodia
  mutica, 786
  pilosa, 732
  pulchella, 732
Tripolium

angustum, 152
frondosum, 153
occidentale, 155
Tripterocalyx
  carneus, 430
    wootonii, 430
  micranthus, 430
  pedunculatus, 430
  wootonii, 430
Trisetum
  canescens, 787
    montanum, 787
  cernvum, 787
    canescens, 787
  montanum, 787
  purpurascens, 774
  Rocky Mountain, 787
  spicatum, 787
  spike, 787
  tall, 787
  wolfii, 787
Triteleia
  grandiflora, 809
Triticum
  aestivum, 788
  cylindricum, 692
  elongatum, 723
  repens, 725
  trachycaulus, 728
Trollius
  albiflorus, 517
  americanus, 517
  laxus, 517
    albiflorus, 517
Tropaeolaceae, 611
Tropaeolum
  family, 611
  majus, 611
  peregrinum, 611
Troximon
  aurantiacum, 137
    purpureum, 138
  glaucum, 138
    dasycephalum, 138
  parviflorum, 138
  pumilum, 138
  taracifolium, 138
Trumpet-vine, 59
Tsuga
  douglasii, 33
  glauca, 33
Tule, 684
Tulip
  common, 810
  empress, 810
Tulipa
  clusiana, 810
  fosteriana, 810
  gesneriana, 810
  praestans, 810
  sylvestris, 810
Tumbleweed, 129
Tumblegrass, 773
Turkey-mullein, 299
Turritis
  brachycarpa, 250
  drummondii, 249
  glabra, 250
  hirsuta, 250
  lasiophylla, 258
  macrocarpa, 250
  ovata, 250
  retrofracta, 251

spathulata, 250
stricta, 249
Twayblade
  broad-leaved, 816
  heart-leaved, 816
  northern, 816
Twinberry, black, 99
Twinflower, 99
Twinpod
  Grand Junction, 280
  Newberry, 280
Twisted-stalk
  clasping, 809
Typha
  angustifolia, 823
    domingensis, 823
  domingensis, 823
  glauca, 823
  latifolia, 823
Typhaceae, 823
Udora
  canadensis, 788
Ulmaceae, 611
Ulmus
  americana, 612
  parvifolia, 612
  procera, 612
  pumila, 612
Umbelliferae, 613
Umbrellawort
  narrowleaf, 428
Uniola
  spicata, 719
Urachne
  micrantha, 754
Uralepis
  pilosa, 732
Uropappus
  linearifolius, 215
Urtica
  breweri, 638
  dioica, 638
    gracilis, 638
    holosericea, 638
    lyallii, 638
    occidentalis, 638
    procera, 638
  gracilis, 638
  holosericea, 638
  lyallii, 638
  procera, 638
  urens, 638
Urticaceae, 637
Utahia
  sileri, 92
Utricularia
  intermedia, 411
  minor, 411
  vulgaris, 411
Uva-ursi
  buxifolia, 297
  patula, 297
  procumbens, 297
  uva-ursi, 297
Uvularia
  amplexifolia, 809
Vaccaria
  pyramidata, 114
  segetalis, 114
Vaccinium
  caespitosum, 298
  erythococcum, 299
  globulare, 298

humifusum, 297
membranaceum, 298
microphyllum, 299
myrtillus, 298
  microphyllum, 299
occidentale, 298
oreophilum, 298
scoparium, 299
Valerian
  family, 638
Valeriana
  acutiloba, 639
    pubicarpa, 639
  arizonica, 639
  capitata, 639
    acutiloba, 639
    pubicarpa, 639
  edulis, 639
  locusta, 639
  occidentalis, 639
  pubicarpa, 639
  puberulenta, 639
  rubra, 638
  utahensis, 639
Valerianaceae, 638
Valerianella, 639
  locusta, 639
Vanclevea
  stylosa, 237
Vaqueros
  espanta, 45
Velvetgrass
  Yorkshire fog, 740
Velvet-leaf, 419
Veratrum
  californicum, 810
Verbascum
  blattaria, 601
  thapsus, 601
  virgatum, 602
Verbena
  bipinnatifida, 640
    latiloba, 641
  bracteata, 641
  bracteosa, 641
  gooddingii, 641
  hastata, 641
  macdougalii, 641
  nodiflora, 640
Verbenaceae, 639
Verbesina
  encelioides, 237
  scaposa, 180
Veronica
  alpina, 603
    wormskjoldii, 603
  americana, 602
  anagallis-aquatica, 602
  arvensis, 602
  beccabunga, 602
    americana, 602
  biloba, 603
  catenata, 603
  hederifolia, 603
  peregrina, 603
  xalapensis, 603
  persica, 603
  serpyllifolia, 603
  wormskjoldii, 603
  xalapensis, 603
Vervain
  blue, , 641
  Dakota, 640

family, 634
Goodding, 641
New Mexico, 641
prostrate, 641
Vesicaria
  alpina, 275
  fendleri, 275
  stenophylla, 275
Vetch
  American, 410
  hairy, 410
  kidney, 339
  Louisiana, 410
Viburnum
  album, 100
  alnifolium, 101
  opulus, 101
  plicatum, 101
  rytidophyllum, 101
Vicia
  americana, 410
    americana, 410
    angustifolia, 410
    minor, 410
    oregana, 410
    truncata, 410
  caespitosa, 410
  dissitifolius, 410
  exigua, 410
  faba, 410
  linearis, 410
  ludoviciana, 410
  oregana, 410
  producta, 410
  sparsifolia, 410
  sylvatica, 410
  thurberi, 410
  trifida, 410
  truncata, 410
  villosa, 410
Viguiera
  annua, 238
  ciliata, 237
  longifolia, 237
  annua, 238
  multiflora, 238
    multiflora, 238
    nevadensis, 238
  nevadensis, 238
  soliceps, 238
Vilfa
  cryptandra, 788
  flexuosa, 788
  depauperata, 749
  filiformis, 749
  richardsonis, 752
  tricholepis, 701
Villarsia
  pumila, 317
Vinca
  major, 50
  minor, 50
Vine
  matrimony, 606
  puncture, 647
Viola
  adunca, 642
    bellidifolia, 642
  arvensis, 642
  beckwithii, 642
    cachensis, 642
    bellidifolia, 642
    valida, 642

    bonnevillensis, 642
  canadensis, 642
    rugulosa, 642
  charlestonensis, 643
  clauseniana, 642
  lingaefolia, 642
  mamillata, 642
  nephrophylla, 642
  nuttallii, 642
    major, 642
    vallicola, 642
    venosa, 643
  odorata, 643
  oxysepala, 642
  palustris, 643
  papilionacea, 643
  praemorsa, 642
  purpurea, 643
    charlestonensis, 643
    venosa, 643
    rugulosa, 642
  tidestromii, 642
  tricolor, 643
  utahensis, 643
  vallicola, 642
  wittrockiana, 643
Violaceae, 641
Violet
  Beckwith, 642
  blue, 642
  bog, 642
  Canada, 642
  dame's, 269
  English, 643
  family, 641
  marsh, 643
  meadow, 643
  Nuttall, 642
  pine, 643
  sweet, 643
Virgilia
  lutea, 382
Virginiacress, 273
Virgins-bower
  purple, 508
  white, 508
Virona
  hirsutissima, 508
Viscaceae, 643
Vitaceae, 645
Viticella
  breviflora, 318
  parviflora, 318
  austinae, 318
Vitis
  arizonica, 645
  labrusca, 646
  vinifera, 646
Vulpia
  bromoides, 734
  microstachys, 734
  myuros, 735
  octoflora, 735
Wahlbergella
  apetala, 108
  kingii, 108
Wahoo, 115
Wallflower
  lesser, 268
  pretty, 268
  spreading, 268
Walnut
  Arizona, 327

black, 327
English, 327
family, 326
Washingtonia
  filifera, 818
Water-clover, 16
Water-crowfoot
  yellow, 514
Waterfern
  family, 24
Waterleaf
  family, 316
  western, 317
Waterlily
  family, 430
  fragrant, 430
Watermelon, 289
Water-milfoil
  family, 314
Water-nymph
  family, 811
Water-parsnip
  cutleaf, 617
  hemlock, 636
Water-pepper, 489
Water-pimpernell, 500
Water-plantain
  family, 650
Water-starwort
  family, 94
  vernal, 94
Waterweed
  squaw, 157
Waterwort
  California, 296
  family, 295
  three-lobed, 296
Wax-flower, 501
Weatherglass
  Poor-mans, 497
Wedgegrass
  prairie, 778
Wedgeleaf, 264
Weed
  dollar, 421
  Jimpson, 605
  Joe-Pye, 193
  pineapple, 164
  poverty, 209
  Telegraph, 204
Wellingtonia
  gigantea, 34
Wet-the-bed, 225
Wheat, 788
Wheatgrass
  annual, 731
  bluebunch, 727
  crested, 693
  fairway, 693
  intermediate, 724
  Scribner, 726
  slender, 728
  spreading, 726
  tall, 723
  thickspike, 725
  western, 727
Whipplea
  utahensis, 558
Whispering-bells, 316
White-sage
Whitetop
  hairy, 257
William
  sweet, 107

Willow
  arctic, 551
  arroyo, 554
  barren-ground, 552
  Bebb, 551
  black, 554
  Booth, 552
  Cascade, 552
  caudate, 554
  coyote, 553
  crack, 553
  desert, 59
  Drummond, 552
  dusky, 553
  family, 546
  Geyer, 553
  glaucous, 553
  grayleaf, 553
  narrowleaf, 551
  peach-leaf, 551
  plainleaf, 555
  red, 554
  Scouler, 555
  whiplash, 554
  Wolf, 556
  yellow, 554
Willowherb
  alpine, 440
  autumn, 441
  Barbey, 441
  glandular, 441
  Hornemann, 442
  Nevada, 442
  northern, 441
  Rocky Mountain, 442
  slender, 442
Willow-weed, 490
Wildrye
  alkali, 727
  beardless, 728
  blue, 724
  Canada, 723
  creeping, 728
  Great Basin, 723
  Russian, 725
  Salina, 726
  thickspike, 725
  Virginia, 728
Windgrass
  Italian, 698
Wiregrass, 793
Windmillgrass
  tumble, 714
Winged-pigweed
Wintercress, 254
Winterfat, 123
Wintergreen
  family, 501
  greenish, , 503
  lesser, 502
  liver-leaf, 502
  pictureleaf, 502
  secund, 502
Wirelettuce
  annual, 230
  desert, 231
  fewflower, 230
  Parry, 230
  slender, 231
  thorn, 231
Wisteria
  floribunda, 411
  Japanese, 411
  macrobotrys, 411

  macrostachya, 411
  Nuttall, 411
  sinensis, 411
Witchgrass, 756
Witch-hazel
  family, 314
Woad
  dyer's, 270
Wolfberry
  Anderson, 605
  pale, 606
Wolffia
  punctata, 800
Wolftail
Wood-anemone, 505
Woodlandstar
  slender, 560
  smallflower, 560
Woodreed
  drooping, 715
Woodrush
  hairy, 798
  millet, 798
  spike, 798
Woodsia
  Oregon, 24
  oregana, 24
  Rocky Mountain, 24
  scopulina, 24
Woodsorrel
  creeping, 449
  erect, 450
  family, 449
Woodyaster
  Cisco, 240
  Cronquist, 239
  Henrieville, 239
  hurtleaf, 240
  smooth, 239
Woollybase
  stemless, 207
Woollyleaf
  gray, 192
  stemless, 207
  Pursh, 192
  Wallace, 192
Wormwood
  biennial, 146
  Carruth, 147
  field, 147
  Louisiana, 148
  Michaux, 148
  Parry, 149
  spruce, 149
  sweet, 146
Woundwort, 336
Wulfenia
  pinnatifida, 601
Wyethia
  amplexicaulis, 238
  arizonica, 238
  scabra, 238
    attenuata, 239
    canescens, 239
    scabra, 239
Wyomingia
  argentata, 184
  vivax, 191
Xanthium
  italicum, 239
  pensylvanicum, 239
  strumarium, 239
Xanthocephalum

  microcephalum, 197
  petradoria, 197
  sarothrae, 197
    pomariense, 197
    sarothrae, 197
Ximenesia
  encelioides, 237
Xylophacos
  amphioxys, 354
  aragaloides, 354
  argophyllus, 355
  brachylobus, 378
  cibarius, 358
  cymboides, 360
  eurekensis, 363
  glareosus, 375
  iodantha, 365
  marianus, 373
  medius, 363
  melanocalyx, 354
  musiniensis, 371
  newberryi, 371
  purshii, 375
  uintensis, 355
  utahensis, 379
  zionis, 380
Xylorhiza
  confertifolia, 239
  cronquistii, 239
  glabriscula, 239
    glabriuscula, 240
    linearifolia, 240
  lanceolata, 240
  tortifolia, 240
    imberbis, 240
    tortifolia, 240
  venusta, 240
Xylosteum
  involucratum, 99
Yabea
  microcarpa, 637
Yam, 241
Yampah, 636
Yarrow
  fernleaf, 137
  milfoil, 137
Yellow-bell, 807
Yellowflower, 194
Yellow-wood, 382
Yerba-santa, 316
Yew
  English, 34
  family, 34
  Japanese, 34
Yucca
  angustissima, 649
    avia, 649
    kanabensis, 650
    toftiae, 650
  arborescens, 649
  baccata, 649
    baccata, 649
    vespertina, 649
  Bailey, 649
  baileyi, 649
  brevifolia, 649
  datil, 649
  draconis, 649
    arborescens, 649
  elata, 650
    utahensis, 650
  filamentosa, 648
  gilbertiana, 649

Harriman, 649
harrimaniae, 649
  gilbertiana, 649
  sterilis, 650
Kanab, 650
kanabensis, 650
narrow-leaved, 649
navajoa, 649
schidigera, 650
standleyi, 649
Toft, 650
toftiae, 650
Utah, 650
  utahensis, 650
Zannichellia
  palustris, 824
Zanichelliacea, 824
Zauschneria
  californica, 448
    latifolia, 448
    garrettii, 448
    latifolia, 448
    garrettii, 448
Zapania
  cuneifolia, 640
Zelkova
  serrata, 613
Zigadenus
  elegans, 810
  gramineus, 811
  paniculatus, 811
  vaginatus, 811
  venenosus, 811
    gramineus, 811
  volcanicus, 811
Zizia
  aptera, 637
Ziziphus
  jujuba, 519
Zoysia
  japonica, 788
  matrella, 788
  tenuifolia, 788
Zuckia
  arizonica, 131
  brandegei, 130
    arizonica, 131
    brandegei, 131
Zygophyllaceae, 646
Zygophyllum
  tridentatum, 646

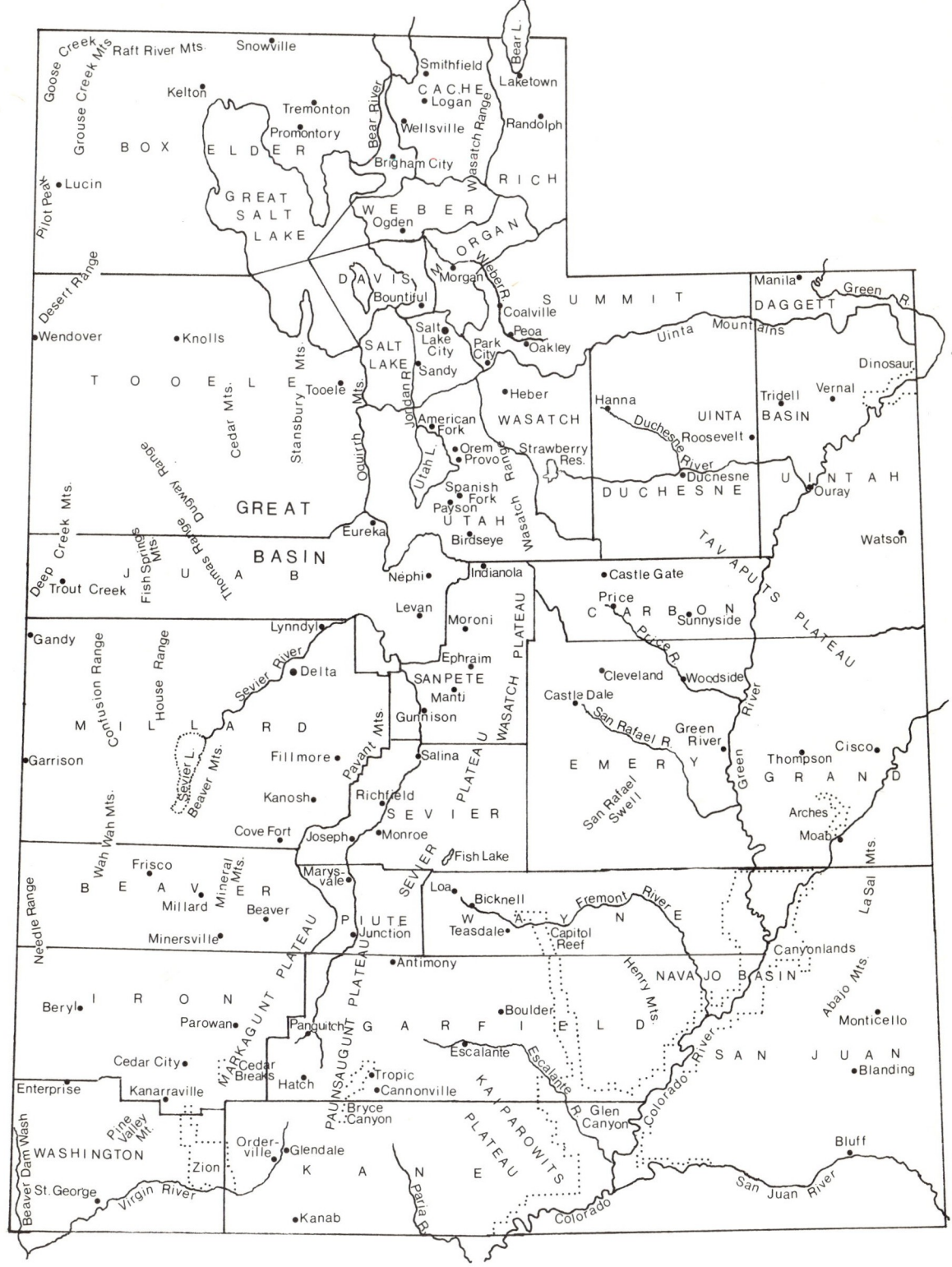